FOUNDATIONS OF ORGANIC CHEMISTRY

FOUNDATIONS OF ORGANIC CHEMISTRY

UNITY AND DIVERSITY OF STRUCTURES, PATHWAYS, AND REACTIONS

David R. Dalton

A JOHN WILEY & SONS, INC., PUBLICATION

Published by John Wiley & Sons, Inc., Hoboken, New Jersey
Published simultaneously in Canada

For general information on our other products and services or for technical support, please contact our Customer Care Department within the United States at (800) 762-2974, outside the United States at (317) 572-3993 or fax (317) 572-4002.

Wiley also publishes its books in a variety of electronic formats. Some content that appears in print may not be available in electronic formats. For more information about Wiley products, visit our web site at www.wiley.com.

Library of Congress Cataloging-in-Publication Data:

Dalton, David R.
  Foundations of organic chemistry : unity and diversity of structures, pathways, and reactions / by David R. Dalton.
    p. cm.
   ISBN 978-0-470-47908-7 (cloth)
   1. Chemistry, Organic–Textbooks.   I. Title.
   QD251.3.D35 2011
   547–dc22

            2010043290

Printed in Singpore

oBook ISBN: 978-1-118-00539-2
ePDF ISBN: 978-1-118-00537-8
ePub ISBN: 978-1-118-00538-5

10 9 8 7 6 5 4 3 2 1

8. Part I. The Reactions of Alcohols, Enols, and Phenols: Oxidation, Reduction, Substitution, Addition, Elimination, and Rearrangement
Part II. Ethers
Part III. Selected Reactions of Alkyl and Aryl Thiols and Thioethers   **562**

Welcome! The study of organic chemistry can be filled with either the joy of discovery of one of the great ongoing intellectual efforts of human beings or the sad drudgery afforded a task seen only as a barrier to "getting on." If you consider it the latter, you will lose the richness of what is offered and will have cheated yourself of a major inheritance.

The organizational tree of organic chemistry, built up over generations, has many branches. Some branches are rich with fruit that should be savored at length; others, fragrant with blossom, invite shorter pause. Still other limbs may be mostly deadwood, on which some new growth is seen, or may be new sprouts from the main trunk ready for a growth spurt. To appreciate this vibrant life takes time; plan now on budgeting enough.

You will probably find that you are expected to master sufficient material from lecture and text to be able to make predictions and to justify observations. Although it is true that you must discover for yourself (if you do not already know) what you must do to learn, the time-tested methods of reading, rereading, doing more problems than assigned, and, when you have thought it through and still cannot quite understand, asking your instructor will work here too. However, I hope you are not intimidated by hard work!

DAVID R. DALTON

Philadelphia, Pennsylvania
March 2011

ACKNOWLEDGMENTS

It is clear that I cannot sufficiently acknowledge the contributions of my family, friends, and colleagues as well as all of the men and women from whom I have learned. However, there are some who, more than others, directly contributed to this volume.

The men and women with whom I worked at Clemson University and, in particular, John Huffman, Karl Dieter, and Melanie Cooper deserve an early mention as their perception of teaching played a significant role. Subsequently, colleagues and students at Temple University struggled with various portions of the manuscript over many revisions and years and the fortitude of Linda Mascavage (Arcadia University), Serge Jasmin (Temple University), Phil Sonnet (USDA, retired), Charlie DeBrosse (Temple University) and Harry Gottlieb (Temple University) in particular, was beyond what any reasonable person could expect.

The staff at John Wiley were left to cope with my efforts and that anything reasonable has come out clearly is the responsibility of Jonathan Rose, Lauren Hilger, Amanda Amunullah, Lisa Morano Van Horn and their colleague Stephanie Sakson.

Everything that you don't like and the blunders and errors are mine alone!

D. R. D.

PART I

BACKGROUND

Nullius addictus iurare in verba magistri ... (I am not bound to swear allegiance to the word of any master ...)

—Horrace, *Epistulae*

INTRODUCTION TO PART I

As is generally true for established subjects, it is difficult to know exactly where to begin learning. If a linear historical route is taken, the advantages gained by subsequent organizational insights are lost. For example, the use of **ethyl alcohol (ethanol,** CH_3CH_2OH, **1**) (**Alcohols**, Chapter 8) as an intoxicant from fermentation of sucrose containing plants (**Carbohydrates**, Chapter 11) is buried in antiquity, preceding the written record, and the substance we call **morphine** (**2**, below) (**Alkaloids**, Chapter 13) has a recorded history of at least 4000 years. Indeed, the use of the former is attributed to the gods themselves! The latter (albeit in crude form) was apparently obtained from the unripe seed pods (as is done today) of what may have been the opium poppy (*Papaver somniferum* L.) by the Sumerians, the early inhabitants of part of what was later called Babylonia and is currently called Iraq. These materials are still being actively studied, and a representation for each of them is shown below. However, to begin to study a well-organized subject with two such diverse materials whose major link appears to be some interesting physiological activity would deny the benefits of that organization.

CH_3CH_2OH

1

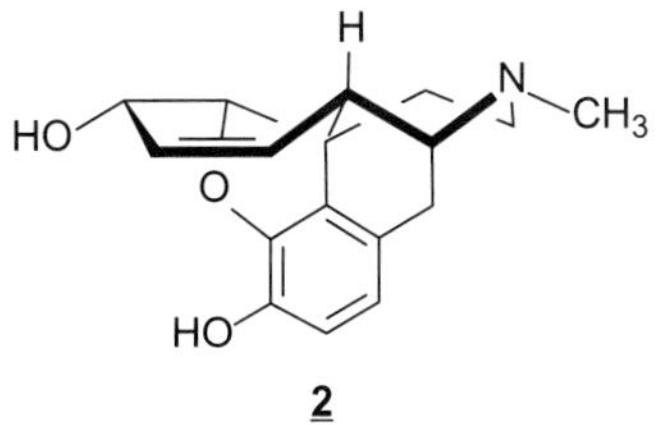

2

Thus, we will take, as is usually done, a mixed approach—selectively using history to suit our needs.

Foundations of Organic Chemistry: Unity and Diversity of Structures, Pathways, and Reactions, First Edition. David R. Dalton.

Only about 2500 years ago, a Greek school led by Demokritos of Abdera (468–370 BCE) laid down ideas we interpret as similar to those of the atomic theory, reformulated in the first decade of the nineteenth century by the English schoolteacher and meteorologist John Dalton.*

The years between Demokritos and Dalton had seen advances in the arts of metallurgy, distillation (heating to cause vaporization followed by condensation of the vapors), wine making, and so on, which gradually advanced with the passing generations. In retrospect, the frustration of that slow pace and the absence of tools with which to accelerate it, to satisfy both curiosity and greed, clearly accompanied mankind's plunge into the Dark Ages. That sad period saw the rise of magic and superstition, which still enslave some today, and the mad search for *al-kimiya*, the Philosopher's Stone. While the alchemist practitioners did slowly advance the tools and techniques of manipulation of materials, that advance was sometimes justified as a search for the "true names" of things, so their transformations could be properly affected.

The progress that might have been anticipated following Dalton was not immediately forthcoming. As we shall see, being able to say how many atoms of any given element are present in a molecule is related to the structure of that molecule as the number of bricks in a pile is related to a building, which might be built with them. Thus, some 30 years after Dalton, we find the "father" of modern dye chemistry, W. H. Perkin,[†] apparently assuming (at age 18) that twice the number of atoms in **p-N-allyltoluidine** ($C_{10}H_{13}N$) (**3**, Chapter 10) plus **oxygen** ($3/2\ O_2$) should be equivalent to **quinine** ($C_{20}H_{24}N_2O_2$) (**4**, Alkaloids, Chapters 12 and 13) plus **water** (H_2O). When he did the experiment, he treated the *p-N*-allyltoluidine ($C_{10}H_{13}N$), **3**, with the known oxidizing agent **potassium dichromate** ($K_2Cr_2O_7$), a dark brown material was obtained, which, although subsequently leading Perkin to dyestuffs (Chapter 10) for wool and cotton, was not the well-defined construction quinine ($C_{20}H_{24}N_2O_2$) (**4**).

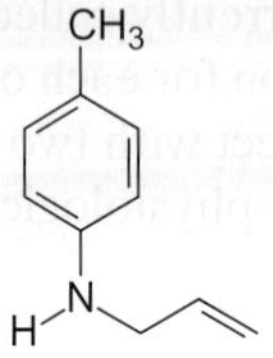

<u>3</u> 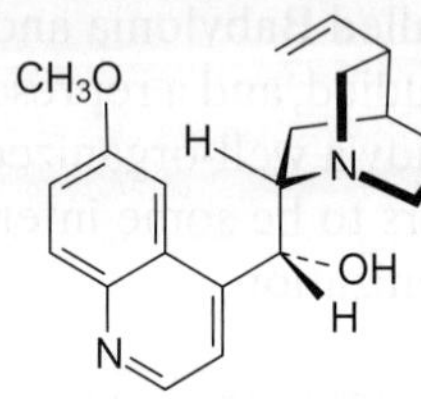   <u>4</u>

These retrospective glimpses clearly require our carrying to them our current ideas and tools. Indeed, it can be argued that the development of tools, which confirmed some ideas and not others, has led to our present understandings. However, the tools of organic chemistry are the same as those used elsewhere, and so it is frequently suggested that the study of organic chemistry might begin with those things that distinguish it from other disciplines.

---

*John Dalton (1766–1844) was a teacher of "natural philosophy" in Manchester, England, where he studied chemistry and physics. He is known for his research into color blindness (called Daltonism) and atomic theory.

[†]William Henry Perkin, FRS (1838–1907), found, among other colors, "mauveine" (today sometimes called simply "mauve") while working with aniline (*vide infra*, Chapter 10) and other aromatic amines.

In that vein, historically, the subject of organic chemistry was divided from other branches of chemistry first by the idea that organic compounds were exclusive to living systems and that, somehow, such materials possessed special characteristics. Second, having disabused ourselves of that idea, the subject was considered as the domain, for the most part, of compounds of carbon and their interrelationships. Third, and most recently, organic chemistry has evolved into what it is that individuals who practice the art choose to call what they do. Nonetheless, it is still generally true that the distinguishing characteristic of the study of organic chemistry is the central role played by compounds containing carbon.

So it is, today, very important that you realize, first, that the enormous, yet incomplete, structure you face, which shares both praise and blame for our current way of life, did not leap into being full grown. The process was slow.

Second, it is continuing to grow. Indeed, the present increasing rate of accretion of information, in part due to a confluence of trends that include, most apparently, a major refinement of our tools as they are interfaced with microprocessors, an ever-increasing population of active workers, and rapid communication of results upon which new building can occur, is leading to a better understanding of our world and our place in it. Nonetheless, much of the early work remains important as it serves as the foundation of our present edifice, and it will serve as a starting place for us.

An Introduction to Structure and Bonding

לכ ל, זמן; ועת לכל-חפץ, תחת חשמים

To every thing there is a season, and a time to every purpose under the sun.

—Ecclesiastes 3

A. THE SOURCES OF CARBON COMPOUNDS

An ever-increasing amount of carbon dioxide (CO_2) is found in the atmosphere and the oceans, apparently as a result of continuing combustion of fossil fuels and normal life processes. Additionally, much of the biosphere contains carbon locked in plants (as carbohydrates [Chapter 11] and related materials about which you will learn). However, most of the compounds of carbon in use today in our technology are obtained either directly from coal or from petroleum—thought by many to have arisen by application of heat and pressure to the decaying products of earlier biosphere inhabitants or by laboratory modification of those materials.

Petroleum and coal consist mainly of compounds that contain only carbon and hydrogen. These materials are called **hydrocarbons**. While the reason for the name should appear clear, it is hoped you might ask how it is known that only hydrogen and carbon are present.

The complete combustion of a hydrocarbon in an oxygen atmosphere generates carbon dioxide (CO_2) and water (H_2O). If all of the carbon is converted to CO_2 and all of the hydrogen to H_2O, and if there are no other elements besides carbon and hydrogen present in the hydrocarbon, then, as matter is conserved, the amount of carbon in the CO_2 plus the amount of hydrogen in the water must equal the amount of carbon and hydrogen, respectively, in the hydrocarbon combusted.

Problem 1.1. A pure material is isolated from petroleum and 5.8 mg of the material is burned in a stream of oxygen to yield 17.6 mg of carbon dioxide (CO_2 and 9.0 mg of water (H_2O).

(a) Show how you know the material isolated from petroleum contains only carbon and hydrogen.

(b) Suggest an empirical formula for the material.

Foundations of Organic Chemistry: Unity and Diversity of Structures, Pathways, and Reactions, First Edition. David R. Dalton.
© 2011 John Wiley & Sons, Inc. Published 2011 by John Wiley & Sons, Inc.

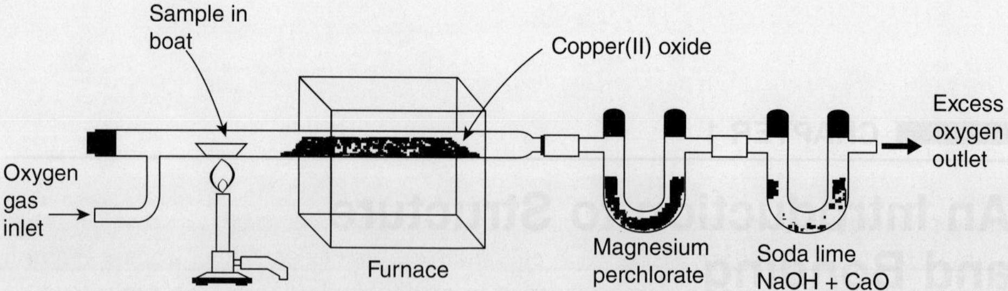

Figure 1.1. A schematic representation of a combustion train for the determination of carbon and hydrogen in combustible substances.

There is a significant amount of technology required to carry out the determination of the carbon dioxide and water produced in the process Problem 1.1 exemplifies. You will note that the material on which the combustion is being carried out is specified as "pure." Second, a weight is given (5.8 mg). Third, a stream of oxygen (O_2) is needed. Fourth, the carbon dioxide (CO_2) must be weighed, as must the water (H_2O). Finally, having solved the problem, it should be clear to you that atomic weights for the elements involved were needed.

We will not deal here with the establishment of atomic weights, save to indicate that current usage assigns the atomic weight of ^{12}C as 12 (exactly) and, for the calculations, it is sufficient to allow ^1H to be 1.00 and ^{16}O to be 16.00. Figure 1.1 is a pictorial representation of combustion analysis equipment.

You will note, as shown in Figure 1.1, that there are two removable sections (at the right-hand side of the combustion train). The first contains magnesium perchlorate [$Mg(ClO_4)_2$] and the second soda lime (a mixture of sodium hydroxide [$NaOH$] and calcium oxide [CaO]). Magnesium perchlorate [$Mg(ClO_4)_2$] readily absorbs water (H_2O), while soda lime ($NaOH + CaO$) absorbs carbon dioxide (CO_2). Thus, weighing the removable tubes before and after combustion will provide the weights of water (H_2O) and carbon dioxide (CO_2), respectively. Purified, dry oxygen (O_2) is currently obtained by distillation from liquefied air (another technology we will not discuss here). However, it should be clear to you that the development of the analytical balance, allowing accurate weights to be taken, is critical to the operation of the system. Finally, there is the problem of the analysis being carried out on pure material.

I. How Do We Know a Material Is Pure?

Classically, pure organic compounds were defined as those that had the correct elemental composition (within a few tenths of 1% of that calculated) and had sharp melting points (mp) or constant boiling points (bp).* However, your exposure to

---

*While these values are called "points," you should keep in mind that they are really "temperatures," and usually, for pure materials, the temperature range over which melting (or boiling) occurs is very small.

general principles of chemistry should be sufficient for you to recognize that, by today's standards, these criteria may be insufficient. Nonetheless, significant basic principles were enunciated with work done on less-than-pure materials (errors often cancel, and frequently, within the limits of the errors, differences could not be detected).

Within the last few decades, powerful tools for purification of materials by separation have been developed. **Distillation** and **crystallization** are two with which you might already be familiar. A third is **chromatography** (from the Greek *chroma*, meaning **color**, and *graph*, meaning **draw** or **record**). The latter, as you might gather from its name, was originally developed as a tool to separate the colored components of a mixture by differences in the way they became distributed between a solid stationary **adsorbing** phase and a liquid **mobile** phase in which they were soluble.

Today, the original concept has been expanded to include materials without absorption in the visible region of the spectrum (color) and any two different phases, for example, gas–liquid and gas–solid, two immiscible liquids, etc.

As an example, consider the case of two gases, ammonia (NH_3) and nitrogen (N_2). Ammonia (NH_3) is soluble in water (H_2O) (at room temperature and 1 atm pressure) to the extent of about 90 g NH_3 per 100 g H_2O. Nitrogen (N_2), under the same conditions, is soluble to the extent of about 4 g N_2 per 100 g H_2O. If we have a mixture of N_2 and NH_3 gases in contact with water, ammonia will distribute (or **partition**) itself mostly into the aqueous phase, whereas nitrogen, being relatively insoluble, will remain in the gas phase. If we now replace the aqueous solution with fresh water, a second **partitioning** will occur, and repetition of this process again and again will lead to relatively pure nitrogen (perhaps contaminated with water vapor) and an aqueous solution of ammonia. This tedious procedure can be replaced so that the process can be carried out with greater simplicity, automatically, and nearly continuously by utilizing **chromatographic columns** that, as originally developed, permit a moving phase containing the materials to be separated to pass over and through particles of a stationary phase that selectively **absorbs** (or, if only on the surface, **adsorbs**) one or more of the materials in the moving phase. Thus, in the example given, if the gaseous mixture were to pass over and through a bed of glass beads, packed in a column and coated with water (H_2O), then, as the mixture of gases moved through the column, it would continuously encounter fresh water (H_2O). At each encounter, ammonia (NH_3) would be selectively absorbed into the water (H_2O) as a function of the amount of water (H_2O) and the various vapor pressures of the gases and water. The gases nitrogen (N_2) and ammonia (NH_3) could be pushed (continuously be adsorbed and desorbed as the vapor mixture changed) over this bed of beads by a **carrier gas**, which is relatively less soluble in water than either of them (e.g., helium [He], which has about half the solubility of nitrogen [N_2] in water). At a suitable detector (which might, e.g., measure changes in the **thermal conductivity** of the effluent gases), the pure helium (the **carrier**) would, as it left the column (or **eluted**), first find itself contaminated with nitrogen (N_2) since it is less soluble in water (H_2O) than ammonia (NH_3) and then with ammonia (NH_3), which was delayed by **its greater absorption.**

Such a process, in which material is partitioned between the gas and the liquid phases, is called **gas–liquid partition chromatography** (GLPC or, sometimes, GLC or GC). A schematic representation of such equipment is shown in Figure 1.2, and,

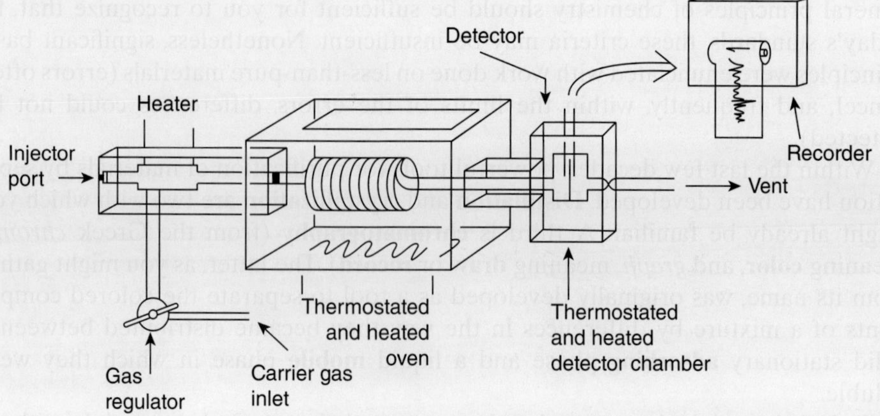

Figure 1.2. A schematic representation of a gas chromatograph.

commonly, organic compounds can be readily separated from each other, **eluted** in pure form and analyzed by combustion. For materials that do not lend themselves to this technique because they are too high boiling or because appropriate solid/liquid phases cannot be found, other methods (which include **column chromatography and high-pressure liquid chromatography**) based on the same principle can be applied.

However, such techniques aside, large quantities of purified volatile materials were obtained, historically, by distillation through long columns packed with glass beads or other materials (to increase the surface area on which **equilibrium** between liquid and vapor could occur). Even today, such distillation columns are part of petroleum refining and most mixtures of fuels are prepared by remixing purified materials to obtain appropriate blends.

The technique of **fractional distillation** is a practical application of **Raoult's law.***

Consider an ideal solution containing two **volatile** components, x (bp 80°C) and y (bp 110°C), both at 1 atm. The total vapor pressure above a solution of the two materials ($\mathbf{P}_t$) is equal to the sum of the partial vapor pressures of all components (Equation 1.1):

$$\mathbf{P}_t = \mathbf{P}_x + \mathbf{P}_y = \mathbf{P}_x{}^\circ\mathbf{N}_x + \mathbf{P}_y{}^\circ\mathbf{N}_y, \tag{1.1}$$

where $\mathbf{P}_x, \mathbf{P}_y$ = the partial vapor pressure of x and y components, respectively, in the solution at the given temperature; $\mathbf{P}_x{}^\circ, \mathbf{P}_y{}^\circ$ = the vapor pressure of pure x and pure y, respectively, at that temperature; and $\mathbf{N}_x, \mathbf{N}_y$ = the mole fraction of x and y, respectively, in the **solution**.

The composition of the **vapor** mixture in terms of the mole fraction in the **vapor state** is given by Equation 1.2 for the mole fraction of x ($\mathbf{N}'_x$):

$$\mathbf{N}'_x = \mathbf{P}_x/\mathbf{P}_t = \mathbf{P}_x{}^\circ\mathbf{N}_x/(\mathbf{P}_x{}^\circ\mathbf{N}_x + \mathbf{P}_y{}^\circ\mathbf{N}_y). \tag{1.2}$$

*François-Marie Raoult (1830–1901), French chemist who long held the chair of chemistry at Grenoble.

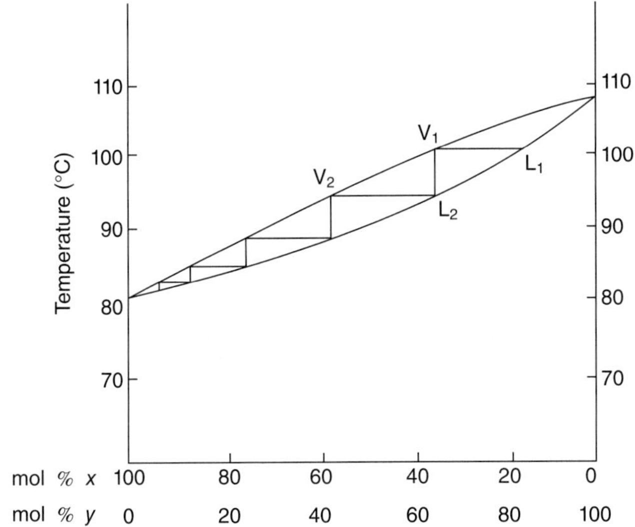

Figure 1.3. Temperature–composition diagram for the mixture of *x* and *y*.

Similarly, for the mole fraction of *y* ($\mathbf{N'}_y$), Equation 1.3 applies:

$$\mathbf{N'}_y = \mathbf{P}_y/\mathbf{P}_t = \mathbf{P}_y{}^{\circ}\mathbf{N}_y/(\mathbf{P}_x{}^{\circ}\mathbf{N}_x + \mathbf{P}_y{}^{\circ}\mathbf{N}_y). \qquad (1.3)$$

The temperature at which "boiling" occurs requires only that the vapor pressure of the mixture above the solution be equal to the external pressure. However, as dictated by Equation 1.1, the total vapor pressure is a function of the composition of the vapor and at lower temperatures, the vapor will be richer in the lower boiling component.

The relationship between boiling point and vapor composition for the mixture of *x* (bp 80°C) and *y* (bp 110°C), both at 1 atm, is shown in Figure 1.3, where the lower curve presents the boiling point of the liquid mixture at any temperature between 80 and 110°C, and the upper curve corresponds to the vapor composition as calculated from Equations 1.2 and 1.3.

As can be seen from Figure 1.3, the vapor, at any given temperature, is always richer in the more volatile component and, if the vapor is liquefied, the liquid corresponding to that vapor will also be richer in that component. For example, consider a liquid mixture composed of 20% *x* and 80% *y*. As shown in Figure 1.3, at 102°C and 1 atm (point **L₁**), the vapor above the liquid will consist of about 40% *x* and 60% *y* (point **V₁**).

Problem 1.2. With reference to Figure 1.3,

(a) if the composition of the vapor at **V₁** is, as stated, about 40% *x* and 60% *y*, what will be the composition of the **liquid** obtained by direct condensation of that vapor?

(b) if the liquid so obtained (part a) is immediately reconverted to vapor, what will be the composition of that vapor?

Repeated condensation and redistillation allows, in this way, separation of the mixture of x and y.

Although it was early found that freezing point lowering of a solvent could be used to determine the number of moles of solute dissolved in the solvent (and hence the molecular weight), Raoult's law applied to solutions of small concentrations of nonvolatile solutes also affords a method of approximately determining the molecular weights of the solutes. This is because the temperature at which the vapor pressure becomes equal to any specified external pressure is higher for the solution than for the pure solvent. Additionally, it should not come as a surprise to you that, other things being equal, it is generally true that materials with lower molecular weights boil at temperatures lower than those with higher molecular weights. So, given the results from Problems 1.1 and 1.2, it should be possible to determine the molecular weight of a **pure**, **volatile** sample by simply measuring the volume that a given mass of the **vapor** of the (pure, volatile) sample occupies at atmospheric pressure. The **Victor Meyer** method accomplishes that end (Problem 1.3).

Problem 1.3. The Victor Meyer (1848–1897, professor of chemistry, University of Heidelberg) method for determination of molecular weights is quite general for low-boiling materials. A diagrammatic representation of the apparatus is shown in Figure 1.4.

Note that the cylindrical bulb with the long stem is filled with dry air and is heated, while still open to the atmosphere, to a constant temperature by a suitable boiling fluid within the hot bath. A small glass vial containing the volatile liquid substance is dropped into the heated bulb and, at the same time, the tube is stoppered at the top of the stem. The glass vial shatters and its liquid content rapidly vaporizes. The vapor drives out of the bulb that quantity of air that occupies the same volume at the temperature of the bulb and at the pressure of the atmosphere. This air is collected in the measuring tube.

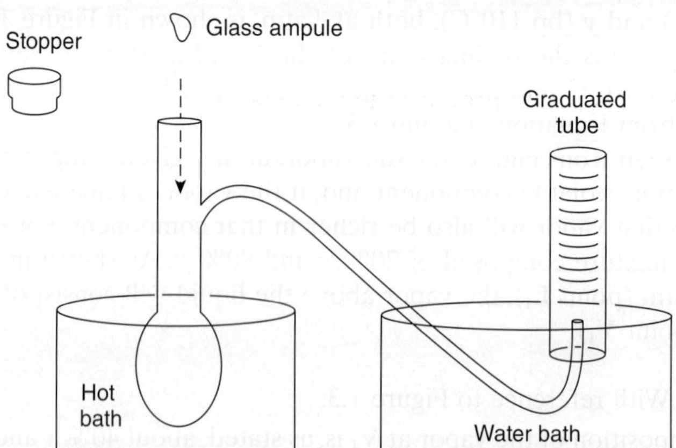

Figure 1.4. A schematic diagram of a Victor Meyer apparatus for the determination of the vapor density of a substance that is volatile at the temperature of the hot bath. The air displaced from the heated bulb by the volatilization of the sample in the bulb is measured in the inverted graduated tube.

Assume that, to a first approximation, the vapor of the organic material analyzed in Problem 1.1 can be considered an ideal gas; that is, it obeys the ideal gas relationship $\mathbf{PV} = \mathbf{nRT}$. If the pressure on the day the experiment was run was 728 torr and a volume of air corresponding to 15.2 cc was displaced from the hot tube at 373 K (but collected at 300 K) by introducing and vaporizing 34.3 mg of the material from Problem 1.1, what must be its molecular weight? ($\mathbf{R} = 0.083\,\mathrm{L\,atm\,mol^{-1}\,K^{-1}}$)

The process of **fractional distillation** is routinely applied to pretreated liquid mixtures of **hydrocarbons** from various sources of petroleum or from liquefaction of coal. The lower boiling fraction (up to about 40°C) is called **gas** and contains a mixture of hydrocarbons with (generally) fewer than five carbon atoms (and, of course, their associated hydrogens since they are hydrocarbons). The next higher boiling fraction, from about 40°C to about 170 or 180°C, is called **gasoline*** and generally contains hydrocarbons with up to 10 carbon atoms. **Kerosene**, which has a boiling temperature range of about 180–230°C, and which is used as a jet airplane fuel as well as a heating fuel, is composed of hydrocarbons with 11 and 12 carbon atoms. As the temperature of the crude oil distillation is raised even further, **light gas oil, heavy gas oil**, and then **asphalt** fractions are obtained. The amount of each of these materials, as well as the nature and amount of the various substances with which they are contaminated, varies with the source of the oil. Frequently, the impurities are not hydrocarbons. The **nonhydrocarbon** impurities contain, in addition to carbon and hydrogen, elements such as oxygen, nitrogen, and sulfur. Such impurities can sometimes be removed by chemical means because hydrocarbons are relatively unreactive materials compared to the others (Chapter 3). Indeed, the major **reaction** hydrocarbons undergo is **oxidation**, and when the oxidation is carried out under controlled conditions, the process is used to power our society—generating electricity as well as heat. Although electricity can, of course, be generated by other means (solar, hydroelectric, nuclear, etc.), the use of oil for this purpose as well as for refinement into gasoline amounted, in 1990 to the consumption of **more than 18 million barrels of crude oil daily** by the United States. Of this amount, the equivalent of about 20% was needed by the chemical industry to both power the industry and as crude feedstock for raw chemicals. Indeed, the equivalent (i.e., considering natural gas, liquids from coal gasification) of more than 1.2 **million barrels of crude oil daily** are needed as feedstock to produce the goods we use.

B. MORE ABOUT HYDROCARBONS

With the availability of relatively pure hydrocarbons from, for example, **fractional distillation**, taking ever-decreasing temperature ranges to get **fractions** as pure as possible (or, more recently, by **gas chromatography**), it becomes possible to attempt to determine how efficient our combustion engines for heat production really are; that is, ignoring for the moment (1) the important issues of economy of production and delivery; (2) social questions of purchase from, and dependency on, nations or

---

*The material purchased at the "gas pump," also called gasoline, contains these and other hydrocarbons and many other materials, called **additives**, which presumably increase the efficiency of combustion.

individuals with whom we might not always agree; as well as (3) the planet-wide danger of increased concentration of greenhouse gases from which **all** combustion and most refining processes suffer, which fuel should we consume to heat our homes, power our automobiles, and generate electricity?

In order to obtain this information, we want to first know how much **heat** is potentially available per unit weight from the hydrocarbons we have; that is, what is their **heat content** or, when *completely* combusted to give *only* carbon dioxide (CO_2) and water (H_2O), how much energy, in the form of heat, can we expect to get?

I. Combustion—Heats of Reaction

At high temperature, all **hydrocarbons** are attacked by **oxygen**. If oxygen is present in excess (see Problem 1.1), complete combustion occurs to give carbon dioxide (CO_2) and water (H_2O). **Methane** (CH_4) is one of the main components of **natural gas**. It is also formed by the action of anaerobic organisms during decomposition of organic matter under water in marshes (from where it gets the name **marsh gas**), during the treatment of sewage sludge, and in the normal digestive processes of cows and other ruminants. The combustion of gaseous methane (indicated by the subscript $_{(g)}$) (Equation 1.4) with the formation of gaseous carbon dioxide ($CO_{2(g)}$) and liquid water ($H_2O_{(l)}$)* is accompanied by the evolution of heat. Careful measurement of the amount of heat evolved with all species adjusted for their standard states, $\Delta H_c°$, (in a calorimeter [Figure 1.5]) during this process has been made and is found to be $890.4\,\text{kJ}\,\text{mol}^{-1}$ ($212.8\,\text{kcal}\,\text{mol}^{-1}$)[†] at 25°C:[‡]

$$CH_{4(g)} + 2O_{2(g)} \rightarrow CO_{2(g)} + 2H_2O_{(l)} \qquad (1.4)$$
$$\Delta H_c° = -890.4\ \text{kJ mol}^{-1}\ (-212.8\ \text{kcal mol}^{-1}).$$

Now, since the same amount of heat is produced whenever a fixed amount of work is done and since this amount is independent of the way the work is done (i.e., the **path**, which may or may not involve intermediate reactions), it is possible to calculate the heat of a specific reaction from measurements of heats of many different reactions. The principle is now called **Hess's law**, after Germain H. Hess (1802–1850), who first published it (Hess, 1840) as **the law of constant heat summation**, and it is quite useful in predicting the heats of reactions, which are otherwise most difficult to carry out. In using Hess's law, the following are accepted **by convention**:

1. **Standard states** will be used. A standard state for a substance is that state in which it is stable at 25°C and 1 atm pressure.

---

*Conversion of 1 mol of liquid water ($H_2O_{(l)}$) to 1 mol of water vapor ($H_2O_{(g)}$) (or the reverse) requires correction for the heat of vaporization of water. For both at 25°C, $43.93\,\text{kJ}\,\text{mol}^{-1}$ ($10.5\,\text{kcal}\,\text{mol}^{-1}$) is required.

[†]The recommended units (Système international d'unités or SI units) for measurement of energy are **joules**. The use of **calories** as a unit of measurement nonetheless remains widespread. Multiply the value in kilocalories by 4.184 to convert it to kilojoules.

[‡]It is important to recognize that the heat measured here is the quantity liberated (hence the negative sign), having begun with the reactants at 25°C and ending with the products at 25°C. The temperature at which the reaction actually occurs does not matter.

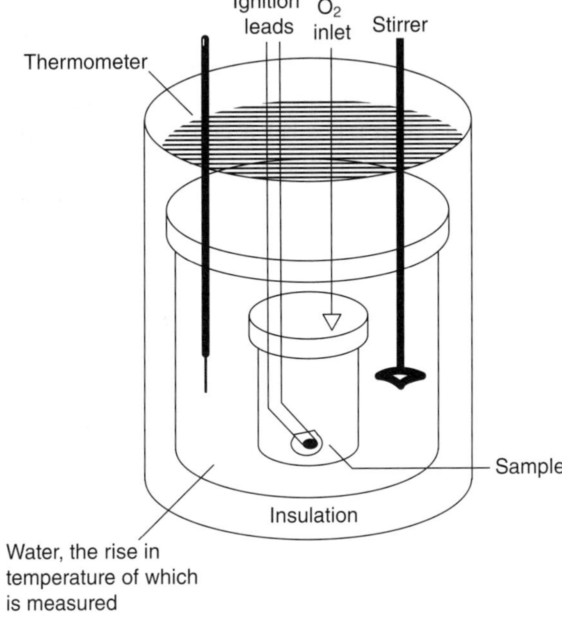

Figure 1.5. A diagrammatic representation of an adiabatic (constant heat) calorimeter. The calorimeter consists of an insulated container filled with water in which the reaction chamber containing the sample and an excess of oxygen gas is immersed. In an exothermic reaction, the heat generated is transferred to the water and measured. Knowing the quantity of heat, the change in temperature, and the heat-absorbing characteristics of the system, the amount of heat evolved in the reaction may be calculated.*

2. **Elements**, in their standard states, have a heat content (**enthalpy**), H, of **zero**.

3. The **standard enthalpy** of any **compound** is thus the heat of the reaction by which it is formed from its elements—with the reactants and products in their standard states at 25°C and 1 atm.

4. The **superscript** ° indicates that the enthalpy is a **standard heat of combustion** (**subscript** $_c$) (i.e., $\Delta H_c°$) or **formation** (**subscript** $_f$) (i.e., $\Delta H_f°$).

5. If **heat is evolved**, $\Delta H°$ is given a negative sign and the reaction is called **exothermic**, while if **heat is absorbed**, $\Delta H°$ is given a positive sign and the reaction is called **endothermic.**

---

*In principle, there are two ways of carrying out such a combustion reaction. In the calorimeter described, the volume of the system is held constant, a pressure change is expected, but no mechanical work is done. It follows that the increase in energy is equal to the amount of heat absorbed by the water. Alternatively, as in the cylinder of an idealized automobile engine, a piston might be driven so that the volume of the system would change while the pressure remained constant. Strictly speaking, the change in heat content or enthalpy (ΔH) equals the heat change at constant pressure—not at constant volume. While it is true that the heat of reaction at constant volume is greater than that at constant pressure (because work is done in the latter case), for reactions involving solids and liquids (most organic compounds are solids or liquids, unlike the example of methane [CH_4] chosen here), the difference is small. For reactions between gaseous species, the difference may be appreciable.

Thus, in the case at hand (Equation 1.4), the value of ΔH_c° is the amount of heat liberated on complete combustion of 1 mol of the hydrocarbon methane (CH_4) when the reactants and products are in their standard states.

In an exactly analogous fashion, the hydrocarbon **ethane** (C_2H_6- or, better, CH_3CH_3 [mp −182.8°C, bp −88.6°C]), also a constituent of natural gas (which is about 85% **methane** [CH_4] and 10% ethane [CH_3CH_3]), undergoes combustion to carbon dioxide (CO_2) and water (H_2O) (Equation 1.5), while carbon* itself (Equation 1.6) and hydrogen (Equation 1.7) similarly burn:

$$CH_3CH_{3(g)} + 7/2\,O_{2(g)} \rightarrow 2CO_{2(g)} + 3H_2O_{(l)} \quad \Delta H_c^\circ = -1558.9 \text{ kJ mol}^{-1} \qquad (1.5)$$
$$(-372.6 \text{ kcal mol}^{-1})$$

$$C_{(s)} + O_{2(g)} \rightarrow CO_{2(g)} \quad \Delta H_c^\circ = -395.4 \text{ kJ mol}^{-1} \qquad (1.6)$$
$$(-94.5 \text{ kcal mol}^{-1})$$

$$H_{2(g)} + 1/2\,O_{2(g)} \rightarrow H_2O_{(g)} \quad \Delta H_c^\circ = -285.8 \text{ kJ mol}^{-1} \qquad (1.7)$$
$$(-68.3 \text{ kcal mol}^{-1}).$$

Application of Hess's law to the thermochemical data provided in Equations 1.5–1.7 then allows calculation of the heat of formation of ethane from carbon and hydrogen (Equation 1.8):

$$2C_{(s)} + 3H_{2(g)} \rightarrow CH_3CH_{3(g)} \quad \Delta H_f^\circ = -89.1 \text{ kJ mol}^{-1} \qquad (1.8)$$
$$(-21.3 \text{ kcal mol}^{-1}).$$

Problem 1.4. Using the data provided in Equations 1.5–1.7, show how the expression of Equation 1.8 is obtained. The most recent widely quoted value, given in reference to Table 1.1 for ΔH_f° for ethane, is −83.7 kJ mol^{-1} (−20.0 kcal mol^{-1}).

We interpret this kind of information to mean that there is more energy stored in the **bonds** holding together the nuclei of the atoms in the starting materials than there is in the bonds holding together the nuclei of the atoms in the products; that is, we have measured the amount of energy held in the methane (CH_4) and oxygen (O_2) molecules as compared to the energy in the carbon dioxide (CO_2) and water (H_2O) molecules, and we have found that there is more in the former than in the latter. We attribute this difference not to the nuclei but rather to **the way the nuclei are held together**, that is, the bonds. Thus, the word "bond" is taken to mean "the force holding nuclei to each other." So it follows that although combustion is critical to energy production, and although heats of combustion, reaction, and formation provide analytical footing to the first law of thermodynamics, it is **bond energies** that have proven to be more generally useful in organic chemistry.

Considering, for the moment, that a bond between nuclei represents only a manifestation of some force holding the nuclei together, it is reasonable to inquire both

---

*Different allotropic forms, for example, graphite, diamond, and the fullerenes (*vide infra*), have slightly different heats of combustion. The value given here is for diamond.

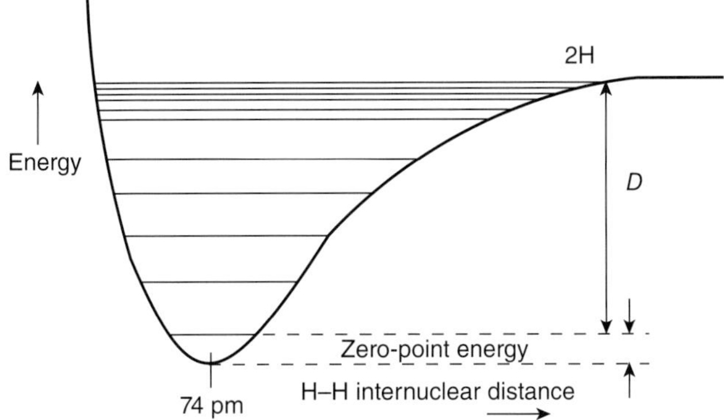

Figure 1.6. A representation of the potential energy curve for H_2.

if the bond can be broken and how much force might be exerted, or energy expended, to accomplish that end. Hopefully, it is clear that, at least in principle, if expenditure of that energy allows a bond to be broken, the same amount of energy can be recouped if the bond is remade (i.e., the process is *microscopically reversible*). Further, as with Hess's law, if bonds in the starting materials can be suitably broken and then new bonds are made in the products and the gain (or loss) in energy for *the overall process* is measured, some indication of whether a net overall gain or loss in enthalpy occurs can be realized. The anticipated value of this exercise, for the moment, will then lie in being able to predict whether a reaction might be **exothermic** or **endothermic** overall. Eventually, other values to understanding this process will become apparent.

Consider the hydrogen molecule (H_2). Figure 1.6 presents a schematic diagram of the potential energy curve for H_2. As thermal (or other) energy is supplied to H_2, the motions of the molecules and the motions of the atoms in the molecules increase. Much of the energy is expended in **translation** (i.e., the molecules move about faster relative to each other), but some increase in energy is manifested as increased vibration of the atomic nuclei in H_2, some quantized values of which are represented by the horizontal lines. At least in principle, when enough energy is supplied, the two atomic nuclei in H_2 can be separated into individual hydrogen atoms. This quantity is called the **bond dissociation energy** (or **BDE**) and is given the symbol D.*

For the dissociation of H_2 into two hydrogen atoms, $\Delta H_d°$ or $DH°$ at 25°C is usually taken as 435.9 kJ mol^{-1} (104.2 kcal mol^{-1}) (Equation 1.9). In an exactly analogous fashion, for the dissociation of O_2 into two oxygen atoms, $\Delta H_d°$ or $DH°$ at 25°C is taken as 498.3 kJ mol^{-1} (119.1 kcal mol^{-1}) (Equation 1.10). Combining these two with the **calorimetric** expression of Equation 1.6 (and adjusting for everything in the vapor phase) then allows the summation shown in Equation 1.11 and the conclusion that, since this $\Delta H_d°$ or $DH°$ at 25°C is for **two** bonds between hydrogen and

---

*The value for D is normally obtained in ways other than that described above. When using dissociation energies with enthalpic calculations, that is, $\Delta H_d°$ or $DH°$, a small correction term is applied, which includes consideration of the dissociation as if it had occurred at 25°C.

TABLE 1.1. Selected Average Bond Strengths (kJ mol^{-1} [kcal mol^{-1}] at 25°C)

Diatomic Molecules

H–H	435.9 (104.2)	H–F	570.3 (136.3)	F–F	158.1 (37.8)
O=O	498.3 (119.1)	H–Cl	431.8 (103.2)	Cl–Cl	242.4 (58.9)
N/N	945.2 (225.9)	H–Br	366.5 (87.6)	Br–Br	192.9 (46.1)
C=O	1076.5 (257.3)^a	H–I	298.3 (71.3)	I–I	151.1 (36.1)

Polyatomic Molecules

C–H	418 ± 4 (100 ± 1)^b	C–C	376 ± 8 (90 ± 2)^c	C–F	452 ± 13 (108 ± 3)
N–H	383 ± 8 (92 ± 2)^d	C=C	720 ± 8 (172 ± 2)^e	C–Cl	306 ± 8 (73 ± 2)^f
O–H	498 ± 4 (119 ± 1)^g	C≡C	962 ± 8 (230 ± 2)	C–Br	293 ± 4 (70 ± 1)
S–H	381 ± 4 (91 ± 1)^h	C–N	297 ± 4 (71 ± 1)^i	C–I	236 ± 4 (56 ± 1)
C–O	277 ± 4 (66 ± 3)^j	C–Pb	239 ± 17 (57 ± 4)^k	O–O	213 ± 4 (51 ± 1)^l
O–I	238 ± 13 (56 ± 3)	O–Cl	251 ± 13 (60 ± 3)	O–Br	234 ± 13 (56 ± 3)

^{a}This value is for carbon monoxide. Use 803.3 (192.0) for C=O in CO_2.
^{b}Found in ethane, propane, and so on. For the first C–H in methane, use 435 (104).
^{c}Found in ethane, propane, butane, and so on. In CF_3CF_3 use 447 ± 4 (107 ± 1).
^{d}Found in N,N-dimethylamine. For the first N–H in ammonia, use 460 ± 8 (110 ± 2).
^{e}Found in ethene (ethylene). For 1,1,2,2-tetrafluoroethene, use 319 ± 13 (76 ± 3).
^{f}Found in carbon tetrachloride (tetrachloromethane).
^{g}Found in water. For methanol use 437 ± 4 (104 ± 1).
^{h}Found in hydrogen sulfide. For methanethiol, use 370 ± 4 (88 ± 1).
^{i}Found in N-methyl benzylamine.
^{j}Found in methyl t-butyl ether.
^{k}Found in tetraethyl lead (tetraethyllead).
^{l}Found in hydrogen peroxide. For dimethyl peroxides, use 157 ± 3 (37 ± 6).
Most of the values in this table are taken from Kerr, J. A. *CRC Handbook of Chemistry and Physics*, 71st edition, CRC Press, Boston, **1990**, pp. 9–95 ff.

oxygen, a single such bond must have an $\Delta H_d°$ or $DH°$ at 25°C of 485.8 kJ mol^{-1} (116.1 kcal mol^{-1}).*

In these and related ways, the **bond energies** of Table 1.1 have been accumulated:

$$H_{2(g)} \rightarrow 2H\bullet_{(g)} \quad \Delta H_d° \text{ or } DH° = 435.9 \text{ kJ mol}^{-1} \tag{1.9}$$
$$(104.2 \text{ kcal mol}^{-1})$$

$$O_{2(g)} \rightarrow 2O\bullet_{(g)} \quad \Delta H_d° \text{ or } DH° = 498.3 \text{ kJ mol}^{-1} \tag{1.10}$$
$$(119.1 \text{ kcal mol}^{-1})$$

$$H_{2(g)} + 1/2O_{2(g)} \rightarrow H_2O_{(g)} \quad \Delta H_c° = -285.8 \text{ kJ mol}^{-1} \tag{1.7}$$
$$(-68.3 \text{ kcal mol}^{-1})$$

$$2H\bullet_{(g)} + O\bullet_{(g)} \rightarrow H_2O_{(l)} \quad \Delta H_c° = -971.1 \text{ kJ mol}^{-1} \tag{1.11}$$
$$(-232.1 \text{ kcal mol}^{-1}).$$

*As shown in Table 1.1, the currently accepted value for the average bond strength for the O–H bond in water is 498 ± 4 kJ mol^{-1} (119 ± 1 kcal mol^{-1}).

It is generally acknowledged, and must be emphasized, that the concept of bond energies (particularly as determined for polyatomic species), while useful, should not be substituted for calorimetric data where the latter is available. However, where there are no other data, it is possible to obtain crude estimates for the heats of formation (ΔH_f°) and combustion (ΔH_c°).

Problem 1.5. Assume that the heat of formation (ΔH_f°) at 25°C for hydrogen bromide (HBr) is not known to be $-36.4\,kJ\,mol^{-1}$ ($-8.7\,kcal\,mol^{-1}$). Using the information given in Table 1.1, provide your own estimate for the value of ΔH_f° at 25°C for HBr. Do you consider the answer "reasonable?"

C. ON THE NATURE OF THE CHEMICAL BOND

I. Ionic and Nonpolar Covalent Bonds

In the most general sense, when we ask about a **chemical bond**, we are inquiring into the nature of the force holding atoms together to form compounds. For example, in methane, CH_4, discussed above, we want to know how it is that the atoms (hydrogen and carbon in this case) are held to (or connected with) each other such that we must put in some energy to cause them to come apart. As we have already seen (e.g., in Table 1.1), different bonds have different energies (as measured by the cost to dissociate them into their component atomic nuclei and associated electrons) and even bonds connecting those same atoms to each other apparently have different energies, depending upon the compounds in which they are found.

Different approaches to describing bonds have been made over the years and, based largely on reactivity and physical properties, two limiting, idealized categories have been popular, viz., **ionic** and **covalent**. Ideally, the latter involves two identical nuclei sharing electrons equally (hence *co*valent), while the former involves the complete transfer of electrons to form ions.

Based largely upon spectroscopic measurements (i.e., the observation of quantized absorption and emission of energy—as in atomic spectroscopy, about which you may have already heard, and other forms of spectroscopy to be discussed in Chapter 2) and the utility of the concept in making successful predictions, we define as an **orbital** that particular volume in space, near a nucleus, in which there is a high probability of finding an electron associated with that nucleus. By the **Pauli exclusion principle**,* no more than two electrons can occupy a given orbital and **if** there are two electrons in the orbital, they must differ with regard to a property called *spin*. This property has only two possible values, frequently symbolized by the notations $\uparrow$ and $\downarrow$ (or sometimes by $+\frac{1}{2}$ and $-\frac{1}{2}$, respectively), and if two electrons have the same values (i.e., $\uparrow\uparrow$ or $\downarrow\downarrow$) for this property, they are said to be *unpaired* and they cannot occupy the same orbital. **Only** paired ($\uparrow\downarrow$) electrons can occupy the same orbital.

---

*Wolfgang E. Pauli (1900–1958), Nobel Prize in Physics, 1945. Pauli spent productive years at the ETH in Zurich and at the Institute for Advanced Studies in Princeton, New Jersey.

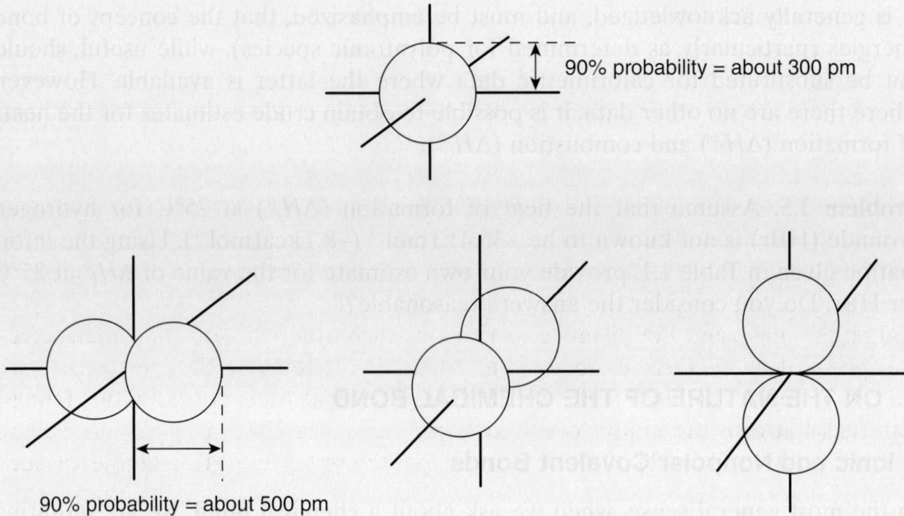

Figure 1.7. Idealized representations of one $2s$ and three $2p$ orbitals with their respective 90% probability radii.

Orbitals are described by specifying their approximate size (even though there may be a finite probability of finding the electron very far from the nucleus), shape, and spatial orientation. Orbitals that are **spherically symmetrical** about the nucleus are called *s orbitals** and by using a numerical prefix, that is, 1, 2, ..., (*the principal quantum number*), their relative energies and the volume of space (of highest probability) they enclose is defined. As measured spectroscopically, an electron in a $1s$ orbital is closer to the nucleus and is lower in energy than an electron in a $2s$ orbital. Additionally, when the principal quantum number is 2, a maximum of six more electrons can be accommodated and the spatial location assigned to them, in pairs, lies along the three normal coordinate axes (x, y, and z). These three $2p$ orbitals (i.e., $2p_x$, $2p_y$, and $2p_z$), each of which can contain no more than two electrons, are found to be of equal energy (and thus *mutually degenerate*). The geometry of the p orbitals differs from that of the s orbitals. In contrast to the spherical symmetry of the latter, the mathematical solution to the wave equation (ψ) for an electron in any of the p orbitals dictates that the region of highest electron probability (ψ^2) is best represented as a pair of slightly distorted spheres, symmetrically disposed about the intersection of the x-, y-, and z-axes.

The set of three p orbitals is thus mutually orthogonal and is drawn to show that there is essentially a zero probability of finding electron density at the nucleus. Figure 1.7 depicts the 90% probability radii of idealized representations of the one $2s$ and the three $2p$ orbitals; the sum of the volumes of the $2p$ orbitals corresponds to a sphere.

---

*The letter s here does not stand for either spherical or symmetrical but rather for *sharp*, a reference to a descriptor assigned on the basis of a spectroscopic transition observed from a particular atomic state of hydrogen. Similarly, the descriptors p, d, and f refer to principal, diffuse, and fundamental spectroscopic lines.

If we make the assumption that occupied (whether with one or two electrons) and unoccupied orbitals, regardless of the nucleus about which they are found, all look pretty much the same, then we can see how, at least in principle, an additional electron (in an appropriate orbital, which still cannot contain more than two) might be associated with one nucleus giving it a net negative charge (measured as its **electron affinity**). Similarly, another atom, having lost an electron (from an appropriate orbital), would then reasonably acquire a net positive charge (measured as its **ionization potential**). The two species, one negative and the other positive, resulting from an idealized complete electron transfer, are now capable of attracting each other and are held together by some electrostatic force, which is proportional to the charge, the mass, and the distance apart from each other the ions find themselves. The physical characteristics that have led to the conclusion that this picture of *ionic bonding* is useful include not only the general high crystallinity of these materials but also the ability of solutions of such substances to carry an electric current. Sodium chloride (NaCl) is often considered a typical example of such material. However, it is clear that the picture of bonding as a result of full transfer of an electron, rather than any sharing, does not accurately describe any diatomic species.

In the idealized picture of *covalent bonding*, on the other hand, commonly thought of as present in most liquids and in many solids and gases, and where electric current is usually not carried by solutions of the substances, there are two popular complementary approaches. For purposes of simplicity, we will first consider both of them using the hydrogen molecule. In more complicated cases, qualitative explanations, sometimes using one, sometimes the other, and, occasionally, both, are common. The first of these is called the **valence bond (VB) method** and is attributed to Heitler and London.* According to the **VB** treatment, a system is imagined in which, at the beginning, each $1s$ orbital on a hydrogen atom has an electron (i.e., $[1s^1]$) and, at the end, each electron in the bonded species, H_2, is associated with both atoms; that is, at the end, one electron **may** be in the $1s$ on one hydrogen atom, and one electron **may** be in the $1s$ orbital on the other hydrogen atom. Alternatively, both electrons **may** be on one hydrogen atom and none on the other. Thus, in this method, electron configurations are combined and, in a case like hydrogen (H_2), *almost* all of the bonding lies between the nuclei but (by the Born–Oppenheimer[†] approximation)

---

*Walter H. Heitler (1904–1981) moved to the University of Bristol as a consequence of the rise of anti-Semitism in Germany, although he did his major work with Fritz London in Zurich on the VB method.

Fritz W. London (1900–1954), who made his major contributions to the weak forces of intermolecular bonding (London dispersion forces), was also forced to leave Germany (University of Berlin) for England, although he eventually enjoyed a professorship at Duke University, Durham, North Carolina.

[†]Max Born (1882–1970), Nobel Prize in Physics, 1954, worked at Göttingen until forced to leave by the Nazis because of his outspoken views. He subsequently held visiting posts at the University of Cambridge and the University of Edinburgh and became a British citizen.

J. Robert Oppenheimer (1904–1967) who held positions at the University of California, Berkley, California Institute of Technology and the Institute for Advanced Studies, Princeton, New Jersey, is more widely known for his work on the development of the atomic bomb and his subsequent efforts toward peace. The approximation that, in essence, separates nuclear and electronic components of the wave function was developed with Max Born in the late 1920s.

since electrons move faster than nuclei, there might be some small contribution to H_2 from something that might look like H^+H^-. In this regard, all of the theories concur in the (warranted) assumption that as long as the two bonded nuclei are identical (as in H_2), such ionlike contributions should be vanishingly small. However, when the two nuclei are not identical, the electrons need not be equally shared. Unequal sharing will be discussed briefly under the heading of Polar Covalent Bonds.

Again, for the case of H_2, as shown in Figure 1.8 (on the lower curve), when the two hydrogen atoms and their respective associated electrons (one electron on each) approach each other from a great distance, the energy of interaction will generally decrease, in the case of paired spins, reaching a minimum at the appropriate average equilibrium bond distance (74 pm or 0.74 Å [where 10 Å = 1 nm] for H_2). Closer approach causes a rapid rise in energy because of internuclear repulsion. The upper curve in Figure 1.8 depicts what happens if the electrons associated with their respective hydrogen atoms have unpaired spins. Thus, in the **VB method**, if two atoms are to form a bond, the atomic orbitals (AOs) associated with those atoms each deliver one electron; two (paired) electrons are engaged in bond making; and the orbitals will tend to overlap with each other because overlap facilitates an exchange of electrons and the total energy of the system is lowered (since the volume in which it is most probable to find the electrons has increased). Regrettably, the VB method underemphasizes the extent to which the energy levels of the final system are modified by covalent bonding.

The alternative picture of covalent bonding, which is of equal merit, is called the **molecular orbital (MO) method** and is attributed to the insight of Pauling and

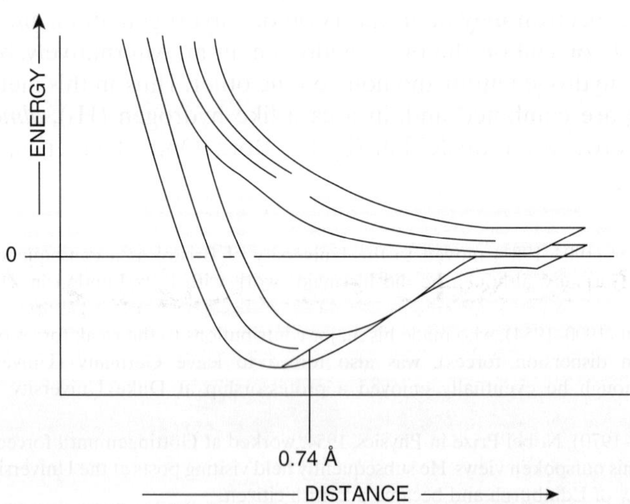

Figure 1.8. An energy diagram for hydrogen (H_2). Zero corresponds to two hydrogen nuclei at a very large distance. The lower curve corresponds to paired electrons and bonding, while the upper curve corresponds to unpaired electrons and antibonding.

Slater,* who independently suggested the same concept in 1931. Just as AOs are associated with one nucleus, a similar probability description can be built up for molecules (MOs). Generally, the situation is more complicated for MOs because as the number of nuclei and their potential arrangements in space increase, so do the possible combinations of wave functions, and the precise solution requires that the electron density be spread out over the entire volume of the molecule (instead of being confined to a particular atom or bond) to minimize the total energy. Therefore, in complicated cases, a limited form of the MO approach (**localized MOs**) resulting from a linear combination of atomic orbitals (**LCAO**) is used.

For example, qualitatively, using the MO method for the relatively simple case of hydrogen (H_2), it is suggested that the $1s$ AOs from the two identical hydrogen atoms (H_A) and (H_B), respectively, be combined so that they might overlap in the volume of space between the two nuclei to create MOs. Because the result of that overlap is symmetric to rotation about the axis joining the nuclei, it is called **sigma** (σ). Further, since orbitals can be neither created nor destroyed, the combination of two orbitals must yield two orbitals and, indeed, the combination of the wave functions (ψ) of the AO $1s$ orbitals to make MO wave functions (Ψ) can occur in two different ways.

Any MO that concentrates electrons between nuclei is called **bonding** and results from an in-phase or constructive overlap of the two AOs, that is, from the adding together of the wave functions for the AOs or, for H_A and H_B, bonding $\Psi_{MO} = (\psi_{1sA} + \psi_{1sB})$. Similarly, the MO formed by combining the same $1s$ AOs of different signs, the result of an out-of-phase or destructive overlap, that is, for H_A and H_B, **antibonding** $\Psi_{MO} = (\psi_{1sA} - \psi_{1sB})$, is also symmetric about the internuclear axis, is called **sigma star** (σ^*), and possesses a nodal plane halfway between the two nuclei. The presence of a nodal plane signifies zero electron density in the plane and requires that the electron density be concentrated near the ends of the molecule. The antibonding MO is less stable than the isolated hydrogen nuclei because the nuclei remain within bonding distance, **but without bonding**, and thus suffer internuclear repulsion. Figure 1.8 (*vide supra*) depicts the relative energies of the bonding (lower curve) and antibonding (upper curve) orbitals to internuclear separation, and Figure 1.9 is a representation showing the relative energies of the bonding and antibonding orbitals at their equilibrium internuclear separation.

Finally, although a number of assumptions have been made, both the **VB** and **MO** methods are currently justified by the fact that, over the years, they have led to predictions confirmed by experiment.

---

*Linus C. Pauling (1901–1994), professor, California Institute of Technology (Nobel Prize in Chemistry, 1954; Nobel Peace Prize, 1962) is the only person to win two *unshared* Nobel prizes. Independently from John Slater, he developed algorithms that describe atomic orbitals.

John C. Slater (1900–1976), Massachusetts Institute of Technology, developed the algorithms that describe atomic orbitals at the same time but independent from Linus Pauling. The use of "Slater-type orbitals" remains useful for depicting the regions in space near a nucleus where finding an electron is probable.

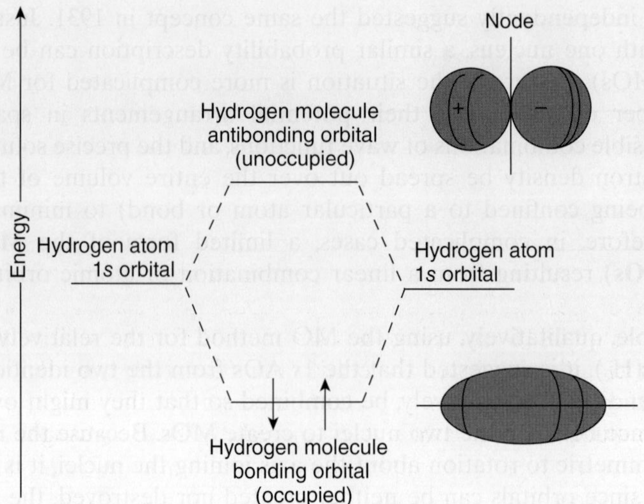

Figure 1.9. A schematic representation of the formation of bonding (σ) and antibonding (σ^*) molecular orbitals of hydrogen (H_2) by the combination of two equivalent $1s$ hydrogen atomic orbitals. The signs (+) and (−) do *not* refer to charges but rather to the sign of the wave function ψ, whose square (ψ^2) gives the probability of finding the electron(s) in the volume shown.

Problem 1.6. Using several schematic representations such as that shown in Figure 1.9 and the usual convention for electrons (i.e., up and down arrows, "↑" and "↓"), show the configuration for

(a) two uncombined hydrogen atoms,

(b) a hydrogen molecule with one electron (H_2^+),

(c) an excited hydrogen molecule with two electrons,

(d) a hydrogen molecule with three electrons (H_2^-), and

(e) a hydrogen molecule with four electrons (H_2^{2-}).

Problem 1.7. There are three different electronic configurations for the excited hydrogen molecule (part c above), all of which conform to the Pauli exclusion principle. Describe them using the notation of Problem 1.6. Which configuration do you expect to be the most stable?

II. Polar Covalent Bonds

It has already been pointed out that the twin concepts of **ionic bonding**, on the one hand, and **covalent bonding**, on the other, are idealized; that is, electrons are neither completely transferred nor (except perhaps for identical nuclei) equivalently shared. In the most general case, as suggested by Lewis* early in this century,

*Gilbert N. Lewis (1875–1946), professor, University of California, Berkeley, was a physical chemist who worked on the reformulation of thermodynamics but is most widely known for his proposals about the distribution of electrons around nuclei as proposed in *J. Am. Chem. Soc.*, **1916**, *38*, 762.

the noble gas configuration represents a particularly stable arrangement and, while such configurations might be arranged by the complete transfer of electrons, they can also be reached by electron sharing. Such sharing need not be equal. Indeed, hydrogen (H_2), which is composed of two identical atoms, and other such diatomic species composed of identical atoms (e.g., F_2, N_2) are atypical. What is done if the bonded elements are not the same? Consider then the case of hydrogen fluoride (HF). The electronic configuration of the highly electronegative* fluorine atom is represented as $[(1s^2)(2s^2)(2p^5)]$ and thus it might be expected to readily accept an electron to yield the anion (F^-) $[(1s^2)(2s^2)(2p^6)]$ from, for example, a hydrogen atom $[(1s^1)]$, and the resultant species H^+F^- would then be expected to be ionic. However, the gas HF does not possess all of the properties typically associated with ionic materials, and thus, following the suggestion of Lewis, we might suppose that the electron is not completely transferred from hydrogen to fluorine, but rather it is shared by both hydrogen and fluorine. Considering the one electron from hydrogen and the seven electrons with the principle quantum number 2 on fluorine, representations such as that shown below can be written as

$$H{:}\ddot{\underset{..}{F}}{:}.$$

From the MO point of view, a first approximation would be to consider the interaction of the $1s$ orbital on hydrogen (in which there is already one electron) and one of the $2p$ orbitals on fluorine (with five $2p$ electrons, i.e., fluorine is $[(1s^2)(2s^2)(2p^5)]$, we assign four of the five as paired in two $2p$ orbitals and the fifth in the last) while keeping in mind that fluorine has a high nuclear charge and that the energy levels around fluorine will not be equivalent to those around hydrogen. If we then take the z-axis as the internuclear line and allow the $1s$(H) and $2p_z$(F) to overlap, the σ bonding combination can be taken as $[1s(\mathrm{H}) + C_1 2p_z(\mathrm{F})]$, where the coefficient C_1 gives the relative weight of the $2p_z$(F) AO in the $\sigma_{\mathrm{bonding}}$ MO. In contrast to a covalent bond between two identical nuclei, the combining valence AO orbitals have different weights in the MO because they have different energies. The more

---

*The term **electronegativity** refers to the relative ability of an atom, in a bonding situation, to attract electrons to itself. A widely applied semiquantitative method is described by Pauling (*vide supra*), who suggested that a scale of eletronegativities could be created by considering that if bonding electrons between two nonidentical nuclei, X and Y, were equal, then it would be reasonable to expect the bond energy for XY to be the mean of $X_2 + Y_2$. Generally, it is found that $X - Y$ is almost always greater than that mean, and thus the extra contribution to that energy must be due to a partial ionic character. Relative values for a scale can thus be arranged. Given the "electronegativity" of hydrogen as 2.2, an abbreviated electronegativity table can be created:

H (2.2)								
Li (1.0)	Be (1.5)			B (2.0)	C (2.5)	N (3.1)	O (3.5)	F (4.1)
Na (0.9)	Mg (1.2)			Al (1.5)	Si (1.7)	P (2.1)	S (2.4)	Cl (2.8)
K (0.8)	Ca (1.0)	Cu (1.9)	Zn (1.6)		Ge (2.0)	As (2.2)	Se (2.5)	Br (2.7)
		Ag (1.9)	Cd (1.7)		Sn (1.7)			I (2.2)
			Hg (1.9)		Pb (1.6)	Bi (1.7)		

Taken in part from Pauling, L. *The Nature of the Chemical Bond*, 3rd edition, Cornell University Press, Ithaca, NY, **1960**, p. 93.

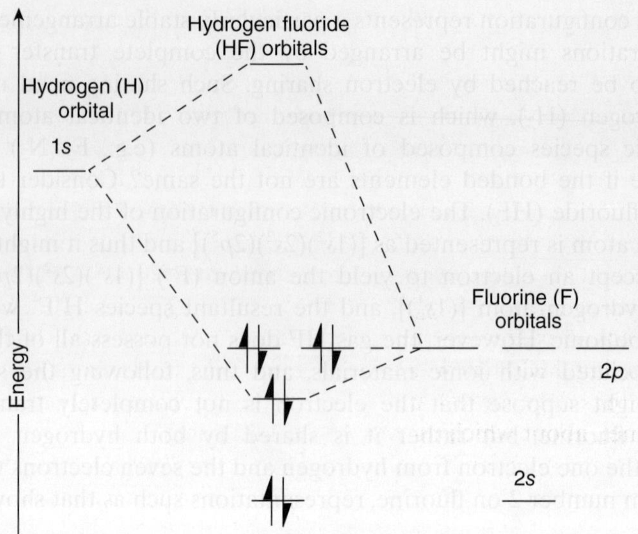

Figure 1.10. A schematic representation of a few of the molecular orbital energies in hydrogen fluoride (HF). The sigma (σ) and pi (π) bonding relative positions are shown, but only the sigma (σ^*) antibonding orbital position is provided.

electronegative fluorine will attract a greater electron "share" in the $\sigma_{bonding}$ orbital. In the same way, the antibonding sigma orbital (σ^*), taken as $[1s(H) - C_2 2p_z(F)]$, where C_2 gives the relative weight of the $2p_z(F)$ AO in the σ^* orbital, will reduce electron density between the nuclei, but with most near the hydrogen. The filled $2p_x$ and $2p_y$ orbitals are also available for bonding, but since there is only an s orbital on hydrogen, there is nothing with which they can overlap. The relative positions for some of the MOs of HF are shown in Figure 1.10.

A somewhat more sophisticated MO picture of a system as complicated as HF can also be created. In this picture, the LCAO is recognized as involving **all** of the electrons in the system. However, since an exact solution to the total energy of the resulting 10-electron system with two nuclei cannot currently be calculated, this self-consistent field (**SCF**) method assumes that the total energy can be approximated as some function that includes the potential energy of the repulsion of each nucleus toward the other, the attraction of each nucleus to each electron, the kinetic energy of each electron, and the potential energy of repulsion of each electron with every other electron. Using the Born–Oppenheimer approximation (*vide supra*), the kinetic energies associated with the nuclei can be ignored and the nuclei considered stationary. Since all of the electrons are moving relative to one another, the electron–electron repulsion energy is approximated in the SCF method by considering each electron to move in the average field of all of the other electrons. The MOs created in this way extend over the entire molecule and the resulting picture, even though restricted to s and p orbitals, is very different from that which contains more traditional-looking localized orbitals.

For HF, then, using the LCAO-MO-SCF method, there are a total of 10 electrons in the system that must be considered (nine from fluorine, originally $[(1s^2)(2s^2)(2p^5)]$, and one from hydrogen, originally $[(1s^1)]$), and they can be accommodated in five MOs (i.e., two electrons per orbital). The lowest-lying occupied orbital in HF appears to be the $1s$ orbital on fluorine and, (perhaps) in part because of the high nuclear charge, little mixing occurs. The next highest MO in HF has about the right energy for the $2s$ orbital in fluorine, but perturbation by the hydrogen $1s$ orbital is present; that is, the contribution of hydrogen to this orbital is small. In traditional terms, this suggests that the orbital is simply $2s^2$ (as shown in the simpler representation of Figure 1.10). The third MO contains a large bonding contribution from a fluorine $2p$ orbital and the hydrogen $1s$ orbital, **as well as some input from the fluorine $2s$ orbital (!)**. This suggests that the bonding from fluorine to hydrogen has mixed s and p characters, about which we shall say more later. The fourth and fifth bonding MOs (now accounting for all 10 electrons) have the same energies (are thus degenerate) and look like the traditional $2p$ orbitals (see http://www.webmo.net, about which more shall be said later).

Now, the ionic picture for HF, which supposes complete electron transfer, fails to account for the properties of the material but has merit on the basis of electronegativity arguments. The idealized Lewis dot structure more closely accounts for the lack of ionic character by suggesting a sharing of the electrons but formally neglects to account for the inequality of that sharing. The simplified MO picture is somewhat more detailed and it not only suggests that the bonding electrons are not equally distributed within the bond, but it also shows that the bond between hydrogen and fluorine can be considered as arising from an overlap of a $2p$ orbital from fluorine with a $1s$ orbital from hydrogen. Finally, the more comprehensive SCF-MO picture suggests that the bond between hydrogen and fluorine may actually involve a mixture of fluorine orbitals.

All of these pictures are usually thought of as complimentary approximations. The advantage to be gained by using one concept rather than another at any given time lies in developing an understanding of experimentally determined (i.e., observed) physical properties and reactivity (i.e., reality) so that predictions about other systems can be made. **It is important to keep in mind that all of the methods dealing with orbitals are approximations, as are the orbitals themselves**.

Rapid and convenient representations of such molecular systems, in which there is an unequal electron distribution in bonds (called **polar** bonds), are frequently represented simply with structures with an arrow drawn over the bond or by using the symbols (delta) δ^+ and δ^- to indicate partial positive and negative character, respectively:

$$+ \rightarrow - \qquad \delta^+ \delta^-$$

$$\text{H-F} \qquad \text{H-F.}$$

When such unlike charges (i.e., as q and $-q$) are separated, they constitute an electric dipole. The dipole moment, μ, is given by qR, where R is the distance between the charges. The dipole moment is a vector quantity. Polar molecules possess permanent dipoles. When the charge q is that on the electron (assumed to be a point charge) whose value is taken as 1.60×10^{-19} Coulombs (C) and the distance R is about $1\,\text{Å}$ ($1\,\text{Å} = 10^{-10}\,\text{m} = 100\,\text{pm}$), then $\mu = 1.6 \times 10^{-29}\,\text{C m}$. The units of

μ are generally given in debye (D)* where $1D = 3.34 \times 10^{-30}$ Cm, and thus $\mu = 4.8$ D for point charges 1 Å (100 pm) apart.

Molecules with permanent electric dipoles have moments that can be measured by applying an electric field to remove the degeneracy seen in their rotational spectra (Chapter 2). This is called the **Stark effect**[†]

Alternatively, an average dipole moment on a bulk sample can be obtained through measurement of the potential difference between charged plates with and without the dielectric present. The latter is, of course, a **macroscopic** property of the sample, which assumes that the microscopic structure, from which the dipole moment comes, does not actually exist; that is, the entire medium is "continuous" and isotropic. However, despite the assumption of continuity, this macroscopic property, ε, the **dielectric constant** of the medium, can be related to the dipole moment, μ, through the density of the medium, the number of molecules between the charged plates, and thus the induced dipole per unit volume (but taking into account an initial contribution of preferential orientation).

The direct measurement of the dielectric constant is straightforward. Indeed, the electrical equipment only requires a variable capacitor (of known capacitance) in parallel with the capacitor whose capacitance is to be determined (i.e., between whose plates is placed a known quantity of material whose dielectric constant is to be found).

III. Orbital Hybridization

In drawing structures for hydrogen (H_2) and hydrogen fluoride (HF), it has been possible to account for the location of all (both) of the nuclei in a linear fashion with bonds that are symmetrical about the internuclear axis (i.e., σ bonds). However, with methane, CH_4, which cannot be linear, a geometry problem arises.

When enough experimental information had been gathered,[‡] it became clear that all four hydrogen nuclei of methane (CH_4) are chemically identical; that is, they are all bonded to carbon and are incapable of being distinguished one from another by any chemical process. Thus, it was also clear that the most obvious way to minimize the total energy of the system would be to arrange the hydrogen nuclei around carbon on the surface of an imaginary sphere so that they were as far apart from each other as possible—to reduce any internuclear repulsions (as shown in Figure 1.11a). The radius of the sphere so created is the average carbon–hydrogen bond length (about 1.09 Å, 109 pm). This same arrangement would assure that the paired electrons **in** the bonds holding the hydrogen nuclei to the carbon would also be as far apart as possible (since once the electrons are paired in filled orbitals, they, too, repel other paired electrons, only two paired electrons being allowed in the bond).

---

*Named such in honor of Peter J. W. Debye (1884–1966), who was awarded the Nobel Prize in 1936 for "contributions to our knowledge of molecular structure through his investigations on dipole moments ..."

[†]Johannes Stark (1874–1957), Nobel Prize in Physics for his work on structural and spectral changes of atoms. His work is well described in *Ann. Phys.*, **1914**, *43*, 965.

[‡]The insights and massive contributions of Louis Pasteur, Friedrich August Kekulé, Archibald Scott Couper, et al. (all in the middle of the nineteenth century) to structural organic chemistry require fuller acknowledgment and will be elaborated upon subsequently.

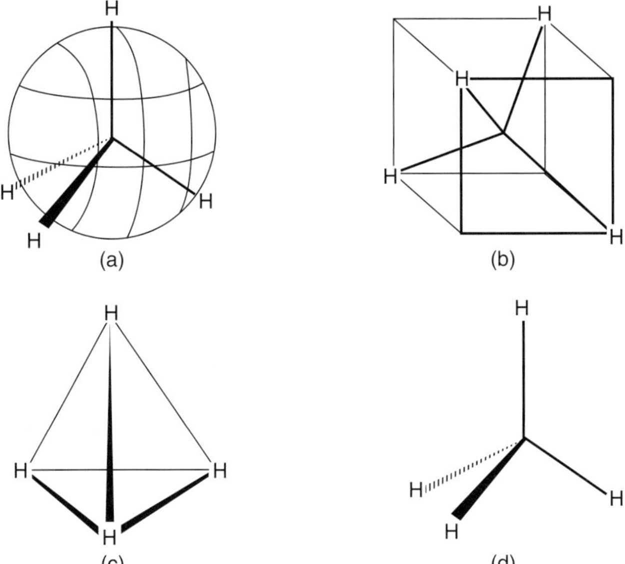

Figure 1.11. Representations of methane, CH$_4$, (a) as a tetrahedron inscribed within a sphere, (b) as a tetrahedral figure within a square, (c) as a tetrahedron, and (d) as four lines drawn to the corners of a tetrahedron. In (a) and (d), the heavy wedged lines ("flying wedge") are meant to indicate that the hydrogen (H) at the terminus lies **above** the plane of the paper. The dashed lines in the same figures are meant to indicate that the hydrogen (H) at the terminus lies **below** the plane of the paper. The two other lines in each figure show the hydrogen (H) at each terminus **in** the plane of the paper, respectively.

Alternative representations (all with the same internal H–C–H angle of about 109.5°) are shown (inscribed in a cube as shown in Figure 1.11b, inscribed in a tetrahedron as shown in Figure 1.11c, and, finally, as a stick figure as shown in Figure 1.11d.)*

The difficulty of imagining how carbon, whose atomic configuration of $[(1s^2)(2s^2)(2p^2)]$ could form four equivalent bonds, was overcome by the independent introduction of the concept of **orbital hybridization** (L. Pauling and J. C. Slater, *vide supra*) and was already introduced above (for HF as orbital mixing). Orbital hybridization involves determining which (if any) combinations of s and p orbitals on a given atom can be utilized to form **hybrid** AOs on that atom. Subsequently, the hybrid orbitals can be used to make bonds between the atoms. The idea is to allow for both better **orbital overlap** and formation of **equivalent** bonds. Better orbital overlap (making stronger bonds) results because orbitals, once hybridized, occupy a greater volume of space. Equivalent bonds are required because, experimentally, the hydrogens are incapable of being distinguished. Implementation of orbital

---

*Bonds are not static like the sticks that are often used to model them. Connecting nuclei with rods is an oversimplification. In each of these representations, the line joining the carbon atom to each of the hydrogen atoms designates a carbon–hydrogen bond, which, being symmetrical to rotation about the internuclear axis, is a σ (sigma) bond. The H–C–H angle so subtended, in each case, is one of the internal angles of a regular tetrahedron, that is, 109°28′.

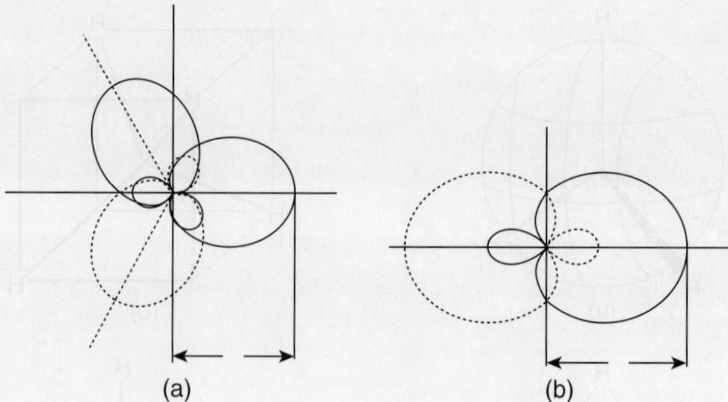

Figure 1.12. Representations of sp^2 (a) and sp (b) hybridized orbitals, respectively.

hybridization for carbon allows it to utilize the one $2s$ orbital and the three $2p$ orbitals to make fixed sets of hybridized orbitals consisting of any of (a) **four** orbitals, each composed of 25% s and 75% p (i.e., 1/4 s and 3/4 p, or $s^{1/4}p^{3/4}$, or sp^3); (b) **four** orbitals, three of which are hybridized orbitals, and each hybridized orbital is composed of 33.3% s and 66.6% p (i.e., 1/3 s and 2/3 p, or $s^{1/3}p^{2/3}$, or sp^2) *and* one unhybridized p orbital left over; and (c) **four** orbitals, two of which are hybridized orbitals, and each hybridized orbital is composed of 50% s and 50% p (i.e., 1/2 s and 1/2 p, or $s^{1/2}p^{1/2}$, or sp) *and* two unhybridized p orbitals left over. Pictorial representations of the three sp^2 orbitals and the two sp orbitals and the relative volumes of space they occupy are presented in Figure 1.12a,b.

The four sp^3 orbitals (as shown in Figure 1.13a) would then also be directed to the corners of a regular tetrahedron and would be available for overlap with the $1s$ orbitals from each of the four hydrogens forming the σ bonds necessary to make methane, CH_4 (Figure 1.13b).*

The power of the concept of localized orbital hybridization, combined with the simple ideas that (a) internuclear repulsion of nonbonded nuclei should be minimized and (b) electron pairs filling orbitals also repel other filled orbitals (sometimes called valence shell electron pair repulsion [**VSEPR**]) is enormous. It allows us (1) to make predictions about the geometry of molecules or, (2) knowing the geometry, to make predictions about areas in space where electrons might be, and (3) to justify, to ourselves, why and how reactions might occur. The following examples are illustrative of these concepts:

(I) The (highly reactive) species BeH_2 should be linear, with an H–Be–H angle of 180°. The hydrogens are equivalent and so the bonds are equivalent. The ground-state configuration of beryllium $[(1s^2)(2s^2)]$ cannot accommodate any bonds at all if we are restricted to using unpaired electrons to

---

*It should be clear that the idea of fixed, localized hybrid orbitals is simply another approximation used to attempt to rationalize experimental observation. The simplistic qualitative picture presented assumes, for example, that the combination of AOs cannot include fractional contributions to the molecular wave function. As a consequence, as will become clear, angular adjustments must be accounted for qualitatively. It should also be clear that, for the moment, all but the most gross symmetry arguments are ignored.

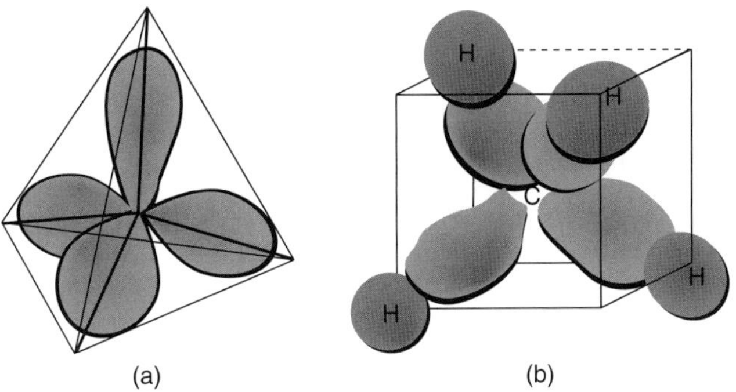

Figure 1.13. (a) A representation of localized sp^3-hybridized orbitals on carbon and (b) a representation of electrons localized in those sp^3 orbitals and bonded to hydrogen (using the $1s$ orbitals of the latter) to form σ (sigma) bonds in methane, CH_4.

make bonds and reach a noble gas configuration. If BeH_2 is to exist, the only way to make bonds (other than using hybridized orbitals) would be to "promote" one $2s$ electron to the $2p$ level (making an excited $[(1s^2)(2s^1)(2p^1)]$ state beryllium) and then to combine this state with the two hydrogens using their respective $1s$ electrons. This would produce an undefined H–Be–H angle and **nonequivalent** hydrogen-to-beryllium bonds. The more appealing picture using the hybridization idea of Slater and Pauling suggests that **equivalent and stronger** bonds would consist of hybridization of the one $2s$ orbital with one of the $2p$ orbitals to produce two sp orbitals (leaving the two other $2p$ orbitals unhybridized). Since the angle formed by two localized sp hybridized orbitals is 180°, if each one of the orbitals was then allocated one of the original $2s$ electrons from beryllium for bonding, it would now be possible to form two equivalent σ bonds using the $1s$ electron from each of the hydrogens. Indeed, this process predicts that all compounds of beryllium in which there are two ligands attached should have the same bond angle (Figure 1.14d). The arrangement can be described as **digonal**.

Problem 1.8. Using appropriate drawings, show why you would predict that $BeCl_2$ might also have a Cl–Be–Cl angle of 180°. Predict the carbon–beryllium–carbon angle in dimethylberyllium, $(CH_3)_2Be$.

Problem 1.9. If, regardless of the nature of the nucleus, appropriate s and p orbitals can be combined (i.e., the principle quantum number can be 2, 3, …), predict the F–Hg–F angle in HgF_2. What is the principle quantum number of the Hg orbitals used in hybridization?

The LCAO-SCF-MO approach, which must also yield a digonal system, suggests a different approximation for BeH_2. Using a total of five orbitals (the two $1s$ orbitals on hydrogen and the $1s$, $2s$, and one $2p$ orbital on beryllium) and the six electrons available, three occupied (containing two

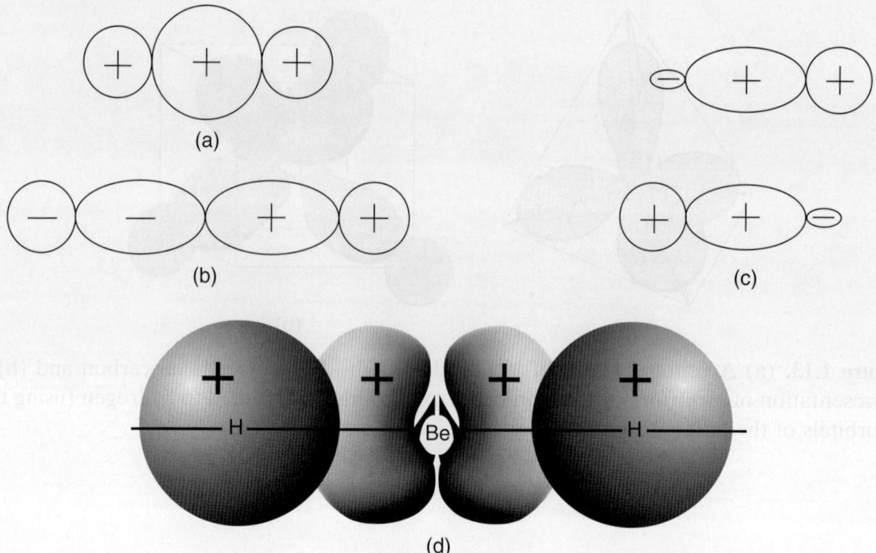

Figure 1.14. Symbolic representations of the SCF-MO bonding orbitals for beryllium hydride (BeH$_2$): (a) Be 2s and positive 1s H contribution, (b) Be 2p and negative 1s contributions, (c) beryllium hydride (BeH$_2$) halves shown as two sp orbitals from Be and two 1s orbitals from H, and (d) beryllium hydride (BeH$_2$).

electrons each) and two unoccupied (or *virtual*) MOs are created. The lowest-lying (occupied) MO is the 1s beryllium core, and this orbital is almost unaffected by the hydrogens. The next highest occupied (bonding) MO derives from the overlap of the beryllium 2s orbital and a positive contribution from *both* 1s orbitals on hydrogen (Figure 1.14a). The third, final, and highest occupied bonding MO arises from an overlap of the beryllium 2p orbital with the *negative* contribution of the two hydrogen 1s orbitals (Figure 1.14b).

These pictures clearly do not individually correspond to two sp hybridized orbitals on beryllium bonding as a result of an overlap with two 1s orbitals on hydrogen. However, if they are equally weighted, their positive and negative contributions give the same final result, which may be viewed as the sum of Figure 1.14c or, more traditionally, as Figure 1.14d.

(II) Considering internuclear repulsion or the repulsion between localized, paired electrons (VSEPR), molecular species, in which there are three ligands bonded to a fourth **and in which there are no other electrons, are presumed to be planar**, for example, boron trifluoride (BF$_3$) and (monomeric) borane (BH$_3$). This is confirmed experimentally. Boron, of course, has a $[(1s^2)(2s^2)(2p^1)]$ ground state. How then can it form three equivalent ligands? Since boron has only three electrons with the principle quantum number 2, in order to form three equivalent bonds, rehybridization to three localized sp^2 orbitals, arranged at angles of 120° to each other, can be effected. The result defines a plane to which the remaining unhybridized

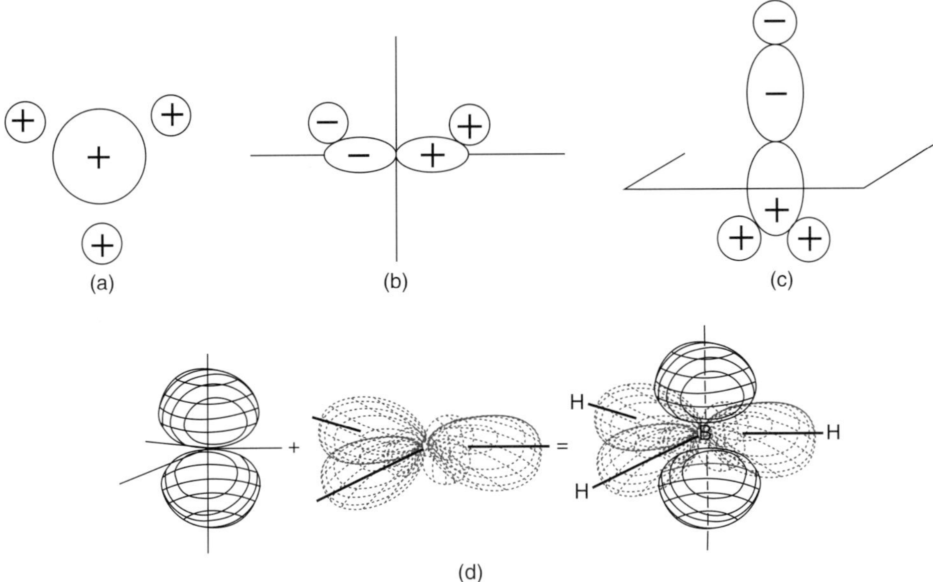

Figure 1.15. Symbolic representations for boron trihydride (monomeric borane), BH_3 of the SCF-MO bonding orbitals: (a) B $2s$ and positive $1s$ H contributions; (b) B $2p$ and one of two degenerate contributions from hydrogen (H); (c) B $2p$ and the other degenerate contribution from hydrogen (H); (d) boron trihydride (monomeric borane), BH_3 where the s-type orbitals on the hydrogens (H) have been omitted for clarity. The boron is in the plane defined by the three hydrogens (H).

$2p$ orbital is orthogonal. Utilizing one electron in each of the sp^2 orbitals makes all three equivalent and hence the bonds formed, using these electrons to pair with those brought in by the ligand, equivalent (Figure 1.15d). Such an arrangement is described as **trigonal**, and the bonds, each symmetrical about the internuclear axis, are sigma (σ). It might further be argued that the availability of a "vacant" $2p$ orbital, into which, in principle, two electrons might somehow be added, would encourage interactions of such trigonal species with electron-rich reactants. We will see that this is the case.

The LCAO-SCF-MO description for (monomeric) borane (BH_3) must also provide a trigonal species. As there are eight electrons (one each from the three hydrogens and five from boron), there will be four occupied MOs. The lowest-lying MO is primarily the $1s$ orbital on boron, only slightly perturbed by the three hydrogens. The second bonding MO is then derived primarily from the $2s$ orbital on boron and enjoys a positive contribution from all three hydrogens (Figure 1.15a). Bonding MOs three and four are of the same energy (degenerate); each has one node, and each is derived from overlap of one of the $2p$ orbitals on boron with a suitable linear combination of hydrogens; the first (Figure 1.15b) is shown with a node going through one of the hydrogens (which is thus not seen), and the second (Figure 1.15c) is at right angles (orthogonal) to the first. The

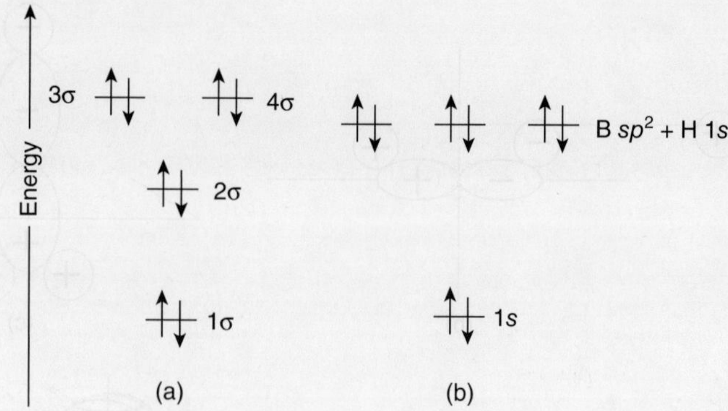

Figure 1.16. Two abbreviated schematic energy diagrams for monomeric boron trihydride (borane, BH_3): (a) delocalized MOs and (b) localized MOs. Bonding orbitals are signified as σ.

conversion of these MOs to localized MOs results in the three sp^2 orbitals given above using the ideas of localized rehybridization (Figure 1.15d). The schematic energy level diagram relating these two different approaches is shown as Figure 1.16.

Problem 1.10. Based on the conclusions drawn concerning monomeric compounds of boron, predict the hybridization in

(a) the methyl cation, CH_3^+;

(b) the methyl radical, $\cdot CH_3$; and

(c) aluminum trichloride, $AlCl_3$.

(III) As already noted, the ground-state configuration of carbon $[(1s^2)(2s^2)(2p^2)]$ requires rehybridization (using the localized two-center MO approach) if four localized, equivalent bonds are to be formed. Thus, four equivalent localized sp^3 orbitals, directed toward the corners of a regular tetrahedron, are expected. One electron can be placed in each of the sp^3 orbitals and sigma (σ) bonding to four equivalent ligands is effected (e.g., using four $1s$ orbitals, one from each of four hydrogen atoms, each with one electron). The geometry is described as **tetrahedral** (Figures 1.11 and 1.13).

The LCAO-SCF-MO delocalized picture for methane (CH_4) begins, as is usual with the basis set of five orbitals on carbon (one $1s$, one $2s$, and three $2p$) and the four hydrogen $1s$ orbitals, which yield nine linear combinations. There are 10 electrons (six from carbon and one each from each of the four hydrogens) that must fit into the nine combinations and, using them pairwise, we can fill the lowest five orbitals. The first and lowest-lying orbital is the carbon core MO consisting primarily of the carbon $1s$ orbital. The second and next highest MO is made up of the carbon $2s$ orbital and a positive combination of the four hydrogen $1s$ orbitals; this is shown in Figure 1.17a. The next three MOs are mutually degenerate. They are found

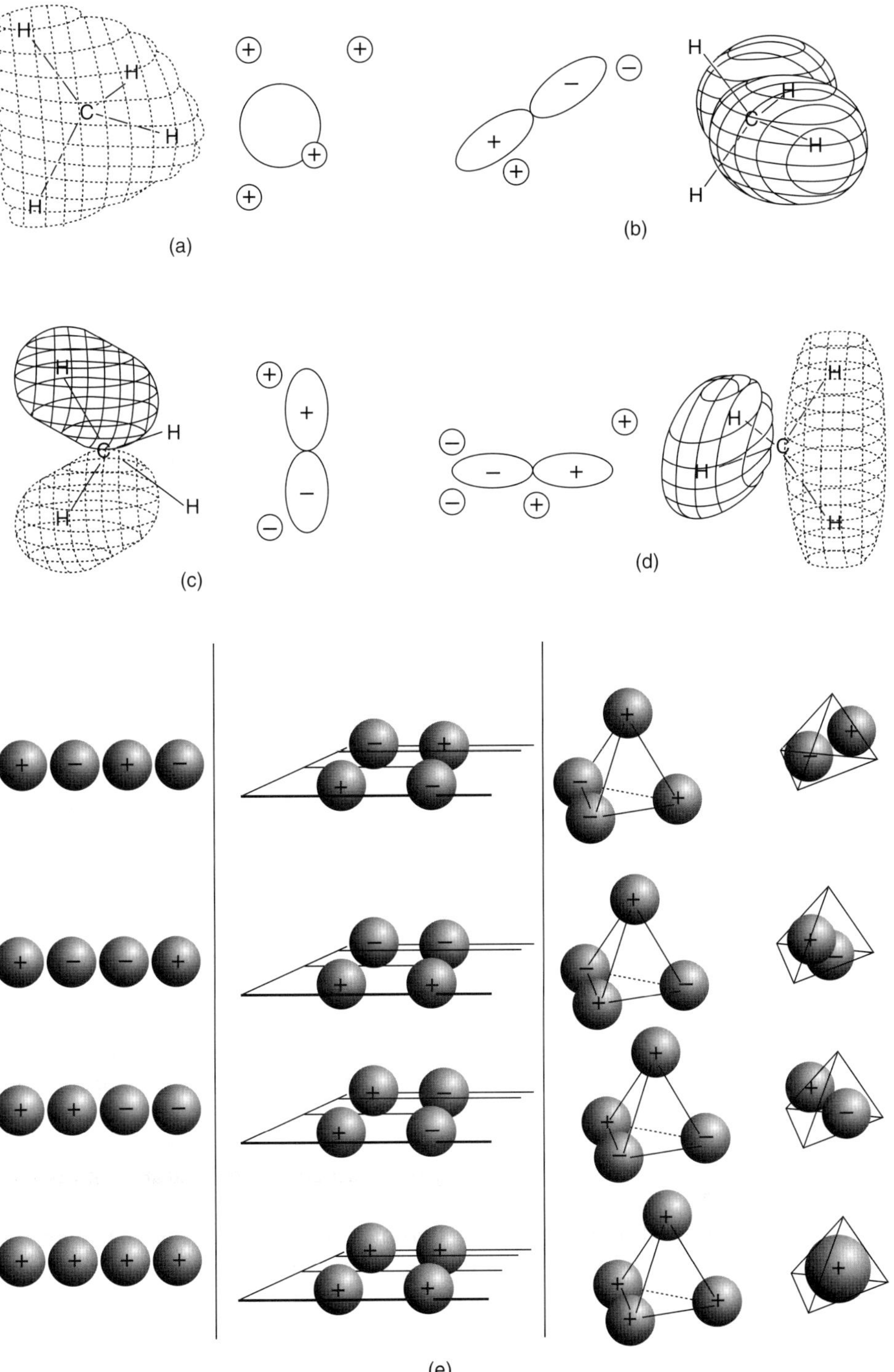

Figure 1.17. (a–d) Schematic representations of four (bonding) SCF-MOs of methane (CH₄). The representations were created using the MOViewer in WebMO version 6.0.002p. (e) Utilization of "unrehybridized" AOs on carbon and hydrogen to create four MOs in methane, CH₄. (See Hamrin, K.; Johansson, G.; Gelius, U.; Fahlman, A.; Nordlling, C.; Siegbahn, K. *Chem. Phys. Lett.*, **1968**, *1*, 613.

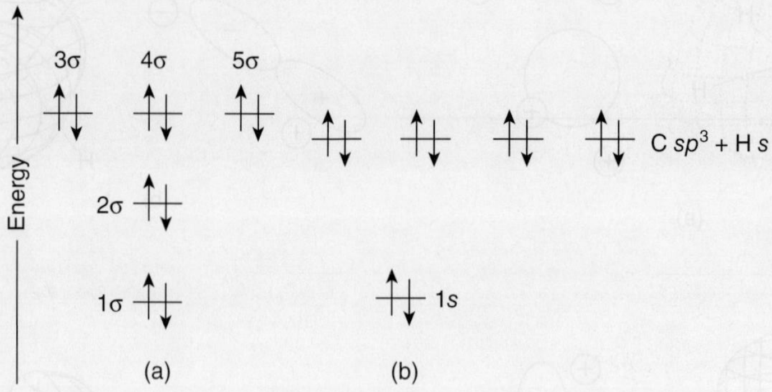

Figure 1.18. Two abbreviated schematic energy diagrams for methane, CH_4: (a) delocalized MOs and (b) localized MOs. Bonding orbitals are signified as σ.

in three mutually orthogonal planes, each with a single node and formed by a combination of each of the carbon $2p$ orbitals, with an appropriate combination of hydrogens. One possible schematic representation for these three is shown in Figure 1.17b–d. These MOs can be transformed into localized MOs that are approximately the same as those four resulting from an interaction of localized sp^3 orbitals from carbon interacting with the four $1s$ orbitals obtained, one each from each hydrogen. As shown in Figure 1.18, the heights of the energy levels using delocalized MOs does not correspond exactly to that from localized MOs, but the overall geometric result is the same.

An alternative way of viewing the same picture is to utilize unrehybridized carbon and four hydrogen (i.e., H_4) MOs made up of a combination of the four hydrogen AOs. If the four hydrogen (H) $1s$ AOs are arranged in a linear fashion, four MOs, increasing in energy (with increasing numbers of nodes) can be written as shown on the left-hand side of Figure 1.17e. An alternative arrangement (in a plane) is shown in the center of Figure 1.17e, and, finally, the same orbitals, now arranged as a tetrahedron (and depicted in relationship to the $2s$ and three $2p$ orbitals of carbon) are shown on the right of Figure 1.17e.

Thus, instead of viewing the system as the rehybridization of one $2s$ and three $2p$ AOs at carbon (ignoring the $1s$ core) to yield four new equivalent sp^3 orbitals at carbon and then permitting a combination of *each* of these with one, *each*, of the $1s$ orbitals of the four hydrogens, the view is one of "unrehybridized" carbon creating suitable MOs with a set of four "rehybridized" hydrogens.

(IV) Nitrogen (N) ground state $[(1s^2)(2s^2)(2p^3)]$ provides three electrons (one each in each of the $2p$ orbitals) and the remaining two electrons with principal quantum number 2 paired in the $2s$ orbital. Thus, nitrogen with three ligands, for example, as in ammonia (NH_3), might be expected to have a geometry in which the H–N–H angle is 90° (Figure 1.19a). However, in ammonia, NH_3, spectroscopic measurements (Chapter 2) show that the equilibrium H–N–H bond angle is about 107°!

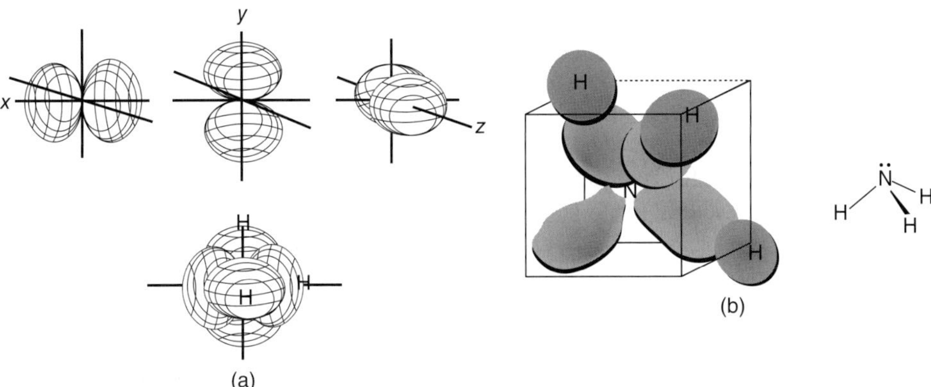

Figure 1.19. Representations for ammonia (NH₃). (a) Assuming H–N–H 90° angles, the bonding is N $2p$ – H $1s$. (b) Assuming H–N–H as nearly tetrahedral, the bonding is N sp^3 – H $1s$.

There are three possible explanations we can consider using localized orbitals. First, we might imagine that since filled, localized bonding orbitals repel each other and since nonbonded nuclei within bonding distance repel each other, these forces make the expected 90° angle (dictated by simply using the unhybridized $2p$ orbitals) wider. Alternatively, for a second choice, we might imagine that rehybridization has been effected to produce three localized sp^2 orbitals, making angles of 120°, with one localized $2p$ orbital, orthogonal to the plane of the three sp^2 orbitals and containing the remaining two electrons. Here we would then argue that repulsion between the pair in the localized $2p$ orbital and the pairs in the bonds holding the hydrogens to nitrogen cause a diminution in the H–N–H bond angle from the expected 120° to the observed 107°. Finally, granting rehybridization, we might imagine that four localized sp^3 orbitals are formed and that three of these, with one electron in each, are used to make σ bonds with the localized $1s$ orbitals of each of the three hydrogens, and that the fourth sp^3 orbital contains the remaining (paired) two electrons (in a *nonbonding* orbital). Here we might anticipate that (in concert with tetrahedral geometry) the H–N–H angle should be about 109.5° but that the angle is diminished by repulsion between the electron pair in the one localized sp^3 orbital with the pairs in the localized orbitals to which the hydrogens are bonded. In each case, with only an s orbital available on hydrogen, the resulting bond would enjoy the same symmetry along the internuclear axis and would be described as σ.

Generally, the explanation invoking sp^3 hybridization is preferred. The preference arises because (1) in geometric terms, 107° is closer to 109.5° than it is to 90° or 120°, so that there will be minimum additional distortion required, and (2) the overlapping ability of sp^3 orbitals is significantly better than that of p orbitals, and somewhat better than sp^2 orbitals, so that stronger bonds will be formed if rehybridization to localized sp^3 orbitals is effected.

Because of the distortion from a tetrahedron, the geometry is described as **trigonal pyramidal** (Figure 1.19b).

The presence of a pair of electrons in an sp^3 orbital directed toward an apex of the trigonal pyramid suggests that the reactivity of ammonia might be governed largely by interactions with electron-poor species, which would thus form a bond with the "nonbonding" pair [e.g., as in the ammonium ion (NH_4^+)].

In the LCAO-SCF-MO approach, ammonia is considered composed of three equivalent hydrogens and a lone pair. Thus, there are the usual five orbitals on nitrogen (one $1s$, one $2s$, and three $2p$ orbitals) and three orbitals, each $1s$, from each of three hydrogens. There are, as there were for methane (CH_4) above, 10 electrons. Again, there are only four orbitals and eight electrons once one pair of electrons has been assigned to the $1s$ core at nitrogen. Interestingly, the next two MOs result from linear combinations of three equivalent hydrogen $1s$ orbitals with the nitrogen $2s$ orbital *and* that $2p$ orbital lying along the threefold rotational axis passing through the vacant axis of the trigonal pyramid! As a consequence, these two MOs contribute both to bonding between nitrogen and hydrogen as well as to the characteristic reactivity of ammonia with electron-poor species. The last two MOs (compare the three highest MOs in methane) are mutually degenerate and are composed of the appropriate combinations using the remaining two $2p$ orbitals from nitrogen and the $1s$ orbitals from the hydrogens. As before, appropriate combinations of these delocalized MOs generate the localized, hybridized MOs typical of sp^3 orbitals.

(V) With oxygen [$(1s^2)(2s^2)(2p^4)$], the situation is very similar to that of nitrogen. The H–O–H bond angle in water is found to be 105°, and a variety of possibilities using hybridized and unhybridized orbitals exist. However, from the previous discussion, it could be concluded that of all the possibilities, the most likely results from hybridization to four localized sp^3 orbitals. Here, the hydrogens would be expected to be identical, forming an H–O–H angle of about 109.5°, and the remaining two pairs of electrons, each pair in one of the remaining two nonbonding, localized sp^3 orbitals, also forming an angle of about 109.5°. It is suggested that this is so because the best overlap, generating the σ bonds of the lowest energy, is obtained when the hydrogens are bonded using localized sp^3 orbitals from oxygen and the s orbitals from hydrogen. That the observed H–O–H angle of 105° is less than the tetrahedral 109.48° is then ascribed to repulsion between the pairs of electrons in the remaining localized, nonbonding sp^3 orbitals (Figure 1.20). Interestingly, the LCAO-SCF-MO approach suggests only a smooth increase in electron density in the plane of the lone pairs and, specifically, the presence of **two** discrete orbitals (not equivalent in energy) and both of which are occupied. Representations of these orbitals are provided in Figure 1.20a.

Problem 1.11. Based on the foregoing discussion,

(a) Draw an orbital picture of ethane (C_2H_6) or, better, H_3CCH_3.

(b) Label the σ bonds with their respective localized orbital notation, for example, a bond between hydrogen and carbon should be labeled with both an s and an sp^3.

(c) In another drawing, replace the orbital lobes with straight lines and draw a representation of ethane emphasizing the tetrahedral nature of each of the carbon atoms.

(VI) The compound ethene (also called ethylene), C_2H_4, is known to consist of a planar array with the geometry shown in Figure 1.21a,b.

As discussed (e.g., with boron trihydride [monomeric borane BH_3] above), localized trigonal hybridization with bond angles *approximating*

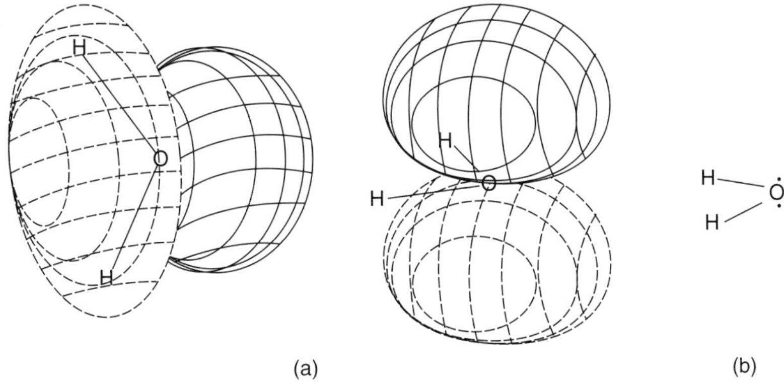

(a)   (b)

Figure 1.20. Representations of water (H_2O). (a) A representation of the occupied nonbonding orbitals and (b) a representation indicating that in addition to the two hydrogens attached to oxygen, there are also two pairs of nonbonded electrons. The representations were created using the MOViewer in WebMO version 6.0.002p.

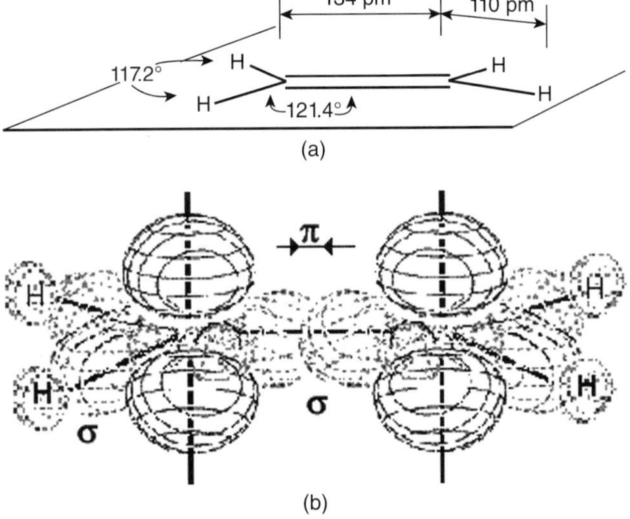

(a)

(b)

Figure 1.21. (a) A pictorial representation of ethene (ethylene), C_2H_4. (b) A cartoon representation of ethene (ethylene, C_2H_4) showing a backbone of sp^2-hybridized carbon and the unhybridized *p*-type orbitals orthogonal to the plane defined by the former. The representations were created using the MOViewer in WebMO version 6.0.002p.

120° suggests sp^2 hybridization. Unlike the case of boron (ground-state configuration $[(1s^2)(2s^2)(2p^1)]$), the ground-state configuration of carbon has one more electron (i.e., $[(1s^2)(2s^2)(2p^2)]$) so that if rehybridization is effected as suggested, the localized $2p$ orbital, which was *vacant* for the compounds of boron, must now contain one electron for this compound of carbon. The bonds formed from carbon to hydrogen, the latter having only the localized $1s$ orbital with which to bond, can be called sigma (σ) in accord with what has gone before.

Furthermore, as the H–C–C angle at *both* carbon atoms is close to 120°, it seems clear that we must argue that the bond formed from one carbon to the next is made by a constructive overlap of an sp^2 orbital from each. Such overlap would result in a bond, which was symmetrical about the internuclear axis and would therefore also be designated as sigma (σ). The $2p$ orbitals, with one electron in each, might now be expected to overlap. However, they cannot overlap end to end to yield a bond that is symmetrical about the internuclear axis; they can only overlap edge to edge (Figure 1.21b). Bonds formed from orbitals that overlap in this way, and which lack the symmetry of the sigma (σ) bonds, are called pi (π) bonds.

This orbital picture of ethene (ethylene, C_2H_4), in which there is to be a pi (π) bond arising from edge on overlap of localized, adjacent unhybridized p orbitals on adjacent sp^2-hybridized carbons, linked by a sigma (σ) framework, is an unusually valuable, but oversimplified, representation. Thus, having formed bonds, the approach argues that the electrons associated with each carbon are localized to their respective orbitals so that there is no mixing or exchange. Even with a "localized picture," it might reasonably be argued that the four electrons lying between the two carbons could be expected to hybridize to some new kind of bonding so as to make them equivalent. However, as we will see, first in Chapter 2 and again later, the approximation that there are *two kinds* of bonds such that either (1) there is no exchange between sigma (σ) and pi (π) electrons or (2) that if there is an exchange, it does not affect the overall energy of the system, works very well. In that vein, if we consider *only* the sigma (σ) and pi (π) bonds lying between the two carbon atoms as a separate subset of bonds in the molecule, then in terms of the traditional MO viewpoint, a picture showing both bonding and antibonding orbitals (Figure 1.22) can be constructed.

(VII) Ethyne (also called acetylene) corresponds to C_2H_2. Its structure is represented as tubelike with H–C–C angles of 180°. Spectroscopic evidence (Chapter 2) allows for the representation shown in Figure 1.23.

As was the case for beryllium, a digonal carbon atom suggests that the localized hybridization is sp. Unlike beryllium $[(1s^2)(2s^2)]$ discussed in (I) above, carbon $[(1s^2)(2s^2)(2p^2)]$ will have two electrons, one each in each localized $2p$ orbital even after hybridization has been effected. Using the two localized sp hybrid orbitals on each carbon (which will form stronger bonds than those that might be made by an overlap of other localized available orbitals), three σ bonds can form: one carbon–carbon bond and two carbon–hydrogen bonds. There are two localized $2p$ orbitals on each carbon available to form two, mutually orthogonal π bonds by sidewise overlap.

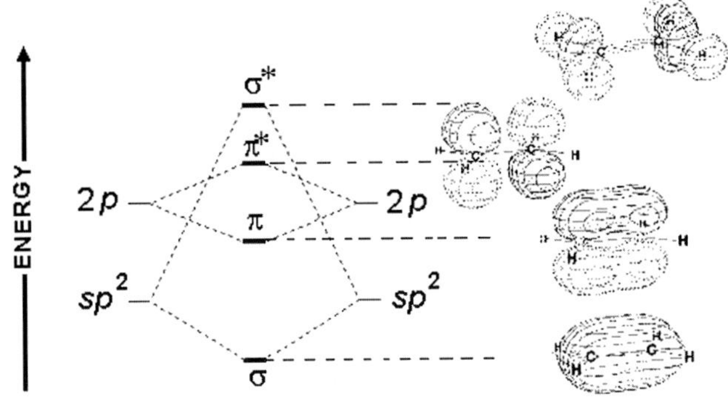

Figure 1.22. A schematic representation of the mixing of $2p$ and sp^2 orbitals in ethene (ethylene, $H_2C\text{–}CH_2$) to form σ and π bonding and antibonding (σ^* and π^*) orbitals and cartoon representations of those orbitals. The representations were created using the MOViewer in WebMO version 6.0.002p.

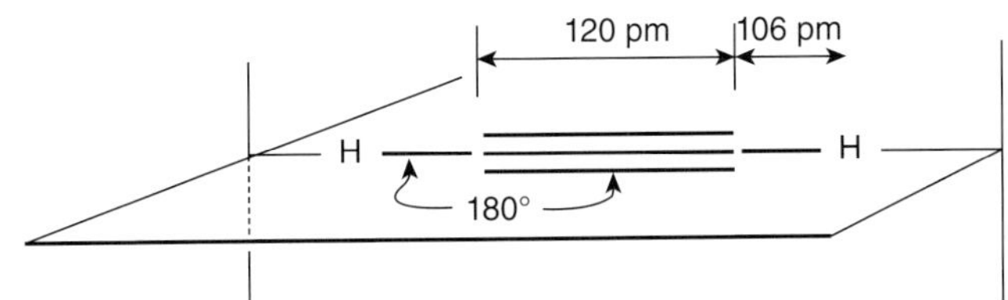

Figure 1.23. A pictorial representation of ethyne (acetylene, C_2H_2).

Again, as was the case with ethene (ethylene), the most widely used (because it seems to work best) approximation avoids rehybridization of the electrons between the carbons into some new kind of bonding in favor of the formation of σ and π bonds, with their corresponding σ^* and π^* antibonding orbitals (unoccupied in the ground-state configuration).

IV. Allotropes of Carbon

As was noted earlier in this chapter, the heat of combustion used for carbon in Hess's law (Equation 1.6) computations is a function of the allotrope of carbon used. The difference in allotropic forms is a function of the bonding present. Thus, as shown in Figure 1.24, the idealized lattice network of flawless diamond is that of a continuous and endless array of sp^3-hybridized carbons about 154 pm apart (ΔH_c° of $C_{(s,\,diamond)}$ is $-94.5\,kcal\,mol^{-1}$ [$-395.4\,kJ\,mol^{-1}$, $-32.95\,kJ\,gm^{-1}$]). In the same Figure 1.24 there is also shown the idealized concept of graphite, an endless array of sheets of planar carbon networks, the planes about 341 pm apart, but the carbon–carbon distance is now 142 pm. Here, in contrast to the idealized sp^3 hybridization found in diamond, it is argued that the hybridization at carbon is sp^2-like and that a continuous π network cloaks each planar sheet (ΔH_c° of $C_{(s,\,graphite)}$ is $-94.1\,kcal\,mol^{-1}$ [$-393.5\,kJ\,mol^{-1}$, $-32.79\,kJ\,gm^{-1}$]). Recently, a new family of carbon compounds, initially observed in interstellar gas clouds by infrared spectroscopy (Chapter 2, Part 3), and which are composed only of closed localized networks, have been isolated by

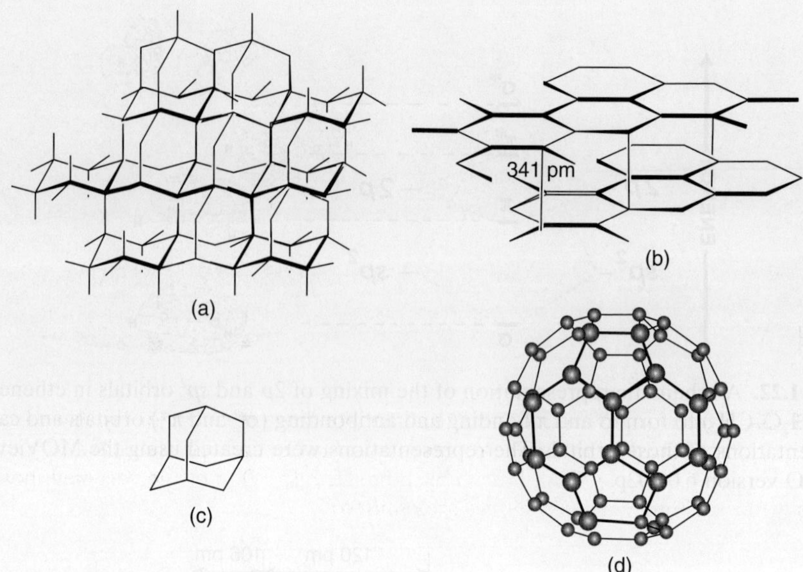

Figure 1.24. Allotropes of carbon. (a) A representation of a portion of the carbon network found in diamond. (b) A representation of a portion of the carbon network found in graphite. (c) A representation of the simplest basic unit of the diamond lattice. If there are hydrogens to fulfill the tetravalency of carbon, that is, producing $C_{10}H_{16}$, then this structure represents the *hydrocarbon* tricyclo[3.3.1.1^{3,7}]decane (whose *trivial* name is **adamantane**). (d) A representation of C_{60}, buckminsterfullerene.

heating graphite rods in an inert atmosphere. The bonding in these compounds, called **fullerenes** (after Buckminster Fuller [1895–1983], designer of geodesic domed structures, *c.f.* Pawley, M. Buckminster Fuller. New York: Taplinger (1991)), is presumed to be very much graphite-like. Figure 1.24 also depicts the structure of the truncated icosahedron C_{60} buckminsterfullerene, the value for the enthalpy of combustion being recently determined (ΔH_c° of $C_{(s,\,fullerene)}$ for C_{60} buckminsterfullerene itself is $-5639.2\,\mathrm{kcal\,mol^{-1}}$ [$-23{,}594.4\,\mathrm{kJ\,mol^{-1}}$, $-32.77\,\mathrm{kJ\,gm^{-1}}$]).*

V. Combination of Ionic and Covalent Bonding

Although there are a number of compounds whose aqueous solutions are characterized by the formation of ions, this discussion will be limited to consideration of sodium carbonate (Na_2CO_3).

Dissolution of sodium carbonate generates sodium cations and the carbonate anion (Equation 1.12):

$$Na_2CO_{3\,(aq)} \rightarrow 2Na^+_{(aq)} + CO_3^{=}{}_{(aq)}. \tag{1.12}$$

We will assume that the state of hydration of the anion can be ignored and that, for the moment, the anion itself can be considered as an isolated system. In that

*I am grateful to Dr. Y. H. Wing for early communication of this value. A later value is given as $25{,}899.1 \pm 14.2\,\mathrm{kJ\,mol^{-1}}$ (Rojas-Aguilar, A. *J. Chem. Thermodyn.*, **2002**, *34*, 1729).

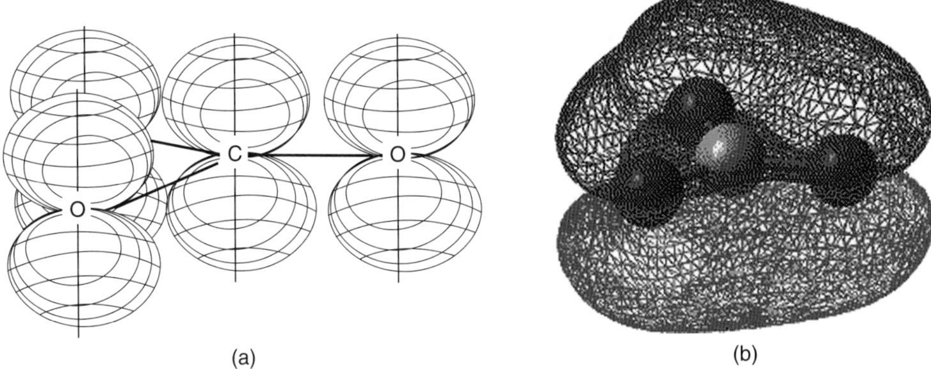

(a)   (b)

Figure 1.25. Pictorial representations of the carbonate anion ($CO_3^=$) showing that the carbon lies in the plane defined by the three oxygen atoms. (a) Orbitals (p) are shown for purposes of illustration only and the charges have been omitted. (b) One of the occupied molecular orbitals of the carbonate anion (charges are omitted).

anion, studies with isotopically labeled oxygen (e.g., ^{16}O, ^{17}O, and ^{18}O), as we shall see, demonstrate that all three of the oxygens are equivalent, and the structure of the anion is best represented by allowing the plane defined by the three oxygens to also contain the carbon (Figure 1.25).

Based upon the foregoing discussion of MOs, it is argued that the carbon atom might best be described with localized sp^2 hybridization (i.e., **trigonal**) and with one electron in the $2p$ orbital orthogonal to the plane of those three sp^2 orbitals. The bonds to each oxygen are symmetrical about the internuclear axis and are described as σ. However, what is to be done with the bonds *from* oxygen to carbon? Without considering hybridization, all three of the oxygens $[(1s^2)(2s^2)(2p^4)]$, presumably using $2p$ orbitals, could participate in forming the σ bonds to carbon (i.e., the σ bond would be sp^2 from carbon and p from oxygen), *but only one* oxygen could use a second one-electron-containing $2p$ orbital with which to form a π bond by sidewise overlap with the occupied localized $2p$ orbital on carbon (as with ethene [ethylene, $CH_2=CH_2$] and ethyne [acetylene, $HC\equiv CH$]). This is because even the π bond cannot contain more than two electrons. Thus, we are left with the two remaining oxygens, each containing one electron available for bonding, but without the ability to form a bond because there are no other orbitals available on carbon. However, this picture, in which two of the oxygens are different from the third, is contrary to the experimental observation that all three oxygens are equivalent.

While the problem can be approximately treated by examining the total MO description for the system in detail (e.g., see Figure 1.25b), that treatment is cumbersome. Further, it is clear that the problem results from the idea that bonding is localized and can only be described in terms of electron pairs. However, bonding electrons can be associated with more than two nuclei, and in such circumstances, stability is enhanced because when electrons can distribute themselves over a greater volume of space, the extent of orbital overlap, and thus bonding, increases. Compare, for example, the volume of space made available using sp^3-hybridized orbitals to the volume using only $2p$ orbitals (e.g., see Figure 1.12). However, it should be very clear that symmetry plays an important role, and the extent to which

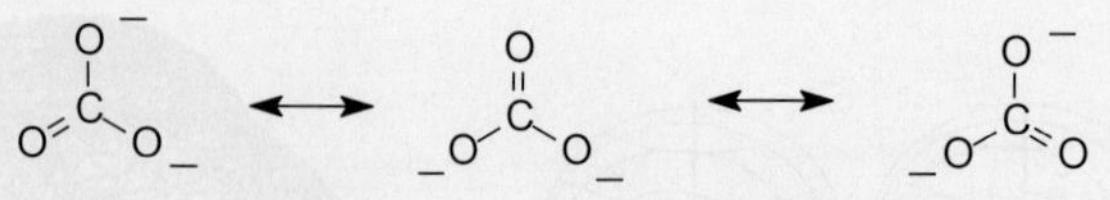

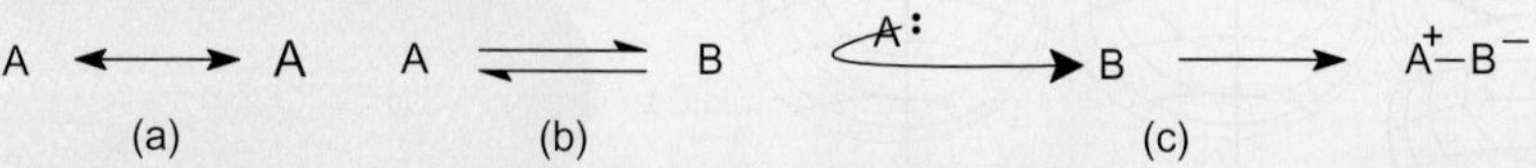

Figure 1.26. Typical resonance forms for the carbonate ($CO_3^=$) anion.

A ⟷ A   A ⇌ B   A: → B   → A⁺–B⁻

(a)   (b)   (c)

Figure 1.27. (a) Double-headed arrow, **used for resonance forms only**. The material "A" on the left has a different-type font than that on the right. They are different representations of the same letter and both are correct. (b) Two arrows, generally used to indicate **equilibrium**. **These two arrows are not to be used for resonance**. "A" and "B" are different from each other. (c) Curved arrows, indicating movement of an electron pair. A pair of electrons originating on A is used to form a bond from A to B. A new compound is formed ("A–B"), and A has become poorer in electrons (they were used to make the bond) and B has become richer.

such bonding occurs will be dramatically dependent upon the geometry or spatial orientation of the participating nuclei and orbitals.

The distribution of electrons over more than two bonds is called **electron delocalization** and the process is described as **resonance**. The qualitative picture of resonance generally uses VB structures such as those shown in Figure 1.26 for the carbonate anion ($CO_3^=$) and, in general, results from an inability to draw a single structure, with lines representing bonds, to depict the object; that is, none of the real pictures in Figure 1.26 can describe the carbonate anion ($CO_3^=$). Rather, the carbonate anion ($CO_3^=$), which nonetheless is real, is a hybrid of all of them. Alternatively, we are describing a real object in terms of unreal representations, all of which are necessary, but none of which is sufficient.

The rules for writing valance bond descriptions of a species like the carbonate anion allow for understanding what constitutes a reasonable set of structures and, with discussion, follow:

1. The individual resonance structures constituting the set are connected by a double-headed arrow (Figures 1.26 and 1.27a) or double-headed arrows. Double-headed arrows are not used for any other purpose and are **not** the same as, nor should they be confused with, two single-headed arrows (Figure 1.27b) that might be used to signify something else, for example, equilibrium.

2. The members of the set of structures have **no** individual reality. They are hypothetical representations of different electron-pairing schemes.

3. A curved arrow (or curved arrows, Figures 1.27c and 1.28) may be drawn on one representation to show "movement" of electrons from one location on (or between) atoms to another location on (or between) atoms, thus creating a new representation. **By convention, electrons "move" in the direction of the arrowhead**. If a curved arrow (or set of curved arrows) is used, (a) the nucleus at the terminus of the arrow (or set) becomes poorer in electrons and that at the head of the arrow (or set) becomes richer in electrons, and (b) an identical

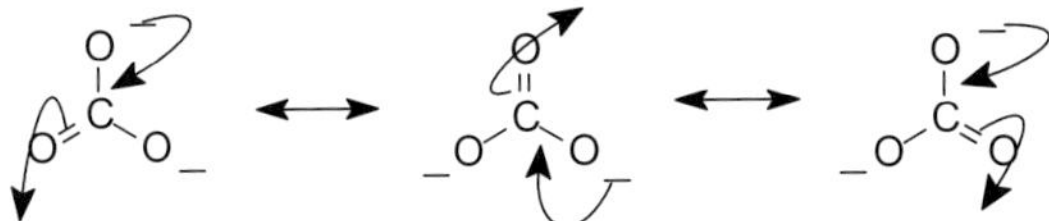

Figure 1.28. Some resonance forms for the carbonate anion ($CO_3^=$) (compare with Figure 1.26) connected by resonance arrows and with curved arrows indicating "movement" of pairs of electrons showing a path from one resonance form to another.

arrow or set of arrows may be drawn on the resultant structure, which returns to the initial representation.

4. All representations of the structures reached by the curved-arrow approach and connected with double-headed arrows **must** have all atoms in the same positions in space. Only electron "motion" is considered.

5. All representations in a set of resonance structures must have the **same number of paired and/or unpaired electrons**.

6. The amount that any specific resonance structure representation actually contributes to the overall hybrid will depend on the presumed energy of that contributor. Generally, the larger the number of identical representations of the lowest-energy contributor, the more stable the structure.

7. Separation of unlike charges on adjacent atoms should be avoided in writing resonance structures since, even though they are acceptable, they are not likely to be major contributors to the minimum energy representation.

8. Finally, it must be understood that the curved arrows are no more than an accounting technique. They must not be mistaken for reality and lent more credence than they are worth lest they be taken seriously as the movement of electrons.

NOTICE TO THE STUDENT

The curved-arrow approach, showing the "movement" of a pair (or pairs) of electrons, will be used subsequently, repeatedly, and relentlessly in other connections. It is worth your time and effort to practice (Problem 1.12).

Problem 1.12. Using the resonance method described, draw structures for NO_3^-, $SO_4^=$, $CH_3CO_2^-$, N_2O, NO_2^-, and N_3^-.

ADDITIONAL PROBLEMS

In the following problems, use $^{12}_{6}C = 12.00$ exactly, $^{1}_{1}H = 1.00$, and $^{16}_{8}O = 16.00$.

Problem 1.13. Benzoic acid ($C_6H_5CO_2H$) is often used as a primary standard for the calibration of thermometers (its melting point is 122°C), for calibration of adiabatic calorimeters (the energy of combustion of benzoic acid is $-26.41\,kJ\,g^{-1}$), such as that shown in Figure 1.5, and by analysts to calibrate their combustion trains (Figure 1.1).

(a) Why would it be necessary to "calibrate" a combustion train?

(b) In calibrating the combustion train, Ms. Anna List weighed out 7.23 mg of the acid and burned it in the oxygen atmosphere as described in Figure 1.1 of the text. What is the increase in weight she expected in the magnesium perchlorate? In the soda lime? To how many decimal points (significant figures) should the results be expressed?

Problem 1.14. In Problem 1.13 and in Table 1.1 (as well as elsewhere in the chapter), numerical values are given with an expressed degree of uncertainty. Why cannot we be more specific? What is the difference (if there is one) between "accuracy" and "precision?"

Problem 1.15. The standard heat of formation (ΔH°_{298}) of methane (CH_4) is frequently given as $-17.87 \pm 0.74\,\text{kcal mol}^{-1}$ ($74.70 \pm 3.10\,\text{kJ mol}^{-1}$). When graphite is sublimed to carbon atoms (i.e., $C_{(s)}$ to $C_{(g)}$), the heat evolved can be considered as the "heat of vaporization" of carbon, and it has been estimated at $170\,\text{kcal mol}^{-1}$ ($711\,\text{kJ mol}^{-1}$).

Using the information in Table 1.1 for the strength of the H–H bond, calculate ΔH° for the reaction $C_{(g)} + 4H_{(g)} = CH_{4(g)}$. Why do you suppose it is frequently argued that one-fourth of this number is significant?

Problem 1.16. In 1951, C. A. Coulson (1910–1974, professor of chemistry, Oxford University) pointed out that "It is the behavior and distribution of the electrons around the nucleus that give the fundamental character of an atom: it must be the same for molecules." Utilizing WebMO (http://www.WebMo.net) and a small basis set (e.g., STO-3G), create the MOs of hydrogen (H_2) and show that they can be approximated as in Figure 1.9. (Note that 1 hartree = $627.51\,\text{kcal mol}^{-1}$ = 27.25 eV.)

Problem 1.17. In 1951, C. A. Coulson (1910–1974, professor of chemistry, Oxford University) pointed out that "It is the behavior and distribution of the electrons around the nucleus that give the fundamental character of an atom: it must be the same for molecules." Utilizing WebMO (http://www.WebMo.net) and a small basis set (e.g., STO-3G), create the MOs of hydrogen fluoride (HF) and show that they can be approximated as in Figure 1.10. (Note that 1 hartree = $627.51\,\text{kcal mol}^{-1}$ = 27.25 eV.)

Problem 1.18. In 1951, C. A. Coulson (1910–1974, professor of chemistry, Oxford University) pointed out that "It is the behavior and distribution of the electrons around the nucleus that give the fundamental character of an atom: it must be the same for molecules." Utilizing WebMO (http://www.WebMo.net) and a small basis set (e.g., STO-3G), create the MOs of monomeric boron hydride (BH_3) and create a diagram of the orbitals such as in Figure 1.10. Note the symmetry in the sets of orbitals and their shapes. (Note that 1 hartree = $627.51\,\text{kcal mol}^{-1}$ = 27.25 eV.)

Problem 1.19. Interestingly, suppose you determined that boron hydride was largely dimeric (how might you do this?) with a structure as shown in the figure

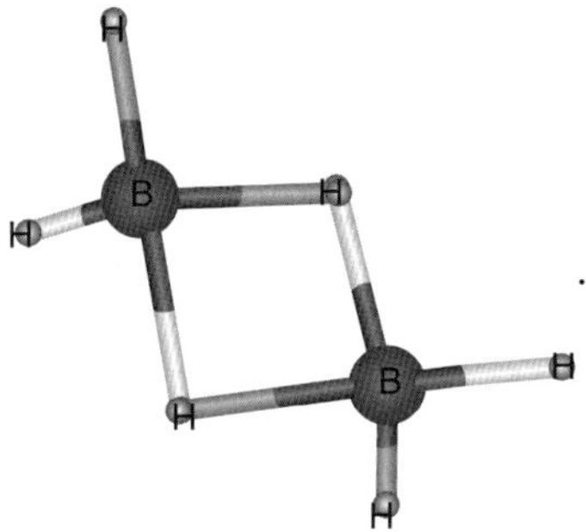

In place of the eight MOs (four bonding occupied orbitals and four unoccupied antibonding orbitals) seen in Problem 1.18, you will now find 16 orbitals. Optimize the structure of diborane at the restricted Hartree–Fock (RHF) 3-21G level (set the charge = 0) and then resubmit the structure at the STO-3G level for the MO picture. What does the structure of MO3 suggest?

Problem 1.20. The sulfate anion (SO_4^{-2}) is, apparently, tetrahedral rather than planar. Using the concepts of the "resonance arrow" and "curved arrow" notations, convince a reader that all four oxygens are identical.

Problem 1.21. The phosphate anion (PO_4^{-3}) is, apparently, tetrahedral rather than planar. Using the concepts of the "resonance arrow" and "curved arrow" notations, convince the reader that all four oxygens are identical.

Utilizing WebMO (http://www.WebMo.net), first draw the phosphate (PO_4^{-3}) anion and optimize the geometry (at the RHF/3-21G level, singlet, charge = −3). Then, with a small basis set (e.g., STO-3G), create the MOs of the phosphate anion (PO_4^{-3}). Note the symmetry in the sets of orbitals and their shapes. In particular, of the 29 MOs, 25 of which are doubly occupied, note the large energy gap between MO 9 and MO 10 and the shape of MO 10. (Note that 1 hartree = 627.51 kcal mol^{-1} = 27.25 eV.)

Problem 1.22. There is a completely different way to look at bonding. Consider the well-known story of *The Wizard of Oz* by L. F. Baum. There are four witches (before, so the story goes, Dorothy's house landed in Oz) on the sphere "Oz" who need to be as far away from each other and from the Wiz (whose throne is at the center of the sphere). L. F. Baum put them at the poles and the equator (east, west, north and south) where the angle subtended between any pair is 90°. At what angle would the separation be greater? After the "wicked witch" was dead, and knowing that three points define a plane, what angle is best? Could one argue that geometry dictates hybridization rather than vice versa? Why or why not?

REFERENCE

Hess, G. H. *Bull. Sci. Acad. Imp. Sci.*, **1840**, *8*, 257.

CHAPTER 2

An Introduction to Spectroscopy and Selected Spectroscopic Methods in Organic Chemistry

> All that we see or seem is but a dream within a dream.
>
> —Poe

A. GENERAL INTRODUCTION

It might have been profitable to the growth of organic chemistry if the advances in the isolation and purification of organic compounds had been coupled with similar progress in the design and creation of methods with which to unambiguously visualize their *structures*. However, in general, the idea of *organized structure* (or the way in which atoms might be connected to each other to form molecules) seems to have originally derived from finding materials that had the same elemental composition but different physical properties (these materials are called *isomers*, and several different kinds of *isomers* will be discussed in the next chapter). The attempts to actually "see" the cause(s) of the differences were frustrated by the lack of the appropriate tools. Thus, structural assignments were *deduced* by reactions and, as challenges due to structural complexity increased because more difficult problems were undertaken, analogies were made to structures that were presumed to be correct. Chemical reactions, which included such processes as oxidation, reduction, and fragmentation (sometimes called *degradation*) of large molecules into smaller ones, were commonly employed. Frequently, when it was possible to tentatively identify small fragments, imaginative constructs as to how they might be reconnected to form the original molecule were developed. Then, the use of additional reactions to combine fragments (the process called *synthesis*) so as to actually recreate what had been oxidized, reduced, or fragmented was explored. The results became incorporated into the framework of the science. Eventually, volumes of information on characteristic properties of various kinds of compounds were compiled. The correlation of that information built the body of organic chemistry, and appropriately, the arts of deductive analysis and synthesis jointly became massive undertakings that only the most ambitious might hope to accomplish. Indeed, as we shall eventually see, the logic involved and the interweaving of ideas and relationships lent much joy and satisfaction to the entire process. Even today, such efforts constitute a major portion of the study of organic chemistry, although much of the required insight and the leaps of imagination required to artfully construct an argument to define molecular structure have been supplemented by instrumental methods, which directly give information about structure.

The problem with structure determination, of course, is that molecules and their constituent atoms cannot be directly seen with the naked eye nor even with the best optical microscope. This is because the wavelengths of visible (VIS) light are much longer than the dimensions of the objects at which one is looking. The use of shorter wavelengths, appropriate for atoms and molecules, requires different detectors than our eyes.* Additionally, much information about bonds and, by inference, to what they are bonding can be obtained from other regions of the electromagnetic spectrum. Again, detectors other than our eyes are needed.

So, in that vein, the purpose of this chapter is to provide an introduction to a "box of tools." The tools and techniques for their use provide consistent (across the array of tools) evidence for structures that we write and the relationships between reactants and products in a variety of processes. **The tools will be used throughout the remainder of this text, singly or in combination, to justify conclusions about structure and reactivity.**

The major tools and techniques to use them currently available to organic chemists for examination of both atoms and bonds make use of only selected portions of the electromagnetic spectrum (Table 2.1). The particular areas have been developed largely because of their relative ease of access. However, almost all of the portions of the spectrum presented in Table 2.1, from wavelengths of meters to tenths of angstroms[†] ($10\,\text{Å} = 1\,\text{nm} = 10^{-9}\,\text{m} = 1000\,\text{pm}$), are used on some occasions and new techniques are being investigated to access the less used regions on a routine basis.

Additionally, in part because of the pervasive intrusion (or timely arrival) of microprocessors, tools for structure determination are constantly being refined and enlarged in scope. Examples include **multidimensional nuclear magnetic resonance (NMR)**, which is now commonly used for large molecules (the first commercial NMR spectrometers were introduced in the late 1950s) and, even more recently, **near-field microscopy**, which uses a lensless technique for VIS spectroscopy and thus sidesteps the normal problems of resolution[‡] by accumulating structural features a little at a time, is being developed (the first compound microscope became available in 1610).[§]

Historically (i.e., over the last half century or so), the general increase in the use of various portions of the electromagnetic spectrum, through creation of reliable

---

*As seen from Table 2.1, shorter wavelengths necessarily mean higher energies. The appropriate wavelengths for direct visualization require energies such that modification of the material at which one is looking might occur.

[†]Named after A. J. Ångström (1814–1874), professor of physics, Uppsala, Sweden. He is famous for his atlas of the solar spectrum.

[‡]The word *resolution* is taken to mean the ability to produce separate images of objects very close together.

[§]Developed and used by Galileo Galilei (1564–1642), Italian astronomer and mathematician, credited with demonstrating Copernican theory was worth dying for.

TABLE 2.1. A Portion of the Electromagnetic Spectrum

Frequency (v) Hz (cps)	Name	Energy (mol⁻¹)	Wavelength (λ)	Transition Studied
10^6 (1 MHz)	Radio	10^{-4} cal	333 m	Nuclear spin states (NMR)
10^7		10^{-3} cal	33.3 m	Molecular rotations
10^8	TV and FM	10^{-2} cal	3.33 m	
10^9 (1 GHz)		10^{-1} cal	33.3 cm	
10^{10}	Microwave	1 cal	3.33 cm	
10^{11}		10 cal	0.33 cm	Molecular rotations and vibrations
10^{12}	Far infrared	10^2 cal	3.33×10^{-2} cm	
10^{13}		1 kcal	3.33×10^{-3} cm	Molecular vibrations
10^{14}	Near infrared Visible	10 kcal Red (800 nm) to blue (400 nm)	3.33×10^{-4} cm	
10^{15}	Near UV	100 kcal 4 eV	3333 Å	Valence electrons $\pi \to \pi^*$ transitions
10^{16}		1000 kcal 40 eV	333 Å	
10^{17}	Far UV	1×10^4 kcal 400 eV	33.3 Å	Middle shell electronic transitions
10^{18}	X-ray	1×10^5 kcal 4 keV	3.33 Å 333 pm	Inner shell electronic transitions
10^{19}		1×10^6 kcal 40 keV	3.33×10^{-1} Å 33.3 pm	
10^{20}	γ-ray	1×10^7 kcal 400 keV	3.33×10^{-2} Å 3.33 pm	

and specific sources and detectors of radiation for those regions, has been recognized as the solution to the problem of molecular visualization. Thus, although the molecules still cannot be "seen," insight into their nature and the nature of the bonds that atoms make to each other to form molecules can be inferred by their interaction with electromagnetic radiation. In some cases, it is even possible to observe, or failing direct observation, deduce, the interactions of molecules or fragments of molecules with each other. The portions of the electromagnetic spectrum currently in general use by organic chemists produce **ultraviolet (UV)**, **VIS**, **infrared (IR)**, and NMR *spectra* (singular *spectrum*).* Because of their utility and widespread availability, these particular spectroscopic methods, as well as **mass spectrometry (MS)**, will be emphasized here and applied throughout the text. Further, although many materials are either noncrystalline or, if crystalline, are nonetheless not amenable to x-ray crystallographic techniques, a brief discussion of **x-ray diffraction spectros-**

*As used here, the word *spectrum* is taken to mean the result of the search for a transition (or transitions) between energy levels as plotted against the frequencies over which the search was conducted.

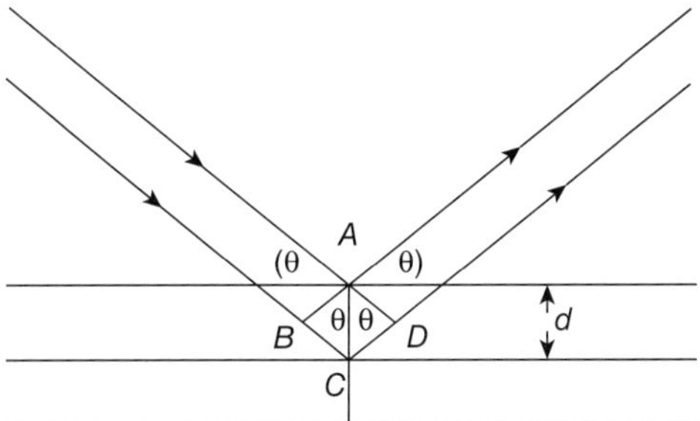

Figure 2.1. Diffraction in a set of parallel planes. For reinforcement of a wave, there must be an integer number of wavelength difference in the path lengths of rays scattered by successive planes. Thus, *BCD* must be a whole number of wavelengths. Since $BC = CD = d\sin\theta$ and $BCD = 2d\sin\theta$, reinforcement requires $n\lambda = 2d\sin\theta$, where n is an integer, λ is the wavelength, d is the distance between successive planes, and θ is the angle between the incident wave and the plane at which diffraction occurs. The statement $n\lambda = 2d\sin\theta$ is known as Bragg's law.*

copy and **Raman, microwave**, and **electron spin resonance (ESR)** techniques will be provided. As appropriate, reference will be made to some of the results derived from using these tools. The high energy and less frequently used methods of electron microscopy, and electron diffraction and neutron diffraction spectroscopies will not be developed here.

B. X-RAY CRYSTALLOGRAPHY

A crystal lattice contains molecules, and thus the atoms of which they are composed, in a structure characterized by a repetitive or periodic three-dimensional lattice, the smallest whole unit of which is called the *unit cell*. In many such crystalline materials, the layers of atoms in the lattice may be treated as a diffraction grating. The spacing in the crystal lattice is appropriate to the wavelength of x-rays.

So, a beam of monochromatic radiation of the appropriate wavelength is allowed to strike the crystal lattice. The radiation is generated by allowing high-energy electrons to impinge on a metal, usually copper or molybdenum, so that an electron is collisionally removed from a low-lying atomic orbital by the impact and photon-producing decay results. Then, because the spacing between adjacent planes of atoms in crystals is of the same order of magnitude as the wavelength of the radiation, the radiation can be diffracted by the electrons associated with the various atomic nuclei (Figure 2.1) and the intensity of the scattered beam measured at a suitable detector. The intensity (as a function of constructive interference) or angular position of the diffracted beam is dependent on the unique positions of the atoms in the crystal and the number of electrons surrounding the different nuclei. Therefore,

---

*Sir William Henry Bragg (1862–1942) and Sir William Lawrence Bragg (1890–1971), a "father and son team," were awarded the Nobel Prize (physics) in 1915 for demonstrating that there were, on the atomic scale, real particles and for showing that both x-ray and neutron diffraction could be used to help determine molecular structure.

it is possible to subsequently construct a three-dimensional electron density map of the time-averaged representation of the contents of all of the unit cells in the crystal and thus to define the structure (both position and atom identity) of the material making up the crystal.

C. PHOTON SPECTROSCOPY

I. General Introduction

The relative energies of bonding and antibonding orbitals and the general shapes of molecules were discussed in Chapter 1. Just as the transfer of electrons from low-lying atomic orbitals to higher-lying atomic orbitals requires energy, so too is energy required to excite electrons from nonbonding and bonding to antibonding molecular orbitals. The process, for both atoms and molecules, is called *excitation*. As with atomic spectroscopy, the absorption of energy by molecules is quantized and **photons** are the units of the (radiant) energy. Further, only certain values, corresponding to the availability of states into which the energy can be put, are available.*

Thus, we may ask how much energy is required to take a molecule (or electrons in a bond in a molecule if they are localized) from one energy state (E_1) to a different energy state (E_2). The answer ($E_2 - E_1$ or ΔE_{21}) clearly depends upon the distance between the states. Further, the answer corresponds to the absorption of a single quantum of radiant energy and is related directly to the frequency, ν, of the absorbed quantum and inversely to its wavelength, λ, by the relationship of Equation 2.1, where h is Planck's[†] constant (taken as 6.62×10^{-27} ergs) (and 1 erg = force of 1 dyne over a distance of $1\,cm = 1\,g\,cm^2\,s^{-2} = 1 \times 10^7$ J, so h, which scales the "mole" world down to the "molecule" world, has the value of 0.662×10^{-33} J s):

$$\Delta E_{21} = h\nu = h(c/\lambda). \tag{2.1}$$

One erg per molecule is equivalent to 1.44×10^{13} kcal mol^{-1}, and c is the speed of light (3.0×10^8 m s^{-1}). Thus, the amount of energy (kilocalorie per mole) involved in the absorption of one quantum by every molecule in a mole (one einstein) is a function of the wavelength (λ) as given by Equation 2.2:

$$\Delta E_{21} = 286,000/\lambda\;(\text{Å})\;\text{kcal mol}^{-1}. \tag{2.2}$$

The relationship among the wavelength, frequency, and velocity of light has been expressed well by Roberts and Caserio (1977, p. 265): "(consider) yourself standing

---

*The situation is much like a multilevel library in which books must be on a certain level, a specific shelf (a given distance from the floor of the level) and on a certain (imagined horizontal) position on that shelf. Such a book could not then be on a level that does not exist nor on a shelf that is not there. Moving such a book from, for example, the third shelf of the first floor to the fourth shelf of the second floor would presumably take more energy than moving it from the third shelf of the first floor to the fourth shelf on the same floor.

[†]Max Planck (1858–1947) was a German physicist who studied thermodynamics but recognized quite early the value of the concept of quantum theory. He was professor at the University of Berlin and received the Nobel Prize in Physics in 1918 for his discovery of "energy quanta."

on a pier watching ocean waves going by. Assuming the waves are uniformly spaced, there will be a uniform distance between the crests, which is λ, the wavelength. The wave crests will pass by at a certain number per minute, which is ν, the frequency. The velocity, c, at which the crests move by you is related to λ and ν by the relationship $c = \lambda\nu$."

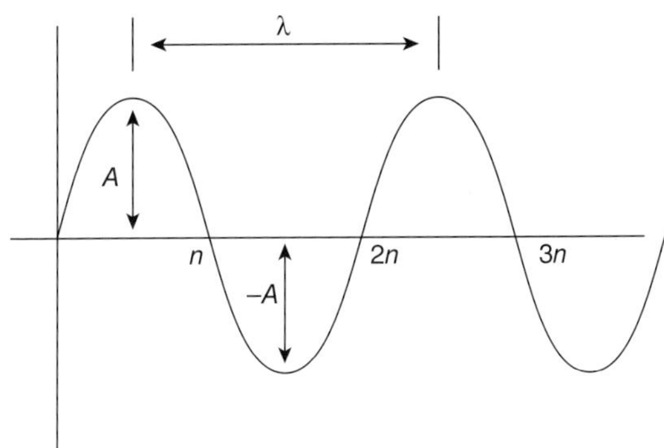

A simple sine wave of wavelength λ and amplitude A.

Molecular *spectra*, which are the outcome of the absorption of energy plotted against the frequencies over which the search for the transitions ΔE_{21} is made, are frequently complex. This complexity is due to the presence of many quantized states within molecules that lie so near to each other as to make them difficult to resolve. Ignoring nuclear and translational energies (the latter referring to the energy involved in entire molecules moving), the energy of a state is given by Equation 2.3; that is, it is the sum of the electronic, vibrational, and rotational energies making up that state:

$$E_{\text{total}} = E_{\text{electronic}} + E_{\text{vibrational}} + E_{\text{rotational}}. \tag{2.3}$$

Electronic transitions are frequently associated with UV and VIS spectroscopies and may involve energies of the order of $1\text{–}10^3\,\text{kcal}\,\text{mol}^{-1}$. With transitions of this magnitude, the population of a large number of vibrational states with energies of the order of $10^{-1}\text{–}10\,\text{kcal}\,\text{mol}^{-1}$ and rotational states with energies of the order of 10^{-4} to $10^{-2}\,\text{kcal}\,\text{mol}^{-1}$ found associated with each electronic state is expected. Similarly, transitions between vibrational states, associated with atoms at the ends of bonds (Chapter 1) moving with respect to each other, are observed by IR spectroscopy and will frequently involve the population of a large number of rotational states (far IR and microwave).

Light of short wavelength (λ) consists of large bundles of energy, and light of long wavelength consists of short bundles of energy.

A not-to-scale schematic diagram representing the relationships outlined above is shown in Figure 2.2.

For some of the portions of the electromagnetic spectrum in which organic chemists work, the frequency (ν) units, hertz (cps), are too small for convenience and thus the numbers for their expression are too large (1 million Hz = 1 MHz = $10^6\,\text{Hz}$) so that **wave numbers**, that is, the number of waves per centimeter, represented often as $\bar{\nu}$, is typically used instead; that is, the number of waves (or wave crests)

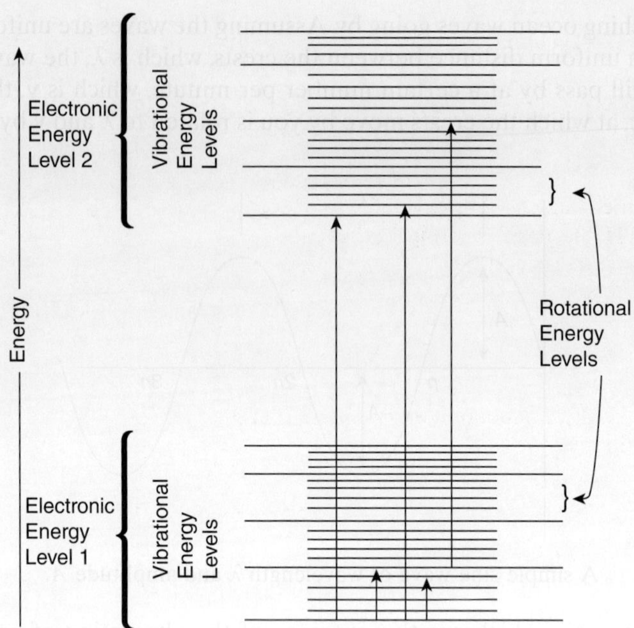

Figure 2.2. A not-to-scale schematic depiction of molecular electronic, vibrational, and rotational energy levels. The horizontal axis is used only to spread out the drawing.

per centimeter is the number of wave crests per second, the frequency (ν), divided by the wavelength (λ) in centimeter, and the units of $\bar{\nu}$ are thus cm^{-1}. So, the units used in optical spectroscopy are meters, centimeters, nanometers, micrometers, as well as hertz (or cps) and wave numbers, $\bar{\nu}$ cm^{-1} and $1\,m = 10^2\,cm = 10^6\,\mu m = 10^9\,nm = 10^{10}\,\text{Å}$, while $1\,Hz = 10^{-6}\,MHz = 3.3 \times 10^{-11}\,cm^{-1}$.

II. UV and VIS Spectroscopy

At the dawn of the nineteenth century, Ritter* announced the discovery of light beyond the violet end of the VIS spectrum. This higher-energy radiation (shorter wavelength) was called UV.

Absorption of energy in the UV region of the spectrum (wavelengths below 400 nm)[†] or in the VIS region (wavelengths between 400 and 800 nm) may be sufficient for the excitation of an electron from a stable, low-lying, bonding orbital to an unstable, higher-lying, antibonding orbital (Chapter 1). Because these energies are of the order of magnitude of those found in bonds themselves (Table 1.1 and Chapter 1), bond breaking can occur and reactions to make new bonds effected. The excitation, occurring from one electronic state of the system to another (Figure 2.2), will be more facile between states that lie closer in energy to each other than

---

*Johann W. Ritter (1776–1810) was an experimentalist who followed early studies on electrochemistry and radiation. He showed that light "beneath" the violet accelerated the rate at which silver metal darkened.

[†]This corresponds to energy greater than about 25,000 cm^{-1} or, since one wave number is equivalent to 2.86 cal mol^{-1}, about 71.5 kcal mol^{-1}.

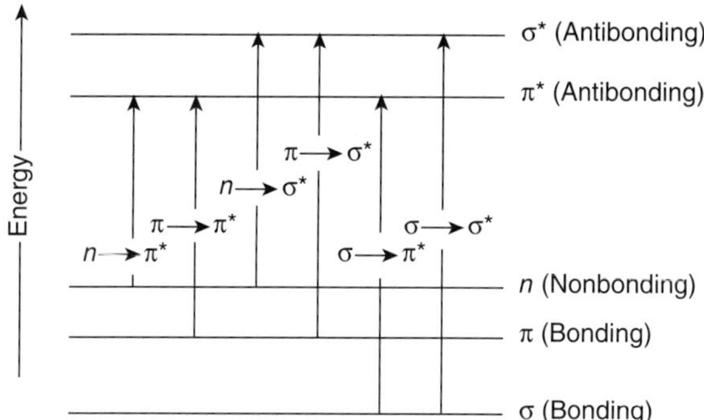

Figure 2.3. An unscaled representation of a reasonable sequence of relative orbital energies of bonding (σ and π), nonbonding (n), and antibonding (σ* and π*) orbitals and different kinds of transitions between them. The representation is drawn to suggest an increase in energy for transitions in going from left to right in the figure.

between states that are far apart (Figure 2.3). Further, as noted above, the energy of the entire manifold of vibrational and rotational states accompanying the lower-lying electronic state is mirrored by another set of vibrational and rotational states in the higher electronic level (just as a set of shelves and books on one floor of a library is mirrored by a similar set on the next higher floor). Then, since the magnitude of the energy changes for electronic transitions is so much greater than that corresponding to vibrational and rotational transitions (i.e., about a factor of 10 for the latter and 10^2 for the former), it is usually not possible to resolve the broad signals (called **lines** or **bands**) of electronic transitions seen in a plot of absorbance (measuring the energy absorbed) versus wavelength (Figure 2.4).

Based on the above discussion, hydrocarbons such as methane (CH_4) and ethane (CH_3CH_3), the bonding in which was discussed in Chapter 1, can only absorb energy in this region of the spectrum through an allowed σ to σ* transition (written as σ → σ*). On the other hand, ethene (ethylene, $CH_2=CH_2$), also discussed in Chapter 1, which, by the linear combination of atomic orbitals (LCAO) method, has a π bond as well as σ bonds, might be expected to undergo a π → π* transition (since corresponding appropriate π and π* levels are closer together than are the σ and σ* levels [Figure 2.3]). Accordingly, the maximum (λ_{max}) for methane (CH_4) (i.e., the region in which the efficiency of the absorption of energy is the greatest) that corresponds to the anticipated σ → σ* transition is found at a lower wavelength (higher energy, $\lambda_{max} \approx 120\,nm$) than that of ethene (ethylene, $CH_2=CH_2$, where a π → π* transition can be effected ($\lambda_{max} \approx 175\,nm$).*

As transitions become more facile, the wavelengths at which absorbance occurs become longer and spectra can be taken in the VIS region. For example, in aqueous solution, both chromate (CrO_4^{-2}), which is yellow, and dichromate ($Cr_2O_7^{-2}$), which is orange, have maxima (λ_{max}) below 400 nm but, both "tail" into the VIS, absorbing

---

*That such a transition is observed is counted as support for that molecular orbital picture (Chapter 1 and Figures 1.21b and 1.22), which suggests that the four electrons lying between the two carbons of ethene (ethylene, $CH_2=CH_2$) can be reasonably approximated with two different kinds of bonding with little electron mixing, that is, an sp^2 hybridized σ framework holding a separate π bond on it.

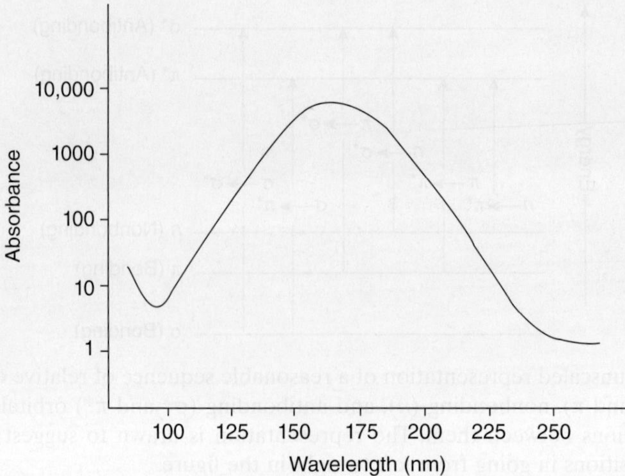

Figure 2.4. A representation of the UV spectrum of gaseous ethene (ethylene [$CH_2=CH_2$]) showing a maximum at about 175 nm corresponding to a $\pi \rightarrow \pi^*$ transition. The intensity of the absorbance (A) is defined as log I_o/I, where I_o is the intensity of the incident light, while I is the transmitted light.*†

in the violet and blue and thus appearing yellow and orange, respectively (Figure 2.5). As we will see, the colors of many organic compounds result from the presence of a group (or groups) of atoms known as *chromophores*. The chromophore, while necessary, is not always sufficient for absorption of light in the **VIS region** of the spectrum. Sometimes, depending upon the system, other attachments, called *auxochromes*, are needed to lower the energy of the system and to move the absorbance out of the UV (Table 2.2).

Finally, the sources of UV radiation in laboratory spectrophotometers are either hydrogen (^1H) or deuterium (^2H) lamps with quartz windows, which produce and transmit much of their useful radiation above 150 nm and below 400 nm. Incandescent sources generate their output in the VIS (>400 nm). Wavelength selection for the UV and VIS is accomplished with ruled gratings or (more rarely today) prisms. Photons coming through the sample in a quartz cell (or cuvette) are counted either with photomultiplier detectors or diode array systems. The resulting voltages are amplified and passed directly either to software for manipulation and then to a

---

*It is often found, within certain concentration limits, that the intensity of the absorption is proportional to both the concentration, c (mole per liter), and the thickness, l (centimeter), of the sample in the beam; that is, the absorbance = εlc, where ε is defined as the molar extinction coefficient. This relationship is referred to as Beer's law or the Beer–Lambert law. The law is named after August Beer (1825–1863), a lecturer in Bonn who studied optics, and Johann H. Lambert (1728–1777), a Swiss mathematician. It has been suggested that Beer's law was initially discovered by the French mathematician Pierre Bouguer (1698–1758). Lambert made reference (with attribution) to it and, much later, Beer extended it to its present form.

†This particular portion (i.e., where $\lambda < $ ca. 200 nm) is often called the "vacuum UV." Special experimental techniques are necessary, including evacuation of the sample chamber, because the common atmospheric gases (i.e., O_2, N_2, and H_2O) have nonbonded electrons and thus their $n \rightarrow \pi^*$ transitions occur and those of the sample might be obscured.

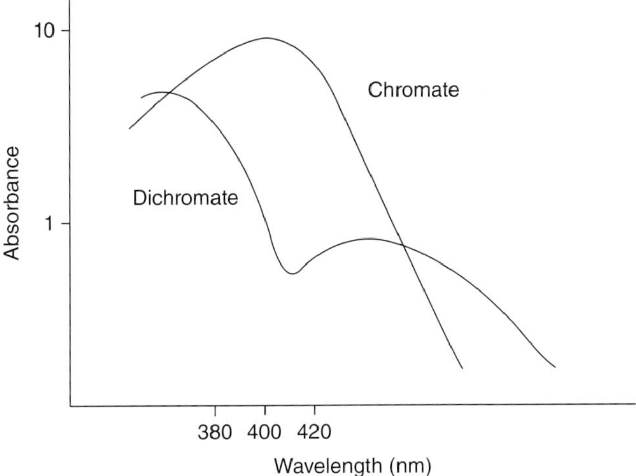

Figure 2.5. An approximation of the absorption spectra of potassium chromate (K_2CrO_4) in dilute aqueous potassium hydroxide and potassium dichromate ($K_2Cr_2O_7$) in dilute aqueous sulfuric acid (H_2SO_4) solutions, respectively.

TABLE 2.2. An Approximate Relationship between Visible Color Absorbed and the Color Seen as a Consequence

Wavelength (λ) (nm)	Color Absorbed	Color Seen
≈ 400	Blue	Yellow
≈ 500	Green	Red
≈ 600	Yellow	Blue
≈ 700	Red	Green

display device or to a digital-to-analog converter and then to a pen for printing on paper.

III. IR Spectroscopy

Although it might be assumed, based on the discussion in Chapter 1 or on the examination of macroscopic models, that bonds between atoms in a molecule are static objects like pieces of wood or plastic, in reality (except perhaps at absolute zero), atoms in molecules are in constant but restricted motion.* This motion, called **vibrational**, is of two types: *stretching* and *bending*.

Absorption of IR radiation (found "beneath" the red by Sir William Herschel[†]) causes transitions between vibrational states.

---

*The nuclei were considered static objects when discussing electron motion as the latter can be treated much more conveniently that way (the Born–Oppenheimer approximation; see Chapter 1).

[†]William Herschel (1738–1822), British Astronomer Royal. Although born in Germany, he emigrated to England where, it has been argued, early musical training, indomitable studiousness, and more than a touch of genius influenced him to become interested in astronomy. In addition to the discovery of the planet Uranus and its moons, double stars, and the movement of our solar system, he found the radiation beneath the red, that is, IR, while searching for solar radiation filters.

Very approximately, a bond between two nuclei can be considered to be like a mechanical spring. By Hooke's law (R. Hooke [1635–1703], British physicist), the two atoms and their connecting bond are treated as a simple harmonic oscillator composed of two masses joined by a spring. The bond vibrates with a stretching frequency, ν, given by Equation 2.4, where c is the velocity of light, k is the force constant (stiffness) of the bond (spring) in its normal equilibrium position and is thus a measure of its "strength," and M, the reduced mass, is given by Equation 2.5 (m_1, m_2 = masses of the objects attached to the ends of the spring):

$$\nu = (2\pi c)^{-1}(k/M)^{1/2} \tag{2.4}$$

$$M = m_1 m_2 / (m_1 + m_2). \tag{2.5}$$

It follows from Equations 2.4 and 2.5 that if m_1 and k are held essentially unchanged and the other mass, m_2, is increased, the frequency will decrease. Alternatively, if we hold the masses constant but increase the stiffness of the spring (i.e., make k larger), the frequency will increase. Thus, for example, if in methane (CH_4) one or more of the protium (^1H) attached to carbon is replaced by deuterium (i.e., from ^{12}C–^1H to ^{12}C–^2H), the frequency at which stretching is observed in the spectrum would be expected to decrease. Alternatively, as the bonds between two carbon atoms change from single to double to triple (e.g., ethane [CH_3CH_3] to ethene [CH_2=CH_2] to ethyne [CH≡CH]), they would presumably get stronger, have respectively larger values of k, and the alkyne would absorb at higher frequency than the alkane with the alkene somewhere in between, as is found.*

Generally, for molecules with n atoms (where $n > 3$), there are $2n - 1$ stretching modes and $2n - 5$ bending modes for a total of $3n - 6$ modes. Generally, also, it is easier to bend a bond (i.e., occurs at lower frequency) than it is to stretch one. While some bending modes are identical to others (because the direction in which a molecule bends may be unimportant), thus simplifying the overall picture, it remains true that for large molecules, spectra are complex. Despite these complexities, the overall features of most spectra in the region between about $4000\,cm^{-1}$ and about $1200\,cm^{-1}$, however rich in detail, can be assigned with some certainty. From about $1200\,cm^{-1}$ to about $400\,cm^{-1}$, absorption bands are frequently associated with complex vibrational *and* rotational phenomena, some of which may be characteristic of the molecule as a whole. This portion of the IR spectrum is called the "fingerprint" region. Although it is certainly true that any two molecules that are identical will have identical fingerprints, it is not always possible to make clear assignments or to analyze this region completely. As an example, although we will be considering peaks of diagnostic value when appropriate, Figure 2.6 presents a portion of the IR spectrum of gaseous propane ($CH_3CH_2CH_3$) (which is the next member of a series beginning with methane [CH_4] and continuing past ethane [CH_3CH_3]). It is clear that C–H stretching can be seen (centered at around $2970\,cm^{-1}$), as is typical for all

---

*Although the general principle outlined is correct, the argument for these *particular* compounds is somewhat disingenuous. IR absorption bands for electrically symmetrical molecules are forbidden. Thus, diatomics such as halogens and O_2 and N_2 do not absorb in the IR, and both ethene (CH_2=CH_2) and ethyne (HC≡CH), along with other symmetrically substituted *alkenes* and *alkynes*, have notoriously weak signals. This is discussed further along with some aspects of Raman spectroscopy.

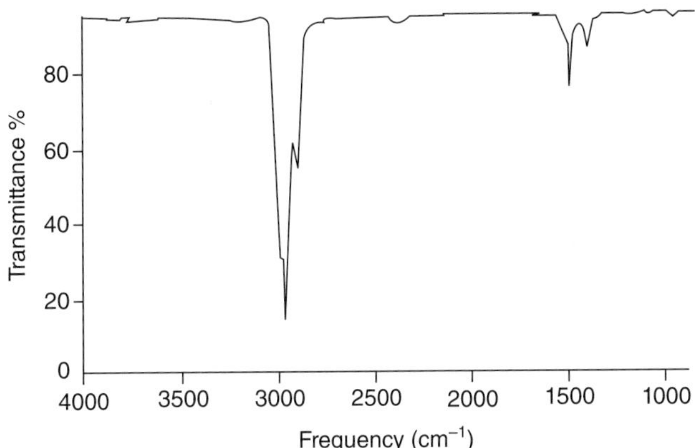

Figure 2.6. A gas-phase infrared spectrum of 8.5 torr of propane in a 10-cm gas IR cell at 25°C. The spectrum was taken on a Digilab FTS-40 FT-IR spectrometer. (see also Figure 2.8).

hydrocarbons (the range for C–H stretching being between about 3200 and 2800 cm^{-1}), and that the other expected peaks for the C–H bend (centered near 1465 cm^{-1}) are also present. The C–C stretch (centered about 910 cm^{-1}) is too weak to be seen. Finally, IR spectra of gases (such as in Figure 2.6), liquids, and solids (in liquid solutions or solid solutions, e.g., in potassium bromide [KBr]) can be obtained.

The source of IR radiation can be any incandescent device but is usually an electrically heated rod of ceramic material prepared by sintering oxides of cerium and zirconium. When hot, these materials emit significant radiation in the 200–10,000 cm^{-1} region and are ideal since the frequencies between about 400 and 4000 cm^{-1} are used for routine analyses. In one method of analysis (called *continuous wave* [CW]), wavelengths needed for analysis are selected with either an appropriately ruled grating (or sets of gratings) or with a prism. However, since silica glasses absorb IR radiation, if a prism is used, it is made from a large, single crystal of salt, cut and polished appropriately. Thermal detectors (such as a blackened piece of gold foil), which respond to heat (as a function of the number of photons received) and the conversion of that energy to an electrical signal (voltage), are common. With the widespread availability of lasers and microprocessors, newer machines use a different method.

Based on the understanding that radiant energy is composed of trains of electromagnetic waves, many frequencies of which are superimposed, it should be possible, in principle, to allow the IR beam to pass through a sample, fall on a suitable detector, and, by plotting the response as a function of time, to obtain all of the information carried by the beam. Measurement and storage of that information before a sample is placed in the beam, remeasurement after a sample is placed in the beam (where now some of the frequencies are missing because they are absorbed by the sample), and ratioing (i.e., taking a ratio) of the results will then dictate which frequencies were absorbed. In practice, however, with frequencies of the order of 10^{13}–10^{14} Hz (Table 2.1), the idea fails because there are no detectors available that can respond fast enough.

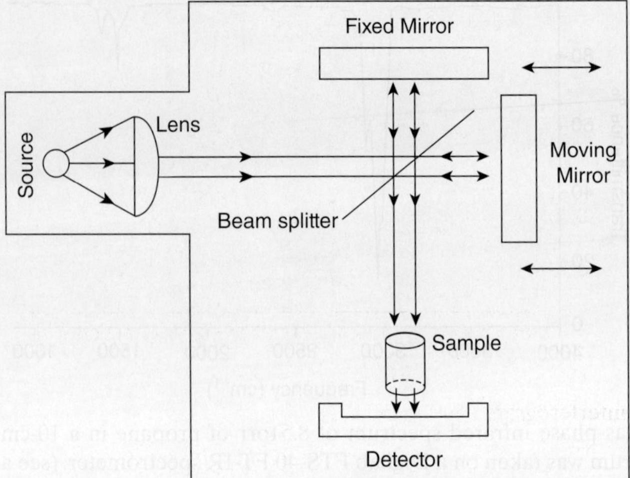

Figure 2.7. A depiction of a Michelson interferometer as used in a Fourier transform (see footnote on page 59) infrared (FT-IR) spectrometer.

The problem has been overcome by the use of a motorized Michelson* interferometer and by the ability to transform the resulting *time domain* spectrum into a *frequency domain* spectrum through an inverse Fourier[†] transform (FT).

While a full discussion is inappropriate for this text, the elegant simplicity of the system is worthy of brief description. As shown in Figure 2.7, radiation from a source is collimated by a lens and then divided into two parts or "arms" of (ideally) equal energy by a beam splitter (in the 400–4000 cm^{-1} region, a very thin film of germanium is commonly used). About 50% of the beam is directed onto a fixed mirror, and the other 50% passes through onto a movable mirror (which mirror is fixed and which is movable is not important). The movable mirror is driven at a constant speed over a distance of a few millimeters (a distance, of course, far greater than the longest wavelength in the IR). The beams then recombine at the beam splitter, where both constructive and destructive interference occurs, the extent of which depends upon the instantaneous position of the moving mirror relative to the stationary one.

A consideration of what has been accomplished can be illustrated by imagining what happens to monochromatic light, that is, a single frequency, but remembering that *all* frequencies in the beam can be treated in the same way. (a) When the two

*Albert A. Michelson (1852–1931) was the first American to win the Nobel Prize in Physics in 1907 "for his optical precision instruments and the spectroscopic and metrological investigations carried out with their aid." He worked and studied at the U.S. Naval Academy, Case Institute of Technology, and, finally, at the University of Chicago. He designed and then developed interferometers and, in 1887, along with Edward Morley ([1838–1923] chemistry professor, Western Reserve College), demonstrated that the speed of light was the same in all directions (the famous Michelson–Morley experiment). Others concluded from those experiments that it was unlikely that there was a "luminiferous ether" to carry waves of light.

[†]Jean Baptiste Joseph Fourier (1768–1830), École Polytechnique, worked on infinite series and showed that some discontinuous functions belonged to these series. He showed that any periodic function could be decomposed into a set of simple oscillating functions, a Fourier series.

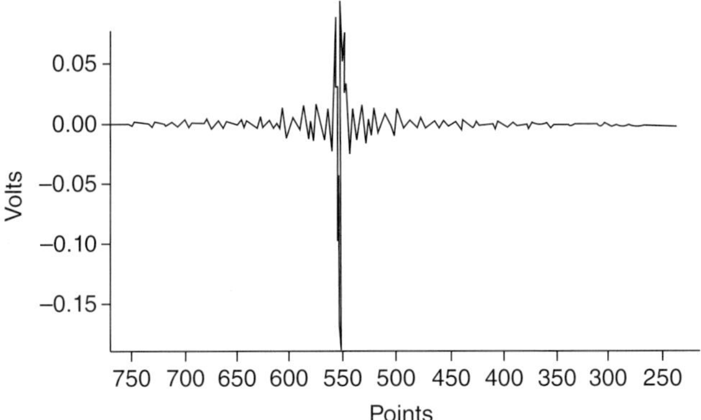

Figure 2.8. The interferogram of propane ($CH_3CH_2CH_3$) from which the spectrum shown in Figure 2.6 was derived.

mirrors are equidistant from the beam splitter, the path length is exactly the same; constructive interference occurs; and the detector observes maximum intensity. (b) When the moving mirror is one-fourth of a wavelength away from the point of zero path difference, the total optical path difference is one-half of the wavelength; the two beams are now 180° out of phase, and destructive interference occurs and no radiation passes. (c) This process repeats for each one-fourth wavelength the mirror moves, and a modulated cosine wave is created (cosine because the amplitude is a maximum at zero deviation). (d) The faster the mirror moves, the higher the frequency of the modulated cosine function. At this point, what has been accomplished is the conversion of *a single frequency into a time-modulated cosine function*. Mathematically, a cosine wave can be described as the Fourier transformation of a single frequency and thus, the interferometer is taking the Fourier sum* of the incoming signal. The same process occurs for every frequency emitted from the IR source, *all* of which will be at a maximum at zero path difference but nowhere else over the path of the moving mirror. Furthermore, the output from such a device, called an interferogram, must contain all of the information of the original radiation, and it can be reconverted from the time domain back to the frequency domain by an inverse FT. Now the detector must respond only to a modulated cosine function across the IR range, that is, 1250 Hz at about 4000 cm^{-1} and 125 Hz at about 400 cm^{-1},

*This process, like many others, can be described either in the **time domain**, by the values of some quantity x as a function of time t, that is, $x(t)$, or in the **frequency domain** by an amplitude (X) as a function of frequency (v), that is, $X(v)$. Here, as well as for NMR spectroscopy (*vide infra*), it is convenient to think of $x(t)$ and $X(v)$ *as two different representations of the same function*. The *FT* equations (Equations 2.6a,b) simply allow one to go back and forth between the representations. The representations of the spectra can be in either form. Thus, the transformation of the interferogram of Figure 2.8 produces the spectrum of Figure 2.6 and vice versa:

$$x(t) = \int_0^\infty X(v)e^{-i2\pi vt}\, dv \tag{2.6a}$$

$$X(v) = \int_0^\infty x(t)e^{i2\pi vt}\, dt. \tag{2.6b}$$

and pyroelectric detectors* are very efficient in this range. The beam (now as a voltage) coming from the interferometer is collected and stored as digital data and the process repeated with a sample. Ratioing the resulting data and performing FT results in a frequency domain spectrum of quality superior to that obtained from CW systems. The superiority derives from features that include (a) increased speed (all frequencies are sampled simultaneously), and the whole scan can be quickly repeated, the results coadded, and thus the signal-to-noise ratio increased (i.e., signals are constant and noise is random), and (b) more radiation gets through to the detector than in other systems because it has not undergone dispersion.

IV. Raman Spectroscopy

Although the absorption of energy from one state to another requires the availability of the state(s), as shown in Figure 2.2, it is nonetheless true that photons can apparently be "captured" momentarily by molecules (in what are sometimes called "virtual" levels). Since the virtual level does not (generally) correspond to a stable position, the molecule must immediately drop back to its ground state.[†] When it does fall back, it may end up in an excited vibrational and/or rotational state higher, lower, or equal to the one from which it was raised, and, in doing so, will emit energy (as scattered light). Generally, the transient absorption reflects a change in the polarizability of the bond, and thus Raman transitions may be observed where IR is not, that is, in molecules that are electrically symmetrical.

Typically, because the exciting radiation can be at any frequency, Raman spectroscopy utilizes monochromatic laser sources and the scattered light is collected at an IR detector.

For example, as noted above, ethene (ethylene, $CH_2=CH_2$) lacks absorption in the carbon–carbon double-bond region of the IR spectrum because of its symmetry. In the Raman, however, ethene (ethylene, $CH_2=CH_2$) absorbs strongly at about $1580\,cm^{-1}$. In more complicated cases, it may be that there are some vibrational modes that affect both the polarizability and the dipole moment (present in molecules not electrically symmetrical), and molecules within which these modes are present may have similar absorbances in both the IR and Raman.

V. Microwave Spectroscopy

Rotational spectra are obtained from gaseous molecules that have permanent dipole moments. Since rotational energy levels are very closely spaced (Figure 2.2 and Table 2.1), low-energy radiation (microwave and far-IR region) is required. Typically, rotational levels are separated by as much as $30–60\,cal\,mol^{-1}$ (or, since $E = h\nu$, about $10–20\,cm^{-1}$).

---

*A typical pyroelectric detector might consist of a small crystal of material with a permanent dipole moment, sandwiched between metal foil electrodes. When the crystal absorbs heat, it undergoes a change in lattice spacing and thus dipole moment. Such a change in dipole moment will then result in a change in the electric charges sensed by the metal foil electrodes.

[†]In the library analogy (*vide supra*), it is as if a user were to pick up a book from a shelf and, instead of setting it on a surface, simply release it. It might fall on a shelf or to the floor. It is unlikely that it would stay suspended in the air.

Examination of such spectra allows one to obtain distances between nuclei and the shapes of molecules from moments of inertia (i.e., $I = m_R r$, where m_R is the reduced mass and r is the internuclear distance). Further, as already pointed out in Chapter 1, application of a constant electric field while the spectrum is being observed causes the energy levels to shift by an amount due to the molecular dipole moment, μ.

Since the measurement of frequencies can be made with high precision, the technique is very valuable. Regrettably, it is only applicable to gases that have a permanent electric dipole moment and so its utility is limited.

Among the more interesting examples of the use of this technique has been the analysis of methanol (methyl alcohol, CH_3OH) in the presence of both ethanol (ethyl alcohol, CH_3CH_2OH) and water (H_2O). The former has a series of bands in the microwave spectrum lying between 29,636.91 and 37,703.67 MHz that are unique, and quantitative measurements of these have been used to determine the amount of methanol (methyl alcohol, CH_3OH) in various samples of wine.*

VI. Magnetic Resonance Spectroscopy

a. NMR. NMR[†] is a very special kind of photon spectroscopy in that the sample is simultaneously subjected to two magnetic fields, one stationary and the other varying—at some radio frequency (RF) (or vice versa). The first field serves to create a specific difference in the energy states of certain nuclei and the second to provide the energy for the transition between them. Thus, at a specific, nucleus-dependent combination of the fields, energy is absorbed. The *absorption* is measured as a signal by an RF detector.

Nuclei that have *both* even mass and even atomic number, such as ^2He, ^{12}C, and ^{16}O, have *no* magnetic properties (spin quantum number $I = 0$). Other nuclei, that is, those with odd-numbered mass and odd or even atomic numbers, including, but not limited to, ^1H, ^3H, ^{13}C, ^{15}N, ^{17}O, ^{19}F, ^{29}Si, and ^{31}P (where I is half-integral), and those with even mass but odd atomic numbers such as ^2H, ^{10}B, and ^{14}N (where I is integral), have a nonzero spin quantum number and thus, since they possess the property called *spin*, simulate tiny magnets that can interact with an externally impressed magnetic field (symbolized as H_o).

Consider, for example, the case of the proton, ^1H ($I = 1/2$). There are two possible quantum orientations (designated as $+1/2$ and $-1/2$) of this magnetic nucleus with respect to the direction of an external field; that is, the nuclear magnets can be aligned either *with* or *against* the field. The two states are not equivalent. The low-energy state is aligned with the field, and the high-energy state is aligned against the field. The NMR experiment measures that difference which, reasonably, must be related not only to the particular nucleus (in this case ^1H) but also to the strength

---

*Kitchin, R. W.; Eillis, R. E.; Cook, R. L. *Anal. Chem.*, **1981**, *53*, 1190.

[†]In contrast to the earlier term "resonance" discussed in Chapter 1, and which had a very special meaning with regard to representations of structures we draw, the word resonance used here means that absorption of energy takes place at specified resonance frequencies. With the exception of x-ray crystallography discussed above (Section B) and MS, discussed in Section D, it should be clear that this whole chapter is dealing with various forms of resonance spectroscopy.

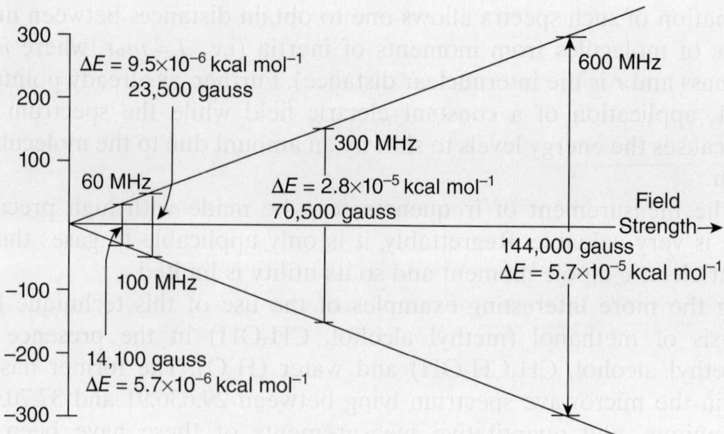

Figure 2.9. A field–frequency diagram depicting the energies of the higher and lower spin states of ^{1}H as a function of magnetic field strength (gauss) (horizontal scale) and radio frequency (megahertz) (vertical scale).

of the magnetic field. The stronger the field (for any given nucleus), the larger the difference between the states (Figure 2.9).*†

Considering a nucleus as a moving magnetic dipole, a magnetic field ($|\mathbf{H}|$) exerts a torque (T) on such a dipole (μ) as given by Equation 2.8, where θ is the angle between the field and the axis of rotation:

$$T = |\mu||\mathbf{H}|\sin\theta. \tag{2.8}$$

This is exactly analogous to a gyroscope or "top," spinning at an angle (θ) with an angular momentum ($|\mathbf{M}|$) and angular velocity ω (Figure 2.11). Thus,

$$T = |\mathbf{M}|\omega\sin\theta \tag{2.9}$$

so that

$$|\mu||\mathbf{H}|\sin\theta = |\mathbf{M}|\omega\sin\theta \tag{2.10}$$

*This will be the case whether one considers the resonance absorption phenomena in either classical or quantum mechanical terms. For the latter, it is sufficient to argue here that two states have been created and that the transition between the states corresponds to absorption (or emission) of radiation when the applied frequency coincides with the natural frequency. Interestingly, and in contrast to UV, IR, and so on, the absence of nearby states suggests that the line should be "infinitely" narrow. That this is not so apparently results from a variety of causes that include, among others, interactions with the remainder of the system (called *spin–lattice relaxation*).

†Examination of an abbreviated form of the classical explanation may be appealing to some. The establishment of an electric current in a loop is accompanied by the presence of a magnetic field that has the shape of a disk (uniformly magnetized) of the same area as the loop (Figure 2.10a). From a distance (Figure 2.10b), the loop looks like a point magnetic dipole. The dipole moment (μ) is given by Equation 2.7, where q is the charge and R the distance between the ends of the dipole:

$$|\mu| = (\text{current})(\text{area}) = qR. \tag{2.7}$$

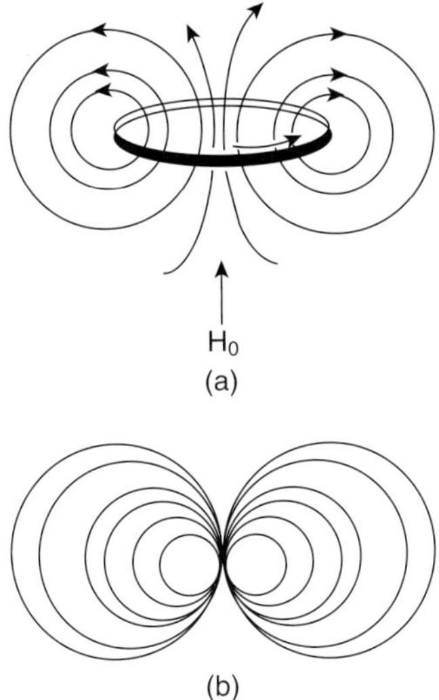

Figure 2.10. (a) A diagrammatic representation of the magnetic field around a ring through which current is flowing. (b) A representation of the magnetic field around a dipole.

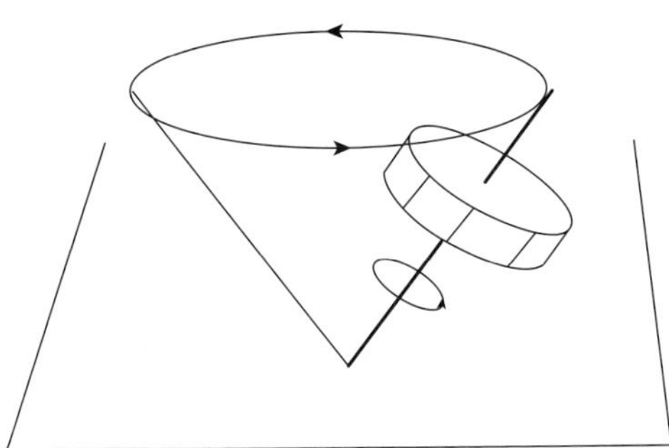

Figure 2.11. A representation of precession in a spinning top.

and

$$\omega = (|\mu|/|\mathbf{M}|)|\mathbf{H}| = \gamma|\mathbf{H}|, \qquad (2.11)$$

where γ is called the *magnetogyric ratio* and is a specific and unique constant for each particle behaving in this fashion. The angular velocity ($\omega/2\pi$) of this precession is called the *Larmor angular frequency*.

Now, suppose a second, smaller magnetic field ($\mathbf{H}_1$ in Figure 2.12) is introduced perpendicular to the original field ($\mathbf{H}_o$ in Figure 2.12) but rotating about that direction. As long as the angular velocity ($\omega_1/2\pi$) is smaller than that of the Larmor angular velocity, its effect will be to exert a torque on the nucleus, which will tip it toward the plane perpendicular to $\mathbf{H}_o$, and this will cause the precession to wobble. However, if the field $\mathbf{H}_1$ is **equal** to the Larmor angular velocity, then the torque, being in the same direction with the same magnitude, will produce large oscillations,

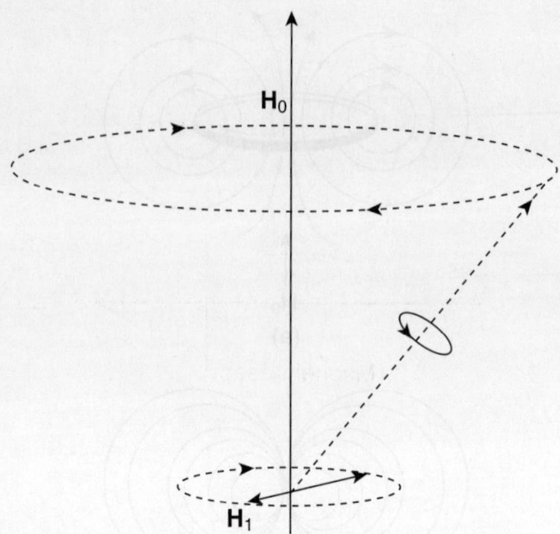

Figure 2.12. Diagrammatic representation of precession of a nucleus with a magnetic moment in a magnetic field (**H**₀).

that is, resonance. Increasing the field **H**₁ further, however, will again diminish the effect.

A larger difference between the states also affects the extent to which the states are populated. At these low energies, there are, comparatively, many fewer nuclei in the higher-energy state (in the magnetic field) than there are in the lower energy state. Roughly (ignoring possible entropy effects) the difference in energy at 60 MHz (5.7×10^{-6} kcal mol^{-1}, Figure 2.9) corresponds to an equilibrium constant $K_{eq} \approx 1.00001$. At 600 MHz, $K_{eq} \approx 1.0001$ (Problem 2.1).

Problem 2.1. Knowing that the speed of light (c) is 3×10^8 m s^{-1} and that $\lambda = c/\nu$, calculate

(a) the wavelength (λ) of the radiation absorbed or emitted in the transition E_{21} at the resonance condition for frequencies of 100 and 300 MHz,

(b) the energy (Δ**E**) of the transition at each of these frequencies (use Equation 2.2), and

(c) since Δ**E** $= -2.303 RT \log_{10} K_{eq}$, K_{eq} for these frequencies at 25°C. (Note that 25°C = 298 K; $R = 1.987$ cal deg^{-1} mol^{-1}.)

Since the energy differences are small, the finely tuned RF oscillator need not be very powerful. However, the frequency must be precisely known, and thus the oscillator must be stable and the detector very sensitive. At lower fields (e.g., 60 and 100 MHz), it is common to hold the frequency constant and slowly change (*sweep*) the magnetic field through the region where absorption of radiation is anticipated (much like CW-IR). The higher frequencies, for which superconducting magnets are used, operate with a pulsed technique (very similar to Fourier transform infrared [FT-IR] but without needing an interferometer). Consider the CW experiment. A diagrammatic representation of such a spectrometer is shown in Figure 2.13. Holding the frequency fixed, the magnetic field,

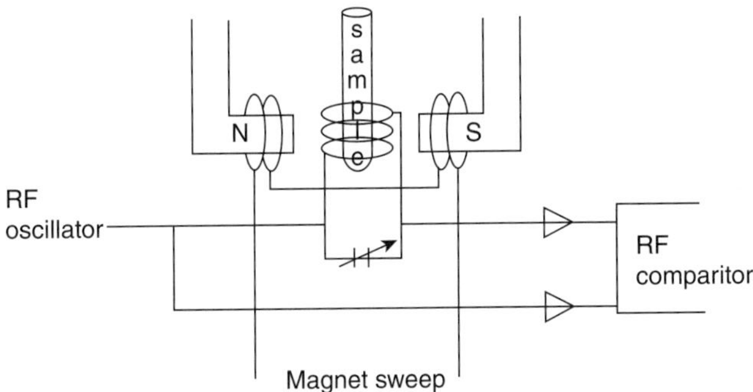

Figure 2.13. A diagramatic representation of a CW-NMR spectrometer.

H_o, is gradually increased until the proton comes into resonance. The sweep is generally from low to high field. At the resonance condition, three pieces of information are obtained: the **chemical shift**, the **multiplicity**, and the **integrated area**. These are discussed below after noting the difference between the CW and the FT experiment.

In the FT-IR experiment, all of the wavelengths (frequencies) of light are present in the source and they are "sorted" by the interferometer. As will be discussed further, in the FT-NMR, those frequencies must be provided and that is done in a single-shaped pulse of (normally) 1- to 10-μs duration of RF radiation. The frequency of the radiation used is, as in the CW experiment, a function of the nucleus to be examined and the magnetic field. This pulse causes **all** the specific nuclei absorbing at that frequency (e.g., as shown in Figure 2.9 at 70,500 gauss the frequency would be ca. 300 MHz) to "flip" into the higher-energy alignment. Now, **all** of the nuclei reemit RF radiation *at their respective resonance frequencies*. This produces an interference pattern (different from but similar to the interferogram seen in FT-IR, *vide supra*) called "free induction decay" (FID). The FID is the result of RF emission versus time. The FT is used to convert the data into emission versus frequency, which is displayed as in the CW experiment.

1. The Chemical Shift. Every nucleus for which NMR spectroscopy is possible (e.g., ^1H, ^{13}C, ^{19}F, …) will come into resonance at a specific field/frequency combination, which is modulated by the local environment. Protons (^1H) are discussed first as both an example of what can be done with NMR and with specific cases in mind because, among other reasons, (a) only limited time and space are available; (b) protons are ubiquitous; and (c) chemical shift and coupling constant information are readily obtained and will serve as good tools to help understand the interplay of structure and dynamics.

In real circumstances, the field actually experienced by a proton bonded to another atom will be *less* than H_o, the externally experienced field. This is because one cannot have an NMR tube containing only protons. Protons are bonded to other nuclei and the electrons (also possessing charge and the property called spin) in the bond **shield** the proton from the full effect of the field by setting up a small magnetic field of their own *opposite to* H_o. This effect is clearly related to the electronic environment in which a particular proton finds itself. For example, protons attached to oxygen in water (H_2O), where there are nonbonded electrons on oxygen, will be in a different environment than those attached to carbon in, for example, methane (CH_4). Similarly, protons in ethane (CH_3CH_3) will be different (although not by much) from those in methane (CH_4) because each carbon to which the protons in ethane (CH_3CH_3) are bonded is *also* attached to *another carbon*.

The magnitude of this small secondary field (produced by the shielding effect of the electrons) is also directly proportional to the field strength of the large magnetic field, H_o, and can be quantified as σH_o. Thus, the field **actually felt at a bonded proton** is given by $H_o - \sigma H_o$ where the constant of proportionality, σ, is called the *shielding parameter*. The greater the shielding, the higher the applied field H_o must be to bring the proton into resonance. Protons that are **more shielded** than others need higher field strengths to come into resonance and are found **upfield**; the less strongly shielded protons are **downfield**; **identical** protons come into resonance at the same field strengths. Further, because many different combinations of field and frequency can be (and are) in use, communication of the information about specific values at which a proton or protons come into resonance can become complex. To simplify the problem, the position at which various protons in molecules come into resonance (i.e., their "shielding value") has been defined *relative to an agreed-upon standard*, tetramethylsilane (TMS) ((CH_3)$_4$Si).* All 12 hydrogens in TMS are chemically identical. The specific position relative to TMS is called the *chemical shift* (symbolized by δ). The chemical shift is intimately connected to the chemical (electronic) environment in which the atom finds itself. The chemical shift (δ) is a dimensionless quantity whose value is given by Equation 2.12:

$$\text{Chemical shift (ppm)} = \delta = (H_{\text{sample}} - H_{\text{reference (TMS)}})/H_{\text{reference (TMS)}}. \qquad (2.12)$$

Alternatively, δ = chemical shift (hertz) relative to TMS $\times 10^6$/spectrometer frequency (hertz).

There is an additional advantage to communication of results as delta (δ) values. As shown in Figure 2.9, a 10-fold increase in operating frequency (e.g., from 60 to 600 MHz) also means a 10-fold increase in magnetic field, H_o, and thus, also, a 10-fold increase in σH_o. As a consequence, protons that differ because of different shielding will have a 10-fold greater difference at 600 MHz than at 60 MHz and will thus be further apart in a plot that is linear in field strength or frequency; that is, protons that might come into resonance 20 Hz apart using a 60 MHz spectrometer

---

*The removal of one hydrogen from methane (CH_4) leaves behind a fragment called "methyl" or "the methyl group" (CH_3-). It is important to note that this is the **name of the fragment** (such names will be discussed in the next chapter when the language of organic chemistry is developed). It is assumed, specifically, that attachment to another ligand is actually present. If another ligand is not present, it is necessary to further identify the fragment by adding to the word **methyl** a further descriptor, for example, the **methyl radical** ($CH_3\cdot$), where the hydrogen *and one electron* have been removed, the **methyl cation** (CH_3^+), where the hydrogen and two electrons (i.e., the hydride ion) have been removed, and the **methyl anion** (CH_3^-), where the hydrogen nucleus only has been removed (i.e., a proton). Thus, TMS [(CH_3)$_4$Si] may be viewed, for the moment, as having arisen from ethane (CH_3CH_3) by replacement of one of the carbon atoms by silicon and the hydrogens that were attached to that carbon (now a silicon, Si) by "methyl groups." TMS is chosen for the NMR standard because the protons attached to the carbon atoms are among the most strongly shielded protons likely to be encountered in routine work. The value of δ for TMS is 0.00 ppm (exactly).

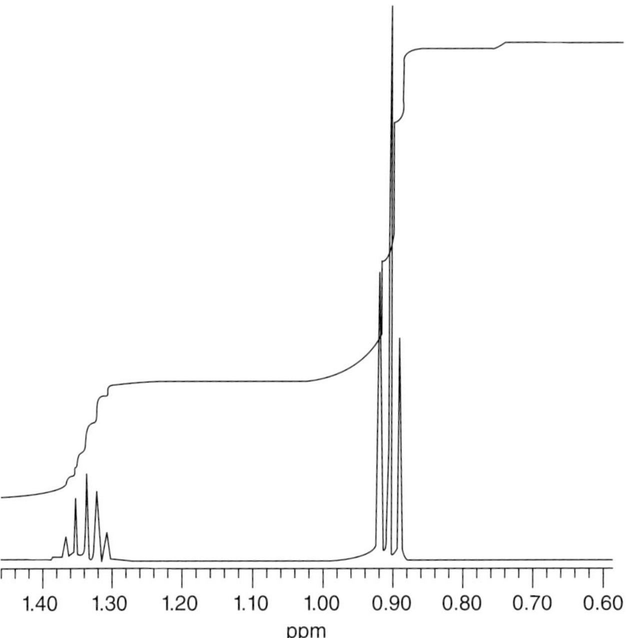

Figure 2.14. The ^{1}H NMR spectrum of propane ($CH_3CH_2CH_3$) in deuterochloroform ($^2HCCl_3$) at 500 MHz.

will now be 200 Hz apart if the frequency of the spectrometer is 600 MHz. Reporting values in parts per million (δ), relative to the same standard, relieves us of the necessity of defining the frequency of the spectrometer and allows rapid communication and comparison of information from various instruments and laboratories.

Thus, one of the major values to such ^{1}H NMR spectroscopy (and for NMR in general) is that structurally different nuclei, irradiated at a given frequency, absorb energy at different field strengths and are distinguishable. Further, although the protons and, indeed, the carbons, within most of the simple hydrocarbons examined so far, for example, methane (CH_4), ethane (CH_3CH_3), ethene (ethylene, $CH_2{=}CH_2$), and ethyne (acetylene, $HC{\equiv}CH$) (but not propane [$CH_3CH_2CH_3$]), respectively, are spectroscopically identical, it should be clear that this will usually not be the case. Figure 2.14 presents the ^{1}H NMR spectrum of propane ($CH_3CH_2CH_3$), the next higher hydrocarbon homologue of the series beginning with methane (CH_4) and continuing past ethane (CH_3CH_3).

Notice in Figure 2.14 that there are two regions of signals, one centered at $\delta \approx 0.90$ ppm and the other at $\delta \approx 1.33$ ppm. The signal farthest upfield (and closest to TMS) is typical of a methyl group ($CH_3{-}$) attached to a methylene ($-CH_2{-}$)*

---

*The removal of two hydrogens from methane (CH_4) leaves behind a fragment called "methylene" or "the methylene group" ($-CH_2{-}$). It is important to note that this is the **name of the fragment** (such names will be discussed in the next chapter when the language of organic chemistry is developed). It is assumed, specifically, that some attachments (which may, or may not, be identical) are actually present. In the particular case of methylene, as we shall eventually discuss, there is some evidence that a species with two electrons and two hydrogens, that is, ($\bullet CH_2\bullet$) may have transient existence.

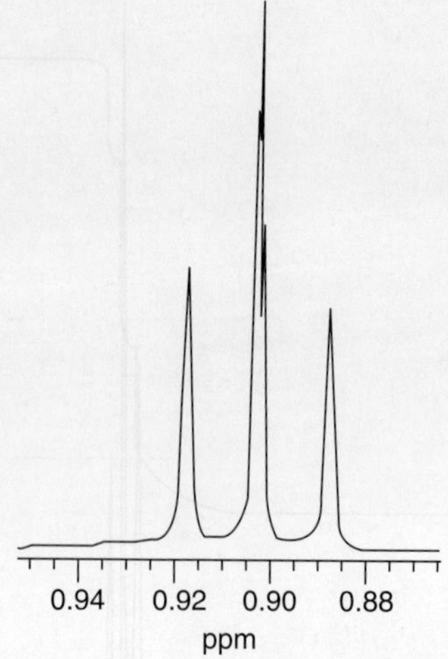

Figure 2.15. An expanded view of the most upfield region of the ^{1}H NMR spectrum of propane ($CH_3CH_2CH_3$) shown in Figure 2.14.

and the more downfield signal (furthest from TMS) that of a methylene ($-CH_2-$) attached to a methyl (CH_3-).

In addition to the chemical shift (δ), which provides information about what kinds of protons are present (e.g., methyl [CH_3-], methylene [$-CH_2-$]), the spectrum in Figure 2.14 also reveals two other important features, routinely obtained, and which contain information useful for structure determination. These features are the multiplicity of signals (due to coupling relationships between protons giving rise to spin–spin splittings) and the integrated areas under the peaks, corresponding to the *relative* numbers of the different kinds of protons as dictated by the chemical shift.

2. Multiplicity (The Coupling Constant—Spin–Spin Splitting). Figure 2.15 presents an expanded view of the most upfield region [the methyl group (CH_3)] of the ^{1}H NMR spectrum of propane ($CH_3CH_2CH_3$) (Figure 2.14), while Figure 2.16 presents an expanded view of the other signal [the methylene group ($-CH_2-$)] for the same compound. The chemical shift values (δ) have identified these as the methyl (CH_3-) and methylene ($-CH_2-$) groups, respectively.

The distance *between the centers of the upfield lines is identical to the distance between the centers of the downfield lines.* This distance is known as the *coupling constant* (symbolized by the letter J) and is the result of the process known as *spin–spin splitting.* The reason that these resonances appear as groups of equally spaced lines, rather than as a single line, is because the hydrogens on the contiguous carbons are **nonidentical**. The *multiplicity* of lines, a **triplet** (three lines) for the methyl group (CH_3-) and a **septet** (seven lines) for the methylene ($-CH_2-$), is produced by the **mutual** interaction of the nonidentical magnetic nuclei. In *this* particular case, $J \approx 7\,Hz.$* Identical protons (i.e., those that come at the same chemical shift) do not couple (or have a coupling constant of zero).

---

*As we will see, the coupling constant can be related to (among other things) the angle subtended between the hydrogens.

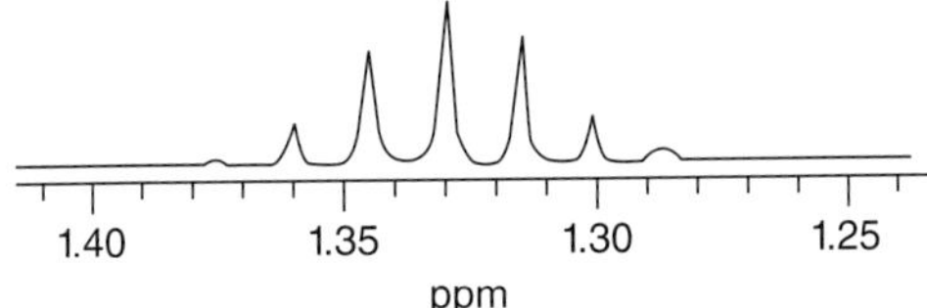

Figure 2.16. An expanded view of the most downfield region of the ^{1}H NMR spectrum of propane ($CH_3CH_2CH_3$) shown in Figure 2.14.

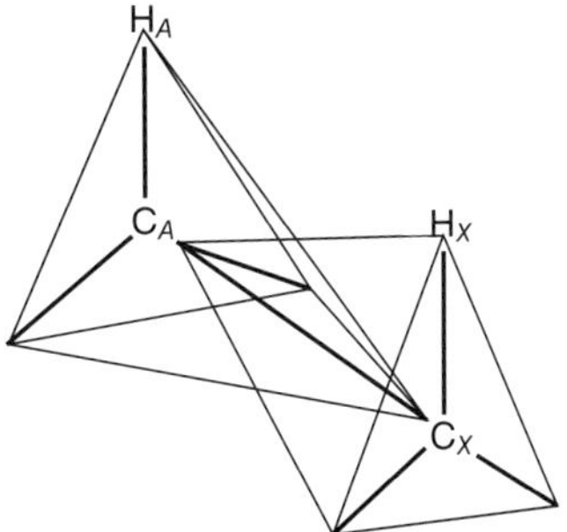

Figure 2.17. A representation of two chemically different protons ($\mathbf{H}_A$ and $\mathbf{H}_X$) on two chemically different contiguous tetrahedral carbon atoms. It is important to recognize that although the additional ligands on each of the carbons are *not* specified, they are, nonetheless, present. Their exact nature, as long as they are not identical, is unimportant to the discussion here.

In contrast to the chemical shift (δ), the absolute value of which is dependent upon the magnetic field strength (or transmitter frequency), such first-order splitting, centered *around* the chemical shift (δ) of a particular proton, is the same for a specific compound regardless of the frequency (or field).

Additionally, as is clear from Figures 2.15 and 2.16, the multiplicity of the lines is ($n + 1$), where n is the number of **identical** protons **on the contiguous carbons**. Thus, both terminal methyl (CH_3-) groups of propane ($CH_3CH_2CH_3$) are identical. Each consists of three identical protons attached to a carbon. The one intervening methylene ($-CH_2-$) group consists of two identical protons attached to one carbon. As a consequence, it is expected that the six protons on the two methyl groups will be split into three lines (a **triplet**) by the two protons on the methylene (i.e., $n + 1$, where $n = 2$). The methylene signal will be split into seven lines (a **septet**) by the six identical protons on the two adjacent methyl groups that flank them (i.e., $n + 1$, where $n = 6$).

Qualitatively, although the origin of spin–spin coupling involves all of the electrons and the nuclei in the coupling system, it may be described as follows: consider two nonidentical protons ($\mathbf{H}_A$ and $\mathbf{H}_X$) on nonidentical contiguous carbons $\mathbf{C}_A$ and $\mathbf{C}_X$, respectively (Figure 2.17). As we sweep through the field (holding the frequency constant) and reach that combination of field and frequency at which $\mathbf{H}_A$ comes into resonance, the absolute sense of the spin of $\mathbf{H}_X$ (which will come into resonance at a *different* combination of field and frequency because $\mathbf{H}_A \neq \mathbf{H}_X$) may be the same as that of $\mathbf{H}_A$ or opposite (there being only two possibilities). Each possibility, $\mathbf{H}_X$ having the same absolute sense of spin or $\mathbf{H}_X$ having the opposite absolute sense of spin as H_A, corresponds to two new energy states for H_A, and thus the line corresponding to H_A is split into a doublet. The same will be true for H_X.

Similarly, if there are *two* identical protons (H_A) on a carbon contiguous to a nonidentical carbon bearing one proton (H_X), as we sweep through H_X, *both* H_A protons may be aligned *with* H_X (symbolized as ↑↑), one of the two H_A protons may be aligned *with* H_X and one *against* (symbolized as ↑↓ and ↓↑; i.e, there are two possibilities) and, finally both H_A protons may be aligned *against* H_X (symbolized as ↓↓). Thus, there are three possibilities (i.e., $n + 1 = 3$ for two identical protons) and their probabilities are in the ratio 1:2:1. In the same way, for three identical neighbors, there will be four sets of possibilities (i.e., set one [↑↑↑], set two [↑↑↓, ↑↓↑, ↓↑↑], set three [↑↓↓, ↓↑↓, ↓↓↑], and set four [↓↓↓]) and thus four lines in the ratios 1:3:3:1. Further, if H_A is coupled to H_X with a coupling constant J_{AX}, then H_X is coupled to H_A with the **same** coupling constant, that is, $J_{AX} = J_{XA}$, so that the relationship of the carbons being contiguous is established.*

Finally, the numerical value of the coupling constant (i.e., its magnitude) depends on the number and kind of intervening chemical bonds and the angles the bonds holding the protons to carbon make with each other.

3. The Integrated Area. The **chemical shift** (δ) provides information relating to particular protons according to the extent to which they are shielded by nearby electrons. The **coupling constant** (J) supplements this information by defining relationships between protons on neighboring atoms. Integration of the area under the peak or peaks (the **integrated area**) gives the *relative numbers* of protons coming into resonance at that particular value of the chemical shift. Thus, in the ^{1}H NMR spectrum of propane ($CH_3CH_2CH_3$) (Figure 2.14), the integrated area for the six methyl ($CH_3–$) protons relative to that of the two methylene ($–CH_2–$) protons is 3:1. It should be clear that, in a similar vein, integration is generally used as a *relative measure*; therefore, integration of the spectrum of a hydrocarbon such as ethane (CH_3CH_3), in which there is only one kind of proton (and thus the spectrum consists of a single line [a **singlet**]), is generally not a worthwhile endeavor.

4. Expansion of the Principle. Given the necessity to maintain close tolerance in the field-to-frequency ratio so that the relatively small energy changes involved in NMR spectroscopy can be accurately and reproducibly measured, it is necessary to provide a means for an NMR spectrometer to adjust to magnetic drift, temperature changes, and so on. This is accomplished by utilizing a frequency-controlling feedback circuit that locks onto a specific resonance frequency and continuously adjusts the magnetic field to keep the reference signal maximized. Historically, the reference lock frequency was set to monitor a nucleus internal to the sample and in ^{1}H NMR, the protons in TMS itself served this function. However, it is now common to use a frequency for the reference lock, which is tuned to a nucleus other than the

---

*The subscripts A and X are chosen, as has become customary, to indicate that the protons bearing those subscripts are significantly different. By that is meant that the chemical shifts of the protons are far enough apart so that a simple first-order treatment (such as that provided) is possible. When the chemical shift difference between the protons ($\Delta\delta$) is comparable to the coupling constant (J), the ($n + 1$) rule frequently breaks down. When this happens, there may be more lines, or lines in different positions with different intensities, than expected.

one being observed. The use of deuterium (^{2}H, $I = 1$) is a popular choice since it is conveniently incorporated, instead of protium (^{1}H), into a variety of commercially available solvents. At a field strength of about 23,500 gauss, ^{2}H comes into resonance at a frequency of about 15.3 MHz, whereas protons come into resonance near 100 MHz. Thus, the NMR system is modified to include two RF transmitters and receivers, one to measure the proton (^{1}H) resonance spectrum and the other, tuned to the deuterium (^{2}H) frequency (in the solvent, thus minimizing solvent interference with the proton spectrum) and part of a feedback circuit, to keep the system stationary.*

Second, just as in the IR experiment, the efficiency of the NMR experiment can be improved upon by exciting all possible resonance frequencies (e.g., of the protons) simultaneously rather than scanning them sequentially.[†]

It will be recalled that in the FT-IR experiment, undispersed IR radiation was passed into an interferometer so that the timescale on which the receiver was to operate could be modified. The results (an interferogram) were subjected to an FT to convert them from the time to the frequency domain.

In the NMR, because the frequencies are now in the radio region of the spectrum (i.e., at wavelengths that can be handled directly), an interferometer is not necessary. Thus, in FT-NMR, the appropriate frequencies over which the nuclei are expected to undergo transition between the energy states (established by the externally applied magnetic field) are supplied all at once. To supply such a burst of energy, a pulse of RF radiation is created. The pulse is centered around the carrier frequency needed to observe the particular nucleus and is of limited duration (the length of the pulse determining the width of the band). A 10 ms (10^{-5} s) pulse might cover, for example, a range of about 100,000 (10^5) Hz. (Note the relationship between *time*, i.e., a *pulse* in seconds, and the *frequency* in hertz, i.e., cycles per second.) The excited nuclei now emit radiation at their characteristic frequencies as they relax back to their equilibrium configuration (Figure 2.18a). This fading response is called *free induction decay* and the signal is referred to as a *FID*. The emission, picked up by a probe coil, contains all the information in a complete spectrum. When all of the nuclei have relaxed back to their original state (frequently in as little as a second), the pulse can be repeated. Again, since noise is random and signals are not, storage of the FIDs and subsequent averaging of the signals results in enhanced signal to noise. Figure 2.18b presents the FID from which the Fourier transformed ^{1}H NMR spectrum of propane ($CH_3CH_2CH_3$) given above as Figure 2.14 was generated.

As was the case with respect to CW-IR and FT-IR, the comparison of CW-NMR and FT-NMR again requires understanding that, with the latter, the amplitude of the signals is being measured in the time domain rather than in the frequency domain. The FT allows the interconversion. What follows are some special features available, as a consequence of using FT-NMR, which have expanded the horizons of the tool.

In FT-NMR, it is neither necessary nor desirable to collect all of the data. Figure 2.14 displays the ^{1}H NMR spectrum of propane collected at 500 **MHz**. Both sets of

---

*It should be clear that, at least in principle, the two could be reversed; that is, the spectrometer could be locked on protons (^{1}H) to observe deuterons (^{2}H).

[†]Sequential scanning of frequencies while holding the field constant, or vice versa, is called **CW** spectroscopy.

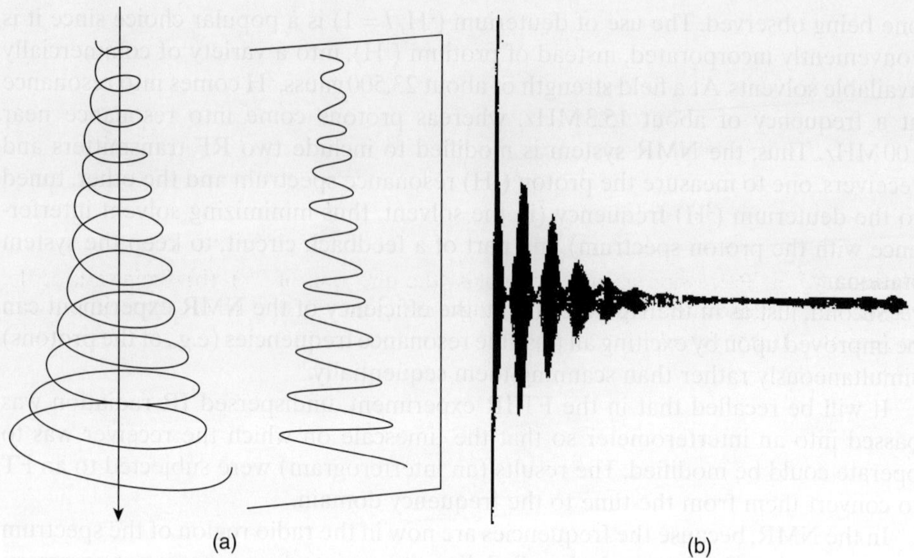

(a)   (b)

Figure 2.18. (a) Decaying precession and the signal observed as the decay proceeds. (b) The FID from which the ^{1}H NMR spectrum of propane (Figure 2.14) is derived.

signals [the set due to the methylene ($-CH_2-$ at $\delta \approx 1.33\,$ppm) and the set for the methyl (CH_3- at $\delta \approx 0.90\,$ppm)] appear below $\delta = 3.0\,$ppm. Indeed, for most samples, signals will appear at *less than* $\delta \approx 10\,$ppm or over about $5000\,\mathbf{Hz}$ (in a $500\,$MHz spectrometer). Therefore, the reference frequency of $500\,$MHz can be subtracted, and only frequencies between 0 and $5000\,$Hz need to be recorded. This is particularly important because the *analog* voltage signal produced must be stored as a *digital* number whose value is proportional to the intensity of the signal. Without this modification, the amount of computer storage space necessary for the entire spectrum could become excessive.

5. NMR in Two Dimensions. NMR in more than one dimension (e.g., two-dimensional [2-D] NMR) refers to the results obtained as a consequence of applying a *second pulse* of RF) before relaxation from the first RF pulse is complete. The use of such sequential pulsing allows the creation of plots of one frequency versus another (two dimensions) with the amplitude in the third dimension, creating what appears to be a topological map. Because coupled nuclei interact (by definition), they affect each other's relaxation and such a topological map more clearly reveals the intimate details of their relationship. The process is straightforward but requires some review of what is accomplished by the FT experiment (besides the advantage of rapid acquisition of spectra).

In the pulsed NMR experiment, a static magnetic field ($\mathbf{H}_o$) is applied to establish two energy levels (parallel and antiparallel to the field), and the transition between them can be stimulated by application of the RF field. At this point, the sample magnetization is precessing (before the RF is applied) about the static magnetic field axis (Figure 2.19a) such that any instantaneous vector quantity can be thought of as composed of two vectors: one along the z-axis and one in the xy plane (Figure 2.19b).

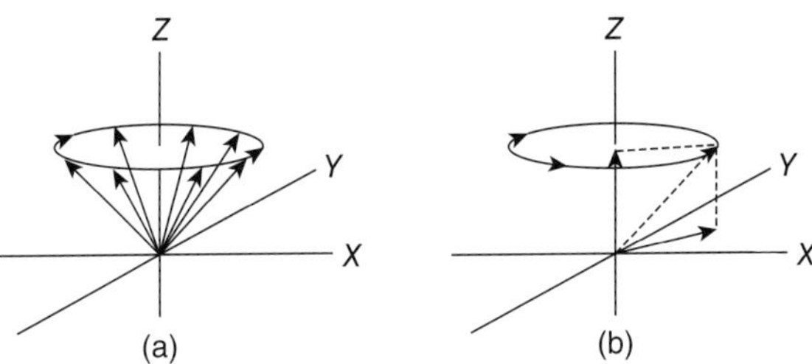

Figure 2.19. (a) Precession around the z-axis (the direction of $\mathbf{H_o}$). (b) Projection of the vectors of an instantaneous precessional angle along the z-axis and in the xy plane of a nucleus at different times.

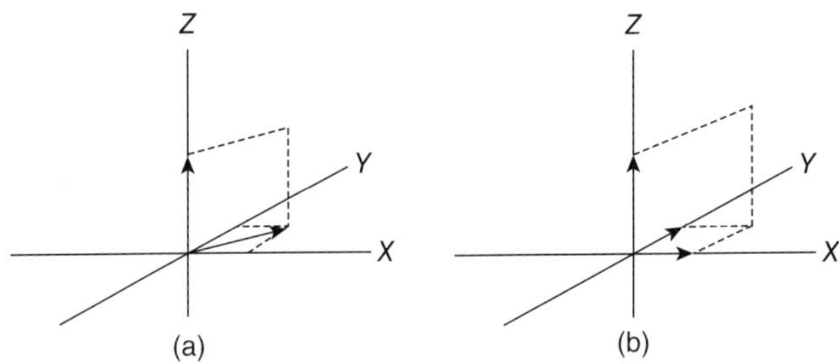

Figure 2.20. (a) Projection of the vector in the xy plane onto the x- and y-axes (and along the z-axis). (b) Final projection of all vectors onto the x, y, and z-axes.

The vector component in the xy plane (Figure 2.20a) can be further described in terms of vectors along x and y or, for the system, along x, y, and z (Figure 2.20b).

The application of the RF pulse (an alternating voltage across the ends of the coil in the NMR probe) induces an alternating magnetic field. This field is, of course, much smaller than the static field (the alternating magnetic field has a frequency of oscillation about the same as the precessional nuclear frequency, which *is* the unique chemical shift). If the RF pulse is, for example, along the x-axis of the coordinate system shown above in Figure 2.19a (which it is designed to be), then it drives the vector *at right angles to it* (e.g., along the z-axis) to rotate toward y, thus affecting the y-component of the magnetization. The length of the pulse determines how far toward y the magnetization is driven.

The decay of the new vector sum just created by the pulse, back to its original condition, is the FID. Only the change in the xy plane, however, is measured in the receiver coil. Thus, a maximum signal appears when the pulse length drives the vector component in the z-direction all the way over to the xy plane (i.e., onto the y-axis, which is 90° or $\pi/2$ radians). The decay of the signal back again along the z-axis is called the *longitudinal relaxation time* (T_1). The magnetization also "spreads" transversely in the xy plane as a consequence of interaction with the local environment. This second form of relaxation is called the *transverse relaxation time* (T_2).

Now, for 2-D NMR, assume some time, t_1 (where $t_1 < T_2$), has passed after the initial pulse is provided. Relaxation is incomplete. The components of magnetization as a consequence of the initial pulse have not yet completely recovered; that is, the vector sum, at the instant t_1, is *not* identical to the original picture and the original magnetization in the *xy* plane has not yet been reached. Thus, the magnitudes of the vectors (M) along the *x*- and *y*-axes that were originally present have not yet been reached. The consequence is that at t_1, the nucleus has precessed to some angle other than what it was originally. If this new precession has occurred over the time t_1 at some frequency ν Hz, then it has passed through an angle $2\pi\nu t_1$ radians, and if the length of the vector in the *xy* plane is M_1, then its component along the *y*-axis is $M_1\cos2\pi t_1$ and along the *x*-axis is described by $M_1\sin2\pi t_1$.

At this instant, the time t_1, a second RF pulse (assume it's of the same magnitude as the first, i.e., 90° or $\pi/2$ radians) is applied along the *x*-axis. Again, as before, the vector along the *z*-axis is driven toward *y*, leaving the value along the *x*-axis unchanged but adding *a larger (–z) component to the final coordinates*. **However, the projection in the *xy* plane, since the magnetization had not fully recovered, is now different**; that is, it is now $M_2\cos2\pi t_1$ along the *y*-axis and $M_2\sin2\pi t_1$ along the *x*-axis. Therefore, the FT transformed spectrum *at this time* is normal except for its magnitude (i.e., $M_1 \neq M_2$).

To complete the picture, imagine sampling the data for every time t_1, from $t_1 = 0$ to $t_1 = T_1$, with the intervals set so that every column of points taken from a complete set of FIDs (normally found in the spectrum of the sample) is subjected to its own FT; that is, there are now two time parameters. There is the acquisition time typical of the regular FID and there is the new time, t_1, the time between two pulses. This yields the 2-D frequency spectrum in which each FID is a function of two times; that is, $f(t_1, t_2)$ transforms to $f(\nu_1, \nu_2)$.

In a coupled system, the second and subsequent pulses (t_2) of the experiment causes magnetization, which arose from one transition during t_1 to be **redistributed among all the other nuclei with which it is coupled**. As a consequence, a line detected in t_2 will have **components of its amplitude** modulated as functions of the frequencies of all of the other lines in t_1, and in a plot of ν_1 versus ν_2, there are cross peaks between coupled systems. The detection of coupling (and its magnitude) is dramatically simplified.

Although trivial, the 2-D NMR spectrum of propane, shown as a contour plot looking down from above the peaks (a now traditional representation), is shown in Figure 2.21.

6. Double Resonance. Finally, in this regard, while spin–spin coupling is valuable for determining relationships among protons, which might be on contiguous carbons, it should be clear that as the numbers of carbons and protons increase, overlapping multiplicities might confuse, rather than clarify, the spectrum. A correlated spectroscopy (COSY) spectrum, such as that shown in Figure 2.21, dramatically clarifies the situation. However, in the absence of the ability to obtain 2-D NMR spectra, the introduction of a second (or third, if different frequencies are used for observe and lock channels) tunable oscillator that can produce an intense RF at selected frequencies alone and impose them on the sample can also dramatically simplify the picture. This technique, called *double resonance* or *spin decoupling*, succeeds by irradiating a given nucleus with an intensity that causes rapid equilibration between the spin states originally established by the external magnetic field. Thus, neighboring nuclei cannot distinguish these states and the multiplicity originally caused by the nucleus (or nuclei) disappears. For example, in propane ($CH_3CH_2CH_3$), while

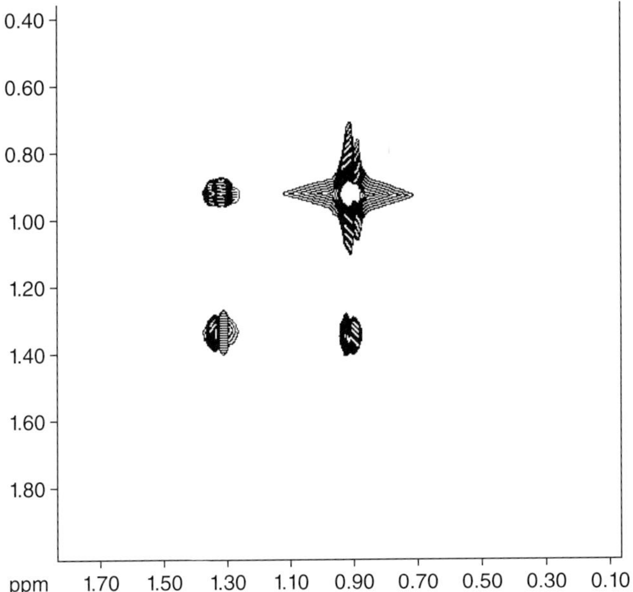

Figure 2.21. The 2-D ^{1}H NMR spectrum of propane. Spectra obtained in this way are said to be derived from COrrelated SpectroscopY (COSY).

observing the spectrum in the region of the methylene group ($-CH_2-$, $\delta \approx 1.33$ ppm) and irradiating at the center of the methyl (CH_3-) triplet, $\delta \approx 0.90$ (Figures 2.14 and 2.15), with a decoupling frequency tuned specifically for the center of the triplet, the collapse of the original septet (Figures 2.14 and 2.16) seen for the methylene ($-CH_2-$) to a singlet (Figure 2.22) is found. The results of the reverse experiment, irradiation of the methylene septet while observing what was the methyl triplet, are shown in Figure 2.23.

7. ^{13}C, ^{19}F, and ^{31}P NMR Spectroscopy. With the exception of the specific relationship between magnetic field strength and frequency at which transitions between energy levels occur, all of the aspects of the methods discussed above with regard to ^{1}H NMR spectroscopy also apply to other nuclei with spin quantum number $I = 1/2$. Indeed, observations similar to those discussed above have been made with many of them. Nonetheless, until recent technical advances made their use routine, many were not examined thoroughly and, even at this writing, major use is made of only a few. These nuclei include ^{13}C, ^{19}F, and ^{31}P. The natural abundance for both ^{19}F and ^{31}P is 100%, and the experimental techniques and results are very similar to those described for protons. By comparison, the natural abundance of ^{1}H is 99.98% and ^{13}C is only 1.11%.

For ^{19}F, the resonance frequency at a field strength of 23,500 gauss, where the resonance frequency for protons is 100 MHz, is 94.2 MHz. Therefore, examination of such spectra can be easily undertaken by providing a frequency generator that can operate either at 100 MHz or at the fairly close frequency of 94.2 MHz. As might be expected, in molecules containing both fluorine and hydrogen, coupling between ^{19}F and ^{1}H can be observed, as well as coupling between nonidentical ^{19}F nuclei.

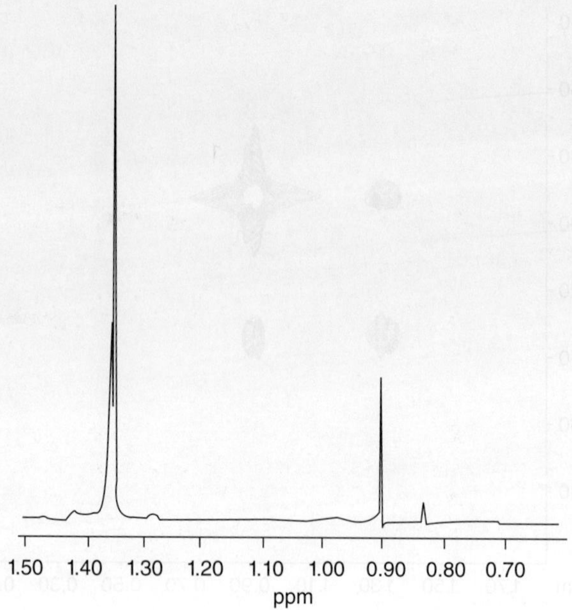

Figure 2.22. Double irradiation while obtaining the ^{1}H NMR spectrum of propane ($CH_3CH_2CH_3$). Irradiation at the methyl (CH_3–) frequency while observing at the methylene (–CH_2–) frequency.

Further, since the range of chemical shifts generally increases with the atomic number, the range for ^{19}F is nearly 600 ppm (protons having a range of about 15 ppm) and differences for different fluorine nuclei are correspondingly magnified. Finally, given the large natural abundance of ^{19}F, CW-NMR experiments were common for some time. FT-NMR is now generally used.

Although the frequency at which ^{31}P nuclei come into resonance at 23,500 gauss is only 40.5 MHz, and thus an entirely different transmitter is necessary to observe this nucleus, all of the experiments described for both ^{1}H and ^{19}F are applicable and have been carried out. Here, the chemical shift range is close to 1000 ppm and, again because of its high natural abundance, ^{31}P NMR is widely exploited.

^{13}C (which also has a range of about 300 ppm much like ^{19}F) has a low natural abundance, *as well as a magnetogyric ratio about 1/4 that of ^{1}H*; the use of this nucleus as a molecular probe awaited widespread availability of FT-NMR machines where many pulses, each containing a complete spectrum, could be accumulated and the other advantages of the method could be capitalized upon.*

In a field $\mathbf{H}_o$ = 23,500 gauss, protons (^{1}H) come into resonance near 100 MHz. In the same field, ^{13}C nuclei come into resonance about 25.2 MHz. Thus, different observing RF transmitters are required for these different nuclei. However, the same locking RF transmitter, if "external," can be employed (e.g., ^{2}H in a deuterated

---

*The problem of low natural abundance for ^{13}C (1.11% ^{13}C, the rest being ^{12}C, while for protons [^{1}H], the natural abundance is 99.98%, the rest being ^{2}H) is exacerbated by its much smaller magnetic moment. The magnetogyric ratio (γ, Equation 2.11) for ^{1}H is 2.68 radians $T^{-1}s^{-1}$ (1 T [or tesla] $\approx$ 1 × 10^4 gauss), while that for ^{13}C is only 0.0675 radians $T^{-1}s^{-1}$ (that for ^{19}F is 2.520 radians $T^{-1}s^{-1}$ and for ^{31}P is 1.086 radians $T^{-1}sec^{-1}$). The intrinsic sensitivity of a nucleus is proportional to γ^3, so that relative to protons, the sensitivity for ^{13}C is $(0.675/2.68)^3(1.11) = 0.0177\%$, that is, a factor of about 5700 relative to ^{1}H.

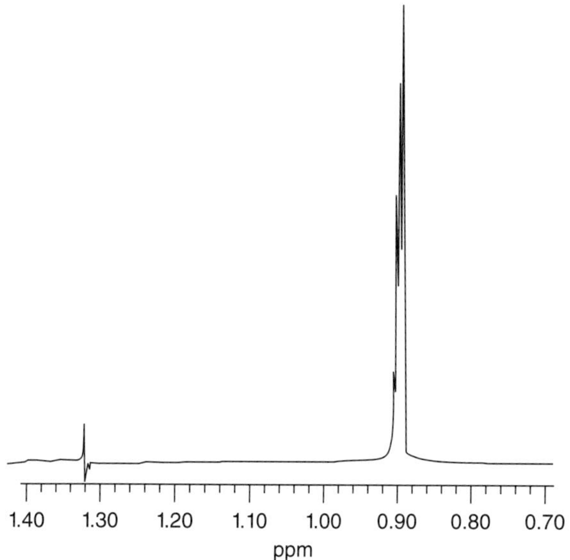

Figure 2.23. Double irradiation while obtaining the ^{1}H NMR spectrum of propane (CH$_3$CH$_2$CH$_3$). Irradiation at the methylene (–CH$_2$–) frequency while observing at the methyl (CH$_3$–) frequency.

solvent since, as noted above, at 23,500 gauss, ^{2}H comes into resonance at 15.3 MHz). Because ^{13}C NMR spectroscopy, with its nearly 500-ppm range, distinguishes between nonequivalent carbon atoms (as ^{1}H NMR, with only about a 15-ppm range, distinguishes between nonequivalent protons), it has become a vital tool in structure determination. However, whereas it is *unlikely* that a proton attached to carbon (e.g., in the hydrocarbon propane [CH$_3$CH$_2$CH$_3$]) would be found attached to a ^{13}C (because only 1.11% of the carbon is ^{13}C), it is *necessary* that each ^{13}C in such a hydrocarbon be found attached to a ^{1}H (because 99.98% of the hydrogen is ^{1}H). Thus, the splitting by ^{13}C of protons in ^{1}H NMR can usually be ignored, but the splitting by ^{1}H of carbons in ^{13}C NMR cannot. Contrarily, again because of the problem of relative natural abundance, whereas coupling between nonidentical protons to give valuable information about the kind and number of nearest neighbors was a major feature of ^{1}H NMR spectra, the low probability of finding (at natural abundance) two adjacent ^{13}C nuclei means that ^{13}C–^{13}C coupling can be ignored in ^{13}C NMR.

The examination of *coupled* (i.e., ^{13}C–^{1}H) ^{13}C NMR spectra is made difficult by the intensity of the coupling and the presence of many long-range couplings. As an example, the proton-coupled ^{13}C NMR spectrum of propane (CH$_3$CH$_2$CH$_3$) is provided as shown in Figure 2.24.* To remove such complexity, various **decoupling**

*In the ^{13}C gas-phase NMR spectrum of propane (CH$_3$CH$_2$CH$_3$), the methyl groups (CH$_3$–) are reported to be found at a chemical shift of $\delta \approx 15.4$ ppm and the methylene (–CH$_2$–) at $\delta \approx 15.9$ ppm (Levy, G. C.; Lichter, R. L.; Nelson, G. L. *Carbon-13 Nuclear Magnetic Resonance Spectroscopy*, 2nd edition, Wiley-Interscience, New York, 1980, p. 52). The spectra shown here are in ^{2}HCCl$_3$, and it is clear that (Figure 2.24) the quartet due to the coupling of the three protons on each of the methyl groups is centered at about $\delta \approx 16.4$ ppm, while the triplet (shaded in Figure 2.24) is centered at about $\delta \approx 16.2$ ppm (i.e., the methyl group is found downfield of the methylene). The decoupled spectrum (Figure 2.25) provides the same information.

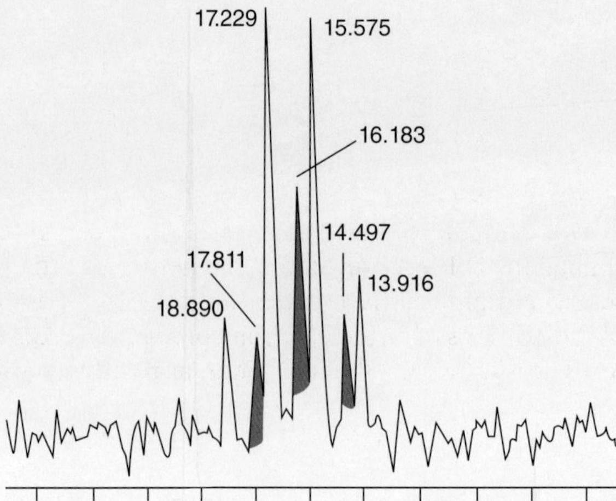

Figure 2.24. The ^{1}H coupled ^{13}C NMR spectrum of propane (CH$_3$CH$_2$CH$_3$). The values on the spectrum are parts per million (δ).

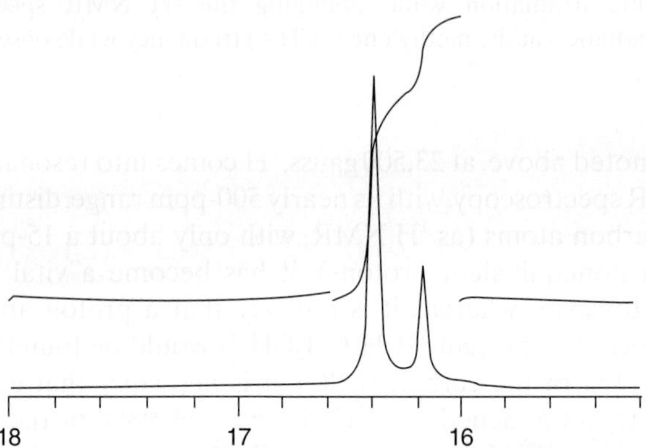

Figure 2.25. The noise-decoupled ^{13}C NMR spectrum of propane (CH$_3$CH$_2$CH$_3$) in ^{2}HCCl$_3$. The ^{13}C resonances are found at $\delta \approx 16.17$ ppm and $\delta \approx 16.39$ ppm, respectively.

techniques are employed. While the principle is similar to that described above for proton–proton decoupling, it is common here to use a wide-band RF oscillator (or in FT spectrometers, a long pulse). The width and intensity of the decoupling RF is such that *all* protons are decoupled simultaneously (called **noise decoupling**).

This is shown for propane (CH$_3$CH$_2$CH$_3$) in Figure 2.25. Such a decoupled spectrum results in a series of lines (**singlets**), *one for each of the unique carbon atoms at its unique chemical shift*; that is, there has been a general collapse of all proton coupling to carbon. Because of the large range of chemical shifts in these spectra, it is rare that two nonidentical carbon atoms would have chemical shifts that could not be distinguished and, again, the use of higher magnetic fields (and correspondingly higher frequencies) can spread out the spectrum so as to minimize the small potential for peak overlap.

b. ESR. ESR spectroscopy provides a way of examining the environment at nuclei bearing **unpaired electrons**, such as are present in **carbon free radicals**, which are relatively rare in organic chemistry.

Since an electron can be considered as a negatively charged particle that, like a proton, possesses the property called spin, it also has a magnetic moment, and the discussion above relating to such particles applies. Also, as was the case with the proton, there are two possible orientations ($+1/2$ and $-1/2$) that define the two energy states in a magnetic field. However, the magnetogyric ratio (γ) (see Equation 2.11) for the electron is significantly greater (almost 10^3 times) than that of the proton (1.76×10^3 radians $T^{-1}s^{-1}$ for the electron compared to 2.68 radians $T^{-1}s^{-1}$ for the proton, where $1\,T$ [tesla] $\approx 1 \times 10^4$ gauss). Thus, at relatively low magnetic fields (about 3600 gauss) compared to those used for routine NMR, the absorption frequency of free electrons is near 10^4 MHz (10^{10} Hz or 10 GHz), which is in the microwave region of the electromagnetic spectrum (Table 2.1).

The basic ESR system is analogous to the NMR system already discussed, in which the frequency is held constant and the field is swept. There is a difference in that at the higher frequencies used in ESR, RF power is better conducted by waveguides than by coaxial cable. Thus, the sample is inserted through a hole in the waveguide where the magnetic vector of the electromagnetic wave is at a maximum (determined by tuning the system with a known sample or standard). To obtain the best signal-to-noise ratio, the spectrometer is designed to plot the first derivative of the curve of absorption versus the magnetic field rather than the absorption curve proper (Figure 2.26).

In organic chemistry, the main function of ESR spectroscopy is to gather information about transient species such as the methyl radical ($\cdot CH_3$), a very reactive substance arising, in principle, by the loss of a hydrogen with an electron (i.e., a hydrogen atom) from methane (CH_4). Fortunately, since such species might be formed in low concentration, the sensitivity of the ESR technique is high, so that under favorable circumstances, concentrations of radicals as low as 10^{-12} M can be detected. The coupling between the "free" electron and nearby hydrogens (called *the hyperfine coupling*) can be measured and used as an identifying tool. Hyperfine

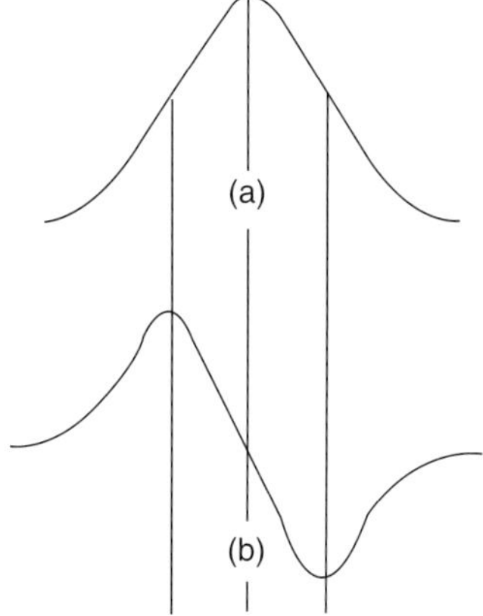

Figure 2.26. Comparison of (a) an absorption curve and (b) its first derivative.

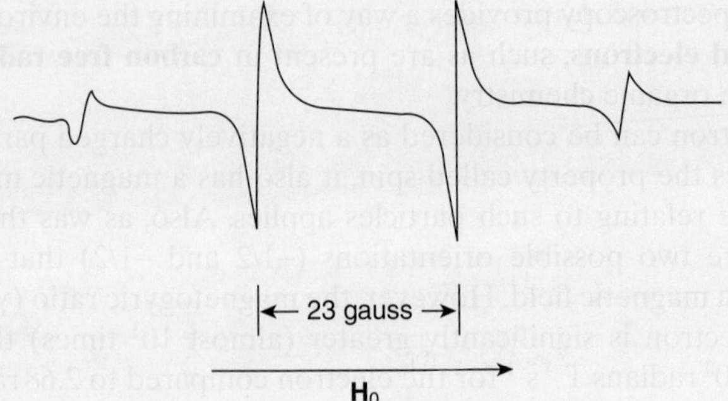

Figure 2.27. A representation of the ESR spectrum of the methyl radical.

splittings generally result in very broad signals, in part due to the fact that lifetimes of electrons in the two available spin states are short and here (as in NMR), the line width is related to the reciprocal of the lifetime. In the methyl radical ($\cdot CH_3$), the four resonance lines, in the ratio 1:3:3:1, resulting from the coupling of the three hydrogens to the electron ($n + 1$), have a coupling constant of about 23 gauss (64.5 MHz), which is dramatically larger than the coupling constants (1–15 Hz) commonly found in the proton NMR for proton–proton coupling (Figure 2.27).*

D. MS

A **mass spectrometer** is a device for creating and sorting gaseous charged molecules according to their mass-to-charge ratio, m/z. Although it uses small quantities of material (frequently in the range of micrograms [10^{-6} g] to nanograms [10^{-9} g]), it stands in contrast to all of the spectroscopic techniques discussed so far in that the sample is destroyed.†

---

*The actual placing of the free electron on carbon (rather than elsewhere) derives, in part, from observing the coupling of that electron to the hydrogens on methyl and the ability to assign, to the transition observed (E_{21} or ΔE), a value proportional to the field H_o and whose constants of proportionality are a function of its location (Equation 2.13). The **Bohr magneton**, β, is a constant, different for each nucleus and charged particle and whose value, for the electron, is 2.216×10^{-24} cal T^{-1}). The **splitting factor**, g, is a dimensionless constant of proportionality, which has a value of 2.0023 for free electrons and a value close to that for carbon-centered free radicals. Were the electron centered on a nucleus other than carbon, the value of the proportionality constant would change (as would β) and the identification of the new location would be ascertained:

$$\Delta E = g\beta H_o. \tag{2.13}$$

†It can be argued that this method of analysis is considered a form of spectroscopy *only* because the usual format for data display is a plot of relative intensity versus mass number and thus resembles the output from the other forms of spectroscopy already discussed.

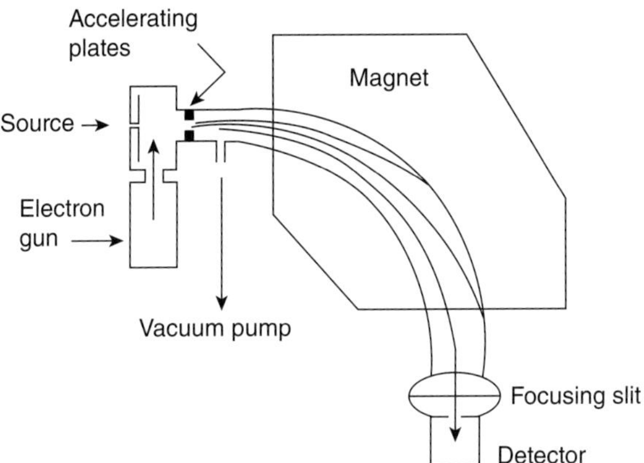

Figure 2.28. A schematic diagram of a 90° (the angle subtended between source and detector) magnetic sector mass spectrometer.

A wealth of structural information as well as a "simple" molecular mass can be gathered from an examination of the mass spectrum of a molecule. Because advances in the technology of creating and manipulating gaseous ions and storing and analyzing the resulting data have been developed, MS has become a routine tool in many laboratories. Indeed, in some quarters, a high-resolution mass spectrum is currently considered equivalent, or superior, to elemental analysis as part of the requirements for structure proof.

Regardless of the details of the various kinds of spectrometers and the various techniques needed for analysis of solids, liquids, and gases, all mass spectrometers have certain common characteristics. Each must have a way to ionize the sample, to accelerate the ions away from their point of creation, to separate the ions according to their masses, and to detect the separated ions. All mass spectrometers manipulate the ions by combinations of magnetic and electric fields, and all of them require that the ions be handled at high vacuum so that collisions with walls, other gaseous species, and so on, may be minimized. Typically, vacuum at least 10^{-5} torr is required, and this can be obtained with a series of vacuum pumps. Again, typically, the ions are accelerated by causing them to be passed between electrically charged plates. A schematic drawing of a mass spectrometer showing these features is presented in Figure 2.28.

I. Creation of Ions in the Mass Spectrometer: The Ionization Chamber

When a material has sufficient vapor pressure so that enough is vaporized in the vacuum of the sample chamber, some of it may be converted to ions by bombardment with a beam of electrons (from an electron gun) whose energies are high enough to dislodge a nonbonding or bonding electron. This method is known as **electron impact** (EI) ionization and, while the energy of the electrons in the stream from the source might be adjusted in almost any range, normally, 50–100 eV is used and 70 eV ($1\,eV = 23.06\,kcal\,mol^{-1}$) is common. As shown in Equation 2.14, loss of an electron from ethane (CH_3CH_3) results in the formation of an ion, the (odd electron) ethyl cation radical* $(CH_3CH_3)^{\bullet+}$:

$$CH_3CH_3 + e^- \rightarrow (CH_3CH_3)^{\bullet+} + 2e^-. \tag{2.14}$$

As an alternative to generating parent ions by bombardment with high-energy electrons, it is possible to form them by the transfer of an electron from the material being analyzed to another electron-deficient species. For example, in the technique called **chemical ionization** (CI), the sample is diluted with a large (about 10^4 times) excess of a reagent gas before being subjected to the electron beam. Under these circumstances, it is much more likely that the electrons from the high-energy source will encounter the reagent gas than the sample so that initially, the ions formed will be those from the reagent gas. Subsequent collisions between the electron-deficient reagent gas and the sample will result in electron transfer. Although the radical ions formed from the sample have the same gross structure as the parent ions formed by EI, they possess much less energy and undergo less initial fragmentation into daughter ions.

A variety of other specialized techniques, particularly applicable to materials of low vapor pressure, have been developed or are currently under development.

II. The Separation of Ions by Mass: The Mass Analyzer

Although there are a number of ways that ions can be separated by their masses, one of the most common mass analyzers (Figure 2.28) makes use of a device called a **magnetic sector**.

As we have already noted, charged particles in a magnetic field move in a circular path. If the charged particles are accelerated *into* the magnetic field (i.e., if they are provided with an accelerating voltage), then the radius of the circle they travel is a function both of their mass (m) and charge (z), as well as the accelerating voltage and magnetic field. Using only a piece of the circle (i.e., 60°, 90°, or even 180°— values common in commercial instruments) and adjusting the voltage and magnetic field, ions of any specific mass/charge (m/z) can then be focused so they will exit the magnetic field centered between the poles. If an appropriate detector is then mounted at the exit (Figure 2.28), scanning an entire range of masses so that they fall, sequentially, on the detector can be accomplished by sweeping either the magnetic field or the voltage. Ions not striking the detector (e.g., because the magnetic field is wrong) will be lost. Although this particular design suffers limitations due to the inability to focus the ions well (sharply defining the boundaries of the mag-

---

*It might be reasonable to ask, considering localized molecular orbitals (Chapter 1), from which orbital the electron was lost. Alternatively, considering the self-consistent field molecular orbital (SCF-MO) picture (Chapter 1), it might be asked if reorganization of bonding levels has occurred. Generally, because the lifetime of the ion is long compared to electron movement, a minimization occurs. Occasionally, the instability of the initially formed radical ion (the **parent ion**) causes it to rapidly fragment (fall apart) so that only pieces ("daughter ions" or "fragments") are observed. Such paths will be discussed.

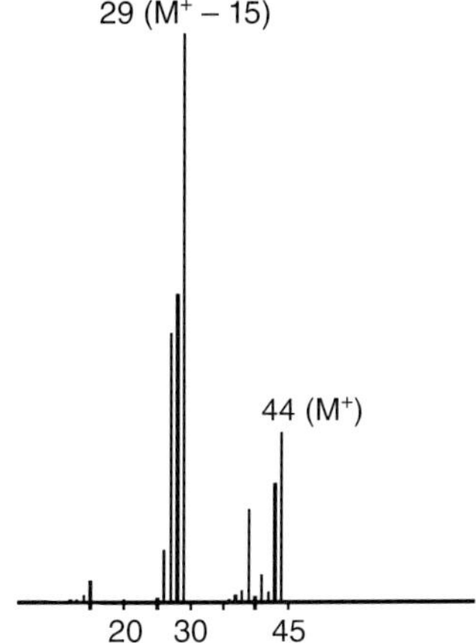

Figure 2.29. The mass spectrum of propane ($CH_3CH_2CH_3$). The molecular ion (M^+) is found at m/z 44. The main fragmentation pattern is to lose the equivalent of 15 mass units (i.e., methyl [$CH_3·$]) to produce the $C_2H_5^+$ cation (m/z 29).

netic field is difficult), the general principle continues, with refinements, to be widely used.

III. Detecting the Ions

The most common (and probably least sensitive) detector is called a *Faraday cup*. This device consists simply of an insulated, cup-shaped conductor, connected directly to an amplifier. The cup shape helps prevent electrons, discharged when the cup is struck by an ion, from being lost.

A typical spectrum (that of propane, $CH_3CH_2CH_3$) is provided in Figure 2.29.

ADDITIONAL PROBLEMS

Problem 2.2. The chapter contains spectroscopic data for propane ($CH_3CH_2CH_3$). As will be subsequently developed, the next higher homologue of the hydrocarbon series of which propane ($CH_3CH_2CH_3$) is a member is butane (sometimes called *n*-butane, $CH_3CH_2CH_2CH_3$). Using any resources available to you, find (or make predictions about) the UV, IR, ¹H NMR, ¹³C NMR, and mass spectrum of *n*-butane ($CH_3CH_2CH_2CH_3$).

Problem 2.3. As will be developed, there is a compound with the same empirical formula as *n*-butane (i.e., C_4H_{10}) that has a different arrangement of the atoms and which is called 2-methylpropane (or sometimes, *iso*-butane). Its structure, provided below, is sometimes abbreviated as ((CH_3)$_3$CH or (CH_3)$_2$CHCH$_3$). Using any resources available to you, find (or make predictions about) the UV,

IR, ^{1}H NMR, ^{13}C NMR, and mass spectrum of *iso*-butane ((CH_3)$_3$CH or (CH_3)$_2$CHCH$_3$).

2-Methylpropane (*iso*-butane)

REFERENCE

Roberts, J. D.; Caserio, M. C. *Basic Principles of Organic Chemistry*, 2nd edition, Benjamin, **1977**.

Structure: The Nomenclature of Hydrocarbons and the Shape of Things to Come

A Chrysanthemum by another name would be easier to spell.

—William J. Johnston

A. INTRODUCTION

The communication of information through drawing structural formulas and writing equations generally permits organic chemists to understand each other. However, it is difficult to converse in structures and equations. So, as organic chemistry developed, what compounds are called became important to the exchange of information. Initially, names of compounds were based on the source of the material. While some of these names remain in current use, and even now newly isolated complex materials are given "trivial" names relating to their sources and/or structures, other systems for naming compounds were gradually developed.

First, commercial and common (or trivial) names for a host of organic molecules abound, and these will be used parenthetically where appropriate here so that you might come to recognize them. However, the preferred system of nomenclature is the definitive set of rules adopted by the International Union of Pure and Applied Chemistry (**IUPAC** rules to be found on the Internet at http://www.chem.qmul.ac.uk/iupac/iupac.html; see Appendix II). While the set is occasionally revised (last in 1979), most of what you learn will remain applicable.

Second, it is typical of organic chemistry that transformations of molecules be examined. This is because the raw materials wrenched from coal or petroleum, which are our current major hydrocarbon sources, are usually not the ones that we want to eventually use. They must be converted to other (desired) products so as to maintain our good health or way of life. Since many of the same starting materials are converted to different products, it is also necessary to have distinct names for the new substances. Indeed, with well over 4 million different organic compounds known (you will learn only a small fraction), major chaos would result without organization. The nomenclature rules for hydrocarbons serve both as a logical point

Foundations of Organic Chemistry: Unity and Diversity of Structures, Pathways, and Reactions,
First Edition. David R. Dalton.
© 2011 John Wiley & Sons, Inc. Published 2011 by John Wiley & Sons, Inc.

of entry to the system, and their names frequently serve as the root from which elaboration can spring.

Third, although this chapter is largely devoted to the systematics of hydrocarbon naming (and some reactions of hydrocarbons), an example of the use of a parent (or root) name and its elaboration is warranted. You already know the name of the compound with one carbon and four hydrogens as **methane** (CH_4).

Replacement of one of the hydrogens (H) of methane by chlorine (Cl) generates **chloromethane** (methyl chloride, CH_3Cl); of two hydrogens by two chlorines, **dichloromethane** (methylene chloride, CH_2Cl_2); of one hydrogen of methane (CH_4) by an –OH group, **methanol** (methyl alcohol, CH_3OH); of two hydrogens by one oxygen, **methanal** (formaldehyde, $H_2C=O$), and so on. It is hoped you will have noticed that the "root" fragment **methan** retained its identity throughout and that it signified that **only one carbon atom was present**.

Finally, **you must carefully watch prefixes and suffixes** as well as spellings. Being nearly correct, for example, using an "a" instead of an "e," such as writing eth**a**ne (C_2H_6 or CH_3CH_3) when you mean eth**e**ne (C_2H_4 or $CH_2=CH_2$) or, perhaps, eth**y**ne (C_2H_2, or $HC\equiv CH$), is being wrong and disaster (as well as diminished examination scores) can result.

B. NOMENCLATURE AND SPECTROSCOPY

I. Alkanes

We begin by naming the alk**ane**s. These are hydrocarbons containing (by the linear combination of atomic orbitals [LCAO] method) *only sp³*-hybridized carbon atoms bonded to hydrogen or other carbon atoms; that is, there are **no** double or triple (i.e., π) bonds and such compounds are called **saturated**. In accord with (approximately) tetrahedral geometry, each carbon bears **four** ligands; these may be either another carbon (or carbons) and/or an appropriate number of hydrogens. In these compounds, carbon–carbon bonds are on the order of 154 pm (1.54 Å) long and carbon–hydrogen bond lengths about 109 pm (1.09 Å).

Broadly, there are two categories of these alk**ane**s: **acyclic** (without **rings**) and **cyclic** (with rings). While both **acyclic** and **cyclic** alk**ane**s are similar, they also have minor differences; we will first continue with **acyclic** alk**ane**s.

a. Acyclic Alkanes. The first four acyclic alk**ane**s have nonsystematic names; they are **methane** (CH_4), **ethane** (CH_3CH_3), **propane** ($CH_3CH_2CH_3$), and **butane** ($CH_3CH_2CH_2CH_3$) and are composed of one, two, three, and four carbons, respectively, and their attendant hydrogens—so that there are **four bonds to carbon**. Beginning with **pentane**, which has five carbon atoms ($CH_3CH_2CH_2CH_2CH_3$), the remaining **continuous-chain** (or **unbranched**) alk**ane**s, that is, those in which all the carbon atoms are linked to **no more than two** other carbon atoms and either three hydrogens (if they are terminal carbons) or two hydrogens (if they are not terminal carbons), are named by adding the appropriate Latin or Greek prefix to the stem **-ane**. Thus, for the compound with 6 carbons (C_6H_{14}), we have **hexane**; for 7 carbons (C_7H_{16}), **heptane**; for 8 carbons (C_8H_{18}), **octane**; for 9 carbons (C_9H_{20}), **nonane**; and for 10 carbons ($C_{10}H_{22}$), **decane**. All of these alk**ane**s correspond to the

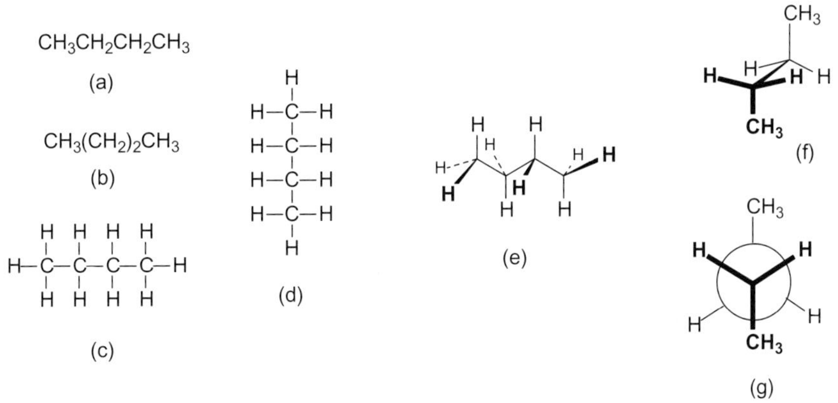

$CH_3CH_2CH_2CH_3$

(a)

$CH_3(CH_2)_2CH_3$

(b)

(c)

(d)

(e)

(f)

(g)

Figure 3.1. Different representations of butane (C_4H_{10}): (a) in-line type set; (b) in-line condensed; (c) expanded horizontally; (d) expanded vertically; (e) expanded three-dimensionally, approximately showing each carbon as a tetrahedron; (f) a "sawhorse" projection looking along the carbon–carbon σ bond connecting the second and third carbon atoms; and (g) a **Newman** (Melvin S. Newman [1908–1993], professor, The Ohio State University) projection in which the σ bond between the second and third carbon atoms is represented as a circle.

TABLE 3.1. IUPAC Names of Unbranched (or Normal) Alkanes (C_nH_{2n+2})

n	Name	Formula	n	Name	Formula
1	Methane	CH_4	11	Undecane	$CH_3(CH_2)_9CH_3$
2	Ethane	CH_3CH_3	12	Dodecane	$CH_3(CH_2)_{10}CH_3$
3	Propane	$CH_3CH_2CH_3$	13	Tridecane	$CH_3(CH_2)_{11}CH_3$
4	Butane	$CH_3(CH_2)_2CH_3$	14	Tetradecane	$CH_3(CH_2)_{12}CH_3$
5	Pentane	$CH_3(CH_2)_3CH_3$	15	Pentadecane	$CH_3(CH_2)_{13}CH_3$
6	Hexane	$CH_3(CH_2)_4CH_3$	20	Icosane	$CH_3(CH_2)_{18}CH_3$
7	Heptane	$CH_3(CH_2)_5CH_3$	30	Triacontane	$CH_3(CH_2)_{28}CH_3$
8	Octane	$CH_3(CH_2)_6CH_3$	40	Tetracontane	$CH_3(CH_2)_{38}CH_3$
9	Nonane	$CH_3(CH_2)_7CH_3$	50	Pentacontane	$CH_3(CH_2)_{48}CH_3$
10	Decane	$CH_3(CH_2)_8CH_3$	100	Hectane	$CH_3(CH_2)_{98}CH_3$

formula C_nH_{2n+2}, where n is the number of carbon atoms. This is true of **all** acyclic alk**ane**s.

In all of these compounds, carbon is approximately tetrahedral and not only must you keep the three-dimensional form in mind but you must also quickly grow accustomed to various ways of presenting information that needs to be communicated. Considering butane ($CH_3CH_2CH_2CH_3$, C_4H_{10}) as an example, the following representations (Figure 3.1) are all simply different ways of **writing** what is meant by "butane," *and they are all equivalent.*

Table 3.1 presents the names of some of the common alkanes and the pattern of nomenclature. While it is not necessary to commit the names of all of the alkanes in Table 3.1 to memory, you should know at least the first ten.

Near the end of the eighteenth century, Berzelius (Jöns J. Berzelius, Stockholm [1779–1848], literally the inventor of "organic chemistry") had already found cases

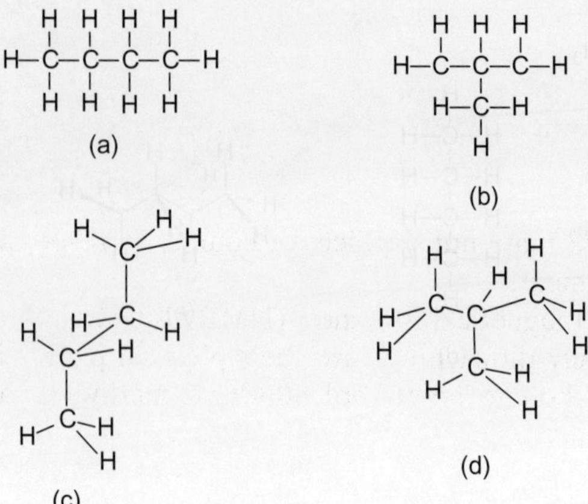

Figure 3.2. Representations of the isomeric hydrocarbons C_4H_{10}: (a) and (c) represent butane (*n*-butane); (b) and (d) represent 2-methylpropane (isobutane).

of different materials with the same elemental composition but exhibiting different physical properties, and had minted the word **isomer** or **isomorph** (of each other) to define the phenomenon.* Beginning with material with four carbon atoms (i.e., C_4H_{10}), it is possible to write *more than one hydrocarbon structure corresponding to this elemental composition*. As shown in Figure 3.2, the compound with the name **butane** (sometimes called "*n*-butane") has two carbons bearing three hydrogens (CH_3-, called a **methyl group**), and each is connected to another carbon bearing two hydrogens ($-CH_2-$, called a **methylene group**), for a total of 4 carbons and 10 hydrogens. It can be considered to have a continuous or **unbranched** chain of carbons (Figure 3.2a,c). Its isomer (sometimes called "isobutane") has three methyl groups, with all of them attached to one carbon bearing only one hydrogen ($-CH-$, called a **methine group**), for a total of 4 carbons and 10 hydrogens. The chain is **branched** at the methine (Figure 3.2b,d). This new constitution of matter is a **positional** or **constitutional** isomer of the original hydrocarbon butane. It should be obvious that as the total number of carbons and hydrogens in such acyclic hydrocarbons increases, the numbers of isomers, as a result of branching, will also dramatically increase. Clearly, some rational system of nomenclature is required. The IUPAC name for the isomer of butane (isobutane) is 2-methylpropane. We arrive at this name systematically, as described below.

Like most alkanes, neither butane (Figure 3.2a) nor its isomer 2-methylpropane (Figure 3.2b) has significant absorption in the 200–400 nm (UV) region of the electromagnetic spectrum. In these compounds, with only σ bonds and no *n* or non-bonded electrons, only $\sigma \rightarrow \sigma^*$ absorption is possible ($\lambda_{max} < 200$ nm). Both isomeric butanes have intense absorption in the 3200–2800 cm^{-1} region of the infrared (IR) (C–H stretching) as well as the weaker C–C stretching and C–H bending modes

---

*His student, Wöhler (1800–1882), described (Wöhler, F. *Ann. Phys. Chem.*, **1928**, *88*, 254) the thermal conversion of the inorganic salt ammonium cyanate ($NH_4^+OCN^-$) to urea (H_2NCONH_2), an organic substance isolated from urine. This work has been considered an important step in discrediting the then current concept of a "vital force" (vitalism) supposedly possessed by living systems and their components and otherwise absent.

at longer wavelengths. The major differences in their IR spectra, allowing them to be uniquely identified, are in the part of the fingerprint region between 1200 and $400\,cm^{-1}$.

In their mass spectra, both butane and 2-methylpropane have small parent ions (M^+) at m/z 58 mass units, although the parent has about four times greater abundance for butane than it does for 2-methylpropane. The most intense (100%) peak for both is at m/z 43 mass units, which, of course, corresponds to a three-carbon eight-hydrogen fragment.[*][†]

The ^1H nuclear magnetic resonance (NMR) spectra of the two isomers of C_4H_{10} are dramatically different[‡] as are their physical properties (butane: melting point [mp] –139°C, boiling point [bp] –0.4°C; 2-methylpropane: mp –160°C, bp –10.2°C).[§]

Problem 3.1. Sketch the ^1H NMR spectrum of the two isomers of butane (n-butane and 2-methylpropane), showing the approximate chemical shifts (relative to tetramethylsilane [TMS]), the integrated areas, and the multiplicities expected. Assume $J = 7.0\,Hz$.

---

[*]For butane, the masses of the observed fragments (and their *relative* intensities) are found at m/z 58 (11), 43 (100), 42 (12), 41 (28), 39 (11), 28 (27), and 27 (28). For 2-methylpropane, the masses of the observed fragments (and their *relative* intensities) are found at m/z 58 (3), 43 (100), 42 (32), 41 (38), 39 (17), 29 (6), 27 (28), and 15 (7).

[†]The base peak in most unbranched alkanes is found at m/z 43, which is followed, in *decreasing* intensity, by peaks in increments of 14 mass units (the mass of a methylene unit [–CH_2–]), which correspond to the cleavage of each carbon–carbon bond in the molecule.

[‡]*For saturated alkanes* in $^2HCCl_3$ solution (the solvent in which many spectra are taken), protons on methyl groups, that is, methyl protons (CH_3–), appear at or near $\delta \approx 0.9$ ppm, while methylene protons (–CH_2–) are somewhat deshielded and appear further downfield near $\delta \approx 1.3\,ppm$. Methine protons (–CH–) are even more deshielded with $\delta \approx 1.5\,ppm$.

[§]Generally, branched alkanes melt lower than unbranched alkanes of the same empirical formula. If, however, the branched compound is dramatically more symmetrical than its unbranched isomer, the melting point will be higher. Determining this by inspection is not easy, so predictions concerning melting point are very difficult to make. (In part, the problem deals with how molecules fit into a crystal lattice.) However, the effect on the boiling point is somewhat more predictable and, again, branching lowers the temperature at which boiling occurs. It is usually argued that branched alkanes have lower boiling points than their unbranched isomers because, in the former, the surface area of the molecule is smaller and thus the area available for intermolecular associations is less. The nature of these interactions in nonpolar molecules, such as these hydrocarbons, has been of interest for a long time, since it was felt necessary to attempt to explain why they could be liquified at all. The classic argument runs about as follows: as two nonpolar molecules (or atoms, for that matter) with filled bonding orbitals approach each other, the interaction between them should steadily become more repelling. At very close distances (as seen with hydrogen nuclei in Chapter 1), that is exactly what happens. At long distances, the same (but certainly smaller) effect might be expected. Indeed, at distances greater than the sum of the *van der Waals* radius for each interacting nonbonded atom or group, they *do* repel each other (J. D. van der Waals, Dutch physicist [1837–1923]). However, there is a small range of distances at which nonbonded atoms must apparently attract each other, or else liquids could not form. These weak attractive forces are called *van der Waals attractive forces* or *London forces* (F. London, Germany [1929]; see also Chapter 1) or *dispersion forces*. It is widely held that they arise as a consequence of temporary dipoles that exist intramolecularly because the electron distribution is not static and these, in turn, create or induce dipoles in neighboring molecules. Thus, the attractive force can be described as an induced-dipole–induced-dipole interaction. Further, although individually these forces must be quite weak and fluctuating, given the large numbers of molecules over which they interact, the total is significant.

Problem 3.2. One of the isomers of pentane (C_5H_{12}), whose IR spectrum contains the ubiquitous C–H absorption in the 3200–2800 cm^{-1} region, has only a singlet (at $\delta \approx 0.9$ ppm in $^2HCCl_3$ solution) in its ^1H NMR spectrum. Its ^{13}C NMR spectrum has only two lines. What structure can be assigned to this isomer?

The systematization for the nomenclature of acyclic alkanes (IUPAC) has the following rules:

(I) Identify the **longest continuous chain** of carbon atoms. The **longest continuous chain** takes the name of the parent hydrocarbon. It is important you particularly note that the **two-dimensional representations**, for example, (a) and (b) in Figure 3.2, can be written left to right or top to bottom as well as twisted. Similarly, the hydrogens associated with any carbon can be written before or after the carbon, and so on. The three-dimensional representations, for example, (c) and (d) in Figure 3.2, should clarify that how the two-dimensional representation of the **tetrahedral** carbon atoms is written is unimportant. This concept is illustrated in Figure 3.3 for the hydrocarbon hexane (C_6H_{14}, *n*-hexane), where **all** of the representations shown are equivalent.

(II) Identify any **substituent groups**, that is, those that are not part of the longest continuous chain but are attached to it. For hydrocarbons, the names of the **substituents themselves** are based on the corresponding hydrocarbons and are assigned by replacing the **-ane** ending of their name with **-yl**. These groups are called **alkyl groups**. Among the simplest and most common are CH_3- (the **methyl** group) and CH_3CH_2- (the **ethyl** group). Examples using one of the isomers of octane, 2,4-dimethylhexane, are found in Figure 3.4.

(III) The parent hydrocarbon is numbered (from either end of the longest continuous chain) in such a way that the substituent(s) are assigned the **lowest**

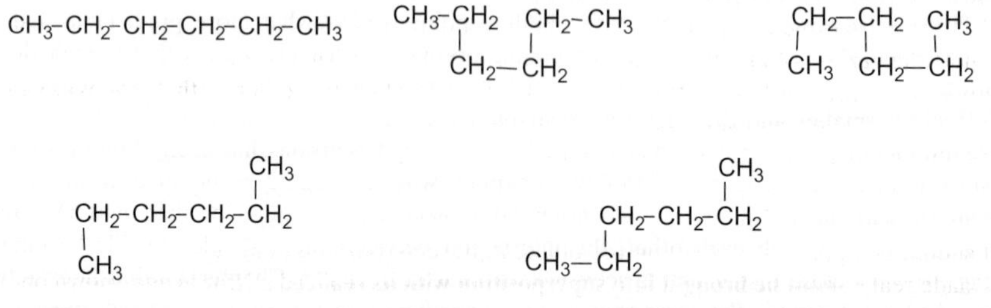

Figure 3.3. Various equivalent representations for the hydrocarbon *n*-hexane (C_6H_{14}).

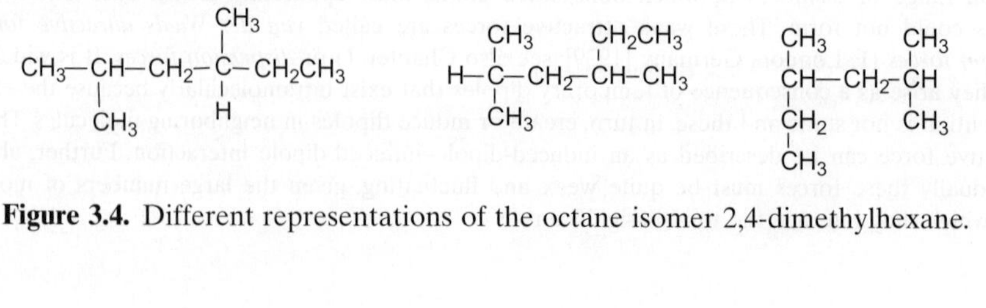

Figure 3.4. Different representations of the octane isomer 2,4-dimethylhexane.

number(s). Thus, as shown in Figure 3.4, the **methyl** substituents are assigned the numbers 2 and 4 rather than 3 and 5. Numbers, specifying the positions of the substituents, are **separated by commas**. Prefixes (di-, tri-, tetra-, penta-, hexa-, etc.) are used to designate **how many identical substituents** lie along the chain. The prefixes are separated from the numerical position identifiers by a **hyphen** and are fused with the name of the substituent. Thus, in the case of the octane isomer 2,4-dimethylhexane, the prefix **di-** (meaning two) follows the numbers but precedes the substituent name **(methyl)**.*

(IV)   When there are two or more different substituents present, they are named **alphabetically** and **without regard to the prefixes** (such as di-, tri). Thus, the undecane isomer shown in the next figure is named 4-**e**thyl-3,3-dimethylheptane:

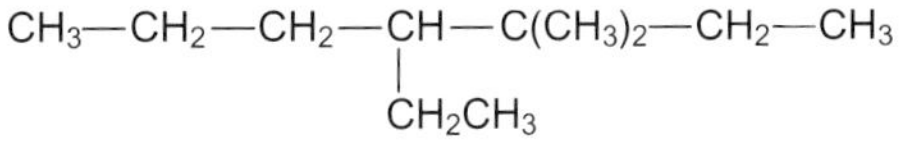

$$CH_3—CH_2—CH_2—CH—C(CH_3)_2—CH_2—CH_3$$
$$|$$
$$CH_2CH_3$$

(V)   If there is a branch **in the substituent** chain, then the entire name of the branch is set off in parentheses and is **numbered beginning with the carbon directly attached to the parent chain**. Thus, the dodecane isomer in the next figure is named 2-methyl-3-(1-methylpropyl)heptane:†

---

*The prefix mono- is not used except conversationally for emphasis. Further, in case of confusion over two possible number systems (i.e., as a result of numbering left to right or right to left), preference is given to that one which has the lowest number **at the first point of difference**. Thus, the nonane isomer shown in the next figure is 2,**3**,5-trimethylhexane and not 2,**4**,5-trimethylhexane:

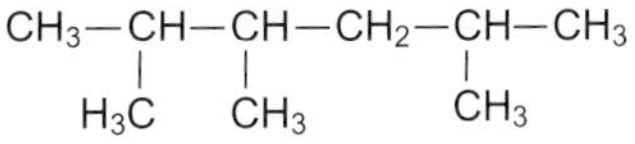

$$CH_3—CH—CH—CH_2—CH—CH_3$$
$$|\qquad|\qquad\qquad|$$
$$H_3C\quad CH_3\qquad CH_3$$

There are other kinds of isomers besides those dependent upon the positions of substituents on the chain. Although more will be said about this topic subsequently, a demonstration of the nature of one of these other kinds of isomerism is appropriate here (by way of introduction).

Remembering that the geometry of each carbon is approximately tetrahedral, carefully examine the situation at the carbon atom number 3 in 2,**3**,5-trimethylhexane. That carbon bears the following substituents: (1) a hydrogen, (2) a methyl group, (3) a carbon to which two other methyl groups are attached, and, finally, (4) a branched chain of four carbons. These are shown in Figure 3.5a in a redrawn version of 2,3,5-trimethylhexane.

If the two-dimensional representation of Figure 3.5a is recast into a three-dimensional representation as Figure 3.5b and is simplified with the abbreviation of Figure 3.5c, the result can be shown as Figure 3.5d. It should be apparent to you that the object represented on the right side of the mirror in Figure 3.5d, if made real, **cannot be brought into superposition with its realized mirror image**, shown on the left of Figure 3.5d.

This property of **handedness** is called **chirality** (from the Greek word *cheir*, meaning "hand" and pronounced as *ki-ral-it-e*), and it is possessed by many objects (such as hands, gloves, shoes, and screws) as well as most organic molecules. Such kinds of isomers are called **stereoisomers** (from the Greek word *stereos*, meaning "solid"). Finally, while it is generally true that any organic molecule in which there is only one carbon atom bearing four **different** ligands may be drawn in a fashion such that its real representation cannot be brought into superposition with that of its realized mirror image, other molecules that lack such a carbon may also be chiral; that is, the property of handedness does not require an asymmetrically substituted carbon atom.

†Normally, of course, a **1-methyl** substituent would simply result in a longer chain; for example, 1-methylbutane is an **incorrect** name for pentane.

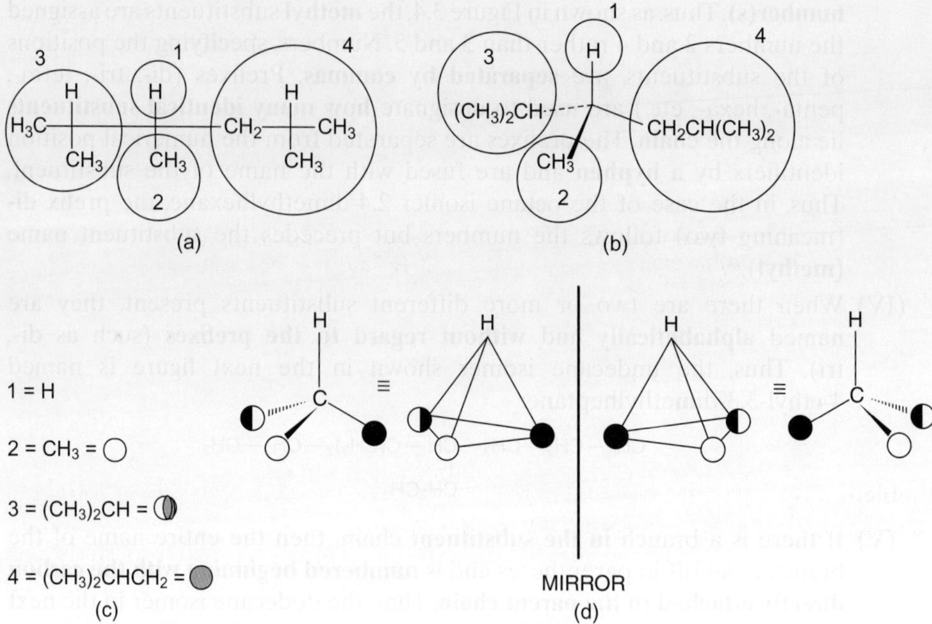

Figure 3.5. (a) A representation of 2,3,5-trimethylhexane showing that the four attachments to the third carbon atom in the chain are different from each other. (b) A redrawing of (a) depicting the (approximately) tetrahedral nature of one particular carbon atom with the four numbered ligands. It is intended that group "2" is above the plane of the paper; "3" is behind the plane of the paper; and "1" and "4" are in the plane of the paper. (c) Abbreviations of the substituents of (a) and (b). (d) A representation of an asymmetrically substituted carbon atom and its mirror image. The realized mirror image representations cannot be brought into superposition without breaking a bond and making a reconnection.

$$H_3C-CH-CH_2-CH_3$$
$$CH_3-CH-CH-CH_2(CH_2)_2CH_3$$
$$CH_3$$

(VI) Finally, any alkyl group attached to **only one other carbon** is called **primary**; if attached to **two** other carbons, then it is **secondary**; to **three** other carbons, **tertiary**; and attached to **four** other carbons, **quaternary**. While these designations are not part of formal IUPAC names, they are part of the systematic description of compounds. Thus, in the earlier example of the isomers of C_4H_{10}, butane and 2-methylpropane, the former may be said to have two primary carbons (in the methyl groups) and two secondary carbons (in the methylene groups), while the latter has three primary carbons (in the methyl groups) and one tertiary carbon (in the methine).

Problem 3.3. For the undecane isomer shown in IV above (4-ethyl-3,3-dimethylheptane), decide which carbon atoms have four different ligands attached. Represent the asymmetrically substituted carbon atoms appropriately and determine if they can be superimposed upon their realized mirror images.

Figure 3.6. Representations of (a) cyclopropane and (b) cyclobutane.

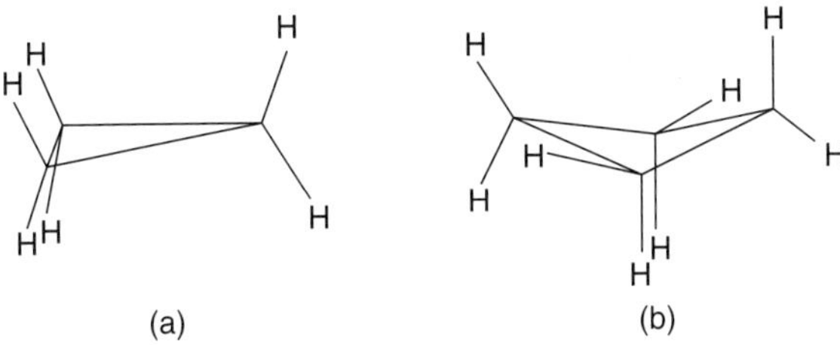

Figure 3.7. Three-dimensional representations of (a) cyclopropane and (b) cyclobutane.

Problem 3.4. For the dodecane isomer shown in V above [2-methyl-3-(1-methylpropyl)-heptane], decide which carbon atoms have four different ligands attached. Represent the asymmetrically substituted carbon atoms appropriately and determine if they can be superimposed upon their realized mirror images.

Problem 3.5. For the undecane isomer shown in IV above (4-ethyl-3,3-dimethylheptane), decide which carbon atoms are primary, secondary, tertiary, and/or quaternary.

Problem 3.6. For the dodecane isomer shown in V above [2-methyl-3-(1-methylpropyl)-heptane], decide which carbon atoms are primary, secondary, tertiary, and/or quaternary.

b. Cyclic Alkanes. Cyclic alkanes, or **cycloalkanes**, are named by adding the prefix **cyclo-** to the name of the corresponding continuous-chain alkane having the same number of carbon atoms as those in the ring. In this way, the **carbocyclic** (a ring composed only of carbon atoms) compound with three carbon atoms and the appropriate number of hydrogens becomes **cyclo**propane (Figure 3.6a), that with four carbons becomes **cyclo**butane (Figure 3.6b), and so on. It will quickly be noted, however, that *in forming the ring, one hydrogen at each terminus of the acyclic system must be removed.* Thus, these compounds with one ring, although composed of σ bonds only (i.e., they remain alkanes), must now correspond to the general formula C_nH_{2n}.

As is the case with the acyclic alkanes, it will be necessary to keep in mind that the two-dimensional representations of Figure 3.6 have a three-dimensional reality, which can be represented, among other ways, as shown in Figure 3.7.

Substituents may also be appended to cycloalkanes. Here, because the way the ring may be drawn is unimportant, the substituents are assigned numbers in accord with their respective positions so as to give them the lowest possible values. Thus, as shown in Figure 3.8a, the addition of one methyl group to cyclopentane produces methylcyclopentane, and the addition of a second methyl to the *same* carbon produces 1,1-dimethylcyclopentane (Figure 3.8b).

In the event a second methyl group is added to cyclopentane at a carbon atom *other* than that already bearing one, there are only two other positions at which it

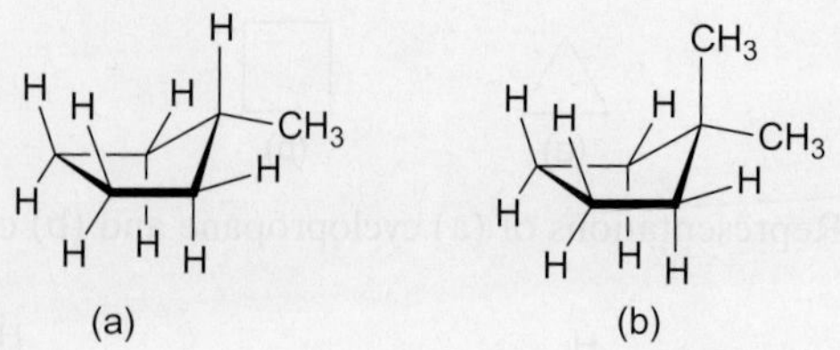

Figure 3.8. Representations of (a) methylcyclopentane and (b) 1,1-dimethylcyclopentane.

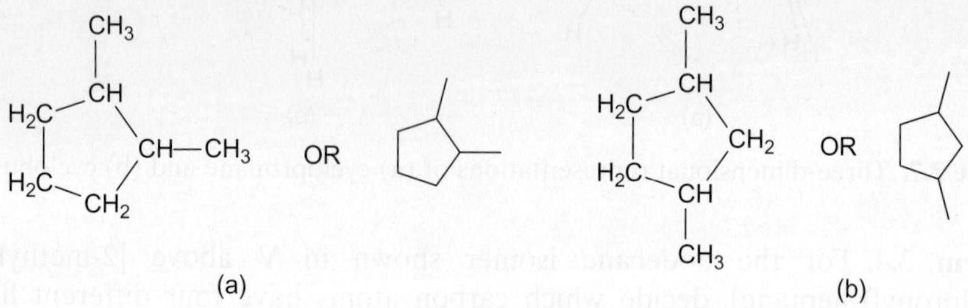

Figure 3.9. Two-dimensional representations of (a) 1,2-dimethylcyclopentane and (b) 1,3-dimethylcyclopentane.

can be placed (since the ring can be numbered clockwise or counterclockwise). **Two-dimensional** (flat) representations of 1,2-dimethylcyclopentane and 1,3-dimethylcyclopentane are depicted in Figure 3.9a,b.

As you might gather from three-dimensional representations such as those of Figures 3.7 and 3.8, the simple two-dimensional picture of Figure 3.9 is incomplete. This is because the two methyl groups can be either on the **same side** (or **face**) **or** on the **opposite sides** (or **faces**) of the five-membered ring. These **stereoisomers**, or **configurational** isomers, are distinguishable from each other and are depicted in Figure 3.10. When the two groups (in this case, methyl groups) are on the same side (or face) of the ring, they are said to be (Z) (from the German *zusammen*, which means "together").* When they are on opposite sides (or faces) they are said to be (E) (from the German *entgegen*, which means "opposed"). The **heavy wedge lines**

---

*Ligands on the same face may also be described as **suprafacial**. The prefixes *syn-* and *cis-* (the latter most commonly) are also used to describe the (Z)-isomer and the *relationship* of the ligands to each other. Ligands on opposite faces may also be described as **antarafacial**. The prefixes *anti-* and *trans-* (the latter most commonly) are also used to describe the (E)-isomer and the relationship of the ligands to each other. It should be clear to you that the (E)- and (Z)-isomers cannot be interconverted without bond breaking followed by suitable reconnection. Assignment of the designation (Z) or (E) is based on the **atomic number** of the atoms at the point of attachment. (The details of this system were developed by Robert Cahn, Sir Christopher Ingold, and Vladimir Prelog [Cahn, R. S.; Ingold, C. K.; Prelog, V. *Angew. Chem. Int. Ed. Engl.*, **1966**, *5*, 385] and will be discussed and developed further in Chapter 4.) Thus, in 1,2-dimethylcyclopentane, at carbon atom number 1 of the cyclopentane ring, the carbon of the methyl group (atomic number 6) takes priority over the hydrogen (atomic number 1). Similarly, at carbon atom number 2 of the cyclopentane ring, the carbon of the methyl takes priority over the hydrogen on the same carbon. If the two groups of highest priority are on the same side, the prefix (Z)- is used; if they are on opposite sides, the (E)- prefix is used. In the case of isotopes, where the atomic number is the same, the masses of the isotopes are used (e.g., ^{2}H, deuterium, would have priority over ^{1}H, protium).

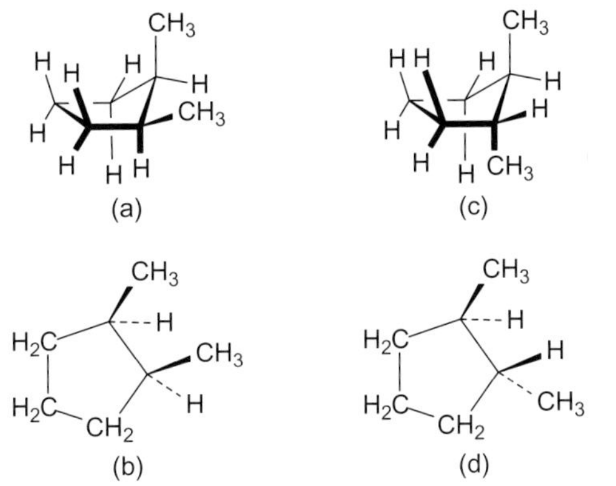

Figure 3.10. Three-dimensional representations of (Z)-1,2-dimethylcyclopentane (a,b) and (E)-1,2-dimethylcyclopentane (c,d).

(Figure 3.10b,d) signify the attachment is *above the plane of the paper*, while the **dashed lines** signify the attachment is *below the plane of the paper.**

Problem 3.7. Cast the two-dimensional representation of 1,3-dimethylcyclopentane (Figure 3.9b) into the appropriate three-dimensional representations of (E)- and (Z)-1,3-dimethylcyclopentane, respectively.

Problem 3.8. Are either (or both, or neither) (E)- or (Z)-1,2-dimethylcyclo**propane** superimposable upon its realized mirror images?

As was the case for acyclic alkanes, cyclic alkanes may also be found as substituents. Again, the **-ane** ending is changed to **-yl**, as in 2-cyclopropylpentane:

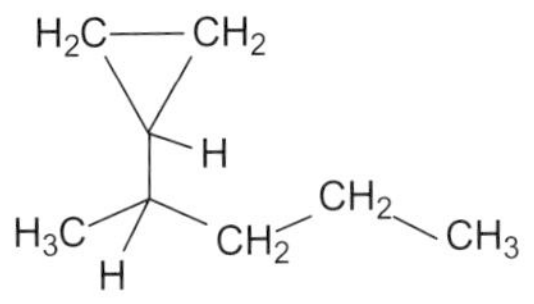

In drawing structural formulas for both alkanes and cycloalkanes, it is customary to **omit** the hydrogens and to draw simply the **carbon framework**. It is your responsibility to know that each carbon has a total of four substituents, arranged in an (approximately) tetrahedral fashion around it and **if nothing else is specified, the ligands are hydrogens**. To help you recognize and grow accustomed to this shorthand notation, the compounds with which you have already dealt in this chapter are recapitulated below, without their IUPAC names, as shown in Figure 3.11a,b.

*In the particular case here, of course, whether the two methyl groups of (Z)-1,2-dimethylcyclocyclopentane are on the top face or at the bottom face of the ring is not important. The same is true for the (E)-isomer. However, as molecular complexity increases, cases will be encountered where the two faces may not be equivalent. A dashed line will still be taken to mean that the substituent is below the plane of the paper and a heavy line that it is above.

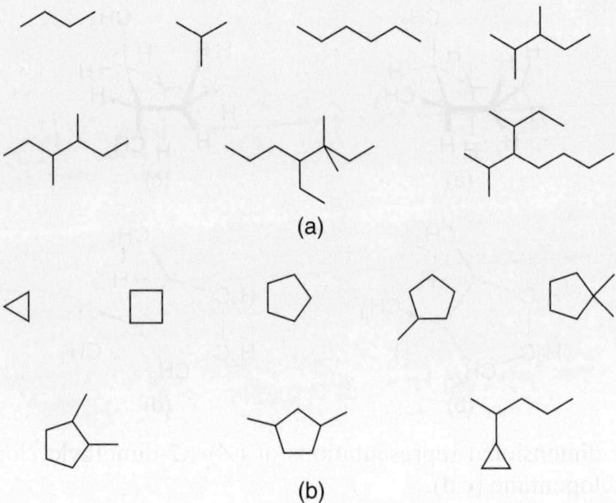

Figure 3.11. (a) Shorthand notation for some acyclic alkanes. (b) Shorthand notation for some cyclic and acyclic compounds introduced earlier.

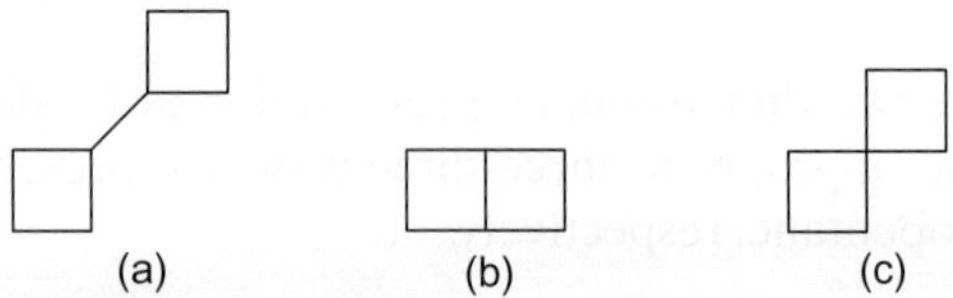

Figure 3.12. The joining of two rings (a) with a single bond (i.e., at one position with **no** carbons in common), generating cyclobutylcyclobutane; (b) with two single bonds (i.e., at more than one position with **two** carbons in common), generating bicycle[2.2.0]hexane; and (c) at **one** carbon, producing spiro[3.3]heptane.

Problem 3.9. Name the structures shown in Figure 3.11a using the IUPAC system.

Problem 3.10. Name the structures shown in Figure 3.11b using the IUPAC system.

Given the ease with which the nomenclature system can be elaborated upon and the ease with which compounds can be created (at least on paper), it should be clear that, just as alkyl fragments can be substituted for hydrogens along alkane chains and as hydrogens on rings can be replaced by alkyl fragments, so rings can be substituted by rings! Although there will be further discussion when additional use is made of such compounds, three different ways to accomplish the joining of two rings are shown in Figure 3.12.

Formulating the names of compounds with two carbons in common, as shown in Figure 3.12b, requires (1) counting the total number of carbon atoms in the rings and assigning to the compound the stem name of the corresponding open-chain hydrocarbon; (2) adding a prefix describing the number of rings (e.g., **bi, tri**); (3) specifying the sizes of the rings present, which is done by *counting how many carbons are in the* **chains that connect the atoms constituting the junctions of the rings** (the "bridgehead" carbons); (4) listing **those numbers, largest first, in brackets,**

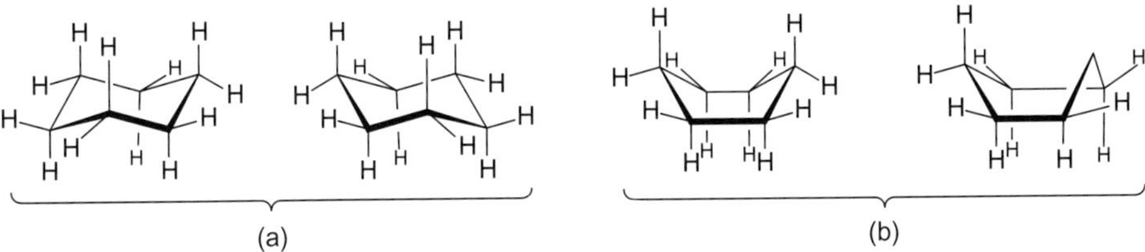

Figure 3.13. Representations of cyclohexane (a) chair and (b) boat conformations.

after the descriptor "cyclo" and before the hydrocarbon name; and (5) separating the numbers by periods.

Finally, in the drawing of three-dimensional representations of organic molecules (e.g., Figures 3.1f,g, 3.2c,d, 3.5d, 3.7, 3.8, and 3.10a,c), attempts are made to provide some perspective to the pictures. Frequently, shading, making some lines heavier than others, and so on, is employed. This is done both to reinforce the spatial relationships to which you must become accustomed and to acquaint you with processes about which you must learn.

For example, as you will learn in Chapter 4 when dynamic processes are considered, the alkanes already discussed, as well as cyclohexane (Figure 3.13), exist, at room temperature, as mixtures of **conformational isomers**,* two of which are shown. These isomers are interconvertible by flexing the ring and, in the case of cyclohexane, are called the "chair" and "boat" conformers.

II. Alkenes, Arenes, and Alkynes

We now continue by naming the alk**ene**s and alk**yne**s. These are hydrocarbons containing (by the LCAO method) sp^2 (for the former) and sp (for the latter) hybridized carbon atoms bonded to hydrogen or other carbon atoms; that is, there are now π bonds between carbon atoms. In alkenes, these bonds are called **double** bonds; alkynes contain **triple** bonds. Such compounds in which π bonds are found are called **unsaturated**. Hopefully, you will recall (Chapter 1) that it is convenient to consider the σ framework independently from the π bonds. For alk**e**nes, in accord with (approximately) trigonal geometry, each carbon of the **double bond** bears **three** ligands; for hydrocarbons, one of these must be the other carbon, the occupied p orbital of which is needed for π bond formation, and the others may be either carbon or hydrogen. Carbon–carbon double bonds (the combination of one σ and one π bond) are typically about 134 pm (1.34 Å) long; carbon–hydrogen bonds for protons bonded to carbon, which is sp^2 hybridized, are about 109 pm (1.09 Å) long; and the length of carbon–carbon bonds from sp^2 to sp^3-hybridized carbons is typically about 150 pm (1.50 Å). For alk**y**nes, which are digonal, each carbon of the **triple bond** bears **two** ligands; one of them must, of course, be the other carbon bearing the mutually

---

*__Conformational isomers__ are isomers that exist as a consequence of temporary spatial arrangements and that can be interrelated by simple flexing, twisting, or rotational motions (**but do not require bond breaking or making for interconversion**). Conformational isomerism generally requires energies much less than those needed for bond breaking. Because the interconversion is rapid at room temperature, some spectroscopic tools (e.g., ¹H NMR) may show the **average** structure.

$$H_2C=CHCH_2CH_3 \qquad\qquad H_3CCH=CHCH_3$$
$$\text{(a)} \qquad\qquad\qquad\qquad \text{(b)}$$

Figure 3.14. "In-line" abbreviated representations for (a) 1-butene and (b) 2-butene.

orthogonal occupied p orbitals, which are needed to form the triple bond. The other may be carbon or hydrogen. Typically, the carbon–carbon triple bond (one σ and two π bonds) is about 121 pm (1.21 Å) long; the carbon–hydrogen bond for a hydrogen bonded to an sp-hybridized carbon is about 106 pm (1.06 Å) long; and the length of the bond from an sp-hybridized carbon to an sp^3-hybridized carbon is about 146 pm (1.46 Å).

As was the case for alk**ane**s, both **acyclic** (without **rings**) and **cyclic** (with rings) alk**e**nes and alk**y**nes are known. While alk**e**nes and alk**y**nes might be considered similar (because of the π bonds), their differences are considerable.

a. Alkenes. Alk**ene**s (which are sometimes called "olefins"—as a double bond is occasionally called an "olefinic linkage") are named using the same roots as alk**ane**s according to the IUPAC system. Thus, the two-carbon alkene is eth**ene** (C_2H_4 or $CH_2{=}CH_2$); that with three carbons is prop**ene** (C_3H_6 or $CH_3CH{=}CH_2$), and so on. The introduction of the double bond means that there are two hydrogens less in the alkene **for each double bond** than there were in the corresponding alkane. Thus, with one double bond, the alkenes must correspond to C_nH_{2n}, where n is an integer greater than two.* If there is **more than one double bond**, the parent name of the alkene is given the prefix corresponding to the number of double bonds present; that is, two double bonds will correspond to a d**i**ene; three to a **tri**ene, and so on, and the formula will be reduced by **two** hydrogens for **each** double bond present.

Just as was the case with alkanes, beginning with the alkene with four carbon atoms, it is possible to write more than one isomer of the formula C_4H_8.

By the IUPAC system, since the double bond must connect contiguous carbons, it is only necessary to specify the location of the **first** carbon of the π system, assign it the **lowest possible number** consistent with its position, and change the **-ane** ending of the corresponding alk**ane** to **-ene**. Double-bond position isomers of butene, that is, 1-butene and 2-butene, are shown in Figure 3.14.

Although it might be anticipated that a $\pi \rightarrow \pi^*$ transition for both of these alkenes could be observed, that absorption happens to occur at wavelengths in the vacuum UV (<200 nm). Further, as 1-butene lacks the symmetry of its isomer, 2-butene, it might be anticipated that, in addition to carbon–hydrogen (C–H) stretching absorption in the region 3100–2800 cm^{-1} of the IR, which both should have, the former would have a band (or at least a more intense band) in the region 1680–1600 cm^{-1} because of the carbon–carbon double bond. The latter would, presumably, be Raman active.

Clearly, however, the ^{1}H NMR spectra of these butene positional isomers should be dramatically different. Indeed, protons on carbon–carbon double bonds (also

---

*The general formula, C_nH_{2n}, also corresponds to that of a one-ring carbocycle as already noted. Formally, therefore, **one site of unsaturation** corresponds to **two missing hydrogens** and can be accounted for by the presence of **one** double bond **or one** ring. Two **sites of unsaturation** might be (a) two rings, (b) one ring and one double bond, (c) two double bonds, or (d) one triple bond. All are possible based on the formula C_nH_{2n-2}.

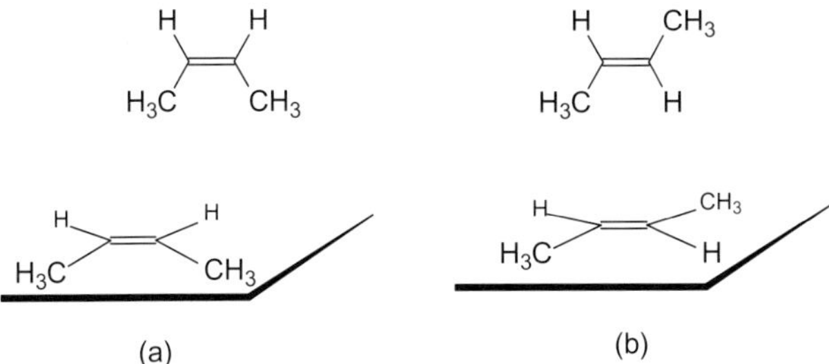

Figure 3.15. Representations of the stereoisomers of 2-butene: (a) (Z)-2-butene (*cis*-2-butene) and (b) (E)-2-butene (*trans*-2-butene).

called **vinylic** protons) are significantly **deshielded** relative to those of alkanes.* Such protons are normally found at a chemical shift around $\delta \approx 5.0$–6.0 ppm. Protons on carbons adjacent to those directly attached to the carbon–carbon double bond (**allylic** protons) are also deshielded, coming into resonance near $\delta \approx 2.5$ ppm.

Even the most cursory examination shows that 1-butene has **three** vinylic protons, while there are only **two** in its isomer. On this basis alone, they are distinguishable. However, there is another complication; it is found that there are **two** 2-butene isomers. It should be apparent that these come about because, in addition to the position isomerism afforded by the location of the double bond, there is also isomerism (stereoisomerism) arising from the different spatial arrangement of the groups attached to the double bond, which **cannot be interconverted without breaking the π bond**.

In one 2-butene isomer, the two methyl groups are on the **same** side of the double bond. This isomer (as was already discussed above for the cyclic alkanes) is called the (Z)- or *cis*-isomer; its stereoisomer is, correspondingly, the (E)- or *trans*-isomer. Representations of (Z)-2-butene (*cis*-2-butene) and (E)-2-butene (*trans*-2-butene) are shown in Figure 3.15.†

There are **three** vinyl protons in 1-butene, while there are only **two** vinyl protons in the each of the isomeric (Z)- and (E)-2-butenes (all of which are expected to appear between about δ 5.0 and 6.0 ppm). Therefore, the integrated ratio of methyl-to-vinyl protons will be 1:1 in the former and 3:1 in the latter. Other features are also noteworthy. Thus, there are two *kinds* of vinyl protons in 1-butene: the two on carbon one (the multiplet of which is centered at about δ 4.90 ppm) and the one on

---

*The deshielding experienced by protons attached to sp^2-hybridized carbon atoms, and **which also extends to protons on nearby** carbon atoms, is presumably due to the presence of a secondary magnetic field arising from the circulation of electrons in the π cloud, as well as from the nature of the electrons in the bonds themselves (i.e., sp^3 localized compared to sp^2-localized orbitals). The circulation of electrons creates both shielding and deshielding zones, and those protons lying **in the plane of the π framework** are in the deshielding zone.

†As indicated earlier, the decision concerning the designation (Z)- or (E)- is based on the **atomic number** of the atoms at the point of attachment. Thus, at carbon atom number 2 of the alkene, the carbon of the methyl group (atomic number 6) takes priority over the hydrogen (atomic number 1). Similarly, at carbon atom number 3 of the alkene, the carbon of the methyl takes priority over the hydrogen on the same carbon. If the two groups of highest priority are on the same side **of the double bond, the prefix (Z)- is used; if they are on opposite sides, the (E)- prefix is used.**

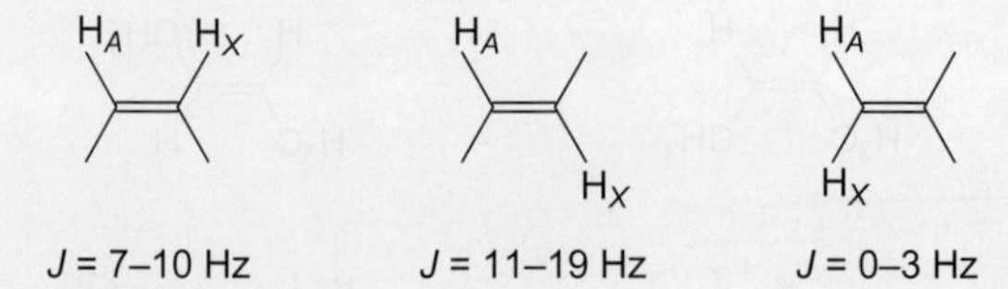

$J = 7–10$ Hz $J = 11–19$ Hz $J = 0–3$ Hz

Figure 3.16. Typical two- and three-bond coupling constants for nonidentical vicinal protons (H_A and H_X). Notice that the nonidentical (**geminal**) protons **on the same** carbon couple with a relatively small coupling constant compared to those on adjacent carbons.

carbon two (the multiplet of which is centered at about δ 5.75 ppm). In both of the 2-butene isomers there is, in each, only **one** kind of vinyl proton, which, for both, is centered at δ 5.45 ppm. There should also be other dramatic differences (e.g., in the methyl region; see Problem 3.11).

Problem 3.11. What multiplicity do you expect for the protons of the methyl group ($CH_3–$) in the ^1H NMR spectra of the 1-butene and the two 2-butene isomers (*cis*- or (Z)-2-butene and *trans*- or (E)-2-butene)?

Additionally, it might be expected that vinyl protons would couple with each other in 1-butene where they are all different but **not** in the 2-butene isomers because the two vinyl protons in (Z)-2-butene are identical to each other, as are the two vinyl protons in (E)-2-butene and identical protons do not couple. Not only is the expectation fulfilled but it is also found that the magnitude of the coupling is **related to the relative location of the hydrogens**. As shown in Figure 3.16, **nonidentical** hydrogens on the same carbon (called **geminal**) of the double bond will couple with each other. The two hydrogens at the first carbon of 1-butene are nonidentical, as one of them is *cis* or (Z) to the ethyl group attached to the other carbon of the double bond, while the other is *trans* or (E) to that same group. Thus, coupling is observed. Nonidentical hydrogens on adjacent carbons that enjoy a *cis* or a *trans* relationship **with regard to each other** will also couple. Coupling constants for these various relationships are different and are usually quite distinctive (Figure 3.16). It is important to note that nonidentical **geminal** protons couple *with each other as well as with those on adjacent carbons.*

In the particular case of 1-butene, coupling constants for the vinyl protons have been assigned (based on decoupling experiments), as shown in Figure 3.17.

Thus, although it is fairly easy to distinguish between 1-butene and its 2-butene double-bond isomers by ^1H NMR, the problem of deciding which 2-butene isomer is which remains. ^{13}C NMR doesn't help either. The methyl groups in (Z)-2-butene are found at $\delta \approx 11.42$ ppm, while those of the (E)-isomer appear at $\delta \approx 16.80$ ppm. The sp^2-hybridized carbons in the former appear at $\delta \approx 124.22$ ppm and those in the latter at $\delta \approx 125.42$ ppm.* Again, although the isomers can be distinguished, there are no definitive characteristics belonging to one and not the other.

In a historical sense, the problem of which isomer of 2-butene was which was resolved, long before the spectroscopic techniques described here were widely avail-

---

*These values, typically between $\delta \approx 100$ and 150 ppm, are taken from de Haan, J. W.; van de Ven, L. J. M. *Org. Mag. Res.*, **1973**, *5*, 147.

$$J_{AM} = 1.9 \text{ Hz}$$
$$J_{AX} = 17.4 \text{ Hz}$$
$$J_{MX} = 10.4 \text{ Hz}$$

Figure 3.17. Coupling constants for vinylic protons in 1-butene (after Alexander, S. *J. Chem. Phys.*, **1958**, *28*, 358.)

able, on the basis of chemical reactions about which you will learn. That information was simply confirmed, however, by noting that (Z)-2-butene must possess a dipole moment (0.33 D), while the (E)-2-butene must not (0 D):

$$\mu = 0.33 \text{ D} \qquad\qquad \mu = 0.0 \text{ D}$$

In longer-chain alkenes, the carbon–carbon double bond is given the lowest possible number and substituents (as in alkanes) are numbered and alphabetized as before. As a substituent, the two-carbon unit, –CH=CH₂, is called the **vinyl** group.

Cyclic alkenes have the double bond(s) within the ring (**endocyclic**) or attached to the ring but outside (**exocyclic**). For systems with one double bond, and for cyclo-propene through cycloheptene, only a *cis-* or (Z)-endocyclic double bond can be accommodated. In cyclooctene and larger rings, both (Z)- and (E)-endocyclic double bonds are allowed. The ring size restriction on geometry arises from the necessity of keeping the angles of sp^2-hybridized carbons near 120° and the bond lengths normal (i.e., ca. 154 pm [1.54 Å] for carbon–carbon single bonds and near 134 pm (1.34 Å) for carbon–carbon double bonds). Thus, a *trans* or (E) double bond simply cannot fit in the smaller rings (nor can your right elbow be held in your right hand ... and for the same reason!).

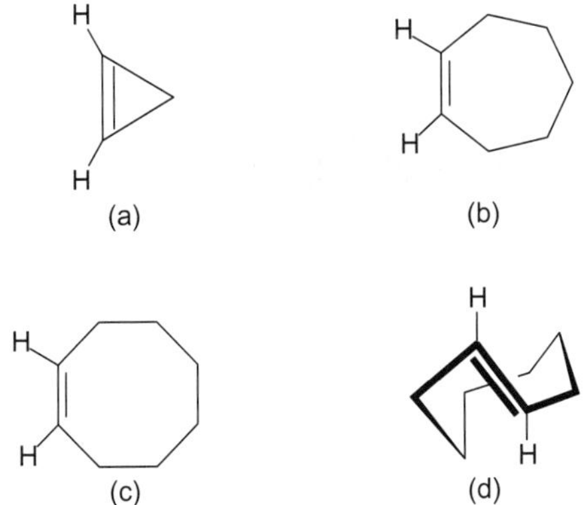

(a) Cyclopropene, (b) cycloheptene, (c) (Z)-cyclooctene, and (d) (E)-cyclooctene

Alkenes containing **more than one** double bond have already been mentioned. Besides utilizing a prefix (di-, tri-, etc.) to indicate how many double bonds are present, you now know that a numerical locator, describing their positions, will be necessary. Using the generic term **pentadiene** (C_5H_8) as an example, you should recognize that the isomeric structures in the next figure correspond to 1,2-pentadiene, 2,3-pentadiene, 1,3-pentadiene, and 1,4-pentadiene, respectively:

$$H_2C=C=CHCH_2CH_3 \qquad CH_3CH=C=CHCH_3 \qquad H_2C=CHCH=CHCH_3 \qquad H_2C=CHCH_2CH=CH_2$$

Compounds, such as 1,2-pentadiene and 2,3-pentadiene, in which there is one carbon atom to which two double bonds are appended, bear the generic name **cumulenes**. Dienes in which there is a single bond interposed between double bonds, as in 1,3-pentadiene, are called **conjugated dienes**. Dienes in which there is at least one methylene unit ($-CH_2-$) lying between the double bonds are called **isolated dienes**.

Isolated dienes (e.g., 1,4-pentadiene) behave much as if the two double bonds were in separate systems. The carbon–carbon double-bond lengths in simple alkenes and in isolated dienes are both about 134 pm (1.34 Å), and there does not seem to be much, if any, interaction between the separated π systems. However, in conjugated dienes, the carbon–carbon single bond lying between the two double bonds is significantly shorter (148 pm [1.48 Å]) than a normal carbon–carbon single bond (154 pm [1.54 Å]). Additionally, the $\pi \rightarrow \pi^*$ UV absorption is shifted about 30 nm to longer wavelength (lower energy) for each double bond placed in conjugation (i.e., separated from its predecessor by a single bond). These observations led to the argument that the two π bonds are interacting (conjugating) across the single bond.

As discussed in Chapter 1, if the π system is considered separately, and now is considered to be composed of a "side on" overlap of **four** $2p$ atomic orbitals (AOs), then **four** molecular orbitals (MOs) can be created to accommodate the **four** electrons. These can be utilized pairwise in making two **bonding** and two **antibonding** orbitals as shown in Figure 3.18 for the case of 1,3-butadiene (chosen for symmetry reasons).

Finally, cumulenes, such as 2,3-pentadiene (Figure 3.19), have the interesting property of containing one carbon atom bearing two mutually orthogonal p orbitals. Similarly, the π bonds must be mutually orthogonal.

Problem 3.12. Carefully examine the drawing of Figure 3.19. Draw the mirror image representation of the 2,3-pentadiene shown in the figure. Is the realized object in Figure 3.19 capable of superposition on its realized mirror image? Identify the asymmetrically substituted carbon atom(s) in 2,3-pentadiene.

Cyclic dienes, except for geometric constraints, already commented upon for the cyclic monoalkenes, present no particular problems. Thus, both 1,3-cyclohexadiene and 1,4-cyclohexadiene clearly fit into their respective categories of conjugated and isolated double-bonded systems. However, as might be expected, cyclic cumulated dienes require larger rings.

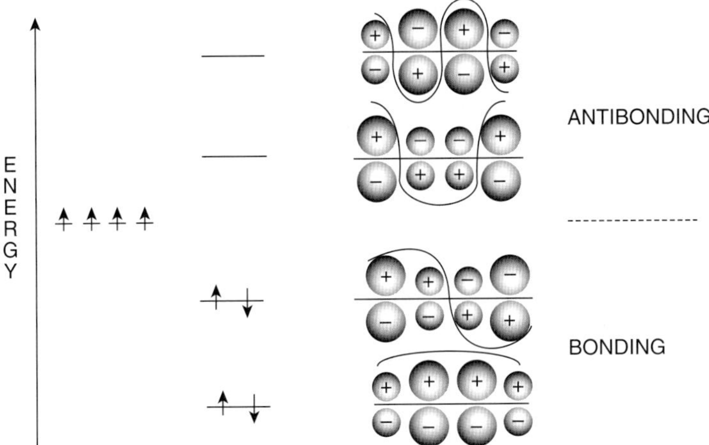

Figure 3.18. Relative energies and a schematic representation of the bonding (π) and antibonding (π^*) molecular orbitals (MOs) of 1,3-butadiene. The four electrons are placed in the bonding π orbitals. The curved lines represent the number of phase changes (nodes) in the MO. The lowest-lying MO has zero nodes, the next highest has one, and so on. The sizes of the orbitals are meant to represent the approximate contribution the orbital makes to the corresponding bonding (antibonding) orbital. The signs represent the sign of the wave function (Chapter 1). Thus, the lowest-lying (bonding) π orbital is bonding "everywhere."

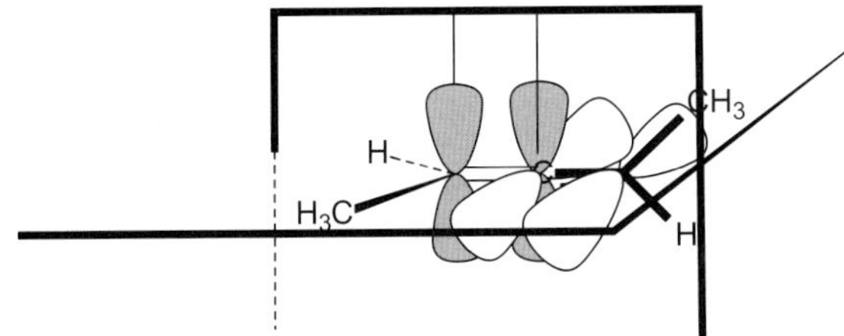

Figure 3.19. A representation of the cumulene 2,3-pentadiene depicting the mutually orthogonal π systems between the second and third carbon atoms and between the third and fourth carbon atoms.

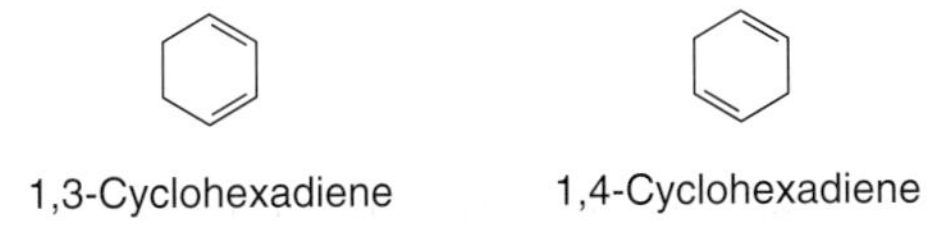

1,3-Cyclohexadiene   1,4-Cyclohexadiene

Both cyclic and acyclic trienes, tetraenes, and so on, also exist, and their nomenclature follows the rules you have already learned. These compounds have properties similar to those already discussed except for the particular case of those cyclic arrays, which are **both** completely conjugated and planar (or nearly so). These compounds form a class of their own and are called **arenes**.

b. Arenes. The extension of the number of double bonds in a cyclic array of six carbons from two (for the known 1,3-hexadiene) to three (for the unknown 1,3,5-hexatriene), although conceptually straightforward, fails.

The failure is not due to the inability to write the structure and, at least in principle, to go from C_6H_8 (which corresponds to C_nH_{2n-4}, a compound with *three* sites of unsaturation) to C_6H_6 (which corresponds to C_nH_{2n-6}, a compound with *four* sites of unsaturation); the failure is to form the "triene!" The compound C_6H_6 can be written and does exist. Its name is **benzene**.

However, this cyclic array is found (by rotation Raman and microwave spectroscopies and by x-ray crystallography) to, surprisingly, have **all** six carbon–carbon bonds of equal length (139 pm [1.39 Å]) and, less surprisingly, all six carbon–hydrogen bonds of equal length (109 pm [1.09 Å]). Additionally, in this planar molecule, all C–C–C bond angles and all H–C–C bond angles are 120°. The 1,3,5-cyclohexatriene structure, with alternating single and double bonds, would be predicted to have single bond lengths about 148 pm (1.48 Å), that is, shorter than 154 pm (1.54 Å) for a normal single bond as noted above for 1,3-butadiene, and double-bond lengths of about 134 pm (1.34 Å). Most interestingly, dramatic spectroscopic differences between benzene, and substituted benzenes (called **arenes**) and simple alkenes (or even alkadienes) also occur.

First, in the UV spectra of most arenes, two bands appear: one fairly intense band at about λ_{max} 203 nm and another weaker, long-wavelength band at $\lambda_{max} \approx 255$ nm. The positions and intensities of these bands are highly dependent upon the substituents on the ring (or rings) of the arenes.

Second, there are several new bands (or sets of bands) in the IR (or Raman) near 3030 cm^{-1} associated with C–H stretch and two bands associated with carbon–carbon stretching near 1600 and 1500 cm^{-1}. While in benzene itself (for symmetry reasons) these bands are fairly weak, substituted arenes frequently have intense absorptions.

Finally, in the ^{1}H NMR, there is a single peak appearing at $\delta \approx 7.3$ ppm for benzene, and in the ^{13}C-NMR, there is also a single peak for benzene at $\delta \approx 128.5$ ppm. In other arenes, similar deshielding is observed. The noticeable **downfield** shift in the NMR spectra, even beyond that typically associated with protons (^{1}H NMR) attached to sp^2-hybridized carbons or such hybridized carbons (^{13}C NMR) themselves, has argued for some special treatment. These spectroscopic observations followed many years after the discovery of the odoriferous (called **aromatic**) compounds described; they are, nonetheless, currently considered major characteristics of that group.*

First, with regard to nomenclature, the parent C_6H_6 hexagonally symmetric hydrocarbon, as you already know, is called **benzene**. Second, trivial nomenclature has become entrenched so that substituents, such as methyl groups, which might replace one or more of the hydrogens and could be named with numerical descriptors to define their positions, are **also**, and equally well, named with common or trivial names. Thus, **methylbenzene** (Figure 3.20b)† is also called **toluene**.

---

*Additional characteristics, related to reactivity differences between aromatic and nonaromatic compounds, which, in an historical sense, long preceded these observations, will be discussed subsequently.
†Although we will deal somewhat more extensively with the notation shortly, it is **very important** that you recall the use of the double-headed resonance arrow from Chapter 1. Note **in particular now** that **all** of the forms shown here **are considered identical, and drawing any one of them is the same as drawing all of them or any other of them.**

(Continued on next page)

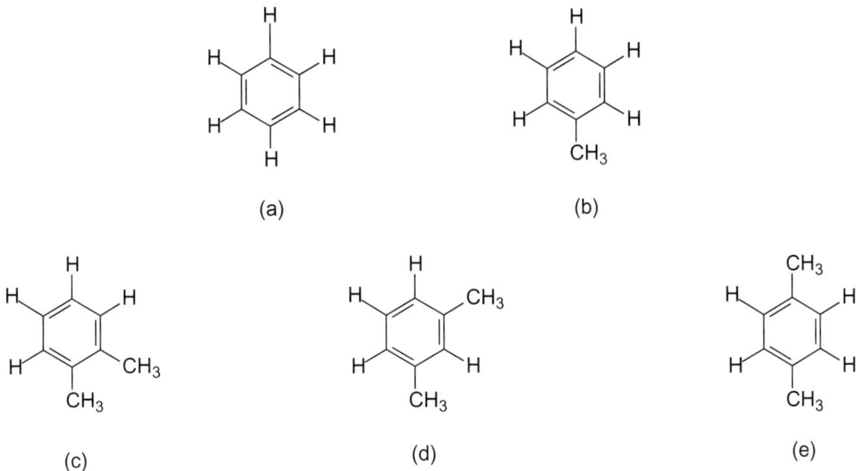

Figure 3.20. Some aromatic compounds. Although drawn with localized double bonds, it is understood that each is only a representative of the resonance forms that could be written for (a) benzene, (b) methylbenzene (toluene), (c) 1,2-dimethylbenzene (*o*-xylene), (d) 1,3-dimethyl-benzene (*m*-xylene), and (e) 1,4-dimethylbenzene (*p*-xylene).

The corresponding dimethylbenzene isomers, also shown in Figure 3.20, are then 1,2-dimethylbenzene, also called *o*-xylene or *ortho*-xylene (1,2-substitution≡*ortho*); 1,3-dimethyl-benzene, also called *m*-xylene or *meta*-xylene (1,3-substitution≡*meta*); and 1,4-dimethylbenzene, also called *p*-xylene (1,4-substitution≡*para*). The properties associated with aromaticity are frequently unaffected by substituent(s).

The spectroscopic properties and, indeed, the chemical properties of the arenes are currently accounted for in terms of **electron delocalization** throughout the system. Both resonance (valence bond [VB]), as suggested by the usage of the double-headed arrow above, and qualitative extension of the LCAO-MO concepts treated in Chapter 1 are considered for benzene and its derivatives.

With regard to the VB treatment, the use of curved arrows (as a shorthand representation for "bookkeeping," as shown in Figure 3.21 with methylbenzene (toluene), serves to interrelate the forms. The various forms are labeled and referred to by the names of those who suggested them, that is, Kekulé (1866) and Dewar (1867).

Remember: the nuclei are stationary—only the electrons "move" in these cartoons.

The MO treatment of benzene suggests that since there are six *p*-type orbitals and six electrons to be considered, and since we must have the same number of MOs as AOs, then there will be six MOs into which the six electrons can be placed.

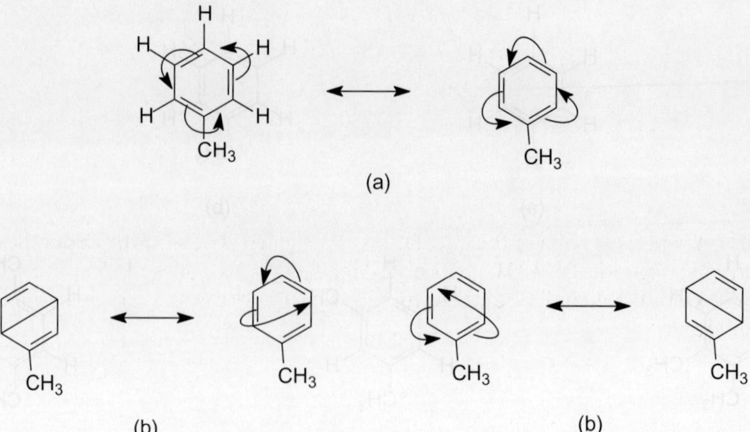

Figure 3.21. Resonance forms (connected by resonance arrows) for methylbenzene (toluene): (a) Kekulé forms *vide infra* and (b) Dewar forms (suggested as early as 1867 by James Dewar as another potential C_6H_6 isomer).

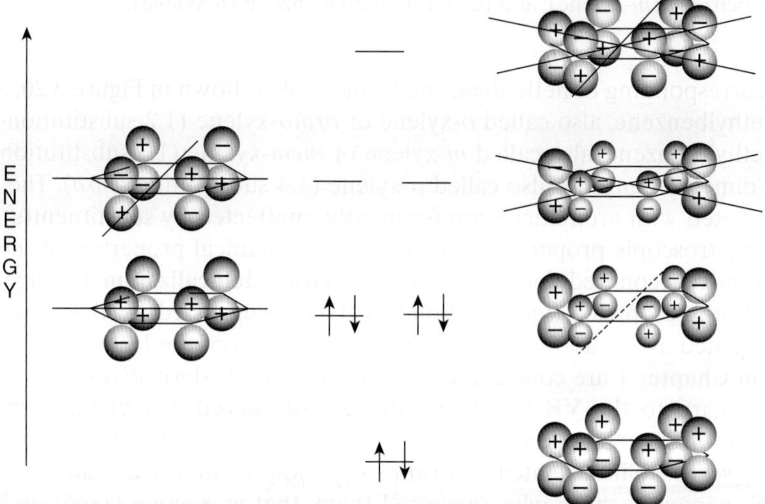

Figure 3.22. A schematic representation of the π MOs of benzene and the relative energies of those orbitals. The three lowest (bonding) orbitals are shown with paired electrons. The antibonding orbitals (the three highest) are vacant. The dashed lines represent the number of nodes or sign changes of the wave function. The lowest MO has zero node, the next two have one node, the next two (now antibonding) have two nodes, and the highest has three nodes. The two highest bonding MOs are mutually degenerate as are the two lowest antibonding MOs. The relative sizes of the individual orbitals are meant to represent the extent to which they contribute to the total molecular orbital.

If this is done pairwise (the Pauli principle requiring not more than two per orbital), then there will be three occupied and three unoccupied MOs. Of the six predicted MOs, three will be bonding (and occupied) and three will be antibonding (and unoccupied). A schematic representation of the π MOs of benzene is presented in Figure 3.22.

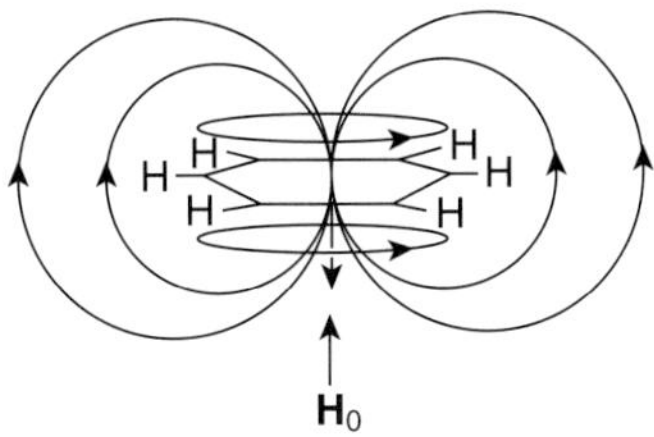

Figure 3.23. A representation of the circulation of π electrons of benzene under the influence of an external magnetic field (H_o). A secondary field is established. The secondary field opposes the external field in the center of the ring but reinforces it along the periphery where the protons lie.

The same kind of picture can be used, in part, to justify the NMR observation that the protons arranged around the periphery of the aromatic ring and, indeed, the ring carbons themselves, have been found to be deshielded. As shown in Figure 3.23, the presence of the π electrons in the cyclic array (a "closed loop") dictates (as discussed in Chapter 1) that a "molecular" magnetic dipole is established. The local magnetic field (when the dipole is orthogonal to the external field) is in the direction **opposite** to the externally applied field, H_o. The applied field in the **center** of the ring is reduced. Protons in the center of the ring (of which there are none) would presumably need a higher than "expected" applied field to come into resonance. Protons around the periphery are affected oppositely since there the return of the lines of magnetic force creates a local field that reinforces the externally applied field. Thus, the protons are deshielded and they come into resonance downfield.

Finally, for arenes, the aromatic ring can also serve as a substituent (i.e., C_6H_5-), in which case it is given the name **phenyl** (rhymes with "kennel") as in 2-phenylpropene or (Z)-1-methyl-2-phenylcyclobutane, and so on.

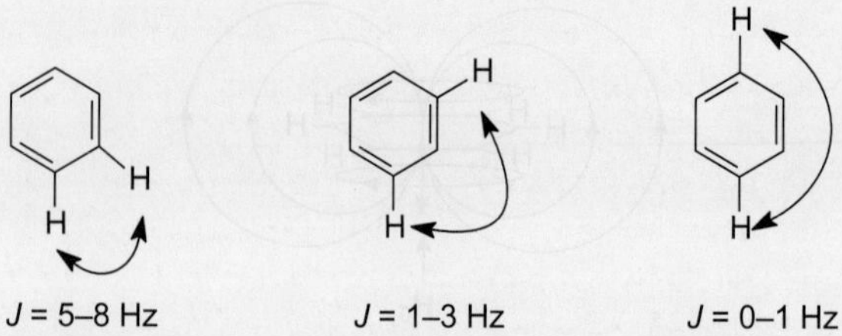

$J = 5–8\ \text{Hz}$ $J = 1–3\ \text{Hz}$ $J = 0–1\ \text{Hz}$

Figure 3.24. Typical coupling constants for aromatic protons.

As shown above, aromatic rings can also be joined to other aromatic rings such as in (a) biphenyl, (b) naphthalene, (c) anthracene, and (d) phenanthrene, as well as to nonaromatic rings such as in (e) 1,2,3,4-tetrahydronaphthalene (tetralin).

As you might expect, the protons on alkyl substituents (i.e., alkyl groups attached to the aromatic rings) are also deshielded because of their proximity to the ring. Whereas a methyl group as part of an alkyl chain produces a signal in the ^{1}H NMR near $\delta \approx 0.9\,$ppm, the methyl group in methylbenzene (toluene) is found at $\delta \approx 2.3\,$ppm. Similarly, methylene units ($-CH_2-$) attached to aromatic rings (e.g., in ethylbenzene) are found downfield ($\delta \approx 2.7\,$ppm) relative to those where no aryl group is present ($\delta \approx 1.3\,$ppm). Furthermore, once the hexagonal symmetry of benzene is destroyed by substitution, the remaining protons are no longer equivalent. While the chemical shift of the aromatic protons remains below about $\delta \approx 7.0\,$ppm, the spectra become dramatically more complex. This results from the protons, which are no longer identical, coupling with each other. Generally, as shown in Figure 3.24, *para* couplings are smaller (about 1 Hz or less) than *meta* couplings (less than about 3 Hz), while *ortho* protons give the largest couplings ($J_{\text{ortho}} \approx 5–8\,$Hz).

In their mass spectra, alkyl benzenes are characterized by an apparent ready fragmentation with the formation of an ion, which corresponds to $C_7H_7^+$ (i.e. m/z 91). This fragment is written either as (a) the *benzyl cation* or (b) an ion with a seven-membered ring (called the *tropylium ion*), represented in the next figure.

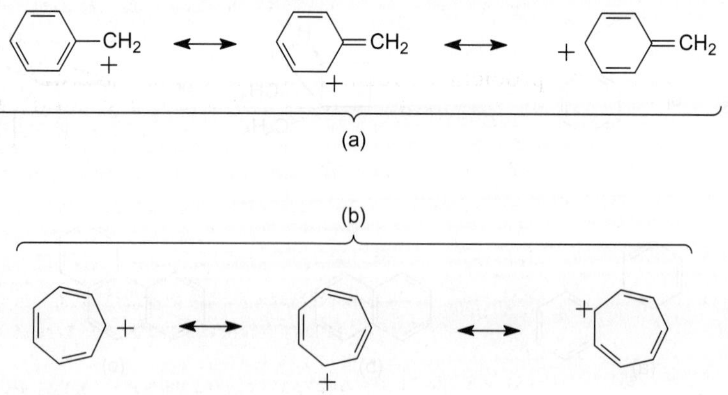

(a)

(b)

c. Alkynes. Alkynes are also known as **acetylenes** after the first member of the class, C_2H_2 ($H–C\equiv C–H$). The nomenclature follows the usual IUPAC rules for hydrocarbons and the **-ane** suffix is replaced by **-yne**. An alternative, commonly used system is based on the addition of alkyl groups to the parent acetylene. Thus, the first member of the series is ethyne (acetylene, HC≡CH), the second member is propyne (methylacetylene, $CH_3C≡CH$), and so on. When there are four or more carbon atoms in the parent chain, the possibility of triple bond isomerism must be considered. The position of the triple bond along the chain in acyclic alkynes is specified by a number as was done with alkenes, the lowest possible number being assigned to the triple bond. Other substituents are then numbered and alphabetized accordingly. Thus, typical compounds include (a) 1-butyne, (b) 2-butyne, (c) 4-methyl-2-hexyne, and (d) 2,5-dimethyl-3-hexyne.

$$H—C≡C—CH_2CH_3 \qquad H_3C—C≡C—CH_3 \qquad H_3C—C≡C—CH(CH_3)CH_2CH_3 \qquad (CH_3)_2CH—C≡C—CH(CH_3)_2$$

(a) (b) (c) (d)

When found attached as a substituent, the function –C≡CH is called the **ethynyl** group, as in ethynylcyclohexane ([a] in the next figure). Further, as you might imagine, incorporation of a digonal linkage into a cyclic structure is attended by considerable difficulty, and the smallest isolable carbocycle in which a triple bond is found has nine carbon atoms (cyclononyne) ([b] in the next figure).

(a) (b)

Like alkenes, the electronic (UV) spectra of simple alkynes is characterized by only $\pi \to \pi^*$ absorbance, which is generally found at wavelengths <200 nm. For example, for propyne, there is a weak absorbance at $\lambda_{max} = 186.5$ nm. In the IR, however, terminal alkynes have a characteristic sharp H–C stretch between 3350 and 3200 cm^{-1}, and, keeping symmetry considerations in mind, a C≡C stretch between 2250 and 2050 cm^{-1}. The latter is particularly important as absorption of radiation in this region is quite characteristic of compounds with triple bonds.

In the ^{1}H NMR, protons directly attached to triply bonded carbon atoms are typically found **upfield** of vinylic protons (between $\delta \approx 2.5$ and 3.0 ppm for the former and between $\delta \approx 5.0$ and 6.0 for the latter). This difference means that, although vinylic protons are apparently deshielded by the electrons in the double bond, acetylenic protons are shielded. It is argued that although changes in the hybridization at carbon from $sp^3 \to sp^2 \to sp$ (i.e., increasing s contribution to the bonding orbital) should be accompanied by a corresponding downfield shift, an "overcompensation" has occurred as a result of the π system, resulting in an extra deshielding of protons attached to the carbons of the double bond and an extra shielding of protons attached to the carbons of the triple bond.*

A similar phenomenon is observed in the ^{13}C NMR. Acetylenic carbons are normally found between $\delta \approx 65–90$ ppm downfield from TMS, while vinylic carbons are generally between $\delta \approx 100–150$ ppm.

Table 3.2 summarizes the spectroscopic information presented for all of the hydrocarbons discussed.

---

*An alternative suggestion, that in general the extent of shielding of any proton from the electrons in the σ bond is related to the amount of p character in the bond; the more p character, the greater the shielding and the same justification involving "overcompensation" is invoked.

TABLE 3.2. Summary of Some Spectroscopic Properties of Hydrocarbons

Hydrocarbon	UV	IR	^{1}H NMR	^{13}C NMR
Alkanes	<200 nm	3950–2850 cm^{-1}		
Methyl			0.9–1.0 ppm	0–30 ppm
Methylene			1.25–1.40 ppm	15–40 ppm
Methine			1.40–1.60 ppm	25–50 ppm
Alkenes	<200 nm (λ_{max} higher if conjugated)	3100–3000 cm^{-1} (C–H) 1680–1620 cm^{-1} (C=C)		
Methyl			1.6–1.8 ppm	10–20 ppm
Methylene			2.0–2.2 ppm	20–35 ppm
Methine			2.4–2.6 ppm	25–40 ppm
Vinyl			4.9–5.2 ppm	110–135 ppm
Alkynes	<200 nm (λ_{max} higher if conjugated)	3320–3310 cm^{-1} (C–H) 2250–2100 cm^{-1} (C≡C)		
–C≡C–CH$_3$ (methyl)			2.0–2.1 ppm	0–5 ppm
–C≡C–CH$_2$ (methylene)			3.0–3.2 ppm	10–20 ppm
–C≡C–CH (methine)			2.4–2.5 ppm	25–27 ppm
–C≡C–H (alkynyl)			2.5–2.7 ppm	65–80 ppm
Arenes	<200 nm and λ_{max} 255 nm, and higher if conjugated	3100–3000 cm^{-1} (C–H) 1600–1450 cm^{-1} (aromatic) 840–680 cm^{-1} (substituent pattern)		
Ar–CH$_3$ (methyl)			2.2–2.3 ppm	17–21 ppm
Ar–CH$_2$ (methylene)			2.6–2.7 ppm	27–32 ppm
Ar–CH (methine)			2.8–2.9 ppm	34–37 ppm
Ar–H (aromatic)			7.0–8.0 ppm	125–150 ppm

The values for ^{13}C were taken from Breitmaier, E.; Haas, G.; Voelter, W. *Atlas of Carbon-13 NMR Data*, Heyden, Philadelphia, **1979**. The following ^{13}C information (from the same source) is also of some interest (δ, ppm, TMS): ethane 5.30$_{(g)}$, 7.26$_{(soln)}$; propane 15.40$_{(g)}$, 15.90$_{(soln)}$; butane 13.10$_{(g)}$, 24.90$_{(soln)}$; 2-methylpropane 24.30$_{(g)}$, 25.00$_{(soln)}$; and 2,2-dimethylpropane 21.40$_{(g)}$, 31.40$_{(soln)}$.

C. PHYSICAL AND CHEMICAL PROPERTIES; OXIDATION AND REDUCTION OF HYDROCARBONS

Alkanes, also known as paraffins (i.e., *little affinity*), are considered unreactive compared to related alk**ene**s, alk**yne**s, and arenes. Indeed, the π systems in the unsaturated compounds dramatically increase the variety of transformations they can undergo and they *serve as the locus for those reactions*; that is, the alkyl portions of the unsaturated compounds, which are "alkane-like," are also relatively unreactive compared to the site of unsaturation. However, all of the hydrocarbons can be oxidized by oxygen, and those with π systems more or less readily undergo reaction with hydrogen to produce alkanes.

While reactions of the unsaturated systems will be discussed subsequently, quantitative evaluation of the results of oxidation and reduction of hydrocarbons provides the basis for much that follows.

I. The Concept of Homology

Groups of compounds, such as the hydrocarbons called "the alkanes" (and each separately, "the alkenes," "the alkynes," and "the arenes") that have similar structures, frequently have similar chemical and physical properties. The members of such structurally related groups are said to constitute a **homologous series** and, given the properties of some of its members, it is frequently possible to estimate the properties of others.

For example, in Table 3.3 the boiling point of octane (C_8H_{18}) is found to be 125.7°C and that of nonane (C_9H_{20}) 150.8°C (both at 760 torr). The difference (25.1°C), presumably due to the increase in the molecular weight of 14 mass units (i.e., one methylene group [$-CH_2-$]), suggests that decane ($C_{10}H_{22}$), which corresponds to the addition of one more methylene group, should have a similar increase in boiling point, that is, $150.8 + 25.1 = 175.9$°C (760 torr). The experimental value is 174.1°C. Similarly, in the other direction, heptane would be expected to have a decrease in boiling point, that is, $125.7 - 25.1 = 100.6$°C. The experimental value is found in Table 3.3. It is reasonable, and is also clear from the table, that (a) for lower-molecular-weight compounds, each additional methylene ($-CH_2-$) unit will make a more important contribution to the difference in boiling points (with a corresponding larger deviation from predictions based on homology), and (b) cycloalkanes should behave in an analogous fashion.

Branching in alkanes, which provides structural variation and alters the effects of intermolecular interactions, does not lend itself to the same kind of analysis. However, in general, branching (in compounds with the same molecular weight) lowers the boiling point.

For alkenes (Table 3.3), alkynes, and arenes, the concept of homology as pertains to addition of *alkyl* substituents also continues to apply, although positional isomerism and stereochemistry play important roles. Addition of further carbon–carbon double bonds, triple bonds, and/or aryl groups, as well as branching, confounds widespread prediction; however, in a closely related series, the principle is maintained.

TABLE 3.3. Boiling Points (bp, °C, 760 torr), Melting Points (mp, °C), and Heats of Combustion ($-\Delta H_c°$, kcal mol^{-1}) to $CO_{2(g)}$ and $H_2O_{(l)}$ for Selected Hydrocarbons (1 kcal = 4.184 kJ)

Hydrocarbon	Molecular Formula	bp	mp	$-\Delta H_c°$
Unbranched Alkanes				
Methane	CH_4	−161.5	−182.5	212.8
Ethane	C_2H_6	−88.6	−183.3	372.8
Propane	C_3H_8	−42.1	−187.7	530.6
Butane	C_4H_{10}	−0.5	−138.4	687.4
Pentane	C_5H_{12}	36.1	−129.7	845.3
Hexane	C_6H_{14}	68.7	−95.3	995.0
Heptane	C_7H_{16}	98.4	−90.6	1151.3
Octane	C_8H_{18}	125.7	−56.8	1307.5
Nonane	C_9H_{20}	150.8	−53.5	1463.9
Decane	$C_{10}H_{22}$	174.1	−29.7	1620.1
Cycloalkanes				
Cyclopropane	C_3H_6	−32.9	−127.4	499.3
Cyclobutane	C_4H_8	13.1	−80.0	650.4
Cyclopentane	C_5H_{10}	49.2	−93.9	786.5
Cyclohexane	C_6H_{12}	80.7	6.6	936.9
Cycloheptane	C_7H_{14}	119.0	−8.0	1098.7
Cyclooctane	C_8H_{16}	150.0	14.0	258.8
Branched Alkanes				
2-Methylbutane	C_5H_{12}	27.9	−159.9	843.4
2,2-Dimethylpropane	C_5H_{12}	9.5	−16.6	839.9
2-Methylpentane	C_6H_{14}	60.3	−153.7	993.6
3-Methylpentane	C_6H_{14}	63.3	−118.0	994.1
2,2-Dimethylbutane	C_6H_{14}	49.7	−99.9	991.4
2,3-Dimethylbutane	C_6H_{14}	58.0	−128.5	992.9
Alkenes				
Ethene	C_2H_4	−103.7	−169.2	337.3
Propene	C_3H_6	−47.7	−185.3	491.8
1-Butene	C_4H_8	−6.3	−185.4	649.3
(*E*)-2-Butene	C_4H_8	0.9	−105.6	646.9
(*Z*)-2-Butene	C_4H_8	3.8	−138.9	647.7
1-Pentene	C_5H_{10}	29.2	−165.2	800.0[a]
(*E*)-2-Pentene	C_5H_{10}	36.4	−140.2	797.3
(*Z*)-2-Pentene	C_5H_{10}	36.9	−151.4	797.5
1-Hexene	C_6H_{12}	63.5	−139.8	955.0[a]
(*E*)-2-Hexene	C_6H_{12}	67.9	−132.9	953.8
(*Z*)-2-Hexene	C_6H_{12}	68.8	−141.1	954.2
(*E*)-3-Hexene	C_6H_{12}	67.1	−113.4	953.6
(*Z*)-3-Hexene	C_6H_{12}	66.4	−137.8	955.3

The boiling point and melting point values presented are largely from Rappoport, Z. *CRC Handbook of Tables for Organic Compound Identification*, 3rd edition, CRC Press, Cleveland, OH, **1977**. Values for $-\Delta H_c°$ (±0.5) are largely from Cox, J. D.; Pilcher, G. *Thermochemistry of Organic and Organometallic Compounds*, Academic Press, New York, **1970**.
[a] Estimated.

II. Oxidation and Reduction

Oxidation and reduction (sometimes called redox) reactions are traditionally studied in terms of electron transfer. The transfer of an electron (or electrons) **to** an atom, resulting in a **net gain** of an electron (or electrons) by that atom, is called **reduction**. Loss of an electron (or electrons), that is, transfer **from** an atom, resulting in a **net loss** of an electron (or electrons) by an atom, is called **oxidation**.*

While the idea of net electron gain or loss is applicable throughout chemistry, reactions at carbon, where the number of ligands at the beginning and end of the reaction are the same, do not easily lend themselves to such analysis. Therefore, a variety of more or less arbitrary schemes have been developed in organic chemistry to attempt to judge if oxidation and reduction have occurred. Generally, these operate by assigning some number to carbon (e.g., "0") and by using an electronegativity scale to determine whether a change of substituents **on** carbon has resulted in a presumed increase, decrease, or no change in the electron density **at** carbon. In many cases, this valuable exercise can be quickly (albeit more qualitatively) replaced by simply noting whether the carbon in question has gained or lost hydrogen or oxygen. Thus, a loss of hydrogen and/or a gain of oxygen **at carbon** will correspond to oxidation, while a gain of hydrogen or loss of oxygen **at carbon** will correspond to reduction.

a. Oxidation. Evidence has already been presented (Chapter 1) that the heat liberated on complete combustion of any organic material (the **heat of combustion**, ΔH_c°) can be utilized, with other information and Hess's law, to obtain estimates of bond energies. It should now be recalled that (1) in **combustion**, hydrocarbons react with oxygen to produce only CO_2, H_2O, and heat (Equation 3.1); (2) any process that produces heat is called **exothermic**; (3) in such an exothermic process, the "heat content" of the products is less than the heat content of the starting materials; and, (4) under such circumstance, ΔH_c° is a negative number:

$$(\text{Hydrocarbon, e.g., } C_n H_{2n+2}, C_n H_{2n}, \ldots) + O_2(\text{excess}) \rightarrow nCO_2 + mH_2O + \text{heat.}$$

$$(3.1)$$

Although the number of molecules for which heats of combustion have been measured is minuscule (given the total number of compounds known), the available data can be put to a good end, as already shown in Chapter 1 and in other ways.

For example, the compound named "isooctane" (2,2,4-trimethylpentane) is arbitrarily used as the standard for comparing combustion properties of fuels burned in internal combustion engines and is assigned an "octane rating" of 100. It is clear that branching in alkanes (for the same molecular weight) lowers the boiling point (Table 3.3) so that you might predict that octane (bp 125.6°C) will boil **higher** than

---

*The gain or loss of an electron (or electrons) by a particular atom in a molecule (as in the picture of resonance in the carbonate anion [Chapter 1, Figure 1.28] or in aromatic hydrocarbons, etc. [Chapter 3, Figure 3.21]) may be matched by the loss or gain of an electron (or electrons) at another atom in the molecule. Such cartoons, for which curved arrows are drawn to portray a path for electron motion to help us understand and remember, correspond neither to reality nor to a change in oxidation state. Further, while one atom in a molecule *might* actually be oxidized, and a second atom *in the same molecule* might actually be reduced, the **net change for the molecule** might be neither oxidation nor reduction.

2,2,4-trimethylpentane (isooctane, bp 113.4°C); that is, at the same pressure, isooctane is more volatile. Second, branching has little effect on the amount of heat liberated on complete combustion to CO_2 and H_2O (Table 3.3). Thus, since $-\Delta H_c^\circ$ for octane is $1307.5\,kcal\,mol^{-1}$, isooctane should be about the same (it is; $-\Delta H_c^\circ$ for 2,2,4-trimethylpentane is $1305.3\,kcal\,mol^{-1}$). Finally, therefore, isooctane will give off just over six times more heat per mole (which is capable of being converted to work) than will methane ($-\Delta H_c^\circ$ for methane is $212.8\,kcal\,mol^{-1}$); however, if combustion is complete in each case, producing the *same amount of heat* generates about 25% *more* CO_2 and water from the larger hydrocarbon.

A second example may be equally important. It is clear from the information in Table 3.3 (which is illustrative rather than exclusive) that, although the effect is small, less heat is obtained from combustion of branched alkanes than from straight-chain alkanes of the same molecular formula (compare, e.g., the isomers of C_5H_{12} **or** those of C_6H_{14}). Since all of the C_5H_{12} isomers, for example, yield the same amount of CO_2 and H_2O (Equation 3.2) when combustion occurs, the amount of heat liberated must be a reflection of their relative stabilities:

$$C_5H_{12} + 8O_2 \rightarrow 5CO_2 + 6H_2O + heat; \qquad\qquad (3.2)$$

that is, the amount of heat released reflects the potential energy stored in the alkane, and since more heat is released from the straight-chain alkane, it must have more energy stored, and thus it must be *less* stable than its branched isomers. It is not obvious, however, that it *should* be so! One explanation for **this experimental observation** is that intramolecular attractive forces favor the branched isomers.

Problem 3.13. Using any suitable compendium in the library (or online), find as many isomers of C_7H_{16} and at least one other alkane of your choice and make a table for those isomers similar to Table 3.3. For the compounds you found, is it true that branching apparently increases stability?

Problem 3.14. Using Table 3.3, compare the heats of combustion of alkenes with cycloalkanes. What can you conclude (if anything) about the relative stabilities of cycloalkanes compared to alkenes?

Problem 3.15. Using Table 3.3, compare the heats of combustion of isomeric alkenes. What can you conclude (if anything) about (a) the effect of the position of the double bond in the chain and (b) the effect of stereochemistry about the double bond?

The data presented in Table 3.3 can also be used to compare cyclic alkanes with their acyclic analogues. For example, using the simple ideas inherent in the LCAO-MO discussion in Chapter 1, it might be reasonably argued that sp^3 hybridization at carbon requires bond angles approaching the tetrahedral value of about 109.5°. In cyclopropane, assuming the bonds lie along the internuclear axis, the internal bond angles must be only 60°. Similarly, if cyclobutane were planar (but see Chapter 4), and the bonds lay along the internuclear lines, the internal angles would need to be 90°. To constrain angles in this fashion requires energy, which could reasonably be expected to be reflected in the heats of combustion in these systems. The difficulty is finding a way to measure how much **strain energy** is present. One common way is to attempt to estimate how much the heat of combustion of alkanes increases *per –CH$_2$– group*, assuming that acyclic alkanes are relatively strain free and that the carbon–hydrogen bonds are equally strong in cyclic and acyclic systems. The first assumption is almost certainly incorrect since the increase in $-\Delta H_c^\circ$ *per –CH$_2$–* in alkanes is somewhat variable (Table 3.3). Compare methane (CH$_4$) to ethane (CH$_3$CH$_3$), where one –CH$_2$– has been added; this gives an increase in $-\Delta H_c^\circ$ of 160 kcal mol^{-1}. Alternatively, one might compare methane (CH$_4$) to 2,2-dimethylpropane [(CH$_3$)$_4$C], where four "–CH$_2$–" units have been added (to CH$_4$) as one was added before. The increase in $-\Delta H_c^\circ$ for the pentane isomer is 627.1 or 156.8 kcal mol^{-1} per –CH$_2$–. While the first (being at the beginning of the series) might be atypical, the second might also be unusual because crowding between methyl groups may also cause unusual effects. Although (Table 3.3) the heats of combustion for the majority of linear alkanes differ from each other by about 158–159 kcal mol^{-1} per –CH$_2$– and such a value might be used, it too is suspect since, for acyclic linear alkanes, most C–C bond angles are about 112.5° (with corresponding diminution of H–C–H angles); that is, while the differences are small, it is not quite certain what value is best. Similar arguments can be made for the strengths of C–H bonds, which could reasonably become stronger (more *s* character) as *sp^3* bonds are stretched. Any increase in C–H bond strength will make the corresponding strain energy apparent in small rings appear smaller (since it is total energy that is measured in heats of combustion).

It is clear that the C–H bonds have been affected, perhaps through rehybridization at carbon, since the protons around the periphery appear **upfield** of normal methylene (–CH$_2$–) protons. Indeed, protons on the methylene (–CH$_2$–) groups of a cyclopropane are typically found in the ^{1}H NMR spectrum near $\delta \approx 0.2$ ppm (i.e., even further *upfield* than a methyl group (–CH$_3$) attached to carbon).*

The above arguments aside, it is clear (Problem 3.14) that, *per –CH$_2$– unit*, small ring carbocycles have greater heats of combustion than linear alkanes. The difference is attributed to the **strain** associated with confining the system, and the original concept, which supposed (incorrectly) that planarity was required for all cyclic systems, is attributed to the German chemist A. Baeyer ([1835–1917] Nobel Prize, 1905) and is called **Baeyer strain**.

---

*Interesingly, it could also be argued that rehybridization at carbon has resulted in a "ring current" (see the above discusion of the NMR spectrum of benzene) and that the protons around the cyclopropyl ring are now in the "shielding" zone. In this vein, it was suggested by A. D. Walsh (1949) some years ago (when detailed calculations of molecular orbitals were unavailable) that since (a) the H–C–H bond angles in cyclopropane are about 120° and (b) the reactivity of cyclopropane bears some similarity to that of alkenes (Chapter 6), it was *reasonable to assume that the hybridization at carbon for this cycloalkane was sp^2*! In this way, two of the localized *sp^2* orbitals at carbon (angle of 120°) could be used for bonding to the *s* orbital of each of the two hydrogens found at that carbon, and the third *sp^2* orbital **is then pointed toward the center of the ring.** Walsh further suggested that these hybridized AOs combine to form three MOs: one bonding, occupied "σ-like" (but not truly σ because the axial symmetry characteristic of this bond type is absent), and two antibonding, unoccupied "σ*-like" orbitals. Finally, he suggested that the remaining *2p* orbitals, one at each carbon, now directed (roughly) toward each other, overlap in a fashion somewhat between the side- or edge-on overlap characteristic of π bonds and the end-on overlap characteristic of σ bonds (Chapter 1). The interaction of these three AOs forms a second set of three MOs, two bonding, occupied "π-like" and one, unoccupied, "π*-like" (Walsh, A. D. *Trans. Faraday Soc.*, **1949**, *45*, 179). Representations for these orbitals and their relative energies are shown in Figure 3.25.

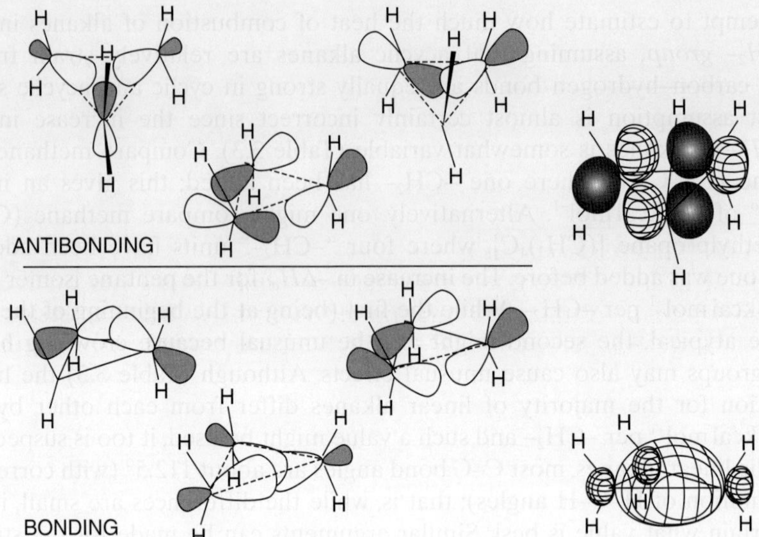

Figure 3.25. Schematic representation of Walsh-type orbitals for cyclopropane.

b. Reduction.

Hydrocarbons containing π bonds (alkenes, alkynes, and arenes) **add*** hydrogen across the double bond and undergo reduction. In a practical sense, when hydrogen is used as the reducing agent (other indirect methods will be discussed subsequently), the reaction between hydrogen and the **unsaturated** system, yielding a **saturated** system, requires the presence of a metal or a metal-containing catalyst[†] in order to proceed at a useful rate.[‡]

As indicated in Table 3.4, the conversion of an alkene to the corresponding alkane is generally thermochemically favorable (i.e., ΔH_r° is negative). It is reasonable that this is so because, as we have earlier seen, π bonds are generally of higher energy than the corresponding carbon–carbon and carbon–hydrogen σ bonds.

The exact details of the path by which hydrogen adds to an alkene, alkyne, or arene are difficult to establish. However, some information is available. First, in the presence of the metal catalyst, the addition is usually suprafacial. Thus, 1,2-dimethylcyclopentene, on hydrogenation over a platinum catalyst, produces (*Z*)-

---

***Addition** reactions to unsaturated (i.e., π) systems, which correspond to the conversion of an sp^2-hybridized carbon to one that is sp^3 hybridized, and its reverse (**elimination**), constitute major pathways for the interconversion of organic compounds. In general, such processes may or may not be reductive (or oxidative). We consider this one process now only because it is reductive.

[†]For the moment, we will consider a **catalyst** as something that facilitates the reaction—whether by increasing the energy of the starting material(s) or by providing a different path (one other than that available in its absence), and so on, linking a specific set of reactants and products. If the catalyst is in a different phase (e.g., solid or liquid) other than that containing the alkene to be reduced (gas, liquid, or solid in solution, etc.), the process is heterogeneous. If both catalyst and unsaturated substrate are in the same phase (e.g., all in solution), the process is homogeneous. Heterogeneous and homogeneous metal-containing catalysts are known.

[‡]Chapter 4 is concerned with the dynamics of processes such as this one. While the thermochemical data are very valuable in helping decide if it is possible that a set of products will form, it is of significant interest to be able to say something about how fast (i.e., at what **rate**) we might expect to see them.

TABLE 3.4. Boiling Points (bp, °C, 760 torr), Melting Points (mp, °C), and Heats of Combustion ($-\Delta H_c^\circ$) and Hydrogenation ($-\Delta H_r^\circ$) (kcal mol^{-1}) for Selected Alkenes and Alkynes (1 kcal = 4.184 kJ)

Alkene	Molecular Formula	bp	mp	$-\Delta H_c^\circ$	$-\Delta H_r^\circ$
Ethene	C_2H_4	−103.7	−169.2	337.3	32.6
Propene	C_3H_6	−47.7	−185.3	491.8	29.9
1,2-Propadiene	C_3H_4	−32.0	−146.0	—	70.5
1-Butene	C_4H_8	−6.3	−185.4	649.3	30.1
(E)-2-Butene	C_4H_8	0.9	−105.6	646.9	27.4
(Z)-2-Butene	C_4H_8	3.8	−138.9	647.7	28.3
1,2-Butadiene	C_4H_6	10.3	−136.3	619.9	—
1,3-Butadiene	C_4H_6	−4.4	−108.9	607.2	56.6
1-Pentene	C_5H_{10}	29.2	−165.2	800.0[a]	29.7[a]
(E)-2-Pentene	C_5H_{10}	36.4	−140.2	797.3	27.2
(Z)-2-Pentene	C_5H_{10}	36.9	−151.4	797.5	28.1
(Z)-1,3-Pentadiene	C_5H_8	44.1	−148.8	768.1	54.1[b]
(E)-1,3-Pentadiene	C_5H_8	42.0	−87.5	761.6	54.1[b]
1,4-Pentadiene	C_5H_8	25.9	−148.3	786.9	60.2
1-Hexene	C_6H_{12}	63.5	−139.8	955.0[b]	30.2
(E)-2-Hexene	C_6H_{12}	67.9	−132.9	—	—
Benzene (an arene)	C_6H_6	80.1	5.5	781.0	49.1
Cyclohexene	C_6H_{10}	82.9	−103.5	896.8	29.1
1,3-Cyclohexadiene	C_6H_8	80.3	−104.8	—	54.9
1,5-Hexadiene	C_6H_{10}	59.5	−140.8	918.8	60.0
1,3,5-Cycloheptatriene	C_7H_8	115.5	−79.5	—	72.1
1,3-Cycloheptadiene	C_7H_{10}	121.5	−110.4	—	50.8
Cycloheptene	C_7H_{12}	114.4	−56.0	1049.9	26.0
1-Heptene	C_7H_{14}	93.6	−119.0	1113.4	29.7
(E)-2-Heptene	C_7H_{14}	98.0	−109.5	—	—
Ethyne	C_2H_2	−84.0	−80.0	—	74.6
Propyne	C_3H_4	−23.2	−102.7	—	69.2
1-Butyne	C_4H_6	8.6	−130.0	620.6	—
2-Butyne	C_4H_6	27.2	−32.3	615.8	65.1

The boiling point and melting point values in this table are largely from Rappoport, Z. *CRC Handbook of Tables for Organic Compound Identification*, 3rd edition, CRC Press, Cleveland, OH, **1977**. Values for $-\Delta H_c^\circ$ (±0.5) are largely from Cox, J. D.; Pilcher, G. *Thermochemistry of Organic and Organometallic Compounds*, Academic Press, New York, **1970**.
[a]Estimated.
[b]Isomers not reported separately.

1,2-dimethylcyclopentane. Second, some alkenes have been shown to interact exothermically with some catalysts, and third, hydrogen (H_2) apparently combines quite readily with many metals to form metal–hydrogen bonds (with cleavage of the hydrogen–hydrogen bond). Although it is clear that not all metal catalysts (Pt, Pd, and Ni are the most commonly used, either alone or in low concentration on a carrier such as finely powdered carbon) and not all alkenes behave the same, the general picture such as that sketched in Figure 3.26 is both vague enough to satisfy the general criteria and clear enough to convey the message.

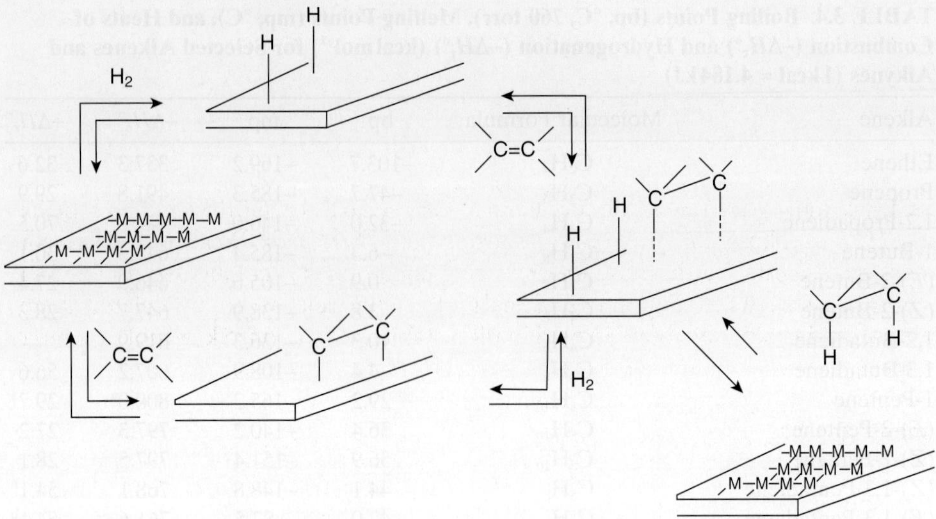

Figure 3.26. A cartoon representing the adsorption of hydrogen and alkene onto a (catalytic) metal surface. Hydrogen and alkene are almost certainly not (if only because they are different) adsorbed in the same way and/or to the same extent. It is presumed that the adsorbed species react and the product is desorbed. Under appropriate conditions, the reaction is known to be reversible for some alkanes and, at least in principle, is reversible for all.

In addition to the utility of a process that allows alkenes and alkynes to be converted to alkanes,* measurement of the heat of hydrogenation ($-\Delta H_r^\circ$) allows the relative stabilities of alkenes to be gauged and relationships to be explored.

Experimental values for the hydrogenation of some selected alkenes and alkynes to the corresponding acyclic and cyclic alkanes are given in Table 3.4.

Generally, an isolated double bond liberates about $30\,\text{kcal}\,\text{mol}^{-1}$ on reduction. Further, the more alkyl groups that are added to the carbons of the double bond, the less the heat evolved; that is, alkyl substitution for hydrogen *on* the double bond produces a stronger bond. Thus, 1-pentene undergoes reduction liberating $1–2\,\text{kcal}\,\text{mol}^{-1}$ more heat than the corresponding 2-pentene isomers. In addition, it is clear that (E)-isomers are more stable than (Z)-isomers (compare (E)-2-butene with (Z)-2-butene and (E)-2-pentene with (Z)-2-pentene; Table 3.4). The amount of heat liberated when a *non*conjugated diene is reduced is about twice that of a single double bond (compare 1-pentene and 1,4-pentadiene; Table 3.4), whereas a conjugated double bond (compare 1,3-pentadiene with 1,4-pentadiene; Table 3.4) imparts significant stability. Cumulated double bonds (e.g., 1,2-propadiene) are apparently much less stable than either conjugated or two isolated double bonds since much more heat is liberated when the former is reduced than either of the latter.

*Formulation of fuels for optimum performance may involve heating mixtures of saturated and unsaturated hydrocarbons over various catalysts at high temperatures and pressures. Under these conditions, structural rearrangements, hydrogenation, and dehydrogenation processes all occur. In the presence of a platinum catalyst, this process has been called *platforming*.

Complete reduction of alkynes involves liberation of about as much heat as compounds with the same number of cumulative double bonds (compare propyne with 1,2-propadiene; Table 3.4).

Medium-ring cyclic alkenes are about as stable as their corresponding acyclic isomers and, again, increasing the number of double bonds in conjugation is stabilizing. It is interesting to compare (Table 3.4) cycloheptene ($-\Delta H_r^\circ =$ 26.0 kcal mol^{-1}) to 1,3-cycloheptadiene ($-\Delta H_r^\circ = 50.8$ kcal mol^{-1}) and cyclohexene ($-\Delta H_r^\circ = 29.1$ kcal mol^{-1}) to 1,3-cyclohexadiene ($-\Delta H_r^\circ = 54.9$ kcal mol^{-1}). The comparison suggests that in these medium-ring compounds, a conjugated double bond contributes between 24.8 and 25.8 kcal mol^{-1}. Thus, relative to an isolated double bond, putting a double bond in conjugation stabilizes by about 5 kcal mol^{-1} (compare 1-pentene with 1,4-pentadiene). Carrying the analogy further, putting a third double bond in conjugation (i.e., 1,3,5-cycloheptatriene [$-\Delta H_r^\circ = 72.1$ kcal mol^{-1}]) increments the heat evolved by 21.3 kcal mol^{-1}. Attempting the same comparison with the homologous six-membered ring isomer shows the dramatic decrease in heat evolution in proceeding to benzene ($-\Delta H_r^\circ = 49.1$ kcal mol^{-1}) from 1,3-cyclohexadiene ($-\Delta H_r^\circ = 54.9$ kcal mol^{-1}); that is, there is a large increase in stability on adding the third "conjugated" double bond.

How much does the aromaticity contribute to stability? Using heats of combustion (comparing butane to any butene and then to 1,3-butadiene and cyclohexane to cyclohexene, etc.) would argue that what would be expected for "1,3,5-cyclohexatriene" might be of the order of $-\Delta H_c^\circ \approx 816$ kcal mol^{-1}. Thus, benzene is about 35 kcal mol^{-1} "more stable than expected." Alternatively, comparing heats of hydrogenation of cyclohexene to cyclohexadiene, it might be expected that the heat of hydrogenation of "1,3,5-cyclohexatriene" might be of the order of $-\Delta H_r^\circ \approx$ 80.7 kcal mol^{-1}). Thus, benzene is about 32 kcal mol^{-1} more stable than expected. Of course, neither answer is "correct." However, it is clear that benzene does have increased stability over what is simply anticipated by utilizing the concept of homology.

Finally, there are some practical considerations. First, it is common to carry out the hydrogenation of most alkenes by dissolving them in an appropriate solvent (of known volume) in the presence of a catalyst, as already noted. If the amount of alkene, the temperature at which the reaction is being carried out (frequently room temperature), and the volume of the system containing hydrogen gas are known, then it is possible to calculate (using the ideal gas law) approximately how much hydrogen (H_2) will be absorbed if the reaction goes to completion. Alternatively, the method may be used to determine *how many* double (and or triple) bonds are present by noting how much hydrogen is absorbed. Separation of alkenes and alkynes from alkanes (should the reduction fail to go to completion), or rapid comparison of materials before and after reaction, can be accomplished with gas–liquid partition chromatography (GLPC) (Chapter 1). It is common to use columns impregnated with silver nitrate (which complexes with the π systems preferentially) to allow the alkane to elute first. That the reaction has gone to completion is clearly demonstrated by examining the ^{1}H NMR spectra of the starting materials (vinylic protons generally appearing between $\delta \approx 4.9$–5.2 ppm) and products. Reduction being complete, there will no longer be vinylic protons. Alternatively, and for alkenes bearing four alkyl substituents, the ^{13}C spectra will be as informative; that is, unsaturated starting materials will exhibit the usual downfield signals (Table 3.2), which will disappear when the unsaturation is gone (to be replaced by additional signals upfield).

Second, despite the generalization above, specific conditions required for reduction may well vary from one alkene to the next. It is particularly worthy to note that most arenes (in accord with their extraordinary stability and as thus expected) can only be reduced with difficulty. In contrast to simple alkenes, which can often be reduced at or near 1 atm of hydrogen in the presence of a Pt or Pd catalyst, benzene requires high pressures and temperatures (e.g., powdered Ni, 40 atm H_2, 200°C) (Equation 3.3). Further, the reduction of alkynes to (Z)-alkenes (addition of hydrogen [H_2] being suprafacial) frequently cannot be stopped at the alkene stage and complete reduction to alkane occurs. Special "poisoned" catalysts have been developed, which frequently succeed in stopping the reaction.* (Equation 3.4).

Finally, a number of noncatalytic reductive methods have been developed. These will be discussed subsequently (Chapter 5).

$$\text{(3.3)}$$

$$\text{(3.4)}$$

$$R = \text{alkyl or aryl}$$

ADDITIONAL PROBLEMS

Problem 3.16. Name the compounds (a)–(e) shown in the figure.

(a)   (b)   (c)   (d)   (e)

Problem 3.17. Presuming that compounds (a)–(e) were obtained, draw a Newman projection of each looking down the bond between C2 and C3.

Problem 3.18. Presuming that each of compounds (a)–(e) were obtained, which might have isomers that were not capable of being brought into superposition with their realized mirror images?

Problem 3.19. Suggest how many signals might be observed in the proton NMR spectrum of

*One such catalyst is frequently prepared by lead salt deactivation of palladium that has been deposited on calcium carbonate ($CaCO_3$). It is called a **Lindlar catalyst** (Lindlar, H. *Helv. Chim. Acta,* **1952**, *35*, 446).

(a) cyclopropane,

(b) methylcyclopropane, and

(c) 1,1-dimethylcyclopropane.

Problem 3.20. The compound whose mass spectrum appears in the figure corresponds to C_5H_{12} and (a) has a single peak in its proton NMR spectrum (at 0.90 ppm) and (b) lacks a parent peak in its mass spectrum where the most intense signal is at M − 15 (shown in the figure). Identify the compound.

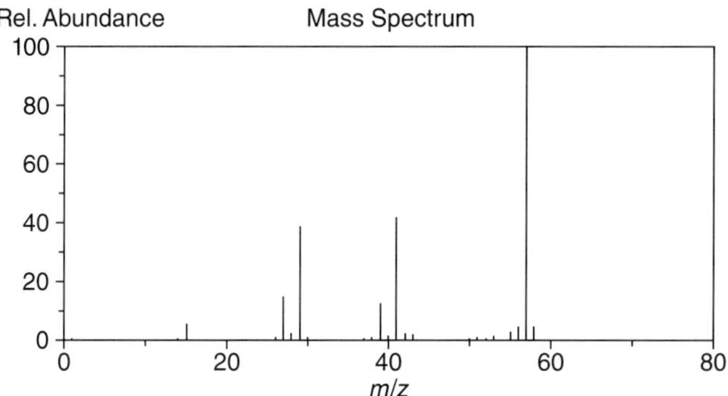

Problem 3.21. In Figure 3.9, two-dimensional (flat!) representations for 1,2-dimethylcyclopentane and 1,3-dimethylcyclopentane were presented and are reproduced in the figure:

(a) Draw a representation for each of these two species with dashed and wedged lines such as in (a) in the next figure.

(b) Draw a representation for each of these two species as an "edge-on" figure such as in (b) in the next figure.

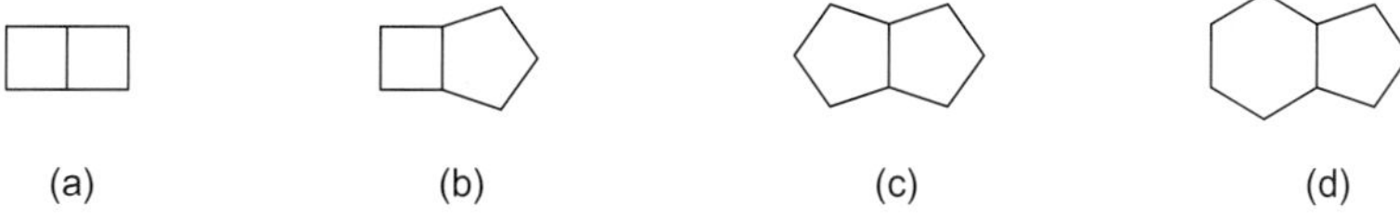

Problem 3.22. Given that (a) in the figure is named bicycle[2.2.0]hexane and (b) is bicycle[3.2.0]heptane, name (c) and (d).

Problem 3.23. Name the following alkenes. Where appropriate, use the designations (*E*) and (*Z*).

(a)   (b)   (c)   (d)   (e)

(f)   (g)   (h)   (i)   (j)

Problem 3.24. Describe the relationships of the vinylic protons in each of the alkenes in Problem 3.23.

Problem 3.25. Describe the differences in the methyl (CH_3–) groups in each of the alkenes in Problem 3.23.

Problem 3.26. Name the following dienes:

(a)   (b)   (c)   (d)   (e)

(f)   (g)   (h)   (i)   (j)

Problem 3.27. Name the following dienes:

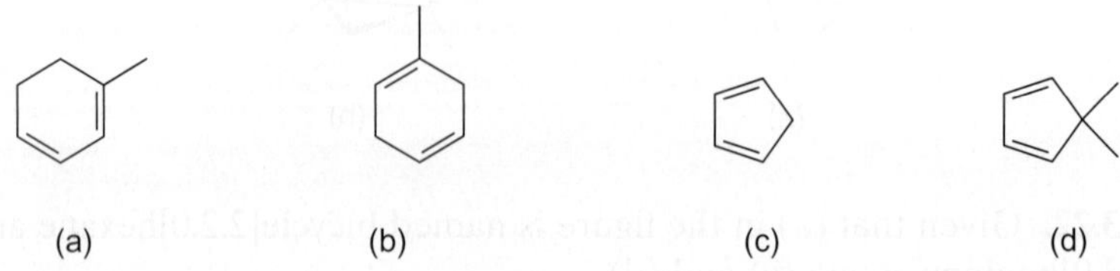

(a)   (b)   (c)   (d)

Problem 3.28. For the dienes (a), (b), and (d) in Problem 3.27, describe *how many* signals (lines) (i.e., what is the multiplicity) you would expect to see in the proton NMR spectrum for the methyl group(s) present.

Problem 3.29. Predict the products of hydrogenation of the following alkenes and the alkyne:

REFERENCES

Dewar, J. *Proc. Royal Soc. Edinburgh*, **1867**, *6*, 82.

Kekulé, A. *Annalen Chem. Pharm.*, **1866**, *137*, 129.

An Introduction to Dynamics

I pass with relief from the tossing sea of Cause and Theory to the firm ground of Result and Fact.

—W. S. Churchill

A. INTRODUCTION

In prelude to actually examining the reactions which make up the body of organic chemistry, this chapter is largely concerned with attempting to provide information about questions that are (or should be) asked in regard to every organic reaction.

First, in the most general sense, it is important to say something about spontaneity. To know why it is that if, for example, the quantities ΔH_c° (a measurement of the heat of combustion) and ΔH_r° (a measurement of the heat of hydrogenation) are negative, in accord with exothermic reactions, the processes fail to occur spontaneously. Some response discussing the nature of energy **barriers** lying between starting materials and products might be expected and will be forthcoming. In the same vein (and at least in principle), every small change on the path over the barrier between starting materials and products can be considered microscopically reversible. Thus, for at least some reactions, starting materials and products might be considered to be in equilibrium. It will clearly be useful to know when this might be the case and how the equilibrium might be affected to produce products.

Second, some comments on the speed with which things happen (or the **rate** at which they occur) will need to be made. After all, while it is (or may be) comforting to know that "all things come 'round to him who will but wait," it is of some useful purpose to consider, before beginning to wait, how long the wait might be.

Third, before we can say too much, except in very general terms, about energy barriers and the crossing of them, it will be useful to attempt to deal with the question, in some detail, about what it is with which we are starting. That is, as reactants approach each other, what can we say about the details of the initial contact? What might the reactants actually best look like in order that the path on which they embark be of minimum difficulty?

Foundations of Organic Chemistry: Unity and Diversity of Structures, Pathways, and Reactions, First Edition. David R. Dalton.
© 2011 John Wiley & Sons, Inc. Published 2011 by John Wiley & Sons, Inc.

These questions deal with quantities of energy. In Chapter 1, you found information about the nature of bonds holding the nuclei together to make molecules and how we might say something about the energy needed to break a bond (and thus measure its "strength"). Generally, the bond strengths were, you will recall, of the order of $\approx 100\,kcal\,mol^{-1}$ (although some were significantly less, few were much more—the **order of magnitude** is about right). Subsequently, in Chapters 2 and 3, more information about organic molecules (some of their names and something of their nature) and some details about the tools that are used to learn what we know about them—by measuring energies of transitions between the states which bonds and nuclei occupy—were provided.

Again, it is important that you keep in mind that at room temperature (taken to be $\approx 300\,K$), spectroscopic tools tell us that not only are molecules moving about ($E_{translation}$) but that bonds between nuclei in the molecules are stretching, bending, and twisting ($E_{vibration}$ and $E_{rotation}$). Indeed, while we can use portions of the electromagnetic spectrum to probe structure, **molecular transformations** that occur with less than about $600\,cal\,mol^{-1}$ ($2.5\,kJ\,mol^{-1}$) cannot be observed at room temperature because they are occurring too fast. Similarly, **molecular transformations** that require $20–30\,kcal\,mol^{-1}$ ($84.6–125.4\,kJ\,mol^{-1}$) (i.e., near the low end of the bond strengths given in Chapter 1) frequently cannot be followed spectroscopically at or near room temperature because the energetic requirements are too high for convenient observation.*

Many exciting processes occur between 0.6 and $20\,kcal\,mol^{-1}$. For example, isomers are considered inseparable if the barrier to their interconversion is less than about $25\,kcal\,mol^{-1}$ ($104.5\,kJ\,mol^{-1}$) and their interconversion might be studied. Thus, to probe what is happening in this region, the low-energy end of the electromagnetic spectrum can be examined.

Next, as molecules undergo reactions, we will need to inquire if more than one species is involved. That is, in each specific case, we will want to know how many species may be undergoing covalent bonding changes as we proceed from starting materials to products. We will also want to know which bonds are being made and which are being broken. (**It is very important to learn early that it is rare for entire molecular systems to fly apart and recombine; it is usually the case that only a few bonds are involved.**) Presumably, because processes are reproducible, some sort of pathway connecting the starting material(s) to product(s) must exist and will usually be followed. Presumably, also, pathways of minimum energy are preferentially taken, and it will be useful to consider what we can and cannot, in general, say about such paths—so that we can make predictions about cases that have not specifically been examined.

The tools to be used to attempt to find answers to some of the questions raised above are just those encountered in Chapters 1 and 2: portions of the electromagnetic spectrum generally available to chemists, our laboratory equipment (for weighing, titrating, separating mixtures, etc.), and our wits.

---

*It is worthwhile noting that the average amount of energy associated with one degree of freedom of molecular motion is (classically) $1/2\,kT$ (where k is the Boltzman constant $= 1.38 \times 10^{-16}\,erg\,deg^{-1}$). This corresponds to $2.981 \times 10^{2}\,cal\,mol^{-1}$ at $300\,K$. Rotational transitions (in the far-infrared and microwave regions of the spectrum) are associated with frequencies of the order of $30,000\,MHz$ ($30\,GHz$), which is about $1\,cm^{-1}$ or $2.86\,cal\,mol^{-1}$ (about $12\,J\,mol^{-1}$).

B. REVIEW OF SOME ENERGY CONSIDERATIONS

The change in internal energy (ΔE) of a system is related to heat changes and work done. That is, systems can be heated or cooled (or can evolve or consume heat), and the system itself either does work or has work done upon it.* Heat changes in a system and work done on or by the system can often be measured by examining its surroundings. That is, since systems can exchange energy with their surroundings, by adjusting the surroundings of different systems to have common properties, all systems can be compared similarly. For example, as described in Chapter 1, the temperature change in the water around the bomb calorimeter (Figure 1.5) is used to determine the heat evolved on combustion of organic compounds. In such cases, because the calorimetry is carried out at constant volume (V), the pressure (P) within the calorimeter changes and no mechanical work is done. As a consequence, the heat evolved (and measured as a temperature increase in the surrounding water) does not *actually* correspond to the enthalpy, ΔH (since the change in heat content is **defined** as the heat change at constant P, not at constant V). As pointed out in Chapter 1, the adjustment is usually minor and **for most organic compounds**, Equation 4.1 can, and is, frequently replaced by Equation 4.2:[†‡]

$$\Delta H = \Delta E + P(\Delta V), \tag{4.1}$$

$$\Delta H \approx \Delta E. \tag{4.2}$$

Of course, the statements of Equations 4.1 and 4.2 (the **first law of thermodynamics**) say nothing about whether a reaction will or will not occur—only that energy is conserved.

The idea that there may be an inherent direction in which any system may tend to go, whether or not the process involved is exothermic, is clearly correlated with its equilibrium constant, K_{eq}. That is, experience dictates that systems tend

---

*In classical terms, $\Delta E = q + w$, where q, if positive, is the heat added to the system and w, if positive, is work done on the system. Various kinds of work can be done. When the work done on the system is that kind of work called pressure–volume work (or P-V work), such as with a gas in a cylindrical chamber, one end of which contains a (movable) piston, two situations can be imagined. First, if the temperature of the gas in the piston is increased and the piston is not allowed to move (i.e., the chamber volume is held constant), then the pressure inside the chamber will increase (this is similar to the adiabatic calorimeter discussed in Chapter 1). This kind of work is abbreviated as $V(\Delta P)$. Alternatively, as the temperature of the gas is increased, the piston is allowed to move so that the volume occupied by the gas is increased too. In this case, the pressure remains constant. This form of work is abbreviated as $P(\Delta V)$. When the work done corresponds to $P(\Delta V)$, then q is ΔH, the **enthalpy**.

[†]For example, $-\Delta H_c^\circ$ for octane (C_8H_{18}) is 1307.5 kcal mol^{-1} (5470 kJ mol^{-1}) and $-\Delta E$ is $\approx$1309.9 kcal mol^{-1} (5481 kJ mol^{-1}).

[‡]Bertholet (M. Berthelot, 1827–1907), who helped complete the overthrow of the "vital force" theory (begun by Wöhler, earlier alluded to, *vide supra*) with the synthesis of some simple organic compounds, played an important role in early thermochemical studies. After becoming convinced that the same laws governed organic and inorganic fields of chemistry, he eventually pursued the study of reaction velocities and, subsequently, thermochemistry. For that discipline, he coined the words *exothermic* and *endothermic*. He apparently believed, at least initially, that a reaction would proceed as written if $\Delta H < 0$. Later, as more reactions were investigated and as experimental techniques improved, the P-V work term became clear.

toward equilibrium; for example, heat flows from hot bodies to those that are colder, wound clocks run down, gases expand into a vacuum, a neat desk at the beginning of the term becomes cluttered as the semester wears on. The direction of such spontaneous changes in isolated systems is always toward increasing disorder and must be accounted for in attempting to determine the total energy change accompanying any process. The increase in randomness or disorder is expressed by a quantity called **entropy** (ΔS) and, like energy (ΔE) and enthalpy (ΔH), is dependent on the initial and final state of the system (i.e., $S_{final} - S_{initial}$); a positive sign indicates an increase in randomness. Interestingly, no process that becomes more orderly (i.e., $-\Delta S$ or lower entropy) can proceed without producing even greater disorder (i.e., $+\Delta S$ or higher entropy) in its surroundings. Thus, although energy may be conserved (the **first law**), entropy continues to increase (the **second law of thermodynamics**).

The relationship (Equation 4.3) that correlates energy change with equilibrium constant involves standard enthalpy (ΔH°), temperature (K), and standard entropy (ΔS°) and is called the **Gibbs standard free energy** (ΔG°):*

$$\Delta G^\circ = \Delta H^\circ - T\Delta S^\circ. \tag{4.3}$$

The equilibrium constant attending a reaction (e.g., Equation 4.4) is thus given in Equation 4.5 and is related to ΔG° through Equation 4.6,

$$A + B \rightleftharpoons C + D, \tag{4.4}$$

$$K_{eq} = \frac{[C][D]}{[A][B]}, \tag{4.5}$$

$$\Delta H^\circ - T\Delta S^\circ = \Delta G^\circ = -2.303RT \log_{10} K_{eq}, \tag{4.6}$$

where $R = 1.987 \, \text{cal} \, \text{deg}^{-1} \, \text{mol}^{-1}$.

Finally, in this regard, it should be clear that if values for ΔG° were readily available, the same would be true for equilibrium constants. However, whereas ΔH° is easily estimated (if not available directly), the same cannot be said for ΔS°. Indeed, it is still found that it is even difficult to prejudge the sign of ΔS for some processes. However, it is generally true that if ΔH° is more negative than about $-15 \, \text{kcal} \, \text{mol}^{-1}$ (about $-63 \, \text{kJ} \, \text{mol}^{-1}$) then $K_{eq} > 1$, while if ΔH° is more positive than about $+15 \, \text{kcal} \, \text{mol}^{-1}$ (about $+63 \, \text{kJ} \, \text{mol}^{-1}$) then $K_{eq} < 1$.

As indicated in Table 4.1, if two materials (e.g., a starting material and a product or two isomers that might be in equilibrium) differ by as little as about $5.5 \, \text{kcal} \, \text{mol}^{-1}$ ($23.0 \, \text{kJ} \, \text{mol}^{-1}$), the more stable isomer will predominate to the extent of about 99.99%. The gas and liquid chromatography techniques discussed in Chapter 1 and most of the spectroscopic tools discussed in Chapter 2, even under routine circum-

---

*As usual, the superscript ($^\circ$) indicates the **standard state**, or the state in which the substance is stable at 25°C and 1 atm pressure. Josiah W. Gibbs (1839–1903), after whom this function is named, published the principles which subsequently found widespread application to the study of chemical equilibria in the rather obscure *Transactions of the Connecticut Academy* in 1878. They were ignored for some years.

TABLE 4.1. The Relationship between Some Values for the Composition of a Mixture of Equilibrating Materials, the Equilibrium Constant Attending that Equilibrium, and the Energy Difference (by Equation 4.6 for $T = 25°C$) Which must be Present between Them

Percent of More Stable Isomer	$\Delta G°$ kcal mol^{-1} (kJ mol^{-1})	Equilibrium Constant K_{eq}
50.00	0.00 (0.00)	1.00
55.00	0.12 (0.50)	1.22
60.00	0.24 (1.00)	1.50
65.00	0.37 (1.55)	1.86
70.00	0.50 (2.09)	2.33
75.00	0.65 (2.72)	3.00
80.00	0.82 (3.43)	4.00
85.00	1.03 (4.31)	5.67
90.00	1.30 (5.44)	9.00
95.00	1.75 (7.32)	19.00
99.00	2.72 (11.38)	99.00
99.99	5.46 (22.84)	9999.00

stances, are able to determine if as little as 0.01% of an isomer (or other impurity) is present.

C. THE BARRIER BETWEEN REACTANTS AND PRODUCTS

In the relationship given in Equation 4.4, which might, for example, be the combustion of methane (CH_4) to carbon dioxide (CO_2) and water (H_2O) (Equation 1.4) or the hydrogenation of (E) or (Z)-2-butene to butane, or indeed, any of a number of other reactions for which $\Delta H°$ is sufficiently negative so that the reaction might be expected to proceed to yield products to the exclusion of starting materials, it becomes necessary to inquire as to why there is no spontaneous reaction. Supposedly, the problem must have something to do with energy and entropy. Indeed, it does seem reasonable to imagine that for species to react, they must not only collide but that the collision must occur with enough energy to overcome some repulsive forces between nonbonded nuclei. The effective collision must also be oriented with some precision, in an ordered way, to allow bonds to break in one molecular species and be made properly in another. These considerations will apply across all reactions, and thus, there will always be a barrier between the way the system we have in hand appears (the "starting materials") and the way it will appear after whatever transformation is effected (the "products"). Clearly, if there were no barrier, we could only have products! Figure 4.1 presents a view of a barrier between starting material(s) and product(s). At the top of the barrier (the **transition state** "‡"), return to starting material is as likely as continuation on to product. You should note that the barrier resembles a pass through a mountain. That is, it is the lowest energy path over the mountain barrier

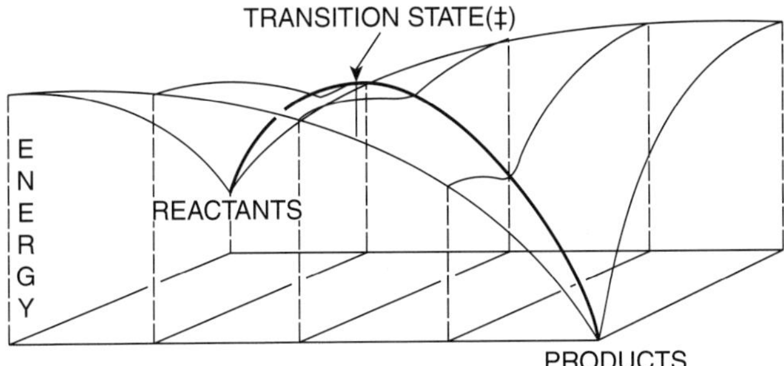

Figure 4.1. A cartoon depiction of a path between some starting reactants and the corresponding products as they pass over a barrier lying between them. This kind of a plot is often called a "More O'Ferrall plot" (Professor R. A. More O'Ferrall, Department of Chemistry, National University of Ireland, Dublin, Ireland).

that divides the promised land of product(s) from the worn land of starting material(s). A further abbreviation of the information in Figure 4.1 is shown in Figure 4.2. Here, it is important to note that, while the line connecting starting material(s) to product(s) is shown, **it is left to your understanding** that it represents a **minimum energy pathway**.*

Finally, in this regard, it should be clear that product(s) of one reaction might be starting material(s) for another. Depending on the amount of energy required, the initial product(s) may neither be isolated nor, indeed, be observed! In such cases, the initial product(s) are described as **intermediate(s)**, and more than one diagram such as that of Figure 4.1 (or Figure 4.2) is needed. Frequently, these are appended one to the next as in Figure 4.3 (abbreviated as Figure 4.4a and redrawn in a different fashion as Figure 4.4b) and the conversion of starting material(s) to product(s) is a multistep process.

*Since the maximum on the curve in Figure 4.1 is also a minimum path "through the space between reactants and products," it is often referred to as a "saddle point." The analogy to a saddle is demonstrated in the cartoon below.

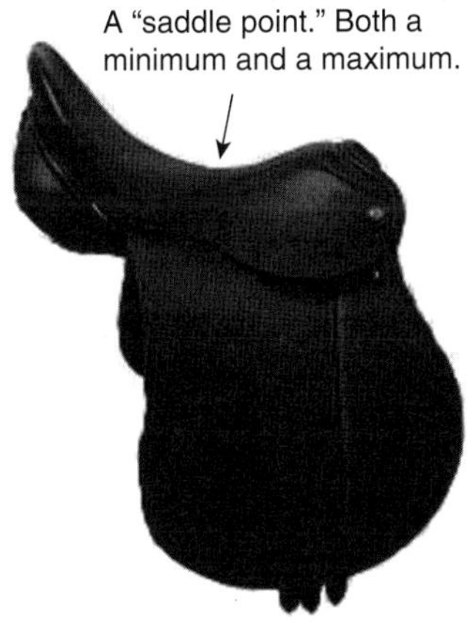

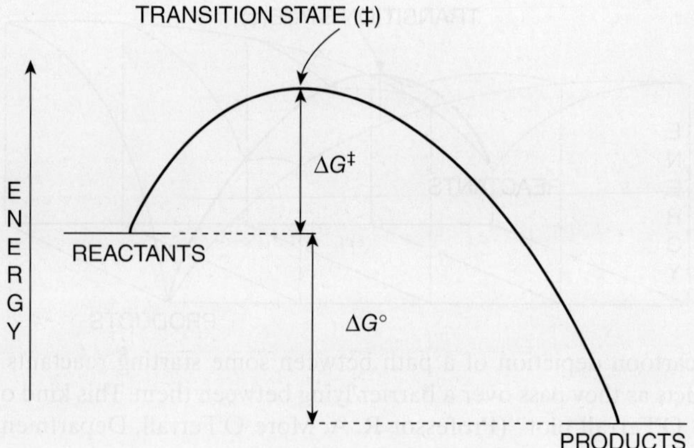

Figure 4.2. An abbreviation of the exothermic reaction of Figure 4.1, in which it is understood that the line represents a minimum energy pathway between reactants and products. The overall energy change (ΔG°) is seen as negative but the energy of activation ($\Delta G^{\ddagger}$) is seen as positive.

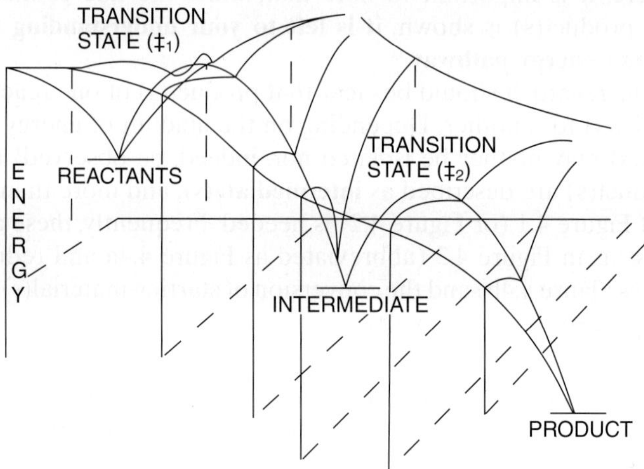

Figure 4.3. A representation of a multistep process with two transition states and an intermediate lying between starting material(s) and product(s).

D. MORE ABOUT THE TRANSITION STATE

Although large collections of molecules* are normally used in chemical processes and thus the physical properties and behaviors reported are average properties of the assembly, the tendency when discussing molecules in organic chemistry is to concentrate on individual species. Thus, for example, as you already know, although the ideal gas law ($PV = nRT$) is widely applicable, assumptions surrounding its use are made. Among them is that the individual molecules behave much like each other

---

*It is sobering to attempt to deal with large numbers. A mole contains about 10^{24} molecules. The cosmos is often estimated to be somewhere between 10 and 20 billion ($\approx 10^{10}$) years or $\approx 10^{17}$ s old!

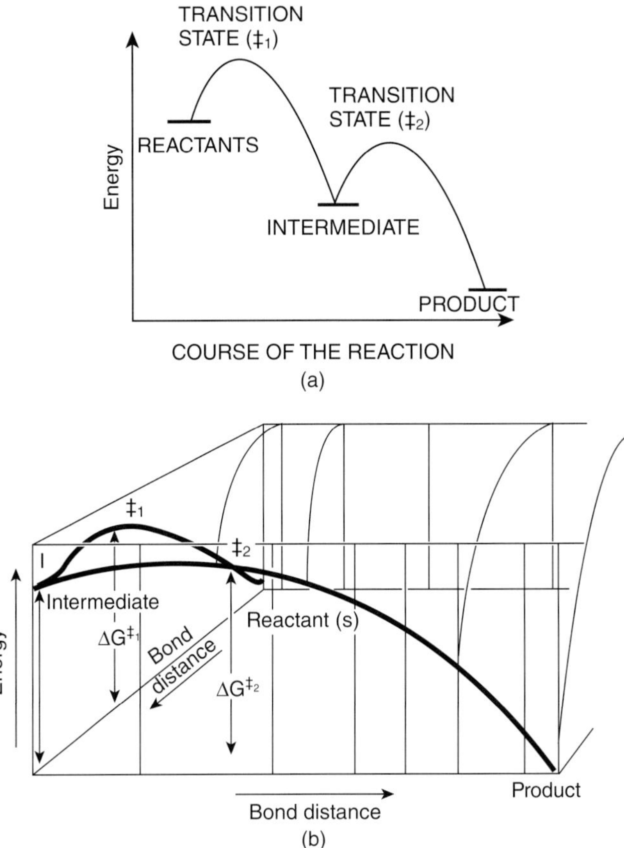

Figure 4.4. (a) An abbreviation of the two step exothermic reaction of Figure 4.3. (b) An alternative representation (to Figure 4.4a) of a multistep process with two transition states and an intermediate lying between starting reactant(s) and product.

or some average. This is done even though we know that there is a distribution of properties.

Largely based on ideas set forth by Boltzmann (L. Boltzmann, 1844–1906), and as depicted in Figure 4.5, we consider that if N is the number of molecules in some specific energy state, then N_i, the number in a state whose potential energy is E_i *above* that state, is given in Equation 4.7,

$$N_i = N\, e^{-(E_i/k_B T)}, \tag{4.7}$$

where k_B, the Boltzmann constant, is the gas constant, R, per molecule, that is, $1.987\,\mathrm{cal\,deg^{-1}\,mol^{-1}}$ ($8.31\,\mathrm{J\,deg^{-1}\,mol^{-1}}$), so ($1.987\,\mathrm{cal\,deg^{-1}\,mol^{-1}}/6.02 \times 10^{23}\,\mathrm{mol^{-1}}$) = $3.30 \times 10^{-24}\,\mathrm{cal\,deg^{-1}} = 1.38 \times 10^{-16}\,\mathrm{erg\,deg^{-1}}$.

Presumably, the first molecules to undergo a reaction and to pass over the barrier (the state at the top of which is called the **transition state**) separating starting materials from products will be that subset N_i with enough energy. That is, not all of the molecules will pass over the barrier simultaneously.

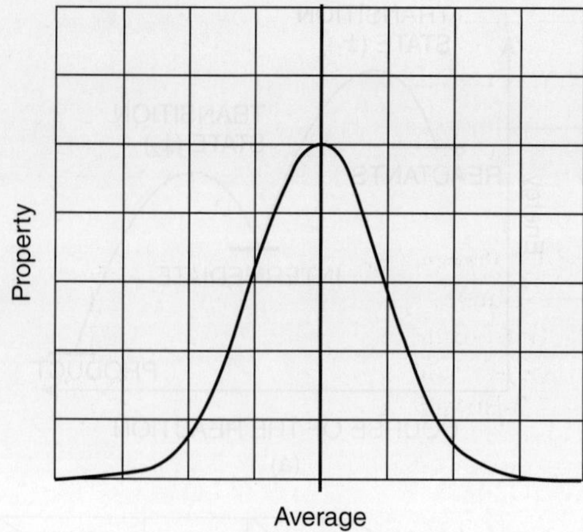

Figure 4.5. A depiction of a distribution of some measured property about an average.

The rate (i.e., the change in observable property per unit time) of any process that must pass over a barrier can be formulated in terms of the amount of energy it takes to reach the pinnacle or top of the barrier and the amount of material (i.e., the concentration) passing over the barrier per second. In this regard, it is frequently assumed that there is a specific arrangement of nuclei (i.e., a discrete species called the **activated complex**) in the transition state. Furthermore, it is assumed that because the activated complex can return to starting material(s) whose concentrations were, at least initially, well known and whose concentrations can be monitored as time goes by, that some equilibrium between the starting material(s) and the activated complex exists, as in the expanded version of Equations 4.4 and 4.5, shown as Equations 4.8 and 4.9, respectively (where the activated complex is denoted as $AB^{\ddagger}$).

$$A + B \rightleftharpoons AB^{\ddagger} \rightleftharpoons C + D \tag{4.8}$$

$$K^{\ddagger} = \frac{[AB^{\ddagger}]}{[A][B]} \tag{4.9}$$

Transition state theory argues that the rate of the reaction (modified by some probability function that the activated complex will go on to product rather than back to starting material) is equal to the concentration of activated complex, that is, $[AB^{\ddagger}]$ *times* the rate at which the complex passes over the barrier. In detail, it is presumed that a bond that was, for example, stretching is now stretched just to the point of being broken, or that a bond that was inhibited from twisting has now overcome that inhibition, and that the event occurs with some frequency v. This frequency, v, is simply E_i/h (i.e., since $E = hv$), where h is Planck's constant (6.62×10^{-27} ergs or 1.58×10^{-34} cals). Furthermore, since the energy, E_i, at which the process occurs is simply $k_B T$, the frequency (v) with which the barrier is overcome is $k_B T/h$, that is,

$$v = k_B T/h. \tag{4.10}$$

Thus, since

$$\text{rate of reaction} = [AB^{\ddagger}] \times (\text{rate of passage over the barrier} = \text{frequency}),$$

then

$$\text{rate of reaction} = [AB^{\ddagger}](k_{\mathrm{B}}T/h) = K^{\ddagger}[A][B](k_{\mathrm{B}}T/h)$$

and, at a given temperature,

$$\text{rate of reaction} = k_{\mathrm{obs}}[A]^{a}[B]^{b}, \tag{4.11}$$

where $k_{\mathrm{obs}} = K^{\ddagger}k_{\mathrm{B}}T/h$.

The **order** of the reaction, **experimentally determined**, is given by the sum of the exponents (a, b) of the concentration terms in the kinetic expression. In the particular case of Equation 4.11, if $a = 1$ and $b = 1$, that is, $[A]^{1}[B]^{1}$, then the order is two. The reaction is said to be **first order in A**, **first order in B**, and **second order overall**.

Furthermore, $K^{\ddagger}$ must be related to the energy of activation, E_{a} (Chapter 1), for the process; since

$$\Delta G^{\mathrm{o}} = -2.303RT \log_{10} K_{\mathrm{eq}} = -RT \ln K_{\mathrm{eq}}, \tag{4.12}$$

it is reasonable that

$$\Delta G^{\ddagger} = -RT \ln K^{\ddagger} \tag{4.13}$$

or

$$K^{\ddagger} = \mathrm{e}^{(-\Delta G^{\ddagger}/RT)}, \tag{4.14}$$

and since $\Delta G^{\mathrm{o}} = \Delta H^{\mathrm{o}} - T\Delta S^{\mathrm{o}}$, separation of the entropy and enthalpy terms gives

$$k_{\mathrm{obs}} = (k_{\mathrm{B}}T/h)\mathrm{e}^{(-\Delta G^{\ddagger}/RT)} = (k_{\mathrm{B}}T/h)\mathrm{e}^{(-\Delta H^{\ddagger}/RT)}\, e^{(-\Delta S^{\ddagger}/R)}. \tag{4.15}$$

Although there are some processes in which the activated complex is highly ordered (i.e., a very specific organization is required), and for these $\Delta S^{\ddagger}$ is positive and the reaction is slowed, in many cases there appears to be enough leeway so that $\Delta H^{\ddagger}$ can be taken as close to the energy of activation (E_{a}).

Thus, in summary, the rough pictures of potential energy diagrams (or of passes [transition states] through mountainous territory [energy barriers] dividing the promised land of products from the swamps of starting materials) can be, through Equations 4.7–4.15, quantified. **While the details of the paths may differ, similarities abound. Recognizing the unity (and the diversity) constitutes much of the body and challenge of organic chemistry.**

E. ROTATION ABOUT SIGMA (σ) BONDS IN ACYCLIC ALKANES, ALKENES, ALKYNES, AND ALKYL-SUBSTITUTED ARENES

I. Alkanes

In addition to the normal infrared (IR) stretching frequencies of hydrocarbons expected in the region of about 2950–2850 cm^{-1} for the C–H sp^{3} hybridized structural unit and typical ^{1}H and ^{13}C nuclear magnetic resonance (NMR) absorbances expected for alkanes, absorption of energy in those portions of the electromagnetic spectrum known as the **far-IR** and **microwave regions**, that is, between about 3000 and 300,000 MHz (3–300 GHz, 0.1–10 cm^{-1} or ≈0.3–30 cal mol^{-1}) is also observed for these materials. Absorption in these regions corresponds to transitions between (closely spaced) rotational energy levels (Figure 2.2).

Spectroscopy in the microwave and far-IR regions of the electromagnetic spectrum did not come into its own until after World War II developments in radar technology became generally available. However, it had been clear for some years before (with the development of statistical mechanics) that calculations of the total energies (E_{total}) of all but the smallest molecules only poorly agreed with experiment, as long as the assumption of completely free rotation about the linear axis of the σ bond was maintained. However, the deviation in E_{total} could be largely accounted for by assuming a **barrier** to such free rotation existed.

As has been subsequently established by microwave, IR, and Raman spectroscopies, the barrier to internal rotation in a species such as ethane (CH$_3$CH$_3$) is sinusoidal in nature. That is, as the methyl groups (CH$_3$–) rotate about the single (σ) bond connecting them, the angle subtended (called the **dihedral angle**) between **any pair of hydrogens on adjacent carbons** varies from 0° to 60°. The same conclusion is reached by holding one methyl group of ethane (CH$_3$CH$_3$) fixed and rotating the other a total of 360°, until the original picture has been returned. During the rotation described, hydrogens attached to the carbon being rotated will pass by those on the carbon being held stationary and all three hydrogens will be eclipsed at 0°, 120°, 240°, and 360°. The forms of ethane (CH$_3$CH$_3$), generated by such rotation, are called **rotational** or **conformational** isomers. The energy required to pass over the barrier connecting them is the **energy of activation** (E_a) for the conversion of one staggered form to another. Figure 4.6 shows both sawhorse and Newman pro-

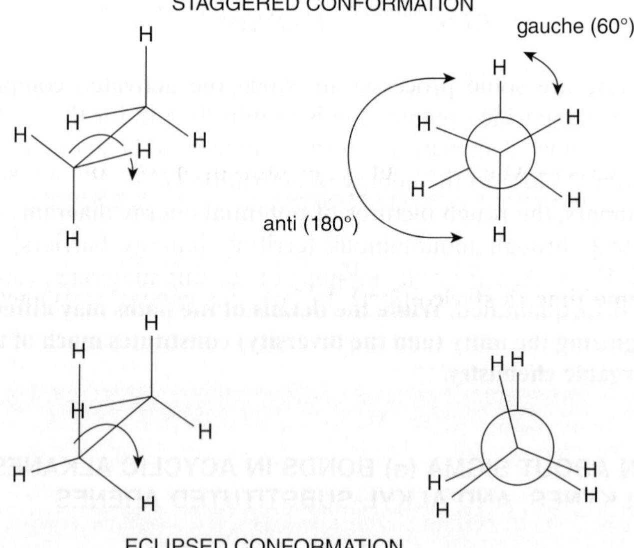

Figure 4.6. Sawhorse and Newman representations for staggered and eclipsed conformers of ethane (CH$_3$CH$_3$).

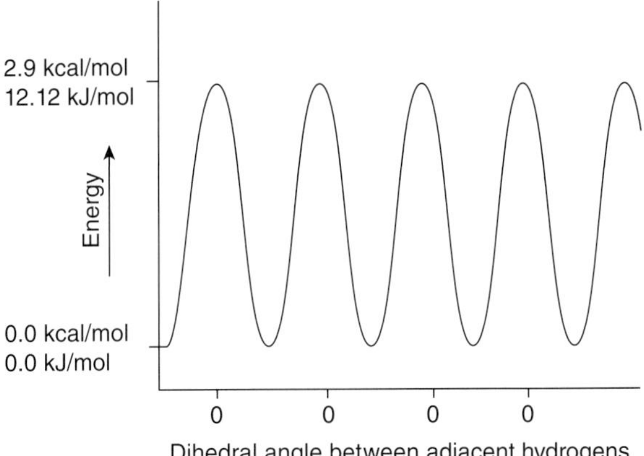

Figure 4.7. A sketch of a potential energy diagram for rotation about the carbon–carbon single bond in ethane (CH_3CH_3). The staggered form with an H–H dihedral angle of 60° is about 2.9 kcal mol^{-1} more stable than the eclipsed form with an H–H dihedral angle of 0°.

jection formulations representing extrema for this process, that is, the more stable **staggered conformation** and the less stable **eclipsed conformation**.

Figure 4.7 shows the variation in potential energy between eclipsed and staggered forms of ethane (CH_3CH_3) (a simple cosine function) assuming that one methyl group is held fixed and the other is permitted to rotate a full 360°. Complete eclipsing of **three** hydrogens on the "front" methyl with **three** hydrogens on the "back" methyl apparently increases the potential energy of the system about 2.9 kcal mol^{-1} (12.1 kJ mol^{-1}). Thus, on the average, it is argued that eclipsing of any **one** hydrogen on the "front" methyl with any **one** hydrogen on the "back" methyl increases the potential energy by about 0.9 kcal mol^{-1} (3.8 kJ mol^{-1}), and thus, generally, such hydrogen–hydrogen eclipsing "costs" about 1 kcal mol^{-1} (4.18 kJ mol^{-1}).

Considering the geometry of ethane, with (approximately) tetrahedral carbon atoms, it should be clear that even in the eclipsed conformation, the hydrogens are pointed away from each other. So, in that conformation, since the carbon–carbon bond is about 154 pm (1.54 Å) long and each C–H bond is about 109 pm (1.09 Å) long, the nonbonded distance between any two eclipsed hydrogens must be approximately 230 pm (2.3 Å)—which is about at the limits of the sum of their van der Waals radii (about 120 pm [1.2 Å] each). Furthermore, when the protons (^1H) are replaced by deuterium (^2H), the barrier to rotation goes down slightly. Thus, the source of the barrier does not seem to be caused by the nuclei attempting to occupy the same space at the same time (a **steric** effect). In fact, this general **resistance to twisting** (called a **torsional** barrier) is not clearly understood. In some quarters, it is currently ascribed to repulsion of the bonding electrons in the filled orbitals holding the hydrogens to the carbons as the angle between those bonds on adjacent carbons diminishes. The ground-state molecular orbital (MO) simply shows that the orbital containing the two electrons is more stable when the hydrogens on adjacent carbons are staggered.

It is clear, however, that if the hydrogens are replaced by methyl groups, the barrier increases in size. Now, it is argued that, in **addition** to a torsional barrier, steric interactions (or van der Waals repulsions) become important. Thus, in butane ($CH_3CH_2CH_2CH_3$), that conformation with the two methyl groups (CH_3–) eclipsed

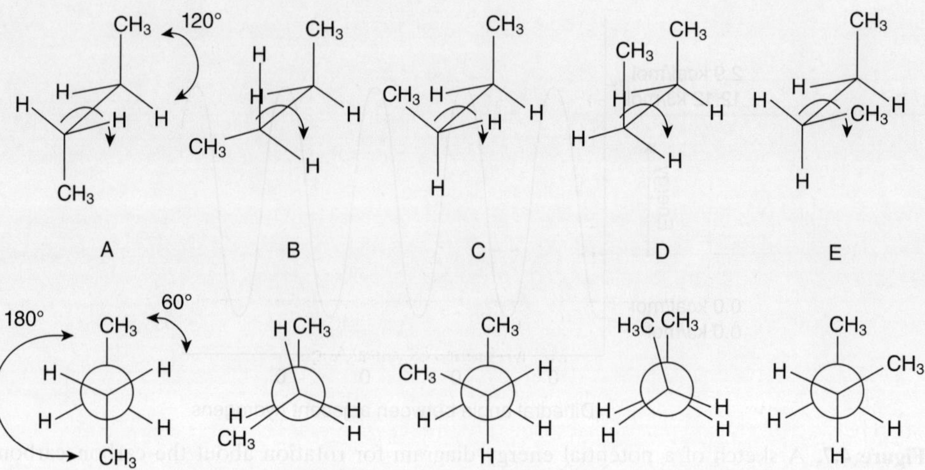

Figure 4.8. Sawhorse (top row) and Newman (bottom row) projections "looking down" or "along" the bond between carbon atoms 2 and 3 of butane ($CH_3CH_2CH_2CH_3$). The angles specified are the dihedral angles as projected onto a flat surface. The dihedral angle between the methyl groups in each projection A–E is, respectively, as follows: A (180°), B (120°), C (60°), D (0°), and E (60°). In A, the methyl groups are described as **anti** (or **trans**). In B, there is hydrogen–hydrogen (H–H) and hydrogen–methyl (H–CH_3) eclipsing. In C and E, the methyl groups are described as **gauche**. Both of these forms are also called **skew**. In D, there is hydrogen–hydrogen (H–H) and methyl–methyl (CH_3–CH_3) eclipsing.

(i.e., a methyl–methyl dihedral angle of 0°) will be higher in energy than any others. Additionally, when the methyl groups are farthest apart (a methyl–methyl dihedral angle of 180°), the system will be at its most stable conformation. Figure 4.8 presents some of the various conformational isomers of butane ($CH_3CH_2CH_2CH_3$). The drawing is an attempt to show what a viewer might see by "looking along" or "looking down" the bond between the second and third carbon atoms.

As shown in Figure 4.9, eclipsing of the methyl groups (CH_3–) and both pairs of hydrogens in butane ($CH_3CH_2CH_2CH_3$) (see D of Figure 4.8) requires passing over a transition whose height is about 6.1 kcal mol^{-1} **above** the most stable configuration (A of Figure 4.8). Since each hydrogen–hydrogen eclipsing (Figure 4.7) results in a torsional strain of about 3.8 kJ mol^{-1} (0.9 kcal mol^{-1}), the contribution to the total due to the methyl–methyl eclipsing must be about 18.0 kJ mol^{-1} (4.3 kcal mol^{-1}). Some of the 18.0 kJ mol^{-1} (4.3 kcal mol^{-1}) is due to steric or van der Waals interactions between the methyl groups; presumably some is also torsional strain. Additionally, hydrogen–methyl (H–CH_3) nonbonded interactions can be examined. Thus, in Figure 4.8 (B), there are two hydrogen–methyl eclipsings and one hydrogen–hydrogen pair eclipsed. This rotational isomer is found to lie 14.2 kJ mol^{-1} (3.4 kcal mol^{-1}) above the methyl–methyl anti-isomer A. Again, assuming that the hydrogen–hydrogen eclipsing can be valued at 3.8 kJ mol^{-1} (0.9 kcal mol^{-1}), each hydrogen–methyl eclipsing must be worth between 5.0 and 5.4 kJ mol^{-1} (1.2 and 1.3 kcal mol^{-1}). Finally, it is worth noting that in both C and E (Figures 4.8 and 4.9), the minimum does not lie at the same low energy as A! Presumably, in these gauche forms, the two methyl groups have some small through-space nonbonded (van der Waals type) interaction (worth about 3.3–3.8 kJ mol^{-1} [0.8–0.9 kcal mol^{-1}]).

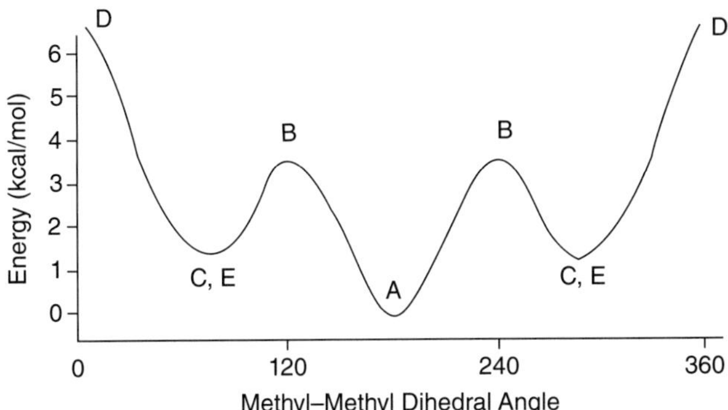

Figure 4.9. A representation of the change in relative energy with torsional angle along the bond between the second and third carbons of butane ($CH_3CH_2CH_2CH_3$). The letter designations A–D indicate the representations with the same letters shown in Figure 4.8.

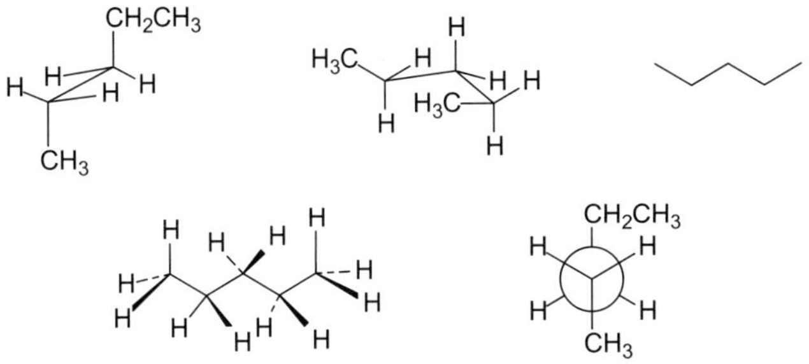

Figure 4.10. Some representations for pentane ($CH_3CH_2CH_2CH_2CH_3$).

Problem 4.1. Sketch a potential energy diagram, such as that in Figures 4.7 and 4.9, for propane ($CH_3CH_2CH_3$). You may use the numbers for H–H and H–CH$_3$ interactions given in the text. Clearly indicate bond angles and, with sawhorse and Newman projections, show which rotational (or conformational) isomer is most stable and which is least stable.

The practice of choosing any given bond and describing the subsequent structure in terms of Newman and/or sawhorse drawings will be used in addition to the line drawings and those with wedges and dotted lines introduced earlier. Figure 4.10 presents some of these representations for pentane ($CH_3CH_2CH_2CH_2CH_3$).

II. Alkenes, Alkynes, and Arenes

The barrier to rotation around the double bond in alkenes, estimated to be about $272.5\,kJ\,mol^{-1}$ ($65\,kcal\,mol^{-1}$) (Table 1.1),* is too great to be observed in uncatalyzed

---

*This value is only approximate and the barrier can apparently be diminished by appropriate substitution, which lowers the electron density in the double bond or causes bulky groups to come too close to each other.

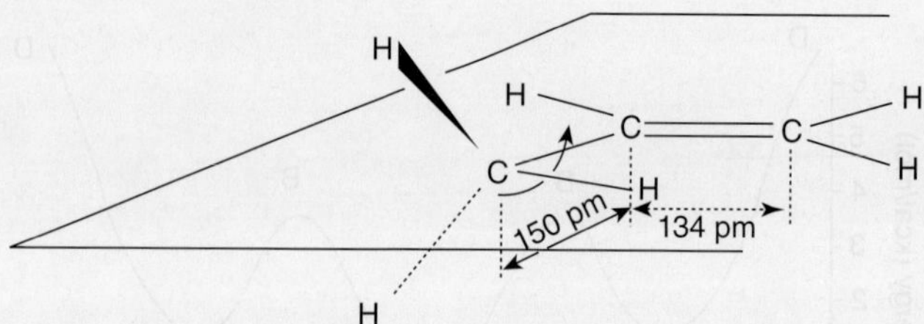

Figure 4.11. A depiction of the rotational isomer of propene, CH_3–CH=CH_2, with a hydrogen of the methyl group (–CH_3) lying in the plane defined by the three carbon atoms.

processes. However, the small barrier to rotation about a σ bond connecting a ligand to an *sp²* hybridized carbon can be measured spectroscopically. Consider propane, $CH_3CH_2CH_3$ (Problem 4.1). Several interesting features are evident. First, the maximum in the barrier to rotation of one of the methyl groups, where three interactions (two H–H and one H–methyl) will be encountered, can be estimated to be about 3 kcal mol^{-1} (2 × 0.9 + 1.2 or 1.3) (about 12.5 kJ mol^{-1}). Second, the minima, in a 360° rotation, will have the methyl as far from the protons on the adjacent carbon as possible, that is, ≈60° (see Problem 4.1). However, in propene, CH_3–CH=CH_2 (Figure 4.11), where different angles obtain, the barrier to rotation of the methyl group appears to be about 8.4 kJ mol^{-1} (2 kcal mol^{-1}), and, second, the minimum appears to be in that rotational isomer (rotomer) with one of the hydrogens of the methyl group *in the plane* defined by the three carbon atoms of the alkene (Figure 4.11).

For propyne (methylacetylene, $HC{\equiv}C$–CH_3), as shown below, where the carbon framework is linear, no barrier to rotation has been detected for either the carbon–carbon sigma (σ) or the carbon–carbon pi (π) system. For aromatic systems such as methylbenzene (toluene, H_3C–C_6H_5), shown below, no barrier to rotation has been detected about the bond connecting the methyl group to the aromatic ring.

propyne

methylbenzene
(toluene)

F. CONFORMATIONAL ANALYSIS OF MEDIUM-RING CYCLIC ALKANES

The relative instability (or greater reactivity) of small (three and four membered) rings, as compared with their medium (five through seven membered) counterparts, is evidenced (Chapter 3) by, for example, the higher heat of combustion per methylene (–CH_2–) for the compounds with the smaller rings. Furthermore, the small-ring compounds cyclopropane and cyclobutane undergo reduction to the corresponding acyclic hydrocarbons (Equations 4.16 and 4.17) under conditions where the medium-ring compounds are stable.

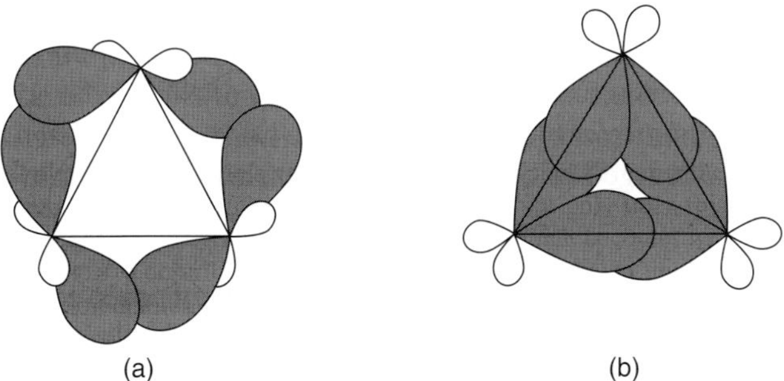

$$\xrightarrow[\text{Pt catalyst}]{\text{H}_2} CH_3CH_2CH_3 \qquad (4.16)$$

$$\xrightarrow[\text{Pt catalyst}]{\text{H}_2} CH_3CH_2CH_2CH_3 \qquad (4.17)$$

That the medium (five through seven) ring carbocycles fail to undergo such **hydrogenolysis** was well known by Baeyer (see Chapter 3), who argued that compression of the carbon–carbon angles to 60° (well below the idealized 109°24″ anticipated for sp^3 hybridization) resulted in steric strain. The generalized concept of such "compression" or "angle" strain, despite the potential for rehybridization (Chapter 3), is still considered a valid concept although (as also pointed out earlier) for rings larger than three, the imposition of planarity, as originally supposed by Baeyer, is now recognized as inappropriate.

The bonding between the three carbon atoms in cyclopropane, which define the plane they occupy, has been considered as any of, or a compromise of, or composite of (1) rehybridization along the model proposed by Walsh (see Chapter 3), (2) preservation of sp^3-type orbitals to minimize interelectronic repulsion of bonded electrons (keeping the interorbital angles near 109.5°) but losing significant orbital overlap, and/or (3) placing the orbitals along the internuclear axes, requiring a 60° angle and thus maximizing overlap but increasing interelectronic repulsion.

The Walsh-type orbitals for cyclopropane were presented in Chapter 3 (Figure 3.25). The compromise solutions for the other two views of bonding in cyclopropane noted above are presented in Figure 4.12.

It should be clear that all hydrogens on each face of the cyclopropane must be eclipsed (Figure 4.13). The same is not true, however, for cyclobutane.

(a)   (b)

Figure 4.12. The carbon framework of cyclopropane showing (a) internuclear lines and bonding orbitals extending beyond those lines. That is, the **interorbital** angle (assumed to be about 105°) is larger than the **internuclear** angle (60°) and (b) the orbitals lying along the internuclear axes. For (b), the **internuclear** and **interorbital** angles are the same (60°).

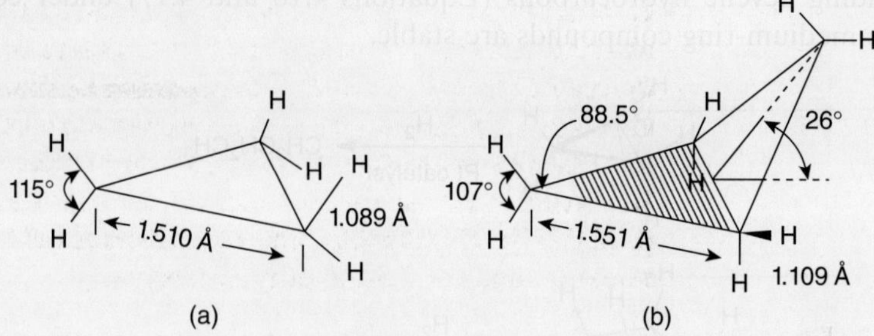

Figure 4.13. Perspective geometric representations for (a) cyclopropane and (b) cyclobutane.

Figure 4.14. Attempted perspective views of cyclopentane: (a) as a flattened "envelope" showing bond lengths and angles, (b) as "envelopes" in equilibrium, and (c) in the "twist" form.

As shown in Figure 4.13, cyclobutane adopts a puckered configuration. While any three of the carbon atoms must occupy a plane, the fourth is not in that plane. Such folding alleviates the eclipsing strain that would result if all four carbons were constrained to be coplanar. In addition to relief of hydrogen–hydrogen eclipsing, which is achieved by avoiding coplanarity, carbon–carbon and carbon–hydrogen bond lengths are significantly longer than in cyclopropane. This bond lengthening also allows a decrease in strain as the filled orbitals representing the bonds between carbon and hydrogen are further apart.

A regular five-sided geometric figure (a pentagon) has internal angles of 108° and, as this is close to 109.5° anticipated for sp^3 hybridization, cyclopentane (C_5H_{10}) might be anticipated to be strain free. However, in such a planar array, the hydrogens at each carbon will be eclipsed with those on the neighboring carbons (leading to an estimated destabilization of about $41.8\,\text{kJ}\,\text{mol}^{-1}$ [$10\,\text{kcal}\,\text{mol}^{-1}$]). Therefore, planar cyclopentane is less stable than its conformational isomers called the "envelope" form (where four contiguous carbons are planar or nearly so) and the "half-chair" or "twist" conformer, where there are only three coplanar carbon atoms (one methylene [$-CH_2-$] unit above and one below the plane defined by the other three). This information is displayed in a perspective geometric fashion in Figure 4.14.

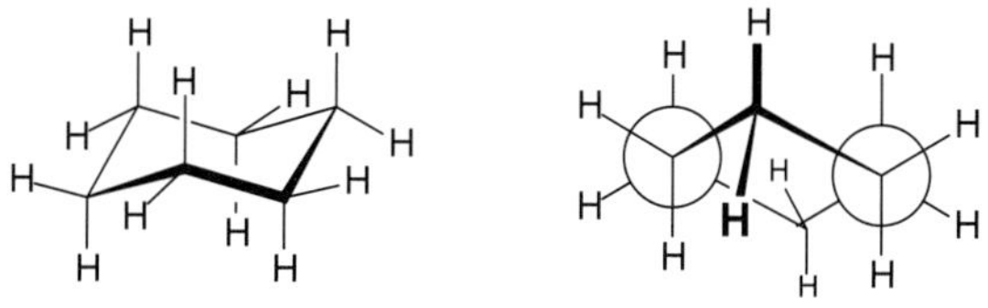

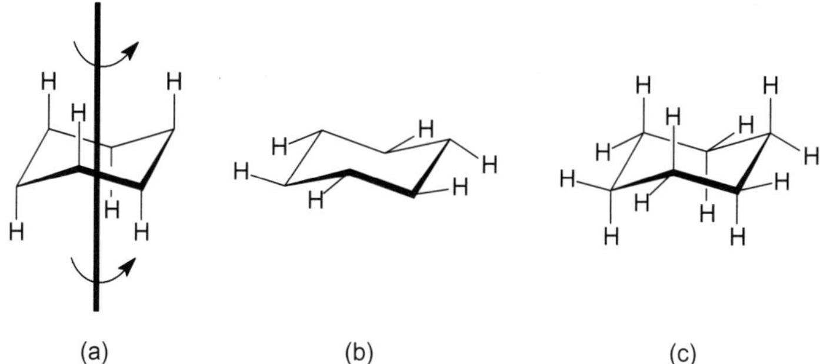

Figure 4.15. Representations of the "chair" form of cyclohexane.

(a)   (b)   (c)

Figure 4.16. Representations of cyclohexane showing (a) axial hydrogens, (b) equatorial hydrogens, and (c) both of them combined.

Cyclohexane (C_6H_{12}) and materials containing this six-membered carbocyclic system occupy a special place in organic chemistry in general because of the widespread natural occurrence of materials containing six-membered rings (which argues for the stability of the arrangement). Furthermore, in part because of the availability of materials containing six-membered rings, studies of carbocycles frequently utilize cyclohexane and compounds derived from it as prototypical examples. In this regard, it is important to realize that in cyclohexane, because the ring possesses a conformation with all bond angles approximately tetrahedral and all carbon bonds, as well as bonds to protons on adjacent carbons, mutually staggered (a particularly stable arrangement which is essentially strain free), its use as a prototype may be inappropriate!

The particular strain-free conformation is (trivially) called the "chair" and two representations for the chair form of cyclohexane are provided in Figure 4.15.

If you can visualize a plane passing through cyclohexane, it should be clear to you that 6 of the 12 hydrogens lie along an axis, which is orthogonal to the plane (Figure 4.16a). The positions these protons occupy, and substituents (such as alkyl groups) that might replace them, are called "axial." The other six protons, and substituents that might replace them (Figure 4.16b), may be thought to lie around the equator (alternately above and below the plane) and are called "equatorial."*

*The perspective drawings here, which you should practice drawing, follow a common convention. The ring is viewed as if the observer is "above" and to the "left." Bonds on opposite sides of the ring are parallel to each other. Axial bonds are all parallel to the "axis" orthogonal to the plane of the ring and thus parallel to each other. Equatorial bonds are drawn to approximate the tetrahedral angle and are parallel to ring bonds once removed from the carbon to which they are attached (i.e., the carbon–carbon ring bonds at the adjacent positions.)

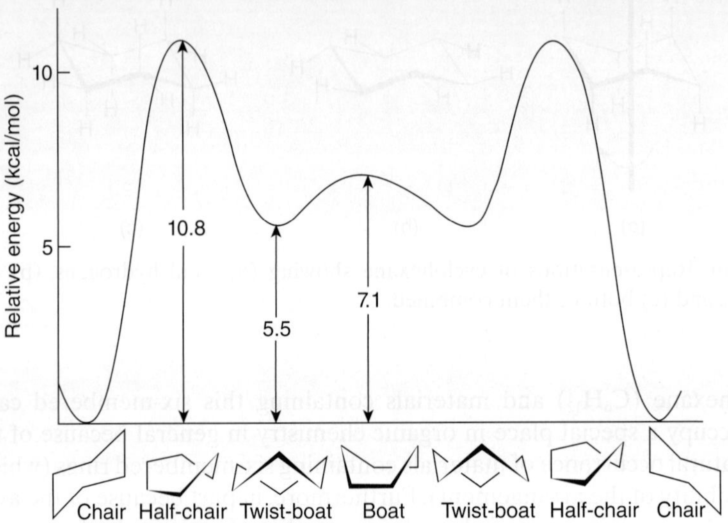

Figure 4.17. The interconversion of two chair representations of cyclohexane. The arrows depict the net motion required to convert one to the other.

Figure 4.18. A representation of the path connecting the two chair forms of cyclohexane and the barriers to be overcome in converting one to the other.

Substituents may thus be placed in an axial or equatorial position. Depending on the particular substituent, a preference may exist for an axial or equatorial orientation. Furthermore, the positions can interchange when the six-membered ring flexes, establishing an equilibrium of axial and equatorial isomers. The flexing of the unsubstituted ring occurs rapidly at room temperature and causes all equatorial positions to become axial and vice versa, as depicted in Figure 4.17. The interconversion of one chair form to the other, as shown in Figure 4.18, requires surmounting of an energy barrier and passing through other more or less flexible conformations called the "boat" and "twist-boat" forms of cyclohexane.

Although the twist-boat and boat forms of Figure 4.18 represent local minima on the energy surface connecting the chair forms of cyclohexane, actual measurement of the equilibration between the two chair forms, an example of which is given in Figures 4.19 and 4.20, usually involves measurement of the height of the **major** (or highest) barrier which, of course, dictates the slowest process. One measurement of the height of the major barrier between the two chair forms of cyclohexane made use of NMR techniques and was independently described by two different groups

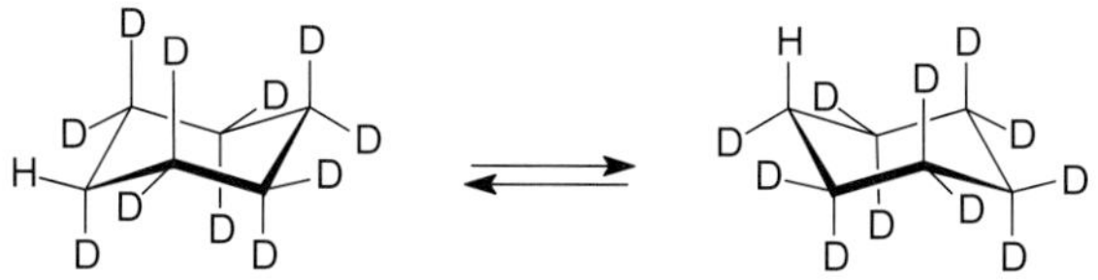

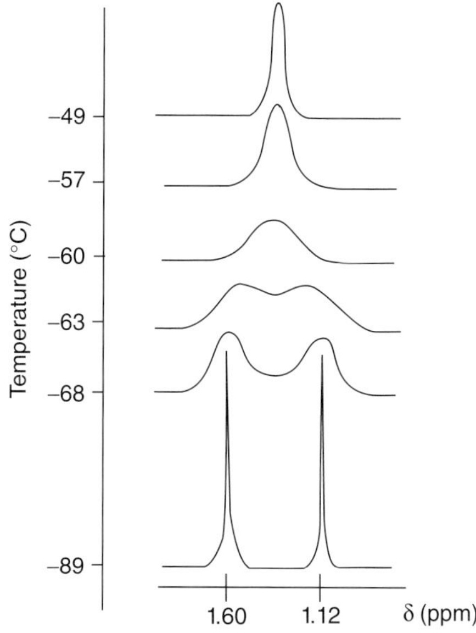

Figure 4.19. Representations of the chair form of cyclohexane in which all but 1 of the 12 protons (^{1}H) have been replaced by deuterium (^{2}H; abbreviated here as "D"). Both conformational isomers are shown: (a) with the ^{1}H equatorial and (b) with the ^{1}H axial.

Figure 4.20. The ^{1}H NMR spectrum of ^{1}H^{2}H$_{11}$-cyclohexane (C$_6$H$_1$D$_{11}$) at 60 MHz as a function of temperature. Coupling between ^{1}H and ^{2}H has been removed by double irradiation (redrawn from Bovey, F. H. *Nuclear Magnetic Resonance Spectroscopy*, Academic Press, New York, **1969**, p. 191).

of workers in the 1960s.* In essence, the experiment involved preparation of cyclohexane (C$_6$H$_{12}$) in which *all but one* of the hydrogen atoms had been replaced by deuterium (^{2}H but abbreviated here as "D," Figure 4.19, C$_6$H$_1$D$_{11}$) and then measurement of the ^{1}H NMR spectrum of the d$_{11}$-cyclohexane while the temperature of the sample was lowered (Figure 4.20).

At room temperature, where the inversion process is rapid (from Figure 4.18, the maximum barrier height is 10.8 kcal mol^{-1}), the proton of ^{1}H^{2}H$_{11}$-cyclohexane appears at a chemical shift (δ) of 1.37 ppm downfield of tetramethylsilane (TMS). However, as the temperature is lowered, the time cyclohexane spends in a given conformation becomes longer, and eventually, the rate of equilibration between the two states (axial and equatorial) is slowed so that the two states (e.g., about 50%

---

*The original work is described in (a) Anet, F. A. L.; Ahmad, M.; Hall, L. D. *Proc. Chem. Soc.*, **1964**, 145 and (b) Bovey, F. A.; Hood, F. P.; Anderson, E. W.; Kornegay, R. L. *Proc. Chem. Soc.*, **1964**, 146.

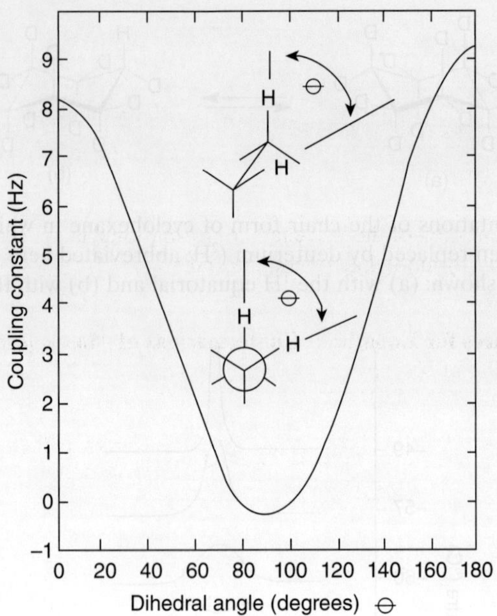

Figure 4.21. A diagrammatic representation of the Karplus relationship showing the variation of coupling constant with dihedral angle subtended between neighboring protons that are coupled to each other.

of the species have the ^{1}H axial and about 50% have the ^{1}H equatorial) are "frozen out" and the proton appears at different chemical shifts ($\delta = 1.12$ ppm for the axial and 1.60 ppm for the equatorial).

Problem 4.2. What experiment might you imagine could be developed to determine which proton is equatorial and which is axial?

The analysis of these data (i.e., rate constant as a function of temperature) leads to the maxima of Figure 4.18.*

Equilibria between cyclohexane conformational isomers as a function of substituent have been reported for a variety of compounds. Generally, as you might expect, equatorial substitution (which is less sterically demanding), as shown in Figure 4.22 for methylcyclohexane, is more stable than axial. For example, in the conformational isomer with the methyl group axial (Figure 4.22a), not only is there interaction between the hydrogens of the methyl group and the axial hydrogens (a 1,3-interaction)

*Because axial and equatorial protons are different, when the flexing of the cyclohexane chair has been "frozen out," the geminal protons (i.e., protons on the **same** carbon) will couple with each other. Furthermore, axial and equatorial protons can also couple with equatorial and axial protons (respectively) on adjacent carbons. Thus, in the experiment described above, all but one proton was replaced by deuterium and the decoupling of deuterium ($I = 1$) from protium effected.

One of the more interesting features to come out of the study of such compounds with fixed angles is the relationship between coupling constant and the dihedral angle subtended between protons on adjacent carbons. Figure 4.21 (known as the Karplus relationship after its originator, M. Karplus [b. 1930, professor, Harvard University]) summarizes the results of many such experiments.

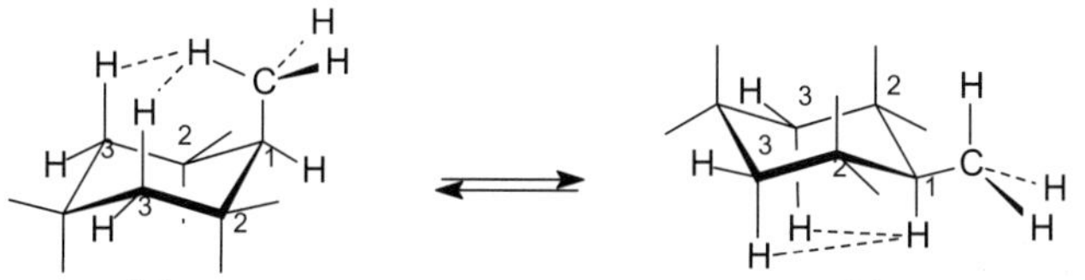

Figure 4.22. A representation of the two chair conformational isomers of methylcyclohexane: (a) methyl group axial; (b) methyl group equatorial.

TABLE 4.2. Preferences for Equatorial Conformations of Monosubstituted Cyclohexanes Expressed as a Function of Free Energy ($-\Delta G°$)

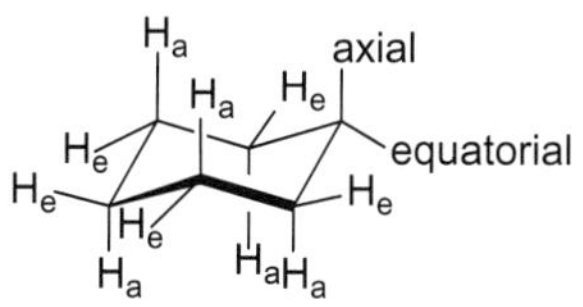

Substituent	$-\Delta G°$ kcal mol^{-1} (kJ mol^{-1})
$-CH_3$	1.8 (7.5)
$-CH_2CH_3$	1.8 (7.5)
$-CH(CH_3)_2$	2.1 (8.8)
$-C(CH_3)_3$	>4.5 (>18.8)
$-CH=CH_2$	1.7 (7.1)
$-C_6H_5$	3.1 (12.9)

The data in this table are taken largely from Allinger, N. L.; Freiberg, L. A. *J. Org. Chem.*, **1966**, *31*, 894.

of the ring but there are also two gauche methyl–methylene interactions, both of which are missing in the conformational isomer with the methyl group equatorial (Figure 4.22b).

Problem 4.3. Using Newman-type projections, show the gauche methyl–methylene interactions for the conformational isomer of methylcyclohexane with the methyl group axial.

Based on experiments similar to that described above for cyclohexane itself, conformational preferences (axial versus equatorial) for a variety of substituents have been determined and some of those known are presented in Table 4.2. Indeed, using this table and the information provided in Table 4.1, it is possible to provide an approximate prediction of the amount by which the preferred isomer will be favored.

According to the information in Table 4.2, the conformational preference for a methyl group ($-CH_3$), an ethyl group ($-CH_2CH_3$), and an isopropyl group [$-CH(CH_3)_2$] are all about the same, and it is not until a *tert*-butyl group [$-C(CH_3)_3$] is encountered that the equilibrium preference for an equatorial substituent becomes overwhelming. Indeed, this preference is such that the *t*-butyl group [$-C(CH_3)_3$] is often referred to as a "conformational anchor."

Problem 4.4. How would you account for the observation that both the ethyl group ($-CH_2CH_3$) and the isopropyl group [$-CH(CH_3)_2$] have about the same $-\Delta G°$ values as the methyl ($-CH_3$) group?

Consider next the attachment of two methyl groups to cyclohexane. With two (or more) substituents present, the previously presented discussion (with regard to ring inversion) must be considered as well as the relationship and interactions between the groups. The (Z)- (or *cis*-) and (E)- (or *trans*-) methylcyclohexane **configurational** isomers cannot be interconverted by simply flexing the ring. However, the **conformational** isomers for the different **configurations** can be interconverted. Furthermore, the different conformational isomers may have different energies (depending on the relative [axial/equatorial] placement of the substituents).

The equilibrium between the conformational isomers of (Z)-1,2-dimethylcyclohexane (Figure 4.23a,b) and (E)-1,2-dimethylcyclohexane (Figure 4.23 c,d) demonstrates the interconversions.

In (Z)-1,2-dimethylcyclohexane, both conformations (Figure 4.23a,b) have one methyl group axial, and thus, ring flexing cannot relieve one axial interaction. However, in (E)-1,2-dimethylcyclohexane (Figure 4.23c,d) *either* both methyl groups are axial (Figure 4.23c) or both are equatorial (Figure 4.23d). Clearly, the latter is preferred, and the equilibrium population of the former (Figure 4.23c) would be expected to be small.

Problem 4.5. Assuming that the "cost" for putting methyl groups axial on a cyclohexane ring is constant for each methyl group, and using the data in Tables 4.1 and 4.2, estimate the equilibrium concentration of the *diaxial* (E)-1,2-dimethylcyclohexane present (at 25°C).

Although the conversion of (Z)-1,2-dimethylcyclohexane (Figure 4.23a,b) into (E)-1,2-dimethylcyclohexane (Figure 4.23c,d) cannot be easily accomplished (i.e., in

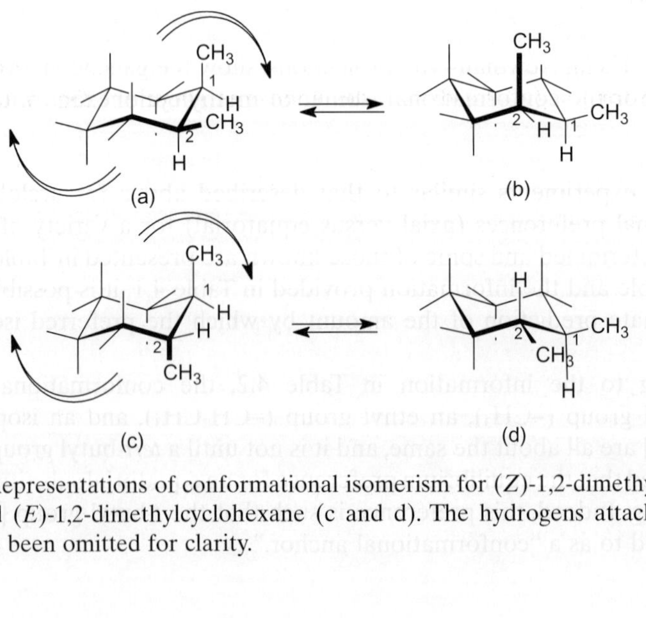

Figure 4.23. Representations of conformational isomerism for (Z)-1,2-dimethylcyclohexane (a and b) and (E)-1,2-dimethylcyclohexane (c and d). The hydrogens attached to C_3–C_6, inclusive, have been omitted for clarity.

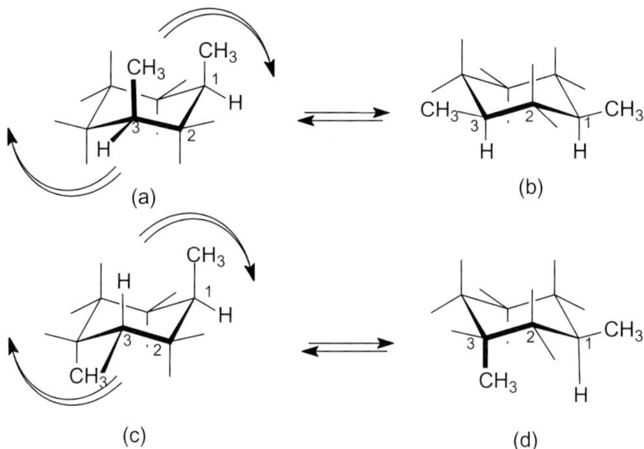

Figure 4.24. Representations of conformational isomerism for (Z)-1,3-dimethyl-cyclohexane (a and b) and (E)-1,3-dimethylcyclohexane (c and d). The hydrogens attached to C_2 and C_4–C_6, inclusive, have been omitted for clarity.

principle, it requires breaking at least one bond, e.g., the carbon–carbon bond of the ring between the carbons bearing the methyl groups, and its reformation), it should be clear to you that of the four conformational isomers shown in Figure 4.23, the one with the two methyl groups equatorial should be the most stable (i.e., Figure 4.23d). Indeed, it can be readily estimated that diequatorial (E)-1,2-dimethylcyclohexane is about $1.8\,\mathrm{kcal\,mol^{-1}}$ ($7.5\,\mathrm{kJ\,mol^{-1}}$) more stable than (Z)-1,2-dimethylcyclohexane. A similar analysis can be performed for the other dimethylcyclohexane isomers (Figures 4.24 and 4.25).*

As above, a comparison can be made between the conformational isomers of (Z)-1,3-dimethylcyclohexane (Figure 4.24a,b) and (E)-1,3-dimethylcyclohexane (Figure 4.24c,d). For the former, it is clear that the conformation with two equatorial methyl substituents (Figure 4.24b) should be much preferred over the corresponding conformer with two axial methyl substituents (Figure 4.24a). Furthermore, as was the case for the (Z)-1,2-dimethylcyclohexane (Figure 4.23a,b), here, with the 1,3-dimethylcyclohexanes, it is the (E)-1,3-dimethylcyclohexane isomer, with one methyl group axial and one equatorial (Figure 4.24c,d) that cannot (e.g., by flexing) relieve one of the methyl groups of its axial interactions. It would be correct to conclude that whether the isomer is (E)- or (Z)-, and whether the alkyl substituents are 1,2- or 1,3- or 1,4- (as shown below in Figure 4.25), the isomer with the two methyl groups **equatorial** will be the most stable. Thus, (Z)-1,3-dimethylcyclohexane is expected to be more stable than the corresponding (E)-isomer by about $1.8\,\mathrm{kcal\,mol^{-1}}$ and a similar value would be expected for the difference between the (Z)- and (E)-isomers of 1,4-dimethylcyclohexane, but now, the (E)- isomer, with the two methyl groups equatorial is preferred (Figure 4.25a).

*The analysis leading to this conclusion is based on an understanding of the nature of the interaction(s) arising when *one* substituent is present. It is worthwhile examining Newman projection formulae of the dimethylcyclohexanes.

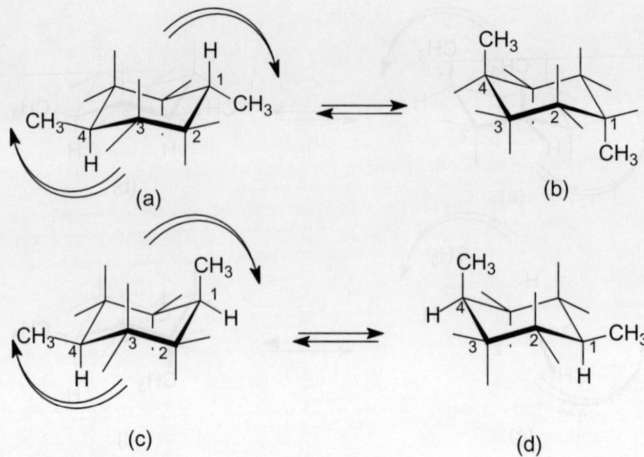

Figure 4.25. Representations of conformational isomerism for (*E*)-1,4-dimethylcyclohexane (a and b) and (*Z*)-1,4-dimethylcyclohexane (c and d). The hydrogens attached to C_2, C_3, C_5, and C_6 have been omitted for clarity.

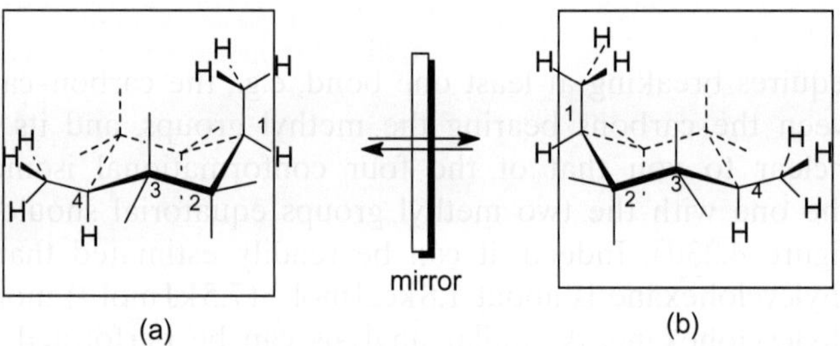

Figure 4.26. Representations of a pair of "frozen out" (*Z*)-1,4-dimethylcyclohexanes. The carbon atoms bearing the numbers 1 and 4 are drawn to be coplanar as they, along with the hydrogens attached to them and the carbons and one of the hydrogens of the methyl groups, are each *bisected* by the planes shown. This relationship is not altered by flexing the ring. Additionally, (a) and (b) are related as mirror images. The hydrogens attached to C_2, C_3, C_5, and C_6 have been omitted for clarity.

The family of dimethylcyclohexanes will also be used as examples in a further discussion of symmetry and chirality (this chapter, *vide infra*).

First, consider (*Z*)-1,4-dimethylhexane (Figure 4.25c,d and redrawn as Figure 4.26a,b). As shown, the two depictions of these conformational isomers (Figure 4.26a,b) are not only related by simply flexing the ring so that the methyl group at C-1, which was axial (in Figure 4.26a), becomes equatorial (in Figure 4.26b), while that at C-4, which was equatorial (in Figure 4.26a), becomes axial (in Figure 4.26b). They are also related as **mirror images** of each other (i.e., the drawing on the left is the mirror image of that on the right—with the exception of the numbers). It should be clear that these mirror image objects, if realized, would be capable of superposition. That is, the (now) three-dimensional object on the right (a two-dimensional representation of which is shown as Figure 4.26b) *can be lifted from the page as a three-dimensional object, rotated 180°* and set onto the realized object on the left

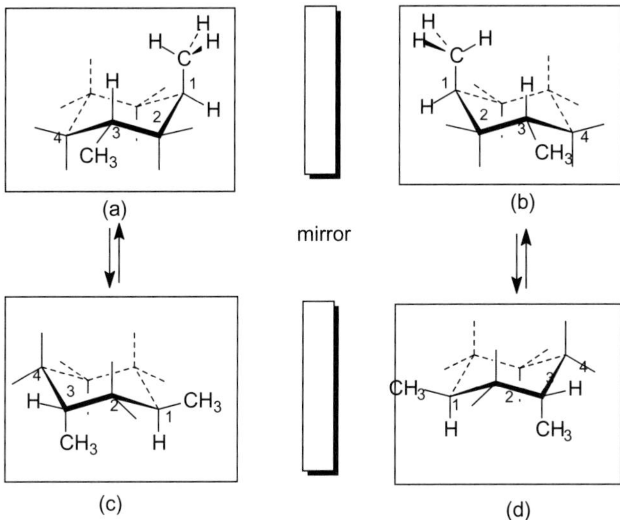

Figure 4.27. Representations of (E)-1,3-dimethylcyclohexanes. The representations (a) and (b), as well as (c) and (d), are, respectively, pairs of nonsuperposable mirror images (enantiomers). They cannot be interconverted by ring flexing, rotation, and so on. However, the pairs (a) and (c), as well as (b) and (d), are conformational isomers. The hydrogens attached to C_2 and C_4–C_6, inclusive, have been omitted for clarity.

(the two-dimensional representation of which is shown as Figure 4.26a), or vice versa, and the figures brought into coincidence; they are identical. Materials that are superposable upon their realized mirror images are **achiral*** (Chapter 3).

Consider next the case of (E)-1,3-dimethylcyclohexane (Figure 4.27). As shown in the drawings at the top (Figure 4.27a,b), not only are the three-dimensional realized objects represented **incapable of being superposed (!)** but they are also **not** related by simply flexing the ring. Here, C-1 and C-3 and the carbons of their respective methyl substituents are not in the same plane. When the realized representation of Figure 4.27b is lifted from the page as a three-dimensional object, rotated 180°, and set on to the realized representation of Figure 4.27a, the methyl substituent at C-3 (which was initially protruding from the plane toward the viewer in the representation of Figure 4.27b) will now be directed behind the plane. The new, realized representation is thus **not identical** to the realized object represented in Figure 4.27a.

The conformational isomers that are related by flexing the ring are also shown; that is, the conformational isomer of the (E)-1,3-dimethylcyclohexane shown in Figure 4.27a is represented in Figure 4.27c and the conformational isomer of the mirror image of the object of Figure 4.27a (i.e., the isomer in Figure 4.27b) is represented in Figure 4.27d. Mirror image objects (Chapter 3) that are incapable of superposition are chiral and are called **enantiomers**. Thus, the realized three-dimensional objects of Figure 4.27 are related **sets** of enantiomeric conformational

---

*It should be clear that while, individually, the diequatorial and diaxial conformers of (E)-1,4-dimethylcyclohexane (Figure 4.25) will similarly be superposable on their respective realized mirror images, the conformer with the two methyl groups axial **cannot** be brought into superposition with that of the one with the methyl groups equatorial.

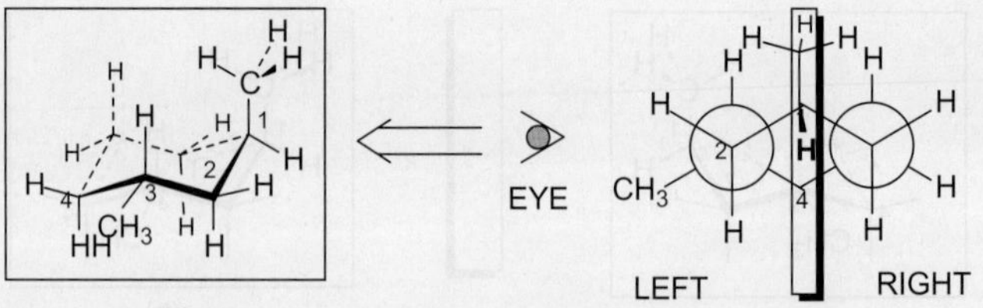

Figure 4.28. (a) One of the enantiomers of (*E*)-1,3-dimethylcyclohexane (originally seen as Figure 4.27a), a representation of the eye of a beholder, and (b) what is seen.

isomers. While conformational isomers may have different energies (and hence, as we shall see, different reactivities), **the interaction of a pair of enantiomers with achiral reactants is identical.** Indeed, except for the sense in which each of the enantiomers, or the pair, interacts with polarized light (*vide infra*), the physical properties of each member of the enantiomeric pair (e.g., boiling point [bp], melting point [mp], IR, NMR) will be identical!

To communicate that these enantiomers can be distinguished one from the other, it is necessary to develop another set of nomenclature rules. This set is largely due to the efforts of R. S. Cahn and Sir Christopher Ingold (University College, London), who first described it in the early 1950s and, as subsequently modified by Vlado Prelog of the Swiss Federal Institute of Technology, was accepted by the International Union of Pure and Applied Chemistry (IUPAC) as the Cahn–Prelog–Ingold rules.* It is also known as the *R,S* system of sequence rules or convention.

Problem 4.6. With reference to Figure 4.27, describe the relationship between the representations of the pair (a) and (d) and the pair (b) and (c).

The Cahn–Prelog–Ingold rules will be outlined using Figure 4.28, at the top left of which Figure 4.27a is redrawn. Added to the drawing (top right) is the representation of an eye, which is placed so as to suggest a view of the representation—which is shown beneath the eye. You should understand that the placement of the viewpoint is arbitrary—the same results will be obtained regardless of where the viewer looks. However, some features are "easier to see"—depending on where you stand (i.e., your point of view)!

To denote the absolute sense of the configuration at carbon atom number one (C-1) (as numbered in Figure 4.28)† using the *R,S* convention, the groups attached to C-1 are **assigned an order of precedence based on the atomic number of the first bonded atom of the attachment**. The higher the atomic number, the higher the precedence.

However, **if the first bonded atoms of the attachments are identical, then precedence is given to the second atom of higher atomic number, and so on.**

*Cahn, R. S.; Ingold, C. K.; Prelog, V. *Angew. Chem. Int. Ed. Engl.*, **1966**, *78*, 413 and Prelog, V.; Helmchen, G. *Angew. Chem. Int. Ed. Engl.*, **1982**, *21*, 567.
†The rules for cyclohexane nomenclature (including numbering) were discussed earlier (Chapter 3). As drawn in this figure, the methyl groups are at C-1 and C-3. Which is C-1 and which is C-3, is not, as will become apparent, material to this discussion. (But *R* takes precedence over *S*!)

Consider now Figure 4.28b. Attached to C-1, there is a methyl (CH_3-) group, a hydrogen (H–) and, both to the left and to the right, methylene ($-CH_2-$) groups. The figure is (hopefully) drawn to show that C-1, the hydrogen attached to it, and the carbon of the methyl group attached to it occupy the plane, which also passes through C-4. In this way, **in an absolute sense**, as drawn, C-2 and C-3 are to the "left" of the observer while C-5 and C-6 are to the "right" of the observer. The methylene groups (C-2 and C-6) are apparently identical. However, the methylene group on the left is **subsequently** attached to a carbon bearing a methyl group and one hydrogen (C-3), while the one on the right is **subsequently** attached to a carbon bearing two hydrogens (C-5).

The assignment of priority ("1" = highest; "4" = lowest) may be then made as follows:

(a) There are four attachments to C-1. Of the four, the point of first attachment for three of the ligands is carbon and for the fourth is hydrogen. The atomic number of hydrogen (1) is smaller than that of carbon (6) and therefore **hydrogen will be assigned the lowest priority**; that is, H– = 4.

(b) Of the remaining three attachments, all are **first** to carbon. However, the methyl group ($-CH_3$) is next (or **second**) attached to *three* hydrogens while each of the methylenes ($-CH_2-$) is attached to *two* hydrogens and *another carbon*. Again, since the atomic number of hydrogen (1) is lower than that of carbon (6) the **methyl group will be assigned the next lowest priority**; that is, CH_3- = 3. That is, the decision about priority is now pushed beyond the first point of attachment. Note that this is done **only** when the first point of attachment is identical (i.e., in this example, there are three carbons as the "first" point).

(c) Considering now only the remaining two attachments, that is, the methylene ($-CH_2-$) groups to the *left* (C-2) and to the *right* (C-6) of the plane passing through C-1, C-4, the carbon of the methyl, and the proton attached to C-1 (Figure 4.28b), it should be clear that methylene ($-CH_2-$) to the right (C-6) is **next** attached to another carbon (C-5) which **bears two hydrogens**, that is, another methylene ($-CH_2-$) and must therefore be of lower priority than the one on the left (i.e., C-6 = 2). Thus, when the "second" point of attachment is identical, one looks at the "third."

(d) The highest priority thus falls to the original substituent with the highest atomic number or, as in this case, to the substituent from which a path beginning with (originally) identical ligands eventually leads to the substituent with the highest atomic number.

Thus, to reiterate, priority goes first to the highest atomic number (if two substituents have the same atomic number, for example, protium [^1_1H] and deuterium [^2_1H], priority goes to the one with the highest mass, i.e., ^2H) and, after that, priorities are assigned at the **first point of difference**. Thus, the drawing of Figure 4.28 at C_1 may now be recast as Figure 4.29.

Having determined the priorities of the various groups, ligands, or subtituents, they must next be ordered, without ambiguity, so as to allow differentiation of realized mirror image objects.

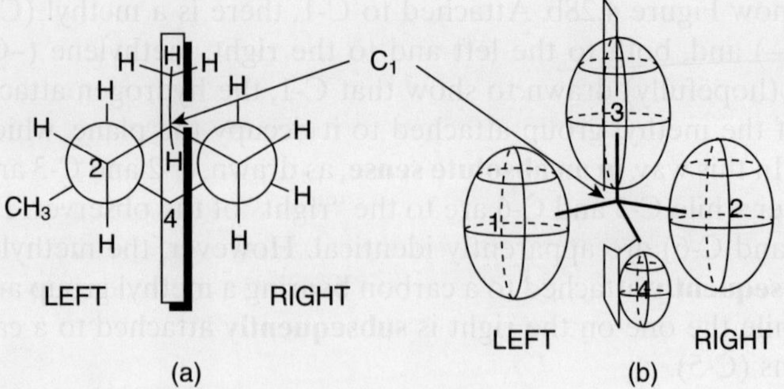

Figure 4.29. (a) A recasting of Figure 4.28 showing (b) the priorities (1–4) of the substituents for the *R,S* system of absolute configuration denotation at C_1.

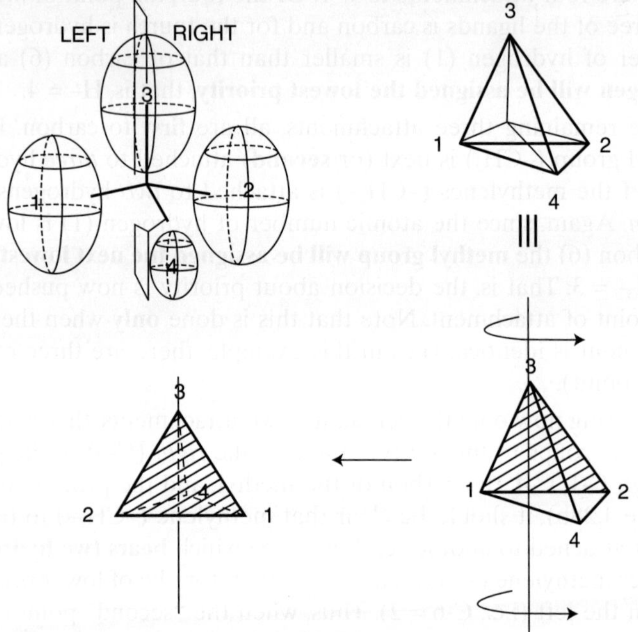

Figure 4.30. A representation of the ordering of priorities at C-1 of one enantiomer of (*E*)-1,3-dimethylcyclohexane as originally drawn in Figure 4.27a.

Thus, a tetrahedron can now be drawn (Figure 4.30), with the priorities arranged as delineated (Figure 4.29). Then, the tetrahedron is turned so that the **face of the tetrahedron opposite the group of lowest priority is toward the viewer**. If, in proceeding from **priorities** 1 → 2 → 3 (Figure 4.31), a clockwise motion is described, the absolute configuration is described as *R* (for *rectus*). If the motion described is anticlockwise (counterclockwise), the absolute configuration is described as *S* (for *sinister*). In this particular case, it should be clear that, as drawn in Figure 4.27a *et seq.*, the configuration at C-1 is *R*.

Now, beginning again with the object shown in Figure 4.28a, examine the configuration at C-3. Convince yourself that the orientation with regard to priorities is as

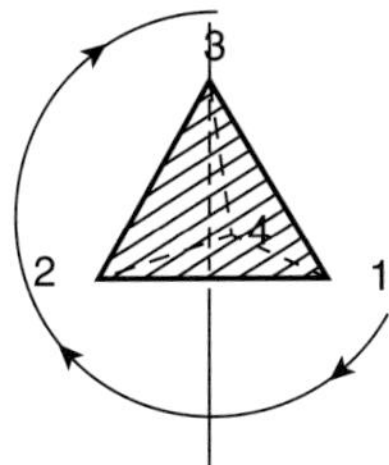

Figure 4.31. The "clockwise" sense of the ordering of priorities at C-1 of one enantiomer of (*E*)-1,3-dimethylcyclohexane as originally drawn in Figure 4.27a.

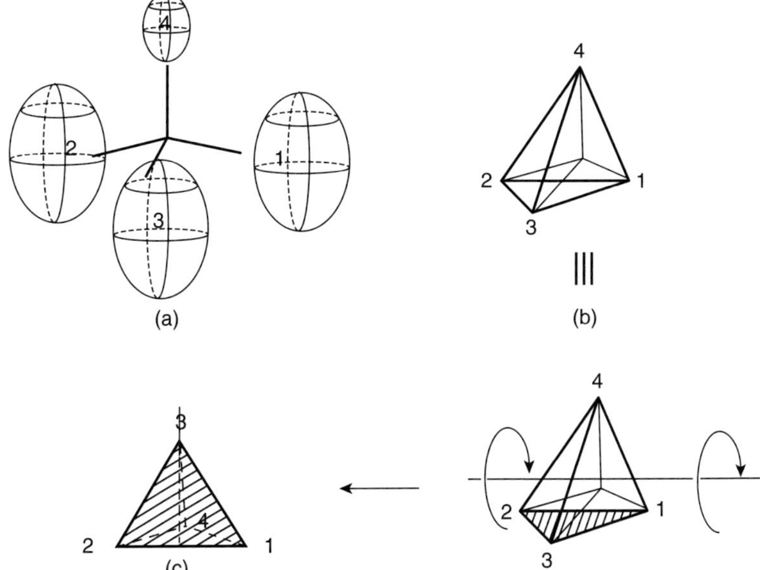

Figure 4.32. Representations at C-3 of the (*E*)-1,3-dimethylcyclohexane of Figure 4.27a.

shown in Figure 4.32a, the corresponding tetrahedral representations as in Figure 4.32b,c are correct, and that, finally, the object represented in Figure 4.27a is (1*R*,3*R*)-(*E*)-1,3-dimethylcyclohexane.

Since the mirror image of *R* must be *S*, it necessarily follows that the mirror image object shown as Figure 4.27b must be (1*S*,3*S*)-(*E*)-1,3-dimethylcyclohexane.* These objects (molecules) are classified as enantiomers, as they have the opposite chirality.

Problem 4.7. Using the 1,2-dimethylcyclohexanes in place of the 1,3-dimethylcyclohexanes, repeat the analysis illustrated above to determine the absolute configuration of the isomers.

Problem 4.8. Draw (1*R*,2*S*)-(*E*)-1-ethyl-2-methylcyclohexane and its mirror image.

*The designations of *R* and *S* precede the IUPAC name, are enclosed in parentheses, and each atom bearing the designation is indicated by a number and labeled *R* or *S*. Commas are used to separate these descriptors as needed.

Chiral materials are pervasive. That is, most of the objects, both throughout nature and hence in this course, lack the symmetry required for superposition on their realized mirror images.

It can be argued that it was originally van't Hoff (J. H. van't Hoff, Utrecht, 1875) and LeBel (J. A. LeBel, Paris, 1874) who related the concept of a tetrahedral carbon atom bearing four different substituents (commonly referred to as "an asymmetric carbon" or "an asymmetrically substituted carbon") with chirality. However, not only they but, indeed, Pasteur (L. Pasteur, 1868), as we will eventually see (Carbohydrates, Chapter 11), had already made observations about the concepts of symmetry, asymmetry, and dissymmetry, which divorced those concepts from the requirement of four different substituents on carbon.*

It should be clear to you that there are many objects that cannot be brought into superposition with their realized mirror images. Thus, as shown in Figure 4.34, placing a double bond in a ring large enough to permit *trans* or (E)-substitution creates a sense of helicity, and alkenes, such as (E)-cyclooctene, are possessed of chirality.

Similarly, the directionality a π bond imposes on substituents also results in a sense of helicity or "twist" when an accumulation of double bonds (as in 2,3-pentadiene, $CH_3CH=C=CHCH_3$) is present (Figure 4.35).

---

*A **symmetric** object is one which can be brought into superposition with its realized mirror image. Such an object *must* possess an n-fold alternating axis of symmetry, that is, an axis such that when the object is rotated around the axis $2\pi/n$ and then reflected across a plane at right angles to the axis, the result is indistinguishable from the original. A **plane of symmetry** is thus equivalent to a onefold axis ($n = 1$), and one-half the object is the mirror image of the other. A **center of symmetry** is a point, lying on a line such that if the line is drawn from any element of the object to this point, extension of the line an equal distance beyond the point will encounter an identical element. An **asymmetric** object lacks all symmetry and cannot be brought into superposition with its realized mirror image. A **dissymmetric** object will have some simple symmetry element (e.g., an axis of symmetry), but it will not be capable of being brought into superposition with its realized mirror image. That is, it is still chiral. Thus, asymmetry is not necessary (but it is sufficient) as dissymmetric objects are also chiral. Dissymmetry, which is less restrictive than asymmetry, is also sufficient. Consider the objects of Figure 4.33 shown below. Keep in mind that the mirror image of R is S. The objects of the figure are all (potentially) isomers of 1,3-di(1-deuterioethyl)-cyclobutane.

Now as shown, for both (a) and (b), these isomers could be brought into superposition with their (respective) realized mirror images because they are (respectively) identical with their respective mirror images. Furthermore, (a) and (b) are related to each other as, respectively, (Z)- and (E)-isomers. However, (c) and its mirror image, represented as (d), are nonidentical mirror image objects and hence, if realized, could not be brought into superposition. Nonidentical mirror image objects are **enantiomers**. However, the (Z)-1,3-di(1-deuterioethyl)cyclobutane represented by (a) is not the mirror image of (c) or (d). Neither is (a) identical with (c) nor (d). Such nonmirror image, nonidentical objects, which are nonetheless related, are called **diastereomers**. The relation of (a) to (c) and (d) (and vice versa) is as diastereomers. They are **diastereomeric**.

Returning to Figure 4.33. Note that (a) possesses a mirror plane passing through its center. Molecules that enjoy such internal symmetry and are thus achiral are called *meso*; one-half is the mirror image of the other. Since the descriptor *meso* refers to any achiral member of a set of diastereomers that also includes at least one chiral member, it should be clear that the (E)-isomer is also *meso*, as it possesses a *point of symmetry*. That is, a line, drawn from the center of the molecule in a given direction for some specific distance, will encounter the identical structural subunit when drawn in the opposite direction at the same distance. Finally, it should also be clear to you that this diastereomer, while **chiral**, is **not asymmetric**; it has an axis of symmetry and asymmetric objects have no symmetry. This diastereomer is **dissymmetric**. (For a thorough discussion see Eliel, E. *Elements of Stereochemistry*, John Wiley & Sons, New York, **1969**.)

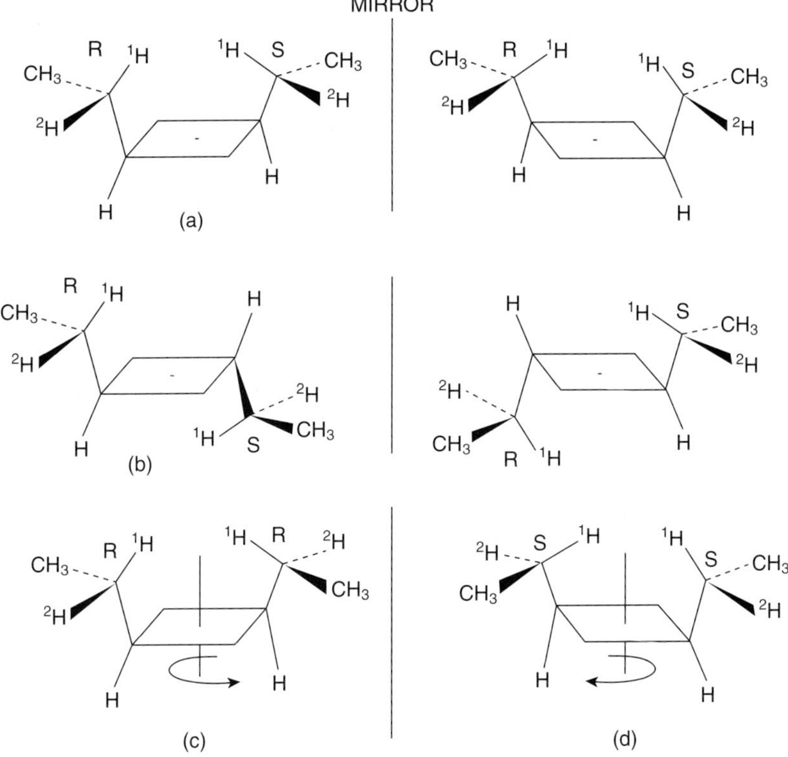

Figure 4.33. Representations of isomers of 1,3-di(1-deuterioethyl)cyclobutane and their mirror images. (For a more complete discussion see Eliel, E. L.; Wilen, S. H. *Stereochemistry of Organic Compounds*, John Wiley & Sons, New York, **1994**, pp. 71*ff*.)

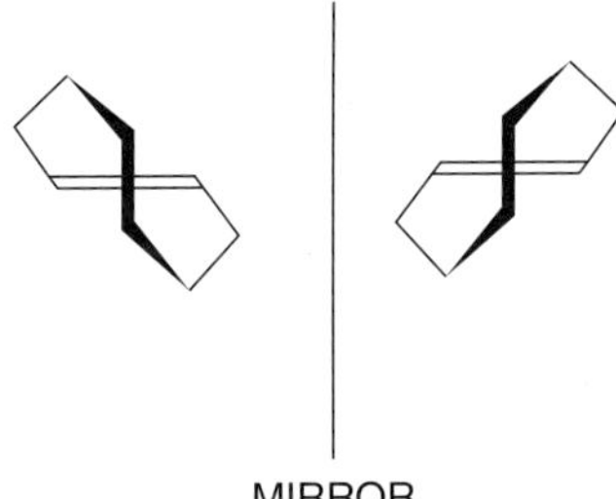

Figure 4.34. Enantiomeric rendering representations of (*E*)-cyclooctene.

The 2,3-pentadiene from the left of Figure 4.35 can be viewed (from end-on) as in Figure 4.36. Since carbon will take priority over hydrogen, the sense of helicity (using the usual priority rules) as shown in Figure 4.36 is clearly anticlockwise and thus the enantiomer represented is *S*.

Thus, it should be clear that any object, even though it might not contain carbon, may lack the symmetry, which would allow it to be brought into superposition with its realized mirror image. Such objects include commonplace items such as shoes (but not socks), pantyhose (but not stockings), gloves, and other items reflecting the

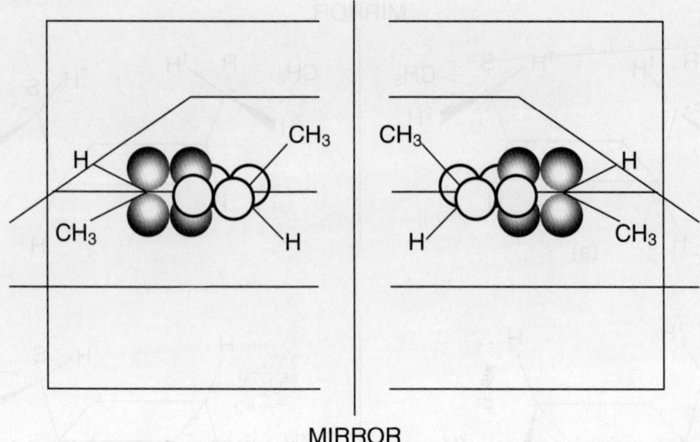

Figure 4.35. A depiction of 2,3-pentadiene showing mirror images incapable of being brought into superposition.

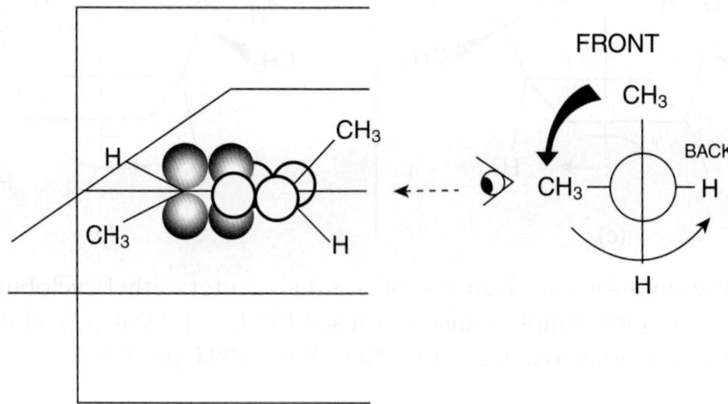

Figure 4.36. A representation of (*S*)-2,3-pentadiene.

chiral (or handed) nature of humans. Nonhuman objects such as nuts and bolts (which might be threaded in a right- or left-handed sense), climbing vines (usually with a right-hand twist), and knots, such as the trefoil of Figure 4.37, also obviously possess chirality.

G. THE CONSERVATION OF SYMMETRY DURING REACTIONS

In addition to the **static** stereochemical features across planes, about axes, and through points (discussed above and seen in realized organic compounds), an understanding of *the stereochemical relationships often found to exist* in the **dynamic** relationship between reactants and products is vital to organic chemistry. Such an understanding allows detailed predictions, for example, of products, to be made!

By way of example, consider the following observations reported in 1958: when (*Z*)-3,4-dimethylcyclobutene is heated, it undergoes decomposition to (*Z*,*E*)-2,4-hexadiene (exclusively!). Furthermore, when (*E*)-3,4-dimethylcyclobutene is heated,

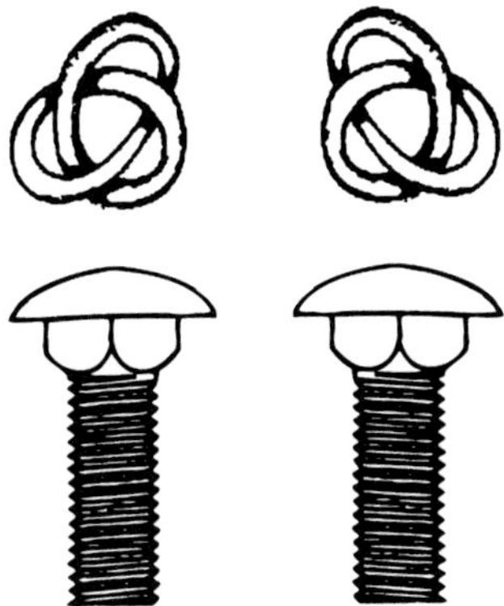

Figure 4.37. The top row shows a trefoil knot and its mirror image. These knots, if realized, are related as mirror images that cannot be brought into superposition. A bolt and its mirror image are more common (bottom), one threaded so that the nut must be turned to the right while the other is threaded so the nut must be turned to the left.

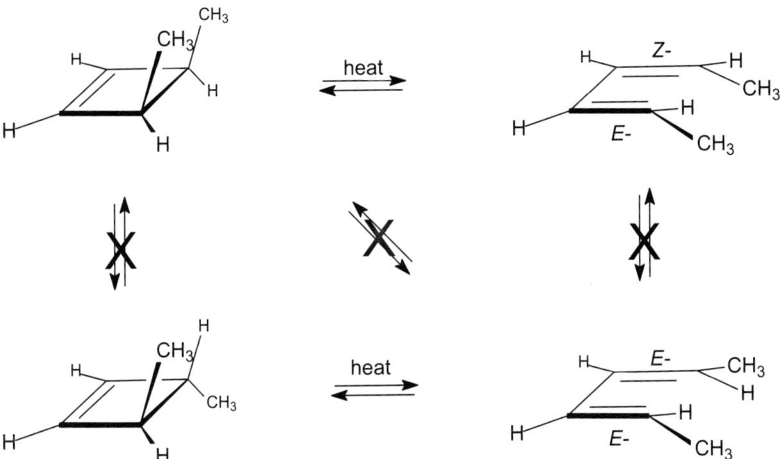

Figure 4.38. A depiction of the thermal conversions of *cis-* or (*Z*)-3,4-dimethylcyclobutene to *cis,trans-* or (*Z,E*)-2,4-hexadiene and *trans-* or (*E*)-3,4-dimethylcyclobutene to *trans,trans-* or (*E,E*)-2,4-hexadiene. The reactants and the products do not interconvert under the conditions of the reaction and *there is no crossover* (see Woodward, R. B.; Hoffmann, R. *The Conservation of Orbital Symmetry*, Verlag Chemie GmbH [Academic Press], New York, **1970**, p. 37*ff*. and references therein).

it undergoes decomposition to (*E,E*)-2,4-hexadiene (exclusively!). These processes are shown in Figure 4.38, respectively.

Problem 4.9. Keeping in mind the coupling constants (Figures 3.16 and 3.17) and chemical shifts (Table 3.2) for vicinal protons on the carbons of the double bond, indicate how one might know which hexadiene was obtained from the specific cyclobutene with which one started.

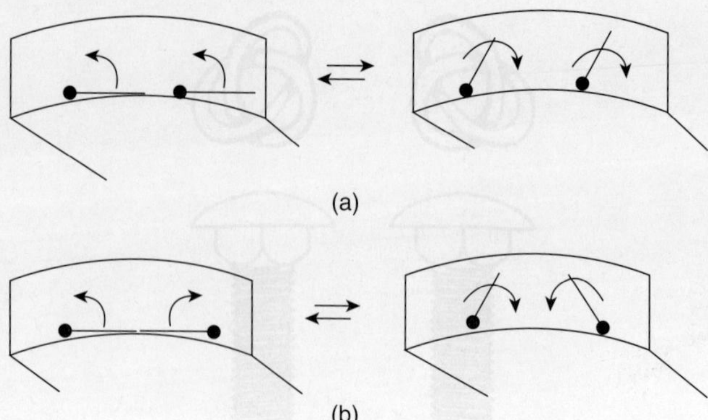

Figure 4.39. (a) Conrotatory windshield wiper motion. (b) Disrotatory windshield wiper motion.

Interestingly, many examples of such reactant–product specificity are known. Their analysis led Woodward and Hoffmann* to propose that these and other **electrocyclic**[†] processes occur in a fashion, which **conserves symmetry**.

Indeed, it is clear that it is the conservation of symmetry that rules the path the reactions take. This powerful predictive tool operates using the following concepts:

1. Overlap of orbitals of like sign corresponds to bonding and overlap of orbitals of unlike sign corresponds to antibonding (see Chapter 1).
2. The symmetry properties[‡] **of the orbitals** undergoing change in the conversion of reactants to products are conserved.
3. The conservation of **orbital symmetry** dictates the products.

There are common manifestations of the conservation of symmetry. Consider, for example, the situation occurring during simultaneous operation of two windshield wipers on the front windshield of an automobile.

On many current or recent vintage U.S. model automobiles, windshield wipers operate together, as shown by the arrows in Figure 4.39a. However, on many older U.S.- and certain non-U.S.-made automobiles, the windshield wipers operate together as shown by the arrows in Figure 4.39b. Hopefully, it is clear to you that these **processes involve fundamentally different symmetry operations**. In the first case (Figure 4.39a), the wipers move in the *same* direction. Such a motion is defined as **conrotatory**. As the tip of the first blade moves toward the tip of the second blade, the latter is **moving away**. (It is as if your mirror image backed away from a mirror as you approached it.) **This process conserves symmetry about an axis lying between the two wipers (and pointing over the hood of the car).**

---

*Woodward, Robert Burns (1917–1979), professor, Harvard University and Hoffmann, Roald (1937–), professor, Cornell University shared the 1965 Nobel Prize in Chemistry for this and other work.

[†]Although it is sometimes expanded upon, the word "electrocyclic" was originally restricted (by Woodward and Hoffmann) to refer to intramolecular cycloadditions where "the formation of a single bond between the termini of a linear system containing k π electrons, and the converse process" was occurring.

[‡]Specifically, mirror and axial symmetry operations of the appropriate orbitals **as they are engaged in bond breaking and making, that is, in a dynamic (as opposed to static) process,** must be examined.

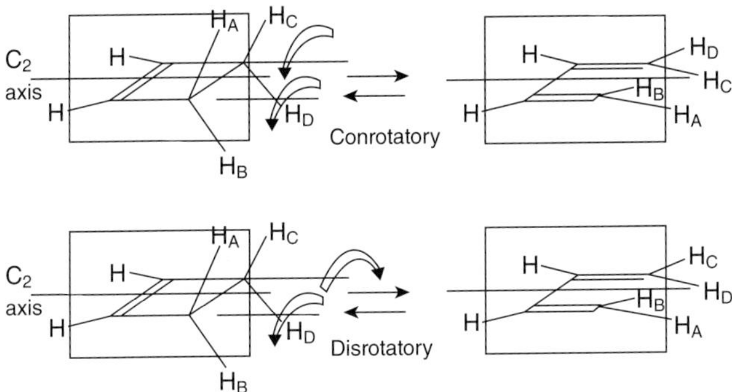

Figure 4.40. A representation of **conrotatory** and **disrotatory** processes, respectively, with the interconversion between cyclobutene and 1,3-butadiene as an example.

In the second case (Figure 4.39b), the wipers move in *opposite* directions ("toward" or "away" from each other). Such a motion is defined as **disrotatory**. As the tip of the first blade moves toward the second blade, the tip of the latter is **moving toward the first**. (It is as if your mirror image approached you as you approached the mirror, that is, the normal case for mirror images.) **This process conserves symmetry about a mirror plane lying between the two wipers (and pointing over the hood of the car)**.

These examples may now be related to the **process** of conversion of cyclobutene to butadiene and the reverse (Figure 4.40). That is, if we suppose the **process** is **conrotatory**, we may ask what kind of symmetry is conserved? Alternatively, if we suppose the **process** is **disrotatory**, what kind of symmetry is conserved? In Figure 4.40, the protons of cylobutene at C-3 and C-4 are labeled with letters so that they can be followed into 1,3-butadiene (and vice versa). Additionally, a mirror plane and a rotational axis, to which the symmetry considerations are referred, are shown. The plane and axis lie between the orbitals that will eventually be considered.

First, examine the conrotatory process shown in Figure 4.40.

For cyclobutene, notice that H_A is the **mirror image** of H_C and H_B is the **mirror image** of H_D. Notice also that **rotation** of cyclobutene 180° **about the axis** shown (this is formally called a C_2 axis) exchanges H_A with H_D and H_B with H_C. Now, examine the 1,3-butadiene. Notice that H_A is the **mirror image** of H_D and that H_B is the **mirror image** of H_C. Notice, too, that rotation about the *same* C_2 axis exchanges H_A with H_D and exchanges H_B with H_C.

Therefore, **the conrotatory process conserved axial symmetry**.

Now, examine the disrotatory process shown in Figure 4.40.

As above, for cyclobutene, H_A is the **mirror image** of H_C and H_B is the **mirror image** of H_D. Notice also that **rotation** of the cyclobutene 180° **about the C_2** axis exchanges H_A with H_D and H_B with H_C. Now, examine the 1,3-butadiene. Notice that now H_A is the mirror image of H_C and that H_B is the **mirror image** of H_D. Notice, too, that now rotation about the same C_2 axis exchanges H_A with H_C and exchanges H_B with H_D.

Therefore, **the disrotatory process conserved mirror symmetry**.

Again, as with the windshield wipers (Figure 4.39a,b), a conrotatory process has conserved axial symmetry and a disrotatory process has conserved mirror symmetry.

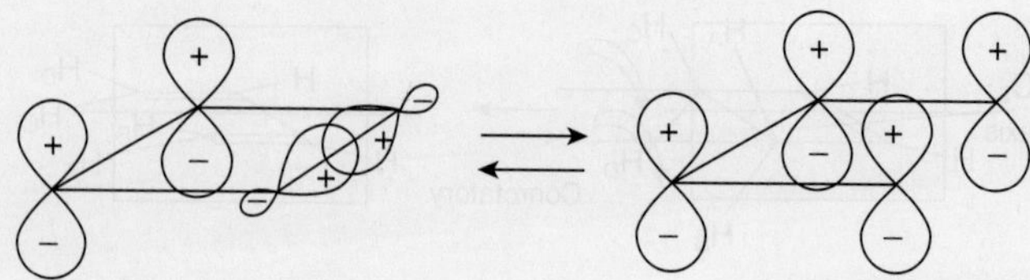

Figure 4.41. A representation of the lowest-lying occupied orbitals (σ and π) in cyclobutene and (π_1) 1,3-butadiene involved in their interconversion.

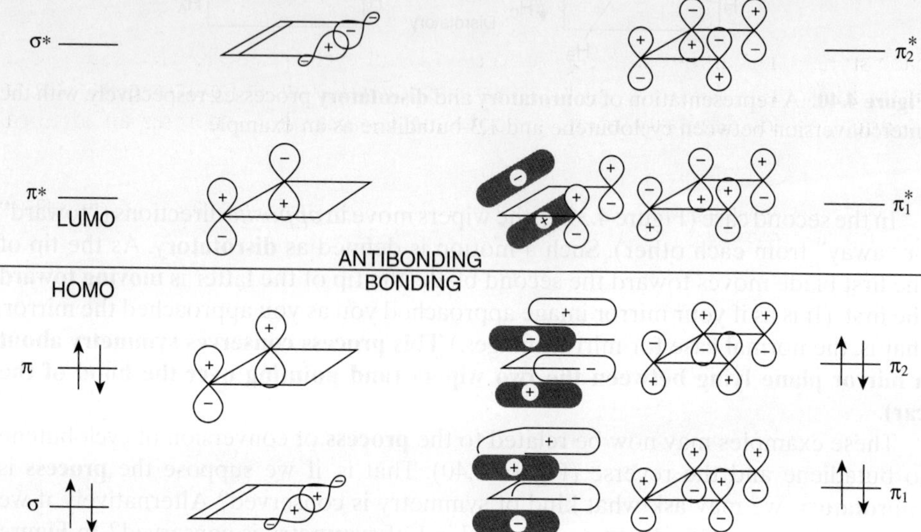

Figure 4.42. A representation of some of the molecular orbitals involved in the interconversion between the C_4H_6 isomers, cyclobutene, and 1,3-butadiene.

However, these general symmetry principles have not answered the question about which process actually occurs nor have they served to account for the experimental observations regarding product specificity with which this section began. To do that, we must consider the orbitals involved. The Woodward and Hoffmann principle is that **orbital symmetry is conserved**. Thus, it is necessary to examine the **path** for the interconversion of cyclobutene to 1,3-butadiene and, in particular, the orbitals involved. Figure 4.41 shows the interconversion again, this time with some orbitals drawn in.* Examination of the figure shows that regardless of the path (i.e., **conrotatory** or **disrotatory**), for the reaction as written from left to right, the π bond between C_1 and C_2 and the σ bond between C_3 and C_4 will be "broken" (involving a total of four electrons). Two π bonds will be "formed" in 1,3-butadiene (involving a total of four electrons). The reaction as written from right to left is the reverse of this process!

From Chapter 3 (e.g., Figure 3.18), it should be clear that the orbitals shown represent the lowest-lying occupied MOs. A more complete description of these

---

*Although individual orbitals are drawn here (and elsewhere), it is important to keep in mind that the **total** orbital symmetry of each ground state or excited state, and so on, must be considered (see Chapter 1).

orbitals, bonding, and antibonding is given in Figure 4.42, where some additional common descriptors are also employed. The HOMO is the highest occupied molecular orbital. The LUMO is the lowest unoccupied molecular orbital. Generally, thermal processes lack sufficient energy to permit or induce occupation of the LUMO at the expense of the HOMO. Thus, thermal processes will generally involve only bonding electrons. Photochemical processes, on the other hand, where promotion of an electron from a bonding (the HOMO) to an antibonding orbital (the LUMO) can occur, for example, in a $\pi \rightarrow \pi^*$ transition (see Figure 2.3), will have other consequences—as dictated by orbital symmetry.*

Next, consider the symmetry (with regard to a mirror plane and the C_2 axis) for each state. For example, with regard to the MO of cyclobutene labeled σ, as shown again in Figure 4.43, it should be clear that reflection in the mirror generates an identical object. Similarly, rotation about the C_2 axis also generates an identical

*A general explanation of the reaction between any two centers (e.g., atoms, molecules) has evolved from arguments combining symmetry with the disruption or *perturbation* of that symmetry as reactions occur. *Perturbation theory* suggests that as the (originally) distant, unperturbed basis set of MOs of the two reacting species approach each other, the overlap between their respective orbitals creates a new interaction as a consequence of orbital mixing. It argues further that the strongest interactions will be between orbitals that are closest in energy to each other. Then, ignoring for the moment the possibility that an orbital might have only one electron (a SOMO, or singly occupied molecular orbital), the simplifying assumption is made that the filled and unfilled orbitals most likely to interact are at the highest energies of each of the reactants. These orbitals are called, collectively, the *frontier orbitals*. If the frontier orbitals are both fully occupied, then the potential gain in energy resulting from overlap to create two (bonding and antibonding) orbitals is offset by the destabilization resulting from creating a fully occupied antibonding orbital and no reaction occurs. Similarly, bonding cannot be effected if the orbitals are both empty. Therefore, it is only the interaction of a filled MO with an unfilled MO that can result in bond formation. The frontier orbitals involved will most likely be the HOMO of one reactant with the LUMO of the other, creating a new, low-lying, filled MO and an unoccupied (higher) antibonding MO. A high-energy HOMO implies that the electrons are weakly held and reactive; a low LUMO implies that its filling will readily occur and that the resulting MO will be particularly stable. The convention is that the high HOMO donates the electrons (the *donor*), while the low LUMO accepts them (the *acceptor*).

Donor–acceptor frontier orbitals. Reactants are (a) and (c). The product is (b).

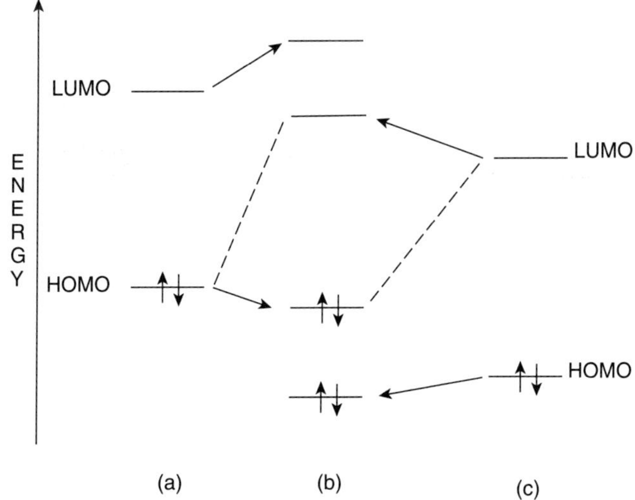

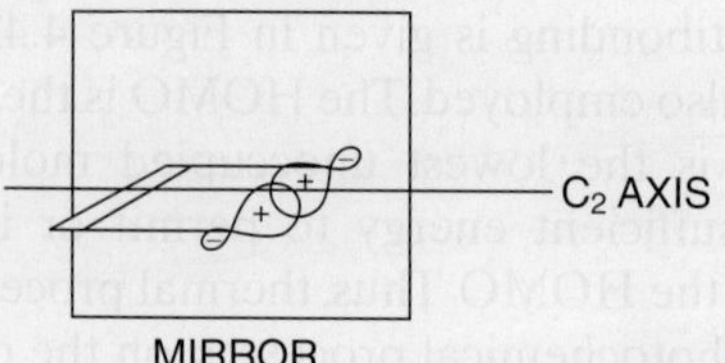

Figure 4.43. A representation of the C_3–C_4 σ bond of cyclobutene and two symmetry elements (a plane and an axis) passing through that bond.

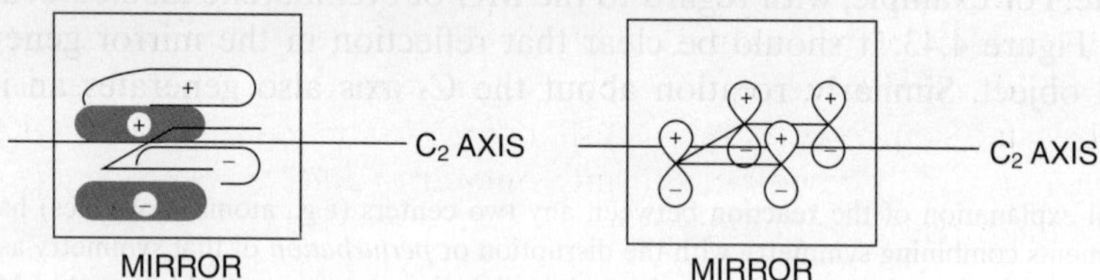

Figure 4.44. A representation of the molecular orbital of 1,3-butadiene labeled π_1 and the same two symmetry elements as shown in Figure 4.43 (a plane and an axis) passing through the representation.

TABLE 4.3. The Results of Consideration of the Symmetry Operations for the Interconversion of Cyclobutene and 1,3-Butadiene[a]

	Cyclobutene		1,3-Butadiene		
	Mirror	C_2 axis	Mirror	C_2 axis	
σ*	Antisymmetric	Antisymmetric	Antisymmetric	Symmetric	π^*_2
π*	Antisymmetric	Symmetric	Symmetric	Antisymmetric	π^*_1
π	Symmetric	Antisymmetric	Antisymmetric	Symmetric	π_2
σ	Symmetric	Symmetric	Symmetric	Antisymmetric	π_1

[a]Remember … A disrotatory process conserves mirror symmetry whilst a conrotatory process conserves C_2 symmetry.

object. Therefore, σ is said to be symmetric with regard to **both** mirror symmetry and with regard to C_2 symmetry. On the other hand, as shown in Figure 4.44 for the MO of 1,3-butadiene labeled π_1, it should be equally clear that, while reflection in the MIRROR generates an identical object, rotation about the C_2 axis produces an object in which the lobes bearing a minus (−) are on the top and those with the plus (+) are on the bottom, not the identical object. Therefore, π_1 is symmetric with regard to mirror symmetry and antisymmetric with regard to C_2 symmetry. In this way, Table 4.3 can be created from the information in Figure 4.42.

It should be clear that, **since the interconversion of cyclobutene and 1,3-butadiene is a thermal process and only bonding orbitals can be considered, symmetry is conserved (with one symmetric and one antisymmetric orbital each) if axial symmetry is followed**. Symmetry is **not** conserved (two symmetric in one and one symmetric and one antisymmetric in the other) if mirror symmetry is followed. Furthermore, as shown in Figure 4.40, **only** a process that is **conrotatory** conserves C_2 symmetry.

Therefore, as shown in Figure 4.38, the thermal conversion of (Z)-3,4-dimethylcyclobutene to (Z,E)-2,4-hexadiene and (E)-3,4-dimethylcyclobutene to (E,E)-2,4-hexadiene cannot *cross over since only a conrotatory process will conserve orbital symmetry.*

In this vein, it is possible to examine how cyclobutene might undergo a photochemical reaction and what the consequences of such a reaction might be. That is, if conservation of orbital symmetry is required, and a $\pi \rightarrow \pi^*$ process is induced by irradiation at the appropriate wavelength, then (examination of Figure 4.42 and Table 4.3 dictates that) the HOMO and the LUMO will now be singly occupied and while conservation of the symmetry of the C_2 axis (a conrotatory process) for photon-excited cyclobutene (σ, symmetric; π, antisymmetric; σ^*, symmetric) would require the same symmetry in the corresponding photon-excited 1,3-butadiene (i.e., π_1, antisymmetric; π_2, symmetric; π^*_1, antisymmetric) *and that symmetry is missing!* Therefore, the process cannot occur.

However, the appropriate symmetry is present if mirror symmetry is conserved. That is, the mirror symmetry for excited cyclobutene is σ, symmetric; π, symmetric; π^*, antisymmetric while for 1,3-butadiene the corresponding states are π_1, symmetric; π_2, antisymmetric; π^*_1, symmetric. Thus, in this excited state cyclobutene has two symmetric and one antisymmetric mirror relationships **as does 1,3-butadiene.** The interconversion of these species in these states conserves symmetry. Since mirror symmetry requires that **disrotatory processes** occur, we can confidently predict that the photochemical process will generate *cis,trans-* or (Z,E)-2,4-hexadiene from *trans-* or (E)-3,4-dimethylcyclobutene and *trans,trans-* or (E,E)-2,4-hexadiene from *cis-* or (Z)-3,4-dimethylcyclobutene with no *cross over since only a disrotatory process will conserve orbital symmetry.*

By definition (*vide supra*), electrocyclic reactions need to be confined to unimolecular systems (i.e., systems in which a *single* molecule is undergoing covalency change in the transition state). However, as early as the end of nineteenth century, reactions between conjugated dienes and alkenes containing only one double bond had been observed (a bimolecular process since *two* molecules are *both* undergoing covalency change in the transition state). Beginning in about 1928, the isolated observations concerning this powerful synthetic tool, in which *acyclic* precursors react to yield *cyclic* products, were correlated and expanded upon by Otto Diels and Kurt Alder in a reaction that now bears their names (the Diels–Alder reaction).[*†‡]

---

[*]Otto Diels (1876–1954), professor, University of Kiel. Kurt Alder (1902–1958), professor, University of Cologne. Alder was a student with Diels and they shared the Nobel Prize for their work in 1950.

[†]So-called name reactions in organic chemistry are common. Not only does this afford recognition to those who discover, or who have worked to explore, a method which becomes widely used, it also serves as a "short-hand" notation to describe that method. Simply stating, for example, that the product arises via a Diels–Alder reaction defines, to those versed in the art, possible reactants, conditions under which the reaction might be carried out, and so on. While you should remember the name and the general outlines of what is occurring, the more exhaustive details are the subject of further study.

[‡]These particular ring-forming reactions, of which the Diels–Alder are only one kind, are generally referred to (in the Woodward and Hoffmann paradigm) as "cycloadditions" and "cycloreversions"—depending on their direction. In the strictest sense given earlier, they are *not* electrocyclic processes. The terms "suprafacial" and "antarafacial," defining the sense of "newly formed or broken bonds" with regard, respectively, to the same or opposite faces of the reacting systems were coined in the same minting.

Problem 4.10. Shown below is a representation of the orbital diagrams for 1,3-cyclohexadiene and 1,3,5-hexatriene (as in Figure 4.42 for cyclobutene and 1,3-butadiene). Create a table (as in Table 4.3) and, keeping in mind that any disrotatory process will conserve mirror symmetry and any conrotatory process will conserve C_2 symmetry, decide what symmetry is conserved for the (a) thermal and (b) photochemical transformations from cyclic to acyclic isomers.

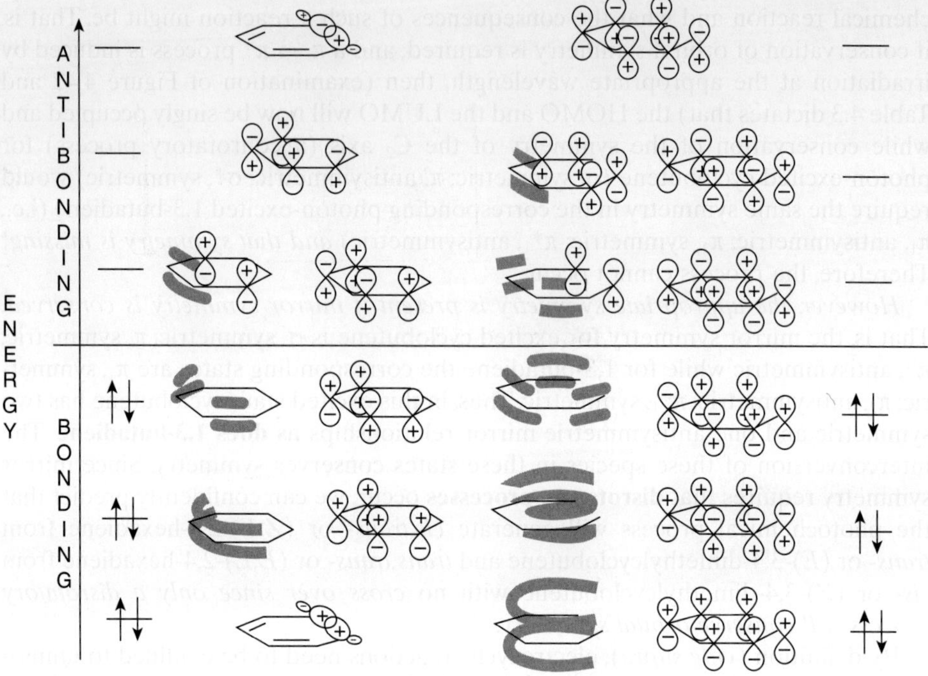

Using 1,3-butadiene and ethene as examples of a conjugated diene and an isolated alkene, respectively (for the purposes of the Diels–Alder reaction, the alkene is called a **dienophile** or "diene lover"), the reaction is illustrated in Figure 4.45.

While the diene portion must contain two double bonds separated by a single bond (an orbital description follows), the dienophile simply requires a π system.

Indeed, in all of these processes, **three** bonds are formed (**two** σ + **one** π) and **three** bonds are broken (**two** π + **one** σ), and it has been argued that in some circumstances, the making and the breaking of these bonds occurs simultaneously. However, in principle, any conjugated diene and any π-containing system, which might reasonably fit together, can be considered reactants in the Diels–Alder

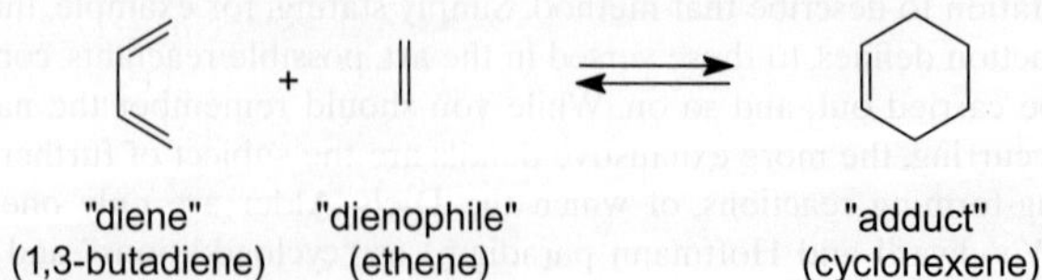

Figure 4.45. A representation of the Diels–Alder reaction between 1,3-butadiene and ethene to produce cyclohexene (and its reverse).

reaction. Thus, 1,3-butadiene can also undergo reaction with ethyne to produce 1,4-cyclohexadiene (Figure 4.46).

It should also be clear, as depicted in Figures 4.47 and 4.48, that there is a geometric constraint on the diene. Thus, the two double bonds **must** lie on the same "side" of the single bond (*s*). Unless the diene is part of a ring (*vide infra*), the *s-trans* diene is generally slightly more stable (and is thus favored at room tempera-

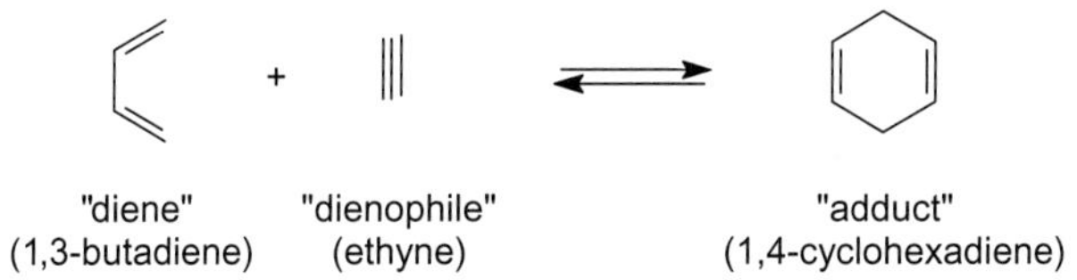

"diene"   "dienophile"   "adduct"
(1,3-butadiene)   (ethyne)   (1,4-cyclohexadiene)

Figure 4.46. A representation of the Diels–Alder reaction between 1,3-butadiene and ethyne to produce 1,4-cyclohexadiene (and its reverse).

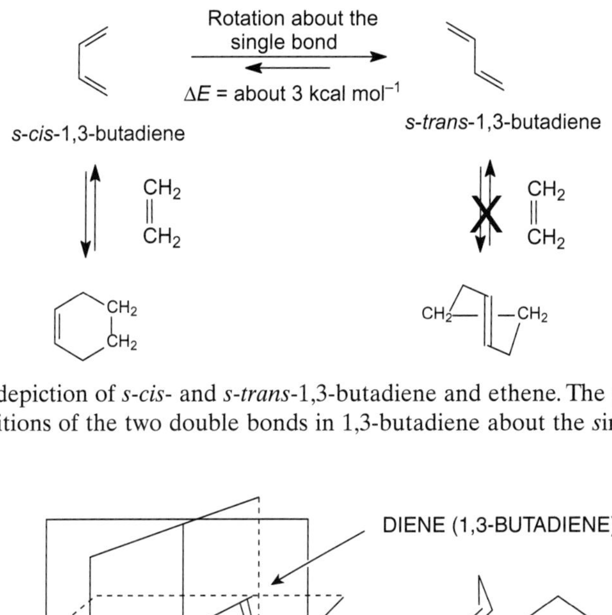

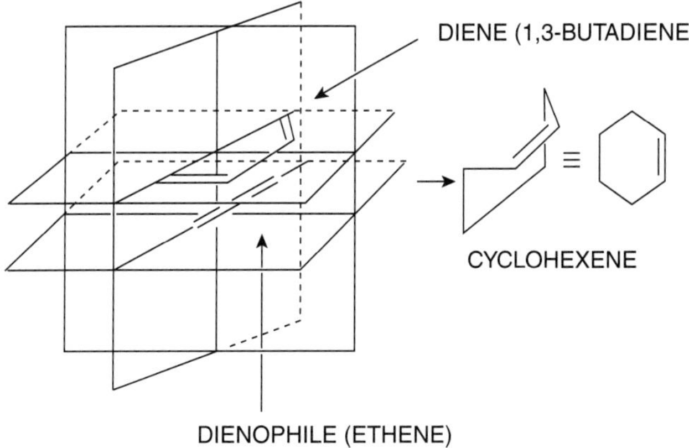

Figure 4.47. A depiction of *s-cis-* and *s-trans*-1,3-butadiene and ethene. The prefix *s* denotes the relative positions of the two double bonds in 1,3-butadiene about the *single* bond.

Figure 4.48. A depiction of the organization (presumably) required during the Diels–Alder reaction between 1,3-butadiene (the diene) and ethene (the dienophile) to produce cyclohexene.

ture) than the *s-cis* diene. However, in the *s-trans* diene, the ends of the double bonds of the conjugated system are too far apart to accommodate ring formation and, even if they were not, **since the geometry of the starting materials is preserved in the products**, should the diene with the *s-trans* geometry react, then the double bond in the cyclohexene product would necessarily be *trans*! Since a double bond cannot be placed *trans* in a ring with only six carbon atoms, the product cannot form. Furthermore, as shown in Figure 4.48, a high degree of order is presumably required to produce the cyclic product.

An orbital symmetry diagram for this process can also be produced (Figure 4.49). The symmetry of the occupied orbitals in products and starting materials (from either direction!) remains the same (two symmetric and one antisymmetric) with regard to the mirror plane bisecting the 2,3-single bond of butadiene and the π cloud of ethene. Thus, symmetry is conserved and the thermal process is allowed. There will certainly be a barrier to the reaction occurring (consider the contribution to the organization required—expressed as ΔS). The process is frequently referred to as a "(4 + 2) cycloaddition" and, as you already expect for such allowed electrocyclic processes, (a) the geometry of the starting materials is reflected in the products and (b) maximum orbital overlap is anticipated.

The first of these observations is reflected in the process depicted in Figure 4.50 and the second in Figure 4.51.

Problem 4.11. Utilize WebMO® to convince yourself that the *endo*-tricyclo[5.2.1.0^{2,6}] deca-3,8-diene is favored over the corresponding *exo*-isomer.

The rigid systems found as a consequence of the Diels–Alder reaction using cyclic dienes have been the subject of extensive study.

Thus, the reaction of cyclopentadiene with ethene (Figure 4.52) results in the formation of the bicyclic alkene bicyclo[2.2.1]heptene (norbornene) which, on

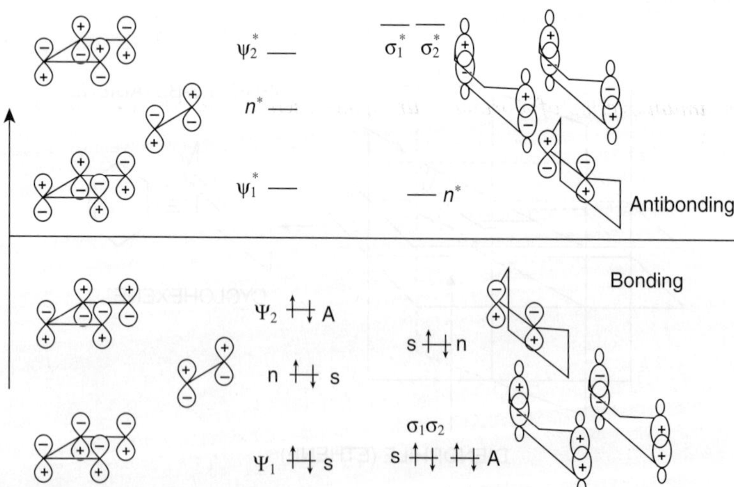

Figure 4.49. A correlation diagram for the Diels–Alder reaction between 1,3-butadiene $(CH_2=CH–CH=CH_2)$ and ethene $(CH_2=CH_2)$ showing it to be a thermally allowed process.

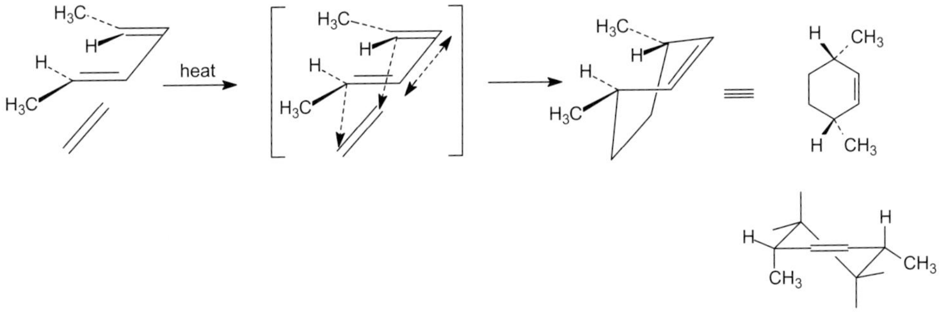

Figure 4.50. A cartoon depiction of the conservation of orbital symmetry during a Diels–Alder reaction between ethene (ethylene, $H_2C=CH_2$) and *trans-*, *trans-*, or (*E,E*)-2,4-hexadiene (CH_3–CH=CH–CH=CH–CH_3). The stereochemistry of the starting materials is conserved.

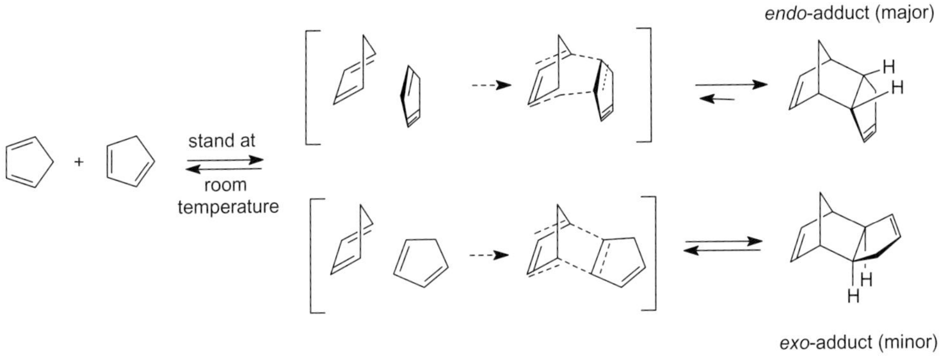

Figure 4.51. A cartoon representation of the Diels–Alder reaction between two cyclopenta-diene equivalents: one of which acts as a diene and the other as a dienophile. Dicyclopentadiene (3a, 4,7,7a-tetrahydro-4,7-methano-1H-indene, tricyclo[5.2.1.0^{2,6}]deca-3,8-diene) is commer-cially available and is thermally decomposed ("cracked") at its boiling point (170°C) to produce cyclopentadiene (bp 46°C). The latter slowly dimerizes at room temperature. Dicyclopentadiene exists almost completely as the *endo* isomer, which, it is argued, forms as result of *maximum overlap of π orbitals during the reaction* (the **Alder Endo rule**). The other potential isomer (the *exo* isomer) is not observed.

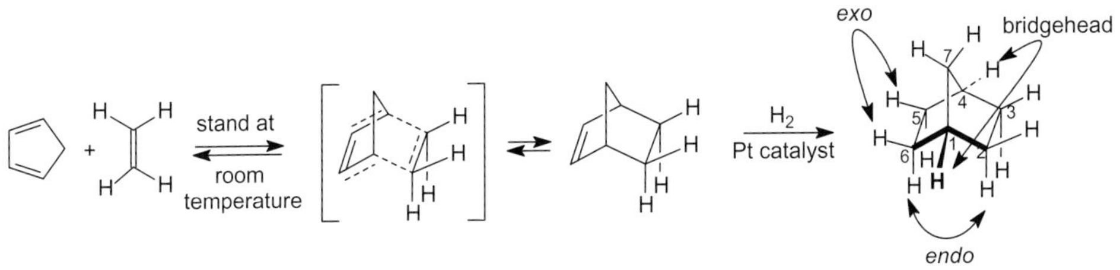

Figure 4.52. A cartoon showing the consequences of the Diels–Alder reaction between cyclopentadiene and ethene to generate bicyclo[2.2.1]heptene (norbornene) and the subse-quent catalytic reduction of the latter to produce bicyclo[2.2.1] heptane (norbornane).

reduction with hydrogen over a suitable catalyst, yields bicyclo[2.2.1]heptane (nor-bornane). In such a system, there are three distinctly different types of carbon atoms (Figure 4.52), that is, the four carbons C_2, C_3, C_5, and C_6 constitute one set, the two *bridgehead* carbons C_1 and C_4 a second set, and the single-carbon C_7 connecting C_1 with C_4, a third. The hydrogens attached to the bridgehead carbons are, reasonably, bridgehead hydrogens. The eight hydrogens attached to the set of four carbons C_2, C_3, C_5, and C_6 are further divided into two groups. Four of the eight are on the *same side* of the six-membered carbocycle as C_7 and are called *exo*; four are on the oppo-site side of C_7, tucked "under," and are called *endo*.

Rigidly fixed systems that are incapable of conformational isomerism and carbon–carbon bond rotations, such as bicyclo[2.2.1]heptane (norbornane), have found an important place in structural studies of organic molecules in solution. Thus, although the *exo* protons at C_2, C_3, C_5, and C_6 are all identical to each other, they are different from the *endo* protons on the carbons bearing them. Geminal coupling is thus expected to be seen in the ^1H NMR spectrum. Furthermore, the *exo* proton at, for example, C_2, will couple with the *endo* proton at C_3 (and vice versa) as well as the bridgehead proton at C_1. The C_1 bridgehead proton, in addition to enjoying coupling to the *different exo* and *endo* protons at C_2, will also be coupled to the protons at C_7 which, as they are equally disposed on opposite sides of the plane defined by the bridgehead proton, C_1, and C_7, are identical. A variety of experiments in which ^2H has been substituted for ^1H (by methods soon to be encountered) has allowed con-struction of Table 4.4, and has demonstrated the presence of long-range couplings (such as the "W" plan in Figure 4.53 between *anti*-H_7 and *endo*-H_2). These couplings follow definite patterns through space and "through" bonds. Figures 4.54a–c present,

TABLE 4.4. Approximate Chemical Shifts (δ, ppm) and Coupling Constants (J, Hz) Found in Bicyclo[2.2.1]Heptane (Norbornane) and Related Compounds

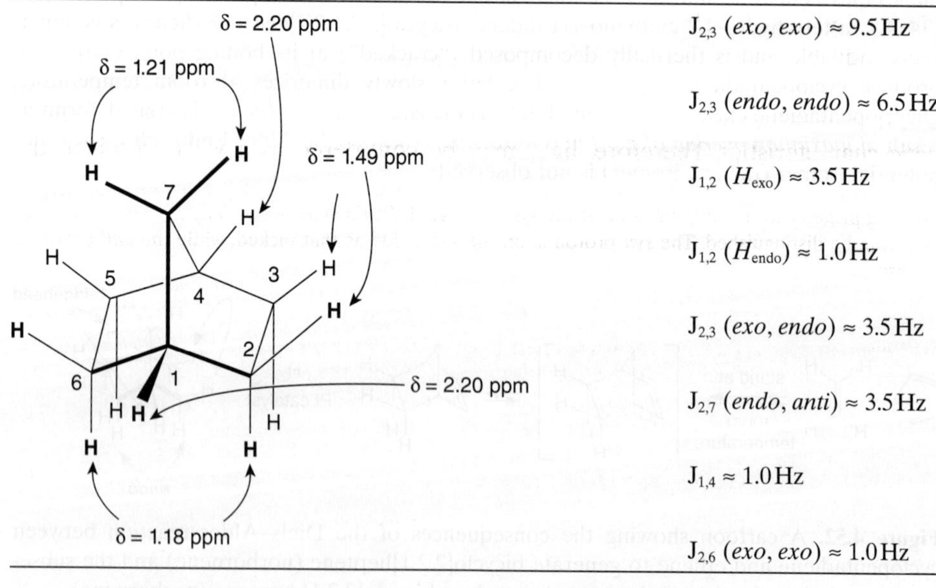

$J_{2,3}$ (*exo, exo*) ≈ 9.5 Hz

$J_{2,3}$ (*endo, endo*) ≈ 6.5 Hz

$J_{1,2}$ (H_{exo}) ≈ 3.5 Hz

$J_{1,2}$ (H_{endo}) ≈ 1.0 Hz

$J_{2,3}$ (*exo, endo*) ≈ 3.5 Hz

$J_{2,7}$ (*endo, anti*) ≈ 3.5 Hz

$J_{1,4}$ ≈ 1.0 Hz

$J_{2,6}$ (*exo, exo*) ≈ 1.0 Hz

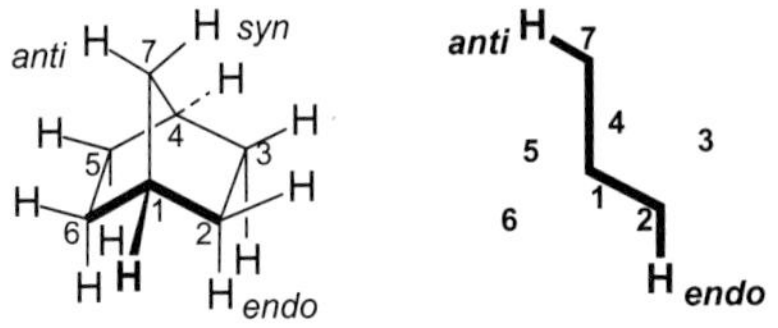

A "W-plan" arrangement between
anti-H$_7$ and *endo*-H$_2$.

Figure 4.53. A representation of bicyclo[2.2.1]heptane (norbornane) labeled to show protons involved in coupling as given in Table 4.4.

respectively, the ^{1}H, the ^{1}H/^{13}C correlated, and the ^{1}H two-dimensional NMR spectra of bicyclo[2.2.1]heptane (norbornane).*

H. THE MEASUREMENT OF CHIRALITY

I. The Wave Nature of Light[†]

Problem 4.12. Identify as many additional "W-plan" arrangements as you can in Figure 4.53.

The wave nature of visible light had been recognized by Huygens (C. Huygens, 1629–1695) who, it is said, got his ideas about wave propagation by watching ripples on a Dutch canal. The principle bearing his name, that each point on a wave front may be regarded as a new source of waves, is particularly easy to demonstrate with water waves. Regardless, although the wave nature of **visible** light was the subject of many experiments, its relationship to other forms of electromagnetic radiation was not established until Maxwell (J. C. Maxwell, professor of physics, Cambridge University, England, 1831–1879)[‡] proposed a theory (1864) based on earlier work of Faraday (M. Faraday, 1791–1867) that related light with electricity. In essence, Maxwell's equations allowed for the predictions of the properties that light waves might have. Because light is electromagnetic radiation, it has both *electric* and *magnetic* characteristics. Therefore, light may be considered as a wave in which the

---

*Having picked (arbitrarily) one side of the system (e.g., the right side—as drawn) then the hydrogens at C$_7$ can be distinguished. The *syn* proton is on the same side as that picked, while the *anti* proton is opposite.

[†]"Does light *really* consist of waves, or of particles? Is the electron *really* a particle or is it a wave?

These questions cannot be answered by one of the two stated alternatives. Light is the name that we have given to a part of nature. The name refers to all of the properties that light has, to all of the phenomena that are observed in a system containing light. Some of the properties … can be described in terms of wavelength. Other properties … resemble those of particles and can be described in terms of a light quantum having a certain amount of energy, hv, and a certain mass, hv/c^2. A beam of light is *neither* a sequence of waves *nor* a stream of particles; it is **both** … . The electron, like the photon, has to be described as having the character both of a particle and of a wave." (Pauling, L. *General Chemistry*, 3rd edition, W.H. Freeman, San Francisco, 1970, p. 80)

[‡]Much of Maxwell's work on electromagnetic theory was done while he was still an undergraduate at Cambridge. His first contributed paper to the Royal Society was in 1846—when he was 15 years old!

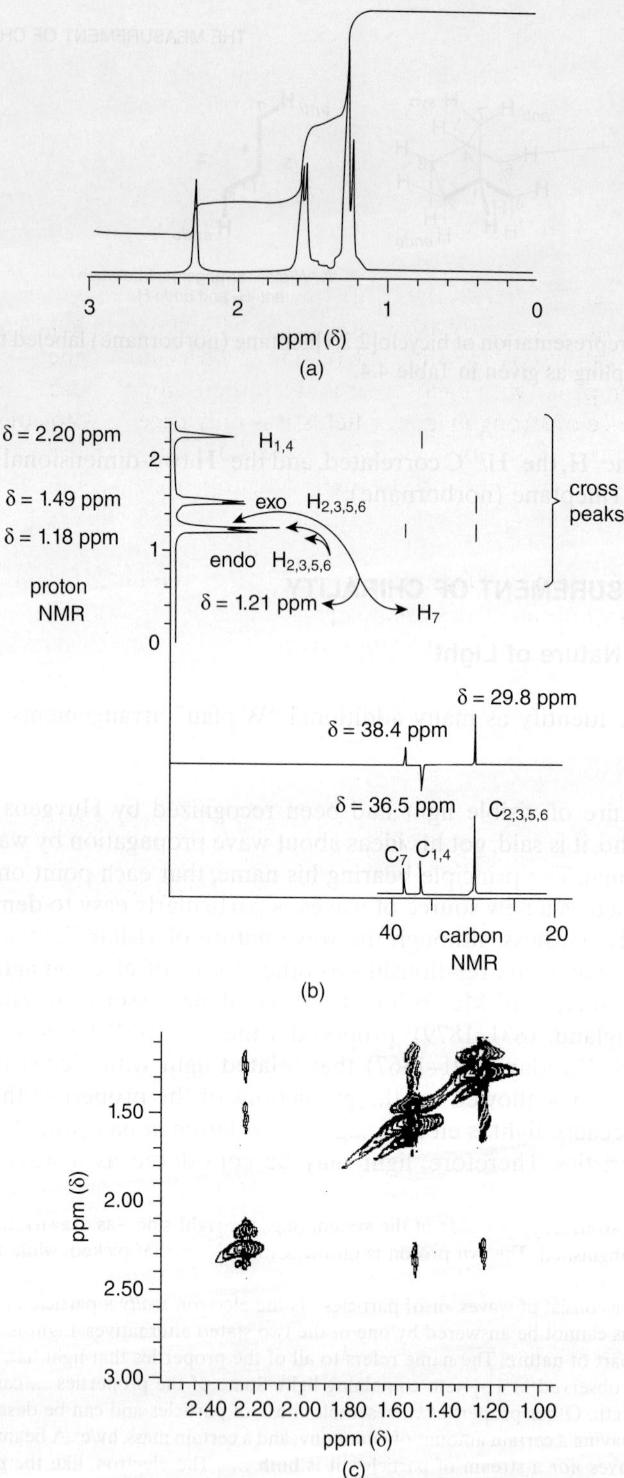

Figure 4.54. (a) The ^{1}H NMR spectrum of bicyclo[2.2.1]heptane (norbornane) at 100 MHz in ^{2}HCCl$_3$. (b) The ^{1}H/^{13}C correlated NMR spectrum of bicyclo[2.2.1]heptane (norbornane). The proton spectrum is plotted along the ordinate and the carbon spectra along the abscissa. The usual (lower) carbon spectrum is the result of complete proton decoupling. The more unusual upper carbon spectrum tests for attached protons. Carbons bearing an odd (one or three) number of protons appear as inverted signals. The positions where the spectra "cross" (the marks in the upper right quadrant of the figure) dictate to which carbons the protons are attached (and vice versa). (c) The two-dimensional ^{1}H/^{1}H NMR spectrum of norbornane.

electric field (**E**) and the magnetic field (**H**) components are perpendicular to each other and to the direction of propagation. Because studies on the polarization of light (*vide infra*) had failed to ever detect *any* vibrations other than those perpendicular to their direction of propagation, and this is what was *required* by Maxwell's relationship, his equations gained currency.

Maxwell's equations allowed differentiation between the vector quantities **E**, the electric field strength, and **H**, the magnetic field strength. As shown in Figure 4.55, for a single-frequency light wave imagined to be a simple planar cosine function, the following restrictions are imposed: (a) the electric and magnetic components of the wave are in phase with each other and (b) their amplitudes are equal.

In the absence of strong magnetic fields, it is only necessary to consider the electric vector component of the wave (**E**, Figure 4.55). In an unpolarized monochromatic (light of a single frequency) beam of light, the electric component is randomly distributed in all directions perpendicular to the direction of propagation. All waves will be moving, each in its own plane, and a point source, looked at "end-on" might appear as in Figure 4.56. Hopefully, it is clear that the electric component of this beam, which is vibrating in all planes (of which only a few are drawn in Figure 4.56a), shows that each vibration (as a vector quantity) consists of two kinds of orthogonal

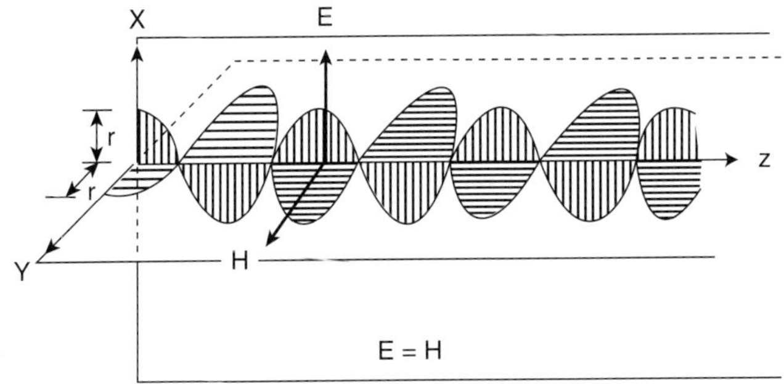

Figure 4.55. A cartoon representation of the distribution of electric (**E**) and magnetic (**H**) vectors (by Maxwell's equations) in a plane-polarized monochromatic light wave.

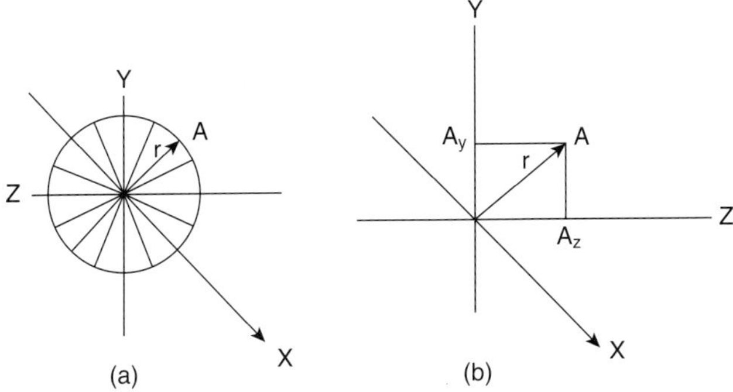

Figure 4.56. (a) A depiction of a point source of light (x, y, z = 0) as seen from the X direction (see Figure 4.57). (b) Depiction of the vectors in the Y and Z directions (of magnitude A_z and A_y) of the beam fragment "A" of (a).

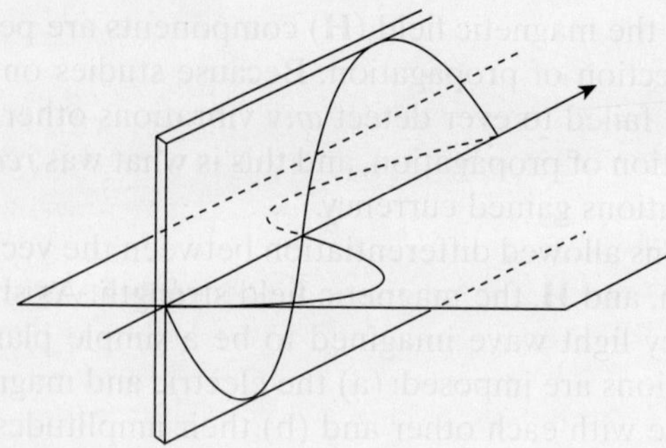

Figure 4.57. The vector A of Figure 4.56b propagating in the x direction in the mutually orthogonal *xz* and *xy* planes.

vibrations *only* (Figure 4.56b). Therefore, two planes are defined (Figure 4.57). Light confined to these planes is called *plane polarized*.

II. Plane-Polarized Light and Handedness

It is important to understand that plane-polarized light can be produced both naturally (calcite and quartz crystals) and synthetically (with aligned polymers [Chapter 6]), and when such polarized light passes through chiral ("handed") materials (or solutions of chiral materials), the plane of polarization is rotated. Historically, this property has been called "optical rotation," and the chiral material was said to be "optically active." Indeed, experimentally, the rotation of the plane of plane-polarized light demonstrates "optical activity" and thus, that the material is chiral.

Huygens (*vide supra*), in 1678, reported on the use of Iceland spar (one particular rhombohedral crystalline habit of calcite, calcium carbonate, $CaCO_3$) to generate, by double refraction,* *two* rays of light. Each ray is plane polarized and the planes are mutually orthogonal (as in Figure 4.57). When he placed a second identical crystal behind the first, and began to rotate it, he found that he either observed the two beams independently (the angle of rotation was 0°) or that the two different beams could be recombined to give the original incident beam (the angle of rotation was 180°).

Subsequently, in 1828, Nicol (W. Nicol 1770–1851, Edinburgh) constructed an optical device (the Nicol prism) made from calcite, cut so as to remove one of the two refracted rays by total reflection. Therefore, only passage of a single beam of plane-polarized light (i.e., light confined to one plane) is allowed through a Nicol prism.

When a beam of plane-polarized light (e.g., leaving a Nicol prism) is directed through a chiral material (a solution containing chiral solute is also chiral), the plane of polarization of the light is rotated and the beam emerges vibrating in a plane other than that in which it entered! An observer watching the beam come through

---

*Double refraction refers to the property possessed by some materials (including calcite [$CaCO_3$] and quartz [SiO_2]) to produce two divergent beams of plane-polarized light from an incident beam of unpolarized radiation.

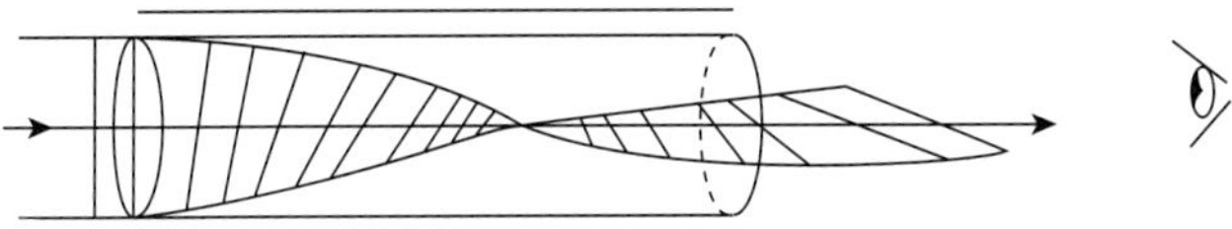

Figure 4.58. A representation of the rotation of the plane of plane-polarized light by a chiral material.

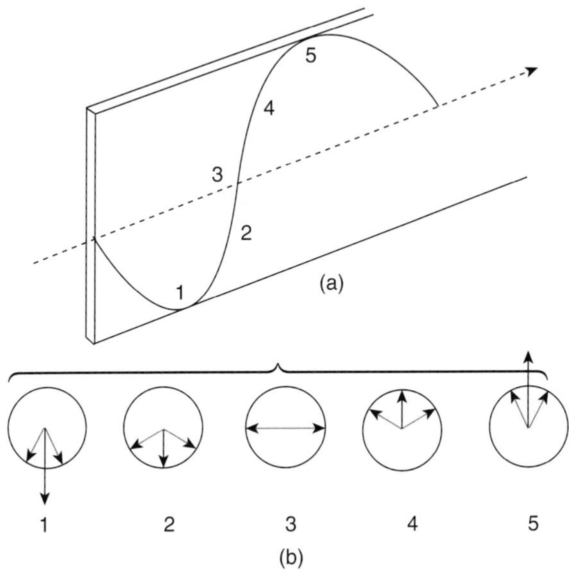

Figure 4.59. (a) The depiction of a plane-polarized wave moving in the "x" direction, the points 1 through 5 are drawn to represent the vectors shown in the pictures of part (b); (b) a depiction of how the harmonic motion of (a) can be the result of two oppositely directed circular motions.

will note (Figure 4.58) that the plane-polarized light has been rotated either to the right or to the left (of the observer). The extent to which the original beam has been rotated can also be noted.

The electric component, **E**, of plane-polarized light, shown above (one of the planes of Figure 4.57 redrawn as Figure 4.59a) is clearly a wave, moving in a given direction, whose amplitude is rising and falling. Any such simple harmonic motion along a straight line may be described as the resultant of **two opposite circular motions.*** This may be seen by considering the rotation of the vector pair in the sequential pattern of Figure 4.59b, which produces the variation in the magnitude of the vector sum. The handed sense of circular polarization, coupled to a direction of propagation is depicted in Figure 4.60.

*It appears that this idea originally came from Fresnel (A. Fresnel, 1788–1827, French engineer), who suggested that plane-polarized light entering a chiral material (or a crystal along its optic axis) may be decomposed into two circularly polarized vibrations rotating in opposite directions with the same frequency (Figure 4.59b).

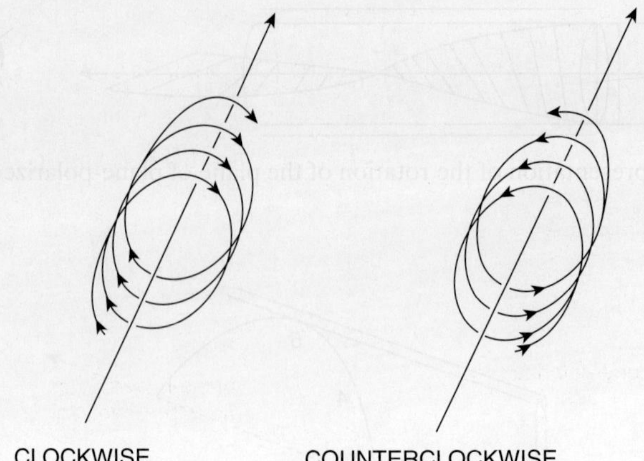

Figure 4.60. A depiction of the senses of circular polarization.

Because a chiral object and its realized mirror image interact oppositely with the circularly polarized light (a perturbation in the bonding electrons if not an absorption), in the presence of one of the enantiomers the velocity of *one* of the circularly polarized components of the light will be effected more than the other so that, on recombination as the beam leaves the material, it will be seen to have had its direction of oscillation, and hence its plane of polarization, rotated relative to its original position.

The extent of such clockwise or anticlockwise polarization is measured in a device called a **polarimeter**, which consists of two Nicol prisms cemented together with a Canada balsam glue. The first, called the polarizer, is used to polarize the light source. The second, called the *analyzer*, is used to measure the extent to which the beam of plane-polarized light has been rotated by the sample. A **polarimeter** is schematically shown as Figure 4.61.

The **specific rotation** $(\alpha)_\lambda^t$ is given in Equation 4.18,

$$(\alpha)_\lambda^t = \alpha/Pc, \tag{4.18}$$

where (α) = specific rotation;

t = temperature (°C) at which the measurement is made;

λ = wavelength of the light used (frequently the D line of sodium [589.3 nm]);

α = observed rotation;

P = path length (dm) of sample through which light passes; and

c = concentration of solute (g mL^{-1}).

Substances that cause the beam to be rotated to the observer's right are called **dextrorotatory (or [+])**. If the beam is rotated to the observer's left, the substances causing the rotation are called **levorotatory (or [−])**. There is no simple relationship between the sense of rotation (+ or −) and the absolute configuration (*S* or *R*). In addition to the above listed parameters, it should also be clear that the magnitude of the **specific rotation** is a function of whether or not the sample is one pure enan-

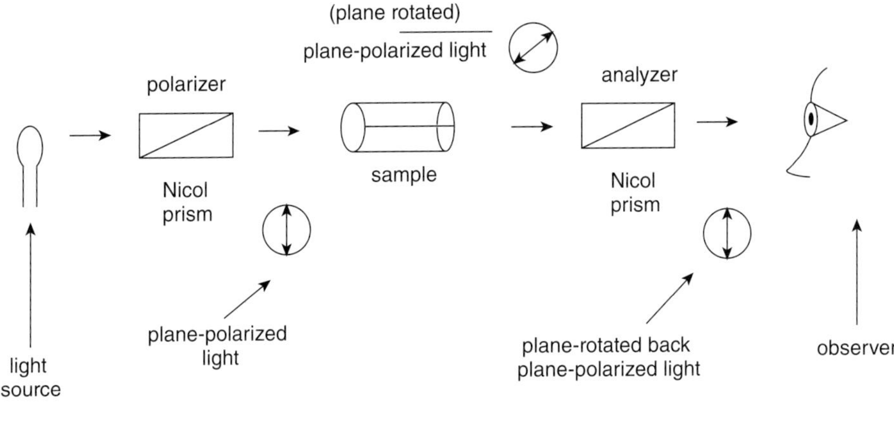

Figure 4.61. A schematic description of a polarimeter. In the absence of a sample, the polarizer and analyzer are aligned so that a maximum amount of light passes. (If they are "crossed" or at 90°, no light passes.) Introduction of a chiral sample between them causes the plane of the plane-polarized light to be tilted, and the extent (the number of degrees) and the sense (clockwise or counterclockwise) the analyzer must be turned to again cause maximum transmission are noted. Generally, the observed rotation is reported as **specific rotation**.

tiomer or a mixture.* If a mixture of enantiomers is present and if the rotation of a pure enantiomer is known, the **enantiomeric excess** (or **ee**) of the major enantiomer can be reported. Finally, in this regard, if there is **exactly a 1:1 mixture** of enantiomers, there is no apparent rotation of the plane of plane-polarized light because the interaction of each of the enantiomers with the plane-polarized light will be equal **and opposite**. Such a mixture is called a **racemate** and the sample itself is said to be **racemic**.

III. Optical Rotatory Dispersion (ORD) and Circular Dichroism (CD)

In the period after about 1860 when Pasteur (Louis Pasteur, 1822–1895) had separated enantiomorphic crystals by hand and shown that their respective solutions possessed equal (in magnitude) and opposite (in sign) specific rotations (cf. Equation 4.18) numerous similar studies on the properties of light and its interaction with solutions containing chiral materials were undertaken. So it was that in 1896, Cotton (Aimé Cotton, 1869–1951) reported that both the sign and the magnitude of the rotation of polarized light was a function of the wavelength (λ, Equation 4.18) of the light used.

Consider, first, plane-polarized light exiting a Nicol prism (e.g., Figure 4.59a for which light of only one wavelength is depicted).

The measurement of optical rotation as a function of wavelength, across the ultraviolet-visible (UV-VIS) region, is referred to as ORD and the resulting plots

---

*A pure liquid enantiomer or a pure solid enantiomer in (homogeneous) solution is said to be **isotropic** (the same throughout). However, when different results are obtained in different directions (as with a crystalline material) the substance is **anisotropic**.

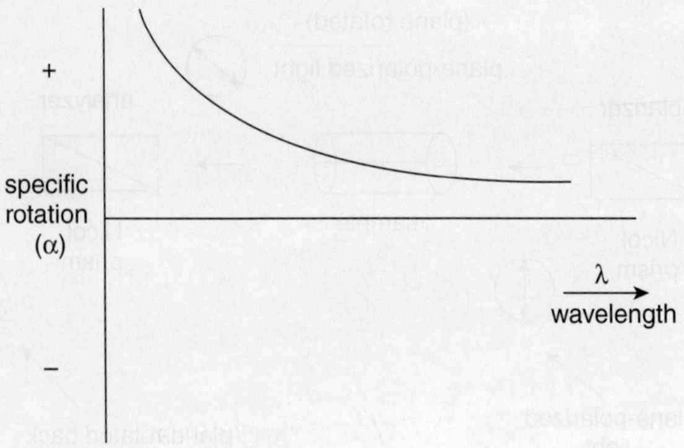

Figure 4.62. A plain negative curve of specific rotation (α) versus wavelength (λ).

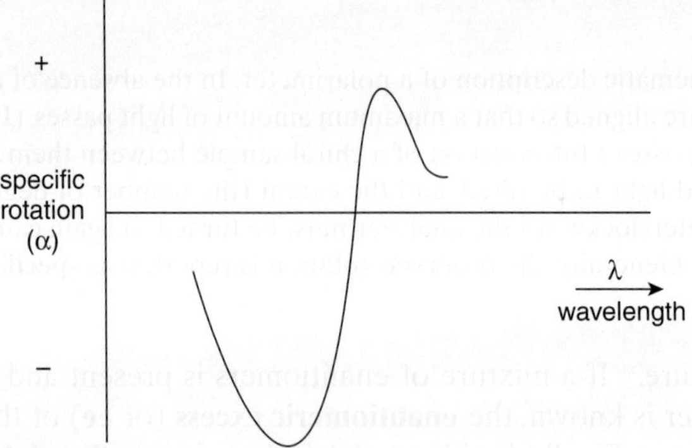

Figure 4.63. A plot of specific rotation (α) versus wavelength (λ) of a chiral material that absorbs in the region over which the rotation is measured. The material is said to have a positive **Cotton effect**.

of (α) versus λ as **ORD curves**. If the curve is not a simple one (called *plain positive* if it rises with increasing λ and *plain negative* if it falls, Figure 4.62), it is usually because the optically active substrate is *absorbing* some of the light. If plane-polarized light is absorbed (as well as rotated), then the plot of (α) against λ takes the form of a (generally) nonsymmetrical (or *anomalous*) curve (Figure 4.63), which has a minimum at short wavelength and a maximum at long wavelength or vice versa with the inflection point at the UV maximum. As the wavelength increases, if the trough comes before the peak, the material is said to exhibit a **positive Cotton effect**; alternatively, if the peak comes before the trough, the **Cotton effect** is **negative**.

Consider, second, what would happen using circularly polarized light (Figures 4.60 and 4.59b).

If the velocity of one of the circularly polarized components is retarded more than the other by some fixed amount (usually one-fourth of a wavelength)*, then the beam be resolved into its left and right circularly polarized components and the

---

*One such device for such resolution is called a **Fresnel rhomb** and depends on correctly choosing angles of a rhomboid to allow internal reflection of the plane-polarized light beam from two glass-air surfaces. More modern devices depend on electrical induction of mechanical stress.

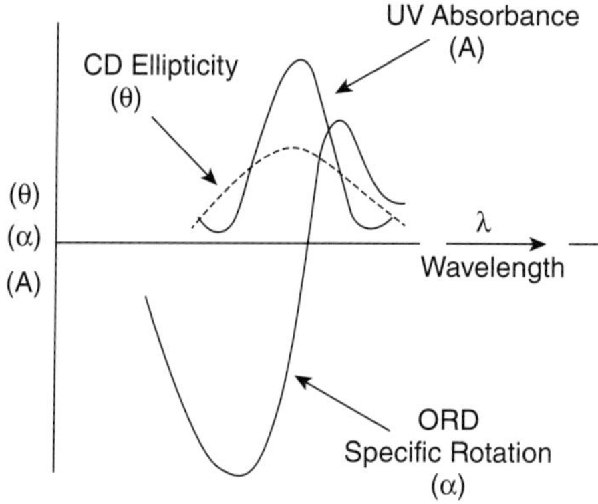

Figure 4.64. A comparison of ORD, CD, and UV spectra.

resultant will be elliptical (rather than circular). Absorption of light, from such an elliptically polarized beam generally defines even otherwise hidden maxima. This process is known as CD. Figure 4.64 presents a comparison that might be made of ORD, CD, and UV spectra.

ADDITIONAL PROBLEMS

Problem 4.13. Draw Newman and sawhorse projections of 2-methylbutane. Identify eclipsed, staggered, gauche, and antirelationships between sets of methyl groups "looking down" the C_2–C_3 bond in your representation of the structure.

Problem 4.14. Draw Newman and sawhorse projections of 2,3-dimethylbutane. Identify eclipsed, staggered, gauche, and antirelationships between sets of methyl groups "looking down" the C_2–C_3 bond in your representation of the structure.

Problem 4.15. Utilizing information in the text, make a plot similar to that of Figure 4.9 (a) for 2-methylbutane "looking down" the C_2–C_3 bond and (b) for 2,3-dimethylbutane also "looking down" the C_2–C_3 bond in your representation of the structure.

Problem 4.16. There is one 1,1-dimethylcyclobutane. How many 1,3-dimethyl-cyclobutanes can be written? What relationship(s) do they have to each other?

Problem 4.17. According to Equation 4.16, hydrogenation of cyclopropane over a platinum catalyst results in "ring opening" and the formation of propane. Suppose the same reaction were carried out on methylcyclopropane. What product(s) might you expect to find?

Problem 4.18. As shown in Table 4.2, the "cost" of putting a phenyl (C_6H_5–) group axial is significantly less than putting a *tert*-butyl [$(CH_3)_3C$–] group axial. How would you account for this observation (use suitable drawings)?

Problem 4.19. Using WebMO® at the STO-3G level, compute the single point energies of *tert*-butylcyclohexane with the *tert*-butyl group equatorial and axial (respectively).

Problem 4.20. The plant kingdom is filled with a variety of products about which you will be learning. Among the materials is a large family of compounds that have a 10-carbon framework. Generically, these compounds are called "terpenes" about which you will learn more in Chapter 11. A typical member of the group has the trivial name "limonene." Limonene rotates plane-polarized light to the left (i.e., it is levorotatory). A representation of the structure of (–)-limonene is shown below. Is it *R* or *S*?

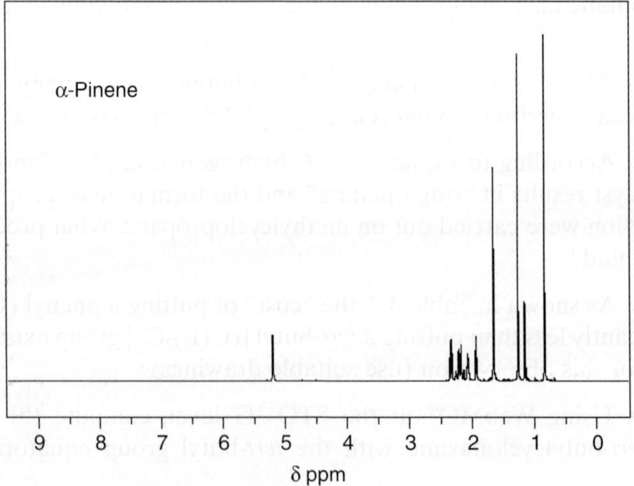

(–)-Limonene

Problem 4.21. The plant kingdom is filled with a variety of products about which you will be learning. Among the materials is a large family of compounds that have a 10-carbon framework. Generically, these compounds are called "terpenes" about which you will learn more in Chapter 11. A typical member of the group has the trivial name "α-pinene." α-Pinene rotates plane-polarized light to the right (i.e., it is dextrorotatory). A representation of the structure of (+)-α-pinene (or (–)-α-pinene) is shown below. How many asymmetrically substituted carbon atoms are present? Is it (are they) *R* or *S* as drawn?

α-Pinene

Problem 4.22. The 400 MHz ^{1}H-NMR spectrum, taken in ^{2}HCCl$_3$ solvent, of α-pinene is shown below. Identify as many of the sets of protons as you can.

Problem 4.23. There are two columns below. There is a set of dienes in one column, and a set of dienophiles in the other. Mix to produce compounds resulting from Diels–Alder reactions. Note: after one Diels–Alder reaction has occurred, the resulting alkene might be considered as a dienophile!

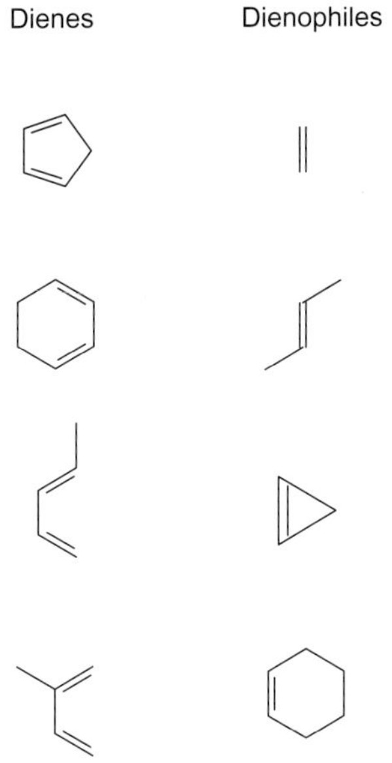

Dienes   Dienophiles

Classes of Organic Compounds—A Survey: An Introduction to Solvents and to Acids and Bases and to Computational Chemistry

"The time has come," the Walrus said,
"To talk of many things:
Of shoes—and ships—and sealingwax—
Of cabbages—and kings—
And why the sea is boiling hot—
And whether pigs have wings"
—L. Carroll

A. INTRODUCTION

This chapter differs from those that have preceded it and those that will follow it in several ways.

First, various kinds (or classes) of organic compounds will be introduced. The distinguishing feature of each class will be seen to be a unit called the **functional group**, which is broadly defined as the structural feature (in what may be a large molecule) that exhibits some specific chemical characteristic that is essentially independent of the molecule in which it is found. Although it can be argued that there is not much theoretical (nor even practical) insight about chemistry to be gained from learning the names of the kinds of compounds bearing functional groups and the names of the functional groups, organic chemistry can be (and has been) considered as "the chemistry of the functional group." So, you will need to develop some sense of familiarity with the names of the various functional groups and the characteristics that distinguish them. For example, as you already know from Chapter 3, alkanes differ from alkenes in that the latter, **regardless of any other structural feature that may be present, must contain at least one carbon–carbon double bond**, and the former, **regardless of any other structural feature that may be present, can contain only single bonds!** Therefore, the **carbon–carbon double bond** (C=C) is the functional group that distinguishes alkenes from alkanes. Similarly, alkynes contain at least one carbon–carbon triple bond (C≡C) as their functional group. Furthermore, you also know that alkenes and alkynes undergo reaction with hydrogen (H_2)—in which the latter *adds* to the site(s) of *unsaturation* in the presence of a catalyst to yield alkanes—whereas alkanes themselves, under the same conditions, cannot react with hydrogen, that is, they are already *saturated.*

In this chapter, all of the major functional groups in organic chemistry that you might reasonably be expected to encounter are represented. While you could choose to attempt to memorize them all now, their presentation here is not (necessarily) for that purpose but rather to provide you with a feeling for the diversity and richness of the banquet you have yet to enjoy and to introduce you to **organization by functional group**.

Again, while it is not anticipated that the information be committed to memory in its entirety, some spectroscopic detail will also be provided for your examination. Spectroscopy is a frequently used tool to distinguish the presence (or absence) of a particular functional group, and further, it provides clear indication that there is some rationale for deciding that the structures drawn are reasonable approximations of reality.

Second, with the exception of some electrocyclic processes at which you have already briefly looked (Chapter 4) and variants of which will be considered again later (see, e.g., Chapter 6), almost all organic reactions occur in a solvent (i.e., the organic reagent undergoing reaction, **the substrate**, and the organic or inorganic reagent with which it is to react, **the reactant**,* are codissolved in another material, **the solvent**). Thus, the solvent is the medium that permits the reaction to occur, only serving (ideally) to bring reactants together. Generally, in organic chemistry, the solvent is not water (although mixtures with water as a component might be suitable). However, the **solvating properties** (i.e., the properties that led us to choose it as a solvent for a particular solute or mixture of solutes) of the organic solvent are also largely imparted by functional groups. Thus, occasionally, solvents are chosen, which may also be reactants. Alternatively, if the functional group (or groups) present in the solvent is sufficiently different in reactivity from those of the reactants themselves then, except for bringing the reactants together, the role of the solvent is more passive. So, it is that the discussion of functional groups, each of which can serve as a locus for reactivity in substrates, is also a key to solvent properties and the interactions between functional groups in solute(s) and solvent(s) must be considered.

Then, we will consider acids and bases. Hopefully, it will become clear that, if definitions are made broadly, and in the absence of water to level acid strengths, almost everything can be viewed as either an acid or a base, depending on that with which it is reacting! While this might suggest that the entire concept of acids and bases is of limited utility, a new viewpoint, that is, that many reactions can be recognized as acid–base processes if the appropriate criteria are applied, is introduced. Eventually, we will find it is not unusual to work with acids (and bases) whose ion-

---

*The substrate is also, clearly, a reactant. The attempt to distinguish between them may frequently appear arbitrary. For example, in the hydrogenation of cyclohexene over a suistable catalyst with gaseous hydrogen, cyclohexane results (Chapter 6). Both cyclohexene and hydrogen are reacting and they are both reactants. In the peculiar market or limiting reagent-related sense in general use in organic chemistry, cyclohexene is considered the substrate while hydrogen is the reactant.

ization constants differ by as much as 10^{60} in a series of organic reactions undertaken on the same substrate. And it will become clear that despite treatment with highly acidic and basic materials, the framework of many organic molecules survives! Therefore, the discussion of acids (and their conjugate bases) will lead to how solvents may interact with acids to enhance or diminish the apparent acidity of the latter, and what is meant by acidity in systems that cannot involve "free" protons, the hydronium ion (H_3O^+) or, indeed, even monomolecular unionized acids!

It is becoming clear that cost (both in time and money) of carrying out what were once rather sophisticated calculations, the results of which might be hoped to accurately describe the structure (and some properties) of molecules, has diminished markedly in only the last few years. There is no sign that the rate at which new innovations occur will slow. Therefore, it is certain that organic chemists will need to appreciate the use of some form of machine-type calculations to interrelate structural and energetic characteristics of molecules. Although calculations from the *first principles* (called ab initio), which are based on quantum mechanics (Chapter 1), can now be performed for simple molecules on desktop workstations, it is more common for organic chemists to utilize programs that contain some experimental parameters. When these *semiempirical methods* are employed to predict relative energies of similar structures (e.g., isomers) and other features (e.g., frequencies of vibrations), different approximations to solve the Schrödinger Equation in Appendix 1 for these larger molecules are used. The approximation(s) used depend on the computational equipment available as well as the level of accuracy desired and the data available for the approximation. Finally, it has become common to utilize an approximation method that largely ignores the electronic structure of the system and concentrates instead on approximations of interactions among nuclei. These approximations are grouped under the general heading of *force fields* and require that different atoms have predefined atom types (depending, among other things on the hybridization given by a defined geometry), sets of Equations (based on classical physics), and sets of parameters that link the predefined atom types to experimental data for a subset of the systems of interest. These *empirical methods* then allow the prediction of the properties of molecules, which either have not been or cannot be measured themselves. Among the more common and widely used techniques for such interrelation is that involving the tool of molecular mechanics (MM), largely developed through the efforts of Allinger (Allinger, N. L., professor, University of Georgia) and his coworkers. Although it can be argued that there are better tools, the development of MM has matured enough so that some discussion of it is warranted and this Chapter will close with an introduction to MM.

B. GENERAL CHARACTERISTICS OF FUNCTIONAL GROUP PLACEMENT

A significant international effort, resulting in the set of rules collectively known as those of the International Union of Pure and Applied Chemistry (IUPAC, Chapter 3) has already been put forth to systematize the names of organic compounds. These rules will be followed here, but trivial (or common) names will also be used because they remain current. In addition to formal nomenclature, it is often convenient to distinguish or signify a particular carbon (or carbons) generically because of, for example, similarities in behavior. There are several ways to do this.

For example, if an sp^3-hybridized carbon is bound to **one** other carbon, we will call it a **primary** carbon and the hydrogens (or any other attached atom) on that carbon will also be called **primary** (abbreviated 1°). Similarly, a **secondary** carbon will be one bound to **two** other carbons and the hydrogens (or other substituents) attached to the **secondary** carbon will also be **secondary** (abbreviated 2°) while a carbon bound to **three** other carbons is **tertiary** and the hydrogen (or other substituent) attached to it will also be **tertiary** (3°). Finally, a carbon bound to **four** other carbons is **quaternary** (4°). Using the compound 2,2,4-trimethylpentane as a template, these designations are shown.

Additionally, the carbons nearby the one being specifically considered can also be identified generically. For example, again considering 2,2,4-trimethylpentane, **assume that the quaternary carbon is specified.** Then, using **the Greek alphabet**, that is, α, β, and γ, one may say that "carbon one is α to the quaternary carbon" or "the one methylene carbon (i.e., $-CH_2-$) is α to the quaternary carbon while the tertiary (or 3°) carbon is β."

The designations used here are α, β, and γ in reference to the quaternary carbon. It should be clear that had a referent other than the quaternary carbon been chosen, the designations α, β, and γ would refer to those carbons bearing the specified relationship to the new referent.

Finally, in this vein, these informal systems that use Greek letters for relative positions and descriptions defining local attachments such as primary (1°), secondary (2°), tertiary (3°), and quaternary (4°) are independent of structure type, functional group, and/or other nomenclature rules and are quite widespread and useful. They will be frequently encountered.

C. THE FUNCTIONAL GROUPS AND THEIR NAMES

I. Hydrocarbons

It will be recalled from Chapter 3 (which can be profitably reviewed at this time) that compounds composed only of hydrogen and carbon are called **hydrocarbons**.

a. Alkanes. **Alkanes** are frequently referred to as **saturated compounds** because they have only carbon–carbon and carbon–hydrogen single bonds. They are also called **aliphatic** compounds (Greek [*aleiphos*, **fat**]) and **paraffins** (Latin [*parum*, **little** + *affinis*, **affinity**]).

When the chains of carbons bearing their appended hydrogens are unbranched (i.e., they are **straight-chain** compounds), the alkanes are occasionally referred to as **normal**; they and their **branched-chain** counterparts are **isomers** (Greek [*isos* + *meros*; **made of the same parts**]) of each other. Thus, in the older literature, **butane** (C_4H_{10}) is often called ***n*-butane** and its isomer, **2-methylpropane** (C_4H_{10}), is referred to as **isobutane**. In the same vein, **pentane** (C_5H_{12}) is called ***n*-pentane** and its isomer **2-methylbutane** (C_5H_{12}) is **isopentane**. The third isomer of C_5H_{12}, **2,2-dimethylpropane**, found some years after the first two, received the name **neopentane**. However, the system using such "trivial" names, demanding, as it does, a new nonsystematic prefix for each new isomer discovered, is clearly wanting and thus the IUPAC (Chapter 3) system has gained currency. However, the older nomenclature remains in use too.

It will be recalled (Chapter 3) that, although **acyclic** alkanes correspond to C_nH_{2n+2} where n is any integer ($n \geq 1$), the presence of rings ($n \geq 3$) (**cyclic** alkanes) will require that the number of hydrogens be reduced by two for each ring present. Thus, for example, while **hexane** (and its isomers) will correspond to C_6H_{14}, cyclohexane must correspond to C_6H_{12}, and bicyclo[2.2.0]hexane to C_6H_{10}.

Problem 5.1. (a) Write the structures for *n*-hexane and as many of its isomers as you can. (b) Draw the structure of cyclohexane, methylcyclopentane, the isomers of dimethylcyclobutane, and isomers of trimethylcyclopropane. Note that they all correspond to C_6H_{12}. (c) Draw the structure of bicyclo[2.2.0]hexane. Are there any other bicyclic isomers corresponding to C_6H_{10}?

Alkanes are characterized spectroscopically by strong infrared (IR) absorption in the C–H stretching region between 3950 and 2850 cm^{-1}, lack of ultraviolet (UV) absorption above 200 nm, ^{1}H nuclear magnetic resonance (NMR) signals below $\delta \approx 2.0$ ppm downfield from tetramethylsilane (TMS) = 0.00, and ^{13}C NMR signals below about $\delta \approx 50$ ppm downfield from TMS = 0.00. Their mass spectra usually involve sequential loss of methylene units ($-CH_2-$, $m/z = 14$).

b. Alkenes. **Alkenes**, it will be recalled (Chapter 3), are **unsaturated** and, if lacking rings and containing only one double bond, correspond to C_nH_{2n} where n is any integer ≥ 2. Their nomenclature follows that of the **alkanes** save for, again, the frequent use of trivial names such as **propylene** (for **propene**) and **isobutylene** (for **2-methylpropene**).

Generically, alkenes are also referred to as **olefins**. The word olefin comes from the early experimental observation that ethene (C_2H_4, bp −107.3°C), even now commonly called ethylene, but then called olefiant gas, reacted with chlorine (Cl_2) to yield an oily compound, 1,2-dichloroethane ($C_2H_4Cl_2$). Thus, it (and by implication all alkenes) was an "oil former" (Latin [*oleum*, **oil** + *facio*, **make**]).

Again, while the subject can be profitably reviewed (Chapter 3), you should recall that more than one double bond can be present in alkenes and thus **dienes**, **trienes**, and so on, are found. The double bonds can be **cumulative** (leading to **allenes**, such as the "symmetrical" allene, **2,3-pentadiene**, or the "unsymmetrical" allene, **3-methyl-1,2-butadiene**), **conjugated** (as in **1,3-pentadiene**), or **isolated** (as in **1,4-pentadiene**).

It is also important to remember that **configurational** isomers are common and that these (Z)- and (E)-isomers have important geometric constraints, including such obvious features as the inability to introduce a strain-free (E) double bond into a ring with fewer than eight carbons. As was the case for alkanes, trivial names also intrude for substituted alkenes. Thus, the **ethenyl group** ($CH_2=CH-$) is commonly called **vinyl** and the **2-propenyl group** ($CH_2=CH-CH_2-$) is called **allyl**.

Problem 5.2. Draw the structures for (a) 2,3-pentadiene; (b) 3-methyl-1,2-butadiene; (c) 1,3-pentadiene; (d) 1,4-pentadiene; (e) vinylcyclohexane [ethenylcyclohexane]; and (f) allylcyclopropane [(2-propenyl)cyclopropane].

Compounds containing isolated carbon–carbon double bonds resemble alkanes in the most commonly used portion of the UV spectrum (200–400 nm) because they have no absorption at wavelengths above 200 nm. However, if a compound contains more than one double bond, and if the double bonds are conjugated, UV absorption above 200 nm is observed and extended conjugation results in absorption at even longer wavelengths. Additionally, the frequency (wavelength) at which the absorption occurs in such conjugated dienes is a function of the relative arrangements of the double bonds and substituents that might be present. A set of empirical rules for such compounds was early worked out by Woodward* and, given the descriptions of the systems below, the rules can be briefly described as follows: (a) the **base value** for a *heteroannular diene* is 214 nm; (b) the **base value** for a *homoannular diene* is 253 nm; (c) a **double bond** *extending conjugation adds* 30 nm to the base value; (d) an **alkyl substituent** (or ring member) adds 5 nm to the base value; (e) if one of the **conjugated double bonds** is *exocyclic*, that, too, *adds* 5 nm to the base value. The following examples serve to define the terms above.

All positions marked (*) are **alkyl substituents** on the conjugated system.

*Woodward, R. B. (*vide supra*) first published these rules in 1941.

Problem 5.3. Based on the values given in the text for Woodward's rules, calculate the expected absorption maxima for the examples shown above. How do the values you have calculated compare with those provided?

In the IR, C–H stretching for protons attached to the double bond is found between 3100 and $3000\,cm^{-1}$, and unless symmetry prevents it (in which case the absorption will be Raman active), C=C absorption occurs between 1600 and $1620\,cm^{-1}$. Again, conjugation moves the absorption attributed to the C=C to longer wavelength (lower frequency).

With regard to NMR, protons directly attached to the carbons of the C=C system generally come into resonance between $\delta \approx 4.9$ and $\delta \approx 5.2\,ppm$ downfield from TMS = 0.00. It is argued that this is because the circulation of electrons in the π-cloud (cf. **aldehydes**) results in the protons in the plane of the double bond lying in a **deshielding** zone. In concert with this argument, in cyclic arrays where protons are forced to lie above (or below) the π-cloud, those protons are, correspondingly, shielded. In the ^{13}C NMR spectrum, the olefinic carbons, similarly deshielded, can generally be found to come into resonance between $\delta \approx 110$ and $\delta \approx 135\,ppm$.

In the mass spectrum, it is not uncommon to find that the C=C unit is difficult to cleave, and fragments containing the $C=C^{+}$ are seen.

c. Alkynes. The functional group characterizing **alkynes** is the carbon–carbon triple bond and, in the absence of rings and other unsaturation, alkynes correspond to C_nH_{2n-2} where $n \geq 2$. The first and simplest member of the group is **ethyne**, $HC{\equiv}CH$, which also enjoys the trivial name **acetylene**. Based on the trivial name, other members of the class are also generally referred to as **acetylenes**. Thus, the names **propyne** ($CH_3C{\equiv}CH$) and **methylacetylene** are synonymous, as are 2-butyne and dimethylacetylene ($CH_3C{\equiv}CCH_3$). As substituents, only **ethynyl** ($HC{\equiv}C-$) and **propynyl** (trivially named propargyl, $CH_3C{\equiv}C-$) are common.

Alkynes frequently absorb weakly in the UV near 255 nm but, if conjugated, the absorption is more intense and shifted to longer wavelength. Since there is a second band normally found below 200 nm in unconjugated compounds, two (or more) bands are frequently seen when the triple bond is placed in conjugation. In the IR, the stretching frequency for hydrogens attached to the carbon–carbon triple bond is usually between 3320 and $3310\,cm^{-1}$ and a clean, sharp band (of modest intensity) in this region is quite characteristic for this rare group. Similarly, the carbon–carbon triple bond, unless symmetrically substituted, has its stretching frequency in the narrow range of $2250–2100\,cm^{-1}$ and weak to modest absorption here is very characteristic of alkynes. Protons attached to carbon bearing the triple bond in alkynes generally come into resonance in the region between $\delta \approx 2.5$ and $\delta \approx 2.7\,ppm$. However, it is important to keep in mind that (1) many other kinds of protons come into resonance in this region and (2) the number of compounds containing a terminal alkynyl function is very small so, unless there is a good reason to suspect the presence of the $H-C{\equiv}C-$ function, the ^1H information is of limited utility. There are many more compounds that contain internal triple bonds and absorption between $\delta \approx 65$ and $\delta \approx 80\,ppm$ in the ^{13}C NMR is diagnostically useful. Interestingly, it is also of value to note that in the ^{13}C NMR, these peaks for internal alkynyl carbon atoms are usually much diminished over others because these carbon atoms **bear no protons**. Therefore, normal (<1 s) delays between pulse sequences in Fourier transform (FT) spectrometers **are frequently too short to allow complete relaxation**, the signal received by the detector is correspondingly diminished, and the resulting peak is small. Increasing the delay between pulses frequently results in increased peak intensity.

Cycloalkynes, given sufficient ring size (Chapter 3), have properties similar to their acyclic counterparts.

d. Arenes. The special character of **arenes** or **aromatic compounds** such as **benzene** (C_6H_6), where the cyclic array results in (1) enhanced stability (measured both in heats of hydrogenation and combustion) as compared with what might be anticipated for less highly conjugated analogues, (2) fundamentally different kinds of reactions, and (3) the demonstrable (by NMR) presence of a ring current, are characteristic of **arenes** in general.

Substituted arenes are common and trivial names abound. Regrettably, replacement of trivial names for aromatic compounds by more systematic ones lags far behind nonaromatic systems. Thus, $C_6H_5CH_3$ is variously called **methylbenzene** or **toluene**, while its dimethyl isomers (Figure 5.1) are referred to as **xylenes**.

Arenes are readily characterized by ^{13}C NMR spectroscopy because the aromatic carbons typically come into resonance between $\delta \approx 125$ and $\delta \approx 150$ ppm, further downfield (as a consequence of the ring current) than all but a few very much more deshielded carbon nuclei. Protons attached to aromatic rings also suffer deshielding and thus are generally found between $\delta \approx 7.0$ and $\delta \approx 8.0$ ppm downfield from TMS in the ^{1}H NMR.

In addition to the expected stretching seen between 3100 and 3000 cm^{-1}, the IR spectra of arenes are characterized by weak-to-moderate intensity absorbance in the 1600–1450 cm^{-1} region for the aromatic ring and, depending on the pattern of substitution, characteristic absorbances in the region 840–680 cm^{-1}. In the UV, simple aromatic compounds not only absorb strongly below 200 nm but they also have two more $\pi \to \pi^*$ bands, one around 200 nm and the other near 255 nm. The longer wavelength absorptions are weak in benzene but are more intense in substituted benzenes. Furthermore, as substituents are added, their presence may alter the basic light-absorbing unit (the **chromophore**) itself. If the added substituents have π electrons that can alter the energy of the molecular orbitals (i.e., the highest occupied molecular orbital [HOMO]–lowest unoccupied molecular orbital [LUMO] distance [Chapter 4]), shifts to longer wavelength can result and such shifts can carry the absorbance into the visible (VIS). Thus, absorbance at about 400 nm (the blue region

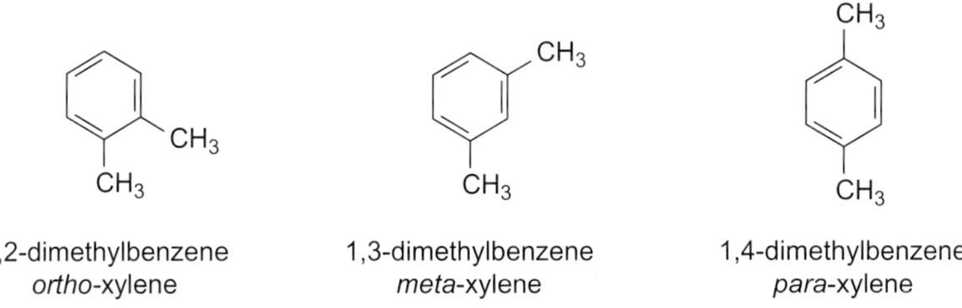

Figure 5.1. A representation of the dimethylbenzene (xylene) isomers.

of the visible spectrum) results in the material appearing yellow, while absorbance near 700 nm (the red region of the visible) results in the material appearing green. Much of the dye industry hinges upon attaching the appropriately substituted aromatic compounds to fabrics and such compounds are even found in inks and foodstuffs as coloring matter.

Finally, it is worthwhile noting that the aromatic ring itself reluctantly undergoes fragmentation in the mass spectrometer.

II. Alkyl and Aryl Halides

In principle, a **halogen** (F, Cl, Br, and I) can replace any **hydrogen** in a hydrocarbon. The **organic halides** thus produced constitute one of the largest groups of organic compounds, their number increased because different halogens may be attached to the same carbon!

Replacement (or **substitution***) of one hydrogen in methane by, for example, chlorine produces **chloromethane** (CH_3Cl, methyl chloride), replacement of two hydrogens, **dichloromethane** (CH_2Cl_2, methylene chloride), replacement of three hydrogens, **trichloromethane** ($CHCl_3$, chloroform), and replacement of all four hydrogens, **tetrachloromethane** (CCl_4, carbon tetrachloride).

Similar nomenclature obtains for **fluoro-**, **bromo-**, and **iodo-**alkanes.

The names of the alkyl halides follow the IUPAC system (Chapter 3 and Appendix II) with the halide being given the lowest number consistent with its position. If there is more than one kind of halogen present, that is, a mixed halide, the halogens are listed in alphabetical order, *which also takes precedent over their position*. Finally, as indicated above, if there is more than one halogen of the same kind (i.e., two bromines, three fluorines), the prefixes **di-** (for two), **tri-** (for three), **tetra-** (for four), and so on, are affixed, *but they are not used when considering the alphabetical ordering*.

In the event there is another **substituent** present, such as an alkyl group, the halogen is considered *equivalent to the alkyl group* and receives the number consistent with its position and alphabetical order and is thus also cited as a prefix (i.e., the halogens, as a group, are regarded as substituents).

In alkyl halides, the carbon bearing the halogen is nearly tetrahedral and is thus nearly sp^3 hybridized. As shown in Figure 5.2, the bond from the halogen atom to the carbon is considered to be made up by overlap of the nearly sp^3 orbital from carbon and an orbital from the halogen, which is mostly p (with some small amount of s character mixed in); the bond distance between the carbon and halogen atom is a function of the particular halogen. Based on this bond being a covalent bond with some polar character (see Polar Covalent Bonds, Chapter 1), it would be expected (and is found) that the smaller, more electronegative fluorine has a shorter bond to carbon than do the other halogens and, indeed, as the size of the halogen increases, the halogen–carbon bond lengthens.

Some typical alkyl, alkenyl, and aryl halides, and their names follow.

---

*The dynamics of the actual **process** called **substitution** will be discussed in (Chapter 6).

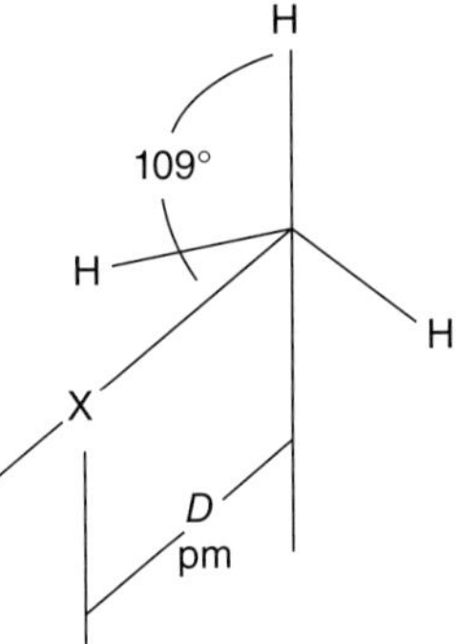

Figure 5.2. A generalized picture of a monohalomethane. The distance D (pm), as well as the X–C–H angle (shown as 109°), varies with the specific halogen, X, as indicated in the text (where X = F, Cl, Br, or I).

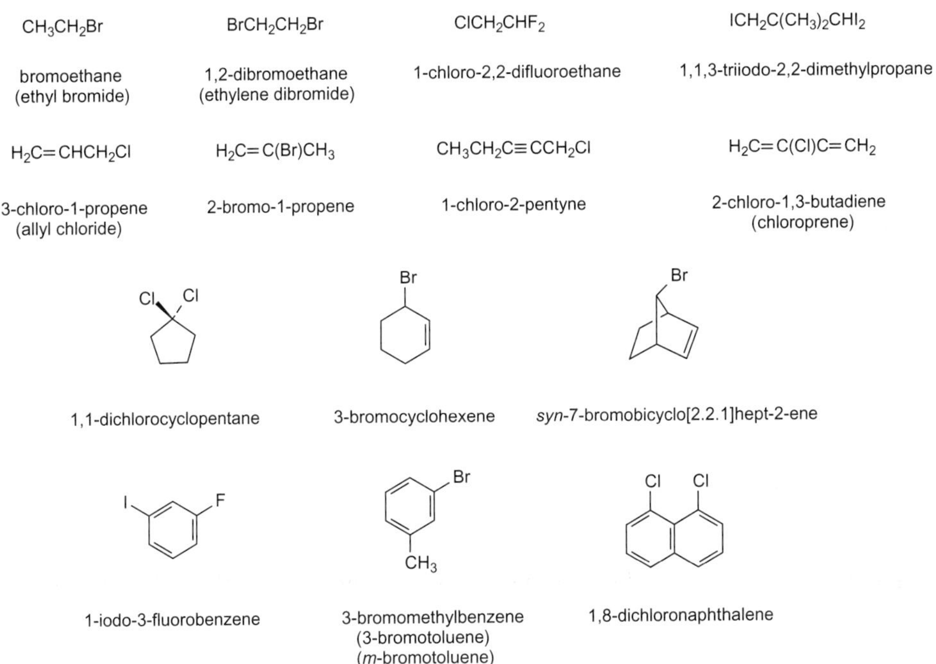

CH_3CH_2Br

bromoethane
(ethyl bromide)

$BrCH_2CH_2Br$

1,2-dibromoethane
(ethylene dibromide)

$ClCH_2CHF_2$

1-chloro-2,2-difluoroethane

$ICH_2C(CH_3)_2CHI_2$

1,1,3-triiodo-2,2-dimethylpropane

$H_2C{=}CHCH_2Cl$

3-chloro-1-propene
(allyl chloride)

$H_2C{=}C(Br)CH_3$

2-bromo-1-propene

$CH_3CH_2C{\equiv}CCH_2Cl$

1-chloro-2-pentyne

$H_2C{=}C(Cl)C{=}CH_2$

2-chloro-1,3-butadiene
(chloroprene)

1,1-dichlorocyclopentane

3-bromocyclohexene

syn-7-bromobicyclo[2.2.1]hept-2-ene

1-iodo-3-fluorobenzene

3-bromomethylbenzene
(3-bromotoluene)
(*m*-bromotoluene)

1,8-dichloronaphthalene

As shown in Table 5.1, the length of the carbon–halogen bond increases with increasing size of the halogen while, correspondingly, the bond strength of the carbon–halogen bond decreases. Because the C–X bond is a polar covalent bond (with the carbon atom at the positive end and the highly electronegative halogen at the negative end [i.e., $C^{\delta+}$–$X^{\delta-}$]), all of the haloalkanes have significant dipole moments.

The IR spectra of alkyl halides also reflect the nature of the halogen. Thus, again using the simple methyl halides as examples, in addition to the anticipated C–H stretching bands around 3300 cm^{-1}, absorbances for C–X (X = halogen) are readily observed as intense bands at relatively long wavelength. As expected from the

TABLE 5.1. Some Structural and Physical Properties of the Methyl Halides

Compound	C–X Bond Length (D pm)	H–C–H Bond Angle (°)	Dipole Moment (μ Debye)	Bond Strength kcal mol^{-1} (kJ mol^{-1})	Boiling Point (°C) (760 torr)
Fluoromethane (CH_3F) (methyl fluoride)	139	110.62	1.82	109 (456)	–78.4
Chloromethane (CH_3Cl) (methyl chloride)	178	110.35	1.94	84 (351)	–24.2
Bromomethane (CH_3Br) (methyl bromide)	193	111.16	1.79	70 (293)	3.6
Iodomethane (CH_3I) (methyl iodide)	214	111.39	1.64	56 (234)	42.4

For purposes of comparison, the boiling point of methane (X = H) is –161.7°C.

TABLE 5.2. Some Selected Spectroscopic Data for the Methyl Halides

Compound	C–X Stretching Frequency ν (cm^{-1})	^{1}H–C Chemical Shift δ (ppm)	^{13}C Chemical Shift δ (ppm)	$^1J_{CH}$ Coupling Constant (Hz)
Fluoromethane (CH_3F) (methyl fluoride)	1475	4.26	74.1	149
Chloromethane (CH_3Cl) (methyl chloride)	1355	3.05	23.8	150
Bromomethane (CH_3Br) (methyl bromide)	1305	2.68	9.3	152
Iodomethane (CH_3I) (methyl iodide)	1255	2.16	–20.5	151

inverse relationship between the masses of the nuclei (Chapter 2) and the IR stretching frequency, **as the atomic weight of the halogen increases, the frequency at which the absorption occurs decreases**. Table 5.2 presents these data.

The ^{1}H and ^{13}C NMR data for the same set of compounds are shown in Table 5.2. As might be anticipated as the substituent becomes **more** electronegative (electronegativity increases in the order I, Br, Cl, F), (i) protons attached to the carbon bearing the halogen are progressively *deshielded* and thus come into resonance further *downfield*, (ii) the carbon bearing the halogen itself is more *deshielded* and thus comes into resonance further *downfield*, and (iii) since there is very little change in hybridization at carbon (as evidenced by the similarities in H–C–H bond angle regardless of substituent), the H–C coupling constant is similar throughout the series.

The deshielding of protons and carbon by substituents has given rise to attempts to predict the chemical shift by application of empirical additivity rules. While these can be useful, they are, at best, only approximate. Thus, as shown in Table 5.3 for the chlorinated derivatives of methane, there is not a simple incremental value to

TABLE 5.3. ^{1}H and ^{13}C Chemical Shifts and J_{CH} Coupling Constants for Chlorinated Derivatives of Methane

Compound	^{1}H-C Chemical Shift δ (ppm)	^{13}C Chemical Shift δ (ppm)	$^1J_{CH}$ Coupling Constant (Hz)
Methane (CH$_4$)	0.23	−2.3	125
Chloromethane (CH$_3$Cl) (methyl chloride)	3.05	23.8	151
Dichloromethane (CH$_2$Cl$_2$) (methylene chloride)	5.33	54.0	178
Trichloromethane (CHCl$_3$) (chloroform)	7.24	77.5	209
Tetrachloromethane (CCl$_4$) (carbon tetrachloride)	—	96.5	—

either the chemical shift (^{1}H or ^{13}C) or to the C–H coupling constant (J_{CH}) for each substitution of a chlorine for a proton.

Fluoromethane (CH$_3$F) and, indeed, all fluorocarbons contain ^{19}F (its abundance is 100%). This nucleus (^{19}F) also has spin 1/2 and, at a field strength of 7.05 T ([Chapter 2] at which protons [^{1}H] come into resonance at 300 MHz and carbon [^{13}C] comes into to resonance at 75.5 MHz); ^{19}F comes into resonance at 282.3 MHz (i.e., within 10% of that of ^{1}H). Most NMR spectrometers that are capable of recording proton (^{1}H) magnetic resonance spectra are also capable of recording fluorine (^{19}F) magnetic resonance spectra. However, since TMS [(CH$_3$)$_4$Si] does not contain fluorine, it cannot be used as an internal (or external) reference. Tetrafluoromethane (CF$_4$) is occasionally used. Thus, a more complete description of the NMR spectra for fluoromethane (CH$_3$F) would also include the ^{19}F chemical shift (and a statement regarding the reference to which it made) as well as the *coupling constant*s between fluorine and hydrogen (J_{HF}) and between fluorine and carbon (J_{CF}).* Figure 5.3 shows the structure of fluoromethane and lists the coupling constants and chemical shifts for the various nuclei.

In the mass spectrum, the parent ion of halogenated compounds is generally quite pronounced as electron impact results in loss of one of the outershell electrons borne by the halogen. Thus, the parent ion (CH$_3$F$^+$) would be expected from fluoromethane. However, the presence of isotopes of both chlorine (i.e., ^{35}Cl and ^{37}Cl) and bromine (i.e., ^{79}Br and ^{81}Br), whose natural relative abundance is well known (^{35}Cl/^{37}Cl $\approx$ 3:1 and ^{79}Br/^{81}Br $\approx$ 1:1) is of major importance in detecting and confirming the presence of these halogens in organic compounds.

---

*It is also interesting to note that both ^{35}Cl and ^{37}Cl possess the property we have called "spin" (both have spin 3/2) and that at 7.05 T they come into resonance at 29.40 and 24.47 MHz, respectively. However, because of a nuclear quadrapole moment, these nuclei relax very fast and, in solution, give rise to very broad lines in all but highly symmetrical compounds. The lines are so broad as to make such spectra essentially useless. Similarly, ^{79}Br and ^{81}Br both of which also have spin 3/2 and ^{127}I whose spin is 5/2, and which at 7.05 T come into resonance at 75.19, 81.05, and 60.04 MHz, respectively, do not produce useful solution spectra.

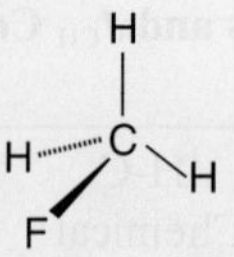

Figure 5.3. Fluoromethane coupling constants and chemical shifts. $J_{HF} = 46.4\,Hz$, $J_{CF} = 157.5\,Hz$, $J_{CH} = 149.1\,Hz$, $\delta_H = 4.26\,ppm$ (TMS = 0.00), $\delta_C = 74.1\,ppm$ (TMS = 0.00), $\delta_F = 210\,ppm$ (CF$_4$ = 0.00).

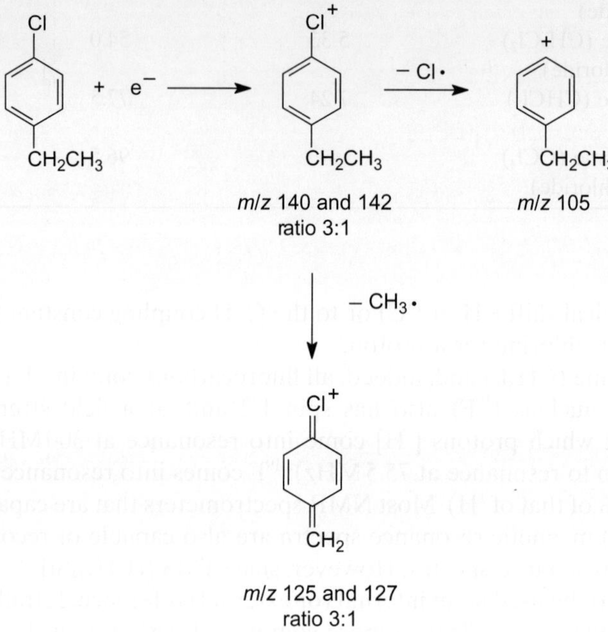

Scheme 5.1. A portion of the fragmentation pathway (showing possible ions) seen in the mass spectrum (Figure 5.4) of 4-chloroethylbenzene.

For example, as shown in Scheme 5.1, electron impact onto 4-chloroethylbenzene produces ions that contain chlorine. There will be, ignoring the presence of small quantities of ^{13}C as a result of its natural abundance of 1.1%, **two** parent ion peaks, one at m/z 140 and the other at m/z 142 (in the ratio 3:1, respectively). The first of these is due to p-CH$_3$CH$_2$C$_6$H$_4$-^{35}Cl and the second to p-CH$_3$CH$_2$C$_6$H$_4$-^{37}Cl. If the chlorine is lost (shown here as a sequential process), the ethylphenyl cation (m/z 105) results. An alternative fragmentation of the parent, involving initial loss of the methyl group (from the ethyl side chain) yields the p-chlorobenzyl cation (or its equivalent). This results in a species that still has chlorine and thus again, *two* ions at m/z 125 and 127 are seen. The mass spectrum of 4-chloroethylbenzene is shown as Figure 5.4.

The ^{1}H and decoupled ^{13}C NMR spectra of 4-chloroethylbenzene are presented as Figures 5.5 and 5.6, respectively.

Problem 5.4. For 4-bromoethylbenzene, sketch (a) the mass spectrum you would expect, (b) the ^{1}H NMR spectrum, and (c) the ^{13}C NMR spectrum.

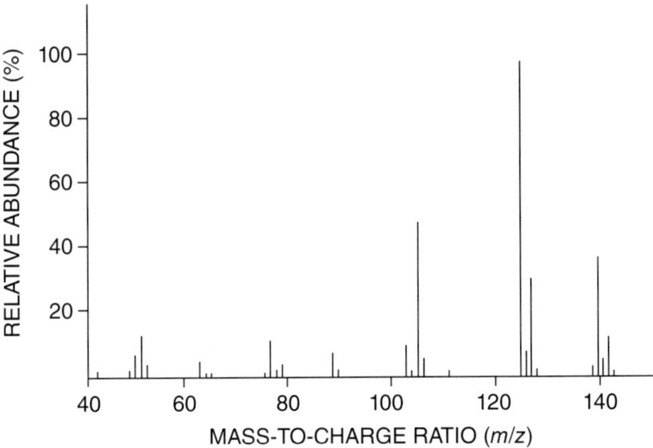

Figure 5.4. A representation of the mass spectrum of 4-chloroethylbenzene.

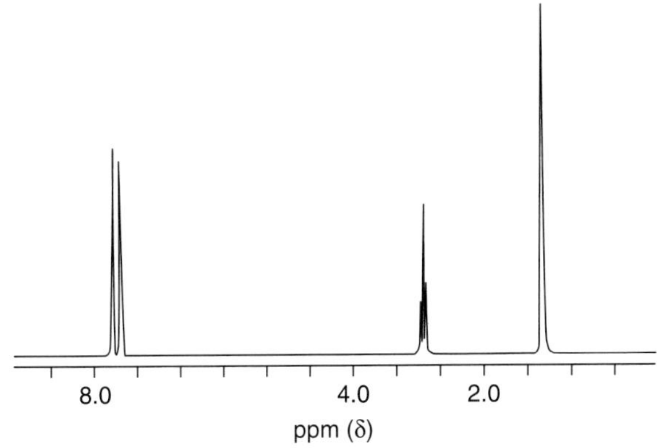

Figure 5.5. The ^{1}H NMR spectrum of 4-chloroethylbenzene in ^{2}HCCl$_3$ solvent (δ TMS = 0.00 ppm).

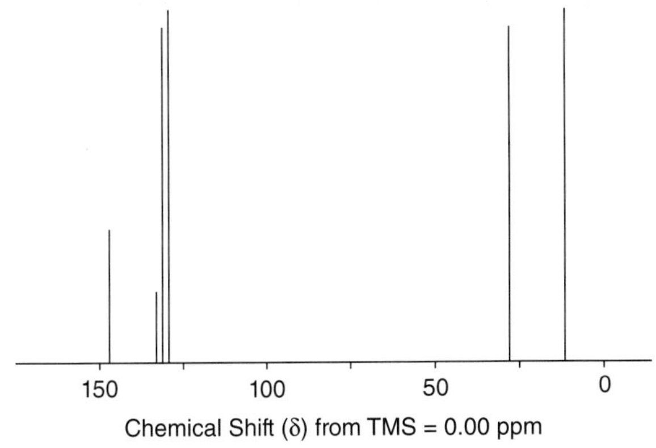

Figure 5.6. The ^{13}C noise decoupled NMR spectrum of 4-chloroethylbenzene in ^{2}HCCl$_3$ solvent.

III. Alcohols and Phenols

In alkanes, the replacement of a **hydrogen (–H)** bonded to carbon by a **hydroxyl (–OH) group** results in the generation of that class of compounds called **alcohols**. Alternatively, one may consider alcohols as arising from the replacement of one **hydrogen (–H)** bonded to the oxygen of **water** (H–O–H) by an **alkyl** group. The former emphasizes the organic nature (and nomenclature) of alcohols, while the latter emphasizes the nature of the functional group. That is, low-molecular-weight alcohols closely resemble water in their reactivity and the properties of the **hydroxyl group (–OH)** are pretty much the same whether the other ligand is carbon or hydrogen.*

Alcohols are conveniently named by replacing the final "e" in the name of the corresponding alkane with the suffix "ol." Thus, the relationship between methane (CH$_4$ or H$_3$C–H) and methanol (H$_3$C–OH or CH$_3$OH) is clear. The names of the other alcohols follow the IUPAC system with the –OH being given the lowest number consistent with its position. Thus, the pentanol isomers 1-pentanol, 2-pentanol, and 3-pentanol differ only by the position of the hydroxyl group along the chain.

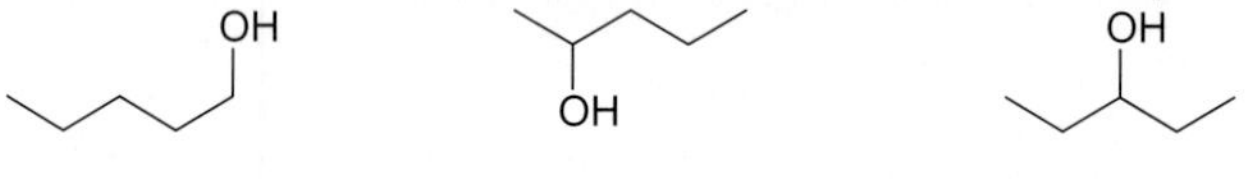

1-pentanol   2-pentanol   3-pentanol

In the event there is another substituent present, for example, a methyl group, the nomenclature follows the usual pattern save that the location of the hydroxyl (–OH) takes precedent. Thus, the series of methyl-1-pentanols is named as follows.

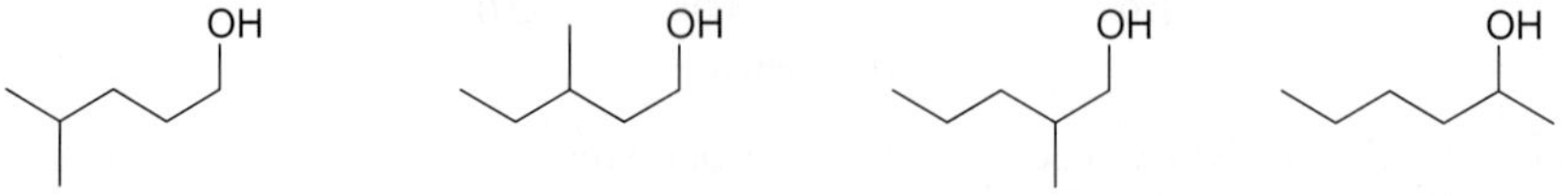

4-methyl-1-pentanol   3-methyl-1-pentanol   2-methyl-1-pentanol   2-hexanol

Additionally, there is a second widespread nomenclature system in use. In this, common for simple alcohols, the alkyl group is named as a **substituent** and followed by the word **alcohol**.

A third system depends on considering alcohols as derivatives of methanol. However, the word "methanol" is replaced by "carbinol."

Finally, the –OH group may be considered as a substituent and its position fixed with a number and the radical name "hydroxy." The use of all four systems is illus-

---

*While the subject will be considered in some detail later, it should be clear to you that simply reaching into an organic compound, grabbing it by a hydrogen, wrenching the latter free, and replacing that hydrogen by a hydroxyl is an unlikely path for the creation of alcohols. Indeed, although not as obvious, the direct oxidation of an alkane to an alcohol is a rather rare reaction in the organic chemistry laboratory and the oxidation of alkanes usually cannot be stopped before completion (i.e., as pointed out in Chapters 1 and 3, the products are CO$_2$ and water).

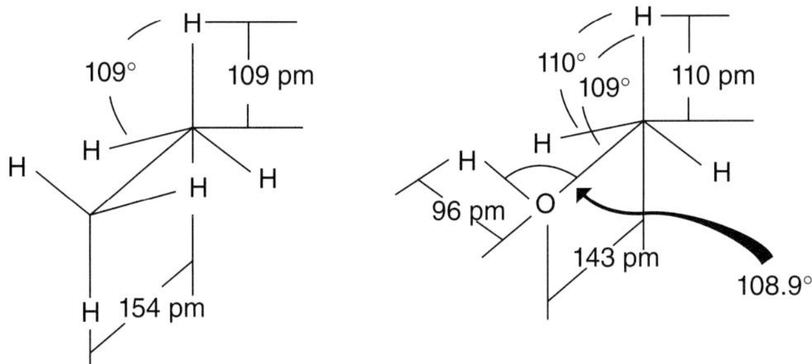

Figure 5.7. A comparison of the structures of ethane and methanol.

trated below for the isomers of C_4H_9OH. Generally, you will find that the first, the IUPAC system, is preferred:

(1) $CH_3CH_2CH_2CH_2OH$:   **1-butanol**, *n*-butyl alcohol, *n*-propylcarbinol, 1-hydroxybutane;

(2) $CH_3CH_2CH(OH)CH_3$: **2-butanol**, *sec*-butyl alcohol,* ethylmethylcarbinol, 2-hydroxybutane;

(3) $(CH_3)_2CHCH_2OH$: **2-methyl-1-propanol**, isobutyl alcohol, isopropylcarbinol, 1-hydroxy-2-methylpropane; and

(4) $(CH_3)_3COH$: **2-methyl-2-propanol**, *tert*-butyl alcohol, trimethylcarbinol, 2-hydroxy-2-methylpropane.

Considering **methanol** (or **hydroxymethane**) as a typical example of primary alcohols (which, in many ways it is not), structural details can be examined. As shown in Figure 5.7, the carbon–oxygen single bond (143 pm [1.43 Å]) (as is typical) is shorter than a similar carbon–carbon single bond in, for example, ethane (154 pm [1.54 Å]), the carbon–hydrogen bonds are about the same in both (110 pm vs. 109 pm, respectively) and the oxygen–hydrogen bond about the same as in water (96 pm vs. 97 pm, respectively).

phenol   an "enol"

The special cases of the hydroxyl bonded to doubly bonded carbon, that is, replacing hydrogen attached to the double bond in arenes and alkenes, leads to **phenols** and vinyl alcohols or **enols** (i.e., phenyl + ol and ene + ol). With few exceptions, enols cannot be isolated. They are usually present, in equilibrium, with the

---

*The prefixes *sec* and *tert* are abbreviations of *secondary* (2°) and *tertiary* (3°). Occasionally, the latter is further abbreviated as "*t-*."

corresponding *carbonyl* (or C=O containing, *vide infra*) isomer (Equation 5.1). Interestingly, compounds with *two* –OH groups attached to the same carbon (i.e., **geminal**) are also usually found to be unstable with respect to the corresponding *carbonyl* isomer (Equation 5.2), a notable exception being **methanal** (formaldehyde) (Equation 5.3). If the –OH (or hydroxyl) groups are attached to *different* carbon atoms, the resulting glycols (**diols**, **triols**, etc., to **polyols** for two, three, and many –OH groups, respectively) are commonly stable and, as we shall see, are frequently naturally occurring (e.g., as **carbohydrates**, Chapter 11).

$$(5.1)$$

2-propenol   propanone

enol   ketone
(ca. 10^{-6}% at 25°C)   (ca. 99.999999% at 25°C)

$$(5.2)$$

2,2-propandiol   propanone

hydrate   ketone
(ca. 0.1% at 25°C)   (ca. 99.9% at 25°C)

$$(5.3)$$

methanediol   methanal (formaldehyde)

hydrate   aldehyde
(ca. 99.9% at 25°C)   (ca. 0.1% at 25°C)

The IR spectra of alcohols are characterized by a very intense O–H stretching absorption band (or bands) in the region between about 3600 and 3200 cm^{-1} and a second band (or group of bands) in the region between about 1200 and 1000 cm^{-1} due to the C–O single-bond stretch. If water is present in the sample, its absorption can be mistaken for that of the alcohol (at the higher frequency). Furthermore, the relatively weak cohesive forces (generally considered to be of the order of about 20 kJ mol^{-1} [5 kcal mol^{-1}]) between the positively polarized proton of a hydroxyl group and the more electronegative, and thus negatively polarized, oxygen of another hydroxyl group routinely results in **hydrogen bond** formation (Figure 5.8), except in very dilute solution (aprotic solvent) or in the gas phase. Thus, the sharp, *free* –OH stretch expected at about 3700 cm^{-1} is frequently absent unless precautions are taken. The IR spectrum of 1-propanol (CH$_3$CH$_2$CH$_2$OH) taken as an undiluted pure liquid film ("neat") is shown in Figure 5.9.

In alcohols without π-bonds, there cannot be an $n \rightarrow \pi^*$ transition in the UV or VIS, and thus, with an –OH as the only chromophore, where only an $n \rightarrow \sigma^*$ transition is allowed, there is no absorption above 200 nm. With phenols and enols,

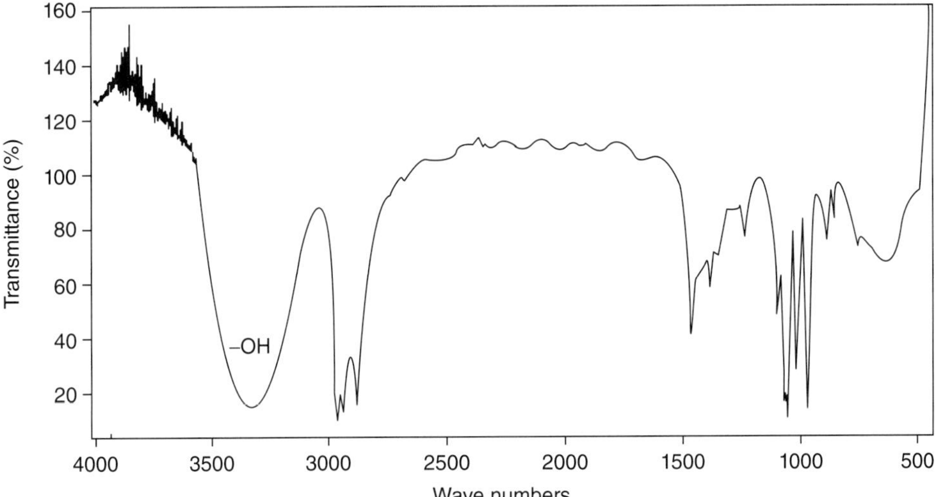

Figure 5.8. A representation of a portion of a three-dimensional intermolecular hydrogen bonding network between protons of hydroxyl groups from one molecule and oxygens from others.

Figure 5.9. The IR spectrum (transmittance) of 1-propanol ($CH_3CH_2CH_2OH$, neat).

however, longer wavelength absorption (even into the VIS) is common. Indeed, phenols are commonly found contributing to coloring matter in plants.

In the ^1H NMR spectrum of both alcohols and phenols, the chemical shift of protons attached to oxygen (because of hydrogen bonding) is frequently found to be solvent, temperature, and concentration dependent, and it is difficult to predict a δ value. However, those protons attached to carbon bearing oxygen, that is, H–C–O–, generally come into resonance between $\delta \approx 3.3$ and $\delta \approx 3.7\,ppm$, while the aliphatic carbon of the alcohol group itself can generally be found between about $\delta \approx 50$ and $70\,ppm$ downfield from TMS in the ^{13}C spectrum.*

---

*Although one of the isotopes of oxygen, that is, ^{17}O, also possesses the property called spin (^{17}O has a spin of $-5/2$) and comes into resonance at $40.7\,MHz$ (in a field of $7.05\,T$, where ^1H is found at $300\,MHz$ and ^{13}C at $75.5\,MHz$), its natural abundance of 3.7×10^{-2} precludes its use in all but the most specialized circumstances. The ^{17}O chemical shifts of alcohols (as for ^1H and ^{13}C) are also a function of substitution pattern with the oxygen of primary alcohols coming into resonance at higher field than secondary and secondary upfield of tertiary relative to water.

Figure 5.10 presents the ^{1}H NMR and ^{13}C NMR spectra of 1-propanol (decoupled and undecoupled) (CH$_3$CH$_2$CH$_2$OH) taken at 300 MHz for ^{1}H and 75.5 MHz for ^{13}C. The sample was dissolved in ^{2}H$_2$O for this particular experiment and some of the hydroxyl protons (^{1}H) may have been exchanged for the deuterium (^{2}H) in the solvent. The ^{17}O spectrum is not shown (*vide infra* for the ^{17}O spectra of **aldehydes** and **carboxylic acids**) only because the resonance assigned to the ^{17}O of the hydroxyl oxygen of 1-propanol was not well separated from the signal assigned to the ^{17}O in the ^{2}H$_2$^{17}O in the solvent. Typically, the ^{17}O of primary alcohols is reported to be found to be about 10 ± 5 ppm downfield of the ^{17}O signal in ^{1}H$_2$^{17}O reference.

Except for methanol (CH$_3$OH) itself, the mass spectra of alcohols are characterized by fragmentation resulting from the loss of water (i.e., M$^+$-18), and depending on the exact circumstances, the signal derived from the parent ion may actually be smaller than the (M$^+$-18) fragment. Additionally, carbon–carbon bond cleavage between the carbon bearing the –OH and an α-carbon, leading to loss of a fragment containing both the carbon and the oxygen of the alcohol and a second fragment, in which they are lacking, is common. Cleavage of the carbon–carbon bond is the major pathway if water cannot be lost (lack of protons on the α-carbons).

The low-molecular-weight alcohols, such as **methanol** (CH$_3$OH), **ethanol** (CH$_3$CH$_2$OH), **1-propanol** (CH$_3$CH$_2$CH$_2$OH), and **2-propanol** [CH$_3$CH(OH)CH$_3$], are completely miscible with water in all proportions and have significantly higher boiling points than alkanes of comparable molecular weight. Thus, at 1 atm, methanol (bp 65°C) boils more than 150°C higher than ethane (bp −88.7°C), while ethanol (bp 78°C) boils about 120°C higher than propane.*

Problem 5.5. (a) How is hydrogen bonding related to the diminishing difference between the temperatures at which alkanes and alkanols (alcohols) boil as their respective molecular weights increase? (b) Boiling points at 1 atm for such high-molecular-weight materials are difficult to find in the literature of organic chemistry. How do you account for this observation? (c) Draw the structure for "racemic 2-nonanol." Do you anticipate the separated enantiomers would have similar or different boiling points?

Methanol (CH$_3$OH, also called wood alcohol) was, at one time, produced by destructive distillation of wood.[†] Since the late 1930s, when its production was considered important for the efforts surrounding WWII, it has been made by the catalytic hydrogenation of carbon monoxide.[‡] Large quantities continue to be made by the catalytic process since it is a valuable solvent for many organic reactions and as material from which other organic compounds are prepared.

*This difference, as well as others that are largely related to hydrogen bonding, diminishes as the hydrocarbon part of the alcohol becomes larger. Thus, for example, at 1 atm, *n*-decane (bp 174°C) boils within 40°C of 1-nonanol (bp 213.5°C) and within 25°C of racemic 2-nonanol (bp 198.2°C).

[†]This process involves heating the dried wood above its decomposition temperature in an atmosphere where combustion is inhibited because oxygen is largely excluded. The liquid evolved is mostly water, which contains only a small percentage (1–4%) of methanol and traces of other materials. The residue is **charcoal** and the gases produced can be used to fuel the process!

[‡]The catalyst for the reaction of CO with 2H$_2$ to produce CH$_3$OH is a mixture of oxides of zinc (ZnO) and chromium (Cr$_2$O$_3$). The reaction is carried out under pressure (200–300 atm) and at high temperature (300–400°C).

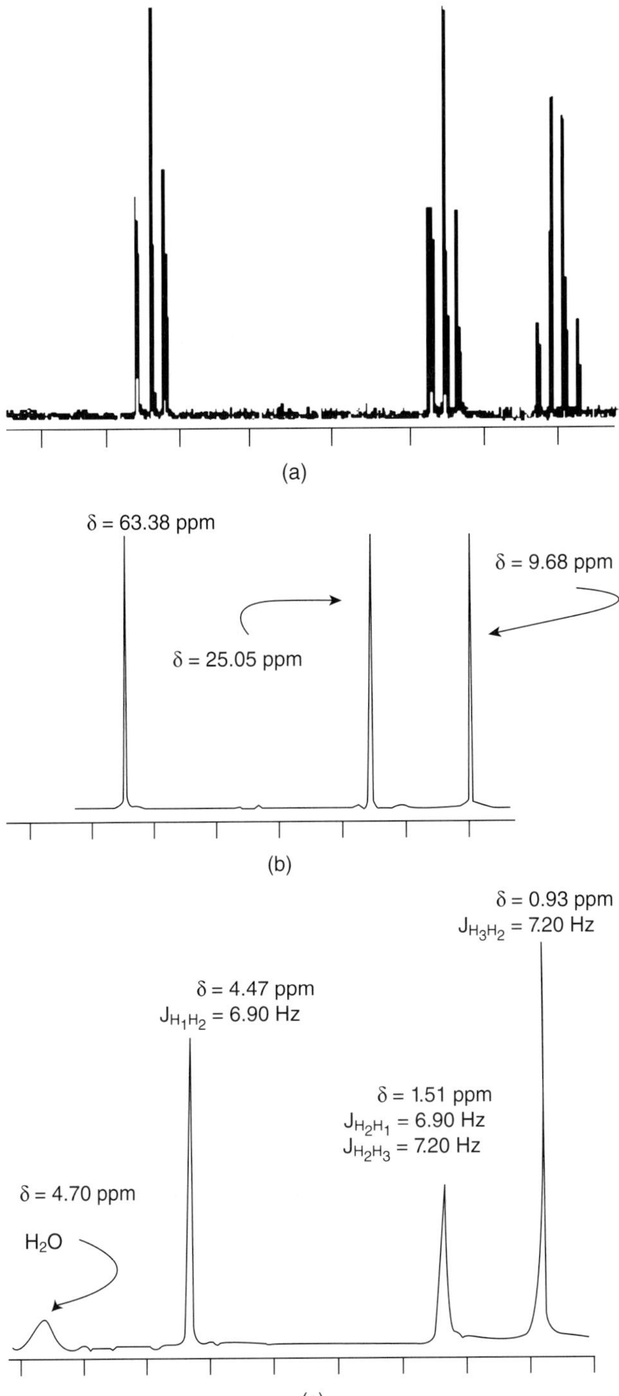

Figure 5.10. The (a) undecoupled (coupled) ^{13}C NMR and (b) proton decoupled ^{13}C NMR spectrum of 1-propanol ($CH_3CH_2CH_2OH$) in ^2H_2O solvent. See text for details. (c) The ^{1}H NMR spectrum of 1-propanol ($CH_3CH_2CH_2OH$) in ^2H_2O solvent. See text for details.

Ethanol (CH₃CH₂OH), or ethyl alcohol, is the "alcohol" to which reference is made as the intoxicating constituent of beverages. Commonly, when used as an intoxicant, it is obtained as a product of the fermentation of carbohydrates (Chapter 11) found in grains (e.g., wheat, rye, rice, and corn) and fruits (e.g., grapes and apples). Ethanol is also produced by the *hydration of* (or, synonymously, *addition of water to*) ethene (CH₂=CH₂, Chapter 6). Depending on demand (economics), ethanol (CH₃CH₂OH) produced by fermentation is usually used in beverages, where other products that accompany the alcohol during distillation are said to enhance flavor. The purer material, when produced from ethene (CH₂=CH₂), lacks the flavoring material and is better suited for industrial purposes, for example, as a fuel. Careful purification of the material obtained by fermentation has allowed its use as a fuel too. When used as an automotive fuel, ethanol is often mixed with gasoline (i.e., gasoline + alcohol = gasohol). 1,2,3-Trihydroxypropane (**glycerol** [HOCH₂CHOHCH₂OH]) is one of the products obtained by the *hydrolysis of* (or, synonymously, *the reaction of water with*) plant and animal fats (Chapter 9) and is commonly used along with 1,2-dihydroxyethane (HOCH₂CH₂OH, **ethylene glycol**), obtained from the oxidation of ethene (CH₂=CH₂), in the plastics industry. Ethylene glycol (HOCH₂CH₂OH), because of its complete miscibility with water, high boiling point, and low melting point, is a major ingredient in commercial anti-freeze solutions. However, because the toxicity of ethylene glycol (HOCH₂CH₂OH) is high compared with, for example, 1,2,3-trihydroxypropane (**glycerol** [HOCH₂ CHOHCH₂OH]), its use is diminishing.

Hydroxybenzene (C₆H₅OH, **phenol**), and substituted hydroxybenzenes (**phenols**), much like the arenes to which they are related, enjoy the same enhanced stability (and other properties) of arenes. As a consequence, although both alcohols and phenols possess the hydroxyl (or **–OH**) functional group, phenols are frequently considered as a separate class of compounds.

hydroxybenzene
(phenol)   4-hydroxytoluene
(*para*-cresol)   1,4-dihydroxybenzene
(hydroquinone)   1,2-dihydroxybenzene
(catechol)

IV. Ethers

As discussed above, **alcohols** may be considered as arising from the replacement of one **hydrogen (–H)** bonded to the oxygen of **water** (H–O–H) by an **alkyl** group, that is, methanol (CH₃OH) may be thought of as resulting from removal of one hydrogen of water and replacement of that hydrogen by a CH₃ group.

The relatively inert group of materials called **ethers** corresponds to that class of compounds that may be thought of as arising by replacement of **both** hydrogens of water by **alkyl (or aryl)** groups. Thus, **methyl ether** (CH₃OCH₃) (notice that the word **ether** is **separate** from the name of the alkyl group) may be thought of as resulting from removal of **two** hydrogens of water (HOH) and their replacement by **two** CH₃ groups. **Methyl ether** (CH₃OCH₃) is occasionally called **dimethyl ether** to emphasize that the two alkyl groups are identical.

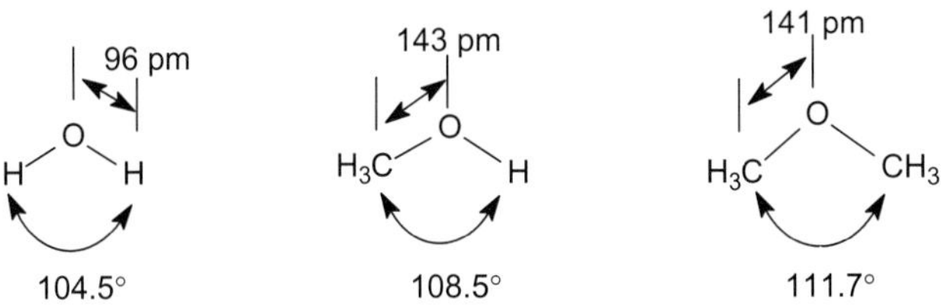

96 pm	143 pm	141 pm
water	methanol (methyl alcohol)	methyl ether (dimethyl ether)

(104.5° — water; 108.5° — methanol (methyl alcohol); 111.7° — methyl ether (dimethyl ether))

The names of the simple **ethers** are created by naming the two groups attached to oxygen (in alphabetical order) and appending the word "ether." If the alkyl groups are identical, they are named only once.

$CH_3CH_2OCH_2CH_3$

ethyl ether
(diethyl ether)
("ether")

$CH_3OC(CH_3)_3$

tert-butyl methyl ether
(*t*-butyl methyl ether)
(2-methoxy-2-methylpropane)

$C_6H_5OCH_3$

methyl phenyl ether
(methoxybenzene)
(anisole)

In the event one of the groups attached to oxygen is "large" and the other is "small," the ether can be named as a derivative of the more complicated attached group. This is done by considering the less complicated group as a derivative of the corresponding alcohol (and thus an **alkoxy** group).*

$CH_3OCH_2CH_2OCH_3$

1,2-dimethoxyethane

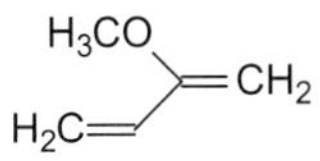

2-methoxy-1,3-butadiene

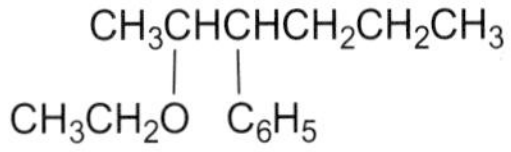

2-ethoxy-3-phenylhexane

The oxygen, since it must be attached to two carbon atoms (in order that it be an ether), can be incorporated into a ring otherwise made up of carbon (or other) atoms. In this way, members of oxygen containing **heterocycles**, corresponding to the similar **carbocycles** encountered earlier (Chapter 4) can be made up. Regrettably, no *simple*[†] system of nomenclature has been developed to deal with these ethers and trivial names are commonly used.

---

*If water, HO–H, loses a proton, the species remaining can be called the **hydroxy** (HO–) group and you will recall that **methanol** (CH_3OH) might be named **hydroxymethane**. By analogy, removal of the proton from the oxygen of (a) **methanol** (CH_3O–H) leaves the **methoxy** (CH_3O–) group; (b) **ethanol** (CH_3CH_2O–H) leaves the **ethoxy** (CH_3CH_2O–) group; (c) **phenol** (C_6H_5O-H) leaves the **phenoxy** (C_6H_5O–) group, and so on, for **alkoxy** groups and thus ethers can be named as **alkoxyalkanes**.

[†]This statement is slightly disingenuous in that it has been suggested that the simple addition of the suffix *oxa* (with, if necessary, a numerical identifier for position) to the carbocyclic name would be sufficient. However, the nomenclature system (commonly called the Hantzsch–Widman system) has gained more currency. Here, the prefix *oxa* is also used, but the ring size is denoted by specific "stem" names such as **ir** (for a three-membered ring), **et** (for a four-membered ring), **ol** (for a five-membered ring), and so on. Finally, the degree of unsaturation is specified by a suffix such as **ane** (for a saturated ring) and **ene** (for a monounsaturated ring). The entire Problem of nomenclature is further complicated by the fact that most of the **heterocyclic** compounds with which one commonly deals also have their own trivial names (A. R. Hantzsch [1857–1935], professor [at 28!] at Zurich and then Wuerzberg and Leipzig [1903]).

Some of the more common cyclic ethers and the trivial names that are most widely used for them, along with more systematic names, are shown below. Generally, the common names will be used.

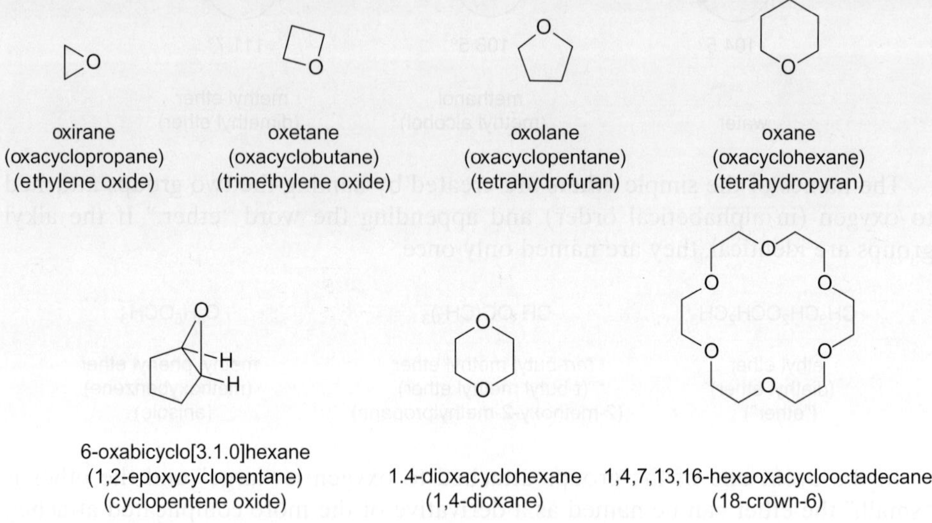

The spectroscopic properties of ethers might be anticipated on the basis of what you already know about the spectroscopic properties of alcohols and alkanes. Thus, the C–O stretching vibrations in the region 1200–1000 cm^{-1} (as seen in alcohols) would be anticipated in the IR. Again, as in alcohols, without unsaturation, there cannot be an $n \rightarrow \pi^*$ transition in the UV or VIS and thus, with the C–O–C as the chromophore, only an $n \rightarrow \sigma^*$ transition is allowed, and there is no absorption above 200 nm. However, variously substituted methylated phenols (i.e., methoxy aryl ethers, ArOCH$_3$) are common plant constituents, absorb strongly in the UV and VIS, and frequently contribute to coloring matter in plants.

The protons attached to the carbon atoms bearing the oxygen in ethers suffer (as was the case in alcohols) significant deshielding by the electrons on the oxygen and thus such protons are commonly found to come into resonance at about $\delta = 3.4$ ppm relative to TMS, while the carbons flanking the oxygen, again as in alcohols, are generally found to come into resonance between about $\delta = 65$ and 75 ppm downfield from TMS in the 13C NMR spectrum. Interestingly, however, as shown in Figure 5.11, the effect of hybridization changes are readily seen in some of the more strained ethers (i.e., oxiranes) and the effect of the deshielding by the electrons on oxygen made apparent when a carbon is flanked by more than one oxygen.

The ^{17}O spectra are similar to those found for alcohols with, for example, the oxygen of diethyl ether (CH$_3$CH$_2$^{17}OCH$_2$CH$_3$) coming into resonance at about 12 ppm downfield of the water signal.

Again, the mass spectra of ethers are very similar to that of alcohols save that loss of water cannot occur from the former. However, to the extent that α-cleavage can occur, it is a commonly found pathway in ethers. Thus, ions are rationalized accordingly (e.g., Scheme 5.2) where initial electron impact causes loss of an

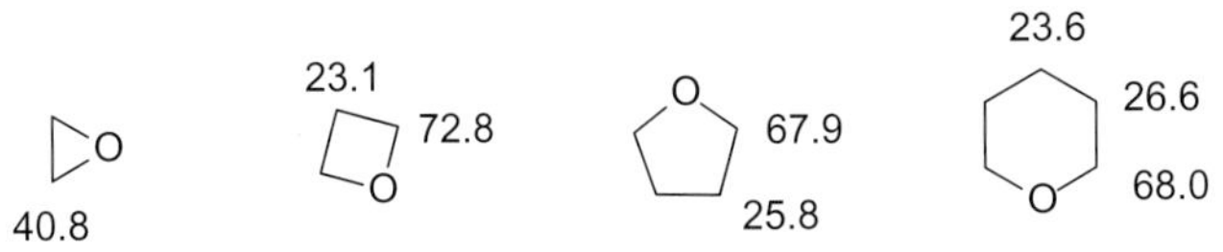

Figure 5.11. The ^{13}C chemical shift values (δ) for some cyclic ethers relative to TMS ($\delta = 0.00$) in ^{2}HCCl$_3$ solution (after Levy, G. C.; Lichter, R. L.; Nelson, G. L. *Carbon-13 Nuclear Magnetic Resonance Spectroscopy*, 2nd edition, Wiley-Interscience, New York, 1980, p. 59).

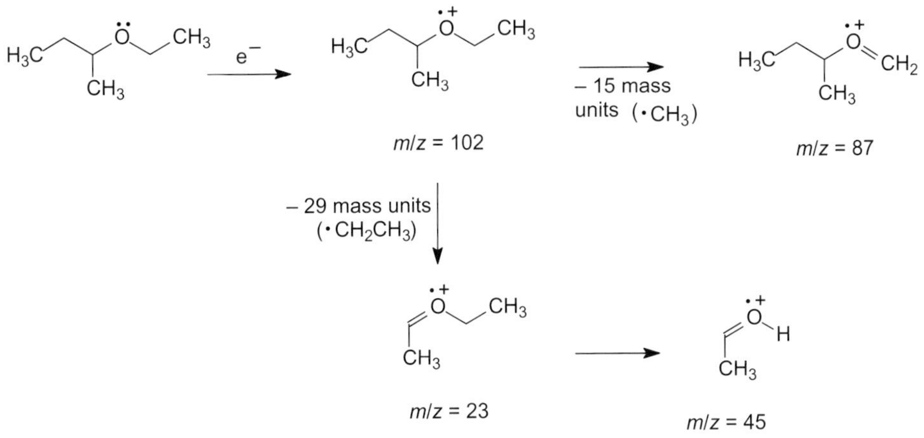

m/z = 102

m/z = 87

− 29 mass units
(·CH$_2$CH$_3$)

m/z = 23

m/z = 45

Scheme 5.2. Some suggestions of ions to account for the masses observed during electron impact mass spectroscopy of 2-ethoxybutane (2-butyl ethyl ether).

electron to give the parent cation (*m/z* 102) and the two different α-cleavage pathways follow, one with loss of a methyl group (102 − 15 = 87) and the other with a loss of an ethyl group (102 − 29 = 73). Interestingly, the latter apparently undergoes the subsequent loss of ethylene (i.e., 28 mass units) while retaining a hydrogen, to provide a fragment at *m/z* 45 (73 − 28 = 45).

Although the electrons on the oxygen of ethers are available for hydrogen bonding, the lack of protons that can be donated from the ethers themselves result in compounds whose melting and boiling points are very similar to those of hydrocarbons of about the same molecular weights. However, water is more soluble in ethyl ether [(CH$_3$CH$_2$)$_2$O, the "ether" which has served as a general anesthetic] (about 7% by weight at room temperature) than in pentane (0.04% by weight), and it is argued that this is because of hydrogen bonding.

The ability to dissolve both polar materials and nonpolar materials while failing to **react with the solutes thus dissolved** has made ethers such as **ethyl ether** [(CH$_3$CH$_2$)$_2$O] and **oxolane (oxacyclopentane, tetrahydrofuran**, *vide supra*), as well as 1,4-dioxacyclohexane (1,4-dioxane), *vide supra*, valuable to organic chemists as media (e.g., solvents [*vide infra*]) in which reactions between two (or more) organic compounds can be carried out. That these materials are also relatively low boiling has made them particularly attractive despite their high flammability and a tendency to form explosive peroxides.*

V. Thiols, Thioethers, Disulfides, and Their Oxides

The **sulfur** analogues of alcohols, ethers, and peroxides are known and are called **mercaptans**. Furthermore, just as **alcohols** may be considered as **derived** by the **substitution** of a **hydrogen (–H)** bonded to carbon by a **hydroxyl (–OH) group** and **ethers** to **substitution** of **both** hydrogens of water by **alkyl (or aryl)** groups, so the replacement of a **hydrogen** bonded to carbon by **hydrosulfide (–SH)** produces a **thiol** (the sulfur analogue of an alcohol) and **thioethers** may be considered to arise by substitution of both hydrogens of hydrogen sulfide (H_2S) by alkyl groups. Thus, for example, **methyl thioether** (also called **methyl sulfide** or **dimethyl sulfide**) (CH_3–S–CH_3) may be thought of as resulting from removal of the **two** hydrogens of hydrogen sulfide and their replacement by **two** CH_3 groups. Then, just as ethers may be named as **alkoxyalkanes**, emphasizing the nature of the **alkoxy** group as a substituent, when the **thiol group** (i.e., R–S–H without the hydrogen) is named as a substituent, the prefix **mercapto-** is used. Finally, in this vein, the (much more stable) sulfur analogue of the **peroxide** (R–O–O–R) functionality, the **disulfide** linkage (R–S–S–R) (common to proteins, Chapter 12), is also widely recognized (i.e., dimethyl disulfide, H_3C–S–S–CH_3).

As shown above for water (H_2O), methanol (H_3COH), and dimethyl ether (H_3COCH_3), the corresponding sulfur-containing compounds (hydrogen sulfide [H_2S], methanethiol [H_3CSH], and dimethyl sulfide [H_3CSCH_3]) can also usefully be compared. Interestingly, the H–S–H, H–S–C, and C–S–C angles in the latter set are, respectively, considerably smaller than the corresponding H–O–H, H–O–C, and C–O–C angles, respectively, in the former. It is usually argued that this implies that there is considerable p character, largely from mixing in of $3p$ orbitals, in the bonds from sulfur to the respective ligands. This suggestion is partially supported by the rather weak, longer wavelength IR absorbance of thiols (near $2600\,cm^{-1}$) as com-

---

*Although noted for their relative lack of reactivity, one of the more common reactions of ethers, albeit undesirable, is the one that occurs with atmospheric oxygen. In this process (which will be discussed [Chapter 8]), the initial formation of a **hydroperoxide**, as shown below, is followed by generation of a **peroxide**. These materials are (explosively) unstable. Commercially available ethers are frequently contaminated by addition of small amounts of materials (**antioxidants**) to inhibit the formation of **hydroperoxides** and **peroxides**. An ether is of the form R–O–R, a hydroperoxide of the form R–O–O–H, and a peroxide of the form R–O–O–R.

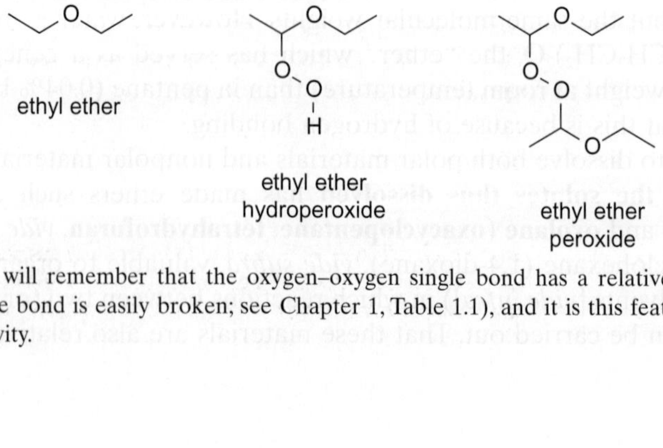

ethyl ether

ethyl ether
hydroperoxide

ethyl ether
peroxide

Hopefully, you will remember that the oxygen–oxygen single bond has a relatively low energy of homolysis (i.e., the bond is easily broken; see Chapter 1, Table 1.1), and it is this feature that accounts for its high reactivity.

pared with alcohols (alcohols, it will be recalled, usually have intense absorption between 3600 and 3200 cm^{-1}).

hydrogen sulfide	methanethiol (methyl mercaptan)	methyl thiomethyl ether (dimethyl sulfide)
133 pm / 92.2°	182 pm / 100.3°	182 pm / 98.9°

There are additional significant differences between sulfur compounds and their corresponding oxygen analogues. For example, while hydrogen sulfide has a higher molecular weight than water, both its boiling point (−60.3°C) and its melting point (−85.5°C) are significantly lower than those of water. The same differences, while diminished, are also found in thiols (ethanethiol, bp 36°C; mp −144.4°C) and alcohols (ethanol, bp 78.3°C; mp −117.3°C). It is usually argued that the differences in boiling point are as a result of diminished hydrogen bonding in the case of thiols. Credence is lent to this suggestion by the additional observations that the boiling and melting points of diethyl sulfide (92 and −102.5°C) are appropriately higher than those of diethyl ether (34.6 and −116.3°C), respectively.

Second, almost all of the low-molecular-weight sulfur compounds have objectionable (to humans) odors. Thus, in defense, the skunk sprays offenders with a mixture of thiols (and other compounds) containing the disulfide, (*E*)-2-buten-1-yl methyl disulfide, and the thiols 1-butane thiol, 3-methylbutane-1-thiol, and (*E*)-2-butene-1-thiol. Additionally, although the mixture of low-molecular-weight hydrocarbons used as "natural gas" is essentially odorless (to humans), contamination of that material with only a few parts per billion (ppb) of a mixture of low-molecular-weight (mostly C$_5$-isomers) thiols is sufficient to allow humans to smell a "gas leak."

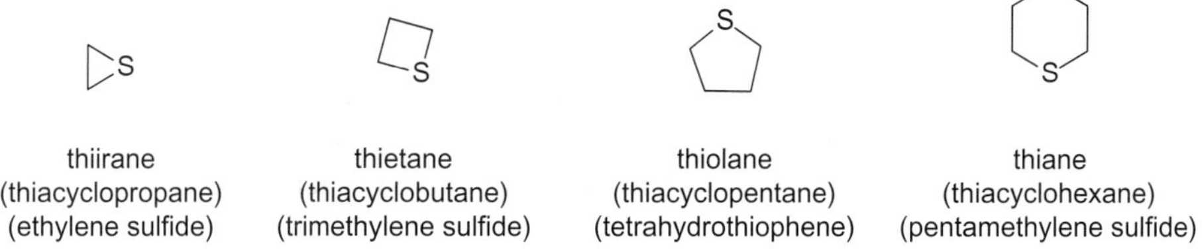

butanethiol	3-methylbutane-1-thiol	(*E*)-2-butene-1-thiol	(*E*)-2-buten-1-yl methyl disulfide

As was the case for the corresponding compounds containing oxygen (i.e., ethers), thioethers (sulfides) occur both in acyclic and cyclic forms. Thus, compounds directly analogous to the common cyclic ethers (*vide supra*) are known, and they are shown in the following illustrations. The trivial names that are most widely used for these thioethers (sulfides), along with more systematic names, are shown. Generally, the common names will be used.

thiirane (thiacyclopropane) (ethylene sulfide)	thietane (thiacyclobutane) (trimethylene sulfide)	thiolane (thiacyclopentane) (tetrahydrothiophene)	thiane (thiacyclohexane) (pentamethylene sulfide)

The ^{1}H NMR spectra of thiols again parallels that of alcohols save that the diminished hydrogen bonding in thiols leads to a very slow exchange rate. Thus, the proton on sulfur (i.e., the **sulfhydryl** proton, –S–H) is usually observed between about δ 1.2–1.6 ppm (TMS = 0.00) for aliphatic thiols (between δ 2.8 and 3.6 ppm for aromatic thiols) and the coupling between the proton on sulfur and that on the carbon bearing the sulfur (i.e., H–C–S–H) of ca. 8 Hz is also seen. Furthermore, as would be anticipated from the suggested electron configuration at sulfur, the protons attached to the carbon bearing the sulfur should be found further upfield than the similar protons in the oxygen analogue. Thus, while the protons on the carbon of alcohols and ethers (i.e., H–C–O) are generally found between δ 3.3 and 3.7 ppm (TMS = 0.00), the corresponding protons in thiols and sulfides (thioethers) are between δ 2.1 and 3.1 ppm.

Similar effects are observed in the ^{13}C NMR spectrum, although the changes are somewhat more dramatic. Thus, while the carbon of alcohols and ethers bearing the oxygen (i.e., C–O) typically comes into resonance somewhere between about δ 50 and 75 ppm (ethers being, generally, further downfield than alcohols), the corresponding carbon in thiols and thioethers or sulfides (i.e., C–S) is generally found between δ 20 and 35 ppm (thioethers being, generally, further downfield than thiols).

In addition to ^{1}H and ^{13}C NMR, ^{33}S (abundance about 1%) would be expected to possess spin and indeed, it has spin 3/2 and a quadrapole moment. Thus, unless the system is highly symmetrical (e.g., the sulfate anion, SO_4^{-2}), the ^{33}S NMR lines for materials in solution are expected to be very broad.

Mass spectrometrically, the contribution of 4.4% of ^{34}S results in a significant peak two mass units above the expected (for ^{32}S) parent (i.e., at $M^+ + 2$) for both thiols and thioethers (sulfides). Furthermore, in all of the daughter ions that contain sulfur, the same isotopic ratio will be expected. Again, as was the case for alcohols and ethers, cleavage of the C–C bond between the carbon bearing the sulfur and the adjacent carbon (i.e., α-cleavage) is a generally observed pattern with the formation of species such as $H_2C=S^+$-R. Additionally, sulfides frequently yield an ion that presumably bears a positive charge on sulfur, for example, R-S$^+$, by direct cleavage of a carbon–sulfur bond.

For example, the mass spectrum of ethyl sulfide (ethyl thioether) $[(CH_3CH_2)_2S]$ would be expected to have the parent ion M^+ at m/z 86 (and another parent ion, M^+, as a result of 4.4% ^{34}S at m/z 88) and daughter ions at m/z 75 (77) and m/z 61 (63) corresponding to ions, which might be written as $(CH_3CH_2–S=CH_2)^+$ and $(CH_3CH_2–S)^+$, respectively.

Finally, there are classes of sulfur compounds that have no oxygen analogues. These materials (mostly oxides) are considered to have formed by oxidation at sulfur (i.e., the sulfur is in a higher oxidation state) and include analogues of sulfuric $[HOS(O_2)OH$ or $H_2SO_4]$ and sulfurous $[HOS(O)OH$ or $H_2SO_3]$ acids as well as some sulfonium salts (R_3S^+) as shown in Table 5.4.

Sulfur-containing compounds similar to those of Table 5.4 are common industrial chemicals. Thus (*vide infra*), the solvent properties of dimethyl sulfoxide (in which many polar organic compounds dissolve and which, nonetheless, is completely miscible with water) have made it very valuable. Additionally, the acid properties (*vide infra*) of many sulfonic acids, which are very strong acids (i.e., they ionize

TABLE 5.4. Some Sulfur Compounds Which Have No Corresponding (Stable) Oxygen Analogues and Which Contain Multiply-Bonded Sulfur

Class	Example	Formula	Structure
Sulfoxide	Dimethyl sulfoxide	$(CH_3)_2SO$	
Sulfone	Dimethyl sulfone	$(CH_3)_2SO_2$	
Sulfinic acid	Methanesulfinic acid	CH_3SO_2H	
Sulfonic acid	Methanesulfonic acid	CH_3SO_3H	
Sulfonium salt	Trimethylsulfonium iodide	$(CH_3)_3S^+\ I^-$	

completely), coupled with their solubility in a large number of organic solvents, have made them suitable where acids that only dissolve in water are of limited utility.

VI. Amines, Hydrazines, and Other Nitrogenous Materials

The group of basic, nitrogen-containing organic compounds which might be considered as derived from the **substitution** of one or more **hydrogens (–H)** bonded to **ammonia (NH₃)** by a carbon ligand are, together, called **amines**. Among these compounds, those that have only one carbon to nitrogen bond (and two hydrogens still bonded to nitrogen) are called **primary amines. Secondary amines** have two carbon-to-nitrogen bonds and one hydrogen bonded to nitrogen, and **tertiary amines** have only carbon–nitrogen bonds, all of the hydrogens having been replaced. By analogy to inorganic ammonium salts, **quaternary ammonium salts** have four carbon-to-nitrogen bonds (with a positive charge on nitrogen).*

*It is worthwhile remembering that this is different from the situation in hydrocarbons and *alcohols*, where the designations **primary (1°), secondary (2°),** and **tertiary (3°)** refer to the number of carbons attached to the one under discussion or, for the latter, bearing the –OH group.

Amines are commonly named as *alkyl* derivatives. Thus, the series in which the hydrogens of ammonia (NH_3) are replaced by, for example, *methyl* groups would be called, respectively, methanamine (methylamine, CH_3NH_2),* *N*-methylmethanamine (dimethylamine, $(CH_3)_2NH$), and *N,N*-dimethylmethanamine (trimethylamine, $(CH_3)_3N$). The positively charged species corresponding to the ammonium cation $(NH_4^+)^\dagger$ would thus be the tetramethylammonium cation $[(CH_3)_4N^+]$.

In this "trivial" nomenclature, when the alkyl (or aryl) substituents on nitrogen are not all identical, the amine is named by using an alphabetical listing of the groups (e.g., butylethylmethylamine $[CH_3CH_2CH_2CH_2N(CH_3)(CH_2CH_3)]$). Alternatively, when there might be some ambiguity because not all of the alkyl groups are unbranched, the longest alkyl chain serves as the base for the nomenclature and the prefix *N* is used to specify the other alkyl groups attached to nitrogen. Thus, $CH_3CH_2CH_2CH_2N(CH_3)(CH_2CH_3)$, which was named as butylethylmethylamine (above), is more clearly defined as *N*-ethyl-*N*-methylbutanamine (*N*-ethyl-*N*-methylbutylamine), and a simple compound such as $CH_3CH_2CH_2CH_2N(CH_3)_2$ is reasonably called **N,N-dimethylbutanamine (N, N-dimethylbutylamine)**.

As already intimated above, a more systematic nomenclature (used by Chemical Abstracts, Appendix II) is closely related to what you have already learned about alkane nomenclature in that the longest chain is used as the parent and the final *e* in the name of the alkane is replaced by *amine*. Thus, $CH_3CH_2CH_2CH_2N(CH_3)_2$ is **N, N-dimethylbutanamine**, while CH_3NH_2 is **methanamine**, $(CH_3)_2NH$ is **N-methylmethanamine**, and so on.

A third (and final) alternative is to use the rules of the IUPAC, where amines are named as substitution derivatives of the longest chain hydrocarbon to which they are attached by using the *prefix* **amino-** for the $-NH_2$ group. In the IUPAC system then $CH_3CH_2CH_2CH_2N(CH_3)_2$ is simply **dimethylaminobutane**, while CH_3NH_2 is **aminomethane**, $(CH_3)_2NH$ is **methylaminomethane**, and so on. The added advantage of naming the amino group as a substituent then permits it to be easily located along a chain or on a ring, as shown below.

Generally, common names are used for the simplest compounds, and as most literature searching is currently done using the Chemical Abstracts Service (SciFinder, Appendix II), this system will be the nomenclature of choice.

Structural data (shown above for water $[H_2O]$, methanol $[H_3COH]$, dimethyl ether $[H_3COCH_3]$, hydrogen sulfide $[H_2S]$, methanethiol $[H_3CSH]$, and dimethyl sulfide $[H_3CSCH_3]$) are also available for ammonia (NH_3) (see also Chapter 1) and many alkanamines (alkylamines), such as methanamine (methylamine; CH_3NH_2) and *N,N*-dimethylmethanamine (trimethylamine; $(CH_3)_2NCH_3$ or $(CH_3)_3N$).

---

*It should be noted that *amine* is used as a suffix that has been added to the name of the alkyl group. That is, unlike the functional group name *alcohol*, which is a separate word, for example, **methyl alcohol**, or the functional group name, *ether*, as in **methyl ether**, the functional group name *amine*, is appended directly to that of the **alkyl** substituent, that is, **methanamine**.

$\dagger$As this is a positively charged species, it should be clear that a negatively charged counterion (a *gegenion*) must also be present. It is frequently simply written as "X^-" unless it is specifically known to be something else (e.g., halide, hydroxide).

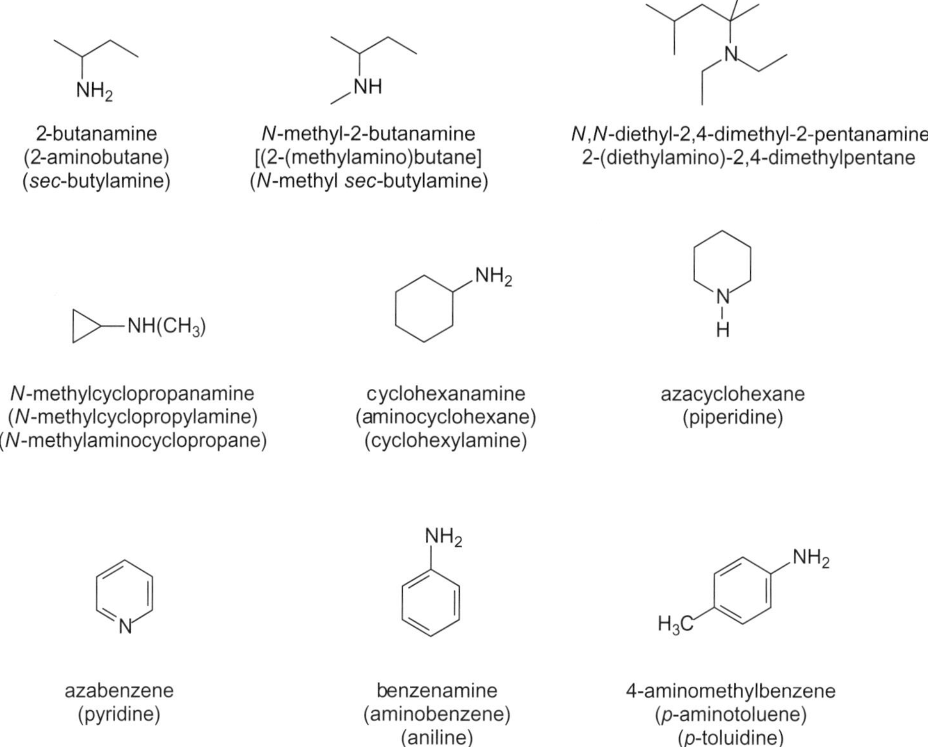

If a comparison is made in bond lengths between ostensibly* sp^3-hybridized carbon–carbon (154 pm), carbon–nitrogen (147 pm), carbon–oxygen (143 pm), and carbon–fluorine (139 pm) systems, it is possible to conclude that as the nuclear charge becomes more concentrated (or as the electronegativity increases), the covalent bond between carbon and that to which it is bonded becomes shorter.

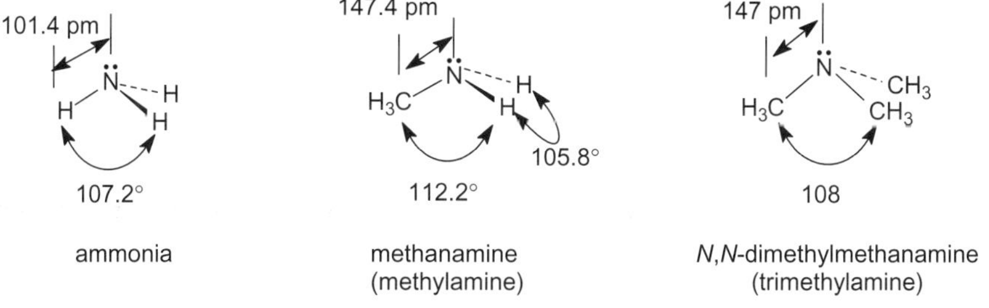

The reduced electronegativity of nitrogen as compared with oxygen makes the extent to which ammonia and amines in general hydrogen bond to themselves arguably less clear than is the case for water and alcohols. Thus, as shown below, in Table 5.5 where (as one measure of the extent of hydrogen bonding) the boiling points of

*Although the exact hybridization (Chapter 1) need not conform to this integral value, the angles observed are very close to those anticipated for such hybridization. Methane thiol is excluded intentionally.

TABLE 5.5. A Comparison of the Change in Boiling Points (°C), at Atmospheric Pressure, with Increasing Substitution of Hydrogen by Alkyl groups for Water, Hydrogen Sulfide, and Ammonia

Compound	Boiling Point (°C)	Compound	Boiling Point (°C)	Compound	Boiling Point (°C)
H_2O	100.0	H_2S	−60.3	H_3N	−33.3
CH_3OH	64.7	CH_3SH	7.6	CH_3NH_2	−6.5
$(CH_3)_2O$	−23.7	$(CH_3)_2S$	38.0	$(CH_3)_2NH$	7.4
CH_4	−161.5			$(CH_3)_3N$	3.5

alkyl derivatives of water (H_2O), hydrogen sulfide (H_2S), and ammonia (NH_3) are compared, the last two are seen as being much more like each other than they are like water. However, it is argued that since primary amines have boiling points intermediate between alkanes and alcohols of about the same molecular weight and since tertiary amines boil at lower temperatures than either primary or secondary amines of the same molecular weight, some hydrogen bonding is affirmed.

Problem 5.6. Based on your understanding of hydrogen bonding in water, alcohols, and primary amines, how do you account for the observation that tertiary amines fail to hydrogen bond to other tertiary amines?

Almost all of the low-molecular-weight amines have objectionable (to humans) odors. Many of the compounds smell "fishy" and, indeed, the odor of rotting fish has been attributed to such materials.* Some diamines produced by the action of bacteria on decaying organic materials enjoy names reflective of their "unpleasant" odors, for example, putrescine [$H_2N(CH_2)_4NH_2$] and cadaverine [$H_2N(CH2)_5NH2$].

As was the case for the corresponding compounds containing oxygen (i.e., ethers) and sulfur (i.e., thioethers or sulfides) and as shown briefly above, amines also occur both in acyclic and cyclic forms. Thus, compounds directly analogous to the common cyclic ethers and thioethers (*vide supra*) are known and they are shown below. The trivial names that are most widely used for these amines, along with more systematic names, are provided. Generally, the common names will be used.

aziridine
(azacyclopropane)
(ethylenimine)

azetidine
(azacyclobutane)
(trimethylenamine)

azolidine
(azacyclopentane)
(pyrrolidine)

perhydroazine
(azacyclohexane)
(piperidine)

The IR spectra of primary and secondary amines are characterized largely by the N–H stretching frequency although those bands (in the 3500–3300 cm^{-1} region) are

*As already mentioned and will be discussed more fully later in this chapter, amines are basic compounds. Bases can be converted to their salts with acids. Use of vinegar (dilute, flavored ethanoic acid [acetic acid; CH_3CO_2H]) and lemon juice (citric acid, *vide infra*) converts the odoriferous amines to their ammonium salts, which have much diminished vapor pressures and, arguably, different odors.

frequently weak for primary amines and very weak for secondary amines (unlike the generally strong –OH bands in the 3600–3200 cm^{-1} region). Fortunately, amines usually also have a medium to strong N–H bending band in the region 1650–1550 cm^{-1}, which can (occasionally) be used to bolster the evidence for the presence of the N–H group. The utility of this absorbance is moderated by the common occurrence of other absorbances in the same region.

The ^{1}H NMR spectra of amines parallels that of alcohols, ethers, thiols, and thioethers. However, since nitrogen is less electronegative than oxygen (but more electronegative than carbon), the chemical shifts of protons on carbons *directly attached* to nitrogen are further downfield from those on carbon bonded to other carbons, but upfield from those on carbon attached to oxygen. Thus, the protons on a methyl group directly attached to nitrogen (CH$_3$–N) come into resonance at about $\delta \approx 2.2$ ppm (for comparison, remember the typical ^{1}H chemical shift values of CH$_3$–C [$\delta \approx 0.9$]; CH$_3$–O [$\delta \approx 3.2$]; CH$_3$–S [$\delta \approx 2.1$]) and, correspondingly, the methylene and methine protons are shifted downfield by about $\delta \approx 0.2$–0.4 ppm. As was the case for alcohols, the proton(s) on the nitrogen itself come into resonance at chemical shifts that are solvent and concentration dependent (but generally between $\delta \approx 0.5$ and 3.5 ppm).

The ^{13}C NMR spectra of amines follow the same pattern. Thus, while the carbon of a methyl group attached to oxygen (CH$_3$–O) comes into resonance significantly downfield from a C-methyl (CH$_3$–C; $\delta \approx 6$ [ethane]*) at about $\delta \approx 49$ ppm, one attached to nitrogen (CH$_3$–N) is found at about $\delta \approx 28$ ppm (somewhat downfield from the analogous sulfur derivative at ca. $\delta \approx 20$ ppm).

In addition to ^{1}H and ^{13}C NMR, both naturally occurring isotopes of nitrogen, viz. ^{14}N (spin = 1; abundance 99.63%; frequency at 7.05 T = 21.7 MHz) and ^{15}N (spin = –1/2; abundance = 0.37%; frequency at 7.05 T = 30.5 MHz) are available for magnetic resonance studies. Since the multiplicity of coupling interactions are reduced with spin 1/2 systems, ^{15}N spectroscopy is somewhat more widely used than ^{14}N spectroscopy despite its lower natural abundance. ^{15}N NMR spectroscopy has become an important tool in the structural elucidation of peptides (Chapter 12) because these materials may be considered (to a first approximation) as polyamides (*vide infra*) and much can be learned about their structures and the conformations they adopt in solution by characterizing the N–H bonds.

The parent ion (M$^+$) in the mass spectra of all aliphatic monoamines (indeed of all amines with an *odd number of nitrogens*) is an odd number. The value of this useful information is diminished by the observation that many aliphatic amines produce a very weak M$^+$ signal. Indeed, the base peak of most aliphatic amines consists of a fragment resulting from carbon–carbon bond cleavage between the carbon bearing the nitrogen and its nearest neighbor (i.e., α, β). In the case of primary aliphatic amines, this fragment (CH$_2$NH$_2^+$) is at *m/z* 30 mass units and it is quite characteristic of these compounds.

Problem 5.7. Briefly justify the observation that the parent ion (M$^+$) will always be odd if the number of nitrogen atoms in the molecule is odd. Show that this need not obtain in the case of either thiols or alcohols.

---

*Comparison to this value may be somewhat disingenuous. Indeed, the methyl groups of 2,2-dimethylpropane [(CH$_3$)$_4$C] come into resonance at $\delta = 31.5$ ppm.

Finally, as was the case with sulfur, there are organic compounds of nitrogen that have no oxygen analogues (although, on a formal basis, symmetrically alkylated **hydrazines**, e.g., 1,2-dimethylhydrazine [$CH_3NHNHCH_3$], might be considered to be like **peroxides** [*vide supra*]). These organonitrogen compounds are, for the most part, oxides and may be considered as derived from nitrous (HONO) and nitric acids ($HONO_2$) as shown below in Table 5.6.

Many nitrogen-containing organic compounds similar to those of Table 5.6 are common industrial chemicals.

VII. Phosphines, Phosphonium Salts, and Other Phosphorus Derivatives

Although phosphorus has the same number of outer shell electrons as nitrogen and bears the same kernel charge, it (along with other second row elements) enjoys the

TABLE 5.6. Some Nitrogen Compounds Which Have No Corresponding (Stable) Oxygen Analogues

Class	Example	Formula	Structure
Hydroxylamine	*N*-methylhydroxylamine	CH_3NHOH	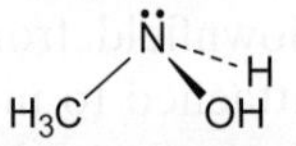
	O-methylhydroxylamine	CH_3ONH_2	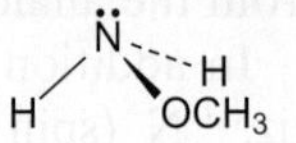
Alkyl nitrate	Methyl nitrate	CH_3ONO_2	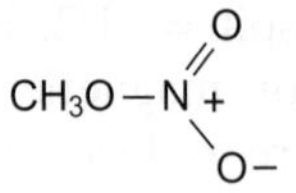
Alkyl nitrite	Methyl nitrite	CH_3ONO	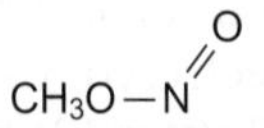
Nitroalkane	Nitromethane	CH_3NO_2	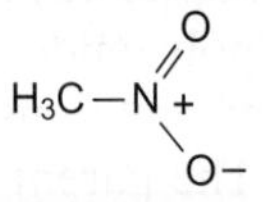
Nitrosoalkane	Nitrosomethane	CH_3NO	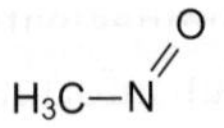
Amine oxide	Trimethylamine oxide	$(CH_3)_3NO$	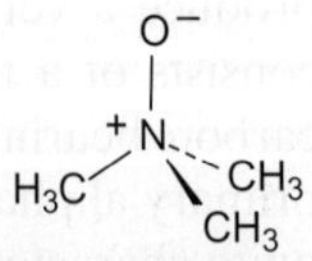
Alkyl azide	Methylazide	CH_3N_3	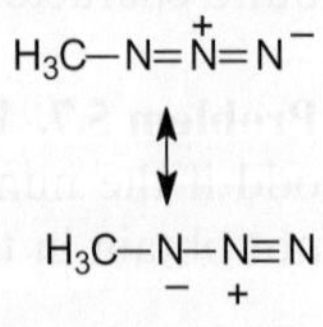

Note that both multiply and singly bonded nitrogen species are present. Some representations can only be written with formally charged or polar bonds.

benefits of d orbitals, which can participate in covalent bonding and thus, like sulfur, (*vide supra*) can form more than four covalent bonds. However, the **phosphines** are the phosphorus analogues of the nitrogenous **amines** and, like amines, are named by adding the suffix (*phosphine*) to the *alkyl* stem name. Phosphine (PH_3) is reported to be a poisonous gas (bp $-87.5°C$ at 1 atm), which is spontaneously flammable. Phosphines in which the hydrogens on the phosphorus of PH_3 have been replaced (substituted) by carbon-containing fragments are less reactive.

As shown below, the H–P–H angle in phosphine (93°) is quite far from tetrahedral and, indeed, the C–P–C angle even in trimethylphosphine, although closer (99°), deviates significantly from the idealized (109.5°).

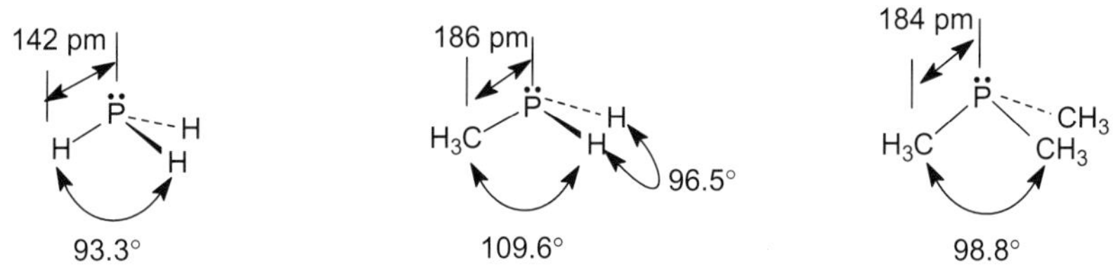

142 pm   186 pm   184 pm

93.3°   109.6° 96.5°   98.8°

phosphine   methylphosphine   trimethylphosphine

Alkyl phosphines readily undergo oxidation to give **oxides**, which may be considered as having either pentacovalent phosphorus or a semipolar phosphorus–oxygen bond, for example, in trimethylphosphine oxide $[(CH_3)_3P{=}O \leftrightarrow (CH_3)_3P^+{-}O^-]$. Similar compounds with carbon replacing oxygen are also known, for example, methylenetriphenylphosphorane (Chapter 10) $[(C_6H_5)_3P{=}CH_2 \leftrightarrow (C_6H_5)_3P^+{-}CH_2^-]$. As a group the compounds with such phosphorus-to-carbon semipolar bonds are known as **ylides** or, occasionally, as **Wittig reagents** (Chapter 10), and they are commonly prepared from the corresponding **phosphonium salts**, for example, methyltriphenylphosphonium bromide $[(C_6H_5)_3P^+CH_3\ Br^-]$.

The IR spectra of compounds with at least one P–H bond are characterized by a medium intensity band in the 2450–$2250\,cm^{-1}$ region. As very few other types of compounds absorb in this region, a band here is generally of diagnostic significance. The NMR spectra of alkyl phosphines are particularly interesting because ^1H, ^{13}C, and ^{31}P nuclei all have spin 1/2 and all of them interact with each other! For these nuclei, in a 7.05 T field, where protons come into resonance at 300 MHz and ^{13}C 75.5 MHz, ^{31}P (natural abundance $= 100\%$) is found at 121.5 MHz. Furthermore, although TMS $[(CH_3)_4Si]$ contains both protons and carbons and can therefore be used a standard for these nuclei, it clearly does not contain phosphorus and, as a consequence, a different standard must be used for the latter. Traditionally, the standard for ^{31}P has been 85% phosphoric acid (H_3PO_4) sealed in an "external" capillary tube and placed within the normal NMR tube holding the sample. As shown in Table 5.7, ^1H, ^{13}C, and ^{31}P NMR spectra for compounds containing all of these nuclei are known. Additionally, the coupling between all of these nuclei can be observed and, generally, the coupling to ^{31}P for both ^1H and ^{13}C is quite large. Most notably (Table 5.7), the chemical shift of ^{31}P is highly dependent on the nature of the substituents it bears and upon its oxidation state.

The mass spectra of phosphines (other than the simple methylphosphines) generally result in the observation of an M^+ peak and fragmentation that occurs produces daughter ions that have come about by loss of an alkene fragment.

TABLE 5.7. Chemical Shifts (^{1}H, ^{13}C and ^{31}P) and Coupling Constants (Homo- and Heteronuclear) for Simple Phosphines

Compound	Formula	^{1}H NMR (300 MHz) (TMS = 0.00 ppm)	^{13}C NMR (75.5 MHz) (TMS = 0.00 ppm)	^{31}P NMR (121.5 MHz) (85% H_3PO_4 = 0.00)
Phosphine	PH_3	H; δ = 1.73[a] J_{H-P} = 179 Hz		P; δ = 238 ppm J_{P-H} = 179 Hz
Methylphopshine	$(CH_3)PH_2$	H-C; δ = 0.98 H-P; δ = 2.63 $J_{H-C} \approx$ 135 Hz J_{H-P} = 186 Hz J_{H-C-P} = 4 Hz $J_{H-C-P-H}$ = 8 Hz	C; δ = 14.3 $J_{C-H} \approx$ 135 Hz J_{C-P} = 12 Hz	P; δ = 163.5 J_{P-H} = 186 Hz J_{P-C} = 12 Hz J_{P-C-H} = 4.1 Hz
Dimethylphosphine	$(CH_3)_2PH$	H-C; δ = 1.06 H-P; δ = 3.12 $J_{H-C} \approx$ 135 Hz J_{H-P} = 191 Hz J_{H-C-P} = 3.6 Hz $J_{H-C-P-H}$ = 8 Hz	C; δ = 14.3 $J_{C-H} \approx$ 135 Hz J_{C-P} = 14 Hz	P; δ = 98.5 J_{P-H} = 191 Hz J_{P-C} = 14 Hz J_{P-C-H} = 3.6 Hz
Trimethylphosphine	$(CH_3)_3P$	H-C; δ = 0.93 $J_{H-C} \approx$ 135 Hz J_{H-C-P} = 2.7 Hz	C: δ = 14.3 $J_{C-H} \approx$ 135 Hz J_{C-P} = 14 Hz	P; δ = 62 J_{P-C} = 14 Hz J_{P-C-H} = 2.7 Hz

Some of the values reported are not confirmed.

Values not obtained by the author have been taken from (a) Emsley, J. W.; Feeney, J.; Sutcliffe, L. H. *High Resolution Nuclear Magnetic Resonance Spectroscopy*, Vols. 1 and 2, Pergamon, Oxford, UK, **1966**; (b) Levy, G. C.; Lichter, R. L.; Nelson, G. L. *Carbon-13 Nuclear Magnetic Resonance Spectroscopy*, 2nd edition, Wiley Interscience, New York, **1980**, and (c) Jackman, L. M.; Sternhell, S. *Applications of Nuclear Magnetic Resonance Spectroscopy in Organic Chemistry*, 2nd edition, Pergamon, Oxford, UK, **1969**.

[a]Gas phase.

Finally, as was the case with sulfur and nitrogen, there are organic compounds of phosphorus that have no oxygen analogues and might be considered as based on the corresponding oxyacids. These organophosphorus compounds belong to two groups, that is, those that formally have either three or five bonds to phosphorus and a representative (but far from inclusive) sampling is provided in Table 5.8.

VIII. An Introduction to Organometallic Compounds

Even the most cursory examination of the periodic Table will show that all of the compounds so far discussed in this Chapter are composed of only a few elements seen there (i.e., C, H, O, S, N, P, and the halogens) and which represent but a fraction of those available. Furthermore, *none of these so far used are, under normal conditions, metals*! Since the vast majority of elements are metals under normal conditions, it might reasonably be asked if there are compounds in which carbon–metal bonds are found.

TABLE 5.8. Some Phosphorus Compounds Which Have No Corresponding (Stable) Oxygen Analogues

Class	Example	Formula	Structure
Phosphine oxide	Trimethylphosphine oxide	$(CH_3)_3PO$	
Phosphite	Trimethylphosphite	$(CH_3O)_3P$	
Phosphinate	Methyl diethylphosphinate	$CH_3OP(O)$ $(CH_2CH_3)_2$	
Phosphonate	Dimethyl ethylphosphonate	$(CH_3O)_2P(O)$ CH_2CH_3	
Phosphonic acid	Methylphosphonic acid	$CH_3PO_3H_2$	
Phosphate	Trimethyl phosphate	$(CH_3O)_3PO$	
Phosphonium salt	Methyltrimethoxyphosphonium halide	$CH_3P^+(OCH_3)_3\ X^-$	

Both multiply and singly bonded phosphorus species are present. Some representations can only be written with formally charged or polar bonds.

There are many compounds that contain carbon–metal bonds. Most **organometallic** compounds you will encounter in this text are highly reactive and can only be used in the absence of oxygen, CO_2, and so on. Furthermore, many of the organometallic compounds are only rarely isolated in a pure state because the impure ("crude") materials can be used by chemists as intermediates in the preparation of other organic compounds, which, the metal having been removed, are less reactive and can be purified by the traditional techniques of distillation, crystallization, chromatography, and so on. However, some organometallic compounds, such as tetraethyllead [$(CH_3CH_2)_4Pb$] and dimethylmercury [$(CH_3)_2Hg$], are stable enough to be environmentally dangerous (because they are toxic and have been shown to accumulate in tissue). Other organic compounds that contain metals and are quite stable are frequently **not** considered to be organometallic compounds because (except perhaps fleetingly) the metal is not bonded to carbon. Examples of these compounds abound, and they include the related families of the iron-containing cytochromes and hemes (Chapters 12–14), cobalt-containing vitamin B_{12} (Chapters 12 and 14), the chlorophylls (Chapter 14), which contain magnesium, and metal salts of carboxylic acids (*vide infra*).

The name of the **metal** is used as the parent in the nomenclature of these compounds. Then, the appropriate organic fragment (alkyl or aryl) *directly bonded to the metal* is placed as a **prefix**. The names are written as one word. Thus, in addition to tetraethyllead $[(CH_3CH_2)_4Pb]$ and dimethylmercury $[(CH_3)_2Hg]$ mentioned above, compounds such as methyllithium $[CH_3Li]$, diethylmagnesium $[(CH_3CH_2)_2Mg]$, TMS $[(CH_3)_4Si]$, and triethylaluminum $[(CH_3CH_2)_3Al]$ are either isolable or presumed to be present in organic reactions with the metal somehow attached to the organic fragment. Additionally, both **mixed metal** (e.g., lithium dimethylcuprate $[(CH_3)_2CuLi]$) and **mixed ligand** (e.g., diethylaluminum chloride $[(CH_3CH_2)_2AlCl]$ and methylmagnesium bromide $[(CH_3)MgBr]$) organometallics are presumed to have at least transient existence since materials behaving as if they had those respective compositions can be prepared.

Having already considered compounds in which carbon is bonded covalently to atoms that are either the same (another carbon) or more (e.g., chlorine) electronegative than it is,* and recognizing that the electrons in the bond are *polarized* toward the **more** electronegative element (leaving carbon slightly electron deficient and thus **positive**), it should come as no surprise that, when carbon is bonded to atoms that are **less** electronegative than it is, the same idea should apply. In most organometallic compounds, the carbon–metal bond is polarized leaving the metal electron deficient and the carbon partially **negative**.

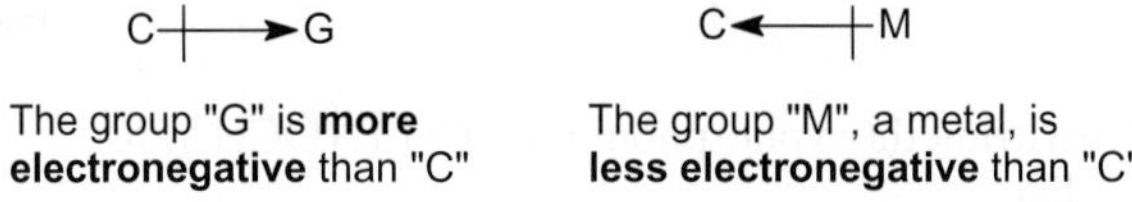

The group "G" is more electronegative than "C"
The group "M", a metal, is less electronegative than "C"

However, if electronegativity were the only factor involved in electron distribution in the bond, then it might be difficult to rationalize the reactivity (to say nothing of the properties) of many of the organometallic compounds. Thus, the difference between the electronegativities of carbon (2.5) and fluorine (4.1) (4.1 − 2.5 = 1.6) is about the same as the difference between electronegativities of carbon (2.5) and lithium (1.0) (2.5 − 1.0 = 1.5). While the gas fluoromethane (methyl fluoride, CH_3F) is quite stable (*vide supra*) with a strong covalent bond between carbon and fluorine, methyllithium (CH_3Li) contains a highly ionic bond, where there is poor overlap between the metal and the carbon orbitals, leaving the electrons largely associated with the carbon. So, the material called methyllithium, and written as CH_3Li, is very reactive and it must be protected from air and moisture. For this reason, it is frequently handled only in solution. However, in solution (where the **reactivity** of the solvent itself will now be important, *vide infra*), it appears that methyllithium (CH_3Li) is no longer monomeric (from the Greek, *monos* = alone or one; *mero*[*s*] = part), that is, aggregates form, and their structures and properties, for what is still written as "CH_3Li" are somewhat solvent (*vide infra*) dependent. For

---

*On the Pauling Electronegativity Scale (Chapter 1), the electronegativities of C, N, O, and F are 2.5, 3.1, 3.5, and 4.1, respectively. Although H is 2.2, P is 2.1, and S is 2.4, most metals lie between 0.8 and 1.7.

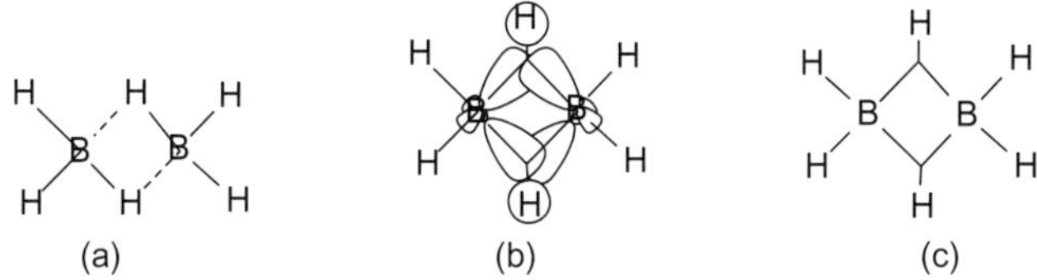

Figure 5.12. Representations of the "two-electron three-center" bond in diborane. (a) A structural representation indicating that the (mostly) *p* orbital from boron is bonded to a proton, which is "formally" bonded to another boron. (b) A representation of orbital overlap of three orbitals to give a molecular orbital that contains two electrons. (c) A shorthand representation of the picture of the molecular orbital made up of three atomic orbitals (containing two electrons).

example, in hydrocarbon solution, methyllithium is frequently found to be tetrameric (i.e., four Li atoms arranged as a tetrahedron with four [tetrahedrally] associated methyl groups—one centered on each face of the Li_4). This material is apparently quite volatile.

As already noted, the relatively poor overlap between the orbitals on the bonding partners in organometallic compounds partially accounts for the high reactivity of these materials, and it should be clear that the bonding in organometallic compounds is not the same as in the more covalent materials considered earlier. Indeed, with the alkaline-earth metals, for example, Be, Mg, Ca, and other metals, for example, B and Al, more electronegative than those in the furthest left column of the periodic table, it appears that the structure of the material (under certain conditions) can only be accounted for if three nuclei are bonded together with two electrons, that is, a *two-electron three-center* bond (Figure 5.12).

Therefore, it should not come as a surprise that, given the variety of particular electronic configurations extant in the metals as a class, bonding between carbon and any specific metal can vary from highly ionic to nearly covalent, with the former exhibiting saltlike properties (sodium chloride and methylpotassium have similar lattice structures) and the latter those more traditionally associated with organic compounds. (The boiling point of tetramethyllead [$(CH_3)_4Pb$, MW $\approx$ 267] is 110°C while that of 2,2-dimethylpropane [$(CH_3)_4C$, MW $\approx$ 72] is about 10°C; alternatively, hydrocarbons of MW $\approx$ 268 such as the isomers of $C_{19}H_{40}$ tend to have boiling points closer to 300°C.)

IX. Compounds Containing Unsaturated Functional Groups

All of the compounds to be considered in this section have multiple (double or triple) carbon-to-heteroatom bonds. The most common contain the carbon–oxygen double bond (the **carbonyl group**).

a. Aldehydes. The **aldehyde** functional group (–CHO) will always begin (or end) a chain in simple hydrocarbon-derived molecules since there must be (by definition) one hydrogen attached to the carbon of the carbonyl. In such simple hydrocarbons, the IUPAC nomenclature requires that the final "e" in the name of the corresponding **alkane** be replaced with the suffix "al." Thus, the relationship between the alkanes, alcohols and aldehydes shown below follows logically.

methane

methanol

methanal
(formaldehyde)

2-methylpentane

4-methyl-1-pentanol

4-methylpentanal

When the **functional group** (–CHO) is appended to a ring or a longer chain, the compound may be named using "–carbaldehyde" as a *suffix*. However, in the event some other group (*vide infra*) must serve as the parent and the above suffix cannot be appended, the *prefix* "formyl" can be used instead. This *prefix* is derived from an alternative nomenclature system, from which another set of trivial names also arises. The latter is based on the relationship between **carboxylic acids** (*vide infra*) and aldehydes. **Carboxylic acids** bear the **functional group** "–CO$_2$H" and can be considered as (and often are) oxidation products of **aldehydes** (as **aldehydes** can be considered as oxidation products of **primary alcohols** which they too often are). Trivial names for carboxylic acids abound (in part because they were among the earliest products obtained when organic chemistry was in its infancy) and, if their names are known, those of the aldehydes can be derived.

Finally, often as a function of industrial use, other systems and/or trivial names may be found. For example, having defined a parent aldehyde, substituents are often located using Greek letters: beginning with the first carbon to which the carbonyl is attached. Thus, 2-chloropropanal (CH$_3$CH(Cl)CHO) is also called α-chloropropanal and the α,β-unsaturated (i.e., the double bond is between the α and β carbons) aldehyde, 2-propenal (CH$_2$=CH-CHO), also enjoys the trivial name **acrolein**.

The low-molecular-weight aldehydes, for example, methanal (formaldehyde, H$_2$C=O) and ethanal (acetaldehyde, CH$_3$CHO), are generally unpleasant "sharp-" smelling, low-boiling liquids (methanal, H$_2$C=O, bp –19°C). A 40% aqueous solution of methanal (H$_2$C=O) is commercially available as "formalin," commonly used in biology as a preservative.

Aldehydes have a variety of acceptable names:

	Methanal	Formaldehyde (from formic acid [methanoic acid] HCO$_2$H)
	Ethanal	Acetaldehyde (from acetic acid [ethanoic acid] CH$_3$CO$_2$H)
	Cyclohexanecarboxyaldehyde	Formylcyclohexane

Although it might be predicted that sp^2 hybridization would be found at the carbon of the carbonyl and that, as a consequence, the hydrogen–carbon–hydrogen (H–C–H) angle and the hydrogen–carbon–oxygen (H–C–O) angles in, for example, methanal (formaldehyde, H_2CO) would be identical and equal to 120°, it is found that these idealized angles are only close (as was also the case with ethene, Chapter 1) to what is actually found. Thus, as shown in Figure 5.13, the H–C–H angle (116.5°) is somewhat diminished, while the H–C–O angle (121.8°) is increased.

A similar picture is also found for ethanal (acetaldehyde, CH_3CHO) and, indeed, for most other aldehydes (Figure 5.14). Interestingly, as shown in Figure 5.14, it appears that the *favored conformation* of aldehydes (by about $1\,kcal\,mol^{-1}$ [$4.18\,kJ\,mol^{-1}$]) is that one with one of the hydrogens on the methyl in an *eclipsed* conformation with the oxygen of the carbonyl. Although the energy difference between eclipsed and staggered forms is small, even propanal (CH_3CH_2CHO) clearly shows a preference for that conformational isomer in which the methyl group is eclipsed with the oxygen (Figure 5.15).

Aldehydes without unsaturation other than the carbon–oxygen double bond have only weak absorption in the portion of the electronic (UV) spectrum readily accessible in the laboratory (i.e., 200–400 nm). This band (generally found between about 270 and 300 nm) is the result of an $n \rightarrow \pi^*$ transition and has a very low extinction coefficient, $\varepsilon \leq 30$ (Chapter 2). It is of little practical value. Unfortunately,

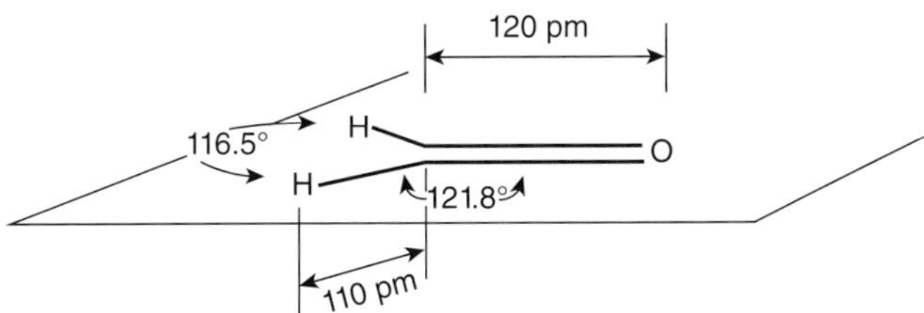

Figure 5.13. A representation of methanal (formaldehyde, $H_2C=O$) showing bond angles and lengths.

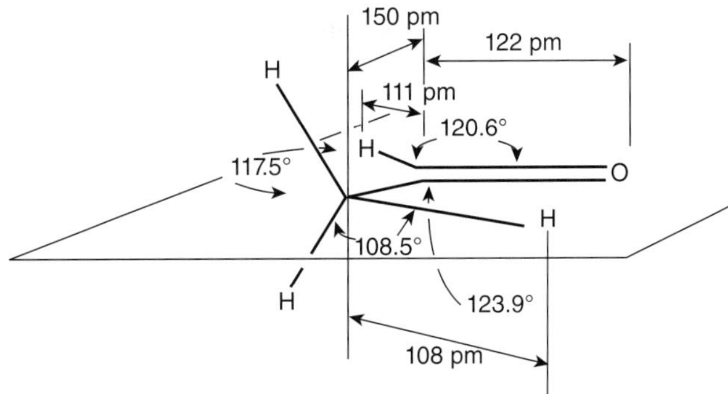

Figure 5.14. A representation of ethanal (acetaldehyde, CH_3CHO) showing bond angles and lengths.

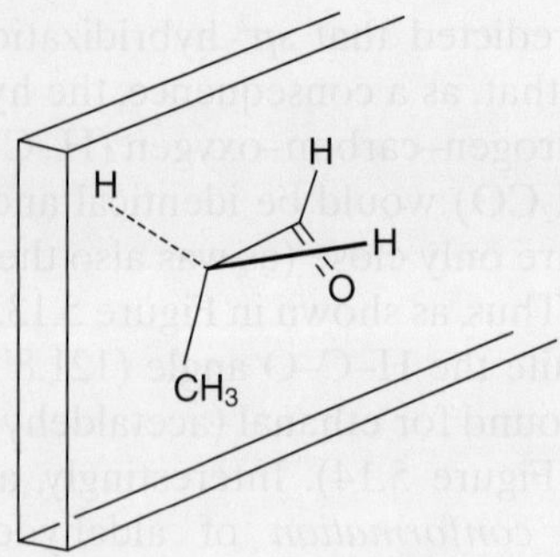

Figure 5.15. A representation of propanal (CH₃CH₂CH=O) showing that the methyl group lies in the plane of the carbon–oxygen double bond.

the two more intense bands ($\varepsilon \geq 1000$), the $\pi \to \pi^*$ transition near 150 nm and the $n \to \sigma^*$ absorption near 190 nm, are at energies that are too high (or wavelengths that are too low) for routine UV spectrometers. However, as already seen for conjugated dienes, if there is at least one double bond conjugated with the carbonyl group, the absorption band is moved to longer wavelength and, normally, an intense $\pi \to \pi^*$ transition ($\varepsilon \geq 10,000$) and a weaker $n \to \pi^*$ transition are observed. The position of the $\pi \to \pi^*$ absorption band can frequently be predicted utilizing Woodward's rules (*vide supra*) for α,β-unsaturated carbonyl compounds. Thus, using the same definitions as earlier (p. 185, this chapter, Section C.I.b) an α,β-unsaturated aldehyde (*partial structure given as* $C^\beta{=}C^\alpha{-}CH{=}O$) if *unsubstituted* will absorb at $\lambda_{max} \approx 208$ nm; α- or β-alkyl substitution shifts the maximum to longer wavelength (a *bathochromic shift*) to about $\lambda_{max} \approx 220$ nm and each additional alkyl substitution causes an additional 10- to 12-nm bathochromic shift in the absorption wavelength. Finally, for every additional double bond placed in conjugation, the absorption maximum shifts to longer wavelength (by about 30 nm) and, if any of the double bonds are **exocyclic**, a further small shift (ca. 5 nm) in the same direction is observed.

In the IR, the C–H of the aldehyde function (i.e., –CH=O) itself produces a sharp, occasionally weak, but quite distinctive band in the region between about 2800 and 2700 cm⁻¹. Additionally, for simple, unconjugated, aliphatic aldehydes (R–CHO) an intense carbonyl (C=O) absorption is routinely observed between 1720 and 1740 cm⁻¹. For example, the carbonyl band in ethanal (acetaldehyde [CH₃CHO]) is found at 1730 cm⁻¹. In concert with what was observed in the UV, conjugation (α,β-unsaturation) causes the IR absorption to be shifted to longer wavelength (1710–1685 cm⁻¹). Figure 5.16 presents the IR spectrum of propanal (CH₃CH₂CHO) taken as an undiluted pure liquid film ("neat").

The ¹H NMR spectra of aldehydes are distinctive, in particular because the proton directly attached to the carbon of the carbonyl is strongly deshielded, and routinely appears very far downfield (between about $\delta = 9.2$ and 10.1 ppm; TMS = 0.00). It is usually argued that the deshielding of aldehydic protons, as well as the protons attached to carbon(s) α (and even β) to the carbonyl, results from induction of a secondary magnetic field (in the polarized π-bond), which reinforces the externally applied magnetic field—thus bringing the protons into resonance at a lower applied field (i.e., downfield) (Figures 5.17 and 5.18a).

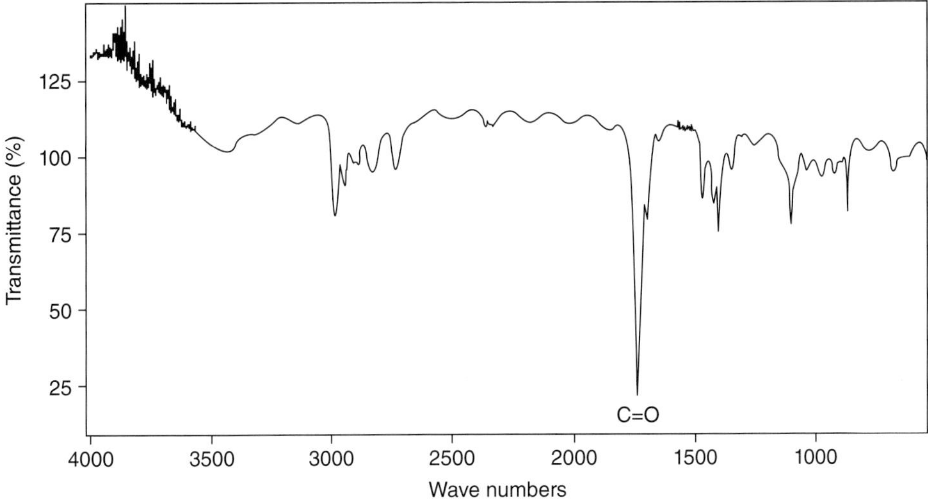

Figure 5.16. The infrared spectrum of propanal (CH$_3$CH$_2$CHO) as a neat liquid film.

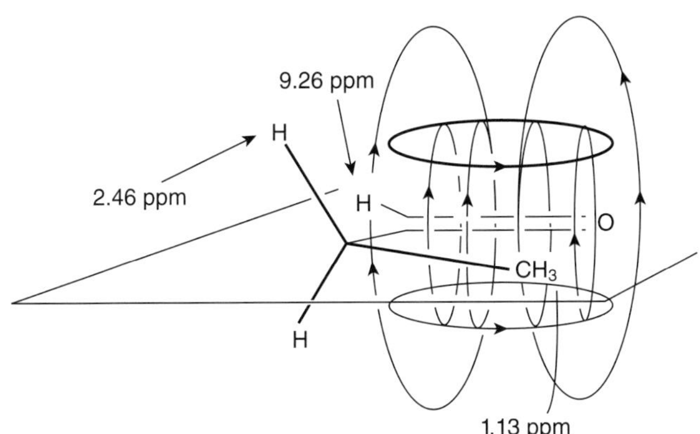

Figure 5.17. A representation of propanal (CH$_3$CH$_2$CH=O) attempting to account for the experimental observation that protons attached to the carbon of the carbonyl, as well as to the α- and β-carbons, are shifted downfield relative to the positions expected when the carbonyl is absent.

The diamagnetic anisotropy induced is also experienced, as must be the case, by the carbon atoms themselves. Thus, the carbon of the carbonyl must be deshielded strongly and, in the ^{13}C NMR spectrum of propanal (CH$_3$CH$_2$CHO), the aldehydic carbon comes into resonance at about 201.8 ppm downfield from TMS. Furthermore, still considering propanal, C-2 (the α-carbon) is less deshielded than the carbon of the carbonyl and is found at 36.7 ppm, while the β-carbon, essentially unaffected by the anisotropy induced by the carbonyl, comes into resonance at 5.2 ppm (all downfield from TMS = 0.00) as shown in Figure 5.18b.

It is reasonable to ask if the ^{17}O resonance (around 40.7 MHz at 7.05 T) is similarly affected, and indeed, as expected, the ^{17}O resonance for the oxygen of aldehydes is

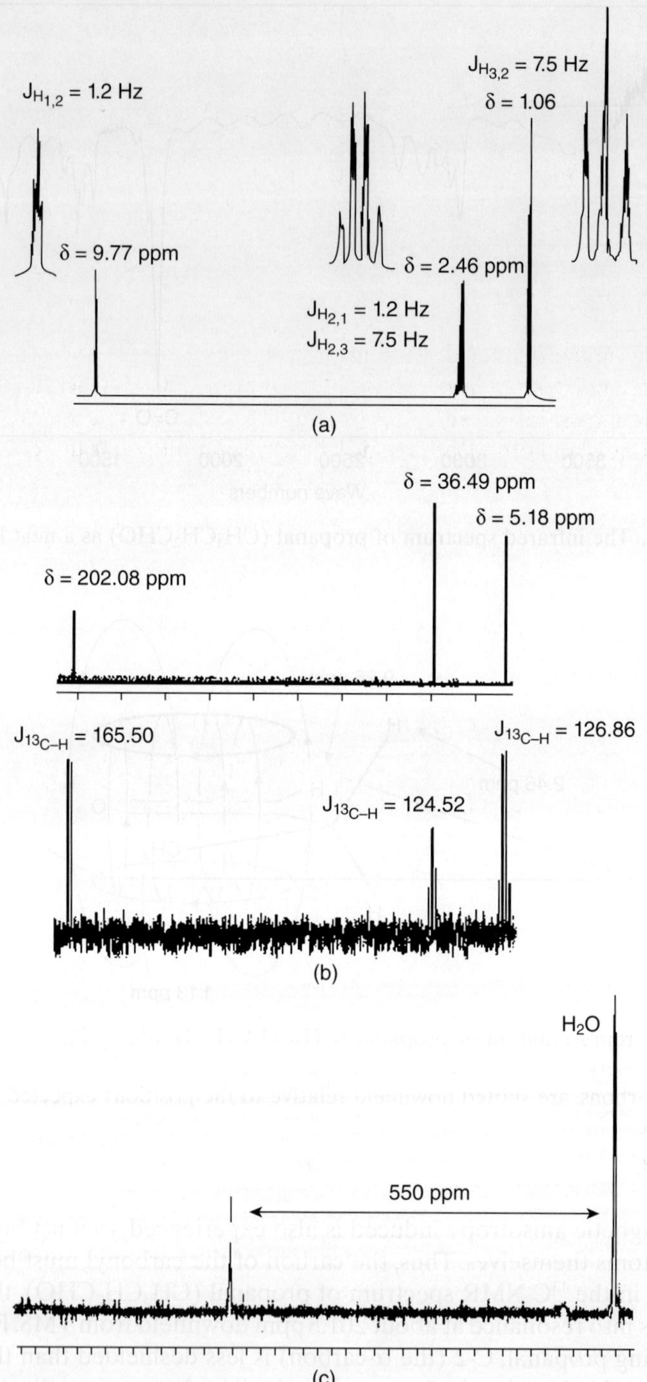

Figure 5.18. NMR spectra of propanal (CH_3CH_2CHO). (a) The ^1H NMR spectrum in $^2HCCl_3$ at 300 MHz. (b) The ^{13}C spectra (both decoupled and undecoupled) in $^2HCCl_3$ at 75.5 MHz. (c) The ^{17}O NMR spectrum in ^2H_2O at 67.8 MHz.

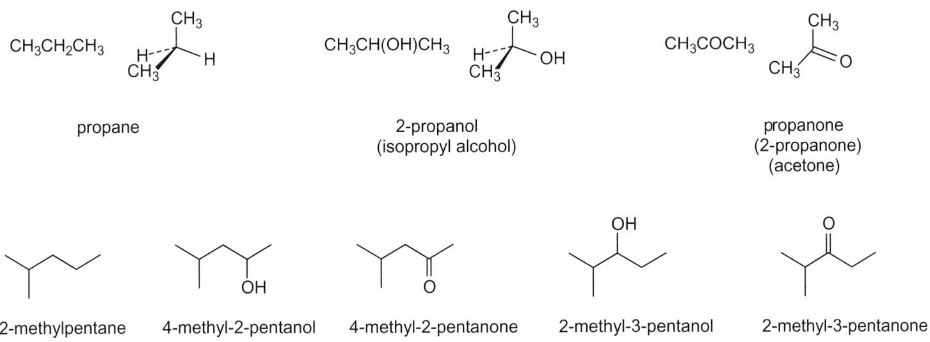

Scheme 5.3. A generic aldehyde producing the products of the McLafferty rearrangement. The arrows are drawn to indicate possible movement of electrons to make and break bonds— thus accounting for the observed products.

generally found much further downfield than that of alcohols and ethers. Thus, the oxygen of propanal comes into resonance at about 550 ppm downfield from the water signal as shown in Figure 5.18c. (It may be recalled [*vide supra*] that the ^{17}O signals for alcohols and ethers are found about 550 ppm further upfield [i.e., the ^{17}O chemical shifts of both alcohols and ethers are generally in the range of 7–60 ppm downfield of $H_2^{17}O$]).

Electron impact mass spectrometry of aliphatic aldehydes usually produces a molecular ion (M^+), an M^+-1 fragment (due to the loss of the hydrogen on the carbon of the carbonyl), and a daughter ion resulting from loss of CHO by a process called **α-cleavage** ($CH=O^+$ itself is found at m/z 29). Interestingly, it has been observed that when there is a hydrogen positioned six atoms away from the oxygen of the carbonyl, an alternative fragmentation pathway becomes available (Scheme 5.3). This process is known as a **McLafferty** rearrangement.*

b. Ketones. Ketones all have a carbonyl carbon (C=O) bonded to two other carbons, that is, RR′C=O, where R may or may not be the same as R′, but both R and R′ form carbon–carbon bonds to the C=O. Thus, the **keto** functional group never begins (nor ends) a chain. In simple hydrocarbons, the IUPAC nomenclature system requires that the final "e" in the name of the corresponding **alkane** be replaced with the suffix "one" and the position of the doubly bonded oxygen provided as a numerical prefix separated from the name by a hyphen. Thus, the relationship between the alkanes, alcohols, and ketones follows logically.

*McLafferty, F. W. (1923–) was born in Evanston, IL, and held a professorship at Purdue University (1964–1968) before moving to Cornell University.

Problem 5.8. Given the apparent structural relationship between secondary alcohols and ketones shown above, write additional "hexanol" isomers that might reasonably be related to other ketones in the same way.

Because some ketones (e.g., propanone and 2-butanone) are excellent, low-boiling solvents, they are important materials of commerce and trivial names are common. Thus, 2-propanone (CH_3COCH_3) is frequently called acetone or dimethyl ketone (DMK), while 2-butanone ($CH_3COCH_2CH_3$) is similarly referred to as ethyl methyl ketone (or methyl ethyl ketone [MEK]). In addition to the IUPAC system, Greek letters are commonly used to define the location of a substituent (or substituents) flanking the carbonyl. As was the case for aldehydes (*vide supra*), lettering begins with the positions **adjacent** to the carbonyl carbon and can proceed in both directions, the less substituted side bearing Greek letters and prime signs.

$$\delta' \quad \beta' \quad \overset{\displaystyle O}{\underset{\gamma' \quad \alpha' \quad \alpha \quad \gamma}{C}} \quad \beta \quad \delta$$

As was the case for aldehydes, ketones without unsaturation other than the carbon–oxygen double bond have only weak absorption in the portion of the UV spectrum above 200 nm. This $n \to \pi^*$ transition, seen between 270 and 300 nm for saturated ketones (but shifted to longer wavelength when the ketone is highly substituted) has a very low extinction coefficient, $\varepsilon \leq 30$ (Chapter 2). Again, as for aldehydes, this absorption is of little practical value, and the more intense bands due to the $\pi \to \pi^*$ and the $n \to \sigma^*$ transitions near 150 and 190 nm, respectively, are at energies that are too high (or wavelengths that are too low) for routine UV spectrometers (i.e., they are in the **vacuum UV**). However, as already seen for conjugated dienes and aldehydes, if there is at least one double bond conjugated with the carbonyl group, the absorption band is moved to longer wavelength, and normally, an intense $\pi \to \pi^*$ transition ($\varepsilon \geq 10,000$) and a weaker $n \to \pi^*$ transition are observed. Woodward's rules (*vide supra*) for α,β-unsaturated carbonyl compounds already discussed for aldehydes are applicable here too. Thus, using the same definitions as earlier (p. 185, this chapter, Section C.I.b), an α,β-unsaturated ketone (*partial structure given as $C^\beta=C^\alpha-C=O$*) if unsubstituted will absorb at $\lambda_{max} \approx 208$ nm; α- or β-alkyl substitution shifts the maximum to longer wavelength (a *bathochromic shift*) to about $\lambda_{max} \approx 220$ nm and each additional alkyl substitution causes an additional 10- to 12-nm bathochromic shift in the absorption wavelength. Finally, for every additional double bond placed in conjugation, the absorption maximum shifts to longer wavelength (by about 30 nm) and, if any of the double bonds are **exocyclic**, a further small shift (ca. 5 nm) in the same direction is observed. For example, 4-methyl-3-penten-2-one (mesityl oxide) would be expected to absorb at 230 nm (220 + 10) and a major band is found at 230.6 nm.*

Largely because ketones that are not conjugated have only weak absorption above about 270 nm (*vide supra*) and because many naturally occurring *and*

---

*This value obtains when the solvent used is a hydrocarbon (such as isooctane). Other solvents (e.g., chloroform and water) apparently cause a further bathochromic shift. The role of solvents, in general, will be discussed subsequently in this chapter.

optically active compounds contain this functional group, **optical rotatory dispersion (ORD)** data for *structurally rigid* systems can be (and has been) correlated with *absolute configuration*! As shown in Figure 5.19, any molecule (or at least a model thereof) containing a cyclohexanone or other rigidly fixed carbonyl group can be "viewed, looking down the carbonyl from oxygen to carbon" and the space around the bonding region divided into four quadrants. Looking in the other direction (i.e., from carbon to oxygen) is usually unnecessary since it is usually an empty space, but, since it might be occasionally necessary, it should be clear that another set of four quadrants is defined in that way. Thus, in total there are **eight** regions of space around the oxygen–carbon double bond and these **octants** may or may not be occupied. They are shown in Figure 5.19a.

In the IR, ketones, like aldehydes (and other carbonyl compounds to be encountered), have intense absorption in the 1870–$1550\,\mathrm{cm^{-1}}$ region. Although the concept of a single band centered at about $1715\,\mathrm{cm^{-1}}$ is widespread, the exact position (within the range given) is found to depend on a variety of factors. For example, as expected, α,β-unsaturation producing a conjugated carbonyl moves the absorbance to longer wavelength ($1650\,\mathrm{cm^{-1}}$ is common), while confining the carbonyl to a small ring

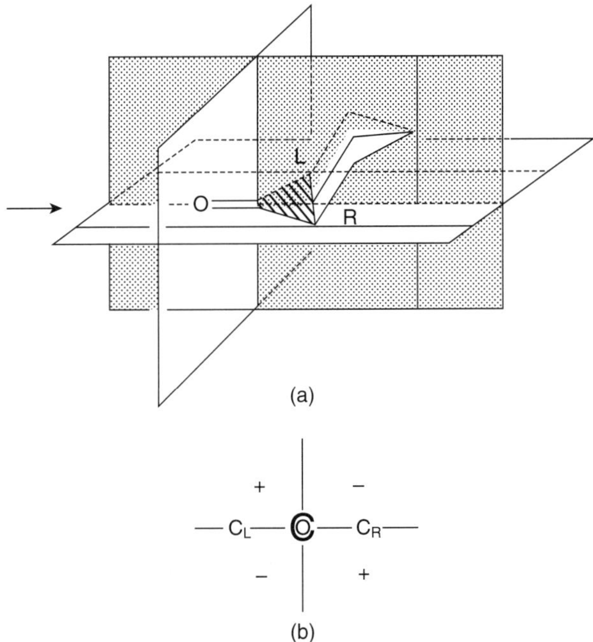

Figure 5.19. The eight octants surrounding a carbonyl group. (a) Cyclohexanone drawn with three planes: one mirror plane bisects the carbonyl group and C-4 and, orthogonal to it; a second plane contains the carbonyl group and the two α-carbons (labeled C_L [to the left of the viewer] and C_R [to the right of the viewer]); the third plane is orthogonal to the other two and is drawn through the carbon of the carbonyl. (b) The four octants labeled with the signs (+ and −) and defined by (a) as seen "looking down" the oxygen–carbon double bond. It is found that, when a single absolute configuration is present and, as a result, one octant is more fully occupied than the others, the sense of the Cotton effect (positive or negative) is given by the most occupied octant.

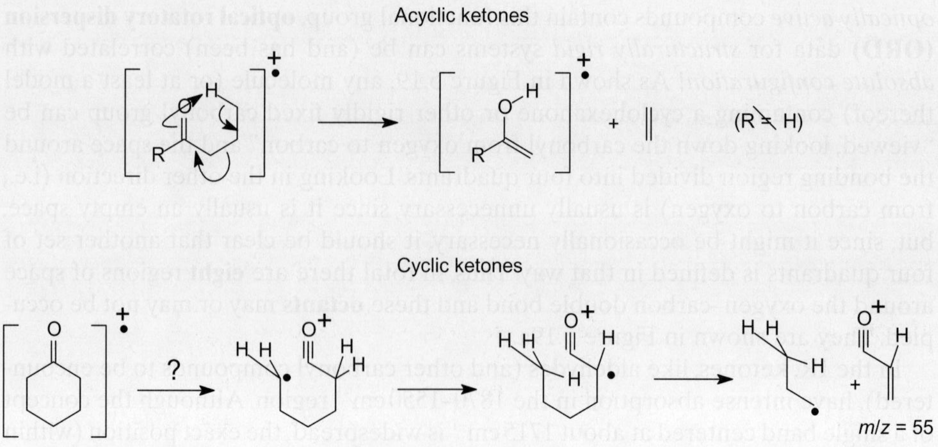

Acyclic ketones

Cyclic ketones

$m/z = 55$

Scheme 5.4. Examples of electron impact fragmentation of cyclic and acyclic ketones.

results in the opposite effect (cyclohexanone, C=O λ_{max} = 1715 cm^{-1}; cyclopentanone, C=O λ_{max} = 1751 cm^{-1}; cyclobutanone, C=O λ_{max} = 1775 cm^{-1}; cyclopropanone, C=O λ_{max} = 1906 cm^{-1}).*

The NMR spectra (^{1}H, ^{13}C, and ^{17}O) of ketones are similar to those of aldehydes and other carbonyl compounds (*vide infra*). Thus, although there can be no proton on the carbon of the carbonyl (as there is for aldehydes), the chemical shift of protons on the α-carbon can be compared. Again, looking at propanal (CH$_3$CH$_2$CHO) as an example, it will be recalled that in the ^{1}H NMR spectrum (Figure 5.18a), the triplet for the methyl group was centered at 1.06 ppm and the methylene quartet (each line of which was further split by coupling to the proton on the aldehydic carbon) at 2.46 ppm. Correspondingly, in the case of 2-butanone (CH$_3$COCH$_2$CH$_3$), the triplet for the protons at C-4 is centered at 1.02 ppm and the methylene quartet at 2.39 ppm. The methylene quartet is, of course, not noticeably split further since there is no proton on the carbon of the carbonyl (the methyl group at C-1 appears as a three-proton singlet at 2.06 ppm).

The same kind of observations also apply to ^{13}C and ^{17}O spectra, that is, small changes, save for the proton on the carbon of the carbonyl, over those already noted for aldehydes.

In the mass spectrum, ketones are also similar to aldehydes. The molecular ion for most ketones is readily observed, and **α-cleavage** also occurs. The charged oxygen bearing daughter fragment is usually that one with the fewer carbon atoms and is often the base peak. Again, as was the case with aldehydes, the McLafferty rearrangement (Scheme 5.3) can be expected if the structure of the ketone allows for the process (i.e., if a cyclic species where a hydrogen is the sixth atom away from the oxygen can be achieved by simple rotation, Scheme 5.4). Cyclic ketones undergo a slightly different fragmentation pathway (Scheme 5.4).

---

*The value for cyclopropanone was obtained on gaseous material. Cyclopropanone is reported to be too reactive (it reacts with itself to yield "polymer") to isolate. The infrared spectra for the other ketones are reported to be for the "neat" (i.e., no solvent) materials. Values not obtained by the author are taken from Silverstein, R. M.; Webster, F. X. *Spectroscometric Identification of Organic Compounds*, 6th edition, Wiley, New York, 1998.

TABLE 5.9. Some Nitrogen, Sulfur, and Phosphorus Derivatives of Ketones and Aldehydes

Ketone or Aldehyde (Name)	Class (Type) of Derivative	Structure (Name)
ethanal (acetaldehyde)	Imine from reaction with methylamine CH_3NH_2	(*E*)-ethanal N-methylimine
propanal	Oxime from reaction with hydroxylamine hydrochloride H_2NOH (HCl)	(*E*)-propanaloxime
propanone (acetone)	Hydrazone from reaction with hydrazine H_2NNH_2	propanone hydrazone
cyclopentanone	Semicarbazone from reaction with semi-carbazine hydrochloride	cyclopentanone semicarbazone
cyclohexanone	Thioketone by reaction with phosphorus pentasulfide (P_2S_5) or other reagents	thiocyclohexanone (unstable)
ethanal (acetaldehyde)	Phosphorane (ylid) or Wittig reagent prepared from the halide not the aldehyde (see Chapter 9)	$(C_6H_5)_3P=CH_2CH_3$ ⇅ $(C_6H_5)_3P—CH_2CH_3$ (+ −) ethylidine triphenylphosphorane

c. Nitrogen, Sulfur, and Phosphorus Analogues of Aldehydes and Ketones.

As amines, thiols, and phosphines may be considered as nitrogen, sulfur, and phosphorus analogues, respectively, of alcohols, it can reasonably be asked if a similar situation occurs with aldehydes and ketones.

As shown in Table 5.9, a variety of nitrogen, sulfur, and phosphorus derivatives of (literally derived from) aldehydes and ketones are known.

d. Carboxylic Acids.

The distinguishing structural feature of that group of compounds known as **carboxylic acids** is that all of them contain the **carboxy group**, variously written as $-CO_2H$ and $-COOH$, as well as where the latter emphasizes that the O–C–O angle is close to 120° (*vide infra*). Although it appears that the **carboxy** group is formally made by combining the carbonyl and hydroxyl groups on one carbon and that therefore the **carboxy** group should have properties of both, the presence of one so modifies the other that while there are some similarities, the differences far outweigh them; indeed, as might be gathered from the name **carboxylic acid**, the defining feature of this function is its **acidity** (*vide infra*).

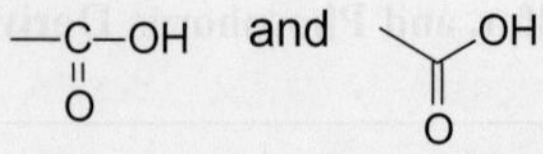

Like aldehydes (of which **carboxylic acids** can be considered oxidation products), the **carboxylic acid** functional group must always begin (or end) a chain in simple hydrocarbon-derived molecules. The IUPAC nomenclature for these compounds requires that the longest chain **beginning** with the **carboxy** group be appropriately named and that the substituents identified (and enumerated) as usual. The acid is then named on the basis of taking the name of the longest chain bearing the carboxy group, dropping the final "e" and replacing it with "oic acid." The IUPAC system also allows (as for **aldehydes**) for the **carboxy** group to be named as a substituent.

Generally, the separate word "acid" ends the name (as in cyclohexanecarboxylic acid [$C_6H_{11}CO_2H$]) and thus is (as for aldehydes) a suffix to the otherwise systematic name. Alternatively, should the situation warrant, "carboxy" for a single appended –CO_2H group can also be used.

As was the case for aldehydes and ketones, the letters of the Greek alphabet can be used to indicate the location of substituents. Again, as earlier (*vide supra*), the lettering starts at the first carbon **adjacent** to the carbonyl.

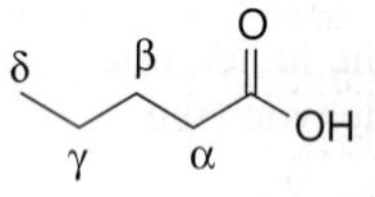

Carboxylic acids occur widely in nature and were among the first compounds isolated—both because of their "acidity" and their stability. It was common knowledge as early as the middle of the seventeenth century that a blue vegetable material called "syrup of violets" turned red with acids—an early working definition of something called an **acid** (*vide infra*).

It is generally found that, as the most highly oxidized form (except for carbon dioxide CO_2) of oxygen-containing organic compounds, carboxylic acids are **relatively** unreactive toward oxygen and may be separated, without change, from accompanying impurities. However, one of the consequences of early isolation is that a plethora of trivial names abound, and indeed, these trivial names have come to serve as the "root" names of many other compounds (cf. **aldehydes**). The relationship among **primary alcohols, aldehydes,** and **carboxylic acids** is shown as follows.

methanol

methanal
(formaldehyde)

methanoic acid
(formic acid)

1-butanol

butanal

butanoic acid
(butyric acid)

hydroxymethyl-
cyclohexane

cyclohexyl-
carboxaldehyde

cyclohexane-
carboxylic acid

Some of the common carboxylic acids and their trivial (and IUPAC) names are provided in Table 5.10.

Although it might be predicted that sp^2 hybridization would be found at the carbon of the carbonyl in, for example, methanoic acid (formic acid, HCO_2H) and that, as a consequence, the oxygen–carbon–oxygen (O=C–O) bond angle and the hydrogen–carbon–oxygen (H–C–O) angles (one to the oxygen of the hydroxyl and the other to the carbonyl) would all be about 120°. As seen in Figure 5.20, there are two different forms of formic acid (at high dilution in the gas phase). They differ in the sense of rotation about the carbon–oxygen single bond.

For this, as most other carboxylic acids, it is usual to find that in the gas phase (at reasonable pressures), as well as in solution and in the solid phase, carboxylic acids are **dimeric**.

Although, as already noted above, carboxylic acids are characterized by their acidity, it is generally found that these materials are among those called "weak acids" as most do not ionize completely in aqueous (or other) solution. A fuller discussion of acidity follows later in this Chapter but, as expected, the carboxylic acids do react with bases to form salts, which are frequently water soluble.

Carboxylic acids (and their salts) without unsaturation other than the carbon–oxygen double bond have only weak absorption in the UV in the 200–400 nm range. However, as was the case for α,β-unsaturated aldehydes and ketones, acids and their salts that also enjoy the presence of a double bond in conjugation with the C=O unsaturation do absorb strongly. Thus, for example, propanoic acid ($CH_3CH_2CO_2H$) has λ_{max} at about 204 nm ($\varepsilon \approx 41$), while propenoic acid (acrylic acid, H_2C=$CHCO_2H$),

TABLE 5.10. Some Common Carboxylic Acids

Compound	Trivial Name	IUPAC Name	Boiling Point (°C, 1 atm)
HCO_2H	Formic acid	Methanoic acid	101
CH_3CO_2H	Acetic acid	Ethanoic acid	118
$CH_3CH_2CO_2H$	Propionic acid	Propanoic acid	141
$CH_3(CH_2)_2CO_2H$	Butyric acid	Butanoic acid	163
$(CH_3)_2CHCO_2H$	Isobutyric acid	2-Methylpropanoic acid	154
$CH_3(CH_2)_3CO_2H$	Valeric acid	Pentanoic acid	187
$CH_3(CH_2)_4CO_2H$	Caproic acid	Hexanoic acid	205
$CH_3(CH_2)_5CO_2H$	Enanthic acid	Heptanoic acid	223
$CH_3(CH_2)_6CO_2H$	Caprylic acid	Octanoic acid	239
$CH_3(CH_2)_7CO_2H$	Pelargonic acid	Nonanoic acid	255
$CH_3(CH_2)_8CO_2H$	Capric acid	Decanoic acid	270
$HOCH_2CO_2H$	Glycolic acid	Hydroxyethanoic acid	dec (mp 78)
$CH_3CH(OH)CO_2H$	Lactic acid	2-Hydroxypropanoic acid	dec (mp 17)
$CH_2=CHCO_2H$	Acrylic acid	Propenoic acid	141
HO_2CCO_2H	Oxalic acid	Ethanedioic acid	dec (mp 180)
$HO_2CCH_2CO_2H$	Malonic acid	Propanedioic acid	dec (mp 135)
$HO_2C(CH_2)_2CO_2H$	Succinic acid	Butane-1,3-dioic acid	dec (mp 183)
$HO_2C(CH_2)_3CO_2H$	Glutaric acid	Pentane-1,4-dioic acid	303
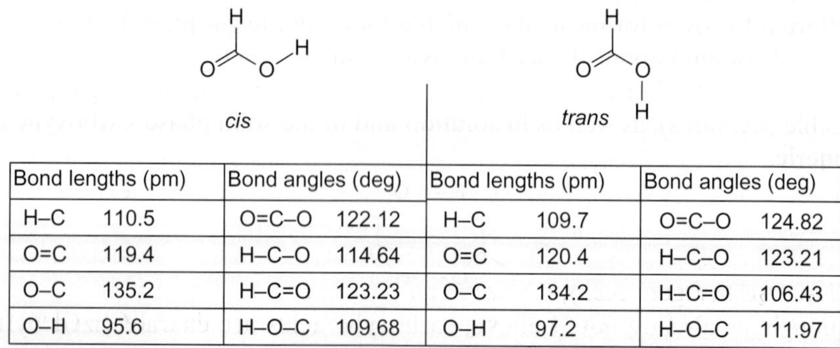	Benzoic acid	Benzenecarboxylic acid	mp 122

The data in this Table were obtained from the *Dictionary of Organic Compounds*, CHEMnetBase, CRC Press, at http://doc.chemnetbase.com.

Bond lengths (pm)		Bond angles (deg)		Bond lengths (pm)		Bond angles (deg)	
H–C	110.5	O=C–O	122.12	H–C	109.7	O=C–O	124.82
O=C	119.4	H–C–O	114.64	O=C	120.4	H–C–O	123.21
O–C	135.2	H–C=O	123.23	O–C	134.2	H–C=O	106.43
O–H	95.6	H–O–C	109.68	O–H	97.2	H–O–C	111.97

Figure 5.20. Representations of *cis*- and *trans*-methanoic acid (formic acid, HCO_2H) showing bond angles and lengths (from microwave spectroscopy). These were obtained in the gas phase at high dilution.

which also absorbs near 200 nm, does so with an $\varepsilon \approx 10,000$! Indeed, the same set of Woodward's rules already briefly outlined for aldehydes and ketones (which can now be profitably reviewed) can also be used for carboxylic acids and their salts.

Carboxylic acids have dramatically distinctive IR spectra. Thus, as shown in Figure 5.21, the broad, intense, hydrogen bonded O–H is spread across the entire

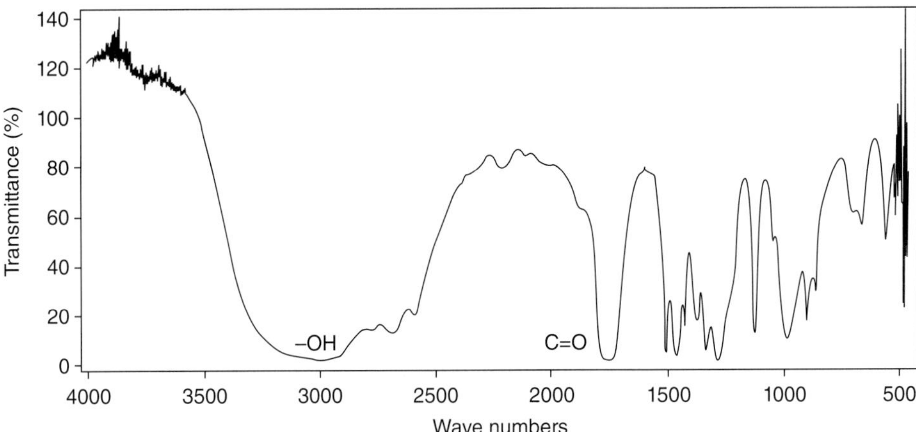

Figure 5.21. The infrared spectrum of propanoic acid ($CH_3CH_2CO_2H$) (transmittance) taken as a thin film ("neat").

region between about 2500 and 3300 cm^{-1} (the "free" –OH near 3520 cm^{-1} is seen only in very dilute solutions in solvents considered incapable of hydrogen bonding), and there is a second intense band, due to absorption by the carbonyl group in the region between about 1705 and 1720 cm^{-1}. In concert with what was observed in the UV, conjugation (α,β-unsaturation) causes the IR absorption to be shifted to slightly longer wavelength (1680–1710 cm^{-1}).

The ^{1}H NMR spectra of carboxylic acids are also distinctive, in particular, because the proton directly attached to the hydroxyl of the carboxyl is strongly deshielded, and routinely appears very far downfield (between about $\delta = 10$ and 13 ppm; TMS = 0.00). However, because this proton is easily exchanged, it is frequently found that it will exchange with solvent protons (depending, of course, on the solvent) or the signal has been broadened to such an extent that it is not observed (Figure 5.22).

The diamagnetic anisotropy induced as a result of electron circulation within and about the carbonyl group of aldehydes and ketones might also be expected to be experienced by the same group in carboxylic acids. Thus, the carbon of the carboxyl group should be deshielded, and, in the ^{13}C NMR spectrum of propanoic acid ($CH_3CH_2CO_2H$) (Figure 5.23), the carboxyl carbon comes into resonance at about 178.4 ppm downfield from TMS. As was the case with propanal (CH_3CH_2CHO) and butanone ($CH_3CH_2COCH_3$), *vide supra*, the carbon atom α to the carboxyl is also deshielded (relative to what the chemical shift would be anticipated to be were the C=O absent). This is observed in Figure 5.23.

It is reasonable to ask, as before, if the ^{17}O resonance (around 40.7 MHz at 7.05 T or 67.8 MHz at 11.75 T) is effected in the same way and to the same extent as was the case with aldehydes and ketones (*vide supra*). Interestingly, the ^{17}O resonance for carboxylic acids is generally found between that of the hydroxyl of alcohols and the carbonyl of aldehydes and ketones! Thus, as shown in Figure 5.24, the ^{17}O resonance for propanoic acid ($CH_3CH_2CO_2H$) in ^2H_2O is found at about 255 ppm downfield from the water signal. This information is taken as evidence that, on the NMR

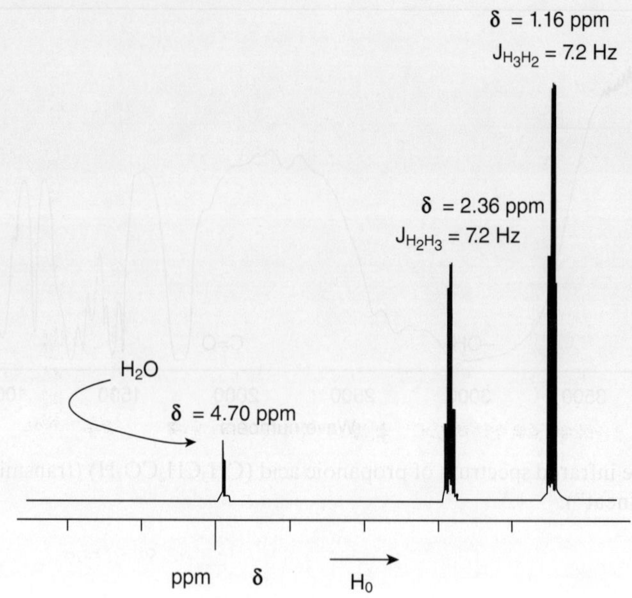

Figure 5.22. The ^{1}H NMR spectrum of propanoic acid ($CH_3CH_2CO_2H$) in ^2H_2O at 300 MHz.

timescale (Chapter 2), the difference between the two oxygens (C=O/C–OH) is averaged out and the oxygens cannot be distinguished.*

Problem 5.9. How would you account for the observation that while the IR spectrum of propanoic acid, Figure 5.21, clearly shows two different kinds of oxygen (i.e., both C=O and –OH groups can be seen), the ^{17}O NMR spectrum of the same material shows that there is only one kind of oxygen? Note: a correct response will probably give some indication of the *difference* in the two time frames.

Electron impact mass spectrometry of carboxylic acids usually produces only a weak molecular ion (M^+). If the carboxylic acid has a total molecular weight below about 150, an M^+-17 fragment (due to the loss of the hydroxyl), as well as an M^+-45 (as a consequence of losing CO_2H), is frequently observed. With the larger carboxylic acids, a McLafferty rearrangement such as the one shown in Scheme 5.5 (which should be compared with Scheme 5.3) intrudes.

e. Carboxylic Acid Derivatives. Although the chemistry of the derivatives of carboxylic acids[†] is generally richer than that of the acids themselves, the **nomenclature**, as well as the **distinguishing structural features**, of the members of the group makes it desirable to temporarily group them together. All the derivatives (except nitriles, *vide infra*) contain a **carboxy group**, $-CO_2H$, in which the –OH has been

---

*You should compare this result with that seen in the IR (Figure 5.21).

[†]The compounds presented were, historically, actually *derived from* carboxylic acids. The concept of a derivative has stayed although many of these materials, prepared in other ways, are precursors to carboxylic acids.

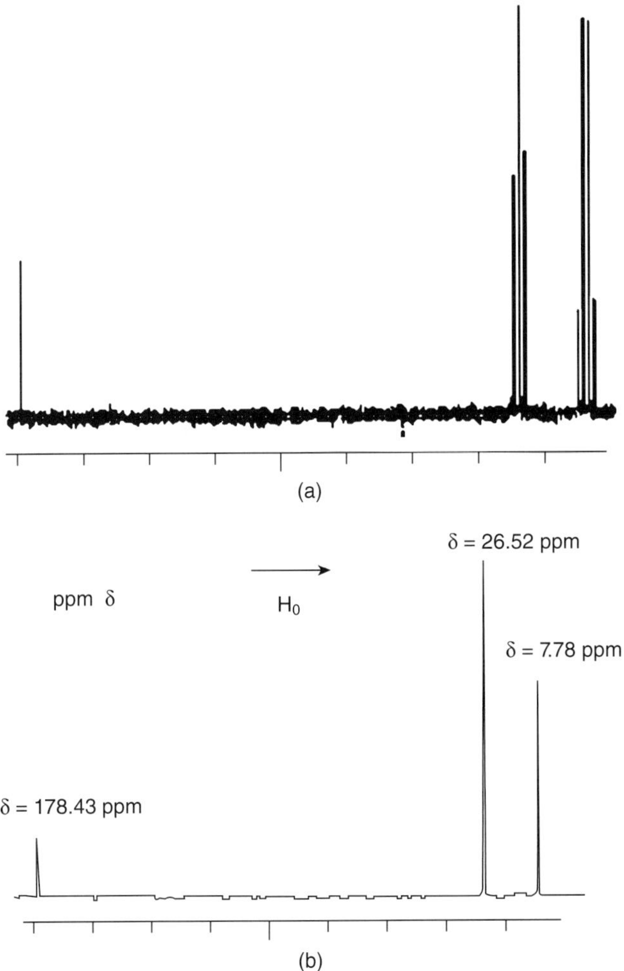

Figure 5.23. The ^{13}C NMR spectrum of propanoic acid ($CH_3CH_2CO_2H$) in $^{2}H_2O$ at 75.46 MHz (a) undecoupled and (b) proton noise decoupled.

replaced by a ligand (L) other than H or carbon (i.e., neither an aldehyde nor a ketone is formed). Table 5.11 contains a sampling of some of these compounds.

1. Carboxylic Acid Halides. The constant portion, that is, the carbon and oxygen of the carbonyl and its appended alkyl or aryl group, RC=O is a functional unit called the **acyl group** and as shown for the acid chloride ethanoyl chloride (acetyl chloride, CH_3COCl) in Table 5.11, the name is derived by simply using the name of the hydrocarbon root with the same number of carbon atoms, dropping the final "e" and replacing it with "oyl." In this way, ethane becomes ethanoyl. The halide (i.e., halogen as "halide") is cited as a *separate word*.

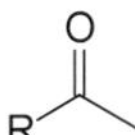

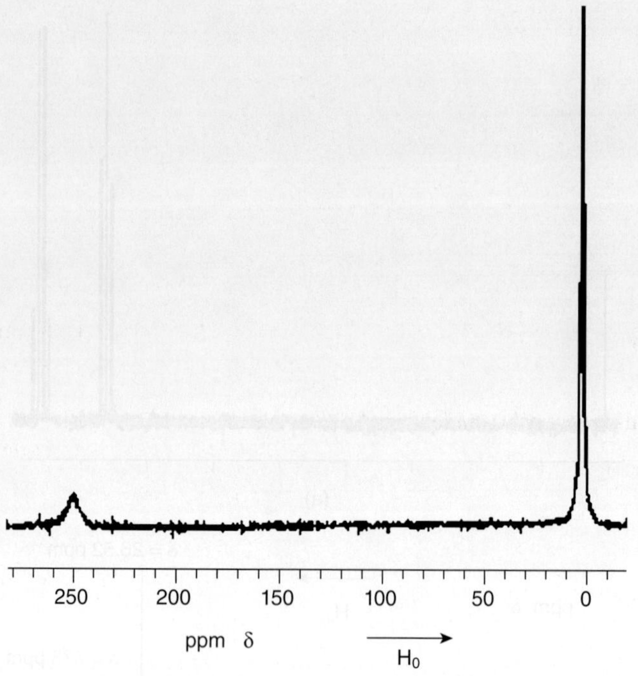

Figure 5.24. The ^{17}O NMR spectrum (at 67.8 MHz) of propanoic acid ($CH_3CH_2CO_2H$) in ^2H_2O solvent. On the "NMR timescale" the oxygens are equivalent.

Scheme 5.5. A carboxylic acid undergoing McLafferty-type rearrangement with fragmentation on electron bombardment in the mass spectrometer.

A common variant, used when the *carbonyl* group can best be considered a substituent (e.g., on a ring), is to simply append the word "carbonyl" to the name of the alkane; for example, the unit **cyclohexanecarbonyl** is

Carboxylic acid chlorides are, generally, reactive compounds. However, they can be important intermediates between carboxylic acids and other materials, and the conversion of an acid to this **derivative** is accompanied, in the IR, by loss of the –OH absorption of the carboxylic acid (2500–3300 cm^{-1}) and a shift of the C=O to shorter wavelength (for the carboxylic acid, the intense C=O absorption is expected

TABLE 5.11. A Sampling of Derivatives of Carboxylic Acids

Ligand (L) (and Generic Name)	$R-\overset{O}{\underset{\|}{C}}-L$	Name
Cl (acid chloride)	$CH_3-\overset{O}{\underset{\|}{C}}-Cl$	Ethanoyl chloride (acetyl chloride)
$\overset{O}{\underset{\|}{C}}$ O—C—CH$_3$ (acid anhydride)	$CH_3-\overset{O}{\underset{\|}{C}}-O-\overset{O}{\underset{\|}{C}}-CH_3$	Ethanoic anhydride (acetic anhydride)
–OR (R = alkyl or aryl) (ester)	$CH_3-\overset{O}{\underset{\|}{C}}-OCH_2CH_3$	Ethyl ethanoate (ethyl acetate)
NR$_2$ (R = H, alkyl or aryl) (amide)	$CH_3-\overset{O}{\underset{\|}{C}}-\underset{H}{\overset{}{N}}-CH_3$	*N*-methylethanamide (*N*-methylacetamide)
Nitriles	$CH_3C{\equiv}N$	Ethanenitrile (acetonitrile)

to be centered about 1760 cm^{-1}, while in the acid chloride, a weaker band in the 1785–1815 cm^{-1} region is frequently observed).

Acid halides are generally handled with good ventilation as they are acrid smelling and frequently *lachrymators*, and while acyl fluorides, bromides, and iodides have been reported, the chloride is the most common derivative.

Although the simplest acyl chloride, methanoyl chloride (formyl chloride, HCOCl) is unstable at room temperature, its microwave spectrum has been reported. The structural data available are outlined in Figure 5.25. It is worthwhile noting that the carbon–oxygen double bond (C=O) in this acid chloride, as well as in ethanoyl chloride (acetyl chloride, CH$_3$COCl), are about 118–119 pm long (about the same as in methanoic acid [formic acid, HCO$_2$H], *vide supra*) and slightly shorter than in methanal (formaldehyde, H$_2$CO) and ethanal (acetaldehyde, CH$_3$CHO) (Figure 5.26).

2. Carboxylic Acid Anhydrides. The replacement of the **proton** attached to the oxygen in the carboxylic acid functional group (–CO$_2$H) by an acyl group produces the derivative called a **carboxylic anhydride**. There are, of course, two possibilities for carboxylic anhydrides of the general form RCO(O)OCR or (RCO)$_2$O since the "R" groups can be the same as each other or they can be different from each other.

$$R-\overset{O}{\underset{\|}{C}}-O-\overset{O}{\underset{\|}{C}}-R \quad \text{same as} \quad (RCO)_2O$$

As noted in Table 5.11, if they are the same, the name of the parent acid is used and the word "acid" is replaced by "anhydride." If they are not the same, each parent acid is cited (alphabetically) as a separate word and the descriptor "anhydride" appended as before.

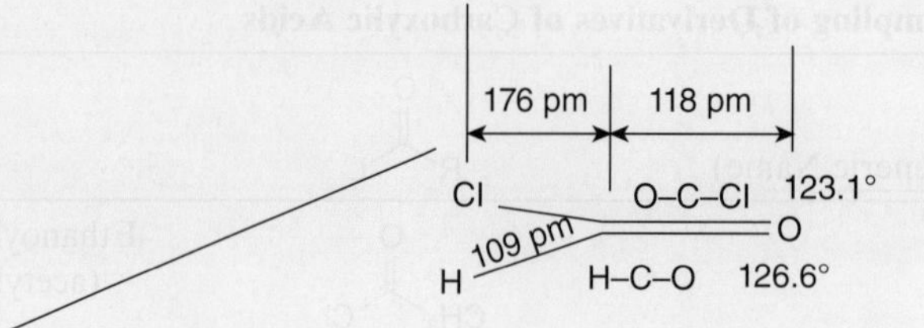

Figure 5.25. A representation of methanoyl chloride (formyl chloride, HCOCl), an unstable molecule, showing bond lengths and angles (as determined by microwave spectroscopy).

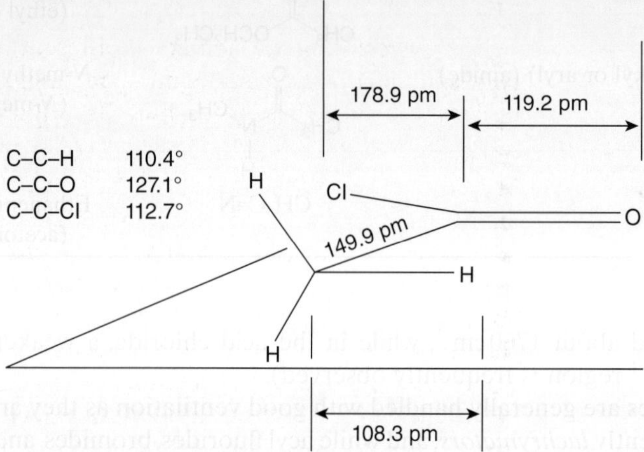

Figure 5.26. A representation of ethanoyl chloride (acetyl chloride, H$_3$CCOCl) showing bond lengths and angles (as determined by microwave spectroscopy).

Interestingly, the IR spectra of anhydrides, both where the carboxylic acid parents are the same and where they are different, show two carbonyl stretching bands (asymmetric and symmetric coupled), generally separated by less than 100 cm^{-1}. For example, as shown in Figure 5.27 for ethanoic anhydride (Table 5.11, acetic anhydride [(CH$_3$CO)$_2$O or CH$_3$COOCOCH$_3$]), the *two* carbonyl stretching bands occur at 1758 cm^{-1} (symmetric) and 1825 cm^{-1} (asymmetric), and, of course, the –OH has disappeared.

As might be anticipated, the ^{1}H and ^{13}C NMR spectra confirm the symmetry with only a singlet observed in the case of the former and, in the undecoupled ^{13}C spectrum of the latter, the anticipated methyl quartet because of ^{1}H-^{13}C coupling. Interestingly, long range (but small) coupling of the carbonyl carbon to the protons on the methyl group can also be seen. The spectra are shown in Figure 5.28. Of course, the ^{17}O NMR spectrum (Figure 5.29) shows that there are two distinctively different types of oxygen present.

In the presence of water, carboxylic anhydrides that are soluble in water are quickly converted back to their corresponding carboxylic acids. Thus, like acid halides, carboxylic anhydrides are reactive compounds and are frequently used in synthesis as intermediates between acids and other materials. Ethanoic anhydride (acetic anhydride [(CH$_3$CO)$_2$O]) and 1,2-benzenedicarboxylic anhydride (phthalic

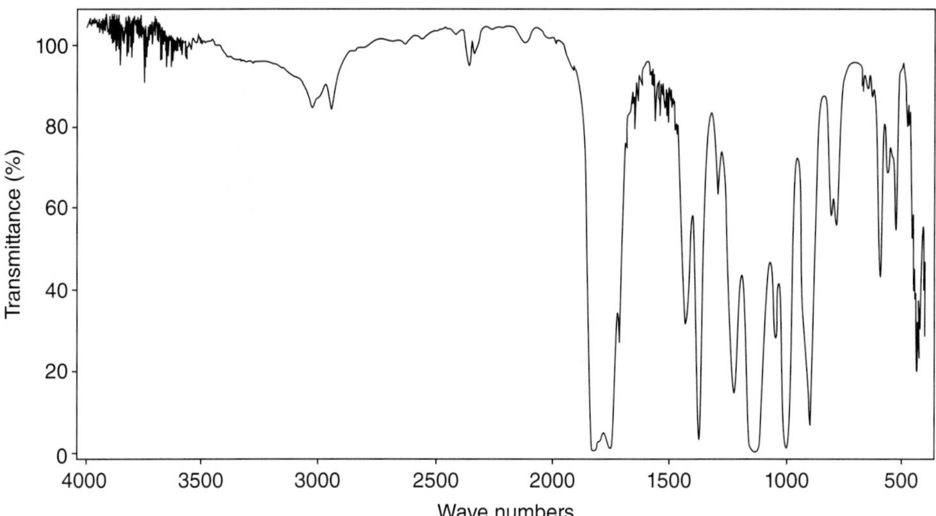

Figure 5.27. The infrared transmittance spectrum (neat) of ethanoic anhydride (acetic any-dride) as a thin film between two sodium chloride (NaCl) windows.

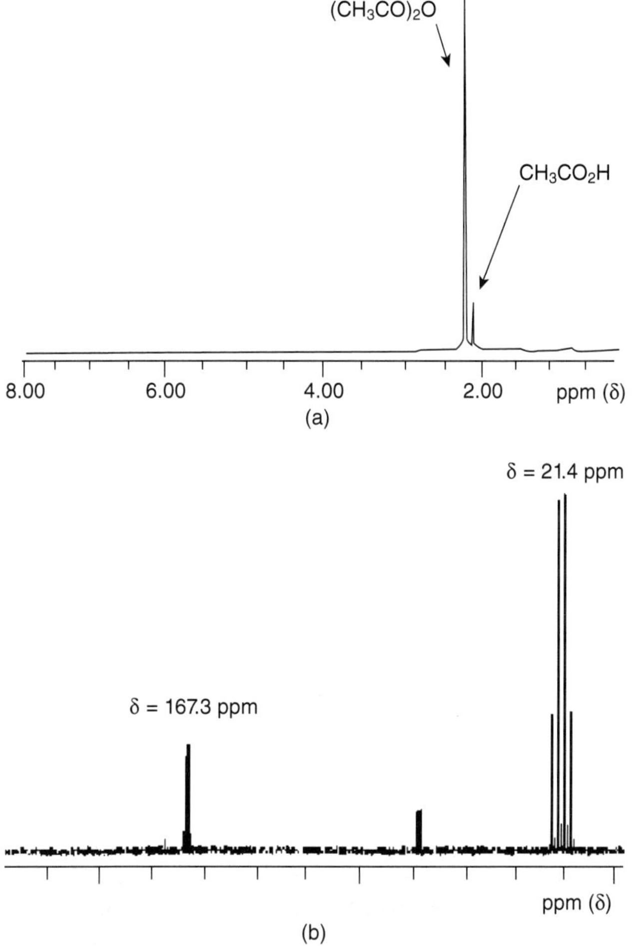

Figure 5.28. (a) The ^{1}H NMR spectrum of ethanoic anhydride [(CH$_3$CO)$_2$O] in ^{2}HCCl$_3$ at 300 MHz; (b) the coupled ^{13}C NMR spectrum of the same compound at 75.5 MHz. In both spectra, a trace amount of acetic acid is seen.

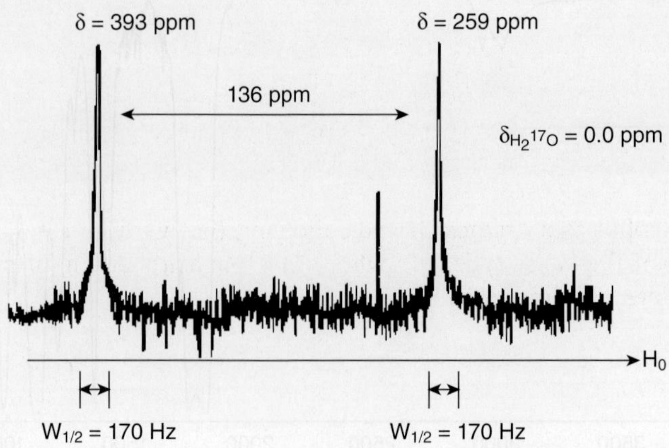

Figure 5.29. The ^{17}O NMR spectrum in $^2HCCl_3$ of ethanoic anhydride [acetic anhydride, $(CH_3CO)_2O$] at 67.8 MHz. The carbonyl oxygen is found at 393 ppm from water.

Scheme 5.6. The equilibrium between propyl propanoate and water and propanoic acid and propyl alcohol.

anhydride [$C_6H_4(CO)_2O$]), shown below, are valuable commercial materials widely used in the plastics industry.

1,2-benzenedicarboxylic acid
(*ortho*-phthalic acid)

1,2-benzenedicarboxylic anhydride
(phthalic anhydride)

Problem 5.10. Of the two isomeric (maleic and fumaric) butenedioic acids ($HO_2CCH=CHCO_2H$), only one of them (maleic acid) forms an intramolecular anhydride. Draw and label the structures of the two acids and the one anhydride.

3. Carboxylic Acid Esters and Lactones. In principle, the replacement of **either** the hydrogen of the generalized carboxylic acid group $–CO_2H$ by some alkyl or aryl group "R" or the replacement of the hydroxyl group of $–CO_2H$ by an alkoxy or aryloxy group "OR," while clearly different operations, lead to an equivalent result, that is, the production of a *carboxylic acid ester* $–CO_2R$ (or, if cyclic, a *lactone*).

Generally, *esters* are actually derived from a reaction between a *carboxylic acid* and an *alcohol* (accompanied by the loss of water) and the reaction of water with an *ester* ("hydrolysis") produces the carboxylic acid and the alcohol. Thus, as shown in Scheme 5.6, the formation of the *ester* **propyl propanoate** and its **hydrolysis** to

1-propanol and **propanoic acid** are connected, in **equilibrium**, which may lie, **depending on the conditions**, on one side or the other!

As might be gathered from the name **propyl propanoate** (*vide supra*), the names of **esters**, in general, are compounded from those of the *alcohol* and *carboxylic acid* from which the specific ester might be considered "derived." The "compounding" consists of replacing the "ol" ending of the IUPAC name of the alcohol by "yl" (i.e., naming the alcohol as a "radical") and following that with, as a second word, the IUPAC name of the carboxylic acid where the "oic acid" ending has been replace by the suffix "ate." Examples follow.

methyl ethanoate   methyl propanoate   ethyl cyclopropanecarboxylate

Interestingly, the salts of carboxylic acids, utilize the same nomenclature system. Thus, conversion of ethanoic acid (acetic acid, CH_3CO_2H) to its sodium salt (with, e.g., aqueous sodium hydroxide, $NaOH$) produces water and **sodium ethanoate** (sodium acetate, $CH_3CO_2^-Na^+$). Furthermore, when the (RCO_2-) group must be named as a substituent, it is called the "acyloxy" or "R-carbonyloxy" group.

1-ethanoyloxycyclohexene
(1-acetoxycyclohexene)

And, it may sometimes be necessary to name the $-CO_2R$ group as a substituent (note: $-CO_2R$ is not the same as RCO_2-, the acyloxy group). The $-CO_2R$ group is called "alkoxycarbonyl."

3-ethyl-1-methoxycarbonylcyclopentene

Cyclic esters (esters derived from **hydroxyl** bearing **carboxylic acids**) are called **lactones**. Two different systems of nomenclature exist for lactones. The first, names lactones as cyclic ethers (*vide supra*) bearing carbonyl groups. The second uses letters of the Greek alphabet to specify the relationship between the position of the hydroxyl and carboxylic acid groups and thus, the ring size. Both nomenclature systems are shown in Table 5.12.

In contrast to what was observed in the microwave spectrum of methanoic acid (formic acid, HCO_2H) (Figure 5.20), where *two forms of the acid* (as a consequence of restricted rotation because of hydrogen bonding) could be seen, the microwave spectrum of methyl methanoate (methyl formate, HCO_2CH_3) contains no evidence for more than the one conformer shown in Figure 5.30.

TABLE 5.12. Some Hydroxycarboxylic Acids and the Lactones Which May Be Derived from Them

Hydroxycarboxylic Acid	Lactone
Hydroxyethanoic acid (hydroxyacetic acid, glycolic acid)	Oxacyclopropanone; α-acetolactone; oxiranone (unstable; not isolable)
3-Hydroxypropanoic acid (β-hydroxypropanoic acid)	Oxacyclobutane-2-one (oxetan-2-one, β-propiolactone)
4-Hydroxybutanoic acid (γ-hydroxybutyric acid)	Oxacyclopentan-2-one (oxolan-2-one; γ-butyrolactone)

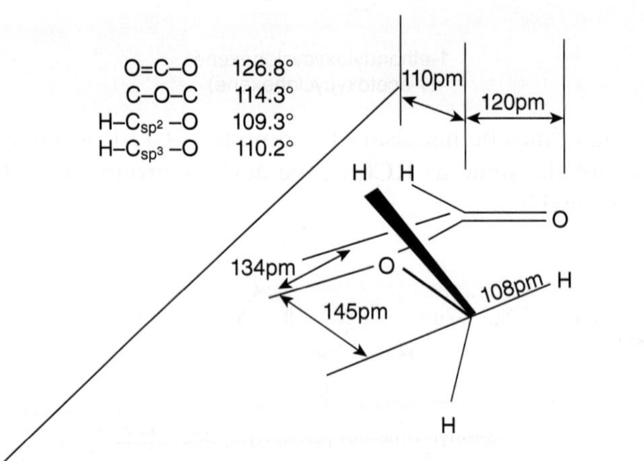

Figure 5.30. A representation of the structure of methyl methanoate (methyl formate, HCO_2CH_3) as determined by microwave spectroscopy.

The UV spectra of saturated (cyclic and acyclic) esters resemble those of the corresponding acids and their simple metal (e.g., sodium) salts. Indeed, as with the acids from which they are "derived," conjugation moves absorption to longer wavelength and with increasing comtensity.

Both normal aliphatic esters and six-membered ring lactones (δ-lactones) have their carbonyl-stretching absorptions in the region between 1735 and 1750 cm^{-1}; the IR spectrum of propyl propanoate, neat liquid, is shown in Figure 5.31.

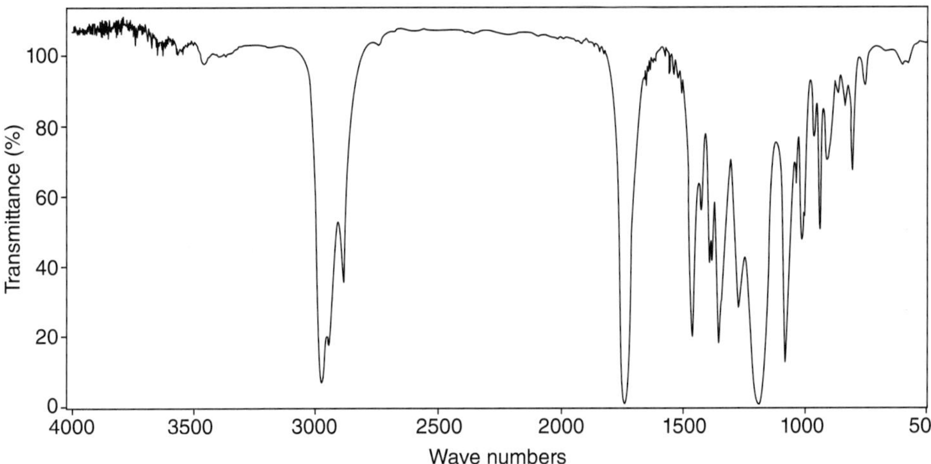

Figure 5.31. The infrared (transmittance) spectrum of propyl propanoate, $CH_3CH_2CO_2CH_2$ CH_2CH_3, neat, as a thin film.

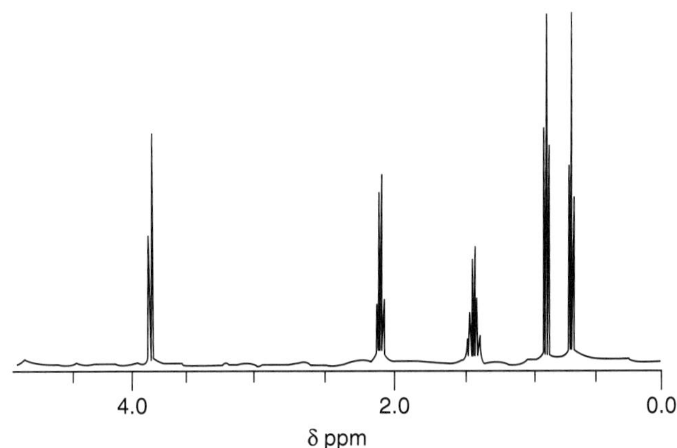

Figure 5.32. The ^1H NMR spectrum of propyl propanoate ($CH_3CH_2CO_2CH_2CH_2CH_3$) in $^2HCCl_3$ at 300 MHz.

The ^1H, ^{13}C (coupled and decoupled), and the ^{17}O NMR spectra of propyl propanoate ($CH_3CH_2CO_2CH_2CH_2CH_3$) are shown as Figures 5.32–5.34, respectively. All of these spectra are, of course, unremarkable in light of what is expected. Thus, analysis of the ^1H NMR spectrum (Figure 5.32) shows typical, characteristic signals for the protons on carbon adjacent to a carbonyl ($\delta = 2.28$ ppm, 2H, q, $J = 7.5$ Hz), attached to a carbon bearing an oxygen ($\delta = 3.49$ ppm, 2H, t, $J = 6.9$ Hz), on the additional methylene ($-CH_2-$) in the alcohol-derived portion and, finally, the terminal methyl (CH_3-) groups (each a triplet because of the adjacent [different] methylenes [$-CH_2-$]). Similarly, in the ^{13}C NMR spectra, the carbonyl carbon stands alone at $\delta = 173.3$ ppm and the one carbon, attached directly to oxygen, is found, about where expected, at $\delta = 65.0$ ppm. Meanwhile, the corresponding methylene ($-CH_2-$) carbons and methyl carbons are at about the chemical shift values expected for their

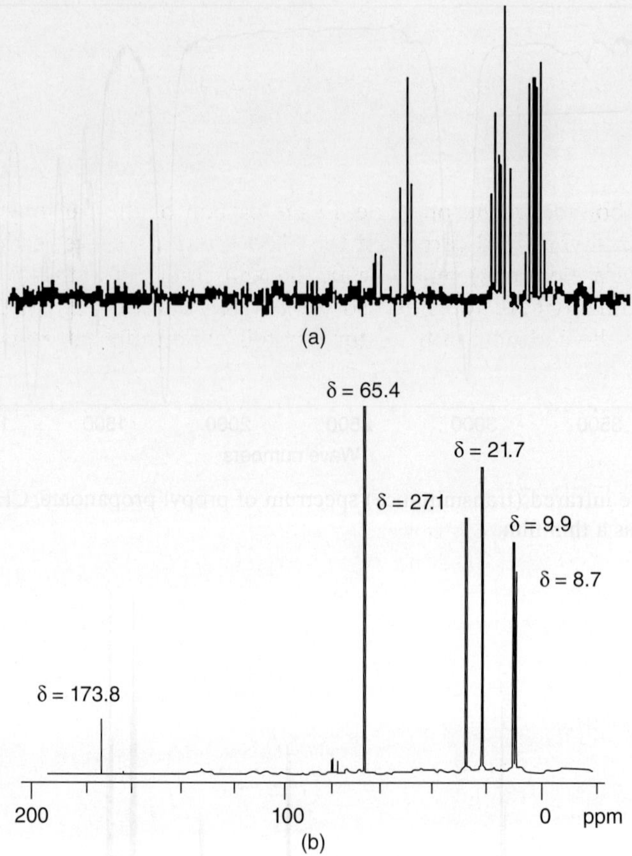

Figure 5.33. The ^{13}C NMR spectra (a, coupled [or undecoupled]; b, decoupled) of propyl propanoate ($CH_3CH_2CO_2CH_2CH_2CH_3$) in $^2HCCl_3$ at 75.5 MHz.

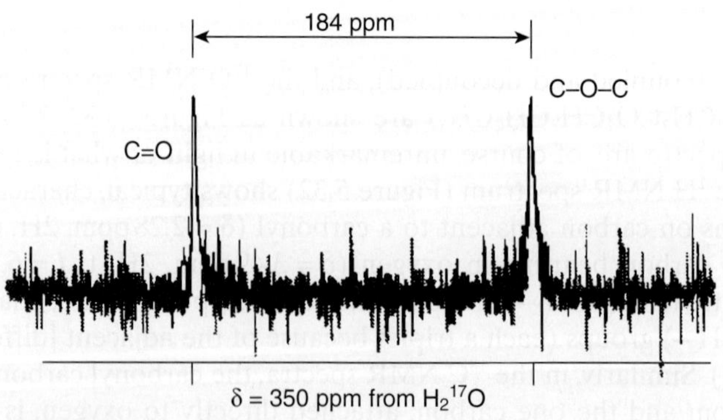

Figure 5.34. The ^{17}O NMR spectrum of propyl propanoate ($CH_3CH_2CO_2CH_2CH_2CH_3$) in $^2HCCl_3$ at 67.8 MHz.

respective types (see Figure 5.33). Of striking interest, however, is the observation that the ^{17}O NMR spectrum (Figure 5.34) shows the presence of two different oxygen atoms (a result that should be contrasted to that for the propanoic acid ($CH_3CH_2CO_2H$) itself (Figure 5.24) and compared with that seen, for example, for ethanoic anhydride (acetic anhydride [$(CH_3CO)_2O$]) (Figure 5.29). It is expected that the doubly bonded carbonyl oxygen is to be found further downfield than the oxygen singly bound, on the one side, to the carbon of the carbonyl and, on the other, to the sp^3-hybridized carbon of the alkyl group (because of the anisotropy induced by the π-electrons) but, as with the anhydride, the other oxygen is also affected! Presumably here, too, it is also the π system carbonyl that is the cause and it suggests that the carbonyl of the ester group is quite different from that of aldehydes and ketones (where the chemical shifts of the carbons and oxygens of these groups are found, respectively, much further downfield and thus more strongly deshielded).*

Esters, generally, are found to be responsible for the "flavor" (odor/taste) of many fruits. Frequently, although the mixture of natural constituents responsible for flavor is very complex, simple ester mimics can be found and these materials are used as artificial (although there is nothing "artificial" about them) flavorings. Some examples of simple esters and the "flavors" they possess are given in Table 5.13.†

As shown in Scheme 5.7, esters, even when a parent ion is found to be present, undergo the (now familiar) McLafferty rearrangement typical of carbonyl compounds where there is a hydrogen atom six atoms removed (across five bonds) from the oxygen. A direct consequence of this is that **all methyl esters lacking α-substitution** will, regardless of other factors, possess a peak (frequently the base peak) at m/z 74!

Interestingly, many esters are found to yield ions resulting from both possible senses of fragmentation depicted in Scheme 5.8. Thus, fragmentation on both sides of the carbonyl group is found and the specific detail of which side preferentially cleaves (as surmised from the fragments found) is a result of the substitution pathway of the parent.

4. Amides, Lactams, Imides, Hydroxamic Acids, and Ureas. Attachment of an acyl group (RCO) to a trigonally substituted nitrogen (as in ammonia and amines), or the equivalent operation of replacement of the hydroxyl function (–OH) of the carboxylic acid group ($-CO_2H$) by trigonally substituted nitrogen, produces **amides**. Amides are further subdivided (as were amines) into those bearing two hydrogens and the acyl group on nitrogen (primary amides); those bearing the acyl group, one hydrogen, and an alkyl (or aryl group) on nitrogen (secondary amides); and finally,

---

*The chemistry to be studied will lead to the same conclusion, that is, that there is a striking difference between these two types of carbonyl groups. Indeed, correlation of the massive volumes of chemical information led to this same conclusion first and further suggested, as foreshadowed in the resonance structures for, for example, the carbonate anion (Chapter 1), that the nonbonded electrons on the (nominally) sp^3-hybridized oxygen helped ameliorate the charge difference between the oxygen and carbon of the carbonyl. This same relief cannot be experienced by aldehydes and ketones but, *vide infra*, would be expected to be enjoyed by amides!

†The subject of how such small molecules interact with an appropriate receptor to produce an electrical signal, which is transported to a remote site where it is subsequently interpreted remains under active investigation.

TABLE 5.13. Some Carboxylic Esters and Their Agreed upon "Flavors"

Compound	Structure	Boiling Point (°C)	Flavor
3-Methylbutyl ethanoate (isoamyl acetate)	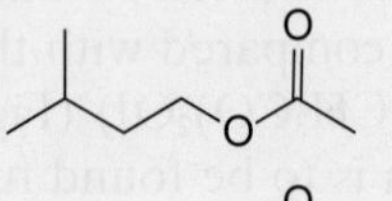	142	Banana
3-Methybutyl-3-methyl-butanoate (isoamyl valerate)	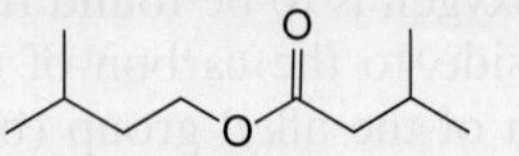	194	Apple
Butyl butanoate	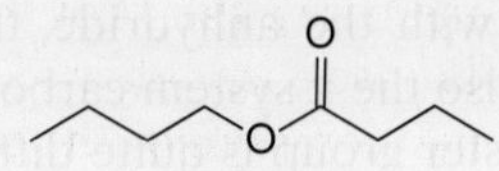	156	Pineapple
2-Methylpropyl propanoate (isobutyl propanoate)	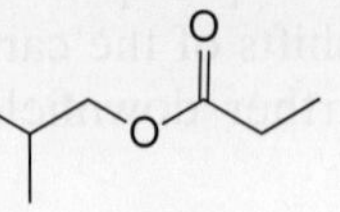	111	Rum
Methyl (2-hydroxy)-benzenecarboxylate (methyl salicylate)	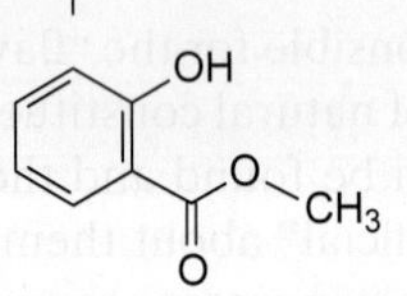	220	Wintergreen oil
5-Heptyloxacyclo-pentan-2-one (γ-undecanolactone)	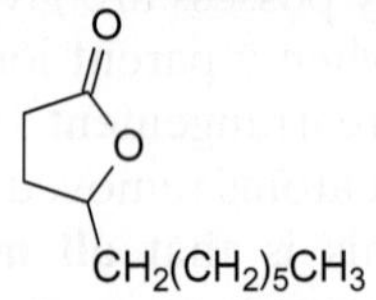	164–166 at 13 torr	Peach

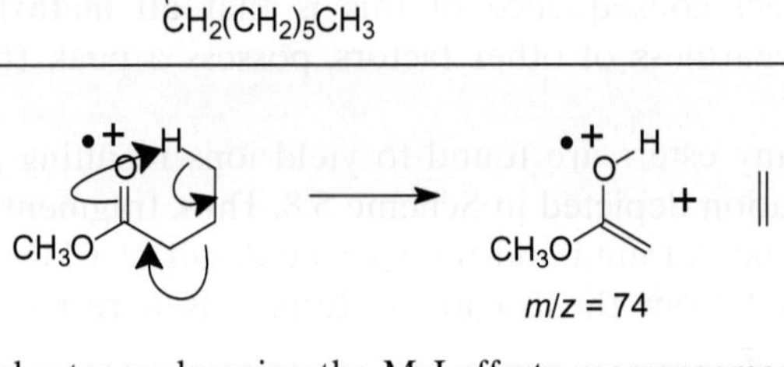

$$m/z = 74$$

Scheme 5.7. A methyl ester undergoing the McLafferty rearrangement and producing the common (to all such methyl esters) m/z 74 ion.

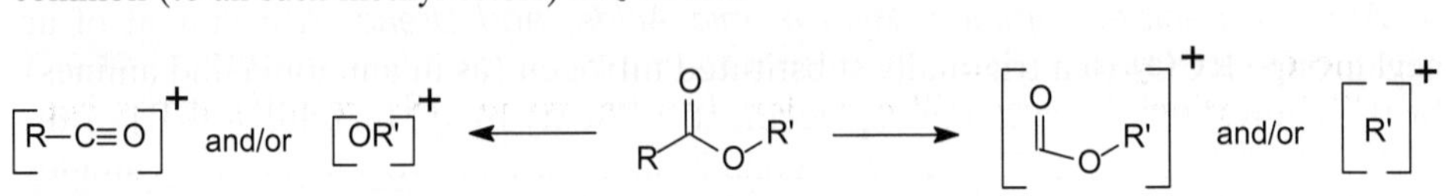

Scheme 5.8. Possible fragments anticipated in the mass spectrum of esters as a result of cleavage on both sides (or either side) of the carbonyl group.

those bearing two alkyl, two aryl, or one alkyl and one aryl group on nitrogen along with the acyl attachment (tertiary amides).

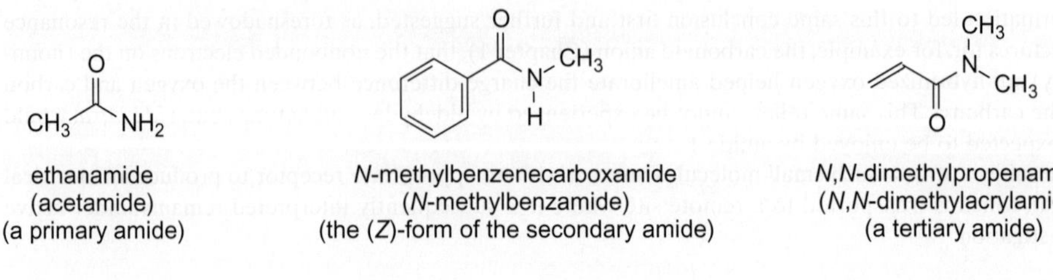

ethanamide
(acetamide)
(a primary amide)

N-methylbenzenecarboxamide
(*N*-methylbenzamide)
(the (*Z*)-form of the secondary amide)

N,N-dimethylpropenamide
(*N,N*-dimethylacrylamide)
(a tertiary amide)

As is thus clear from these few examples, the names of **amides** can be considered to be derived from those of the parent **alkanes, alkenes**, and so on, by simply removing the final "e" and replacing it with the suffix "amide." Alternatively, reinforcing their relationship to **carboxylic acids**, the names of **amides** can be considered to be derived from those of the corresponding **carboxylic acids** by removing the suffix "oic" and the added word "acid" and replacing them by the suffix "amide."

However, as noted above in the examples shown, if the **nitrogen** atom of the **amide** bears substituents, they are (a) located by placing an *N* for **each** substituent on **nitrogen** as a hyphen-separated prefix along with (b) the name, as a radical ending in "yl," of the **substituent**, or substituents. If more than one substituent is present, they are named in alphabetical order if nonidentical, or with the further prefix "di" if they are identical.

As lactones are **cyclic esters**, so **lactams** are **cyclic amides**. Again, as was the case with lactones, the nomenclature of the lactams is based on the use of Greek letters to show ring size, substitution of nitrogen for carbon in a cyclic array (thus using the prefix "azacyclo"), or the Hantzsch–Widman system (*vide supra*). Some examples follow in Table 5.14.

Imides are compounds that have **two** acyl groups attached to the same nitrogen and, thus are, formally, the nitrogen analogues of anhydrides. The nomenclature follows accordingly, with the nitrogen analogue of, for example, **phthalic anhydride**, being **phthalimide** and that of **succinic anhydride**, **succinimide**.

TABLE 5.14. Some Aminocarboxylic Acids and the Lactams Which May Be Derived from Them

Aminocarboxylic Acid	Lactam
Aminoethanoic acid (aminoacetic acid, glycine)	Azacyclopropanone (azaridinone)
3-Aminopropanoic acid (β-aminoalanine)	Azacyclobutan-2-one (2-azetidinone, β-propiolactam)
4-Aminobutanoic acid (γ-aminobutyric acid, GABA)	Azacyclopentan-2-one (2-pyrrolidinone; γ-butyrolactam)

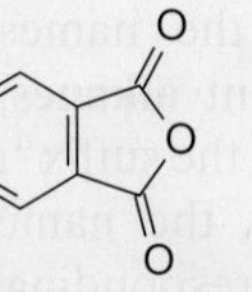

phthalic anhydride

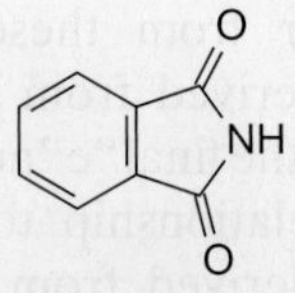

phthalimide

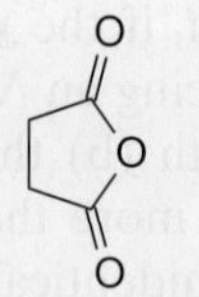

succinic anhydride
(oxacyclopentane-2,5-dione,
oxalane-2,5-dione)

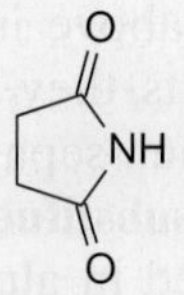

succinimide
(azacyclopentan-2,5-dione,
pyrrolidin-2,5-dione)

Replacement of the hydroxyl (–OH) of the carboxylic acid group (–CO$_2$H) with an *N*-hydroxyl group (i.e., –NH–OH) generates that small group of compounds known as hydroxamic acids, for example,

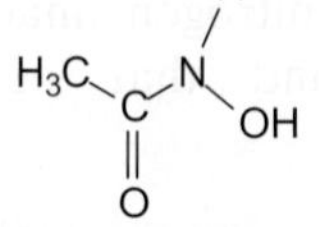

N-hydroxyethanamide

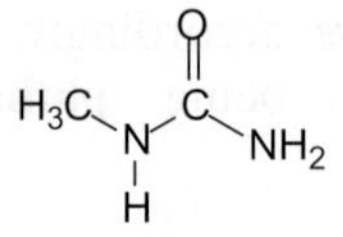

N-methylurea

while the group of compounds known as **ureas**, named after the parent "bisamide" H$_2$NCONH$_2$, originally produced by heating ammonium cyanate (NH$_4$$^+OCN^-$, p. 88) are simply derived by substitution at nitrogen, for example, *N*-methylurea (CH$_3$NHCONH$_2$).

Amides (and selected related compounds) are generally quite stable and, except for methanamide (formamide, HCONH$_2$) the *N*-unsubstituted amides are all crystalline solids. The simple *N*-methyl derivatives of the one- and two-carbon carboxylic acids (methanoic acid [formic acid, HCO$_2$H] and ethanoic acid [acetic acid, CH$_3$CO$_2$H]) as well as selected other derivatives are important materials of commerce. Indeed, materials such as *N,N*-dimethylmethanamide (*N,N*-dimethylformamide [HCON(CH$_3$)$_2$], DMF), *N*-methylethanamide (*N*-methylacetamide [CH$_3$CONH(CH$_3$)], NMA), and *N,N*-dimethylethanamide (*N,N*-dimethylacetamide [CH$_3$CON(CH$_3$)$_2$], DMA) are all high boiling (the temperatures at which these materials distill [at about 1 atm] are, respectively, 153, 204, and 165°C), relatively inert, and excellent solvents (*vide infra*). It is interesting, furthermore, to note that although *N,N*-dimethylmethanamide (*N,N*-dimethylformamide [HCON (CH$_3$)$_2$], DMF) and *N*-methylethanamide (*N*-methylacetamide [CH$_3$CONH(CH$_3$)], NMA) are isomeric amides (both being C$_3$H$_7$NO) they distill some 50°C apart and that the derivative of the latter with one more methyl group viz., *N,N*-dimethylethanamide (*N,N*-dimethylacetamide [CH$_3$CON(CH$_3$)$_2$], DMA) distills about 40°C lower! It is argued that the hydrogen on the nitrogen atom of the *N*-

mono-methylamide (N-methylethanamide [N-methylacetamide {CH₃CONH-CH₃}, NMA]) is available for hydrogen bonding to the oxygen of a neighboring *N*-methylamide and that such a network of hydrogen bonding (Figure 5.35) produces the increase in distilling temperature. In concert with this idea, removal of the proton and its replacement by a methyl group, while raising the molecular weight, produces *N,N*-dimethylethanamide (*N,N*-dimethylacetamide [CH₃CON(CH₃)₂]), an amide that lacks an N–H and therefore cannot hydrogen bond, thus distilling at a lower temperature. Further discussion of solvent properties of these and similar cases will be undertaken later in this chapter.

The details of the structures of simple amides are well known (largely from microwave and electron diffraction spectroscopy), and it is clear that they are planar (or nearly so, Figure 5.36), and that there is some barrier to rotation about the carbonyl-to-nitrogen single bond. The barrier is larger than the few (≤ca. 5) kcal mol⁻¹ found in the corresponding carbonyl-to-carbon single bond in an aldehyde or ketone but smaller than a typical carbon-carbon double bond (≥ca. 80 kcal mol⁻¹). Generally, the barrier in amides lies between about 12 and 20 kcal mol⁻¹ as can be seen by NMR spectroscopy in suitable cases. The details of the structure of methanamide (formamide [HCONH₂]) are provided in Figure 5.36.

As indicated in Figure 5.36, the (*E*)- and (*Z*)-protons on nitrogen are not identical (the former lying on the *same side* of the C–N single bond as the carbonyl oxygen

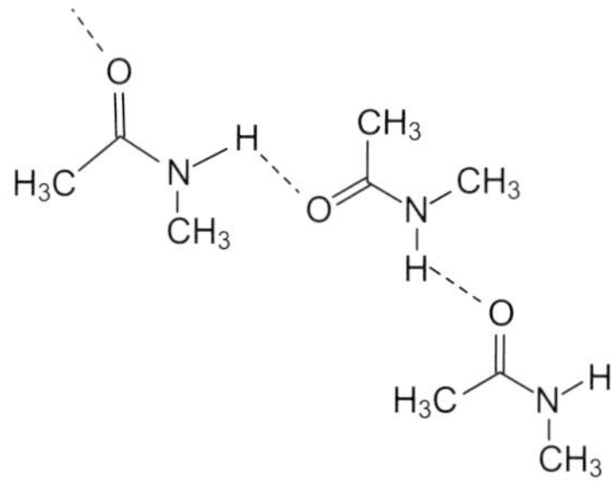

Figure 5.35. A depiction of possible hydrogen bonding in *N*-methylethanamide (*N*-methyl-acetamide, NMA).

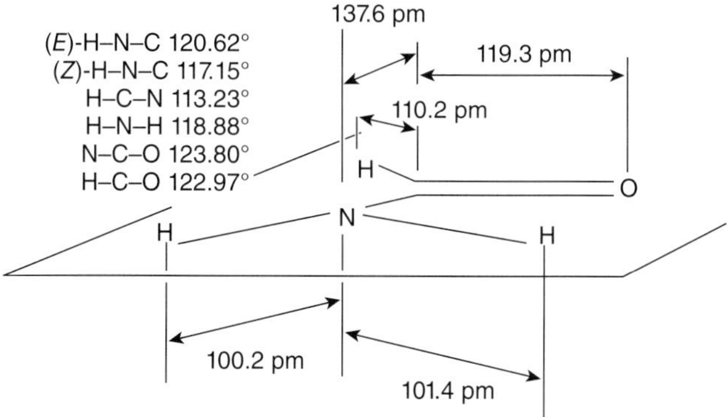

Figure 5.36. A representation of methanamide (formamide [HCONH₂]) showing bond lengths and angles as determined by microwave spectroscopy.

and the latter lying on the *opposite side* of this same bond). The barrier to rotation is ca. 20 kcal mol^{-1}. This observation is mirrored in the corresponding *inseparable* (because the barrier is too small) *N*-methyl derivative(s) (*E*)- and (*Z*)-*N*-methylmethanamide ((*E*)- and (*Z*)-*N*-methylformamide [(*E*)- and (*Z*)-HCONHCH$_3$]) as shown in Figure 5.37.

As with esters (and lactones), α,β-unsaturated amides (and lactams) absorb in the UV with λ$_{max}$ between 200 and 220 nm and an ε near 10^4. Also, as with esters (and lactones), if the α,β-unsaturation (extending the conjugation of the carbonyl) is lacking, absorption above 200 nm is absent; an additional double bond (e.g., γ, δ) moves the absorption to longer wavelength and usually increases its intensity.

All amides and lactams have intense carbonyl (C=O) absorption in the IR in the region 1650–1700 cm^{-1}. It is generally argued that the major reason that the absorption occurs at somewhat longer wavelength (lower frequency) than that for other carbonyl compounds is because there is a smaller force constant in these carbon–oxygen double bonds. The diminished force constant here is thought to arise from electron donation to the carbon of the carbonyl by the nonbonded electrons on nitrogen. In resonance terms, a picture such as that shown in Figure 5.38 is thought to obtain. It might be argued that these pictures suggest that the real structure of an amide would have a carbon–nitrogen single bond *shorter* than one in an amine and a carbon–oxygen double bond *longer* than in an aldehyde or ketone. Problem

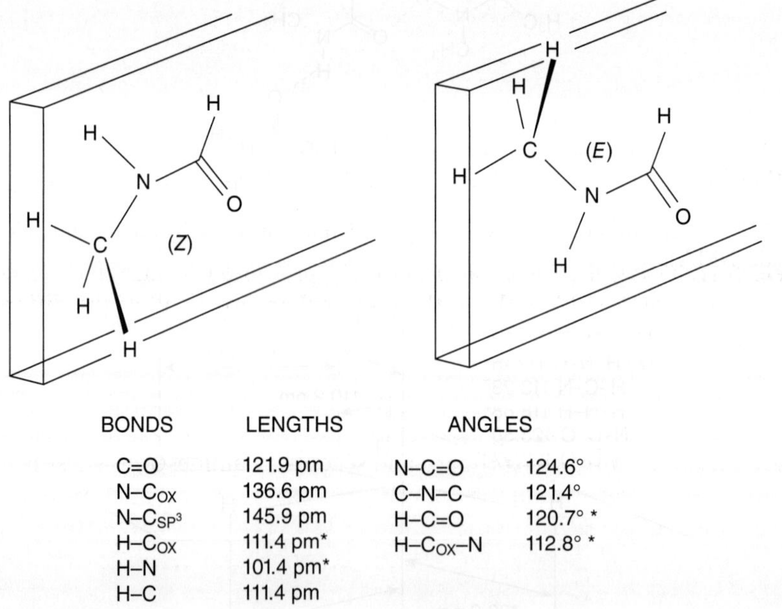

BONDS	LENGTHS	ANGLES	
C=O	121.9 pm	N–C=O	124.6°
N–C$_{OX}$	136.6 pm	C–N–C	121.4°
N–C$_{SP^3}$	145.9 pm	H–C=O	120.7° *
H–C$_{OX}$	111.4 pm*	H–C$_{OX}$–N	112.8° *
H–N	101.4 pm*		
H–C	111.4 pm		

Figure 5.37. A representation of the (*Z*)- and (*E*)-isomers of *N*-methylmethanamide (*N*-methylformamide). Although the isomer mixture is readily detectable, they are inseparable because they interconvert too rapidly at room temperature. The numbers marked with an asterisk (*) are the result of calculation; the abbreviation C$_{ox}$ is used to refer to the carbon of the carbonyl (C=O) group.

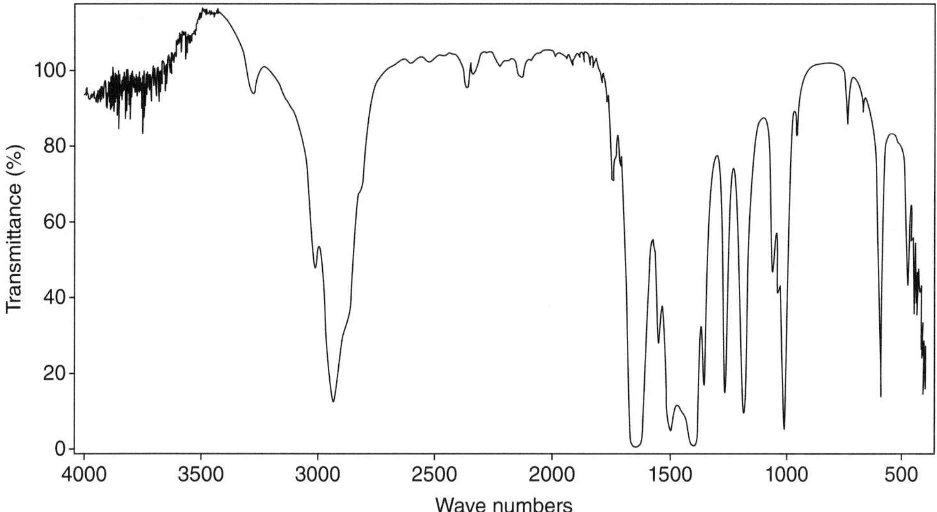

Figure 5.38. A standard depiction of resonance forms for an amide.

Figure 5.39. The transmittance IR spectrum of *N,N*-dimethylethanamide (*N,N*-dimethylacetamide, DMA [$H_3CCON(CH_3)_2$]) (neat) as a thin film.

5.11 deals with what is known about the carbon–oxygen double bond, and it is clear that the bond between the carbon of the carbonyl and the nitrogen of the amide is significantly shorter than a "normal" carbon–nitrogen bond in amines (the data show that in amines, the C–N bond is typically 145–148 pm, while in amides, it is typically 135–139 pm).

Problem 5.11. Using the information already presented in this chapter, compare the carbon–oxygen bond lengths in the carbonyl groups of all of the different carbonyl containing compounds. What conclusion, if any, do you reach?

As was the case with alcohols (*vide supra*), and as has already been pointed out in the discussion about boiling points of both amides and alcohols, amides that possess a proton on nitrogen (primary and secondary amides) show distinct, hydrogen bonded, N–H stretching in the region 3520–3060 cm^{-1}. The position and intensity of the band is a function of concentration as well as the exact structure (i.e., what else is connected to the nitrogen and/or carbon of the carbonyl) of the amide. Interestingly, in dilute solution, *primary amides* actually show *two* moderately strong N–H bands (near 3320 and 3400 cm^{-1}), which are attributed to different stretching modes. However, the N–H bands in both primary and secondary amides are weaker than that due to the carbonyl group (frequently the most intense band in the amide's spectrum) and also weaker than the very intense –OH absorption seen in alcohols (but stronger than the N–H seen in amines). Figure 5.39 presents the IR transmittance spectrum of *N,N*-dimethylethanamide (*N,N*-dimethylacetamide, DMA). As expected, there is no N–H band.

NMR spectroscopy has been extensively used in the study of amides because (a) the substituents on nitrogen, for example, hydrogen (H) or methyl (CH_3–) have different chemical shifts because of their different environments (i.e., (*E*) or (*Z*) to the oxygen of the carbonyl) and (b) the process that interconverts the (*E*)- and (*Z*)-isomers (rotation about the [carbonyl] carbon-to-nitrogen bond) can be conveniently measured in the NMR as the temperature of the sample is varied over some reasonable range (ca., 25–100°C). It appears to require about 12–20 kcal mol^{-1} (about 50–85 kJ mol^{-1} to overcome the rotational barrier).

When there is a hydrogen on nitrogen (i.e., a **primary** or **secondary** amide), it is generally found that a broad signal somewhere in the range of about δ 5–8 ppm is seen. This broad signal results, in part, from the exchange of these protons with others and, in part, because the most abundant isotope of nitrogen (^{14}N; 99.63%) also has a quadrupole moment ($I = 1$), which results in rapid nuclear relaxation of the proton to which it is bound and significantly broadened lines.*

As might be anticipated, however, as all amides contain protons, carbon, nitrogen, and oxygen, all of these nuclei might be examined at their appropriate respective frequencies. With simple amides, the ^{14}N, ^{15}N, and ^{17}O information that is obtained is of limited value and, unless there are protons on nitrogen (the N–H would be seen in the IR too) with which coupling can be observed (^{15}N has a spin of 1/2) each provides a single signal. For example, for *N,N*-dimethylethanamide (*N,N*-dimethylacetamide, DMA [$H_3CCON(CH_3)_2$]), single lines are seen for all three of these nuclei. However, the ^{1}H and ^{13}C spectra are much more informative.

Figure 5.40 presents the ^{1}H spectrum of *N,N*-dimethylethanamide (*N,N*-dimethyl-acetamide [$H_3CCON(CH_3)_2$]) as taken in ^{2}HCCl$_3$ at 300 MHz. The methyl group on the carbon of the carbonyl has a chemical shift of $\delta = 1.98$ ppm while the two methyl groups attached to nitrogen are found at $\delta = 2.84$ and 2.92 ppm. It is argued that the methyl group farthest downfield is (*Z*) that is, on the same side of the single bond as the oxygen.

As anticipated, the proton decoupled ^{13}C NMR spectrum at 75.46 MHz has only four lines. It is clear that the carbon of the carbonyl comes into resonance furthest downfield ($\delta = 170.4$ ppm) and that the methyl of the acyl unit furthest upfield ($\delta = 21.5$ ppm). Largely on the basis of solvent shifts (i.e., which methyl group is most affected by a change in solvent), it has been decided that the methyl group at $\delta = 35.0$ ppm (the one furthest upfield) is on the same side of the C–N single bond as the oxygen. The spectrum is shown as Figure 5.41.

As already noted for amines, the presence of *one nitrogen* (or any odd number of nitrogens) requires that the parent ion be an *odd* number. The same must, of course, be the case for any amide that contains only one nitrogen. Also, as might be anticipated considering that amides are derivatives of carbonyl compounds which have a penchant, as seen earlier in this section, for undergoing the McLafferty

---

*The lifetime of the excited state is inversely related to the linewidth. The return to equilibrium of the nuclear magnetism following a perturbation, as noted in Chapter 2, takes some time—called the *relaxation time*. There are different kinds of relaxation times, usually called T_1 (spin-lattice relaxation) for magnetism parallel to H_o and T_2 (spin-spin relaxation) for magnetism perpendicular to H_o. The linewidth at half-height (W_2) is given by $(\pi T_2)^{-1}$.

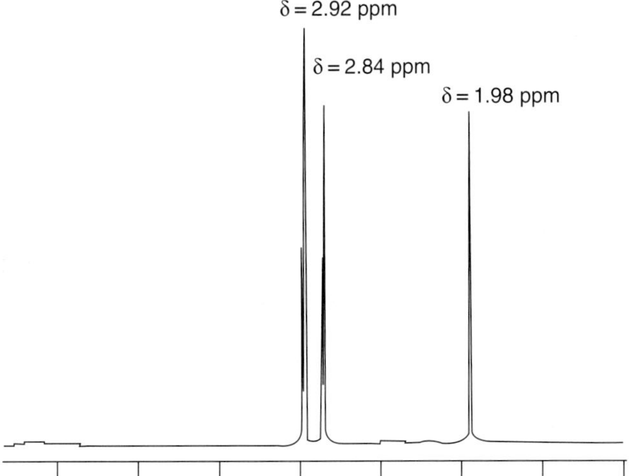

Figure 5.40. The ^{1}H NMR spectrum at 300 MHz of *N,N*-dimethylethanamide (*N,N*-dimethylacetamide, DMA [H$_3$CCON(CH$_3$)$_2$]) in ^{2}HCCl$_3$. The chemical shift assignment is given in the text.

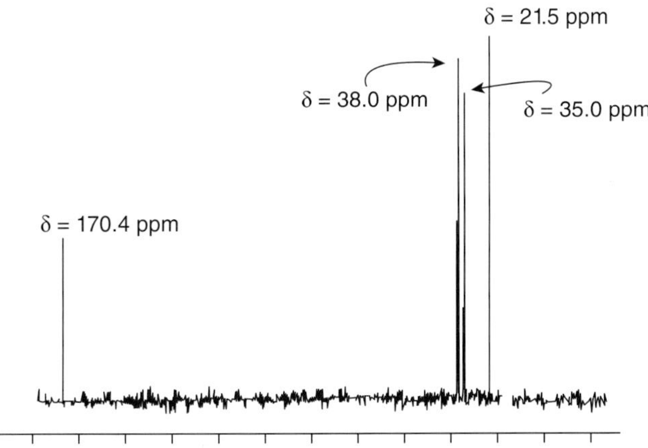

Figure 5.41. The ^{13}C NMR spectrum at 75.48 MHz of *N,N*-dimethylethanamide (*N,N*-dimethylacetamide, DMA [H$_3$CCON(CH$_3$)$_2$]) in ^{2}HCCl$_3$. The chemical shift assignment is given in the text.

rearrangement that, structure permitting, the same process would occur here. Indeed, as seen in Scheme 5.9, when there is a simple substitution pattern on nitrogen, the McLafferty process is the major course and will generally lead to the base ion. However, with sizeable groups on nitrogen, an alternative pathway is common (Scheme 5.10) and even the fragment from Scheme 5.10 can undergo further fragmentation (Scheme 5.11).

Of the few kinds of truly large molecules (macromolecules) common to all life (nucleic acids, polysaccharides, and proteins) about which more will be said later (Chapters 12–14), it is important to point out now that the last of these, proteins, are **amides**.

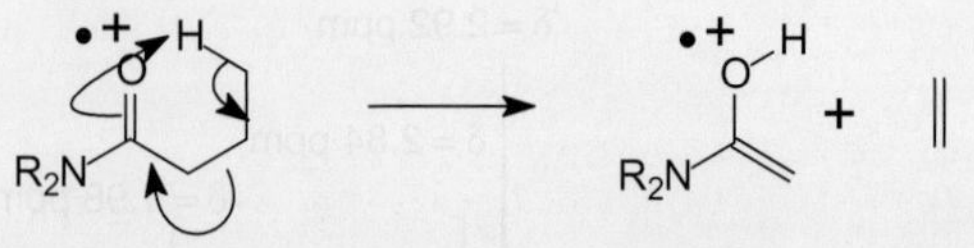

Scheme 5.9. The McLafferty rearrangement occurring on an amide ($R' = H$ or CH_3).

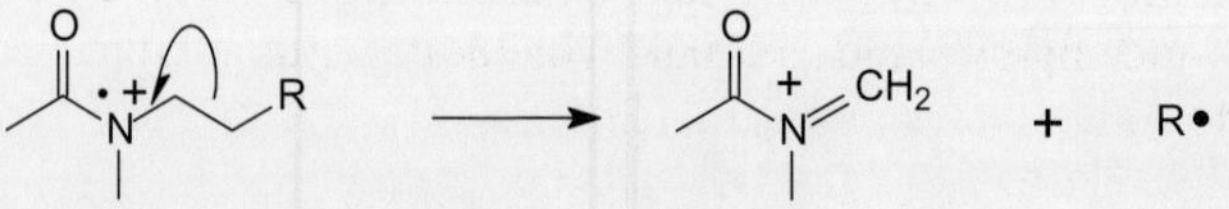

Scheme 5.10. A mass spectrometric fragmentation pattern for an N-alkyl-substituted amide, which may occur along with a McLafferty-type rearrangement.

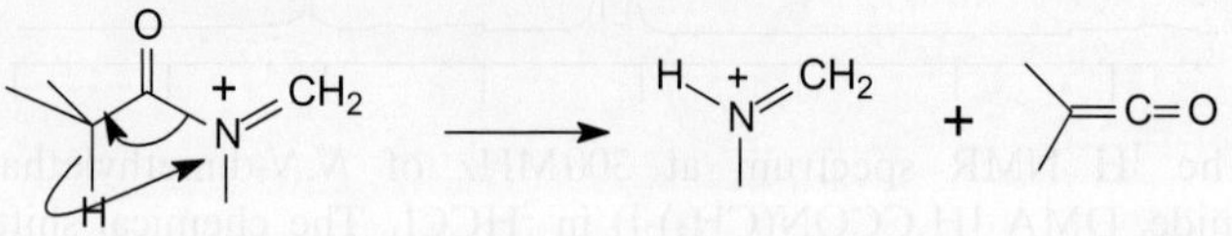

Scheme 5.11. A mass spectrometric fragmentation pattern for an intermediate produced from an N-alkyl-substituted amide undergoing fragmentation by a non-McLafferty-type rearrangement.

Indeed, proteins may be considered that set of polyamides (many amides linked as amides) arising from a selected subset of carboxylic acids bearing amino groups at C-2 (or the α-carbon). Consider, for example, the simple **amino acid** aminoethanoic acid (aminoacetic acid, glycine [$H_2NCH_2CO_2H$], Gly, G).

If two molecules of aminoethanoic acid (aminoacetic acid, glycine, [$H_2NCH_2CO_2H$], Gly, G) are allowed to combine (with loss of water) in such a way that the amino group of one becomes attached to the carbonyl of the other, then a *dimer* (called a **dipeptide**) results.

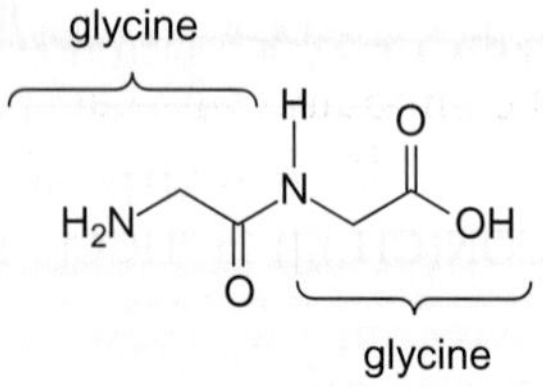

As we will explore in some detail later (Chapter 12), there is a group of some 20 so-called standard amino acids (of which aminoethanoic acid [glycine, Gly, G, $H_2NCH_2CO_2H$] is one), which are common to almost all **peptides** in living systems. These amino acids may be viewed as arising by removal of one of the α-hydrogens of aminoethanoic acid (aminoacetic acid, glycine, Gly, G, [$H_2NCH_2CO_2H$]) and its replacement by some other group (methyl, ethyl, etc.). Long peptide chains are called **polypeptides** and materials with one or more polypeptide chains are called **proteins**. These **polypeptides** and **proteins** consist of different sets, in different order, of about 50–4000 amino acids **all linked as amides and all chosen from the "standard" 20 but in different sequences**. The subtle changes in bond angles, conformation, and the relationships between the various side chains branching from the multitude of α-carbons, as well as features arising from the presence of **basic amino groups** and **acidic carboxylic acid groups** in the same molecule (!), among these **chiral** materials, lead to their various shapes and exciting physical and chemical properties (Chapter 12).

The breakdown of proteins during digestion and reuse of the components that result permit living systems to continue to function. However, excess nitrogen may be produced and many land animals (including humans) excrete it as urea (H_2NCONH_2).

5. Nitriles. Organic compounds that contain the $-C\equiv N$ function are, as a group, called **nitriles.** The IUPAC nomenclature for this group is based on simply adding the **suffix** *nitrile* to the appropriate hydrocarbon chain name. The number of carbons used in deciding upon the appropriate hydrocarbon name *includes* the carbon of the nitrile itself. Examples are the following:

$CH_3C\equiv N$	ethanenitrile (acetonitrile),
$(CH_3)_3CCH_2C\equiv N$	3,3-dimethylbutanenitrile, and
$C_6H_5C\equiv N$	benzenenitrile (benzonitrile, cyanobenzene).

In a formal sense, the nitrile group can be considered to be derived from a carboxylic acid ($-CO_2H$) via a primary amide ($-CONH_2$), which has "lost the elements of water (H_2O)." That is, "$-CONH_2$" – "H_2O" = "$-C\equiv N$."

Interestingly, the **dehydration** (loss of water) of amides can *actually* be effected and some nitriles are prepared in this way.

The structural details available for ethanenitrile (acetonitrile, $CH_3C\equiv N$) are given in Figure 5.42.

As was the case with other similar carboxylic acid derivatives already discussed, nitriles do not absorb in the UV unless they have α,β-unsaturation. With only one α,β-double bond, most nitriles absorb with a λ_{max} in the range 210–220 nm ($\varepsilon_{max} \approx 10,000$); additional conjugation moves the absorption maximum to longer wavelength.

It may be remembered that compounds with carbon–carbon triple bonds ($-C\equiv C-$, i.e., alkynes, acetylenes) absorbed IR radiation in the region around 2200 cm^{-1}. It might then be expected that nitriles, because of the carbon–nitrogen triple bond, would absorb similarly. Indeed, compounds that contain the $-C\equiv N$ group do absorb in the same region although they usually have a modestly intense peak over a

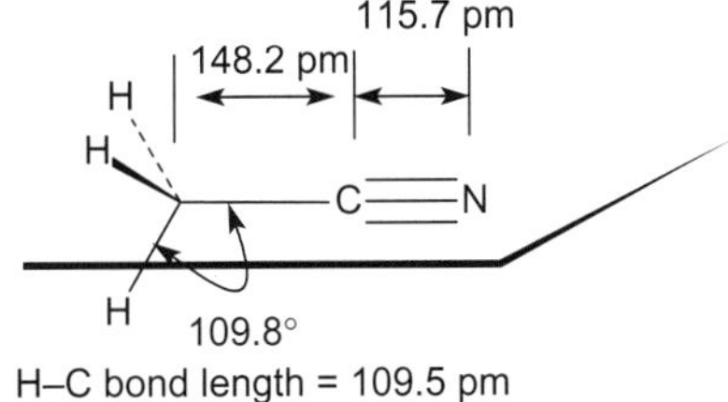

Figure 5.42. Ethanenitrile (acetonitrile, $CH_3C\equiv N$).

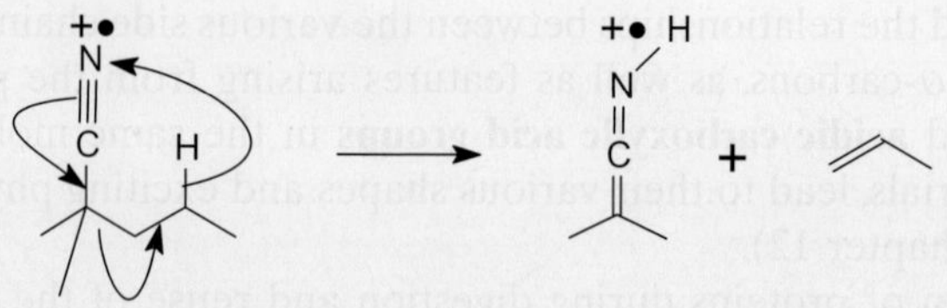

Scheme 5.12. The McLafferty rearrangement occurring on a nitrile.

somewhat narrower range than the alkynes (between 2220 and 2260) and, of course, the nitriles, which must end the chain, lack a terminal hydrogen.

Because there are no protons directly associated with the $-C \equiv N$ group (except for the unique case of hydrogen cyanide, HCN), ^1H NMR spectra are similar to those of other compounds bearing nonhydrogen containing substituents, that is, the analysis of the spectra is primarily concerned with the protons on the α-carbon, as those are the ones most affected. Indeed, it is argued that nitriles are very much like alkynes and while, for example, the protons on the methyl of propyne ($H_3C-C \equiv CH$) appear at about $\delta = 1.80\,ppm$, those of ethanenitrile (acetonitrile, $H_3C-C \equiv N$) are found at about $\delta = 1.98\,ppm$. Methylene and methine protons are similarly affected.

However, in the ^{13}C NMR, where the $-C \equiv C-$ (alkyne) carbons come into resonance in the range of $\delta \approx 65–80\,ppm$, carbons in nitriles ($-C \equiv N$) are typically found between about $\delta \approx 114–125\,ppm$ downfield from TMS. Both the ^{14}N and ^{15}N spectra are known for simple nitriles.

It is common to find that although the molecular ion of many nitriles is missing, an $(M-1)^+$ peak is seen. It is argued that this peak arises by loss of an α proton. It is also clear that, as with other acid derivatives, a McLafferty-type rearrangement can occur. A potential path is shown in Scheme 5.12.

D. AN INTRODUCTION TO SOLVENTS

As pointed out in the Introduction to this chapter, most reactions that will be encountered in organic chemistry occur in **solution**. Typically, solutions are considered made up of a minor component called a **solute**, which, for our purposes here, is frequently a **reactant** and thus will be undergoing some change (i.e., reacting) in a process called a **reaction**, and a **solvent** (or mixture of solvents) which constitutes the bulk of the mixture. The purpose of using a solvent (if it is not a reactant itself!) may be to moderate a reaction which, in the absence of the solvent, is too vigorous or, perhaps, the solvent might simply serve to bring reactants together which might not otherwise be miscible. Frequently, the choice of solvent is determined by the **reactants** and/or the **reaction** that is being carried out. Although solvents, often the simple organic compounds already discussed in this chapter, are usually divided into categories as a function of their properties (about which there has already been some discussion as the various groups of compounds were being introduced), there are a few overriding considerations that are important regardless of the specific class of compound chosen as solvent.

First, and rather obvious, it is generally desirable for the solvent to be a liquid over the entire temperature range in which the reaction is occurring.

Second, and also rather obvious, is the general observation (occasionally flaunted with good reason) that the **reactants** (or at least one of them) **should be soluble** (or at least partially so) in the solvent. Largely, it is the solubility of the reactant(s) over an appropriate temperature range that determines the choice of solvent. As noted above, this is because it is usually the function of the solvent to allow reactants to come into contact so that they might react as well as to moderate reaction by intruding itself between reacting species. Furthermore, in the same vein, not only are the reactants expected to be soluble, but (again generally) some **appreciable solubility** is sought. Although the concept of "appreciable solubility" is vague, it is important to recognize that it is frequently possible to adjust the quantity of solvent (relative to solute) so that the solubility (often expressed either as gram [of solute] per milliliter [of solvent] or as gram [of solute] per 100 g [of solvent]) may not be a limiting concern.

Third, **unless the solvent is to also be a reactant** (a process called *solvolysis* which will be discussed), **it will generally have a different kind of functional group (if any) present than any of the reactants**. For example, methanol (CH_3OH) is completely soluble, in all proportions, in ethanol (CH_3CH_2OH). However, a reaction involving ethanol and some other reagent would almost certainly also occur between that reagent and methanol (and, of course, vice versa). Therefore, they would **not** be chosen as solvents one for the other. This is in concert with the idea that the hydrocarbon chain attached to the **functional group** plays a lesser role in reactivity than does the functional group. Fourth, unless it is the intended outcome, the product of the reaction should not contain the solvent. Alternatively, the solvent, while the exact nature of its participation in the reaction might not be known, should not have been incorporated into the product.

Finally, the time worn construct that **like dissolves like** will continue to prove valid, and, in order to determine what kinds of material are "alike" and what "makes them alike," it may be necessary to reexamine the various functional groups just introduced.

I. Protic and Aprotic Solvents

Of the functional groups discussed, there are clearly those in which at least one proton is shown as attached directly to oxygen, an atom whose *electronegativity* (Chapter 1) is second only to fluorine. These –OH-containing compounds, as already seen, form **hydrogen bonds** through partial proton donation (sharing) to other electronegative atoms and, therefore, would be expected to dissolve solutes whose solubility would be increased by hydrogen bonding—that is, materials just like them! Solvents (as well as solutes) that function by hydrogen bonding are called **protic**. Solvents (as well as solutes) whose primary mode of interaction is not through hydrogen bonding are called **aprotic**. Aprotic solvents can have protons in their framework, but the protons are not used to the same extent for hydrogen bonding as they are with protic solvents.

Protic solvents include such obvious candidates as alcohols (R-OH) and carboxylic acids (RCOOH) as well as water (HOH), while aprotic solvents include materials such as tetrachloromethane (carbon tetrachloride, CCl_4), ether (diethyl ether [$(CH_3CH_2)_2O$]), benzene (C_6H_6), and deuteriochloroform ($^2HCCl_3$). It should be clear that there must be many more aprotic solvents than protic ones.

II. Polar and Nonpolar Solvents

In addition to the difference in hydrogen bonding ability, a second criterion can be applied to solvating ability (i.e., the ability to act as a solvent for a particular solute or set of solutes).

As pointed out in Chapter 1, the force of attraction (F) between two charges (e.g., q and q') separated by some distance, r, is a function, ε (called the **dielectric constant**) of the (isotropic) medium (Equation 5.4):

$$\varepsilon = qq'/Fr^2. \tag{5.4}$$

Thus, the **dielectric constant** is a measure of how much the strength of an electric field is reduced by a homogeneous medium around a particle (i.e., ions or molecules) compared with what would be found for the same particle in a vacuum ($\varepsilon = 1$ for a vacuum). Some liquids have small dielectric constants over a wide temperature range. Other liquids have large dielectric constants that (usually) decrease with increasing temperature. Because the force between charged particles varies inversely with the dielectric constant (Equation 5.4) solvents with high dielectric constants (Table 5.15) are said to have *high ionizing power*, that is, ions are more easily

TABLE 5.15. Dielectric Constants (ε), Dipole Moments (μ, debyes), and Boiling Points (°C) of Some Common Solvents

Compound	Structure	Dielectric Constant, ε	Dipole Moment μ (Debyes)	Boiling Point (°C)
Methanamide (formamide)	$\text{O} \parallel \text{H–C–NH}_2$	110	1.66	211
Water	H_2O	79	1.64	100
Methanoic acid (formic acid)	$\text{O} \parallel \text{H–C–OH}$	59	1.82	100
Ethanenitrile (acetonitrile)	CH_3CN	38	3.4	82
Methanol	CH_3OH	33	2.9	65
Ethanol	CH_3CH_2OH	25	1.7	78
Propanone (acetone)	$\text{O} \parallel \text{H}_3\text{C–C–CH}_3$	21	2.7	56
Ethanoic acid (acetic acid)	$\text{O} \parallel \text{H}_3\text{C–C–OH}$	6.1	1.67	118
Trichloromethane (chloroform)	$HCCl_3$	4.8	1.03	61
Ether (diethyl ether)	$(CH_3CH_2)_2O$	4.3	1.2	35
Tetrachloromethane (carbon tetrachloride)	CCl_4	2.2	0	77
Hexane	$H_3C(CH_2)_4CH_3$	1.9	0.09	69

The data in this Table were taken from Riddick, J. A.; Bunger, W. B. Organic Solvents, in Weissberger, A. (ed.), *Techniques of Chemistry*, Vol. II, 3rd edition, Wiley-Interscience, New York, **1970**.

separated in such solvents and the solvent is thought to surround and hold them in a three-dimensional cage-like structure (occasionally called a "solvent cage") and to thus **solvate** the ions. These materials are **called polar solvents. Nonpolar** solvents are those with small dielectric constants (ε) and these generally do not separate ions well.

Interestingly, as shown in Table 5.15, the solvents in that Table with the highest **dielectric constants**, viz., methanamide (formamide, $HCONH_2$, $\varepsilon = 110$) and water (H_2O, $\varepsilon = 79$) do not have the highest **dipole moments**! From this it should be clear that the "polarity" of an individual molecule (as measured by the **dipole moment**) might be modified by identical surrounding molecules so that, when measured in bulk, a quite different result (the dielectric constant) obtains. In this vein, it is also clear that structure cannot always be a reliable guide. Consider, for example the case of methanoic acid (formic acid, HCO_2H) and ethanoic acid (acetic acid, H_3CCO_2H), information for both of which is included in Table 5.15. Both are carboxylic acids, the latter differs from the former by "replacement" (or "substitution") of a nonpolar methyl group ($-CH_3$) for a nonpolar hydrogen ($-H$) and, as therefore expected, both have similar dipole moments. Yet in bulk, methanoic acid (formic acid [HCO_2]) is a polar solvent capable of dissolving compounds that will ionize. Ethanoic acid (acetic acid [CH_3CO_2H]), in bulk, in incapable of dissolving those materials because it is a much less polar medium.

III. Polarizability

In addition to describing species as **protic** or **aprotic** and considering the likelihood of hydrogen bonding, examining the permanent dipoles to attempt to understand how **internal charge separation** (μ) will effect the clustering of molecules, and weighing the effect the **dielectric constant** (ε) might have on the degree to which ionic substances might be associated, it is frequently important to consider the **polarizability** or extent to which the distortion of electrons within a molecule (or assembly of molecules) can be effected. Generally, the forces (called **dispersion forces**) that arise as a consequence of molecules approaching one another (already mentioned in Chapter 1) are small compared with the others listed but, in molecules with non-bonded electrons, for example, ether (diethyl ether [$(CH_3CH_2)_2O$]), the approach of charged species can induce sizeable effects which, in turn, aid in their solubility. Induced (in contrast to permanent) effects such as this are grouped together under the general heading of **polarizability**.

IV. Choosing a Solvent

Although it might be argued that the easiest way to choose a solvent for some solute is by trial and error (and, on occasion, that may be the method of choice), a more logical approach begins with knowing something about the solute and what is to be done with it. Although the literature of chemistry (Appendix II) is an excellent guide, a few examples here will illustrate how decisions can be reached.

a. Solvents for Spectroscopy. Consider first the choice of solvents that might be utilized for the various spectroscopic methods with which you have already become familiar. It should be clear that not only should the solute be soluble, but it should not react with the solvent and, of course, it should not interfere with the spectra to be obtained; that is, either it should not itself absorb in the region to be examined or its absorbances should be so well known that they can be "subtracted" from the spectrum of the solute in the solvent.

1. Solvents for UV Spectroscopy. Because the transitions of diagnostic interest generally occur above 200 nm, solvents that absorb UV in the region of interest should be avoided. Fortunately, both very polar (e.g., water and methanol) and nonpolar (e.g., hexane) solvents are available so that a wide variety of solutes can be accommodated. This is aided by the sensitivity of the technique.

2. Solvents for IR and Raman Spectroscopy. Many IR spectra can be obtained on gaseous materials in an IR gas cell, or as a film simply placed in the IR beam, or even as a "neat" liquid, that is, without solvent, by simply placing a drop of the material on a polished salt plate (using the plate, or an identical one, without the drop as reference). Other IR spectra, on crystalline material, can be obtained as a solid solution in potassium bromide (KBr). The latter by grinding the sample with the potassium bromide (KBr) and then compressing the mixture so that a "window" of KBr containing the solute and suitable for placement directly in the IR beam is formed. However, solution IR and Raman spectra are often either necessary or desirable for many purposes (such as measuring the extent of hydrogen bonding in alcohols or removing traces of crystal habit).

Without special reason to do otherwise, solvents for IR and Raman spectroscopy are chosen that do not absorb in the region of interest anticipated for the solute. Thus, if it is anticipated that there is a carbonyl group in the solute, solvents such as propanone (acetone, $[(CH_3)_2CO]$) would be avoided. Furthermore, for the IR, since solution cells generally have windows made of material that does not absorb in the region of interest (from about 4000 to 400 cm^{-1}) such as sodium chloride (NaCl) or potassium bromide (KBr) it is necessary to avoid water (and aqueous solutions). While other window material is available, the general practice is to seek an acceptable solvent, for example, trichloromethane (chloroform, $HCCl_3$), when halogen-containing materials are not being examined.

3. Solvents for NMR Spectroscopy. Generally, for NMR spectroscopy in solution (in contrast to gas-phase NMR or solid-state NMR), solvents that lack the nucleus being examined are preferred. For ^1H NMR spectroscopy, this usually means that materials such as tetrachloromethane (carbon tetrachloride, CCl_4, Table 5.14, $\mu = 0$ debye) or sulfur dioxide (SO_2, $\mu = 1.6$ debye) are needed. However, as might be anticipated, the former is of limited utility because not too many materials will dissolve in it, and the latter is reactive enough to change many solutes and require extensive precautions for routine handling. Generally, therefore, deuterated (i.e., ^2H containing rather than ^1H containing) solvents are needed. Fortunately, the demand for such materials is high and many, for example, trichlorodeuteriomethane (deuteriochloroform [$^2HCCl_3$]), perdeuteriomethanol ($^2H_3CO^2H$), deuterioacetone [$(^2H_3C)_2CO$], and deuteriobenzene ($^2H_6C_6$), whose respective dipole moments and dielectric constants span a respectable range are commercially available. It is usually possible to find a suitable solvent among these. However, as might be anticipated, NMR spectra obtained for other nuclei could be somewhat more complicated as, by analogy, finding suitable solvents for organic solutes that lack carbon for ^{13}C NMR, oxygen for ^{17}O NMR, phosphorus for ^{31}P NMR, nitrogen for ^{14}N or ^{15}N NMR, or fluorine for ^{19}F NMR could prove to be difficult. Fortunately, the analogy fails. Thus, if a solvent lacking the nucleus to be examined cannot be found, it is nonetheless possible to obtain spectra since the chemical shifts for nuclei other than protons occur over a much wider range than do those for protons and they usually will not interfere. Indeed, it is frequently possible to use the chemical shift of the solvent as the reference peak against which the chemical shifts of the solute are reported. Thus, the ^{17}O chemical shifts for alcohols are reported relative to the ^{17}O chemical shift of water, which is often acceptable as a solvent.

Thus, it is clear that the choice of solvent has derived from the particular purpose to which it was put, and it remains necessary to find a solvent in which the solute has suitable solubility. Frequently, the literature of organic chemistry will hold examples of compounds similar to the one of interest and the results of investigators who have already undertaken solubility studies will be available. However, lacking that, the traditional rule of thumb "like dissolves like" is a reasonable place to start.

b. Immiscible Liquids. The societal imperative "alcohol and gasoline don't mix" should have its chemical counterpart. After all (Chapter 2), gasoline is a particular distillation cut of (largely) hydrocarbons, which (by observation) "floats on water," while the alcohol to which reference is made is ethanol (H_3CCH_2OH), which, as we already know, is completely soluble in water in all proportions. How then can the mixture of ethanol and gasoline called "gasohol" be a homogeneous mixture? The amount of ethanol in gasohol fluctuates,* but it remains homogeneous because of the addition of other components (e.g., some compounds of phosphorus) called **additives** whose presence enhances the solubility of ethanol in the hydrocarbon mixture. These **solubilizing agents** might frequently be used in commercial products, but, in the laboratory, gasoline would be considered an inappropriate solvent for ethanol.

Problem 5.12. If ethanol dissolves in gasoline (for the commercial product), what is wrong with using it as a solvent in the laboratory?

Problem 5.13. The all-too-frequent crude oil "spills" resulting from drilling, shipping, and so on; operations; and accidents common to our current way of life demonstrate that oil and water do not mix. Yet, ways are found to remove grease and oil stains from fabrics and contaminated surfaces. Why is spilt oil such a problem then?

c. Organic Compounds that Dissolve in Water. Consider next the crystalline, chiral carbohydrate (Chapter 11) called (+)-sucrose (table sugar, β-D-fructo-furanosyl-α-D-glucopyranoside [$C_{12}H_{22}O_{11}$]),[†] a structure for which is shown.

---

*Fluctuations occur, in part, because most of the ethanol is produced by the fermentation of corn not destined for human consumption but for which other markets (e.g., feeding other animals) exists and, in part, because the price of crude oil fluctuates. Other noneconomic factors play a role too. The internal combustion engine can be more or less successfully "tuned" to meet the available fuel.

[†]Although the subject will be discussed more fully later, this is only one example of the class of "hydrates of carbon" (or carbohydrates). It should be noted that the molecular formula for (+)-sucrose might be (and at one time was!) written as $C_{12}(H_2O)_{11}$. Other carbohydrates were also written this way to emphasize the class similarity.

β-D-fructofuranosyl-α-D-glucopyranoside[(+)-sucrose]

Examination of the structure shows that for the 12 carbon atoms, not only are there 11 oxygens (i.e., almost a 1:1 ratio) but of the 11 oxygens, 8 are hydroxyl groups! Based on the earlier observation that hydroxyl groups engage in hydrogen bonding and that hydrogen bonding to the solvent should increase the solubility of the solute in the solvent, it should be predicted that sucrose would be water soluble. On the other hand, it should also be predicted that sucrose would be insoluble in gasoline. Both "predictions" are, of course, as we already knew, true! Of somewhat more interest is the work suggested in Problem 5.14.

Problem 5.14. Using the material to be found in a library or online (Appendix II), find the experimental values for the solubility of sucrose in at least two other solvents. Briefly justify the observations reported in the literature.

The conversion of a material such as (+)-sucrose from the crystalline solid form to its solvent-shell-surrounded form as a solute in solution requires destruction of the crystal lattice, a highly organized state which, it could be argued, profits not only from a decreased entropy but also from the intermolecular hydrogen bonding. However, on dissolution, while the crystalline order is lost, water molecules in the solvent have now become associated with the hydroxyl groups of the carbohydrate and are more highly organized than before.

Many commercially available packaged food products proudly announce on their respective labels that they contain small quantities of "sodium benzoate" as a preservative (not more than one part *per* thousand). The sodium salt of ben-zenecarboxylic acid (benzoic acid), much like the salts of other acids (e.g., sodium chloride is the sodium salt of hydrochloric acid) to be discussed later in this chapter, has nearly twice the solubility in water (about 1 g/1.8 mL at 25°C) as that of its "parent" carboxylic acid (about 1 g/2.5 mL). In general, compounds, such as salts, that can ionize in solution are better carried into solution by solvents that can support charges. Such solvents are thought to play an active role in solvating the ions by forming shells of solvent around them and shielding them from their oppositely charge counterparts. The solvent then serves to spread out the charge, which was concentrated on the ion. Both hydrogen bonding by protic solvents on the one hand, and electron donation by polarized solvents on the other, can aid in reducing the extent to which the charge might be localized. Since the energy of solvation of an ion in a polar solvent may be of the same order of magnitude as the energy of a covalent bond, it should be anticipated that reactions that involve making and breaking bonds might be effected in solvents that allow ionization (presuming that the course of the reaction is not otherwise changed!).

Thus, in summary, the nature of the solvent will play an important role in reactions and no discussion of a process is complete unless the solvent is expressly considered.

d. Phase Transfer Catalysts. When reactants are incapable of being brought into contact, for example, when they are in different immiscible phases, no reaction can occur. So, a reaction between an ionic species, soluble in water, and a second species that is not ionic and thus insoluble in water (but soluble in organic solvents that are not water miscible) should fail. In some cases, the Problem can be overcome by using solvents that are partially miscible with each other or by changing the solvent system in other ways. In other cases, small compounds that do not themselves enter into reactions but are soluble to a limited extent in both phases can be employed to transfer reagents from one phase to another. These materials are called **phase transfer catalysts**. Quaternary ammonium salts, such as tetrabutylammonium hydroxide $[(CH_3CH_2CH_2CH_2)_4N^+ \ OH^-]$ and N,N,N-trimethylhexadecylammonium bromide [cetyltrimethylammonium bromide, $CH_3(CH_2)_{14}CH_2N(CH_3)_3^+ \ Br^-$], that have limited solubility both in water and in organic solvents can move their anions (OH^- and Br^-, respectively) from water into the immiscible second solvent such as benzene (C_6H_6) or dichloromethane (methylene chloride, CH_2Cl_2). Furthermore, after a reaction has occurred, it is often possible to move a product out of the phase in which it has been generated into the other phase, taking advantage of Le Châtelier's principle to push the process to completion.

It should thus be clear that the process functions on a continuous basis and only a small amount of phase transfer catalyst is necessary. It also follows that the phase transfer agent functions as a true **catalyst** in that it lowers the activation energy for the reaction by bringing reactants into the same space and in this way facilitates the reaction while, at least in principle, being capable of recovery unchanged.

E. ACIDS AND BASES

The development of indicators for acids and bases is interwoven into the fabric of organic chemistry. Thus, since the middle of the seventeenth century, when the "acidifying principle" was still considered to be oxygen (Saurestoff), it has been appreciated that the blue vegetable matter, syrup of violets, turned red with acids and green with bases.* At the time, all materials classified as acids and bases fit this description. Indeed, depending on the particular materials called acid or base and the concentration of the "syrup of violets," acids and bases might be described as "strong" or "weak" and a scale (or scales) of "strength" could be established. However, today, we recognize that such a scale that might be developed has unacceptable limits. The recognition derives from having somewhat more elaborate techniques of measurement, purer materials of more certain composition, and an appreciation of an expanded understanding or definition of what constitutes "acids," "bases," and "strength."

---

*It appears that Robert Boyle (1627–1691) first codified the observation in *The Sceptical Chymist* published in 1661.

Before continuing in this vein, however, it is important to recognize that acids and bases are an inseparable pair. That is, acids can only be defined in terms of their reaction with a base and vice versa. We recognize this concept by pointing out for some **acid** HX (e.g., Equation 5.5 where the nature of X is not limited) that will be losing a proton, for example, to water, that X^- can be called the **conjugate base of the acid (HX)** while H_3O^+ is the **conjugate acid of water (acting as a base)**.

Also, we now recognize, for example, that the "strength" of an acid (or base) may depend on the solvent in which it is dissolved. Thus, from the previous discussion of solvents, it might be imagined that if a solvent can engage in hydrogen bonding, acids resulting from proton donation, such as hydrogen chloride (HCl), hydrogen bromide (HBr), and hydrogen iodide (HI), that could also participate in such hydrogen bonding (perhaps by ionization), might have their respective "strengths" **leveled** by such participation. Thus, we would not be surprised to learn that in aqueous solution the reaction shown in Equation 5.5, which produces the same hydronium ion (H_3O^+) from

$$H\text{-}X + H_2O \rightarrow H_3O^+ + X^- \ (X = Cl, Br, I), \tag{5.5}$$

each yields the result that all of these acids, which ionize to the same extent, are equally "strong" and from this derive the general principle that no acid stronger than the conjugate acid of a solvent can exist in that solvent.

However, if a hydrocarbon (such as benzene, C_6H_6) is substituted for water, then when gaseous hydrogen chloride (HCl) and gaseous hydrogen bromide (HBr) are, respectively, passed into the solvent, neither of them ionize appreciably, the acid strengths are no longer leveled, and the latter behaves as the "stronger" acid. Therefore, it is necessary to attempt to determine not only what constitutes "strength" and how it might be measured under various conditions but, in the absence of ionization, what is meant by "strength."

With regard to the concept of "strength," it should be clear that some "base" other than H_2O (as that is the material capturing the proton in water) should be used in the hydrocarbon solvent and, indeed, various scales for the measurement of acidity are established in just that way. Care must still be taken because the conjugate acid of whatever solvent is used might exercise a leveling effect.

I. Brønsted Acids and Bases

In 1923, Brønsted,* following in the footsteps of Arrhenius[†] who, apparently, was only concerned with the proton as the acidifying principle, proposed that *an acid is a proton donor* and *a base is a proton acceptor*. While we recognize that this is dramatically oversimplified given what we now know about solvent participation, it nonetheless remains a useful construct. Indeed, for materials that ionize, completely or otherwise, the **acid strength** or **acidity** is referred to as *the position of equilibrium* in the reaction of the acid with a base. Thus, for the generalized acid "HA" ionizing in water (H_2O) Equation 5.6 can be written as

---

*J. N. Brønsted (1879–1947), Denmark.
[†]S. Arrhenius (1859–1927), Sweden.

$$HA + H_2O = H_3O^+ + A^-. \tag{5.6}$$

And we may then write that the **equilibrium constant, K_a, is a measure of the acid strength** and is given as the

$$\text{eqilibrium constant} = K_a = [H_3O^+][A^-]/[HA] \tag{5.7}$$

where (a) the concentration of water (H_2O) is omitted from the denominator because it is considered large and invariant, thus becoming part of K_a, and (b) the use of brackets [] denotes concentrations (usually in mol L^{-1}) so that [HA] is the concentration of unionized HA, [A^-] is the concentration of the anion of the acid HA, and [H_3O^+] is the concentration of the hydronium ion.

Thus,

$$-\log_{10} K_a = -\log_{10}[H_3O^+] + \log_{10}\{[HA]/[A^-]\}. \tag{5.8}$$

But

$$-\log_{10} K_a/pK_a \tag{5.9}$$

and

$$-\log_{10}[H_3O^+]/pH, \tag{5.10}$$

so

$$pK_a = pH + \log_{10}\{[HA]/[A^-]\}. \tag{5.11}$$

Problem 5.15. Based on Equations 5.6 through 5.11, justify the observation that acids that dissociate completely and are stronger acids than H_3O^+ will have pH's that are essentially indistinguishable.

While the participation of water in forming the hydronium ion and allowing expression of pH is generally acceptable and it is used as the standard solvent for acidity expressions, we will find (as noted above) that it may be inconvenient to be too strongly tied to this medium since many organic compounds are not soluble in water. Thus, a significant effort has been expended to construct a scale of acidity, which can be transferred across a variety of solvents. While the details of the various scales that have been established are beyond the scope of this text, the principle of comparing various acids both in water and in other solvents by choosing solvents that are either more acidic or more basic than water and reference indicators that respond in some region of the electromagnetic spectrum (not necessarily the visible)* has been quite successful. As organic chemistry has sought ever more acidic reagents for protonating even the weakest bases materials called "superacids" have been developed. It is difficult to measure the acidity of mixtures of hydrogen fluoride—antimony pentafluoride (HF-SbF$_5$) when there is evidence that this

---

*Without considering any detail, one might imagine, for example, monitoring the extent of protonation of the carbonyl of a ketone, rather than the oxygen of water.

mixture is capable of protonating the π electrons of benzene (C_6H_6) to produce what is apparently the benzenium ion (Equation 5.12) and, of course, requires that benzene is acting as a **base**:

$$C_6H_6 + HF\text{-}SbF_5 \rightarrow C_6H_7^+ + SbF_6^-. \tag{5.12}$$

So, acidity requires basicity for its expression, and as it was true for acids, so it is for bases. Thus, just as acids stronger than H_3O^+ cannot exist to any extent in water, so as bases stronger than HO^- similarly cannot exist to any appreciable extent in water. The amide anion (H_2N^-) is a typical example. As shown in Equation 5.13, the reaction of potassium metal (K) with anhydrous ammonia (NH_3) results in the nonequilibrium formation of hydrogen (H_2), which is lost from the reaction mixture, and potassium amide (KNH_2).* (This is similar, of course, to the extraordinarily vigorous [explosive] reaction between potassium [K] metal and water [H_2O] to produce hydrogen [H_2] and potassium hydroxide [KOH].) Potassium amide may be stored (apparently) indefinitely in the absence of water. However, when exposed to water, as shown in Equation 5.14, potassium amide reacts to produce ammonia and potassium hydroxide, a weaker base (because water is a stronger **acid** than ammonia):

$$K + NH_3 \rightarrow 1/2\ H_2 + KNH_2, \tag{5.13}$$

$$KNH_2 + H_2O \rightarrow KOH + NH_3. \tag{5.14}$$

Given, then, the idea that any proton-containing material can be considered an acid (when treated with a strong enough base) or, contrarily, any material might be a base when treated with a strong enough acid, and the ability to measure the extent to which a proton can be transferred even outside an aqueous environment, it is possible to construct a table (Table 5.16) with a range of pK_a's of some ca. 70 units (corresponding therefore to a range of acidity constants [K_a's] of about 10^{70}!).

II. Lewis Acids and Bases

G. N. Lewis proposed a different definition of acids and bases in 1923.[†]

He defined an **acid** as an **electron pair acceptor**, a definition the proton (H^+) and, indeed, any positively charged species meets. A Lewis **base** is then an **electron pair donor**, a definition clearly met by any species bearing a negative charge (e.g., the conjugate base of any of the acids of Table 5.16).

In a historical sense, but one which continues, the expansion of the *concept* of an acid to nonprotonaceous materials has had important effects. These include the idea that the interactions between electron-rich species and those with "empty orbitals" might be described in terms of "acids and bases" and that equilibrium constants

---

*This is an unfortunate lapse in nomenclature in organic chemistry. The anion generated by deprotonation of ammonia (i.e., NH_2^-) is referred to as the amide anion and the metal associated with this anion is named first. By its usage, it is seldom confused with the amide group (–N–C=O). Occasionally, reference to materials produced from amines and metals will simply be called **metal salts of amines**, which, although more cumbersome than the briefer "metal amide," gains in clarity.

[†]G. N. Lewis (1875–1946), professor, Massachusetts Institute of Technology (1905–1912), University of California, Berkeley (1912–1946).

TABLE 5.16. Some Acids (and Their Conjugate Bases) (or Bases [and Their Conjugate Acids]) Typically in Use in Organic Chemistry

Compound	Acid Form	Base Form	pK_a
2-Methylpropane	$(CH_3)_3CH$	$(CH_3)_3C^-$	≥ 51
Ethane	CH_3CH_3	$CH_3CH_2^-$	≈ 42
Methane	CH_4	CH_3^-	≈ 40
Benzene	C_6H_6	$C_6H_5^-$	≈ 37
Ethene	$H_2C=CH_2$	$H_2C=CH^-$	≈ 36
Ammonia	NH_3	NH_2^-	≈ 36
Hydrogen	H_2	H^-	≈ 35
N-ethylethanamine	$(CH_3CH_2)_2NH$	$(CH_3CH_2)_2N^-$	≈ 33
Triphenylmethane	$(C_6H_5)_3CH$	$(C_6H_5)_3C^-$	≈ 31
Ethyne	$HC\equiv CH$	$HC\equiv C^-$	≈ 23
Propanone	CH_3COCH_3	$CH_3COCH_2^-$	≈ 20
2-Methyl-2-propanol	$(CH_3)_3COH$	$(CH_3)_3CO^-$	19
Methanol	CH_3OH	CH_3O^-	16
Water	H_2O	HO^-	15.7
N,N,N-triethylammonium ion	$(CH_3CH_2)_3NH^+$	$(CH_3CH_2)_3N:$	10
Pyridinium ion	(pyridinium N–H structure)	(pyridine N structure)	5
Ethanoic acid	CH_3CO_2H	$CH_3CO_2^-$	4.8
Trifluoroethanoic acid	CF_3CO_2H	$CF_3CO_2^-$	0.2
Methanesulfonic acid	CH_3SO_3H	$CH_3SO_3^-$	−1.2
Hydronium ion	H_3O^+	H_2O	−1.7
Sulfuric acid	H_2SO_4	HSO_4^-	≈ -5.2
Hydrogen chloride	HCl	Cl^-	≈ -6
Benzenesulfonic acid	$C_6H_5SO_3H$	$C_6H_5SO_3^-$	≈ -6.5
Flurosulfonic acid	FSO_3H	FSO_3^-	≤ -20

Many pK_a values (particularly those measured far outside the range from ca. 0 to 15) shown in the Table are strongly solvent dependent and should be considered approximate.

might be defined for such processes. Consider, for example, the reaction between **Lewis acid** trimethylboron $[(CH_3)_3B]$ and **Lewis base** N,N-dimethylmethanamine (trimethylamine$[(CH_3)_3N:]$) (Equation 5.15).* Although N,N-dimethylmethanamine (trimethylamine $[(CH_3)_3N:]$ is a stronger base (by about 1 pK_a unit) than ammonia in water the dissociation constant is smaller for the former, that is, the complex between trimethylboron and the ammonia is stronger (by about a factor of 10) because the methyl groups on the boron and on the nitrogen interfere with each other. Thus, as might be expected, if the steric requirements are made even more demanding, no compound should form and, indeed, when trimethylboron is treated with N,N-diethylethanamine (triethylamine $[(CH_3CH_2)_3N:]$), there is no reaction! (That is, it is argued that one arm of one ethyl group so hinders the electron pair on nitrogen that no bond can form.)

*Review, if necessary, the convention of the curved arrow, already defined in Chapter 1. Note that the arrow is always used to show that a pair of electrons is moving from the tail of the arrow to the head; the tail becomes, correspondingly, poorer in electrons while the head becomes richer; and the arrow actually) becomes the bond holding the two species together!

$$(CH_3)_3B \quad :N(CH_3)_3 \quad \rightleftharpoons \quad (CH_3)_3\overset{-}{B}—\overset{+}{N}(CH_3)_3 \qquad (5.15)$$

Other important Lewis acids include a host of electron-deficient metal halides (e.g., aluminum trichloride [$AlCl_3$], ferric chloride [$FeCl_3$], zinc chloride [$ZnCl_2$], boron trichloride [BCl_3], and stannic chloride [$SnCl_4$] as well as organometallic "mixed" halide derivatives such as diethylaluminum chloride [$(CH_3CH_2)_2AlCl$]).

Although lists of such compounds can be made, it is more important to understand that the Lewis definition is broad enough to include almost any species as acids or bases—depending on what is to be reacted! Consider, for example, the observation that hydrogen cyanide (HCN) is an acid (at least in the Brønsted sense, $pK_a = 9.1$) and that nitrogen is more electronegative than carbon. So, in the unionized material, a dipole in the direction shown below would be expected.

$$H \rightarrow C \equiv N$$

In the same way, if each of the hydrogens in ethene (ethylene, $H_2C=CH_2$) is replaced by a cyano group ($C \equiv N$) the *electron-deficient* (i.e., a Lewis acid!) double bond in the tetracyanoethene (tetracyanoethylene [$(N \equiv C)_2C=C(C \equiv N)_2$]), which results might be expected to react with electron-rich (i.e., Lewis base) species. Thus, it comes as little surprise that, an *electron-rich* **aromatic hydrocarbons**, such as benzene (C_6H_6) serve to donate electrons to tetracyanoethene (tetracyanoethylene [$(N \equiv C)_2C=C(C \equiv N)_2$]) in what is known as a "**donor–acceptor pair**" or a "**charge–transfer complex**" with π electrons serving both as "acid" and as "base." The evidence for reaction is striking as both benzene (C_6H_6) and tetracyanoethene (tetracyanoethylene [$(N \equiv C)_2C=C(C \equiv N)_2$]) are colorless while a bright orange color (presumably a material formed by π electrons mixing in the acid–base reaction) is seen. The material is frequently written as shown in Figure 5.43.

Not only can a given compound be—depending on that with which it is treated—both an acid and a base (e.g., water, H_2O, acting as an acid, can lose a proton to a stronger base while, acting as a base, it can donate its electrons to any suitable acid) but different acid and base groups can also be present in the same molecule! For example, both acid and base functional groups are present in amino acids such as

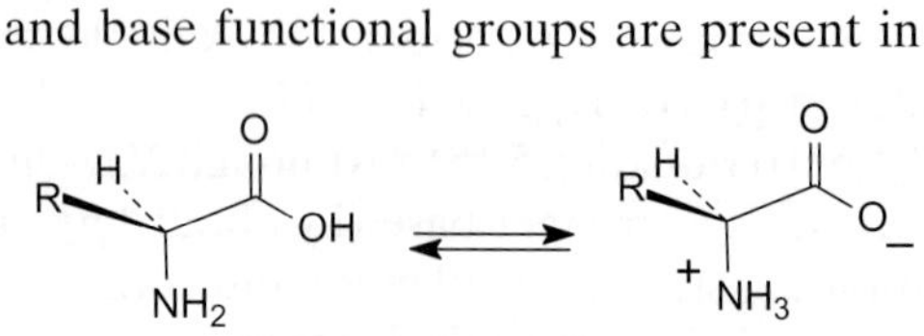

$$(5.16)$$

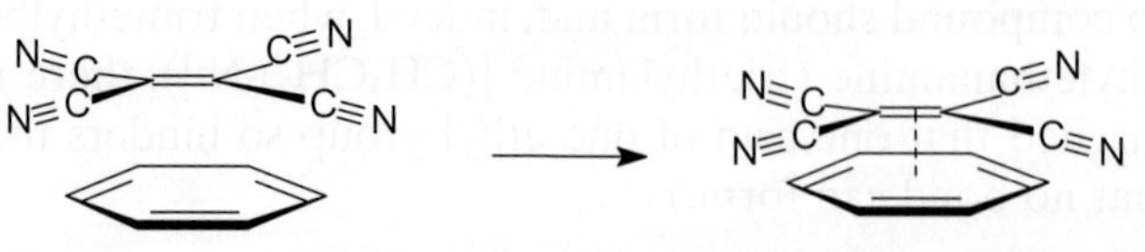

Figure 5.43. A depiction of a "charge–transfer complex" between tetracyanoethene (tetracyanoethylene [$(N \equiv C)_2C=C(C \equiv N)_2$]), which, in this case, is acting as an electron acceptor (Lewis acid) and benzene (C_6H_6), which, in this case, is acting as an electron donor (Lewis base).

those (α-amino acids) shown below, where, except for R = H, the chiral, substituted (R = H, alkyl, aryl) system contains the basic amino function, which can be thought of as "having been protonated by" or "reacted with" the acidic carboxylic acid. Frequently, the "neutral" amino acids are written as the bis-ionic species (called **zwitterions**) as shown on the right-hand side of the equilibrium of Equation 5.16 in order to reflect the relative basicities of the amino ($-NH_2$) and carboxylate ($-CO_2^-$) functions (or the acidities of the ammonium [$-NH_3^+$] and carboxylic acid [$-CO_2H$] functions). It should be clear, however, that, as a function of the pH in acidic aqueous solution, both the amino and carboxylic acid functions are expected to be protonated, and, in basic aqueous solution, neither is expected to be protonated.

Finally, in this vein, those groups of unsaturated compounds that also contain the hydroxyl functionality (called phenols and enols and already introduced in the discussion of alcohols, *vide supra*), along with their saturated counterpart alcohols, are also acidic—but somewhat more so (Table 5.17). Thus, as noted earlier, enols

TABLE 5.17. Some Carbonyl Compounds, Alcohols, and Their Conjugate Bases and pK_a's

Compound (Conjugate Acid)[a]	Approximate pK_a	Compound (Conjugate Base)[a]
R–O–C(=O)–CH$_3$	25	R–O–C(=O)–CH$_2^-$ ↔ R–O–C(–O$^-$)=CH$_2$
CH$_3$–C(=O)–CH$_3$	20	CH$_3$–C(=O)–CH$_2^-$ ↔ CH$_3$–C(–O$^-$)=CH$_2$
cyclohexanone (=O)	17	cyclohexanone enolate (↔ –O$^-$)
R–O–C(=O)–CH$_2$–C(=O)–O–R	13	R–O–C(=O)–CH$^-$–C(=O)–O–R ↔ R–O–C(=O)–CH=C(–O$^-$)–O–R
R–O–C(=O)–CH$_2$–C(=O)–CH$_3$	11	R–O–C(=O)–CH$^-$–C(=O)–CH$_3$ ↔ R–O–C(=O)–CH=C(–O$^-$)–CH$_3$
CH$_3$–C(=O)–CH$_2$–C(=O)–CH$_3$	9	CH$_3$–C(=O)–CH$^-$–C(=O)–CH$_3$ ↔ CH$_3$–C(=O)–CH=C(–O$^-$)–CH$_3$
CH$_3$CH$_2$OH	18	CH$_3$CH$_2$O$^-$
HOH	16	HO$^-$
C$_6$H$_5$–OH	10	C$_6$H$_5$–O$^-$ ↔ (cyclohexadienyl =O) ↔ (cyclohexadienyl =O)

[a] R = any alkyl or aryl group.
The rates for conversion of carbonyl and enol forms for many compounds are slow enough to be measured by ^{1}H NMR spectroscopy. Thus, although ethanol and cyclohexanone are acids of comparable strength, the rate at which the equilibrium between the acids and their respective conjugate bases is attained differs. A C–H bond is broken more slowly than an O–H bond.

are in equilibrium with their corresponding carbonyl derivative(s). Indeed, in general, compounds that possess protons on carbons adjacent (or α-) to the carbon of the carbonyl *can be considered acidic* to a host of different bases. Much of the chemistry of the carbonyl group (Chapters 8 and 9) is dependent on loss of the proton α- to the carbonyl in an equilibrium as shown in Figures 5.44 and 5.45 (and discussed further in Problems 5.13 and 5.14). In Figure 5.44, the loss of the proton α to the carbonyl to some generalized base, B⁻, is shown as an equilibrium process. As an equilibrium, Figure 5.44 can be read from either direction, but in Figure 5.44, the *curved arrows* are used to indicate the "flow of electrons" and the eventual disposition of the proton in what thus appears to be conversion of a carbonyl compound to its enol (according to the direction of the arrows). The arrows are the reverse of those shown in Figure 5.45.

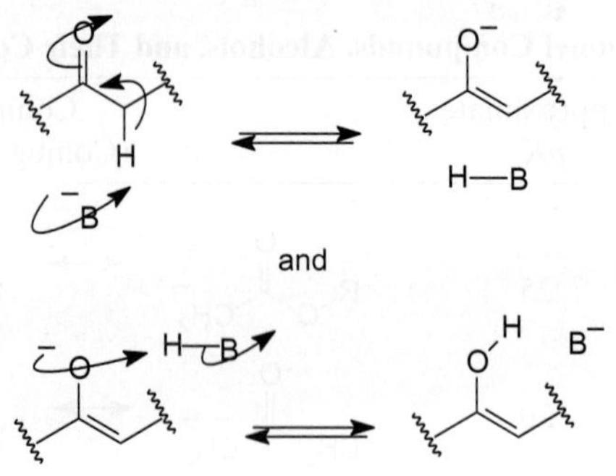

Figure 5.44. Keto-enol equilibrium. The curved arrows are used to depict electron motion connecting the starting ketone to the ending enol. Although an equilibrium, the path depicted by the curved arrows is intended to show enol formation.

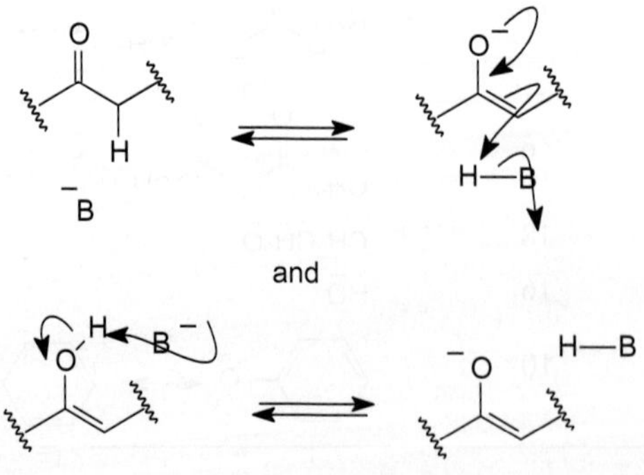

Figure 5.45. Keto-enol equilibrium. The curved arrows are used to depict electron motion connecting the starting enol to the ending ketone. Although an equilibrium, the path intended to be depicted by the curved arrows runs from the enolate anion at the lower left to the keto isomer on the upper left.

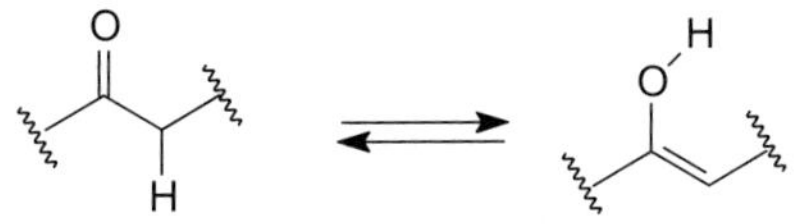

Figure 5.46. The generalized keto-enol tautomerism.

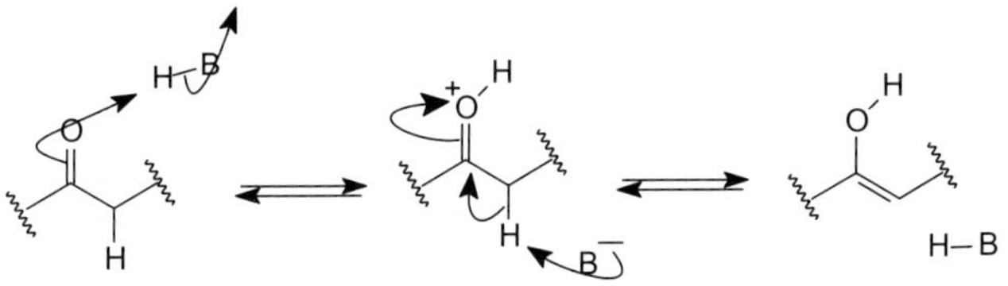

Figure 5.47. Protonation of the carbonyl of a ketone and formation of the corresponding enol.

This particular process of isomerization in which the idealized resonance forms of the enolate anion are protonated either on oxygen or on carbon is called **proton tautomerization**. The keto and enol forms are thus **tautomers**.*

In Figure 5.45, the curved arrows are now written to show loss of the protons attached to a hydroxyl group to some generalized base B⁻. Read from the lower left to the upper left, the *curved arrows* are again used to indicate the "flow of electrons" and the eventual disposition of the proton and to consummate the movement of nuclei in the reversal of the process of Figure 5.44.

The abbreviated, generalized proton tautomerism processes connecting the keto- and enol forms is shown in Figure 5.46.

Problem 5.16. Using propanone (acetone) and any other compound listed in Table 5.16, write an equilibrium expression for the generation of the enolate anion.

It should be clear that since the oxygen of the carbonyl is a Lewis base it can be protonated and keto-enol tautomerism is therefore also promoted by acid! Thus, as shown in Figure 5.47, protonation of the oxygen of the carbonyl and subsequent proton loss from the carbon α to the carbonyl results in enol formation (and, because this is an equilibrium process, vice versa).

Problem 5.17. Using propanone (acetone) and any other compound listed in Table 5.16, write an equilibrium expression for the generation of the enol of propanone through protonation (as in Figure 5.47).

*In addition to **proton tautomers** shown here, other **tautomers** (broadly defined as positional isomers created by minor nuclear shifts and electronic reorganizations over relatively low energy barriers) are known. **Valence tautomers** include, for example, the bicyclo[4.2.0]octatriene and cyclooctatetraene C_8H_8 isomers which are related via an electrocyclic process (Chapter 3).

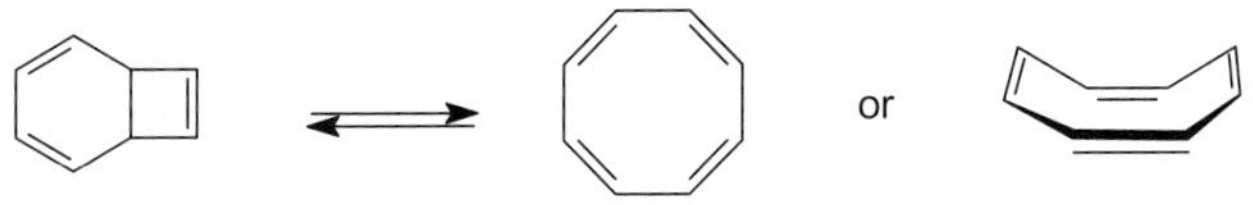

or

Thus, reasonably, it might be predicted that the acidity of protons α to a carbonyl group should depend, at least in part, on the extent to which the carbonyl group succeeds in attracting electrons from the carbon–hydrogen bond of the proton to be lost. Therefore, two carbonyl groups flanking the carbon bearing that proton (i.e., a 1,3-dicarbonyl compound, Table 5.17) should increase the acidity of the proton relative to the case where only one carbonyl is present. Indeed, because the resulting anion is also resonance stabilized (electron delocalized), its stability will be increased further and the kinetic acidity (the rate at which proton loss occurs) will be higher. Table 5.17 contains data on the acidity of some carbonyl derivatives.

Problem 5.18. From the data in Table 5.17, it is clear that protons α to the carbonyl of esters are less acidic than those α to the carbonyl of ketones. How might you account for this observation?

III. Hard and Soft Acids and Bases (HSABs)

The measurement of the position of equilibrium (K_{eq}) for reactions in which proton transfer is the common feature permits (as noted above) a definition of acid "strength." The same concept cannot, of course, be applied to Lewis acids and bases because proton transfer is no longer the unifying element. How then can a decision be made about what constitutes acid or base strength? A number of attempts to correlate the wealth of chemical reactions that involve Lewis acids and bases have been made and among the more recent is that of Pearson (Pearson, R. G., Professor, University of California, Santa Barbara) who has urged the division of acids and bases into categories called **hard** and **soft**. By this concept, hard acids are generally (but not necessarily) positively charged species that hold their charge on small atoms whose outer electrons (if present) are not easily excited (i.e., the gap between the HOMO and the LUMO is large). Soft acids (which might also be positively charged—but less so) are larger, more easily polarized, and have outer electrons that are more easily excited (i.e., a smaller HOMO–LUMO gap). Similarly, hard bases consist of electron donors of high electronegativity, low polarizability, and filled (or nearly so) low-energy orbitals, while soft bases are more highly polarizable and have unfilled low-energy orbitals. Generally, soft bases are easily oxidizable while hard bases can be oxidized only with difficulty.

Having thus divided materials by oxidation state, polarizability, and so on, Pearson set forth the HSAB principle, where hard acids prefer to bind to hard bases while soft acids prefer to bind to soft bases.

For example, boron trifluoride (BF_3), a hard acid, forms strong complexes with amines, such as ammonia (NH_3) and methanamine (methyl amine, H_3CNH_2), both of which are hard bases. Phosphorus (which lies just below nitrogen in the periodic table) forms compounds similar to nitrogen, and it might be expected that boron trifluoride (BF_3) would also form complexes with phosphine (PH_3) and trimethylphosphine, $(CH_3)_3P$. Now, although the complexes between the Lewis acid boron trifluoride (BF_3) and both amine and phosphine Lewis bases form, those between boron trifluoride (BF_3) and the nitrogenous bases are stronger (as measured by their respective equilibrium constants). On the other hand, when the (soft) Lewis acid Ag^+ is used in place of boron trifluoride (BF_3) with the same amines, the situation is reversed and it is concluded that the phosphorus containing Lewis bases must be soft.

TABLE 5.18. Some Hard and Soft Acids and Bases (after Pearson, R. G. *Science*, 1966, *151*, 172 ff.)

Hard Acids	Soft Acids	Hard Bases	Soft Bases
H^+, Li^+, Na^+, K^+ Mg^{+2}, Ca^{+2}, Al^{+3}, Fe^{+3} $AlCl_3$, etc.	Cu^+, Ag^+, Hg^+, BH_3, I^+ Br^+, I_2, Br_2, etc.	H_2O, OH^-, F^-, Cl^- ROH, NH_3, etc.	I^-, RS^-, CN^-, H^-, $CH_2{=}CH_2$, etc.

Although hydrogen bonding, charge–transfer complexes, and the role of metals and metal surfaces can be correlated by the principle, the role of the solvent can moderate its effects, some nuclei and groups must be considered borderline, and the volume of data has proved daunting for its routine application. Nonetheless, the correlation has proved useful, and Table 5.18 contains some of the material that will allow predictions to be made.

F. COMPUTATIONAL METHODS

As pointed out in the introduction to this chapter, and as used in Chapter 1, there are a variety of computational methods that allow simulation of structures and prediction of properties to be made. In many cases, where measurements have been made subsequent to the predictions, these computational methods have proven their worth.

The broad and growing number of such methods has derived strength from the speed at which new computer hardware and software is introduced and developed. Indeed, it is now clear that quality calculations can be carried out on widely available desktop equipment. Broadly, the current tools can be described as (a) *ab initio* (first principles) methods, (b) *semiempirical* methods, and (c) *empirical* methods. *Ab-initio* type calculations require some constants (such as the speed of light) and the use of quantum mechanics alone. Generally, *ab initio* calculations are time-consuming and applicable to only the smallest molecular systems. Semiempirical methods use some experimental data to approach approximate solutions to the Schrödinger Equation (see Appendix I) and, depending on the desired outcome and the particular approximations (their *level*) as expressed as the *basis set* used and as already seen in Chapter 1, can be valuable for comparing properties of sets of molecules. Empirical methods, which utilize (i) predefined atom types such as the presumed hybridization of carbon as *sp*, *sp*^{2}, or *sp*^{3} and its connectivity; (ii) Equations derived from classical physics that predict interactions among nuclei as if they were "hard spheres"; and (iii) sets of parameters to fit the Equations and atom types to experimental data, have become very popular. With regard to the latter, while it might seem that such empiricism would lead to results of limited utility, the more empirical methods have proved their value (as very good approximations) by their successes. Among the more successful of the latter is the MM *force field* due largely to the efforts of Allinger* and coworkers.

*Allinger, N. L., Professor of Chemistry, University of Georgia. Various versions of MM have been (and are being) developed.

I. MM

As seen already, MM calculations are used to "clean up" structures prior to more "sophisticated" calculations at the semiempirical or ab initio level in WebMO and are generally performed assuming that each molecule is isolated from its neighbors. Electron diffraction, X-ray diffraction, microwave, NMR, and IR spectroscopic data are used. Then, parameters (such as "stretching constants" for bonds, *vide infra*) are derived iteratively so that calculations using them reproduce experimentally determined values for as large a number of systems as possible. These trial-and-error optimized parameters are often useful directly so that one molecule can be compared with another. In MM, the total energy (E_{total}) is given by an expression such as that of Equation 5.17, each term of which is discussed further below:

$$E_{\text{total}} = E_{\text{stretch}} + E_{\text{bend}} + E_{\text{stretch-bend}} + E_{\text{van der Waals}} + E_{\text{torsional}} + E_{\text{dipole}}. \qquad (5.17)$$

a. Stretching Energy Contribution (E_{stretch}). The overall Equation to describe the stretching energy is given by an expression such as that in Equation 5.18, where K_s is the *stretching force constant parameter* developed to work in the equation, and r and r_o are, respectively, the distance between the nuclei and the equilibrium distance between the nuclei taken, initially, from relevant experimental data. In principle, once a set of constants K_s and r_o have been defined, subsequent calculations for E_{total} are routine. The constant (71.94) adjusts the units from millidyne $(\text{Å})^{-1}$ to kilocalorie per mole:

$$E_{\text{stretch}} = 71.94 \, K_s (r - r_o)^2 [1 - 2.00(r - r_o)]. \qquad (5.18)$$

b. Bending Energy Contribution (E_{bend}). The parameters, K_b, the *bending force constant* (in millidyne Å rad^2) and σ_o (degrees), the *equilibrium bond angle*, are used to determine the bending energy. The force constant is usually obtained from IR spectroscopy, while angles come from electron diffraction (or microwave) spectroscopy. Iteration to fit Equation 5.19 is then undertaken:

$$E_{\text{bend}} = (2.194 \text{ s } 10^{-2}) K_b (\sigma - \sigma_o) [\{r(1) - r_o(1)\} + \{r(2) - r_o(2)\}]. \qquad (5.19)$$

c. Stretch-Bend Energy Contribution ($E_{\text{stretch-bend}}$). A potential function called stretch bend is also included in most calculations. This function is designed to cause bond lengths to stretch when angles are compressed and to shrink when angles are made wider. In Equation 5.20, K_{sb} is the *stretch-bend constant* (millidyne rad^{-1}) and the angle effected has the equilibrium value σ_o (degrees) between nuclei separated from the atom about which they may be considered to pivot by equilibrium bond lengths $r_o(1)$ and $r_o(2)$:

$$E_{\text{stretch-bend}} = 2.51124 \, K_{sb} (\sigma - \sigma_o) [\{r(1) - r_o(1)\} + \{r(2) - r_o(2)\}]. \qquad (5.20)$$

d. van der Waals Energy Contribution ($E_{\text{van der Waals}}$). Nonbonded interaction energies (excluding 1,3-interactions) are calculated with a force-field Equation that considers both distance between the nuclei and their relative "hardness" as a van der Waals constant. In Equation 5.21, the term $p = r^*/r$, where r^* is the sum of the van der Waals radii of the two atoms [(1) and (2)] while r is the distance between the two atoms. The constant K_{vdw} is defined as $[\varepsilon(1)\varepsilon(2)]^2$ where $\varepsilon(1)$ and $\varepsilon(2)$ are the van der Waals constants for the atoms 1 and 2, respectively:

$$E_{\text{van der Waals}} = K_{vdw}[(2.90\times10^5)(e^{-12..50/p}) - 2.25p^6]. \tag{5.21}$$

e. Torsional Energy Contribution ($E_{\text{torsional}}$). The torsional (Equation 5.22) contribution is given by a series of *one-*, *two-*, and *threefold torsional constants*, v_1, v_2, and v_3 (kcal mol^{-1}) which "modify" the cosine function of the torsion angle (ω). As a cosine function, the energy will be at a maximum when the torsion angle is a minimum:

$$E_{\text{torsional}} = (v_1/2)(1+\cos\omega) + (v_2/2)(1+\cos2\omega) + (v_3/2)(1+\cos3\omega). \tag{5.22}$$

f. Dipole Interaction Energy and Dipole Moment Contribution (E_{dipole}). Bond moments for polar bonds are needed to obtain dipole moments and dipole interaction energies in MM. Generally, bond moments are found by trial and error. Thus, some reasonable value is taken (based on precedent), and it is then refined until the experimental dipole moment is reproduced. Equation 5.23 is used to this end. In Equation 5.23, R is taken as the distance between the midpoints of the two bonds from a central atom to two substituents "A" and "B," μ_A and μ_B are their respective bond moments (in Debye units), χ is the angle between the bond vectors, α_A and α_B are the angles between the bond axes and the line along which R is measured, and D is the effective dielectric constant. The energy is converted to kilocalorie per mole by use of the conversion factor k (erg molecule^{-1} to kcal mol^{-1} whose value is given as 14.39418):

$$E_{\text{dipole}} = (k\mu_A\mu_B(\cos\chi - 3\cos\alpha_A\cos\alpha_B)/R^3D. \tag{5.23}$$

Generally, programs that use MM for structural calculations have tables of parameters, atom types, and so on, as integral to their function. Usually, the user is asked if modification of any of the input data is desired. Failure to modify the data when that for atom types being calculated is absent results in use of default values built into the program or a request for new values. If the appropriate values are missing, the approximations lose much of their value.

Finally, many MM programs contain terms in addition to those of Equation 5.17. Among the more popular is that of Equation 5.24 which attempts to account for out-of-plane (E_{oop}) effects. For example, in a molecule such as cyclobutanone, the carbonyl group (C=O), normally considered to be sp^2 hybridized should form angles of 120° with the carbons flanking it. However, as the carbon atoms of cyclobutanone are constrained to a four-membered ring, the internal angles are diminished and the C–C–O angle (found to be close to 133°) is made wider (Figure 5.48). Although, in principle, some strain energy might be relieved by forcing the oxygen out of the plane defined by the carbon of the carbonyl and the carbons flanking it, this would adversely affect the orbital overlap in the π system. Thus, a correction is needed:

$$E_{\text{oop}} = \frac{1}{2}k(d^2). \tag{5.24}$$

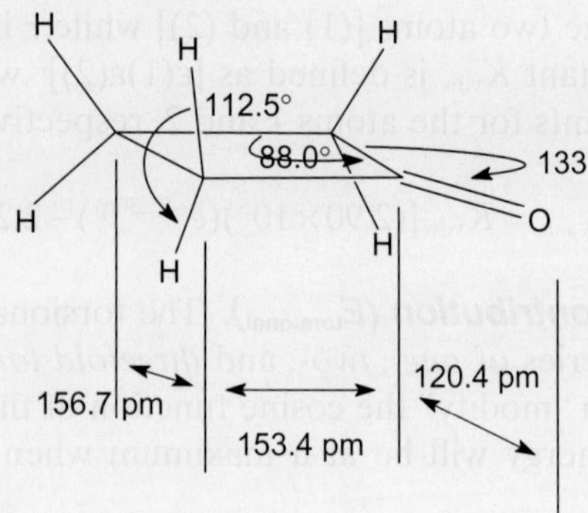

Figure 5.48. A representation of cyclobutanone.

In Equation 5.24, k is the out-of-plane bending constant (in kcal/Å^2) and d is the height of the central atom above the plane of its substituents (in Å).

However, even without knowing the details, and even in the absence of accurate constants, valuable **comparisons** between two (or more) molecules can be obtained as long as it is remembered that the results are no better than the data and approximations warrant. Clearly, the "energy" of a molecule found this way (as distinct from, e.g., its heat of formation) is strictly a relative term.

ADDITIONAL PROBLEMS

Problem 5.19. Name the following compounds and identify the positions of the substituents relative to the functional group both with the appropriate numbers and with the appropriate letters, for example,

2-methylcyclohexanone and α-methylcyclohexanone

(a)   (b)   (c)   (d)

Problem 5.20. Based on systematic nomenclature and Woodward's rules (this chapter, p. 185) (a) name the following compounds, (b) predict something about the wavelength at which the UV maximum might be found for each compound, and (c) look up literature values for the positions of the maxima you have predicted.

(a)   (b)   (c)

(d)

β-Carotene or (all *E*-)-1,1′-(3,7,12,16-tetramethyl-1,3,5,7,9,11,15,17-octadecanonaene-1,18-diyl)bis[2,6,6-trimethylcyclohexene]

Problem 5.21. Draw the structures for the following compounds:
(a) 1,1-dichlorocyclohexane,
(b) 1-bromo-2-chloropentane,
(c) Z-1,2-dichlorocyclopropane,
(d) 3-bromo-2-methylcyclohexene,
(e) 3-iodo-1,4-pentadiene, and
(f) 2,4,6-trinitrotoluene (TNT).

PART II

MIDDLEGROUND

Unity and Diversity, and never one without the other ...
—A. Camus, Le Partie de notre génération, *Demain*, **1957**

INTRODUCTION TO PART II

Organic chemistry largely remains involved in the preparation of compounds of carbon and transformations between them. While the techniques for carrying out and observing reactions as they occur continue to grow in sophistication, there are many unresolved questions about most reactions, and skill and art both play important roles.

Generally, the same (or similar) processes are used to prepare specific compounds as well as to modify those prepared earlier (and thus, perhaps, commercially available) or those found in Nature. Not surprisingly, the overall sense of any particular transformation is limited (as are the processes), but the variation in detail is frequently very great. Among the particular transformations we will consider *for every functional group* are these: *oxidation, reduction, substitution, addition, elimination*, and *rearrangement*.

In formulating the transformations of *starting materials to products* for *all* of the functional groups, evidence has accumulated that the pathways by which the former are converted to the latter frequently involve the intermediacy of the same types of transient, highly reactive species. These intermediate species include *carbocations, carbanions, free radicals*, and *carbenes*.

Finally, it is necessary to inquire if the overall energetics, which will include the formation and subsequent utilization of the intermediates on the path linking starting materials to products, is favorable (a question of *thermodynamics*), and even if favorable, whether the time frame in which the products might be formed is reasonable (a question of *kinetics*)?

Then, knowing the **process** (*oxidation, reduction*, etc.), as well as something about the **pathway** (*kinetics* and *thermodynamics*) and about any **intermediates** (does it involve carbocations, free radicals, etc.) that might be present, it may be possible to predict what could reasonably happen in a case similar to that examined. Indeed, although in principle it might be possible to simply memorize the outcome of every

Foundations of Organic Chemistry: Unity and Diversity of Structures, Pathways, and Reactions,
First Edition. David R. Dalton.
© 2011 John Wiley & Sons, Inc. Published 2011 by John Wiley & Sons, Inc.

possible reaction that has been reported (trusting that the details have been properly examined with the tools now available), it should be clear that an understanding of the general principles might be applied to unfamiliar cases. To that end, the intermediates common to *all* of the transformations will be briefly described here and, in the forthcoming chapters, the transformations listed above will be discussed for *each* of the functional groups. As dictated by space and time limitations, some small portion of the accumulated evidence for what is presented will also be made available. However, it is important to recognize that in each case, *only a few examples of the multitude known will be provided.*

A. THE INTERMEDIATES

Single bonds that are broken in processes that occur either stepwise or without symmetry control (Chapter 4) can either (a) fragment so that each piece gets one of the two electrons in the bond, a *homolytic process* (illustrated for the case of ethane in Equation II-A), or (b) fragment in such a way that one fragment gets the electron pair that was the bond, while the other fragment loses all contact with that pair, a *heterolytic process* (again illustrated for the case of ethane in Equation II-B).

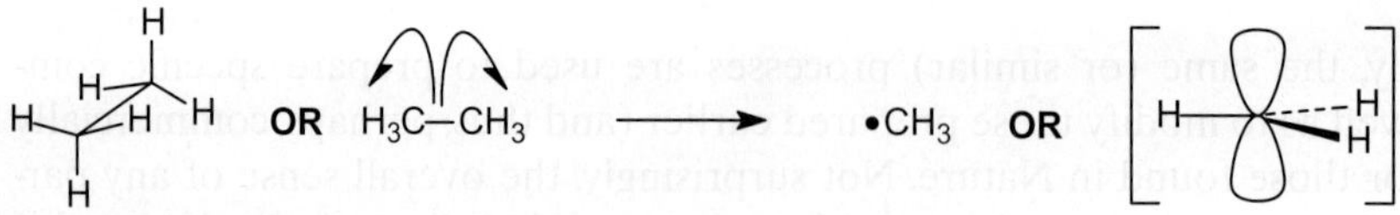

$$H_3C-CH_3 \longrightarrow \bullet CH_3 \quad \text{OR} \tag{II-A}$$

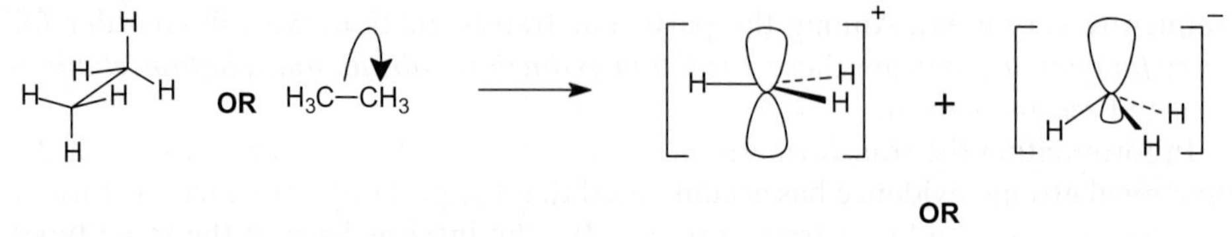

$$H_3C-CH_3 \longrightarrow \tag{II-B}$$

$$CH_3^+ \quad + \quad CH_3^-$$

Most organic reactions are carried out in solution rather than in the gas phase. Although reactions in solution can occur by both homolytic and heterolytic processes, it is common that solvents (Chapter 5) promote the latter by helping to ameliorate the local buildup of charge and by solvating the ions that result. Nonetheless, the methyl carbocation (CH_3^+)* and methyl carbanion (CH_3^-) of Equation II-B are such high-energy species (so reactive) that the equation should

---

*Positively charged species such as the methyl carbocation have gone through several "name" changes in recent years. Thus, there is a lengthy period (and thus a significant body of literature) in which CH_3^+ and related species are known as "carbonium ions" and a shorter period when, because "'onium ions" such as NH_4^+ and CH_5^+ (the latter in "super acids") were clearly related, the electron-deficient species R_3C^+ were called "carbenium ions." The name carbocation has been generally accepted and has been in use for a sufficiently long period so that it seems likely to last.

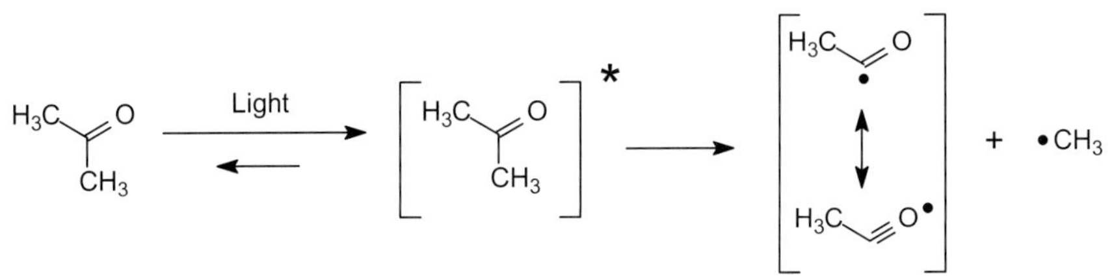

Scheme II-A. A representation of the photochemical decomposition of propanone (acetone [CH$_3$COCH$_3$]) to carbon monoxide (CO) and the methyl radical (CH$_3$•). The species in brackets represents a photochemically excited molecule in which an antibonding orbital is occupied.

only be considered as illustrative of a type of process. Indeed, in the absence of some external agent to cause the bond to break, the only source of energy for the process to occur is provided by collisions between the molecules themselves and those of the solvent. At reasonable temperatures (e.g., in the vicinity of temperatures up to about 100°C), bonds with energies much greater than ca. 35 kcal mol^{-1} (146.3 kJ mol^{-1}) are unlikely to be broken without intrusion of a *catalyst** or direct solvent participation.

Thus, the specific process of Equation II-B is not observed in solution as the solvation energies required are too great, nor is it seen under conditions usually available for gas-phase reactions (although such species can occasionally be observed or their presence inferred at high enough energies).

Leaving aside, for the moment, the cationic and anionic species of Equation II-B and understanding that the reaction of Equation II-A is illustrative only of a process, the fragments (CH$_3$•) are called **radicals** (in this case, the *methyl radical*). Although the descriptor "radical" has previously been taken to mean (in the nomenclature of, e.g., hydrocarbons [Chapters 3 and 5]) a group of distinguishable atoms, it is here taken to mean a species with *an unpaired electron. A diradical* would therefore have *two unpaired electrons*.

While it is true that, in principle, the methyl radical (CH$_3$•) might be produced by the breaking of the carbon–carbon bond of ethane (*DH*° = 90 ± 2 kcal mol^{-1} [376 ± 8 kJ mol^{-1}]; Table 1.1, Chapter 1), this high-energy route is usually avoided by finding and using some alternative process that produces the same result starting from materials that already have higher energy (i.e., raising the ground state). One such alternative is shown in Scheme II-A. Thus, when propanone (acetone [CH$_3$COCH$_3$]) is irradiated with light at about 25,370 nm, an $n \rightarrow \pi^*$ absorption occurs. The excited propanone, in addition to reemitting some of the radiation, also

*The height of the barrier over which the reactants must travel to become products (the distance between reactants and the rate-determining transition state for the reaction [Chapter 4]) can be effected by materials called *catalysts* (Chapter 3). As pointed out there, these materials facilitate the conversion of starting materials to products by diminishing the energy gap between the reactants and the rate-determining transition state. It will be recalled that, in principle, this is accomplished by raising the energy of the ground state of the reactants and/or lowering the height of the transition state lying between the reactants and the products. Many more cases will be encountered where catalysts intrude and, as is usually the case, catalysts having been used are then regenerated so that their overall effect, as seen in the kinetic expression for the process, is not in proportion to their stoichiometric concentration.

(a)   (b)

Figure II-A. Representation of the methyl radical (CH$_3\cdot$) showing (a) a p orbital at right angles to the plane defined by the three hydrogens and containing the carbon atom and (b) the transference of the sense of electron "spin" through the bonds as a path for accounting for the electron's interaction with a nucleus.

efficiently undergoes *photolytic decomposition* or *photolysis* to give acetyl radicals (CH$_3$CO$\cdot$) and methyl radicals (CH$_3\cdot$) (Scheme II-A). The former continue, in a second decomposition, to carbon monoxide (CO) and another equivalent of methyl radical (CH$_3\cdot$).

Although not as easily studied as more highly substituted radicals,* the electron spin resonance (ESR, Chapter 2) spectrum of the methyl radical (CH$_3\cdot$) has been obtained and the hyperfine coupling constant of the electron both with the hydrogens on carbon and with the carbon (as ^{13}C) itself determined. Both ($a_H \approx -23$ gauss for ^{1}H and $a_C \approx 41$ gauss for ^{13}C) are in accord with a **planar** species; in general, this and other similar carbon-centered radicals are considered to have the **odd electron in a p orbital**, that is, at right angles to the plane defined by the three hydrogens and containing the carbon (Figure II-A).†

*The introduction of one or more aromatic rings into conjugation with the carbon bearing the radical allows for electron delocalization (Chapter 1) and thus increased stability. For example, in the benzyl radical (see next figure), electron deocalization involving the aromatic ring is detected and the hyperfine coupling to the ortho-, meta-, and para-positions (Chapter 3) observed (the coupling constants a_H are provided).

Some representations of the benzyl radical

†The discussion of the combination of atomic orbitals to produce localized molecular orbitals was presented at some length in Chapter 1. In that vein, localized orbitals of (CH$_3\cdot$) might more easily be used to account for the interaction of the "odd electron" with the hydrogen nuclei (i.e., Figure II-A can be a representation of a high-lying localized orbital).

There are two possible pictures:

(I) The first corresponds to bonding orbitals between the three 1s hydrogen orbitals and the one 2s orbital on carbon and two of the 2p orbitals, respectively, leaving the third 2p orbital unhybridized and yielding a planar structure. These bonding orbitals are shown in the next figure as a–d, respec-

(Continued on next page)

In addition to the odd electron species called *radicals* described above, it is also possible to have species with **two** electrons that lack ligands. There are two possibilities:

(I) If the species has the unliganded electrons on different atoms that are insulated from direct interaction by, for example, saturated carbon, they are called diradicals (e.g., the 1,3-cyclopentanyl diradical shown in the next figure). As shown, however, it is possible that the diradical species could have the electrons with spins that are either paired, called the "singlet state," or with the spins unpaired, called the "triplet state" (the **state** being named as a function of the number of "unpaired electons $n + 1$"). If the number of "unpaired" electrons is zero, because the electrons are paired, then $n = 0$, $(n + 1) = 1$, and the state is a "singlet"; if they are unpaired, then $n = 2$, $(n + 1) = 3$, and the state is a "triplet."

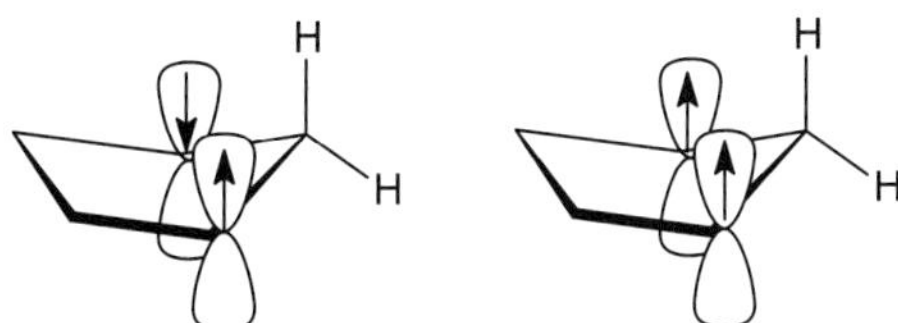

Spin paired, singlet   Spin unpaired, triplet

It might be supposed that either of the species above could be obtained by simply breaking the single bond in bicyclo[2.1.0]pentane as shown in

tively (the unoccupied antibonding orbitals are not shown). The representations were created using the MOViewer in WebMO version 6.0.002p.

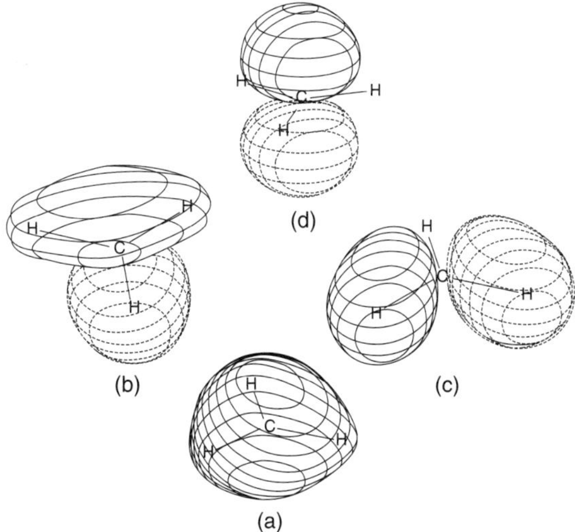

(II) The other possibility results in a pyramidal methyl radical, and the localized orbitals are produced by a combination of the 2s orbital with one 2p orbital to generate a σ-type orbital that points *away* from the concave set of three hydrogens and a σ-type orbital pointing *into* the concave set.

(Continued on next page)

Equation II-C. However, as pointed out earlier (Chapter 4), such electrocyclic processes occur with conservation of symmetry, and the process (and its reverse) is *thermally disrotatory* to the singlet. The triplet might then be produced by a process that occurs in a stepwise fashion, such as the low-temperature photochemical decomposition of the azo compound 2,3-diazabicyclo[2.2.1]hept-2,3-ene, shown in Equation II-D.

$$\text{(II-C)}$$

$$\text{(II-D)}$$

(II)   If the diradical species has the electrons on the **same** atom, that is, results from the formal removal of two singly bonded ligands with one of their associated electrons with each, and if the atom from which the ligands is removed is carbon, then the resulting diradical is called a carbene; if the atom from which the ligands are removed is nitrogen, the diradical is a *nitrene*.

Combination of the σ *in-pointing* orbital with the 1*s* hydrogen orbitals produces an occupied bonding orbital (cf.[e] in the next figure) and the corresponding unoccupied (and not shown) antibonding orbital. The other two occupied bonding (and corresponding unoccupied antibonding) orbitals as well as the hybrid orbital containing the "odd" electron (i.e., [f–h] in the figure) are strikingly similar to b–d (except that the carbon atom does not lie in the plane of the three hydrogen nuclei; i.e., it is pyramidal).

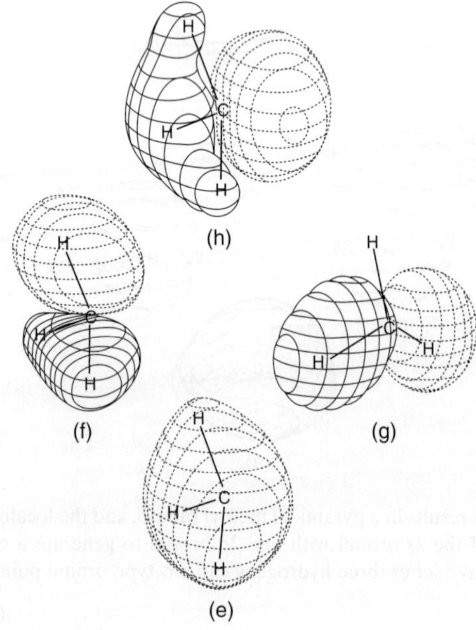

Carbenes may be generated, among other ways, by the thermal or photochemical decomposition of diazomethane (CH_2N_2) (Equation II-E) and **nitrenes** by the thermal decomposition of a suitable azide (R-N_3) such as phenyl azide (azidobenzene [$C_6H_5N_3$]) (Equation II-F).

$$CH_2\overset{+}{=}N\overset{-}{=}N \xrightarrow[\text{light}]{\text{Heat or}} :CH_2 \ + \ N_2 \qquad\qquad \text{(II-E)}$$

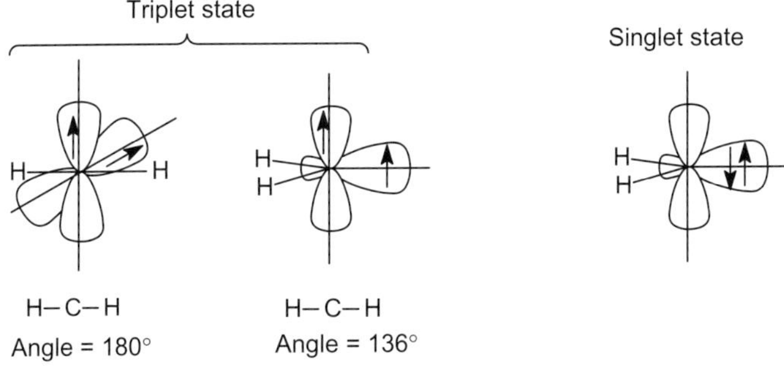

$$\text{(II-F)}$$

A **carbene** such as *methylene* (methylidene, :CH_2, Equation II-E) can also have the electrons *paired* (a *singlet* [$^1CH_2$:]) or *unpaired* (a *triplet* [$^3CH_2$:]). Since the nitrogen (N_2) produced in the photolytic decomposition of diazomethane (CH_2N_2) is in the singlet state, the **principle of spin conservation** argues that the carbene is also a singlet. This is the excited-state methylene since its decay to a triplet has been observed and thus it is argued that the triplet state for methylene itself (:CH_2) is apparently the ground state. However, not all **carbenes** are ground-state triplets.

It was originally argued that triplet methylene ($^3CH_2$:) should be linear, that is, that the H—C—H angle should be 180° with the electrons localized in mutually orthogonal *p*-type orbitals as shown in the next figure. However, it is found that the angle is closer to about 136°. Interestingly, the singlet state ($^1CH_2$:), which is apparently about 9 kcal mol^{-1} (about 36 kJ mol^{-1}) higher in energy than the triplet state ($^3CH_2$:), has a somewhat smaller H—C—H angle, in concert with the idealized concept of the paired electrons occupying an sp^2-like orbital and a *p*-type orbital remaining vacant.*

Triplet state   Singlet state

H—C—H   H—C—H

Angle = 180°   Angle = 136°

*Although the occupancy (and hence the energy) of the hybrid orbitals differs for the singlet and triplet states, the general shapes of those orbitals (shown in the next figure in order of increasing energy a → f) must be about the same if the H—C—H angles are about the same.

These alternative representations of the bonding and antibonding orbitals of the singlet and triplet states in the figure are of slightly different energies (but approximately the same shapes) for both of the states. The representations were created using the MOViewer in WebMO version 6.0.002p.

(Continued on next page)

While it is uncommon now, the singlet carbene ($^1CH_2$:) was, at one time, written as $CH_2\pm$ to emphasize that a vacant orbital should be associated with a positive charge and a filled orbital with a negative charge. Electron-deficient carbon species (and thus bearing positive charges) are **carbocations**; those with negative charges (and a filled but not bonded orbital) are **carbanions**.

Although **carbocations** can be observed spectroscopically in superacid/SO_2 media* at low temperature, it was by other methods that their presence in organic

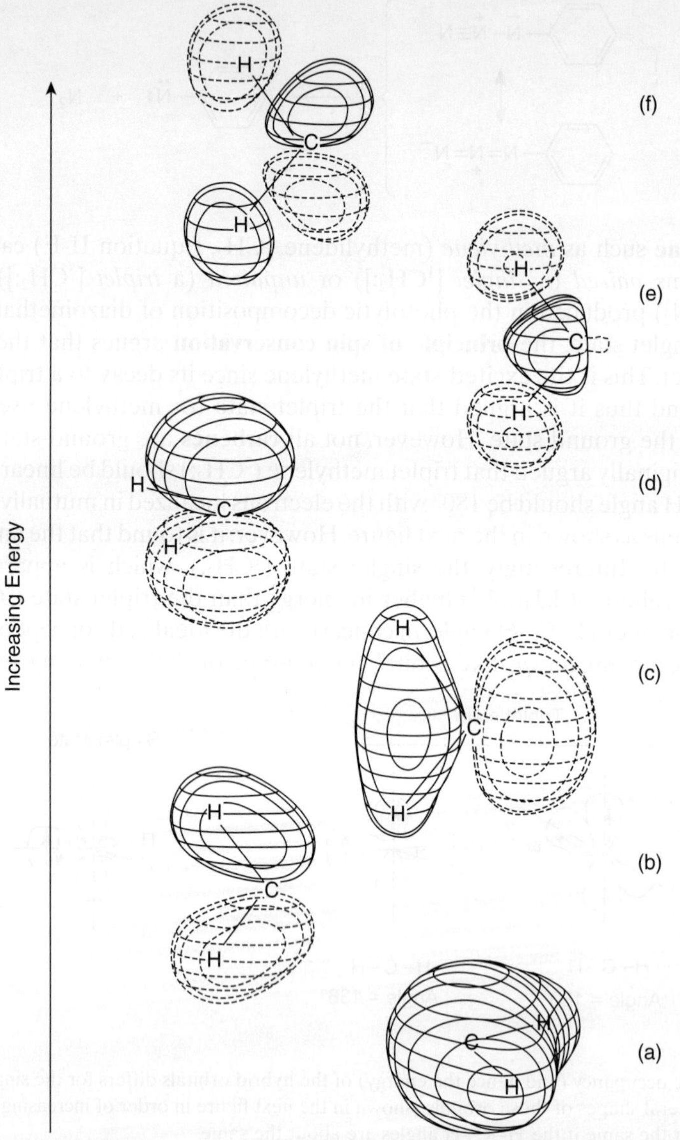

*Much of this work was done by G. Olah and his coworkers (Nobel Prize, 1994, University of Southern California). The ^{13}C chemical shift is particularly sensitive, as might be anticipated, as electrons shielding the nucleus are missing. For example, C2 in 2-methylpropane [$(CH_3)_3CH$] is normally found at $\delta \approx$ 25.0 ppm, while the chemical shift for the same carbon in the corresponding 2-methylpropyl carbocation $(CH_3)_3C^+$ is $\delta \approx$ 335.2 (at $-80°C$ in SO_2ClF-SbF_5 where it can be observed).

reactions as intermediates was originally inferred. Thus, it was early found that some highly substituted alkyl halides such as chlorotriphenylmethane (triphenylmethyl chloride, trityl chloride [$(C_6H_5)_3C–Cl$], Equation II-G) apparently underwent ionization in polar medium (e.g., liquid SO_2) spontaneously to produce solutions that could conduct current. As it was appreciated that organic molecules in solution generally did not conduct current (being covalent rather than ionic), it was concluded that chlorotriphenylmethane (triphenylmethyl chloride, trityl chloride [$(C_6H_5)_3C–Cl$]) was ionizing (Equation II-G).*

$$\text{(structure)} \xrightarrow{\text{Liquid } SO_2} \left[\text{(structure)} \right]^+ \;+\; Cl^- \qquad \text{(II-G)}$$

In addition to spontaneous ionization (in ionizing solvent), freezing point depression (as well as boiling point elevation) has long been used to establish the concentration of solute particles in solvents with suitable melting (boiling) points. A large volume of work was done with alcohols in sulfuric acid (H_2SO_4) because its melting point is convenient (pure H_2SO_4, mp 10.49°C) and also because the ions produced do not form stable covalent products with each other. Thus, as shown in Equation II-H, the dissolution of triphenylmethanol (triphenylcarbinol [$(C_6H_5)_3C–OH$]) in sulfuric acid results in the conversion of a single solute particle into four solute particles—with observation of the corresponding melting point depression.

$$\text{(structure)}–OH \;+\; 2H_2SO_4 \longrightarrow \left[\text{(structure)} \right]^+ \;+\; 2HSO_4^- \;+\; H_3O^+$$

$$\text{(II-H)}$$

Subsequently, numerous additional carbocations have been detected either spectroscopically or by other means, or their presence has been inferred from other arguments.

In contrast to carbocations, the ionization of carbon acids to generate **carbanions** has been somewhat more difficult to demonstrate. This appears to be due largely to the difficulty of finding solvents of suitable dielectric constant to support ionization but yet not too acidic (allowing proton transfer) or otherwise too reactive. Thus, while diethyl ether (ether [$(CH_3CH_2)_2O$]) and other similar solvents are, in principle,

---

*It will be appreciated that, a priori, it could not be known if the ionization shown in Equation II-G to produce the chloride anion and the trityl cation is correct or if the chloride cation and the trityl anion were being generated. The former was more in concert with other evidence, some of which was only established subsequently.

suitable from the point of view of reactivity, they do not easily support ionization. Nonetheless, the reaction of, for example, chlorotriphenylmethane (triphenylmethyl chloride, trityl chloride $[(C_6H_5)_3C\text{—}Cl]$) with sodium metal in ether (Equation II-I) produces a material that behaves as if it were triphenylmethyl sodium (triphenylmethanide).

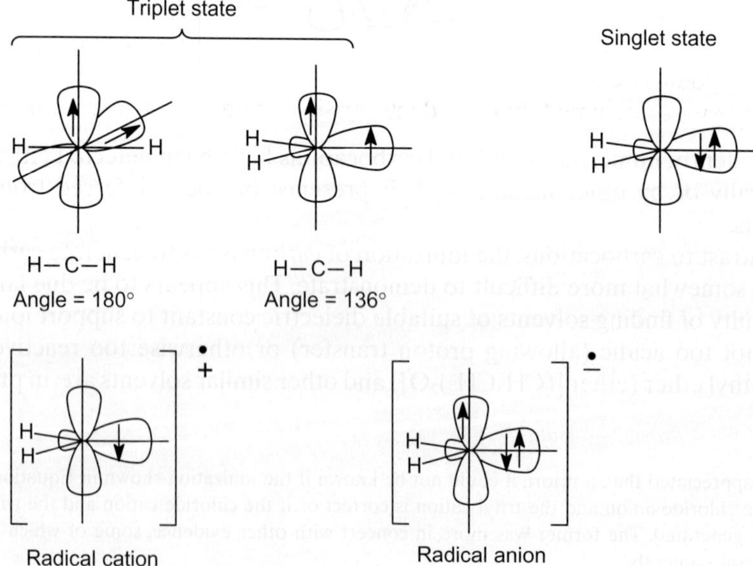

In general, it has been argued (as will become apparent) that when the negative charge is not delocalized over an adjacent unsaturation (for the triarylmethyl anion shown in Equation II-I), the protons on the rings *ortho*- to the central carbon so interfere with each other that planarity cannot be attained and the anion retains its geometry. Thus, the electrons in the occupied but unliganded orbital are kept at a maximum distance from those in the adjacent occupied (ligand bearing) orbitals. Further, it will be found that unless such a delocalization exists, the hydrocarbon is not sufficiently acidic to allow easy removal of the proton and that reaction of an alkyl halide with a metal is necessary to produce a carbanion. Examples illustrative of this concept are found in Chapter 5 (Section E, Acids and Bases).

Returning again to the picture of the singlet and triplet carbenes discussed above and shown again below, removing one of the electrons from either the singlet or triplet results in the formation of a **radical cation**: a species that possess, in principle, both a vacant orbital *and* an "odd electron." Alternatively, the addition of one electron to either the singlet or the triplet produces a **radical anion**.

B. THE TRANSFORMATIONS

Rearrangements, already encountered in Chapter 4, are processes in which the number of nuclei (e.g., carbon, hydrogen, and oxygen) in the framework of the reactant and product remain the same, but the framework itself has been changed. Occasionally, rearrangements are accompanied by the processes of **addition, elimination**, and/or **substitution**, to be discussed further below; however, in the three examples of rearrangements that follow, only the skeleton changes.

The first rearrangement example (Equation II-J) is the reversible thermal interconversion of (Z)-3,4-dimethylcyclobutene into (Z,E)-2,4-hexadiene (as seen earlier [Chapter 4, **Figure 4.38**]). It should be noted that although the relationships between carbons and attendant hydrogens have been retained the numbers of sigma (σ) and pi (π) bonds has clearly changed, as has the relationship between carbon atoms 3 and 4 (adjacent in (Z)-3,4-dimethylcyclobutene) and carbon atoms 2 and 5 (at opposite ends of the conjugated double bonds in (Z,E)-2,4-hexadiene).

$$\text{(II-J)}$$

Second, in Equation II-K, a carbocation, which is **secondary**, is transformed into a carbocation that is **tertiary** by the migration of a **methyl group with its pair of electrons**. The process is called a **methide shift**. Since the stability of carbocations is increased by electron donation from attached substituents, it is reasonable that, generally, tertiary (abbreviated as 3°) carbocations are found to be more stable than secondary (abbreviated as 2°) and 2° more stable than primary (abbreviated as 1°). The **shift** of a substituent (the **methide shift** here), which allows the conversion of 1° to 2° carbocations and 1° and 2° carbocations to 3° carbocations, can be of an alkyl group, an aryl group, and, indeed, even a hydride itself. In the example of Equation II-K, where the **methide** shift occurred, as well as in numerous other cases, it was the products derived from the reaction in which the generation of the 2° carbocation was presumed but products arising from the 3° carbocation were obtained that provided the initial impetus for the rationale behind the statement concerning the relative stabilities of the carbocations.

$$\text{(II-K)}$$

A 2° carbocation
undergoing a
methide shift
rearrangement

A 3° carbocation

Third, in Equation II-L, it is seen that structures in which substituents can be placed as far apart as possible and which lie at minima along reaction coordinates are the favored products when equilibrium between isomers can be achieved. Here (in Equation II-L), 1,2,3-trimethylbenzene, when allowed to stand in the presence of the Lewis acid catalyst aluminum chloride (aluminum trichloride, $AlCl_3$), undergoes rearrangement to its more symmetrical isomer 1,3,5-trimethylbenzene.

$$\text{(1,2,3-trimethylbenzene)} \xrightarrow{\ AlCl_3\ } \text{(1,3,5-trimethylbenzene)} \qquad \text{(II-L)}$$

Addition reactions and their formal reverse, **elimination** reactions, can be considered as related processes that (largely) in one direction convert compounds with double and triple bonds to those that have *more* substituents on carbon, while in the other direction produce compounds that have *fewer* substituents on carbon. Thus, when **elimination** occurs, **unsaturation increases** and when **addition** occurs, **unsaturation decreases**. These related processes are shown in Equation II-M.

Substitution involves the replacement of one ligand (or substituent) on carbon by another. The replacement of hydrogen in methane (CH_4) by chlorine in a process (descriptively) called a **free radical substitution** is shown in Equation II-N.

$$H_2C{=}CH_2 \ + \ H_2O \ \underset{\text{Elimination}}{\overset{\text{Addition}}{\rightleftarrows}} \ CH_3CH_2OH \qquad \text{(II-M)}$$

$$CH_4 \ + \ Cl_2 \ \xrightarrow{\text{Light}} \ CH_4Cl \ + \ HCl \qquad \text{(II-N)}$$

As pointed out in Chapter 3, the processes of **oxidation** and **reduction** may be considered, for the former, as an increase in the number of oxygen atoms or a decrease in the number of hydrogens and, for the latter, as a decrease in the number of oxygens or an increase in the number of hydrogens. This definition is useful in many cases to make a quick decision about an oxidation/reduction reaction, but its utility is limited by noting that many organic reactions do not involve changes in the numbers of oxygen or hydrogen nuclei but do involve gain or loss of electrons. As you probably have already learned, an atom is said to be "oxidized" if, as a consequence of its reaction, it has suffered a net loss of electrons. Since each reaction must be balanced, each oxidation must be accompanied by a corresponding reduction, and another atom (or group of atoms) has enjoyed a net gain of electrons and is said to be "reduced." In order to decide which has "gained" and which has "lost," some arbitrary rules are used and they are coupled to the concept of electronegativity (Chapter 1). Briefly, the idea is to find out whether a specific carbon in a specific reaction may be thought of as having gained electrons, lost electrons, or neither gained nor lost.

1. Elemental carbon is arbitrarily assigned the oxidation state "zero."
2. Each atom bonded to carbon that is *more* electronegative than carbon (e.g., O, N, S, and halogens) is assigned a relative oxidation state of +1 with doubly and triply bonded atoms in this group being counted twice and thrice (respectively).

3. Each atom bonded to carbon that is *less* electronegative than carbon (e.g. H, Li, Na, and Mg) is assigned a relative oxidation state of −1.

4. A carbocation, already having a + charge on carbon, changes the oxidation state of that carbon by +1. A carbanion, on the other hand, changes the oxidation state of the carbon bearing it by −1. A radical leaves the oxidation state of carbon unchanged.

Consider the following examples:

$$\Sigma = -4 \qquad \Sigma = -2 \qquad \Sigma = -3$$

$$\Sigma = -2$$
At one carbon
$$\Sigma = +2$$
At one carbon
$$\Sigma = 0$$
Overall

$$\Sigma = -2$$
At each carbon
$$\Sigma = -4$$
Overall

$$\Sigma = -3$$
At one carbon
$$\Sigma = +1$$
At one carbon
$$\Sigma = -2$$
Overall

Based on the idea that a gain of oxygen (or loss of hydrogen) is an oxidation, it should be clear that chloromethane (methyl chloride, CH_3Cl) is more highly oxidized (it has less hydrogen) than methane. However, it is somewhat less clear what might be made of methylmagnesium chloride (CH_3MgCl). This **organometallic compound** (Chapter 5), **a Grignard reagent**, (Chapter 7), formed by the reaction of chloromethane with magnesium metal in ether (Equation II-O), is an **oxidation** at magnesium and a **reduction** at carbon, although neither oxygen nor hydrogen has been gained or lost.

$$CH_3Cl \ + \ Mg \ \xrightarrow[\text{Solvent}]{\text{Ether}} \ CH_3MgCl \qquad\qquad \text{(II-O)}$$

More traditionally, the conversion of 1,1-dichloroethene (CH_2=$CHCl_2$) to 1,1-dichloroethane (CH_3CHCl_2), where there has been a gain of one hydrogen atom at each carbon, clearly corresponds to a reduction and its reverse to an oxidation.

The Reactions of Hydrocarbons: Oxidation, Reduction, Substitution, Addition, Elimination, and Rearrangement

> A man will turn over half a library to make one book.
> —S. Johnson

A. INTRODUCTION

Hydrocarbons are molecules composed only of hydrogen and carbon. Some of the chemistry associated with acyclic and cyclic alkanes, alkenes, and alkynes, as well as the aromatic compounds (arenes) has already been touched upon briefly, for example, combustion (in Chapter 1), electrocyclic processes (in Chapter 4), and acidity (in Chapter 5).* These processes, as well as others to be encountered here, can be cast into the traditional groupings of *oxidation and reduction, addition and elimination,* and *substitution and rearrangement.*

When there are **intermediates** involved that are capable of detection (or of being inferred from the observed outcome of the reactions), the paths leading through the groups of reactions will be seen to involve the intermediates called *free radicals, carbanions, carbocations,* and *carbenes.* Of course, since our information is limited, the details of most reactions will remain obscure and inferences from a few well-studied cases will be used to extend our reach. The examples provided arise from the knitting together of only a few reactions and a few intermediates. They can be expanded upon by the vast numbers of substrates and reactants available and the subtle interplay of steric and electronic effects as reactions occur.

---

*Hydrocarbons are the weakest acids (and thus, their conjugate bases are the strongest bases). The "best estimates" for the pKa's of saturated hydrocarbons, that is, $R_3C-H \rightarrow R_3C^- + H^+$ (relative to water) are of the order of ≈ 50. A proton on the carbon–carbon double bond of an alkene is some 4–5 pKa units *more* acidic (and that is about the same for *unsubstituted* arenes, i.e., benzene itself has a reported pKa of ≈ 37). Alkynes are considerably *more* acidic. The pKa of ethyne (acetylene $HC \equiv CH$]) is about 23.

Foundations of Organic Chemistry: Unity and Diversity of Structures, Pathways, and Reactions,
First Edition. David R. Dalton.
© 2011 John Wiley & Sons, Inc. Published 2011 by John Wiley & Sons, Inc.

Therefore, even among the relatively simpler compounds composed only of carbon and hydrogen, there is much excitement.

B. ALKANES

I. Oxidation

As pointed out earlier (Chapter 1) all hydrocarbons are attacked by oxygen (O_2) when the temperature of the system is high enough. Although cyclic and acyclic alkanes, lacking unsaturation in the form of double and triple bonds, as well as other functional groups, have "little affinity" for other reactants, their ready conversion to carbon dioxide and water has proved efficient enough to (temporarily) power society. Nonetheless, while heat evolved in the process of burning natural gas, gasoline, oil, and other hydrocarbon products, and producing carbon dioxide (CO_2) and water (H_2O) (Equation 6.1) is desirable (to heat our homes and power our engines), the "cost" is high. Thus, the process is seldom complete, resulting in polution of the atmosphere (e.g., lower oxidation states such as carbon monoxide [CO] are emitted too), and even the huge quantities of carbon dioxide (CO_2) and water, both of which are "greenhouse" gases, are inimical to normal life as they intrude on the reemission of heat from the earth.*

$$C_nH_{(2n+2)} + (3n+1)/2\, O_2 \rightarrow n\, CO_2 + (n+1)\, H_2O \qquad (6.1)$$

The introduction of unsaturation (i.e., the loss of hydrogen) is also an oxidation and is an alternative to the concept of oxidation involving the gain of oxygen. Thus, as pointed out in Chapter 3, the presence of a ring in a cycloalkane or a double bond in an alkene can be accommodated by "losing" two hydrogen atoms from the corresponding acyclic saturated alkane. Such "sites of unsaturation" are, formally, oxidations since the "addition" of hydrogen (H_2) could (and in some cases does) result in ring opening with formation of the corresponding acyclic alkane (*vide infra*).

Hydrocarbons may be slowly **oxidized** by standing in the presence of oxygen (a ground state triplet, Figure 6.1) and light to produce hydroperoxides (Equation 6.2). The process is called *autoxidation* and is generally avoided by keeping hydrocarbons under an inert atmosphere and away from sunlight. **Autoxidation** should be avoided because hydroperoxides are unstable and decompose to produce an intractable mixture of materials.

---

*It has been estimated that about 10^{10} tons of CO_2 and 10^7 tons of SO_2 along with significant amounts of NO_x are produced annually by the combustion of fossil fuels. All of these "more complete" products of combusion may contribute to global warming. For example, the CO_2 absorption band at about $2350\,cm^{-1}$ means that radiation of this frequency does not readily escape the atmosphere. Thus, normal radiative cooling is diminished and, while entrapment of some heat by CO_2 and H_2O moderates surface conditions to make life possible, too much may result in climate changes, which we find less desirable.

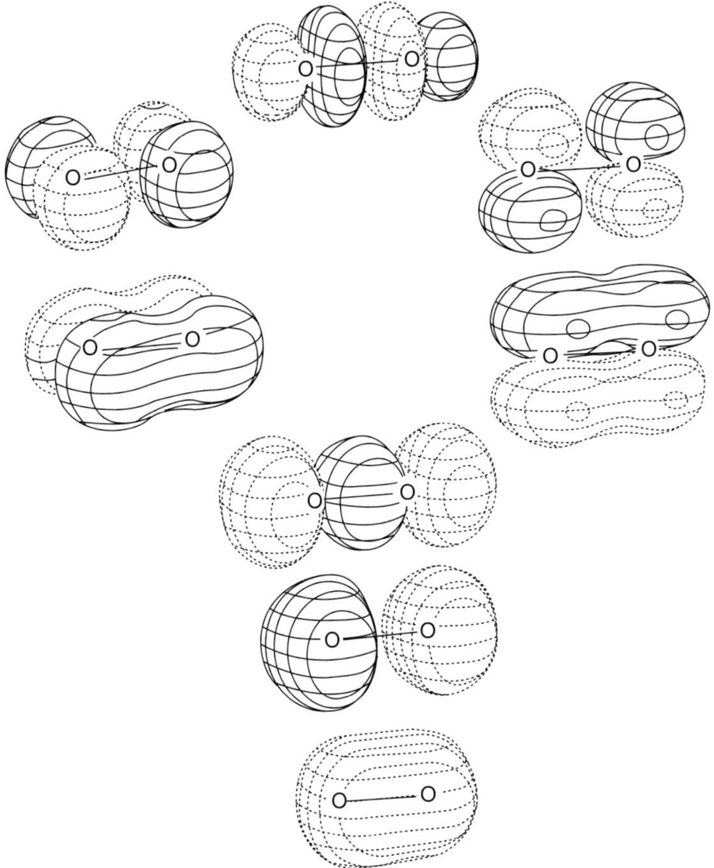

Figure 6.1. A representation of the molecular orbitals of oxygen ($^3O_2$). The representations were created using the MOViewer in WebMO version 6.0.002p. (WebMO, LLC. www.webmo .net).

As might be expected on the basis of bond strengths (Chapter 1), tertiary hydrogens are abstracted more readily than secondary hydrogens and others.*

$$\text{(6.2)}$$

The details of the process are not known, but it is believed that they are similar to what occurs in other *substitution* processes where an atom *other than oxygen* is found to **replace** (or *be substituted for*) a hydrogen attached to carbon and, in principle, involve either (a) the removal of the hydrogen followed by its replacement

---

*Although it would be worthwhile to quickly review homolytic bond strengths (which have probably not changed much since the last overview), the necessary values are briefly noted here. The homoytic bond dissociation energy (DH°) for a primary C–H (as in ethane [CH_3CH_3]) is about 422 kJ mol^{-1} (101 kcal mol^{-1}), for a secondary C–H (as in propane [$(CH_3)_2CH_2$]) is about 410 kJ mol^{-1} (98 kcal mol^{-1}), and for a tertiary C–H (as in 2-methylpropane [$(CH_3)_3CH$]) is about 401 kJ mol^{-1}. (96 kcal mol^{-1}) For methane (CH_4) itself, breaking the *first* C–H bond requires about 434 kJ mol^{-1} (104 kcal mol^{-1}).

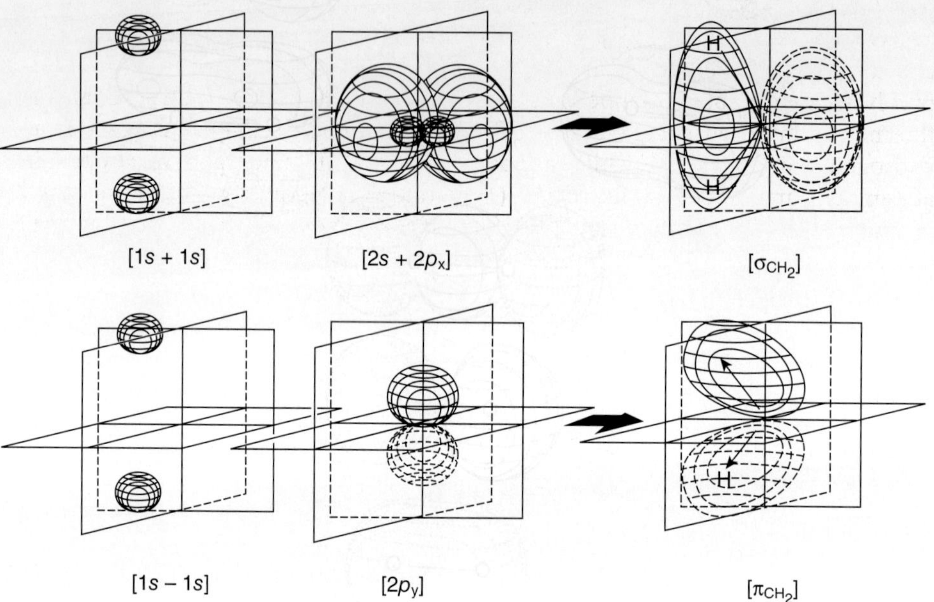

Scheme 6.1. A cartoon depicting a possible path for the removal of a hydrogen from an alkane by oxygen, producing a carbon radical and the hydrogen peroxide radical, followed by recombination of the radicals to yield an alkylhydroperoxide.

Scheme 6.2. Bonding MOs in a $-CH_2-$ fragment available for further reaction. The representations were created using the MOViewer in WebMO version 6.0.002p.

with oxygen and others, or (b) the reaction of the electron-rich species at carbon followed by loss of the hydrogen.

Scheme 6.1 presents a potential pathway, using arrows showing the *movement of single electrons*, which produces a carbon-free radical and a hydrogen peroxide radical. A combination of these two radicals generates the product of Equation 6.2.

An alternative possibility allows for reaction directly at carbon. Consider the $-CH_2-$ fragment as shown in Scheme 6.2. Ignoring other attachments, it can be argued (as for methylene [$:CH_2$] itself, *vide supra*) that the $-CH_2-$ unit contains *bonding* molecular orbitals (MOs) that include (i) a σ_{CH2} bonding MO (made up of positive overlap between the orbital composed of the two hydrogen $1s$ orbitals [i.e., $1s + 1s$] and an orbital composed of the carbon $2s$ and one carbon $2p$ [$2p_x$] pointed *into* the bonding hydrogen $1s + 1s$ MO) and (ii) a π_{CH2} bonding MO (made up of positive overlap between the other hydrogen MO [i.e., $1s - 1s$] and the appropriate carbon $2p$ [$2p_y$] orbital). Insertion of the π_{CH2} orbital into the σ^* O_2 (Chapter 1) orbital then produces a fragment that, on hydrogen migration, yields hydroperoxide. Although in principle either path can be considered, the direct abstraction of the hydrogen by oxygen to generate a radical is frequently considered more favorable for steric reasons.

Problem 6.1. Using the data in Table 1.1 (Chapter 1) calculate $-\Delta H^\circ$ for the process of Scheme 6.1 at 25°C and 1 atm.

The introduction of unsaturation (an oxidation) by dehydrogenation of an alkane either directly with high heat or over a suitable catalyst at a lower temperature produces alkenes and hydrogen. Because of the variety of products formed, laboratory techniques for such dehydrogenations are uncommon. However, industrially, they are important for **reforming** hydrocarbon mixtures (to produce gasoline with the appropriate *properties)* and for production of some aromatic compounds. There is a large body of evidence that shows that the larger alkanes begin to become pyrolytically unstable at temperatures above about 300°C. The smaller alkanes are thermally more robust, and methane (CH_4), for example, decomposes to carbon and hydrogen at about 1000°C while propane ($CH_3CH_2CH_3$) undergoes decomposition at temperatures above about 500°C (Equation 6.3), giving two different sets of products, only one of which actually corresponds to an *oxidation* in that hydrogen has been lost and a double bond introduced. Both the *oxidation* and the other process could also be called *fragmentations* as carbon–carbon bonds are broken or fragmented. In the absence of purposefully added catalysts (*vide supra*), the walls of the reaction vessels can serve as sites to help catalyze the bond breaking processes.

$$H_3C-CH=CH_2 + H_2 \longleftarrow H_3C-CH_2-CH_3 \longrightarrow CH_2=CH_2 + CH_4 \quad (6.3)$$

Further, and not surprisingly, cyclohexane (C_6H_{12}) undergoes dehydrogenation to benzene (C_6H_6) and three equivalents of hydrogen (H_2) (Equation 6.4) at a lower temperature (300–350°C) in the presence of a metal catalyst (e.g., Pt, Pd, Ni) or by using sulfur (S) or selenium (Se) (Equation 6.5) to pick up the hydrogen. While intermediate stages between the alkane and the aromatic system are doubtlessly present, they have not been detected.

$$\text{(cyclohexane)} \underset{300\text{–}350°C}{\overset{\text{metal catalyst}}{\rightleftharpoons}} \text{(benzene)} \quad + \quad 3\,H_2 \qquad (6.4)$$

$$\text{(cyclohexane)} + \text{S (or Se)} \xrightarrow{\text{heat}} \text{(benzene)} \quad + \quad 3\,H_2S \text{ (or } H_2Se) \qquad (6.5)$$

Problem 6.2. Calculate ΔH° from the bond energies in Table 1.1 (Chapter 1) for the conversion of cyclohexane to benzene and hydrogen as shown in Equation 6.4 at 25°C and 1 atm. From this, with ΔS as $-100\,JK^{-1}mol^{-1}$, and Equation 4.3 estimate ΔG° under the same conditions. In order for the reaction to proceed as written (without a catalyst) to what temperature must it be heated?

II. Reduction

Although acyclic alkanes are devoid of unsaturation, longer chain compounds can be converted to those with shorter chains in the presence of hydrogen (H_2) at high temperatures and pressures in the presence of suitable catalysts. However, the small ring cyclic derivatives are more labile. Thus, under relatively mild conditions (just above room temperature) and in the presence of a Ni, Pt, or Pd catalyst,* cyclopropane reacts with hydrogen (H_2) to produce propane (Equation 6.6). Since hydrogen has been added, the process is, reasonably, called a **reduction**. The **reduction** of cyclobutane to produce butane (Equation 6.7) requires more forcing conditions (a temperature of about 120°C), and the larger rings (viz. cyclopentane and cyclohexane) are, like the acyclic alkanes themselves, resistant enough to be used as solvents for certain other catalytic reductions (which will be encountered).[†]

$$\xrightarrow[\text{catalyst}]{H_2} \quad CH_3CH_2CH_3 \tag{6.6}$$

$$\xrightarrow[\text{catalyst}]{H_2} \quad CH_3CH_2CH_2CH_3 \tag{6.7}$$

III. Substitution

Reactions of acyclic and cyclic alkanes in which one or more hydrogens are replaced and a different substituent introduced at carbon *in place of the hydrogen* are frequently referred to as **substitution** reactions. The subtle distinction as to whether the reaction is actually an oxidation rather than a substitution (as based on a system of oxidation numbers for carbon such as that discussed at the end of the introduction to this part) is usually avoided. The **halogenation of alkanes** falls into this category. Thus, replacement of hydrogen by halogen (e.g., chlorine, as in Equation 6.8) can be thought of as an oxidation at carbon since the halogens are more electronegative than carbon. The alternative point of view that the reaction is a **substitution** ignores the electronegativity difference and, based upon the presumed pathway by which the reaction proceeds, suggests that electrons are neither gained nor lost—as would be expected in an oxidation–reduction process.

$$CH_4 \; + \; Cl_2 \quad \xrightarrow[\text{light}]{\text{heat or}} \quad CH_3Cl \; + \; HCl \tag{6.8}$$

When methane (CH_4) is mixed with chlorine (Cl_2), nothing happens. Indeed, although **(Problem 6.3)** it is clear that the reaction should be exothermic, it is not until the mixture of gases is heated to temperatures above 300°C or irradiated with visible (violet) or ultraviolet (UV) light that the reaction occurs (explosively).[‡]

---

*Catalysts** were discussed in the introduction to Part II as well as in Chapter 3. The use of catalysts will be encountered again.

[†]The relatively higher reactivity of the smaller cyclic hydrocarbons can be justified on the basis of their increased strain energy (Chapter 4).

[‡]Assuming that the purpose of the heat or light is to cause the **homolysis** or **heterolysis** of a bond (one of the C–H or Cl–Cl bonds), it is possible to ask about the potential pathways for the reaction. The ionization energy of chlorine is about 1254 kJ mol⁻¹ (300 kcal mol⁻¹) and its electron affinity is 359.5 kJ mol⁻¹ (86 kcal mol⁻¹), making the separation of the ions Cl⁺ and Cl⁻ unstable relative to atoms by about 890.3 kJ mol⁻¹ (213 kcal mol⁻¹). On the other hand, the homolytic bond strength of the Cl–Cl bond is 242 kJ mol⁻¹ (58.9 kcal mol⁻¹). It is argued that absorption of light is required in order to effect bond cleavage (i.e., an n → σ* transition). Further, it can be supposed that it requires the absorption of a single

(Continued on next page)

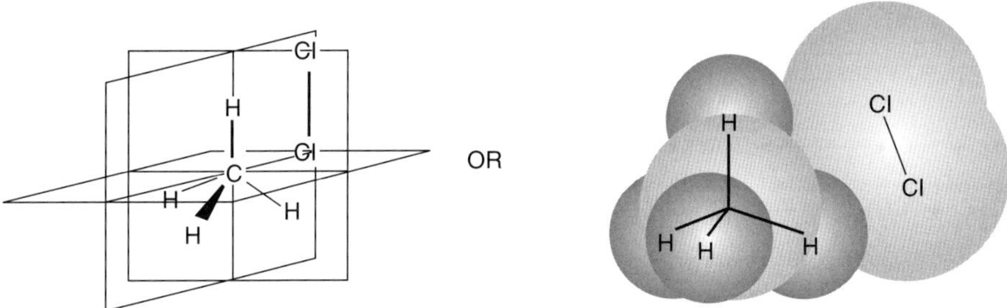

Figure 6.2. Cartoon representations concerning the attempted concerted addition of chlorine (Cl_2) to methane (CH_4).

Problem 6.3. Using the information presented in Table 1.1 (Chapter 1) calculate $\Delta H°$ for the process represented in Equation 6.8.

Although direct orbital interaction (an alternative suggested for the oxidation of hydrocarbons by oxygen) might apply here too, the steric requirement of forcing a large chlorine into the area occupied by the orbitals on carbon suggests (Figure 6.2) that, as before, such pathway is unlikely. Further, the precise orientation of the approach of molecular chlorine (Cl_2) to methane (CH_4), followed by a "four-center" reaction with the simultaneous (or nearly so) breaking (of C–H and Cl–Cl) and making (of C–Cl and H–Cl) of four σ-bonds in concert is **neither** in accord with the relative reactivities of different kinds of C–H bonds (i.e., primary [1°], secondary [2°], and tertiary [3°] C–H's react at different rates) **nor** other reaction characteristics typical of free radicals.

If a radical pathway is supposed for this process, it might reasonably be argued that the yellow green gas, chlorine (Cl_2), absorbs some UV or near-UV light (a σ → σ* transition) and some Cl–Cl bond breaking occurs yielding chlorine atoms (Cl•) as the first step in a multistep process (Scheme 6.3).

Then, continuing with Scheme 6.3, the *initiation* of the reaction having been effected, the process continues to *propagate* by subsequent (slower) abstraction of a hydrogen by a *chlorine atom* (Cl•) with the formation of a methyl radical ($H_3C•$) and hydrogen chloride (HCl). Further *propagation* by (slow) reaction of the methyl radical ($H_3C•$) with another molecule of chlorine (Cl_2) now produces the product, chloromethane (methyl chloride, H_3CCl), *and another chlorine atom* (Cl•).

quantum of light (of the appropriate energy) *per* molecule to affect the transition and thus, eventually, bond cleavage (if reemission, etc., do not intrude). A quantum of **quanta** is *1 Einstein* and has energy E = *Nh*v (where *N* is Avogadro's number). Thus,

$$E = h\nu = hc/\lambda = 2.859 \times 10^5 / \lambda \, (\text{Å})$$

and light of ca. 286 nm (2860 Å) corresponds to an energy of ca. 100 kcal Einstein^{-1} (418 kJ Einstein^{-1}), clearly enough to cause cleavage of Cl–Cl *and/or C–H* bonds (Table 1.1). However, the energy is not delivered all at once. Thus, a 1000-watt light bulb only provides about 5×10^{-2} watts of energy as visible light over a 10 Å region (in a given direction). This is of the order of 10^{17} quanta s^{-1} and for 1 mol (ca. 6×10^{23} quanta) it would thus take 6×10^6 s (ca. 10 weeks). Intense UV sources are much more efficient and, *an entire mole of quanta of light is not necessary* because, as discussed, the consumption of one radical may occur with the production of another.

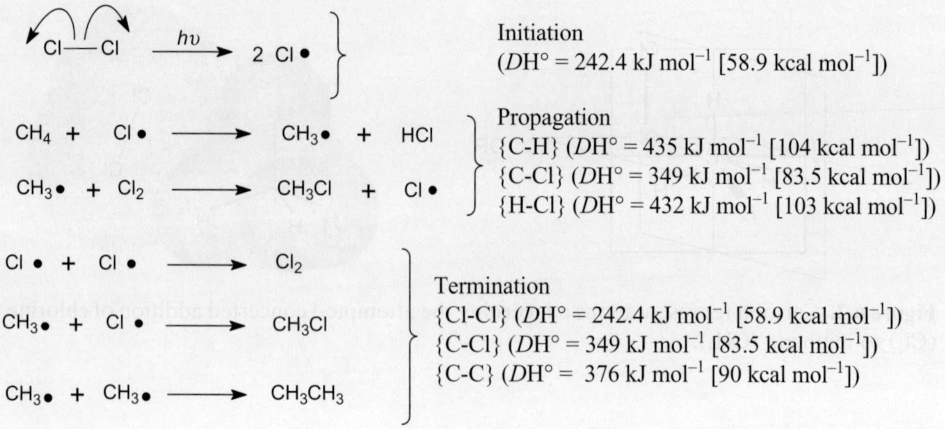

$$Cl\text{—}Cl \xrightarrow{h\upsilon} 2\ Cl\bullet$$

Initiation
($DH° = 242.4$ kJ mol^{-1} [58.9 kcal mol^{-1}])

$$CH_4\ +\ Cl\bullet\ \longrightarrow\ CH_3\bullet\ +\ HCl$$

$$CH_3\bullet\ +\ Cl_2\ \longrightarrow\ CH_3Cl\ +\ Cl\bullet$$

Propagation
{C-H} ($DH° = 435$ kJ mol^{-1} [104 kcal mol^{-1}])
{C-Cl} ($DH° = 349$ kJ mol^{-1} [83.5 kcal mol^{-1}])
{H-Cl} ($DH° = 432$ kJ mol^{-1} [103 kcal mol^{-1}])

$$Cl\bullet\ +\ Cl\bullet\ \longrightarrow\ Cl_2$$

$$CH_3\bullet\ +\ Cl\bullet\ \longrightarrow\ CH_3Cl$$

$$CH_3\bullet\ +\ CH_3\bullet\ \longrightarrow\ CH_3CH_3$$

Termination
{Cl-Cl} ($DH° = 242.4$ kJ mol^{-1} [58.9 kcal mol^{-1}])
{C-Cl} ($DH° = 349$ kJ mol^{-1} [83.5 kcal mol^{-1}])
{C-C} ($DH° = 376$ kJ mol^{-1} [90 kcal mol^{-1}])

Scheme 6.3. A representation of the overall path for the light initiated chlorination of methane. The enthalphy change ($\Delta H° = -99.5$ kJ mol^{-1} [-23.8 kcal mol^{-1}]) shows the reaction is exothermic. Since the reaction is a **chain** reaction, only a small number of chlorine atoms need to be formed initially. Second, although the combination of atoms and radicals is fast, the mean free paths of all of the gases is relatively small (of the order of 20–30 nm [at 1 atm and room temperature]) and the relative concentrations of radical and atomic species is low (about a factor of 10^{-10}) compared with that of reactants.

Clearly this process of hydrogen atom (H•) abstraction by atomic chlorine (Cl•) and subsequent reaction of the methyl radical (H$_3$C•) with molecular chlorine (Cl$_2$) could, in principle, continue as long as the reacting species could find each other. Near-room temperature, the reactants are moving at about 10^3 m s^{-1} and undergoing about 10^{10} collisions s^{-1}. Not all collisions result in a reaction and the concentration of chlorine atoms (Cl•) and methyl radicals (H$_3$C•) is small relative to their respective parents. However, the reaction proceeds to produce chlorinated product.

As shown in Scheme 6.3, the process is *terminated* by the combination of (a) chlorine atoms with each other, regenerating molecular chlorine (Cl$_2$); (b) chlorine atoms (Cl•) with methyl radicals (CH$_3$•), forming chloromethane (H$_3$CCl); and (c) methyl radicals (CH$_3$•) with each other, forming ethane (CH$_3$CH$_3$). Additionally, deactivation of any of the radical species by a wall reaction serves to end the reaction.

An additional complication is shown in Scheme 6.4. As the concentration of chloromethane (methyl chloride, CH$_3$Cl) grows, it can serve as a reactant in place of methane (CH$_4$) (the concentration of which is diminishing). In this way, additional sets of reactions occur in parallel with formation of chloromethane (methyl chloride, CH$_3$Cl). These arise from (a) the abstraction of a hydrogen from chloromethane (methyl chloride, ClCH$_3$) by a chlorine atom (Cl•) producing the chloromethyl radical (ClCH$_2$•) and hydrogen chloride (HCl) followed by (b) reaction of the chloromethyl radical (ClCH$_2$•) with chlorine (Cl$_2$) to yield dichloromethane (CH$_2$Cl$_2$). Repetition of this process on the dichloromethane (CH$_2$Cl$_2$) product then generates the dichloromethyl radical (•CHCl$_2$) and thence trichloromethane (chloroform, CHCl$_3$). Finally, another repetition, this time on trichloromethane (chloroform, CHCl$_3$) as the substrate undergoing reaction with chlorine atoms (Cl•) produces the

Chloroalkane Product

	bp (°C)	δ (ppm; TMS = 0.0)	
		^{1}H	^{13}C
$CH_4 + Cl_2 \longrightarrow CH_3Cl + HCl$	−24	3.05	24.9
$CH_3Cl + Cl_2 \longrightarrow CH_2Cl_2 + HCl$	40	5.30	54.0
$CH_2Cl_2 + Cl_2 \longrightarrow CHCl_3 + HCl$	61	7.27	77.5
$CHCl_3 + Cl_2 \longrightarrow CCl_4 + HCl$	77		96.5

Scheme 6.4. The products of further chlorination of methane (CH_4). For purposes of reference, the boiling point of methane (CH_4) is −161.5°C and the ^{1}H and ^{13}C chemical shifts (relative to δ = 0.00 ppm for TMS) are δ = 0.24 ppm and δ = −2.3 ppm, respectively.

trichloromethyl radical (·CCl_3). Reaction of the latter with chlorine (Cl_2) yields tetrachloromethane (carbon tetrachloride, CCl_4) and a chlorine atom (Cl·), continuing the chain.

As pointed out earlier in this chapter, the interesting consequence of the fact that different C–H bonds have different dissociation energies is to produce nonstatisical product distributions when such hydrocarbons react with atomic chlorine (Cl·) at temperatures below those that would produce leveling of reactivity. Keeping in mind that while the dissociation energy ($DH°$) associated with the homolytic C–H bond cleavage for methane (CH_4) is 435 kJ mol^{-1} (104 kcal mol^{-1}), it is noteworthy that other primary C–H bonds are generally of lower energy, for example, the C–H bond energy is closer to 422 kJ mol^{-1} (101 kcal mol^{-1}) in ethane (CH_3CH_3) and secondary C–H bonds in, for example, propane [$(CH_3)_2CH_2$], have $DH°$ of about 410 kJ mol^{-1} (98 kcal mol^{-1}). Tertiary C–H bonds, as in, for example, 2-methylpropane (isobutane [$(CH_3)_3CH$]), which has $DH°$ 401 kJ mol^{-1} (96 kcal mol^{-1}) are the most fragile. This is emphasized in Scheme 6.5 and Problem 6.3.

Problem 6.4. These questions refer to the material in Scheme 6.5. (a) At higher temperatures, all of the hydrogens in 2-methylbutane react with *equal* facility. What ratio of products A–D are expected under such conditions? (b) Sketch the ^{1}H nuclear magnetic resonance (NMR) spectrum of *each* of the products A–D and show how you would, therefore, know which isomer corresponded to which letter. (c) Which of the isomers A–D are, at least in principle, capable of resolution into enantiomers?

Because homolysis of the bromine–bromine bond in Br_2 is *more facile* ($DH°$ = 193 kJ mol^{-1} [46.1 kcal mol^{-1}]) than the bond between chlorine atoms in Cl_2 ($DH°$ = 242.4 kJ mol^{-1} [58.9 kcal mol^{-1}]), it should come as no surprise that when bromine is allowed to react with 2-methylbutane, **only** 2-bromo-2-methylbutane

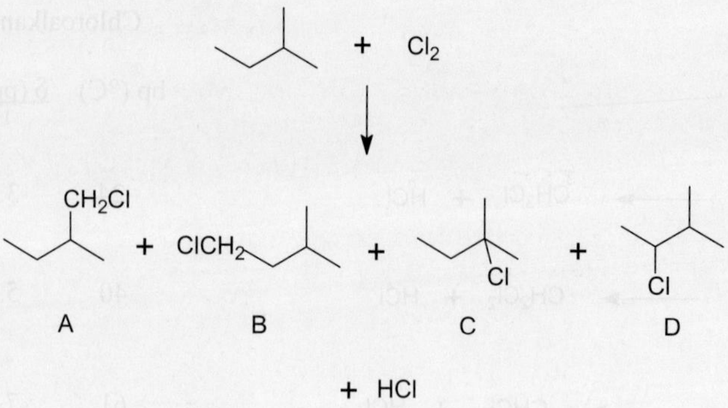

Scheme 6.5. The monochlorination products of 2-methylbutane at 300°C. Hydrogen chloride (HCl) is also obtained. The products are separable. As shown by gas chromatography, the monochlorination products 1-chloro-2-methylbutane (A), 1-chloro-3-methylbutane (B), 2-chloro-2-methylbutane (C), and 2-chloro-3-methylbutane (D) are *actually* obtained in the following percent yields: 30% A; 15% B; 22% C; and 33% D.

(93%) and 2-bromo-3-methylbutane (7%) are formed (Equation 6.9) and that I_2 ($DH° = 151\,kJ\,mol^{-1}$ [36.1 kcal mol^{-1}]) fails to react at all.

$$+ \; Br_2 \longrightarrow \qquad + \qquad + \; HBr \qquad (6.9)$$

At high temperature (above about 400°C) alkanes undergo nitration with nitric acid (HONO$_2$). The reaction has all of the characteristics of the radical processes already discussed and, except for methane (CH$_4$) (Equation 6.10), the large number of products produced from most alkanes due to carbon–carbon bond cleavage and introduction of more than one nitro group (–NO$_2$) limits the utility of the process.

$$H_3C-H \; + \; HONO_2 \longrightarrow H_3C-\overset{+}{N}\overset{\displaystyle O}{\underset{-O}{\big\Vert}} \; + \; H_2O \qquad (6.10)$$

In the same vein, a free radical reaction between alkanes and sulfur dioxide (SO$_2$) and chlorine (Cl$_2$) produces chlorosulfonates (RSO$_2$Cl) (Equation 6.11) while highly reactive carbenes, for example, methylene (:CH$_2$) generated photochemically from diazomethane (CH$_2$N$_2$), *vide supra* insert into carbon–hydrogen (C–H) bonds indiscriminately (Equation 6.12) and less reactive carbenes insert with a preference for weaker bonds first (i.e., tertiary > secondary > primary). Details of the paths by which the processes of Equations 6.10–6.12 occur suggest that radicals may be involved although, for singlet carbenes (*vide supra*), direct insertion has not been ruled out.

$$CH_4 \; + \; SO_2 \; + \; Cl_2 \longrightarrow H_3C-\overset{\displaystyle O}{\underset{\displaystyle O}{\overset{\big\Vert}{\underset{\big\Vert}{S}}}}-Cl \; + \; HCl \qquad (6.11)$$

$$CH_4 \; + \; :CH_2 \longrightarrow CH_3CH_3 \qquad (6.12)$$

IV. Rearrangement

Largely as a consequence of the observation that branching of alkanes appears to increase the efficiency with which an internal combustion engine functions, that is, the "octane number" (Chapter 3) increases with increased branching, evidence has accumulated that certain catalysts (e.g., Pt) affect the rearrangement of the skeletal backbone of alkanes. In the petroleum industry, the process by which the crude petroleum gasoline fraction is reformulated with the use of platinum (Pt) is referred to as "platforming." Although the exact details may vary with the particular circumstances, it appears that the overall process involves free radicals, and in addition to *rearrangement* of the skeleton, *dehydrogenation* (as pointed out earlier) may also occur. The interconversion of 1,2-dimethylcyclopentane and methylcyclohexane, in which a methyl group apparently "inserts" into the five-membered ring, is an example of such a rearrangement (Equation 6.13).

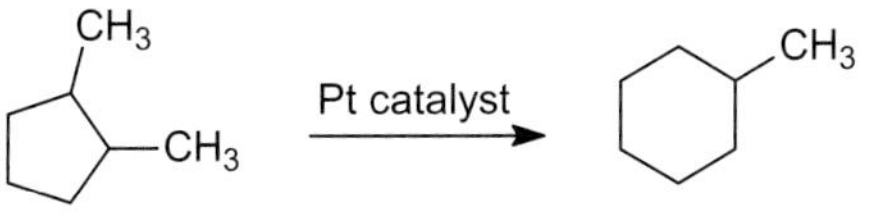

(6.13)

Additional metal-catalyzed alkane rearrangements, such as the photochemically initiated mercury-catalyzed dimerization of alkanes, are also known (Equation 6.14), and it is again argued that the process involves free radicals (as well as dehydrogenation).

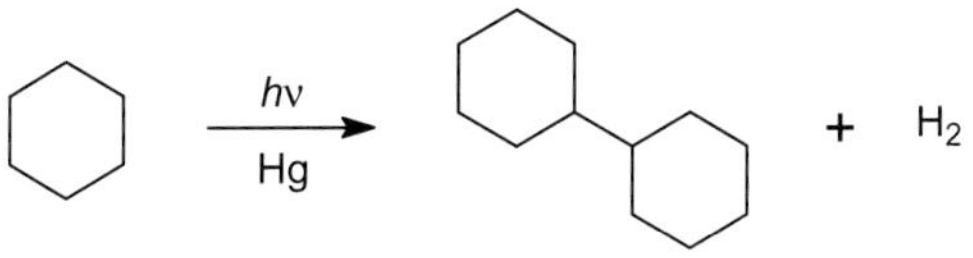

(6.14)

C. ALKENES

I. Oxidation

As was the case for alkanes (*vide supra*), alkenes are also attacked by oxygen. Indeed, while the anticipated radical (or radical-like) reaction of *triplet*—or ground state—oxygen with alkenes occurs with preferential (but not exclusive) abstraction of benzylic and allylic hydrogens to yield resonance stabilized species, its synthetic utility is limited. However, in contrast to that process, the reaction between *singlet oxygen** and alkenes bearing an allylic hydrogen to produce allylic hydroperoxide

*Singlet oxygen can be prepared by photosensitized excitation, that is, the process in which some donor (e.g., Rose Bengal [RB], *vide infra*) is photochemically excited and then, presumably by collision, transfers its energy to oxygen. In this way, high-energy (or excited) electronic states (Chapters 1 and 2), not otherwise readily available, can be populated. Thus, for example, the $\pi \rightarrow \pi^*$ absorbance for ground state singlet to excited state singlet transition for RB is at about λ_{max} 558 nm (Equation 6.15a). Then, it is presumed that singlet-excited RB crosses over or *decays* by some nonradiative process to the triplet state (Equation 6.15b). As a triplet, it reacts with ground state oxygen ($^3O_2$) to produce singlet-state oxygen

(Continued on next page)

is of practical value. In this process, for example, the terpene* S-(-)-limonene[†] produces (exclusively) a chiral hydroperoxide (Scheme 6.6). As shown, and in contrast to the reaction of triplet oxygen ($^3O_2$), which appears to involve free radicals, singlet oxygen ($^1O_2$) seems to add directly to yield a "pre-peroxide" by what is called a

($^1O_2$) and singlet (ground state) RB (Equation 6.15c). The singlet-state oxygen is about $92\,kJ\,mol^{-1}$ ($23\,kcal\,mol^{-1}$) above the ground state.

Rose Bengal (potassium salt) = RB

$$^1RB \xrightarrow{h\nu} {}^1RB^* \tag{6.15a}$$

$$^1RB^* \longrightarrow {}^3RB^* \tag{6.15b}$$

$$^3RB^* + {}^3O_2 \longrightarrow {}^1O_2^* + {}^1RB \tag{6.15c}$$

*As will be discussed more fully later (Chapter 11, *et seq.*), it was found very early in the examination of the world around us that many plants produced fragrant odors (such as those in cloves, pepermint, sandalwood, and lemons) and that these volatile "essential oils" could be, formally, dissected into "C_5-units" called "isopentenyl" or "isoprene" units. Isoprene is the trivial name given to 2-methyl-1,3-butadiene [C_5H_8] and the branched hydrocarbons of two "isoprene units" are called **terpenes** (a word derived, etymologically, from the terebinth tree, *Pistaca terebinthus*, which provides a fragrant terpene-rich exudate) and possess the general composition $C_{10}H_{16}$. **Diterpenes** and **triterpenes** correspond to compounds with twenty (20) and thirty (30) carbon atoms, respectively, and the very common plant products with fifteen (15) carbon atoms, the **sesquiterpenes**, correspond to "one and one-half terpenes." The visual dissection of these large compounds into "isoprene" units gave impetus to the establishment of "biogenetic" relationships that provided suggestions for actual experiments and, eventually, the linking of acetate units (C_2) through (R)-mevalonic acid ($C_6 = 3 \times C_2$) and its decarboxylation (or loss of CO_2 [C_6–$C_1 = C_5$]) to the "isoprene units" of the **terpenes** (C_{10}) and the **sesqui-** (C_{15}), **di-** (C_{20}), and **triterpenes** (C_{30}). As will be discussed later (Chapter 11), much of this work was carried out in the laboratories of Sir John Cornforth (b. 1917, National Institute of Medical Research, Great Britain; Nobel Prize 1975) and coworkers.

(R)- mevalonic acid

2-methyl-1,3-butadiene
("isoprene")

[†]Although isoprene itself has been implicated as part of a strategy some plants have for coping with heat and apparently contributes to the bluish haze hanging over some wooded hills in summer, "condensation" products of isoprene, for example, limonene abound. S-(-)-Limonene (boiling point [bp] 175–177°C) can be obtained from citrus (e.g., *Citrus limon*). Considered simply as an achiral, unconjugated

(Continued on next page)

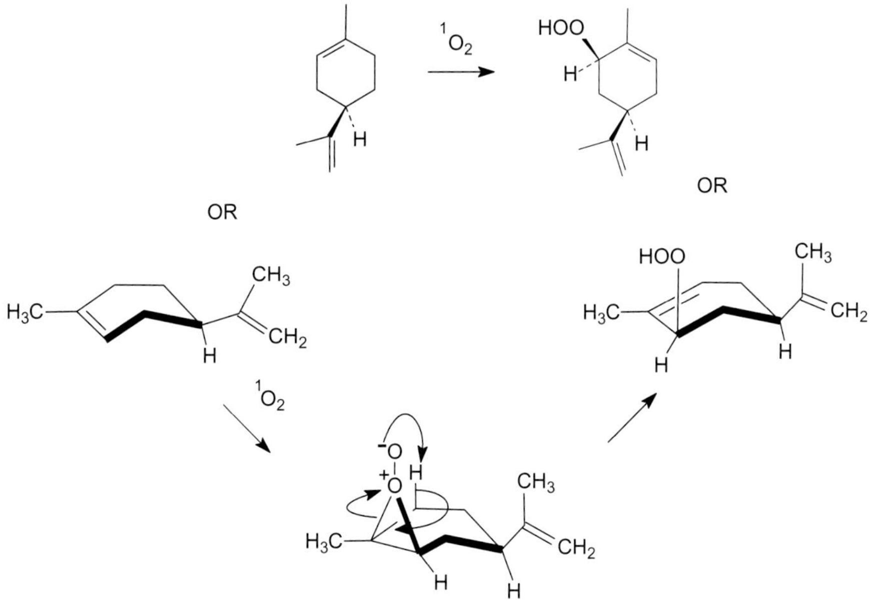

OR

Scheme 6.6. A possible pathway for the reaction of S-(-)-limonene with singlet oxygen ($^1O_2$) yielding a chiral hydroperoxide.

*cheletropic** process, and this product subsequently rearranges to the observed hydroperoxide. Finally, in this vein, (a) the most highly substituted (i.e., electron-rich) double bond reacts preferentially and (b) singlet oxygen can be produced in nonphotochemical ways, for example, by the reaction between hydrogen peroxide and sodium hypochlorite (Equation 6.16).

$$HOOH \ + \ NaOCl \ \longrightarrow \ ^1O_2 \ + \ H_2O \ + \ NaCl \qquad (6.16)$$

If the alkene is a **conjugated** diene, then $^1O_2$ adds in an electrocyclic (Diels–Alder) fashion (Scheme 6.7).

In the presence of a palladium catalyst [commonly Pd(II)Cl$_2$] oxygen (O$_2$) reacts with alkenes to yield carbonyl compounds. This process (the *Wacker Process*) utilizes the plant growth regulator ethene (ethylene, CH$_2$=CH$_2$) as the *exclusive* raw material for the commercial production of ethanal (acetaldehyde, CH$_3$CH=O). The overall reaction is shown in Scheme 6.8 and it is important to note that only a trace

diene, it can be viewed as a **Diels–Alder** (Chapter 4) adduct between two equivalents of "isoprene," *viz.*

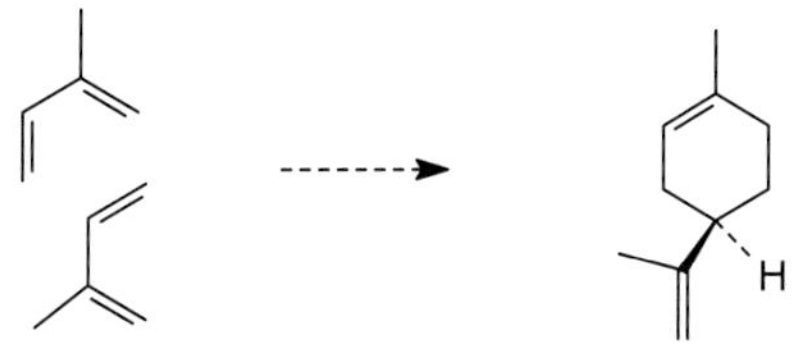

*Cheletropic (from the Greek *chele*, for *claw*) processes are related to the Diels–Alder reaction (Chapter 4). The difference is that here, two σ bonds *terminating in a single atom* are being made simultaneously or in concert.

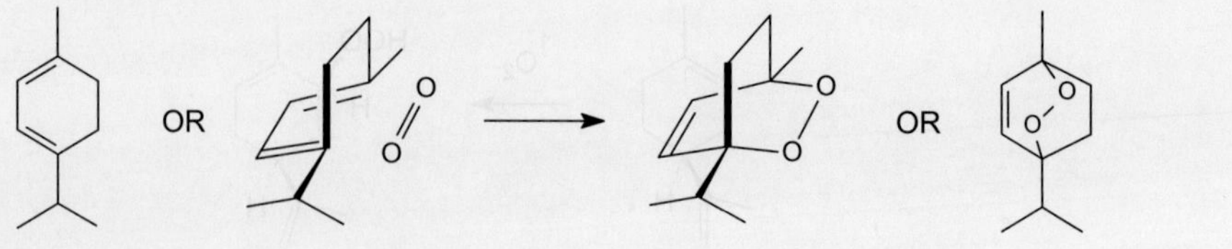

Scheme 6.7. A cartoon-like depiction of a possible path for the reaction of singlet oxygen ($^1O_2$) with α-terpinene, a conjugated terpenoid diene.

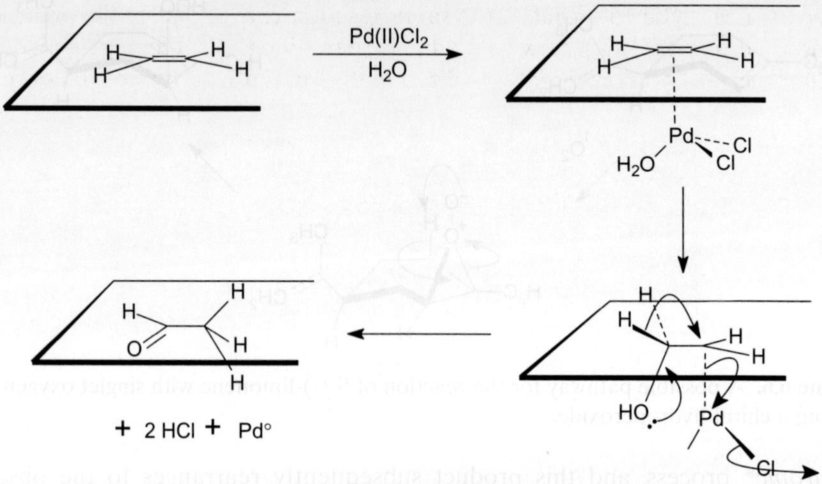

$$Pd^{\circ} + 2\ CuCl_2 \longrightarrow Pd(II)Cl_2 + 2\ CuCl$$

$$2\ CuCl + 1/2\ O_2 \longrightarrow 2\ CuCl_2 + H_2O$$

Scheme 6.8. An abbreviated cartoon depiction of what might be occurring in the oxidative conversion of ethene ($CH_2{=}CH_2$) to acetaldehyde ($CH_3CH{=}O$) by the *Wacker Process*. Unspecified ligands to Pd might include H_2O, Cl, and H.

of Pd(II) is necessary because a cycle, involving Cu(I) and Cu(II), to regenerate it from Pd° also operates, as shown at the end of Scheme 6.8. An expression for the overall reaction is given in Equation 6.17.

$$H_2C{=}CH_2 + \tfrac{1}{2}O_2 \xrightarrow[\text{Cu(II)Cl}_2]{\text{Pd(II) Cl}_2} CH_3CHO \tag{6.17}$$

Terminal alkenes (e.g., R–CH=CH_2) yield methyl ketones rather than aldehydes and numerous small-scale synthetic applications have been developed.

Treatment of alkenes with other metals in their respective high oxidation states [e.g., Cr(VI)] has met with mixed results and, generally, under conditions sufficient to promote oxidation, only low yields of oxidized products are isolable. However, there are notable exceptions.

For example, under carefully controlled alkaline conditions, a low yield of vicinal (from the Latin, *vicinus*, near) **diols** can be obtained from alkenes on treatment with potassium permanganate ($KMnO_4$). The vicinal diols are usually reactive enough so that, unless care is taken, they are further oxidized. However, as shown in Scheme 6.9, an alkene such as cyclohexene (C_6H_{10}) yields the *cis*-1,2-cyclohexanediol in modest yield, presumably via a cyclic manganate ester.

Of somewhat greater utility is the related reaction between alkenes and osmium tetroxide (OsO_4) (Scheme 6.10). Although the process is strikingly similar to that shown for permanganate, the reaction between the alkene and osmium tetroxide is much more easily controlled and high yields of the cyclic osmate ester intermediate are frequently isolable. Osmium is a heavy metal poison and thus its use is largely limited to a catalytic role, that is, the same reaction carried out in the presence of hydrogen peroxide (H_2O_2). Now, the hydrogen peroxide (H_2O_2) converts the reduced osmium back to its more highly oxidized state for reuse in hydroxylation of the alkene.

When a stream of oxygen (or air) is passed through a silent electric discharge, low yields (about 5% using air and 15% using oxygen [O_2]) of the powerful oxidant ozone (O_3) (Figure 6.3) are generated (Equation 6.18).

$$ ^3/_2\,O_2 \;\rightleftharpoons\; O_3 \tag{6.18} $$

When the stream of oxygen (O_2) containing ozone is passed through a methylene chloride (dichloromethane [CH_2Cl_2]) solution of an alkene at low temperature, the double bond undergoes oxidation because it is attacked by the ozone (O_3) (the *Criegee* mechanism, Scheme 6.11); the initial complexation of ozone (O_3) with the double bond is followed by formation of a *primary ozonide* (sometimes called a *molozonide*) by a **"1,3-dipolar addition"** process. Although the primary ozonide is unstable, the presence of such 1,2,3-trioxolanes has been confirmed by microwave spectroscopy. It is argued that the primary ozonide decomposes with cleavage of

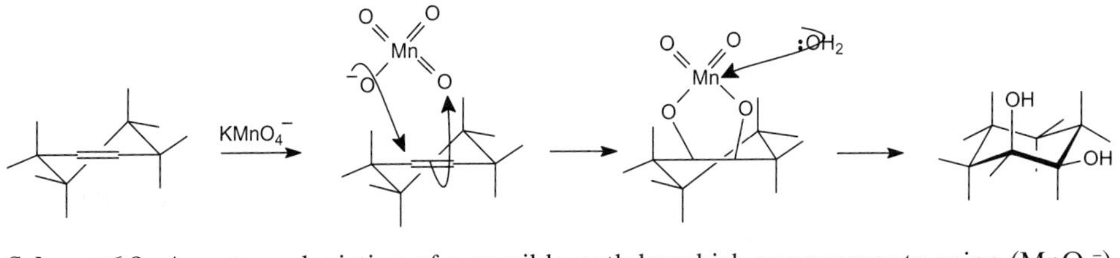

Scheme 6.9. A cartoon depiction of a possible path by which permanganate anion (MnO_4^-) in basic aqueous medium yields (*Z*)- or *cis*-1,2-dihydroxycyclohexane (*cis*-1,2-cyclohexanediol). Note that the oxidation at carbon is accompanied by a reduction at manganese.

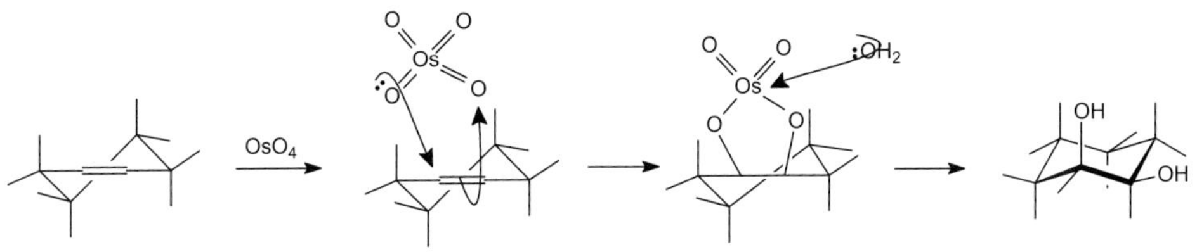

Scheme 6.10. A depiction of a possible path by which osmium tetroxide (OsO_4) yields (*Z*)- or *cis*-1,2-dihydroxycyclohexane (*cis*-1,2-cyclohexanediol). Note that the oxidation at carbon is accompanied by a reduction at osmium. Note further that the reduced osmium can be reoxidized by hydrogen peroxide (H_2O_2) (which in turn is reduced to water [H_2O]).

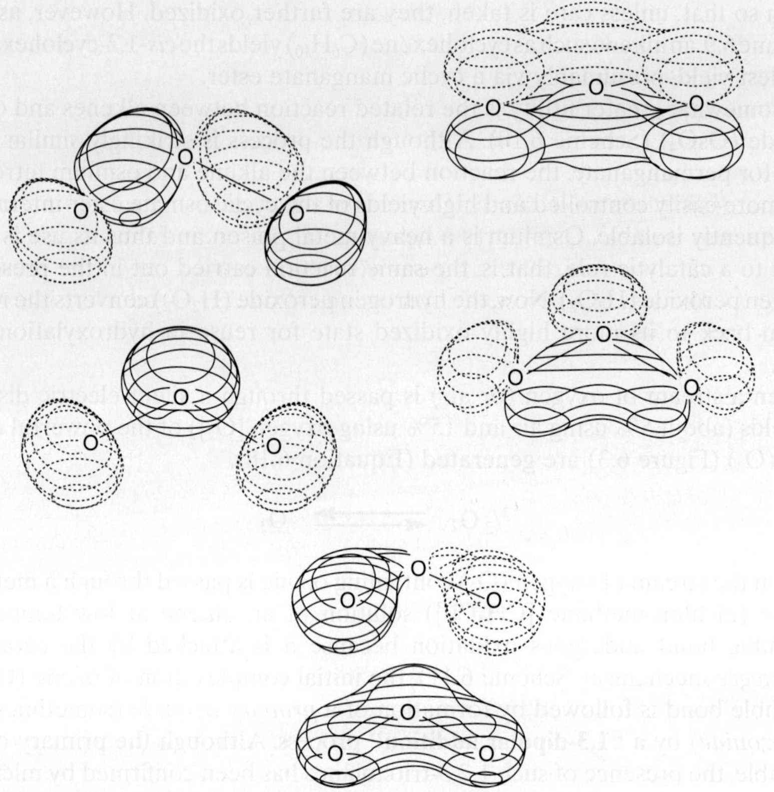

Figure 6.3. Representations of the molecular orbitals of ozone (O_3). The representations were created using the MOViewer in WebMO version 6.0.002p.

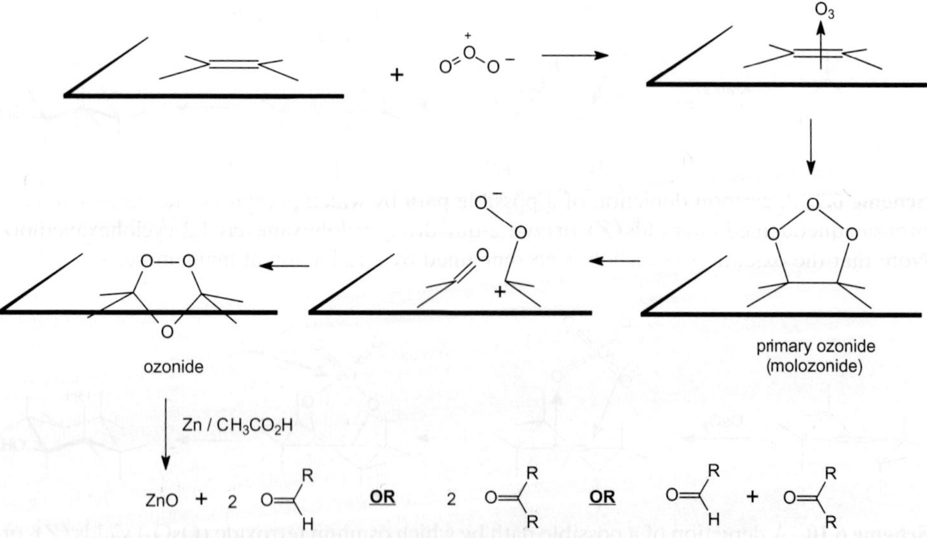

Scheme 6.11. A proposed pathway (after Criegee) for the ozonolysis (reaction with ozone followed by reductive hydrolysis) of alkenes. $R \neq H$ so that two equivalents of aldehyde, or two equivalents of ketone, or one equivalent of each, depending upon the substitution pattern of the alkene, results. (See Criegee, R. *Angew. Chem. Int. Edit.*, **1975**, *87*, 745.)

the carbon–carbon bond to a carbonyl compound (an aldehyde or ketone, depending upon the substitution at the carbon atoms of the double bond) and a dipolar dioxygen ion. In the next step the carbonyl species and the dipolar ion recombine to yield an *ozonide* (a 1,3,4-trioxolane or 1,3,4-trioxacyclopentane).

Although in some cases the ozonide can be isolated, these compounds are generally fragile and decompose explosively. Usually, the ozonides are reductively decomposed immediately after formation with zinc (Zn) in acetic acid (ethanoic acid, CH_3CO_2H) or dimethyl sulfide $[(CH_3)_2S]$ to give two equivalents of aldehyde, two equivalents of ketone, or one equivalent of each (depending upon the structure of the alkene). The overall process of ozonization coupled with reductive hydrolysis is known as *ozonolysis*.

Dienes (as well as polyenes) can also undergo ozonolysis and the products obtained reflect cleavage at all fragile double bonds. Generally, it has been observed that when electron-donating substituents are attached to the double bond the reaction with ozone is faster than when electron-withdrawing substituents are present. Thus, it is common to find that *more highly substituted alkenes react faster than their less highly substituted analogues*. The following examples (Equations 6.19a–6.19g) are illustrative:

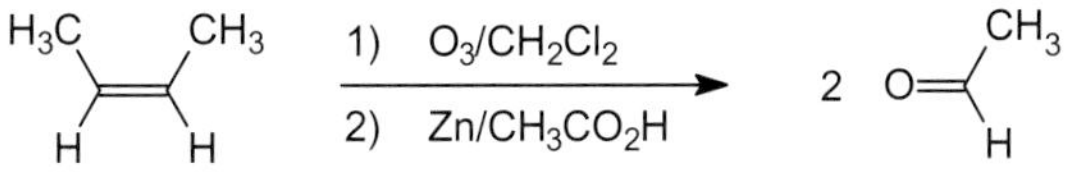

$$(6.19a)$$

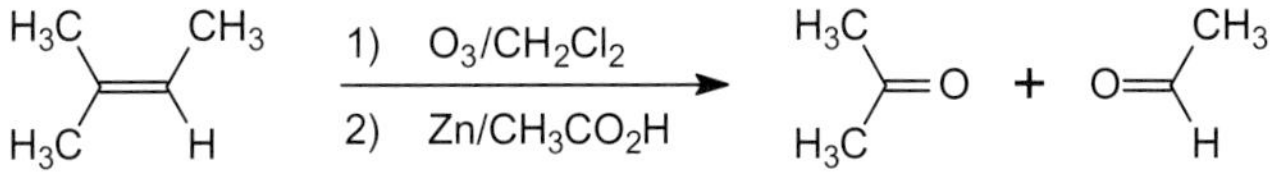

$$(6.19b)$$

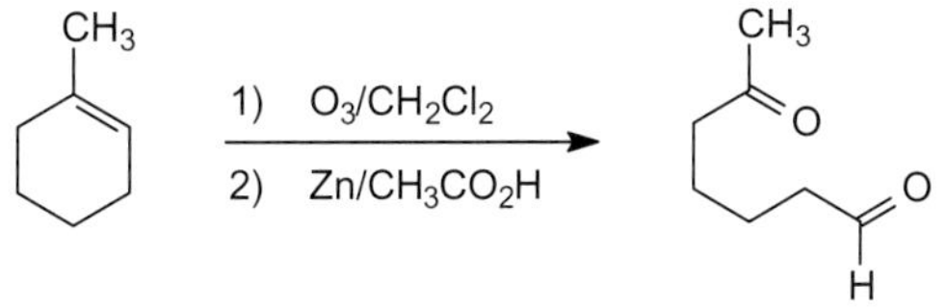

$$(6.19c)$$

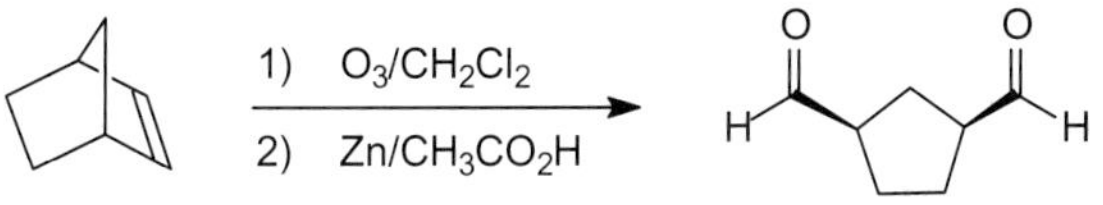

$$(6.19d)$$

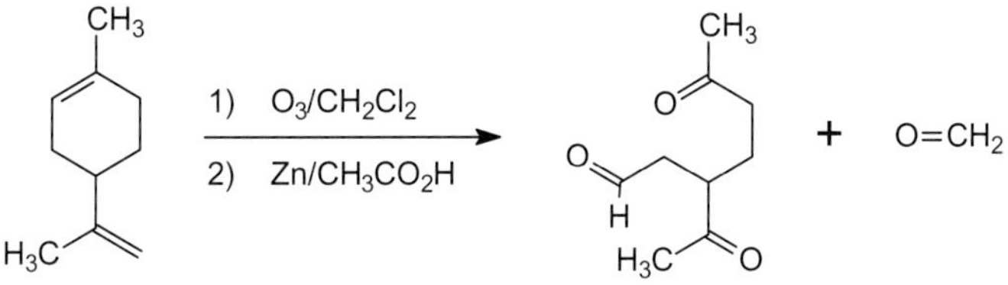

$$(6.19e)$$

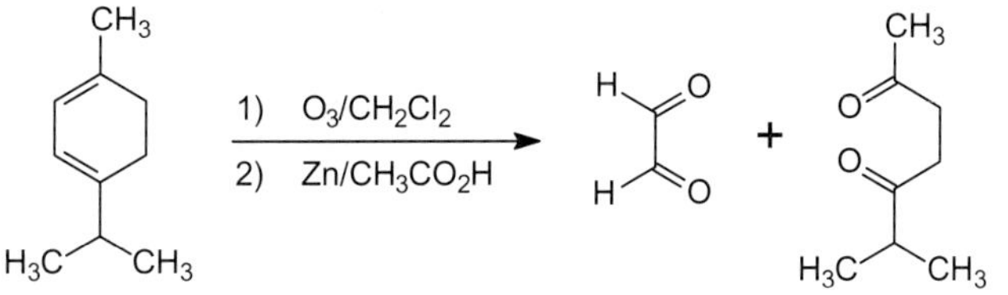

$$(6.19f)$$

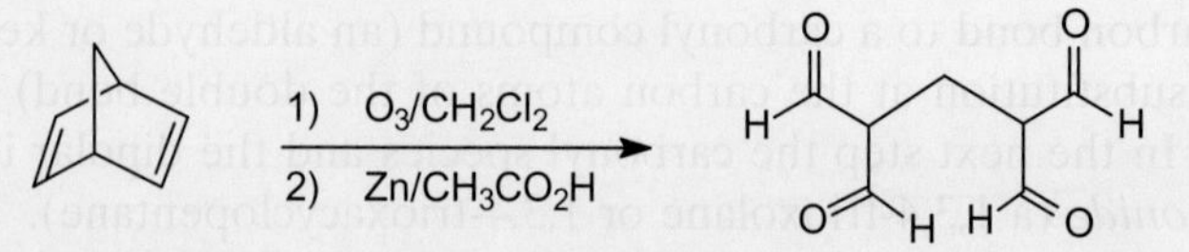

$$\text{(6.19g)}$$

Problem 6.5. Predict the products obtained on ozonolysis of the following alkenes:

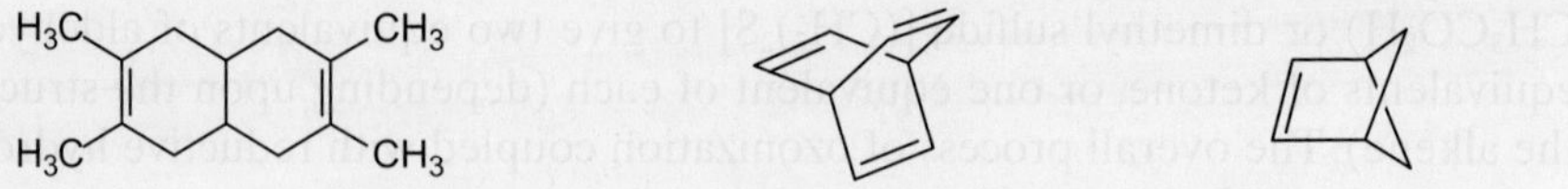

Alkenes (including dienes) also react with per(carboxylic) acids such as performic acid (HCO_3H), peracetic acid (CH_3CO_3H), trifluoroperacetic acid (CF_3CO_3H), and (because it is relatively stable and commercially available) *meta*-chloroperbenzoic acid ($m\text{-}ClC_6H_5CO_3H$) to yield **epoxides** or **oxiranes** (Scheme 6.12).

Although it will be discussed in more detail later, it is important to note that treatment of oxiranes with aqueous acid (H_3O^+) results in protonation of the oxygen of the epoxide and **ring opening** in an "*anti-*" or an antarafacial fashion with production of the corresponding diol. Thus, *in contrast to the results of permanganate (MnO_4^-) and osmium tetroxide (OsO_4) oxidation of alkenes, this two-step process* (epoxidation followed by aqueous acid hydrolysis) yields the *(E)-* or *trans*-diol (Equation 6.20**).** The ^1H NMR and Fourier transform infrared (FT-IR) spectra for cyclohexene and the *(E)-* and *(Z)-*diols are shown in Figure 6.4.

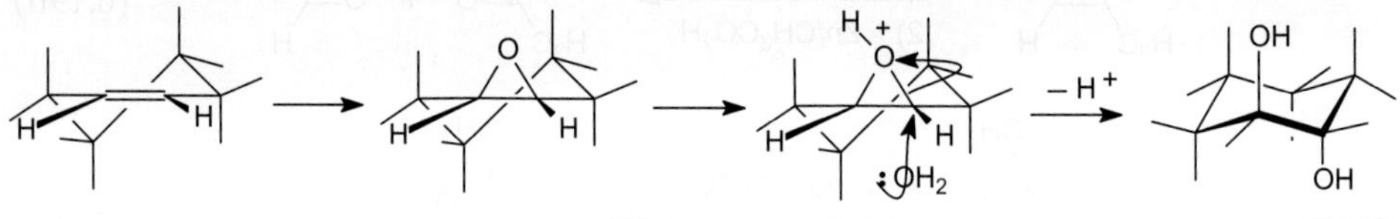

$$\text{(6.20)}$$

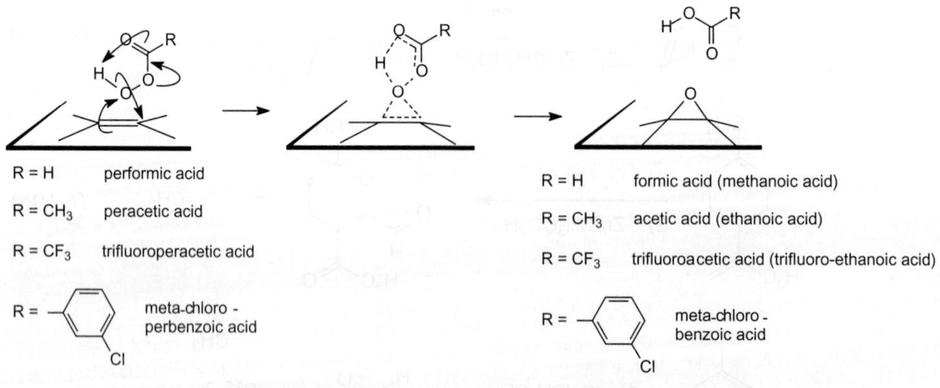

R = H   performic acid

R = CH_3 peracetic acid

R = CF_3 trifluoroperacetic acid

R =   meta-chloro -
           perbenzoic acid

R = H   formic acid (methanoic acid)

R = CH_3 acetic acid (ethanoic acid)

R = CF_3 trifluoroacetic acid (trifluoro-ethanoic acid)

R =   meta-chloro -
           benzoic acid

Scheme 6.12. A depiction of a possible path by which an alkene reacts with a percarboxylic acid to transfer an oxygen, producing an epoxide (oxirane) and the corresponding carboxylic acid.

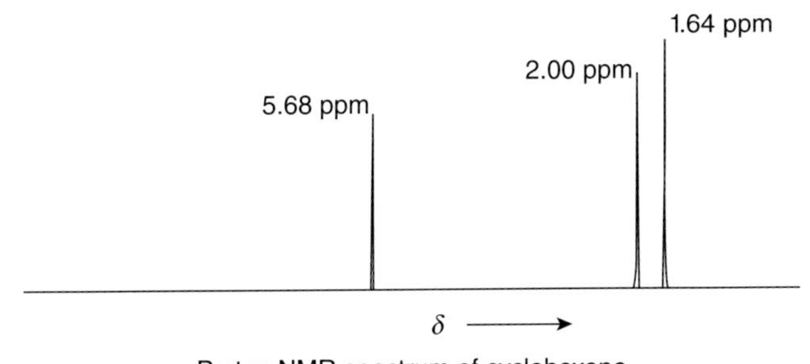

Proton NMR spectrum of cyclohexene
in deuteriochloroform at 300 MHz

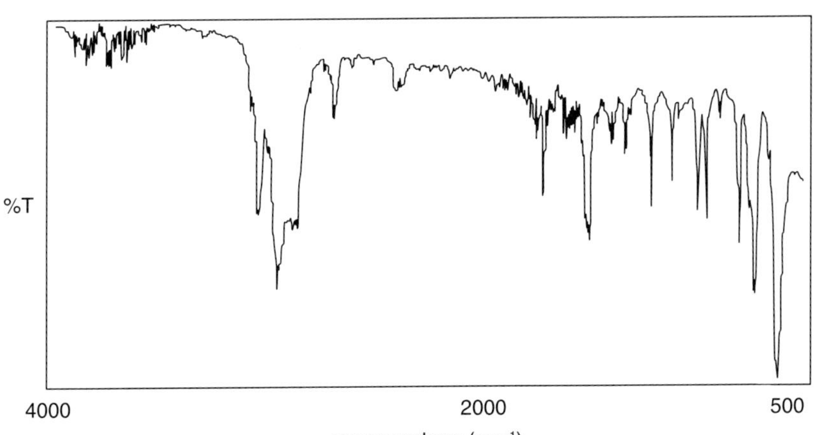

Infrared spectrum of cyclohexene (neat liquid)

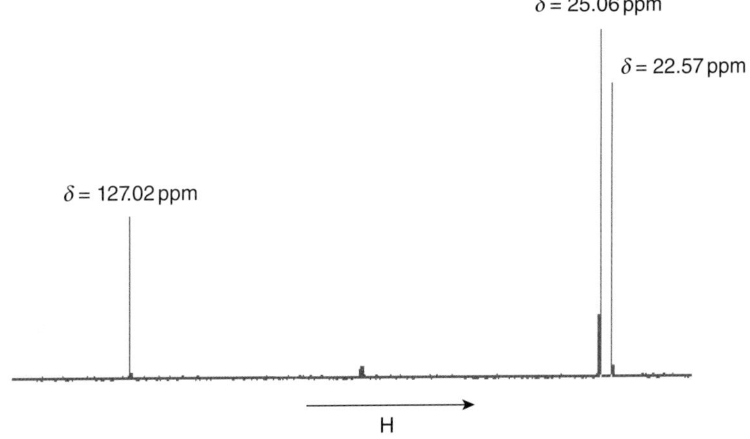

Carbon NMR spectrum of cyclohexene in deuteriochloroform at 75 MHz

Figure 6.4. The ^{1}H, ^{13}C, and FT-IR spectra of cyclohexene.

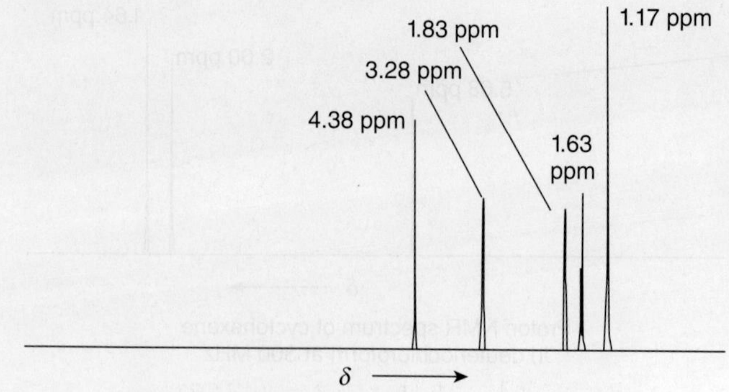

Proton NMR spectrum of trans-1,2-cyclohexanediol
in deuteriochloroform at 300 MHz

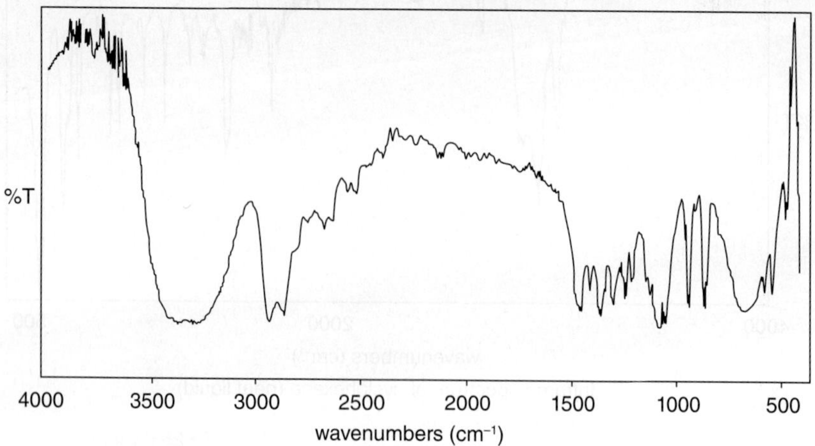

Infrared spectrum of trans-1,2-cyclohexanediol in KBr

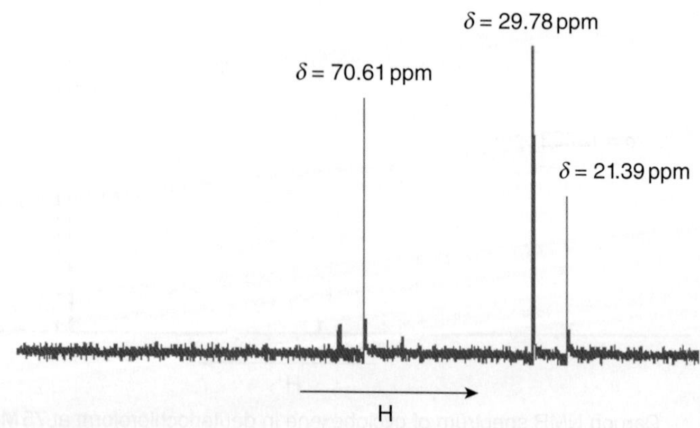

Carbon NMR spectrum of trans-1,2-cyclohexanediol in deuteriochloroform at 75 MHz

Figure 6.4. *Continued.*

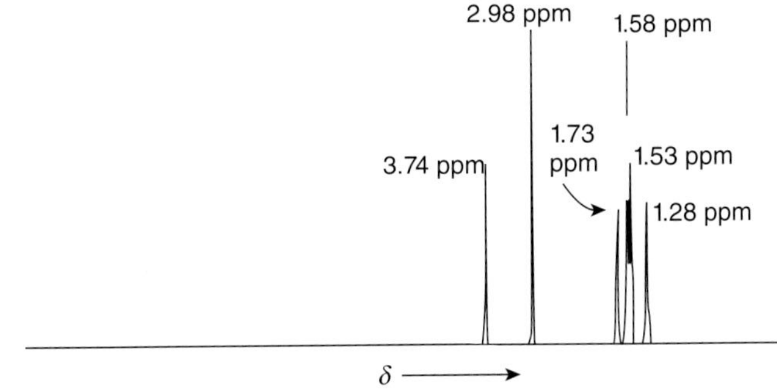

Proton NMR spectrum of cis-1,2-cyclohexanediol
in deuteriochloroform at 300 MHz

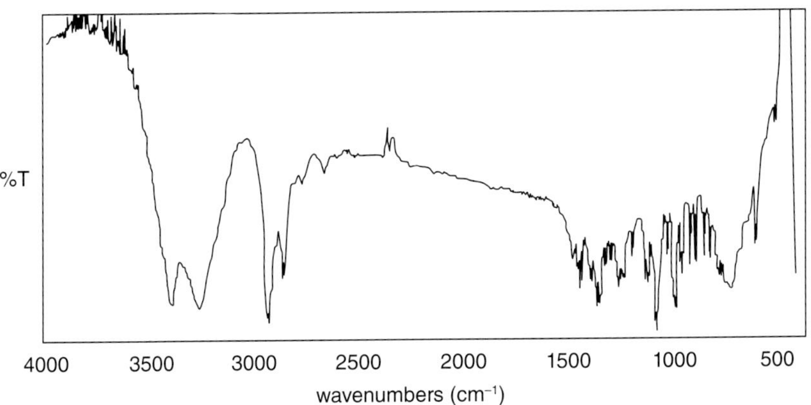

IR spectrum of cis-1,2-cyclohexanediol as KBr pellet

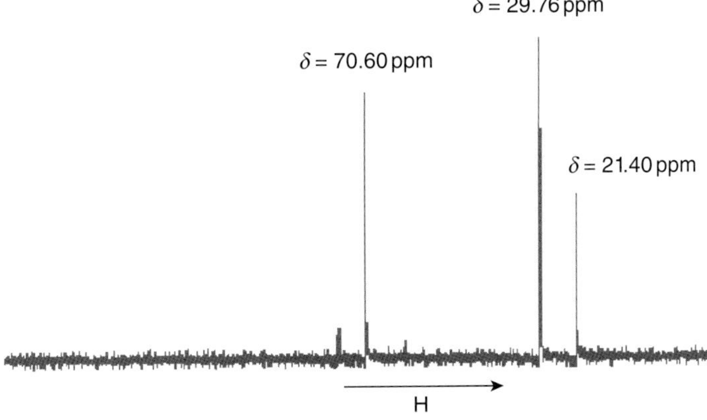

Carbon NMR spectrum of cis-1,2-cyclohexanediol in deuteriochloroform at 75 MHz

Figure 6.4. *Continued.*

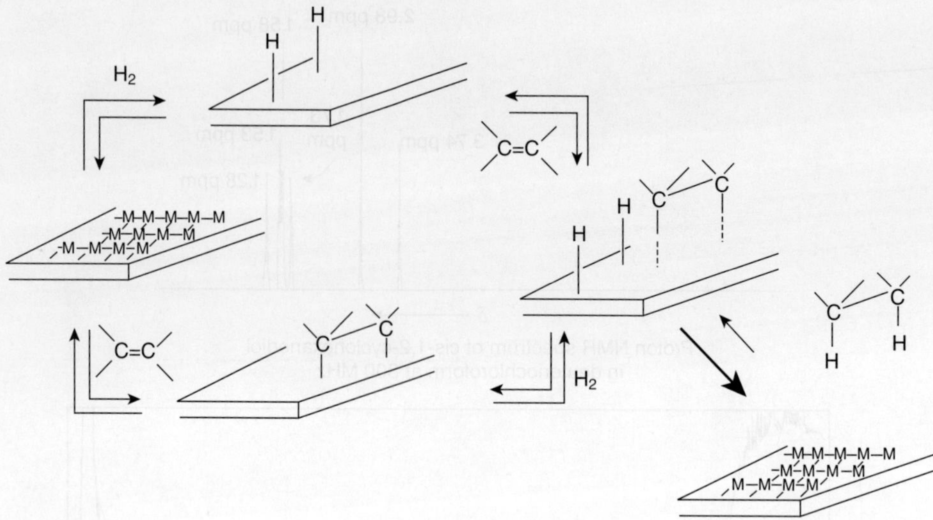

Figure 6.5. A cartoon representing the adsorption of hydrogen and alkene onto a (catalytic) metal surface. Hydrogen and alkene are almost certainly not (if only because they are different) adsorbed in the same way and/or to the same extent. It is presumed that the adsorbed species react and the product is desorbed. Under appropriate conditions, the reaction is known to be reversible for some alkanes and, at least in principle, is reversible for all.

II. Reduction

As pointed out in Chapter 3 the exact details of reduction of alkenes in the presence of catalysts are unclear. Further, just as **oxidation** can be considered as **addition** of oxygen to the double bond so **reduction** can be thought of as **addition** of **hydrogen** to an alkene. Overall, of course, the process is one that converts an unsaturated to a saturated system (with one equivalent of H_2 being consumed for every site of unsaturation—double bond or ring).

Although catalytic reduction on an active metal surface is most common, a variety of methods for adding hydrogen across the double bond is available. As shown in Chapter 3 (Figure 3.26) and in Figure 6.5, over a suitable catalyst (e.g., Pt on carbon or reduced platinum oxide; *Adams' catalyst* [PtO_2]*), hydrogen adds in a suprafacial (*cis-*, *syn-* or *Z-*) manner. While both hydrogen and alkene are adsorbed on the metal, it is probably true that they are adsorbed to different extents and in different ways.[†]

As shown in Scheme 6.13, if the catalyst for the reduction is changed from Pt to Pd, *the stereochemistry of the reduction is apparently altered*. However, once it is recognized that Pd on alumina (aluminum oxide [Al_2O_3]) causes *isomerization*—a **rearrangement** (*vide infra*) from one alkene to another—faster than reduction occurs, the result is explicable. Thus, (Scheme 6.13) 1,2-dimethylcyclohexene

---

*Roger Adams (1889–1971), Professor, University of Illinois (1916–1971).
[†]Not all of the sites on a surface are identical. Further, broadly, *chemisorption* involves strong interactions between adsorbant and surface. This type of adsorption may involve interactions whose strengths approximate those of true chemical bonds. At the opposite extreme, *physical adsorption*, apparently involves weaker bonding to the surface.

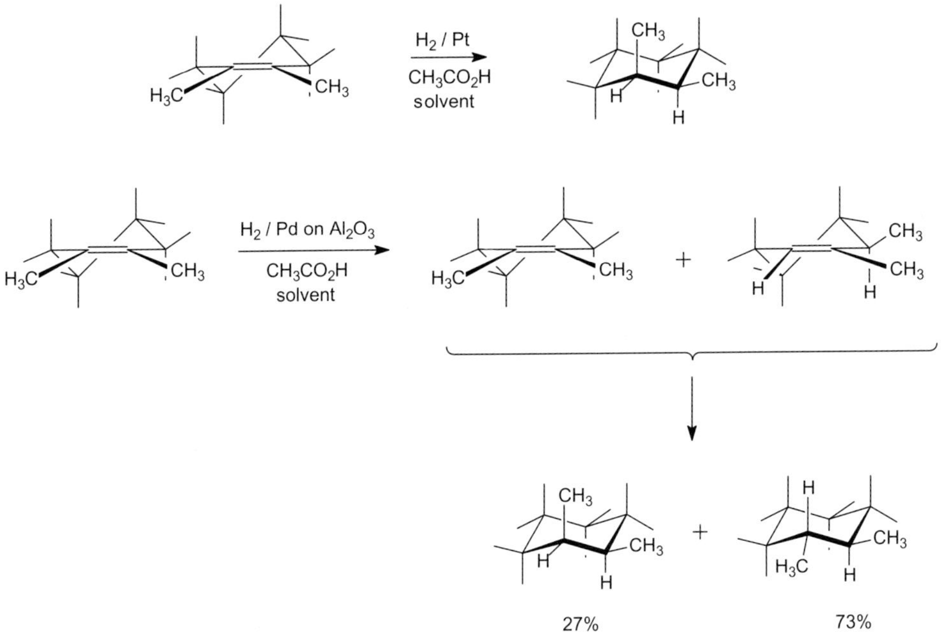

Scheme 6.13. Reduction of 1,2-dimethylcyclohexene to 1,2-dimethylcyclohexane over Pt and Pd catalysts. Different results are obtained as a function of catalyst.

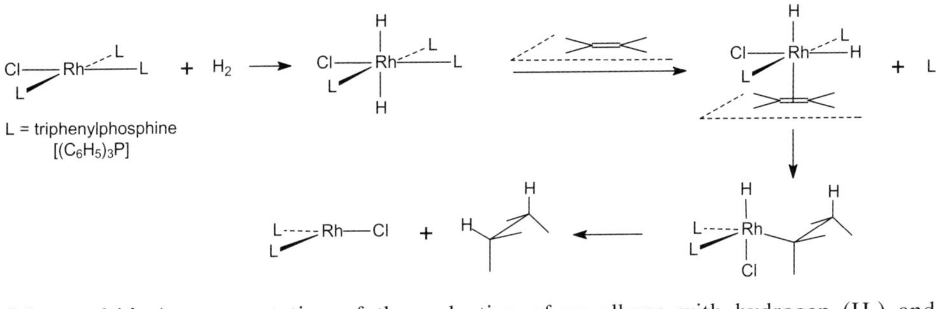

L = triphenylphosphine
[(C$_6$H$_5$)$_3$P]

Scheme 6.14. A representation of the reduction of an alkene with hydrogen (H$_2$) and chlorotris(triphenylphosphine)rhodium {ClRh[(C$_6$H$_5$)$_3$P]$_3$}.

yields *cis*-1,2-dimethylcyclohexane over Pt and a mixture of *cis*- and *trans*-1,2-dimethylcyclohexanes over Pd (on alumina).

Depending upon the unsaturated system and other special needs, a variety of other **insoluble** (or **hetereogeneous**) catalysts (i.e., metals such as Ni, Fe, Rh, and Ru) might be used. However, it is becoming common to attempt to utilize catalysts that are **soluble** (or **homogeneous**) in the reaction medium because they are easier to prepare reproducibly and they are more selective (if two or more differently substituted double bonds are present, e.g., in limonene, Scheme 6.6). The transition metal complex known as *Wilkinson's catalyst*, chlorotris(triphenylphosphine)-rhodium {ClRh[(C$_6$H$_5$)$_3$P]$_3$}, has proved very effective in many cases and some details of its mechanism have been established (Scheme 6.14). Interestingly, although an intermediate can be isolated and thus the addition of hydrogen (H_2) proceeds in *two* steps, the net result remains *suprafacial* (*syn, cis,* or *Z*-).

Although a variety of additional reduction techniques that do not specifically utilize hydrogen gas (H_2) have been employed for the conversion of specific alkenes to alkanes (using reagents such as [a] lithium metal [Li] and alkyl amines [RNH_2], as well as [b] sodium metal [Na] and ethanol [ethyl alcohol, CH_3CH_2OH]), there are three that have proved quite general. The first of these uses hydrogen transfer from (*Z*)-diimide [(*Z*)-H–N=N–H] which, itself, can be produced by the metal-catalyzed dehydrogenation of hydrazine (H_2NNH_2) (Scheme 6.15) and, by what is presumably an allowed electrocyclic process, results in overall suprafacial addition of hydrogen across the double bond. Alternatively, *transfer hydrogenation* can be effected. In this technique, the catalyst (heterogeneous or homogeneous) promotes transfer of hydrogen from another organic molecule, added specifically for that purpose, and which is oxidized in the process. For example (Equation 6.21), the double bond in the straight chain alkene, 1-decene [CH_2=$CH(CH_2)_7CH_3$] is conveniently reduced by Pd-catalyzed hydrogen transfer from cyclohexene (C_6H_{10}), which is correspondingly oxidized to benzene (C_6H_6). Finally, *reductive hydroboration* can be employed. Reductive hydroboration is a *two-step* process that involves initial *addition of borane* across the double bond followed by replacement of the boron bonded to carbon by hydrogen. Although shown in Equation 6.22, the details of the process are reserved for consideration in the next section (*vide infra*) of this chapter.

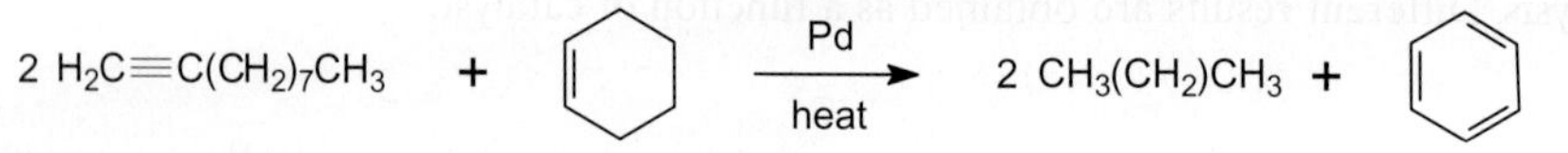

$$2\ H_2C{=}C(CH_2)_7CH_3 \;+\; \bigcirc \xrightarrow[\text{heat}]{\text{Pd}} 2\ CH_3(CH_2)CH_3 \;+\; \bigcirc \tag{6.21}$$

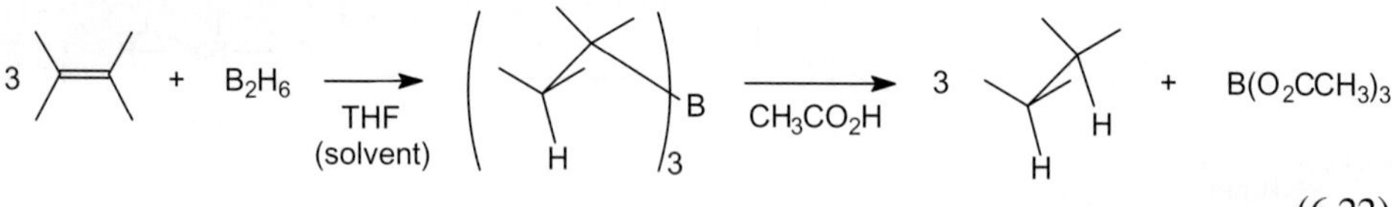

$$(6.22)$$

As pointed out earlier, there is *less* heat evolved on hydrogenation of alkyl bearing double bonds than there is on hydrogenation of double bonds that are

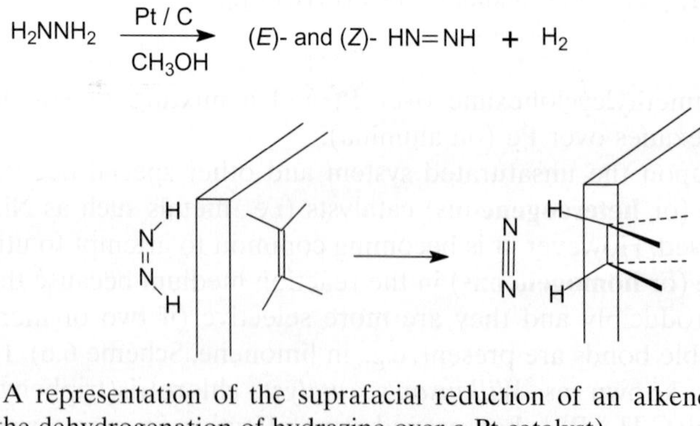

$$H_2NNH_2 \xrightarrow[\text{CH}_3\text{OH}]{\text{Pt / C}} (E)\text{- and } (Z)\text{- HN{=}NH} \;+\; H_2$$

Scheme 6.15. A representation of the suprafacial reduction of an alkene by diimide (as generated by the dehydrogenation of hydrazine over a Pt catalyst).

unsubstituted. Thus, alkyl substitution stabilizes the carbon–carbon double bond. Additionally, it is worth remembering that, based on heats of hydrogenation: (a) *(E)*-isomers of alkyl-substituted alkenes are more stable than the corresponding *(Z)*-isomers; (b) the amount of heat liberated when a *non*conjugated diene is reduced is about twice that of a single double bond whereas a conjugated double bond imparts significant stability; and (c) cumulated double bonds [e.g., allene (1,2-propadiene, $H_2C=C=CH_2$)] are less stable than either two conjugated or two isolated double bonds, since much more heat is liberated when the former are reduced than either of the latter.

Problem 6.6. It has been reported that, if the reduction of a *conjugated* diene (e.g., 1,3-butadiene over Pd) is interrupted when only one equivalent of hydrogen (H_2) has been used up, a mixture of alkenes (e.g., 1-butene and *(E)*- and *(Z)*-2-butene) is found. How would you account for this observation?

The biological reduction of the carbon–carbon double bond can also be brought about. Commonly this is accomplished by transfer of hydrogen (as hydride) from reduced nicotinamide adenine dinucleotide (NADH) and subsequent (or simultaneous) proton transfer, that is,

X = H; nicotinamide adenine dinucleotide $(NAD)^+$ (NADH)

X = PO_3^{-2}; nicotinamide adenine dinucleotide phosphate $(NADP)^+$ (NADPH)

III. Addition

Addition reactions to carbon–carbon double bonds are of major importance to the fabric of organic chemistry because they allow for the introduction of functional groups into hydrocarbons. In this way, the carbon–carbon double bond is elaborated upon and new and valuable compounds are prepared.

It should be clear that the spectroscopic characteristics of the disappearance of a double bond might be matched by those of the appearance of new functionality—as appropriate to the addition. While the latter will clearly vary, for the double bond: (a) the ^{13}C signals (at δ 110 to 135 ppm relative to tetramethylsilane [TMS]) should disappear; (b) in the ^1H NMR, protons on the carbon–carbon double bond—if any are present—are normally found between about δ 4.9 and 5.2 ppm and the resonance signals due to them should no longer be seen; (c) the medium-to-weak IR absorbance (conjugation absent) between about 1600 and 1620 cm^{-1} should disappear; and (d) for conjugated dienes that had been demonstrated to have UV absorbance, removal of the double bond(s) (because of addition) should result in a shift to higher energies (shorter wavelengths).

The kinds of addition reactions to alkenes are legion. However, the fundamental process only requires that one of the carbon–carbon double bonds be broken and that two new bonds to carbon be formed, that is, Equation 6.23 (where X and Y may be the same or different). This broad generality then allows that (it can be argued) the oxidation and reduction reactions already discussed are just examples of such addition processes, with oxidation only an example of the addition of oxygen ("X" = "Y" = O; "XY" = O_2) and reduction the addition of hydrogen "X" = "Y" = H; "XY" = H_2).

$$(6.23)$$

Generally, such addition processes are thermodynamically favorable. For example, consider the case where "X" = "Y" = Cl (i.e., "XY" = Cl_2) and the alkene is ethene (ethylene, $CH_2{=}CH_2$) to produce 1,2-dichloroethane (CH_2ClCH_2Cl) and, for comparison, the reaction of chlorine (Cl_2) with ethane (CH_3CH_3). From the data in Table 1.1 (Chapter 1), the homolytic bond strength of the carbon–carbon single bond in ethane (CH_3–CH_3) is about 376 kJ mol^{-1} (90 kcal mol^{-1}) and the carbon–carbon double bond in ethene (ethylene, $CH_2{=}CH_2$) is about 720 kJ mol^{-1} (172 kcal mol^{-1}). Thus, the cleavage of one-half the carbon–carbon double bond should need about 259 kJ mol^{-1} (62 kcal mol^{-1}) and addition reactions to the double bond will be significantly more exothermic than the cleavage of a carbon–carbon single bond. Using the data in Table 1.1 (Chapter 1):

for CH_3–CH_3 + Cl_2 → $2CH_3Cl$;
$\Delta H° = [(90) + (59) = (149) - 2(73)] = 3$ kcal mol^{-1} (12.5 kJ mol^{-1})

for $CH_2{=}CH_2$ + Cl_2 → $ClCH_2CH_2Cl$;
$\Delta H° = [(62) + (59) = (121) - 2(73)] = -25$ kcal mol^{-1} (105 kJ mol^{-1})

Thus, it might be concluded that the second reaction should be paticularly facile and, indeed, the reaction between chlorine (Cl_2) and ethene (ethylene, $H_2C{=}CH_2$)

to produce 1,2-dichloroethane (CH_2ClCH_2Cl) occurs rapidly in solution *and in the dark*.

To accomplish the generalized addition process shown in Equation 6.23, it should be clear that either the bonds from carbon to "X" and "Y" and the double bond between the two carbons are either (a) formed (and broken) at the same time or nearly so, with no discrete separate intermediate steps, in a simultaneous fashion or in concert, or (b) in clearly separate steps and thus consecutively.

The idea of the reactions occurring in concert has already been briefly discussed in Chapter 4 where some stereochemical consequences of some orbital-symmetry-allowed processes (e.g., the Diels–Alder reaction) were introduced. Further discussion of these and related (e.g., cheletropic, *vide supra*) reactions will be included here.

However, it is clear from countless experimental reports that even when "X" and "Y" (Equation 6.23) are the same ("X" = "Y"), reactions in which bonds are broken and made sequentially are far more common than those involving simultaneity. For the sequential reactions, there are two possibilities, viz (a) the "X–Y" bond breaking occurs **homolytically** with "X" and "Y" each retaining one of the electrons originally in the bond (thus generating a pair of radicals). Then, either "X•" (or "Y•") reacts with the carbon–carbon double bond to produce a new radical (or radicals) (Equation 6.24) whose eventual combination yields product; or (b) the bond breaking occurs heterolytically with either "X" or "Y" (depending upon relative electronegativity) retaining the pair of electrons (an **anion**), and the other original partner, stripped of its share, is now a **cation**. Since such heterolytic bond breaking is energetically unfavorable without *simultaneous bond making*, solvent assistance and/or the reaction with the double bond must occur and, indeed, *may lead the process*.

$$\text{(6.24)}$$

With both negative (*nucleophilic*—or nucleous seeking) and positive (*electrophilic*—or electron seeking) species present, the double bond could, in principle, be attacked by the nucleophile or the electrophile. Both are known but the latter are much more common and are collectively organized under the rubric: **electrophilic addition**.

a. Electrophilic Addition. The reactions grouped under this heading are extraordinarily widespread and the ease with which they appear to occur has encouraged the argument that the relatively high electron density present in the region between the two carbon atoms in an alkene serves as a locus for the attack on the system by those "electron-seeking" reagents. Reasonably, species with incomplete outer electron shells, bearing a positive charge (e.g., H^+) or neutral (e.g., BF_3), would be expected to interact strongly with the pi (π) orbital electrons. Then, the electron-deficient, high energy (generally) unstable result of such an interaction should be available for attack by electron-rich nucleophiles *in a discrete second step*, the latter consummating the addition process (Equation 6.25).

$$(6.25)$$

The formalism of Equation 6.25 is common among **electrophilic addition** reactions. Generally, the *curved arrow* is drawn *from the double bond* (showing the movement of a *pair* of electrons) to the *electron-deficient* species, thus forging a bond to carbon. *The curved arrow then becomes the bond.* The consequence is the formation of an electron-deficient *intermediate* (shown here as a *carbocation*), which is then attacked by an electron-rich *nucleophile*. Again, the curved arrow, showing the movement of a pair of electrons, *becomes the bond.**

Thus, the generalized process of electrophilic addition can involve numerous electrophilic reagents (Table 6.1) interacting with any alkene (diene, etc.). Although each different alkene and different electrophilic reagent must give a unique product (or set of products) under whatever specific set of reaction conditions are used, a similar set of questions can be asked of *all* of them: viz (1) With what **stereochemistry** does the reaction proceed? That is, are "X" and "Y" added suprafacially (*cis-* or *syn-*), antarafacially (*trans-* or *anti-*), or both (equally or nearly so); (2) With what **regiochemistry** does the reaction proceed? That is, if "X" ≠ "Y" and the carbon atoms at the double bond termini are *not* equivalent, is there a preference for the sense of the direction in which "X" and "Y" are added?; and (3) With what kinetics does the reaction occur? That is, are the exponents of the concentration terms in the rate equation (determined experimentally for the rate determining step) the same numbers as those given by the stoichiometry in the balanced equation?

Finally, it is reasonable to ask what might occur as a second step in electrophilic addition should no reasonably reactive nucleophile be present.

A discussion of all but a few examples is clearly beyond this text.

1. The Stereochemistry of Electrophilic Addition. Consideration of the electron-rich area between the carbons of the double bond where initially (a) the four ligands attached to the double bond are also coplanar with the carbons of the double bond and hence with each other; (b) the plane of the π-cloud is orthogonal to the plane containing the substituents; and (c) the initial approach of the electrophilic reagent is into the area of highest electron density (i.e., the π-cloud) leads to the suggestion expressed in Figure 6.6. That is, the attack into the vacant orbital of the electrophile by the electrons of the alkene would result in bonding and the initial (high-energy) species formed would be composed of electrophilic reagent and alkene. It is called a "π-complex." There is some spectroscopic evidence that such "π-complexes" are formed readily and, in a few cases, under very carefully controlled conditions, they have actually been isolated. It is not clear that the π-complex is on the reaction path to product.

---

*The formal pathway shown is somewhat arbitrary but is commonly used and, if adhered to, will generally provide good results. An alternative view, using sets of single electrons, is usually more difficult for beginning students. Further, the correlation between the curved arrows already associated with resonance (Chapter 1) and these, shown here in Equation 6.25, leading to products, is mutually reinforcing.

Reagent Added to C=C of Alkene	Product
1 HX (X = Cl, Br)	
2 H_2SO_4/H_2O	
3 H_2SO_4/ROH	
4 H_2SO_4/RCO_2H	
5 H_2SO_4/RSH	
6 $Hg(O_2CCH_3)_2/H_2O$	
7 RX /AlCl$_3$ (X = Cl, Br)	
8 X_2 (Cl$_2$, Br$_2$, BrCl, IBr) (I$_2$ and F$_2$ require special conditions)	
9 HOX (X = Cl, Br, I)	
10 $RSO_2Cl/CuCl$	
11 NOCl	

(Continued)

Reagent Added to C=C of Alkene	Product
12 IN$_3$	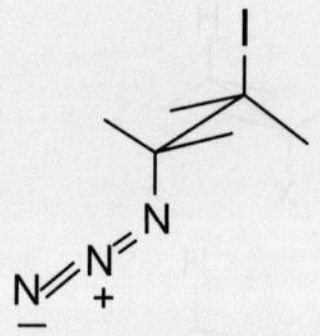
13 INCO	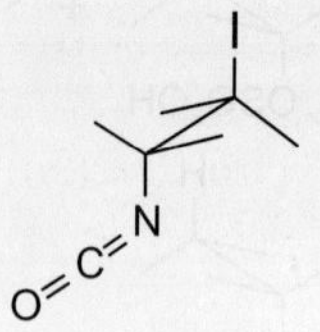
14 RCOCl (with AlCl$_3$)	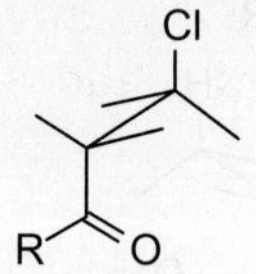
15 O$_3$	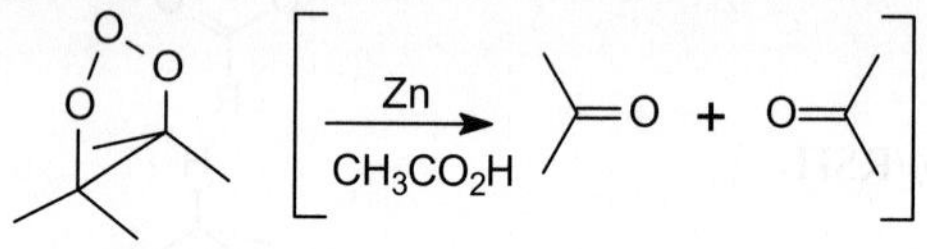
16 RCO$_3$H	
17 (BH$_3$)$_2$/THF 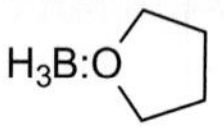 H$_3$B:O	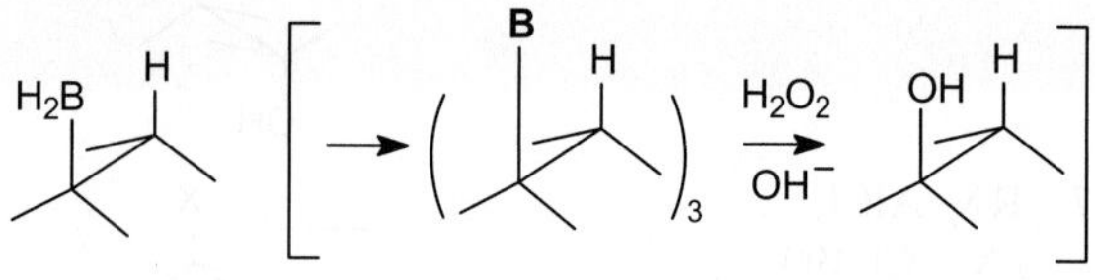
18 MH$_3$ (M = Al, Sn)	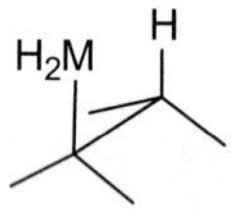
19  / H$_2$SO$_4$	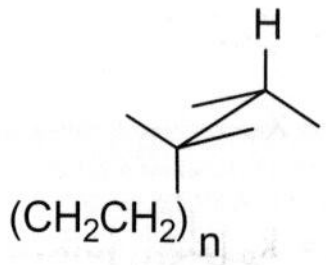
20 I$_2$/RCO$_2^-$ Ag$^+$	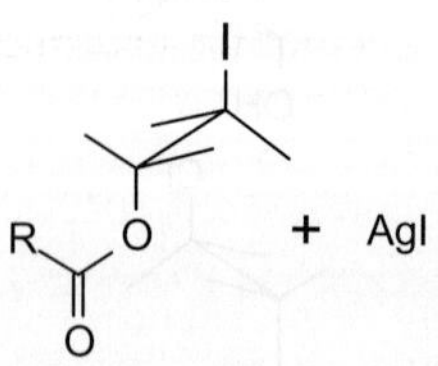

The 20 examples in the table are an incomplete (but representative) set of electrophiles and examples of the kinds of products that result from their respective reactions with a representative generic alkene. The representations of product stereochemistry in the table are incomplete. Reactions of double bonds with nucleophilic reagents and electrocyclic and related processes are intentionally excluded and will be considered subsequently.

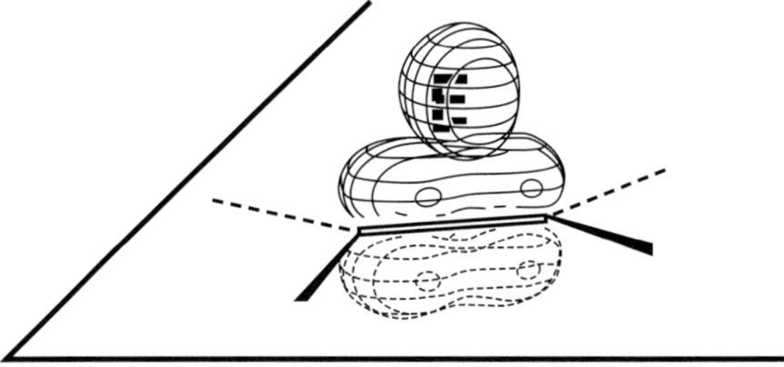

Figure 6.6. A representation of an initial π-complex formed by attack of the π electrons of an alkene with a vacant orbital of an electrophile (**E**). The representation was created using the MOViewer in WebMO version 6.0.002p.

For many electrophiles, under ionic conditions (particularly when **E** has non-bonded electrons [Table 6.1, examples 6 and 8–13]), a so-called "*'onium*"* ion (Figure 6.6, **E** = **E⁺**) is apparently formed. This ion is distinct from the initial π-complex since here a bond within the electrophile has broken and the 'onium ion bears a positive charge. However, as with the π-complex, the ion apparently has one face blocked by the electrophile and so subsequent attack by a nucleophile prefer-entially occurs *from the opposite face*. The net result is *antarafacial* (or *anti-* or *trans-*) addition across what was the carbon–carbon double bond. For example, consider the case of addition of bromine (Br_2) (Table 6.1, example 8) in carbon tetrachloride (tetrachloromethane, CCl_4) solution to the isomers (*Z*)- and (*E*)-2-butene [(*Z*)- and (*E*)-$CH_3CH=CHCH_3$] (Scheme 6.16).

The following observations have been made: (a) UV/Visible (UV/VIS) spectros-copy (following e.g., disappearance of the color of bromine [Br_2]) suggests that a π-complex is formed; (b) in what is apparently the rate-determining step, the electron-rich π-cloud attacks the bromine molecule (Br_2) forming a **bromonium ion** (the *'onium* ion) and bromide anion (Br^-, or its equivalent[†]). The bromine atom in the **bromonium ion** apparently bonds equivalently to both carbon atoms of what was the symmetrical alkene,[‡] forming a "*bridged ion*"; (c) bromide anion (Br^-, or its

---

The nomenclature of "'onium*" ions derives from positively charged hypervalent ions such as ammo-*nium* (NH_4^+) and hydr*onium* (H_3O^+).

[†]Examination of the **kinetics** (*vide infra*) shows that at low bromine concentrations the rate of product formation is given by an expression approximately first order in bromine (Br_2) and first order in alkene, that is,

$$\text{rate} = k_o \, [Br_2]^1 \, [\text{alkene}]^1$$

where k_o is the observed rate constant. At higher bromine (Br_2) concentrations the kinetics become much more complicated and there is some evidence that the tribromide anion (Br_3^-) is involved as the nucleo-philic species.

[‡]With nonsymmetrical alkenes, the species called a "bromonium ion" is almost certainly nonsymmetrical. In extreme cases (*vide infra*) it appears that an α-bromocarbocation, *viz.*

(where R ≠ H and is electron donating)

is either in equilibrium with the bromonium ion or is preferred.

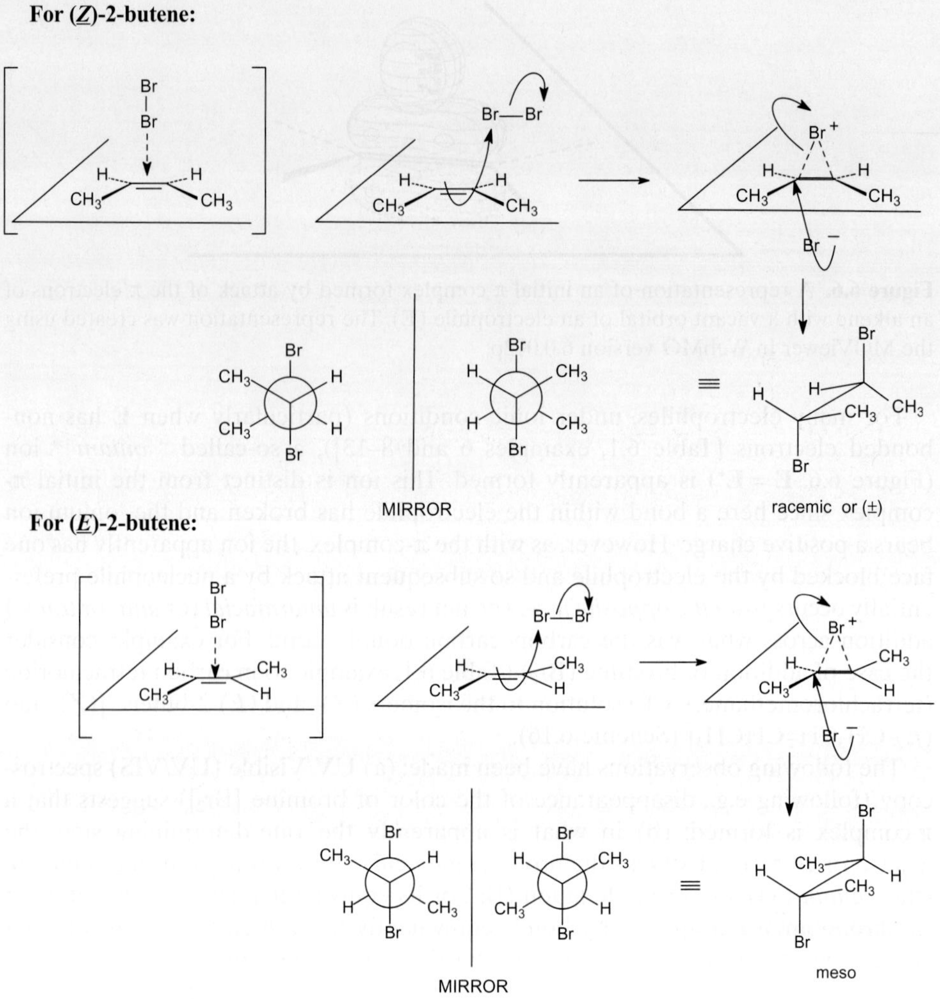

For (_Z_)-2-butene:

MIRROR

racemic or (±)

For (_E_)-2-butene:

MIRROR

meso

Scheme 6.16. Representations of the course of electrophilic addition of bromine (Br_2) to _cis_- or _(Z)_- and _trans_- or _(E)_-2-butene to give, respectively, _racemic_ or (±)-2,3-dibromobutane and _meso_-2,3-dibromobutane.

equivalent) attacks the three-membered ring at carbon from the side opposite to the bromine already present in the bromonium ion.

Now, (Scheme 6.16) having begun with _cis_- or _(Z)_-2-butene, the net result will be to form _racemic_ or (±)– or a 1:1 mixture of the enantiomers (2R,3R)-2,3-dibromobutane and (2S,3S)-2,3-dibromobutane (the original attack occurring with equal probability from the "top" or "bottom" face). Interestingly, each of these enantiomers has a twofold rotating axis of symmetry and, although chiral, is thus not asymmetric. They are dissymmetric.

Alternatively, having begun with _trans_- or _(E)_-2-butene, the net result will be to form _meso_ or (2R,3S)-2,3-dibromobutane [or (2S,3R)-2,3-dibromobutane], a molecule clearly possessing a plane of symmetry.

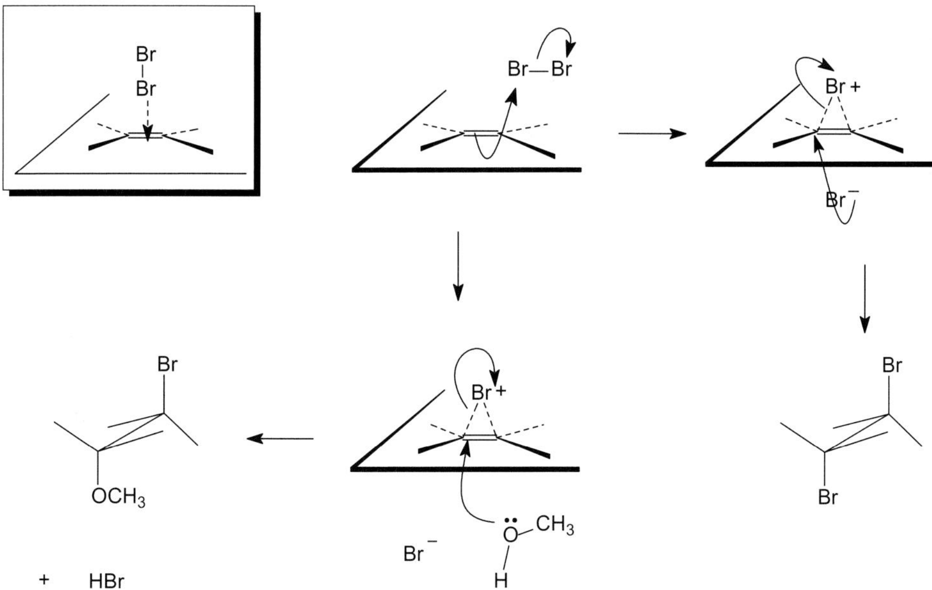

Scheme 6.17. A representation of the intrusion of methanol (methyl alcohol, CH₃OH) into the electrophilic addition of bromine (Br₂) across a double bond to form an α-bromo methyl ether.

The idea that the process requires at least two steps, that is, formation of an ion (shown as the bromonium ion) as a first step followed by its subsequent destruction (in a second separate step), is supported by many and diverse observations. For example, if the same bromination reaction is attempted in methanol (methyl alcohol, CH₃OH) rather than in carbon tetrachloride, (tetrachloromethane, CCl₄) the solvent (with its [nucleophilic] electron-rich oxygen) intrudes into the reaction and some O-methyl ether is obtained (Scheme 6.17).

2. The Regiochemistry of Electrophilic Addition. Since a symmetrical addend (e.g., Br₂, Cl₂) does not permit a decision to be made regarding which carbon atom in a nonsymmetrical alkene is attacked first (unless solvent or another nucleophile intrudes in the second step), only nonsymmetrical addends such as those in Table 6.1 entries 1–7 and 9–14 (or a symmetrical addend with intrusion of a "different" nucleophile) can be used to define which end of the double bond (*region*) receives/ bears which piece of addend.

The first five entries in Table 6.1 involve a mineral acid; either one of the hydrogen halides HX (X = F, Cl, Br or I) or sulfuric acid (H₂SO₄) and another reactant is incorporated into the final product. For reactions in solution, there appear to be numerous similarities for all of these "proton acids" although, for example, the additions of hydrogen fluoride (HF) and hydrogen iodide (HI) do not generally occur under conditions that succeed for hydrogen chloride (HCl) and hydrogen bromide (HBr).

However, the differences are easily understood. Thus, while hydrogen fluoride (HF) is capable of fulfilling its role as addend, the addition is reversible and the presence of a small amount of base is apparently required to inhibit the reversibility. The problem with hydrogen iodide (HI) is different; here the reaction between this hydrogen halide and alkenes suffers from the ready oxidation of hydrogen iodide (HI) to iodine (I_2), which then intrudes into the reaction and consumes alkene. Indeed, even with hydrogen bromide (HBr), care must be taken to insure that the reaction occurs under ionic conditions so that radical process do not intrude (*vide infra*).

That nonsymmetrical alkenes do *not* yield equal amounts of both possible products has been known for some time and, indeed, after a careful study, V. Markownikoff* in 1870, reported that when HX adds to a nonsymmetrically substituted alkene, the hydrogen (H^+) attaches to the carbon with the fewer alkyl substituents and the X^- to the carbon with more alkyl substituents. Why should this be so?

Addition of HX to an unsymmetrical alkene could, in principle, give two different products. The *ionic* reaction between hydrogen bromide (HBr) and 2-methylpropene (isobutylene [$(CH_3)_2CH=CH_2$]), where both 1-bromo-2-methylpropane (isobutyl bromide [$BrCH_2CH(CH_3)_2$]) and 2-bromo-2-methylpropane (*t*-butyl bromide [$(CH_3)_3CBr$]) *could have* formed is shown in Equation 6.26.

2-methylpropene   1-bromo-2-methylpropane   2-bromo-2-methylpropane

$$(6.26)$$

In this case *only* 2-bromo-2-methylpropane is found.

As shown in Figure 6.7, both the ^1H and ^{13}C NMR spectra of the C_4H_9Br isomers are significantly different. Thus, as anticipated, all nine hydrogens and three of the four carbon atoms of 2-bromo-2-methylpropane [*t*-butylbromide, $(CH_3)_3CBr$] are identical while there are three different kinds of carbon atoms and three different kinds of protons in 1-bromo-2-methylpropane [isobutylbromide, $(CH_3)_2CHCH_2Br$] and it is clear, therefore, which is which.

Further, as shown in Equations 6.27 and 6.28, similar reactions are observed when the addend is hydrogen chloride [(HCl) either gas or liquid] or the hydronium ion [(H_3O^+) from, *for example*, water (H_2O) and sulfuric acid (H_2SO_4)].

$$(CH_3)_2C=CH_2 + HCl_{(aq)} \longrightarrow (CH_3)_3CCl \quad \textbf{NOT} \quad (CH_3)_2CHCH_2Cl \quad (6.27)$$

$$(CH_3)_2C=CH_2 + H_2O \xrightarrow[\text{catalyst}]{H^+} (CH_3)_3COH \quad \textbf{NOT} \quad (CH_3)_2CHCH_2OH \quad (6.28)$$

In order to predict which products will form in these reactions it is necessary to realize, first, that *the initial step apparently involves attack of the double bond onto*

*Markownikoff, V. *Liebigs Ann. Chem.*, **1870**, *153*, 228.

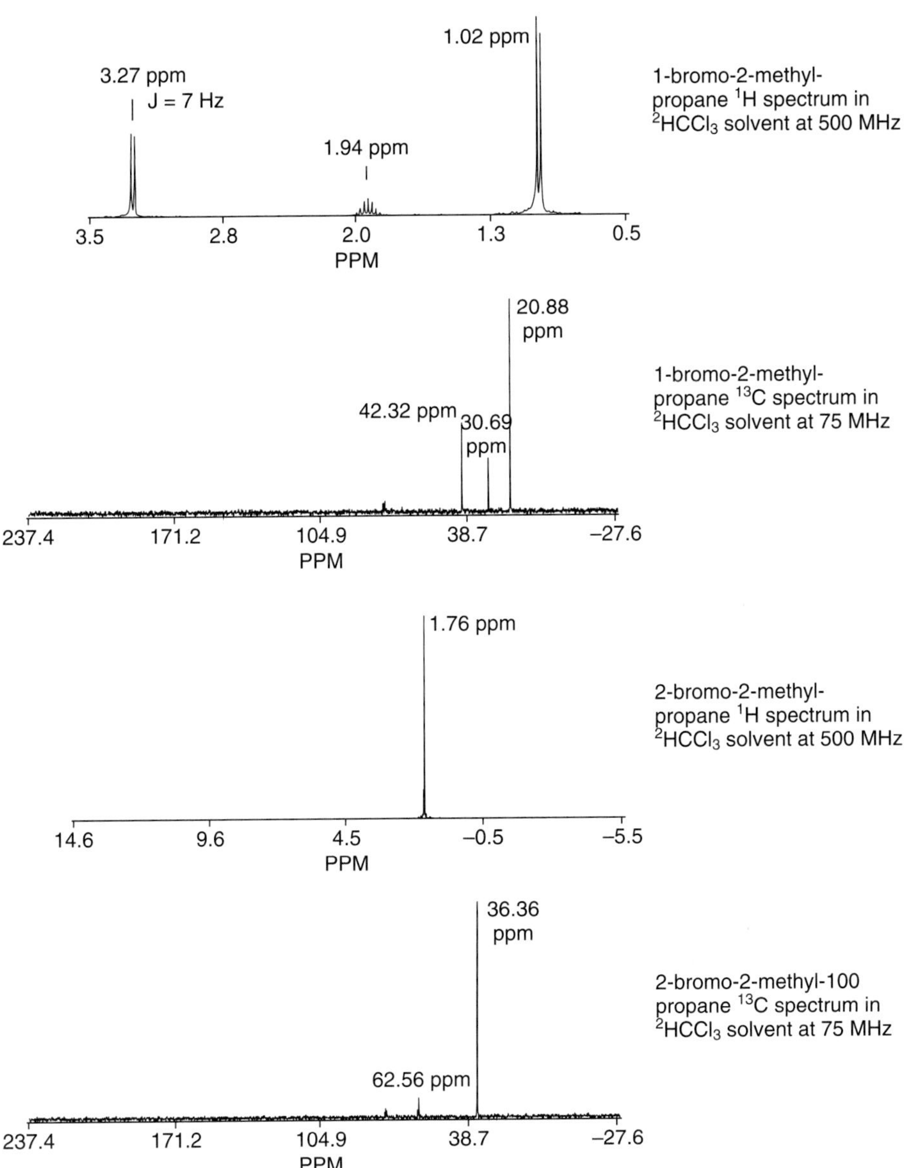

Figure 6.7. The ^{1}H and ^{13}C spectra of 1-bromo-2-methylpropane [isobutyl bromide, (CH$_3$)$_2$CHCH$_2$Br] and 2-bromo-2-methylpropane [*t*-butyl bromide, (CH$_3$)$_3$CBr].

an electron-deficient species—in these cases that species is a proton. That is, the electrophile *is* the proton (regardless of which nucleophile accompanies it).

Second, since alkyl groups are *electron-donating* relative to hydrogen, carbocations that have three alkyl groups attached to the carbon bearing the positive charge (*tertiary* [or 3°]) will be preferred to those that have only two alkyl groups attached to the carbon bearing the positive charge (secondary [or 2°]). Thus, carbocations that have only one alkyl group attached to the carbon bearing the positive charge (primary [or 1°]) should be the least stable substituted carbocations but they should be more stable than the simple methyl cation (CH_3^+). That is, in order of decreasing stability 3° > 2° > 1° > CH_3^+.*

Third, it is important to know whether the reaction is carried out under conditions such that the products are in *equilibrium* with the starting materials and each other (*thermodynamic conditions*) or, alternatively, conditions where the product formed fastest (the first formed product) is the product isolated (*kinetic conditions*). In the first case, the stabilities of the starting materials and products (as measured by, e.g., heats of formation ($\Delta H°$) and entropy factors ($\Delta S°$)] can be used to estimate the equilibrium constants (K_{eq}) (Chapter 4) for each. Alternatively, conditions might be found such that the equilibrium constant between the different products can be measured directly. Reactions run under *thermodynamic control*, or *equilibrium control* will be reversible so that the same product/starting material mixture will be obtained beginning from either starting materials or reaction products after sufficient time has passed. The alternative, *kinetic control*, requires that the conditions are such that the product/starting material ratio (as well as the ratio of the different products) at any time is determined by the rate(s) of formation of the product(s) and/or the rate(s) of disappearance of the starting material(s) and equilibrium is not achieved.

It is very important to understand that the ratio of products or, indeed, the products themselves need not be the same for reactions run under the different conditions. In the particular case of the primary and tertiary bromides, 1-bromo-2-methylpropane [isobutyl bromide, $(CH_3)_2CHCH_2Br$] and 2-bromo-2-methylpropane (*t*-butyl bromide, $(CH_3)_3CBr$], respectively, the experimentally determined equilibrium constant $K_{eq} = [(CH_3)_3CBr]/(CH_3)_2CHCH_2Br] \approx 4.5$ but, within experimental error, the exclusive product is the tertiary bromide. Thus it is clear that the reaction is *not* being carried out under thermodynamic control and it is worthwhile to consider the paths by which the product may be formed (Scheme 6.18a,b). Indeed, since the products and starting materials are not equilibrating, the product actually isolated should be that which follows the path of lowest energy and it is reasonable that the most stable carbocation (i.e., a tertiary [3°] carbocation rather than a primary [1°] carbocation) should be formed *as an intermediate* (i.e., before being attacked by nucleophile) (Chapter 4) on the way to the final product.[†] This is depicted in Figure 6.8.

As shown in Figure 6.8a, since the numbers and kinds of atoms and bonds in the C_4H_9Br isomers 1-bromo-2-methylpropane [isobutyl bromide, $(CH_3)_2CHCH_2Br$]

---

*While the order provided is correct for simple alkyl-substituted carbocations, it should be clear that special circumstances might cause changes. For example, if alkyl groups are replaced by electron-rich substituents, carbocations that bear them enjoy enhanced stability. Alternatively, if a carbocation is so constrained by other structural features that it *cannot* become planar and thus profit from the alkyl groups it bears, then its relative stability will suffer.

[†]The link between intermediate stability and raction rate is made through a device called the **Hammond Postulate** (G. Hammond, (1921–2005), Iowa State University, California Institute of Technology and University of California, Santa Cruz), which argues that the structure of the species to be found at the maximum in the pass through the mountainous potential energy surface (i.e., the **transition state**) over which reactants must pass on their way to product, resembles the structure of the nearest stable species. Thus, the transition state leading to tertiary carbocation will be lower than that leading to primary carbocation. (See Hammond, G. S. *J. Am. Chem. Soc.*, **1955**, 77, 334.)

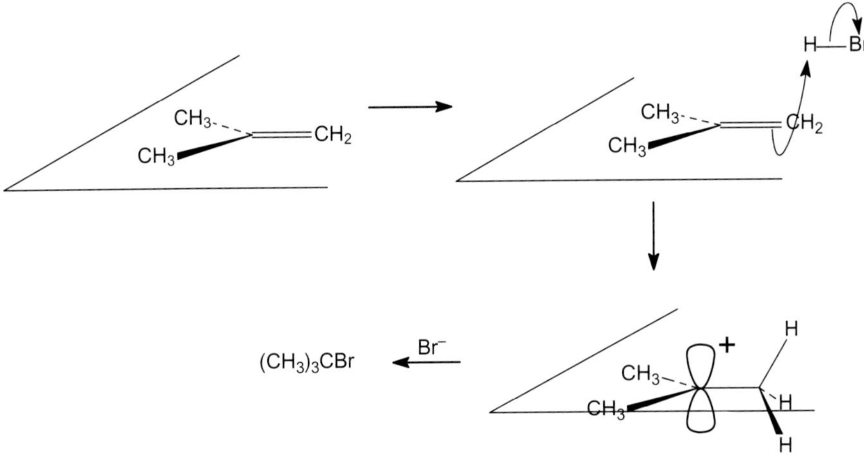

Scheme 6.18a. A depiction of a potential **tertiary** carbocation lying on the path from the reactants hydrogen bromide (HBr) and 2-methylpropene [isobutylene, $(CH_3)_2C=CH_2$] to the product 2-bromo-2-methylpropane [t-butylbromide, $(CH_3)_3CBr$]. The latter is the *only* product formed in the reaction under ionic conditions.

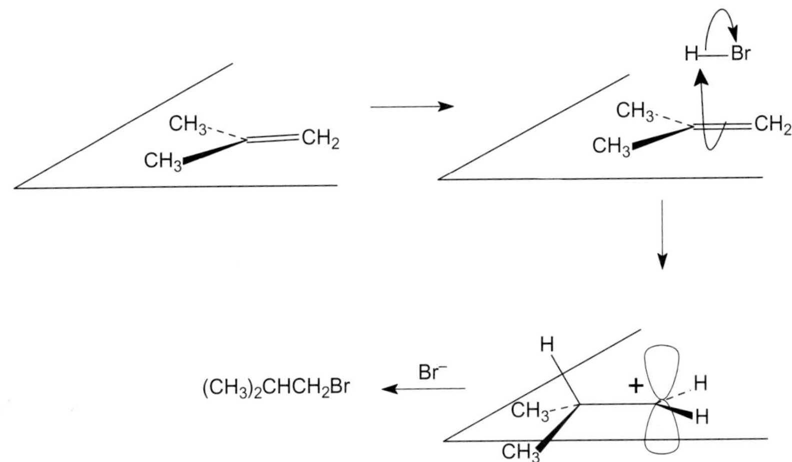

Scheme 6.18b. A depiction of a potential **primary** carbocation lying on the path from the reactants hydrogen bromide (HBr) and 2-methylpropene [isobutylene, $(CH_3)_2C=CH_2$] to 1-bromo-2-methylpropane [isobutylbromide, $(CH_3)_2CHCH_2Br$]. The latter is *not found* in the reaction.

and 2-bromo-2-methyl-propane [t-butylbromide, $(CH_3)_3CBr$] are very similar, their energies (as measured by, e.g., heats of formation) should also be very similar. Thus, having begun with the same alkene, and ending at nearly the same place, the overall energetics should be about the same. Indeed, that they are not suggests that the determining feature is the *height of the rate determining transition state*($\ddagger_1$).

Despite the difference in energies of the rate determining transition states, it should be clear that given enough time and a normal distribution of energies, the

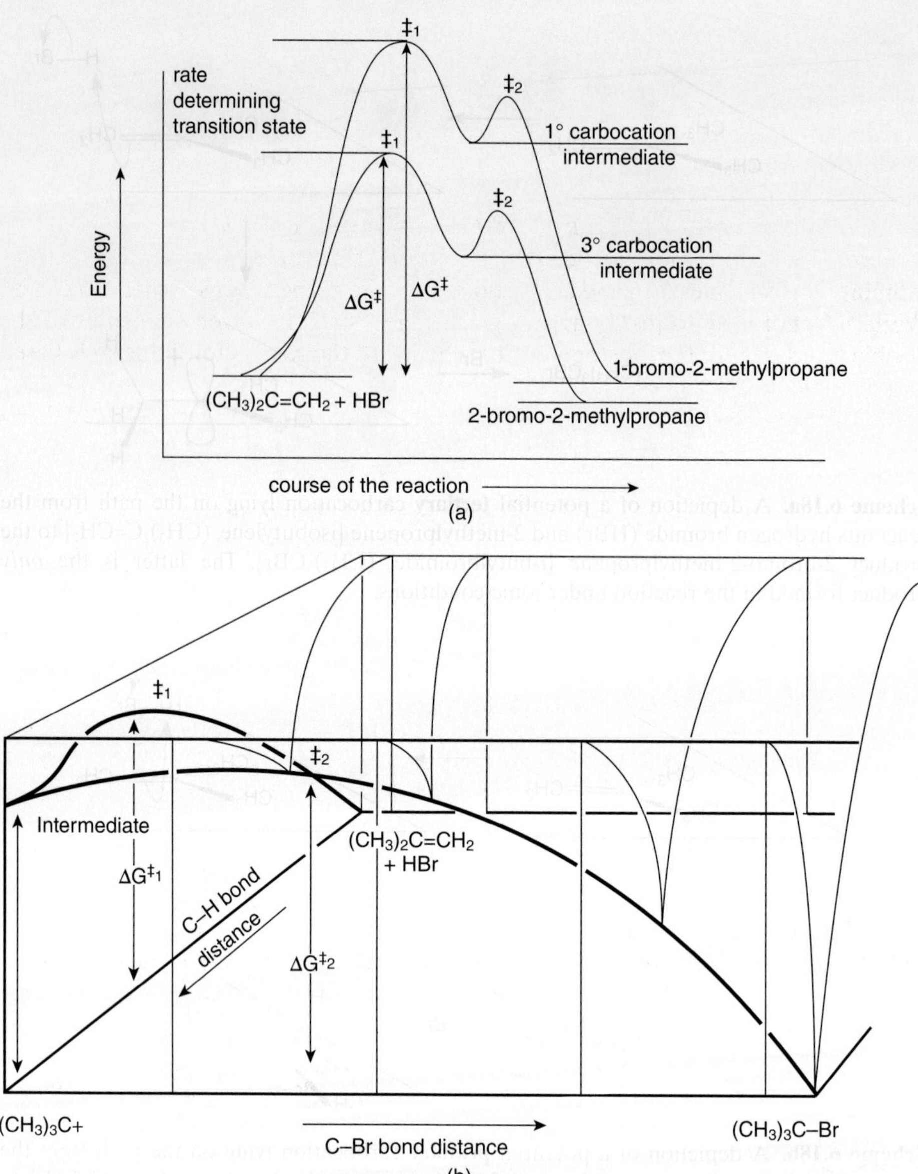

Figure 6.8. (a) A plot of total energy *versus* the course of the addition of hydrogen bromide (HBr) to 2-methylpropene [isobutylene, $(CH_3)_2C=CH_2$]. In accord with the **Hammond Postulate** the upper curve shows that the transition state ($‡_1$) leading to a (high-energy) *primary* (1°) carbocation intermediate is postulated to be higher in energy than the transition state ($‡_2$) leading to a (lower energy) *tertiary* (3°) carbocation intermediate. A different representation (showing some of the potential surface) of the lower curve is shown in Figure 6.8b. (b) A representation of a potential energy surface for the addition of hydrogen bromide (HBr) to 2-methylpropene [isobutylene, $(CH_3)_2C=CH_2$] to yield 2-bromo-2-methylpropane [*t*-butyl bromide, $(CH_3)_3CBr$]. The lower curve of Figure 6.8a is a "flattened" representation of the path the reaction takes as shown on this surface.

path traversed by starting materials to products *must also be followed in the reverse direction.* This principle of *microscopic reversibility,* which argues that all of the infinite numbers of intermediate states linking reactants to products are the same in either direction generally does not intrude into the isolation of the kinetic product because for that product, the reaction is interrupted before equilibrium is achieved. However, as we will see, advantage can be taken of the different products formed under conditions of *kinetic control* and *thermodynamic control.*

It will not have escaped notice that a bridged structure was written for the bromonium ion (Schemes 6.16 and 6.17) and not for the carbocations (Scheme 6.18a,b). While it is not easy to justify avoiding a bridged structure over the center of the double bond (where the electon density is highest), the reason for doing so is based largely on (1) the result of calculations, which indicate that the difference in energy of the gas phase species (bridged protonium ion *vs.* nonbridged carbocation) is small; and (2) the *experimental* observations that when the addend has nonbonding electrons *and is capable of bridging,* the only addition products are those of antarafacial (or *trans-* or *anti-*) addition. The sense of addition following protonation (a proton having no nonbonding electrons) is more commonly dependent upon the exact conditions of the reaction. Consider the following example: when the bicyclodecene isomer (bicyclo[4.4.0^{1,6}]decene) shown in Equation 6.29 is treated with bromine in carbon tetrachloride (CCl_4) or a variety of other solvents, the exclusive product is the dibromide resulting from antarafacial (or *trans-* or *anti-*) addition. However, when hydrogen chloride (HCl) in dichloromethane (CH_2Cl_2) solution is added to the same alkene (Equation 6.30), the result is quite different from when it is added in diethyl ether [($CH_3CH_2)_2O$] (Equation 6.31). Presumably, in dichloromethane (CH_2Cl_2), the carbocation cannot separate as well from the anion (Cl^-) as it can in ether [($CH_3CH_2)_2O$].* Further, carbocations (in contrast to bromonium ions) are frequently found to have *rearranged* before a nucleophile adds. Thus, as shown in Scheme 6.19, protonation of 3,3-dimethyl-1-butene [($CH_3)_3CCH=CH_2$] would be expected to lead to a *secondary* carbocation with which the nucleophile (X^-) might react. However, the major product (the exact quantity depending upon the nuclophilicity of X^-) is derived from the *tertiary carbocation* arising from *methide* (CH_3^-) *migration.*

$$(6.29)$$

$$(6.30)$$

---

*It is very important to understand that this discussion is for purposes of illustration only. Indeed, it suffers most notably from the fact that kinetics for the reactions shown have not been reported.

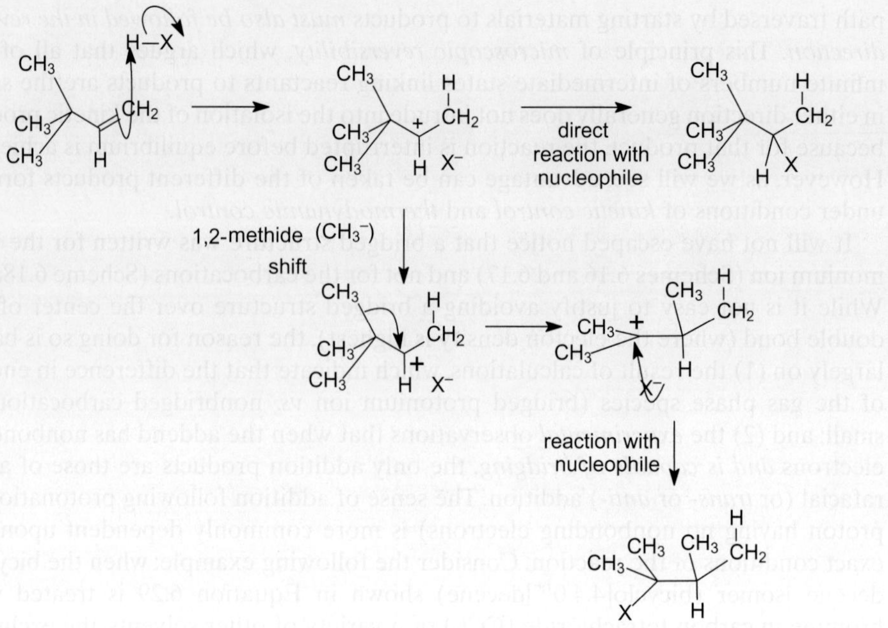

Scheme 6.19. A representation of the protonation of 3,3-dimethyl-1-butene [(CH$_3$)$_3$CCH=CH$_2$] that leads to a *secondary* (2°) carbocation with which the nucleophile (X$^-$) reacts, **as well as** to rearrangement by 1,2-shift of a methide (CH$_3^-$) to produce a *tertiary* (3°) carbocation—which is then attacked by the nucleophile.

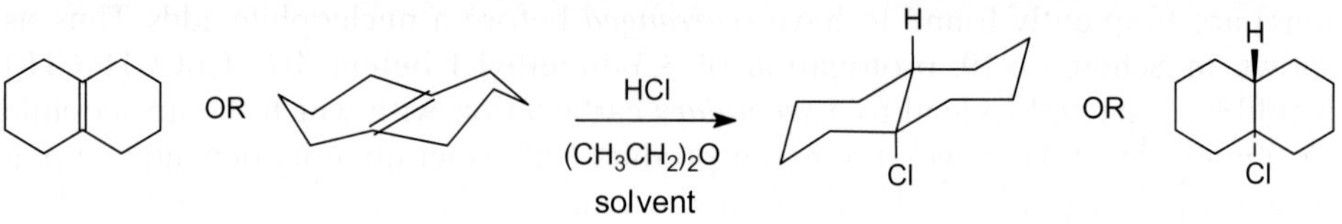

$$(6.31)$$

In addition to complications introduced by the skeletal rearrangement shown in Scheme 6.19 and related processes such as hydride (H$^-$) shifts leading to more stable carbocations and thus initially unanticipated products, the early literature of organic chemistry even contained disagreement on the regiochemistry of the addition of hydrogen bromide (HBr) to simpler alkenes. In some circumstances the rule first adumbrated by Markownikoff (*vide supra*) and now recognized as a statement regarding the relative stability of the first formed (kinetic) carbocation seemed to apply while in others it failed. It was recognized that the problem largely dealt with terminal alkenes *and only with hydrogen bromide (HBr) and not hydrogen chloride (HCl) or hydrogen iodide (HI)*. In 1930, M. Kharasch (University of Chicago) showed that under *ionic* conditions the anticipated Markownikoff sense was observed while in the presence of certain impurities [shown to be traces of peroxides (RO)$_2$] the *opposite regiochemistry* obtained because a *free radical reaction* intruded. Thus, as shown in Scheme 6.20 for **anti-Markownikoff** addition, hydrogen abstraction *from* hydrogen bromide (HBr) occurs and the bromine atom generated then

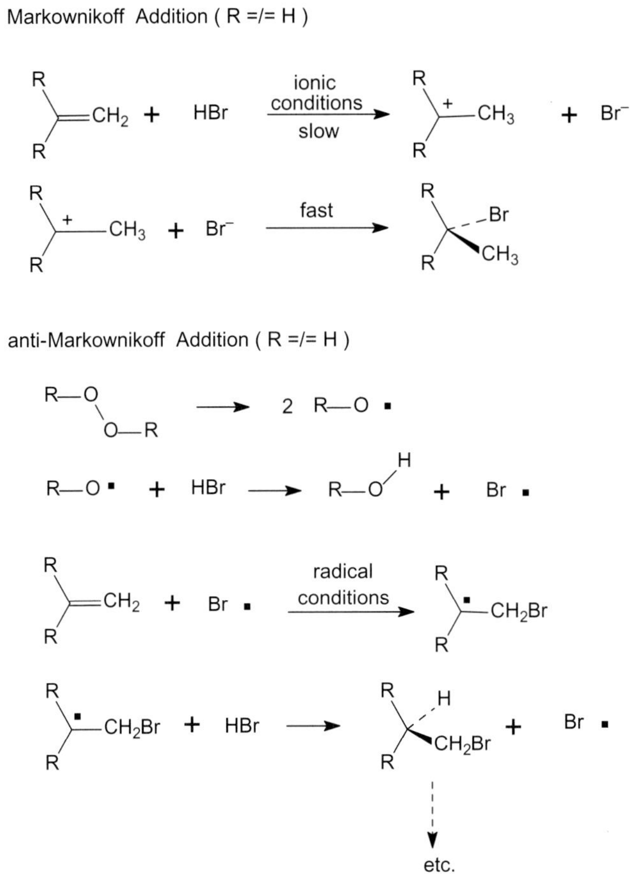

Markownikoff Addition (R =/= H)

anti-Markownikoff Addition (R =/= H)

etc.

Scheme 6.20. A representation of the ionic and radical pathways for reaction of hydrogen bromide (HBr) with a generic terminal alkene [$R_2CC=CH2$] showing both "Markownikoff" and "anti-Markownikoff" products.

reacts at the exposed terminus, generating a tertiary (3°) radical [more stable than a primary (1°) radical] that subsequently abstracts a hydrogen atom from hydrogen bromide (HBr) with formation of another bromine atom. Thus, **anti-Markownikoff** (or "abnormal" or "peroxide effect") product can be formed in a radical process that continues by the generation of one bromine atom for each that is consumed—in a chain of reactions that, once initiated, grows as long as reagents are present to be consumed.

Returning to 3,3-dimethyl-1-butene [$(CH_3)_3CCH=CH_2$] (Scheme 6.19), it is now possible to consider what products are actually obtained from reaction of this alkene with hydrogen bromide (HBr) as the conditions are permitted to vary from ionic (where both unrearranged, i.e., 2-bromo-3,3-dimethylbutane [$CH_3CH(Br)C(CH_3)_3$] and rearranged i.e., 2-bromo-2,3-dimethylbutane [$(CH_3)_2C(Br)CH(CH_3)_2$] products are obtained) to radical (producing 1-bromo-3,3-dimethylbutane [$BrCH_2CH_2C(CH_3)_3$]). These products, all of which can be obtained in the same reaction mixture, are shown in Scheme 6.21.

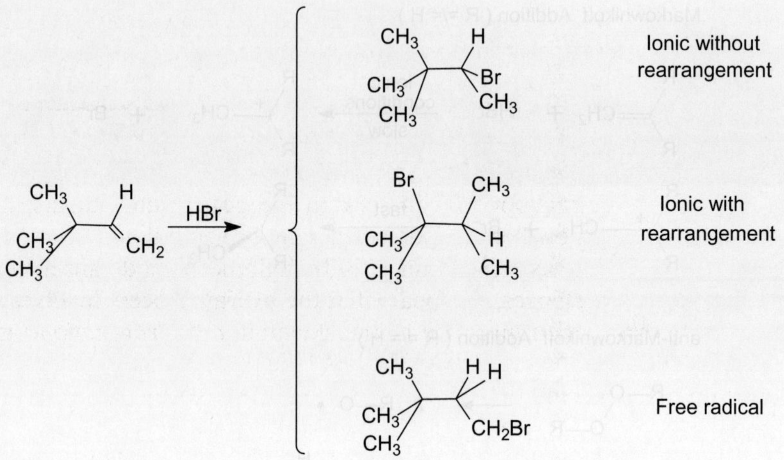

Scheme 6.21. A summary representation of the products obtained on reaction of 3,3-dimethyl-1-butene [$(CH_3)_3CCH=CH_2$] with hydrogen bromide (HBr).

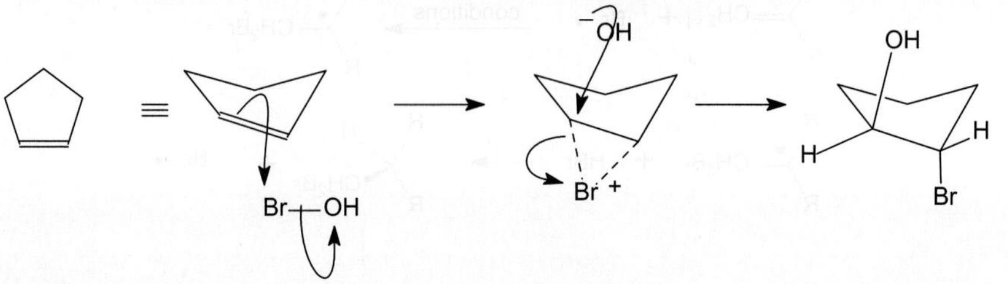

Scheme 6.22. A representation of the path by which hypohalous acids in general and hypobromous acid (HOBr) in particular reacts with alkenes in general and cyclopentene (cyclo-C_5H_8) in particular. Note that the addition is antarafacial.

Problem 6.7. Presume that a set of conditions intruded such that treatment of 3,3-dimethyl-1-butene [$(CH_3)_3CCH=CH_2$] with hydrogen bromide (HBr) was found to actually yield the *three* possible bromoalkane products, viz. 2-bromo-3,3-dimethylbutane [$CH_3CH(Br)C(CH_3)_3$], 2-bromo-2,3-dimethylbutane [$(CH_3)_2C(Br)CH(CH_3)_2$], and 1-bromo-3,3-dimethylbutane [$BrCH_2CH_2C(CH_3)_3$]. Presume further that the *three* bromoalkanes were cleanly separable by gas chromatography. Sketch the ^1H NMR spectra for the *three* bromoalkanes and the alkene starting material paying particular attention to regions that allow them to be distinguished from each other.

The regiochemistry of addition for the other examples in Table 6.1 is also consistent with formation of the most stable carbocation.

Thus, for the addition of the hypohalous acids (Table 6.1, entry 9, HOX; X = Cl, Br, I) the halogen adds as X^+, forming the expected halonium ion and this is attacked by HO^- (or its equivalent) as shown in Scheme 6.22. The same process obtains for the addition reactions of nitrosyl chloride (NOCl), iodine azide (IN_3), and iodine isocyanate (INCO) (Table 6.1, entries 11, 12, and 13, respectively) where the more nucleophilic end of the ambident nuclophile preferentially attacks the incipent carbocation. Indeed, even the addition of borane [diborane $(BH_3)_2$], although *suprafacial*, follows the same pattern as it is the vacant orbital on boron that initially interacts with the electron-rich alkene π-cloud.

The hydration of alkenes with aqeuous acid (Table 6.1, entry 2; Scheme 6.23) or, more conveniently in the laboratory, by oxymercuration (Table 6.1, entry 6) and reduction (Scheme 6.24), can be usefully compared with oxidative hydroboration (Table 6.1, entry 17; Scheme 6.25) as alternative synthetic techniques for preparation of alcohols.

The use of borane (diborane, B_2H_6) and its derivatives (*vide infra*) for **reduction** (Equation 6.22) as well as for **oxidation** (Scheme 6.25) of alkenes is noteworthy. In this vein it is worthwhile recognizing that, as shown in Scheme 6.26, borane in tetrahydrofuran ($H_3B{\cdot}THF$) behaves as if it were monomeric rather than dimeric (B_2H_6) (Chapter 5). Second, although in both Equation 6.22 and in Scheme 6.25 only one ligand to boron was specified in order to emphasize the *reduction* and

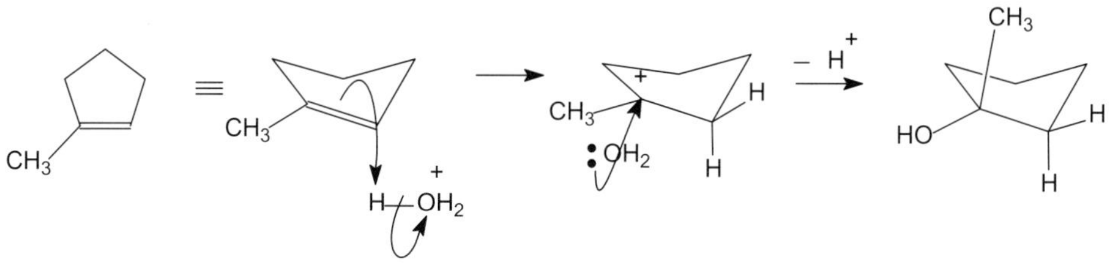

Scheme 6.23. A representation of the path by which aqueous acid (represented as the hydronium ion) in general reacts with alkenes in general and 1-methylcyclopentene in particular. Note that the addition occurs via the most stable carbocation (regiospecifically) but without regard to facial selectivity.

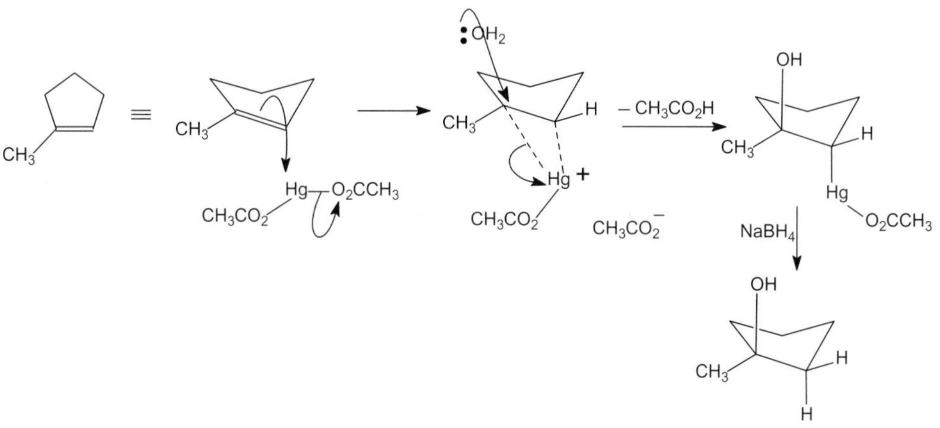

Scheme 6.24. A representation of oxymercuration of alkenes (in general) and 1-methylcyclopentene in particular. Note the regio- and stereochemistry of the *two*-step process. The net result is the addition of water (H–OH) across the double bond in a "Markownikoff sense."

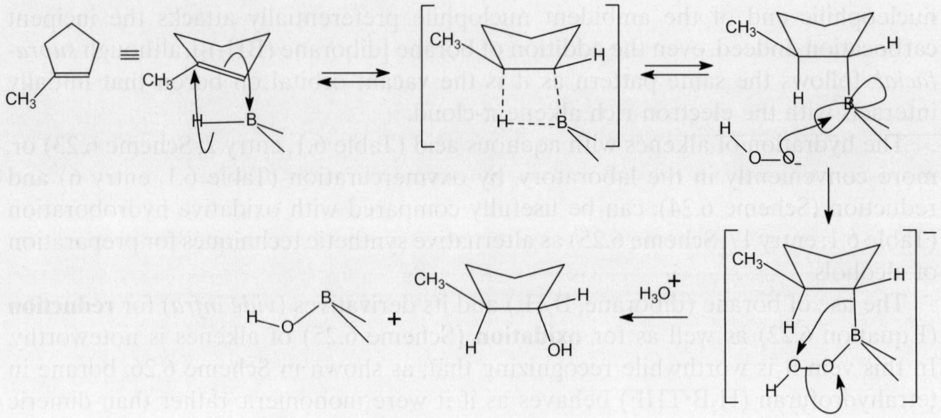

Scheme 6.25. A representation of the *oxidative hydroboration* of 1-methylcyclopentene (as a "typical" alkene). The addition of boron and hydride (H⁻) is suprafacial with the double bond attacking the boron so as to build up charge on that carbon of the double bond that would be the most stable carbocation; it is to this carbon of the double bond that the hydride (H⁻) adds. In the oxidation step, the anion of hydrogen peroxide (HO₂⁻) attacks the boron. Subsequent rearrangement of carbon to oxygen with oxygen–oxygen bond cleavage produces the alcohol. *Only one ligand to boron is shown to emphasize the oxidation process.*

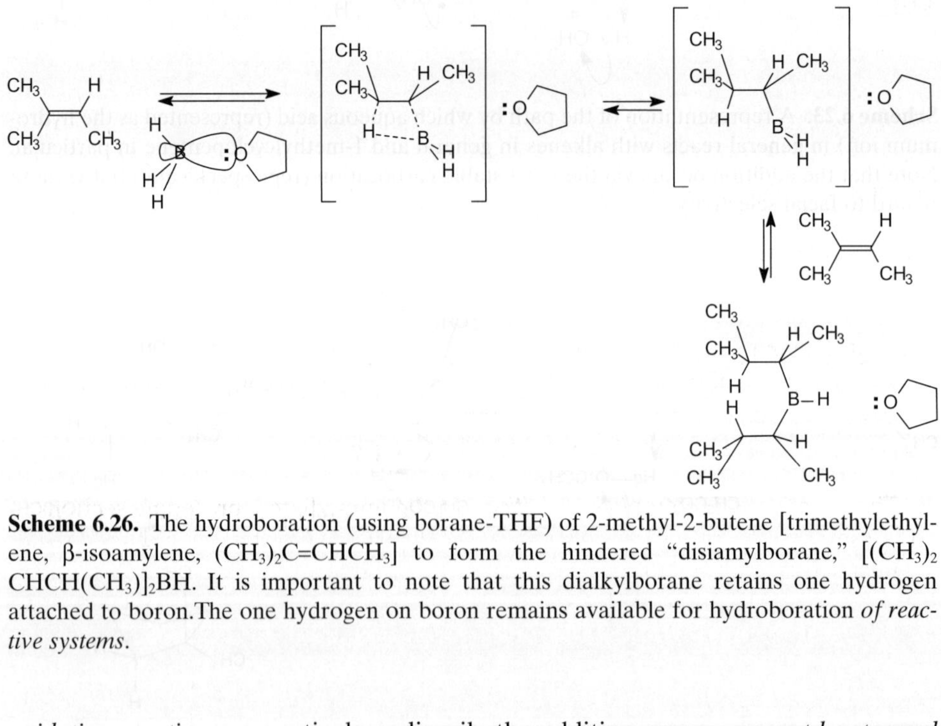

Scheme 6.26. The hydroboration (using borane-THF) of 2-methyl-2-butene [trimethylethylene, β-isoamylene, $(CH_3)_2C=CHCH_3$] to form the hindered "disiamylborane," $[(CH_3)_2CHCH(CH_3)]_2BH$. It is important to note that this dialkylborane retains one hydrogen attached to boron. The one hydrogen on boron remains available for hydroboration *of reactive systems.*

oxidation reactions, respectively, ordinarily the addition process *cannot be stopped with the addition of one alkene.* Thus, the RBH_2 (R ≠ H) reacts with a second alkene to produce R_2BH (R ≠ H) and, unless hindered sterically, with a third alkene to produce R_3B (R ≠ H). Thus, in general, in both reduction (Equation 6.22) and oxidation (Scheme 6.25) reactions, *the remaining ligands on boron are second and third addition products to alkene reactant.*[*]

A variety of useful materials, hindered at boron and yet capable of delivery of hydride, can be prepared in this way. Although their utility will become clearer as additional functional groups with which they can interact are discussed, both disiamylborane $\{[(CH_3)_2CHCH(CH_3)]_2BH\}$ (Scheme 6.26) and 9-borabicyclo[3.3.1] nonane (9-BBN) (Equation 6.32) form dimers that readily dissociate so that they behave as monomeric species in the presence of reactive alkenes. Experimental determination of the kinetics (*vide infra*) of hydroboration are much simplified in this way.

$$(6.32)$$

Further, it should be clear that if a *chiral* alkene such as the terpene α-pinene (Scheme 6.27) is used, and if steric effects are important, then the specificity of addition would result *in a chiral borane reagent.* Use of the *chiral borane reagent,* in, for example, oxidative hydroboration, would then produce a chiral alcohol.

Finally, because the adduct of borane with an alkene is formed *reversibly,* when the temperature is raised, borane is lost and *the least sterically hindered alkene results.* As shown in Scheme 6.28, the initially formed adduct, at temperatures above *ca.* >150°C, produces (reversibly) an alkene, isomeric with the original. The readdition and elimination reoccurs until the double bond has migrated to the terminus. Three possibilities then intrude, *viz* (a) borane is removed and the isomeric alkene is isolated; (b) oxidative hydroboration is effected to produce a primary alchol; or (c) reduction can be produced by addition of a carboxylic acid.[†]

3. The Kinetics of Electrophilic Addition.
In the immediately preceding sections (1 and 2) some of the details of the stereo- (e.g., suprafacial vs. antarafacial) and regiochemistry (e.g., more highly *vs.* less highly substituted terminus) *actually* resulting from the attack of the relatively electron-rich (and yet weak) π-bond between two carbon atoms on a sampling of some electrophilic addends have been enumerated.

[*]A significant amount of the chemistry of alkyl boranes was pioneered by Professor H. C. Brown (1912–2004), former Wetherill Research Professor Emeritus at Purdue University who received the Nobel prize for his work in 1979. See, for example, Brown, H. C.; Chandrasekharan, J.; Wang, K. K. *Pure and Appl. Chem.,* **1983**, *55*, 1387.
[†]Addition of a carboxylic acid in which the acidic proton has been replaced by deuterium (^2H) i.e., $CH_3CO_2{}^2H$, then allows production of the corresponding mono-deuterated alkane!

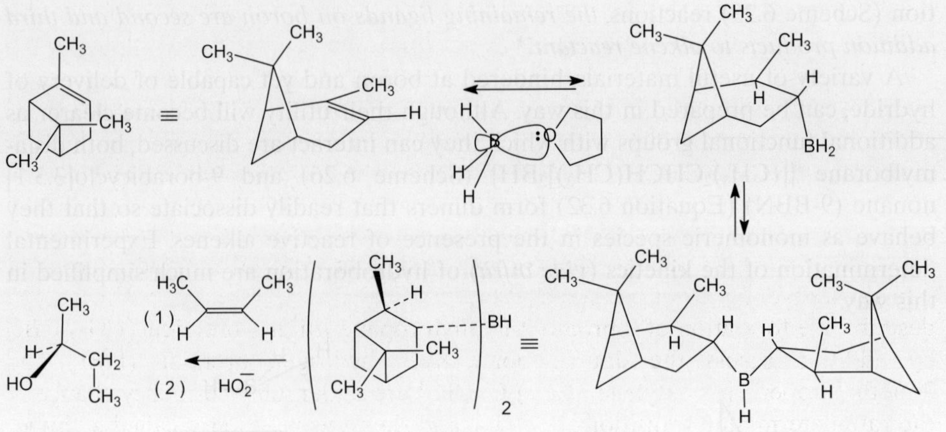

Scheme 6.27. A representation of the reaction of borane [$^1/_2(B_2H_6)$] with α-pinene to produce a chiral hyborating reagent which, on reaction with 2-butene, generates a chiral 2-butanol. The extent of enantiomeric excess (ee), as well as the specific enantiomer of 2-butanol produced, depends on the conditions of the reaction and the sense of chirality of the α-pinene.

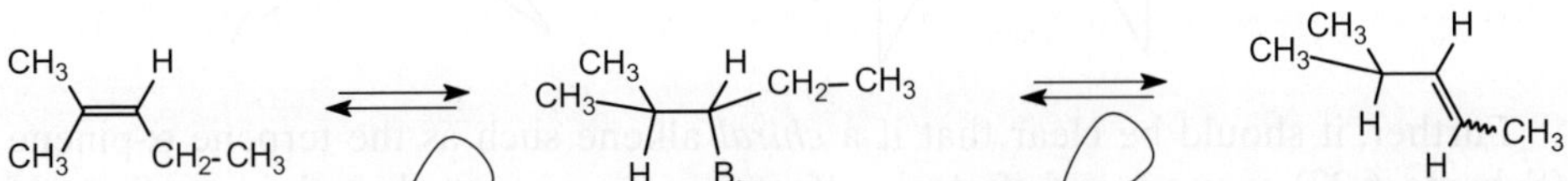

Scheme 6.28. A representation of the hydroboration of 2-methyl-2-pentene at elevated temperature where hydroboration is reversible. The net result is the migration of the boron to the least crowded carbon atom from which alkene, alcohol, and alkane (depending upon subsequent reactions) can be obtained.

Some evidence (from e.g., ^{1}H and/or ^{13}C NMR and/or IR spectroscopies) was provided to lend credence to the reported products (rather than some others) being actually obtained. In addition, pathways involving "curved arrows," designed to represent the "movement of electrons," were shown to "justify" the products obtained and some suggested "rules" or "postulates" were adumbrated.

All of the drawings with arrows that connect starting materials to products are representations of the potential "mechanisms" by which the former are converted to the latter. The ideas that show **the numbers of molecules necessarily undergoing covalency change in the rate-determining transition state** lead to the concept of **"molecularity."** For example, in the pictures of Scheme 6.18a and Figure 6.8b, which describe the formation of 2-bromo-2-methylpropane [*t*-butyl bromide, $(CH_3)_3CBr$] by addition across the double bond, **both** hydrogen bromide (HBr) and 2-methylpropene [*iso*butylene, $(CH_3)_2C=CH_2$] are undergoing covalency change in the rate-determining transition state ($\ddagger_1$), and thus, this reaction is called a **"bimolecular addition."** This is theory.

However, as pointed out in very general terms in Chapter 4, the **experimental** rate at which any reaction occurs *must* be a function of the number of species (ions, complexes, molecules, etc.) passing over that transition state (the *rate-determining transition state*) lying between starting materials and products during some time periods. Therefore, consider (as shown below) the combination of parts of Scheme 6.18b and of Figure 6.8a, redrawn as Scheme 6.29, for the *ionic* reaction of hydrogen bromide (HBr) with 2-methylpropene [*iso*butylene, $(CH_3)_2C=CH_2$] to yield *only* the tertiary bromide, 2-bromo-2-methylpropane [*t*-butyl bromide, $(CH_3)_3CBr$]. As

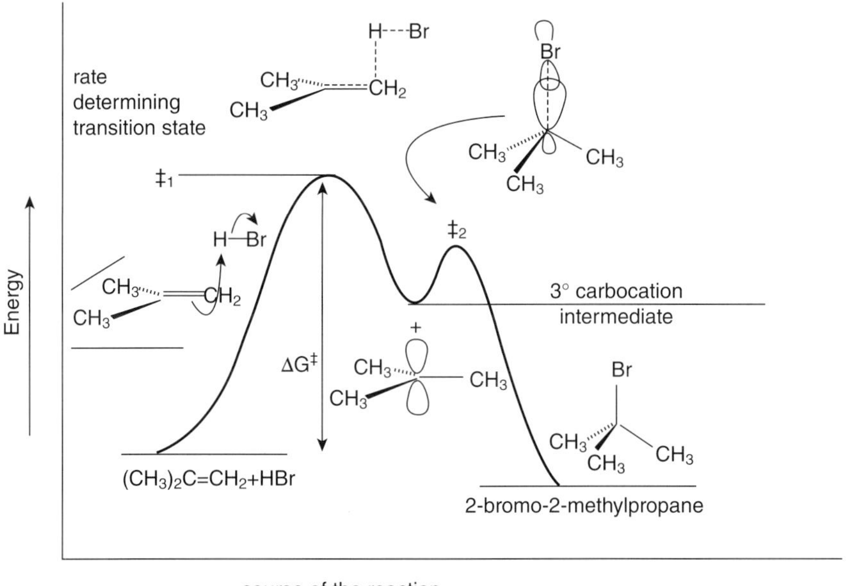

Scheme 6.29. A representation of pathway for the reaction between 2-methyl-2-propene [$(CH_3)_2C=CH_2$] and hydrogen bromide (HBr) to form 2-bromo-2-methylpropane (*t*-butyl bromide [$(CH_3)_3CBr$]) that shows representations for the transition states and intermediates along the way.

shown in Scheme 6.29, the highest transition state ($\ddagger_1$) lying on the path between the starting materials, alkene [2-methylpropene, *iso*butylene, $(CH_3)_2C{=}CH_2$] and hydrogen bromide (HBr), and product haloalkane [2-bromo-2-methylpropane, *t*-butyl bromide, $(CH_3)_3CBr$] contains *both* alkene and hydrogen bromide. It can be imagined that at this point ($\ddagger_1$) the conversion of the starting materials to transition state is half-way consummated. Thus, the carbon–carbon double bond can be thought of as now being half-way between a double bond, with which we began, and the single bond in the symmetrically substituted carbocation of the intermediate; at the same time, the bond from carbon to the proton, the latter originally committed to the bromine, is now half-way to beccomming a carbon–hydrogen bond, and finally, half-way to staying bonded to the bromine, that is, the bond between bromide anion and the proton to which it was originally attached is half-broken.*

To the extent that the picture (Scheme 6.29) represents what is actually happening in the reaction it should be clear that *both* hydrogen bromide (HBr) and 2-methylpropene [*iso*butylene, $(CH_3)_2C{=}CH_2$] are *present in the rate-determining transition state*. Therefore, both of the reactants should be represented in the expression for the rate (Equation 6.33) and, experimentally, this can be determined by measuring the rate at some specific temperature at which product forms (or *either* starting material disappears) as a function of the concentrations of the individual reactants. For the Equation 6.33, *et seq* as usual, the expressions in brackets are concentration terms (mol L^{-1}); k_o is observed rate constant; and the exponents "n" and "m" are *experimentally* determined values whose sum (n + m) is the "order" of the reaction.

$$-d[HBr]/dt = -d[alkene]/dt = +d[product]/dt = rate = k_o[HBr]^n[alkene]^m \quad (6.33)$$

Interestingly, examination of the contents of Table 6.1 might lead to the conclusion that all (or almost all) of the reactions shown there might be broadly described as "bimolecular" and, perhaps, thus "second order." At the risk of being repetitious, "molecularity" is a **concept** to which all (or almost all) of the reactions of Table 6.1 might subscribe. "Order" is the sum of the exponents in the concentration terms of the rate equation. Concepts aside, "order" is found as the result of an **experiment** or series of experiments. It is the difference between dreams and reality.

Now, assume that the generalized alkene of Table 6.1 is called "A" and a generalized electrophile (all of the addends of Table 6.1) is called "E" and the generalized product resulting from the electrophilic addition reaction (all of the products of Table 6.1) is called "P."

Then, for the general case, and following Equation 6.33,

$$-d[E]/dt = -d[A]/dt = +d[P]/dt = rate = k_o[E]^n[A]^m \quad (6.34)$$

*Although these bonds can be considered as "half-way committed" and "half-way uncommitted" to their "new bonding partners," it should be recognized that this is for illustration only and, in that vein, a "half-made" carbon–hydrogen (C–H) bond is not the same as a "half-made" bromine–hydrogen (Br–H) bond. Thus, the electronegativities of, and electron distribution around, carbon and bromine are quite different and thus, bonds to them could be more or less made or broken in the real transition state. However, for discussion purposes, it is convenient to think of the transition state as resembling the one pictured.

It is common to determine "n" and "m" by increasing the amount of one of the two reactants to such an extent that its concentration remains "unchanged" over the course of the reaction and then undertaking to observe the effect on the rate of the reaction produced by changes in the concentration of the other reactant. A doubling in concentration for one of the reagents for a reaction that is first order in that reagent results in a doubling of the rate. If, however, the reaction is second order in that reagent, the rate increases as the square of the concentration term.

Now, assuming that $n = m = 1$ (a reaction that is first order in each component is second order overall) so that for each equivalent of alkene (a $mol\,L^{-1}$ of A) that is consumed one of electrophile (e $mol\,L^{-1}$ of E) is also consumed and one equivalent of product (p $mol\,L^{-1}$ of P) is formed then Equation 6.34 can be rewritten as Equation 6.35, that is,

$$+dp/dt = rate = k_o(e-p)(a-p) \tag{6.35}$$

Although it need not generally be true, it is often the case that the starting concentrations of alkene and electrophile are set the same so that

$$(e-p) = (a-p) = c \tag{6.36}$$

and

$$-dc/dt = k_o c^2 \tag{6.37}$$

and, separating the variables,

$$\int_{c_0=0}^{c=c} dc/c^2 = -k_o \int_{t=0}^{t=t} dt \tag{6.38}$$

which leads to

$$1/c = 1/c_o = k_o t \tag{6.39}$$

and a plot of $1/c$ versus t is linear, the slope being k_o. Further, since $c = (a - p)$ and $c_o = a$,

$$\{1/(a-p)\} - \{1/a\} = k_o t \tag{6.40}$$

Finally, multiplying numerator and denominator by $(a - p)$ and rearranging

$$p/a(a-p) = k_o t \tag{6.41}$$

Of course, it is possible that the initial concentrations (a $mol\,L^{-1}$ of A and e $mol\,L^{-1}$ of E) might not be set equal. Then, returning to Equation 6.35, the variables can be separated

$$dp/\{(a-p)(e-p)\} = k_o dt \tag{6.42}$$

which can be broken up into partial fractions

$$k_o \int dt = \int dp/\{(a-p)(e-p)\} = \int dp/\{(a-p)(e-a)\} + \int dp/\{(e-p)(a-e)\} \tag{6.43}$$

and integrated to yield

$$[\ln(a-p) - \ln(e-p)]/(a-e) = k_o t + \text{constant} \tag{6.44}$$

and, when $t = 0$, $p = 0$, and thus const $= \ln[(a/e)/(a-e)]$, so

$$1/(a-e)\ln[e(a-p)/a(e-p)] = k_o t \tag{6.45}$$

While using the above expressions, it is necessary to keep in mind that as the concentration changes the reaction mechanism may change (and, indeed, any mechanism may only be useful for a few half-lives, if only because as the starting materials are used up, the likelihood of reactants encountering each other diminishes). In addition, nothing has been said about *how* the concentrations are measured (i.e., How might you determine the concentration of {a} hydrogen bromide [HBr]; {b} 2-methylpropene [$(CH_3)_2C=CH_2$]; and {c} 2-bromo-2-methylpropane [$(CH_3)_3CBr$] in ways that do not alter them as they are being measured) or about the solvent in which this reaction is being run (and how that might effect the reaction). Thus, it is important to realize that considerable effort has been expended to determine some of the details of the mechanism of each of the reactions in Table 6.1. Indeed, although much is known and is generally in concert with what has been presented in Table 6.1, as more is learned, the answers may change and significant gaps yet remain.

For the material presented in Table 6.1, it is generally true that, at least over some of the concentration ranges examined, the reactions appear to be first order in alkene and first order in addend. Thus, in the addition of hydrogen bromide (HBr) to 1-methylcyclohexene (in acetic acid solvent), products can be accounted for as shown in Scheme 6.30.

It is of interest to examine the kinetics of hydroboration of the same alkene as shown in Scheme 6.30. It has generally not proved possible to use diborane itself for kinetic measurements. However, as shown in Scheme 6.31, the dimer of 9-borabicyclo[3.3.1]nonane (9-BBN), (i.e., [(9-BBN)$_2$]) can be used. As predicted, the reaction is half-order in (9-BBN)$_2$ and first order in alkene.

4. Cationic Polymerization: Electophilic Addition in the Absence of a Reactive Nucleophile. In the presence of catalytic amounts of acidic materials (protic as well as nonprotic) lacking nucleophilic counterions, the attack of the electron-rich alkene on the electrophile appears to occur with the formation of a carbocation as expected but, as a nucleophilic anion is missing, another *alkene* attacks the carbocation (Table 6.1, entry 19). The result is the formation of a **new carbocation**, larger than the original by *one alkene unit* (Scheme 6.32 R ≠ H). The process continues until termination is effected (e.g., by proton loss).

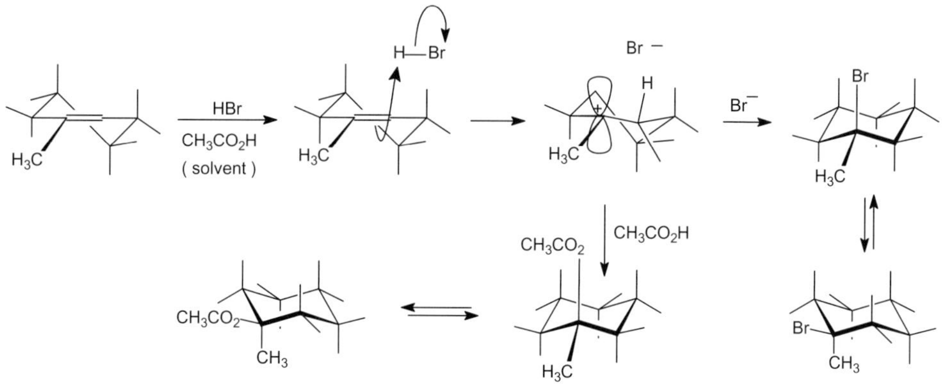

Scheme 6.30. A representation of the addition of hydrogen bromide (HBr) to 1-methylcyclohexene in acetic acid (CH_3CO_2H), that is, under acidic conditions, showing that the solvent intrudes into the electrophilic addition reaction. The intrusion of solvent will complicate the kinetic arguments. Initially, the formation of 1-bromo-1-methylcyclohexane is first order in alkene and first order in hydrogen bromide (HBr) only at low (HBr). The kinetics become much more complicated as the reaction progresses.

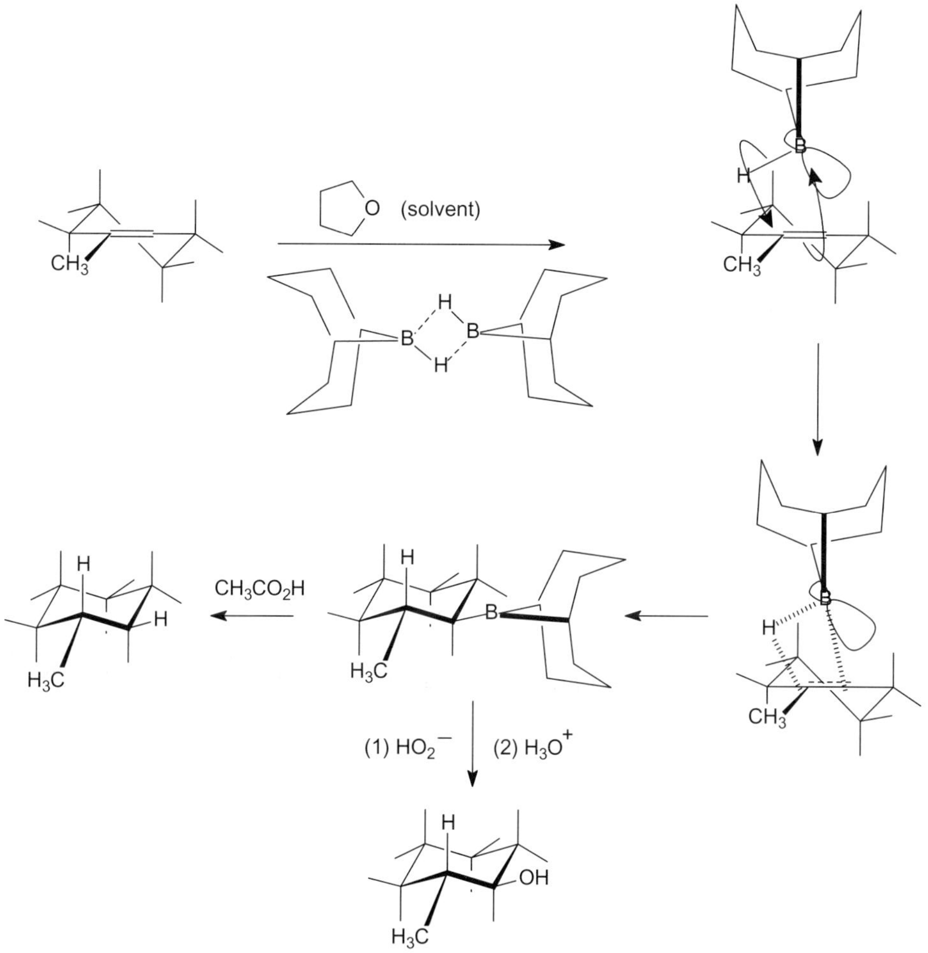

Scheme 6.31. A representation of the pathway for hydroboration of 1-methylcyclohexene with the dimer of 9-borabicyclo[3.3.1]nonane (9-BBN) (i.e., [9-BBN]$_2$), which yields (a)(*E*)- or (*trans*)-2-methylcyclohexanol oxidatively (hydrogen peroxide anion, HO_2^-), or (b) methylcyclohexane, reductively (acetic acid, CH_3CO_2H). The kinetics of hydroboration are in accord with rapid, reversible formation of 9-BBN monomer from the dimer *prior* to the rate-determining step. (See Brown, H.C.; Chandrasekharan, J.; Wang, K. K. *Pure Appl. Chem.*, **1983**, *55*, 1387.)

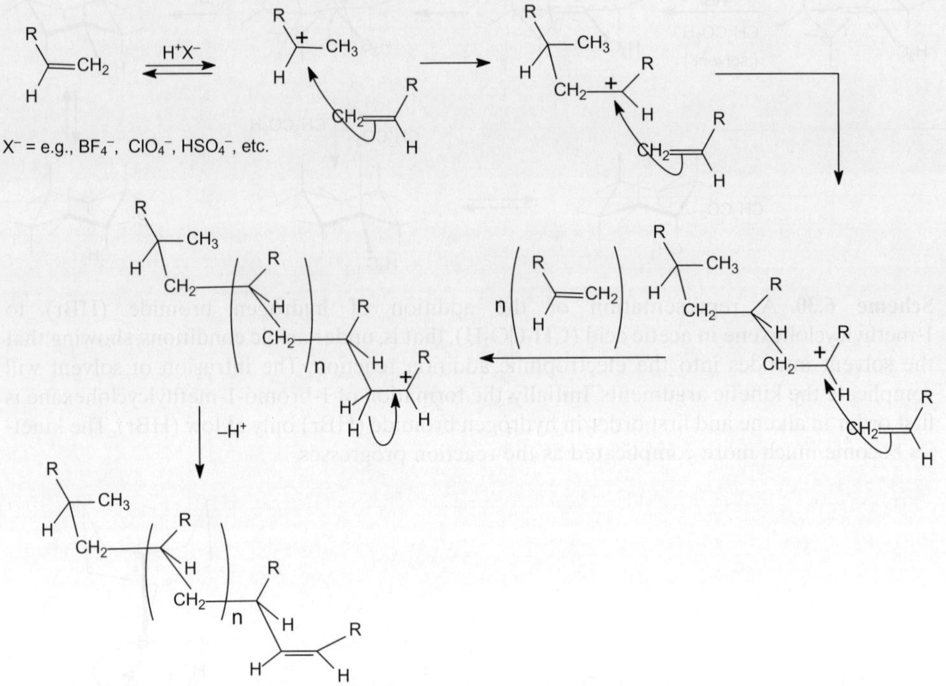

Scheme 6.32. Polymerization of an alkene: R ≠ H; X^- is non-nucleophilic; n = a large number that is concentration- and condition-dependent.

Generally, polymerization can only be effected when the double bond is substituted with alkyl or other groups that stabilize carbocations. This is because the barriers to both the initial protonation and the growing polymer chain bearing the carbocation are lower for substituted alkenes. Thus, ethene (ethylene, $CH_2=CH_2$) and alkenes bearing electron-withdrawing groups (EWGs) (such as halogen) cannot generally be polymerized with acid (**radical** and **anionic** polymerizations, *vide infra*, do not suffer the same drawback).

When a single alkene (**monomer**) is polymerized, there are identical repeating units. This is a **homopolymer**. When a mixture of different monomers is polymerized, a **copolymer** results. Depending upon the relative quantities of the different monomers used, different polymers with different properties can be realized.

With substituted alkenes such as 2-methylpropene [*iso*butylene, $(CH_3)_2C=CH_2)$] and a catalyst such as sulfuric acid under anhydrous conditions, long polymer chains can be formed. In this way 2-methylpropene [*iso*butylene, $(CH_3)_2C=CH_2$] forms a sticky polymer (*poly*isobutylene) with a very high molecular weight.

Interestingly, when aqueous sulfuric acid (≈60% sulfuric acid by weight) rather than concentrated sulfuric is used, this four-carbon alkene, 2-methylpropene [*iso*butylene, $(CH_3)_2C=CH_2$], begins to *poly*merize but, before the *third* equivalent of alkene can react, a proton is lost. Thus, a mixture of *di*mers (two eight carbon alkenes), viz. 2,4,4-trimethyl-1-pentene [$(CH_3)_3CCH_2C(CH_3)=CH_2$] and 2,4,4-trimethyl-2-pentene [$(CH_3)_3CH=C(CH_3)_2$]), results. The catalytic (Pt, Ni) hydrogenation (H_2) of the mixture of octenes yields 2,2,4-trimethylpentane ["*iso*octane," $(CH_3)_3CCH_2CH(CH_3)_2$] (Scheme 6.33).

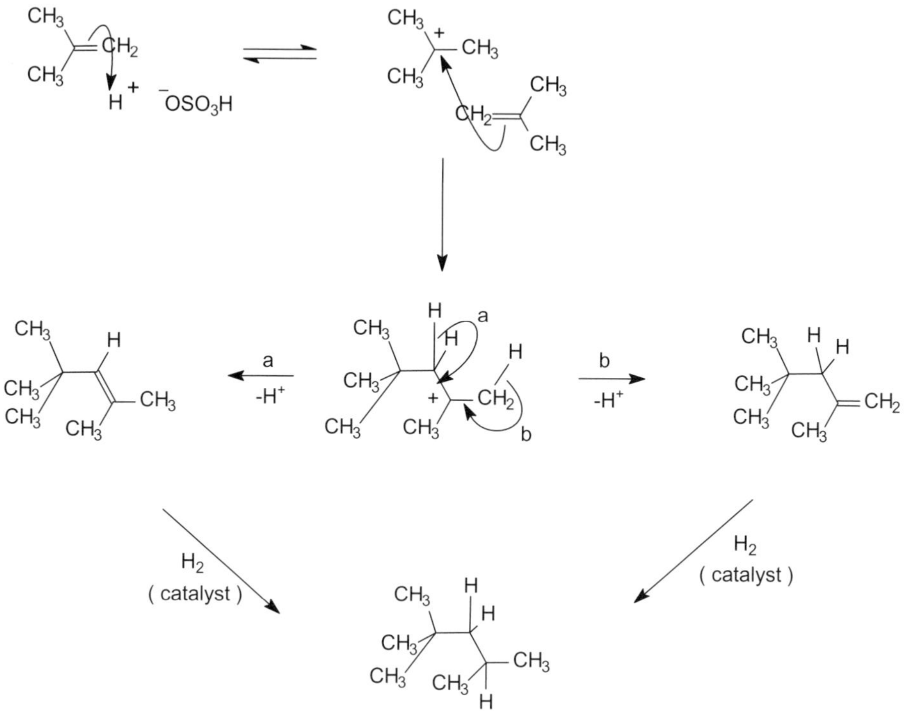

Scheme 6.33. A representation of the path of the dimerization of 2-methylpropene [*iso*butylene $(CH_3)_2C=CH_2$] to form a carbocation that loses a proton (H^+) to give the two alkenes, (path a) 2,4,4-trimethyl-2-pentene $[(CH_3)_3CH=C(CH_3)_2]$ and (path b) 2,4,4-trimethyl-1-pentene $[(CH_3)_3CCH_2C(CH_3)=CH_2]$. The catalytic (Pt, Ni, etc.) reduction (hydrogenation) of the alkene mixture yields 2,2,4-trimethylpentane [isooctane, $(CH_3)_3CCH_2CH(CH_3)_2$].

Problem 6.8. As shown in Scheme 6.33, the dimerization of 2-methylpropene [*iso*-butylene $(CH_3)_2C=CH_2$] results in the formation of 2,4,4-trimethyl-1-pentene $[(CH_3)_3CCH_2C(CH_3)=CH_2]$ and 2,4,4-trimethyl-2-pentene $[(CH_3)_3CH=C(CH_3)_2]$. These isomers can be separated by gas chromatography. The separated isomers were examined by ^1H NMR spectroscopy, which quickly showed which was which. Sketch the spectra that were obtained.

As might be anticipated, aprotic acids such as titanium tetrachloride $(TiCl_4)$, ferric chloride $(FeCl_3)$, aluminum chloride $(AlCl_3)$, and some of the organometallic materials (alkylaluminum dihalides, dialkylaluminum halides, etc.) as well as mixtures of these materials can also be used to catalyze polymerization. The literature of organic chemistry is replete with trials of different materials used under different reaction conditions to produce different polymers. However, at the beginning of the second half of the twentieth century, K. Ziegler and G. Natta* were able to revolutionize the entire subject when they found that catalysts (now called *Ziegler–Natta*

---

*K. Ziegler (Max Plack Institute for Coal Research) and G. Natta (Polytechnic Institute of Milan); Nobel Prize 1963 for the discovery and use of titanium-based catalysts in stereoregular polymerization processes.

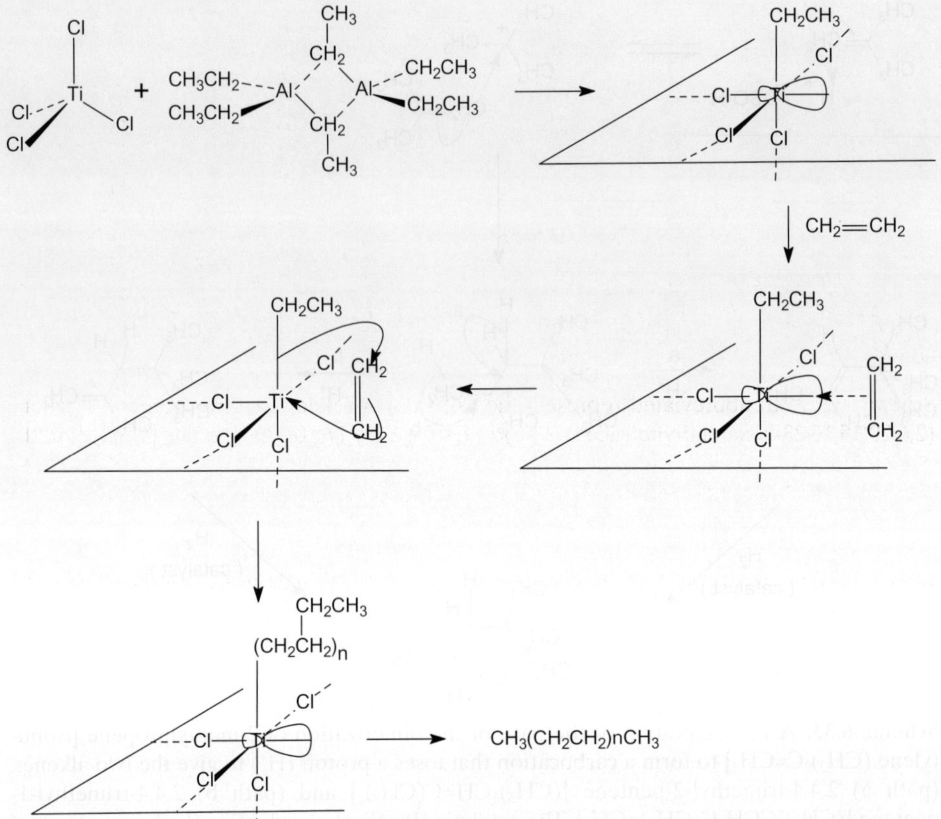

Scheme 6.34. A representation of the polymerization of ethene (ethylene, $CH_2=CH_2$) by a catalyst composed of the Lewis acid mixture of titanium tetrachloride ($TiCl_4$) and triethylaluminum [$(CH_3CH_2)_3Al$]: a *Natta–Ziegler Catalyst*. The actual catalytic species is (presumably) an alkyl surface-substituted octahedral titanium trichloride ($TiCl_3$) derivative with a vacant site capable of coordination with ethene (ethylene, $CH_2=CH_2$) and thence, by linear addition of alkene units, "polyethylene" [$CH_3(CH_2CH_2)_nCH_3$].

catalysts) could be prepared from a *mixture* of a transition metal halide (such as titanium tetrachloride [$TiCl_4$]) and a metal alkyl such as triethylaluminum [$(CH_3CH_2)_3Al$].

The catalyst itself is apparently a form of titanium trichloride ($TiCl_3$, commonly octahedral), which is formed on reduction of the tetrachloride by the alkyl aluminum. Some surface sties of the solid catalyst bear an alkyl group and a vacant site (Scheme 6.34). The polymer that forms is an unbranched linking of ethylene units.

5. Electrophilic Addition to Dienes and Polyenes. The tools utilized to examine the reactions of alkenes in Sections 1–3, viz. stereochemistry, regiochemistry, and kinetics, as well as the particular reaction of an alkene with an electrophile in the absence of a reactive nucleophile (i.e., polymerization), can now be applied to the reactions of the same reagents with dienes. In this regard, it should be clear that for isolated dienes or polyenes, that is, alkenes with more than one double bond and in

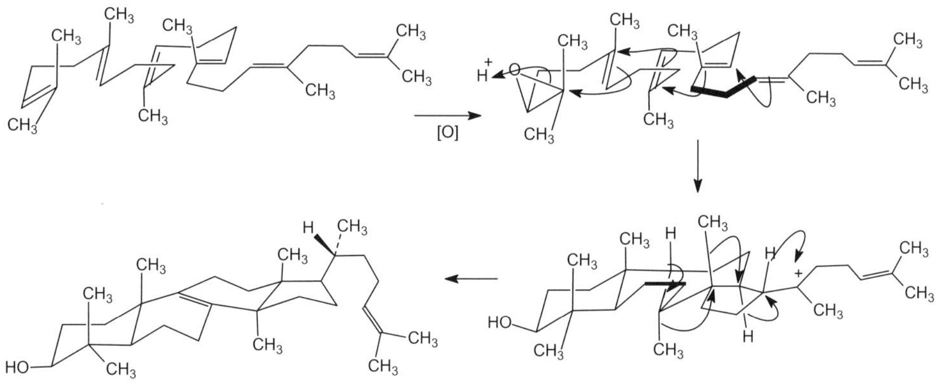

Scheme 6.35. An abbreviated representation of the conversion of the triterpene squalene (2,6,20,15,19,23-hexamethyltetracosa-2,(E)6, (E)10, (E)14, (E)18, 22-hexaene [$C_{30}H_{50}$]) to the plant sterol, lanosterol. This is included here to illustrate the cooperation in reactivity between "isolated" double bonds. See Problem 6.9 for a laboratory example of a similar process.

which at least one methylene (CH_2) group lies between the double bonds, the double bonds may be thought of as behaving independently from each other—although, of course, as noted below, when they are held in the appropriate geometry they do interact with dramatic consequences.

For example, as shown in Scheme 6.35, in the presence of the appropriate catalyst(s) it is known (Sir John Cornforth, et al., *vide supra*) that the triterpene squalene (2,6,20,15,19,23-hexa-methyltetracosa-2,(E)6,(E)10,(E)14,(E)18, 22-hexaene [$C_{30}H_{50}$]) undergoes oxidation (forming an epoxide) and then via a zipper-like cascade, with one double bond adding to the next, cyclization and rearrangement with migration of hydride and methide (or their equivalent) follow to produce lanosterol.

Cumulative double bonds apparently undergo electrophilic addition as expected to form the most stable carbocation. Although relatively little work has been done to examine the details of the reaction with electrophiles (because compounds with cumulative double bonds are rare), it is nevertheless interesting that addition of hydrogen chloride (HCl) to 1,2-propadiene (allene [$CH_2=C=CH_2$])* produces 2-chloropropene [$CH_2=C(Cl)–CH_3$] rather than the isomeric 3-chloro-1-propene [$CH_2=CH–CH_2Cl$] (Scheme 6.36). Thus, it appears that a secondary vinylic carbocation ($CH_2=C^+CH_3$) is preferred to a primary allylic carbocation ($CH_2^+CH=CH_2$) where the empty orbital is out of phase with the filled π-system.

Problem 6.9. Given that the Lewis Acid stannic chloride [tin (IV) chloride {$SnCl_4$}] can induce dehydration of alcohols to produce carbocations (much as protonation of alkenes can accomplish the same end), write arrows on the (presumed) tertiary carbocation produced from the alcohol below to account for the product observed. (See, Fish, P. V.; Johnson, W. S. *J. Org. Chem.*, **1994**, *59*, 2324 and references therein.)

*In 1,2-propadiene (allene, $CH_2=C=CH_2$) the carbon–carbon double bonds are *shorter* (131 pm) than those in isolated alkenes (134 pm in ethene [ethylene, $CH_2=CH_2$]) but *longer* than in alkynes (120 pm in ethyne [acetylene, HCCH]). 1,2-Propadiene (allene, $CH_2=C=CH_2$) is about 6.25 kJ mol^{-1} (1.5 kcal mol^{-1}) *less* stable than propyne (methylacetylene, $CH_3C\equiv CH$).

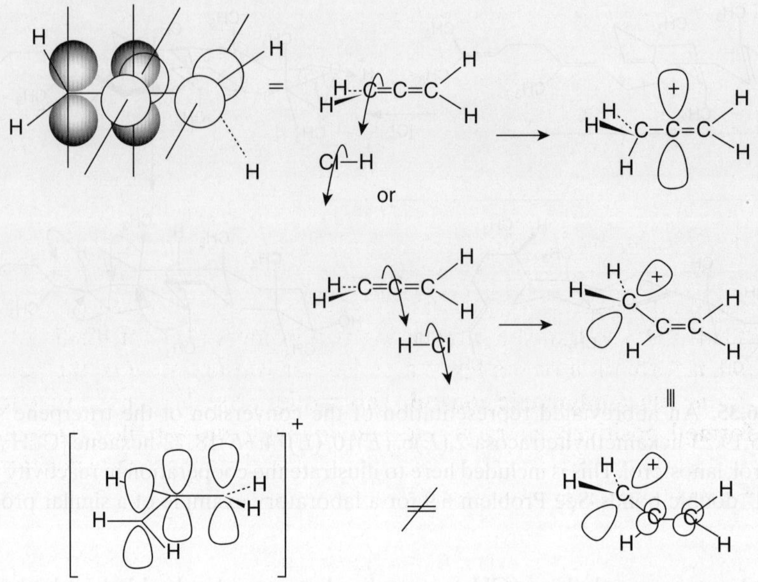

Scheme 6.36. A representation of the protonation of 1,2-propadiene (allene,$CH_2=C=CH_2$) that can lead to either a secondary carbocation or a primary carbocation. The latter is *not* equivalent to an allylic carbocation with orbital overlap across three carbons because the "empty *p*-orbital" is out of phase with the "filled π-orbital" and thus, there is no orbital overlap. Indeed, it has been estimated that the barrier to rotation is about $138\,kJ\,mol^{-1}$ ($33\,kcal\,mol^{-1}$).

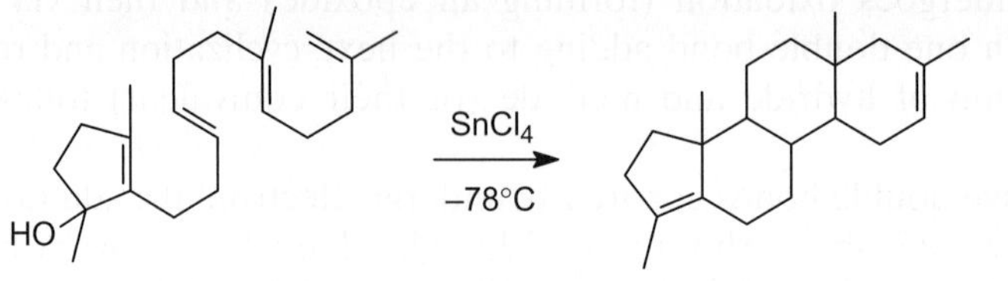

However, in contrast to compounds with either isolated or cumulative double bonds, electrophilic addition reactions with alkenes that contain conjugated π-systems routinely produce products resulting from the interaction of both double bonds. While hydroboration appears to be an exception, with each double bond reacting separately (and the second more rapidly than the first), conjugated dienes, such as 1,3-butadiene ($CH_2=CH-CH=CH_2$), suffer addition across each double bond ("1,2-addition") *as well as across the entire conjugated system* ("1,4-addition"). Commonly, the products ("1,2-addition" and "1,4-addition") are formed concurrently.

As pointed out earlier (Chapter 3), the carbon–carbon double bonds in 1,3-butadiene ($CH_2=CH-CH=CH_2$) are about the same length (134 pm) as normal carbon–carbon double bonds, for example, 134 pm in ethene (ethylene, $CH_2=CH_2$), but the carbon–carbon single bond between carbon atoms C2 and C3 is *shorter* (148 pm) than a normal carbon–carbon single bond, for example, 154 pm in ethane (CH_3CH_3) and this is attributed to both (a) orbital overlap in the lowest occupied MO and (b) the fact that the C2–C3 σ bond may be thought of as made up by overlap of two (shorter) sp^2-type orbitals rather than two (longer) sp^3-type orbitals.

The orbital overlap, it can be argued, is related to the fact that *two* sets of products are formed on electrophilic addition to such conjugated dienes. Thus, for example, when a solution of bromine (Br_2) in hexane (C_6H_{14}) is added to a cold ($-15°C$) solution of 1,3-butadiene (CH_2=CH–CH=CH_2) in the same solvent *both* 3,4-dibromo-1-butene [CH_2=CHCH(Br)CH_2Br] *and* (*E*)-1,4-dibromo-2-butene [(*E*)-BrCH$_2$CH=CHCH$_2$Br] in a 4:1 ratio (respectively) are formed. Interestingly, as expected, the latter is marginally more stable than the former and, at equilibrium, (*E*)-1,4-dibromo-2-butene [(*E*)-BrCH$_2$CH=CHCH$_2$Br], is favored by about 4.14 kJ mol^{-1} (1 kcal mol^{-1}) so that the ratio of the two isomers is about 15:85.

The *initial* formation of the *less* stable isomer is illustrative of the paradigm of **kinetic** *versus* **thermodynamic control.** This model provides for the reaction of the initally formed, positively charged, intermediate with nucleophile (in a second step) at positions whose relative reactivities differ, with the more reactive position leading to the less stable product and *vice versa*.

The concept is illustrated in Schemes 6.37a,b and 6.38.

The general principle illustrated in these different representations (Schemes 6.37a,b and 6.38) of the same process is that, in reactions in which two (or more) products form simultaneously and reversibly, the ratio of the products actually isolated is a function of solvent, temperature, as well as *time*. Thus, the ratio of products *at equilibrium* may be different because the *rates* of attack in nonrate-determining steps at different sites are different.

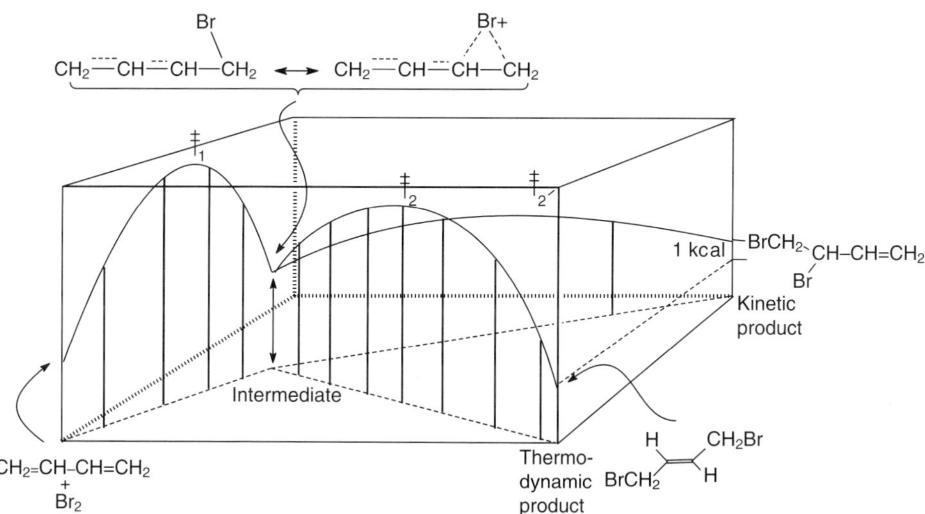

Scheme 6.37a. A representation of the course of the reaction between bromine (Br_2) and 1,3-butadiene (CH_2=CH–CH=CH_2) showing the potential surface and the rate-determining lowest-energy transition state ($\ddagger_1$) leading to a resonance stabilized intermediate. The reaction of the intermediate with bromide anion (Br$^-$) over the lowest-energy transition state ($\ddagger_{2'}$) leading from the intermediate leads to the kinetic product. The transition state leading to the thermodynamic product ($\ddagger_2$) is lower than the rate-determining transition state but higher than that leading to the kinetic product.

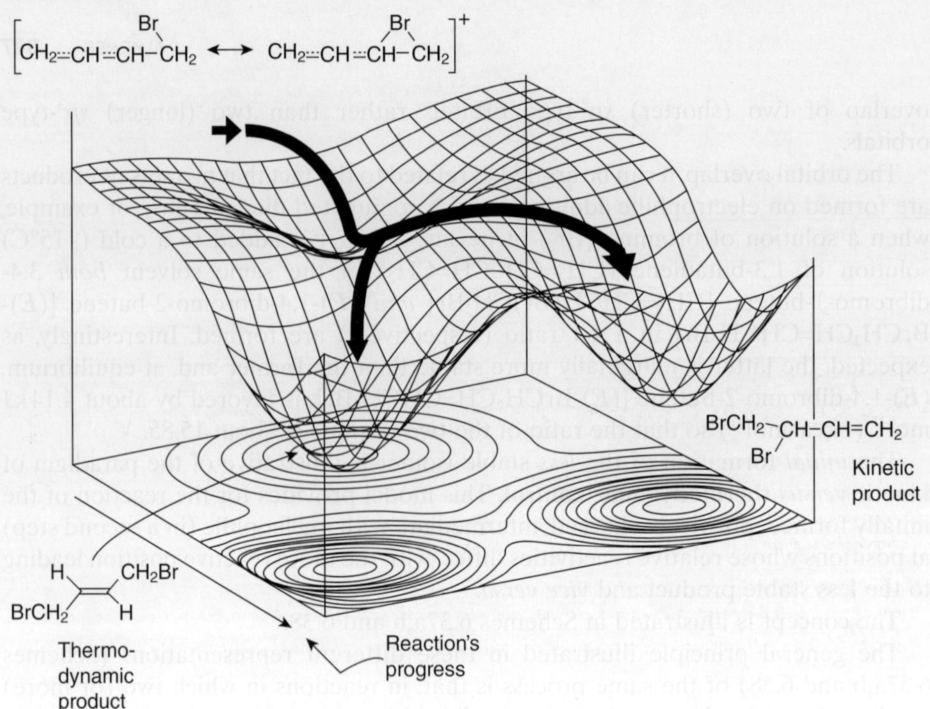

Scheme 6.37b. A representations of the course of the reaction between bromine (Br_2) and 1,3-butadiene (CH_2=CH–CH=CH_2) showing the potential surface and the rate-determining transition state ($\ddagger_1$) leading to a resonance stabilized intermediate. The reaction of the intermediate with bromide anion (Br^-) over the lowest transition state ($\ddagger_{2'}$) leading from the intermediate leads to the kinetic product. The transition state leading to the thermodynamic product ($\ddagger_2$) is lower than the rate-determining transition state but higher than that leading to the kinetic product.

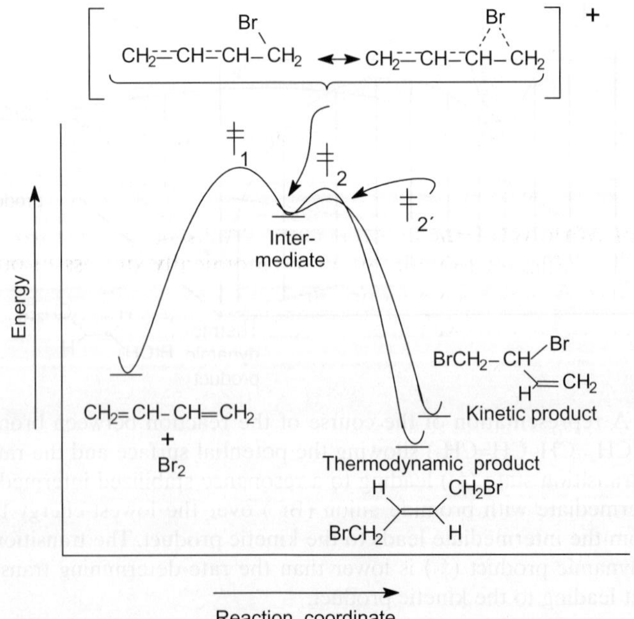

Scheme 6.38. An alternative representation of the reaction shown in Scheme 6.37a and 6.37b. The paths themselves, shorn of supporting framework and flattened to two dimensions, are shown.

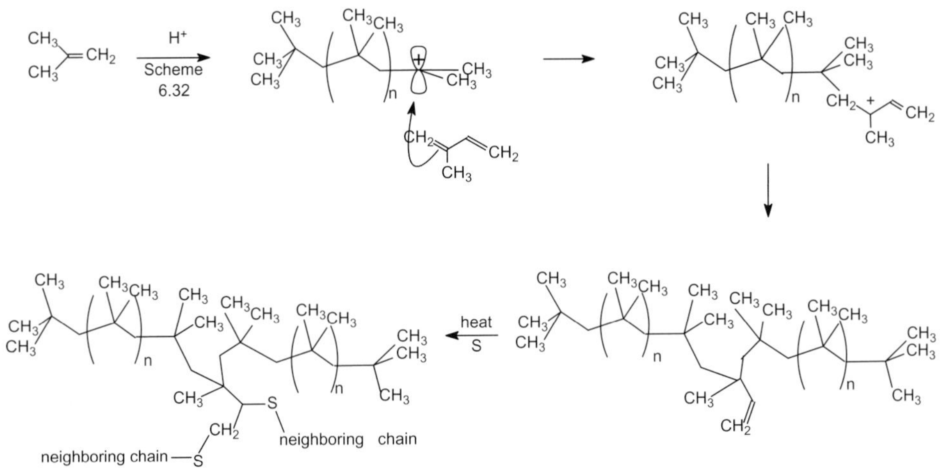

Scheme 6.39. A representation of acid-catalyzed cross polymerization of 2-methylpropene [*iso*butylene, $(CH_3)_2C=CH_2$] with a trace of 2-methyl-1,3-butadiene [isoprene, $CH_2=C(CH_3)-CH=CH_2$] followed by cross-linking chains with sulfur (part of the process of Vulcanization).

The polymerization of dienes is economically important and although it is possible to write pathways using acid catalysis to produce such polymers, anionic (nucleophilic) and, particularly, free radical polymerization (*vide infra*) are much more common. However, acid-catalyzed polymerization is often utilized when dienes are a minor component of *copolymers* of alkene with a diene—to provide "handles" for "cross-linking" the chains.*

For example, as shown in Scheme 6.39 the sulfuric acid-catalyzed polymerization of 2-methylpropene (*iso*butylene [$(CH_3)_2C=CH_2$]) discussed above, (Scheme 6.32) when carried out in the presence of a few percent of 2-methyl-1,3-butadiene (isoprene, [$CH_2=C(CH_3)-CH=CH_2$]) results in a material that retains enough double bonds to allow "chains" of the hydrocarbons to be "cross-linked." The vulcanized material is called "butyl rubber."

6. Special Cases: The Oxo and Ritter Reactions. There are two cases of electrophilic addition that warrant special consideration because of their historic and economic value.

At the end of World War I, the German chemical industry was deprived of quantities of petroleum sufficient to maintain an economically successful posture, so the synthesis of hydrocarbons became a priority.

The resulting **Fischer–Tropsch process** led to both alkanes and alkenes from the reduction of carbon monoxide (CO) by hydrogen (H_2) over a cobalt catalyst (which is a mixture of dicobalt octacarbonyl [$Co_2(CO)_8$] and cobalt tetracarbonyl hydride [$HCo(CO)_4$] and which is used at temperatures over 120°C and pressures above 200 atm), that is,

---

*The cross-linking of chains by heating the polymer with sulfur, graphite, and other additives to modify its properties to produce a material suitable for specific purposes is part of the art of **Vulcanization** of rubber. The details of **Vulcanization** vary from polymer to polymer, depending upon, for example, the specific monomer(s) used in the polymerization, the extent of polymerization, and the properties desired.

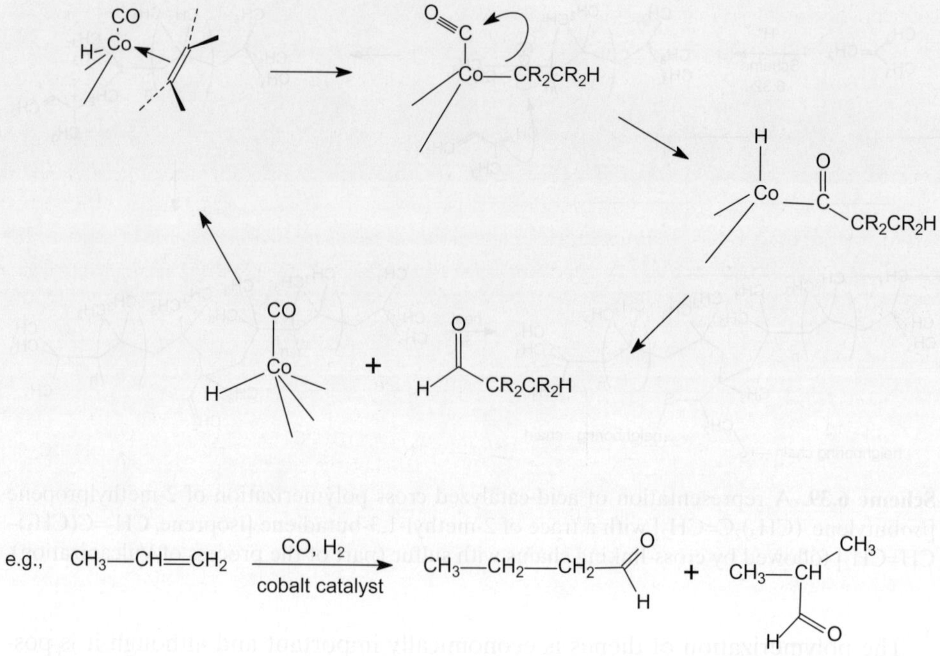

e.g., $CH_3-CH=CH_2$ $\xrightarrow[\text{cobalt catalyst}]{CO, H_2}$ $CH_3-CH_2-CH_2-CHO$ + $(CH_3)_2CH-CHO$

Scheme 6.40. A general representation of the Oxo reaction showing involvement of the cobalt catalyst and a specific example of the reaction of propene ($CH_3CH=CH_2$) under the conditions of the Oxo reaction to produce butanal ($CH_3CH_2CH_2CHO$) and 2-methylpropanal [$(CH_3)_2CHCHO$]. The ratio of the aldehydes in the case of propene can be changed by altering the temperature, pressure, and others.

$$n\,(CO) \; + \; (2n+1)\,H_2 \; \xrightarrow[\text{catalyst}]{\text{cobalt}} \; C_nH_{(2n+2)} \; + \; n\,(H_2O) \qquad (6.46)$$

$$n\,(CO) \; + \; (2n)\,H_2 \; \xrightarrow[\text{catalyst}]{\text{cobalt}} \; C_nH_{(2n)} \; + \; n\,(H_2O) \qquad (6.47)$$

Later, it was found that alkenes underwent reaction with carbon monoxide (CO) and hydrogen (H_2) in the presence of dicobalt octacarbonyl [$Co_2(CO)_8$] catalysts to produce aldehydes (i.e., the alkene has been *hydroformylated*). Presumably (Scheme 6.40) the process involves the coordination of the alkene to the metal (with loss of carbon monoxide, CO) followed by transfer of the carbonyl group to the alkene (with carbon cobalt bond formation) and subsequent reduction. As will be noted later (Chapter 9), aldehydes formed this way can be reduced to alcohols. Thus, additional important chemical intermdiates arise from this, the **Oxo Reaction**.*

In another vein, addition of sulfuric acid to an alkene in the presence of a nitrile (R–C≡N) or hydrogen cyanide (HC≡N), either of which can react with the carbocation as a nucleophile, apparently leads to "nitrilium ions." These ions, on hydrolysis, produce amides. As amides can be hydrolyzed in acid or basic medium to amines and carboxylic acids, the overall process (the **Ritter** reaction)[†] of carbocation

*Cornils, B.; Herrmann, W. A.; Rasch, M. *Angew. Chem. Internat. Ed.*, **1994**, *33*, 2144.
[†]Ritter, J. J.; Minieri, P. P. *J. Am.Chem. Soc.*, **1948**, *70*, 4045 *et seq.*

formation, reaction with nitrile nucleophile, and hydrolysis, as shown in Scheme 6.41, is widely used to prepare both amines (Chapter 10) and "unusual" amides (Chapters 9 and 10) for incorporation into peptides (Chapter 12).

b. Nucleophilic Addition to Alkenes, Dienes, and Polyenes.

Even though the carbon–carbon double bond is electron-rich, suitable nucleophiles might be able to attack it if, for example, appropriate substitution is present on the double bond carbons (Scheme 6.42). Presumably, arguing by analogy to electrophilic addition, the first step would produce a **carbanion** (rather than a carbocation) and, in a second step, the anion would attack some electrophile. Again, by analogy (and given the

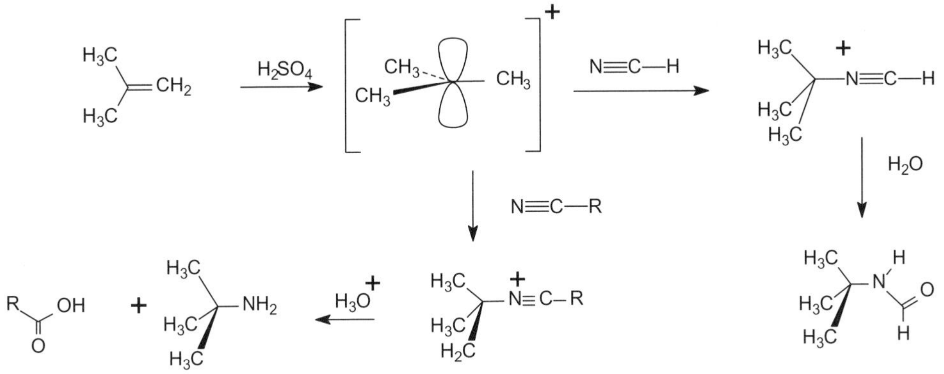

Scheme 6.41. A representation of the Ritter reaction. Hydrolysis of the iminium ion produces either an amide (as shown for reaction with hydrogen cyanide [HCN] producing the formamide of 2-amino-2-methylpropane [*N-tert*-butylformamide (CH₃)₃CNHCHO]) or, further, past the amide to the amine (2-amino-2-methylpropane [2-methyl-2-propanamine, *t*-butylamine, (CH₃)₃CNH₂] and the corresponding (unspecified and thus unnamed) carboxylic acid (R-CO₂H).

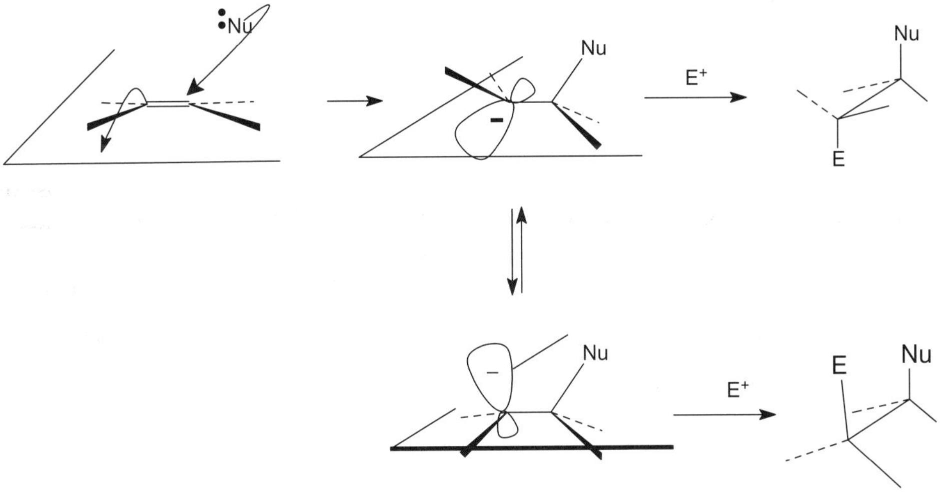

Scheme 6.42. A schematic representation of the addition of a nucleophilic reagent to a carbon–carbon double bond.

need to attack an area of relatively high electron density), the first step might have the highest barrier and thus be rate determining and, because carbanions apparently retain their tetrahedral geometry (in contrast to carbocations, which, barring bridging or other geometric restrictions, tend toward planarity), some preferred geometry might be expected to result.

It should be clear, however, that there might be difficulty in producing a nucleophile (Nu:) both capable, on the one hand, of attacking the "electron-rich" alkene and, on the other hand, resisting reaction with the electrophile necessary to consummate the reaction.

Indeed, these constraints have dictated that *nucleophilic addition to simple, unsubstituted alkenes occurs rarely* (and then under very special circumstances).

Thus, as shown in Table 6.2, there are generally three types of nucleophilic addition reactions to alkenes that succeed. First, there are those alkenes to which are attached *inductively withdrawing substituents*, such as fluorine (F) atoms, for example, tetrafluoroethene (tetrafluoroetylene, CF_2CF_2), and to which potent, but weakly acidic, nucleophiles such as ethanethiol (CH_3CH_2SH) in alcohol solvents add (Equation 6.48). Second, there are those alkenes that have *conjugative EWGs or substituents*. These substituents include, among others, the nitro- (NO_2) and cyano- (CN) groups and, in particular, carbonyl-containing functions (aldehydes, ketones, esters, amides, etc.). Although from Equation 6.49 it appears that the addition has simply occurred across the double bond, it is clear (*inter alia*, from the

TABLE 6.2. Representative Addition Reactions between Nucleophiles and Alkenes

Reactant	Reagent	Catalyst	Product
$F_2C=CF_2$	CH_3CH_2SH	Base to generate anion	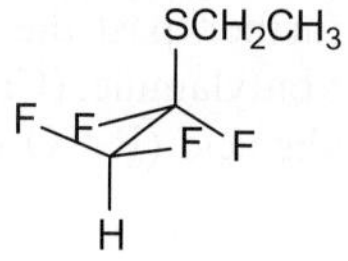
$H_2C=CH_2$	$^-C{\equiv}N$	$Fe(CO)_4$	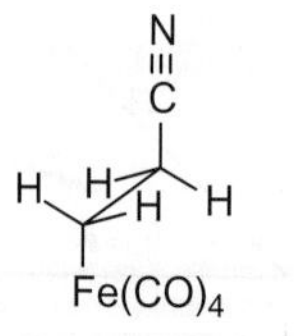
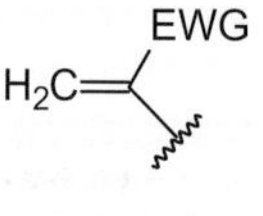 $H_2C=$ EWG	nucleophile (Nu:)		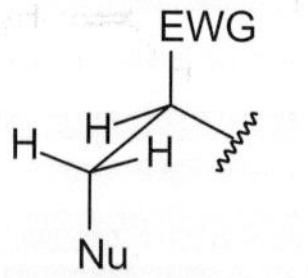

The examples shown are incomplete but representative of those few nucleophilic reactions at the carbon–carbon double bond. The abbreviations Nu: and EWG refer, respectively, to any nucleophile (Nu:) capable of reaction with the alkene bearing a suitable electron-withdrawing group (EWG). Typically, as will be discussed in greater detail (**Chapter 9**), the Nu: species will be carbanionic species and EWG might be any of the group including, but not limited to, cyano (CN), Nitro (NO_2), and any carbonyl-bearing function (aldehyde, ketone, ester, etc.) directly attached to the sp^2-hybridized carbon of the alkene. The latter is known as **Michael Addition**.

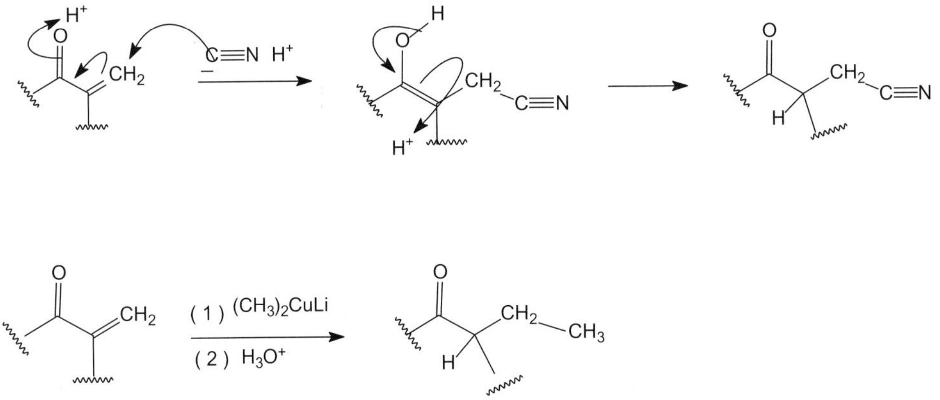

Scheme 6.43 diagrams and Equation 6.48/6.49 figures appear here.

Scheme 6.43. Two examples of nucleophilic addition reactions to alkenes bearing electron-withdrawing carbonyl groups. These addition reactions both result in the formation of new carbon–carbon bonds. The second addition reaction involves use of an **organometallic reagent** (a dialkyl lithium cuprate, sometimes called a **Gilman Reagent** (Chapter 7).

isolation of intermediates) that the addition actually occurs across the conjugated system with subsequent proton transfer (Scheme 6.43). This mode of addition, the **Michael Condensation,*** particularly important with carbanionic nucleophiles because new carbon-carbon bonds are created, will be discussed subsequently (Chapter 9).

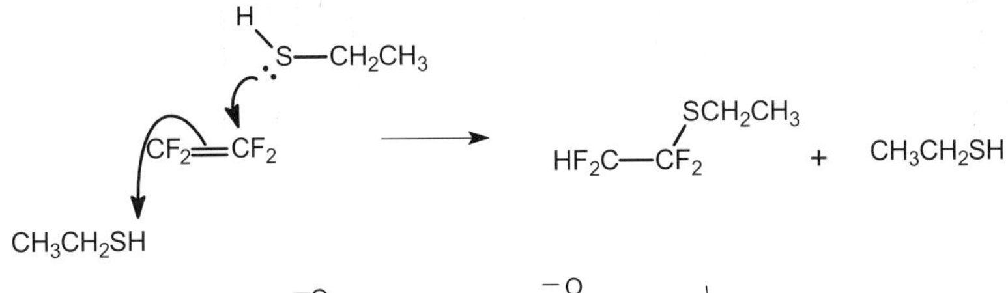

$$(6.48)$$

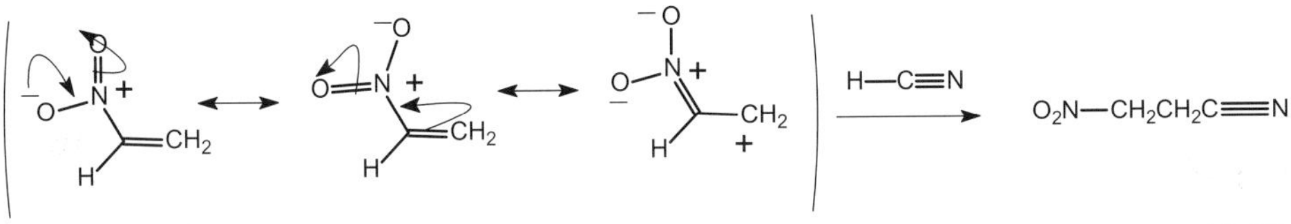

$$(6.49)$$

Finally, complexation of the π-system with empty, low-lying metal orbital(s) presumably induces sufficient electron defficiency at the carbon atoms of the alkene to allow attack by reactive nucleophiles. Again, this is particularly important for the creation of new carbon–carbon bonds (Equation 6.50).

*A. Michael (1853–1942), Professor, Harvard University, (1912–1936).

$$(6.50)$$

When EWGs such as cyano (-CN) are attached to the carbon–carbon double bond (e.g., cyanoethene [acrylonitrile, CH_2=CH-C≡N]) anionic polymerization, initiated by, for example, sodium amide ($NaNH_2$) or butyllithium ($CH_3CH_2CH_2CH_2Li$), can be effected. This polyacrylonitrile (now better made by radical processes, *vide infra*) has been used to make fibers, and materials prepared from them are sold under the trade name Orlon™.

c. Radical Addition to Alkenes, Dienes, and Polyenes.

As already discussed, addition of hydrogen bromide (HBr) to alkenes under ionic (Markownikoff) and radical (anti-Markownikoff) conditions, as shown in Schemes 6.20 and 6.21, respectively, gives different (regiochemistry) products. And yet, for each, the same argument, that is, formation of the most **stable** intermediate, be it carbocation or radical, is made. Although none of the other hydrogen halides (hydrogen fluoride [HF], hydrogen chloride [HCl], and hydrogen iodide [HI]) succeed in radical additions,* a number of other reagents (Table 6.3) provide products by what appear to be one electron transfer processes. Additionally, *free radical polymerization*, that is, the initiation of polymerization of alkenes by free radical initiators such as peroxides (R–O–O–R), has proven to be a commercially important pathway to numerous materials with desirable (i.e., marketable) properties.

Almost all of the reactions in Table 6.3 require the addition of an initiator or the use of light (usually in the UV region of the spectrum) of sufficient energy to induce homolysis of the most fragile bond in the system. Thermal homolysis, when peroxides (RO–OR) are used, is possible only because of the lability of the oxygen–oxygen bond (see Table 1.1, Chapter 1).

Interestingly, chlorine (Cl_2) and bromine (Br_2), which undergo ionic electrophilic addition (Table 6.1), also undergo radical (atom) addition to the double bond in the presence of UV light as an initiator. However, if one or more carbon atoms *next to the double bond*, that is, *allylic*, have at least one hydrogen, and if the stereochemistry permits, it is much more common to observe hydrogen abstraction from the allylic position and subsequent allylic halogenation (Scheme 6.44). To this end, and because ionic addition often competes with the radical process, reagents have been

---

*The production of bromine atoms (Br) from hydrogen bromide (HBr) via hydrogen (H•) abstraction by another radical (e.g., RO•) is exothermic. The difference in energy between breaking the carbon–carbon double bond and forming a carbon–bromine bond is small and the subsequent formation of the carbon–hydrogen bond exothermic, thus making the overall process favorable. However, the hydrogen fluorine bond in hydrogen fluoride (HF) (about $564\,kJ\,mol^{-1}$ [$135\,kcal\,mol^{-1}$]) is too strong to be easily broken by the alkoxy radical and, for hydrogen iodide (HI), the bond formed, eventually, between carbon and iodine, is too weak. Hydrogen chloride (HCl) seems to be more equally balanced thermochemically but free radical reactions, if they occur at all, are rare.

TABLE 6.3. Representative Reactions between Free Radicals and Alkenes

Reagent	Product
HBr (free radical initiator)	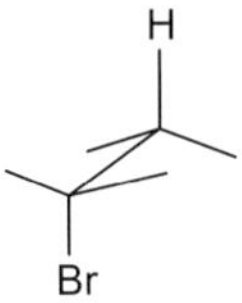
H_2S	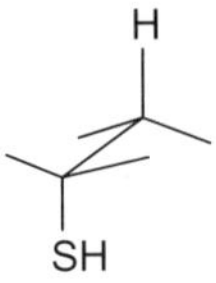 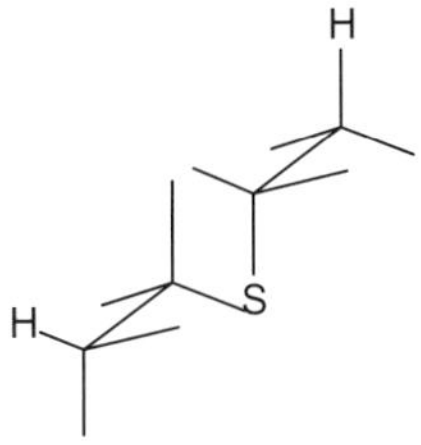
X_2 (X = F, Cl, Br, I) (with UV light)	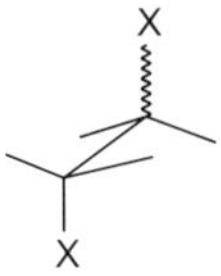
CX_4 (X= F, Cl, Br, I) (free radical initiator)	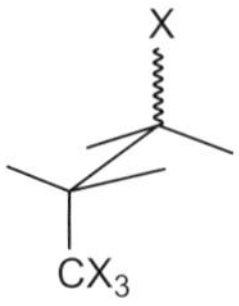
RO–OR (trace) + 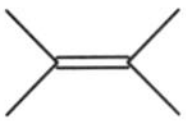	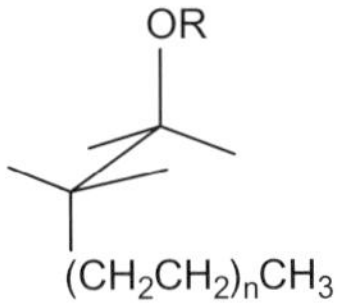

The examples in the table are an incomplete (but representative) set of free radical reactions with alkenes.

sought to provide very small concentrations of bromine (Br_2) or even bromine atoms (Br•) and, as shown in Scheme 6.44, N-bromo- and N-chlorosuccinimide can be used.

Therefore, it would be correct to assume that the reaction of bromine (Br_2) and chlorine (Cl_2) with alkenes enjoys a subtlety that results from the dual mode of reactivity (ionic and radical) in the *addition* reaction *and* the possibility of an allylic *substitution* reaction when there are allylic hydrogens (α-hydrogens) suitably placed. Interestingly, it appears that addition is competitive with substitution over a large concentration range and that the role of the solvent, π-complexes and others, is important in deciding which set of products will predominate. However, with special attention to details, particularly concentration of bromine (Br_2) or chlorine (Cl_2) and their respective sources (such as N-bromosuccinimide [NBS] and N-chlorosuccinimide) and radical initiator, electrophilic addition can be minimized when α-halogenation (substitution) is sought.

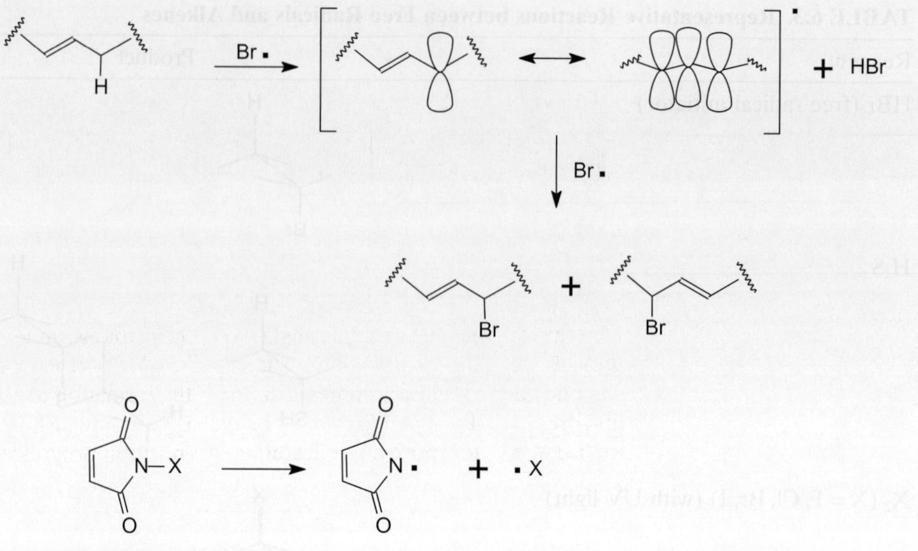

X = Cl (N-chlorosuccinimide)
X = Br (N-bromosuccinimide)

Scheme 6.44. A generalized representation of *allylic* halogenation. Although photochemical generation of halogen can be effective, it is more common to use N-halosuccinimides.

In this vein, the allyl radical ($H_2C=CH-CH_2\cdot$) itself (similar to the allylic radical of Scheme 6.44), which can be generated from, for example, 1,2-propadiene (allene, $H_2C=C=CH_2$) by addition of a hydrogen atom ($H\cdot$) or from propene ($CH_2=CHCH_3$) by hydrogen atom abstraction has been calculated to have a barrier to rotation of about $62.7\,kJ\,mol^{-1}$ ($15\,kcal\,mol^{-1}$) (which is in accord with the difference (ca. $58.5\,kJ\,mol^{-1}$ [$14\,kcal\,mol^{-1}$] in C–H bond dissociation energy).

Free radical polymerization, another example of a chain reaction (this chapter), is quite common and, for many alkenes and dienes, is the preferred method of polymer formation. Typically, as shown in Table 6.3 and Scheme 6.45, the initiator of the free radical process is a peroxide (such as di-*tert*-butylperoxide [$(CH_3)_3C-O-O-C(CH_3)_3$]). In Scheme 6.45a, the radical polymerization of ethylene (ethene, $CH_2=CH_2$) normally carried out at high pressure ($>10^3\,atm$) is shown, while as shown in Scheme 6.45b, the radical polymerization of a diene, 2-chloro-1,3-butadiene [chloroprene, $CH_2=C(Cl)CH=CH_2$], produces the *all trans* or (Z)-polymer called "neoprene."

The polymerization of 2-methyl-1,3-butadiene (isoprene [$CH_2=C(CH_3)CH=CH_2$]) by radical processes produces a "rubber-like" material. Indeed, a synthetic rubber very close to natural rubber (all *cis*-1,4-polyisoprene, Figure 6.9) can be prepared from 2-methyl-1,3-butadiene (isoprene [$CH_2=C(CH_3)CH=CH_2$]) with a Ziegler–Natta catalyst (*vide supra*).

Interestingly, the all *trans*-1,4-polyisoprene (Figure 6.9) is distinctly different from its *cis*-isomer. The *trans*-isomer (called **gutta-percha**) is hard and brittle. Although it softens when heated, it cools to its original consistency and for some years was used in dentistry as a material for "temporary fillings." Although occasion-

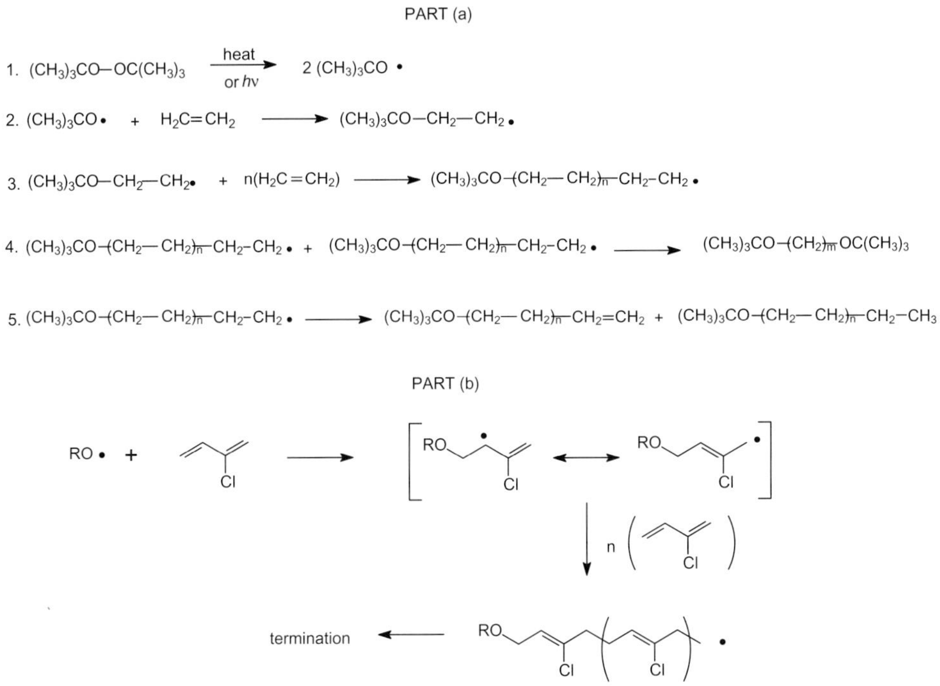

PART (a)

1. $(CH_3)_3CO-OC(CH_3)_3 \xrightarrow[\text{or } h\nu]{\text{heat}} 2\ (CH_3)_3CO\ \bullet$

2. $(CH_3)_3CO\bullet\ +\ H_2C=CH_2 \longrightarrow (CH_3)_3CO-CH_2-CH_2\bullet$

3. $(CH_3)_3CO-CH_2-CH_2\bullet\ +\ n(H_2C=CH_2) \longrightarrow (CH_3)_3CO\text{-}(CH_2-CH_2)_{\overline{n}}\text{-}CH_2\text{-}CH_2\bullet$

4. $(CH_3)_3CO\text{-}(CH_2-CH_2)_{\overline{n}}\text{-}CH_2\text{-}CH_2\bullet\ +\ (CH_3)_3CO\text{-}(CH_2-CH_2)_{\overline{n}}\text{-}CH_2\text{-}CH_2\bullet \longrightarrow (CH_3)_3CO\text{-}(CH_2)_{\overline{m}}OC(CH_3)_3$

5. $(CH_3)_3CO\text{-}(CH_2-CH_2)_{\overline{n}}\text{-}CH_2\text{-}CH_2\bullet \longrightarrow (CH_3)_3CO\text{-}(CH_2-CH_2)_{\overline{n}}\text{-}CH_2=CH_2\ +\ (CH_3)_3CO\text{-}(CH_2-CH_2)_{\overline{n}}\text{-}CH_2\text{-}CH_3$

PART (b)

Scheme 6.45. Free radical polymerization catalyzed by di-*tert*-butyl-peroxide [$(CH_3)_3$ C–O–O–C$(CH_3)_3$]) of **(a)** ethene (ethylene [$CH_2=CH_2$]) at high pressure to yield polyethylene. **Step 1** is termed "initiation" as the initial species needed for polymerization have been formed in this step. **Steps 2** and **3** are called "propagation" since a new radical forms for each that is used. The chain continues. **Steps 4** and **5** are examples of "termination." All steps that destroy radicals and do not create new ones terminate the reaction. In Part b 2-chloro-1,3-butadiene [chloroprene, $CH_2=C(Cl)CH=CH_2$] undergoes a similar polymerization to yield "neoprene" rubber.

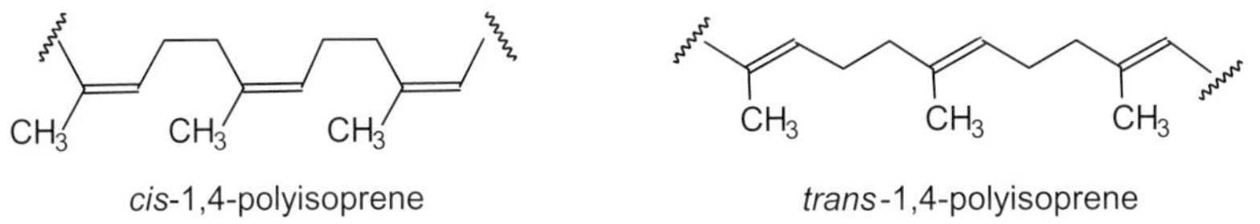

cis-1,4-polyisoprene

trans-1,4-polyisoprene

Figure 6.9. Two polymers of 2-methyl-1,3-butadiene [isoprene, $CH_2=C(CH_3)CH=CH_2$]. The first, all *cis*-1,4-polyisoprene, is close to natural rubber (wrenched from the naturally occurring latex obtained by tapping *Hevea brasiliensis*, the rubber "tree") and can be obtained by Ziegler–Natta-catalyzed polymerization of the monomer. The second, all *trans*-1,4-polyisoprene, is also obtained naturally (from hybrids of *Palaquium*).

ally still pressed into use for that kind of work, it has largely been supplanted by more modern materials.

In Table 6.4 there is a representative set of dienes that have been polymerized, either alone or (copolymerized) with another diene or some suitable alkene to produce elastomers of commercial value. In many cases the details of the process(es) by which the polymerization has occurred remain obscure and/or unreported.

TABLE 6.4. Representative Polymers (Elastomers) Made from Dienes by Free Radical and Related Polymerization Processes

Example	Monomers	Catalyst(s)	Products
1	1,3-butadiene + 1-phenylethene (styrene)	RO• $\left(\begin{array}{l}\text{ROOH} + \text{Fe}^{+2} \rightarrow \\ \text{RO•} + \text{OH} + \text{Fe}^{+3}\end{array}\right)$	[both (*E*)- and (*Z*)-isomers] / C_6H_5 — Units found in the polymer.
2	1,3-butadiene	Natta–Ziegler	All *cis*-polybutadiene
3	2-methyl-1,3-butadiene (isoprene)	Natta–Ziegler	All *cis*-polyisoprene
4	$3H_2C{=}CH_2 + 2H_2C{=}CHCH_3$ ethene (ethylene) propene (propylene)	Natta–Ziegler	CH_3 — Units found in the polymer
5	1,3-butadiene + $H_2C{=}CHC{\equiv}N$ cyanoethene (acrylonitrile)	RO•	[both (*E*)- and (*Z*)-isomers] / CN — Units found in the polymer.

Nonetheless, the general principles as outlined in the earlier discussion of the Natta–Ziegler catalyst system (*vide supra*) and the role played by free radicals permit some predictions to be made, which, in general, are substantiated by experiment.

Other types of polymers, for example, cellulose, will be discussed subsequently (Chapter 11).

d. Intermolecular Cheletropic and Other Cycloaddition Reactions.

As indicated earlier (Chapter 4), it has become common to group reactions that occur through reorganization of electron pairs within a closed (or nearly so) loop *while conserving orbital symmetry* under the rubric "**pericyclic.**"

Cheletropic* and other **cycloaddition**[†] reactions conserve orbital symmetry and can occur both *intra*molecularly as well as *inter*molecularly. Both versions are **pericyclic** reactions. The *inter*molecular set is discussed here (since reactions in that set may be viewed as *addition* reactions *between* two molecules). The *intra*molecular reactions, which can be considered **rearrangements,** will be treated in that section of this chapter.

In all **pericyclic** processes, the bonds are presumed to form simultaneously (or nearly so) but *not necessarily* to the same extent at the same time. Thus, they *are* occurring in *concert* (and the process is "*concerted*"). However, the process *need not be synchronous*.

Electrocyclic reactions (Chapter 4), in which rings are opened and closed *within* a single molecule through σ and π bond interconversions (also introduced in Chapter 4), and **sigmatropic** rearrangements, in which bonds migrate over conjugated systems (similar to carbocation rearrangements seen earlier in this chapter), are additional types of **pericyclic** processes and they, as well as *intra* molecular **cheletropic** and **cycloaddition** reactions, will be considered subsequently (see Section IV, Rearrangements, of this chapter).

The ring forming reactions described in this section, with the exception of the reaction of *triplet* methylene (carbene, $^3{:}CH_2$), the second example in Table 6.5, follow the symmetry rules outlined in Chapter 4. Thus, the simplest example of an allowed[‡] (2 + 2) addition is shown the first item in Table 6.5.

Although halogenated carbenes such as dichlorocarbene ($:CCl_2$) are common and can be generated (among other ways) by base treatment of the corresponding trihalomethane, for example, chloroform (trichloromethane, $HCCl_3$), with base, the discussion of that and related processes and those compounds is reserved until

---

***Cheletropic** (Greek: *chele*, claw) reactions are defined as "those processes in which two σ bonds that terminate at a single atom are made, or broken, in concert."

[†]**Cycloaddition** reactions, in general, can be defined as "those processes in which two or more molecules condense to form a ring by electron transfer from π-type bonds to newly made σ-type bonds." Woodward, R. B.; Hoffmann, R. *The Conservation of Orbital Symmetry*, **1970**, Verlag Chemie GmbH [Academic Press, Inc., NY].

[‡]A reaction is *thermally* allowed (it occurs without the necessity of involving high-energy antibonding orbitals) when the bonding pairs of electrons of the reactant(s) can be transferred into the bonding orbitals of the product(s) (and *vice versa*) with the overall conservation of symmetry (i.e., *conrotatory* processes conserve *axial* symmetry and *disrotatory* processes conserve *mirror* symmetry, cf. (Chapter 4). Interestingly, *photochemical* processes, in which antibonding orbitals are populated, are also allowed— but with a reversal of symmetry over the corresponding thermal process. Finally, "allowed" can be read as "favored."

TABLE 6.5. Representative Cycloaddition Reactions

Addend	Alkene Example	Electrons from Addend	Electrons from Alkene	Product(s)
¨ $^1CH_2$ (singlet)	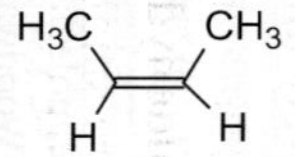	2	2	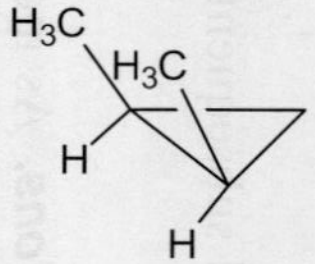
¨ $^3CH_2$ (triplet)	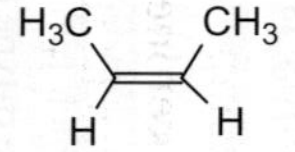	2	2	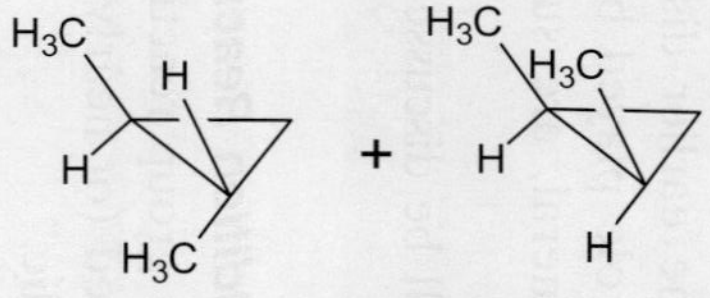
$F_2C=CCl_2$	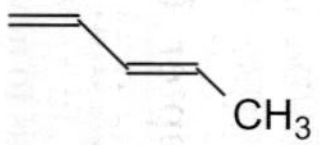	2	2	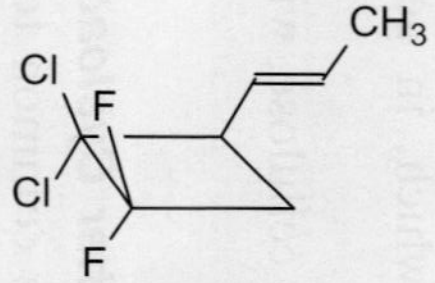
$H_2C=C=CH_2$	$H_2C=C=CH_2$	2	2	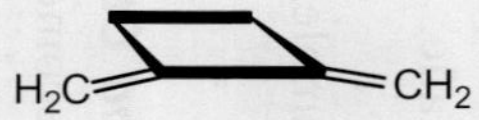
$H_2C=CH_2$	$H_2C=CH_2$	2	2	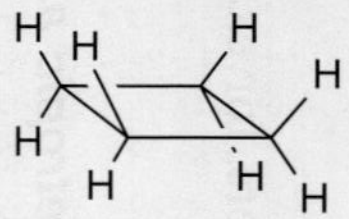
$O=\overset{+}{O}-O^-$	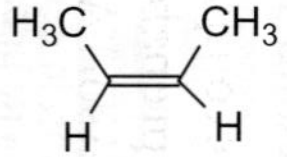	4	2	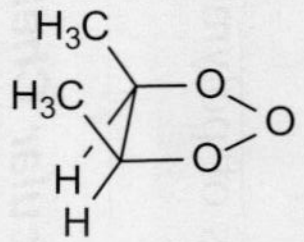
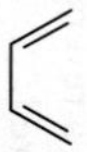	$CH_2=CH_2$	4	2	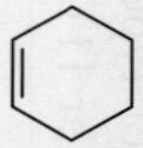
SO_2	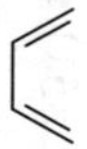	2	4	

The table contains examples of the set of cheletropic and other cycloaddition reactions to be considered here.

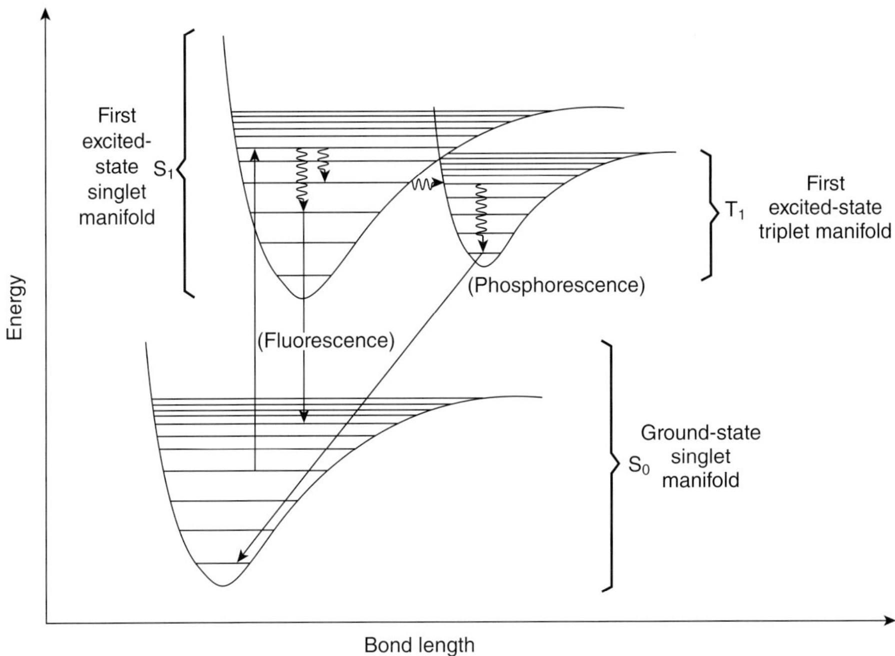

Scheme 6.46. An idealized representation of nondissociative excitation of a diatomic molecule. Quantized absorption of a photon from an upper vibrational state of a singlet (paired spins) ground state (S_0) manifold into an upper vibrational state of a singlet excited state (S_1) leads the process. If there is no photochemical reaction, rapid nonradiative decay to a lower vibrational S_1 level, which preceeds the reemission called **fluorescence**, can occur. The bond lengths between the nuclei of the diatomic in the S_1 state may be longer than in the S_0 state but, as the electrons remain paired, it is not necessarily so. The slow (forbidden) intersystem crossing (paired electrons → unpaired electrons) from $S_1 \rightarrow T_1$ is an alternative to decay to S_0. The energy levels of the S_1 and T_1 states should match. If there is no reaction from the triplet manifold, nonradiative vibrational decay to a level from which **phosphorescence** to the singlet ground state (S_0) (with spin inversion) can occur. The bond length in the triplet state diatomic is (presumably) longer than that in the singlet state.

Chapter 7. On the other hand, methylene ($:CH_2$) and alkyl substituted methylenes (carbenes, $:CHR$; $:CR_2$; $R \neq H$) can be readily prepared either by the thermal (or photochemical) decomposition of the corresponding diazo compounds (Equation 6.51) themselves prepared from an amine (Chapter 10) or ketone (or aldehyde) (Chapter 9) precursor. Alternatively, for a species that behaves as if it were singlet methylene ($:CH_2$), the treatment of methylene iodide (diiodomethane, I_2CH_2) with copper (Cu)-activated zinc (Zn) metal (Equation 6.52) can be used (the *Simmons–Smith reaction**).

$$\underset{R}{\overset{R}{\diagdown}}C\overset{+}{=}\overset{-}{N}=N \longleftrightarrow \underset{R}{\overset{R}{\diagdown}}\overset{-}{C}-\overset{+}{N}\equiv N \xrightarrow[\text{light}]{\text{heat or}} \underset{R}{\overset{R}{\diagdown}}C: \ + \ N\equiv N \qquad (6.51)$$

*Simmons, Howard E., E. I. du Pont Co., Inc., Wilmington, DE.

$$I_2CH_2 \ + \ Zn \ \xrightarrow{\text{Cu}} \ I—Zn–CH_2—I \ \longrightarrow \ H_2C\text{:} \ + \ ZnI_2 \tag{6.52}$$

It is important to recognize the distinction between the reaction of the singlet carbene (e.g., ^{1}:CH$_2$) formed directly either thermally or photochemically and triplet carbene (e.g., ^{3}:CH$_2$) formed from either (a) the singlet state by crossing over to the triplet, or (b) sensitized excitation.* *Singlet carbene undergoes electrocyclic addition with retention of stereochemistry*. On the contrary, *triplet carbene adds as a diradical and loss of stereochemistry* (Scheme 6.47). The stereochemical evidence lends credence to the presumed path for the addition.

Analysis of the three other 2 + 2 reactions in Table 6.5 (examples 3, 4, and 5) follows a similar course. Since the simplest reaction, the dimerization of ethylene (ethene [H$_2$C=CH$_2$]) with itself is thermally forbidden ($_\pi2_s + {_\pi2_s}$) and photochemically allowed ($_\pi2_s + {_\pi2_a}$) (Table 6.5, example 5) (Scheme 6.48a,b), it is reasonable to ask how the other examples might occur. Example 3, the cyclization of 1,1-dichloro-2,2-difluoroethene (Cl$_2$C=CF$_2$) with *trans*- or (*E*)-1,3-pentadiene to produce the 2,2-dichloro-1,1-difluoro-3-[1-(*E*)-propenyl]cyclobutane, appears to be a diradical process (Scheme 6.49) and thus nonconcerted, while the allene (1,2-propadiene, H$_2$C=C=CH$_2$) dimerization makes use of the central sp-hybridized carbon to allow a reaction to occur either via radicals (nonconcerted) or by a suprafacial–antarafacial (i.e., $[_\pi2_s + {_\pi2_a}]$ concerted) process (Scheme 6.50).

The reaction of 1,1-dichloro-2,2-difluoroethene (Cl$_2$C=CF$_2$) with *trans*- or (*E*)-1,3-pentadiene (Scheme 6.49) is particularly interesting since while it appears that minor perturbation of the respective highest occupied molecular orbital (HOMO) and lowest lying unoccupied molecular orbital (LUMO) (Chapter 4) orbitals might

---

*As shown in Scheme 6.46 and noted earlier with regard to Rose Bengal (RB) photosensitized oxidation, intersystem crossing between singlet and triplet states can occur. Consider Scheme 6.46, which is intended to depict a photochemical event resulting from the absorption of a photon by an idealized diatomic molecule in the singlet ground state (S$_0$) (i.e., paired electrons) and which has a variety of vibrational states available to it in that manifold. The quantized absorption of energy from a specific vibrational level of the singlet ground-state manifold into a vibrational level of the first excited state occurs without inversion of spin (S$_0 \to$ S$_1$). That is, the paired electrons remain paired even though one of them is now in an upper level vibrational state of the excited-state manifold (S$_1$). Generally, the molecule in the excited singlet (S$_1$) vibrational state begins to relax internally to a lower vibrational level (but still within S$_1$ manifold) and the relaxation continues until a transition *back to the ground state* S$_0 \leftarrow$ S$_1$ occurs. This reemission of energy will be at a different wavelength from that absorbed (note the different vibrational level) and it is called **fluorescence**.

However, *there are other possibilities*. For example, a chemical reaction involving the molecule in its excited singlet state can occur. In the case of electrocyclic reactions, such photochemical processes occur with a symmetry opposite to those of the thermal reactions, that is, a (thermally) conrotatory process will occur in a (photochemically) disrotatory way and *vice versa* since antibonding orbital(s) are now occupied. The dimerization of ethene (ethylene, H$_2$C=CH$_2$), a thermally disallowed 2$_\pi$ + 2$_\pi$ reaction, is one example that will be encountered later in in this chapter. Other photochemical processes will be examined when carbonyl compounds (Chapter 9) are considered.

Alternatively, energy might be transferred in a collision. The photosensitized (RB) oxidation reactions of hydrocarbons (*vide supra*) is among these.

Further, as shown in Scheme 6.46, if the excited singlet state has lived long enough (longer than about 10^{-6} s), the forbidden (but energetically favorable) transition to the triplet (T$_1$) state (i.e., S$_1 \to$ T$_1$), called **intersystem crossing**, occurs. The chemistry of triplet state systems, as noted previously, is that resembling diradicals. However, if the triplet state fails to undergo reaction, decay may take place to the (singlet) ground state (S$_0$) by emission of energy (called **phosphorescence**).

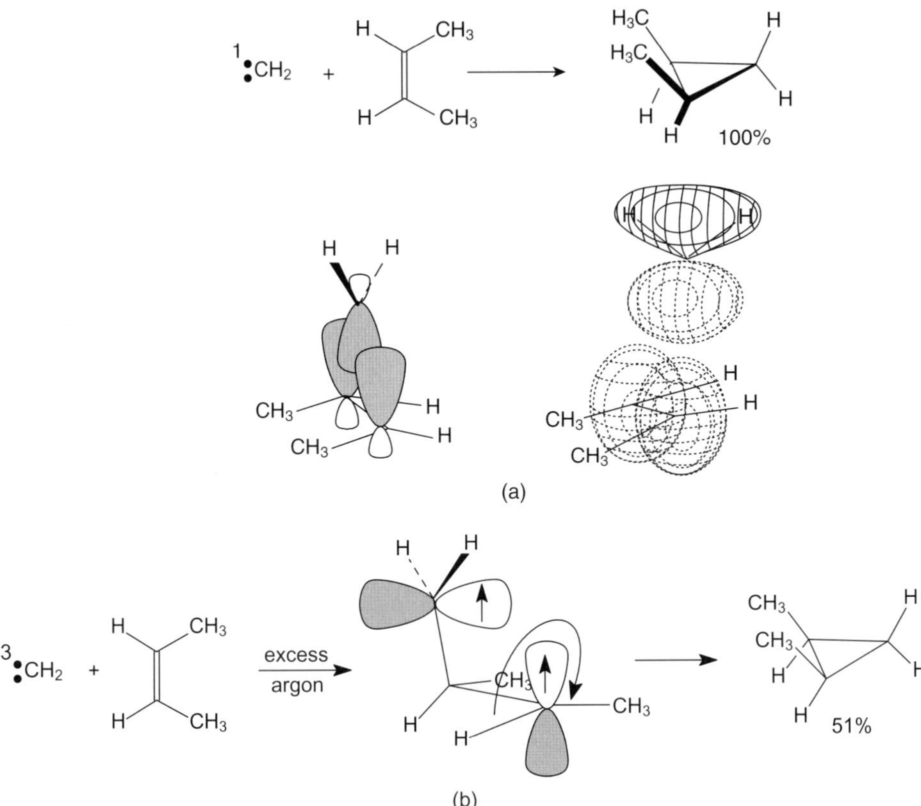

(a)

(b)

Scheme 6.47. The reaction of singlet methylene (^{1}:CH$_2$) and triplet methylene (^{3}:CH$_2$) with an alkene. In (a), the singlet methylene (^{1}:CH$_2$) adds across the double bond with retention of alkene stereochemistry, that is, the methyl groups in (Z)-2-butene are on the same side of the double bond while in the product, (Z)-1,2-dimethylcyclopropane, they are on the same side of the three-membered ring. Alternatively, in (b), triplet methylene (^{3}:CH$_2$) reacts with the loss of alkene stereochemistry. The orbital representations were created using the MOViewer in WebMO version 6.0.002p.

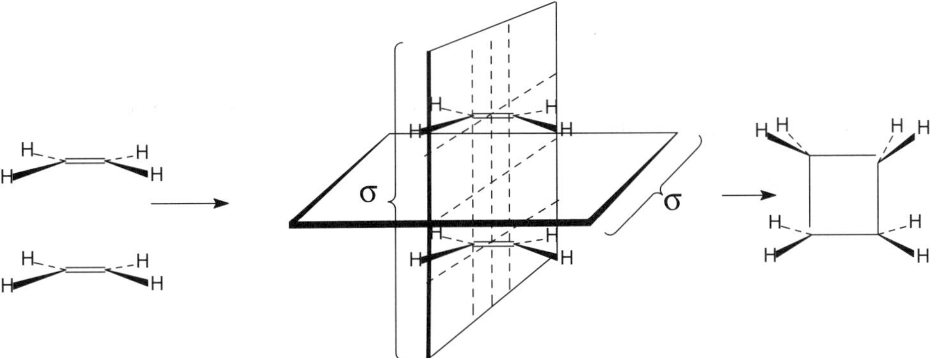

Scheme 6.48a. A representation of the dimerization of ethene (ethylene, H$_2$C=CH$_2$) to form cyclobutane (C$_4$H$_8$). The two mirror planes (σ and σ') that contain the symmetry elements that must be conserved in the dimerization are also presented. The process is thermally forbidden as described in Scheme 6.48b.

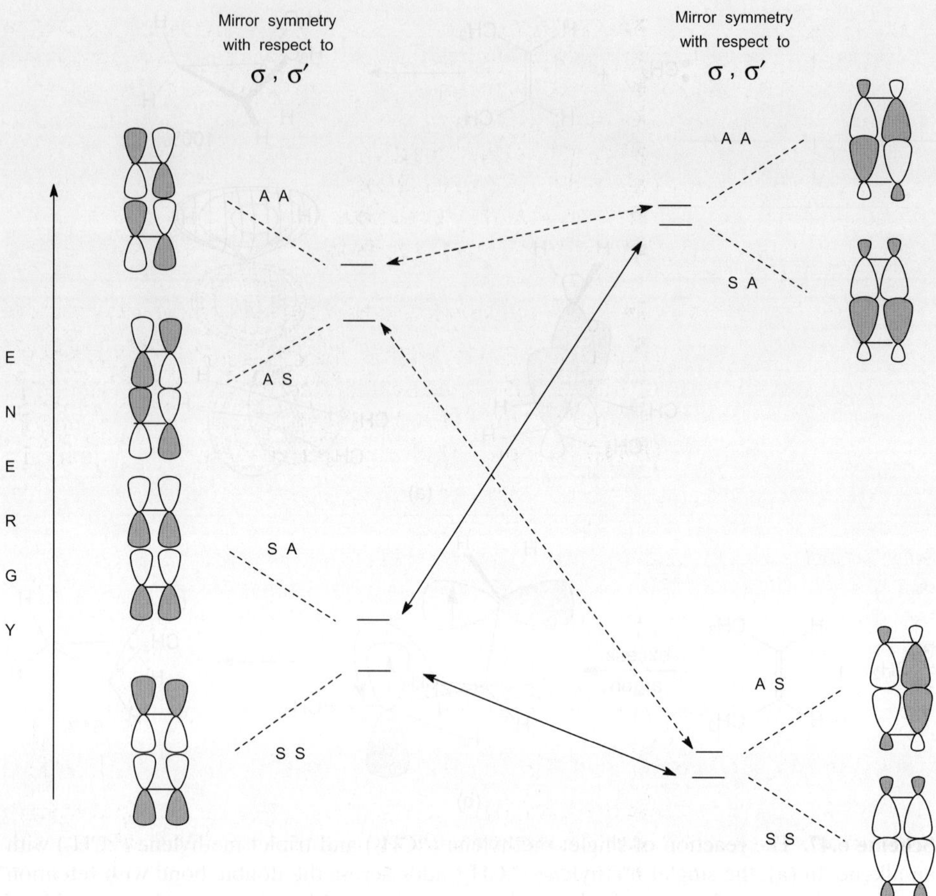

Scheme 6.48b. An approximation of an orbital correlation diagram that uses *p*-type orbitals (on the left) whose overlap defines localized π bonds for ethene ($H_2C=CH_2$) and *sp^3*-type orbitals (on the right) whose overlap defines localized σ bonds in cyclobutane. The diagram should be used in conjunction with the center drawing of Scheme 6.48a where it is intended to show that a reaction between two ethenes has begun. The symmetry of the *p*-type orbitals with regard to the two planes (σ and σ′) and their *relative* energies (for the two ethenes) is given on the left side of this scheme. On the right side of this scheme, the symmetry of the bonding σ-type orbitals of cyclobutane, with regard to the two planes (σ and σ′), and their *relative* energies are provided. It is important to note that while the symmetry of the lowest lying set of orbitals with regard to each element is the same in both the starting material and product, the same *is not true* for the next highest (and *also occupied*) set of orbitals. Thus the preservation of orbital symmetry would require that *an antibonding orbital be occupied*. For this system, the antibonding orbitals will be occupied in photochemical, not thermal, processes.

account for the result if it is assumed that the process is concerted, an alternative explanation is also available. Thus, it is appreciated that if an alkene has electron-withdrawing substituents, its LUMO is lowered (relative to ethene [ethylene, $CH_2=CH_2$]) because it can more readily accept electrons. That observation is in concert with the attack of nucleophiles on alkenes bearing EWGs (discussed earlier). It is also in accord with the observation that the product of the reaction between

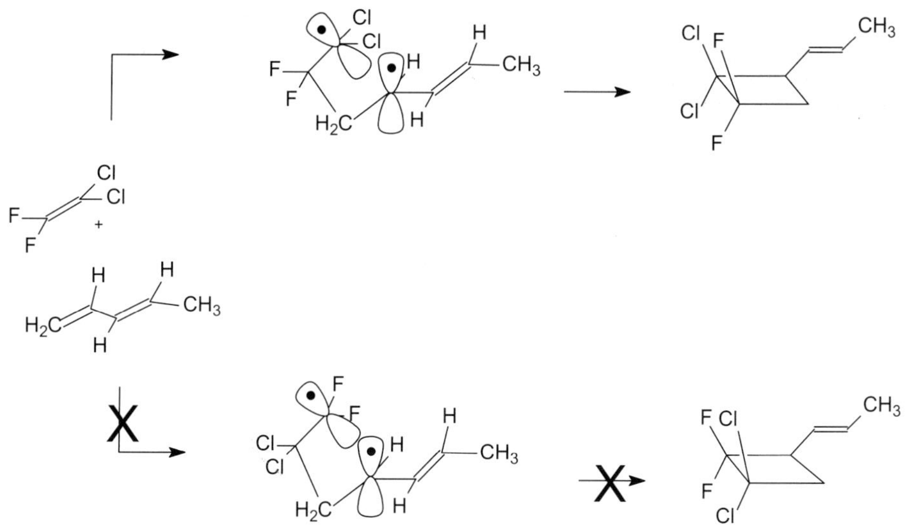

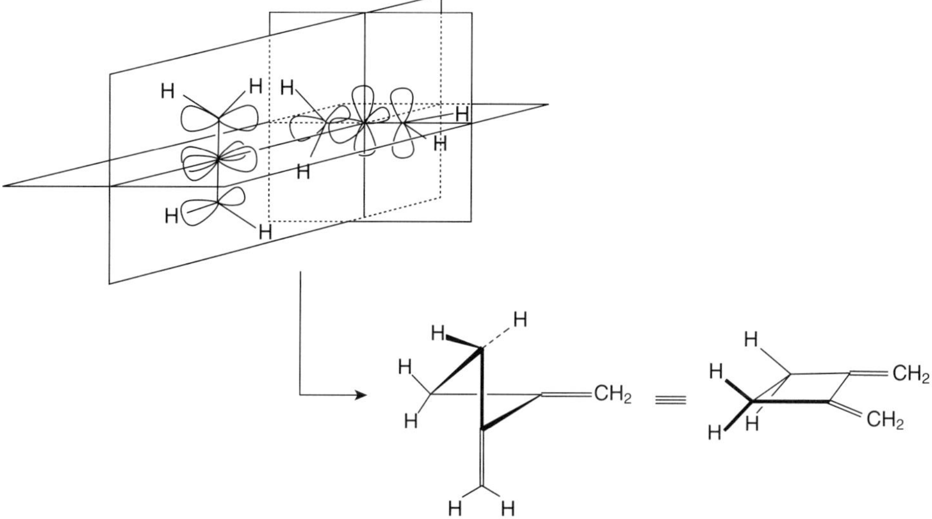

Scheme 6.49. A representation of the radicals that might be involved in the reaction of 1,1-dichloro-2,2-difluoroethene with (*E*)-1,3-pentadiene. The justification (in part) for suggesting that the reaction involves radicals is the observation that the product is 2,2-dichloro-1,1-difluoro-3-[1-(*E*)-propenyl]cyclobutane rather than the isomeric 1,1-dichlor-2,2-difluoro-3-[1-(*E*)-propenyl]cyclobutane.

Scheme 6.50. A representation of the dimerization of allene ($H_2C=C=CH_2$) to produce 1,2-dimethylenecyclobutane. The process, as a $_{\pi}2_s + _{\pi}2_a$ concerted electrocyclic reaction is thermally allowed. The more general case of dimerization of substituted allenes also appears to occur through a radical pathway.

1,1-dichloro-2,2-difluoroethene ($Cl_2C=CF_2$) with (*E*)-1,3-pentadiene produces 2,2-dichloro-1,1-difluoro-3-[1-(*E*)-propenyl]cyclobutane rather than the isomeric 1,1-dichlor-2,2-difluoro-3-[1-(*E*)-propenyl]cyclobutane *if it is assumed* that the reaction occurs with the involvement of radicals in a stepwise process. This is because reaction at the fluorinated end of 1,1-dichloro-2,2-difluoroethene ($Cl_2C=CF_2$) produces the more stable radical at the dichlorinated end of that fragment and, on the other fragment, reaction at C-1 of (E)-1,3-pentadiene produces the (more stable) allylic radical. Finally, it appears that allylic radicals, enjoying added stability from electron delocalization across all three carbons (lowering the lowest lying bonding MO), resist "single-bond" rotation, insuring that the stereochemistry of the double bond is retained in the product.

Problem 6.10. Presuming that without both isomers A and B available for examination, spectroscopic methods might not allow you to be *certain* which isomer was which; suggest a *chemical* reaction that might allow you to decide.

(a)   (b)

The addition of ozone (O_3) to alkenes to give a primary ozonide (molozonide), which rearranges to an ozonide and eventually leads, on reduction, to carbonyl compounds (aldehydes and/or ketones), has already been mentioned and the reaction itself is shown in Scheme 6.11. However, it is important to recognize that this is only one example of a $4\pi + 2\pi$ electrocyclic addition and that orbital overlap for many sets of these reactions dictates their courses as well. Thus, to show the similarity of some of these "dipolar $3 + 2$ addition reactions" Equations 6.53–6.56 are provided. Although any alkene might be used as an example, (Z)-2-butene is used in each to emphasize that all of them occur with *retention of stereochemistry* and, in the first (Equation 6.53), the reaction with ozone to form the primary ozonide (molozonide) is presented again (i.e., see Scheme 6.11). In a similar way, with a suitable azide, $R–N_3$, readily prepared from an alkyl halide (Chapter 7), the same alkene forms a triazoline (Equation 6.54) and with nitrous oxide (N_2O) the heterocycle (Chapter 13) *cis*-4,5-dimethyl-Δ^2-1,2,3-oxadiazoline (*cis*-4,5-dihydro-4,5-dimethyl-1,2,3-oxadiazole) (Equation 6.55). Finally, with a nitrile oxide, such as the oxide derived from ethanenitrile (acetonitrile [$CH_3C\equiv N$]), the same alkene yields a different heterocycle*, the dihydroisoxazole, 3,4,5-trimethyl-4,5-dihydroisoxazole (Equation 6.56).

$$\tag{6.53}$$

$$\tag{6.54}$$

*Most simply, a heterocyclic compound (contrast *homo*cyclic) is one having at least two kinds of atoms in a cycle or ring. Most commonly, one or two oxygen, nitrogen, sulfur, or phosphorus atoms are found in an otherwise carbocyclic system.

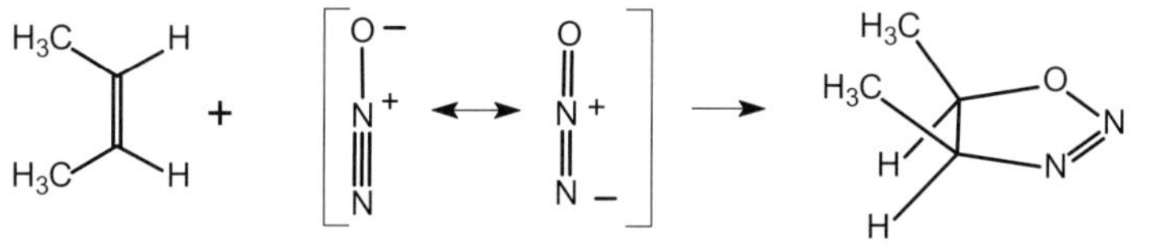

$$(6.55)$$

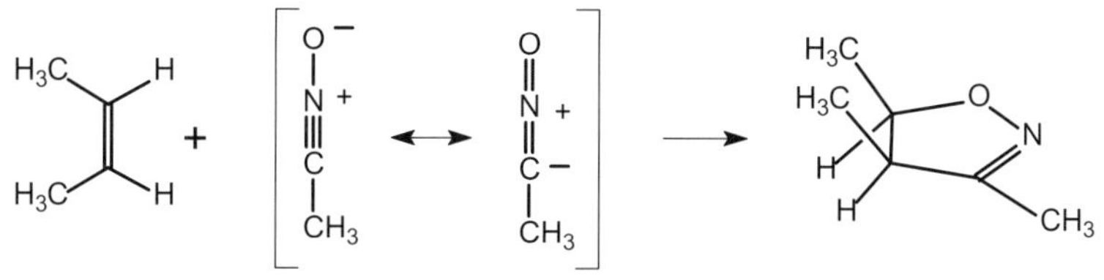

$$(6.56)$$

The penultimate item in Table 6.5 is, of course, the Diels–Alder reaction, discussed earlier in Chapter 4. As pointed out there, the process is considered to be concerted and the stereochemistry in the diene and dienophile (here taken as the alkene) is retained. In expanding upon those themes, it is now worthwhile examining substituted dienes and dienophiles. However, in doing so, the simplifying assumption will be made that despite the perturbations introduced by substituents (e.g., *overall* symmetry may be destroyed), the principles requiring the conservation of *orbital* symmetry are maintaned *and* the analyses are similar. Consider the substitution of an electron acceptor (A) on the alkene and an electron donor (D) on the diene (Scheme 6.51).

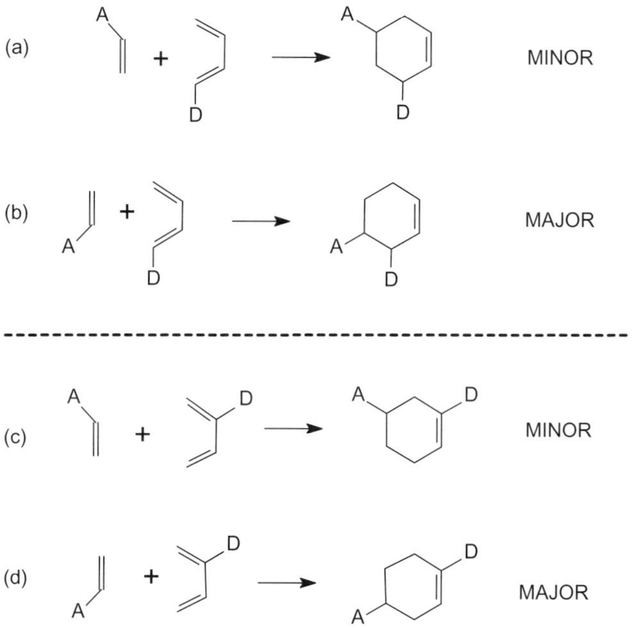

(a) MINOR

(b) MAJOR

(c) MINOR

(d) MAJOR

Scheme 6.51. Potential products arising from the Diels–Alder reaction between a diene bearing an electron-donating group (D) and a dienophile (an alkene) bearing an electron-accepting group (A). In addition to confirming the stereochemical outcome associated with concerted processes, the general experimental observations regarding the "major" and "minor" products are followed for a host of donating and accepting groups. The preponderance of "1,2"- and "1,4"- products has been justified on the basis of HOMO–LUMO interactions.

As noted earlier, EWGs (acceptors) lower the LUMO of the alkene (dienophile) and, in doing so, encourage attack by nucleophilic reagents—suggesting that the electron density at the carbon *adjacent* to the one bearing the substituent is reduced. Electron-donating groups on the diene raise the HOMO and, in particular, when the donor is at C-1 (Scheme 6.51a and b), the electron density at C-4 enjoys the greatest increase. Thus, although the reaction between diene and dienophile is concerted, the carbon with the lowered electron density in the dieneophile is receptive to that with the higher electron density in the diene and the major product has the donor and acceptor groups adjacent to each other (i.e., **b** instead of **a** in Scheme 6.51). Alternatively, when the diene has the donor group at C-2 (Scheme 6.51c and d), it is C-1 of the diene that enjoys the greatest electron enrichment and thus preferentially interacts with that carbon of the dienophile (alkene) of lower electron density, leading to the product with the donor (D) and acceptor (A) groups farthest apart (i.e., **d** instead of **c** in Scheme 6.51).

Finally, the last entry in Table 6.5 is, like the first, another example of a cheletropic reaction. This time, in contrast to the first, a diene (rather than a monoene) and sulfur (IV) dioxide (rather than the more reactive carbene) are undergoing an orbital symmetry allowed ($_\pi 4_s + _\pi 2_s$) pericyclic process. This allowed reaction occurs with retention of symmetry since (although not shown in Table 6.5) the geometry of the substituted diene is retained in the product. Thus, as shown in Equation 6.57, the diene, (2*E*,4*E*)-hexadiene (cf. Figure 4.42), enters the reaction in a suprafacial fashion and, as expected for a disrotatory process (Chapter 4), produces the *cis*-2,5-dihydrothiophene-1,1-dioxide*

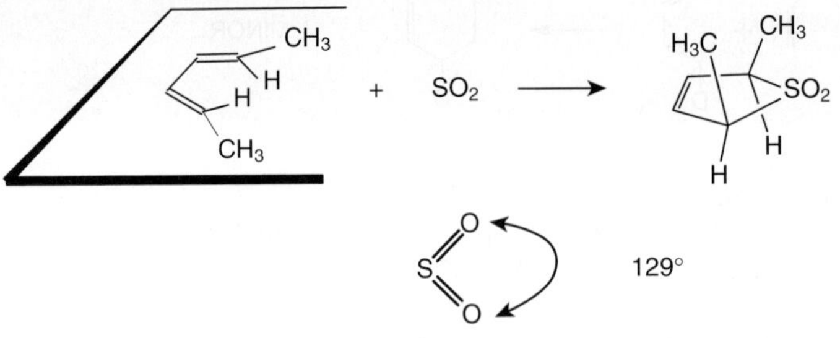

$$\text{(6.57)}$$

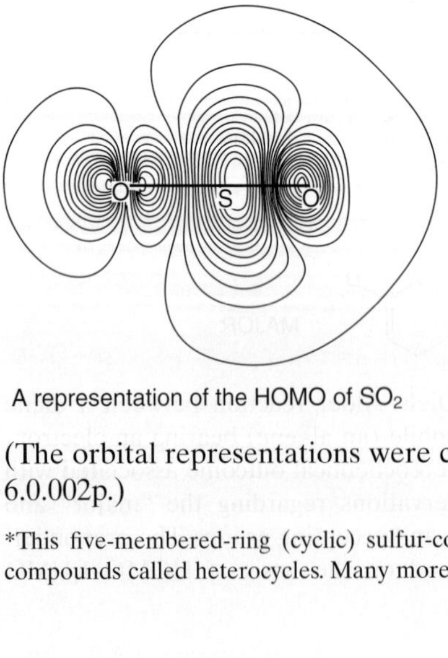

A representation of the HOMO of SO_2 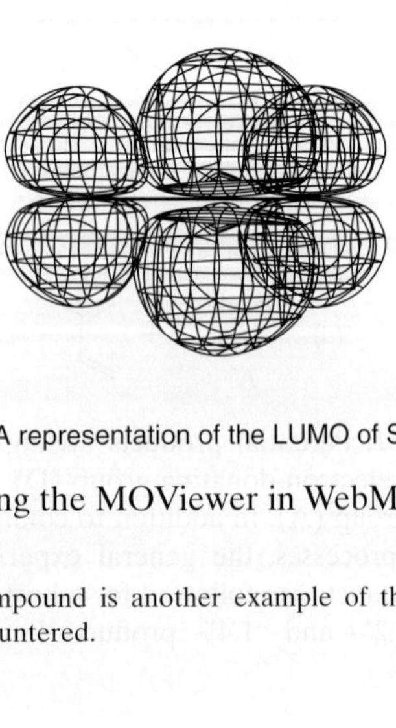   A representation of the LUMO of SO_2

(The orbital representations were created using the MOViewer in WebMO version 6.0.002p.)

*This five-membered-ring (cyclic) sulfur-containing compound is another example of that group of compounds called heterocycles. Many more will be encountered.

IV. Substitution

The *direct* substitution of hydrogen attached to the carbon–carbon double bond by, for example, halogen, (as was the case for alkanes) cannot usually be effected because addition across the double bond or substitution at the allylic position intrudes. One rationale considers that the vinyl carbons are (formally) sp^2-hybridized and thus the C–H bonds have "additional" s character compared with the C–H (formally sp^3-hybridized) bonds in alkanes. Indeed, while a typical alkane C–H bond has a homolytic dissociation energy (DH^o) of ca. 418 kJ mol⁻¹ (100 kcal mol⁻¹ [Table 1.1]), that of a typical vinylic C–H bond (i.e., =C–H) is closer to ca. 452 kJ mol⁻¹ (ca.108 kcal mol⁻¹) and hydrogens in an allylic position (C=C–CH₂-) are even more reactive (DH^o is about 372 kJ mol⁻¹ [89 kcal mol⁻¹]) (Figure 6.10). A path for the process of allylic halogenation as shown in Equation 6.58 has already been presented in Scheme 6.44.

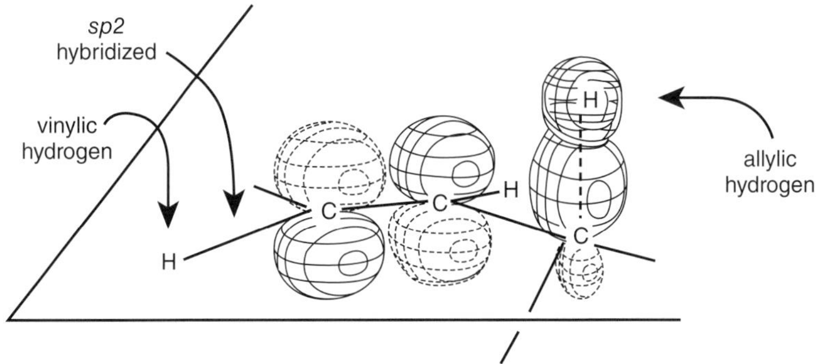

$$(6.58)$$

It should not be surprising that reaction at the allylic position is not restricted to halogenation. Thus, although the carbon–carbon double bond is easily oxidized (Part I of this chapter), an oxidizing agent such as selenium dioxide (SeO_2) preferentially directs oxidation to the allylic position and the process by which this is thought to occur (Scheme 6.52) is also indicative of the *indirect* path by which *vinylic* substitution can be accomplished.

In the same way, vinylic substitution can be accomplished in two steps. The first is an addition across the double bond. The second is an elimination *of part of what*

Figure 6.10. A representation of a hydrocarbon fragment with vinylic and allylic hydrogens. The localized allylic C–H bond is shown as coplanar with the localized π bond to suggest partial orbital overlap. The representations were created using the MOViewer in WebMO version 6.0.002p.

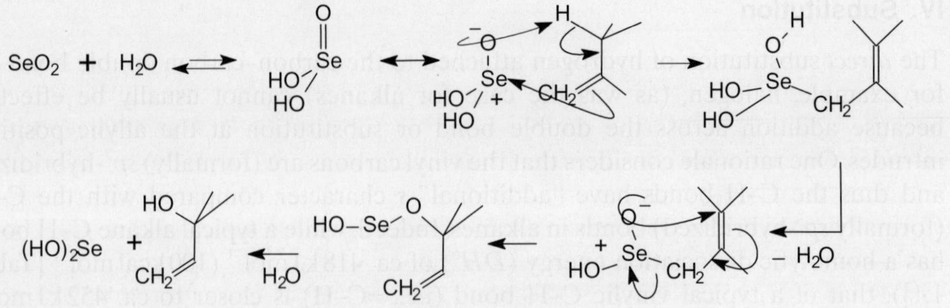

Scheme 6.52. A representation that accounts for allylic hydroxylation on selenium dioxide (SeO₂) oxidation of alkenes bearing an allylic hydrogen.

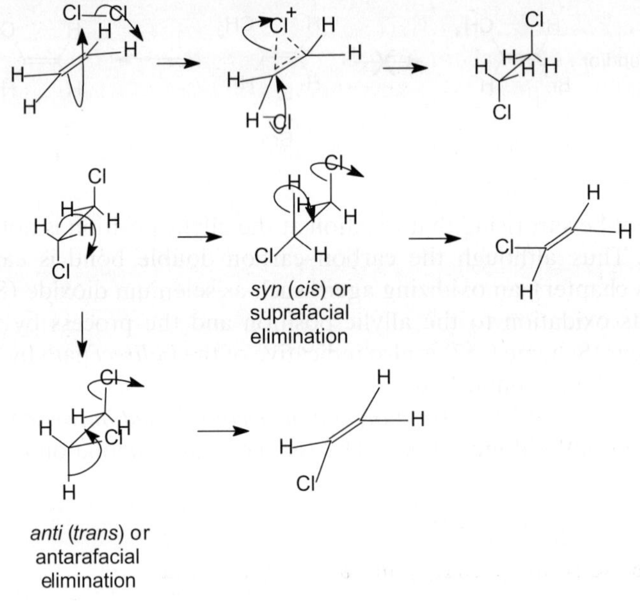

syn (cis) or
suprafacial
elimination

anti (trans) or
antarafacial
elimination

Scheme 6.53. The reaction of chlorine (Cl₂) with ethene (ethylene, H₂C=CH₂) to produce 1,2-dichloroethane (ClH₂CCH₂Cl) followed by elimination of hydrogen chloride (HCl) from a different rotational conformer in either a *syn*-coplanar (or *cis* or suprafacial) or an *anti*-coplanar (or *trans* or antarafacial) sense to produce chloroethene [Cl(H)C=CH₂].

was added. Consider the *two* steps of Scheme 6.53 *carried out separately.* In the first step, chlorine (Cl₂) is added to ethene (ethylene, H₂C=CH₂) in a typical electrophilic addition forming 1,2-dichloroethane (ethylene dichloride, ClH₂CCH₂Cl). Then, separately, an **elimination reaction** (in principle, and sometimes in practice, the *microscopic reverse* of an addition reaction) is carried out. However, the conditions are arranged so that *hydrogen chloride (HCl), not chlorine (Cl₂), is eliminated; the product is the vinyl halide, chloroethene [ethylene chloride, H₂C C(Cl)H].*

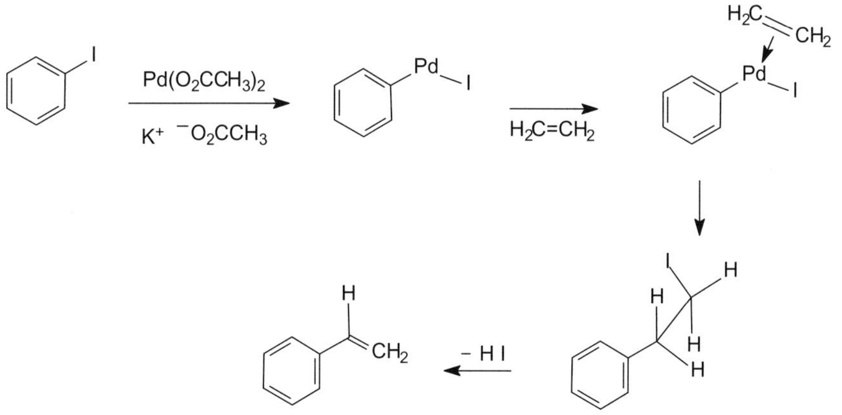

Scheme 6.54. A representation of a potential path accounting for the products of the Heck reaction.

Also, as shown in the scheme, there is evidence that it is convenient to look at the *elimination reaction** as being best (most easily; over the lowest transition state) carried out when the orbitals containing the species to be eliminated (e.g., **H** and **Cl**) are coplanar (sometimes called *peri*planar [Greek "peri" for "about"]) (or nearly so). Thus, the **H** and **Cl** are either on what is destined to become the *same* face of the double bond (*syn-coplanar*) or what will be the *opposite* faces (*anti-coplanar*).

The haloalkenes, for example, chloroethene [ethylene chloride, $Cl(H)C=CH_2$], prepared in this and related ways and discussed in more detail with the other organic halides (Chapter 7) serve as starting material for a host of other types of compounds. However, additional substitution reactions (also apparently occurring through "addition–elimination" pathways) might be effected. Typical of these is the "apparent" arylation of an alkene in which an aryl group is "substituted" for a hydrogen attached to the double bond. The process is known as the *Heck reaction* and is shown in Scheme 6.54.[†]

In the Heck reaction, an "arylpalladium" reagent is formed by treatment of an aryl halide, such as iodobenzene (C_6H_5I) (*vide infra*) with a palladium salt such as palladium acetate [$Pd(O_2CCH_3)_2$] and a base (e.g., potassium acetate, $CH_3CO_2^-K^+$) to produce the reagent, written, in this case, as "ArPdI." Addition of the alkene such as ethene (ethylene, $H_2C=CH_2$) results in formation of an adduct that then undergoes elimination of HI to form ethenylbenzene (styrene, $C_6H_5CH=CH_2$), a product of "substitution" of phenyl (C_6H_5-) for hydrogen ($H-$) on the carbon–carbon double bond.

V. Rearrangement

As pointed out earlier, the *process* of structural interconversions while retaining the same numbers of nuclei (in the same overall oxidation state) is accomplished by the

---

*A fuller discussion of **Elimination Reactions** is postponed until Chapter 7 because the kinetics with which many of them occur, as well as intermediates that may be involved, resemble those also found in **nucleophilic substitution**.

[†]Heck, R. F.; Nolley, Jr., J. P. *J. Org. Chem.*, **1972**, *37*, 2320. Richard Heck, Ei-ichi Negishi and Akira Suzuki shared the Chemistry Nobel Prize in 2010.

movement of bonds from one place to another and is called, simply, **rearrangement**. Experience dictates that successfull understanding of what is occurring and the ability to predict what will occur in previously unseen circumstances depends upon pattern recognition and the *understanding* that these interconversions are not random. Indeed, they *all* follow clearly defined regio- and stereochemical pathways. The patterns of electron flow, shown by "curved arrows," on the one hand, and symmetry considerations in electrocyclic processes, on the other, as already established *in this chapter* will continue to be followed. Thus, the conversion of one carbocation (a secondary one) into another carbocation (a tertiary one) by movement of a methyl group (CH_3–) *with its pair of electrons* (a *methide*) from one carbon *to an adjacent one* (a "1,2-methide shift") was shown in Scheme 6.19. Further, the *reversible* nature of **hydroboration** was shown in Scheme 6.28 to lead to alkene isomerization since *addition* of the elements of R_2BH across a double bond in one direction and *elimination* of R_2BH in a different direction produces an isomer.*

Table 6.6 lists the rearrangements to be considered here. The unifying feature that these have in common is that all begin with and end with molecules that have at least one double bond, *and no other functionality* (e.g., –*OH, C O*) in their respective frameworks. The functionalized systems will be considered separately later— every family has its rearrangements.

The first example in Table 6.6 involves an exchange reaction between two different alkenes (metathesis), and is elaborated upon in Scheme 6.55. This is a particularly important process that has gained value with the introduction of very specific catalysts that even allow rings to form with expulsion of small alkenes (i.e., **Ring Closing Metathesis [RCM]**) using Grubbs[†] catalysts.

It is important to note that the alkyl groups do not simply fly off and become reconnected since then random reattachment would be expected and the substituents are not randomly attached. Second, a "reasonable" pathway in which there is reversible metal catalyzed formation of substituted cyclobutanes (put together one way and taken apart another) is probably wrong because, when substituted cyclobutanes are subjected to the conditions of the reaction, it is reported that they *do not* undergo conversion to alkenes!

The next four items (examples 2–5, inclusive) in Table 6.6 are all electrocyclic reactions, clearly related to the cycloadditions and others already discussed earlier in this chapter and the symmetry controlled processes of Chapter 4. Example 2, a conrotatory four-electron $(2\pi + 2\pi = 2\pi + 2\sigma)$ process relating *trans* or (*E*)-3,4-dimethylcyclobutene to *trans, trans* or (*2E,4E*)-hexadiene conserves C2 symmetry as shown in Figure 4.41 and again here in Equation 6.59. Examples 3, 4, and 5 are six-electron disrotatory processes.

---

*It would be correct to argue that "one hydrogen *has been substituted* for another" and to suggest that this process of reversible hydroboration is really another example of "substitution" rather than rearrangement. The point is moot. As shown in Scheme 6.28, the isomerization process results in repetitive additions and eliminations as the carbon–carbon double bond "moves down the chain."

[†]Grubbs, R. H., (California Institute of Technology, Pasadena, CA.) and others have developed a series of catalysts based on ruthenium rather than tungsten. These are highly selective and have the general formulation of $Cl_2L_2Ru=CHR$, where R is any alkyl or similar substitutent and "L" corresponds to various ligands, such as $(C_6H_5)_3P$. Cf. Grubbs, R. H. *Handbook of Metathesis*, Wiley-VCH GmbH, Weinheim, Germany, **2003**.

TABLE 6.6. Representations of Some Examples of Alkene Rearrangement Reactions

Example	Name	Reaction
1	Olefin methasis	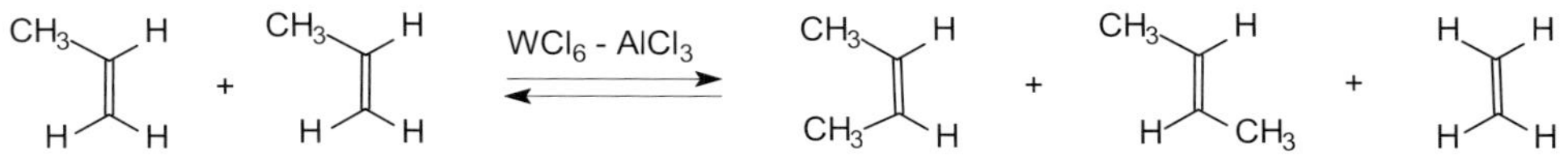
2	Electrocyclic reaction	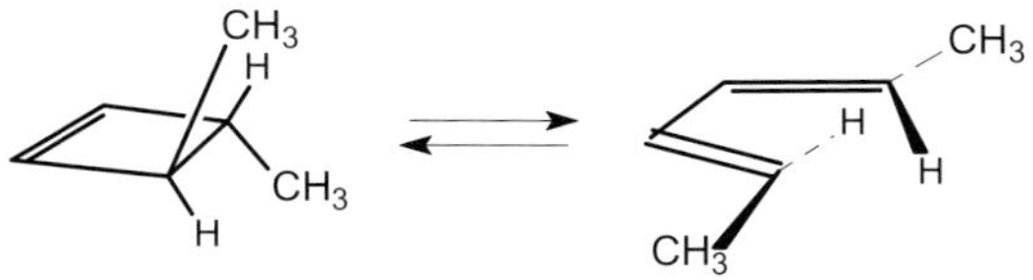
3	Ene reaction	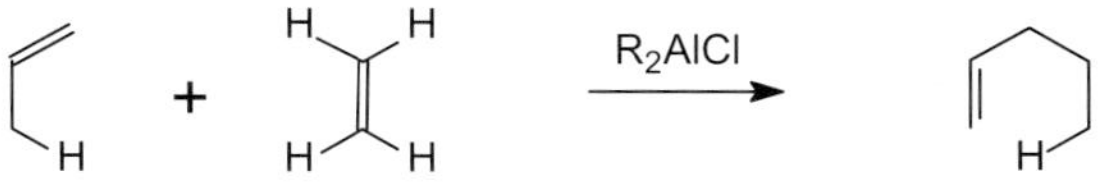
4	Cope rearrangement	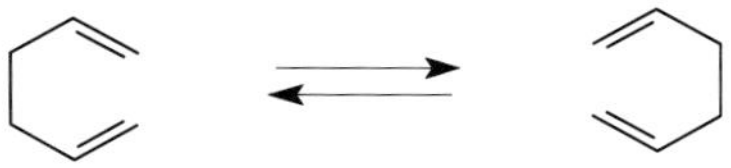
5	Intramolecular cycloaddition	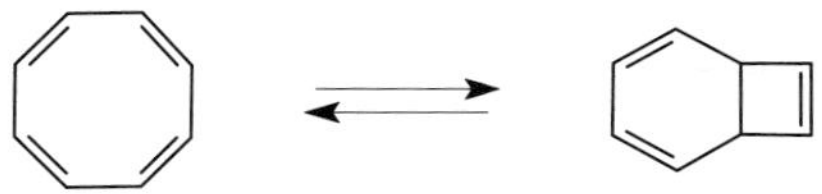
6	Carbene insertion	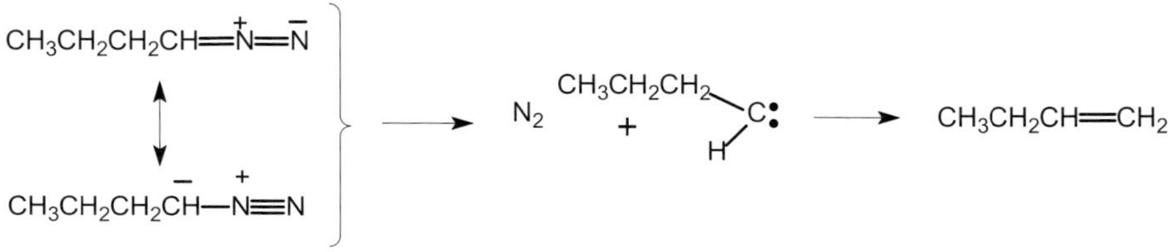

(Continued)

374

TABLE 6.6. *Continued.*

Example	Name	Reaction
7	Protonation-deprotonation	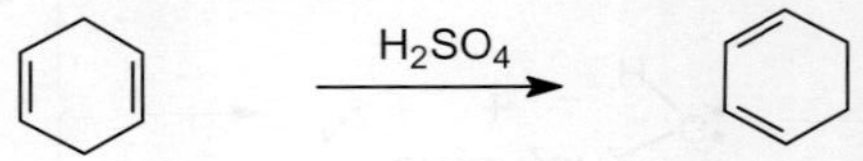
8	Di-pi-methane	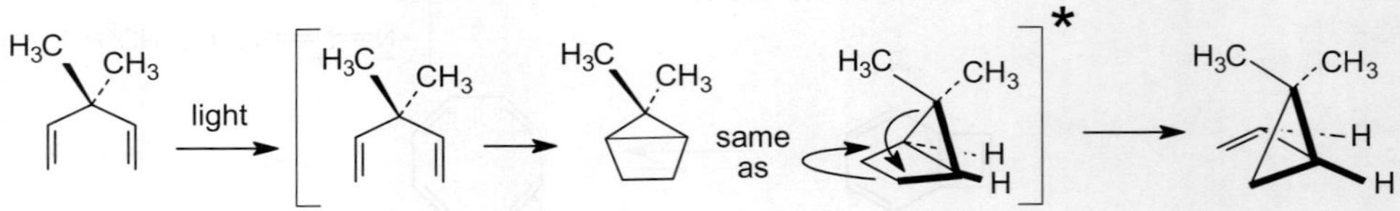
9	Methide shift (a sigmatropic rearrangement)	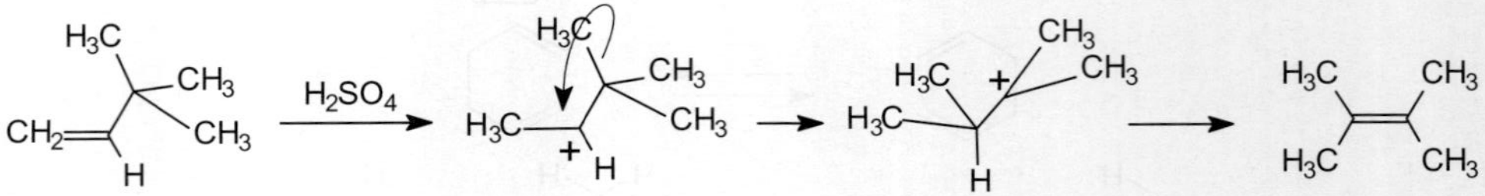
10	Reversible hydroboration	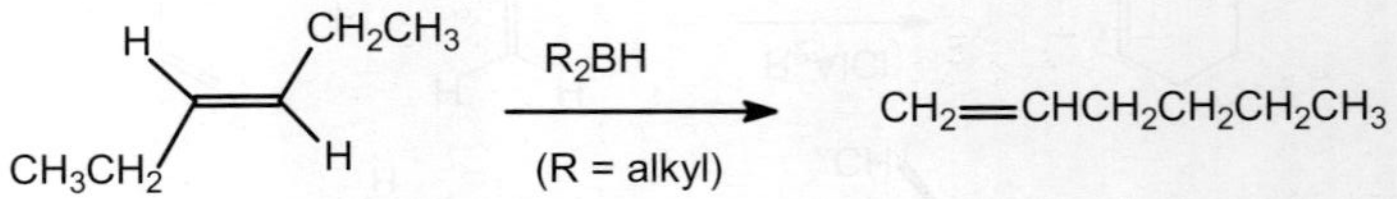
11	Reversible hydrozirconation	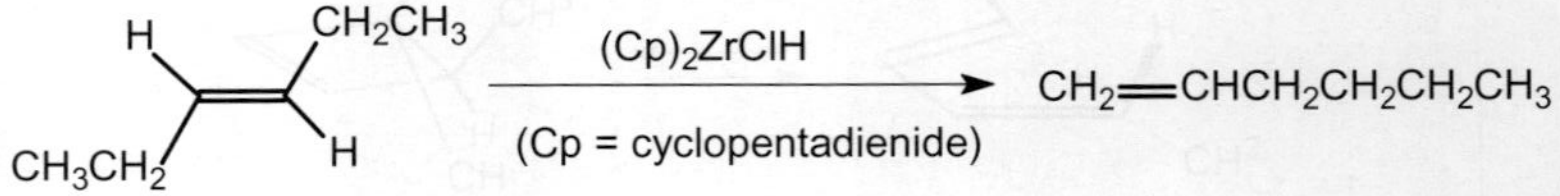 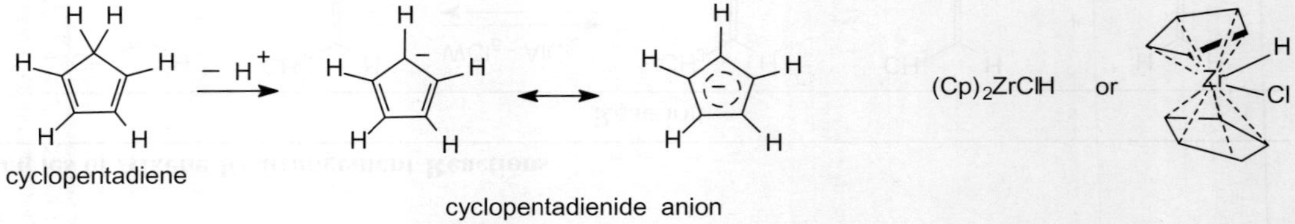

cyclopentadiene

cyclopentadienide anion

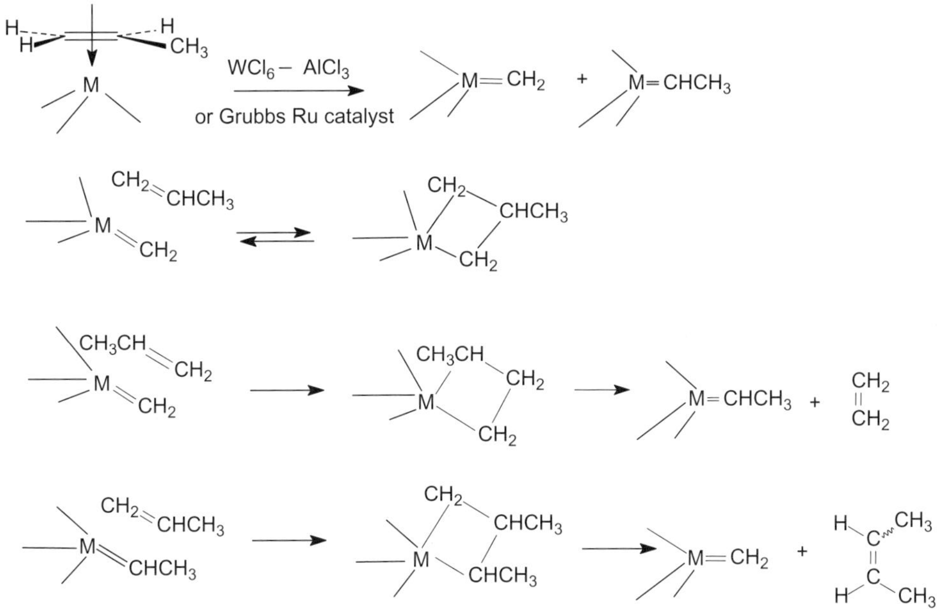

Scheme 6.55. A representation of a possible pathway using metal (M = Ru or W) carbene-like intermediates to account for alkene metathesis.

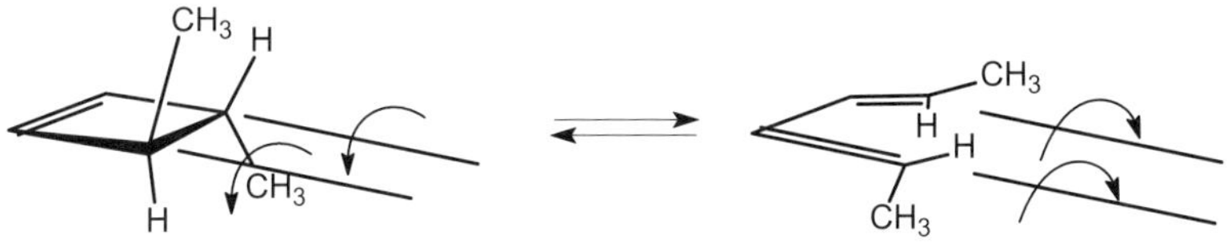

$$(6.59)$$

Example 3, the "ene reaction" ($2\sigma + 2\pi + 2\pi = 2\pi + 2\sigma + 2\sigma$), has been observed as an allowed suprafacial intramolecular process, as well as the intermolecular reaction between propene ($CH_3CH=CH_2$) and ethene ($CH_2=CH_2$) shown in Equation 6.60. Oxygen and nitrogen analogues are also known. Similarly, the **Cope* reaction** (a $2\sigma + 2\pi + 2\pi = 2\pi + 2\pi + 2\sigma$ narcissistic[†] process) is shown in Equation 6.61 for 1,5-hexadiene (Example 4) and a similar reaction also occurs with involvement of heteroatoms. The oxygen and nitrogen analogues, which give rise to the "oxycope" and "azacope" reactions, respectively, will be encountered in later chapters. The last of the set, the electrocyclic ($2\pi + 2\pi + 2\pi = 2\pi + 2\pi + 2\sigma$) process shown in Example 6 and Equation 6.62, which converts cyclooctatetraene into the *cis* or (Z)-2,4,7-bicyclo[4.2.0]octatriene, is clearly related to the 1,3,5-hexatriene = cyclohexadiene rearrangement shown in Problem 4.10.

Although all of these (and many related) processes follow the predictions of the rules of orbital symmetry, it has not yet proved possible to exclude the involvement of radical or radical-like intermediates.

---

*Cope, A. C. (1909–1966), Professor, Massachusetts Institute of Technology, Boston, MA
[†]A reaction in which an object is converted into its mirror image and named after the youth who, in Greek Mythology, was said to have pined away for love of his own reflection.

$$(6.60)$$

$$(6.61)$$

$$(6.62)$$

The sixth example in Table 6.6, carbene insertion into an adjacent C–H sigma (σ) bond, is representative of the gas-phase reaction of carbenes. Indeed, since this intramolecular reaction effectively competes with the intermolecular addition of carbenes to alkenes seen earlier, the intermolecular process (Scheme 6.47) is best examined when no α-hydrogens are available in the carbene.

Rearrangements in which a carbon–carbon double bond "moves into conjugation," as in the seventh example in Table 6.6, are common. Indeed, since conjugated dienes are generally found to be about 12–17 kJ mol^{-1} (3–4 kcal mol^{-1}) more stable than their nonconjugated isomers (based on heats of hydrogenation and/or combustion),* it is clear that at equilibrium (Table 4.1) there will be essentially no nonconjugated diene. This is most easily seen by taking the UV spectrum of the isomers since the $\pi \rightarrow \pi^*$ transition will be much more facile for the conjugated alkene (Problem 6.11). In addition, in this particular case, symmetry of the 1,4-diene (two mutually orthogonal planes) dictates that the ^{13}C and ^{1}H NMR spectra for this isomer will enjoy a simplicity lacking in the spectra of its conjugated isomer. The NMR spectra are given in Figure 6.11.

Problem 6.11. Based upon Equation 2.2 and the observation that 1,3-cyclohexadiene has a $\pi \rightarrow \pi^*$ at 263 nm while that for its nonconjugated isomer is about 190 nm, calculate the difference in energies of the excited states.

The path for the rearrangement shown in Scheme 6.56 and Figure 6.12 is in concert with the available evidence.

The di-π-methane rearrangement (Table 6.6, example 8) is an example of a large family of processes that occur photochemically. As expected, excitation of the nonconjugated π bonds imparts significant energy to the system and, as a direct consequence, bond lengths (as well as angle distortions) generally unatainable thermally become possible. The process shown in Table 6.6 (example 8) is written in Scheme 6.57 as occuring *both* in a concerted disrotatory manner from an excited singlet state (reached by direct photochemical excitation as in Scheme 6.46) *and* as a diradical

---

*As indicated in Table 3.4, two double bonds in 1,4-cyclohexadiene that do not interact should have about twice the heat of hydrogenation [ΔH_r] of cyclohexene, that is, about 243.3 kJ mol^{-1} (58.2 kcal mol^{-1}) or (2 × 29.1 kcal mol^{-1}). The heat of hydrogenation [ΔH_r] of the conjugated isomer, 1,3-cyclo-hexadiene, is 229.5 kJ mol^{-1} (54.9 kcal mol^{-1}) and thus, it is reckoned that the conjugated isomer is 13.8 kJ mol^{-1} (3.3 kcal mol^{-1}) more stable.

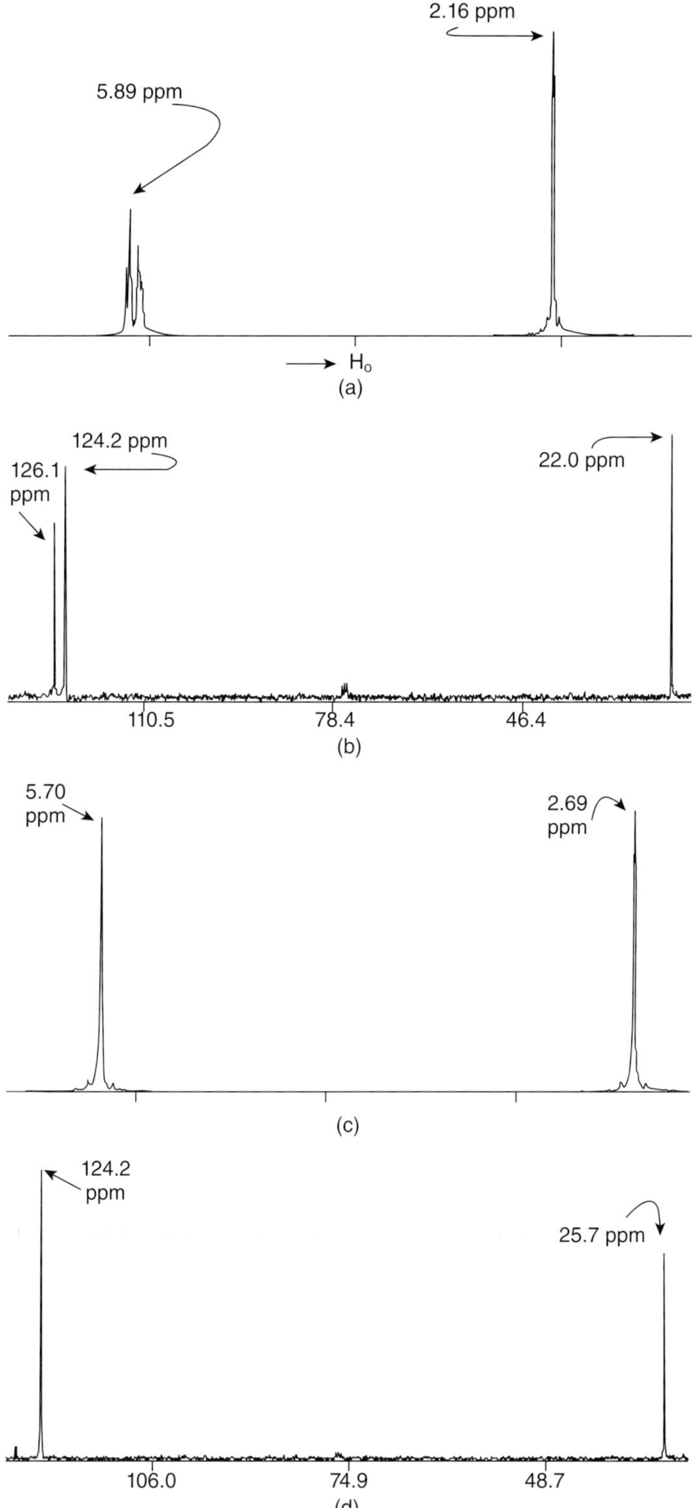

Figure 6.11. (a) The ^1H NMR spectrum of 1,3-cyclohexadiene in $^2HCCl_3$ at 300 MHz. (b) The ^{13}C NMR spectrum of 1,3-cyclohexadiene in $^2HCCl_3$ at 75 MHz. (c) The ^1H NMR spectrum of 1,4-cyclohexadiene in $^2HCCl_3$ at 300 MHz. (d) The ^{13}C NMR spectrum, of 1,4-cyclohexadiene in $^2HCCl_3$ at 75 MHz.

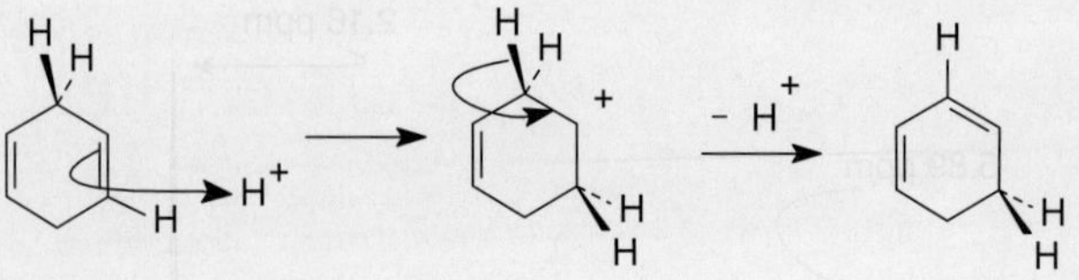

Scheme 6.56. A path for the acid-catalyzed isomerization of a nonconjugated alkadiene (1,4-cyclohexadiene) to its conjugated isomer (1,3-cyclohexadiene). Note that although the reaction is reversible, that is, the protonated intermediate could have lost the same proton originally acquired to return to the nonconjugated isomer, it is shown as occurring in one direction (to the most stable isomer).

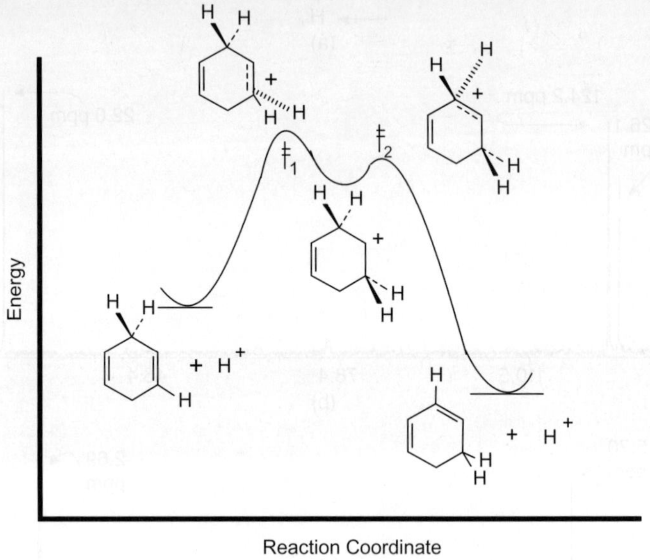

Figure 6.12. A representation of a slice through the three-dimensional reaction surface for the acid (H^+)-catalyzed isomerization of the unconjugated diene, 1,4-cyclohexadiene, to its conjugated isomer, 1,3-cyclohexadiene.

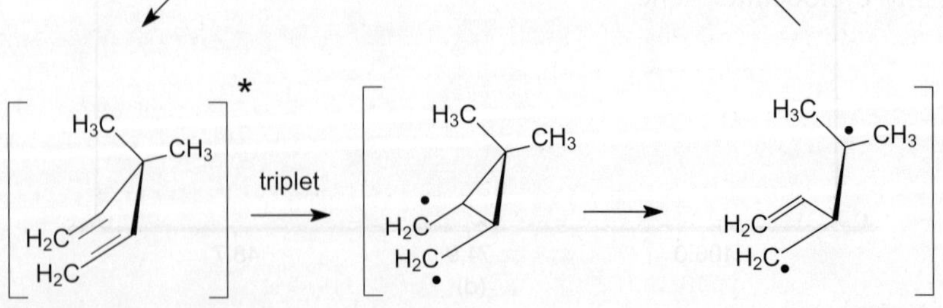

Scheme 6.57. The di-π-methane of 3,3-dimethyl-1,4-pentadiene to ethenyl-2,2-dimethylcyclopropane. In principle the reaction shown could occur from a singlet or triplet state. In this particular case both states give rise to the same product.

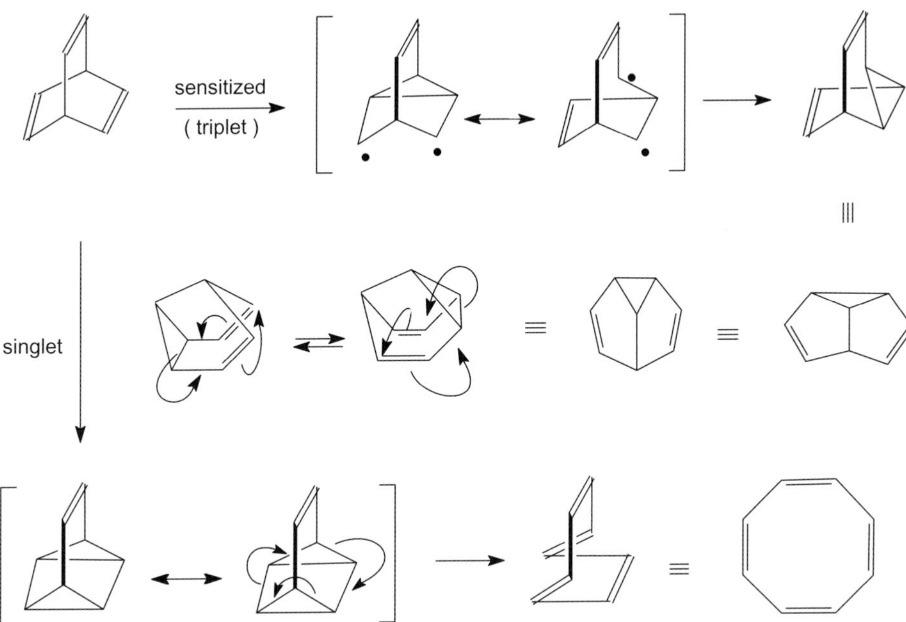

Scheme 6.58. Representations for the photochemical (both triplet [sensitized] and singlet [unsensitized]) reactions of bicyclo[2.2.2]octatriene ("barrelene"). The former leads to tricyclo[5.1.0.0^{4,8}]octa-2,5-diene (semibulvalene) while the latter produces cyclooctatetraene. See Luo, J.; Ihmels, H.; Deiseroth, H.-J.; Schlosser, M.*Can. J. Chem.*, **2009**, *87*, 619 and references therein.

reaction (i.e., a triplet process). These are both shown in the scheme since, as a function of the substituents on the double bond, both can occur. Alternatively, in some systems singlet and triplet processes give *different* products *from the same starting material.* In Scheme 6.58 the result of photochemical rearrangement of bicyclo[2.2.2] octatriene (whose trivial name is "**barrelene**" because, when the π electron clouds are included, some argue that the system resembles a barrel) is considered. In this case, the triplet product, obtained from the sensitized photochemical reaction, is the fluxional* molecule tricyclo[5.1.0.0^{4,8}]octa-2,5-diene ("semibullvalene"). It is believed that this forms stepwise via the diradicals shown. The concerted process, on the other hand, results from the unsensitized (and thus presumably singlet) photochemical transformation. In the latter (also in Scheme 6.58), barrelene is converted into cyclooctatetraene.

The rearrangement of one carbocation to another, as seen in example 9 of Table 6.6 is a common phenomenon and, indeed, it can be argued that example 7, the conversion of 1,4-cyclohexadiene to its conjugated isomer 1,3-cyclohexadiene, falls into the same group. However, it will be recalled that in the latter, a proton was added, forming a carbocation and, subsequently, a different proton was lost, leading to the rearranged diene. In the case of the sigmatropic rearrangement of which the

---

*The molecules grouped as **fluxional** are those in which rapid tautomeric interconversion between systems with identical structures is in evidence. Scrambling of double bonds (and thus the carbon atoms) can be found by suitable labeling experiments.

conversion of 3,3-dimethyl-1-butene into 2,3-dimethyl-2-butene is an example (Table 6.6, example 9), it is clear that protonation, followed by a *movement of a methyl group with its pair of electrons* (i.e., *a methide*) from one carbon to its immediately adjacent neighbor and then loss of a proton, is needed.

A wide variety of such structural rearrangements, which include hydride, methide, and other shifts from carbon to carbon have been recognized. Although these rearrangements have commonly been observed when substituents, other than the unadorned carbon–carbon double bond, are present, some are also seen with alkenes.

The reactions needed to account for interconversions observed in bicyclic terpenes which will be dealt with in some detail in Chapter 11. Schemes 6.59 and 6.60 are shown as a sampling of the field, rich with such rearrangements, served to demonstrate that these carbon-to-carbon migrations occurred with *retention of stereochemistry in the migrating group* as well as to stimulate a generation of chemists to examine these processes in detail.

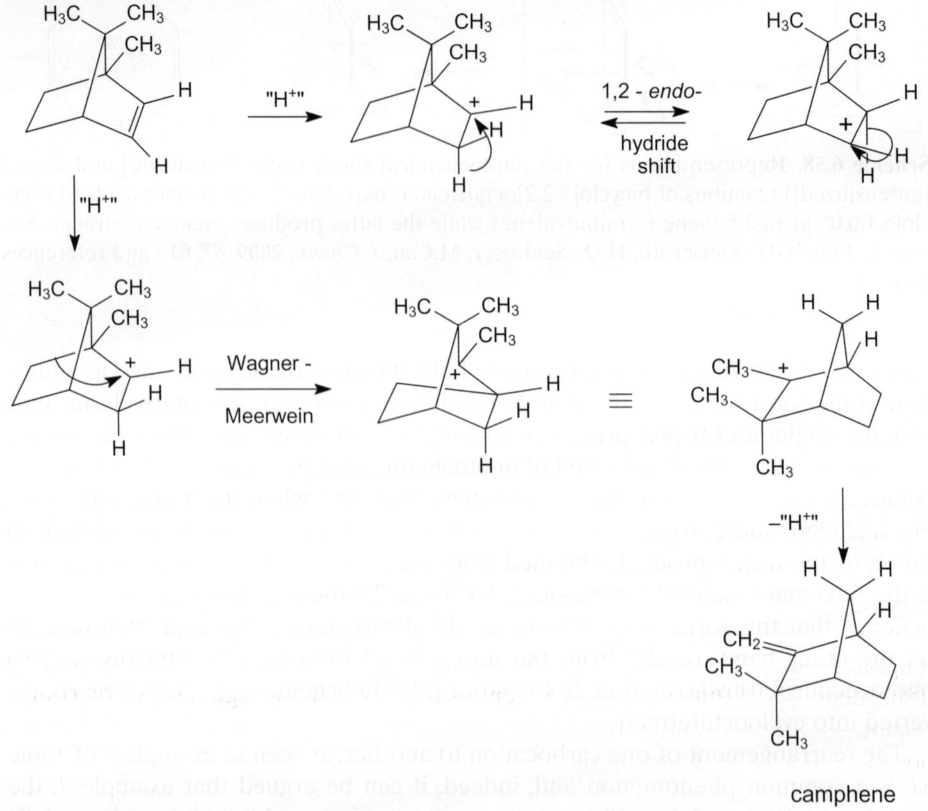

Scheme 6.59. Representations of the 1,2-hydride migration and Wagner–Meerwein* rearrangements on a bicyclic carbocation.

*G. Wagner (1849–1903), Professor, Warsaw University and H. L. Meerwein (1879–1965), Professor, Bonn University. The early work was published in the German and Russian literature around the beginning of the twentieth century. An interesting review article is available. Birladeanu, L. *J. Chem. Ed.*, **2000**, *77*, 858. But see Wagner, G. *J. Russ Phys. Chem Ges.*, **1859**, *31*, 680; Meerwein, H.; van Emster, K., *Chem. Ber.*, **1922**, *55*, 2500; Streitweiser, Jr., A. *Chem. Rev.*, **1956**, *56*, 571.

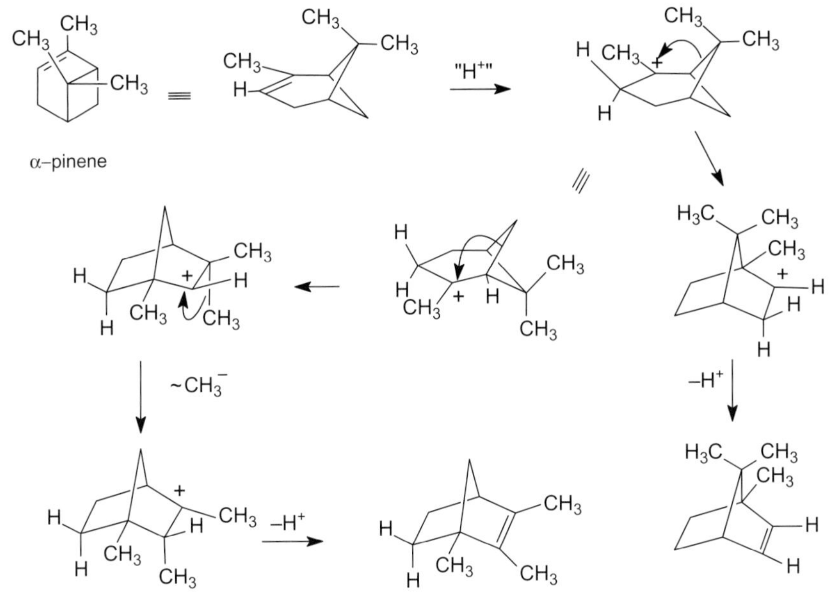

Scheme 6.60. Representation of two different Wagner–Meerwein rearrangements (and a methide shift) that account for products derived from α-pinene.

Thus, in Scheme 6.59, 1,7,7-trimethylbicyclo[2.2.1]hept-2-ene ("bornene") is protonated to yield a secondary carbocation,* capable of undergoing 1,2-*endo*-hydride migration and/or the migration of a bond in a Wagner–Meerwein rearrangement to a "new" bicyclic (but now **tertiary**) carbocation and thence, by proton loss, to 3,3-dimethyl-2-methylenebicyclo[2.2.1]heptane ("camphene").

Then in Scheme 6.60, 2,6,6-trimethylbicyclo[3.1.1]hept-2-ene ("α-pinene") can produce a carbocation on protonation that is shown to rearrange *from* a tertiary carbocation *to* a secondary carbocation *while relieving ring strain*. That is, formation of the carbocation allows a rearrangement from a substituted cyclobutane ring to a substituted cyclopentane ring. Since either bridge can migrate, there are two possibilities. First, there is that one in which the carbon bearing the two methyl groups migrates and that subsequently leads to 1,7,7-trimethylbicyclo[2.2.1]hept-2-ene ("bornene") and that, as seen in Scheme 6.59, can also undergo a set of rearrangements. Alternatively, migration of the other one carbon bridge, followed by proton loss, leads to 1,2,3-trimethylbicyclo[2.2.1]hept-2-ene.[†]

As already outlined in Scheme 6.28 (*vide supra*), not only is hydroboration (Table 6.6, example 10) reversible but oxidation, reduction, and isomerization of alkenes

---

*Generally the carbocations shown in Schemes 6.59 and 6.60 are *not* produced by protonation. As will be seen subsequently, these carbocations arise from the corresponding alcohols and/or alkyl halides by solvolysis and/or elimination reactions.

[†]It is noteworthy that these alkene products are, however, seldom observed as the carbocations produced, rather than losing a proton as shown, readily react with solvent or adventitious moisture. Thus, some of these rearrangements will be seen again when alcohol rearrangements are considered (Chapter 8).

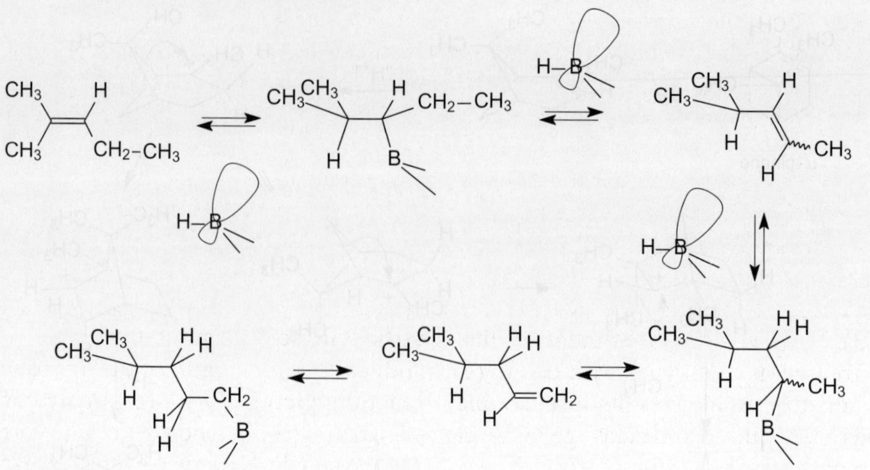

Scheme 6.61. The use of the hydroboration reaction to effect alkene isomerization.

using diborane [$(BH_3)_2$] and its derivatives can also be effected. However, as noted in Scheme 6.28 and reiterated in Scheme 6.61, if isomeriztion rather than oxidation or reduction is desired, it is sufficient to simply remove the alkyl borane (by distillation) and to isolate the rearranged alkene. Since more highly substituted alkenes are more reactive toward diborane than less highly substitutued alkenes, the latter will be produced.

The final example in Table 6.6, reversible hydrozirconation, resembles reversible hydroboration. Hydrozirconation has the advantage that crystalline materials are often obtained and that the zirconcene hydridochloride can be recovered for reuse. However, hydrozirconation is used less and thus is more expensive than hydroboration. As shown at the bottom of Table 6.6, the cyclopentadienyl π-type ligands, lying in the coordination sphere of the metal, are not coplanar. It appears (by x-ray crystallography) that when the interplanar distance between the cyclopentadienyl planes is small (in other metalocines), the ligands are close to parallel. However, with a metal atom with a larger radius, the interplanar distance increases and the planes of the ligands are no longer parallel. This presents an open "face" to the alkene for bonding to the metal and increases the possibility for additional transformations (such as the introduction of substituents onto carbon) as well as the isomerization shown in the table.

D. ALKYNES

I. Oxidation

As was true for alkanes and alkenes (*vide supra*), alkynes are attacked by oxygen. Indeed, one of the earliest uses of the parent or simplest alkyne, ethyne (acetylene, $HC{\equiv}CH$), was to burn it in air for purposes of illumination. As an illuminant, when mixed with the appropriate amount of air and using special gas jet tips,* ethyene (acetylene, HC≡CH) burns with an intense white flame.

Terminal alkynes (R–C≡C–H; R = H, for ethyne [acetylene, HC≡CH], but, in general, 1-alkynes where R = alkyl or aryl) and internal alkynes (R–C≡C-R'; where R, R' ≠ H but R may or may not be the same as R' and either or both could be alkyl or aryl) behave in slightly different ways toward oxidizing agents. Thus, gentle treatment with reagents such as neutral potassium permanganate (KMnO$_4$) (Scheme 6.62) and sodium periodate (NaIO$_4$) in the presence of ruthenium (IV) dioxide (RuO$_2$) (to reoxidize the reduced periodate) will successfully convert *internal* alkynes to α-diketones and, occasionally, terminal alkynes to α-ketoacids (Scheme 6.63). More vigorous treatment results in carboxylic acid formation.

Treatment of alkynes with ozone (O$_3$) under the same conditions as those used for alkenes usually results in incomplete reaction. Generally, more forcing conditions are required and cleavage of the carbon–carbon triple bond occurs with formation of the corresponding carboxylic acids (Scheme 6.64). However, even with excess ozone (O$_3$), α-diketones are occasionally obtained if particularly unreactive alkynes (as a function of the substitutents, R and R') are encountered.

The attempted epoxidation of alkynes to form the corresponding unsaturated epoxide (oxirene) fails (Equation 6.63) to produce a stable compound and, when oxidation occurs, a mixture of more highly oxidized compounds usually results.

---

*In part, the special tips are necessary because the temperature of the flame (around 1500°C) is so high. This very hot flame causes carbon particles produced to be heated to white incandescence. The heat is such that oxyacetylene flame is used for welding and cutting iron and steel. The heat liberated in this reaction reflects the instability of such unsaturated compounds relative to their saturated analogues and, indeed, to the elements themselves. The values for the bond strengths (Chapter 1, Table 1.1), of carbon–carbon single bonds (376 ± 8 kJ mol^{-1} [90 ± 2 kcal mol^{-1}] for ethane, CH$_3$CH$_3$), carbon–carbon double bonds (720 ± 8 kJ mol^{-1} [172 ± 2 kcal mol^{-1}] for ethene [ethylene, H$_2$C=CH$_2$]), and carbon–carbon triple bond (962 ± 8 kJ mol^{-1} [230 ± 2 kcal mol^{-1}] for ethyne [acetylene, HC≡CH]) can be exploited to demonstrate that one of the bonds in an alkene or alkyne "must" be weaker than the others. The use of this idea, that is, that *one* of the two bonds of a double bond (172–90 = 82 kcal mol^{-1} [343 kJ mol^{-1}]) or *one* of the three bonds of a triple bond (230–172 = 58 kcal mol^{-1} [242 kJ mol^{-1}]) is particularly susceptable to breaking, leads to the prediction that reactions at carbon–carbon double and triple bonds will be 40–120 kJ mol^{-1} (10–30 kcal mol^{-1}) *more exothermic* than the reactions at the carbon–carbon single bond. Further, both ethene (ethylene, H$_2$C=CH$_2$) and (as already noted) ethyne (acetylene, HC≡CH) are unstable with respect to formation of carbon and hydrogen, that is,

$$\text{H-C} \equiv \text{C-H}_{(g)} \rightarrow 2\text{C}_{(s)} + \text{H}_{2(g)} \quad -\Delta H° = -234 \text{ kJ mol}^{-1} (56 \text{ kcal mol}^{-1}).$$

As a consequence, beginning with ethyne's (acetylene, HC≡CH) formation from calcium carbide

$$3\ \text{C}_{(s)} + \text{CaO} \xrightarrow{\text{heat}} \text{CaC}_2 + \text{CO}$$

$$\text{CaC}_2 + 2\ \text{H}_2\text{O} \longrightarrow \text{HC} \equiv \text{CH} + \text{Ca(OH)}_2$$

this colorless gas (bp –85°C) is neither liquified nor handled under pressure unless it is mixed with inert gases (such as nitrogen) or it is dissolved in a diluent (typically propanone [acetone, CH$_3$COCH$_3$]) and free space around the solution miminized. Often, diatomaceous earth or clay pipes under about 30 mm in diameter (with smaller pipes in larger ones) are used as absorbent. Further, it is recommended that contact with copper and all its alloys and salts (which appear to promote its explosive decomposition) be carefully avoided.

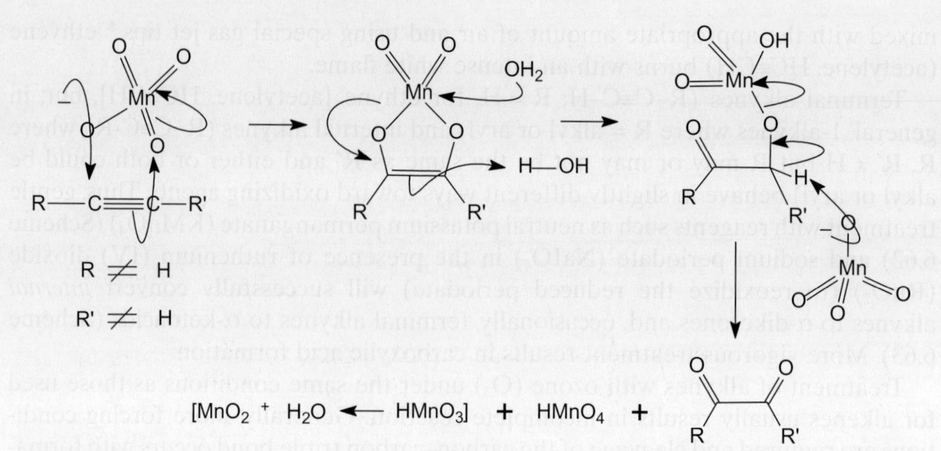

Scheme 6.62. A representation of the conversion of an internal (i.e., nonterminal) alkyne to an α-diketone by oxidation with neutral permanganate (MnO_4^-). Although two equivalents of permanganate (MnO_4^-) are shown, the second equivalent is not known to be necesary to remove the proton. The manganese in $HMnO_3$ eventually undergoes disproportionation and manganese dioxide (MnO_2) separates.

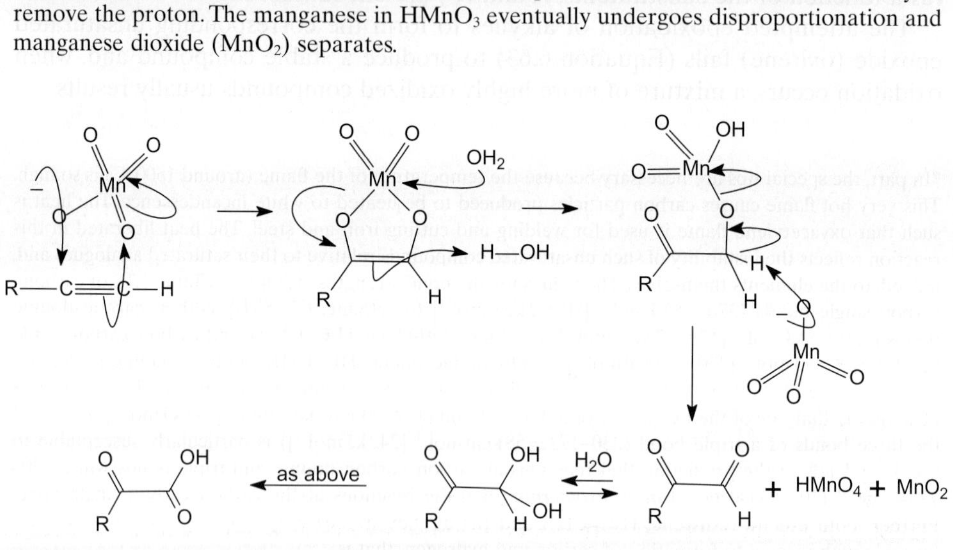

Scheme 6.63. A representation of the conversion of a terminal alkyne to an α-ketocarboxylic acid by oxidation with permanganate under neutral conditions (MnO_4^-). The oxidation of the aldehyde (via the hydrate) requires a second equivalent of permanganate (MnO_4^-) as described for the oxidation of alcohols (Chapter 8).

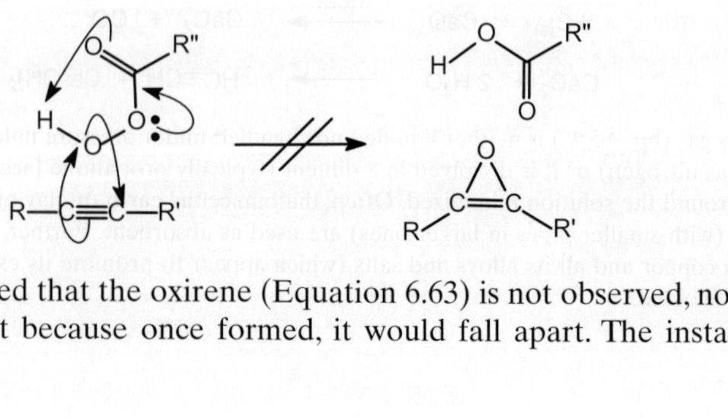

$$(6.63)$$

It has been argued that the oxirene (Equation 6.63) is not observed, not because it fails to form, but because once formed, it would fall apart. The instability, the

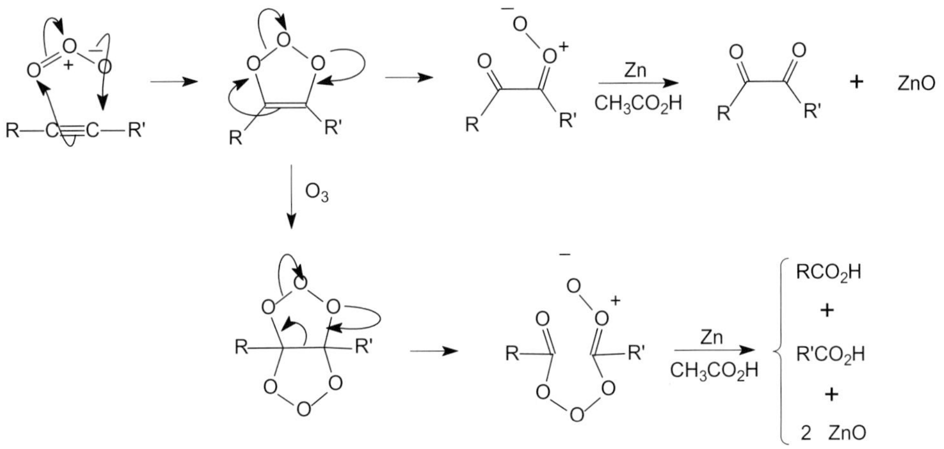

$$\downarrow O_3$$

Scheme 6.64. A representation of the ozonolysis of a nonsymmetrically substituted internal alkyne to yield an α-diketone and, with sufficient ozone (O_3), carboxylic acids. The quantity of ozone (O_3) added to a reaction mixture in, for example, dichloromethane (H_2CCl_2) solvent is difficult to measure as the air or oxygen stream is passed through the silent electric discharge and then into the solution. As a consequence, mixtures of carboxylic acids and α-diketones are the usual products.

argument continues, results from the overlap between the nonbonding electrons on oxygen and the π-electrons of the double bond, unavoidably coplanar, producing a system in which the bonding and antibonding MOs would both be fully occupied.

As was the case for alkenes, other reactions in which an "oxidation" will be seen to have occurred (e.g., oxidative hydroboration) are also found for alkynes. Following the same pattern, however, dictates that those reactions be found under the general heading of "**Addition**" (*vide infra*).

II. Reduction

The catalytic reduction of alkynes converts the carbon–carbon triple bond to a carbon–carbon single bond with the addition of two equivalents of hydrogen (H_2) across the triple bond (Equation 6.64, where R and R′ may be hydrogen or alkyl or aryl). Presumably this reaction proceeds via addition of one equivalent of H_2 to produce the alkene and then, in a separate step, a second equivalent of H_2 adds. Interestingly, alkynes are more easily reduced than the corresponding alkenes, in concert with the bond strengths as discussed above and also because, apparently, alkynes are more easily adsorbed on the catalyst than are alkenes.

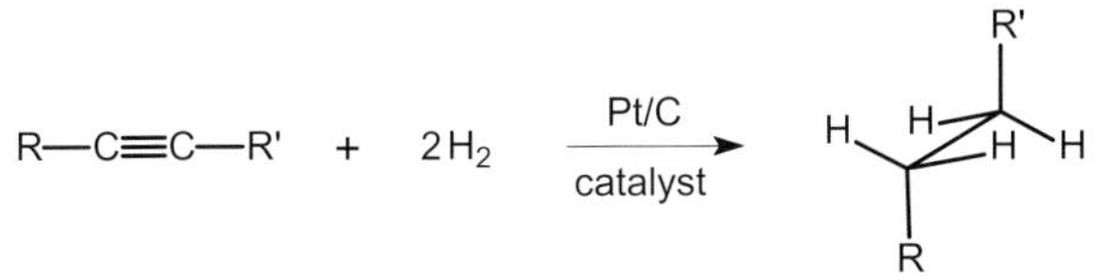

$$R—C≡C—R' \ + \ 2H_2 \ \xrightarrow[\text{catalyst}]{\text{Pt/C}} \quad\quad\quad\quad (6.64)$$

In accord with the idea that the reduction of alkenes is in general less facile, a catalyst that fails to reduce alkenes to alkanes nonetheless succeeds in reducing alkynes to alkenes. The reduction thus stops at the alkene stage and the addition has, as expected, been suprafacial [*cis-* or *syn-* or (*Z*)-]. The special lead (Pb/CaCO$_3$) "poisoned" catalyst (Equation 6.65 with R and R' as above) is known as a **Lindlar* catalyst**.

$$R—C≡C—R' \xrightarrow[\text{CaCO}_3/\text{PbSO}_4]{\text{H}_2} \quad (6.65)$$

Lindlar catalyst

Also, as was the case for alkenes, the carbon–carbon triple bond can be reduced by active metals (e.g., sodium, Na, lithium, and Li) in the presence of a proton donor, such as ethanol (ethyl alcohol, CH$_3$CH$_2$OH). Interestingly, the reduction appears to occur in a series of one-electron transfer steps (first to a radical or radical-like species and then to an anion) from the active metal to the alkyne in liquid ammonia (NH$_{3(l)}$) solvent followed by the transfer of the proton. The alkene produced arises from net antarafacial [or *trans-*, *anti-*, or (*E*)-] hydrogen addition (Equation 6.66).

$$R—C≡C—R' \xrightarrow[\text{CH}_3\text{CH}_2\text{OH}]{\text{Na/NH}_{3\,(l)}} \quad (6.66)$$

Again, as seen with alkenes, both borane (B$_2$H$_6$) (Equation 6.22) and diimide (H–N=N–H) (Scheme 6.15) can be used to reduce alkynes. Indeed, the reaction of internal alkynes with borane is apparently more facile than that with the alkene that results from consummation of the reduction. Further, as would be expected, suprafacial addition of hydrogen and boron obtains and *Z*- (or *cis-*) alkene is the only product. The use of deuterated boron compounds (e.g., the hindered 9-^{2}H-9-borabicyclo[3.3.1]nonane [9–^{2}H–9–BBN]), commercially viable since it contains only one deuterium (^{2}H), followed by use of deuterated acetic acid (CH$_3$CO$_2$^{2}H, i.e., reductive workup) produces *Z*- (or *cis-*)-dideuteroalkene of high stereospecificity and in high yield (Scheme 6.65).

III. Addition

Addition reactions to the carbon–carbon triple bond resemble those of the carbon–carbon double bond, continuing the trend already noted for **Oxidation** and **Reduction** reactions. However, alkynes are significantly more rare than alkenes and terminal alkynes (i.e., compounds bearing the functional group –C≡C–H) are among the least common. Therefore, the utility of the presence (or absence) of the sharp carbon–hydrogen IR band (between 3320 and 3310 cm^{-1}) found in these compounds is limited. On the other hand, the typical carbon–carbon triple bond band (between 2250 and 2100 cm^{-1}), while weak to moderate, is particularly valuable (unless the compound is symmetrical) since so few absorbances are seen in that region and its disappearance during the course of a reaction is easily monitored.

The ^{1}H resonance for the proton on the terminal carbon of a 1-alkyne (δ = 2.5–2.7 ppm) also has limited utility because terminal alkynes are so rare and the ^{13}C

**H. Lindlar, F. Hoffmann-LaRoche & Co. Ltd., Basel, Switzerland.*

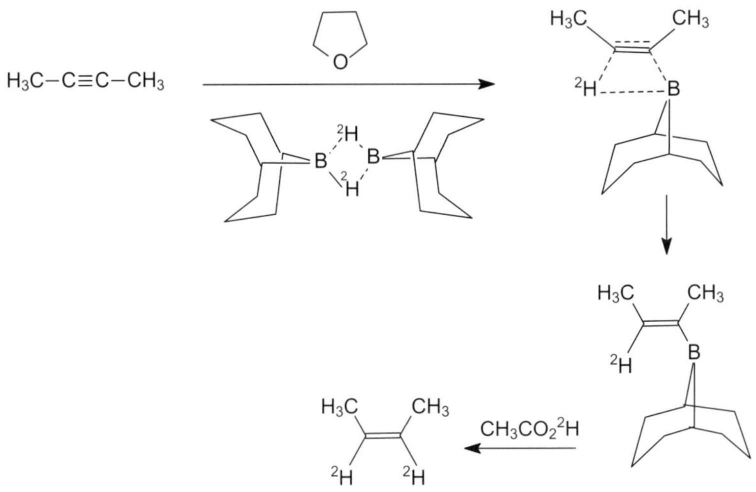

Scheme 6.65. A potential pathway for the production of Z-2,3-[^{2}H$_2$]-2-butene by reductive hydroboration of 2-butyne.

resonances (found between $\delta = 65$ ppm and $\delta = 80$ ppm) for the alkyne carbons occur at the same place in the ^{13}C spectrum as those for many other carbons. Again, however, the change between the initial spectrum and that seen after a reaction can prove particularly valuable.

Depending upon the specific conditions and reagents, reactions at the triple bond can be more or less facile than those at the corresponding double bond. As already pointed out, hydrogenation of the triple bond occurs *more* readily than does hydrogenation of the double bond. Further, it appears that although the triple bond is *less* reactive toward some **electrophilic** reagents (e.g., bromine, Br$_2$), it is *more* reactive toward others (e.g., hydrogen chloride, HCl).

While nucleophilic addition reactions appear to occur with greater facility at the carbon–carbon triple bond than they do at the corresponding double bond, the reactions of carbenes and radicals, as well as electrocyclic processes, are comparable.

a. Electrophilic Addition. Because the electron density between the two carbon atoms at the termini of the carbon–carbon triple bond is higher than that in a corresponding alkene, it would appear reasonable that "electron-seeking reagents" (i.e., electrophiles) should more readily attack the triple bond (Figure 6.13) than the double bond.

Interestingly, however, despite that logical prediction, and the experimental observation that, indeed, some reactions with electrophiles are more facile with alkynes than with alkenes, the principle is frequently violated. Thus, it appears that *when a bridged intermediate* is required, the alkyne is significantly less reactive while in reactions that appear to involve proton (H$^+$) transfer the reactivities are at least comparable. It has been argued that this is so because the *vinylic* carbocation (Figure 6.14) generated in the reaction of the triple bond with an electrophile must remain *sp*-hybridized (or nearly so). Alternatively, a bridged ion, which might be

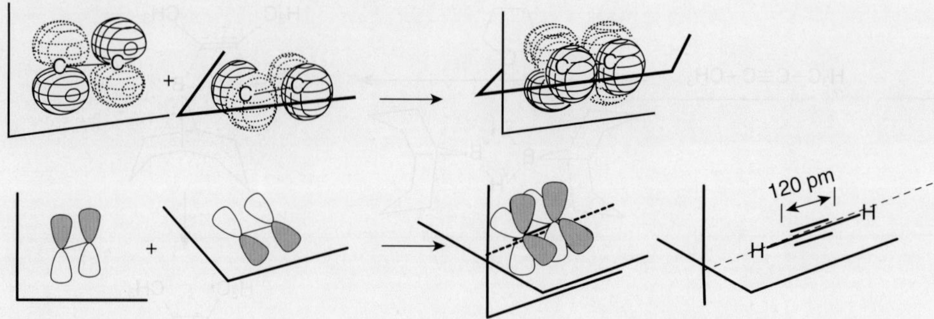

Figure 6.13. Representations of the overlap of *p*-type orbitals to produce the carbon–carbon triple bond. The representations were created using the MOViewer in WebMO version 6.0.002p.

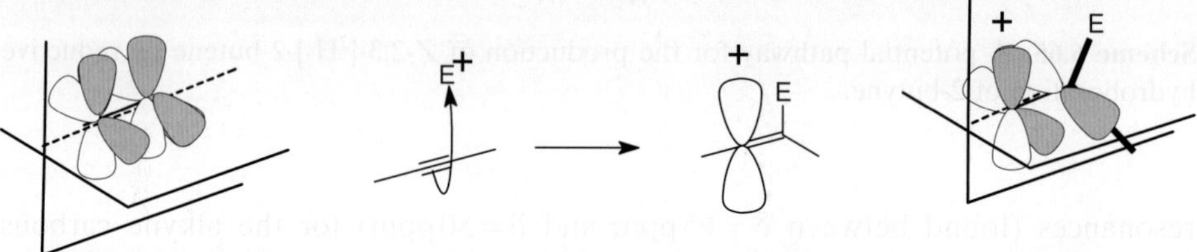

Figure 6.14. A representation of the formation of a carbocation by addition of an electrophile to an alkyne.

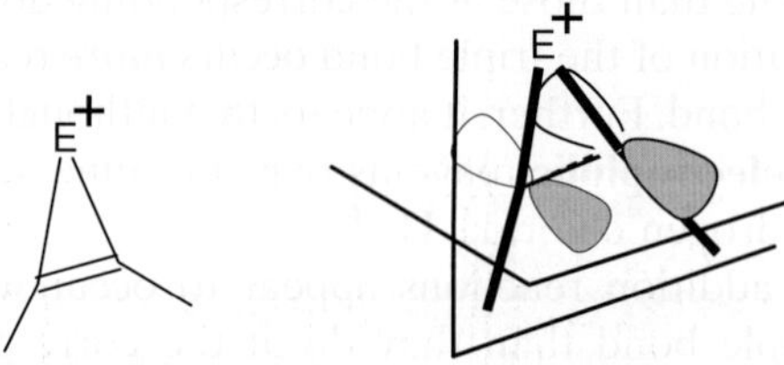

Figure 6.15. A representation of a bridged " 'onium" ion generated by electrophilic addition to an alkyne.

perceived as benefiting from the nearness of the carbon termini, would formally resemble a destabilized cyclopropenium ion (Figure 6.15).

Further, when double and triple bonds are both present in the same substrate (*vide infra*), their relative reactivity appears to depend upon what is being added. Thus, generalizations regarding the *relative* reactivities of alkenes and alkynes can be made only within narrow limits.

In a reaction that is similar to that for alkenes, all four hydrogen halides, hydrogen fluoride (HF), hydrogen chloride (HCl), hydrogen bromide (HBr), and hydrogen iodide (HI), can be added in an ionic fashion across the carbon–carbon triple bond as shown in the first example of Table 6.7. In each case the carbon–carbon triple bond (seen between 2250 and 2100 cm^{-1} in the IR) vanishes and a signal at $\delta \approx 5.5$ ppm appears in the ^{1}H NMR spectrum.

TABLE 6.7. Representative Addition Reactions between Electrophiles and Alkynes

Number	Reagent	Product
1	HX (X = Cl, Br)	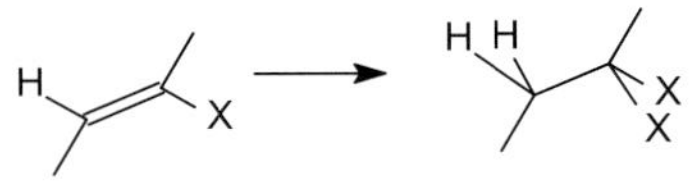
2	$H_2SO_4/H_2O/HgSO_4$	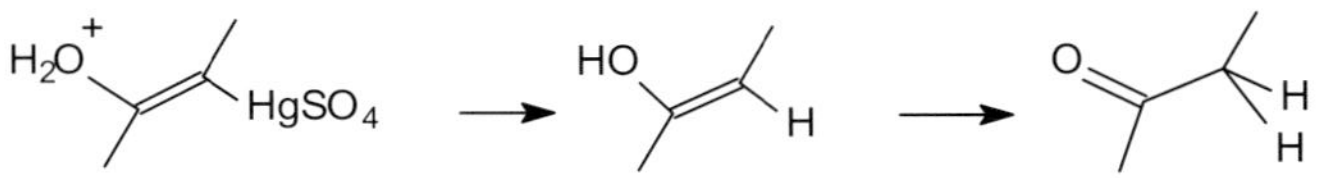
3	$HgSO_4/RCO_2H$	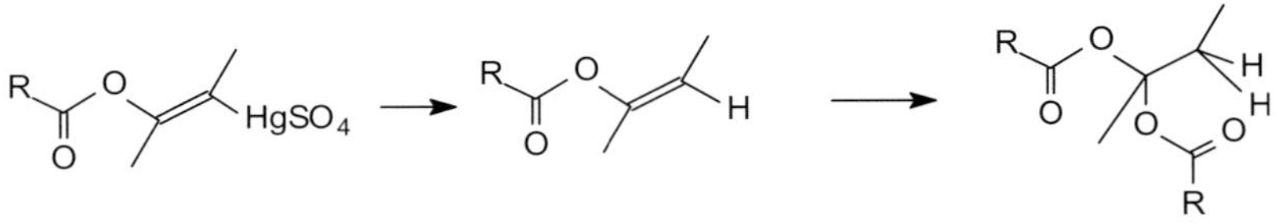
4	HOX (X = Cl, Br)	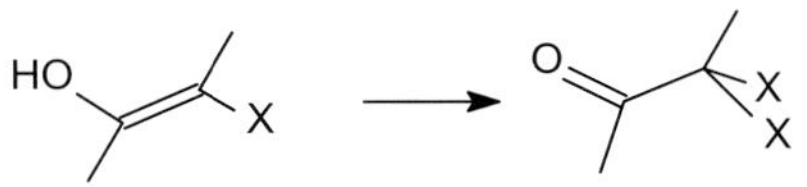
5	X_2 (X = Cl, Br)	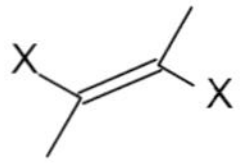
6	R_2BH	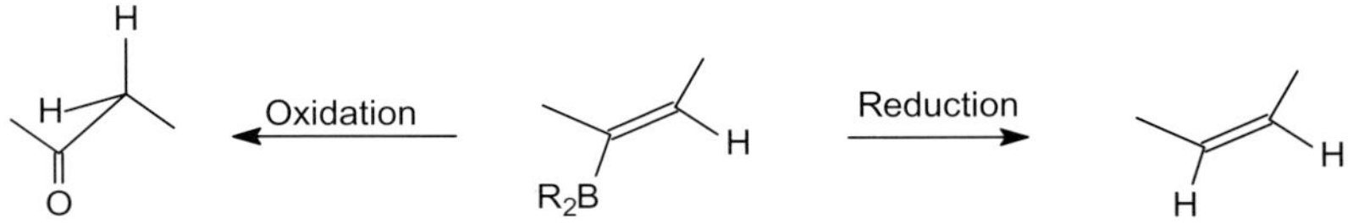

An incomplete but representative list of addition reactions between electrophiles and alkynes.

Under proper conditions it is possible to add one or two equivalents of acid producing, first, the vinyl halide and then, because a cation on carbon bearing a halogen is generally more stable than a simple secondary vinyl carbocation, a *geminal* (rather than a *vicinal*) dihalide (Scheme 6.66). For nonsymmetrical alkynes (e.g., 2-pentyne) *both* sets of vinyl and geminal halides are obtained and, if the solvent (e.g., methanol [CH_3OH], or acetic acid [CH_3CO_2H]) is capable of partici- pating, solvent-incorporated products (as was the case with alkenes) are also found. Finally, the addition appears to produce vinyl halide product that is the result of *anti-* [or (*E*)-, *trans-*, or antarafacial] addition.

The generalized scheme for the addition of hydrogen halides to alkynes in solu- tion, as shown for the specific case of 2-butyne in Scheme 6.66, depicts a bimolecular process. Therefore, it might be expected to be, experimentally, first order each in alkyne and hydrogen halide (and thus second order overall) as shown in Equation 6.67.

The actual process (at least in solution) appears more complicated.

$$-d[\text{alkyne}]/dt = -d[\text{HCl}]/dt = +d[\text{vinyl halide}]/dt = k_o[\text{HCl}]^1[\text{alkyne}]^1 \qquad (6.67)$$

The acid-catalyzed hydration of alkynes (Table 6.7, example 2) is commonly carried out using mercury (II) salts, such as mercuric sulfate ($HgSO_4$), as catalysts. The addition (Scheme 6.67) appears to involve a bridged mercurinium ion, which, for unsymmetrical cases such as 1-alkynes other than ethyne (acetylene [$HC\equiv CH$]), is subsequently attacked by water (H_2O) at the carbon that best supports a positive charge. The regiochemistry of **Markownikoff** addition, seen with alkenes, is followed.

With unsymmetrical internal alkynes, for example, 2-pentyne, the mercurinium ion appears to be attacked equally well by water at both carbon termini.

It is argued that the intermediate bearing both mercury (Hg) and a hydroxyl group (OH) attached to the double bond then loses mercury (Hg^{+2}) to produce the enol, which is generally nonisolable and which rapidly undergoes proton tautomer-

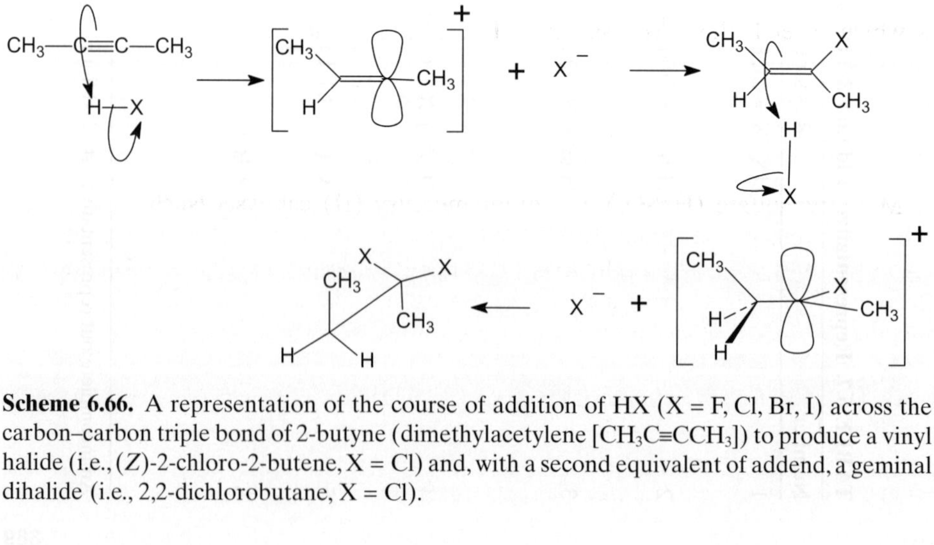

Scheme 6.66. A representation of the course of addition of HX (X = F, Cl, Br, I) across the carbon–carbon triple bond of 2-butyne (dimethylacetylene [$CH_3C\equiv CCH_3$]) to produce a vinyl halide (i.e., (*Z*)-2-chloro-2-butene, X = Cl) and, with a second equivalent of addend, a geminal dihalide (i.e., 2,2-dichlorobutane, X = Cl).

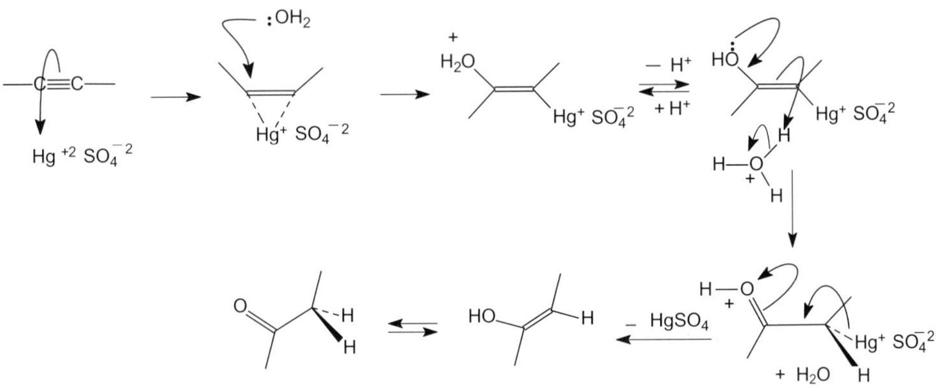

Scheme 6.67. A potential pathway for the mercuric sulphate-catalyzed hydration of an alkyne. The enol form of the carbonyl containing product is usually not isolable.

ization to acetaldehyde (if the starting material was ethyne [acetylene, $HC \equiv CH$]) or ketone(s). Some spectroscopic evidence for the intermediacy of an enol has been reported but, regardless, it is clear that in the products, the IR band assigned to the carbon–carbon triple bond in the alkyne starting material (between 2250 and $2100 \, cm^{-1}$) has been replaced by a (frequently intense) carbonyl (C=O) absorption due to the aldehyde ($1730 \, cm^{-1}$) or the ketone (around $1715 \, cm^{-1}$).

Because the sense of the addition reaction is in the "**Markownikoff**" direction terminal alkynes, with the exception of ethyne (acetylene, $HC \equiv CH$), will always generate methyl ketones.

Problem 6.12. What are the products obtained on the mercury (II)-catalyzed hydration of (a) ethyne (acetylene, $HC \equiv CH$), (b) propyne (methylacetylene, $CH_3C \equiv CH$), and (c) 2-butyne (dimethylacetylene, $CH_3C \equiv CCH_3$)? Because ethyne (acetylene, $HC \equiv CH$) and 2-butyne (dimethyl-acetylene, $CH_3C \equiv CCH_3$) are symmetrical, the carbon–carbon triple bond stretching frequency is IR inactive (Raman active) and ^1H and ^{13}C NMR spectroscopies are necessary to determine which starting material is which. Sketch the NMR spectra (^1H and proton-decoupled ^{13}C) for all three alkynes.

Problem 6.13. Sketch the ^1H and the proton-decoupled ^{13}C NMR spectra of the three products obtained in Problem 6.12.

Mercuric sulfate ($HgSO_4$) and other mercury (II) catalysts {such as mercuric ethanoate (mercuric acetate [$Hg(O_2CCH_3)_2$])} can also serve to help produce enol esters or esters of carbonyl hydrates (acylals) (Table 6.7, example 3) as shown in Scheme 6.68. The pathway is analogous to that shown in Scheme 6.67 save that a second equivalent of carboxylic acid adds to the double bond formed.

As shown in the fourth example of Table 6.7, hypochlorous and hypobromous acids (HOCl and HOBr, respectively) react at the carbon–carbon triple bond to produce the corresponding α,α-dihaloketone (or dihaloethanal or acetaldehyde, CH_3CHO) derivative in the case of ethyne (acetylene, $HC \equiv CH$, itself). It is not clear whether the α-halo enol initially formed rapidly adds a second equivalent of hypo-

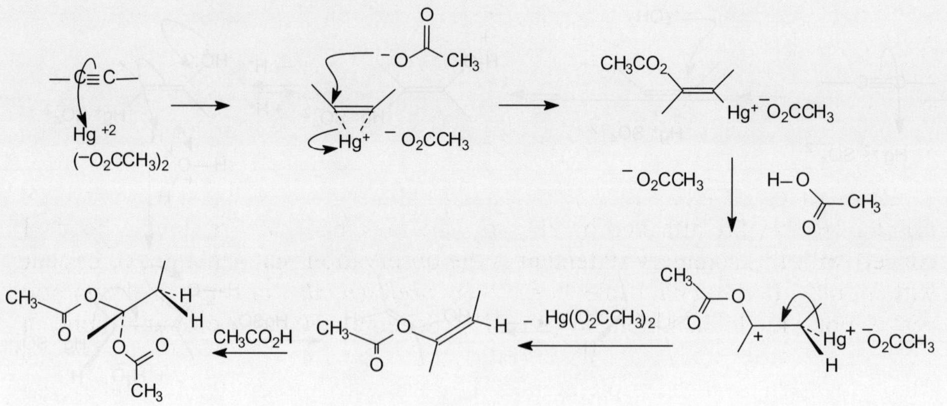

Scheme 6.68. A potential pathway for the mercuric ethanoate (mercuric acetate [Hg(O$_2$CCH$_3$)$_2$]) catalyzed electrophilic addition of ethanoic acid (acetic acid [CH$_3$CO$_2$H]) to an alkyne to form an enol ethanoate (enol acetate) and then, with excess acid, the corresponding diester (an acylal) of ethanal (acetaldehyde, CH$_3$CHO) or ketone hydrate.

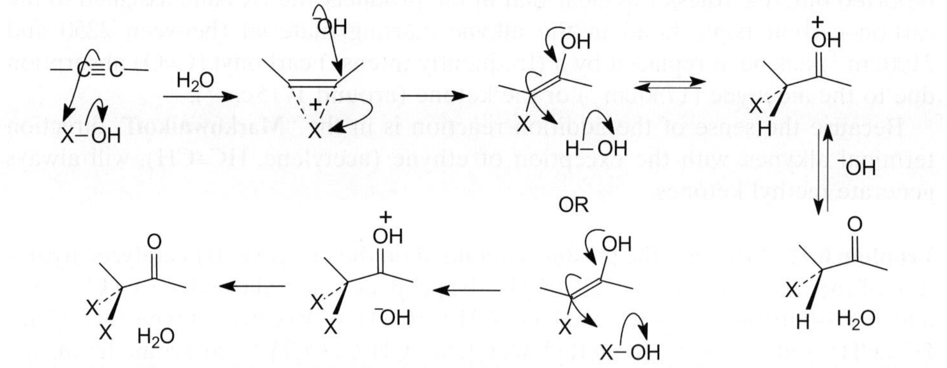

Scheme 6.69. A representation of a potential pathway for the reaction between an alkyne and hypohalous acid (X = Cl, Br). Although it is possible that an α-halo carbonyl derivative (aldehyde or ketone) might form, the result of enol–keto equilibrium will insure that the α,α-dihalo carbonyl derivative results (as long as sufficient hypohalous acid is present).

halous acid or if the equilibrium between α-halo enol and α-halo ketone intrudes. It is assumed that the initial addition is *anti-* [or *trans*, or (E), or antarafacial] by analogy with other additions where rearrangement is less facile. The potential sets of reactions involved are shown in Scheme 6.69.

As a general rule, and as shown in the fifth example in Table 6.7, bromine (Br$_2$) and chlorine (Cl$_2$) in, for example, carbon tetrachloride (tetrachloromethane, CCl$_4$) solution, add across the carbon–carbon triple bond to give *trans-* [antarafacial, *anti-*, or (E)] dibromoalkene. Also, as a general rule (albeit with a very limited number of comparisons), these addition reactions occur more slowly with alkynes than they do with alkenes and a second equivalent of halogen can add to produce the tetrahalide (Equation 6.68).

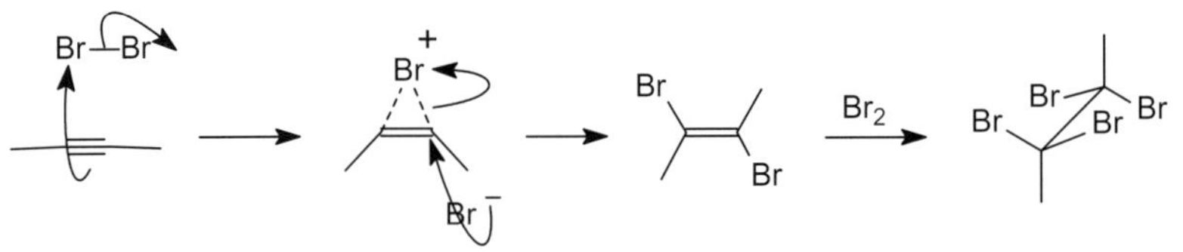

$$(6.68)$$

Examination of many such addition reactions indicates that triple bonds are, *usually*, less readily attacked by electrophilic reagents than are double bonds. In concert with this summary statement is the observation that generally, compounds with double bonds *and* triple bonds will *preferentially* undergo hydroboration with 9-borabicyclo[3.3.1]nonane (9-BBN) (Scheme 6.70) at the double bond *but* will *preferentially* undergo hydroboration with bis(2,4,6-trimethylphenyl)borane (dimesitylborane) at the triple bond. (The former hydroborating reagent is more reactive than the latter, which is more sterically demanding.)

Although not specified in the table, electrophilic addition to conjugated enynes (as in Scheme 6.70) appears to provide both 1,2- and 1,4-addition simultaneously. Thus, as shown in Scheme 6.71, the addition of hydrogen chloride across the conjugated enyne, but-3-en-1-yne (vinylacetylene, $HC{\equiv}C{-}CH{=}CH_2$) produces both 2-chloro-1,3-butadiene (chloroprene [$H_2C{=}C(Cl)CH{=}CH_2$]) and 4-chloro-1,2-butadiene ($H_2C{=}C{=}CHCH_2Cl$).

Finally, as noted in the last example in Table 6.7, triple bonds can undergo *mono*-hydroboration to produce vinylic boranes (both regiochemical adducts) and the latter oxidized with basic hydrogen peroxide to aldehydes (terminal alkynes) or ketones (nonterminal alkynes) or reduced, with acetic acid, to *Z*-alkenes (Scheme 6.72).

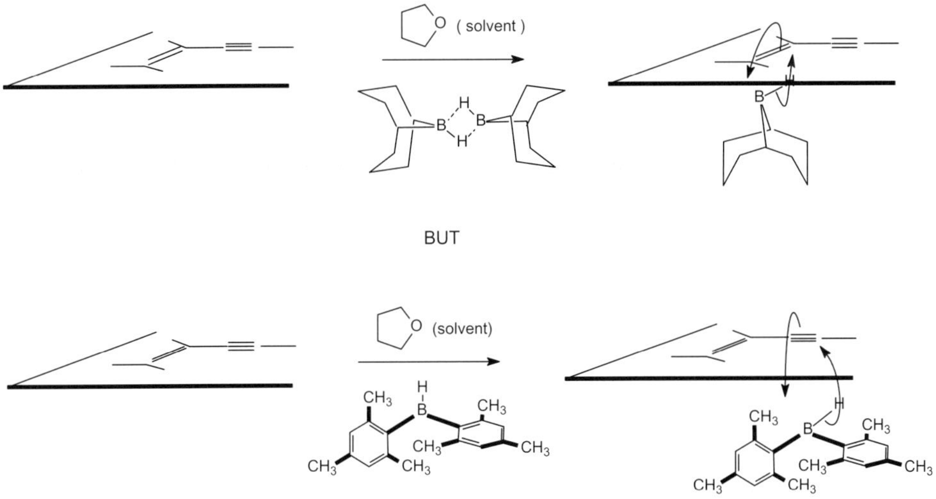

BUT

Scheme 6.70. The hydroboration of an enyne showing that 9-borabicyclo[3.3.1]-nonane (9-BBN) preferentially undergoes hydroboration at the carbon–carbon double bond but that the more sterically demanding bis(2,4,6-trimethylphenyl)borane (dimesitylborane) preferentially reacts at the carbon–carbon triple bond.

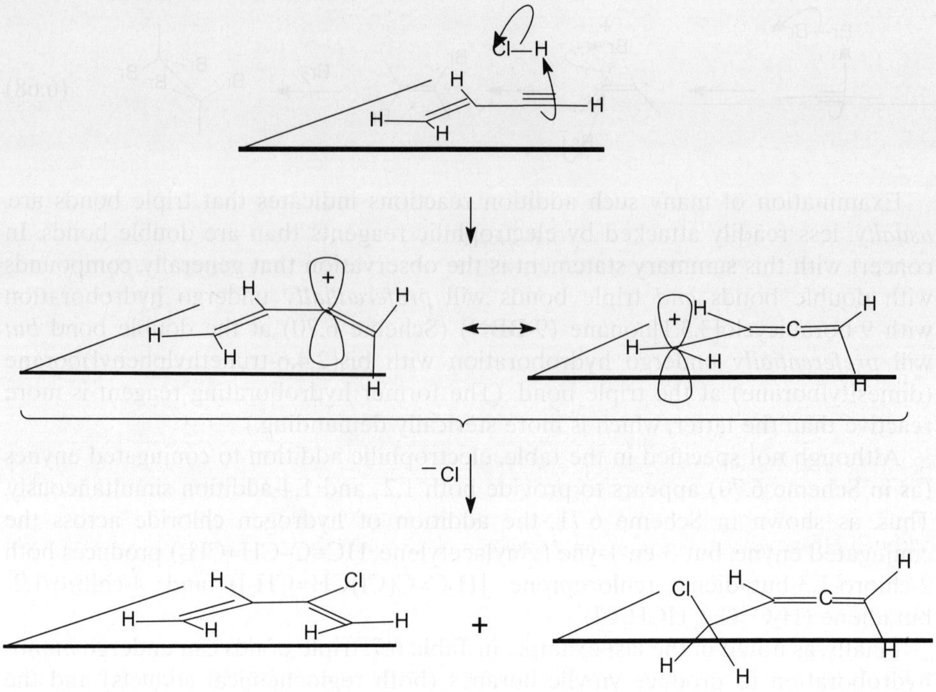

Scheme 6.71. A representation of the addition of hydrogen chloride to but-3-en-1-yne (vinylacetylene [HC≡C–CH=CH$_2$]) to produce 2-chloro-1,3-butadiene (chloroprene [H$_2$C=C(Cl)CH=CH$_2$]) and 4-chloro-1,2-butadiene (H$_2$C=C=CHCH$_2$Cl).

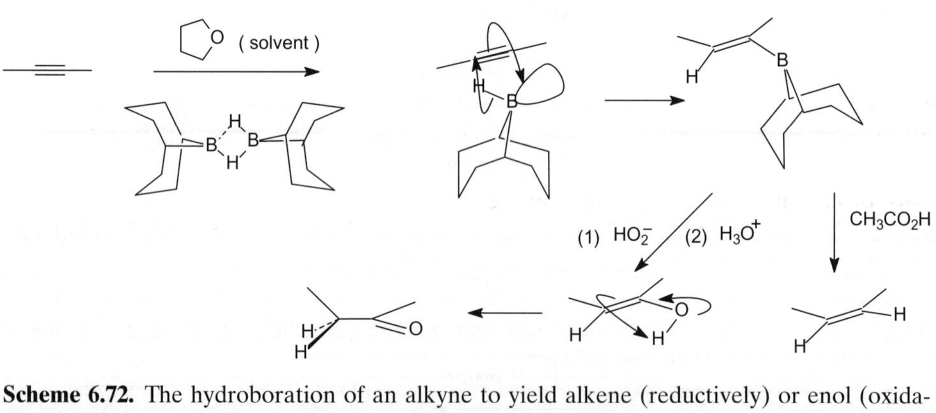

Scheme 6.72. The hydroboration of an alkyne to yield alkene (reductively) or enol (oxidatively). The latter usually tautomerizes to ketone or aldehyde.

b. Nucleophilic Addition to Alkynes and Conjugated Enynes.

Despite the higher electron density between carbon atoms held together by a triple bond, the addition of electron-rich nucleophiles to alkynes is generally more facile than the same addition to alkenes. It has been argued that the developing sp^2-type carbanion enjoys advantages over what would be the corresponding sp^3 species in an

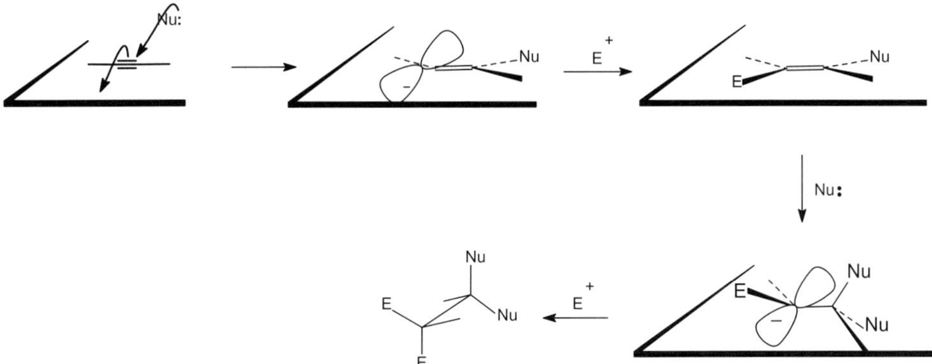

Scheme 6.73. A schematic representation of nucleophilic addition to an alkyne. The sequence of addition of nucleophile and (subsequently) electrophile (to consummate the reaction) is the opposite of that in electrophilic addition.

alkene. However, it is clear, as expected, that electron-withdrawing substituents on the unsaturated system, carbon–carbon triple or double bond, help the attack of nucleophile.

As shown in Scheme 6.73, it is the timing of the steps and the nature of the attacking species that determines the addition is nucleophilic rather than electrophilic. The nature of the process is further confirmed by the examples in Table 6.8.

Thus, as shown in the first and second examples in Table 6.8, the addition of ethanethiol (thioethane, ethyl mercaptan, CH_3CH_2SH) as well as ethanol (CH_3CH_2OH) across the triple bond of ethyne (acetylene, $HC{\equiv}CH$) is catalyzed by the corresponding metal salts and requires high temperature (at which high pressures will develop). These reactions are *not* generally observed in the presence of acid.

The third example in Table 6.8 is particularly interesting since the "product" shown is an intermediate in a general method for the synthesis of various trisubstituted alkenes. Thus, the alkylcopper–magnesium bromide species (the **Normant* reagent**) adds to a primary alkyne so that the alkyl and metal addends are *Z*- (or *syn*-, or *cis*- or suprafacial) with the metal at C-1. In a second (and thus separate) step *and* with stereochemical retention (i), a second alkyl group can be added—as an alkyl iodide, for example, iodoethane (ethyl iodide, CH_3CH_2I)—which then *replaces* the metal; (ii) a carboxylic acid can be formed by treatment of the copper reagent with carbon dioxide (CO_2); or (iii) iodine (I_2) itself can be used so that a vinyl iodide results (Scheme 6.74).

It is generally argued that as shown in example 4, Table 6.8, the addition of hydrogen cyanide (HCN) to ethyne (acetylene, $HC{\equiv}CH$) is a nucleophilic rather than an electrophilic process because the reaction is made more facile by the presence of electron-withdrawing substituents attached to the carbon of the triple bond. Indeed, although more reluctantly (thus necessitating more active catalysts), even alkenes with electron-withdrawing substituents (see Table 6.2) undergo the same reaction. The specific example in Table 6.8, propenenitrile (acrylonitrile, $CH_2{=}CH{-}CN$), has been prepared commercially this way.

*Normant, J., Professor, University P.& M. Curie, Paris, France.

TABLE 6.8. Representative Addition Reactions between Nucleophiles and Alkynes

Number	Reactant	Reagent	Catalyst	Product
1	$H-\!\!\equiv\!\!-H$	CH_3CH_2SH	$CH_3CH_2S^- K^+$ (160°C)	$CH_2=C(SCH_2CH_3)(H)$
2	$H-\!\!\equiv\!\!-H$	CH_3CH_2OH	$CH_3CH_2O^- Na^+$ (160°C)	$CH_2=C(OCH_2CH_3)(H)$
3	$CH_3-\!\!\equiv\!\!-H$	$CH_3Cu(MgBr_2)$		$[(CH_3)(CH_3)C=C(H)(Cu(MgBr_2))]$
4	$H-\!\!\equiv\!\!-H$	$H-C\equiv N$	$CuCl$	$(H)(H)C=C(H)(C\equiv N)$
5	$H-\!\!\equiv\!\!-H$	$H-\!\!\equiv\!\!-H$	$CuCl/NH_4Cl$	$H-C\equiv C-CH=CH_2$

These are representative examples of nucleophilic addition to the triple bond of alkynes. If the triple bond bears an electron-withdrawing group the nucleophilic addition is much more facile (as was pointed out for the same reaction on the carbon–carbon double bond, *q.v.*)

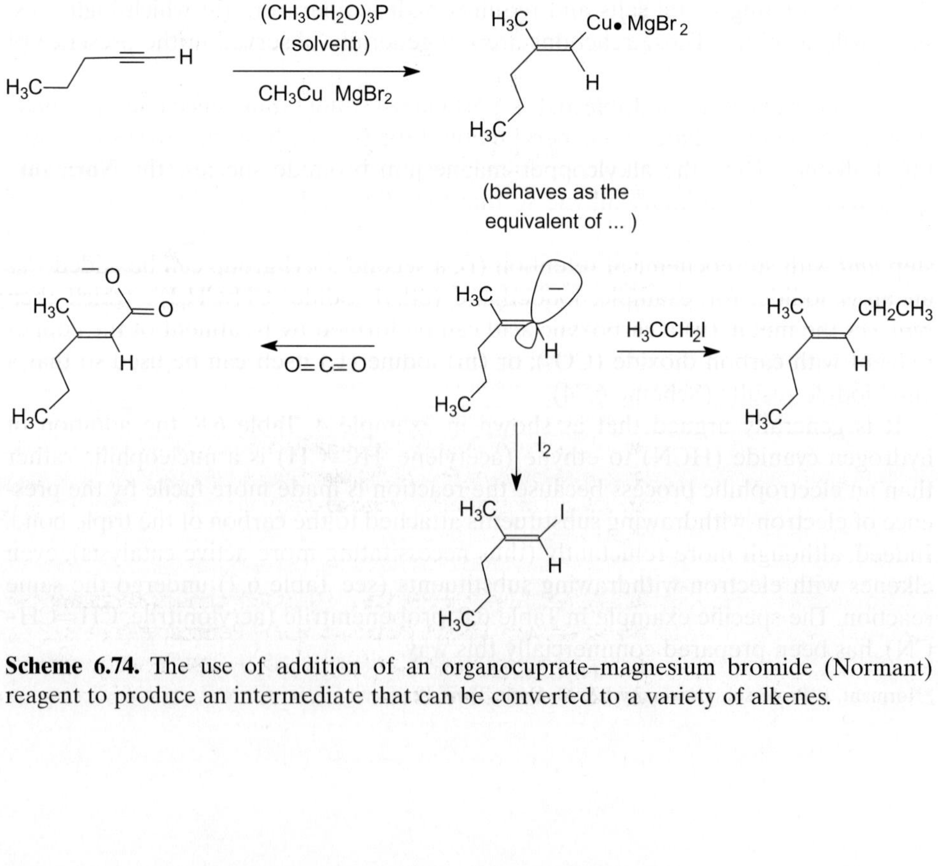

Scheme 6.74. The use of addition of an organocuprate–magnesium bromide (Normant) reagent to produce an intermediate that can be converted to a variety of alkenes.

Finally, but-3-en-1-yne (vinylacetylene, HC≡C–CH=CH$_2$) can be prepared (Table 6.8, example 5) by the catalyzed dimerization of ethyne (acetylene, HC≡CH) itself. Interestingly, this is only one of the possiblities of reaction of this alkyne (and substituted alkynes) with itself (or themselves). For example, as shown in Scheme 6.75, treatment of a disubstituted alkyne, such as 2-butyne (dimethylacetylene, CH$_3$C≡CCH$_3$) with zirconcene hydridochloride ([Cp]$_2$ZrHCl, where Cp = cyclopentadienyl [see, Table 6.6]) apparently causes the formation of a vinylic zirconium intermediate. When the intermediate is treated with methyllithium (CH$_3$Li) and then a second alkyne (or a second equivalent of the original) added, and this is followed by aqueous acid, a conjugated diene results.

Through an apparently less complicated process that nonetheless requires complexation with an organometallic catalyst, ethyne (acetylene, HC≡CH) itself undergoes trimerization in the presence of dicobalt octacarbonyl [Co$_2$(CO)$_8$] (Equation 6.69) to benzene and (Equation 6.70) tetramerization in the presence of nickel(II) cyanide [Ni(CN)$_2$] to cyclooctatetraene.

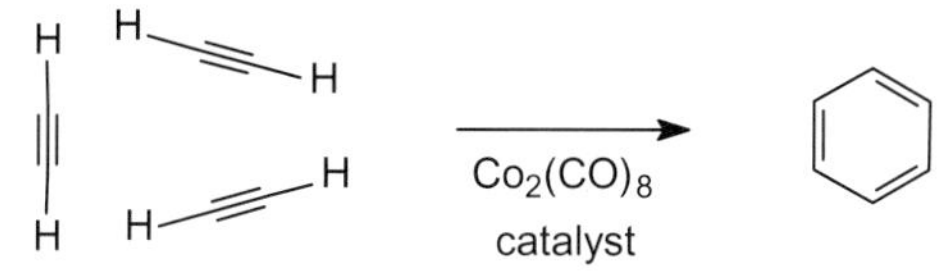

$$(6.69)$$

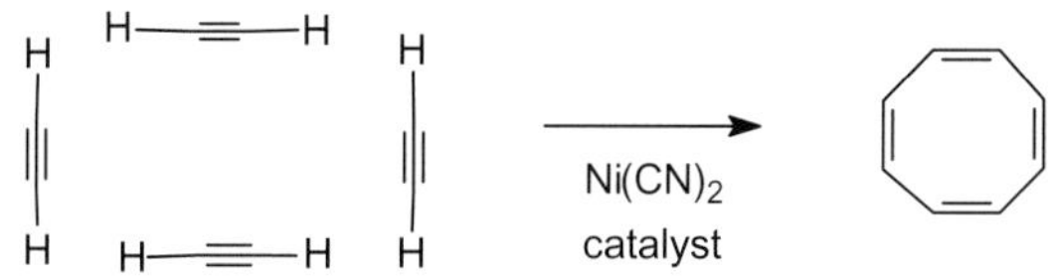

$$(6.70)$$

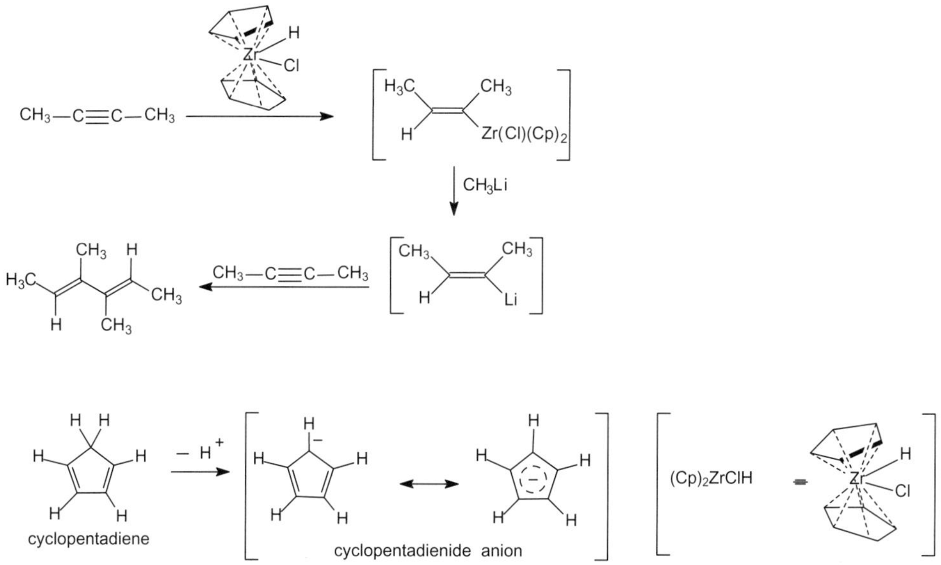

Scheme 6.75. The formation of (2*E*,4*E*)-3,4-dimethyl-2,4-hexadiene by the zirconecene hydridochloride-catalyzed dimerization of 2-butyne and its use in the dimerization of 2-butyne.

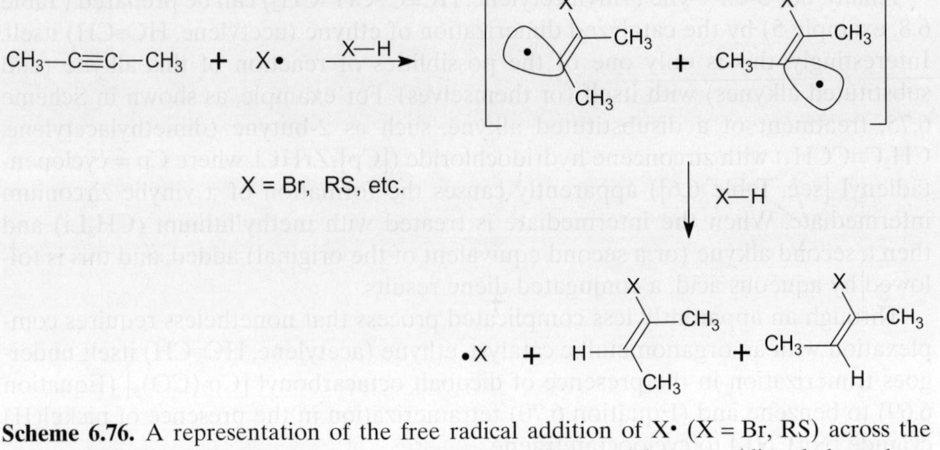

Scheme 6.76. A representation of the free radical addition of X· (X = Br, RS) across the carbon–carbon triple bond of a symmetrical alkyne, 2-butyne (dimethylacetylene, $CH_3C{\equiv}CCH_3$).

c. Radical Addition to Alkynes.

The addition of free radical species (Br·, RS· [R = alkyl]) to alkynes has been investigated much less than has the corresponding reaction with alkenes. However, it appears that, in general, radicals add less readily to alkynes than to similar alkenes (although electron-withdrawing substituents facilitate the reaction for both). Additionally, the vinylic radicals formed from reaction with alkynes (Scheme 6.76) appear to enjoy less stabilization from attached unsaturated substituents than did the corresponding alkyl radicals resulting from addition to alkenes.

d. Intermolecular Cheletropic and Other Cycloaddition Reactions.

As was the case with alkenes (*qv*), simple alkynes react with carbenes (example 1, Table 6.9, **as the dienophilic** component in the Diels–Alder reaction (example 2, Table 6.9), as well as in [2 + 2] cycloaddition reactions and with 1,3-dipolarophiles (examples 3 and 4, respectively, Table 6.9). The final reaction in Table 6.9 (example 5) involves the cyclization of enediynes to produce transient diradical species. This reaction, now called the **Bergman*** cyclization is important in converting naturally occurring (and synthetic) antibiotics (Chapters 8 and 11) found to contain the enediyne group (Figure 6.17) into an active form.

The first item in Table 6.9 provides an example of the addition of two equivalents of a carbene, methylene (:CH_2), an electrophilic species, to the alkyne, 2-butyne (dimethylacetylene, $CH_3C{\equiv}CCH_3$) to provide the bicyclobutane (1,3-dimethylbicyclo[1.1.0]butane). With ethyne (acetylene, $HC{\equiv}CH$) itself, as shown in Equation 6.71, the expected cyclopropene cannot be isolated because (apparently) it rearranges to 1,2-propadiene (allene, $H_2C=C=CH_2$).

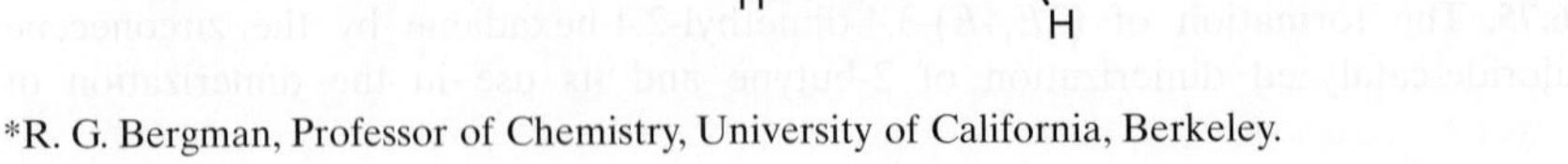

$$H{-}{\equiv}{-}H \;+\; {:}CH_2 \longrightarrow \qquad \longrightarrow H_2C{=}C{=}CH_2 \qquad (6.71)$$

*R. G. Bergman, Professor of Chemistry, University of California, Berkeley.

TABLE 6.9. Some Examples of Cycloaddition Reactions of Alkynes

Number	Addend	Alkyne	Product(s)
1	$^{1}:CH_2$	$CH_3-C\equiv C-CH_3$	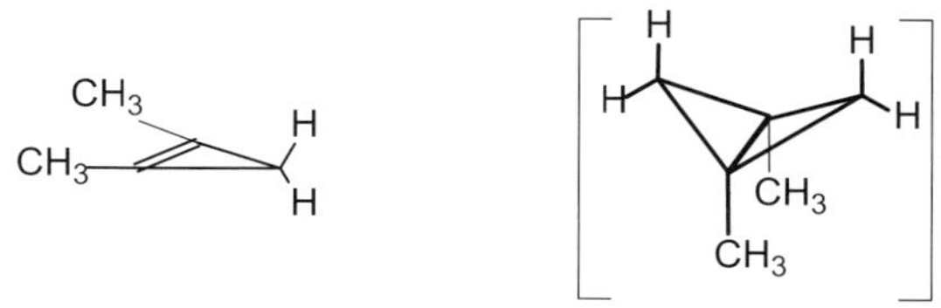
2	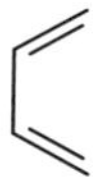	$CH_3-C\equiv C-CH_3$	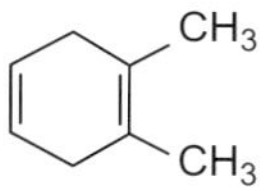
3	$CF_2=CF_2$	$HC\equiv C-CH=CH_2$	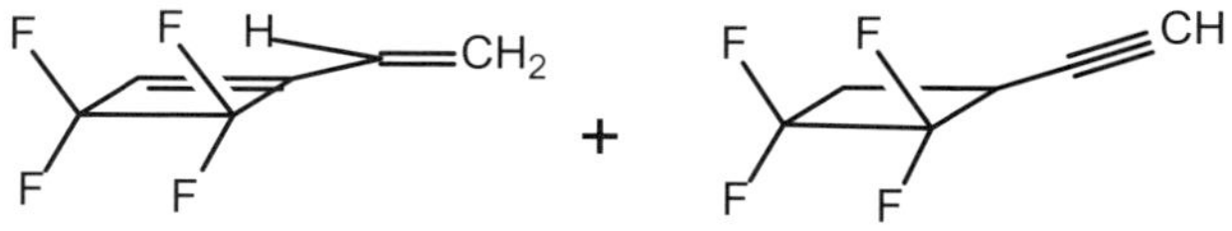
4	$[C_6H_5-N=\overset{+}{N}=\overset{-}{N} \leftrightarrow C_6H_5-\overset{-}{N}-\overset{+}{N}\equiv N]$	$CH_3-C\equiv C-CH_3$	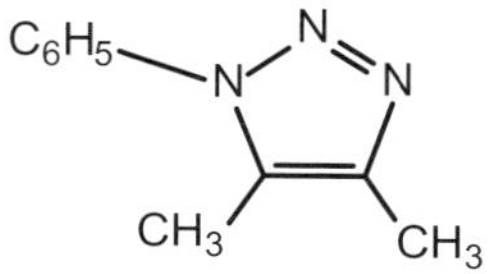
5		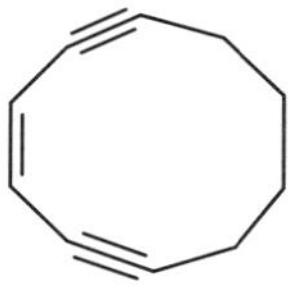	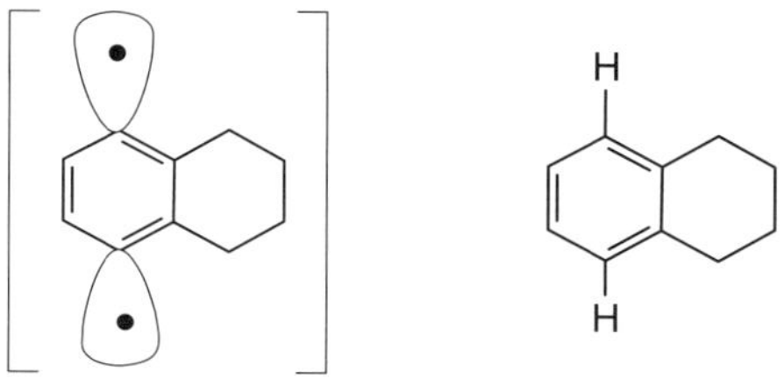

The second example in Table 6.9 is the (electrocyclic [4 + 2]) Diels–Alder reaction similar to that already discussed for alkenes and while ethyne (acetylene, HC≡CH) can serve, it is generally found that electron-withdrawing substituents (as was also true for alkenes, *qv*) on the triple bond facilitate the reaction. Conjugated alkynes (or eneynes), which lack the appropriate "cisoid angles," do not serve as the "diene" component (but see example 5, Table 6.9).

In example 3 in Table 6.9, the [2 + 2] addition reaction of 1,1,2,2-tetrafluoroethene to *both* the double and triple bonds of but-3-en-1-yne (vinylacetylene, HC≡C–CH=CH$_2$) is shown. Both addition reactions appear to occur with about equal facility (a 1:1 mixture is obtained) and the process may involve free radicals.

The self-dimerization of ethyne (acetylene, HC≡CH) would give rise to the "4n" antiaromatic compound cyclobutadiene and, although the metal-catalyzed trimerization and tetramerization (*vide supra*) occur, the reaction (Equation 6.72) is not observed. Interestingly, cyclobutadiene and substituted cyclobutadienes have been prepared by several more circuitous routes and one of the ways of producing cyclobutadiene (albeit temporarily) is shown in Scheme 6.77.

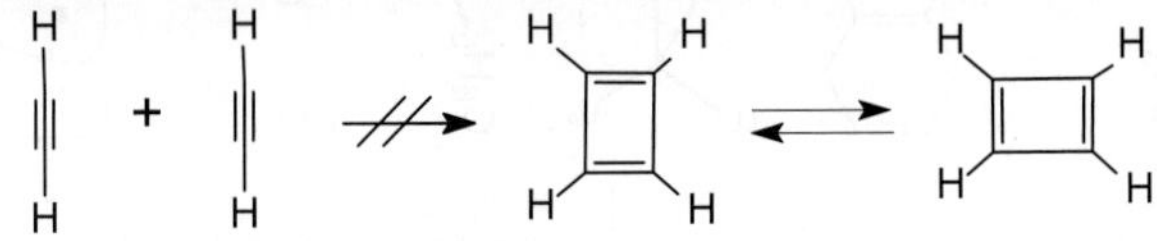

$$\text{(6.72)}$$

When the lactone (Chapter 9) 2H-pyran-2-one (2-oxo-2H-pyran, α-pyrone) is irradiated below 290 nm (with a pyrex filter to form a bicyclic lactone and, subsequently, without a filter to let in higher energy radiation) while frozen in an argon (Ar) matrix at 8 K, carbon dioxide (CO$_2$) is eliminated and a species is formed whose IR spectrum (with only C–H and C–C bands and a weak band at 1523 cm^{-1}) is in concert with a *rectangular* (not square planar) structure. On warming, decomposition to "dimer" (*endo*-tricyclo[2.0^{1,4}.0^{5,8}]octa-2,6-diene) and ethyne (acetylene, HC≡CH) occurs.

In a successful effort to stabilize the cyclobutadiene, advantage was taken of the repulsion expected between *adjacent* *t*-butyl [-C(CH$_3$)$_3$] groups to induce deviation from planarity and bond stretching. Thus, using the photochemical reactions shown

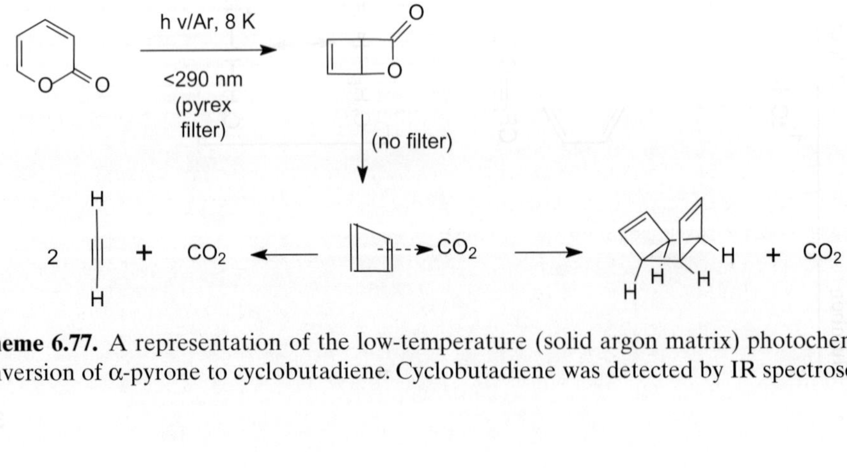

Scheme 6.77. A representation of the low-temperature (solid argon matrix) photochemical conversion of α-pyrone to cyclobutadiene. Cyclobutadiene was detected by IR spectroscopy.

in Scheme 6.78, 2,3,4,5-tetra(*t*-butyl)-cyclopentadienone was induced to lose carbon monoxide (CO) and to form the 1,2,3,4-tetra(*t*-butyl)cyclobutadiene. It has been suggested that the distortion from square planar symmetry, which would have required mutually degenerate MOs and thus (Chapter 1) (by the **Aufbau** and **Pauli Exclusion Principles**) *un*paired electrons, is necessary for the stability of the system. This idea is further elaborated upon in Figure 6.16 in which the orbital occupancy for mutually degenerate orbitals expected for a square planar system is compared with the occupancy expected for its rectangular isomer.

In contrast to the difficulty in producing cyclobutadiene by the "simple" [2 + 2] cycloaddition of alkynes, the addition of 1,3-dipolar species across the carbon–carbon triple bond of alkynes is quite facile. Thus, in addition to the fourth example of Table 6.9 showing that alkyl azides can add across the triple bond, reactions such as that of ethyne (acetylene, H–C≡C–H) with isocyanic acid (fulminic acid,

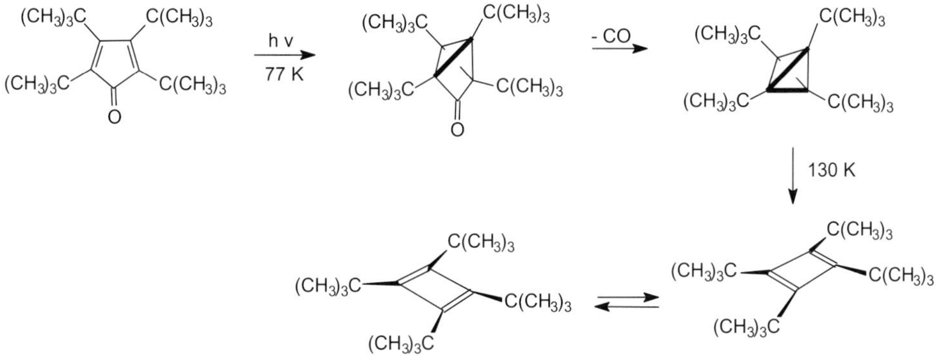

Scheme 6.78. A representation of the low-temperature (liquid nitrogen) photochemical conversion of 2,3,4,5-tetra(*t*-butyl)cyclopentadienone to 1,2,3,4-tetra(*t*-butyl)cyclobutadiene.

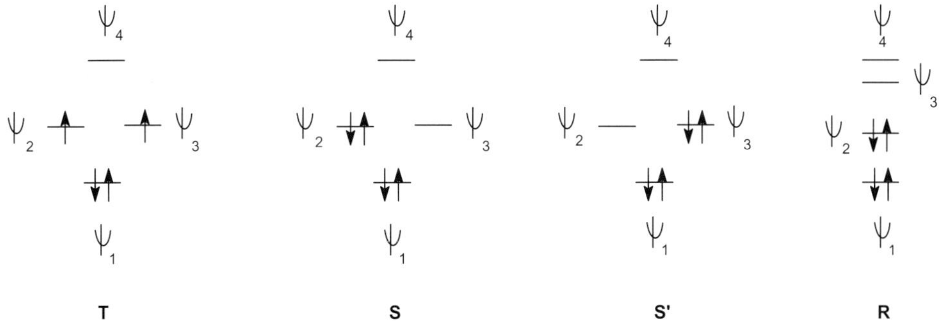

Figure 6.16. Valence bond representation for electron distribution in "cyclobutadiene." **T** is a presumed triplet state that corresponds to filling the lowest lying MO (ψ_1) with two π electrons and then placement of the remaining two π electrons in the mutually degenerate (equal energy) ψ_2 and ψ_3 orbitals (**Pauli Principle**). Because the material isolated in an argon matrix at 7 K has none of the properties of a triplet, the more energetic singlet states (**S** and **S'**), with paired electrons but with mutually degenerate, **Principle** violating, orbitals that are *not fully occupied* might obtain if cyclobutadiene were *forced to be a square*. However, distortion to the rectangular system, represented by **R**, avoids the problem.

H–C≡N⁺–O⁻) (Equation 6.73) to produce oxazolines (Chapter 10) have been observed.

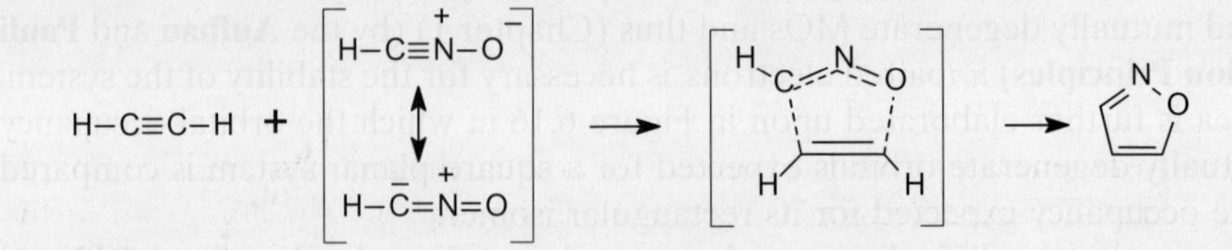

$$H-C≡C-H \;+\; \left[\begin{matrix} H-C≡N-O \\ \updownarrow \\ H-C=N=O \end{matrix}\right] \longrightarrow \left[\begin{matrix} \end{matrix}\right] \longrightarrow \qquad (6.73)$$

The final example in the table (example 5, Table 6.9) is one of a class of similar cyclizations found in several families of antibiotics. For example, it has been argued that the activity of antibiotics such as Dynemicin A (Figure 6.17) results from initial binding (intercalation) of the antibiotic into the minor groove of deoxyribonucleic acid (DNA; Chapter 14) followed by cycloaromatization of the enediyne system to produce a diradical intermediate. The diradical then irreversibly links to the DNA by hydrogen abstraction, damaging it permanently.

Interestingly (Figure 6.18), it appears that the distance between the ends of the enediyne conjugated system must be between 320 and 330 pm if the cyclization is to occur readily. In the 10-membered ring of example 5 in Table 6.9, the distance is 325 pm. If the distance is made shorter by removing one of the methylene units (–CH₂–) to make a nine-membered ring (the distance is reduced to 284 pm) then

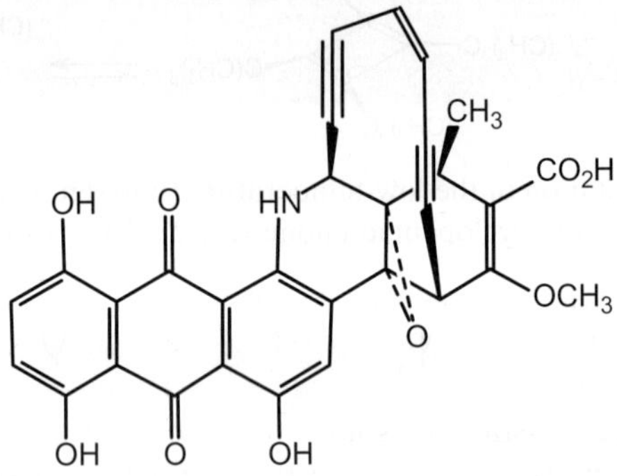

Figure 6.17. A representation of Dynemicin A.

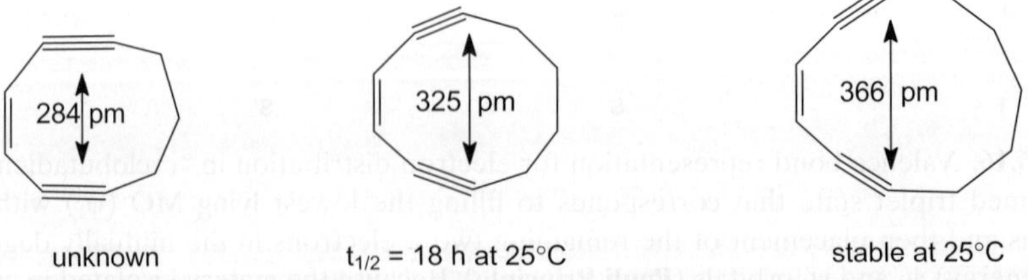

Figure 6.18. Representations of (3$\underline{Z}$)-cyclonon-3-en-1,5-diyne, (3$\underline{Z}$)-cyclodec-3-en-1,5-diyne, and (3$\underline{Z}$)-cycloundec-3-en-1,5-diyne showing the approximate distance between the termini of the respective conjugated "enediyne" systems (after Nicolaou, K. C. and Smith, A. L., in Stang, P. J. and Diederich, F. (eds.), *Modern Acetylene Chemistry*, VCH, New York, **1995**).

the enediyne is not isolable. It has been reported that the corresponding 11-membered ring enediyne (where the distance is now 366 pm) does not undergo cyclization.

E. ARENES AND AROMATICITY: SPECIAL INTRODUCTION

That group of benzenoid hydrocarbons called arenes, already introduced in Chapter 3 and again in Chapter 5, might be thought of as compounds needing "no further introduction." However, the *the reactions undergone by arenes are distinctly different from those typical of the other hydrocarbons (alkanes, alkenes, and alkynes) already considered.*

The frequently sweet-smelling, sometimes called "aromatic," compounds of carbon and hydrogen, known as arenes, burn (undergo oxidation) with a characteristically "smokey" flame and such compounds were constituents of fluids used for heating and light for some time. Indeed, it is commonly held that benzene (C_6H_6) itself was first isolated by Michael Faraday* in 1820 from a mixture of liquids called "illuminating oil" (prepared by the thermal decomposition of whale and other oils) from which it separated spontaneously. Although arenes do undergo oxidation and reduction, there is a signal difference in that, as pointed out earlier (Chapter 3), both of these processes occur with much more difficulty than with their "related" unsaturated compounds. This difference is attributed to the property of **aromaticity** (which, in turn, is related to the presence of a pair of mutually degenerate, i.e., energetically equivalent, doubly occupied bonding orbitals, *vide infra*).

The early work on benzene (C_6H_6) (which was also obtained from heating coal in the absence of air)[†] led to some confusion because combustion analysis dictated that it had to be an unsaturated compound[‡] and yet, in the laboratory, it was almost as stable to oxidizing agents such as permanganate (MnO_4^-) and solutions of the halogens (Br_2) as the saturated alkanes. Therefore, it was clear that structures corresponding to six carbons and six hydrogens (i.e., C_6H_6) containing unsaturation in the form of "normal" double and triple bonds, such as $H_2C=C=C=CH–CH=CH_2$, were not consistent with the properties of benzene. Despite some very inventive ideas, such as that due to Ladenburg[§] (Figure 6.19), which would account for a C_6H_6 lacking unsaturation, a satisfactory solution was not immediately obvious.

Subsequently, the problem was compounded by the isolation of additional materials, related to benzene (C_6H_6), from the fractionation of *coal tar (vide supra).*

---

*The research of Michael Faraday (1791–1867; Professor, University of Vienna) was largely concerned with physics although he may be remembered for the electrochemistry law (and constant), which bears his name. He is commonly acknowledged as the first person to liquefy chlorine (Cl_2) and, with characteristic caution, is said to have reported his research work on benzene (C_6H_6) in 1825, some 5 years after its isolation.

[†]The process of heating one or more of the varieties of coal to ca. 1000°C is called *destructive distillation.* The noncondensable gas is *coal gas*; the liquid condensate is *coal tar*; and the residue is *coke*, used for *smelting* iron in the steel industry.

[‡]A compound with six carbons should, of course, if it is saturated, have 14 hydrogens. Thus, benzene (C_6H_6) must have four sites (rings and/or double and triple bonds) of unsaturation.

[§]Albert Ladenburg (1842–1911) was a student of Kekulé (*vide infra*) who showed, by chemical reactions, the equivalence of the hydrogens of benzene. Ladenburg eventually became Professor at University of Breslau.

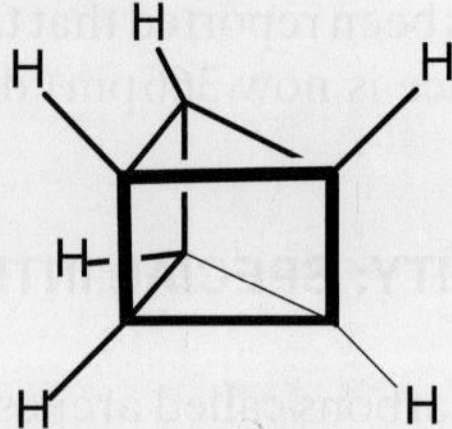

Figure 6.19. A representation, proposed by Ladenburg, of a C_6H_6 hydrocarbon with no carbon–carbon double or triple bonds.

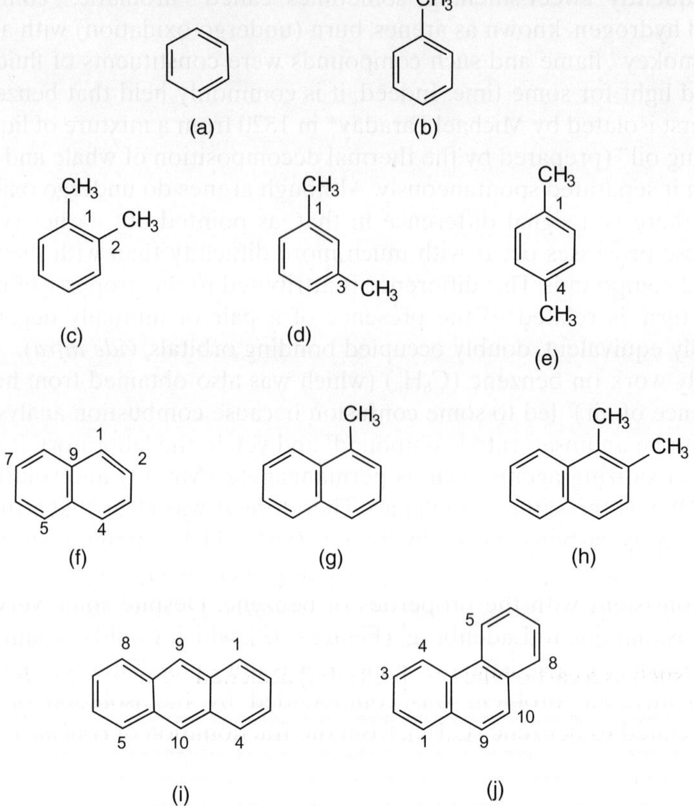

Figure 6.20. Examples of arenes and substituted arenes:(a) benzene; (b) methylbenzene (toluene); (c) 1,2-dimethylbenzene (*ortho*-xylene); (d) 1,3-dimethylbenzene (*meta*-xylene); (e) 1,4-dimethylbenzene (*para*-xylene); (f) naphthalene; (g) 1-methylnaphthalene (α-methylnaphthalene); (h) 1,2-dimethylnaphthalene (α,β-dimethylnaphthalene); (i) anthracene; (j) phenanthrene.

These included (Figure 6.20) alkyl substituted benzene derivatives such as methylbenzene (toluene, CH_3–C_6H_5, Figure 6.20b) and dimethylbenzenes {xylenes, $(CH_3)_2C_6H_4$, Figure 6.20c–e} as well as *fused* aromatics such as naphthalene ($C_{10}H_8$), Figure 6.20f, (and its methyl, Figure 6.20g, and dimethyl, Figure 6.20h, derivatives),

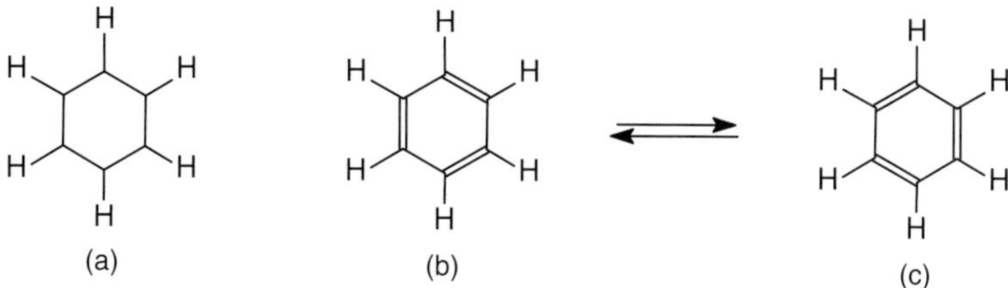

Figure 6.21. Early representations of C_6H_6. (a) A hexagonal representation that fails to account for tetravalency. (b) and (c) Hexagonal representations of C_6H_6 that are intended to show rapid equilibration between isomers of 1,3,5-hexatriene in which the double and single bond lengths are alternating too quickly for isolation.

and the isomers anthracene ($C_{14}H_{10}$), Figure 6.20i, and phenanthrene ($C_{14}H_{10}$), Figure 6.20j.

In the early part of the nineteenth century, the structures provided in Figure 6.20 were not, of course, known and the valence structures proposed by Kekulé in 1859, while successful for the alkanes and alkenes already considered in this chapter, could not accommodate the different behavior of aromatic compounds. Then, in 1865, Kekulé* and, independently, Couper[†] proposed that the structure of benzene (C_6H_6) was that of a regular hexagon with a carbon and one hydrogen at each corner (Figure 6.21a). However, since this violated the tetravalency of carbon, it was further (but subsequently) proposed that the correct structure involved a rapidly oscillating or equilibrating (and thus inseparable) pair of double bond isomers of 1,3,5-cyclohexatriene (Figure 6.21b,c).

The major difficulty of the "diminished" reactivity of benzene (C_6H_6) relative to that expected for alkenes alluded to earlier (*vide supra*) could not be resolved by this (or any other equally inventive) suggestion. Indeed, until the tools of MO theory and experimental spectroscopy were integrated into the mainstream of organic chemistry, aromaticity was largely defined in terms of "unexpected stability" as measured by heats of hydrogenation ($-\Delta H_r^\circ$, Chapter 3) and (with somewhat circular reasoning) by finding reactions that were "characteristic of aromatic compounds" (*vide infra*).

Indeed, it has only recently become clear that the planar, closed, overlapping set (or ring) of what look like "occupied *p-type* atomic orbitals" could be experimentally probed.[‡] As already discussed in Chapter 2, the π-electron system of benzene (C_6H_6) can be treated as a circular loop, in which an induced current results in dia-

*Friedrich August Kekulé (1829–1896) was a professor at the Universities of Ghent and, later, Bonn. Interestingly, before becoming involved in the problems of the structures of organic compounds he studied architecture.

[†]Archibald Scott Couper (1831–1892), a Scottish chemist, was among the first to publish structures comparable to those used today. He apparently stopped work early in his career (before he was 30) because of illness.

[‡]The set of monocyclic compounds corresponding to $(CH)_n$ are collectively known as *annulenes*. Cyclobutadiene (C_4H_4), *vide infra*, is thus [4]annulene; benzene (C_6H_6) is [6]annulene; and *cyclooctatetraene (C_8H_8) is [8]annulene*.

magnetic anisotropy,* which is observable in the NMR spectra of aromatic systems by its effect on the protons lying around the periphery of the ring (and, indeed, above and below the ring). The diamagnetic anisotropies and shielding/deshielding effects *calculated* by (among the first) J. Pople[†] has now confirmed and extended the thermochemical work as well as the studies of the reactions undergone by aromatic compounds. The experimentally observed anisotropy is definitive of aromaticity.

In a historical sense then, the stability of benzene was quantified, roughly, as follows: the heat of hydrogenation (or reduction) $[-\Delta H_r^\circ]$ of cyclohexene (C_6H_{10}) to cyclohexane (C_6H_{12}) (Figure 6.22) (Table 3.4) is $121.8\,kJ\,mol^{-1}$ ($29.1\,kcal\,mol^{-1}$) and it is reasonable to use this number (which is similar to that for other cyclic alkenes, Table 3.4) as a basis for extrapolating a value for the heat of hydrogenation expected for benzene (C_6H_6). Benzene (C_6H_6) is also reduced (without isolation of any intermediates) to cyclohexane (C_6H_{12}). So, if the heat liberated on reduction of one double bond is $121.8\,kJ\,mol^{-1}$ ($29.1\,kcal\,mol^{-1}$) then three times that amount or $365.3\,kJ\,mol^{-1}$ ($87.3\,kcal\,mol^{-1}$) would be expected from the reduction of benzene (C_6H_6) to cyclohexane (C_6H_{12}). Table 3.4 gives the heat of reduction $[-\Delta H_r^\circ]$ found for benzene (C_6H_6) to cyclohexane (C_6H_{12}) as $205.4\,kJ\,mol^{-1}$ ($49.1\,kcal\,mol^{-1}$), which is thus $159.8\,kJ\,mol^{-1}$ ($38.2\,kcal\,mol^{-1}$) *lower* than expected. Thus, benzene (C_6H_6) enjoys this much added stability over a hypothetical "1,3,5-cyclohexatriene" that might have been "constructed" by "replacing" alternate single bonds in cyclohexane (C_6H_{12}) with double bonds.

However, as pointed out earlier (Chapter 5), conjugated alkenes possess an enhanced stability over that of their "isolated double bond" isomers and thus, it can be argued, it is not surprising that benzene (C_6H_6) is "more stable" than what is "expected" when it is compared with cyclohexene (C_6H_{10}). A conjugated system would make a better choice for the comparison.

The heat of hydrogenation (or reduction) $[-\Delta H_r^\circ]$ of 1,3-cyclohexadiene (C_6H_8) (Figure 6.22) (Table 3.4) is $229.7\,kJ\,mol^{-1}$ ($54.9\,kcal\,mol^{-1}$), some $14.6\,kJ\,mol^{-1}$ ($3.3\,kcal\,mol^{-1}$) *lower* in energy than anticipated in the absence of interaction between the double bonds. While placing yet a third double bond into conjugation (forming "1,3,5-cyclohexatriene") cannot be accomplished [benzene (C_6H_6) forms instead], it can be argued that it is reasonable to predict that introduction of the additional double bond might be expected to add about $108.8\,kJ\,mol^{-1}$ ($26\,kcal\,mol^{-1}$) (i.e., a $[-\Delta H_r^\circ]$ [Figure 6.22] of about $338.5\,kJ\,mol^{-1}$ [ca. $80.9\,kcal\,mol^{-1}$]). Using this value and comparing it with the heat liberated on reduction ($205.4\,kJ\,mol^{-1}$; $49.1\,kcal\,mol^{-1}$; Table 3.4) of benzene (C_6H_6) results in a "stabilization" that is now some $133\,kJ\,mol^{-1}$ ($31.8\,kcal\,mol^{-1}$) more than expected.

---

*From the work of L. Pauling (and others in the 1930s) on the diamagnetism of aromatic compounds and advances in the MO theory of such cyclic systems by E. Hückel (about the same time). Until his recent death, Linus Pauling (1901–1995) was Professor of Chemistry at California Institute of Technology and Stanford. He received two Nobel Prizes: 1954 (Chemistry); 1963 (Peace). Additionally, he was a staunch advocate of the need for large daily doses of vitamin C in the human diet. Erich Hückel (1896–1980) was Professor of Physics at Stuttgart and Marburg Universities in Germany.

[†]John Pople (1925–2004), Professor, Carnegie-Mellon University and Northwestern University, educatd at Cambridge (Nobel Prize, 1998).

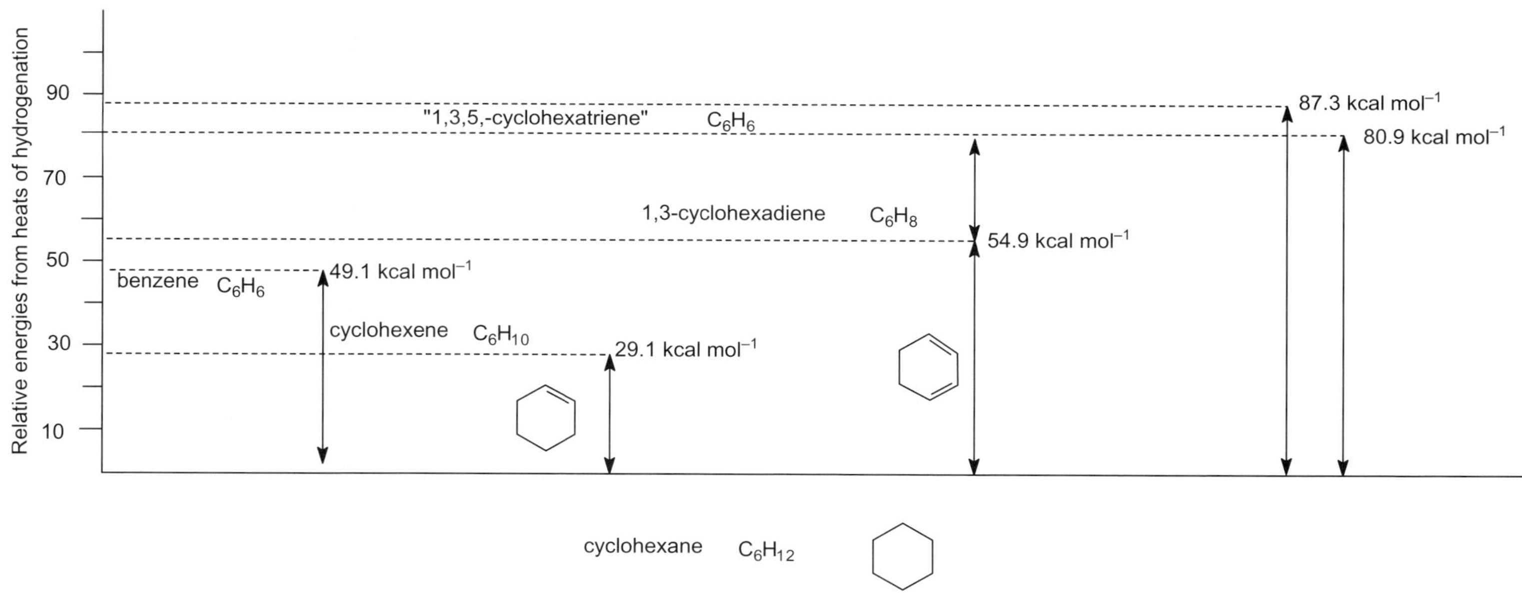

Figure 6.22. A diagramatic representation of the heats of hydrogenation ($-\Delta H^\circ_r$) relative to cyclohexane (C_6H_{12}) for cyclohexene (C_6H_{10}), 1,3-cyclohexadiene (C_6H_8), benzene (C_6H_6), and a hypothetical "1,3,5-cyclohexatriene" (C_6H_6).

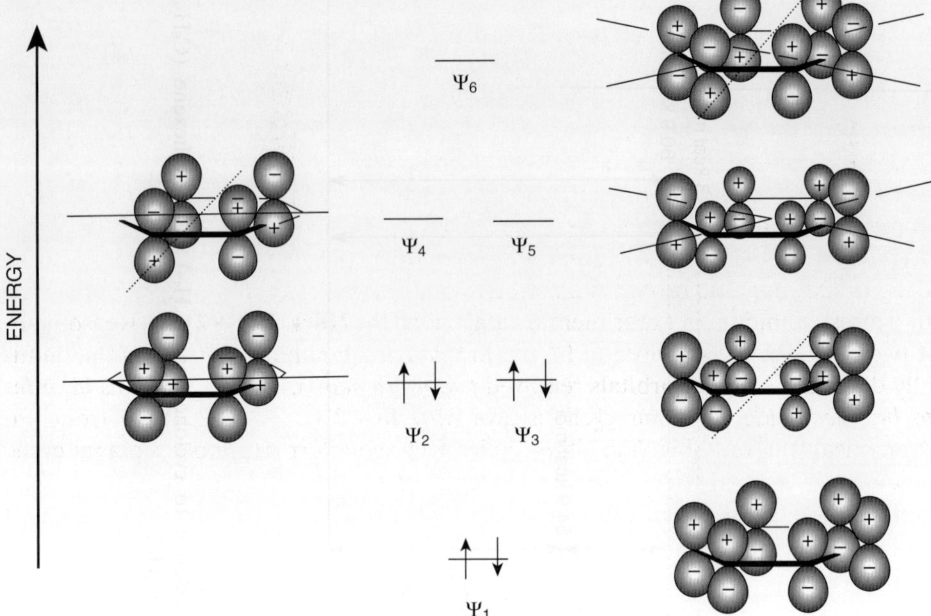

Figure 6.23. A schematic representation of the π MO's of benzene (C_6H_6) and the relative energies of those orbitals. MOs of equal energy (ψ_2, ψ_3; bonding, occupied, and ψ_4, ψ_5; antibonding, unoccupied) are *degenerate* because they are of equal energy and each has the same number of nodes (one and two, respectively). Each MO can contain a maximum of two (a pair) electrons. The bonding orbitals (ψ_1, ψ_2, ψ_3) are occupied.

Therefore, benzene (C_6H_6) enjoys added stability somewhere between about 32 and 38 kcal mol^{-1}. The value generally agreed upon is 150.6 kJ mol^{-1} (36 kcal mole^{-1})* and this value is called the "**resonance energy**" of benzene (C_6H_6).

The MO treatment of the π-type orbitals in benzene was discussed in Chapter 3 and Figure 3.22, is seen again here as Figure 6.23, and is also used to justify the additional stability found in benzene (C_6H_6) and other arenes.

Thus, the planar, cyclic array depicted in Figure 6.23 is supposed to show the results of the creation of six MOs (the lowest of which is bonding everywhere and has zero nodes) from the combination of six *p*-type AOs. The combination results in enhanced stability for the bonding orbitals and instability for the antibonding orbitals relative to the *p*-type orbitals from which the combination was made. Then, with the orbitals in place, consideration can be given to their occupancy by electrons—of which six are available and, subject to the idea that the repulsion between electrons will prohibit occupancy of any single member of a degenerate set of orbitals by two electrons as long as any of the set is unoccupied (**Hund's rule**), the six electrons can be distributed among the lowest lying (bonding) MOs. Now, since only the π-MOs (ψ_1, ψ_2, ψ_3) that are bonding are filled, and these are the lowest lying orbitals, the stability of benzene (C_6H_6) and related arenes is accounted for.

*Since heats of hydrogenation are experimental values that have been obtained at different times and by different workers with more or less precision an agreed upon value is necesssary.

The shift in paradigm from the *two* arrows signifying equilibrium (Figure 6.21) to the *one* double-headed arrow of Figure 6.24 signifying resonance is of fundamental importance. The resonance concept argues that *neither* "classical" structure accounts for benzene (C_6H_6), that *neither* accounts for the experimental observation that all carbon-carbon bonds are of equal length (139 pm), as are all carbon-hydrogen bonds (108 pm) and all bond angles are (120°), and thus that *neither* representation is correct. Indeed, *both* or, perhaps, something else should be written. Some other, less popular, choices are also shown in Figure 6.24.

Interestingly, the filled, mutually degenerate *pair* of orbitals *above* the lowest lying (filled) bonding orbital is illustrative of a simple, general principle (supported by a massive amount of experimental data) called the **Hückel 4n + 2 rule**. Reasonably, if two electrons are required to fill the lowest lying bonding orbital then the mutually degenerate higher orbitals required *two more pair* (i.e., *four*) *electrons in order to be filled*. Indeed, planar cyclic arrays with $4n + 2$ ($n = 0, 1, 2...$) electrons are dramatically more stable than either their acyclic counterparts or other planar cyclic arrays with $4n$ or any other number of electrons. Further, such planar cyclic systems with $4n$ π-electrons are apparently *destabilized*. They have become known as *antiaromatic*.

In this vein, a variety of cyclic systems related to benzene (C_6H_6), *although not arenes* (as *arenes* are derivatives of benzene), have been examined. Consider cyclopropene (C_3H_4) and the ions derived from it as shown in Figure 6.25. If a hydride ion (H^-) is removed, the cyclopropenyl cation ($C_3H_3^+$), a planar array with $4n + 2$ electrons ($n = 0$) results, while if a proton (H^+) is removed and the electron pair left behind, the cyclopropenide anion, a planar array with $4n$ ($n = 1$) electrons can be

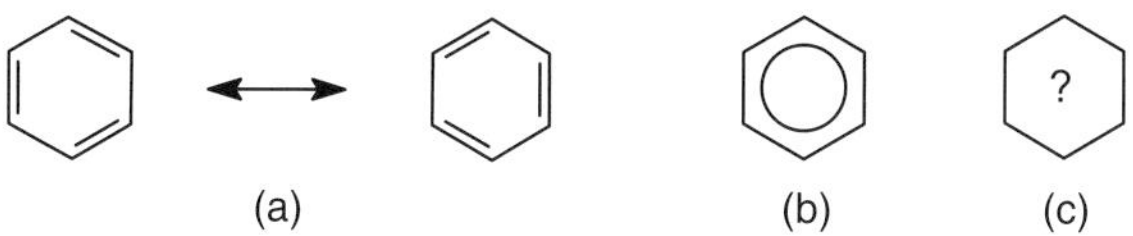

(a)　　　　　　　　(b)　　(c)

Figure 6.24. (a) Traditional resonance forms representing benzene (C_6H_6); (b) and (c) alternative representations for benzene (C_6H_6).

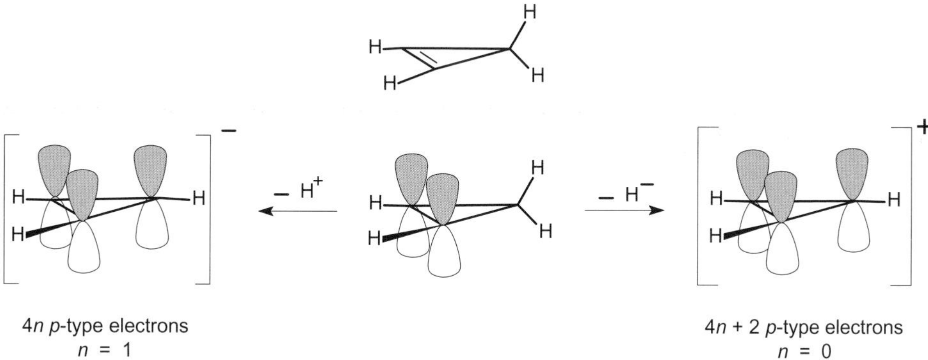

Figure 6.25. A depiction of the cyclopropenyl cation and the cyclopropenide anion. The former, with $4n + 2$ electrons ($n = 0$) is particularly stabilized and is considered aromatic while the latter, with $4n$ electron ($n = 1$) is considered antiaromatic.

imagined. The anion is unknown while the cation is readily observed and, with suitable substituents and the proper counter-ion, is even isolable.

As pointed out earlier in this chapter, cyclobutadiene (C_4H_4) (Figure 6.26), a $4n$ π-electron system ($n = 1$), exists only fleetingly and, even then, it apparently prefers a rectangular (not square planar) form (Equation 6.72). This is in concert with the concept that it is destabilized by "antiaromaticity." On the other hand, the cyclopentadienide anion (C_5H_5) (Equation 6.74) with $4n + 2$ π-electrons ($n = 1$) is readily formed by deprotonation of cyclopentadiene (C_5H_6), an *abnormally acidic* hydrocarbon ($pK_a = 16$). Further, as shown in Figure 6.27, the chemical shifts of the protons, nonequivalent in the parent, all become equivalent in the sodium salt and the normally expected upfield shift due to shielding by the "added" electron (negative charge) is ameliorated by the distribution of charge around the ring.

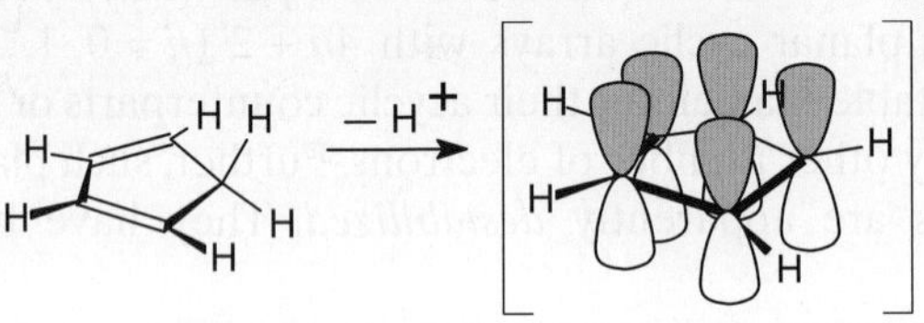

$$\tag{6.74}$$

Cycloheptatriene (C_7H_8) (Figure 6.28) is not planar. Indeed, it is clear from its ^{1}H NMR spectrum that there are four (4) different sets of protons (Figure 6.29). However, the corresponding $4n + 2$ ($n = 1$) π-electron cycloheptatrienyl cation clearly enjoys planarity (all of the protons are identical and *deshielded*) and, with the appropriate counter ion, its salts are isolable (Equation 6.75). Both the enhanced stability and the dramatic downfield shift of the protons (Figure 6.29) (due to both the positive charge and the induced diamagnetic anisotropy resulting from a ring current) confirm its aromaticity.

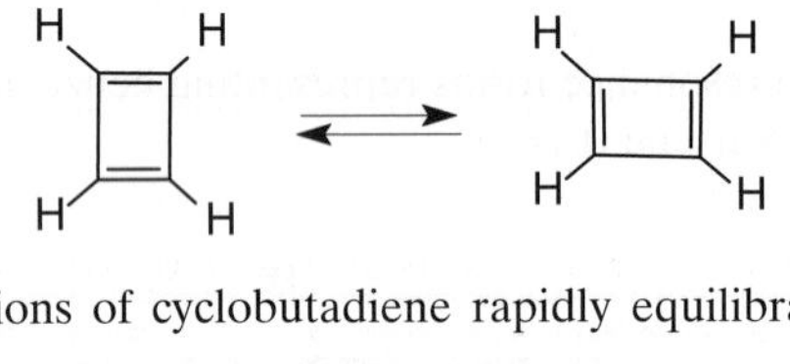

Figure 6.26. Representations of cyclobutadiene rapidly equilibrating as bonds alternately shorten and lengthen.

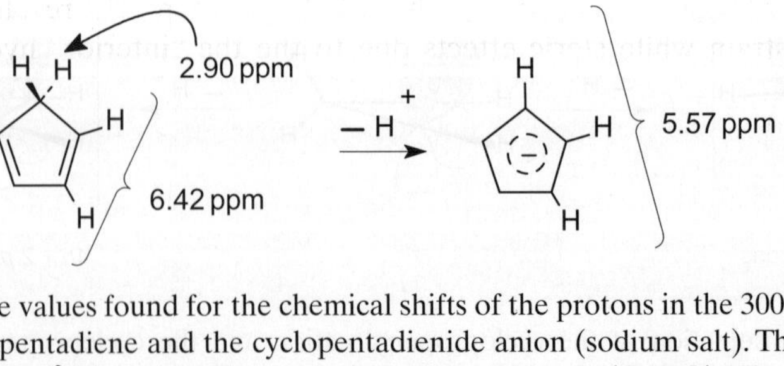

Figure 6.27. The values found for the chemical shifts of the protons in the 300 MHz ^{1}H NMR spectra of cyclopentadiene and the cyclopentadienide anion (sodium salt). The former spectrum was taken in ^{2}HCCl$_3$ and the latter in tetrahydrofuran (C_4H_8O). The chemical shift values in ppm are relative to TMS ($\delta = 0.00$ ppm).

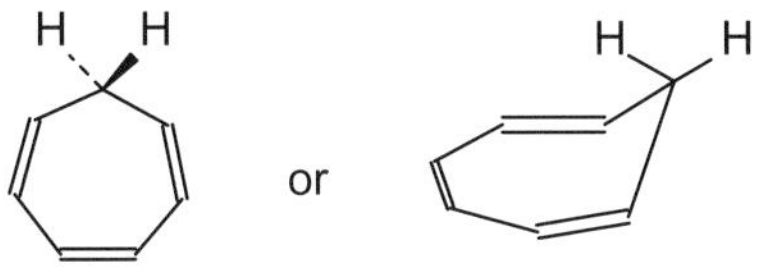

or

Figure 6.28. Representations of cycloheptatriene.

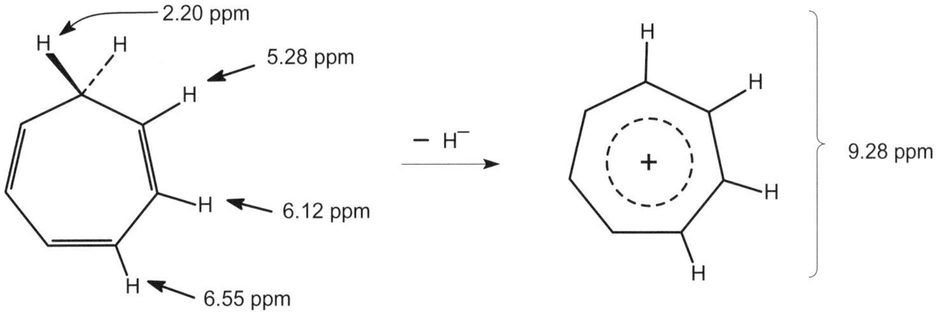

Figure 6.29. The values found for the chemical shifts of the protons in the 300 MHz ^{1}H NMR spectra of cycloheptatriene and the cycloheptatrienyl cation. The former spectrum was taken in ^{2}HCCl$_3$ and the latter in liquid sulfur dioxide (SO$_2$). The chemical shift values in ppm are relative to TMS ($\delta = 0.00$ ppm).

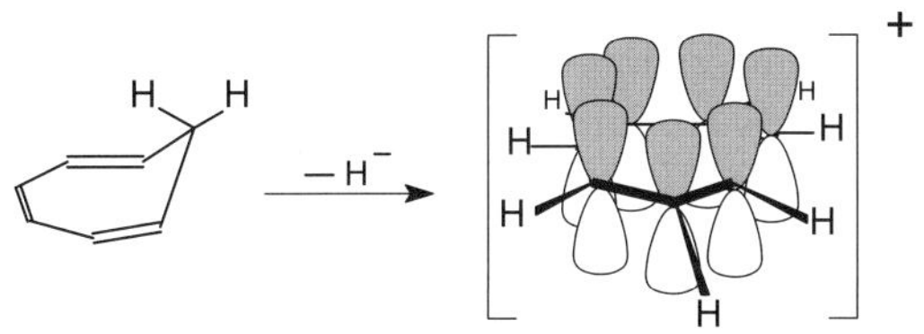

$$\tag{6.75}$$

Although the Hückel $4n + 2$ rule is confined to monocyclic systems and requires degeneracy of the HOMO for $n > 0$ so that four electrons (in two pairs) are specified, it can be frustrated by other constraints. For example, consider two possibilities for the monocyclic 10-carbon, 10 π-electron ($4n + 2$; $n = 2$) hydrocarbon cyclodecapentaene (C$_{10}$H$_{10}$) (Figure 6.30). The internal angles of the planar, almost circular, all-(Z)-isomer (Figure 6.30a) are significantly larger than 120°, resulting in destabilizing steric strain while steric effects due to the the "interior" hydrogens in the isomeric (Z,Z,E,Z,E)-cyclodecapentaene (C$_{10}$H$_{10}$) (Figure 6.30b) that intrude into each other's space in the intramolecular cavity prevent planarity. Cyclodecapentaene (C$_{10}$H$_{10}$) is a reactive, alkene-like hydrocarbon.

Additional aromatic compounds that, however, are not benzenoid (i.e., lack the fundamental six-carbon hexagonal structure characteristic of benzene and its "derivatives") are also known. The bright blue crystalline, planar, $4n + 2$ π-electron nonbenzenoid hydrocarbon azulene (C$_{10}$H$_8$), while not enjoying the same "diminished reactivity" characteristic of its benzenoid isomer naphthalene (C$_{10}$H$_8$), nonetheless clearly enjoys a similar ring current (Figure 6.31).

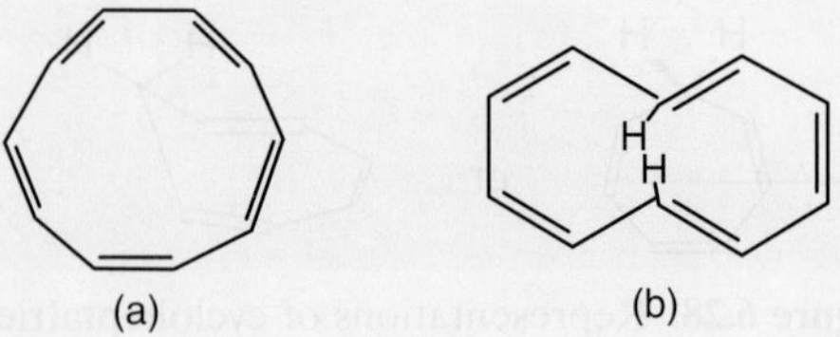

Figure 6.30. Isomers of cyclodecapentaene ($C_{10}H_{10}$). (a) The internally strained all-(Z)-isomer with angles greater than 120°. (b) The (Z,Z,E,Z,E)-isomer, strained because of internal hydrogen steric repulsion.

Figure 6.31. The proton chemical shifts (relative to $\delta = 0.00$ ppm for TMS) for the arene naphthalene ($C_{10}H_8$) and its blue, nonbenzenoid-but-aromatic isomer azulene ($C_{10}H_8$).

It is to be expected (and has been long known) that the UV spectra (Chapter 2) of benzenoid and other aromatic (nonbenzenoid) compounds are similar and that these compounds possess a chromophore that is distinguishable from the more simply conjugated analogues. Thus, benzene absorbs strongly ($\varepsilon \approx 47{,}000$) in the vacuum UV ($\lambda_{max} = 184$ nm) and into the UV more weakly at λ_{max} 202 nm ($\varepsilon = 7000$) and has a series of small bands ($\varepsilon \approx 200$) in the region between λ_{max} 230 nm and λ_{mas} 270 nm. The band near 200 nm is presumably associated with a $\pi \rightarrow \pi^*$ transition that is similar to that found in conjugated dienes (Chapter 5) and polyenes except that the band in benzene (C_6H_6) is at lower frequency (higher energy) (a larger HOMO-LUMO gap). However, the second series of bands (called the *benzenoid bands*) appears to be characteristic of aromatic compounds since analogous structure is not seen in the spectra of conjugated alkenes. These benzenoid bands also must correspond to $\pi \rightarrow \pi^*$ transitions but (a) they are of lower energy (longer wavelength) and (b) they are weak. Weak bands are frequently associated with forbidden transitions and, in this case, the transition is forbidden because the π^* state enjoys the same electronic symmetry as the ground state and transitions between states of the same symmetry are normally forbidden. That they are seen here means that vibrations of the ring must be causing sufficient distortion at any given instant to partially destroy the symmetry. In concert with this is the effect of solvent on these bands. They are significantly broadened and in some cases dampened to the point of extinction by choice of solvent. Interestingly, the same long wavelength series of fine-structured bands is characteristic of nonbenzenoid and benzenoid compounds but the solvent caused dampening effect is typical of aromatic systems.

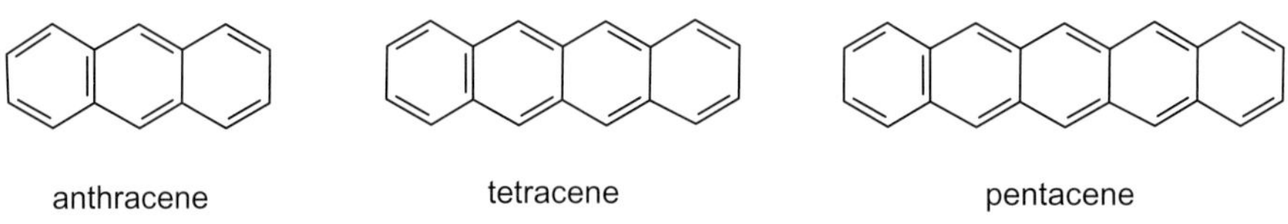

Figure 6.32. Colorless, yellow, and blue arenes, respectively.

Finally, in this vein, while the benzenoid bands in benzene are centered around λ_{max} 255 nm (ε = 230), they are found near λ_{max} 380 (ε = 7900) for colorless anthracene ($C_{14}H_{10}$) (Figure 6.32) near λ_{max} 480 (ε = 11,000) for yellow napthacene (tetracene, $C_{18}H_{12}$) (Figure 6.32) near λ_{max} 580 (ε = 12,600) for blue pentacene ($C_{22}H_{14}$) (Figure 6.32) and between 605 and 699 nm (ε = 300) for azulene (Figure 6.31).

Arenes also absorb characteristic IR frequencies. In addition to C-H absorption near 3030 cm^{-1}, four weak-to-moderate bands are present in the tradititonal "carbon-carbon double bond" region (between about 1600 and 1400 cm^{-1}). These are attributed to the carbon-carbon double bond in-plane vibrations and, although they may be obscured by other more intense absorptions, if they are absent, it is unlikey that the compound is aromatic. Finally, in this vein, there are a number of (generally) weak bands (usually between two and four) in the region between 2000 and 1600 cm^{-1} as well as in the fingerprint region (1000 to 650 cm^{-1}), which, if not obscured, are diagonistic of the substitution pattern on the ring.

I. Oxidation

a. Oxidation of the Aromatic Ring. As already pointed out, arenes burn (combining with oxygen [O_2]) with a characteristic yellow "sooty" flame. The residue is reported to contain additional polynuclear aromatic hydrocarbons and is considered evidence for the difficulty with which these graphatic materials combine with oxygen. Nonetheless, as with conjugated dienes *(vide supra)*, some aromatic compounds such as anthracene ($C_{14}H_{10}$) (Table 6.10, item 1) combine with oxygen (O_2) to give peroxides and, in the presence of enzyme systems (Chapter 12) classified as dehydrogenases or oxidases (dioxygenases and monooxygenases);* other aromatic nuclei incorporate oxygen as epoxides (oxiranes), which may subsequently rearrange to **phenols** (Ar-OH) (Table 6.10, item 2) *(vide infra)* or to oxepine.[†]

Additionally, as shown in Table 6.10 (items 5 and 6), strong oxidizing agents applied under vigorous conditions can result in the destruction of the aromatic ring

---

*These enzymatic systems do not apparently involve the immediate participation of either singlet or triplet molecular oxygen (O_2). Dioxygenases catalyze the incorporation of two oxygen atoms while monooxygenases (also called *hydroxylases*) catalyze the incorporation of only one oxygen; the other oxygen is going to produce an equivalent of water (H_2O). Some monooxygenases are capable of hydroxylation of aliphatic as well as aromatic compounds. In the case of the latter, they produce phenols (Ar-OH).

[†]Oxepine is the name given to the parent, fully unsaturated seven-membered ring containing one oxygen (and six carbons). It is also called oxacycloheptatriene.

TABLE 6.10. Oxidation Reactions of Arenes

Number	Arene	Reactant	Product(s)
1	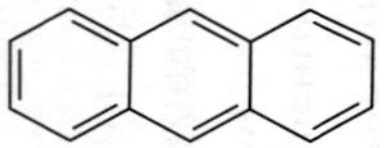 (anthracene)	O_2	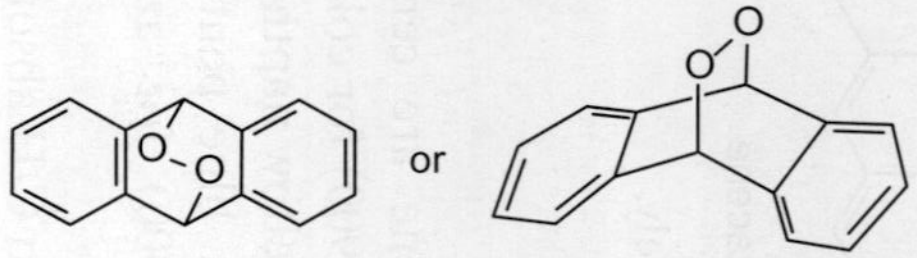 (endoperoxide) or (endoperoxide isomer)
2	 (benzene)	O_2 Monooxygenase enzyme	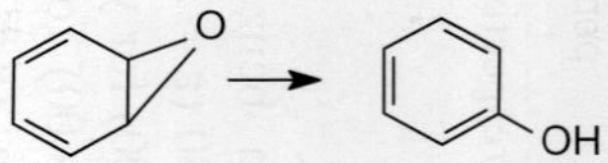 (arene oxide) → (phenol, OH)
3	 (benzene)	CF_3CO_2H $(CF_3CO_2)_4Pb$	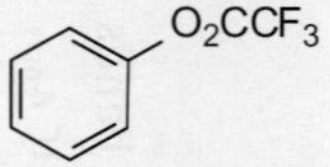 (phenyl O_2CCF_3)
4	 (benzene)	Electrochemical Oxidation (O_2)	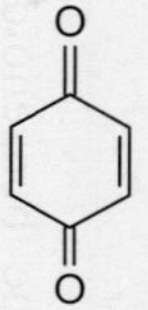 (1,4-benzoquinone)
5	 (benzene)	V_2O_5 Air	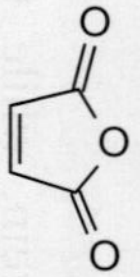 (maleic anhydride)
6	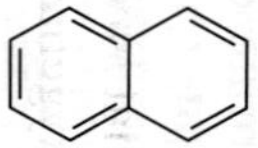 (naphthalene)	Ruthenium tetroxide NaOCl	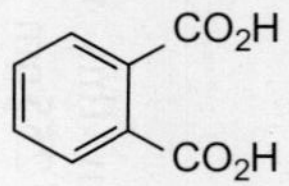 (phthalic acid, CO_2H / CO_2H)
7	 (benzene)	H_2O_2 and $FeSO_4$ "Fenton's Reagent"	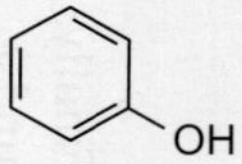 (phenol, OH)

A sample of some oxidation reactions of the aromatic ring. Most of the reactions shown do not proceed in good yield.

while electrochemical oxidation (item 4) can produce members of that class of compounds called **quinones** (*vide infra*).

Finally, in this vein, oxidation by ozone (O_3) followed by reductive hydrolysis (ozonolysis) (Equation 6.76) can be effected on aromatic compounds and, while it occurs less readily than with alkenes, produces results that suggest that the Kekulé structures are real.

$$ \tag{6.76} $$

Problem 6.14. Experimentally, the products of the ozonolysis of 1,2-dimethylbenzene (*o*-xylene)(Equation 6.76), viz. (a) 2,3-butanedione [biacetyl; $(CH_3CO)_2$], (b) 2-oxopropanal (methylglyoxal; CH_3COCHO), and (c) ethanedial [glyoxal; $(HCO)_2$] are actually formed in different quantities. What is the ratio of products (a):(b):(c) anticipated if they are in fact formed in amounts that reflect the resonance forms of 1,2-dimethylbenzene (o-xylene)?

In general, the oxidation of aromatic rings (except in biological systems) is not particularly facile. As the processes are difficult to carry out cleanly in the laboratory, very little is known about the details of when carbon-oxygen bonds are actually made and carbon-hydrogen bonds get broken, whether triplet ($^3O_2$) or singlet ($^1O_2$) oxygen (the latter about 92–100 kJ mol^{-1} [ca. 22–24 kcal mol^{-1}] higher in energy than the former) is involved, and whether the process is even radical or ionic.*

Animals that have the digestive organ known as the "liver" appear to utilize the liver monooxygenases to convert benzene (C_6H_6) and other arenes to epoxides (Table 6.10, item 2) on the way to phenols (Table 6.10, item 2) and quinones (Table 6.10, item 4) that might then be excreted or further oxidized (ultimately to carbon

---

*In many of the processes (particularly in living systems) it is believed that triplet oxygen ($^3O_2$) is reduced ($E_o = -0.33$ V) to the superoxide radical ion ($\cdot O_2^-$) that, among other things, can subsequently pick up a proton (from, e.g., adventitious moisture) to produce the hydrogen peroxide radical ($HO_2\cdot$), a potent oxidizing agent. Alternatively, the superoxide radical ion ($\cdot O_2^-$) might undergo disproportionation to peroxide anion (O_2^{-2}) and oxygen (O_2). Many other possibilities also exist and, despite long-term and ongoing efforts, the details, even in the simplest cases, are not yet clear.

dioxide, CO_2, and water, H_2O).* Arene epoxides are readily attacked by a variety of functional groups present in both DNA and ribonucleic acids (RNA) (Chapter 14), resulting in the irreversible modification of those carriers of genetic information and also serving as sites from which malignancies can begin. Many industrial organizations have banned the use of *unsubstituted arenes*[†] (such as benzene, C_6H_6) in the workplace.

Quinones (Table 6.10, item 4) are formed only with difficulty from benzene (C_6H_6) itself. However, as shown in Equations 6.77 and 6.78, on oxidation in solution with, for example, ammonium cerium(IV) sulfate [ceric ammonium sulfate, $Ce(NH_4)_4(SO_4)_4$], naphthalene ($C_{10}H_8$) and anthracene ($C_{14}H_{10}$) yield *p*-quinones and phenanthrene ($C_{14}H_{10}$), an isomer of the latter (Equation 6.79), yields an *o*-quinone much more readily. It has been argued that maintaining aromaticity in portions of each of these polynuclear systems aids in making the oxidation reaction much more facile. Similar oxidation patterns are common in naturally occurring quinones.

$$(6.77)$$

$$(6.78)$$

*The pathway by which this occurs was studied in some detail at the National Institutes of Health (NIH). Interestingly, during the oxidation of the aromatic ring to a phenol, there is apparently a hydride (H^-) shift **on** the aromatic ring so that the *hydrogen*, initially on the carbon that winds up bearing the oxygen, moves over to an adjacent carbon. This process (schematically represented here), where ^3H is tritium, is called the **NIH shift** (after the laboratories in which it was found).

[†]As will be seen in Part b of this section, alkyl-substituted arenes are apparently both more readily oxidized than their unsubstituted congeners and more readily excreted as normal metabolic waste products.

$$\text{(6.79)}$$

Phenols (e.g., phenol itself [C_6H_5-OH or Ar-OH], Table 6.10, item 2) and their esters (e.g., the trifluoroacetate ester of phenol [C_6H_5-O_2CCF_3 or ArO_2CCH_3], Table 6.10, item 3) have been oxidized with air and oxygen (O_2), in neutral and alkaline solutions, with and without ionic and/or radical catalysts and/or irradiation and in a variety of solvents. Enzymes (this chapter and Chapter 12) from a wide variety of sources have also been used. Frequently, oxidation of aromatic systems to phenols cannot be stopped before quinones and products of ring fragmentation occur and numerous, sometimes ill-defined, products result. Thus, as shown in Equation 6.80, oxidation of the polynuclear hydrocarbon *chrysene* with ammonium cerium(IV) sulfate [ceric ammonium sulfate, $Ce(NH_4)_4(SO_4)_4$] is reported to produce 6*H*-benzo[*d*]naphtho[1,2-*b*]pyran-6-one (8% yield)* and a quinone (23% yield). The remainder of the product(s) (69%) was unidentified.

---

*The nomenclature for polynuclear systems such as this one depends upon using letters (*a*, *b*, *c* ...) for the sides of the *heterocyclic* ring. As shown here, a compound that might be called "oxabenzene," in which an oxygen atom replaces one carbon (and its associated hydrogen), would necessarily bear a positive charge. This compound is called the "pyrylium ion" and some salts of substituted pyrylium ions are well-characterized compounds. The oxygen containing heterocycle corresponding to cyclohexadiene clearly contains a "reduced pyrylium" nucleus and is called a "pyran." However, since the position next to the oxygen bears two hydrogens instead of the one present in the pyrylium ion (befitting a "reduced" species), that carbon is specified by adding the *prefix 2H- to that proton tautomer* (the numbering beginning with the heteroatom). The structure of 4*H*-pyran is also shown.

pyrylium ion   2*H*-pyran   4*H*-pyran   dihydropyran

If (a) 2*H*-pyran is redrawn with the ring oxygen at the lower right corner, and (b) the ring sides lettered then when (c) a "naphthalene" is "added" across the "*b*" side of the ring (and the 1,2-positions of the napthalene) and (d) a "benzene" is "added" across the "*d*" side a **benzo[*d*]naptho[1,2-*b*]pyran** system is produced. Finally, (e) the carbon bearing the two hydrogens is oxidized to a ketone (producing a pyranone) and, keeping the lettering system in the order used, it is a 6*H*-6-pyranone rather than a 2*H*-2-pyranone and 6*H*-benzo[*d*]naphtho-[1,2-*b*]pyran-6-one results.

(a)   (b)   (c)   (d)   (e)

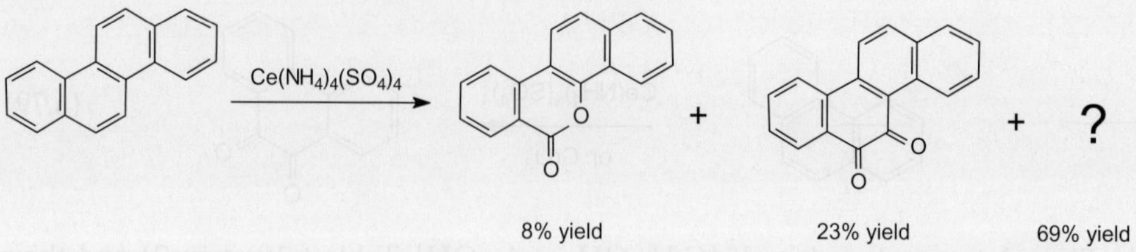

8% yield   23% yield   69% yield

$$(6.80)$$

Quinones, as oxidation products of aromatic systems, are capable of being reduced back to aromatic systems and some quinones are used particularly for this purpose. For example, as shown in Scheme 6.79, 2,3-dichloro-5,6-dicyano-1,4-benzoquinone (**DDQ**) can be used to oxidize 1,2,3,4-tetrahydronaphthalene (tetralin, $C_{10}H_{12}$) to naphthalene ($C_{10}H_8$) while it is reduced to the corresponding phenol, 2,3-dichloro-5,6-dicyanohydroquinone (**DDQ·H₂**). Further, while reduced to the same product, DDQ is also capable of oxidizing 1,3,5-cycloheptatriene to the corresponding, aromatic cycloheptatrienyl cation and, if the oxidation is carried out in perchloric acid, the perchlorate of the cation is an isolable salt (Scheme 6.79).

The last example (item 7) in Table 6.10, the replacement of a hydrogen of the aromatic ring by a hydroxyl, may be formally considered as an oxidation reaction (as the product has more oxygen than the starting material). Alternatively, it may be considered a substitution because the path appears to involve the •OH (as a radical) substituting for the •H (as an atom).

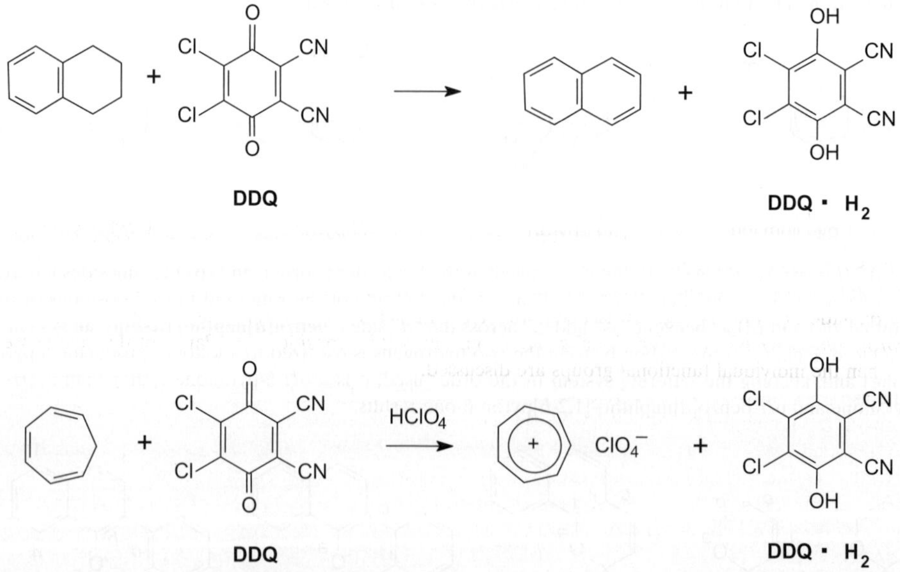

Scheme 6.79. Processes demonstrating the ability of 2,3-dichloro-5,6-dicycano-1,4-benzoquinone (**DDQ**) to effect oxidation of hydrocarbons and, in the process, undergo reduction to the corresponding phenol.

b. Oxidation of Alkyl Substituents on the Aromatic Ring*. A variety of oxidizing agents that include oxygen, permanganate, and dichromate are capable of converting primary and secondary alkyl chains attached to aromatic rings to the corresponding carboxylic acid(s) (Equation 6.81);[†] tertiary substituents (which would require oxidation at a quaternary carbon) react with much more difficulty. Indeed, while the oxidation of methyl groups is most common, substituents consisting of longer chains, saturated and unsaturated, are also accommodated in the general expression of Equation 6.81.

$$\underset{R\ =\ alkyl}{\text{(Ar)}-R} \xrightarrow{\ MnO_4^{-1}\ \text{or}\ Cr_2O_7^{-2}\ } \text{(Ar)}-CO_2^{-} \tag{6.81}$$

With side chains longer than methyl, oxidation with air (or oxygen) apparently involves a hydroperoxide radical and the reaction appears to occur, at least initially, at the benzylic position (i.e., α to the aromatic ring) as shown in Equation 6.82. Interestingly, as shown in Equation 6.83, careful adjustment of oxidizing reagent, solvent, and reaction conditions will even allow the conversion of methyl groups to aldehydes.[‡]

The lability of the aldehyde function under oxidizing conditions can be ameliorated by carrying out the oxidation under conditions where the aldehyde group is protected (*vide supra*) (Chapters 8 and 9) as shown in Equation 6.84. Here the acylal (an ester) produced can be hydrolyzed with aqueous acid or base to the corresponding aldehyde (Chapter 9).

$$\text{(Ar)}-CH_2R \xrightarrow{\ O_2\ } \text{(Ar)}-\overset{O}{\overset{\|}{C}}-R \ +\ H_2O \tag{6.82}$$

---

*Other substituents or groups on the aromatic ring that can also undergo oxidation (e.g., hydroxyl [–OH], amino [–NH$_2$], and thio [–SH]) will be considered when the chemistry associated with those functional groups is encountered. However, in general, since these *functional groups* are **more easily oxidized** than the alkyl groups, they must be *masked* or **protected** from the influence of the oxidizing agent(s) while the oxidation of the alkyl group is being effected. The **art** of *protecting* (*masking*) and subsequently *deprotecting* (*unmasking*) a functional group or groups will be considered in Part 3 of this text as well as when the individual functional groups are discussed.

[†]The astute reader will immediately recognize that this is another unbalanced equation. What has happened to the remainder of "R?" What has happened to the chromium, the manganese, and the others? The remainder of "R" has presumably been oxidized to CO$_2$ and, perhaps, acetic acid (ethanoic acid, CH$_3$CO$_2$H) since methyl groups are generally more resistant to oxidation than methylene and methine groups. In a laboratory experiment, it would be appropriate to assume that the chromium(VI) of dichromate is reduced to Cr(III) while the Mn(VII) will be found as Mn(IV), that is, purple MnO$_4^-$ generating brown MnO$_2$.

[‡]This is particularly interesting because aldehydes (Chapter 9), lying (in a redox sense) between primary alcohols and carboxylic acids, are generally too labile to be isolated in the presence of oxidizing agents. Indeed, as will be encountered later (Chapter 9), many aldehydes spontaneously oxidize (in the presence of air) to the corresponding carboxylic acids.

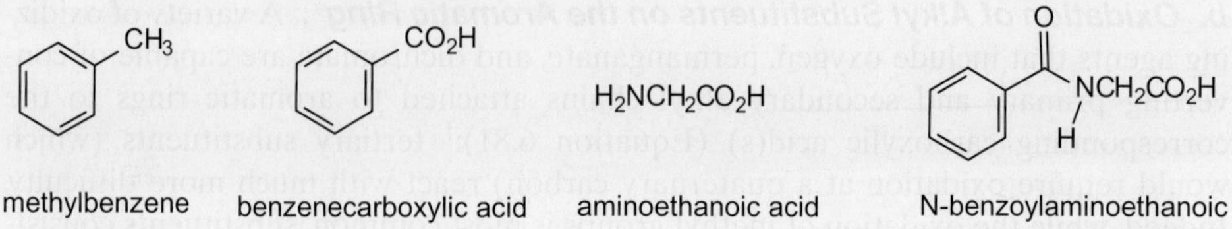

methylbenzene   benzenecarboxylic acid   aminoethanoic acid   N-benzoylaminoethanoic

(toluene)   (benzoic acid)   (glycine)   (hippuric acid)

Figure 6.33. An example of an alkyl-substituted arene that can be converted into a material capable of excretion.

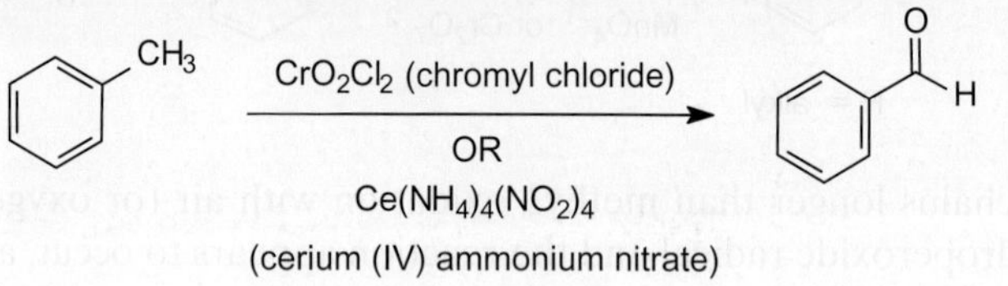

$$\text{(6.83)}$$

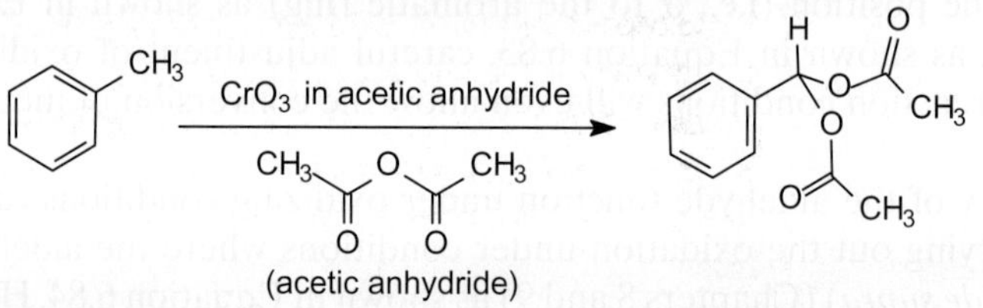

CrO₂Cl₂ (chromyl chloride)

OR

Ce(NH₄)₄(NO₂)₄

(cerium (IV) ammonium nitrate)

$$\text{(6.84)}$$

CrO₃ in acetic anhydride

(acetic anhydride)

The ease of oxidation of the side chain attached to the aromatic ring, compared with the relative resistance to oxidation of the ring itself, is reflected in the ability of liver enzymes (in *all* animals that have livers) to oxidize the side chain (almost regardless of length) to the corresponding benzenecarboxylic acid. For example, using the compounds drawn in Figure 6.33, it is already appreciated that oxidation converts methylbenzene (toluene, $C_6H_5CH_3$) to benzenecarboxylic acid (benzoic acid, $C_6H_5CO_2H$). Then, by pathways that will be broadly considered in Chapter 12, the reaction of benzenecarboxylic acid (benzoic acid, $C_6H_5CO_2H$) in the living system with the aminoacid, aminoethanoic acid (glycine, Gly, G, $H_2NCH_2CO_2H$) produces N-benzoylaminoethanoic acid (hippuric acid, *N*-benzoylglycine, $C_6H_5CONHCH_2CO_2H$), which is excreted in the urine. Thus, in contrast to the unsubstituted arenes, which produce oxiranes that can intrude into replication of genetic information, the alkyl-substituted arenes can be oxidized to excretable materials. Methylbenzene (toluene, $C_6H_5CH_3$) has largely replaced benzene (C_6H_6) wherever possible in manufacturing processes and other uses where contact with animals is possible (e.g., glues and paints) as required by enlightened providers.

II. Reduction

Arenes are reduced with more difficulty than alkenes and alkynes so that if double and/or triple bonds are present in side chains attached to the ring, hydrogenation in the presence of a catalyst will cause them to be reduced first. Further, the conditions for catalytic reduction of the aromatic ring are vigorous enough to insure that intermediates are not isolated. That is, reduction of the "first" double bond is the slow step in the process. As a consequence, reduction of one equivalent of benzene (C_6H_6) with one equivalent of hydrogen (H_2) produces one-third of an equivalent of cyclohexane and two-thirds of an equivalent of benzene remain unreacted. Typical reduction conditions involve Pt or Pd in acetic acid (CH_3CO_2H) at about 100°C and high pressure (ca. 100 atm [1 atm = 14.69 lb/in^2 at sea level]). Alternatively, Ni at higher temperatures or 5% Ru or Rh on carbon have all been used. There are relative trade-offs in catalyst cost and other expenses that encourage one method over another from time-to-time. In concert with what was observed with alkenes and alkynes (*vide supra*) and as shown in Equation 6.85, catalytic reduction appears to be a suprafacial process. That is, reduction of 1,2-dimethylbenzene (*o*-xylene) generates *mostly* (*Z*)- or *cis*-1,2-dimethylcyclohexane. The origin of the (*E*)- or *trans*-1,2-dimethylcyclohexane (as much as 10% under some conditions) is not clear, although since oxidation (dehydrogenation) of cyclohexane yields benzene (Equations 6.4 and 6.5) and, thus, to at least some extent, the reduction is reversible, some isomerization can be justified.

$$\text{(6.85)}$$

As with alkynes (Equation 6.66), arenes can be reduced with alkali metals in liquid ammonia and amines in the presence of an alcohol as a proton source. The reaction, known as a **Birch* reduction**, presumably also occurs via a series of one electron transfer process as it does with alkynes. Thus, the initial process, as shown in Scheme 6.80, presumably involves transfer of an electron from the metal to the LUMO of the arene to produce a radical anion. Then the radical anion abstracts a proton from the alcohol, generating the alkoxide and an aryl radical. After a second electron transfer, the aryl anion once again abstracts a proton to produce the dihydroarene, a 1,4-cyclohexadiene. It is important to note that the **unconjugated** diene is produced. Second, if there is an **electron-donating substituent** (such as a methoxyl [$-OCH_3$] group) attached to the arene, *one of the double bonds is found in conjugation* with the substituent (Equation 6.86) while if there is an **electron-withdrawing substituent** (such as a carboxylic acid [$-CO_2H$]) attached, *neither of the double bonds is in conjugation*, that is, reduction occurs at the carbon atom bearing the carboxylic acid (Equation 6.87).

$$\text{(6.86)}$$

$$\text{(6.87)}$$

*A. J. Birch, Professor, Australian National University. See, for example, Birch, A. J. *Pure Appl. Chem.*, **1996**, *68*, 553.

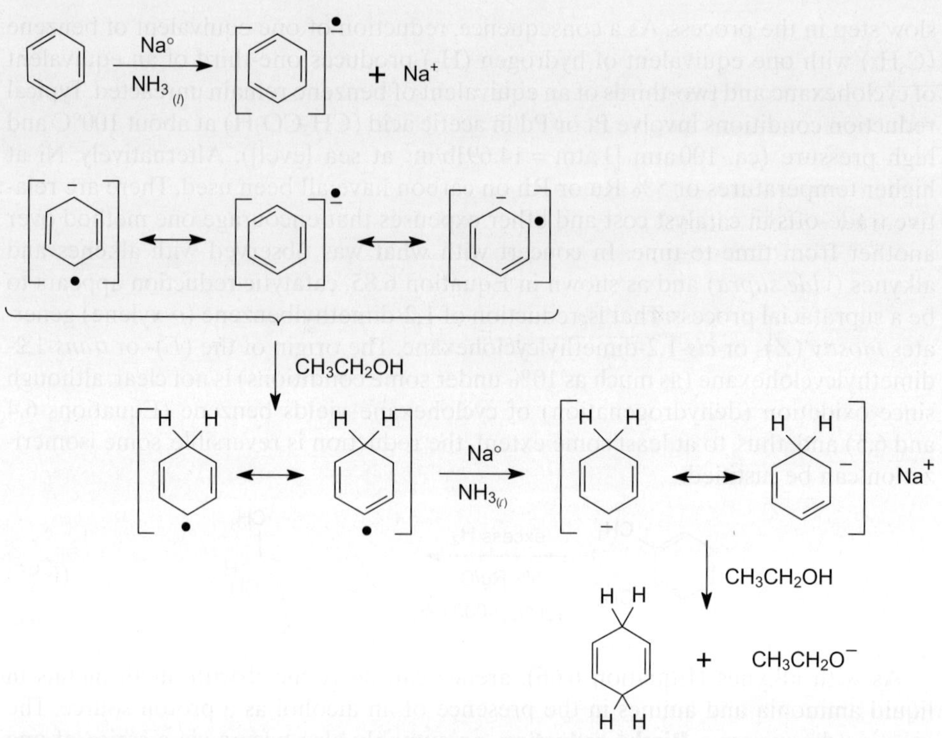

Scheme 6.80. A potential pathway for the reduction of benzene with sodium in liquid ammonia in the presence of ethanol as a proton source. This is an example of the **Birch reduction**.

Finally, in this regard, if the reduction of the aromatic ring is carried out with calcium (Ca) rather than sodium (Na) in the presence of an amine rather than ammonia, it is possible to isolate cyclohexene rather than cyclohexadiene. Thus, all three double bonds can be reduced, two of them, or, only one (albeit to the *unconjugated* diene).*

III. Addition

For benzene itself, it is most common for addition to be followed by elimination (i.e., the process of *substitution, vide infra,* involves *an addition coupled to an elimination*). However, there are a few common reactions in which addition occurs without subsequent elimination.

For example, there remain some specalized needs for potent insecticides, many of which were developed when their potential for general harm to the environment was not widely recognized. Among this class of materials is found the mixture of

*As discussed earlier, conjugated dienes are, generally, more stable than their unconjugated isomers. Therefore, other things being equal, it is often possible to isomerize the unconjugated to the conjugated isomer. This can be accomplished with acid or base but care must be taken that other groups (e.g., the enol ether in the 1-methoxy-1,3-cyclohexadiene, Equation 6.86) are not destroyed in the process.

compounds obtained when benzene is chlorinated *under free radical conditions* (in the presence of light). Under these conditions, chlorine does *not* undergo addition-elimination to produce chlorobenzene, the product of substitution (*qv*), but rather produces a mixture of the isomers of hexachlorocyclohexane. This mixture, marketed under trade names such as Lindane™ and Hexachlor™ (among others), contains about 40% (by weight) of the γ-isomer (Equation 6.88) as its major and most lethal component.

$$\text{excess Cl}_2 \quad (\text{light}) \qquad + \quad \text{isomers} \qquad (6.88)$$

γ -1,2,3,4,5,6-hexa-
chlorocyclohexane

A second type of addition reaction to benzene (C_6H_6) itself is found with the more reactive carbenes, which apparently add in a cheletropic process to the aromatic ring as well as insert into the carbon-hydrogen bonds. Thus, carbene ($:CH_2$) itself, generated by the decomposition of diazomethane (CH_2N_2), produces the bicyclo[4.1.0]heptadiene, norcaradiene (C_7H_8), which undergoes a disrotatory electrocyclic ring opening (*vide supra*) to cycloheptatriene (C_7H_8) and methylbenzene (toluene, C_7H_8) as shown in Equation 6.89.

$$+ :CH_2 \quad \longrightarrow \qquad + \qquad \longrightarrow \qquad (6.89)$$

Finally, extended arenes such as anthracene ($C_{14}H_{10}$) undergo other cycloaddition reactions. Thus, as shown in Equation 6.90, anthracene ($C_{14}H_{10}$) undergoes a thermally allowed [4 + 2] cycloaddition (Diels–Alder) reaction with dienophiles such as *Z*- or *cis*-butenedioic anhydride (maleic anhydride, $C_4H_2O_3$) and a photochemically allowed [4 + 4] cycloaddition reaction with itself (Equation 6.91).

$$+ \qquad \xrightarrow{\text{heat}} \qquad (6.90)$$

$$+ \qquad \xrightarrow{\text{light}} \qquad (6.91)$$

IV. Substitution

It will be remembered that the replacement of one group or atom by another is called **substitution**. Arenes, as well as alkanes, alkenes and alkynes, undergo substitution reactions. For all of these *hydrocarbons*, the atom being replaced (or "substituted") is hydrogen and in that vein, this section deals with the substitution of hydrogenic groups (protium [^{1}H], deuterium [^{2}H], and tritium [^{3}H]) attached to an aromatic ring by other electrophiles. This means that the production, for example, of phenols (Ar-OH) and arylamines (Ar-NH$_2$), which rely upon substitution of, for example, halogen, will not be discussed here. That information can be found in later chapters. It also means that reactions of the substituents, once they have replaced the hydrogen will, for the most part, be considered subsequently.*

As pointed out in Part III, aromatic compounds, unlike alkenes and alkynes, only reluctantly undergo addition reactions and it is often found when forcing conditions are used that substitution occurs instead. Presumably, it is the destruction of aromaticity that accounts for the initial reluctance to react and its reestablishment, resulting in substitution rather than addition, that allows consummation of the reaction.

Although the attack on (or by) the aromatic ring that leads to substitution can be brought about by free radical, electron-rich (nucleophilic) or electron-poor (electrophilic) reagents, the surfeit of electrons associated with aromaticity provides a richness that dictates that the latter should be most common. Nonetheless, all three substitution pathways will be considered. Indeed, the three that will be examined are benzene (C$_6$H$_6$) and the polycyclic arenes (naphthalene [C$_{10}$H$_8$], anthracene [C$_{14}$H$_{10}$], and phenanthrene [C$_{14}$H$_{10}$]).

Finally, much of the wealth of information for radical, nucleophilic, and electrophilic substitution results from the exploration of a *second* (or further) substitution reaction of the aromatic ring. For example, as shown in Equation 6.92, the reaction of some electrophile, E_1^+, with benzene (C$_6$H$_6$), where all six positions are equivalent, can, at some rate under some specified reaction conditions of solvent and temperature, produce a substitution compound, where E_1 has replaced (been substituted for) a proton (H$^+$).†

$$\text{(6.92)}$$

Then, as shown in Scheme 6.81, where the rate for substitution by the electrophile E_1^+ is **rate$_1$,** a second substitution, by essentially the same process, with some other

---

*The astute reader will note that oxidation of the methyl group (–CH$_3$) of methylbenzene (toluene, C$_6$H$_5$CH$_3$) has already been discussed and that the methyl group of methylbenzene (toluene, C$_6$H$_5$CH$_3$) *is* a substituent attached to the aromatic ring of benzene (C$_6$H$_6$) that has been substituted for a hydrogen. The oxidation is only one of the possible reactions of the methyl group; others will be considered subsequently.

†In concert with what has gone before and despite the confusion that the choice of words might cause, it will be seen that the paths written here will use the curved arrow to indicate (in the traditional sense) the "movement and direction" of electrons. Thus, whether the electrophile "attacks" the aromatic ring or *vice versa*, the electron flow will be **from** the aromatic ring **to** the electrophile.

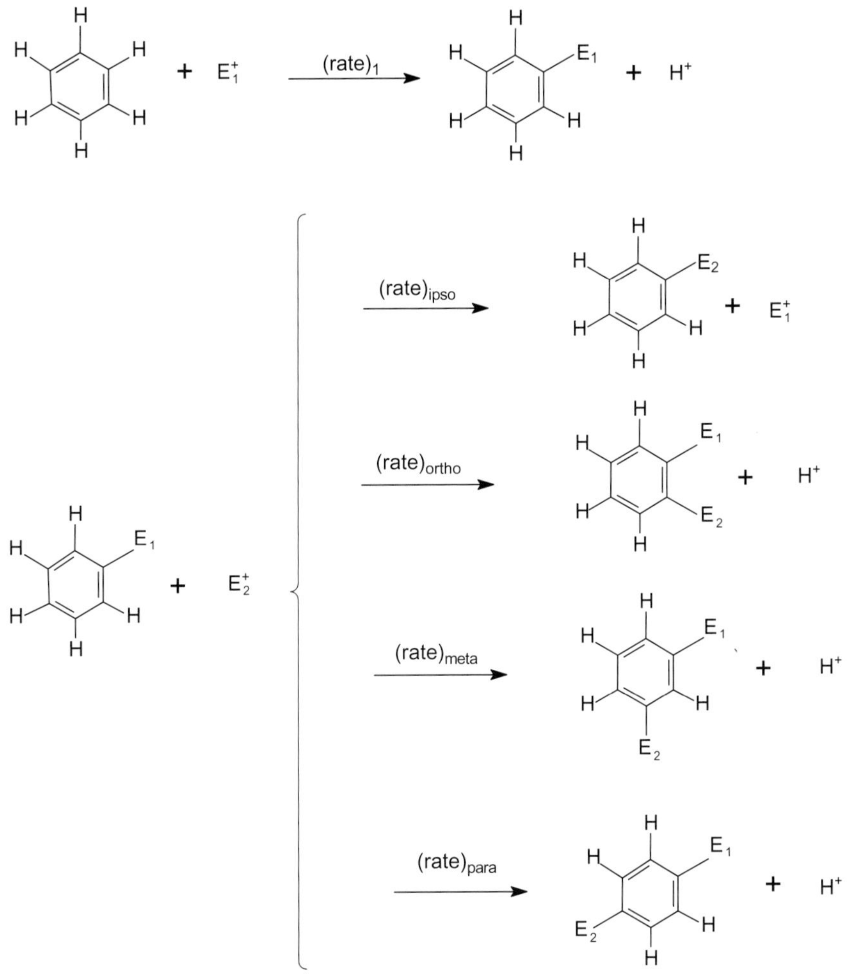

Scheme 6.81. The substitution reaction of an electrophile (E_1) with benzene (C_6H_6) at any one of the six equivalent positions on the ring. (E_1) has been substituted for a proton (H^+). In a second reaction (E_2), an electrophile that might or might not be the same as (E_1) reacts with the product of the initial substitution reaction. There are four possible products, those corresponding to *ipso, ortho, meta,* and *para* substitution. Not all products are formed with equal facility but each forms at its own ***rate***.

electrophile E_2^+ or, perhaps the same electrophile E_1^+ but under more or less forcing conditions, can now be studied to provide an insight into whether E_1: (a) is replaced by E_2 ($E_2 \neq E_1$) (*ipso* substitution); or (b) a proton is replaced to provide *ortho- (o-), meta- (m-),* or *para- (p-)* product. Indeed, the *relative rates* of o- ($\text{rate}_{\text{ortho}}$), m- ($\text{rate}_{\text{meta}}$), and p-product ($\text{rate}_{\text{para}}$) formation can be measured and conclusions about the nature of E_1 as a "substituent" and as an electrophile, as well as about the process by which substitution occurs, can be reached.

a. Electrophilic Aromatic Substitution. There is an enormous amount of evidence that all electrophilic aromatic substitution reactions almost regardless of the electrophile proceed by the same path. The genealized mechanism is much like the reactions of alkenes with electrophiles in that there are **two steps**. In the first step, which is often (but not always) rate determining, the bond between electrophile and aromatic ring is forged, leading (in the case of benzene, C_6H_6) to a benzenium ion or, more generally, to an **arenium ion intermediate** (Figure 6.34 and Scheme 6.82). The arenium ion (also called a σ-complex), which is considered to be a resonance-stabilized intermediate ("**I**" in Figure 6.34), is no longer aromatic. Both the proton (H^+) due to leave and the incoming electrophile (E^+) due to stay are bonded to the same carbon in the σ-complex. Both the proton (H^+) and the incoming electrophile (E^+) can be considered bonded, although the bond for each will be different (unless $E^+ = H^+$). The positive charge that was originally associated with the electrophile (E^+) is now distributed (unevenly) into the ring. If the electrophile were originally accompanied by or bonded to some nucleophilic partner, that partner would now be a spectator and would accompany the positively charged arenium ion. Thus, the fate of the resonance-stabilized (Scheme 6.82) intermediate ("**I**" in Figure 6.34) is quite different from that previously seen with alkenes (compare Scheme 6.29) in that the second step (over the nonrate-determining transition state) *does not involve addition of the nucleophile but rather loss of a proton (H^+) to regain aromaticity.*

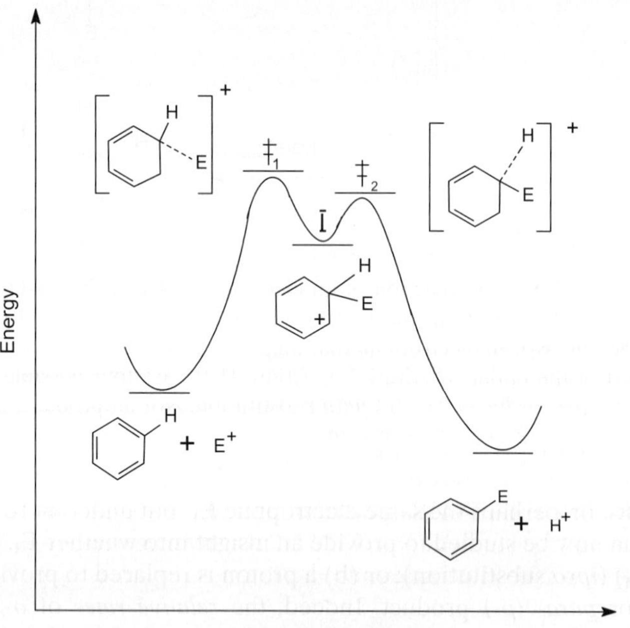

Figure 6.34. A representation of a slice through the three-dimensional reaction surface depicting a pathway for **electrophilic aromatic substitution**. The formation of an arenium ion intermediate ("**I**") in the addition of the electrophile to the aromatic ring (the first step) over the rate-determining transition state ($\ddagger_1$) is followed by loss of a proton (the second step, $\ddagger_2$) from that ion to regain aromaticity.

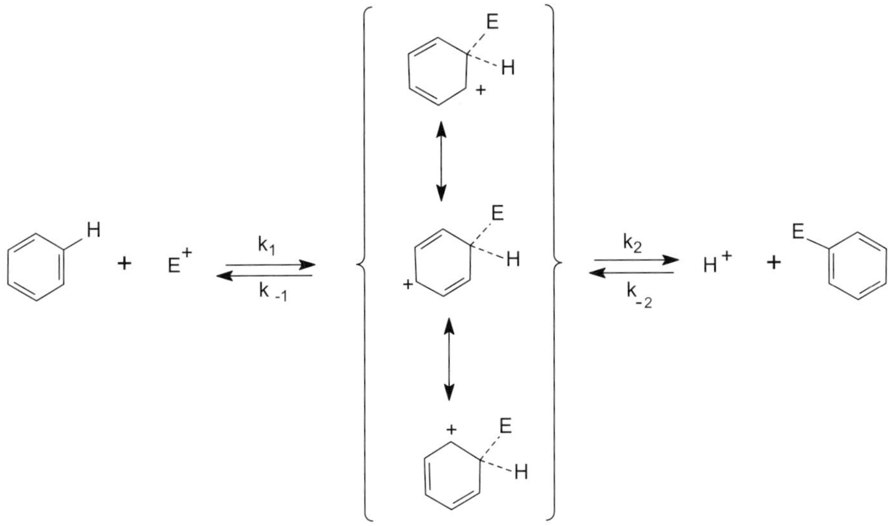

Scheme 6.82. The addition of an electrophile (E^+) to benzene (C_6H_6) showing the formation of a resonance stabilized σ-complex (arenium [benzenium, $C_6H_6E^+$]) intermediate.

In many cases there is evidence that a π-complex (Equation 6.93) or some other "encounter complex" may form before the σ-complex (arenium ion). However, it is generally not clear if such complexes actually lie on the path to product. What is clear is that the species labeled "**I**" (or something very much like it) in Figure 6.34 and for which the resonance-stabilized representations are given in Scheme 6.82 lies between the starting materials and the products.

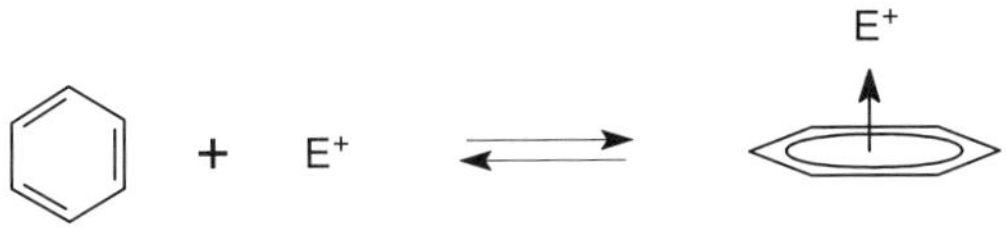

$$\text{(6.93)}$$

The evidence for the first step of Figure 6.34 being rate determining and the intermediacy of the resonance-stabilized (Scheme 6.82) σ-complex or arenium ion following that transition state comes from (a) the *isolation* of some suitably substituted arenium ion intermediates that then go on to product, (b) low-temperature studies of the NMR spectra of arenium ions, and (c) the presence or lack of a primary isotope effect when the carbon-hydrogen bond is broken. If the bond-breaking step is rate determining then an isotope effect on the rate will be seen.

With regard to the first, when 1,3,5-trimethylbenzene (mesitylene) Equation 6.94 is allowed to react with fluoroethane (ethyl fluoride [CH_3CH_2F]) in the presence of boron trifluoride (BF_3) at –80°C, a crystalline material, mp –15°C, forms. This material analyzes for the tetrafluoroborate salt of the 1-ethyl-2,4,6-trimethylbenzenium ion, which, as shown in Equation 6.94, on warming produces 1-ethyl-2,4,6-trimethylbenzene and tetrafluoroboric acid.

Where this carbocation is described more fully below....

$$(6.94)$$

With regard to spectroscopy, as shown in Table 6.11, when benzene (C_6H_6) is dissolved in a "super acid" (a mixture of SbF_5, FSO_3H, SO_2ClF, and SO_2F_2 is often used) at a low temperature (about $-140°C$), both ^1H and ^{13}C NMR spectra in concert with the presumed benzenium ion ($C_6H_7^+$) are found. Many related systems produce similar results.

Finally, isotope effects have been used to show that the formation of the arenium (specifically benzenium ion) is often the rate-determining step in the reaction of arene and electrophile. Table 6.12 provides a sample of electrophilic reactions using a variety of electrophiles and benzene (C_6H_6) as the "common" substrate. In the first example it is seen that when benzene (C_6H_6) is treated with deuteriosulfuric acid ($^2H_2SO_4$) under relatively mild conditions (i.e., at or near room temperature) so as to avoid sulfonation (Table 6.12, example 4), the sequential, equilibrium controlled, substitution of each of the protons (^1H) by deuterium (^2H) occurs. Eventually all of the protons are replaced producing hexadeuteriobenzene ($C_6{}^2H_6$) even though only one such replacement is shown in the electrophilic substitution reaction of Table 6.12 (example 1) and Equation 6.95. Since a carbon-deuterium bond ($C-{}^2H$) is stronger than a carbon-protium bond ($C-{}^1H$),* it would be anticipated that hexadeuteriobenzene ($C_6{}^2H_6$) should undergo electrophilic substitution by other electrophiles (E^+) more slowly than benzene (C_6H_6) itself *if breaking the carbon-hydrogen*

---

*Although a detailed discussion of isotope effects is beyond the scope of this text, the fundamental idea is straightforward. Generally, isotopic substitution (e.g., ^1H for or by ^2H and ^{13}C for or by ^{12}C) does not change much the potential energy surface for a reaction. Further, what changes occur must do so as a result of the change of **mass** in going from one isotope to another. Changes in **mass** largely cause **vibrational changes** in the bonds to which the masses are attached and are thus easily detected in the IR spectra of the isotopically substituted compounds. A simple calculation based on vibrational changes can provide an approximation of what might be expected as the result of a change of ^1H for or by ^2H **in a C-H bond that was being made or broken during a reaction**.

Clasically, vibrational frequencies are given by $\upsilon = (1/2\pi)\,(k/m)^{1/2}$, where m is the reduced mass, that is, $m = [m_1 m_2/(m_1 + m_2)]$

Since the $C-{}^2H$ frequency should be smaller than the $C-{}^1H$ by about $1/1.41$ and the $C-{}^1H$ frequency is found at about $3000\,cm^{-1}$, a difference in activation energies resulting in a difference of rate constants for protium (k_H) and deuterium (k_D) containing materials (k_H/k_D) *where the carbon-hydrogen bond is being broken in the rate-determining step* is expected to be of the order of about 6 (at or near room temperature).

TABLE 6.11. Data for the Low-Temperature ^{1}H and ^{13}C NMR Spectra of the Benzenium Ion

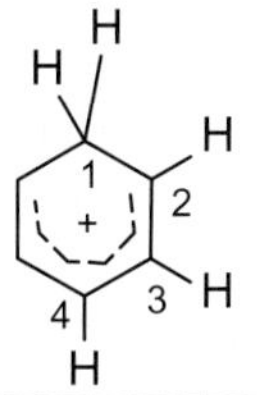

^{1}H NMR			^{13}C NMR	
Position	ppm (TMS = 0.00)		Position	ppm (TMS = 0.00)
1	3.6		1	52.2
2	8.8		2	186.6
3	7.2		3	136.9
4	7.5		4	178.1

Presented in a different format in Olah, G. A.; Staral, S.; Asencio, G.; Liang, G.; Forsyth, D. A.; Mateescu, G. D. *J. Am. Chem. Soc.*, **1978**, *100*, 6299.

TABLE 6.12. A Sample of Some Electrophilic Substitution Reactions on Benzene as a Substrate. Further Substitution Reactions on the Products May Occur

Number	Arene	Electrophile	Product(2)
1		^{2}H$_2$SO$_4$	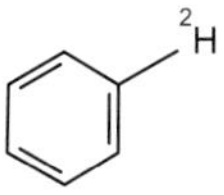 ^{2}H-benzene + polydeuterated benzenes
2		H$_2$SO$_4$ + HONO$_2$ (H$_3$O$^+$/HSO$_4^-$/NO$_2^+$)	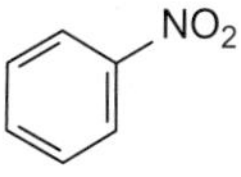 NO$_2$-benzene
3		Br$_2$/FeBr$_3$	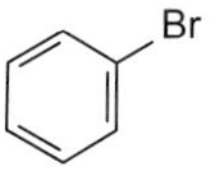 Br-benzene + H$^+$ / FeBr$_4^-$
4		SO$_3$/H$_2$SO$_4$	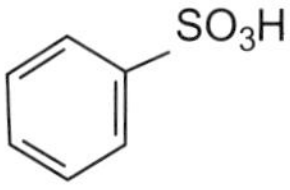 SO$_3$H-benzene
5		CH$_3$CH$_2$Cl/AlCl$_3$	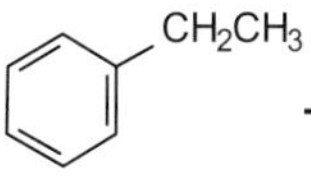 CH$_2$CH$_3$-benzene + H$^+$ / AlCl$_4^-$
6		CH$_3$CH=CH$_2$/H$_3$PO$_4$	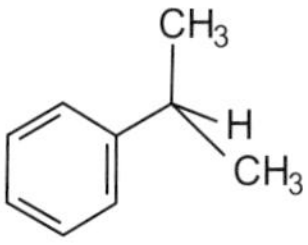 (CH$_3$)$_2$CH-benzene
7		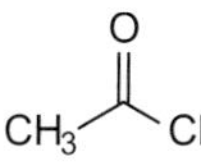 CH$_3$C(=O)Cl / AlCl$_3$	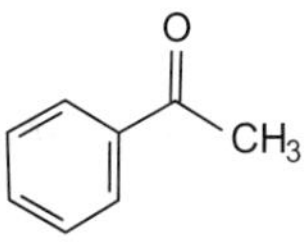 C(=O)CH$_3$-benzene + H$^+$ / AlCl$_4^-$

bond occurred in the rate-determining step. Ususally, there is no deuterium isotope effect or, where there is one, it is much smaller than expected. The small effect is accounted for on the basis of the closeness of the values of the rate constants k_{-1} and k_2 (Scheme 6.82).

Equation 6.95 describes the substitution of ^2H for ^1H in the reaction between deuteriosulfuric acid ($^2H_2SO_4$) and benzene (C_6H_6).

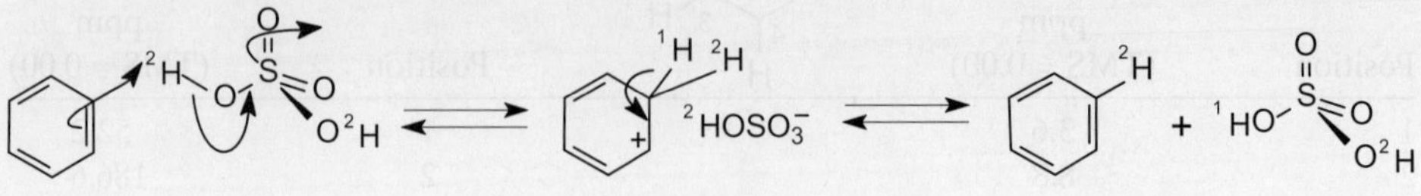

$$(6.95)$$

The second example of Table 6.12, the nitration of benzene (C_6H_6) in sulfuric acid (H_2SO_4) (Equation 6.96) is a widely investigated reaction. It is clear that the nitronium ion (NO_2^+) is the

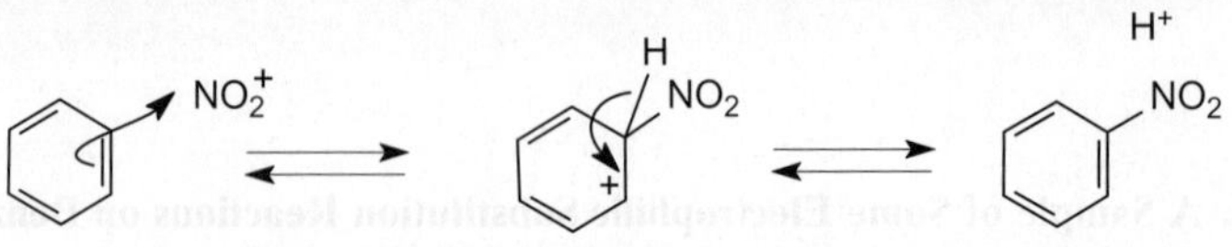

$$(6.96)$$

nitrating species and inorganic salts such as nitronium tetrafluoroborate (NO_2BF_4) as well as the less expensive mixture of nitric ($HONO_2$) and sulfuric [$(HO)_2SO_2$] acids are effective. However, if the mixture of sulfuric (H_2SO_4) and nitric (HNO_3) is used, it is necessary to avoid the presence of water (H_2O) because, as shown in Scheme 6.83, it will inhibit the reaction.

Problem 6.15. Using the information present in Scheme 6.83, account for the observation that water (H_2O) is deleterious to the progress of nitration of arenes with nitric ($HONO_2$) and sulfuric [$(HO)_2SO_2$] acids.

$$HONO_2 + H_2SO_4 \rightleftharpoons H_2ONO_2^+ + HSO_4^-$$

$$H_2ONO_2^+ \rightleftharpoons H_2O + NO_2^+$$

$$H_2O + H_2SO_4 \rightleftharpoons H_3O^+ + HSO_4^-$$

$$HONO_2 + 2\,H_2SO_4 \rightleftharpoons H_3O^+ + NO_2^+ + 2\,HSO_4^-$$

Scheme 6.83. A suite of reactions that accounts for the production of the nitronium ion (NO_2^+) by the sulfuric (H_2SO_4) acid promoted dehydration of nitric (HNO_3) acid. It is important to notice the stoichiometry of the summed reaction since cryoscopic measurements identify four ionic species.

Among the more interesting features of the nitration reaction are the observations that arenes *more* reactive than benzene (C_6H_6) [and benzene (C_6H_6) itself under certain conditions] undergo nitration at a rate that is **independent** of the concentration of arene as long as nitric acid ($HONO_2$) is in constant excess. This is accounted for by arguing that the production of the nitronium ion (NO_2^+) is rate determining (Equation 6.97). Similarly, arenes that are *less* reactive than benzene (C_6H_6) [and benzene (C_6H_6) itself under certain conditions] undergo nitration at a rate that is **first order** in arene with nitric acid ($HONO_2$) in constant excess.* This is accounted for by arguing that the formation of the intermediate "**I**" (Figure 6.34) is rate determining (Equation 6.98).

$$\text{rate} = k_{\text{observed}} \tag{6.97}$$

$$\text{rate} = k'_{\text{observed}}[\text{arene}]^1 \tag{6.98}$$

Finally, with regard to nitration with mixed nitric ($HONO_2$) and sulfuric (H_2SO_4) acids, it should be recognized that this is a potent oxidizing mixture and not only must care be exercised but its successful use argues for the stability of arenes.

When bromine (Br_2) or iodine (I_2) is added to benzene (C_6H_6), there is an instantaneous color change. It has been argued that this is due to the formation of some sort of π-complex [see Equation 6.93, where for bromine (Br_2), $E^+ = Br^+$ and the counter ion is Br_3^-, and similarly for iodine (I_2)]. However, unlike the situation for alkenes and alkynes, addition across the double bond does not occur. Indeed, benzene (C_6H_6) and iodine (I_2) or bromine (Br_2) can be separated and the starting materials recovered unchanged. However, reaction with halogens does occur in the presence of a catalyst as shown in the third example in Table 6.12.

As was the case with other hydrocarbons, the order of reactivity of arenes with halogens in the presence of the appropriate catalysts is $F_2 > Cl_2 > Br_2 > I_2$. Generally, fluorine (F_2) is too reactive to control and, even with newer methods such as the use of xenon difluoride (XeF_2), the preparation of aryl fluorides requires alternatives to direct fluorination (*vide infra*). At the other end of the reactivity spectrum, iodine (I_2) is the least reactive and generally iodination with iodine (I_2) will not succeed in producing iodobenzene (C_6H_5I) *unless* there is an oxidizing agent present to produce significant quantities of reactive iodinating species (i.e., I^+ or its equivalent). However, chlorination and bromination, in the presence of metal salt catalysts, appear quite facile. Thus, as shown in Scheme 6.84, X = Br, bromination in the presence of iron(III) bromide (ferric bromide, $FeBr_3$) or chlorination in the presence of zinc(II) chloride ($ZnCl_2$) or iron(III) chloride ($FeCl_3$) (Scheme 6.84, X = Cl) succeeds in the production of, respectively, bromobenzene (C_6H_5Br) and chlorobenzene (C_6H_5Cl). Finally, in this regard, it will become clear that if there are (so called) "activating substituents" (Scheme 6.81) on the arene, the electrophilic substitution reactions of halogens may become so facile that catalysts are not required.

The sulfonation of benzene (C_6H_6) to produce benzenesulfonic acid ($C_6H_5SO_3H$) is given (in Table 6.12) as the fourth example of electrophilic aromatic substitution. Because the reaction has been extensively studied (in part, as will be discussed

---

*Although not yet discussed, the first substitutent added (viz. Scheme 6.81) plays a role that activates or deactivates an arene toward further electrophilic, nucleophilic, or radical substitution. The details of the reactions of such substituted arenes toward further substitution are discussed later in this section.

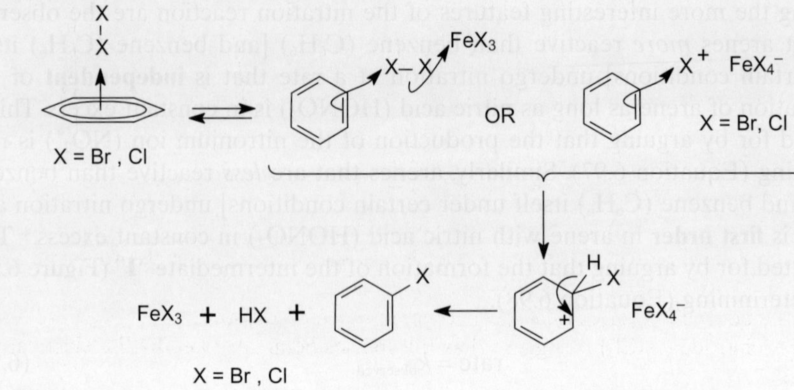

Scheme 6.84. A representation of iron(III)-catalyzed bromination and chlorination (X = Br, Cl) of benzene (C_6H_6).

subsequently, because salts of alkyl substituted benzensulfonic acids are commercially available as **detergents**), it has become clear that the exact nature of the sulfonating reagent is strongly dependent upon the concentration of sulfur trioxide (SO_3) in the sulfuric acid. However, as shown in Equation 6.99, if the concentration of sulfur trioxide (SO_3) is high enough, it is likely that most of the reaction involves pyrosulfuric acid ($H_2S_2O_7$) and/or its protonated form ($H_3S_2O_7^+$). Pyrosulfuric acid ($H_2S_2O_7$) is the adduct formed between sulfuric acid (H_2SO_4) and sulfur trioxide (SO_3).

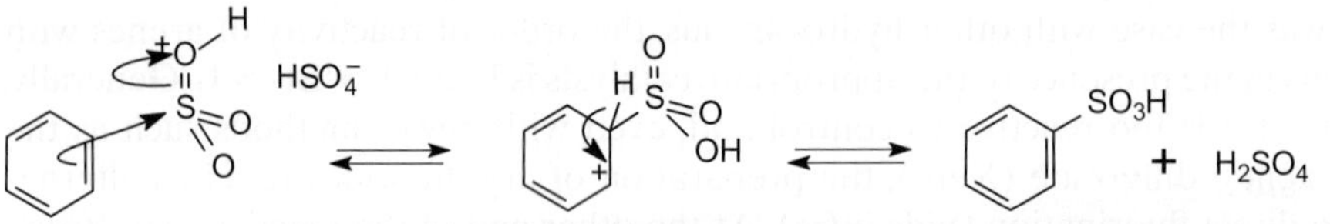

$$\text{(6.99)}$$

In the fifth, sixth, and seventh examples in Table 6.12, a **new carbon-carbon bond is forged.** These three processes and related others are variations on the theme of transformations called the **Friedel-Crafts*** reactions.

As seen in the fifth example of Table 6.12, and in more detail in Scheme 6.85, when a cold solution of benzene (C_6H_6) is treated with chloroethane (ethyl chloride, CH_3CH_2Cl) in the presence of a Lewis acid (such as aluminum trichloride, $AlCl_3$), phenylethane (ethylbenzene, $C_6H_5CH_2CH_3$) is formed.

Although the chemistry of alkyl halides will be discussed in some detail in the next chapter (Chapter 7), it should be clear even now that complexation of the

---

*Charles Friedel (1832–1899), Professor, Sorbonne, where he first worked with C. A. Wurtz, successor to Dumas. Among his other accomplishments (with J. M. Crafts) was the first reported investigation of the process of transesterification (Chapter 9).

James M. Crafts (1839–1917), Professor, Cornell University and then Massachusetts Institute of Technology (MIT). After a lengthy leave (about 17 years) during which he worked at the Sorbonne in the laboratory of Wurtz, where he collaborated with Friedel, he returned to MIT. At MIT he moved into the administration, eventually becoming president (1897–1900). Subsequently, he resigned and reascended to the faculty.

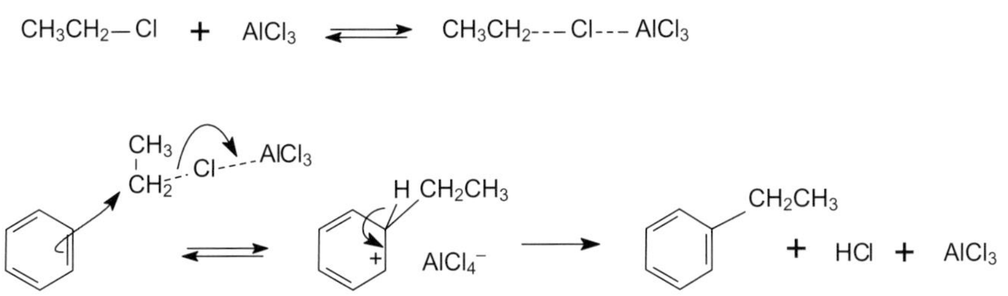

$$CH_3CH_2\!-\!Cl \;+\; AlCl_3 \;\rightleftharpoons\; CH_3CH_2\text{---}Cl\text{---}AlCl_3$$

Scheme 6.85. A representation of the Friedel–Crafts reaction between chloroethane (ethyl chloride, CH_3CH_2Cl) and benzene (C_6H_6) in the presence of aluminum trichloride ($AlCl_3$) to produce ethylbenzene ($C_6H_5CH_2CH_3$). A complex between chloroethane (CH_3CH_2Cl) and aluminum trichloride ($AlCl_3$) is suggested as the species being attacked by the electron-rich aromatic ring.

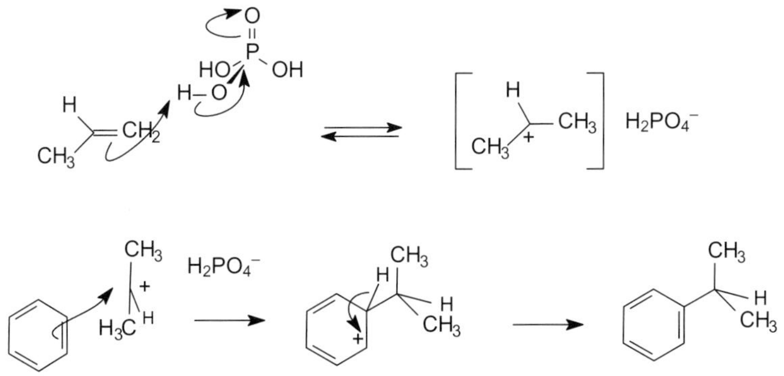

Scheme 6.86. A representation of a pathway for the formation of 2-phenylpropane by the reaction of benzene (C_6H_6) with propene ($CH_3CH=CH_2$) in the presence of phosphoric acid (H_3PO_4). The formation of a secondary cation or its equivalent by phosphoric acid protonation of propene ($CH_3CH=CH_2$) presumably proceeds the attack by benzene (C_6H_6).

electron-rich chlorine of chloroethane (ethyl chloride, CH_3CH_2Cl) with the Lewis acid aluminum trichloride ($AlCl_3$) would present an electron-deficient primary carbon to the electron-rich aromatic ring.

Indeed, it is this pattern that is typical of **Friedel-Crafts** reactions.

Thus, as shown in Scheme 6.86, protonation of propene ($CH_3CH=CH_2$) by phosphoric acid (H_3PO_4) should reasonably lead to the corresponding secondary carbocation that, were there a nucleophilic anion present, would be consumed in a more typical addition reaction (*vide supra*, this chapter, Section III, Part a). However, as the only ions present are those in equilibrium with the dihydrogen orthophosphate anion ($H_2PO_4^-$), a weak nuclophile, the electron-rich ring attacks either the cation or, what is less likely, the more highly substituted carbon atom while the alkene is undergoing protonation. Subsequent loss of a proton then regenerates the aromatic ring. The formation of 2-phenylpropane [(1-methyl)ethylbenzene] has clearly followed the "Markownikoff" pattern (most stable carbocation) of substitution.

The idea of a carbocation participating in these acid-catalyzed processes is reenforced as shown in Scheme 6.87, where a pathway for the reaction between

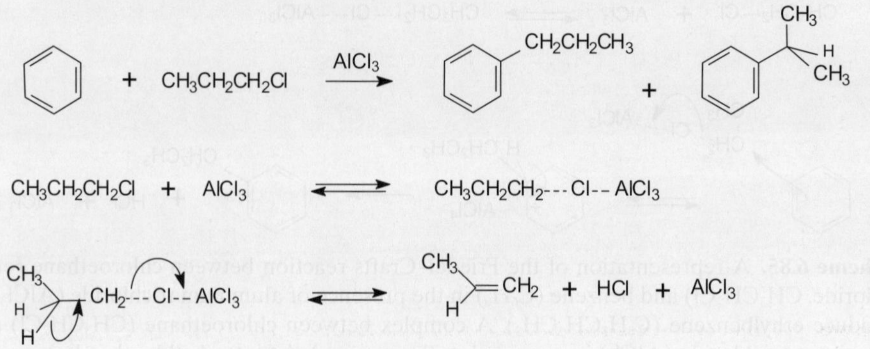

Scheme 6.87. A representation of the reaction between benzene (C_6H_6) and 1-chloropropane ($CH_3CH_2CH_2Cl$) to produce both 1-phenylpropane and 2-phenylpropane. It is suggested that the former is generated by the reaction of an aluminum trichloride ($AlCl_3$) complex with the alkyl halide. Simultaneously, dehydrohalogenation of the halide leads to alkene that, in the presence of the Lewis acid, undergoes reaction to produce the latter.

1-chloropropane (propyl chloride, $CH_3CH_2CH_2Cl$) and benzene (C_6H_6) in the presence of aluminum trichloride ($AlCl_3$) is presented. If this reaction is carried out in dichloromethane (methylene chloride, CH_2Cl_2) solution at about −5°C then both 1-phenylpropane and 2-phenylpropane are obtained in a 3:2 ratio. However, at about +35°C, while the same two isomers are found, the ratio is now 2:3. From this result and other similar experimental findings, it is argued that the initial complexation of 1-chloropropane with aluminum chloride leads to a species that is capable of being attacked by benzene (C_6H_6) and that 1-phenylpropane forms as a result of that process. And, at the same time, the alkyl chloride (1-chloropropane, $CH_3CH_2CH_2Cl$) is reacting with aluminum trichloride ($AlCl_3$) to produce rearranged alkyl chloride (i.e., 2-chloropropane, [$(CH_3)_2CHCl$]) or simply alkene (i.e., propene, $CH_3CH=CH_2$), hydrogen chloride (HCl), and aluminum trichloride ($AlCl_3$).

Another variation on the Friedel-Crafts reaction is to utilize an acyl halide (rather than an alkyl halide). Although the chemistry of acyl halides and other derivatives of carboxylic acids is discussed in Chapter 9, it is appropriate to point out that a hydrogen can be substituted by an acyl group ($CH_3C=O$) as shown in Scheme 6.88.

In the same vein, Gatterman* showed that carbon monoxide (CO) and hydrogen chloride (HCl) would react with benzene (C_6H_6) in the presence of aluminum trichloride ($AlCl_3$) at high pressure so as to substitute an aldehyde functional group for a hydrogen. The actual electrophile attacked by the aromatic ring may be the acylium ion shown in Scheme 6.89 complexed with the aluminum tetrachloride anion.

Although not shown in Table 6.12, aryl metallation by reaction with mercury(II) (Equation 6.100) and thallium(III) (Equation 6.101) salts is also known. The mercuration is commonly effected with the perchlorate [$Hg(ClO_4)_2$] to give the corresponding mercurial salt. Interestingly, unlike other electrophilic substitution reactions discussed above, a deuterium isotope effect indicative of *rate-determining*

*Ludwig Gatterman (1860–1920), Professor, Heidelberg.

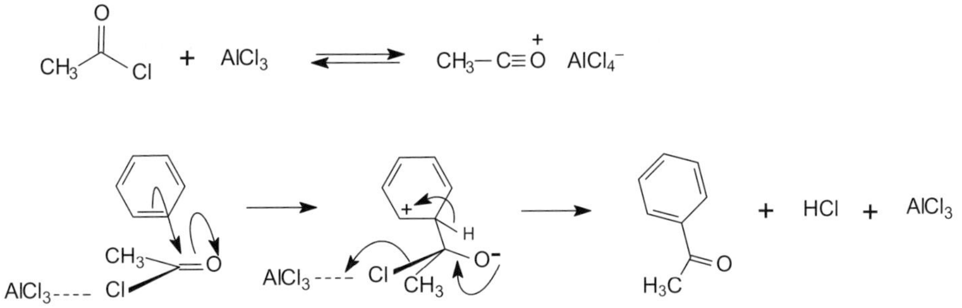

Scheme 6.88. A representation of a pathway to substitute an acyl group (CH_3C=O) onto benzene (C_6H_6) ring in place of hydrogen. The production of 1-phenylethanone (acetophenone, $C_6H_5COCH_3$) could occur by attack of the aromatic ring onto an aluminum chloride ($AlCl_3$) complex with the acyl chloride or by attack of the aromatic ring on a preformed carbocation generated by dehalogenation of the acyl chloride.

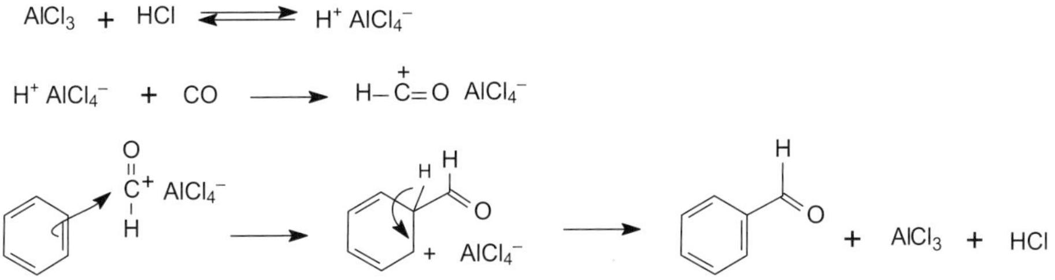

Scheme 6.89. A potential pathway for the formation of benzenecarbaldehyde (benzaldehyde C_6H_5CHO) by the Gatterman Reaction among benzene (C_6H_6), carbon monoxide (CO), hydrogen chloride (HCl), and aluminum chloride ($AlCl_3$) at high pressure. The process is formally similar to the Friedel–Crafts reaction.

proton loss (i.e., $k_H/k_D \approx 6$, *vide supra*) is observed here. Generally, it has proved possible to replace the mercury with other metals, such as palladium, and to produce highly active catalysts for addition and related reactions.

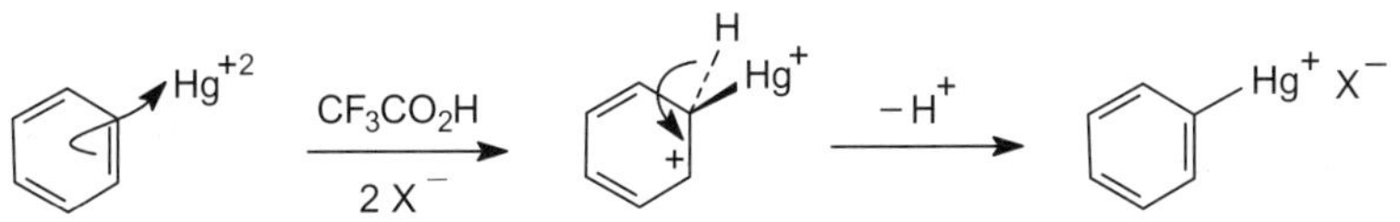

$$\text{(6.100)}$$

The arylthallium bistrifluoroacetate can be converted to many other compounds (e.g., phenols and aryl iodides) and its formation (and conversion to the aryl iodide) is included here since at least some evidence can be found that it is formed in an electrophilic substitution reaction. There is additional evidence, however, that its formation may involve a one-electron-transfer process.

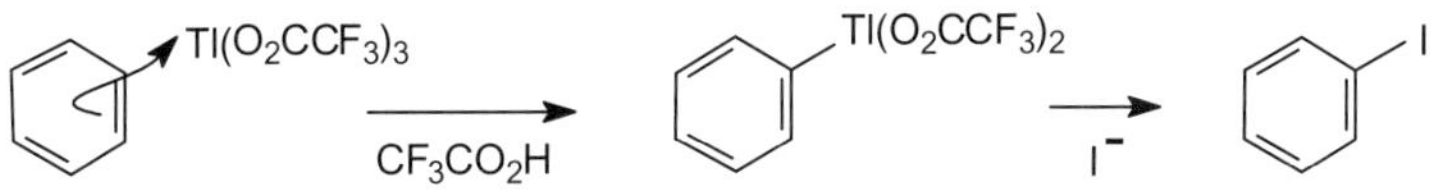

$$\text{(6.101)}$$

TABLE 6.13. A Sample of Some Electrophilic Substitution Reactions on Naphthalene as a Substrate. While More Than One Product is Shown in Each Case, the Ratio of Products Is Seldom 1:1 and Is Strongly Dependent Upon the Exact Conditions under Which the Reaction is Carried Out. The Products Shown Frequently Continue to React to Give Further Substitution

Number	Arene	Electrophile	Product(s)
1	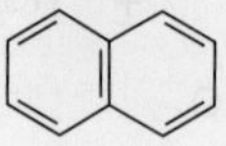	$^2H_2SO_4$	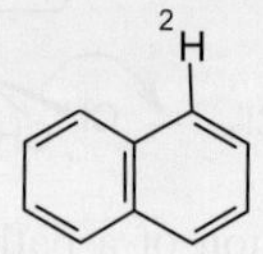 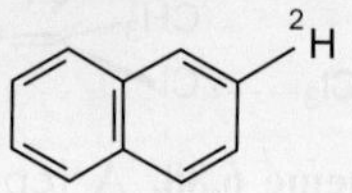
2	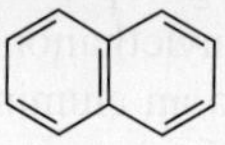	$Br_2/CH_3CO_2H/H_2O$	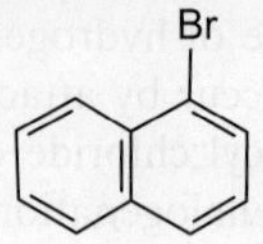 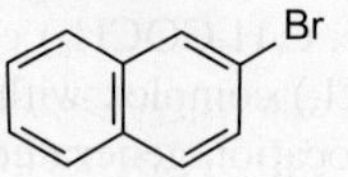
3	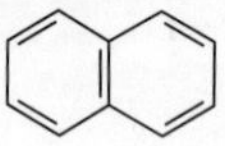	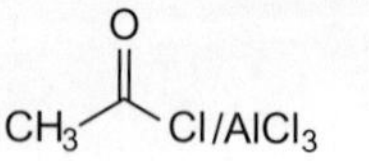 CH_3—CO—$Cl/AlCl_3$	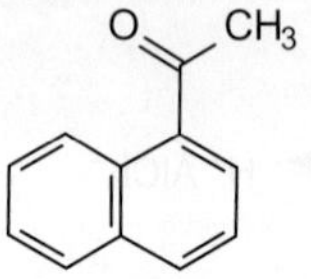 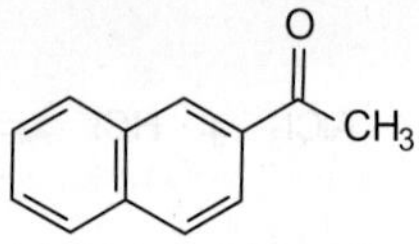
4	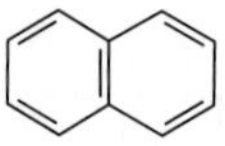	H_2SO_4 heat	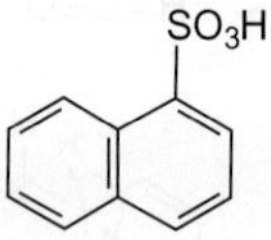 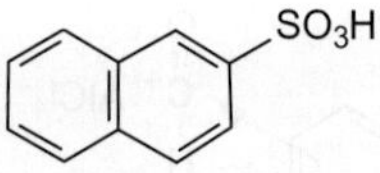
5	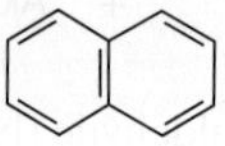	$H_2SO_4{}^+HONO_2$ $H_3O+/HSO_4{}^-/NO_2{}^+$	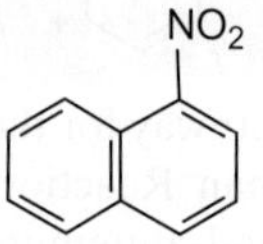 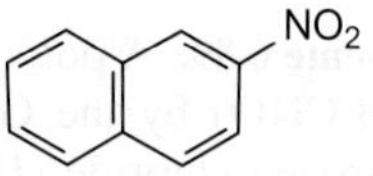

All of the results of substitution of various groups for hydrogen on benzene (C_6H_6) outlined above also apply to alkyl-substituted benzenes. They apply as well to other aromatic systems. In both cases, the symmetry of benzene (C_6H_6) has been perturbed and thus it is common for more than one product to be obtained. Consider the case of naphthalene ($C_{10}H_8$), the "first" of the class of polynuclear aromatic hydrocarbons. A sampling of some of the reactions of naphthalene ($C_{10}H_8$) is provided in Table 6.13 and, when comparing the reactions to those of Table 6.12, it is important to note that it is *common to find simultaneous formation of more than one substitution product.*

The exchange of the protons at the C-1 (C-α) and C-2 (C-β) positions of naphthalene (example 1, Table 6.13) has been measured and the rate has been found to depend upon the acidity of the medium in which the exchange occurs. However, regardless of the rate at which exchange occurs, it is clear that reaction at C-1 (C-α) occurs about 8 to 10 times faster than it does at C-2 (C-β).*

Possible pathways are shown in Scheme 6.90 for exchange of tritium (^3H) at C-1 and C-2. It is common to argue that the carbocation intermediate for the exchange

*The exchange reaction was actually carried out in separate experiments beginning with tritium (^3H) C-1 (C-α) and C-2 (C-β)-labeled naphthalene and the loss of ^3H to the solution measured.

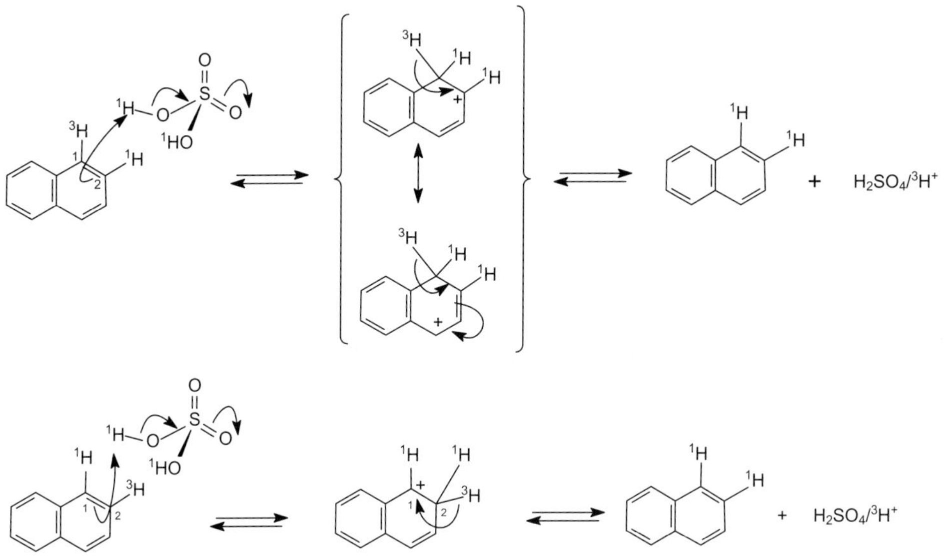

Scheme 6.90. Potential pathways for the exchange of protons at C-1 (C-α) and C-2 (C-β) of naphthalene. The exchange of tritium (^{3}H) at C-1 is 8 to 10 times faster than at C-2.

at C-1 is more stable than that at C-2 because the former is stabilized by *allylic* conjugation that does not involve "intruding" on the aromaticity of the ring (i.e., it is allylic as well as benzylic while the latter is only benzylic) and thus the transition state leading to it is lower. The same argument was presented earlier in relation to carbocation stability during addition to alkenes (*vide supra*).

The nitronium ion (NO_2^+), generated as indicated earlier from the action of sulfuric acid (H_2SO_4) on nitric acid ($HONO_2$)(Scheme 6.83), introduced in the form of nitronium tetrafluoroborate ($NO_2^+BF_4$) or formed *in situ* from the reaction of nitric acid ($HONO_2$) with ethanoic anhydride (acetic anhydride [$(CH_3CO)_2$]) (Equation 6.102) also reacts with napththalene (Table 6.13, example 2) to give substitution at *both* C-1 (C-α) and C-2 (C-β) as

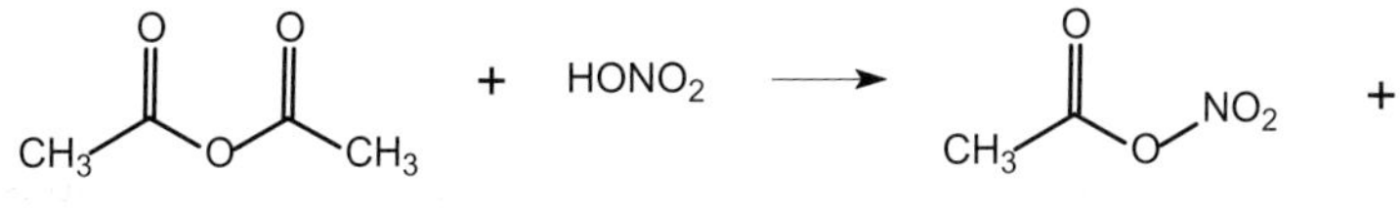

$$(6.102)$$

shown in Scheme 6.91. Although under very gentle conditions nearly exclusive substitution at C-1 (C-α) can be effected, using sulfuric (H_2SO_4) and nitric ($HONO_2$) acids as before produces a mixture that is about 9:1 in favor of C-1 (C-α) substitution. The rationale is the same as that provided above for proton exchange.

As indicated in the third example in Table 6.13, the bromination of naphthalene ($C_{10}H_8$) produces a mixture (99:1) of C-1 (C-α) and C-2 (C-β) products even in the absence of iron(III) catalysts. Indeed, as shown, bromination can be effected in aqueous acetic acid (ethanoic acid, CH_3CO_2H). There is no reason to believe that the brominating agent is not bromine (Br_2) and that the process is not directly

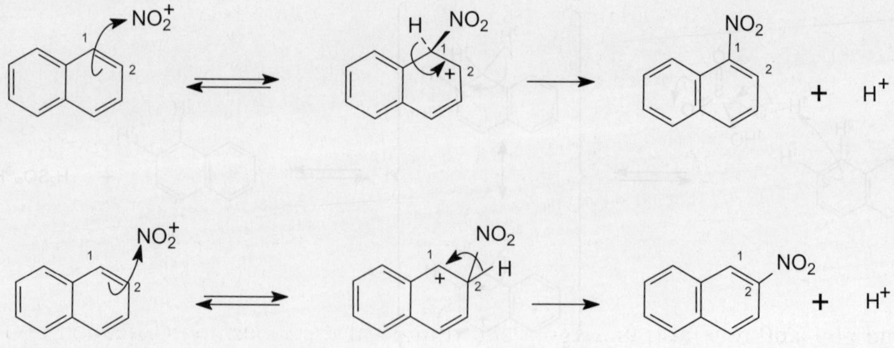

Scheme 6.91. Representations for the reaction of the nitronium ion (NO_2^+) with naphthalene ($C_{10}H_8$) to produce both C-1 (C-α) and C-2 (C-β)-substituted nitronaphthalene. Using the nitrating mixture from sulfuric (H_2SO_4) and nitric (HNO_3) acids at or near room temperature produces about a 9:1 mixture of α-nitronaphthalene:β-nitronaphthalene.

analogous to that shown for nitration (Scheme 6.91) and proton exchange (Scheme 6.90). Presumably the process is similar to that depicted in Equation 6.103.

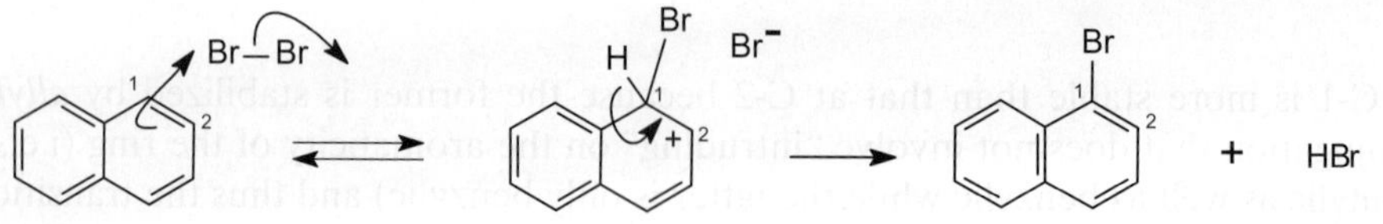

$$\text{(6.103)}$$

In contrast to nitration (Scheme 6.91) and bromination (Equation 6.103), the products of sulfonation of naphthalene ($C_{10}H_8$) with sulfuric acid (Table 6.13, example 4) (Equation 6.104) are different at different temperatures. Indeed, low temperature sulfonation has been demonstrated to be reversible and below 120°C only 1-naphthalenesulfonic acid is isolated (i.e., α-substitution). Above 160°C only 2-naphthalenesulfonic acid (β-substitution product) is isolable. The low-temperature product (1-naphthalenesulfonic acid) can be isomerized by heating to yield its isomer (2-naphthalenesulfonic acid). This is another case of kinetic and thermodynamic control providing different products (as discussed earlier for addition to conjugated alkenes [this chapter]).

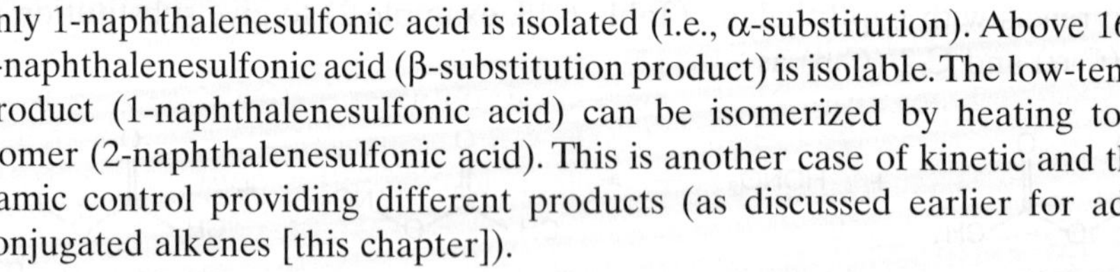

$$\text{(6.104)}$$

The last example in Table 6.13 is the Friedel-Crafts acylation of naphthalene ($C_{10}H_8$). Substitution is found at both C-1 (C-α) and C-2 (C-β). Presumably, these products form by pathways that are similar to those already examined. However, dramatic solvent effects on the product ratio have been reported and if the acylation is carried out in solvents that can form complexes with ethanoyl chloride (acetyl chloride, CH_3COCl), then most of the substitution is found at C-2 (C-β). It has been argued that this is because C-2 (C-β) is less hindered than C-1 (C-α).

It has not yet been found possible to find conditions under which naphthalene ($C_{10}H_8$) can be cleanly alkylated to provide simple products. Thus, alternative methods that involve reduction of more highly oxidized substituents to alkyl groups have been used when specific alkyl-substituted naphthalenes are sought.

Finally, with regard to polynuclear arenes and as might be anticipated from the results of oxidation discussed earlier, these compounds frequently undergo substitution *so as to maximize the number of aromatic rings that are unaffected.* Thus, the bromination of anthracene provides 9-bromoanthracene (Equation 6.105) and phenanthrene also undergoes substitution at the 9-position (Equation 6.106). Careful addition of bromine (Br_2) in both cases even permits isolation of the respective unstable 9,10-dibromides. This demonstrates that in these cases the bimolecular addition process that involves bromide anion (generating dibromide) and the unimolecular process involving proton loss to produce the rearomatized arene must pass over transition states of similar heights.

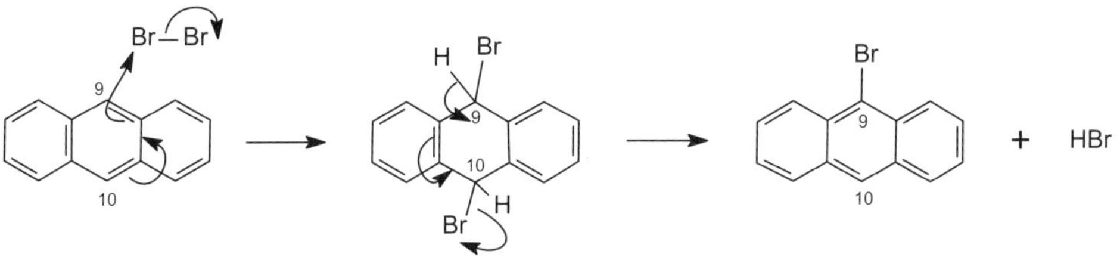

$$(6.105)$$

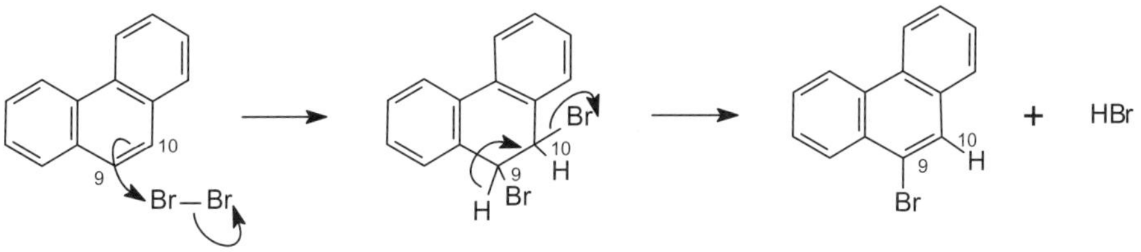

$$(6.106)$$

When electrophilic substitution is carried out on an aromatic hydrocarbon that already bears *another alkyl (or aryl) group,** it is found that the group present **directs the position(s) into which the incoming electrophile enters**.

---

*Since the subject of this chapter is **"Hydrocarbons,"** the discussion of the directing effect(s) of other substituents (e.g., Cl, NO_2, and OCH_3) will be dealt with (only) obliquely here. Those topics are covered more substantively in sections in subsequent chapters dealing with halogen, nitrogen, ethers, and others. Although not specifically addressed earlier, benzene is obtained by the catalytic reforming (platinum catalyst) of *light naptha* (see Chapter 1) petroleum distillate. Generally, benzene (C_6H_6), toluene (methylbenzene, $C_6H_5CH_3$), dimethylbenzenes (xylenes, [$C_6H_4(CH_3)_2$]), and a mixture of other aromatic compounds (called "C_9 aromatics") are obtained. Thus, methylbenzene (toluene, $C_6H_5CH_3$) is not generally obtained by the Friedel-Crafts methylation of benzene (C_6H_6). Indeed, it is general that *more* methylbenzene (toluene, $C_6H_5CH_3$) is produced during the reforming process than is necessary for commerce and it is converted to benzene and methane by further treatment with hydrogen and a cobalt catalyst at high temperature (ca. 600°C) and high pressure (ca. 1000 psi).

Consider the Friedel-Crafts alkylation reaction (Table 6.12, example 5). The addition of aluminum trichloride ($AlCl_3$) to, for example, a chloroalkane in benzene (C_6H_6) solution as in Scheme 6.85 or Scheme 6.87, produces an alkyl-substituted benzene. Interestingly, the product of the reaction, the alkylated benzene, undergoes further alkylation; the first alkylation apparently activates the ring toward the subsequent reaction. Products of a second and third alkylation are frequently more common than those where only mono-alkylation has occurred. However, the details of the competition for the specific site (*ortho* [or 1,2-], *meta* [or 1,3-], and *para* [or 1,4-]) into which the next substitution occurs is obscured because the alkyl groups *rearrange on the aromatic nucleus* to produce the most stable isomer (i.e., the one most likely to resist further rearrangement). The process can be observed with the dimethylbenzenes {xylenes [$C_6H_4(CH_3)_2$]}, obtained from petroleum hydrocarbons (see footnote p. 439) and purified by distillation.

Thus, 1,2-dimethylbenzene (*ortho*-xylene [$1,2\text{-}C_6H_4(CH_3)_2$] (Figure 6.35), 1,3-dimethylbenzene (*meta*-xylene [$1,3\text{-}C_6H_4(CH_3)_2$] (Figure 6.36), and 1,4-dimethylbenzene (*para*-xylene [$1,4\text{-}C_6H_4(CH_3)_2$] (Figure 6.37) equilibrate in the presence of aluminum trichloride ($AlCl_3$) to produce a mixture enriched in the *meta*-isomer (Scheme 6.92).

Problem 6.16. Based upon the relationships described in the legend of Figure 6.36a and the expanded picture ($\delta = 7.53$, 7.51, 7.50, 7.48, 7.35, and 7.32 ppm) of the aromatic region of that figure, identify the aromatic protons of 1,3-dimethylbenzene (*m*-xylene).

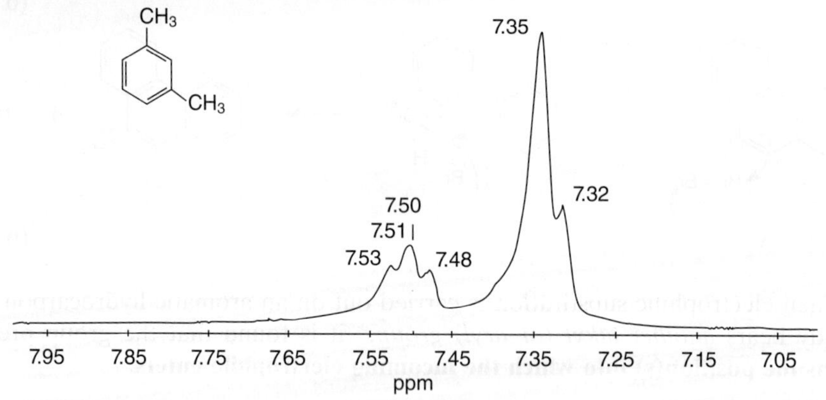

In contrast to the rearrangements that accompany the further Lewis acid-catalyzed alkylation of an already alkylated arene, electrophilic substitution reactions with NO_2^+ (i.e., *nitration*) and Br_2 (i.e., *iron(III)-catalyzed* bromination) appear somewhat less complicated. In these cases too, it is clear that any alkyl (or aryl) substituent, being more electron-rich than the hydrogen it has replaced, activates the arene toward further electrophilic substitution. Thus, in a competition between benzene (C_6H_6) and methylbenzene (toluene, $C_6H_5CH_3$) for a limited amount of nitronium ion (NO_2^+) the latter succeeds in reacting while the former remains unchanged. Interestingly, positions *ortho-*, *meta-*, and *para-* are all activated, with the *ortho-* and *para-* positions being activated more than the *meta-* (Equation 6.107).

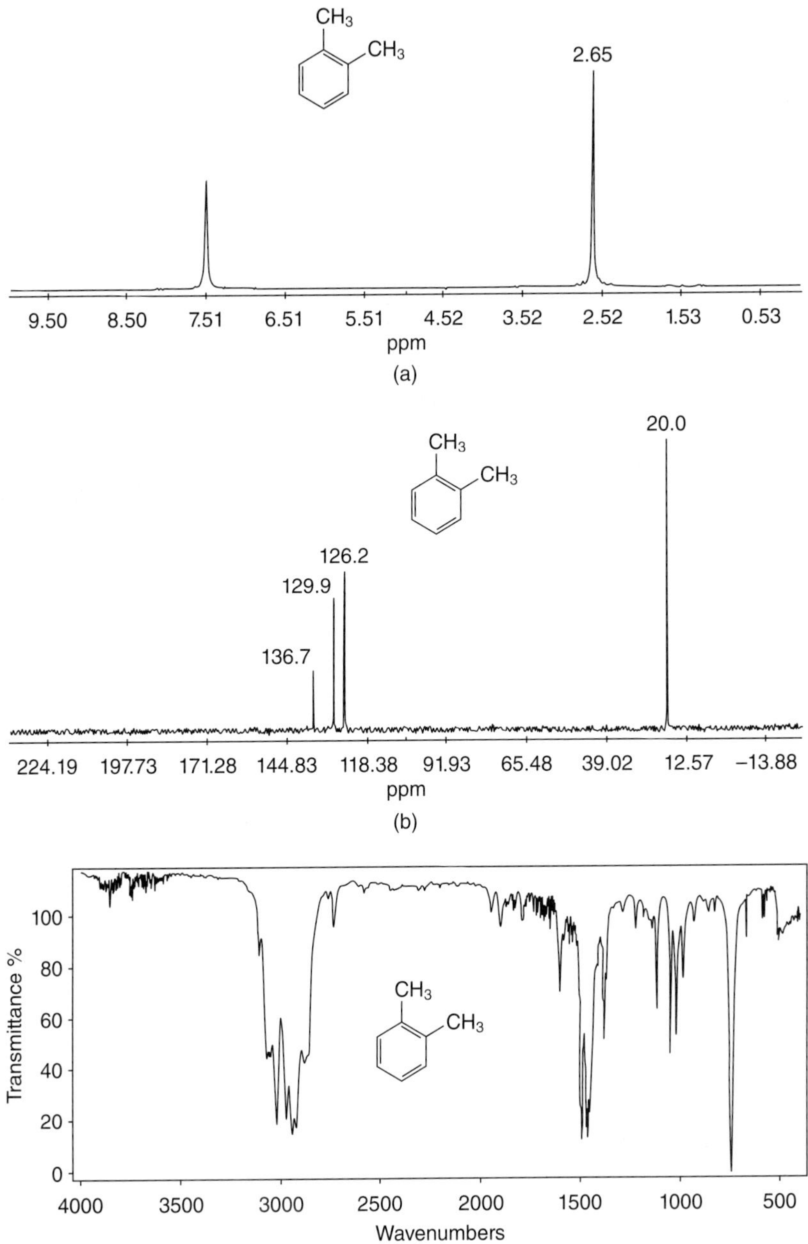

Figure 6.35. (a) The ^{1}H NMR spectrum of 1,2-dimethylbenzene (*ortho*-xylene) in ^{2}HCCl$_3$ at 300 MHz. The chemical shift of the methyl groups (CH$_3$–) at δ = 2.65 ppm is very close to those of methyl groups in its isomers (*vide infra*). (b) The ^{13}C NMR spectrum of 1,2-dimethylbenzene (*ortho*-xylene) in ^{2}HCCl$_3$ at 75.5 MHz. The carbons (C-1 and C-2) bearing the methyl (CH$_3$–) groups come into resonance at δ = 136.7 ppm, the carbons that flank the methyl (CH$_3$–) groups (C-3 and C-6) at δ = 129.9 ppm, the remaining aromatic carbons (C-4 and C-5) being the furthest upfield (least deshielded) at δ = 126.2 ppm. The identical methyl (CH$_3$–) groups are found at δ = 20.0 ppm downfield from TMS. (c) The FT-IR (neat) spectrum of 1,2-dimethylbenzene (*ortho*-xylene). As is typical of aromatic hydrocarbons, evidence for the *ortho*- methyl groups is seen in the pattern of low-intensity bands in the region 2000–1650 cm^{-1} and in the intense signal near 800 cm^{-1}. These features, which are quite characteristic of the substitution pattern, should be compared with those found in the IR spectra of 1,3-dimethylbenzene (*meta*-xylene), Figure 6.36c, and 1,4-dimethylbenzene (*para*-xylene), Figure 6.37c.

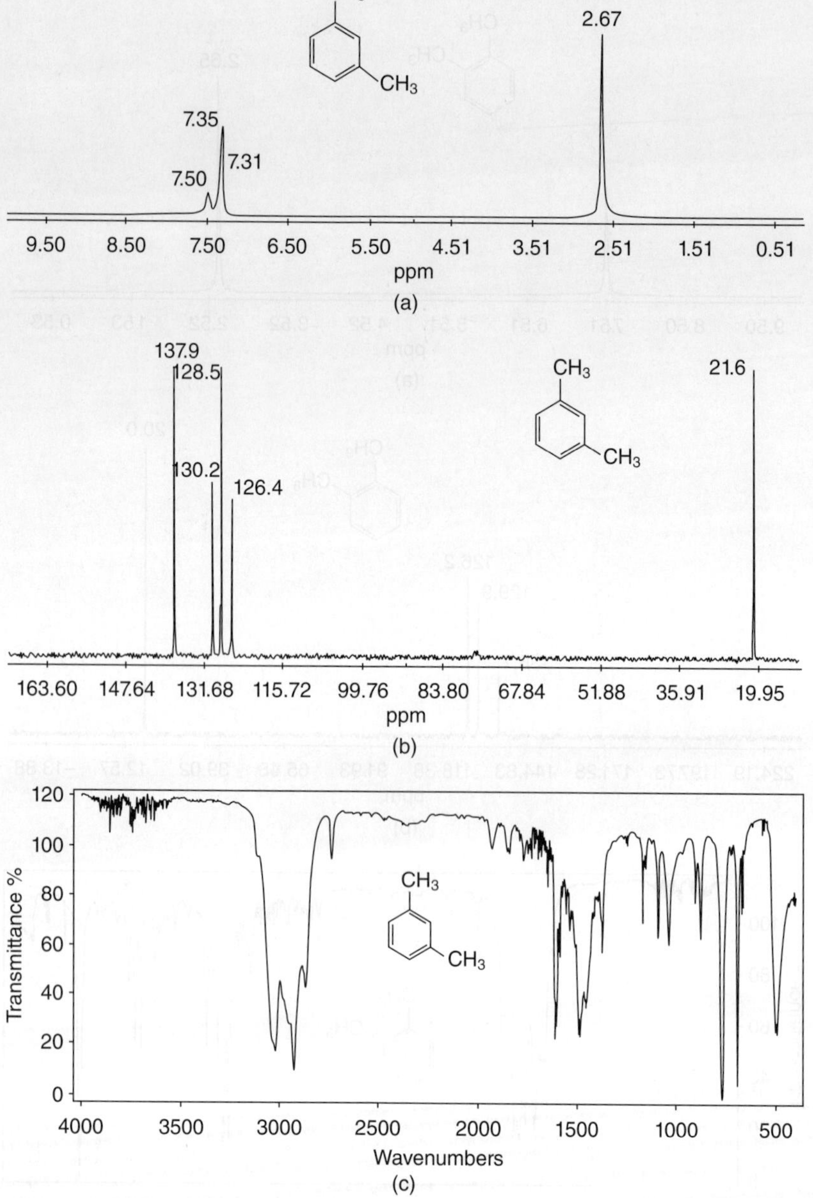

Figure 6.36. (a) The ^{1}H NMR spectrum of 1,3-dimethylbenzene (*meta*-xylene) in ^{2}HCCl$_3$ at 300 MHz. The chemical shift of the methyl groups (CH$_3$–) at δ = 2.67 ppm is very close to those of the methyl groups in its isomers (*vide infra*). It will be noted that the aromatic protons are different and coupled to each other. Nonidentical adjacent protons (1,2- or *ortho*-) have coupling constants in the range of 5–8 Hz while nonidentical protons bearing a 1,3- (or *meta*-) relationship have coupling constants in the range of 1–3 Hz. The coupling of aromatic protons bearing a 1,4-relationship (or *para*-) is in the range of 0–1 Hz. All of these relationships are present here. (b) The ^{13}C NMR spectrum of 1,3-dimethylbenzene (*meta*-xylene) in ^{2}HCCl$_3$ at 75.5 MHz. The carbons (C-1 and C-3) bearing the methyl (CH$_3$–) groups come into resonance at δ = 137.9 ppm, the carbon lying between them (C-2) at δ = 130.2 ppm, which flank the methyl (CH$_3$–) groups (C-4 and C-6) at δ = 126.4 ppm, the remaining aromatic carbon (C-5) being at δ = 128.3 ppm. The identical methyl (CH$_3$–) groups are found at δ = 21.6 ppm downfield from TMS. (c) The FT-IR (neat) spectrum of 1,3-dimethylbenzene (*meta*-xylene). As is typical of aromatic hydrocarbons, evidence for the *meta*- methyl groups is seen in the pattern of low-intensity bands in the region 2000–1650 cm^{-1} and in the intense signals between 800 and 500 cm^{-1}. These features, which are quite characteristic of the substitution pattern, should be compared with those found in the IR spectra of 1,2-dimethylbenzene (*ortho*-xylene), Figure 6.35c, and 1,4-dimethylbenzene (*para*-xylene), Figure 6.37c.

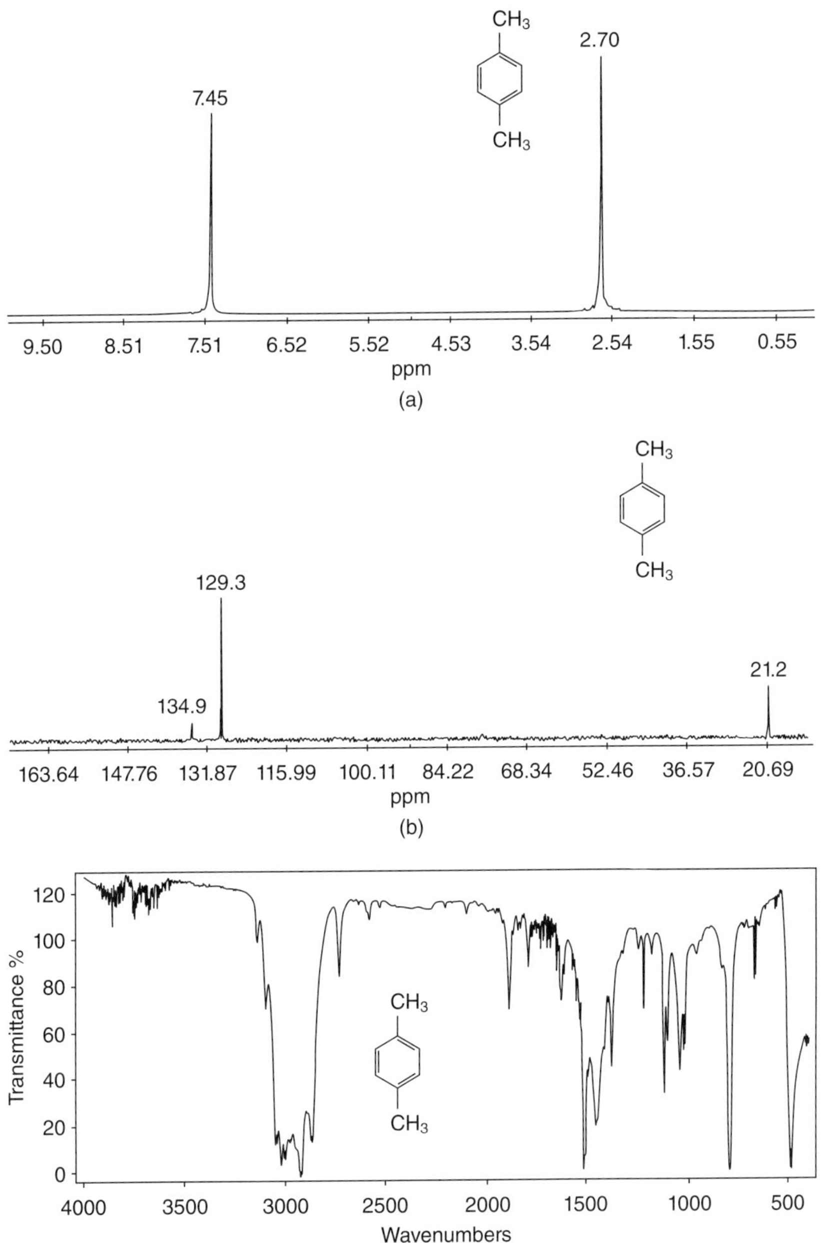

Figure 6.37. (a) The ^{1}H NMR spectrum of 1,4-dimethylbenzene (*para*-xylene) in ^{2}HCCl$_3$ at 300 MHz. The chemical shift of the methyl groups (CH$_3$–) at δ = 2.70 ppm is very close to those of the methyl groups in its isomers (*vide supra*). (b) The ^{13}C NMR spectrum of 1,4-dimethylbenzene (*para*-xylene) in ^{2}HCCl$_3$ at 75.5 MHz. The carbons (C-1 and C-4) bearing the methyl (CH$_3$–) groups come into resonance at δ = 134.9 ppm and the other four (identical) carbons at δ = 129.2 ppm. The identical methyl (CH$_3$–) groups are found at δ = 20.9 ppm downfield from TMS. (c) The FT-IR (neat) spectrum of 1,4-dimethylbenzene (*para*-xylene). As is typical of aromatic hydrocarbons, evidence for the *para*- methyl groups is seen in the pattern of low-intensity bands in the region 2000–1650 cm^{-1} and in the intense signal near 800 cm^{-1}. These features, which are quite characteristic of the substitution pattern, should be compared with those found in the IR spectra of 1,2-dimethylbenzene (*ortho*-xylene), Figure 6.35c, and 1,3-dimethylbenzene (*meta*-xylene), Figure 6.36c.

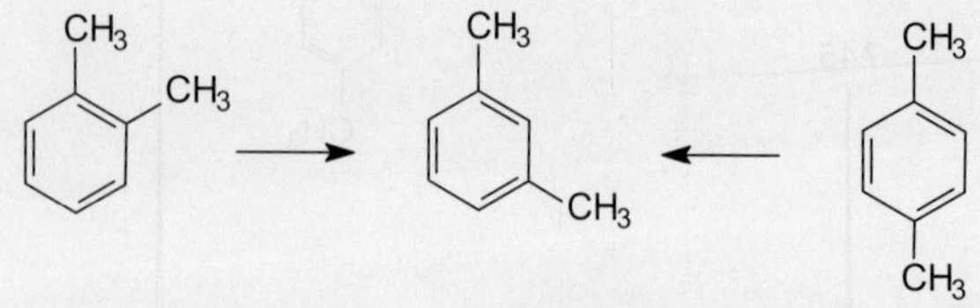

Scheme 6.92. A representation of 1,2-dimethylbenzene (*ortho*-xylene) and 1,4-dimethylbenzene (*para*-xylene) undergoing rearrangement to 1,3-dimethylbenzene (*meta*-xylene). The path followed for this process presumably involves both intramolecular and intermolecular movement of the methyl (CH_3-) groups in the presence of a (not shown) Lewis acid (e.g., $AlCl_3$ and BF_3).

Therefore, alkyl substituents on the aromatic ring are known as *ortho—para directing groups; and they are* **activating**.

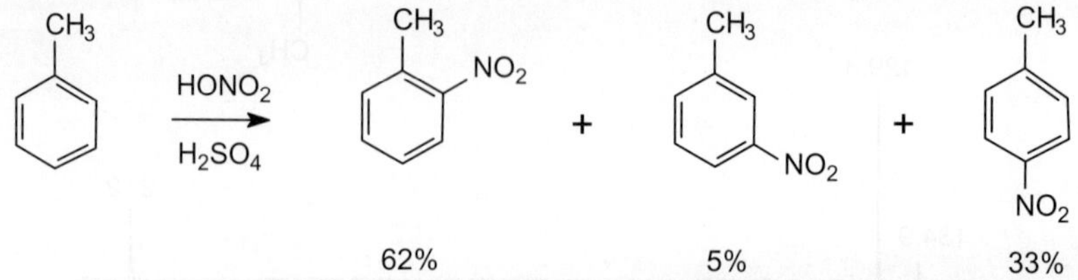

$$\tag{6.107}$$

$$62\% \qquad 5\% \qquad 33\%$$

It has been argued that the delocalization of positive charge onto the ring carbon bearing the alkyl group more readily occurs when the incoming electrophile attaches itself to the *ortho-* and *para-* positions of the aromatic ring than when it is *meta-*. The resonance structures of Scheme 6.93 support the argument.

The predilection for *initial* attachment of the incoming electrophile to one position (i.e., *ortho-* or *para-*) rather than another (i.e., *meta-*) is thought to arise from overall electron donation from the alkyl group to the ring. The methyl group is **activating**. The activation has been thought of as coming from a (a) conjugative effect in which sigma (σ) electrons from the C-H bond(s) participate in resonance with the aromatic ring and (b) "through-the-σ-bond" electron push in which the relatively (compared with the hydrogen it replaced) electron-rich methyl (CH_3-) group *adds electrons to the ring, facilitating* the attack. The idea of using a π_{CH2}-bonding MO (made up of positive overlap between the appropriate carbon and hydrogen orbitals as outlined in the beginning pages of this Chapter) has not received much attention.

As shown in Scheme 6.94, a set of resonance hybrids can be written in which delocalization of the C-H electrons (**as is traditional for resonance structures the nuclei do not move**) into the ring would cause the incoming electrophile to choose *ortho-* and *para-* over *meta-* substitution. This explanation has variously been called "no-bond-resonance" and the theory of "hyperconjugation." Clearly, the same idea could also be used to account for the stability of tertiary carbocations over their secondary and primary analogues. One difficulty with the widespread acceptance of the idea of no-bond-resonance has been the observation that the nitration of other alkyl substituted benzenes such as (1,1-dimethylethyl)benzene (*t*-butylbenzene) (Figure 6.38), which is incapable of participating in the same way, is also *enhanced* relative to benzene (C_6H_6). Indeed, although (1,1-dimethylethyl)benzene (*t*-

$$HONO_2 + H_2SO_4 \rightleftharpoons H_2ONO_2^+ + HSO_4^- \xrightarrow{H_2SO_4} H_3O^+ + NO_2^+ + 2\,HSO_4^-$$

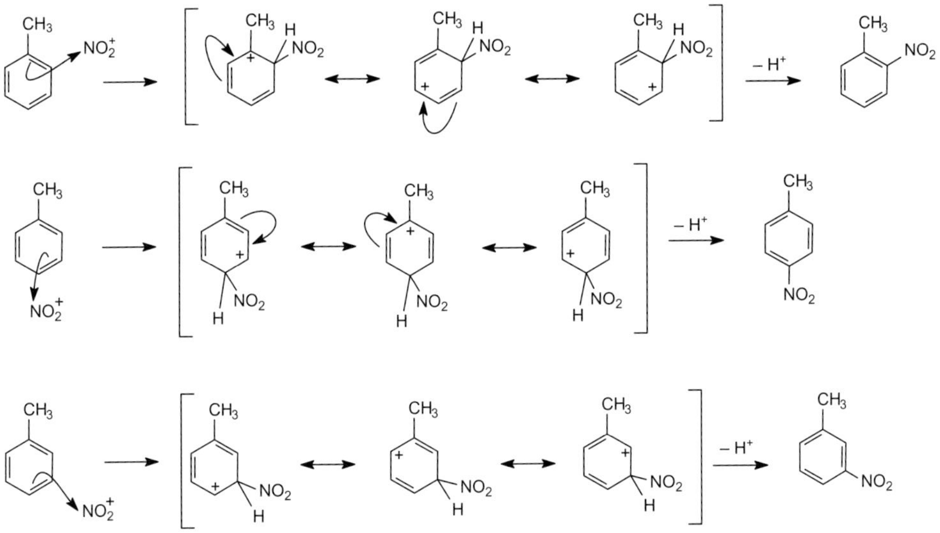

Scheme 6.93. The formation of the nitronium ion (NO_2^+) from nitric ($HONO_2$) and sulfuric (H_2SO_4) acids in the presence of methylbenzene (toluene [$C_6H_5CH_3$]) and the subsequent reaction of NO_2^+ to produce 1-methyl-2-nitrobenzene (*ortho*-nitrotoluene), 1-methyl-4-nitrobenzene (*para*-nitrotoluene), and 1-methyl-3-nitrobenzene (*meta*-nitrotoluene). Using the curved arrow to indicate the movement of a pair of electrons (as usual) there is a resonance form in which the positive charge is formally placed on the ring carbon bearing the methyl (CH_3-) group *for the first two*. No such form can be written for the last.

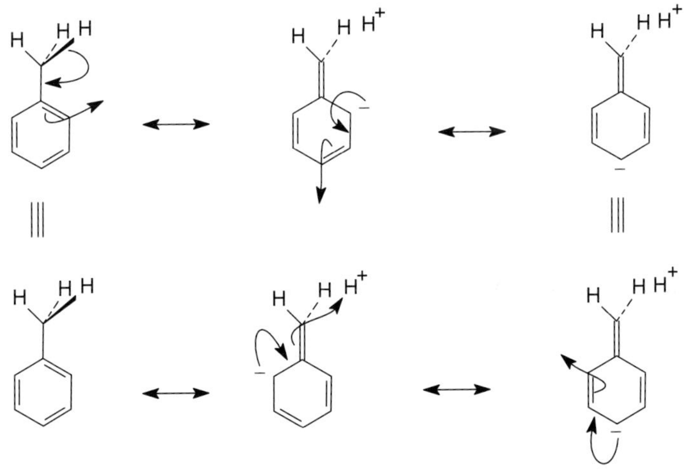

Scheme 6.94. Resonance forms for methylbenzene (toluene) suggesting that *ortho*- and *para*-positions may be particularly attractive for an incoming electrophile. The drawing is intended to illustrate the principle called "no-bond-resonance" or the theory of hyperconjugation. It is suggested that the σ-bonding C–H electrons participate in resonance. It is important to recongize that the nuclei do not move.

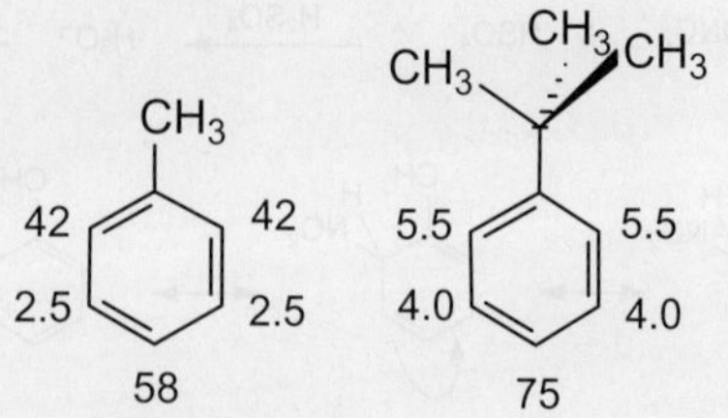

Figure 6.38. A representation of the relative "rates of attack on the individual nuclear positions" of methylbenzene (toluene) and (1,1-dimethylethyl)benzene (*t*-butylbenzene). After Ingold, C. K. *Structure and Mechanism in Organic Chemistry*, Cornell University Press, Ithaca, NY, **1953**.

butylbenzene) is sterically inhibited from undergoing nitration (or other substitution) at the positions *ortho*- to the 1,1-dimethylethyl *(t*-butyl $[(CH_3)_3C–]$) substituent, its reactivity at *meta*- and *para*- positions can be compared with methylbenzene (toluene). For example, it is reported that (1,1-dimethylethyl)benzene (*t*-butylbenzene) is *less* reactive toward nitration than methylbenzene (toluene) under the same conditions in the ratio 64:100. But, as anticipated, the diminished reactivity appears associated only with the reduced reactivity at the *ortho*-positions (for steric reasons) *not* at the *meta*- and *para*-positions that are otherwise activated Figure 6.38.

Although most of the methylbenzene (toluene, $C_6H_5CH_3$) isolated from light naphtha is either used as an octane enhancer in gasoline or it is converted to benzene (by the acid-catalyzed reverse of the Friedel-Crafts reaction), some small amount is diverted to other ends as demanded by local economy and needs. Among these is its exhaustive nitration (Equation 6.108) to 2,4,6-trinitrotoluene (TNT), which continues to be used as an explosive.

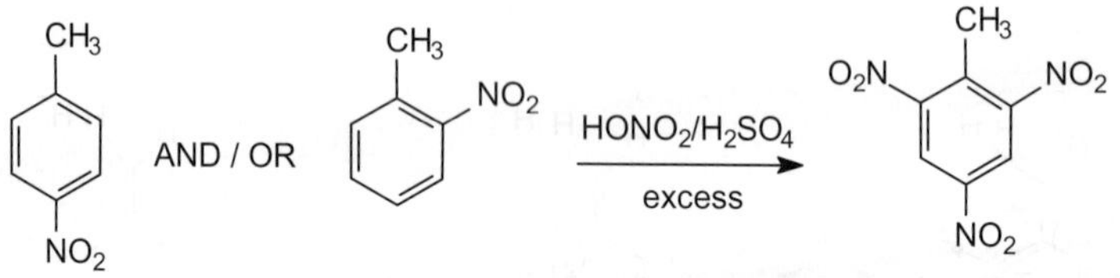

$$\text{(6.108)}$$

With respect to bromination, in the presence of iron(III) salts or iodine (I_2) as catalyst *and in the dark* the addition of bromine (Br_2) to methylbenzene (toluene, $C_6H_5CH_3$) occurs almost exclusively *ortho*- and *para*- to the methyl group. 2-Bromo-1-methylbenzene (2-bromotoluene) and 4-bromo-1-methylbenzene (4-bromotoluene) (Equation 6.109; compare Equation 6.107) are produced. However, *in the sunlight* (or with NBS) the outcome is different (Equation 6.110). Under these different conditions bromomethylbenzene (benzyl bromide, $C_6H_5CH_2Br$) results. In contrast to the ionic electrophilic substitution, the reaction in the presence of sunlight (or NBS) apparently involves free radicals and the intermediacy of the phenylmethyl (benzyl) radical $[C_6H_5CH_2•]$ (*vide infra*).

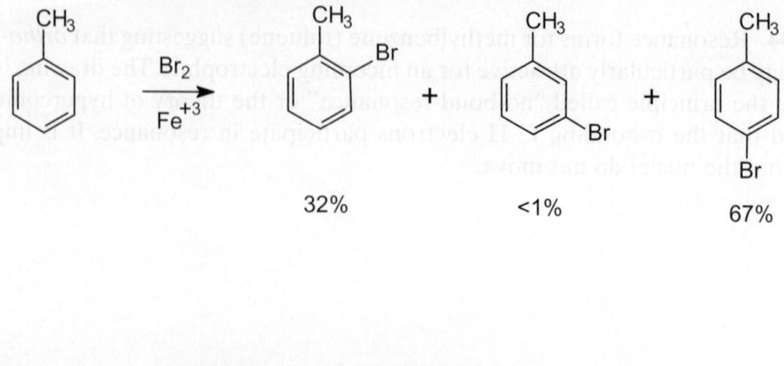

$$\text{(6.109)}$$

32%   <1%   67%

(6.110)

The presence of the benzyl-"type" radical ($C_6H_5CH_2\cdot$) is also suggested by the observation that ethylbenzene ($C_6H_5CH_2CH_3$) undergoes side-chain bromination at the benzylic carbon under essentially the same conditions (NBS or $Br_2/h\nu$) to produce the corresponding 1-bromo-1-phenylethane [$C_6H_5CH(Br)CH_3$] (Equation 6.111). Interestingly, in this vein, although the benzyl ($C_6H_5CH_2\cdot$) radical has been observed by electron spin resonance (ESR) spectroscopy (Chapter 2), the observation was made under different conditions and it has not yet proved possible to see it during the bromination reaction.

(6.111)

b. Nucleophilic Aromatic Substitution. The attack of an electron-rich nucleophile onto the electron-rich aromatic ring of a _hydrocarbon_ in order to substitute the nucleophile for a hydrogen* (which is thus presumably required to leave _with_ its electron pair, i.e., as a hydride [H^-] ion), is very rare.

Treatment of arenes such as benzene (C_6H_6) and naphthalene ($C_{10}H_8$) with alkyl lithium reagents (such as n-butyllithium, BuLi, $CH_3CH_2CH_2CH_2Li$, or its equivalent} can, depending upon the conditions, result in the removal of a proton to produce an arene _anion_.[†‡] However, alkylation, which is thus _effectively_ a substitution of a hydride ion, occasionally occurs (depending upon the conditions under which the reaction is run) and the alkylated arene is obtained as a minor product (Equation 6.112).

---

*The substitution of one nonhydrogenic group (e.g., halogen) for another is a reaction that is much more common (particularly if the aromatic ring also bears other substituents that can withdraw electrons from the site being attacked). That substitution reaction and related processes will be discussed in Chapter 7.

[†]It will be recalled (Chapter 5, Table 5.16) that alkyl anions (which are not monomeric although usually written as such) are generally considered to be the strongest bases known. Although measurements and comparisons are difficult to make, the alkyllithium reagents are generally considered to be about 4–6 pK units _more basic_ than their aryl counterparts. Thus, the deprotonation of the arene is simply an acid-base reaction. As will be seen in Chapter 7, preparation of the aryllithium reagent is better effected from an aryl halide (_vide infra_).

[‡]The anion obtained in this way is not the same as the radical anion obtained by, for example, addition of an alkali or alkaline earth metal to a solution of arene in liquid ammonia (NH_3)$_{liq}$, a reaction discussed earlier. In the present case, the anion corresponds, empirically, to $C_6H_5\ ^-M^+$. The radical anion created by adding an electron to the aromatic system (_vide supra_) corresponds to $C_6H_6\cdot^-\ M^+$.

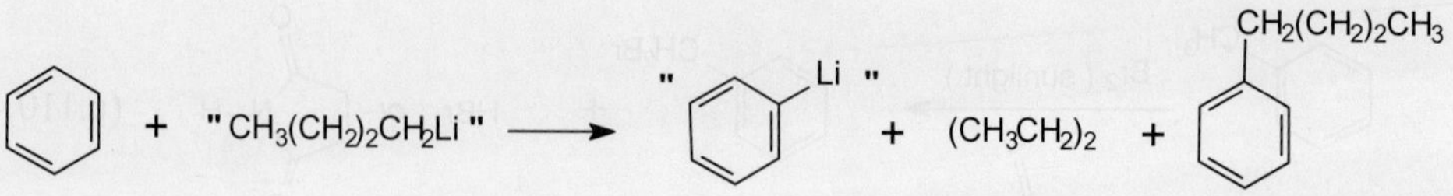

$$\text{(6.112)}$$

It also appears possible to alkylate nitrobenzene with the displacement of hydride (or its equivalent) by the use of the anion obtained from deprotonation of methylsulfinylmethane (dimethyl sulfoxide [$(CH_3)_2SO$]), viz. the methylsulfinylmethide anion (dimsyl anion [$CH_3SO(CH_2)^-$]. Although the details are not known, the reaction is particularly useful because Friedel-Crafts alkylation of nitrobenzene usually fails. Indeed, the reaction of nitrobenzene with Lewis acids and alkyl (or acyl) halides is so poor that nitrobenzene has occasionally been used as a solvent for the reaction of other arenes with these substrates. It is argued that nitrobenzene is so unreactive because the nitro substituent withdraws electrons, making them unavailable for an incoming electrophile, that is, the ring is deactivated toward electrophilic substitution by the electron-withdrawing substituent.

A representation of a possible pathway to the product is provided in Scheme 6.95.

Problem 6.17. Based upon the pathway for *para-* alkylation shown in Scheme 6.95 show a similar scheme for alkylation to produce 2-nitromethylbenzene (**ortho**-nitrotoluene).

c. Free Radical Substitution. As already pointed out in the first part of this section [Chapter 6, Section E, Part Ia], the *substitution* of a hydroxyl group as a free radical (i.e., •OH) for a hydrogen atom (i.e., •H), a free radical hydroxylation, can be considered an oxidation as well as a substitution. Thus, as shown below in Scheme 6.96 (as well as in Table 6.10, example 8), iron(II) sulfate (ferrous sulfate, $FeSO_4$) is oxidized to iron(III) sulfate (ferric sulfate $Fe_2(SO_4)_3$) while an equivalent of hydrogen peroxide (H_2O_2) is reduced to hydroxide ion (OH^-) and the hydroxyl radical (•OH) (**Fenton's Reagent**). *That process is the oxidation-reduction reaction* and it is rate determining.

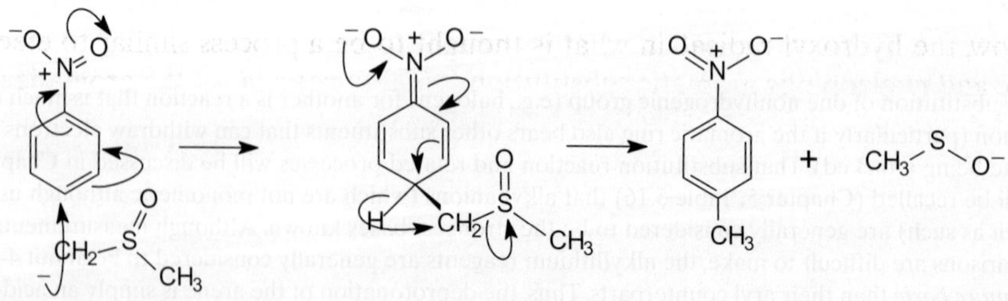

Scheme 6.95. A possible pathway for the alkylation of nitrobenzene by the methylsulfinylmethide anion [$CH_3(SO)CH_2^-$]. Although only *para-* alkylation to produce 4-nitromethylbenzene (*para*-nitrotoluene) is shown, *ortho-* alkylation also occurs. A similar pathway can be written for the latter.

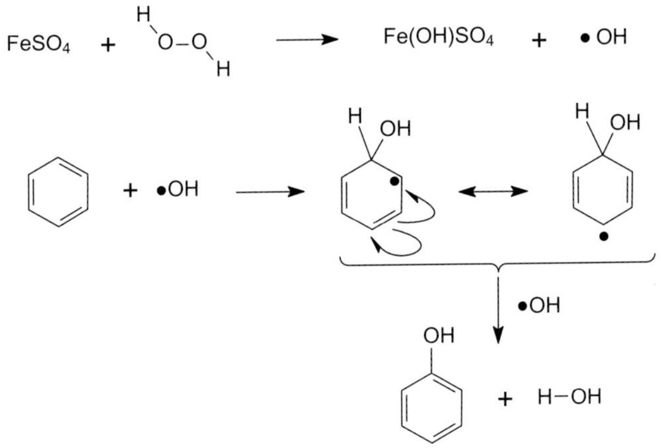

Scheme 6.96. A representation of the *free radical substitution* of hydroxyl (•OH) for a hydrogen atom (•H). Note the use of two half-headed arrows to indicate the movement of *single electrons*. This abbreviation is only rarely used.

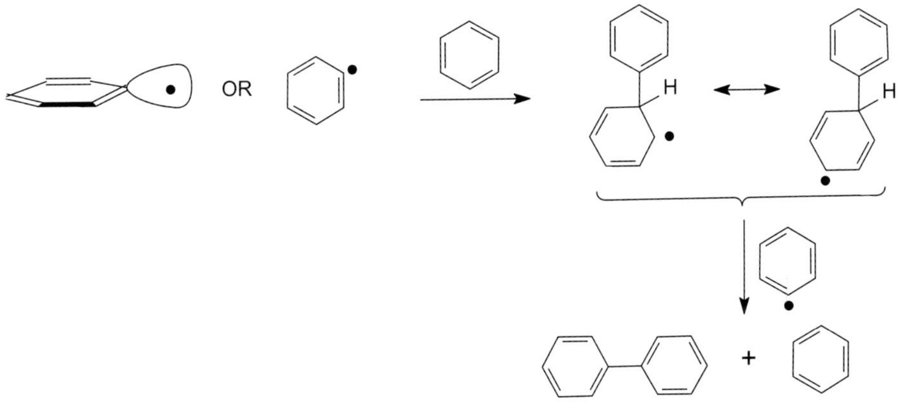

Scheme 6.97. A potential pathway for the formation of biphenyl by the reaction of the phenylradical with benzene.

Now, the hydroxyl radical, in what is thought to be a process similar to electrophilic and nucleophilic aromatic substitution (*vide supra*) *adds* to the aromatic ring, generating a resonance stabilized *radical* species that, in a subsequent step, loses a hydrogen atom (•H) to, for example, another hydroxyl radical (•OH), forming hydroxybenzene (phenol, C_6H_5OH) and water (H_2O).

Aryl radicals, that is, •C_6H_5, which have been generated **thermally** by the decomposition of, for example, aryl diazonium salts (ArN_2^+) in the presence of copper(I) salts (Equation 6.113) and the decomposition of diacyl peroxides (Equation 6.114), or **photolytically** from aryl iodides (e.g., iodobenzene, C_6H_5I) (Equation 6.115) and arylthallium trifluoroacetates as shown in Equation 6.116, react with benzene (C_6H_6) to give biphenyl (C_6H_5-C_6H_5) (Scheme 6.97).

$$(6.113)$$

$$(6.114)$$

$$(6.115)$$

$$(6.116)$$

ADDITIONAL PROBLEMS

Problem 6.18. Given the information shown in Scheme 6.5, predict the products of monobromination and, separately, monochlorination of:

(a) 2-methylpropane

(b) methylcyclopropane

(c) 1,2-dimethylcyclohexane.

Problem 6.19. The alkenes shown can be readily oxidized by potassium permanganate or osmium tetroxide to yield diols as shown in Schemes 6.9 and 6.10. Write the products expected using either of those reagents.

(a)   (b)   (c)

Problem 6.20. Write the product(s) of the ozonolysis of the alkenes shown.

(a)   (b)   (c)

(d)   (e)   (f)

Problem 6.21. The alkenes shown in Problem 6.19 can be epoxidized to yield the corresponding epoxides (oxiranes) with *m*-chlorperbenzoic acid in chloroform solution. Write the product(s) expected from such epoxidation.

Problem 6.22. The products of epoxidation in Problem 6.21 can be induced to undergo acid-catalyzed hydrolysis to yield diols. Draw the structures of the diols obtained that way.

Problem 6.23. The reduction of alkenes can be accomplished in a number of ways. The methods include using hydrogen (H_2) with (a) Adam's Catalyst and (b) Wilkinson's Catalyst as well as with (c) diimide and (d) borane. Using 1,2-dimethylcyclopentene as a substrate, compare and contrast these methods.

Problem 6.24. Using 1-methylcyclopentene as a substrate write the product(s) of the reaction of the reagents listed in Table 6.1 with that alkene. Pay close attention to the regio- and stereochemistry of the addition.

The Reactions of Alkyl, Alkenyl, and Aryl Halides: Oxidation, Reduction, Substitution, Addition, Elimination, and Rearrangement

Lend me the stone strength of the past and I will lend you the wings of the future.
—Robinson Jeffers, *To the Rock That Will be a Cornerstone*

A. INTRODUCTION

As already pointed out in the Survey of organic compounds (Chapter 5), halogen (fluorine [F], chlorine [Cl], bromine [Br], and iodine [I]) can, at least in principle, replace any hydrogen (H) in a hydrocarbon to produce the corresponding alkyl, alkenyl, alkynyl, or aryl halide. In practice, not all hydrogens are easily "replaced," and some are more easily replaced than others. Furthermore, as indicated in the last chapter, the halogenation might be accomplished by a radical (Schemes 6.3–6.5 and Equation 6.9) or ionic substitution reaction as well as by an electrophilic (Tables 6.1 and 6.7) or radical (Table 6.3) addition reaction. The latter might be followed by an elimination reaction (involving either ionic or radical species). Finally, more than one halogen might be added.

Although there are distinctive differences between the different alkyl and aryl halides as a function of the particular halogen, there are also striking similarities. As a consequence, it is frequently possible to abbreviate their structures; for example, a primary alkyl halide is written as RCH_2X where X = F, Cl, Br, or I. However, as might be expected from (a) their electronegativities (Chapter 1, II) and (b) their formation from hydrocarbons (Chapter 6), the halides containing chlorine (Cl), bromine (Br) and iodine (I) are used more frequently as intermediates in organic synthesis than are the fluorides (R–F). On the other hand, the relative stability (lack of reactivity) of many of the fluorine-containing compounds has encouraged their development for a host of specialized applications.

It is of some interest that in the search for naturally occurring (in contrast to "man-made") materials that serve as palliatives to our illnesses, we have found that marine plants and animals have produced a rich variety of chlorine (Cl)-, bromine

---

Foundations of Organic Chemistry: Unity and Diversity of Structures, Pathways, and Reactions,
First Edition. David R. Dalton.
© 2011 John Wiley & Sons, Inc. Published 2011 by John Wiley & Sons, Inc.

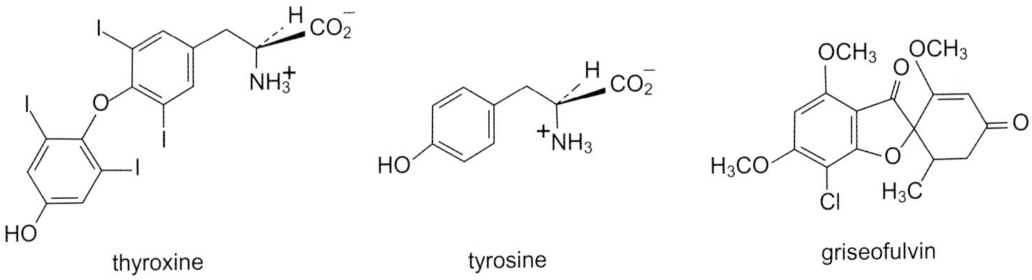

thyroxine   tyrosine   griseofulvin

Figure 7.1. Molecules from terrestrial species that contain halogen (and a precursor to one of them).

(Br)-, and iodine (I)-containing natural products (Chapters 11–14). It is presumed that this reflects the aqueous environment since only a few terrestrial species have been discovered that normally produce fluorine (F), chlorine (Cl), or iodine (I) compounds. In that vein, normal humans are a major iodine (I)-containing terrestrial species as they contain the thyroid hormone thyroxine (Figure 7.1).*

Other examples of terrestrial halides include the antibiotic griseofulvin (Figure 7.1), from the mold *Penicillium griseofulvum* Dierckx, and fluoroethanoic acid (fluoroacetic acid, FCH_2CO_2H), the toxic principle of *Gastrolobium grandiflorum* (a range fodder), which has been known to cause livestock loss in free-ranging cattle.

Information on some (more) typical organic halides is found in Table 7.1.

It should be remembered (Chapter 5), although not included in the table that ^{19}F nuclear magnetic resonance (NMR) spectroscopy is a commonly employed tool for structural elucidation of fluorine-containing compounds. However, in addition to obtaining ^{19}F spectra,[†] in which the coupling between ^{14}C and ^{19}F as well as ^1H and ^{19}F and between ^{19}F and ^{19}F (if more than one fluorine is present) can be used to map the structure, it should be clear that the reverse is coupling with those nuclei. Although coupling between ^{19}F and ^{13}C is very useful and widely practiced, its further treatment here, except for the example of fluorobenzene (C_6H_5F) in Figure 7.2, which follows (to give some feeling for the magnitude of the coupling constants), will not be pursued.

The natural abundance of ^{35}Cl (spin 3/2) is 75.5% while that of ^{37}Cl (spin 3/2) is 24.5% so that either might be, in principle, suitable for the NMR experiment (they should come into resonance at about 29.4 and 24.5 MHz, respectively, at 7.05T). However, because of the presence of a quadrupole moment, these nuclei relax very quickly and ^{35}Cl and ^{37}Cl solution spectra of alkyl chlorides suffer from unacceptable

*This hormone is apparently formed *in vivo* by a series of enzymatic reactions that begin with the electrophilic iodination of the phenolic ring (Chapter 6) of the amino acid tyrosine (Chapter 12) (Figure 7.1). Thyroxine (Figure 7.1) is required for stimulating the intracellular synthesis of numerous metabolic enzymes. The condition known as hypothyroidism is found among iodine-deprived individuals. During fetal development, iodine deprivation can result in mental (and physical) retardation with the development of a syndrome called cretinism. If a normal, but more mature, individual is deprived of iodine (even in the trace quantities added—in the developed world—to table salt, i.e., "iodized salt"), their thyroid gland becomes enlarged and the condition known as goiter develops.

[†]As noted earlier, ^{19}F, with natural abundance of 100%, has spin 1/2 and comes into resonance at 282.3 MHz at a field strength of 7.05T, that is, the field strength at which protons (^1H) come into resonance at 300 MHz and carbon (^{13}C) at 75.5 MHz.

TABLE 7.1. Some Physical and Spectroscopic Properties of Alkyl, Alkenyl, and Aryl Halides

	mp (°C)	bp (°C)	^{1}H (δ ppm)	^{13}C (δ ppm)
Alkyl halides				
Fluoromethane [CH_3F] (methyl fluoride)	−142	−78	4.26	74.1
Chloromethane [CH_3Cl] (methyl chloride)	−97	−24	3.05	23.8
Bromomethane [CH_3Br] (methyl bromide)	−94	4.5	2.69	9.0
Iodomethane [CH_3I] (methyl iodide)	−67	42.5	2.16	−21.7
Tetrafluoromethane [CF_4] (carbon tetrafluoride)	−184	−129		117.5
Tetrachloromethane [CCl_4] (carbon tetrachloride)	−23	76.7		95.4
Tetrabromomethane [CBr_4] (carbon tetrabromide)	92	190		−29.7
Tetraiodomethane [CI_4] (carbon tetraiodide)	168	(dec)		−293.5
Fluoroethane [CH_3CH_2F] (ethyl fluoride)	−143	−38	1.24; 4.36	14.6; 79.3
Chloroethane [CH_3CH_2Cl] (ethyl chloride)	−138	12	1.48; 3.57	18.7; 39.9
Bromoethane [CH_3CH_2Br] (ethyl bromide)	−28	38	1.66; 3.37	20.3; 28.3
Iodoethane [CH_3CH_2I] (ethyl iodide)	32	78	1.88; 3.16	21.6; −0.2
Alkenyl halides				
Fluoroethene [$CH_2{=}CHF$] (vinyl fluoride)	−161	−72	6.31 (*gem*); 4.39 (*cis*); 3.09 (*trans*)	
Chloroethene [$CH_2{=}CHCl$] (vinyl chloride)	−160	−14	6.28 (*gem*); 5.47 (*cis*); 5.31 (*trans*)	C1 126.0; C2 117.3
Bromoethene [$CH_2{=}CHBr$] (vinyl bromide)	−137	15.8	6.32 (*gem*); 5.68 (*cis*); 5.83 (*trans*)	C1 115.5; C2 122.0
Iodoethene [$CH_2{=}CHI$] (vinyl iodide)		56		C1 85.3; C2 130.4
Aryl halides				
Fluorobenzene [C_6H_5F]	−42	85	7.57 *ortho*; 7.29 *meta*; 7.49 *para*	163.3 ipso; 115.6 o; 129.9 m; 124.0 p
Chlorobenzene [C_6H_5Cl]	−45	132	7.25 *ortho*; 7.33 *meta*; 7.31 *para*	134.7 ipso; 128.9 o; 129.8 m; 126.6 p
Bromobenzene [C_6H_5Br]	−31	156	7.05 *ortho*; 7.40 *meta*; 7.30 *para*	123.0 ipso; 131.9 o; 130.2 m; 126.9 p
Iodobenzene [C_6H_5I]	−31	189	6.87 *ortho*; 7.53 *meta*; 7.30 *para*	96.5 ipso; 131.9 o; 130.2 m; 129.5 p

Alkynyl halides are omitted as they are less common, and their values, where available, are less certain. Additional information for fluoromethane (methyl fluoride, CH_3F) can be found in Figure 5.3 and for chloromethane in Table 5.3. The ^{13}C NMR shift for fluoromethane (methyl fluoride, CH_3F) has also been reported as 65.7 ppm and is a function of the state (gas/solution) of the sample. All NMR values are reported relative to tetramethylsilane (TMS) (δ = 0.00 ppm). Values for the spectroscopic data not obtained by the author have been taken from (a) Emsley, J. W.; Feeney, J.; Sutcliffe, L. H. *High Resolution Nuclear Magnetic Resonance Spectroscopy*, Vols. 1 and 2, **1966**, Pergamon, Oxford, UK; (b) Levy, G. C.; Lichter, R. L.; Nelson, G. L. *Carbon-13 Nuclear Magnetic Resonance Spectroscopy*, 2nd edition, **1980**, Wiley Interscience, New York; and (c) Jackman, L. M.; Sternhell, S. *Applications of Nuclear Magnetic Resonance Spectroscopy in Organic Chemistry*, 2nd edition, **1969**, Pergamon, Oxford, UK. Physical data were taken from the *Dictionary of Organic Compounds*, CHEMnetBase, CRC Press, at http://doc.chemnetbase.com.

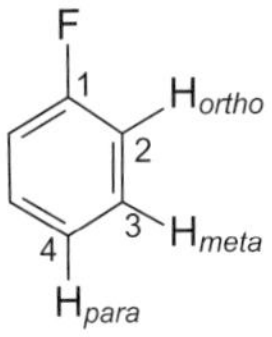

	J_{CF} (Hz)	J_{HF} (Hz)
	C_1–F –244.7	F–H$_{ortho}$ 9.4
	C_2–F 20.98	F–H$_{meta}$ 5.8
	C_3–F 7.81	F–H$_{para}$ ± 0.5
	C_4–F 3.18	

Figure 7.2. Fluorobenzene. The coupling constants J_{CF} and J_{HF} were measured on 300 MHz ^{1}H and 75.5 MHz ^{13}C spectra, respectively.

line broadening. Indeed, the relaxation of these nuclei is so fast and the lifetime thus so short that potential coupling between attached ^{13}C and neighboring ^{1}H vanishes. The same phenomenon attends the organic bromides (^{79}Br; spin = 3/2; abundance = 50.5%; frequency = 75.2 MHz at 7.05 T; and ^{81}Br; spin = 3/2; abundance = 49.5; frequency = 81.0 MHz at 7.05 T) as well as iodides (^{127}I; spin = 5/2; abundance = 100%; frequency = 60.0 MHz at 7.05 T).

On the other hand, mass spectroscopy is a particularly useful tool for those cases where more than one halogen isotope is present.

Although abundance of ^{19}F and ^{127}I isotopes is 100% each, the mass spectra of alkyl fluorides and iodides are quite different. Thus, the presence of ^{19}F generally results in weak molecular ion peaks and a peak at (M-1)$^+$, believed to result from loss of a hydrogen on the carbon-bearing fluorine. On the other hand, the presence of ^{127}I produces an intense molecular ion peak as well as daughter ions that result, if there is a hydrogen on the α-carbon, from loss of hydrogen iodide (HI), that is, (M-128)$^+$, and α-cleavage (e.g., cleavage of the bond from the carbon-bearing iodide to an adjacent carbon so that R–CH$_2$I would produce the CH$_2$ = I$^+$ ion).

Alkyl, alkenyl, and aryl chlorides and bromides, in addition to fragmentation patterns that are similar to the corresponding iodides, also each have more than one isotope. This feature is of importance in structure elucidation.

Consider the case of chlorine-containing compounds. There are two isotopes (^{35}Cl and ^{37}Cl) that are normally present in the ratio of (roughly) 75:25 (or 3:1). Thus, a compound that contains one chlorine will have two molecular ion (or parent) peaks (at M$^+$ and [M + 2]$^+$), and the second one, two mass units higher than the first, will be approximately one-third its intensity. A compound with two chlorine atoms will have three molecular ion peaks corresponding to molecules containing ^{35}Cl^{35}Cl, ^{35}Cl^{37}Cl, and ^{37}Cl^{37}Cl at M$^+$, (M + 2)$^+$, and (M + 4)$^+$ in the ratio approximately 10:6.5:1. Similarly, a compound with bromine, again with two isotopes (^{79}Br and ^{81}Br) but this time in the ratio of (about) 1:1 provides a spectrum with two molecular ion (or parent) peaks (at M$^+$ and [M + 2]$^+$), and they are in the ratio of 1:1. For molecules with two bromines, there will be three molecular ion peaks, corresponding to molecules containing ^{79}Br^{79}Br, ^{79}Br^{81}Br, and ^{81}Br^{81}Br at M$^+$, (M + 2)$^+$, and (M + 4)$^+$ in the approximate ratio 1:2:1.

Problem 7.1. Given the relative abundance of ^{35}Cl:^{37}Cl is 3:1 and that a major fragment of trichloromethyl-containing species (i.e., R–CCl$_3$) corresponds to an ion that can be written as CCl$_3^+$, predict the relative intensities of the four peaks found at m/z 117, 119, 121, and 123.

Finally, it should be clear from the above discussion that there are many organic compounds that contain more than one halogen and, indeed, mixed organic halides containing, for example, fluorine (F) and chlorine (Cl), are quite common (*vide infra*, this Chapter, B [I]). Additionally, it will be recalled that the reaction of bromine (Br$_2$) and chlorine (Cl$_2$), as well as their mixed halides (Table 6.1), with monoalkenes produces, respectively, dibromides, dichlorides, and mixed vicinal dihalides, while the free radical halogenation of hydrocarbons with chlorine (Cl$_2$) or bromine (Br$_2$) (the latter more selectively, Chapter 6) consecutively introduces halogen in place of hydrogen (H) as a function of the respective reactivities of the position, that is, in the order tertiary, secondary, and primary. Multiple introduction of halogen atoms can, in principle, continue until all of the hydrogen atoms have been replaced, for example, in methane (Scheme 6.4), chlorination successively produces chloromethane (methyl chloride, CH$_3$Cl), dichloromethane (methylene chloride, CH$_2$Cl$_2$), trichloromethane (chloroform, CHCl$_3$), and tetrachloromethane (carbon tetrachloride, CCl$_4$).

B. FLUOROCARBONS

As pointed out briefly in Chapter 6 (Table 6.1), the ionic addition of hydrogen fluoride (HF) across the carbon–carbon double bond of alkenes produces alkyl fluorides. Generally, the addition is effected by treating a solution of alkene in an ethereal solvent (e.g., tetrahydrofuran [THF]) with a pyridine solution of HF, which must thus contain pyridinium fluoride (Equation 7.1).

$$\text{(7.1)}$$

Interestingly (example eight in Table 6.1), fluorination of alkenes with fluorine (F$_2$) also requires special conditions. This is because fluorine (F$_2$) attacks other bonds too and (*vide infra*) will even attack saturated carbon to replace (substitute for) hydrogen.*

Thus, to add fluorine (F$_2$) across the carbon–carbon double bond (C=C) the temperature of a solution of alkene in an inert solvent [such as Xenon (Xe)] is lowered nearly to –100°C and a dilute solution of fluorine (F$_2$) in nitrogen (N$_2$) or argon (Ar) is added. If the solution of fluorine (F$_2$) in nitrogen (N$_2$) is added to an alkane (rather than an alkene) and a metal catalyst is present, reasonable yields of completely fluorinated alkanes (perfluoroalkane) can be produced. The same results are obtained with other fluorinating agents too. In this way, using chlorine trifluoride (ClF$_3$), a modest yield (ca. 60%) of perfluoroheptane (C$_7$F$_{16}$) can be obtained by the fluorination of heptane (C$_7$H$_{16}$). The details of the pathway are unknown, but it is

---

*As given in Table 1.1, DHo for fluorine (F$_2$) is only 158.1 kJ mol^{-1} (37.8 kcal mol^{-1}), but formation of a carbon–fluorine (C–F) bond is about 452 ± 13 kJ mol^{-1} (108 ± 3 kcal mol^{-1}).

thought unlikely that radicals are involved and, specifically, the reaction does not have the characteristics of a radical-chain process, such as that seen in chlorination of alkanes (Chapter 6). Because the abundance of ^{19}F is 100%, the mass spectra of fluorine (F)-containing compounds are fairly straightforward. Intervals between peaks and peaks charateristic of complicated mixtures such as perfluorokerosene (PFK) (which produce a series of regularly spaced m/z peaks) have convinced many mass spectroscopists to utilize them as reference and calibration standards.

In the infrared (IR), strong absorption below about 1350 cm^{-1} and no absorption at frequencies that are above this (but below 4000 cm^{-1}) and used for identification of other functional groups, encouraged spectroscopists to utilize perfluoro-oils as a medium for spectra of insoluble materials. Fluorolube™, in which solid samples were ground and the resulting paste put in the IR beam on a salt (NaCl) plate, has largely been replaced by potassium bromide (KBr) as a matrix for crystalline solids. However, many (older) reference spectra of solids are still to be found in a Fluorolube medium.

It will be remembered that NMR spectra of ^{19}F-containing systems is routine (Chapter 2, VI [7] and Chapter 5, C.II). Indeed, the value to structure determination of fluorine-containing compounds should be clear. However, it is important to remember (Chapter 5, C.II) that ^{19}F has spin 1/2 and will thus couple with protons (^{1}H) and carbon (^{13}C). Furthermore, while noise decoupling of protons to yield ^{13}C spectra in which different carbons appear as single lines at their different chemical shifts is common, ^{1}H decoupling will not, of course, remove coupling between ^{19}F and ^{13}C.

Commercially interesting fluorocarbons fall, more or less, into three groups or categories of compounds, each of which is discussed more fully below.

I. Freons and Halons

The (now nearly discontinued) low-molecular-weight, hydrogen-free, so-called chlorofluorocarbons (CFCs or freons) are manufactured by the reaction between HF and tetrachloromethane [carbon tetrachloride (CCl$_4$)] in the presence of antimony pentachloride (SbCl$_5$) (Equation 7.2).

$$2CCl_4 + 3HF \xrightarrow{\text{SbCl}_5} CCl_2F_2 + CCl_3F + 3HCl \tag{7.2}$$

These compounds were (and in some places still are) widely used as refrigerants, foam-blowing agents, solvents for cleaning circuit boards, and propellants for aerosols. The commercialized materials are usually named using three digits. The first digit is the number of carbon (C) atoms present less 1 (i.e., ΣCn − 1); the second digit is the number of hydrogen (H) atoms present increased by 1 (i.e., ΣHn + 1); the third digit is the number of fluorines (F) (i.e., ΣFn). All other atoms are chlorines (Cl). Thus, CCl$_2$F$_2$ is CFC-12 (i.e., $1 - 1 = 0$; $1 + 0 = 1$; $2 = 2$) and CCl$_3$F and CCl$_2$FCClF$_2$ are, respectively, CFC-11 and CFC-113.

Apparently, since their development in the early 1930s by the E.I. duPont Company, the increasing use of these materials has proved to be deleterious to the ozone (O$_3$) present in the upper atmosphere. It is now widely held that the photochemical destruction of the CFCs results in the production of chlorine atoms (•Cl) and chlorofluoromethyl radicals (e.g., •CClF$_2$) (Equation 7.3) and the former, among other reactions, interacts with ozone (O_3), to produce chlorine monoxide (ClO) and oxygen (O_2) (Equation 7.4).

$$CCl_2F_2 \xrightarrow{\text{UV light}} \bullet Cl + \bullet CClF_2 \tag{7.3}$$

$$Cl\bullet \ + \ O{=}O\overset{+}{}_{\diagdown O^-} \longrightarrow ClO \ + \ O_2 \tag{7.4}$$

Among the suite of reactions available to chlorine monoxide (ClO) in the presence of polar stratospheric ice clouds (PSCs) are those leading to species such as hydrogen chloride (HCl) and chlorine nitrate ($ClONO_2$). These species are apparently reactive enough (on the PSC ice surfaces in the presence of ultraviolet [UV] radiation) to (re)produce chlorine atoms (Cl). In this way, the chlorine atoms (Cl) are again available to destroy additional ozone (O_3), forming oxygen (O_2) and chlorine monoxide (ClO) (Equation 7.4) again. Thus, a catalytic cycle of ozone (O_3) destruction, once established, continues. By international treaty, production of CFCs has been banned and substitutes are to be used. Regrettably, the CFCs have become pervasive and much of the world relies on refrigeration, for which these materials serve so well.

A wide search for alternative materials has resulted in the duPont Company suggesting that CFCs be replaced by hydrochlorofluorocarbons (HCFCs) (Equation 7.5).

$$2HF + CHCl_2 \xrightarrow{SbF_5} HCClF_2 + 2HCl \tag{7.5}$$

When a chlorine (•Cl) or fluorine (•F) in the CFC is replaced by a hydrogen, HCFCs result.* Reports of widespread testing of HCFCs appear to indicate that their decomposition in the lower atmosphere will preserve the upper atmosphere ozone (O_3). Of course, "full-scale" use of HCFCs may have unforeseen consequences.

Halons™, which contain fluorine (•F) and bromine (•Br) and occasionally chlorine (•Cl), were commonly (and again occasionally still are) used as fire retardants. For example, although substitutes continue to be sought, fire extinguishers aboard aircraft continue (at this writing) to contain Halons as no good substitutes are yet available.

Halons are named in a way similar to that used for CFCs and HCFCs. That is, they are named with the word Halon and a series of digits. The first digit is the number of carbons (C), the second is the number of fluorines (F), the third is the number of chlorines (Cl), and the fourth is the number of bromines (Br). Thus, CF_2BrCl is Halon 1211.

Research continues to find suitable replacements for both "groups" (Freons and Halons) of materials, which, given what will inevitably be worldwide use, will be benign or, at least, less deleterious to the environment than our previous choices.

---

*The nomenclature for the hydrogen-containing CFCs is the same as that discussed for the CFCs that lack hydrogen. Thus, for example, $CHClF_2$ is HCFC-22 and CF_3CHCl_2 is HCFC-123 (when isomers are possible, the more symmetrical isomer is named with numbers only while a lowercase letter [a or b] is used for the less symmetrical analogues). The same is true for compounds that lack chlorine, the hydrofluorocarbons or HFCs. Thus, HCF_2CHF_2, is HFC-134 but the less symmetrical CH_2FCF_3 is HFC-134a.

II. Polymers of Highly Fluorinated Monomers

It is reported that largely as a result of a need to protect metals from the corrosive effects of uranium hexafluoride ($^{235}UF_6$ and $^{238}UF_6$) during separation of those isomers from each other by gaseous diffusion, it was necessary to develop lubricants and films. Once again, the E.I. duPont Company stepped in to use organofluorine technology to solve the problem. As shown in Equation 7.6, the dehydrohalogenation of HCFC-22 (chlorodifluoromethane, $HCClF_2$) at 700–900°C presumably produces singlet difluorocarbene ($F_2C:$), which dimerizes to tetrafluoroethene (tetrafluoroethylene, $F_2C=CF_2$).

$$HCClF_2 \xrightarrow{700\text{–}900°C} HCl + F_2C\!: \longrightarrow F_2C=CF_2 \qquad (7.6)$$

The polymers, thus simply based on $(CF_2)_n$ that subsequently come from that monomer are actually produced by the free radical polymerization of the alkene (tetrafluoroethene; tetrafluoroethylene, $CF_2=CF_2$). They are known, generically, as Teflon™.*

III. Use of Fluorocarbons to Carry Oxygen

Finally, although many low-molecular-weight fluorine-containing compounds are toxic, others are completely innocuous and, in particular, some nontoxic fluorocarbon derivatives dissolve large quantities of oxygen (O_2). Reports of proprietary fluorocarbon compounds and mixtures being used as blood replacement materials in nonhuman animals surface from time to time along with a flurry of interest in developing emergency blood substitutes for humans. Interesting perfluoroethers such as the perfluoro-18-crown-6 (the hydrogen-containing analogue of which was discussed in Chapter 5, C.IV) apparently form stable emulsions, which show promise for these practical uses.

C. OXIDATION

Halons (*vide supra*) and related compounds provide evidence that the presence of halogen reduces the ease of combustion. It is held that radical-chain inhibition by halogen atoms may play a role. Since the presence of halogens inhibits oxidation,

---

*As might be expected from entry 3 in Table 6.5 and as shown below, in the absence of free radical initiators, the alkene dimerizes to produce tetrafluorocyclobutane. This cycloaddition reaction may well occur as indicated in Chapter 6 with the involvement of radical species, which fail to polymerize (because the chain length is limited), or a completely different pathway may intrude.

the disposal of unwanted halogenated materials by combustion, unless it is effected at very high temperature, has not succeeded (but see Reduction).

It is possible to consider that elimination of hydrogen halide (H–X; X = F, Cl, Br, I) from an alkyl or alkenyl halide constitutes "oxidation" as the alkene (or alkyne) that results contains less one proton less (i.e., oxidation corresponding to the loss of hydrogen). However, while alkenes (or alkynes) are doubtlessly in a higher oxidation state than the corresponding alkanes (to which they can be reduced) they are, nonetheless, in the same oxidation state as the alkyl halide from which they were produced—as measured by electron loss and gain (Introduction to Part II).

Although particularly sensitive to subtle structural and electronic effects, it is found that some primary alkyl halides (RCH_2X; X = Cl, Br, I) will react with methylsulfinylmethane (dimethyl sulfoxide [DMSO] [$(CH_3)_2SO$]) in the presence of a weak base such as sodium carbonate (Na_2CO_3) or sodium hydrogen carbonate (sodium bicarbonate, $NaHCO_3$) to produce the corresponding aldehyde.

In this way, as shown in Scheme 7.1, a (halomethyl)benzene (benzyl halide, $C_6H_5CH_2X$; X = Cl, Br, I) can be converted to benzenecarbaldehyde (benzaldehyde, C_6H_5CHO). It will be noted that the reaction is thought to involve an initial substitution of the halogen by the methylsulfinylmethane (DMSO [$(CH_3)_2SO$]), which occurs with loss of halide. This is followed by a separate, second step, of proton loss and elimination of dimethyl sulfide [$(CH_3)_2S$] to actually effect the oxidation. Two possible pathways are shown for conversion to the aldehyde by proton loss. The first involves direct abstraction of the proton on the carbon-bearing oxygen and appears to be followed when this proton is the most acidic. The alternative pathway, rapid and reversible loss of one of the protons on a carbon attached to the positively charged sulfur followed by a slower intramolecular proton loss (to lead to aldehyde) has been identified by deuterium exchange reactions where the proton on the carbon-bearing oxygen is less acidic than one on the methyl group.

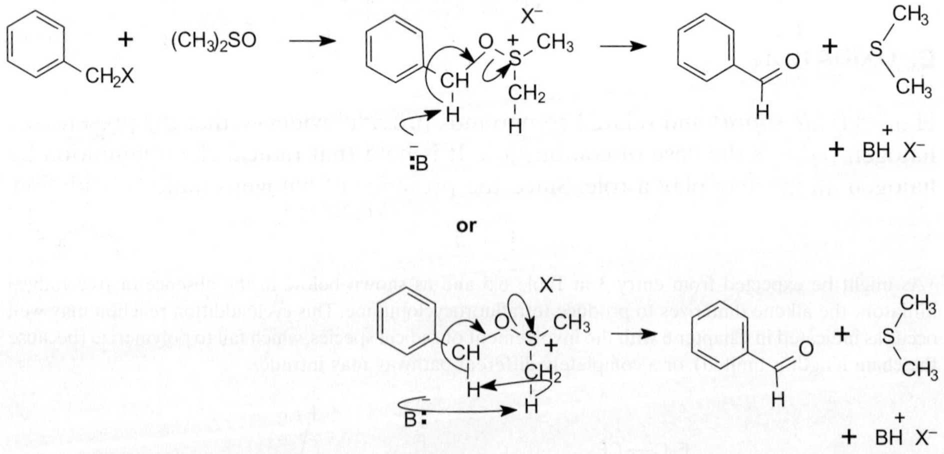

Scheme 7.1. Representation of two identified pathways from a halomethylbenzene (benzyl halide, $C_6H_5CH_2X$; X = Cl, Br, I) to benzenecarbaldehyde (benzaldehyde, C_6H_5CHO) by reaction with methylsulfinylmethane (dimethyl sulfoxide, DMSO [$(CH_3)_2SO$]) in the presence of a base B:$^-$ (commonly sodium bicarbonate, $NaHCO_3$).

The use of methylsulfinylmethane (DMSO [$(CH_3)_2SO$]) in the presence of sodium bicarbonate ($NaHCO_3$) has become a common reagent for the conversion of halides to the corresponding carbonyl compounds. While being used by the chemical community, the reaction was found to have (among others) the following characteristics: with all other parameters constant, (1) iodides react more rapidly than bromides or chlorides; (2) the efficiency of conversion of primary alkyl chlorides and bromides to the corresponding aldehydes is increased by the addition of iodide ion to the reaction; and (3) primary alkyl halides react much more rapidly (and "cleanly") than secondary halides, while tertiary alkyl halides (which could not give carbonyl compounds anyhow) undergo elimination reactions. These observations are related to the first step in the reaction, which, as noted above, involves a substitution reaction and will be discussed more fully under that heading (*vide infra*).

Before leaving the topic of oxidation of alkyl halides, it is worthwhile noting that benzylic ($C_6H_5CH_2X$; X = Cl, Br, I) and allylic ($RCH=CH–CH_2X$; X = Cl, Br, I) halides and related compounds that form stabilized carbocations, have long been known to undergo oxidation by heating them with hexamethylenetetramine (1,2,5,8-tetraazatricyclo[3.3.1.1^{3,7}]decane) in water.*

While aryl halides are particularly resistant to oxidative destruction,[†] alkenes, in which there is at least one halogen on one of the carbons of the double bond, that is, vinylic (Equation 7.7), undergo oxidation with a variety of oxidants (e.g., potassium permanganate, $KMnO_4$) in aqueous solution to produce the corresponding carboxylic acids (presumably via the acid halide).

As shown in Equation 7.7, trifluoroacetic acid (CF_3CO_2H) can be prepared by oxidation of the 2(Z)-perfluoro-2,3-dichloro-2-butene (although the 2(E)-isomer is equally reactive).

---

*The details of the process shown in this footnote are not known. However, it is possible that partial hydrolysis of the adamantane (tricyclo[3.3.1.1^{3,7}]decane, Chapter 1) analogue hexamethylenetetramine (1,2,5,8-tetraazatricyclo[3.3.1.1^{3,7}]decane) to formaldehyde (methanal, CH_2O) and ammonia (NH_3), from which it was formed by a condensation reaction (see Chapter 10), provides an oxidizing agent (methanal).

[†]Typical aryl halides include the polychlorinated biphenyls (PCBs). Generally, the commercial PCBs constitute the mixture of compounds obtained, as shown, on chlorination of biphenyl in the presence of iron and iron(II) chloride at high temperature (see Chapter 6, Table 6.12). This mixture has been in more or less continuous use since the early 1930s, and, although they are now considered environmental hazards, they remain in restricted use since substitutes to replace them as insulating fluids in electrical transformers (where resistance to oxidation in important) have not been found. Because of their presumed deleterious role as environmental pollutants, their use as hydraulic fluids, softeners for plastics, and so on, has been discontinued. However, they persist because (as expected) they have resisted oxidative

(*Continued on next page*)

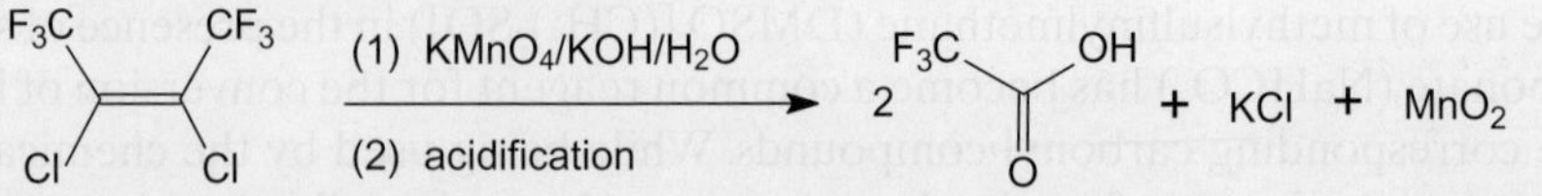

$$\text{(F}_3\text{C)(Cl)C}=\text{C(CF}_3\text{)(Cl)} \xrightarrow[\text{(2) acidification}]{\text{(1) KMnO}_4/\text{KOH/H}_2\text{O}} 2\ \text{F}_3\text{C}-\text{C(=O)OH} + \text{KCl} + \text{MnO}_2 \quad (7.7)$$

If the halogen is neither vinylic nor allylic, its reactions (except for reduction as described in the following section of this chapter) and those of the double bond are largely independent.

D. REDUCTION OF ALKYL, ALKENYL, AND ARYL HALIDES

I. Dehalogenation and Reductions at Carbon

In the reduction of alkyl, alkenyl, and aryl halides, the halogen (X = F, Cl, Br, or I) rarely leaves without one or more of the electrons in the bond accompanying it (i.e., it rarely leaves as X^+). Thus, even though the halogen generally leaves as an anion or a radical, there are four major reductive pathways that will be considered here. The first three differ from the fourth since, in them, the halogen is directly replaced by hydrogen. The fourth method is more circuitous; there is electron transfer from a metal to the alkyl, alkenyl, or aryl halide and then, in a true second step, hydrolysis of the metal–carbon bond.

First, catalytic reduction (frequently referred to as *hydrogenolysis*) can be used. Depending on the conditions, the unsaturation in alkenes, alkynes, and even arenes (if they are present) can be effected since the reduction follows the same general path of adsorption on the surface of the catalyst and then transfer of hydrogen seen in the reduction of those unsaturated groups (Chapter 6).

degradation. It is of some interest that 2,3,7,8-tetrachlorodibenzofuran (dioxin), a minor impurity in the PCB mixture, is reported to be among the most toxic chemicals (to humans). Presumably it results from chlorination of dibenzofuran (which might be present when the initial chlorination was effected) or it has been introduced oxidatively during or, more slowly, subsequent (in incineration) to the chlorination.

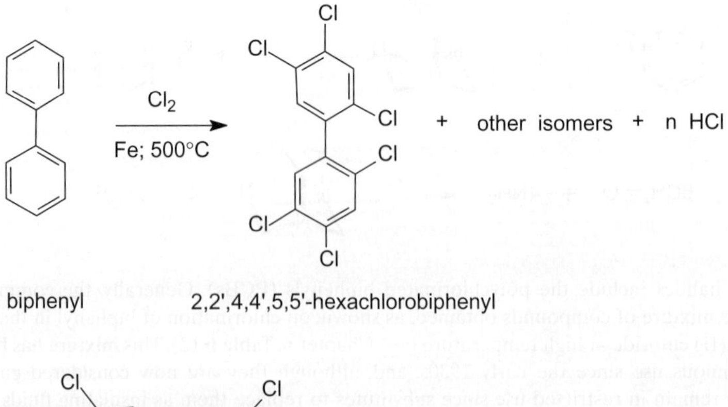

biphenyl   2,2',4,4',5,5'-hexachlorobiphenyl

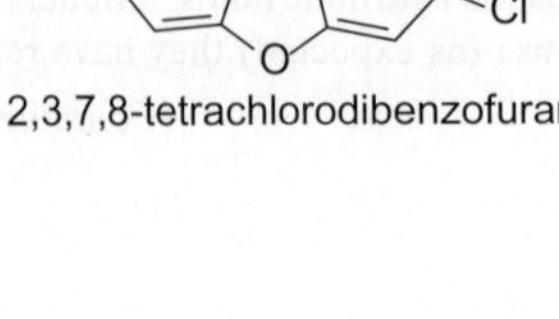

2,3,7,8-tetrachlorodibenzofuran

Second, substitution by hydride (H^-) for halide (X^-) can be carried out. Although substitution reactions will be discussed subsequently in this chapter, when many more nucleophiles will be introduced, the replacement of halogen by the nucleophile hydride (H^-) can also be thought of as a reduction, and it is thus briefly considered here too.

Third, there are a few radical processes in which a hydrogen atom (H) enters a site previously occupied by halogen. Presumably, the halogen (X) was abstracted by some radical or atomic species first and a chain reaction similar to that already discussed (Chapter 6) for halogenation results. This time, the reaction is a dehalogenation and hydrogen replaces halogen (i.e., it is the reverse of that discussed in Chapter 5).*

Fourth, differently, and with diminishing ease, alkyl iodides (R–I), bromides (R–Br), chlorides (R–Cl), and (rarely) fluorides (R–F), also react with a variety of metals (M). The conditions of the reaction can be adjusted either to isolate only the reduced haloalkane or to produce some organometallic intermediary species (frequently written as R–M–X) species. In the organometallic compounds, electron transfer from the metal has generally been accomplished so that hydrolysis produces the corresponding hydrocarbon.

a. Hydrogenolysis. The majority of catalytic methods of reduction generally appear simple: a solution of substrate in alcohol (or aqueous alcohol) solvent is stirred with the catalyst (Pd, Pt, Ni, etc.) in a hydrogen (H_2) atmosphere. Since hydrogen halide (H–X) will be generated, some base (such as sodium carbonate, Na_2CO_3) is added to combine with it as it forms and, presumably, protect the catalyst from deactivation by halogen. These conditions are similar to those required for reduction of carbon–carbon double (C=C) and triple (C≡C) bonds (Chapter 6), and it is generally found that those sites of unsaturation are also reduced during hydrogenolysis. Since aromatic rings are more resistant to reduction, hydrogenolysis of aryl halides may occasionally be effected without reduction of the ring as shown (Equation 7.8) for the conversion of *ortho*-chlorocinnamic acid to 3-phenylpropanoic acid.

$$\begin{array}{ccc} & \xrightarrow[\text{(2) acidification}]{\text{(1) } H_2/Pt/K_2CO_3} & \end{array} \qquad (7.8)$$

b. Substitution of Hydride for Halide. Although the details (including stereochemistry and kinetics) of substitution reactions, a group of processes which also includes substitution of hydride (H^-) for halide (X^-), for all of the alkyl, alkenyl, and aryl halides will be discussed later in this chapter, the process is also mentioned here since the net result is reduction. Generally, the ease of reduction (R–I > R–Br > R–Cl >>> R–F) follows from the expected bond strength and an easily replaced halide, such as a secondary alkyl iodide, will react with sodium borohydride ($NaBH_4$) in a dipolar aprotic solvent, such as (methylsulfinyl)methane (DMSO [$(CH_3)_2SO$]), and following a work-up with aqueous acid to destroy

---

*It will be found in your study of organic chemistry that a reaction and its reverse are both studied—but at different times and in different ways. Indeed, it will be in your interests to consider the reverse of every reaction examined.

unreacted borohydride, the alkane will result (Equation 7.9). If the halide, such as the aryl bromide, 1-bromonaphthalene (Equation 7.10), requires more forcing conditions, lithium aluminum hydride ($LiAlH_4$) in oxacyclopentane (THF) solvent at reflux (Equation 7.10, with the same kind of work-up) would better serve as the "source of hydride (H^-) ion" since it is a more powerful reducing agent (i.e., experience dictates that $LiAlH_4$ succeeds where $NaBH_4$ fails).*

$$CH_3(CH_2)_5CHCH_3 \ + \ NaBH_4 \ \xrightarrow[\text{(solvent)}]{(CH_3)_2SO} \ CH_3(CH_2)_6CH_3 \qquad (7.9)$$

$$\qquad (7.10)$$

c. Radical Replacement of Halogen by Hydrogen.

Replacement of a halide by a hydrogen atom donor is a radical process and has been investigated most thoroughly using a low concentration of radical initiator such as azo-bis-isobutyronitrile ($[(CH_3)_2C(CN)N=N(CN)C(CH_3)_2]$ AIBN)[†] to start the reaction and tri-1-butyltin hydride (tri-n-butyltin hydride, $[CH_3CH_2CH_2CH_2]_3SnH$)[‡] as the reducing agent. The order of reactivities of alkyl halides generally reflects the strengths of the carbon–halogen bond, that is, R–I > R–Br > R–Cl > R–F, and thus, as shown in Scheme 7.2, for the reduction of 1-bromo-2-fluorocyclohexane by reaction in solution with tributyltin hydride (tri-n-butyltin hydride, $[CH_3CH_2CH_2CH_2]_3SnH$), it is expected that bromine (Br) would be abstracted in preference to fluorine (F). The limiting reagent is presumably 1-bromo-2-fluorocyclohexane, and the tin hydride ($[CH_3CH_2CH_2CH_2]_3SnH$) can be used in excess. Thus, the termination step yields only small quantities of dicyclohexyldifluoride. Furthermore, it appears that hydride abstraction occurs at the diffusion rate. Thus, radical rearrangements are avoided.

In another vein, it is frequently found that if more than one halogen is present on the same carbon, it is possible to adjust the conditions (e.g., by using a limited amount of tri-1-butyltin hydride [tri-n-butyltin hydride {$CH_3CH_2CH_2CH_2$}$_3SnH$]) so

*It is very important to recognize that reduction with these group III metal hydrides almost certainly does not involve a "free" hydride (H^-) ion. That is, the potent reducing agent lithium aluminum hydride ($LiAlH_4$), which may be dimeric in some solvents, functions through species that appear to involve lithium (Li) complexes to the halogen due to depart and hydride donation from a tetrahedral aluminum hydride (AlH_4^-) anion. Indeed, the gentler agent, sodium borohydride ($NaBH_4$) probably functions in a similar way with sodium (Na) complexing the leaving group (less tightly than lithium [Li]) and the tetrahedral borohydride anion (BH_4^-) donating hydride. Interestingly, the "simpler" hydrides, such as sodium hydride (NaH) and calcium hydride (CaH_2), scarcely function as reducing agents at all!

[†]The decomposition of azobisisobutyronitrile ($[(CH_3)_2C(CN)N=N(CN)C(CH_3)_2]$ AIBN), the initiator for the reaction, as shown in Scheme 7.2, occurs to produce the stabilized 2-cyanopropyl radical and nitrogen. The preparation of this compound and related materials is discussed more fully in Chapter 10.

[‡]The formation of the reducing agent tri-1-butyltin hydride (tri-n-butyltin hydride, $[CH_3CH_2CH_2CH_2]_3SnH$) is discussed below under Grignard reagents an organometallic reagent prepared by exchange of one metal for another.

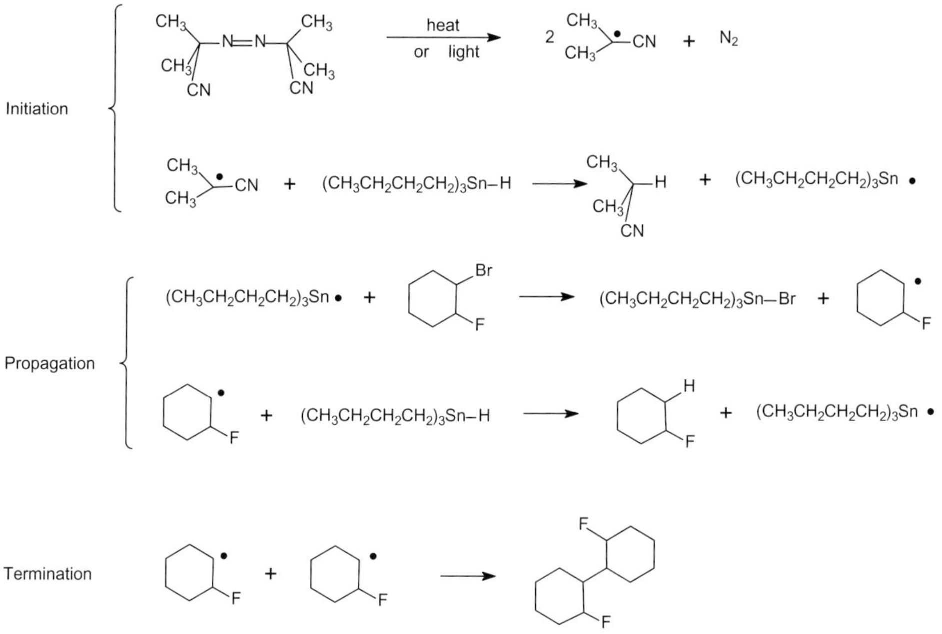

Initiation

Propagation

Termination

Scheme 7.2. A representation of the radical reduction of 1-bromo-2-fluorocyclohexane with tri-1-butyltin hydride (tri-*n*-butyltin hydride, [CH$_3$CH$_2$CH$_2$CH$_2$]$_3$SnH) in a radical reaction initiated by azobisisobutyronitrile ([(CH$_3$)$_2$C(CN)N=N(CN)C(CH$_3$)$_2$], AIBN). It is important to note that the bromine (Br) is replaced in preference to the fluorine and that the amount of the one termination product shown (the dicyclohexanedifluoride) is minor.

as to remove only one of them. Presumably, this is because the radical is stabilized by the presence of the other halogen. Thus, as shown in Equation 7.11, 1,1-dibromo-2,2,3-trimethylcyclopropane yields an unequal mixture of (*E*)- and (*Z*)-1-bromo-2,2,3-trimethylcyclopropanes when only 1 equivalent of tri-1-butyltin hydride (tri-*n*-butyltin hydride, [CH$_3$CH$_2$CH$_2$CH$_2$]$_3$SnH) is used.

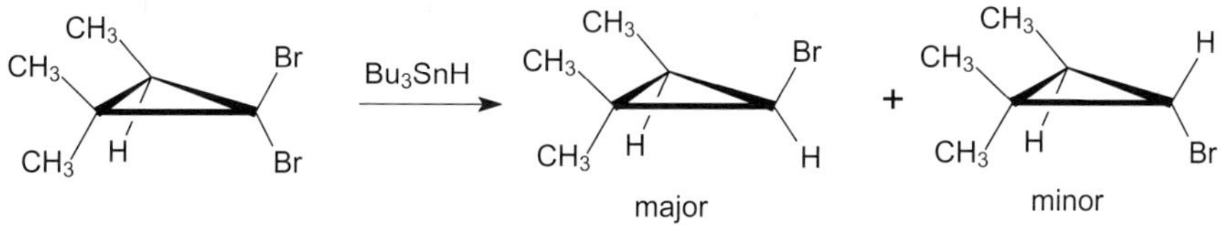

$$\tag{7.11}$$

Problem 7.2. Predict the product(s) of the reduction of each of the following with tri-1-butyltin hydride (tri-*n*-butyltin hydride, [CH$_3$CH$_2$CH$_2$CH$_2$]$_3$SnH; Bu$_3$SnH).

(a)   (b)

It is argued that the reduction of alkyl iodides with HI is also a free radical process (Equation 7.12).

$$\text{(cyclohexane with CH}_3\text{ and I)} + \text{HI} \longrightarrow \text{(cyclohexane with CH}_3\text{)} + \text{I}_2 \tag{7.12}$$

***d. Reaction of Alkyl, Alkenyl, and Aryl Halides with Metals.**[*] The history of organic chemistry is replete with the reactions of metals with organic halides. Indeed, in 1855, Wurtz[†] reported on the preparation of alkanes by the coupling reaction between 2 equivalents of primary alkyl bromide (Equation 7.13) on treatment with sodium (Na) metal.

$$\text{CH}_3\text{---}\text{---}\text{Br} \xrightarrow{2\text{Na}} \text{CH}_3(\text{CH}_2)_6\text{CH}_3 + 2\text{NaBr} \tag{7.13}$$

This is among the earliest mention of reactions of alkyl halides with sodium or potassium, and, indeed, it follows fairly closely the announcement, by Humphry Davy[‡] (in the Bakerian lecture in 1807) of the production of sodium (Na) and potassium (K) by electrolysis of their alkali salts.

In that vein, it has been subsequently shown (by reactions which most probably involve a series of one electron transfer processes) that halogens can be "stripped" from almost any alkyl, alkenyl, or aryl halide by treatment with lithium (Li) or sodium (Na) in the presence of a proton donor such as (commonly used) 2-methyl-2-propanol (*t*-butanol [$(\text{CH}_3)_3\text{C--OH}$], *t*-BuOH). A solvent such as oxacyclopentane (THF) helps to moderate the reaction. Rearrangements of the carbon skeleton are common. Because it usually lacks selectivity, the method is only rarely used (unless the goal is destruction of the halogenated compound).

It is also possible to reductively dehalogenate more reactive alkyl halides with either zinc (Zn) dust or a copper–zinc alloy (Cu–Zn) in the presence of an alcohol and protic acid, but this method also suffers from lack of specificity and is also frequently attended by skeletal rearrangement.

The more selective methods of reduction, which also minimize rearrangement reactions of the reacting alkyl halide (substrate), generally involve isolation (and thus at least partial purification) of an organometallic intermediate. Using one or more of these methods has enabled alkyl, alkenyl, and aryl halides, in the presence of an appropriate (usualy ethereal or hydrocarbon) solvent, to react with a variety of metals to produce organometallic compounds.

[*]The Heck reaction (the reaction among an aryl iodide [e.g., iodobenzene, $\text{C}_6\text{H}_5\text{I}$], palladium acetate [$\text{Pd}(\text{O}_2\text{CCH}_3)$], and an alkene [e.g., ethene, $\text{H}_2\text{C=CH}_2$] to yield an aryl-substituted alkene [e.g., ethenylbenzene, styrene, $\text{C}_6\text{H}_5\text{CH=CH}_2$]) has already been described in Chapter 6 and Scheme 6.54. In contrast to what is discussed here, the organometallic (palladium) intermediate was not isolated.
[†]Charles A. Wurtz (1817–1894), successor to J. Dumas, was the first occupant of the chair of organic chemistry established at the Sorbonne (in 1875). The Wurtz coupling reaction is only one of the many reactions upon which he worked as he explored the many ways sodium (Na) and potassium (K) could be used in organic chemistry.
[‡]Sir Humphry Davy (1778–1819) was, arguably, among the premiere chemists of the early nineteenth century. His study of chlorine (Cl_2) was instrumental in demonstrating that it was an element and comparable to his studies on the preparation of sodium (Na) and potassium (K) metals.

In addition to these direct processes, many organometallic compounds are found to undergo exchange reactions with other (less reactive) metals, and, in this way, it has proved possible to produce organic derivatives of almost every metal. Despite this, only a few groups of organometallic compounds have been explored in depth and, of that few, all of them enjoy the ability to undergo a rich variety of useful transformations including conversion to hydrocarbons.

Although the differences in reactivity and end use* among the various organometallic compounds that can be prepared ultimately dictate which particular organic halide and which metal will be used, it is probably still true that organomercurial, organomagnesium (Grignard reagents),[†] organolithium, organolithium cuprates (Gilman reagents),[‡] and organotin compounds receive the most attention.

1. Organomercurials. Both alkyl and aryl mercury compounds can be prepared by reaction of the corresponding halide with sodium amalgam (NaHg) (Equation 7.14). The necessity for sodium suggests that even though the organomercurial separates as the stable product, the electron density of the reducing medium needs to be increased, and, perhaps, small quantities of the organosodium compound may be necessary to induce the reaction to begin. Although the organomercury products are usually stable in water (it has been argued that alkyl mercurials are industrial aqueous pollutants and that occasional outbreaks of heavy metal poisoning in fish and thence humans results), they can be converted to the corresponding hydrocarbon(s) by treatment with acid (e.g., hydrochloric acid [HCl]) (Equation 7.15) as well as to other derivatives (such as a different halide) by treatment with the appropriate halogen.

$$\text{PhBr} \xrightarrow[\substack{C_6H_6/CH_3CO_2CH_2CH_3 \\ (\text{solvent})}]{2.7\%\ \text{Na in Hg}} \text{Ph–Hg–Ph} \qquad (7.14)$$

$$\text{Ph–Hg–Ph} \xrightarrow{\text{aq. HCl}} \text{PhH} + \text{Ph–HgCl} \xrightarrow{\text{aq. HCl}} \text{PhH} + \text{HgCl}_2 \qquad (7.15)$$

*In the context of reduction of alkyl, alkenyl, and aryl halides, replacement of halogen by hydrogen through the intermediacy of an organometallic compound can only be more or less facile from one organometallic compound to the next. While curiosity might be a sufficient rationale for the preparation and study of organometallic compounds, the variety available provides a spectrum of reactivities toward many other compounds too. Thus, the organometallic compounds described here have largely been developed for other purposes. Their uses for those other purposes will be described as warranted but, in that vein, it is worth noting that (it has been estimated) in the years (ca. 1900–1935) between the first publications regarding the reactions of organomagnesium halides by Victor Grignard and his death, more than 6000 papers had appeared in the literature making use of Grignard reagents. As noted below, Grignard received the Nobel Prize for its development.

[†]Victor Grignard (1871–1935) was professor of chemistry at University of Nancy and then the University of Lyon. He received the Nobel Prize in Chemistry in 1912 (jointly with Sabatier, the latter for catalytic hydrogenation studies).

[‡]Henry Gilman (1893–1986) was Professor of chemistry at Iowa State University.

2. Organomagnesium Compounds (Grignard Reagents). Alkyl, alkenyl, alkynyl, and aryl halides (particularly bromides [R–Br] and iodides [R–I]) react with magnesium (Mg) in diethyl ether [$(CH_3CH_2O)_2$] or other suitable solvents (such as oxacyclopropane [THF]) to produce compounds that are generally written as R–Mg–X but may have significant quantities of MgX_2 and R_2Mg present in equilibrium with the alkylmagnesium halide (Equation 7.16).

$$CH_3CH_2\text{---}Br + Mg \xrightarrow{\text{ether}} CH_3CH_2\text{---}Mg\text{---}Br \rightleftharpoons (CH_3CH_2)_2Mg + MgBr_2$$

$$(7.16)$$

Regardless of the actual structure of the Grignard reagent, which may depend on temperature, solvent, and particular substrate and even source of magnesium metal, it is generally written in the form R–Mg–X (or Ar–Mg–X) with the justification that it "appears to behave that way."

The rigid of exclusion moisture is necessary during the reaction because, as shown in Equation 7.17, for the formation of the Grignard reagent phenylmagnesium iodide (C_6H_5MgI) and its reaction with water (H_2O), the arylmagnesium halide is converted to hydrocarbon (C_6H_6) and magnesium salt [Mg(OH)I] when it is exposed to a proton source. Indeed, addition of water (or other proton source) to any of the Grignard reagents constitutes a route to reduction of the precursor halide to the corresponding hydrocarbon!

$$(7.17)$$

The path for the formation of the Grignard reagent is believed to involve transfer of an electron from the "sea" of electrons present in the metal into the lowest unoccupied molecular orbital (LUMO) of the alkyl (aryl) halide. The radical anion/surface-metal cation can continue with the transfer of a second electron or, depending on steric and electronic circumstances, an alkyl (aryl radical) and halogen anion can be produced. Coupling and elimination products occasionally accompany reduction and those products can be rationalized as being derived from free radicals. The halide anion remains associated with the surface as it does with most reactions in solvents that cannot support ionic materials. The alkyl radical would be free from that constraint.

After transfer of a second electron, the Grignard reagent itself ("RMgX") apparently forms by removing a magnesium atom from the metal surface, leaving a "hole" or pit etched in the metal. Pitted surfaces appear most reactive, and both exposure of fresh surface in this way and clean surfaces (by scratching away oxidized areas) appear to enhance the reaction. A generalized picture for the Grignard reaction using 1-bromobutane ($CH_3CH_2CH_2CH_2Br$) is presented in Scheme 7.3.

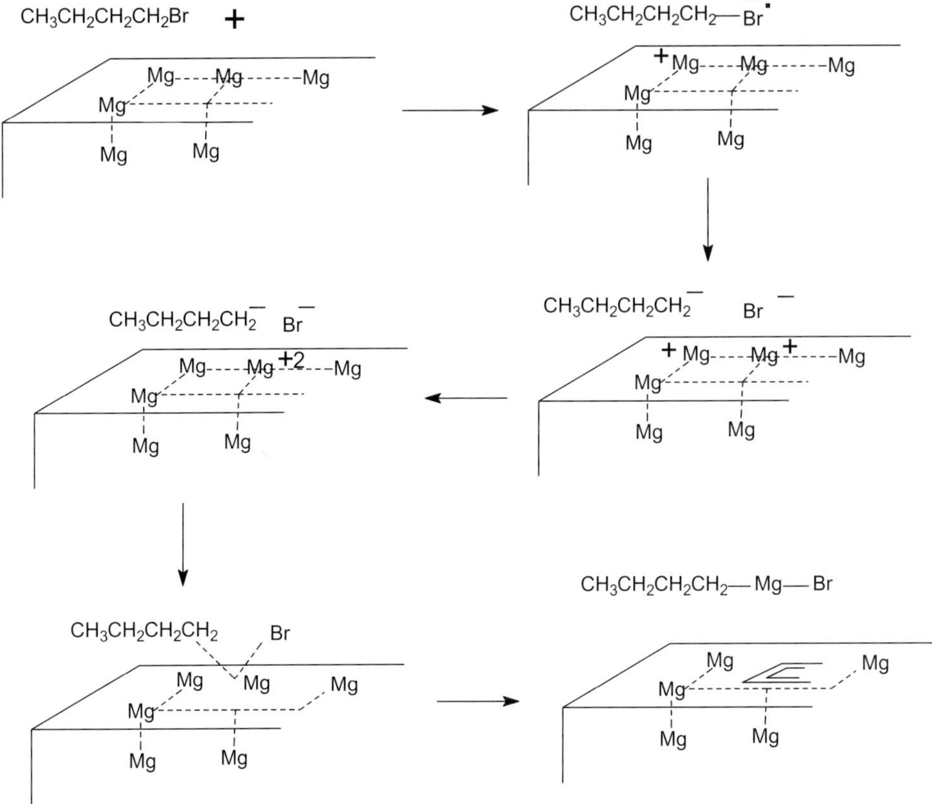

Scheme 7.3. A representation (note that solvent is involved but not specifically shown) of the conversion of 1-bromobutane ($CH_3CH_2CH_2CH_2Br$) to the corresponding Grignard reagent by reaction with magnesium (Mg) metal in ether ($(CH_3H_2)_2O$).

Problem 7.3. When bromocyclohexane is treated with magnesium metal in ether and the Grignard reagent formed is allowed to stand for a while before adding water, both cyclohexene and a cyclohexane dimer are isolated. How can one account for these additional products?

Although alkenyl halides such as bromoethene (vinyl bromide, $CH_2=CHBr$) can be converted to the corresponding Grignard reagent in oxacyclopentane (THF) as shown in Equation 7.18 and the product hydrolyzed to produce ethene (ethylene, $CH_2=CH_2$), this would doubtlessly be considered a wasteful exercise—given the volume of ethene (ethylene, $CH_2=CH_2$) available from other sources. However, consider the processes depicted in Equation 7.19 where (E)-1-bromopropene is converted to the Grignard reagent and the product treated with ^2H_2O so that the bromine (Br) is replaced with deuterium (^{2}H) rather than protium (^{1}H). Although the evidence that vinyl carbanions retain their geometry argues that the deuterium will be found *trans* to the methyl group, it is not always easy to maintain conditions that allow retention of configuration. The difficulty, of course, is that to maintain the geometry, the temperature must be kept low, while to induce reaction, the mixture must be heated.

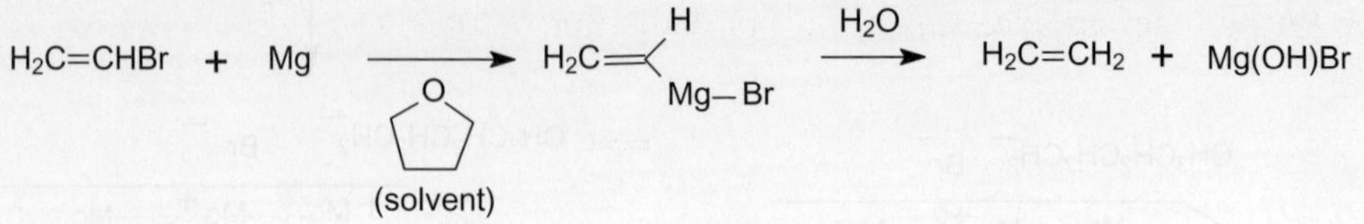

$$\text{(7.18)}$$

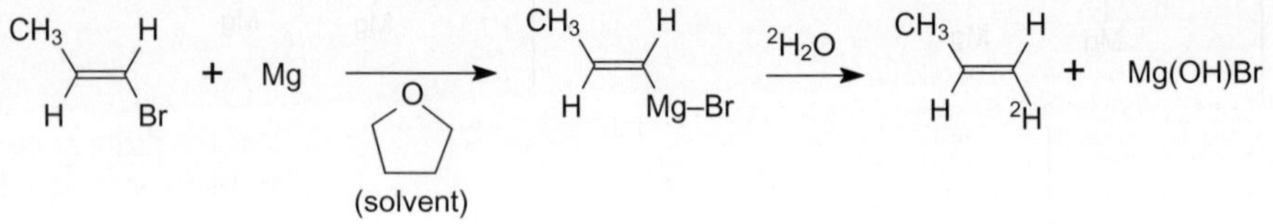

$$\text{(7.19)}$$

Problem 7.4. Start with either (*E*)-1-bromopropene or (*Z*)-1-bromopropene and assume that conditions are found to produce the Grignard reagent. Assume that it retained its configuration while reacting with ^{2}H$_2$O. Explain how you might know (what experiments would you do to determine) (a) if ^{2}H was actually incorporated, (b) if the ^{2}H is bound to the carbon from which the bromine (Br) left, and (c) if the geometry of the product was (or was not) in concert with the ^{2}H replacing the bromine stereospecifically.

Thus, for vinyl carbanions, it is more common to use the corresponding alkyl lithium derivatives (*vide infra*), which can be formed under less forcing conditions. Similarly, consider the case of *R*- (or *S*) 2-bromobutane (as an example) as shown in Scheme 7.4. To the extent that the carbanion retains uninverted tetrahedral geometry, the treatment of the Grignard reagent should provide the corresponding enantiomerically pure 2-[^{2}H$_1$]butane. In the event, conditions have not yet been found that allow the Grignard reagent to be converted to either enantiomer to the exclusion of the other (i.e., neither inversion nor retention is found; the product is

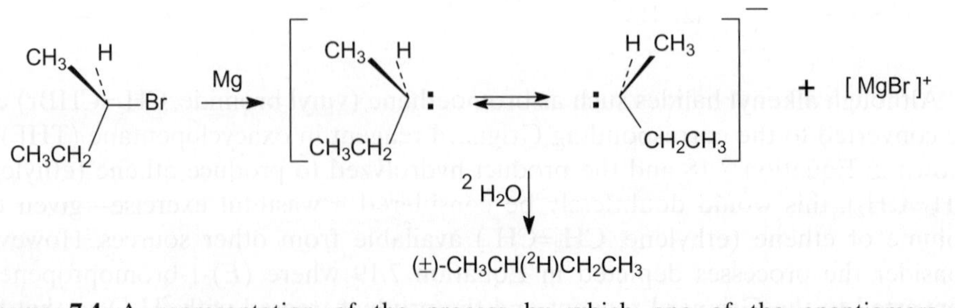

Scheme 7.4. A representation of the process by which one of the enantiomers of 2-bromobutane provides racemic 2-[^{2}H$_1$]butane.

racemic). It is not clear why this should be the case but it has been suggested (a) that the inversion barrier in the anion (through carbon) is low enough to account for the experimental observation or (b) two, symmetrically bridged, dianions (to two MgX^+) are present.

In a similar vein, allylic and benzylic halides also present a problem. Here, these reactive, conjugated halides preferentially react with the Grignard reagent already formed rather than magnesium. In this way, coupling to produce dimer (much like the Wurtz reaction, *vide supra*) occurs (Equation 7.20).

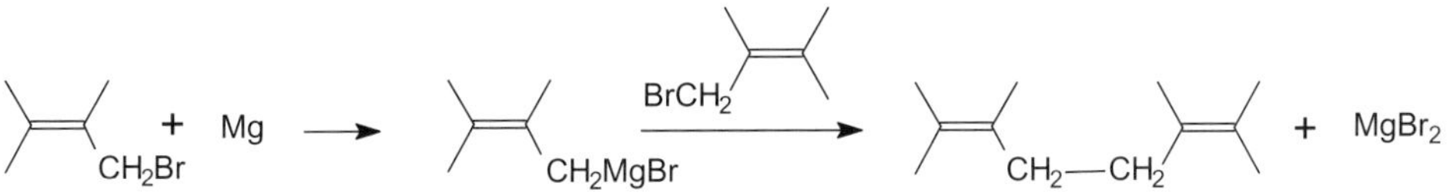

$$(7.20)$$

Although (in principle) alkynyl halides can be converted to Grignard reagents just as their more highly saturated analogues are, the acidity of the terminal hydrogen in primary alkynes is such that a simple exchange reaction is sufficient (Equation 7.21; R = H, alkyl, alkenyl, or aryl). Again, generation of 1-[$^2H_1$]-substituted alkynes is readily effected in this way.

$$R\!\!-\!\!\!\equiv\!\!\!-\!\!H \xrightarrow{\;CH_3CH_2MgBr\;} R\!\!-\!\!\!\equiv\!\!\!-\!\!MgBr + CH_3CH_3 \qquad (7.21)$$

Grignard reagents are occasionally used to prepare other organometallic compounds too. For example, as shown in Scheme 7.5, the reaction of 1-butylmagnesium bromide (*n*-butylmagnesium bromide, $CH_3CH_2CH_2CH_2MgBr$) with tin(IV) chloride (stanic chloride, $SnCl_4$) in hydrocarbon solvent produces tetra-1-butyltin [tetra-*n*-butyltin, $(CH_3CH_2CH_2CH_2)_4Sn$] and magnesium chloride ($MgCl_2$). When the tetraalkyltin is subsequently treated with 1 equivalent of chlorine (Cl_2), 1-chlorobutane (*n*-butyl chloride, $CH_3CH_2CH_2CH_2Cl$) and tri-1-butyltin chloride [tri-*n*-butyltin chloride, $(CH_3CH_2CH_2CH_2)_3SnCl$] result. The latter, on reduction

$$CH_3CH_2CH_2CH_2Cl \;+\; Mg \longrightarrow CH_3CH_2CH_2CH_2MgCl$$

$$4CH_3CH_2CH_2CH_2MgCl \;+\; SnCl_4 \longrightarrow (CH_3CH_2CH_2CH_2)_4Sn \;+\; 4MgCl_2$$

$$(CH_3CH_2CH_2CH_2)_4Sn \;+\; Cl_2 \longrightarrow (CH_3CH_2CH_2CH_2)_3SnCl \;+\; CH_3CH_2CH_2CH_2Cl$$

$$4(CH_3CH_2CH_2CH_2)_3SnCl \;+\; LiAlH_4 \longrightarrow 4(CH_3CH_2CH_2CH_2)_3SnH \;+\; LiAlCl_4$$

Scheme 7.5. A series of reactions utilized to convert an alkyl halide (1-chlorobutane) to the corresponding alkyltin hydride. The alkyl halide, after reaction with magnesium to produce a Grignard reagent can be hydrolyzed to produce the alkane. The trialkyltin hyride can be used to convert other alkyl halides to their respective alkanes.

with lithium aluminum hydride (LiAlH$_4$) produces tri-1-butyltin hydride [tri-*n*-butyltin hydride, (CH$_3$CH$_2$CH$_2$CH$_2$)$_3$SnH], the reagent of choice for radical reduction of alkyl halides (*vide supra*).

Although 1,2-dihalides undergo elimination reactions on treatment with magnesium (Mg), it has been reported that if the halogens are different from each other and at least three carbon atoms apart, selective conversion to Grignard reagents is possible.

3. Alkyl-, Alkenyl-, and Aryllithium Reagents. Alkyl, alkenyl, and aryl organolithium compounds are readily prepared by reaction of lithium metal (Li) with the corresponding halide in hydrocarbon solvents. As shown in Equation 7.22 an equivalent of lithium halide is produced simultaneously. Alternatively (Equation 7.23), exchange between a dialkyl mercurial (R$_2$Hg) and lithium metal (Li) can be effected. The advantage of the latter being that halide is no longer present to contaminate the reaction mixture.

Interestingly, lithium NMR (^{7}Li has spin 3/2 and an abundance of >92%) has been explored to investigate the nature of alkyllithiums and although spectra can be difficult to interpret (not the least because the nature of the bond between carbon and lithium is not well understood), it is generally agreed that the structure of the organolithium reagent is (among other things) a function of solvent and, frequently, the method of its preparation. Usually, regardless of the way it is formed, reaction with protic solvents results in, overall, substitution of hydrogen for (originally) halogen. Indeed, the reaction of alkyllithium reagents with moisture is frequently violent and hydrocarbon solutions of some of these reagents for example, *t*-butyllithium [(CH$_3$)$_3$C$^-$Li$^+$] are known to spontaneously ignite and burn with a bright flame on exposure to air.

In contrast to what is seen with Grignard reagents, organolithium compounds prepared from vinyl halides in hydrocarbon solvents appear to retain their configuration (see Problem 7.4).

$$CH_3CH_2CH_2CH_2Cl + 2Li \longrightarrow CH_3CH_2CH_2CH_2-Li + LiBr \qquad (7.22)$$

$$(CH_3CH_2CH_2CH_2)_2Hg + 2Li \longrightarrow 2CH_3CH_2CH_2CH_2-Li + Hg \qquad (7.23)$$

Although not frequently used directly for reduction to hydrocarbons, aryl iodides (e.g., iodobenzene, C$_6$H$_5$I) on treatment with activated copper (Cu) apparently form arylcuprates, which, in a subsequent step, react with a second equivalent of aryl iodide to produce a biaryl. Equation 7.24 is an example of the production of biphenyl by this coupling process (known as the Ullman reaction).*

$$(7.24)$$

*Fritz Ullmann, Universities of Zurich, Geneva and Berlin Technische Hochschulle.

If an alkyl copper is treated with an alkyllithium, a lithium dialkylcopper reagent can be formed. Alternatively, treatment of a copper halide (e.g., cuprous iodide, CuI) with an alkyllithium (e.g., n-butyllithium, $CH_3CH_2CH_2CH_2Li$) produces the same reagent (Equation 7.25).

$$2CH_3CH_2CH_2CH_2Li + CuI \longrightarrow (CH_3CH_2CH_2CH_2)_2CuLi + LiI \qquad (7.25)$$

The lithium diorganocoppers (also called Gilman reagents)* are stable in ethereal solvents (i.e., they are less reactive than the alkyllithium from which they are made). Additionally, unlike the less selective alkyl sodium (Wurtz coupling) and aryl copper (Ullman coupling) reagents already mentioned, solutions of Gilman reagents can be prepared (Scheme 7.6a) and subsequently used for high-yield coupling reactions with with alkyl, alkenyl, and even aryl halides (Scheme 7.6b).[†]

As will be discussed later in this chapter (Section E, 7), vinyl iodides such as (E)-1-iodopentene (Scheme 7.6c) also react with alkenyl boron reagents (Chapter 6, Scheme 6.26) in a palladium-catalyzed process. The Suzuki reaction (cf. the Heck reaction, Chapter 6, Scheme 6.54 and footnote page 371) also results in the formation of a new carbon–carbon bond (Scheme 7.6c), and, generally, the iodides are more reactive than the bromides, and they, in turn, are more reactive than the vinyl chlorides.

II. Reductions at Halogen

Based on the values of the first ionization potentials of the halogens,[‡] it might be concluded that, except for iodine, cationic species are unlikely, and indeed, as seen here (*vide infra*), compounds prepared from aryl iodides (Ar–I) are the only positive halogen compounds with carbon–halogen bonds both stable enough for routine work and reactive enough to fulfill their function.

However, the formation of other compounds of the halogens with other elements in which the sharing of electrons leads to more positive oxidation states for the halogen has served organic chemistry well. Examples of such compounds with positive oxidation states for the halogens were provided in Chapter 6 (cf. Table 6.1), and they include hypochlorous acid (HOCl) nitrosyl chloride (NOCl), iodine azide (IN_3), and iodine isocyanate (INCO). It may be recalled that it was argued there that the electron-rich double bond attacked the electrophilic halogen to lead the process of electrophilic addition.

Such positive halogen compounds (and others, including sodium hypochlorite [NaOCl], sodium chlorite [$NaClO_2$], and sodium chlorate [$NaClO_3$]) can serve to

---

*Henry Gilman (1893–1986), professor, Iowa State University.
[†]Gilman and related lithium organocopper reagents are also capable of undergoing selective addition reactions (as discussed in Chapter 9).
[‡]Accepted values (in electron volts [$1\,eV = 23.05\,kcal$]) are as follows: fluorine, 17.42; chlorine, 13.01; bromine, 11.84; and iodine, 10.44 (and for purposes of comparison, carbon, 11.26 and hydrogen, 13.59). It can be recalled that the ionization potential is an experimental value. It is the amount of energy necessary to effect the removal of the most loosely bound electron from a gaseous atom. The ionization potential can be used as a measure of the ability of an element (or compounds containing that element) to behave as a reducing agent.

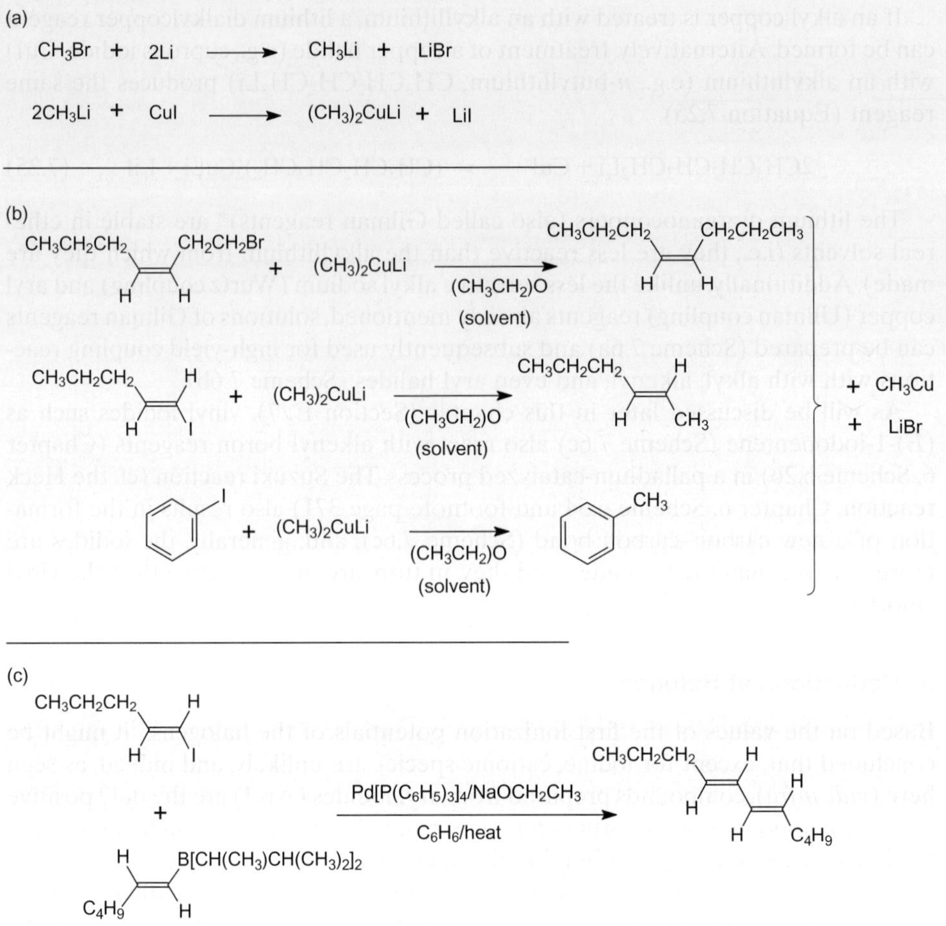

(a)

$$CH_3Br + 2Li \longrightarrow CH_3Li + LiBr$$

$$2CH_3Li + CuI \longrightarrow (CH_3)_2CuLi + LiI$$

(b)

[chemical scheme: (Z)-1-bromo-3-heptene + (CH₃)₂CuLi, (CH₃CH₂)O (solvent)]

[chemical scheme: (E)-1-iodopentene + (CH₃)₂CuLi, (CH₃CH₂)O (solvent)]

[chemical scheme: iodobenzene + (CH₃)₂CuLi, (CH₃CH₂)O (solvent)]

$$+ CH_3Cu$$
$$+ LiBr$$

(c)

[chemical scheme: (E)-1-iodopentene + alkenyl borane reagent B[CH(CH₃)CH(CH₃)₂]₂, Pd[P(C₆H₅)₃]₄/NaOCH₂CH₃, C₆H₆/heat → diene product]

Scheme 7.6. (a) A representation of the preparation of methyllithium (CH$_3$Li) from the reaction of bromomethane (methyl bromide, CH$_3$Br) with lithium metal (Li) and its conversion to the Gilman reagent, lithium dimethyl copper [(CH$_3$)$_2$CuLi]. (b) Representative coupling reactions with the Gilman reagent, lithium dimethyl copper [(CH$_3$)$_2$CuLi], with a primary alkyl halide [(Z)-1-bromo-3-heptene], a vinyl halide [(E)-1-iodopentene], and an aryl halide (iodobenzene), where, in each case, the halogen is successfully replaced with a methyl group. The yields in these reactions are routinely reported to be above 70% of theory. (c) A representation of a palladium (Pd)-catalyzed (Suzuki) coupling reaction of the same vinyl iodide [(E)-1-iodopentene] of part (b) with an alkenyl borane reagent made from addition of disiamyl borane (H-B[CH(CH$_3$)CH(CH$_3$)$_2$]$_2$) to 1-hexyne (HC≡CCH$_2$CH$_2$CH$_2$CH$_3$).

oxidize a variety of organic substrates,* and some organic compounds containing halogen can serve the same role.

Organic compounds that contain halogen and can effect such oxidations fall into two groups. Representatives of the first group are given in Figure 7.3. In this group, the halogen is attached to either oxygen or nitrogen. These compounds are not the

---

*As already noted in Chapter 6 B.III, substitution of chlorine for hydrogen, for example, in the chlorination of methane (CH$_4$), can be considered an oxidation at carbon and a reduction at chlorine. However, use of positive halogen bleaching powders and solutions to cause oxidative decomposition of organic contaminants in drinking water and on clothing, and so on, constitute more common uses.

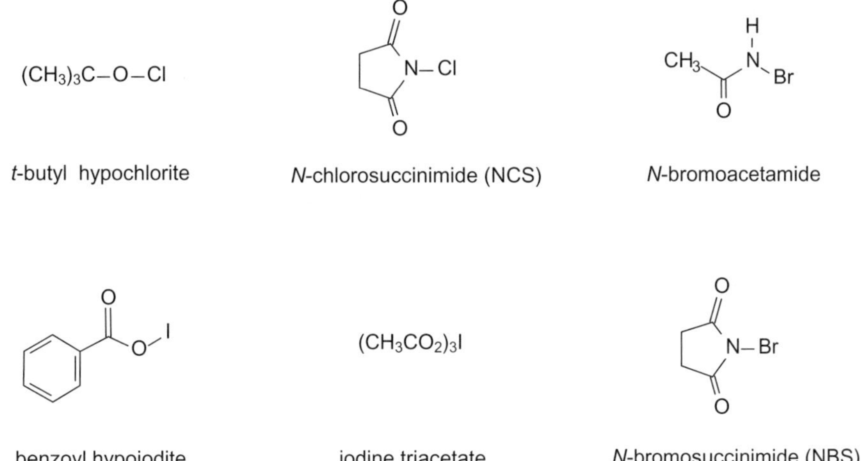

t-butyl hypochlorite   N-chlorosuccinimide (NCS)   N-bromoacetamide

benzoyl hypoiodite   iodine triacetate   N-bromosuccinimide (NBS)

Figure 7.3. Some representative examples of organic compounds that serve as sources of positive halogen (some of these are discussed further in Chapter 10).

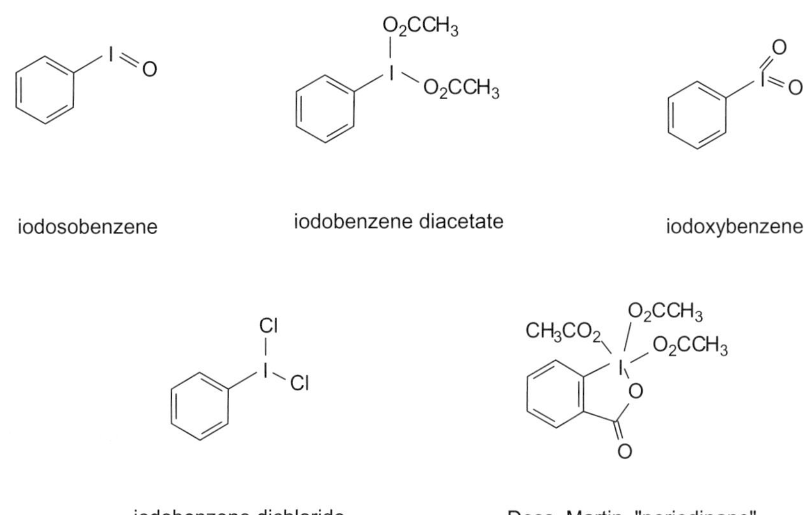

iodosobenzene   iodobenzene diacetate   iodoxybenzene

iodobenzene dichloride   Dess–Martin "periodinane"

Figure 7.4. A representative group of iodoarenes that serve as oxidizing agents. The iodine is reduced as a consequence of the oxidation reaction.

subject of this chapter and, while useful as oxidants in organic chemistry, will be considered further here in Chapter 10.

Representatives of the second group, which consists of iodine-containing organic compounds with a carbon-to-iodine bond, are provided in Figure 7.4. Some of these compounds can be obtained commercially, and they are generally used as specific and mild oxidizing reagents, soluble in organic solvents, and readily measured for quantitative use. Most often, they are called for to convert alcohols to aldehydes (Chapter 8) or alkenes to diols in place of metal salts of osmium or manganese oxides (Chapter 6).

E. NUCLEOPHILIC SUBSTITUTION

The earliest struggles to place the study of organic chemistry into a position where it was possible to predict the outcome of a reaction, by understanding the path between a starting material and product, involved substitution reactions.

In 1854, a year after his death, the publication of the landmark work of Auguste Laurent,* which codified and extended the ideas of Dumas,[†] set firmly in place the concept of the retention of a central, unchanging molecular core and attached replaceable substituents.

About the same time (1850), Wilhelmy[‡] undertook to measure the rate at which acid-converted cane sugar to fruit sugar.[§] This study is arguably the first in which the mathematics for a first-order process (Chapter 4) was deduced.

Although it appears that the kinetic work of Wilhelmy was largely ignored, substitution reactions, converting one material into another, continued to

*Auguste Laurent (1807–1853), after receiving a degree as a mining engineer, began work in a technical college founded by Dumas. Laurent is, arguably, the first to clearly distinguish among the terms equivalents, atoms, and molecules, as part of putting organic chemistry on a systematic basis.

[†]Jean Baptiste André Dumas (1800–1884) was a teacher, chemist, and French public servant whose name is still associated with analytical methods for both the determination of nitrogen by combustion analysis and vapor density.

[‡]Ludwig Ferdinand Wilhelmy (1812–1864) was born in Poland but studied in Berlin and Heidelberg. He was a skilled businessman who devoted himself to chemistry and physics.

[§]The chemistry of the conversion of the disaccharide sucrose ($[\alpha]_D^{20} = +66.5$) to the monosaccharide fructose ($[\alpha]_D^{20} = -92.4$) and an equilibrium mixture of α-D-glucopyranose (α-D-glucose) and its anomer, β-D-glucopyranose (β-D-glucose) ($[\alpha]_D^{20} = +52.7$ for the mixture), the structures of which were unknown at the time, will be discussed in Chapter 11. The positive rotation of the mixture of glucose epimers is less than the high negative rotation of fructose.

This inversion of sign during the hydrolysis played an important role in the generalized concept of "inversion." The hydrolysis is often carried out with catalysis by the enzyme (Chapter 11) sucrase (which is also called "invertase"), and the resulting mixture is called invert sugar.

CH$_2$OH

HO
HO
OH

β-D-glucopyranose

$[\alpha]_D^{20} = +18.7°$

HOCH$_2$ OH   CH$_2$OH
O
HO   OH

fructose

$[\alpha]_D^{20} = -92.4°$

CH$_2$OH

HO
HO
OH

sucrose

$[\alpha]_D^{20} = +66.5°$

CH$_2$OH

HO
HO
OH

α-D-glucopyranose

$[\alpha]_D^{20} = +112°$

($[\alpha]_D^{20} = +52.7°$ for the mixture at equilibrium)

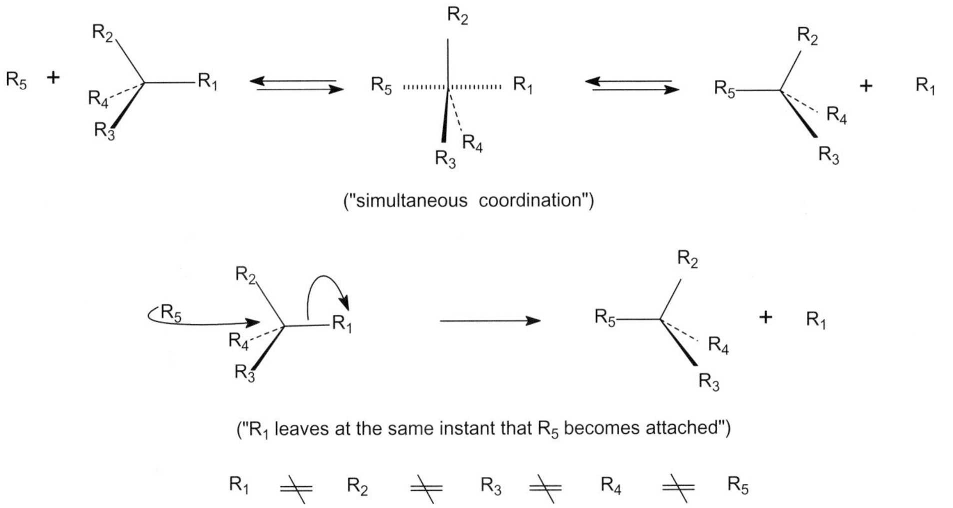

Figure 7.5. A representation of the Walden inversion. In his 1923 publication *Valence and the Structure of Atoms and Molecules*, Lewis states "There seems to be but one possible way of accounting for this peculiar behavior … Let us consider a carbon atom attached to the four radicals R_1, R_2, R_3, and R_4, and let us assume that a fifth group, R_5, becomes temporarily attached to the carbon atom near the face of the tetrahedron that is opposite R_1. A slight shift of the kernel might make it now the center of a new tetrahedron with corners at R_2, R_3, R_4, and R_5, while R_1 would become detached from the molecule. Then if the radical R_5 in the new molecule were to be replaced by the radical R_1, the resulting molecule would be the mirror image of the one with which we started. In this explanation, it is not necessary to assume that the five radicals are attached to the carbon for any appreciable time; indeed, it might be assumed that the R_1 leaves at the same instant that the R_5 becomes attached to the carbon atom."

prove a fertile field, which Wurtz (*vide supra*) and Gerhard* et al. pursued vigorously.

Then, in 1895, in an event that has been called "the most astonishing observation which had been made in the field of optical activity since the fundamental investigations of Pasteur," Walden[†] described a remarkable series of transformations in which one enantiomer of an α-aminocarboxylic acid (Chapter 9) could be converted into either enantiomer of an α-bromocarboxylic acid (Chapter 9) (i.e., the amino [–NH2] group was replaced by a bromine [–Br]). Shortly thereafter, he showed that the same α-bromocarboxylic acid could be converted into its own enantiomer. The principle, inversion of configuration by substitution, is diagramatically represented in Figure 7.5.

---

*Charles Gerhard (1816–1856) had worked with Laurent but was recognized subsequently as attempting to "take credit" for Laurent's efforts. This, coupled with a, reportedly, dogmatic manner succeeded in earning him the emnity of his contemporaries. He apparently alienated numerous leading figures of the time.

†Paul Walden (1863–1957) was born in Russia and educated in Germany, eventually becoming a professor at the University of Rostock and then Tubingen. Foreshadowing current trends, it is reported that he was

(Continued on next page)

The general process continues to be called Walden inversion.*

In addition to the possibility of substitution with inversion as a result of simultaneous entry of the nucleophile and departure of the leaving group, the alternative, that is, prior dissociation followed by readdition, also needed to be considered. The latter was suggested (in electronic terms) shortly thereafter and within the decade,

guaranteed pension "for life" by three different European governments but, after the second war (WWII), while in his 80s, he found it difficult to collect from any of them. The quotation, attributed to Emil Fischer (who is credited with the discovery of the configuration of glucose, Chapter 11), is from Ingold, C. K. *Structure and Mechanism in Organic Chemistry*, Cornell University Press, Ithaca, NY, 1953, p. 373. Emil Fischer (1852–1919) was professor of organic chemistry and director of organic chemical research at the University of Berlin. The work that he guided across the canvas of the chemistry of amino acids, carbohydrates, indoles, proteins, and purines, as well as his contributions to dye chemistry and stereochemistry, earned him the Nobel Prize in Chemistry in 1902.

*One of the series of reactions used by Walden and reported (in part) in *Chemische Berichte*, **1895**, *28*, 1287, is shown. Additional discussion is postponed until Chapter 10 when more of the chemistry of the functional groups will have been described, but it is important to understand that even now, some of the interconversions described by Walden are considered subtle and would not be chosen to demonstrate a principle.

However, in 1895, working with naturally occurring materials (such as 2-hydroxysuccinic acid [malic acid, $HO_2CCH_2CH(OH)CO_2H$], *vide infra*) of (then) unknown structure was typical of the best chemists. Optical activity, based on the work of Huygens, Maxwell, Fresnel, et al., as outlined in Chapter 4 Section H, was a well-entrenched tool and, along with the source from which the material was obtained, its appearance, physical state, melting point and/or boiling point, and combustion analysis, was a commonly provided analytical characteristic. However, the complexity of the starting materials masked the elegant simplicity of the discovery and the passage of time (some 20 years), and the insightful genius of G. N. Lewis was required to even suggest the possibility of substitution with inversion. (Gilbert Newton Lewis [1875–1946], Ph.D., Harvard University, Professor MIT [1905–1912], University of California [1912–1946].) The 1895 example of Walden follows:

S-(–)-2-hydroxysuccinic acid (malic acid)

S-(+)-2-aminosuccinic acid (aspartic acid)

S-(–)-2-bromosuccinic acid

dimethyl S-(–)-2-hydroxysuccinate

dimethyl S-(–)-2-bromosuccinate

R-(+)-2-hydroxysuccinic acid (malic acid)

dimethyl R-(+)-2-bromosuccinate

R-(+)-2-hydroxysuccinic acid (malic acid)

S-(–)-2-bromosuccinic acid

(Continued on next page)

Ingold* and coworkers, albeit working with less-than-pure materials, had experimental evidence in hand to propose two limiting pathways: one involving initial dissociation and the other concerted entry and departure.

Both kinds of reactions were in accord with the incoming species being electron rich and thus with an electron being transferred from the "substituting agent" to the nucleus originally holding the "expelled" species. The result would be that the incoming "nucleophile" (L. *nucleus*, kernel; Gr. *philos*, loving) would be, formally, one electron poorer and the "leaving" substituent one electron richer when the process had been consummated. It was "postulated" that these heterolytic reactions would differ in that one of them would have "only one stage, in which two molecules simultaneously undergo covalency change" while the other would have "two stages: a slow heterolysis ... followed by a rapid coordination between carbonium ion[†] and the substituting agent ... with only one molecule undergoing covalency change ... in the rate determining step."

The concept of two molecules simultaneously undergoing covalency change in the rate-limiting step (abbreviated as S_N2, which stands for substitution, nucleophilic, bimolecular) can be compared with and contrasted with the concept of only one molecule undergoing covalency change (similarly abbreviated as S_N1, for substitution, nucleophilic, unimolecular) in the rate-limiting step. However, in that

The overall process, the conversion of S-(–)-malic acid to R-(+)-malic acid through the corresponding halosuccinic acids was subsequently repeated using phosphorus pentachloride (PCl_5). As shown below, this is an abbreviated version of the original set of transformations:

*Sir Christopher Ingold (1893–1970) was professor at Leeds (1924–1930) and then at University College London (1930–1970) until his death. The words in quotations that follow are taken from Ingold, C. K. *Structure and Mechanism in Organic Chemistry*, Cornell University Press, Ithaca, NY, 1953.

[†]At the time this was written by Ingold (1953, *vide supra*), the suffix "onium" was used for this species. However, it has subsequently been (largely) agreed to reserve "onium" for positively charged species with filled valence shells (e.g., ammonium [NH_4^+], oxonium [R_3O^+]). Current use has the words "carbonium ion" for the species R_3C^+ replaced by the single word "carbocation."

conceptual exercise, it is very important to understand the difference between the *idea* of S_N1 and S_N2 and the *experimental reality* of reaction order.

I. Nucleophiles and Nucleophilicity

Both S_N1 and S_N2 reactions involve substitution of a nucleophile for a leaving group. Although this act may be construed as constituting a definition (i.e., that species substituting for a leaving group is a nucleophile), there is little illuminated in that definition about those properties that define nucleophilicity.

First, those electron-rich species involved in donating electrons (or an electron) in these heterolytic reactions and which are called "nucleophiles" clearly cannot be imagined as being interested in nuclei, as such. Rather, for the most part, because they are electron rich and may even bear a negative charge, they are attracted to areas of low electron density.

Second, as originally proposed by G. N. Lewis in his generalized theory of acids and bases (Chapter 5), any substance that can donate an electron pair to form a covalent bond is a base. Thus, if chloride ion (Cl^-) or water (H_2O:) acting as "nucleophiles" donate electrons to, for example, the methyl cation (CH_3^+, a Lewis acid) so as to form a bond, they are also Lewis bases. Indeed, any proton acceptor (Brønsted base, Chapter 5) can be classified as a nucleophile!

What then are the differences between a base and a nucleophile and between an acid and an electrophile?

As pointed out earlier (Chapter 5) base strength (or basicity) and acid strength (or acidity) are defined in terms of an equilibrium in an acid–base reaction (generally involving proton transfer). On the other hand, nucleophilicity (and electrophilicity) is defined in terms of rates of reactions (generally involving carbocations, not protons). Thus, base (acid) strength deals with the overall energy change in a reaction (from starting material[s] to product[s]) (Equation 4.6, Chapter 4; i.e., $\Delta G^\circ = -RT \ln K_{eq}$), while nucleophilicity deals with the activation energy to reach the rate-determining transition state for the reaction. For the latter, the observed rate constant, k_{obs}, is given by an expression such as that of Equation 7.26 (and as discussed previously in Chapter 4, Equation 4.15),

$$\begin{aligned} k_{obs} &= (k_B T/h)e^{(-\Delta G^\ddagger/RT)} \\ &= (k_B T/h)e^{(-\Delta H^\ddagger/RT)}e^{(-\Delta S^\ddagger/R)}, \end{aligned} \tag{7.26}$$

where k_B is the Boltzman constant ($1.38 \times 10{-}16\,erg\,K^{-1}$), h is Planck's constant ($6.62 \times 10{-}27\,ergs$), and T is the temperature in K.

It is nonetheless true that many strong bases, frequently as a function of solvent, are good nucleophiles since, whether charged (e.g., [OH^-]) or not (e.g., [$(CH_3)_3N$:]), an electron pair is available for bond formation. However, many good nucleophiles (e.g., [I^-], [Br^-], [Cl^-], [$(CH_3)_2S$:], [R_3P:], [R_2O:]) are not strong bases and their nucleophilicity is frequently a function of how well they are solvated by the solvent in which the heterolytic reaction is being carried out. Indeed, a "good nucleophile" is one with which, in a particular solvent (and frequently under a particular set of nonequilibrium circumstances), reaction with a particular electrophile occurs "rapidly" (a kinetic judgment).

Finally, nucleophilic substitution reactions are ubiquitous. As will become apparent, alcohols (and their ester derivatives), ethers, and ammonium salts among others (in addition to alkyl halides) can also be used as substrates for nucleophilic substitution reactions.

II. S_N1

Under the appropriate conditions, tertiary alkyl halides, benzylic halides, and allylic halides, that is, species capable of generating stable carbocations (Table 7.2), are found to undergo reactions that are generally classified as S_N1.

Broadly, the following characteristics (some of which are discussed more fully below) are part of the picture of unimolecular nucleophilic substitution:

(a) The initial rate of disappearance of alkyl halide (i.e., $-d[\text{R–L}]/dt$) should be found to fit (overall) first-order kinetics (Chapter 4). This is presumed to indicate rate-determining heterolytic cleavage of the alkyl-halide bond to yield halide anion and a stabilized carbocation. Since the nucleophile should not be present in the rate-determining step of a unimolecular substitution, its nature as well as its concentration (because the reaction is first order in substrate) should be irrelevant.

TABLE 7.2. Halide Substrates in the S_N1 Reaction

Type	Example	Carbocation	Leaving Group
Tertiary alkyl halide	2-Chloro-2-methylpropane (*tert*-butyl chloride) 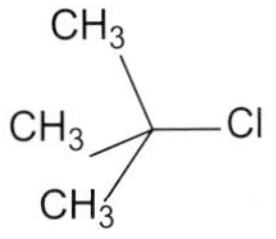	2-Methylpropyl cation (*tert*-butyl cation) 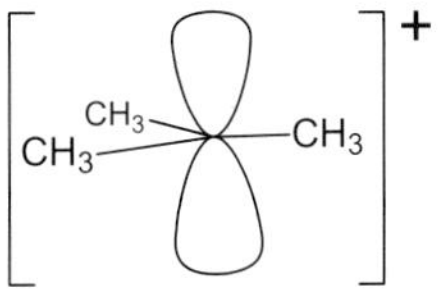	Chloride anion (Cl^-)
Benzylic halide	*Racemic* 1-bromo-1-phenylethane 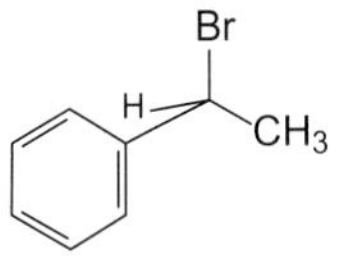	1-phenylethyl cation 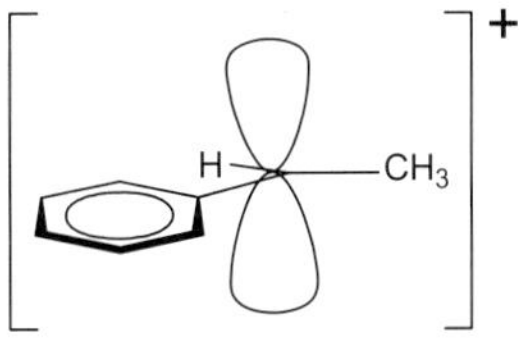	Bromide anion (Br^-)
Allylic halide	*Racemic* 3-bromocyclo-hexane 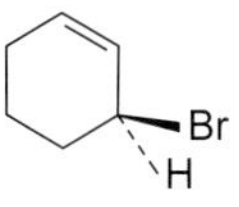	Cyclohexenyl cation 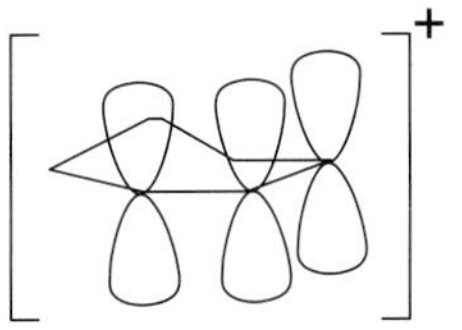	Bromide anion (Br^-)

A nucleophile is unspecified; it is not in the rate-determining step. Initial kinetics are used to determine the order in reacting substrate. Solvents in which such reactions are carried out are chosen to support ions (see text for details).

(b) Large electronegativity differences between the species that is leaving (i.e., the leaving group, L) and the carbon to which it is initially attached in the substrate should facilitate the process. Furthermore, the presence of Lewis acids (such as $AlCl_3$, BF_3), which can form complexes with the leaving group (L, particularly if it is a halide), might be expected to promote the reaction.

(c) Substituents on the carbon to which the leaving group (L) is attached that stabilize the carbocation (e.g., alkyl groups, aryl groups, conjugated double bonds) are expected to accelerate the reaction.

(d) Solvents capable of disbursing and/or carrying charge should promote the reaction. These solvents can also react with the carbocation to yield solvent-incorporated products (in which case the process is called solvolysis).

(e) If the substrate is optically active because the carbon bearing the leaving group (L) is "asymmetrically substituted," the product (forming from a planar carbocation) is expected to be (largely) racemic.

(f) If an ion corresponding to the leaving group (L), that is, L^-, is added to the reaction mixture, the rate of disappearance of the starting material is expected to diminish. That is, to whatever extent the initial bond breaking might be reversible, increasing the concentration of one of the products is expected to shift that equilibrium back toward the starting material (the common ion effect). Alternatively, addition of a non-nucleophilic anion (e.g., ClO_4^-), which increases the ion-carrying capacity of the solvent, should facilitate the reaction.

a. The Kinetics. Between about 1933 and 1937, Ingold and coworkers measured the rates of, and thus established the kinetics associated with, a number of substitution reactions on a variety of substrates and, from these, developed the broad outline of the concept of molecularity. It is clear that they recognized that those reactions organized under the concept of S_N1 did not always proceed with first-order kinetics.

Consider, for example, a nucleophilic substitution reaction on 2-chloro-2-methylpropane (*t*-butyl chloride [$(CH_3)_3CCl$], Table 7.2, example 1) in an ionizing solvent and in the presence of some nucleophile (Nu:). The starting material is going to disappear and the product is going to form. These events will occur over some measureable period of time and the object is to determine the rate at which the disappearance of starting material and/or the appearance of product occurs. This can be expressed as follows:

$$(CH_3)_3CCl \underset{k_{-1}}{\overset{k_1}{\rightleftharpoons}} (CH_3)_3C^+ + Cl^-, \tag{7.27}$$

$$(CH_3)_3C^+ + Nu: \overset{k_2}{\longrightarrow} (CH_3)_3CNu^+, \tag{7.28}$$

where k_1 is the rate constant associated with the disappearance of 2-chloro-2-methylpropane (*t*-butyl chloride [$(CH_3)_3CCl$]) to form the 2-methyl-2-propyl cation (*t*-butyl cation [$(CH3)_3C^+$]) and chloride anion (Cl^-), k_{-1} is the rate constant associated with the reverse reaction, and k_2 is the rate constant associated with product formation.

The rate of product formation is

$$\text{rate} = +d[(CH_3)_3C\text{–}Nu^+]/dt. \tag{7.29}$$

Then, since product can only be formed from the carbocation,

$$+d[(CH_3)_3C\text{–}Nu^+]/dt = k_2[(CH_3)_3C^+][Nu{:}], \tag{7.30}$$

but $[(CH_3)_3C^+]$, the intermediate (**I** in Figures 7.6 and 7.7) is a short-lived species partitioned between the starting material, 2-chloro-2-methylpropane (*t*-butyl chloride [$(CH_3)_3C\text{–}Cl$]) and the product, $(CH_3)_3C\text{–}Nu^+$.

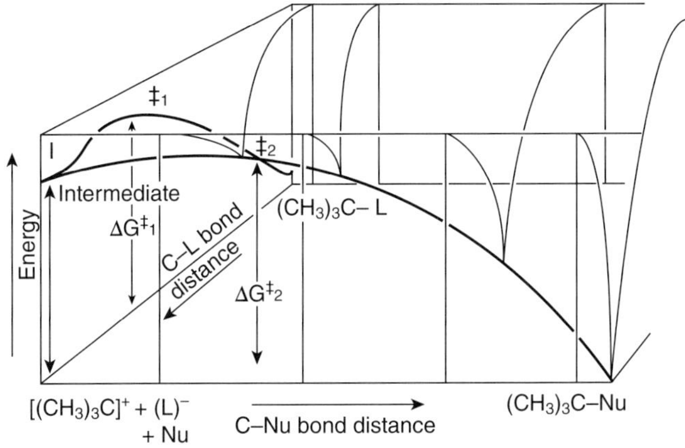

Figure 7.6. A representation of an energy diagram for the S_N1 reaction of 2-chloro-2-methylpropane. In the example, L, the leaving group, is chlorine (Cl). This three-dimensional figure can be transposed into two dimensions as shown in Figure 7.7.

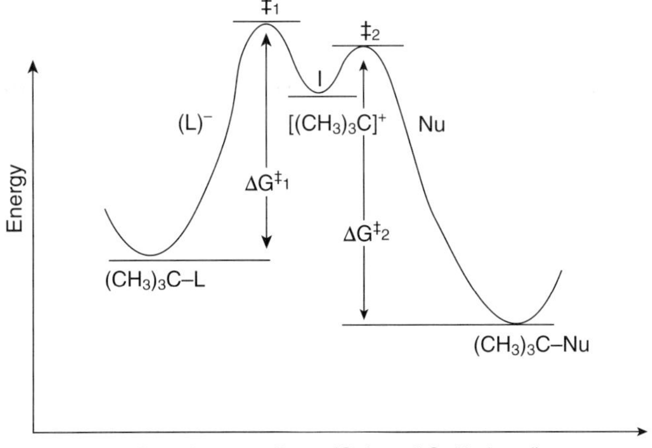

Figure 7.7. A two-dimensional representation of an energy diagram for the S_N1 reaction of 2-chloro-2-methylpropane. In the example, L, the leaving group, is chlorine (Cl). This two-dimensional figure can be represented in three dimensions as shown in Figure 7.6.

Thus, it is commonly taken that there is no change in the instantaneous concentration of the carbocation intermediate—it is used as quickly as it is formed! This assumption is called the steady-state approximation, that is,

$$d[(CH_3)_3C^+]/dt \approx 0 \tag{7.31}$$

and

$$d[(CH_3)_3C^+]/dt \approx 0 = k_1[(CH_3)_3Cl] - k_{-1}[(CH_3)_3{}^+][Cl^-] - k_2[(CH_3)_3{}^+][Nu:], \tag{7.32}$$

so

$$k_{-1}[(CH_3)_3C^+][Cl^-] + k_2[(CH_3)_3{}^+][Nu:] = k_1[(CH_3)_3Cl] \tag{7.33}$$

or

$$[(CH_3)_3C^+] = k_1[(CH_3)_3Cl]/\{k_{-1}[Cl^-] + k_2[Nu:]\} \tag{7.34}$$

and

$$d[(CH_3)_3C\text{–}Nu^+]/dt = \{k_1k_2[(CH_3)_3Cl][Nu:]\}/\{k_{-1}[Cl^-] + k_2[Nu:]\}. \tag{7.35}$$

Then, dividing the numerator and denominator by $k_2[Nu:]$,

$$d[(CH_3)_3C\text{–}Nu^+]/dt = k_1[(CH_3)_3Cl]/\{1 + (k_{-1}[Cl^-]/k_2[Nu:]) \tag{7.36}$$

so that if $k_2[Nu:] \gg k_{-1}[Cl^-]$,

$$d[(CH_3)_3C\text{–}Nu^+]/dt = k_1[(CH_3)_3Cl], \tag{7.37}$$

and the reaction is first order in substrate. However, it should be clear that if either $[Cl^-]$ is large or if $k_2[Nu:]$ is small compared with $(k_{-1}[Cl^-])$, then the reaction cannot be cleanly first order.

Experimentally, it is best if both the product and starting material can be detected and their concentrations measured at any time during the reaction. Frequently gas chromatography (Chapter 1) or one of the spectroscopic tools (Chapter 2) is found suitable. Thus, for the first-order disappearance of 2-chloro-2-methylpropane (t-butyl chloride [$(CH_3)_3Cl$]), if the initial concentration that is, at time $t = 0$, is b, then at some later time, $t = t$, after the reaction has progressed for a while, the concentration b will have been reduced by formation of the product p to $(b - p)$ so that

$$+d[(CH_3)_3C\text{–}Nu^+]/dt = +d[p]/dt = k_1[b - p] \tag{7.38}$$

and

$$d([p]/[b - p]) = k_1dt \tag{7.39}$$

so

$$\int_{p=0}^{p=p} d[p]/[b-p] = \int_{t=0}^{t=t} k_1 dt \qquad (7.40)$$

and because when $t = 0, p = 0$,

$$\ln[b/(b-p)] = k_1 t \qquad (7.41)$$

so that a plot of $\ln[b/(b-p)]$ versus t should give a straight line whose slope is k_1. Finally, in this vein, since $(b-p)$ is the amount of 2-chloro-2-methylpropane (*t*-butyl chloride $[(CH_3)_3C\text{–}Cl]$) remaining at any given time, the first-order rate constant k_1 can be obtained with any two concentrations and the corresponding times, that is,

$$\ln\{[(CH_3)_3C\text{–}Cl]_{t=t1}\}/\{[(CH_3)_3C\text{–}Cl]_{t=t2}\} = k_1(t_2 - t_1). \qquad (7.42)$$

b. Electronegativity Differences. The values of the relative ability of an atom, in a bonding situation, to attract electrons to itself (its electronegativity, Chapter 1) for the elements in the second row of the periodic table, that is, C (2.5), N (3.1), O (3.5), and F (4.1), suggest that leaving group ability should increase in that order (from left to right along the row). However, without appropriate additional substitution, none of these are particularly good as a leaving group. Indeed, it can be argued that since the orbitals forming the sigma (σ) bond holding the leaving group to the carbon from which it must depart are similar in size to those on the carbon itself, there should be good orbital overlap, the corresponding antibonding orbitals (σ^*), which have the appropriate symmetry for bond breaking, will be distant, and thus, as is found, none of them should be good leaving groups.

However, as will be seen subsequently, protonation of alcohols (Chapter 8) or their conversion into esters (Chapter 9) of strong acids and conversion of amines (Chapter 10) into tetraalkylammonium salts, changes the nature of the bonding sufficiently so that the C–O and C–N bonds more easily undergo heterolysis. Fluorine, however, cannot be aided in the same way, and even Lewis acids, which might serve to complex with the fluorine and help it to leave, suffer from the drawback that most solvents that promote ionization have nonbonded electrons with which the Lewis acid might preferentially complex. Thus, fluorine remains a poor leaving group.

Although electronegativity decreases in the halides as one proceeds down the periodic column (F [4.1], Cl [2.8], Br [2.7], I [2.2]), the nature of the bonding also changes, with each succeeding halide being further removed from the carbon to which it is attached. Additionally, the energies of the antibonding (σ^*) orbitals also decrease in the same way, making them more accessible. Thus, as anticipated, other factors being constant, the ease with which they leave should increase in the same way, with iodide (I^-) being superior to bromide (Br^-) and bromide superior to chloride (Cl^-). This is frequently found to be the case. In the same way, it is clear that leaving group ability will also, approximately, correlate with acid strength, since the latter is a measure of the ease of ionization of a proton away from the anion. The stronger H–L is as an acid, the better "L" will be as a leaving group. HF is a relatively weak acid (pKa ≈ 3.2) and fluoride is a poor leaving group, while hydrogen chloride (HCl) (pKa ≈ −6.5), hydrogen bromide (pKa ≈ −8.5), and HI (pKa ≈ −9) are all significantly stronger and their relative leaving group abilities follow the trend of their acidities.

c. The Structure of the Alkyl Group.

As anticipated, substituents on the carbon atom to which the leaving group is attached that donate electrons or provide delocalization so as to stabilize the positive charge will facilitate the ability of the leaving group to leave. Since alkyl groups and substituents that can donate electrons to the developing positive charge (e.g., aromatic rings and double bonds) will perform this function better than hydrogen, tertiary alkyl (abbreviated as 3° or R_3C-L), benzyl ($ArCH_2-L$), and allyl ($RCH=CH-CH_2-L$) substrates undergo S_N1 substitution reactions with greater ease than do simple secondary (abbreviated as 2° or R_2CH-L) and the 2° are, in turn, more reactive than primary (abbreviated as 1° or RCH_2-L). With the possible exception of positively charged nitrogen (e.g., in diazo compounds such as the methyldiazonium ion [$CH_3-N_2^+$, Chapter 10]), leaving groups attached to a simple methyl fail to leave in S_N1 processes.

As shown qualitatively in Figure 7.8, since a tertiary (3°) carbocation is more stable than its secondary (2°) carbocation counterpart, the transition state leading to it should be lower. Furthermore, since the ionization process in which the leaving group leaves is rate determining, it is reasonable that the relative rates of S_N1 reactions should reflect the stability of the carbocations Thus, it is not surprising to find that in water, at 50°C, 2-bromo-2-methylpropane (t-butyl bromide [$(CH_3)_3C-Br$])

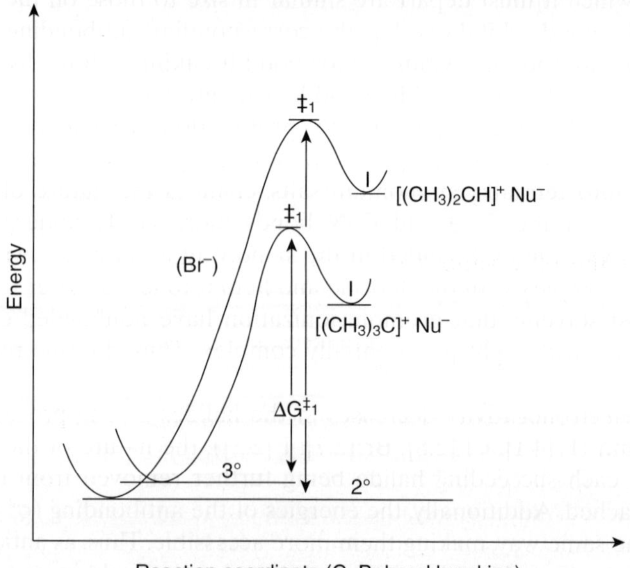

Figure 7.8. A qualitative representation of the differences in ionization energies between tertiary and secondary alkyl bromides (2-bromo-2-methylpropane [$(CH_3)_3CBr$] and 2-bromopropane [$(CH_3)_2CHBr$]) leading to tertiary and secondary carbocation intermediates (respectively). The events occurring after formation of the intermediates are not represented as they are not part of the rate-determining S_N1 process.

TABLE 7.3. A Comparison of the Relative Rates of Solvolysis of Tertiary Alkyl Halides in Methanol (CH_3OH) Solution

Compound	Structure	Relative Solvolysis Rate
2-Chloro-2-methylpropane	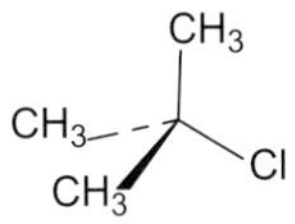	1
2-Chloro-2,4,4-trimethylpentane	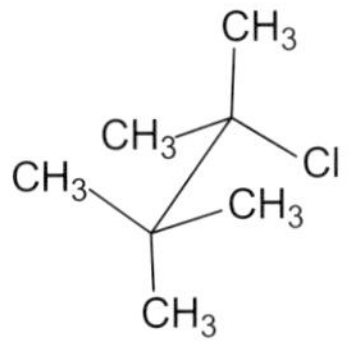	20
4-Chloro-2,2,4,6,6-pentamethylheptane	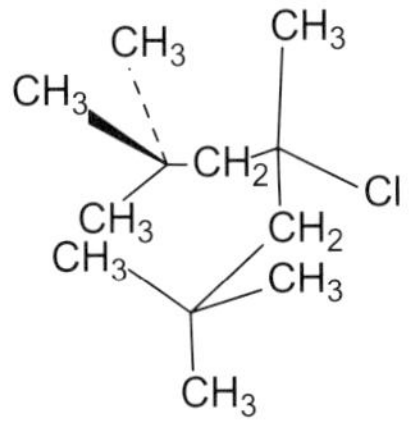	500

It is argued that strain resulting from steric encumbrance "behind" the leaving group ("B-strain") can be used to justify increased rates (after Takeuchi, K.; Takasuka, M; Shiba, E.; Tokunaga, H.; Endo, T.; Ushino, T.; Tokunega, K.; Okazaki, T.; Kinoshita, T.; Ohaga, Y. *J. Phys. Org. Chem.*, **2001**, *14*, 229).

dissociates approximately 105 times faster than 2-bromopropane (isopropyl bromide [$(CH_3)_2CHBr$]), which apparently reacts about a factor of 10 faster than bromoethane (ethyl bromide [CH_3CH_2Br]) itself.*

It is also clear that intramolecular strain (sometimes called "back strain" or "B-strain"), which results from steric repulsion between groups attached to the carbon bearing the leaving substituent, also plays an important role in accelerating the reaction of haloalkenes. Such ground-state destabilization is, for example, used to account for the observations that, under the same conditions, 2-chloro-2,4,4-trimethylpentane undergoes solvolysis about 20 times faster than 2-chloro-2-methylpropane and 4-chloro-2,2,4,6,6-pentamethylheptane is nearly 500 times faster (Table 7.3)!

d. The Role of the Solvent. With the rate-determining transition state in the S_N1 reaction requiring the separation of a positively charged carbocationic species from a leaving group that is departing with an electron pair (either as an anion or as a neutral species, with halides yielding the former), it should be clear why solvents capable of supporting charge (i.e., ionizing solvents, Chapter 5) facilitate the reaction. Solvents capable of hydrogen bonding (e.g., water, alcohols, carboxylic acids)

*It is not clear that the reactions of the secondary and primary bromides are completely or, for the latter in particular, even largely, S_N1. Indeed, as pointed out earlier, except in the idealized limit, all nucleophilic substitution reactions probably involve a mixture of both S_N1 and S_N2 processes with tertiary following a larger amount of the former and primary a significantly larger amount of the latter.

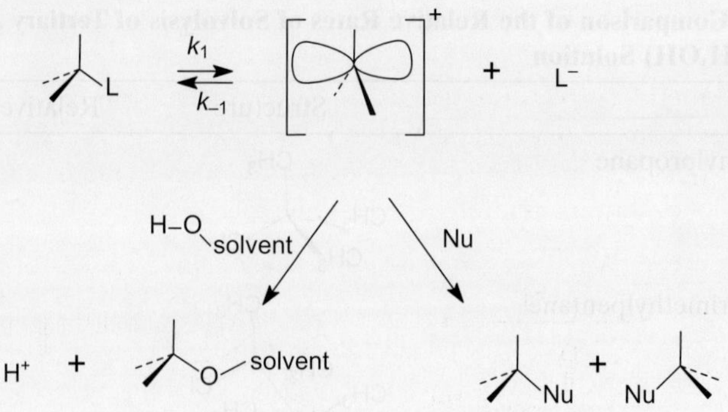

Scheme 7.7. A representation of the partitioning of a carbocation intermediate between the solvolysis product and that of nucloephilic substitution.

are frequently chosen for S_N1 reactions, although care must be exercised since these solvents (a) may also solvolyze the nucleophile so well that its reactivity is diminished and (b) may themselves act as nucleophiles since they possess the required electron pair(s) to qualify as nucleophiles. When the solvent reacts as the nucleophile with the intermediate carbocation, the process is called *solvolysis* (Scheme 7.7).

As shown in Scheme 7.7, a nonbonded electron pair on a solvent heteroatom participates by reacting with the cationic species. However, the best solvents for the S_N1 reaction will be those that, in addition to solvating the cation, also solvate the anion and make it easy to separate the ions once they are formed. To satisfy the latter, a high dielectric constant medium is desirable (Chapter 5, D [II]). However, the necessity to solvate both cation and anion means that not only should nonbonded pairs be available (for the cation) but hydrogen bonding (for the anion) is also important.

To this end, water (H_2O), alcohols (e.g., methanol [methyl alcohol, CH_3OH], ethanol [ethyl alcohol, CH_3CH_2OH], and 2,2,2-trifluoroethanol, CF_3CH_2OH), as well as carboxylic acids (e.g., methanoic acid [formic acid, HCO_2H] and ethanoic acid [acetic acid, CH_3CO_2H]) and mixtures of these solvents, have been used in S_N1 reactions. Dipolar aprotic solvents such as methanamide (formamide, H_2NCOH), methylsulfinylmethane (DMSO [$(CH_3)_2SO$], and ethanenitrile (acetonitrile, $CH_3C{\equiv}N$]) are found to be (as expected) less effective in the S_N1 reaction, despite their high dipole moments, since the anionic leaving group is not as well solvated.

Finally, despite use of solvents (or mixtures of solvents) to solvate the intermediate carbocation and the anion, the structure of the substrate continues to play the critical role and, for example, while, as noted above, at 50°C, 2-bromo-2-methylpropane (*t*-butyl bromide [$(CH_3)_3C–Br$]) dissociates approximately 105 times faster than 2-bromopropane (isopropyl bromide [$(CH_3)_2CHBr$]) in water at 50°C, the difference is only slightly diminished (from 105 to 104) in 60% ethanol : 40% water (v/v) at 55°C.

e. The Substrate Stereochemistry Attending the S_N1 Reaction. When optically active (+ or −), 1-chloro-1-phenylethane (α-phenyethyl chloride [$C_6H_5CH(Cl)CH_3$]) is allowed to react with water (H_2O) (i.e., to undergo hydrolysis) in aqueous propanone (acetone, CH_3COCH_3) and in the presence of added alkali, the corresponding alcohol, 1-phenylethanol [$C_6H_5CH(OH)CH_3$] (Equation 7.43), forms at a rate that is independent of the concentration of the added alkali. Thus, the rate-determining step (*vide supra*) is formation of the carbocation. And, depending on the ratio of propanone (acetone, CH_3COCH_3) to water (H_2O), the product is 89–96% racemized (with the remaining 4–11% inverted).*

$$\text{(7.43)}$$

It is argued that complete racemization, producing equal amounts of both of the enantiomers shown in Equation 7.43, does not occur even though the "free" carbocation should be planar (or nearly so) as a consequence of "dissymmetric solvation" of the cation. That is, the solvent shell surrounding the carbocation as it forms is not symmetrically arranged (if only because the leaving group cannot get completely out of the way before the shell begins to collapse to yield solvated product).

The necessity for the production of a planar carbocation was recognized quite early as a consequence of the observation (shown in Scheme 7.8) that apocamphyl chloride (1-chloro-7,7-dimethylbicyclo[2.2.1]heptane) did not give a precipitate (of silver chloride, AgCl) on boiling for as long as 48 h with aqueous, alcoholic silver nitrate ($AgNO_3$ in water [H_2O]–ethanol [CH_3CH_2OH] mixtures), whereas the acyclic analogue 3-chloro-3-ethyl-2,2-dimethylpentane reacted rapidly in the cold with the same reagent (an estimated rate difference of ca. 10^9)! Similarly, although bromoalkanes are more reactive than their chloro analogues and as shown in Scheme 7.9, 1-bromonorbornane (1-bromobicyclo[2.2.1]heptane) only partially reacts with aqueous silver nitrate ($AgNO_3$) at 150°C (in a sealed tube) at 48 h, the less strained 1-bromobicyclo[2.2.2]octane produces a precipitate of silver bromide (AgBr) in aqueous ethanolic silver nitrate ($H_2O/CH_3CH_2OH/AgNO_3$) at room temperature in about 4 h.

The use of chiral substrates has allowed the importance of solvent and nucleophile-associated cation to be probed. For example, as shown in Scheme 7.10, when chiral 1-chloro-1-phenylethane is heated in ethanoic acid (acetic acid, CH_3CO_2H]) at 50°C in the presence of potassium ethanoate (potassium acetate, $CH_3CO_2^-$ K^+), the ethanoic acid (acetic acid, CH_3CO_2H) ester of 1-phenylethanol is obtained. In this weakly nucleophilic system, about 15% excess inverted product is found (i.e., 85% of the product is racemic and the substitution reaction occurred with about 57.5% inversion and 42.5% retention). Presumably, the carbocation formed first and then

---

*The expression of percent racemization with percent excess inversion or retention is quite common. Thus, a reaction producing products with 50% inversion and 50% retention would be said to occur "with racemization." A reaction proceeding with 55% inversion and 45% retention then occurs with "90% racemization and 10% inversion," and one with 55% retention and 45% inversion with "90% racemization and 10% retention." It is also common to express results as occurring with a specific amount of "enantiomeric excess" (or ee). In this way, a reaction that produced products with "90% racemization and 10% net inversion" (or 55% inversion and 45% retention) could be said to yield 10% ee.

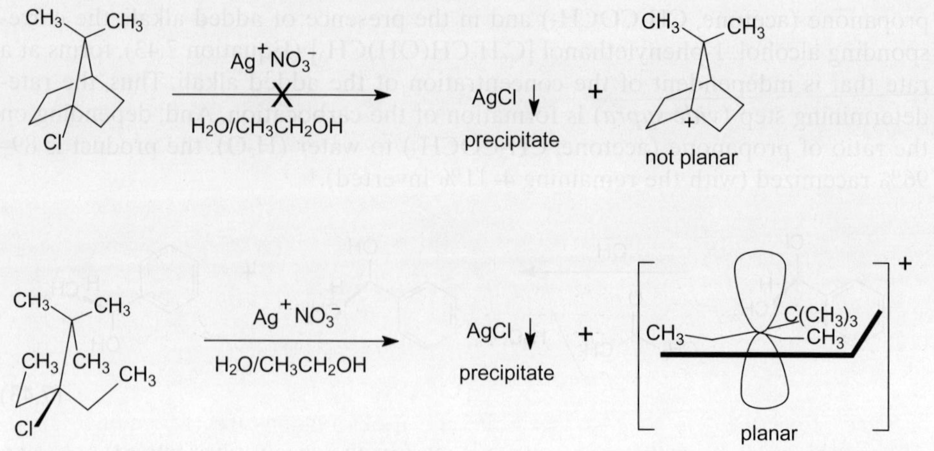

Scheme 7.8. Representations of the experimental observations that apocamphyl chloride (1-chloro-7,7-dimethylbicyclo[2.2.1]heptane) does not give a precipitate (of silver chloride [AgCl]) on boiling for as long as 48 h with aqueous, alcoholic silver nitrate [AgNO$_3$ in water (H$_2$O)–ethanol (CH$_3$CH$_2$OH) mixtures], whereas the acyclic analogue, 3-chloro-3-ethyl-2,2-dimethyl-pentane, reacts rapidly in the cold with the same reagent. This is explained by arguing that the strain accompanying the necessity for planarity in the bicyclic system (as it is converted to the carbocation) is too large to permit the process to occur. The general failure to introduce such strain (which would include attempting to place an sp2 hybridized carbon, as in an alkene, in the same position) is often referred to as Bredt's rule after the German chemist J. Bredt who proposed that alkenes such as 1-norbornene (bicyclo[2.2.1]hept-1-ene) are too strained to exist (except, perhaps, fleetingly).

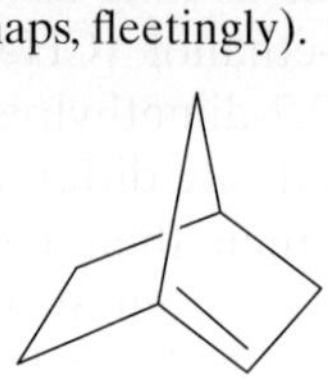

1-norbornene
(too strained to exist?)

Bredt's rule, first announced in 1924, was based on the examination of a large amount of chemistry explored, as noted earlier (Chapter 6, Section C) as a consequence of isolation of numbers of naturally occurring C10 compounds called terpenes (discussed more fully in Chapter 11) because of the derivation of some of them from the resinous exudate of the terebinth tree, *Pistacia terebinthus*. Some of these bicyclic compounds are shown (see Bredt, J. *Liebigs Ann. Chem.*, **1924**, *437*, 1).

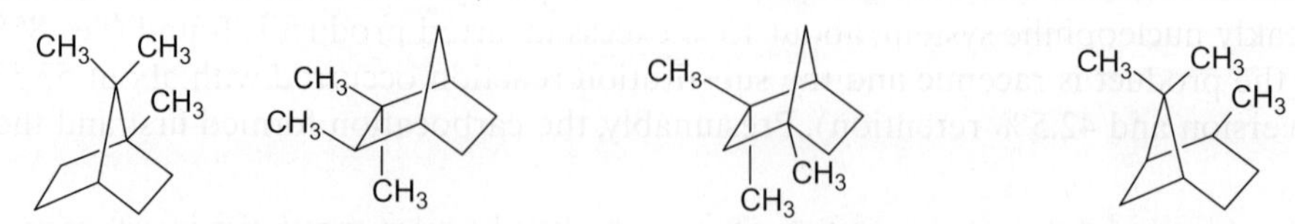

bornane

1,7,7,-trimethylbicyclo-
[2.2.1]heptane

camphane

exo-2,2,3-trimethylbi-
cyclo[2.2.1]heptane

fenchane

1,3,3-trimethylbicyclo-
[2.2.1]heptane

pinane

6,7,7-trimethylbicyclo-
[3.1.1]heptane

It is clear that when the rings are large enough, as in the bicycle[3.3.1]non-1-ene, the strain associated with a double bond at a bridgehead can be accommodated.

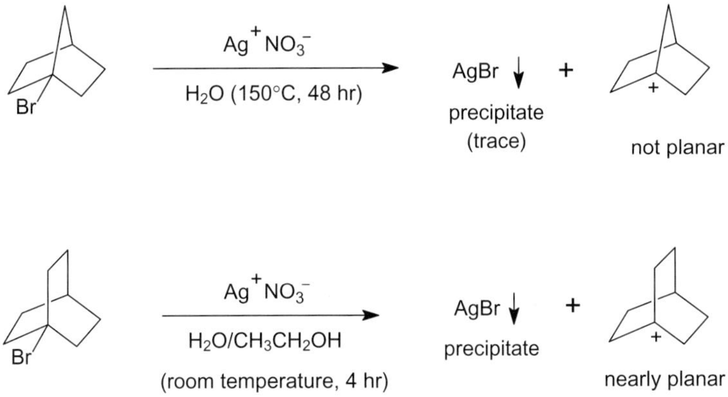

Scheme 7.9. A representation of the experimental observations that 1-bromonorbornane (1-bromobicyclo[2.2.1]heptane) only partially reacts with aqueous silver nitrate ($AgNO_3$) at 150°C (in a sealed tube) at 48 h, while the less strained 1-bromobicyclo[2.2.2]octane produces a precipitate of silver bromide ($AgNO_3$) in aqeuous ethanolic silver nitrate ($H_2O/CH_3CH_2OH/AgNO_3$) at room temperature in about 4 h. The tertiary carbocation produced from the latter is arguably less strained than that anticipated from the former although the analogous alkene would still appear to violate Bredt's rule (*vide supra*) concerning placing a double bond at the bridgehead and, in practice, it is not isolable.

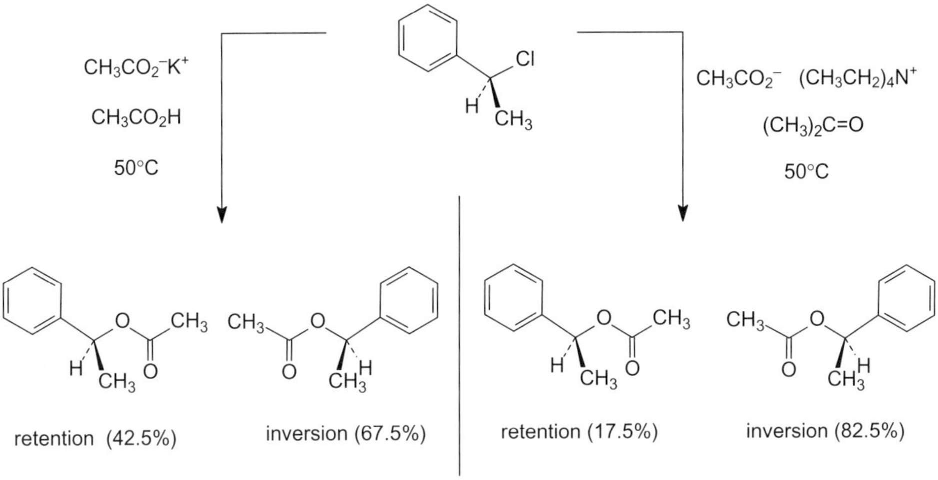

Scheme 7.10. A representation of inverted and retained product distribution obtained from optically active 1-phenyl-1-chloroethane as a function of solvent (and counterion) (See Swain, C. G.; Dittmer, D. C.; Kaiser, L. E. *J. Am. Chem. Soc.*, **1955**, *77*, 3737 and references therein.).

was attacked subsequently from either side but, because the leaving group could not easily get away, there is an excess of inversion over retention and racemization is incomplete. It might be reasonably argued that if a better ionizing solvent is chosen to facilitate the ability of the leaving group (chloride anion, Cl^-) to leave, the extent of racemization should increase. However, in addition to improving the ability of the leaving group to leave, there should also be a similar improvement in

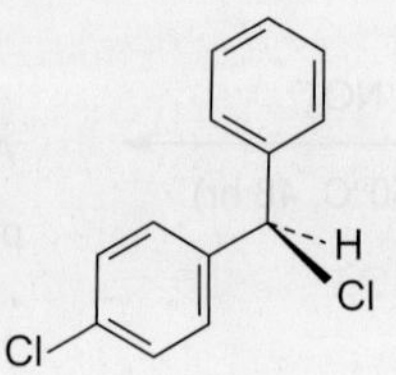

Figure 7.9. *para*-Chlorophenylphenylchloromethane.

the function of the nucleophile. In such cases, the kinetics must be reevaluated. Thus, if the solvent in the above reaction is changed to propanone (acetone, CH_3COCH_3) and the cation to tetraethylammonium $[(CH_3CH_2)_4N^+]$ instead of potassium (K^+), then the extent of racemization might be expected to increase. However, the extent of inversion increased instead (65% inversion), and the rate was given by an expression involving two terms (Equation 7.44):

$$\text{rate} = k_1[C_6H_5CHClCH_3] + k_2[C_6H_5CHClCH_3][CH_3CO_2^-]. \qquad (7.44)$$

Additionally, this dramatic departure from expectation signaled that the leaving of the leaving group might actually occur in stages. That is, while the bond between the leaving group and the carbon to which it was attached could be broken in the rate-determining step, the resulting cation and anion would remain confined to a "solvent shell." Only later, when each was separately solvated, might the nucleophile intrude.*

This interesting idea was tested in a slightly different system, Chloro(p-chlorophenyl)phenylmethane [1-chloro-4(chlorophenylmethyl)benzene] (Figure 7.9), chosen because it emphasizes what is expected. For this chloride, it was found that recovered starting material was undergoing racemization (i.e., ionizing to carbocation and recombining with chloride anion) about 30 times faster than it was undergoing solvolysis!

III. S$_N$2

In contrast to what was found for the generalized S$_N$1 reaction, the S$_N$2 reaction is most facile on methyl halides and unhindered primary alkyl halides, significantly less so on secondary halides and uncommon for tertiary halides (i.e., essentially the opposite order of the S$_N$1 reaction). These halides do not generally form stable carbocations. They do fit the overall idea that there is relatively little charge at the reaction site as the nucleophile enters and the leaving group leaves. A few examples of substrates, solvents, and nucleophiles for the S$_N$2 reaction are listed in Table 7.4 and more of the latter in Table 7.5.

The following characteristics (some of which are discussed more fully below) are part of the overall picture of bimolecular nucleophilic substitution (S$_N$2):

---

*The ideas involved in collapse of the solvent shell, ethanoic acid (acetic acid, CH_3CH_2OH) compared with propanone (acetone, CH_3COCH_3) and the use of solvent to participate in the nucleophilic substitution to yield product (in the first case) and intermediates that do not lie on the path to product (in the second case) are beyond the scope of the present discussion.

TABLE 7.4. Some Examples (Including Conditions) of the S_N2 Reaction

Starting Alkyl Halide	Conversion to	Reaction Conditions	Product(s)
(1) [structure: CH₃–C(H)(CH₂CN)–Br] 2-bromopropanenitrile	Alkyl halide	NaI in propanone (acetone, [structure: CH₃–CO–CH₃])	[structure: CH₃–C(CH₂CN)(H)–I] 2-iodopropanenitrile
(2) CH_3Br bromomethane (methyl bromide)	Alcohol	NaOH in water (H_2O)	CH_3OH methanol (methyl alcohol)
(3) [structure: trans-2-chlorocyclopentanol with HO, H, Cl, H] *trans*-2-chlorocyclopentanol	Oxirane (epoxide)	NaOH in water (H_2O) and ethanol (CH_3CH_2OH) mixture	[structure: bicyclic epoxide with O, H, H] bicyclo[3.1.0²,⁶] oxacyclohexane (cyclopentene epoxide)
(4) CH_3CH_2Br bromoethane	Ester	[structure: Na⁺ ⁻O–CO–CH₃] sodium ethanoate (sodium acetate) in propanone (acetone, [structure: CH₃–CO–CH₃])	[structure: CH₃CH₂O–CO–CH₃] (ethyl ethanoate)

(a) The initial rate of disappearance of alkyl halide (i.e., $-d[\text{R–L}]/dt$) should be linked to the rate of disappearance of the nucleophile (i.e., $-d[\text{Nu:}]/dt$) and, ideally, will be found to fit (overall) second-order kinetics: first order in alkyl halide and first order in nucleophile (Chapter 4). This is presumed to indicate that the rate-determining step involves both the leaving of the leaving group (heterolytic cleavage of the alkyl-halide bond) and simultaneous formation of a bond to the incoming group. There is no carbocation intermediate.

(b) In order to avoid the departure of the leaving group, the incoming nucleophile should approach the site of reaction (i.e., the "seat" at which reaction occurs) from the opposite direction (180° away). The S_N2 reaction will thus occur with inversion of configuration. Furthermore, substituents on the carbon to which the leaving group (L) is attached that hinder the approach of the nucleophile (steric hindrance) should retard the reaction. Thus, the order of reactivity for halogen-substituted alkyl substrates is expected to be methyl > primary > secondary >> tertiary. Alkenyl (vinyl), alkynyl (acetylenic), and aryl halides, where the incoming nucleophile is not attacking an

TABLE 7.5. Some Additional Examples of the S_N2 Reaction

(a) Formation of hydroperoxides:

CH_3CH_2Br + ^-OOH ⟶ CH_3CH_2—OOH + ^-Br

(b) Formation of thiols:

CH_3CH_2Br + ^-SH ⟶ CH_3CH_2—SH + ^-Br

(c) Formation of thiocyanates:

CH_3CH_2Br + { ^-S—$C{\equiv}N$ ⟷ $S{=}C{=}N^-$ } ⟶ CH_3CH_2—S—$C{\equiv}N$ + Br^-

(d) Formation of isocyanates:

CH_3CH_2Br + { ^-O—$C{\equiv}N$ ⟷ $O{=}C{=}N^-$ } ⟶ CH_3CH_2—$N{=}C{=}O$ + Br^-

(e) Formation of nitrates and nitrites:

CH_3CH_2Br + { ^-O—$\overset{\cdot\cdot}{N}{=}O$ ⟷ $O{=}\overset{\cdot\cdot}{N}$—$O^-$ } ⟶ CH_3CH_2—$\overset{+}{N}\overset{O}{\underset{O_-}{}}$ + CH_3CH_2—O—$N{=}O$ + Br^-

(f) Formation of alkynes:

CH_3CH_2Br + CH_3—$C{\equiv}C^-$ ⟶ CH_3CH_2—$C{\equiv}C$—CH_3 + Br^-

(g) Formation of nitriles and isonitriles:

CH_3CH_2Br + $C{\equiv}N^-$ ⟶ CH_3CH_2—$C{\equiv}N$ + CH_3CH_2—$\overset{+}{N}{\equiv}\overset{-}{C}$ + Br^-

(h) Formation of ammonium salts:

CH_3CH_2Br + $:N(CH_3)_3$ ⟶ CH_3CH_2—$\overset{+}{N}(CH_3)_3$ + Br^-

(i) Formation of alkyl azides:

CH_3CH_2Br + $^-N{=}\overset{+}{N}{=}N^-$ ⟶ CH_3CH_2—$\overset{+}{N}{=}N{=}\overset{-}{N}$ + Br^-

(j) Formation of phosphonium salts:

CH_3CH_2Br + $P(C_6H_5)_3$ ⟶ CH_3CH_2—$\overset{+}{P}(C_6H_5)_3$ Br^-

A representation of some nucleophilic displacement reactions utilizing bromoethane (ethyl bromide, CH_3CH_2Br) as a prototypical primary alkyl substrate upon which the various nucleophiles act. The display of synthetic versatility in the S_N2 reaction producing the many varieties of compounds illustrates carbon–oxygen, carbon–sulfur, carbon–nitrogen, carbon–phosphorus, and carbon–carbon bond formation and should not be taken to exclude other possibilities. The details of the conditions might be specified as in Table 7.4 for each case but are omitted here for brevity.

sp^3-hybridized carbon atom (and different geometric constraints apply, *vide infra*) should not undergo the S_N2 reaction.

(c) The reaction should be most facile if the LUMO assigned to the halogen-alkyl bond of the substrate is close in energy to the highest occupied molecular orbital (HOMO) containing the nonbonded electrons (which will become bonding) of the nucleophile. Thus, although electronegativity differences might dictate the reverse, the leaving group ability is generally I > Br > Cl.

(d) Since, unlike the S_N1 reaction, the presence of the nucleophile in the rate-determining step is required, the factors that define and govern nucleophilicity are important.

(e) The solvent must be capable solvating the nucleophile and substrate, but the need to disburse charge is less important than in the S_N1 reaction since a carbocation is not formed. The solvent itself can act as a nucleophile and might be incorporated into the product(s) (in which case the process is called *solvolysis*).

a. The Kinetics. As noted earlier (this chapter, Section E.II.a) during the 4-year period between 1933 and 1937, in a series of 10 papers, Ingold and coworkers established the kinetics and stereochemistry associated with a number of substitution reactions on a variety of substrates. Cases that fit the kinetics expected to be associated with both the S_N1 and S_N2 reactions were identified.

Consider, for example, the production of ethyl azide by the reaction of bromoethane with azide anion (N_3^-) (Table 7.5i).* A representation of the pathway over which the reactants must pass, the transition state between reactants and products, and the products themselves is provided in Figures 7.10 and 7.11.

The overall reaction of Equation 7.46 is given by the rate expression of Equation 7.47, and experimentally, this can be determined by measuring the rate at some specific temperature at which product forms (or either starting material disappears) as a function of the concentrations of the individual reactants:

$$CH_3CH_2Br + N_3^- \xrightarrow{k_0} CH_3CH_2N_3 + Br^- \tag{7.45}$$

$$\begin{aligned} \text{rate} &= -d[CH_3CH_2Br]/dt = -d[N_3^-]/dt = +d[CH_3CH_2N_3]/dt \\ &= +d[Br^-]/dt = k_o[N_3^-]^n[CH_3CH_3Br]^m, \end{aligned} \tag{7.46}$$

where the expressions in brackets are concentration terms (mol L^{-1}); k_o is the observed rate constant; and the exponents n and m are experimentally determined values (presumably "1" each) whose sum ($n + m$) is the "order" of the reaction (presumably "2" or "second-order overall").

---

*This particular example is chosen for discussion because azide (N_3^-) is an excellent nucleophile in a variety of solvents (*vide infra*) and the reaction is readily monitored by spectroscopic means. For example, the C–Br stretch for most alkyl bromides is seen as an intense absorption in the region 690–515 cm^{-1}. Thus, a plot of absorbance versus concentration having been made for the appropriate peak area, the rate at which the compound disappears (the area will diminish as the reaction progresses), will correspond to $-d$[reactant]/dt.

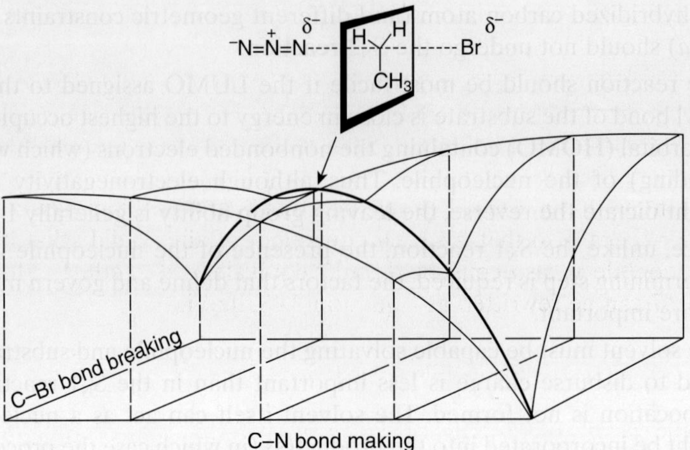

Figure 7.10. A representation of an energy diagram for the S_N2 reaction of 2-bromoethane (ethyl bromide, CH_3CH_2Br) with azide anion (N_3^-). This three-dimensional figure can be transposed into two dimensions as shown in Figure 7.11.

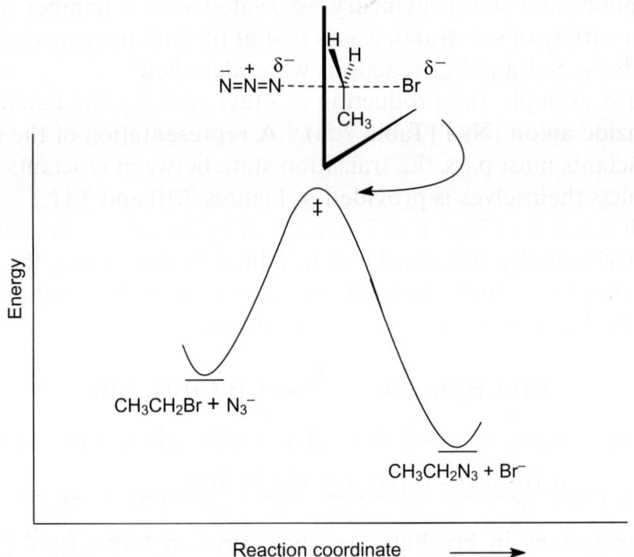

Figure 7.11. A two-dimensional representation of the energy diagram of Figure 7.10. In the example, bromoethane (ethyl bromide [CH_3CH_2Br]) undergoes reaction with azide anion (N_3^-) producing ethyl azide ($CH_3CH_2N_3$) and bromide (Br^-) anion.

Then, for the general case, and following Equation 7.46,

$$-d[E]/dt = -d[A]/dt = +d[P]/dt = \text{rate} = k_2[E]^n[A]^m, \qquad (7.47)$$

where "E" is the abbreviation for ethyl bromide (CH_3CH_2Br), "A" is the abbreviation for azide anion (N_3^-), and "P" is the abbreviation for the product (ethyl azide, $CH_3H_2N_3$). As noted earlier, it is common to determine n and m by increasing the amount of one of the two reactants to such an extent that its concentration remains "unchanged" over the course of the reaction and observing the effect on the rate of the reaction produced by changes in the concentration of the other reactant. A doubling in concentration for one of the reagents for a reaction which is first order in that reagent results in a doubling of the rate. If, however, the reaction is second order in that reagent, the rate will increase as the square of the concentration term!

Now, assuming that $n = m = 1$ (a reaction that is first order in each component is second-order overall) so that for each equivalent of alkyl halide that is consumed one of nucleophile is also consumed and 1 equivalent of product is formed, then Equation 7.47 can be rewritten as Equation 7.48, that is,

$$+dP/dt = \text{rate} = k_2(E_o - P)^1(A_o - P)^1, \tag{7.48}$$

where E_o and A_o are the initial concentrations of ethyl bromide (CH_3CH_2Br) and azide anion (N_3^-), respectively, and P is the concentration of product formed. Although it need not generally be true, it is often the case that the starting concentrations of alkyl halide and nucleophile are set the same so that

$$(E_o - P) = (A_o - P) = C \tag{7.49}$$

and

$$-dC/dt = k_2C^2 \tag{7.50}$$

and, separating the variables,

$$\int_{C_0=c}^{c=c} dC/C^2 = -k_o \int_{t_0=0}^{t=t} dt, \tag{7.51}$$

which leads to

$$1/C - 1/C_o = k_o t, \tag{7.52}$$

and a plot of $1/C$ versus t is linear, the slope being k_o. Furthermore,

$$C = (A_o - P) \text{ and } C_o = A_o. \tag{7.53}$$

Finally, multiplying the numerator and denominator by $(A_o - P)$ and rearranging,

$$P/[A(A_o - P)] = k_o t. \tag{7.54}$$

Of course, it is possible that the initial concentrations ($A_o \, mol \, L^{-1}$ of A and $E_o \, mol \, L^{-1}$ of E) might not be set equal. Then, returning to Equation 7.50, the variables can be separated

$$dP/[(A_o - P)(E_o - P)] = k_o dt, \tag{7.55}$$

which can be broken up into partial fractions

$$k_o \int dt = \int dP / [(A_o - P)(E_o - P)]$$

$$= \int dP / [(A_o - P)(E - A)] + \int dP / [(E_o - P)(A - E)] \qquad (7.56)$$

and integrated to yield

$$[\ln(A_o - P) - \ln(E_o - P)]/(A - E) = k_o t + \text{const} \qquad (7.57)$$

and, when $t = 0$, $P = 0$ and thus const $= \ln[(A/E)/(A - E)]$, so

$$[1/(A - E)]\ln[E(A_o - P)/A(E_o - P)] = k_o t. \qquad (7.58)$$

As before, it is necessary to keep in mind that as the concentration changes, the reaction mechanism may change (as the starting materials are used up, the likelihood of reactants encountering each other diminishes).

b. The Stereochemistry of the S$_N$2 Reaction. Over a period of 3 years (between 1935 and 1938), three examples of reactions between optically active alkyl halides (where the halogen-bearing carbon atom was the only reason for chirality) and metal halides (M$^+$ X$^-$) in which the halogen was a radioactive isotope of the leaving group were reported (Equations 7.59–7.61, in which the element bearing the asterisk [*] is radioactive).*

$$(7.59)$$

$$(7.60)$$

$$(7.61)$$

Consider the reaction of Equation 7.59 in which unlabeled S-(+)-2-iodooctane [CH$_3$CH(I)CH$_2$(CH$_2$)$_4$CH$_3$] is converted to the corresponding labeled R-(−)-

---

*Although, normally, bromine is a ≈1:1 mixture of ^{79}Br and ^{81}Br and with the advent of mass spectroscopy, it is possible to use enriched mixtures of one or the other isomer to solve the same problem, at the time, radioactive isomers were a reasonable choice. For example, ^{77}Br, a β- and γ-emitter, has a half-life of 57 h, while ^{82}Br has a half-life of 35.7 h. Similarly, although the natural abundance of iodine ^{127}I is 100%, ^{125}I, a γ-emitter, has a half-life of 56 days (See, Hughes, E. D., *et al.*).

enantiomer by reaction with radioactive iodide anion (e.g., $^{125}I^-$ and shown here as *I^-). In the idealized S_N2 process, each radioactive iodide that is introduced results in an inversion of configuration at the carbon atom attacked. Furthermore, in any real assembly of molecules where the number of reacting species approaches Avogadro's number, it is unlikely that a radioactively labeled and inverted product molecule will (early in the reaction) encounter another radioactive iodide anion before the anion encounters an unlabeled and uninverted reactant. Thus, each inversion results in racemization of two molecules; one of which contains radioactive iodide and is inverted and the other of which does not contain the radioactive anion and is an unreacted original. The demonstration that the initial rate of label incorporation, occurring with second-order kinetics, is equal to the rate of racemization serves to show that the S_N2 process obtains.

It is not necessary to have optically pure material upon which to carry out the reaction and, indeed, the early studies on the reactions of chiral S-(+)-2-bromooctane by Ingold and coworkers were carried out on less than pure material. However, it is necessary to know with what enantiomeric composition one begins in order to determine the extent of change.

With regard to the reactions shown in Equations 7.59–7.61, it was found that substitution with the respective radioactive nuclide did occur with second-order kinetics at rates identical with those of the respective racemization processes. It is now accepted that the products shown form in the S_N2 reaction.* An explanation for backside attack, required to obtain inverted product, in terms of a simple two-electron–two-orbital interaction diagram is given in Figure 7.12.

As shown in Figure 7.12, the back lobe of the bonding σ at the sp^3-hybridized carbon is of the appropriate sign (unshaded) for a bonding interaction with the HOMO of a nucleophile bearing nonbonded (n) electrons. However, effective bonding requires that the orbitals grow together. So, as the sp^3 bond holding the

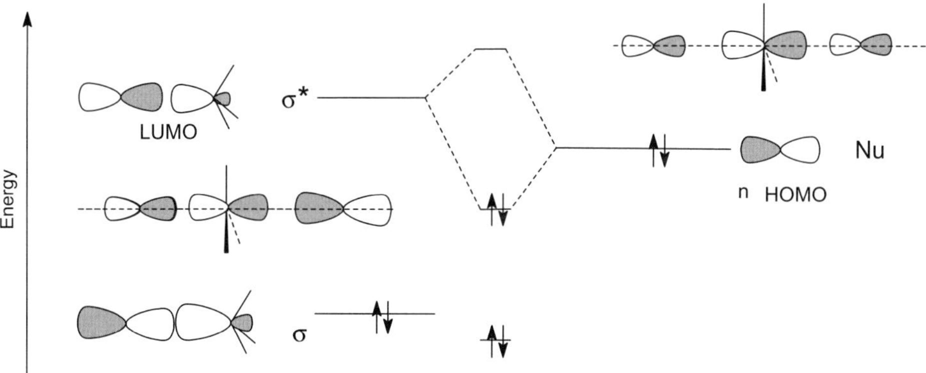

Figure 7.12. An abbreviated representation of an orbital picture of the σ-bonding orbital of the halogen-alkyl bond, the LUMO (σ*) associated with it, the HOMO of the nucleophile, and the occupied, low-lying, inversion pathway.

*As will become clear, S_N2 substitution (like S_N1 substitution) is frequently accompanied by elimination. Thus, the expected nucleophilic substitution product can be accompanied by alkene(s) as well as products of solvolysis.

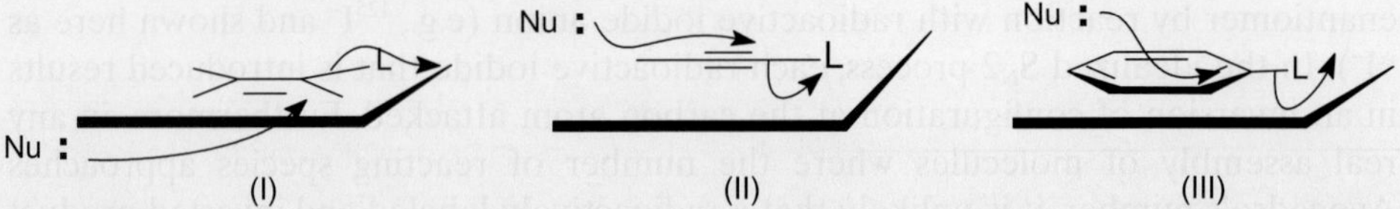

Figure 7.13. A representation depicting the difficulty expected for attack by a nucleophile opposite to the direction from which the leaving group (L) must leave in (I) alkenyl, (II) alkynyl, and (III) aryl halides (L = halogen). The drawing attempts to illustrate the encumbrance entailed in a presumed S_N2 process.

leaving group to the carbon increases in antibonding character (breaking the bond!) with the electron pair departing with the leaving group, the bonding to the nucleophile begins.

Frontside attack of the electrons in the *n*-orbital of the nucleophile (i.e., into the nodal surface of the carbon–halogen bond) is unfavorable.

Since the S_N2 reaction is characterized by "backside" attack on the substrate by the nucleophile, it is argued that halogen attached to unsaturated carbon will be less readily displaced because the incoming (electron rich) nucleophile must pass through an encumbered area. Furthermore, in addition to geometrically unfavorable constraints (Figure 7.13), the bonds to the leaving group (halide) in alkenyl, alkynyl, and aryl are generally stronger (more *s* character) than the corresponding bonds in alkyl halides and this too should retard the reaction. However, despite these constraints, nucleophiles do react with some alkenes bearing suitable leaving groups.

There appear to be at least three processes by which S_N2 substitution at unsaturated sites can occur. When there are sufficient electron-withdrawing substituents on the carbon–carbon bond (as well as the leaving group), the nucleophile can add to the double bond and, in a separate step, the leaving group can be lost. This pathway is known as *addition–elimination* and such pathways will be encountered again when they will be discussed in greater detail.

The second process appears to involve only very specific leaving groups such as the trifluoromethanesulfonate (triflate; $CF_3SO_2–$) derivatives of enols (Chapter 8) rather than halides.*

In-plane, backside, vinylic nucleophilic substitution of an appropriate halide leaving group can also be effected. Thus, the (*E*)-β-alkylvinyl(phenyl)iodonium salt, which can be prepared from the corresponding trimethylsilyl alkene and iodosobenzene, undergoes both S_N2 substitution to produce the inverted chloroalkene and elimination to the corresponding alkyne (Scheme 7.11).

c. The Nature of the Leaving Group. Although, in general, the ability of a leaving group to accommodate the electron pair that it acquires as a consequence of the displacement process is a question of stability, it is not always easy to decide

---

*As will be discussed again in Chapter 8, when the leaving group is excellent, for example, the trifluoromethanesulfonate (triflate) anion ($CF_3SO_3^-$) and the solvent is appropriate, for example, 2,2,2-trifluoroethanol (TFE), the departing ion appears to separate (but not very far) from the carbon to which it is attached (solvated ion pair) and then, one face encumbered by the departing solvated triflate, addition of solvent (to form the ether) occurs on the face opposite that of the departing leaving group. The result is inversion of configuration.

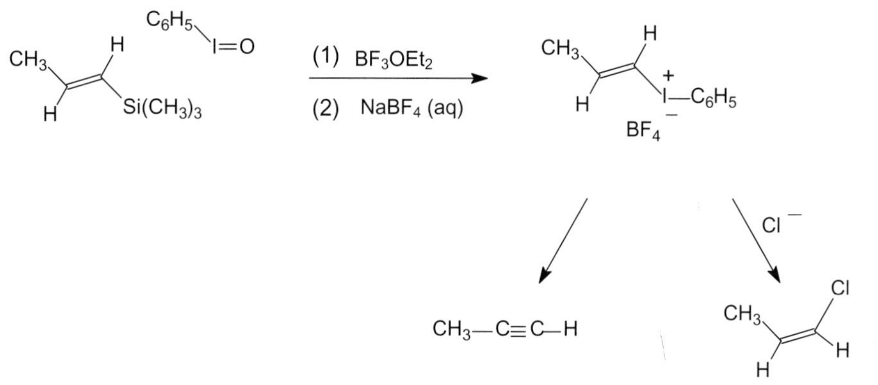

Scheme 7.11. The formation of a vinylic aryl iodonium tetrafluoroborate and its S_N2 displacement by chloride.

how to weigh the various contributing factors (such as solvation and electron delocalization). However, among the halides, it appears that leaving group ability in S_N2 reactions is about the same as it was in the S_N1 reaction (*vide supra*) and that the simplest correlation, as pointed out earlier, is with the strength of the corresponding protic acid H–L (L = halogen). Additionally, as noted above, halonium salts make excellent leaving groups and even modest nucleophiles can then be successfully utilized.

d. The Nature of the Nucleophile. The total number of electrons at the site of a nucleophilic displacement reaction remains the same as the nucleophile enters and the leaving group departs. The process of replacing the leaving group with the nucleophile may thus be considered as measuring the respective electron affinities of the incoming and outgoing ligands. Alternatively, nucleophilicity can be thought of as a measure of the ability of any species to donate an electron pair to carbon. Regrettably, both of these fairly straightforward ideas are complicated by the observation that a number of interacting, occasionally opposing effects influence the outcome of the nucleophilic displacement process.

For example, in protic solvents such as water (H_2O) and water–ethanol (ethyl alcohol, CH_3CH_2OH) mixtures, large ions that are more polarizable and may more readily provide an electron pair for the displacement process are also more poorly solvated than small ions. Thus, with the same substrate (e.g., bromoethane [ethyl bromide, CH_3CH_2Br]) as shown in Table 7.5, nucleophilicity increases with increasing atomic number in any one group (column) of the periodic table (i.e., $I^- > Br^- > Cl^- > F^-$) in such solvents. In this vein, it should be noted that this means that nucleophilicity does not parallel basicity since, as noted earlier, the order of basicity in aqueous solution is $Cl^- > Br^- > I^-$.

However, a change in solvent to one which is polar, yet aprotic, such as methylsulfinylmethane (DMSO, $(CH_3)_2SO$) dramatically improves the nucleophilicity of the smaller anions. The improved nucleophilicity of, for example, Cl^-, is apparently the consequence of better solvation of the cation (that originally accompanied the anion), which, in turn, has the result of leaving the nucleophilic anion (relatively) unencumbered and ready for reaction. In concert with the above observations, it is important to note that (a) sulfur lies below oxygen in the periodic table and (b) HS^- is a weaker base than HO^-, so it would be correctly predicted that with regard to its ability to serve as a nucleophile that $HS^- > HO^-$ (Table 7.5b). In the same vein, a phosphine (R_3P) (Table 7.5j) is a better nucleophile than an amine ($R_3N:$) (Table 7.5h).

Interestingly, nucleophiles in which two adjoined atoms have nonbonded electrons are particularly reactive as nucleophiles. Thus, a nucleophile, such as the peroxide anion (HOO^-) (Table 7.5a), is about 102 times more reactive than the corresponding hydroxide anion (HO^-) and the azide anion ($^-N=N^+=N^-$) (Table 7.5i) is particularly effective. In the same vein, bidentate (literally "two toothed") anions such as thiocyanate (isothiocyanate) ($^-S–C\equiv N \leftrightarrow S=C=N^-$) (Table 7.5c) and cyanate (isocyanate) ($^-O–C\equiv N \leftrightarrow O=C=N^-$) (Table 7.5d) are highly reactive with the former yielding mostly thiocyanate ($S > N$) and the latter mostly isocyanate ($N > O$) product.

e. The Nature of the Solvent.

As already noted, the S_N1 reaction is facilitated by solvents that have relatively high dielectric constants (which increase the ionizing ability of the substrate) and solvents that are able to ameliorate the localized charges by surrounding the separated ions and delocalizing the charge over a wide area (solvation). Solvents such as water (H_2O) and alcohols ($R–OH$) are particularly efficacious.

While the S_N2 reaction does not require the solvent to facilitate ionization of the substrate, it does require that cations associated with nucleophilic anions separate from them and, additionally, that the solvent only poorly solvate the electron-rich nucleophile. Generally, solvents that have protons attached to electronegative atoms (i.e., protic solvents) solvate electron-rich species (through proton donation) particularly well and thus are not good in promoting the S_N2 reaction. However, solvents such as methylsulfinylmethane (DMSO [$(CH_3)_2SO$]), N,N-dimethylmethamide (N,N-dimethylformamide [$(CH_3)_2NCOH$]), ethanenitrile (acetonitrile [$CH_3C\equiv N$]), propanone (acetone [$(CH_3)_2C=O$]), and nitromethane (CH_3NO_2), which have relative high dielectric constants ($\varepsilon = 49, 37, 38, 21,$ and $41\,D$ [at $25°C$], respectively), solvate cations well and anions poorly. These dipolar aprotic solvents thus serve to facilitate the S_N2 reaction.

IV. The S_N2' Reaction

When there is unsaturation α,β to the carbon bearing the halogen undergoing substitution (e.g., in allylic halides) both S_N1 (*vide supra*) and S_N2 reactions can occur. The normal S_N1 reaction produces a carbocation or carbocations, which can lead to a set of products different from those found in the S_N2 reaction (see Problem 7.5).

The normal S_N2 reaction is accompanied by a specific rearrangement involving the attack of the nucleophile at the more remote terminus of the carbon–carbon double bond and loss of the leaving group, accompanied by synchronous (or nearly so) migration of the double bond. This process, shown in Scheme 7.12, is known as the S_N2' reaction. As also shown in Scheme 7.12, the attack of the nucleophile is preferentially onto the same face of the unsaturated system (i.e., *syn* or *cis* or suprafacial) from which the leaving group departs. The specificity of the attack limits the number of possible compounds that might be produced. (It has been suggested that there may be no S_N2' reaction! See Bordwell and Mecca, 1972.)

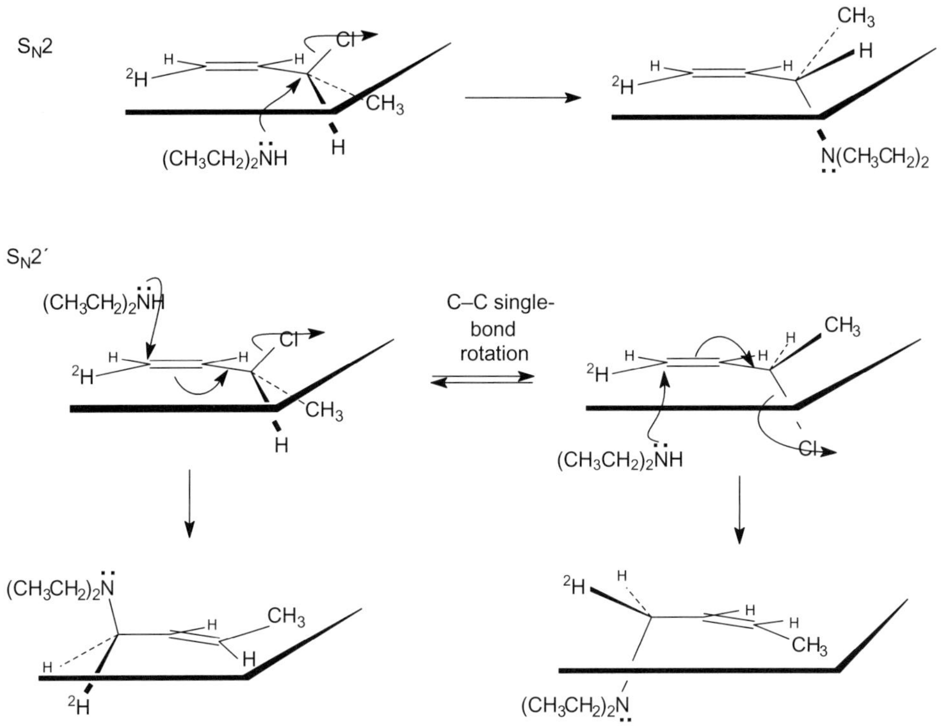

Scheme 7.12. An illustration of the S_N2 and S_N2' reaction of *N,N*-diethylamine (*N*-ethylethanamine [$(CH_3CH_2)_2NH$]) with (*Z*)-3*R*-1-[$^2H_1$]-3-chloro-1-butene. The former yields product of inverted configuration but with retained geometry about the carbon–carbon double bond and the isotopic label remaining at C-1. The latter leads to two products with the double bond between C-2 and C-3 and the isotopic label attached to the carbon-bearing nitrogen. The two products arise from the C–C single-bond rotational isomers of the starting alkene and suprafacial substitution.

Problem 7.5. (*Z*)-3*R*-1-[$^2H_1$]-3-chloro-1-butene undergoes S_N2 and S_N2' reactions with formation of the specific products shown in Scheme 7.12, thus helping to define those reaction pathways. Interestingly, at least in principle, (*Z*)-*R*-1-[$^2H_1$]-3-chloro-1-butene is also capable of undergoing the S_N1 reaction. Write the products you believe should form as a consequence of the S_N1 reaction of (*Z*)-3*R*-1-[$^2H_1$]-3-chloro-1-butene using *N,N*-diethylamine (*N*-ethylethanamine [$(CH_3CH_2)_2NH$]) as the nucleophile. Note all products that are different from those arising from the S_N2 and S_N2' reactions.

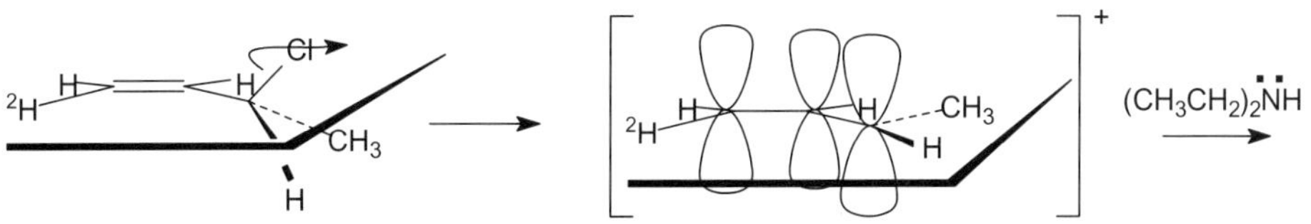

V. Nucleophilic Aromatic Substitution

Halogens on aromatic rings can be replaced by a variety of other substituents. There are two general pathways (and a few more which are rare and will not be considered here) by which the substitution can be effected. The first involves loss of halogen and a proton on the adjacent carbon (an *elimination* reaction, *vide infra*) to produce an intermediate (called *benzyne*), which subsequently *adds* a nucleophile. The overall result is substitution.

The second pathway generates an (occasionally isolable) intermediate but requires electron attracting ("withdrawing") substituents in positions *ortho* and/or *para* to the carbon bearing the halogen to be replaced. These electron attracting groups make the carbon bearing the halogen susceptible to attack by the incoming nucleophile.

a. The Elimination-Addition Pathway (Benzyne).

The fusion of organic halides with caustic soda (sodium hydroxide [NaOH]) was among the earliest reactions examined in organic chemistry. It was clear that the halogen could be removed and a hydrocarbon recovered (for a more complete discussion of this elimination process, see this chapter, Section G). When this reaction is performed on alkyl-substituted aryl halides, a mixture of phenols results. For example, as shown in Equation 7.62, heating *para*-chlorotoluene (4-chloro-1-methylbenzene) with molten sodium hydroxide (NaOH) produces a 1:1 mixture of 3- and 4-methylphenols (3- and 4-methylbenzenols, respectively).

$$\text{(7.62)}$$

It subsequently became clear that the substitution reaction could be carried out even if no other substituents were present on the aromatic ring and that even less vigorous conditions could be used. As shown in Equation 7.63, bromobenzene (C_6H_5Br) on treatment with lithium diethylamide (*N,N*-diethylaminolithium [$LiN(CH_2CH_3)_2$])* in *N,N*-diethylamine (*N*-ethyl-ethanamine [$(CH_3CH_2)_2NH$]) solvent results in *substitution* of the *N,N*-diethylamino group for the bromine.

$$\text{(7.63)}$$

*Lithium salts of organic amines, powerful organic bases, whose corresponding amines have pKa's > 30, are commonly used for proton abstraction from carbon. Some of these, such as lithium diisopropylamide (LDA, [$(CH_3)_2CH]_2NLi$) have become commercially available and their use in other connections will be discussed subsequently. In addition, there is a nomenclature problem. Metal salts of amines are called "amides." The amine derivatives of carboxylic acids in which there is a carbonyl-to-nitrogen bond, that is, O=C–N, are also called "amides." The remainder of the name and/or the context in which the terms are used will suffice to distinguish which is meant. However, formal resolution of the ambiguity has not yet been promulgated.

In a further development, it was found (Equation 7.64) that substituents already present could affect the outcome of the reaction either by electron donation (making some positions less likely to be attacked by a nucleophile), electron withdrawal, and/or exertion of a steric effect to block the incoming species.

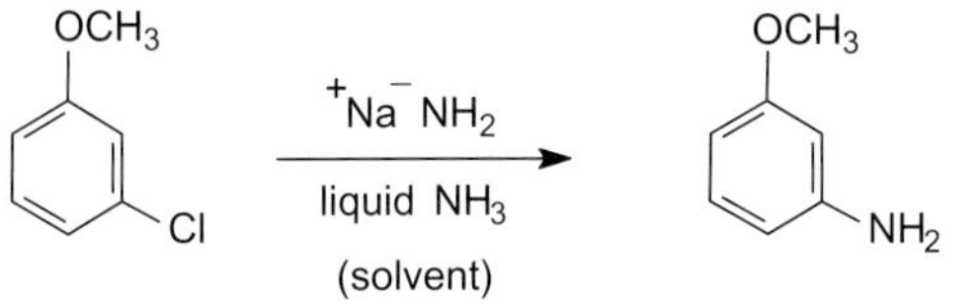

$$\text{(7.64)}$$

Then, using radiolabeled (^{14}C) bromobenzene (C_6H_5Br), it was shown that the aniline (aminobenzene, $C_6H_5NH_2$) resulting from the reaction with sodium amide in liquid ammonia was a 1:1 mixture of ipso- and ortho-[$^{14}C_1$]-anilines (aminobenzenes) (Equation 7.65).

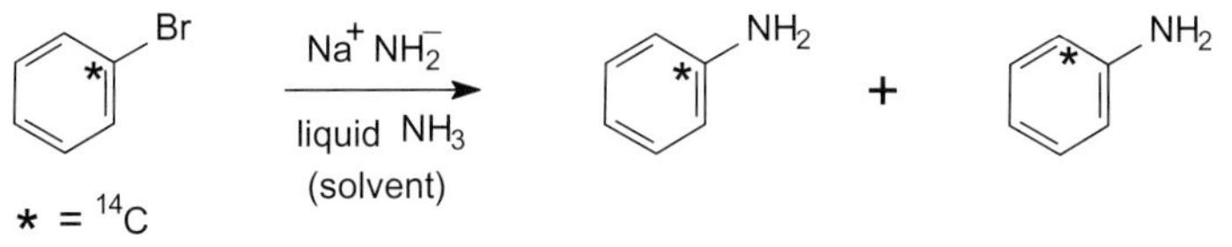

$$\text{(7.65)}$$

A pathway that accounts for the observations illustrated in Equations 7.62–7.65 is provided in Scheme 7.12. It is argued that hydrogen halide is lost (in an elimination reaction) to provide an intermediate, which is written with a carbon–carbon triple bond in the six-membered ring. Given the observation that stable compounds do not result from attempted introduction of a triple bond into a six-membered ring (but lacking other unsaturation, e.g., cyclohexyne), it is not surprising that the intermediate (called *benzyne*) is not isolable. As an intermediate, it will lie in a minimum on the reaction pathway surface and, both termini being identical, might be expected to react at either terminus so that the incoming nucleophile need not be at the carbon atom from which the leaving group left. Reasonably, substituents already present could also exert some influence on the incoming nucleophile since the termini would then not be identical (Scheme 7.13).

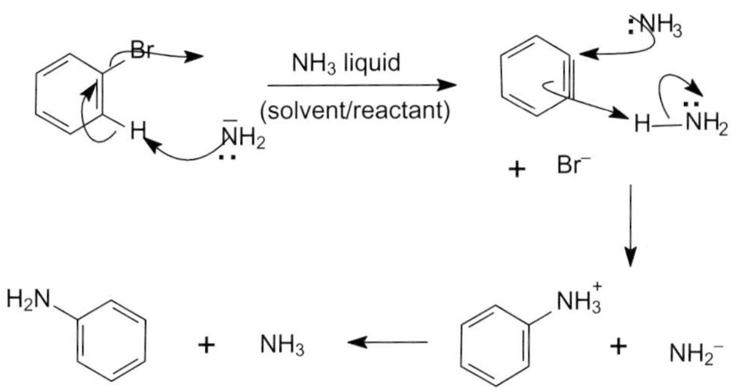

Scheme 7.13. A representation of the conversion of bromobenzene (C_6H_5Br) to aniline (benzeneamine, $C_6H_5NH_2$]) by treatment of the former with sodium amide (Na$^+$ NH$_2^-$) in liquid ammonia (NH$_3$) solvent. Benzyne is shown as an intermediate.

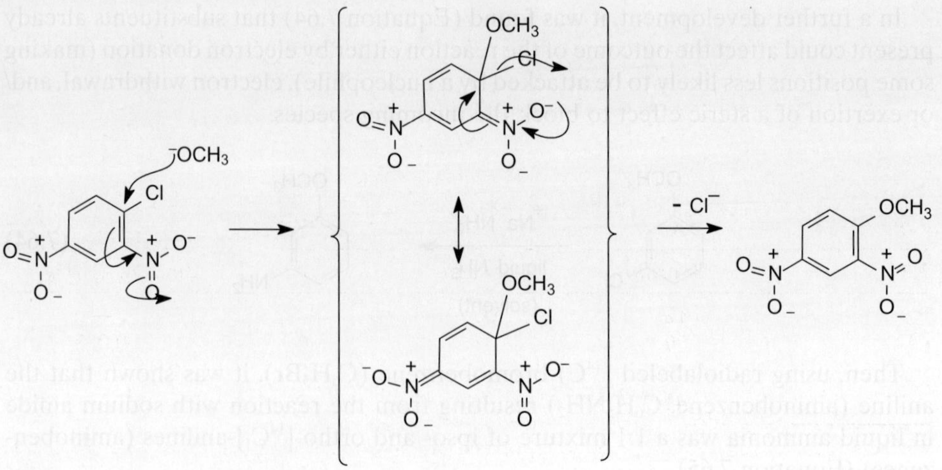

Scheme 7.14. A representation of an aromatic nucleophilic substitution reaction (S_NAr) involving the intermediacy of a *Meisenheimer complex*. The curved arrows on the one resonance structure show a way that chloride ion might be lost.

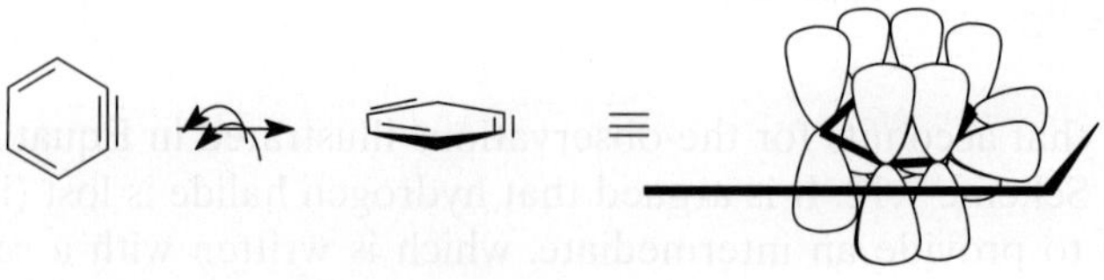

Representation of benzyne

b. The Addition-Elimination Pathway (S_NAr Substitution).

When the aromatic halide bears electron-withdrawing substituents (e.g., nitro [$-NO_2$] groups), nucleophilic substitution of halogen can be effected. The reaction occurs with, overall, second-order kinetics (first order in aromatic substrate and first order in nucleophile). The reaction was found early in the twentieth century by Meisenheimer,* and the (frequently) isolable complexes formed between nucleophile and substrate are referred to as *Meisenheimer complexes* (Scheme 7.14).

Interestingly, kinetic studies have shown that, depending on the nucleophile, the leaving group and even the solvent, *either* the first step (the addition of the nucleophile) or the second step (the loss of the leaving group) can be rate determining.

VI. Electrophilic Aromatic Substitution

Electrophilic aromatic substitution can be effected on aryl halides. In contrast to nucleophilic substitution (*vide supra*), electrophilic substitution does not generally result in substitution of the electrophilic reagent for the halogen. Instead, the incoming electrophile is substituted for a proton on the aromatic ring and, as shown in Table 7.6 for the nitration of halobenzenes, the substitution occurs largely *ortho* and *para* to the halogen.

---

*Jacob Meisenheimer (1876–1938) was professor of chemistry at the Universities of Berlin and Tübingen.

TABLE 7.6. The Products (and Their Relative Yields) Obtained on Nitration of the Monohalobenzenes with a Mixture of Nitric and Sulfuric Acids

X =	% ortho	% meta	% para	Relative Rate Benzene $(C_6H_6) = 1$
F	12	0	87	—
Cl	29	0	68	0.033
Br	36.5	1.2	62.4	0.030
I	41	—	59	—

The relative rates for chlorobenzene (C_6H_5Cl) and bromobenzene (C_6H_5Br) are relative to benzene (C_6H_6) itself under the same conditions. (After Bird, M. L.; Ingold, C. K. *J. Chem. Soc.*, **1936**, 918. See also Roberts, J. D.; Sanford, J. K.; Sixma, F. L. J.; Cerfontain, J.; Zagt, R. *J. Am. Chem. Soc.*, **1954**, 76, 4525.)

Furthermore, also as shown in Table 7.6, the rate of product formation is diminished relative to benzene (C_6H_6). From this, it is clear that, relative to hydrogen, halogen substituents deactivate the ring toward electrophilic substitution. This is in contrast to what was seen for the alkyl substituents (Chapter 5), which, it will be recalled, while also *ortho-* and *para*-directing, were activating toward further electrophilic substitution.

It is sometimes argued that the *ortho/para*-directing influence can be separated from the deactivating influence. The latter presumably arises as a consequence of the electronegative halogen's ability to withdraw electrons from the ring and to thus discourage the approach of *any* electrophile to *any* position on the ring. Furthermore, the deactivating effect is most pronounced at the positions meta to the halogen since the intermediates that would be formed do not enjoy any stabilization by electron delocalization of the nonbonded (p-type) electrons on the halogen (Figure 7.14). Indeed, it is just the stabilization seen in the intermediates for *ortho-* and *para*-substitution, which serve to suggest that the transition states leading to those intermediates will favor them.

Finally, in this vein, and as noted earlier for electrophilic substitution in alkyl-substituted arenes, a variety of substitutions can be effected. Among these is the Friedel–Crafts-type acylation of chlorobenzene (C_6H_5Cl) by 2,2,2-trichloroethanal (trichloroacetaldehyde, chloral $[CCl_3CHO]$), which, as shown in Scheme 7.15, results in the formation of 1,1,1-trichloro-2-[di(*p*-chlorophenyl)]ethane (dichlorodiphenyltrichloroethane [DDT]).

VII. Substitution by Carbon

The formation of new carbon–carbon bonds through the coupling of organometallic reagents (this chapter, Section D, Part 1, d) is commonly regarded (and was discussed in that vein) as a series of one-electron transfer processes resulting in radicals, which subsequently join together. However, the specificity (both stereochemical and regiochemical) that attends some of these processes suggests that complexation

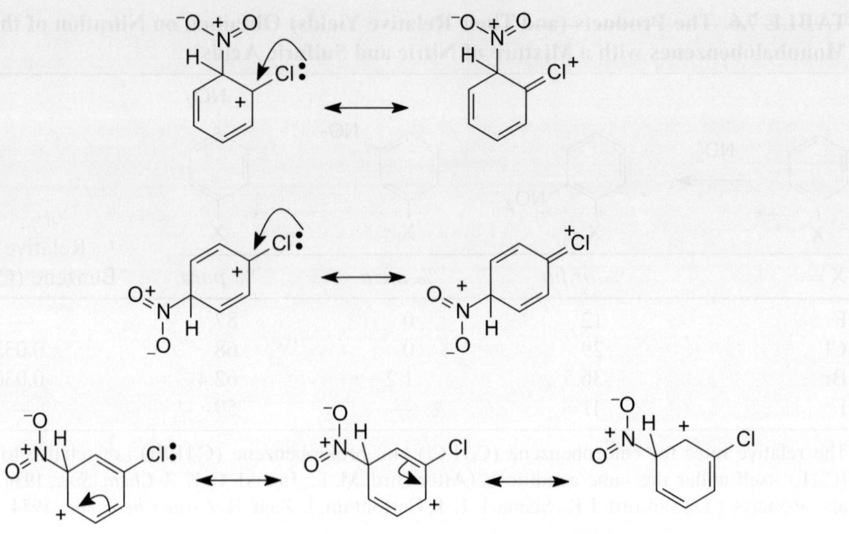

Figure 7.14. Representations of resonance forms for *ortho-*, *para-*, and *meta*-intermediates anticipated for the nitration (with NO_2^+) of chlorobenzene (C_6H_5Cl). For both *ortho-* and *para*-nitration, nonbonded electrons associated with chlorine are capable of participation in stabilization of the positive charge produced in the aromatic ring as a consequence of the attack of the electrophile. Similar stabilization is not available for *meta*-nitration.

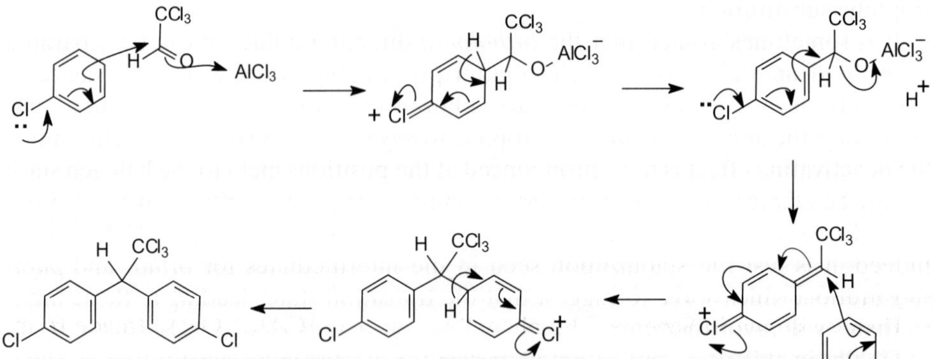

Scheme 7.15. A representation of a possible pathway for the production of 1,1,1-trichloro-2-[di(*p*-chlorophenyl)]ethane (dichlorodiphenyltrichloroethane [DDT]) by the Friedel–Crafts aluminum trichloride-catalyzed reaction of chlorobenzene (C_6H_5Cl) with chloral (2,2,2-trichhloroethanal, Cl_3CCHO).

with metals such as palladium (Pd) may preclude the formation of "free" radicals and that reaction occurs directly from the complexes. The same idea was already encountered in the brief discussion of the Heck reaction (Chapter 6, Scheme 6.54) and the mention of the *Suzuki* reaction (this chapter, Scheme 7.6). Thus, these processes may more closely resemble substitution reactions.

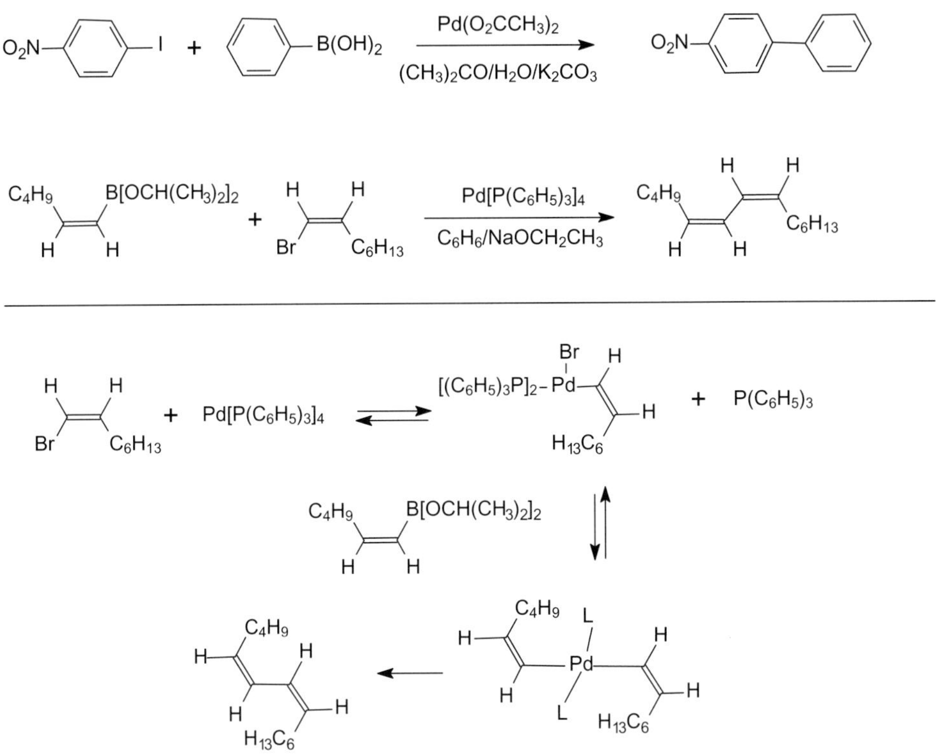

Scheme 7.16. Some examples of the *Suzuki reaction*. The process permits creation of carbon–carbon bonds by coupling of alkyl, alkenyl, and aryl halides (I > Br >> Cl) with organoboron compounds. Akira Suzuki, Ei-ichi Negishi and Richard Heck shared the Chemistry Nobel Prize in 2010.

The Suzuki reaction permits the palladium (Pd)-catalyzed substitution of a halogen (I > Br >> Cl) attached to an alkyl, alkenyl, or aryl group by a second alkyl, alkenyl, or aryl group, originally attached to boron and with *retention of stereochemistry* (see Miyaura et al., 1979).

Thus, as shown in Scheme 7.16, phenylboronic acid [$C_6H_5B(OH)_2$] (made from, e.g., the reaction between phenyl magnesium bromide [$C_6H_5Br + Mg \rightarrow C_6H_5MgBr$] and trimethylborate [$B(OCH_3)_3$] followed by hydrolysis) undergoes reaction with p-nitroiodobenzene in aqueous acetone containing some potassium carbonate and in the presence of palladium acetate [$(CH_3CO_2)_2Pd$] to yield p-nitrobiphenyl (4-nitrobiphenyl [p-C_6H_5-C_6H_4-NO_2]) in about 85% yield. Furthermore, demonstrating retention of stereochemistry at the carbon–carbon double bond, (Z)-1-hexenyldiisopropylborate reacts with (Z)-1-bromooctene in the presence of tetrakis(triphenylphosphine)palladium(0) ($Pd[P(C_6H_5)_3]_4$) and sodium ethoxide and in benzene (C_6H_6) solvent to give (5Z,7Z)-tetradeca-5,7-diene in >99% yield. These reactions and a possible pathway that might account for the products are shown (Scheme 7.16).

Processes similar to the Suzuki coupling reaction have been observed with alkynes. Thus, the *Stephens–Castro reaction* between unsaturated iodides (aryl, vinyl) and copper(I) acetylides in pyridine solvent results in the formation of new

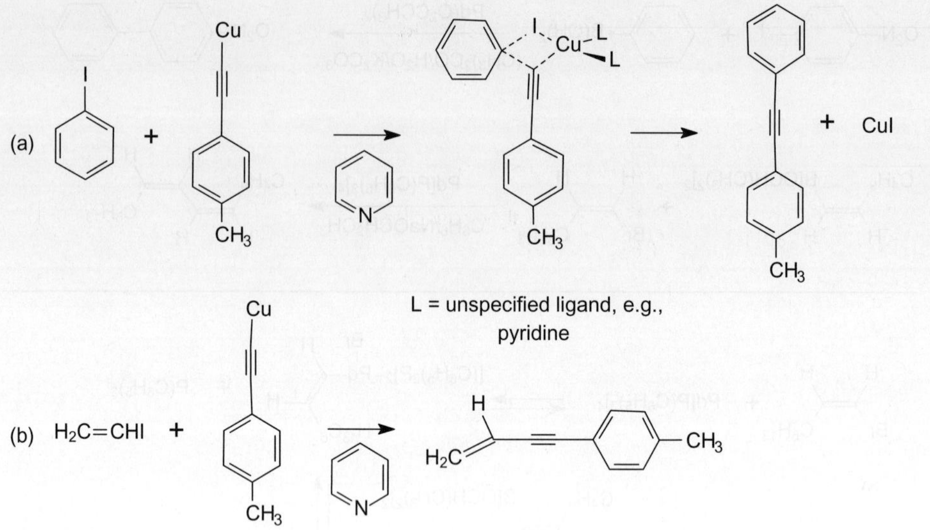

L = unspecified ligand, e.g., pyridine

Scheme 7.17. Examples of the *Stephens–Castro* coupling reaction. In reaction (a), coupling between an aryl iodide (iodobenzene) and the organometallic (copper) alkyne [copper(*p*-methylphenyl)ethyne] produces the diarylacetylene. In reaction (b), the same process, this time starting with a vinyliodide (iodoethene), generates a vinylacetylene [4-(*p*-methylphenyl)-1-buten-3-yne].

carbon–carbon bonds and yields, respectively, diarylalkynes and vinylacetylenes (Scheme 7.17) (see Stephens and Castro, 1963).

Some years later, Sonogashira and coworkers (1975) showed that both bromo- and iodoarenes and alkenes could be brought into reaction with terminal alkynes (in a direct extension of the *Suzuki* and *Stephens–Castro* reactions) in the presence of catalytic quantities of palladium(II) (e.g., bis(triphenylphosphine)palladium dichloride, $([C_6H_5]_3P)_2PdCl_2$, copper(I), e.g., cuprous iodide [CuI]), salts, and a primary (e.g., *n*-butylamine [$CH_3CH_2CH_2CH_2NH_2$]) or secondary (e.g., diethyl-amine [$(CH_3CH_2)_2NH$]) amine (Chapter 10). The details (Scheme 7.18) of this mild process (which are beyond the scope of this work) are thought to resemble the *Heck* (Chapter 6, Scheme 6.54) reaction in that a π-cloud-metal complex may play an intimate role.

Finally, there is the *Stille* reaction (Scheme 7.19). Here, vinyl-substituted orga-nostannanes (e.g., trinbutylvinylstannane [$(Bu)_3SnCH=CH_2$]) prepared from tri-*n*-butyltin chloride and the corresponding Grignard reagent and unsaturated halides (or sulfonates, Chapter 8) couple in the presence of a palladium catalyst (e.g., tris(dibenzylideneacetone)palladium [$Pd_2(dba)_3$]) to forge new carbon–carbon bonds.

VIII. Photochemically Induced Substitution of Vinyl and Aryl Halides

Irradiation of both aryl and vinyl halides (bromides are commonly used) induces homolysis of the carbon–halogen bond and produces aryl radicals from aryl halides and (apparently) both vinyl radicals and vinyl cations from vinyl halides.

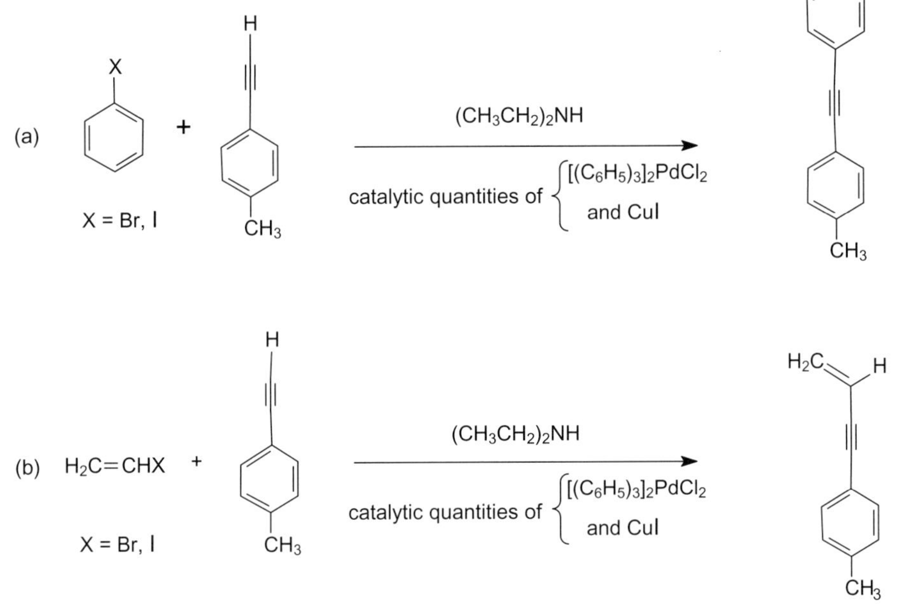

Scheme 7.18. Examples of the *Sonogashira* coupling reaction. In reaction (a), an aryl halide (X = Br, I) is coupled to the terminus of the (*p*-methylphenyl)ethyne. In reaction (b), the coupling is to a haloethene, producing a substituted vinylacetylene derivative. In both cases, catalytic (not stoichiometric) quantities of palladium and copper catalysts are used.

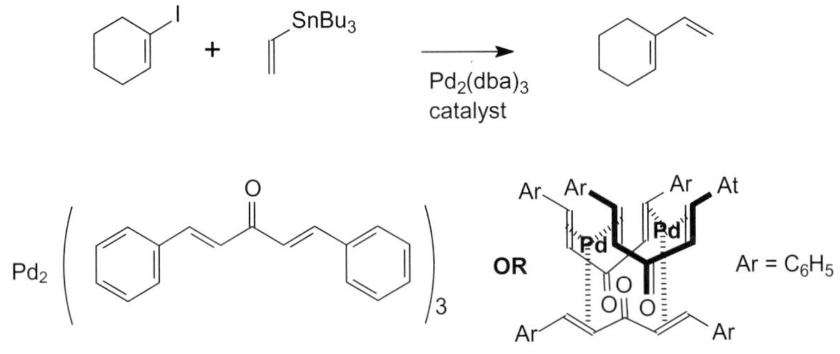

tris(dibenzylideneacetone)dipalladium(0)

Scheme 7.19. An example of the Stille coupling reaction (see, e.g., Stille, J. K. *Angew. Chem. Int. Ed. Engl.*, **1986**, 25, 508).

Thus, photolysis of 1,2,2-triphenylbromoethene at 350 nm in ethanol (CH_3CH_2OH) yields products from both substitution (from vinyl cation) and reduction (from vinyl radical). These products are formed under conditions where, in the absence of light, no substitution occurs.

Although details are obscure, it is presumed that weakening of the carbon–bromine bond ($n \rightarrow \pi^*$ absorption) precedes heterolysis to a stabilized vinyl cation and bromide anion (Br^-). The cation then reacts with ethanol (CH_3CH_2OH) and a

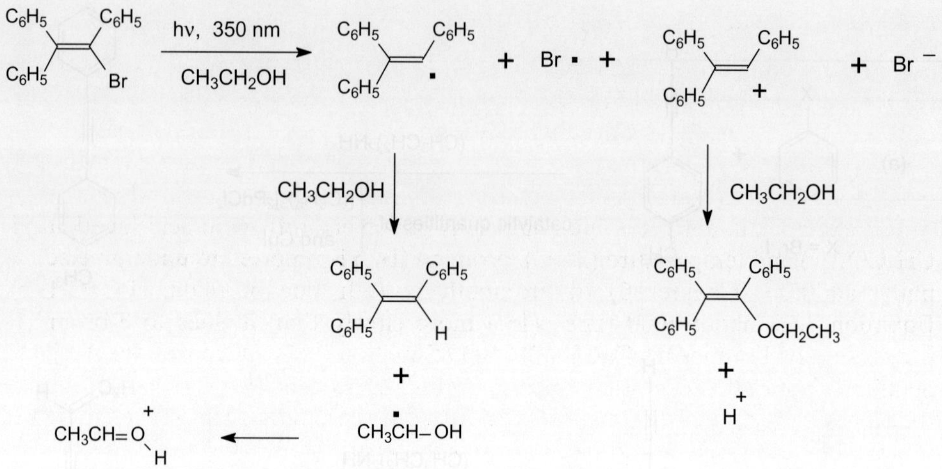

Scheme 7.20. Photochemically initiated substitution reactions of vinyl halides.

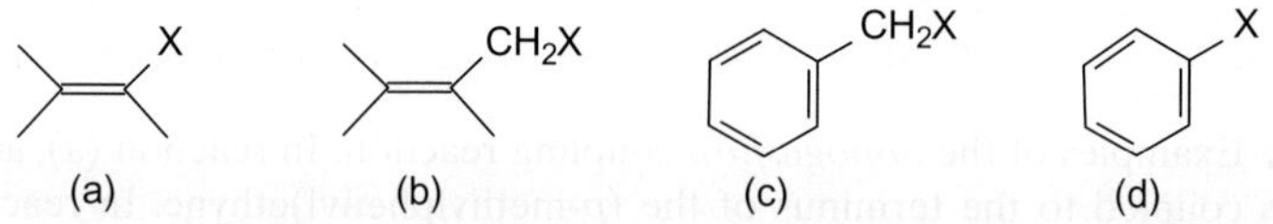

Figure 7.15. Representations of (a) vinyl, (b) allyl, (c) benzyl, and (d) aryl halides. In these compounds, the halogen affects addition reactions to the site(s) of unsaturation. X = halogen (F, Cl, Br, I).

proton (H^+) is lost. Alternatively, σ-bond cleavage to a bromine atom (Br•) and a vinyl radical is followed by hydrogen abstraction by the vinyl radical from ethanol (CH_3CH_2OH) (which is then oxidized to ethanal [acetaldehyde, CH_3CHO] while the bromine atom [Br•] is reduced to bromide ion [Br^-]) (Scheme 7.20).

F. ADDITION REACTIONS

Addition reactions to alkyl and aryl halides fall into two broad categories. First, there are those alkyl (alkenyl and alkynyl) and aryl halides in which the carbon bearing the halogen is isolated from the unsaturation by at least one methylene ($-CH_2-$) or methine ($-CH$) unit. In these compounds, it is generally found that the presence of the halogen is irrelevant to reaction at the site of unsaturation.* Clearly, none of vinyl (Figure 7.15a), allyl (Figure 7.15b), benzyl (Figure 7.15c), and phenyl (Figure 7.15d) halides fit this group.[†]

---

*There are exceptions to this generalization. Some of the exceptions are discussed later in this chapter under the heading of Rearrangements and, as is usually the case for "exceptions," they add flavor and spice to the study of what might otherwise be routine.

[†]Very few alkynyl halides (i.e., $-C{\equiv}C-X$; X = halogen [F, Cl, Br, I]) have been prepared and thoroughly characterized. They have not been investigated as fully as those other groups discussed here.

I. Addition Reactions to Vinyl Halides

The general observation that replacement of a hydrogen or carbon attached to the carbon–carbon double bond by halogen results in retardation of electrophilic addition reactions is justified by arguing that there is decreased electron density in the double bond. That is, halogen inductively withdraws electrons better than hydrogen or carbon (electronegativity, Chapter 1).

For example, under the same conditions of solvent (ethanoic acid [acetic acid, CH_3CO_2H]) and temperature (25°C), bromine (Br_2) is reported to undergo electrophilic addition (Chapter 6) to bromoethene (ethylene bromide, $H_2C=CHBr$), Equation 7.66, about 1000 times (10^3) more slowly than it does to 3-bromo-1-propene (allyl bromide, $H_2C=CH–CH_2Br$) (Equation 7.67). Of course, for the justification to apply, it is necessary to assume that the path of electrophilic addition is the same for both and that the only difference is the nearness (or connectivity) of the bromine to the double bond. Interestingly, however, other factors intrude and it is clear from appropriate labeling experiments that the carbocation expected to be present as an intermediate in 3-bromo-1-propene (allyl bromide, $H_2C=CH–CH_2Br$]) is stabilized by the adjacent bromine! The stabilization results in lower energy and, presumably, a more readily surmounted transition state (cf. the Hammond postulate; see Figure 6.8 and Scheme 6.29) leading to that intermediate (Scheme 7.21).

$$H_2C=CHBr + Br_2 \xrightarrow[\text{(solvent)}]{CH_3CO_2H} BrCH_2—CHBr_2 \qquad (7.66)$$

$$H_2C=C(CH_2Br)H + Br_2 \xrightarrow[\text{(solvent)}]{CH_3CO_2H} BrCH_2—C(Br)(CH_2Br)H \qquad (7.67)$$

Interestingly, bromination under radical conditions appears to occur with about equal facility to bromoethene (ethylene bromide [$H_2C=CHBr$]) and 3-bromo-1-propene (allyl bromide [$H_2C=CH–CH_2Br$]).

Problem 7.6. Provide a justification for the observation that vinyl (C=C–X; X=Cl, Br) and allyl (C=C–C–X; X = Cl, Br) halides undergo free radical addition reactions to the carbon–carbon double bond with about equal facility.

As pointed out earlier (Chapter 6, Section 2), nonsymmetrical alkenes undergo addition of nonsymmetrical addends with regiochemical preference dictated by carbocation stability.* Thus, as shown in Scheme 7.22, chloroethene (ethylene chloride [$H_2C=CHCl$]) reacts with hydrogen chloride (HCl) to produce 1,1-dichloroethane (H_3CCHCl_2) rather than the isomeric 1,2-dichloroethane (H_2CClCH_2Cl) because

---

*Examples of alkenes bearing halogen on one of the carbons of the double bond were among the first studied (by V. Markownikoff, 1870) and gave rise to the Markownikoff rule (Chapter 6, Part 2), *viz.* the hydrogen of H–X attaches to the carbon with the fewer nonhydrogen substituents and the X to the carbon with the larger number of nonhydrogen substituents. As pointed out earlier, the rule is now justified by noting that differences in carbocation stability produce the required result.

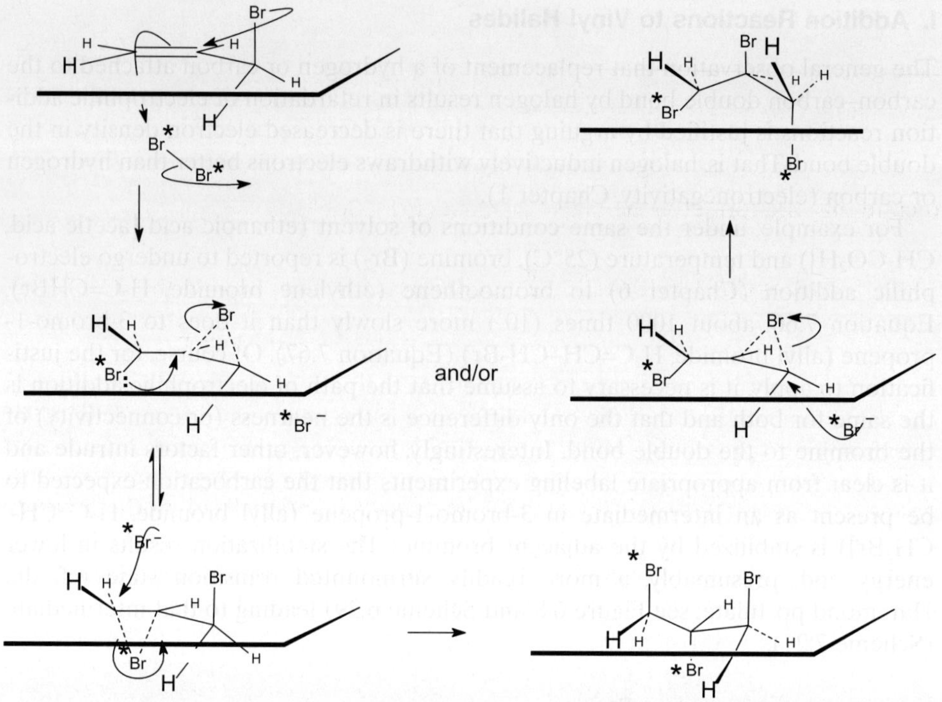

Scheme 7.21. A representation of the addition of radiolabeled bromine (indicated as *Br$_2$) to 3-bromo-1-propene (allyl bromide, H$_2$C=CHCH$_2$Br). It is important to note that the initially formed bromonium ion is capable of being "opened" by internal (backside) attack from the bromine atom on the neighboring carbon. This is an example of the process called neighboring group participation (this chapter, Rearrangement). In this particular case, the newly formed bromonium ion is, except for the position of the label (*Br), the mirror image of the bromonium ion from which it was generated. Each of these bromonium ions is capable of being attacked by exogenous labeled bromide anion. Thus, depending on the specific bromonium ion, the final relationship of the two isotopically unique bromine atoms in the product (1,2,3-tribromopropane) will be either 1,2 or 1,3.

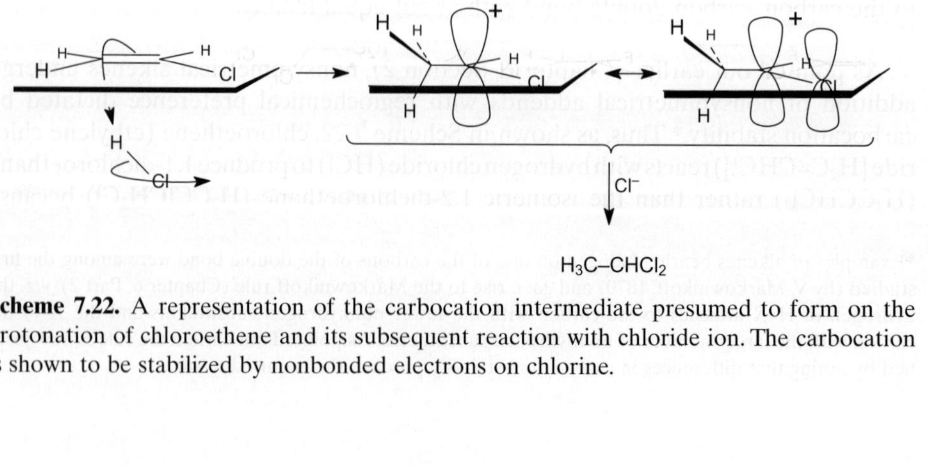

$$H_3C-CHCl_2$$

Scheme 7.22. A representation of the carbocation intermediate presumed to form on the protonation of chloroethene and its subsequent reaction with chloride ion. The carbocation is shown to be stabilized by nonbonded electrons on chlorine.

the chlorine stabilizes the carbocation through electron donation of its nonbonded electrons.

Reaction with hydrogen bromide (HBr), which, unlike hydrogen chloride (HCl) can undergo both free radical and ionic addition reactions, produces both 1-bromo-2-chloroethane ($BrCH_2CH_2Cl$) and 1-bromo-1-chloroethane ($CH_3CHBrCl$). Under free radical conditions (i.e., in the presence of peroxide catalyst), steric effects dictate that the former is the major product (Equation 7.68).

$$H_2C{=}CHCl + HBr \xrightarrow{\text{peroxide}} BrCH_2{-}CH_2Cl + CH_3{-}CHBrCl \qquad (7.68)$$

$$\text{major} \qquad\qquad \text{minor}$$

Although it is true (Chapters 4 and 6) that concerted, thermal $\pi2s + \pi2s$ processes are forbidden and $\pi2s + \pi2a$, while allowed, are not generally observed as they are sterically difficult, alkenes bearing fluorine on the double bond often dimerize to cyclobutanes (Scheme 7.23). The result is rationalized on the grounds that the observed lack of stereospecificity means that orbital symmetry considerations can be ignored. Indeed, the rules of orbital symmetry may be violated (or circumvented) if the process resulting in dimer formation is not concerted. Thus, both stepwise radical and ionic reactions can, in principle, be used to produce cyclobutane-containing products. When dealing with simple alkenes, these processes are rarely used since four-membered rings are formed only with difficulty (Chapter 4) and other reactions of the required ionic and/or radical intermediates intrude. However, as shown in Scheme 7.23, 1-chloro-1,2,2-trifluoroethene undergoes nonstereospecific dimerization to (*E*)- and (*Z*)-isomers of 1,2-dichloro-1,2,3,3,4,4-hexafluorocyclobutane. Presumably, this thermal process involves a biradical intermediate.

Interestingly, a similarly specific cyclization, which may also be occurring by a radical pathway, has been reported with the fluoroalkyne 1-fluoro-3,3-dimethyl-1-butyne (*t*-butylfluoroacetylene [$FC{\equiv}CC(CH_3)_3$]). As shown in Scheme 7.24, this

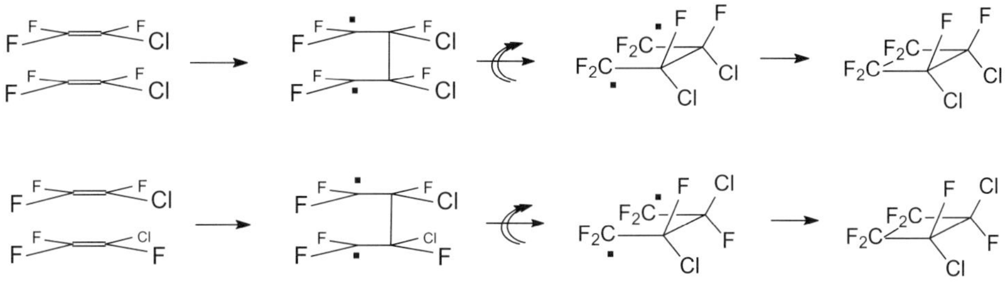

Scheme 7.23. A representation of the radical intermediates presumed to lie on the path between 1-chloro-1,2,2-trifluoroethene and the dimers (*E*)- and (*Z*)-1,2-dichloro-1,2,3,3,4,4-hexafluorocyclobutane formed from it (see Atkinson, B.; Stedman, M. *J. Chem. Soc.* **1962**, 512).

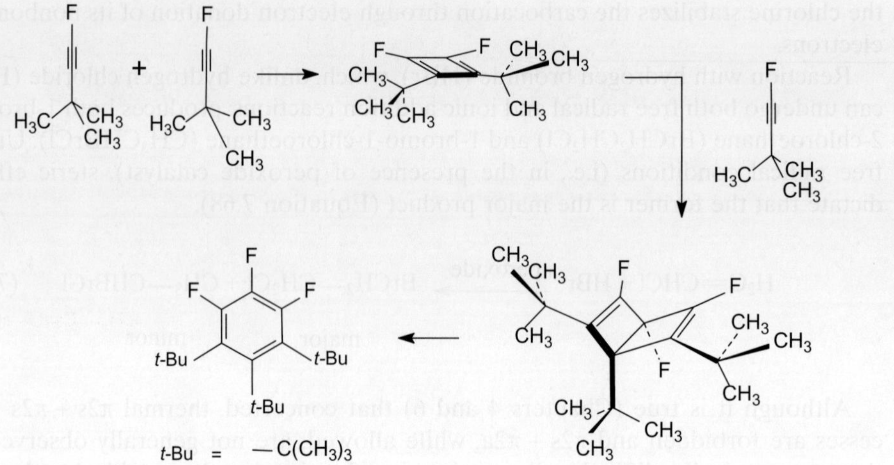

Scheme 7.24. The trimerization of 1-fluoro-3,3-dimethyl-1-butyne [*t*-butylfluoroacetylene, F–C≡CC(CH₃)₃] to 1,2,3-trifluoro-4,5,6-tri(1,1-dimethylethyl)benzene. The final step is an allowed six-electron ring opening. The cyclobutadiene intermediate is presumably formed in a diradical process (see Viehe, H. G.; Merenyi, R.; Oth, J. F. M.; Valange, P. *Angew. Chem. Int. Ed. Engl.*, **1964**, *76*, 888).

alkyne apparently undergoes an initial dimerization to cyclobutadiene,* subsequent addition of another equivalent of alkyne and a final electrocyclic ring opening to 1,2,3-trifluoro-4,5,6-tri(1,1-dimethylethyl)benzene.

In concert with the general observation that electron withdrawing groups in the dienophilic portion of the Diels–Alder electrocyclic π4s + π2s addition reaction facilitate the process (raising both the HOMO and LUMO of the alkene), haloalkenes can serve in this capacity. As shown in Equation 7.69, 1-bromoethene (ethylene bromide, H₂C=CHBr) serves as a dienophile in the reaction with cyclopentadiene to provide both endo-2-bromobicyclo[2.2.1]heptane and the corresponding *exo*-isomer. As expected on the basis of the Alder Endo rule (Chapter 4), the former predominates in the uncatalyzed reaction.

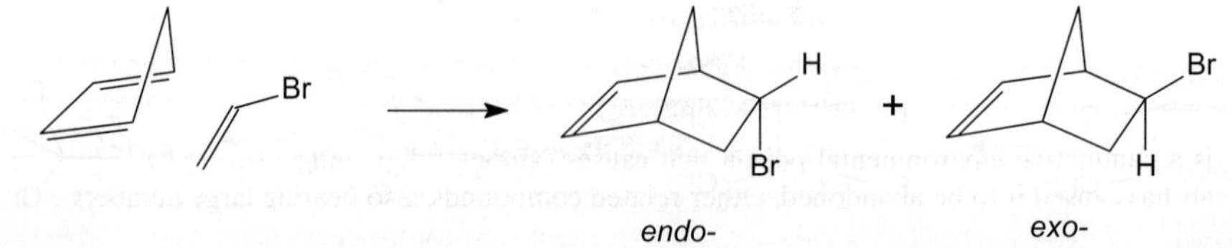

$$(7.69)$$

*As pointed out earlier (Chapter 6) unsubstituted cyclobutadiene, which has only fleeting existence, is apparently best represented as an equilibrating rectangular pair of isomers rather than a single square-like one. Substituted cyclobutadienes, such as the 1,4-difluoro-2,3-di(1,1-dimethylethyl)-1,3-cyclobutadiene shown in Scheme 7.24 appear to be stabilized by the strongly electron-withdrawing fluorine atoms (which reduce the electron density in the ring) and the bulky *t*-butyl substituents (which inhibit the approach of other reactants).

Problem 7.7. Draw and name the products expected in the Diels–Alder reaction between cyclopentadiene and 3-chloro-1-propene (allyl chloride [$H_2C=CH–CH_2Cl$]).

In contrast to the expression of Equation 7.69, electron-withdrawing groups on the diene portion of the reacting pair in the Diels–Alder reaction generally inhibit the reaction (which, it will be recalled [Chapter 4], is controlled by a dominant HOMO [diene]—LUMO [dienophile] interaction). However, if the diene bears a significant array of electron-withdrawing substituents, the HOMO–LUMO roles can be exchanged (reverse-demand) and the Diels–Alder reaction proceeds as expected. In this way (Equation 7.70), hexachlorocyclopentadiene serves as the diene component in the Diels–Alder reaction with cyclopentadiene (as the dienophile) to yield an alkene, which can be further chlorinated to a mixture of (largely) octachlorotricyclodecanes called "chlordane" (or "chlordan").*

$$(7.70)$$

Problem 7.8. Based on the path shown for the synthesis of chlordan(e), provide a reasonable synthesis for aldrin and deldrin.

G. ELIMINATION REACTIONS OF ALKYL AND ALKENYL HALIDES

Although it is possible (this Chapter, Section C) to consider elimination of the elements of hydrogen halide (H–X; X = F, Cl, Br, I) as an oxidation–reduction reaction or (this chapter, Section E, V [a]) as part of a substitution process by which halogen is replaced in aryl halides (Scheme 7.13), the concern of this section is to attempt

---

*The name "chlordan" is registered with the U.S. Patent Office as a common name for the mixture of compounds produced in this reaction. While the crude oily mixture was commonly used as a potent insecticide against (largely) ticks, fleas, lice, and so on, for a number of years, the eventual recognition that it is a cumulative environmental poison that causes (among other things) severe liver damage in mammals has caused it to be abandoned. Other related compounds, also bearing large numbers of relatively labile chlorine atoms, such as aldrin and deldrin, have suffered the same fate.

aldrin

dieldrin

TABLE 7.7. Alkyl Halides

Type of Elimination	Example
1,1-Elimination (α-elimination)	trichloromethane (chloroform) $\xrightarrow{-\,HCl}$ dichlorocarbene
1,2-Elimination (β-elimination)	chloroethane (ethyl chloride) $\xrightarrow{-\,HCl}$ ethene (ethylene)
1,3-Elimination (γ-elimination)	1,3-dibromopropane $\xrightarrow[\text{(e.g., as ZnBr}_2)]{-\,2\,Br^-}$ cyclopropane

Substrates and generalized pathways for the elimination of HX and X_2 from haloalkanes. Mechanistic information will be discussed subsequently. No presumption is intended about the path from the drawings provided.

to understand the pathways by which alkyl and alkenyl halides lose H–X or X_2 (X = F, Cl, Br, I) with the introduction of unsaturation.

There are five different types of substrates that must be considered. These are shown in Tables 7.7 and 7.8 along with a brief illustration of what product might be expected from each. The first example in Table 7.7 is that of 1,1- or α-elimination, that is, both the proton and halogen that are going to be lost are on the *same* carbon. The second example, perhaps the most thoroughly studied, is called β- or 1,2-elimination and covers those cases where the proton and halogen that are eliminated are on *adjacent* carbons. And then, there are those compounds in which only halogen (i.e., X_2) is eliminated. These reactions generally require the presence of a metal and while reactions that begin with two halogen atoms (a) on the same carbon, (b) on adjacent carbons, and (c) on carbons separated by a methylene (–CH$_2$–) unit are all known and will be discussed, only an example of the last is provided in Table 7.7. Arene halides undergo elimination of HX (X = Cl, Br) to produce benzynes. This yields substitution products as noted above (this chapter, Section E, V [a]) and will not be considered further here.

Examples of elimination of hydrogen halide (HX, X = Br) from sp^2-hybridized carbon are provided in Table 7.8. Reactions in which other hydrogen halides have been eliminated are also known but, in general, they have been less extensively studied. The first example in Table 7.8 illustrates a thermal 1,1- (or α) elimination to produce a carbene, which inserts into an adjacent C–H bond to produce ethyne (acetylene, H–C≡C–H). This reaction is first order in substrate and commonly occurs

TABLE 7.8. Alkenyl Halides

Type of Elimination	Example
1,1-Elimination (α-elimination)	bromoethene (ethylene bromide) $\xrightarrow{-\,HBr}$ → H—C≡C—H, ethyne (acetylene)
1,2-Elimination (β-elimination)	bromoethene (ethylene bromide) $\xrightarrow{-\,HBr}$ H—C≡C—H, ethyne (acetylene)
1,2-Elimination (β-elimination)	2-bromopropene $\xrightarrow{-\,HBr}$ 1,2-propadiene (allene)

Substrates and generalized pathways for the elimination of HX from haloalkenes. Mechanistic information will be discussed subsequently. No presumption is intended about the path from the drawings provided.

in the gas phase. Interestingly, the reaction in the second example of Table 7.8, β- (or 1,2) elimination, occurs at the same time under the same conditions—but to a lesser extent. In solution, on the other hand, 1,2- (or β) elimination appears to occur to the exclusion of the vinylcarbene. The third example of Table 7.8 is another case of 1,2- (or β) elimination but, as shown, serves to generate a cumulative diene rather than an alkyne.

In principle (and in some cases, in practice), gas-phase and low-temperature (ca. 5 K), elimination reactions can be photochemically induced. Such processes are of restricted utility and limited application. Some gas-phase reactions can also be induced thermally. While pyrolytic decomposition of halides has been investigated on some small molecules in detail, most elimination studies have been carried out in solution where the loss of HX or X_2 (X = F, Cl, Br, I) with the introduction of unsaturation is generally effected, for the former, by the addition of base and, for the latter, by using suitably reactive metals.

The introduction of unsaturation by the loss of HX (X = F, Cl, Br, I) will be considered first.

I. α-Elimination (1,1-Elimination)

a. α-Elimination of HX (X = Cl, Br) from Alkyl and Alkenyl Halides. In the
1950s a series of kinetic experiments, which also included trapping of the products,

showed that haloforms (HCX_3; $X = Cl, Br$) underwent reaction with base in aqueous solution to produce the corresponding dihalomethylene (carbene, $:CX_2$, $X = Cl, Br$). As shown for trichloromethane (chloroform [$HCCl_3$]) in Scheme 7.25, the first step is rapid and reversible proton removal and this is followed by a slow (rate determining) step involving loss of chloride anion (Cl^-) to generate dichlorocarbene ($:CCl_2$). Accordingly, the reaction is found to be first order in base, first order in trichloromethane (chloroform, $HCCl_3$), and thus second-order overall.

Although dihalocarbenes do not generally insert into carbon–hydrogen bonds, dichlorocarbene ($:CCl_2$) and dibromocarbene ($:CBr_2$) generated with base (e.g., potassium t-butoxide [$K^+ {}^-O–C(CH_3)_3$]) in solution (Equation 7.71) in this way have been shown to add across the carbon–carbon double bond of various alkenes with retention of stereochemistry (Scheme 7.26). Since triplet carbenes (Chapter 6) do not preserve the stereochemistry of their reaction partners (Scheme 7.27), it is argued that α-elimination of HX ($X = Cl, Br$) from trichloromethane (chloroform, $HCCl_3$) or tribromomethane (bromoform, $HCBr_3$) must produce singlet state carbene.

$$H–CBr_3 + K^+ {}^-OC(CH_3)_3 \xrightarrow[\text{(solvent)}]{(CH_3)_3COH} :CBr_2 + KBr \tag{7.71}$$

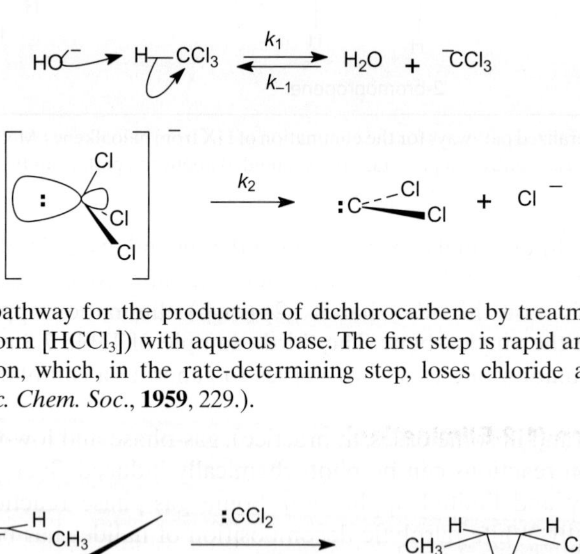

Scheme 7.25. A pathway for the production of dichlorocarbene by treatment of trichloromethane (chloroform [$HCCl_3$]) with aqueous base. The first step is rapid and reversible and produces carbanion, which, in the rate-determining step, loses chloride anion (Cl^-) (See, Wagner, W. J. *Proc. Chem. Soc.*, **1959**, 229.).

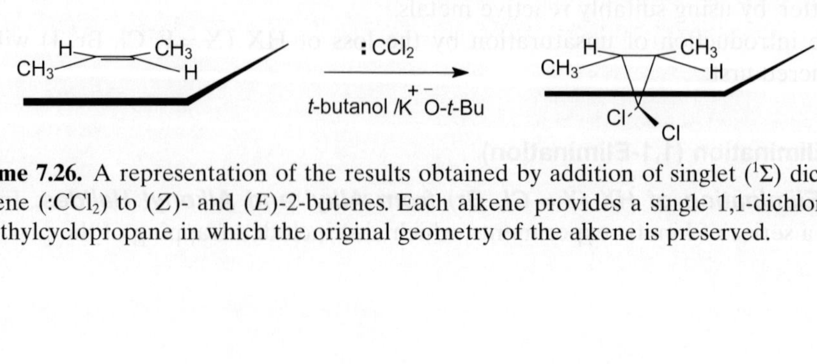

Scheme 7.26. A representation of the results obtained by addition of singlet ($^1\Sigma$) dichlorocarbene ($:CCl_2$) to (Z)- and (E)-2-butenes. Each alkene provides a single 1,1-dichloro-2,3-dimethylcyclopropane in which the original geometry of the alkene is preserved.

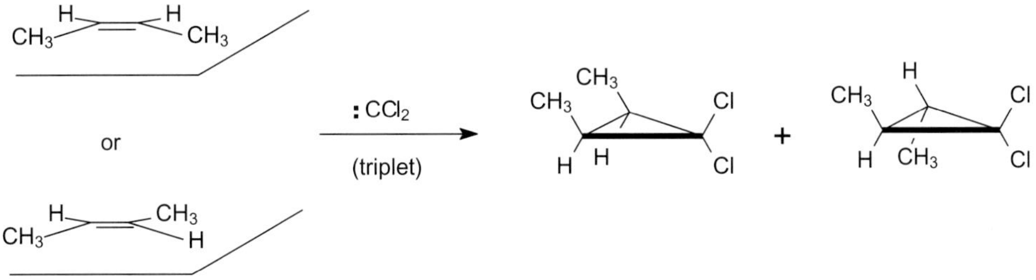

Scheme 7.27. A representation of the results obtained by addition of triplet (^{3}T) dichlorocarbene (:CCl$_2$) produced photochemically to (Z)- and (E)-2-butenes. Both alkenes provide a mixture of 1,1-dichloro-2,3-dimethylcyclopropanes. The original geometry of the alkene is lost and the reaction is presumed to occur via a radical process (see Wescott, L. D.; Skell, P. S. *J. Am. Chem. Soc.*, 1965, *87*, 1721 and Doering, W. von E.; Hoffmann, A. K. *J. Am. Chem. Soc.*, **1954**, *76*, 6162).

The symmetry of the LUMO of the singlet carbene and the HOMO of an alkene (and the HOMO of the carbene and LUMO of an alkene) is shown in Figure 7.16. The overlap of orbitals of like sign shows that the reaction to form cyclopropane is allowed.

b. α-Elimination of X$_2$ (X = Cl) from Alkyl Dihalides. At low temperatures, an exchange reaction between organometallic compounds such as butyllithium (CH$_3$CH$_2$CH$_2$CH$_2$Li) and geminal di- and polyhalides that lack acidic protons (e.g., tetrachloromethane [carbon tetrachloride, CCl$_4$]) can be effected to produce α-halo organolithium compounds. As the temperature is increased, these compounds lose metal halide, forming carbenes (Equation 7.72).

$$CCl_4 + BuLi \longrightarrow BuCl + Cl_3CLi \xrightarrow{-LiCl} :CCl_2 \qquad (7.72)$$

$$CH_3CH_2CH_2CH_2- \ = Bu$$

II. β-Elimination (1,2-Elimination)

a. β-Elimination of HX (X = F, Cl, Br, I) from Alkyl and Alkenyl Halides. The β-elimination of hydrogen halides from the corresponding alkyl or alkenyl halide frequently accompanies nucleophilic subsitution.* When a nucleophile is strongly basic, elimination may be made the major reaction pathway.†

As early as 1935, Ingold (*vide supra*) and coworkers recognized and codified two, *idealized*, limiting elimination pathways. By analogy to the nucleophilic substitution processes, S$_N$1 and S$_N$2 (discussed above), these were called E1 and E2 (for

---

*The solvolysis of optically active 2-bromooctane under alkaline conditions in 60% aqueous ethanol, early examined by Ingold et al. (in the period 1935–1938) produced only small amounts of (inverted) alcohol (2-octanol) and ether (ethyl 2-octyl ether) and major amounts of alkenes.

†As will be seen, even with nucleophile and substrate held constant, reaction conditions (e.g., solvent and temperature) can be optimized to favor elimination instead of substitution.

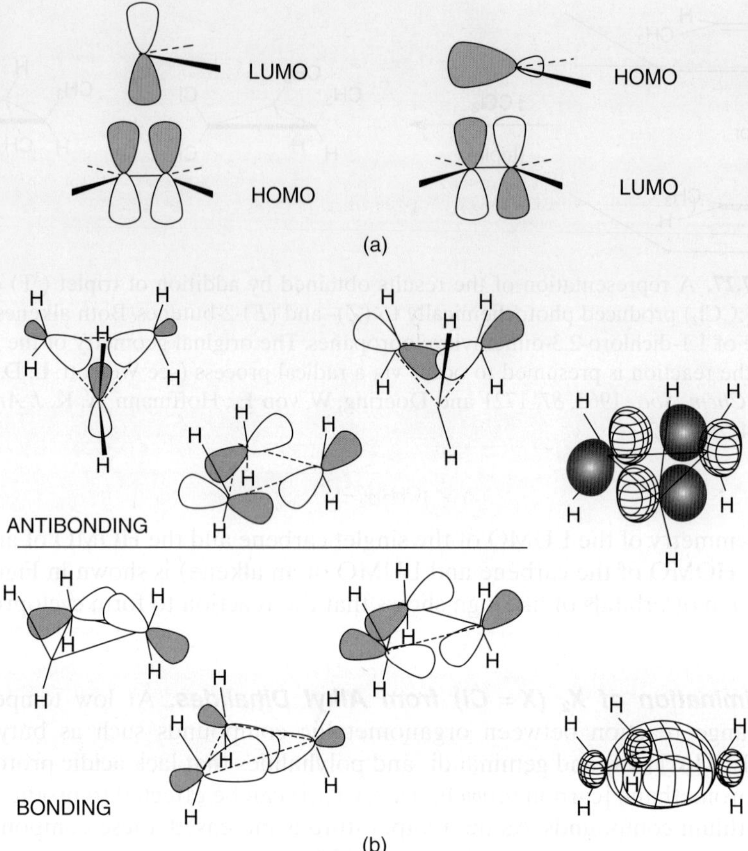

Figure 7.16. (a) A representation of the HOMO and LUMO orbitals of a singlet carbene and an alkene, showing potential overlap that leads to cyclopropane product. (b) A representation of the HOMO and LUMO orbitals of cyclopropane.

unimolecular and bimolecular elimination), respectively. Later, a third process, that is, elimination from a "preformed" carbanion, was suggested. The carbanion is the "conjugate base" of the substrate. This process is called the E1cb because, although it is second-order overall (first order in base, which is required to produce the anion, and first order in substrate), the rate-determining step involves only one species—the carbanion.

1. Elimination: Unimolecular (E1). For the E1 (elimination, unimolecular) process, it was supposed that the halogen (the leaving group, "L"), by analogy to the S_N1 reaction (*vide supra*), would leave. The leaving of "L" with its pair of electrons would be rate determining and a carbocation would remain behind. At least in principle, the carbocation could be the same ion as that produced in the S_N1 reaction! Subsequently, in a process that was not rate determining, a proton would be lost. The kinetics accompanying this process should be identical with those of the S_N1 reaction since the rate-determining step is the same. An idealized representation of a simplified version of the process is provided in Scheme 7.28.

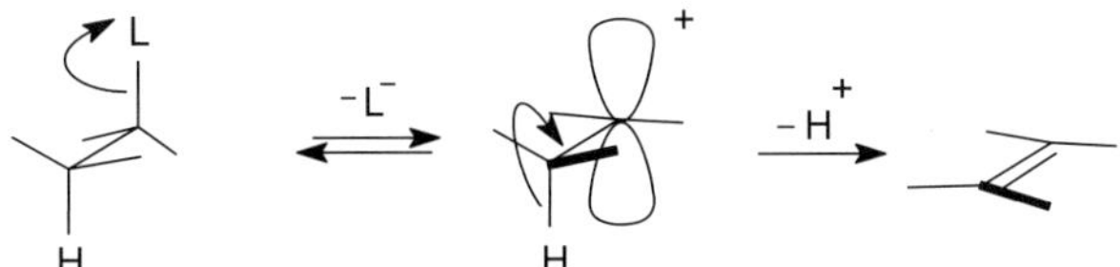

Scheme 7.28. A representation of the formation of a carbocation through the leaving of a leaving group ("L") followed by loss of a proton to produce an alkene.

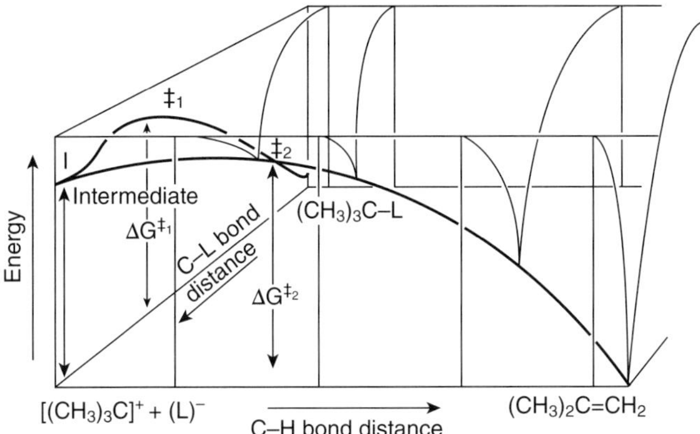

Figure 7.17. A representation of an energy diagram for the E1 reaction of 2-halo-2-methylpropane (L = halogen) to yield 2-methylpropene. This three-dimensional figure can be transposed into two dimensions as shown in Figure 7.18.

As anticipated on the basis of the idea that the rate-determining step in the E1 reaction is the same as that in the S_N1 reaction, representations such as that of Figures 7.17 and 7.18, shown for the path of E1 elimination from a 2-halo-2-methyl substrate (L = halogen) to the corresponding alkene product (i.e., 2-methylpropene [$(CH_3)_2C=CH_2$]), will be strikingly similar to that seen for an S_N1 reaction on the same substrate. Thus, except for the product-determining step, it would be anticipated that the paths seen in Figures 7.6 (or 7.7) and 7.17 (or 7.18) would be the same.

Furthermore, again in concert with the idea that the intermediate carbocation in the S_N1 and E1 reactions is the same and that the product-determining step is not rate determining, the same effects resulting from changes in the substrate structure should be observed in both unimolecular nucleophilic substitution and in unimolecular elimination. Thus, substituents on the carbon atom to which the leaving group is attached that donate electrons or provide delocalization to stabilize the positive charge should facilitate the E1 as they did for the S_N1 (*vide supra*). As pointed out earlier, the order of carbocation stability (3° > 2° > 1°) suggests (Hammond postulate [Chapter 6, Figure 6.8]) that the E1 reactions (like the S_N1 reaction) should be significantly more facile for tertiary alkyl halides than for either secondary or primary alkyl halides. Generally, the suggestion is supported by experimental observation.

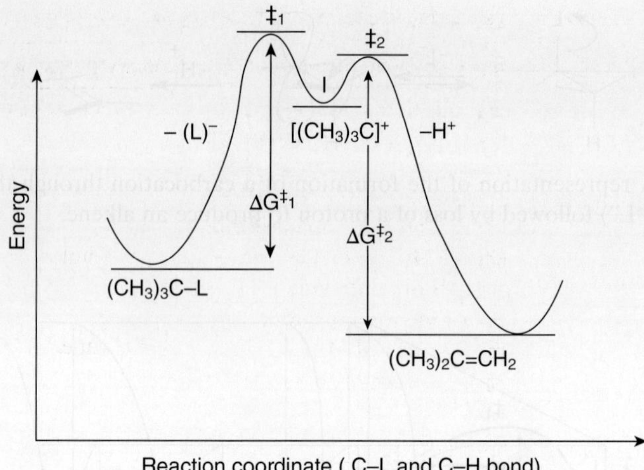

Figure 7.18. A two-dimensional representation of an energy diagram for the E1 reaction of a 2-halo-2-methylpropane (L = halogen) to produce 2-methylpropene. This two-dimensional figure can be represented in three dimensions as shown in Figure 7.17.

In the same vein, the role of the solvent should also be similar in the S_N1 and E1 reactions. Solvents that facilitate the separation of charge (i.e., ionizing solvents) will aid the reaction much more than nonionizing solvents. Reasonably, elimination will be promoted by efficient solvation of the leaving group and any other nucleophiles that might be present. Solvation of the leaving group is important to help it leave and to suppress its "(internal) return." Solvation of the nucleophile is important to retard its attack on substrate, and solvents of low nucleophilicity (to inhibit attack of the solvent on the carbocation [i.e., solvolysis]) are desirable.

Finally, given the similarities between the S_N1 and E1 reactions, the choice as to which might preferentially occur frequently rests on (1) the temperature at which the reaction is carried out and (2) the basicity of the nucleophile.

i. THE EFFECT OF TEMPERATURE ON THE E1 REACTION. As early as 1948, Ingold (*vide supra*) and coworkers pointed out that chlorides, bromides, and iodides all produce higher proportions of alkene as the temperature is increased under conditions where the nature of the leaving group is unimportant. Thus, under ostensibly first-order conditions and with the same base and solvent (alkali in "60% ethanol"), 2-bromopropane produces 53.2% alkene at 45°C and 63.6% alkene at 100°C. They concluded that in all the cases they examined, the elimination reactions must have the higher activation energy (relative to substitution) as shown in Figure 7.19. Indeed, it is largely because substitution (by the S_N1 pathway) is more facile than elimination (by the E1 pathway); that is, the ratio (S_N1/E1) of products derived from these reactions even under the most favorable conditions is usually greater than 1, and that the E1 reaction is not widely used to produce alkenes.

Problem 7.9. Examine the equations (Equations 7.29–7.42) describing the kinetics associated with the S_N1 reaction and, either *de novo* or by suitable modification, write a set that is in concert with what might be expected for the E1 reaction.

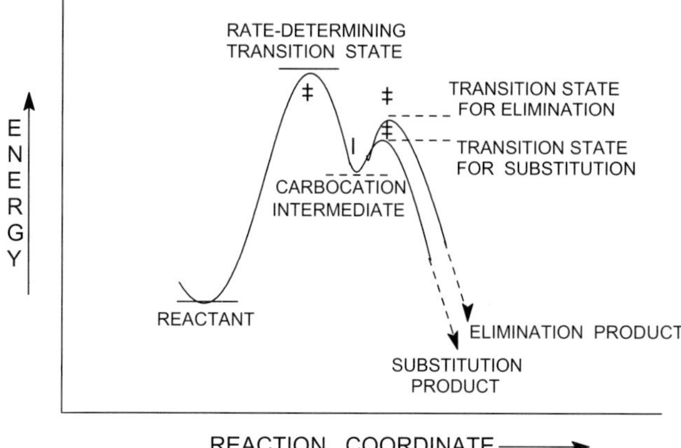

Figure 7.19. A representation of the course of a generalized $S_N1/E1$ reaction. The rate-determining transition state leads to a carbocation that subsequently partitions between products of substitution and elimination. The transition state for elimination *from the common intermediate* is generally higher than that for substitution.

ii. THE BASICITY OF THE NUCLEOPHILE. Since a proton (H^+) is being lost from a carbocation, even relatively weak bases such as hydroxide anion (HO^-), which normally could not abstract a proton from unactivated carbon, are sufficient to promote elimination. However, it is important to recognize that measurement of basicity is frequently difficult in non-nucleophilic solvents and base strengths need not follow the order established under other conditions.

iii. THE REGIOCHEMICAL SENSE OF E1 ELIMINATION. In principle (and in practice), the E1 elimination might occur to produce a variety of alkenes from a carbocation that is not symmetrical. Generally, the high-energy carbocation loses a proton to preferentially form the most stable alkene, which is usually the most highly substituted alkene (the Saytzeff* product) as shown in Scheme 7.29. It is argued that preferential formation of the most stable alkene results from a transition state in which partial electron donation from alkyl groups facilitates proton loss.

Interestingly, as shown in Table 7.9, the ratio of more highly substituted (more stable) alkene to less highly substituted (less stable) alkene produced in alcohol–water solvent mixture at 25°C appears to also be a function of the steric bulk of the attached substituents.

The same sense of elimination to form the more stable alkene (the Saytzeff-type product) can also be observed in cyclic systems. For example, as shown in Scheme 7.30, the unstable tertiary iodide 1,2-dimethyliodocyclohexane undergoes elimination to a mixture of three alkenes, *viz.* 1,2-dimethylcyclohexene, 2,3-dimethylcyclohexene, and 2-methyl-1-methylenecyclohexane, the ratio of which (66:31:3) clearly represents some kinetic partitioning between tetrasubstituted:

---

*Alexander Saytzeff (University of Kazan) formulated a rule in 1875 concerning elimination based on the observation that in the elimination of hydrogen halides (chlorides and bromides) from alkyl halides, the preferred product is the alkene with the greatest number of alkyl groups attached to the double bond.

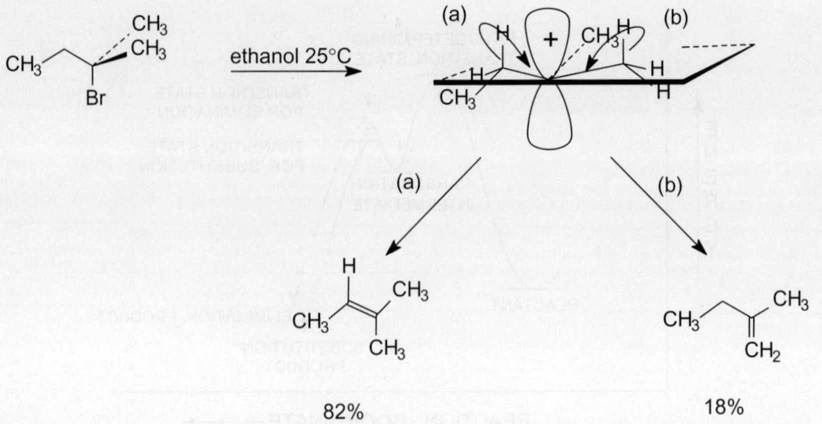

82%   18%

Scheme 7.29. A representation of two potential pathways for proton loss from a carbocation intermediate on the S_N1–E1 reaction surface. Both pathways are followed. It is supposed that proton loss via the pathway labeled (a), which results in the most highly substituted alkene (the Saytzeff product), occurs preferentially because that is the product formed in highest yield. The alkene resulting from pathway (b), the Hofmann product, is also formed. A. W. Hofmann (1818–1895) was a German chemist who was professor of chemistry at the Royal College of Chemistry in London (1845–1864) and then accepted a post as professor at the University of Berlin. Most of Hofmann's work dealt with amines (Chapter 10). Hofmann found, in contrast to Saytzeff, that the least highly substituted alkene is formed when the elimination is carried out on amine quaternary salts (so-called *onium salts*). This is, in part, presumably due to the close association between the base and the positively charged "onium salt" as well as to the removal of the proton in the rate-determining step (cf. the E2 reaction). (Note: the 82:18 ratio of products shown here should be considered identical, within experimental error, to the 79:21 ratio of Table 7.9.)

TABLE 7.9. The Variation in Yield of More-Substituted-Alkene/Less-Substituted-Alkene Products and Total Alkene as a Function of Steric Bulk on the Carbon from Which the Proton Is Lost

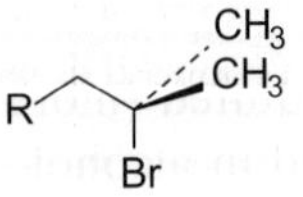	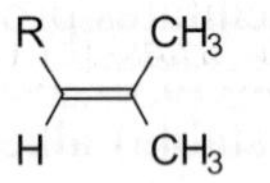 $R\overset{CH_3}{\underset{H}{\diagdown}}\overset{CH_3}{\diagup}CH_3$	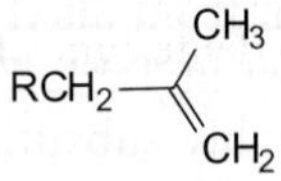 $RCH_2\overset{CH_3}{\underset{CH_2}{\diagdown}}$	Total Alkene
R = CH$_3$	79	21	27
CH$_2$CH$_3$	71	29	32
CH(CH$_3$)$_2$	59	41	46
C(CH$_3$)$_3$	19	81	57

Within experimental error, the first entry (2-bromo-2-methylbutane) in the table provides results the same as those of Scheme 7.23. Interestingly, 2-bromo-2,4,4-trimethylpentane (the last entry in the table) undergoes E1 elimination to yield a larger amount of the less-substituted alkene, which equilibration studies show to be the more stable (see Saunders, W. H. Jr.; Cockerill, A. F. *Mechanisms of Elimination Reactions*, **1973**, Wiley, New York).

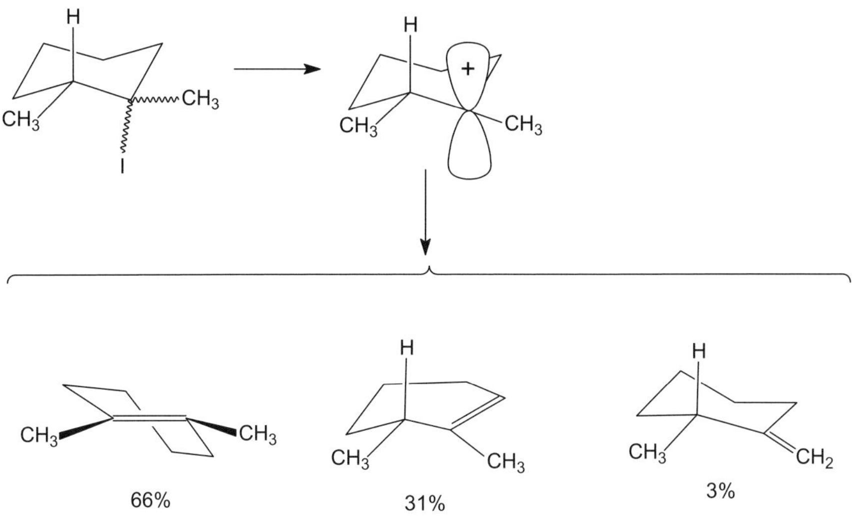

Scheme 7.30. The products of E1 elimination from the unstable 1-iodo-1,2-dimethylcyclohexane to a mixture of alkenes. The most stable alkene is preferred by about 2:1. At equilibrium, it is preferred nearly 5:1.

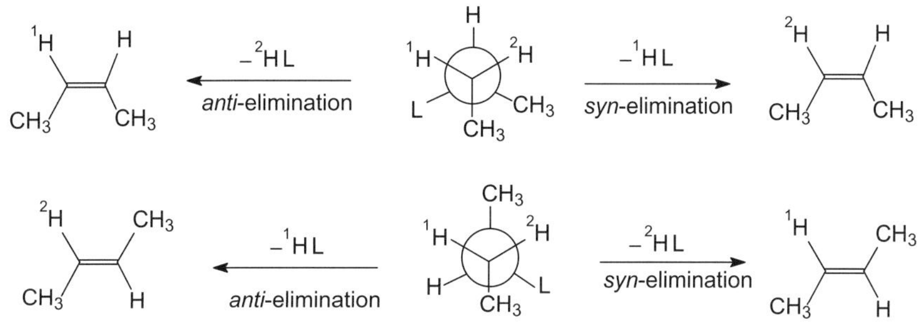

Scheme 7.31. The elimination of ^{1}H–L and ^{2}H–L (where L = halogen) from a *threo*-2-[^{2}H$_1$]-3-halobutane diastereomer showing the results of *anti*- and *syn*-elimination. As noted in the text, *anti*-elimination is apparently preferred since, for the most part, the *cis*- [or (*Z*)] 2-butene produced does not contain deuterium (^{2}H), while the *trans*- [or (*E*)] 2-butene produced does not contain protium (^{1}H). Thus, ignoring a deuterium isotope effect, *syn*-elimination is apparently avoided.

trisubstituted:disubstituted alkenes. The equilibrium mixture, reflecting thermodynamic partitioning, is about 85:15:0. Thus, in the elimination reaction, there is clearly a nonexclusive preference for the most stable alkene.

iv. THE STEREOCHEMICAL SENSE ATTENDING PROTON LOSS IN THE E1 REACTION. The stereochemistry of the alkenes produced in the E1 reaction in both acyclic and cyclic systems has been examined in order to compare the results with those of the E2 reaction (*vide infra*) on similar systems.

First, as shown in Scheme 7.31, it appears that in the most ionizing solvents, an acyclic carbocation preferentially loses a proton opposite to the direction (so-called *anti* [or antarafacial—the opposite "face"]) from which the leaving group departs.

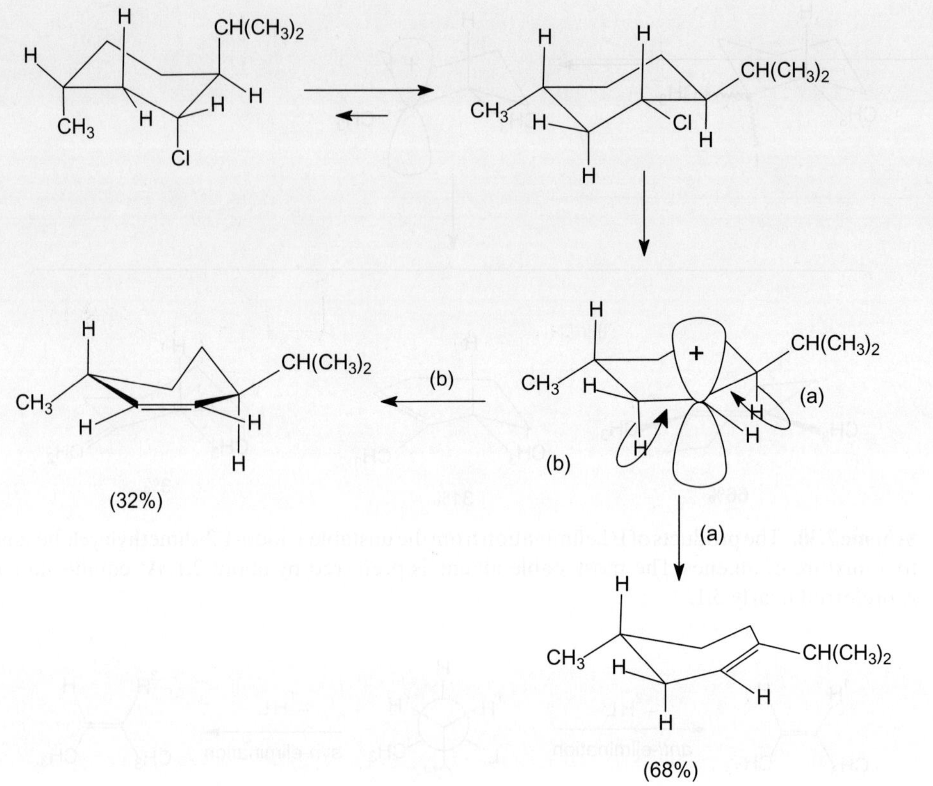

Scheme 7.32. The E1 elimination as effected on (1*R*,3*R*,4*S*)-3-chloro-4-(1-methylethyl)methylcyclohexane [(+)-menthyl chloride]. It is important to note that there are no protons *anti* (or antarafacial) to the leaving group (Cl) on the carbons β to it (see, e.g., Barton, D. H. R.; Head, A. J.; Williams, R. J. *J. Chem. Soc.*, **1952**, 453 and Bamkole, T. O.; Maccoll, A. *J. Chem. Soc. B*, **1970**, 6, 1159).

This process leads to both *cis*- [or (*Z*)] and *trans*- [or (*E*)] alkenes and a preference for the final geometry [(*Z*)/(*E*)] does not seem to be expressed!

Second, while elimination of the proton *anti* (or antarafacial) to the leaving group is apparently preferred, *syn* (or suprafacial) elimination can occur. Consider the case of (1*R*,3*R*,4*S*)-3-chloro-4-(1-methylethyl)methylcyclohexane [(+)-menthyl chloride], Scheme 7.32, in which, in the most stable all equatorial conformation, there is no proton *anti* to the leaving halogen (Cl) on either of the β-carbons. When this chloroalkane is subjected to the conditions of the E1 reaction, the proton lost can be from either of the two adjacent carbons, none of the protons being *anti* (or antarafacial) and all making angles of ca. 60° with the chlorine. The major product (68%), 4-methyl-(1-methylethyl)cyclohexene (3-menthene) (the Saytzeff [more highly substituted alkene] product) is derived from loss of a proton which was *cis* (or suprafacial) to the chlorine. The minor product (32%), 3-(1-methylethyl)-6-methylcyclohexene (2-menthene) (the Hofmann [less highly substituted alkene] product) (a 2:1 product ratio) may be derived from loss of a proton that was, initially, either *cis* [or (*Z*)] or *trans* [or (*E*)] to the chlorine—but neither of which was *anti* (or antarafacial)!

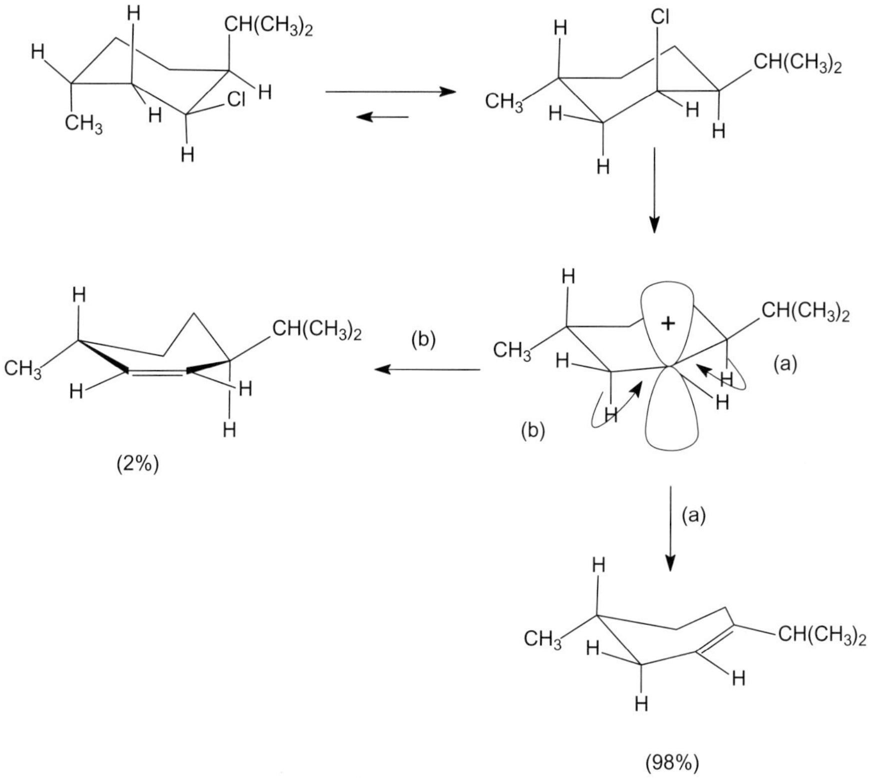

(2%)

(b)

(a)

(b)

(a)

(98%)

Scheme 7.33. The E1 elimination as effected on (1*R*,3*S*,4*S*)-3-chloro-4-(1-methylethyl)methylcyclohexane [(+)-neomenthyl chloride]. It is important to note that there are two protons *anti* to the leaving group (Cl) on the carbons β to it.

It is important to note that when E1 elimination is effected on the epimeric chloro-derivative [neomenthyl chloride, (1*R*,3*S*,4*S*)-3-chloro-4-(1-methylethyl) methylcyclohexane], Scheme 7.33, which has the chlorine axial and the two alkyl substituents equatorial, and where, as a consequence of that arrangement, there are now two *anti* (or antarafacial) and axial protons situated β to the chlorine (Cl), the same product mixture is obtained, but now, with *anti*-elimination possible, the ratio of 3-menthene:2-menthene is 98:2.

2. Elimination, Unimolecular, Conjugate Base (E1cb). * In a unimolecular elimination reaction from the conjugate base (E1cb), a proton on the carbon β to the leaving group (halogen) is rapidly and reversibly lost to the base. The resulting carbanion (whose concentration is dependent on the initial equilibrium creating it) is the *single* species in the rate-determining transition state, which undergoes loss of the leaving group (L = halogen) to produce alkene. Thus, the E1cb process occurs with second-order kinetics (first order in substrate and first order in base) but is nonetheless a unimolecular elimination. Reactions thought to occur via the E1cb process require a particularly acidic proton on the carbon β to that bearing the

*See Breslow, R. *Tetrahedron Lett.* **1964**, 399.

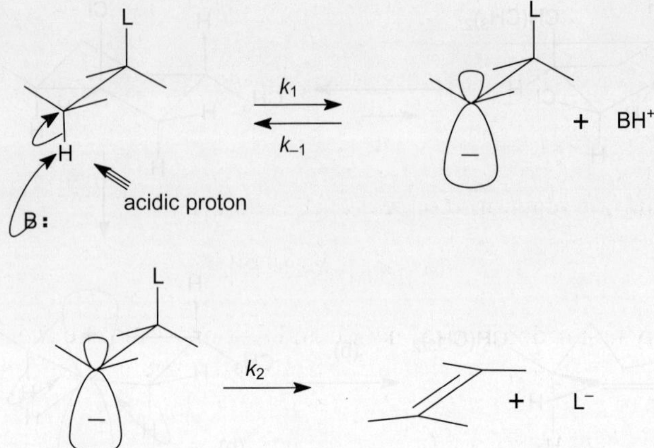

Scheme 7.34. A representation of an E1cb process. In the first step, an acidic proton is removed from the β-carbon by the exogenous base to generate the conjugate base of the substrate (a carbanion). The product-determining step contains only the carbanion.

leaving group (Scheme 7.34) and therefore usually carry other electron-withdrawing groups on that carbon too. As will become apparent later, these reactions are thought to lie at an extreme of possible E2 processes.

In a process such as that shown in Scheme 7.34, the rate of product formation (where A^- represents the carbanion or conjugate base) is

$$\text{rate} = +d[\text{alkene}]/dt = -d[A^-]/dt = k_2[A^-]. \tag{7.73}$$

Then, since product can only be formed from the carbanion (or conjugate base),

$$d[A^-]/dt = k_1[\text{A-H}][\text{B:}] - k_{-1}[A^-][\text{BH}^+] - k_2[A^-], \tag{7.74}$$

where A-H is the alkyl halide and B: is the base.

Now, if it is argued that there is no change in the instantaneous concentration of the carbanion (conjugate base) intermediate, that is, it is a reactive intermediate that, in its short lifetime, either reacts with a proton source and returns to starting material or goes on to product by loss of the leaving group, then its concentration remains small and constant and the steady-state approximation applies, so that

$$d[A^-]/dt = 0 = k_1[\text{A-H}][\text{B:}] - k_{-1}[A^-][\text{BH}^+] - k_2[A^-] \tag{7.75}$$

and

$$k_1[\text{A-H}][\text{B:}] = [A^-](k_2 + k_{-1}[\text{BH}^+]) \tag{7.76}$$

or

$$[A^-] = k_1[\text{A-H}][\text{B:}]/(k_2 + k_{-1}[\text{BH}^+]) \tag{7.77}$$

and, from Equation 7.73,

$$\text{rate} = k_2 k_1 [\text{A-H}][\text{B:}]/(k_2 + k_{-1}[\text{BH}^+]). \tag{7.78}$$

Then, if $k_2 \gg k_{-1}[\text{BH}^+]$, the carbanion proceeds to alkene faster than it returns to the starting material, $k_{-1}[\text{BH+}]$ can be ignored, k_2 in the numerator and denominator cancel, and the reaction reduces to Equation 7.79:

$$\text{rate} = k_1 [\text{A-H}][\text{B:}]. \tag{7.79}$$

This is clearly first order in alkyl halide and first order in base (i.e., second-order overall). It should also be clear that if $k_{-1}[\text{BH}^+] \gg k_2$, the process is still first order in alkyl halide and first order in base, but, in addition, there is an inverse dependence on the conjugate acid of the base!

3. Elimination: Bimolecular (E2). In an E2 reaction, two species (the base and substrate only since, by analogy with substitution processes, consideration of solvent molecules is expressly omitted) are expected to be undergoing covalency change in the rate-determining transition state. This is directly analogous to the S_N2 process, and, as seen in the S_N2 reaction (*vide supra*), the rate of a reaction which is first order in each of two reactants is second-order overall. Equation 7.80, where k_o is the observed rate constant and the other meanings are clear, describes such an elimination reaction:

$$-d[\text{substrate}]/dt = -d[\text{base}]/dt = +d[\text{alkene}]/dt = k_o[\text{base}]^1[\text{substrate}]^1. \tag{7.80}$$

In the E2 process, the leaving of the leaving group ("L" = Cl, Br, I), with its pair of electrons, occurs in concert with removal of a β-proton (Equation 7.81). It is expected that the developing σ-π framework of the alkene will be enhanced and promoted by arranging coplanarity between the proton to be removed and the leaving group to be lost. In producing this arrangement, the dihedral angle between these two is effectively limited to the values of $0°$ (Figure 7.20; synperiplanar) or $180°$ (Figure 7.20; antiperiplanar). It should also be clear that the plane containing the proton to be lost and the leaving group also contains the two carbon atoms, which will eventually be doubly bonded and thus the σ sp^3-bonding orbitals by which the proton and the leaving group are bonded to the termini of what is to become the alkene will be "replaced" by the p-orbitals of the π-bond. It is often convenient to think of the electron pair liberated as base removes a proton as "helping" the leaving group leave (with its electron pair).

L

:B = base

L = leaving group

(Cl, Br, I)

H

:B

$$+ \quad \text{B} \!\!-\!\! \text{H} \quad + \quad \text{L}^- \tag{7.81}$$

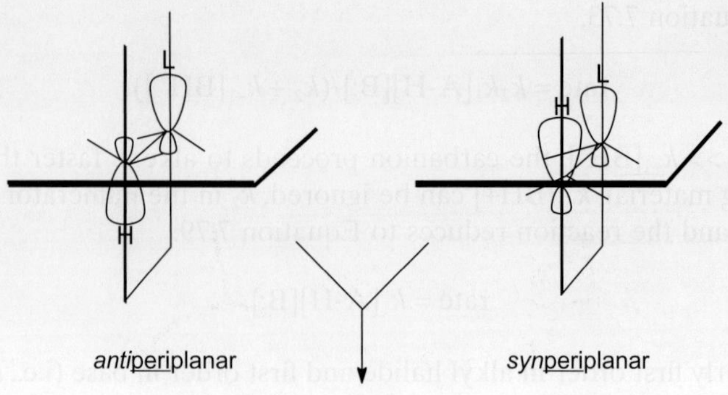

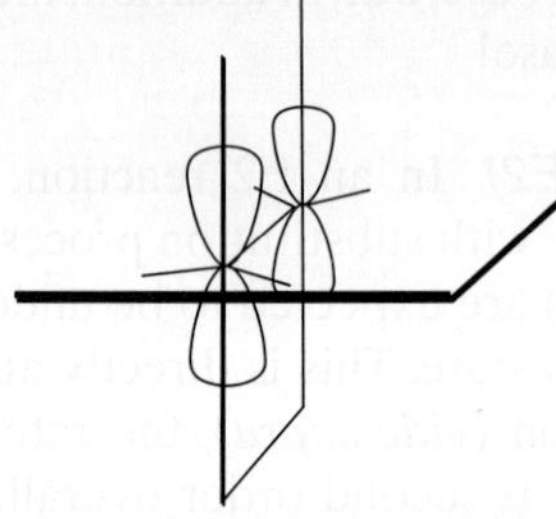

Figure 7.20. A representation of antiperiplanar and synperiplanar relationships between a leaving group ("L"; L = Cl, Br, I) and a proton to be lost to some suitable base "B:" giving rise to an alkene. Such relationships appear to be preferred in the E2 elimination (see Ingold, C. K. *Structure and Mechanism in Organic Chemistry*, 2nd edition, Cornell University Press, Ithaca, NY, **1969**, p. 688 ff.).

In an idealized case for an E2 reaction, it can be argued that the bond-making and bond-breaking steps occur completely in concert and synchronously so that in the transition state the bond from the base to the proton on the β-carbon will be half-made, the bond between the β-carbon and the proton will be half-broken, the carbon–carbon double bond will be half-made, and the bond between carbon and the leaving group "L" (L = Cl, Br, I) will be half-broken. Such an idealized picture is shown in Figures 7.21 and 7.22.

Generally, however, bond breaking might lead to bond making or vice versa since a completely symmetrical case is almost certainly idealized. This is accommodated by the creation of the "variable transition state theory of elimination reactions" (see Bunnett, 1969, p. 53 ff.).

The necessity for the creation of the concept of a variable E2 process arises because of the number of possible variations that can intrude in the overall elimination process. For example, it is well appreciated that some bases "B:" will be "stronger" than others or (in other words) some alkyl halides will be "more acidic" toward those bases than others.

Then, there is the possibility that the leaving group will more readily leave from one alkyl halide than from another. As shown in Figure 7.23, the E2 reaction is considered to span the distance from a process in which the proton on the β-carbon is completely lost (the E1cb, *vide supra*) all the way to a process in which the leaving group completely leaves before the proton is lost (the E1, *vide supra*).

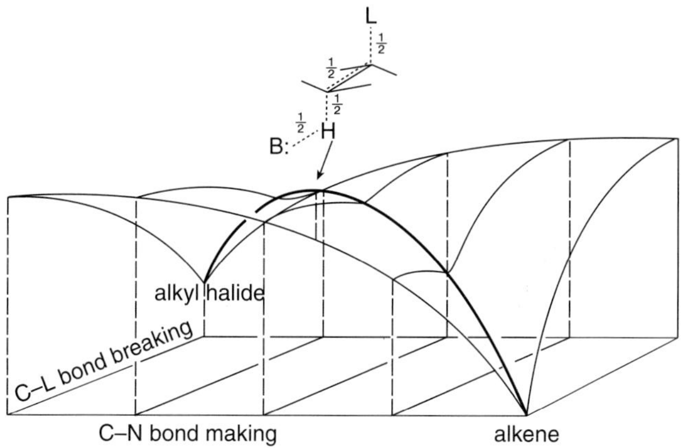

Figure 7.21. A representation of an idealized E2 reaction in which the bond-making and bond-breaking steps occur in concert, synchronously and to the same extent. Thus, in this idealized transition state, each bond is formed (broken) to the same extent at the same time, that is, exactly halfway! A two-dimensional representation of this process is shown in Figure 7.22.

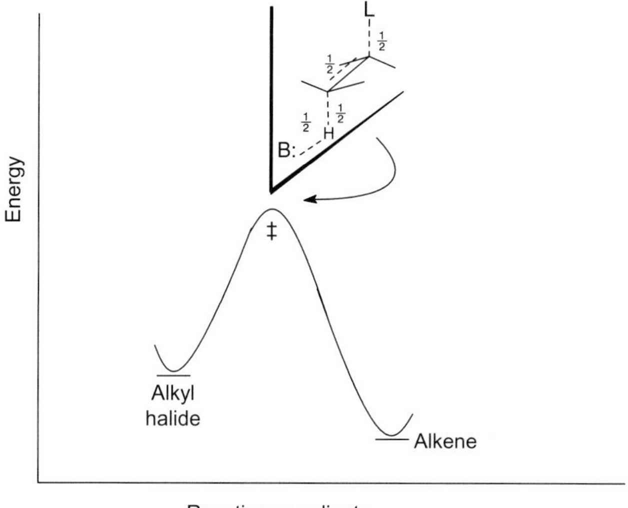

Figure 7.22. A representation of an idealized E2 reaction in which the bond-making and bond-breaking steps occur in concert, synchronously and to the same extent. Thus, in this idealized transition state, each bond is formed (broken) to the same extent at the same time, that is, exactly halfway! A three-dimensional representation of this process is shown in Figure 7.21.

It should be clear that whether the E2 process is "nearly E1cb" or "nearly E1" will be a function of the same set of variables already encountered. Thus, while second-order kinetics may apply except at the extreme of the "pure" E1 the nature of the base, the substrate, the leaving group, and the solvent, as well as the temperature at which a reaction is run will all play a role in defining where on the continuum any particular reaction lies.

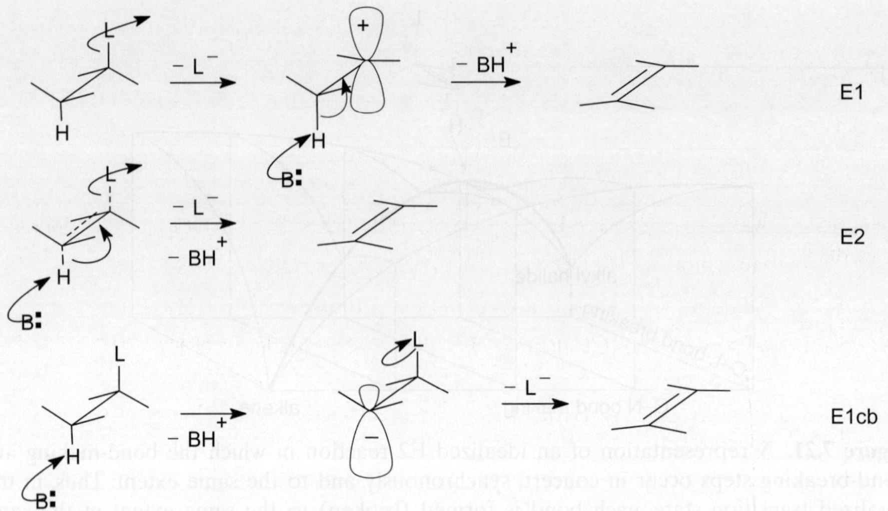

Figure 7.23. 1,2-Elimination extremes. The E1 and E1cb processes can be thought of as extreme examples of the symmetrical, concerted, synchronous, E2 reaction. Thus, the E1 reaction is, in the limit, an E2 reaction in which bond breaking between carbon and the leaving group occurs "completely" before bond breaking to the hydrogen on the β-carbon "has begun." Similarly, the E1cb reaction is, in the limit, an E2 reaction in which bond breaking between carbon and the hydrogen on the β-carbon occurs "completely" before bond breaking between carbon and the leaving group has begun.

i. THE NATURE OF THE BASE IN THE E2 ELIMINATION REACTION. Alkoxide anions (the conjugate bases of alcohols [Chapter 8]) are frequently used in solutions of the corresponding alcohols as the proton-abstracting base in E2 reactions. These bases are commonly prepared (*in situ*) by reaction of alkali metals with the alcohol (Chapter 8) as shown here in Equation 7.82.

$$R{-}\overset{O}{\diagup}{}_{H} \;+\; M \quad \xrightarrow[\text{solvent}]{\text{ROH}} \quad R{-}\overset{-}{O}\,M^{+} \;+\; 1/2\;H_2 \qquad (7.82)$$

R = alkyl

M = Na, K

It is currently held that as the bulk of the alkyl groups surrounding the negatively charged oxygen increases, the ability of the anion to be solvated by neighboring proton donors decreases. Thus, *t*-butoxide (2-methyl-2-propoxide [$(CH_3)_3CO^-$]), the alkoxide anion prepared from *t*-butanol (2-methyl-2-propanol [$(CH_3)_3COH$]) in *t*-butanol, is a stronger base than 2-propoxide [$(CH_3)_2CHO^-$] in 2-propanol [$(CH_3)_2CHOH$], ethoxide ($CH_3CH_2O^-$) in ethanol (CH_3CH_2OH), and methoxide (CH_3O^-) in methanol (CH_3OH), in that order. A comparison of the base strengths is given in Table 7.10.

The strength of the base plays an important role in product distribution in the E2 reaction. As shown in Scheme 7.35, the treatment of 2-chlorohexane [$(\pm)$-$CH_3(CH_2)_3CH(Cl)CH_3$] with methoxide (CH_3O^-) in methanol (CH_3OH) at 100°C yields three alkene products, *viz.* 1-hexene [$H_2C{=}CH(CH_2)_3CH_3$] (33%),

TABLE 7.10. A Comparison of the Acidities of Alcohols, the Anions (Alkoxides), of Which in the Corresponding Alcohol Solution Are Commonly Used as Bases ("B:" or "B⁻") in the E2 Reaction

	Compound (alcohol)	pKa (in water)	Alkoxide		
Decreasing acidity	Water	HO–H	15.7[a] Hydroxide	HO⁻	Increasing basicity
	Methanol	CH_3O–H	16 Methoxide	CH_3O^-	
	Ethanol	CH_3CH_2O–H	18 Ethoxide	$CH_3CH_2O^-$	
	2-Propanol (*iso*-propanol)	$(CH_3)CHO$–H	18 2-Propoxide (*iso*-propoxide)	$(CH_3)_2CHO^-$	
	2-Methyl-2-propanol (*t*-butanol)	$(CH_3)_3CO$–H	19 2-Methyl-2-propoxide (*t*-butoxide)	$(CH_3)_3CO^-$	

[a]Based on $K_w = 10^{-14}$ and $[H_2O] = 55.5\,M$.

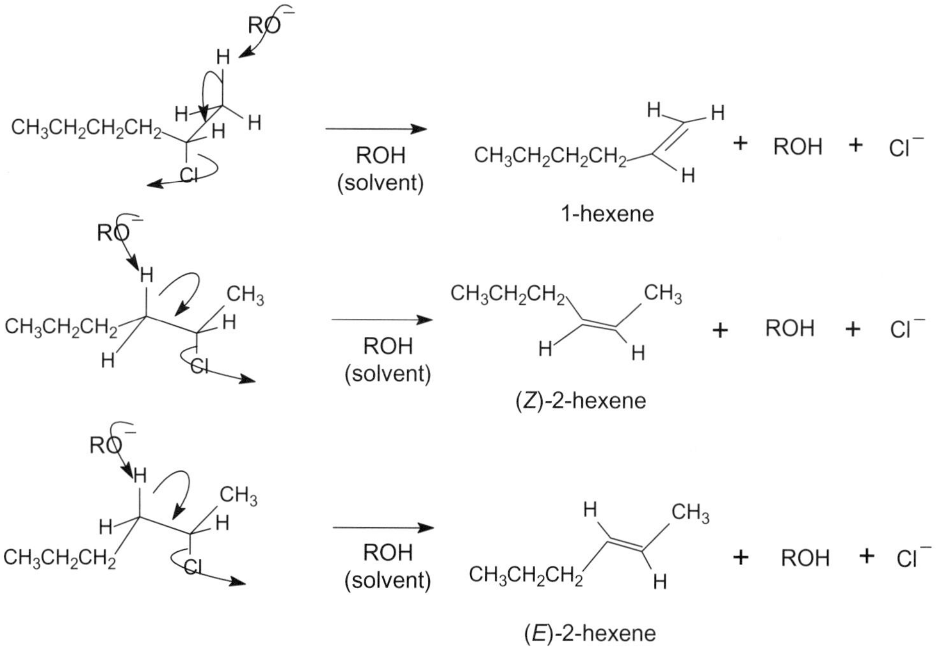

Scheme 7.35. A representation of bimolecular elimination pathways for the loss of HCl from (±)-2-chlorohexane [$H_3C(Cl)CH(CH_2)_3CH_3$] to produce 1-hexene [$H_2C=CH(CH_2)_3CH_3$] and (*E*)- and (*Z*)-2-hexene [(E,Z)-$CH_3CH=CH(CH_2)CH_3$]. The base "RO⁻" is an alkoxide anion. The ratio of products is a function of the alkoxide in the particular alcohol from which it is derived. Only antiperiplanar elimination is shown (see Bartsch, R. A.; Bunnett, J. F. *J. Am. Chem. Soc.*, **1968**, *90*, 408 and Bartsch, R. A.; Bunnett, J. F. *J. Am. Chem. Soc.*, **1969**, *91*, 1376).

trans- or (*E*)-2-hexene [(*E*)-CH$_3$CH=CH(CH$_2$)$_2$CH$_3$] (50%), and *cis-* or (*Z*)-2-hexene [(*Z*)-CH$_3$CH=CH(CH$_2$)$_2$CH$_3$] (17%). The same three products are found when the same substrate is treated with *t*-butoxide (2-methyl-2-propoxide [(CH$_3$)$_3$CO$^-$]) in *t*-butanol ([(CH$_3$)$_3$COH] 2-methyl-2-propanol) at 99°C. Now the ratio is 88% 1-hexene, 7% (*E*)-2-hexene, and 6% (*Z*)-2-hexene! The reactions are shown in Scheme 7.35 (see Bartsch and Bunnett, 1968, 1969).

As discussed above, the Saytzeff (more highly substituted alkene) product predominates in the first case, while the Hofmann (less highly substituted alkene) product is the major result of the second. Both of the reactions follow second-order kinetics. However, the stronger, less well solvated, more bulky base apparently abstracts a proton from the terminal methyl group more readily than it does from an internal carbon. To the extent that the proton abstraction leads the leaving of the chloride in the E2 process, the reaction becomes more E1cb like.

ii. THE NATURE OF THE LEAVING GROUP. Along the same lines, and as might be anticipated, if the leaving group (halogen) is made poorer, proton abstraction might also get ahead of the departure of the leaving group. Thus, again, a symmetrical E2 process such as that depicted in Figures 7.21 and 7.22 would not be found and a shift toward an E1cb-like process would be expected. In accord with this idea and as shown in Table 7.11, it is found that, holding everything else constant, changing the leaving group from iodine to bromine to chlorine and then to fluorine (each halogen worse than its predecessor) produces an increase in the amount of Hofmann product (less substituted alkene) generated. The same trend, despite a dramatic change in the sense of elimination is observed with *t*-butoxide (2-methyl-2-propoxide [(CH$_3$)$_3$CO$^-$]) when the solvent is changed from *t*-butanol (2-methyl-2-propanol [(CH$_3$)$_3$COH] to methylsulfinylmethane (DMSO [(CH$_3$)$_2$SO]), a dipolar aprotic solvent. Reasonably, acids whose conjugate bases have highly localized charges, that is, they cannot distribute the negative charge on the anion internally (by resonance), are much less acidic in dipolar aprotic solvents (where intermolecular hydrogen bonding is unavailable) than in water. The strength of the conjugate base is correspondingly increased.

TABLE 7.11. A Comparison of the Yields of Alkenes Obtained from 2-Halohexanes as a Function of Leaving Group (Halogen) and Base

2-Halohexane (Halogen)	Base/Solvent	Temperature (°C)	1-Hexene	Percent Yield (*E*)-2-Hexene	(*Z*)-2-Hexene
I	CH$_3$O$^-$/CH$_3$OH	100	19	63	18
Br	CH$_3$O$^-$/CH$_3$OH	100	28	55	18
Cl	CH$_3$O$^-$/CH$_3$OH	100	33	50	17
F	CH$_3$O$^-$/CH$_3$OH	100	70	21	9
I	(CH$_3$)$_3$CO$^-$/(CH$_3$)$_3$COH	99	69	20	11
Br	(CH$_3$)$_3$CO$^-$/(CH$_3$)$_3$COH	99	80	12	8
Cl	(CH$_3$)$_3$CO$^-$/(CH$_3$)$_3$COH	99	88	7	6
F	(CH$_3$)$_3$CO$^-$/(CH$_3$)$_3$COH	99	97	1	1

See Scheme 7.35 and compare Table 7.12. See Bartsch, R. A.; Bunnett, J. F. *J. Am. Chem. Soc.*, **1967**, *91*, 1376.

TABLE 7.12. A Comparison of the Yields of Hexenes Obtained from 2-Halohexanes as a Function of Leaving Group (Halogen) with *t*-Butoxide (2-Methyl-2-Propoxide [(CH₃)₃CO⁻]) in Methylsulfinylmethane (Dimethyl Sulfoxide [(CH₃)₂SO])

2-Halohexane (Halogen)	Base/Solvent	Temperature (°C)	1-Hexene	(E)-2-Hexene	(Z)-2-Hexene
I	$(CH_3)_3CO^-/(CH_3)_2CSO$	51	35	55	10
Br	$(CH_3)_3CO^-/(CH_3)_2CSO$	51	47	44	9
Cl	$(CH_3)_3CO^-/(CH_3)_2CSO$	51	59	34	7
F	$(CH_3)_3CO^-/(CH_3)_2CSO$	51	81	5	13

Compare with Table 7.11. See Scheme 7.35. See Bartsch, R. A.; Bunnett, J. F. *J. Am. Chem. Soc.*, **1967**, *91*, 1382.

As shown in Table 7.12 and in concert with what is expected as the leaving group is changed, the amount of Saytzeff product (more highly substituted alkene) diminishes as the leaving group changes in the order I, Br, Cl, and F. The poorer leaving group produces a larger amount of the less highly substituted alkene (the Hofmann product). However, in contrast to what was seen in Table 7.11, this strong base more closely resembles methoxide (CH_3O^-) in methanol (CH_3OH) than it does *t*-butoxide in *t*-butanol as regards the relative yields of alkene products. It is possible that this result can be accounted for by noting that in methylsulfinylmethane (DMSO [$(CH_3)_2SO$]), where hydrogen bonding to the solvent is unavailable, *t*-butoxide (2-methyl-2-propoxide [$(CH_3)_3CO^-$]) is unaccompanied by solvent molecules and thus its "effective" steric bulk, as "seen" by the proton to be abstracted, is diminished (relative to that in *t*-butanol [2-methyl-2-propanol, $(CH_3)_3COH$]). Reasonably, this would cause the preference for abstraction of the proton from the primary carbon to yield the Hofmann product (the less substituted alkene) to diminish too. The IR, ¹H NMR, and ¹³C NMR spectra of 1-hexene are presented in Figure 7.24a–c. The IR, ¹H NMR, and ¹³C NMR spectra of (Z)-2-hexene are presented in Figure 7.25a–c, and the IR, ¹H NMR, and ¹³C NMR spectra of (E)-2-hexene are presented in Figure 7.26a–c. It should be clear from these data that once the alkenes are separated (by gas chromatography), their identities are easily established.

Problem 7.10. Assume that the products from the elimination reactions with (±)-2-chlorohexane and the various bases discussed above could be separated by gas chromatography and isolated. The experimental data of Figures 7.24–7.26 provide the actual spectra one might expect to obtain on the purified samples. Define, in as much detail as you can, how it is known that the figures have not been mislabeled.

iii. THE STEREOCHEMISTRY OF THE E2 ELIMINATION REACTION. *Preliminary comments*: there are two facets of stereochemical interest in every E2 elimination reaction. First, the elimination may occur with either synperiplanar (*syn* or suprafacial) or antiperiplanar (*anti* or antarafacial) loss of the leaving group (halogen) and β-proton (Figure 7.16). Second, the unsaturated product may, in either case, be an (*E*)- or *trans*-alkene or a (*Z*)- or *cis*-alkene. Both facets can be examined simultaneously. To do so requires the following digression into stereochemical analysis and nomenclature.

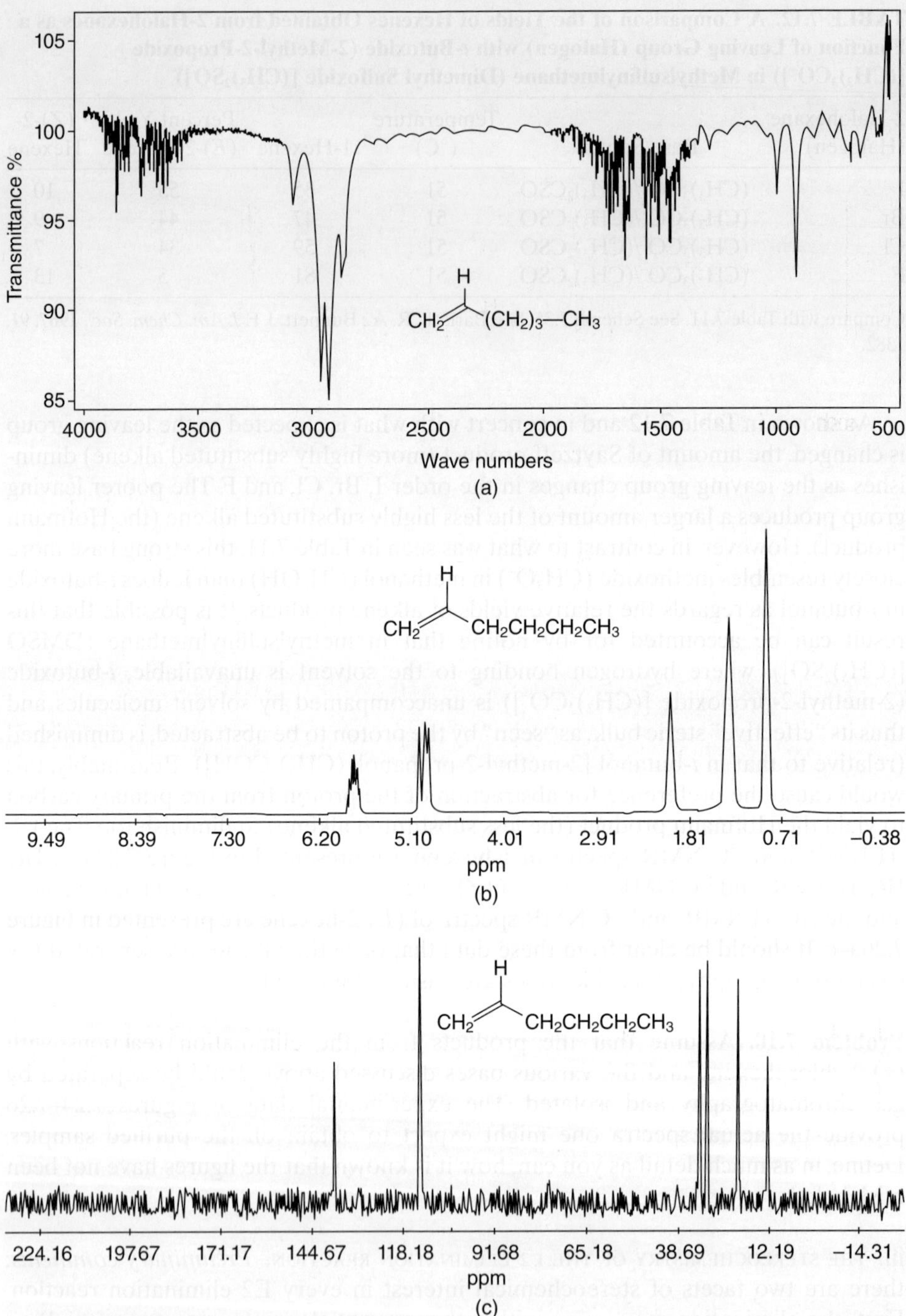

Figure 7.24. (a) The Fourier transform infrared (FT-IR) spectrum of neat 1-hexene between two NaCl slate plates. The noisy regions of the spectrum between 4000 and 3500 cm⁻¹ and between 1800 and 1400 cm⁻¹ are due to water vapor. (b) The ¹H NMR spectrum (at 300 MHz) of 1-hexene in ²HCCl₃. (c) The ¹³C NMR spectrum (at 75 MHz) of 1-hexene in ²HCCl₃.

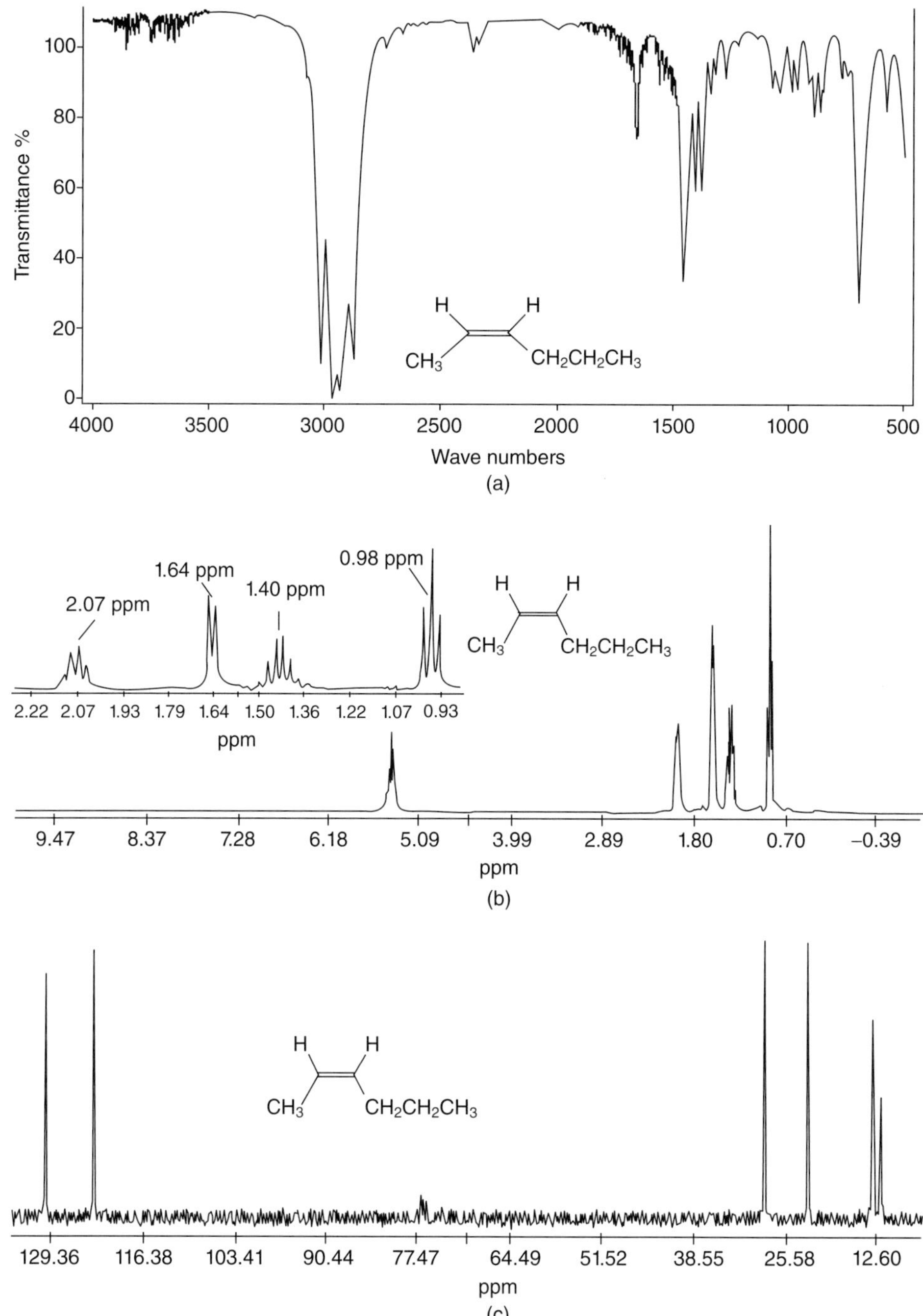

Figure 7.25. (a) The Fourier transform infrared (FT-IR) spectrum of (*Z*)-2-hexene neat liquid between NaCl windows. (b) The ^{1}H NMR spectrum (at 300 MHz) of (*Z*)-2-hexene in ^{2}HCCl$_3$. The inset is an expanded view of the upfield portion of the spectrum. (c) The ^{13}C NMR spectrum (at 75 MHz) of (*Z*)-2-hexene in ^{2}HCCl$_3$.

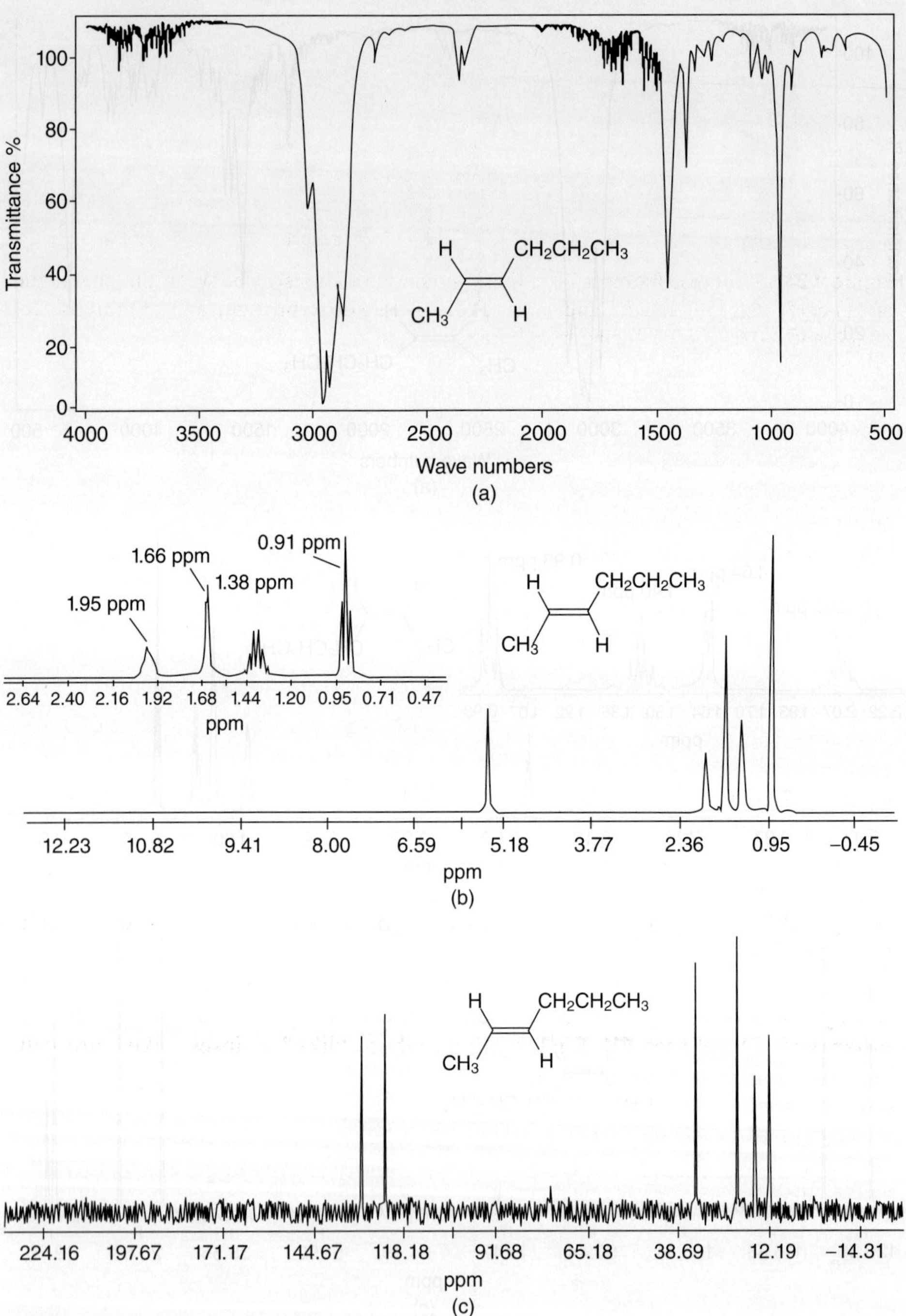

Figure 7.26. (a) The Fourier transform infrared (FT-IR) spectrum of (*E*)-2-hexene neat liquid between NaCl windows. (b) The ^{1}H NMR spectrum (at 300 MHz) of (*E*)-2-hexene in ^{2}HCCl$_3$. The inset is an expanded view of the upfield portion of the spectrum. (c) The ^{13}C NMR spectrum (at 75 MHz) of (*E*)-2-hexene in ^{2}HCCl$_3$.

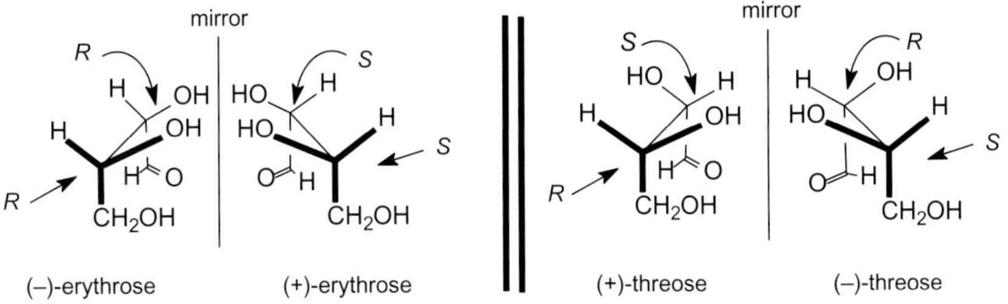

Figure 7.27. A representation of the diastereomeric relationship between the enantiomers of (±)-erythrose [(–)-2*R*,3*R*- and (+)-2*S*,3*S*-2,3,4-trihydroxybutanal] and (±)-threose [(+)-2*R*,3*S*- and (–)-2*S*,3*R*-2,3,4-trihydroxybutanal].

Historically, a significant body of chemistry was developed along with the analysis of structure and stereochemistry by examination of natural products. In the period between about 1875 and 1888, Emil Fischer* and his assistants patiently unraveled the structures and interrelationships among various "simple" sugars called "mono-saccharides."† Among these compounds, which are often written as acyclic polyhy-droxylated aldehydes (i.e., as a linear array of carbons bearing many hydroxyl [–OH] groups [Alcohols, Chapter 8] and at least one aldehyde [O=C–H] functional-ity at a chain terminus), are the two, four-carbon diastereomers (Chapter 4) erythrose and threose. (–)-Erythrose (2*R*,3*R*-2,3,4-trihydroxybutanal) and its enan-tiomer (+)-erythrose (2*S*,3*S*-2,3,4-trihydroxybutanal) are diastereomeric with the enantiomeric pair of (+)-threose (2*R*,3*S*-2,3,4-trihydroxybutanal) and (–)-threose (2*S*,3*R*-1,2,3-trihydroxybutanal) as shown in Figure 7.27. Although the chemistry of these and related materials will be discussed in more detail later (Chapter 11), they are introduced now because (a) a trivial nomenclature has arisen based on these structures and its understanding will make analysis of elimination reactions more facile, and (b) it is now possible to introduce the use of another form of projection formula representation.

Examine the stereochemistry about the C2–C3 bond in (±)-erythrose. Note that the hydroxyl (–OH) groups eclipse each other, the hydrogens (–H) eclipse each other, and C1 eclipses C4. Such systems, where "like" eclipses "like," are called *erythro*. Those resembling (±)-threose, where "like groups" are not eclipsed are, generically, *threo*. Additional examples are provided in Figure 7.28.

The communication of the stereochemical information elucidated for these and related carbohydrates (Chapter 11) is conveniently conveyed graphically by use of the Fischer projection formulas.

In a Fischer projection, each asymmetrically substituted carbon atom is placed in the plane of the paper and is symbolized by crossed vertical and horizontal lines.

---

*A brief biography of Emil Fischer is given earlier in this chapter, Section E in connection with Walden inversion.

†Monosaccharides are one of a class of compounds generically known as carbohydrates. As implied by the name, carbohydrates are, approximately, "hydrates of carbon," Cn(H₂O)n. They will be discussed further in Chapter 11.

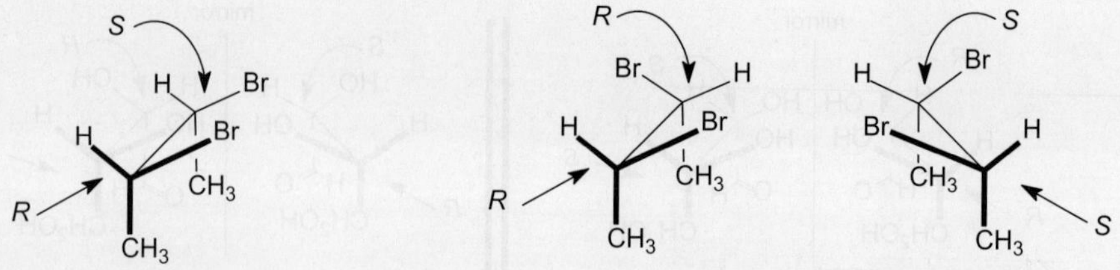

meso-2,3-dibromobutane

(*erythro*-2,3-dibromobutane)

(±)-2,3-dibromobutane

(*threo*-2,3-dibromobutane)

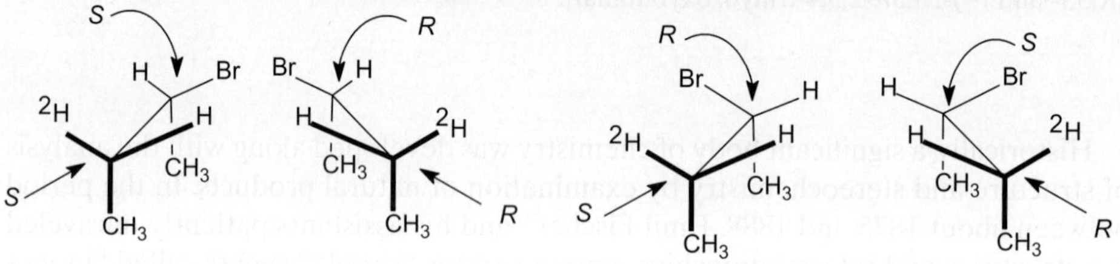

(±) 2S,3S- and 2R,3R-

2-bromo-3-deuteriobutane

(*threo*-2-bromo-3-deuteriobutane)

(±) 2S,3R- and 2R,3S-

2-bromo-3-deuteriobutane

(*erythro*-2-bromo-3-deuteriobutane)

Figure 7.28. A representation of the use of *erythro* and *threo* nomenclature with several different compounds. It is important to note that the use of these descriptors is independent of rotational isomerism and that the eclipsed conformers are shown only to emphasize similarities and differences.

The letter "C" is specifically omitted. The four substituents are then placed at the extrema of each line understanding that those substituents on the vertical lie beneath the plane of the paper and those on the horizontal lie above the plane of the paper. The structures of the enantiomers of erythrose and threose and their corresponding Fischer projections are shown in Figure 7.29.

Problem 7.11. Examine the enantiomers and diastereomers shown in Figure 7.29. Then, as with the example shown for C2 of (−)-erythose, define the absolute stereochemistry as *R* or *S* at each asymmetrically substituted carbon in the figure using the Cahn–Prelog–Ingold rules.

As shown in Figure 7.29 and Problem 7.11, the Fischer projection can only be rotated in the plane of the paper 180° without changing its meaning. A rotation of 90° or 270° inverts the absolute configuration. Therefore, it is frequently desirable to translate such two-dimensional representations into a recognizable three-dimensional shape (such as a tetrahedron viewed "edge-on") before attempting to determine the order in which the substituents must be placed.

Now, consider the diastereomeric *erythro*- and *threo*-5-chloro-6-deuteriodecanes shown in Schemes 7.36 and 7.37, respectively. As shown, when these diastereomers are, individually, treated with with *t*-butoxide (2-methyl-2-propoxide [$(CH_3)_3CO^-$])

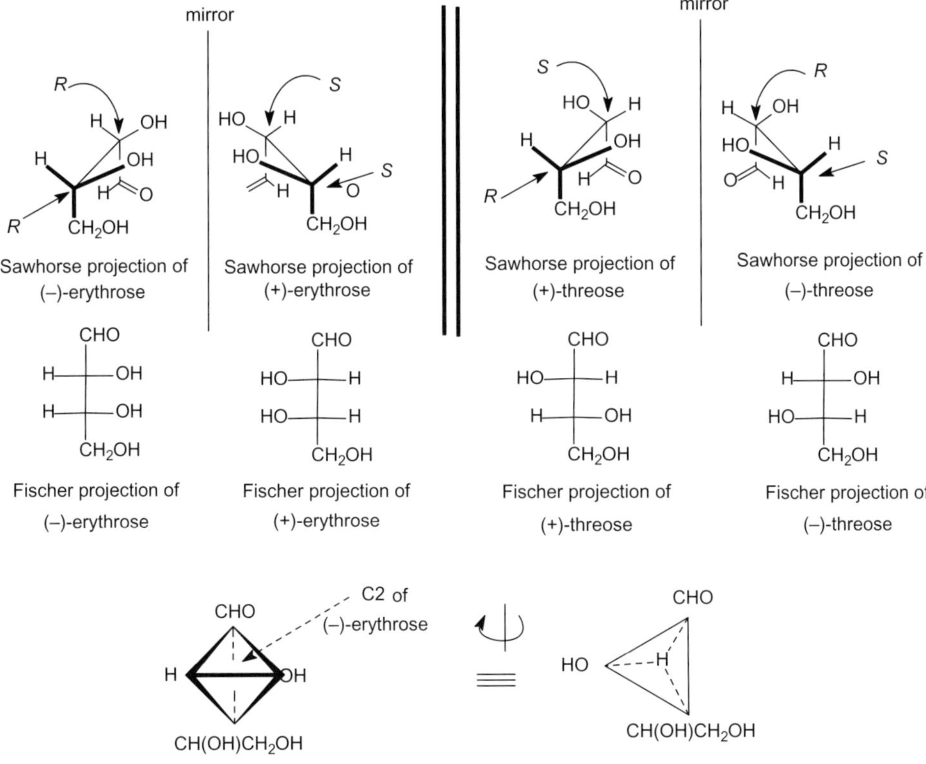

Figure 7.29. A set of representations for the isomers of erythrose and threose. They are shown first as sawhorse representations (**as in** Figure 7.27) and then as Fischer projections. An analysis of the stereochemistry at C2 of (–)-erythose, showing that it is R is also presented. See Problem 7.11.

both ^{1}HCl and ^{2}HCl are lost. Then, since the (E)- and (Z)-alkenes are separable, the sense of elimination can be defined!

As shown in Scheme 7.36 for the *erytho*-isomer, *syn*- (*cis* or suprafacial) elimination produces (Z)- (or *cis*) 5-decene **without** ^{2}H and (E)- (or *trans*) 5-decene **with** ^{2}H. Alternatively, *anti*- (*trans* or antarafacial) elimination produces (Z)- (or *cis*) 5-decene **with** ^{2}H and (E)- (or *trans*) 5-decene **without** ^{2}H. Then, since (Z)- (or *cis*) alkene can contain deuterium (^{2}H) if (and only if) the course of the elimination process is *anti* (*trans*- or antarafacial) and (E)- (or *trans*-) alkene can contain deuterium (^{2}H) if (and only if) the course of the elimination process is *syn* (*cis* or suprafacial), gas chromatographic separation of the isomeric alkenes and their examination by mass spectrometry permits determination of the path of elimination or, as is seen, the extent to which each path is followed. Careful product analysis shows that in methylsulfinylmethane (DMSO [$(CH_3)_2SO$]) solvent, about 6–7% of both the (E)- and (Z)-alkenes are produced by *syn*- (*cis* or suprafacial) elimination. However, when the solvent is benzene (C_6H_6), about 60% of the (E)- or *trans*-product and 40% of the (Z)- or *cis*-product are produced by *syn*- (*cis* or suprafacial) elimination. It is argued that in the less polar solvent (benzene,

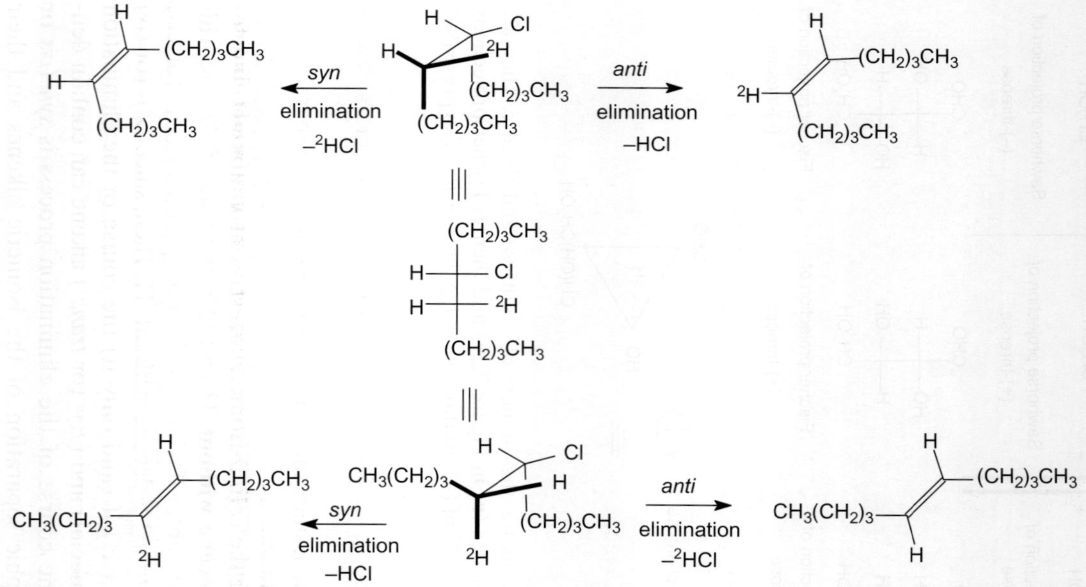

Scheme 7.36. The results of E2 elimination of ^{2}HCl and ^{1}HCl from *erythro*-5-chloro-6-deuteriodecane with *t*-butoxide (2-methyl-2-propoxide $[(CH_3)_3CO^-]$) in methylsulfinylmethane (dimethyl sulfoxide, DMSO $[(CH_3)_2SO]$) solvent. Note that *syn*-elimination provides (*Z*)-alkene that has lost deuterium (^{2}H) and (*E*)-alkene that has lost protium (^{1}H). Thus, examination of the separated alkenes by mass spectroscopy provides definitive information about the path and the results must be complementary to those in Scheme 7.37. The work reported above and in Schemes 7.36 and 7.37 is taken from Sicher, J.; Zavada, J.; Pankova, M. *Collect. Czech. Chem. Commun.*, **1971**, *36*, 3140 and Pankova, M.; Svoboda, M.; Zavada, J. *Tetrahedron Lett.*, **1972**, 2465. The interpretation of the results is subject to two caveats. First, elimination also occurs to produce the enantiomers of (*E*)- and (*Z*)-6-deuterio-4-decene. The 4-decene and 5-decene isomers are spectroscopically distinguishable and, at least in principle, chromatographically separable. Second, since a carbon–deuterium bond is stronger than a carbon–protium bond (the mass of deuterium [^{2}H] is greater than that of protium [^{1}H]; see Chapter 2, D.III for a brief discussion of the effect of nuclear mass on bond strength), breaking the ^{2}H–C bond will be more difficult than breaking the ^{1}H–C bond. The problem of carbon–hydrogen bond strength differences is exacerbated by the extent to which the reaction is E1cb like. As indicated above, the use of the *t*-butoxide (2-methyl-2-propoxide $[(CH_3)_3CO^-]$) in methylsulfinylmethane (dimethyl sulfoxide, DMSO $[(CH_3)_2SO]$) solvent causes the E2 reaction to have E1cb characteristics.

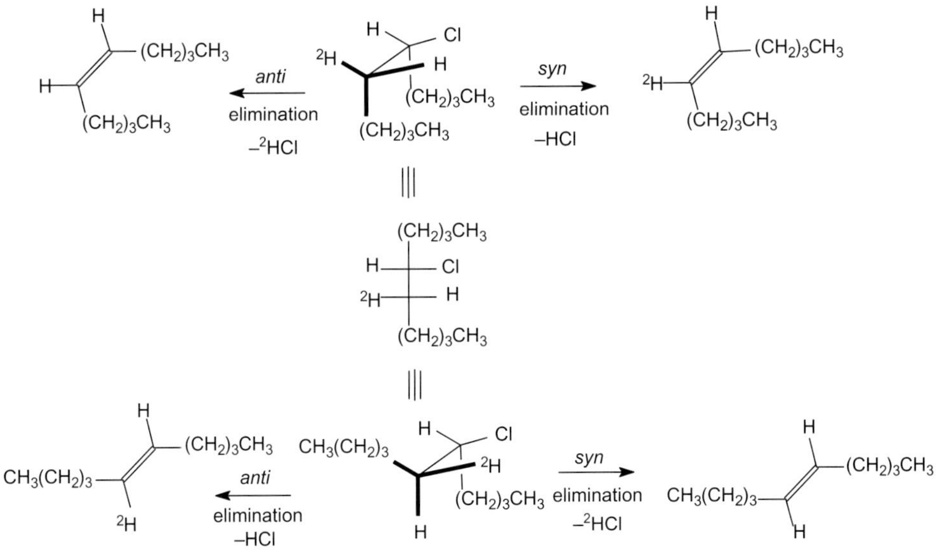

Scheme 7.37. The results of E2 elimination of ^{2}HCl and ^{1}HCl from *threo*-5-chloro-6-deuterio-decane with *t*-butoxide (2-methyl-2-propoxide [$(CH_3)_3CO^-$]) in methylsulfinylmethane (dimethyl sulfoxide, DMSO [$(CH_3)_2SO$]) solvent. Note that *syn*-elimination provides (*E*)-alkene that has lost deuterium (^{2}H) and (*Z*)-alkene that has lost protium (^{1}H). Thus, examination of the separated alkenes by mass spectroscopy provides definitive information about the path, and the results must be complementary to those in Scheme 7.36. (See Scheme 7.36 for reference information).

C_6H_6), the negatively charged base and its associated cation are closer to the more polar chlorine during the elimination process (i.e., in a nonpolar solvent, separation of charges is more difficult) and that this accounts for the change in direction of elimination. As shown in Scheme 7.37, work with the *threo*-isomer is exactly complementary.

Problem 7.12. Assuming that medium resolution mass spectroscopy was available for examination of the products of elimination, suggest what you might find for results of the various elimination pathways of ^{1}HCl and ^{2}HCl shown in Schemes 7.36 and 7.37.

Interestingly, in six-membered rings where the double bond must be *cis*, it is clear that E2 elimination preferentially occurs in an *anti* (*trans* or antarafacial) fashion. For example, as shown in Figure 7.30, the isomeric pair of (*Z*)- and (*E*)- (or *cis* and *trans*) 4-bromo-(1,1-dimethylethyl)cyclohexane [4-bromo-*t*-butylcyclohexane] both undergo E2 elimination with *t*-butoxide (2-methyl-2-propoxide [$(CH_3)_3CO^-$]) in *t*-butanol (2-methyl-2-propanol [$(CH_3)_3COH$]) to yield the corresponding cyclohexene. Remembering that the *t*-butyl group acts as a bulky conformational anchor, thus strongly preferring the equatorial position (Chapter 4, Table 4.2), it is not surprising that the (*Z*)- (or *cis*) isomer, with the bromine axial and two β-diaxial protons arrayed antiperiplanar to the bromine, reacts about 500 times faster than the (*E*)- (or *trans*) isomer.

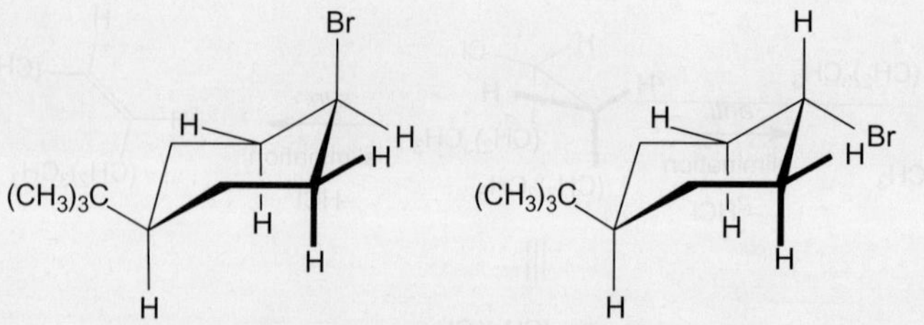

Figure 7.30. (*Z*)- and (*E*)- (or *cis* and *trans*) 4-bromo-(1,1-dimethylethyl)-cyclohexane (4-bromo-*t*-butylcyclohexane). Both undergo E2 elimination with *t*-butoxide (2-methyl-2-propoxide [$(CH_3)_3CO^-$]) in *t*-butanol (2-methyl-2-propanol [$(CH_3)_3COH$]). The (*Z*)-isomer, with the bromine axial and two anticoplanar protons, reacts about 500 times faster than its isomer.

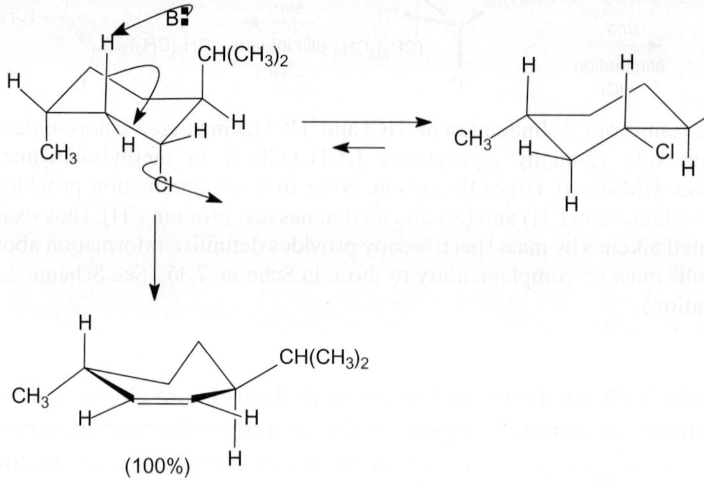

Scheme 7.38. A representation of the course of the E2 reaction on (1*R*,3*R*,4*S*)-3-chloro-4-(1-methylethyl)methylcyclohexane (menthyl chloride). The base B: is hydroxide and the solvent is ethanol.

Similarly, elimination of hydrogen chloride (HCl) from (1*R*,3*R*,4*S*)-3-chloro-4-(1-methylethyl)methylcyclohexane (menthyl chloride) and (1*R*,3*S*,4*S*)-3-chloro-4-(1-methylethyl)methylcyclohexane (neomenthyl chloride) with sodium hydroxide in ethanol (conditions favoring the E2 reaction) has been examined (Schemes 7.38 and 7.39).*

As shown in Scheme 7.38 for menthyl chloride [(1*R*,3*R*,4*S*)-3-chloro-4-(1-methylethyl)methyl-cyclohexane], the isomer in which all of the substituents are (preferentially) equatorial, antiperiplanar elimination can only occur when the ring flexes into the less stable conformation where all groups are axial. As shown in the scheme, there is only one proton that is both axial and β to the chlorine. Thus, there is only one E2 elimination product on reaction with sodium hydroxide in ethanol,

*Hughes, E. D.; Ingold, C. K.; Rose, J. B. *J. Chem. Soc.*, **1953**, 3839.

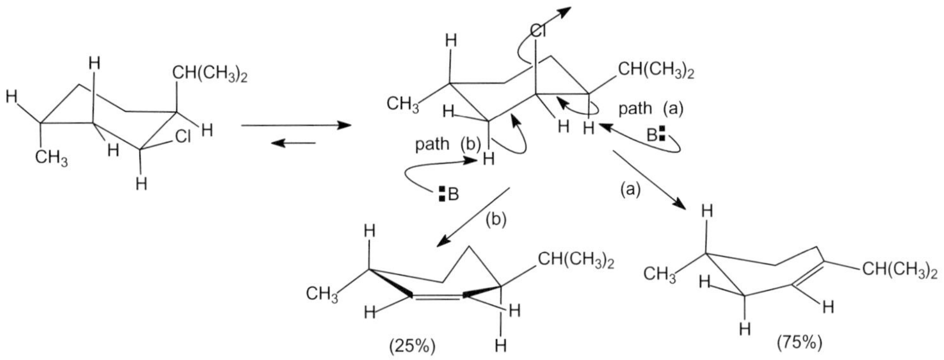

Scheme 7.39. A representation of the course of the E2 reaction on (1*R*,3*S*,4*S*)-3-chloro-4-(1-methylethyl)methylcyclohexane (neomenthyl chloride). The base B: is hydroxide and the solvent is ethanol.

that is, 2-menthene [3-(1-methylethyl)-6-methylcyclohexene], the less substituted (and thus thermodynamically less stable) alkene (the Hofmann product). The results of this E2 reaction (100% Hofmann product) should be compared with those of the E1 reaction on the same compound (Scheme 7.32) where the ratio of Hofmann to Saytzeff products was 1:2 (i.e., only about 33% of the product was 2-menthene).

On the other hand, with neomenthyl chloride [(1*R*,3*S*,4*S*)-3-chloro-4-(1-methylethyl)methylcyclohexane] (Scheme 7.39), where the chlorine is axial and the two alkyl substituents are equatorial and where, as a consequence of that arrangement, there are now two *anti-* (or antarafacial) and axial protons situated β to the chlorine (Cl), two products are possible and, as expected, the Saytzeff product is preferred. As before, the results of this E2 reaction (3:1 in favor of the Saytzeff [more highly substituted alkene]) should be compared with those of the E1 reaction on the same compound (Scheme 7.33) where the ratio of the Saytzeff to Hofmann products was 98:2.

Table 7.13 contains a short list of the almost limitless number of (variable) E2 reactions from which examples might be provided. Indeed, although the table holds only alkyl and alkenyl halides, in concert with the substance of this chapter, the nature of the leaving group can be altered to include derivatives of such common functional groups as alcohols (Chapter 8), thiols (Chapter 8), and amines (Chapter 10). As more work is done, the list grows.

The first example of Table 7.13 illustrates that preferential *anti-* (*trans* or antarafacial) elimination is encountered with ethoxide in ethanol from an acyclic substrate. Only one product is reported from the (shown) racemic *threo*-dibromide [(±)-1,2-dibromo-1,2-diphenylethane (*cis*-stilbene dibromide)] as well as from the (not shown) diastereomeric *erythro*-dibromide [*meso*-1,2-dibromo-1,2-diphenylethane (*trans*-stilbene dibromide)] in the first step of the elimination. Interestingly, both isomers also undergo a second elimination [shown from (*Z*)-1-bromo-1,2-diphenylethene] to 1,2-diphenylethyne (diphenylacetylene).

In the second example, a hindered alkoxide base and high-boiling alkyl-substituted arene were used to insure elimination. There are two interesting principles (Scheme 7.40) illustrated by this example. First, loss of the bridgehead proton (the proton at

TABLE 7.13. Some Additional E2 Reactions of Alkyl and Alkenyl Halides

Example	Substrate	Base/Solvent	Product(s)
1	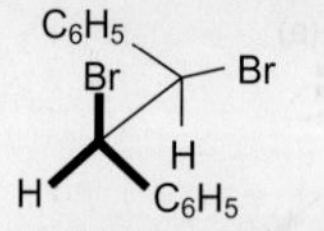 (±)-*threo*-stilbene dibromide	$CH_3CH_2O^- K^+$ in CH_3CH_2OH	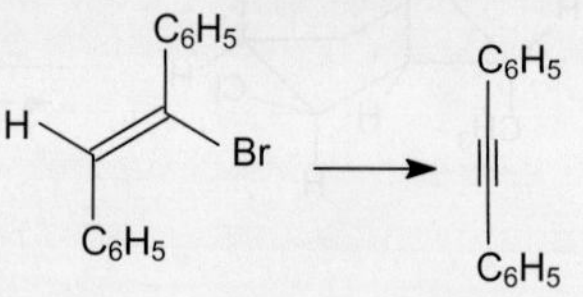
2	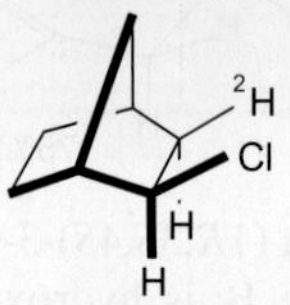 (±)-*exo*-2-chloro-*exo*-3-(deuterionorbornane)	$(CH_3CH_2)_2(CH_3)CO^-$ in *para*-cymene [4-(1-methylethyl)toluene]	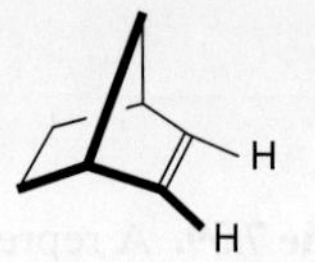
3	$(CH_3)_3C$—Br 2-bromo-2-methylpropane (*tert*-butyl bromide)	$CH_3CH_2O^- K^+$ in CH_3CH_2OH	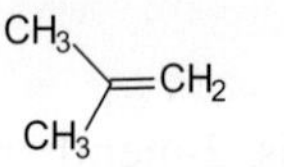
4	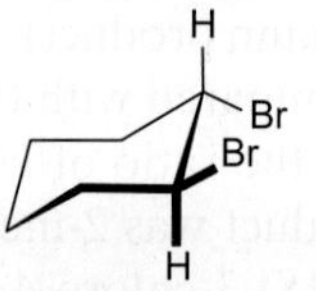 (*E*)-1,2-dibromocyclohexane (*trans*-1,2-dibromocyclohexane)	$(CH_3)_2CHO^-$ in $(CH_3)_2CHOH$	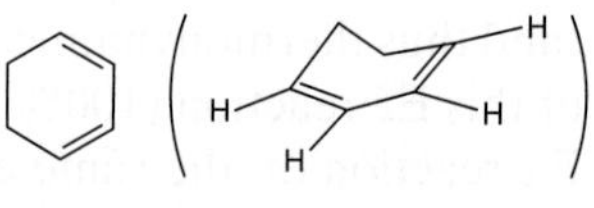

For these and additional examples, see Saunders, W. H. Jr.; Cockerill, A. F. *Mechanisms of Elimination Reactions*, **1973**, Wiley, New York.

C1) is not observed. Introduction of a double bond at a bridgehead of a bicyclo[2.2.1] system is "forbidden."*

Second, since neither rotation about the C2–C3 bond nor ring flexing motion can be effected and the β-endo hydrogen (H) at C3 cannot be forced to a 180° dihedral angle with respect to chlorine (Cl) at C2, only *syn* (*cis* or suprafacial) is possible.

The third example is the E2 elimination of hydrogen bromide (HBr) from *t*-butyl bromide (2-bromo-2-methylpropane [$(CH_3)_3CBr$]) by ethoxide in ethanol and is unremarkable save that it serves to emphasize that only elimination and no substitution occurs. Thus, *t*-butyl ethyl ether [$(CH_3)_3COCH_2CH_3$] cannot be prepared this way. Although not shown, bromoethane (CH_3CH_2Br) primarily undergoes

---

*The introduction of a double bond to the bridgehead of such small bicyclic systems is "forbidden" according to a rule named after the German chemist J. Bredt, who first reported (1935) that the proton at the bridgehead was apparently not lost on enolization of bicyclo[2.2.1]heptan-2-one (Chapter 9). Further investigation and popularization of, as well as limitations to, "Bredt's rule" was undertaken by V. Prelog (Chapter 4) some two decades later. Prelog pointed out that the limit of applicability of Bredt's rule lay between systems with seven- and eight-membered rings. Thus, a bridgehead double bond is readily accommodated in, for example, bicyclo[5.3.1]undecane. It should be clear that the question of stability requires specification of conditions.

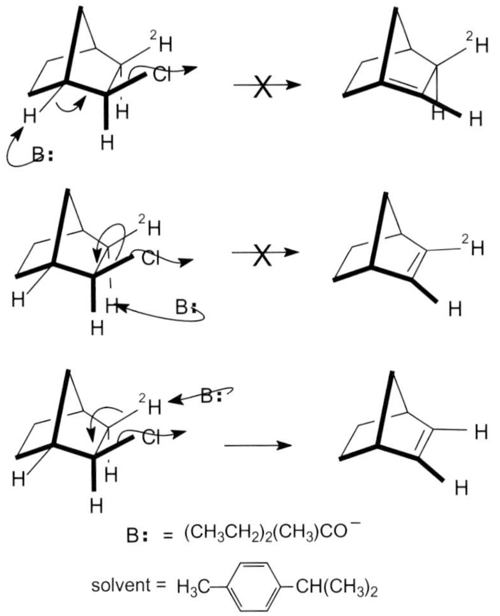

Scheme 7.40. Elimination reaction attempts on *exo*-2-chlorobicyclo[2.2.1]heptane. The first elimination fails because of failure to attain coplanarity (Bredt's rule is violated). The second elimination fails since chlorine cannot be arranged antiperiplanar to any proton. The last process is synperiplanar elimination.

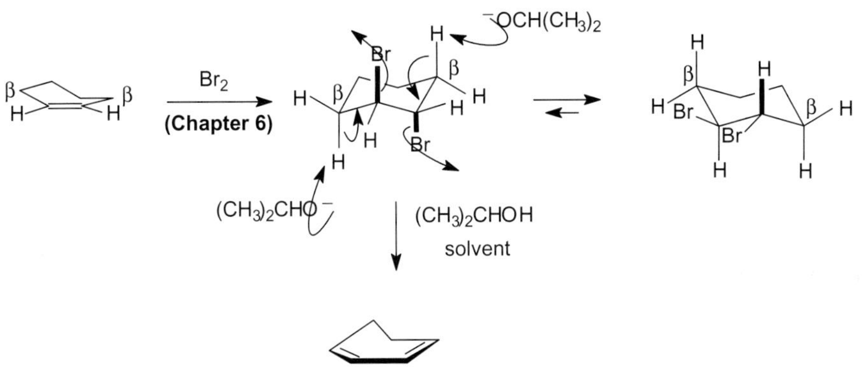

Scheme 7.41. A representation of the elimination of 2 equivalents of hydrogen bromide (HBr) from (*E*)-1,2-dibromocyclohexane. The formation of this isomer by bromination of cyclohexene is also shown.

substitution under the same conditions, and 2-bromopropane [$(CH_3)_2CHBr$] yields a mixture of elimination and substitution products.

The final example in Table 7.13 shows that two elimination reactions can occur on the same substrate. Although it appears that elimination occurs from the 1,2-diequatorial isomer, it is almost certainly true that conformational change to the diaxial isomer (prepared, e.g., by addition of bromine [Br_2] to cyclohexene) precedes the elimination as shown in Scheme 7.41.

4. 1,2-Elimination and the Primary Deuterium Isotope Effect. Figure 7.23 is **repeated** below to emphasize the difference between the rate-determining steps for the E1, the E2, and the E1cb reactions.

The rate-determining step of the idealized E1 reaction involves only the departure of the leaving group. The proton is lost to the base in a subsequent, rapid step, that is,

$$\text{rate} = -d[\text{substrate}]/dt = +d[\text{product}]/dt = k[\text{substrate}]. \tag{7.83}$$

In contrast, in the E2 reaction, proton loss occurs in the rate-determining step, that is,

$$\text{rate} = -d[\text{substrate}]/dt = -d[\text{base}]/dt = k[\text{substrate}][\text{base}]. \tag{7.84}$$

Now, the isotope of hydrogen (^{1}H), which bears a single neutron (along with the proton) in its nucleus is deuterium (^{2}H) and the mass of the neutron, is about the same as that of the proton. Thus, there is almost a doubling of mass when a proton is replaced by a deuteron and the difference in masses results in a difference in the zero point energies of the C–^{1}H and C–^{2}H bonds with the latter being stronger than the former. It is more difficult to break a carbon–deuterium bond than it is to break a carbon–hydrogen bond.

The difficulty in breaking the bond to carbon is reflected in the rate constants for a reaction in which the breaking of the bond is in the rate-determining step. When protium (^{1}H) is replaced by deuterium (^{2}H) and the effect is seen, it is reported as a *primary deuterium isotope effect*. Largely as a consequence of the difference in zero point energies for bond stretching, the ratio is expected to be

$$k_{\text{H}}/k_{\text{D}} \approx 7. \tag{7.85}$$

On the other hand, there should be no primary deuterium isotope effect if, as in the E1 reaction, the removal of the proton is not involved in the rate-determining step, the rate constants would be equal, and the ratio unity.

If a reaction occurs by a mixture of the two processes, a result greater than 1 but less than 7 would be anticipated.

Finally, for the E1cb process, if formation of the carbanion is rate determining and the carbanion, once formed, more rapidly goes on to product than it returns to starting material, the E1cb "reduces" to the E2 process and an E2 isotope effect would be expected. However, if the rate of return to reactant from carbanion is comparable with the rate of product formation or if the rate of return to reactant is faster than the rate of product formation, then the kinetic expression becomes much more difficult to evaluate and it is not clear that an E1cb mechanism can be identified.

Interestingly, if the starting material is recovered before an elimination reaction has gone to completion, the amount of deuterium originally present should be reduced if a carbanion is formed as an intermediate in a protic (^{1}H) solvent. This is because (re)formation of reactant (substrate) from carbanion in such a solvent would be unlikely to result in reincorporation of a deuteron (^{2}H). However, evidence of the exchange of ^{1}H for ^{2}H in recovered starting material does not necessarily dictate that the E1cb pathway is being followed since the carbanion necessary for the exchange may not be the same carbanion that lies on the pathway for elimination.

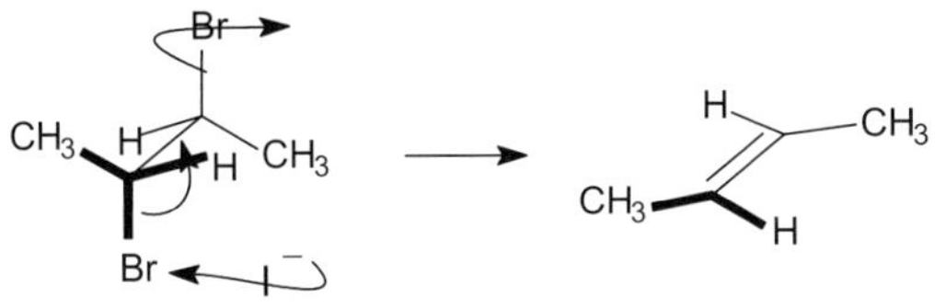

erythro-2,3-dibromobutane

(*meso*)

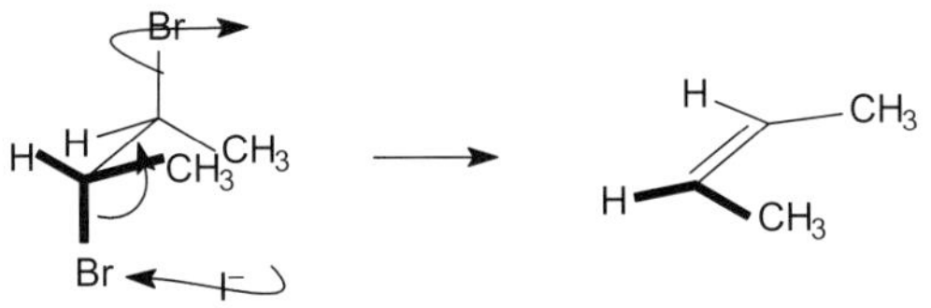

threo-2,3-dibromobutane

(racemic)

Scheme 7.42. A representation of α,β-dehalogenation from *erytho-* and *threo*-2,3-dibromobutanes on treatment with iodide anion (I⁻) in ethanol (CH₃CH₂OH).

b. 1,2- or α,β-Elimination of X₂ (X = Cl, Br) from Alkyl and Alkenyl Dihalides.

Elimination of halogen from α,β-dihalides in a variety of circumstances has been observed and can be used to prepare alkenes and alkynes from vicinyl dihalides. The process is usually of limited utility since the dihalides themselves are often prepared from the unsaturated compounds. As shown in Scheme 7.42, both *erythro* (*meso*) and *threo* ($\pm$)-2,3-dibromobutane diastereomers undergo stereospecific elimination of bromine to their respective (*E*)- (or *trans*) and (*Z*)- (or *cis*) 2-butenes on treatment with iodide anion (I⁻) in ethanol (ethyl alcohol, CH₃CH₂OH).

Metal-promoted elimination of any pair of halogens (except fluorine [F]) has also been used to produce double and triple bonds. The use of zinc (Zn) metal (with formation of zinc(II) dibromide [ZnBr₂]) is common and, occasionally, salts of other metals, for example, tin(II) chloride [SnCl₂], have also been used as shown in Scheme 7.43. There is some evidence that when these Lewis acids are used, the products form by single-electron transfer or radical process.

III. γ-Elimination (1,3-Elimination) and δ-Elimination (1,4-Elimination)

a. γ-Elimination of HX (X = Cl, Br, I) from Alkyl and Alkenyl Halides.

Although a discussion of the rich chemistry of ketones will not be undertaken until Chapter 9, where substitution (S$_N$1 and S$_N$2) and elimination (E1 and E2) reactions at the carbon α to the carbonyl will be considered in some detail, the peculiar reaction reported by the Russian chemist A. Faworsky* early in the twentieth century and

---

*A. Faworsky (A. Favorskii), a Russian chemist, who simultaneously held positions and had coworkers at the University of St. Petersburg and the Frauen-hochschule in the same city, published a clear study of the rearrangement reaction in *J. Prakt. Chemie*, **1913**, *88*, 641.

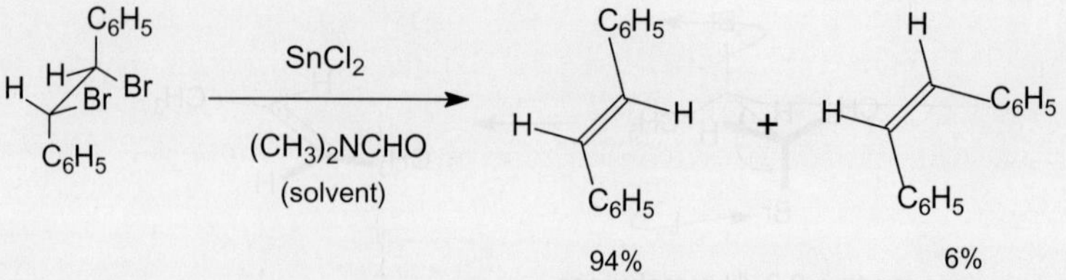

Scheme 7.43. A representation of the use of tin(II) chloride (stannous chloride, $SnCl_2$) in the dehalogenation of racemic *threo*-1,2-dibromo-1,2-diphenylethane [(±)-stilbene dibromide] to (mostly) (*E*)- (or *trans*) stilbene (94%) and a trace (6%) of (*Z*)- (or *cis*) stilbene. There is some evidence that this and related reactions may proceed through a process of one electron transfer reactions.

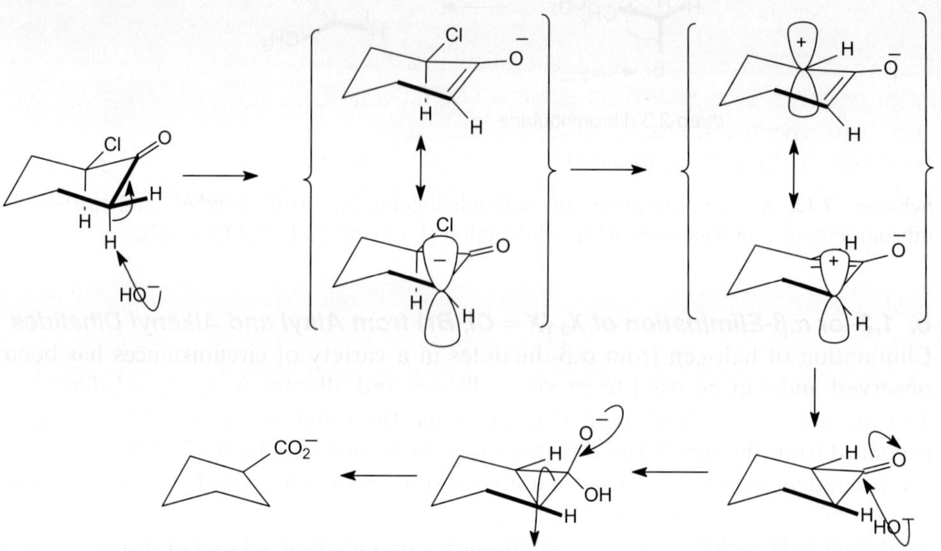

Scheme 7.44. A representation of a pathway to cyclopentanecarboxylic acid anion from 2-chlorocyclohexanone (the Faworsky [Favorskii] rearrangement). The path is presumed to take place via γ- (or 1,3) elimination of hydrogen chloride (HCl) through a substituted cyclopropanone intermediate.

which continues to be actively investigated will be initially considered here. The Favorskii rearrangement is generally accepted as one in which 1,3 (or γ) loss of HX (X = Cl, Br, I) occurs to produce an intermediate (a cyclopropanone), where, under the same conditions (base) required for its creation, it continues to react, producing (in aqueous medium) the anion of a carboxylic acid (Chapter 9)! As shown in Scheme 7.44, when the Favorskii rearrangement is carried out on 2-chlorocyclohexanone with aqueous base, the product is the salt of cyclopentanecarboxylic acid (which yields the carboxylic acid itself on treatment with mineral acid). The pathway shown requires initial proton loss from a carbon α to the carbonyl to produce a resonance-stabilized carbanion. Then, the (now) allylic halide departs in an S_N1-like process producing a second resonance-stabilized "zwitterion" (Chapter 1), which collapses to the cyclopropanone. The latter is attacked at the carbon of the carbonyl by the base (hydroxide anion, HO⁻) forming a tetrahedral intermediate, which opens (in either direction) to generate the carboxylic acid anion. Different bases (e.g., methoxide, CH_3O^-) give other products (i.e., methoxide $[CH_3O^-]$ yields the corresponding methyl ester).

Numerous examples of the Favorskii (Faworsky) rearrangement, including some with labeled carbon (see Problem 7.13), are known and the reaction will be encountered again.

Problem 7.13. When 2-chlorocyclohexanone-2-^{14}C is treated with sodium methoxide in methanol, labeled methyl cyclopentanecarboxylate results. Careful degradation of the carboxylic acid ester (Chapter 9) showed that the location of the position of the labeled carbon was restricted to three carbon atoms. (a) On which three carbons is the label found? (b) What is the distribution of the label (i.e., what percent of the label is found at each carbon)?

b. γ-Elimination of X_2 (X = Cl, Br, I) from Alkyl Halides. A number of specific cases of elimination from 1,3-dihalides are known. Those that occur most readily appear to have the specific geometry that minimizes the number of conformations available to the two sites across which reaction takes place. It has been argued that the dehalogenation occurs in two steps and if conformational mobility is restricted, unwanted reactions are minimized. Thus, as shown in Equation 7.86 both the (*E*)- (or *trans*) and (*Z*)- (or *cis*) isomers of 1-bromo-3-chlorocyclobutane react to produce the same bicyclo[1.1.0]butane with about equal facility.

$$\text{Cl}\!-\!\langle\rangle\!-\!\text{Br} \xrightarrow[\substack{\text{1,4-dioxacyclohexane} \\ \text{(dioxane)}}]{\text{Na metal}} \langle\!\ominus\!\rangle \qquad (7.86)$$

(solvent)

c. δ-Elimination of X_2 (X = Cl, Br, I) from Alkenyl Halides. When a double or triple bond lies between two halogens (i.e., 1,4-dihalide), loss of both halogens with introduction of another bond to extend conjugation across the system can often be effected (Equation 7.87). Indeed, as shown in Equation 7.88, such 1,4-elimination can be used to prepare even highly strained systems.

$$BrCH_2-C\equiv C-CH_2Br + Zn \longrightarrow CH_2{=}C{=}C{=}CH_2 + ZnBr_2 \qquad (7.87)$$

$$\underset{\text{CH}_2\text{Br}}{\overset{\text{CH}_2\text{Br}}{\bigcirc}} \xrightarrow{\text{Zn}} \bigcirc\!\square \;+\; ZnBr_2 \qquad (7.88)$$

H. REARRANGEMENT REACTIONS OF ALKYL AND ALKENYL HALIDES

Rearrangements, it will be recalled, involve nonrandom structural interconversions in which the same number of nuclei is present in the substrate and the product but their connectivity has changed. For example, in addition to undergoing S_N1 and

S_N2 reactions, allylic halides can react with rearrangement of the carbon–carbon double bond.

Thus, in systems such as (*Z*)- (or *cis*) 3-chloro-1-deuterio-1-butene exogenous chloride reacts stereospecifically, with the incoming chloride (which might be radio-labeled) entering *syn* (or *cis* or suprafacially) to the leaving group. Other nucleophiles behave in the same way, leading to rearranged carbon skeleton and the process, which frequently occurs with second-order kinetics, is referred to as the S_N2' reaction (Scheme 7.45).

Many of the rearrangements of alkyl chlorides were first observed in the exploration of the chemistry of the naturally occurring group of hydrocarbons called terpenes (see Chapter 6, Section I, Oxidation, for an introduction to this family of C10 hydrocarbons). Although similar rearrangements were known before the twentieth century began, their more detailed examination was not undertaken until about 1920.*

It was clearly understood by about 1940 (Scheme 7.46) in one of the first experiments to use radiolabeled chlorine, that the product of addition of hydrogen chloride (HCl) to (+)-camphene (Figure 7.31) [(+)-3,3-dimethyl-2-methylenebicyclo[2.2.1] heptane; from "oil of camphor" isolated from *Lauraceae* spp.] produced camphene hydrochloride (*exo*-2-chloro-2,3,3-trimethylbicyclo[2.2.1]heptane). The addition product not only spontaneously lost hydrogen chloride (HCl) while standing in chloroform ($CHCl_3$) solution, reforming camphene, but it also slowly rearranged (in a reversible Wagner–Meerwein process) to produce isobornyl chloride (*exo*-2-chloro-1,7,7-trimethylbicyclo[2.2.1]heptane).

(*E*)-1*R*-1-chloro-1-deuterio-2-butene

+

(*Z*)-3*R*-3-chloro-1-deuterio-1-butene

(*Z*)-1*S*-1-chloro-1-deuterio-2-butene

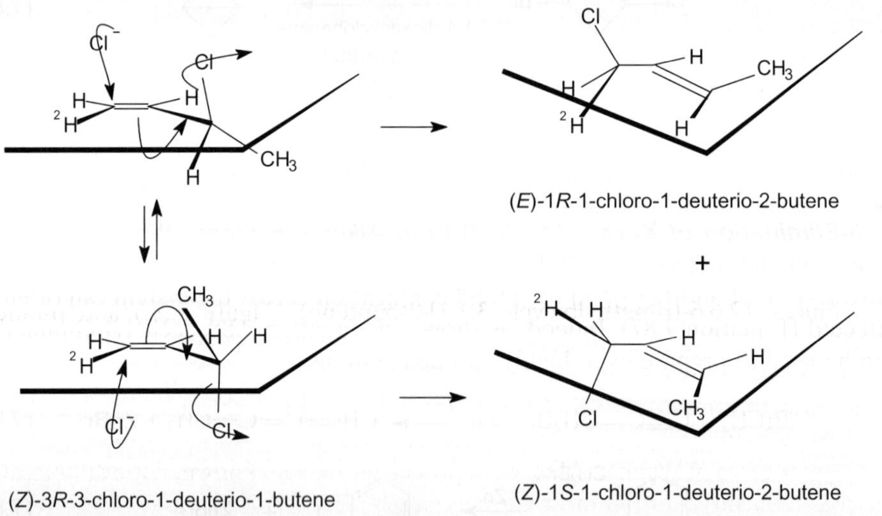

Scheme 7.45. A representation of the rearrangement of an allylic chloride [(*Z*)-3*R*-3-chloro-1-deuterio-1-butene], demonstrating that the reaction (S_N2') occurs with stereospecificity.

*Some rearrangements of hydrocarbon skeletons were originally reported in 1899 (in Russian) by A. Wagner in work that was subsequently elaborated upon by H. Meerwein and coworkers more than a decade later. The family of reactions involving alkyl group migrations have collectively become known as Wagner–Meerwein rearrangements. H. Meerwein, at the University in Bonn, was a prolific chemist. More of his work will be encountered.

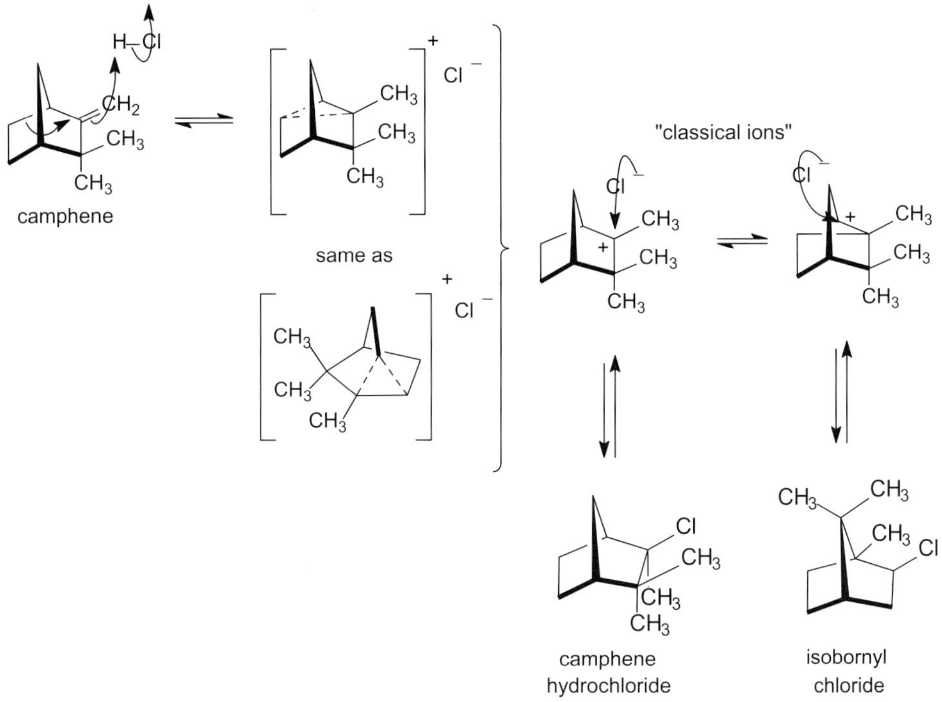

Scheme 7.46. The formation of camphene hydrochloride (*exo*-2-chloro-2,3,3-trimethylbicyclo[2.2.1]heptane and its rearrangement to isobornyl chloride (*exo*-2-chloro-1,7,7-trimethylbicyclo[2.2.1]heptane). The addition of a proton to the carbon–carbon double bond of camphene 3,3-dimethyl-2-methylenebicyclo[2.2.1]heptane is shown as accompanied by σ-bond migration to produce a single ion with partial bonding to two sites (called, variously, "nonclassical" or "bridged") or a pair of "rapidly equilibrating" ions. The "classical" ions are shown, leading to the observed products. Debate raged over a period of years about the nature of the ion or ions lying between the starting materials and products. Additional discussion is provided in Chapter 8.

α-Pinene (2,6,6-trimethylbicyclo[3.3.1]-2-heptene) (Figure 7.32) and β-pinene (2-methylene-6,6-dimethylbicyclo[3.3.1]heptane) (Figure 7.33) both provide pinene hydrochloride (*exo*-2-chloro-2,6,6-trimethylbicyclo[3.3.1]heptane), which then undergoes Wagner–Meerwein rearrangement to bornyl chloride (*endo*-2-chloro-1,7,7-trimethylbicyclo[2.2.1]heptane) (Scheme 7.47).

Simpler cases of Wagner–Meerwein processes are also known. For example, alkyl migration during addition of HX to alkenes (Chapter 6, Scheme 6.19) has already been noted. Similarly, to the extent that the same carbocations are generated during nucleophilic substitution reactions, the same processes occur! Indeed, almost identical rearrangments will be encountered again in Chapter 8 in the discussion of derivatives of alcohols as they are here with alkyl halides.

For example (Scheme 7.48), it might be supposed that in an ionizing solvent, 3-iodo-2,2-dimethylpentane, following an S_N1 pathway, undergoes solvolysis with loss of iodide anion and formation of secondary carbocation. Then, in a subsequent step, a 1,2-methyl shift (a Wagner–Meerwein rearrangement) from the initially

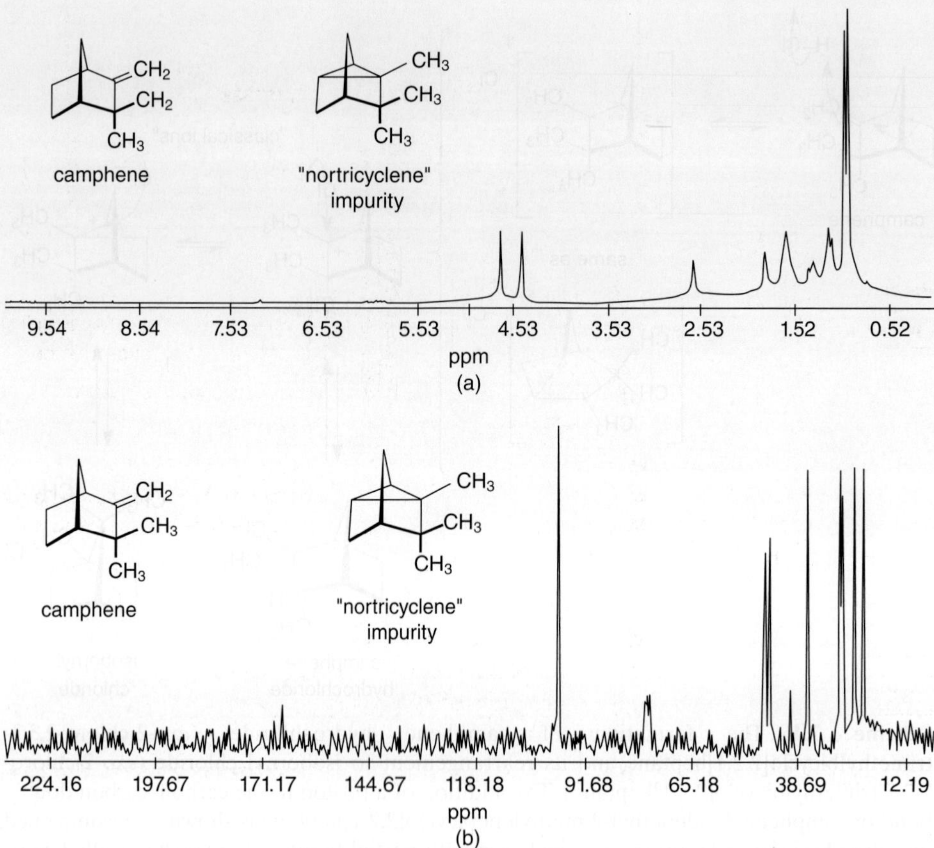

Figure 7.31. (a) The ^{1}H NMR spectrum of (+)-camphene in ^{2}HCCl$_3$ at 300 MHz. There is about 10% impurity of "nortricyclene" present. (b) The ^{13}C NMR spectrum of (+)-camphene in ^{2}HCCl$_3$ at 75 MHz. There is about 10% impurity of "nortricyclene" present.

generated (secondary) cation to a tertiary cation would be expected and return of the leaving group would provide some 2-iodo-2,3-dimethylpentane. In the event, of course, the tertiary halide would undergo solvolysis itself, precluding its isolation, but products arising from the methide shift would be expected.

As also shown in Scheme 7.48, the alternative possibility of migration of the methide at the same time the iodide is leaving, is indistinguishable from the formation of a secondary arbocation followed by migration of the methide (on the basis of the evidence presented).

Problem 7.14. As shown in Scheme 7.48, there are two paths that can account for rearranged product in the solvolysis of 3-iodo-2,2-dimethylpentane. In the first, the iodide leaves and then the methyl moves. In the second, the movement of the methide is simultaneous with the iodide leaving. Assume, in each case, the step described (involving the breaking of the carbon–iodine bond) is rate determining. Assume further that (a) radiolabeled iodide and carbon were available and (b) 3-iodo-2,2-dimethypentane could be resolved into its respective enantiomers and

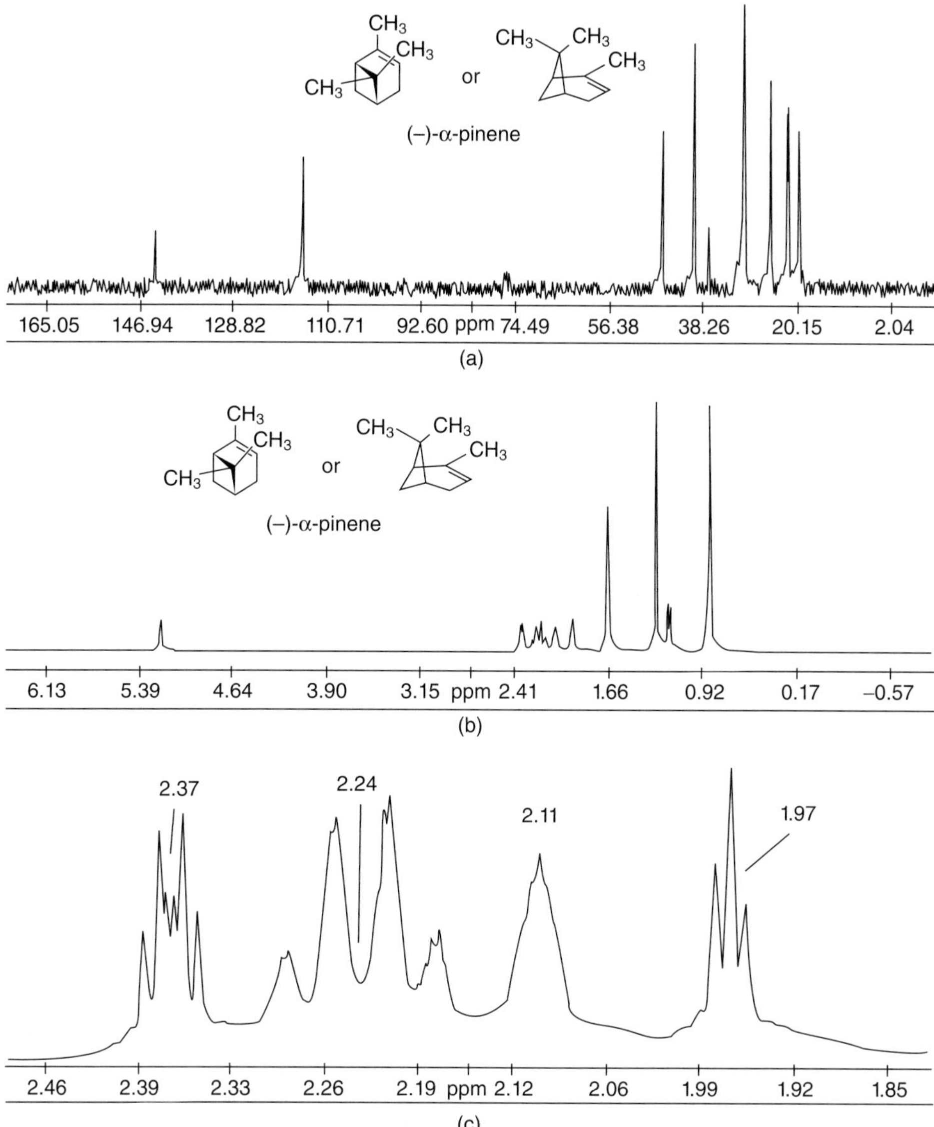

Figure 7.32. (a) The ^{13}C NMR spectrum of α-pinene in ^{2}HCCl$_3$ at 75 MHz. (b) The ^{1}H NMR spectrum of α-pinene in ^{2}HCCl$_3$ at 300 MHz. (c) The region of the ^{1}H NMR spectrum of α-pinene (in ^{2}HCCl$_3$ at 300 MHz) from δ = 2.0–2.5 ppm shown in an expanded view.

they were available. Suggest what might be done in order to attempt to resolve the issue of whether the migration of the methyl occurred after the iodide left or was simultaneous with its departure. What assumptions might be made in addition to those specified above (justify them briefly)?

There is one additional rearrangment to be considered here. An essentially identical process will be encountered again with derivatives of alcohols (Chapter 8)

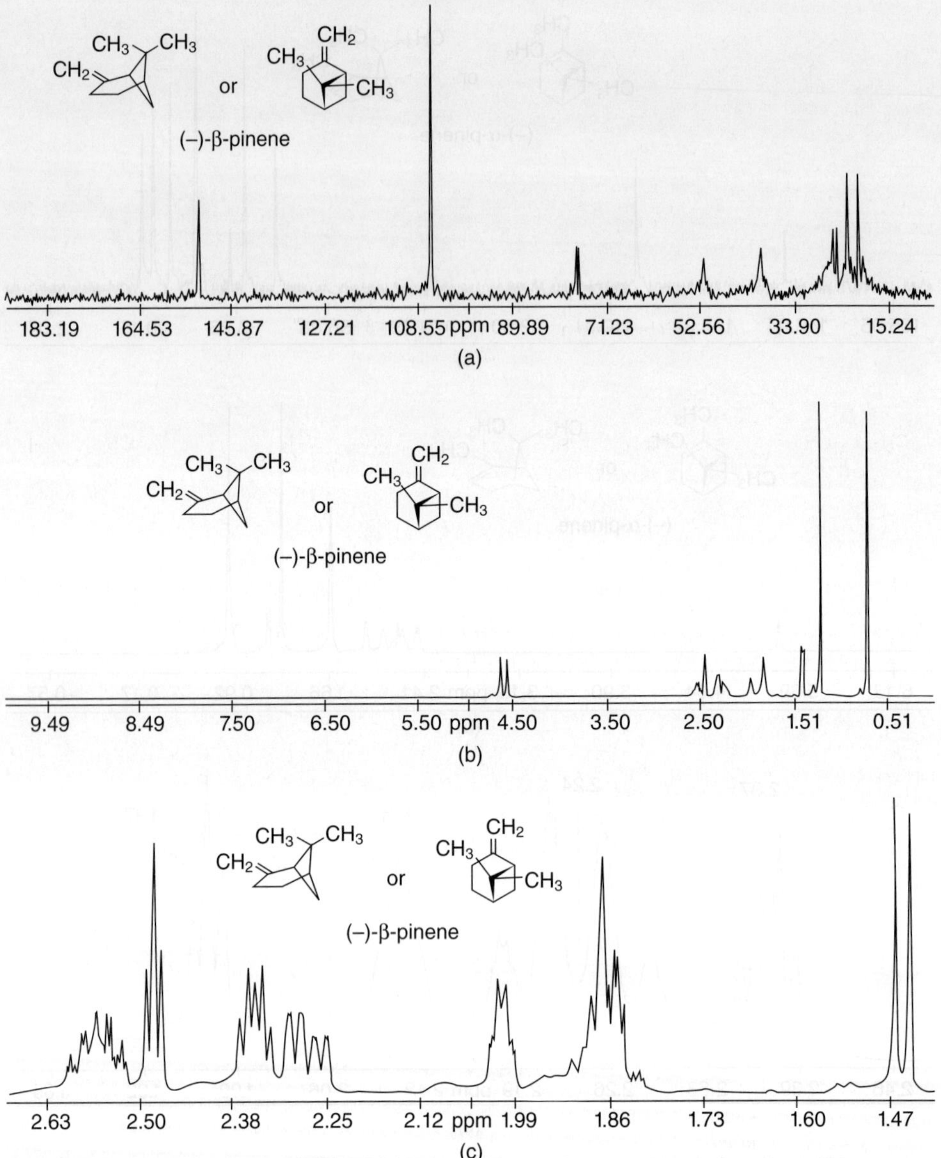

Figure 7.33. (a) The ^{13}C NMR spectrum of β-pinene in ^{1}HCCl$_3$ at 75 MHz. (b) The ^{1}H NMR spectrum of β-pinene in ^{1}HCCl$_3$ at 300 MHz. (c) The region of the ^{1}H NMR spectrum of β-pinene (in ^{1}HCCl$_3$ at 300 MHz) from δ = 1.4–2.7 ppm shown in an expanded view.

(where it was most thoroughly studied), but its presence must be acknowledged wherever carbocations are formed (as in the S_N1 reaction) or backside displacement observed (as in the S_N2 reaction).

When there is a phenyl group (C_6H_5–) on the carbon adjacent to the one bearing the leaving group, then the phenyl "neighboring group" is seen to participate in the departure of the leaving group. Although the subject will be addressed again in more

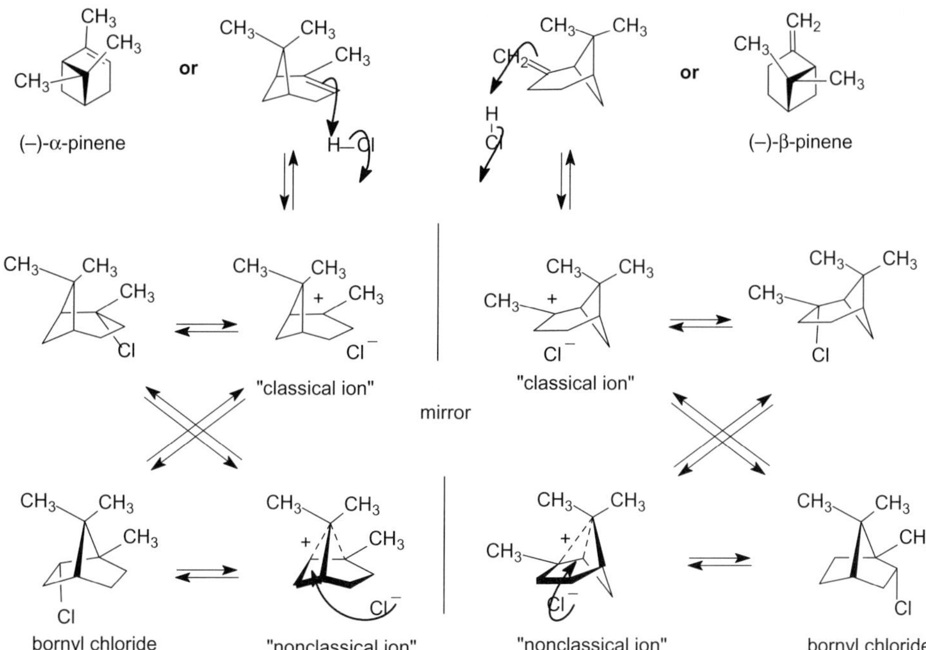

Scheme 7.47. A pathway using "classical ions" and the Wagner–Meerwein rearrangement as well as "nonclassical ions" to depict the formation of pinene hydrochloride (*exo*-2-chloro-2,6,6-trimethylbicyclo[3.3.1]heptane) from α-pinene (2,6,6-trimethylbicyclo[3.1.1]-2-heptene) and β-pinene (2-methylene-6,6-dimethylbicyclo[3.3.1]heptane) and the subsequent interconversion of pinene hydrochloride (*exo*-2-chloro-2,6,6-trimethylbicyclo[3.3.1]heptane) and bornyl chloride (*endo*-2-chloro-1,7,7-trimethylbicyclo[2.2.1]heptane).

Two step:

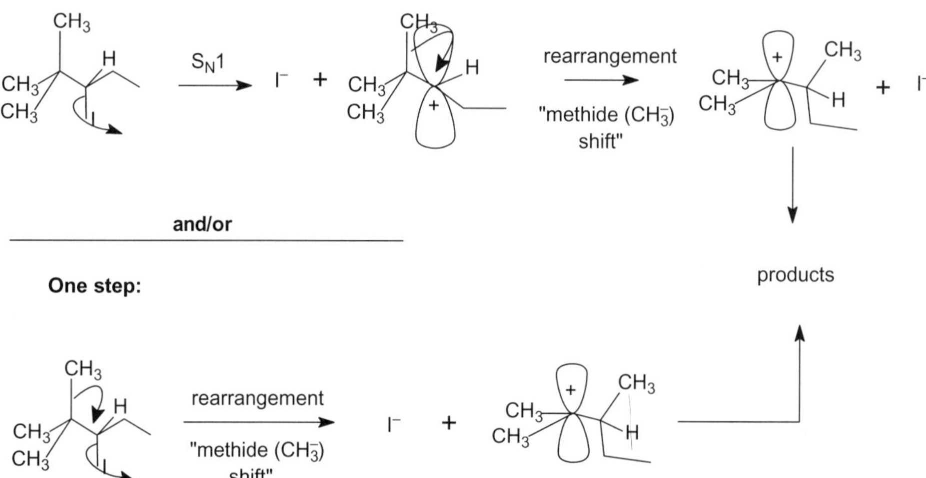

Scheme 7.48. Two alternatives to account for methide migration. The first is a two-step process, where the leaving group departs in the first step and the methyl migrates in the second step. The alternative is a one-step process, where the methide shifts in concert with the leaving of the leaving group. The rearrangement is generally of the Wagner–Meerwein type.

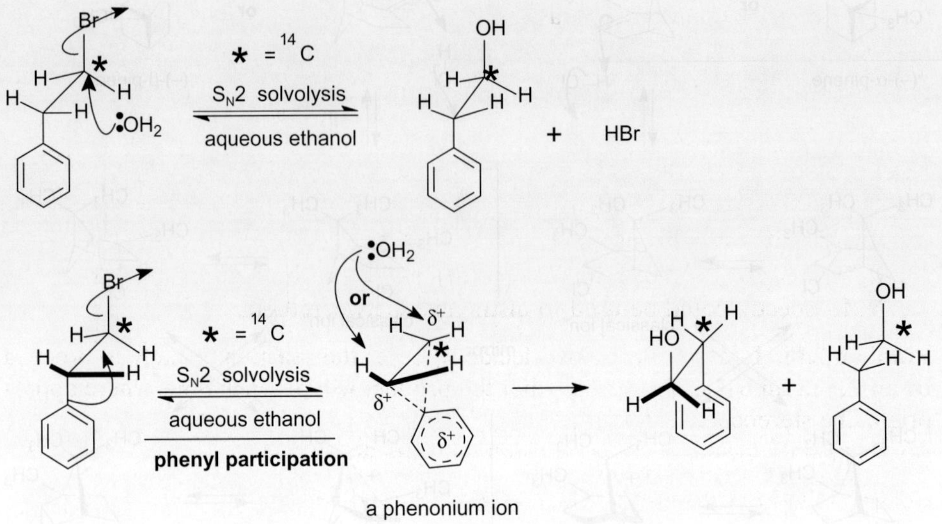

* = ¹⁴C
S_N2 solvolysis
aqueous ethanol
+ HBr

* = ¹⁴C
S_N2 solvolysis
aqueous ethanol
phenyl participation
or
δ⁺
a phenonium ion

Scheme 7.49. A rearrangement during solvolysis in which a phenyl group (C₆H₅–) migrates from one carbon to another. The rearrangement appears to occur along with the normal solvolysis (in this case). The scrambling of label is accounted for by involving a "phenonium ion" intermediate (but see Brown, H. C.; Kim, C. J. *J. Am. Chem. Soc.*, **1968**, *90*, 2082 and references therein).

detail in the next chapter, where the evidence to justify the argument is very rich (most of the experimental work having been performed with 2-phenylethanol derivatives and some homologues), a simple case is presented here. As shown in Scheme 7.49, the solvolysis, in aqueous solution, of 1-bromo-2-phenyl-(1-¹⁴C)-ethane results in the formation of alcohol in which some scrambling of the ¹⁴C label has occurred. It is argued that, in addition to normal S_N2 displacement, there is neighboring group participation by the phenyl, which results in the intermediacy of a "phenonium ion." Attack of solvent on the (symmetrical) phenonium ion leads to material in which the ¹⁴C has been "scrambled."

ADDITIONAL PROBLEMS

Problem 7.15. Allylic halides such as 3-chloropropene undergo S_N1-type reactions faster than their corresponding saturated isomers. How would you explain the observation that chloromethyl ethyl ether (ClCH₂OCH₂CH₃) also rapidly undergoes substitution reactions and does so much faster than methyl (2-chloroethyl) ether (CH₃OCH₂CH₂Cl)?

Problem 7.16. Write a suitable reaction (or several suitable reactions) using curved arrows between the following pairs of compounds. Assume that the solvent is aqueous ethanol unless otherwise specified:

(a) bromoethane and sodium azide

(b) allyl chloride (3-chloropropene) and sodium ethoxide

(c) (−)-2-bromopentane and potassium cyanide

(d) sodium *t*-butoxide (sodium 2-methyl-2-propoxide) and methyl iodide

(e) sodium methoxide and 2-iodo-2-methylpropane (*t*-butyl iodide).

Problem 7.17. Assume that an E1 elimination reaction was carried out on (*Z*)-1-bromo-1,2-dimethylcyclopropane. Three products are isolated by gas chromatography:

(a) Identify the three products.

(b) Describe (using a sketch if appropriate) how they would differ so that their ^{1}H NMR spectra could be used to distinguish between them.

Problem 7.18. In each case below, decide whether the reaction is likely to proceed by an S_N1 or an S_N2 pathway and what the product will be (including, where appropriate, the stereochemistry):

(a) 3-phenyl-1-bromopropane and sodium azide in ethanol

(b) (*S*)-3-phenyl-1-bromobutane and sodium azide in ethanol

(c) (*R*)-1-phenyl-1-bromobutane and sodium azide in ethanol.

REFERENCES

Bartsch, R. A.; Bunnett, J. F. *J. Am. Chem. Soc.*, **1968**, *90*, 408.

Bartsch, R. A.; Bunnett, J. F. *J. Am. Chem. Soc.*, **1969**, *91*, 1376.

Bordwell, F. G.; Mecca, T. G. *J. Am. Chem. Soc.*, **1972**, *94*, 5829.

Bunnett, J. F. *Survey of Progress in Chemistry*, Vol. 5, Academic Press, New York, **1969**.

Crowdrey, W. A.; Hughes, E. D.; Nevell, T. P.; Wilson, C. L. *J. Chem. Soc.*, **1938**, 209.

Hughes, E. D.; Juliusberger, F.; Masterman, S.; Topley, B.; Weiss, J. *J. Chem. Soc.*, **1935**, 1525.

Hughes, E. D.; Juliusberger, F.; Scott, A. D.; Topley, B.; Weiss, J. *J. Chem. Soc.*, **1936**, 1173.

Markownikoff, V. *Liebigs Ann. Chem.*, **1870**, *153*, 228.

Miyaura, N.; Yamada, K.; Suzuki, A. *Tetrahedron Lett.*, **1979**, 3437.

Sonogashira, K.; Tohda, Y.; Hagihara, N. *Tetrahedron Lett.*, **1975**, 4467.

Stephens, R. D.; Castro, C. E. *J. Org. Chem.*, **1963**, *28*, 3313.

Part I. The Reactions of Alcohols, Enols, and Phenols: Oxidation, Reduction, Substitution, Addition, Elimination, and Rearrangement
Part II. Ethers
Part III. Selected Reactions of Alkyl and Aryl Thiols and Thioethers

And much as wine has play'd the Infidel
And robb'd me of my Robe of Honor - Well
I wonder often what the Vintners buy
One half so precious as the stuff they sell.

—E. Fitzgerald, *Quatrain XCV*, 5th Edition, 1889,
Rubáyát of Omar Khayyám of Naishápúr

SPECIAL INTRODUCTION TO CHAPTER 8

A brief discussion of both the nomenclature and some of the properties of those classes of organic compounds (a) containing the hydroxyl (–OH) functional group and known as **alcohols**, **enols**, and **phenols** (R–OH; Ar–OH; R ≠ H) (Figure 8.1); (b) in which both protons of water [H–O–H] (or the remaining proton attached to oxygen in an alcohol, [R–O–H; R ≠ H]) can be considered to "have been replaced" by alkyl, alkenyl, and/or aryl substituents and are called **ethers** (R–O–R′; R, R′ ≠ H); and (c) that correspond to the organic analogues of hydrogen sulfide (H–S–H) called thiols (R–SH; R ≠ H) and thioethers (R–S–R′; R, R′ ≠ H) was provided in Chapter 5. Since it will not be reiterated, it can be profitably reviewed.

For the most part, and as the hydroxyl (–OH) functional group is more common, alcohols will receive greater emphasis in this chapter. However, to the extent that similarities and differences between, for example, alcohols and thiols, and alcohols and ethers can be drawn upon for emphasis, and to the extent that some of, in par-

---

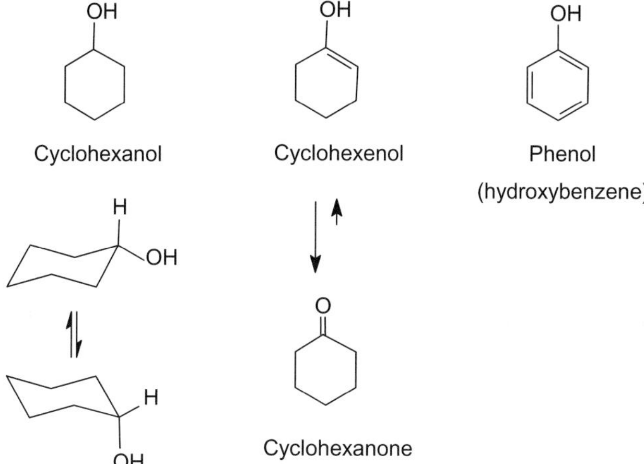

Figure 8.1. An alcohol (cyclohexanol), enol (cyclohexenol), and phenol (phenol [hydroxy-benzene]). Cyclohexenol is shown in equilibrium with its tautomer cyclohexanone and, at room temperature, is present in low (ca. $10^{-5}\%$) concentrations relative to the ketone.

ticular, the ethers (R–O–R′) are of significant commercial importance, their chemistry will also be discussed.

The use of alcohols (R–OH; R ≠ H) and ethers (R–O–R′; R, R′ ≠ H) as solvents and (for the former) reactants was also pointed out in Chapters 5–7. Additionally, as has already been mentioned and as will be discussed further, alcohols (R–OH; R ≠ H) and ethers (R–O–R′; R, R′ ≠ H), much like water (R–OH; R = H), can be protonated (they are basic) by strong acid and, for the former, deprotonated (they are acidic) by strong bases.

As shown in Table 8.1, the preparation of some alcohols (with one or more hydroxyl [–OH] groups) has been provided earlier. Table 8.2 reminds the reader that (a) enols are in equilibrium with their corresponding carbonyl tautomers (Chapter 5); (b) acid-catalyzed addition of alcohols (R–OH; R ≠ H) and thiols (R–SH; R ≠ H) to, for example, alkenes can be used to produce ethers (R–O–R′; R, R′ ≠ H) and thioethers (R–S–R′; R, R′ ≠ H), respectively (Chapter 6); (c) ethers are readily formed by nucleophilic substitution of halogen (X = Cl, Br, I) by alkoxide (RO⁻) (Chapter 7) under the appropriate (S_N1 or S_N2) conditions; and (d) substitution of a halogen on an aromatic ring (Ar–X, X = Cl, Br) can be effected by fusion with molten alkali (e.g., NaOH) (Chapter 7).

As this chapter develops, it will be useful to keep in mind that in addition to the simple compounds containing one or more hydroxyl groups (–OH) such as the alcohols in Table 8.1, hydroxyl groups (–OH) and alkoxyl (–OR) containing organic compounds are literally the building blocks of nature; that is, carbohydrates (i.e., "hydrates of carbon" [Chapter 11] corresponding, approximately, to $C_n(H_2O)_n$ ($n \geq 3$)) are essential components of living organisms and are arguably the most abundant class of biological molecules. Carbohydrates contain at least one and generally more than one hydroxyl group (–OH). Carbohydrates are not only found in all nucleic acids (Chapters 13 and 14) but they are also the main structural component of cellulose and starch (Chapter 11). In the nucleic acids and elsewhere,

TABLE 8.1. Examples of Some of the Reactions That Produce Alcohols and That Have Been Encountered in Previous Chapters

Chapter 5

$$CO \;+\; H_2 \xrightarrow[\text{Catalyst}]{ZnO-Cr_2O_3} CH_3OH$$

$$C_6H_{12}O_6 \xrightarrow{\text{Yeast}} 2CO_2 \;+\; 2CH_3CH_2OH$$

α- and β-anomers of glucose

Chapter 6

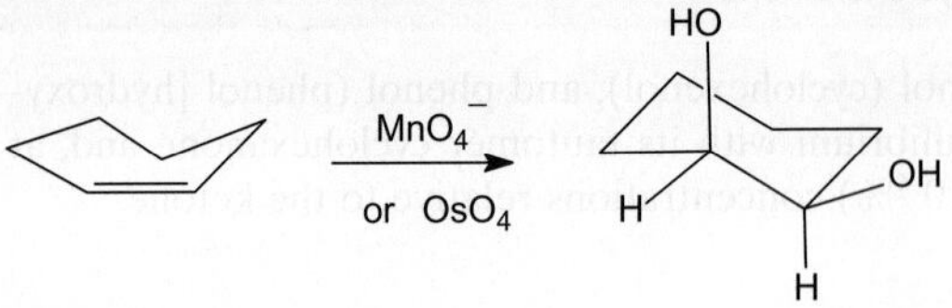

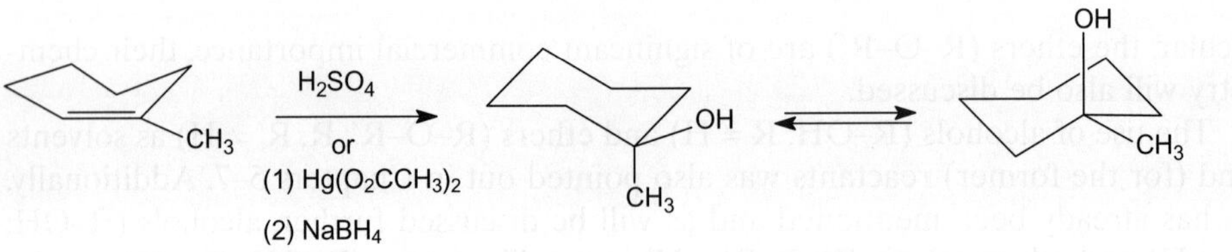

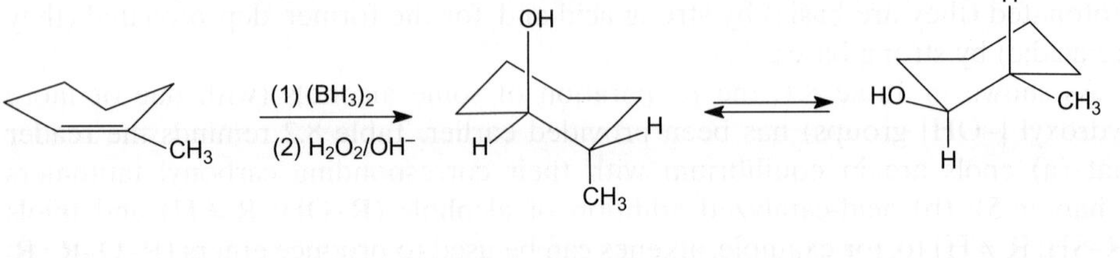

Chapter 7

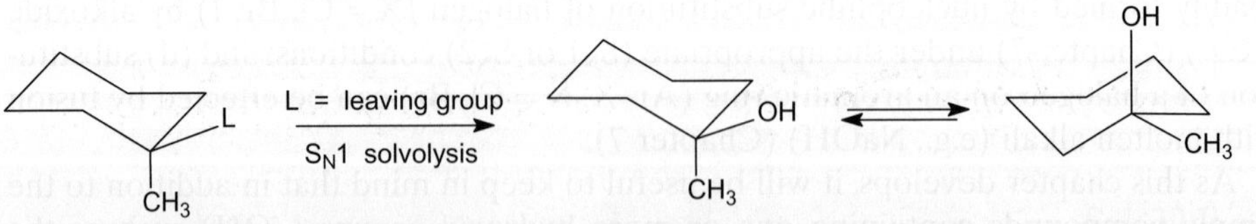

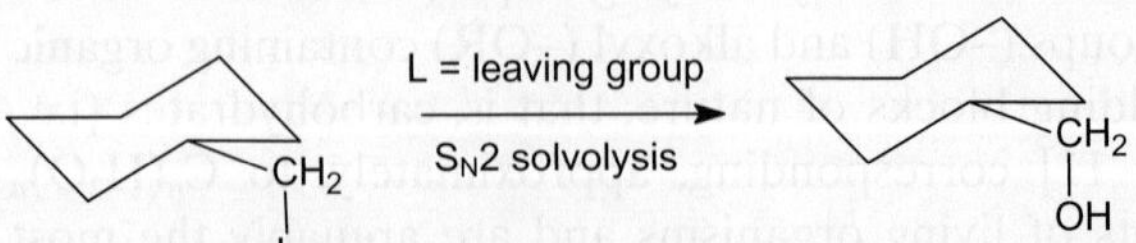

TABLE 8.2. Reactions Previously Encountered That Produce (1) an Enol, (2) an Ether by the Acid-Catalyzed Addition of an Alcohol to an Alkene, (3) an Ether via an S_n2 Solvolysis in an Alcohol, (4) an Ether via an S_n1 Reaction, and (5) a Phenol

1.

Propanone → 2-Propenol

(a ketone)   an enol (ca. 10^{-6}% at 25°C)

2.

Cyclohexene $\xrightarrow[\text{H}_2\text{SO}_4]{\text{CH}_3\text{OH}}$ Cyclohexyl methyl ether (methoxycyclohexane)

3.

Benzyl bromide $\xrightarrow[\text{CH}_3\text{O}^-]{\text{CH}_3\text{OH}}$ (S_N2) Benzyl methyl ether

4.

Chlorodiphenylmethane $\xrightarrow[\text{S}_N1]{\text{CH}_3\text{OH}}$ Diphenylmethyl methyl ether

5.

Chlorobenzene $\xrightarrow{\text{Molten KOH}}$ Phenol (hydroxybenzene)

carbohydrates are generally found as cyclic acetals or ketals (compounds formed as a result of the reaction of alcohols with aldehydes and ketones—to be discussed further in Chapter 9), bristling with an array of hydroxyls, whose stereochemical relationships are part of a recognition pattern for their use. Additionally, a number of amino acids (Chapter 12) and many products of the further metabolism of the simple building blocks of nature, such as steroids (Chapter 11) and alkaloids (Chapter 13), are graced with a complement of hydroxyl (–OH) and alkoxyl (–OR) groups.

Regarding the chemistry of alcohols (ROH) and thiols (RSH), it should be clear that any discussion of the reactions of the carbon-bound hydroxyl (C–O–H) or sulfhydryl (C–S–H) group must take into account *both* the chemistry of the oxygen–hydrogen (sulfur–hydrogen) system that, it will be seen, closely resembles what is found in water (H–OH) and hydrogen sulfide (H–SH), as well as the processes that affect the carbon–oxygen (C–O) or carbon–sulfur (C–S) bond.

For all of these related compounds (alcohols, enols, phenols, ethers, thiols, and thioethers), some properties of which are provided in Table 8.3, it might also be anticipated that the nonbonding electrons (Chapter 1) on oxygen or sulfur could participate in reactions occurring *elsewhere in the organic framework*, and, indeed, it is common to find that processes taking place remote to the *heteroatom* are also affected by its presence.

Finally, the higher oxidation states of sulfur (Chapter 5) are also important. The brief treatment here is supplemented by materials that will be developed as parts of other chapters.

PART I. ALCOHOLS, ENOLS, AND PHENOLS

A. ACIDITY AND BASICITY

Current ideas about acids and bases as applied to alcohols, enols, and phenols, as noted earlier (Chapter 5), are divided into the main categories of proton donors/acceptors (Brønsted acids/bases) and electron acceptors/donors (Lewis acids/bases). Because alcohols, enols, and phenols all bear protons on oxygen, which can be donated *and* have sufficient electron density on oxygen to permit protonation, they readily fit the Brønsted criteria. However, since the electron density on oxygen can also be "donated" to nonprotonaceous species (e.g., the "empty orbital" on BF_3 [Chapter 1]), alcohols, enols, and phenols also qualify as Lewis bases.

As before (Chapter 5), the acid strength of alcohols is measured as a function of an ionization constant, K_a, which is defined in terms of an equilibrium between protonated (alcohol, enol, or phenol) and deprotonated (alkoxide, enolate anion, or phenoxide) species (Equations 8.1 and 8.2).

$$R\text{—}O\text{—}H \quad \overset{K_a}{\rightleftharpoons} \quad R\text{—}O^- \;+\; H^+ \tag{8.1}$$

so

$$K_a = [RO^-][H^+]/ROH, \tag{8.2}$$

where R = alkyl, alkenyl, or aryl.

The process of Equation 8.1, as shown, is meant to be recognized as idealized. "Free" protons are not generated and the degree of solvation of the protons, as well as of the alkoxide anion, changes with the nature of the alkyl group.

Similarly, the base strength of an alcohol, enol, or phenol can, in principle, be measured by examination of the corresponding *di*protonated species (Equation 8.3).

$$R\text{—}O^+H_2 \quad \overset{K_a'}{\rightleftharpoons} \quad R\text{—}O\text{—}H \;+\; H^+ \tag{8.3}$$

TABLE 8.3. Some Representative Examples of (a) Alcohols and Phenols, (b) Ethers, and (c) Thiols and Thioethers[a]

Compound	Name	mp °C	bp °C	d^{20}	^{1}H NMR Data (δ, ppm TMS = 0.00)	^{13}C NMR Data (δ, ppm TMS = 0.00)
(a)						
CH_3OH	Methanol	−97.8	65.0	0.7914	3.39 (CH_3, s), 4.14 (OH)	49.0
CH_3CH_2OH	Ethanol	−114.7	78.5	0.7893	1.18 (CH_3, t), 3.59 (CH_2, q) 4.40 (OH)	17.6 (CH_3), 57.0 (CH_2)
$CH_3CH_2CH_2OH$	1-Propanol	−126.5	97.4	0.8035	0.93 (CH_3, t), 1.53 (β–CH_2, m), 3.49 (α–CH_2, t), 4.3 (OH)	10.0 (CH_3), 25.8 (β CH_2), 63.6 (α CH_2)
$CH_3CH(OH)CH_3$	2-Propanol	−89.5	82.4	0.7855	1.29 (CH_3, d), 3.99 (−CH, m), 2.90 (OH)	25.10 (CH_3), 63.40 (CH)
$CH_3(CH_2)_2CH_2OH$	1-Butanol	−90.2	117.3	0.8098	0.91 (CH_3, t), 1.10–1.70 (CH_2, m), 3.52 (CH_2), 4.11 (OH)	13.60 (δ CH_3), 19.10 (γ CH_2), 35.00 (β CH_2), 61.40 (α CH_2)
$CH_3CH_2CH(OH)CH_3$	2-Butanol	−114.7	99.5	0.8063	0.90 (CH_3, t), 1.11 (CH_3, d), 1.38 (CH_2, m), 3.62 (CH, m), 3.99 (OH)	9.90 (CH_3), 22.60 (CH_3), 32.00 (CH_2), 68.70 (CH)
$(CH_3)_2CHCH_2OH$	2-Methyl-1-propanol		108.1	0.8019	0.89 (CH_3, d), 1.67 (CH, m), 3.27 (CH_2, d), 3.98 (OH)	18.90 (CH_3), 30.80 (CH), 68.90 (CH_2)
$(CH_3)_3COH$	2-Methyl-2-propanol	25.5	82.2	0.7887	1.20 (CH_3), 2.40 (OH)	31.30 (CH_3), 68.40 (C)
$CH_3(CH_2)_3CH_2OH$	1-Pentanol	−79.0	138.0	0.8144	0.91 (CH_3, t), 1.09–1.80 (CH_2's, m), 3.50 (CH_2), 4.41 (OH)	13.80 (ε CH_3), 22.60 (δ CH_2), 28.20 (γ CH_2), 32.50 (β CH_2), 61.80 (α CH_2)
$CH_3(CH_2)_4CH_2OH$	1-Hexanol	−47.0	158.0	0.8136	0.90 (CH_3, t), 1.35 (CH_2's, m), 3.50 (CH_2), 3.67 (OH)	14.20 (ζ CH_3), 22.80 (ε CH_2), 32.00 (δ CH_2), 25.80 (γ CH_2), 32.80 (β CH_2), 61.93 (α CH_2)
$CH_3(CH_2)_6CH_2OH$	1-Octanol	−16.0	195	0.8249	0.88 (CH_3), 1.29 (CH_2's, m), 3.49 (CH_2), 3.90 (OH)	13.90 (θ CH_3), 22.80 (η CH_2), 32.10 (ζ CH_2), 29.60 (ε CH_2), 29.70 (δ CH_2), 26.10 (γ CH_2), 32.90 (β CH_2), 61.90 (α CH_2)
$CH_3(CH_2)_8CH_2OH$	1-Decanol	6.4	231	0.8292	0.90 (CH_3), 1.30 (CH_2's, m), 3.40 (CH_2), 3.50 (OH)	14.00 (κ CH_3), 22.80 (ι CH_2), 32.20 (θ CH_2), 29.60 (η CH_2), 29.90 (ζ CH_2), 29.80 (ε CH_2), 29.80 (δ CH_2), 26.10 (γ CH_2), 32.90 (β CH_2), 61.90 (α CH_2)

(Continued)

TABLE 8.3. Continued

Compound	Name	mp °C	bp °C	d^{20}	^{1}H NMR Data (δ, ppm TMS = 0.00)	^{13}C NMR Data (δ, ppm TMS = 0.00)
(phenol structure)	Phenol (hydroxybenzene)	42.0	183	1.072	OH 6.11; ring H 6.65–7.32	154.90; 115.40; 129.70; 121.00
(1-naphthol structure)	1-Naphthol	94.0	278	1.103[b]	OH 8.92; a. 7.01, b. 7.22–7.60, c. 7.72, d. 8.41	1. 153.98, 2. 108.81, 3. 126.93, 4. 119.82, 5. 128.17, 6. 126.78, 7. 125.27, 8. 122.87, 9. 125.81, 10. 135.75
(catechol structure)	Catechol (1,2-dihydroxybenzene)	105.0	246	1.371	7.09	117.30; 122.10; 145.00
(4-methylphenol structure)	4-Methylphenol (p-cresol) (4-methylhydroxybenzene)	36.0	202	1.0347	2.20 (H_3C); OH 6.45; 6.92 H; 6.69 H	130.50; 20.60 (H_3C); 152.6; 130.20; 115.30
(b) (oxirane structure)	Oxacyclopropane (oxirane)	−111.7	10.7	0.897	2.54 (CH_2)	39.50 (CH_2)
(ethyl vinyl ether structure)	Ethyl vinyl ether (ethoxyethene)	−115.8	35.75	0.7589	6.31 H; H 3.81; 1.21 CH_3CH_2–O; H 3.99; 3.66	153.30; 85.20; 13.8 CH_3CH_2–O; 54.30

$CH_3CH_2OCH_2CH_3$	Ethyl ether (diethyl ether)	−116.3	34.6	0.714	1.13 (CH_3, t), 3.38 (CH_2, q)	17.10 (CH_3), 67.40 (CH_2)
(oxacyclopentane ring)	Oxacyclopentane (tetrahydrofuran)		65.0	0.889	3.71 (α CH_2), 1.81 (β CH_2)	68.60 (α CH_2), 26.20 (β CH_2)
(1,4-dioxane ring)	1,4-Dioxane (1,4-dioxacyclohexane)	11.8	101.4	1.234	3.63 (CH_2)	67.4 (CH_2)
$CH_3OCH_2CH_2CH_3$	1,2-Dimethoxyethane (ethylene glycol dimethyl ether		84.7	0.8665	3.29 (CH_3), 3.40 (CH_2)	59.00 (CH_3), 72.00 (CH_2)
$CH_3CH_2OCH_2CH_2OH$	2-Ethoxyethanol		135.0	0.9297	1.18 (CH_3), 3.51 (CH_2's, OH)	72.00 66.50 $HOCH_2CH_2OCH_2CH_3$ 62.50 15.00
(phenyl)$-CH_2CH_2OCH_3$	Methyl phenethyl ether					

(c)

CH_3SH	Methanethiol (methyl mercaptan)	−123.0	5.96	0.859	2.07 (CH_3), 1.24 (SH)
CH_3CH_2SH	Ethanethiol (ethyl mercaptan)	−144.4	36.0	0.831	1.31 (CH_3, t), 2.52 (CH_2, q), 1.17 (SH)
$CH_3CH_2CH_2SH$	1-Propane-thiol (propyl mercaptan)	−113.8	67.5	0.8047	0.98 (CH_3, t), 1.67 (β CH_2), 2.52 (α CH_2), 1.31 (SH)
$CH_3CH(SH)CH_3$	2-Propanethiol	−130.7	56.0	0.8142	1.31 (CH_3, d), 3.07 (CH), 1.40 (SH)
$HSCH_2CH_2SH$	1,2-Ethanedithiol	−41.0	147.0	1.119	2.72 (CH_2), 1.57 (SH)

(Continued)

TABLE 8.3. *Continued*

Compound	Name	mp °C	bp °C	d^{20}	^{1}H NMR Data (δ, ppm TMS = 0.00)	^{13}C NMR Data (δ, ppm TMS = 0.00)
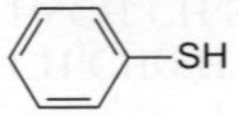 (phenyl–SH)	Thiophenol (benzenethiol) (mercaptobenzene)	−14.9	172.5	1.078	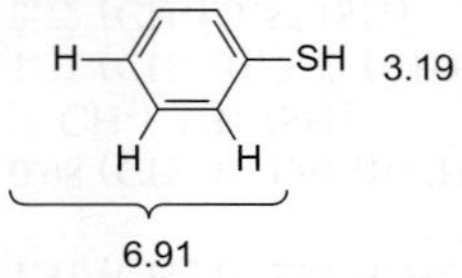 SH 3.19; ring H 6.91	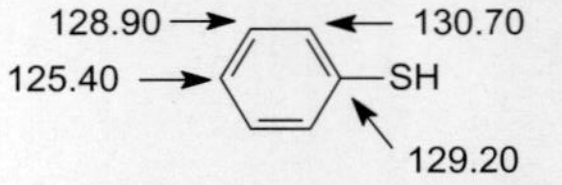 128.90; 125.40; 130.70; 129.20
$CH_3CH_2SCH_2CH_3$	Diethylsulfide	−102.1	92.0	0.8368	1.23 (CH_3, t), 2.47 (CH_2, q)	
$ClCH_2SCH_3$	Chloromethyl methyl sulfide		107.0		2.30 (CH_3), 4.71 (CH_2)	
(thiirane ring)	Thiirane (thiacyclopropane, ethylene sulfide)		55.0	1.005	2.27 (CH_2)	60.12 (CH_2)
 (thiolane ring)	Thiolane (thiacyclopropane, tetrahydrothiophene, tetramethylene sulfide)	96.17	121.2	0.999	2.77 (α CH_2), 1.90 (β CH_2)	31.40 (α CH_2), 31.20 (β CH_2)
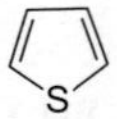 (thiophene ring)	Thiophene	−38.3	84	1.06	7.10 (α CH_2), 6.90 (β CH_2)	125.40 (α CH_2), 127.20 (β CH_2)
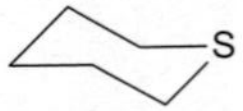 (thiane ring)	Thiane (thiacyclopropane, pentamethylene sulfide)	16	170.5		2.58 (α CH_2), 1.30–2.10 (β and γ CH_2's)	

Values for the spectroscopic data in the table not obtained by the author have been taken from (a) Emsley, J. W.; Feeney, J.; Sutcliffe, L. H. *High Resolution Nuclear Magnetic Resonance Spectroscopy*, Vols. 1 and 2, Pergamon, Oxford, **1966**; (b) Levy, G. C.; Lichter, R. L.; Nelson, G. L. *Carbon-13 Nuclear Magnetic Resonance Spectroscopy*, 2nd edition, Wiley Interscience, New York, **1980**; and (c) Jackman, L. M.; Sternhell, S. *Applications of Nuclear Magnetic Resonance Spectroscopy in Organic Chemistry*, 2nd edition, Pergamon, Oxford, **1969**.

Physical data were taken from tables in the *CRC Handbook of Chemistry and Physics*, 71st edition, CRC Press, Boston, **1990**, or from the *Dictionary of Organic Compounds*, CHEMnetBase, CRC Press, at http://doc.chemnetbase.com

[a]The information presented is the result of experiment and is subject to change.

[b]This value is at 101°C.

TABLE 8.4. Selected (and approximate) pK_a Values for Some Compounds Bearing O–H Bonds

Conjugate Acid	pK_a	Conjugate Base
1. H–Cl	≈ -6.5	Cl^-
2. protonated phenol (C_6H_5–OH_2^+)	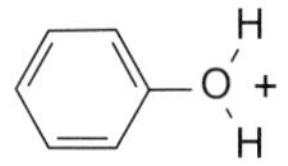 -7	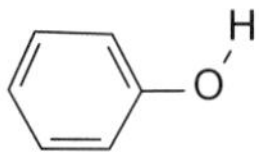 phenol (C_6H_5–OH)
3. protonated acetone ($(CH_3)_2C$=OH^+)	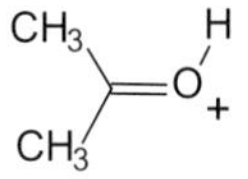 -3	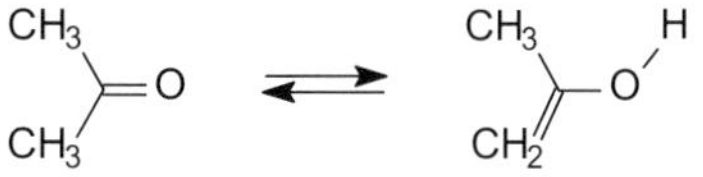 acetone ($(CH_3)_2C$=O) $\rightleftharpoons$ enol of acetone (CH_3(C)=CH_2–OH)
4. protonated methanol (CH_3–OH_2^+)	-2	methanol (CH_3–OH)
5. hydronium ion (H–OH_2^+)	-1.7	water (H–OH)
6. phenol (C_6H_5–OH)	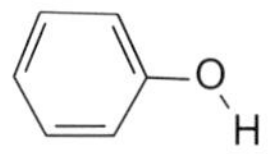 10	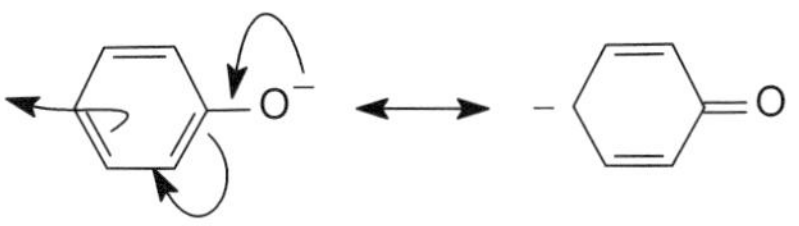 phenoxide anion (C_6H_5–O^-) (resonance-delocalized)
7. H–OH (water)	15.7	H–O^-
8. CH_3–OH (methanol)	16	CH_3–O^-
9. acetone ($(CH_3)_2C$=O) $\rightleftharpoons$ enol (CH_3(C)=CH_2–OH)	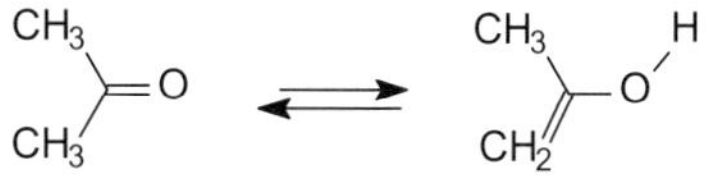 19^a	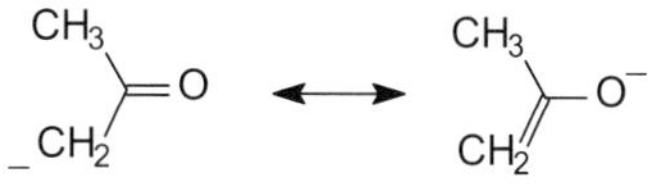 enolate anion (CH_3(C)=$O \cdots CH_2^-$) (resonance-delocalized)
10. NH_3	32	$^-NH_2$
11. H_2	35	H^-
12. CH_4	≈ 50	$^-CH_3$

Approximate values for hydrogen chloride (HCl) and the amide (NH_2^-), hydride (H^-), and methide (CH_3^-) ions are also provided for comparison purposes. Data in this Table has been largely taken from Lowry, T.; Richardson, K. S. "Mechanism and Theory in Organic Chemistry", 3rd edit, Harper & Row, N. Y. **1987** and references therein.

[a]The value given is for the loss of the proton from carbon. The enol has been reported to have a pK_a of ca. 14. See, Chiang, Y.; Kresge, A. J., Tang, Y. S.; Wirz, J. *J. Am. Chem. Soc.*, **1984**, *106*, 460.

Approximate pK_a values for some protonated alcohols, enol precursor (i.e., carbonyl compounds bearing α-protons), alcohols, protonated phenol, and phenol are provided in Table 8.4.

As shown in Table 8.4, protonated phenol (entry 2) is a very strong acid. It can be argued that the increase in stabilization on proton loss (resonance stabilization involving the electron pair on oxygen is possible in phenol but not when it is protonated) is largely responsible for this. Further, protonated alcohols (R–OH_2^+), ketones (R_2C=OH^+), and the hydronium ion (H_3O^+) are all about equally acidic (pK_a's are all between about -2 and -5).

At the other end of the scale, phenol (entry 6) itself is acidic and forms the resonance-stabilized phenoxide anion on proton loss. Then, as already noted (Table 7.12), alcohols are (in alcohol solution) about as acidic as water. Presumably, the position of the equilibrium (Equation 8.1) lies further to the side of the alcohol because (at least in part) the larger alkoxide anions are more poorly solvated than the smaller ones. Thus, their acidity decreases as their size increases (i.e., the order of acidity of various alcohols relative to water is H_2O > primary (RCH_2OH) > secondary (R_2CHOH) > tertiary (R_3COH), and the corresponding conjugate base strength of the *alkoxide anions* is tertiary (R_3CO^-) > secondary (R_2CHO^-) > primary (RCH_2O^-) > hydroxide (HO^-)).*

Water (H_2O) reacts rapidly with alkali metals (e.g., sodium [Na] and potassium [K]) to produce hydrogen (H_2) and the corresponding alkali hydroxides (e.g., NaOH). While methanol (methyl alcohol, CH_3OH) reacts similarly, the reaction is significantly moderated by increasing the size of the alkyl group (Equation 8.4). The more highly substituted alcohols, for example, 2-methyl-2-propanol (*t*-butanol [$(CH_3)_3COH$]), react sluggishly with sodium (Na) at room temperature to produce the corresponding alkoxide. Very strong bases such as sodium amide ($NaNH_2$) or lithium amide ($LiNH_2$) (sources of the amide anion NH_2^-) (Equation 8.5), sodium hydride (NaH), or other hydrides (sources of the hydride anion H^-) or metal alkyls such as methyllithium (CH_3Li) and butyllithium ($CH_3CH_2CH_2CH_2Li$ [BuLi]) are all strongly hydroscopic and react vigorously with water. By analogy, they would also be predicted to react with alcohols and, indeed, they do so to form the corresponding metal alkoxides and ammonia (NH_3), hydrogen (H_2), or alkane (methane [CH_4] or butane [$CH_3(CH_2)_2CH_3$]), respectively.

$$(CH_3)_3C\!-\!O\backslash_H \quad + \quad Na \quad \longrightarrow \quad (CH_3)_3C\!-\!O^- \; Na^+ \quad + \quad \tfrac{1}{2}H_2 \qquad (8.4)$$

$$CH_3CH_2\!-\!O\backslash_H \quad NH_2\,Na^+ \quad \longrightarrow \quad CH_3CH_2\!-\!O^- \; Na^+ \quad + \quad NH_3 \qquad (8.5)$$

The solvent for the reaction with the alkali metal, metal hydride, and so on, is often the alcohol itself. However, on occasion, such as when the alcohol is not a liquid or when other functional groups with which the alkoxide anion might react are present, ethers or hydrocarbons are used as solvents. Commercially available butyllithium reagents are sold as solutions in hydrocarbon solvents.

Although a more robust discussion is postponed until the next chapter, it is important to understand now that preparation of enolate anions is a particularly sensitive process. Thus, abstraction of the (acidic)[†] proton α to the carbonyl group (Table 8.4, entry 9) leads to an enolate anion, and that anion is capable of *addition to the carbon of the carbonyl group* of a second equivalent of carbonyl compound or another carbonyl compound. Further, as already noted in the caption to the table, the enol, bearing a proton on oxygen, is *more* acidic ($pK_a \approx 14$) than the corresponding ketone. It can be argued that the proton is more rapidly lost from oxygen than from carbon.

---

*Interestingly, the order of acidity of the ions in the *gas phase* seems to be reversed. It is argued that this is so because solvation is lacking and the additional alkyl groups ("R") stabilize the anionic oxygen by electron polarization (i.e., as there is a larger number of sp^3-type orbitals, it is easier to polarize the electrons).

[†]It may be recalled (Chapter 5 and see, e.g., Figure 5.44) that protons on the carbon atoms to which the carbon of the carbonyl are attached are acidic. One possible rationale is that the carbon of the carbonyl is the positive end of the polarized carbon–oxygen bond and is thus electron "attracting."

Problem 8.1. With an alcohol of your own choice (which you should name and for which you should draw the structure), write the product of the reaction of that alcohol with

(a) methylmagnesium iodide [CH_3MgI] in ether,

(b) lithium hydride in oxacyclopentane (tetrahydrofuran [THF]), and

(c) *n*-butyllithium in hexane.

Finally, as anticipated from Table 8.4, alcohols are protonated by strong Brønsted acids to form oxonium salts (Equation 8.6, producing ethyl oxonium chloride), and coordination of the electron pair on oxygen can also be effected by the reaction with Lewis acids (Equation 8.7, forming ethyl oxonium aluminum dichloride).

$$CH_3CH_2{-}\overset{..}{\underset{H}{O}} \;+\; H{-}Cl \;\longrightarrow\; CH_3CH_2{-}\overset{H}{\underset{H}{O^+}} \;+\; Cl^- \qquad (8.6)$$

$$CH_3CH_2{-}\overset{..}{\underset{H}{O}}{:}AlCl_3 \;\longrightarrow\; CH_3CH_2{-}\overset{AlCl_3}{\underset{H}{O^+}} \;\rightleftarrows\; CH_3CH_2{-}\overset{..}{\underset{AlCl_2}{O}} \;+\; HCl \qquad (8.7)$$

B. OXIDATION OF ALCOHOLS, ENOLS, AND PHENOLS

I. Introduction

Complete combustion of alcohols with oxygen at high temperature results, as with hydrocarbons, in the formation of carbon dioxide (CO_2) and water (H_2O). In principle, therefore, alcohols might be used to replace such hydrocarbon-based products derived from petroleum. Furthermore, the production of methanol (methyl alcohol, CH_3OH) from carbon monoxide (CO) and hydrogen (H_2) at high pressure (Chapter 6) and of ethanol (ethyl alcohol, CH_3CH_2OH) by fermentation of glucose ($C_6H_{12}O_6$) (Chapter 11) (Table 8.1) are both processes using "renewable resources." As a consequence, these low-molecular-weight alcohols are potentially attractive as alternatives to petroleum-derived gasoline.

In that regard, although methanol (methyl alcohol, CH_3OH), which is quite toxic, is occasionally used in small quantities as a gasoline additive (often with other alcohols), the use of pure methanol as a fuel remains experimental. Indeed, most methanol is used to produce formaldehyde (methanal, $H_2C{=}O$) or as an intermediate in other commercial endeavors (*vide infra*).

On the other hand, ethanol (ethyl alcohol, CH_3CH_2OH), which is less toxic, and which is largely made for industrial purposes by the acid-catalyzed hydration* of ethene (ethylene, $H_2C{=}CH_2$, Chapter 6), has generated greater interest as a fuel for

*The commercial process uses high temperature and a phosphoric acid (H_3PO_4) on alumina (Al_2O_3) catalyst and produces ethanol (CH_3CH_2OH), which contains about 4.4% water (H_2O) by weight. This material, which is a constant boiling mixture (an *azeotrope*) of water and ethanol (bp 78.1°C), is called 95% ethanol. Pure ethanol (*absolute* ethanol, bp 78.5°C), which is 100% ethanol, cannot be isolated by distillation from this mixture. Absolute ethanol can be produced from 95% ethanol by removing the water, and this "dehydration" is commonly accomplished by the addition of benzene (bp 80.0°C).

(Continued on next page)

automobiles. Most of the alcohol currently destined for automotive use is produced by fermentation of the sugars present in corn and is mixed with gasoline (forming "gasohol"). The urge to use this alternate fuel apparently derives from the need to provide farmers with a market for surplus corn. Fermentation of corn mash (as well as barley, grape juice, etc.) serves and has (historically) served as the source for beverage alcohol.

The heats of combustion (Chapter 3) of methanol (methyl alcohol [CH_3OH], $-\Delta H_c^\circ = 171\,kcal\,mol^{-1}$ [$714.8\,kJ\,mol^{-1}$]) and ethanol (ethyl alcohol [CH_3CH_2OH], $-\Delta H_c^\circ = 328\,kcal\,mol^{-1}$ [$1371\,kJ\,mol^{-1}$]) are about half that of hydrocarbons of comparable molecular weight (Table 3.3) and, indeed, the amount of heat liberated on combustion of ethanol is significantly less than that from gasoline.* The rationale for the use of ethanol and ethanol–gasoline mixtures is therefore largely economic and is a function of both the price of crude oil and the crop surplus.

II. Oxidation at the Hydroxyl-Bearing Carbon

a. Chemical Oxidation of Alcohols. The methods for oxidation of alcohols to the corresponding carbonyl compounds (aldehydes and thence carboxylic acids from primary alcohols or ketones from secondary alcohols) are legion. For each process, regardless of detail, it is important to note that (a) tertiary alcohols (R_3C–OH, R ≠ H) cannot easily undergo oxidation at carbon and, under forcing conditions, carbon–carbon bond cleavage occurs; (b) secondary alcohols (R_2CHOH; R ≠ H) yield ketones (R_2C=O, ≠H); and (c) primary alcohols (RCH_2OH) yield aldehydes (RCH=O), which are often so labile that oxidation to the corresponding carboxylic acid (RCO_2H) intrudes and the aldehyde is not isolated (Figure 8.2).

As already noted above, most of the methanol (methyl alcohol, CH_3OH) produced by the reduction of carbon monoxide is reoxidized to methanal (formaldehyde, H_2C=O). The latter is largely used in making thermosetting plastics. Industrially, the oxidation is accomplished by heating (450–900°C) methanol (CH_3OH) over a silver (or silver oxide) catalyst in the presence of air for a short time (ca. 0.01 s). The oxidation of methanol (CH_3OH) in the laboratory to produce methanal (formaldehyde, H_2CO) is generally avoided as both materials are readily available commercially.

A few of the more common laboratory processes for other alcohols are provided in Table 8.5.

The first item in Table 8.5 shows the use of potassium permanganate ($KMnO_4$) to oxidize 1-propanol ($CH_3CH_2CH_2OH$) to propanal (CH_3CH_2CHO). Potassium permanganate ($KMnO_4$) is a powerful oxidizing agent, which was previously encountered in Chapter 6, where it was shown to be useful in the preparation of vicinal

---

Benzene forms a ternary azeotrope with water and ethanol (18.5% ethanol, 7.4 % water, and 74.1% benzene), which distills first.

$$H_2C{=}CH_2 \;+\; H_2O \;\xrightarrow[\text{Alumina 300°C}]{H_3PO_4}\; CH_3CH_2OH$$

*Fuel alcohol (denatured ethanol, which contains impurities such as methanol so that it is not potable) produces about 11,500 Btu lb^{-1} (1 Btu or British thermal unit is the amount of heat required to raise the temperature of 1 lb of water at its maximum density, 1°F, and is equal to about 252 cal). Gasoline produces about twice as many British thermal units.

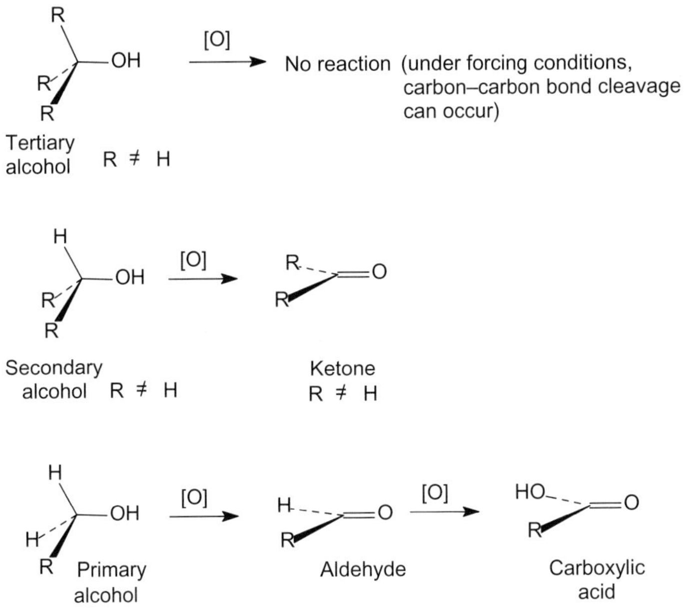

Figure 8.2. A schematic representation of the results of oxidation of tertiary, secondary, and primary alcohols. Generally, the "R" groups shown can be alkyl, aryl, and so on, but, *except for the primary alcohol (R=H is methanol, CH₃OH)*, not hydrogen (H). Aldehydes are easily oxidized to carboxylic acids. The abbreviation [O] is meant to indicate that oxidation occurs. The details are omitted.

diols from alkenes. As noted there, the diol product was susceptible to overoxidation. Indeed, the laboratory use of potassium permangate (KMnO₄) as an oxidizing agent is often limited by its lack of specificity and its ability to oxidize many functional groups. Its use is further complicated by the fact that the reduction product of permanganate (MnO₄⁻), which is purple, is manganese dioxide (MnO₂), which is dark brown. Consequently, it is difficult to observe the course of the reaction.

A balanced equation for the oxidation of some unspecified alcohol to some unspecified carbonyl compound (aldehyde or ketone) is provided in Equation 8.8.

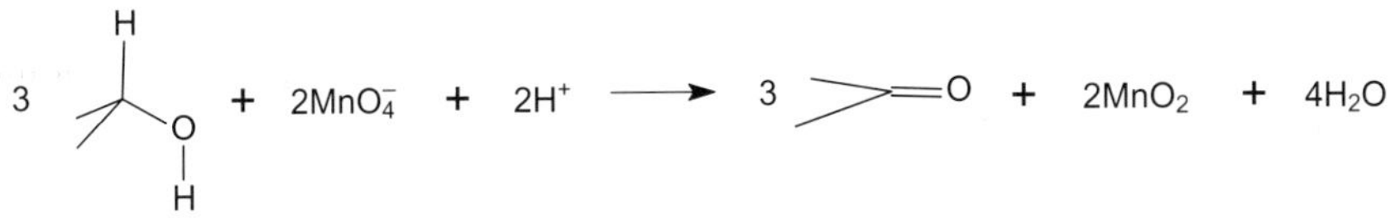

$$3 \; \text{(alcohol)} + 2MnO_4^- + 2H^+ \longrightarrow 3 \; \text{(carbonyl)} + 2MnO_2 + 4H_2O \tag{8.8}$$

It is important to note that 2 equivalents of permanganate are required for each three of alcohol and that protons are "consumed" (or base "generated") as the reaction progresses. Since the conversion of aldehyde to carboxylic acid frequently proceeds through the hydrate of the aldehyde (Chapter 9), the formation of which is catalyzed by both acid *and* base, stopping the oxidation at the aldehyde stage is often difficult and direct conversion to the carboxylic acid occurs. Scheme 8.1 provides a pathway linking 1-propanol with propanal and shows that permanganate (manganese(VII)) is initially reduced to manganate (manganese(VI)), which

TABLE 8.5. A Few of the Many Different Oxidation Reactions That Have Been Carried Out on Alcohols

	Alcohol	Reagent/Conditions	Products(s)
1.	$CH_3CH_2CH_2OH$ 1-Propanol	Alkaline $KMnO_4$	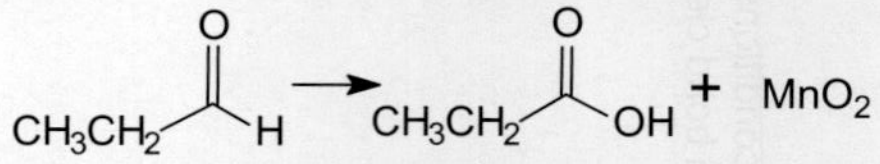 Propanal → Propanoic acid $+ MnO_2$
2.	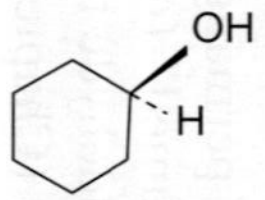 Cyclohexanol	CrO_3/H_2SO_4 (Jones Oxidation)	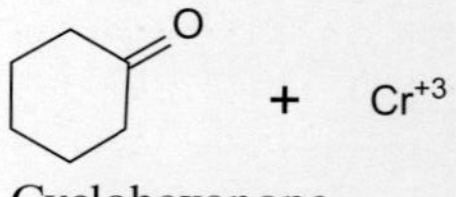 Cyclohexanone $+ Cr^{+3}$
3.	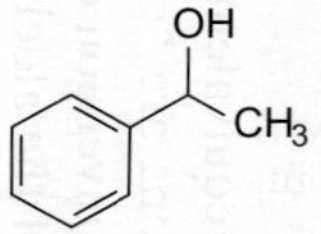 1-Phenylethanol	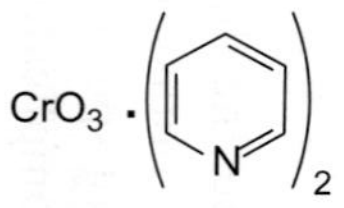 $CrO_3 \cdot$ (pyridine)$_2$ (Collins reagent)	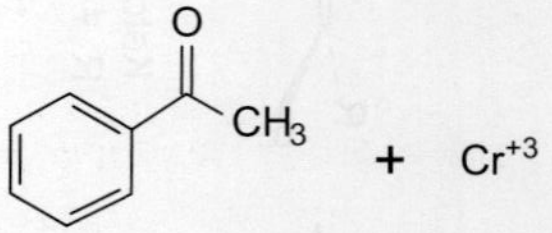 Phenylethanone (acetophenone) $+ Cr^{+3}$
4.	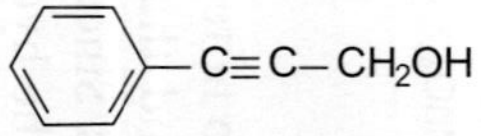 1-Hydroxy-3-phenyl-2-propyne (3-phenylpropargyl alcohol)	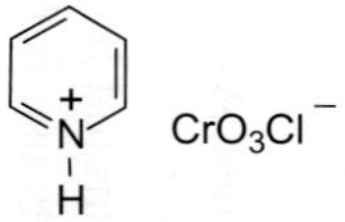 Pyridinium chlorochromate (PCC)	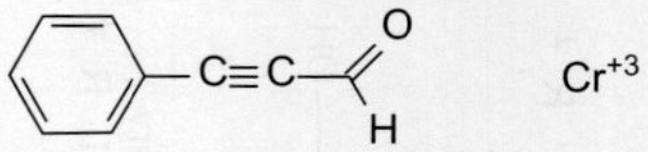 3-Phenylbutynal Cr^{+3}

5. Cyclooctanol

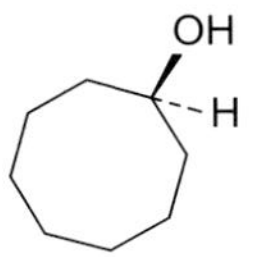

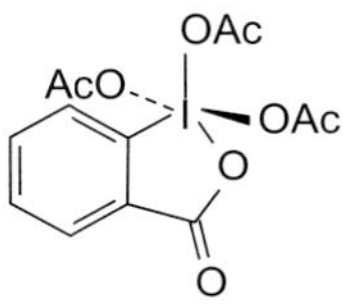

1,1,1-Triacetoxy-1, 1-dihydro-
1,2-benziodoxol-3(1*H*)-one
(Dess-Martin triacetoxyperiodinane)

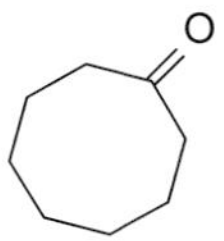

Cyclooctanone

6. Cyclopentylmethanol
(hydroxymethylcyclopentane)

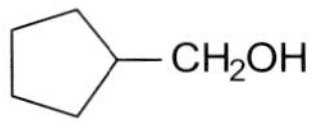

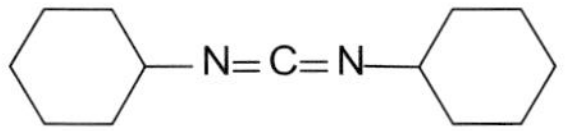

Dicyclohexylcarbodiimide

H_3PO_4 /

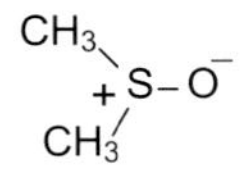

(Moffatt oxidation)

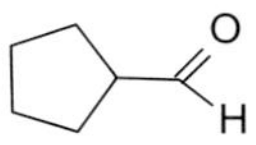

Cyclopentane carboxaldehyde $+$ $(CH_3)_2S$ $+$ dicyclohexylurea

7. $H-C{\equiv}C(CH_2)_2CH_2OH$
4-Pentyne-1-ol

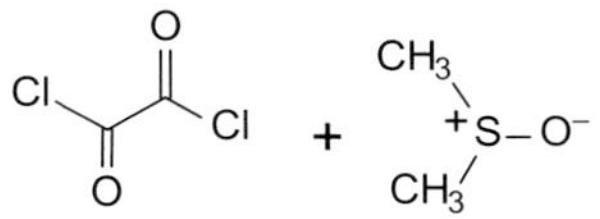

Swern oxidation

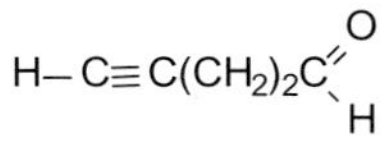

O_2/PtO_2

$H-C{\equiv}C(CH_2)_2C$ $+$ $(CH_3)_2S$

4-Pentynal

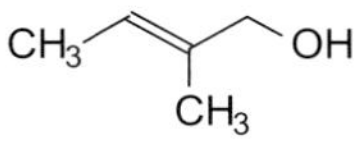

8. (*E*)-2-methyl-2-buten-1-ol

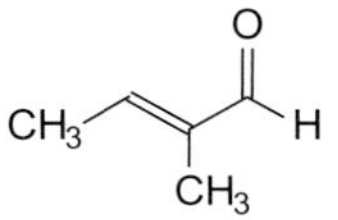

(*E*)-2-Methyl-2-butenal $+$ H_2O

(Continued)

TABLE 8.5. *Continued*

Alcohol	Reagent/Conditions	Products(s)
9. 3-Hydroxycyclohexene	$Al[OCH(CH_3)_2]_3$ (catalyst) (Oppenauer oxidation) (Meerwein–Ponndorf–Verley Reduction)	3-Cyclohexenone + $(CH_3)_2CHOH$
10. 1-Phenylpropanol	$N=N$ diethyl azodicarboxylate (OCH_2CH_3)	Phenylpropanone
11. *endo*-Bicyclo[2.2.1]heptan-2-ol (*endo*-norbornanol)	$(CH_3CH_2CH_2)_4N^+ RuO_4^-$ tetra-*n*-Propylammonium perruthenate (TPAP); N-Methylmorpholine oxide (NMO)	Bicyclo[2.2.1]heptan-2-one (norbornanone)
12. 2-Propanol (isopropanol)	Halogen, X_2 (X = Cl_2, Br_2, I_2)	Propanone (acetone)

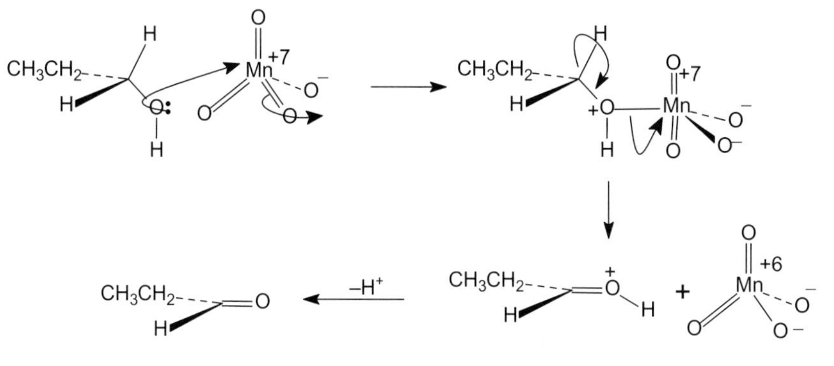

$$3MnO_4^{-2} \; + \; 2H_2O \; \longrightarrow \; 2MnO_4^- \; + \; MnO_2 \; + \; 4OH^-$$

Scheme 8.1. A representation of a path showing the oxidation of 1-propanol to propanal by permanganate [manganese(VII)]. The equilibrium between manganate [manganese(VI)] and both manganese dioxide [(MnO$_2$), manganese(IV)] and permanganate [manganese(VII)], which allows for regeneration of the latter during oxidation, is also shown.

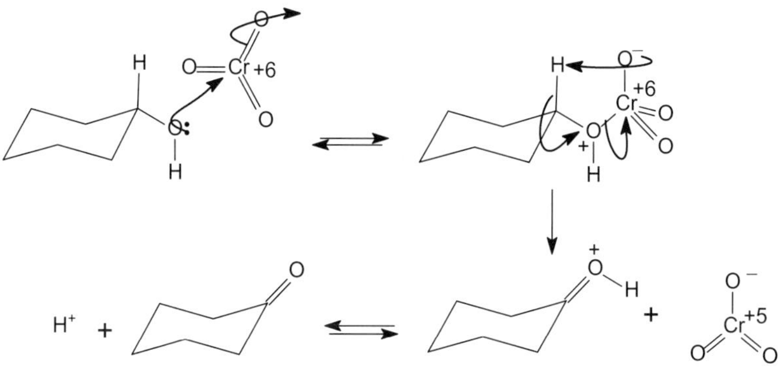

Scheme 8.2. A representation of a pathway for the oxidation of cyclohexanol to cyclohexanone by chromic anhydride.

undergoes disproportionation to regenerate permanganate and manganese dioxide (MnO$_2$). As noted in the table, it is common to employ a strong alkaline medium to inhibit this equilibrium (which regenerates oxidizing agent and may contribute to overoxidation).

The second item in Table 8.5 shows the oxidation of cyclohexanol to cyclohexanone with chromic anhydride (chromate, chromium(VI)). A wide variety of subtly different methods, largely substrate dependent, utilize the reduction of yellow orange chromium(VI) to green chromium(III) to affect oxidation. Among these is the **Jones oxidation**, which involves the treatment of alcohol substrate with a stochiometric quantity of chromic anhydride (CrO$_3$, chromium(VI)) in aqueous sulfuric acid (H$_2$SO$_4$). A potential pathway for the conversion is shown in Scheme 8.2. The disproportionation of Cr^{+5} to Cr^{+3} and Cr^{+6} follows much the same pattern as that for manganese (Mn^{+6} to Mn^{+4} and Mn^{+7}) as shown in Scheme 8.1.

Although there is conformational mobility at room temperature so that *both* axial and equatorial conformers are present, it is clearly possible to observe the overall course of the oxidation by examination of the infrared (IR) spectra of the starting material and product (Figure 8.3). As anticipated, the broad –OH stretching band centered at $3368\,cm^{-1}$ disappears and is replaced by an intense carbonyl (C=O) band centered at $1716\,cm^{-1}$ as the reaction progresses.*

Given a potentially limitless number of alcohols and the possible variation in functionality that **might** be present *in addition to the hydroxyl group*, it is not surprising that a variety of reagents based on chromic anhydride have been developed

---

*As shown some years ago (Eliel, E. L.; Schroeter, S. H.; Brett, J.; Biros, F. J.;Richer, J.-C. *J. Am. Chem. Soc.*, **1966**, *88*, 3327), (*Z*)- or *cis*-4-*t*-butylcyclohexanol, where the hydroxyl (–OH) is axially disposed (the *t*-butyl group serving as a conformational "anchor," which remains equatorial; Chapter 4), undergoes oxidation about three times more rapidly than the (*E*)- or *trans*-isomer, where the hydroxyl (–OH) is equatorial. It is argued that the energy difference in the ground state (estimated as $0.7\,kcal\,mol^{-1}$ [$3\,kJ\,mol^{-1}$]) is diminished in the rate-determining transition state because if the ground-state difference were maintained the two isomers should react at the same rate. The reaction of chromic anhydride with the hydroxyl oxygen in both cases (to form the respective chromate esters) is probably not rate determining since the axial hydroxyl is more sterically encumbered than the equatorial hydroxyl, and thus the (*Z*)- or *cis*-isomer should react more slowly. Proton abstraction from the **carbon** bearing the hydroxyl is another alternative, and that step might be rate determining because relief of steric strain should be greater for the (*Z*)- or *cis*-isomer than for the (*E*)- or *trans*-isomer. Both isomers yield the same product (4-*t*-butylcyclohexanone).

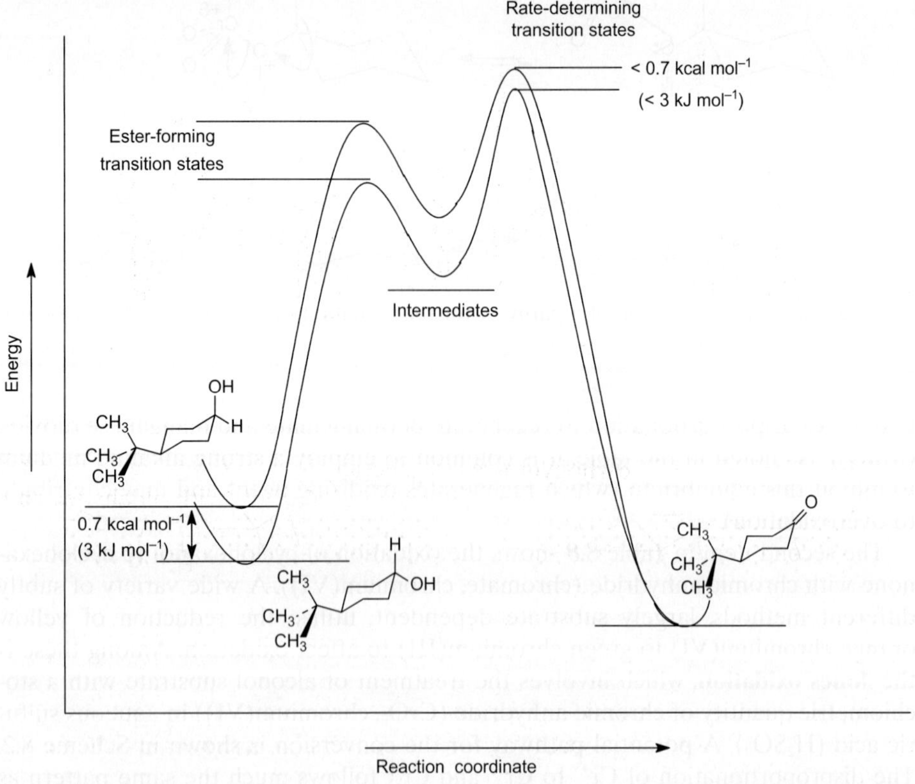

(Continued on next page)

for use in different solvents and capable of functioning at different pH values. Thus, oxidation of 1-phenylethanol to phenylethanone (acetophenone) and of the propargyl alcohol (1-hydroxyl-3-phenyl-2-propyne, 2-phenylpropargyl alcohol) to 3-phenylbutynal, the third and forth items, respectively, in Table 8.5, also make use of the reduction of chromium(VI) to chromium(III). However, **Collins** reagent (dipyridine chromium(VI); CrO_3·$(C_5H_5N)_2$) is not acidic and is soluble enough in dichloromethane (CH_2Cl_2) to allow its use as solvent. Pyridinium chlorochromate (**PCC**, $(CrO_3Cl)^-$·$(C_5H_5NH)^+$) is similar to Collins reagent* and is somewhat easier to prepare (Collins reagent requires the addition of chromic anhydride (CrO_3) to pyridine (C_5H_5N), a process that must avoid overheating), while to prepare **PCC**, chromic anhydride (CrO_3) is added to a solution of pyridine (C_5H_5N) in hydrochloric acid (HCl).

Many reactions of halogens (F_2, Cl_2, Br_2, and I_2) are related to their oxidizing abilities. However, they are often difficult to handle and control in the laboratory environment and overoxidation is common. Some organic compounds that contain iodine in its higher oxidation states are relatively easy to work with, and the development of periodinanes, in particular, has shown the way to clean oxidation of primary and secondary alcohols to aldehydes and ketones, respectively. Among the more useful reagents is that shown in item 5 in Table 8.5, the **Dess–Martin** reagent[†] (1,1,1-triacetoxy-1,1-dihydro-1,2-benziodoxol-3(1H)-one,   triacetoxyperiodinane). Although the details of the process by which the oxidation is effected have not been established, it is clear that the iodine is reduced as the alcohol is oxidized. Scheme 8.3 presents a reasonable pathway.

In the search for specific and gentle oxidizing agents for alcohols that leave other functional groups unaffected, it has been found that reduction of the common solvent, methylsulfinylmethane (dimethyl sulfoxide [DMSO] [$(CH_3)_2SO$]) to dimethyl sulfide can be very useful. Thus, both items 6 and 7 in Table 8.5, the **Moffatt** oxidation[‡] and the **Swern** oxidation,[§] respectively, which can be performed with exquisite selectivity at low to moderate temperatures, have found wide use.

The figure shown next attempts to illustrate the sterically encumbered axial isomer relative to the less crowded equatorial isomer.

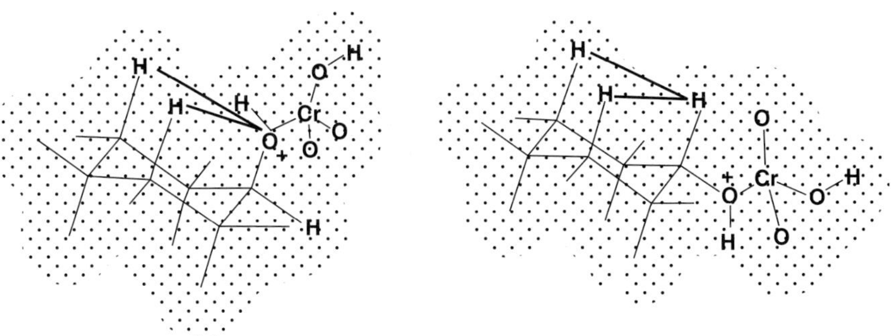

Axial chromate ester   Equatorial chromate ester

*Collins, J. C.; Hess, W. W.; Frank, F. J. *Tetrahedron Lett.*, **1968**, 3363.

[†]Dess, P. B.; Martin, J. C. *J. Am. Chem. Soc.*, **1978**, *100*, 300.

[‡]Pfitzner, K. E.; Moffatt, J. G. *J. Am. Chem. Soc.*, **1963**, 85, 3027.

[§]Mancuso, A. J.; Huang, S.-L.; Swern, D. *J. Org. Chem.*, **1978**, 43, 2480.

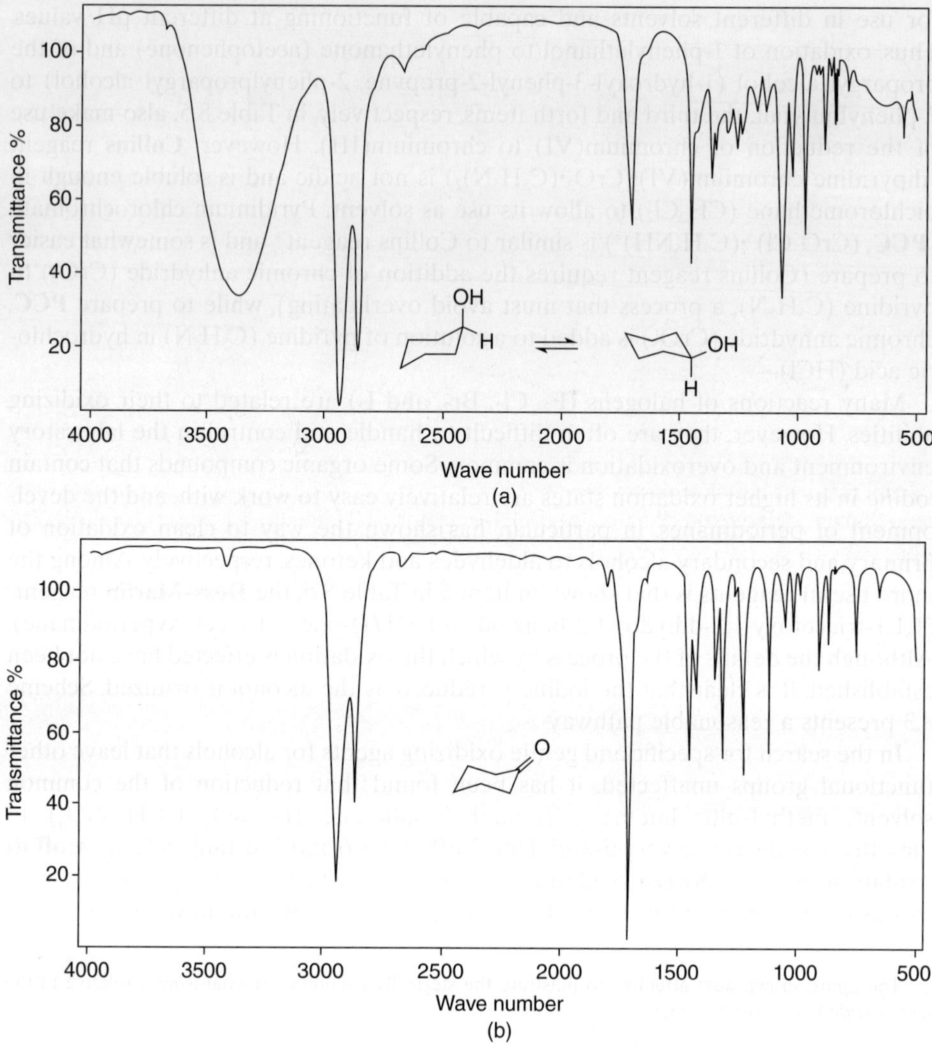

Figure 8.3. (a) The (neat) IR spectrum of cyclohexanol containing a small quantity of cyclohexanone (presumably formed by air oxidation of the alcohol). (b) The (neat) IR spectrum of cyclohexanone from which all of the cyclohexanol has been removed by column chromatography.

As shown in Scheme 8.4, oxygen is "transferred" from methylsulfinylmethane (DMSO, $[(CH_3)_2SO]$) to dicyclohexylcarbodiimide ($C_6H_{11}N=C=NC_6H_{11}$) to yield dicyclohexylurea $[(C_6H_{11}NH)_2C=O]$ and a sulfoxonium salt of the alcohol substrate. The non-nucleophilic dihydrogen phosphate anion, produced from phosphoric acid (H_3PO_4) when the latter protonates the diimide, does not compete for the protonated diimide with the more nucleophilic sulfoxide in forming the trigonal intermediate shown in the scheme. It is this intermediate that is attacked, *at sulfur*, by the alcohol substrate, resulting in the transfer of dimethyl sulfide $[(CH_3)_2S]$ to the alcohol to be oxidized. Then, reductive loss of dimethyl sulfide $[(CH_3)_2S]$ and a

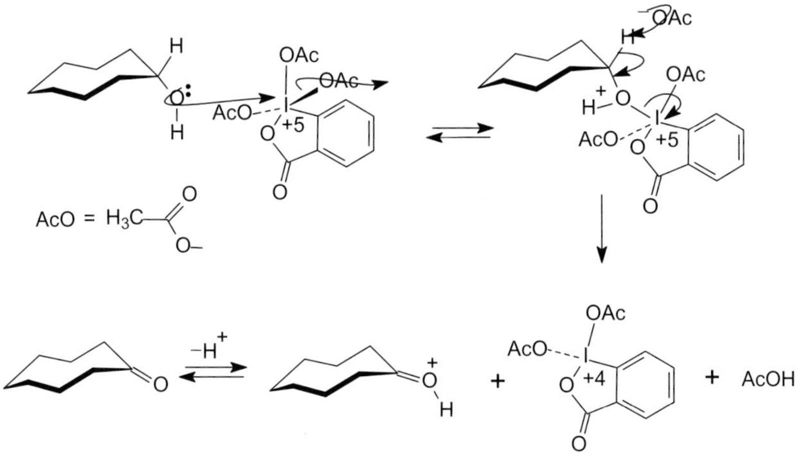

AcO = H_3C—C(=O)—O⁻

Scheme 8.3. A possible pathway for the oxidation of cyclooctanol to cyclooctanone by the Dess–Martin triacetoxyperiodinane [1,1,1-triacetoxy-1,1-dihydro-1,2-benziodoxol-3(1*H*)-one].

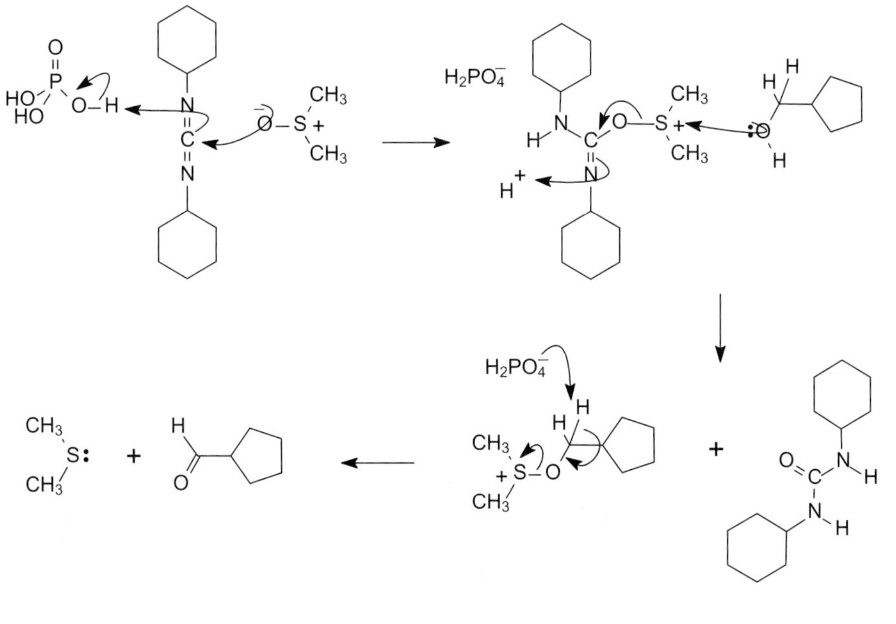

Dicyclohexylurea

Scheme 8.4. A representation of the Moffatt oxidation showing the conversion of dicyclohexylcarbodiimide to the corresponding urea while methylsulfinylmethane (dimethyl sulfoxide [DMSO] [$(CH_3)_2SO$]) is reduced to dimethyl sulfide [$(CH_3)_2S$] and cyclopentylmethanol is oxidized to cyclopentane carboxaldehyde.

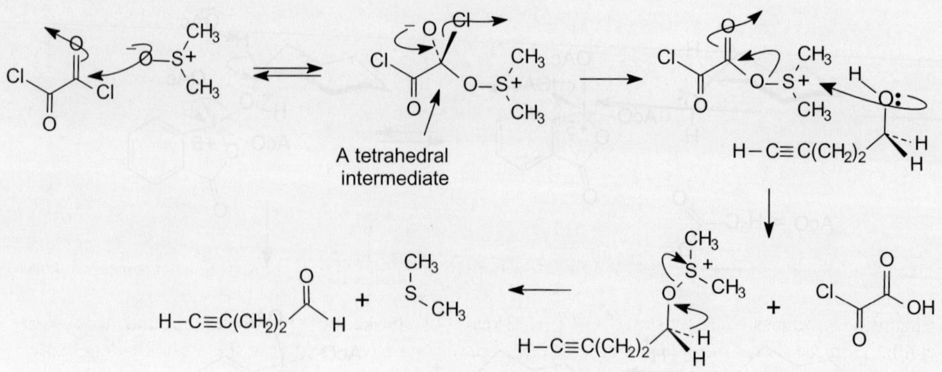

A tetrahedral
intermediate

Scheme 8.5. A representation of the Swern oxidation showing the conversion of 4-pentyne-1-ol to 4-pentynal using oxalyl chloride (ClCOCOCl) and methylsulfinylmethane (dimethyl sulfoxide [DMSO] [$(CH_3)_2SO$]). The sulfoxide is reduced to dimethyl sulfide [$(CH_3)_2S$], while the primary alcohol is oxidized to the corresponding aldehyde.

proton (H^+) from the sulfoxonium salt produces the oxidized product. The *overall* process of the Moffatt oxidation suffers from the necessity to separate the final (oxidized) product (aldehyde or ketone) from the by-product dicyclohexylurea [$(C_6H_{11}NH)_2C=O$], a process that can be difficult.

In an exactly analogous fashion, the bis acid chloride of oxalic acid (oxalyl chloride [ClCOCOCl]) serves to activate methylsulfinylmethane (DMSO [$(CH_3)_2SO$]) by formation of a tetrahedral intermediate. This kind of intermediate, common in the reaction of carbonyl compounds (Chapter 9), is formed as a result of attack at the carbon of the carbonyl by the oxygen of the methylsulfinylmethane (DMSO [$(CH_3)_2SO$]). Although reversion to starting materials is possible, loss of chloride anion with reformation of the trigonal carbonyl center produces an activated complex. When the complex is attacked by alcohol, *the same sulfoxonium salt found in the Moffatt oxidation results* along with the unstable, half-acid chloride of oxalic acid (HO_2CCO_2H) (Chapter 9). As was the case before, decomposition of the sulfoxonium salt produces dimethyl sulfide [$(CH_3)_2S$] and the corresponding aldehyde.

A pathway for the formation of 4-pentynal from 4-pentyne-1-ol (Table 8.5, item 7) by the Swern oxidation is shown in Scheme 8.5.

Acid chlorides other than oxalyl chloride (ClCOCOCl) and even acid anhydrides (such as trifluoroacetic anhydride [$(CF_3CO)_2O$] can be used in the Swern oxidation. The choice is often dependent upon what other functional groups are present.

The eighth item in Table 8.5, the oxidation of the primary alcohol (*E*)-2-methyl-2-buten-1-ol to the corresponding aldehyde, (*E*)-2-methyl-2-butenal, by oxygen (O_2) in the presence of platinum oxide (PtO_2) is particularly interesting. It has already been noted, here and in preceeding chapters, that oxygen (O_2) reacts with compounds of carbon and hydrogen to produce carbon dioxide (CO_2) and water (H_2O). So it must be that, under the specified conditions for item 8 in Table 8.5, the platinum dioxide (PtO_2) catalyzed oxidation by oxygen (O_2) is "incomplete." Further, at the beginning of Chapter 6, the reaction of oxygen (O_2) with alkenes was discussed. Since (*E*)-2-methyl-2-buten-1-ol clearly contains a carbon–carbon double bond, it is also necessary to assume that the conditions specified preclude reaction at the double bond.

The selectivity imparted to organic reactions by the presence of catalysts, coupled to the observation (Chapter 4) that "small" changes in the energy of activation ($\Delta G^{\ddagger}$) can result in apparent "large" changes in product distribution, serves to enrich organic chemistry. Although the pathway by which this "dehydrogenation" occurs (presumably on the surface of the catalyst) is not known, the fact that it occurs is only one example of such enrichment.*

*It will be recalled from the earlier discussion (Chapter 6) that the ground state of oxygen is a triplet (i.e., as shown in the next figure, two electrons with spins parallel occupy degenerate π orbitals) [$^3O_2$]. Excitation produces a singlet state [$^1O_2$] (ca. 23 kcal mol^{-1} [96 kJ mol^{-1}] above the ground state), which can be pictured as resulting from the spin inversion of one electron; that is, the mutually degenerate π orbitals are each singly occupied by spin paired electrons. Pairing the electrons in *one orbital* removes the degeneracy (raising the energy) of the singlet, $^1O_2$ (which is now ca. 37 kcal mol^{-1} [159 kJ mol^{-1}], above the ground state). As pointed out earlier, excited, singlet-state oxygen can be produced by the irradiation of some donor (e.g., rose bengal, fluorescein, and methylene blue, called a *sensitizer*), which undergoes intersystem crossing as a result of, for example, collision from *its* initially produced singlet state to a triplet state. The triplet-state sensitizer then transfers its energy (it is *quenched*) to oxygen. Spin is conserved (the asterisk [*] indicating an excited state); that is,

$$^1(\text{Sensitizer}) + h\nu \rightarrow {}^1(\text{sensitizer})^*$$

$$^1(\text{Sensitizer})^* \rightarrow {}^3(\text{sensitizer})^*$$

$$^3(\text{Sensitizer})^* \,[\uparrow\uparrow] + {}^3O_2\,[\downarrow\downarrow] \rightarrow {}^1(\text{sensitizer})\,[\uparrow\downarrow] + {}^1O_2{}^*\,[\uparrow\downarrow]$$

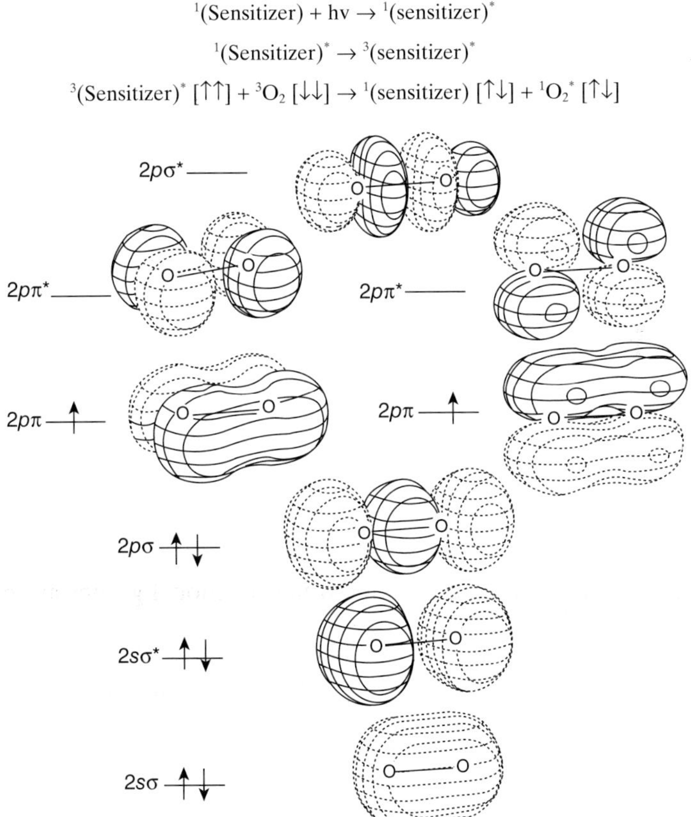

A representation of the molecular orbitals (MOs) (excluding 1s) of ground-state (triplet [$^3O_2$]) oxygen. The representations were created using the MOViewer in WebMO version 6.0.002p.

(Continued on next page)

The ninth item in Table 8.5 is of particular interest as it involves the catalytic conversion of an alcohol to the corresponding carbonyl compound while another carbonyl compound is being converted to an alcohol. Indeed, the **Oppenauer** oxidation,* which uses, for example, aluminum isopropoxide $\{Al[i\text{-}OPr]_3, Al[OCH(CH_3)_2]_3\}$ as a catalyst, is the microscopic reverse of the **Meerwein–Ponndorf–Verley** reduc-

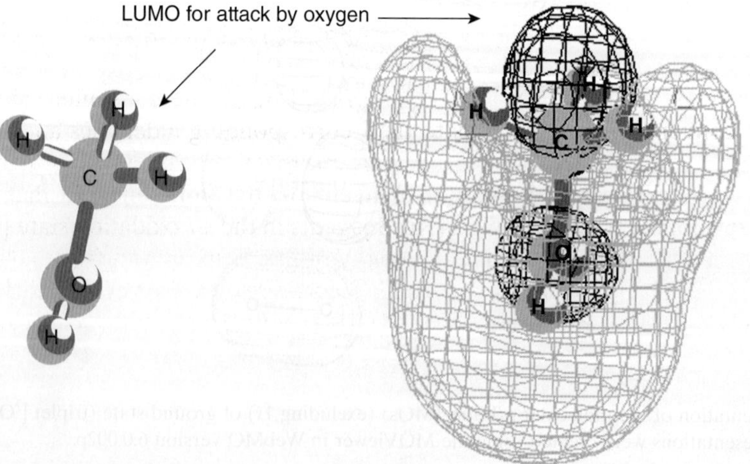

X = Y = Z = H: fluorescein

X = H: Y = Z = Br: eosin yellow

X = H, Y = Br, Z = NO$_2$: eosin I blue

X = Cl, Y = Z = I: rose bengal

Methylene blue

Benzophenone

Interestingly, in contrast to what is shown in Table 8.5 for the platinum catalyzed (PtO$_2$) oxidation of an alcohol to an aldehyde, *singlet oxygen ($^1O_2$) apparently reacts with alcohols to produce hydroperoxides*, for example,

$$\text{(12–25\%)}$$

It is possible that insertion of the oxygen $^1O_2$* into the LUMO of an alcohol, such as methanol (shown in the next figure), followed by hydrogen atom migration from carbon to oxygen might apply here as it did in the Oxidation of Alkanes (Chapter 6, Section B, I). Such a pathway would directly account for the hydroperoxide product.

*Oppenauer, R. V. *Recl. Trav. Chim. Pays Bas*, **1937**, 56, 137.

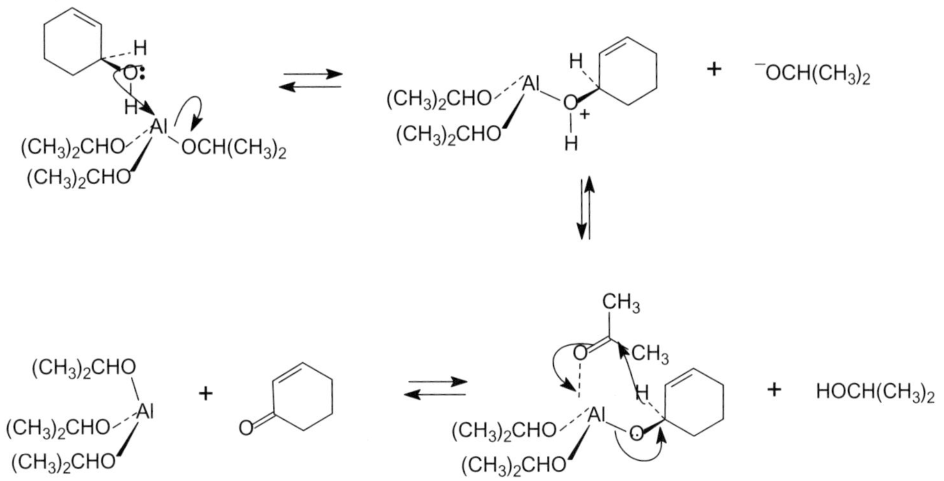

Scheme 8.6. A representation of the Oppenauer oxidation. Note that a hydride ion (H⁻) is transfered between the substrate and solvent, the latter being **reduced**.

tion* and will be encountered in that form again in Chapter 9 when the reduction of carbonyl compounds is discussed. As shown in Scheme 8.6, it is presumed that the first step is the reversible displacement of the aluminum (Al) of an alkoxide (*iso*propoxide anion [$(CH_3)_2CHO^-$, (i-OPr⁻)]) by the alcohol to be oxidized. Then, with the carbonyl compound to be reduced (the *solvent*) also coordinated to oxygen, *hydride (H^-) is transferred from the carbon bearing the hydroxyl in the substrate to the carbon of the carbonyl in the solvent.* Release of the carbonyl group-bearing substrate consummates the oxidation. Since the solvent is in vast excess, the reaction is driven to completion.

The overall acceptance of hydrogen by a reducing reagent is demonstrated by item 10 of Table 8.5. Here, the diethyl ester of diazodicarboxylic acid (diethyl azodicarboxylate [DEAD]) serves to oxidize 1-phenylpropanol to phenylpropanone while it is reduced to diethyl hydrazo-dicarboxylate in the process. Although the details are sketchy, a possible pathway is shown in Scheme 8.7.

Although the variety of oxidizing agents already considered provides a flavor of the richness explored by chemists in attempting to find convenient and specific ways to convert primary alcohols to their corresponding aldehydes and secondary alcohols to ketones in high yield, there is an apparently unending stream of new reagents and new ways to apply old reagents to the same ends. In that vein, the recent development of ruthenium (Ru) reagents in the +7 oxidation state [Ru(VII)] serves as a striking example of what can be done. Thus, although ruthenium (Ru) in the +6 [Ru(VI)] oxidation state had been known to convert primary alcohols (RCH_2OH) to carboxylic acids (RCO_2H) and secondary alcohols to the correspond-

---

*(a) Meerwein, H.; Schmidt, R. *Liebigs Ann. Chem.*, **1925**, *444*, 221; (b) Ponndorf, W. *Angew. Chem.*, **1926**, 39, 138; (c) Verley, A. *Bull. Soc. Chim. Fr.*, **1925**, 37, 537.

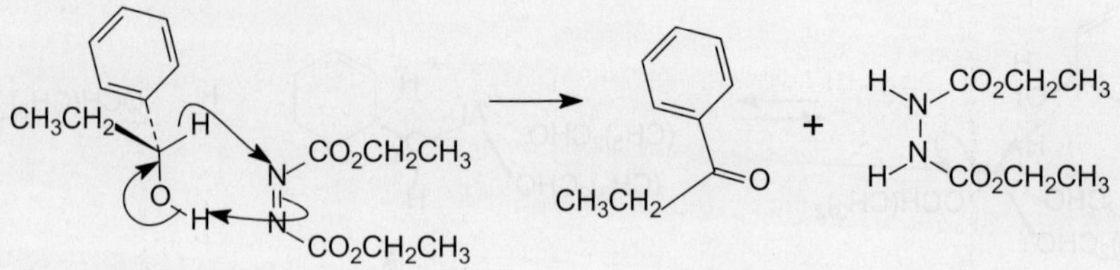

Scheme 8.7. A representation of a possible pathway for the oxidation of a secondary alcohol (1-phenylpropanol) to the corresponding ketone (phenylpropanone) with diethyl azodicarboxylate (DEAD) (see Mitsumobu, O.; Yamada, M. *Bull. Chem. Soc. Jpn.*, **1967**, *40*, 2380 and Mitsunobu, O. *Synthesis*, **1981**, 1).

ing ketones as part of a catalytic cycle with other oxidants in aqueous base, by-products of overoxidation were common in some cases. Ruthenium (Ru) in the +8 [Ru(VIII)] oxidation state (as RuO_4), on the other hand, is too powerful an oxidant. However, it was found after extensive studies that ruthenium (Ru) could be converted to the unusual "perruthenate" ion [Ru(VII)] by the reaction of ruthenium tetroxide (RuO_4) with (among others) tetra-*n*-propylammonium hydroxide $[(CH_3CH_2CH_2)_4N^+OH^-, n\text{-}Pr_4N^+OH^-]$ to provide a salt (tetra-*n*-propylammonium perruthenate [TPAP], $(nPr_4)N^+RuO_4^-$), soluble in organic solvents (such as dichloromethane [methylene chloride, CH_2Cl_2]), tolerant of a wide variety of other functional groups and which cleanly oxidizes both primary and secondary alcohols (with a few exceptions) to their respective aldehydes and ketones. There are two other features of value too.

First, on completion of the reaction, the ruthenium reagent is easily removed by filtration through a thin pad of silica gel and, second, in the presence of a co-oxidant, such as *N*-methylmorpholine *N*-oxide, only a catalytic amount of ruthenium reagent $((n\text{-}Pr)_4N^+RuO_4^-, TPAP)$ is needed. The role of the co-oxidant is to regenerate the Ru(VII) after its reduction as a consequence of alcohol oxidation. While the details of the exact pathway are unknown, Scheme 8.8 provides a potential pathway to account for the oxidation shown in item 11 in Table 8.5.

The final item in Table 8.5, the oxidation of 2-propanol to propanone with halogen (chlorine [Cl_2], bromine [Br_2], and iodine [I_2]), is a process scarcely used in the laboratory since it lacks specificity. Thus, as noted in Chapter 7, halogens *add* to compounds with carbon–carbon double and triple bonds, and thus the presence of such groups would preclude their use as a specific oxidant for alcohols. Nonetheless, it is important to realize that positive halogen ($X^+ = Cl^+$, Br^+, I^+, or higher oxidation states) *can* be used to effect oxidation and, if a primary or secondary hydroxyl group is present, conversion to aldehyde (and carboxylic acid) or ketone will occur. Bleaching agents, for example, hypochlorous acid (HOCl), may also undergo addition across unsaturated systems. Hypochlorous acid (HOCl), for example, is a source of Cl^+ and will serve to convert primary and secondary alcohols to their respective carbonyl compounds (frequently on the way to more complete degradation). A potential pathway to accomplish such oxidation is shown in Scheme 8.9.

In addition to the variety of reagents (some examples of which are provided in Table 8.5) that are generally available for the oxidation of alcohols, specific reagents

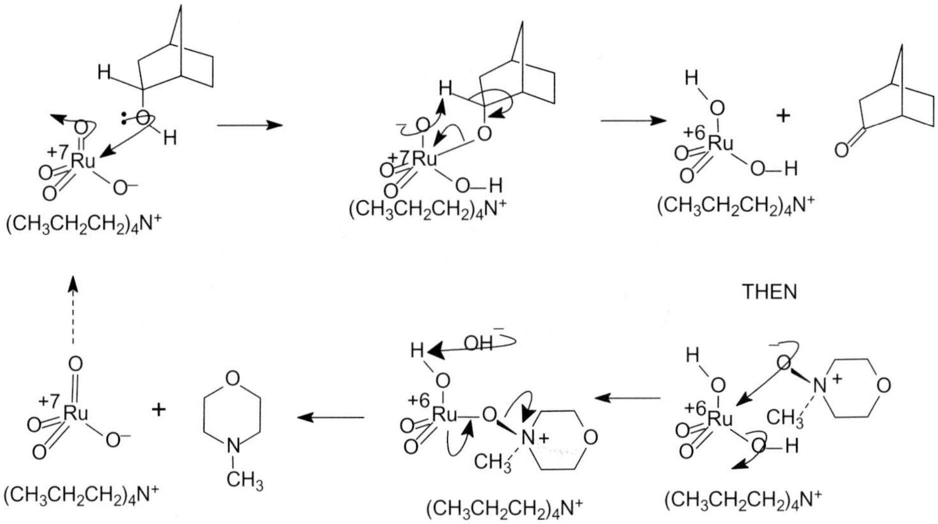

Scheme 8.8. A representation of the oxidation of *endo*-bicyclo[2.2.1]heptan-2-ol to the corresponding ketone (bicyclo[2.2.1]heptan-2-one) with tetra-*n*-propylammonium perruthenate (TPAP) in the presence of *N*-methylmorpholine *N*-oxide. Despite what is shown, all four ruthenium–oxygen (Ru–O) bonds in TPAP are equivalent and the process may actually involve a series of one-electron rather than two-electron redox steps. Curved arrows representing two-electron processes are used for convenience. The actual path is not yet known in detail (see Ley, S. V.; Griffith, W. P. *J. Chem. Soc. Chem. Commun.*, **1978**, 1625).

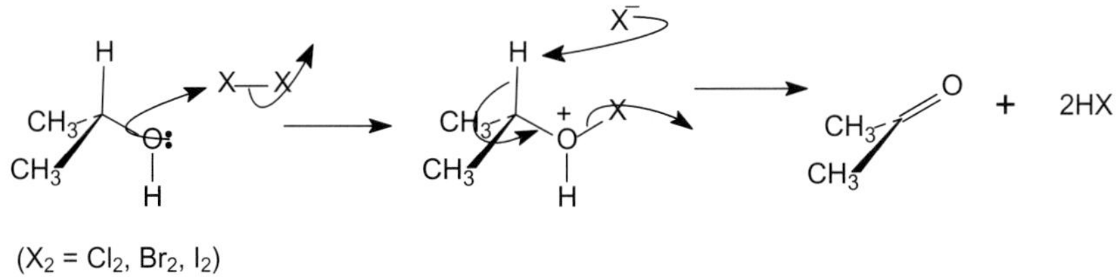

$(X_2 = Cl_2, Br_2, I_2)$

Scheme 8.9. A pathway for the oxidation of a secondary alcohol (2-propanol, $(CH_3)_2CHOH$) by halogen to produce a ketone (propanone, $(CH_3)_2C=O$).

have been developed for selective oxidation of some alcohols. Among these is *activated* manganese dioxide (MnO_2),* which, as shown in Equation 8.9, is especially useful for the oxidation of α,β-unsaturated (i.e., allylic) alcohols to the corresponding aldehydes or ketones. As illustrated in item 9 of Table 8.5, it is not always *necessary* to use such specialized reagents.

*Different preparations of activated manganese dioxide (MnO_2) are reported in the chemical literature. Generally, the commercially available material destined for catalyst use has suffered particle size adjustment (through grinding), acid (or base) washing to add or subtract surface protons, and heat treatment to remove adsorbed water. It is not uncommon to find that different "batches" of commercially obtained material (even from the same supplier) have different activites.

$$\text{(8.9)}$$

If more than one hydroxyl group is present in the substrate, as will be the case with that large group of compounds called carbohydrates (Chapter 11), many of the oxidizing reagents already discussed can successfully be used, but difficulties can be encountered because of lack of specificity. With only two hydroxyl (–OH) groups, the situation is somewhat simpler, and, if both hydroxyl groups (–OH) are the same and the oxidation reaction is performed carefully, that is, choosing a suitable solvent, using the stoichiometrically correct amount of oxidant and maintaining appropriate temperature control, a dialdehyde or diketone can be obtained (Equation 8.10).

$$\text{(8.10)}$$

However, it should be clear that different hydroxyl (–OH) groups may oxidize differently. Thus, since tertiary hydroxyl groups (R_3C–OH; R ≠ H) resist oxidation, compounds that contain tertiary hydroxyl groups (R_3C–OH; R ≠ H), along with those that are primary (RCH_2OH) or secondary (R_2CHOH; R ≠ H), will only suffer oxidation of the last two. Although there are exceptions, the order of ease of oxidation of different kinds of hydroxyl groups is allylic (HO–C–C=C) > saturated secondary (R_2CHOH; R ≠ H) > saturated primary (RCH_2OH) >> tertiary (R_3C–OH; R ≠ H). Again, as will be seen later in the discussion of carbohydrates (Chapter 11), specific conditions for oxidation of primary hydroxyl (RCH_2OH) groups in the presence of secondary hydroxyl (R_2CHOH; R ≠ H) have been identified.

The special cases of geminal diols (i.e., both hydroxyl groups [–OH] on the same carbon atom) and vicinal diols (i.e., hydroxyl groups on adjacent carbon atoms) are dealt with below.

The first hydrate of carbon dioxide (CO_2) (Equation 8.11) corresponds to carbonic acid (H_2CO_3). The hydration of carbonic acid (H_2CO_3) or the formation of a dihydrate ($2H_2O$) of carbon dioxide (CO_2) would then have four hydroxyl (–OH) groups on the same carbon atom. It is often argued that the absorption of carbon dioxide (CO_2) into water (H_2O) produces carbonic acid (H_2CO_3), and the dissociation constants of H_2CO_3 to $H^+ + HCO_3^-$ (4.16×10^{-7}) and HCO_3^- to $H^+ + CO_3^{-2}$ (4.84×10^{-11}) are frequently cited. However, these values assume that all of the CO_2 dissolved is present, initially, as H_2CO_3. It is probably true that most of the CO_2 dissolved is only loosely hydrated and that H_2CO_3 may actually be a "stronger" acid than suggested.

$$\text{(8.11)}$$

The hydrate of methanoic acid (formic acid [HCO_2H], Equation 8.12) can be described as having three hydroxyl groups on one carbon. This hydrate is formed only with difficulty as carboxylic acid dimers (Chapter 9) tend to predominate.

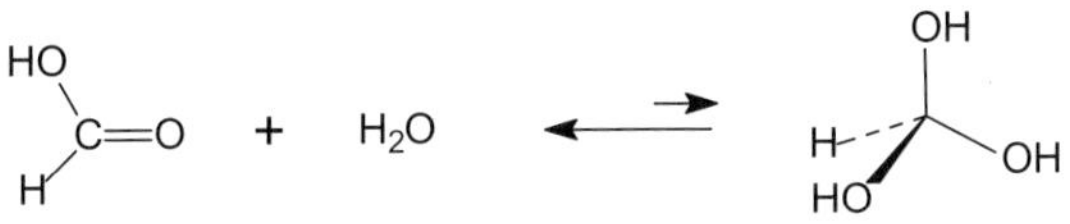

$$(8.12)$$

Hydrates of ketones and aldehydes are also known and, in general, as already noted in Chapter 5, they too are unstable with regard to the carbonyl compound. Exceptions occur when there are electron-withdrawing substituents on the adjacent carbon (aldehydes) or carbons (ketones) and, as shown by the large equilibrium constant, some of the hydrates can be isolated (Figure 8.4).

When two hydroxyls (–OH) are on adjacent carbon atoms (vicinal), it is possible to cleave the carbon–carbon bond between them to produce two aldehydes, an aldehyde and a ketone, or two ketones, depending upon the substituents on the diol. In principle, although many oxidizing agents could be used, in practice it is largely

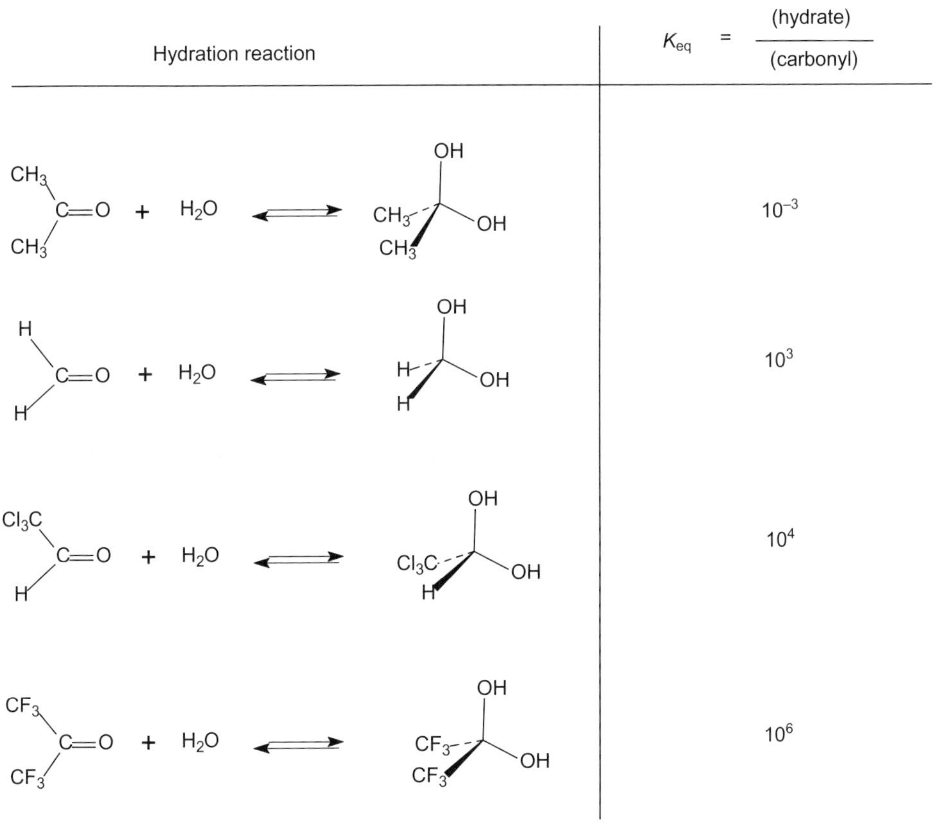

Hydration reaction	$K_{eq} = \dfrac{\text{(hydrate)}}{\text{(carbonyl)}}$
(acetone)	10^{-3}
(formaldehyde)	10^{3}
(chloral)	10^{4}
(hexafluoroacetone)	10^{6}

Figure 8.4. Some hydrates of aldehydes and ketones and equilibrium constants for their formation. (See, Lowry, T., Richardson, K. S. *Mechanism and Theory in Organic Chemistry*, 3rd ed., Harper & Row, New York, **1987** and references therein.)

either lead tetraacetate $[Pb(O_2CCH_3)_4]$ or periodic acid $(HIO_4 \cdot 2H_2O$ or $H_5IO_6)$, which serves the purpose.

In either case, it is necessary that a cyclic intermediate be formed. This requires, therefore, that the hydroxyl groups be close enough together to permit such a structure. A typical reaction with lead tetraacetate $[Pb(O_2CCH_3)_4]$ is shown in Equation 8.13.

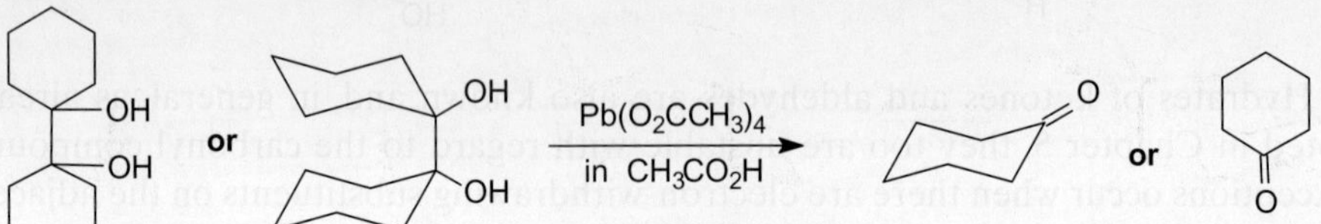

$$\text{(8.13)}$$

Since it is necessary to form a cyclic intermediate, not all diols can be used. Consider the set of diasteriomeric diols of Figure 8.5 (enantiomers not shown). It should be clear to you that the dihedral angle subtended between the hydroxyl groups on carbon atoms 2 and 3 in the decalin diols* "A," "C," "D," "E," and "E'" are all about 60°, while in "B," and "D′," the angle is about 180°. It should also be

*A note about nomenclature. The compounds shown in Figure 8.5 are all derivatives of bicyclo[4.4.0] decane but are not generally named as such. Historically, the aromatic hydrocarbon naphthalene $(C_{10}H_8)$ was reduced to produce decahydronaphthalene (decalin, $C_{10}H_{18}$, the *cis*- and *trans*-isomers of which are shown next), and thus the compounds of Figure 8.5 are, by some systems of nomenclature (including *Chemical Abstracts*) considered derivatives of decahydronaphthalene (decalin, $C_{10}H_{18}$). Second, as noted earlier (Chapter 5), *trans*-decalin cannot undergo ring inversion as that would require the ring fusion to be diaxial. *cis*-decalin, with one axial and one equatorial fusion, can, on the other hand, undergo such inversion, that is,

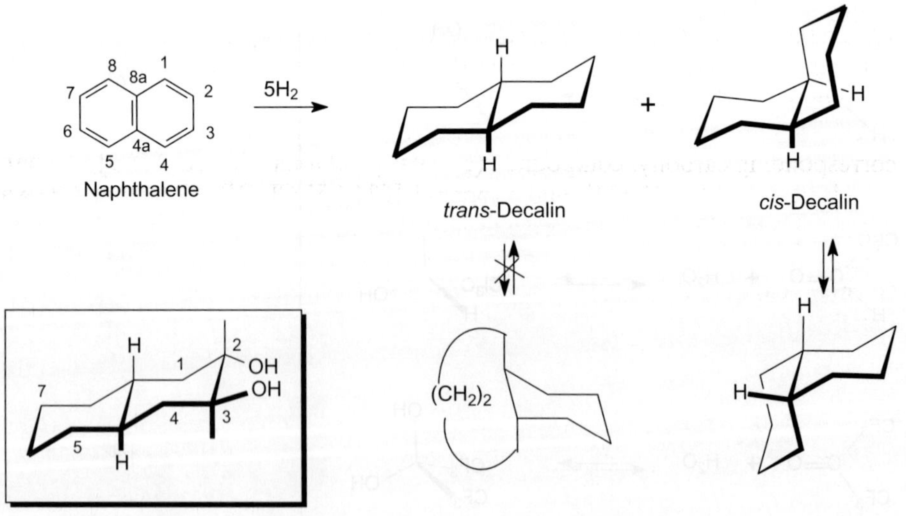

2α, 3β, 4aα, 8aβ-Decahydro-2,3-naphthalenediol

(Continued on next page)

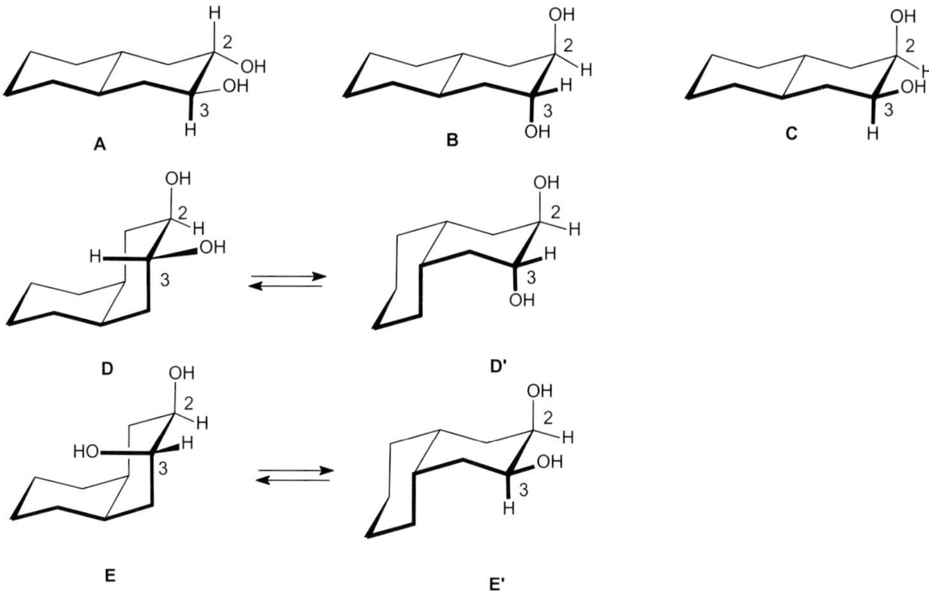

Figure 8.5. Representations of the *trans*-decalin (A, B, and C) and *cis*-decalin (D and E) *cis* and *trans*-2,3-diols.

clear that, as shown in Scheme 8.10, unless a *cyclic intermediate* forms as shown, carbon–carbon bond cleavage cannot occur. Thus, the isomers represented by the drawings labeled B and D′ are incapable of undergoing ring cleavage. However, since D (2β,3α,4aβ,8aβ-decahydro-2,3-naphthalenediol) and D′ (2α,3β,4aα,-8aα-decahydro-2,3-naphthalenediol) (Figure 8.5) are clearly related by ring inversion, it should be clear that *it is only* B (2β,3α,4aα,8aβ-decahydro-2,3-naphthalenediol) that will fail to undergo the cleavage reaction.

If there are three contiguous hydroxyl (–OH) groups on adjacent carbon atoms, as shown (Equation 8.14) for 2,3,4-hexanetriol, oxidation occurs to produce the corresponding carbonyl compounds (ethanal [acetaldehyde, CH_3CHO], methanoic acid [formic acid, HCO_2H], and propanal [CH_3CH_2CHO]). Note that C3, which bears a hydroxyl group (–OH), is flanked by two carbon atoms, each of which also bears a hydroxyl group, and that *C3 is excised* as the carbon in methanoic acid (formic acid [HCO_2H]) as a consequence of *two* carbon–carbon bond cleaving reactions.

Consider, next, compound "A," shown in the illustration and in Figure 8.5. Compound A is clearly a *trans*-decalin and, further, a decahydro-2,3-naphthalenediol. *By convention*, if the first substituent (the hydroxyl [–OH] at C2) is arranged on the same side of the ring system as the proximal ring juncture hydrogen (H), that is, the hydrogen at C8a, then the descriptor β is appended. Alternatively, the substituent is designated as α if it is opposite to the proximal ring juncture hydrogen. Thus, the hydroxyl at C2 in A is α. Continuing around the ring in "A," since the hydroxyl at C3 is opposite to that at C2, the hydroxyl at C3 is β and so on. The ring juncture hydrogens at 4a and 8a follow the same pattern. Compound A is thus 2α,3β,4aα,8aβ-decahydro-2,3-naphthalenediol and the other isomers are named similarly.

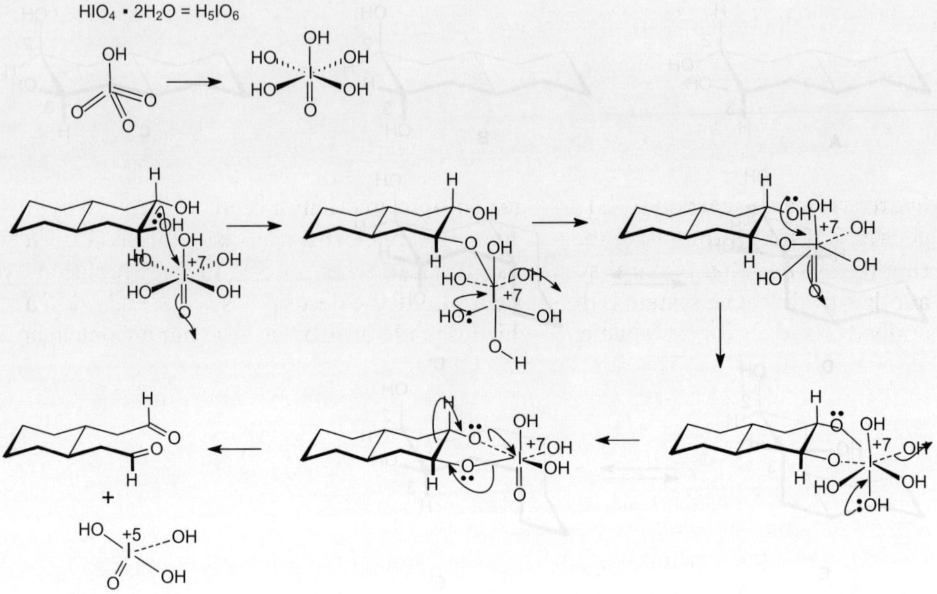

$HIO_4 \cdot 2H_2O = H_5IO_6$

Scheme 8.10. A representation for a pathway showing the periodic acid (H_5IO_6) cleavage of the C2–C3 bond in the decalin diol, 2α,3β,4aα,8aβ-decahydro-2,3-naphthalenediol. The representation demonstrates the necessity of a small dihedral angle between the hydroxyl groups that coordinate to periodic acid (H_5IO_6). Note that the overall result of double-bond hydroxylation (e.g., with OsO_4) followed by periodic acid (H_5IO_6) oxidation, as shown, corresponds to the processes of ozonolysis in that the same dialdehyde is expected.

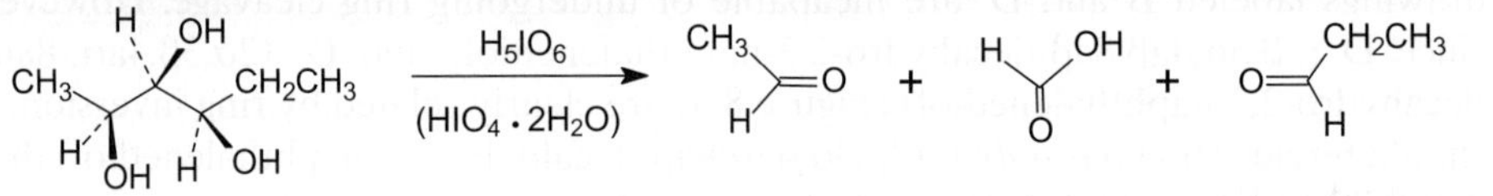

$$\text{(8.14)}$$

b. Biological Oxidation of Alcohols.

Although there are many examples of the oxidation of alcohols in living systems and more will be encountered later in this chapter and elsewhere (see Chapter 11), it should not be surprising that such systems have managed to avoid the extreme conditions and reagents (with the obvious exception of oxygen [O_2]) already discussed for laboratory transformations. Additionally, oxidations (and the reductions to which they are linked) in biological systems usually occur with stunning efficiency and exquisite stereospecificity (i.e., only one of a pair of enantiomers reacting) compared to those in the laboratory.

The example of the reversible oxidation of ethanol (CH_3CH_2OH) to ethanal (acetaldehyde, CH_3CHO) by *yeast alcohol dehydrogenase* (**alcohol: NAD$^+$ oxidoreductase** [EC 1.1.1.1])* is one such (now classical) process and is discussed below. The oxidoreductase uses nicotinamide adenine dinucleotide (NAD$^+$) (oxidized form) as its coenzyme, and it is in the pyridine (azabenzene, C_5H_5N) ring that the reduction is clearly seen.

*EC stands for Enzyme Commission numbers and their meaning are described in detail in Chapter 11. See also, the website of the International Union of Pure and Applied Chemistry (www.iupac.org).

However, in order to demonstrate the specificity of the oxidation, the material on chirality (handedness), discussed earlier (Chapter 4), requires extension. This is because it has been demonstrated that *the (chiral) enzyme can distinguish between chemically identical nuclei.*

Consider the case of the enantiomers of 1-deuterioethanol shown in Figure 8.6.

The two hydrogen nuclei (one protium [^{1}H] and one deuterium [^{2}H]) on the hydroxyl-bearing carbon (C-1) are **enantiotopic**. In ethanol itself, where both hydrogens at C-1 are protium [^{1}H], the two hydrogen nuclei may be distinguished by referring to the *idea* that a chiral system is produced when one of them is replaced by another nuclide. The system is thus **prochiral**, and the descriptors *pro-R* and *pro-S* are applied. The decision as to which is which is made by arbitrarily assigning one ligand at the prochiral site a higher priority than the other. If the resulting system is *R*, then the selected ligand is *pro-R*. Subscripts are used at the appropriate nuclei, that is,

In some systems, when the reversible enzymatic oxidation occurs with the **yeast alcohol dehydrogenase**, *only the pro-R hydrogen is lost* (Equation 8.15).

$$(8.15)$$

Additionally, the *two faces of the ethanal (acetaldehyde [CH$_3$CHO]) produced are also distinguishable*; that is, because they bear a mirror image relationship to each other, the faces are **enantiotopic**. Thus, as shown in Figure 8.7, the plane defined by the carbonyl oxygen, the carbon of the methyl group, and the hydrogen nucleus attached to the carbon of the carbonyl is examined. Then, the priority is assigned based on the Cahn–Prelog–Ingold sequence rules (Chapter 4), and the two faces are determined to be *re* (from *rectus*) or *si* (from *sinister*) if the substituents *on that face* appear, respectively, in a clockwise or counterclockwise order of decreasing priority.

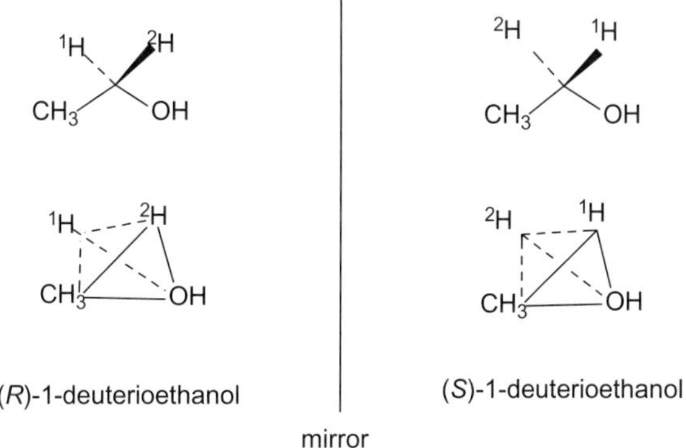

Figure 8.6. The enantiomers of 1-deuterioethanol.

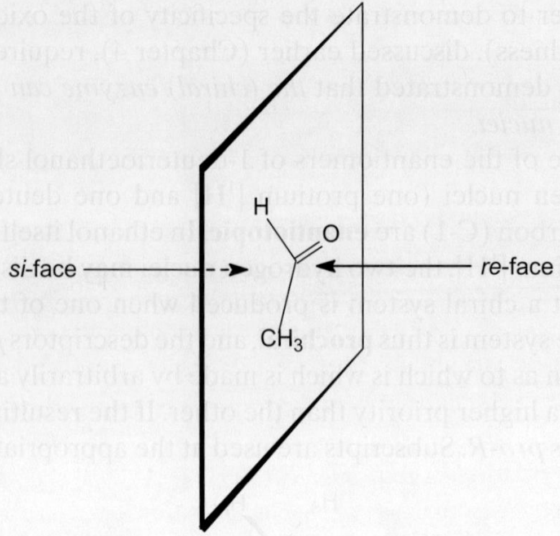

Figure 8.7. The *re*- and *si*-faces of ethanal (acetaldehyde, CH₃CHO).

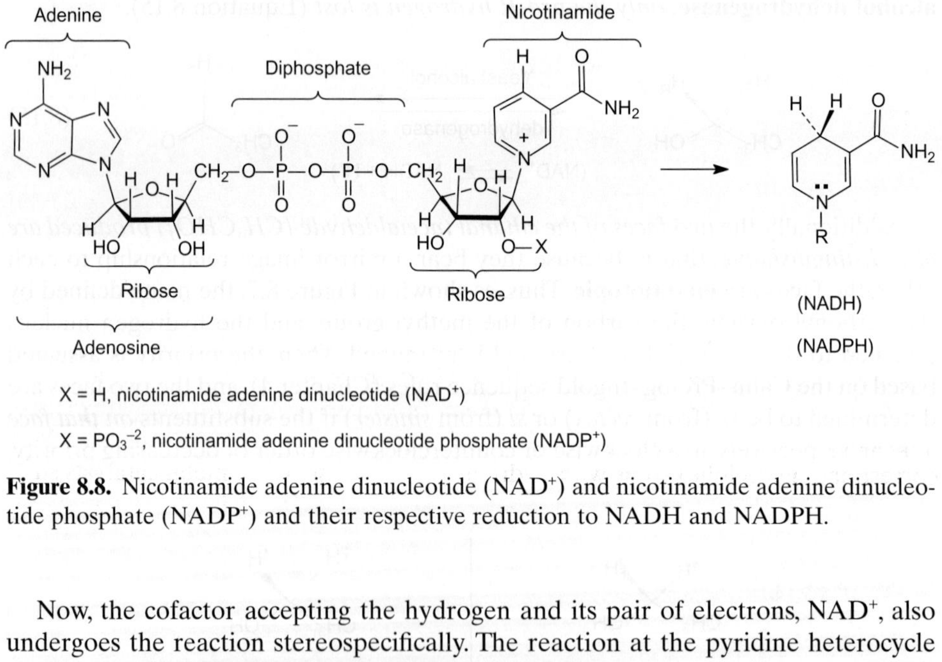

Figure 8.8. Nicotinamide adenine dinucleotide (NAD⁺) and nicotinamide adenine dinucleotide phosphate (NADP⁺) and their respective reduction to NADH and NADPH.

Now, the cofactor accepting the hydrogen and its pair of electrons, NAD⁺, also undergoes the reaction stereospecifically. The reaction at the pyridine heterocycle occurs with the hydrogen being transferred to the *re*-face, becoming *pro-R* in the process (Figure 8.8 and Scheme 8.11).

III. Oxidation at Sites That Do Not Bear Hydroxyl

a. Oxidation of Enols. As noted earlier, in the absence of special circumstances, enols are in equilibrium with their proton tautomer carbonyl isomers (aldehydes and ketones) and the equilibrium lies far on the side of the latter (e.g., see item 1 in Table 8.2). Thus, oxidation of enols generally results in a change in oxidation state

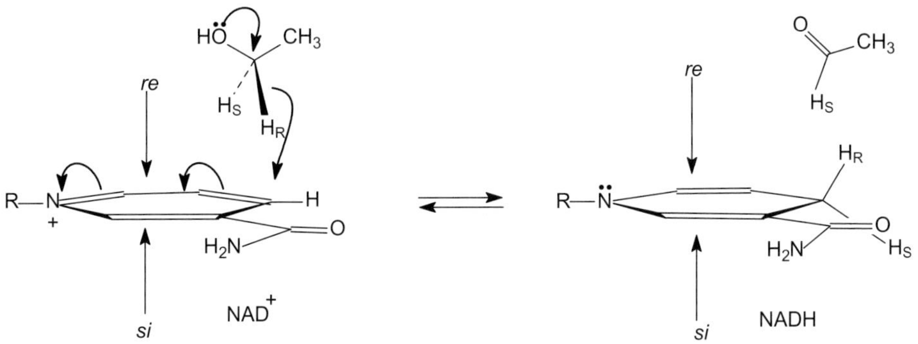

Scheme 8.11. A cartoon representation of the oxidation of ethanol (ethyl alcohol, CH_3CH_2OH) to ethanal (acetaldehyde, CH_3CHO) with the cofactor nicotinamide adenine dinucleotide (NAD^+). The curved arrows depict the loss of the *pro-R* C-1 hydrogen from the former as it is being added to the *re*-face of the latter. The reverse reaction, loss of the C-4 *pro-R* hydrogen from NADH to ethanal (acetaldehyde, CH_3CHO), where it becomes the *pro-R* hydrogen of ethanol (ethyl alcohol, CH_3CH_2OH), is indicated by the arrows connecting the structures on the right and left. It is important to note that *it is the same hydrogen removed in the oxidation process and which is returned in the reduction process.*

at a carbon (the α-carbon) immediately attached to the hydroxyl-bearing carbon (Equation 8.16)

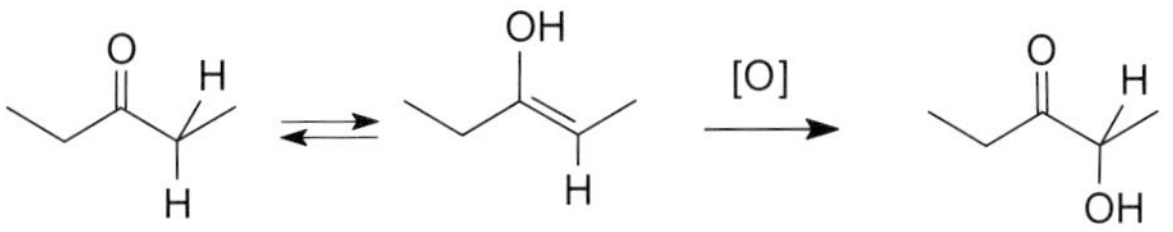

$$(8.16)$$

Thus, as shown in Equation 8.17, the oxidation of the diarylenol, 2-(2′,4′,6′-trimethyl-phenyl)-2-phenyl-1-ethenol with lead tetracetate in warm ethanoic acid (acetic acid, CH_3CO_2H) yields the corresponding acetoxyaldehyde.

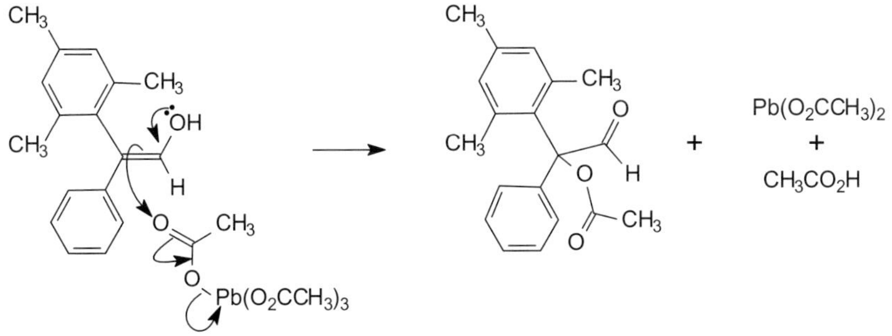

$$(8.17)$$

More recent work has shown that direct hydroxylation of enolate anions can be achieved with a molybdenum peroxide reagent, which consists of MoO_5 as a complex with pyridine (C_5H_5N) and hexamethylphosphoric acid triamide (hexameth-ylphosphoramidate [HMPA], $O=P[N(CH_3)_2]_3$). As shown in Scheme 8.12, phenylcy-clohexanone is first converted to the corresponding anion (e.g., with *n*-butyllithium, BuLi, $CH_3CH_2CH_2CH_2Li$) and the latter is treated with the molybdenum complex

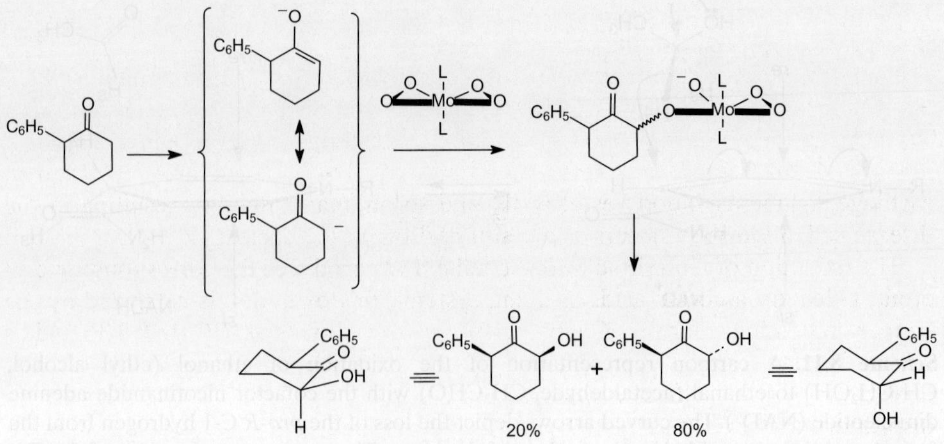

20%   80%

Scheme 8.12. A partial representation of a pathway by which the enolate anion of 2-phenylcyclohexanone is converted to a mixture of (*E*)- and (*Z*)-2-hydroxy-6-phenylcyclohexanones by oxidation with MoOPH. "L" is a suitable ligand, for example, pyridine (e.g., see Vedejs, E.; Engler, D. A.; Telschow, J. E. *J. Org. Chem.*, **1978**, *43*, 188).

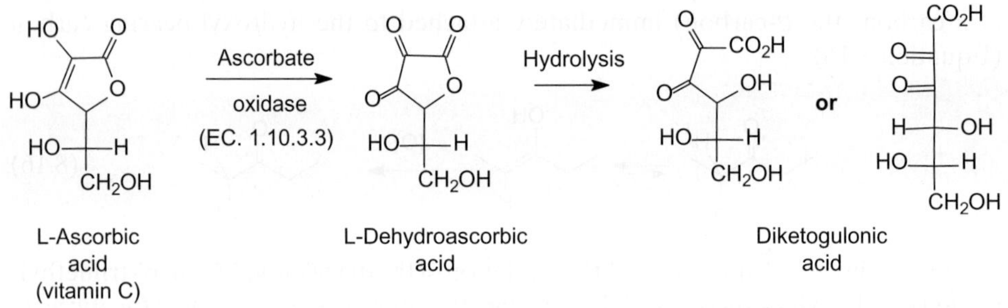

Ascorbate
oxidase
(EC. 1.10.3.3)

Hydrolysis

or

L-Ascorbic
acid
(vitamin C)

L-Dehydroascorbic
acid

Diketogulonic
acid

Scheme 8.13. A representation of the oxidation of L-ascorbic acid (vitamin C) to L-dehydroascorbic acid and then to diketogulonic acid.

(called **MoOPH**) at low temperature. After quenching the reaction, a mixture of stereoisomeric α-hydroxyketones ((*E*)- and (*Z*)-2-hydroxy-6-phenylcyclohexanones) results.

Problem 8.2. The enolate anion of Scheme 8.12 is only one of two possible enolate anions that might have been formed. Although the details of which anion forms under what conditions will be discussed in Chapter 9, you should be able to guess what the "other" enolate anion might look like. Draw at least one resonance form for an enolate anion resulting from loss of a proton other than the one shown. Presuming that the enolate anions might be isolated, which do you suppose might be more stable?

The bis-enol, L-ascorbic acid (vitamin C) (Scheme 8.13), is easily oxidized. Therefore, in addition to its function as a vitamin (*vide infra*), L-ascorbic acid and related materials (derivatives of L-ascorbic acid) are often used as "antioxidants," the L-ascorbic acid being oxidized *instead* of the material with which it is mixed, thus "protecting" it from oxidation.

L-ascorbic acid is produced by normal carbohydrate metabolism (Chapter 11) in plants and in most animals (except for primates and guinea pigs). When it is missing from the diet of humans for a prolonged period, **collagen** formation is impared. Since collagen is the major stress-bearing protein of connective tissue found in cartilage, ligaments, blood vessel walls, and so on, major debility, resulting in the disease called **scurvy**,*† occurs as a result of vitamin C deficiency.

The oxidation of L-ascorbic acid (Scheme 8.13) produces the corresponding diketone L-dehydroascorbic acid. In living systems, this oxidation is catalyzed by the copper-containing enzyme (Chapter 12) *ascorbate oxidase* (EC 1.10.3.3) and is linked to the reduction of oxygen (O_2) to water (H_2O). Interestingly, the enzyme produces a radical (called *semidehydroascorbate*) first, and this species apparently disproportionates back to ascorbic acid and on to the dione (dehydroascorbic acid). Under physiological conditions, dehydroascorbic acid is *irreversibly* hydrolyzed to diketogulonic acid (Scheme 8.13), thus providing the rationale for dietary L-ascorbic acid (vitamin C).

Problem 8.3. Assume that any ketone can be simply represented by the carbonyl fragment shown in the figure and any enol by the corresponding two carbon fragment with which the former is in equilibrium. Using the bond energies of Chapter 1 (Table 1.1), estimate the relative stabilities (expressed as the difference in energy [ΔE]) of the two proton tautomers.

Finally, acid-catalyzed enolization of aldehydes and ketones in the presence of selenium dioxide (SeO_2) also results in oxidation α- to those carbonyl functionalities via the corresponding enol. As shown in Scheme 8.14, the reaction of the enol with SeO_2 followed by the loss of selenium(II) oxide (which subsequently

---

*Before the need for vitamin C was recognized, scurvy was widespread among seafaring nations. Sailors on long voyages, deprived of fresh fruits and vegetables, would suffer bleeding gums, weakness, anemia, and so on. The introduction of citrus (e.g., lemons and limes), rich in vitamin C, into the regular diet of British seamen led to them being referred to as "limeys."

†Collagen polypeptides (Chapter 12) are composed of a large amount of the amino acid proline, which is hydroxylated (to 4-hydroxyproline) after the peptide has been assembled. It is the latter that imparts structural rigidity to collagen. The enzyme (Chapter 12) *prolyl hydroxylase* (EC 1.14.11.2), which affects the conversion, requires L-ascorbic acid to function.

L-proline   4-Hydroxyproline

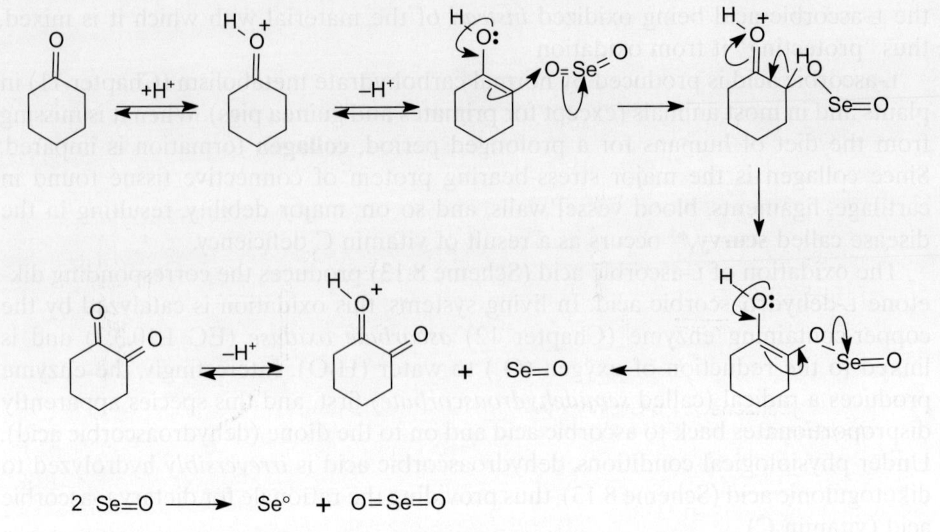

$$2\ Se=O \longrightarrow Se° + O=Se=O$$

Scheme 8.14. A representation of the acid-catalyzed oxidation of a ketone (cyclohexanone) with selenium dioxide to the corresponding α-diketone (1,2-cyclohexanedione) (e.g., see Singh, K. J.; Anand, S. N. *J. Indian Chem. Soc.*, **1979**, *56*, 363).

disproportionates to metallic selenium [Se°] and selenium dioxide [SeO$_2$]) produces the corresponding diketone.

b. Oxidation of Phenols. When double bonds are placed in the molecular framework so that the enolization of a carbonyl group to form the corresponding enol results in systems with extended conjugation, the enol can be significantly stabilized. Hydroxybenzenes (phenols), compounds in which the hydroxyl is attached directly to the aromatic ring, can be thought of as examples of such compounds (although there is really no "double bond" attached to the carbon bearing the hydroxyl).

Interestingly, the results of oxidation (and other reactions) provide some evidence that carbonyl compounds are in equilibrium with the hydroxybenzenes (phenols). Indeed, in extreme cases such as 1,3,5-trihydroxybenzene (phloroglucinol) (Figure 8.9), some reactions occur as if the tricarbonyl compound were present as the major tautomer. However, it is generally the case that for phenols, as with enols, oxidation occurs either at the adjacent α (or *ortho*) position or at the more remotely conjugated γ (or *para*) position.

The oxidation of phenols in the laboratory as well as in living systems is generally accomplished as a series of one-electron processes through phenoxy radicals that are subsequently converted to *ortho*- or *para*-quinones. A wide variety of oxidizing agents can be used (Scheme 8.15 and Figure 8.10).

An adequate dietary intake of the particular quinones known as K vitamins ("K" from the Danish word *koagulation*) is required for the synthesis (in the liver) of prothrombin (a propeptide that leads to the peptide [Chapter 12] thrombin) and other "factors" needed for blood clotting. The metabolic cycle of the K vitamins includes the corresponding hydroquinones. The oxidation of vitamin K$_3$ hydroquinone to vitamin K$_3$ (menadione) and its reduction is shown in Figure 8.11. (The

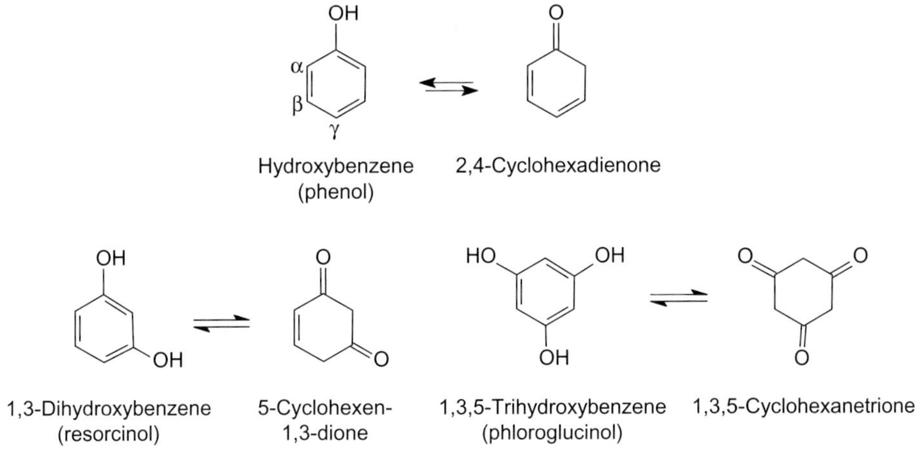

Figure 8.9. Proton tautomers of hydroxybenzene (phenol), 1,3-dihydroxybenzene (resorcinol), and 1,3,5-trihydroxybenzene (phloroglucinol).

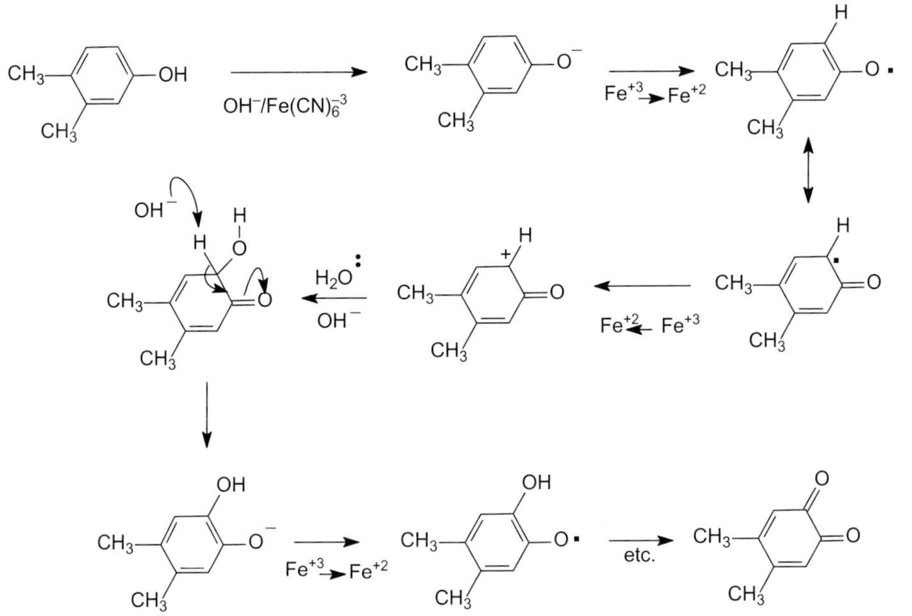

Scheme 8.15. A representation of one-electron oxidation of a hydroxybenzene (3,4-dimethylhydroxybenzene [3,4-dimethylphenol]) to an *ortho*-quinone.

other K vitamins are substituted with long alkenyl chains α or *ortho* to the methyl group.)

It is widely held that enzyme-catalyzed oxidative coupling of variously substituted phenols is a common occurrence in nature.* The process is thought to involve

*A very interesting early exposition of this point of view is found in Barton, D.H.R.; Cohen, T. *Festschrift Arthur Stoll*, Birkhäusesr Verlag, Basel, Switzerland, **1957**, p. 117ff. Although apparently complicated, the formation of naturally occurring **usnic acid** is among the simpler examples available. Usnic acid is an antibacterial agent isolated from varieties of *Usnea barbata* (lichens). Although many possible oxidation

(Continued on next page)

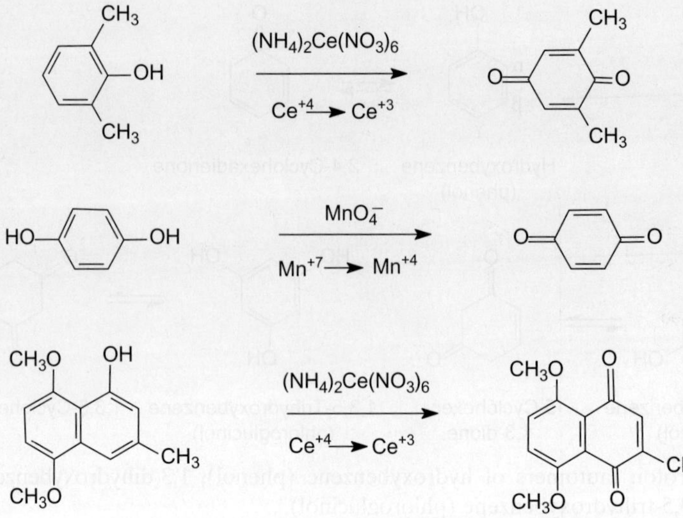

Figure 8.10. Representative oxidations of phenols to the corresponding quinones.

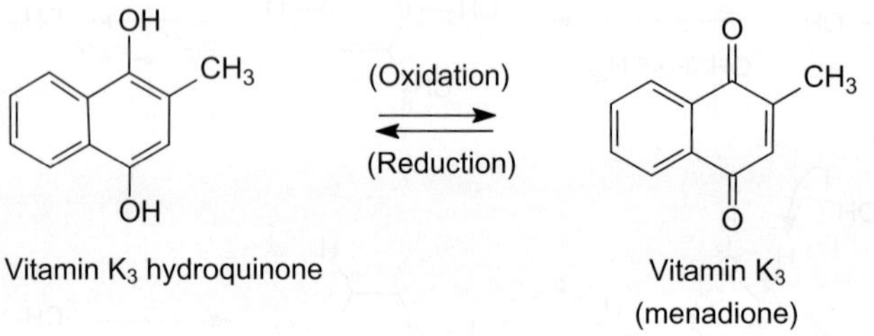

Vitamin K_3 hydroquinone

Vitamin K_3
(menadione)

Figure 8.11. A representation of part of the cycle involving the reduction of vitamin K_3 (menadione) to the corresponding hydroquinone.

pathways and products can be written, *usnic acid* may be thought of as being derived from two molecules of 2,4,6-trihydroxy-3-methylacetophenone as shown in the figure:

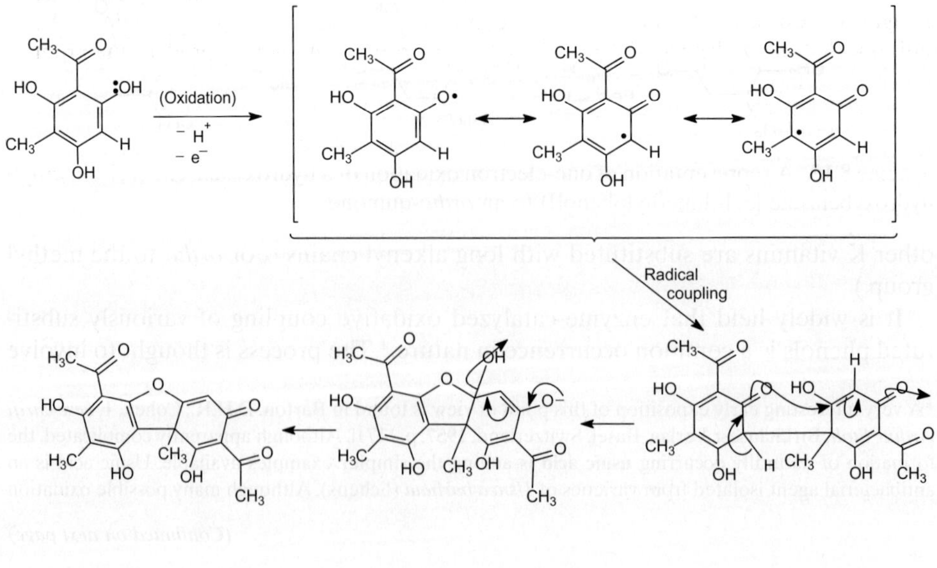

Radical
coupling

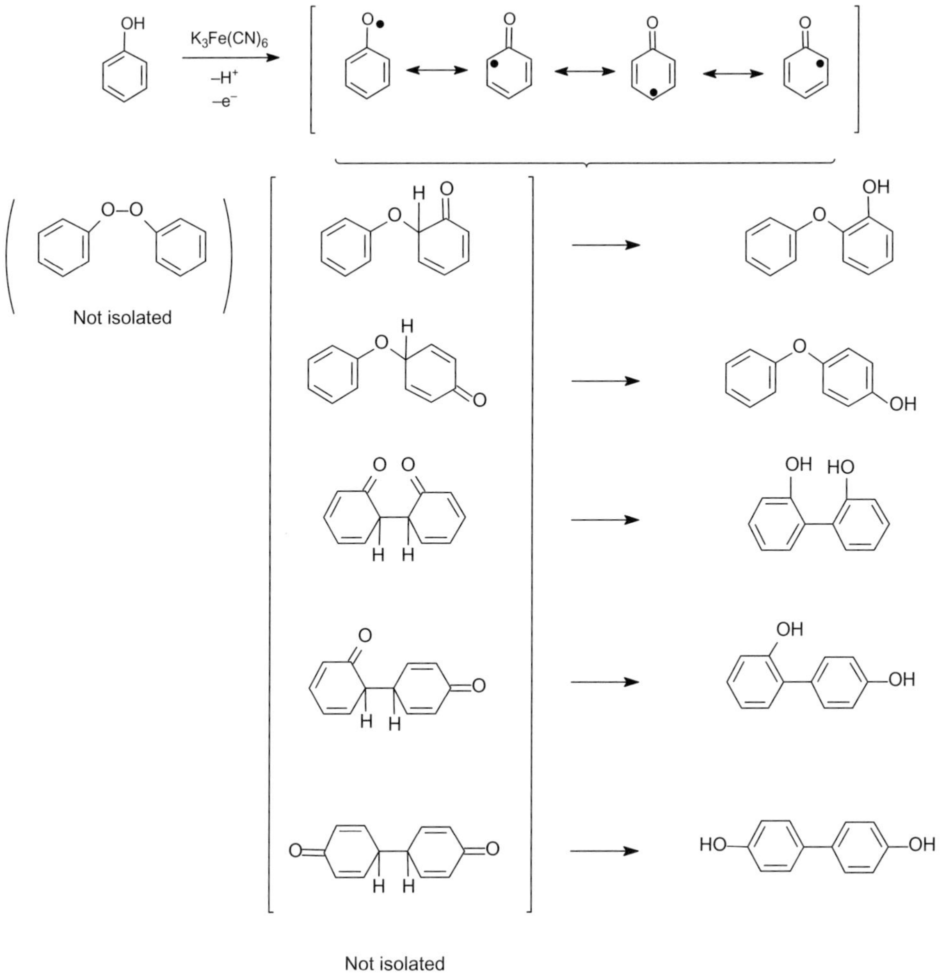

Not isolated

Not isolated

Scheme 8.16. The results of oxidation of phenol by basic iron(III) cyanide. Ferricyanide is thought to oxidize the phenol to the corresponding radical, and coupling reactions of the radical with itself at electron-rich sites followed by proton tautomerism produce the product mixture shown. The peroxide, resulting from "oxygen–oxygen" coupling, is not obtained nor is it expected to be stable under the reaction conditions.

one-electron oxidation to produce "phenoxy" radicals, which then proceed to add to each other. Subsequent rearomatization, through proton tautomerism, then accounts for the products.

In the laboratory, treatment of phenol with potassium ferricyanide (potassium iron(III) cyanide, $K_3Fe(CN)_6$) results in a product mixture that can be viewed as if the radical generated by proton and electron loss from phenol was allowed to react and couple with another like itself *at all those sites predicted by resonance considerations to be electron rich.* The product mixture obtained and the suggested intermediates to that mixture are shown in Scheme 8.16.

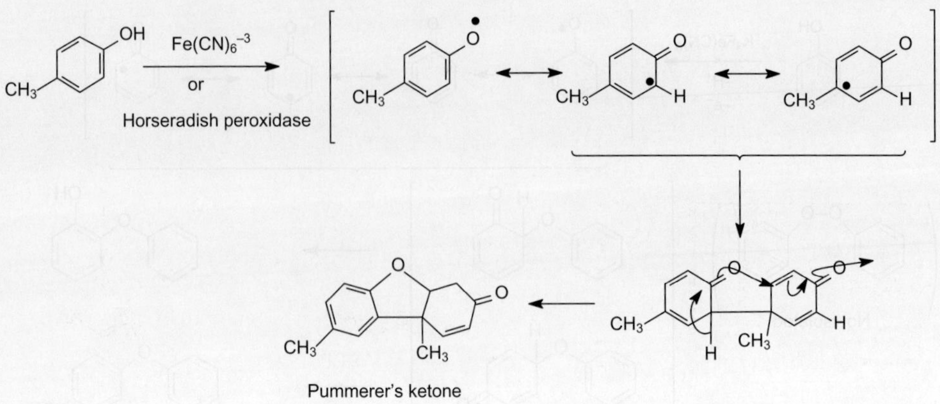

Pummerer's ketone

Scheme 8.17. A pictorial representation of a pathway to Pummerer's ketone by the oxidation of 4-hydroxytoluene with basic iron(III) cyanide (ferricyanide, $Fe(CN)_6^{-3}$). The material prepared in the laboratory is racemic.

Similarly, as shown in Scheme 8.17, **Pummerer's ketone*** can be isolated from *p*-cresol (4-hydroxytoluene) if either basic iron(III) cyanide or the enzyme *horseradish peroxidase* is used.

Interestingly, despite the expenditure of a large amount of work, the pathway shown for this and other oxidative coupling reactions, common in nature, remains speculative.

c. Oxidation at the Double Bond of Allylic Alcohols.

Generally, hydroxyl groups (–OH) of primary (1°) and secondary (2°) alcohols are more easily oxidized than alkenes. Despite this, many of the oxidation reactions at the carbon–carbon double bond (Chapter 6, Section C) can be effected whether or not there is a hydroxyl group on a flanking carbon (i.e., allylically disposed [–CH=CH–C(OH)]). For example, epoxidation of most allylic alcohols with 3-chloroperoxybenzene carboxylic acid (*m*-chloroperbenzoic acid) yields the corresponding oxirane at what was the carbon–carbon double bond and leaves the alcohol intact (producing an "epoxy–alcohol"). Additionally, osmium(VIII) tetroxide (OsO₄) oxidation succeeds at the carbon–carbon double bond of many allylic alcohols, introducing two hydroxyl (–OH) groups in the process and produces a triol (–C(OH)–C(OH)–C(OH)–).

Despite its apparent lack of involvement, it is clear from stereochemical studies that an allylic hydroxyl group can have a major influence on the result of reactions at the carbon–carbon double bond. Indeed, the hydroxyl group of the allylic alcohol can be used to guide the outcome of the oxidation process.

Consider the peracid epoxidation of 3-hydroxycyclohexene as shown in Scheme 8.18. Since the oxirane (epoxide) product is mostly (97%) the result of oxidation on the *same* side as the hydroxyl group [i.e., *cis*-, *syn*-, (*Z*)- or suprafacial] it is

*Endo, Y.; Shudo, K.; Okamoto, T. *Chem. Pharm. Bull.* **1983**, 31, 3769.

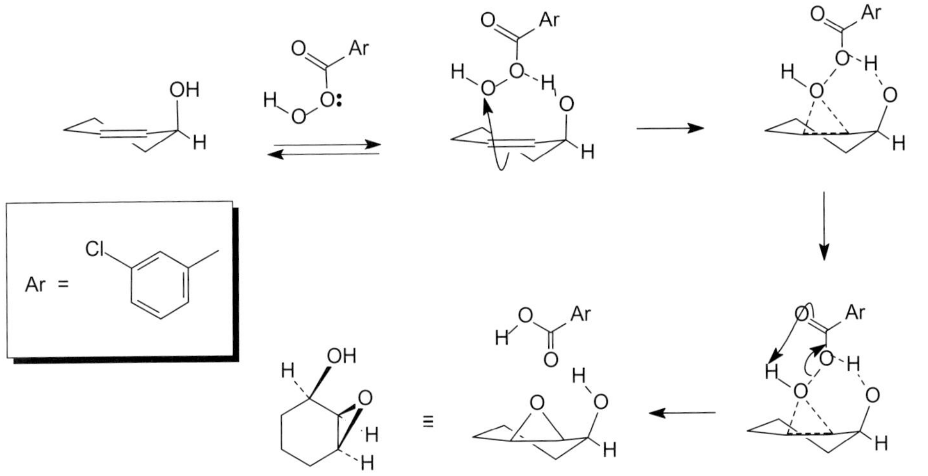

Scheme 8.18. A representation of a pathway by which an allylic alcohol (3-hydroxycyclohexene) produces an oxirane on epoxidation with a peracid (3-chloroperoxybenzene carboxylic acid [*m*-chloroperbenzoic acid]). It is suggested that the hydroxyl group coordinates to the peracid.

argued that complexation of the peracid with the hydroxyl group directs the incoming oxygen.

However, the stereochemistry of the carbon–carbon double bond may intrude upon the expected complexation. Thus, for (*Z*)- (or *cis-*) 3-hydroxycyclooctene, epoxidation with 3-chloroperoxybenzenecarboxylic acid (*m*-chloroperbenzoic acid) produces the *trans*-epoxide (99.8%), while the (*E*)- (or *trans-*) 3-hydroxycyclooctene yields (84%) the *cis*-epoxide (Figure 8.12).

The recognition that the hydroxyl group might be used to direct the incoming oxidizing agent has recently led to the exciting technique of **stereospecific** oxidation of allylic alcohols where the **chirality developed in the product is dependent on the chirality present in the oxidizing reagent.*** Many naturally occurring materials, including (but not limited to) carbohydrates (Chapter 11), have a series of hydroxyl-bearing carbons embedded in their structures. Almost all of these natural products are chiral, and thus a synthesis, which produces an excess of one enantiomer (*enantiomeric excess* [ee]) or diastereomer (*diastereomeric excess* [de]),[†] is of particular value.

---

*The process is called the **Sharpless oxidation**, after Sharpless, K. B., Professor, Massachusetts Institute of Technology (MIT). The first descriptons of the process appeared in 1980 (Katsuki, T.; Sharpless, K. B. *J. Am. Chem. Soc.*, **1980**, *102*, 5974).

[†]The following definitions are taken from Eliel, E. L.; Wilen, S.H. *Stereochemistry of Organic Compounds*, Wiley, **1994**, pp. 1996–1997: "**Enantiomer excess (ee)**. In a mixture of a pure enantiomer (*R* or *S*) and a racemate (*RS*), ee is the percent excess of the enantiomer over the racemate. **Diastereomer excess (de)**. In a diastereoselective reaction producing two (and only two) diastereomers in amounts A and B, de = 100 (|A − B|)/(A + B)."

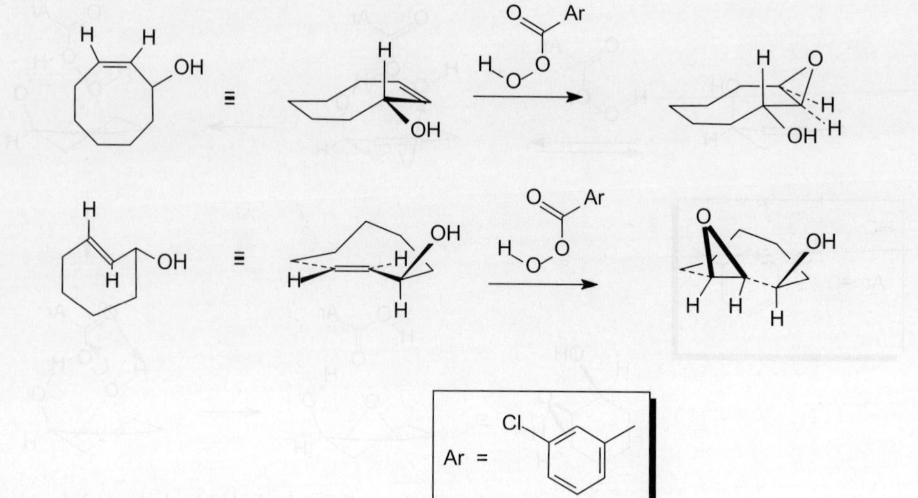

Ar =

Figure 8.12. The major products resulting from the epoxidation of (*Z*)- and (*E*)-3-hydroxycyclooctene with 3-chloroperoxybenzene carboxylic acid (*m*-chloroperbenzoic acid), showing that the interplay between the geometry of the double bond and the hydroxyl group often determines the product.

Indeed, if the *absolute* sense of the oxidation can be predicted beforehand, structural elucidation as well as synthesis will profit. The Sharpless oxidation (*vide supra*) accomplishes this goal by (a) making use of the *chirality present* in the readily available esters (Chapter 9) of (2*R*,3*R*)-2,3-dihydroxybutane-1,4-dicarboxylic acid ([L-(+)-tartaric acid] $HO_2CCH(OH)CH(OH)CO_2H$) and its D-(-)- mirror image; (b) involving titanium(IV) tetra*iso*propoxide $\{Ti[OCH(CH_3)_2]_4\}$ as a *complexing reagent* on which the oxidation will occur to hold the reacting species appropriately; and (c) invoking the gentle *oxidizing agent*, 2-methyl-2-hydroperoxypropane (*t*-butyl hydroperoxide [$(CH_3)_3COOH$]) as the source of the oxygen destined to form the ring.

A cartoon representation of the process is provided in Scheme 8.19 and an overall result in Scheme 8.20.

As shown in Scheme 8.19 for diethyl (2*R*,3*R*)-2,3-dihydroxybutane-1,4-dicarboxylate ([L-(+)-diethyl tartrate] where Et = CH_2CH_3), two 2-hydroxypropyl (*iso*propoxide, [$-OCH(CH_3)_2$]) ligands, L, attached to the titanium(IV) reagent, titanium(IV) *iso*propoxide [$Ti(-OCH(CH_3)_2]$, are expected to be displaced by two hydroxyl groups of two different dihydroxydicarboxylic acid esters. The resulting **dimer** (a 10-membered ring) has *one face (shown in Scheme 8.19 as the bottom face) encumbered by the carboxylate groups of the ethyl esters.*

Then, the replacement of a third *iso*propoxide [$-OCH(CH_3)_2$] by the *t*-butylhydroperoxide [$H-O-O-C(CH_3)_3$] oxidizing agent and the introduction of the allylic alcohol [$C=C-C-OH$] with the loss of the fourth *iso*propoxide [$-OCH(CH_3)_2$] produces an intermediate in which a "simple" transfer of oxygen from the *t*-butyl-

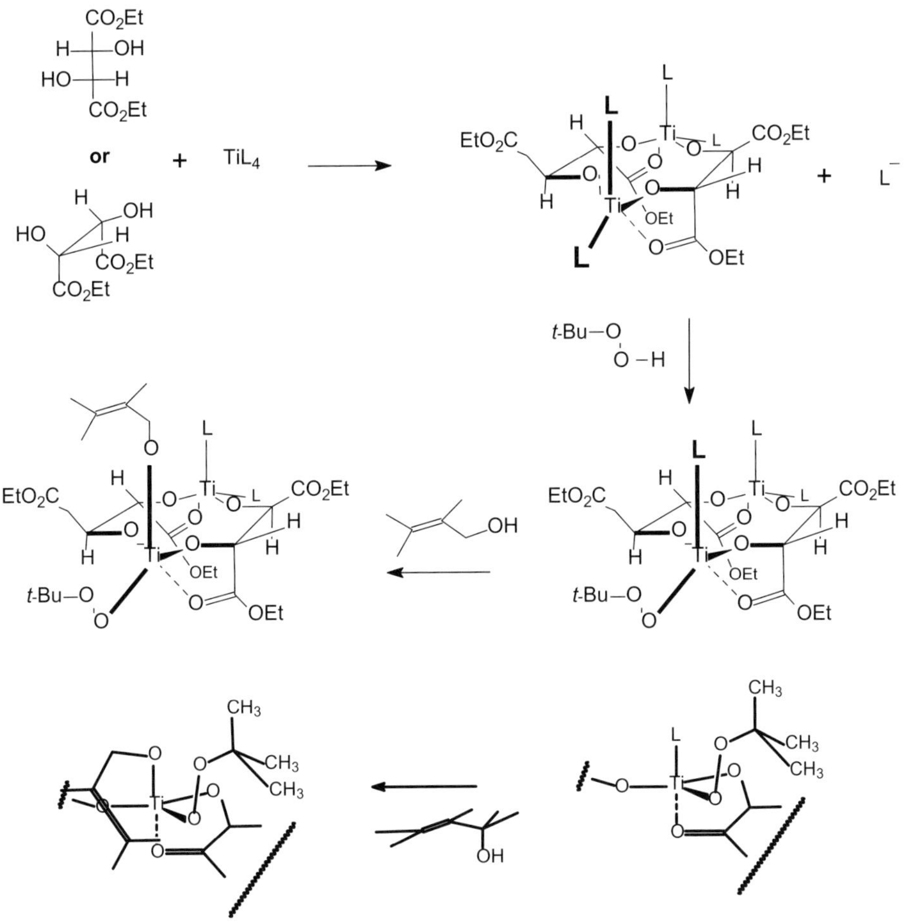

Scheme 8.19. A cartoon representation of the process by which stereospecific introduction of an oxygen (as an oxirane) is accomplished in the Sharpless oxidation. Diethyl (2R,3R)-2,3-dihydroxybutane-1,4-dicarboxylate [L-(+)-diethyl tartrate], Et = CH_2CH_3 forms a complex with titanium(IV) *iso*propoxide, TiL_4, where L = $OCH(CH_3)_2$. In the process, the two hydroxyl groups at C2 and C3 of the tartrate displace (replace) two of the four *iso*propoxy groups originally attached to titanium (see Katsuki, T.; Sharpless, K. B. *J. Am. Chem. Soc.*, **1980**, *102*, 5974).

hydroperoxide to the carbon–carbon double bond can occur to produce the oxirane. Although not shown, the oxygen *remote from* the tertiary carbon of the *t*-butylhydroperoxide (the oxygen bonded to the titanium in the scheme) is apparently the oxygen that is in the three-membered (oxirane) ring at the end of the reaction. Finally, it is important to realize that the picture of the process shown in Scheme 8.19 is in concert with what evidence is available, but the overall process should not be taken as established in detail. Kinetic evidence, which would help limit the mechanistic possibilities, is (at this writing) unavailable. But, regardless of the details, the outcome is that the sense of chirality of the process is defined by **the face of the**

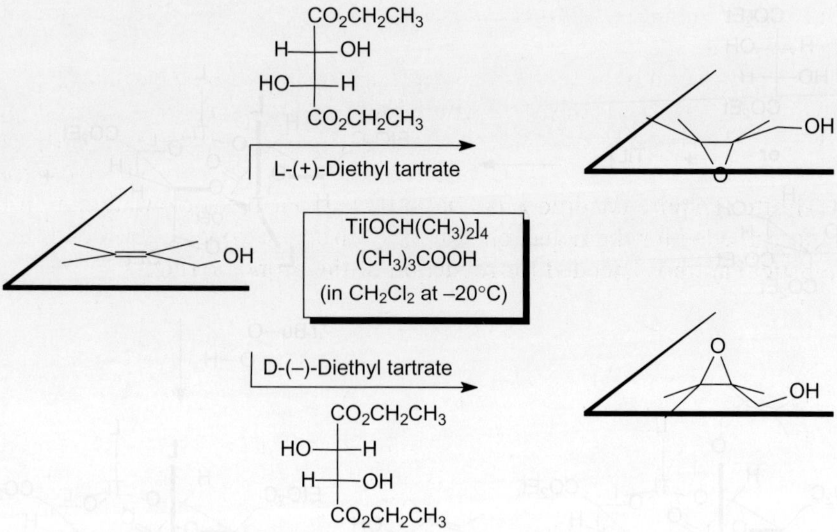

Scheme 8.20. A representation of the overall stereopecific conversion of allylic alcohol to oxirane as a function of the absolute stereochemistry of the diethyl tartrate used to form a complex with titanium(IV) isopropoxide.

intermediate that is blocked as a consequence of which isomer of diethyl tartrate is used to begin.*

C. REDUCTION OF ALCOHOLS, ENOLS, AND PHENOLS

I. Reduction of Alcohols

As a general rule, the reduction of alcohols (R–OH) to hydrocarbons (R–H) cannot be achieved directly. Indeed, the catalytic reduction of other functional groups with hydrogen and metals such as platinum (Pt) or palladium (Pd) at or near room tem-

---

*The idea of oxidizing an unsymmetrically substitued carbon–carbon double bond to a diol with control of stereochemistry *even in the absence* of the hydroxyl group of the allylic alcohol has provoked a significant amount of work. The principle should, of course, work. The dissymmetry (Chapter 4) present in the unsymmetrically substituted double bond coupled with a potentially dissymmetric (or even asymmetric) oxidizing agent could, as a key fits a lock, produce only a single enantiomeric (or diastereomeric) diol. The years of effort have currently reached a reasonable plateau (i.e., good-to-excellent ee's or de's **can confidently be predicted** based upon the **reagent** used). This note digresses into current practice.

Oxidation of alkenes with osmium(VIII) tetroxide (OsO_4) (Chapter 6, Scheme 6.10) involves a cyclic intermediate and results in the addition of two hydroxyl groups to the same face [*cis-, syn-, (Z)-* or suprafacial] of the reacting double bond. However, it is usually desirable to avoid stoichiometric concentrations of expensive, heavy metal poisons. Therefore, the observation that *catalytic* quantities of osmium(VIII) tetroxide (OsO_4) could effectively be used in the presence of a less noxious oxidizing agent (the N-oxide of N-methylmorpholine, NMO) was particularly important. The function of the latter then is to convert the osmium(VI) trioxide (OsO_3), produced when the osmium(VIII) tetroxide is reduced (as it performs the oxidation), **back to osmium(VIII) tetroxide** (OsO_4), that is,

(Continued on next page)

perature, 1 atm pressure, and pH values close to neutrality is often carried out in alcohol solvents.

However, the family of compounds related to phenylmethanol (benzyl alcohol, $C_6H_5CH_2OH$) and its variously substituted alkyl and aryl derivatives are a major exception to that general rule. These compounds can be reduced to the corresponding methylbenzene (toluene, $C_6H_5CH_3$) analogues by processes (called *hydrogenolysis*) that often require conditions (Equation 8.18) that are somewhat more forcing than those needed for the reduction of typical carbon–carbon double bonds but are less drastic than those needed for reduction of the aromatic ring.

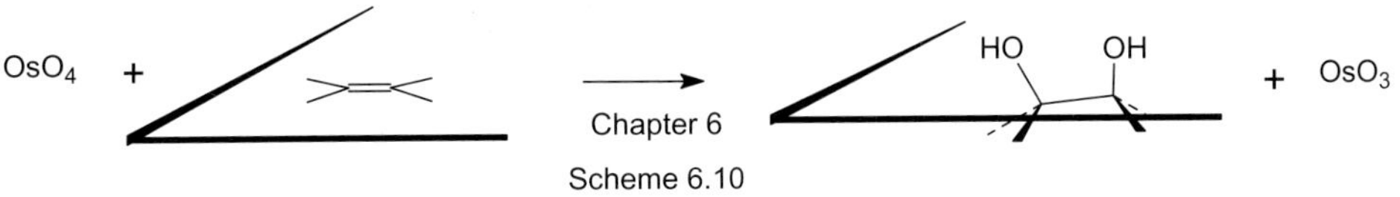

$$OsO_4 \; + \qquad \longrightarrow \qquad + \; OsO_3$$

Chapter 6

Scheme 6.10

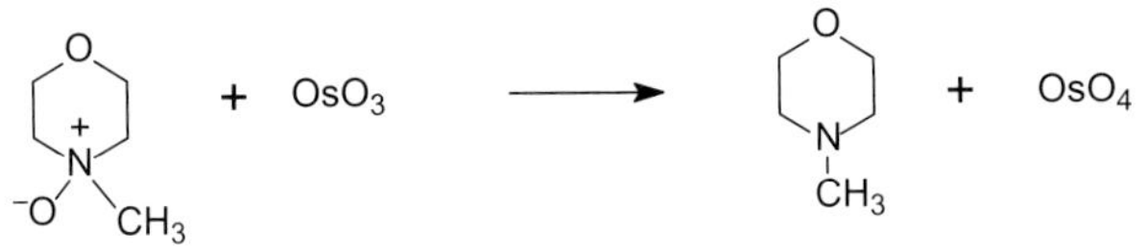

Now, the second feature of this process is to **hold the osmium in a chiral environment throughout its oxidation–reduction cycle.**

Because osmium(VIII) tetroxide (OsO_4) is a Lewis acid, a diastereomeric base pair was sought to insure that either member of an enantiomeric (or diastereomeric) pair of diols can be prepared, that is,

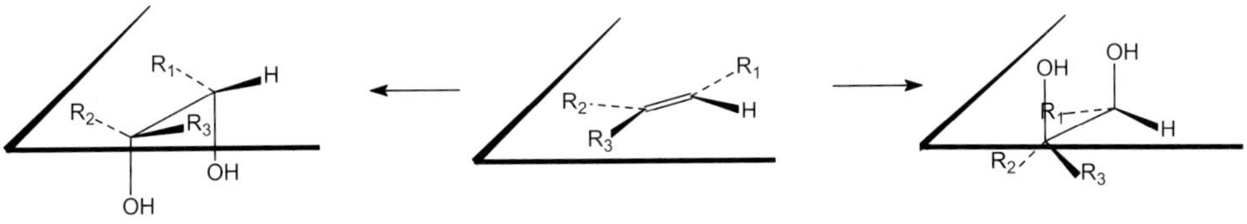

Although many possible base pairs can be imagined and a number of those have been implemented, the naturally co-occurring epimeric **cinchona** alkaloids (Chapter 13) quinine and quinidine is one diastereomeric pair that, when converted to their respective dihydrobases by the reduction of the double bond in the side chain, have proved to be a popular, useful, and commercially successful pair.

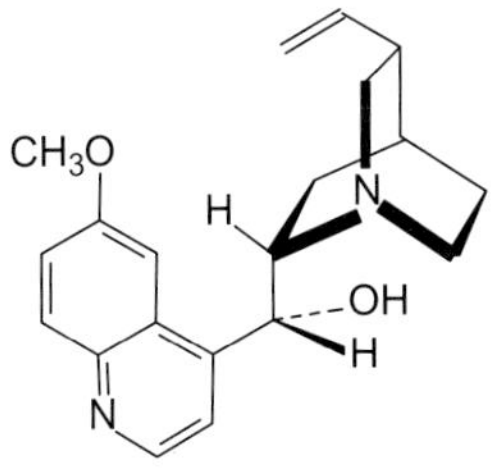

Quinine

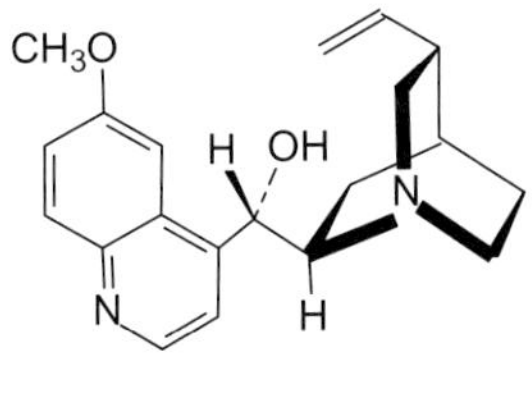

Quinidine

Then, two equivalents of dihydroquinine (**DHQ**) (on the one hand) and dihydroquinidine (**DHQD**) (on the other), respectively, are converted to the corresponding phthalizine derivatives (from phthalic anhydride and hydrazine [Chapter 10]) and the complexing agents (DHQ)$_2$-PHAL and (DHQD)$_2$-PHAL are produced.

(Continued on next page)

$$\text{Ph}-CH_2OH \xrightarrow[\substack{\text{1,2-Dimethybenzene}\\ (\textit{ortho}\text{-Xylene) solvent}\\ 100-125°C}]{H_2 \text{ (Ni catalyst)}} \text{Ph}-CH_3 \qquad (8.18)$$

The failure of direct catalytic reduction of alcohols has spawned a variety of more circuitous routes. Most of these other methods require the conversion of the

Bis(dihydroquinine)phthalizine [(DHQ)$_2$-PHAL]

Bis(dihydroquinidine)phthalizine [(DHQD)$_2$-PHAL]

Now, when the (Lewis acid) osmium(VIII) tetroxide (OsO$_4$) is added to a solution containing either (DHQ)$_2$-PHAL or (DHQD)$_2$-PHAL, the metal oxide bridges the two basic nitrogens in the azabicyclo[2.2.2]octyl systems so that only a narrow channel is available into which the unsymmetrically substituted alkene will fit. Thus, *one face of the alkene is more readily oxidized than the other*.

Finally, therefore, the determining feature of *which* face of the alkene (*re* or *si*) is oxidized is solely a function *only* of the hindered oxidizing agent, and the same alkene will produce either dihydroxy isomer simply by changing the alkaloid from quinine to quinidine.

(*Continued on next page*)

hydroxyl group into a more labile function and then a replacement of the latter by a hydrogen. For example, if a primary or secondary alcohol is allowed to react with the acid chloride of methanesulfonic acid (methanesulfonyl chloride, CH_3SO_2Cl) at low temperature in the presence of a base such as pyridine (azabenzene, C_5H_5N), the corresponding methanesulfonate ester (trivially called a "mesylate") can frequently be isolated. Then, the addition of a reducing agent such as lithium aluminum hydride ($LiAlH_4$) results in the reduction of the sulfonate to alkane (Scheme 8.21).

Tertiary alcohols can be reduced with mixtures of metals (such as zinc [Zn] or tin [Sn]) and aqueous hydrochloric acid (HCl) or red phosphorus (P) with hydrogen iodide (HI). The pathway by which such reductions occur is not clear.

Additionally, the reduction process of **Barton** and **McCombie** has been developed into a useful technology. The **Barton–McCombie*** procedure (Scheme 8.22) involves the conversion of an alcohol to a thiocarbonate (or dithiocarbonate) derivative (usually in two steps) and the reaction of the latter with tributyltin hydride and azobisisobutyronitrile (AIBN) (bis(1-cyano-1-methylethyl)diazine, $[(CH_3)_2C(CN)N=NC(CN)C(CH_3)_2]$). As shown in Scheme 8.22, decomposition of AIBN produces a radical, which, in turn, abstracts hydrogen from tributyltin hydride, generating the tributyltin radical. The reaction of the latter with sulfur and the subsequent decomposition of that intermediate produces a deoxygenated,

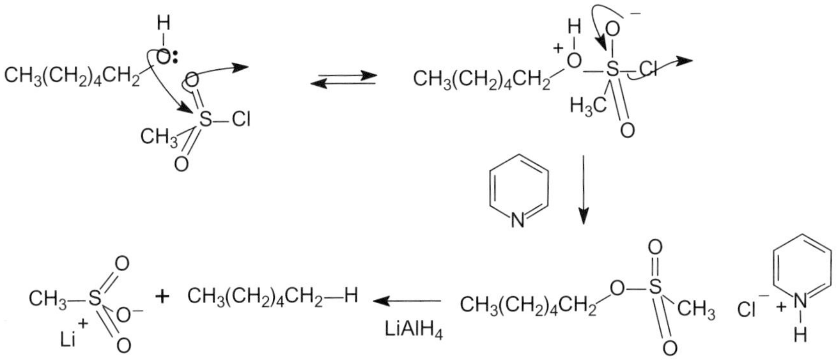

Scheme 8.21. A representation of a way to reduce a primary alcohol (1-hexanol) to the corresponding hydrocarbon (hexane) by the initial formation of the corresponding 1-hexyl methanesulfonate ester in the presence of azabenzene (pyridine) followed by reduction of the ester with lithium aluminum hydride ($LiAlH_4$).

Thus, for example, 2-methyl-2-heptene yields 3S-2,3-dihydroxy-2-methylheptane using $(DHQ)_2$-PHAL (95% ee) and the 3R enantiomer (98% ee) with $(DHQD)_2$-PHAL in 1:1 t-butanol : water at 0°C with a catalytic amount of OsO_4 and 1.1 eqivalents of NMO.

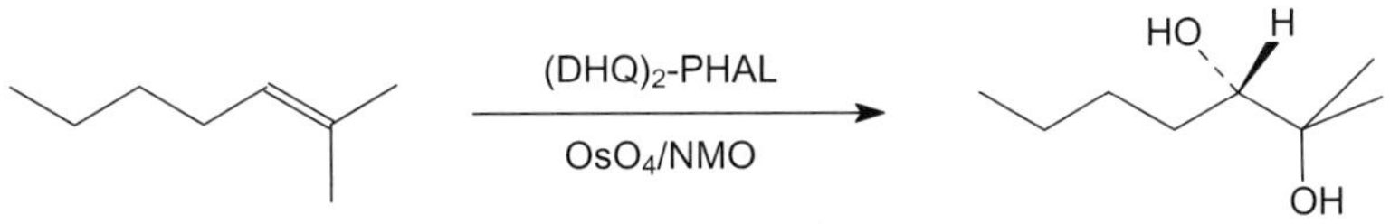

*Barton, D. H. R.; McCombie, S. W. *J. Chem. Soc., Perkin Trans. 1*, **1975**, *16*, 1574.

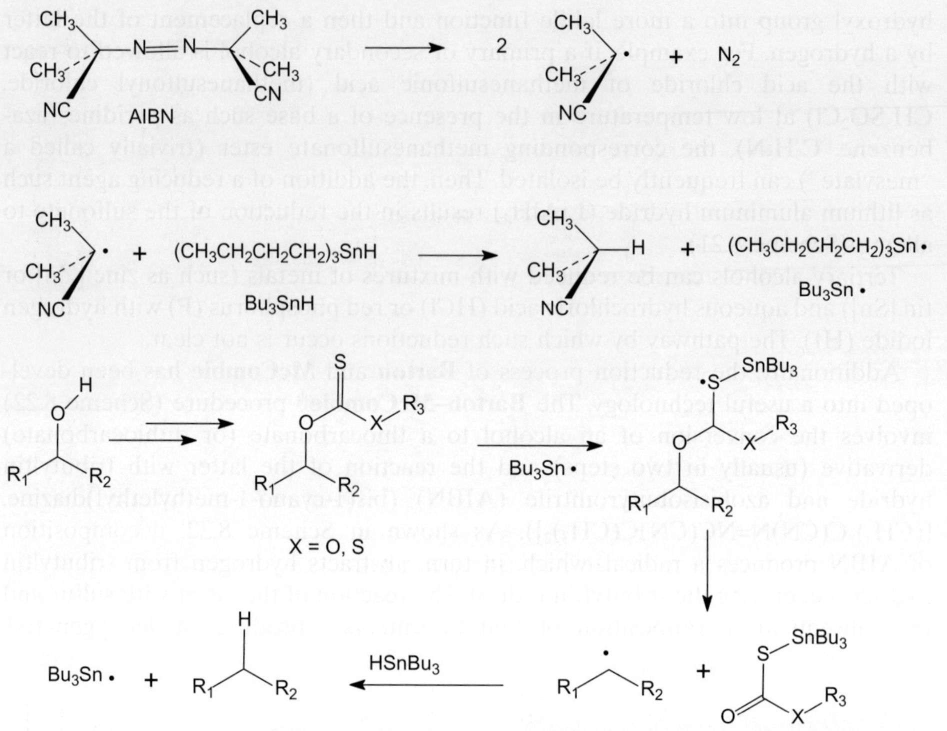

Scheme 8.22. A generic example of the Barton–McCombie procedure for the reduction of alcohols. The reaction is normally carried out in a solvent such as oxacyclopentane (THF) and is thought to involve, as a first step, the decomposition of a catalytic quantity of azobi-sisobutyronitrile (AIBN) (bis(1-cyano-1-methylethyl)diazine, $[(CH_3)_2C(CN)N=NC(CN)C(CH_3)_2]$) to generate the 2-cyanopropyl radical and nitrogen. The radical then abstracts hydrogen from tributyltin hydride, generating, in turn, the corresponding tributyltin radical, which reacts at the sulfur of the thiocarbonate. Decomposition of the latter intermediate produces the corresponding tin-substituted thiocarbonyl derivative and an alkyl radical. The reaction is then consummated by the alkyl radical abstracting a hydrogen atom from a second equivalent of tributyltin hydride, producing the corresponding alkane and regenerating the tributyltin radical.

carbon-centered radical, which, with additional tin hydride, yields hydrocarbon and another equivalent of trialkyltin radical. The process is clearly a **chain reaction**. An excess of hydride is employed.

Other free radical processes can be harnessed to the same end. For example, as shown in Scheme 8.23, when a primary alcohol is treated with tri-*n*-butyl-phosphine $[(CH_3CH_2CH_2CH_2)_3P, Bu_3P]$ and *ortho*-nitrophenyl selenocyanate in a suitable solvent (such as oxacyclopentane [THF]), the corresponding primary alkylselenide is formed. Cleavage of the carbon–selenium bond by tributyltin hydride (Bu_3SnH) in the presence of a radical initiator such as AIBN $((CH_3)_2C(CN)N=NC(CN)C(CH_3)_2$, bis(1-cyano-1-methylethyl)diazine) produces a radical, which, by hydrogen transfer and chain propagation as shown in Scheme 8.22, results in the replacement of the selenium by hydrogen and the generation of the hydrocarbon.

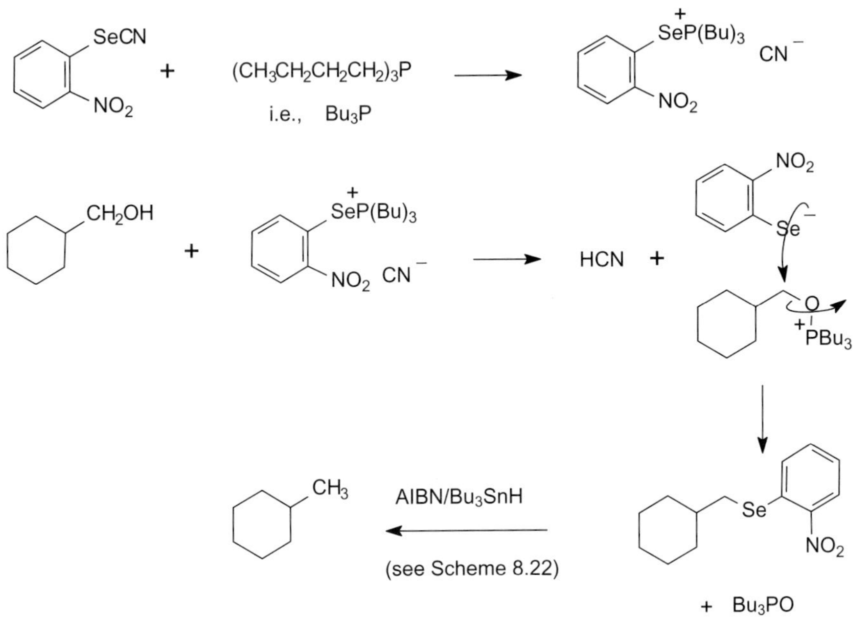

Scheme 8.23. Reduction of hydroxymethylcyclohexane, as an example of a primary alcohol, by the replacement of the hydroxyl with an arylselenium and the subsequent cleavage of the carbon–selenium bond with tri-n-butyltin hydride.

II. Reduction of Enols and Phenols

Since enols are generally in tautomeric equilibrium with the corresponding carbonyl compounds, the methods that apply to produce hydrocarbons are those used for the reduction of aldehydes and ketones (Chapter 9) and are discussed there. For a few stable enols (*vide supra*), catalytic methods, which correspond to a direct reduction of the carbon–carbon double bond as with alkenes, yield the corresponding alcohols.

Phenols (hydroxybenzenes) and alkyl-substituted phenols can be reduced to the corresponding cyclohexanols with hydrogen in the presence of metal catalysts such as platinum (Pt) or nickel (Ni), and it is often possible to isolate more than one isomer. Thus, as shown in Equation 8.19, the reduction of 2-methylbenzenol (2-methylphenol, *ortho*-cresol) with hydrogen (H_2) in the presence of platinum (Pt) under acidic conditions yields (Z)- (or *cis*-) 2-methylcyclohexanol, while under neutral conditions, the (E)- (or *trans*-) isomer predominates.

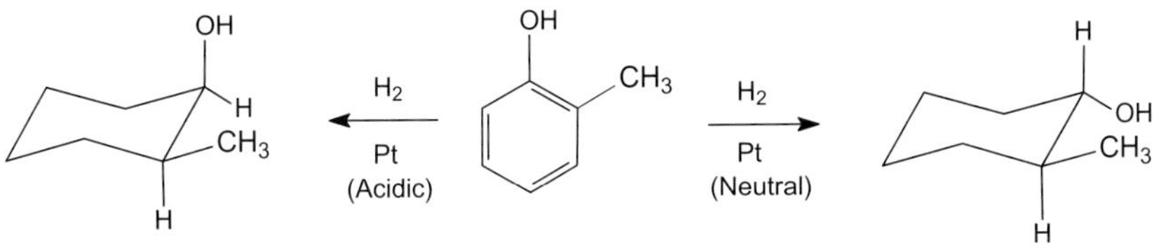

$$(8.19)$$

The reduction of the aromatic ring has proved valuable in the synthesis of some highly hindered cyclohexanols. For example (as will be discussed more fully

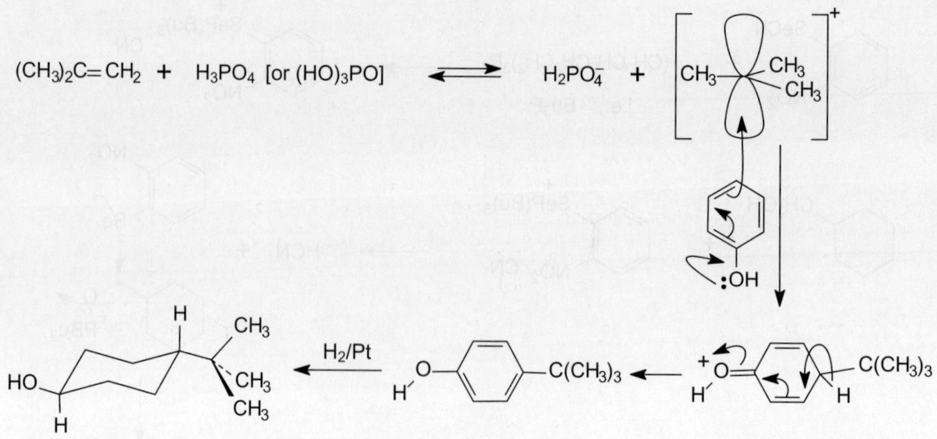

Scheme 8.24. The reaction of 2-methylpropene (*iso*butylene [$H_2C=C(CH_3)_2$]) with phosphoric acid to produce the *t*-butyl cation followed by the reaction of the latter with phenol (hydroxybenzene). Substitution occurs in the *para*-position. The resulting 4-(1-methylethyl) hydroxybenzene (4-*t*-butylphenol) is reduced with hydrogen (H_2) in the presence of a platinum catalyst to produce (*E*)-(or *trans*-) *t*-butylcyclohexanol [(*E*)- or (*trans*)-4-(1-methylethyl) cyclohexanol)].

later in this chapter), the aromatic electrophilic substitution (**Friedel–Crafts**) reaction (Chapter 6, Table 6.12) occurring between 2-methylpropene (*iso*butylene, $H_2C=C(CH_3)_2$) and hydroxybenzene (phenol) results (largely) in the substitution of a *t*-butyl group [2-methylethyl, $-C(CH_3)_3$] for the hydrogen *para* to the hydroxyl group. Then, catalytic reduction of the aromatic ring produces 4-(*t*-butyl)cyclohexanol [4-(1-methylethyl)cyclohexanol], a compound that has been widely used to study reactions of cyclohexanols (and compounds derived from them) where the ring is (essentially) incapable of chair–boat–chair interconversions (Chapter 4). Scheme 8.24 describes the processes discussed.

Reductive removal of the hydroxyl group, leaving an aromatic ring (i.e., the substitution of hydrogen [H] for hydroxyl [OH]), is not easily accomplished. Methods of direct reduction often produce significant amounts of cyclohexanols and indirect methods are generally poor. However, one of the more recently developed techniques, used in alkaloid synthesis (Chapter 13),* involves the initial conversion of the phenol to a tetrazole (Chapter 10) derivative and then the hydrogenolysis of the resulting product (Scheme 8.25).

---

*Historically, the proof of structure of a naturally occurring material was only considered complete when the natural product had been synthesized (or assembled by more or less well-understood processes from materials whose structures were certain). The methods of synthesis learned in these studies were later put to use when the interest in modifying the structures of naturally occurring materials to enhance their desirable properties or to remove or limit their undesirable properties was required. Many new reactions were discovered in these studies. On occasion, although an undesirable product was obtained, ways were found to modify the result to produce what was originally sought. Hydroxyl group activation (toward substitution) of an aromatic ring is discussed later in this chapter. You are called upon to recognize now that *after* the hydroxyl group is used to direct an incoming substituent, *its removal might subsequently be desirable*.

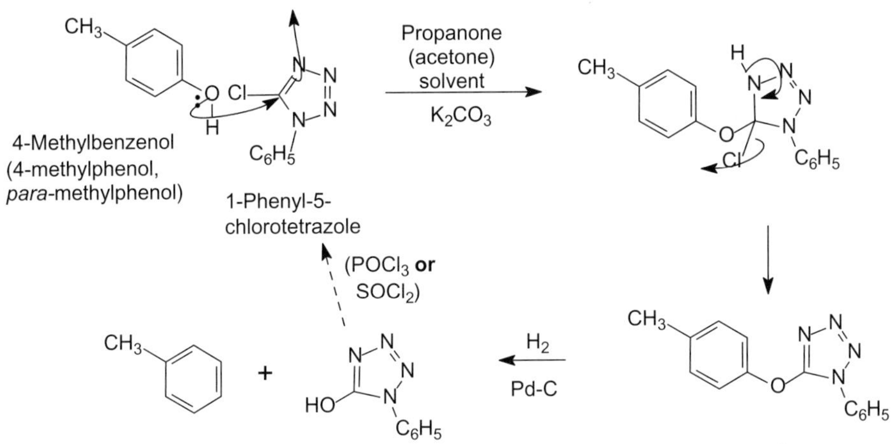

Scheme 8.25. A path for hydrogenolysis of a phenol to an aromatic hydrocarbon. The *para*-methylphenol first reacts (via addition–elimination) with 1-phenyl-5-chlorotetrazole to produce a dehydrohalogenated adduct. Catalytic reduction with hydrogen (H_2) and palladium on carbon powder (Pd-C) as the catalyst then produces methylbenzene (toluene) and hydroxytetrazole (which can be recycled to chlorotetrazole by treatment with phosphoryl chloride or thionyl chloride) (see Musliner, W. J.; Gates, J. W. Jr. *Org. Chem. Bull.*, **1967**, *39*, 1).

D. SUBSTITUTION REACTIONS OF ALCOHOL, ENOLS, AND PHENOLS

I. Introduction

As already noted (in Chapter 7), nucleophilic substitution reactions at carbon (both S_N1 and S_N2) occur most readily when the leaving group is the conjugate base of a strong acid (i.e., when the leaving group is a weak base) and with nucleophiles that are electron rich (so that they can easily capture a carbocation (S_N1) or they can help the leaving group depart [S_N2]). Both of these features prove to be difficulties when dealing with alcohols.

The first problem is the nature of the leaving group. If the leaving group, in the case of alcohols, enols, and phenols, is to be the hydroxyl (–OH), it will leave as hydroxide (OH⁻) anion and that ion is the conjugate base of the weak acid water (H–OH). Hydroxide (OH⁻) is a poor leaving group.

Second, alcohols, enols, and phenols are, like water, protic (i.e., they are proton donors) and electron-rich nucleophiles would be expected to cause deprotonation at oxygen, resulting in alkoxide formation. The alkoxide anion itself is a nucleophile (as well as a base).

Further, as noted below, "substitution" could mean (particularly for phenols) reaction elsewhere than at the oxygen or the carbon-bearing oxygen.

When substitution occurs **at** oxygen, the alcohol, enol, or phenol is acting as the nucelophile and the proton on oxygen is replaced (e.g., C–O–H → C–O–G [G ≠ H]) by the electrophile. These reactions are generally referred to as **addition reactions** (the alcohol having "added" to something).

Some addition reactions (or substitutions at oxygen) involving replacement of the proton are "temporary"; that is, the **addition** (or substitution) product is eventually reconverted to the original alcohol, enol, or phenol. Further, some products of such additions to (substitutions at) oxygen are considered to be "derivatives" of alcohols.*

The reactions, in which only the proton is replaced and which take advantage of the nucleophilicity of oxygen, despite their apparent variety, are all very similar. An electron pair on oxygen is considered to undertake an attack on an electrophilic site, and the proton, originally on oxygen, is lost. Of course, the alternative, that is, loss of the proton first and then reaction of the oxide anion, is, formally, equivalent.

When substitution occurs at carbon, there are two possibilities. Either the carbon–oxygen bond is broken and the entire hydroxyl (–OH) group is replaced by the group being substituted for it (e.g., a halogen; C–OH $\rightarrow$ C–X [X=F, Cl, Br, I]) or substitution occurs elsewhere (as on the aromatic ring of a phenol or at a site distant from the hydroxyl [–OH] group of the alcohol).

The reactions in which the carbon–oxygen bond is broken or another substitution process occurs are rich in both kind and pathway. However, they all involve converting a "poor" leaving group into a "good" leaving group, and all fall into two basic catagories. The first catagory is simply protonation on oxygen. This process converts the leaving group from hydroxide (OH^-), which is a poor leaving group (the conjugate acid is the weak acid H_2O), to water (OH_2), which is a good leaving group (the conjugate acid being the hydronium ion (H_3O^+). The second catagory is conversion of the hydroxyl (–OH) group into a derivative (*vide supra*). The derivative may or may not be isolated before subsequent substitution occurs. Thus, the hydroxyl group (–OH) is converted from a poor leaving group to a good leaving group by replacing the proton and, in the process, generating a leaving group, which will be the conjugate base of a strong acid.

Then, in the second step, the substitution reaction at carbon is effected.

As was the case for alkenyl and aryl halides (Chapter 7), S_N1 and S_N2 (or S_N1-like and S_N2-like) reactions do not appear to take place for enols and phenols either. However, along with addition–elimination processes, which occur with enols and phenols as with halides (e.g., see Chapter 7, Figure 7.11), truly exceptional leaving groups *can be created* from simple enols and phenols by forming suitable derivatives (as is the case with alcohols). Further, there is some evidence that actual S_N1 (or S_N1 like) processes to form vinyl cations ($R_2C=CR^+$) may occur from such derivatives!

Catalytic processes involving compounds of (largely) palladium (Pd) and/or nickel (Ni) along with carbanionic species based on magnesium (Mg), copper (Cu),

---

*In organic chemistry, interrelationships between compounds allow the concept that one kind of compound may be considered "derived from" another. Historically, *derivatives* were entire classes of compounds that were often crystalline (for easy isolation) or very stable (for easy purification). The concept remains useful and, on occasion, the derivation actually occurs. For example, as already indicated above and as will be discussed more fully within, alcohols may be converted to the corresponding methanesulfonate esters (mesylates) by reaction with methanesulfonyl chloride in the presence of base (to pick up the proton). Mesylates can (generally) be *reconverted to their corresponding alcohols* by hydrolysis. To the extent that (at least in principle) any alcohol can be converted to its corresponding mesylate, mesylates, as a class, are *derivatives* of alcohols. On the other hand, although both an alcohol and its derived mesylate can be converted to an alkyl halide and, in principle, the alkyl halide can be reconverted to alcohol (and mesylate), the halides are not generally considered as derivatives of either the alcohol or the mesylate. Neither is a necessary precursor of the halide.

or tin (Sn) reagents have been invented for reactions of the trifluorosulfonate and other derivatives of both enols and phenols. These processes allow replacement of oxygen by a suitable nucleophilic fragment resulting in the formation of new carbon–carbon bonds (see Chapter 7, Section E). For example, the reaction of the trifluorosulfonate (often called triflate) derivative of the enol of 4-(2-methylethyl) cyclohexanone (4-*t*-butylcyclohexanone) with an aryl tributyltin compound in the presence of a palladium (Pd) catalyst produces good yields of the corresponding arylcyclohexene derivative (Equation 8.20).

$$(8.20)$$

Electrophilic aromatic substitution reactions are common with phenols and in these processes, the hydroxyl group is retained. However, as pointed out above, preparation of suitable derivatives can induce loss of that hydroxyl group too.

II. Substitution Reactions of Alcohols, Enols, and Phenols at Oxygen

As noted above, most of the reactions that might be listed as substitution at oxygen are to be considered as addition reactions in this work. However, it is worthwhile emphasizing a few of those reactions here.

When simple primary alkyl halides are permitted to react with alkoxide anions, substitution of the alkoxide for the halide occurs or, alternatively, the alkyl group has been substituted for the hydrogen which was on oxygen. Variants of this reaction, the **Williamson ether synthesis**,* have been seen before as a **substitution** reaction at carbon (of the alkyl halide) and will be seen again as an **addition** reaction of alcohols.

Here, the reaction between methyl iodide and the anion of 2-methyl-2-propanol (*tert*-butoxide) to form *tert*-butyl methyl ether is used to demonstrate the process in which a methyl group has been substituted for the proton originally on the oxygen of 2-methyl-2-propanol (*tert*-butanol) (Equation 8.21). The example of Equation 8.21 is also used to emphasize that the ether (*tert*-butyl methyl ether) **cannot** be prepared from the reaction between methoxide anion and 2-iodo-2-methylpropane (or indeed any 2-halo-2-methylpropane) because **elimination** would intrude (Equation 8.22).

$$(8.21)$$

---

*Named after, A. W. Williamson (1824–1904), who found the reaction and who was appointed professor of chemistry at University College, London in 1849. See Williamson, A. W. *Liebigs Ann. Chem.*, **1851**, 77, 37.

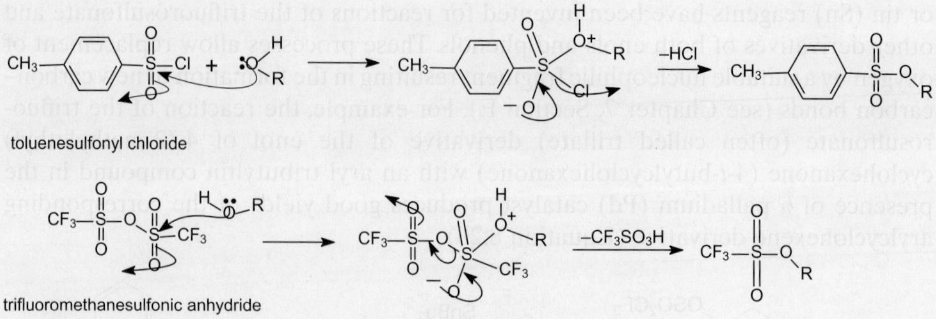

toluenesulfonyl chloride

trifluoromethanesulfonic anhydride

Scheme 8.26. Representations of substitution (addition) at oxygen of an alcohol (ROH) to produce a toluenesulfonate (*tosylate*) of the alcohol and a trifluoromethanesulfonate (*triflate*). Although not shown, it is common to add a base (such as pyridine [azabenzene]) to react with the acid (HCl and trifluoromethanesulfonic acid, respectively) so that it is removed from the reaction mixture.

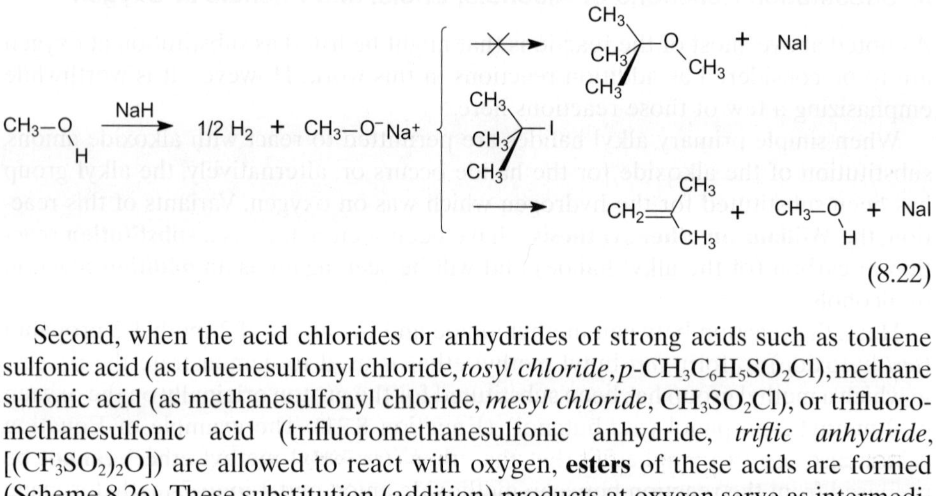

$$(8.22)$$

Second, when the acid chlorides or anhydrides of strong acids such as toluene sulfonic acid (as toluenesulfonyl chloride, *tosyl chloride*, p-$CH_3C_6H_5SO_2Cl$), methane sulfonic acid (as methanesulfonyl chloride, *mesyl chloride*, CH_3SO_2Cl), or trifluoro-methanesulfonic acid (trifluoromethanesulfonic anhydride, *triflic anhydride*, $[(CF_3SO_2)_2O]$) are allowed to react with oxygen, **esters** of these acids are formed (Scheme 8.26). These substitution (addition) products at oxygen serve as intermediates for further substitution since (a) the proton on oxygen has been removed and is unavailable for reaction with electron-rich nucleophiles and (b) the hydroxyl "leaving group" is now the anion of a strong acid.

III. Substitution Reactions of Alcohols at Carbon

a. Formation of Alkyl Halides.
The replacement of the hydroxyl group of an alcohol by any of the substituents already discussed for alkyl halides (Chapter 7) can be accomplished by *converting the alcohol to the corresponding alkyl halide*.

In general, inorganic acids react with alcohols with loss of water and the formation of alkyl halide (either through direct nucleophilic substitution or by an elimination–addition process). The facility with which this substitution occurs varies

Tertiary alcohols react rapidly...

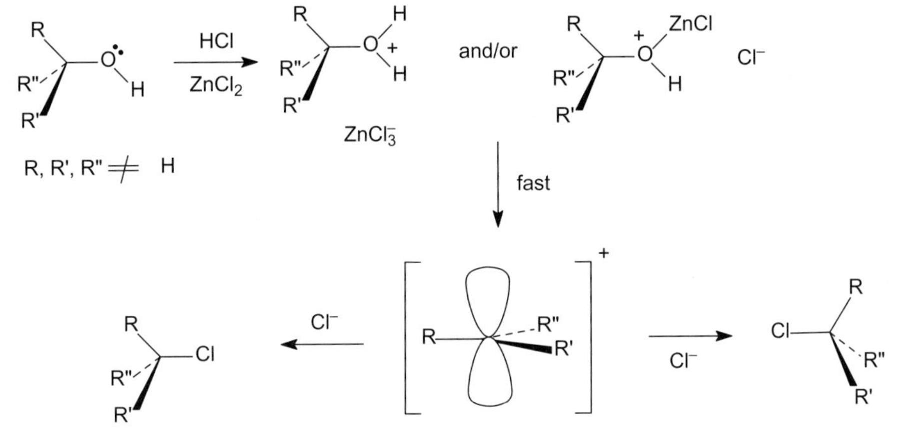

R, R', R" ≠ H

and/or

fast

Secondary alcohols react more slowly...

R, R' ≠ H

and/or

slow

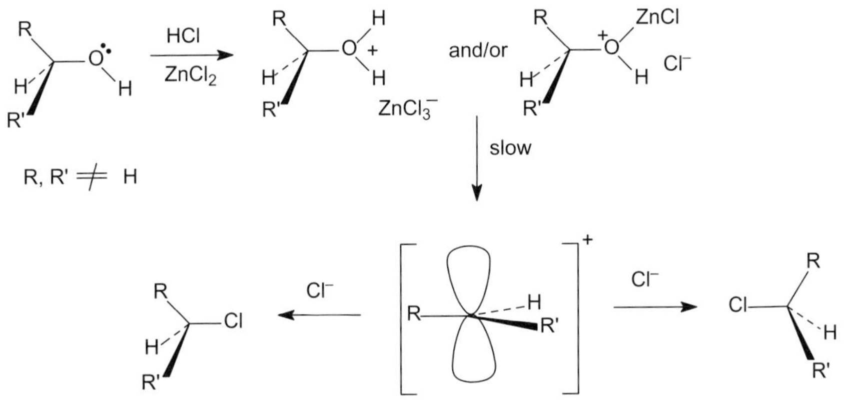

Primary alcohols reaction slowly or they do not react at all.

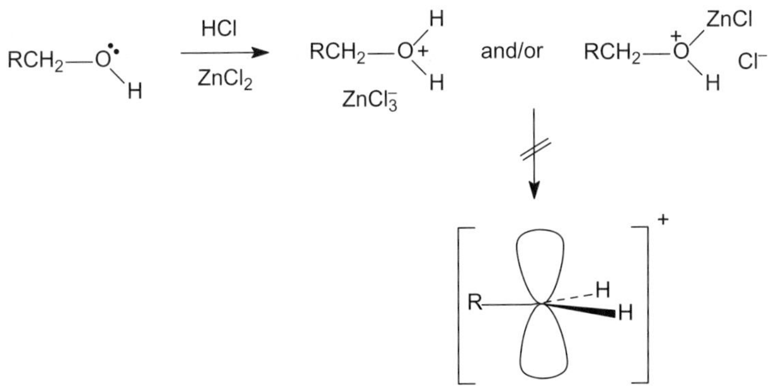

and/or

Scheme 8.27. A representation of the Lucas test showing that in a mixture of aqueous HCl and ZnCl$_2$ tertiary alcohols react quickly to provide tertiary alkyl chlorides, secondary alcohols react more slowly and primary alcohols not at all. The alkyl halide separates from the aqueous solution as a distinct phase.

with the type of alcohol, the particular acid, and the ease with which rearrangements, and so on, might intrude. Generally, for example, primary alcohols will react with hydrogen iodide to produce the corresponding iodide much more readily than with hydrogen bromide (HBr) to yield the bromide and fail to react with hydrochloric acid. On the other hand, tertiary alcohols will react with concentrated aqueous hydrochloric acid to yield the corresponding chlorides. Secondary alcohols are intermediate in their reactivity.

Prior to the widespread availability of spectroscopic tools (Chapter 2), the relative reactivities of primary, secondary, and tertiary alcohols toward acid gave rise to a qualitative test (called the Lucas test),* which might, generally, distinguish between isomeric alcohols. Although there are many cases for which the simple test fails (see Problem 8.4), tertiary alcohols react with the Lucas reagent almost immediately at room temperature to produce the corresponding alkyl chlorides (which separate as a second phase in the reaction mixture); secondary alcohols react (often with gentle warming) after several minutes; and primary alcohols fail to react. As indicated in Scheme 8.27, the test clearly relies upon carbocation formation. Thus (as shown there), it should be clear that the hydroxyl group (–OH) is being *converted* from a poor leaving group into a good leaving group by protonation from the acidic medium.[†]

---

*The Lucas test was first published by Professor H. Lucas, California Institute of Technology, in 1930. It consisted then, as it does now, of using hydrochloric acid ($HCl_{(aq)}$)in the presence of the Lewis acid zinc chloride ($ZnCl_2$). See Lucas, H. *J. Am. Chem. Soc.*, **1930**, *52*, 802.

[†]It is important to understand that the loss of the hydroxyl (–OH) group under acidic conditions here and in the subsequent discussion, either through protonation or after substitution of the hydrogen of the hydroxyl by some other function (*vide infra*), simply converts the oxygen into a good (or better) leaving group. Therefore, the cartoons shown below (and already seen in Chapter 7 for S_N1 and S_N2 reactions, respectively, with alkyl halides) apply (parts a and b for the S_N1 reaction and parts c and d for the S_N2).

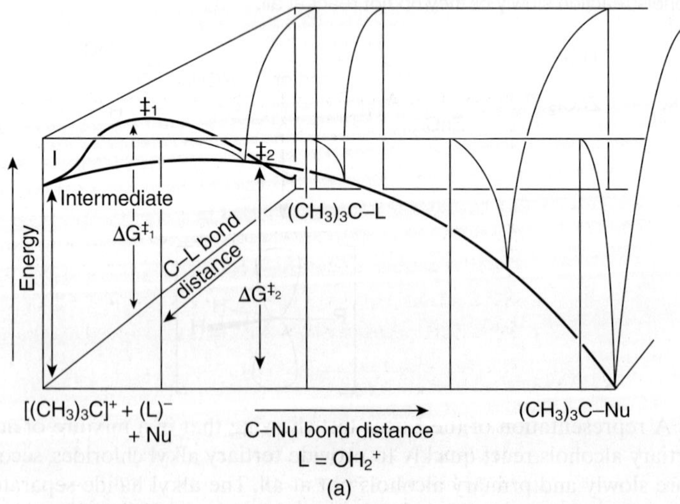

(Continued on next page)

Problem 8.4. Phenylmethanol (benzyl alcohol, $C_6H_5CH_2OH$) and 2,2-dimethyl-1-propanol (neopentyl alcohol [$(CH_3)_3CCH_2OH$]) are among those alcohols that fail to give "appropriate" responses in the Lucas test. Although both of these alcohols are primary and would thus be expected to remain unreacted, chlorine-containing products are formed. Using curved arrows to show what bonds are broken and made, predict what each of these might provide.

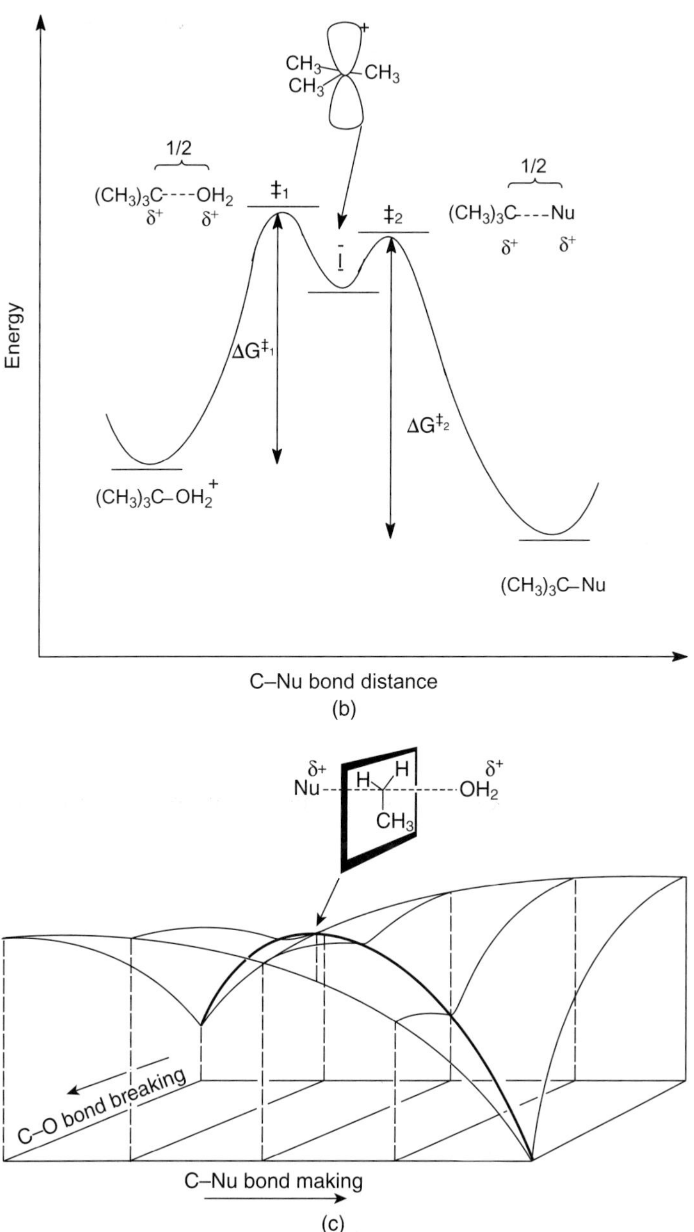

(Continued on next page)

Acid derivatives* from which water has been removed often serve as alternatives to the use of protic acid (hydrogen halide [HX], X=Cl, Br, I) for the formation of the corresponding alkyl halide because they are both more soluble in the organic solvent medium in which the reactions are commonly carried out and they are higher energy reactants (thus requiring less forcing conditions to surmount the barrier lying between reactants and products).

Although such methods have long been used to prepare alkyl halides, alternatives are available. Thus, as shown in Scheme 8.28, the reaction between triphenylphosphine [$(C_6H_5)_3P$] and tetrachloromethane (carbon tetrachloride, CCl_4) results in the formation of the corresponding chlorotriphenylphosphonium trichloromethide salt [$(C_6H_5)_3P^+Cl\ CCl_3^-$]. In carbon tetrachloride (CCl_4) solution, the addition of (*Z*)- or *cis*-4-*t*-butylcyclohexanol [(*Z*)- or *cis*-4-(1-methylethyl)cyclohexanol] apparently results in the formation of an intermediate in which a new bond has been formed between oxygen and phosphorus (by a displacement reaction at phosphorus). The chloride ion so liberated is now able to displace oxygen in a (mostly) S_N2 process leading to triphenylphosphine oxide and (*E*)- or *trans*-4-*t*-butyl-1-chlorocyclohexane [(*E*)- or *trans*-4-(1-methylethyl)-1-chlorocyclohexane]. The opposite stereochemical result is expected if one begins with the (*E*)-or *trans*-4-*t*-butylcyclohexanol [(*E*)- or *trans*-4-(1-methylethyl)cyclohexanol].[†]

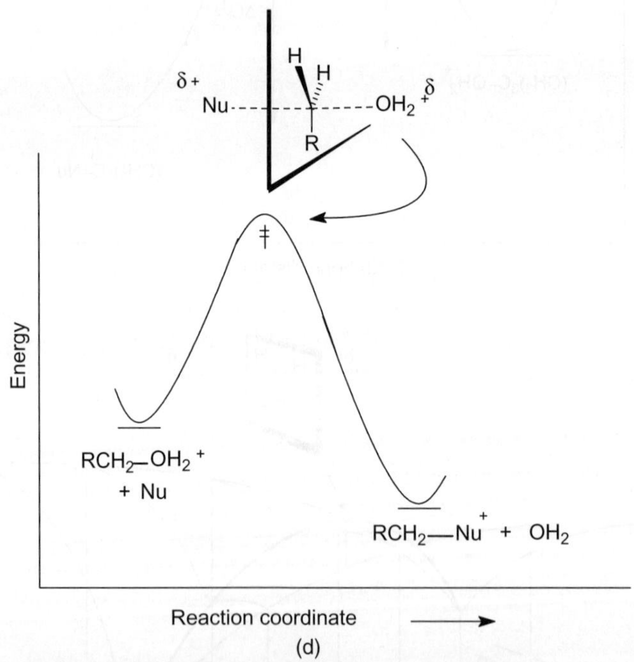

*In this vein, sulfuryl chloride (SO_2Cl_2) is an *acid chloride* derivative of sulfuric acid [H_2SO_4 or $SO_2(OH)_2$]; phosphorous oxychloride ($POCl_3$) is an acid chloride derivative of phosphoric acid [H_3PO_4 or $PO(OH)_3$], and so on.

[†]To the extent that the transition state for S_N2 substitution of 4-*t*-butylcyclohexyl systems requires the substituents at C-1 adapt coplanarity and is the *same* when approached from either direction, it should be clear that the approach of a nucleophile is generally less encumbered when an axial substituent

(Continued on next page)

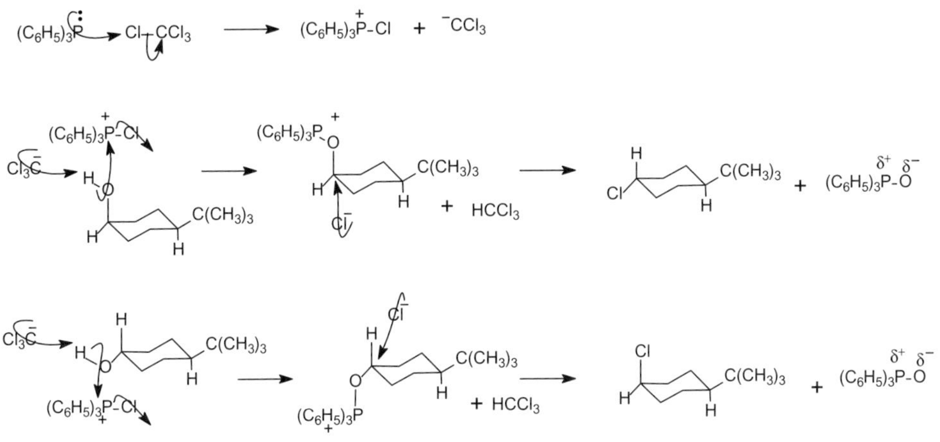

Scheme 8.28. A representation of the formation of alkyl chloride from the reaction between an alcohol and triphenylphosphine in carbon tetrachloride. The 4-t-butylcyclohexanols yield mostly the inverted product, although the reaction of the *cis*-isomer is cleaner than that of the *trans*-isomer (see Appel, R. *Angew. Chem. Int. Ed.*, **1975**, *14*, 801).

Hydrogen fluoride (HF) fails to convert alcohols to the corresponding fluorides. Sulfur tetrafluoride or its derivative, diethylaminosulfur trifluoride (DAST $(CH_3CH_2)_2NSF_3$),* is required for the conversion of alcohols to the corresponding fluorides. As with analogous proton-catalyzed reactions and as shown in Scheme 8.29, alkyl fluoride formation results from mixtures of S_N1 and S_N2 processes.

Interestingly, in the presence of a scavenger for HX (such as triethylamine [$(CH_3CH_2)_3N$] or azabenzene [pyridine, C_5H_5N], which acts by forming an insoluble salt, for example, triethylamine hydrochloride [$(CH_3CH_2)_3NH^+ Cl^-$] or pyridinium hydrochloride [$C_5H_5NH^+ Cl^-$], and thus removes halide from the reaction system), the sulfur and phosphorus halides produce the esters (Scheme 8.30) of the corresponding acids, for example, dimethyl sulfate [$(CH_3O)_2SO_2$] and trimethyl phosphate [$(CH_3O)_3PO$] from methanol, and sulfuryl chloride (SO_2Cl_2) and phosphorous oxychloride ($POCl_3$), respectively.

b. Replacement of the Hydroxyl (–OH) Functional Group by Other Substituents.

Esters of methanesulfonic acid (CH_3SO_3H), trifluoromethanesulfonic acid (CF_3SO_3H), and p-toluenesulfonic acid (p-$CH_3C_6H_4SO_3H$) have proven

(usually of higher initial energy) is being displaced than when an equatorial substituent (usually of lower initial energy) is being displaced, that is,

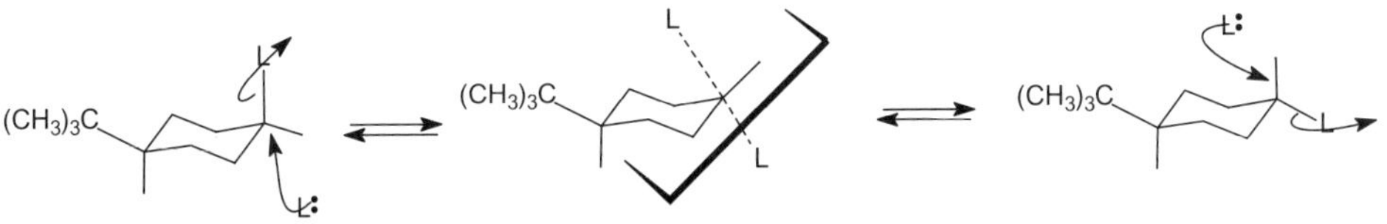

*See Markovskij, L. N.; Pashinnik, V. E.; Kirsanov, A. V. *Synthesis*, **1973**, 787 and Middleton, W. J.; Bingham, E. M. *J. Org. Chem.*, **1980**, *45*, 2883.

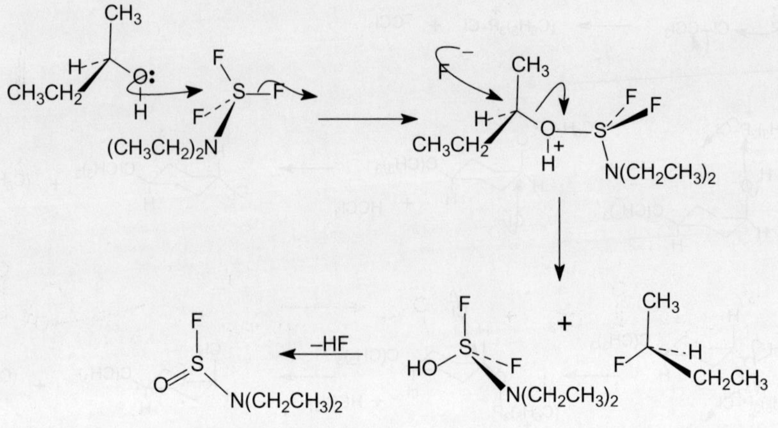

Scheme 8.29. The use of diethylaminosulfur trifluoride (DAST, $(CH_3CH_2)_2NSF_3$) to convert 2-butanol to the corresponding fluoride (2-fluorobutane).

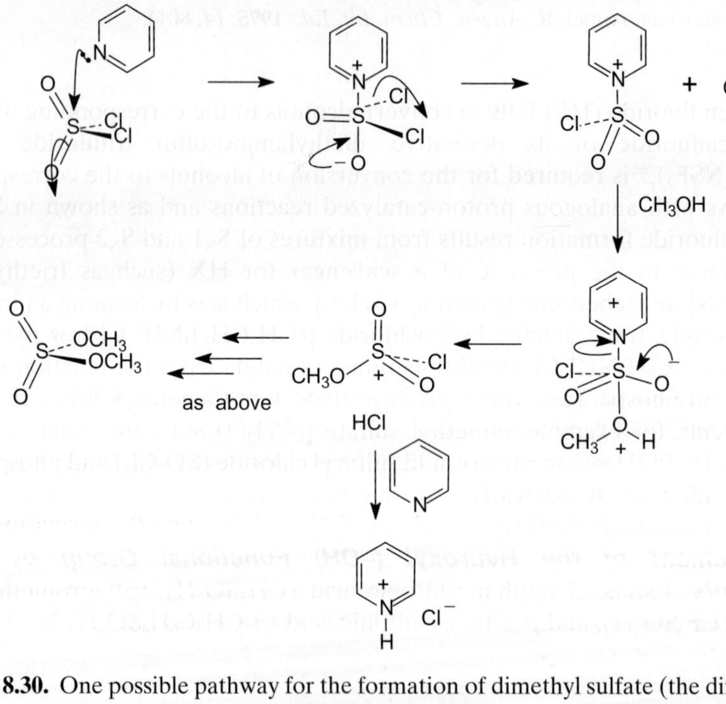

Scheme 8.30. One possible pathway for the formation of dimethyl sulfate (the dimethyl ester of sulfuric acid) by the reaction of sulfuryl chloride (SO_2Cl_2) with methanol (CH_3OH) in the presence of azabenzene (pyridine, C_5H_5N).

particularly useful in effecting the replacement of the hydroxyl group thus esterified by various nucleophiles. Converting the hydroxyl group into the ester of a carboxylic acid (a topic covered in greater detail in Chapter 9) generally makes the hydroxyl a better leaving group, but the particular acids mentioned are superior since their anions are the conjugate bases of very strong acids. Thus, as shown in Scheme 8.31, the reaction of the anhydride of trifluoromethanesulfonic acid ("triflic anhydride")

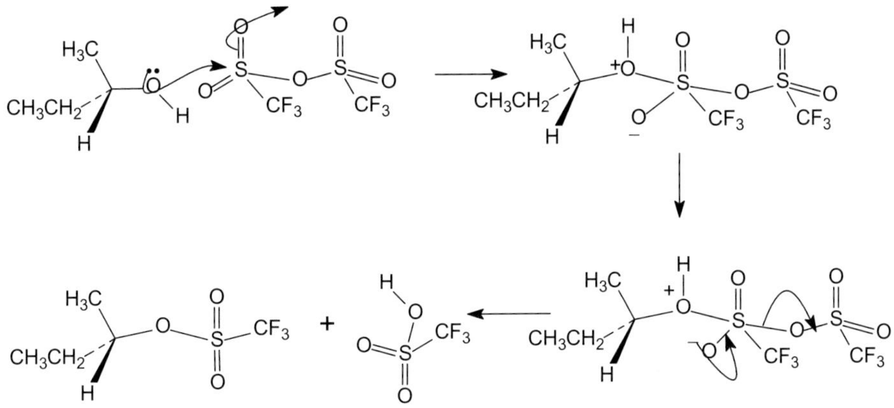

Scheme 8.31. The formation of the trifluoromethanesulfonate derivative (triflate ester) of S-(+)-2-butanol from the alcohol and trifluoromethanesulfonic anhydride (triflic anhydride).

with an alcohol (such as S-(+)-2-butanol [$CH_3CH(OH)CH_2CH_3$]) produces the corresponding ester and the latter, with a nucleophile such as chloride (Cl^-), cleanly undergoes S_N2 substitution (Equation 8.23).

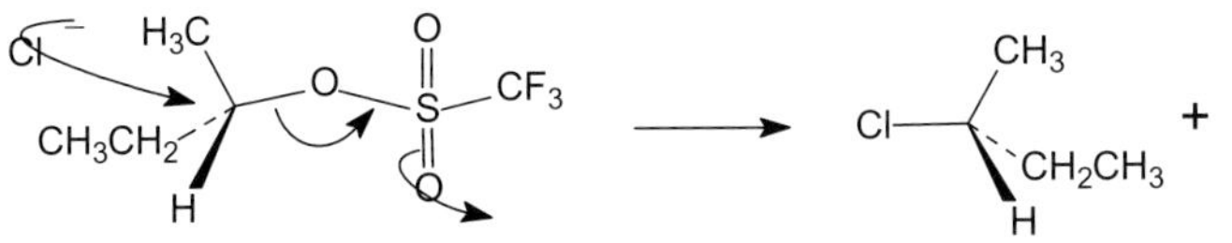

$$\qquad\qquad (8.23)$$

Since the formation of esters (RCO_2R') is reversible (i.e., esters can be **recon**verted to alcohols ($R'OH$) and acids (RCO_2H, RSO_3H)) by hydrolysis (reaction with water—frequently base and/or acid catalyzed as discussed more fully in Chapter 9), a reaction sequence which might *convert* a specific enantiomer of one *alcohol into its mirror image* could involve the S_N2 displacement of an ester leaving group by an acid anion or its equivalent and the subsequent hydrolysis of the resulting ester. Such a process is depicted in Scheme 8.31, which shows the formation of the trifluoromethanesulfonate ("triflate") ester of S-(+)-2-butanol, and Scheme 8.32, which shows the S_N2 displacement (with inversion) of the trifluoromethanesulfonate anion by the anion of benzoic acid (benzoate) and the subsequent hydroysis of the benzoate ester of R-(+)-2-butanol.

Although the replacement of hydroxyl **by** hydroxyl can often be achieved in just the way shown in the sum of Schemes 8.31 and 8.32, other reactions can intrude. Indeed, elimination to form an alkene often occurs when the nucleophile is too basic (Equation 8.24).

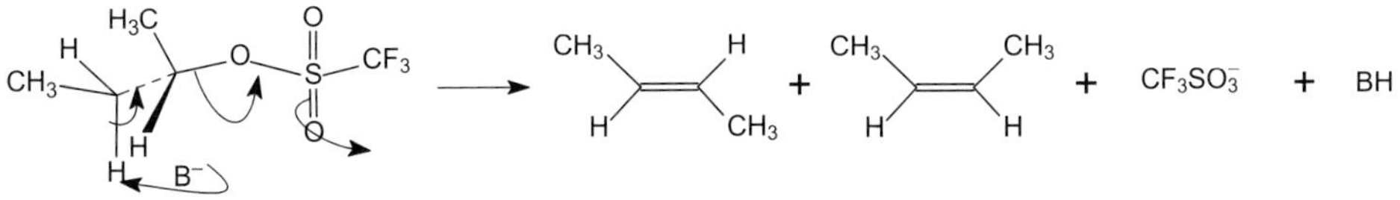

$$\qquad\qquad (8.24)$$

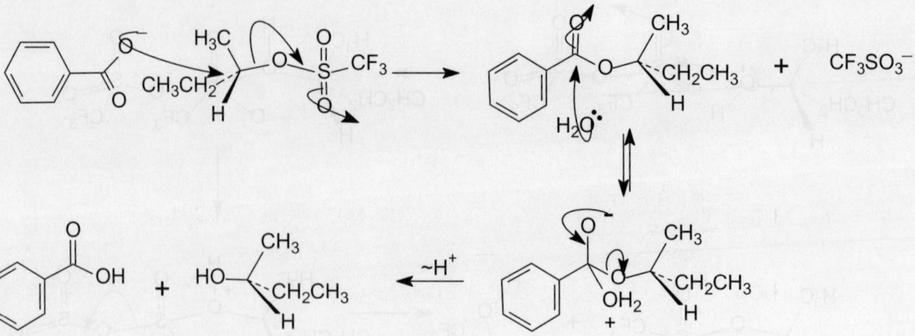

Scheme 8.32. A representation of the reaction between the triflate derivative of S-(+)-2-butanol and the benzoate anion showing displacement of the trifluoromethylsulfonate anion and its replacement, with inversion of configuration, by benzoate. The hydrolysis of the benzoate ester generates benzoic acid and the alcohol with **inverted configuration**.

One of the more successful processes developed for the substitution of the hydroxyl group of an alcohol by another hydroxyl group *with (usually) an inversion of configuration at the carbon bearing the hydroxyl,* and which often minimizes concurrent elimination, is known as the **Mitsunobu reaction.***† The overall process is shown in Equation 8.25.

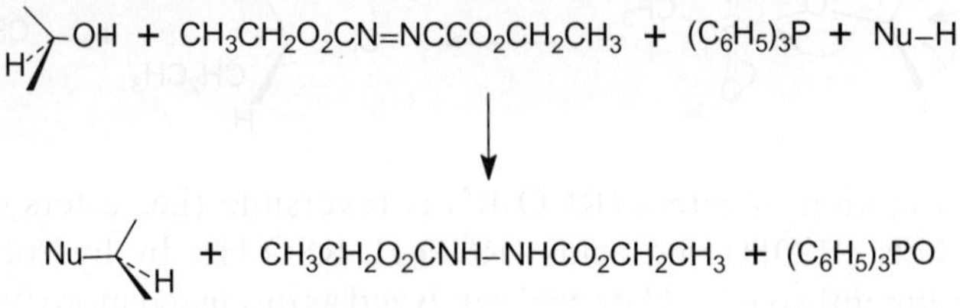

$$\text{H}\!\!>\!\!\text{OH} + CH_3CH_2O_2CN=NCCO_2CH_2CH_3 + (C_6H_5)_3P + Nu–H$$

(8.25)

$$Nu\!\!<\!\!\text{H} + CH_3CH_2O_2CNH–NHCO_2CH_2CH_3 + (C_6H_5)_3PO$$

Although the details of the **Mitsunobu protocol** remain under investigation, it is widely held that much of the reaction proceeds in three steps (as shown in Scheme 8.33). The first step is thought to be between a phosphene (shown here as triphenylphosphine [$(C_6H_5)_3P$], an ester of diazodicarboxylic acid (shown here as diethyl diazodicarboxylate) and a proton bearing nucleophile (shown here simply as H–Nu). In a second step, the phosphorus–nitrogen intermediate thus formed is proposed to undergo reaction with the alcohol substrate to produce the reduced diazodicarboxylic acid ester and a second intermediate. The latter contains a phosphorus–oxygen bond (thus having converted the poor –OH leaving group into an activated species). In the final step of Scheme 8.33, the displacement of the phosphine oxide (shown

---

*Named after the late Professor Oyo Mitsunobu, Aoyama Gakuin University, Tokyo, Japan, who initially discovered and then explored the reaction in some detail (see Mitsumobu, O.; Yamada, M. *Bull. Chem. Soc. Jpn.*, **1967**, *40*, 2380 and Mitsunobu, O. *Synthesis*, **1981**, 1).
†Because (in part) of its expansion to include a variety of nucleophiles that allow for carbon–carbon, carbon–halogen, carbon–nitrogen, and carbon–sulfur **as well as** carbon–oxygen bond formation and because (in part) of the variety of subtle changes that are employed to effect the desired overall reaction and because (in part) the changes can be carried out in one or more of the steps in the process, the overall technique has become known as the Mitsunobu protocol.

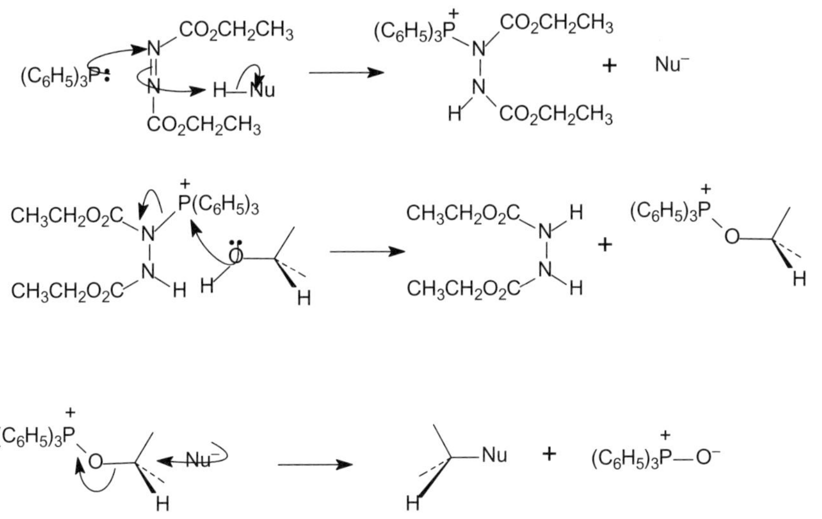

Scheme 8.33. The Mitsunobu protocol. Conversion of an alcohol (shown here as secondary) to an enantiomeric species by the displacement of the (modified) hydroxyl group. There are three steps shown. The first activates the phosphine (shown here as triphenylphosphine) for reaction with the alcohol; the second produces a good leaving group as the alcohol reacts with phosphorus; and, in the third, the activated hydroxyl group is lost.

here as triphenylphosphine oxide) by an incoming nucleophile with an overall net inversion consummates the process. Of course, as noted earlier, if the nucleophile is a carboxylic acid anion, hydrolysis to the alcohol itself will be required.

c. Replacement of the Hydroxyl (–OH) Functional Group by Carbon: An Example from Nature.

Geranyl diphosphate (Chapter 11) is a metabolic intermediate lying between acetate-derived mevalonate* (Chapter 6, Section C, I and Chapter 11) and cholesterol (Chapter 11).

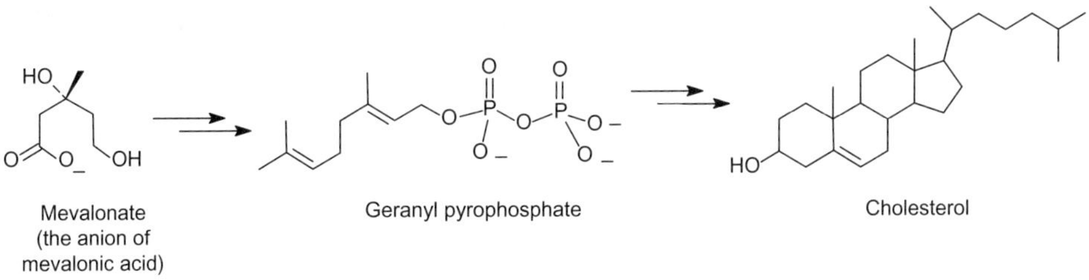

Mevalonate
(the anion of
mevalonic acid)

Geranyl pyrophosphate

Cholesterol

*When mevalonate, written below as the anion of (R)-mevalonic acid (Chapter 6, Section C, I and Chapter 11), is acted upon by the enzyme mevalonate-5-phosphotransferase, a phosphate "unit" is transferred from adenosine triphosphate (ATP) to the primary alcohol of mevalonate. Adenosine diphosphate (ADP) and phosphomevalonate are produced. Although enzyme mediated, this process may be viewed as the reaction of an activated derivative of phosphoric acid (an "anhydride") reacting with an alcohol to produce an "ester" of phosphoric acid [H_3PO_4 or $O{=}P(OH)_3$].

(Continued on next page)

The intermediate, geranyl diphosphate, is a C-10 hydrocarbon, which is formed by the enzyme-catalyzed (the enzyme is named *prenyltransferase*) reaction between the C-5 isomers dimethylallyl diphosphate and isopentenyl diphosphate (C-5 + C-5 = C-10). The combination of these fragments apparently involves an S_N1

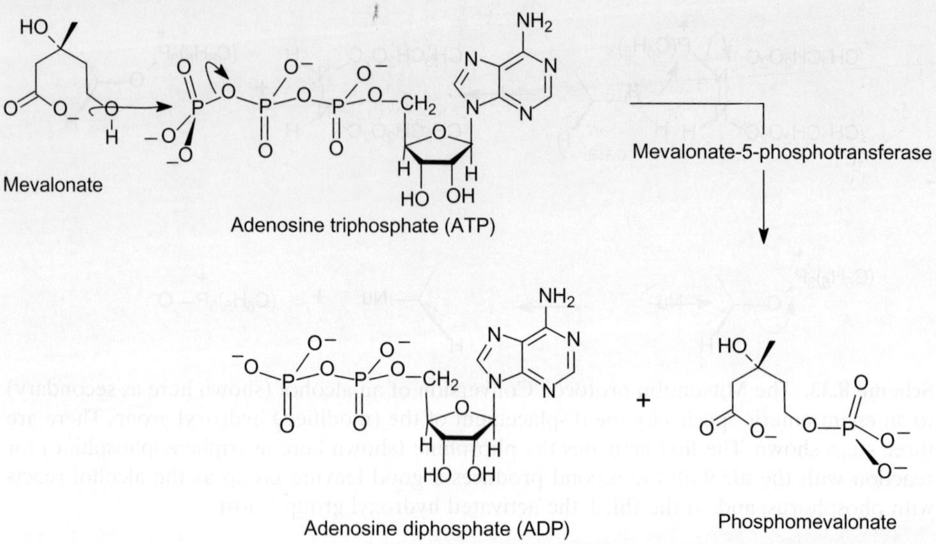

Mevalonate

Adenosine triphosphate (ATP)

Mevalonate-5-phosphotransferase

Adenosine diphosphate (ADP)

Phosphomevalonate

Then, in an analogous process, when phosphomevalonate is acted upon by the enzyme phosphormevalonate kinase (Chapter 12), a second equivalent of ATP is used and a second phosphate residue is transferred. The result is the formation of 5-pyrophosphomevalonate (and another equivalent of ADP).

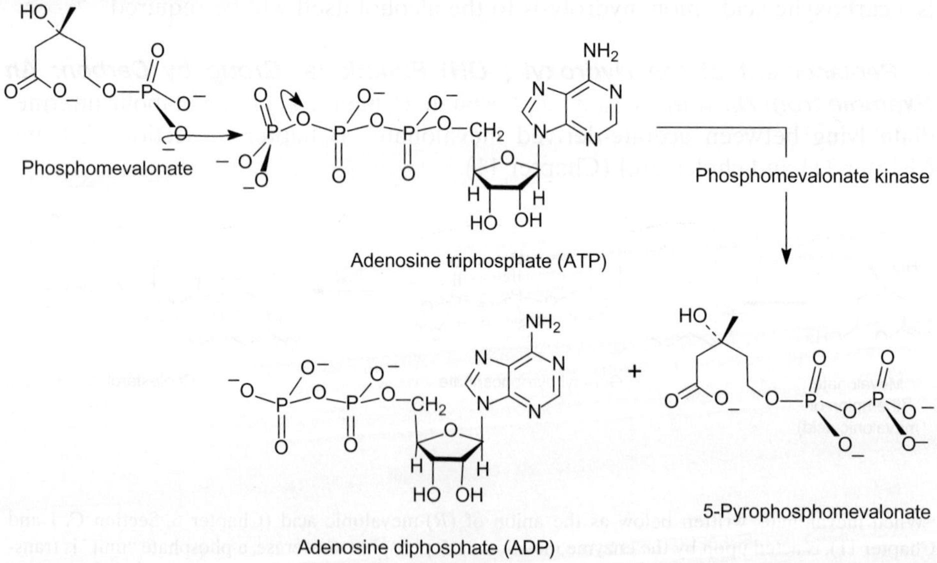

Phosphomevalonate

Adenosine triphosphate (ATP)

Phosphomevalonate kinase

Adenosine diphosphate (ADP)

5-Pyrophosphomevalonate

Action of the enzyme pyrophosphomevalonate decarboxylase on 5-pyrophosphomevalonate results in the consumption of a third equivalent of ATP and the simultaneous dehydration and decarboxylation (loss of carbon dioxide, CO_2) of the pyrophosphomevalonate. Isopentenyl diphosphate results.

(Continued on next page)

process in which the dimethylallyl diphosphate loses the diphosphate leaving group forming an allylic carbocation. In addition to whatever inhibition to rotation is enforced by the prenyl transferase enzyme, it can also be argued that rotation of the cation is inhibited by an overlap with the adjacent unsaturation (shown here as a resonance structure).

In a subsequent step, the allylic carbocation condenses with isopentenyl diphosphate, forming a new carbocation, which then loses the H_R (this chapter) proton to form the product. Alternatively, geranyl diphosphate can form if (as shown in the figure) proton loss is concomitant with condensation.

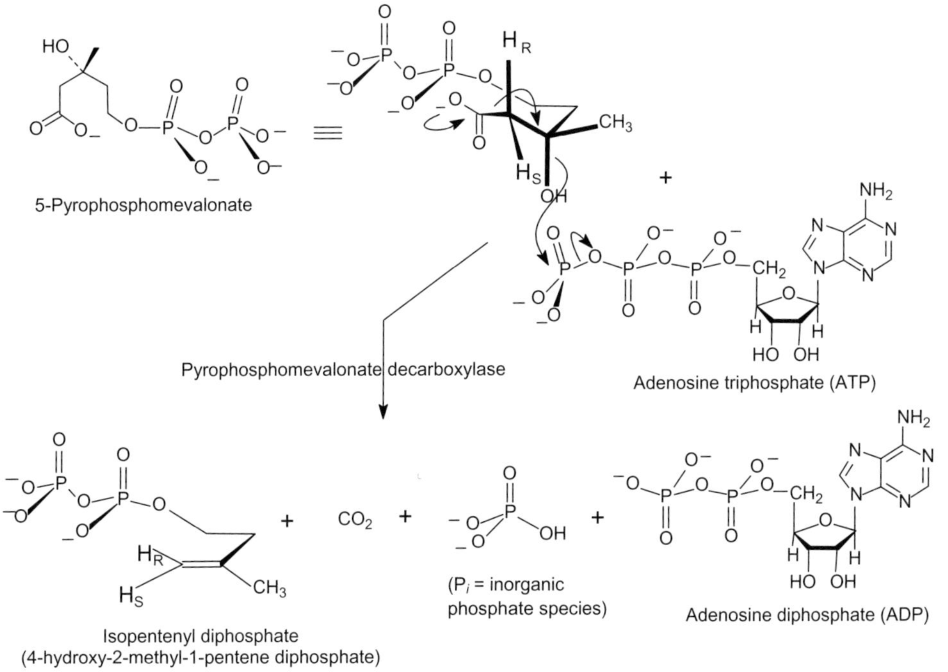

5-Pyrophosphomevalonate

Pyrophosphomevalonate decarboxylase

Adenosine triphosphate (ATP)

Isopentenyl diphosphate
(4-hydroxy-2-methyl-1-pentene diphosphate)

$+$ CO_2 $+$

(P_i = inorganic phosphate species)

Adenosine diphosphate (ADP)

The isomerization of isopentenyl diphosphate to dimethylallyl diphosphate is catalyzed by the enzyme isopentenyl pyrophosphate isomerase. In this process, the 2R hydrogen is lost and another hydrogen is picked up on the opposite face. Thus, the original methyl group is now *cis* or (*Z*)- to the methylene bearing the diphosphate group, that is,

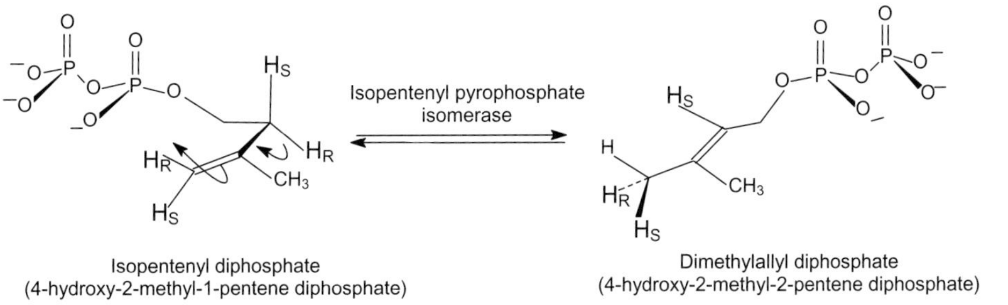

Isopentenyl pyrophosphate isomerase

Isopentenyl diphosphate
(4-hydroxy-2-methyl-1-pentene diphosphate)

Dimethylallyl diphosphate
(4-hydroxy-2-methyl-2-pentene diphosphate)

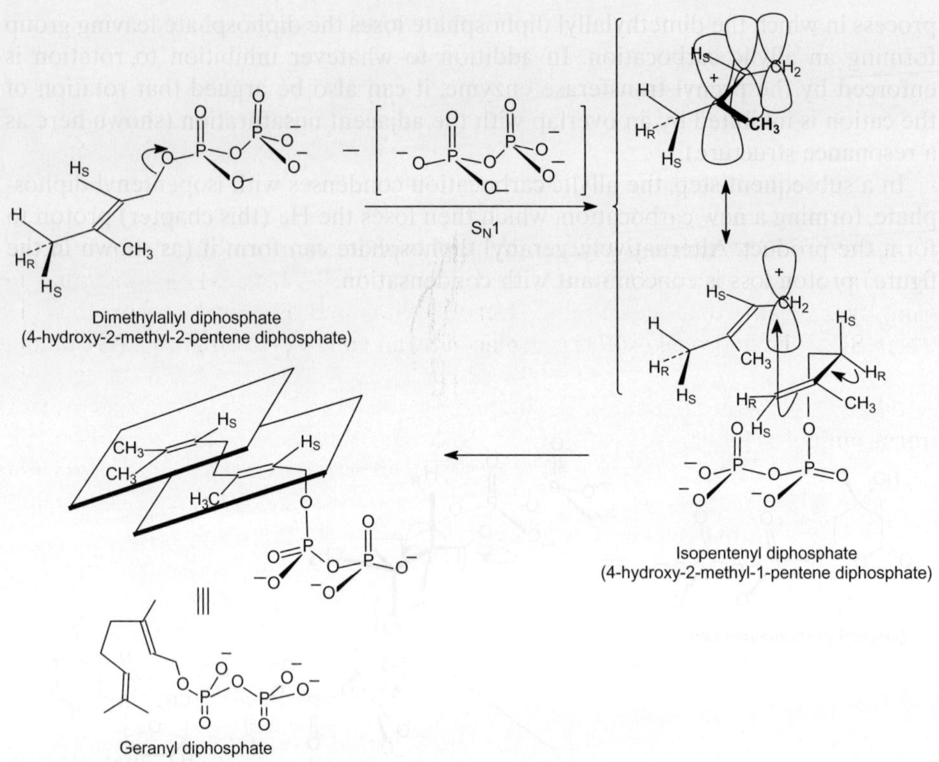
Dimethylallyl diphosphate
(4-hydroxy-2-methyl-2-pentene diphosphate)
S_N1
Isopentenyl diphosphate
(4-hydroxy-2-methyl-1-pentene diphosphate)
Geranyl diphosphate

IV. Substitution Reactions of Enols and Phenols at Carbon

a. Substitution at the Carbon-Bearing Oxygen. Esters of **enols** can also be induced to undergo substitution. Thus, as outlined in Scheme 8.34, allowing 1-(*o*-methoxyphenyl)-2-methylpropen-1-yl tosylate to stand in 2,2,2-trifluoroethanol (CF_3CH_2OH) (in the presence of a base [triethylamine, $(CH_3CH_2)_3N$] to react with the acid formed) produces the corresponding 2,2,2-trifluoroethyl ether (95%) and

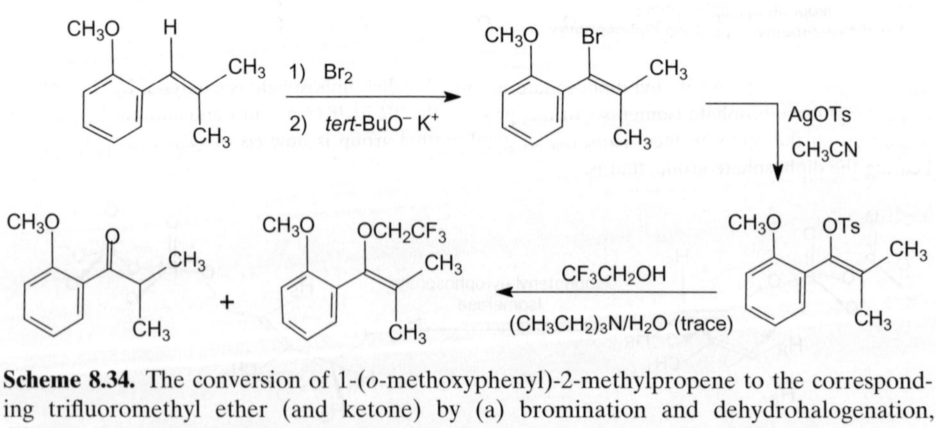
1) Br_2
2) *tert*-BuO⁻ K⁺
AgOTs
CH_3CN
CF_3CH_2OH
$(CH_3CH_2)_3N/H_2O$ (trace)

Scheme 8.34. The conversion of 1-(*o*-methoxyphenyl)-2-methylpropene to the corresponding trifluoromethyl ether (and ketone) by (a) bromination and dehydrohalogenation, (b) silver ion-assisted substitution of bromine by tosylate (presumably via the vinylic cation), and (c) S_N1 substitution of trifluoroethanol for tosylate.

some ketone (ca. 5%) by a reaction that appears to be largely S_N1. Presumably, the leaving group leaves, an "ion pair" forms, and then the solvent (2,2,2-trifloroethanol) reacts.

Problem 8.5. Write a pathway for the formation of the ketone product shown in Scheme 8.34. Hint: the amount of water used in the reaction is important.

Substitution of hydroxyl groups attached to aromatic rings can also be effected. If the hydroxyl group (–OH) is attached either α- or β- (1- or 2-) on a naphthalene ring system and that naphthol is treated with ammonia and sodium bisulfite (NaHSO₃); the hydroxyl (–OH) is replaced by an **amino** (–NH₂) group. As outlined in Scheme 8.35, the process apparently involves the addition of the bisulfite to the aromatic ring, tautomerization of the phenolic hydroxyl to its corresponding keto form, aminolysis of the ketone, and subsequent elimination leading to rearomatization. As noted earlier, this pattern of **addition–elimination** resulting in **substitution** is common in unsaturated systems. The overall process of naphthol to aminonaphthalene is called the **Bucherer reaction.***

Problem 8.6. Draw arrows on the intermediate structures of Scheme 8.35 similar to those on the first and last structures in that scheme to account for the movement of electrons and nuclei linking the reactants to the products.

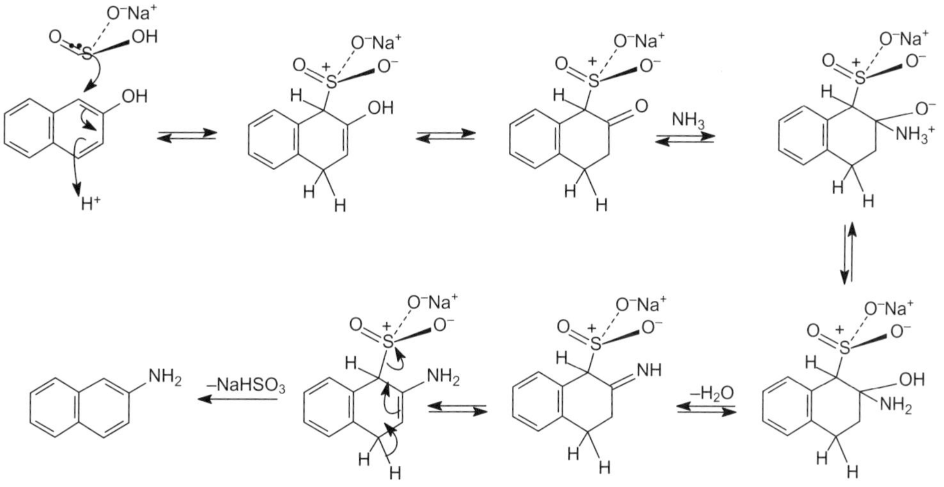

Scheme 8.35. A possible path for the aminolysis of β-naphthol (the Bucherer reaction or the **Bucherer–LePetit** reaction). It is suggested that bisulfite adds to the aromatic ring, the phenolic hydroxyl tautomerizes, an addition reaction of ammonia to the carbonyl carbon occurs with loss of water, and, finally, the imine so generated tautomerizes, bisulfite is lost, and the aminonaphthalene is produced.

*Hans T. Bucherer, Professor, Munich Technische Hochschule, who, it is reported, was the only member of the Bavarian Polytechnical Society to resign (in 1933) when the Society (under some pressure) pledged its fealty to Nazi Chancellor A. Hitler (see Bucherer, H. T. *J. Prakt. Chem.*, **1904**, *69*, 49).

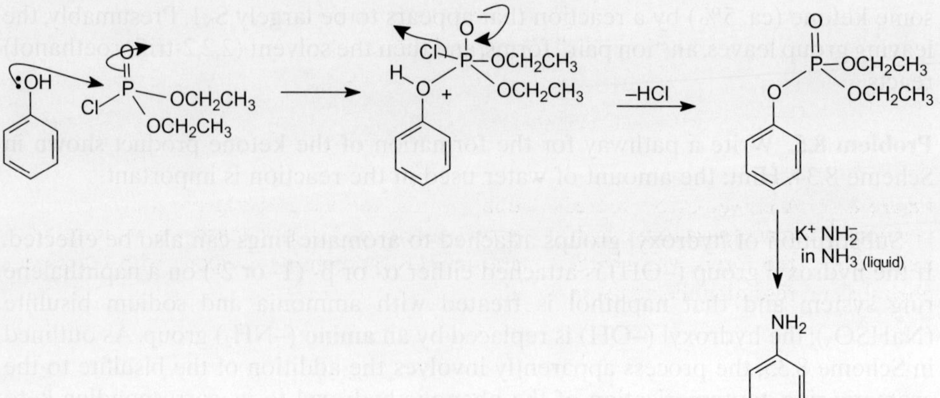

Scheme 8.36. The conversion of hydroxybenzene (phenol, C_6H_5OH) to aminobenzene (aniline, $C_6H_5NH_2$) via a substitution reaction involving a phosphate intermediate.

Interestingly, if the hydroxyl group (–OH) of phenol (C_6H_5OH) is converted to the corresponding aryl diethylphosphate ester by the reaction with diethylchlorophosphate [$ClP(O)(OCH_2CH_3)_2$], substitution of nitrogen with the phenolic oxygen, producing aminobenzene (aniline, $C_6H_5NH_2$), can be brought about. In this case, liquid ammonia (NH_3) is used as the solvent and the nucleophile is the amide anion (NH_2^-) obtained from, for example, potassium amide [$K^+NH_2^-$] (Scheme 8.36).

Additionally, as pointed out earlier (Chapter 7, Section D7) with regard to vinyl halides, the Stille (*vide supra*) reaction affords substitution of the leaving group by a carbon nucleophile at the site of unsaturation. Thus, as shown in the Equation 8.26, when the enol trifluorosulfonate (triflate, $–OSO_2CF_3$, –OTf) of 4-*tert*-butylcyclohexanone (Chapter 9) is treated with tri-*n*-butylphenylstannane in the presence of a palladium catalyst (tris(dibenzylideneacetone)palladium, $Pd_2(dba)_3$) in a polar aprotic solvent (1-methyl-2-pyrrolidinone, *N*-methylpyrrolidinone [NMP]), substitution is readily affected.

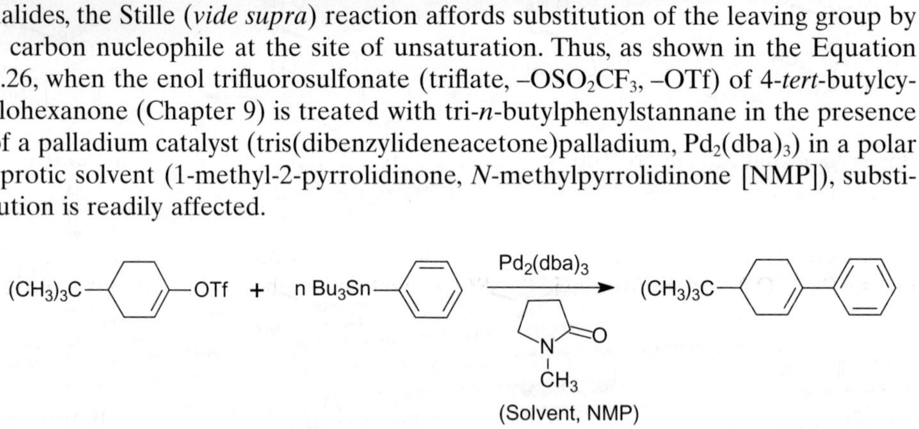

$$(8.26)$$

b. Electrophilic Aromatic Substitution of Phenols.

As already noted (Scheme 8.24), reaction of phenol with 2-methylpropene (*iso*-butylene) in the presence of an acid catalyst results in the substitution of the *t*-butyl group for the *para*-hydrogen in the phenol. The *electrophilic substitution process* shown there (Scheme 8.24) is typical of the reaction of electrophiles with phenol (and phenols in general).

Oxygen is more electronegative than carbon. It is thus reasonable to expect the oxygen substituent on the aromatic ring to be (inductively) electron withdrawing. However, the experimentally observed *acceleration* of electrophilic substitution

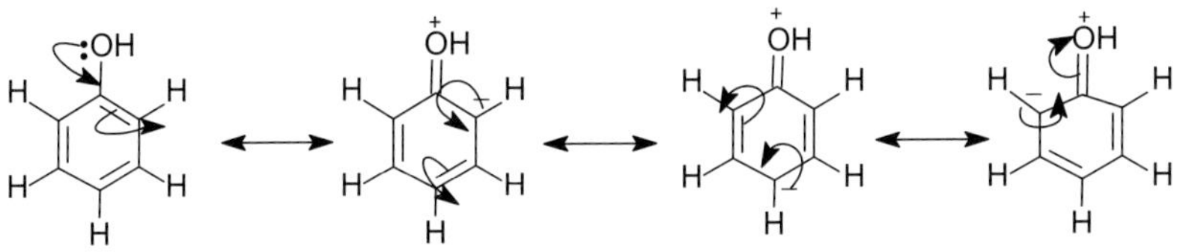

Figure 8.13. A curved-arrow representation showing the increased electron density at the positions *ortho* and *para* to the hydroxyl group of phenol. The resonance arrows are used to indicate that the forms with charge separation must be considered along with the parent phenol as representing the actual material.

$$H_2SO_4 \; + \; HONO \; = \; HSO_4^- \; + \; H_2ONO^+$$

$$H_2ONO^+ \; = \; H_2O \; + \; N{=}O^+$$

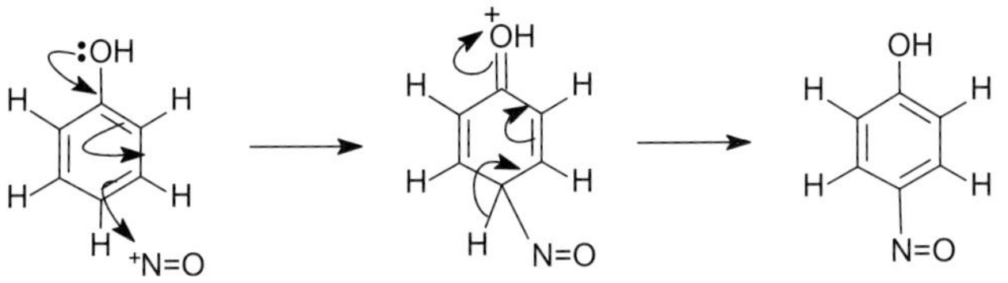

Scheme 8.37. Representation of a pathway that accounts for the *para*-nitrosation of phenol.

reactions of phenol relative to benzene (a factor of about 10^3 for the reaction with a mixture of nitric and sulfuric acids leading to replacement of a hydrogen by a nitro [NO_2] group, *vide infra*) argues that the electron-rich oxygen donates electrons to the aromatic ring so as to facilitate reaction with an incoming electrophile (E^+).

Further, to account for the multitude of observations that electrophilic substitution occurs primarily at the positions *ortho* and *para* to the hydroxyl-bearing carbon, it is argued that the electron density resulting from the participation of the electrons on oxygen must be highest at those positions. The typical resonance picture for phenol (benzenol, hydroxybenzene) is shown in Figure 8.13.

Thus, as shown in Scheme 8.37, nitrous acid (HNO_2) produced *in situ* from sodium nitrite ($NaNO_2$) and sulfuric acid (H_2SO_4) reacts with phenol in aqueous solution to produce nitrosophenol. Substitution occurs *para* to the hydroxyl.

The same pathway can be written for the nitration of phenol with a mixture of nitric ($HONO_2$) and sulfuric (H_2SO_4) acids (which serve to generate the nitronium ion [NO_2^+]), and, indeed, in dilute aqueous mixtures of these acids, mononitration occurs to produce *p*-nitrophenol. However, when concentrated acids are used and there is a large excess of the nitronium ion (NO_2^+) present, di- and trinitrophenols are formed. Picric acid (2,4,6-trinitrophenol), a material reported to decompose explosively, is formed in this way (Scheme 8.38).

Formation of the nitronium ion (NO_2^+) *in situ* from the action of sulfuric acid (H_2SO_4) on nitric acid ($HONO_2$), as shown in Scheme 8.38, can be avoided by the use of nitronium tetrafluoroborate ($NO_2^+BF_4^-$), perchlorate ($NO_2^+ClO_4^-$), acetate ($NO_2^+[CH_3CO_2^-]$), and so on, thus precluding the need for strong protic acids.

When phenol is treated with concentrated sulfuric acid (98% H_2SO_4) alone, sulfonation occurs. As shown in Scheme 8.39, if the substitution reaction is carried out

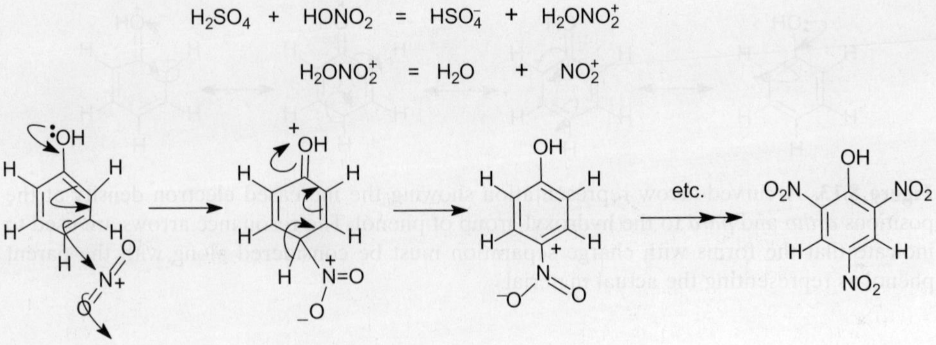

$$H_2SO_4 \; + \; HONO_2 \; = \; HSO_4^- \; + \; H_2ONO_2^+$$

$$H_2ONO_2^+ \; = \; H_2O \; + \; NO_2^+$$

Scheme 8.38. Representation of a pathway that accounts for the nitration of phenol to produce picric acid (2,4,6-trinitrophenol).

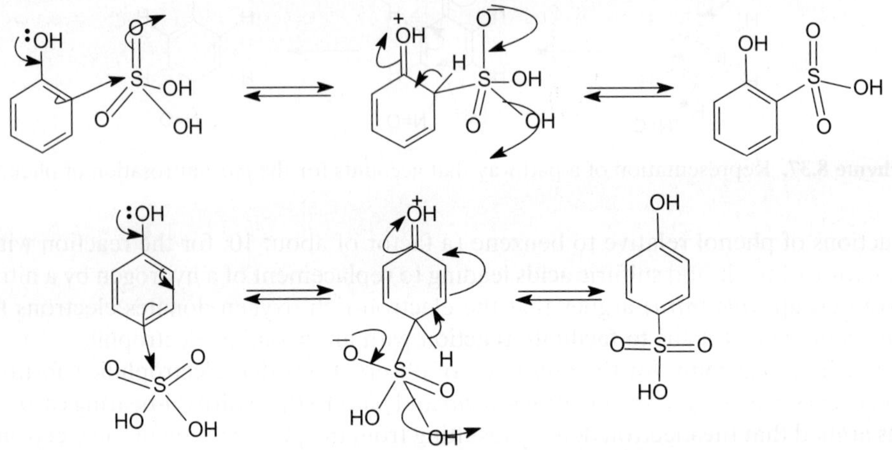

Scheme 8.39. A representation of the reversible sulfonation of phenol to provide both *ortho*- and *para*-hydroxybenzenesulfonic acids. At 20°C with 98% sulfuric acid (H_2SO_4), the ratio of *para*- to *ortho*-substitution is about 1:1. At 100°C, the ratio of *para* to *ortho* is about 9:1.

at room temperature, about a 1:1 mixture of *ortho*- and *para*-hydroxybenzenesulfonic acid is obtained. However, at about 100°C, the *para*-isomer is the major product. It is argued that the sulfonation reaction is reversible (with a proton replacing the sulfonic acid group in the reverse reaction) and that the *p*-hydroxybenzenesulfonic acid is the thermodynamically favored product.

Sulfonation to produce disulfonic acids can also be carried out if fuming sulfuric acid (sulfuric acid [H_2SO_4] containing sulfur trioxide [SO_3]) is used. The second sulfonic acid group is introduced into the unfilled *ortho*- or *para*-position.

As already noted in Scheme 8.24, substitution of hydrogen by an alkyl group can also be effected. Generally, such Friedel–Crafts alkylations with phenols require catalysis by **Lewis acids** just as they did with unsubstituted and alkyl-substituted arenes (Chapter 6). Despite the fact that the nonbonded electrons on the hydroxyl

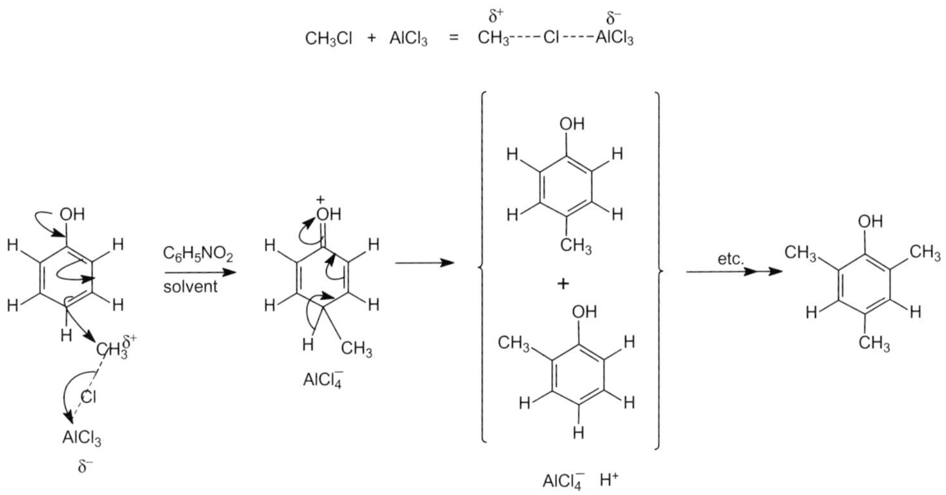

Scheme 8.40. A representation of the Friedel–Crafts alkylation of phenol by chloromethane (CH_3Cl) in nitrobenzene solvent in the presence of aluminum trichloride ($AlCl_3$).

group of the phenol are expected to complex with the Lewis acid (leading to an equilibrium between complexed and uncomplexed phenols), alkylations occur *ortho* and *para* to the hydroxyl substituent. For example, as shown in Scheme 8.40, introducing a solution of chloromethane (methyl chloride, CH_3Cl) in a solvent such as nitrobenzene (a solvent inert to Friedel–Crafts reactions, Chapter 9)* into a nitrobenzene solution of phenol containing aluminum trichloride ($AlCl_3$) results in alkylation to produce a mixture of *ortho-* and *para*-methylphenols. Further alkylation to the 2,4,6-trimethylphenol can be effected if an excess of chloromethane (CH_3Cl) is used.

As suggested by the alkylation of *p*-methylphenol to 2,4,6-trimethylphenol in Scheme 8.40, further electrophilic substitution of *p*-methylphenol occurs *ortho* to the hydroxy group (–OH) rather than *ortho* to the alkyl group. Thus, although *both* the alkyl and hydroxyl groups are *ortho/para* directing groups (i.e., they *direct further incoming substitution into positions* ortho *and* para *to the ones they currently [respectively] occupy*), the hydroxyl group does so "more powerfully" than does the alkyl group. For example, nitration (with a mixture of nitric and sulfuric acids) of *p*-methylphenol produces mainly 2-nitro-4-methylphenol rather than 3-nitro-4-methylphenol (Scheme 8.41).

Problem 8.7. Draw a series of resonance structures corresponding to those expected to stabilize the intermediate shown in Scheme 8.41.

---

*The Friedel–Crafts reaction was discussed earlier (Chapter 7) and will be encountered again. The work, modified continuously since its inception with different Lewis acids and conditions, was reported initially before the twentieth century began by Friedel, C.; Crafts, J. M. *Compt. Rend.*, **1877**, *84*, 1392 and 1450, and it was quickly expanded upon by Bœdtker, E. *Bull Soc. Chim. Fr.*, **1901**, *25*, 845.

$$H_2SO_4 + HONO_2 = HSO_4^- + H_2ONO_2^+$$

$$H_2ONO_2^+ = H_2O + NO_2^+$$

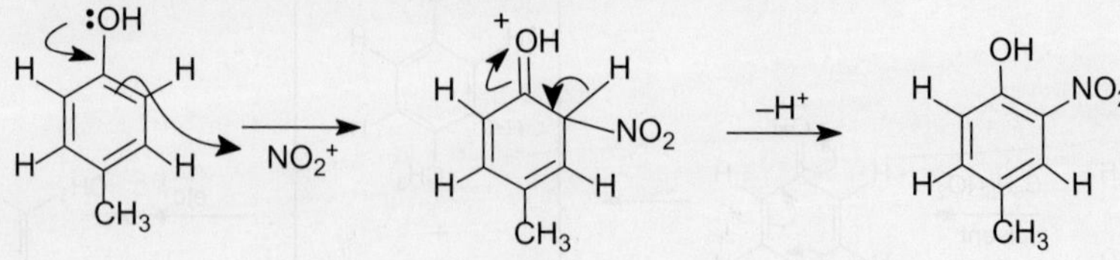

Scheme 8.41. A representation of the formation of 2-nitro-4-methylphenol by the nitration of 4-methylphenol.

Halogen substitution for hydrogen in phenols can also be effected. Halogens (Cl, Br, and I) enter *ortho* and *para* with respect to the phenol hydroxyl (–OH). Chlorination with excess chlorine (Cl_2) in the presence of iron(III) chloride (ferric chloride, $FeCl_3$) at 70–75°C produces 2,3,4,6-tetrachlorophenol (Equation 8.27).

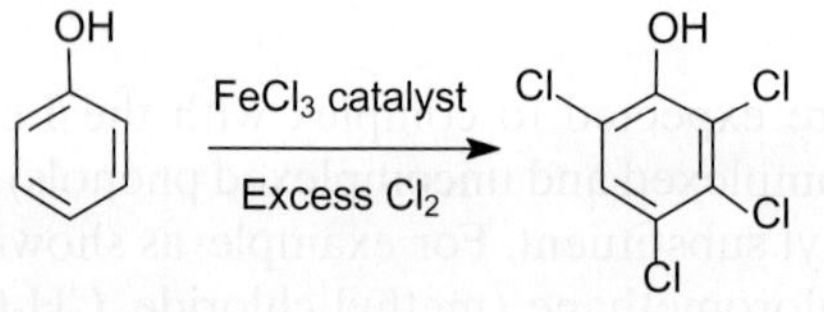

$$(8.27)$$

In aqueous solution, bromine (Br_2) reacts with phenol to provide 2,4,6-tribromophenol and isolation of intermediate 2- and 4-bromophenols cannot be effected. However, in nonpolar solvents, where proton transfer to the solvent is inhibited, substitution occurs less readily and both 2- and 4-bromophenols can be obtained and separated from each other (Equation 8.28).

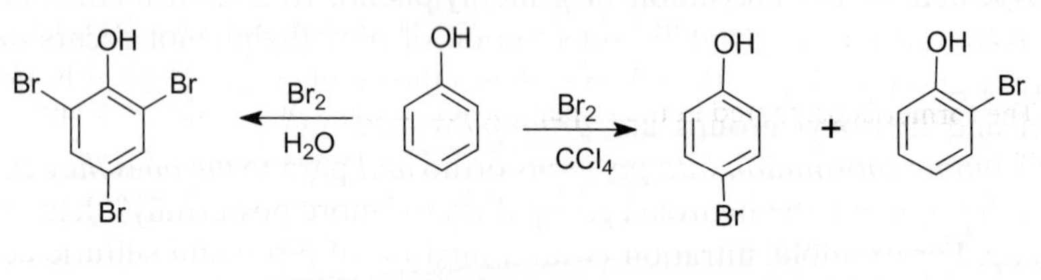

$$(8.28)$$

Problem 8.8. How would you account for the observation concerning the effect of solvent as noted in Equation 8.28?

There are a number of substitution reactions of phenols in which the hydrogen (for which substitution is being made) is replaced by carbon. The Friedel–Crafts alkylation already mentioned (Scheme 8.40 above) has an acylation counterpart, which, in some cases, is complicated by an initial acylation of the phenol *on oxygen* ("O-acylation") and a subsequent rearrangement to the ring substitution products. Indeed, in practice, initial preparation, isolation, and purification of the phenol *ester* (Chapter 9) resulting from O-acylation, followed by treatment of the ester with a Lewis acid such as aluminum trichloride ($AlCl_3$), often provides a cleaner overall

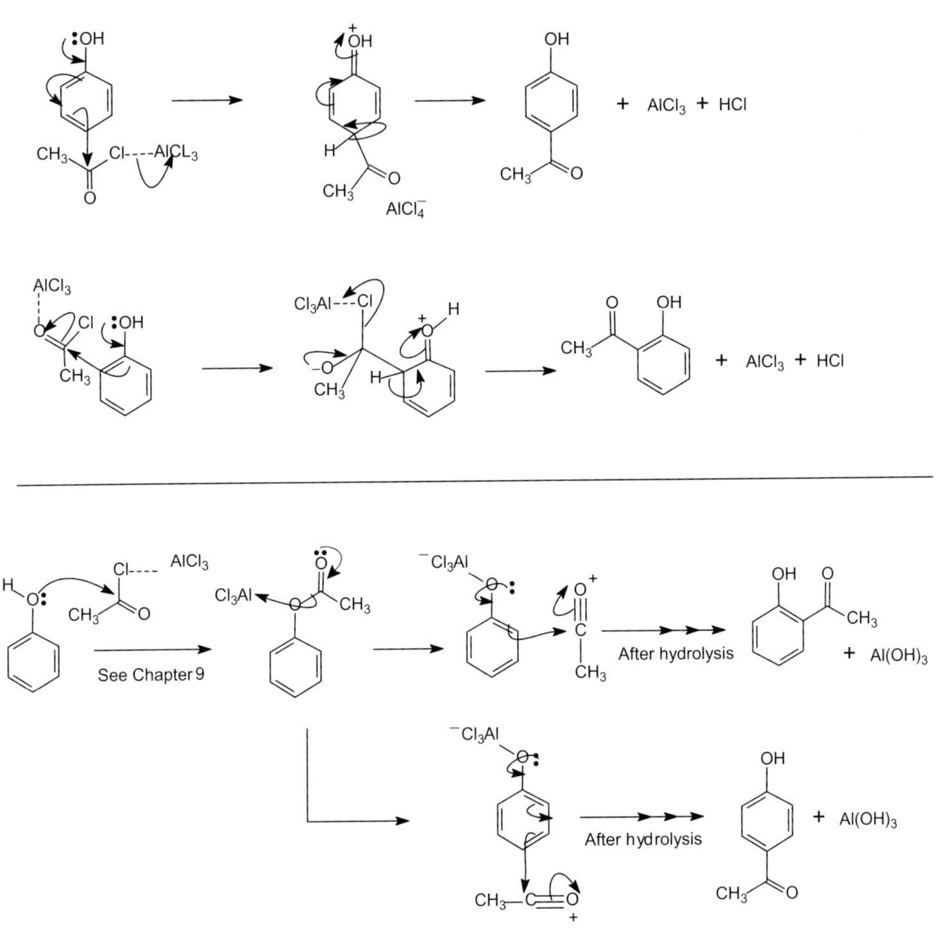

Scheme 8.42. Electrophilic aromatic substitution of phenol with an acyl group. The acylation of phenol with acetyl chloride (CH_3COCl) in the presence of aluminum trichloride ($AlCl_3$) can apparently occur via a direct addition of the aluminum trichloride complexed acetyl chloride (in competition with O-acylation) or by a subsequent rearrangement of O-acylated phenol. The former is presented in the upper portion of the scheme, while the latter is shown in the lower portion.

reaction. The conversion of the ester to the products of substitution as shown in Scheme 8.42 is called the **Fries rearrangement.***

A related process is shown in Scheme 8.43. Here, phenoxide anion is permitted to react (as a nucleophile) in a typical S_N2 (or S_N2') process with an allylic halide such as 3-bromopropene (allyl bromide). The product is phenyl allyl ether (2-propenyloxybenzene). Heating this ether above 200°C results in substitution by

*See Fries, K.; Finck, G. *Chem. Ber.*, **1908**, *41*, 4271 and Fries, K.; Pfaffendorf, W. *Chem. Ber.*, **1910**, *43*, 212.

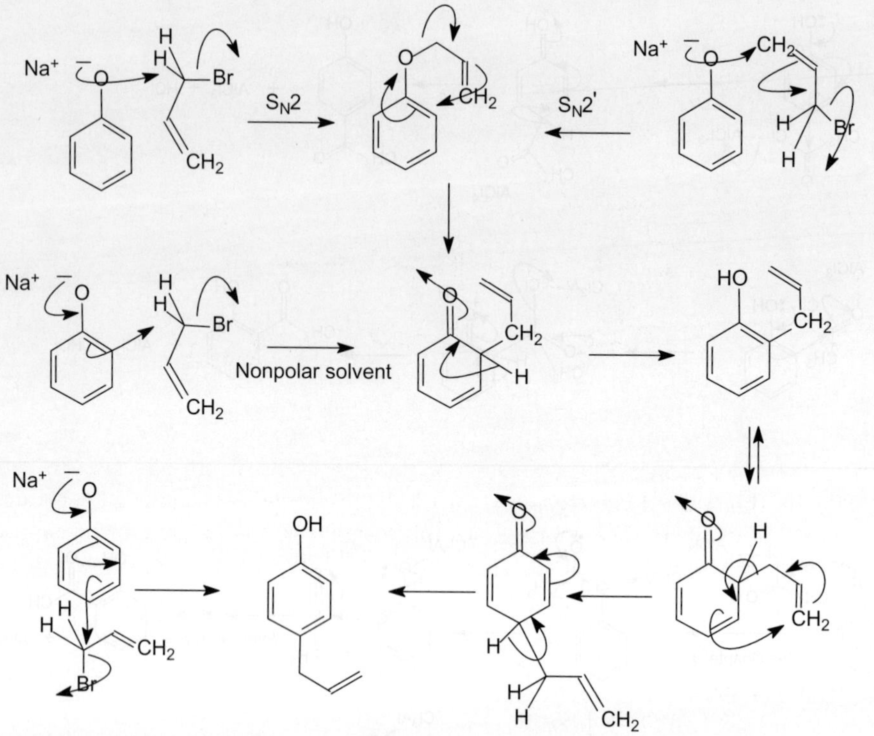

Scheme 8.43. Aromatic substitution *ortho* and *para* to the phenolic hydroxyl via the Claisen rearrangement. Two pathways are shown. The first involves nucleophilic substitution (S_N2 and/or S_N2') at the oxygen of the phenol yielding an ether, which subsequently rearranges. The second involves a direct substitution of the vinyl group with a hydrogen on the aromatic ring.

rearrangement to produce the corresponding 3-(*o*-hydroxyphenyl)-1-propene. This electrocyclic rearrangement process is an example of the **Claisen rearrangement**.[*]

As also shown in Scheme 8.43, the *same* product can be obtained in what may be a direct substitution reaction by heating the sodium phenoxide and the allyl bromide together in a nonpolar solvent (where the S_N2 process is inhibited).

Carbon–carbon bond formation as a consequence of substitution on the phenolic aromatic ring can be effected in a variety of other ways too. For example, if the sodium salt of phenol is heated at high pressure with carbon dioxide to 125–250°C, *ortho*-substitution occurs (Equation 8.29); at 250–300°C, the *para*-hydroxy benzoic acid results (Equation 8.30). The carboxylation process is known as the **Kolbe–Schmitt** reaction.[†]

---

[*]The Claisen rearrangement (which will be encountered again) was found and partially developed (there are many variants now) by L. Claisen in the early part of the last century. See Claisen, L. *Chem. Ber.*, **1912**, *45*, 3157 and Claisen, L.; Tietze, E. *Chem. Ber.*, **1925**, *58*, 275; Claisen, L.; Tietze, E. *Chem. Ber.*, **1926**, *59*, 2344.

[†]See Kolbe, J. *J. Prakt. Chem.*, **1875**, *120*, 151 and Schmidt, R., *J. Prakt. Chem.*, **1885**, *31*, 397.

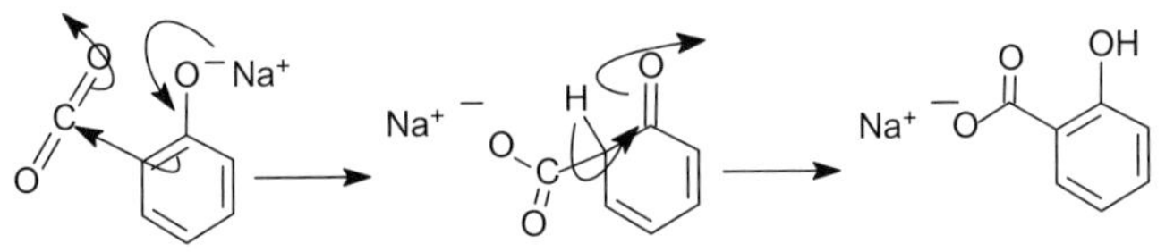

(8.29)

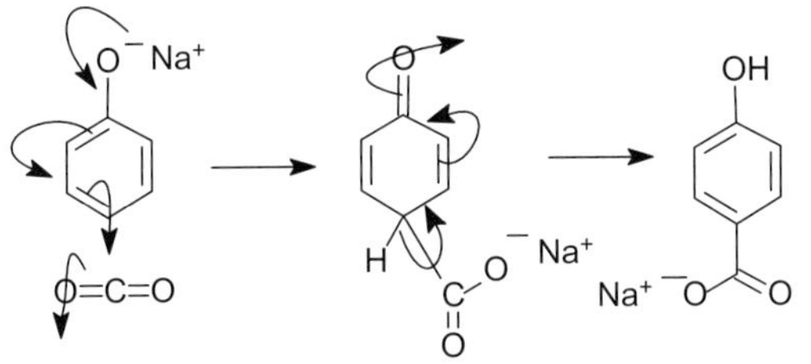

(8.30)

When phenol is carefully treated under basic conditions with formaldehyde (H_2CO), substitution occurs at positions *ortho* and *para* to the phenolic hydroxyl (–OH). In this way, hydroxybenzyl alcohols (Scheme 8.44) can be produced. Acidification results in the generation of the respective stabilized benzylic carbocations (an **E1 reaction**, Chapter 7), which can then react with unreacted phenol and/

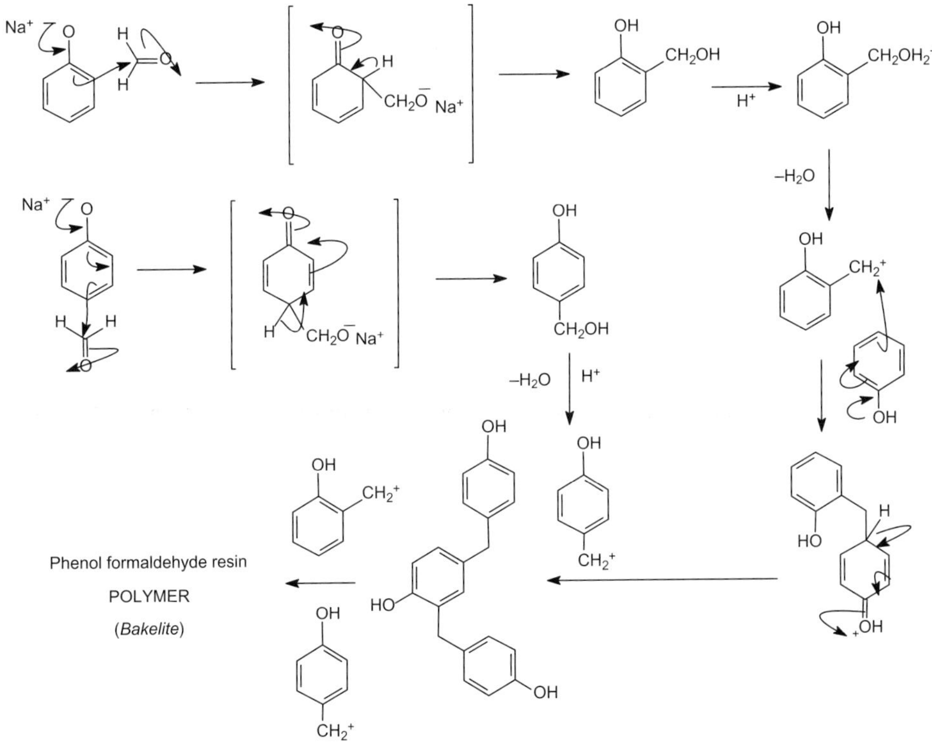

Scheme 8.44. The reaction of formaldehyde with the phenoxide anion to produce *ortho*- and *para*-hydroxybenzyl alcohols. Acidification leads to benzylic carbocations (an E1 elimination), which react with more formaldehyde, phenols, and substituted phenols to produce a formaldehyde–phenol resin-type polymer.

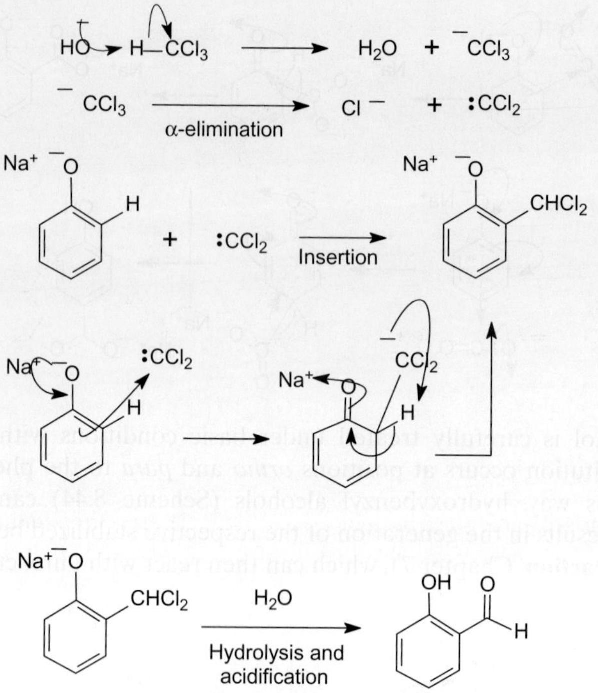

Scheme 8.45. An example of the Reimer–Tieman reaction. Dichlorocarbene is known to be generated from the reaction of trichloromethane (chloroform, $HCCl_3$) with, for example, aqueous sodium hydroxide. Introduction of sodium phenoxide solution results in the formation of *ortho*-hydroxybenzaldehyde. Substitution via either carbene insertion or electrophilic addition results.

or other intermediates to produce a heavily cross-linked phenol-form aldehyde resin polymer called *Bakelite*.

Substitution of a **carbene** with a hydrogen can also occur. In the **Reimer–Tieman** reaction,* sodium phenoxide is allowed to react with dichlorocarbene, the latter generated *in situ* from the reaction of chloroform ($HCCl_3$) with base. Thus, as shown in Scheme 8.45, chloroform ($HCCl_3$) loses a proton to the base to form the trichloromethyl carbanion, which, on α-elimination of chloride anion, produces dichlorocarbene. If the phenoxide anion is present when the carbene is generated in this fashion, the carbene inserts into the *ortho*-carbon–hydrogen bond (or the electrophilic carbene is attacked by the aromatic ring and there is subsequent proton migration). The corresponding *ortho*-dichloromethylphenol is formed and this produces *o*-hydroxybenzaldehyde on hydrolysis.

As already anticipated by the acylation of phenol in the presence of Lewis acid catalysts (Scheme 8.42, *vide supra*) and the substitution with formaldehyde (Scheme 8.44), other similar substitution reactions can be brought about. As will be seen later (Chapter 9) in the study of carbonyl compounds, the carbon of the carbonyl group

*Reimer, K. *Chem. Ber.*, **1876**, *9*, 423; Reimer, K.; Tiemann, F., *Chem. Ber.*, **1876**, *9*, 824; Reimer, K.; Tiemann, F., *Chem. Ber.*, **1876**, *9*, 1285.

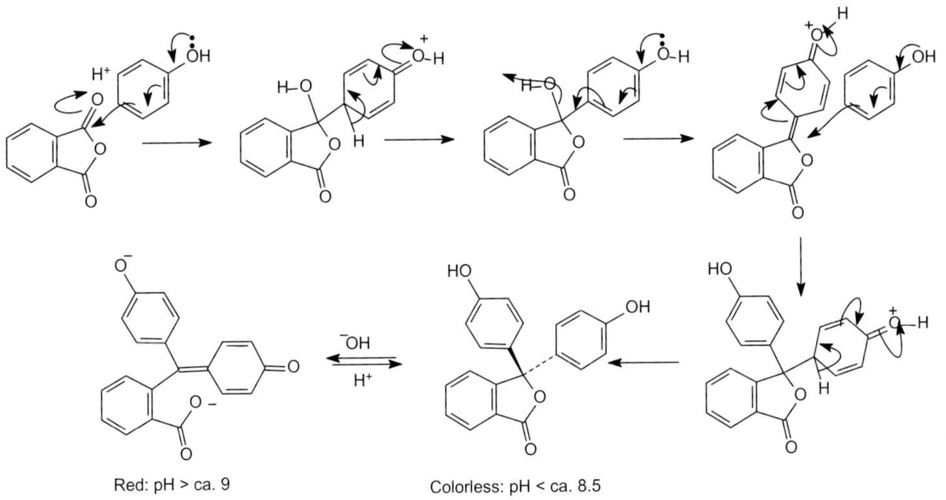

Red: pH > ca. 9   Colorless: pH < ca. 8.5

Scheme 8.46. An accounting for the reaction of phthalic anhydride with 2 equivalents of phenol to produce the acid–base indicator phenolphthalein. Acid catalysis is necessary for the substitution reactions shown.

is the positive end of a dipole (while the oxygen is the negative end), and the reactivity encountered above is typical of such functionality. In the same vein and as shown in Scheme 8.46, when phenol is treated with the anhydride of phthalic acid (i.e., phthalic anhydride) in the presence of zinc chloride ($ZnCl_2$) or sulfuric acid (H_2SO_4), 2 equivalents of phenol react at *one* carbonyl group to produce an internal ester (a lactone) called "phenolphthalein." Phenolphthalein is widely used in chemical laboratories as an acid–base indicator.

Problem 8.9. Write arrows to account for the reversible transformation between the forms of phenophthalein shown as the pH is changed from base to acid (and vice versa).

The substitution reactions shown are, of course, not limited to phenol itself. If 1,3-dihydroxybenzene (a **bis**phenol, resorcinol) is treated under the same conditions as phenol with phthalic anhydride, the intensely yellow green fluorescent dye "fluorescein" is obtained (Equation 8.31).

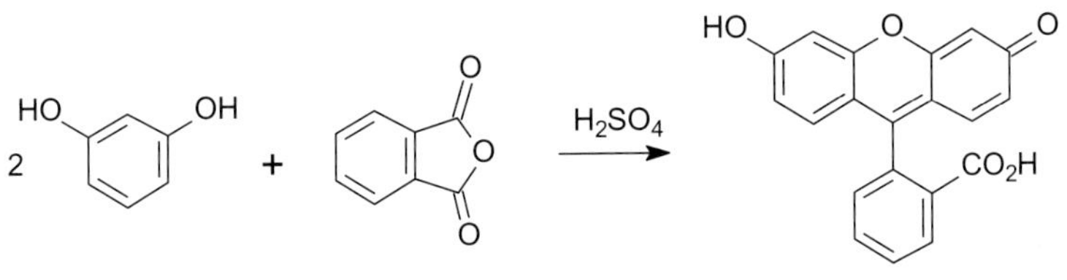

$$\text{(8.31)}$$

Problem 8.10. Write a scheme, such as that shown in Scheme 8.46, which accounts for the formation of fluorescein from phthalic anhydride and 2 equivalents of resorcinol.

Similarly, naphthols and other phenolic aromatics also undergo substitution with the same set of electrophilic reagents as phenol and simple substituted phenols. Thus, as shown in Equation 8.32, 1-naphthol (α-naphthol) reacts with the potent electrophilic benzenediazonium ion (generated from aminobenzene [aniline] and sodium nitrite [$NaNO_2$] in acid solution and is written here as the tetrafluoroborate [BF_4^-] salt) to produce the corresponding brightly colored phenyldiazonaphthalenol, in which substitution has occurred *para* to the hydroxyl.

$$(8.32)$$

E. ADDITION REACTIONS OF ALCOHOLS, ENOLS, AND PHENOLS

I. Introduction

One of the major questions to be faced in the reactions of organic compounds is formally encountered here for the first time.

When considering the interaction between two (or more) organic reagents, the point of view as to which is the reactant and which is the substrate can become clouded. Generally, as the astute observer will have noted, it is the limiting reactant (reagent) on which reactions are carried out that is considered important. The idea of the limiting reagent serving as the substrate derives from the observation that the reactant present in limited quantity is the more valuable and is being acted upon by other reagents (reactants), which are only employed to perform the desired transformation. The transforming reactant, having been used, can be recycled, discarded or, if need be, replaced by other suitable reagents. Thus, for example, in the oxidation of alcohols (this chapter, Part I, B), the focus is on the conversion of the alcohol to an aldehyde, ketone, and so on, and it makes little difference *in theory* which oxidizing agent is employed. After the substrate alcohol is oxidized, the (now reduced) oxidizing agent is simply separated by some suitable means and the oxidized product used to the end it is intended.*

The problem is particularly acute in the subject of addition reactions of alcohols, enols and phenols; that is, there are two kinds of processes that might be imagined. First, there is the *addition **of** the electron-deficient reagent **to** the electron-rich oxygen of the alcohol, enol, or phenol*. Enols also undergo addition reactions with reagents that **add** to the carbon–carbon double bond.

---

*In practice, of course, the use of a particular oxidizing agent may be critical to the overall outcome of a process. While the issue has not been examined in major detail here, it should also be clear that in the event that *more than one functional group is present in the substrate*, the choice of a reagent for a particular process will depend upon the relative reactivities of the various functional groups and their interactions. It is this interplay between *specific* reactants that lends much interest to the art of organic synthesis.

Second, there is *addition of the electron-rich oxygen of the alcohol, enol, or phenol to an electron-deficient center*. Enols can also react by the addition at a carbon to which the oxygen is conjugated (i.e., the α-carbon).

Since the protocol for diagramatically representing reactions with curved arrows shows bond formation by the use of the electron pair of the electron-rich species attacking the corresponding electron-poor species, all of the reactions shown here are of the second type.

II. Addition of the Oxygen of Alcohols to Carbon (with Loss of Hydrogen)

Addition reactions of alcohols to alkenes and alkynes have already been provided (Chapter 6) and a few more are shown in Table 8.6. Further, although many more addition reactions will be encountered later in the text (when the functional group with which the alcohol, enol, or phenol reacts is discussed in more detail), an introduction to some of the more common reactions is provided in Table 8.6. The table presents some of the partners of addition reactions with alcohols, enols, and phenols and the products they form together. The reactions shown are typical and, as such, are briefly discussed.

In the first example of Table 8.6, a proton is transfered from some mineral acid to a typical alcohol (ethanol, CH_3CH_2OH). It will be recalled (from earlier in this chapter) that alcohols are, generally, somewhat less acidic than water (Table 8.4) and that they become weaker acids as more substituents are added to the carbon bearing the hydroxyl group. Conversely, basicity increases and the more highly substituted alcohols are more easily protonated. The basicity of alcohols toward Lewis acids such as boron trifluoride (BF_3) and aluminum chloride ($AlCl_3$) (Table 8.6, examples 2 and 3) results, eventually, in the replacement of all of the halogens attached to, respectively, boron and aluminum. The replacement is stepwise and the compounds that eventually result are esters of boric acid (trimethylborate, $[B(OCH_3)_3]$) and (amphoteric) aluminum hydroxide (aluminum isopropoxide $[Al(O\text{-}iPr)_3]$).

Problem 8.11. Using the curved-arrow approach and remembering that there are nonbonded electrons on oxygen and a "vacant" *p*-type orbital on boron, show how boron trifluoride might be converted to trimethyl borate and HF.

The fourth item in the table, phenol (hydroxybenzene), is alkylated on oxygen, forming an ether, methoxybenzene (anisol), with the powerful alkylating agent trimethyloxonium tetrafluoroborate $[(CH_3O)_3{}^+ BF_4{}^-]$. Other alkoxonium tetrafluoroborates are also commercially available and can be used to the same end with phenols, enols, and alcohols, forming aryl ethers, enol ethers, and dialkyl ethers, respectively. In contrast to dialkyl, diaryl, and aralkyl ethers, which are quite inert and are often used as solvents, enol ethers are capable of acid-catalyzed hydrolysis to produce ketones (or their equivalent enol) and the alcohol from which the enol ether is formed (Scheme 8.47).

Oxonium tetrafluoroborates are powerful alkylating agents that are also used to alkylate the carbonyl oxygen of esters and other derivatives of carboxylic acids

TABLE 8.6. Some Selected Addition Reactions of Alcohols, Enols, and Phenols

	Substrate (alcohol, enol, or phenol)	Reagent added to the substrate	Result of the addition of a reagent to substrate alcohol
1.	CH_3CH_2OH	"H^+"(X^-)	$CH_3CH_2\overset{+}{O}H_2 \ X^-$
2.	CH_3OH	BF_3	$\left[CH_3\overset{+}{O}(H)BF_3^- \right] \longrightarrow B(OCH_3)_3 + HF$
3.	$(CH_3)_2CH-OH$	$AlCl_3$	$\left[(CH_3)_2CH\overset{+}{O}(H)AlCl_3^- \right] \longrightarrow Al[OCH(CH_3)_2]_3 + HCl$
4.	phenol	$CH_3\overset{+}{O}(CH_3)CH_3 \ BF_4^-$	anisole (OCH_3) $+ CH_3OCH_3 + HBF_4$
5.	CH_3CH_2OH	$(CH_3)_3C$–Si$(CH_3)_2$–Cl	$(CH_3)_3C$–Si$(CH_3)_2$–$OCH_2CH_3 + HCl$
6.	cyclohexenone enolate $O^- Na^+$	$(CH_3)_3SiCl$	cyclohexenone silyl enol ether $OSi(CH_3)_3 + HCl$
7.	$CH_2=C(CH_3)O^- Na^+$	$NaOCl$ (excess)	$CH_3CO_2^- Na^+ + CHCl_3$
8.	$CH_2=C(OH)O$–PBr_2	Br_2/red phosphorus	$BrCH_2C(=O)OH$
9.	CH_3CH_2OH/H^+	$CH_2=CH_2$	$(CH_3CH_2)_2O$
10.	CH_3OH/OH^-	$CH_3-C\equiv C-CH_3$	$CH_3{\sim}C=C(CH_3)OCH_3$
11.	CH_3OH	benzaldehyde (C_6H_5CHO)	$C_6H_5CH(OCH_3)(OH) \longrightarrow C_6H_5CH(OCH_3)_2$
12.	$CH_2(OH)$–$CH_2(OH)$	cyclohexanone	hemiketal HO–O–$OH \longrightarrow$ cyclic ketal (dioxolane)
13.	open-chain aldose	—	cyclic pyranose forms (α and β anomers)

(Continued)

TABLE 8.6. *Continued*

	Substrate (alcohol, enol, or phenol)	Reagent added to the substrate	Result of the addition of a reagent to substrate alcohol
4.	HO–(cyclohexyl)	CH₃CH₂C(=O)Cl	cyclohexyl–O–C(=O)CH₂CH₃ + HCl
5.	HO–(cyclohexyl)	CH₃CH₂C(=O)OH, Mineral acid catalysis	cyclohexyl–O–C(=O)CH₂CH₃ + H₂O
6.	HO–(cyclohexyl)	CH₃C(=O)–O–C(=O)CH₃	cyclohexyl–O–C(=O)CH₃ + CH₃CO₂H
7.	HO–(cyclohexyl)	CH₃C(=O)–O–CH₃	cyclohexyl–O–C(=O)CH₃ + CH₃OH
8.	HO–(cyclohexyl)	CH₂=C=O	cyclohexyl–O–C(=O)CH₃
9.	HO–(cyclohexyl)	C₆H₅–N=C=O	C₆H₅–N(H)–C(=O)–O–cyclohexyl
20.	CH₃OH	dihydropyran, Acid catalysis	2-methoxytetrahydropyran (O–OCH₃)
21.	HO–(cyclohexyl)	ROCH₂Cl [R = CH₃] [R = C₆H₅CH₂]	ROCH₂O–cyclohexyl
22.	CH₃OH	propylene oxide (epoxide), Acid catalysis	CH₃CH(OH)CH₂OCH₃ + CH₃CH(OCH₃)CH₂OH
23.	(cyclohexenyl)–O⁻ Li⁺	CH₃I	2-methylcyclohexanone (CH₃) + Li⁺ I⁻
24.	cyclohexanone ⇌ HO–(cyclohexenyl)	CH₂=CH–C(=O)CH₃	octahydronaphthalenone + H₂O
25.	CH₃OH	SOCl₂	CH₃O–S(=O)–OCH₃ + HCl
26.	CH₃OH	POCl₃	(CH₃O)₂P(=O)–OCH₃ (CH₃O–P(=O)(OCH₃)OCH₃) + HCl

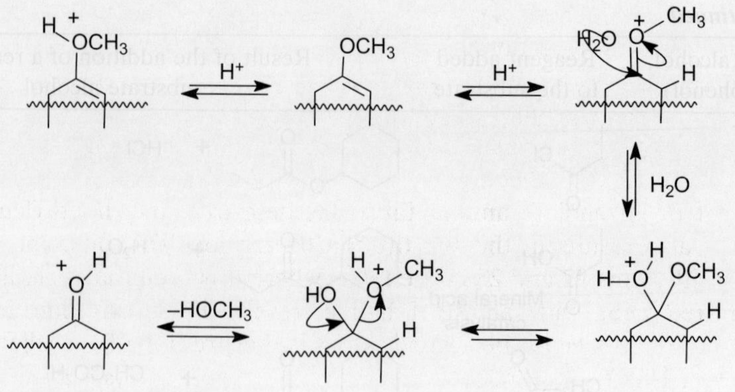

Scheme 8.47. A pathway for the acid-catalyzed hydrolysis of enol ethers.

(Chapter 9) in order to increase their reactivity. We will also encounter the use of oxonium salts in other circumstances where alkylation is required.*

The protection of hydroxyl groups against change in the event functional groups elsewhere in a molecule are undergoing reaction might be accomplished by their conversion to alkyl ethers as noted above. However, except for particular cases where their need is warranted, such ethers are avoided and hydroxyl protection is accomplished instead as shown in the fifth item of the Table 8.6. Here, treatment of ethanol (CH_3CH_2OH) with the silylating reagent, *tert*-butyldimethylchlorosilane [$(CH_3)_3C(CH_3)_2SiCl$], in the presence of a base (such as triethylamine [$(CH_3CH_2)_3N$]) to absorb the hydrogen chloride (HCl) produces an alkyl silyl ether [$R–O–Si(R')_3$]. The advantage of silylation (to produce silyl ethers, $R–O–Si(R')_3$) over alkylation (to produce alkyl ethers, $R–O–R'$) lies in the ease of removal of the silyl group. Thus, while it is generally true that liberation of the "masked" hydroxyl can be reversed

***Alkylation at Oxygen: a Biological Example. The Use of a Sulfonium Salt.**

As noted, oxonium salts are commonly used in the laboratory to effect the addition of an alkyl group onto oxygen. In biological systems, methylation can be effected using an 'onium salt, and it has been shown for many cases that the methylating species is a sulfonium salt, *S-adenosylmethionine* (**SAM**).

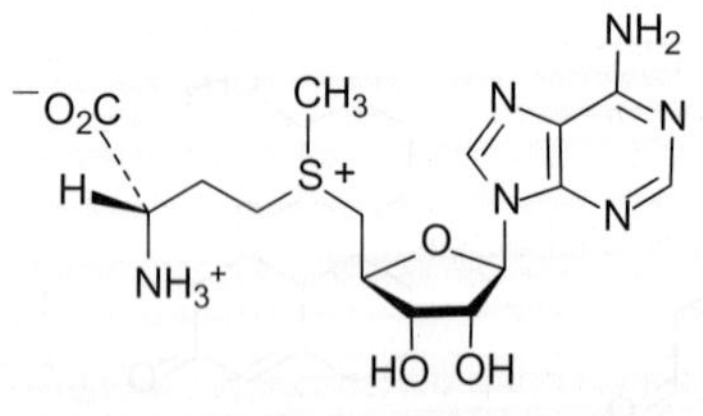

S-adenosylmethionine (SAM)

For example (one of many possible), hydroxylated cinnamic acids are ubiquitous in nature (largely in plants). Frequently, the hydroxyl groups are methylated. Investigations have shown that the process by which the methylation occurs here, as in many other cases, involves the transfer of the methyl group from SAM to the phenol. The process is shown to be, essentially, an S_N2 substitution by oxygen of the phenol on the methyl group of SAM with inversion of configuration at the methyl. The experiments were

(Continued on next page)

only with difficulty when alkylated,* such regeneration of the hydroxyl function is readily accomplished by treatment of the silyl ether with a source of fluoride ion.[†]

In addition to ease of removal after serving to protect a hydroxyl group from unwanted reaction, placement of trialkylsilyl groups on oxygen serves the added function of decreasing the polarity of the (former) hydroxyl group. This is a major aid in chromatography of alcohols.

Treatment of an enolate anion (Table 8.6, item 6) with trimethylchlorosilane $[(CH_3)_3SiCl]$ also results in the substitution on oxygen. This outcome should be contrasted with items 22 and 23 in the table, where it is seen that a similar enolate anion, on treatment with an alkyl halide, results in substitution on carbon. Interestingly, not only can it be argued that it is the strength of the silicon–oxygen bond ($D° = 88.2\,kcal^{-1}$ [$369.0\,kJ\,mol^{-1}$]) relative to the strength of the silicon–carbon bond ($D° = 69.3\,kcal^{-1}$ [$289.9\,kJ\,mol^{-1}$]) that helps direct this choice but there also appears to be a role played by the cation with which the enolate anion is associated.

The seventh item in Table 8.6 is the sodium salt of the enolate ion of acetone. When any methyl ketone (i.e., a ketone of the form $R–CO–CH_3$) or any secondary alcohol that can be oxidized to a methyl ketone (i.e., an alcohol of the form $R-CH(OH)CH_3$) is treated with excess sodium hypochlorite (or any hypohalite), the **haloform** reaction occurs, resulting in the loss of the methyl group as chloroform (trichloromethane, $CHCl_3$) (or any haloform) and the formation of the salt of the

carried out using SAM, in which two of the hydrogens on the methyl were replaced by heavier isotopes of protium, that is, ^1H, ^2H, and ^3H substituted methyl. The details of how this was done and the outcome became known will be elaborated upon in Chapter 13.

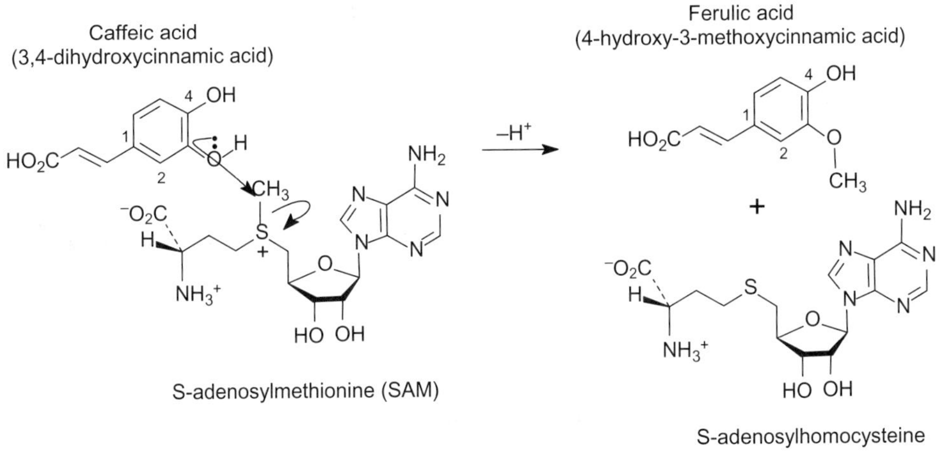

Caffeic acid
(3,4-dihydroxycinnamic acid)

Ferulic acid
(4-hydroxy-3-methoxycinnamic acid)

S-adenosylmethionine (SAM)

S-adenosylhomocysteine

*This clearly excludes alkylating agents that yield acetals or other reactive functionalities. The use of alkylating agents such as chloromethyl methyl ether ($ClCH_2OCH_3$) or chloromethyl benzyl ether ($ClCH_2OCH_2C_6H_5$), which produce acetals similar to that from tetrahydropyran (Table 8.6, item 20), is also provided in Table 8.6 (item 21).

[†]The ready cleavage of the silicon–oxygen bond ($D° = 88.2\,kcal^{-1}$ [$369.0\,kJ\,mol^{-1}$]) (only slightly stronger than the carbon–oxygen bond [$D° = 84.0\,kcal^{-1}$ {$351.4\,kJ\,mol^{-1}$}]) can be attributed to the extraordinary strength of the silicon–fluorine bond ($D° = 129.3\,kcal^{-1}$ [$540.0\,kJ\,mol^{-1}$]) relative to the carbon–fluorine bond ($D° = 105.4\,kcal^{-1}$ [$441.0\,kJ\,mol^{-1}$]) The values in this note are from Pauling, L. *Nature of the Chemical Bond*, 3rd edition, Cornell University Press, Ithaca, NY, **1960**. The value given for the carbon–fluorine bond is within the error limits expressed in Chapter 1 and in Table 1.1.

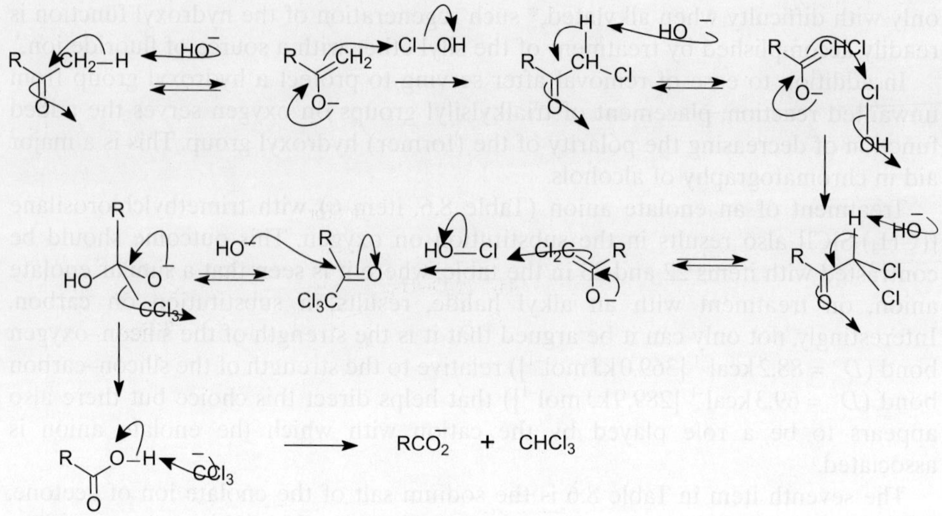

Scheme 8.48. A representation of the pathway for the formation of chloroform and a carboxylic acid from the treatment of a methyl ketone with basic hypochlorite solution (the haloform reaction).

corresponding carboxylic acid as the other product. It is clear that this process occurs through the enolate ion, which undergoes reaction with positive halogen at the terminus. Repetition of the process three times yields the corresponding trihalomethyl ketone, which then undergoes carbon–carbon bond cleavage after the carbonyl of that ketone is attacked by the base. The resulting trihalomethyl carbanion is protonated from the aqueous solvent and the trihalomethane separates from water. The use of alkaline hypochlorite with a dilute aqueous solution of enolizable ketone containing iodide (I^-) results in the generation of iodine and, presumably, some hypoiodite (OI^-), which produces iodoform (CHI_3), a yellow crystalline solid (mp ca. 120°C). The process was originally widely used to determine the presence of enolizable methyl ketones or secondary alcohols that might be oxidized to such compounds, but it has largely been replaced as a diagnostic tool for this functional group by ^1H and ^{13}C nuclear magnetic resonance (NMR) spectroscopies. A schematic representation of what is believed to occur is shown in Scheme 8.48. Interestingly, and in accord with the scheme and the concept that the enol is required for the reaction, the rate of formation of trihalomethane is *independent* of which halogen (Cl, Br, or I) is used and is *dependent* on the rate of enolization.

For many years, the reaction between carboxylic acids and bromine in the presence of red phosphorus was known to produce the corresponding α-bromocarboxylic acid. The process (the **Hell–Volhard–Zelinsky** reaction)* can also be effected with phosphorus trihalides too, and so it is believed that bromine first reacts with phosphorus to produce phosphorus tribromide (PBr_3), and the latter converts

*Hell, C. *Chem. Ber.*, **1881**, *14*, 891; Vollhard, J. *Liebigs Ann. Chem.*, **1887**, *242*, 141; Zelinsky, N. *Chem. Ber.*, **1887**, *20*, 2026.

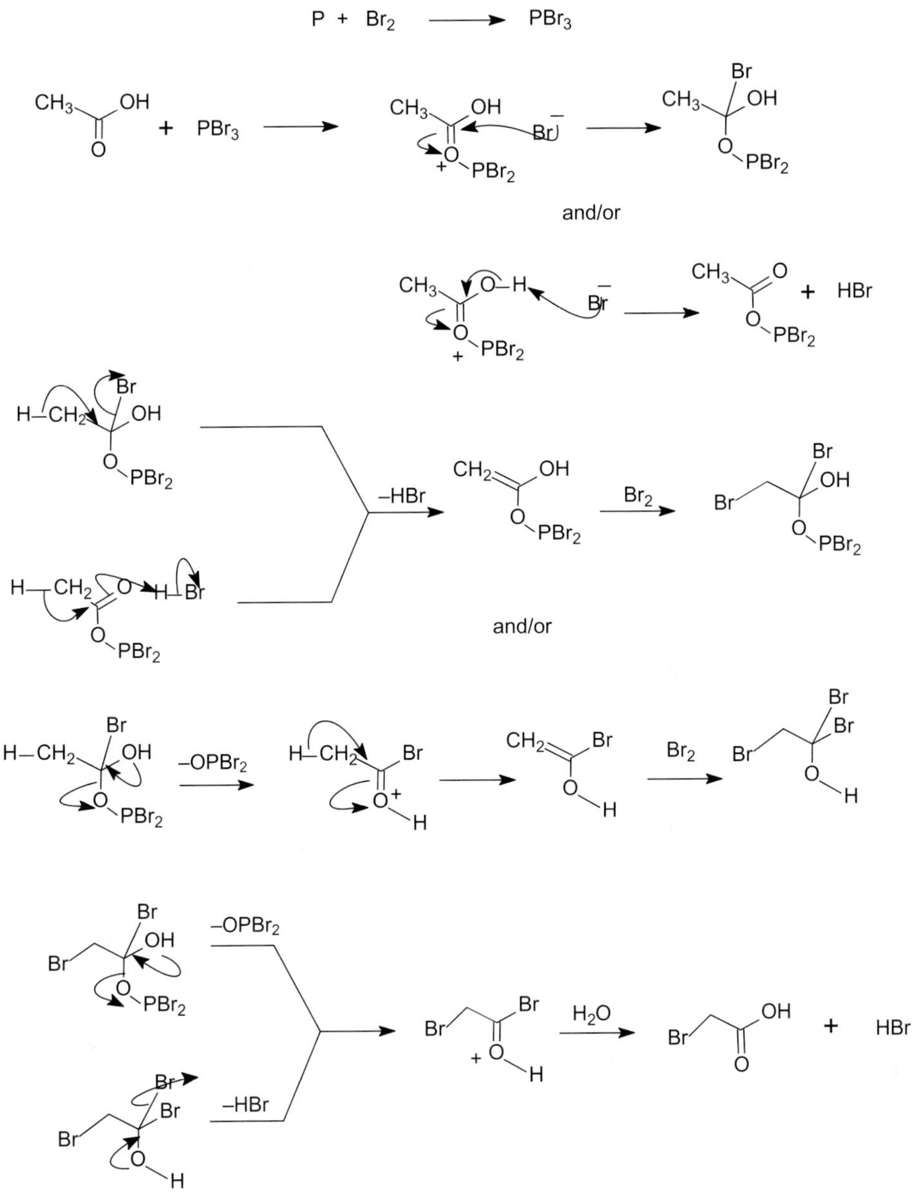

Scheme 8.49. A variety of possibilities to account for the formation of α-bromoacetic acid (bromoethanoic acid [$BrCH_2CO_2H$]) from acetic acid (ethanoic acid [CH_3CO_2H]) with red phosphorus and bromine (the Hell–Volhard–Zelinsky reaction).

the carboxylic acid to the corresponding acid bromide. Then, as shown in Scheme 8.49, enolization of the acid bromide is followed by the reaction with bromine to introduce the halogen α to the carbon-bearing oxygen. Reformation of the carbonyl group is accompanied by loss of halogen. Aqueous workup of the acid bromide results in hydrolysis to produce the α-bromo-carboxylic acid and hydrogen bromide (HBr).

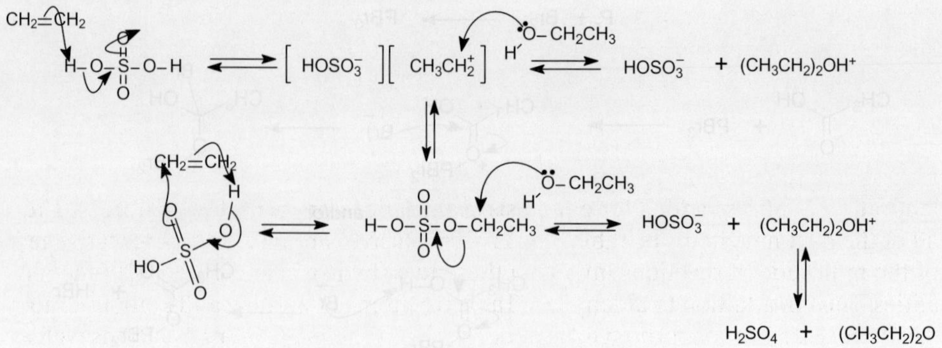

Scheme 8.50. Pathways to diethyl ether (ethyl ether [$(CH_3CH_2)_2O$]) by the addition of ethanol (ethyl alcohol, CH_3CH_2OH) to ethene (ethylene, $CH_2=CH_2$) in the presence of sulfuric acid (H_2SO_4).

As already noted in Chapter 6 (Table 6.1, item 3), alcohols undergo addition to carbocations formed by the reaction of alkenes with protic acids. Alcohols also react with the adducts formed by the addition of protic acids to alkenes (i.e., via S_N1 and S_N2 reactions). Thus, reaction of, for example, ethanol (ethyl alcohol, CH_3CH_2OH) with ethene (ethylene, $CH_2=CH_2$) (Table 8.6, item 9, and Scheme 8.50) results in the formation of ethyl hydrogen sulfate ($CH_3CH_2OSO_3H$) and, eventually, diethyl ether (ethyl ether [$(CH_3CH_2)_2O$]). Presumably, the process begins with protonation of the alkene to generate, however transiently, a (complexed) carbocation. The carbocation can react with ethanol (ethyl alcohol, CH_3CH_2OH) to produce the ether (ethyl ether [$(CH_3CH_2)_2O$]) and/or with the hydrogen sulfate anion (HSO_4^-) to produce ethyl hydrogen sulfate ($CH_3CH_2OSO_3H$). Direct formation of the latter (avoiding carbocations) is also shown in Scheme 8.50 and, however formed, the ethyl hydrogen sulfate ($CH_3CH_2OSO_3H$) also leads to diethyl ether (ethyl ether [$(CH_3CH_2O)_2$]) on reaction with ethanol (CH_3CH_2OH). Interestingly, alcohols add to alkynes (Table 8.6, item 10) more readily (albeit at high temperature) under basic conditions than they do under acidic conditions (Chapter 6, Table 6.8). It will be recalled that protonation of alkynes is generally more sluggish than protonation of alkenes since a high-energy vinyl cation [$RHC=CR^+$] must form in the former case. In contrast, the addition of alkoxide in the former case, while requiring sufficient energy (in the form of heat) to overcome the initial energy barrier bringing the reactants together, finally produces an enol ether (after proton transfer), in which a double bond remains and a new carbon–oxygen bond has been forged (Equation 8.33).

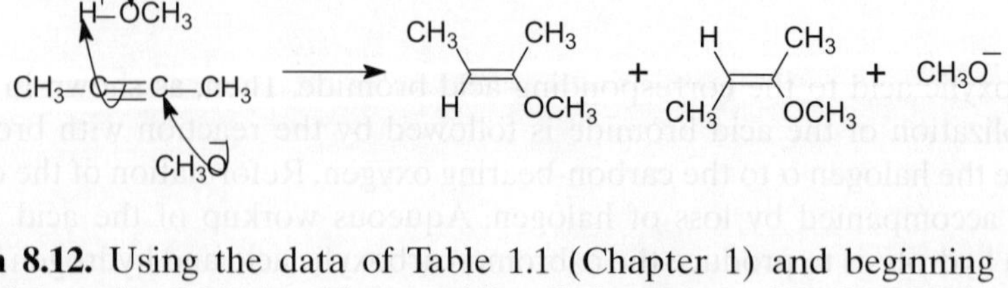

$$CH_3-C\equiv C-CH_3 \longrightarrow \; \text{(enol ethers)} \; + \; CH_3O^- \quad (8.33)$$

Problem 8.12. Using the data of Table 1.1 (Chapter 1) and beginning with the alkyne of Equation 8.33, viz., 2-butyne (dimethylacetylene, $CH_3C\equiv CCH_3$), show how it is possible to account for the (overall) favorable addition of methoxide (CH_3O^-) and subsequent protonation. Apply the same method to the (hypothetical) addition of methoxide (CH_3O^-) to either (Z)- or (E)-2-butene.

The addition of alcohols to the carbon of variously functionalized carbonyl ($C=O$) groups is among the most common reactions in nature as well as in the chemistry laboratory (a phrase added for emphasis as the laboratory is part of "nature"). Thus, all of the next nine items of Table 8.6 (11–19 inclusive) are only a representative few of the multitude of reactions involving these two (hydroxyl and carbonyl) functionalities. Thus, in addition to aldehydes (Table 8.6, items 11 and 13) and ketones (Table 8.6, item 12), a host of carboxylic acid derivatives (Table 8.6, items 14–18) as well as one nitrogen reactant (Table 8.6, item 19) are represented. In all of these cases, the general principle of attack on the positive carbon of the carbonyl by a nonbonded electron pair on the oxygen of the alcohol leads the process. In the case of aldehydes (Table 8.6, items 11 and 13) and ketones (Table 8.6, item 12) as well as carboxylic acids (Table 8.6, item 15) and esters (Table 8.6, item 17), the positive nature of the carbon of the carbonyl is enhanced by mineral acid catalysis, which serves (in a rapidly established equilibrium) to protonate the oxygen of the carbonyl group, thus raising the energy of the ground state. In contrast, acid chlorides (Table 8.6, item 14), anhydrides (Table 8.6, item 16), ketene (Table 8.6, item 18), and isocyanates (Table 8.6, item 19), the substrate with which the alcohol is reacting, are high enough in energy to preclude the necessity for initial protonation.

In the acid-catalyzed reaction of methanol (CH_3OH) with benzenecarbaldehyde (benzaldehyde, C_6H_5CHO) (Table 8.6, item 11), typical of reactions of alcohols with aldehydes, the first step (Scheme 8.51) is equilibrium protonation of the aldehyde.

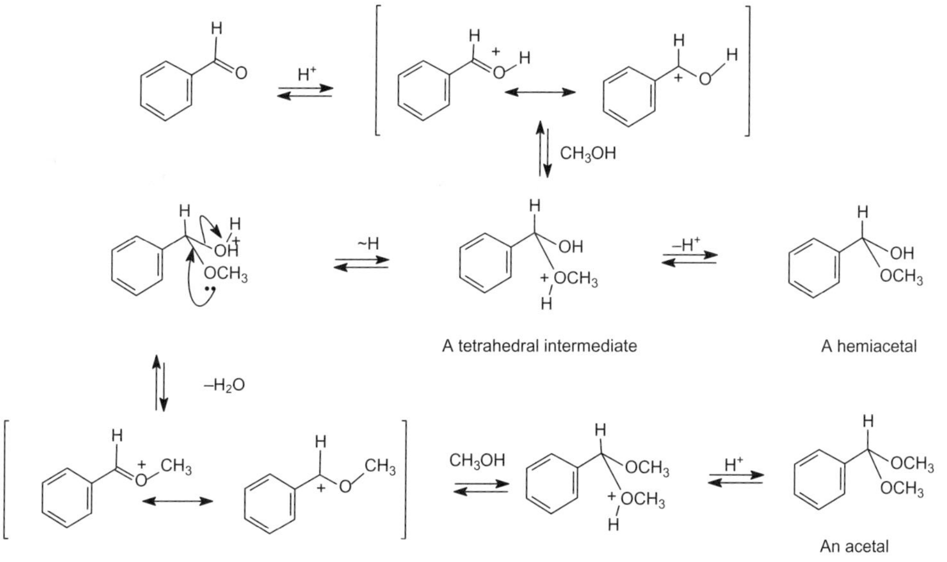

Scheme 8.51. The formation of, first, a hemiacetal and, second, an acetal on the acid-catalyzed addition of methanol (methyl alcohol, CH_3OH) to benzenecarbaldehyde (benzaldehyde, C_6H_5CHO).

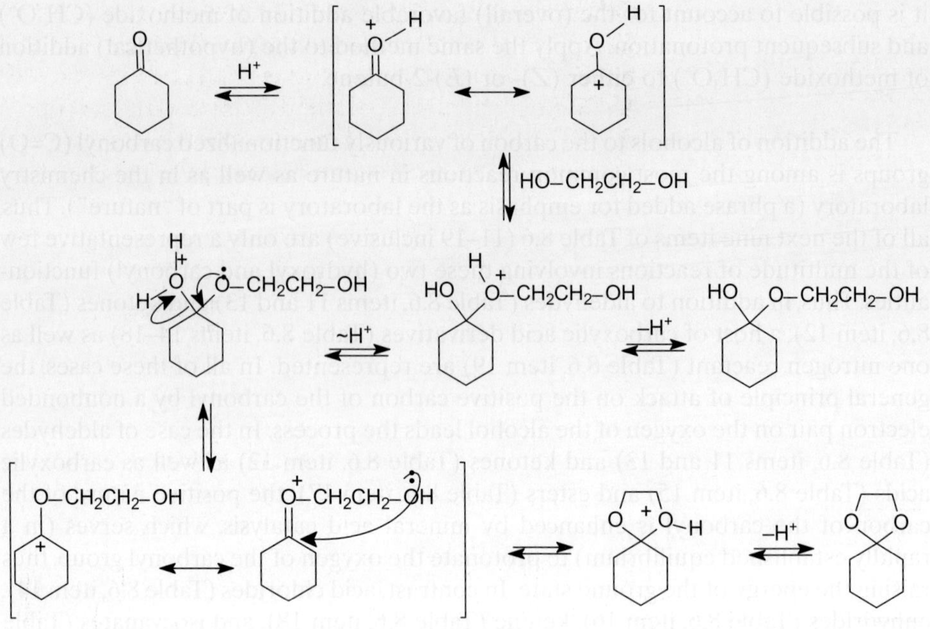

Scheme 8.52. The formation of, first, a hemiketal and, second, a (cyclic) ketal as a result of the acid-catalyzed addition of 1,2-dihydroxyethane (ethylene glycol, HOCH$_2$CH$_2$OH) to cyclohexanone.

Then, attack by the alcohol (methanol, CH$_3$OH) and proton loss produces the tetrahedral *hemiacetal* product. In the event (as is common) excess alcohol (methanol, CH$_3$OH) is present (usually, it is the solvent), proton transfer results in water loss and formation of an oxonium ion, which is susceptible to attack by a second equivalent of alcohol (methanol, CH$_3$OH). The product is an *acetal*.

The formation of *hemiketals* and *ketals* from alcohols and ketones is exactly analogous (with somewhat more difficulty as ketones are generally less reactive than aldehydes; Chapter 9). In the reaction of a ketone, such as cyclohexanone (Table 8.6, item 12) with a 1,2-diol, such as 1,2-dihydroxyethane (ethylene glycol [HOCH$_2$CH$_2$OH]) (Scheme 8.52), the "second equivalent" of alcohol is part of the initial alcohol substrate and the cyclic product results.

Finally, in this vein, it is important to note that the carbohydrates (Chapter 11) composing significant amounts of our biosphere commonly exist as hemiacetals and hemiketals, and thus the "cyclic forms" of glucose (a pair of diasteromeric pyrans) (Table 8.6, item 13) predominate over the "open form," which lies in equilibrium between them (Scheme 8.53). Indeed, shorn of its elaborate extra functionality, the cyclization simply represents the same sort of hemiacetal formation seen with methanol (CH$_3$OH) and benzenecarbaldehyde (benzaldehyde, C$_6$H$_5$CHO) (Table 8.6, item 11) shown above (Scheme 8.51).

Primary and secondary alcohols (e.g., cyclohexanol) add to the carbon of the carbonyl of carboxylic acids and compounds derived from carboxylic acids to form

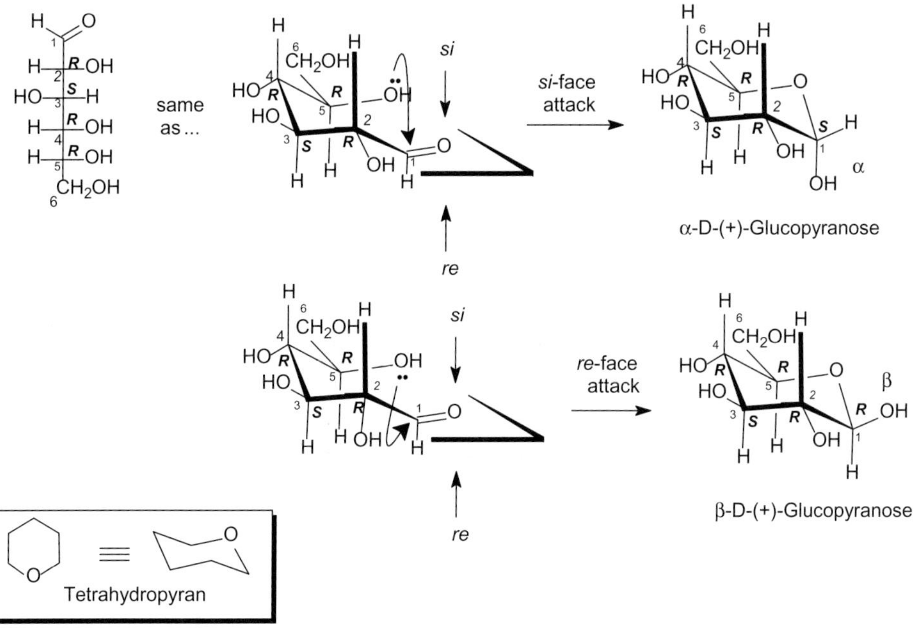

Scheme 8.53. The formation of the "cyclic" or **hemiacetal** forms (the pyranose) of D-(+)-glucose ($2R, 3S, 4R, 5R$-2,3,4,5,6-pentahydroxyhexanal—as the "open" chain form). Note that if cyclization occurs by the attack of the hemiacetal-forming oxygen (from C-5) onto the aldehyde from the *si*-face, the newly formed hydroxyl will be forced downward, on the opposite side of the –CH$_2$OH substituent (at C-5) and thus will generate α-D-(+)-glucopyranose, while the attack from the *re*-face pushes the newly formed hydroxyl into the equatorial position, yielding β-D-(+)-glucopyranose.

esters (Table 8.6, items 14–18). Thus, as shown in Scheme 8.54, cyclohexanol reacts with the acid chloride of propanoic acid (propanoyl chloride, CH_3CH_2COCl) to produce the corresponding ester, cyclohexyl propanoate, and hydrogen chloride (HCl). For practical reasons, the reaction is commonly carried out in the presence of a tertiary amine such as pyridine, which both "activates" the acid chloride* and reacts with the hydrogen chloride (HCl) generated, removing the chloride ion (a potential competing nucleophile as the reaction proceeds and more is formed) from the reaction mixture.[†]

---

*As noted earlier (Chapter 3) a reaction can be induced to proceed more efficiently (and often with fewer unwanted "side products") if the barrier lying between the reactants and products can be made smaller. Also, as pointed out there, it is possible to effect this change by raising the ground state (as well as by lowering the transition state). As shown in Scheme 8.54, pyridine reacts with the acid chloride, forming a more electrophilic acylpyridinium species, higher in energy than the acid chloride. Both the acid chloride and the acylpyridinium intermediate can react with the alcohol to produce the ester; the latter is more facile.

[†]As the reaction proceeds (in the absence of pyridine), the hydrogen chloride (HCl) generated can protonate the alcohol and the ester. Chloride ion, thus liberated, is available to compete with alcohol for unreacted acid chloride and (unwanted) acid-catalyzed alcohol reactions can intrude (*vide supra*).

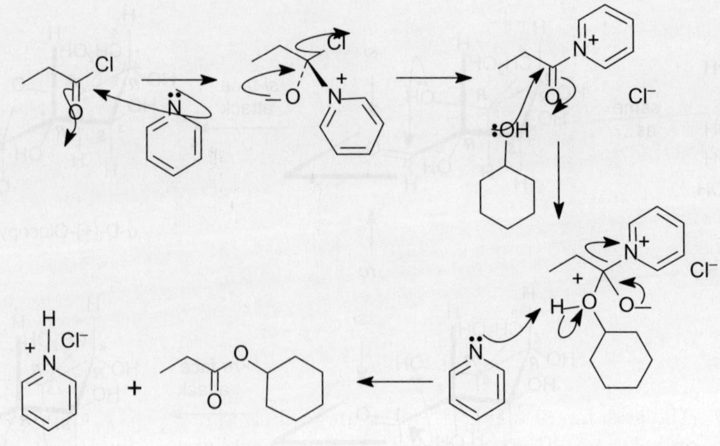

Scheme 8.54. The reaction of propanoyl chloride with cyclohexanol in the presence of pyridine to form cyclohexyl propanoate.

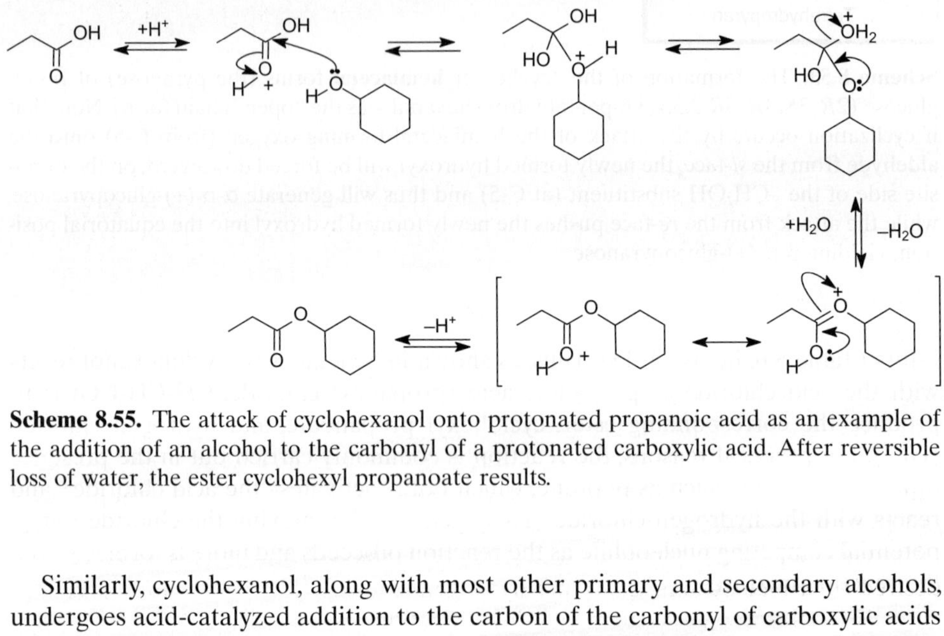

Scheme 8.55. The attack of cyclohexanol onto protonated propanoic acid as an example of the addition of an alcohol to the carbonyl of a protonated carboxylic acid. After reversible loss of water, the ester cyclohexyl propanoate results.

Similarly, cyclohexanol, along with most other primary and secondary alcohols, undergoes acid-catalyzed addition to the carbon of the carbonyl of carboxylic acids such as propanoic acid ($CH_3CH_2CO_2H$) (Table 8.6, item 15), and the tetrahedral intermediate so formed (reversibly) loses water to produce an ester (here, cyclohexyl propanoate). The conditions are generally more vigorous than those required for the addition reaction of alcohols to acid chlorides (*vide supra*). Commonly, the alcohol and carboxylic acid are heated at reflux in the presence of the acid (e.g., a mineral acid such as hydrogen chloride [HCl] or toluene sulfonic acid [$CH_3C_6H_4SO_3H$]) (Scheme 8.55) or, alternatively (and less commonly), the carboxylic acid is converted to the corresponding acylium ion, for example, with sulfuric acid (H_2SO_4), by protonation and dehydration, and the acylium ion is then treated with the alcohol (Scheme 8.56).

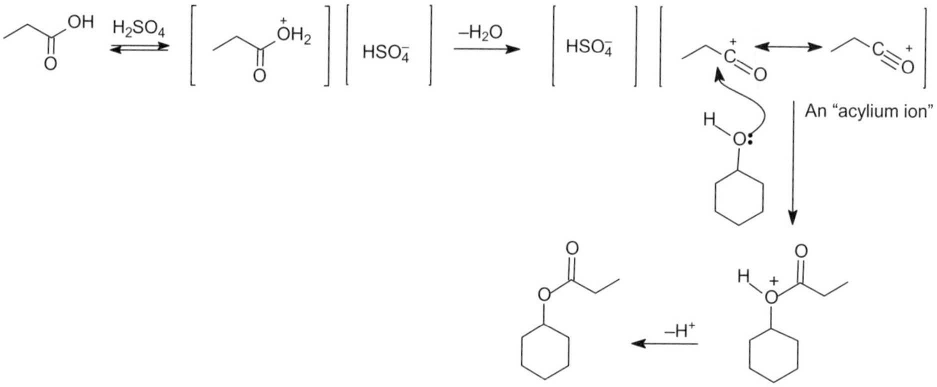

Scheme 8.56. The addition of cyclohexanol to an acylium ion derived from propanoic acid. The acylium ion is produced by treatment of propanoic acid with sulfuric acid.

Problem 8.13. As a general rule, neither of the acid-catalyzed addition reactions of primary and secondary alcohols with carboxylic acids described and shown in Schemes 8.55 and 8.56 (using propanoic acid and cyclohexanol as illustrative examples) works well with carboxylic acids and tertiary alcohols. How might you account for this observation?

The use of a carboxylate anion as a leaving group, as is the case when alcohols react with carboxylic acid anhydrides (Table 8.6, item 16), is common when the carboxylic acid is sterically unencumbered or when the acid is too labile to convert to the corresponding acid halide with, for example, thionyl chloride ($SOCl_2$). Mixed methanoic (formic)–ethanoic (acetic) anhydride (HCO_2OCCH_3) and other simple anhydrides are readily formed or are commercially available (e.g., ethanoic anhydride [acetic anhydride ($CH_3CO)_2O$] is, from year to year, among the top 100 industrial chemicals produced worldwide; Chapter 10). As shown in the reaction between ethanoic anhydride (acetic anhydride ($CH_3CO)_2O$) and cyclohexanol (Scheme 8.57), the addition process is frequently catalyzed by a trace of acid (to further polarize the carbon of the carbonyl and to make it more electrophilic). In the laboratory, the formation of ethanoate (acetate) esters of alcohols commonly is promoted by the addition of a small amount of ethanoyl chloride (acetyl chloride, CH_3COCl) to the reaction mixture along with the anhydride. This produces the desired ester (see Scheme 8.54) and hydrogen chloride (HCl), the latter serving as the acid catalyst for the remainder of the reaction.

As seen in Scheme 8.55, alcohols react, **reversibly**, with carboxylic acids to produce esters and water. As will be discussed later (Chapter 10), but should be clear from the scheme, the *hydrolysis* (reaction with water) of esters produces carboxylic acids and alcohols. In a similar vein, if an ester (such as methyl ethanoate [methyl acetate, $CH_3CO_2CH_3$], Table 8.6, item 17) is treated with an excess of alcohol (such as cyclohexanol, Table 8.6, item 17) in the presence of an acid catalyst and (generally) heated at a temperature above the distillation temperature of the alcohol with which the acid is esterified, **ester interchange** can be effected (Scheme 8.58) through the addition of the higher boiling alcohol to the carbon of the carbonyl followed by expulsion of the lower boiling one.

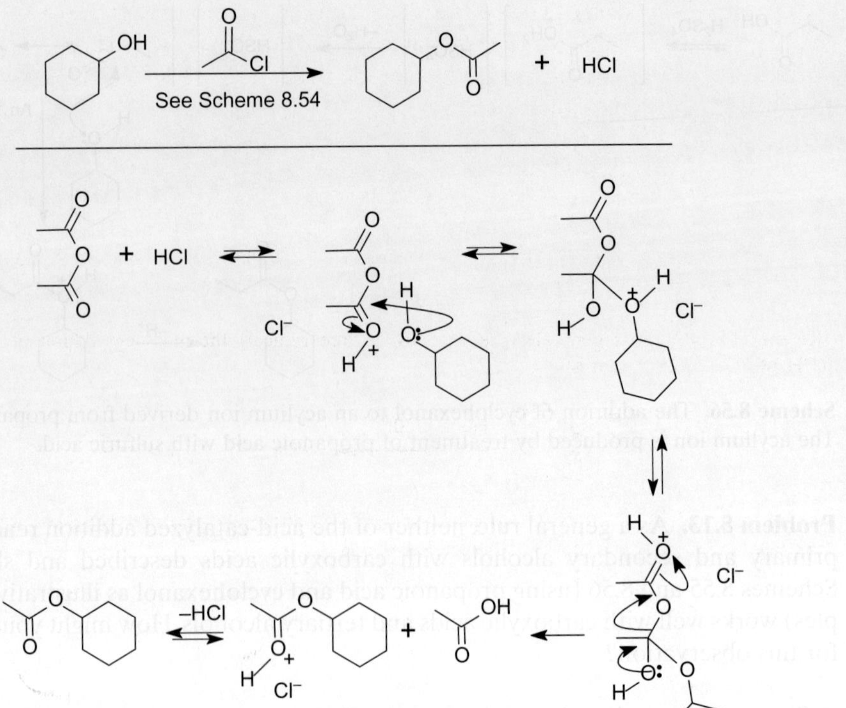

Scheme 8.57. The addition of cyclohexanol to ethanoic anhydride (acetic anhydride) with catalysis from hydrogen chloride (HCl). The catalyst is generated in the reaction between ethanoyl chloride (acetyl chloride, CH_3COCl), which also (Scheme 8.54) produces cyclohexyl ethanoate (cyclohexyl acetate) at the same time.

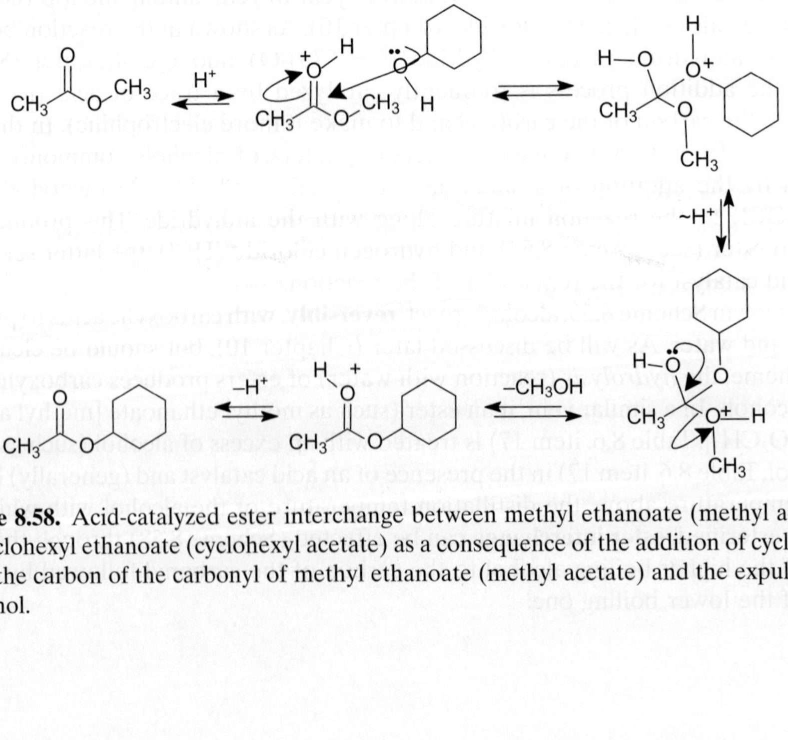

Scheme 8.58. Acid-catalyzed ester interchange between methyl ethanoate (methyl acetate) and cyclohexyl ethanoate (cyclohexyl acetate) as a consequence of the addition of cyclohexanol to the carbon of the carbonyl of methyl ethanoate (methyl acetate) and the expulsion of methanol.

Ketene (CH_2=C=O) may be thought of as an ethanoic acid (CH_3CO_2H, acetic acid) (Chapter 10) that has "lost" water (H–OH) and, as such, it reacts vigorously with hydroxyl-containing species such as alcohols (Table 8.6, item 18), which add to it to generate esters (Equation 8.34).*

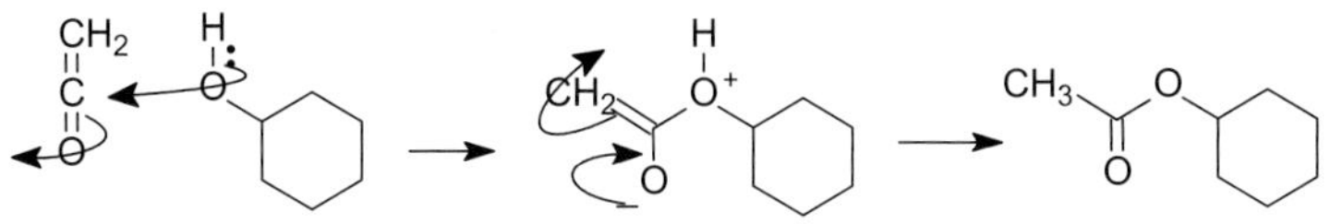

$$(8.34)$$

*Ketene (CH_2=C=O) is generally prepared by (a free radical) thermal degradation of propanone [$(CH_3)_2$C=O, acetone], that is,

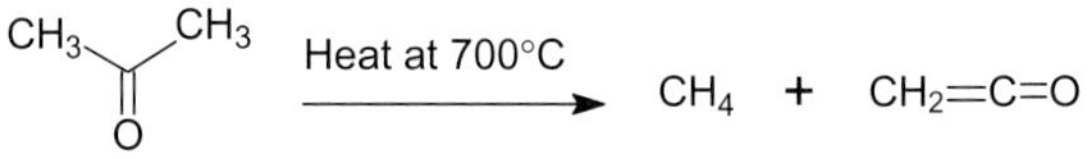

As expected, ethanoic acid (CH_3CO_2H, acetic acid) readily adds to ketene (CH_2=C=O) to produce ethanoic anhydride (acetic anhydride) (Chapter 10), that is,

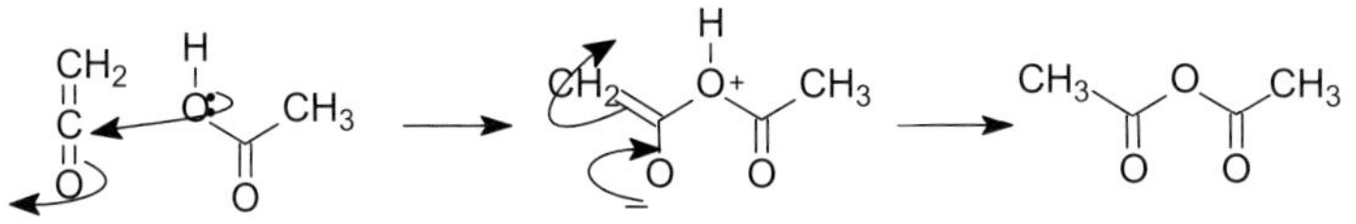

Substituted ketenes are usually prepared by the dehalogenation of α-halocarboxylic acid halides with tertiary amines (Chapter 10) or zinc dust, that is,

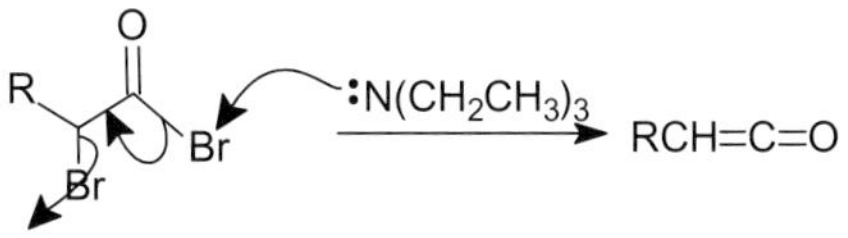

If storage of ketene (CH_2=C=O), bp –56°C, is attempted by confining it at high pressure, it dimerizes (in an allowed $2\pi + 2\pi$ cycloaddition reaction) to produce "diketene" (4-methylidenoxacyclobutan-2-one).

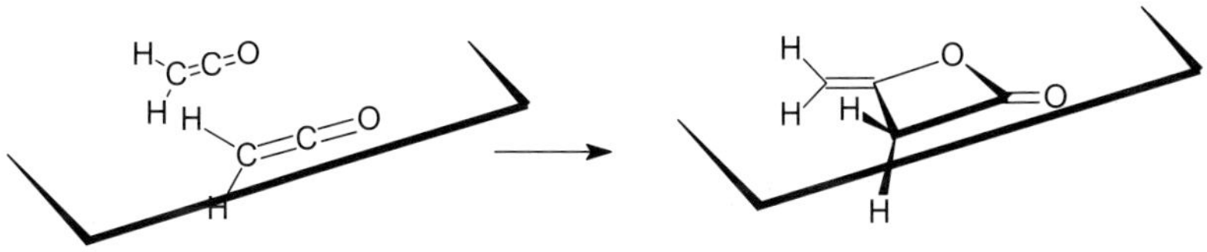

Diketene itself is a useful synthetic tool since alcohols add to it to produce esters of β-ketocarboxylic acids (Chapter 10), that is,

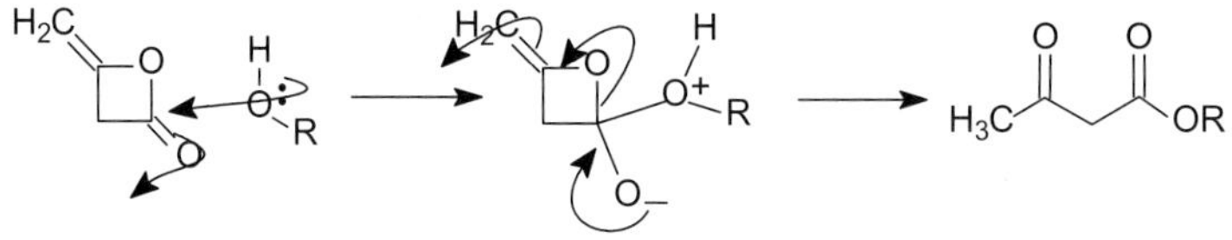

Interestingly, the structure of diketene was not firmly established until almost 50 years after it was initially prepared.

Problem 8.14. How might you employ one or more of the spectroscopic tools (IR, UV, and NMR) frequently used by organic chemists to distinguish between some of the "possible" isomeric structures of diketene shown in the next figure?

In a similar fashion, alcohols undergo addition to isocyanates (R–N=C=O), which are nitrogen analogues of carbenes, to produce esters of carbamic acids (called **urethanes**). Thus, as shown in Table 8.6 (item 19) and in Scheme 8.59, the addition of cyclohexanol ($C_6H_{11}OH$) to phenylisocyanate ($C_6H_5N=C=O$) produces the cyclohexyl ester of phenylcarbamic acid.* Interestingly, as shown in Scheme 8.59, it appears that the proton found on the nitrogen of the urethane is transferred directly to nitrogen from an alcohol dimer (or trimer).

As pointed out above (Table 8.6, items 11–13), alcohols undergo reversible acid-catalyzed addition to the carbonyl group of aldehydes and ketones to form

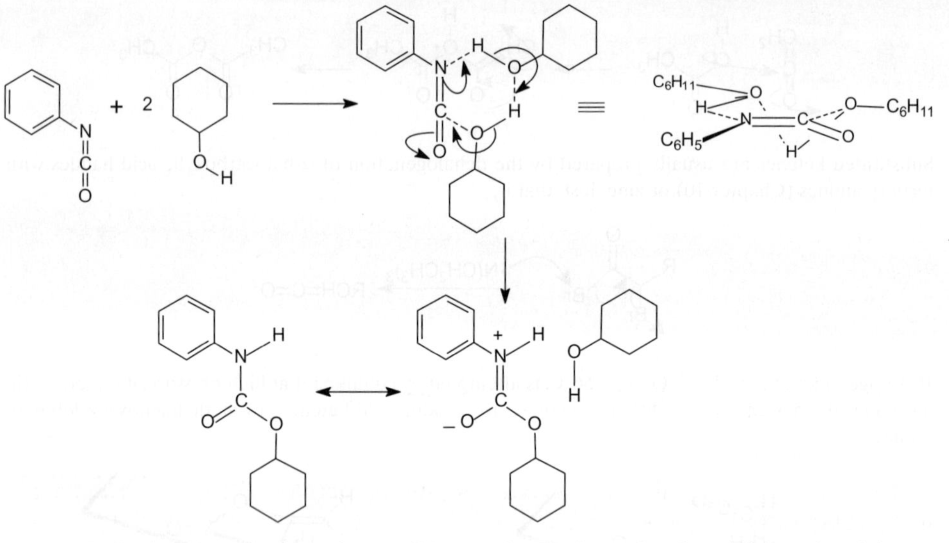

Scheme 8.59. A representation of a possible reaction pathway for the addition of cyclohexanol to phenylisocyanate to produce the corresponding cyclohexyl phenylcarbamate (a urethane).

*It is worthwhile noting that if **di**ols add to **di**isocynates, polymeric carbamic acid esters (<u>poly**urethanes**</u>) result. Polyurethanes, such as the idealized one shown in the next figure, have found considerable industrial use (e.g., as insulating materials).

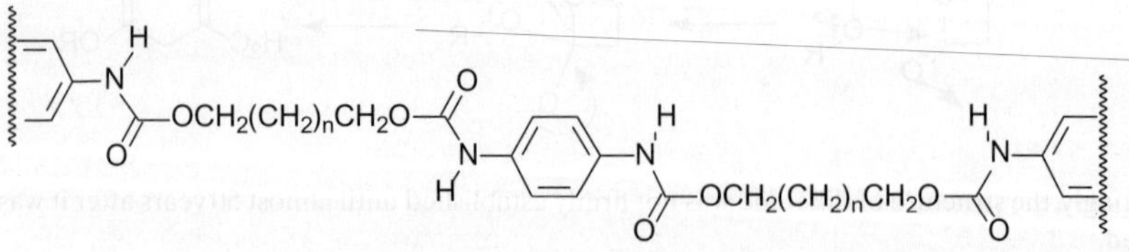

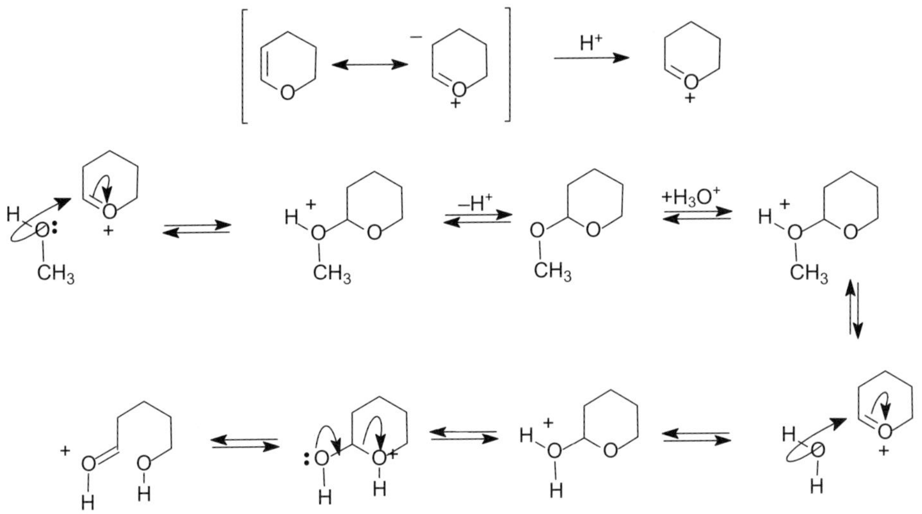

Scheme 8.60. The formation (and subsequent hydrolysis) of 2-methoxyoxacyclohexane by the acid-catalzyed addition of methanol to 3-oxacyclohexene (dihydropyran).

hemiacetals and hemiketals, respectively. With a second equivalent of alcohol, acetals and ketals, respectively, are generated. Acetals and ketals serve to "protect" the carbonyl group from attack by basic reagents (Chapter 9) but, as shown in the reversible processes of Schemes 8.51 and 8.52, these protecting groups are not useful under acidic conditions because of their ready hydrolysis to free alcohol and aldehyde (ketone). Of course, the "protection" of the hydroxyl group has been effected at the same time.

An alternative method of masking the hydroxyl group of alcohols involves acetal formation by a nonaqueous acid-catalyzed addition of alcohol to the carbon–carbon double bond of 3-oxacyclohexene (dihydropyran) (Table 8.6, item 20). The product, an acetal, will also undergo acid-catalyzed hydrolysis, regenerating the alcohol and the hydrate of 3-oxacyclohexene (dihydropyran), 2-hydroxyoxacyclohexane (the cyclic form of 5-hydroxypentanal), as shown in Scheme 8.60.

The "protection" of alcohols through addition reactions with oxygen, and again with acetal formation, is also easily effected through nucleophilic attack by oxygen at the chlorine-bearing carbon of α-chloroethers (Table 8.6, item 21). Thus, methoxychloromethane (chloromethyl methyl ether, CH_3–O–CH_2Cl) and phenylmethylenoxychloromethane (benzyl chloromethyl ether, $C_6H_5CH_2$–O–CH_2Cl), materials that can be prepared by the reaction of the corresponding aldehydes (under anhydrous conditions) with gaseous hydrogen chloride (HCl), react with alcohols (Scheme 8.61) to yield the corresponding acetals of methanal (formaldehyde, $H_2C{=}O$). *It is important to know that these and related compounds, for example, chloromethoxychloromethane (bischloromethyl ether, $ClCH_2OCH_2Cl$) must be used with extreme care as they are strongly suspected of being potent chemical carcinogens.* As shown in the scheme, these acetals are readily hydrolyzed by aqueous acid to yield the corresponding alcohol and aldehyde fragments.

Although, as a rule, most simple ethers (R–O–R; R = alkyl, aryl) are resistant to cleavage at the C–O single bond (in contrast to acetals and ketals, where geminal

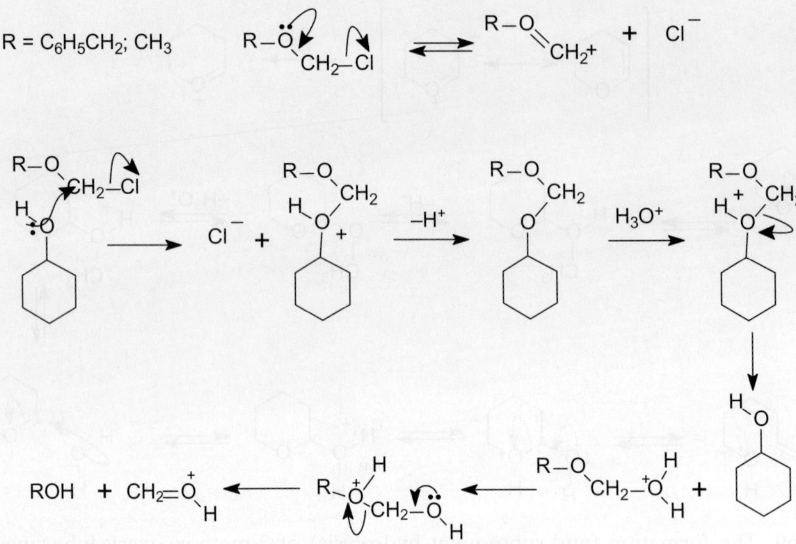

$$R = C_6H_5CH_2;\ CH_3$$

Scheme 8.61. The reaction of an alcohol (cyclohexanol) with a chloromethyl ether, RO-CH$_2$Cl (R=CH$_3$, chloromethyl methyl ether; R=C$_6$H$_5$CH$_2$, benzyl chloromethyl ether) to produce the corresponding acetal of methanal (formaldehyde).

oxygen substitution alters reactivity, *vide infra*), the strain present in small-ring compounds such as oxacyclopropanes (oxiranes, epoxides) dictates that they serve as exceptions to the rule. As will be discussed later in this chapter, the three-membered ring is opened under very mild conditions by a variety of reagents to yield a host of products. Indeed, the many materials of commerce produced from oxacyclopropane (oxirane, ethylene oxide) itself have made it a material of commercial importance (ca. 5.6 billion pounds produced in 1980).

Although the process is general, for the specific addition process shown in Scheme 8.62, that is, the addition of methanol (methyl alcohol, CH$_3$OH) to 2-methyloxacyclopropane (2-methyloxirane, propylene oxide) (Table 8.6, item 22), it is seen that the acid-catalyzed reaction occurs by a mixture of what appears to be (substrate dependent) S$_N$1 and S$_N$2 processes with the protonated oxygen of the oxirane serving as the "leaving group." As shown, both 2-methoxy-1-propanol and 1-methoxy-2-propanol are obtained.

The next two entries in Table 8.6 (items 23 and 24) involve the addition of enolate anions to alkylating reagents (although there is a complicating second step in item 24). While additional discussion of such systems will be provided when carbonyl compounds themselves are discussed (Chapter 9), these examples are inserted here to demonstrate how enols, closely related to alcohols (yet derived from and in equilibrium with carbonyl compounds), exhibit both similar and different reactivites. There are several unifying features to all of these reactions.

First, addition of the alkylating group occurs at the terminus of the carbon–carbon double bond away from the end bearing oxygen (C-alkylation). The carbon-bearing oxygen is, of course, the original carbonyl group. Second, alkylation can occur on oxygen (O-alkylation) as well as on carbon. The product of the second process is an enol ether. Third, the subsequent process may mask what has occurred.

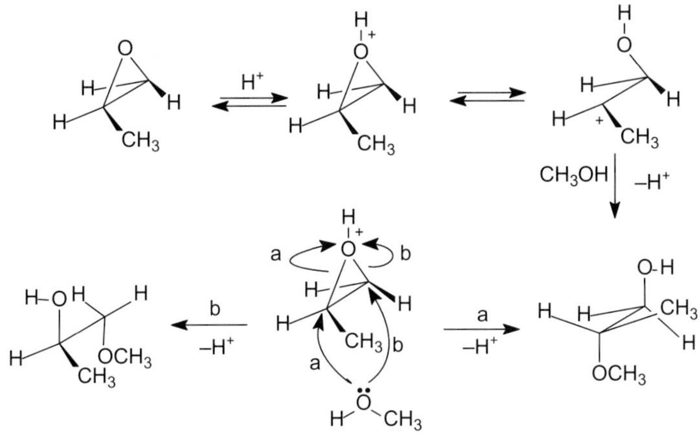

Scheme 8.62. A representation of the addition of methanol (methyl alcohol, CH_3OH) to 2-methyloxacyclopropane (2-methyloxirane, propylene oxide). The acid-catalyzed process appears to occur by a mixture of S_N1 and S_N2 processes where the protonated oxirane oxygen is the "leaving group" to produce both 2-methoxy-1-propanol and 1-methoxy-2-propanol.

In the first reaction (Table 8.6, item 23), alkylation of the anion of cyclohexanone is not reversible. Hence, the question of O-alkylation (Equation 8.35) versus C-alkylation (Equation 8.36) is a **kinetic** problem ($\Delta H° < 0$ for both O- and C-alkylation, Table 1.1) and, in principle, the enolate anion of cyclohexanone could alkylate on carbon and/or oxygen.

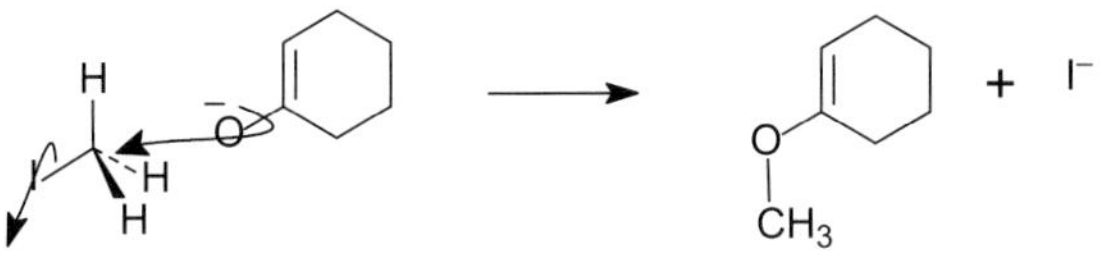

$$\text{(8.35)}$$

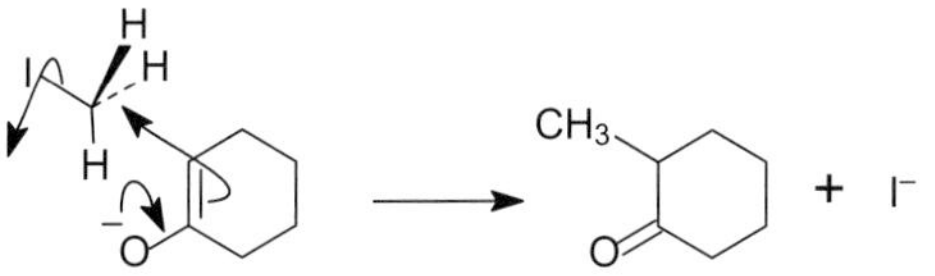

$$\text{(8.36)}$$

Although it is not easy to predict whether alkylation will occur at carbon or oxygen, it appears that the *less* associated the anion is with its cation (Li^+ in item 23 in Table 8.6, but Na^+, K^+, Mg^{2+}, etc., have all been examined), the *more* likely alkylation is to occur on oxygen. Therefore, to maximize the amount of C-alkylation, it is common to find that the lithium salt is used (because there is a high degree of association between anion and cation) in a solvent, such as oxacyclopentane (THF), where ionization is limited. However, even under what might be the most favorable circumstances, mixed O- and C-alkylation can continue to co-occur because the temperature and the group lost from the carbon (but see below) also play important roles.

The second example in this set (Table 8.6, item 24) involves a series of reactions often carried out *in one flask*. Initially, there is a formation of an enolate anion,

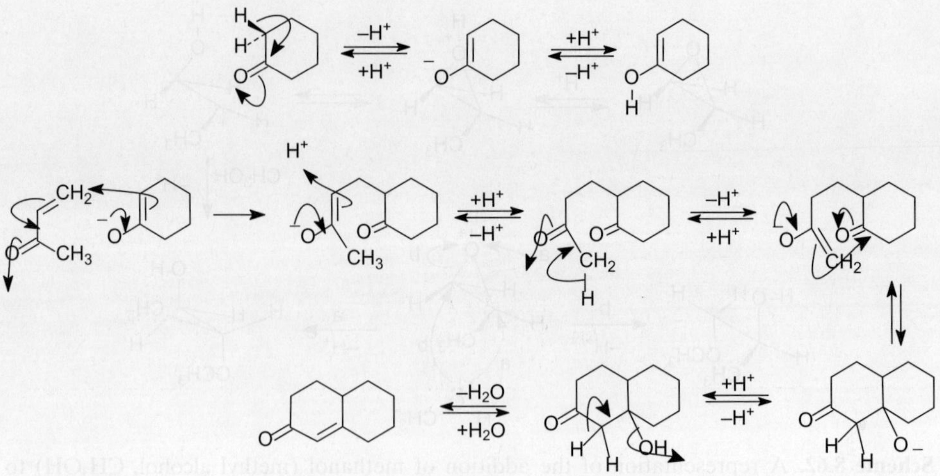

Scheme 8.63. A representation of the Robinson annulation process. The initial addition of the enolate anion to the α,β-unsaturated ketone represents a Michael reaction. Subsequent cyclization with loss of water is an aldol reaction.

which then, in a second step, undergoes addition to the terminus of an α,β-unsaturated ketone. This generalized process is called the **Michael reaction*** and will be encountered again (Chapter 9). Then, the product of that reaction, a ketone, enolizes and that enol subsequently adds to the carbon of the carbonyl. Finally, dehydration occurs. This sequence of reactions (Scheme 8.63), which also will be encountered again (Chapter 9), is called the **Robinson annulation**.[†]

The final two entries in Table 8.6 (items 25 and 26) involve the addition of alcohol to nuclei other than carbon (heteroatoms). It is hoped that the similarity between the two reactions shown, viz., the addition of methanol (CH_3OH) to thionyl chloride ($O{=}SCl_2$), to the "acid chloride" of sulfurous acid [$O{=}S(OH_2)$] (item 25), and to phosphorus oxychloride ($O{=}PCl_3$), the "acid chloride" of phosphoric acid [$O{=}P(OH)_3$] (item 26), will be seen as analogous to the addition of cyclohexanol to the acid chloride of propanoic acid (item 14) (Scheme 8.54). Indeed, the evidence appears to indicate that the processes as shown in Scheme 8.64 are not only similar to each other but also to that shown in Scheme 8.54 (although the addition reaction must occur twice for thionyl chloride [$O{=}SCl_2$] and three times for phosphorus oxychloride [$O{=}PCl_3$]).

---

*The addition of a nucleophile, not only an enolate, to the carbon at the terminus of the α,β-unsaturated carbonyl system is called a Michael reaction. It is named after Professor Arthur Michael (1853–1942), who explored it and who was educated in Heidelberg, Berlin, and Paris. After serving as professor at Tufts University, he moved to Harvard in 1912 and stayed until his retirement in 1936 (see Michael, A. *J. Prakt. Chem.*, **1894**, *49*, 20).

[†]Sir Robert Robinson (1996–1975), who developed this sequence, received his DSc at the University of Manchester under the supervision of C. Perkin and eventually returned there after posts in Sidney, Australia, and Liverpool, England, to serve as professor. In 1930, he accepted the professorship at the University College London. He received the Nobel Prize in Chemistry in 1947 (see Rapson, W. S.; Robinson, R. *J. Chem. Soc.*, **1935**, 1285).

The word "annulation" is derived from the Latin *annulus*, meaning "ring."

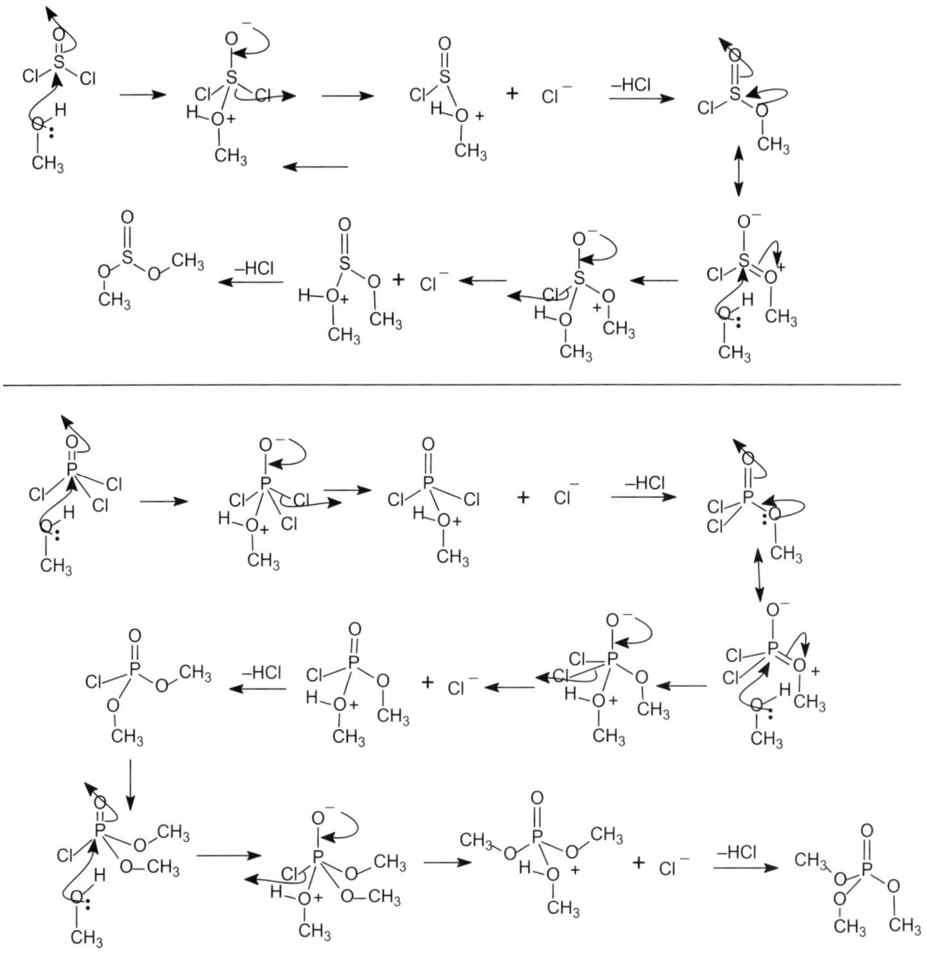

Scheme 8.64. Representations of the addition of methanol (methyl alcohol, CH_3OH) to thionyl chloride ($O{=}SCl_2$), the acid chloride of sulfurous acid (upper portion of the scheme), and phosphorus oxychloride ($O{=}PCl_3$), the acid chloride of phosphoric acid (lower portion of the scheme). Comparison should be made to the closely analogous reaction with carboxylic acid chlorides and other derivatives (Scheme 8.54, et. seq.)

F. ELIMINATION REACTIONS OF ALCOHOLS, ENOLS, AND PHENOLS

I. Introduction

Except for the rare cases of spontaneous loss of water from geminal diols (a carbonyl forming elimination reaction [Figure 8.4 and Equation 8.37]), and similar situations where one of the –OH groups of the geminal diol is replaced by SH or NH, the simple loss of water from an otherwise stable compound fails.

$$(8.37)$$

Thus, under general laboratory conditions, elimination of the hydroxide anion (OH^-), hydroxyl radical ($\cdot OH$), or the hydroxyl cation (OH^+) from an alcohol, enol, or phenol, leaving behind a carbocation, carbon radical, or carbanion, respectively, appears to be unknown. Indeed, even the loss of water, with the OH and H coming from adjacent atoms and, in this way, introducing unsaturation (or further unsaturation) into a compound in the laboratory and without involvement of an acid, base,* or acidic, basic, or neutral metal catalyst is rare.

Nonetheless, elimination does occur.

Alcohols and derivatives of alcohols (e.g., esters) undergo both catalzyed and thermal elimination of the elements of water to form alkenes (Equation 8.38).

$$(8.38)$$

In some cases, ethers are also formed (Equation 8.39) and elimination of water in that reaction may precede carbon–oxygen (ether) bond formation.

$$2RCH_2OH \xrightarrow{-OH_2} RCH_2OCH_2R \qquad (8.39)$$

Although direct elimination of water from phenols and enols is unknown, those compounds and some of their derivatives do formally lose water in the sense that an addition–elimination reaction series can replace the hydroxyl unit (see Chapter 7, Part 5b).

Further, 1,2-diols can undergo elimination of both oxygens to generate alkenes, and α-haloalcohols are known to eliminate the equivalent of hypohalous acid (HOCl, HOBr, etc.) to the same end.

---

*The pK_a of a typical alcohol is in the range 16–19, while that for a typical alkane is closer to 35–40. Therefore, unless the acidity of the proton on the carbon is increased dramatically, loss of water under basic conditions should not occur, the proton being lost to the base from the hydroxyl in preference to that from the carbon. Interestingly, as will be seen when the aldol and related reactions are disussed (Chapter 9), for water to be lost under *basic conditions*, it is sufficient to have the proton to be lost on a carbon α- to *at least one* carbonyl group on one side (pK_a about 13–20) while also on a carbon α- to that bearing the hydroxyl group on the other, that is,

$$H \longleftarrow pK_a = \text{ca. }13\text{–}20$$

$$H \longleftarrow pK_a = \text{ca. }16\text{–}19$$

II. Acid-Catalyzed Elimination of Water

Most alcohols undergo elimination of water in ionizing solution in the presence of either Lewis (e.g., Scheme 8.27) and/or Brønsted acid catalysts. For alkyl alcohols, the rate of dehydration generally follows carbocation stability, that is, as in the E1 reactions of alkyl halides (Chapter 7): tertiary > secondary > primary.

Elimination of water from alcohols can occur under E1 and E2 conditions and these reactions are completely analogous to those already discussed for alkyl halides in Chapter 7 except that the leaving group, "L," is OH_2^+ (Figures 8.14–8.16).

However, when acid-catalyzed elimination of water occurs, and *in contrast to reactions of alkyl halides*, it is far *more* common to find (a) Saytzeff (Chapter 7,

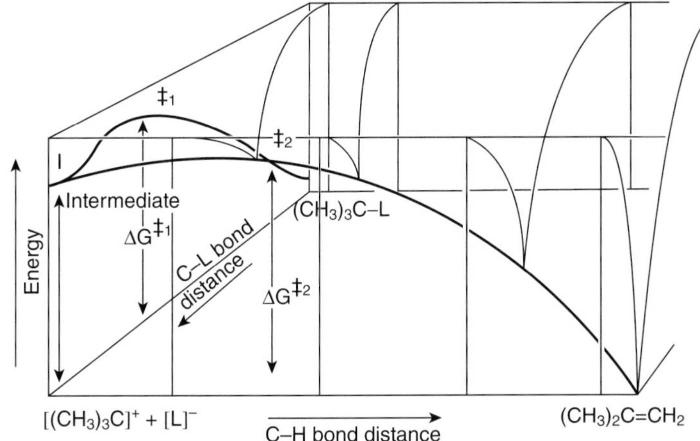

Figure 8.14. A representation of an energy diagram for the E1 reaction of protonated 2-methyl-2-propanol ($L = OH_2^+$) to yield 2-methylpropene. This three-dimensional figure can be transposed into two dimensions as shown in Figure 8.15. As is usual, the superscript symbol ‡ refers to the transition state.

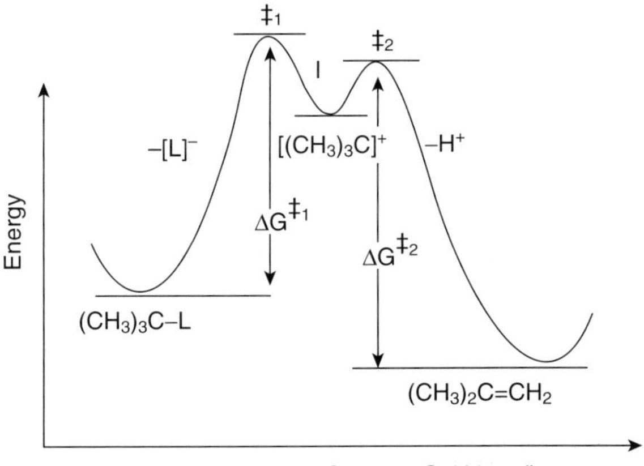

Figure 8.15. A two-dimensional representation of an energy diagram for the E1 reaction of protonated 2-methyl-2-propanol ($L = OH_2^+$) to produce 2-methylpropene. This two-dimensional figure can be represented in three dimensions as shown in Figure 8.14. As is usual, the superscript symbol ‡ refers to the transition state.

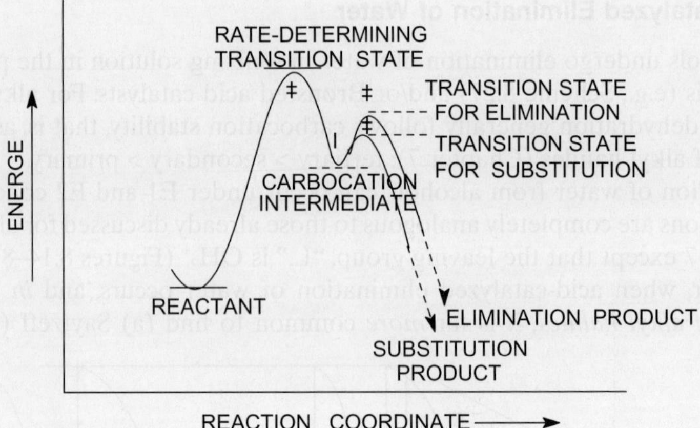

Figure 8.16. A representation of the course of a generalized $S_N1/E1$ reaction. The reactant is a protonated alcohol. The rate-determining transition state leads to a carbocation that subsequently partitions between products of substitution and elimination. The transition state for elimination *from the common intermediate* is generally higher than that for substitution.

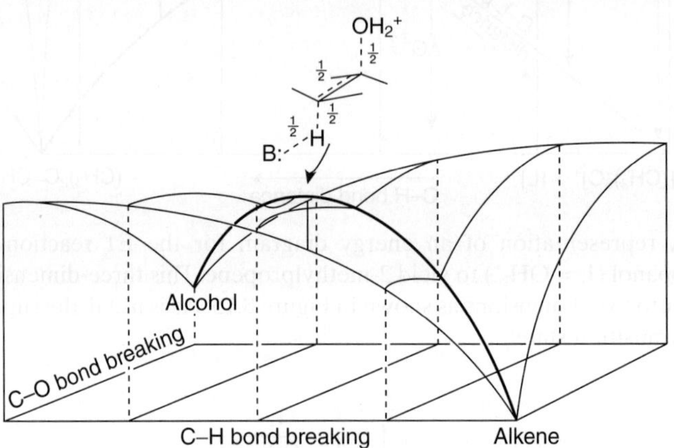

Figure 8.17. An idealized three-dimensional representation of an E2 elimination reaction of a protonated alcohol in which the breaking of the C–O bond and the H–C bond of a hydrogen on the carbon adjacent to that from which the OH₂ is leaving along with the formation of the C–C double bond are all occurring to about the same extent at the same time. A two-dimensional representation of the same process is shown in Figure 8.18.

Section G, 2), that is, more highly substituted, rather than Hofmann (Chapter 7, Section G, 2), that is, less highly substituted, alkene products and (b) that carbocation rearrangements have occurred.

Both of these observations are considered to arise from the ability of the alkene product(s) to undergo reprotonation and subsequent proton loss under the conditions of the reaction. Thus, since (a) the more highly substituted alkenes are thermodynamically favored and (b) rearrangements arise either for thermodynamic reasons (formation of more stable carbocations) or simple equilibration between ions of identical or near identical energies, both of these processes intrude (Figures 8.17).

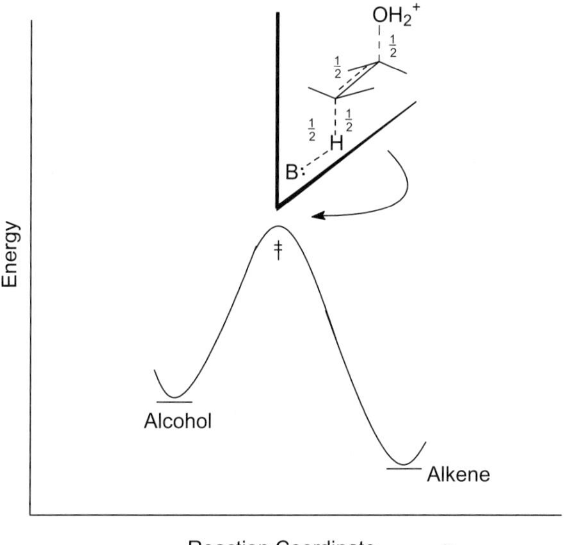

Figure 8.18. An idealized three-dimensional representation of an E2 elimination reaction of a protonated alcohol in which the breaking of the C–O bond and the H–C bond of a hydrogen on the carbon adjacent to that from which the OH₂ is leaving along with the formation of the C–C double bond are all occurring to about the same extent at the same time. A three-dimensional representation of the same process is shown in Figure 8.17.

Scheme 8.65. A representation of the acid-catalyzed elimination of water from *cis*- or (*Z*)-2-phenylcyclohexanol showing that there is a minimum amount of rearrangement and suggesting that the elimination is largely *anticoplanar*.

As shown in Figure 8.18 (*vide supra*), there is usually a clear preference for antarafacial [*anti*-, *trans*-, or (*E*)-] elimination (with the proton and water being coplanar) where it can occur. Thus, for example, for (*Z*-) or *cis*-2-phenylcyclohexanol (Scheme 8.65), acid-catalyzed (85% phosphoric acid) elimination of water produces predominately the Saytzeff product with a minimum of rearrangement. However, when antarafacial [*anti*-, *trans*-, or (*E*)-] elimination is geometrically frustrated, as in the

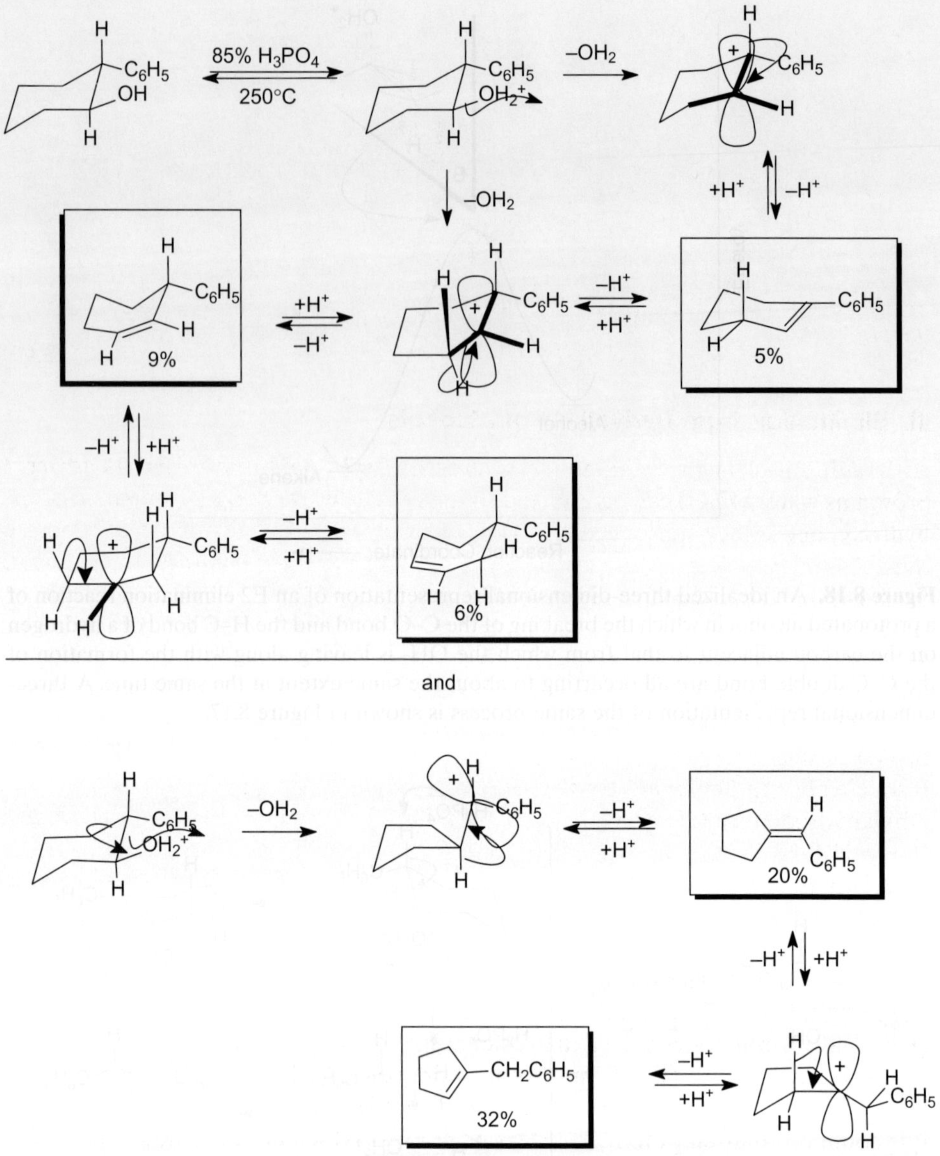

Scheme 8.66. A representation of the acid-catalyzed elimination of water from *trans*- or (*E*)-2-phenylcyclohexanol showing many of the products derived from carbocation rearrangements that intrude and cartoon representations of paths to those products.

corresponding (*E*-) or *trans*-isomer (Scheme 8.66), treatment under the same conditions gives rise to a multitude of carbocation-derived rearrangement products.

Interestingly, although the products shown in Scheme 8.66 represent most of the material, even that number does not complete the material balance. Further, utilizing labeled material (Equation 8.40, where the symbol "*" represents a ^{13}C or ^{14}C-labeled carbon atom), it is clear that the phenyl group participates in carbocation stabilization via a phenonium ion (see Chapter 7, Scheme 7.44, and this chapter).

$$(8.40)$$

III. Elimination from Derivatives of Alcohols

Although it is clear that elimination of the hydroxyl (–OH) group and an adjacent proton as water (H_2O) can be effected by acid catalysis (through protonation of the hydroxyl and either subsequent E2 or E1 processes), it is much more common to utilize ester (Chapter 9) intermediates to the same end. In some cases, the esters are not even isolated before they go on to form alkenes.

The esters of strong acids are among the most commonly used for elimination. Thus, as shown in Equation 8.24 (*vide supra*), an attempted substitution reaction following the formation of the trifluoromethanesulfonate (triflate) ester of 2-butanol (i.e., $CH_3CH_2CH(CH_3)OSO_2CF_3$) leads to the Satyzeff elimination mixture of *trans-* or (*E*)- and *cis-* or (*Z*)-2-butene ((*E*)-$CH_3CH=CHCH_3$ and (*Z*)-$CH_3CH=CHCH_3$, respectively) rather than to the expected elimination product. Similarly, the ethyl hydrogen sulfate ($CH_3CH_2OSO_3H$) ester of the primary alcohol ethanol (CH_3CH_2OH) is readily formed. Ethene (ethylene, $CH_2=CH_2$) and diethyl ether (ethyl ether, $(CH_3CH_2)_2O$) result from its decomposition, the former by E2 elimination and the latter by a combination of S_N2 substitution and acid-catalyzed addition of ethanol (CH_3CH_2OH) to ethene (ethylene, $CH_2=CH_2$). While it might have been anticipated that a simple acid-catalyzed elimination of water (H_2O) could have occurred, the fact that the alcohol is primary precludes ready carbocation formation (E1), and careful study shows that ester formation from ethanol (CH_3CH_2OH) is faster than water loss (E2) from protonated alcohol (Scheme 8.67).

Since sulfuric acid is not particularly soluble in higher-molecular-weight alcohols, it is common to attempt to form esters from acids, which both promote elimination (or substitution, *vide supra*) and which enjoy such solubility. To this end, and as shown in Equation 8.41, even the acetate (ethanoate) ester (formed from the reaction of alcohol—in this case cyclopentanol—with, e.g., acetic acid [ethanoic acid, CH_3CO_2H] or acetic anhydride [ethanoic anhydride, $(CH_3CO)_2O$)] or acetyl chloride [ethanoyl chloride, CH_3COCl]), by processes that are discussed further in Chapter 9, can be used to good effect to produce *cis-* or *suprafacial* elimination of the elements of acetic acid (ethanoic acid).

$$(8.41)$$

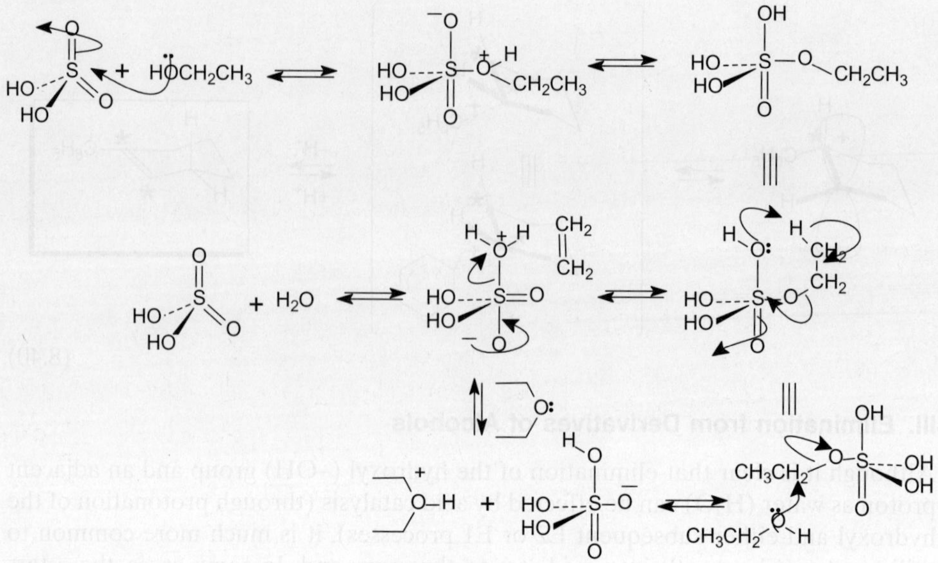

Scheme 8.67. The formation of ethene (ethylene, CH₂=CH₂) and diethyl ether (ethyl ether, (CH₃CH₂)₂O) from the reaction of ethanol (CH₃CH₂OH) with sulfuric acid (H₂SO₄). The reaction proceeds through the hydrogen sulfate ester of ethanol, which is apparently formed faster than the dehydration of the protonated ethanol can occur.

Further, other esters, such as those formed from alcohols with trifluoromethanesulfonic acid (CF₃SO₃H) (e.g., Equation 8.24), methanesulfonic acid (CH₃SO₃H), (e.g., Scheme 8.21), and, in particular, toluenesulfonic acid (Scheme 8.68) are even more commonly used. Indeed, the toluenesulfonate ("tosylate") ester is often employed since it is usually stable enough to be isolated and purified (at low temperature) but will undergo elimination to alkene and toluenesulfonic acid as the temperature is raised.

Certain esters (Chapter 9) are specifically designed to produce alcohol derivatives that are more labile than the alcohols themselves. The special case of the xanthate ester (the **Chugaev** reaction,* Scheme 8.69) has often been used to effect elimination when temperatures for simple ester pyrolysis are too high and other processes intrude.

There are many examples of loss of water to form alkenes in biological systems and in Chapter 12, some will be discussed further as part of metabolic processes. Among the products of metabolism, there are numerous long-chain carboxylic acids ("fatty acids") (Chapter 9) that occur in biological membranes (and elsewhere) that contain double bonds. The acids are often found as esters (Chapter 9) of the triol 1,2,3-trihydroxypropane [1,2,3-propanetriol, glycerol, HOCH₂CH(OH)CH₂OH] and consist of chains with an even number of carbons (12, 14, 16, etc.) arrayed in a linear fashion (Figure 8.19). These triesters of carboxylic acids are called "fats" or "glycerides." The carboxylic acids can be separated from the 1,2,3-propanetriol (glycerol) by hydrolysis of the esters (Chapter 9).

---

*Although the original work was done on terpenes (cf. Chapter 12), the process is much more general. See Chugaev, L. (Tschugaeff, L.) *Chem. Ber.*, **1900**, *33*, 3118.

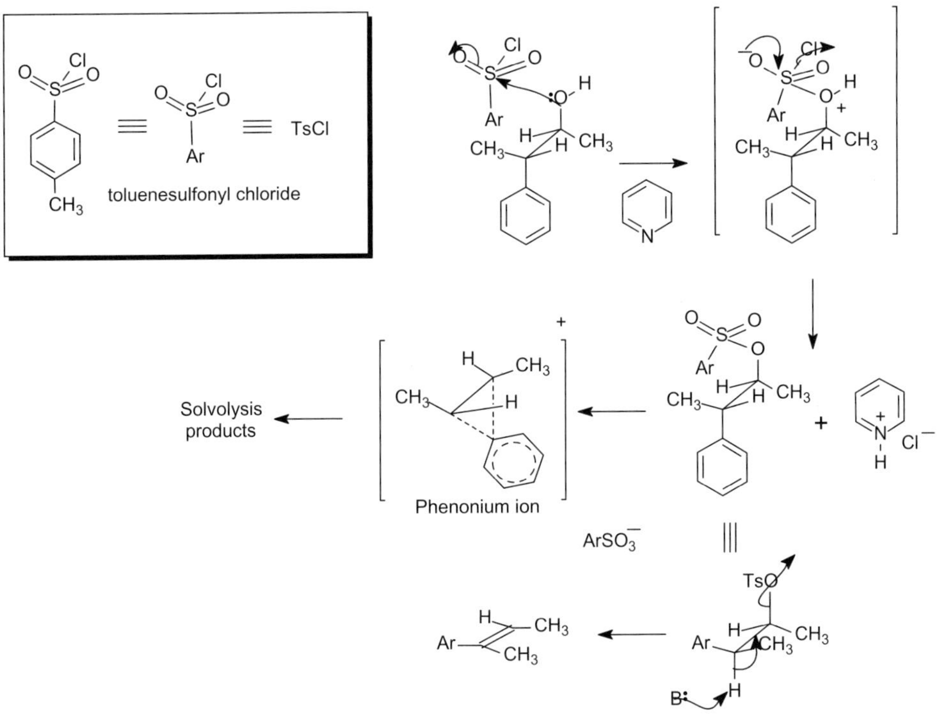

Scheme 8.68. Formation of the toluenesulfonate (tosylate) ester of *erythro*-3-phenyl-2-butanol and its decomposition (with phenyl participation via a phenonium ion) to produce the *erythro*-solvolysis product as well as elimination (after bond rotation) to produce the dehydration product, (Z)-2-phenyl-2-butene.

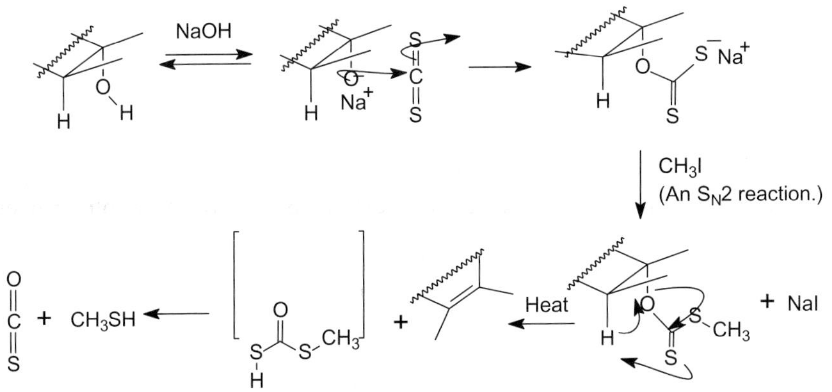

Scheme 8.69. The Chugaev reaction. The salt of an alcohol is allowed to react with carbon disulfide, producing the sodium salt of a dithiocarbonate derivative, which, on S_N2 reaction with methyl iodide (CH_3I), produces the corresponding xanthate ester. Heating the xanthate (pyrolysis) results in loss of carbon oxysulfide (COS), methane thiol (thiomethane), and the corresponding alkene.

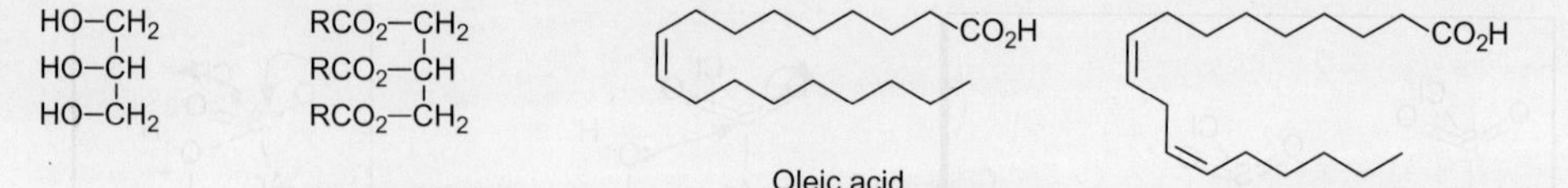

HO–CH₂   RCO₂–CH₂
HO–CH   RCO₂–CH
HO–CH₂   RCO₂–CH₂

1,2,3-Propanetriol   A triglyceride (fat)
(glycerol)   (The "R" groups can be different.)

Oleic acid
(*Z*)-9-Octadecenoic acid

Linoleic acid
(9*Z*,12*Z*)-9,12-octadecadinoic acid

Figure 8.19. Representations of glycerol, a glyceride (R = fatty acid), (*Z*)-9-octadecenoic (oleic) acid, and (9*Z*,12*Z*)-9,12-octadecadinoic (linoleic) acid.

When there are **no** double bonds in the carboxylic acids, they are called **saturated fatty acids**. **Unsaturated fatty acids** have at least one double bond present and this unsaturation results from the elimination of water from an alcohol (dehydration). (*Z*)-9-octadecenoic acid (oleic acid) and (9*Z*,12*Z*)-9,12-octadecadinoic acid (linoleic acid) are examples of the latter (Figure 8.19).

There are two points to note. First, both of these acids contain an **even number** (18) of carbon atoms and, second, the double bonds are *cis* or (*Z*).

A significant amount of investigation as to the origin of these and related natural products (Part 3) has been undertaken, and it now seems clear that a family of enzymes is required for the assembly of such compounds. Further, despite the wide diversity of acids (and related compounds) across numerous plant **and** animal families, these products of metabolism (metabolites) appear to be derived from related multifunctional enzyme complexes called *polyketide synthases* (PKSs). Different PKSs produce different structural arrays by the assembly of two (or, what is more rare, three) carbon units each of which terminates in an acyl (R–C=O; R=CH₃ or CH₃CH₂) unit originally attached to *coenzyme A* (Scheme 8.70). An abbreviated and somewhat simplified version* of the process by which fatty acids are formed is shown in Scheme 8.70.

As shown in Scheme 8.70, the acyl unit is transferred from coenzyme A by an *acyltransferase* component of the *PKS* family to an *acyl carrier protein* where condensation using a *ketosynthase* component of the *PKS* family puts the two components together. Then, reduction by a *ketoreductase* component held next in line in the *PKS* family yields a keto alcohol from which elimination of water is effected by the next *PKS* component, a *dehydratase*. If the coding of the *PKS* requires reduction, the next enzyme in the complex will be an *enoyl reductase*. If the double bond is to remain, the *reductase* member of the *PKS* family will be missing and, in either case, if additional two (or more rarely, three) carbon fragments are to be added, the entire growing chain is passed to the next *acyltransferase* to continue the process and to produce a long chain. In different cases the different *dehydratase* enzymes can produce either (*Z*) or (*E*) unsaturation in the growing chain. The actual details of the dehydration are enzyme (Chapter 11) dependent and, despite great effort, somewhat obscure, although it is apparent that for such thioesters (Chapter 9), as those shown in Scheme 8.70, the proton abstracted is *syn* to the hydroxyl.

*The simplification derives, in part, from the current lack of clarity about the details (at the molecular level) of the processes described and, in part, from the necessity of avoiding discussion of the use of malonyl-CoA (HO₂CCH₂COCoA) rather than the use of acetyl-CoA (CH₃COCoA), and so on, *at this point in the text.*

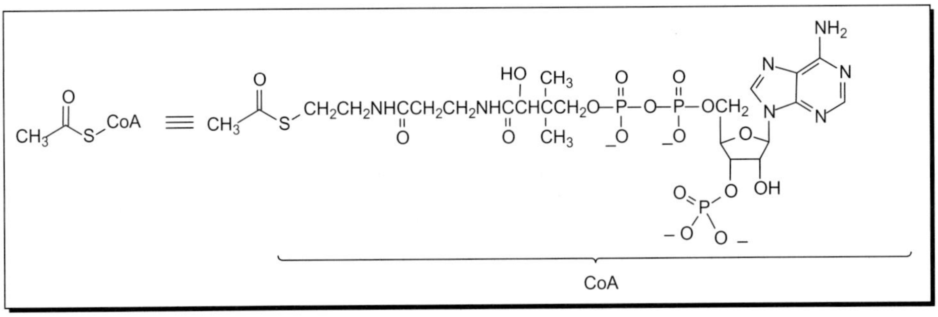
CoA

(Simplified) Example of the Assembly of Fatty Acid from Polyketide Synthase (PKS)

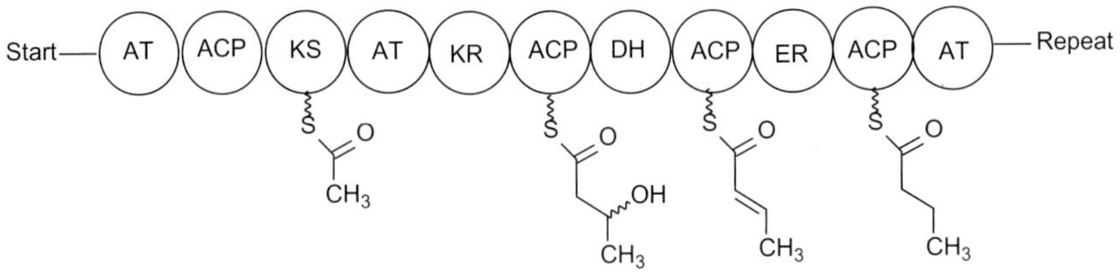
Start
AT
ACP
KS
AT
KR
ACP
DH
ACP
ER
ACP
AT
Repeat

AT = acyl transferase (from acetyl-CoA to the PKS)

ACP = acyl carrier protein

KS = ketosynthase

KR = ketoreductase

DH = dehydratase

ER = enoyl reductase

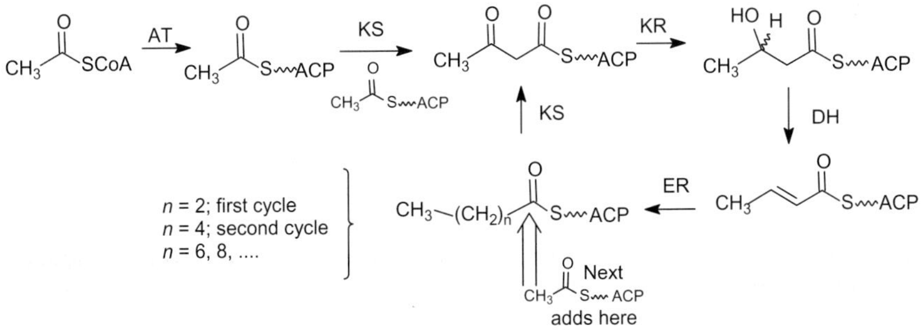
AT
KS
KR
DH
ER
KS
n = 2; first cycle
n = 4; second cycle
n = 6, 8, ...
Next
adds here

Scheme 8.70. A cartoon representation of the formation of saturatured (and unsaturated) fatty acids utilizing a polyketide synthase (PKS) bundle of enzymes.

In some other cases in nature when elimination of a hydroxyl group (–OH) is to be effected, activation of the hydroxyl is through phosphorylation of the hydroxyl to a phosphate (i.e., R–OH to R-OPO_3^{-2}) using an enzyme (a *kinase*) to catalyze the process. Conversion of the hydroxyl to the phosphate group does not utilize phosphoric acid but rather, as noted earlier (and see Chapter 14) relies on the cleavage of an oxygen phosphorus bond in a polyphosphate, that is, conversion of ATP to ADP.

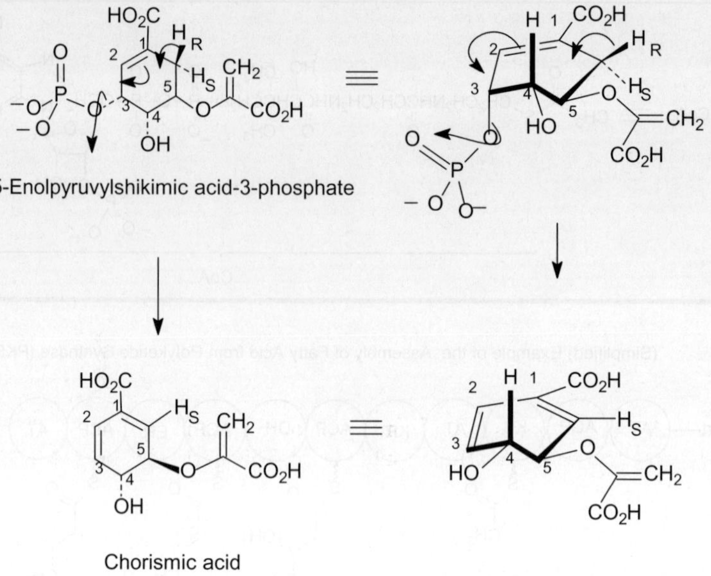

5-Enolpyruvylshikimic acid-3-phosphate

Chorismic acid

Scheme 8.71. A representation of the elimination of phosphate from 5-enolpyruvylshikimic acid-3-phosphate to form chorismate. The elimination is a conjugate (or 1,4-) elimination.

An example of elimination from such a phosphate ester, important in the biosynthesis of aromatic amino acids (Chapter 12), involves not only the loss of water but also the simultaneous migration of a carbon–carbon double bond. This kind of elimination is referred to as *conjugate elimination* (or 1,4-elimination).

As shown in Scheme 8.71, a carbohydrate* derived precursor to the aromatic amino acids phenylalanine and tyrosine (Chapter 12) called 5-enolpyruvylshikimic acid-3-phosphate undergoes elimination of the C-6 *pro*-R hydrogen with concomitant loss of the phosphate leaving group to produce the intermediate chorismic acid.

Having broached the subject of polyols by reference to carbohydrates and by the introduction of 1,2,3-trihydroxypropane (glycerol), it is worthwhile noting that unless the hydroxyls are geminal (generally an unstable arrangement, *vide supra*) or vicinal, they frequently behave independently. When they are vicinal (in a *diol*), *both* can be eliminated with the introduction of a carbon–carbon double bond between the carbons previously bearing the hydroxyls. The elimination is effected by forming the thiocarbonate derivative (the **Corey–Winter** reaction) (Scheme 8.72).

Despite the specificity of the Corey–Winter process,[†] it is worthwhile reemphasizing that most of the chemical methods for elimination of the hydroxyl group are

---

*Much of the chemistry of that group of compounds referred to as carbohydrates (compounds of carbon corresponding to, approximately, $C(H_2O)_n$) is a function of the numerous hydroxyl groups that adorn their structures. However, despite the view that they be considered as polyhydroxy derivatives of carbon (polyols), their rich chemistry must include reference to reactions of the carbonyl group (aldehydes, ketones, and carboxylic acids). Therefore, elaboration of the chemistry of carbohydrates is postponed to Chapter 11.

[†]Elias J. Corey received the Nobel Prize in Chemistry for his elegant developmental work in synthetic chemistry. Careful planning and execution of those plans has been the hallmark of the work reported. For the Corey–Winter reaction, see Corey, E. J.; Winter, R. A. E. *J. Am. Chem. Soc.*, **1963**, *85*, 2677.

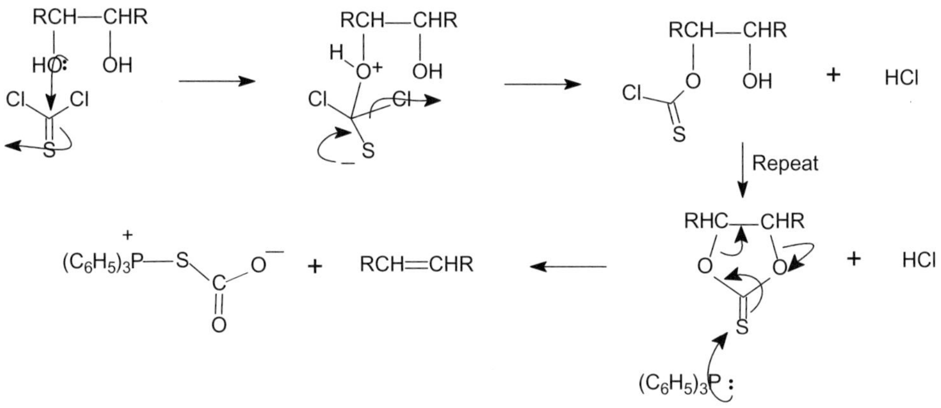

Scheme 8.72. A representation of the elimination of two vicinal hydroxyl (–OH) groups with the iintroduction of a carbon–carbon double bond. This particular process is known as the Corey–Winter reaction.

quite vigorous and often involve heating esters or other alcohol derivatives so that rearrangement reactions and/or equilibration of products to produce the thermodynamically favored isomers results. As a consequence, the Saytzeff rather than the Hoffmann product is often obtained from alcohols and their derivatives. There are three additional cases to be considered.

First, there is the coenzyme B_{12}-dependent (Chapter 12 and 14) catalyzed dehydration of 1,2,3-trihydroxypropane (glycerol) to 3-hydroxypropanal (Scheme 8.73) by what is apparently a free radical reaction. Such a free radical process is apparently related to the similar oxidation of 1,2-hydroxyethane (ethylene glycol) by iron(II) in the presence of hydrogen peroxide (the **Fenton reagent**),* also shown as part of Scheme 8.73.

Second, among the newer methods developed to effect the elimination to produce the *least substituted* alkene is one that involves converting the alcohol to a **selenium** derivative. In this procedure, a primary alcohol is treated with *o*-nitrophenyl selenocyanate in a suitable solvent such as (THF, oxacyclopentane) in the presence of a phosphine (such as tri-*n*-butylphosphine [$(CH_3CH_2CH_2CH_2)_3P$]) to produce the primary alkyl selenide (Scheme 8.74). Then, in a second step, the primary alkyl *o*-nitrophenyl selenide is oxidized with hydrogen peroxide to yield the corresponding selenoxide, which readily undergoes elimination to the alkene. Scheme 8.74 shows the application of the sequence of reactions described above to cyclohexylmethanol and the resulting formation of the *exo*-methylenecyclohexane.

The process using *o*-nitrophenyl selenocyanate was described by Clive (1978) and has been elaborated upon as described by Reich (1979).

The final elimination reaction to be considered here is quite old and, although commercially useful, not well studied. It will be encountered again along with other similar processes when the chemistry of carbohydrates (Chapter 11 and *vide supra*)

---

*The Fenton reagent remains widely used as a general way to produce hydroxyl radicals (HO•) for a variety of purposes. See Fenton, H. J. H. *J. Chem. Soc. Trans.*, **1894**, *65*, 899.

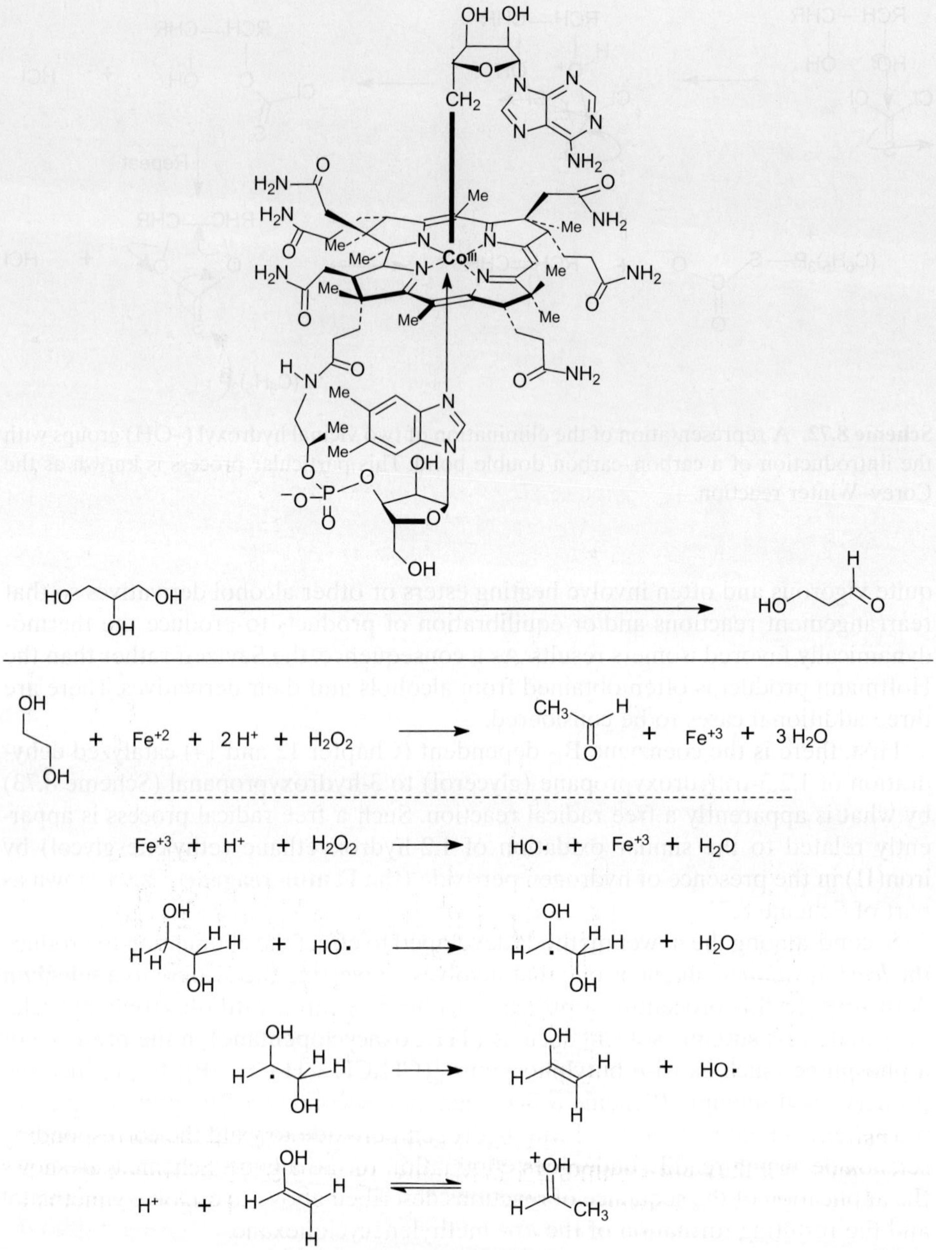

Scheme 8.73. The oxidation of 1,2,3-propanetriol (glycerol) to 3-hydroxypropanal utilizing B$_{12}$ (the details are not known) and the similar oxidation of 1,2-dihydroxyethane (ethylene glycol) to acetaldehyde by acidic hydrogen peroxide in the presence of Fe(II) (Fenton reagent).

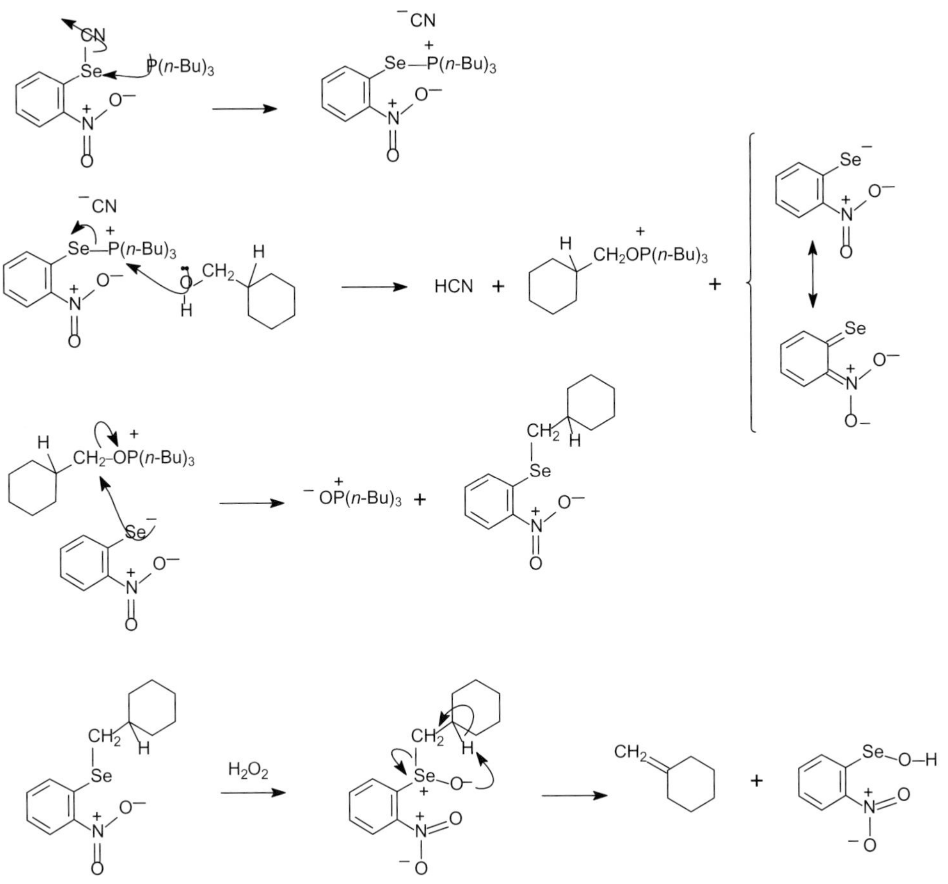

Scheme 8.74. The formation of a terminal alkene by an elimination reaction sequence that begins with the conversion of the oxygen of cyclohexylmethanol to a good leaving group through a reaction with tri-*n*-butylphosphine and *o*-nitrophenyl selenocyanate to form the corresponding selenide (where the oxygen of the starting alcohol has been replaced with the aryl-substituted selenium). Subsequent oxidation to a selenoxide is followed by a rapid elimination to the desired alkene (methylenecyclohexane).

is considered in more detail. Carbohydrates containing five carbon atoms are, generically, **pentoses**; they are common in plants and, on acid treatment, they are converted, in bulk, to commercially valuable furan 2-carboxaldehyde (furfural). In pentoses, each carbon atom bears oxygen. As shown in Scheme 8.75, if the pentose is an **aldo**pentose, one of the terminal carbons is an *aldehyde* (and all of the other carbons bear hydroxyl groups). If the pentose is a **keto**pentose, one of the penultimate carbons is a *ketone* (and all of the other carbons bear hydroxyl groups). The ketohexose and aldohexose can be brought into equilibrium through the common enol (Scheme 8.75).

As shown in Scheme 8.75 (the details of which should be taken as suggestive rather than proven), treatment of such pentoses with sulfuric acid results in the elimination of water and cyclization. The product, furan 2-carboxaldehyde (furfural), is necessarily formed by a combination of a cyclization process and several

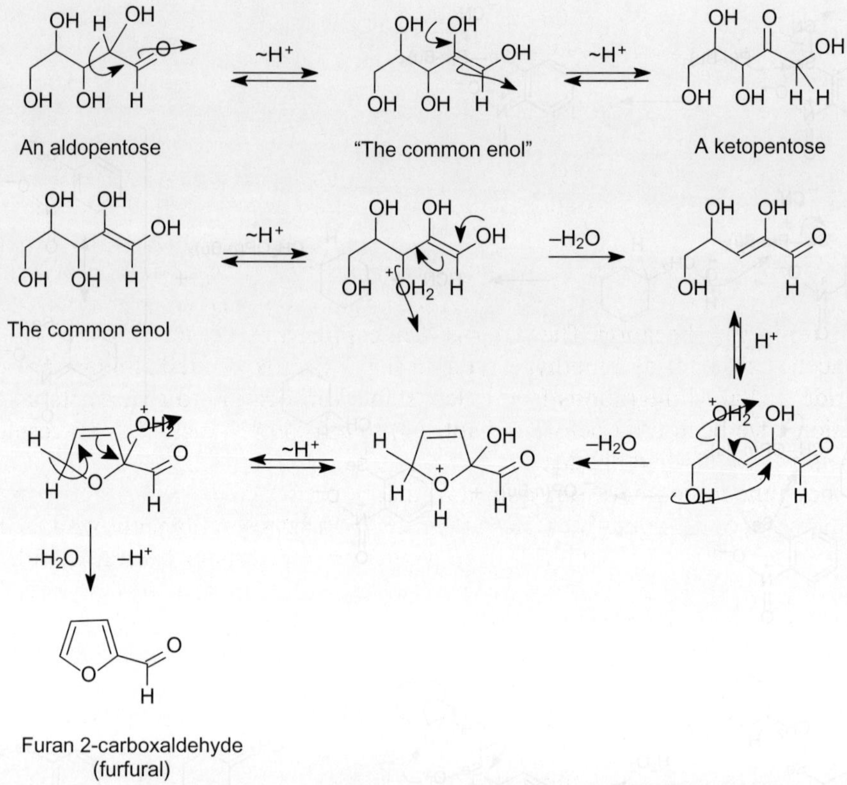

An aldopentose   "The common enol"   A ketopentose

The common enol

Furan 2-carboxaldehyde
(furfural)

Scheme 8.75. A representation of a potential pathway for the formation of furan 2-carboxaldehyde (furfural) as the result of treating aldopentoses and ketopentoses with strong acid (H_2SO_4). The elimination processes shown must occur to generate the product. The details (stereochemistry, timing, etc.) should be considered as reasonable rather than certain.

elimination reactions, although the details of the order in which these processes occur are obscure.

G. REARRANGEMENT REACTIONS OF ALCOHOLS, ENOLS, AND PHENOLS

I. Introduction

Rearrangement reactions of alcohols, enols, and phenols fall into several marginally different catagories or groups. First, there are those rearrangements in which the hydroxyl group (–OH) or its equivalent (as a derivative, such as an ester, *vide supra*) leaves (having been protonated, or otherwise activated) forming a carbocation. The carbocation then undergoes some skeletal change to a new carbocation. The latter is then captured by water or its equivalent. In this way, a new alcohol, enol, or phenol of rearranged structure is formed.

Another group of rearrangement reactions of alcohols, enols, and phenols results in the formation of compounds in which the hydroxyl group (–OH) or its equivalent has been replaced by a carbonyl group (C=O) because the substitution pattern on and/or the hybridization of the *carbon to which the hydroxyl (–OH) was originally attached has changed.*

Some of these reactions are presented in Table 8.7 and are discussed subsequently.

The first item in Table 8.7 is the toluenesulfonic acid ester (toluenesulfonate, tosylate, –OTs) of 3-^{2}H$_1$-3-methyl-2-butanol, which is permitted to undergo solvolysis in an acetic acid solvent. As might be anticipated (and as shown in Scheme 8.76), the secondary tosylate group leaves and the secondary carbocation suffers a hydride migration (in this case, a deuteride, i.e., ^{2}H rather than ^{1}H) to generate the more stable tertiary carbocation. The latter is then captured by the nucleophilic solvent. The acetic acid ester of 2-methyl-2-butanol (i.e., 2-acetoxy-2-methylbutane) results. Hydride and alkyl migrations from a less stable carbocation to a more stable carbocation have been used before (Chapter 7) to explain product mixtures resulting from solvolysis of alkyl halides.

Among the more interesting questions that might be asked, however, is whether the migration of the deuteride (^{2}H) (Scheme 8.76) occurs simultaneous with the loss of the tosylate (i.e., the deuteride is moving from the tertiary carbon toward the secondary carbon **while** the tosylate is leaving) or whether the tosylate leaves first, the carbocation forms, and, in a second step, uncoupled from the first, the deuteride moves from the tertiary carbon to the secondary carbocation and the tertiary carbocation is created.

One way to examine this problem has already been addressed in the previous chapter. Thus, if the migration of the hydrogen (^{2}H) is part of the rate-determining process, then the reaction with ^{1}H should be faster than with ^{2}H (breaking a carbon–hydrogen bond is easier than breaking a carbon–deuterium bond) and the ratio of

TABLE 8.7. Some Rearrangement Reactions of Alcohols, Enols, Phenols, and Their Derivatives

Item	Substrate	Product(s) of Rearrangement	Comments
1.	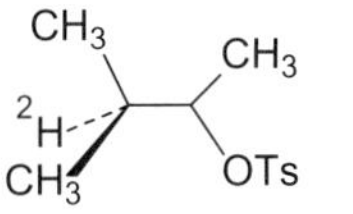 CH$_3$, CH$_3$, ^{2}H, CH$_3$, OTs	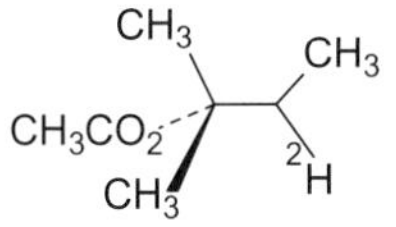 CH$_3$, CH$_3$, CH$_3$CO$_2^-$, CH$_3$, ^{2}H	Product of solvolysis of the toluenesulfonate (tosylate) ester in ethanoic acid (acetic acid, CH$_3$CO$_2$H)
2.	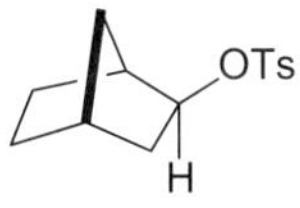 OTs, H	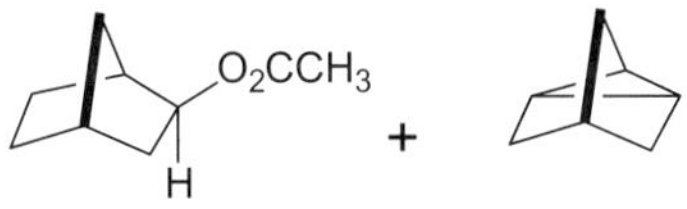 O$_2$CCH$_3$, H +	Products of solvolysis of the toluenesulfonate (tosylate) ester in ethanoic acid (acetic acid, CH$_3$CO$_2$H); Chiral reactant yields racemic product
3.	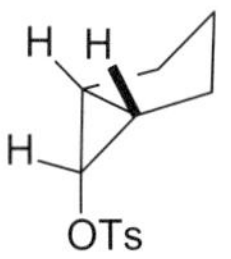 H H, H, OTs	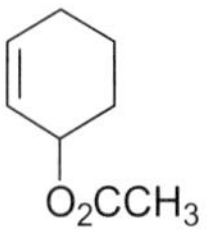 O$_2$CCH$_3$	Product of solvolysis of the toluenesulfonate (tosylate) ester in ethanoic acid (acetic acid, CH$_3$CO$_2$H)

(Continued)

TABLE 8.7. *Continued*

Item	Substrate	Product(s) of Rearrangement	Comments
4.	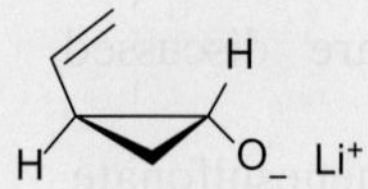	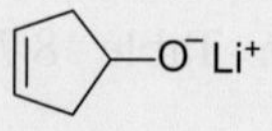	The lithium salt rearranges about 10^5 times faster than the alcohol.
5.	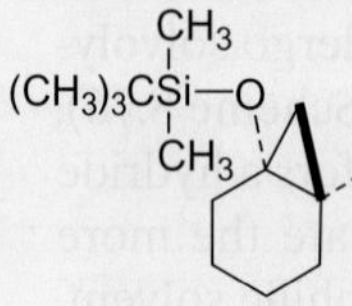	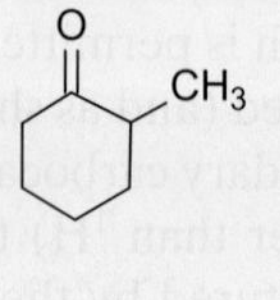 CH_2I_2/Zn	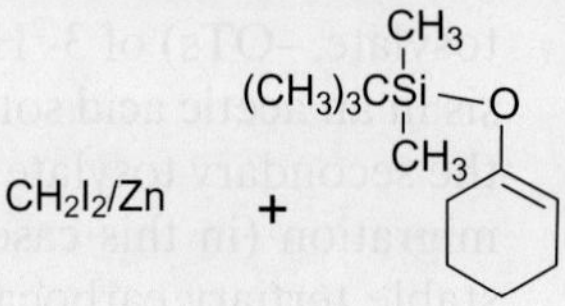
6.	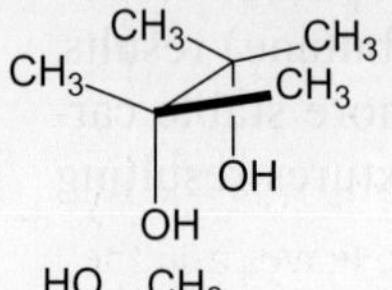	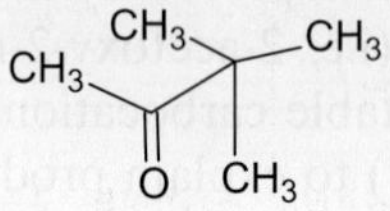	Acid catalysis (H_2SO_4) required; The *pinacol rearrangement*
7.	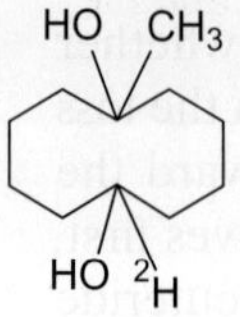	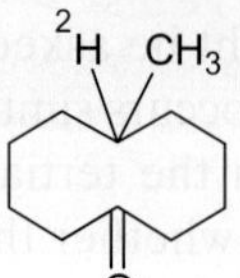	Acid catalysis (H_2SO_4) required
8.	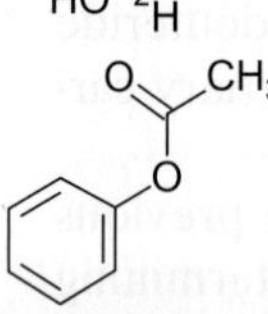	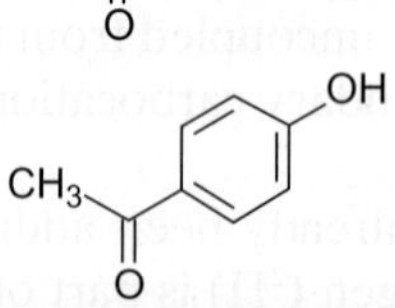	Catalysis by $AlCl_3$ or other Lewis acid required
9.	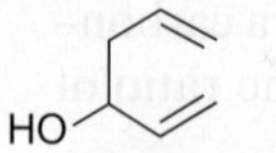	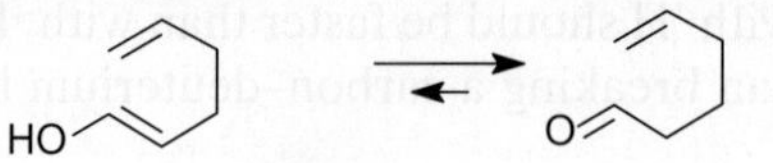	The lithium salt rearranges about 10^{10} times faster.
10.	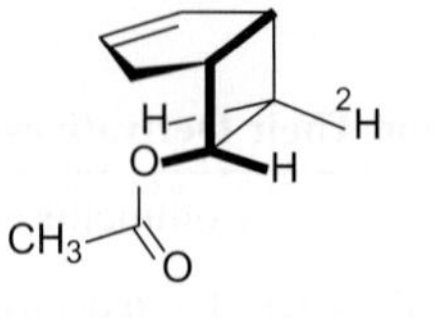	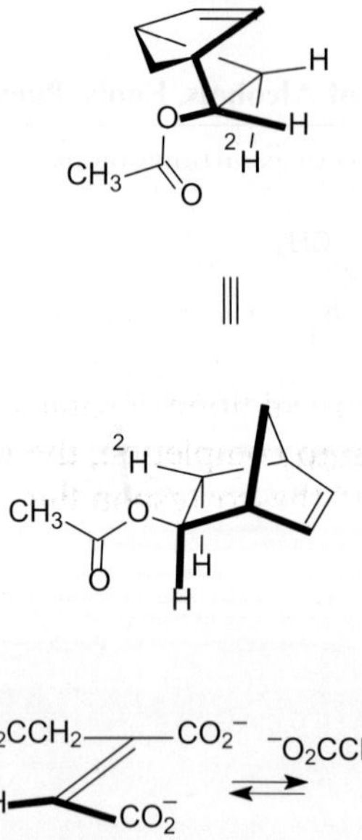	Thermal (ca. 300°C)
11.	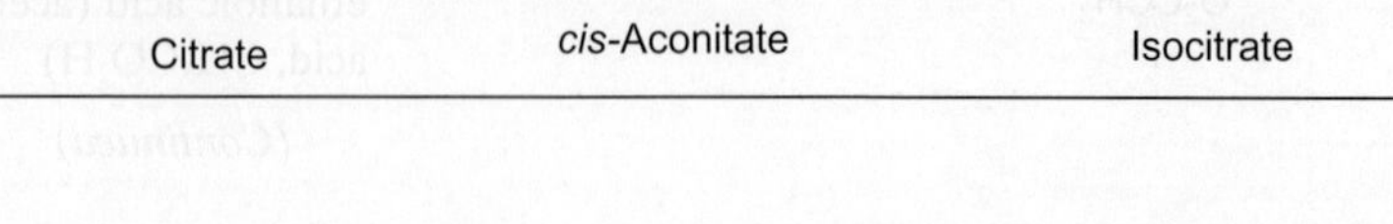		Catalysis is by the enzyme *aconitase*

Citrate   *cis*-Aconitate   Isocitrate

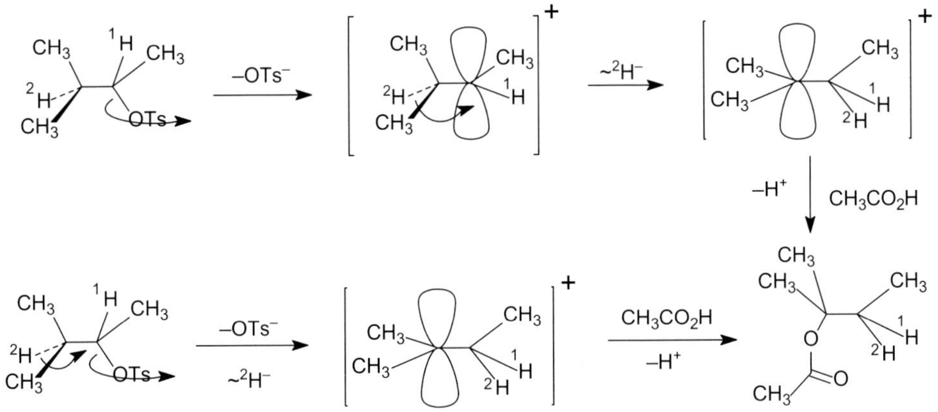

Scheme 8.76. A representation of two possible pathways from the tosylate of 3-^{2}H$_1$-3-methyl-2-butanol to the acetate of 3-^{2}H$_1$-2-methyl-2-butanol. In the first, the tosylate leaves and then the deuteride moves. In the second, the deuterium moves while the tosylate leaves. The same tertiary carbocation is generated and subsequently captured by acetate.

the rate constants should be greater than one; that is, $k_H/k_D > 1$ (where H = ^{1}H and D = ^{2}H).

Alternatively, if the rate-determining step is more E1-like (i.e., the tosylate leaves to form the carbocation in the slow step and the migration of the hydrogen [either ^{1}H or ^{2}H] is fast), then the rates should be identical and the ratio of the rate constants unity.

The second example of Table 8.7 has proved to be a major contentious problem in organic chemistry and, although investigted with many physical methods and debated with vigor (and unusual acrimony), is considered to remain unresolved in all its details. As outlined in Scheme 8.77, when the toluenesulfonate ester (tosylate), or any one of a number of similar esters of *exo*-bicyclo[2.2.1]heptan-2-ol (*exo*-2-hydroxybicyclo[2.2.1]heptane, *exo*-norbornanol) is allowed to undergo solvolysis in acetic acid, the corresponding *exo*-2-acetate ester (along with a small amount of *tricyclic* elimination product {tricyclo[2.2.1.0^{2,6}]heptane, nortricyclene}) is isolated. If the starting chiral alcohol has been resolved and the chiral tosylate is used in the reaction, the acetate product is racemic. In addition, if some of the tosylate is recovered before the reaction has proceeded to completion, the recovered tosylate has also begun to undergo racemization. If the corresponding optically active *endo*-tosylate is used as starting substrate, again, only the racemic *exo*-acetate (along with some norticyclene) is isolated. However, the *endo*-tosylate undergoes reaction more slowly than the *exo*-tosylate, and thus the latter is then considered to react in an accelerated fashion (or, conversely, the *exo*-tosylate undergoes reaction faster than the *endo*-isomer because the latter reacts too slowly).*

---

*There is little argument as to which undergoes reaction faster (or more slowly). However, there is some debate as to which is "normal"; is the *exo*-tosylate solvolysis faster than expected, that is, accelerated (the *endo*-isomer being normal), or is the *exo*-isomer undergoing solvolysis at the normal, unaccelerated rate while the *endo*-isomer is reacting more slowly than normal?

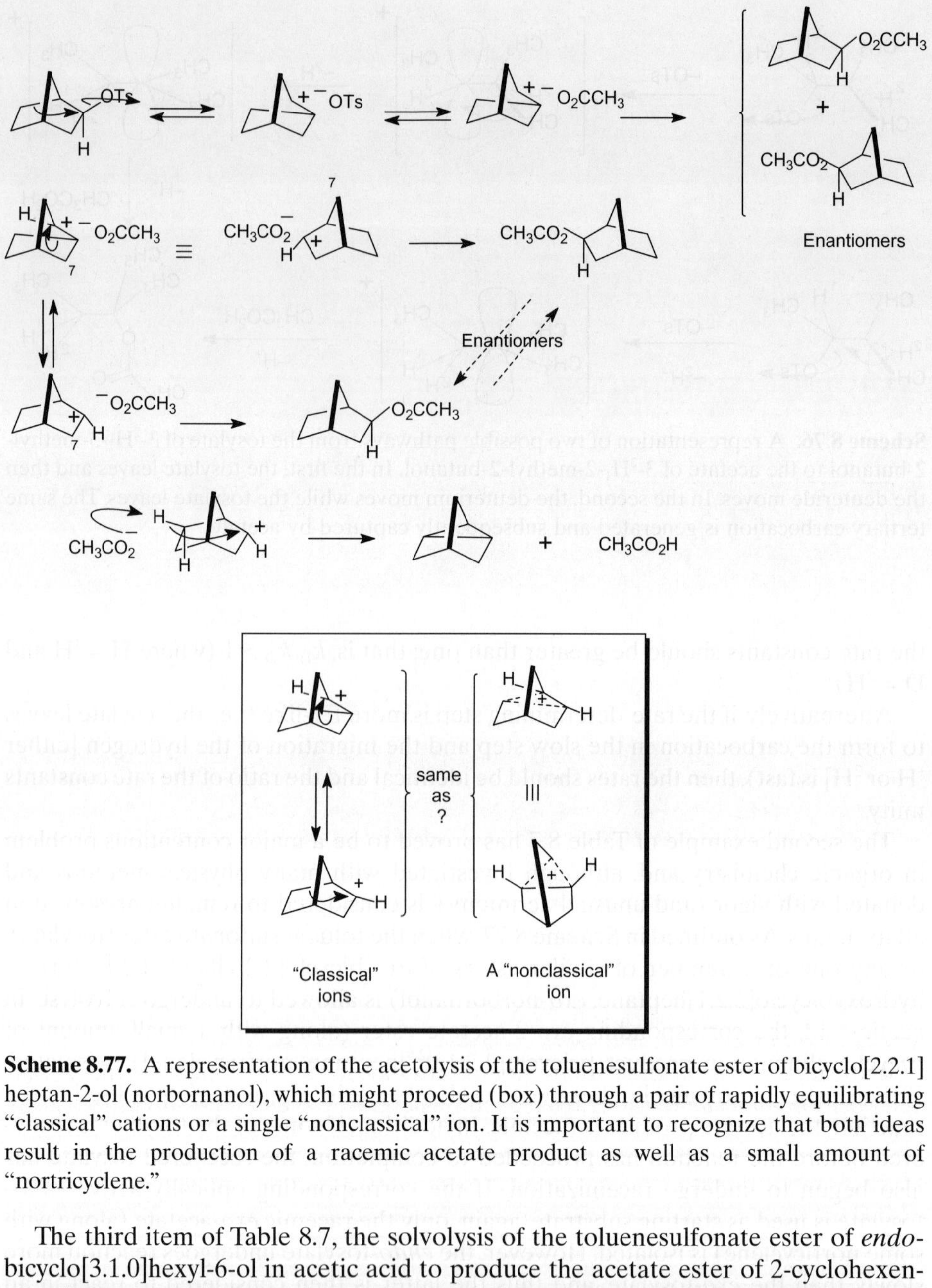

Scheme 8.77. A representation of the acetolysis of the toluenesulfonate ester of bicyclo[2.2.1] heptan-2-ol (norbornanol), which might proceed (box) through a pair of rapidly equilibrating "classical" cations or a single "nonclassical" ion. It is important to recognize that both ideas result in the production of a racemic acetate product as well as a small amount of "nortricyclene."

The third item of Table 8.7, the solvolysis of the toluenesulfonate ester of *endo*-bicyclo[3.1.0]hexyl-6-ol in acetic acid to produce the acetate ester of 2-cyclohexen-1-ol (Scheme 8.78), is of particular interest since it requires the opening of the cyclopropane ring in an allowed electrocyclic reaction (Chapter 4, Section G). In this case, the σ bond of the three-membered ring opens in a disrotatory fashion (conserving mirror symmetry) to generate the allyl cation.*

*As shown below, with only two electrons available over the three carbons of the allyl cation, only the lowest-lying MO is occupied, that is,

(Continued on next page)

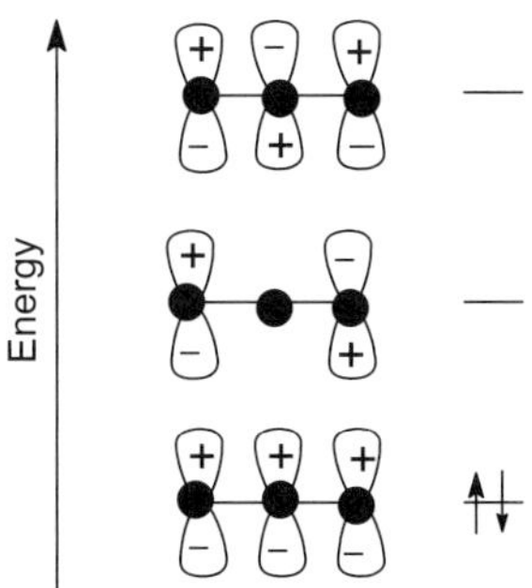

Scheme 8.78. A representation of the (allowed) disrotatory ring opening accompanying solvolysis of *endo*-bicyclo[3.1.0]hexyl-2-ol tosylate to produce cyclohexene-3-ol acetate (see Schoellkopf, U. *Chem. Ber.*, **1966**, *99*, 3391 and Schoellkopf, U.; Fellenberger, K.; Patsch, M.; Schleyer, P. V. R.; Su, T.; Van Dine, G. W. *Tetrahedron Lett.*, **1967**, 3639).

Presumably, the cation is generated with the departure of the tosylate leaving group. It is important to recognize that there are, in principle, **two** disrotatory pathways: *inward* and *outward*. However, these pathways have different outcomes.

In the latter case, the *outward* motion places the electrons originally in the σ bond in a position *opposite* to the leaving group where they are in a position to help the

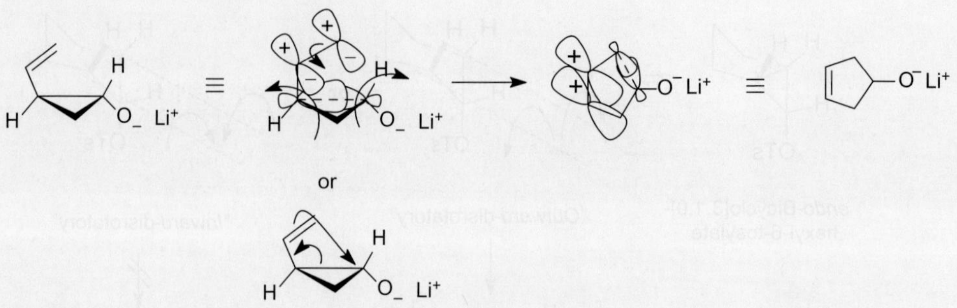

or

Scheme 8.79. A representation of the allowed thermal conversion of the lithium salt of *trans*-2-vinylcyclopropanol to the lithium salt of 4-hydroxycyclopentene.

leaving group (tosylate) depart. At the same time, the hydrogens on the carbons of that σ bond are put on the *outside* of the resulting six-membered ring (Scheme 8.78). In this way, the *endo*-bicyclo[3.1.0]hexyl-6-tosylate undergoes ring opening to the allylic cation, which subsequently reacts with acetate to yield the acetate ester of 2-cyclohexen-1-ol. The opposite motion would put the hydrogens *inward*, which is sterically difficult.

Interestingly, attempted solvolysis *under the same conditions* of the *exo*-bicyclo[3.1.0]hexyl-6-tosylate (Scheme 8.78), where the hydrogens *must* be put inward if the ring opening is to assist the departure of the leaving group, apparently undergoes reaction about 10^6 times more slowly. Indeed, it can be argued that the reaction of the *exo*-bicyclo[3.1.0]hexyl-6-tosylate to produce the product does not involve the ring-opening process at all but that a combination of normal S_N2 displacement (to give *endo*-bicyclo[3.1.0]hexyl-6-tosylate, which then undergoes ring opening with σ bond participation) and S_N1 solvolysis (to give a carbocation, which subsequently undergoes a ring-opening process) are necessary for reaction.

The fourth item of Table 8.7, the thermal conversion of a vinyl cyclopropane to a cyclopentene, is, again, an electrocyclic process. The rearrangement can be viewed as a 1,3-sigmatropic process. Alternatively, an internal $[2\pi + 2\sigma]$ sigmatropic cyclo-addition can be written and even a diradical intermediate can be envisioned. The presence of the hydroxyl group is not necessary for the thermally allowed skeletal change to occur, although, as noted, the alkoxide anion undergoes an accelerated conversion, reacting at or near room temperature (in contrast to the higher temperatures required in its absence). In concert with an allowed process, antarafacial reorganization of the π–σ system with inversion at the migrating carbon is shown in Scheme 8.79.

As shown in item 5 of Table 8.7, when a silylated enol of cyclohexanone is treated with the **Simmons–Smith reagent** ($[CH_2I_2/Zn]$; Chapter 6, Equation 6.52) cyclopro-panation is effected (Scheme 8.74). On reaction with base, the silylated cyclopropa-nol undergoes ring opening (via a presumed tetrahedral intermediate) to produce siloxide and 2-methylcyclohexanone (Scheme 8.80).

A much more common rearrangement (the *pinicol rearrangement*) is depicted as item 6 in Table 8.7. The particular case shown in item 6 bears the overall responsibil-ity for the generic type of reaction (i.e., 2,3-dimethylbutan-2,3-diol, the *trivial name* for which is **pinacol**, is converted to 3,3-dimethyl-2-butanone, the trivial name for

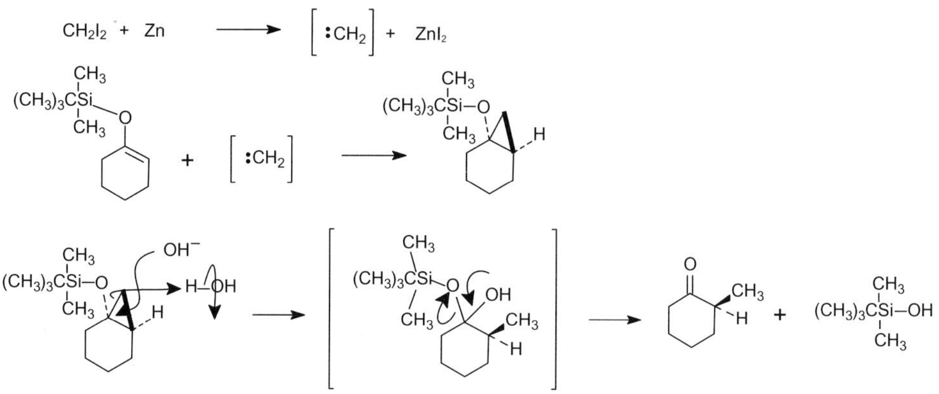

Scheme 8.80. A representation of the pathway from the *t*-butyldimethylsilyl ether of the enol of cyclohexanone to 2-methylcyclohexanone via the addition of "methylene" from the Simmons–Smith reagent and rearrangement of the cyclopropane formed (see Simmons, H. E.; Smith, R. D. *J. Am. Chem. Soc.*, **1959**, *81*, 4256).

which is **pinacolone**) but, in general, any such vicinal diol can be converted to the corresponding ketone with concomitant hydrogen or alkyl shift.* As shown in Scheme 8.81, initial protonation on oxygen is followed by loss of water with simultaneous migration of the adjacent methide to produce the corresponding protonated ketone (3-methyl-2-butanone [*pinacolone*]) in one step. **Alternatively,** dehydration to form a carbocation might precede the methide migration and the overall process can be considered as occurring in two distinct steps; that is, first, water is lost and then the migrating group moves.[†]

The seventh item in Table 8.7 is a *transannular* (across the ring) pinacol–pinacolone rearrangement in which it can be argued that a carbocation forms and that the hydride (^{2}H, deuteride) shift occurs with participation of the oxygen after the tertiary hydroxyl is lost. As shown in Scheme 8.82, the transannular hydride (^{2}H; deuteride can be quite close to the carbocation (or incipient carbocation if loss of water is accompanied by hydride (^{2}H, deuteride) migration.

---

*Although the subject will be discussed in Chapter 9, the *preparation* of such symmetric diols is tied to the early theories of valence. **Moses Gomberg** (1866–1947), University of Michigan, was interested in organometallic compounds (having shown that triarylmethyl halides could produce "stable" triaryl-methyl radicals) and, along with his colleague Werner E. Bachmann (1901–1951), University of Michigan, accomplished the bimolecular reduction of dialkyl and aryl ketones to vicinal diols with metals. For acetone (propanone), it has been found that amalgamated magnesium [Mg(Hg)] routinely produces pinacol (as a hydrate) in about 50% yield.

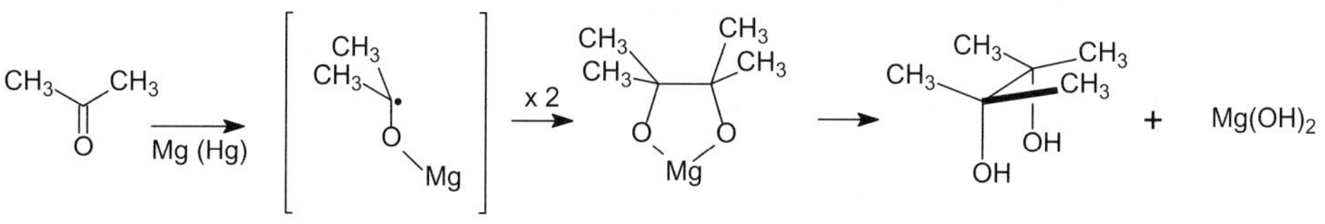

[†]Not surprisingly, it appears that the degree of participation is a function of the particular migrating group with an electron-rich phenyl group participating more than alkyl and alkyl more than hydrogen.

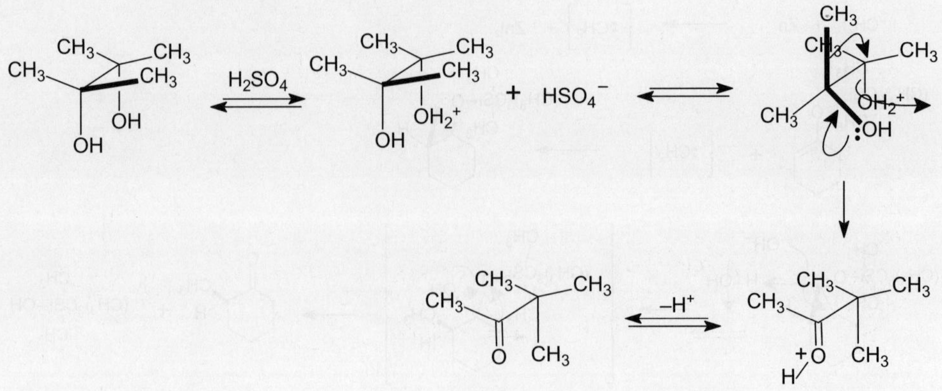

Scheme 8.81. A representation of the pinacol–pinacolone rearrangement. Pinacol (2,3-dimethlbutan-2,3-diol) on treatment with acid (sulfuric acid, H_2SO_4) is protonated, loses water, and a methyl group migrates (as methide) to the carbon from which water was lost with participation of the remaining hydroxyl group. Proton loss from the resulting protonated ketone generates pinacolone (3-methyl-2-butanone).

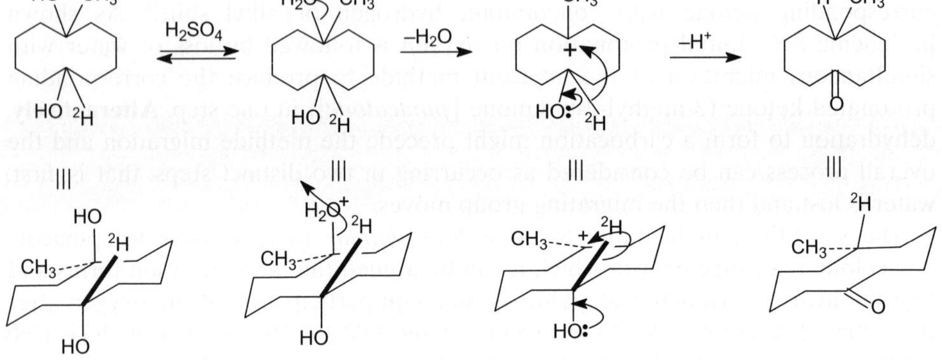

Scheme 8.82. A representation of a pinacol-type rearrangement involving a transannular 1,6-hydride (^2H, deuteride) migration in a 10-membered ring.

The eighth item in Table 8.7 is an example of the Fries rearrangement (after K. Fries, *vide supra*, Scheme 8.42). In this interesting reaction, evidence has accumulated that carboxylic acid esters of phenols, on treatment with aluminum trichloride, **reversibly** dissociate into two fragments (presumably the aluminum-substituted phenolate and acetyl chloride). The fragments then recombine into the corresponding *ortho-* and *para*-substituted phenolic ketones (Scheme 8.83).

The following is relevant: first, the *ortho-* and *para*-substituted phenolic ketones can be interconverted; that is, they are in equilibrium. At low temperature, the *para*-isomer predominates, but, as the temperature is raised, the *ortho*-isomer becomes the major product. Second, if two different esters are used, cross products are observed. Third, the rearrangement is *only* into the positions *ortho* and *para* to the hydroxyl group. If there is a substituent (such as a chlorine or methyl) *meta* to the phenol, it remains unaffected, but the product mixture seldom contains

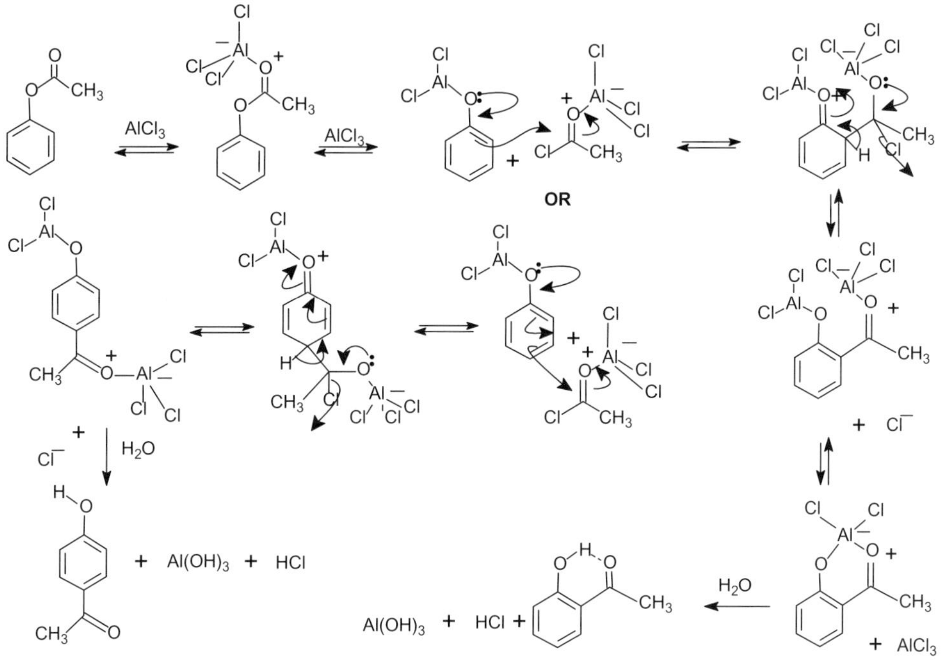

Scheme 8.83. Representations of the pathway of the Fries rearrangement by which the acetyl ester of phenol is converted to the *ortho-* and *para*-hydroxyacetophenones in the presence of aluminum chloride ($AlCl_3$). It is important to note that the reactions are reversible (save for the final hydrolysis) so that both acetophenones can be reconverted to starting ester. It is argued that the *ortho*-hydroxyacetophenone is stabilized by the aluminum chloride complex between the two oxygen atoms.

1,2,3-substituted products (for steric reasons) (Problem 8.15). As shown in Scheme 8.83, it is suggested that the *ortho*-product is stabilized by formation of an aluminum complex between the carbonyl oxygen and the phenolic hydroxyl group.

Problem 8.15. Given Scheme 8.83, account for the observations that

(a) *meta*-product is rarely (if ever) observed and

(b) that the acetate ester of 3-methylphenol (*meta*-cresol) produces 2-hydroxy-4-methylacetophenone and 3-methyl-4-hydroxyacetophenone but *not* 2-hydroxy-6-methylacetophenone under typical Fries rearrangement conditions with an aluminum trichloride ($AlCl_3$) catalyst (see Fries and Finck, 1908; Fries and Pfaffendorf, 1910).

The ninth item in Table 8.7 is called the *oxy-Cope rearrangement*. It is closely related to the Cope rearrangement of the parent hydrocarbon, 1,5-hexadiene, which, it will be recalled, is capable of an allowed $2\pi + 2\pi + 2\sigma$ electrocyclic transformation in Chapter 6 (e.g., see Equation 6.61) into itself. The electrocyclic process shown here (Scheme 8.84) enjoys an advantage over its parent in that (a) the conversion of the alcohol, 3-hydroxy-1,5-hexadiene, to the aldehyde, 5-hexenal, is readily detectable and (b) the alcohol-to-aldehyde conversion is a synthetically useful process

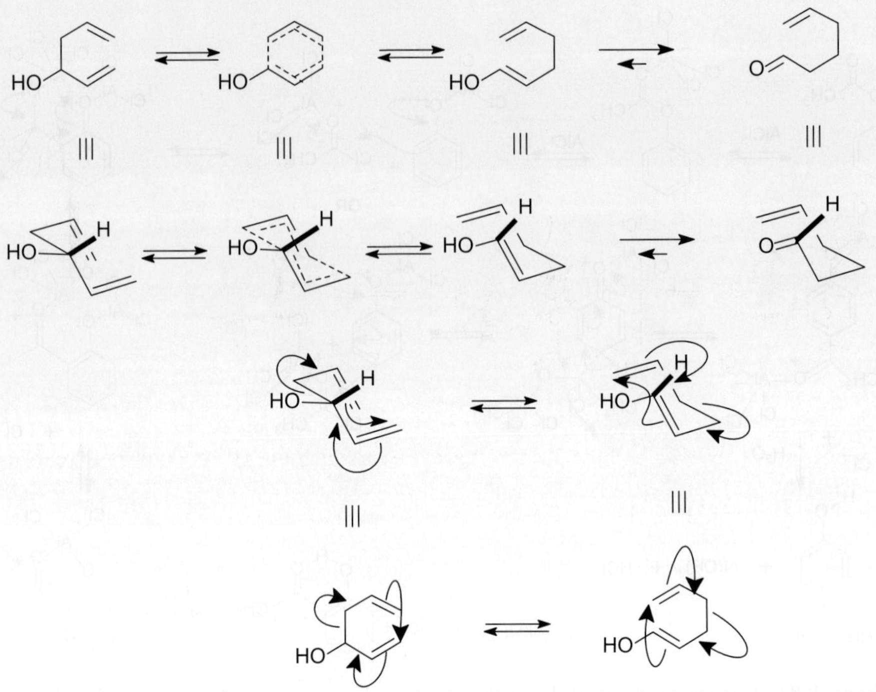

Scheme 8.84. A representation of the *oxy-Cope* rearrangement of 3-hydroxy-1,5-hexadiene to 5-hexenal. An allowed $2\pi + 2\pi + 2\sigma$ electrocyclic transformation (see Fujita, Y.; Onishi, T.; Nishida, T. *J. Chem. Soc. Chem. Commun.*, **1978**, 972 and Evans, D. A.; Nelson, J. V. *J. Am. Chem. Soc.*, **1980**, *102*, 774).

(Chapter 9). Interestingly, as noted in the table, the alkoxide anion (as, e.g., lithium salt) undergoes rearrangement about 10^{10} times more rapidly than the alcohol itself. The anion leads to the enolate anion of the aldehyde, which tautomerizes to the aldehyde on protonation.

The penultimate item (item 10) in Table 8.7, the acetate ester of *trans-7-exo*-[^{2}H$_1$]-bicyclo[3.2.0]hept-2-en-6-ol, enjoys a special place in the study of rearrangement reactions of alcohols and their derivatives. As noted in the table and as shown in Scheme 8.85, the deuterium ([^{2}H$_1$]) substituted bicyclo[3.2.0]heptene acetate, in which the deuterium ([^{2}H$_1$]) **begins** *trans* [or *anti*- or (*E*)-] to the acetate is thermally converted to the isomeric deuterium ([^{2}H$_1$]) substituted bicyclo[2.2.1]heptene acetate, in which the deuterium ([^{2}H$_1$]) **finishes** *cis* [or *syn*- or (*Z*)-] to the acetate. The conversion shown in Table 8.7, item 10, was the **first** test of the *Woodward–Hoffmann* (Chapter 4) prediction that a [1,3]-sigmatropic rearrangement is **required** to proceed with **inversion** of configuration at the migrating carbon. Interestingly, although not shown here, other cases have subsequently been found where the inversion is less than complete. In such cases, it is believed that the reaction takes a different pathway (i.e., radicals are formed) rather than that the rules have failed.

The last example shown in Table 8.7 is a rearrangement common in living systems. Both eukaryotes (cells with a membrane-enclosed nucleus that encapsulates their DNA [deoxyribonucleic acid]) and prokaryotes (simpler, unicellular structures

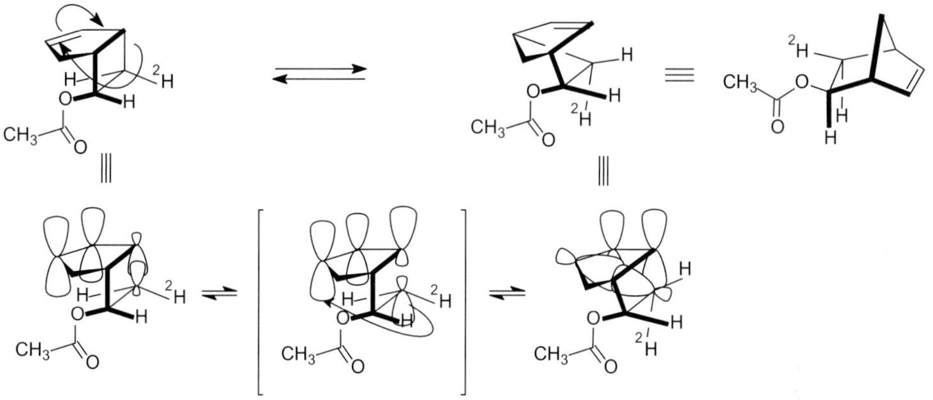

Scheme 8.85. The rearrangement of *trans*-7-*exo*-[$^2H_1$]-bicyclo[3.2.0]hept-2-en-6-yl acetate to *cis*-6-*exo*-[$^2H_1$]-bicyclo[2.2.1]hept-2-en-5-yl acetate in which the [1,3]-sigmatropic rearrangement is shown to proceed with inversion of configuration at the carbon-bearing deuterium ([$^2H_1$]) (see Berson, J. A. *Acc. Chem. Res.*, **1968**, *1*, 152 and references therein).

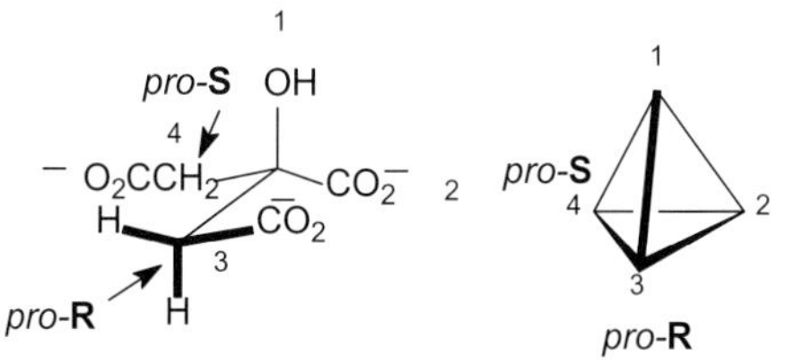

Figure 8.20. The *pro*-R and *pro*-S carboxymethylene arms of citrate.

that lack the membrane-enclosed nucleus) commonly utilize the **citric acid cycle** (also known as the **tricarboxylic acid [TCA] cycle** and the **Krebs cycle** [Chapter 11]) to account for most of the oxidation of fatty acids (Chapter 11), carbohydrates (Chapter 11), and amino acids (Chapter 12). Although the overall series of reactions occurs in a "cycle" (and therefore cannot have a true beginning or end), it is possible to examine portions of it by blocking reactions that allow subsequent parts of the cycle to function (and thus "shutting them down") or by actual isolation of specific enzymes and cofactors that perform the specific reactions within the cycle.

Among the eight enzymes used in the citric acid cycle of more or less comon organic reactions needed to convert the acetyl (CH_3CO) group of acetyl coenzyme A (acetyl-CoA; e.g., see this chapter and Chapter 11) into two molecules of carbon dioxide (CO_2) is that one called *aconitase*. Aconitase has been carefully examined and it has been determined that, as shown in Table 8.7, the enzyme catalyzes the reversible isomerization of citric and isocitric acids through the intermediacy of *cis*-aconitic acid.

The enzyme (aconitase), being itself chiral, can distinguish between the two enantiotopic carboxymethylene groups (one *pro*-**R** and the other *pro*-**S**) of citrate (Figure 8.20). Then, as shown in Scheme 8.86, the *pro*-R proton from the *pro*-R carboxymethylene is abstracted and the hydroxyl is lost in an *antiperiplanar*

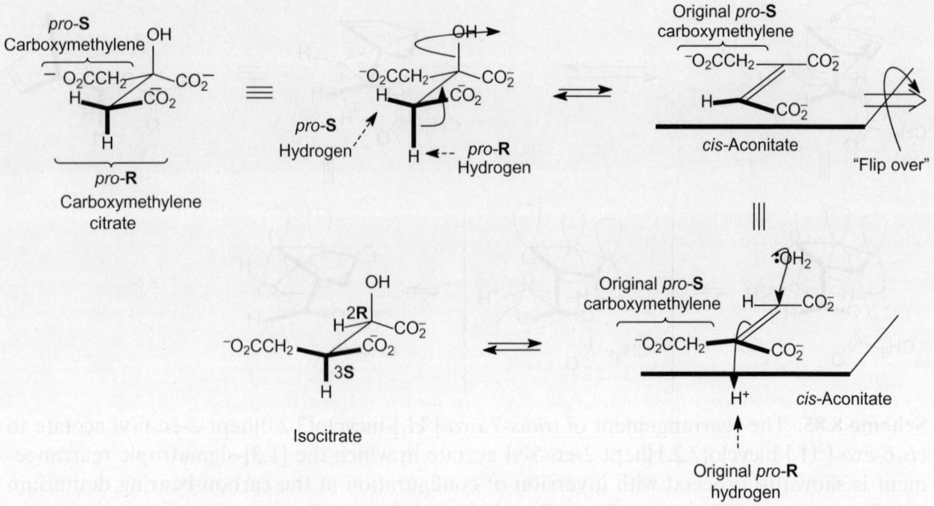

Scheme 8.86. A cartoon representation of the equilibrium between citrate and isocitrate, occurring through *cis*-aconitate and as catalyzed by the enzyme *aconitase* (see Glusker, J. P. Aconitase, in Boyer, P. D. (ed.), *Enzymes*, Vol. 5, 3rd edition, Academic Press, New York, **1971**, 413).

elimination.* Then, the enzyme catalyzes the rehydration of the *cis*-aconitate to produce **only** (2R,3S)-isocitrate. It was particularly interesting to find that the oxygen of the hydroxyl group at C-2 in the (2R,3S)-isocitrate was **not** the same oxygen that was lost from citrate but that the *pro*-**R** proton at C-3 of isocitrate **was** the same proton lost in the previous step. As shown, this suggests that the *cis*-aconitate "flips over" before water attacks and a proton is picked up (Scheme 8.86).

PART II. ETHERS

A. INTRODUCTION

As already pointed out at the beginning of this chapter (Chapter 8) and in Chapter 5, Section IV, the group of compounds called **ethers** may be considered as arising from the replacement of **both** of the hydrogens of water (H_2O) by alkyl and/or aryl groups. In this way, compounds such as those of Table 8.2 and in Chapter 5, Section IV, as well as some additional examples shown in Figure 8.21, all fit into the family.

As pointed out earlier (Chapter 5), ethers, although polar molecules, have physical properties very similar to hydrocarbons of about the same molecular weight and structure. For example, diethyl ether (ethyl ether, $CH_3CH_2OCH_2CH_3$), with a molecular weight of 84, boils at 34.6°C, while pentane ($CH_3CH_2CH_2CH_2CH_3$), with a

*The reader will appreciate that information in addition to that presented here is available in the literature and in some advanced texts. Indeed, it is difficult to decide exactly what level of sophistication can be expected here. At this writing, several of the amino acids in the active site of the enzyme have been identified through a combination of X-ray analysis and site-directed mutagenesis.

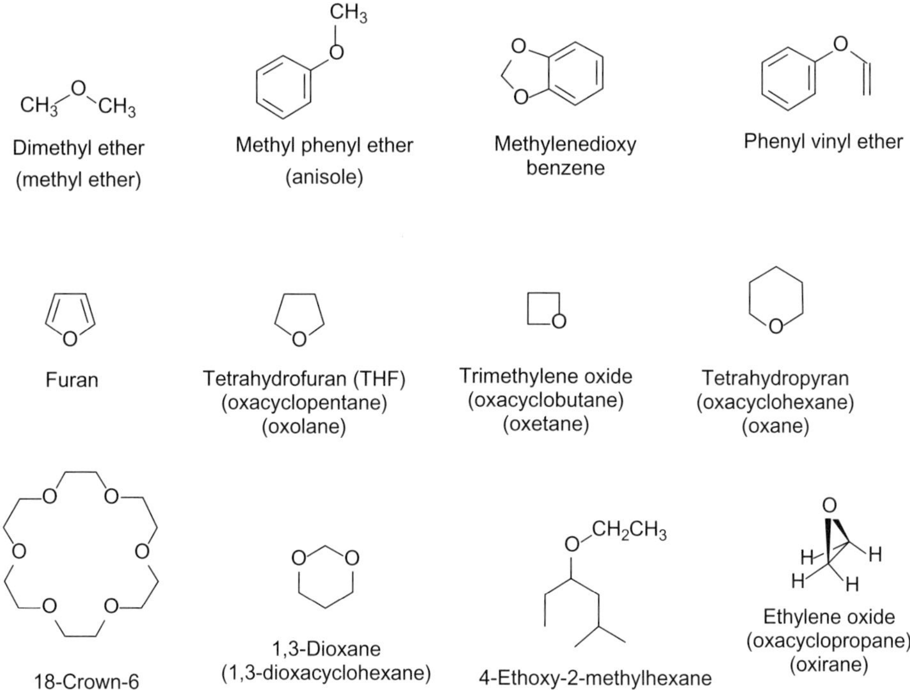

Figure 8.21. Some examples of cyclic and acyclic ethers.

molecular weight of 82, boils at 36.1°C. Not unexpectedly (because of hydrogen bonding), 1-butanol, isomeric with diethyl ether, boils at a significantly higher temperature (117.6°C). However, despite this similarity, ethers are dramatically different from hydrocarbons because of the availability of the electrons on oxygen.

Therefore, it is not surprising that low-molecular-weight ethers have a greater solubility in water (hydrogen bonding) than the hydrocarbons of about the same molecular weight and, what is significantly more important to their chemistry, ethers are both capable of protonation (Brønsted bases) and reaction with Lewis acids. Although alcohols, R–OH (see this chapter, Part I), enjoy the same capability with respect to acids (because they too have nonbonded electrons on oxygen), it will be recalled that much of their reactivity is dictated by their ability to lose the proton on oxygen, a feature unavailable to ethers. So, in contrast to alcohols, solutions of Lewis acids in ethers can serve as vehicles for delivery of the Lewis acid into reaction mixtures. Nonetheless, it is important to keep in mind that since the acids can also react with ethers to give products resulting from covalent bond making/breaking reactions, the delivery process frequently requires low temperatures.

Thus, for the most part, ethers are relatively unreactive compounds. However, they have a particular susceptability to oxidation in air. This proclivity has already been alluded to in Chapter 5 where it was pointed out that storage of low-molecular-weight ethers (such as diethyl ether [$(CH_3CH_2)_2O$]) is considered dangerous as they form unstable peroxides (Equation 8.42).

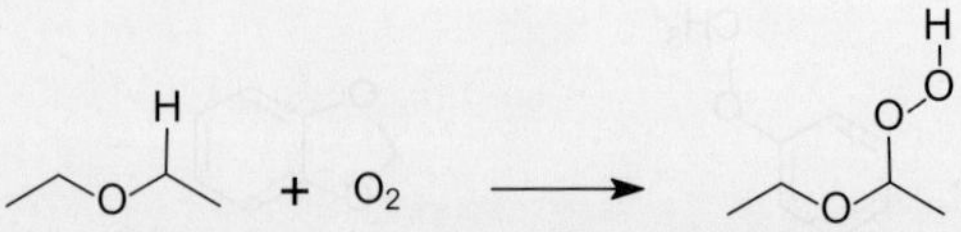

$$\text{(8.42)}$$

The process by which hydroperoxide is formed is not well understood, and it may be that traces of other impurities are necessary to initiate the reaction. There is evidence that peroxide formation can be minimized by storing ethers in the dark and that, as shown in Equation 8.42, oxidation preferentially occurs on the carbon bearing the oxygen of the ether linkage.

Other oxidation reactions at the same carbon atom can be induced with oxidizing agents too. For example, as shown in Equation 8.43, oxidation at the α-carbon to produce a carbonyl group can be brought about by ruthenium tetroxide (RuO$_4$).

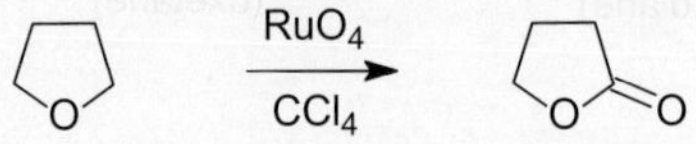

$$\text{(8.43)}$$

The details of the pathway by which this oxidation occurs are not known.

B. THE REACTIONS OF ETHERS

Hydroboration of alkenes (Chapter 6) is frequently effected by use of solutions of borane (B$_2$H$_6$) in THF, the ether (THF)–borane (BH$_3$) complex (H$_3$B•THF) serving to deliver the borane to the alkene Scheme 8.87).

As might be expected from the ability of THF to complex with borane (*vide supra*), reaction of ethers with electron-deficient species is common and, indeed, the reactions of ethers with electron-deficient species defines most of their chemistry.

All ethers react with Brønsted and Lewis acids. Under forcing conditions, most ethers can be induced to fragment into an alkyl halide (or similar) component and alcohol, enol, or phenol. This is the formal reverse of a substitution reaction between

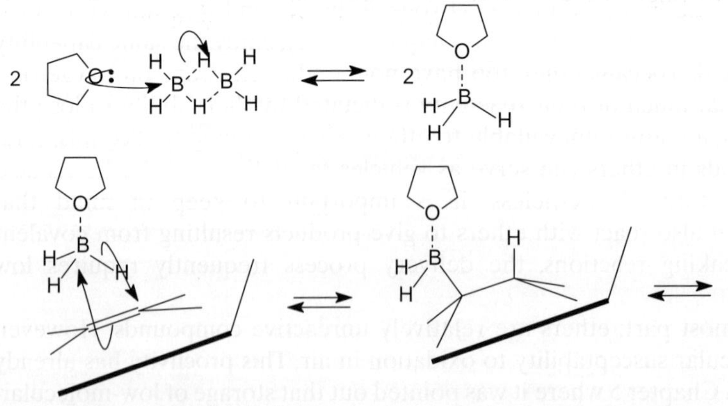

Scheme 8.87. A representation of use of the borane–tetrahydrofuran (BH$_3$•THF) complex in the hydroboration of an alkene.

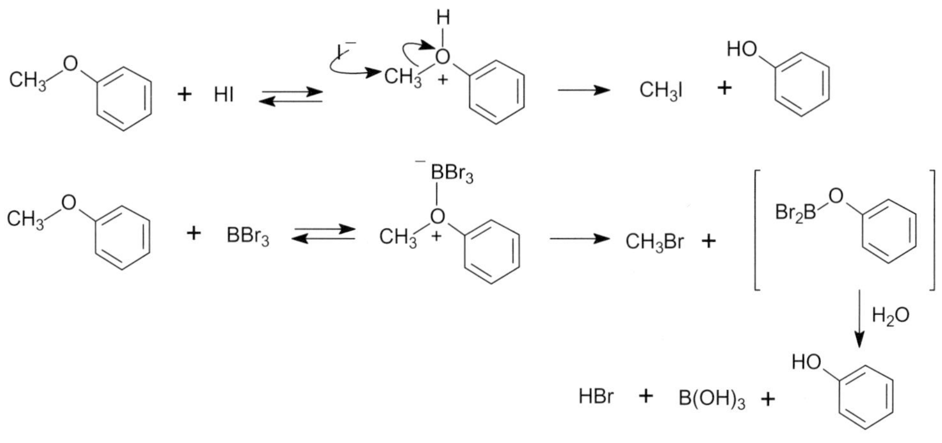

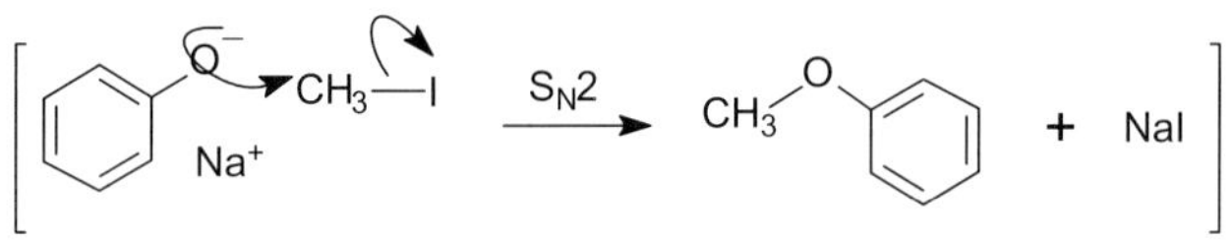

Scheme 8.88. A representation of the acid-catalyzed fragmentation of methyl phenyl ether (anisole) by the reaction with both hydrogen iodide (HI) and boron tribromide (BBr$_3$). In the first case, iodomethane (methyl iodide, CH$_3$I) and phenol are formed. In the second case, bromomethane (methyl bromide, CH$_3$Br) and phenol result. The formation of methyl phenyl ether (anisole) from the reaction between methyl iodide and sodium phenoxide is also shown.

the alkyl halide and the alcohol, enol, or phenol. For example, as shown in Scheme 8.88, methyl phenyl ether (anisole) can be induced to undergo cleavage to methyl iodide and phenol by a reaction with hydrogen iodide* or to methyl bromide and phenol by a reaction with boron tribromide. Also, as shown in the same scheme, methyl phenyl ether (anisole) can be prepared by an S$_N$2 reaction between phenoxide anion and methyl iodide. The cleavage reactions (Scheme 8.88) do not produce phenyl iodide (or bromide) and methanol; that is, although in general *dialkyl* ethers will undergo the cleavage reaction to produce two alkyl iodides and two alcohols as a consequence of the two possible displacement reactions by iodide on the alkyl groups bearing the positively charged oxygen (and subsequent hydrolysis), with *alkaryl* ethers, displacement by iodide is exclusively by an attack on the alkyl group.

Problem 8.16. In our present "spectroscopic" era, how might you determine the presence of an ether functionality in methyl phenyl ether? Be as specific as you can.

---

*In the prespectroscopic era, the reaction of methyl phenyl ethers, or indeed, any ether that can be cleaved by heating with aqueous hydroiodic acid (HI) to produce a volatile alkyl iodide, was used to demonstrate the presence of these functional groups. The test (called the **Zeisel** test) depended upon a color change observed when the hot vapors of alkyl iodide impinged upon a piece of filter paper moistened with (colorless) mercury(II) nitrate solution and the resulting reaction in which (orange) mercury(II) iodide is formed.

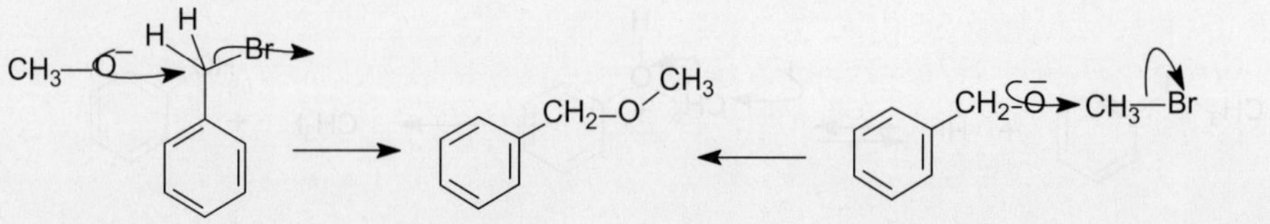

Scheme 8.89. The Williamson ether synthesis of benzyl methyl ether by the reaction of the anion of benzyl alcohol with methyl iodide (iodomethane, CH_3I) and by the reaction of the anion of methanol with benzyl iodide (*vide supra*; see Williamson, A. W. *Liebigs Ann. Chem.*, **1851**, *77*, 37.)

The S_N2 reaction between alkoxide and alkyl halide (Schemes 8.88 and 8.89) is known as the **Williamson ether synthesis**.

The Williamson synthesis cannot, of course, be utilized to prepare ethers under all circumstances. For example, as might be expected, methyl *t*-butyl ether is readily prepared by the reaction of *t*-butoxide anion $[(CH_3)_3C-O^-]$ with methyl iodide (CH_3I) (Equation 8.44), but the same ether cannot be prepared by the reaction of methoxide anion (CH_3O^-) and 2-iodo-2-methylpropane (*t*-butyl iodide).

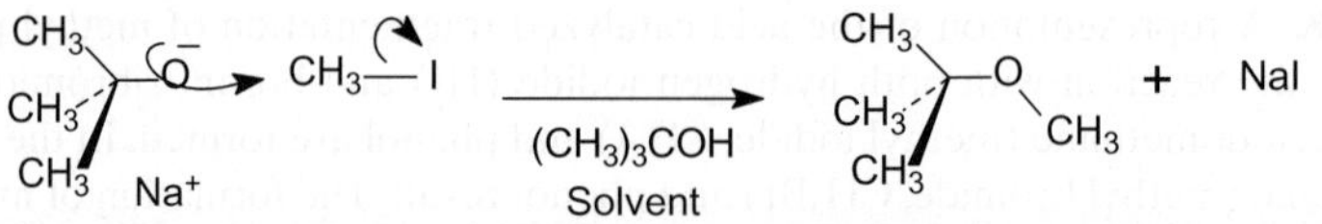

$$\text{(8.44)}$$

Problem 8.17. How would you account for the experimental observation that the reaction between methoxide anion (CH_3O^-) and 2-iodo-2-methylpropane (*t*-butyl iodide) fails to yield methyl *t*-butyl ether? What do you suppose does happen when these reagents are combined?

Despite the previous examples of ether cleavage by acid, it is generally true that, except for oxiranes (oxacyclopropanes, epoxides) and certain other ethers such as acetals (and ketals) (Chapter 9), orthoesters (Chapter 9), and enol ethers (this chapter), all of which are discussed next, reactions of alkyl, aryl, and alkaryl ethers with mineral acids are limited to rapid and reversible protonation. Thus, *it is important to remember that the reactions resulting in carbon–oxygen bond cleavage generally occur slowly and only under forcing conditions.*

On the other hand, as might be anticipated, oxiranes (oxacyclopropanes, epoxides) react with mineral acids such as aqueous hydrochloride acid (HCl) under ionic conditions to produce the corresponding halohydrins. With oxiranes (oxacyclopropanes, epoxides) in which one of the carbon termini is primary and the other secondary, protonation is usually followed by an attack at the primary carbon, yielding a secondary alcohol (Equation 8.45).

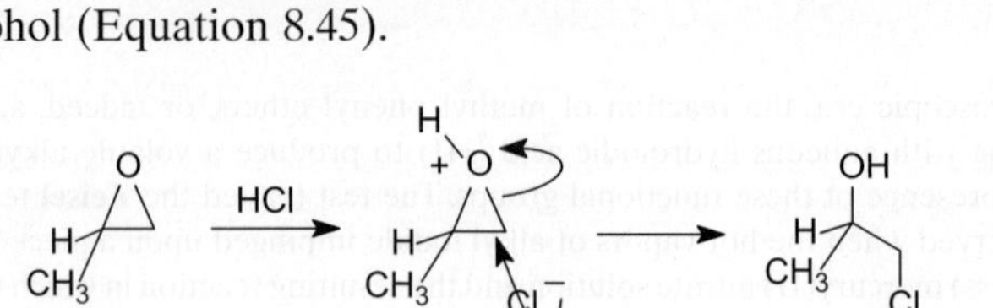

$$\text{(8.45)}$$

However, if one of the termini is tertiary and the other is primary, aqueous acid generally causes the oxirane (oxacyclopropane, epoxide) to open in the opposite way, yielding a primary alcohol (Equation 8.46).

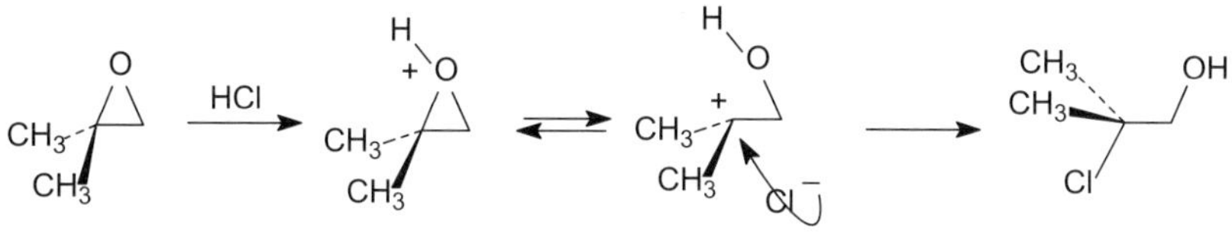

$$(8.46)$$

Further, if an aqueous acid with a non-nucleophilic gegenion (e.g., perchlorate [ClO_4^-] or hydrogen sulfate) [HSO_4^-]) is used, the oxirane (oxacyclopropane, epoxide) is attacked by water in a *trans-* (or *anti-* or antarafacial) fashion to produce a *trans*-diol (Equation 8.47).

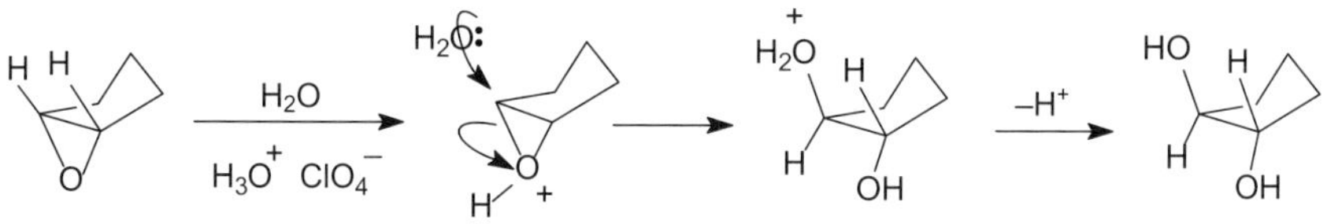

$$(8.47)$$

In reactions directly analogous to the above, enol ethers can be converted to the corresponding carbonyl compounds (Scheme 8.90) and acetals (and ketals) (Chapter 9) to the corresponding aldehydes (and ketones) via the intermediate hemiacetals (hemiketals) (Scheme 8.91). Similarly, orthoesters (Chapter 9) generate esters and/ or the corresponding alcohols and carboxylic acids (Scheme 8.92).

The ring-opening process of Equation 8.45 is, of course, simply the reverse of the process by which oxiranes (oxacyclopropanes, epoxides) are formed from halohydrins (e.g., see item 3, Table 7.6). Further, as written, the processes shown in Schemes 8.90–8.92 are reversible and thus, at least in principle, carbonyl compounds can be converted to enol ethers, acetals (and ketals), and orthoesters. However, while acetals and ketals readily form from alcohols and acids under dehydrating conditions (Chapter 9) and esters undergo exchange reactions with alcohols in the

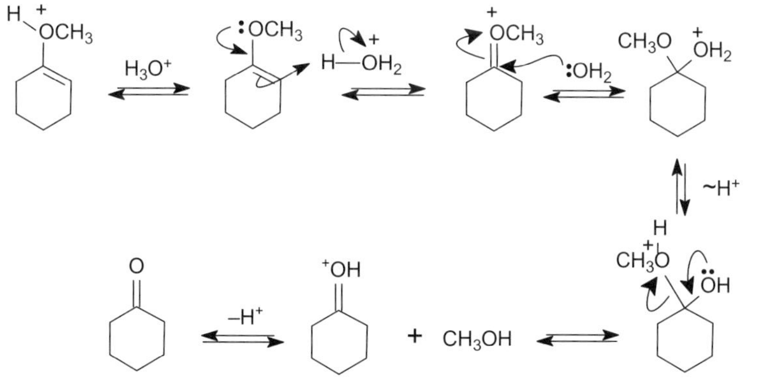

Scheme 8.90. A representation of a pathway for the acid-catalyzed hydrolysis of the methyl enol ether of cyclohexanone.

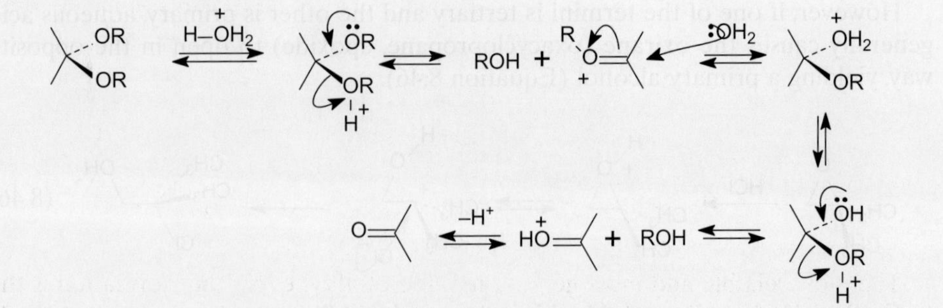

Scheme 8.91. A representation of a pathway for the formation of a carbonyl compound (aldehyde or ketone) by the acid-catalyzed hydrolysis of an acetal (to aldehyde) or ketal (to ketone) via the corresponding, respective hemiacetal or hemiketal.

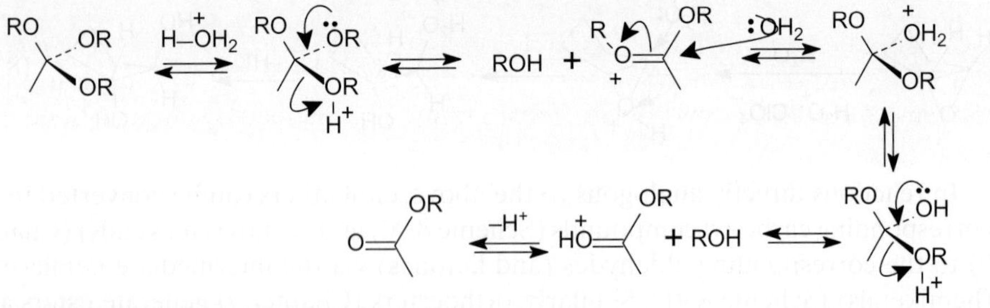

Scheme 8.92. A representation of a pathway for the formation of an ester (Chapter 9) by the acid-catalyzed hydrolysis of an orthoester. Subsequent hydrolysis of the ester to alcohol and carboxylic acid is also possible and is discussed more fully in Chapter 9.

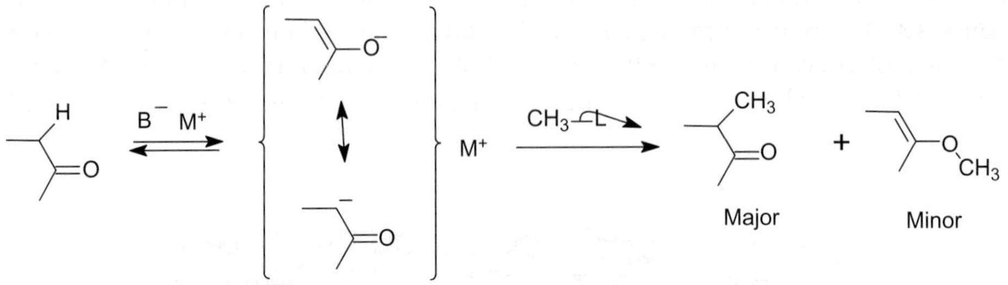

Major

Minor

Scheme 8.93. A generalized depiction of the possiblities of C- and O-alkylation of an enolate anion. C-alkylation is generally preferred, although varying amounts of O-alkylation can be produced by judicious choice of metal–base, leaving group combinations.

presence of acid catalysts (Chapter 9), the formation of enol ethers and orthoesters by the reverse of the forward processes of Schemes 8.90 and 8.92 is not particularly facile.

Indeed, even attempts at the formation of enol ethers by O-alkylation of enols (or enolate anions) is generally frustrated by the intrusion of C-alkylation (Scheme 8.93; see Chapter 9), although some adjustment of the ratio of C- to O-alkylation can be effected by choice of metal–base (e.g., *t*-butyllithium [$(CH_3)_3C^-Li^+$] rather than sodium ethoxide [$CH_3CH_2O^-Na^+$]) and leaving group (e.g., toluemesulfonate rather than iodide). The reaction of ketones with diazomethane (CH_2N_2) (Equation 8.48) does provide O-methyl enol ethers, but the yields are (generally) not very good.

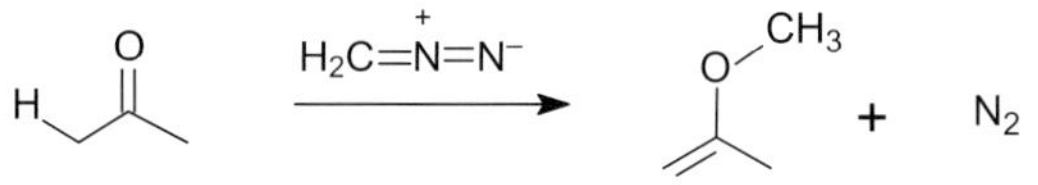

$$(8.48)$$

The most reliable and most general method of alkyl enol ether formation is the halogenation of an alkene in alcohol solvent and the subsequent elimination of HX (dehydrohalogenation) (Scheme 8.94).

Although discussed more fully in Chapter 9 when reactions of aldehydes and ketones are examined, it is worthwhile noting here that silyl enol ethers are easily prepared by the reaction between a carbonyl compound bearing an acidic α-proton with an alkyl-, alkaryl-, or aryl-substituted chlorosilane (e.g., trimethylchlorosilane [$(CH_3)_3SiCl$] and *t*-butyldimethylchlorosilane [$(CH_3)_3C(CH_3)_2SiCl$]) in the presence of a base (whose basicity is adjusted to fit the acidity of the proton to be abstracted) (Scheme 8.95). When silyl enol ethers are treated with aqueous solutions of fluoride ion (e.g., from tetra-*n*-butylammonium fluoride [n-$Bu_4N^+F^-$]), they revert to the corresponding carbonyl compounds via a displacement of oxygen from silicon by fluoride.

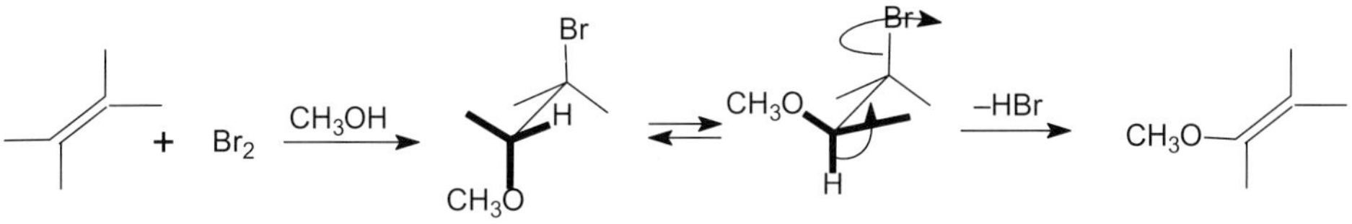

Scheme 8.94. A representation of the reaction of a "generic" alkene with bromine (Br_2) in methanol (CH_3OH) (Chapter 6) to produce a β-bromo methyl ether (the O-methyl ether of a bromohydrin) followed by elimination of hydrogen bromide (shown as an E2 process) to yield a (generic) enol ether.

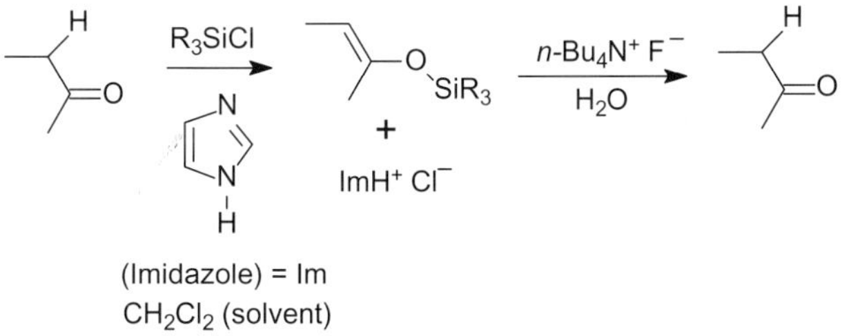

Scheme 8.95. The formation of a generic silyl enol ether (R = alkyl, alkaryl, aryl, etc.) from a generic carbonyl compound (aldehyde or ketone) with imidazole as base in methylene chloride solvent. It is argued that the silylation occurs preferentially on oxygen because of the strength of the silicon–oxygen bond. The silyl enol ether can be *reconverted* to the corresponding aldehyde or ketone by treatment with tetra-*n*-butylammonium fluoride (n-$Bu_4N^+F^-$).

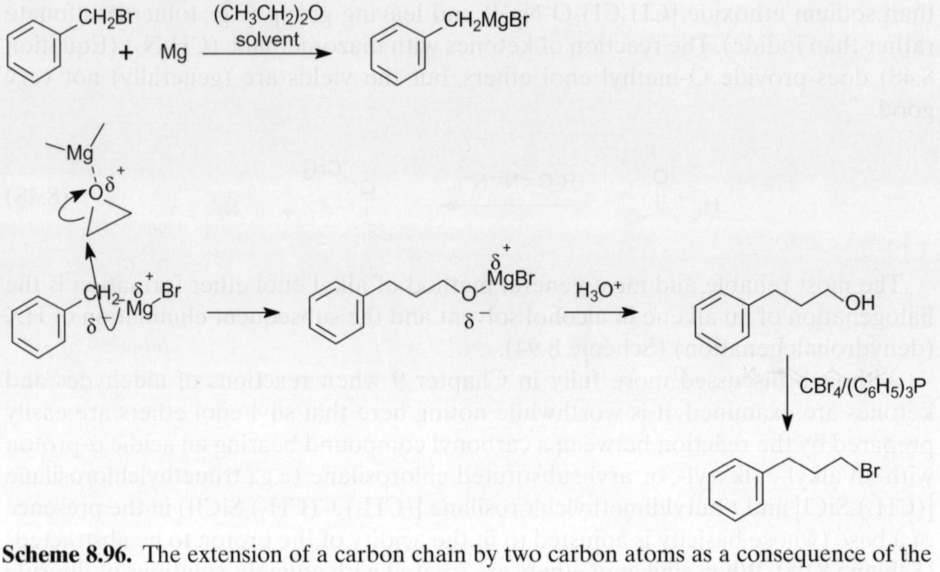

Scheme 8.96. The extension of a carbon chain by two carbon atoms as a consequence of the ring opening of an oxirane (epoxide, ethylene oxide) by the reaction with a representation of an organometallic reagent (benzylmagnesium bromide, a Grignard reagent), hydrolysis, and conversion of the resulting alcohol (3-phenyl-1-propanol) to the corresponding alkyl bromide (1-bromo-3-phenylpropane).

Ethers are generally stable to organometallic reagents, such as Grignard reagents (RMgX, Chapter 7) and, indeed, Grignard reagents (RMgX) are often prepared in ethyl ether. However, epoxides (oxiranes, oxacyclopropanes), acetals, ketals, and orthoesters react with such organometallic compounds with the formation of new carbon–carbon bonds. Thus, as shown in Scheme 8.96, reaction of a Grignard reagent (benzylmagnesium bromide) with ethylene oxide (oxirane, oxacyclopropane) produces, after hydrolysis, an alcohol (3-phenyl-1-propanol) with a carbon chain *two carbon atoms longer* than the original. Since alcohols (this chapter) can be converted to alkyl halides, the process can be seen to serve to extend a carbon chain by two units. Other organometallic reagents react similarly to cause ring opening of the oxirane (oxacyclopropane, epoxide) or to convert acetals and ketals to alcohols and orthoesters to ketals.

Problem 8.18. Using Scheme 8.96 as a template, write the reactions expected between benzylmagnesium bromide (in ether) with

(a) an acetal of your choice,

(b) a ketal of your choice, and

(c) an orthoester of your choice.

Among the more interesting developments in ring-opening reactions of epoxides (oxiranes, oxacyclopropanes) has been the observation that the process can be made enantioselective by the use of suitable Lewis acid–base pairs. Thus, with silicon tetrachloride in the presence of HMPA [$(Me_2N)_3P{=}O$], chlorohydrins can be generated. Then, if one enantiomer of a chiral phosphoramidate (such as the bis-

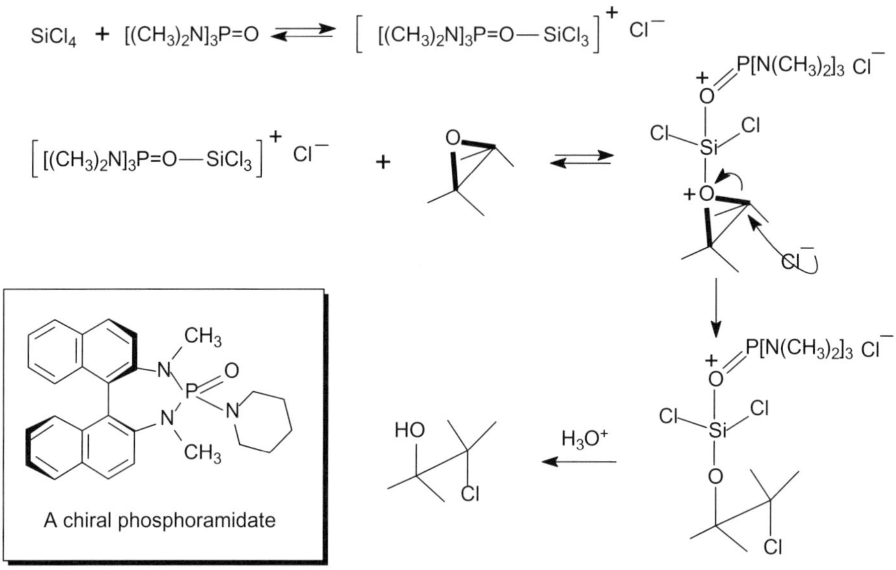

$$SiCl_4 + [(CH_3)_2N]_3P=O \rightleftharpoons \left[[(CH_3)_2N]_3P=O—SiCl_3 \right]^+ Cl^-$$

$$\left[[(CH_3)_2N]_3P=O—SiCl_3 \right]^+ Cl^- +$$

A chiral phosphoramidate

Scheme 8.97. Epoxide opening to chlorohydrin through the action of silicon tetrachloride ($SiCl_4$) and hexamethylphosphoramidate (HMPA, $[(CH_3)_2N]_3P=O$). The chiral phosphoramidate shown induces enantioselectivity in the opening process (Denmark, S. E.; Barsanti, P. A.; Wong, K.-T.; Stavenger, R. A. *J. Org. Chem.*, **1998**, *63*, 2428).

naphthaline derivative shown in Scheme 8.97) is employed in place of the HMPA, the stereochemical outcome of the opening is biased, and enantiomerically enriched chlorohydrin is produced. A proposed pathway is outlined in Scheme 8.97.

Phosphines, for example, triphenylphosphine $[(C_6H_5)_3P]$ (Scheme 8.98), also react with oxiranes (epoxides, oxacyclopropanes). In this process, an alkene and triphenylphosphine oxide are generated. Interestingly, where rotation about the carbon–carbon bond of the ring opened species is possible, the alkene resulting from deoxygenation has the geometry opposite to that of the alkene originally subjected to epoxidation. As shown in Scheme 8.98, epoxidation (e.g., with trifluoroperacetic acid $[CF_3CO_3H]$) of (*Z*)-2-butene produces an oxirane (epoxide, oxacyclopropane) in which the two methyl groups are on the same face of the ring. Then, on treatment with triphenylphosphine $[(C_6H_5)_3P]$, the heterocyclic ring opens, rotation occurs, and suprafacial (*cis*) elimination of triphenylphosphine oxide $[(C_6H_5)_3P=O]$ to produce (*E*)-2-butene follows.

Reduction of a variety of functional groups (e.g., see Chapter 9) can be accomplished with lithium aluminum hydride ($LiAlH_4$) as reductant in etherial solvents, for example, diethyl ether $((CH_3CH_2)_2O$ or THF [oxacyclopentane]). However, as shown in Equation 8.49, oxiranes (oxacyclopropanes, epoxides), unlike other ethers, react with lithium aluminum hydride ($LiAlH_4$) to produce alcohols (after aqueous workup). With nonsymmetrical oxiranes (epoxides, oxacyclopropanes), lithium aluminum hydride ($LiAlH_4$) reduction generally produces the more highly substituted alcohol (hydride having apparently added to the less highly substituted carbon).

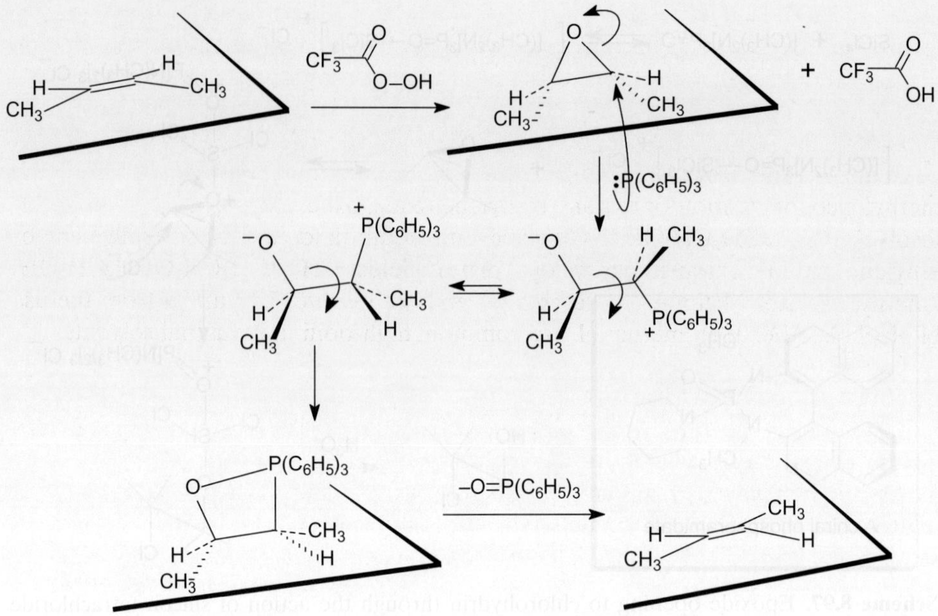

Scheme 8.98. A representation of the reaction of (Z)-2-butene with trifluoroperacetic acid to produce the corresponding oxirane (epoxide, oxacycloproapne) and the opening of the latter with triphenylphosphine to yield (E)-2-butene and triphenylphosphine oxide.

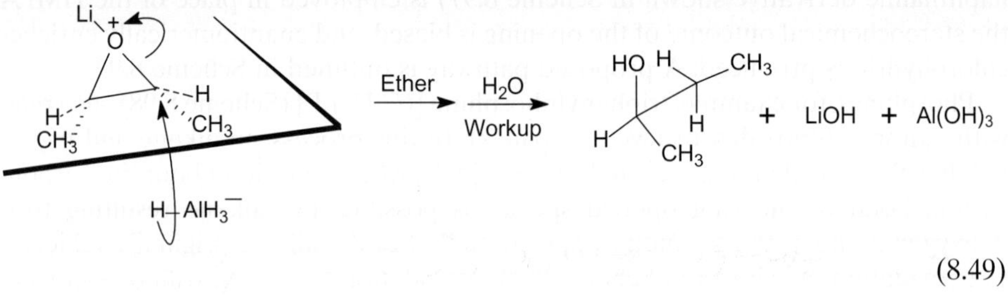

(8.49)

Although oxirane itself (oxacyclopropane, ethylene oxide) can be converted to 1,4-dioxacyclohexane (dioxane) with acid (Equation 8.48), the best results are apparently obtained

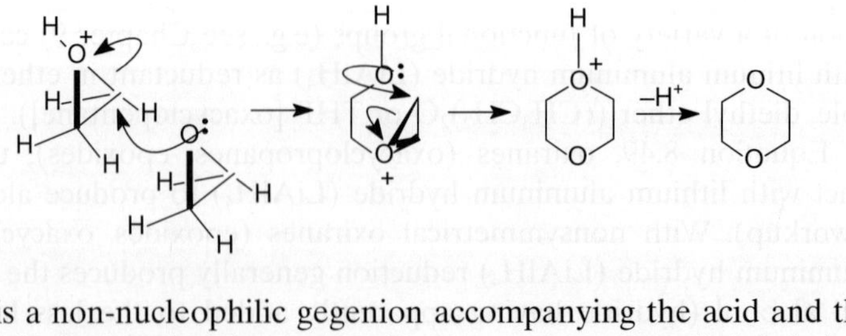

(8.50)

when there is a non-nucleophilic gegenion accompanying the acid and the system is water free (anhydrous). When water is present, 1,2-ethanediol (ethylene glycol,

HOCH$_2$CH$_2$OH) is formed. The addition of more ethylene oxide (oxirane, oxacyclopropane) goes on to produce diethylene glycol (3-oxa-1,5-pentanediol, HOCH$_2$CH$_2$OCH$_2$CH$_2$OH). A third such reaction produces triethylene glycol ([HO(CH$_2$CH$_2$O)$_n$CH$_2$CH$_2$OH]; $n = 2$) and many such further additions yield "polyethylene glycol" (n = "many").

If, in place of water, the acid with the non-nucleophilic gegenion is added to a methyl alcohol solution of oxirane (ethylene oxide, oxacyclopropane), "methyl cellosolve" [CH$_3$OCH$_2$CH$_2$OH] is formed and addition of a second equivalent of ethylene oxide (oxirane, oxacyclopropane) yields CH$_3$OCH$_2$CH$_2$OCH$_2$CH$_2$OH, "methyl carbitol." Families of "carbitols" and "cellosolves," resulting from the use of alcohols other than methanol, are common, high-boiling industrial solvents.

Before considering the chemistry of the corresponding four- (oxetane, oxacyclobutane), five- (oxolane, oxacyclopentane, THF), and six- (oxane, oxacyclohexane, tetrahydropyran [THP]) membered heterocycles and some of their close relatives, a brief digression to unsaturated ethers is warranted.

As might have been anticipated on the basis of the Cope rearrangement of hydrocarbons, which was extended to alcohols and enols earlier in this chapter (e.g., see Scheme 8.84), alkyl and aryl ethers also enjoy the ability to rearrange (a Claisen rearrangement, *vide supra*, Scheme 8.43; see Claisen and Tietze, **1925, 1926**) to, respectively, isomeric ethers (Scheme 8.99) or ketones (aldehydes) (Scheme 8.100) and phenols (Scheme 8.101).

As anticipated by the formation of oxiranes (oxacyclopropanes, epoxides) from halohydrins, the four-membered oxygen-containing ring, oxetane (oxacyclobutane), can be prepared by heating 3-chloropropanol (HOCH$_2$CH$_2$CH$_2$Cl) with potassium hydroxide (KOH) (Equation 8.51).

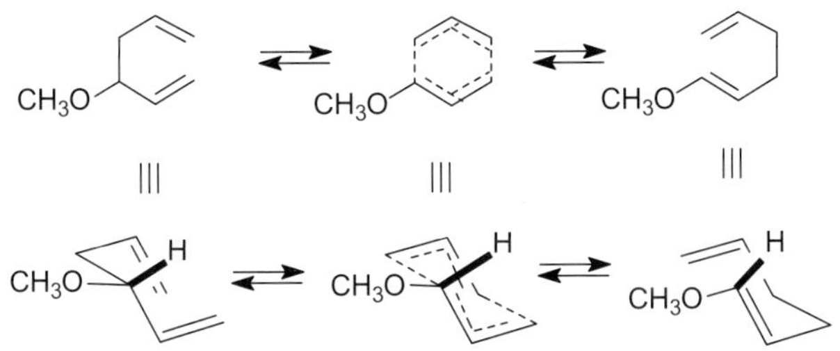

Scheme 8.99. An example of a Claisen rearrangement in which isomeric methyl ethers are seen to interconvert.

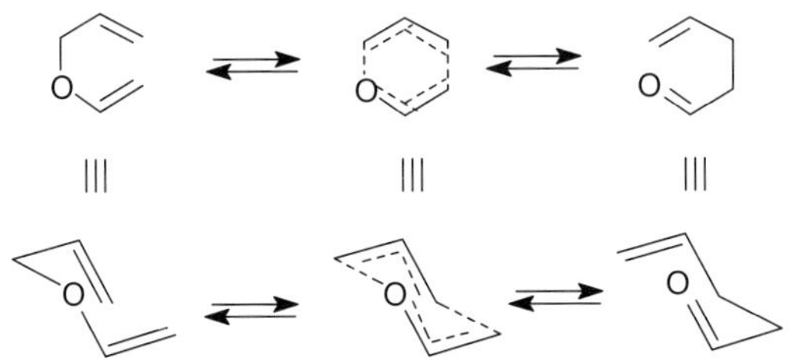

Scheme 8.100. An example of a Claisen rearrangement in which ether is converted to an isomeric aldehyde or ketone. This material will be discussed again in Chapter 9.

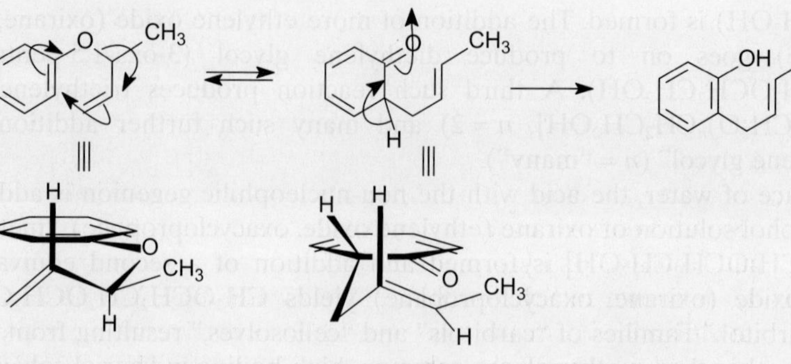

Scheme 8.101. An example of a Claisen rearrangement in which aryl ether is converted to an isomeric phenolic alkene.

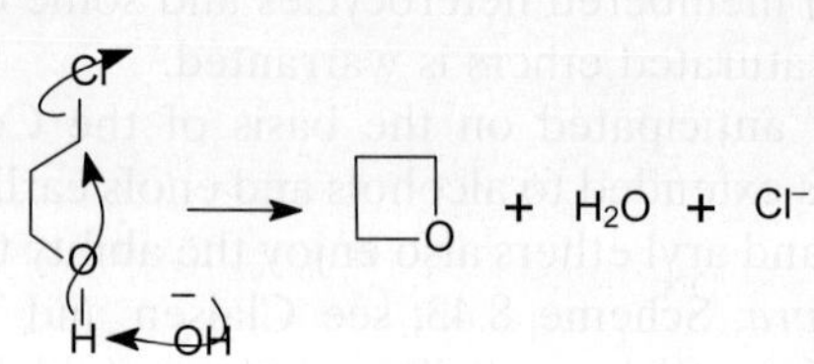

$$+ \; H_2O \; + \; Cl^- \tag{8.51}$$

Interestingly, many substituted oxetanes (oxacyclobutanes) can be prepared by an allowed $_\pi 2_s + _\pi 2_s$, generally stereospecific, photochemical addition of aldehydes (or ketones) to alkenes (Equation 8.52).

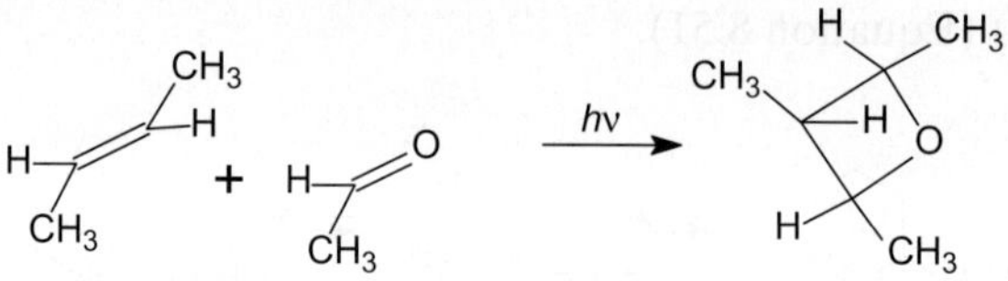

$$\tag{8.52}$$

Oxetanes (oxacyclobutanes) react much like oxiranes (epoxides, oxacyclopropanes), but much more vigorous conditions are required. Thus, oxetane (oxacyclobutane) reacts with hydrogen bromide (HBr) under ionic conditions to provide 1,3-dibromopropane. Presumably, the ring-opening reaction occurs first to yield 3-bromo-1-propanol and then, under the more vigorous conditions, protonation and S_N2 displacement of water follows (Equation 8.53) to produce the 1,3-dibromide.

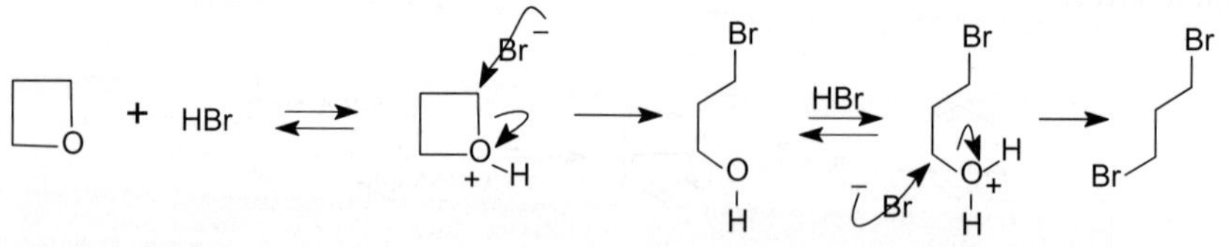

$$\tag{8.53}$$

Oxalane (oxacyclopentane, THF), a widely used solvent, can be obtained by the dehydration of 1,4-pentanediol (Equation 8.54).

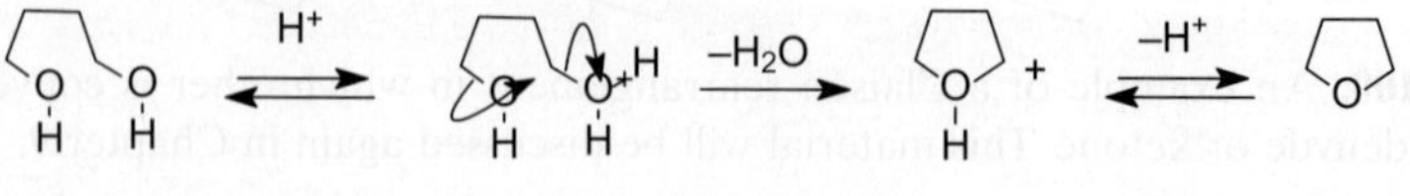

$$\tag{8.54}$$

However, it was originally prepared (over a century ago) from pentoses (five-carbon carbohydrates; see Chapter 11), which, on sulfuric acid treatment, produce furfural (furan 2-carboxaldehyde) (Scheme 8.75), and which, in turn, can be easily oxidized to the corresponding carboxylic acid (2-furoic acid). The acid (2-furoic acid) is then readily decarboxylated (i.e., loss of CO_2) to furan (oxa-2,4-cyclopentadiene) and the resulting product is catalytically reduced with hydrogen (H_2) (Scheme 8.102) over a Ni catalyst to yield the desired five-membered heterocyclic compound (oxalane, oxacyclopentane, THF).

Interestingly, as shown in Scheme 8.103, if furfural (furan 2-carboxalehyde) is reduced with hydrogen (H_2) in the presence of a Ni catalyst, the corresponding primary alcohol, tetrahydrofurfuryl alcohol, results. Then, if the that primary alcohol is heated to 350°C in the presence of an alumina (aluminum oxide, Al_2O_3) catalyst, rearrangement and dehydration occur to yield dihydropyran (oxahex-2-ene).

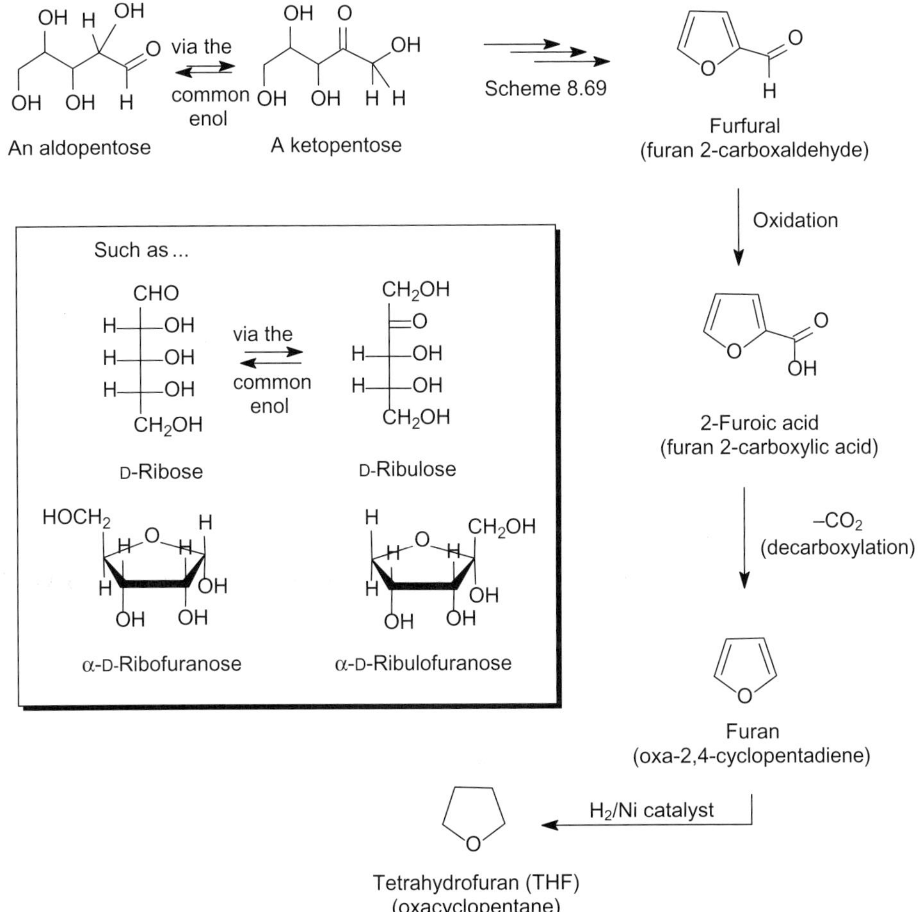

Scheme 8.102. The formation of furfural (furan 2-carboxaldehyde) from pentoses (typical examples, D-ribose and its proton tautomer D-ribulose, are shown) on treatment with sulfuric acid (Scheme 8.75). The conversion to furan 2-carboxylic acid, furan (oxa-2,4-cyclopentadiene), and tetrahydrofuran (THF, oxacyclopentane, oxolane) follows.

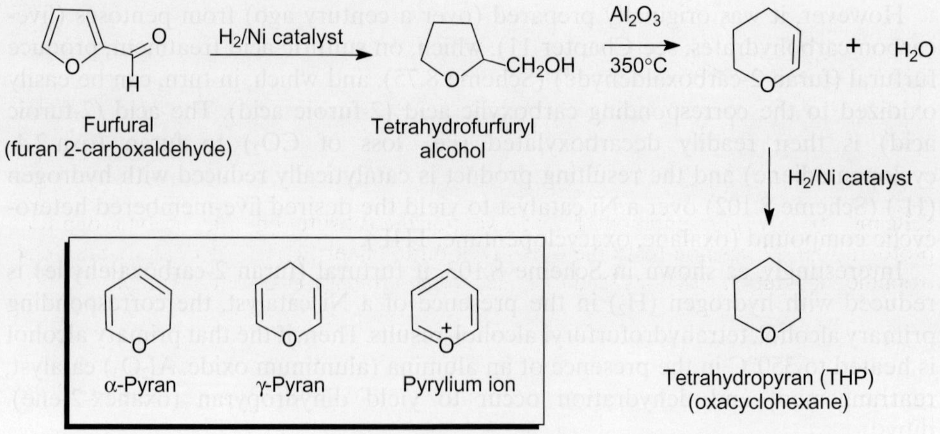

Scheme 8.103. The formation of tetrahydropyran (THP) (oxacyclohexane) from tetrahydrofurfuryl alcohol by catalytic rearrangement and dehydration followed by reduction.

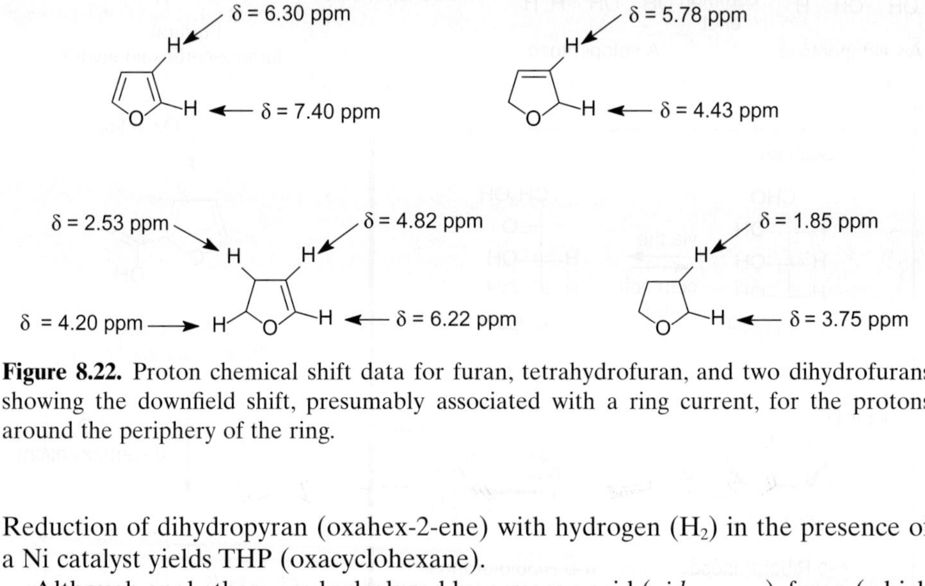

Figure 8.22. Proton chemical shift data for furan, tetrahydrofuran, and two dihydrofurans showing the downfield shift, presumably associated with a ring current, for the protons around the periphery of the ring.

Reduction of dihydropyran (oxahex-2-ene) with hydrogen (H_2) in the presence of a Ni catalyst yields THP (oxacyclohexane).

Although enol ethers are hydrolyzed by aqueous acid (*vide supra*), furan (which might be considered a bis enol ether) is more resistant than many others.

The resistance of furan to hydrolysis can be attributed to the observation that furan possesses *aromatic* character, as expected from combining the four *p*-type orbitals (one each on each of the four carbon atoms) with a fifth (*p*-type) orbital from oxygen, to give, among others, a low-lying (bonding) π-type MO. Distribution of the six electrons (two from each of the π-bonds and two from oxygen) into the three low-lying (bonding) MOs can then be accommodated. Although lacking the symmetry of benzene (C_6H_6), the result of such redistribution nonetheless increases the stability of the system and lends aromatic character. Thus, one criterion of aromaticity, downfield shifts in the proton spectrum that can be attributed to the presence of a ring current (Chapter 2), is found and, as shown in Figure 8.22, the protons

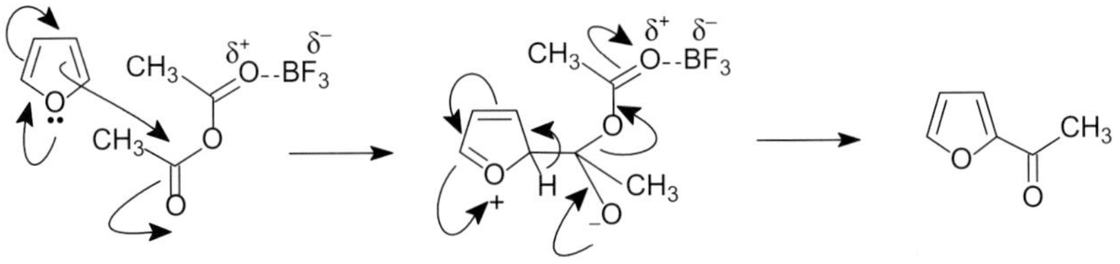

Scheme 8.104. A representation of a pathway for the substitution of an acetyl group for a proton at the C2 position (α to the oxygen) of furan under Lewis acid catalysis. Such electrophilic substitution reactions lend credence to the aromatic character of furan.

on the furan ring are shifted further downfield than protons on the corresponding dihydrofurans.

A second criterion of aromaticity, electrophilic aromatic substitution, is also seen in the reaction between furan and acetic anhydride (in the presence of a boron trifluoride-etherate Lewis acid catalyst) Scheme 8.104.

Nonetheless, furan does react with bromine in methanol to produce the corresponding **addition compounds**, (E)- and (Z)-2,5-dimethoxy-2,5-dihydrofuran (Equation 8.55) and "aromaticity" can also be lost through a Diels–Alder reaction of furan and substituted furans with reactive dienophiles such as the diethyl ester of acetylenedicarboxylic acid (Equation 8.56)

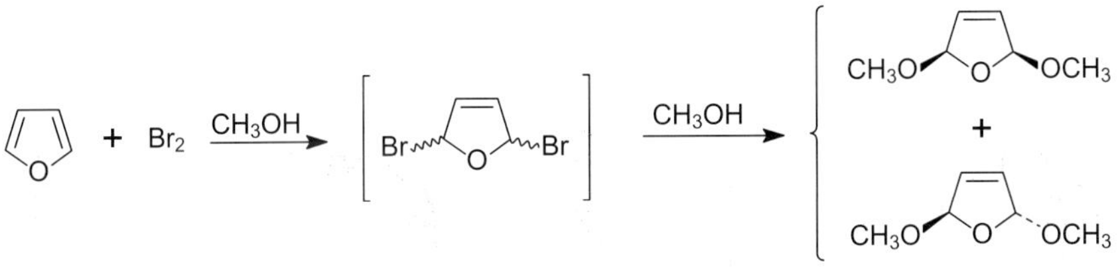

$$ (8.55) $$

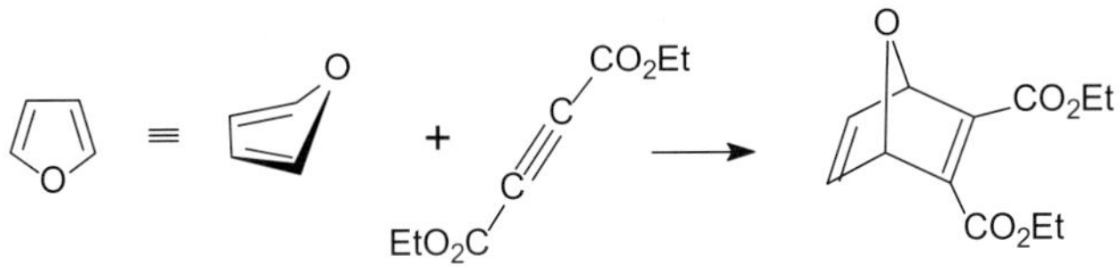

$$ (8.56) $$

Thus, there is evidence (a ring current and some substitution reactions) that furan is aromatic, and it is argued that the stabilization resulting from electron delocalization produces compounds that "possess aromatic character." As might be expected, if the oxygen in the heterocyclic ring of furan is replaced by other nuclei that bear nonbonded electrons such as nitrogen (pyrrole, Chapter 10) or sulfur (thiophene, *vide infra*) or even a carbon with an electron pair (i.e., the cyclopentadienide anion), approximately the same conclusion regarding aromaticity can be reached (Figure 8.23).

When low-molecular-weight aldehydes such as methanal (formaldehyde, $CH_2{=}O$) (Chapter 9) or ethanal (acetaldehyde, $CH_3CH{=}O$) (Chapter 9) are treated with catalytic quantities of acids with non-nucleophilic gegenions (e.g., HBF_4), cyclization to trimers (e.g., $(CH_2O)_3$) (Table 8.8) and tetramers (e.g., $(CH_3CHO)_4$) (Table 8.8)

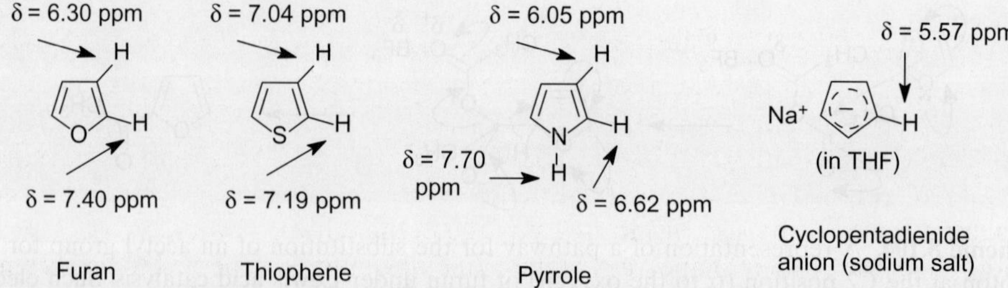

Figure 8.23. The proton chemical shifts of species with a low-lying MO and in which a ring current (denoting aromaticity) can be adduced.

TABLE 8.8. Some Representative Examples of Macrocyclic Polyethers

Compound	Structure	Comments	
1,3,5-trioxacyclohexane (1,3,5-trioxane)		Prepared by the trimerization of methanal (formaldehyde)	
		$3 \; \text{H}_2\text{C}{=}\text{O} \xrightarrow{BF_3/HF}$	
1,3,5-Trioxa-2,4,6-trimethylicyclohexane (paraldehyde)		Prepared by the trimerization of ethanal (acetaldehyde)	
		$3 \; \text{CH}_3\text{CH}{=}\text{O} \xrightarrow{BF_3/HF}$	
		Host Cavity (size: pm)	Cation Guest
1,4,7,10-tetraoxa-cyclodecane ([12]-crown-4)		120–140	Li⁺
1,4,7,10,13-pentaoxa-cyclopentadecane ([15]-crown-5)		170–220	Na⁺
1,4,7,10,13,16-hexaoxa-cyclooctadecane ([18]-crown-6)		260–320	K⁺

occurs to give polyoxaheterocyclic compounds in which the ring consists of alternating carbon and oxygen atoms. On heating, these materials can often be induced to decompose to their starting aldehydes (Chapter 9).

It might be expected that oxacycloalkanes as oxirane (ethylene oxide) would undergo a similar reaction and, indeed, a new chapter in the chemistry of ethers is being written by exploring the chemistry of that process. Thus, recognizing that the oxygen, even when in a cyclic ether, would retain its basic (or nucleophilic) characteristics and realizing that such oxygens might reasonably be expected to interact with electron-deficient species, such as metal cations, it was not surprising to learn that treatment of oxirane (ethylene oxide) with a trace of BF_3/HF in the presence of a metal cation induced cyclization to an oxygen containing ether. Within the structural limits of the size of the cavity of the cyclic ether (now known as "crown ethers" because of their shape), any suitable metal would be expected to be accommodated. If the crown ether is the "host," the bound species is the "guest."

Now, as shown in Table 8.8, depending upon whether it is "puckered" or "flattened," the cavity diameter of [12]-crown-4*† lies between 120 and 150pm. Since the ionic diameter of a lithium cation (Li^+) is about 120pm, it would be expected to fit in the cavity. Neither the sodium cation (Na^+), diameter = 180pm, nor the potassium cation (K^+), diameter = 266pm, could fit.

Crown ethers have found a niche in the organic chemistry laboratory as "phase transfer" reagents. Thus, although potassium permangante ($KMnO_4$) is insoluble in benzene (C_6H_6), the addition of a trace of dibenzo[18]-crown-6 produces a purple solution ("purple benzene") and permits transfer of the permanganate ion (MnO_4^-) into a medium, for example, for oxidation, in which it would otherwise be insoluble.

Further, the crown ethers distinctly resemble that group of oxygen-rich, macrocyclic, naturally occurring ion transporters called *ionophores*, such as nonactin.

Nonactin

Cell walls in biological systems contain lipid (Chapter 11) materials that impart hydrocarbon character to the wall. Transportation of ions (e.g., sodium [Na^+] and potassium [K^+]) across these walls in order to keep the ionic balance in the more

---

*It will be remembered that the nomenclature of these compounds derives first from the number of atoms, n, in the ring (expressed in leading brackets [n]), separated from the following word "crown" by a hyphen and terminating, after another hyphen, with the number of oxygens in the heterocycle.

†For the discovery and exploration of the properties of these and similar materials, the Nobel Prize in Chemistry was awarded to Donald J. Cram, Jean-Marie Lehn, and Charles J. Pedersen in 1987. D. J. Cram (1919–2001) was professor of chemistry at the University of California, Los Angeles; J.-M. Lehn is a professor of chemistry at The Université Louis Pasture, Strasbourgh, France, and the College de France in Paris. C. J. Pederson (1904–1989) was a chemist working for the E. I. du Pont Company, which he joined in 1927 and from which he retired in 1969.

polar intra- and extracellular environments appropriate for cellular activity occurs through channels in the walls and cell death can occur if the balance is impaired.

It is not surprising therefore that ionophores can act as antibiotics. They form tight complexes with metal cations (nonactin complexes with potassium [K^+]) and disturb the bacterial ionic balance and lead to cell death.

PART III. THIOLS, THIOETHERS, AND SOME PRODUCTS OF THEIR OXIDATION

Sulfur lies directly below oxygen in the periodic table so that series of sulfur-containing compounds analogous to those containing oxygen might be anticipated. The general concept of periodicity further suggests that similar properties might also be anticipated for the similar compounds with the *exception* that, since third period elements are more polarizable and can utilize hybridized s, p, **and** d orbitals to accommodate ligands surrounding them, some differences should intrude.

As reported in Chapter 5, thiols (mercaptans [from the Latin word *mercurium captans* or "seizing mercury"], RSH, or ArSH) and sulfides (thioethers, R_2S, ArSR, ArSAr) bear a deceptively simple relationship to their corresponding respective oxygen derivatives, viz., alcohols (ROH), phenols (ArSH), and aryl and alkyl ethers (R_2O, ArOR, ArOAr). As also pointed out in Chapter 5, one of the major differences between simple oxygen and sulfur compounds lies in the smaller C–S–H and C–S–C bond angles and longer H–S and C–S bonds in thiols and thioethers relative to alcohols and ethers. Spectroscopic differences were also emphasized in Chapter 5 and again at the beginning of this chapter. Also, as already noted, most humans find the odors of most thiols offensive and/or irritating. Typical odoriferous materials commonly encountered (along with other compounds) are *n*-propyl mercaptan [$CH_3CH_2CH_2SH$] (onions), allyl disulfide [$(CH_2{=}CHCH_2)_2S_2$] (garlic), *n*-butyl mercaptan [$CH_3CH_2CH_2CH_2SH$] (skunks), and dimethyl sulfide [$(CH_3)_2S$] (odorant added to natural gas, methane, CH_4, allowing it to be detected). Mustard gas, β-chloroethyl sulfide [$(ClCH_2CH_2)_2S$], was a material of warfare during the First World War (WWI), and it has been suggested that major stockpiles were prepared by combatants during the Second World War (WWII) but were not used.

Just as hydrogen sulfide (H_2S) is a stronger acid (i.e., it is morecompletely ionized) than water (H_2O), so it is that thiols are more acidic than alcohols. Indeed, ethanethiol (CH_3CH_2SH, $pK_a = 10.6$) is essentially completely converted to the corrsponding thiolate anion ($CH_3CH_2S^-$) by hydroxide (HO^-) in water ($pK_a = 15.7$), while ethanol (CH_3CH_2OH, $pK_a = 15.9$) remains largely protonated.

The nucleophilicity of sulfur anions, for example, the hydrosulfide anion (HS^-), the hydrogen disulfide anion (HS_2^-), and thiolate anions (RS^-), appears to be significantly superior to the corresponding oxygen derivatives (i.e., the hydroxide anion [OH^-], the hydrogen peroxide anion [HO_2^-], and alkoxide anions [RO^-]) under similar conditions, and both the polarizability of sulfur and the extent of solvation of the respective oxyanions (but not the thioanions) have been implicated.

In that vein, thiols are routinely prepared by S_N2 substitution reactions with the hydrogen sulfide anion (HS^-) serving as a nucleophile to displace halogen (frequently iodine) from the corresponding alkyl halide, as in the reaction of 1-iodobutane to produce butanethiol (Equation 8.57).

$$\text{CH}_3\text{CH}_2\text{I} + \text{ }^-\text{SH} \xrightarrow[\text{S}_\text{N}2]{\text{CH}_3\text{CH}_2\text{OH}} \text{ }\text{SH} + \text{I}^- \qquad (8.57)$$

Interestingly, the hydrogen sulfide anion (HS$^-$) is a strong enough base to deprotonate the thiol formed (Equation 8.58).

$$\text{SH} + \text{ }^-\text{SH} \underset{\longleftarrow}{\overset{\text{CH}_3\text{CH}_2\text{OH}}{\longrightarrow}} \text{S}^- + \text{H}_2\text{S} \qquad (8.58)$$

The resulting thiolate anion is then capable of displacing halogen from a second equivalent of alkyl halide. This produces the corresponding thioether (dibutyl sulfide, Equation 8.59). In practice, thiol synthesis via the S$_\text{N}$2 process is best undertaken using a large excess of hydrosulfide anion so that the *relative* concentration of thiolate anion is minimized and thioether formation (Equation 8.59) is repressed. However, even under the best of circumstances, elimination competes with substitution. Alkenes can even be the major product and with tertiary alkyl halides, it is common to find that *only* alkenes result.

$$\text{I} + \text{S}^- \xrightarrow[\text{S}_\text{N}2]{\text{CH}_3\text{CH}_2\text{OH}} \text{S} + \text{I}^- \qquad (8.59)$$

In addition to the oxygen analogues noted above, other compounds containing sulfur have already been introduced in this work; there are the sulfonium salts (R$_3$S$^+$), such as S-adenosylmethionine, and the simpler trimethylsulfonium iodide [(CH$_3$)$_3$S$^+$ I$^-$], the thioester, acetyl choline (CH$_3$COSAc, Scheme 8.70), as well as the toluenesulfonate esters (p-CH$_3$C$_6$H$_5$SO$_3$R), which are used to convert a poor leaving group (–OH) to a good leaving group (–OTs) as discussed earlier. Some of these and other examples are shown below in Figure 8.24.

Disulfides, such as the diallyl disulfide shown in Figure 8.24, can be obtained by the reaction of sodium disulfide (Na$_2$S$_2$), itself prepared by the dissolution of an equivalent of sulfur in a concenctrated sodium sulfide (Na$_2$S) solution, with the corresponding alkyl halide. Reduction of the disulfide (e.g., with zinc dust in acetic acid) results in cleavage of the sulfur–sulfur bond and thiol formation (Scheme 8.105).*

Alternatively, as might be expected, thiols are rather easily oxidized (e.g., by hypohalites [NaOCl]) and the products are disulfides (Equation 8.60).

$$\text{SH} + \text{NaOI} \longrightarrow \text{S}-\text{S} + \text{H}_2\text{O} + \text{NaI} \qquad (8.60)$$

Indeed, gentle oxidation of thiols with a variety of reagents, including air (O$_2$), manganese(IV) dioxide (MnO$_2$), and even hydrogen peroxide (H$_2$O$_2$) (Equation

---

*Disulfide linkages within cysteine (HSCH$_2$CH(NH$_3^+$CO$_2^-$)-rich proteins (such as hair and wool) help determine the structure and function of the protein (Chapter 12). Interestingly, the one-time-common fashion of "permanent waving" of hair was based upon the reduction of disulfides making up the fibrous proteins in hair to cause their cleavage to the corresponding hydrosulfides. Then, when the "reduced hair" was held in place with suitable mechanical devices or was "set" and allowed to stand or actively oxidized with some suitable reagent, new disulfide bonds formed and removal of the holding devices resulted in hair in which the "wave" was "set."

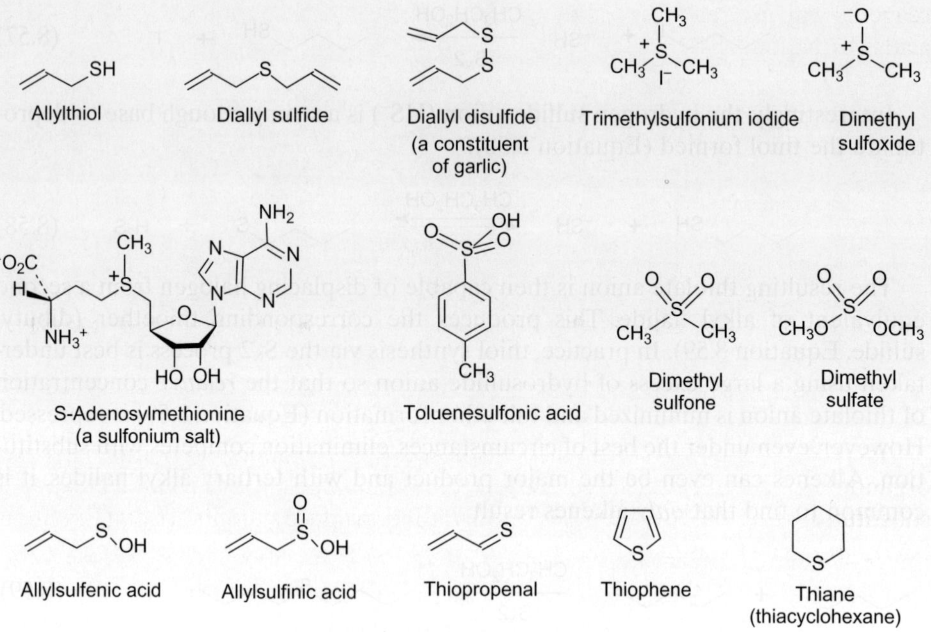

Allylthiol

Diallyl sulfide

Diallyl disulfide
(a constituent
of garlic)

Trimethylsulfonium iodide

Dimethyl
sulfoxide

S-Adenosylmethionine
(a sulfonium salt)

Toluenesulfonic acid

Dimethyl
sulfone

Dimethyl
sulfate

Allylsulfenic acid

Allylsulfinic acid

Thiopropenal

Thiophene

Thiane
(thiacyclohexane)

Figure 8.24. Some typical sulphur-containing compounds.

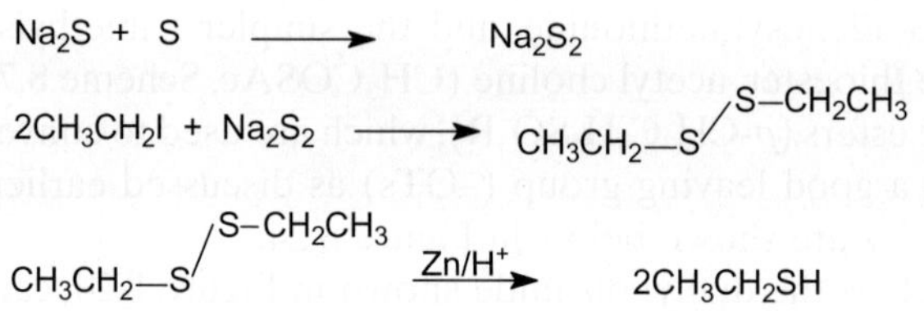

$$Na_2S + S \longrightarrow Na_2S_2$$

$$2CH_3CH_2I + Na_2S_2 \longrightarrow CH_3CH_2-S{\overset{S-CH_2CH_3}{\diagup}}$$

$$CH_3CH_2-S{\overset{S-CH_2CH_3}{\diagup}} \xrightarrow{Zn/H^+} 2CH_3CH_2SH$$

Scheme 8.105. The formation of diethyl disulfide by the reaction of sodium disulfide (itself prepared from sodium sulfide and sulfur) with iodoethane.

8.61) produces disulfides, while more vigorous oxidation with, for example, nitric acid (HNO_3) induces further oxidation to, ultimately, the corresponding sulfonic acid (Equation 8.62)

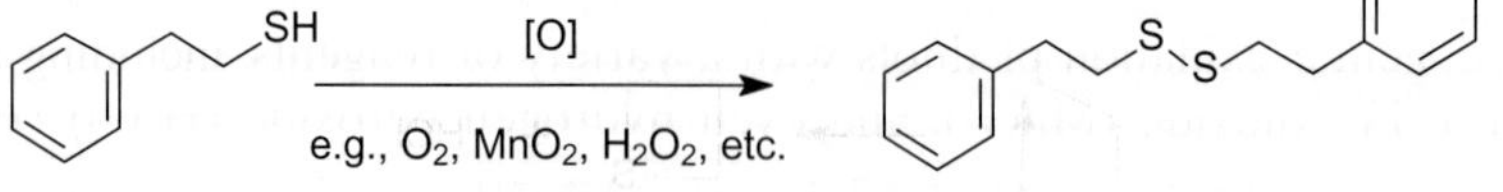

$$(8.61)$$

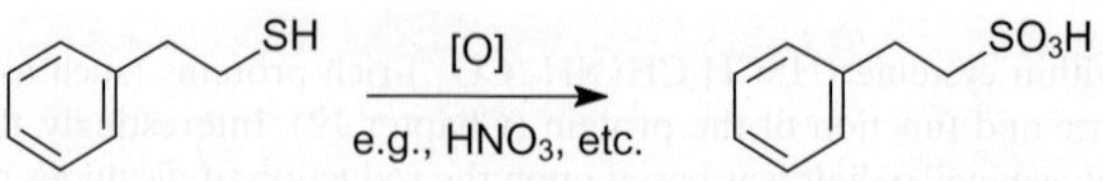

$$(8.62)$$

In addition to their generation by substitution reactions, thiols can also be produced directly by the reaction of organometallic compounds, such as Grignard reagents, with sulfur (S_8) (Equation 8.63). Not surprisingly, the same Grignard reagents will also react with sulfur dioxide (SO_2) to generate, on hydrolysis, sulfinic acids (Equation 8.64).

$$(8.63)$$

$$(8.64)$$

As already noted above, thiols, produced by nucelophilic substitution or from organometallic reagents, can be converted to thioethers by a second nucleophilic substitution reaction. These sulfides, on alkylation, produce sulfonium salts (Equation 8.65), which can be used or further alkylation reactions (e.g., see SAM; Figure 8.24).

$$(8.65)$$

Cyclic sulfides (cyclic thioethers), such as thiirane (which is less stable than oxirane), can also be made by such displacement reactions. However, it is often more convenient to prepare thiirane and substituted thiiranes by thermal decomposition of thiocarbonates, as shown in Equation 8.66.

$$(8.66)$$

Interestingly, the relatively unstable four-membered-ring analogue, thietane (thiacyclobutane), can also be obtained by cyclization of the corresponding chlorothiol (Equation 8.67).

$$(8.67)$$

Thiophene (thiocyclopentadiene), the sulfur analogue of furan, is routinely obtained from coal tar and shale oil as a by-product in petroleum production.*

*It has been reported that its discovery in 1882 (by Victor Meyer [1948–1897], professor of chemistry at the University of Heidelberg) was the result of the use of benzene derived from coal tar rather than purer benzene derived by decarboxylation of calcium benzoate.

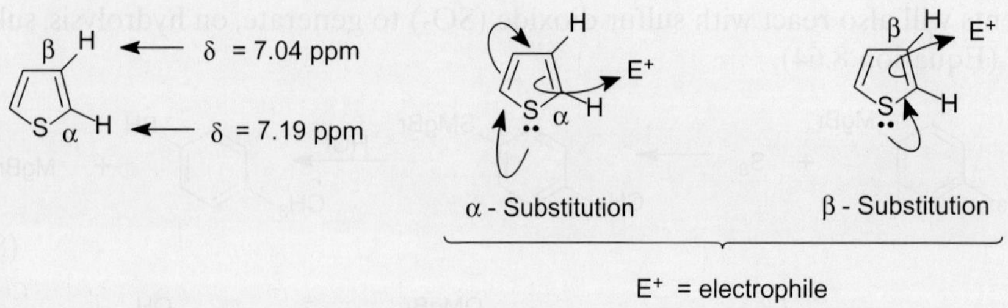

$$\beta \xleftarrow{H} \quad \delta = 7.04 \text{ ppm}$$

$$\text{S} \; \alpha \; H \xleftarrow{} \quad \delta = 7.19 \text{ ppm}$$

α - Substitution

β - Substitution

E^+ = electrophile

Figure 8.25. The ^{1}H-chemical shifts of the protons on the α- and β-positions of thiophene (thiacyclopentadiene) and generalized schemata for electrophilic aromatic substitution at each position.

Thiophene, which possesses aromatic character (Figure 8.25), undergoes substitution reactions at both the α- and β-positions (although there is some preference for the former). Thus, sulfonation (with 90% sulfuric acid [H_2SO_4]) produces approximately a 2:1 mixture of both thiophene-2- and thiophene-3-sulfonic acids (Equation 8.68).

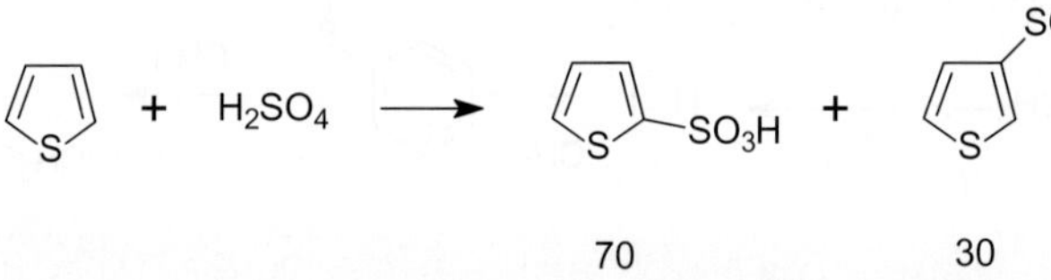

$$\text{thiophene} + H_2SO_4 \longrightarrow \text{thiophene-2-SO}_3\text{H} + \text{thiophene-3-SO}_3\text{H} \tag{8.68}$$

70   30

For short reaction times and at low temperatures (e.g., 0°C, 1 h), oxidation of sulfides (R_2S) with, for example, trifluoroperacetic acid (CF_3CO_3H) results in the formation of sulfoxides, $R_2S^+O^-$ (Equation 8.69), while under more vigorous conditions (30°C, several hours, 2 equivalents of trifluoroperacetic acid (CF_3CO_3H)), further oxidation to sulfones (R_2SO_2) occurs (Equation 8.70).

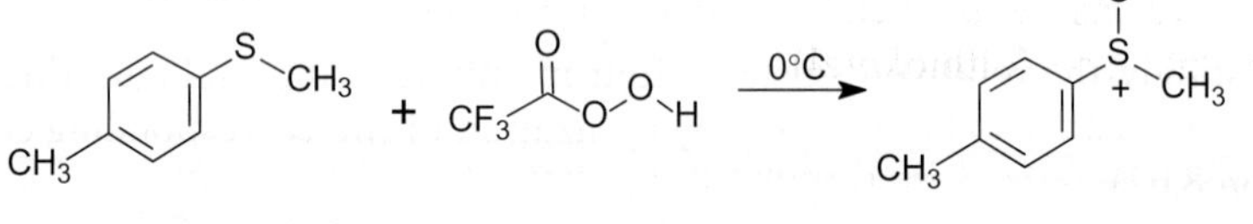

$$\text{CH}_3\text{-C}_6\text{H}_4\text{-S-CH}_3 + CF_3CO_3H \xrightarrow{0°C} \text{CH}_3\text{-C}_6\text{H}_4\text{-S}^+(\text{O}^-)\text{-CH}_3 \tag{8.69}$$

$$\text{CH}_3\text{-C}_6\text{H}_4\text{-S-CH}_3 + 2\,CF_3CO_3H \xrightarrow{30°C} \text{CH}_3\text{-C}_6\text{H}_4\text{-SO}_2\text{-CH}_3 \tag{8.70}$$

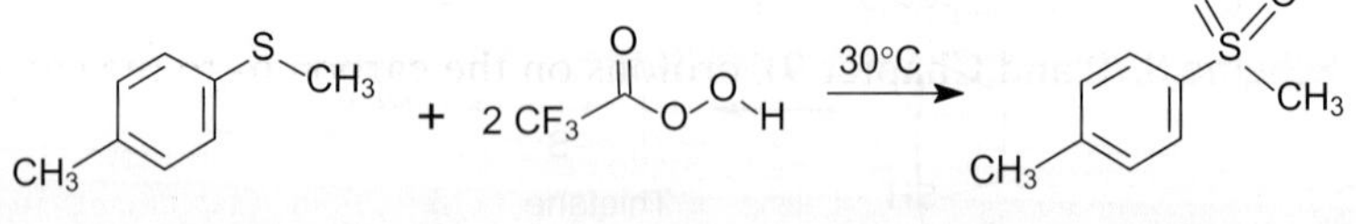

Although many different reagents for the oxidation of sulfides (R_2S) to sulfoxides ($R_2S^+O^-$) and sulfones (R_2SO_2) have been reported (reagents including air, manganese dioxide, and MnO_2), it is particularly interesting to note that use of t-butyl peroxide in the presence of titanium tetraisopropoxide [$Ti(O\text{-}i\text{-Pr})_4$] and diethyl (2$R$,3$R$)-tartrate (or its enantiomer) generates **chiral** sulfoxide. These are the same conditions utilized for the Sharpless oxidation (this chapter) to produce chiral oxiranes (epoxides). As earlier, the sense of induced chirality is dictated by the handedness of the tartrate (Equation 8.71).*

$$(8.71)$$

Further, as noted much earlier (Chapter 6, Table 6.5, item 8), sulfones can also be produced by a cheletropic reaction between, for example, 1,3-butadiene and sulfur dioxide (Equation 8.72). The resulting alkene on reduction yields the corresponding dioxide (thiacyclopentane-S,S-dioxide, sulfolane), which has found use as a commercial solvent (Equation 8.72).

$$(8.72)$$

Sulfolane

Despite their ease of oxidation, thiols have formed part of a valuable technology for the conversion of ketones and aldehydes (Chapter 9) to the corresponding hydrocarbons. Thus, it should not come as a surprise that by analogy with the reaction of 1,2-ethanediol (ethylene glycol, $HOCH_2CH_2OH$) with aldehydes and ketones (Chapter 9) as shown earlier in this chapter (Table 8.6, item 12, and Scheme 8.52) to produce the corresponding 1,3-dioxycyclopentane (acetal or ketal, respectively), and as shown in Scheme 8.106, 1,2-ethanedithiol ($HSCH_2CH_2SH$) under similar conditions reacts with cyclohexanone to produce the cyclohexane fused-1,3-dithiocyclopentane (dithioketal)

As also shown in Scheme 8.106, the sulfur in dithioketals (and dithioacetals) can be reductively removed (and replaced by hydrogen, a **hydrogenolysis reaction**) by treatment with hydrogen gas (H_2) in the presence of **Raney nickel.**[†]

Disulfides also undergo reaction with carbonyl compounds. As already noted (e.g., see Scheme 8.70 and Chapter 9), protons on the carbon α- to the carbonyl are

---

*The nature of the intermediate generated between the tartrate ester and the titanium tetraisopropoxide has already been illustrated earlier in this chapter (in Scheme 8.19). The specific handedness of this complex is presumed to dictate the chirality of the sulfoxide.

[†]Raney nickel is finely divided nickel powder already saturated with hydrogen. It is prepared by dissolving the aluminum from a nickel–aluminum alloy with aqueous sodium hydroxide. Raney nickels of various "activities" are generated by using different base concentrations. The catalyst is generally stored under water and the more active forms are known to be pyrophoric when dry. Interestingly, a U.S. patent (filed May 14, 1926) with no references was granted on May 10, 1927 (patent number 1,628,190) to Murray Raney of Chattanooga, Tennessee, for the production of this family of materials.

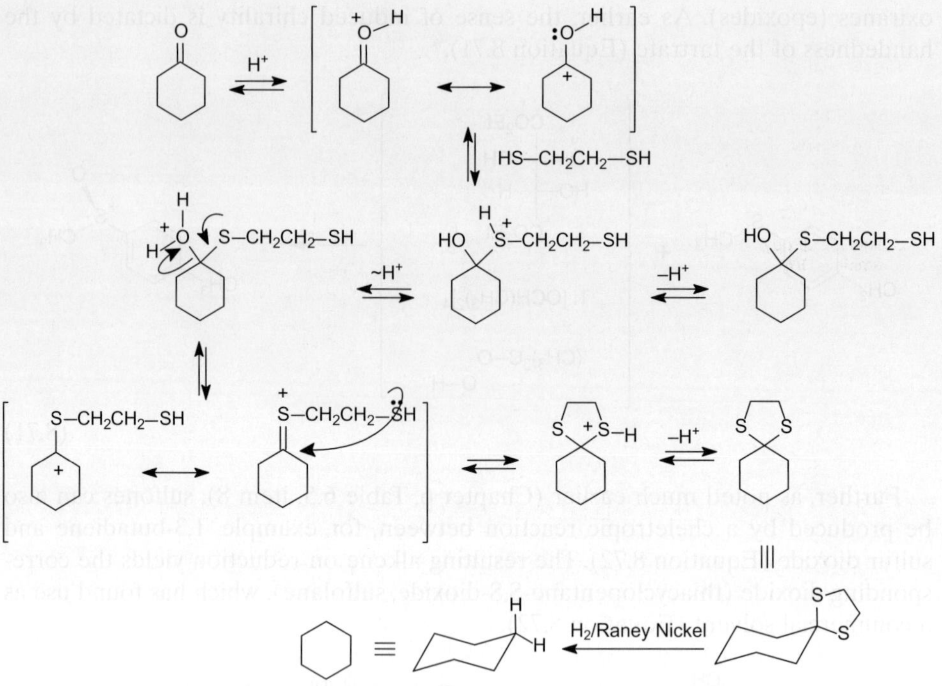

Scheme 8.106. The reaction of 1,2-ethanedithiol with cyclohexanone to produce the corresponding dithioketal and reduction of the latter with Raney nickel to cyclohexane.

acidic enough to be removed to produce the corresponding enolate anion. Nucleophilic attack by this anion on a disulfide (e.g., dibutyldisulfide) (Scheme 8.107) introduces a sulfur on the carbon α- to the carbonyl.* Then, if the carbonyl group is reduced (Chapter 9) to the corresponding alcohol and the alcohol is induced to undergo dehydration, a thioenol ether results. Hydrolysis of the thioenol ether produces the thiol and the corresponding carbonyl compound. However, it will be noted that the oxygen is *now* attached to a carbon atom adjacent to the one to which it was originally bonded.

Problem 8.19. Show the detail of each step hidden in the multiple arrows in the last process of Scheme 8.107.

As pointed out earlier (Chapter 7, Scheme 7.24), where there is a choice of direction of elimination to form alkenes and all other factors are the same, neutral leaving groups generally lead to the *more* highly substituted alkene (the **Saytzeff** product), while 'onium salts lead to the *less* highly substituted alkene (the **Hofmann product**). Thus, as shown in Equation 8.73, elimination of dimethylsulfide and a proton from the 2-substituted butane produces more 1-butene than 2-butene (mixture). Interestingly, sulfoxides also undergo elimination, the oxygen attached to sulfur

*As will be discussed in Chapter 9, when the reactions of carbonyl compounds are more fully elaborated, selenium (using e.g., diphenyldiselenide [$(C_6H_5Se)_2$] is often preferred.

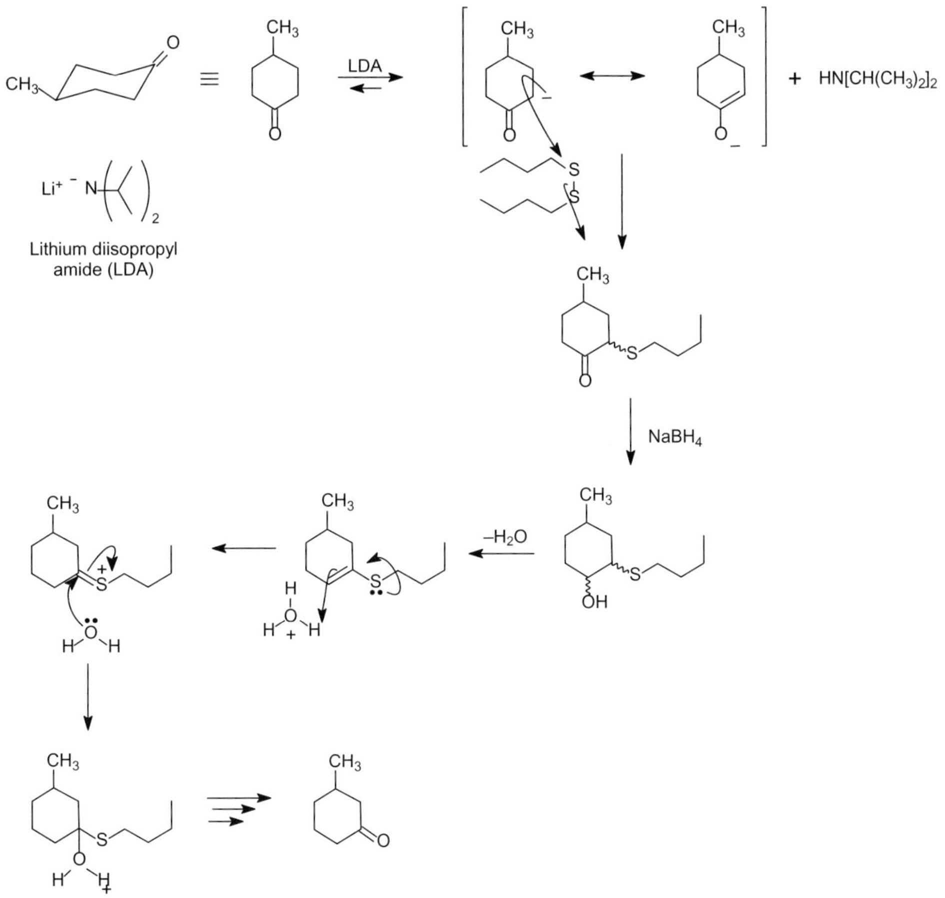

Scheme 8.107. Utilization of butanedisulfide (a dithioether) to convert a ketone (4-methylcyclohexanone) to a thioenol ether and thence to the corresponding 3-methylcyclohexanone by hydrolysis.

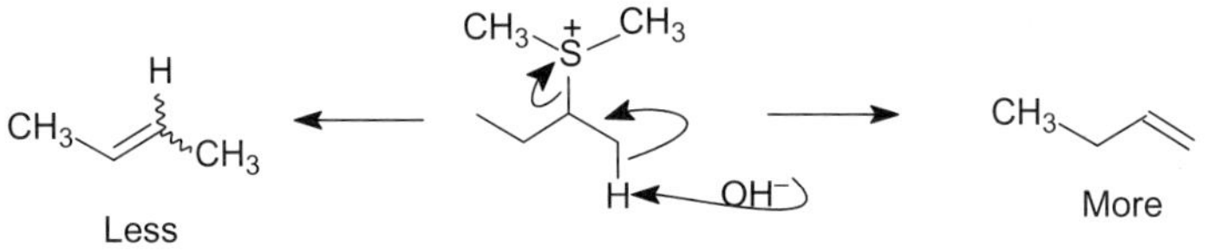

$$(8.73)$$

serving to remove the proton on the α-carbon generating alkene and methyl sulfinic acid (CH_3SOH). Steric effects, which promote cyclic transition states that allow proton removal more easily, appear important (Equation 8.74).

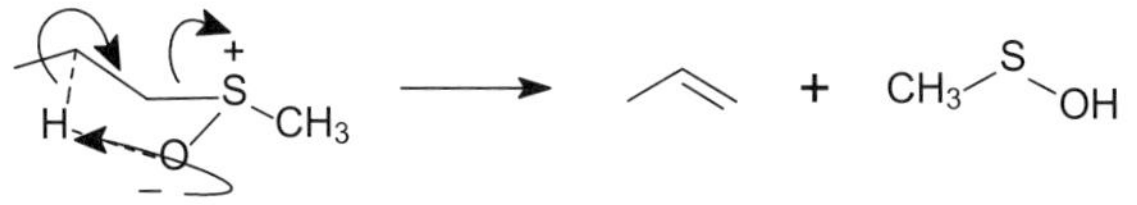

$$(8.74)$$

Sulfonium salts ($R_3S^+O^-$) and sulfoxides (R_2SO) are also involved in several well-characterized rearrangement reactions. As shown in Scheme 8.108, when there is an

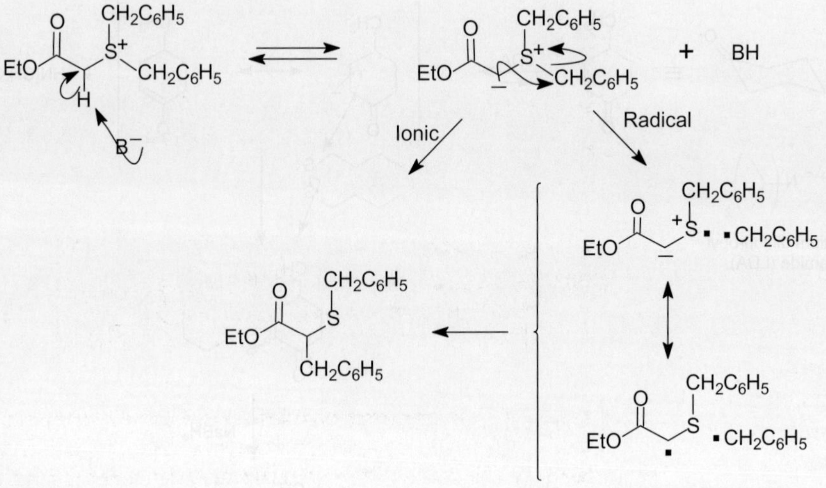

Scheme 8.108. An example of the Stevens rearrangement. The mechanistic details appear to suggest that, depending upon the substituents, the pathway from reactants to products can involve either ions or radicals (See Stevens, R. S.; Creighton, E. M.; Gordon, A. B.; MacNicol, M. *J. Chem. Soc.* **1928**, 3193).

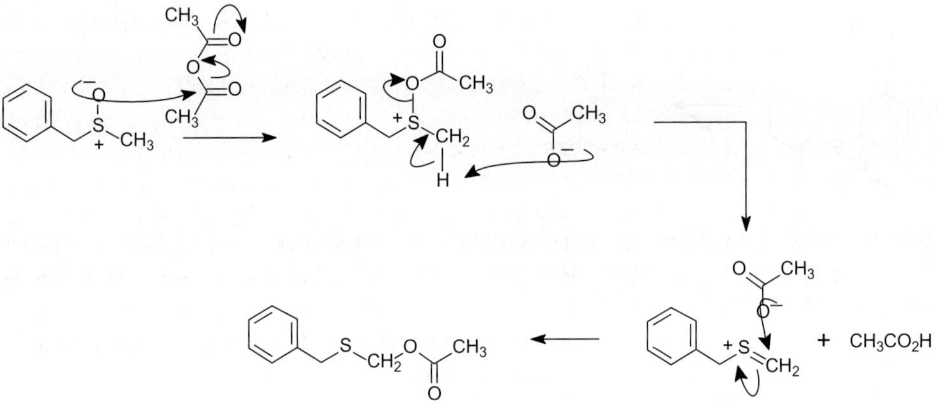

Scheme 8.109. An example of the Pummerer rearrangement in which sulfoxides bearing α-protons that are acidic enough to be abstracted by acetate anion yield a rearranged product rather than the expected sulfoxide derivative (See Pummerer, R. *Chem. Ber.* **1909**, *42*, 2282 and Pummerer, R. *Chem. Ber.* **1910**, *43*, 1401).

acidic proton on the carbon to which the sulfur is attached, treatment with base may result in migration of an alkyl group from sulfur to carbon (the *Stevens rearrangement*). Sulfoxides, on the other hand, undergo the *Pummerer rearrangement* when the proton on the carbon to which the sulfur is attached is removed and an acetylating reagent (such as acetic anhydride [$(CH_3CO)_2O$]) is present (Scheme 8.109).

Rearrangements in aromatic compounds (both sulfoxides and sulfones attached to aromatic rings) are also known. Thus, as shown in Scheme 8.110, the phenolic

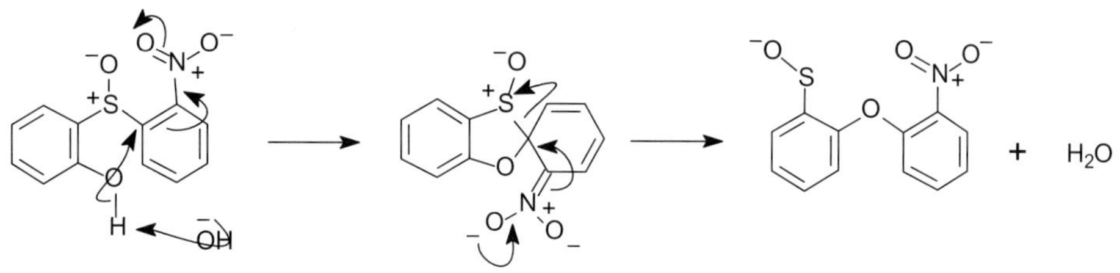

Scheme 8.110. A representation of the *Smiles rearrangement* in which a sulfoxide is converted to a sulfinic acid. (See Levy, A. A.; Rains, H. C.; Smiles, S. *J. Chem. Soc.* **1931**, 3264).

sulfoxide is transformed (by a *Smiles rearrangement*) into the anion of a sulfinic acid on treatment with base.

Finally, it will be remembered that (a) the "strong" organic acids such as benzenesulfonic acid and related aromatic sulfonic acids can be prepared by the sulfonation of the aromatic ring (Equation 8.75) through electrophilic aromatic substitution and that (b) acid chlorides of these acids (Equation 8.76) on reaction with alcohols will produce aryl sulfonate (e.g., toluenesulfonate) esters (Equation 8.77) used earlier (Chapter 8) to effect alkene formation from alcohols.

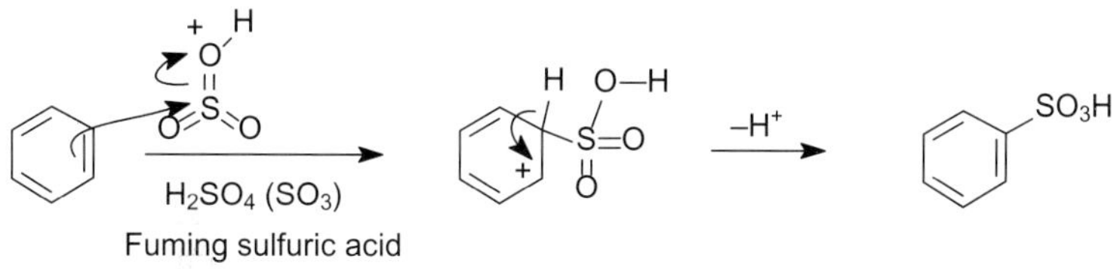

$$(8.75)$$

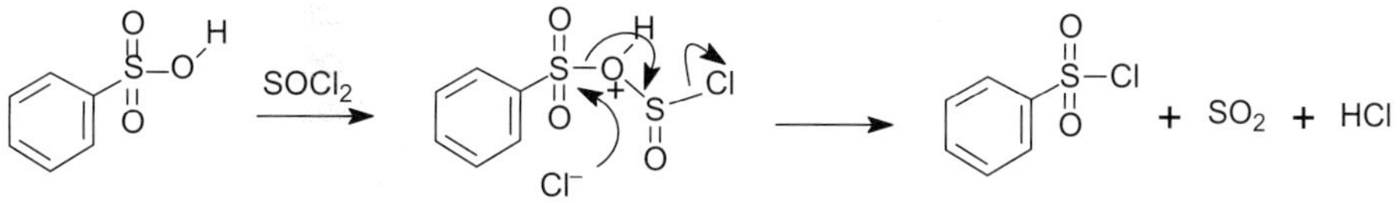

$$(8.76)$$

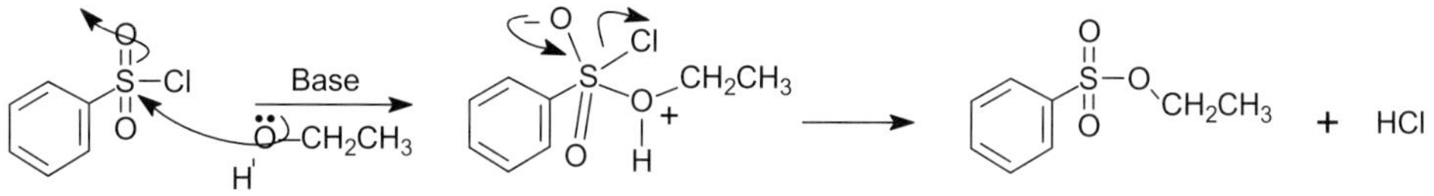

$$(8.77)$$

ADDITIONAL PROBLEMS

Problem 8.20. A representation for a pathway for the conversion of the phenyl diethylphosphate to aminobenzene (aniline, $C_6H_5NH_2$) is provided in Scheme 8.36. It has been proposed that the strong base (potassium amide) removes a proton *ortho* to the phosphate linkage of the aromatic bond and that benzyne is formed. Benzyne then reacts with the ammonia solvent to produce the product. An alternative process might involve, for example, direct displacement of the diethylphosphate anion. Write out at least these two suggested processes and propose what you might do to determine which is occurring (if either).

Problem 8.21. Electrophilic aromatic substitution reactions of phenol generally produce *ortho-* and *para*-products. Write a pathway using curved arrows to account for the bromination of phenol as shown in Equation 8.28.

REFERENCES

Claisen, L.; Tietze, E. *Chem. Ber.*, **1925**, *58*, 275.
Claisen, L.; Tietze, E. *Chem. Ber.*, **1926**, *59*, 2344.
Clive, D. L. J. *Tetrahedron*, **1978**, *34*, 1049.
Fries, K.; Finck, G. *Chem. Ber.*, **1908**, *41*, 4271.
Fries, K.; Pfaffendorf, W. *Chem. Ber.*, **1910**, *43*, 212.
Levy, A. A.; Rains, H. C.; Smiles, S. *J. Chem. Soc.*, **1931**, 3264.
Pummerer, R. *Chem. Ber.*, **1909**, *42*, 2282.
Pummerer, R. *Chem. Ber.*, **1910**, *43*, 1401.
Reich, H. J. *Acc. Chem Res.*, **1979**, *12*, 22.
Stevens, R. S.; Creighton, E. M.; Gordon, A. B.; MacNicol, M. J. *Chem. Soc.*, **1928**, 3193.

Part I. The Reactions of Aldehydes and Ketones: Oxidation, Reduction, Addition, Substitution, and Rearrangement
Part II. The Reactions of Carboxylic Acids and Their Derivatives: Oxidation, Reduction, Addition, Substitution, Elimination, and Rearrangement

"There are more things in heaven and
earth, Horatio,
Than are dreamt of in your philosophy."

Hamlet, Act I, Scene 5
W. Shakespeare

A. INTRODUCTION

The carbonyl group (Figure 9.1) can be considered the hinge upon which a door turns. That passage links the simpler reactions of carbon nuclei bearing one heteroatom substituent (e.g., one halogen and one oxygen.) with the more elaborate systems in which carbon bears two or more, alike or dissimilar, heteroatom substituents. As shown in Figure 9.1, reduction of the simplest aldehyde, methanal (formaldehyde, $H_2C=O$), leads to the one carbon alcohol, methanol (methyl alcohol, CH_3OH), while oxidation of methanal (formaldehyde, $H_2C=O$) leads to the one carbon carboxylic acid, methanoic acid (formic acid, HCO_2H).

Further reduction of methanol (methyl alcohol, CH_3OH) then produces methane (CH_4) while further oxidation of methanoic acid (formic acid, HCO_2H) generates carbon dioxide (CO_2) (Figure 9.1).

A brief discussion of both the nomenclature and some of the properties of the functional groups containing the carbonyl group is provided in Chapter 5. That

Foundations of Organic Chemistry: Unity and Diversity of Structures, Pathways, and Reactions,
First Edition. David R. Dalton.
© 2011 John Wiley & Sons, Inc. Published 2011 by John Wiley & Sons, Inc.

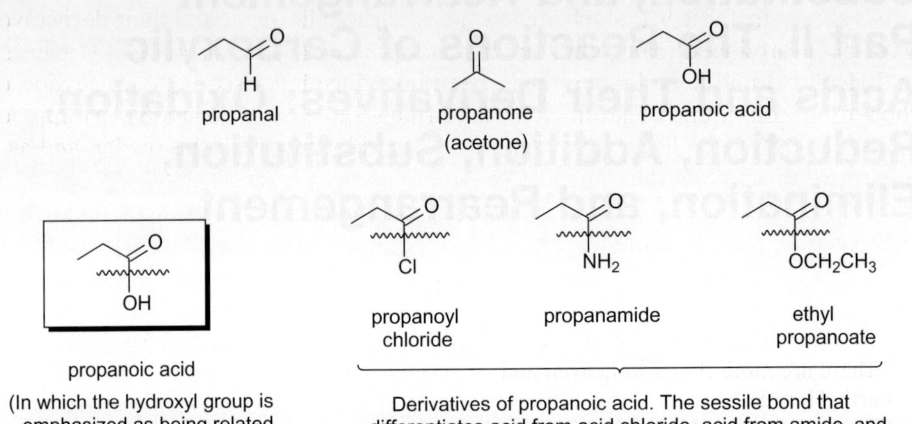

methane   methanol   methanal   methanoic acid   carbon dioxide
                                   (formaldehyde)   (formic acid)

Figure 9.1. The relationship between methane, methanol, methanal, methanoic acid, and carbon dioxide as a redox series.

propanal   propanone   propanoic acid
                  (acetone)

propanoic acid

(In which the hydroxyl group is emphasized as being related to those derivatives that follow)

propanoyl chloride   propanamide   ethyl propanoate

Derivatives of propanoic acid. The sessile bond that differentiates acid from acid chloride, acid from amide, and acid from ester, respectively, is shown crossed by a wavy line.

Figure 9.2. An aldehyde (propanal, CH_3CH_2CHO), ketone (propanone, acetone, dimethyl ketone, DMK, CH_3COCH_3) and carboxylic acid (propanoic acid, propionic acid, $CH_3CH_2CO_2H$). An acid chloride (propanoyl chloride, propionyl chloride, CH_3CH_2COCl), amide (propanamide, propionamide, $CH_3CH_2CONH_2$), and ester (ethyl propanoate, ethyl propionate, $CH_3CH_2CO_2CH_2CH_3$), typical carboxylic acid derivatives, are also shown.

material can be profitably reviewed, and it will be noted, in particular, that (Figure 9.2):

Aldehydes (abbreviated as RCH=O even though the hydrogen is singly bonded to carbon and the oxygen is doubly bonded to carbon, Figures 8.2 and 9.2) contain the carbonyl functional group, and at least *one of the other bonds to the carbon of the carbonyl is from hydrogen.*

Ketones (abbreviated as R_2C=O, Figures 8.2 and 9.2) are compounds (cyclic and acyclic) in which *both of the other bonds to the carbon of the carbonyl are from carbon.*

Carboxylic acids (abbreviated as RCO_2H, Figures 8.2 and 9.2) possess one hydroxyl group (–OH) bonded to the carbon of the carbonyl. For these compounds, if the hydroxyl is replaced by any nonhydrogen or noncarbon substituent (e.g., –OR, NH_2, and halogen.), the resulting material is considered a **carboxylic acid derivative**. The rationale for this linkage of names was provided earlier (Chapter 5).

The use of aldehydes and ketones as well as carboxylic acids and their derivatives is discussed in Chapters 5–8, where their reactions with other species (such as alcohols and alkyl halides) is introduced.

The common theme that unites ALL of the reactions of compounds containing carbonyl groups stems from the polarity of the carbonyl group itself. The two nuclei, carbon and oxygen, (the latter more electronegative than the former [Chapter 1]), lying at the ends of the electron rich double bond, are different. Relative to the oxygen, the carbon of the carbonyl is positive and thus reacts with electron rich species. Relative to the carbon, the oxygen of the carbonyl is negative and thus reacts with electron poor species.*

As shown in Table 9.1, aldehydes, ketones, carboxylic acids, and their derivatives have been produced from other organic compounds (hydrocarbons, alkyl halides, and alcohols) and have been already been introduced as part of the discussion of those other compounds. Therefore, the material of Table 9.1 serves **both** as a reminder of reactions already discussed and thus, presumably, known to the reader, and as a device to foreshadow that which is yet to come. Thus, in addition to a plethora of new reactions that will expand "the philosophy" of "things in heaven and earth," it remains important to remember that the organic compounds already discussed can themselves produce or be produced from aldehydes, ketones, and/or carboxylic acids (and or their derivatives).

As this chapter develops, it will be useful to keep in mind that in addition to the simple compounds containing one or more carbonyl groups (C=O), such as most of the aldehydes, ketones, and carboxylic acids (and/or their derivatives) of Table 9.1, carbonyl-containing organic compounds are, along with their reduced alcoholic relatives, literally the building blocks of nature.

Thus, as shown for some of the compounds of Table 9.1, as seen in the previous chapter and as will be discussed in detail in subsequent chapters, carbohydrates can contain **both** carbonyl (aldehyde, ketone, and carboxylic acid) as well as hydroxyl functionalities while, as expected, amino acids contain the amino ($-NH_2$) as well as carboxylic acid ($-CO_2H$) groups.

Further, with regard to carbohydrates, not only are they the main structural component of cellulose and starch (Chapter 11), the matter of the plant world, but they are also present on the surface of cells and in ribonucleic acids (RNA) and deoxyribonucleic acids (DNA). It is also now clear that the complexity provided by the rich assortment of possibilities that arise as a consequence the presence of carbohydrates on the surfaces of cells that a means of cell-cell recognition is provided. This allows living systems to recognize and differentiate self from nonself. In the nucleic acids and elsewhere, carbohydrates are generally found as aldehyde-derived cyclic hemiaminals Additionally, amino acids (Chapter 12) bear the amino ($-NH_2$) and carboxylic acid ($-CO_2H$) functionality, and the amide (O=C–N) backbone forged from the linkage of one amino acid to the next produces the peptides as directed by the genome.

*The pedant will argue that use of the word "ALL" broadens the concept too much. It can, however, be argued in return that when the direction of addition of electrophilic and nucleophilic reagents is reversed, it is because the polarity at the carbon (and oxygen or its equivalent) (an **umpolung** process, *vide infra*) has been reversed and therefore the general principle has not been violated.

TABLE 9.1. Some of the Reactions That Produce Aldehydes, Ketones and Carboxylic Acids (or Their Derivatives) and Have Been Encountered in Previous Chapters

Chapter 5

hydrate (~0.1% at 25°C)
[2,2-propanediol]

ketone (~99.9% at 25°C)
[propanone]

Chapter 6

$CH_2{=}CH_2 + 1/2\ O_2 \xrightarrow[\text{Cu (II) Cl}_2]{\text{Pd (II) Cl}_2}$ (Both aldehydes and ketones can be prepared this way.)

1) O_3 / CH_2Cl_2
2) Zn / CH_3CO_2H (Both aldehydes and ketones can be prepared this way. Note the use of ethanoic acid [acetic acid] in the reductive step.)

I_2 / $CH_3CO_2^-$ Ag^+ α-haloesters prepared by electrophilic addition to alkenes

CO, H_2 / cobalt catalyst $CH_3CH_2CH_2{-}CHO$ +

$R{-}C{\equiv}C{-}R'$ $R, R' \neq H$ $\xrightarrow[\text{H}_2\text{O}]{\text{MnO}_4^-}$

$R{-}C{\equiv}C{-}H$ $\xrightarrow[\text{HgSO}_4]{\text{H}_2\text{SO}_4 / \text{H}_2\text{O}}$

$R{-}C{\equiv}C{-}H$ 1) B_2H_6 2) H_2O_2 / HO^-

$\xrightarrow{h\nu}$ $\longrightarrow$ $CO_2 + H{-}C{\equiv}C{-}H$

a lactone

$\xrightarrow{[O]}$ Oxidation conditions vary as a function of substituent and result desired.

$\xrightarrow{[O]}$ Oxidation conditions vary as a function of substituent and result desired.

$\xrightarrow{[O]}$ $\xrightarrow{[O]}$ Oxidation conditions vary as a function of substituent and result desired.

$\xrightarrow{\text{AlCl}_3}$ + HCl

Chapter 7

S-(+)-2-aminosuccinic acid
(aspartic acid; 2-aminobutanedioic acid)

S-(−)-2-bromosuccinic acid
(2-bromobutanedioic acid)

Chapter 8

L - ascorbic acid
(Vitamin C)

L - dehydroascorbic
acid

diketogluconic
acid

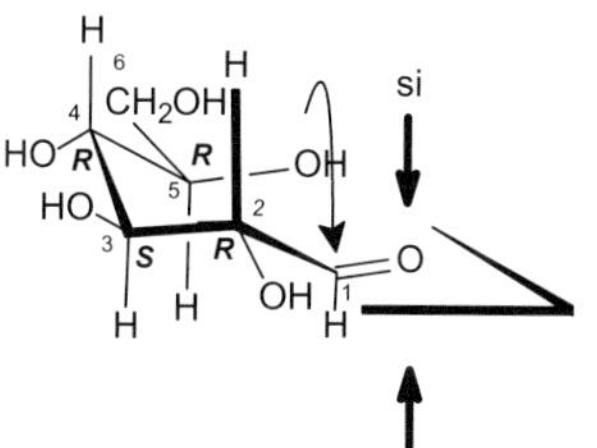

same
as...

si face
attack

α-D-(+)-glucopyranose

re face
attack

β-D-(+)-glucopyranose

TABLE 9.1. *Continued*

AT = acyl transferase (from acetylCoA to the PKS); ACP = acyl carrier protein; KS = keto synthase

KR = keto reductase; DH = dehydratase; ER = enoyl reductase

D-ribose furfural furoic acid

citrate *cis*-aconitate isocitrate

Acid catalysis generally required

Finally, of course, many natural products such as steroids (Chapter 11), alkaloids (Chapter 13), and the building blocks of the genetic code (Chapter 14) are also graced with a compliment of carbonyl groups such as aldehydes, ketones, and carboxylic acids or their derivatives.

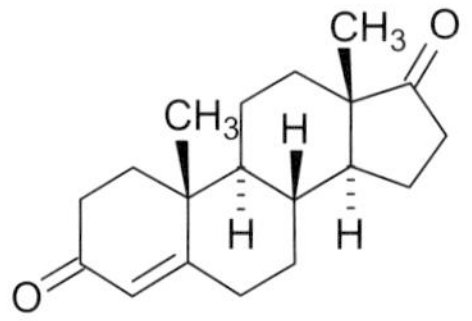

Δ^4 - androstenedione
(a diketosteroid)

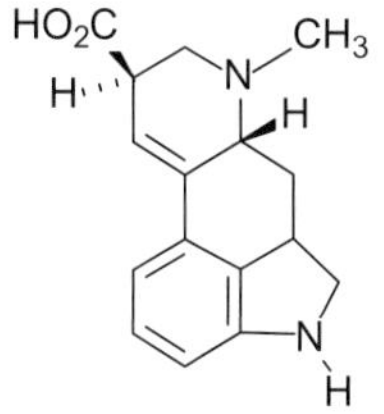

lysergic acid
(a carboxylic acid
hydrolysis product from
ergot alkaloids)

For all of these kinds of carbonyl-containing compounds (aldehydes, ketones, carboxylic acids, etc.), some properties of which are provided for some of the more simple members in Table 9.2, it might be anticipated that protonation of the non-bonded electrons (Chapter 1) on oxygen would increase the electron deficiency at the carbon of the carbonyl and facilitate attack at that carbon by nucleophiles. It might further be expected that, since the carbon of the carbonyl is positive, protons on the carbon adjacent to the carbon of the carbonyl (i.e., the α-carbon) would be particularly acidic (relative to protons on carbon *not* in such a position). In addition, it might be anticipated that the positive carbon of the carbonyl would exert an influence on sites of unsaturation (both double and triple bonds) that were conjugated with the carbonyl (i.e., α,β-unsaturated).

Such expectations, anticipations, and forecasts are to be realized.

PART I. ALDEHYDES AND KETONES

A. OXIDATION OF ALDEHYDES AND KETONES

Aldehydes are more easily oxidized than ketones, but, as with other organic compounds, complete combustion in the presence of oxygen produces carbon dioxide (CO_2) and water (H_2O) (assuming, of course, that halogen, etc., are absent).

In the presence of a suitable initiator, which might be triplet oxygen ($^3O_2$) or a radical source, the controlled oxidation of aldehydes by oxygen results in the abstraction of the hydrogen attached to the carbon of the carbonyl to produce an acyl radical (Equation 9.1)*

$$\text{(9.1)}$$

In • = "initiator"

---

*The experimental bond dissociation energy ($D°$) of most carbon–hydrogen bonds where the carbon is double bonded to oxygen (i.e., H–C=O) is, generally, about 21–63 kJ mol⁻¹ (5–15 kcal mol⁻¹) less than similar carbons lacking the doubly bonded oxygen. It can be argued that this is a consequence of the polarization of the carbon–oxygen double bond. Thus, as the carbon of the carbonyl is positive ($\delta+$) relative to oxygen of the carbonyl (which is negative, $\delta-$) the electrons in the σ bonds from the carbon of the carbonyl to its neighbors will be attracted toward the carbon and the σ bond correspondingly weakened.

TABLE 9.2. Representative Examples of Aldehydes, Ketones, Carboxylic Acids, and Derivatives of the Latter[a]

Compound	Name	mp °C	bp °C	d^{20}	^{1}H NMR data [δ, ppm TMS = 0.00][b]	^{13}C NMR data [δ, ppm TMS = 0.00]
CH_3CHO	Ethanal (acetaldehyde)	−123.5	20.2		2.18 (CH_3, d, J = 3); 9.64 (CH, q, J = 3)	30.7 (CH_3); 199.8 (C=O)
$(CH_3CHO)_3$	Paraldehyde[c]	12.6	124		1.26 (CH_3, d, J = 8); 4.92 (CH, q, J =9)	20.4 (CH_3); 98.1 (O–CH–O)
$(CH_3CHO)_4$	Metaldehyde[c]	115–246			1.28 (CH_3, d, J = 6); 4.98 (CH, q, J = 6)	22.5 (CH_3); 98.2 (O–CH–O)
$H_2C=O$	Methanal[d] (formaldehyde)	−91	−21		9.6 (5% in TMS)	199 (5% in TMS)
$(H_2CO)_3$	1,3,5-trioxane[c]	61 62	115		5.00	94.6
	paraformaldehyde[c]	160–165			5.17 (–O–CH_2–O) 5.04–CH_2OH	90.5 (internal O–CH_2–O); 90.13 (penultimate O–CH_2–O); 86.9 (terminal CH_2–OH)
$HOCH_2$ - $(OCH_2)_nOCH_2OH$						
HCO_2H	Methanoic acid (formic acid)	8.4	100.7	1.22	8.99 (CH); 3.95 (OH)	166.3
CH_3CO_2H	Ethanoic acid (acetic acid)	16.6	118.2	1.05	2.08 (CH_3); 8.85 (OH)	20.48 (CH_3); 177.14 (C=O)
$CH_3CO_2COCH_3$	Ethanoic anhydride (acetic anhydride)	−73	140	1.08	2.34 (CH_3)	21.85 (CH_3); 167.15 (C=O)
CH_3CONH_2	Ethanamide (acetamide)				2.02 (CH_3, t, J = 2.6); 6.33 (NH_2, q, J = 2.6)	23.97 (CH_3) 23.97;175.90 (C=O)

$CH_3CONHCH_3$	N-methylethanamide (N-methylacetamide)	26–28	204–206	0.957	1.85 (C–CH_3); 2.76 (N–CH_3); 5.36 (N–H)	22.20 (C–CH_3); 26.00 (N–CH_3); 173.00 (C=O)
$CH_3CON(CH_3)_2$	N,N-dimethylethanamide (N,N-dimethylacetamide)		164–166	0.938	2.01 (C–CH_3); 2.89 (s–Z–N–CH_3, q, J = 14); 2.93 (s–E–N–CH_3, q, J = 14)	21.3 (C–CH_3); 34.99 (s–Z–N–CH_3); 38.01 (s–E–N–CH_3); 169.82 (C=O)
C_6H_5CHO	Benzenecarbaldehyde (benzaldehyde)		180–183	1.047	7.45 (*meta*-aromatic–CH, m); 7.54 (*para*-aromatic–CH, m); 7.81 (*ortho*-aromatic–CH, m); 9.94 (H–C=O)	129.08 (*meta*-aromatic–C); 129.74 (*ortho*-aromatic–C); 134.24 (*para*-aromatic–C); 136.28 (*ipso*-aromatic–C); 191.90 (C=O)
$C_6H_5CO_2H$	Benzenecarboxylic acid (benzoic acid)	122			7.44 (*meta*-aromatic–CH, m); 7.51 (*para*-aromatic–CH, m); 7.97 (*ortho*-aromatic–CH, m); 9.39 (–OH)	129.01 (*meta*-aromatic–C); 130.08 (*ortho*-aromatic–C); 130.50 (*ipso*-aromatic–C); 133.76 (*para*-aromatic–C); 168.57 (C=O)
$C_6H_5CONH_2$	Benzenecarboxamide (benzamide)	128–129	288		5.75 (NH_2); 7.45 (*meta*-aromatic–CH, m); 7.51 (*para*-aromatic–CH, m); 7.95 (*ortho*-aromatic–CH, m)	127.52 (*ortho*-aromatic–C); 128.56 (*meta*-aromatic–C); 131.34 (*para*aromatic–C); 133.90 (*ipso*-aromatic–C); 168.32 (C=O)
$C_6H_5C{\equiv}N$	Benzonitrile (cyanobenzene)	–13	191	1.007	7.44 (*meta*-aromatic–CH, m); 7.51 (*ortho*-aromatic–CH, m); 7.56 (*para*-aromatic–CH, m)	112.54 (*ipso*-aromatic–C); 119.10 (C/N); 129.18 (*meta*-aromatic–C); 132.08 (*ortho*-aromatic–C); 132.79 (*para*-aromatic–C)

TABLE 9.2. *Continued*

Compound	Name	mp °C	bp °C	d^{20}	^{1}H NMR data [δ, ppm TMS = 0.00][b]	^{13}C NMR data [δ, ppm TMS = 0.00]
$C_6H_5COCH_3$	Phenylethanone (acetophenone)	19–20	201–203	1.030	2.56 C–CH_3); 7.37 (*meta* - aromatic–CH, m); 7.46 (*para*-aromatic–CH, m); 7.87 (*ortho*-aromatic–CH, m)	24.88 (CH_3); 128.26 (*meta* - aromatic–C); 128.45 (*ortho* - aromatic–C); 131.79 (*para*-aromatic–C); 136.93 (*ipso*-aromatic–C); 196.30 (C=O)
Cyclohexanone	Cyclohexanone	–26	154–156	0.947	1.32 (*ax*-H, m, C_4); 1.83 (*ax*-H, m, C_3 and C_5); 1.83 (*eq*-H, m, C_4); 1.87 (*eq*-H, m, C_3 and C_5); 2.22 (*ax*-H, m, C_2, and C_6); 2.29 (*eq*-H, m, C_2 and C_6)	24.94 (C_4), 27.09 (C_3 and C_5); 41.72 (C_2 and C_6); 210.58 (C=O)
CH_3CH_2CHO	Propanal (propionaldehyde)	–81	48–49		CH_3 (1.13, t, J = 7); CH_2 (2.46, m, J = 7,1.5); HC=O (8.89, t, J = 1.5)	CH_3 (6.00); CH_2 (37.30); HC=O (202.80)
CH_3COCH_3	Propanone (2-propanone) (acetone)	–95	56		CH_3 (2.00)	CH_3 (30.66); C=O (206.65)
$CH_3CH_2CO_2H$	Propanoic acid (propionic acid)	–20.8	141	1.38	1.16 (CH_3, t, J = 7.8); 2.36 (CH_2, q, J = 7.8)	8.95 (CH_3), 27.63 (CH_2), 180.40 (C=O)
CH_3CH_2COCl	Propanoyl chloride (propionyl chloride)		80	1.06	1.27 (CH_3, t, J = 7.23); 2.96 (CH_2, q, J = 7.23);	7.28 (CH_3); 39.80 (CH_2) 174.42 (C=O)
$CH_3CH_2COCH_3$	2-butanone	–86.4	82	0.80	0.87 (CH_3, t, J = 7); 2.10 (CH_3, s); 2.42 (CH_2, q, J = 7)	7.47 (CH_3 [C–4]); 28.37 (CH_3 [C-1]); 36.00 (CH_2); 207.57 (C=O)

Formula	Name	mp	bp	density	^{1}H NMR (ppm)	^{13}C NMR (ppm)
$CH_3CH_2CH_2CHO$	Butanal (butyraldehyde)	−97.1	74.7		0.97 (CH_3, t, J = 7); 1.67 (β–CH_2, m); 2.42 (α–CH_2, d[t]); 9.54 (O=CH, t, J = 1.5)	13.63 (CH_3); 15.90 (β–CH_2); 45.83 (α–CH_2); 202.13 (HC=O)
$CH_3CH_2CH_2CO_2H$	Butanoic acid (butyric acid)	−5.5	162.5	0.958	1.00 (CH_3, t, J = 7); 1.69 (β–CH_2, m); 2.31 (α–CH_2, t, J = 7); 11.59 (O–H);	13.30 (CH_3); 18.57 (β–CH_2); 36.28 (α–CH_2); 179.77 (–CO_2H)
$CH_3CH_2CH_2COCl$	Butanoyl chloride (butryl chloride)		101		0.96 (CH_3, t, J = 7.4); 2.05 (β–CH_2, m); 2.80 (α–CH_2, t, J = 7.0)	13.00 (CH_3); 18.70 (β–CH_2); 48.90 (α–CH_2); 173.70 (C=O)
$(CH_3CH_2CH_2CO)_2$	Butanoic anhydride (butyric anhydride)		198		0.94 (CH_3, t, J=7.25); 1.63 (β–CH_2, m); 2.36 (α–CH_2, t, J=7.25)	13.30 (CH_3); 18.20 (β–CH_2); 37.20 (α–CH_2); 169.6 (C=O)
$CH_3CH_2CH_2CO_2CH_3$	Methyl butanoate (methyl butyrate)	−84	102	0.898	0.98 (C=CH_3, t, J = 7.25); 1.65 (β–CH_2, m); 2.22 (α–CH_2, t, J = 7.25); 3.71 (O–CH_3, s)	13.77 (C–CH_3); 18.70 (β–CH_2); 35.80 (α–CH_2); 51.51 (O–CH_3); 173.13 (C=O)
$HCONH_2$	Methanamide (formamide)	2.5	193		7.09 (NH_2, d, 7.1); 8.21 (H–C=O, t, 7.1)	171.2 (C=O)

[a]The information on this table are derived from experiment and subject to change. The bp and mp values prese nted are largely from Rappoport, Z. *CRC Handbook of Tables for Organic Compound Identification*, 3rd edition, CRC Press, Cleveland, OH, **1977**.

[b]The following abbreviations are used: d = doublet; t = triplet; q = quartet; m = multiplet. Data not obtained by the author are from (a) Levy, G. C.; Lichter, R. L.; Nelson, G. L. *Carbon-13 Nuclear Magnetic Resonance Spectroscopy*, 2nd edition, Wiley-Interscience, New York, **1980**, and; (b) Jackman, L. M.; Sternhell, S. *Applications of Nuclear Magnetic Resonance Spectroscopy in Organic Chemistry*, 2nd edition, Pergamon, Oxford, UK, **1969**.

[c]Acetals such as these are commonly used as progenitors of the corresponding aldehydes. A brief discussion has already been provided (Chapter 8).

[d]An aqueous solution of formaldehyde (ca. 37% CH_2O) is commonly available as "formalin."

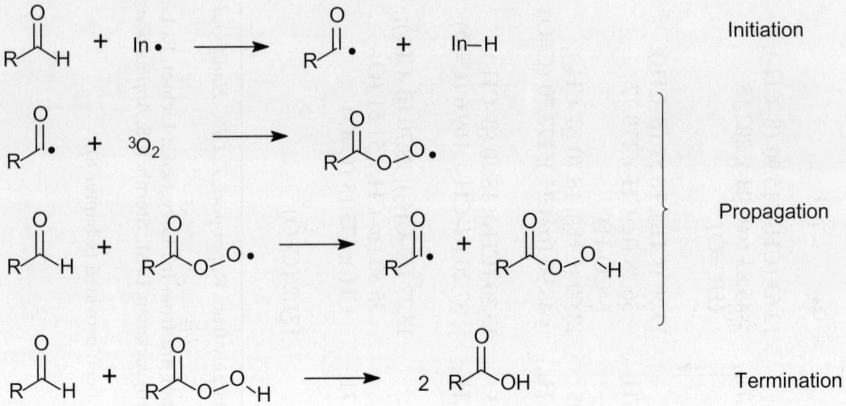

Initiation

Propagation

Termination

Scheme 9.1. A representation for the oxidation of an aldehyde by oxygen to the corresponding carboxylic acid. It will be noticed that the *initiation* step is identical to the reaction shown in Equation 9.1. The last step shown is a variation of the Baeyer–Villiger reaction (typically, the oxidation of an aldehyde or ketone by percarboxylic acid and which will be discussed later in this chapter).

Now, as shown in Scheme 9.1, Equation 9.1 is but the first step in the air-oxidation of aldehyde to carboxylic acid. Thus, after the hydrogen attached to the carbon of the carbonyl group is abstracted, forming the acyl radical, reaction of the latter with $^3O_2$ produces an acylperoxy radical. This step, combined with the next (the use of this just formed acylperoxy radical to effect the abstraction of the acyl hydrogen of another aldehyde producing a new acyl radical and a *percarboxylic* acid, RCO_3H), continues to propagate the chain reaction. Then, termination of the process occurs when another equivalent of aldehyde is oxidized (a form of the *Baeyer–Villiger* reaction* [this chapter]) by the percarboxylic acid. The termination reaction results in the formation of two equivalents of carboxylic acid (one from *oxidation* of the aldehyde and the other from *reduction* of the percarboxylic acid) (Scheme 9.2).

The oxidation of aldehydes by oxygen is so facile that samples of aldehyde left exposed to the air[†] are often contaminated by the corresponding carboxylic acids. An interesting example of this can be found by observing that benzenecarbaldehyde (benzaldehyde, C_6H_5CHO, Table 9.2) left in an open bottle is slowly converted to benzenecarboxylic acid (benzoic acid, $C_6H_5CO_2H$, Table 9.2), which is found in crystalline form (melting point [mp] 121–122°C) around the rim of the bottle.

The variety of oxidizing agents already encountered in the previous chapter [e.g., chromic anhydride (CrO_3), potassium permanganate ($KMnO_4$), hydrogen peroxide (H_2O_2) and sodium hypochlorite ($NaOCl$)] can also be used to oxidize aldehydes to carboxylic acids, and, in many cases, aldehyde oxidation with these reagents appears to occur through the hydrate of the aldehyde (Scheme 9.3).

However, these reagents cannot generally be used in the presence of the other functional groups they oxidize if the *only* desired reaction is aldehyde oxidation.

---

*Baeyer, A.; Villiger, V. *Chem. Ber.*, **1899**, *32*, 3625. A review has appeared; see Krow, G. R. *Org. React.*, **1993**, *43*, 251 ff.
[†]The mole fraction of oxygen in dry air is about 0.21 at sea level.

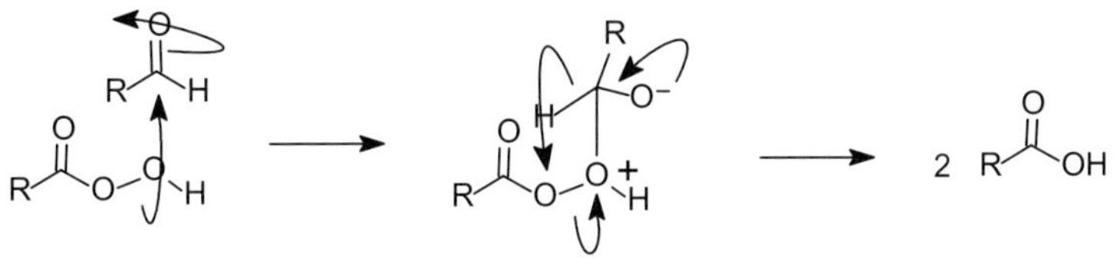

Scheme 9.2. One of the variations of the Baeyer–Villiger reaction. A discussion of the Baeyer–Villiger reaction can be found later in this chapter.

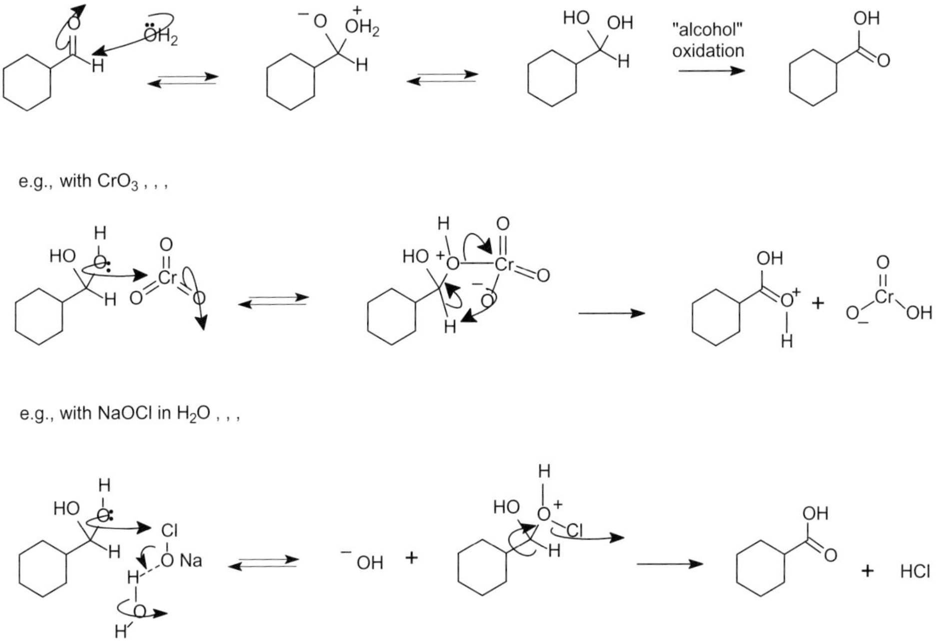

e.g., with CrO_3 , , ,

e.g., with NaOCl in H_2O , , ,

Scheme 9.3. Oxidation of cyclohexanecarbaldehyde via the hydrate. The examples, that is, chromic anhydride (CrO_3) and sodium hypochlorite (NaOCl), are illustrative. All reagents capable of alcohol oxidation can be utilized.

More selective oxidizing agents such as manganese dioxide (MnO_2), sodium chlorite ($NaClO_2$), and even silver oxide (AgO)* occasionally cause problems when other functional groups are present. However, silver oxide (AgO) can be used in the presence of most other oxidizable functionality and it has traditionally been used to demonstrate the presence of an oxidizable function (an aldehyde or hemiacetal as discussed in Chapter 8) in "reducing sugars"[†] (Chapter 11) and, with such sugars, has been used to form silver mirrors on clean glass. Indeed, although almost any aldehyde (i.e., R–CHO undergoing oxidation to RCO_2H) can be used to produce such silver (Ag^{+1} or Ag^{+2} reduced to Ag^{o}) mirrors, the "reducing sugar" glucose (Equation 9.2 and Chapter 8) is not only easily obtainable in pure form, that is, it

---

*Silver oxide (AgO) is generally prepared *in situ* from silver nitrate ($AgNO_3$) and base (NaOH).
[†]"Reducing sugars" derive their names from their ease of oxidation. These sugars possess aldehyde or masked aldehyde (hemiacetal but *not acetal*) functionality.

is inexpensive, but it is also water soluble as is the gluconic acid (Equation 9.2) product so that both unreacted organic starting material and organic product can be readily removed after the reaction that produces the silver mirror. In the pre-spectroscopic era, aqueous ammoniacal (NH_3) solutions of silver(I) (Ag^+) and the production of such mirrors had a place in the organic chemistry qualitative scheme as a test for reducing sugars (the **Tollens** reagent). In the same period, the reduction of copper(II) (Cu^{+2}) salts to Cu(I) (Cu_2O) (**Benedict's** solution or **Fehling's** test) was also used to "test for reducing sugars" or (and this was the problem) any easily oxidizable function, which might or might not be an aldehyde.*

Generally:
$$R\text{-CHO} \xrightarrow{AgO} R\text{-CO}_2H + Ag^0\downarrow$$

D-glucose
(open [acyclic]
representation)

si face
attack

α-D-(+)-glucopyranose

gluconic acid

$$(9.2)$$

Interestingly, aldehydes can serve as their own oxidizing agents too. Thus, when aldehydes *lacking α-protons*[†] (e.g., benzenecarbaldehyde [benzaldehyde]) are

*__Benedict's reagent__ consists of a solution containing copper(II) sulfate ($CuSO_4$), citric acid, and sodium carbonate (Na_2CO_3), while **Fehling's** solution contains copper sulfate ($CuSO_4$), tartaric acid, and sodium hydroxide (NaOH). Presumably, for the latter, L-(+)-tartaric acid is preferentially used since it is most easily obtained. However, in principle, its enantiomer or its diastereomer (*meso*-tartaric acid) might also be used.

$$RCHO + 2Cu^+_2 + 4OH^- \longrightarrow RCO_2H + Cu_2O + 2H_2O$$

citric acid

L-(+)-tartaric acid

[†]Protons α- to the carbonyl group are acidic (viz. enol formation, Chapter 8) and, as a consequence, other reactions (to be discussed later) intrude.

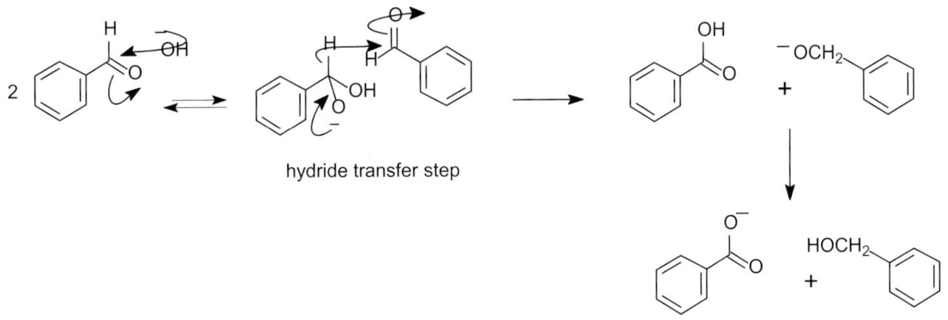

hydride transfer step

Scheme 9.4. The Cannizzaro reaction. The oxidation–reduction of benzenecarbaldehyde (benzaldehyde) to benzenecarboxylic acid (benzoic acid) and phenylmethanol (benzyl alcohol). Note the transfer of a hydride (H⁻) ion from one carbon to another, accounts for the oxidation (hydrogen, H⁻, loss), and reduction (hydrogen, H⁻, gain). (See Cannizzaro, S. *Liebigs Ann. Chem.*, **1853**, *88*, 129.)

treated with base they undergo self-oxidation–reduction (the **Cannizzaro*** reaction, Scheme 9.4).

Problem 9.1. Predict the product(s) of the treatment of furan-2-carbaldehyde (furfural) with base under the conditions of the Cannizzaro reaction.

A complication intrudes when the aldehyde functional group (CHO) in benzene-carbaldehyde (benzaldehyde) *also* has a hydroxyl group (–OH) present in either the position *ortho-* or the position *para-* to the aldehyde functional group (–CHO) and the oxidizing agent is basic hydrogen peroxide (OH⁻ and H_2O_2). In this case (the **Dakin**† reaction) the aldehydic carbon is lost (as methanoic acid, formic acid, HCO_2H), and it is replaced by a hydroxyl group (Scheme 9.5).

Before leaving the subject of aldehyde oxidation, it is worthwhile considering a biologically relevant case. The conversion of glucose (*vide supra*, Table 9.1, Scheme 9.6, and Chapter 11), a 6-carbon sugar, into two equivalents of the three carbon ($2 \times 3 = 6$) ketocarboxylic acid, pyruvic acid (2-ketopropanoic acid) (Equation 9.3) also produces two equivalents of adenosine triphosphate (ATP) from adenosine diphosphate (ADP) (Chapter 12) per equivalent of glucose used. The process involves a series of 10 enzymatically mediated steps (see glycolysis, Chapter 11) and plays a crucial role in supplying energy to many organisms.

About half-way in the process (called glycolysis, i.e., the overall conversion of glucose to pyruvic acid [Equation 9.3 and Chapter 11]), glucose has been converted to two equivalents of glyceraldehyde-3-phosphate (Equation 9.4). It is at this point that oxidation (and phosphorylation) appears to occur.

The overall process is shown in Equation 9.5. (The phosphorylation, using "inorganic phosphate" [PO_4^{-3} or its equivalent, i.e., P_i], converts the thioester [Scheme 9.6], already at the oxidation state of the carboxylic acid, to the phosphoric acid

---

*Stanislao Cannizzaro (1826–1910) studied at Pisa and was a professor of Chemistry at the universities of Genoa, Palermo, and Rome at various times during his career.
†Dakin, H. D. *Am. Chem. J.*, **1909**, *42*, 477.

anhydride of that acid. With water, instead of "inorganic phosphate" [PO_4^{-3} or its equivalent, i.e., P_i], the acid itself is formed.)

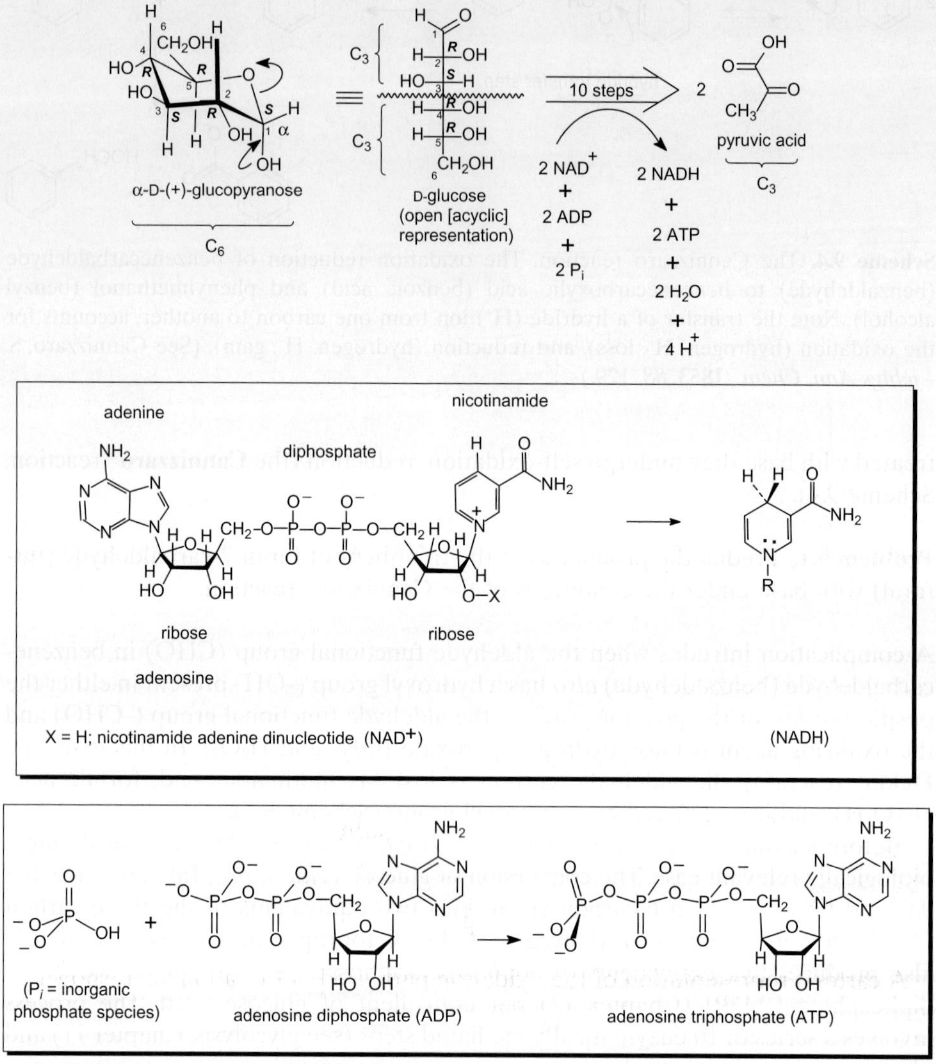

X = H; nicotinamide adenine dinucleotide (NAD$^+$)   (NADH)

(9.3)

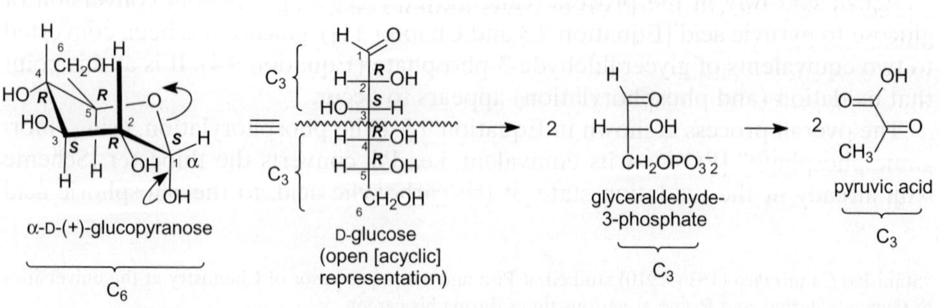

(9.4)

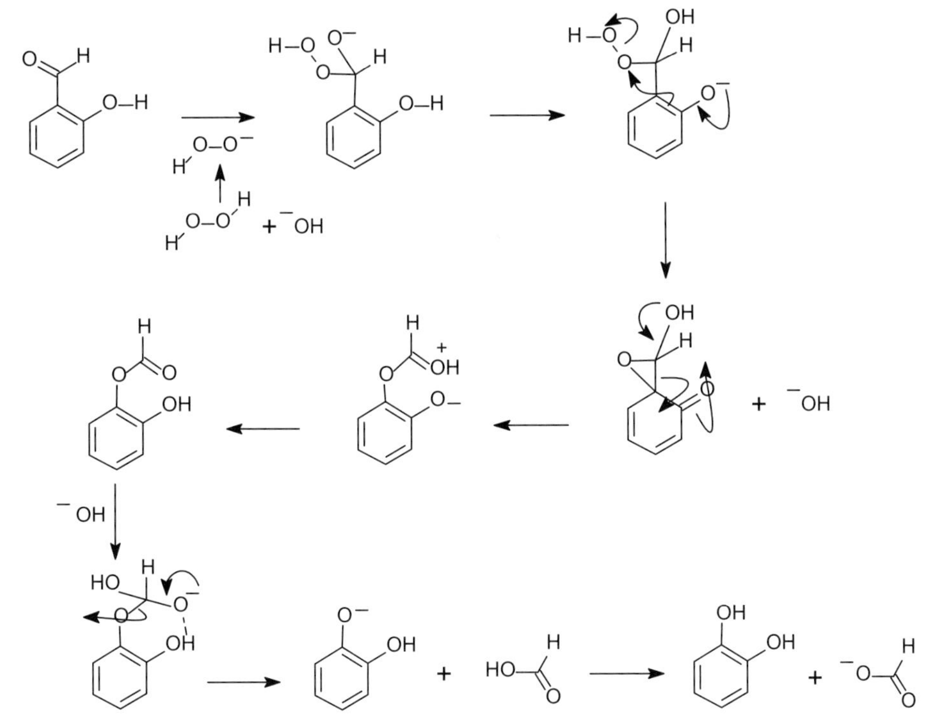

Scheme 9.5. A representation of the **Dakin** reaction, which converts an *ortho-* or *para-*hydroxyl-substituted benzenecarbaldehyde (benzaldehyde) into a bisphenol. The particular case shown generates 1,2-dihydroxybenzene (*o*-catechol).

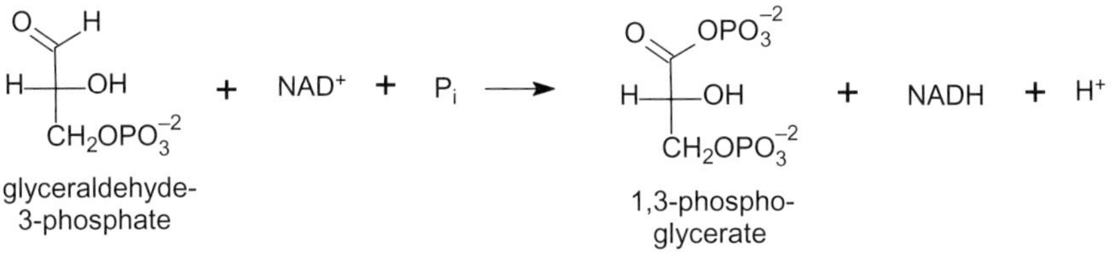

$$\text{glyceraldehyde-3-phosphate} + NAD^+ + P_i \longrightarrow \text{1,3-phosphoglycerate} + NADH + H^+ \tag{9.5}$$

glyceraldehyde-
  3-phosphate

1,3-phospho-
  glycerate

A cartoon representation of the oxidation path used in Equation 9.5 is shown in Scheme 9.5. The process is catalyzed by the enzyme glyceraldehyde-3-phosphate dehydrogenase (EC 1.2.1.12). A large body of evidence has been accumulated about the role of the enzyme and, in particular, that the catalytic site on the enzyme that is used to help the oxidation occur bears a thiol (-SH, Chapter 8) group (from the amino acid cysteine, Chapter 12). Further, it has been shown that hydride (H$^-$) transfer to nicotinamide dinucleotide (NAD$^+$) (see Chapter 12), a required cofactor that accounts *for the transfer of the hydride anion and, thus, the oxidation itself*, is also involved.

The overall proposed pathway is shown in Scheme 9.6.

The difficulties surrounding the oxidation of ketones stand in stark contrast to the ease of oxidation of aldehydes. Although the major observable difference between aldehydes and ketones is the replacement of the hydrogen–carbon bond at the carbon of the carbonyl with a carbon–carbon bond and although (Table 1.1) C–H bonds are **stronger** than carbon–carbon bonds, ketones fail to undergo

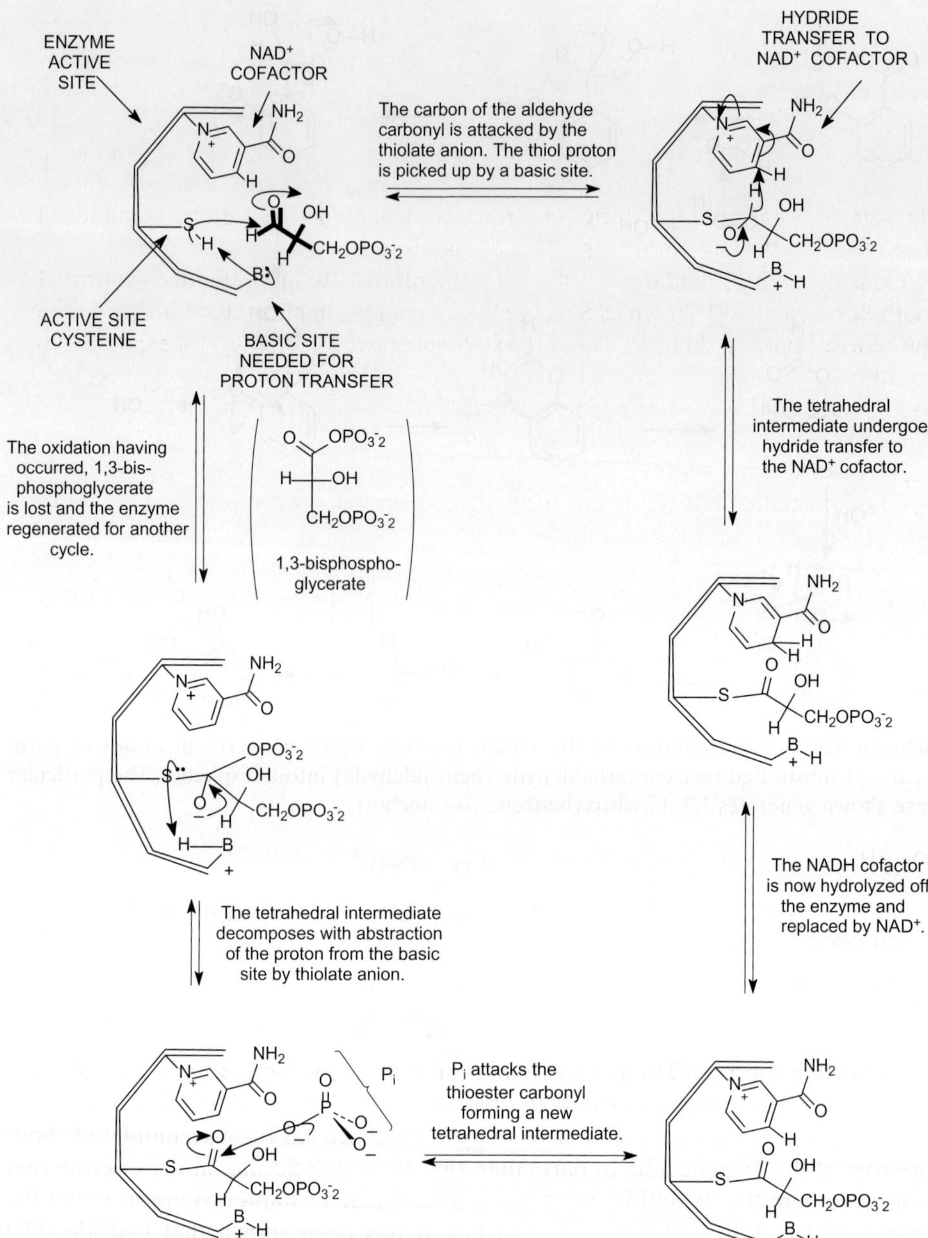

Scheme 9.6. A cartoon representation of the enzyme (named glyceraldehyde-3-phosphate dehydrogenase, EC 1.2.1.12) catalyzed **oxidation** of the aldehyde glyceraldehyde-3-phosphate to the carboxylic acid–phosphoric acid mixed anhydride, 1,3-bisphosphoglycerate (which can be hydrolyzed to the corresponding carboxylic acid and phosphoric acid). Start in a clockwise direction from the upper left.

oxidation under the relatively gentle conditions required for aldehydes. Some of the difference in ease of oxidation can be explained by noting (1) that the attack of oxidizing agents, initially at the carbon of the carbonyl in both cases, will suffer less hindrance with aldehydes, which tend to undergo addition reactions (*vide infra*) much more rapidly than do ketones, and (2) hydride (H⁻) transfer to the species being reduced is generally much more facile than alkalide (RC⁻) transfer. Hydride (H⁻) transfer has minimal steric requirements, while alkalide anions, being tetrahedral or nearly so, have more stringent demands.*

Consider the oxidation of D-(+)-menthol {(+)-[1S,2R,5S]-2-isopropyl-5-methylcyclohexanol} shown as Scheme 9.7. In *cold* permanganate (MnO$_4^-$) solution, the corresponding ketone (+)-[2R,5S]-2-isopropyl-5-methylcyclohexanone [(+)-menthone] is cleanly produced in high yield. In principle the reaction could be used preparatively. However, in *warm* permanganate (MnO$_4^-$), the keto acid 3,7-dimethyl-6-oxooctanoic acid, acetone, the dicarboxylic acids, 3-methylhexanedioic acid (β-methyladipic acid) and 2-methylbutanedioic acid (methylsuccinic acid), and additional smaller fragments are produced. A similar outcome with other ketones and other oxidizing agents is common.

However, other oxidation reactions of ketones do give cleaner results.

Both aldehydes (RCHO) and ketones (R$_2$CO) undergo the **Baeyer–Villiger**[†] reaction.

In this process, a peracid (RCO$_3$H) is allowed to react with the carbonyl compound and the product of addition of that peracid to the carbonyl undergoes rearrangement to an ester (RCO$_2$R).

As shown in Equation 9.6, propanal (CH$_3$CH$_2$CHO) on treatment with perethanoic acid (peracetic acid, CH$_3$CO$_3$H) produces propanoic (CH$_3$CH$_2$CO$_2$H) and ethanoic (acetic, CH$_3$CO$_2$H) acids, the former the result of oxidation and the latter the result of reduction. Similarly, as shown in Equation 9.7, propanone (acetone [(CH$_3$)$_2$C=O]) under the same conditions produces methyl ethanoate (methyl acetate, CH$_3$CO$_2$CH$_3$) and ethanoic (acetic, CH$_3$CO$_2$H) acids.

$$(9.6)$$

*Such ingenious arguments aside, ΔH_f° for propanal (CH$_3$CH$_2$CHO) is **found experimentally to be** $-188.7\,\mathrm{kJ\,mol^{-1}}$ ($-45.09\,\mathrm{kcal\,mol^{-1}}$), while ΔH_f° for propanone [(CH$_3$)$_2$CO] is **found experimentally to be** $-218.5\,\mathrm{kJ\,mol^{-1}}$ ($-52.23\,\mathrm{kcal\,mol^{-1}}$). Thus, the ketone is simply more stable.

[†]Johann Friedrich Wilhelm Adolf von Baeyer (1835–1917) studied under Bunson and Kekulé at the University of Heidelberg and Hofmann at the University of Berlin (from which he received the doctorate). Victor Villiger was von Baeyer's student. As an interesting footnote, although this work was first published in 1899, it received little attention presumably because suitable peracids were not widely available. Baeyer, A.; Villiger, V. *Chem. Ber.*, **1899**, *32*, 3625. A review has appeared; see Krow, G. R. *Org. React.*, **1993**, *43*, 251 ff.

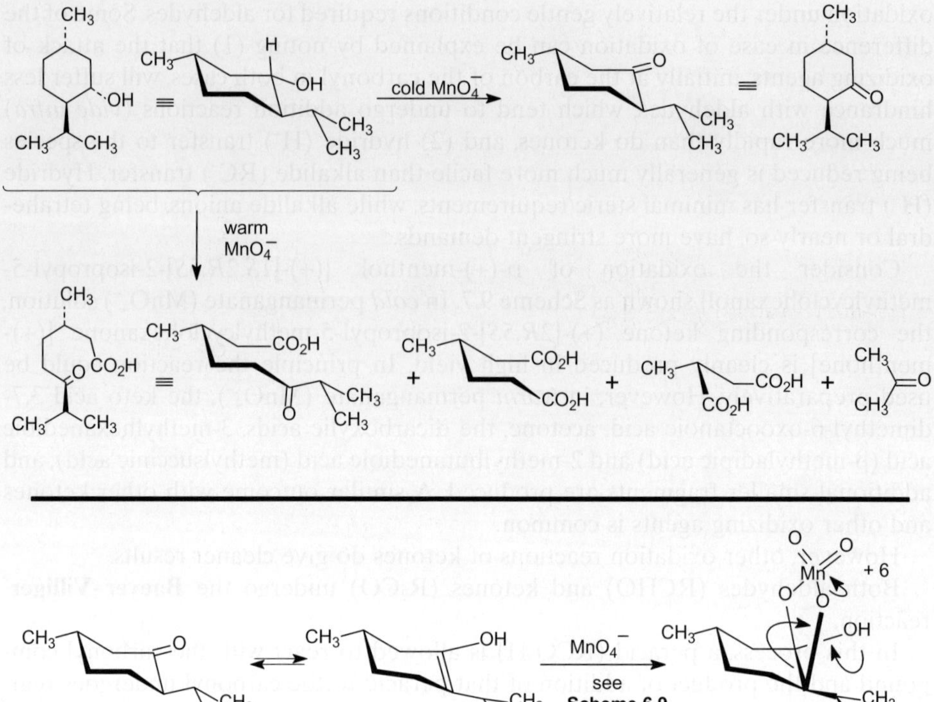

Scheme 9.7. The oxidation of (+)-[1*S*,2*R*,5*S*]-2-isopropyl-5-methylcyclohexanol [(+)-menthol] to the corresponding ketone (+)-[2*R*,5*S*]-2-isopropyl-5-methylcyclohexanone [(+)-menthone] with cold aqueous permanganate (MnO_4^-) solution and its cleavage to many fragments by warm aqueous permanganate (MnO_4^-). A potential pathway accounting for the initially formed fragmentation product is also shown.

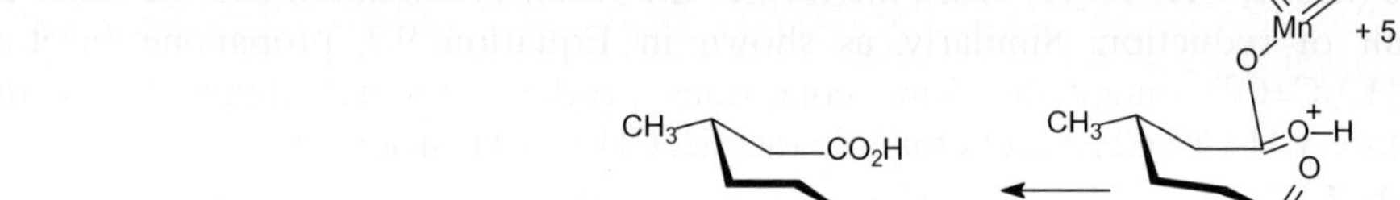

$$(9.7)$$

When the Baeyer–Villiger reaction is carried out on ketones that are not symmetrical, it is generally found that one of the two possible products forms preferentially.

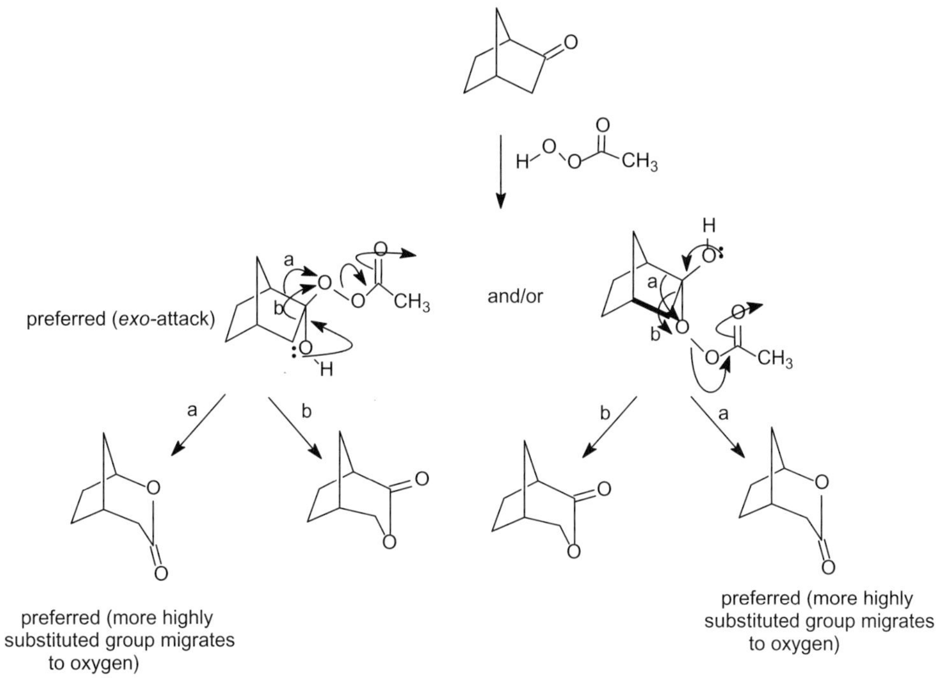

preferred (*exo*-attack)

and/or

preferred (more highly
substituted group migrates
to oxygen)

preferred (more highly
substituted group migrates
to oxygen)

Scheme 9.8. A representation of the Baeyer–Villiger reaction as carried out on bicyclo[2.2.1]-heptan-2-one showing that, along with *exo*-attack by peracid, preferential migration of the most highly substituted alkyl group occurs. (After Meinwald, J.; Frauenglass, E. *J. Am. Chem. Soc.*, **1960**, *82*, 5235.)

For example as shown in Scheme 9.8, bicyclo[2.2.1]heptan-2-one can be attacked by perethanoic acid (peracetic acid) from either the *exo*- or the *endo*-direction. Generally, *exo*-attack is preferred because the *exo*-face is less sterically encumbered. After addition of peracid to the carbonyl group, there are two possible pathways (shown as "a" and "b") by which rearrangement might occur. In pathway "a," the bond from the bridgehead carbon (a tertiary carbon) migrates to the oxygen to provide the bicyclo-[3.2.1]-heptyllactone in which the oxygen is connected to the bridgehead carbon. In pathway "b," the bond from the nonbridgehead carbon (a secondary carbon) migrates to give a different bicyclo[3.2.1]-heptyllactone in which the carbon of the carbonyl group is connected to the bridgehead carbon.

In general, the ability of a group to migrate in the Baeyer–Villiger reaction (the "migratory aptitude") appears to increase in the series:

$$CH_3 < \text{primary alkyl} < \text{secondary alkyl. phenyl} < \text{tertiary alkyl} < H.$$

In addition to the Baeyer–Villiger process, there is a small group of oxidation reactions that can be successfully applied to ketones that bear α-hydrogens. These reactions, in contrast to what was found for potassium permanganate ($KMnO_4$) and chromic anhydride (CrO_3) (*vide supra*), do not generally cause extensive degradation into the smaller fragments found there. However, other functional groups (e.g., alkenes, alkynes, alcohols, and thiols) may be affected.

Problem 9.2. When cyclohexyl methyl ketone is treated with trifluoroperethanoic acid (trifluoroperacetic acid, CF_3CO_3H), the Baeyer–Villiger reaction occurs to yield a product, the ^1H and ^{13}C spectra of which are shown. Given these results and the anticipated migratory aptitude expressed in the series shown above, decide which of the two possible products, cyclohexyl ethanoate (cyclohexyl acetate) or methyl cyclohexanecarboxylate, is obtained.

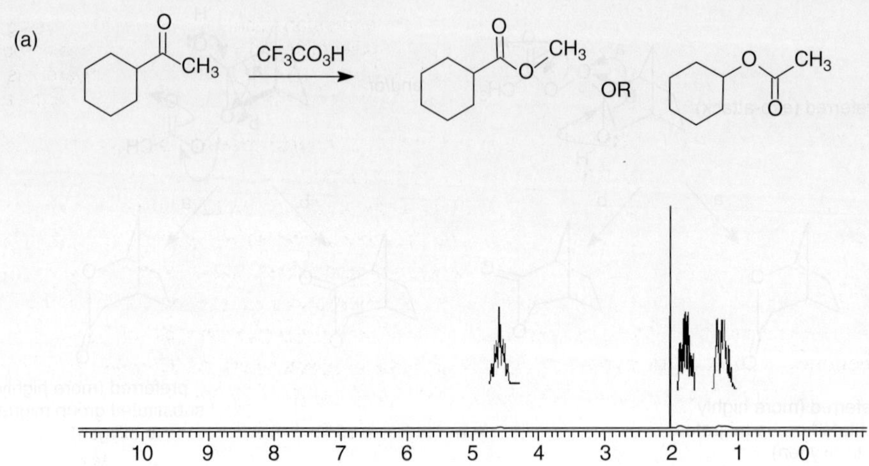

The ^1H NMR spectrum of the product in Problem 9.2.

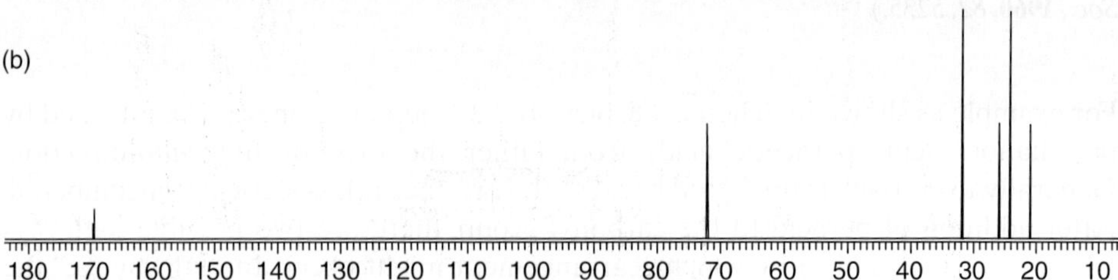

The ^{13}C NMR spectrum of the product in Problem 9.2.

This additional group includes (a) *ortho*-iodosobenzoic acid, which, in the presence of base (Equation 9.8), leads to the corresponding α-hydroxyketones (acyloins) via the corresponding enolate anion; (b) selenium dioxide (SeO_2), which in the presence of acid, produces the corresponding α-diketone (or α-ketoaldehyde) via the enol of the ketone (or aldehyde) (Equation 9.9); (c) ammonium polysulfide (NH_4S_x) or morpholine and sulfur (S_8), which, after hydrolysis of the thioamide formed in the reaction, produces carboxylic acids that result from apparent migration of the carbonyl group to the end of the chain (the **Willgerodt** reaction)* (Equation 9.10); (d) basic hypohalites (e.g., sodium hypochlorite [NaOCl].), which

*Willgerodt, C. *Chem. Ber.*, **1887**, *20*, 2467; *ibid*, **1888**, *21*, 534.

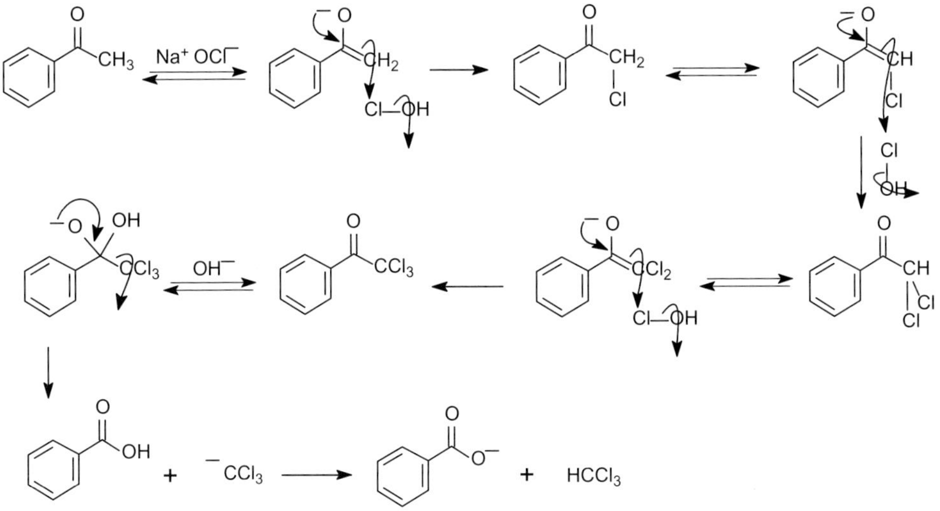

Scheme 9.9. A representation of the pathway for the reaction of methyl phenyl ketone (ace-tophenone) with sodium hypochlorite (NaOCl) in aqueous base to produce chloroform (CHCl₃) and benzenecarboxylic acid (benzoic acid). An example of the **haloform reaction**.

convert methyl ketones (O=C–CH₃) and compounds that can be oxidized to methyl ketones to the corresponding carboxylic acids and haloforms, the latter generated from the terminal methyl (–CH₃) group (Scheme 9.9); and (e) the use of molybdenum pentoxide (MoO₅) in the presence of pyridine and hexameth-ylphosphoramide (HMPA, [(CH₃)₂N]₃P=O; the **MoOPh** reagent, Chapter 8), which, again, produces acyloins (α-hydroxyketones) via the corresponding enolate anion (Equation 9.11).

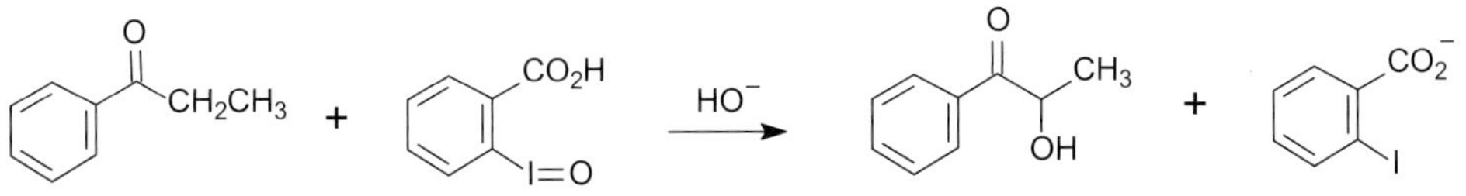

$$(9.8)$$

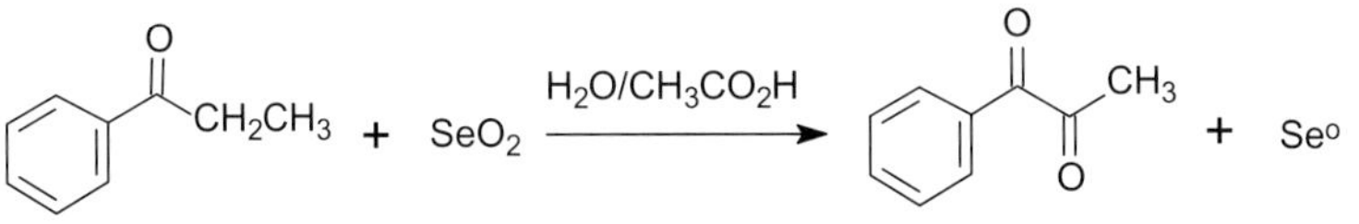

$$(9.9)$$

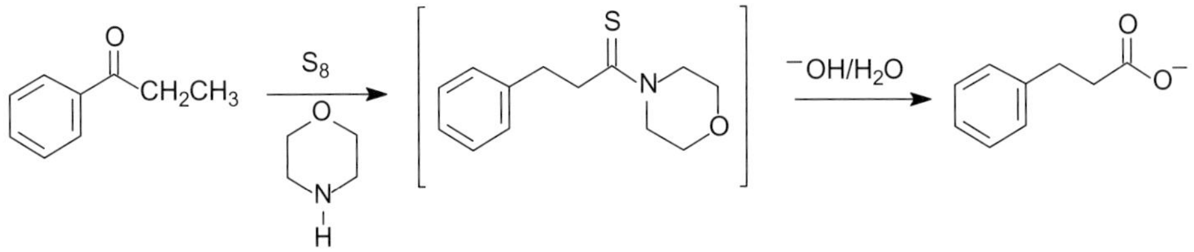

$$(9.10)$$

$$\text{(9.11)}$$

Problem 9.3. The following experimental observations have been made with respect to the representation of the **haloform reaction** shown in Scheme 9.9 and other such base-catalyzed haloform processes: (a) the rate of disappearance of starting ketone is independent of **both** the concentration and nature (Cl, Br, or I) of the hypohalite used, and (b) as long as the solution remains basic and there is sufficient halogen present, the reaction will continue until all of the starting ketone is consumed. Formulate a rate expression for what you believe to be the rate-determining step in Scheme 9.9 and that is consonant with the observations.

To the extent that the oxidation reactions depicted in Equations 9.8–9.11 and Scheme 9.9 can be accounted for by the presence of the enol or enolate anion as a reacting intermediate, which is obligatory on the path from reactant to product, the processes clearly resemble electrophilic addition to alkenes. Indeed, as will become apparent, there are **two** major defining reactive processes of carbonyl-containing compounds: (1) those that involve **addition** to the carbonyl group, and (2) **reactions at the α-carbon** (i.e., reactions that occur at the carbon to which the carbonyl group is attached). For the latter, enols and enolate anions are common, and their generation under conditions of acid and base catalyses, respectively, are vital to the process in which they occur. These *addition* reactions will be considered following a discussion of reduction processes.

However, before leaving this brief introduction to some of the oxidation reactions of aldehydes and ketones, it is important to point out (as already seen in some of the examples used, e.g., in Table 9.1, Equation 9.3 *et seq*) that other functional groups, along with the carbonyl, are often found in aldehydes and ketones. As already noted, the more vigorous oxidizing agents such as permanganate (MnO_4^-) and chromic anhydride (CrO_3) will often react with these more labile accompanying groups. But there are a number of other possibilities. Two of the more common, viz. epoxidation of α,β-unsaturated aldehydes and ketones* and the Baeyer–Villiger oxidation of α-diketones are addressed now.

In order to avoid the Baeyer–Villiger oxidation while introducing an epoxide function into cyclohexen-3-one (Scheme 9.10) epoxidation (*an addition reaction occurring at the carbon–carbon double bond*) must be carried out under basic conditions with the hydrogen peroxide anion (HO_2^-). Here, although the anion might add reversibly to the carbonyl group, the **Michael**[†] type addition at the β-carbon occurs preferentially and the epoxide results.

---

*It will be recalled (Chapter 8) that epoxides (oxiranes) can be induced to undergo acid-catalyzed ring opening to diols. Thus, formation of an epoxide (oxirane) in place of the conjugated double bond of the α,β-unsaturated ketone puts three oxygen-substituted carbon atoms in adjacent positions.

[†]Arthur Michael (1853–1942) was born in the United States (Buffalo, NY) but studied at the Universities of Heidelberg and Berlin in Germany and the Ecole de Médecine in Paris, France. Early in his career,

(*Continued on next page*)

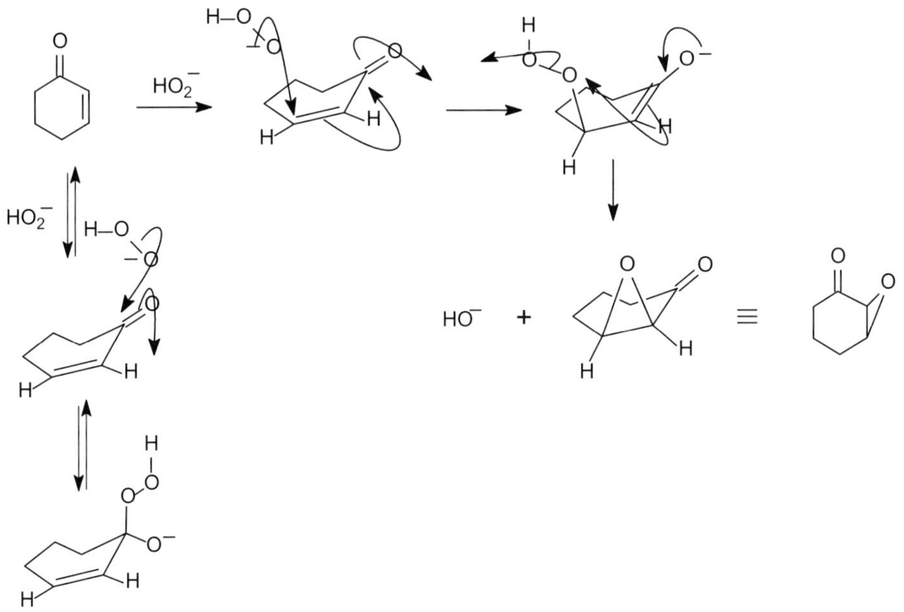

Scheme 9.10. Oxidation (epoxidaton) of cyclohexen-3-one with basic hydrogen peroxide (HO_2^-) to produce the corresponding epoxide (oxirane), avoiding the Baeyer–Villiger reaction.

When the Baeyer–Villiger reaction is carried out on α-diketones, the bond between the two carbonyl groups is broken (i.e., the carbonyl group migrates to oxygen) and the corresponding carboxylic anhydride is obtained. The formation of the anhydride (and the alternate possibility, the formation of α-ketoester) is depicted in Scheme 9.11.

B. REDUCTION OF ALDEHYDES AND KETONES

I. Introduction

Both alkanes **and** alcohols can be produced by the reduction of aldehydes and ketones. However, reactions that produce alcohols generally do not produce alkanes and vice versa. This means that the alcohols do not lie on the reductive pathway to alkanes. Of course, as noted in Chapter 8, alcohols produced by the reduction of aldehydes or ketones can be subsequently converted to alkanes, but the reductive methods for alkanes from the aldehydes and ketones are different from those for alcohols to alkanes.

Second, the variety of conditions applied to the reduction of the carbonyl groups of aldehydes and ketones, both in the presence and in the absence of other substituents, is immense. Further, many studies involving the reduction of the carbonyl

he was Professor of Chemistry at Tufts University and later at Harvard University (1912–1936). Many more examples of this *very useful reaction* have been encountered before and will be encountered again later when addition reactions are discussed in more detail.

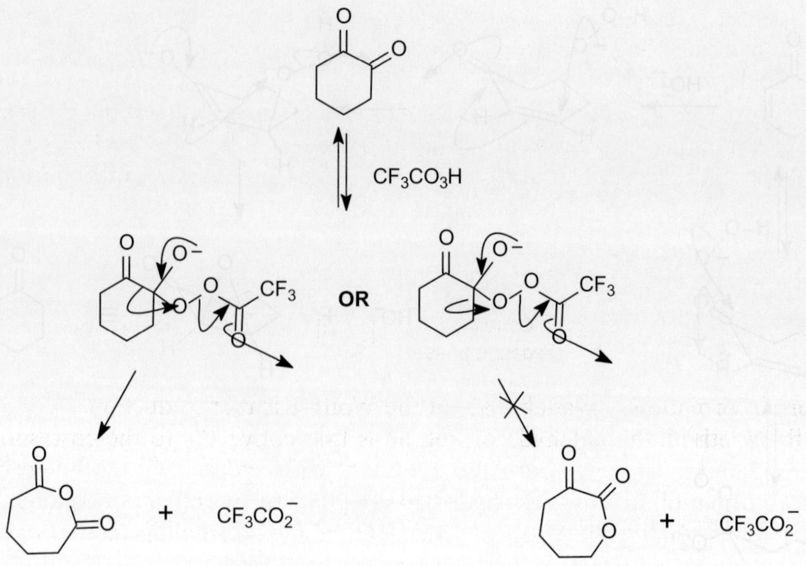

Scheme 9.11. The Baeyer–Villiger reaction on cyclohexane-1,2-dione showing that the anhydride of hexane-1,6-dioic acid (adipic acid [$HO_2C(CH_2)_4CO_2H$] rather than the α-ketolactone of α-keto-ω-hydroxyhexanoic acid.

group of various aldehydes and ketones to alcohols have been undertaken to study the stereochemistry of the reduction (the *re* or *si* face of the carbonyl being preferentially attacked), the relative potency of various reducing agents (which differ in some well-defined way), the role of the solvent, and so on. Common themes have been actively sought.

Finally, it is clear that under almost every set of conditions examined for the reduction of aldehydes and ketones, aldehydes are more easily reduced than ketones.

II. Reduction of Aldehydes and Ketones to Hydrocarbons

There is a small number of reactions that convert aldehydes and ketones directly to hydrocarbons.

Perhaps among the most common is the classical Clemmensen* reduction, a reaction in which the aldehyde or ketone is treated with amalgamated[†] zinc and aqueous hydrochloric acid and which, on consummation, produces the alkane and zinc salts (Equation 9.12). In cases where the aldehyde or ketone is not soluble in

---

*E. C. Clemmensen (1876–1941), a Danish PhD (U. Copenhagen), founded the Clemmensen Corporation in the United States.

[†]An **amalgam** is an alloy of mercury metal ($Hg°$) with another metal. Silver ($Ag°$) amalgam is used in dental fillings. The zinc ($Zn°$) amalgam is a solution of zinc ($Zn°$) metal in mercury ($Hg°$) that is prepared either by treatment of $Zn°$ with small quantities of $Hg°$ directly or by treatment of $Zn°$ with an aqueous solution of mercury(II) chloride ($HgCl_2$). In the latter, since $Zn°$ has a higher reduction potential, it is oxidized to Zn^{+2}, while the Hg^{+2} is reduced to $Hg°$ and both Zn^{+2} and unused Hg^{+2} ions move into the aqueous solution, leaving amalgamated zinc.

the aqueous solution, a nonaqueous second phase (such as toluene) in which both the carbonyl reactant and hydrocarbon product are soluble can often be used. Interestingly, although the Clemmensen reduction has been known and studied off and on by various groups of chemists for nearly a century,* and although it is known that an intermediate alcohol is not involved (the corresponding alcohols cannot be reduced to hydrocarbons under the conditions of the reaction), little more about the pathway from substrate to product is well defined.

$$\text{(9.12)}$$

More information is available about the Wolff–Kishner[†] reduction.

In this method, the aldehyde or ketone is first converted to the corresponding hydrazone derivative (*vide infra*) by treatment with hydrazine[‡] and the derivative is treated with molten potassium hydroxide (KOH, mp 360°C) or potassium hydroxide (KOH) in refluxing diethylene glycol ([HOCH$_2$CH$_2$]$_2$O, boiling point [bp] 204°C) or triethylene glycol ([HOCH$_2$CH$_2$]$_3$O, bp 280–290°C) (the **Huang–Minlon** modification). The process is described in Scheme 9.12.

Dithioketals and dithioacetals can be induced to undergo reductive desulfurization on heating alcoholic solutions of those dithio-derivatives in the presence of moist Raney nickel (see, e.g., page 746). These sulfur compounds are prepared by treating the aldehyde or ketone with 1,2-ethanedithiol (**CAUTION: STENCH**) in the presence of an acid catalyst (an **addition** to the carbon of the carbonyl with loss of water, a process that will be discussed at length, *vide infra*) (Scheme 9.13).

Finally, when ketones and aldehydes are treated with *p*-toluene sulfonyl hydrazine (*p*-CH$_3$C$_6$H$_4$SO$_2$NHNH$_2$, Chapter 10) in the presence of an acid catalyst, addition to the carbon of the carbonyl occurs (*vide infra*) with the formation of the corresponding *p*-toluene sulfonyl hydrazone by a process formally identical with that of the addition of hydrazine (H$_2$NNH$_2$) itself (Scheme 9.12). Reduction of the hydrazone with lithium aluminum hydride (LiAlH$_4$) produces the corresponding hydrocarbon (Scheme 9.14).

III. Reduction of Aldehydes and Ketones to Alcohols

Catalytic hydrogenation of both aromatic and aliphatic aldehydes and ketones with hydrogen gas (H$_2$) over platinum (Pt) and nickel (Ni) catalysts to produce the cor-

---

*Clemmensen, E. C. *Chem. Ber.*, **1913**, *46*, 1837.

[†]The Wolff–Kishner reduction appears to have been discovered independently by L. Wolff (1857–1919), who received his Ph.D. from Strasburg and taught at the University of Jena, Germany, and N. M. Kishner (1867–1935), who received his Ph.D. from the University of Moscow (he was a student of Markownikoff). Kishner eventually also assumed a professorial post at the University of Moscow.

[‡]The **addition** of hydrazine (H$_2$NNH$_2$) across the carbon–oxygen double bond is considered in the next section in more detail. In contrast to what is shown in Scheme 9.12, this and similar addition reactions at the carbonyl are generally *acid catalyzed*, reaching a maximum rate at pH's less than 7.0. While there is clearly some reaction under basic conditions (as shown here) the high temperature and the *irreversible* liberation of nitrogen combine, in this case, to drive the reaction to completion.

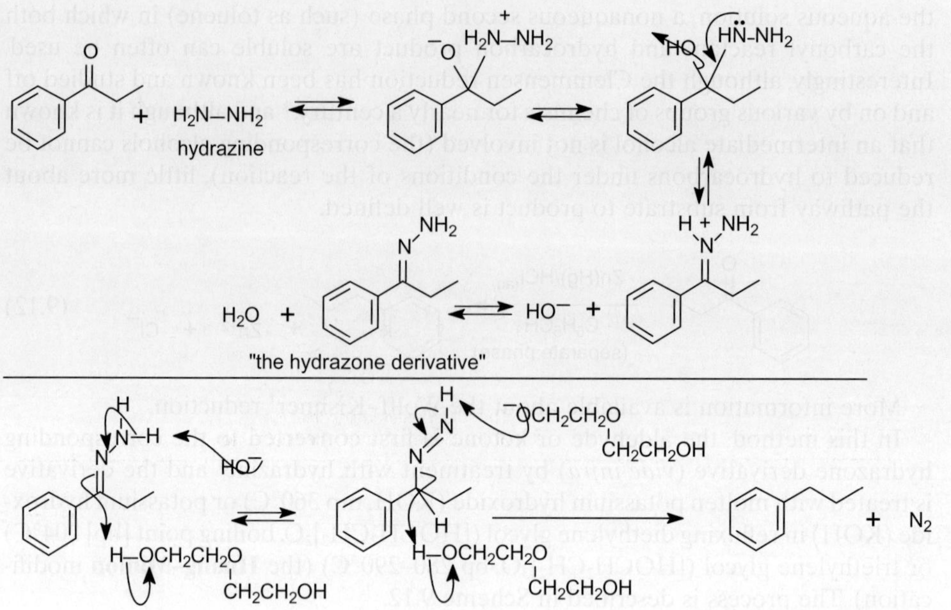

Scheme 9.12. In the top half, the formation of a hydrazone derivative from ethyl phenyl ketone and hydrazine is shown. In the lower half, the decomposition of the hydrazone in basic diethylene glycol to alkane and nitrogen is presented.

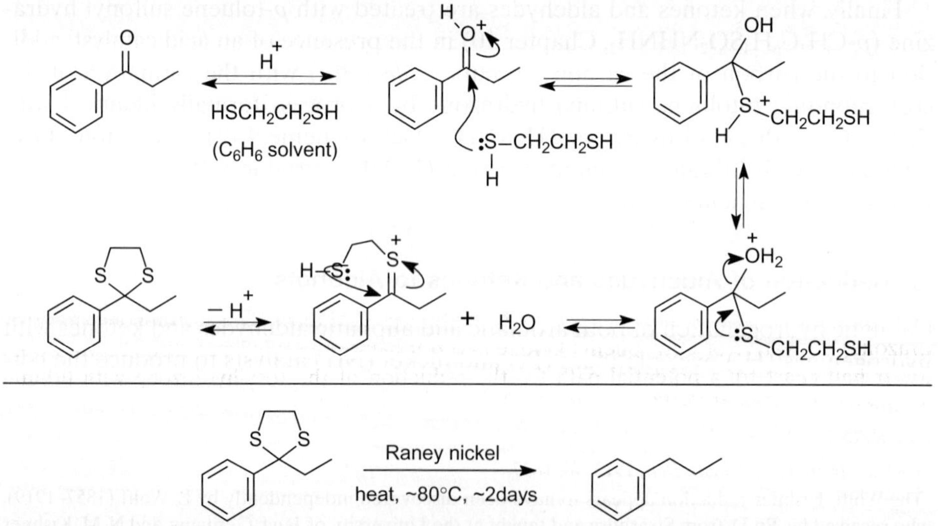

Scheme 9.13. In the top half, the formation of a dithioketal derivative from ethyl phenyl ketone and 1,2-ethanedithiol is shown. In the lower half, the reduction of the dithioketal by Raney nickel is depicted.

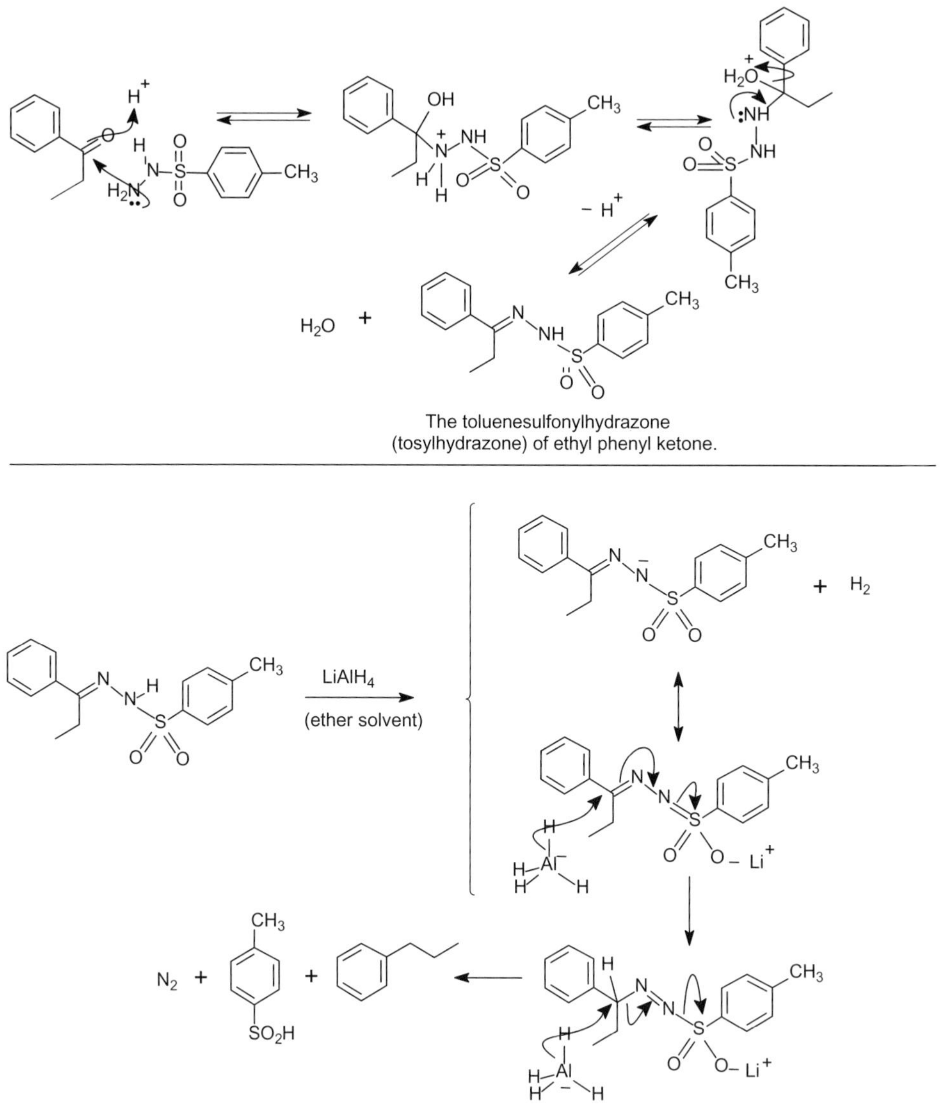

The toluenesulfonylhydrazone
(tosylhydrazone) of ethyl phenyl ketone.

Scheme 9.14. In the top half (Part a), the formation of a toluenesulfonylhydrazone (tosylhydrazone) derivative of ethyl phenyl ketone and p-toluenesulfonylhydrazine is shown. In the lower half (Part b), a potential path for the reduction of the tosylhydrazone with lithium aluminum hydride (LiAlH$_4$) in ether solvent is presented (and the products shown are obtained after hydrolysis).

responding alcohols can often be accomplished. However, since carbon–carbon double bonds, if present, might also be reduced and since other processes (e.g., hydrogenolysis of benzylic alcohols, *vide infra*) might also intrude, catalytic reduction is generally limited to specific cases using well-defined parameters. For example as shown in Equation 9.13, the reduction of propanone (acetone, CH$_3$COCH$_3$) to 2-propanol [(CH$_3$)$_2$CHOH] can be carried out with hydrogen (H$_2$) at between one and three atmospheres pressure over a Raney nickel catalyst at 25–30°C in less than half an hour.

$$
\underset{CH_3}{\overset{CH_3}{>}}\!\!=\!O \;+\; H_2 \xrightarrow{\text{Pt or Ni}} \underset{CH_3}{\overset{CH_3}{>}}\!\!<\!\underset{H}{\overset{OH}{}} \tag{9.13}
$$

Methyl phenyl ketone (acetophenone, $C_6H_5COCH_3$) can be reduced to 1-phenylethanol [$C_6H_5CH(OH)CH_3$], Equation 9.14, under the same conditions (in ethanol, CH_3CH_2OH, solvent) in 2.5 h. However, with a platinum (Pt) catalyst in ethanoic acid (acetic acid, CH_3CO_2H), the reduction of methyl phenyl ketone (acetophenone, $C_6H_5COCH_3$) is very fast and most of the product is phenylethane, *hydrogenolysis** having occured (Equation 9.15).

$$
\underset{C_6H_5}{\overset{CH_3}{>}}\!\!=\!O \;+\; H_2 \xrightarrow[\substack{CH_3CH_2OH \\ (\text{solvent})}]{\text{Pt or Ni}} \underset{C_6H_5}{\overset{CH_3}{>}}\!\!<\!\underset{H}{\overset{OH}{}} \tag{9.14}
$$

$$
\underset{C_6H_5}{\overset{CH_3}{>}}\!\!=\!O \;+\; H_2 \xrightarrow[\substack{CH_3CO_2H \\ (\text{solvent})}]{\text{Pt}} \underset{C_6H_5}{\overset{CH_3}{>}}\!\!<\!\underset{H}{\overset{H}{}} \tag{9.15}
$$

Although a variety of chemical methods that include reagents such as sodium hydrosulfite ($Na_2S_2O_4$), sodium metal (Na) dissolved in liquid ammonia ($NH_{3\,(l)}$) in the presence of an alcohol (which is the ultimate proton donor), and even sodium (Na° and the other alkali metals [Li°, K° and Rb°]) in alcohol (e.g., ethanol, CH_3CH_2OH) itself have been used over the years to reduce ketones to alcohols (Scheme 9.15). Newer methods appear to be preferred for their selectivity, ease of use, and success in avoiding unwanted side products.

Currently, the most widespread methods of reduction of the carbonyl group of aldehydes and ketones involve the use of variously substituted complex hydrides of boron and aluminum, compounds in which the hydrogen–boron and hydrogen–aluminum bonds are polarized with the negative charge residing on the more electronegative hydrogens.

These complex hydrides balance basicity with nucleophilicity, and the counterions (e.g., Li^+) or the hydride itself (e.g., BH_3) strongly coordinate with the oxygen of the carbonyl so that addition of hydride (H^-) to the carbon of the carbonyl is preferred to proton abstraction from the carbon α to the carbonyl. Indeed, the simpler hydrides such as sodium hydride (NaH) are generally too basic to affect reduction. Thus, the simpler hydrides are used for proton abstraction to produce enolate anions, while the complex hydrides are used for reduction.

Further, the development of a multitude of methods based on these families of compounds has been rapid because reduction of one functional group in the presence of another is often a major synthetic challenge and the different reagents can

---

*Hydrogenolysis refers to the reductive removal of a substitutent (–OH, –Cl, etc.) and its replacement by hydrogen (–H). Catalytic hydrogenolysis of benzylic substituents is common.

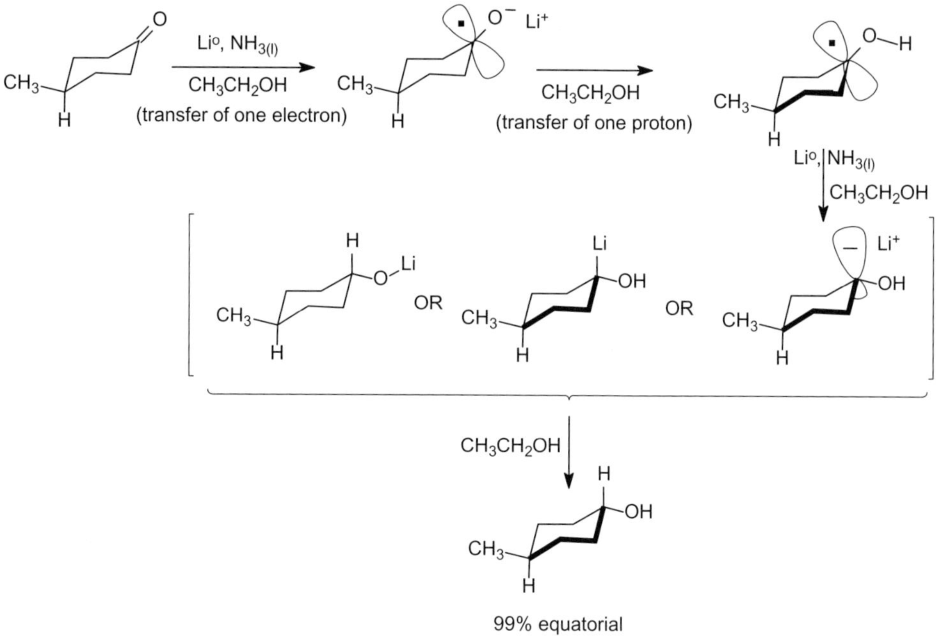

Scheme 9.15. A representation of the reduction of 4-methylcyclohexanone with lithium metal (Li°) in liquid ammonia (NH₃ ₍ₗ₎) and with ethanol (CH₃CH₂OH) as the proton source. The product is almost exclusively *trans*-4-methylcyclohexanol, and it is held that a series of one-electron transfer processes occurs in conversion of the starting material to product. (See Huffman, J. W.; McWhorter, W. W. *J. Org. Chem.*, **1979**, *44*, 584).

often be made to react quite specifically under fixed conditions. The reducing reagents include (a) sodium borohydride (NaBH₄), a gentle and mild reagent capable of being used in alcohol (and water) solvents at pH's≅7, although it is clearly a "hydride (H⁻) donor"; (b) diborane (B₂H₆), a second gentle reducing agent that is used in ether solvents (commonly oxacyclopentane, tetrahydrofuran [THF]) and which is also a "hydride (H⁻) donor" but which functions by a different pathway; (c) lithium aluminum hydride (LiAlH₄), a potent and highly reactive family member that can only be used in ether solvents and reacts violently with water; and (d) a host of other related compounds, including, for example, sodium bis(2-methoxyethoxyaluminum)hydride [NaAlH₂(OCH₂OCH₂CH₃)₂], whose reactivity is intermediate between sodium borohydride (NaBH₄) and lithium aluminum hydride (LiAlH₄) but which has markedly different steric requirements.

Given the observation that all of these reagents are capable of, at least, reducing aldehydes and ketones, a variety of investigations have been undertaken to study how changes in the structure of the carbonyl compound and the nature of the solvent affect the outcome of the reaction.

However, since the question of steric effects and geometry of addition is conveniently addressed in cyclic systems, it has been common to employ cyclic ketones in these studies.

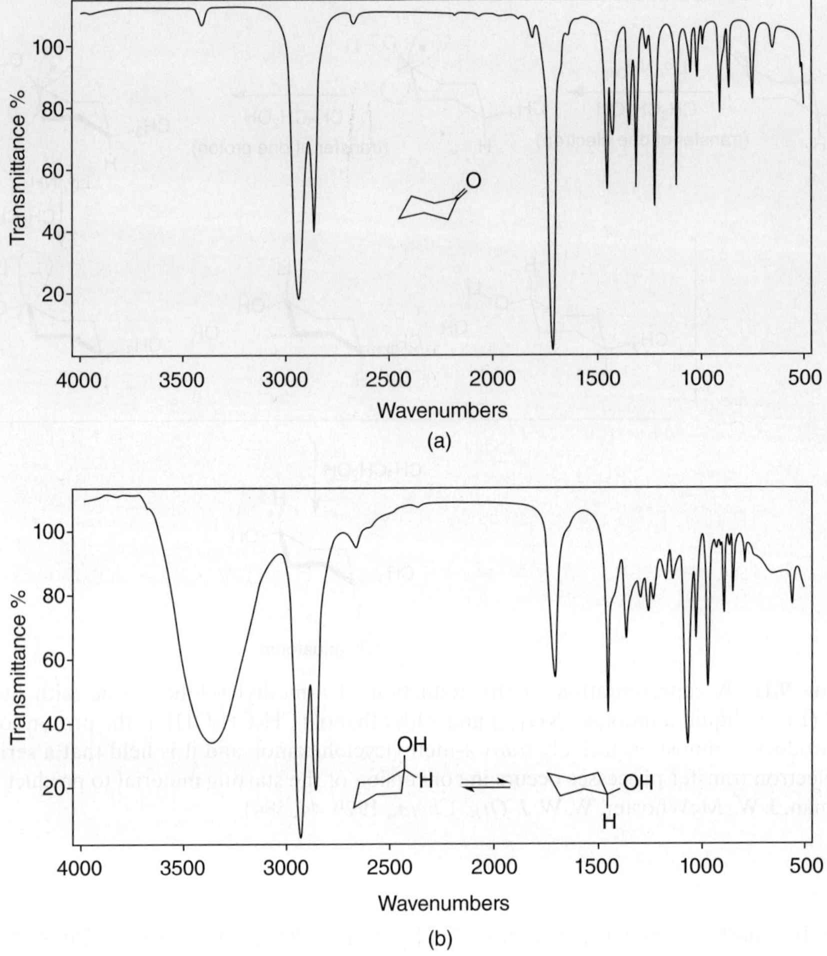

Figure 9.3. The infrared spectra (neat) of (a) cyclohexanone and (b) cyclohexanol.

Consider the case of cyclohexanone (Equation 9.16). All of the reducing agents listed earlier, in their respective solvents, rapidly and cleanly reduce cyclohexanone to cyclohexanol. The infrared spectra are given in Figure 9.3a,b.

$$\text{(9.16)}$$

An accepted pathway for the reduction of cyclohexanone with borane (B_2H_6) in oxacyclopentane (THF) is presented in Scheme 9.16. As shown there, it is presumed that, initially, the borane·THF complex is converted to a carbonyl group·borane complex (i.e., $C{=}O{\cdot}BH_3$) and that hydride donation from the complexed borane to the carbon of the carbonyl leads to a monoalkoxyborane.

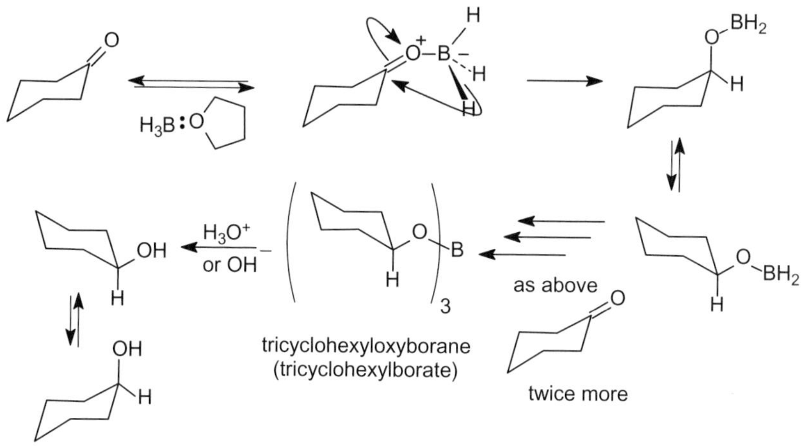

Scheme 9.16. An accepted pathway for the reduction of cyclohexanone to cyclohexanol with borane (B_2H_6). The intermediate tricyclohexylborate is also shown.

Successively, two more equivalents of aldehyde or ketone are reductively added to the same boron to lead, ultimately, to a trialkoxyborane (a trialkylborate), which is then hydrolyzed to boric acid and the corresponding alcohol.

In contrast, reduction of the same ketone with lithium aluminum hydride ($LiAlH_4$), sodium borohydride ($NaBH_4$) or related hydrides (*vide supra*) is presumed to begin by complexation of the oxygen of the carbonyl with the cation (Li^+, Na^+, or K^+) accompanying the hydride-bearing metal (aluminum or boron). Following such complex formation, attack by the tetrahydroaluminate, tetrahydroboronate, and so on, anion occurs and, in a subsequent step, the original complexing cation (Li^+, Na^+, etc.) is replaced by the aluminum (or boron) to form an alkoxytrihydroaluminate (or boronate) ion. It is this alkoxytrihydroaluminate or boronate ion that then goes on to react with other metal-cation-coordinated carbonyl species to, eventually, produce a tetralkoxyaluminate (or boronate) anion. Finally, the alkyloxyaluminate or alkyloxyboronate anion is hydrolyzed to generate the corresponding alcohol and the amphoteric aluminum hydroxide [$Al(OH)_3$] or boric acid [$B(OH)_3$] species (Scheme 9.17).

The structure of the carbonyl compound being reduced plays an important role in determining the structure of the alcohol produced. Historically, there have been two alternative view points that have been promoted as being important to consider. These are expressed as **steric approach control** and **product development control**.*

Application of the product development control suggests, reasonably, that the direction of approach of the hydride delivering species is a function, at least in part, of the encumbrances around the carbonyl group.

Application of the latter is more controversial. The principle is invoked to explain the formation of **the more stable product,** even though it is not clear that one product-forming transition state is necessarily more stable (lower in energy) than

---

*The concepts of "steric approach control" and "product development control" were put forth, respectively, by Dauben, W. G.; Fonken, G. J.; Noyce, D. S. *J. Am. Chem. Soc.*, **1956**, *78*, 2579, and Wheeler, D. M. S.; Wheeler, M. M. *J. Org. Chem.*, **1962**, *27*, 3796. An alternative idea that as new bonds form, new steric effects develop and the static picture changes has been put forth. See Richer, J.-C. *J. Org. Chem.*, **1965**, *30*, 324.

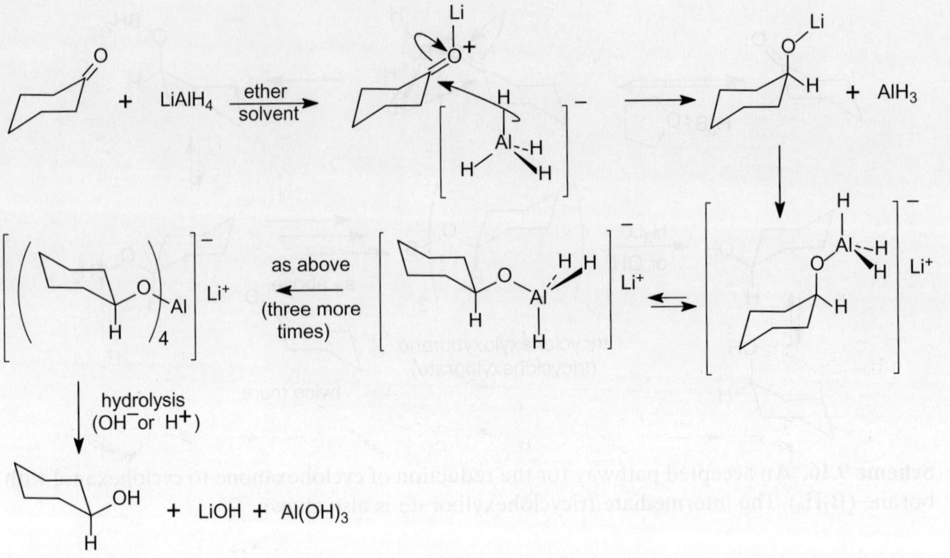

Scheme 9.17. A representation of the reduction of cyclohexanone with lithium aluminum hydride (LiAlH$_4$) in ether and the hydrolysis of the aluminum alkoxide product to produce cyclohexanol.

another. Consider, for example the case of 4-*tert*-butylcyclohexanone on reduction in ether with lithium aluminum hydride (LiAlH$_4$).*

As shown in Scheme 9.18, the approach of the tetrahydroaluminate anion (AlH$_4^-$) to the lithium complexed carbonyl of 4-*tert*-butylcyclohexanone can reasonably be expected to occur from either direction (Figure 9.4). Debate might center around whether the hydrogens on the carbons immediately adjacent to the carbonyl are more of a hindrance to the approach of the reducing species than those two carbon atoms away, but, in any event, it is not clear that a specific preference should be expressed. Nonetheless, the outcome is quite dramatic, with more than 90% of the product resulting from the attack by the hydride-delivering reagent from the axial direction (producing equatorial alcohol).

Some effort has been expended in attempting to determine the effect of changes in steric bulk around the site of reduction. As shown in Equation 9.17, the reduction of bicyclo[2.2.1]heptan-2-one with both borane (B$_2$H$_6$) in oxacyclopentane (THF) and lithium aluminum hydride (LiAlH$_4$) in ether (CH$_3$CH$_2$OCH$_2$CH$_3$) yields a mixture of *endo*- and *exo*-bicyclo[2.2.1]heptan-2-ols. Despite the different pathways (*vide supra*) by which borane (B$_2$H$_6$) and lithium aluminum hydride (LiAlH$_4$) function, the *endo*-product predominates in both cases because delivery of hydride from the *exo*-direction (leading to *endo*-alcohol) is sterically preferred.

---

*It may be recalled (Table 4.2) that the equatorial:axial preference for a *tert*-butyl group in cyclohexane itself is >18.8 kJ mol^{-1} (>4.5 kcal mol^{-1}), and although there will probably be a smaller difference in cyclohexanone (why?), it is still anticipated that a strong preference to keep the *tert*-butyl group equatorial will be present. Further, it is estimated that in polar solvents an axial hydroxyl group (–OH) on cyclohexane (i.e., cyclohexanol) has about the same stability as an equatorial hydroxyl group (–OH).

trans-4-*tert*-butylcyclohexanol

(*E*)-4-*tert*-butylcyclohexanol

trans : *cis* = 90 : 10

cis-4-*tert*-butylcyclohexanol

(*Z*)-4-*tert*-butylcyclohexanol

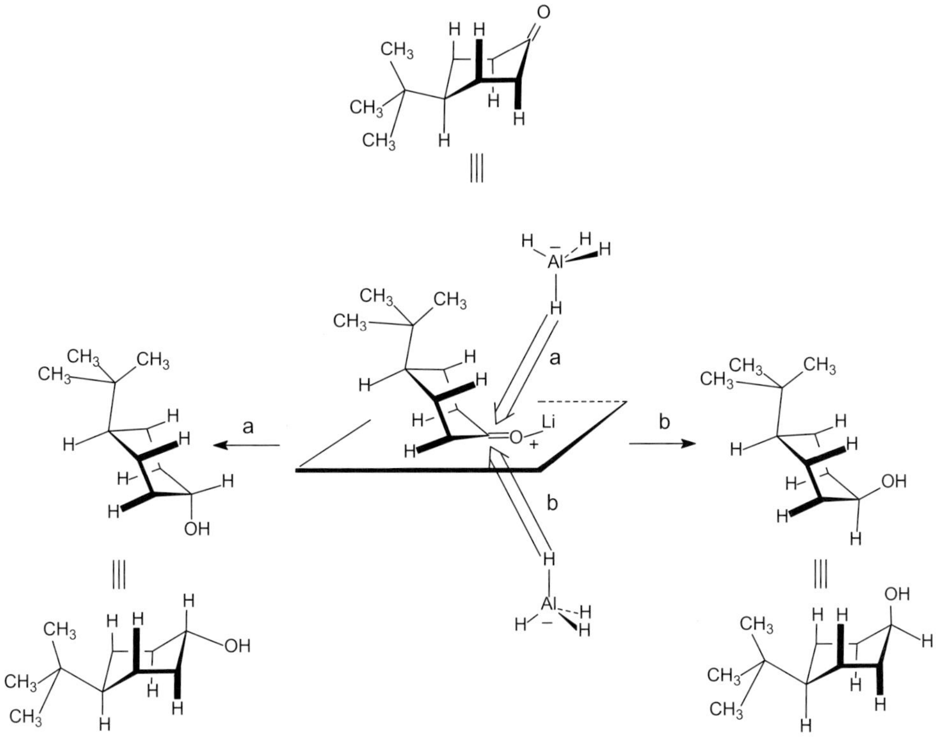

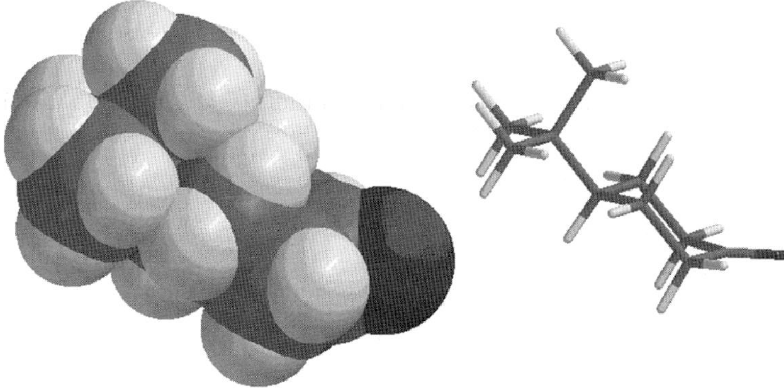

Scheme 9.18. The results of the reduction of 4-*tert*-butylcyclohexanone with lithium aluminum hydride ($LiAlH_4$) in ether showing that the diequatorial product is preferred, although it is unclear if one face of the carbonyl is less encumbered than the other. (See Dauben, W. G.; Fonken, G. J.; Noyce, D. S. *J. Am. Chem. Soc.* **1956**, *78*, 2579.)

Figure 9.4. Space filling and stick representations of 4-*tert*-butylcyclohexanone endeavoring to show that both faces are approximately equally encumbered.

$$\text{(9.17)}$$

endo-bicyclo[2.2.1]heptan-2-ol   *exo*-bicyclo[2.2.1]heptan-2-ol

		isomer ratio	
reducing agent and solvent	$H_3B:O$ (THF)	98	2
	$LiAlH_4/(CH_3CH_2)_2O:$	90	10

In contrast, as shown in Equation 9.18, the reduction of 1,7,7-trimethylbicyclo[2.2.1]-heptan-2-one with both borane (B_2H_6) in oxacyclopentane (THF) and lithium aluminum hydride ($LiAlH_4$) in ether ($CH_3CH_2OCH_2CH_3$) also yields a mixture of *endo*- and *exo*-1,7,7-trimethylbicyclo[2.2.1]heptan-2-ols but in very different ratios. It is argued that the different pathways (*vide supra*) by which borane (B_2H_6) and lithium aluminum hydride ($LiAlH_4$) function play a critical role.* Borane, which delivers hydride **after** forming a complex with the carbonyl group continues to deliver hydride from the *exo*-face (but with greater difficulty because of the hindrance resulting from the methyl groups), while lithium aluminum hydride ($LiAlH_4$), which delivers hydride from the larger aluminum hydride anion (AlH_4^-), is much more sterically retarded and is directed toward the *endo*-face of the ketone, producing *exo*-alcohol.

$$\text{(9.18)}$$

endo-1,7,7-trimethyl-bicyclo[2.2.1]heptan-2-ol   *exo*-1,7,7-trimethyl-bicyclo[2.2.1]heptan-2-ol

		isomer ratio	
reducing agent and solvent	$H_3B:O$ (THF)	50	50
	$LiAlH_4/(CH_3CH_2)_2O:$	10	90

The effects of changes in solvent and reducing agent are also profound. This is illustrated in Equation 9.19, where reduction of 3,3,5-trimethylcyclohexanone to the corresponding axial alcohol (from equatorial hydride addition) and equatorial alcohol (from axial hydride addition) is shown.

*For the details, see Brown, H. C.; Muzzio. *J. Amer. Chem. Soc.*, **1966**, *88*, 2811, and Brown, H. C.; Dickason, W. D. *J. Am. Chem. Soc.*, **1970**, *92*, 709.

reducing agent and solvent	isomer ratio	
LiAlH$_4$/(CH$_3$CH$_2$)$_2$O	50	50
LiAlH$_4$/THF	75	25
NaBH$_4$/CH$_3$OH	95	5

(9.19)

Now, since (at least in principle) all four hydrides of the tetrahydroaluminate anion (AlH$_4^-$) and all three of the hydrides of borane (BH$_3$) can be transferred during a reaction, it should be clear that the details of the reduction process are more complex than those shown in Schemes 9.16–9.18.

Further, as already noted, in the drive to create reagents that produce particular products under particular circumstances, there is a wide variety of reducing agents from which to choose to help dictate (a) the direction of addition to the carbonyl (the stereochemistry); (b) the selectivity (what other functional groups might be present); and (c) the ease of use (can the reduction be carried out without resorting to anhydrous conditions, etc.). Additional data regarding the different reducing agents and the consequent stereochemical outcome are shown in Equation 9.20.*

reducing agent and solvent	isomer ratio	
NaBH$_4$/CH$_3$OH	75	25
LiAlH$_4$/THF	76	24
LiAl(O-tBu)$_3$H/THF	65	35
LiB(s-Bu)$_3$H/CH$_3$OH	2	98

(9.20)

Because the metal (e.g., Pt, Pd, Ni) catalyzed reduction of a carbon–carbon double bond is more facile than that of the carbon–oxygen double bond, unsaturated alde-

---

*The information in this table is taken from Brown, H. C.; Dickason, W. D. *J. Am. Chem. Soc.*, **1970**, *92*, 709; Rickborn, B.; Wuesthoff, W. T. *J. Am. Chem. Soc.*, **1970**, *92*, 6894; Ashby, E. C.; Sevenair, J. P.; Dobbs, F. R. *J. Org. Chem.*, **1971**, *36*, 197; Brown, H. C.; Krishnamurthy, S. *J. Am. Chem. Soc.*, **1972**, *94*, 7159.

hydes and ketones can frequently be converted to their saturated analogues by hydrogenation reactions that leave the carbonyl group unaffected.

If the double bond is *conjugated* with the carbonyl (i.e., an α,β-unsaturated aldehyde, or ketone), reduction of the double bond only, yielding the saturated carbonyl compound, can also be effected by the same metal catalysts with continuous careful monitoring so that only one equivalent of hydrogen is consumed before the reaction is stopped. In contrast, with complex hydrides, it is only the reduction of the carbonyl that occurs and the corresponding alcohol, with the double bond remaining intact, is generally isolated from the reaction mixture.

Similarly, reduction of both aromatic and aliphatic aldehydes and ketones where there is a chlorine (Cl) or bromine (Br) present elsewhere in the carbonyl-containing compound succeeds with sodium borohydride ($NaBH_4$) and, occasionally, with lithium aluminum hydride ($LiAlH_4$) at low temperatures. Catalytic reduction frequently leads to hydrogenolysis of the halogen, producing the halogen-free alcohol.

In a different vein and as already pointed out in Chapter 8 (Scheme 8.6), the **Meerwein–Ponndorf–Verley** reduction is the reverse of the **Oppenauer** oxidation of aldehydes and ketones, and *it is only a change of solvent that dictates whether the reaction that occurs is an oxidation or a reduction.* The same catalyst is used. Scheme 9.19 is the reverse of Scheme 8.6. Thus, it is now suggested that the carbonyl oxygen of cyclohexen-3-one displaces an isopropoxy group (2-propoxy [$^-OCH(CH_3)_2$]) from the catalyst, aluminum isopropoxide [$Al(O\text{-}iPr)_3$]. Then, after intramolecular hydride transfer, propanone (acetone, CH_3COCH_3) is lost by displacement from aluminum by the solvent, 2-propanol (isopropanol [$CH_3CH(OH)CH_3$]), and finally, the cyclo-

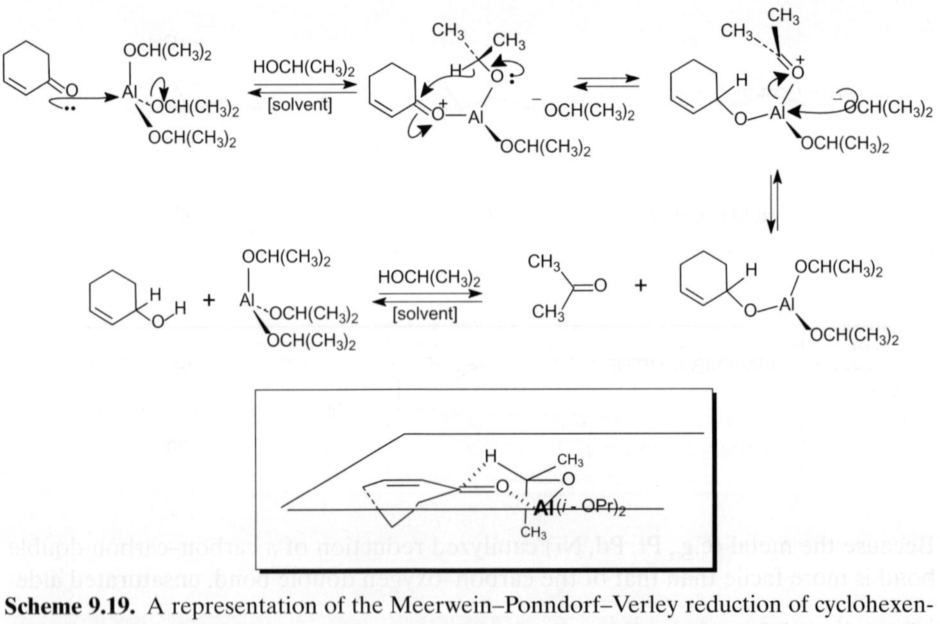

Scheme 9.19. A representation of the Meerwein–Ponndorf–Verley reduction of cyclohexen-3-one using aluminum isopropoxide catalyst. The reaction is the reverse of that seen as the Oppenauer Oxidation (Scheme 8.6). See (a) Meerwein, H.; Schmidt, R. *Liebigs Ann. Chem.*, **1925**, *444*, 221; (b) Ponndorf, W. *Angew. Chem.*, **1926**, *39*, 138; (c) Verley, A. *Bull. Soc. Chim. Fr.*, **1925**, *37*, 537, and Oppenauer, R. V. *Recl. Trav. Chim. Pays-Bas*, **1937**, *56*, 137.

hexenol product is also displaced (by another equivalent of solvent) from the aluminum, thus regenerating the catalyst. Since the solvent, 2-propanol (isopropanol [CH$_3$CH(OH)CH$_3$]), is in vast excess, the reduction of the ketone continues until it is all consumed and an equivalent amount of propanone (acetone, CH$_3$COCH$_3$) has been generated.

Although reduction of ketones with alkali metals, for example, sodium (Na°) in ethanol (CH$_3$CH$_2$OH), to the corresponding alcohol, as already noted, has been replaced with complex hydrides, a minor product in the original reaction, the result of **coupling** two ketones at their respective carbonyl carbons, was occasionally found on workup of the reaction mixture. The diol (a **pinacol**) thus formed could be made the major product if the reaction was carried out with magnesium (Mg°) or amalgamated aluminum (Al–Hg) instead of sodium (Na°). The process, which doubtlessly involves a series of one-electron transfer reactions, is shown in Equation 9.21.

$$\text{(9.21)}$$

55%

In Chapter 8, the biological oxidation of aldehydes and ketones was seen as the **reverse** of the enzymatic reduction of alcohols by yeast alcohol dehydrogenase (**alcohol:NAD$^+$ oxidoreductase** [EC 1.1.1.1]). In the discussion, it was pointed out that it was the oxidized form of the cofactor, nicotinamide adenine dinucleotide (NAD$^+$), that served to remove the hydrogen **with its pair of electrons**, that is, hydride (H$^-$), from the carbon bearing the hydroxyl. Now, in contrast to the case there, it is the reduced form of the cofactor (NADH) that donates a hydrogen, **with its pair of electrons**, that is, hydride (H$^-$), to the carbon of the carbonyl group undergoing reduction and being converted to a carbon bearing a hydroxyl group. Equation 9.22 was seen before as Equation 8.15, but now we consider the equation as proceeding from right to left rather than from left to right.

$$\text{(9.22)}$$

Second, as noted in Chapter 8, Figure 8.7, and as repeated here as Figure 9.5, the two faces of ethanal (acetaldehyde, CH$_3$CHO), and indeed, any aldehyde other than methanal (formaldehyde, O=CH$_2$), can be distinguished from each other by whether the substituents attached to carbon (following the Cahn–Prelog–Ingold order of priorities) appear in a clockwise (*re*) or counterclockwise (*si*) order.

In the case of reduction of the carbonyl group by the yeast enzyme (alcohol:NAD$^+$ oxidoreductase [E.C. 1.1.1.1]), the hydrogen is placed on the *re*-face of the aldehyde, becoming the *pro-R* hydrogen of the alcohol (Scheme 9.20, which has been seen previously as Scheme 8.11).

Interestingly, laboratory results that also occur with high enantioselectivity can be achieved by using a chiral catalyst. As shown in Scheme 9.21, phenylethanone (methyl phenyl ketone, acetophenone, C$_6$H$_5$COCH$_3$) on treatment with borane

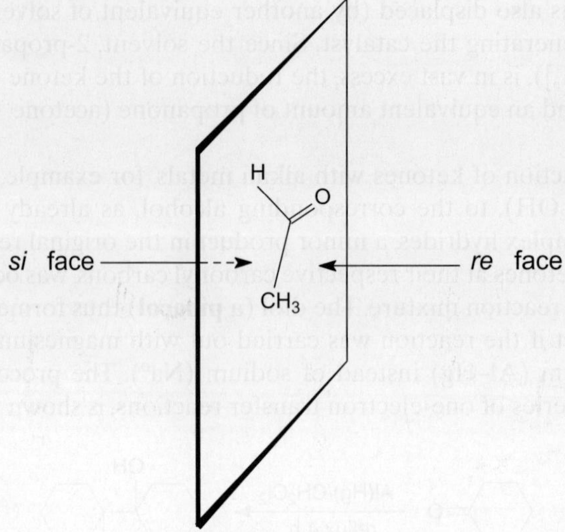

Figure 9.5. The *re*- and *si*- faces of ethanal (acetaldehyde, CH₃CHO).

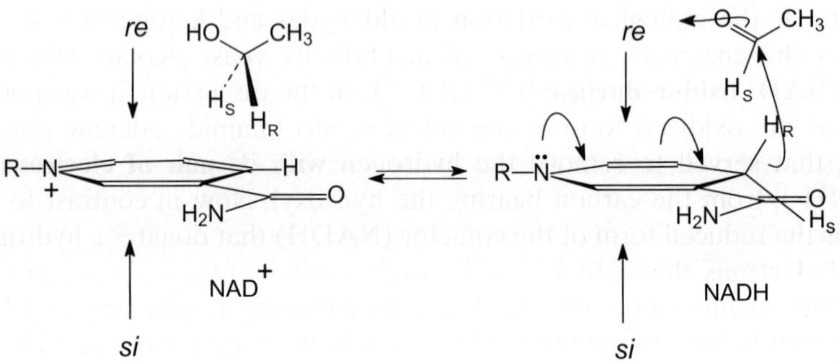

Scheme 9.20. A different representation of Scheme 8.11. A cartoon representation of the reduction of ethanal (acetaldehyde, CH₃CHO) with the cofactor, reduced nicotinamide adenine dinucleotide (NADH). The curved arrows depict the loss of a hydrogen, with its electron pair, from the nicotinamide portion of the cofactor as it is being added to the *re*-face of the aldehyde, thus becoming the *pro-R* hydrogen of the resulting alcohol. The hydrogen added in the reduction is *the same hydrogen as that removed in the oxidation*.

(BH₃) in THF (presumably present as the borane·THF complex) in the presence of about 0.1 equivalents of chiral oxazaborolidine generates a >95% yield of the corresponding chiral alcohol in >95% enantiomeric excess.

C. ADDITION TO ALDEHYDES AND KETONES

I. Introduction

In general, "addition" to aldehydes and ketones is understood to mean that *either* the **oxygen of the carbonyl group is replaced by some other functionality** (usually

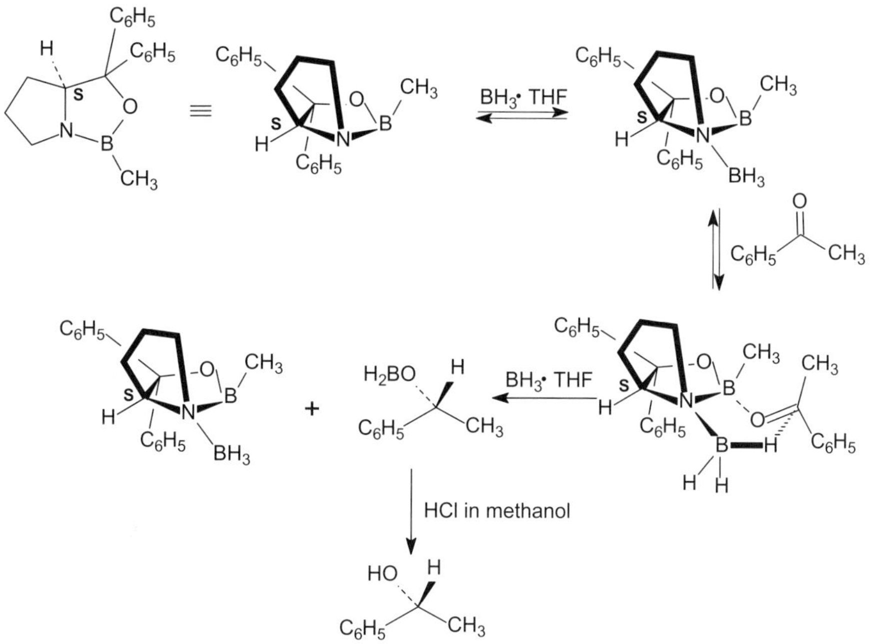

Scheme 9.21. A representation of the reduction of phenylethanone (methyl phenyl ketone, acetophenone, $C_6H_5COCH_3$) to *(R)*-1-phenylethanol with borane (BH_3) in oxacyclopentane (tetrahydrofuran, THF) with an *(S)*-oxazaborolidine. (See Corey, E. J.; Bakshi, R. K. *Tetrahedron Lett.*, **1990**, 611, and Rao, A. V. R.; Gurjar, M. K.; Shamma, P. A.; Kaiwar, V. *Tetrahedron Lett.*, **1990**, 2341.)

through transient formation of a tetrahedral intermediate) *or* the **number of groups (including oxygen) attached to carbon increases from three to four** in the isolated product.

Therefore, it should be clear that some addition reactions (e.g., to alkenes and alkynes) occurring in some position far removed from the carbonyl functional group (C=O) of an aldehyde or ketone, while formally "addition reactions," are not generally considered here. Of course, some reactions remote from the carbonyl group (C=O) might nonetheless affect it. In the event such addition **does not** affect the carbonyl group (C=O), it will be ignored. In the event it **might** affect the carbonyl group (C=O), it is important to note that the carbonyl group (C=O) can be "protected" or "masked" prior to the transformation. For example, reversible acetal (ketal) formation (Chapter 8 and this Chapter) effectively converts the carbonyl group into an "ether." Thus, as shown in Scheme 9.22, reaction of the carbonyl group (C=O) of the aldehyde with the oxidizing agent bromine (Br_2) can be avoided by first "masking" the carbonyl (C=O) as an acetal [$C(OR)_2$], effecting the addition at the double bond, and then "unmasking" the carbonyl group (C=O). Second, as will be discussed within, the special case of α,β-unsaturated carbonyl (C=C–C=O) compounds requires close attention because, depending on the conditions of the addition and the addend, these compounds can undergo addition **either** at the carbonyl carbon (C=O) or at the double bond (C=C).

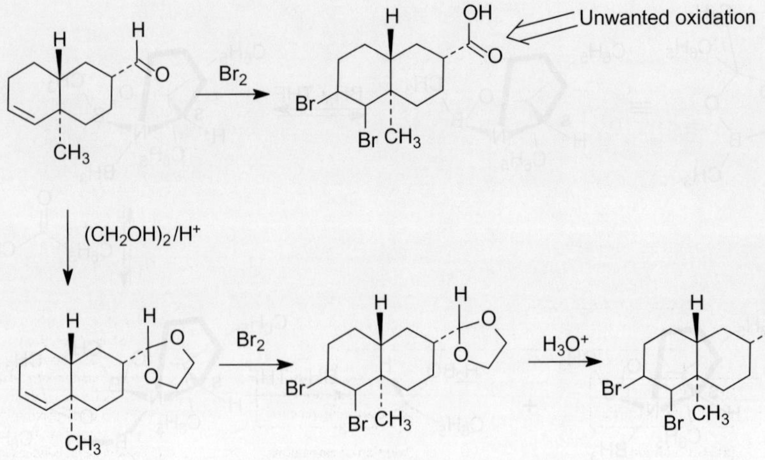

Scheme 9.22. A representation of the bromination of a carbon–carbon double bond in the presence of an aldehyde. "Masking" the aldehydic carbonyl group as an acetal avoids its oxidation to a carboxylic acid during addition of bromine to the remote carbon–carbon double bond.

Problem 9.4. Show the details for each step in Scheme 9.22. Note that the stereochemistry of addition to the carbon–carbon double bond should be shown.

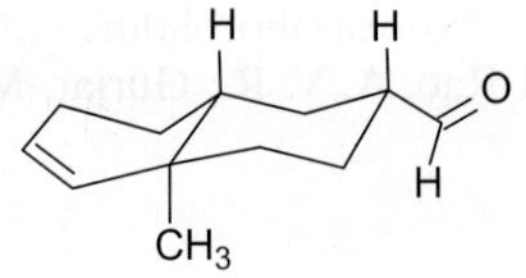

Third, the attentive reader will not have failed to note that *both* Section A **(Oxidation of Aldehydes and Ketones)** *and* Section B **(Reduction of Aldehydes and Ketones)** in this chapter involve *addition* to the carbonyl group (C=O). Thus, as shown in Section A, processes that result in oxidation of the carbonyl group (C=O) proceed from initial attack at the sp^2-hybridized carbon of the carbonyl (C=O) by an electron rich species. The sp^3-hybridized intermediate resulting from that attack is ultimately consumed when the carbon–hydrogen aldehydic (O=C–H) **or** the carbon–carbon (O=C–C) ketonic bond breaks. Similarly, in reduction (Section B), the processes involving hydride ion transfer are shown as attack of hydride (H⁻) donating species onto the sp^2-hybridized carbon of the carbonyl (C=O) with rehybridization at carbon to sp^3 and eventual protonation at oxygen.

Thus, considering only reactions at the carbonyl group (C=O) and ignoring for the moment α,β-unsaturated carbonyl compounds (i.e., compounds containing the structural fragment corresponding to C=C–C=O) and other remote processes (e.g., addition reactions at remote sites and substitution of aromatic rings bearing aldehydic and/or ketonic groups directly attached to the ring, Ar–C=O), there are two broad categories of addition reactions to be considered (Figure 9.6). The first group is that in which the oxygen of the carbonyl (C=O) remains attached to the carbon while in the second group, the oxygen is lost. The hybridization at the carbon of the carbonyl (C=O) changes from sp^2 to sp^3 if only transiently, in both groups. So, before

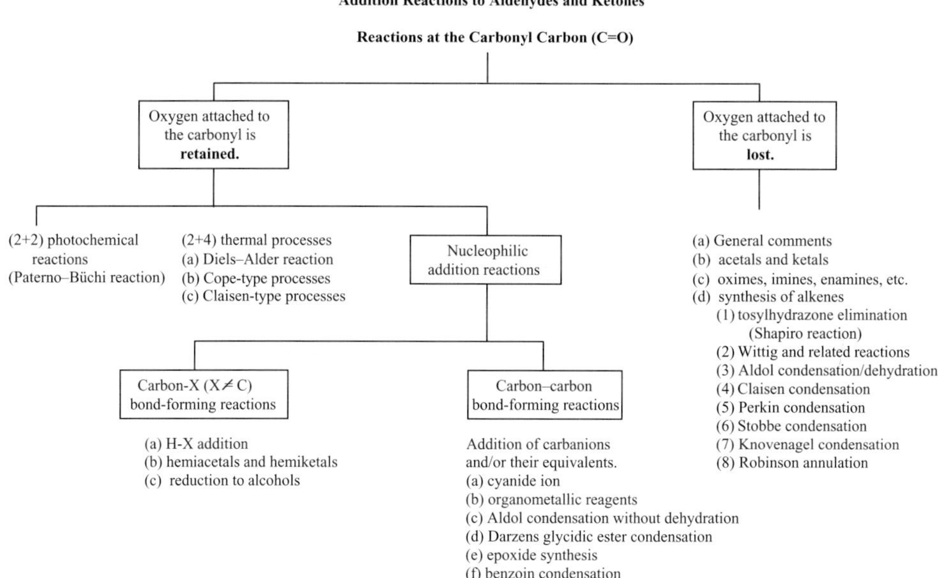

Figure 9.6. Substitution reactions at the carbon of the carbonyl group in aldehydes and ketones.

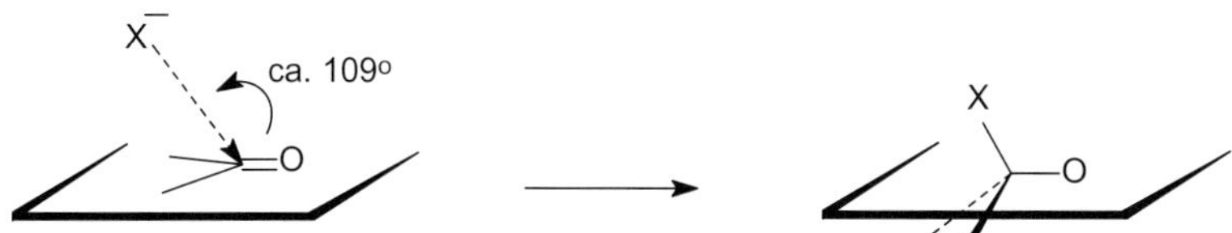

Figure 9.7. A depiction of the presumed favored direction (the "Bürgi–Dunitz trajectory") of attack of an electron rich species on the carbon of the carbonyl (C=O) (Bürgi, H. B.; Dunitz, J. D.; Shefter, E. *J. Am. Chem. Soc.*, **1973**, *95*, 5065).

considering the specific processes of Figure 9.6 in more detail, it is important to recognize that there are stereochemical issues involved in the conversion of the trigonal carbonyl to the tetrahedral species.

First, as shown in Figure 9.7, it is widely recognized that the favored trajectory (the "Bürgi–Dunitz trajectory") for addition of an electron rich species (X⁻) to the carbonyl (C=O) proceeds by a path that requires minimum subsequent distortion to produce the appropriate tetrahedral angle (or nearly so). In particular, this process, which is of paramount importance in ring closure reactions, introduces the necessity of asking whether the *re* or *si* face of an asymmetrically substituted carbonyl group (C=O) is attacked.

Second, as shown in Figure 9.8, the steric influence of neighboring groups can dictate the preference of attack at one face or another. Experimental observations of such preferences have resulted in the establishment of rules, such as **Cram's rule***

---

*Donald J. Cram, (1919–2001), Professor of Chemistry, University of California at Los Angeles, Nobel Prize, Chemisty, 1987 (Cram, D. J.; Abd Elhafez, F. A. *J. Am. Chem. Soc.*, **1952**, *74*, 5828).

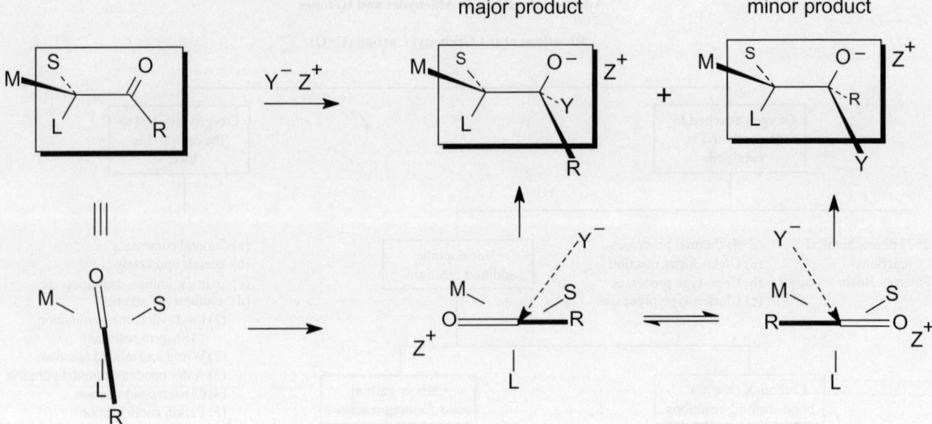

Figure 9.8. A pictorial representation of **Cram's rule** (the **Felkin–Anh model**), which suggests that attack at the carbonyl group (along the "Bürgi–Dunitz trajectory") is strongly influenced by the steric effects introduced by neighboring substituents.

(Figure 9.8). The empirical **Cram's rule,** later put on a more firm footing by Felkin[*] and Anh,[†] predicts that if a species Y^-X^+ attacks the carbon of the carbonyl group (C=O) in a compound where the carbon α to the carbonyl (C=O) bears small (**S**), medium (**M**), and large (**L**) groups, the major product will result from attack of the nucleophile at the side nearest the small group.

It should be clear from Figure 9.8 that the **Felkin–Anh model** requires that the carbonyl (C=O) oxygen will preferentially lie (in the reacting conformation) between the medium (**M**) and large (**L**) groups rather than between the small (**S**) and large (**L**) groups and that the Bürgi–Dunitz trajectory be followed. The latter insures that the electron rich nucleophile approach the carbon of the carbonyl (C=O) from the direction of the smallest substituent.

II. Photochemical Reactions of Aldehydes and Ketones

a. Nonconjugated Carbonyl Compounds.

Most carbonyl compounds have an absorption band near 3200 Å, which corresponds to an $n \rightarrow \pi^*$ absorption (Equation 9.23).[‡]

$$\underset{}{\text{C=O:}} \xrightarrow{h\nu} \text{C}\cdot\cdot\text{O:} \tag{9.23}$$

As shown in Scheme 6.46, energy absorption initially produces a singlet (electrons paired) excited state. The excited state has a variety of paths by which chemical change can occur.

---

[*]Hugh T. Felkin, Institut de Chimie des Substances Naturelles, Gif-sur-Yvette (Cherest, M.; Felkin, H.; Prudent, N. *Tetrahedron Lett.*, **1968**, 4199).
[†]Nguyên T. Anh, Laboratoire de Chimie Théorique, Orsay (Ahn, N. T. *Top. Curr. Chem.*, **1980**, *88*, 145).
[‡]Pyrex glass is transparent at 3200 Å. Mercury arc lamps were used for much of the early photochemical work on ketones. These lamps put out ca. 5 % of their energy at 3130 Å.

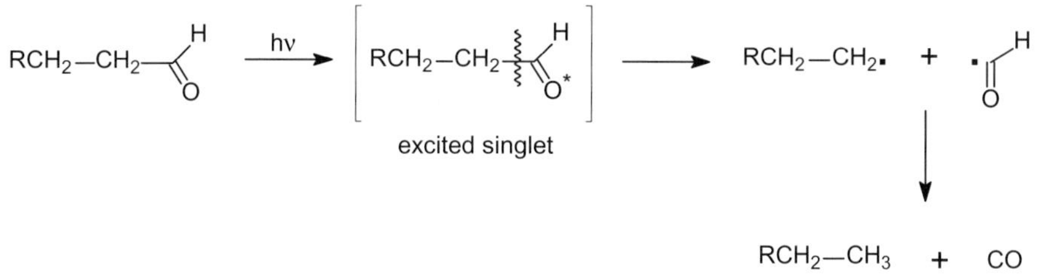

Scheme 9.23. A representation of the fragmentation of an aldehyde to an alkane and carbon monoxide (CO) on photochemical excitation.

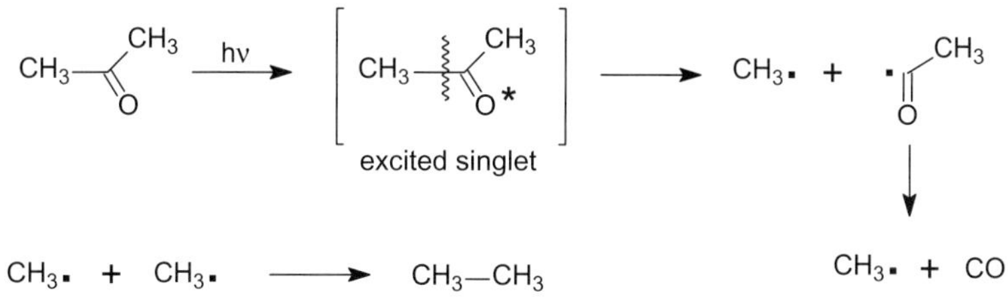

Scheme 9.24. A representation of the photochemical decomposition of propanone (acetone, CH_3COCH_3) to yield carbon monoxide (CO) and ethane (CH_3CH_3).

First, the singlet excited state (Scheme 6.46) can decay back to the ground state of the starting carbonyl compound, releasing the energy absorbed as light (**fluorescence**). Alternatively, absent another reactant (e.g., in the gas phase) and without **intersystem crossing** to the triplet state, the highly energetic singlet can decompose.

In nonconjugated aldehydes and ketones, decomposition from the singlet state commonly results in carbon–carbon bond cleavage to produce free radicals (Schemes 9.23 and 9.24).

Interestingly, ketones containing at least one alkyl group attached to the carbon of the carbonyl, which is three or more carbon atoms long and with at least one hydrogen on the γ-carbon, can decompose by an intramolecular process involving hydrogen transfer from the side chain and carbon–carbon bond cleavage (Scheme 9.25).

Second, if another reactant, such as an alkene, is present during the photochemical excitation of ground state aldehyde or ketone to the excited singlet state, a four-membered cyclic ether (an oxetane or oxacyclobutane) can form (the **Paterno–Büchi reaction**).* Usually, oxetane (oxacyclobutane) formation is stereoselective with the *more stable adduct* being formed. This implies that the singlet excited state undergoes intersystem crossing to the triplet (diradical) state (Scheme 6.46, Chapter 6) and the diradical is the reacting species (Scheme 9.26).

If the alkene is a cumulative diene, that is, allene (1,2-propadiene [$CH_2=C=CH_2$]) or a substituted allene, addition can occur once and/or twice (Equation 9.24),

---

*Paternnò, E.; Chieffi, C. *Gazz. Chim. Ital.*, **1909**, *39*, 341, and Büchi, G.; Inmann, C. G.; Lipinsky, E. S. *J. Am. Chem. Soc.*, 1954, *76*, 4327.

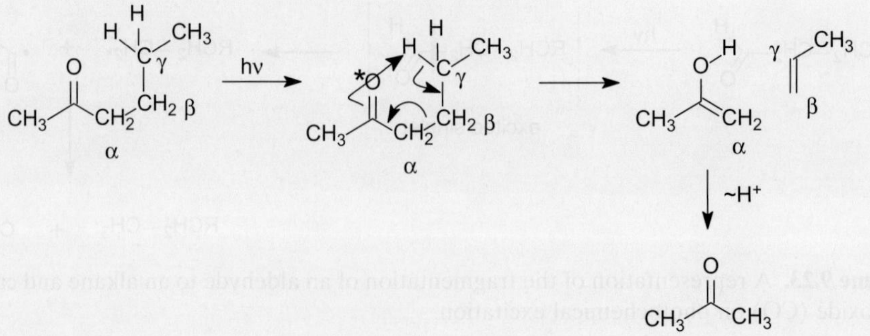

Scheme 9.25. A representation of the pathway by which 2-hexanone undergoes photochemical decomposition to propanone (acetone, CH_3COCH_3) and propene ($CH_3CH=CH_2$) by internal abstraction of a γ-hydrogen and cleavage of the β-γ-carbon–carbon bond.

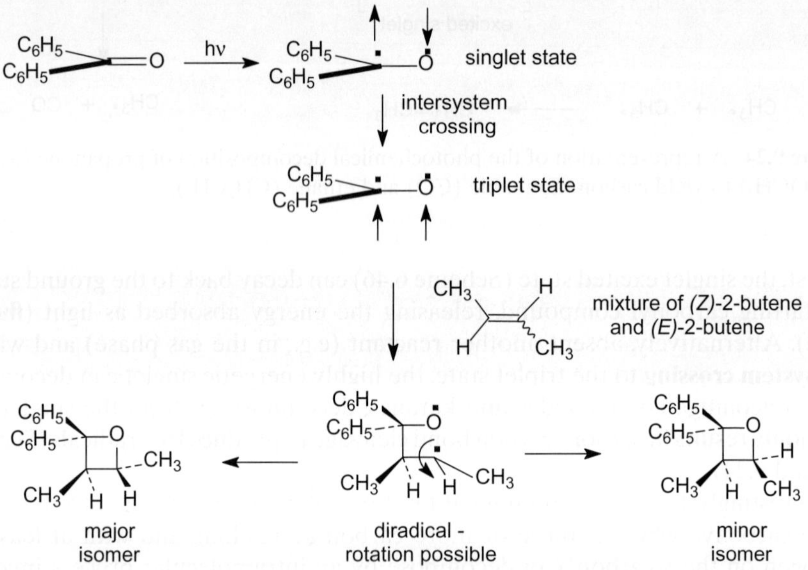

Scheme 9.26. An example of a photochemically induced 2 + 2 cycloaddition reaction between benzophenone (diphenylketone [$(C_6H_6)_2C=O$]) and 2-butene [an *(E)*- and *(Z)*-mixture of isomers]).

producing alkylideneoxetane (alkylideneoxacyclobutane) and dioxaspirobutane (spirodioxacyclobutane) isomers.

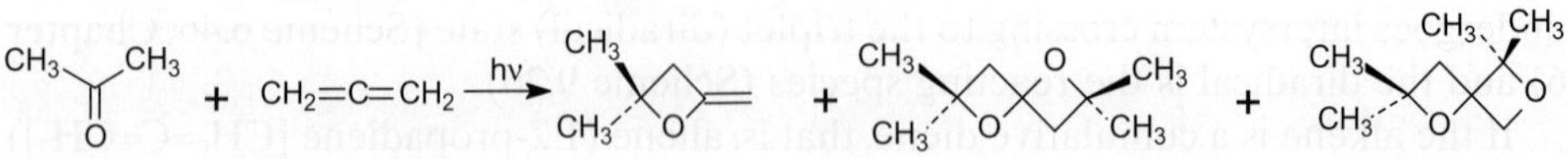

$$(9.24)$$

b. Conjugated Carbonyl Compounds. Carbonyl compounds with α,β-unsaturation frequently have absorption maxima at longer wavelengths (lower energy) than the corresponding saturated analogues. Absorption of energy thus results in an n $\rightarrow \pi^*$ transition of lower energy in the former. Nonetheless, photo-excitation of acyclic α,β-unsaturated ketones that have one or more hydrogens on the γ-carbon results in hydrogen atom transfer to yield the corresponding dienol. The dienol can then be induced to undergo base-catalyzed proton transfer to a β,γ-unsaturated ketone. The result is a ketone in which the "double bond has been deconjugated" (Equation 9.25).

$$\tag{9.25}$$

If, however, the α,β-unsaturation is in a medium size ring where the hydrogen on the γ-carbon cannot be abstracted for geometric reasons, the rearrangement cannot occur. Under these circumstances, either the enone dimerizes (Equation 9.26) or, if an alkene is present, addition across the carbon–carbon double bond (C=C) of the alkene occurs (Equation 9.27).[†]

$$\tag{9.26}$$

$$\tag{9.27}$$

*For this and other similar photochemical processes, see Eaton, P. E. *Acc. Chem. Res.*, **1968**, *1*, 50.
[†]Although the situation is somewhat more complicated (in that the six-membered ring is a *heterocycle*), absorption of energy into the α,β-unsaturated carbonyl system found in the **thymine** residue of DNA (deoxyribonucleic acid) occurs with radiation whose wavelength lies between 200 and 300 nm. As shown in the DNA representations in this footnote where two thymine residues are adjacent, this results in the formation of a four-membered ring between adjacent **thymine** residues on the same DNA strand (the result is called an **intrastrand thymine dimer**). These dimers distort the DNA strand, and thus there is a failure to pass on the correct transcriptional information when the DNA is copied. Fortunately, the process is reversible.

Interestingly, in the absence of available external addend, and when there are substituents at C-4 (i.e., 4,4-disubstituted-2-cyclohexenones), the eneone undergoes internal cycloaddition processes, which observe the **Woodward–Hoffman rules** (Chapter 4, Section G) for photochemical processes. Thus, as shown in Equation 9.28, irradiation (through pyrex) of an alcohol solution of (+)-4-methyl-4-*n*-propyl-2-cyclohexenone produced about equal amounts of the two bicyclo[3.1.0]hexan-2-one "lumi" isomers, a somewhat larger quantity of the cyclopentanone, and some recovered starting material. As shown, all of the products are produced with high enantioselectivity (i.e., only one enantiomer was detected) **and** the recovered starting material was of the same optical purity as before the reaction began.*

$$(9.28)$$

Although it might be argued (as it was before the experiments on chiral material were conducted) that the process by which this rearrangement took place involved initial fragmentation of the β,γ-carbon–carbon single bond to produce a pair of radicals, which then recombined to generate the products shown (the center panel in Scheme 9.27), the observed results with chiral material can only be accounted for if the rules of orbital symmetry are followed (Chapter 4, Section G.)

Problem 9.5. Based on the information presented in Scheme 9.27 and Chapter 4, Section G, predict which of the possible products in the photochemical transformation shown below is actually obtained.†

In the case of the cross-conjugated dienone, 4,4-diphenylcyclohexa-2,5-dienone, irradiation in dioxane-water produces 6,6-diphenylbicyclo[3.2.1]-hex-3-one-2. The formation of this material can be accounted for either through application of the electrocyclic process outlined earlier or by postulating that initial photoexcitation is followed by intersystem crossing to a triplet diradical, which rearranges and then decays back to ground state singlet rearranged product as shown in Scheme 9.28.

*See Schuster, D. I.; Brown, R. H.; Resnick, B. M. *J. Am. Chem. Soc.*, **1978**, *100*, 4505.
†See Chapman, O. L.; Sieja, J. B.; Welstead, W. J. *J. Am. Chem. Soc.*, **1966**, *88*, 161.

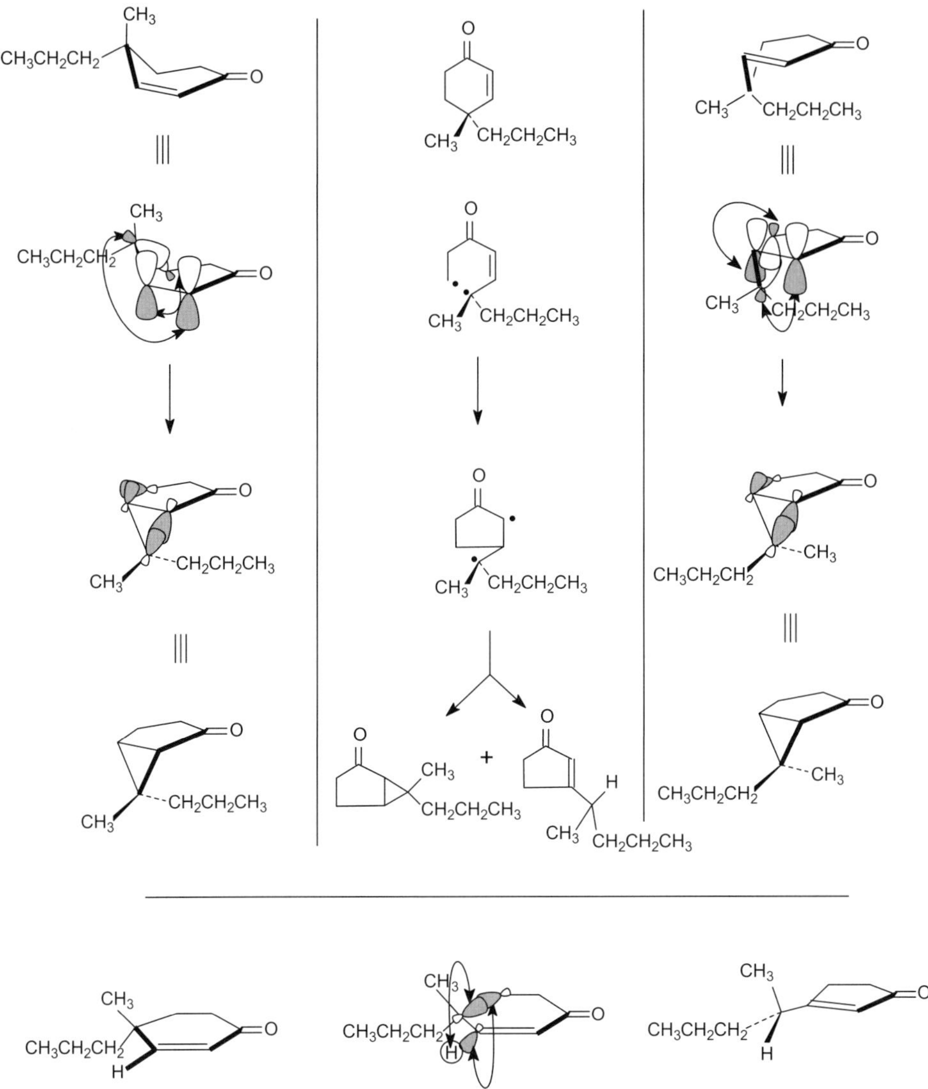

Scheme 9.27. The results of the photochemically induced rearrangement of (+)-4-methyl-4-*n*-propyl-2-cyclohexenone that demonstrate that fragmentation of the β,γ-carbon–carbon bond to a radical pair fails to account for the observed specificity seen in the $_\pi 2_a + _\sigma 2_a$ and $_\sigma 2_a + _\sigma 2_a$ processes. (After Schuster, D. I, Brown, R. H., Resnick, B. M. *J. Am. Chem. Soc.*, **1978**, *100*, 4505.)

Interestingly, as shown in Scheme 9.29, further irradiation of 6,6-diphenylbicyclo[3.1.0]hex-3-en-2-one yields phenolic and acidic material. As shown in Scheme 9.29, the results can be accounted for by postulating a triplet diradical similar to one seen in Scheme 9.28 and which also leads to a zwitterion. Phenyl migration produces the observed phenols. The carboxylic acid product presumably comes about by a process involving attack of hydroxide on the carbonyl (*vide infra*).

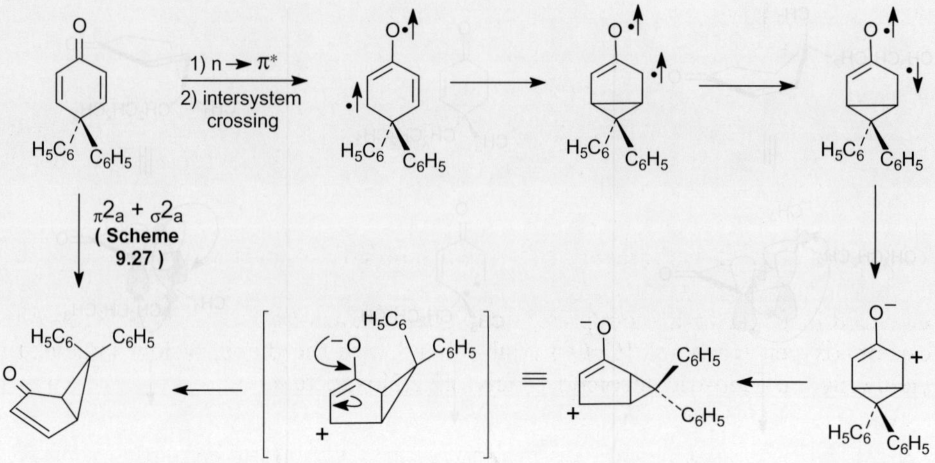

Scheme 9.28. The photochemical conversion of 4,4-diphenylcyclohexa-2,5-dienone to produce 6,6-diphenylbicyclo[3.1.0]hex-3-en-2-one. Two paths are indicated, one via an electrocyclic process and the other via a radical process. (See Zimmerman, H. E.; Swenton. J. S. *J. Am. Chem. Soc.*, **1967**, *89*, 906).

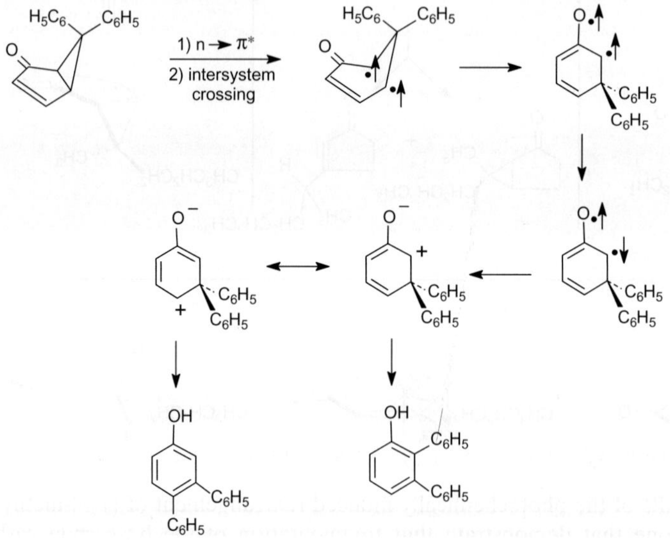

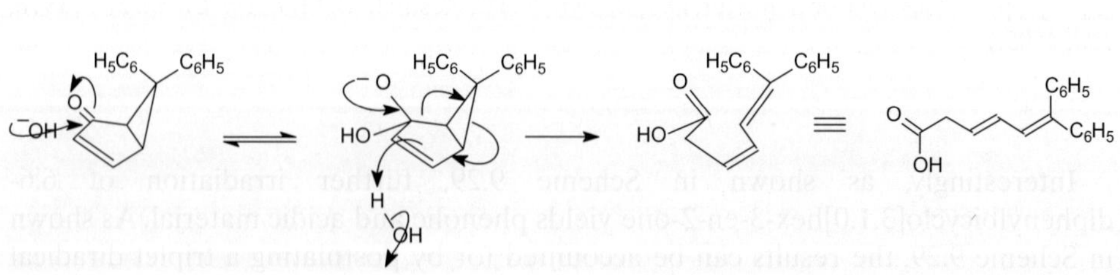

Scheme 9.29. The photochemical conversion of 6,6-diphenylbicyclo[3.1.0]hex-3-en-2-one to two phenols and the conversion of the same starting material to 6,6-diphenylhexa-3,5-dienoic acid on treatment with hydroxide ion. (See Zimmerman, H. E.; Swenton. J. S. *J. Am. Chem. Soc.*, **1967**, *89*, 906).

III. Thermal Electrocyclic and Related Reactions of Aldehydes and Ketones

a. Nonconjugated Carbonyl Compounds. Although, in general, aldehydes and ketones react only poorly as the dienophilic partner in cycloaddition reactions with dienes to generate oxacyclohexyl-3-ene (5,6-dihydro-2*H*-pyran) products, the use of Lewis acid catalysts and highly reactive dienes can induce formation of cyclic products. Further, it is clear that the $4\pi + 2\pi$ process is enhanced if the aldehyde or ketone has electron-withdrawing substituents directly attached to the carbon of the carbonyl. For example, as shown in Equation 9.29, trichloroacetaldehyde (chloral, Cl_3CCHO) acting as a dienophile, undergoes a Diels–Alder reaction across the carbon–oxygen (carbonyl [C=O]) double bond with the diene, cyclohexadiene, to yield only the *endo*-trichloromethylbicyclo[2.2.2]oxaoctene.

$$(9.29)$$

Similarly, ketones with electron-withdrawing substituents (such as the carboxylic acid ester [–CO$_2$R], *vide infra*) can react as dienophiles. Thus, 1,3-butadiene reacts with diethyl ketomalonate (diethyl 2-oxopropan-1,3-dicarboxylate [(CH$_3$CH$_2$O$_2$C)$_2$CO]) to produce the expected diethyl oxacyclohex-3-ene-6-dicarboxylate (diethyl 5,6-dihydro-2*H*-pyran-6-dicarboxylate) as shown in Equation 9.30.

$$(9.30)$$

Finally, in this vein, if the diene reactant is sufficiently electron-rich, then it is no longer necessary to have electron-withdrawing substituents on the aldehyde (ketone). Thus, as shown in Equation 9.31, 1-methyoxy-3-trimethylsilyloxy-1,3-butadiene (the **Danishefsky** diene*) undergoes reaction with benzene carboxyaldehyde (benzaldehyde [C$_6$H$_5$CHO]) in the presence of zinc(II) chloride with the *regioselectivity* shown.

$$(9.31)$$

*Named after Professor S. Danishefsky, Columbia University, in whose research group the diene was developed and initially used in natural product synthesis. See Danishefsky, S.; Kitahara, T. *J. Am. Chem. Soc.*, **1974**, *96*, 7807.

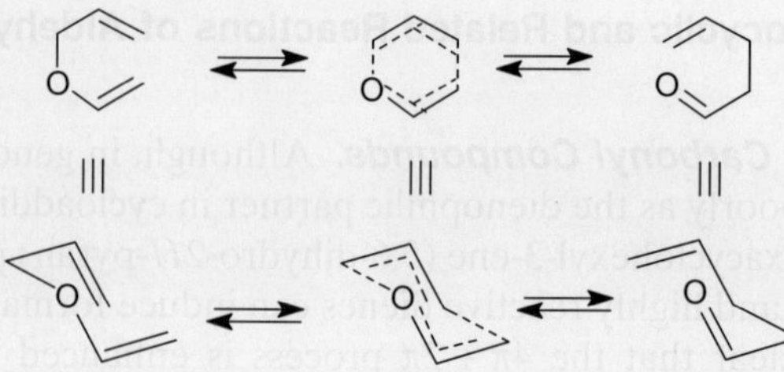

Scheme 9.30. Repetition of Scheme 8.100 depicting equilibrium between allylvinyl ether and 4-pentenal. The process is an example of the Claisen rearrangement. (See Tarbell, D. S. *Org. Reactions*, **1944**, *2*, 1.)

Just as the carbocyclic Diels–Alder reaction has its heterocyclic counterpart, so does the **Cope** rearrangement (Chapter 6). It will be recalled that in the archtypical **Cope** process (without heteroatoms), the narcissistic conversion of 1,5-hexadiene into "itself" is observed (Equation 9.32).

$$\tag{9.32}$$

The "**oxy-Cope**" is different in that the "diene" is aldehydic at one end and, indeed, it is the *enol* of the aldehyde which appears to be the reactive species. It is generally the case that the allylic alcohol (the 3-hydroxy-1,5-hexadiene on the right in Equation 9.33) is synthesized (*vide infra*)

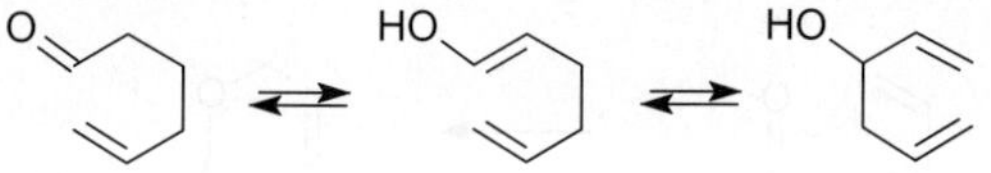

$$\tag{9.33}$$

and specifically converted to the enolate anion, which rearranges to the aldehyde (the 5-hexenal in Equation 9.33). Thus, the reaction is **used** in the sense opposite to the one shown in Equation 9.33. The anion of the enol undergoes the conversion to aldehyde faster than does the enol itself.

Finally, as mentioned in Chapter 8, aliphatic allylvinyl ethers are in equilibrium with the corresponding 4-pentenals (the parent is shown in Scheme 8.100 and the scheme is repeated here as Scheme 9.30). The position of the equilibrium can be adjusted catalytically.

It has been argued that the enzyme *chorismate mutase* (EC 5.4.99.5) promotes a Claisen-type rearrangement in converting **chorismate** to **prephenate** as shown in Scheme 9.31 (see also Scheme 8.71 and Chapter 12 for more information about **chorismate** and **prephenate**).

b. Conjugated Carbonyl Compounds. The 4 + 2 thermal dimerization of the simplest conjugated aldehyde, 2-propenal (acrolein, $CH_2{=}CH{-}CHO$), as well as similar dimerization of the simplest conjugated ketone, buten-3-one (methyl vinyl ketone, $CH_2{=}CH{-}CO{-}CH_3$) (Equation 9.34), to produce the corresponding 3,4-dihydro-2*H*-pyrans (6-acyl substituted oxacyclohex-2-enes) has been known since about 1940. The principle having been established, many other examples are also known, and the regiochemistry seen in Equation 9.34 is generally observed.

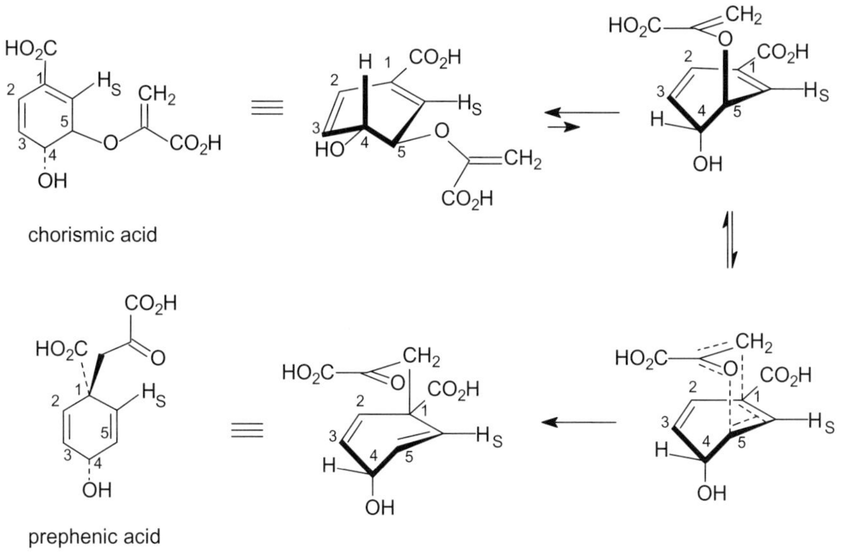

chorismic acid

prephenic acid

Scheme 9.31. A representation of the conversion of **chorismate** to **prephenate** utilizing a Claisen-type rearrangement. The conversion is catalyzed by the enzyme *chorismate mutase* and the representation shown may not be the path followed. (See Mandal, A.; Hilvert, D. *J. Am. Chem. Soc.*, **2003**, *125*, 5598 and references therein.)

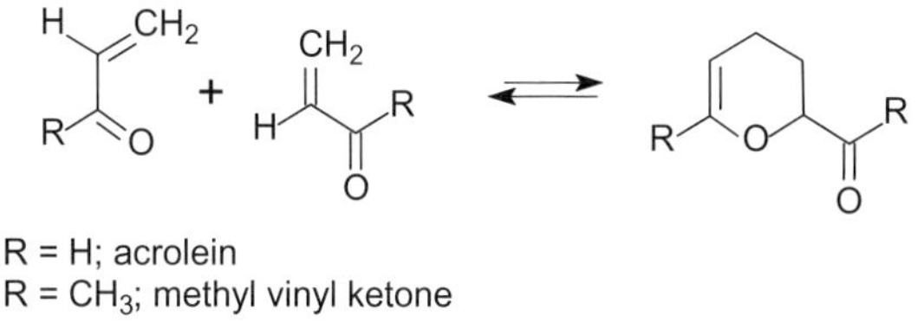

$$(9.34)$$

R = H; acrolein
R = CH$_3$; methyl vinyl ketone

Finally, and as pointed out in Scheme 8.101 (Chapter 8), there is an aromatic version of the Claisen rearrangement. As the "ketone" intermediate is conjugated (prior to its tautomerization to the corresponding phenol), the process may be considered an addition to an unsaturated carbonyl compound. Equation 9.35 represents the aromatic Claisen rearrangement, which, as shown from left to right, allows the conversion of an arylvinyl ether into a substituted phenol.

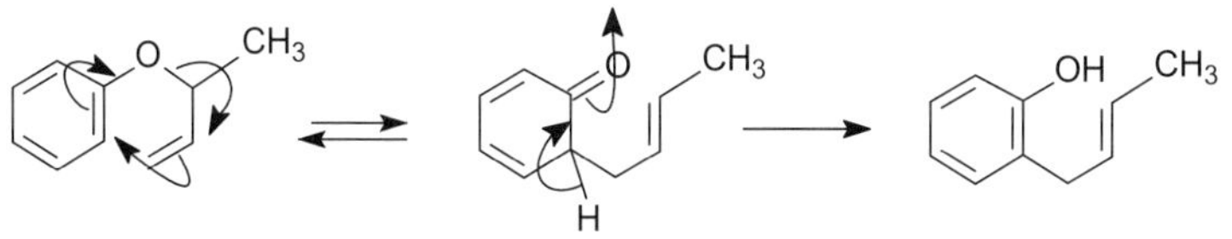

$$(9.35)$$

c. The Carbonyl "ene" Reaction. The carbocyclic "ene" reaction is discussed in Chapter 6 (Table 6.6) and the corresponding carbonyl reaction is among a variety that have been encountered. Indeed, in the presence of appropriate catalysts, as will be seen (*vide infra*), this reaction provides a pathway for the construction, in high enantiomeric excess, of chiral α-hydroxyesters. The generalized **Lewis acid-catalyzed**

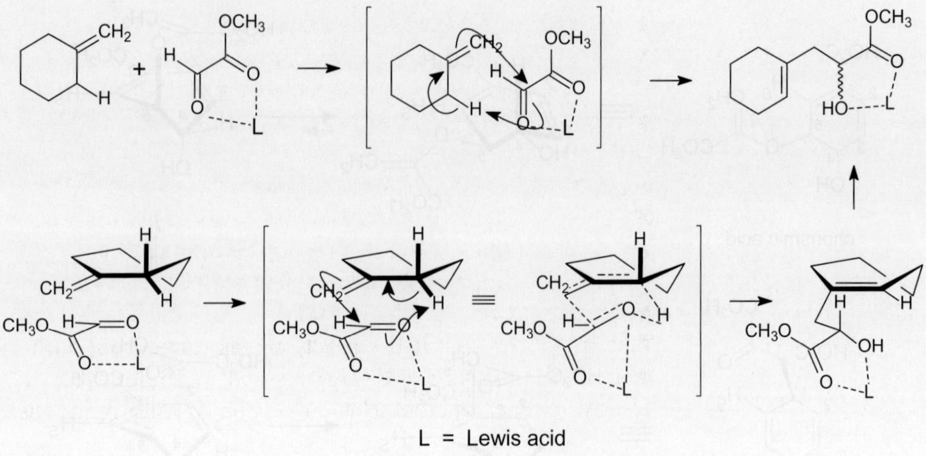

L = Lewis acid

Scheme 9.32. The carbonyl "ene" reaction between methylenecyclohexane and methyl glyoxalate (CH$_3$O$_2$CCHO) producing methyl 3-(1-cyclohexenyl)-2-hydroxypropanoate. Since addition can occur on either the *re* or the *si* face of the carbonyl group, the product is racemic. **The details of the Lewis acid catalyst "L" have been omitted for clarity**.

electrocyclic process corresponding to that in the carbocyclic system (Chapter 6) is shown in Scheme 9.32 (where the details of any specific **Lewis acid "L" have been omitted for clarity**).

In the presence of a chiral Lewis acid catalyst or if chirality is introduced into the ester by esterifying the aldehydic acid with a chiral alcohol (a "chiral auxilary"), high enantiomeric excess of one alcohol over its enantiomer can be achieved. Examples of both of these processes, generating α-hydroxyesters, are shown in Schemes 9.33 and 9.34.

IV. Nucleophilic Addition Reactions Retaining the Carbonyl Oxygen

a. General Comments.
The addition of nucleophiles to the carbon of the carbonyl group (C=O), both where the oxygen is retained and where the oxygen is subsequently lost, is commonly found to be a catalyzed process. *Both acid and base catalyses have been observed.*

In any given case, whether the catalysis is best effected by acid or base is a function of the species being added to the carbonyl. However, for some addends both acid and base catalysis have been observed.

In base catalysis, the role of the base generally appears to be to convert a proton-bearing nucleophile (i.e., Nu–H) into an anionic species (Nu⁻) by rapid, reversible proton abstraction. The negatively charged nucleophile can then add, in a slow (rate determining) step, to the carbon of the carbonyl (C=O) group, and, in a final fast step, the adduct thus formed can abstract a proton from the protonated base, regenerating the base and the fully formed adduct. These steps are illustrated in Scheme 9.35.

In acid catalysis, on the other hand, the initial step appears to be protonation of (Brønsted acid) or complex formation with (Lewis acid) the oxygen of the carbonyl

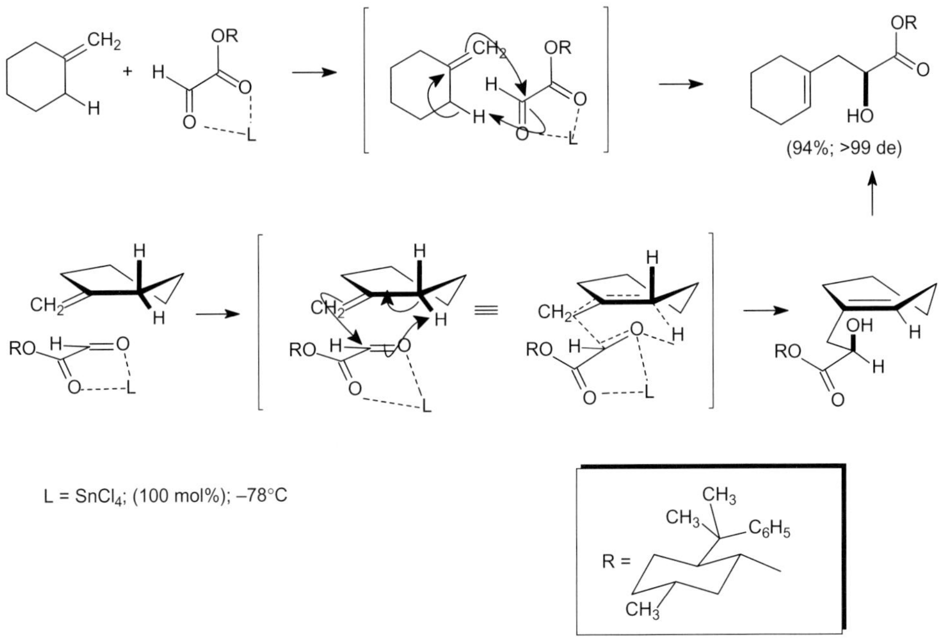

(94%; >99 de)

L = SnCl₄; (100 mol%); –78°C

Scheme 9.33. The formation of a diasteromer of a 2-(1-methyl-1-phenyl)ethyl-4-methylcy-clohexyl 3-(1-cyclohexenyl)-2-hydroxypropanoate from tin(IV) chloride catalyzed "ene" reaction between methylenecyclohexane and the corresponding glyoxalate ester.

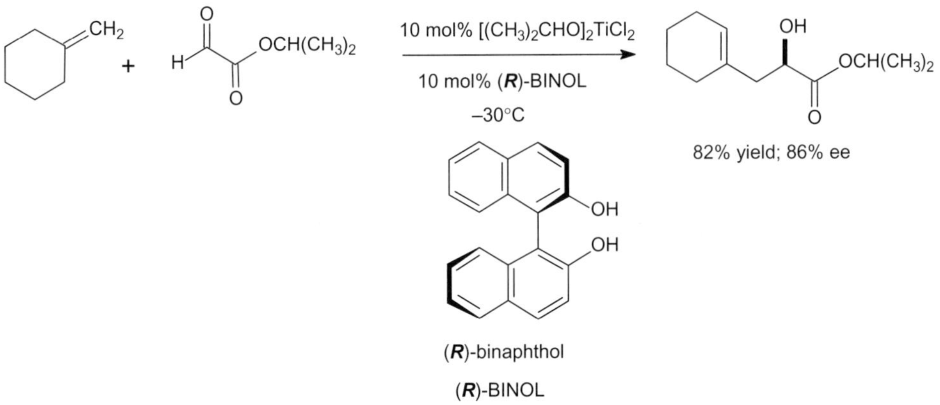

10 mol% [(CH₃)₂CHO]₂TiCl₂

10 mol% (**R**)-BINOL

–30°C

82% yield; 86% ee

(**R**)-binaphthol

(**R**)-BINOL

Scheme 9.34. The formation of isopropyl 3-(1-cyclohexenyl)-2-hydroxypropanoate in high enantiomeric excess (ee) by titanium(IV) Lewis acid catalyzed-"ene" reaction in the presence of (**R**)-BINOL. (See Evans, D. A.; Burgey, C. S.; Paras. N. A.; Vojkovsky, T.; Tegay, S. W. *J. Am. Chem. Soc.*, **1998**, *120*, 5824.)

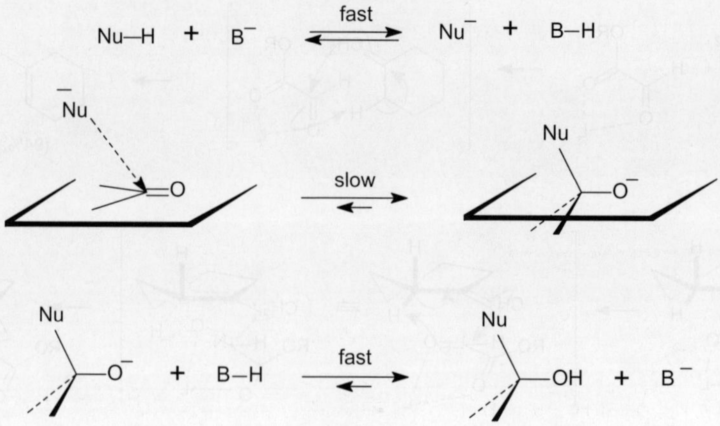

$$\text{Nu–H} + \text{B}^- \xrightleftharpoons{\text{fast}} \text{Nu}^- + \text{B–H}$$

Scheme 9.35. A representation of base catalysis of nucleophilic addition to the carbonyl of an aldehyde or ketone in which the oxygen is not lost. Nu–H is a protonated nucleophile; B is a generalized (Brønsted or Lewis) base.

(C=O) in a fast reversible step, allowing the nucleophile to attack the (now more) electrophilic carbon of the carbonyl. These steps are illustrated in Scheme 9.36.*

b. Addition of H–X. Table 9.3 presents a variety of species of the general form H–X that add across the carbonyl (C=O) carbon–oxygen double bond and in which the oxygen originally present in the carbonyl (C=O) group is retained. Many of these reactions have been encountered before and all of them follow the general patterns already outlined. Thus, in the absence of contradictory evidence, it is assumed that the Bürgi–Dunitz trajectory is used by the incoming nucleophile and the products reflect the individual processes.

The first item in Table 9.3 is simply the familiar hydration (see, e.g., Figure 8.4) of aldehydes and ketones. As previously discussed, these "geminal diols" are normally in equilibrium with the corresponding carbonyl (C=O) compound and, with few exceptions (e.g., methanal [formaldehyde, H_2CO]; 2,2,2-trichloroacetaldehyde [chloral, Cl_3CCHO], see Chapter 8), the equilibrium lies far on the side of the aldehyde or ketone. The same general ideas obtain with regard to the next three items in Table 9.3, except that hemiacetals and hemiketals in cyclic systems (e.g., carbohydrates, Chapter 11), are more common than their "open chain" forms.

The addition of HCl across the carbon–oxygen double bond yields an α-chlorohydrin which is unstable relative to the starting carbonyl compound, and,

---

*Acids and bases vary in "acid strength" or "base strength," that is, the extent of proton transfer or the position of the equilibrium during proton transfer. This variation can be a function of the solvent into which the acid is placed as well as the stability of the conjugate base/acid, and so on. Differences in the extent of proton transfer from the acid or to the base in some carbonyl (C=O) reactions can be observed. For example, with "strong" acids where ionization of the acid is complete (or nearly so), the proton from the acid B-H will be completely transfered (or nearly so) to the carbonyl (C=O) oxygen ***before*** the nucleophile adds to the carbonyl (C=O) carbon as shown in Scheme 9.36. Such processes are said to function under **specific acid catalysis**. The alternative, where the nucleophile adds to the carbon of the carbonyl (C=O) while the proton is partially bonded to the (weaker) acid and to the oxygen of the carbonyl (C=O) simultaneously so that proton transfer is consummated ***while*** the nucleophile is in the process of becoming bonded to carbon is known as **general acid catalysis**. The same comments apply to **specific base** and **general base** catalysis. That is, is the proton completely removed ***before*** an electron pair becomes available, or is the removal of the proton simultaneous with use of the electron pair elsewhere?

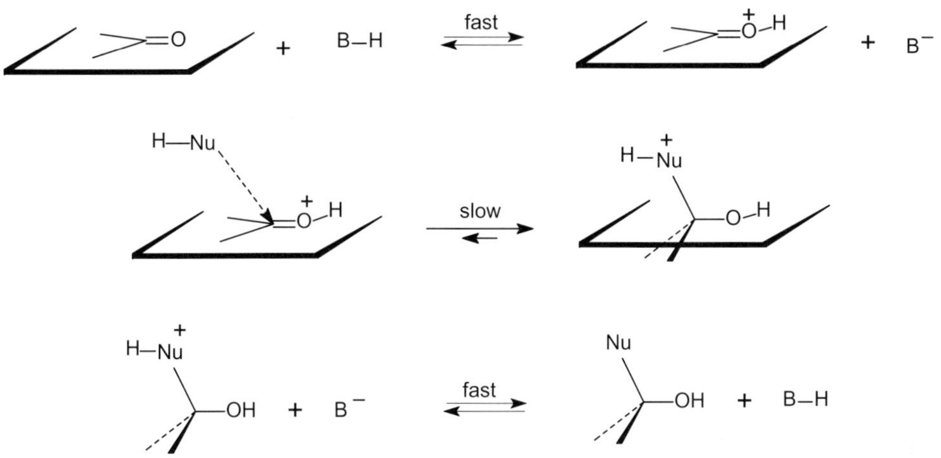

Scheme 9.36. A representation of (Brønsted) acid catalysis of nucleophilic addition to the carbonyl of an aldehyde or ketone in which the oxygen is not lost. H–Nu is a nucleophile that is capable of proton loss to consummate the reaction; B–H is a generalized (Brønsted) acid. The use of a Lewis acid is not shown.

finally, except for the addition of sodium bisulfite ($NaHSO_3$) to aldehydes, all of the other reactions in Table 9.3 have already been discussed in the context of reduction of the carbonyl (C=O) group.

It is argued that, for steric reasons, the addition of sodium bisulfite ($NaHSO_3$) to aldehydes and ketones generally succeeds only with the former. Traditionally, the addition has been used to produce crystalline derivatives of liquid aldehydes so that they may be more easily purified. The carbonyl group is regenerated on base treatment.

c. Addition of Carbon Nucleophiles

1. Hydrogen Cyanide. The reversible addition of hydrogen cyanide to ketones and aldehydes to give hydroxy nitriles (cyanohydrins) is base catalyzed (Equation 9.36).

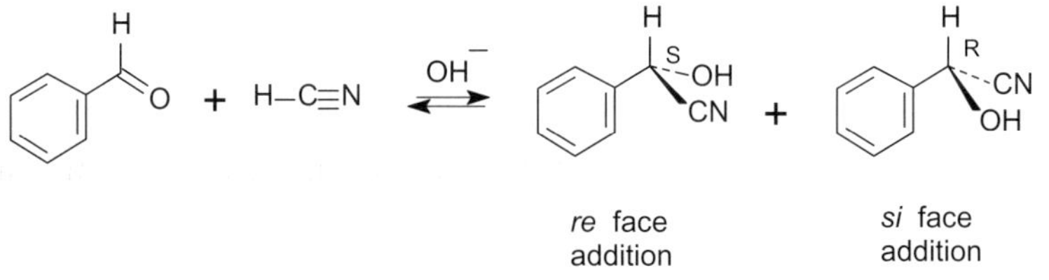

$$\text{(9.36)}$$

It is argued that the weak acid (hydrogen cyanide, HCN, $pK_a = 9.1$) is converted to its conjugate base by hydroxide anion (itself capable of reversible addition to the carbonyl [C=O] group) and, in the rate-determining step, the cyanide anion thus formed adds to the carbon of the carbonyl (C=O) group. The resulting intermediate alkoxide anion then abstracts a proton from another hydrogen cyanide (HCN), producing the cyanohydrin and another equivalent of cyanide anion (Scheme 9.37).

As shown in Scheme 9.37 the Bürge–Dunitz pathway is presumed and, with aldehydes and unsymmetrical ketones, addition of cyanide to either (both) the *re* and *si* faces of the carbonyl is (are) possible.

TABLE 9.3. Addition of Reagents (H–X) to the Carbonyl Carbon

Reagent	Adduct to aldehyde or ketone	Conditions and comments
H–OH	(OH / OH)	Hydrates are generally unstable in the absence of electron withdrawing groups. See **Chapter 8** (**Figure 8.4**) for some geminal diols.
H–OR	(OR / OH)	Acid-catalyzed addition of one equivalent of an alcohol to an aldehyde produces hemiacetals and to a ketone produces hemiketals. See **Chapter 8** and carbohydrates, **Chapter 11**.
H–SH (SR)	(SH (SR) / OH)	Acid and/or base catalyzed addition of one equivalent of hydrogen sulfide (H_2S) or a thiol (HSR) to an aldehyde produces thiohemiacetals and to a ketone produces thiohemiketals. See **Chapter 8**.
H–Cl	(Cl / OH)	A generally unstable adduct.
$H-O-\overset{O}{\underset{O}{S}}{}^{-}Na^{+}$	($SO_3^-Na^+$ / OH)	The crystalline bisulfite adducts of aldehydes are often produced in an effort to purify aldehydes. Ketones react slowly, if at all.
$LiAlH_4$	(H / $O-AlH_3^-Li^+$)	The intermediate formed in the lithium aluminum hydride ($LiAlH_4$) reduction of aldehydes and ketones. Hydrolysis produces the corresponding alcohol. See this **Chapter (Reduction)**.
$NaBH_4$	(H / $O-BH_3^-Na^+$)	The intermediate formed in the sodium borohydride ($NaBH_4$) reduction of aldehydes and ketones. Hydrolysis produces the corresponding alcohol. See this **Chapter (Reduction)**.
B_2H_6	(H / O—B)$_3$	The intermediate formed in the borane (B_2H_6) reduction of aldehydes and ketones. Hydrolysis produces the corresponding alcohol. See this **Chapter (Reduction)**.
$Al[OCH(CH_3)_2]_3$	(H / $O-Al[OCH(CH_3)_2]_2$)	The intermediate formed in the aluminum isopropoxide ($Al[OCH(CH_3)_2]_3$) (Meerwein–Pondorf–Verley) reduction of aldehydes and ketones. Hydrolysis produces the corresponding alcohol. See this **Chapter (Reduction)**.
H_2/catalyst	(H / O—H)	The direct product of catalytic reduction of aldehydes and ketones.

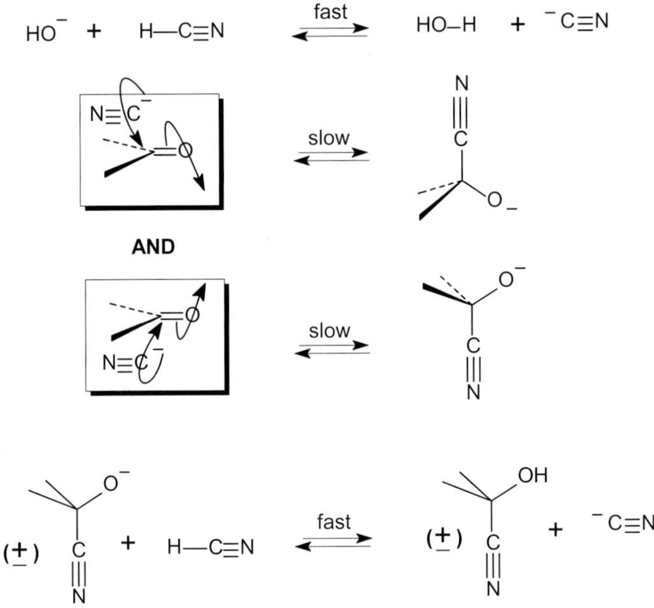

Scheme 9.37. A representation of the steps in base-catalyzed addition of cyanide (CN⁻) anion across the carbonyl (C=O) group. Addition to either face (*re* or *si*) of the carbonyl is possible.

The addition of cyanide to the aldehydic carbon of polyhydroxyaldehydes (see **Carbohydrates**, Chapter 11, and the **Kiliani–Fischer** synthesis) and simple aldehydes in the presence of ammonia (*vide infra*, and see also, **Amino Acids**, Chapter 12, and the **Strecker** synthesis) is of some historical interest. Recently, asymmetric versions of the latter have become important and more information about the **Strecker** synthesis will be provided later in this chapter.

Cyanide is a potent inhibitor of cytochrome oxidase, the terminal respiratory enzyme in aerobic organisms and, therefore, most aerobic organisms have (a) some cyanide present in a relatively small concentration and (b) a pathway by which excess cyanide can be eliminated. When the elimination pathway is overwhelmed, death of the organism results.

Historically, it appears that cyanide-containing plant materials have been recognized as poisonous since antiquity. In both the few plants and animals containing indigenous cyanide in any sizeable concentration, the cyanide is present in the form of cyanohydrins. Two examples, one each from the plant and animal kingdoms, will be considered.

First, **cassava** (also known as manioc, tapioca, and yuca but only *Manihot esculanta* Crantz) is a human food crop. The plant is a short-lived shrub that produces root tubers (resembling potatoes), which are harvested as a starch source in much of the tropical world. Global production in 1990 was estimated in excess of 100 million tons. The toxicity of "incorrectly" processed (insufficiently leached with water) tubers has been known for at least 400 years. The major toxic constituents are reported to be the two cyanogenic glycosides, linemarin, and lotaustralin, whose structures are shown below.

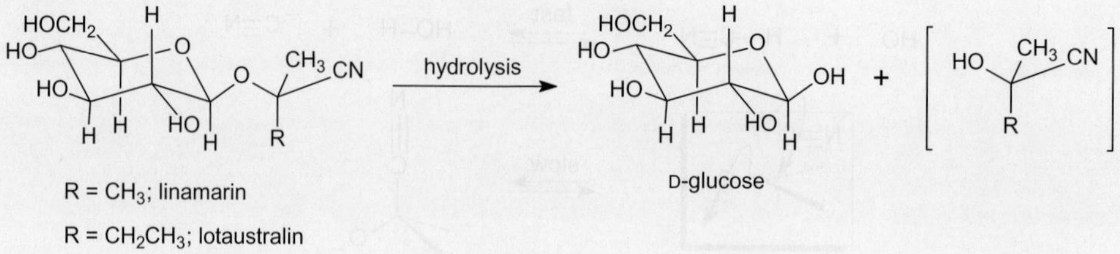

R = CH$_3$; linamarin

R = CH$_2$CH$_3$; lotaustralin

D-glucose

Now, as shown, when hydrolysis occurs, linamarin produces D-glucose and 2-hydroxy-2-methylpropanenitrile (the cyanohydrin of propanone [acetone, R = CH$_3$, [CH$_3$COCH$_3$]]), while lotaustralin generates D-glucose and 2-hydroxy-2-methylbutanenitrile (the cyanohydrin of butanone [methyl ethyl ketone, R = CH$_2$CH$_3$, [CH$_3$COCH$_2$CH$_3$]]). Both of the cyanohydrins decompose to hydrogen cyanide (HCN) and the respective corresponding ketones.

Second, the phylum *Arthropoda*, which contains the classes Chilopoda (centipedes), Diplopoda (millipedes), and Insecta (insects), appears to have the only groups of animals that produce cyanide as a defensive weapon. For example the millipede *Harpaphe haydeniana* produces and **stores** the 2-hydroxy-2-phenylethanenitrile (the cyanohydrin of benzenecarboxaldehyde [benzaldehyde, C$_6$H$_5$CHO]) shown in Equation 9.36. When disturbed, the millipede explosively ejects a drop of liquid that apparently contains benzenecarboxaldehdye (benzaldehyde, C$_6$H$_5$CHO) and hydrogen cyanide (HCN), enzyme-catalyzed hydrolysis having occurred to produce this mixture from the cyanohydrin along with enough energy for an explosive ejection.

Interestingly, for both aldehydes and ketones, if there is a double bond in conjugation with the carbonyl (C=O) group (i.e., an α,β-unsaturated aldehyde or ketone), the product of addition of cyanide is that which results from addition to the β-carbon (i.e., **conjugate** or **Michael addition**) rather than to the carbonyl (C=O) carbon (Equations 9.37 and 9.38). It is argued that this is so because the transition state for conjugate addition has the (presumed) negative charge delocalized (and thus of lower energy) than the transition state for addition to the carbonyl (Scheme 9.38).

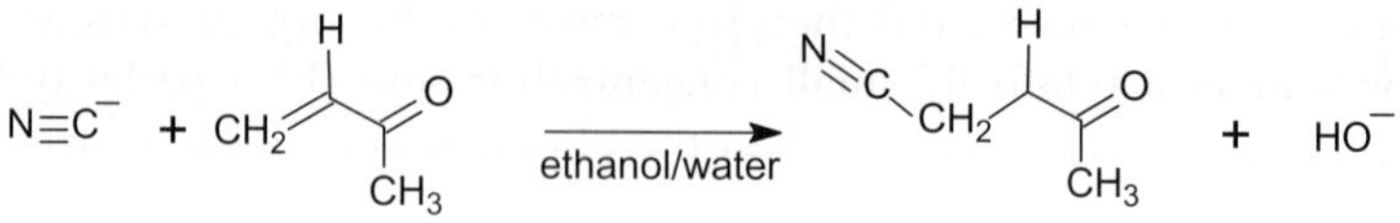

$$\text{N}\equiv\text{C}^- + \text{CH}_2=... \xrightarrow{\text{ethanol/water}} ... + \text{HO}^- \qquad (9.37)$$

compared with

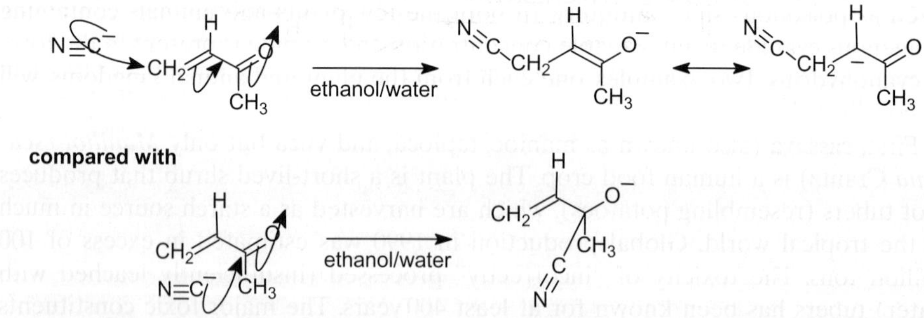

Scheme 9.38. A comparison of the initial results of addition of cyanide anion (CN-) to the β-carbon of an α,β-unsaturated ketone (but-3-en-2-one) and to the carbon of the carbonyl (C=O) of the same ketone.

$$\text{(9.38)}$$

2. Organometallic Reagents. Some organometallic reagents (viz. Grignard reagents ["RMgX"], organolithium reagents ["RLi"], Gilman reagents [R_2CuLi]) were introduced in Chapter **7, Section D, Part IV**) where it was noted that they were produced by reaction of organic halides with metals.

All of these reagents have been important to synthetic efforts where formation of new carbon–carbon bonds *at the carbon of the carbonyl (C=O)* or at the carbon β- to the carbon of the carbonyl (i.e., in conjugate addition) were required.

Consider, as a typical process, the addition of a Grignard reagent, methylmagnesium iodide ("CH_3MgI") to methanal (formaldehyde, $H_2C=O$). Although, for convenience, the overall process can be shown with just the two reacting species (upper part of Scheme 9.39), there is evidence that it is more complicated. First, as with any organometallic species, the process may be one where single electron transfer (see Chapter **7, Section D, Part IV**) intrudes along with (or to the exclusion of) any "two electron" pathways. Further, it appears that "dimeric" and/or solvated dimeric Grignard reagent may be generally involved (lower part of Scheme 9.39), and even these observations probably represent situations that are oversimplified.*

Despite the lack of specific detail and the necessity for simplification that lack introduces, the use of the Grignard reagent with formaldehyde *produces a primary alcohol that is* **one** *carbon longer than the halide* from which the alkylmagnesium reagent was produced. Given the observation (Chapters **7** and **8**) that alkyl halides can be produced from alcohols, a sequence of reactions to lengthen a chain bearing a primary alcohol at its terminus by one carbon atom can be imagined. Such a sequence is shown in Scheme 9.40.

Given the structural variation possible in alkyl halides and in aldehydes and ketones, a wide variety of products can be obtained using the Grignard reaction. Some typical processes are shown in Equations 9.39 (an aldehyde) and 9.40 (a ketone).

$$\text{(9.39)}$$

---

*The astute reader, remembering what has gone before, will be suspicious of the simplicity of this process. First, he or she will have remembered that the species called methylmagnesium iodide (CH_3MgI) prepared in diethyl ether (ethyl ether [$(CH_3CH_2)_2O$]) from methyl iodide (CH_3I) and magnesium metal (Mg) is, in part, a solvate of what may be a mixture of dimethylmagnesium [$(CH_3)_2Mg$], magnesium iodide (MgI_2), methylmagnesium iodide (CH_3MgI), and, in part, a mixture of related but unknown species. Second, he or she will have recalled that methanal (formaldehyde [$H_2C=O$]) is a gas at room temperature and is generally available only as a hydrate (formalin) or as a trimer (1,3,5-trioxocyclohexane [$(CH_2O)_3$]). Nonetheless, the process of yielding the primary alcohol as shown in Scheme 9.39 is written for convenience.

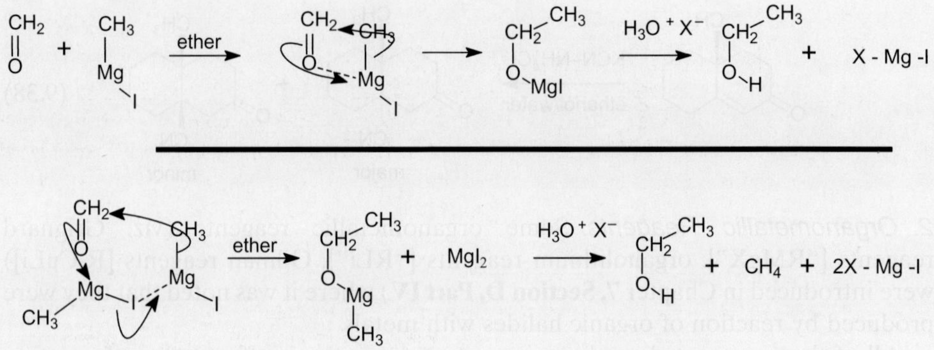

Scheme 9.39. Two idealized representations of the reaction between the Grignard reagent, methylmagnesium iodide (CH₃MgI), and methanal (formaldehyde, H₂C=O) followed by hydrolysis. In both, the primary aldehyde is converted to a primary alcohol. The difference is in the molecularity of the Grignard reagent.

$$R{-}CH_2OH \ + \ HI \ \longrightarrow \ H_2O \ + \ R{-}CH_2I$$

$$R{-}CH_2I \ + \ Mg \ \xrightarrow[\text{ether}]{\text{anhydrous}} \ R{-}CH_2MgI$$

$$R{-}CH_2MgI \ + \ O{=}CH_2 \ \xrightarrow[\text{ether}]{\text{anhydrous}} \ R{-}CH_2CH_2OMgI$$

$$R{-}CH_2CH_2OMgI \ + \ H_3O^+ \ \longrightarrow \ R{-}CH_2CH_2OH \ + \ Mg(OH)I$$

Scheme 9.40. A sketch of a series of reactions that might be used to lengthen a carbon chain by **one** carbon atom, beginning and ending with a primary alcohol.*

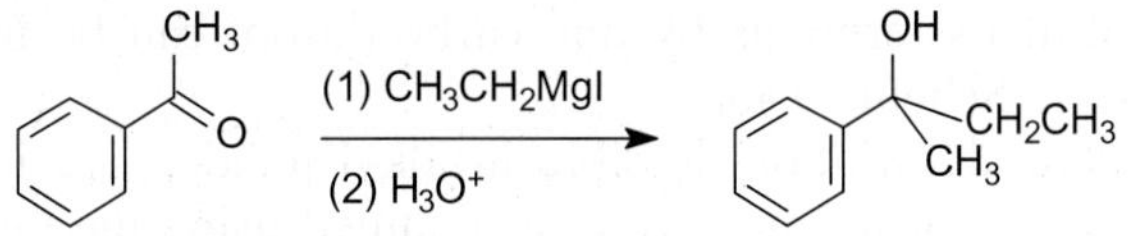

$$\text{(9.40)}$$

The following generalities also apply:

(a) Formation of the Grignard reagent (Chapter 7) is more facile from alkyl and aryl iodides and bromides than from chlorides.

(b) Formation of Grignard reagents from primary and secondary alkyl iodides and bromides generally gives fewer side (additional and unwanted) products.

(c) Reaction of Grignard reagents with methanal (or its equivalent) produces primary alcohols one carbon longer than the starting alkyl halide. Other aldehydes with the same Grignard reagent yield secondary alcohols, while ketones yield tertiary alcohols.

*Once again, the astute reader will recognize the provision of a tool complementary to the one provided in the previous chapter (cf. Scheme 8.76), which allowed the ring opening of oxiranes, particularly oxacyclopropane (ethylene oxide) by reaction with a Grignard reagent, to *add two carbons to the terminus of an alkyl chain* and producing, in that way, a primary alcohol.

(d) Since amphoteric magnesium hydroxide $[Mg(OH)_2]$ is obtained on hydrolysis of the initial addition product, a weak acid (such as ammonium chloride $[NH_4Cl, pK_a = 9.2)$ is commonly used instead. Strong acids are avoided in order to minimize the possibility of elimination of water from the product.

(e) There are two complicating "side reactions" about which to be vigilant. The first of these alternative processes intrude when a hindered ketone is brought into reaction with a Grignard reagent that possesses a hydrogen on the carbon β- to that to which the magnesium is attached (Equation 9.41). The second complication can arise with any carbonyl compound that bears an *acidic* proton on the carbon α- to the carbon of the carbonyl (C=O) (Equation 9.42).

$$(9.41)$$

$$(9.42)$$

Interestingly, α,β-unsaturated aldehydes and ketones generally react with Grignard reagents at the carbonyl. Reactions in which the major product is derived from a process involving the double bond and the carbonyl (conjugate addition) are rare.

Organolithium compounds, which usually occur as multimolecular aggregates of the form $(RLi)_n$ in the nonpolar solvents in which they are used, are less commonly used as addends to the carbonyl (C=O) group than are Grignard reagents. Nonetheless, and despite the fact that alkyllithium compounds are generally stronger bases than Grignard reagents and thus often react by proton abstraction, it is possible to add alkyllithium reagents across the carbonyl group. The results are similar to those found with Grignard reagents. For example, methyllithium adds to the carbonyl of 4-(1,1-dimethylethyl)cyclohexanone (4-*t*-butylcyclohexanone) to produce both axial and equatorial isomers of 1-methyl-4-(1,1-dimethylethyl)cyclohexanol (4-*t*-butyl-1-methylcyclohexanol) (Equation 9.43) in about the same ratio as the corresponding Grignard reagent (methylmagnesium bromide, "CH₃MgBr"). In both cases, the axial alcohol (presumably resulting from favored equatorial attack) predominates.*

$$(9.43)$$

*See Ashby, E. C.; Laemmle, J. T. *Chem. Rev.*, **1975**, *75*, 521.

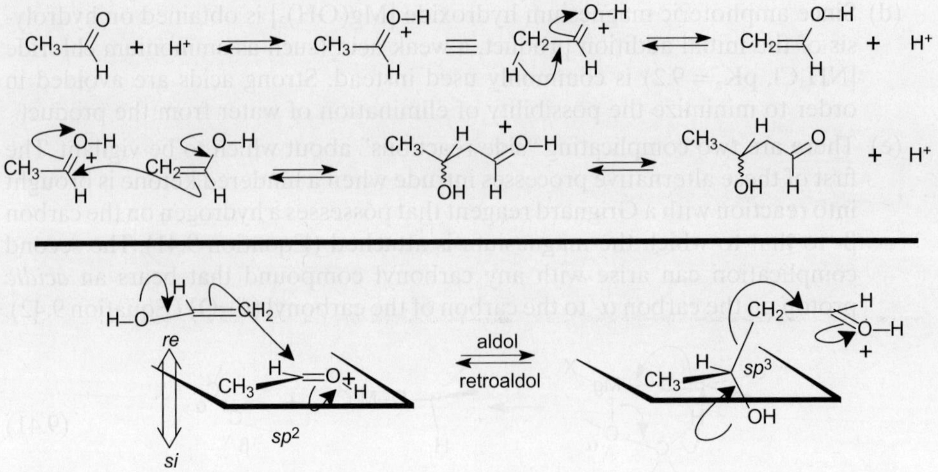

Scheme 9.41. A representation of the acid-catalyzed aldol condensation reaction (using a proton [H$^+$] as a "typical" acid) between two equivalents of ethanal (acetaldehyde [CH$_3$CHO]). The steps, both protonation–deprotonation and carbon–carbon bond forming are reversible (the **retroaldol** reaction). The carbon–carbon bond forming step is shown as occurring on the *re* face of the aldehyde and with an approximate Bürgi–Dunitz trajectory. Attack at the *si* face is equally likely and the product is racemic.

In contrast to the Main Group (Mg and Li) organometallic compounds, transition metal oganocuprates of the general form R$_2$CuLi (Gilman reagents) and higher mixed valence copper compounds (introduced in Chapter 7) have been found to undergo conjugate addition to α,β-unsaturated ketones (Equation 9.44).

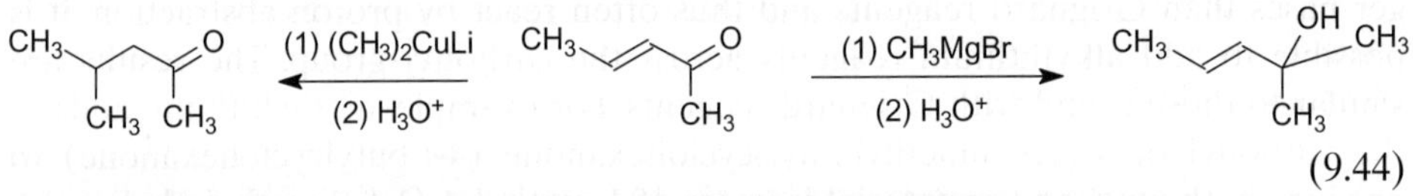

$$(9.44)$$

3. The Aldol Reaction (Without Dehydration). In the presence of acid or base catalysis, ethanal (acetaldehyde, CH$_3$CHO) undergoes rapid, reversible self-condensation to the β-hydroxyaldehyde 3-hydroxybutanal ("aldol"). Equation 9.45 presents a view of the overall process while Scheme 9.41 provides more detail of the acid-catalyzed process while Scheme 9.42 does the same for that catalyzed by base. Although the

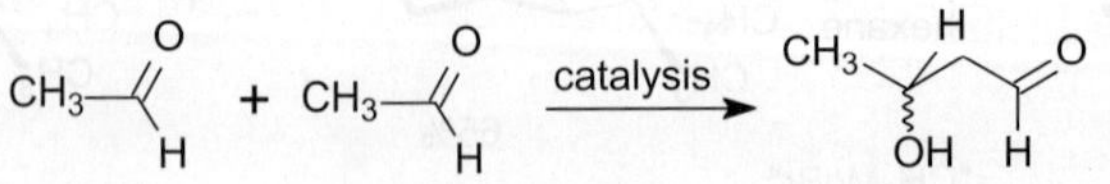

$$(9.45)$$

subtle details of the acid- and base-catalyzed aldol condensation reactions are important, it is the common features of these processes and many related addition reactions that should be noted; that is, their unity rather than their diversity is thread to follow. That unity lies in the formation of a *new carbon–carbon bond*. The bond

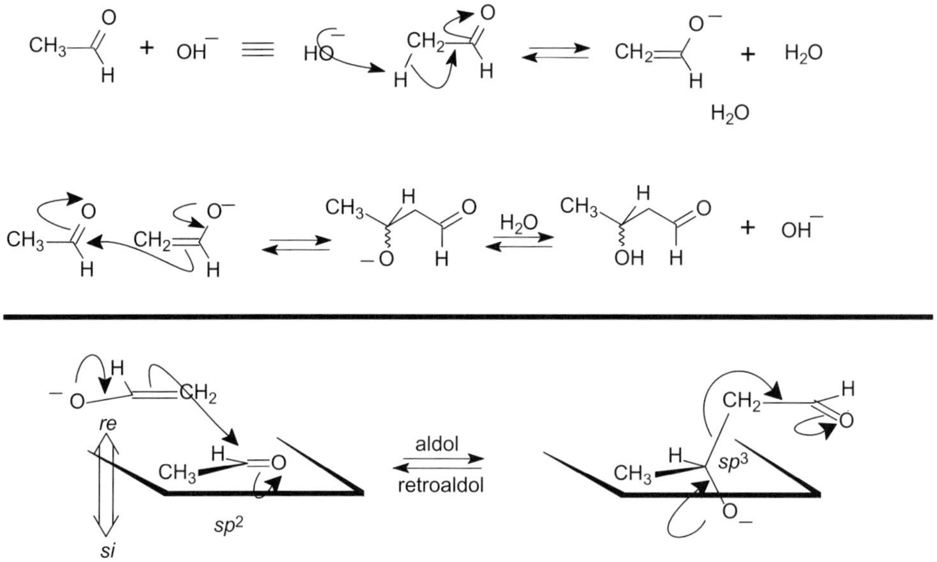

Scheme 9.42. A representation of the base-catalyzed aldol condensation reaction (using the hydroxide ion [OH⁻] as a "typical" base) between two equivalents of ethanal (acetaldehyde [CH₃CHO]). The steps, both deprotonation–protonation and carbon–carbon bond forming are reversible (the **retroaldol** reaction). The carbon–carbon bond forming step is shown as occurring on the *re* face of the aldehyde and with an approximate Bürgi–Dunitz trajectory. Attack at the *si* face is equally likely and the product is racemic.

is forged between the two carbonyl-bearing reactants (or intramolecularly in one with two carbonyl groups) and lies, in the product, between the carbon of what was the carbonyl group (rehybridization from sp^2 to sp^3) in one reactant and the carbon α- to the carbonyl group of the other.

Now, formation of an enolate anion (base catalysis) or the enol itself (acid catalysis) is generally fast and reversible so that the *rate-determining step* is that one in which a σ bond is made between the two carbon atoms.

Second, in the bond-making process, the carbon of the carbonyl, as already noted, changes from sp^2 to sp^3 hybridization, the carbon which has just become tetrahedral will generally be asymmetrically substituted, and a new chiral "center" will have been formed. While attack from either face could, in principle, have occurred, it will become clear that directing such attack is a challenge that has been met and continues to be developed.*

Third, dehydration to introduce a double bond conjugated with the carbonyl group by loss of the proton (H⁺) α to the carbonyl (C=O) and the hydroxyl (OH⁻)

---

*It has been argued that the specificity of attack on one or the other face of a carbonyl group in natural processes under enzymatic control (some of which have already been encountered) is a characteristic that distinguishes the macroscopic world of the bench chemist from the microscopic world of the cell. Control so far developed by the bench chemist pales in comparison with that of the cell, but work in the laboratory of the former continues.

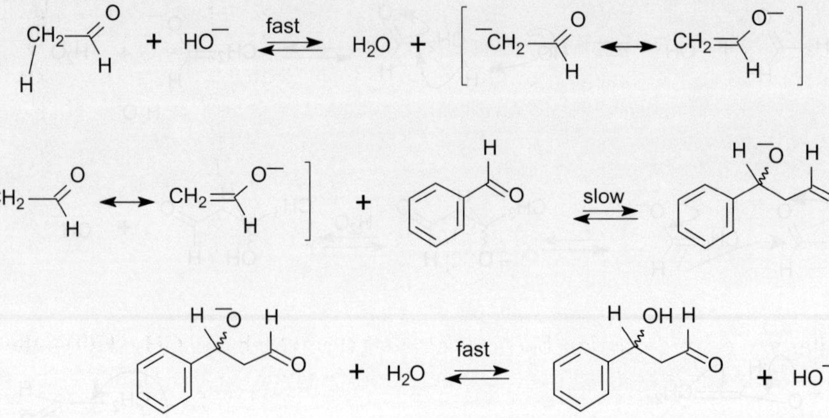

Scheme 9.43. Steps involved in the mixed aldol condensation between benzenecarboxaldehyde (benzaldehyde, C_6H_5CHO) and ethanal (acetaldehyde, CH_3CHO) to produce racemic 3-hydroxy-3-phenylpropanal. Only ethanal (acetaldehyde, CH_3CHO) possesses protons α- to the carbonyl.

β to the carbonyl is common and will be discussed in the next section of this chapter (see Figure 9.6). It is important to note that the oxygen in the water (H_2O) that is lost when dehydration occurs is originally from one of the two reactants. When water (H_2O) is lost, the reaction is no longer as readily reversible.

Fourth, although generalized as the "aldol" reaction, the product need not be aldol (3-hydroxybutanal). That is, two different carbonyl-bearing compounds may be brought into reaction, the reaction will be called a "mixed aldol" condensation, and a variety of different products will be obtained (Problem 9.6).

Problem 9.6. Write the products derived from the mixed aldol condensation between two equivalents of 2-butanone under acidic conditions. Assume that enol formation can occur with equal facility from *either* α-position and that dehydration does not occur. *Note:* Generally, with nonsymmetrically substituted carbonyl groups, the most hightly substituted enol would be expected to predominate. Why?

Thus, for successful mixed aldol condensations, it is common to attempt to utilize at least one component that lacks α-hydrogens. For example the base-catalyzed mixed aldol condensation between benzenecarboxaldehyde (benzaldehyde, C_6H_5CHO)—the component lacking α-hydrogens—and ethanal (acetaldehyde, CH_3CHO)—the component bearing α-hydrogens—can yield only one aromatic ring containing (a β-hydroxyaldehyde) product (Scheme 9.43).

Problem 9.7. As shown in Scheme 9.43, the anion generated by proton abstraction from the carbon α to the carbonyl in ethanal (acetaldehyde, CH_3CHO) is resonance stabilized. The picture in the scheme encourages the argument that there is negative charge on the α-carbon *and* on the oxygen. To the extent that the charge is "shared" and since oxgyen is more electronegative than carbon (Chapter 1, **Part II**; C = 2.5; O = 3.5), it might be argued that there should be "more" negative charge on oxygen than on carbon and that the oxygen, not the carbon, of the anion should attack the carbon of the carbonyl of benzenecarboxaldehyde (benzaldehyde), resulting in the formation of the hemiacetal anion shown below. Protonation of the latter would then produce the corresponding hemiacetal, *which is not detected*. How do you account for its absence?

Similarly, the condensation between ethanal (acetaldehyde, CH_3CHO)—the component bearing α-hydrogens—and methanal (formaldehyde, H_2CO) proceeds as expected (Scheme 9.44) and, in the presence of calcium hydroxide [$Ca(OH)_2$] as the base catalyst, the process terminates with a Cannizzaro reaction (Scheme 9.4), the tetraol, 2,2-dihydroxymethyl-1,3-propanediol (pentaerythritol) and methanoic acid (formic acid, HCO_2H) result.

Because selective formation of carbon–carbon bonds producing functionalized species, for example, those containing β-hydroxycarbonyl units, subsequently capable of transformation into many useful compounds, is important, a significant effort has been expended to control the mixed aldol reaction between aldehydes and/or ketones.

This effort has several components. First, the problem of more than one kind of α-proton (either in different reactants or on opposite sides of the same carbonyl in one reactant) can be controlled by preforming an enol, enolate anion, or enol derivative of one of the carbonyl components before the other is added.

The problem of stereoselection, that is, to which face of the carbonyl component the preformed enol, enolate anion, or enol derivative will add, can then be analyzed as a steric problem.

An analysis of the first of these factors is begun with Equation 9.46.

$$(9.46)$$

The enolate anion on the least substituted carbon is the least stable (more highly substituted being the most stable) and the ratio of (E)-enolate to (Z)-enolate is a function of their method(s) of preparation, the temperature at which their preparation is effected, and the length of time between preparation and use. Thus, it appears that the enolate anions are related as "kinetic" and "thermodynamic" products. It is common to find that the (Z)-enolate is the kinetic product.

The reaction of the lithium salt of the preformed (Z)-enolate anion with, for example, an aldehyde, is thought to take place through a *cyclic transition state* where the metal cation is coordinated to *both* the anion and the oxygen of the carbonyl

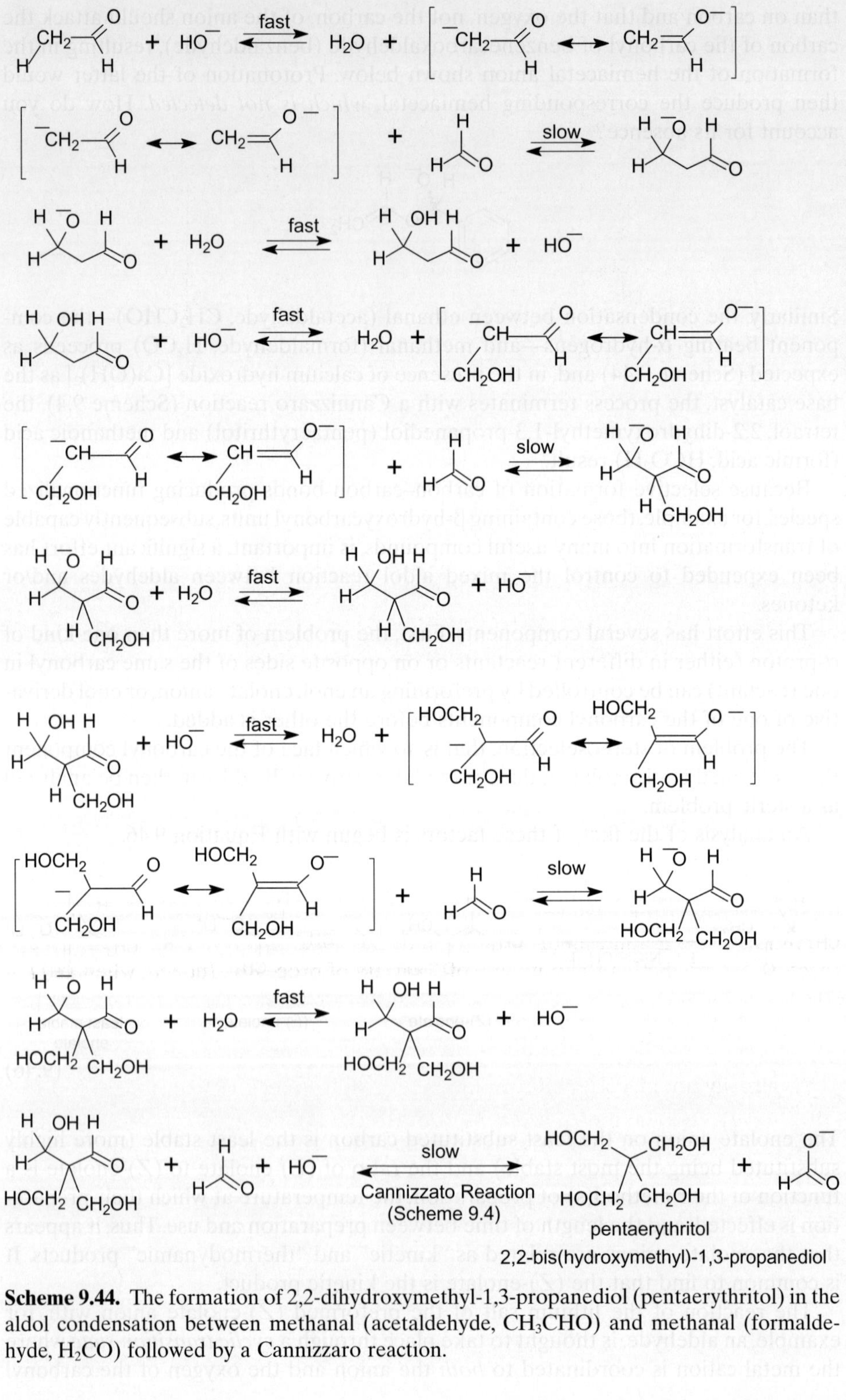

Scheme 9.44. The formation of 2,2-dihydroxymethyl-1,3-propanediol (pentaerythritol) in the aldol condensation between methanal (acetaldehyde, CH_3CHO) and methanal (formaldehyde, H_2CO) followed by a Cannizzaro reaction.

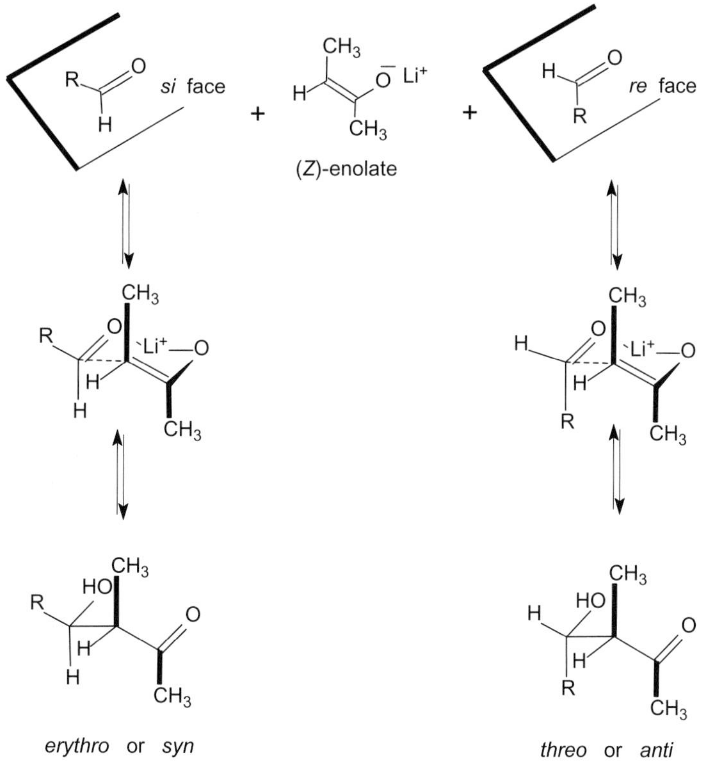

Scheme 9.45. A correlation of enolate geometry and relative configuration of the aldol adduct using the Zimmerman–Traxler transition state. The "R" substituent of the aldehyde would presumably prefer an equatorial position, thus encouraging attack at the *si* face of the aldehyde and producing the *erythro* or *syn* β-hydroxy ketone product on the left.

group to which the enolate will add. This picture (Scheme 9.45) is referred to as the **Zimmerman–Traxler** model.*

Although the Zimmerman–Traxler model is clearly over simplified it has been successful in predicting the outcome of a variety of processes. Indeed, when an (*E*)-enolate is prepared (as enolates of cyclic systems must be), the *threo* or *anti* product is both predicted and found (Scheme 9.46). However, the diastereoselection exhibited by (*Z*)-enolates is generally superior to that of (*E*)-enolates.

As already noted, the aldol reaction is reversible (Schemes 9.41 and 9.42, *et seq*) but it is possible to adjust the conditions (low temperature, rapid workup) so that the kinetic enolate anion (frequently (*Z*)-) leads to a kinetic product. Alternatively, higher temperatures and longer reaction times lead to the thermodynamic product. As shown in Scheme 9.47, when the enolate anion of 2,2-dimethyl-3-pentanone is

---

*The model was initally formulated when Professor Howard Zimmerman (University of Wisconsin) was at Northwestern University in the latter part of the decade ending in 1960. Marjorie Traxler was an undergraduate student working with him. See Zimmerman, H. E.; Traxler, M. D. *J. Am. Chem. Soc.,* **1956,** *79,* 1920.

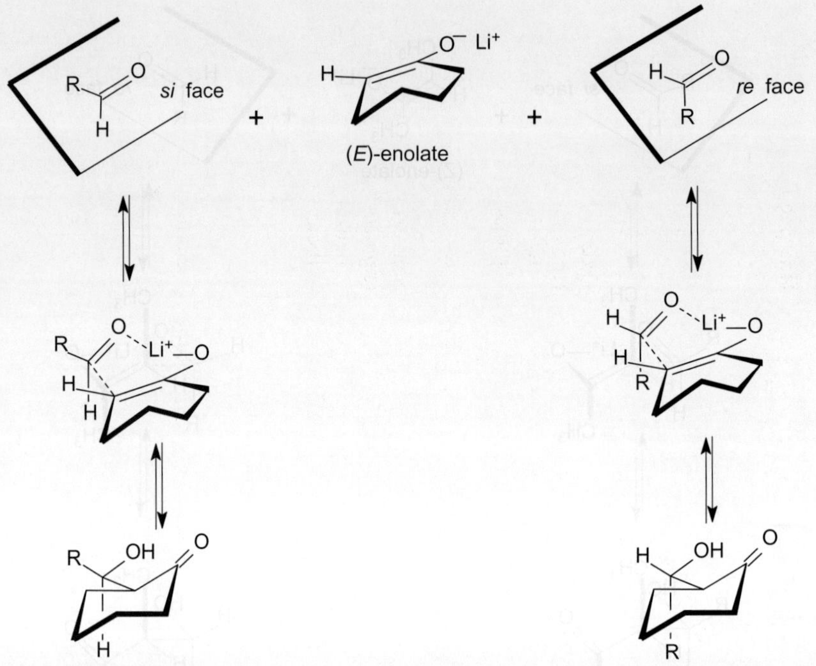

Scheme 9.46. A correlation of enolate geometry and relative configuration of the aldol adduct using the Zimmerman–Traxler transition state with an (*E*)-enolate. The crowded area around the ring is suggested to force the"R" substituent of the aldehyde to assume a pseudoaxial position, thus encouraging attack at the *re* face of the aldehyde and producing the *threo* or *anti* β-hydroxy ketone product on the right.

prepared at low temperatures and allowed to react with benzenecarboxaldehyde (benzaldehyde), *erythro*-5-hydroxy-2,2,4-trimethyl-5-phenyl-3-pentanone is the predominate product. However, under equilibrating conditions (at 25°C), the corresponding *threo*-ketone, which has the bulky groups farthest apart, forms.

In order to improve the stereoselectivity of the aldol process even further, metal salts of enolate anions other than those bearing lithium have been examined. For example, both magnesium and boron enolates have been prepared. Magnesium enolates are very much like lithium enolates in their stereoselectivity, while boron enolates, where there are relatively short metal–oxygen bonds, give improved selectivity. For the boron enolates, the (*Z*)-enolate is generally more stable than its (*E*)-isomer, and *erythro*- or *syn*-products are developed.

The aldol reaction to produce specific β-hydroxyketones was dramatically advanced by the recognition that silyl enol ethers in the presence of Lewis acid catalysts could be used in the carbon–carbon bond-forming process (**Mukaiyama* aldol reaction**) (Scheme 9.48). Thus, although it had been known for some time that silyl enol ethers could be prepared by the reaction between ketones and trialkylsilyl

---

*T. Mukaiyama, Professor, Science University of Tokyo. See Mukaiyama, T.; Inomata, K.; Muraki, M. *J. Am. Chem. Soc.*, 1973, *95*, 967; Mukaiyama, T. *Aldrichimica Acta*, **1996**, *29*, 59.

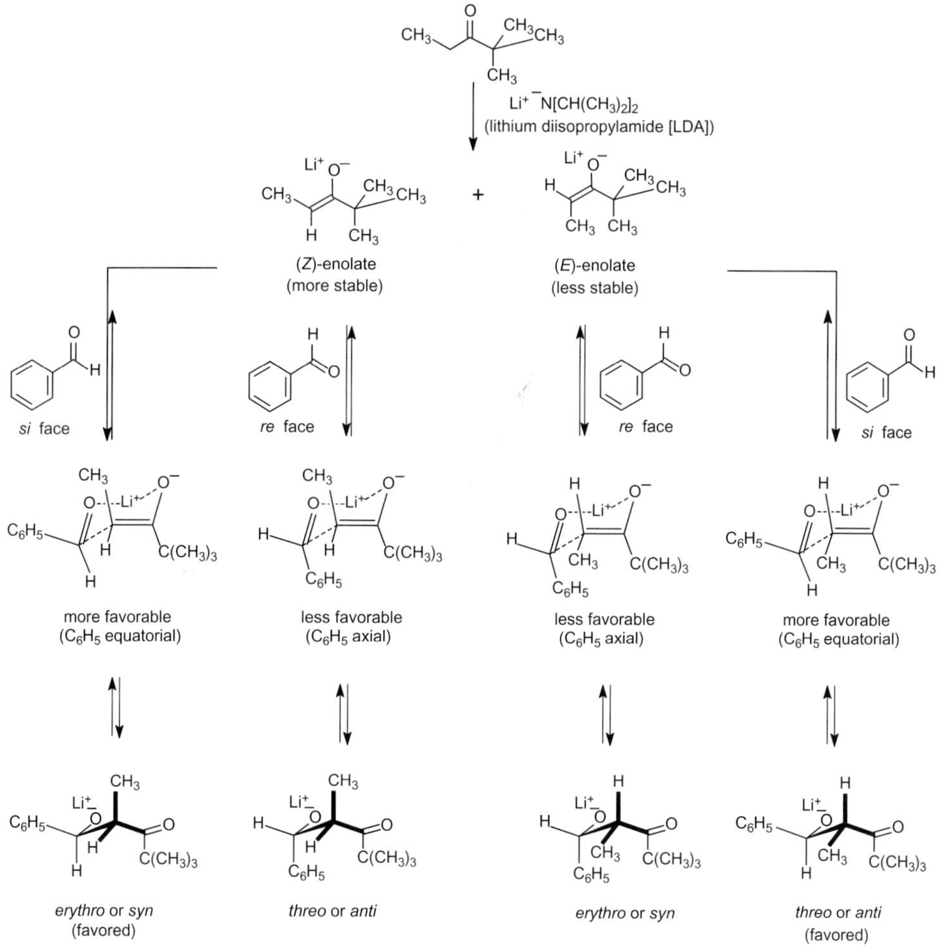

Scheme 9.47. A schematic representation of the formation of the enolate anion of 2,2-dimethyl-3-pentanone and its reaction at low temperatures with benzenecarboxaldehyde (benzaldehyde) to form *erythro*-5-hydroxy-2,2,4-trimethyl-5-phenyl-3-pentanone. Under equilibrating conditions (at 25°C), the corresponding *threo*-ketone, which has the bulky groups farthest apart and thus is most stable, forms.

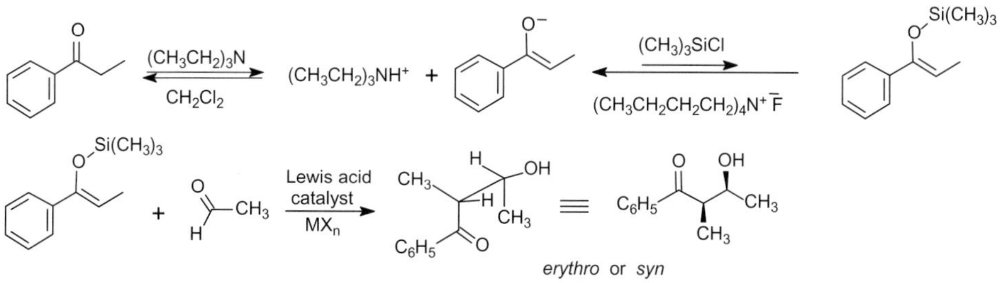

Scheme 9.48. The use of a (Z)-silyl enol ether derivative of ethyl phenyl ketone in the aldol condensation reaction with ethanal (acetaldehyde, CH_3CHO) in the presence of a Lewis acid (MX_n). The **Mukaiyama aldol reaction**.

chlorides in the presence of base and that they could be isolated and fully characterized, their use in the aldol reaction (except for fluoride-catalyzed reversion to
enolate anions) failed because they are relatively poor nucleophiles. The realization
that Lewis acid catalysis would promote the aldol process came later and, most
recently, chiral Lewis acids have been employed to induce stereodifferentiation with
the induction of chirality. Similar effects can be obtained by introducing chirality
into the enol and/or into the aldehyde with which it is to react. Although the details
are beyond the scope of this text, the complications can be sorted out and, when
the selectivity for the reaction is complementary, specific diastereomers can be
preferentially formed (Figure 9.9).

Several examples can serve to illustrate the use of chiral auxiliaries. First,
the chiral, heterocyclic, amino-acid-derived lactam, (**S**)-4-(phenylmethyl)-
2-oxazolidinone,* is capable of conversion into an anion on treatment with
butyllithium [$CH_3(CH_2)_2CH_2Li$]. Acylation at nitrogen with propanoyl chloride
produces chiral propionamide.

As shown in Scheme 9.49 (after Gage, J.R.; Evans, D.A. *Org. Synth. Coll.*, **1993**,
VIII, 339), the boron enolate of the chiral *N*-propionyloxazolidinone is generated
by treatment of the amide with di-*n*-butylboron triflate (i.e., di-*n*-butylboron trifluorosulfonate {[$CH_3(CH_2)_2CH_2]_2BOSO_2CF_3$}) and triethylamine [$(CH_3CH_2)_3N$]. Then,
addition of benzaldehyde (benzenecarboxaldehyde, C_6H_5CHO) results in directed
aldol-type condensation to the oxazolidinone-protected (2*S*, 3*S*)-3-hydroxy-3-
phenyl-2-methylpropanoic acid, and hydrolysis regenerates the chiral oxazolidinone
and liberates the carboxylic acid.

A second example is presented in Scheme 9.50, where the dihexafluoroantimonate [2 $(SbF_6)^{-2}$] complex with (*S,S*)-diphenyl[bis(oxazolinyl)pyridine]Cu(II)
is seen to coordinate with benzyloxyacetaldehyde (2-benzyloxyethanal,

---

*The (**S**)-4-(phenylmethyl)-2-oxazolidinone is readily derived (see Gage, J. R.; Evans, D. A. *Org. Synth.
Coll.*, **1993**, *VIII*, 538) from the amino acid (Chapter 12) (**S**)-phenylalanine. As shown below,
the carboxylic acid group (this Chapter) of (**S**)-phenylalanine can be reduced to the corresponding
alcohol by a borane derivative [$BH_3{\cdot}S(CH_3)_2$] without causing racemization at the asymmetrically
substituted carbon or reduction of the aromatic ring. Then, the resulting amino alcohol can be
converted to an oxazolidinone by reaction with diethyl carbonate [$(CH_3CH_2O)_2C{=}O$]. "Oxazolidinone"
is the name given to a cyclic five-membered compound containing both oxygen and nitrogen as
"hetero" (in contrast to "carbo") members of the ring (it is a *heterocyclic* compound). Further, the
ring must contain a carbonyl group (it is an "-*one*"), and here the carbonyl is connected, on one side,
to the oxygen (an "ester" linkage—"*oxa*") and on the other, to the nitrogen (an "amide"
linkage—"*aza*").

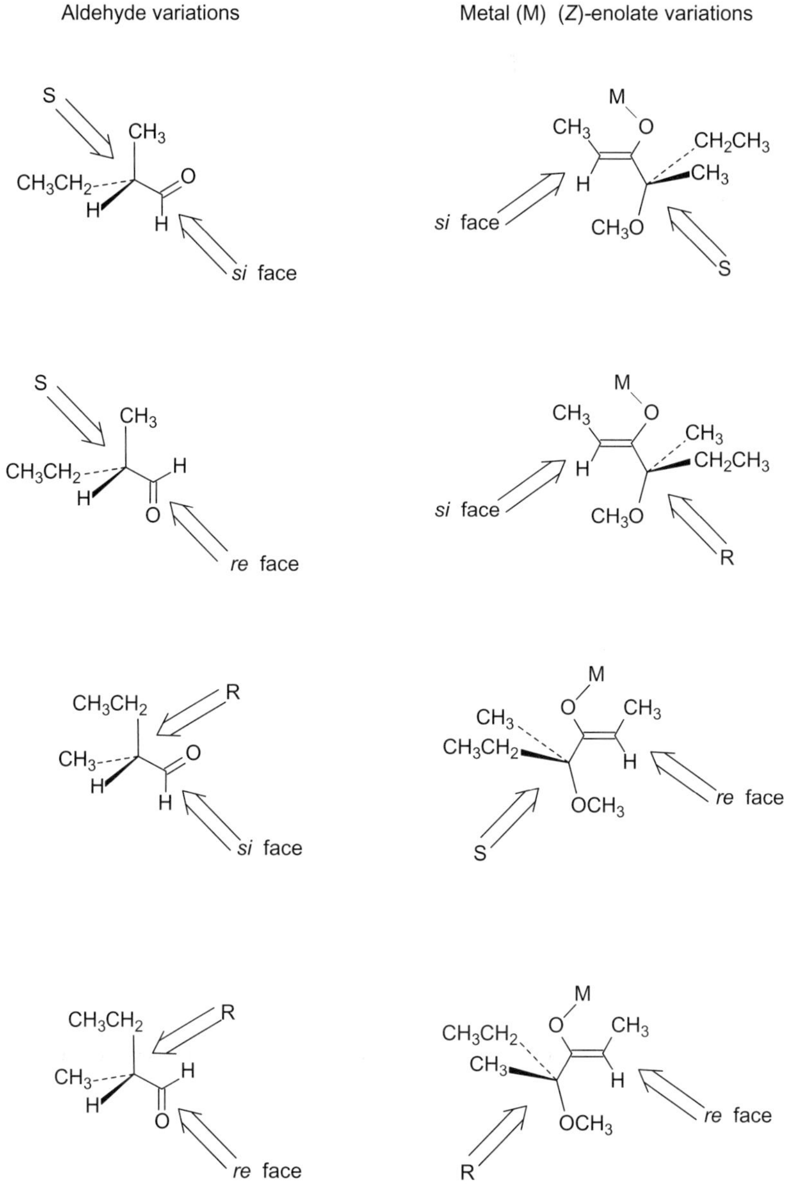

Figure 9.9. The introduction of chirality into enolate anions and into an aldehyde with which it is to react. Fitting together the appropriate anion with the appropriate face of the aldehyde (which also bears a chiral substituent) produces one product preferentially.

$C_6H_5CH_2OCH_2CHO$) in such a way as to allow only one face of the aldehyde to be exposed to an incoming nucleophile (ethyl trimethylsilylketene acetal), and which thus proceeds to react in 99% yield with 98% enantiomeric excess.

Finally, it will be recalled (Chapter 6) that when a second double bond lies in conjugation with the first, reactions across the entire unsaturated system can occur.

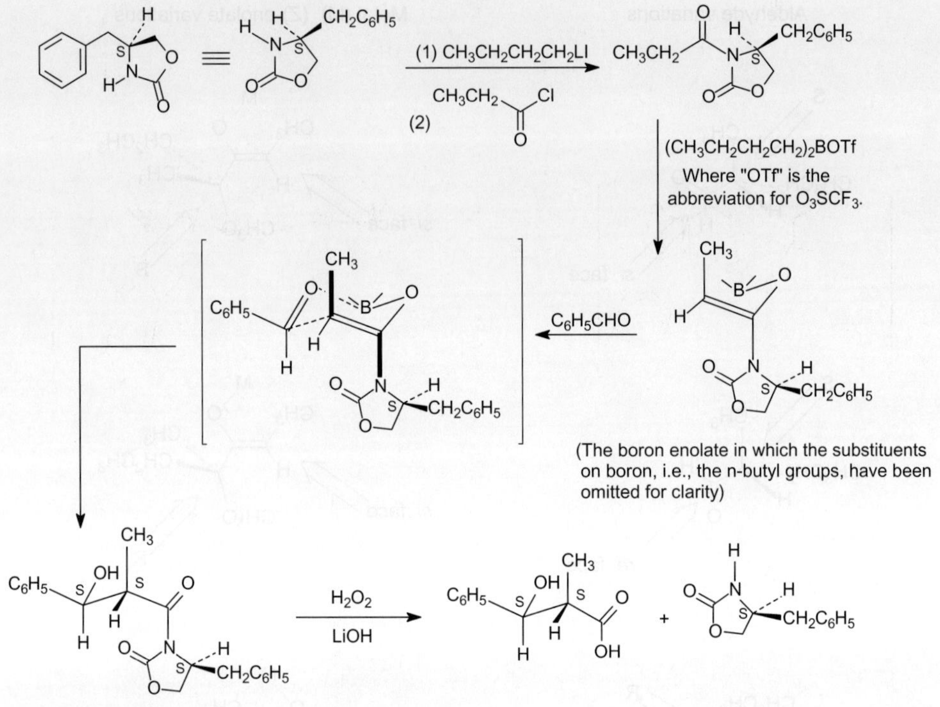

Scheme 9.49. The use of a chiral oxazolidinone and boron enolate to effect a diasteroselective aldol condensation (after Gage, J. R.; Evans, D. A. *Org. Synth. Coll.*, **1993**, *VIII*, 339).

Thus, the *vinylogous aldol* is no exception, and, as shown in Scheme 9.51, use of the *same* oxazoline-derived catalyst and the same benzyloxyacetaldehyde (2-benzyloxyethanal, $C_6H_5CH_2OCH_2CHO$) on 1-methoxy-1,3-bis(trimethylsilyloxy)-1,3-butadiene results in a 97% yield (of a 15:1 mixture of *anti:syn*) of vinylogous aldol products (after hydrolysis).

Many α,β-unsaturated carbonyl compounds (aldehydes and ketones as well as esters and amides [this chapter]) undergo aldol reactions on the α-carbon of the unsaturated partner with the carbonyl of a second partner. This interesting reaction (the Baylis–Hillman* reaction) depends upon the *temporary addition* of a hindered base (1,4-diazabicyclo[2.2.2]octane, DABCO, is commonly used) to the β-carbon of the α,β-unsaturated system rather than the *deprotonation* of the α-carbon. The enolate anion, α to the carbonyl of what was the α,β-unsaturated system, then adds to the other reactant and subsequent elimination provides the condensation product and the base is eliminated. The process is shown in Scheme 9.52 for the reaction between ethanal (acetaldehyde) and the α,β-unsaturated ester ethyl propenoate (ethyl acrylate).

*The Baylis–Hillman reaction was apparently discovered by those after whom it is named (A. Baylis and M. Hillman) while they were employed at the Celanese Corporation in New York in 1972 (Baylis, A. B.; Hillman, M. E. D. U.S. Patent 1973, 3,743,669).

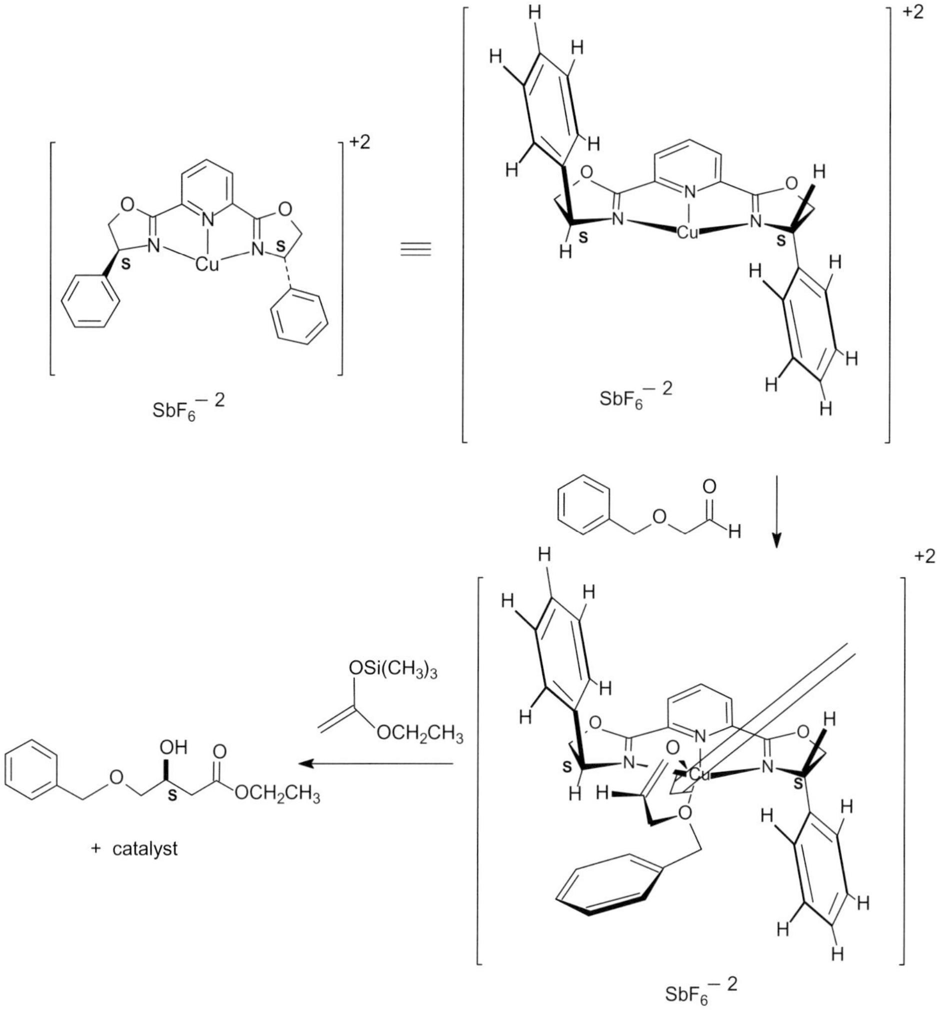

Scheme 9.50. A representation of the reaction (in the direction shown by the arrow) between dihexafluoroantimonate [2 (SbF$_6$)$^{-2}$] (*S,S*)-diphenyl[bis(oxazolinyl)pyridine]Cu(II) with benzyloxyacetaldehyde (2-benzyloxyethanal [C$_6$H$_5$CH$_2$OCH$_2$CHO]) (after Evans, D. A.; Kozlowski, M. C.; Murry, J. A.; Burgey, C. S.; Campos, K. R.; Connell, B. T.; Staples, R. J. *J. Am. Chem. Soc.*, **1999**, *121*, 669).

4. The Darzens Glycidic Ester Condensation.* When an aldehyde (or ketone) is allowed to react with the carbanion formed by proton abstraction from a halogen-bearing carbon α to the carbonyl of an ester, RCO$_2$R′ (this Chapter), an oxirane (epoxide) results. These compounds, containing an oxirane (epoxide) α,β to the carbon of the carbonyl of the ester, are known as "glycidic esters." An accepted

---

*The reaction was explored initially by Darzens (Darzens, G. *Compt. Rend.*, **1906**, *214*, and references to his early work in that paper) but has proved to be very useful. A lengthy review has appeared (Ballester, M., *Chem. Rev.*, **1955**, *55*, 283.)

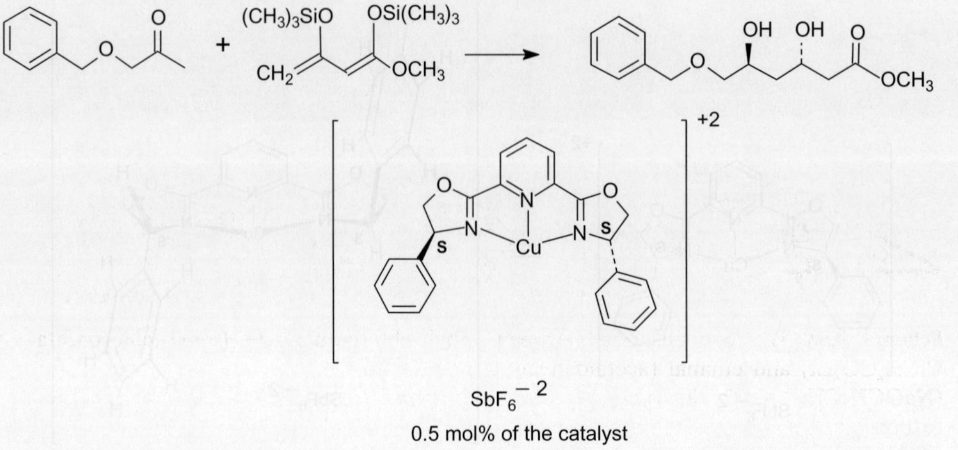

0.5 mol% of the catalyst

Scheme 9.51. The result of the copper–oxazoline-derived catalyst to promote the reaction between 2-benzyloxyacetaldehyde (2-benzyloxyethanal, $C_6H_5CH_2OCH_2CHO$) and 1-methoxy-1,3-bis(trimethylsilyloxy)-1,3-butadiene. The reaction results in a 97% yield of methyl (3S,5S)-6-benzyloxy-3,5-dihydroxyhexanoate, the vinylogous aldol product (after hydrolysis). See Evans, D.; Woerpel, K. A.; Scott, M. I. *Angew. Chem. Int. Ed.*, **1992**, *31*, 430.

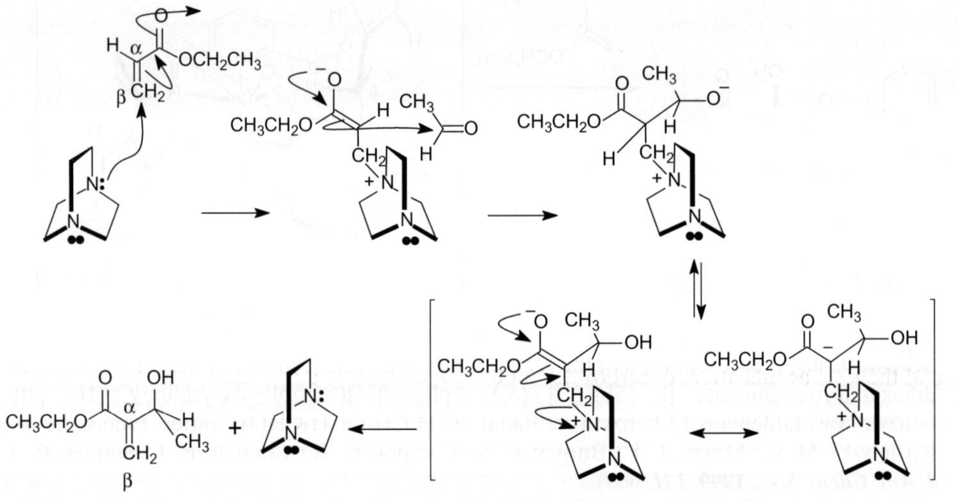

Scheme 9.52. An example of the Baylis–Hillman reaction—a variation of the aldol condensation—in which 1,4-diazabicyclo[2.2.2]octane (DABCO) catalyzes the reaction between ethanal (acetaldehyde, CH_3CHO) and ethyl propenoate (ethyl acrylate, $CH_2{=}CHCO_2CH_2CH_3$).

pathway for the formation of a typical glycidic ester from the haloester ethyl chloroethanoate (ethyl chloroacetate) and ethanal (acetaldehyde) is shown in Scheme 9.53.

When the glycidic ester undergoes hydrolysis (hydrolysis of esters is dicussed in this Chapter) to the corresponding carboxylate anion and conversion to the corresponding acid is attempted by acidification, carbon dioxide (CO_2) is lost, and an

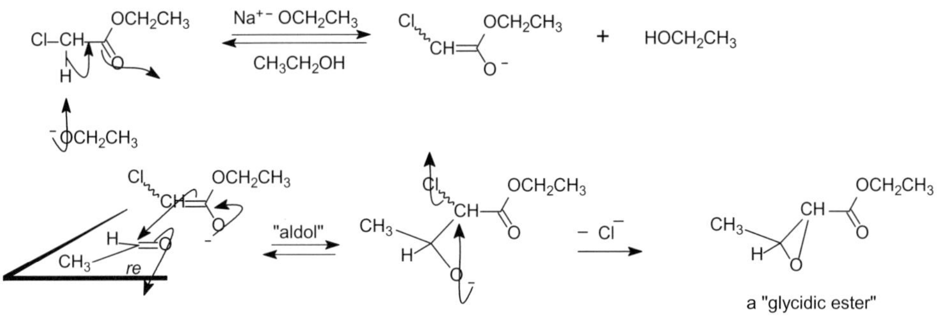

a "glycidic ester"

Scheme 9.53. The condensation between ethyl chloroethanoate (ethyl chloroacetate, $ClCH_2CO_2Et$) and ethanal (acetaldehyde, CH_3CHO) in the presence of sodium ethoxide ($NaOCH_2CH_3$) to produce the glycidic ester isomers ethyl (E)- and (Z)-3-methyl-2-carboxyoxirane.

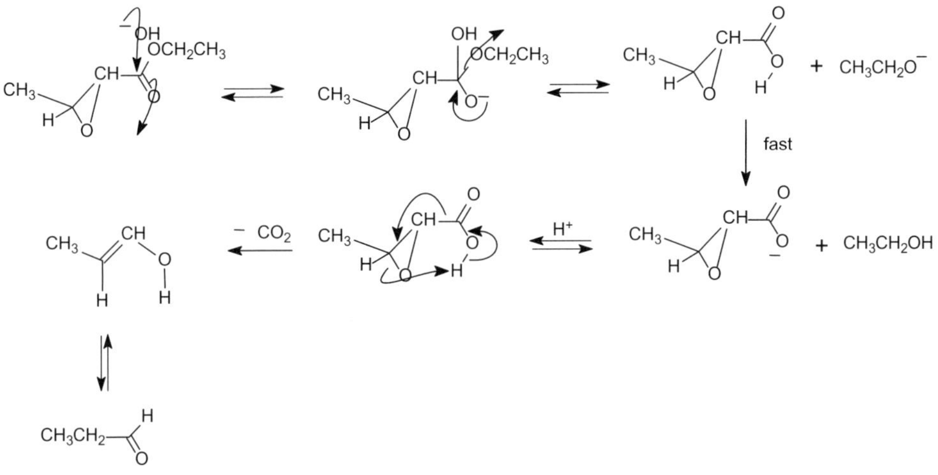

Scheme 9.54. A representation of a pathway by which a glycidic ester undergoes hydrolysis and then, on acidification, decarboxylation to produce a new aldehyde. The backbone of the product aldehyde is one carbon atom longer than the original aldehyde undergoing the Darzens glycidic ester condensation.

aldehyde, one carbon longer than the aldehyde with which the reaction was begun, is formed (Scheme 9.54). Thus, this method serves as a pathway for extending the carbon–carbon chain of aldehydes.

Problem 9.8. What is the product obtained if 2-butanone is allowed to react with ethyl chloroethanoate anion and the resulting glycidic ester is hydrolyzed and acidified?

5. Epoxide Syntheses. Both methylsulfinylmethane [dimethyl sulfoxide, DMSO, $(CH_3)_2SO$] and dimethyl sulfide [$(CH_3)_2S$] react with methyl iodide (CH_3I). The latter reacts at sulfur to produce trimethylsulfonium iodide $[(CH_3)_3S^+\ I^-]$, which, on treatment with base (e.g., the sodium salt of the dimethylsulfoxide anion $[Na^+\ ^-CH_2S(O)CH_3]$ in methylsulfinylmethane (dimethyl sulfoxide, DMSO, $(CH_3)_2SO)]$, produces dimethylsulfonium methylide $[(CH_3)_2S^+CH_2^-]$(Equation 9.47).

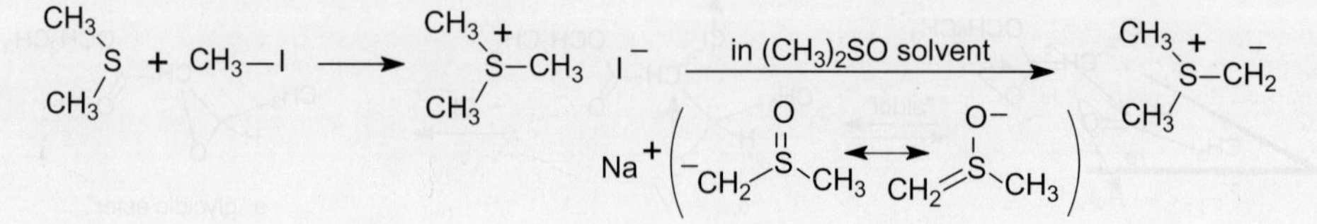

$$(9.47)$$

Methylsulfinylmethane [dimethyl sulfoxide, DMSO, $(CH_3)_2SO]$ reacts with methyl iodide (CH_3I) to produce the trimethylsulfoxonium iodide $[(CH_3)_3SO^+\ I^-]$.* The sulfoxonium iodide can be induced to eliminate hydrogen iodide (HI) by treatment with sodium hydride (NaH) in methylsulfinylmethane [dimethyl sulfoxide, DMSO, $(CH_3)_2SO]$ to generate the corresponding dimethylsulfoxonium methylide $[(CH_3)_2S^+(O)CH_2^-]$ (Equation 9.48).

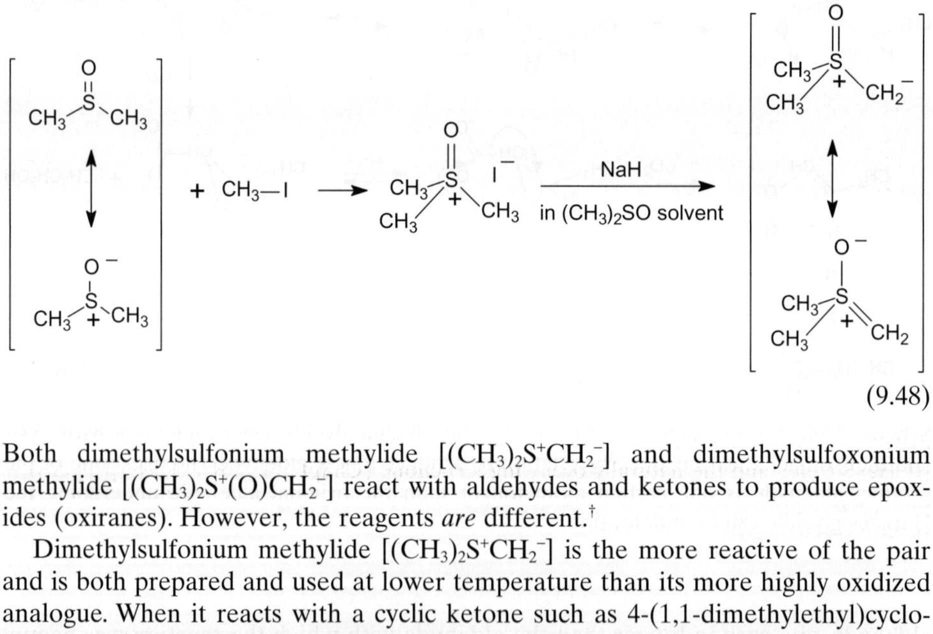

$$(9.48)$$

Both dimethylsulfonium methylide $[(CH_3)_2S^+CH_2^-]$ and dimethylsulfoxonium methylide $[(CH_3)_2S^+(O)CH_2^-]$ react with aldehydes and ketones to produce epoxides (oxiranes). However, the reagents *are* different.[†]

Dimethylsulfonium methylide $[(CH_3)_2S^+CH_2^-]$ is the more reactive of the pair and is both prepared and used at lower temperature than its more highly oxidized analogue. When it reacts with a cyclic ketone such as 4-(1,1-dimethylethyl)cyclohexanone (4-*t*-butylcyclohexanone), addition occurs primarily (*ca.* 80:20) from the axial direction, and subsequent elimination of dimethyl sulfide $[(CH_3)_2S]$ leads to the corresponding epoxide (oxirane) (Scheme 9.55).

---

*The reaction at sulfur generally occurs in preference to that at oxygen, although in the presence of silver oxide the latter can be effected.
†The differences in pathways and other interesting aspects have been explored in detail. See Aggarwal, V. K.; Winn, C. L. *Acc. Chem. Res.*, **2004**, *37*, 611; Corey, E. J.; Chaykovsky, M. *J. Am. Chem. Soc.*, **1965**, *87*, 1353, and earlier papers; Aggarwal, V. K.; Alonso, E.; Bae, I.; Hynd, G.; Lydon, K. M.; Palmer, M. J.; Patel, M.; Porcelloni, M.; Richardson, J.; Stenson, R. A.; Studley, J. R.; Vasse, J.-L.; Winn, C. L. *J. Am. Chem. Soc.*, **2003**, *125*, 10926.

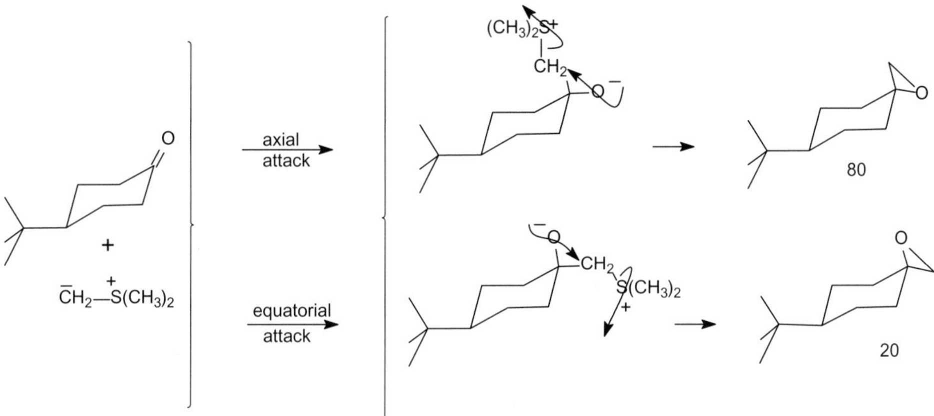

Scheme 9.55. A representation of the reaction of 4-(1,1-dimethylethyl)cyclohexanone (4-*t*-butylcyclohexanone), with dimethylsulfonium methylide [(CH$_3$)$_2$S$^+$CH$_2^-$]. Addition occurs primarily (ca. 80:20) from the axial direction and elimination of dimethyl sulfide [(CH$_3$)$_2$S] leads to the corresponding epoxide (oxirane).

Scheme 9.56. A representation of the reaction between dimethylsulfonium methylide [(CH$_3$)$_2$S$^+$CH$_2^-$] and the naturally occurring terpenone (Chapter 11), R-(−)-5-isopropenyl-2-methylcyclohex-2-en-1-one [R-(−)-carvone]. The reaction occurs to produce the oxirane (epoxide).

Second, in the event that a carbon–carbon double bond (C=C) is conjugated (i.e., α,β) with the carbonyl group (C=O), as in the case of the naturally occurring terpenone (Chapter 11), R-(−)-5-isopropenyl-2-methylcyclohex-2-en-1-one [R-(−)-carvone], the reaction occurs to produce the oxirane (epoxide) (Scheme 9.56).

In contrast, dimethylsulfoxonium methylide [(CH$_3$)$_2$S$^+$(O)CH$_2^-$] reacts with α,β-unsaturated ketones such as R-(−)-5-isopropenyl-2-methylcyclohex-2-en-1-one [R-(−)-carvone] to produce the cyclopropane product of conjugate addition (Scheme 9.57), 4-isopropenyl-1-methylbicyclo[4.1.0]heptan-2-one, and addition to 4-*t*-butyl-cyclohexanone produces the product resulting from equatorial attack (exclusively) (Scheme 9.58).

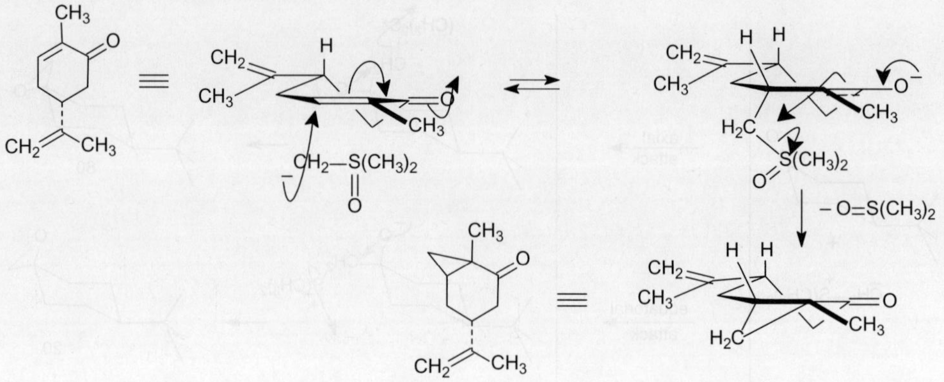

Scheme 9.57. A representation of the reaction between dimethylsulfoxonium methylide [(CH$_3$)$_2$S$^+$(O)CH$_2^-$] and the naturally occurring terpenone (Chapter 11), R-(−)-5-isopropenyl-2-methylcyclohex-2-en-1-one [R-(−)-carvone] to produce the corresponding cyclopropane. (See Corey, E. J.; Chaykovsky, M. *J. Am. Chem. Soc.*, **1962**, *84*, 3782.)

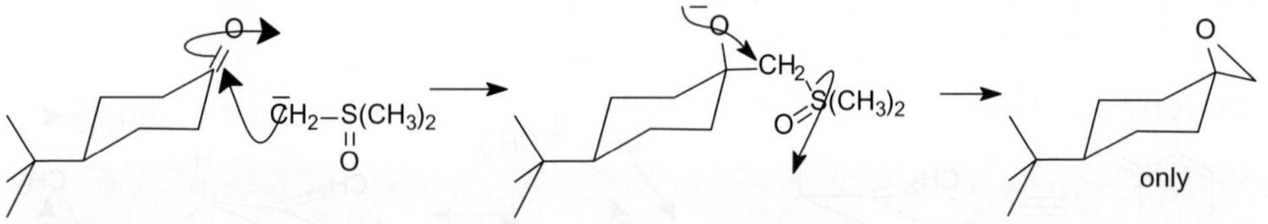

Scheme 9.58. A representation of the reaction of 4-(1,1-dimethylethyl)cyclohexanone (4-*t*-butylcyclohexanone), with dimethylsulfoxonium methylide [(CH$_3$)$_2$S$^+$(O)CH$_2^-$]. Addition occurs exclusively from the equatorial direction and elimination of dimethyl sulfoxide [(CH$_3$)$_2$SO] leads to the corresponding epoxide (oxirane).

Although aldehydes and ketones can also be converted to epoxides by reaction with diazomethane (CH$_2$N$_2$), it is not uncommon to find that methylene insertion occurs at the same time (often preferentially) as shown in Scheme 9.59.

The ring expansion reaction (methylene insertion) with diazomethane (CH$_2$N$_2$) appears to involve the same intermediate as that formed from the corresponding primary amine on treatment with nitrous acid (HONO) in the **Tiffeneau–Demjanov reaction.***

6. The Benzoin Condensation. Treatment of benzenecarboxaldehyde (benzaldehyde, C$_6$H$_5$CHO) and similar aldehydes with sodium cyanide (Na$^+$ CN$^-$) in ethanol (ethyl alcohol, CH$_3$CH$_2$OH) as solvent at reflux for brief periods results in the formation of a dimer, 2-hydroxy-1,2-diphenylethanone (benzoin) (Equation 9.49). A mechanism has been proposed (Scheme 9.60) that accounts for the following observations:

*Although (apparently) first worked on by Demjanov (Demjanov, N. J. *J. Russ. Phys. Chem Ges.*, **1903**, *35*, 375, the work was elaborated on and explored by Tiffeneau (Tiffeneay, Ml.; Weill, P.; Tchoubar, B. *Compt Rend Séances Acad. Soc.*, **1937**, *205*, 54.

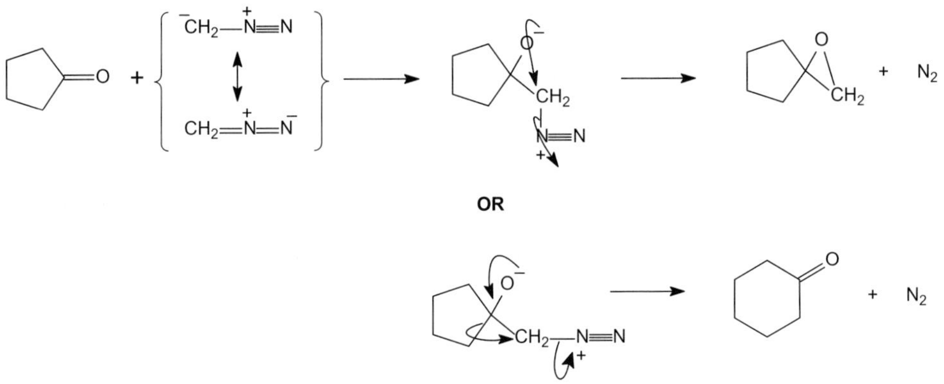

Scheme 9.59. Attack at the carbon of the carbonyl of cyclopentanone by diazomethane [CH$_2$N$_2$] producing oxirane as well as ring-expanded product (cyclohexanone).

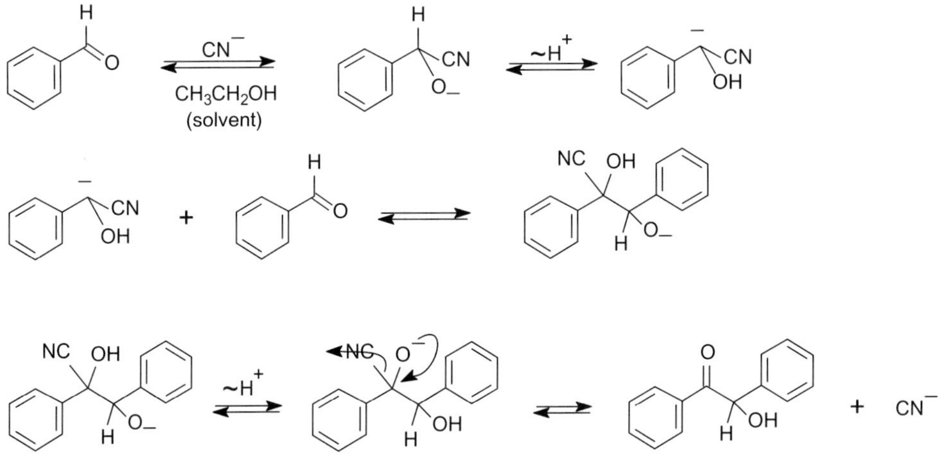

Scheme 9.60. A proposed mechanism for the benzoin condensation which is in concert with the observed rate of formation of product, the effect of electron-donating and electron-withdrawing substituents, and the reversibility of the overall reaction.

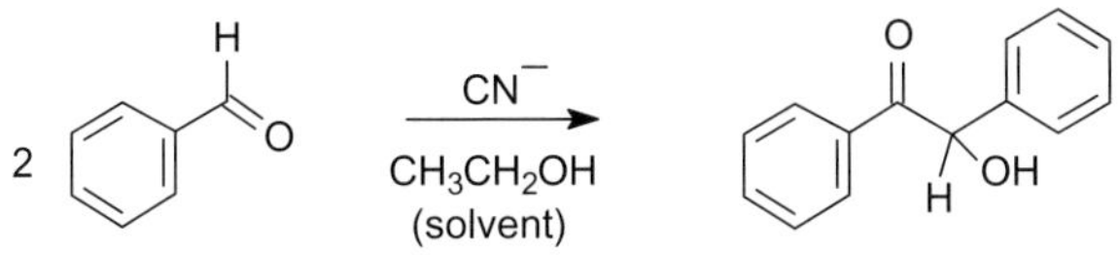

$$\text{(9.49)}$$

1. The rate of product formation is second-order in benzenecarboxyaldehyde (benzaldehyde, C$_6$H$_5$CHO) and first-order in cyanide ion (CN$^-$), requiring that the transition state for the rate-determining step have two equivalents of aldehyde and one of cyanide present, that is,

$$\text{Rate} = +d[\text{product}]/dt = k_{\text{observed}}[C_6H_5CHO]^2[CN^-]$$

2. Both electron-withdrawing (e.g., $-NO_2$) and electron-donating (e.g., $-OCH_3$) substituents on the aromatic ring of a (substituted) benzencarboxaldehyde (benzaldehyde, C_6H_5CHO) **retard** the conversion of (substituted) aldehyde to (substituted) benzoin. It is argued that the electron-donating substituents inhibit the first step in Scheme 9.60 (the attack on the carbonyl group by hydroxide $[HO^-]$) by making the carbonyl group less electron attracting and the electron-withdrawing substituents stabilize the carbanion formed after proton migration from carbon to oxygen subsequent to attack by cyanide (CN^-). A more stable carbanion (lower in energy) will react more slowly with the second equivalent of aldehyde.

3. Finally, product formation is reversible. Thus, treatment of 2-hydroxy-1,2-diphenyl-ethanone (benzoin) with other aromatic aldehydes in the presence of cyanide results in the formation of "mixed" benzoins (Equation 9.50).*

$$(9.50)$$

V. Nucleophilic Addition Reactions with Loss of the Carbonyl Oxygen

a. General Comments. As pointed out in the introduction to Part IV (Nucleophilic Addition Reactions Retaining the Carbonyl Oxygen), base- and acid-catalyzed addition of nucleophiles to the carbon of the carbonyl group (C=O) **both** where the oxygen is retained and where the oxygen is subsequently lost are common. The role of these catalysts in the addition process to the carbonyl (C=O) has been described. However, in those cases where the oxygen originally bonded to the carbon of the carbonyl is **lost**, the acid or base catalyst frequently plays an additional role.

Thus, for example it will be found that, following protonation of the oxygen of the carbonyl by an acid catalyst and subsequent attack by a nucleophile, water is lost and a second bond to the nucleophile formed. Alternatively a second nucleophile can attack, a double bond form, or a ring contract. All of these possibilities produce a rich palette from which to draw the multitude of possibilities presented.

b. Formation of Acetals, Ketals, and Thioketals. The acid-catalyzed reaction of alcohols and thiols with aldehydes and ketones has already been examined both in this Chapter (Part IVb) as well as in Chapter 8. Thus, as shown in Scheme 9.61 (a repetition of what was previously seen as Scheme 8.51), the reaction of methanol (methyl alcohol, CH_3OH) with benzenecarbaldehyde (benzaldehyde, C_6H_5CHO) with acid catalysis produces, first, the hemiacetal and then, following a second proton transfer, loss of water (H_2O) and addition of a second equivalent of methanol (methyl alcohol, CH_3OH) to produce the corresponding acetal.

---

*The benzoin condensation is, by some, called the "Lapworth–Benzoin condensation" as it was originally worked upon by Lapworth. See Lapworth, A. *J. Chem. Soc.*, **1903**, 995.

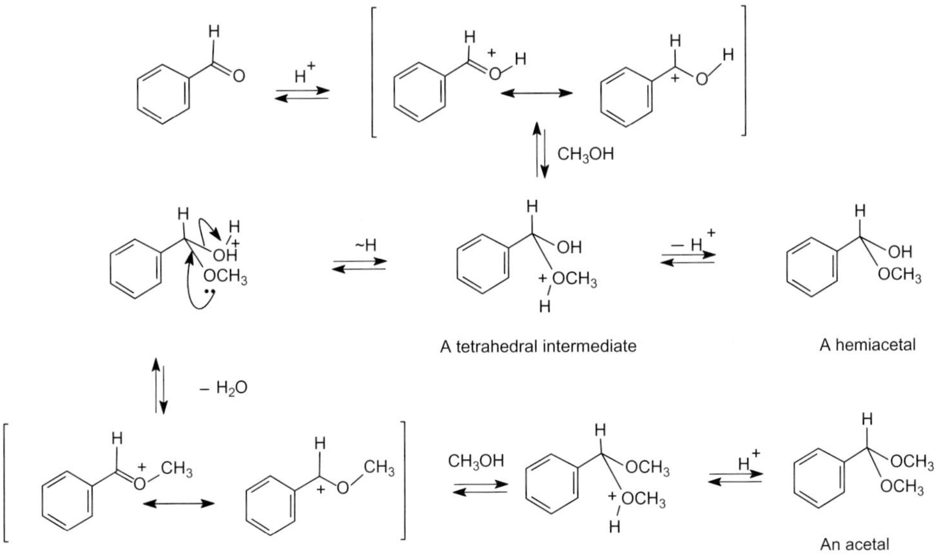

Scheme 9.61. (A repetition of Scheme 8.51) The formation of, first, a hemiacetal and, second, an acetal on acid-catalyzed addition of methanol (methyl alcohol [CH_3OH]) to benzenecarbaldehyde (benzaldehyde [C_6H_5CHO]).

Similarly, as shown in Scheme 9.62 (a repetition of what was previously seen as Scheme 8.52), the reaction of 1,2-dihydroxyethane (ethylene glycol [$HOCH_2CH_2OH$]) with cyclohexanone (again with acid catalysis) produces, first, the hemiacetal and then, following a second proton transfer, loss of water and addition of the second hydroxyl of 1,2-dihydroxyethane (ethylene glycol [$HOCH_2CH_2OH$]) to produce the corresponding cyclic ketal.

Finally, in this vein, as pointed out earlier (Chapter 8), thiols behave much the same as alcohols and thus 1,2-ethanedithiol reacts with cyclohexanone to produce the corresponding dithioketal (via the corresponding thiohemiketal) as shown in Scheme 9.63. Reductive desulfurization of the dithioketal with, for example, Raney nickel, in the presence of hydrogen (H_2), yields the alkane (Chapter 8, Scheme 8.106).

The equilibrium existing between products of addition of H–X nucleophiles to the carbonyl group (C=O) and the reactants from which the products form has been utilized widely in carbohydrate chemistry (Chapter 11). As already pointed out (Scheme 8.53), the cyclic hemiacetal (pyranose) form of glucose is in equilibrium with the acyclic aldehydic form. However, reasonably, the hemiacetal as either the α- or β-glucopyranose can, in an alcohol medium, be converted to an acetal.

For example (Scheme 9.64), on protonation in methanol (methyl alcohol, CH_3OH), the glucopyranose shown undergoes loss of water (H_2O) to generate an oxonium ion that, reacting with the solvent, produces the corresponding ketal. In the absence of further information, it is assumed that attack on the oxonium ion could occur from either face. Incidently, it should be clear (and will be emphasized

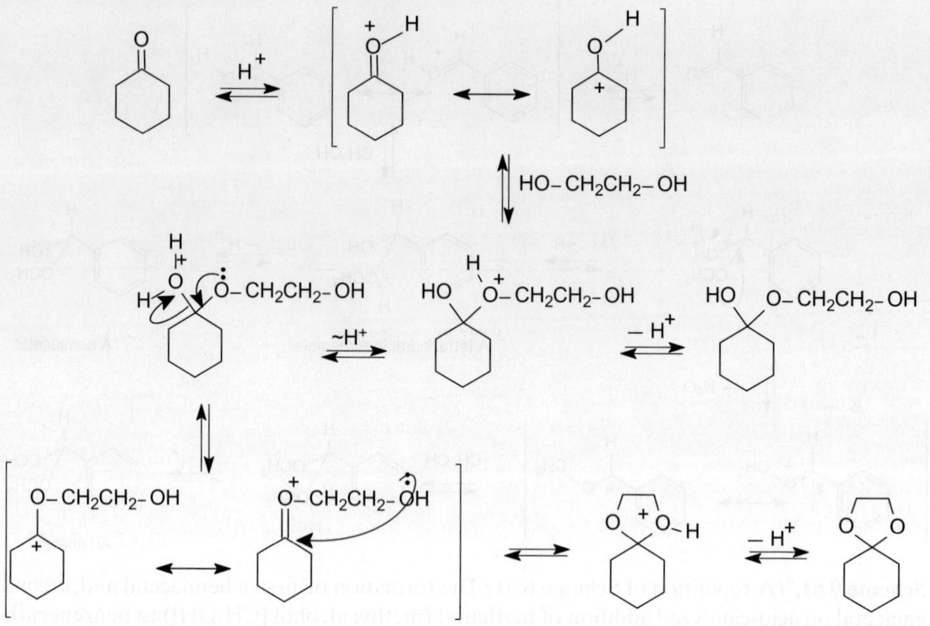

Scheme 9.62. (A repetition of Scheme 8.52) The formation of, first, a hemiketal and, second, a ketal on acid-catalyzed addition of 1,2-dihydroxyethane (ethylene glycol, $HOCH_2CH_2OH$) to cyclohexanone.

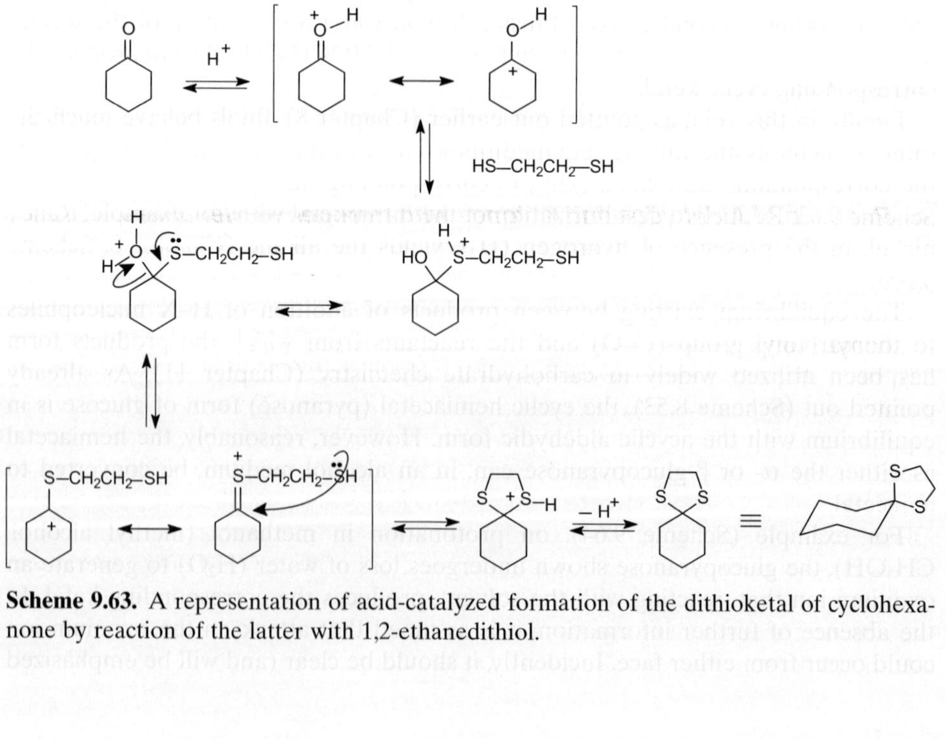

Scheme 9.63. A representation of acid-catalyzed formation of the dithioketal of cyclohexanone by reaction of the latter with 1,2-ethanedithiol.

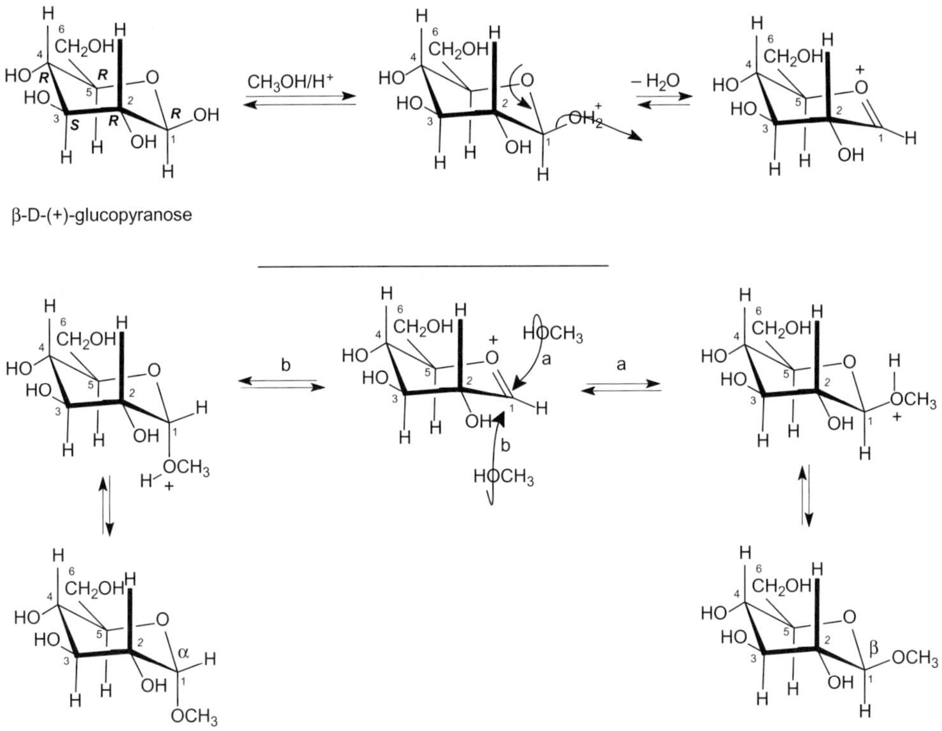

β-D-(+)-glucopyranose

Scheme 9.64. The conversion of β-D-(+)-glucopyranose to an oxonium ion by acid-catalyzed dehydration and subsequent reaction of that ion with methanol (methyl alcohol [CH_3OH]) to produce the corresponding β- and α-methylglycosides.

again in Chapter 11) that in this process one of the five original primary and secondary hydroxyl groups has been replaced by a methoxy and **only** that oxygen originally present as the aldehydic oxygen has been lost.

c. Reaction of Aldehydes and Ketones with Nitrogen Nucleophiles.

A variety of nitrogenous, electron-rich species react reversibly with carbonyl compounds by addition of the nitrogen to the carbon of the carbonyl, producing a tetrahedral intermediate that (almost always) subsequently loses the oxygen of the original carbonyl group (as water, H_2O). Most of these reactions are acid catalyzed and require initial reaction of the catalyst with the oxygen of the carbonyl group (C=O) to increase the positive nature of the carbon of the carbonyl (C=O) at the beginning of the reaction and to convert the leaving group (–OH) from a poor one to a good one (–OH_2) at the end of the reaction as shown in Scheme 9.65. Both Lewis and Brønsted acid catalysts have been used. Since the electron-rich nitrogen nucleophiles (e.g., G–NH_2, where G = alkyl, aryl, H_2N, and HO in Table 9.4) react with acids, it is common to find that the overall rate of product formation is pH dependent (Figure 9.10).

The path from reactants to products (Scheme 9.65) has been studied in some detail for some of the species G-NH_2 and some aldehydes and ketones because

TABLE 9.4. A Representative Group of Nitrogenous Nucleophiles (G–NH₂) and Their Reactions with Aldehydes and Ketones

Group "G" [G–NH₂]	Typical Nuclophilic Reactant	Typical Carbonyl Reactant	Typical Product	Product Class (Name of Typical Product)
1. H	NH_3 (ammonia)	$H_2C{=}O$ (methanal [formaidehyde])	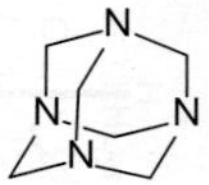	Polymer (hexamethylenetetramine) (1,3,5,8-tetrazatri cyclo-[3,3,1,1$^{3.7}$]decane)
2a. R, Ar (alkyl, aryl) **Primary amine**	CH_3NH_2 (methanamine [methyl amine])	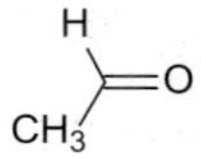 (ethanal [acetaldehyde])	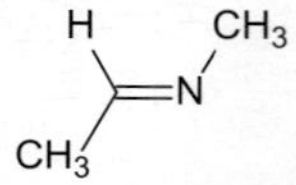 (*Z*) and (*E*) isomers	Imine (the methylimine of ethanal [acetaldehyde])
2b. **Secondary amine**	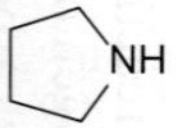 (azacyclopentane [pyrrolidine])	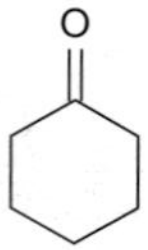 (Cyclohexanone)	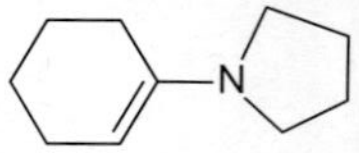	Emamine (the pyrrolidine enamine of cyclohexanone)
3a. H_2N	H_2NNH_2 (hydrazine)	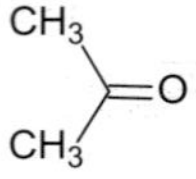 (propanone [acetone])	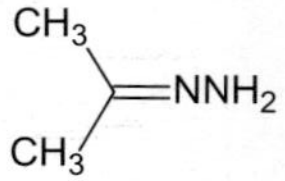	Hydrazone (the hydrazone of propanone [acetone])
				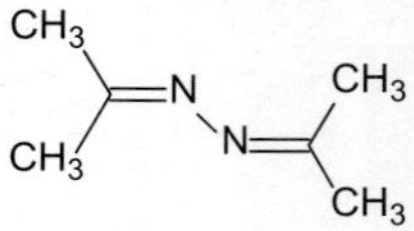 Azine (the azine of propanone [acetone])
3b. RNH (R = alkyl, aryl)	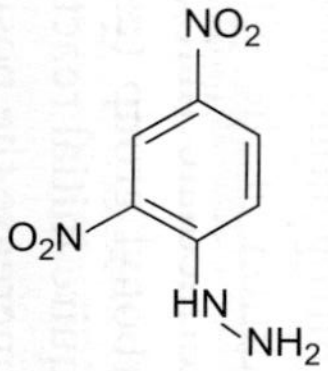 (2,4-dinitrophenylhydrazine)	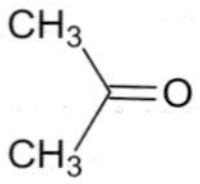 (propanone [acetone])	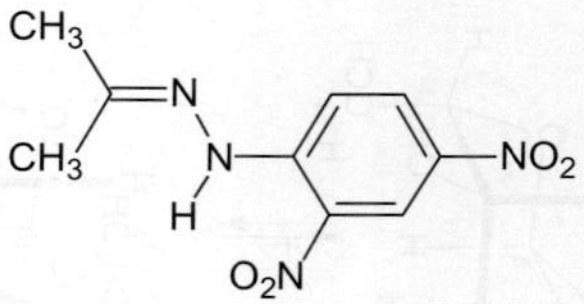	Hydrazone (the 2,4-dinitrophenyl-hydrazone of propanone [acetone])

3c.	RNH R = alkyl, aryl)	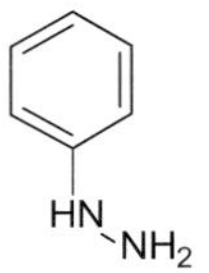 (phenylhydrazine)	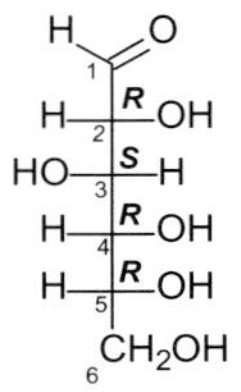 [D-(+)-glucose]	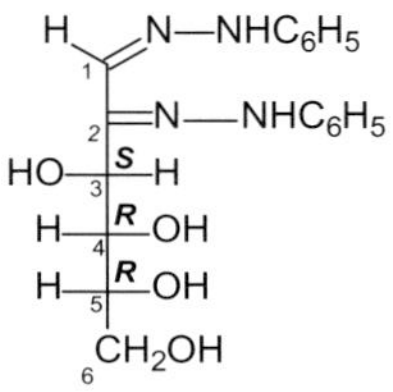	Osazone (the phenylosazone of glucose)
4.	HO	$H_2N–OH$ (hydroxylamine)	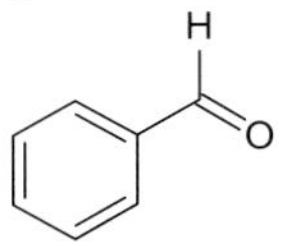 (benzenecarboxaldehyde [benzaldehyde])	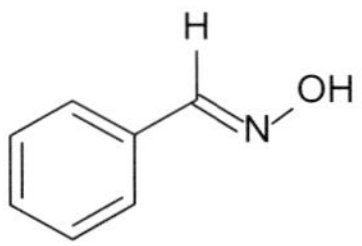 (E)- and (Z)-isomers	Oxime (the oxime of benzenecarboxaldehyde [benzaldehyde])
5.	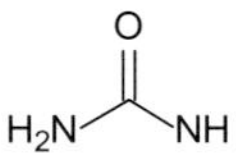	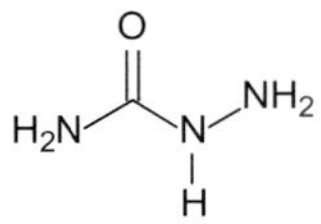 (semicarbazine)	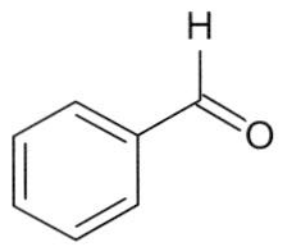 (benzenecarboxaldehyde [benzaldehyde])	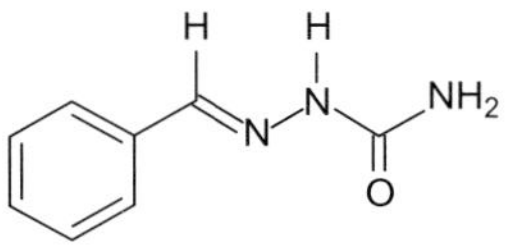	Semicarbazone (*semicarbazone of* benzenecarboxaldehyde [benzaldehyde])
	$(CH_3)_3N^+CH_2CONH$ Cl^-	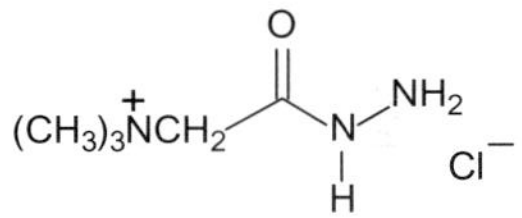 [(carboxymethyl)tri-methylammonium chloride hydrazide] ("Girard's Reagent T")	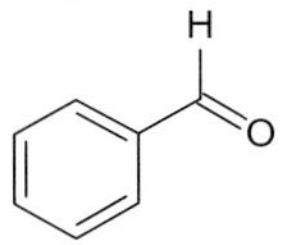 (benzenecarboxaldehyde [benzaldehyde])	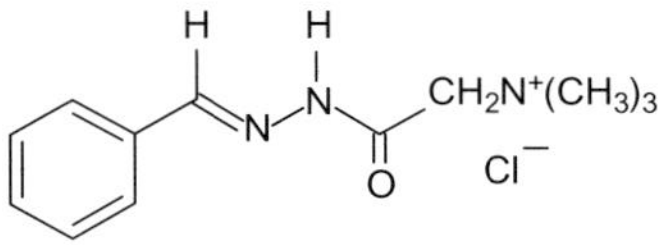	A water-solubtle semicarbazone

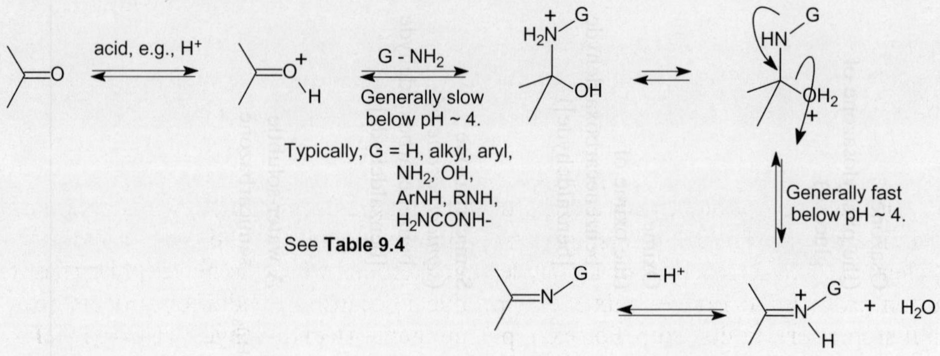

Scheme 9.65. A representation of the reaction between a nitrogenous, electron-rich nucleophile (GNH$_2$) and a carbonyl-containing molecule (such as an aldehyde or ketone) under the influence of acid catalysis. Water is lost and a carbon–nitrogen doubly bonded species results. A representation of some of the various groups ("G") permitted on nitrogen can be found in Table 9.4.

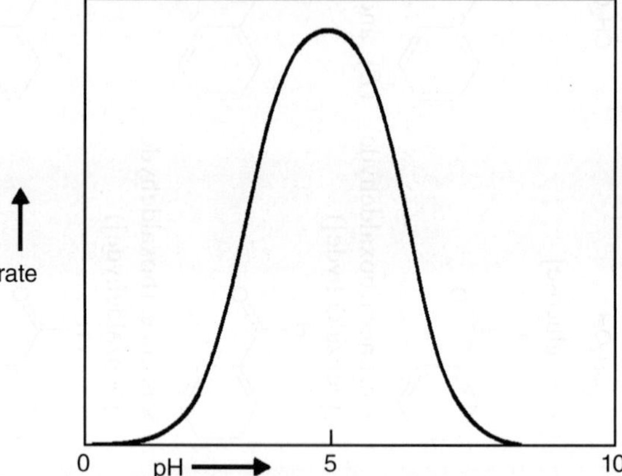

Figure 9.10. A representation of the variation in the rate of product formation with pH in the reaction between an aldehyde or ketone and a basic nucleophillic species G–NH$_2$ (G=H, R, Ar, etc.) to produce the corresponding derivative (Table 9.4).

transamination* (Chapter 12), although mediated by enzymes, can be related to the steps of Scheme 9.65 and Figure 9.10.

Interestingly, however, it appears that the rate-determining step in the process shown in Scheme 9.65 is often a function of the particular nitrogenous nucleophile. Strongly basic nucleophiles such as hydroxylamine (G=OH; H$_2$NOH) and primary

---

*As suggested by the name, the process of transamination effectively removes the amino group of one aminoacid and replaces it with a carbonyl group while affecting the opposite change on its reacting partner, that is,

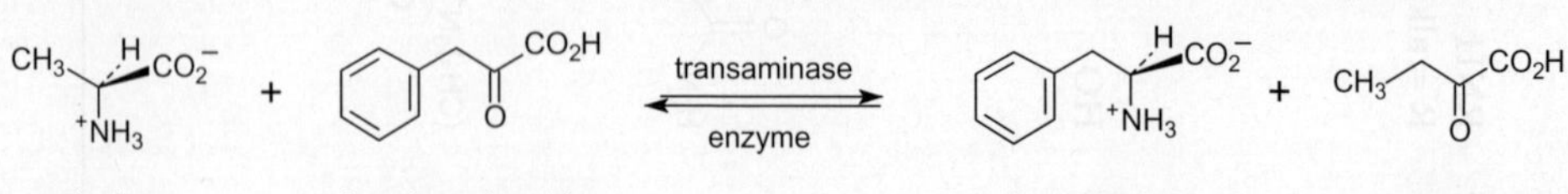

L-alanine + phenylpyruvic acid $\rightleftharpoons$ (transaminase enzyme) L-phenylalanine + pyruvic acid

alkyl amines (G = alkyl, R–NH$_2$) reach a maximum rate at higher pH's (close to pH 7), where the loss of water (H$_2$O) from the tetrahedral intermediate hemiaminal (or carbinolamine) is rate determining. On the other hand, weakly basic nucleophiles, protonated less readily, reach a maximum rate at more acidic pH's (close to pH 4), where the formation of the hemiaminal (or carbinolamine) is rate determining (Figure 9.10).

The reaction of aldehydes and ketones with ammonia (NH$_3$) itself is shown as the first item in Table 9.4. Despite the apparent simplicity of the process shown in Scheme 9.65, the reaction of aldehydes and ketones with this nucleophile (base) often produces a complex mixture of products because the addition process does not stop after the first step. For example, methanal (formaldehyde, H$_2$C=O) yields hexamethylenetetramine (Table 9.4) and ethanal (acetaldehyde, H$_3$CCHO) provides the corresponding C-methylated trimer. However, diarylketones (Ar$_2$C=O) often give stable imines (Equation 9.51).

$$\text{(9.51)}$$

Problem 9.9. The reaction of methanal (formaldehyde, H$_2$C=O) with ammonia (NH$_3$) is used to prepare 1,3,5,8-tetrazatricyclo-[3.3.1.1^{3,7}]decane (hexamethylenetetramine) (Table 9.4). Write a plausible series of reactions between these reagents that would have the desired product as the outcome. Account for the different result to produce the trimer (shown below) with ethanal (acetaldehyde) and ammonia.

Problem 9.10. It was found quite early that treatment of 1,3,5,8-tetrazatricyclo-[3.3.1.1^{3,7}]decane (hexamethylenetetramine) with nitric acid (HNO$_3$) produced the corresponding explosive trinitrotrimethylenetriamine (trivial name "cyclonite"), formaldehyde, and ammonia. Write a plausible series of reactions to account for the observed product.

In contrast, both ketones and aldehydes, the former more slowly than the latter, generally react with aromatic and aliphatic *primary* amines (Table 9.4, 2a) to produce the corresponding hemiaminals (carbinolamines), which are usually not isolated as they readily lose water to produce imines (sometimes called *Schiff's bases*). Thus, the reaction of benzenamine (aniline, $C_6H_5NH_2$) with cyclohexanone shown in Equation 9.52 produces the benzenimine of cyclohexanone. A second equivalent of benzenamine (aniline, $C_6H_5NH_2$) can add to the imine to form a diamine.

$$\tag{9.52}$$

Interestingly, as shown in Table 9.4, when unsymmetrical ketones or aldehydes are allowed to react with primary amines (RNH_2) the imines that are formed are capable of *(E)–(Z)* isomerism and both products can co-occur. However, the barrier to isomerization between these geometric isomers can be low enough to make them inseparable at room temperature.

The reaction between amino group ($-NH_2$) and a carbonyl group ($C=O$) elsewhere in the same molecule has been used in the synthesis of many heterocyclic compounds. For example as shown in Equation 9.53, substituted quinolines can be produced by acid-catalyzed cyclization of the appropriate aminoketones (formed during the *Friedlander synthesis*, in this instance between propanone and *ortho*-aminobenzaldehyde, in an aldol-type reaction with loss of water; this chapter, *vide infra*), and even more than one reaction can be induced to occur. Thus, in Equation 9.54, 1,2-benzenediamine (*ortho*-aminoaniline) undergoes acid-catalyzed addition to 2,3-butane-dione to produce 2,3-dimethyl-1,3-benzopyrazine (2,3-dimethylquinoxaline).

$$\tag{9.53}$$

$$\tag{9.54}$$

While the early part of the acid-catalyzed path to form an imine (i.e., production of a protonated hemiaminal) appears the same for secondary amines (Table 9.4, 2b) and primary amines (Table 9.4, 2a), the absence of a proton on nitrogen denies that intermediate the ability to lose water as shown in Scheme 9.65. Therefore, if there is a proton on the adjacent carbon, water is lost with the formation of a nitrogen-bearing, carbon–carbon double bond species called an **enamine** (Scheme 9.66).*

Although it will be discussed later in this chapter when addition to the carbon α (*alpha*) to the carbonyl ($C=O$) is separately considered[†], it is worthwhile noting here

---

*In the event that there is no hydrogen to lose on the α-carbon a second equivalent of secondary amine has been known to add to the carbon of what was the carbonyl. This kind of product ($(R_2N)_2CR_2$) is called an **aminal** (Scheme 9.66).

[†]It will not have gone unnoticed by the alert reader that the reactions of enols and enol ethers as well as the aldol-type reaction (with retention of the carbonyl oxygen) already discussed are, effectively, of just this kind.

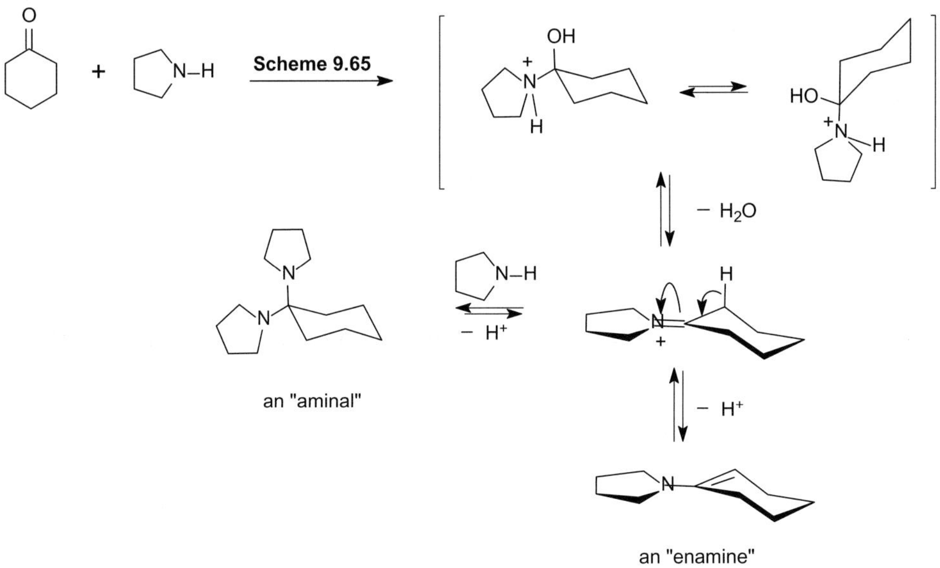

Scheme 9.66. A representation of the reversible formation of both the aminal and the enamine derived from the acid-catalyzed reaction between cyclohexanone and azacyclopentane (pyrrolidine). Under dehydrating conditions and with only one equivalent of azacyclopentane (pyrrolidine), only the enamine is isolated.

that reaction of enamines with electrophiles occurs at the carbon from which the proton was lost.

The third example in Table 9.4, that is, 3a–c, where G = NH_2, describes the addition of hydrazine (H_2NNH_2) and substituted hydrazines ($ArNHNH_2$) to the carbonyl (C=O) to produce the corresponding hydrazones ("hydrazine derivatives") of the aldehyde or ketone with which the reaction began. It will be remembered that the hydrazone derived from hydrazine (H_2NNH_2) itself has already been encountered during the discussion of the reduction of ketones to alkanes (Scheme 9.12) as part of the *Wolff–Kishner reaction* sequence (*vide supra*).

Now, it is worth noting (Table 9.4, 3a) that when the carbonyl group (C=O) in the aldehyde or ketone is unhindered, a second equivalent of carbonyl compound can be added to the hydrazone with the formation of an azine. The aryl-substituted hydrazines, for example, phenylhydrazine ($C_6H_5NHNH_2$) and in particular 2,4-dinitrophenylhydrazine ("2,4-DNPH") (Table 9.4, 3b) frequently form crystalline hydrazones. For many years (in a largely "pre-spectroscopic" era), it had been accepted that if the melting points of two derivatives (e.g., 2,4-DNP and oxime) of an unknown ketone or aldehyde corresponded to values found in published tables or lists of those previously reported, then the ketone or aldehyde had been identified. Generally, the method was successful but clearly dependent on ready access to the appropriate tables or lists.

The use of such tables and lists as well as direct comparison of samples of these highly crystalline materials was also prevalent in the general field of carbohydrate chemistry (a topic more fully dealt with in Chapter 11). Interestingly α-hydroxyaldehydes, such as D-(+)-glucose, and α-hydroxyketones react with *excess*

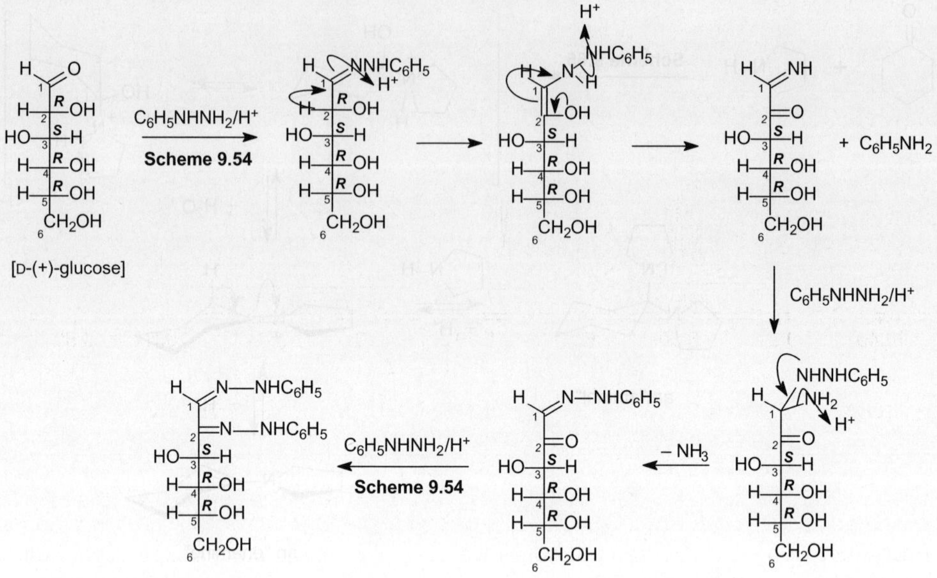

Scheme 9.67. A proposed pathway for the formation of the osazone of D-(+)-glucose.

phenylhydrazine ($C_6H_5NHNH_2$) to yield osazones (Table 9.4, 3c). It is clear that an oxidation must have occurred and, since an excess of phenylhydrazine ($C_6H_5NHNH_2$) is used, it has been proposed that the reduction occurs at the initially formed phenylhydrazone. The latter, it is supposed, is reduced to ammonia (NH_3) and aminobenzene (aniline, $C_6H_5NH_2$). Thus, as shown in Scheme 9.67, the α-hydroxyaldehyde reacts with phenylhydrazine ($C_6H_5NHNH_2$) to the corresponding phenylhydrazone. Then, enolization to the corresponding enol–enamine, followed by protonation on nitrogen in the acidic solution and ketone formation from the enol results in the loss of aminobenzene (aniline, $C_6H_5NH_2$) and the production of an α-ketoaldimine. Subsequent protonation of the imine and attack by additional phenylhydrazine ($C_6H_5NHNH_2$), first at the aldehydic carbon (less hindered) and then at the ketonic carbon, generates the osazone (Scheme 9.67).

In this vein, it should be clear that both D-(+)-glucose (Table 9.4, 3c) and its diasteriomer, D-(+)-mannose, must give the same osazone (Equation 9.55).

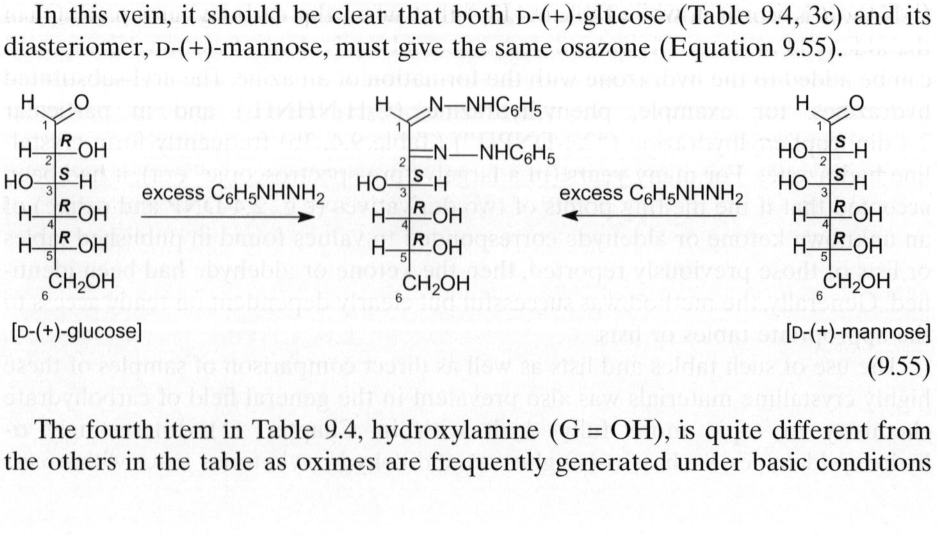

$$(9.55)$$

The fourth item in Table 9.4, hydroxylamine (G = OH), is quite different from the others in the table as oximes are frequently generated under basic conditions

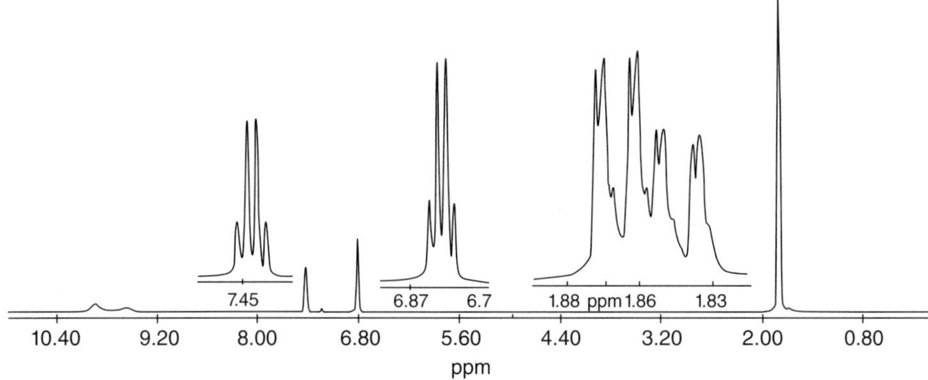

Figure 9.11. The ^{1}H NMR spectrum at 500 MHz in ^{2}HCCl$_3$ of a mixture of the *(Z)*- and *(E)*-oximes of ethanal (acetaldehyde, CH$_3$CHO).

from the readily available crystalline hydroxylamine hydrochloride (H$_2$NOH·HCl). Further, it is common to find that both the *(E)*- and the *(Z)*-oximes can be separated, are stable crystalline materials, and do not interconvert in the solid phase. Many, placed in solution, will interconvert to an equilibrium mixture of *(E)*- and *(Z)*-isomers. The ratio of the isomers and the rate of interconversion are dependent not only on the specific oxime structure but also on the nature of the solvent in which they are dissolved and the presence or absence of catalysts. Both photochemical and thermal isomerization pathways are also present. Of course, the *(E)*- and *(Z)*-isomers have different physical properties and chemical properties and reactivities.*

The proton (^{1}H) NMR spectrum of the mixture of *(E)*- and *(Z)*-oximes of ethanal [*(E)*- and *(Z)*-acetaldoximes; CH$_3$CH=NOH], obtained on treatment of ethanal (acetaldehyde, CH$_3$CHO) with hydroxylamine hydrochloride (H$_2$NOH·HCl), is given in Figure 9.11. It is important to note (in the expanded areas shown) that the proton on the carbon bearing nitrogen comes into resonance at two widely different chemical shifts. The downfield signal (a quartet because of coupling with the methyl [–CH$_3$] group) is attributed to the proton on the *same side as the hydroxyl* (i.e., the *(E)*-isomer), while the upfield signal (also a quartet because of coupling with the methyl [–CH$_3$] group) is attributed to its geometric isomer. The methyl group is correspondingly coupled to the hydrogen on the carbon bearing the nitrogen.

The last items in Table 9.4 make use of the carboxyamino (amido)-substituted hydrazine called "semicarbazine" to form highly crystalline semicarbazides and carboxymethyl-substituted derivatives of hydrazine, one of which {(carboxymethyl)

---

*Although the subject will come up again later (Chapter 10), it is worth noting, for example, that the *(Z)*-isomers of most aldoximes more readily undergo loss of water to produce nitriles than do their *(E)*- counterparts.

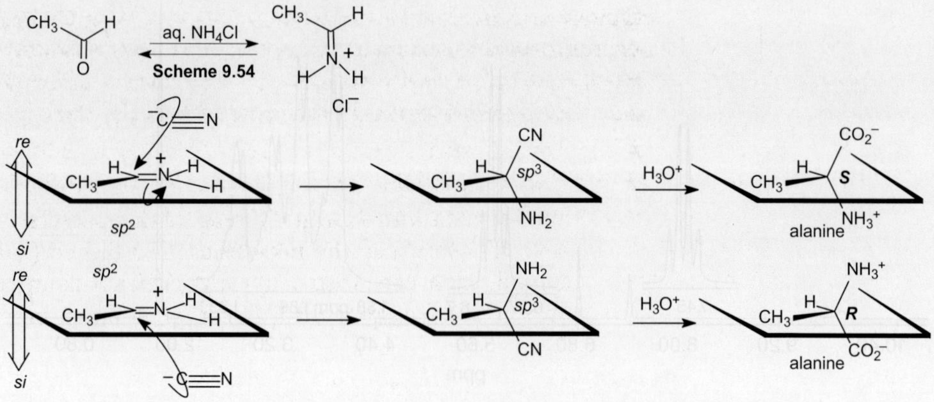

Scheme 9.68. A representation of the **Strecker synthesis** of amino acids.

trimethylammonium chloride hydrazide [$(CH_3)_3N^+CONHNH_2$; Gerard's reagent T]}, is shown. Reagents of the latter type prove valuable both in characterizing the carbonyl compounds with which they are reacted (because they are crystalline) and as tools of isolation since they are often soluble in water.

Finally, although not shown in the table, the material that is shown there can be combined with other reactions. The **Strecker amino acid synthesis**, which is discussed again in Chapter 12, is among the most valuable, while the **Shapiro Olefination** and the related **Bamford–Stevens reaction** have proved useful in producing alkenes from aldehydes and ketones.

When an aldehyde such as ethanal (acetaldehyde, CH_3CHO, Scheme 9.68) is brought into reaction with ammonium cyanide (NH_4CN) (or sodium cyanide [$NaCN$]/ammonium chloride [NH_4Cl]), the product obtained results from capture of the iminium ion intermediate by cyanide (CN^-) ion. As noted in Scheme 9.68, capture can occur from either the *re* or *si* face of the aldehyde, thus creating a chiral α-aminonitrile. As we will learn (this Chapter, *vide infra*) acid-catalyzed hydrolysis of nitriles produces carboxylic acids so that if the nitrile is hydrolyzed to the corresponding carboxylic acid, the **Strecker amino acid synthesis*** will have been accomplished.

Similarly, when an aldehyde or ketone is treated with *p*-toluenesulfonylhydrazine (p-$CH_3C_6H_4NHNH_2$), a *p*-toluenesulfonylhydrazone (tosylhydrazone) (Scheme 9.69) is formed. Now, when this stable, easily purified carbonyl derivative is treated with at least two equivalents of a strong base (such as methyllithium, CH_3Li), proton loss on the carbon adjacent to the sp^2-hybridized center (i.e., the α-carbon) occurs and an organolithiated alkene results. Hydrolysis of the latter results in alkene formation and the consummation of the **Shapiro reaction**.[†]

---

*The synthesis, originally developed by Strecker, has been of increasing value as methods for the direct synthesis of *chiral* amino acids have been sought. See Strecker, A. *Ann. Chem. Pharm.*, **1850**, *75*, 27; *ibid*, **1854**, *91*, 349, and for more recent work, Kuethe, J. T.; Gauthier, Jr. D. R.; Beutner, G. L.; Yasuda, N. *J. Org. Chem.*, **2007**, *72*, 7469.

[†]The Shapiro reaction is described in two publications (although it has also been used successfully by others). See Shapiro, R. H.; Heath, M. J. *J. Am. Chem. Soc.*, **1967**, *89*, 5734, and Shaprio, R. H.; Lipton, M. F.; Kolonko, K. J.; Buswell, R. S.; Capuano, L. A. *Tetrahedron Lett.*, **1975**, 1811.

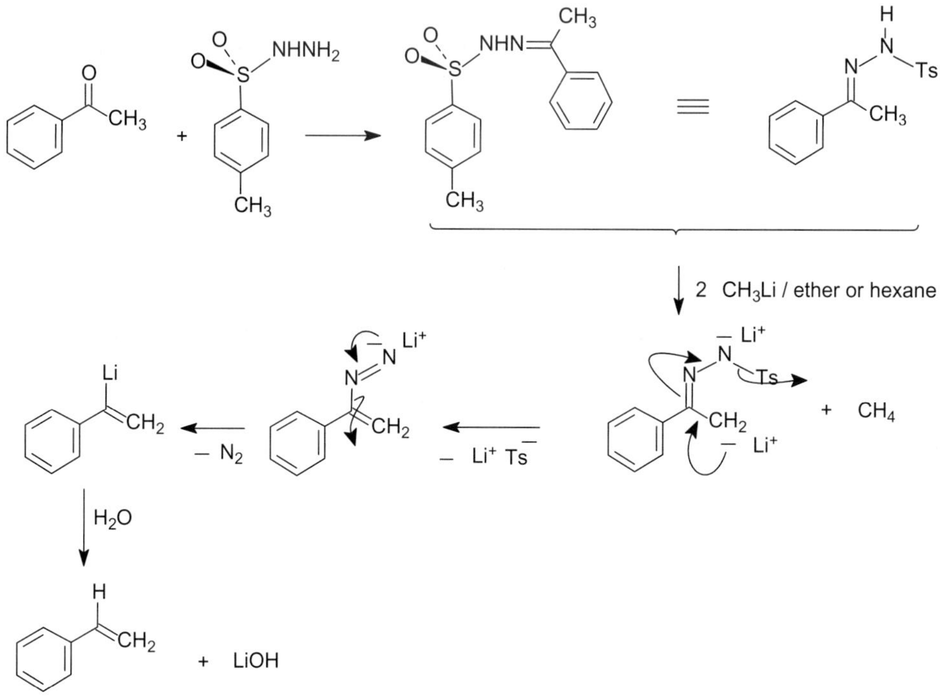

$$\downarrow\ \ 2\ \ CH_3Li\ /\ ether\ or\ hexane$$

$$+\ \ CH_4$$

$$+\ \ LiOH$$

Scheme 9.69. A representation of the **Shapiro reaction**. The formation of the tosylhydrazone derivative of acetophenone ($C_6H_5COCH_3$) and its decomposition on treatment with methyllithium to 1-phenylethene (styrene).

Alternatively, for the **Bamford–Stevens reaction**,* a sulfonylhydrazone, formed in the same way, can be treated **either** with sodium metal (Na) in the presence of a proton source (typically, 1,2-dihydroxyethane [ethylene glycol, $HOCH_2CH_2OH$] is used and only enough sodium metal is added to react with one of the two hydroxyl groups) and a carbocation is generated **or** with sodium metal (Na) in the absence of a proton source in an aprotic solvent such as methylbenzene [toluene, $C_6H_5CH_3$] where a carbene forms (Scheme 9.70).

d. Replacement of the Carbonyl Oxygen by Halogen and Sulfur. As already noted (Table 9.3), when ketones and aldehydes are treated with hydrogen halides, ready reversion of the 1,1- or α,α-halohydrin (fourth item in Table 9.3) to HX and the carbonyl parent preclude its isolation (Equation 9.56). However, in the presence of an alcohol, it is often possible to isolate the corresponding 1,1- or α,α-haloether (Scheme 9.71).

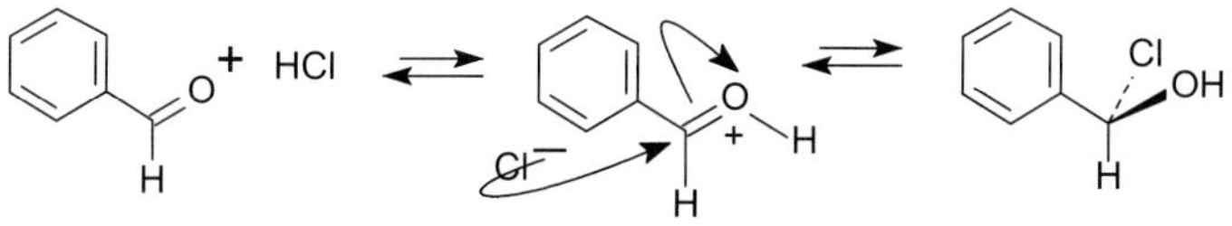

$$\tag{9.56}$$

---

*Bamford, W. R.; Stevens, T. S. J. Chem. Soc., **1952**, 4735. The process has been modified by Caglioti and Shapiro. See Caglioti, R. *Tetrahedron Lett.*, **1962**, 1261, and Shapiro, R. H. *ibid*, **1968**, 345.

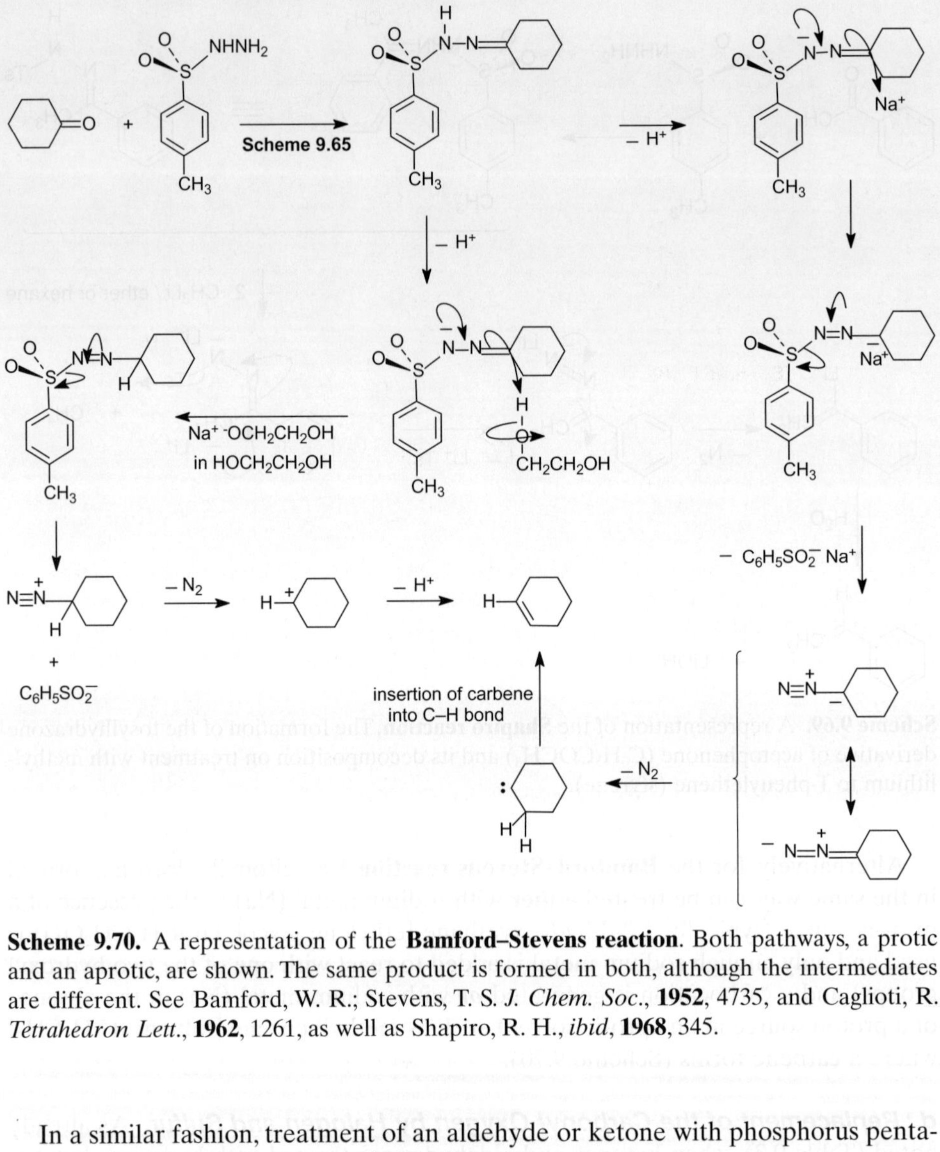

Scheme 9.70. A representation of the **Bamford–Stevens reaction**. Both pathways, a protic and an aprotic, are shown. The same product is formed in both, although the intermediates are different. See Bamford, W. R.; Stevens, T. S. *J. Chem. Soc.*, **1952**, 4735, and Caglioti, R. *Tetrahedron Lett.*, **1962**, 1261, as well as Shapiro, R. H., *ibid*, **1968**, 345.

In a similar fashion, treatment of an aldehyde or ketone with phosphorus penta-chloride (PCl_5) in diethyl ether [$(CH_3CH_2)_2O$] results in replacement of the oxygen of the carbonyl (C=O) by two chlorines (Scheme 9.72).

Replacement of the oxygen of the carbonyl group (C=O) with fluorine to gener-ate a geminal difluoride can also be accomplished. Sulfur tetrafluoride (SF_4), molyb-denum hexafluoride (MoF_6), and most recently, diethylaminosulfur trifluoride [DAST; $(CH_3CH_2)_2NSF_3$] have all been used. Diethylaminosulfur trifluoride [DAST; $(CH_3CH_2)_2NSF_3$], because of its relative stability, is now widely used for such fluo-rinations. The pathway for the introduction of halogen is held to be similar to that for chlorination.

Replacement of the oxygen of the carbonyl (C=O) with sulfur by treatment of the aldehyde or ketone with hydrogen sulfide (H_2S) is rarely successful. It appears that the failure is due to the trimerization of the expected product (Scheme 9.73)

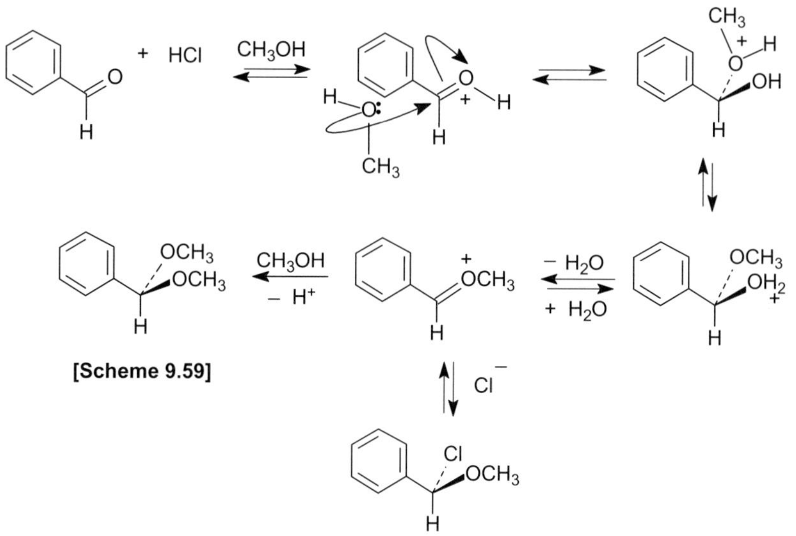

[Scheme 9.59]

Scheme 9.71. A pathway to chlorophenylmethyl methyl ether via an intermediate which can revert to benzenecarboxaldehyde (benzaldehyde) or proceed to the dimethyl acetal of benzenecarboxaldehyde (benzaldehyde via Scheme 9.46 [Scheme 9.50]).

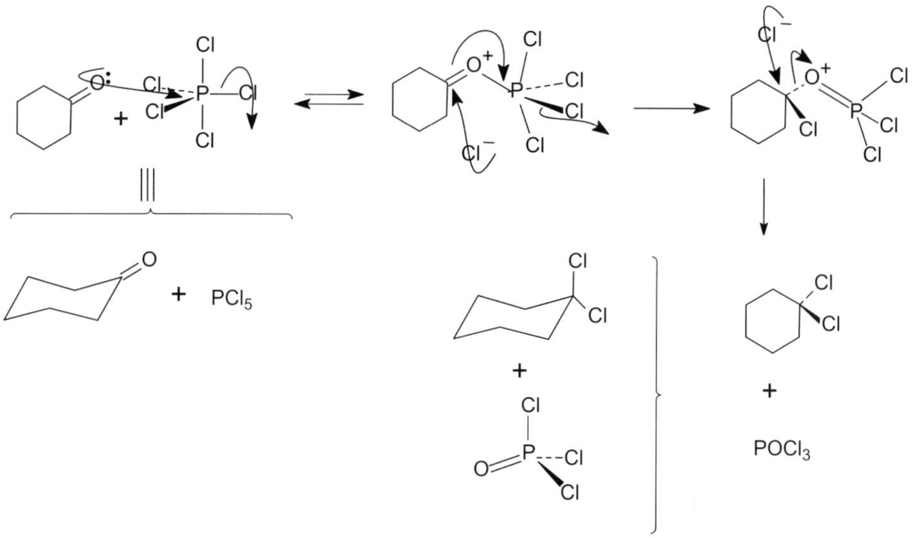

Scheme 9.72. A representation of a pathway for the replacement of the oxygen of the carbonyl group (C=O) of cyclohexanone by two chlorines (2 Cl) through the use of phosphorus pentachloride (PCl_5) in diethyl ether [$(CH_3CH_2)_2O$]. Phosphorus oxychloride ($POCl_3$) accompanies the 1,1-dichlorocyclohexane product.

under the mildly acidic conditions (the first K_a of H_2S is reported to be about 10^{-7}). However, as already noted (Scheme 9.13), aldehydes and ketones react with alkyl sulfides (RSH) or disulfides (e.g., 1,2-ethanedithiol, $HSCH_2CH_2SH$) to generate thioacetals or thioketals, respectively.

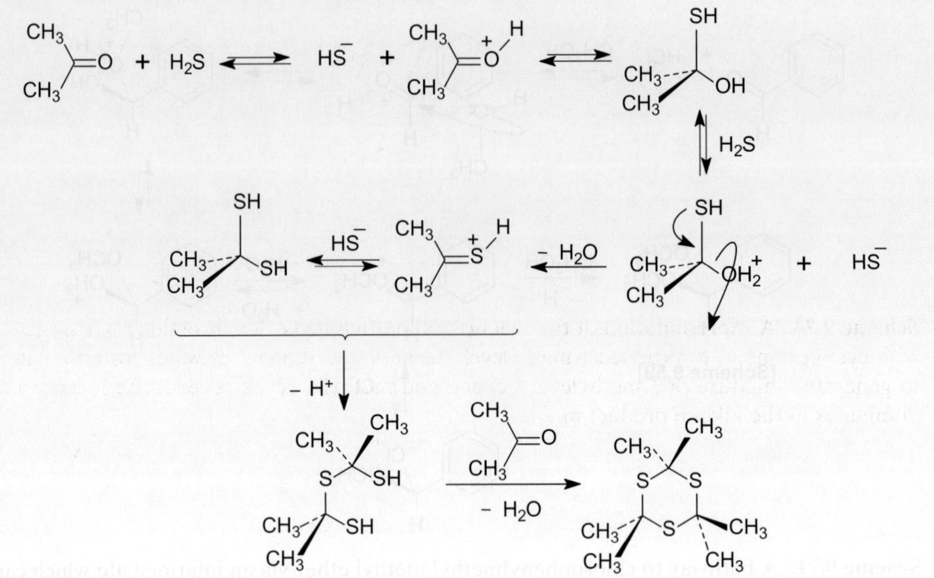

Scheme 9.73. A possible pathway for the formation of the trimer (2,2,4,4,6,6-hexamethyl-1,3,5-trithiocyclohexane) of propanethione [$(CH_3)_2C=S$]. The trimer is generated from the reaction between propanone [$(CH_3)_2C=O$] and excess hydrogen sulfide (H_2S).

The successful replacement of the oxygen of the carbonyl (C=O) of aldehydes and ketones (as well as amides [–NC=O], *vide infra*) can be effected through the use of a specialized reagent, viz. 2,4-bis(4-methoxyphenyl)-1,3,2,4-dithiadiphosphetane-2,4-disulfide (Lawesson's reagent)* as shown in Equation 9.57.

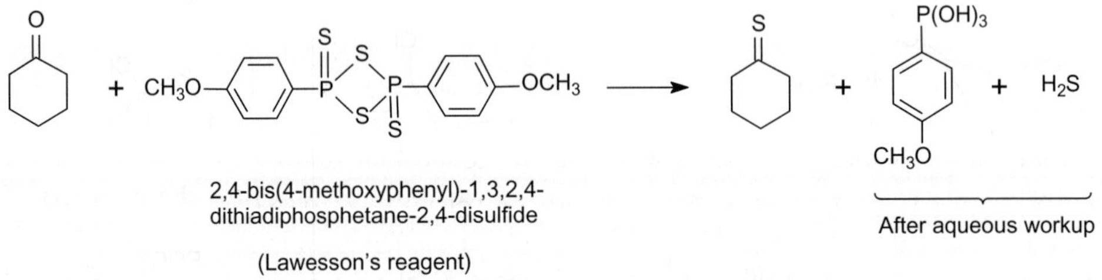

2,4-bis(4-methoxyphenyl)-1,3,2,4-
dithiadiphosphetane-2,4-disulfide

(Lawesson's reagent)

After aqueous workup

(9.57)

e. Replacement of the Oxygen of the Carbonyl by Carbon. As already noted, reaction of organometallic compounds (this chapter, **Part c2**, Schemes 9.39 and 9.40, Equations 9.39–9.44) such as **Grignard reagents** (e.g., "CH_3MgBr," methylmagnesium bromide) with aldehydes or ketones results in the formation of alcohols. Further, as pointed out in Chapter 8 (**Part 3, Elimination**), loss of water ("H_2O") from alcohols and compounds derived from alcohols can be effected under a wide

*For the early work with Lawesson's reagent, see Pedersen, B. S.; Scheibye, S.; Nilsson, N. H. Lawesson, S.-O. *Bull. Soc. Chim. Belg.*, **1978**, *87*, 223.

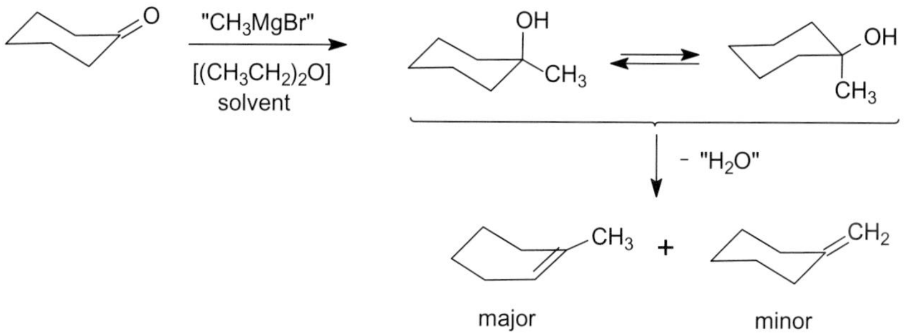

Scheme 9.74. A representation of the reaction of methylmagnesium bromide ("CH$_3$MgBr") with cyclohexanone to produce 1-methylcyclohexanol and the loss of water from the latter to generate a mixture of 1-methylcyclohexene and methylenecyclohexene. The former predominates in the alkene product mixture.

variety of conditions, and it should be clear that the oxygen lost was originally derived from the carbonyl group. In this circuitous way, for example, the reaction between methylmagnesium bromide ("CH$_3$MgBr") and cyclohexanone in ether, followed by dehydration, will produce (Scheme 9.74) a mixture of 1-methylcyclohexene (major) and methylenecyclohexane (minor) products. But the latter can not be obtained this way without the former as the isomer with the double bond endocyclic is the more stable.

Problem 9.11. (a) In Scheme 9.74, the loss of water (–H$_2$O) was placed in quotation marks (" ") because the dehydration process can take a variety of forms, most of which would not actually involve loss of H$_2$O but rather some derivative of the alcohol. This material was discussed in the last chapter. By way of review, suggest at least three dehydration techniques. (b) Why is the Grignard reagent put in quotation (" ") marks?

While special cases might be imagined where dehydration could favor one alkene rather than another, the situation depicted for cyclohexanone in Scheme 9.74 is typical. Aldehydes generally behave similarly with the potential for dehydration of the alcohol generated by the addition routinely realized. However, utilizing β-acyloxysulfones (the **Julia** synthesis) regioselective double bond introduction is possible.

In the **Julia alkenation reaction** (Scheme 9.75) an α-lithiosulfone is allowed to condense with an aldehyde or ketone at low temperatures (e.g., –90°C) in a ethereal solvent, for example, oxacyclopentane (THF). The resulting alcohol is acetylated (this chapter and Chapter 8, Table 8.6, item 16) and the acetoxysulfone is reduced with amalgamated sodium [Na(Hg)] and disodium hypophosphate (Na$_2$HPO$_4$) in methanol (CH$_3$OH). The resulting alkene forms with *trans-* or *(E)*-selectivity.*

---

*It is important to note that α-halosulfones are **inappropriate** starting materials for α-lithiosulfones, which should be prepared from the sulfone itself. The α-halosulfone undergoes base-catalyzed

(Continued on next page)

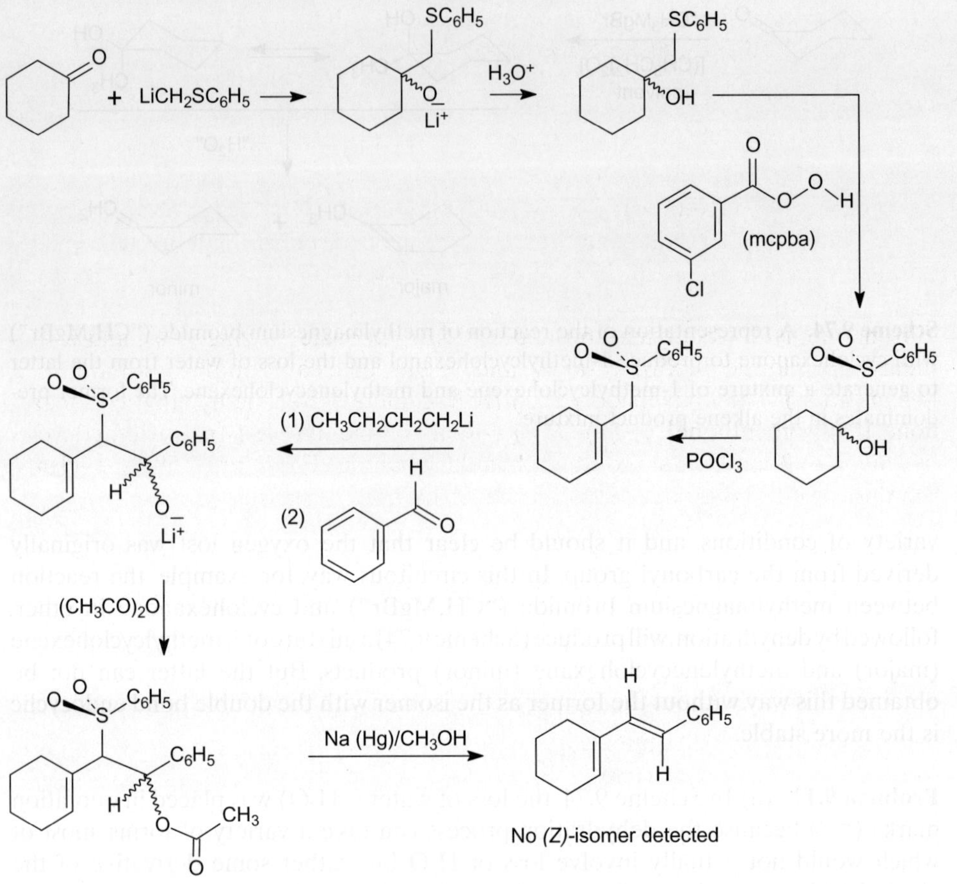

Scheme 9.75. A generalized description of the **Julia synthesis** in which *trans-* or *(E)*-alkenes are preferentially formed. In this case, cyclohexanone is converted into a substituted alkene, which eventually produces an anion capable of attacking benzenecarboxaldehyde (benzaldehyde). Although both diasteriomers are (apparently) produced, elimination generates only the *(E)*-alkene.

Problem 9.12. Write curved arrow "mechanisms" that account for the products shown in each step in the **Ramberg–Bäcklund** process.

elimination of haloacid to produce a thiirane dioxide, which then undergoes extrusion of SO_2 to generate an alkene (the **Ramberg–Bäcklund reaction**), that is,

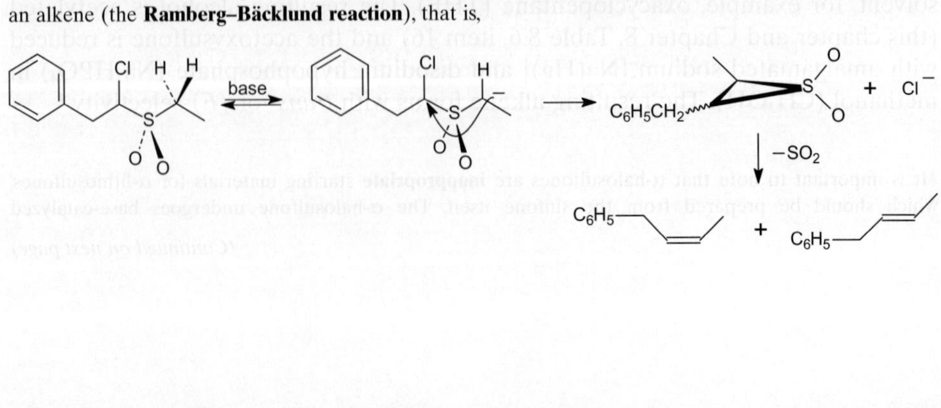

Problem 9.13. Write curved arrow "mechanisms" that account for the products shown in each step in the **Julia alkenation**.*

Interestingly, it appears that a reaction reminiscent of the **Simmons–Smith** reaction (Chapter 6, Equation 6.52), which, it will be recalled, was useful in constructing cyclopropyl alkanes by addition of diiodomethane (CH_2I_2) derived methylene (:CH_2) across the alkene carbon–carbon double bond (C=C), can also be used to replace the carbonyl oxygen of a ketone with a methylene (CH_2) group. (Aldehydes are also reported to work, but the yields are generally found to be lower and other product(s) accompany the one desired.) This variation requires the use of dibromomethane (methylene bromide, CH_2Br_2), powdered zinc (Zn), and a titanium tetrachloride ($TiCl_4$) catalyst, and results in clean addition (without, e.g., racemization). The conversion of (+)-*(R,R)*-5-methyl-2-(1-methylethyl)cyclohexanone [(+)-isomenthone] to (+)-*(R,R)*-4-methyl-2-methylene-1-(methylethyl)cyclohexane [(+)-3-methylene-*cis*-*p*-menthane] shown in Equation 9.58 is a classic example (see Lombardo, L. *Org. Syntheses*, **1967**, *65*, 81) of this process.

$$\text{(9.58)}$$

Similarly, the reagent (the **Tebbe reagent**)[†] generated by the reaction of bis(cyclopentadienyl)titanium dichloride with trimethylaluminum has been found to transfer a methylene group (CH_2) to replace the oxygen of the carbonyl (C=O) (Scheme 9.76). The **Tebbe reagent**[†] will successfully replace the oxygen of the carbonyl groups (C=O) of esters (*vide infra*) as well as aldehydes and ketones, while the reagent produced from dibromomethane (methylene bromide; CH_2Br_2) will not.

Finally, although both of these methods (and some others less common) are useful, the **Wittig reaction**[‡] (and its variants) is/are most general method(s) of replacing the oxygen of the carbonyl group (C=O) of an aldehyde or ketone by a methylene fragment *or any other alkenyl unit*.

The Wittig reaction makes use of the ylide[§] derived by removal of hydrogen halide from a halophosphonium salt. Thus, as shown in Equation 9.59, treatment of

---

*For the Julia alkenation, see Julia, M.; Paris, J. M. *Tetrahedron Lett.*, **1973**, 4833, and Kocienski, P. J.; Lythgoe, B.; Ruston, S. *J. Chem. Soc., Perkin Trans. 1*, **1978**, 829. For more information on the Ramberg–Bäcklund process, see *Ark. Kern. Mikeral Geol.*, **1940**, *13A*, 50, and Paquette, L. A. *Acc. Chem. Res.*, **1968**, *1*, 209.

[†]For the background work on the Tebbe reagent, see Tebbe, F. N., Parshal, G. W.; Reddy, G. S. *J. Am. Chem. Soc.*, **1978**, *100*, 3611; Howard, T. R.; Lee, J. B.; Grubbs, R. H. *J. Amer Chem. Soc.*, **1980**, *102*, 6876, and Francl, M. M.; Hehre, W. J. *Organometallics*, **1983**, *2*, 457.

[‡]Staudinger, H.; Meyer, J. *Helv. Chim. Acta*, **1919**, *2*, 619 and Wittig, G.; Rieber, M. *Liebigs Ann. Chem.*, **1949**, *562*, 187.

[§]An **ylide** (pronounced "*ill-id*") is a species in which adjacent nuclei bear opposite charges as well as an octet of electrons. In the case at hand, both structures shown below are accepted.

$$(C_6H_5)_3\overset{+}{P}-\overset{-}{C}H_2 \quad\longleftrightarrow\quad (C_6H_5)_3P{=}CH_2$$

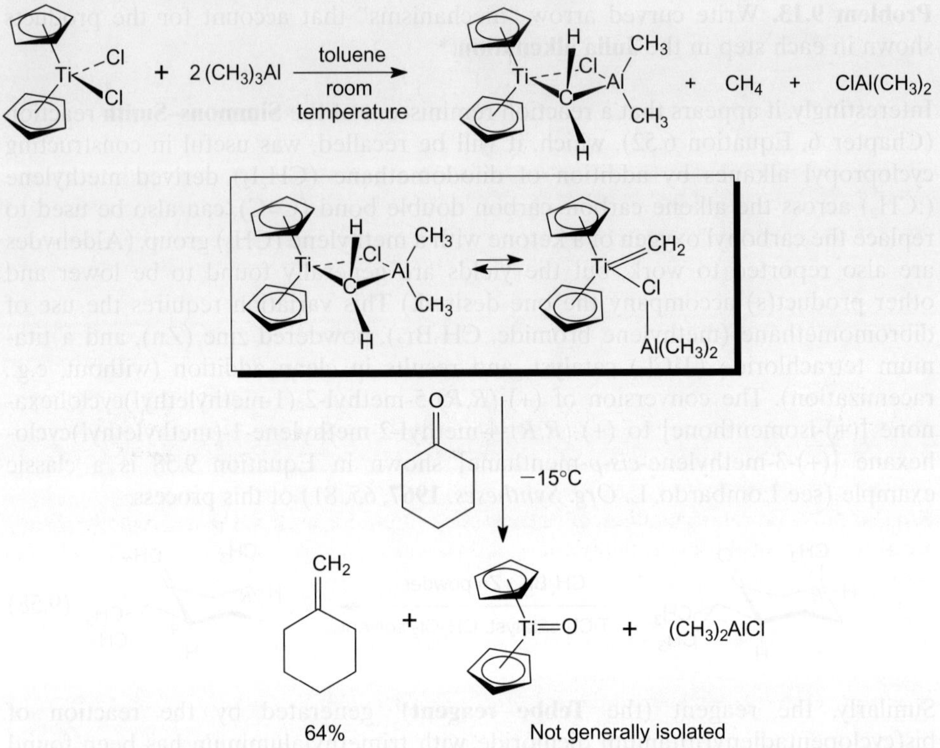

Scheme 9.76. The formation [from bis(cyclopentadienyl)titanium dichloride and trimethyl-aluminum] and use [to replace the oxygen of the carbonyl group (C=O) in cyclohexanone] of the **Tebbe reagent**.

triphenylmethylphosphonium bromide $[(C_6H_5)_3PCH_3^+ \; Br^-]$ with butyllithium $(CH_3CH_2CH_2CH_2Li)$ results in the formation of butane $[CH_3(CH_2)_2CH_3]$, lithium bromide (LiBr), and triphenylphosphinemethylene $[(C_6H_5)_3P^+–CH_2^-$ or $(C_6H_5)_3P=CH_2]$.

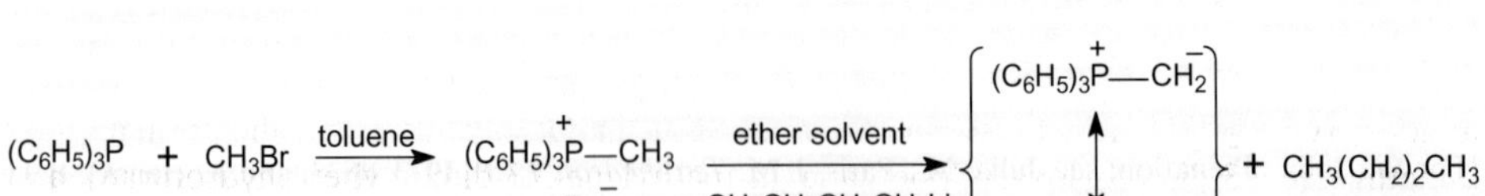

$$(9.59)$$

Now, when cyclohexanone is added to the solution in which the reagent has been generated, reaction occurs to produce methylenecyclohexane and triphenylphosphine oxide (Scheme 9.77). There remains some debate about the details of the process by which the products form, although there is general agreement that a four-membered ring containing phosphorus and oxygen (a phosphetane) is an intermediate. The debate is with regard to formation of the phosphetane. There are two

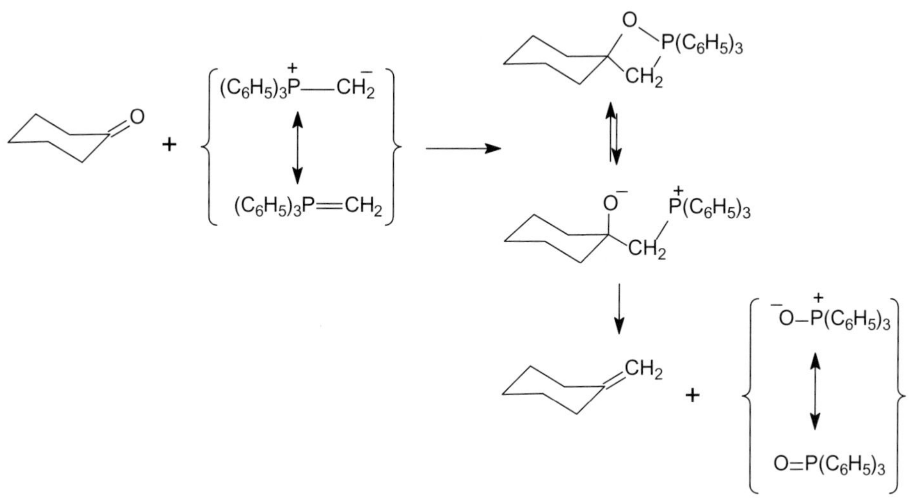

Scheme 9.77. A representation of the reaction between cyclohexanone and the Wittig reagent, triphenylphosphinemethylene (methylenetriphenylphosphorane), generated from treatment of triphenylmethylphosphonium bromide with butyllithium to produce methylenecyclohexane.

choices: either the new carbon–carbon bond and the phosphorus–oxygen bond form at the same time, or they form sequentially (with either carbon–carbon or phosphorus–oxygen leading). Thus, while there are reports that phosphetane intermediates are occasionally capable of detection at low temperatures, little is known about the transition state leading to that intermediate.

First, a variety of phosphorus ylides, from alkyl [$RCH=P(C_6H_5)_3$] and aryl [$ArCH=P(C_6H_5)_3$] and alkoxymethyl [$ROCH=P(C_6H_5)_3$] to carbomethoxy [$(C_6H_5)_3P=CHCO_2CH_3$] substituted species, can be generated for use in the Wittig reaction with aldehydes and ketones. The use of the method for replacing the oxygen of the carbonyl ($C=O$) group has become widespread, and although not completely understood, a variety of principles have been developed. As shown in Equation 9.60, ylides that are not stabilized by an adjacent carbonyl or other electron-withdrawing group (EWG) yield predominately *(Z)*-alkenes.

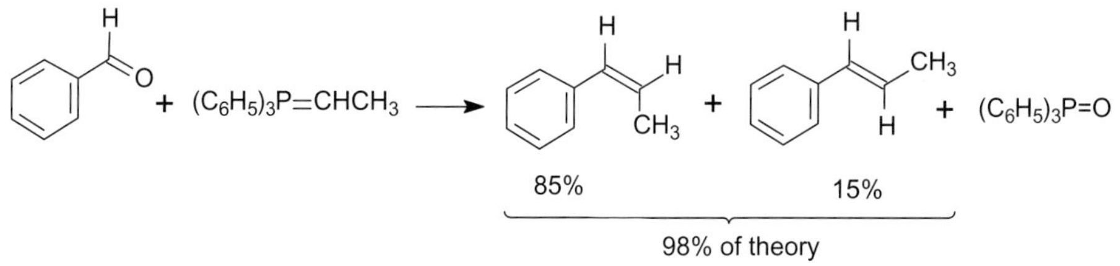

$$85\% \qquad\qquad 15\%$$

$$98\% \text{ of theory}$$

$$(9.60)$$

Second, the ratio of *(Z)*- to *(E)*-isomer, while generally as noted previously, depends on the base that is used to produce the Wittig reagent from the

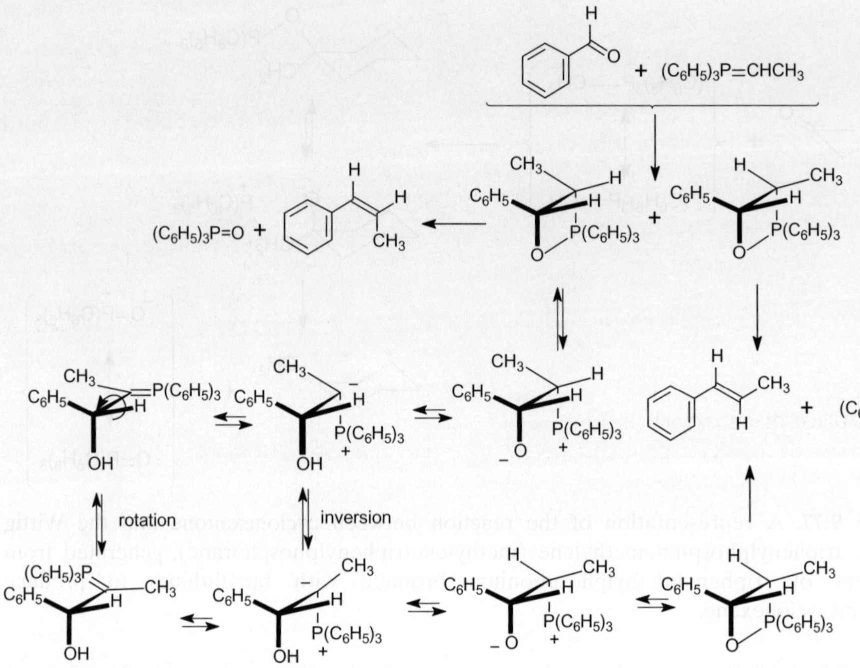

Scheme 9.78. Representation of a pathway that accounts for a change in the ratio of *(Z)*- to *(E)*-1-phenylpropene by change in the reaction conditions (lower temperature and additional strong base, sometimes called the *Schlosser modification of the Wittig reaction*). Early phosphetane fragmentation to alkene is avoided and rotation (or carbanion inversion) to change the isomer ratio is allowed. An excellent discussion of various permutations can be found in Schlosser, M.; Strunk, S. *Tetrahedron*, **1989**, *45*, 2649.

phosphonium salt, the solvent in which the reaction is carried out, and the conditions under which the reaction is run. Indeed, by adjusting conditions so that the phosphetane does not immediately fragment to alkene and phosphine oxide, it may (usually with additional base) undergo ring opening and rotation (or inversion) before fragmentation, thus reversing the generally preferred *(Z)*-geometry (Scheme 9.78).

Third, stabilized Wittig reagents produce almost exclusively the *(E)*-isomer and, indeed, in the example shown (Equation 9.61), only one product was detected.

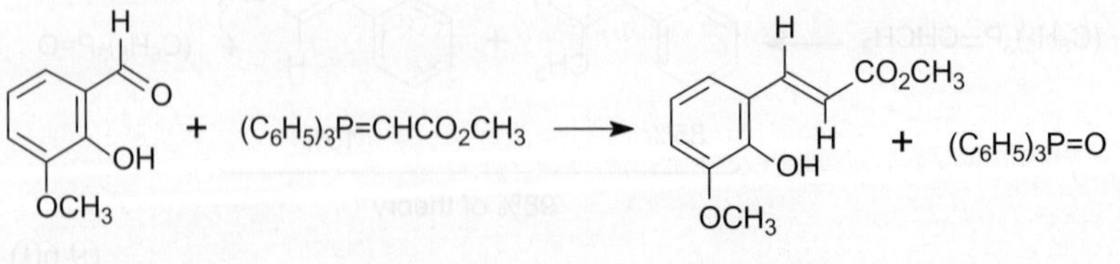

$$(9.61)$$

Of additional interest is the observation that substituents on phosphorus can also be changed and, as a consequence of such changes, different stereochemical outcomes can be effected. These modifications include:

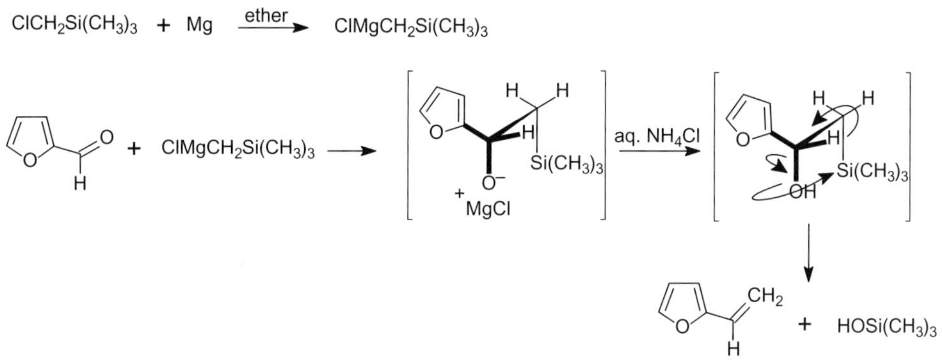

$$ClCH_2Si(CH_3)_3 \; + \; Mg \xrightarrow{\text{ether}} ClMgCH_2Si(CH_3)_3$$

Scheme 9.79. A representation of the steps in the pathway for the introduction of a methylene in place of an oxygen in furan-2-carboxaldehyde (furfural) to generate 2-vinylfuran through use of the **Peterson olefination** process. The Peterson olefination was first reported by Peterson, D. J. *J. Org. Chem.*, **1968**, *33*, 780, and has been modified by Corey, E. J.; Enders, D.; Bock, M. G. *Tetrahedron Lett.*, **1976**, 7.

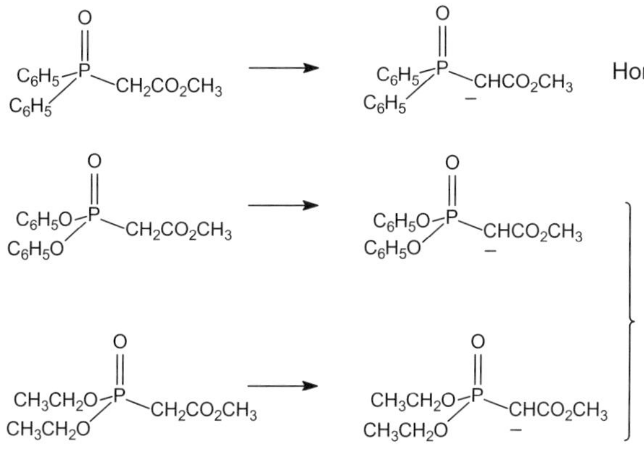

Finally, in this vein, the phosphorus (P) itself can be replaced by a silicon (Si) (Scheme 9.79). Thus, a lithium or magnesium (**Grignard** reagent) coordinated nucleophile can be formed by reaction of an α-halosilane, for example, chloromethyltrimethylsilane [ClCH$_2$Si(CH$_3$)$_3$], with the metal in an appropriate solvent. Reaction with the carbonyl group (C=O) of an aldehyde or ketone produces a β-hydroxysilane, which, in a separate step, undergoes elimination of hydroxytrimethylsilane [(CH$_3$)$_3$SiOH] to generate the alkene. The stereochemistry of the alkene obtained by this *silyl-Wittig* (or ***Peterson olefination***) can also be adjusted by varying the conditions of the reaction.

In addition to these "olefination" reactions, it should be clear that the aldol reaction (*vide supra*) ***when accompanied by dehydration*** also accomplishes the goal of replacing the oxygen of the carbonyl (C=O) by carbon. Such dehydration is common when the aldol reaction is carried out with acid or strong basic catalysts. Further, the generalized aldol reaction can be expanded to include a host of processes for which the unifying factor is ***proton loss*** α ***to the carbon of the carbonyl followed by addition to the carbon of another carbonyl group.***

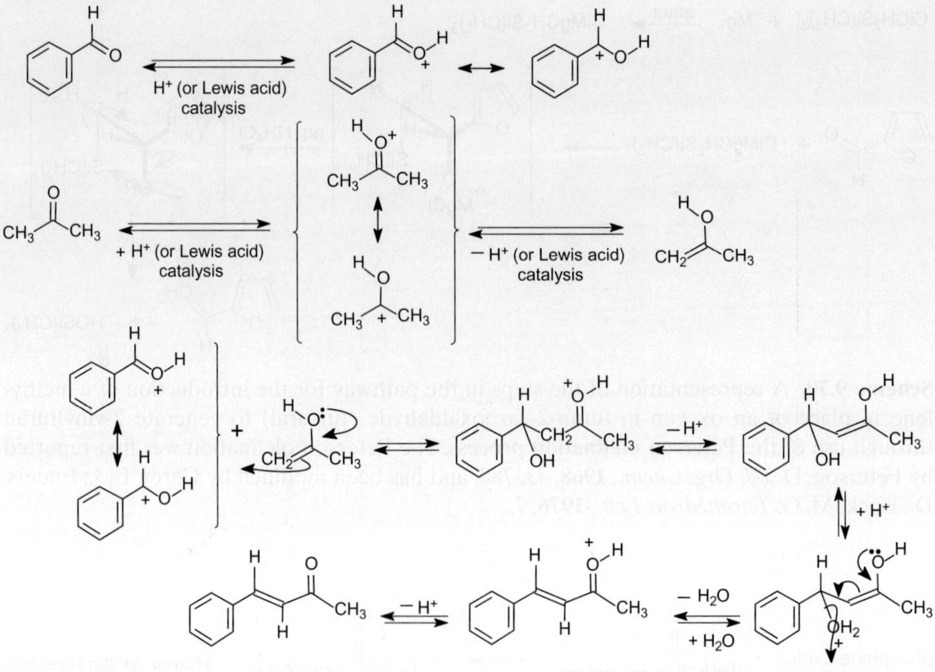

Scheme 9.80. A representation of a set of processes in an aldol condensation reaction resulting in dehydration and the formation of 1-phenyl-1-butene-3-one from benzaldehyde (benzenecarboxaldehyde) and acetone (propanone) in the presence of acid.

The history of organic chemistry is replete with variations upon the generalized theme expressed above. The resulting "name" reactions are generally addend specific, as is seen in Table 9.5.

The first item in Table 9.5 is the aldol condensation itself. As shown in Scheme 9.80, the typical aldol condensation, *when acid is present*, can be followed by further protonation on the hydroxylic oxygen and then loss of water. As is also noted in the scheme, the entire process is reversible—even to the addition of water (in a **Michael**, *vide infra*, fashion).

The second item in Table 9.5 is the Claisen condensation (not to be confused with the Claisen rearrangement discussed in Chapter 8, cf. Scheme 8.99). Here, an aldehyde, lacking α-protons, is allowed to react with an ester of a carboxylic acid (*vide infra*) in the presence of an alkoxide base. The ester must bear protons on the carbon α- to the carbonyl and the (alkoxide) base is chosen to be the conjugate base of the alcohol with which the carboxylic acid was esterified to make that ester.* Thus, as

*Although discussed in more detail (*vide infra*) subsequently, esters are capable of reacting with alkoxide anions at the carbon of the carbonyl group to permit exchange (through a "tetrahedral intermediate") of one alkoxide group for another ("ester interchange"). Thus, it is common to use an alkoxide anion identical to the one that would be leaving should exchange occur. In such a narcissistic process, the exchange, if it occurs, is rendered invisible.

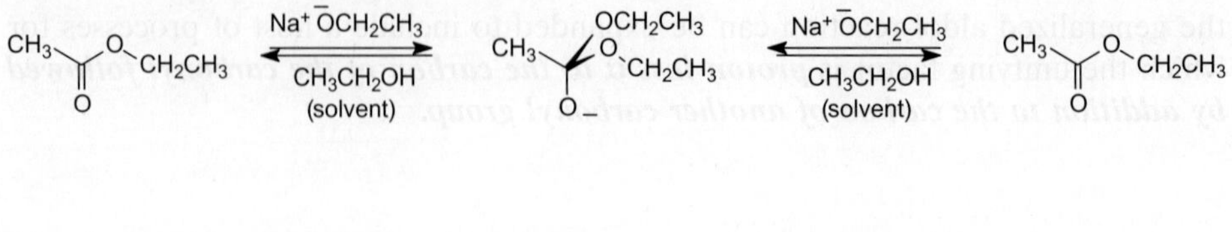

TABLE 9.5. Some "Aldol"-Type Processes

Number	Name of Reaction	Example of Substrate	Example of Addend	Catalyst	Example of Product
1.	Aldol condensation	Aldehyde or ketone 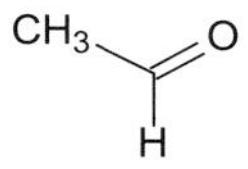 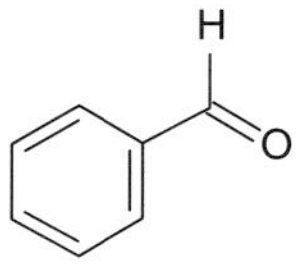	Aldehyde or ketone 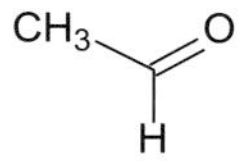 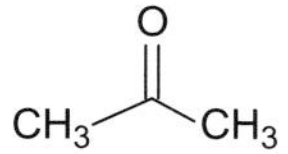	Acid or strong base	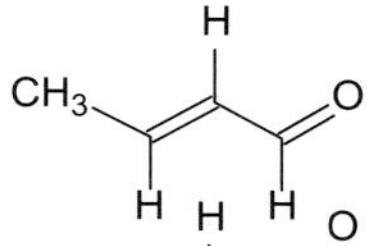 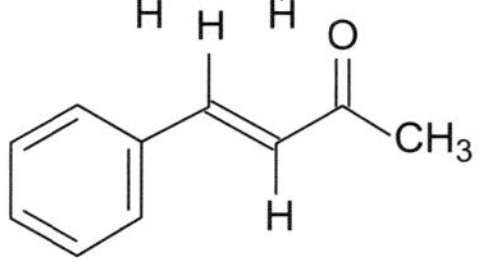
2.	Claisen condensation	Aldehyde 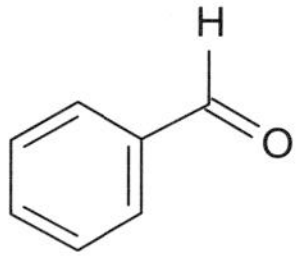	Ester 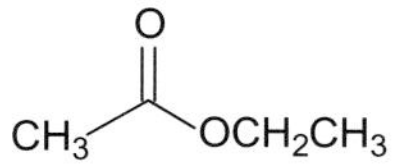	Sodium (Na) metal in ethanol	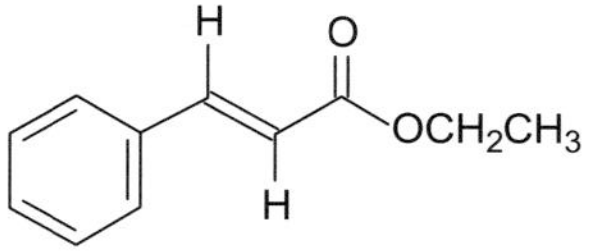
3.	Stobbe condensation	Aldehyde or ketone 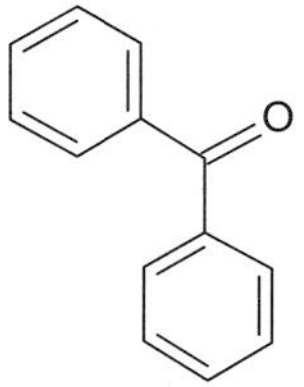	Dialkyl succinate (diethyl butane-dicarboxylate) 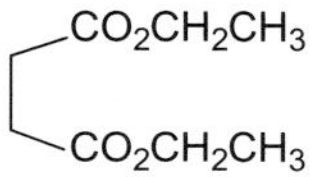	Potassium *tert*-butoxide [KOC(CH$_3$)$_3$]	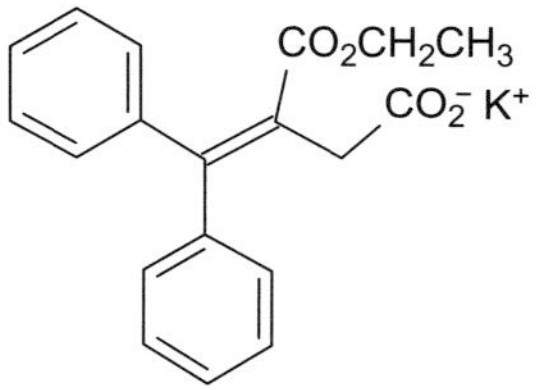

TABLE 9.5. *Continued*

Number	Name of Reaction	Example of Substrate	Example of Addend	Catalyst	Example of Product
4.	Perkin condensation	Aldehyde 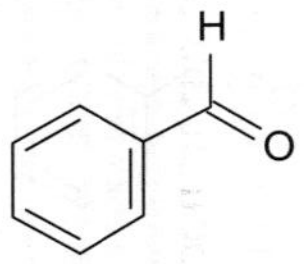	Anhydride 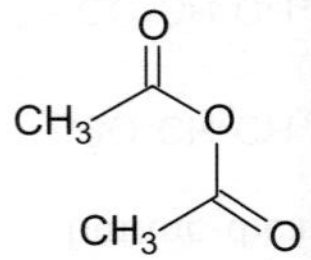	Sodium acetate (CH_3CO_2Na)	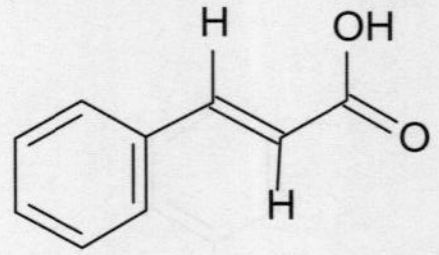
5.	Knovenagel condensation	Aldehyde or ketone 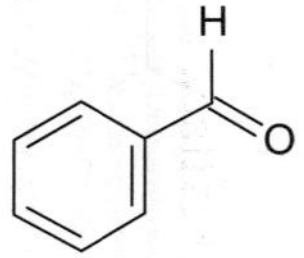	Malonic ester (diethyl propane-dicarboxylate) 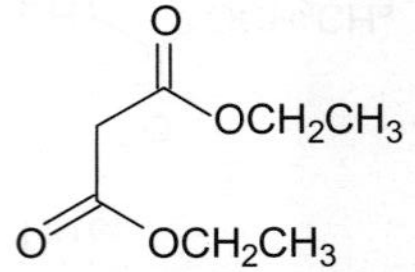	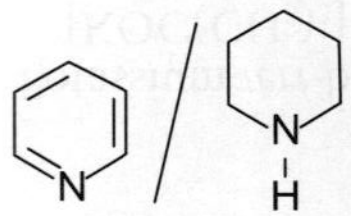 Pyridine and piperidine mixture	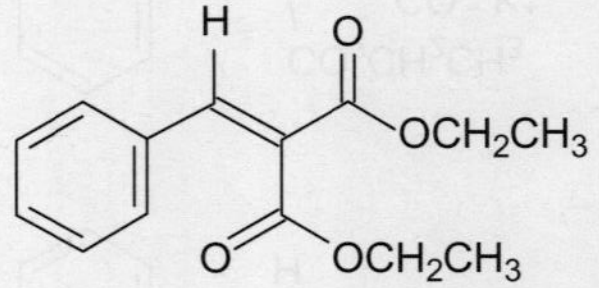
6.	Nitroalkane	Aldehyde or ketone 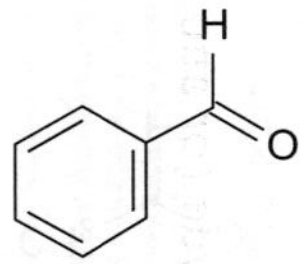	Nitromethane $CH_3–NO_2$	Sodium acetate (CH_3CO_2Na)	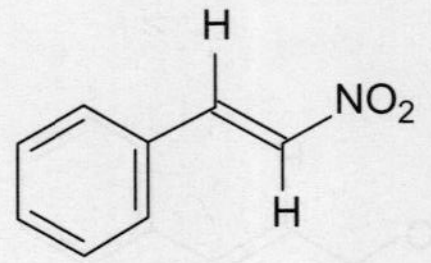
7.	Robinson annulation	Ketone 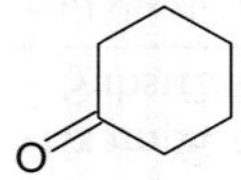	Methyl vinyl ketone (1-butene-3-one) 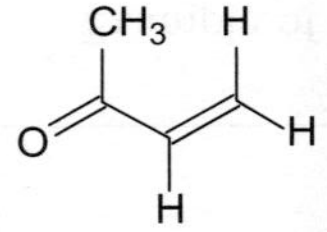	Sodium ethoxide ($NaOCH_2CH_3$)	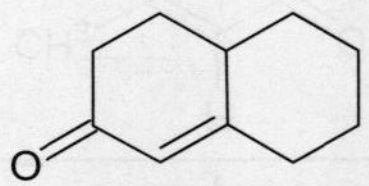

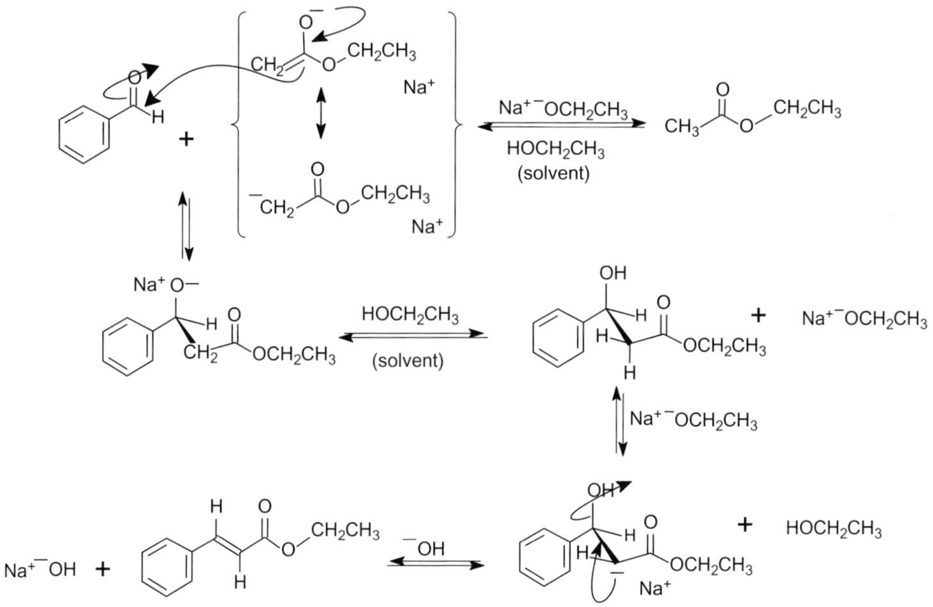

Scheme 9.81. A representation of the Claisen condensation. The reaction benzenecarboxaldehyde (benzaldehyde, C_6H_5CHO) and ethyl ethanoate (ethyl acetate, $CH_3CO_2CH_2CH_3$) in the presence of sodium ethoxide ($NaOCH_2CH_3$) is shown to produce ethyl *(E)*-cinnamate [ethyl *(E)*-3-phenylpropanoate]. See Claisen, L.; Lowman, O. *Chem. Ber.*, **1887**, *20*, 651, and the earlier papers to which reference is therein made, and especially to Geuther, A. *Liebigs Ann. Chem.*, **1855**, *231*, 197.

shown in Scheme 9.81, the addition of alkoxide to ethyl acetate (ethyl ethanoate, $CH_3CO_2CH_2CH_3$) can not only result in (invisible) exchange of one ethoxide for another but can also result in the loss of a proton from the methyl group of the ester. The resulting resonance-stabilized carbanion is then capable, on the one hand, of being reprotonated and, on the other, of attacking the carbon of the carbonyl group of the aldehyde (or another ester).* Attack at the carbon of the carbonyl of the aldehyde is followed by proton transfer from the solution and subsequent loss of hydroxide anion (a weaker base than ethoxide anion) or a second proton transfer and loss of water to produce the conjugated ester (ethyl *(E)*-cinnamate). Since the reaction has traditionally been carried out by addition of sodium metal to the alcohol, there is some suggestion that one electron processes may also be involved.

The third item in Table 9.5 is the Stobbe condensation of diphenylketone (benzophenone [$(C_6H_5)_2C{=}O$]) with diethyl succinate (diethyl butanedicarboxylate [$(CH_3CH_2O_2CCH_2)_2$]). Although the original procedure called for sodium ethoxide

---

*Generally, the carbon of the carbonyl of aldehydes (and ketones) is more readily attacked by nucleophiles than those of esters (and amides). This is rationalized in terms of steric hinderance (aldehydes being less hindered since one of the substituents on the carbon of the carbonyl must be hydrogen) and electron density. For the latter argument, the relative differences in the ability of hydrogen and oxygen to donate electrons to the electron-deficient carbon of the carbonyl are mooted.

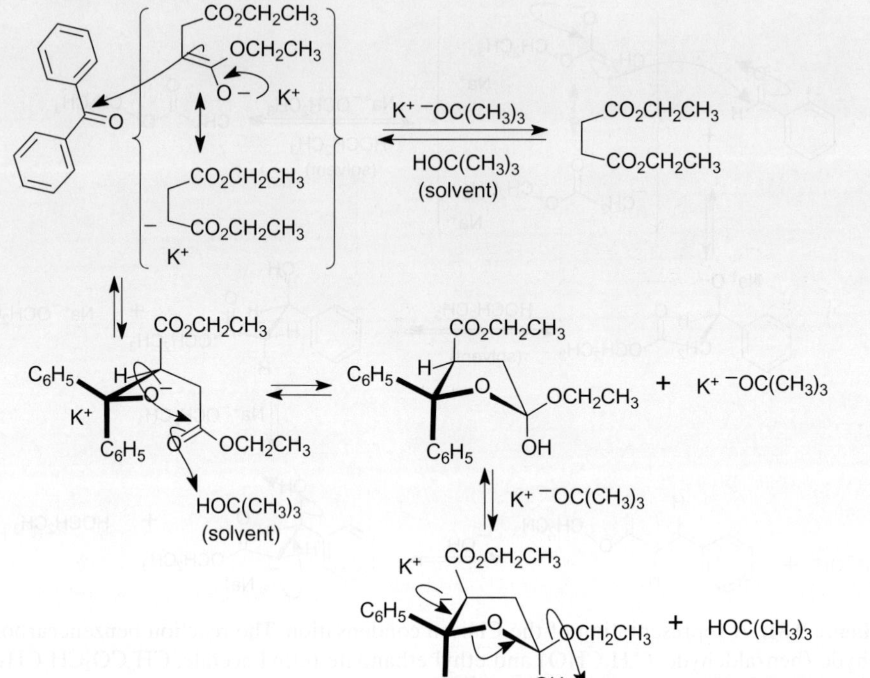

Scheme 9.82. The Stobbe condensation between diphenyl ketone (benzophenone) and diethyl succinate (diethyl butanedicarboxylate) in the presence of potassium *tert*-butoxide [KOC(CH$_3$)$_3$] and in 2-methyl-2-propanol (*tert*-butyl alcohol) solvent. The condensation is shown as occurring via a cyclic intermediate. (See Stobbe, H. *Chem Ber.*, **1893**, *26*, 2312, and, for a particularly insightful use, see Johnson, W. S.; McCloskey, A. L.; Dunnigan, D. A. *J. Am.Chem.Soc.*, **1950**, *72*, 514.

(NaOCH$_2$CH$_3$) in ethanol (CH$_3$CH$_2$OH) as the base–solvent combination, it has been found that a stronger base such as potassium *tert*-butoxide [KOC(CH$_3$)$_3$] in *tert*-butyl alcohol [HOC(CH$_3$)$_3$] gives higher yields and cleaner products. Further, since under either set of conditions the ester group distal to the site of condensation is hydrolyzed and the one proximal is not, the process involving formation of the heterocyclic ring (which has occasionally been isolated) as part of the pathway to product, as shown in Scheme 9.82, has been postulated.

The condensation of aromatic aldehydes with an anhydride in the presence of a salt of the acid corresponding to the anhydride to serve as a catalyst is the **Perkin**

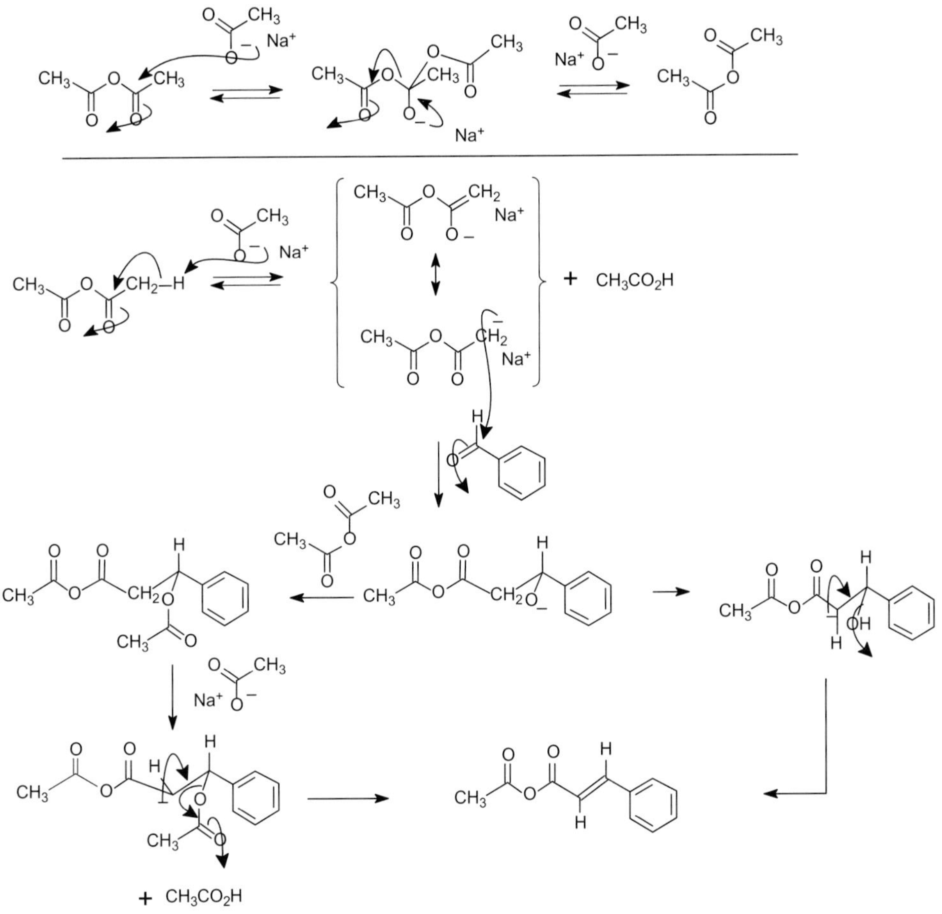

Scheme 9.83. A representation of the Perkin reaction between ethanoic anhydride (acetic anhydride, $CH_3CO_2COCH_3$) and benzenecarboxaldehyde (benzaldehyde, C_6H_5CHO) in the presence of sodium ethanoate (sodium acetate, CH_3CO_2Na) to yield the mixed acetic cinnamic anhydride. The upper part of the scheme indicates "nonproductive" exchange of ethanoate anion (acetate anion, $CH_3CO_2^-$) with the anhydride.

reaction* and in shown as the fourth item in Table 9.5. As shown in Scheme 9.83, it is generally held that the function of the carboxylate salt catalyst is to remove a proton from the carbon of the anhydride α to the carbonyl (C=O) producing an anion which is then available to add to the aldehyde carbonyl. Since the salt of the acid can also react with the anhydride itself to exchange one carboxylate anion for another (*vide infra*), the rationale for using the *same* carboxylate salt is clear. Finally, as shown in the scheme, water is lost either as H_2O or, after additional anhydride (generally present in excess as, e.g., the solvent) reacts with the intermediate alcohol, as acetic acid in the elimination step. A mixed (cinnamic–acetic) anhydride aldol product is obtained. Subsequent hydrolysis of the mixed anhydride yields the corresponding acids.

*Perkin, W. H. *J. Chem. Soc.,* **1868**, *21*, 53. See, also, Leake, P. H. *Chem. Rev.,* **1956**, *56*, 27.

Problem 9.14. Variously ring-substituted benzenecarboxaldehydes (benzaldehydes) have been examined in their reaction with ethanoic anhydride (acetic anhydride). Generally, substituents that withdraw electrons (i.e., EWGs, e.g., $-NO_2$) on the aromatic ring *increase* the yield of the corresponding cinnamic acid, while those that donate electrons (i.e., electron-donating groups [EDGs], e.g., CH_3O-) on the aromatic ring have the opposite effect. Given the pathway of Scheme 9.83, how would you account for this trend?

In a similar way, the condensation of aldehydes and some ketones with diesters of propanedicarboxylic acid (malonic acid) (methylene proton $pK_a \approx 16.5$) and related compounds produces the corresponding alkylidine diester malonates in an aldol-type process called the Knovenagel condensation. Interestingly, it has been known for some time that the preferred base for this process is piperidine (azacyclohexane, $C_5H_{11}N$, pK_a for the conjugate acid, $C_5H_{12}N^+ \approx 10$, *vide infra*) and the preferred solvent, as shown in Scheme 9.84, is pyridine (azabenzene, C_5H_5N). Further, (1) the choice of piperidine (azacyclohexane, $C_5H_{11}N$) appears to have been found (by Knovenagel in 1898) empirically; (2) the reaction with aliphatic aldehydes is often complicated by addition of a second equivalent of malonic diester to the terminus of the conjugated carbon–carbon double bond of product shown in Scheme 9.84 (in a Michael addition, *vide infra*); and (3) the simple esters of malonic acid do not appear to react well with ketones (for what may, in part, be steric reasons) and other derivatives such as ethyl cyanoethanoate (ethyl cyanoacetate [$CH_3CH_2O_2CCH_2C\equiv N$], *vide infra*) are required.

The penultimate item in Table 9.5 is illustrative of the more general aldol-type condensation process. The addition of nitromethane (CH_3NO_2) to benzenecarboxyaldehyde (benzaldehyde, C_6H_5CHO) demonstrates that addition to the carbonyl group of an aldehyde only requires a sufficiently nucleophilic addend.

Interestingly, nitromethane (CH_3NO_2) undergoes aldol condensation with methanal (formaldehyde, H_2CO) in the presence of base to produce the nitrotriol, 2-hydroxymethyl-2-nitro-1,3-propanediol (Problem 9.15). The process by which it forms is presumed to be analogous to that seen in the formation of pentaerythritol (2,2-dihydroxymethyl-1,3-propanediol), Scheme 9.44, from the reaction between ethanal (acetaldehyde, CH_3CHO) and methanal (formaldehyde, H_2CO).

Problem 9.15. Propose a pathway for the generation of 2-hydroxymethyl-2-nitro-1,3-propanediol from nitromethane (CH_3NO_2) and methanal (formaldehyde, H_2CO) in the presence of a base catalyst.

The final example in Table 9.5, the **Robinson annulation**, actually involves two steps. In the first step, conjugate addition (a **Michael reaction**) occurs.*

---

*Conjugate addition, the **Michael reaction**, has been encountered before in this chapter. In Equations 9.37 and 9.38 as well as Scheme 9.38, addition of cyanide anion (^-CN) to α,β-unsaturated ketones was set forth. Then, in **Equation 9.44**, it was shown that a **Gilman reagent**, lithium dimethylcuprate [$(CH_3)_2CuLi$], could be used to effect addition of a methyl group to the β-position of an α,β-unsaturated ketone. Finally, in Scheme 9.57, dimethylsulfoxonium methylide [$(CH_3)_2S^+(O)CH_2^-$] was seen to react (initially) β- to the carbonyl of an α,β-unsaturated ketone to produce, subsequently, a three-membered ring (a cyclopropane) across what was the double bond of the original unsaturation.

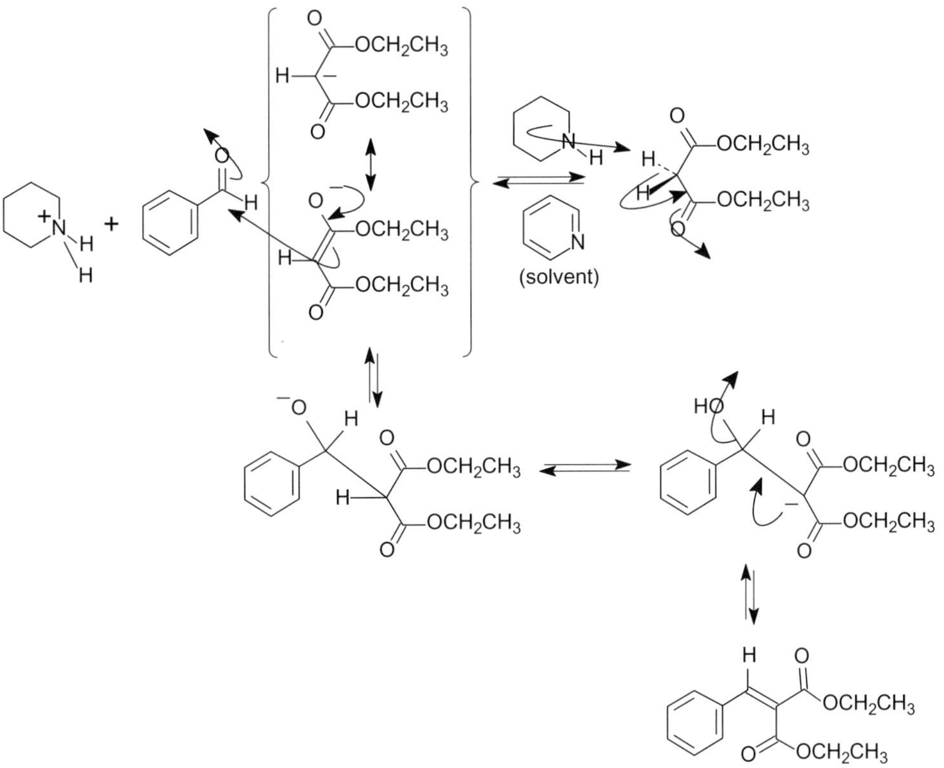

Scheme 9.84. A representation of the Knovenagel condensation. The reaction shown is the reaction between benzenecarboxaldehyde (benzaldehyde, C_6H_5CHO) and diethyl propanedicarboxylate (diethyl malonate [$CH_2(CO_2CH_2CH_3)_2$]) in the presence of azacyclohexane (piperidine, $C_5H_{11}N$) and azabenzene (pyridine, C_5H_5N). The early work was reported by Japp, F. R.; Streatfield, F. W. *J. Chem. Soc.*, **1883**, *43*, 27, and only later by Knovenagel, who developed the reaction further. See Knovenagel, E. *Chem. Ber.*, **1898**, *31*, 735 and references therein.

In the Robinson annulation (as shown in Scheme 9.85), it is argued that, initially, ethoxide anion ($^-OCH_2CH_3$) abstracts a proton α to the carbonyl of cyclohexanone. The enolate anion thus produced affects the **Michael reaction** by addition to the terminus (the β-position) of the α,β-unsaturated ketone (1-butene-3-one) producing a new enolate anion. Then, perhaps following tautomerization to the ketone, a new enolate anion, generated by removal of a proton α to the carbonyl group (on the methyl), is produced (in equilibrium with other enolate anions) and aldol-type condensation of the latter with the original carbonyl group of the cyclohexanone occurs. This step produces a *new six-membered ring*. Finally, dehydration of the bicyclic aldol product yields the observed bicyclo[4.4.0]-$\Delta^{1,8a}$-decene-2-one.

f. Addition to the Carbon Alpha (α) to the Carbonyl (C=O). Hopefully, it is clear that the various aldol-type and related condensation reactions seen in the previous section (*vide supra*, Table 9.5), where addition of an enol or enolate anion

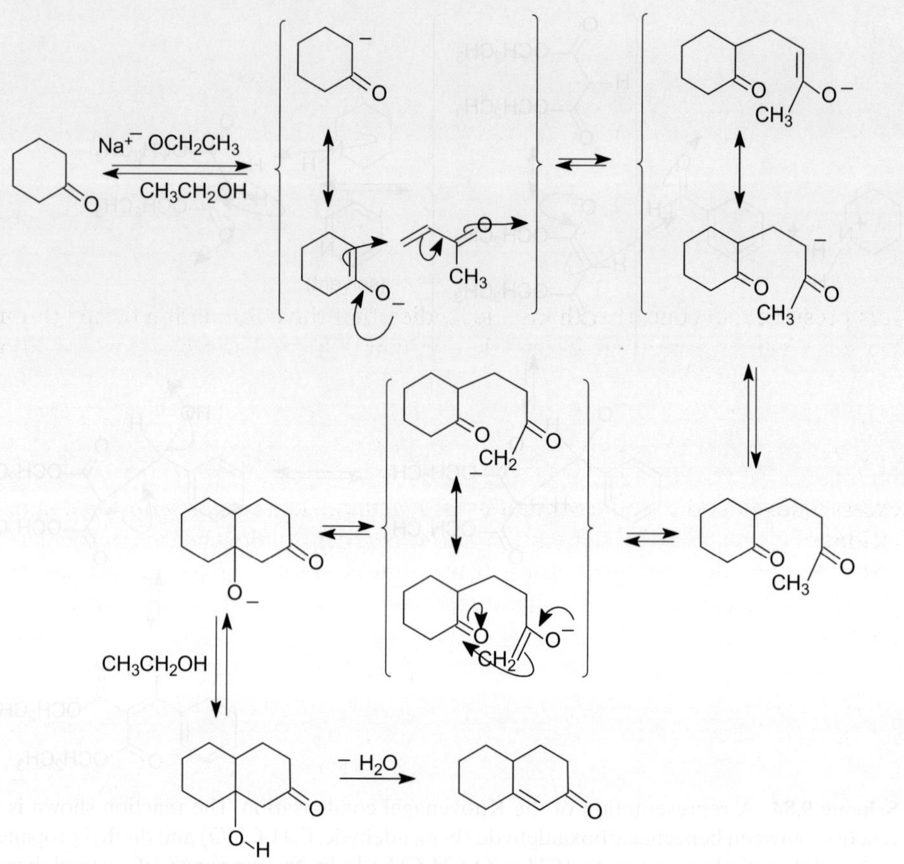

Scheme 9.85. The **Robinson annulation**. The reaction of cyclohexanone with 1-butene-3-one in the presence of ethoxide anion to produce the corresponding bicyclo[4.4.0]-$\Delta^{1,8a}$-decene-2-one. See Rapson, W. S.; Robinson, R. *J. Chem. Soc.*, **1935**, 1285.

to the carbon of a carbonyl group (C=O) eventually led to product, not only involved addition to the carbon of the carbonyl but that it also involved addition of the enol or enolate. That is, carbon–carbon bond formation α- to the carbon of the carbonyl (albeit addition to another carbonyl group) occurred.

In the absence of simple proton exchange (e.g., where the alkylating species possesses an acidic proton), addition α- to the carbon of the carbonyl (C=O) is an important synthetic tool.

I. HALOGENATION. As already noted (Scheme 9.9), halogens such as bromine (Br$_2$) and chlorine (Cl$_2$) in a basic medium react readily with aldehydes and ketones to produce α-halo-derivatives. However, the ease with which aldehydes are oxidized often results in the isolation of carboxylic acid rather than aldehydic products from these starting materials (Scheme 9.3), thus limiting the utility of the halogenation reaction with aldehydes.

With ketones, such as cyclohexanone, monochlorination can be effected (Equation 9.62).

$$\text{(9.62)}$$

It is presumed, in concert with kinetic studies, that the chlorination occurs through the enol, whose formation is rate determining. Under basic reaction conditions, where the enolate anion is formed, it is difficult to avoid polyhalogenation and halogens accumulate on the *same* carbon atom since the presence of a halogen on a carbon increases the acidity of the remaining protons (see the **haloform reaction** [this chapter]) on the halogen-bearing carbon. The course of the same reaction with excess halogen under acidic conditions is, in general, less certain with some ketones yielding halogens on the same atom and others with halogens symmetrically disposed around the carbonyl group. Thus, dibromination of propanone (acetone, $[(CH_3)_2C=O]$) in water (where hydrogen bromide [HBr] production accompanies bromoketone formation and the acidity of the reaction medium increases as the reaction progresses) yields principally 1,1-dibromo-2-propanone (Equation 9.63), while in dilute ethanoic acid (acetic acid, CH_3CO_2H), cyclohexanone generates predominately 2,6-dibromocyclohexanone (Equation 9.64) under similar conditions.

$$\text{(9.63)}$$

$$\text{(9.64)}$$

II. ALKYLATION. The generalized reaction resulting in alkylation α- to the carbonyl group enjoys both the subtleties of nucleophilic substitution processes and reactions of bidentate anions.

As shown in Scheme 9.86, a nonsymmetrically substituted carbonyl group can, in principle, produce four different enolate anions (or their corresponding enols). Again, in principle, each of these enolate anions (or enols) could undergo alkylation on carbon or oxygen leading to an intractable mixture of isomers (Problem 9.16). Fortunately, not all isomers form with equal facility.

Problem 9.16. With reference to Scheme 9.86: (a) Assume that R = R′, and it is achiral and remains so throughout. Under these conditions which, if any, of the 12 products in the scheme are identical? (b) Assume that R, R′, and R″ are all different and that they are achiral (and remain so throughout). Under these conditions which, if any, of the 12 products in the scheme are identical?

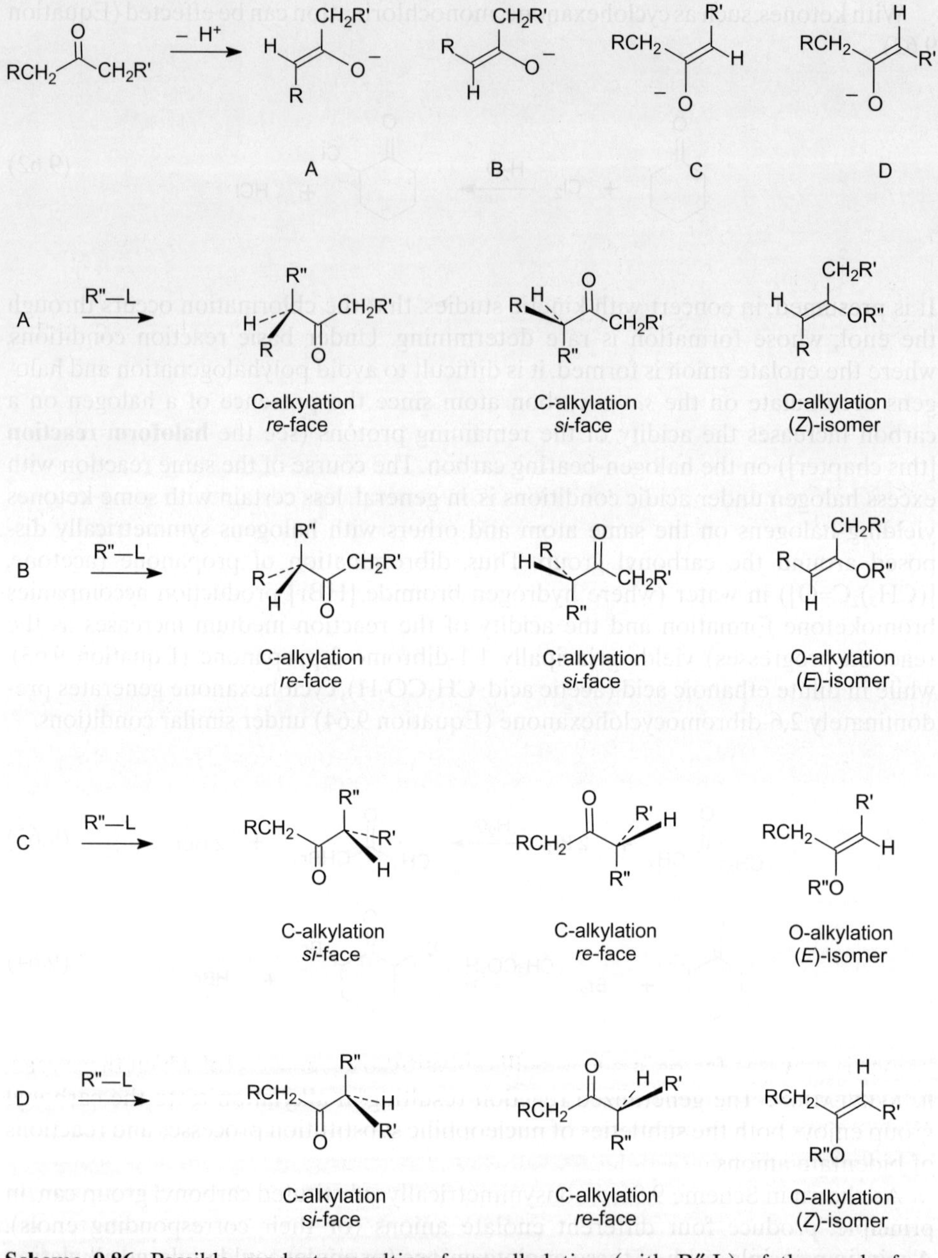

Scheme 9.86. Possible products resulting from alkylation (with R″-L) of the enol of an unsymmetrically substituted ketone (RCH₂COCH₂R′).

III. C-ALKYLATION VERSUS O-ALKYLATION. Since oxygen is more electronegative than carbon (3.5 vs. 2.4, Chapter 1, **Part II**), it is reasonable to suppose that most of the electron density of an enolate anion would reside on oxygen and O-alkylation would predominate. However, since charge density on a nucleophile (ambident enolate

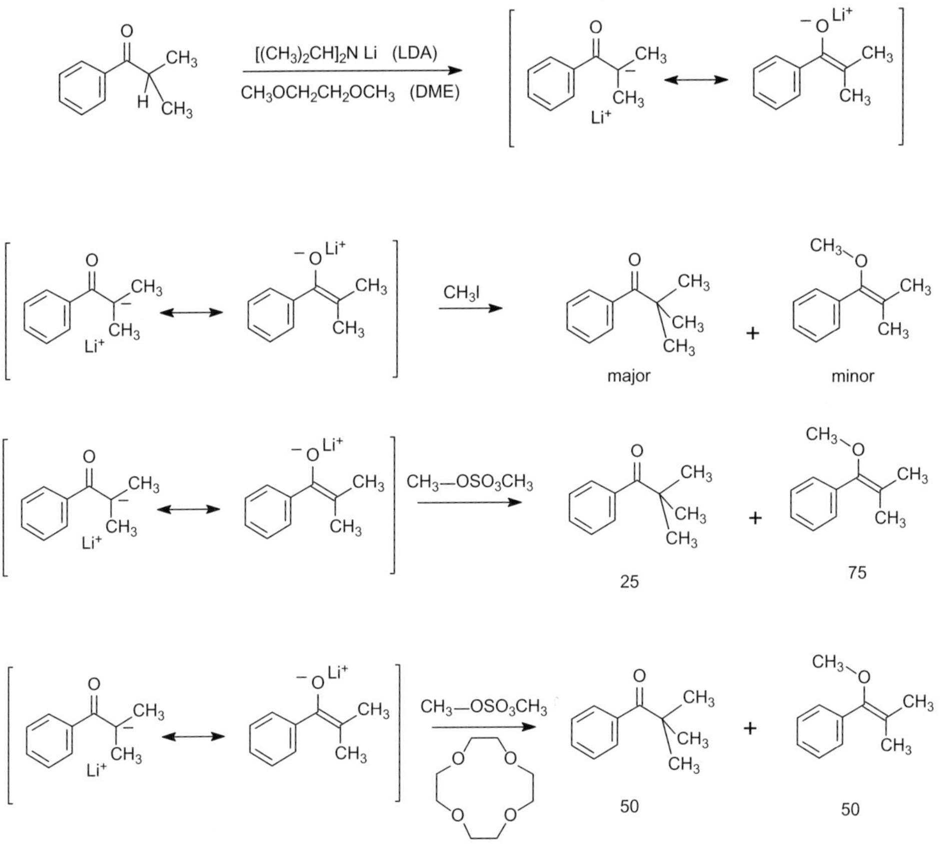

Figure 9.12. O- versus C-alkylation of isopropyl phenyl ketone. (See Jackman, L. M.; Lange, B. C. J. *Am. Chem. Soc.*, **1981**, *103*, 4494.)

anion) is only one factor in nucleophilic substitution reactions (of which these reactions may be considered a subset), conditions can usually be found to insure that C-alkylation predominates.

As a general rule-of-thumb, O-alkylation occurs when the anion is unencumbered by a counter ion or solvent (i.e., a "naked anion") such as hexamethylphosphoramide (HMPA, $[(CH_3)_2N]_3P=O$) and when the leaving group of the alkylating species also contains oxygen (e.g., sulfate [as in dimethyl sulfate $(CH_3)_2SO_4$)], sulfinate [as in methyl methylsulfinate $(CH_3O_3SCH_3$), etc.). For example (Figure 9.12), in DME ($CH_3OCH_2CH_2OCH_3$) solvent, the lithium enolate of isopropyl phenyl ketone $[C_6H_5COCH(CH_3)_2]$ undergoes largely C-alkylation when the alkylating agent is methyl iodide (as would be expected for a poorly solvated, strongly coordinated enolate anion). The addition of [12]-crown-4 (see Chapter 8, Table 8.8) in which the lithium cation (Li^+) can serve as a guest and which should liberate the oxygen (although solvation remains poor) produces mainly C-alkylation. However, when, under the same conditions, the leaving group is changed from iodide (I^-) to methyl sulfate ($CH_3OSO_3^-$) (i.e., the alkylating species is changed from methyl iodide (CH_3I) to dimethyl sulfate [$(CH_3O)_2SO_2$]), the O- to C-alkylation ratio becomes about 3:1, and even when [12]-crown-4 is added, it only drops to 1:1. Thus, most generally, when alkylations are carried out with alkyl halides in ether solvents (e.g., oxacyclopentane [THF] or DME ($CH_3OCH_2CH_2OCH_3$), C-alkylation will predominate.

Interestingly, however, despite changes in cation and solvent, it is often stereochemical demands that dictate the product formed. For example, in the intramolecular cyclizations shown in Scheme 9.87, treatment of 5-bromo-3,3-dimethyl-2-pentanone with lithium diisopropylamide ($[(CH_3)_2C]_2N^-$ Li^+, LDA) in ether results in O-alkylation with the formation of 2-methylene-3,3-dimethyloxacyclopentane (2-methylene-3,3-dimethyltetrahydrofuran), despite the fact that carbon–oxygen bonds tend to be shorter (141 pm) than carbon–carbon bonds (154 pm). On the other hand, the analogous 6-bromo-3,3-dimethyl-2-hexanone, under the same reaction conditions, produces 2,2-dimethylcyclohexanone.

As shown in the Scheme 9.87, it is argued that orbital overlap of the π-system, necessary for enolate stability, cannot accommodate the geometric requirement of backside displacement needed for the formation of a five-membered carbocycle. The heterocycle, with greater geometric latitude, forms instead. However, addition of an extra methylene unit gives the system sufficient freedom to produce the carbocycle.

IV. REGIOSELECTIVITY. Treatment of the symmetrical ketone cyclohexanone (or any such symmetrical ketone) in an etherial solvent such as THF (oxacyclopentane) or dimethoxyethane ($[CH_3OCH_2CH_2OCH_3]$, DME) with an appropriate base such as sodium amide ($NaNH_2$), the anion of dimethyl sulfoxide (the *dimsyl anion* [$CH_3SOCH_2^-$]), LDA ($[(CH_3)_2CH]_2NLi$), or lithium hexamethyldisilazide [$([CH_3]_3Si)_2NLi$] serves to generate the enolate anion, and subsequent addition of methyl iodide (CH_3I) results in the formation of 2-methylcyclohexanone (Equation 9.65).

$$(9.65)$$

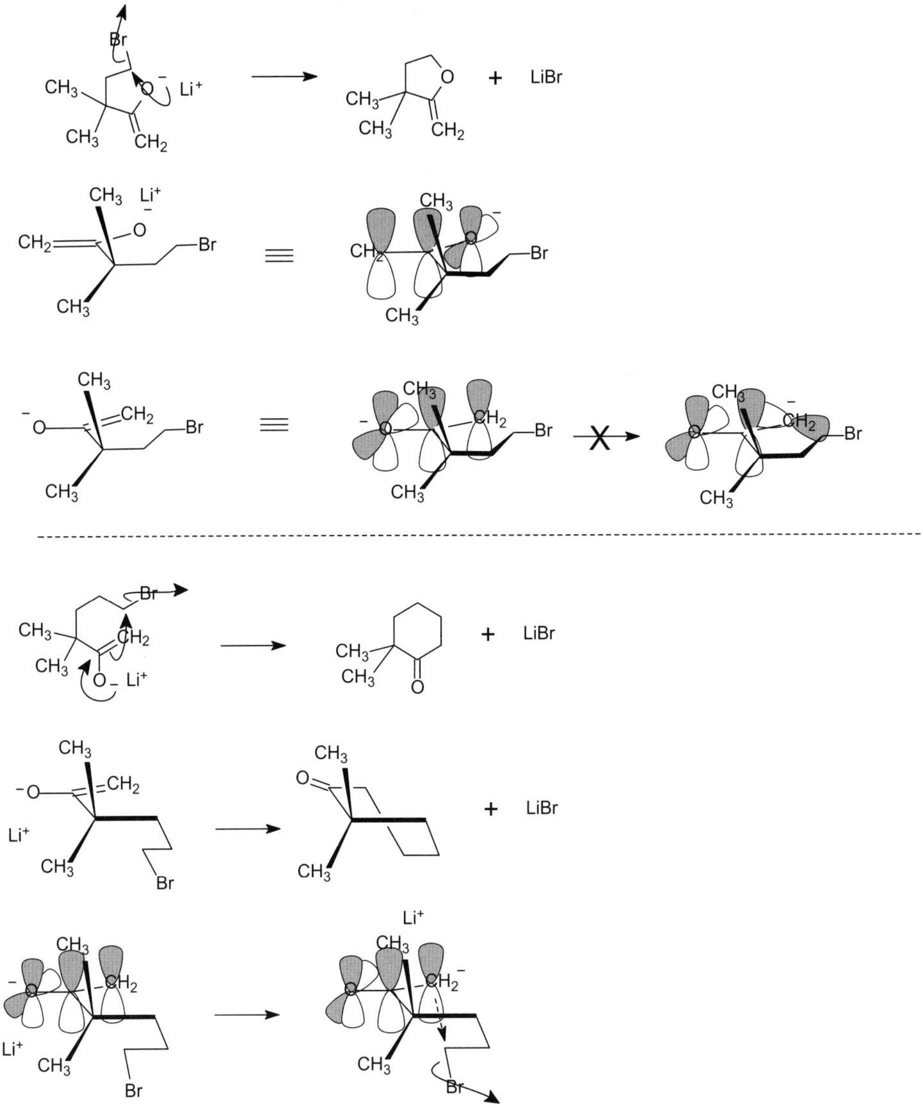

Scheme 9.87. The reaction of 5-bromo-3,3-dimethyl-2-pentanone with lithium diisopropyl-amide ($[(CH_3)_2C]_2N^-$ Li^+, LDA) in ether resulting in O-alkylation and the formation of 2-methylene-3,3-dimethyloxacyclopentane (2-methylene-3,3-dimethyltetrahydrofuran) and the analogous reaction of 6-bromo-3,3-dimethyl-2-hexanone under the same reaction conditions to produce 2,2-dimethylcyclohexanone. (See Baldwin, J. E.; Kruse, L. I. *J. Chem. Soc., Chem. Commun.*, **1977**, 233.)

However, when 2-methylcyclohexanone is treated under the same conditions, two regioisomeric cyclohexanone enolates are expected (Equation 9.66). The issue of regioselectivity is extant in enol formation from any unsymmetrically substituted ketone and must be addressed in order to control the site of the alkylation in the next step.

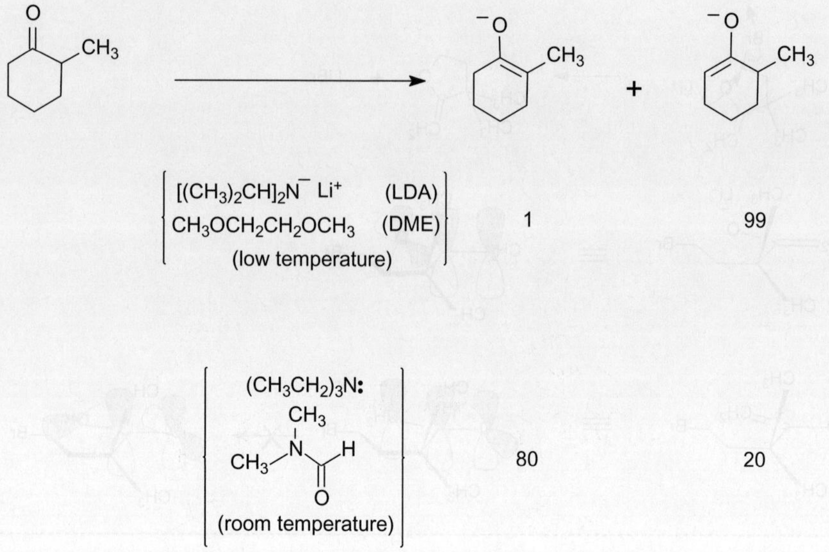

Figure 9.13. The ratio of enolate anions formed from 2-methylcyclohexanone as a function of their method of formation. (See House, H. O.; Kramar, V. *J. Org. Chem.*, **1963**, *28*, 3362, and Caine, D. *J. Org. Chem.*, **1964**, *29*, 1868.)

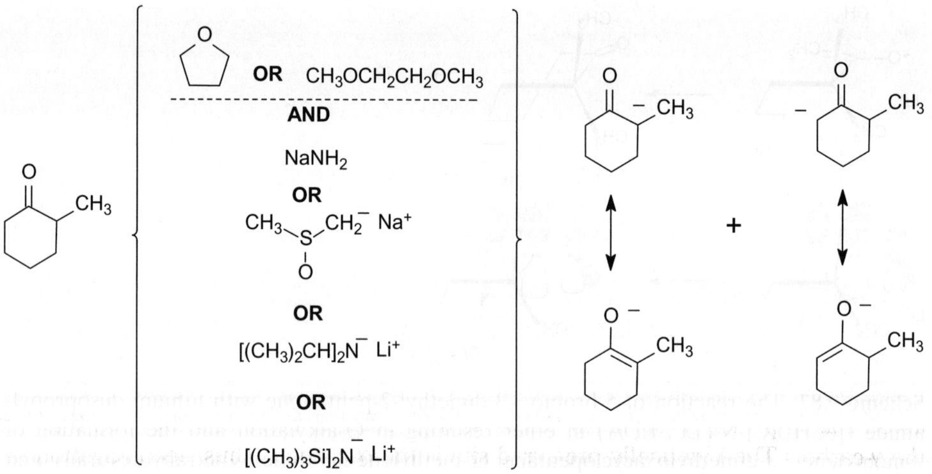

$$(9.66)$$

Expectedly, since more highly substituted alkenes are more stable than less (Chapter 6), if equilibration of the enolate anions (through reprotonation and subsequent deprotonation) is permitted, alkylation will occur on the more highly substituted carbon (Figure 9.13). This is said to be the ***thermodynamic product*** since equilibration has occurred. Alternatively, when the rates of proton abstraction from one side or the other govern the product ratio (because the enolate anions do not have ***time*** to equilibrate before reacting), the product mixture reflects this and the ***kinetic product*** is obtained (Figure 9.13).

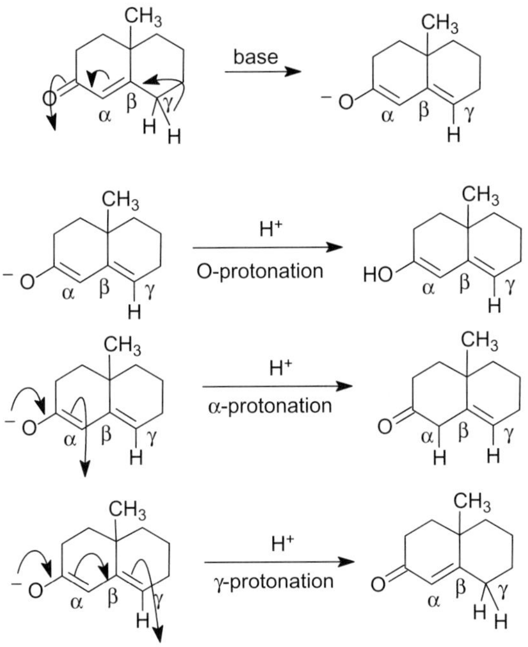

Figure 9.14. Deprotonation and reprotonation (or alkylation) of an α,β-unsaturated 2-decalone showing that reaction can occur on oxygen or β- or γ- to the carbon of the carbonyl. The kinetically preferred site is β-, thus leading to deconjugation of the conjugated ketone. (See Stork, G.; Rosen, P.; Goldman, N.; Coombs, R. V.; Tsujii, J. *J. Am. Chem. Soc.,* **1965,** *87,* 275.)

Thus, as shown in Figure 9.13 for 2-methylcyclohexanone, use of LDA ([(CH₃)₂CH]₂NLi) in DME (CH₃OCH₂CH₂OCH₃) *at low temperatures* produces a 99:1 ratio of less-to-more substituted enolate anions, while a 4:1 mixture is produced using triethylamine [(CH₃CH₂)₃N] as the base in N,N-dimethylformamide [(CH₃)₂NCHO] solvent (Figure 9.13).

Interestingly, if there is an α,β-double bond in conjugation with the carbon of the carbonyl (Figure 9.14), then the proton γ- to the carbonyl can be removed by the base. In such circumstances, protonation (or alkylation) of the resulting enolate anion can, in principle, occur on the oxygen (producing the enol), the α-carbon, or the γ-carbon. The kinetically preferred site is the α-carbon. Thus, deconjugation in the event of protonation can be effected.*

v. STEREOSELECTIVITY. As already shown in Scheme 9.86 and Problem 9.16, both symmetrically and unsymmetrically substituted ketones can lead to stereochemically diverse enolate anions.

---

*It is reasonable to inquire why one of the two *other* protons α- to the carbonyl is not considered acidic enough for removal and thus capable of producing either enol (or enolate anion) or alkylated product. Generally, it has been found that such cross-conjugated systems, although composed of the same number of double bonds, are thermodynamically less stable than their linearly conjugated counterparts.

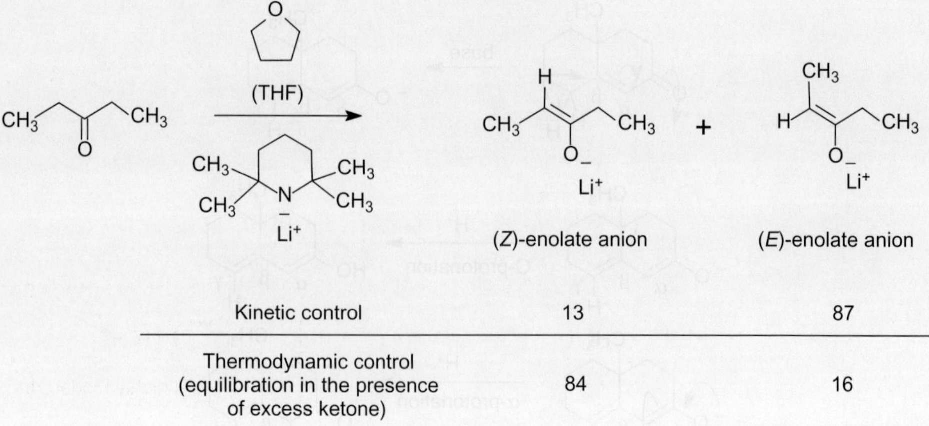

	(Z)-enolate anion	(E)-enolate anion
Kinetic control	13	87
Thermodynamic control (equilibration in the presence of excess ketone)	84	16

Figure 9.15. Formation of the (*E*)- and (*Z*)-enolates from 3-pentanone under both equilibrating and nonequilibrating conditions in THF (oxacyclopentane) with lithium 2,2,6,6-tetramethylpiperidide as the base. (See Fataftah, Z. A.; Kopka, I. E.; Ratke, M. W. *J. Am. Chem. Soc.*, **1980**, *102*, 3959, and Balamraju, Y.; Sharp, C. D.; Gammill, W.; Manue, N.; Pratt, L. M. *Tetrahedron*, **1998**, *54*, 7357.)

Consider, for simplicity, the case of the symmetrical ketone 3-pentanone (diethylketone [(CH$_3$CH$_2$)$_2$C=O]) (Figure 9.15) where the (*Z*)-enolate predominates under conditions of thermodynamic control (equilibration) and the (*E*)-enolate is the major isomer under conditions of kinetic control (first formed) when the base is lithium 2,2,6,6-tetramethylpiperidide and the solvent is THF (oxacyclopentane). Interestingly, each specific case must be examined for base and solvent. Thus, while it would be predicted that in unsymmetrical cases steric effects should play an important role in establishing the thermodynamic isomer, it does not follow that the thermodynamic isomer is not also the isomer first formed.

VI. ENEAMINE-ASSISTED ALKYLATION OF KETONES. As shown in Figure 9.13 (*vide supra*) the kinetic enolate of 2-methylcyclohexanone (formed at low temperature) is the enolate anion of the *less substituted* alkene and, reasonably, alkylation with, for example methyl iodide will yield 2,6-dimethylcyclohexanone. Regrettably, however, the yield of alkylated product from the kinetic enolate is difficult to maximize since even the smallest perturbation (e.g., rise in temperature and presence of excess ketone.) tends to drive the system toward equilibrium, resulting in the more highly substituted enolate anion. Enamines can be employed to overcome this tendency to produce alkylation on the more substituted side.

It will be recalled (Scheme 9.66) that reaction of ketones with secondary amines (such as azacylopentane [pyrrolidine]) produces enamines. Now, as shown in Scheme 9.88, when the unsymmetrically substituted ketone (2-methylcyclohexanone) forms an enamine, that specific enamine which is less sterically encumbered is favored and alkylation occurs on the previously unsubstituted side producing 2,6-dimethylcyclohexanone.

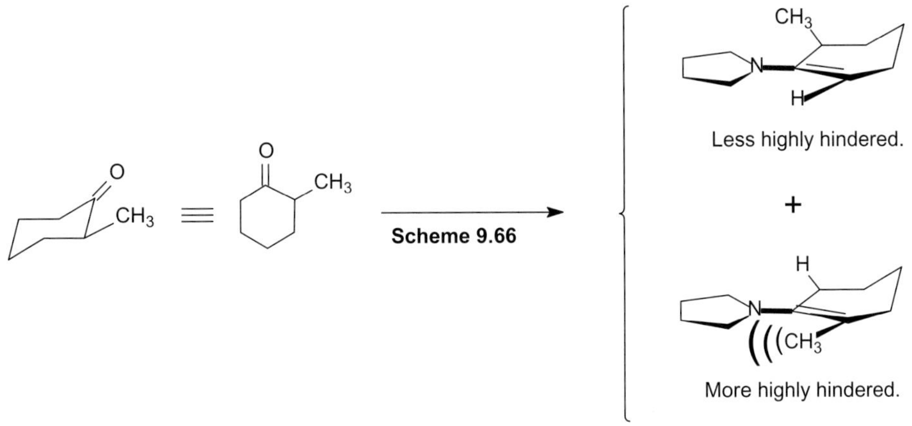

Scheme 9.88. The reaction of pyrrolidine (azacyclopentane) with 2-methylcyclohexanone (Scheme 9.55) to form two different enamines. The less hindered enamine is the thermodynamically favored and will, via alkylation on the *less* highly substituted carbon, produce 2,6-substituted cyclohexanone. (See Gurowitz, W. D.; Joseph, M. A. *J. Org. Chem.*, **1967**, *32*, 3289.)

Problem 9.17. Given the relative electronegativities of nitrogen and carbon (Chapter 1), it might be supposed that alkylation on nitrogen **and/or** carbon might occur. What products would result from alkylation of the enamines in Scheme 9.87 *on nitrogen* instead of carbon with methyl iodide (CH_3I)? What factors might favor or preclude nitrogen alkylation?

Interestingly, if ammonium chloride ($NH_4^+Cl^-$) is used in the presence of methanal (formaldehyde, H_2CO), the iminium ion thus formed ($H_2C=NH_2^+Cl^-$), lacking α-protons and thus incapable of forming an enamine, is nonetheless capable of condensing α- to the carbonyl of ketones that do have α-protons. The two-step process apparently occurs via the enol of the α-proton-bearing carbonyl component adding to the carbon of the iminium ion to eventually produce a primary amine. This is the **Mannich reaction**. The process is sketched in Scheme 9.89 for methyl phenyl ketone (acetophenone, $C_6H_5COCH_3$).

D. SUBSTITUTION REACTIONS PRODUCING ALDEHYDES AND KETONES

I. Introduction

In contrast to previous sections on **substitution** where hydrogen (Chapter 6), halogen (Chapter **7**) or hydroxyl (Chapter 8) attached to carbon was replaced by some other (*substituting*) species, this section is devoted to the introduction of the carbonyl group where it will be shown to replace a hydrogen.

This material was already discussed (Chapter 6) and, indeed, reference will be made to that chapter and the more limited examples provided there. Here, with the reader somewhat more understanding, it will be recognized that not all of the desired information could have been conveyed earlier. Further, it will be seen later

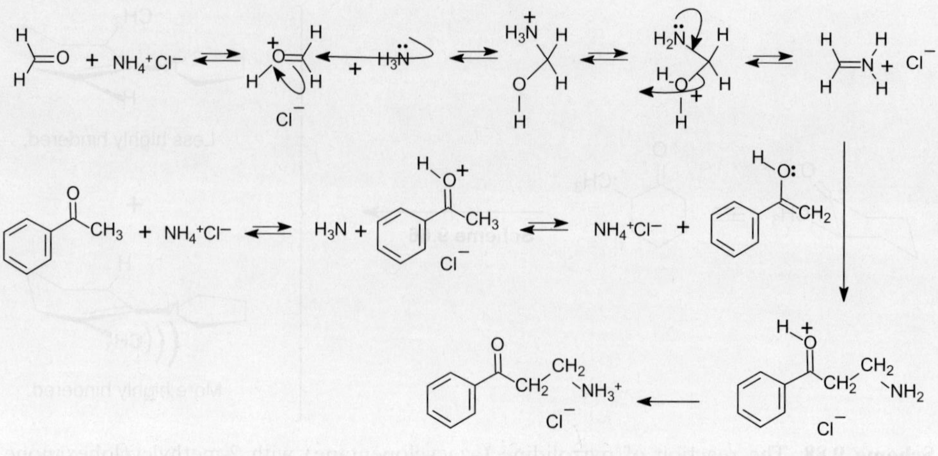

Scheme 9.89. The **Mannich** reaction. The reaction between methanal (formaldehyde, H_2CO) and ammonium chloride ($NH_4^+Cl^-$) produces the corresponding iminium ion ($H_2C=NH_2^+Cl^-$), which cannot form an enamine (it is lacking α-protons). The iminium ion is capable of condensing α- to the carbonyl (shown occurring via the enol) of methyl phenyl ketone (acetophenone, $C_6H_5COCH_3$). (See Mannich, C.; Schumann, P. *Chem. Ber.*, **1936**, *69*, 2299, and Schreiber, J.; Maag, H.; Hashimoto, N.; Eschenmoser, A. *Angew. Chem. Int. Ed.*, **1971**, *10*, 330.)

in this chapter, when the discussion has turned to carboxylic acids and their derivatives, that *replacement of the carbonyl group of the carboxyl* does not occur without oxidation. This has already been briefly outlined (see, e.g., Scheme 9.5).

II. Reimer–Tiemann Synthesis

The observation that an aldehyde group can be substituted for a hydrogen *ortho*- or *para*- to a phenolic hydroxyl (an HO– on an aromatic ring) by treatment of the hydroxybenzene (phenol) with trichloromethane (chloroform, $CHCl_3$) in the presence of concentrated aqueous sodium hydroxide ($NaOH_{(aq)}$) was reported by Reimer and Tiemann in 1876 (*vide supra* and *Chem. Ber.*, **1876**, *9*, 824). Thus, as shown in Equation 9.67, hydroxybenzene (phenol) undergoes conversion to

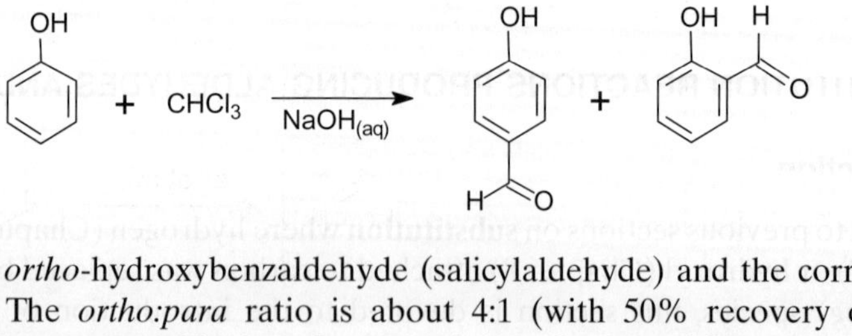

$$\text{(9.67)}$$

a mixture of *ortho*-hydroxybenzaldehyde (salicylaldehyde) and the corresponding *para*-isomer. The *ortho:para* ratio is about 4:1 (with 50% recovery of starting material) on treatment of hydroxybenzene (phenol) with trichloromethane (chloroform, $CHCl_3$) at 50–60°C in the presence of 40% aqueous sodium hydroxide ($NaOH_{(aq)}$).

There are two possible pathways by which the product can form. First, as shown in Scheme 9.90, it is possible that the phenoxide anion, produced by proton removal

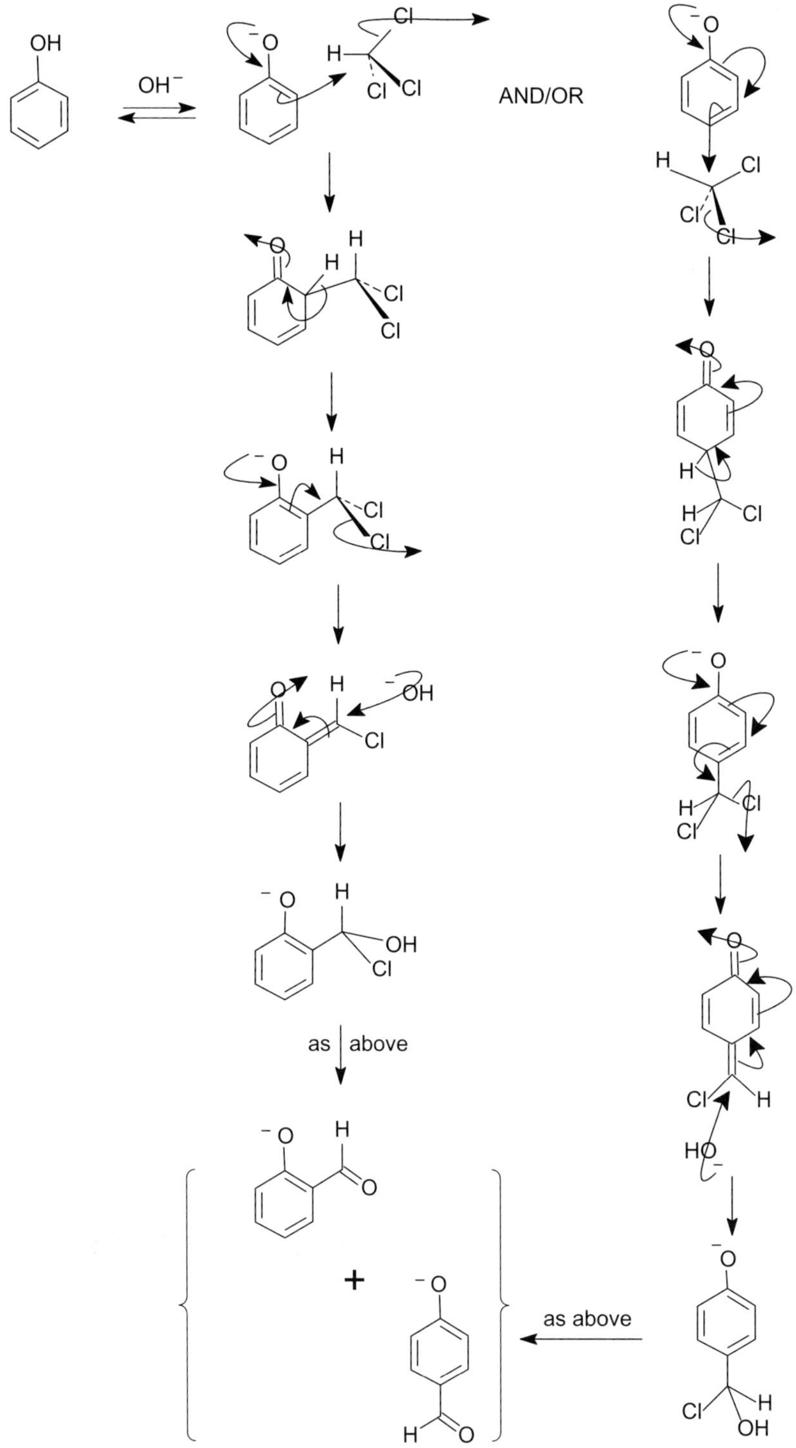

Scheme 9.90. One possible pathway (involving nucleophilic displacement of chloride anion by phenoxide anion at a position *ortho-* or *para-* to the original hydroxyl [–OH] group) for the Reimer–Tiemann conversion of a phenol to a mixture of phenolic aldehydes. The hydroxybenzene (phenol) is treated with trichloromethane (chloroform, $CHCl_3$) and aqueous sodium hydroxide ($NaOH_{(aq)}$).

by hydroxide from hydroxybenzene (phenol) [$pK_a \approx 10$], attacks the trichloromethane (chloroform, $CHCl_3$) *with carbon–carbon bond formation ortho-* or *para-* to produce the corresponding dichloromethyl–phenoxide species, which subsequently undergoes hydrolysis.

Alternatively, as shown in Scheme 9.91, initial proton loss from trichloromethane (chloroform, $CHCl_3$) [$pK_a \approx 15$] is followed by α-elimination (see, e.g., Scheme 7.20) of chloride anion (Cl^-) to produce dichlorocarbene (:CCl_2). As noted earlier (Chapter 7), dichlorocarbene (:CCl_2) generated by base treatment of trichloromethane (chloroform, $CHCl_3$) is generally an electrophilic singlet carbene that reacts rapidly with electron-rich species (e.g., alkenes). So, it is presumed that the dichlorocarbene (:CCl_2) reacts with the aromatic ring at the positions *ortho-* and *para-* to the hydroxyl group, that is, the positions of highest electron density.

Subsequent hydrolysis produces the corresponding aldehydes.

Problem 9.18. Consider what experiments might be carried out to attempt to differentiate the pathways of Schemes 9.90 and 9.91.

III. Gatterman–Koch (Friedel–Crafts) Formylation

The reaction between acyl halide and an aromatic system to produce aromatic aldehydes in the presence of aluminum chloride ($AlCl_3$) or other Lewis acid catalyst is formally analogous to the Friedel–Crafts acylation process already mentioned (Chapter 6, see, e.g., Scheme 6.88). The Friedel–Crafts acylation will be discussed again in this chapter as the reaction between **carboxylic acid chlorides (and related derivatives)** is particularly important (*vide infra*). However, for the formylation (i.e., the addition of the elements HC=O) to occur, formyl chloride (methanoyl chloride, HCOCl) would be required. Although there is good evidence that methanoyl chloride (formyl chloride, HCOCl) can be prepared at low temperatures (ca. –60°C), it is not stable under the conditions required for the reaction. Thus, the Gatterman–Koch process,* which utilizes aluminum chloride ($AlCl_3$) and copper(I) chloride [cuprous chloride, Cu_2Cl_2] as co-catalysts in the presence of hydrogen chloride for the reaction between an aromatic system, such as benzene (Equation 9.68) or methylbenzene (toluene) (Equation 9.69), and carbon monoxide (CO) has served as a viable alternative. While it is presumed that the catalyst coordinates with the carbon monoxide (CO) as well as the aromatic system during the substitution reaction, details remain sketchy.

$$\text{benzene} + CO \xrightarrow[HCl]{AlCl_3/Cu_2Cl_2} \text{benzaldehyde} \tag{9.68}$$

$$\text{toluene} + CO \xrightarrow[HCl]{AlCl_3/Cu_2Cl_2} \text{p-tolualdehyde} \tag{9.69}$$

*Gatterman, L.; Koch, J. A. *Chem. Ber.*, **1897**, *30*, 1622.

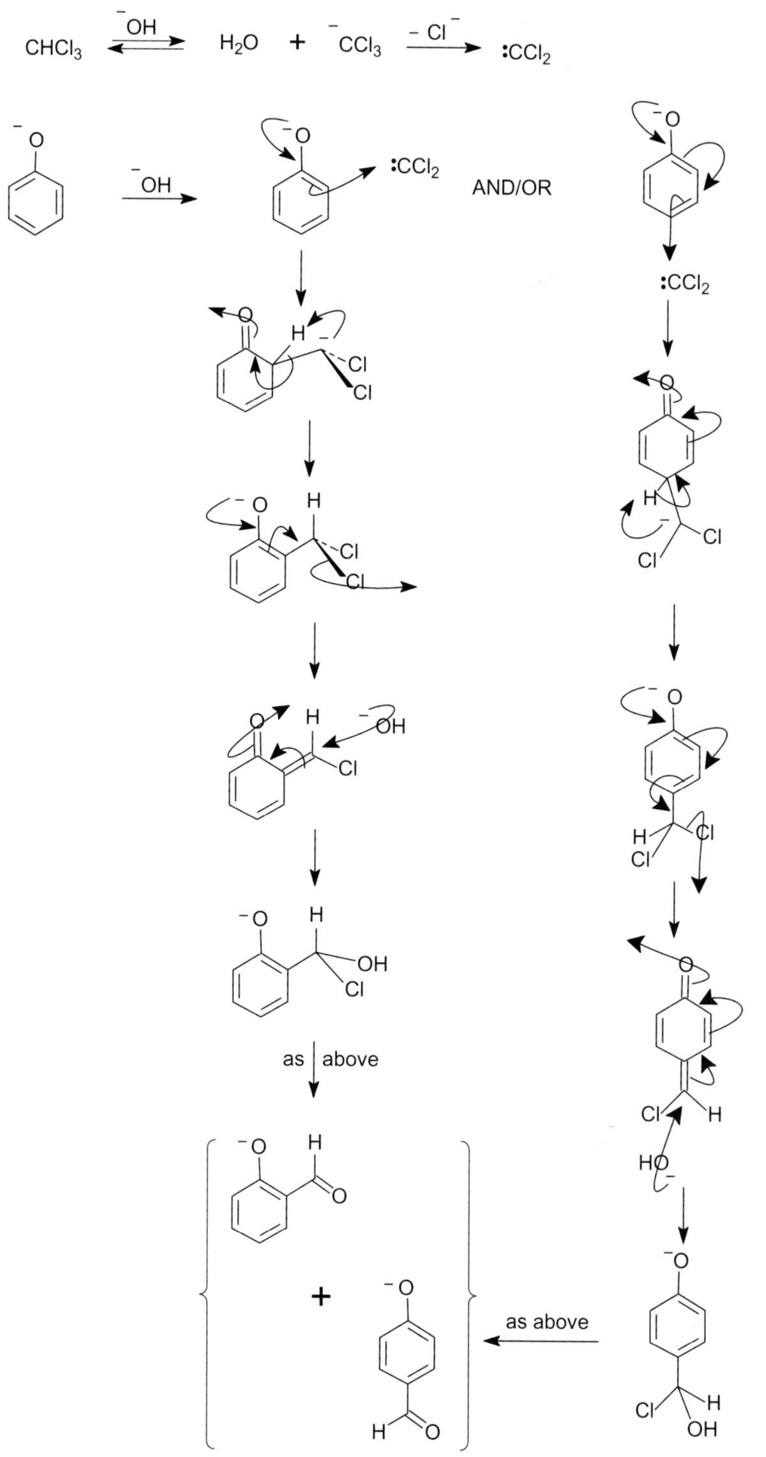

Scheme 9.91. One possible pathway (involving α-elimination of hydrogen chloride [HCl] from chloroform [HCCl₃] to generate singlet dichlorocarbene, which is then attacked by the phenoxide anion *ortho-* or *para-* to the original hydroxyl [–OH] group) for the Reimer–Tiemann conversion of a phenol to a mixture of phenolic aldehydes. The hydroxybenzene (phenol) is treated with trichloromethane (chloroform, CHCl₃) and aqueous sodium hydroxide (NaOH(aq)).

IV. The Pauson–Khand Reaction

The Pauson–Khand* reaction is a recent addition to the armamentarium of synthetic organic chemistry. In this process, a carbonyl group (provided initially as a ligand attached to dicobalt octacarbonyl [$Co_2(CO)_8$]) bridges an alkyne to an alkene to produce, finally, a cyclopentenone. While many of the details of the process remain sketchy, a general outline of the process is provided in Scheme 9.92.

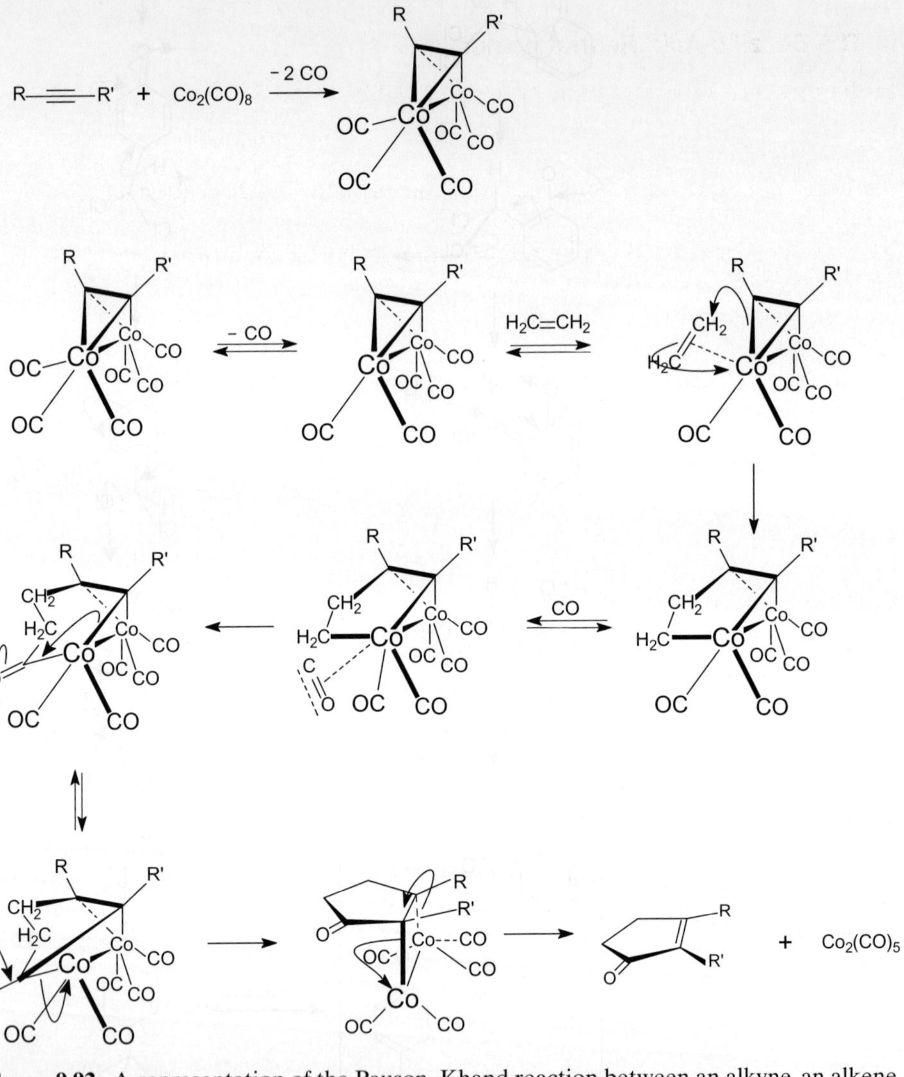

Scheme 9.92. A representation of the Pauson–Khand reaction between an alkyne, an alkene, and carbon monoxide (CO) in the presence of cobalt octacarbonyl [$Co_2(CO)_8$].

*Pauson, P. L. and Khand, I. U., University of Strathclyde, Glasgow, Scotland. See Khand, J. U.; Knox, G. R.; Pauson, P. L.; Watts, W. E. *J. Chem. Soc., Perkin Trans.*, **1973**, *1*, 975, and Pauson, P. L. *Tetrahedron*, **1985**, *41*, 5855.

E. REARRANGEMENT REACTION OF ALDEHYDES AND KETONES

I. Introduction

The carbon of the carbonyl is positive and thus represents an electron-deficient center toward which electron rich bonds migrate. The rearrangements encountered with aldehydes and ketones are generally acid or base catalyzed and thus other processes, such as condensation reactions (*vide supra*) may intrude.

II. The Benzilic Acid Rearrangement*

In the presence of strong base, aromatic 1,2-diketones are converted to salts of α-hydroxycarboxylic acids (Scheme 9.93).

It is important to note the similarities and differences between the benzylic acid rearrangement and the Cannizzaro reaction (Scheme 9.4). The aryl group migrates in the benzylic acid rearrangement. The hydride migrates in the Cannizzaro reaction. Further, the Cannizzaro reaction is **inter**molecular (second-order in aldehyde and first-order in base), while the benzylic acid rearrangement is **intra**molecular and first-order both in base and in diketone.

Aliphatic diketones rearrange similarly. Thus, 1,2-cyclohexanedione is converted to 1-hydroxy-1-cyclopentanecarboxylic acid (Equation 9.70).

$$(9.70)$$

Problem 9.19. Write a pathway, involving curved arrows, to account for the transformation depicted in Equation 9.70.

In a related process, acyloins (α-hydroxyketones and α-hydroxyaldehydes) are transformed into isomeric acyloins with dilute alcoholic sulfuric acid (H_2SO_4) as shown in Scheme 9.94.

III. The Dienone–Phenol Rearrangement

In α,β-unsaturated dienones (Scheme 9.95), migration of a γ-substituent can result in conversion of the dienone to a phenol.[†]

The dieneone–phenol rearrangement is also observed in numerous ring expansion processes (cf. Equation 9.71).

---

*Although Liebig himself (Liebig, J. *Liebigs Ann. Chem.*, **1838**, *25*, 1) found the rearrangement of benzil to hydroxycarboxylic acid (Scheme 9.93), it was not until almost 100 years later that Sir Christopher Ingold proposed the currently accepted mechanism (Ingold, C. K. *Ann. Rept. Chem. Soc.*, **1928**, *25*, 124).
[†]Interestingly, the first example of the dienone–phenol rearrangement was found in the sesquiterpene **santonin** ($C_{15}H_{18}O_3$). Santonin was initially reported to be isolated from the dried flower heads of wormseed (*Artemisia martima* L.) in 1830. It can still be obtained from that source and now it can be used as a "folk" medicine to combat intestinal worms. In cold aqueous sulfuric acid (H_2SO_4), santonin yields

(Continued on next page)

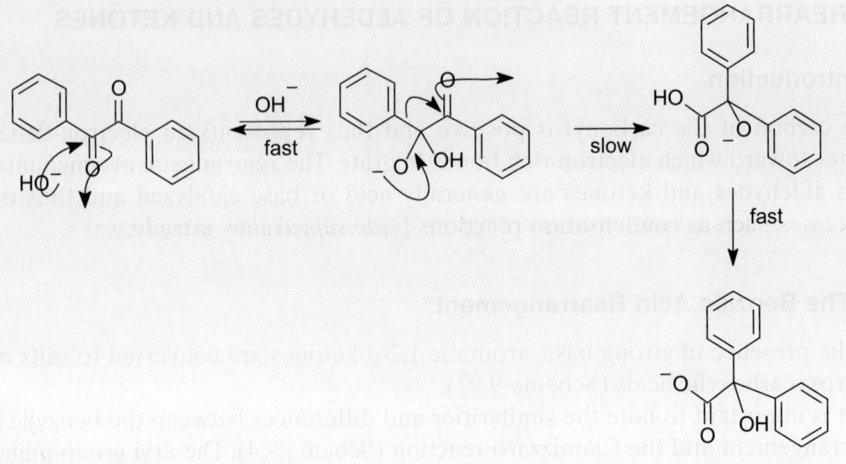

Scheme 9.93. The benzylic acid rearrangement involving the migration of a phenyl group and production of an α-hydroxycarboxylic acid.

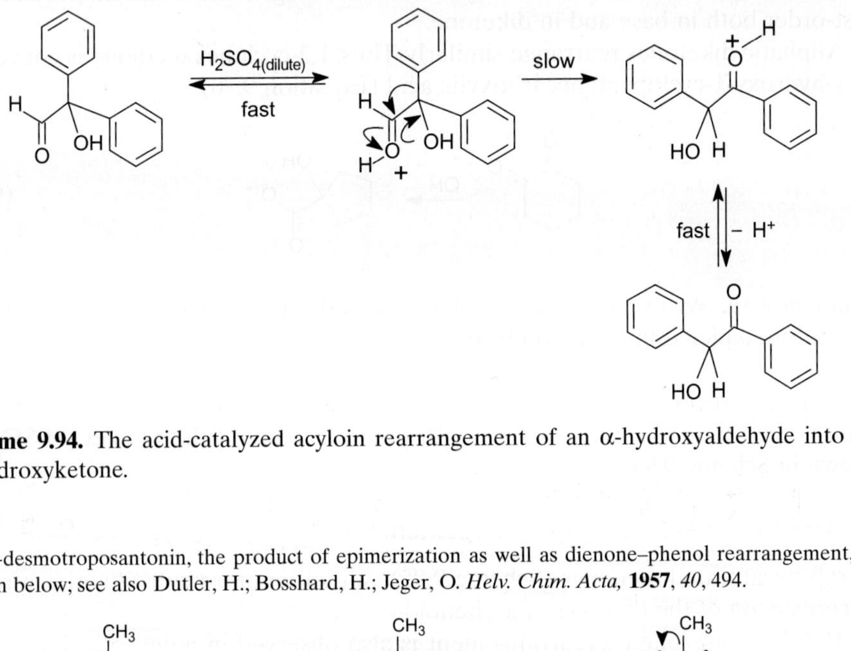

Scheme 9.94. The acid-catalyzed acyloin rearrangement of an α-hydroxyaldehyde into an α-hydroxyketone.

(–)-α-desmotroposantonin, the product of epimerization as well as dienone–phenol rearrangement, as shown below; see also Dutler, H.; Bosshard, H.; Jeger, O. *Helv. Chim. Acta*, **1957**, *40*, 494.

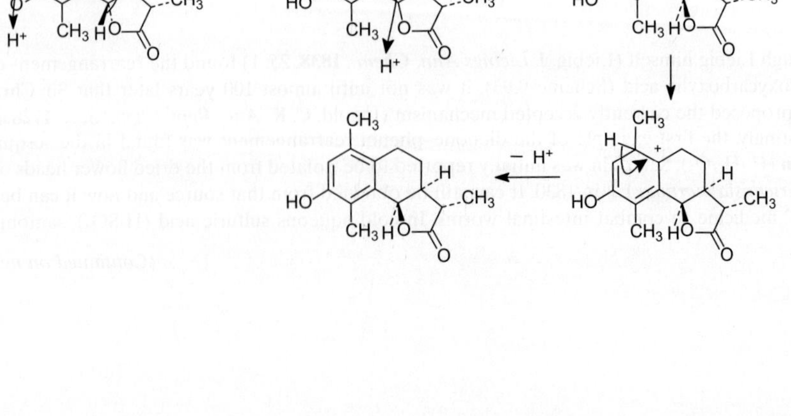

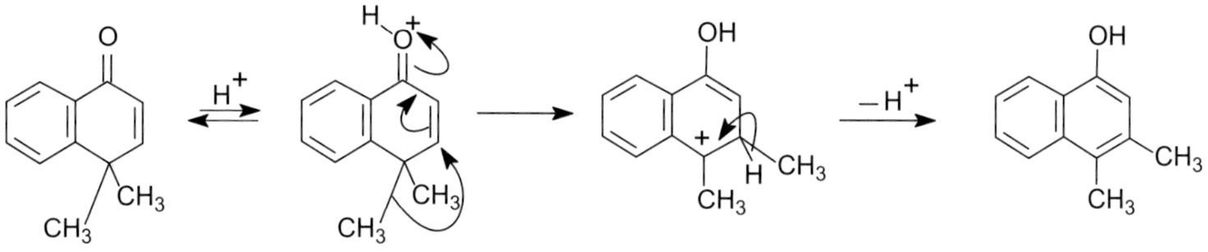

Scheme 9.95. An example of the dieneone–phenol rearrangement.

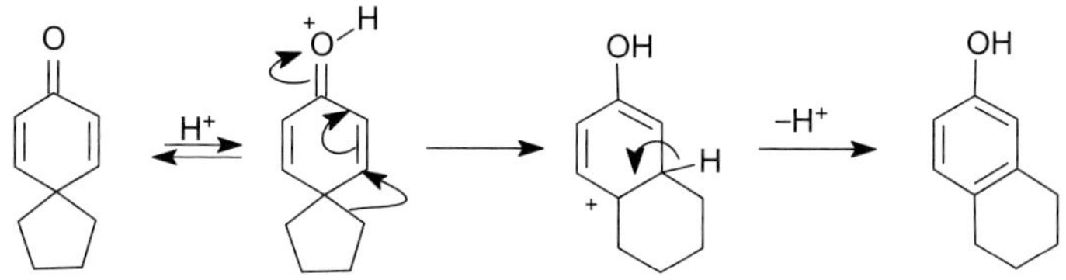

$$(9.71)$$

IV. Anionic Rearrangements

Rearrangements that result when an anion α- to the carbonyl or a similar negatively charged species is produced are common. Thus, the **Stevens*** rearrangement can be effected on both ammonium (Equation 9.72) and sulfonium (Equation 9.73) salts to produce α-aminoketones and α-ketothioethers, respectively.

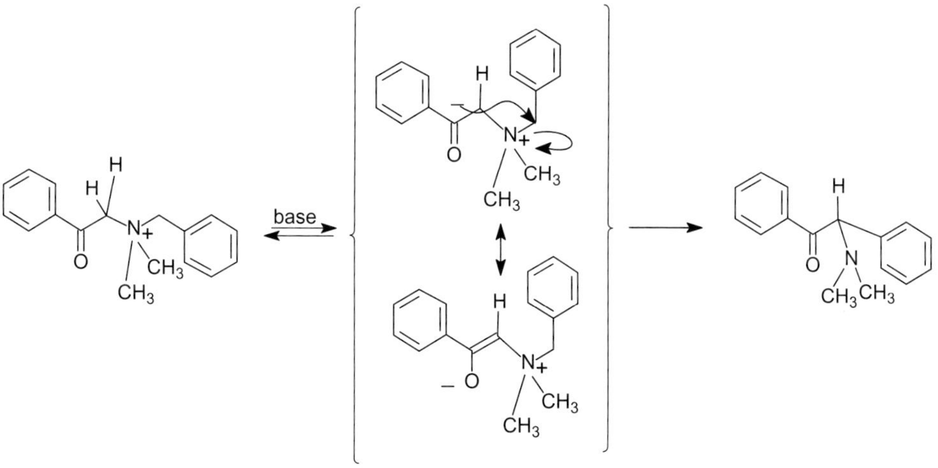

$$(9.72)$$

*Stevens, T. S., Creighton, E. M.; Gordon, A. B.; MacNicol, M. *J. Chem. Soc.*, **1928**, 3193, *et seq.*

$$(9.73)$$

Similarly, α-haloketones undergo rearrangement (the **Favorskii** rearrangement*) to esters (*vide infra*) on treatment with alkoxide in a process that can be written as one involving a simple migration of a group attached to the carbonyl (Equation 9.74). Interestingly, the Favorskii rearrangement may occur by other pathways (Schemes 9.96–9.98).

$$(9.74)$$

Thus, the Favoraskii rearrangement carried out on ^{14}C-labelled-α-chlorocyclo-hexanone results in a label distribution suggesting that a three-membered ring is involved. It is not clear whether an intramolecular S_N2 type displacement occurs (Scheme 9.96) or an α-elimination to generate a carbine, which then inserts (Scheme 9.97). However, while these both might be possible, even in lieu of addition to the carbonyl followed by migration of the group attached to the carbonyl (Equation 9.74), it is quite clear that such migratory processes also occur since, as shown in Scheme 9.98, a Favorskii rearrangement of 1-chloro-1-benzoylcyclohexane, a species lacking any protons α to the carbonyl nonetheless also occurs!

Problem 9.20. Presume that the Favorskii rearrangement of 2-^{14}C-labelled 2-chlorocyclohexanone could occur along a pathway similar to Equation 9.74 (Scheme 9.98). What labeled product(s) would be expected?

Although the subject will arise again in Chapter 10, it is worthwhile considering at this point the rearrangement of oximes derived from aldehydes and ketones. It will be recalled (item 4, Table 9.4) that the reaction of hydroxylamine hydrochloride (HONH$_2$·HCl) with these carbonyl compounds yielded the possibility of (*E*)- and

*Favoraskii, A.J. *Russ. Phys. Chem. Soc.*, **1905**, *37*, 643.

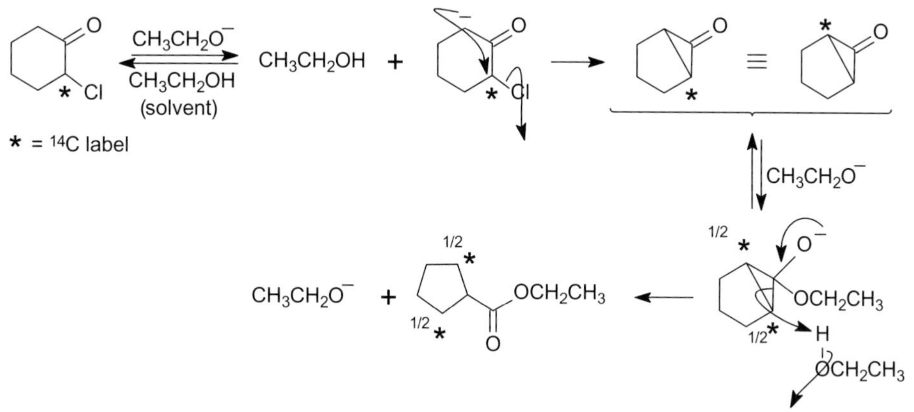

Scheme 9.96. One possible pathway for the Favorskii rearrangement of 2-^{14}C-labelled 2-chlorocyclohexanone to a 1:1 mixture of ethyl 2-^{14}C-cyclopentanecarboxylate and ethyl 5-^{14}C-cyclopentanecarboxylate. Compare with Scheme 9.97. See Loftfield, R. B. *J. Am. Chem. Soc.*, **1950**, *72*, 632; **1951**, *73*, 4707.

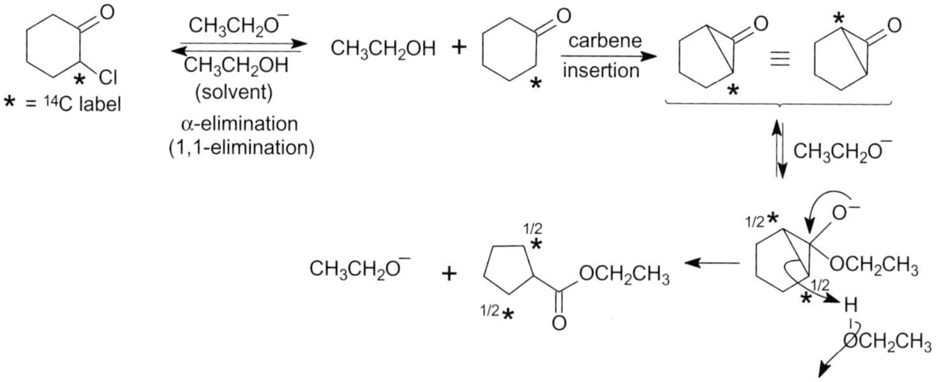

Scheme 9.97. One possible pathway for the Favorskii rearrangement of 2-^{14}C-labelled 2-chlorocyclohexanone to a 1:1 mixture of ethyl 2-^{14}C-cyclopentanecarboxylate and ethyl 5-^{14}C-cyclopentanecarboxylate. Compare with Scheme 9.96. See Loftfield, R. B. *J. Am. Chem. Soc.*, **1950**, *72*, 632; **1951**, *73*, 4707.

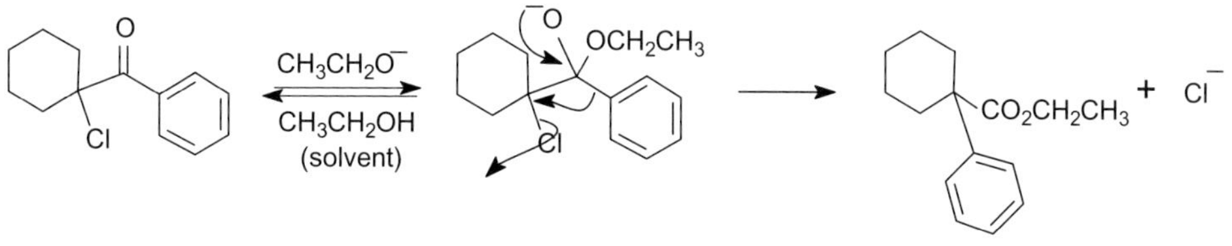

Scheme 9.98. One possible pathway for the Favorskii rearrangement (a specific example of the general expression of Equation 9.74) where there are no hydrogens α- to the carbonyl and thus *neither* a cyclopropanone (Scheme 9.96) nor a carbene (Scheme 9.97) intermediate can obtain in the formation of ethyl 1-phenylcyclohexanecarboxylate.

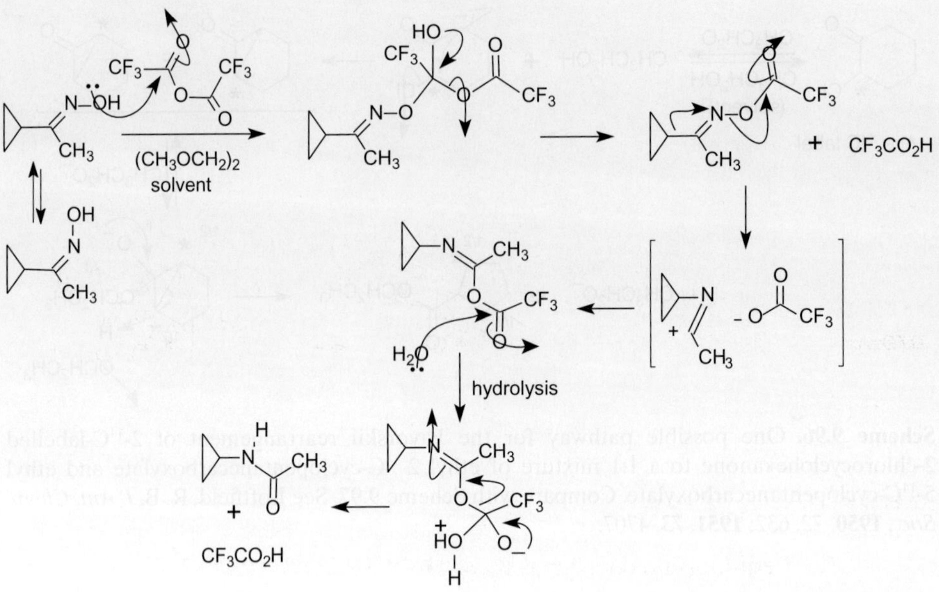

Scheme 9.99. A representation of an example of the **Beckmann rearrangement**. The oxime of (*E*)-1-cyclopropylethanone (cyclopropyl methyl ketone) is treated with trifluoroethanoic anhydride (trifluoroacetic anhydride [(CF$_3$CO)$_2$O]) in 1,2-dimethoxyethane (CH$_3$OCH$_2$CH$_2$OCH$_3$). It is presumed that the intermediate ester undergoes cyclopropyl-to-nitrogen migration (as the cyclopropyl group is *anti*- to the leaving group) and the resulting carbocation is captured by the trifluoroacetate anion. Hydrolysis leads to the corresponding acetamide. First investigated by Beckmann, E. *Chem. Ber.*, **1886**, *19*, 988; *ibid*, **1887**, *20*, 1507 and widely used subsequently. See, Popp, F. D.; McEwen, W. E. *Chem. Rev.*, **1958**, *58*, 370.

(*Z*)-oximes. It is not uncommon to find that, at least for aldehydes and unsymmetrically substituted ketones, one isomer, usually with the (*E*)-geometry for steric reasons, predominates; and often to the exclusion of the other isomer. It is possible to isomerize the (*E*)- to the (*Z*)-isomer via the hydrochloride salt of the oxime and the **Beckmann rearrangement** can, in principle, be effected from either. As shown in Scheme 9.99 for the oxime of (*E*)-1-cyclopropylethanone (cyclopropyl methyl ketone), when the oxime is treated with trifluoroethanoic anhydride (trifluoroacetic anhydride [(CF$_3$CO)$_2$O]) in an etherial solvent such as 1,2-DME (CH$_3$OCH$_2$CH$_2$OCH$_3$), rearrangement to the corresponding amide occurs. As will be seen in Chapter 10, the amide can undergo hydrolysis to the corresponding amine.

PART II. CARBOXYLIC ACIDS AND THEIR DERIVATIVES

A. GENERAL INTRODUCTION

While it is possible to divide this material on oxidation, reduction, addition, substitution, elimination, and rearrangement of carboxylic acids (RCO$_2$H) into sections corresponding to carboxylic acids themselves and then, further, to consider in the same way each of the major groups of derivatives* to be discussed, that is, ([R = alkyl or aryl] acid halides [RCOX; X = halogen], anhydrides [$(RCO)_2O$], esters [RCO_2R'], amides [$RCONR_2$], and nitriles [RCN] as well as some analogues where sulfur replaces oxygen), the similarities would result in excessive repetition.

Throughout the chapter it is important to keep in mind that carboxylic acids are, clearly, acidic. Generally, carboxylic acids are "weak acids" in that proton donation to water is incomplete and thus carboxylic acids are not completely ionized. So, as pointed out earlier (Chapter 5), the **acid strength** or **acidity** is referred to *as the position of equilibrium* in the reaction of the acid with a base. Thus, for the generalized acid "RCO_2H" ionizing in water (H_2O), Equation 9.75 can be written:

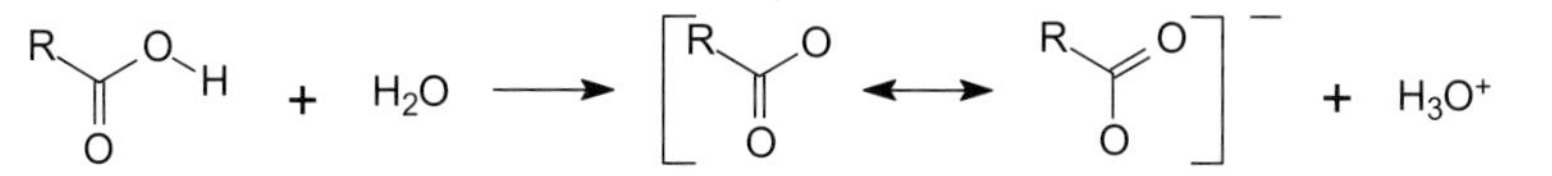
(9.75)

We may then write that the **equilibrium constant, K_a, is a measure of the acid strength** of the carboxylic acid, RCO_2H, and is given by:

$$\text{equilibrium constant} \equiv K_a = [H_3O^+][RCO_2^-]/[RCO_2H] \quad (9.76)$$

where (a) the concentration of water [H_2O] is omitted from the denominator because it is considered large and invarient, thus becoming part of K_a, and (b) the use of brackets [] denotes concentrations (usually in mol L^{-1}) so that [RCO_2H] is the concentration of un-ionized carboxylic acid, RCO_2H, [RCO_2^-] is the concentration of the anion of the acid RCO_2H, and [H_3O^+] is the concentration of the hydronium ion.

Thus,

$$-\log_{10} K_a = -\log_{10}[H_3O^+] + \log_{10}\{[RCO_2H]/[RCO_2^-]\} \quad (9.77)$$

and since

$$-\log_{10} K_a \equiv pK_a \quad (9.78)$$

and

$$-\log_{10}[H_3O^+] \equiv pH \quad (9.79)$$

therefore

$$pK_a = pH + \log_{10}\{[RCO_2H]/[RCO_2^-]\} \quad (9.80)$$

*The reader is encouraged to review Chapter 5, **Part C, IX, d** and **e,** for a discussion of the concept of "derivatives" as used in this context. Indeed, as pointed out there and in Chapter 5**, Part E**, many compounds related to carboxylic acids (in the sense of "derivatives") were, historically, actually obtained **from** carboxylic acids. However, many of the same compounds were (and can be) obtained in other ways and then converted to carboxylic acids from which they remain considered as "derivative."

So, as is fully anticipated, and as shown in Table 9.11, carboxylic acids enjoy a wide range of acidities since their relative ability to ionize must clearly be a function of substituents in the ubiquitious "R" group. That is, those "R" groups that increase the positive nature of the carbon of the carbonyl through, for example, electron withdrawal, will increase the ability of the proton to ionize; those in which the positive nature of the carbonyl is diminished will produce the opposite effect.

B. OXIDATION

As would be expected, under vigorous conditions, carboxylic acids can be oxidized to carbon dioxide and water. However, controlled oxidation (with e.g., 90% hydrogen peroxide [H_2O_2] and sulfuric acid [H_2SO_4]) can be effected to convert carboxylic acids (RCO_2H) to the corresponding peroxycarboxylic acid (peracid [RCO_3H]) (Equation 9.81 and Chapter 6, Scheme 6.12, for uses to which the peracid can be put).

$$(9.81)$$

While, along with other such processes, the reaction of acyl chlorides (e.g., acetyl chloride, CH_3COCl), *vide infra*, with sodium peroxide (Na_2O_2) might be considered a substitution reaction, it nevertheless results in the formation of the corresponding diacyl peroxide [$CH_3CO_2COCH_3$] (Equation 9.82), which is clearly richer in oxygen (an acknowledged result of oxidation).

$$(9.82)$$

An alternative method for the preparation of diacylperoxides involves the autoxidation of aldehydes (Scheme 9.1). The diacylperoxide results from the combination of the acyl radical and a peroxyacyl radical (Equation 9.83), both generated as indicated in the Scheme 9.1 (repeated below).

$$(9.83)$$

Under appropriate conditions, passing an electric current (a **Kolbe*** electrolysis) through an aquous solution of a salt of a carboxylic acid results in the formation of

*Kolbe, H. *Liebigs Ann. Chem.*, **1849**, *69*, 257.

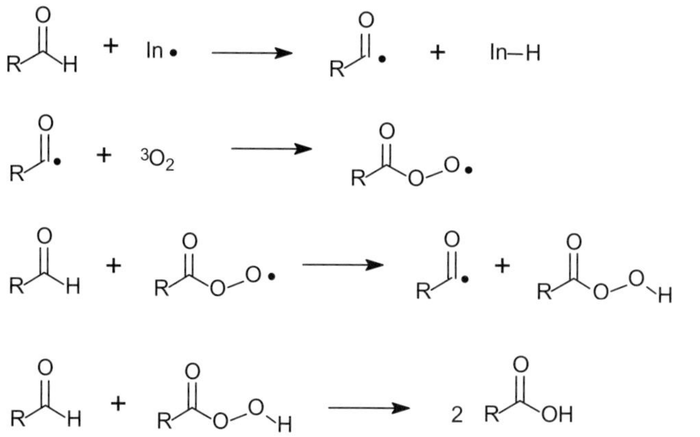

Initiation

Propagation

Termination

Scheme 9.1 (repeated). A representation of the oxidation of an aldehyde by oxygen to the corresponding carboxylic acid [In• represents an unspecified initiator; which might be triplet oxygen ($^3O_2$)].

hydrocarbons (and carbon dioxide [CO_2] and hydrogen [H_2]). Thus, as expected for a negatively charged species, the anion of the carboxylic acid migrates to the anode, where it gives up an electron (an oxidation) and is reduced to a carboxyl radical (RCO_2•). The reduction half reaction, at the cathode, converts water (H_2O) to the hydroxide anion (OH^-) and hydrogen ($2\ H_2$), for which the standard reduction potential is $-0.83\,V$.

Decomposition of the carboxyl radical (RCO_2•) to carbon dioxide (CO_2) and an alkyl radical (R•) follows and, eventually, the alkyl radical dimerizes or loses hydrogen to another alkyl radical (Scheme 9.100). It is important to note that hydrogen abstraction from water by the hydrocarbon radical is retarded by the need to break the strong hydrogen–oxygen bond.

In a similar vein, the silver (Ag^+) salts of carboxylic acids undergo decarboxylation in the presence of bromine (Br_2) to produce silver bromide and bromoalkane (the **Hunsdiecker*** reaction). The Hunsdiecker reaction also appears to involve radicals as shown in Scheme 9.101, where, in the first step, silver bromide precipitates from the reaction mixture with formation of an acyl hypobromite (RCO_2Br). Then, homolysis of the oxygen–bromine bond generates bromine atoms and the carboxyl radical, seen here as the same radical generated in Scheme 9.100. Following loss of carbon dioxide (CO_2), it is held that the alkyl radical is captured by the bromine atom (Br•) to produce alkylbromide (1-bromoethane [CH_3CH_2Br]).

The Hunsdiecker process has the disadvantage of requiring *dry* silver salts. As a consequence, other methods have been sought, and it has been found that some carboxylic acids undergo oxidative decarboxylation on formation of mercury (Hg), lead (Pb), or copper (Cu) salts. For example, cyclopropane carboxylic acid, on treatment with bromine (Br_2) in the presence of red mercury(II) oxide (HgO), yields hydrogen bromide (HBr), carbon dioxide (CO_2), and bromocyclopropane (cyclopropyl bromide) as shown in Equation 9.84.

*Hunsdiecker, H.; Hunsdiecker, C. *Chem. Ber.*, **1942**, *75*, 291.

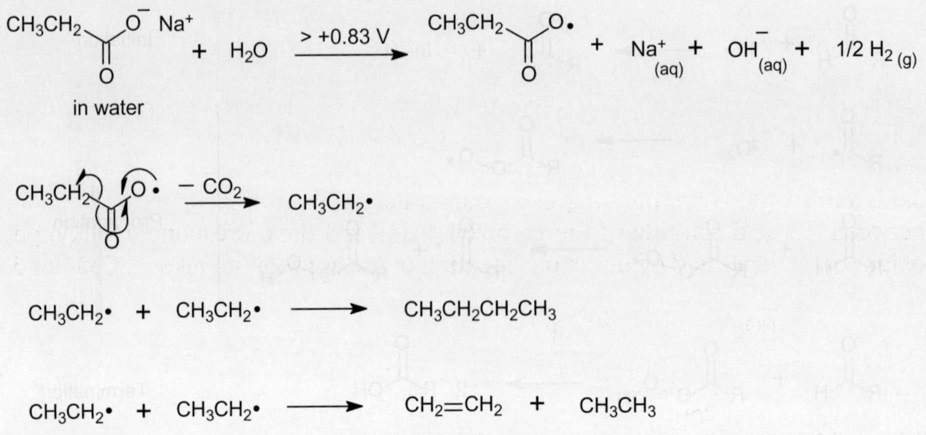

Scheme 9.100. A representation of the **Kolbe** electrolysis of the sodium salt of propanoic acid. The voltage for the electrolysis must be greater than that required for the reduction of water to hydroxide anion (OH^-) and hydrogen (H_2) which occurs at the cathode. It is generally accepted that one electron is lost by the carboxylate anion at the anode to generate a carboxyl radical. The arrows shown on the carboxyl radical (with half-heads) are drawn to account for the apparent movement of one electron (in contrast to the usual cartoons showing two electron arrows) at a time to produce carbon dioxide (CO_2) and the ethyl radical ($CH_3CH_2\cdot$). The latter dimerizes to butane or loses a hydrogen to a second radical, producing ethane and ethene.

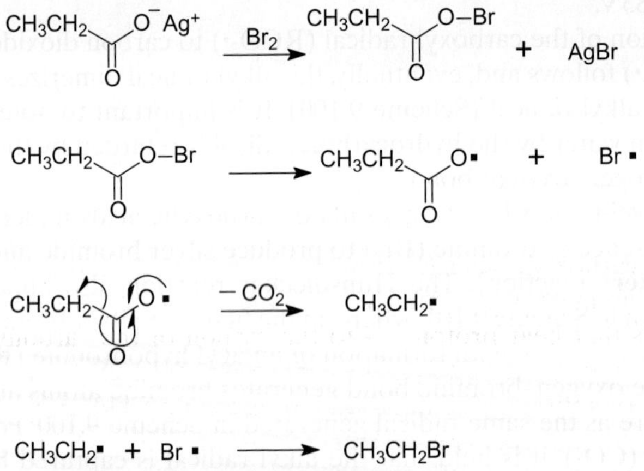

Scheme 9.101. A representation of the **Hunsdiecker** reaction of the silver salt of propanoic acid. It is generally accepted that the precipitation of silver bromide (AgBr) is accompanied by formation of an acylhypobromite and the homolysis of the oxygen bromine bond generates a bromine atom and the carboxyl radical. The arrows shown on the carboxyl radical (with half-heads) are drawn to account for the apparent movement of one electron (in contrast to the usual cartoons showing two electron arrows!) at a time to produce carbon dioxide (CO_2) and the ethyl radical ($CH_3CH_2\cdot$). The latter reacts with bromine to yield bromobutane (CH_3CH_2Br).

$$(9.84)$$

Interestingly, the same kind of carboxyl radical can be generated from the thermal decomposition of esters (usually the *t*-butyl ester as its reactivity is more easily controlled) of peroxycarboxylic acids. In this case, the peroxyester is *not* formed by oxidation of the corresponding ester but rather (Equation 9.85) by the reaction between the acid chloride of the carboxylic acid and the corresponding alkyl peroxide (often generated by direct oxygenation of the appropriate alkane, Chapter 5).

$$(9.85)$$

Not surprisingly, thio acids (thiolocarboxylic acids, Equation 9.86), prepared by reaction of acyl chlorides with hydrogen sulfide (H_2S), are easily oxidized (e.g., on standing in the air) to disulfides (Equation 9.87).

propanoyl
chloride

thionopropanoic
acid

thiolopropanoic
acid

$$(9.86)$$

thionopropanoic
acid

thiolopropanoic
acid

$$(9.87)$$

When esters of carboxylic acids ($R'CO_2R$) that bear protons α- to the carbon of the carbonyl, such as methyl propanoate (Scheme 9.102), or amides ($R'CONR_2$) of carboxylic acids that bear protons α- to the carbon of the carbonyl *and* that *lack* protons on nitrogen, such as N,N-dimethylpropanamide (Scheme 9.102), are converted to their corresponding anions with, for example, LDA, in THF, and the anion allowed to react with 2-sulfonyloxaziridine, hydroxylation α- to the carbonyl occurs. Thus, as shown in Scheme 9.102, the anion is thought to attack the oxygen of the oxaziridine, causing the three-membered ring to open and to produce a compound with a sulfur–nitrogen double bond, which subsequently decomposes with the expulsion of the α-hydroxylated species. It is worth noting that the 2-sulfonyloxaziridine can be obtained as a chiral compound and is capable of stereoselection in the oxidation, making it possible to obtain one enantiomer of the α-hydroxylated product in excess over the other.

Problem 9.21. As indicated in Scheme 9.102, an ester or an N,N-dialkyl (or aryl) amide must be used. Why is it necessary to use the dialkyl (or aryl) amide or ester rather than the carboxylic acid itself?

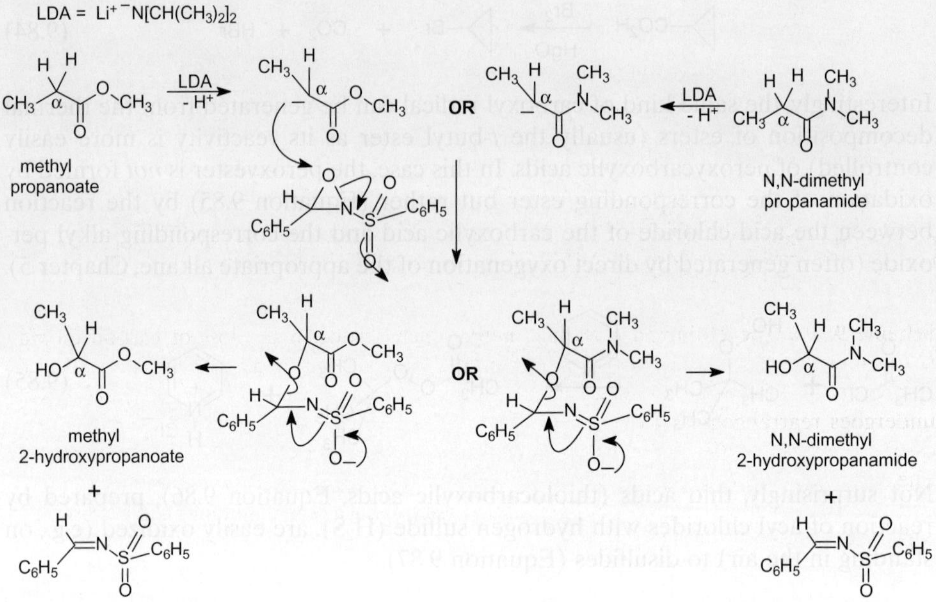

LDA = $Li^+ {}^-N[CH(CH_3)_2]_2$

methyl propanoate

N,N-dimethyl propanamide

methyl 2-hydroxypropanoate

N,N-dimethyl 2-hydroxypropanamide

Scheme 9.102. The use of 2-sulfonyloxaziridine to oxidize carbon α- to the carbonyl group of the carboxylic acid derivative (ester or amide) producing the corresponding α-hydroxycarboxylic ester or amide. See Davis, F. A.; Billmers, J. M.; Gosciniak, D. J.; Towson, J. C.; Bach, R. D. *J. Org. Chem.*, **1986**, *51*, 4240.

In a similar vein, when an amide with two hydrogens on nitrogen (Scheme 9.103) is treated with an oxidizing agent such as chlorine (Cl_2) or bromine (Br_2), the corresponding *N*-chloro- or *N*-bromo-compound is often readily formed. Then, as shown in Scheme 9.103 treatment of this *N*-haloamide derivative with base results in skeletal rearrangement (the **Hofmann rearrangement**) to produce an isocyanate that on reaction with water suffers **the loss of carbon dioxide** and generates the corresponding amine (Chapter 10).

As already noted (Chapter 8, Scheme 8.49, reproduced here as Scheme 9.104), oxidation of a carboxylic acid at the carbon α- to the carbonyl can be effected by treatment of the carboxylic acid with red phosphorus (P) and bromine (Br_2). The α-bromocarboxylic acid produced in this way can serve as a source from which to generate a variety of α-substituted derivatives of carboxylic acids. Further, through elimination of HBr, the α-bromocarboxylic acid, can also serve as a source of α,β-unsaturated carboxylic acids.

Enzyme-mediated (biological) oxidation of carboxylic acids is being studied at an increasing pace and screening of newborn infants for normal and abnormal metabolites of such processes is becoming common.

While (at this writing) the details are not known, it appears that during the anaerobic growth of certain *Escherichia coli* bacteria, an "electron transport chain" of one electron oxidation–reduction processes involving quinones (cf. Chapter 8, Figure 8.11, *et. seq.*), the (required) trace element selenium (Se), as

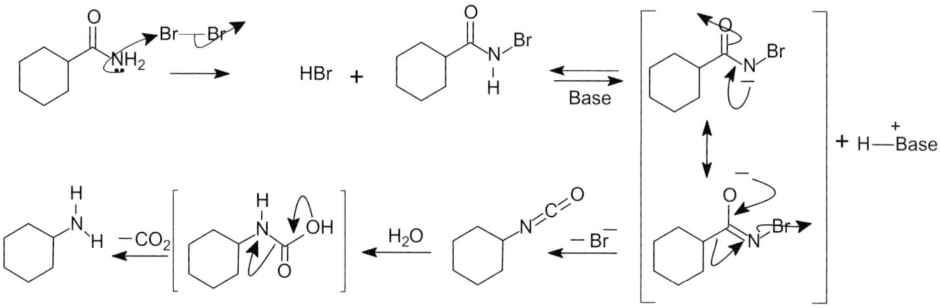

Scheme 9.103. The **Hofmann rearrangement**, which results in the loss of carbon dioxide (CO_2) and the production of an amine from oxidation of an amide with bromine (Br_2) or chlorine (Cl_2). The oxidation produces the *N*-halo-compound, which, on treatment with base, undergoes rearrangement to generate the N-alkylisocyanate (cyclohexylisocyanate in this case). Treatment of the *N*-alkylisocyanate with water, produces the transient **carbamic acid**. The carbamic acid undergoes spontaneous decarboxylation in water to carbon dioxide (CO_2) and cyclohexylamine (Chapter 10). Details of the **Hofmann rearrangement** will be discussed subsequently. But see Hofmann, A. W. *Chem. Ber.*, **1881**, *14*, 2725.

well as iron–sulfur (e.g., Fe_4S_4) clusters associated with the enzyme **formate dehydrogenase** (EC 1.2.1.2) contribute to the oxidation of formate (HCO_2^-) to carbon dioxide (CO_2) and the transport of protons (H^+) across cellular membranes (Equation 9.88).

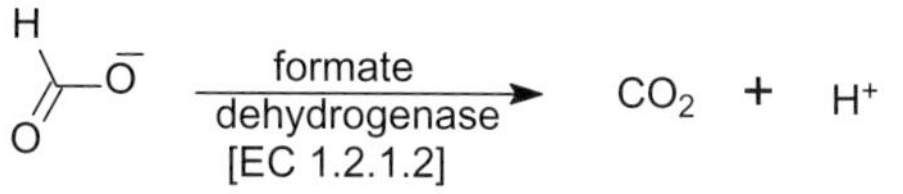

$$\text{formate} \xrightarrow[\text{[EC 1.2.1.2]}]{\text{dehydrogenase}} CO_2 + H^+ \qquad (9.88)$$

It is, however, better known that flavoenzymes (i.e., enzymes utilizing the flavin adenine dinucleotide [$FAD:FADH_2$] redox system) mediate the introduction of α,β carbon–carbon double bonds into carboxylic acids and into acetyl Coenzyme A (acetyl CoA) thioesters of long-, medium-, and short-chain fatty acids. In carboxylic acids, such as those of the tricarboxylic acid (**citric acid, TCA,** or **Krebs**) cycle (Chapter 11) the oxidation is affected by the enzyme **succinate dehydrogenase** (**fumerate reductase**—EC 1.3.99.1), which utilizes the cofactor flavin adenine dinucleotide (FAD) The latter is reduced to $FADH_2$ and an (*E*)-double bond is introduced. The process shown in Scheme 9.105, for the conversion of succinate (1,4-butanedioic acid) to fumerate [(*E*)-1,4-butenedioic acid], is a fragment of the tricarboxylic acid (**citric acid, TCA,** or **Krebs**) cycle (Chapter 11), which is the pathway commonly utilized for oxidative degradation of acetate to carbon dioxide.

It is important to note that as shown in Scheme 9.105 one *pro*-R and one *pro*-S hydrogen are removed during the conversion of succinate to fumerate (FAD to $FADH_2$) and that the cofactor to the **succinate dehydrogenase**, that is, FAD, appears to be bound to a histidine residue (Chapter 12) *on the enzyme* through (what was)

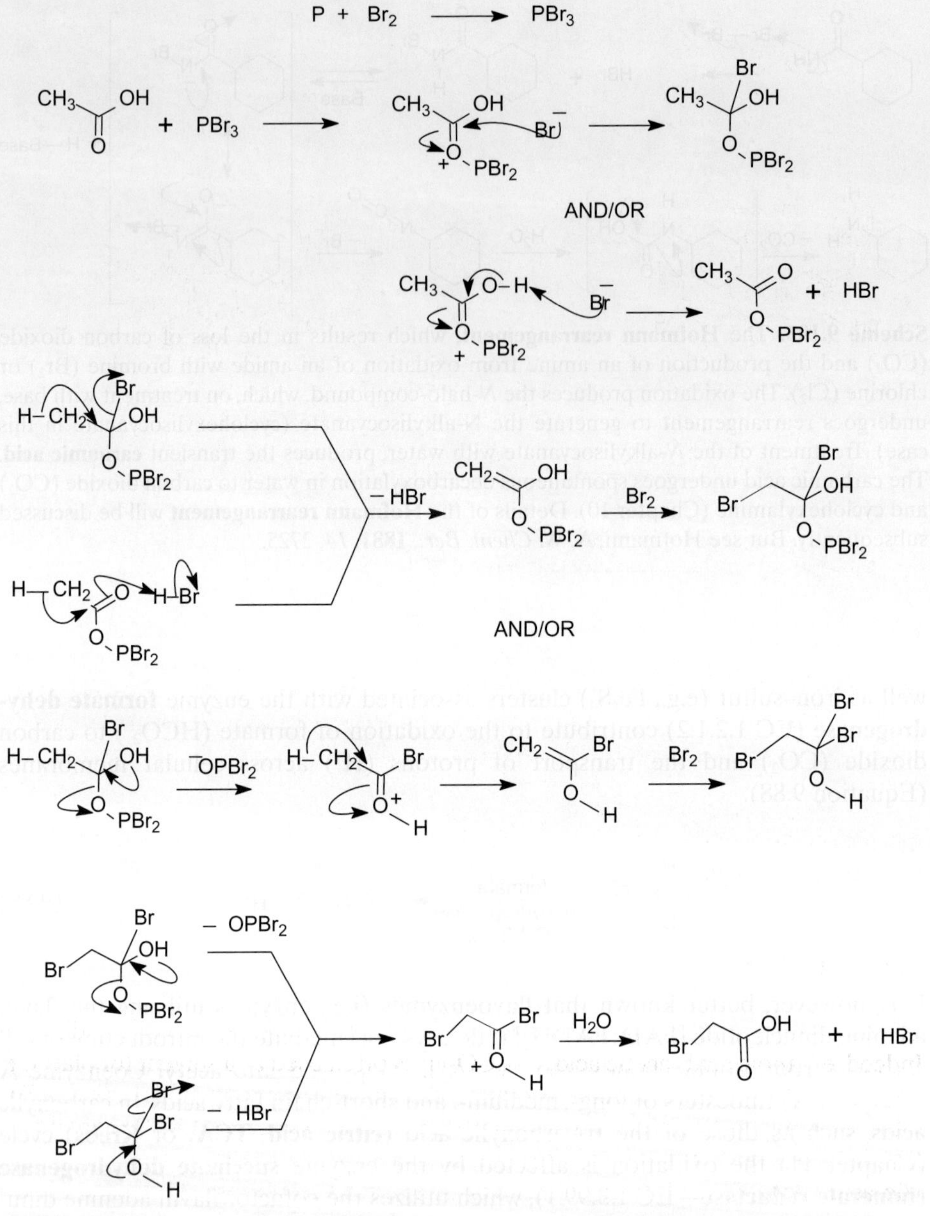

$$P + Br_2 \longrightarrow PBr_3$$

AND/OR

AND/OR

AND/OR

Scheme 9.104. (Scheme 8.49 repeated). A variety of possibilities to account for the formation of α-bromoacetic acid (bromoethanoic acid) by bromination of acetic acid (ethanoic acid) with bromine (Br_2) in the presence of red phosphorus (P).

one of the methyl groups on the aromatic ring, thus requiring reoxidation through the "electron transport chain."

Interestingly, when the dehydrogenation of coenzyme-bound *thio*ester (Scheme 8.70, *et seq*)—in the growing chain of acetyl CoA thioesters of long-, medium-, and short-chain fatty acids generated in the polyketide synthase cassette of reactions—occurs, it still appears that the FAD:$FADH_2$ cofactor is involved, but there are

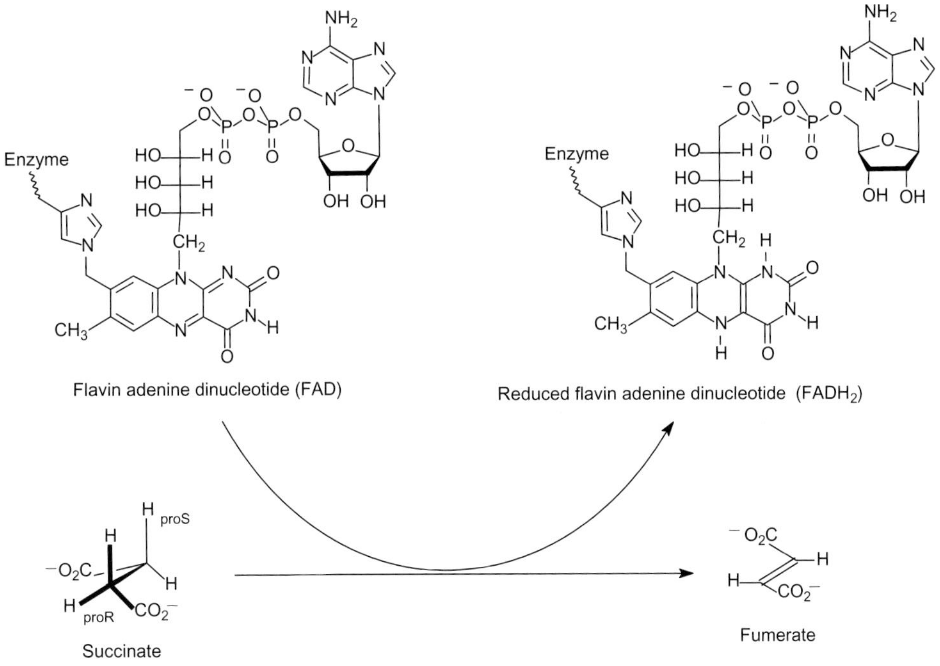

Flavin adenine dinucleotide (FAD)

Reduced flavin adenine dinucleotide (FADH$_2$)

Succinate

Fumerate

Scheme 9.105. A representation of the oxidation of succinate (1,4-butanedioic acid) to fumerate (1,4-butenedioic acid) with the concomitant reduction of enzyme bound flavin adenine dinucleotide.

specific short-, medium-, and long-chain dehydratase enzymes.* Further, the double bond is found to be (*Z*)- rather than (*E*)-.

C. REDUCTION

The direct reduction of a carboxylic acid (RCO_2H) to an alkyl group is very difficult. Indeed, ethanoic acid (acetic acid, CH_3CO_2H) is often used as a solvent for platinum (Pt) and palladium (Pd) catalyzed hydrogenation of other functional groups. The same difficulty is encountered in the reduction of esters (RCO_2R'), anhydrides (RCO_2COR), and acid halides ($RCOCl$, etc.), except that esters of benzenecarboxylic acid (benzoic acid, $C_6H_5CO_2R$) can be reduced to methyl-substituted arenes with hydrogen (H_2) and a copper chromite catalyst ($CuCr_2O_4$) at high temperatures (200–300°C) and pressures (100–300 atm H_2) (Equation 9.89). The reduction of the esters of benzenecarboxylic acids (benzoic acid, $C_6H_5CO_2R$) succeeds similarly because the initial reduction of the ester generates the corresponding alcohol ($C_6H_5CH_2OH$), which (Chapter 8, Equation 8.18) undergoes hydrogenolysis to the methyl group (and water).

*Fatty acid oxidation generally increases when glucose levels fall. Failure of these dehydratase enzymes, which lead the way in degradation of fatty acids, has been implicated in a variety of early childhood illnesses.

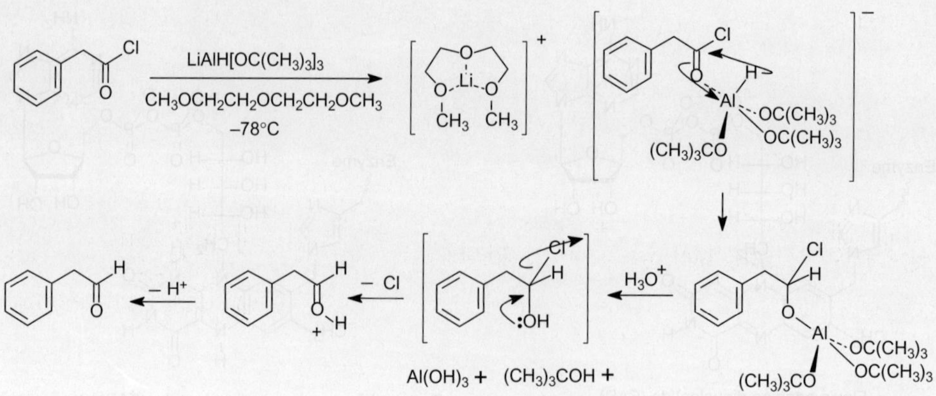

Scheme 9.106. A cartoon representation of the reduction of the acid chloride of 2-phenylethanoic acid (2-phenylacetic acid, α-phenylacetic acid, $C_6H_5CH_2CO_2H$) to the corresponding aldehyde 2-phenylethanal (2-phenylacetaldehyde, α-phenylacetaldehyde, $C_6H_5CH_2CHO$) with lithium aluminum tri-*t*-butoxyhydride ($LiAlH[OC(CH_3)_3]_3$) in bis(2-methoxyethyl)ether [diglyme, $(CH_3OCH_2CH_2)_2O$], at −78°C.

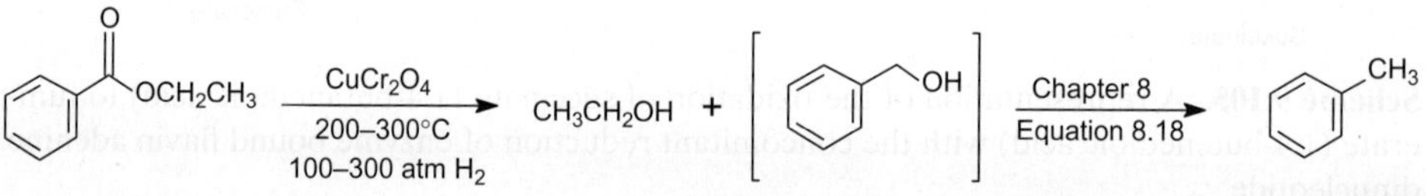

$$(9.89)$$

However, as anticipated on the basis of the reactions of other carbonyl-containing compounds, carboxylic acids (RCO_2H) and the other derivatives (i.e., esters, anhydrides, and acid halides) can be reduced to the corresponding alcohols in a variety of ways, and, under carefully controlled conditions, it is often possible to stop the reduction at the aldehyde ($RCHO$) stage.

For example, reduction of acid chlorides ($RCOCl$), prepared from carboxylic acids by reaction between the acid and, for example, thionyl chloride (*vide infra*), with hydrogen over a barium sulfate ($BaSO_4$) poisoned palladium (Pd) catalyst (the **Rosenmund** reduction),* can often be used to produce the corresponding aldehyde ($RCHO$). The same product can more easily be obtained from the same starting material by using commercially available lithium aluminum tri-*t*-butoxy hydride ($LiAlH[OC(CH_3)_3]_3$)[†] in an ether solvent, such as bis(2-methoxyethyl)ether [diglyme, $(CH_3OCH_2CH_2)_2O$], at −78°C (Scheme 9.106).

In a similar way, carboxylic acids (RCO_2H) (Scheme 9.107) and esters (RCO_2R') (Scheme 9.108, L = OR′) themselves can often be reduced to aldehydes ($RCHO$), rather than reduced completely to the corresponding alcohol (*vide infra*), with

---

*Rosenmund, K. W. *Chem. Ber.*, **1918**, *51*, 585, but see Saytzeff, M. *J. Prakt Chem*, **1873**, *6*, 130.
[†]A variety of reducing agents of similar structure have been developed and, depending on the particular needs of a system, can be utilized. See Kim, S. *Pure Appl. Chem.*, **1987**, *59*, 1005, and Walker, E. R. H. *Chem. Soc. Rev.*, **1976**, *5*, 23.

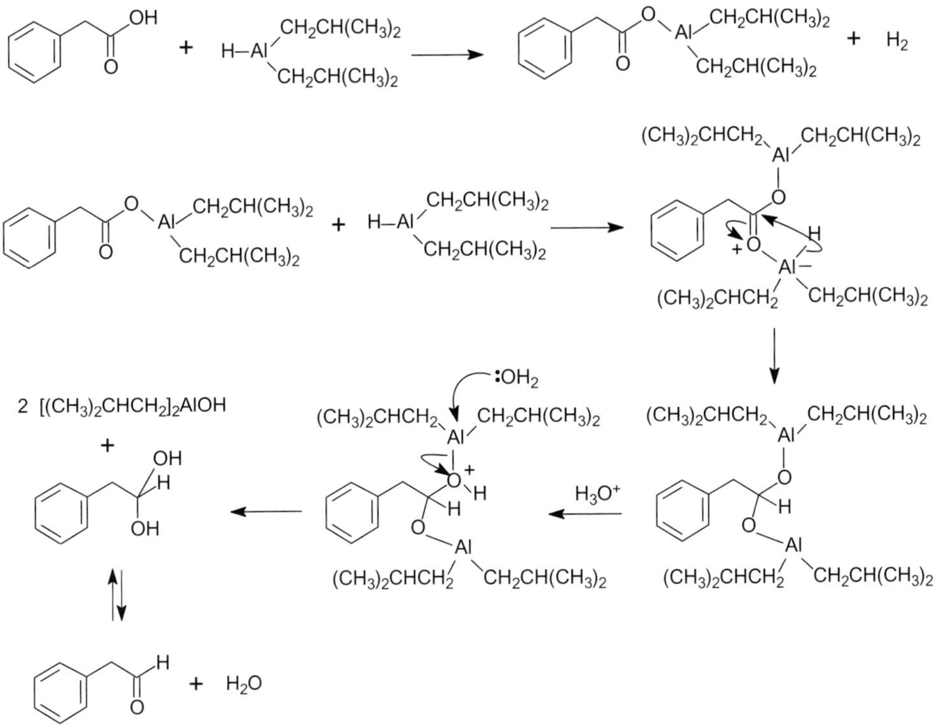

Scheme 9.107. A cartoon representation of the reduction of 2-phenylethanoic acid (2-phenylacetic acid, α-phenylacetic acid, $C_6H_5CH_2CO_2H$) to the corresponding aldehyde 2-phenylethanal (2-phenylacetaldehyde, α-phenylacetaldehyde, $C_6H_5CH_2CHO$) with diisobutyl aluminum hydride (DIBAL-H, $[(CH_3)_2CHCH_2]_2AlH$) in methylbenzene [toluene $(C_6H_5CH_3)$] solution at low temperatures.

L = leaving group
For esters, L = OR; R = alkyl or aryl
For *N,N*-disubstituted amides, L = NR_2; R = alkyl or aryl

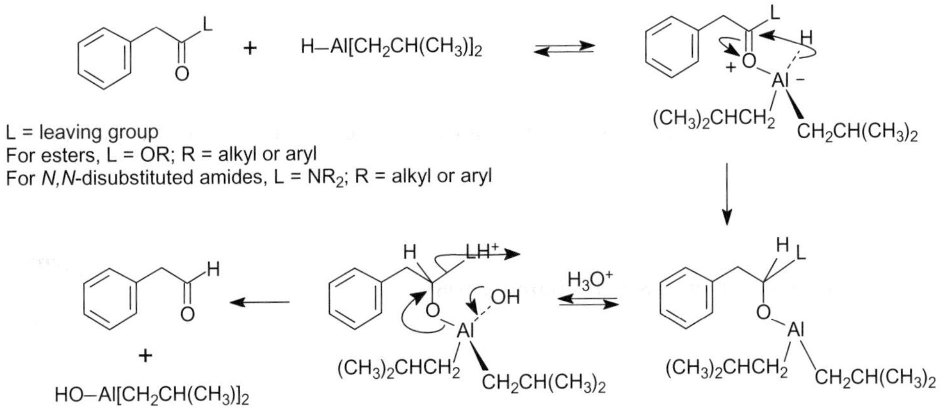

Scheme 9.108. A cartoon representation of the reduction of an ester (L = OR′) and an *N,N*-disubstituted amide (L = NR_2) of 2-phenylethanoic acid (2-phenylacetic acid, α-phenylacetic acid, $C_6H_5CH_2CO_2H$) to the corresponding aldehyde 2-phenylethanal (2-phenylacetaldehyde, α-phenylacetaldehyde, $C_6H_5CH_2CHO$) with diisobutyl aluminum hydride (DIBAL-H, $[(CH_3)_2CHCH_2]_2AlH$) in methylbenzene [toluene $(C_6H_5CH_3)$] solution at low temperatures.

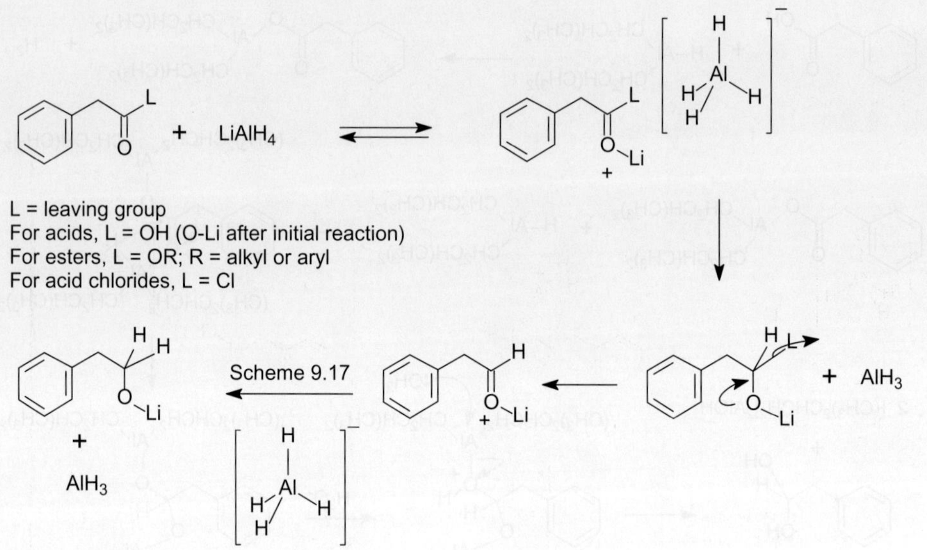

L = leaving group
For acids, L = OH (O-Li after initial reaction)
For esters, L = OR; R = alkyl or aryl
For acid chlorides, L = Cl

Scheme 9.17

Scheme 9.109. A cartoon representation of the reduction of a carboxylic acid (L = OH), an ester (L = OR′), and an acid chloride (L = Cl) of 2-phenylethanoic acid (2-phenylacetic acid, α-phenylacetic acid, $C_6H_5CH_2CO_2H$) to the corresponding alcohol, 2-phenylethanol ($C_6H_5CH_2CH_2OH$) with lithium aluminum hydride ($LiAlH_4$) in ether [$(CH_3CH_2)_2O$] solution.

diisobutyl aluminum hydride (DIBAL-H, [$(CH_3)_2CHCH_2]_2AlH$) in methylbenzene (toluene, $C_6H_5CH_3$) solution at low temperatures. It is generally held that the effects of low temperature and hindered hydride combine to effect the limited reduction. Interestingly, N,N-disubstituted amides (Scheme 9.108, L = NR_2) can also be reduced with diisobutyl aluminum hydride (DIBAL-H, [$(CH_3)_2CHCH_2]_2AlH$) in methylbenzene (toluene, $C_6H_5CH_3$) solution at low temperatures to the corresponding aldehyde, and, under favorable conditions, even the more potent and less hindered reducing agent, lithium aluminum hydride, ($LiAlH_4$) will yield the same product when the amide is in excess. However, generally, amides (*vide infra*) yield amines (Chapter 10) with lithium aluminum hydride ($LiAlH_4$) in excess or, even with diisobutyl aluminum hydride (DIBAL-H, [$(CH_3)_2CHCH_2]_2AlH$), alcohols when the aldehyde is labile.

Lithium aluminum hydride ($LiAlH_4$), a reactive, pyrophoric solid, is commonly used in ether [$(CH_3CH_2)_2O$] solvents to reduce acids (RCO_2H), esters ($RCO_2R′$), anhydrides (RCO_2COR), and acid chlorides ($RCOCl$) to alcohols (Scheme 9.109). Amides, with the exception of some N,N-disubstituted compounds (*vide supra*), and nitriles ($R–C{\equiv}N$) generally yield amines with this reagent (Scheme 9.110)

In the same vein, it is also common, as was the case for aldehydes and ketones, that borane (diborane, B_2H_6) can be used for the reduction of primary ($RCONH_2$), secondary ($RCONHR′$), and tertiary ($RCONR′_2$) amides to the corresponding amines.

Prior to the widespread use of metal hydrides, it was common to reduce (or attempt to reduce) esters with active metals (e.g., Na, Zn) in suitable solvents. Thus,

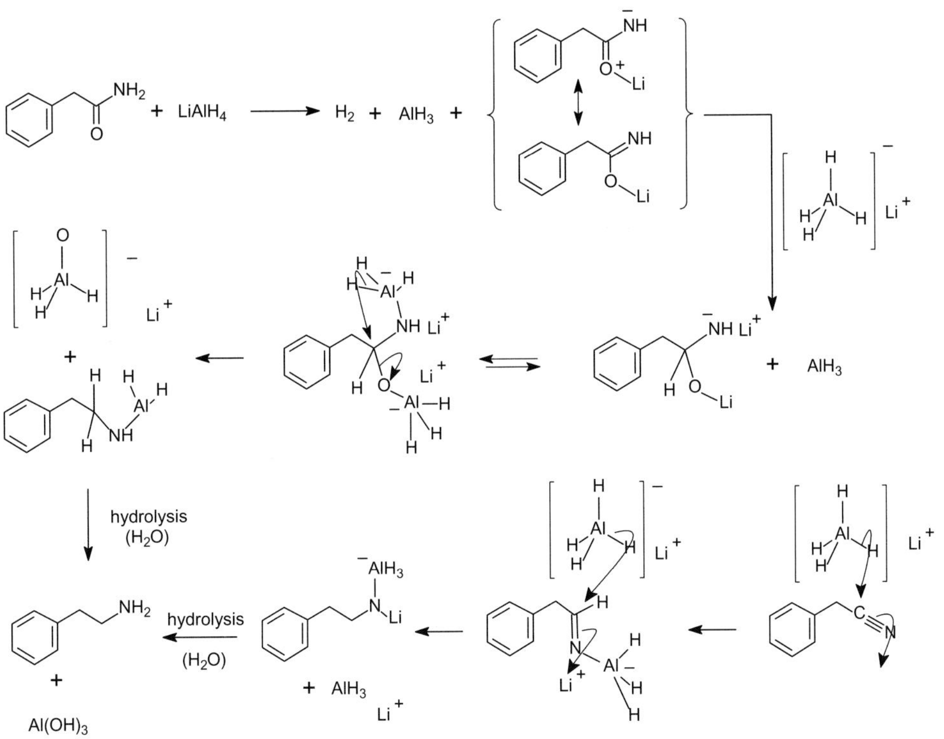

Scheme 9.110. A cartoon representation of the reduction of the primary amide 2-phenylethanoamide ($C_6H_5CH_2CONH_2$), that is, the amide from 2-phenylethanoic acid [2-phenylacetic acid, α-phenylacetic acid, $C_6H_5CH_2CO_2H$] and ammonia, *vide infra*, and also of the nitrile (2-phenylethanonitrile [2-phenylacetonitrile, α-phenylaceto-nitrile, $C_6H_5CH_2C{\equiv}N$]) to the corresponding amine, 2-phenylethylamine, with lithium aluminum hydride ($LiAlH_4$) in ether [$(CH_3CH_2)_2O$] solution.

sodium (Na) in ethanol (CH_3CH_2OH) (the **Bouvealt–Blanc** method)* (Scheme 9.111), was often effective for carboxylic acid esters but *not* for the corresponding carboxylic acids. As shown in Scheme 9.111, when the *trans*-β-(2-carbomethoxycyclohexane)propanoic acid is treated with sodium (Na) metal and ethanol (CH_3CH_2OH) in liquid ammonia ($NH_{3\,(l)}$) solvent, the corresponding *trans*-β-(2-hydroxymethylcyclohexane)propanoic acid is obtained after aqueous workup. Presumably, the carboxylic acid is initially converted to a carboxylate anion, which, relative to the ester, is resistent to accepting an electron from sodium (Na) metal. Thus, the initial electron transfer produces a radical anion at the carbonyl group (C=O) of what was the ester (see, e.g., Scheme 9.15), which, after proton (H^+) transfer, is further reduced to the corresponding carbanion by donation of an electron to the radical from a second equivalent of sodium (Na) metal. A second proton (H^+) transfer directly produces a hemiacetal (Chapter 8), from which the loss of methanol (CH_3OH) generates an aldehyde. Then, by analogy to the reduction shown in Scheme 9.15, the aldehyde (but, again, not the remaining carboxylate anion) is

*Bouveault, L.; Blanc, G. *Compt. Rend. Séances Acad. Sci.*, **1903**, *136*, 1676.

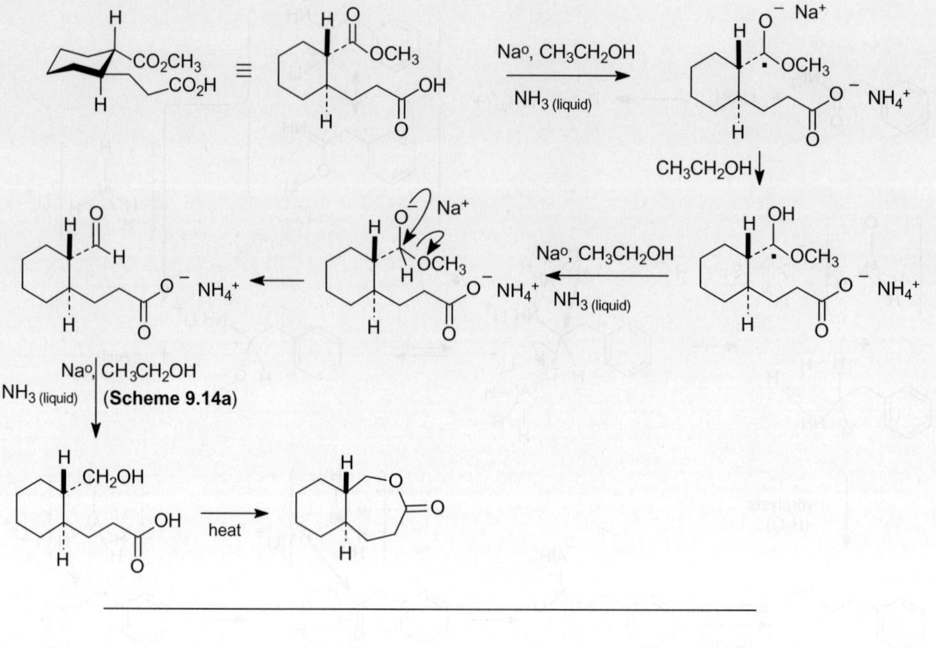

Scheme 9.111. The reduction of *trans*-β-(2-carbomethoxycyclohexane)propanoic acid. The upper part of the scheme shows reduction with sodium (Na) in ethanol (CH_3CH_2OH) and liquid ammonia ($NH_{3\ (l)}$), where the ester is reduced in preference to the carboxylic acid. The lower part of the scheme shows reduction with borane in THF (oxacyclopentane) where the opposite result obtains. For the preparation and reduction of *trans*-β-(2-carbomethoxycyclohexane)propanoic acid, see Paquette, L. A.; Nelson, N. A. *J. Org. Chem.*, **1962**, *27*, 2272.

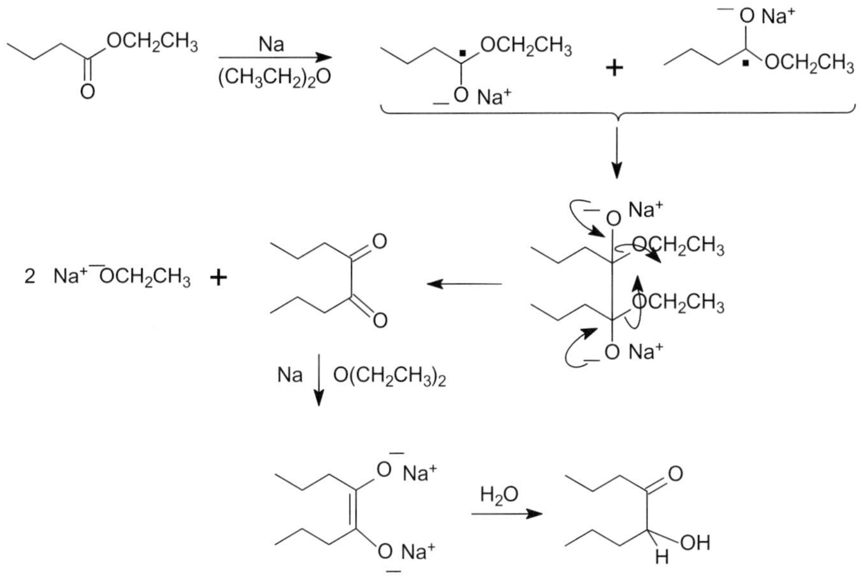

Scheme 9.112. A representation of the reduction of ethyl butanoate by sodium metal (Na) in the absence of a hydrogen donor. Two equivalents of ester condense to produce 4-hydroxy-5-octanone (an α-hydroxyketone or **acyloin**). The process is also known as the **acyloin condensation**.*

reduced by sodium (Na) in ethanol (CH_3CH_2OH) to the corresponding primary alcohol, *trans*-β-(2-hydroxymethylcyclohexane)propanoic acid. As shown in the scheme, heating the hydroxy carboxylic acid producs a lactone, *trans*-3-oxa-4-oxo-bicyclo[5.4.0]undecane.

Interestingly, as also shown in Scheme 9.111, diborane (B_2H_6) in oxacyclopentane (THF, i.e., H_3B·THF) preferentially reacts with the carboxylic acid portion of the same *trans*-β-(2-carbomethoxycyclohexane)propanoic acid. Presumably this is because the bond between oxygen and boron is more easily formed and a proton can be lost to, for example, solvent. Thus, it is predicted that reduction under these circumstances preferentially leads to *trans*-3-(2-carbomethoxycyclohexane)-1-propanol, which then leads to *trans*-3-oxa-2-oxobicyclo[5.4.0]undecane, a lactone isomeric with that in the first example of the scheme.

Further, if the ester reduction with sodium (Na) metal is carried out in the absence of a hydrogen donor, **two equivalents of ester condense with each other** to produce an α-hydroxyketone (an acyloin) (Scheme 9.112). As shown in Scheme 9.112, when the **acyloin condensation** reaction is carried out on ethyl butanoate with sodium (Na) metal in diethyl ether [$(CH_3CH_2)_2O$] at reflux, 4-hydroxy-5-octanone is produced (in about 70% yield) by what appears to be a series of one-electron transfer processes.

*Early work on the acyloin condensation (which can also be carried out between separate ester reactants) was effected by Bouveault, L.; Locquin, R. *Compt. Rend. Séances Acad. Sci.*, **1903**, *136*, 1676, and, in an attempt to establish the mechanism by subsequent investigators, for example, Lynn, J. W.; English, J. *J. Am. Chem. Soc.*, **1951**, *73*, 4284.

Similarly, diesters form cyclic α-hydroxyketones (acyloins) (Equation 9.90),

$$(9.90)$$

The value of the cyclization process to form α-hydroxyketones (acyloins) shown in Equation 9.90 is diminished by the intrusion of a **Dieckmann cyclization**.* (Equation 9.91) and when the acidity of a proton α-to the carbonyl is sufficient for it to be abstracted by traces of hydroxide ($^-$OH) or alkoxide ($^-$OR) that might be present or formed during the reaction.

$$(9.91)$$

As noted earlier (*vide supra*) and as shown in Equation 9.92, reduction of anhydrides with lithium aluminum hydride ($LiAlH_4$) in diethyl ether [$(CH_3CH_2)_2O$] results in the formation of two equivalents of alcohol (Equation 9.92) by a process presumed similar to that shown in Scheme 9.109. In the case of an anhydride such as that from propanoic acid ($CH_3CH_2CO_2H$), that is, propanoic anhydride [$(CH_3CH_2CO)_2O$] shown in Equation 9.92, the leaving group "L" of Scheme 9.109 would, at least initially, be the propanoate anion ($^-O_2CCH_2CH_3$) and, subsequently, it would also be reduced to 1-propanol (CH_3CH_2OH).

$$(9.92)$$

Interestingly, with hindered cyclic anhydrides, such as the cyclohexene derivative (*cis*-1-methyl-3-oxa-2,4-dioxobicyclo[4.3.0]-7-nonene, *cis*-1-methylcyclohex-4-ene-1,3-dicarboxylic acid anhydride) shown in Scheme 9.113, only one of the two (the most hindered) carbonyl groups is reduced and the lactone (cyclic ester) *cis*-1-methyl-3-oxa-4-oxobicyclo[4.3.0]-7-nonene (*cis*-1-methyl-1-hydroxymethylcyclohex-4-ene carboxylic acid γ-lactone) is produced. As shown in the scheme, it is argued that the initial lithium aluminum hydride ($LiAlH_4$) attack is on the *less* hindered carbonyl and intramolecular transfer of hydride (H^-) to the *more* hindered carbonyl follows later.

More generally, reduction of cyclic anhydrides such as phthalic anhydride (benzene-1,2-dicarboxylic acid anhydride) with zinc (Zn) in ethanoic acid (acetic acid, CH_3CO_2H) results in the formation of cyclic esters (lactones) (Scheme 9.114). As shown, the zinc (Zn) metal is oxidized (to zinc oxide [ZnO_2]), while the

---

*The early work on the Dieckmann cyclization was carried out by Dieckkmann, W. *Chem. Ber.*, **1894**, 27, 102, 965, and the significant amount that followed has been reviewed (Thyagarajan, B. S. *Chem. Rev.*, **1954**, 28, 90).

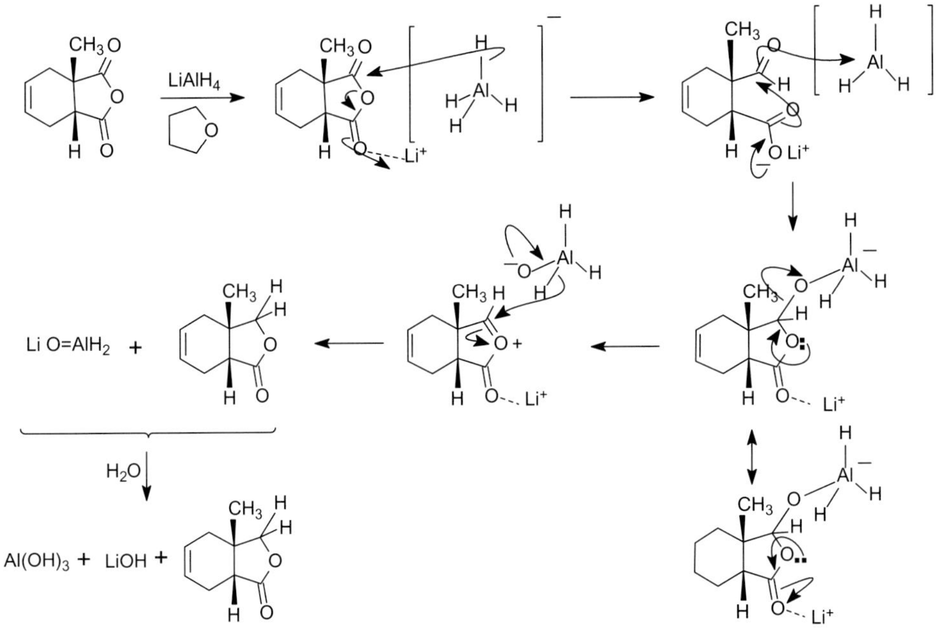

Scheme 9.113. A representation of the reduction of the cyclohexene derivative (*cis*-1-methyl-3-oxa-2,4-dioxobicyclo[4.3.0]-7-nonene, *cis*-1-methyl-cyclohex-4-ene-1,3-dicarboxylic acid anhydride) with lithium aluminum hydride (LiAlH$_4$) to produce the corresponding lactone, *cis*-1-methyl-3-oxa-4-oxobicyclo[4.3.0]-7-nonene (*cis*-1-methyl-1-hydroxymethylcyclohex-4-ene carboxylic acid γ-lactone).

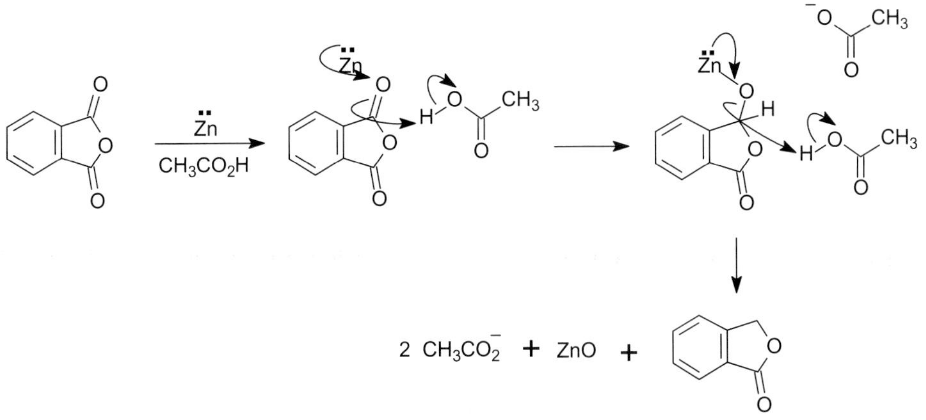

Scheme 9.114. A representation of the reduction of phthalic anhydride (benzene-1,2-dicarboxylic acid anhydride) to the corresponding lactone with zinc (Zn) in ethanoic acid (acetic acid).

anhydride is reduced and the ethanoic acid (acetic acid) serves as the source of protons.

Problem 9.22. The starting material for the reduction reaction shown in Scheme 9.113 is the product of a Diels–Alder reaction between what two reactants?

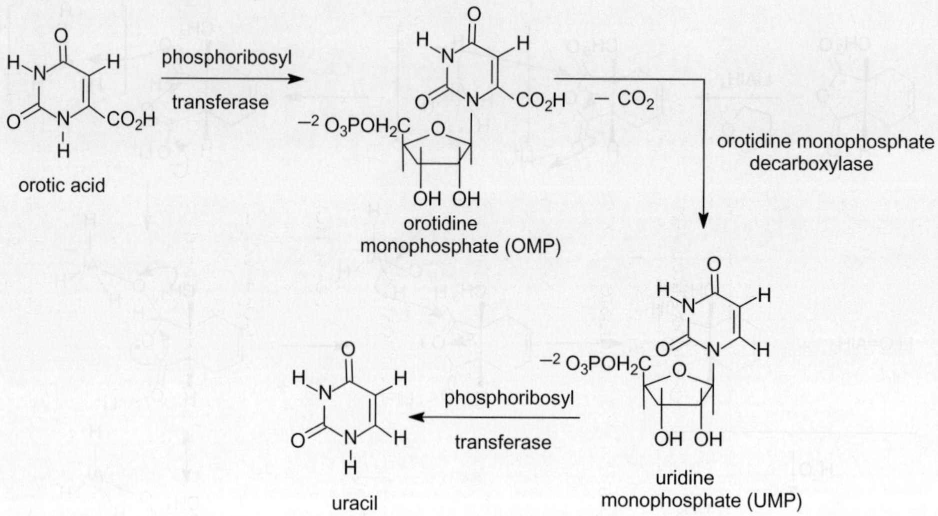

Scheme 9.115. A representation of the enzyme-catalyzed decarboxylation of orotic acid to yield uracil.

D. SUBSTITUTION: ADDITION AND ELIMINATION

The process of substitution undertaken on carboxylic acids and the derivatives of carboxylic acids (anhydrides, acid halides, esters, amides, and nitriles) generally involves a **series** of replacement processes. Thus, individually, substitution may involve replacement of: (a) the proton attached to oxygen of the –OH group (i.e., ionization of the acid); (b) the hydroxyl (–OH) portion of the carboxylic acid (or derivative) (e.g., esterification); (c) the carbonyl oxygen *and* the hydroxyl (–OH) (e.g., orthoester formation, *vide infra*); (d) the entire carboxylic acid functionality (e.g., the Hunsdiecker reaction, already discussed Scheme 9.101) and the decarboxylation of orotic acid (as orotidine monophosphate) to uracil (as uridine monophosphate)—catalyzed by the enzyme orotidine monophosphate decarboxylase (Scheme 9.115); or (e) the protons (if any) on the carbon to which the carboxylic acid functional group is attached (e.g., the Dieckman cyclization, already discussed earlier, cf. Equation 9.91). Indeed, processes already discussed (i.e., reduction and oxidation) have also accomplished some of these ends. Some additional substitutions for the carboxylic acid group itself are presented in Table 9.6, while other substitutions for derivatives of carboxylic acids are shown in Tables 9.7–9.10 and discussed subsequently.

Direct displacement of the leaving group from the sp^2-hybridized carbon of the carbonyl, if it occurs at all, is rare. Generally the substitution process occurs by one of two pathways. First, elimination of the leaving group can be followed by addition of the nucleophile to the resulting positively charged oxonium species (Equation 9.93). Here, the replacement is the result of an **elimination followed by an addition**.

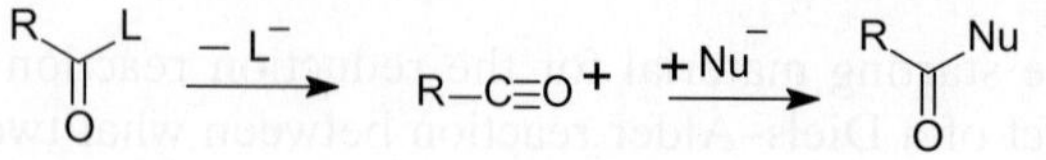

$$R-C(=O)L \xrightarrow{-L^-} R-C\equiv O^+ \xrightarrow{+Nu^-} R-C(=O)Nu \tag{9.93}$$

TABLE 9.6. Representative Carboxylic Acids and Their Derivatives

	Alkyl (RCH_2CO_2H) or Aryl ($ArCO_2H$) Carboxylic Acid	On Reaction with (Species and/or Process)	Generates Substituted Acid "Derivative"
1.	$CH_3CH_2CO_2H$ propanoic acid	$NH_4^+OH^-$ ammonium–hydroxide (acid–base reaction) (substitution of NH_4^+ for H^+)	$CH_3CH_2CO_2^-\ NH_4^+ + H_2O$ ammonium propanoate
2.	benzenecarboxylic acid (benzoic acid)	CH_3CH_2OH/H^+ ethanol and acid catalyst (substitution of OCH_2CH_3 for OH)	ethyl benzenecarboxylate (ethyl benzoate)
3.	cyclohexanecarboxylic acid	$SOCl_2$ (thionyl chloride) or $ClCOCOCl$ (oxalyl chloride) (substitution of Cl for OH)	cyclohexanecarbonyl chloride

TABLE 9.6. *Continued*

Alkyl (RCH_2CO_2H) or Aryl ($ArCO_2H$) Carboxylic Acid	On Reaction with (Species and/or Process)	Generates Substituted Acid "Derivative"
4. CH_3CO_2H ethanoic acid (acetic acid)	$CH_3CH_2NH_2$ aminoethane (ethyl amine)	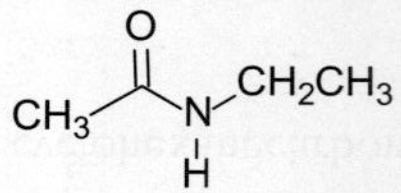 N-ethyl ethanoamide (N-ethyl acetamide)
	$C_6H_{11}N=C=NC_6H_{11}$ or 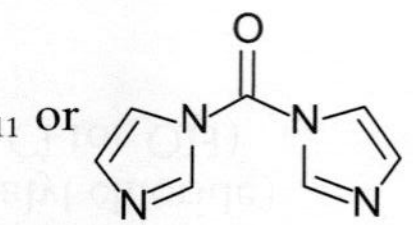 (dicyclohexylcarbodiimide) (carbonyl diimidazole) (substitution of aminoalkyl for OH)	+ $(C_6H_{11}NH)_2C=O$ N,N′-dicyclohexylurea or 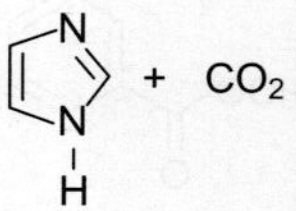 imidazole $+ CO_2$
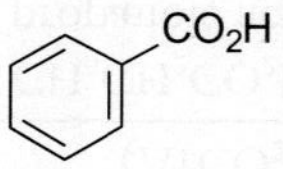 5. benzenecarboxylic acid (benzoic acid)	$Na^+N_3^-/H_2SO_4$ sodium azide / sulfuric acid (a source of hydrazoic acid [HN_3]) (substitution of N_3 for OH)	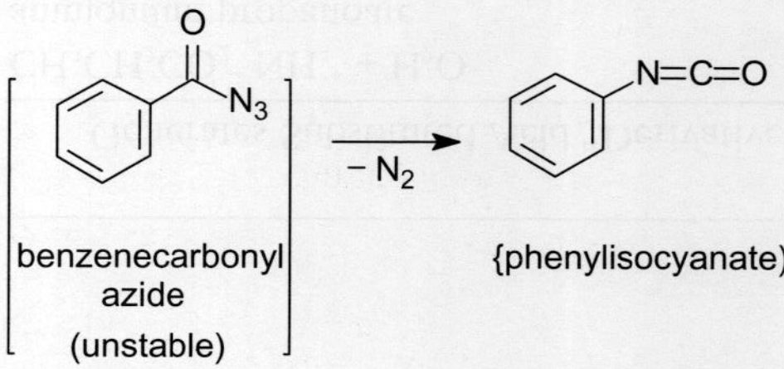 benzenecarbonyl azide (unstable) $\xrightarrow{-N_2}$ {phenylisocyanate}
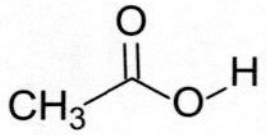 6. ethanoic acid (acetic acid)	$CH_2=C=O$ ketene (substitution of CH_2CO for hydrogen)	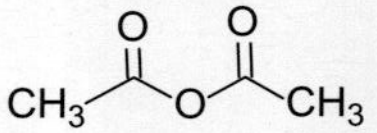 ethanoic anhydride (acetic anhydride)

7.

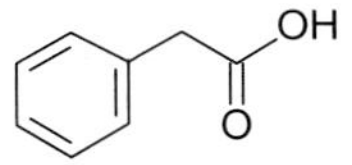

phenylethanoic acid
(phenylacetic acid)

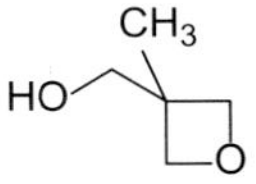

3-(hydroxymethyl)-3-methyloxetane

$C_6H_{11}N=C=NC_6H_{11}$ or

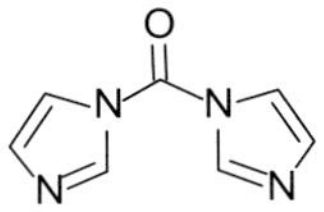

(dicyclohexylcarbodiimide) (carbonyl
  diimidazole)
(substitution of an alcohol for
  carboxylic acid OH)

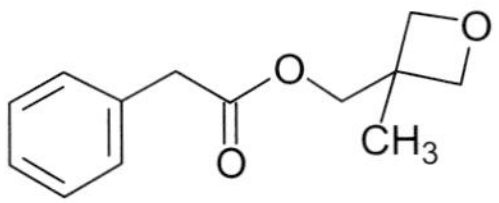

$(C_6H_{11}NH)_2C=O$
N,N′-dicyclohexylurea
or

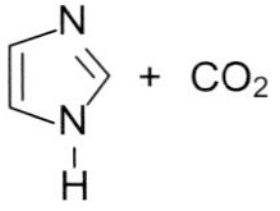

+ CO_2

imidazole

8.

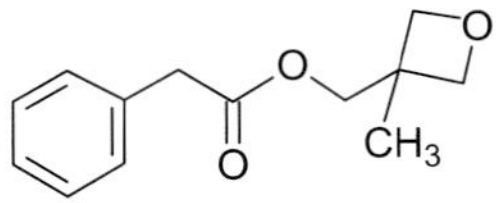

[3-(hydroxymethyl)-3-methyloxetyl]-
phenyl ethanoate

$BF_3{\cdot}O(CH_2CH_3)_2$
borontrifluoride etherate
(a Lewis acid)
(substitution at the C=O)

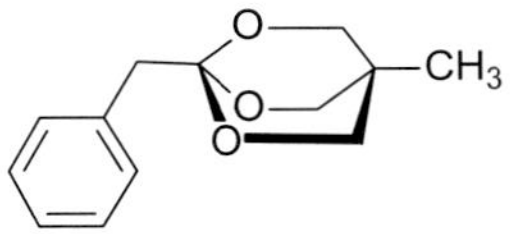

1-benzyl-4-methyl-2,6,7-trioxabicyclo[2.2.2]octane

9.

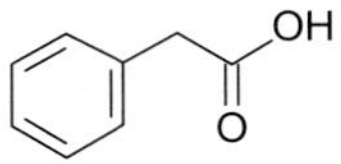

phenylethanoic acid
(phenylacetic acid)

$Na^+N_3^-/H_2SO_4$
sodium azide/ sulfuric acid
(substitution of N_3 for OH) the
  Schmidt reaction

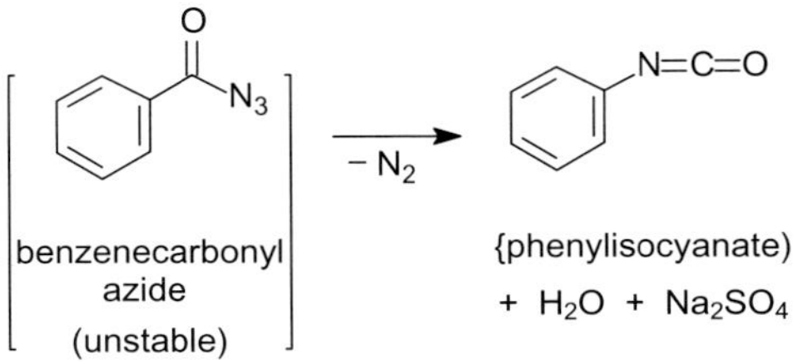

benzenecarbonyl
azide
(unstable)

− N_2

{phenylisocyanate)
+ H_2O + Na_2SO_4

TABLE 9.7. Representative Carboxylic Acid Chlorides and Their Derivatives

Alkyl (RCH$_2$COCl) or Aryl (ArCOCl) Carboxylic Acid Chloride	On Reaction with Species and/or Process)	Generates Substituted Acid "Derivative"
1. CH$_3$CH$_2$COCl propanoic acid chloride (propanoyl chloride)	H$_2$O water hydrolysis (substitution of OH for Cl)	CH$_3$CH$_2$CO$_2$H + HCl propanoic acid
2. benzenecarboxylic acid chloride (benzoyl chloride)	CH$_3$CH$_2$OH / pyridine ethanol and base catalyst [azabenzene (pyridine)] (substitution of OCH$_2$CH$_3$ for Cl)	ethyl benzenecarboxylate (ethyl benzoate) + pyridinium Cl$^-$
3. cyclohexanecarboxylic acid chloride (cyclohexanecarbonyl chloride)	Na$^+$ $^-$O$_2$CH sodium formate (sodium methanoate) (substitution of O$_2$CH for Cl)	cyclohexanecarboxylic formic anhydride + Nacl
4. CH$_3$COCl ethanoyl chloride (acetyl chloride)	CH$_3$CH$_2$NH$_2$ aminoethane (ethyl amine) (substitution of NHCH$_2$CH$_3$ for Cl)	N-ethyl ethanoamide (N-ethyl acetamide) + CH$_3$CH$_2$NH$_3$$^+Cl^-$

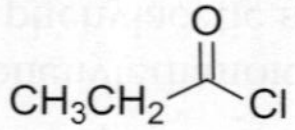

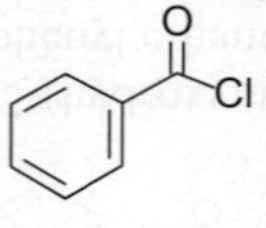

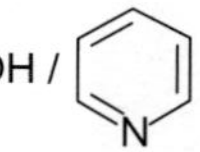

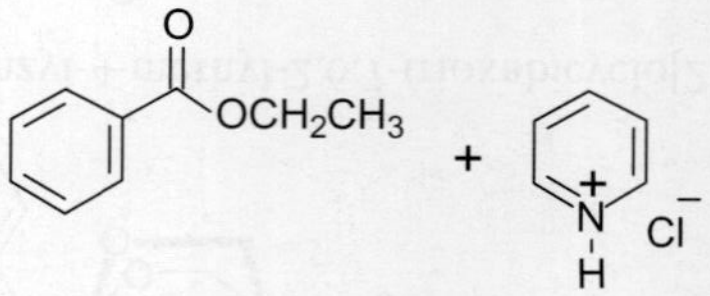

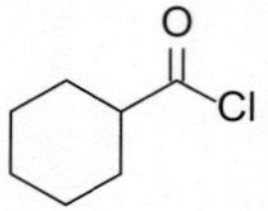

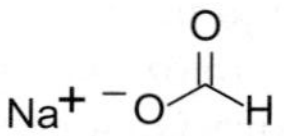

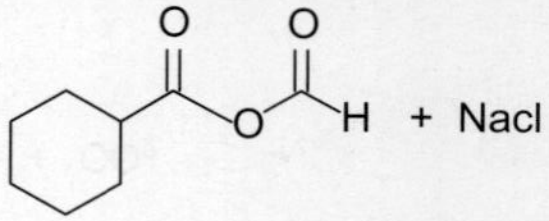

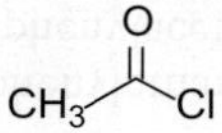

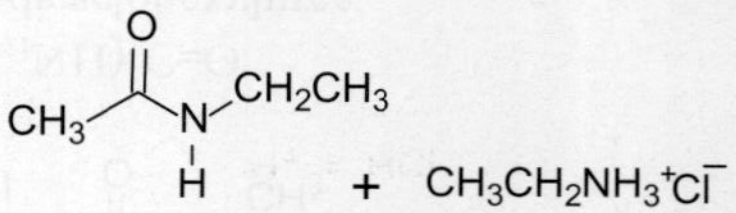

#	Starting material	Reagent	Product
5.	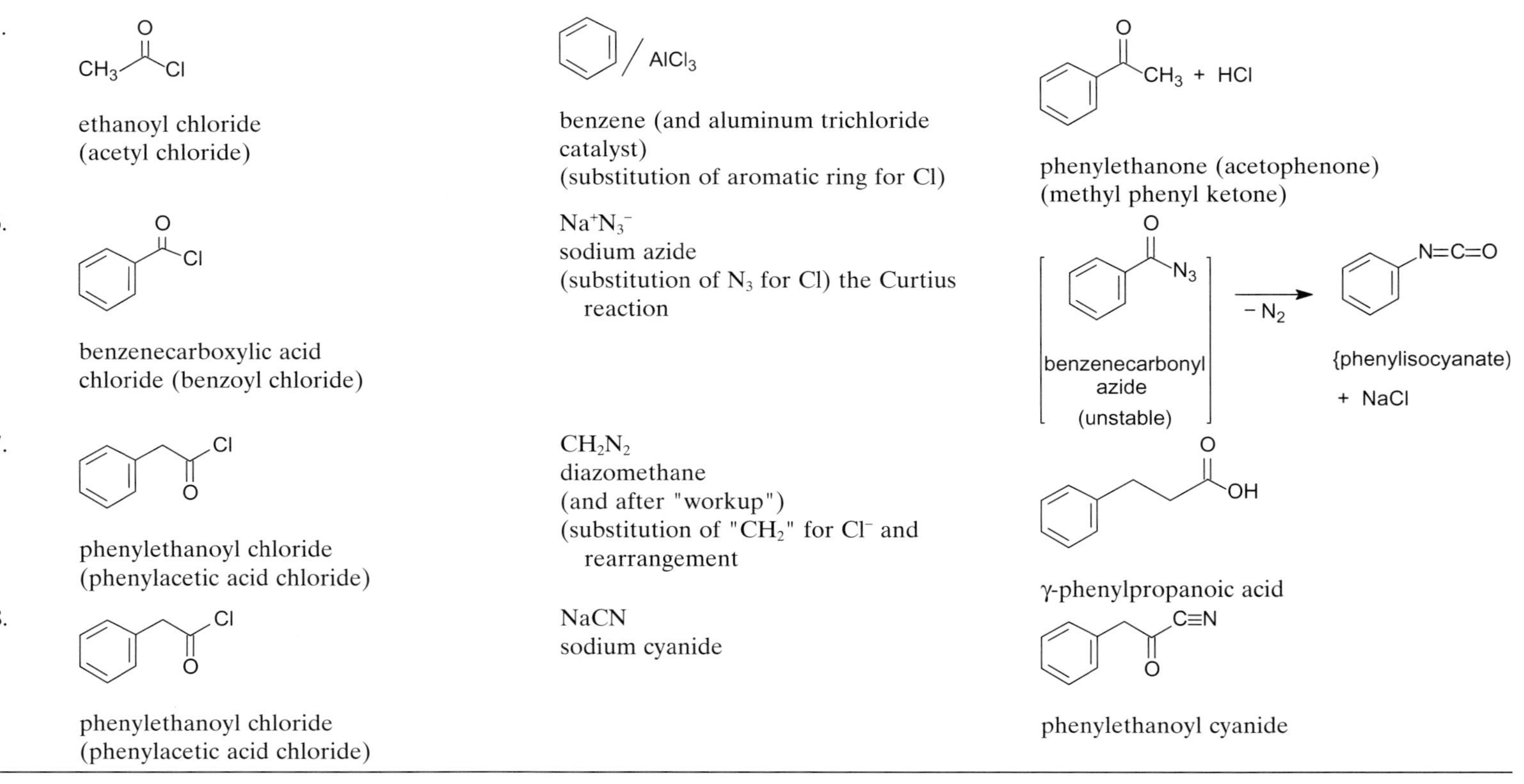 ethanoyl chloride (acetyl chloride)	benzene (and aluminum trichloride catalyst) (substitution of aromatic ring for Cl)	phenylethanone (acetophenone) (methyl phenyl ketone)
6.	benzenecarboxylic acid chloride (benzoyl chloride)	$Na^+N_3^-$ sodium azide (substitution of N_3 for Cl) the Curtius reaction	benzenecarbonyl azide (unstable) $\xrightarrow{-N_2}$ {phenylisocyanate) + NaCl
7.	phenylethanoyl chloride (phenylacetic acid chloride)	CH_2N_2 diazomethane (and after "workup") (substitution of "CH_2" for Cl^- and rearrangement	γ-phenylpropanoic acid
8.	phenylethanoyl chloride (phenylacetic acid chloride)	NaCN sodium cyanide	phenylethanoyl cyanide

TABLE 9.8. Representative Carboxylic Acid Anhydrides and Their Derivatives

Alkyl (RCH_2—...—CH_2R)

or

Aryl (Ar—...—Ar)

Carboxylic Acid Anhydrides	On Reaction with (Species and/or Process)	Generates Substituted Acid "Derivatives"
1. ethanoic anhydride (acetic anhydride)	H_2O water (hydrolysis) substitution of –OH for acetate	CH_3CO_2H ethanoic acid (acetic acid)
2. benzene carboxylic anhydride (benzoic anhydride)	NaN_3 sodium azide substitution of N_3 for benzoate	benzenecarbonyl azide $\rightarrow$ (phenylisocyanate) $+ N_2$ + sodium benzenecarboxylate (sodium benzoate)
3. ethanoic anhydride (acetic anhydride)	$CH_3CH_2NH_2$ ethanamine (ethylamine; aminoethane) substitution of ethanamine for acetate	N-ethylacetamide (N-ethylethanamide) $+ CH_3CO_2^- CH_3CH_2NH_3^+$ ethylammonium acetate (ethylammonium ethanoate)

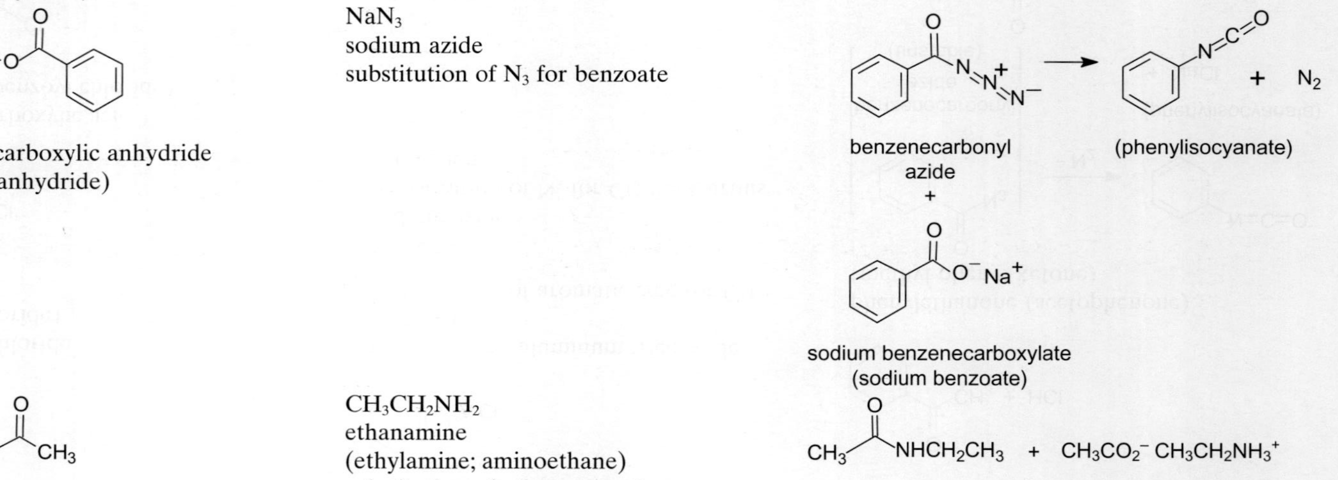

4.

CH_3CH_2—C(=O)—O—C(=O)—CH_2CH_3

propanoic anhydride

CH_3CH_2OH
ethanol
(ethyl alcohol)
substitution of ethanol for acetate

CH_3CH_2—C(=O)—O—CH_2CH_3 + CH_3CH_2—C(=O)—OH

ethyl propanoate   propanoic acid

5.

succinic anhydride
(butane-1,4-dioic anhydride)

benzene and aluminum trichloride
(catalyst)/Friedel-Crafts reaction
substitution of aromatic ring for
  acetate
NH_2NH_2
hydrazine
substitution of nitrogen
for oxygen

4-phenyl-4-oxobutanoic acid

6.

1,2-benzenedicarboxylic anhydride
(phthalic anhydride)

2,3-diaza-2,3-dihydro-1,4-napthaquinone + H_2O

TABLE 9.9. Representative Carboxylic Acid Esters and Their Derivatives

Alkyl (RCH_2CO_2R') or Aryl ($ArCO_2R'$) Carboxylic Acid Esters	On Reaction with (Species and/or Process)	Generates Substituted Acid "Derivative"
1. $CH_3CH_2CO_2CH_2CH_3$ ethyl propanoate	M^+OH^- (aq) metal (e.g., Na^+, K^+, etc.) hydroxide aqueous solution (hydrolysis—substitution of –OH for –OCH_2CH_3)	$CH_3CH_2CO_2^-M^+$ + CH_3CH_2OH metal (e.g., sodium) ethanol propanoate (ethyl alcohol)
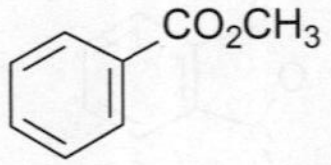 2. (methyl benzenecarboxylate structure, CO_2CH_3) methyl benzenecarboxylate (methyl benzoate)	CH_3CH_2OH/H^+ ethanol and acid catalyst (substitution of OCH_2CH_3 for OCH_3)	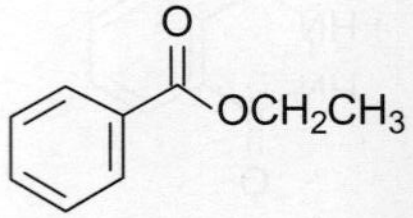 (benzoate ester structure, OCH_2CH_3) + CH_3OH ethyl benzenecarboxylate methanol (ethyl benzoate) (methyl alcohol)
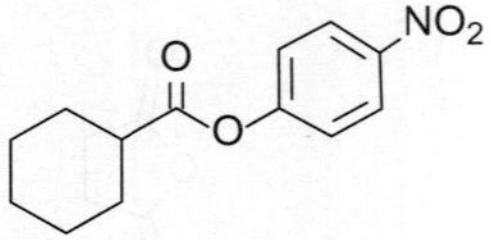 3. (p-nitrophenoxy cyclohexanecarboxylate structure, NO_2) p-nitrophenoxy cyclohexanecarboxylate	$C_6H_5CH_2NH_2$ benzylamine (substitution of benzylamine for p-nitrophenoxy)	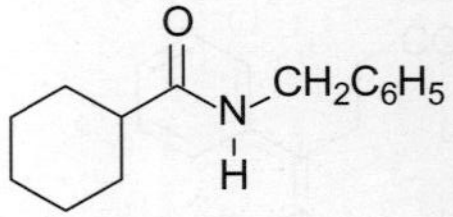 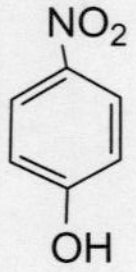 (amide structure, N-$CH_2C_6H_5$, H) + (p-nitrophenol structure, NO_2, OH) N-benzylcyclohexane- p-nitrohydroxy- carboxamide benzene (p-nitrophenol)
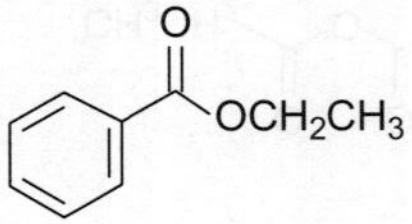 4. (ethyl benzenecarboxylate structure, OCH_2CH_3) ethyl benzenecarboxylate (ethyl benzoate)	CH_3MgBr or CH_3Li methylmagnesium bromide or methyllithium (substitution [and addition] of methyl for OCH_2CH_3)	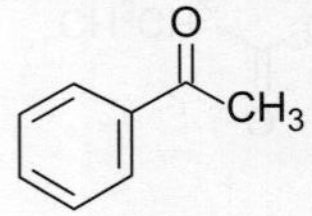 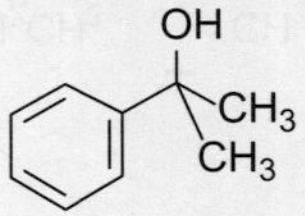 (ketone structure, CH_3) + (alcohol structure, CH_3, CH_3, OH) + magnesium, halide and/or lithium salts phenylethanone (after hydrolysis) (acetophenone) 2-phenyl-2-propanol (methyl phenyl ketone)

5.

ethyl benzenecarboxylate
(ethyl benzoate)

$CH_3CO_2CH_2CH_3$ / $CH_3CH_2O^-Na^+$
ethyl ethanoate (ethyl acetate) and sodium
  ethoxide
(substitution of ethyl ethanoate [ethyl acetate]
  for OCH_2CH_3)
(the Claisen condensation)

ethyl 3-phenyl-
3-oxopropanoate

ethyl 3-oxobutanoate

6.

diethyl
  1,6-hexanedicarboxylate
(diethyl adipate)

$CH_3CH_2O^-$ M^+
metal (e.g., Na^+, K^+ etc)
solution in ethanol (CH_3CH_2OH)
(substitution of carbon for $-OCH_2CH_3$)
(an acyloin condensation)

ethyl 2-oxocyclopentane-
carboxylate

ethanol
(ethyl alcohol)

7.

methyl
  cyclohexanecarboxylate

$[(CH_3)_2CH]_2N^-$ Li^+
lithium diisopropylamide (LDA)
followed by CH_3I (iodomethane [methyl iodide])
substitution of proton α- to carbonyl with a
  methyl (CH_3)
(an alkylation)

methyl 1-methylcyclo-
hexanecarboxylate

LiI
lithium
iodide

$[(CH_3)_2CH]_2NH$
diisopropylamine

8.

ethyl benzenecarboxylate
(ethyl benzoate)

the lithium enolate of ethanal (acetaldehyde)
(an aldol-type condensation)

3-phenyl-3-oxopropanal

lithium ethoxide

TABLE 9.9. *Continued*

Alkyl (RCH$_2$CO$_2$R′) or Aryl (ArCO$_2$R′) Carboxylic Acid Esters	On Reaction with (Species and/or Process)	Generates Substituted Acid "Derivative"
9. methyl cyclohexanecarboxylate	the Tebbe reagent titanium μ-chlorobis(η5-2,4-cyclopentadien-1-yl)-(dimethylalminum)-μ-methylene-(9Cl)	1-cyclohexyl-1-methoxyethene
10. ethyl benzenecarboxylate (ethyl benzoate)	Lawesson's reagent 2,4-bis(4-methoxyphenyl)-2,4-dithioxo-1,2,3,4-dithiadiphosphetane	ethyl benzenethiocarboxylate (ethyl thiobenzoate)
11. ethyl benzenecarboxylate (ethyl benzoate)	H$_2$NNH$_2$/H$_2$SO$_4$ hydrazine/sulfuric acid substitution of hydrazine for –OCH$_2$CH$_3$ a Curtius rearrangement	phenylisocyanate + (NH$_4$)$_2$SO$_4$
12. ethyl benzenecarboxylate (ethyl benzoate)	NH$_2$OH/HCl hydroxylamine hydrochloride a Lossen rearrangement	a hydroxamic acid → phenylisocyanate + H$_2$O

TABLE 9.10. Representative Carboxylic Acid Amides and Their Derivatives

Alkyl (RCH_2CONR_2) or Aryl ($ArCONR2$) Carboxylic Acid Amides (where R = H, alkyl, and/or aryl) AND Alkyl (R–CN) or Aryl (Ar-CN) Nitriles	On Reaction with (Species and/or Process)	Generates Substituted Acid "Derivative"
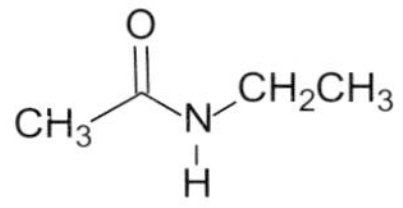 1. N-ethyl ethanoamide (N-ethyl acetamide)	H_2O/H^+ aqueous acid catalyst (hydrolysis of the amide) (substitution of OH for $HNCH_2CH_3$)	CH_3CO_2H + ethanoic acid (acetic acid) $CH_3CH_2NH_3^+$ ethyl ammonium ion
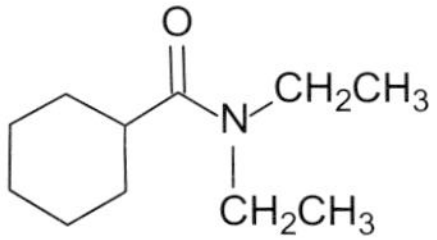 2. N,N-diethyl cyclohexanecarboxamide	CH_3MgBr methylmagnesium bromide (Grignard reagent) (substitution of a methyl for N,N-diethyl)	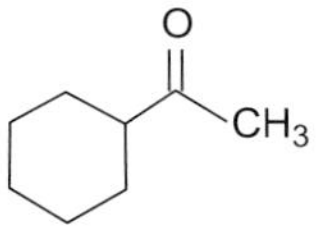 1-cyclohexylethanone (cyclohexyl methyl ketone) + $(CH_3CH_2)_2NMgBr$
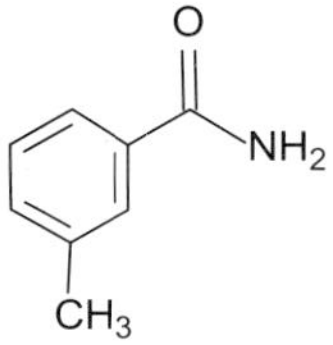 3. 3-methylbenzenecarboxamide (*m*-methylbenzamide)	bromine (Br_2) followed by base (e.g., KOH) (substitution of NH_2 for carbon) (the Hofmann degradation)	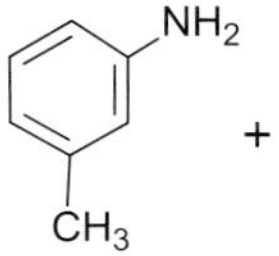 3-methylbenzenamine (3-methylaniline; *m*-toluidine) + KBr + CO_2
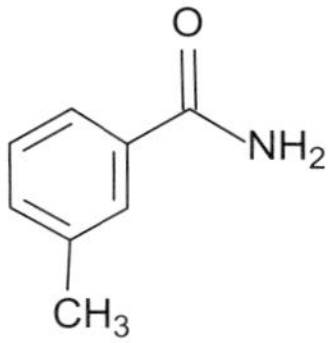 4. 3-methylbenzenecarboxamide (*m*-methylbenzamide)	P_4O_{10} (substitution of a triple bond for water—a dehydration)	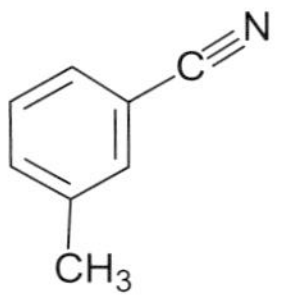 3-methylbenzonitrile (*m*-methylbenzonitrile)

TABLE 9.10. *Continued*

Alkyl (RCH$_2$CONR$_2$) or Aryl (ArCONR2) Carboxylic Acid Amides (where R = H, alkyl, and/or aryl) AND Alkyl (R–CN) or Aryl (Ar-CN) Nitriles	On Reaction with (Species and/or Process)	Generates Substituted Acid "Derivative"
5.	HONO (substitution on nitrogen)	R$_1$=R$_2$=H yields R–COOH + N$_2$ + H$_2$O; R$_1$ or R$_2$=H yields (N-nitroso derivative); neither R$_1$ nor R$_2$=H yields no reaction
6. 3-methylbenzonitrile (*m*-methylbenzonitrile)	H$_2$O / H$^+$ aqueous acid catalyst (hydrolysis of the nitrile) (substitution of O,OH for N)	3-methylbenzenecarboxylic acid (*m*-methylbenzoic acid) + NH$_4^+$ ammonium ion
7. 3-methylbenzonitrile (*m*-methylbenzonitrile)	CH$_3$CH$_2$OH / H$^+$ ethanol (ethyl alcohol) and acid catalyst (substitution of OCH$_2$CH$_3$ for N) (the Pinner reaction)	ethyl 3-methylbenzoic acid ortho ester

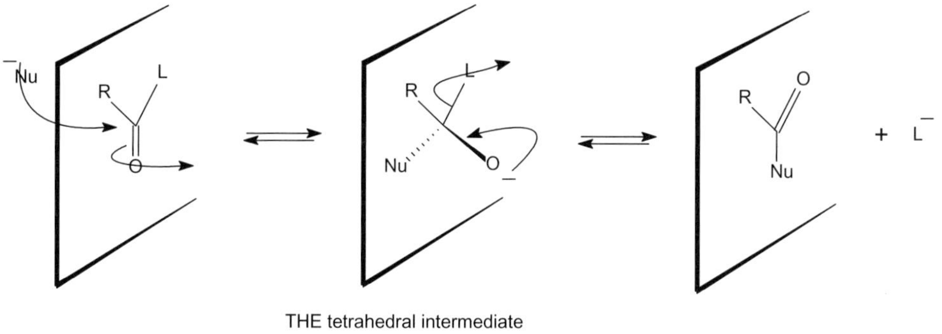

THE tetrahedral intermediate

Scheme 9.116. A representation of the formation of THE tetrahedral intermediate, which results in a substitution reaction via an addition–elimination process.

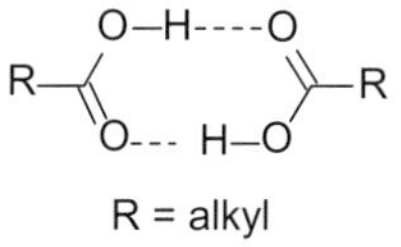

R = alkyl

Figure 9.16. A representation of a carboxylic acid dimer.

The second path involves addition of the nucleophile to the carbon of the carbonyl, generating a quaternary intermediate, which then loses the leaving group (Scheme 9.116). This replacement is the result of an **addition followed by an elimination**.

The first item in Table 9.6, the reaction between propanoic acid ($CH_3CH_2CO_2H$) and ammonium hydroxide, corresponds to an **acid–base** reaction. Carboxylic acids are, as noted from the pK_a's* in Table 9.11, acidic and while, in the gas phase, the low molecular weight carboxylic acids contain significant quantities of dimer (Figure 9.16), it is likely that solvated monomer is the major species (at least in solvents capable of hydrogen bonding, e.g., H_2O).

With simple alkyl groups attached to the α-carbon of the carboxylic acid (the acids in Table 9.11) the pK_a's are all about the same (as measured in aqueous ethanol). However, as seen with the chloroethanoic (chloroacetic) acids (Table 9.11), the effect of electron-withdrawing groups is to increase the acidity (decrease the pK_a) of the carboxylic acid. As expected, a correlation can also be effected with the corresponding downfield shifts of the chlorine bearing carbon atom in the respective ^{13}C NMR spectra of the chloroethanoic (chloroacetic) acids.

Other substituents, such as hydroxyl groups, appear to play a minor role in acidity enhancement. However, where there is a second carboxylic acid group ($-CO_2H$) attached to the carbon of the carbonyl of a carboxylic acid group (CO_2H) (only one such compound, ethanedioic [oxalic] acid, can exist), the acidity increases, and it is argued that the mutually electron-withdrawing carbonyl groups are the cause. Interestingly, there is a second exception. In the case of *cis-* or (*Z*)-2-butenedioic

---

*The subject of the pKa and its relation to pH, as well as acidity in general, was introduced in Chapter 5 (Part E) and can usefully be reviewed at this time.

TABLE 9.11. Representative Examples of Carboxylic Acids[a]

Compound	Name	mp °C	bp °C	pK_a	^{1}H NMR data [δ, ppm TMS = 0]	^{13}C NMR data [δ, ppm TMS = 0]
HCO_2H	Methanoic acid (formic acid)	8.4	101	3.77	8.99 (CH); 3.95 (OH)	166.3 (C=O)
CH_3CO_2H	Ethanoic acid (acetic acid)	17	118	4.76	2.08 (CH_3); 8.85 (OH)	20.48 (CH_3); 177.14 (C=O)
$CH_3CH_2CO_2H$	Propanoic acid	−22	141	4.88	1.16 (CH_3, t, J = 7.8); 2.36 (CH_2, q, J = 7.8); (OH) 7.82	8.95 (CH_3); 27.63 (CH_2); 180.40 (C=O)
$(CH_3)_2CHCO_2H$	2-Methylpropanoic acid (isobutyric acid)	−46	154	4.85	1.20 (CH_3, d, J = 6.9); 2.58 (CH, m, J = 6.9); (OH) 11.88	18.79 (CH_3); 34.06 (CH); 184.00 (C=O)
$(CH_3)_3CCO_2H$	2,2-Dimethylpropanoic acid (privalic acid)	35.4	163	5.05	1.23 (CH_3); 11.49 (OH)	27.01 (CH_3); 38.67 (C); 185.90 (C=O)
$CH_3CH_2CH_2CO_2H$	Butanoic acid (butyric acid)	−5	163	4.82	0.98 (CH_3, t, J = 7.0); 1.67 (CH_2, m, J = 7.0); 2.33 (CH_2, m, J = 7); 11.51 (OH)	13.65 (CH_3); 18.36 (CH_2); 36.18 (CH_2); 180.66 (C=O)
$CH_3(CH_2)_3CO_2H$	Pentanoic acid (valeric acid)	−35	187	4.86	0.93 (CH_3, t, J = 6.6); 1.39 (CH_2, m); 1.62 (CH_2, m), 2.35 (CH_2, t, J = 7.0); 11.96 (OH)	13.71 (CH_3); 22.32 (CH_2); 26.89 (H_2); 33.99 (CH_2); 180.85 (C=O)
$CH_3(CH_2)_4CO_2H$	Hexanoic acid (caproic acid)	−2	205	4.85	0.90 (CH_3, t, J = 6.6); 1.33–1.64 (3 CH_2, m); 2.35 (CH_2, t, J = 7.0); 11.59 (OH)	13.91 (CH_3); 22.42 (CH_2); 24.50 (CH_2); 31.36 (CH_2); 34.23 (CH_2); 180.78 (C=O)
$CiCH_2CO_2H$	Chloroacetic acid	63	Dec	2.87	4.15 (CH_2, s); 11.26 (OH)	40.58 (CH_2); 173.72 (C=O)
Ci_2CHCO_2H	Dichloroacetic acid	10	198	1.26	5.89 (CH, s); 11.20 (OH)	63.75 (CH); 170.20 (C=O)
Ci_3CCO_2H	Trichloroacetic acid	55	196	0.52	11.36 (OH)	89.00 (C); 166.5 (C=O)
$HOCH_2CO_2H$	Hydroxyethanoic acid (glycolic acid)	79	Dec	3.9	4.29 (CH_2, s), 9.38 (2 OH)	60.16 (CH); 177.04 (C=O)
$(\forall)–CH_3CH(OH)CO_2H$	(racemic)-2-hydroxy-propanoic acid (lactic acid)	17		3.87	1.40 (CH_3, d, J = 6.88); 4.33 (CH, q, J = 6.88), 8.26 (2 OH)	20.15 (CH_3); 66.55 (CH); 177.45 (C=O)
$CH_2=CHCO_2H$	Propenoic acid (acrylic acid)	13	141	4.26	5.96 (E–H3, dd, J = 10.5, 1.66); 6.14 (H2, dd, J = 10.5, 17.2); 6.52 (Z–H3), dd J = 1.66, 17.2); 12.0 (OH)	128.14 (CH); 133.15 (CH_2); bbg 171.95 (C=O)

Structure	Name				^{1}H NMR	^{13}C NMR
HO_2CCO_2H	Ethanedioic acid (oxalic acid)	180	d	1.27[b]	7.85 (OH)	162.31 (C=O)
$HO_2CCH_2CO_2H$	Propanedioic acid (malonic acid)	135	d	2.85[b]	3.26 (CH_2); 12.2 (OH)	42.00 (CH_2); 169.60 (C=O)
$HO_2CCH_2CH_2CO_2H$	Butanedioic acid (succinic acid)	183	235d	4.21[b]	2.42 (CH_2); 12.0 (OH)	29.50 (CH_2); 175.00 (C=O)
$HO_2C(CH_2)_3CO_2H$	Pentanedioic acid (glutaric acid)	98	303	4.33[b]	1.90 (CH_2, m, J = 7.0); 2.43 (2 CH_2, t, J = 7.0); 9.89 (OH)	20.43 (CH_2); 33.59 (CH_2); 178.53 (C=O)
$HO_2C(CH_2)_4CO_2H$	Hexanedioic acid (adipic acid)	153	338	4.43[b]	1.50 (2 β-CH_2, m); 2.21 (2 α-CH_2, m); 12.0 (OH)	24.00 (β-CH_2); 33.35 (α-CH_2); 174.28 (C=O)
	tartaric acid (2,3-dihydroxybutanoic acid)	174		3.02[b]	4.51 (CH, s); 10.62 (OH)	73.06 (CH); 174.81 (C=O)
	Maleic acid [(Z)-2-butenedioic acid]	133		1.91[b]	7.06 (CH, s); 11.63 (OH)	130.04 (CH); 166.68 (C=O)
	Fumaric acid [(E)-2-butenedioic acid]	276		3.10[b]	6.80 (CH, s); 11.63 (OH)	133.96 (CH); 166.10 (C=O)
	Benzoic acid (benzenecarboxylic acid)	122	249	4.20	7.44 (meta-aromatic-CH, m); 7.51 (para-aromatic–CH, m); 7.97 (ortho-aromatic–CH, m); 9.39 (–OH)	129.01 (meta-aromatic–C); 130.08 (ortho-aromatic–C); 130.50 (ipso-aromatic–C); 133.76 (para-aromatic–C); 168.57 (C=O)
	Benzene-1,2-dicarboxylic acid (ortho-phthalic acid)	207		2.95[b]	6.72 (OH); 7.64 (meta[para]-aromatic–CH, m, J = 7.83, 7.19, 1.25); 7.82 (ortho-aromatic–CH, m, J = 7.83, 1.25, 0.57)	128.23 (ipso-aromatic–C); 130.36 (ortho-aromatic–C); 132.80 (meta[para]-aromatic–C); 168.78 (C=O)

TABLE 9.11. *Continued*

Compound	Name	mp °C	bp °C	pK$_a$	^{1}H NMR data [δ, ppm TMS = 0]	^{13}C NMR data [δ, ppm TMS = 0]
CO$_2$H, OH	2-hydroxybenzenecarb-oxylic acid (o-hydroxybenzoic acid) (salicylic acid)	158	256	2.94	7.02 (CH[5]-aromatic, m); 7.09 (CH[3]-aromatic, m); 7.50 (CH[4]-aromatic, m); 8.01 (CH[6]-aromatic, m); 9.76 (2 OH) (J's = 0.43, 1.17, 1.78, 7.74, 7.98, 8.24)	112.77, C$_1$; 116.94, C$_3$; 118.77, C$_5$; 130.28, C$_6$; 135.25, C$_4$; 161.51, C$_2$; 172.17 (C=O)
CO$_2$H, NH$_2$	2-aminobenzenecarboxylic acid (o-aminobenzoic acid) (anthranilic acid)	147		4.80	6.69 (CH[5]-aromatic, m); 6.71 (CH[3]-aromatic, m); 7.26 (CH[4]-aromatic, m); 7.46 (CH[6]-aromatic, m); 7.57 (OH, NH) (J's = 0.47, 1.1; 1.5; 7.38; 8.0; 8.19)	109.54, C$_1$; 114.48, C$_3$; 116.22, C$_5$; 131.07, C$_6$; 133.59, C$_4$; 151.38, C$_2$; 169.48 (C=O)
CO$_2$H (E)-cinnamic structure	(E)-3-phenylpropenoic acid [(E)-cinnamic acid]	133	300	4.43	6.70 (α-CH, d, J = 15.8); 7.30 (meta-aromatic, m); 7.40 (para-aromatic, m); 7.53 (ortho-aromatic); 7.90 (β-CH, d, J = 15.8); 8.90 (OH)	119.18 (α-C); 128.84 (ortho-aromatic); 129.13 (meta-aromatic); 130.17 (para-aromatic); 134.56 (ipso-aromatic); 157.73 (β-C); 169.04 (C=O)
CO$_2$H, H$_3$C	p-Methylbenzenecarboxylic acid (p-methylbenzoic acid) (p-toluic acid)	179	273	4.37	2.42 (CH$_3$, m, J = 0.34, 0.68); 7.27 (CH[3,5]-meta-aromatic, m of d, J = 8.34); 8.01 (CH[2,6]-ortho-aromatic, m of d, J = 8.34); 9.40 (OH)	21.75 (CH$_3$); 126.58 (ipso-aromatic); 129.15 (meta-aromatic); 130.21 (ortho-aromatic); 144.58 (para-aromatic); 172.45 (C=O)

Structure	Name				
	p-Nitrobenzenecarboxylic acid (p-nitrobenzoic acid)	241		3.43	8.21 (CH[2,3,5,6], m, J = 0.43, 1.67, 2.28, 8.51); 9.39 (OH)
					126.83 (meta-aromatic); 129.52 (ortho-aromatic); 134.49 (ipso-aromatic); 149.60 (para-aromatic); 167.78 (C=O)
	p-Chlorobenzenecarboxylic acid (p-chlorobenzoic acid)	243		3.98	7.54 (CH[3,5]-meta-aromatic, d, J = 8.43, m, J = 0.42, 2.44); 8.02 (CH[2,6]-ortho-aromatic, d, J = 8.43, m, J = 0.42, 2.44); 9.39 (OH)
					128.35 (meta-aromatic); 129.54 (ipso-aromatic); 130.94 (ortho-aromatic); 138.07 (para-aromatic); 166.57 (C=O)
	p-Methoxybenzene-carboxylic acid (p-methoxybenzoic acid) (p-anisic acid)	184	275	4.47	3.89 (CH₃); 6.96 (CH[3,5]-meta-aromatic, d, J = 9.54); 8.07 (CH[2,6]-ortho-aromatic, d, J = 9.54); 9.39 (OH)
					55.57 (CH₃); 114.00 (meta-aromatic); 122.77 (ipso-aromatic); 132.05 (ortho-aromatic); 163.74 (para-aromatic); 168.66 (C=O)
	2,4,6-Trinitrobenzene-carboxylic acid (2,4,6-trinitrobenzoic acid)	228		0.65	4.17 (OH); 8.57 (CH–aromatic)
					125.78 (meta-aromatic); 127.10 (ipso-aromatic); 144.01 (para-aromatic); 146.53 (ortho-aromatic); 169.54 (C=O)

[a]The information in this table is derived from experiment and subject to change.
[b]The pK$_a$ values are based on the dissociation of the first carboxylic acid group.
[c]The following abbreviations are used: d = doublet; t = triplet; q = quartet; m = multiplet.

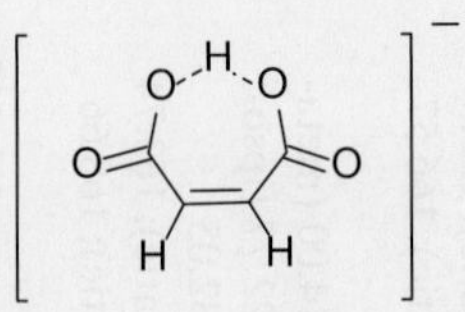

Figure 9.17. A representation of hydrogen-bonded anion from (*Z*)-2-butenedioic (maleic) acid.

(maleic) acid, it has been suggested that the increased acidity can be accounted for by the ability of the second carboxylic acid group to stabilize the anion through sharing the remaining proton (Figure 9.17) and the implication that such an internally hydrogen-bonded structure may be (partially) present prior to initial proton loss, which, in effect, loosens the proton that is not internally bonded.

The substituted benzenecarboxylic (benzoic) acids enjoy a special place in the discussion of carboxylic acids in general. This is because they were (and continue to be) used in an effort to quantitate the relationship between structure and reactivity.

Beginning in the 1930s, Louis Hammett* and coworkers began to examine the correlation between acidity (recorded as K_a) and groups that are attached to the aromatic ring of benzenecarboxylic (benzoic) acids. By that time, it was clear that functional groups on the aromatic ring could be grouped into those that were thought to withdraw electrons (i.e., an **EWG**) from the ring. The EWG could accomplish this either through induction or resonance. The nitro ($-NO_2$) group serves as an example. On the other hand, groups that, were thought to donate electrons (i.e., an **EDG**) to the aromatic ring, such as the methoxy ($-OCH_3$) group. Of course, as noted earlier (Chapter 4, Equation 4.6), K_{eq} (which is K_a here for these acids) is related to the free energy change ΔG° accompanying the process. Hammett then went on to point out " ... It is almost obvious and easily demonstrated that two series of rate or equilibrium constants which are both linearly related to a third series are linearly related to each other ... (and) the ionization constants of benzoic acids in water at 25°C are convenient because so many accurate values are available ..." "If one then defines a substituent constant σ_i characteristic of a substituent *i* by the relation

$$\sigma_i = \log K_i - \log K_o \tag{9.94}$$

where K_o is the acidity constant of benzoic acid and K_i is the constant of the substituted benzoic acid, the linear free-energy relationship[†] takes the form

$$\log k_{ij} - \log k_{oj} = \rho_j \sigma_i \tag{9.95}$$

where k_{ij} is the rate (and K_{ij} would be the equilibrium) constant for the reaction *j* when the substituent *i* is present and k_{oj} (and K_{oj}) is the corresponding constant in the absence of any substituent. The parameter ρ_j is called the **r**eaction constant for

*Louis P. Hammett (1894–1987) was Professor of Chemistry at Columbia University. Hammett, Louis P. J. *Amer. Chem. Soc.*, **1928**, *50*, 2660. Hammett, L. *Physical Organic Chemistry. Reaction Rates, Equilibria and Mechanisms*, McGraw-Hill, New York, 1970.

[†]The principle involved is based on the idea that, as a first approximation, molecules can be divided into a "reacting group, X" and a "nonreacting residue, R." Further, the quantities to be measured result from free-energy (ΔG°) differences, which are related to changes in "R."

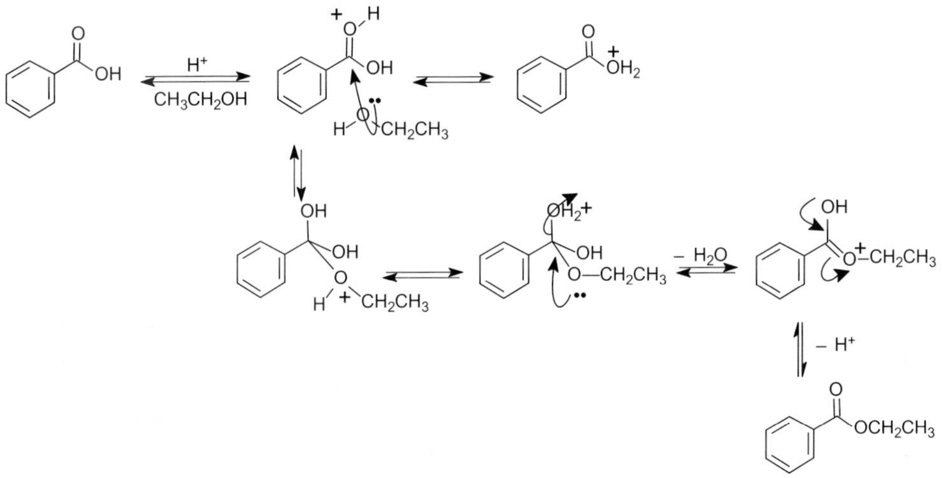

Scheme 9.117. A representation of the esterification of benzenecarboxylic (benzoic) acid by ethanol (ethyl alcohol, CH_3CH_2OH) in the presence of a protic catalyst. In the reversible process, initial protonation of the carboxylic acid is followed by attack on the carbonyl group by the alcohol or, in the reverse direction, by water.

the reaction j…and is set equal to 1 for the ionization of benzoic acids." So, if $\rho_j = 1$ for the ionization of benzoic acids, then it is possible to define a set of σ (substituent) constants …

$$\begin{aligned}
\sigma_x &= \log\left(K_a\text{X-}C_6H_4CO_2H / K_a C_6H_5CO_2H\right) \\
&= pK_a(C_6H_5CO_2H) - pK_a(\text{X-}C_6H_4CO_2H)
\end{aligned} \tag{9.96}$$

Although a large amount of data on substituted benzenecarboxylic acids (benzoic acids) supports the Hammett relationship, it should be clear even on the basis of the limited examples in Table 9.11 that a positive σ is found for carboxylic acids with EWGs (the acid ionizes more completely than benzenecarboxylic acid [benzoic acid] and is thus a "stronger" acid), while a negative σ results when the acid is a "weaker" acid and ionizes less completely than benzenecarboxylic acid (benzoic acid).

Additional consideration will be given to linear free-energy relationships when reactions at the carbonyl group (e.g., ester hydrolysis, *vide infra*) are considered and reaction constants (i.e., ρ) can be developed. Meanwhile, it is important to recognize that the principle of comparing reactivity (and other attributes) of substituted and unsubstituted species to allow predictions to be made has been extended with varying degrees of success to a wide variety of aromatic and non-aromatic systems. It is also important to understand that the difficulty of accurate prediction arises, in part, from the inability to cleanly separate inductive, resonance, and steric effects.

The second example in Table 9.6 is the esterification of benzenecarboxylic (benzoic) acid by ethanol (ethyl alcohol) in the presence of an acid catalyst. As shown in Scheme 9.117, the available experimental data indicate that, in equilibrium, initial protonation of the carboxylic acid is followed by attack on the carbonyl group by the alcohol. Subsequent proton transfer and elimination of water consummates

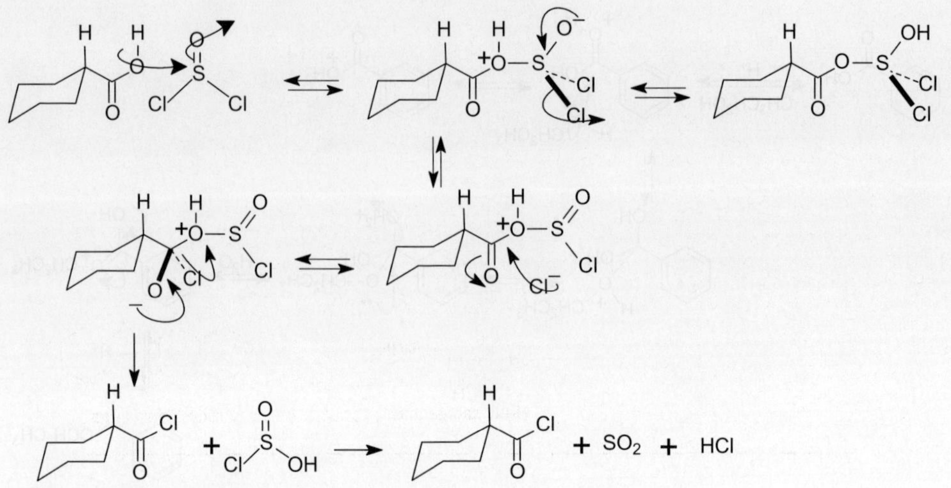

Scheme 9.118. A representation of the conversion of cyclohexanecarboxylic acid into the corresponding acid chloride on reaction with thionyl chloride ($SOCl_2$).

the reaction. It is important to note the reversibility of the process as acid-catalyzed ester hydrolysis.

In a similar vein, the conversion of a carboxylic acid to the corresponding acid chloride can also be shown to proceed through a tetrahedral intermediate. Item 3 in Table 9.6 introduces this acid chloride as a result of the reaction between cyclohexanecarboxylic acid and thionyl chloride ($SOCl_2$) (Scheme 9.118).

Amines (Chapter 5, Part **VI**, and Chapter 10) are basic. Thus, their reactions with carboxylic acids produce the corresponding salts rather than amides (Equation 9.97). Although it is possible for some amine salts of carboxylic acids to be converted to the corresponding amides by heating them to high temperatures (*pyrolysis*), the fragile nature of many compounds prohibits such drastic treatment.

$$CH_3CO_2H + CH_3CH_2NH_2 \longrightarrow CH_3CO_2^- \; {}^+H_3NCH_2CH_3 \qquad (9.97)$$

Thus, it is much more general to carry out the reaction in the presence of a dehydrating reagent that first reacts with the carboxylic acid, activating it for subsequent reaction with the amine and initiating the process that eventually leads to amide and loss of water. Many reagents have been developed to this end. Two of them, dicyclohexylcarbodiimide and carbonyldiimidazole, are shown in the process of item 4 in Table 9.6 and in Scheme 9.119.

In what appears to be a dramatically different process, carboxylic acids that are resistant to attack by sulfuric acid (H_2SO_4) can undergo substitution on the carbon of the carbonyl of the hydroxyl group (–OH) by the azido group (–N_3) (in a process called the **Schmidt** reaction). While it is occasionally possible to isolate the acid azide, rearrangement to the corresponding isocyanate is common. A schematic representation in concert with available experimental data for the process of azide formation and subsequent rearrangement is shown in Scheme 9.120 for the case of benzenecarboxylic acid (benzoic acid). It is particularly important to note that in contrast to the "typical" formation of a tetrahedral intermediate so common for the

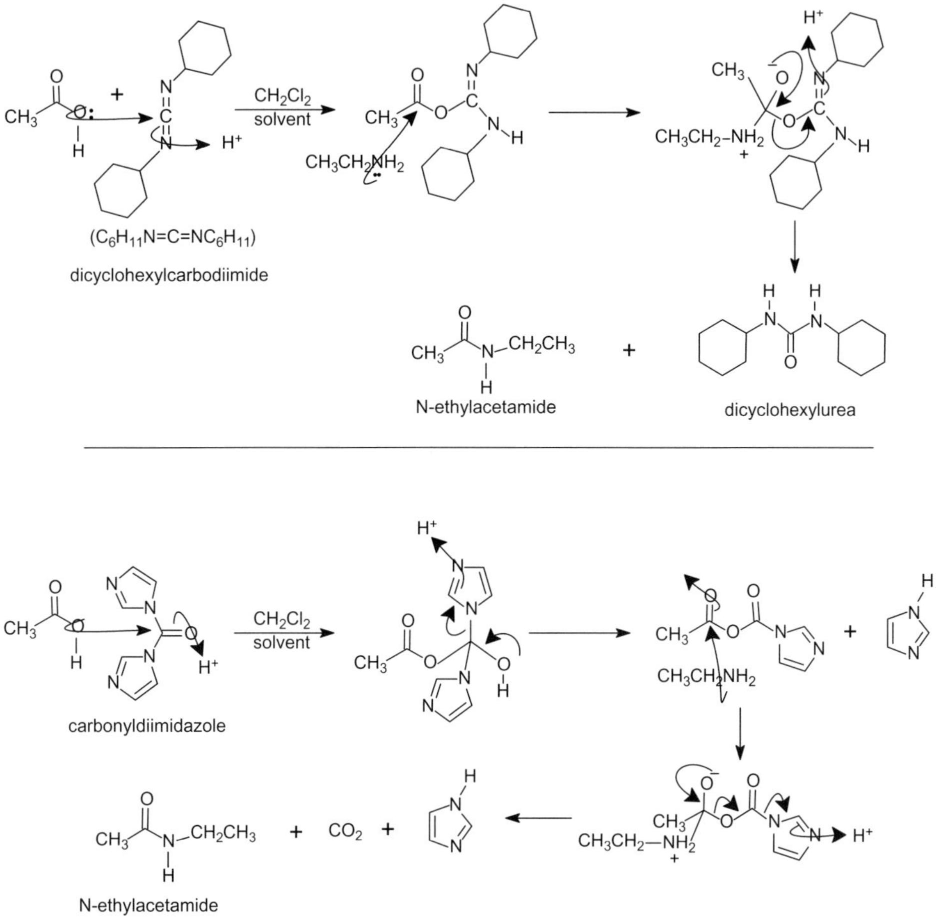

Scheme 9.119. The formation of N-ethylethanonamide (N-ethylacetamide) by reaction of acetic acid with N-ethylamine in the presence of cyclohexylcarbodiimide (top) and carbonyldiimidazole (bottom). In the first case, dicyclohexylurea is formed along with the amide; in the second case, the amide is accompanied by carbon dioxide and two equivalents of imidazole.

reaction of carbonyl compounds, it appears here that the sulfuric acid actually protonates the carboxylic acid, water is lost, and an "acylium" ion generated. The acylium ion is then captured by the nucleophilic azide anion, resulting in the overall substitution process. The rearrangement involving migration of the aryl group follows.

A final point of interest for rearrangement of acylazide to isocyanate, which is not apparent in the example chosen, is that *if the carbon atom of the migrating group bearing the carbonyl is asymmetrically substituted, the migration occurs with retention of configuration.* The process is shown in Equation 9.98 for the rearrangement of the isocyanate derived from (1*R*,3*S*)-3-methylcyclohexanecarboxylic acid.

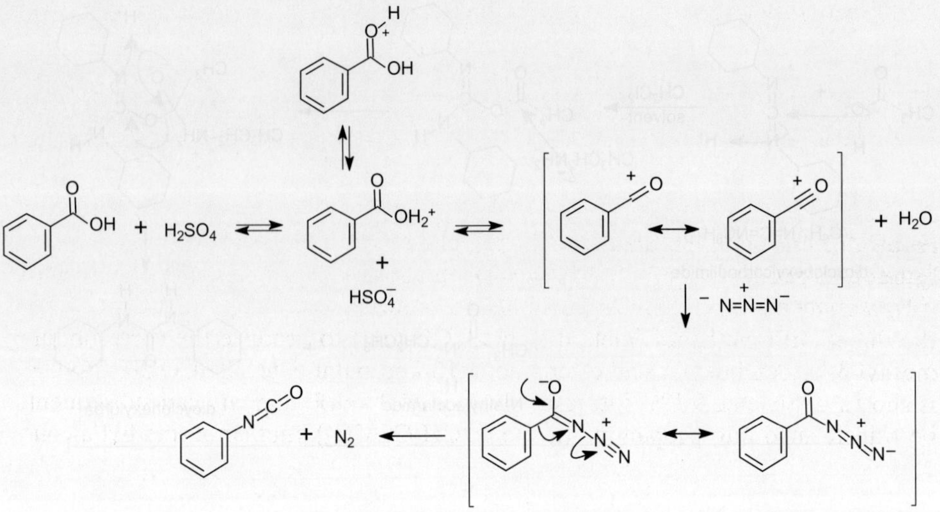

Scheme 9.120. A representation of a pathway for the generation of phenylisocyanate from the reaction of benzenecarboxylic acid (benzoic acid) with sulfuric acid (H_2SO_4) in the presence of azide anion followed by rearrangement.

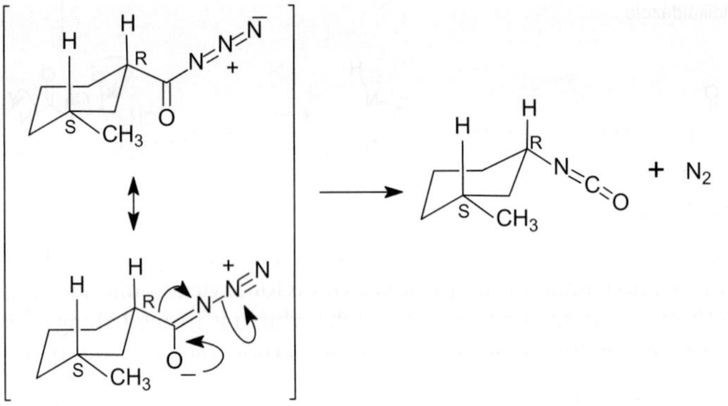

$$(9.98)$$

The sixth item in Table 9.6 illustrates the reaction between ketene (CH_2=C=O) and ethanoic acid (acetic acid, CH_3CO_2H). As shown in Equation 9.99, ketene (CH_2=C=O) can be prepared by the dehydration of ethanoic acid (acetic acid, CH_3CO_2H), and the isolation of ketene is complicated because of its ready dimerization to diketene (4-methylideneoxacyclobutane-2-one).

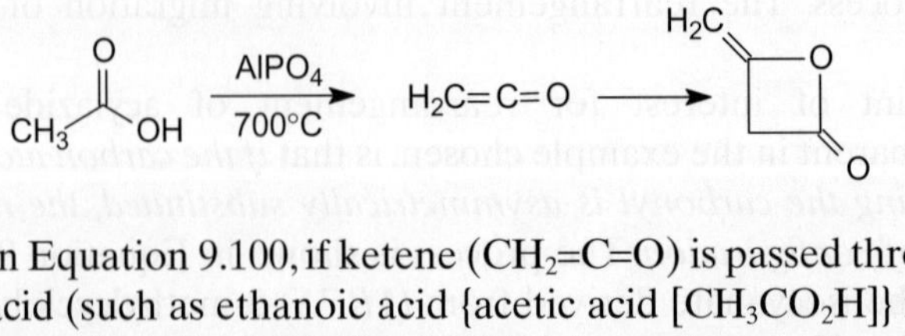

$$(9.99)$$

Then, as shown in Equation 9.100, if ketene (CH_2=C=O) is passed through a solution of a carboxylic acid (such as ethanoic acid {acetic acid [CH_3CO_2H]} itself) reaction, the corresponding anhydride ($CH_3CO_2COCH_3$) is produced. Clearly, the ethanoic acid (acetic acid) anhydride of any carboxylic acid might be prepared in just this way.

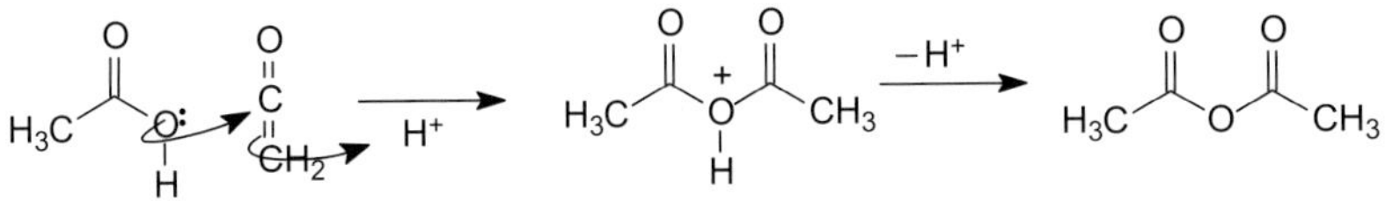

$$\text{(9.100)}$$

Finally, in this vein, the diketene (4-methylideneoxacyclobutane-2-one) produced by the dimerization of ketene (CH_2=C=O) (Equation 9.99) can also undergo reaction with a variety of nucleophiles (alcohols and amines, for example). The reaction of the dimer with methanol (methyl alcohol, CH_3OH) to produce the corresponding methyl β-ketocarboxylic acid ester (methyl 3-ketobutanoate, $CH_3COCH_2CO_2CH_3$) is shown in Scheme 9.121. Interestingly, methyl 3-ketobutanoate can subsequently be transformed into propanone (acetone, CH_3COCH_3), methanol (methyl alcohol,

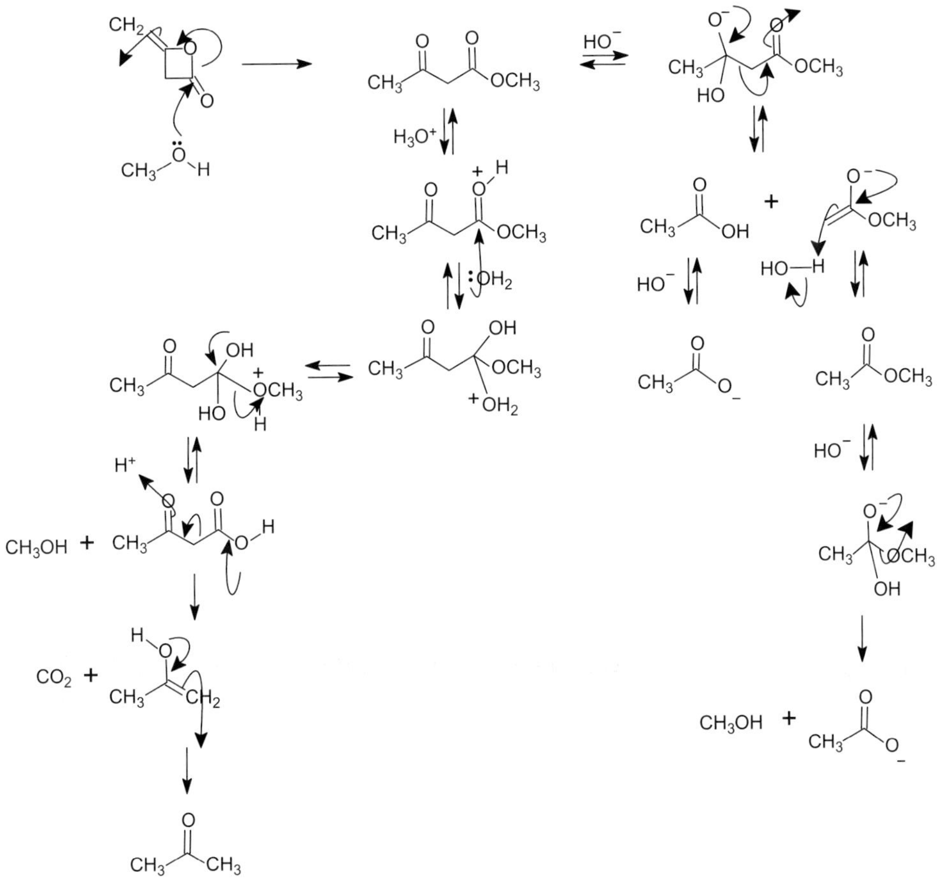

Scheme 9.121. A representation of the formation and subsequent decomposition of methyl 3-ketopropanoate. It is formed by the ring opening of ketene dimer on reaction with methanol. Decomposition in the presence of aqueous acid and aqueous base, separately, produces ketone in the former and fragmentation in the latter.

CH_3OH), and carbon dioxide (CO_2) by *acid hydrolysis* or two equivalents of ethanoic acid (acetic acid, CH_3CO_2H) and methanol (methyl alcohol, CH_3OH) after *basic hydrolysis* and acidification (Scheme 9.121). Other products can also be made from families of versatile keto esters (*vide infra*).

The last two items of Table 9.6 are related. The seventh item shows that the techniques already discussed that employ dicyclohexylcarbodiimide and carbonyldiimidazole to activate a carboxylic acid for attack by primary and secondary amines in order to forge the bond between the carbon of the carbonyl and the nitrogen of the amine to yield amide (*vide supra*, Scheme 9.119) can also be employed in ester formation (Scheme 9.122). The particular alcohol chosen, 3-(hydroxymethyl)-3-methyloxetane, although typical in its reactivity for primary alcohols in general and thus capable of serving as a prototype for ester, was also chosen because, as shown in item 8 of Table 9.6, it can subsequently be used to convert the ester to an *orthoester* of the corresponding carboxylic acid (Scheme 9.123). Such orthoesters are used to protect the carboxylic acid function and preserve it, while manipulations

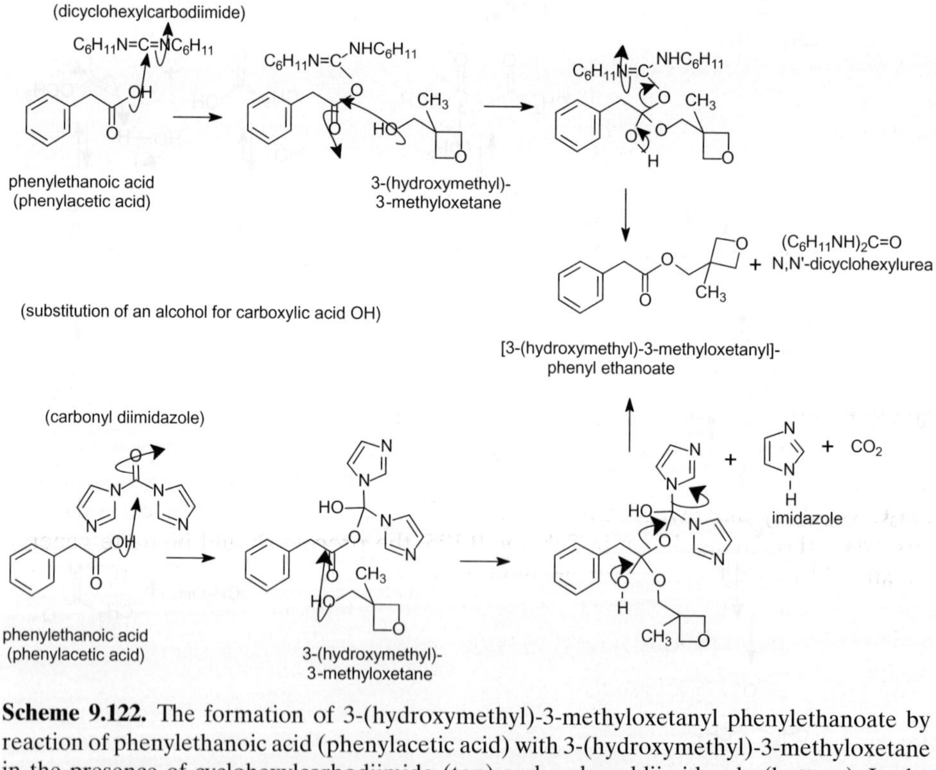

Scheme 9.122. The formation of 3-(hydroxymethyl)-3-methyloxetanyl phenylethanoate by reaction of phenylethanoic acid (phenylacetic acid) with 3-(hydroxymethyl)-3-methyloxetane in the presence of cyclohexylcarbodiimide (top) and carbonyldiimidazole (bottom). In the first case, dicyclohexylurea is formed along with the ester; in the second case, the ester is accompanied by carbon dioxide and two equivalents of imidazole.

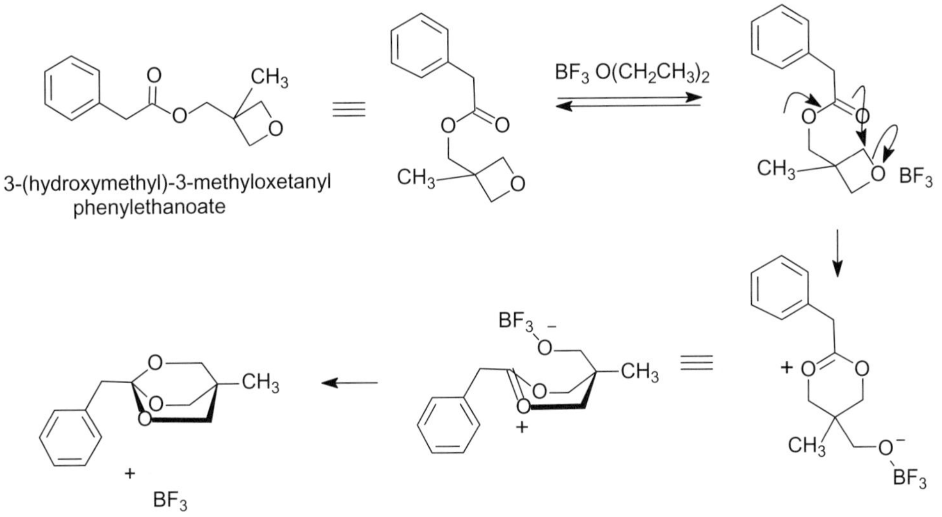

3-(hydroxymethyl)-3-methyloxetanyl
phenylethanoate

Scheme 9.123. A representation depicting the formation of 1-benzyl-4-methyl-2,3,7-trioxabicyclo-[2.2.2]octane by boron trifluoride (etherate) catalyzed rearrangement of 3-(hydroxymethyl)-3-methyloxetanyl phenylethanoate.

that might affect an unprotected carboxylic acid (or derivative in which the carbonyl group is retained) are carried out elsewhere on the molecule. Hydrolysis of orthoester with aqueous acid (H_3O^+) liberates the carboxylic acid and produces the water soluble triol 2,2-bishydroxymethyl-1-propanol.

Some reactions of the first class of compounds to be considered as derived from carboxylic acids, the **acid chlorides**, are presented in Table 9.7. The formation of acid chlorides from carboxylic acids is shown, specifically for a cyclohexanecarboxylic acid, in Scheme 9.118.

However, the reaction is quite general and other reagents (such as oxalyl chloride, ClCOCOCl, or phosphorus oxychloride, $POCl_3$, Scheme 9.124) might also be used. While occasional use of acid bromides (prepared from the corresponding phosphorus oxybromide as in Scheme 9.107) has been made, they are relatively uncommon.

The first item in Table 9.7 corresponds to the aqueous hydrolysis of an acid chloride (propanoyl chloride) to regenerate the carboxylic acid (propanoic acid, $CH_3CH_2CO_2H$) and hydrochloric acid, HCl) from which the acid chloride was formed. Although, as shown in Scheme 9.125, the process should be quite general for all acid chlorides, it is not uncommon that the limited solubility of the acid chloride in water (when the hydrocarbon residue is large enough) retards the reaction to such an extent as to make it appear that the hydrolysis fails. However, since the hydrolysis is not usually desired, such retardation is valuable.

The second item in Table 9.7 is similar to the first in that, again, the chloride of the acid chloride is replaced by an oxygen nucleophile. It would be presumed that the ethyl alcohol (ethanol, CH_3CH_2OH) simply adds to the carbon of the carbonyl (sufficient solubility is expected) with loss of chloride (Cl^-). However, pyridine (azabenzene, C_5H_5N) is present, and it is clear that, in addition to removing the

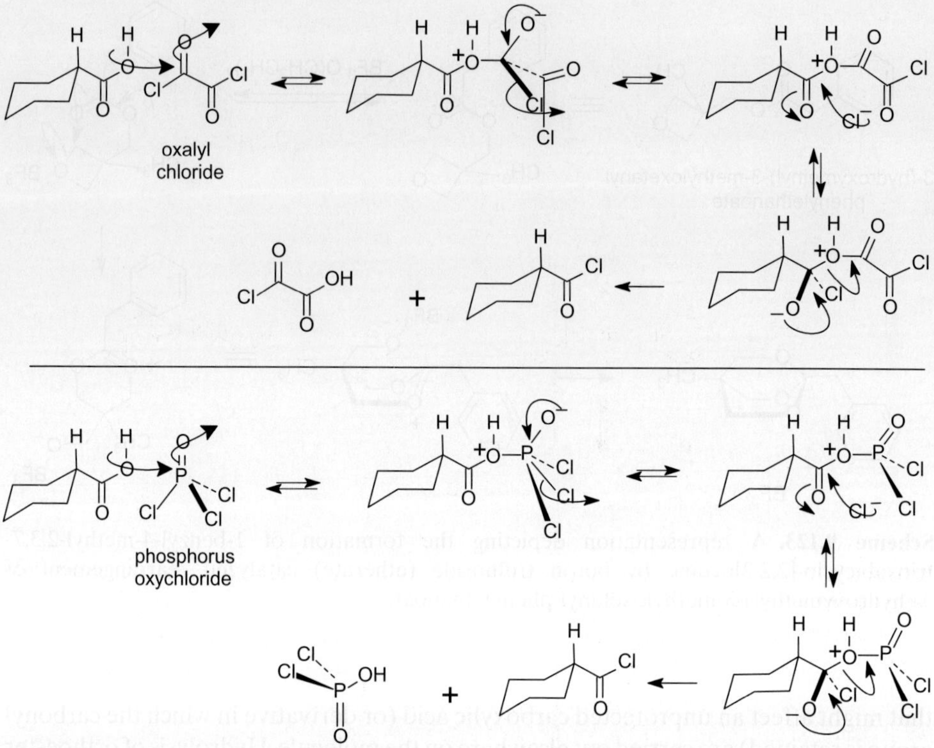

oxalyl
chloride

phosphorus
oxychloride

Scheme 9.124. A representation of the conversion of cyclohexanecarboxylic acid into the corresponding acid chloride on reaction with (upper half) oxalyl chloride (ClCOCOCl) and phosphorus oxychloride, POCl$_3$ (lower half).

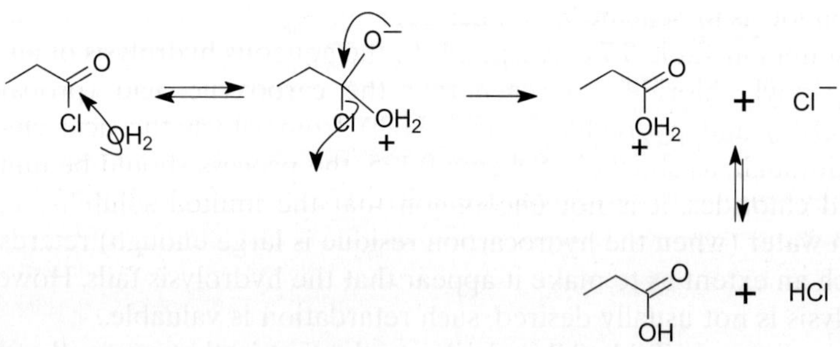

Scheme 9.125. A representation of the hydrolysis (H$_2$O) of propanoyl chloride (CH$_3$CH$_2$COCl) to propanoic acid (CH$_3$CH$_2$CO$_2$H) and hydrogen chloride (HCl).

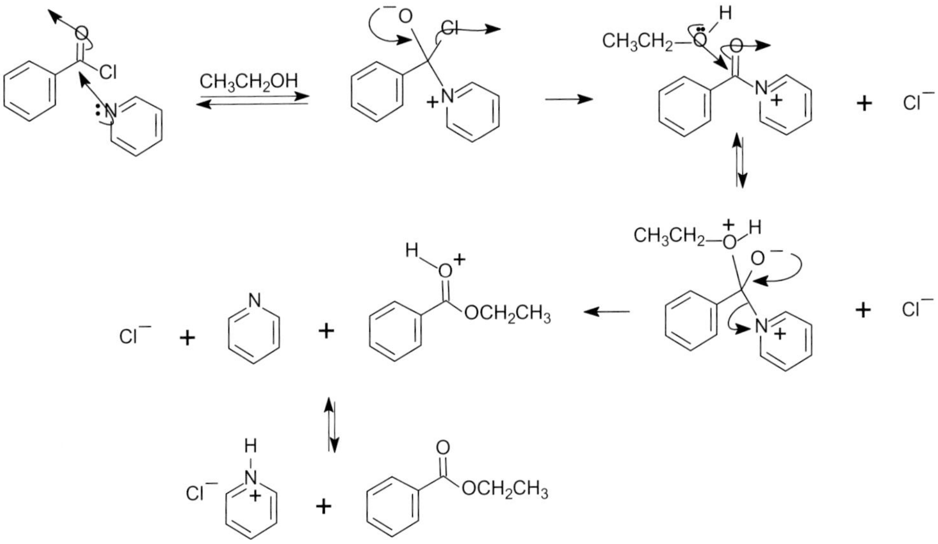

Scheme 9.126. A description of a pathway for the formation of ethyl benzenecarboxylate (ethyl benzoate) by reaction of the acid chloride of benzoic acid (benzoyl chloride) with ethanol (ethyl alcohol, CH_3CH_2OH) in the presence of azabenzene (pyridine).*

hydrogen chloride that forms (through protonation at nitrogen with formation of pyridinium hydrochloride), it plays a role in activation of the acid chloride for substitution. A process accounting for product formation is depicted in Scheme 9.126.

The third item in Table 9.7, the formation of a mixed anhydride from reaction of an acid chloride with the salt of a carboxylic acid, is illustrated in Equation 9.101. Formate mixed anydrides

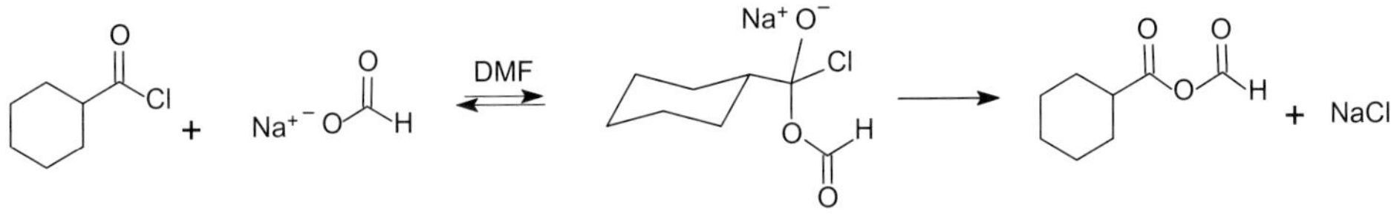

$$(9.101)$$

are generally among the more reactive anhydrides (*vide infra*, Table 9.8). However, their formation requires a suitable solvent (such as dimethylformamide, DMF [$(CH_3)_2NCHO$]) for the addition–elimination reaction, and removal of the solvent is not always simple.

*There is an interesting aside to the information presented in Scheme 9.124 (top). It has recently (Haiss, P.; Zeller, K.-P. *Angew. Chem. Intt. Ed.*, **2003**, *42*, 303) been found that treatment of ^{18}O-enriched benzenecarboxylic acid (benzoic acid, $C_6H_5CO_2H$) with excess oxalyl chloride (ClCOCOCl) produces acid

(Continued on next page)

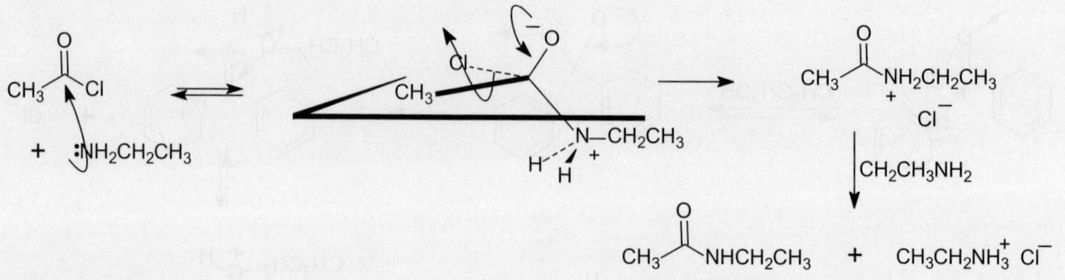

Scheme 9.127. A representation of a pathway for the formation of N-ethyl ethanoamide (N-ethylacetamide, $CH_3CONHCH_2CH_3$) from the reaction of ethanoyl chloride (acetyl chloride, CH_3COCl) with ethylamine (ethanamine, aminoethane, $CH_3CH_2NH_2$). The chlorine on the acyl function has been replaced by a nitrogen.

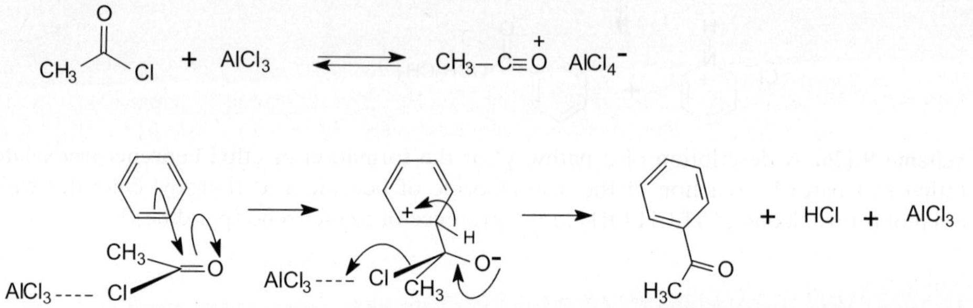

Scheme 9.128. A representation of a pathway to effect the substitution of an aromatic ring for a chlorine atom on an acyl function. In principle, the reaction could occur by attack of the aromatic ring on the carbon of the carbonyl of the acyl halide *or* by reaction of the acyl halide with aluminum trichloride ($AlCl_3$) to produce an "acylium ion," which is then captured by the aromatic ring. See Scheme 6.88.

On the other hand, as shown in Scheme 9.127, acid chlorides (ethanoyl chloride, acetyl chloride, CH_3COCl, in the case of the fourth item in Table 9.7) react readily (often in ether solvents) with excess amine (such as ethaneamine, ethylamine, $CH_3CH_2NH_2$, as shown) to produce the corresponding amides. Generally, after aqueous workup and washing the organic phase with aqueous base and aqueous acid, only the amide (the only neutral species present) remains in the organic phase; thus, this reaction is often very clean.

As already pointed out in Chapter 6 (shown as the seventh item in Table 6.12 and described in Scheme 6.88), acyl halides undergo the Friedel–Crafts reaction with aromatic systems in the presence of aluminum trichloride ($AlCl_3$). Thus, ethanoyl chloride (acetyl chloride, CH_3COCl) reacts with benzene under these conditions

chloride in which much of the ^{18}O has been lost. The authors suggest that the experimental data can be accounted for by the formation of oxetane as shown in the equilibrium below.

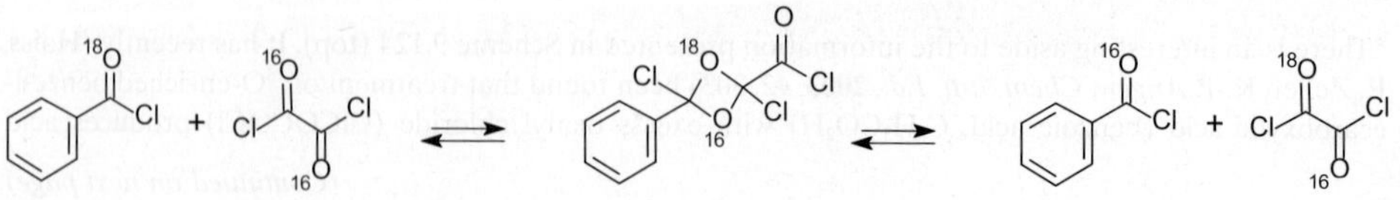

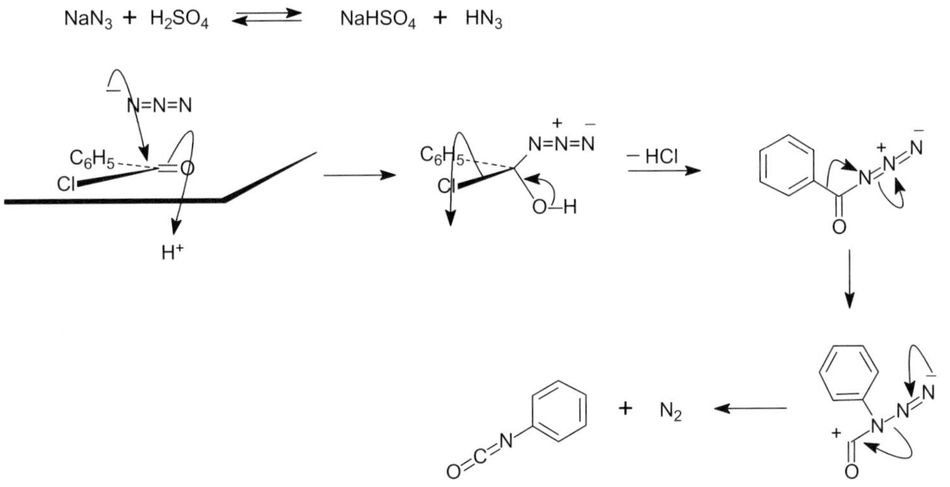

$$NaN_3 + H_2SO_4 \rightleftharpoons NaHSO_4 + HN_3$$

Scheme 9.129. A representation of the formation of phenylisocyanate (C_6H_5NCO) from the reaction between the acid chloride of benzenecarboxylic acid (benzoyl chloride, C_6H_5COCl) and sodium azide (NaN_3) followed by rearrangement of the benzoylazide ($C_6H_5CON_3$) (the **Curtius** reaction)*.

(the fifth item in Table 9.7) to produce the corresponding ketone (phenylethanone, acetophenone, $C_6H_5COCH_3$) as shown in Scheme 9.128 (which repeats what is shown in Scheme 6.88).

In contrast, the sixth item in Table 9.7 requires a more direct attack on the carbonyl group by an excellent nucleophile (the azide anion, N_3^-) and would, nonetheless, not be expected to be particularly facile in aprotic solvents, where the azide anion (N_3^-) is relatively insoluble. As a consequence, this substitution reaction (the **Curtius** reaction)† is frequently carried out in a strong acid (e.g., sulfuric acid [H_2SO_4]) medium where both the nucleophile (now present as hydrazoic acid, HN_3, in equilibrium with its anion) and protonated acyl halide are soluble. As shown in Scheme 9.129, the initial formation of benzenecarbonyl azide (benzoyl azide, $C_6H_5CON_3$) is followed by nitrogen loss and rearrangement (**migration of the aryl ring from carbon to nitrogen**) to produce the corresponding isocyanate (phenylisocyanate, $C_6H_5N=C=O$).

The seventh item in Table 9.7, the reaction of phenylethanoyl chloride (the acid chloride of phenylacetic acid, $C_6H_5CH_2CO_2H$) with diazomethane (CH_2N_2), followed by hydrolytic (using H_3O^+) workup, creates a new carboxylic acid. When this reaction is combined with conversion of the carboxylic acid to the acid chloride (e.g., with thionyl chloride, $SOCl_2$, as in **item 3** of Table 9.6, *vide supra*) the two steps allow for increasing by one the number of carbons (as "methylene units [$-CH_2-$]") between

---

*The decomposition of isocyanate to amine and carbon dioxide had been observed by Wurtz (Wurtz, A. *Ann. Chim.* (Paris), **1854**, *42*, 43) some years before Curtius showed that the reaction with azide succeeded (Curtius, T. *Chem. Ber.*, **1890**, *23*, 3023).

†The family of reactions, one of which is the **Curtius** reaction (the others being **Hofmann, Lossen,** and **Schmidt** reactions), serve to transform such acyl derivatives (acid halides, acids, amides) into compounds in which the group *originally attached to the carbon of the carbonyl is subsequently attached to nitrogen*. These rearrangements will be considered again later in this chapter.

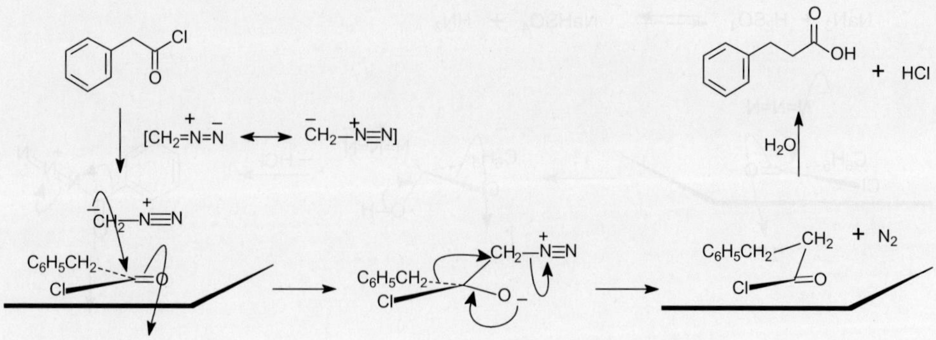

Scheme 9.130. A representation of a pathway to convert phenylethanoyl chloride (the acid chloride of phenylethanoic acid, phenylacetic acid, $C_6H_5CH_2CO_2H$) to the next higher homologue by treatment with diazomethane (CH_2N_2) followed by hydrolysis.

the carboxylic acid functionality (CO_2H) and the aromatic ring, as shown in Scheme 9.130. This kind of processes, the change by one $-CH_2-$ unit, is a **homologization** reaction, and the new carboxylic acid is "homologous with its predecessor." Homologues can, of course, be larger than or smaller than their predecessors, and entire homologous series are created in just this way.

The final item of Table 9.7, the reaction of phenylethanoyl chloride (the acid chloride of phenylethanoic acid, phenylacetic acid, $C_6H_5CH_2CO_2H$) with sodium cyanide (NaCN) adds a one-carbon unit (as did the reaction with diazomethane [CH_2N_2] noted earlier) to the carbon of the carbonyl. Since the cyano group [$-CN$] (a nitrile when connected to an organic group) can undergo hydrolysis to the corresponding carboxylic acid ($-CO_2H$) (via an amide, *vide infra*), this procedure, as outlined in Scheme 9.131, produces the phenethanoyl cyanide (or nitrile), which can subsequently be used to generate an α-ketocarboxylic acid. In the particular reaction shown, the α-ketocarboxylic acid formed on hydrolysis of the nitrile is phenylpyruvic acid ($C_6H_5CH_2COCO_2H$).

As already pointed out, mixed carboxylic acid anhydrides can be obtained from reaction of a carboxylic acid chloride with the salt of a carboxylic acid (Equation 9.101), and this technique can be applied to, in particular, anhydrides of methanoic acid (formic acid, HCO_2H). However, in general, it is the more symmetric anhydrides that are commonly found, and, for example, ethanoic anhydride [acetic anhydride, $(CH_3CO)_2O$], the first item in Table 9.8, remains an important material of commerce (it continues to be used as a raw material in the manufacture of cellulose acetate, cf. this chapter). To that end, it is interesting to note that significant amounts of ethanoic acid (acetic acid, CH_3CO_2H) and ethanoic anhydride [acetic anhydride, $(CH_3CO)_2O$] are produced by oxidation of ethanal (acetaldehyde, CH_3CHO) with oxygen in the presence of a manganese(II) (as the acetate [$Mn(O_2CCH_3)_2$]) or the corresponding cobalt(II) derivative as catalyst. The process apparently involves the intermediacy of *per*ethanoic acid (peracetic acid, CH_3CO_3H), and while ethanoic acid (acetic acid, CH_3CO_2H) can be thought of as arising from a **Baeyer–Villiger reaction** (this chapter) of perethanoic (peracetic acid, CH_3CO_3H) with acetaldehyde, it is also possible to rationalize the formation of the ethanoic anhydride [acetic anhydride, $(CH_3CO)_2O$] by a similar process. The reactions to form the perethanoic

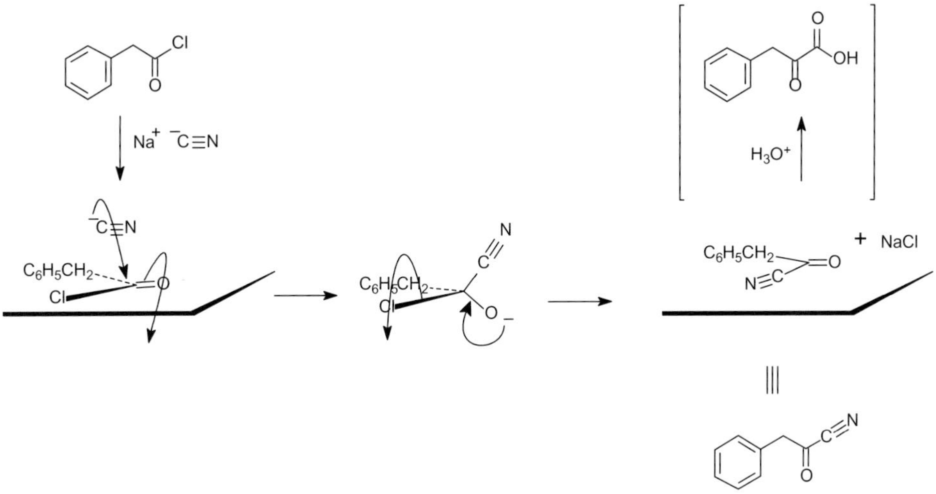

Scheme 9.131. A representation of the reaction between phenylethanoyl chloride (the acid chloride of phenylethanoic acid [phenylacetic acid, $C_6H_5CH_2CO_2H$]) and sodium cyanide to produce the corresponding α-ketonitrile which, on hydrolysis, forms the corresponding α-ketocarboxylic acid (in this case, phenylpyruvic acid).

acid (peracetic acid, CH_3CO_3H) are believed to begin with a sequence of one-electron transfer processes and might, as outlined in Scheme 9.132, continue with ionic species subsequently.

Interestingly, it is often convenient during the manufacturing process of ethanoic anhydride [acetic anhydride, $(CH_3CO)_2O$] to generate the ethanal (acetaldehyde, CH_3CHO) itself on the same site in a separate, palladium(II) chloride ($PdCl_2$), catalyzed oxidation of ethene (ethylene, $CH_2{=}CH_2$) as shown in Scheme 9.133 (the **Wacker** process).*

Finally, in this vein, and despite efforts to find and optimize the catalysts used in the systems just described, it is currently much more common to simply pyrolyze ethanoic acid (acetic acid, CH_3CO_2H) to produce ethenenal (ketene, $H_2C{=}C{=}O$), which then reacts with ethanoic acid (acetic acid, CH_3CO_2H), to generate the anhydride (Equation 9.102).

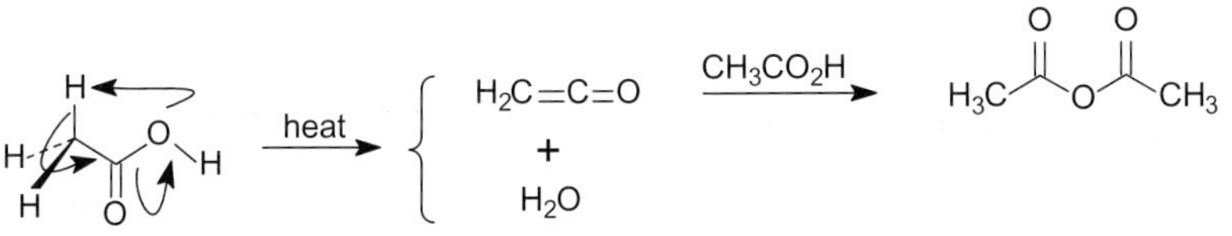

$$\text{(9.102)}$$

A representation describing the hydrolysis of ethanoic anhydride [acetic anhydride, $(CH_3CO)_2O$], the reaction shown first in Table 9.8, is presented in Equation 9.103.

*The Wacker process was developed in part at the German chemical company Wacker–Chemie during the rebuilding the German chemical industry after World War II. A review has been written by Jira, R. *Angew. Chem. Int. Ed.*, **2009**, 48, 9034, and a laboratory method based on the industrial one has been developed; see Tsuji, J.; Nagashima, H.; Hori, K. *Tetrahedron Lett.*, **1982**, 2679.

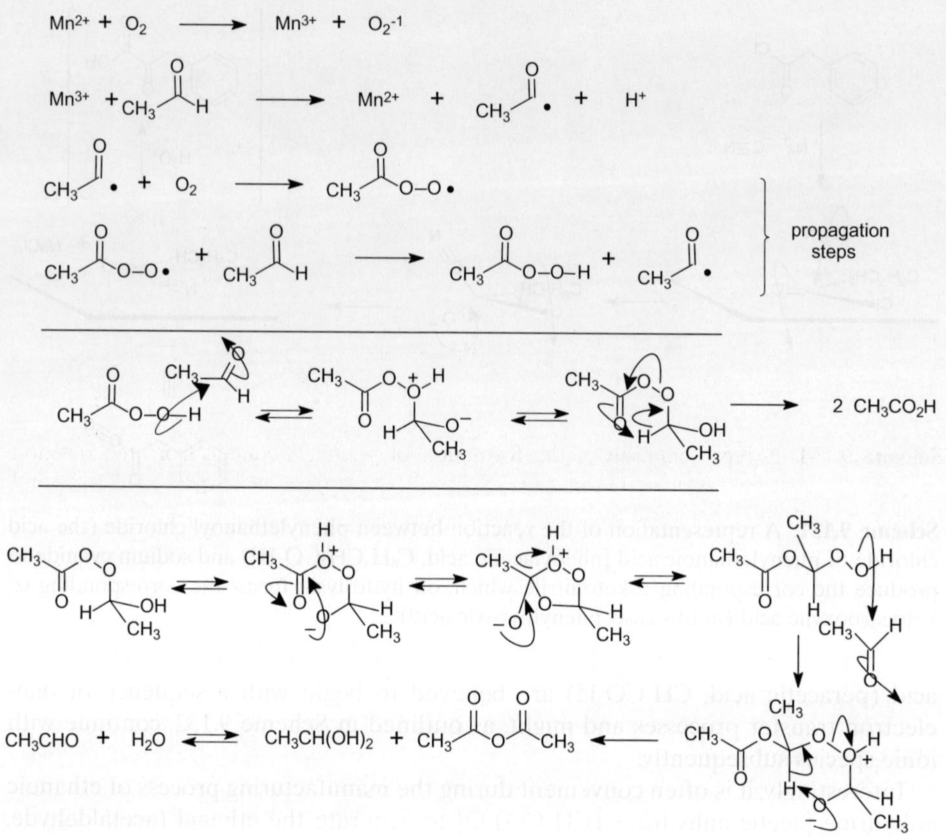

propagation
steps

Scheme 9.132. A representation of the Mn^{+2}-catalyzed oxygen (O_2) oxidation of ethanal (acetaldehyde, CH_3CHO) to perethanoic acid (peracetic acid, CH_3CO_3H) and the subsequent transformation of the peracid and additional ethanal (acetaldehyde, CH_3CHO) to ethanoic acid (acetic acid, CH_3CO_2H), via a Baeyer–Villiger transformation, or to ethanoic anhydride [acetic anhydride, $(CH_3CO)_2O$].

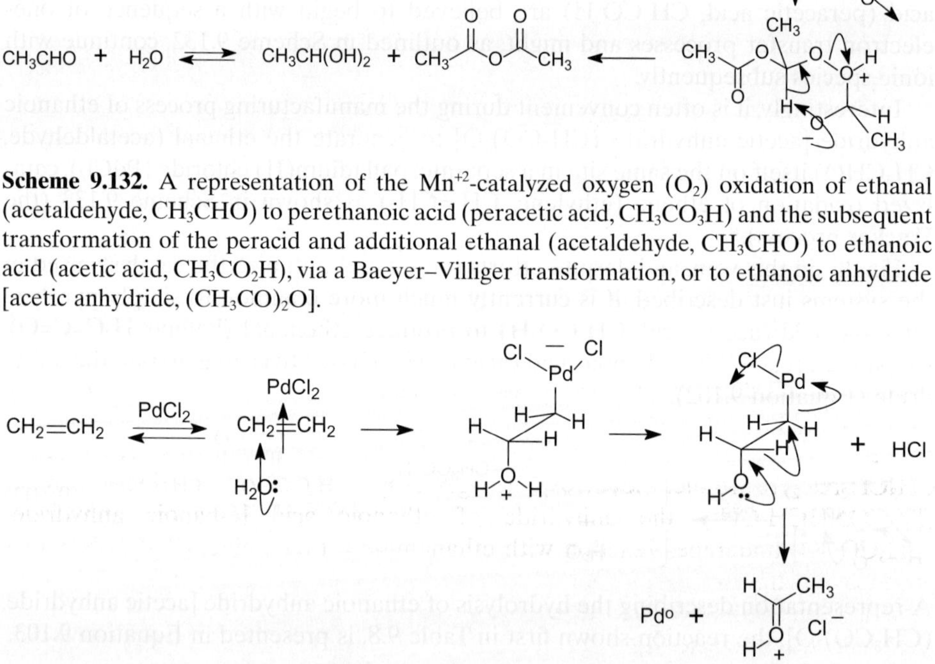

Scheme 9.133. A representation of the palladium(II) chloride-catalyzed oxidation of ethene (ethylene, $CH_2=CH_2$) to ethanal (acetaldehyde, CH_3CHO), the **Wacker process**. Generally, a copper(II) catalyst (not shown) is also employed to effect the reoxidation of the palladium. Interestingly, it is probable that a 1,2-hydride shift occurs as the palladium is lost (rather than elimination to the corresponding enol and then rearrangement to the aldehyde) since use of ^2H_2O (deuterium oxide) in place of its proton analogue (water, ^1H_2O) yields aldehyde in which there is no deuterium (^2H) incorporation.

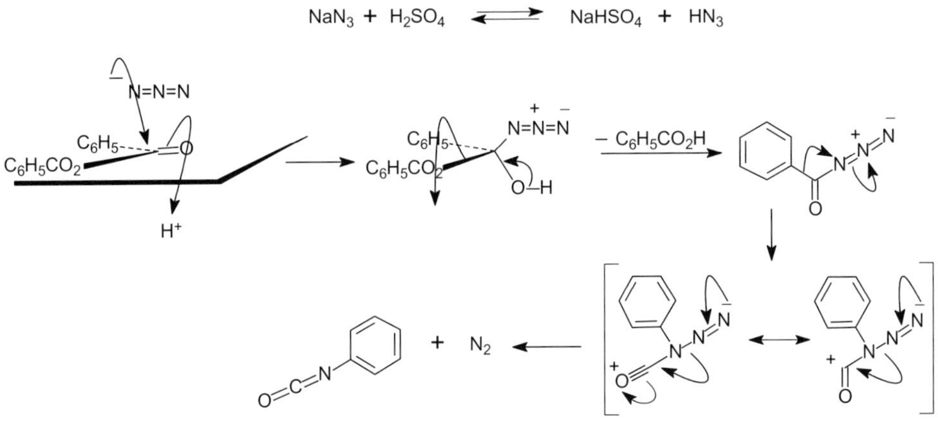

$$NaN_3 + H_2SO_4 \rightleftharpoons NaHSO_4 + HN_3$$

Scheme 9.134. A representation of the formation of phenylisocyanate from the reaction between the acid anhydride of benzenecarboxylic acid [benzoic anhydride, $(C_6H_5CO)_2O$] and sodium azide (NaN_3) followed by rearrangement of the benzoylazide (a **Curtius**-type reaction, *vide supra*).

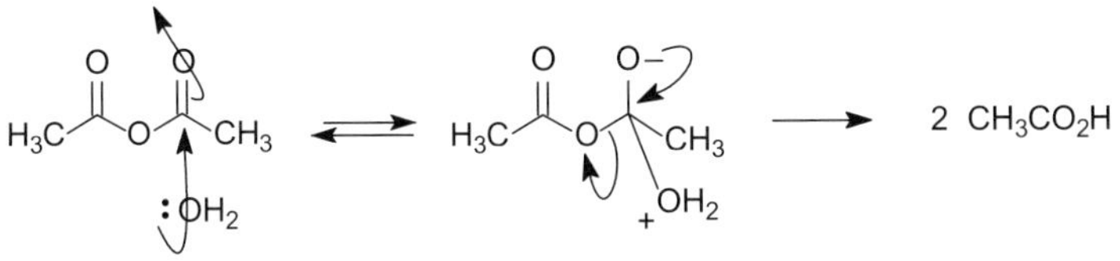

$$2\ CH_3CO_2H \qquad (9.103)$$

The second item in Table 9.8 is analogous to the sixth item in Table 9.7. Thus, by analogy with the material shown in Scheme 9.129 for the formation of phenylisocyanate from the reaction between acid chloride of benzenecarboxylic acid (benzoyl chloride, C_6H_5COCl) and sodium azide (NaN_3), the reaction between the bis-anhydride of benzenecarboxylic acid [benzoic acid anhydride, $(C_6H_5CO)_2$] follows the same path. The difference, of course, lies in the leaving group [chloride (Cl^-) in the former and benzenecarboxylate (benzoate, $C_6H_5CO_2^-$) in the latter]. This process is shown in Scheme 9.134.

Again, in a fashion similar to that shown in Scheme 9.127 for the reaction between the acid chloride of ethanoic acid (acetic acid, CH_3CO_2H), that is, the reaction between ethanoyl chloride (acetyl chloride, CH_3COCl) and ethanamine (ethyl amine, $CH_3CH_2NH_2$) to produce the corresponding N-ethylethanoamide (N-ethylacetamide, $CH_3CONHCH_2CH_3$), the anhydride of ethanoic acid [ethanoic anhydride, $(CH_3CO)_2O$] undergoes reaction with ethanamine (ethyl amine, $CH_3CH_2NH_2$) to produce the same product. A representation of this process is shown in Scheme 9.135.

In a manner analogous to that shown (Scheme 9.126) for the acid chloride of benzenecarboxylic acid (benzoyl chloride, C_6H_5COCl), which is item 2 in Table 9.7, ethanol (ethyl alcohol, CH_3CH_2OH) undergoes reaction with anhydrides, such as propanoic anhydride (item 4 in Table 9.8), to yield the corresponding carboxylic acid ester, ethyl propanoate ($CH_3CH_2CO_2CH_2CH_3$). A representation of the process is shown in Scheme 9.136. Generally, the need for a base such as azabenzene (pyridine, C_5H_5N) as depicted in Scheme 9.126 is unnecessary with anhydrides. In part this is due to the fact that a weak carboxylic acid, propanoic acid ($CH_3CH_2CO_2H$), leaves in this case, rather than hydrochloric acid (HCl) as in the former case.

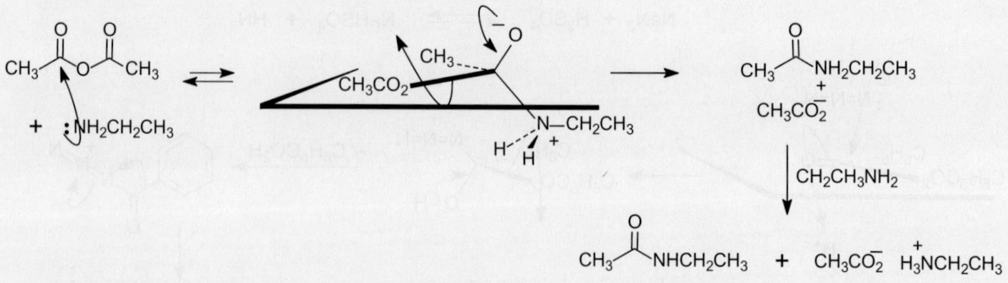

Scheme 9.135. A representation of a pathway for the formation of N-ethyl ethanamide (N-ethylacetamide, $CH_3CONHCH_2CH_3$) from the reaction of ethanoic anhydride [acetic anhydride, $(CH_3CO)_2O$] with ethanamine (aminoethane, ethyl amine, $CH_3CH_2NH_2$). The nitrogen of the ethanamine ($H_2NCH_2CH_3$) has been substituted for an acetate ($CH_3CO_2^-$) functionality [attached to an acyl (CH_3CO-) group].

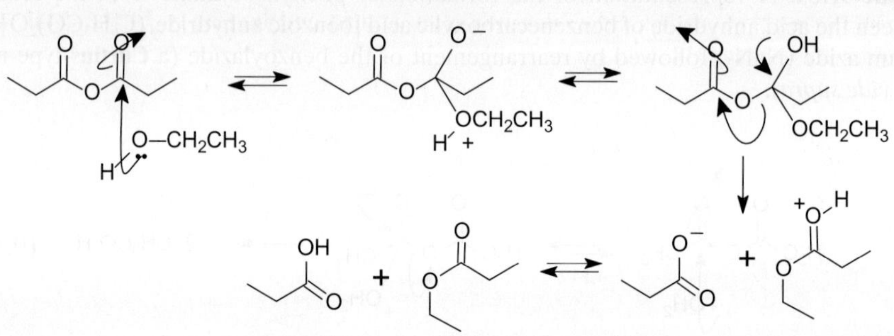

Scheme 9.136. A representation of a pathway for the formation of ethyl propanoate ($CH_3CH_2O_2CCH_2CH_3$) by reaction between propanoic anhydride [$(CH_3CH_2CO)_2O$] and ethanol (ethyl alcohol, CH_3CH_2OH).

The penultimate item in Table 9.8 corresponds to a Friedel–Crafts acylation reaction between benzene and butane-1,4-dioic anhydride (succinic anhydride), which, again, finds analogy with a similar acylation effected under similar conditions between benzene (C_6H_6) and ethanoyl chloride (acetyl chloride, CH_3COCl) as represented in Scheme 9.128. In contrast to the reaction with the acyl halide, it is clear that the Lewis acid catalyst, aluminum chloride ($AlCl_3$), cannot form the corresponding tetrachloride ($AlCl_4^-$) anion. However, it is equally clear that as a Lewis acid it is capable of bonding to either the oxygen of a carbonyl group or the bridging oxygen between the two carbonyl groups and, in this way, increasing the electrophilicity of the carbon of the carbonyl. With increased electrophilicity (in comparison to the case without catalyst), the attack on the carbonyl carbon by the aromatic ring is facilitated. This process is schematically represented in Scheme 9.137.

The final item in Table 9.8 is an additional case of substitution for oxygen by nitrogen at the carbon of the carbonyl of 1,2-benzenedicarboxylic anhydride (phthalic anhydride). In this case, the nitrogen species is bidentate (hydrazine, H_2NNH_2) and undergoes reaction twice, once at each carbonyl group of the cyclic anhydride starting material. The process is depicted in Scheme 9.138.

Esters, introduced in Chapter 5, have already been considered in this chapter as products of substitution of –OR (R = alkyl or aryl) for the –OH group in carboxylic

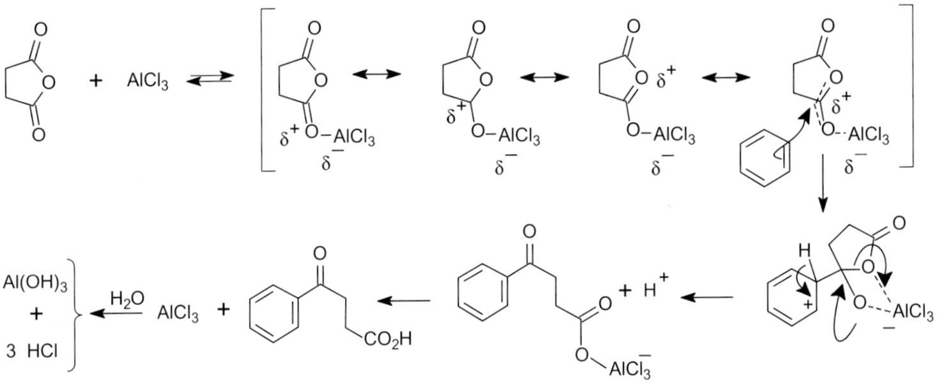

Scheme 9.137. A representation of the Friedel–Crafts acylation of benzene (C_6H_6) by butane-1,4-dioic anhydride (succinic anhydride) in the presence of an aluminum trichloride ($AlCl_3$) catalyst. Generation of the 4-phenyl-4-oxobutanoic acid and hydrolysis of the catalyst are shown as the last step.

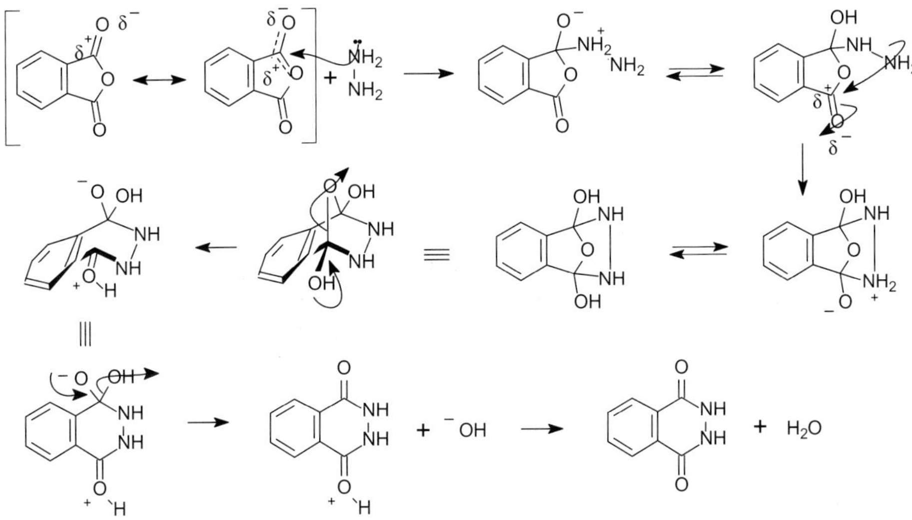

Scheme 9.138. A representation of the formation of 2,3-diaza-2,3-dihydro-1,4-napthaquinone from the reaction between hydrazine (H_2NNH_2) and 1,2-benzenedicarboxylic anhydride (phthalic anhydride).

acids (Table 9.6, item 2), for the halogen (e.g., chlorine) in acyl halides (Table 9.7, item 2), and for a carboxylate (Table 9.8, item 4) leaving group in carboxylic acid anhydrides. There are a host of cases, illustrated by the material in Table 9.9, in which the alkoxide equivalent (–OR, R = alkyl or aryl) present in the esters of carboxylic acids are themselves subjects of substitution reactions. For example, the common thread in the fabric of reactions at the carbon of the carbonyl group is again demonstrated by, among others, the first item in Table 9.9, that is, the simple base promoted hydrolysis of ethyl propanoate. As shown in Scheme 9.139, the unifying concept of attack of the substituting species at the carbon of the carbonyl results in

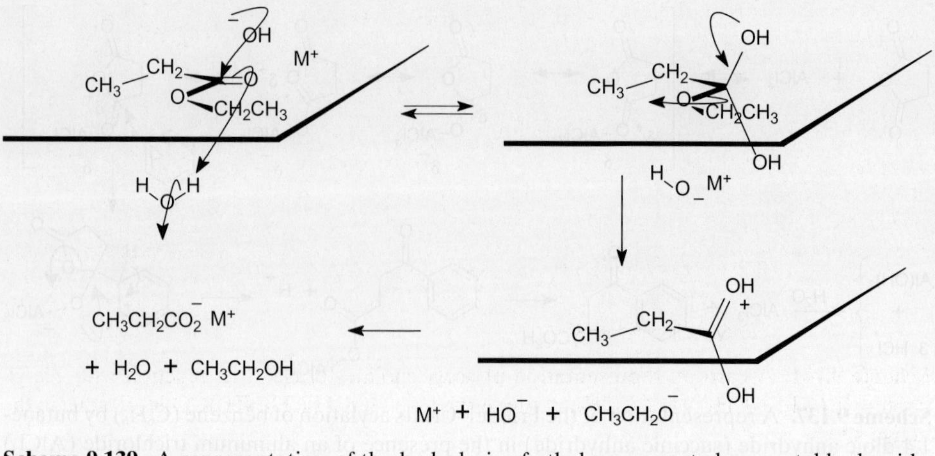

Scheme 9.139. A representation of the hydrolysis of ethyl propanoate by a metal hydroxide. The reaction is depicted as the substitution of an oxygen (forming the salt of the carboxylic acid) for the alkoxy group (ethoxy).

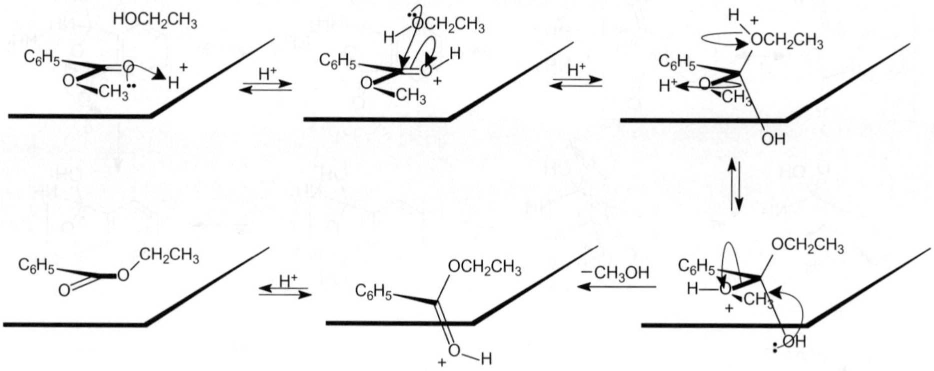

Scheme 9.140. A representation of the process of *ester interchange.* The methoxy group of methyl benzenecarboxylate (methyl benzoate) is replaced by an ethyoxy group from ethanol (ethyl alcohol, CH_3CH_2OH). The reaction is depicted as the acid-catalyzed substitution of an ethoxy for a methoxy.

the formation of the *tetrahedral intermediate.* As should now be clear, the *tetrahedral intermediate* can return to reactant(s) or proceed to product(s). In this particular case, since the substitution of oxygen (to generate, in basic solution, the carboxylate anion) occurs in aqueous solution where there is an excess of metal hydroxide [e.g., sodium hydroxide (NaOH) and potassium hydroxide (KOH)] and the alkoxide ion is too basic [the pK_a of ethanol (ethyl alcohol, CH_3CH_2OH) is about 17 under conditions where the pK_a of water (H_2O) is about 15 and typical carboxylic acids are about 5] to survive in water, the reaction is essentially irreversible. Liberated alkoxide ion abstracts a proton from water, generating alcohol and additional hydroxide.

In contrast, the second item in Table 9.9 involves acid rather than base. Here, as shown in Scheme 9.140, the oxygen of the carbonyl group of methyl benzenecarboxylate (methyl benzoate) is protonated, the tetrahedral intermediate is generated,

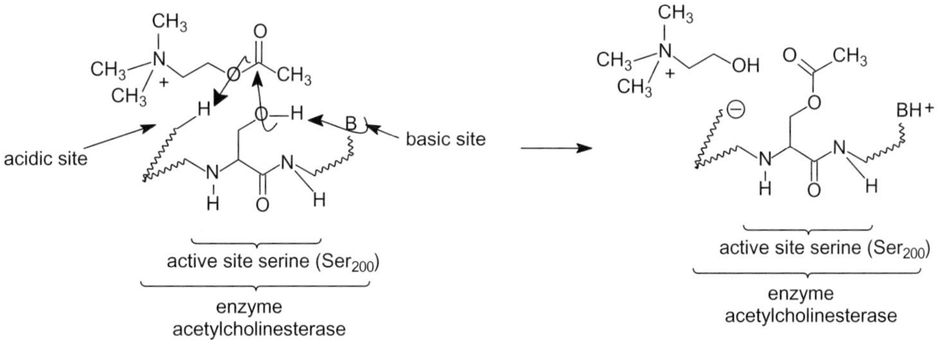

Scheme 9.141. A cartoon representation of acetylcholinesterase in its reaction with acetylcholine [$^+(CH_3)_3NCH_2CH_2O_2CCH_3$] at the serine [($S$)-$HOCH_2CH(NH_2)CO_2H$] residue in the active site.

and substitution of one oxyalkyl group [methoxy ($-OCH_3$)] by another [ethoxy ($-OCH_2CH_3$)] is effected. The reaction is finished by subsequent proton loss. Commonly, these substitution reactions are driven to completion by utilizing an excess of reacting alcohol (ethanol, CH_3CH_2OH, in this case), which also serves as the solvent for the reaction. Overall, such a process is referred to as an *ester interchange*.

While the hydrolysis of esters can be effected by aqueous basic metal hydroxides (Scheme 9.139) and by aqueous acids [Scheme 9.140, water (H_2O) in place of ethanol (ethyl alcohol, CH_3CH_2OH)], biological systems accomplish the same end in a much more subtle fashion. The enzymes accomplishing the hydrolysis of esters are known as "**esterases**" unless they fulfill their function in hydrolyzing esters of glycerol [1,2,3-trihydroxypropane, 1,2,3-propanetriol, $HOCH_2CH(OH)CH_2OH$] when they are called "**lipases**" (triacylglycerol hydrolases).

For example, acetylcholinesterase, an enzymatic esterase common to all animals, is the agent used to discharge the neurotransmitter acetylcholine [$^+(CH_3)_3NCH_2CH_2O_2CCH_3$] after the latter has carried its signal across the space between neuron and muscle (the *neuromuscular junction*). The enzyme bears a serine [(S)-$HOCH_2CH(NH_2)CO_2H$] residue (the "active site serine") to which the acetylcholine [$^+(CH_3)_3NCH_3CH_2O_2CCH_3$] binds and, via the tetrahedral intermediate, is acetylated with release of choline [$^+(CH_3)_3NCH_2CH_2OH$] (Scheme 9.141). Subsequent hydrolysis removes the acetyl group from the enzyme and prepares the enzyme to receive another (signaling) acetylcholine. Blocking the "active site serine" by forming an ester of phosphoric acid (phosphorylation) (Chapter 10) or in some other way inhibits the processes. For example, organophosphorus poisons, such as sarin [isopropyl methylphosphonofluoridate, $(CH_3)_2CHOP(O)(F)CH_3$], convert the hydroxyl group of the "active site serine" to an ester of methyl phosphonic acid [$CH_3P(O)(OR)_2$] and therefore block it. Such blocking (Scheme 9.142) causes the muscle to remain enervated and can lead to death.

The commercial introduction of esterases and lipases to carry out chiral separations has recently (the end of the 20th century) become common. Thus, since the esterase maintains a chiral environment, its use (e.g., by immobilization on a column down which a solution of racemic ester is passed) allows preferential hydrolysis of

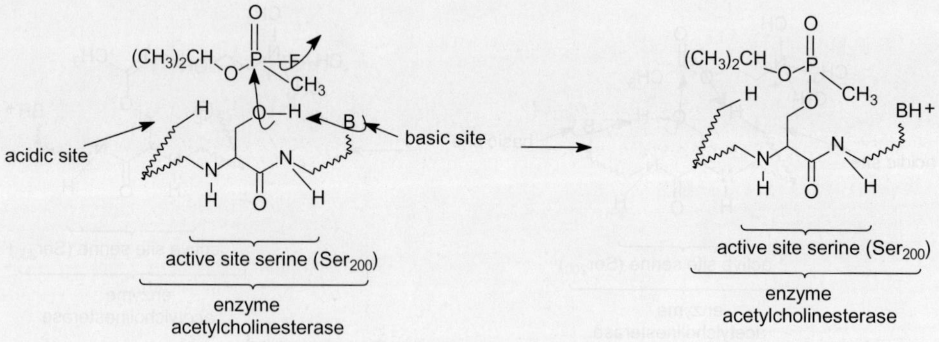

active site serine (Ser$_{200}$)

enzyme
acetylcholinesterase

active site serine (Ser$_{200}$)

enzyme
acetylcholinesterase

Scheme 9.142. A cartoon representation of acetylcholinesterase in its reaction with the poison "nerve" gas, isopropyl methylphosphonofluoridate [sarin $(CH_3)_2CHOP(O)(F)CH_3$] at the serine [(S)-$HOCH_2CH(NH_2)CO_2H$] residue in the active site. The phosphono ester is resistant to hydrolysis.

one enantiomer. It is common to find that some esterases perform better than others. For example, with *Candida rugosa* lipase, menthol esters can be selectively hydrolyzed (Equation 9.104).*

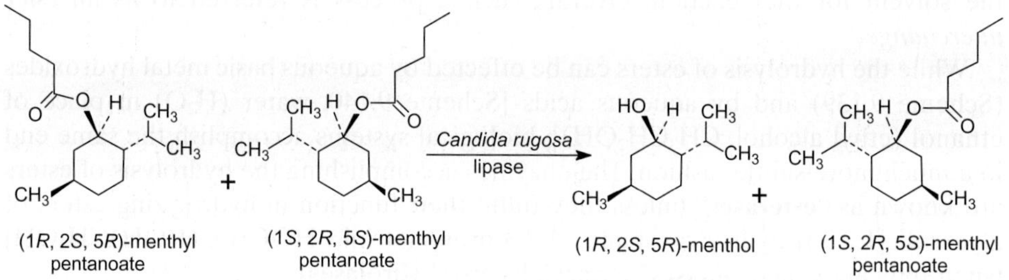

(1*R*, 2*S*, 5*R*)-menthyl pentanoate

(1*S*, 2*R*, 5*S*)-menthyl pentanoate

Candida rugosa lipase

(1*R*, 2*S*, 5*R*)-menthol

(1*S*, 2*R*, 5*S*)-menthyl pentanoate

(9.104)

Substitution of an alkoxy (–OR) group attached to the carbon of the carbonyl (i.e., in an ester) by an amino (–NHR) group is not normally a very facile process. That is, while substitution of an amino group (–NHR) at a carbonyl for a hydroxyl (–OH) leaving group can be effected with carboxylic acids (in the presence of the appropriate coupling reagents [Table 9.6, item 4]) and will also occur with halogen or a carboxylate as leaving group (the former with acid halides [Table 9.7, item 4] and the latter with carboxylic acid anhydrides [Table 9.8, item 3]), esters are significantly more resistant. This appears to be related to the basicity of the leaving group (Chapter 7). However, when the leaving group is aryloxy and particularly if the aryl is substituted with electron-withdrawing substituents, as in the case of item 3 in Table 9.9, then its ability as a leaving group is enhanced and these aryloxy esters do react with amines to generate amides. Thus, as shown in Table 9.9 and in Scheme 9.143, heating the *p*-nitrophenoxy cyclohexanecarboxylate in the presence of benzylamine produces a reaction yielding the corresponding *N*-benzylcyclohexanecarboxamide and *p*-nitrohydroxybenzene (*p*-nitrophenol).

*See Lagrand, G.; Baratti, J. Buono, G.; Triantaphylides, C. *Tetrahedron Lett.*, **1986**, 29.

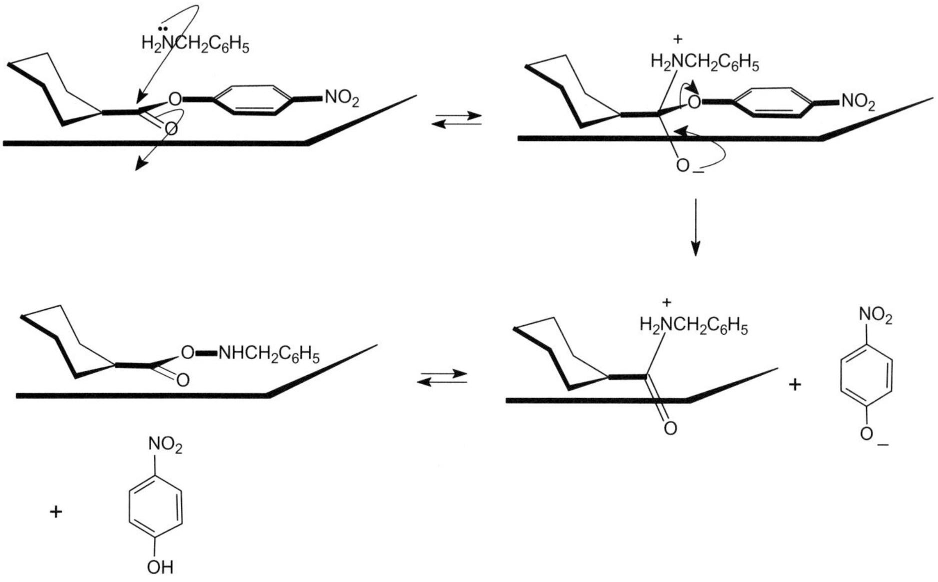

Scheme 9.143. A representation of the reaction of *p*-nitrophenoxy cyclohexanecarboxylate with benzylamine to produce the corresponding *N*-benzylcyclohexanecarboxamide and *p*-nitrohydroxybenzene (*p*-nitrophenol) via the tetrahedral intermediate.

In contrast to the reactions so far shown, it is possible to substitute the leaving alkoxide by *carbon*. To effect this irreversible substitution, either an alkyllithium ("RLi") or an alkylmagnesium halide ("R-MgX") reagent is used.* With the former, it is occasionally possible to stop the reaction after the first equivalent has added (i.e., at the ketone stage), while with the latter, unless the ester is highly hindered, it is common for the ketone to react more rapidly than the ester and thus the tertiary alcohol forms. For the ethyl ester of benzenecarboxylic acid (ethyl benzoate) undergoing reaction with either methyllithium (CH_3Li) or methylmagnesium bromide (CH_3MgBr), item 4 in Table 9.9, 2-phenyl-2-propanol is the major product (small quantities of the ketone, phenylethanone [acetophenone], can occasionally be isolated). Scheme 9.144 depicts a representation of the reaction. It will be recalled from the discussion in Chapter 7 and Schemes 9.39 and 9.40 earlier in this chapter that the representation of the organometallic compounds such as methylmagnesium bromide (CH_3MgBr) as unsolvated and monomeric is a vast simplification, which is presented as shown only to more clearly depict the outcome. The details of the process are not yet well understood to provide a simple explication.

As shown in item 5 of Table 9.9, and similar to what has appeared before in reactions of aldehydes and ketones, anion (or enolate anion) of a carbonyl compound, such as an aldehyde, a ketone, or an ester, is capable of addition to another aldehyde, ketone, or ester. The reaction can become quite complicated if both carbonyl com-

*It will be recalled that although the alkyllithium ("RLi") and magnesium ("RMgX," i.e., Grignard reagents) are written as monomeric, it is likely that they are actually aggregates through most of the reaction.

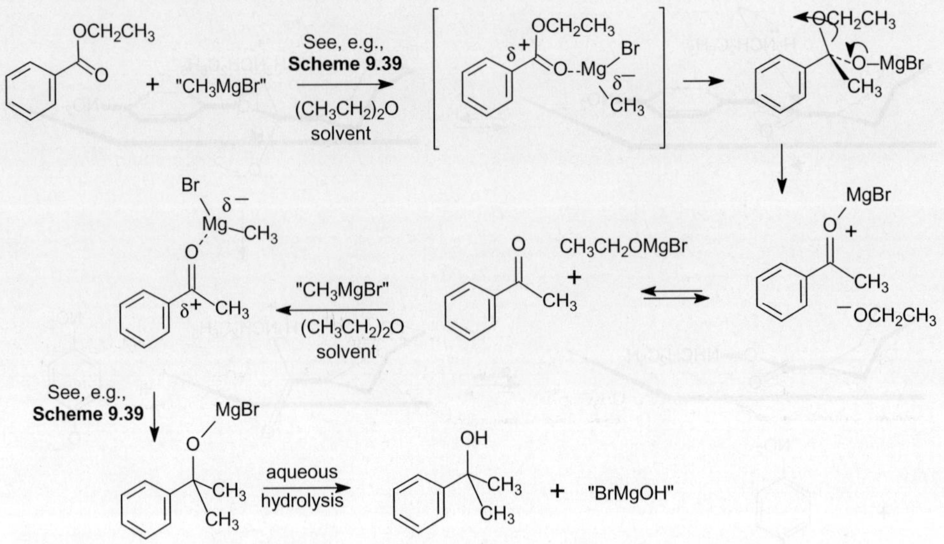

Scheme 9.144. A simplified representation of the reaction between ethyl benzenecarboxylate (ethyl benzoate, $C_6H_5CO_2CH_2CH_3$) and methylmagnesium bromide (CH_3MgBr) in diethyl ether [$(CH_3CH_2)_2O$] solvent.

ponents have protons on the carbons α- to the carbonyl. Thus, as shown in item 5 of Table 9.9, ethyl benzencarboxylate (ethyl benzoate, $C_6H_5CO_2CH_2CH_3$), which has no protons on the carbon bearing the carbonyl group, can only act as substrate. In the presence of ethoxide ion (from sodium ethoxide, $NaOCH_2CH_3$), unprofitable and unobserved ester interchange might occur (the base-catalyzed equivalent of the process shown in Scheme 9.140) with both ethyl benzenecarboxylate (ethyl benzoate, $C_6H_5CO_2CH_2CH_3$) and ethyl ethanoate (ethyl acetate, $CH_3CO_2CH_2CH_3$). However, only ethyl ethanoate (ethyl acetate, $CH_3CO_2CH_2CH_3$) has protons on the carbon atom (the methyl group) α- to the carbonyl. So, in the presence of ethoxide, a proton is removed from the methyl group, generating a carbanion. Attack of the latter at the carbonyl of ethyl benzenecarboxylate (ethyl benzoate, $C_6H_5CO_2CH_2CH_3$) and ethyl ethanoate (ethyl acetate, $CH_3CO_2CH_2CH_3$) can both occur.

In the first case, ethyl 3-phenyl-3-oxopropanoate results, while in the second ethyl 3-oxobutanoate (ethyl acetoacetate, $CH_3COCH_2CO_2CH_2CH_3$) is produced. The processes (called the **Claisen condensation**) are depicted in Scheme 9.145.

Problem 9.23. It is possible to adjust the conditions of the **Claisen condensation** to favor the formation of ethyl 3-phenyl-3-oxopropanate. Predict what would need to be done.

It is not unreasonable to expect that a **Claisen**-type condensation might also occur *intramolecularly*. That is, both the anion α- to a carbonyl group and another carbonyl group might be present in the **same** molecule. As shown in item 6 of Table 9.9 and Scheme 9.146, when the diethyl ester of 1,5-hexanedicarboxylate (diethyl adipate $CH_3CH_2O_2C(CH_2)_4CO_2CH_2CH_3$), is treated with ethoxide ($-OCH_2CH_3$)

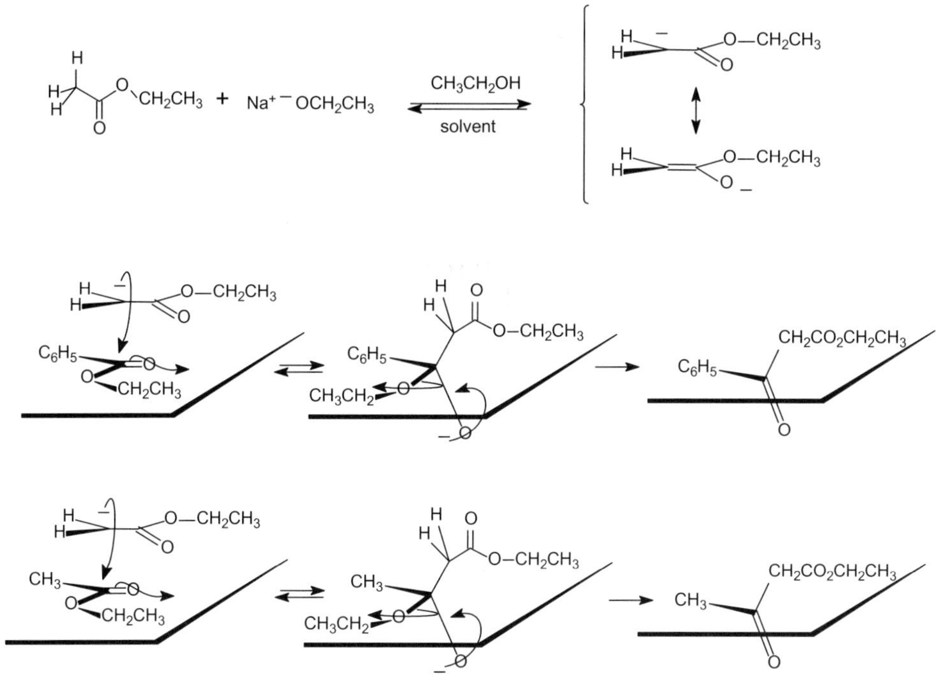

Scheme 9.145. A representation of proton removal from the methyl group of ethyl ethanoate (ethyl acetate, $CH_3CO_2CH_2CH_3$) generating a resonance-stabilized carbanion. Attack of the latter at the carbonyl of ethyl benzenecarboxylate (ethyl benzoate, $C_6H_5CO_2CH_2CH_3$) and ethyl ethanoate (ethyl acetate, $CH_3CO_2CH_2CH_3$) to yield, respectively, ethyl 3-phenyl-3-oxopropanoate and 3-oxobutanoate (ethyl acetoacetate, $CH_3COCH_2CO_2CH_2CH_3$). An example of the **Claisen condensation**.

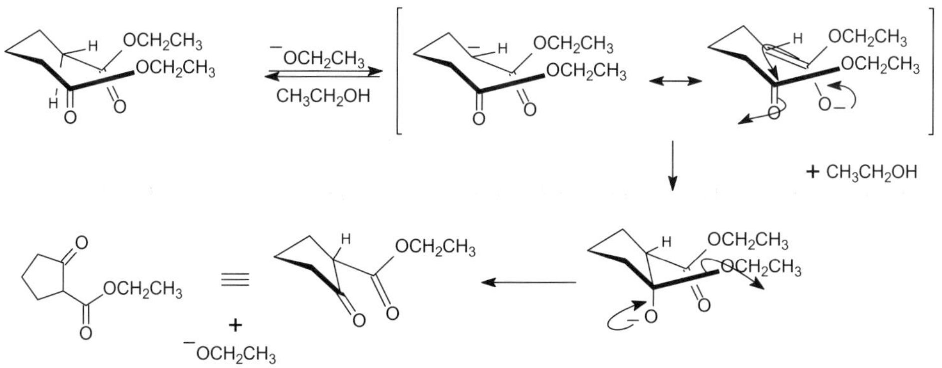

Scheme 9.146. A representation of an intramolecular **Claisen**-type condensation (an **acyloin condensation**) in which the diethyl ester of 1,6-hexanedicarboxylic acid (diethyl adipate, $CH_3CH_2O_2C(CH_2)_4CO_2CH_2CH_3$)) undergoes cyclization to yield ethyl 2-oxocyclopentanecarboxylate in the presence of ethoxide anion ($^-OCH_2CH_3$) in ethanol (ethyl alcohol, CH_3CH_2OH) solvent.

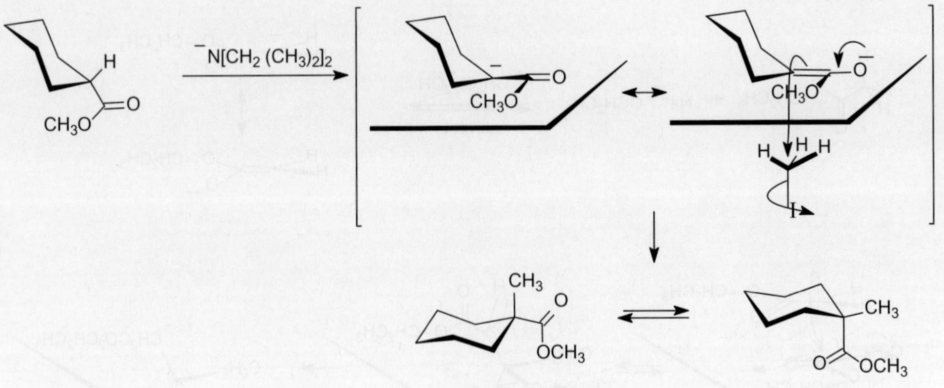

Scheme 9.147. A representation of the reaction between the methyl ester of cyclohexanecarboxylic acid (methyl cyclohexanecarboxylate) with lithium diisopropylamide (LDA) to generate a carbanion on the carbon α- to the carbonyl. The carbanion so formed then acts as a nucleophile toward methyl iodide (CH_3I) to yield methyl 1-methylcyclohexanecarboxylate, lithium iodide, and recovered base, diisopropylamine $\{[(CH_3)_2CH]_2NH\}$.

anion in ethanol the anion, α to one carbonyl, attacks the carbon of the carbonyl five carbon atoms away forming the (now expected) tetrahedral intermediate. Loss of ethoxide anion ($-OCH_2CH_3$) in the substitution process leads to a keto ester (ethyl 2-oxocyclopentanecarboxylate in this case). This intramolecular substitution process version of the Claisen reaction is called the **acyloin condensation**.

Substitution processes focused **_around_** the carbonyl group as well as **_at_** the carbonyl group are, of course, also possible. Consider the case depicted in item 7 of Table 9.9. As noted immediately above for the intermolecular and intramolecular versions of the **Claisen condensation**, success depends upon generation of an anion α- to the carbon of the carbonyl. Generation of such anions, particularly at fairly high dilution (where reaction *between* esters is less likely) with hindered bases, followed by addition of an electrophilic species to the reaction medium, results in overall substitution of the electrophilic species for the proton that was removed. In item 7 of Table 9.9, as shown in Scheme 9.147, the methyl ester of cyclohexanecarboxylic acid (methyl cyclohexanecarboxylate) **does not react with the hindered base (LDA) at the carbon of the carbonyl**. Rather, the base removes the proton on the carbon α- to the carbonyl and the carbanion so formed then acts as a nucleophile toward methyl iodide (CH_3I). Substitution yields methyl 1-methylcyclohexanecarboxylate, lithium iodide, and recovered base, diisopropylamine $\{[(CH_3)_2CH]_2NH\}$.*

*As shown below in this note, it is appropriate to consider two possibilities. First, alkylation at carbon, which is found to occur, produces (Scheme 9.147) the observed product (methyl 1-methylcyclohexanecarboxylate). Alkylation at oxygen, to produce the methoxyether, bis-2′-dimethoxymethylidenecyclohexane, **did not occur**. Generally, while electron density is higher at oxygen, that is, the electronegativity of oxygen is greater than carbon (resulting in, e.g., silylation and acylation of enols

(Continued on next page)

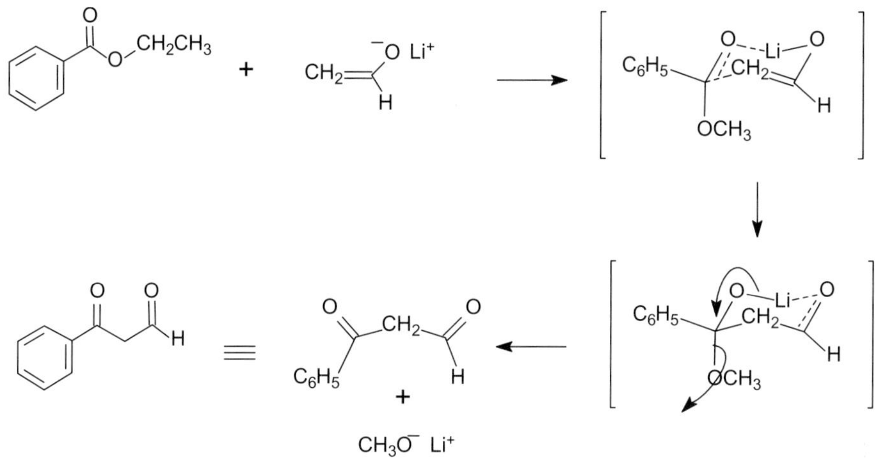

Scheme 9.148. A representation of a pathway accounting for an "aldol-type process" resulting in the substitution of an "acetaldehyde equivalent" for ethoxide. The reaction shown is between the ethyl ester of benzenecarboxylic acid (ethyl benzoate) and the ethanal (acetaldehyde, CH_3CHO) enolate anion. The product is 3-phenyl-3-oxopropanal.

In the same vein, a preformed enolate anion from, for example, ethanal (acetaldehyde, CH_3CHO), as in item 8 of Table 9.9, can substitute for an alkoxide anion. A pathway accounting for this "aldol-type process" between the ethyl ester of benzenecarboxylic acid (ethyl benzoate) and the ethanal (acetaldehyde, CH_3CHO) enolate anion to produce 3-phenyl-3-oxopropanal is shown as Scheme 9.148.

Substitution of carbon for the oxygen of the carbonyl is relatively rare. The **Tebbe reagent**, titanium μ-chlorobis(η5-2,4-cyclopentadien-1-yl)(dimethylaluminum)-μ-methylene-(9Cl), prepared from dicyclopentadienyltitanium dichloride and trimethylaluminum accomplishes this by an exchange process as shown for the reaction with methyl cyclohexanecarboxylate (Scheme 9.149 and item 9 in Table 9.9). The Scheme 9.149 also incorporates suggestions for the initial formation of the **Tebbe reagent** and a possible pathway for the transfer of the methylene group and loss of oxygen to produce 1-cyclohexyl-1-methoxyethene. As noted, **two** equivalents of trimethylaluminum are required to prepare the reagent, and while the stoichiometry affirms the information in the Scheme 9.149, it does not argue for any mechanistic detail.

and enolate anions), substitution reactions require orbital overlap, which is better from the carbon than from the oxygen.

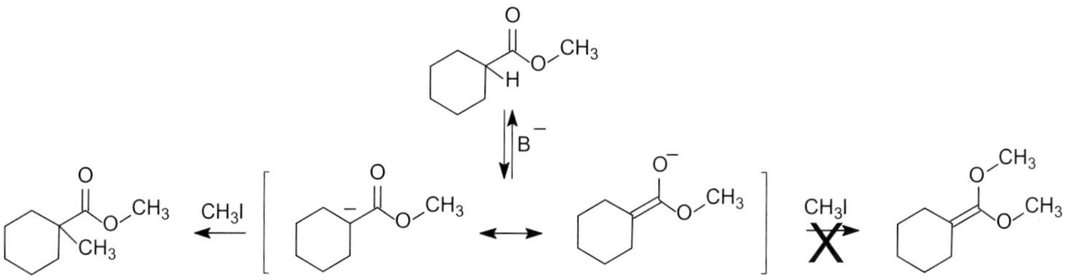

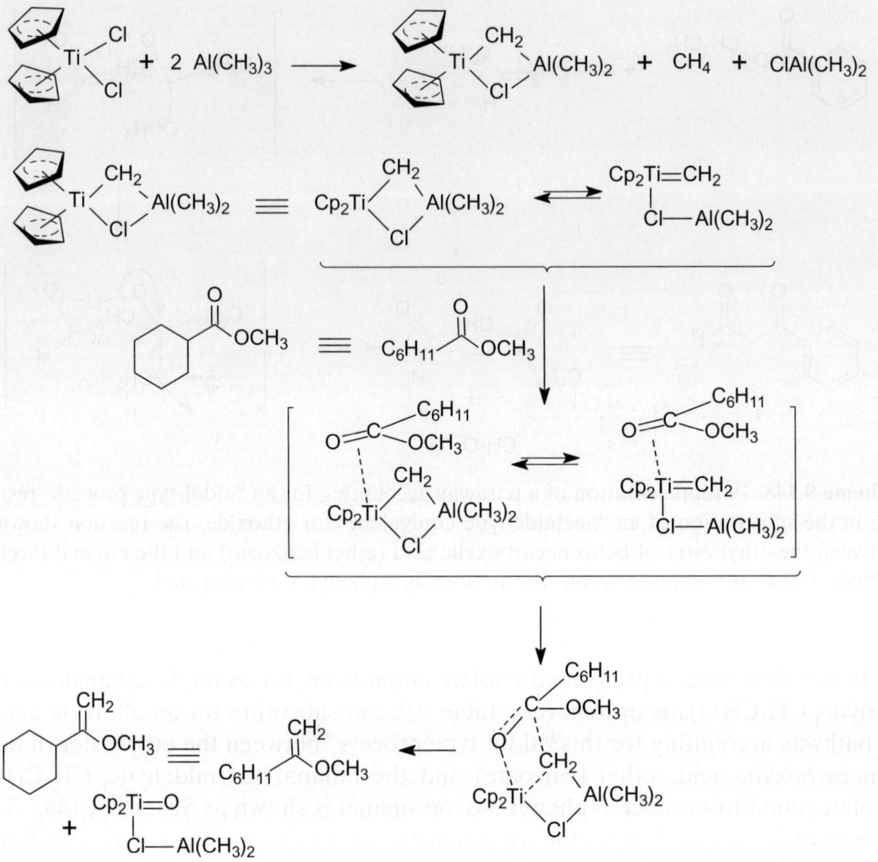

Scheme 9.149. A representation of the formation of the **Tebbe reagent**, titanium μ-chlorobis(η5-2,4-cyclopentadien-1-yl)(dimethylaluminum)-μ-methylene-(9Cl), from dicyclopentadienyltitanium dichloride (Cp$_2$TiCl$_2$) and trimethylaluminum Al(CH$_3$)$_3$, and its reaction with methyl cyclohexanecarboxylate to produce 1-cyclohexyl-1-methoxyethene.

   In a formally similar fashion, a substitution reaction of sulfur for oxygen can also be effected by using **Lawesson's reagent** [2,4-bis(4-methoxyphenyl)-1,2,3,4-dithiadiphosphetane 2,4-disulfide]. Here, it is thought that an initial equilibrium in which diphosphetane decomposes into monomeric species is followed by rate-determining exchange of sulfur for oxygen. A four-membered ring might be involved in the conversion of esters such as ethyl benzenecarboxylate (ethyl benzoate) to thioesters such as ethyl benzenethiocarboxylate (ethyl thiobenzoate), Scheme 9.150 and item 10 in Table 9.9.*

---

*Interestingly, **Lawesson's reagent** (*vide supra*) is simply prepared by heating methoxybenzene (anisole, C$_6$H$_5$OCH$_3$) with phosphorus sulfide (P$_4$S$_{10}$). The reaction involves the generation of hydrogen sulfide (H$_2$S) and should be carried out in an efficient hood.

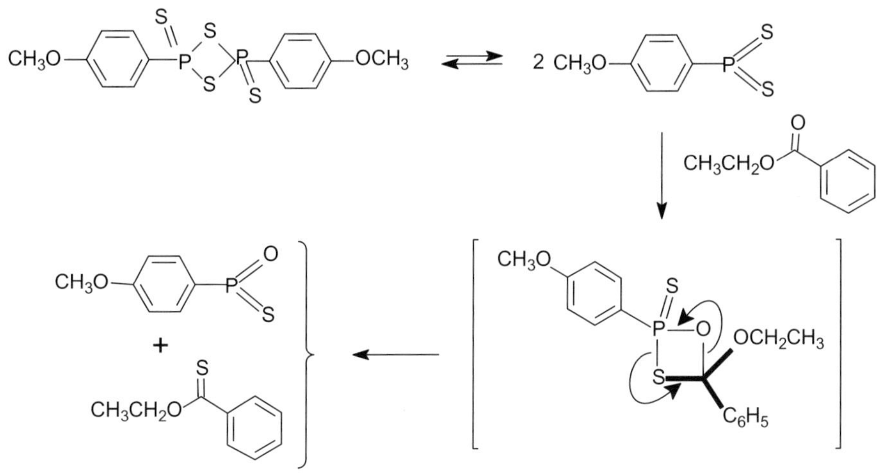

Scheme 9.150. A representation of the use of **Lawesson's reagent** [2,4-bis(4-methoxyphenyl)-1,2,3,4-dithiadiphosphetane 2,4-disulfide] to convert ethyl benzenecarboxylate (ethyl benzoate) to ethyl benzenethiocarboxylate (ethyl thiobenzoate).

In the presence of sulfuric acid, esters of carboxylic acids react in ways similar to the carboxylic acids themselves. Thus, it is not surprising that ethyl benzenecarboxylate (ethyl benzoate, $C_6H_5CO_2CH_2CH_3$) is protonated in the presence of sulfuric acid (H_2SO_4)—as was the corresponding carboxylic acid (benzenecarboxylic (benzoic acid, $C_6H_5CO_2H$). These protonated species are highly susceptible to attack at the carbonyl by nucleophilic agents such as, in this case, hydrazine (H_2NNH_2), which, although protonated by sulfuric acid (H_2SO_4), is also available, in equilibrium, with the unprotonated form. Hydrazine is substituted for the ethoxy ($-OCH_2CH_3$) leaving group. The resulting acylhydrazide undergoes a **Curtius rearrangement** producing the corresponding isocyanate [in this case, phenylisocyanate ($C_6H_5N=C=O$)]. The process is listed as item 11 in Table 9.9, and a representation using curved arrows is shown in Scheme 9.151.

The final item in Table 9.9 resembles that of Scheme 9.151. However, whereas hydrazine (H_2NNH_2) was the species that substituted for the ethoxy group ($-OCH_2CH_3$) in that scheme, hydroxylamine ($HONH_2$) is used in item 12 of Table 9.9 to generate the hydroxamic acid.

The similarity continues in that in both items 11 and 12 a strong acid is used. In the former, it was sulfuric acid (H_2SO_4), whereas in the latter, hydrogen chloride (HCl) is chosen. Then, as depicted in Scheme 9.152, the processes are presumed to be strikingly similar. Thus, after initial protonation, attack by hydroxylamine results in the loss of ethanol ($HOCH_2CH_3$) and the production of benzenehydroxamic acid. The latter rearranges with loss of water to the same phenylisocyanate of the previous example.

As was the case for esters, amides were introduced in Chapter 5 and have already been considered in this chapter. Amides are also derivatives of amines (Chapter 10); that is, they are products of substitution at the carbonyl with $-H-NH_2$ (amides "derived from" ammonia), $-NHR$ (R = alkyl or aryl, amides "derived from" primary

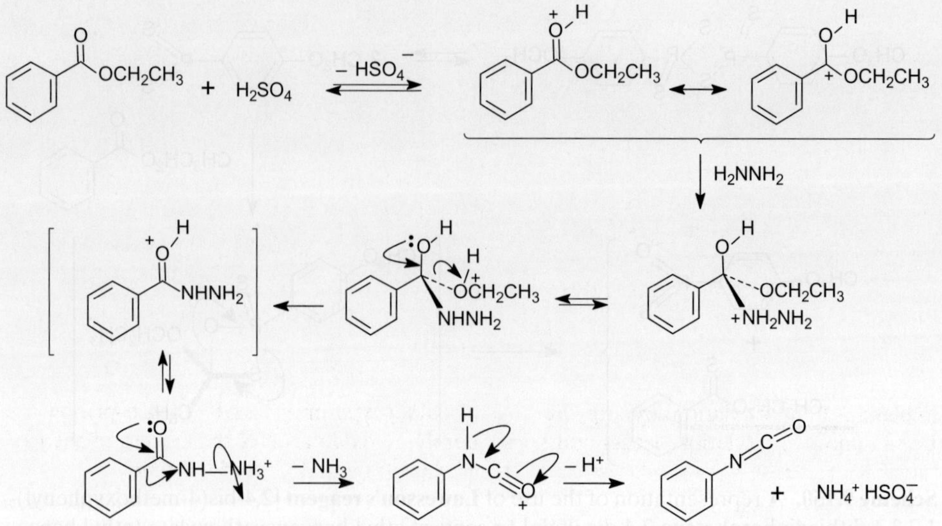

Scheme 9.151. A representation of the reaction of ethyl benzenecarboxylate (ethyl benzoate, $C_6H_5CO_2CH_2CH_3$) with hydrazine (H_2NNH_2) in the presence of sulfuric acid (H_2SO_4). Hydrazine is substituted for the ethoxy ($-OCH_2CH_3$) leaving group and the resulting acylhydrazide undergoes a **Curtius rearrangement,** producing phenylisocyanate ($C_6H_5N=C=O$).

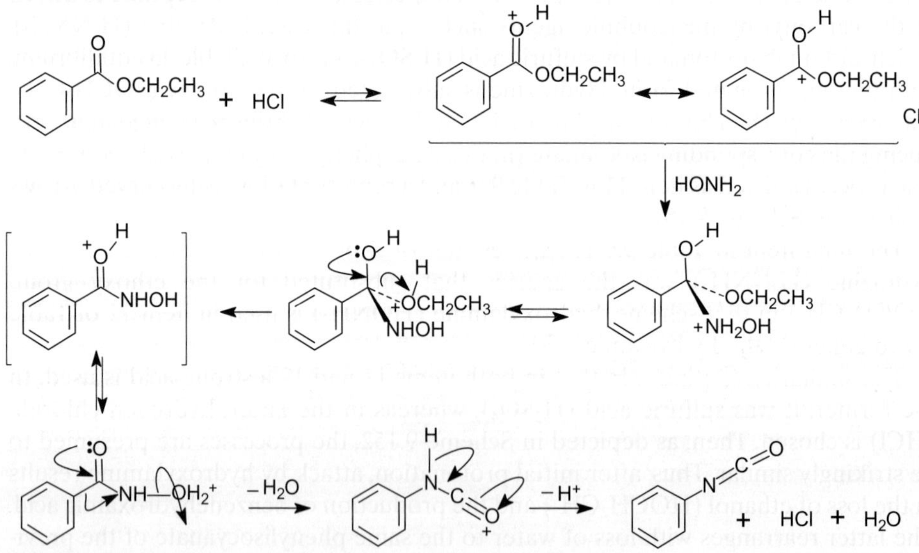

Scheme 9.152. A representation of the reaction of ethyl benzenecarboxylate (ethyl benzoate, $C_6H_5CO_2CH_2CH_3$) with hydroxylamine (H_2NOH) in the presence of hydrochloric acid (HCl). Hydroxylamine is substituted for the ethoxy ($-OCH_2CH_3$) leaving group and the resulting hydroxamic acid undergoes a **Curtius rearrangement,** producing phenylisocyanate ($C_6H_5N=C=O$).

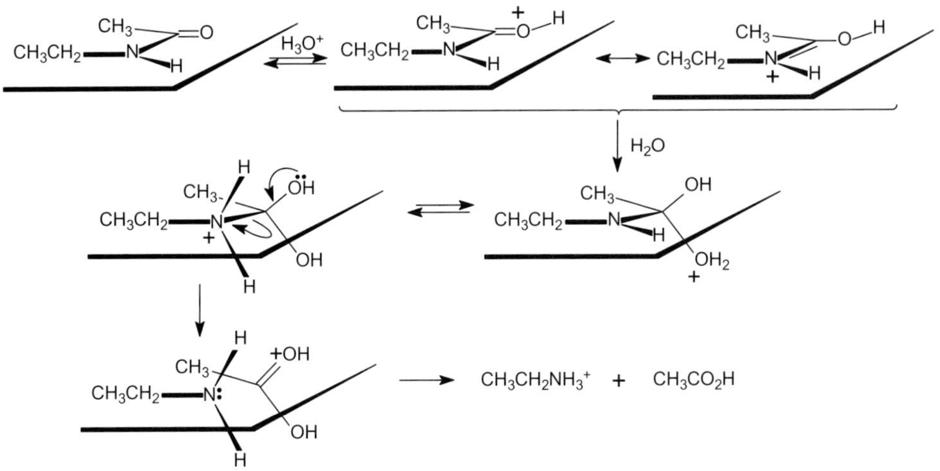

Scheme 9.153. A representation of the hydrolysis of N-ethylethanamide (N-ethylacetamide).

amines), or, finally, $-NR_2$ (R = alkyl or aryl, amides "derived from" secondary amines). These compounds are "amino" substitution products for the $-OH$ group in carboxylic acids (Table 9.6, item 4), for the halogen (e.g., chlorine) in acyl halides (Table 9.7, item 4), for a carboxylate (Table 9.8, item 3) leaving group in carboxylic acid anhydrides, and alkoxy groups (Table 9.9, item 3) in the case of esters. There are a number of instances, illustrated by some of the material in Table 9.10, in which the amine equivalent ($-NH_2$, $-NHR$, $-NR_2$; R = alkyl or aryl) present in the amides of carboxylic acids are themselves subjects of substitution reactions.

As expected, the common thread in the fabric of reactions at the carbon of the carbonyl group is again demonstrated by, among others, the first item in Table 9.10, that is, the simple acid promoted hydrolysis of N-ethylethanamide (N-ethylacetamide, $CH_3CH_2NHCOCH_3$). As shown in Scheme 9.153, the unifying concept of attack of the substituting species ($:OH_2$) at the carbon of the carbonyl results in the formation of the **tetrahedral intermediate**. As should now be clear, the **tetrahedral intermediate** can return to reactant(s) or proceed to product(s). In this particular case, since the substitution of oxygen for nitrogen (to generate the carboxylic acid, ethanoic acid [acetic acid, CH_3CO_2H], and the ethylammonium cation, $CH_3CH_2NH_3^+$) occurs in aqueous solution where there is an excess of aqueous acid (H_3O^+), the reaction is irreversible (Scheme 9.153).

Although it may be supposed from the discussion of item 4 of Table 9.9 (where an ester, reacted with a **Grignard reagent** [methylmagnesium bromide, "CH_3MgBr"], first underwent substitution and subsequently reacted further to yield a tertiary alcohol) that amides such as N,N-diethyl cyclohexanecarboxamide (item 2 in Table 9.10) might behave the same, it falls out that they do not generally do so.*

---

*It is important to recognize that reaction of methyllithium ("CH_3Li") or methylmagnesium halide ("CH_3MgBr") with an amide bearing a proton on nitrogen, that is, RCO-N<u>H</u>$_2$ or RCON<u>H</u>R′, results in generation of the anion as, respectively, the lithium or magnesium salt and the evolution of methane (CH_4). This is a "classic" test for "active" hydrogen.

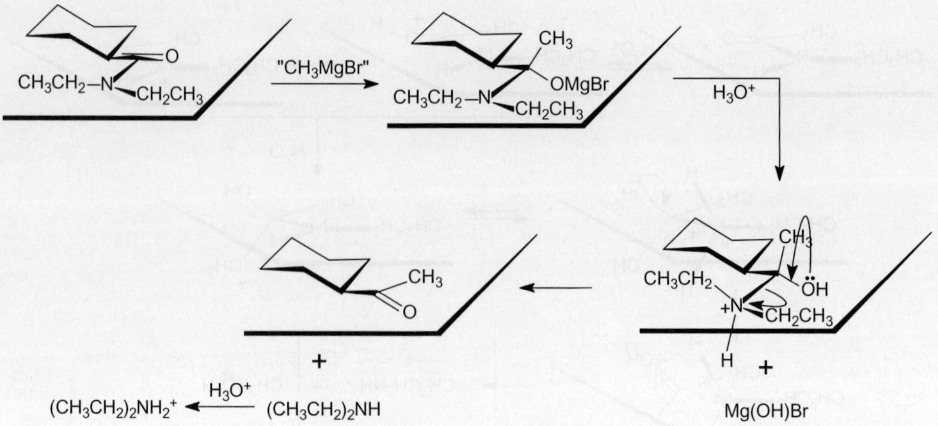

$$(CH_3CH_2)_2NH_2^+ \xleftarrow{H_3O^+} (CH_3CH_2)_2NH$$

Mg(OH)Br

Scheme 9.154. A representation of the addition of methylmagnesium bromide ("CH$_3$MgBr") to the N,N-diethylamide of cyclohexanecarboxylic acid. The reaction stops after the addition until hydrolysis with aqueous acid (H$_3$O$^+$) is effected and cyclohexylmethyl ketone and the N,N-diethylammonium species result.

Thus, as shown in item 2 of Table 9.10 and Scheme 1.154, the addition of what is written as methylmagnesium bromide ("CH$_3$MgBr") to the N,N-diethylamide appears to yield an intermediate that cannot easily go on to eliminate the metallic complex of N,N-diethylamine. So, the reaction stops and the complex remains until the reaction mixture is hydrolyzed with aqueous acid (H$_3$O$^+$), at which stage further reaction is not possible and the corresponding ketone results.

Among the more interesting substitution reactions that amides undergo is one that involves loss of the carbonyl (C=O) and migration of the alkyl or aryl substituent originally attached to the carbonyl over to the nitrogen of the amide. The **Hofmann degradation**, shown as item 3 in Table 9.10, generally involves only those amides in which there are two protons on nitrogen and use of halogen [chlorine (Cl$_2$) or bromine (Br$_2$)] in basic solution (where hypohalite can form). The initial reaction between the amide and halogen in basic solution (Scheme 9.155) results in the formation of an N-halo compound (in the case of item 3 in Table 9.10, N-bromo-3-methylbenzenecarboxamide is generated). Then, α-elimination (Chapter 7) of hydrogen bromide (HBr) produces a "nitrene" (see, e.g., **carbenes**, Chapter 7) which is unstable. Movement of the aryl group in item 3 of Table 9.10 (or, generally, whatever the substituent on the acyl function) to nitrogen results in generation of an isocyanate (Chapter 10) which, on hydrolysis, generates carbon dioxide (CO$_2$) and amine (3-methylbenzamine [3-methylaniline, m-toluidine] in the case of item 3 of Table 9.10).

Interestingly, as will be discussed further in Chapter 10, the migrating group moves with retention of configuration. Thus, if the carbon bearing the carboxylic acid is asymmetrically substituted (i.e., the molecule is chiral) and the migration occurs with retention of configuration at the migrating site, chirality is maintained. In item 3 of Table 9.10, the aromatic ring migrates in such a way that the carbon atom originally bearing the carbonyl group is now attached to the nitrogen. That is, the process is one of substitution.

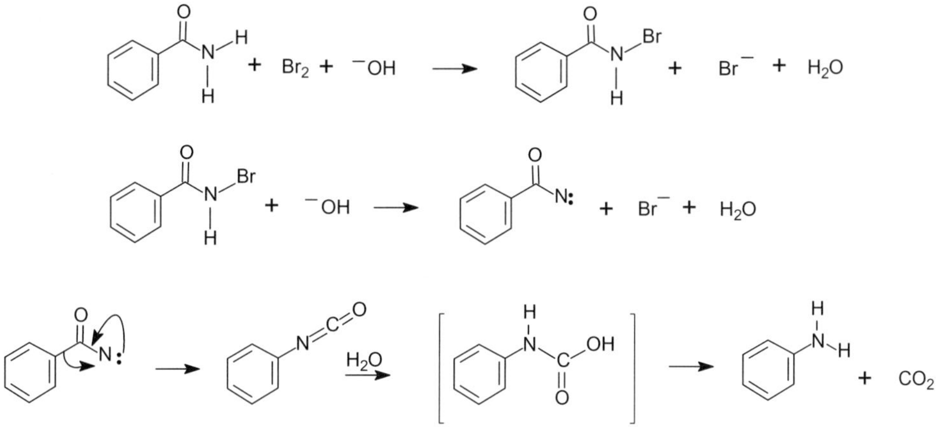
O
N
H
H
+ Br2 + ⁻OH
Br
+ Br⁻ + H2O
+ ⁻OH
N:
N:
N=C=O
H2O
OH
C
+ CO2
an acylnitrene
an isocyanate
a carbamic acid

Scheme 9.155. A representation of the Hofmann degradation of 3-methylbenzenecarboxamide. The amino group attached to the aromatic ring has been substituted for the entire carboxy function. The Hofmann degradation is sometimes referred to as the "Hofmann rearrangement." See Hofmann, A. W. *Chem. Ber.*, **1881**, *14*, 2725.

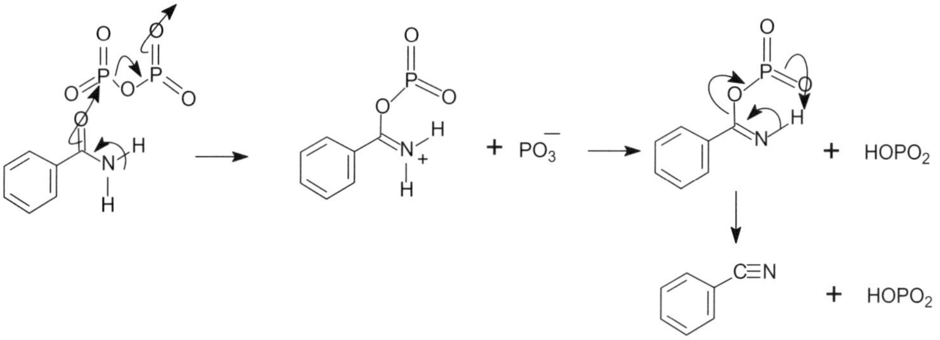
O
P
N
H
+ PO3⁻
+ HOPO2
C≡N
+ HOPO2

Scheme 9.156. A representation of the dehydration of 3-methylbenzenecarboxamide with phosphorus pentoxide (P_2O_5 or P_4O_{10}) to the nitrile (3-methylbenzonitrile).

When the same starting amide, that is, 3-methylbenzenecarboxamide, is treated with phosphorus pentoxide (P_2O_5 or P_4O_{10}), a triple bond is "substituted" for the amide and a nitrile (3-methylbenzonitrile) results. Although many schemes might be written to account for the dehydration of the amide to the nitrile (item 4 of Table 9.10), that depicted in Scheme 9.156 is frequently seen.

In an era before the widespread dissemination of spectroscopic tools, it was important to determine if amides were primary, secondary, or tertiary. That is, if there were two, one, or zero hydrogens on nitrogen, respectively. The use of nitrous acid to **substitute nitric oxide (NO) for hydrogen** was widely employed to this end. More recently, compounds in which an NO replaces a hydrogen on amines (i.e., N-nitroso compounds) have received attention as materials that can lead to cancer formation in susceptible tissues. Such N-nitroso compounds are widespread in the environment since nitrites (e.g., sodium nitrite, $NaNO_2$), which give rise to nitrous acid (HONO), are used in meat preservation. and NO, which can react directly with amides (as in **peptides**, see Chapter 12), is produced by the reaction between nitrogen and oxygen at high temperatures in gasoline-powered internal combustion engines. So, as shown in Scheme 9.17, nitrous acid (HONO) reacts with a generic primary amide, $RCONH_2$, to produce a carboxylic acid (RCO_2H), water (H_2O), and nitrogen (N_2). Nitrous acid also reacts with a generic secondary amide ($RCONHR_1$ or $RCONHR_2$) to produce an N-nitroso derivative (Equation 9.105). However, nitrous acid (HONO) fails to yield an isolable compound with a generic teriary amide ($RCONHR_1R_2$). Presumably, with regard to the latter, some initial reaction might occur, but the resulting compound is unstable and reverts to starting materials on attempted isolation (Equation 9.106).

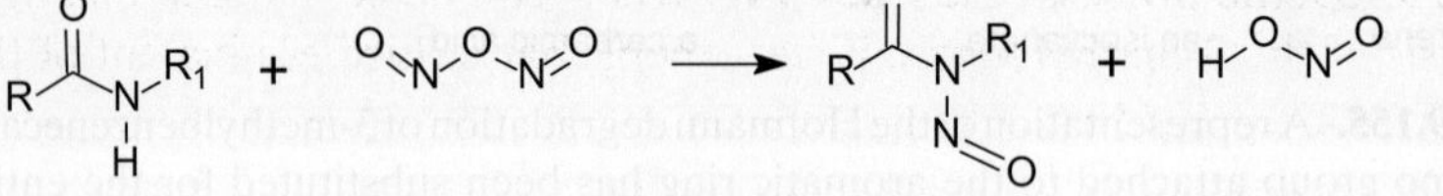

$$(9.105)$$

Problem 9.24. Using Scheme 9.157 as a model, account for the results of Equation 9.105 by drawing curved arrows and keeping track of the cations and anions.

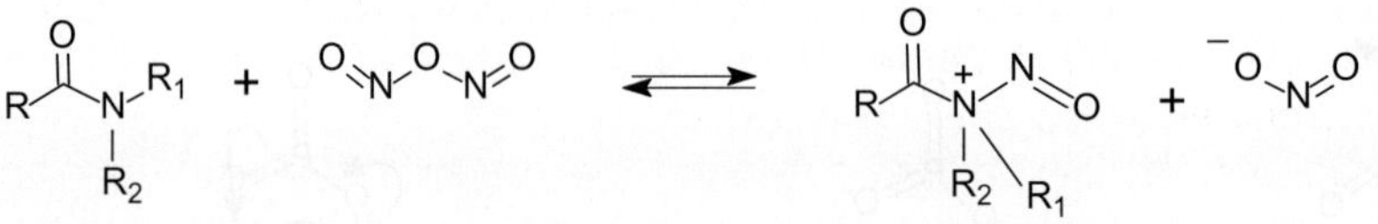

$$(9.106)$$

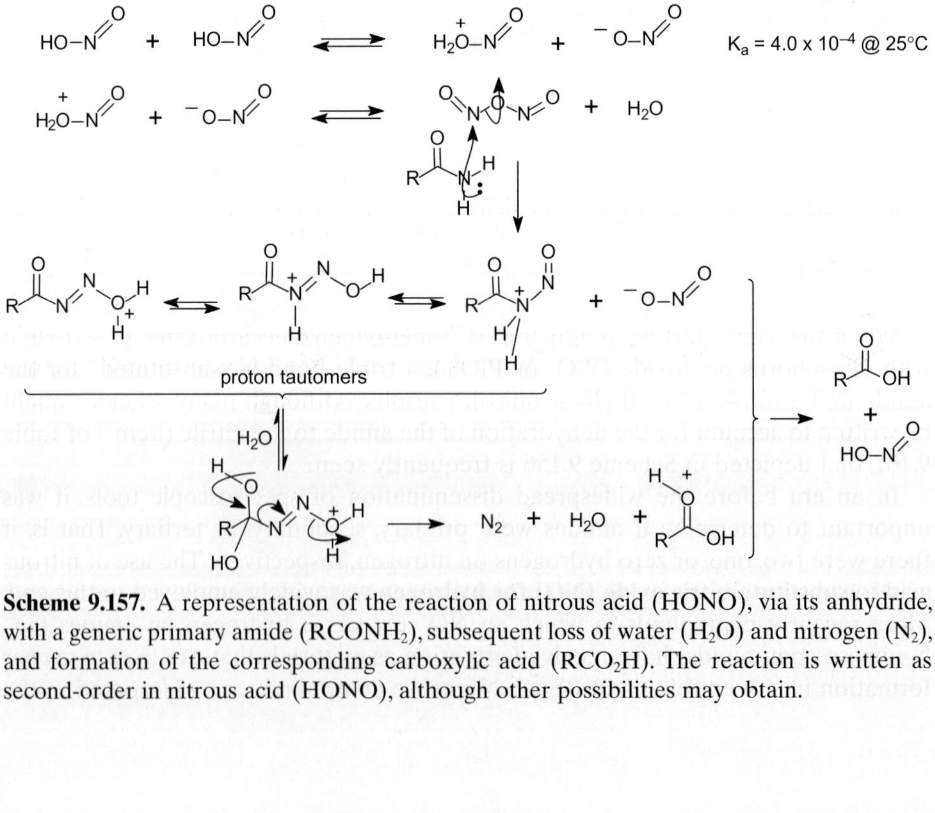

$$K_a = 4.0 \times 10^{-4} \text{ @ } 25°C$$

proton tautomers

Scheme 9.157. A representation of the reaction of nitrous acid (HONO), via its anhydride, with a generic primary amide ($RCONH_2$), subsequent loss of water (H_2O) and nitrogen (N_2), and formation of the corresponding carboxylic acid (RCO_2H). The reaction is written as second-order in nitrous acid (HONO), although other possibilities may obtain.

The penultimate item in Table 9.10, item 6, is again a hydrolysis. Here, a nitrile, 3-methylbenzonitrile, produced as item 4 in Table 9.10 by dehydration of the corresponding amide, 3-methylbenzenecarboxamide, is allowed to react with water in the presence of acid. As amides are often hydrolyzed under the same conditions (item 1 in Table 9.10), it is not generally possible to stop the hydrolysis of the nitrile at the amide stage. A pathway representing the process of item 6 in the table is shown as Scheme 9.158, where it should be noted that since acid is present, the ammonia generated is found in protonated form (as the ammonium cation) and as such is accompanied by whatever gegenion originally accompanied the proton.

Finally, if anhydrous ethanol (ethyl alcohol, CH_3CH_2OH) is used in the presence of a nonaqueous acid catalyst then substitution of the nitrogen of the nitrile by the ethanol can be affected. As depicted in the last item (item 7) in Table 9.10 and in Scheme 9.159, the process (the **Pinner reaction**) can be considered analogous to the hydrolysis of the 3-methylbenzonitrile, except that a third equivalent of the alcohol is utilized and the corresponding triethoxy "orthoester" (ethyl 3-methylbenzoic acid orthoester) results.

It is important to recognize that the reactions in Tables 9.6–9.10 (and the corresponding schemes) are provided only to allow anticipation of reactions not seen or discussed. That is, the principles are general. For example, combination of an acyloin-type process (Table 9.9, item 6, Scheme 9.145) with the chemistry of the nitriles adumbrated in Table 9.10, items 6 and 7, and Schemes 9.158 and 9.159 would lead to an understanding of the **Thorpe–Ziegler** reaction. Thus, as shown in Scheme 9.160, 2,6-dicyano-2-methylhexane on treatment with base, followed by acid

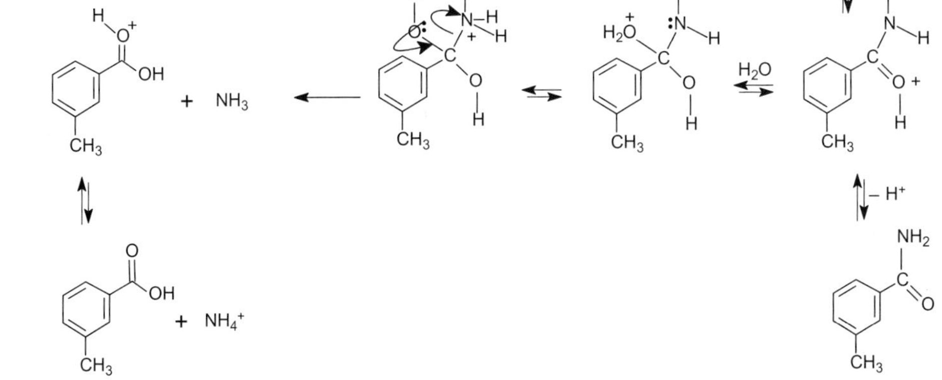

Scheme 9.158. A representation of the pathway for the hydrolysis of 3-methylbenzonitrile under acidic conditions.

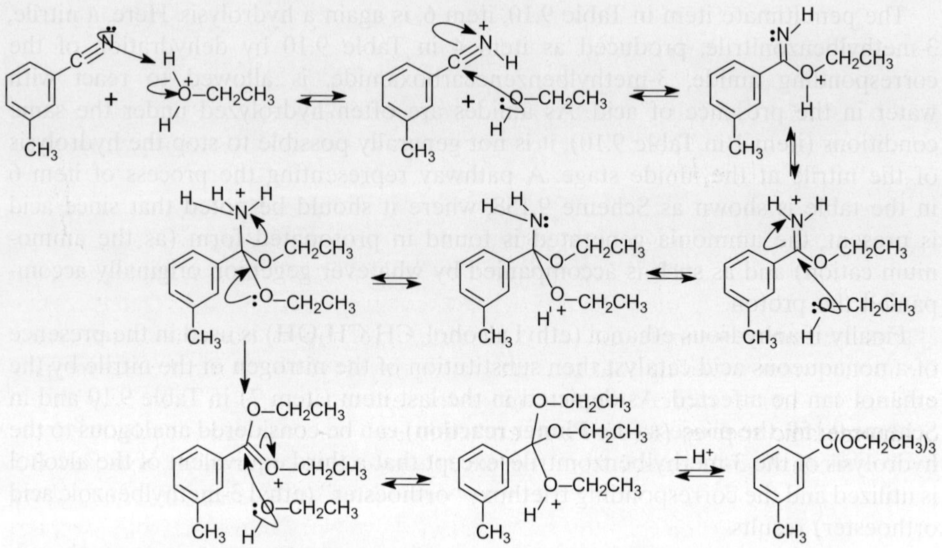

Scheme 9.159. A representation of the pathway for the ethanolysis of 3-methylbenzonitrile under acidic conditions to provide the corresponding orthoester (the **Pinner reaction**). There are some circumstances under which the reaction can be stopped at the imino ether stage where the choice is then available as to whether or not to finish making the orthoester. See Pinner, A. *Chem. Ber.*, **1883**, *16*, 356, and earlier papers.

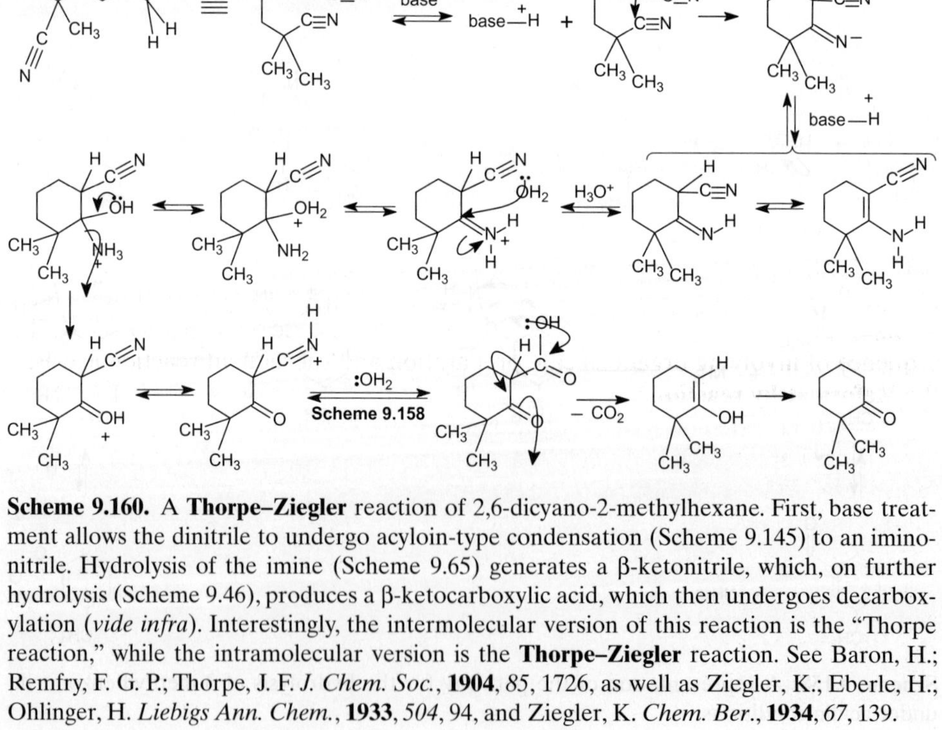

Scheme 9.160. A **Thorpe–Ziegler** reaction of 2,6-dicyano-2-methylhexane. First, base treatment allows the dinitrile to undergo acyloin-type condensation (Scheme 9.145) to an iminonitrile. Hydrolysis of the imine (Scheme 9.65) generates a β-ketonitrile, which, on further hydrolysis (Scheme 9.46), produces a β-ketocarboxylic acid, which then undergoes decarboxylation (*vide infra*). Interestingly, the intermolecular version of this reaction is the "Thorpe reaction," while the intramolecular version is the **Thorpe–Ziegler** reaction. See Baron, H.; Remfry, F. G. P.; Thorpe, J. F. *J. Chem. Soc.*, **1904**, *85*, 1726, as well as Ziegler, K.; Eberle, H.; Ohlinger, H. *Liebigs Ann. Chem.*, **1933**, *504*, 94, and Ziegler, K. *Chem. Ber.*, **1934**, *67*, 139.

hydrolysis, yields 2,2-dimethylcyclohexanone. Scheme 9.160 shown for this process incorporates many of the individual steps shown earlier.

Problem 9.25. Based on the examples prior to that seen in Scheme 9.160, predict the nature of the base utilized to abstract the proton α- to the cyano group in the first step of Scheme 9.160.

Similarly, the understanding that esters react with Grignard reagents (e.g., methylmagnesium bromide, "CH$_3$MgBr") and alkyllithium reagents (e.g., methyllithium, "CH$_3$Li"), as shown in Table 9.9, item 4, and Scheme 9.144, in processes that are not always amenable to simple analysis is important. Its importance derives, first, from the further understanding that substitution of the alkyl group for the leaving alkoxide is apparently more difficult than subsequent reaction of the ketone with the organometallic species (resulting in an "addition" reaction and formation of the tertiary alcohol) and, second, from the hope that new questions might be asked. For example, it might be wondered that since α-haloesters are easily prepared (e.g., via the **Hell–Volhard–Zelinsky** reaction, Chapter 8, Scheme 8.49, and this chapter, Scheme 9.104, and subsequent esterification, Table 9.6, item 2, Scheme 9.117), what would happen if one attempted to prepare an organometallic reagent with magnesium (Mg) or lithium (Li) metal from, for example, ethyl bromoethanoate (ethyl bromoacetate, BrCH$_2$CO$_2$CH$_2$CH$_3$). It might also be wondered if it is possible to find a way to add only one alkyl species and have the reaction *stop* before the resulting carbonyl compound reacts further.

Generally, organometallic reagents cannot be prepared from ethyl bromoethanoate (ethyl bromoacetate, BrCH$_2$CO$_2$CH$_2$CH$_3$) and similar α-haloesters. It appears that as the organometallic reagent begins to form, it reacts at the carbon of the carbonyl of ester substrate, which has not yet formed organometallic species but which retains the active halogen, and after several such steps an intractable mixture results. However, if zinc (Zn) metal is substituted for magnesium (Mg), it is often possible to form the corresponding organozinc derivative, which, being less reactive than the mangesium (Mg) species, does not react with unreacted ester and can be induced, subsequently, to react with (more reactive) aldehydes and ketones such as cyclohexanone (Scheme 9.161). It is common to disguise the ignorance about the path by which such organometallic species form with the usual curved arrows. However, it is much more likely that the organometallic reagents form by a series of one-electron transfer steps such as those shown in Chapter 7 (Scheme 7.3). The sequence of involving organozincate preparation and subsequent reaction is called the **Reformatsky reaction.***

The second problem, that of addition of an alkylating species to the carbon of the carbonyl of the carboxylic acid but with only substitution occurring and without further reaction, can occasionally be accomplished by converting the carboxylic acid (rather than an ester of the acid) to the corresponding lithium salt. Although this salt is written as a monomeric species, it is probably present as aggregates in the

---

*The Reformatsky reaction was first reported in 1887 (Reformatsky, S. *Chem. Ber.*, **1887**, *20*, 1210), and its utility was emphasized in subsequent reviews, the most recent of which (Rao, Y. S. *Chem. Rev.*, **1964**, *64*, 353) is nonetheless not current. More current information can be found in Orsini, F.; Sello, G. *Curr. Org. Chem.*, **2004**, *1*, 111.

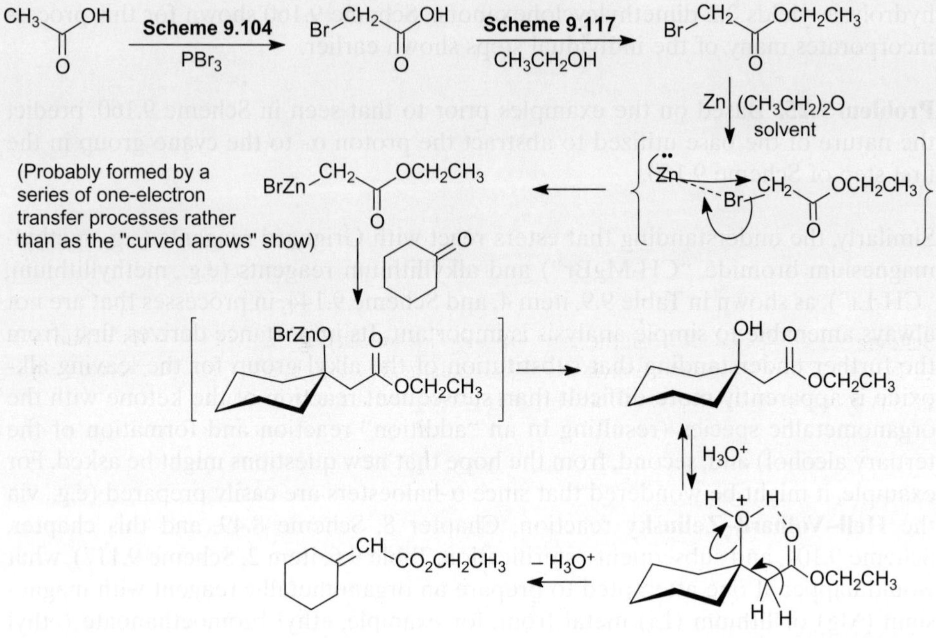
Scheme 9.104
PBr3
Scheme 9.117
CH3CH2OH
Zn (CH3CH2)2O
solvent
(Probably formed by a
series of one-electron
transfer processes rather
than as the "curved arrows" show)
H3O+
– H3O+

Scheme 9.161. The Reformatsky reaction. A representation of the conversion of an α-haloester to the corresponding zincate, which, less reactive than the corresponding magnesium or lithium reagents, does not react with the ester functional group. The reaction with a ketone (cyclohexanone) of the typical ethyl bromo-zincethanoate (ethyl bromozincacetate, $BrZnCH_2CO_2CH_2CH_3$) is shown.

etherial solvent (such as THF [oxacyclopentane]) in which the reaction is carried out. Addition of alkyllithium (again written as monomeric species) to the carbon of the carbonyl produces a product which can be written as the dilithium salt. The addition reaction stops and subsequent hydrolysis produces the ketone. An example of such a process is shown in Scheme 9.162 for a reaction between the lithium salt of benzene carboxylic acid (benzoic acid) and methyllithium.

Discussion of such addition processes would be incomplete without noting that carbon dioxide (CO_2) is a carbonyl compound. So, it should not be surprising that addition reactions can also be effected between carbon dioxide (CO_2) and organometallic compounds such as **Grignard reagents**. Thus, as shown in Scheme 9.163, addition of phenylmagnesium bromide (C_6H_5MgBr), prepared in ether, $(CH_3CH_2)_2O$, by reaction between magnesium metal (Mg) and bromobenzene (C_6H_5Br) and written as the monomeric species, to powdered dry ice ($CO_{2\,(s)}$) or passage of gaseous carbon dioxide (CO_2) into the solution produces the magnesium bromide salt of benzenecarboxylic acid (benzoic acid), which, on addition of mineral acid (e.g., aqueous hydrochloric acid [HCl]), liberates the carboxylic acid itself.

Catalytic addition of aryl halides to carbon monoxide (CO) can also be effected. Thus, in the presence of a palladium catalyst, an alcohol, and base (to remove hydrogen halide) esters of carboxylic acids are produced. The process presumably involves insertion of the catalyst into the carbon halogen bond and subsequent reaction with palladium-coordinated carbon monoxide. A representation of the process is pre-

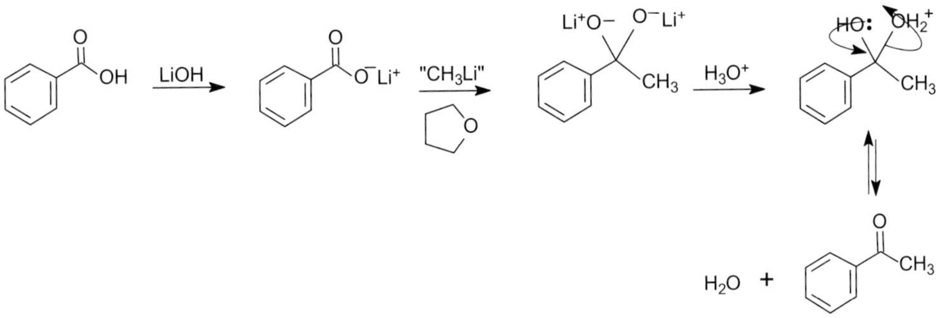

Scheme 9.162. A pathway for the "substitution" of an alkyl group for an –OH group of a carboxylic acid (benzenecarboxylic acid [benzoic acid]). The lithium salt of the carboxylic acid is thoroughly dried prior to attempted reaction with methyllithium.

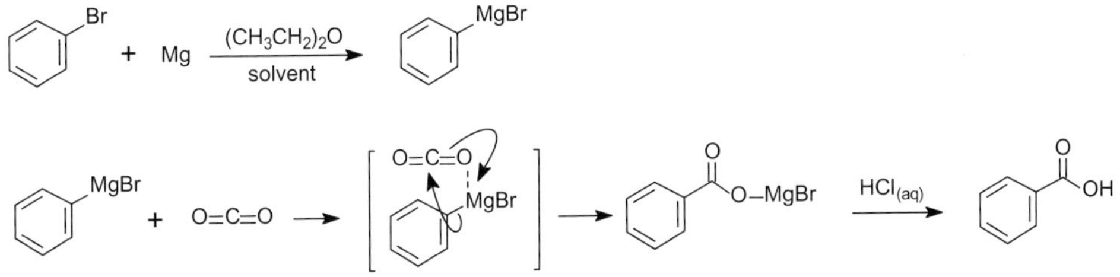

Scheme 9.163. A representation of the formation of benzenecarboxylic acid (benzoic acid, $C_6H_5CO_2H$) by reaction of the **Grignard reagent** [formed from bromobenzene (C_6H_5Br) and magnesium metal (Mg) in ether, $(CH_3CH_2)_2O$ solution] with carbon dioxide (CO_2).

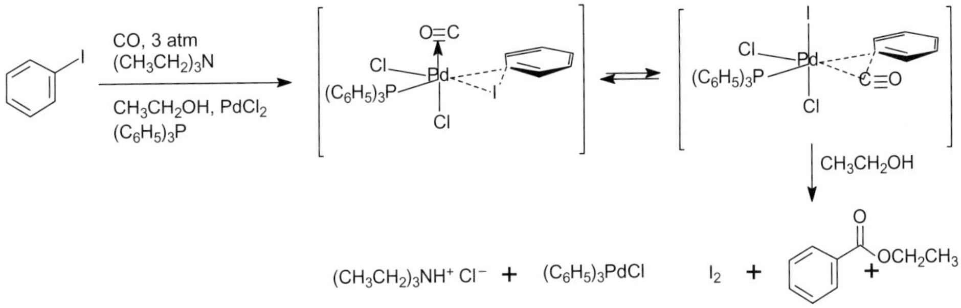

Scheme 9.164. A representation of the conversion of an aryl halide (iodobenzene) into the ester of a carboxylic acid (ethyl benzenecarboxylate [ethyl benzoate]) by reaction with carbon monoxide (CO) in the presence of alcohol (ethanol, ethyl alcohol, CH_3CH_2OH), triethylamine [$(CH_3CH_2)_3N$], and palladium(II) chloride ($PdCl_2$). See Kayaki, Y.; Noguchi, Y.; Iwasa, S.; Ikariya, T.; Noyori, R *Chem. Comm.*, **1999**, 1235.

sented as Scheme 9.164, which shows the reaction utilizing iodobenzene as the substrate and the formation of ethyl benzenecarboxylate (ethyl benzoate) as the product. The details remain vague.

No discussion of the reactions of carbon dioxide (CO_2) would be complete without what can arguably be the most important enzymatically catalyzed process

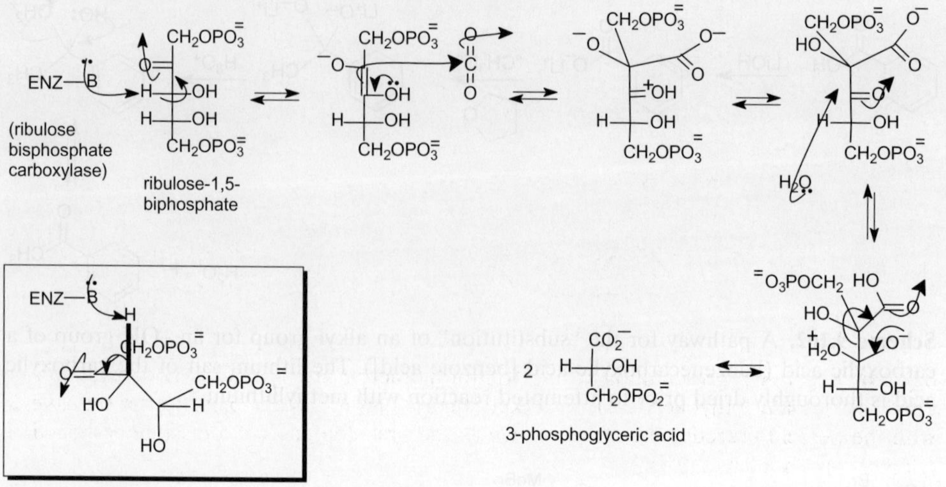

(ribulose bisphosphate carboxylase)

ribulose-1,5-biphosphate

3-phosphoglyceric acid

Scheme 9.165. A portion of the **Calvin cycle** showing the use of carbon dioxide in the enzyme-catalyzed reaction with ribulose-1,5-biphosphate.

and without which life as we know it could not exist. As will be discussed in more detail in Chapter 11, the turning of the **Calvin cycle*** requires the entry of carbon dioxide (CO_2) in the presence of the enzyme ribulose bisphosphate carboxylase to catalyze the reaction with ribulose-1,5-bisphosphate. The process is thought to involve enolization of the ribulose-1,5-bisphosphate followed by addition of the enolate anion to the carbon of the carbonyl of carbon dioxide (CO_2) and subsequent reaction with water to cause fragmentation of the adduct into two three-carbon fragments. A representation of the process is shown in Scheme 9.165.

E. ADDITIONAL REACTIONS AND REARRANGEMENTS OF ESTERS AND β-DICARBONYL COMPOUNDS

It is now necessary to consider three more topics broadly grouped under **addition and elimination**. First, there is the topic of **conjugate addition** for α,β-unsaturated carbonyl compounds; second, although arguably a substitution process, there is the topic of **"addition" occurring α to the carbon of the carbonyl**; and third, there are the special cases of additions occurring between the carbonyl groups of **β-dicarbonyl compounds**, that is, α to **two** carbonyl groups.

Some examples of conjugate addition have already been encountered. For example, the **Michael addition** of cyanide anion (CN^-) to 3-buten-2-one (methyl vinyl ketone, $CH_2CHCOCH_3$) was illustrated in Equation 9.37. Another **Michael addition** example was given in Equation 9.38. Additionally, in Equation 9.44, the

---

*A cycle, by definition, can have no beginning and no end. So, materials are added and withdrawn tangentially as the cycle turns. Many of the features of the **Calvin cycle** were worked out by Melvin Calvin (Nobel Prize 1961), James Bassham, and Andrew Benson (Harvard University) in the decade immediately following World War II. More information about the **Calvin cycle** will be presented in Chapter 11. The "classic" report is Bassham, J.; Benson, A.; Calvin, M. *J. Biol. Chem.*, **1950**, *185*, 781.

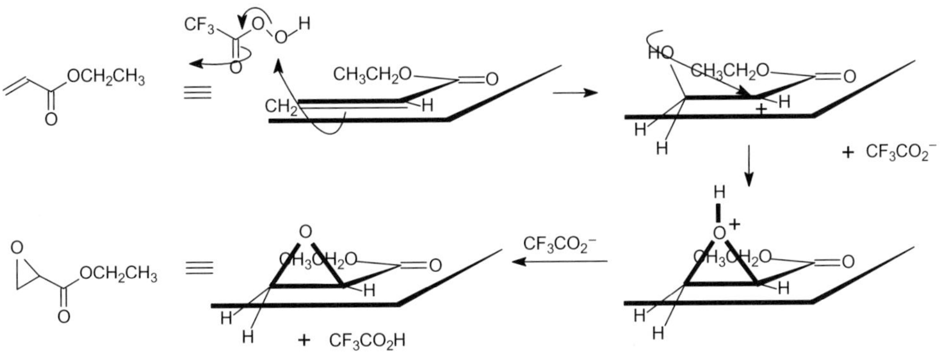

Scheme 9.166. A representation of the trifluoroperacetic acid (CF$_3$CO$_3$H) epoxidation of ethyl propenoate (ethyl acrylate). It is likely that (a) an initial complexation of the alkene with the peracid precedes the attack shown as the first step and (b) a discrete carbocation (shown) may not actually form.

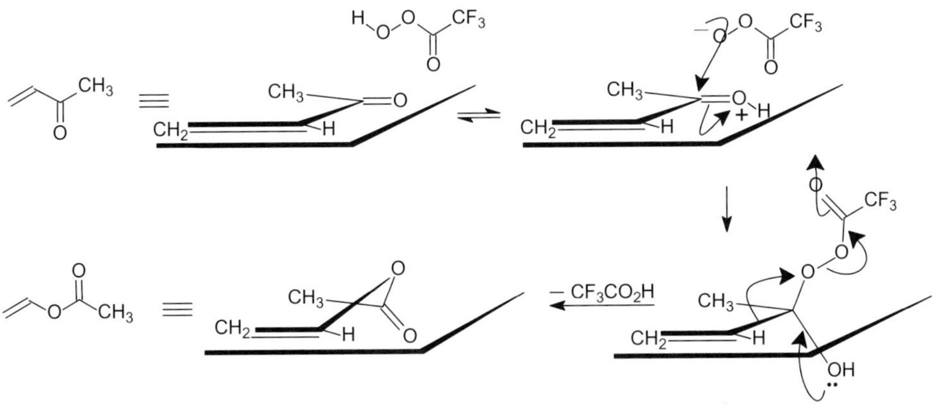

Scheme 9.167. A representation of the trifluoroperacetic acid (CF$_3$CO$_3$H) promoted Baeyer–Villiger oxidation of 3-buten-2-one (methyl vinyl ketone, CH$_2$=CHCOCH$_3$).

difference between **Gilman** and **Grignard reagents** with regard to α,β-unsaturation was illustrated with the observation that the former yielded significant amounts of conjugate addition while the latter yielded predominately direct addition to the carbonyl. Additional examples, for example, Scheme 9.57 and 9.52, serve to emphasize reactivity differences. Now, it is also important to notice that α,β-unsaturated aldehydes and ketones are *different* from α,β-unsaturated esters since the additional oxygen in the latter diminishes the extent to which the carbon of the carbonyl bears a positive charge. Thus, although conjugation with the carbonyl in the case of α,β-unsaturated esters reduces the reactivity of the double bond toward epoxidation with peracids (Chapter 6, Table 6.1, item 16, and Equation 6.63), the reaction succeeds (Scheme 9.166). In contrast, α,β-unsaturated ketones undergo the **Baeyer–Villiger oxidation** (Equation 9.6) preferentially (Scheme 9.167); thus epoxidation of the latter requires a **nucleophilic** reagent for success and, typically, the anion of hydrogen peroxide (HOO$^-$) is utilized (Schemes 9.10 and 9.168). Interestingly, in

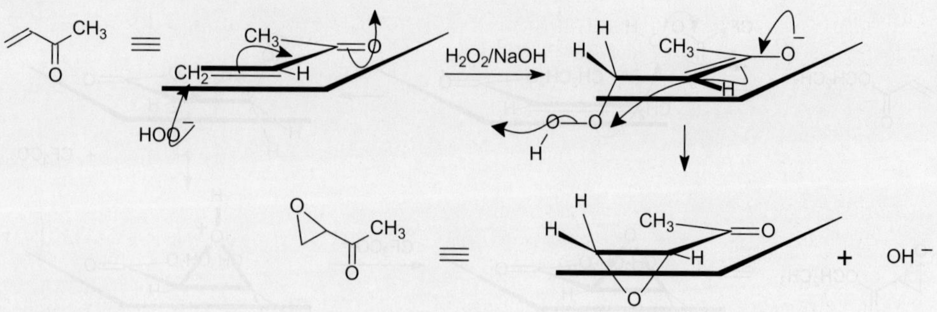

Scheme 9.168. A representation of the epoxidation of 3-butene-2-one (methyl vinyl ketone, CH₂=CHCOCH₃) with basic hydrogen peroxide (a source of HOO⁻).

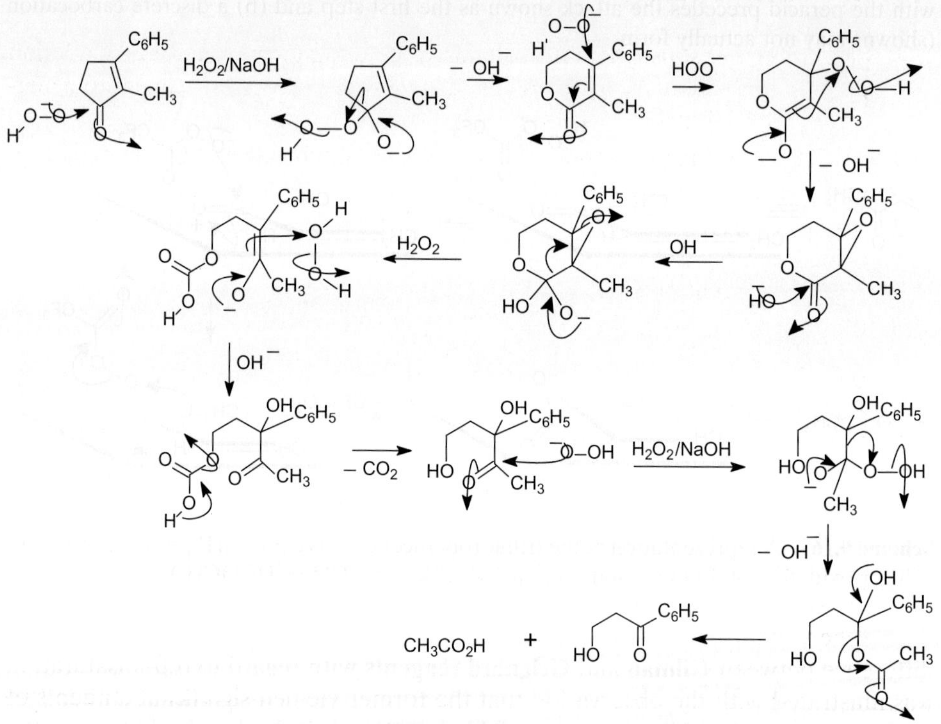

Scheme 9.169. A representation of the basic hydrogen peroxide (H₂O₂/NaOH) oxidation of 2-methyl-1-phenylcyclopentene-3-one involving epoxide formation as well as Baeyer–Villiger oxidation and elimination reactions to produce carbon dioxide (CO₂), ethanoic acid (acetic acid, CH₃CO₂H), and 1-phenyl-3-hydroxypropanone.

some cases (Scheme 9.169) it appears that both epoxidation and Baeyer–Villiger oxidation can occur even with the hydrogen peroxide anion (HOO⁻).

Problem 9.26. The treatment of 3-buten-2-one (methyl vinyl ketone, CH₂=CHCOCH₃) with *m*-chloroperbenzoic acid in methylene chloride (CH₂Cl₂) actually results in the formation of the epoxide shown below. Given the information in Scheme 9.167 and your general knowledge of the material on epoxidation (from Chapter 6), how do you account for the product?

Many reactions beginning with initial proton abstraction α- to the carbonyl have already been seen earlier in this chapter, Additional examples (as shown below) also abound.

It is important that the reader understand the principle of removal of the α-proton and the subsequent nucleophilic nature of the carbanion (enolate anion) so generated.

Some specific examples already discussed and which should be remembered and reviewed if necessary include:

1. Formation and subsequent destruction of an enolate anion through oxidation with the molybdenum pentoxide/pyridine/HMPA (MoOPh) reagent to generate acyloins (Equation 9.11).
2. The **Baylis–Hillman** reaction which results in alkylation α- to the carbonyl in an α,β-unsaturated system (Scheme 9.52).
3. The use of **Grignard reagents**, in limited quantity to remove acidic protons α- to the carbonyl rather than adding to the carbonyl group as might be expected (Equation 9.42).
4. The **Darzens glycidic ester condensation** (Scheme 9.54).
5. The use of enamines to secure a specific α-alkylation site (Scheme 9.66).
6. The entire manifold of **aldol-type** condensation reactions. The aldol condensation itself (Equation 9.45, *et. seq*) and all of those in Table 9.5 (viz. the **Claisen, Stobbe, Perkin,** and **Knovenagel condensation** reactions and the **Robinson annulation**).
7. The internal alkylation seen in the **Favorskii** reaction (Scheme 9.96).
8. In Chapter 8, Scheme 8.49, halogenation α- to the carbonyl, which is repeated (albeit under different conditions) in the **Hell–Volhard–Zelinsky** reaction in Scheme 9.104.

With regard to the special cases of carbanions (enolate anions) lying between two carbonyl groups, it is important to understand that while one carbonyl group dramatically increases the acidity of protons attached to carbons flanking the carbonyl, perhaps by as much as 20 pKa units; that is, the pKa of methane, CH_4, is estimated at ca. 40 and that of propanone (acetone, CH_3COCH_3) at ca. 20. Two carbonyl groups have a smaller but quite profound similar effect. The proton on the methylene ($-CH_2-$) flanked by two carbonyl groups in diethyl 1,3-propandioate [diethyl malonate, $(CH_3CH_2O_2C)_2CH_2$] has a pKa of about 13, that of ethyl 3-ketobutanoate (ethyl acetoacetate, $CH_3COCH_2CO_2CH_2CH_3$) is about 11, and that of 2,4-pentanedione ($CH_3COCH_2COCH_3$), under the same conditions, is about 9. Thus, the process of the **Knovenagel condensation** (item 5 in Table 9.5 and Scheme 9.84), an early example of condensation between an aldehyde and diethyl propane-1,3-dicarboxylate (diethyl malonate) can be considered as a beginning.

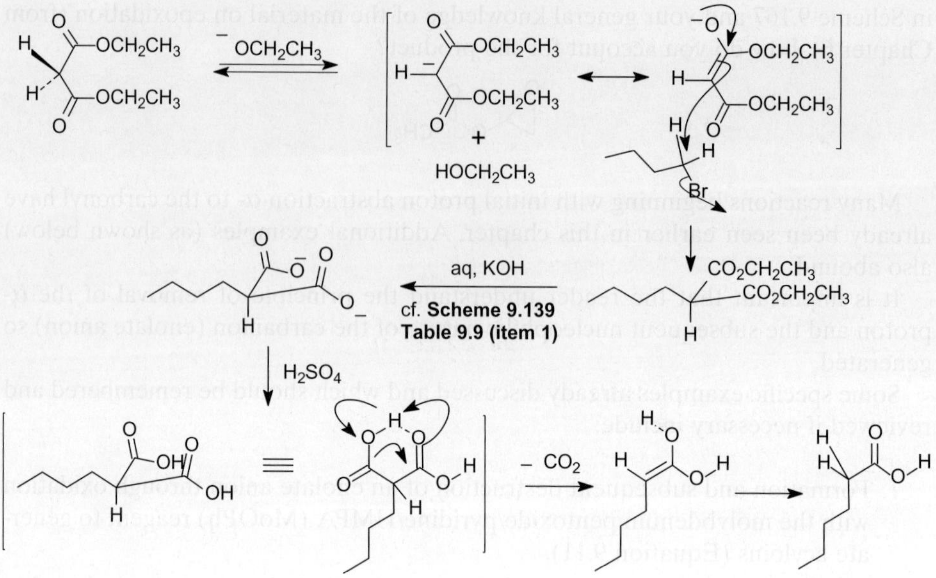

Scheme 9.170. A representation of the alkylation of diethyl propane-1,3-dicarboxylate (diethyl malonate) with 1-bromobutane ($CH_3CH_2CH_2CH_2Br$) in an S_N2-type process using sodium ethoxide in ethanol solution to generate the anion of the diester. The diethyl heptanedioate on hydrolysis undergoes decarboxylation when the dianion is neutralized.

Diethyl propane-1,3-dicarboxylate (diethyl malonate) reacts with bases such as sodium ethoxide in ethanol solution to generate the corresponding anion. The anion is not only able to add to the carbon of carbonyl groups, but it is *also capable of serving as an electrophile in the S_N2 type reaction*. Thus, the anion can be alkylated and, as shown in Scheme 9.170, reaction with 1-bromo-butane produces the corresponding diethyl heptanedioate. The value of this **general process of alkylation of malonate diesters** to the chemical community is increased by noting that basic hydrolysis of the diester to the salt of the dicarboxylic acid, followed by acidification and gentle warming (often the heat evolved in acidification is sufficient), results **in decarboxylation (loss of CO_2)** and formation of, in this case, hexanoic acid (Scheme 9.170). Thus, the synthesis of specific carboxylic acids (and other compounds of value, *vide infra*) is made easy.

The corresponding alkylation of the dinitrile [malonodinitrile, dicyanomethane, $CH_2(CN)_2$] is also easily effected, but, of course, the base must be changed to one that will not attack the cyano functionality and, as shown in Equation 9.107, sodium hydride (NaH) in dimethyl sulfoxide [DMSO, $(CH_3)_2SO$] is frequently the base (and solvent) of choice.

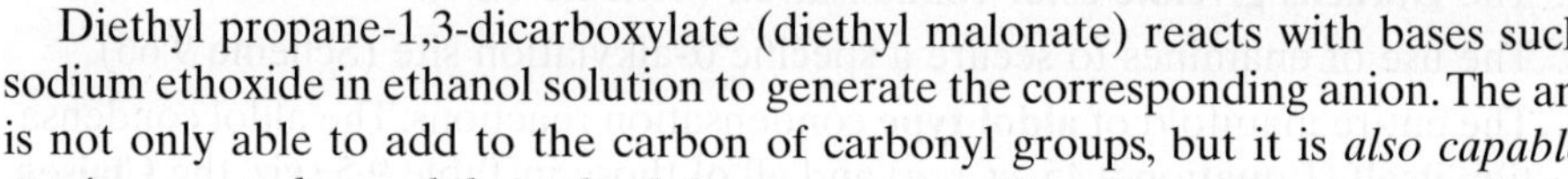

$$(9.107)$$

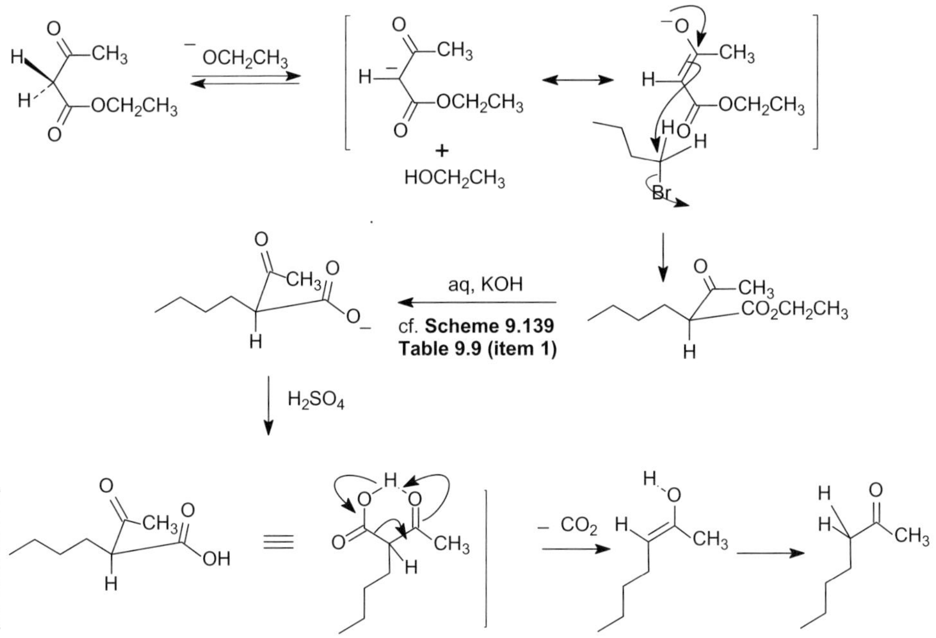

Scheme 9.171. A representation of the alkylation of ethyl 3-ketobutanoate (ethyl acetoacetate) with 1-bromobutane (CH₃CH₂CH₂CH₂Br) in an S_N2-type process using sodium ethoxide in ethanol solution to generate the anion of the ester. The ethyl ester of 2-acetylhexanoate (3-carboethoxy-2-heptanone) on hydrolysis undergoes decarboxylation when the carboxylate anion is neutralized.

Problem 9.27. Sodium hydride (NaH) reacts with dimethyl sulfoxide [(CH₃)₂SO] to produce hydrogen (H₂) and the dimsyl anion sodium salt (Na⁺ ⁻CH₂SOCH₃). Presuming that the proton abstracting species is the dimsyl anion, write a pathway with curved arrows (cf. Scheme 9.170) for the reaction of malonodinitrile [dicyanomethane, CH₂(CN)₂] with 1-bromobutane.

Of course the decarboxylation does not require a **di**carboxylic acid and, in general *any electron accepting functional group β to the carboxylic acid promotes decarboxylation*. Thus, as shown in Scheme 9.171, if the initial alkylation shown in Scheme 9.170 were effected with ethyl 3-ketobutanoate (ethyl acetoacetate), then the ethyl ester of 2-acetylhexanoate (3-carboethoxy-2-heptanone) would be generated. Hydrolysis followed by decarboxylation yields the corresponding 2-heptanone itself. Again, the process is a useful general synthetic method in chemistry and is widely used for the preparation of ketones. However, because base hydrolysis is often complicated by the **retroaldol**-type reaction (Scheme 9.172) occurring simultaneously with the saponification, it is much more common to use acid hydrolysis of the ester. Decarboxylation occurs spontaneously in the acid solution. Additional examples are provided in Schemes 9.173–9.175.

As a consequence of their ready enolization, β-ketoesters can undergo oxidative fragmentation as well as conversion to α,β-unsaturated γ-ketoesters. Thus, when the

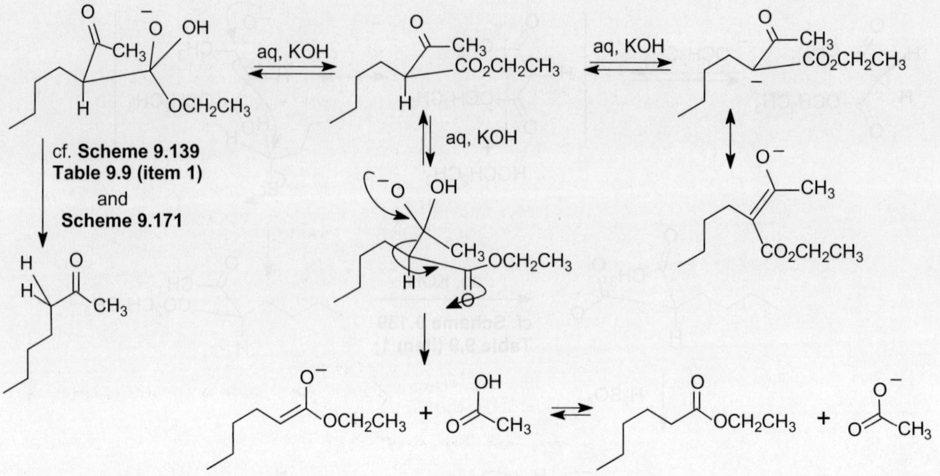

Scheme 9.172. A representation of the retroaldol-type process accompanying alkylation of ethyl 3-ketobutanoate (ethyl acetoacetate) with 1-bromobutane ($CH_3CH_2CH_2CH_2Br$) (Scheme 9.171).

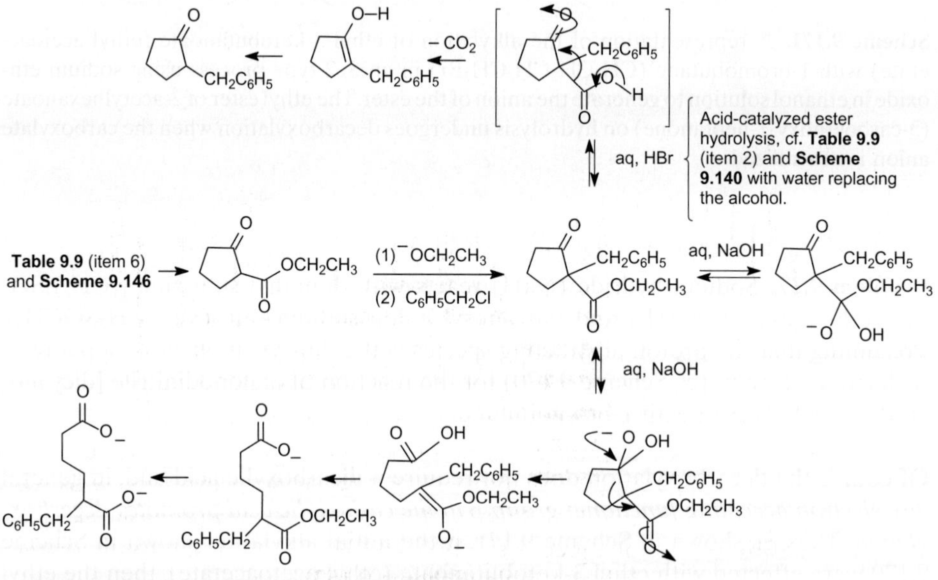

Scheme 9.173. A representation of a comparison between base-catalyzed hydrolysis and decarboxylation (accompanied by a retroaldol-type process), and acid-catalyzed hydrolysis and decarboxylation. Both are shown following alkylation of ethyl 2-carboxycyclopentanone with benzyl chloride.

ethyl ester of cyclohexanone-2-carboxylic acid is treated with permethanoic acid (peracetic acid, CH_3CO_3H), the labile epoxide of the corresponding enol is formed and quickly converted to the corresponding hydroxyketone. With additional permethanoic acid (peracetic acid, CH_3CO_3H) the Baeyer–Villiger Reaction (cf. Scheme 9.8 and Equation 9.6) occurs, the hydroxylactone opens, and the final product is thus

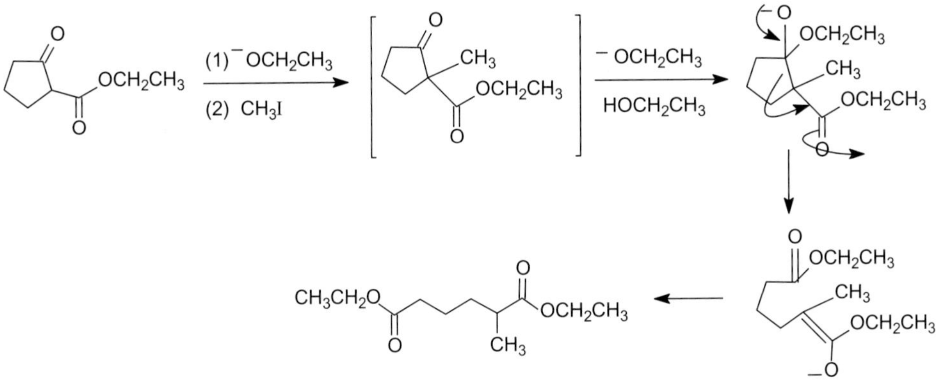

Scheme 9.174. Alkylation accompanied by retroaldol-type process. After the α-proton has been removed, alkylation occurs and subsequently alkoxide attacks the carbon of the carbonyl. The retroaldol type process occurs to produce fragmentation of the ring.

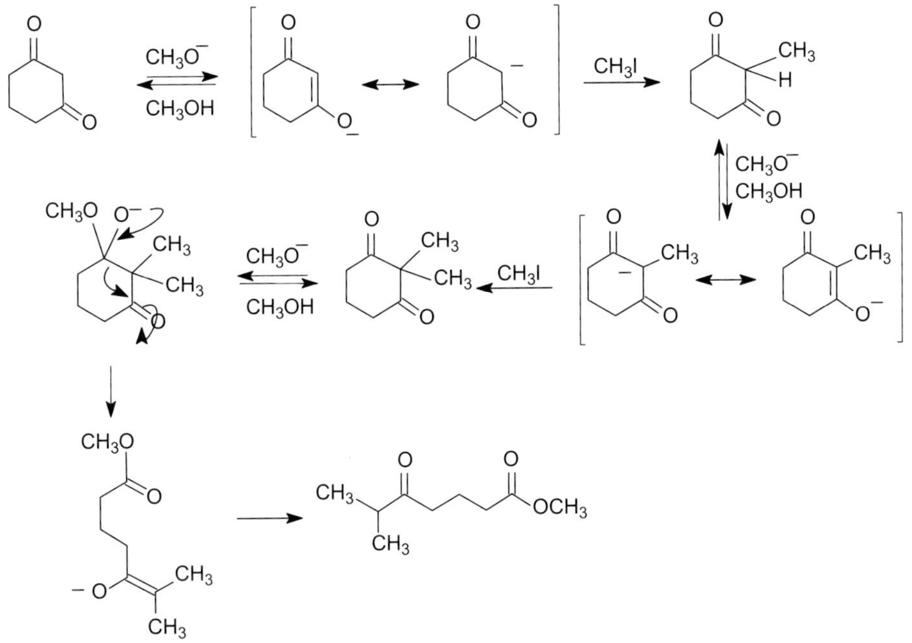

Scheme 9.175. Alkylation accompanied by retroaldol-type process on 1,3-cyclohexanedione. After the α-protons have both been removed and substituted by methyl groups, alkoxide attacks the carbon of the carbonyl and the retroaldol-type process intrudes to produce fragmentation of the ring. The decomposition occurs during the alkylation process.

the half-ester of a new α-ketoacid. The series of reactions described appears in Scheme 9.176.

Similarly, when a β-ketoester, such as ethyl 3-ketobutanoate (ethyl acetoacetate, $CH_3COCH_2CO_2CH_2CH_3$) is treated with the Simmons–Smith reagent (cf. Equation 6.56, item 5 in Table 8.7, and Scheme 8.80), it might be anticipated that the enol of

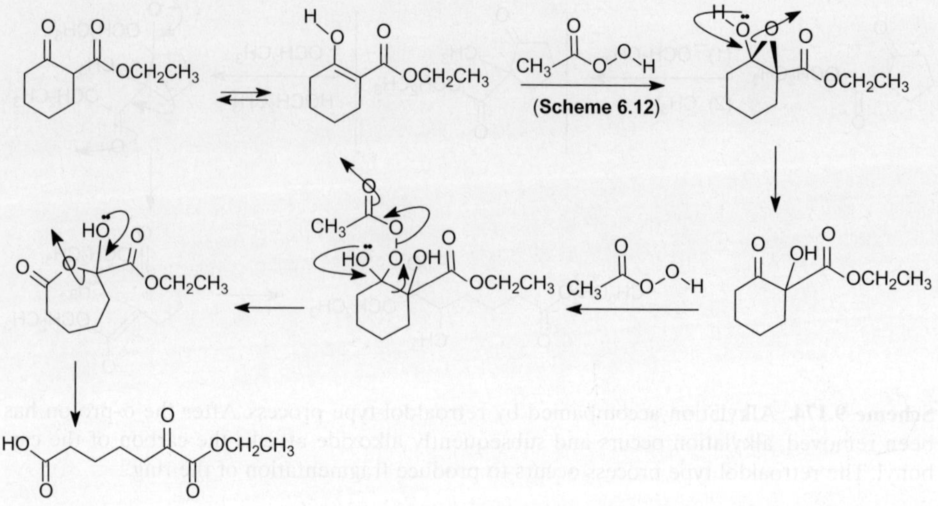

Scheme 9.176. A representation of the reaction of the ethyl ester of cyclohexanone-2-car-boxylic acid with excess permethanoic acid (peracetic acid, CH_3CO_3H). The labile epoxide presumed to be produced from the corresponding enol is expected to rearrange to a ketone, which undergoes the Baeyer–Villiger reaction.

the β-ketoester would yield the corresponding cyclopropane derivative. The fragile cyclopropane does appear to form, but, as with the epoxide (*vide supra*), it apparently undergoes fragmentation and, in the presence of iodine (I_2), small quantities of the iodide, ethyl 2-iodo-4-ketopentanoate, can be realized. If, however, base is added (the base of choice appears to be "DBU," 1,8-diazabicyclo[5.4.0]undec-7-ene) elimination occurs and reasonable quantities of ethyl 4-keto-2-penteneoate are obtained. A representation of a possible pathway accounting for the products is shown in Scheme 9.177.

Interestingly, dicarbonyl compounds that have their carbonyl groups in a 1,5- or 1,6-relationship have been shown to undergo intramolecular pinacol formation (cf. item 6, Table 8.7, and Scheme 8.81 as well as Equation 9.21 for an example showing the preparation of a pinacol). Thus, in the presence of tri-*n*-butyltin hydride [$(CH_3CH_2CH_2CH_2)_3SnH$ or Bu_3SnH] and a free radical initiator such as di(1-cyano-1-methylethyl)diazene (azobisisobutyronitrile, **AIBN**), 1,5-pentanedial (glutaric dialdehyde) undergoes cyclization (via the corresponding 1,3-dioxa-2-stannolane) to the (largely) *cis*-1,2-dihydroxycyclopentane. A representation of a possible pathway from reactants to products is shown in Scheme 9.178.

Problem 9.28. It has recently been pointed out (Yasuda, M.; Chiba, K.; Ohigashi, N.; Katoh, Y.; Baba, A. *J. Am. Chem. Soc.*, **2003**, *125*, 7291) that of the four possible reaction patterns between enolates and α,β-unsaturated carbonyl compounds, as shown below, only the last ("Type D" below) is rare. Why should "Type D" be disfavored relative to the other three?

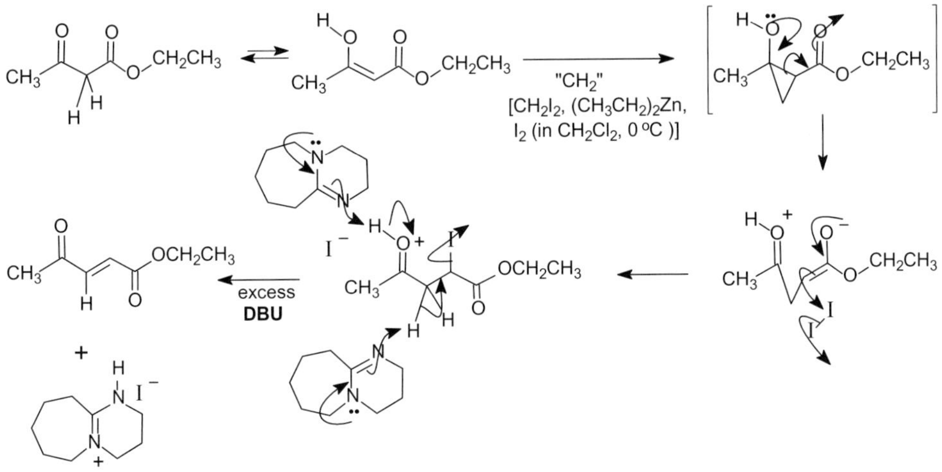

"CH2"
[CH2I2, (CH3CH2)2Zn, I2 (in CH2Cl2, 0 °C)]
excess
DBU

Scheme 9.177. A representation of the conversion of ethyl 3-ketobutanoate (ethyl acetoacetate, $CH_3COCH_2CO_2CH_2CH_3$) to ethyl 4-keto-2-penteneoate. It is presumed that the enol of 3-ketobutanoate proceeds to the product via a cyclopropane intermediate and an α-iodoester, which undergoes elimination of hydrogen iodide (HI) on treatment with DBU (1,8-diazabicyclo[5.4.0]-undec-7-ene).

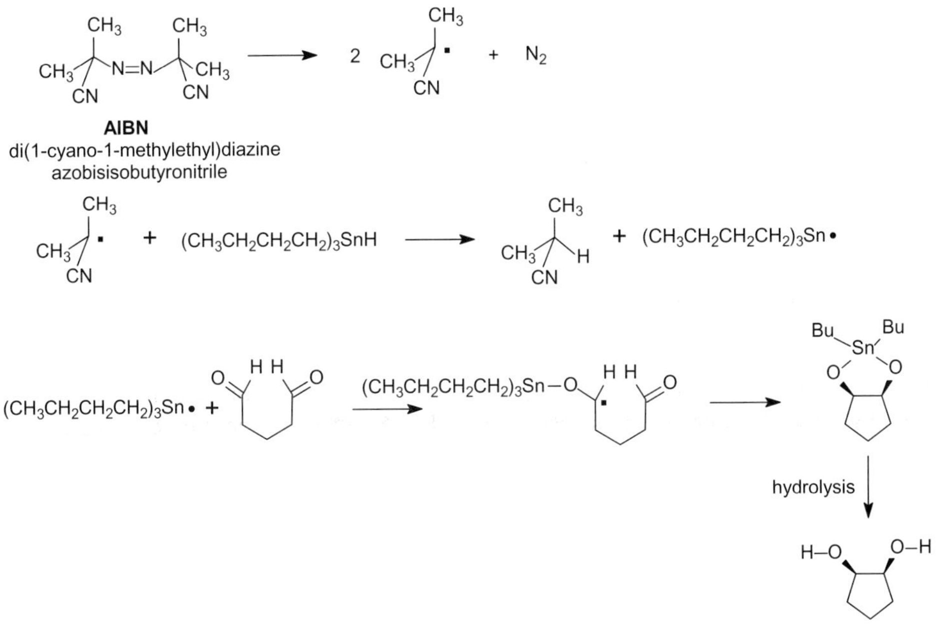

AIBN
di(1-cyano-1-methylethyl)diazine
azobisisobutyronitrile
hydrolysis

Scheme 9.178. A representation of the radical cyclization of 1,5-pentanedial (glutaric dialdehyde) to *cis*-1,2-dihydroxycyclopentane on treatment with tributyltin hydride (catalyzed by AIBN).

Type A

Type B

Type C

Type D

ADDITIONAL PROBLEMS

Problem 9.29. In the discussion of Equation 9.28, it is noted (in particular) that recovered starting material is of the same "optical purity" as when the reaction began. What is the significance of this observation?

Problem 9.30. As pointed out in Scheme 9.74, methylenecyclohexane is the minor product of the dehydration of 1-methylcyclohexanol while 1-methylcyclohexene is the major product. Propose a synthesis of the former in which the latter might not be found at all.

Problem 9.31. As noted in Problem 9.30, the product with the endocyclic double bond (1-methylcyclohexene) is more stable than its exocyclic isomer (methylene–cyclohexane). Using any computational program at your disposal and a small basis set (to conserve machine time), determine the difference in energy between these isomers. Presume that they can be brought into equilibrium. Predict the equilibrium constant.

CHAPTER 10

Part I. The Reactions of Amines: Oxidation, Reduction, Addition, Substitution, and Rearrangement
Part II. Some Organophosphorus Chemistry
Part III. Some Organosilicon Chemistry

He who shall teach a child to doubt, The rotting grave shall ne'er get out.
—William Blake, *Auguries of Innocence,* **1803**

PART I. THE REACTIONS OF AMINES

A. INTRODUCTION

A brief introduction to the material here has been provided in Chapter 5 (Sections VI and VII). That material should be reviewed at this time.

It will be recalled that the basic compounds of nitrogen called amines can be considered as derived from a parent, ammonia (NH_3), by replacement of the hydrogens attached to nitrogen with alkyl and/or aryl groups. Primary amines have only one alkyl or aryl group attached to nitrogen (i.e., RNH_2) while secondary (R_2NH) and tertiary (R_3N) amines have two and three, respectively, such groups attached to nitrogen. Subject to steric and electronic of effects, replacement of hydrogen by alkyl groups results in increased basicity. The groups can be identical or different. The attachment of four groups to the nitrogen produces quaternary ammonium salts ($R_4N^+X^-$), species enjoying the same tetrahedral geometry found in alkanes.

Beyond this, it is worth noting that although there are obvious "electronic" dissimilarities between carbon and nitrogen (the latter having one more electron), which result in major chemical reactivity differences, there are some further structural similarities between the two. Thus, although nitrogen cannot form the long chains and stable rings found for carbon, some simple compounds are analogous.

Foundations of Organic Chemistry: Unity and Diversity of Structures, Pathways, and Reactions,
First Edition. David R. Dalton.
© 2011 John Wiley & Sons, Inc. Published 2011 by John Wiley & Sons, Inc.

For example, nitric oxide (N=O), an odd electron species, resembles carbon monoxide (CO) for which a simple structure satisfying our predilections for single, double, and triple bonds cannot be drawn. Hydroxylamine (H$_2$N–OH) can be thought to resemble methanol (methyl alcohol, CH$_3$OH) despite their major chemical differences, in that one hydrogen of ammonia in the former is replaced by an –OH and, in the latter, one hydrogen of methane is similarly replaced. And so, hydrazine (H$_2$NNH$_2$) and methanamine (methyl amine, CH$_3$NH$_2$) can be viewed in a similar light.

On the other hand, the chemistries of silicon (lying below carbon in the periodic table) and phosphorus (lying beneath nitrogen) are strikingly different. Phosphorus, like nitrogen, can be expected to be found as phosphine (PH$_3$), and as with amines, substituted phosphines and phosphonium salts should be capable of formation. While it might be (correctly) anticipated that with more electrons the chemistry of phosphorus would be more complicated, it might also be (correctly) anticipated that some similarities could be found.

Both phosphines and amines possess some nucleophilic character and while the latter are generally considered to be more basic (nucleophilicity toward protons), the former appear more easily oxidized. Indeed, it is reported that on contact with oxygen, phosphine (PH$_3$) tends to decompose (to oxides and, eventually, to phosphoric acid, H$_3$PO$_4$) and that if the concentration of phosphine (PH$_3$) is much above about 2% in the oxidizing medium, the process can be explosive. Phosphine is reported by some to have the odor of decaying fish at low concentration but others have suggested that the odor is closer to that of garlic (see Chapter 8, Thiols). As noted before (Chapter 5), low-molecular-weight amines are similarly odoriferous.

Amines with sufficient vapor pressure are reported to have unpleasant odors. Some low-molecular-weight amines are reported to have an odor resembling ammonia by some and simply "fishy" by others. Putrefaction of proteins produces toxic materials (ptomaines), and the bis-primary amines 1,4-diaminobutane (H$_2$NCH$_2$CH$_2$CH$_2$CH$_2$NH$_2$, putrescine) and its homologue, 1,5-diaminopentane [H$_2$N(CH$_2$)$_5$NH$_2$, cadaverine], are materials whose trivial names announce their presence in the decaying matter and whose odors are considered so unpleasant that most humans avoid putrescent materials.

In contrast to the case for carboxylic acids (see Chapter 5, Section E and Chapter 10, Table 10.1) and, as recapitulated below, where ionization to the carboxylate ion and the hydronium ion (or a proton) in equilibrium with unionized carboxylic acid defined the equilibrium constant (K_a), in the case of amines, it is the reaction with water to generate the protonated amine and hydroxide ion, which, in equilibrium with the parent amine, defines the equilibrium constant K_b.

So, for carboxylic acids, RCO$_2$H,

$$\text{RCO}_2\text{H} + \text{H}_2\text{O} \xrightleftharpoons{K_a} \text{H}_3\text{O}^+ + \text{RCO}_2^- \qquad (10.1)$$

$$\text{equilibrium constant} \equiv K_a = [\text{H}_3\text{O}^+][\text{RCO}_2^-]/[\text{RCO}_2\text{H}], \qquad (10.2)$$

where as usual, (a) the concentration of water [H$_2$O] is omitted from the denominator because it is considered large and invariant, thus becoming part of K_a, and (b) the use of brackets [] denotes concentrations (usually in mol L^{-1}). Here, obviously, [RCO$_2$H] is the concentration of unionized RCO$_2$H, [RCO$_2^-$] is the concentration of the anion of the acid RCO_2H, and $[H_3O^+]$ is the concentration of the hydronium ion. So

$$-\log_{10}K_a = -\log_{10}[H_3O^+] + \log_{10}\{[RCO_2H]/[RCO_2^-]\} \qquad (10.3)$$

since

$$-\log 10K_a \equiv pK_a \qquad (10.4)$$

and

$$-\log_{10}[H_3O^+] \equiv pH; \qquad (10.5)$$

therefore,

$$pK_a = pH + \log 10\{[RCO_2H]/[RCO_2^-]\}. \qquad (10.6)$$

And, similarly for amines, RNH_2 (R_2NH, and tertiary, R_3N are equally applicable):

$$RNH_2 + H_2O \xrightleftharpoons{K_b} RNH_3^+ + HO^- \qquad (10.7)$$

$$\text{equilibrium constant} \equiv K_b = [RNH_3^+][HO^-]/[RNH_2]; \qquad (10.8)$$

therefore,

$$-\log_{10}K_b = -\log_{10}[HO^-] + \log_{10}\{[RNH_2]/[RNH_3^+]\} \qquad (10.9)$$

since

$$-\log 10K_b \equiv pK_b \qquad (10.10)$$

and

$$-\log 10[HO^-] \equiv pOH, \qquad (10.11)$$

$$pK_b = pOH + \log_{10}\{[RNH_2]/[RNH_3^+]\}. \qquad (10.12)$$

This argument, involving pK_b and pOH, produces a set of numbers analogous to those for pK_a and pH, respectively. While tractable, this requires keeping two sets of processes in mind and, presumably, continued need for assertion of the specific equilibrium constant with which one is dealing. For this and related reasons, as well as the clear understanding that there is a relationship between K_a and K_b, it is generally found that K_a of the conjugate acid of the amine is used to represent its acidity rather than K_b, its basicity (Table 10.1).

Thus,

$$RNH_3^+ + H_2O \xrightleftharpoons{K_a} RNH_2 + H_3O^+ \qquad (10.13)$$

$$\text{equilibrium constant} \equiv K_a = [RNH_2][H_3O^+]/[RNH_3^+]. \qquad (10.14)$$

TABLE 10.1. Representative Examples of Amines[a]

Compound	Name	mp (°C)	bp (°C)	pK_a[b] (K_b)	^{1}H NMR data (δ, ppm TMS = 0.00)[c]	^{13}C and ^{15}N NMR data [δ, ppm TMS = 0.00 (C): NH_3 = 0.00 (N)]
CH_3NH_2	Methanamine (methylamine)	−92.5	−21	10.64 (4.37×10^{-4})	2.47 (CH_3, s, 3H); 0.45 (NH)	28.05 ($^{13}CH_3$); [1.3 (^{15}N)]
$(CH_3)_2NH$	N-methylmethaneamine (dimethylamine)	−96.0	7	10.73 (5.37×10^{-4})	2.21 (CH_3, s, 6H); 0.798 (NH)	38.75 ($^{13}CH_3$); [9.1 (^{15}N)]
$(CH_3)_3N$	N,N-dimethylmethanamine (trimethylamine)	−117	3	9.81 (6.46×10^{-5})	2.05 (CH_3, s, 9H)	47.72 ($^{13}CH_3$); [13.0 (^{15}N)]
$CH_3CH_2NH_2$	Ethanamine (ethylamine)	−80.6	16	10.75 (5.62×10^{-4})	1.12 (CH_3, t, J = 7.04); 2.73 (CH_2, q, J = 7.04); 4.41 (NH_2)	18.83 ($^{13}CH_3$); 36.69 ($^{13}CH_2$); [24.8 (^{15}N)]
$(CH_3CH_2)_3N$	N,N-diethylethanamine (triethylamine)	−115	89	10.64 (4.37×10^{-4})	0.97 (CH_3, t, J = 7.13); 2.43 (CH_2, q, J = 7.13)	11.73 (CH_3); 46.46 (CH_2)
▷—NH_2	Cyclopropylamine (aminocyclopropane)	−50	50		0.197 (CH, m, 2H Z-); 0.322 (CH, m, 2H, E-); 1.59 NH_2; 2.25 (CH, m, C1)	23.90 (CH); 7,30 (CH_2)
NH_2 (cyclohexyl)	Cyclohexylamine (aminocyclohexane)	−17.7	134	10.68 (4.78×10^{-4})	0.88 (H2$_{ax}$, m, 2H); 1.2–1.3 (H4, m, 2H); 1.49 (H3$_{ax}$, m, 2H); 1.52 (H2$_{eq}$, m, 2H); 1.57 (H3$_{eq}$, m, 2H); 2.58 (H1, m, 1H); 2.58 NH_2	24.8 (C3, 2CH_2); 26.6 (C4, CH_2); 36.5 (C2, 2CH_2); 50.3 (C1, CH)
▷N—H	Azacyclopropane (aziridine)		57	8.01 (1.02×10^{-6})	2.60 (CH_2); 0.88 (NH)	18.2 (CH_2)
◇N—H	Azacyclobutane (azetidine) (trimethyleneimine)		63	11.29 (1.94×10^{-3})	3.63 (m, CH_2, 4H); 2.33 (m, CH_2, 2H); 2.08 (s, NH)	48.24 (CH_2); 22.22 (CH_2)

Structure	Name			pK_a (K_a)	^{1}H NMR	^{13}C NMR
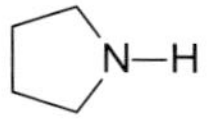	Azacyclopentane (pyrrolidine)		89	11.27 (1.86×10^{-3})	2.64 (m, CH$_2$, 4H); 1.71 (m, CH$_2$, 4H); 0.1 (S, NH)	47.22 (CH$_2$); 25.88 (CH$_2$)
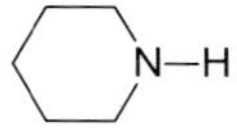	Azacyclohexane (piperidine)	−16	106	11.123 (1.33×10^{-3})	2.78 (m, CH$_2$, 4H); 2.18 (m, NH, 1H); 1.53 (m, CH$_2$, 6H)	47.25 (CH$_2$, 4H); 27.24 (CH$_2$, 2H); 25.21 (CH$_2$, 2H); [^{15}N 37.7 (in C$_6$H$_6$)]
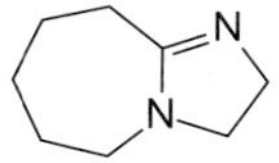	1,8-Diazabicyclo[5.4.0] un-dec-7-ene (DBU)		274	12.8 (6.31×10^{-2})	3.28 (m, C$_9$H$_2$, 2H); 3.21 (m, C^{11}H$_2$, 2H); 3.19 (m, C$_{2,3}$H$_2$, 4H); 2.38 (m, C$_6$H$_2$, 2H); 1.68–1.63 (C$_{4,5}$H$_2$, 4H); 1.58 (C$_{10}$H$_2$, 2H)	161.45 (C7); 52.86 (C11); 48.39 (C2); 44.09 (C9); 37.21 (C6); 29.73 (C5); 28.52 (C3); 26.00 (C4); 22.51 (C10) [190.5 (^{15}N1); 94.7 (^{15}N8)]
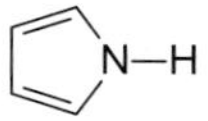	Pyrrole	−23	131	−0.27 (5.37×10^{-15})	7.21 (m, 1H, NH); 6.33 (m, 2H, CH); 6.14 (m, 2H, CH)	117.94 (C1); 107.93 (C2)
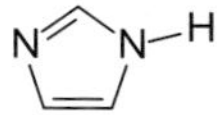	Imidazole	89	256	6.95 (8.91×10^{-8})	7.729 (m, C2, 1H); 7.129 (m, C4,C5, 2H); 1.60 (m, 1H, NH)	135.35 (C2); 121.88 (C4,5)
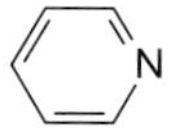	Azabenzene (pyridine)	−42	116	5.25 (1.87×10^{-9})	8.93 (m, H2,6, 2H); 7.62 (m, H4, 1H); 7.23 (m, H3,5, 2H)	149.94 (C2,6); 135.89 (C4); 124.75 (C3,5) [317.2 (^{15}N)]
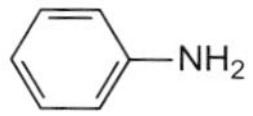	Benzenamine (aniline)	−6	184	4.60 (3.98×10^{-10})	7.12 (m, H3,5, 2H); 6.73 (m, H4, 1H); 6.64 (m, H2,6, 2H); 3.55 (s, NH$_2$)	146.51 (C1); 129.26 (C3,5); 118.39 (C4); 115.07 (C2,C6)
	1,2-Ethanediamine ethylenediamine	8.5	115	9.93 (6.85) (8.51×10^{-5}) (7.08×10^{-8})	2.645 (in D$_2$O)	44.87

(Continued)

TABLE 10.1. *Continued*

Compound	Name	mp (°C)	bp (°C)	$pK_a{}^b$ (K_b)	^{1}H NMR data (δ, ppm TMS = 0.00)^c	^{13}C and ^{15}N NMR data [δ, ppm TMS = 0.00 (C): NH$_3$ = 0.00 (N)]
	Quinoline	−15.6	239	4.81 (6.46×10^{-10})	8.92 (m, H2, 1H); 8.15 (m, H8, 1H); 8.12 (m, H4, 1H); 7.82 (m, H5, 1H); 7.72 (m, H7, 1H); 7.55 (m, H6, 1H); 7.39 (m, H3, 1H)	150.26 (C2); 148.24 (C8a); 135.90 (C4); 129.35 (C7,C8); 128.22 (C4a); 127.72 (C5); 126.43 (C6); 120.95 (C3)
	Indole	52	253	−2.70 (1.99×10^{-17})	7.81 (NH); 7.64 (m, H4, 1H); 7.23 (m, H7, 1H); 7.18 (m, H6, 1H); 7.12 (m, H5, 1H); 7.05 (d, H2, 1H); 6.52 (d, H3, 1H)	135.65 (C7a); 127.73 (C3a); 124.26 (C2); 121.85 (C5); 120.63 (C4); 119.74 (C6); 111.10 (C7); 102.22 (C3)
	Pyrazine	54–56	115	0.65 (−5.78) (4.47×10^{-14}) (1.70×10^{-20})	8.59 (s, 4H)	145.22 (4C)
	Pyrimidine	21	123	1.31 (2.04×10^{-13})	9.27 (m, H2, 1H); 8.78 (m, H4,6, 2H); 7.38 (m, H5, 1H)	159.08 (C2); 156.92 (C4,6); 121.61 (C5); [295.4 (^{15}N)]
	Cytosine	>300		10.28 (4.60) (1.95×10^{-4}) (3.98×10^{-10})	7.50 (H6, d, J = 5.3, 1H); 5.90 (H5, d, J = 5.3, 1H); 8.39 (N1, 1H); 7.69 (NH$_2$, 2H)	168.10 (C4); 159.91 (C2); 144.05 (C6); 95.79 (C5)
	Uracil	316		9.20 (1.58×10^{-5})	11.02 (s, N3, 1H); 10.82 (d, N1, J = 5.7, 1H); 7.41 (dd, C6, J = 5.7, 7.5, 1H); 5.47 (d, C5, J = 7.5,	165.09 (C4); 152.27 (C2); 142.89 (C6); 101.01 (C5)

Structure	Compound	mp (°C)		pK_a (K_a)	^1H NMR	^{13}C NMR
	Thymine	215		9.84 (6.92×10^{-5})	11.0 (s, N1, 1H); 10.6 (s, N3, 1H); 7.28 (s, C6, 1H), 1.75 (s, CH_3, 3H)	164.87 (C4); 151.46 (C2); 137.63 (C6); 107.66 (C5); 11.72 (CH_3)
	Adenine	>300		2.95 (8.91×10^{-12})	12.8 (s, NH, N7); 8.14 (s, C2, 1H); 8.11 (s, C6, 1H); 7.09 (s, NH_2, 2H)	156.36 (C4); 153.41 (C2); 151.71 (C7a); 140.30 (C6); 119.07 (C4a)
	Guanine	>300		9.40 (3.33) (2.51×10^{-5}) (1.74×10^{-11})	10.30 (s, NH, N3); 8.16 (m, NH, N7); 8.16 (m, NH_2, 2H); 8.15 (s, C6, 1H)	168.80 (C4); 162.20 (C7a); 160.10 (C2); 150.10 (C6); 119.60 (C4)
	Morpholine	−7	129	8.33 (2.14×10^{-6})	3.69 (m, J = 11.4, 4.7, 3.71, H2,6, 4H); 2.86 (m, J = 11.1, 4.7, 1.12, H3,5, 4H); 2.59 (s, NH)	68.16 (C2,6); 46.61 (C3,5)
	1,4-Diazabicyclo[2.2.2]-octane (Dabco)	158			4.1 (s, 12H)	55.89 (6C) [11.6 (^{15}N)]
	2-Nitroaminobenzene (2-nitroaniline, *ortho*-nitro-aniline)	71		−0.26 (5.50×10^{-15})	8.06 (m, H3, 1H); 7.43 (NH_2); 7.30 (m, H5, 1H); 6.79 (m, H6, 1H); 6.65 (m, H4, 1H) J_o = 8.5; J_m = 1.5; J_p = 0.5	144.50 (C1); 135.20 (C5); 131.70 (C2); 125.60 (C3); 118.50 (C4); 116.50 (C6)

(Continued)

TABLE 10.1. *Continued*

Compound	Name	mp (°C)	bp (°C)	$pK_a^{\ b}$ (K_b)	^{1}H NMR data (δ, ppm TMS = 0.00)^c	^{13}C and ^{15}N NMR data [δ, ppm TMS = 0.00 (C): NH$_3$ = 0.00 (N)]
H$_2$N—C$_6$H$_4$—NO$_2$	3-Nitroaminobenzene (2-nitroaniline, *meta*-nitro-aniline)	114		$2.47\ (2.95 \times 10^{-12})$	7.43 (m, H2, 1H); 7.411 (NH$_2$); 7.39 (m, H4, 1H); 7.18 (m, H5, 1H); 6.91 (m, H6, 1H) J$_o$ = 8.0; J$_m$ = 2.0; J$_p$ = 0.3	150.05 (C1); 150.00 (C3); 130.69 (C5); 121.27 (C6); 112.17 (C4); 109.00 (C2)
H$_2$N—C$_6$H$_4$—NO$_2$	4-Nitroaminobenzene (4-nitroaniline, *para*-nitro-aniline)	147		$1.00\ (1.00 \times 10^{-13})$	7.72 (m, H3,5, 2H); 6.67 (m, H2,6, 2H); 5.72 (m, NH$_2$); J$_o$ = 9.2; J$_m$ = 2.5; J$_p$ = 0.5	156.85 (C1); 137.32 (C4); 127.55 (C3,5); 113.84 (C2,6)
Isoquinoline	Isoquinoline	26	243	$5.42\ (2.63 \times 10^{-9})$	9.22 (m, H1, 1H); 8.50 (m, H3, 1H); 7.86 (m, H8, 1H); 7.72 (m, H5, 1H); 7.60 (m, H6, 1H); 7.55 (m, H4, 1H); 7.51 (m, H7, 1H)	152.44 (C1); 142.93 (C3); 135.69 (C4a); 128.63 (C8a); 127.50 (C8); 127.15 (C7); 126.37 (C5); 120.38 (C4)

[a]The information presented in this Table is derived from experiment and from the the *CRC Handbook of Tables for Organic Compound Identification*. Z. Rappoport, Compiler, CRC Press, Cleveland, OH **1964**.

[b]The pK_a data are for the corresponding conjugate acid in aqueous solution. $pK_a = -\log K_a = 14 + \log K_b$.

[c]The following abbreviations are used: d = doublet; t = triplet; q = quartet; m = multiplet. Significant portions of the proton (^{1}H) and carbon (^{13}C) data are taken from the SDBS, Integrated Spectral Data Base System for Organic Compounds, National Institute of Advanced Industrial Science and Technology (AIST), Tsukuba, Ibaraki, Japan (http://riodb01.ibase.aist.go.jp/sdbs/cgi-bin/direct_frame_top.cgi). The ^{15}N data shown are taken from Rao, N. S.; Rao, G. B.; Murthy, B. N.; Das, M. M.; Prabhakar, T.; Lalitha, M. *Spectrochim. Acta A*, **2002**, *58*, 2737.

But, as seen earlier

$$\text{equilibrium constant} \equiv K_b = [RNH_3^+][HO^-]/[RNH_2]. \tag{10.15}$$

And

$$K_a K_b = K_w = 1 \times 10^{-14}, \tag{10.16}$$

so

$$K_w = \{[RNH_2][H_3O^+]/[RNH_3^+]\}\{[RNH_3^+][HO^-]/[RNH_2]\} = [H_3O^+][OH^-]. \tag{10.17}$$

Therefore, it is possible to utilize the corresponding protonated amines to define basicity.

The conjugate acids of the weaker bases are more completely ionized and thus have larger K_a's and smaller pK_a's ($-\log_{10}K_a$). The stronger bases have conjugate acids that are less completely ionized so they have smaller K_a's and larger pK_a's. Alternatively, it can be argued that stronger bases hold protons more tightly and therefore the pK_a's of their conjugate acids are larger (the acids, RNH_3^+, are weaker), and weaker bases hold protons less tightly and therefore the pK_a's of their conjugate acids are smaller. For the latter, the acids are "stronger," that is, more fully ionized.

Finally, it has recently been pointed out* that "With a production on million-ton scale per year, aliphatic amines are among the important bulk and fine chemicals in the chemical and pharmaceutical industry. A number of amines, enamines, and imines are useful as pharmaceutically and biologically active substances, dyes, and fine chemicals."

Methods for the preparation of such compounds continue to be developed and will be considered as a part of the material in this chapter.

B. SOME COMMENTS ON THE PREPARATION OF AMINES

It will be recalled that methane (CH_4) can be converted to nitromethane (H_3CNO_2) by the action of nitric acid on methane at high temperature (Chapter 6, Equation 6.10). The nitration of other alkanes by the same process can also be effected, but separation of the multitude of isomers that can be formed (from all but the simplest alkanes) can be difficult. Nitroalkanes, $R–NO_2$ (sometimes accompanied by their corresponding nitrite isomers, $R–ONO$), can also be prepared (as shown in Chapter 7, Table 7.5e) by treatment of alkyl halides with nitrite anion (ONO^-). Reduction of nitroalkanes ($R–NO_2$) with hydrogen (H_2) in the presence of a platinum (Pt) catalyst or lithium aluminum hydride ($LiAlH_4$) in ether produces the corresponding amine (Equation 10.18).

$$CH_3NO_2 + LiAlH_4 \xrightarrow[\text{(2) aqueous hydrolysis}]{\text{(1) ether (solvent)}} CH_3NH_2 + LiOH + Al(OH)_3 \tag{10.18}$$

*Ahmed, M.; Seayad, M. A.; Jackstell, R.; Beller, M. *J. Am. Chem. Soc.*, **2003**, *125*, 10311.

Nitration of arenes (such as benzene) to the corresponding nitro derivatives (such as nitrobenzene, $C_6H_5NO_2$) was introduced during the discussion of electrophilic aromatic substitution (Chapter 6, Section E[IV] (a)). As with nitroalkanes, reduction of the nitro group can be effected to generate the corresponding aryl amines. **However**, unlike nitroalkanes, lithium aluminum hydride ($LiAlH_4$) **cannot be used** since if reduction is attempted with this reagent, complicated mixtures of partially reduced compounds are obtained. Reduction methods that are effective in producing amines from nitroarenes include

(a) reduction with hydrogen in the presence of a platinum or nickel catalyst (Equation 10.19) with the choice of catalyst frequently being either convenience or sensitivity of other groups that may be present to the reduction process.

$$\text{(10.19)}$$

H$_2$/Pt catalyst (or H$_2$/Ni catalyst)

(b) reduction with hydrazine (H_2NNH_2) in the presence of a platinum catalyst (Equation 10.20). The reduction presumably involves the catalytic dehydrogenation of the hydrazine (H_2NNH_2) to diimide ($HN=NH$) and hydrogen (H_2). Diimide ($HN=NH$) is subsequently oxidized to nitrogen (N_2) and the nitro group reduced to amine. Water (H_2O) is also formed.

$$\text{(10.20)}$$

H$_2$NNH$_2$/Pt catalyst CH$_3$OH solvent

(c) reduction with metal (e.g., zinc [Zn], tin [Sn], iron [Fe]) and acid, such as HCl (Equation 10.21). In the process, the metal is oxidized and the acid serves as the source of hydrogen. The details of the pathway are not yet clear.

$$\text{(10.21)}$$

Zn/HCl

Although, generally, production of aminobenzene (aniline) and related *N*-arylamines by nitration and subsequent reduction is so common as to far outweigh other methods of formation of such compounds, special circumstances (such as a need to avoid the acidic conditions normally associated with nitronium ion [NO_2^+] formation) occasionally prevail and alternative methods need to be used.

It may be remembered (Chapter 7, Scheme 7.13) that treatment of aryl halides, such as bromobenzene (C_6H_5Br), with liquid ammonia ($NH_{3(l)}$) (Equation 10.22) results in the formation of benzyne (C_6H_4), and the capture of the latter by ammonia (NH_3) produces aminobenzene (aniline).

$$\text{(10.22)}$$

Similarly, **Friedel–Crafts** acylation (Chapter 6 cf. Scheme 6.88) of aromatic systems produces carbonyl compounds, which can be converted to the corresponding oximes with hydroxylamine hydrochloride ($H_2NOH \cdot HCl$) (see, e.g., Scheme 9.65, Table 9.4, item 4). So, acylation of benzene to form phenylethanone (acetophenone, methylphenylketone, $C_6H_5COCH_3$) can be followed by oxime formation.

As shown in Scheme 10.1, reaction between phenylethanone (acetophenone, methylphenylketone, $C_6H_5COCH_3$) with hydroxylamine hydrochloride ($H_2NOH \cdot HCl$) produces *two* oximes: the (*Z*)-isomer, with the hydroxyl on the same side of the carbon–nitrogen bond as the aryl group, and the (*E*)-isomer whose arrangement of substituents is opposite. These isomers are separable in the event both are formed and it is possible to effect their interconversion under acidic conditions. Now, **Beckmann rearrangement** (Scheme 9.99) carried out on the (*E*)-isomer by reaction with trifluoroacetic anhydride $[(CF_3CO)_2O]$ leads to the corresponding acetamide (*N*-phenyltrifluoroacetamide, $C_6H_5NHCOCF_3$), which, on hydrolysis, produces aminobenzene (aniline, $C_6H_5NH_2$) and trifluoroethanoic acid (trifluoroacetic acid, CF_3CO_2H).

Problem 10.1. Predict the products resulting from the Beckmann rearrangement of the (*Z*)-isomer of the oxime derived from reaction of hydroxylamine ($H_2NOH \cdot HCl$) with phenylethanone (acetophenone, methylphenylketone, $C_6H_5COCH_3$).

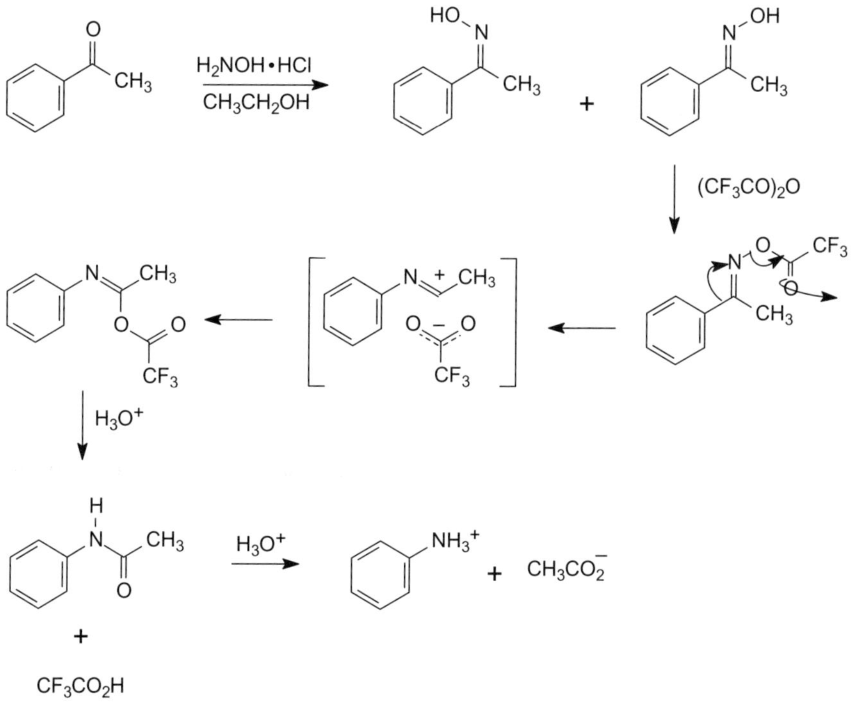

Scheme 10.1. A representation of oxime formation from phenylethanone (acetophenone, methylphenylketone, $C_6H_5COCH_3$) on reaction with hydroxylamine hydrochloride. ($H_2NOH \cdot HCl$), followed by Beckmann rearrangement of the (*Z*)-isomer to produce aminobenzene (aniline, $C_6H_5NH_2$) after hydrolysis.

Interestingly, a host of aryl amines, including diphenylamine (Equation 10.23) and alkyl-substituted aryl amines can now readily be prepared from the corresponding aryl halides on reaction of the latter with amine using a palladium catalyst such as tri-*t*-butylphosphinepalladium(II) in toluene/water with potassium hydroxide (KOH) and an *N,N,N*-trimethylcetylammonium bromide $[C_{16}H_{33}N(CH_3)_3{}^+Br^-]$ phase transfer catalyst (Chapter 5).*

$$\text{(10.23)}$$

Clearly, therefore, the nitro group ($-NO_2$) is not the only source of nitrogen for the production of the amino group ($-NH_2$). Indeed, reduction of azides (RN_3), which might be produced, for example, in nucleophilic substitution reactions using the azide anion (N_3^-) as nucleophile (Table 7.7), with lithium aluminum hydride ($LiAlH_4$), tin (Sn), or zinc (Zn) and acid or even with hydrogen H_2 in the presence of a catalyst (Equation 10.24) also generates the corresponding amine (RNH_2).

$$CH_3CH_2N=N=N \quad \xrightarrow[\substack{\text{Sn/HCl or} \\ \text{Pt/H}_2 \\ \text{after workup}}]{\text{LiAlH}_4 \text{ or}} \quad CH_3CH_2NH_2 + N_2 \qquad \text{(10.24)}$$

Both alkyl ($R-C\equiv N$) and aryl ($Ar-C\equiv N$) nitriles (Chapter 9, Scheme 9.110) (Equation 10.25) and amides ($RCONH_2$ and $ArCONH_2$) (Chapter 9, Scheme 9.110) (Equation 10.26) can also be reduced with lithium aluminum hydride ($LiAlH_4$) or diborane (B_2H_6) in oxacyclopentane (tetrahydrofuran [THF]) or another ether as solvent to produce amines.

$$\text{(10.25)}$$

$$\text{(10.26)}$$

Amides, of course, can also undergo the Hofmann rearrangement (see, e.g., Scheme 9.103) so that amines bearing one carbon less than that obtained on simple reduction can be generated (Equation 10.27). The same isocyanate intermediate has also been seen in Table 9.6 (item 5) and Scheme 9.120, the **Schmidt rearrangement**, Table 9.7 (item 6), and the **Curtius rearrangement**, and the same result (i.e., amine formation) was obtained on hydrolysis.

*Kuwano, R.; Utsunomiya, M.; Hartwig, J. F. *J. Org. Chem.*, **2002**, *67*, 6479.

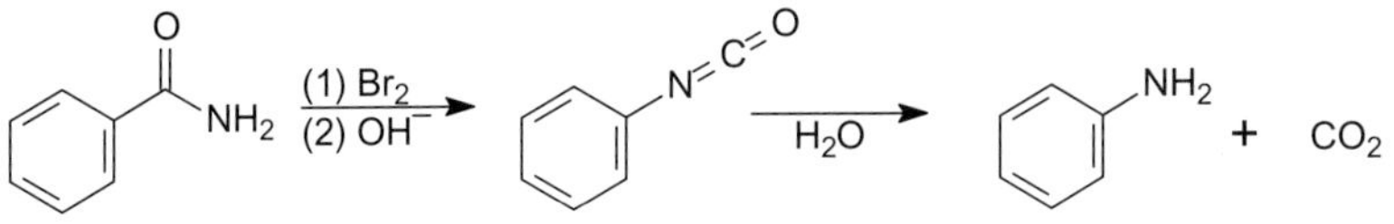

$$\text{(10.27)}$$

Additionally, in this vein, the host of derivatives of carbonyl compounds presented in Chapter 9, Table 9.4, viz. aldimines and ketimines, enamines, hydrazines, and oximes, are all capable of reduction to amines, whose substitution patterns are defined by these precursors.

Finally, it might be suggested (in part by perusal of Table 9.4 as recommended above) that amines could be produced by allowing alkyl halides (R–X, where X is any suitable leaving group) to react with ammonia (NH_3) in an S_N1- or S_N2-type process. Generally, those reactions are unsatisfactory since, for the S_N1-type processes that might occur, ammonia (NH_3) is a strong enough base to remove a proton adjacent to the carbocation and elimination results. On the other hand, for the S_N2-type process, the extent of alkylation of the ammonia (NH_3) is not easily controlled since alkyl amines are, generally, stronger bases and better nucleophiles than ammonia (NH_3). Therefore, mixtures of primary (RNH_2), secondary (R_2NH), tertiary (R_3N), and even quaternary salts ($R_4N^+ X^-$) (X^- now representing any anion associated with the quaternary ammonium species) result.

To avoid this problem, several specialized methods have been developed to introduce alkyl groups selectively to amino groups. Among them are two reactions involving alkylation on nitrogen anions attached to electron-withdrawing species, viz. the **Gabriel synthesis** (Scheme 10.2) and alkylation of sulfonamides produced in the **Hinsberg reaction** (Scheme 10.3).

Thus, as shown in Scheme 10.2, in the **Gabriel synthesis**,* the anion of phthalimide [1,2-benzenedicarboximide or 1H-isoindole-1,3(2H)-dione] generated by treatment of the imide with sodium ethoxide ($Na^+ {}^-OCH_2CH_3$) in ethanol (CH_3CH_2OH) is

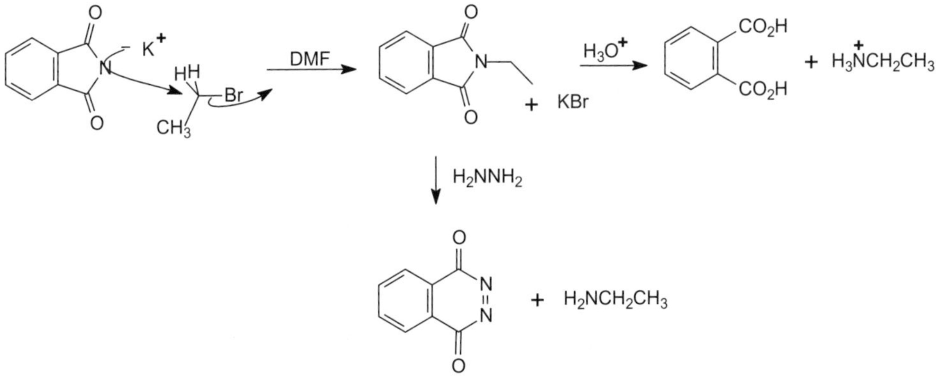

Scheme 10.2. The **Gabriel synthesis** using bromoethane (CH_3CH_2Br) on reaction with phthalimide [1,2-benzenedicarboximide or 1H-isoindole-1,3(2H)-dione] in the presence of base to produce ethanamine. The solvent is shown as dimethylformamide (DMF).

*The original work: Gabriel, S. *Chem. Ber.*, **1887**, *20*, 2224. A review is found at Mitsunobu, O. *Comp. Org. Syn.*, **1991**, *6*, 79.

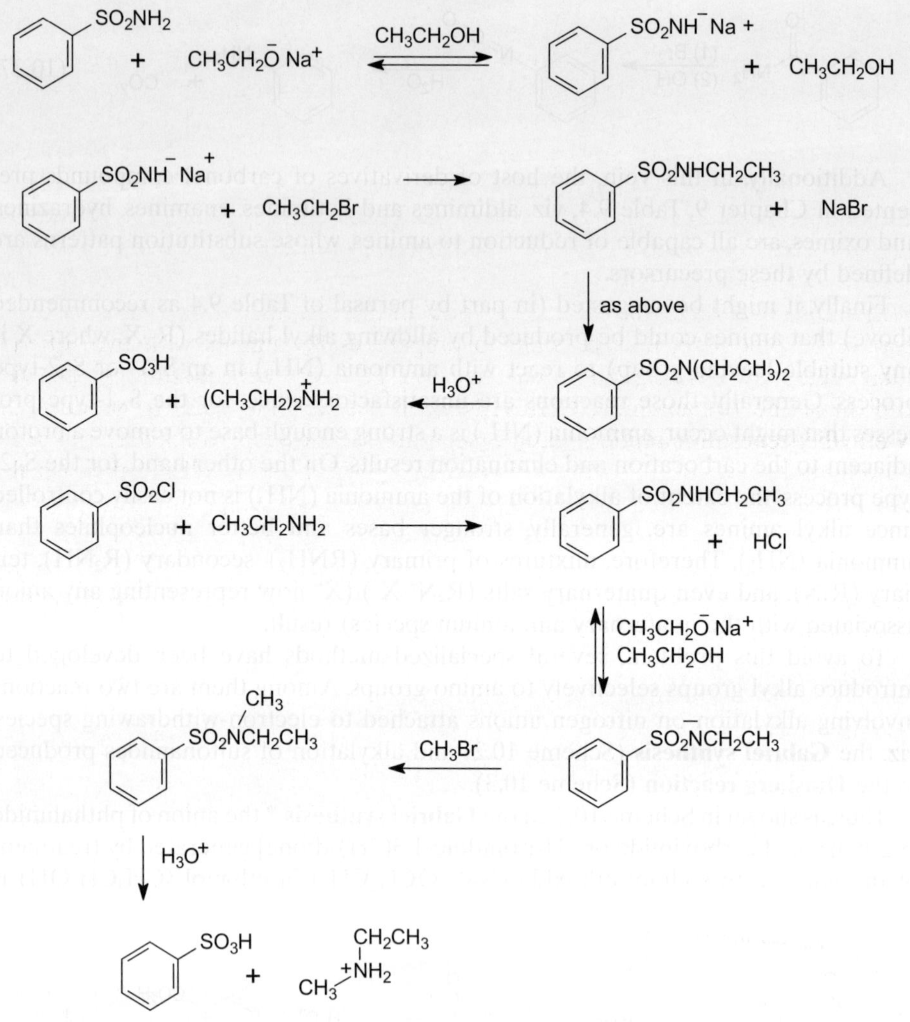

Scheme 10.3. The **Hinsberg reaction**. Alkylation at nitrogen of benzenesulfonamide with bromoethane (CH_3CH_2Br) produces *N*-ethylbenzenesulfonamide and, on repetition of the process, *N,N*-diethylbenzenesulfonamide. An alternative process, the reaction of benzenesulfonyl chloride with ethanamine ($CH_3CH_2NH_2$) to produce *N*-ethylbenzenesulfonamide and subsequent alkylation of the latter via formation of the sulfonamide anion and reaction of that anion with bromomethane (CH_3Br) is also provided.

allowed to react with an alkyl halide (in the scheme, the example is bromoethane but, as expected from S_N2-type processes, methyl > primary > secondary) to effect a displacement of the halide and formation of alkylated phthalimide. It should, of course, be clear that any suitable leaving group "L" could be used in place of halide, for example, triflate (L = $-OSO_2CF_3$) (Scheme 8.32) or even alcohol (L = $-OH$), as with the Mitsonobu protocol (Scheme 8.33). Then, hydrolysis of the *N*-substituted phthalimide [*N*-ethyl-1,2-benzenedicarboximide or 2-ethyl-1H-isoindole-1,3(2H)-dione] produces *o*-phthalic acid and the corresponding ethaneammonium salt (ethylammonium salt, $CH_3CH_2NH_3^+$), which is converted to ethaneamine (ethyl-amine, $CH_3CH_2NH_2$) on basification. Alternatively, with hydrazine (H_2NNH_2), the corresponding 1,4-phthalazinedione is produced and the free base (ethanamine, ethylamine, $CH_3CH_2NH_2$) is isolated directly.

In a similar way, the **Hinsberg reaction*** (Scheme 10.3) produces amines via the anion derived from benzenesulfonamide and reaction of the latter with an alkyl halide (again shown in the scheme as bromoethane, CH_3CH_2Br) generates the substituted product. In contrast to the **Gabriel synthesis** (above), however, a second alkylation is possible here (since there were initially two protons on nitrogen) and so N,N-diethylamine can be formed. Interestingly, as shown in the lower part of Scheme 10.3, it is possible to prepare dialkylamines with two *different* substituents by utilizing an initial reaction between benzenesulfonyl chloride and an amine (ethanamine, $CH_3CH_2NH_2$, in the Scheme 10.3) to produce the N-ethylbenzenesulfonamide and, in a separate second step, to alkylate the anion generated from N-ethylbenzenesulfonamide with a different alkyl halide (in this case bromomethane) to produce N-ethyl-N-methylbenzenesulfonamide. Hydrolysis of the sulfonamide produces benzenesulfonic acid and N-methylethanamine ($CH_3NHCH_2CH_3$).

Representative samples of some amines either prepared by the methods provided above or isolated from natural materials are provided in Table 10.1.

C. OXIDATION OF AMINES

I. Oxygen and Peroxide Oxidations

Oxidation at nitrogen is generally considered to be a fairly facile process. It is argued that with oxygen or peracids the first step in such oxidation processes is the formation of the corresponding N-oxide. For example, as shown in Equation 10.28, treatment of pyridine (azobenzene) with peracetic acid (generated *in situ* from hydrogen peroxide and acetic acid) generates the corresponding pyridine N-oxide. Substituted pyridines behave similarly.

$$\text{(10.28)}$$

Similar oxidation reactions also occur with aliphatic amino compounds. Thus, with tertiary amines, the corresponding N-oxides are formed, and it is common for these materials to be highly crystalline. As shown in Equation 10.29, trimethylamine forms the corresponding hygroscopic N-oxide.

$$(CH_3)_3N + CH_3CO_3H \longrightarrow (CH_3)_3N^+\!-O^- + CH_3CO_2H \tag{10.29}$$

*This is but one of the reactions investigated by Hinsberg (Hinsberg, O. *Chem. Ber.*, **1890**, *23*, 2962) and was not more fully investigated until some years after the reaction was found (Hinsberg, O.; Kessler, J. *Chem. Ber.*, **1905**, *38*, 906).

With secondary amines, R_2N–H, it appears that similar oxidation initially results in formation of the *N*-oxide, and this is often followed by a tautomeric rearrangement to generate the corresponding *N,N*-disubstituted hydroxylamine (Equation 10.30). However, with primary amines the product mixture is generally not easily analyzed and extensive decomposition (except in the case of aryl amines such as aminobenzene [aniline], *vide infra*) may intrude (although, as shown in Scheme 10.4, it is common to find some hydroxamic acid [generically, RCONH(OH)] among the products).

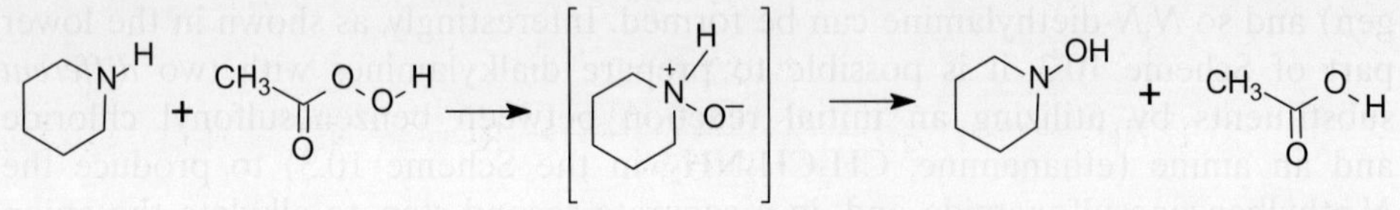

$$(10.30)$$

Additional oxidation pathways for primary amines to other species with a variety of oxidizing agents (as well as oxygen) are also available, and reports of nitrile (R–C≡N) group production as well as the generation of nitro compounds (R–NO$_2$) from primary amines (RNH$_2$) are occasionally made and much success is reported to attended oxidation of amines with ruthenium (Ru) on alumina (Al$_2$O$_3$) catalysts. However, historically, the oxidation of amines with oxygen (O$_2$) is not routine. It is much more general to obtain dark intractable materials from air and/or peroxide oxidation of primary alkyl amines than reasonable quantities of clean material.

On the other hand, bacteria, yeast, plants, and mammals all possess copper-containing proteins, called "copper amine oxidases" or CAOs, which serve to

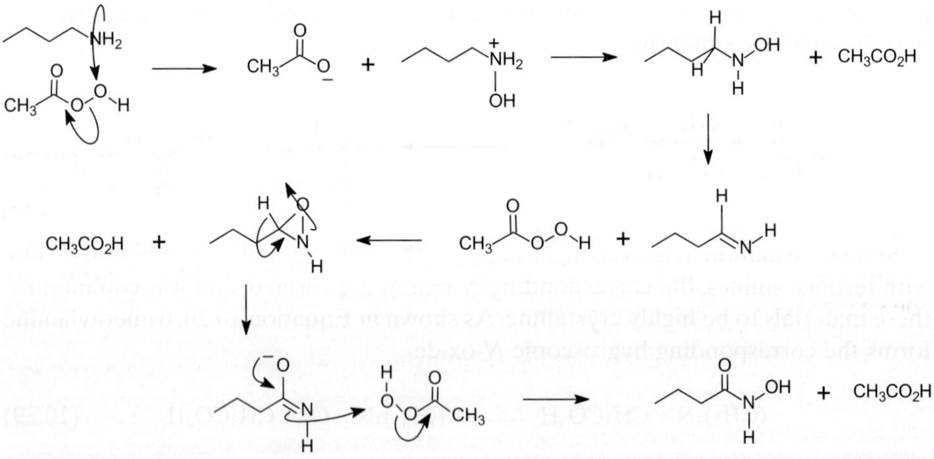

Scheme 10.4. A representation of the formation of *N*-hydroxybutanamide (the hydroxamic acid of butanoic acid) from 1-aminobutane (butanamine, butylamine) and perethanoic acid (peracetic acid). The exact details of the process are obscure and the hydroxamic acid yield is generally low (and variable).

regulate amine levels in the living organism. Work on these proteins, designed to elucidate their mode of action, has been going on for some years, and recent advances in both chemistry and biochemistry are providing significant information on their modes of action. Generally, their function is to convert amines (or their bound equivalents) to aldehydes (or their bound equivalents) and ammonia (or its bound equivalent) as shown in Equation 10.31. This process can be of great importance since the aldehydes produced are very reactive. For example, *lysyl oxidase* (LAO)* is the enzyme responsible for the formation of α-aminoadipic-ω-semialdehyde (Equation 10.32),

$$RCH_2NH_3^+ + O_2 + H_2O \xrightarrow{\substack{\text{copper amine} \\ \text{oxidase (CAO)} \\ \text{enzyme}}} RCHO + NH_4^+ + H_2O_2 \quad (10.31)$$

where H_2O_2 and NH_3 are shown in quotation marks because these species are not formed "free". The α-aminoadipic-ω-semialdehyde is instrumental in forming cross-links in the proteins *collagen* and *elastin*, connective tissues required for normal growth.

$$(10.32)$$

As might be expected, copper (Cu) is needed to maintain normal **CAO** levels as some is lost in normal metabolism. So in, for example, poultry, copper deficiency results in changes in elastin and death (largely) due to aortic ruptures.

More generally, the oxidative breakdown of amino acids results in the formation of ammonia, urea, and/or (Chapter 13) uric acid (Equation 10.33).

$$(10.33)$$

Glutamate ($[^-O_2CCH_2CH_2CH(NH_4^+)CO_2^-]$ Scheme 10.5), through the action of the enzyme *glutamate dehydrogenase* (EC 1.4.1.3),[†] is oxidized to the corresponding imine $[^-O_2CCH_2CH_2C(=NH)CO_2^-]$. *Glutamate dehydrogenase* utilizes the cofactor nicotinamide adenine dinucleotide (NAD$^+$) or the corresponding phosphate derivative (NADP$^+$) and that cofactor is reduced (to NADH or NADPH, respectively)

---

*See, for example, Kagan, H. M.; Williams, M. A.; Williamson, P. R.; Anderson, J. M. *J. Biol. Chem.*, **1984**, *259*, 11203 and earlier papers for leading references on lysyl oxidase (EC 1.4.3.13).
†Glutamic acid dehydrogenase (EC 1.4.1.3) was crystallized and characterized in the early 1950s (Olson, J. A.; Anfinsen, C. B. *J. Biol. Chem.*, **1952**, *197*, 67).

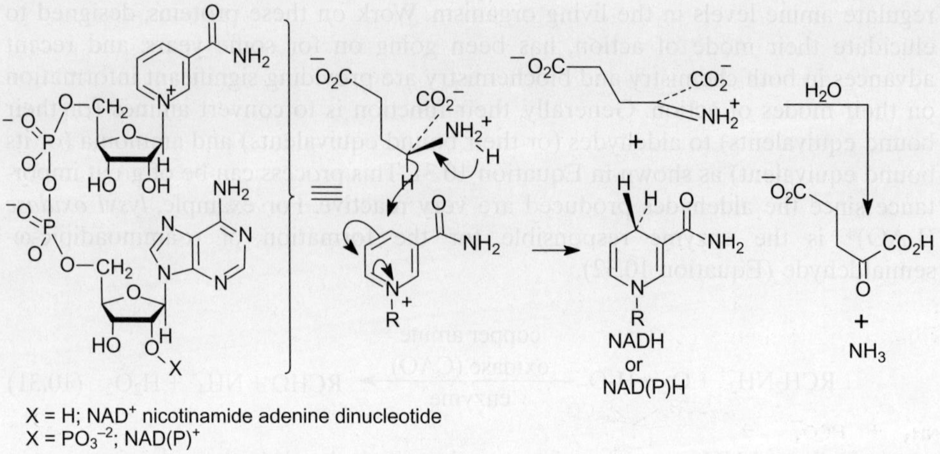

X = H; NAD$^+$ nicotinamide adenine dinucleotide
X = PO$_3^{-2}$; NAD(P)$^+$

Scheme 10.5. A representation of the oxidation of glutamate using the enzyme *glutamate dehydrogenase* (EC 1.4.1.3). The cofactor nicotinamide adenine dinucleotide (NAD$^+$) or the corresponding phosphate derivative (NADP$^+$) is reduced (to NADH or NADPH, respectively) as the glutamate [$^-$O$_2$CCH$_2$CH$_2$C(NH$_4^+$)CO$_2^-$] is oxidized to the imine, α-iminoglutarate [$^-$O$_2$CCH$_2$CH$_2$C(=NH)CO$_2^-$]. The latter undergoes hydrolysis to ammonia and α-ketoglutaric acid [$^-$O$_2$CCH$_2$CH$_2$C(=O)CO$_2^-$] (see Olson, J. A.; Anfinsen, C. B. *J. Biol. Chem.*, **1952**, *197*, 67).

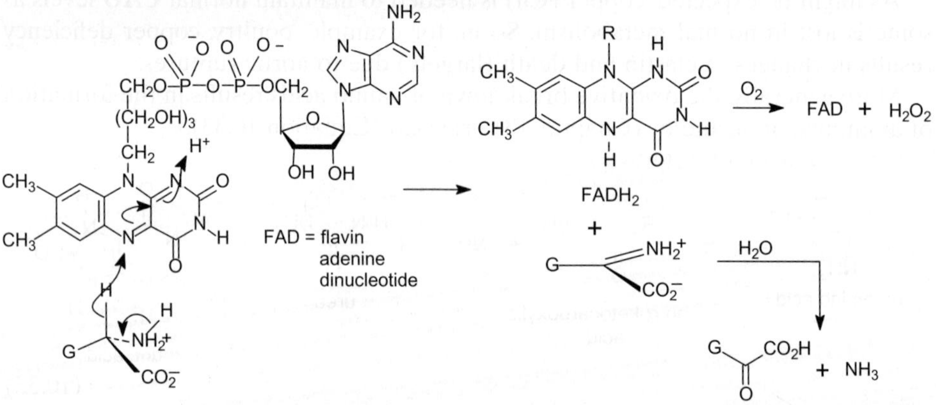

FAD = flavin adenine dinucleotide

Scheme 10.6. The oxidation of an amino acid (G = alkyl or alkaryl) by an oxidase utilizing flavin adenine dinucleotide (FAD). The latter is reduced to FADH$_2$ as the amino acid is oxidized to imine. The imine is hydrolyzed to ammonia and α-ketocarboxylic acid. The reduced dinucleotide (FADH$_2$) is reoxidized by oxygen (O$_2$) to FAD. The reduction product in the second reaction is hydrogen peroxide, H$_2$O$_2$ (or its equivalent).

as the glutamate ($[^-O_2CCH_2CH_2CH(NH_4^+)CO_2^-]$ is oxidized to the imine, α-iminoglutarate $[^-O_2CCH_2CH_2C(=NH)CO_2^-]$. The latter is said to undergo hydrolysis to ammonia and α-ketoglutaric acid $[^-O_2CCH_2CH_2C(=O)CO_2^-]$ (Scheme 10.5).

Interestingly, there are also nonspecific amino acid oxidases that utilize flavin adenine dinucleotide (FAD) to produce $FADH_2$ (Scheme 10.6). The reduced dinucleotide ($FADH_2$) is reoxidized by oxygen (O_2) to FAD. The reduction product in the second reaction is hydrogen peroxide, H_2O_2 (or its equivalent).

Lastly, in this regard, urea (H_2NCONH_2) results from ammonia and aspartate (for the nitrogens) and bicarbonate (HCO_3^-) for the carbon. The aspartate yields fumerate (!) as shown in Equation 10.34 and discussed as part of the **urea cycle** in Chapter 12 (Scheme 12.7).

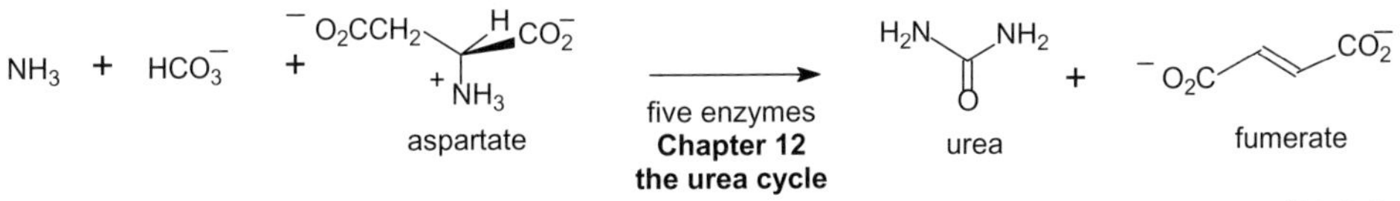

$$NH_3 \;+\; HCO_3^- \;+\; \text{aspartate} \;\xrightarrow[\substack{\textbf{Chapter 12} \\ \textbf{the urea cycle}}]{\text{five enzymes}}\; \text{urea} \;+\; \text{fumerate}$$

$$(10.34)$$

In contrast to the relatively intractable laboratory oxidation products primary amines form on air and/or peroxide oxidation, the more stable oxides of tertiary amines have provided a platform for a significant amount of interesting chemistry. Among the reactions that have been explored in some detail are (1) the allylic $N \rightarrow O$ rearrangements of suitably substituted N-oxides (Scheme 10.7), (2) the **Cope** (or **Hofmann**) type elimination reactions (*vide infra* and see Chapter 9) (Scheme 10.8), and (3) the **Polonovski** reaction (Scheme 10.9), of particular interest because of its generality.

Similarly, the secondary amine oxides formed with oxygen and/or peroxide, having rearranged to the corresponding hydroxylamines are subsequently oxidized further with, for example, silver oxide (Ag_2O) in ether, to produce stable free

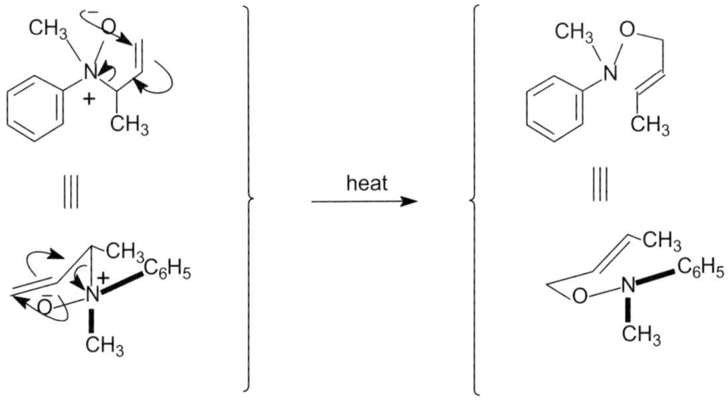

Scheme 10.7. A thermally induced allylic $N \rightarrow O$ migration of the butenyl group of N-methyl-N-phenyl-3-amino-1-butene oxide. See Moriwaki, M.; Yamamoto, Y.; Oda, J.; Inouye, Y. *J. Org. Chem.*, **1976**, *41*, 300.

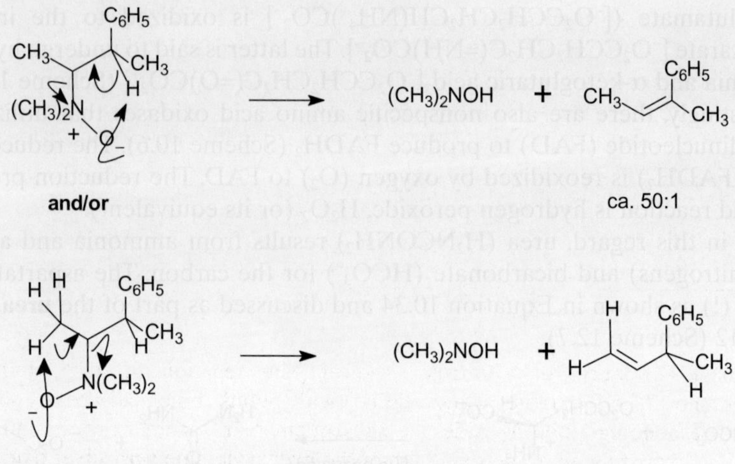

and/or

$(CH_3)_2NOH$ +

ca. 50:1

$(CH_3)_2NOH$ +

Scheme 10.8. A representation of a **Cope**-type elimination of N,N-dimethylhydroxylamine from the N-oxide of N,N-dimethyl-2-amino-3-phenylbutane showing that the *more* highly substituted alkene is obtained in greater quantity than the less highly substituted isomer. See Cope, A. C.; Ross, D. L. *J. Am. Chem. Soc.*, **1961**, *83*, 3854.

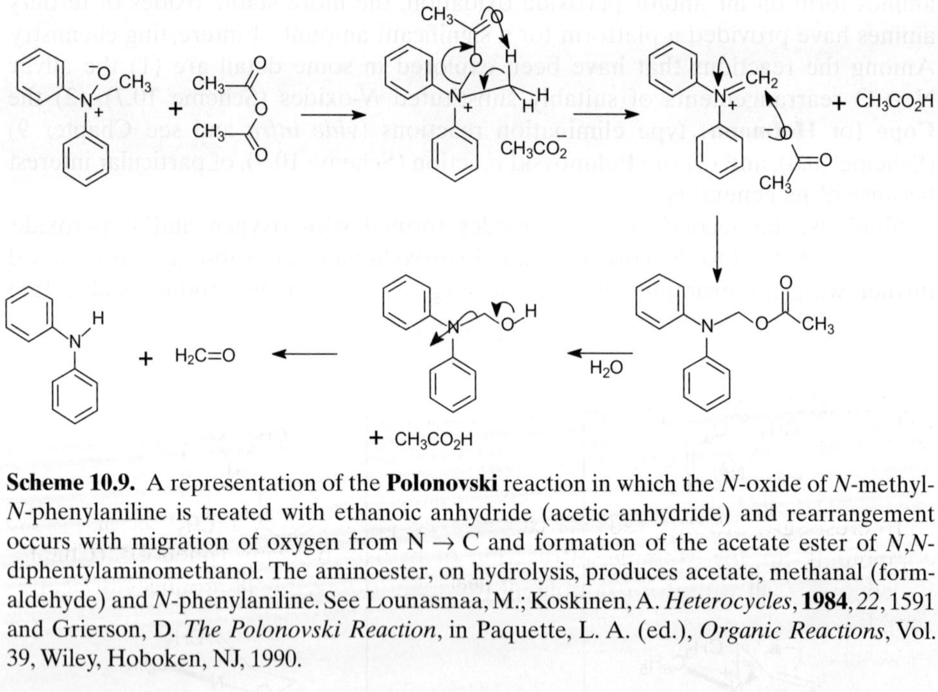

+ CH_3CO_2H

+ CH_3CO_2H

Scheme 10.9. A representation of the **Polonovski** reaction in which the N-oxide of N-methyl-N-phenylaniline is treated with ethanoic anhydride (acetic anhydride) and rearrangement occurs with migration of oxygen from $N \rightarrow C$ and formation of the acetate ester of N,N-diphentylaminomethanol. The aminoester, on hydrolysis, produces acetate, methanal (formaldehyde) and N-phenylaniline. See Lounasmaa, M.; Koskinen, A. *Heterocycles*, **1984**, *22*, 1591 and Grierson, D. *The Polonovski Reaction*, in Paquette, L. A. (ed.), *Organic Reactions*, Vol. 39, Wiley, Hoboken, NJ, 1990.

radical species (called "nitroxyls"). These one-electron species, also available by direct oxidation of secondary amines, have proved valuable as structural probes. The formation of the nitroxyl radical derived from 2,2,5,5-tetramethylpyrrolidine by oxidation of the corresponding hydroxylamine, *or by direct oxidation* with hydrogen peroxide in the presence of sodium tungstate, is shown in Scheme 10.10.

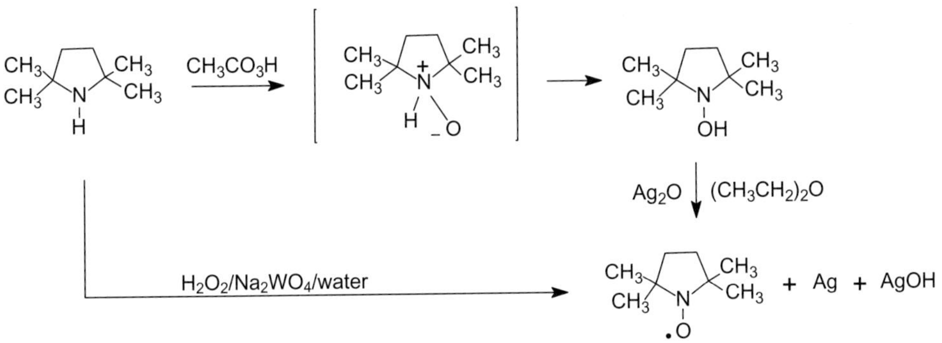

Scheme 10.10. Formation of a nitroxyl radical. The reaction of 2,2,5,5-tetramethyl-azacyclopentane (2,2,5,5-tetramethylpyrrolidine) with perethanoic acid (peracetic acid) to form the corresponding labile *N*-oxide, the subsequent rearrangement of the latter to the corresponding hydroxylamine, and oxidation of the hydroxylamine with silver oxide (Ag_2O) to the stable nitroxyl radical. The alternative oxidation of the same starting material to the same radical with hydrogen peroxide in the presence of sodium tungstate is also shown. See Utsumi, H.; Takeshita, K.; Ichikawa, K.; Matsumoto, K.-I. *Curr. Top. Biophys.*, **1996**, *20*, 41 and references therein.

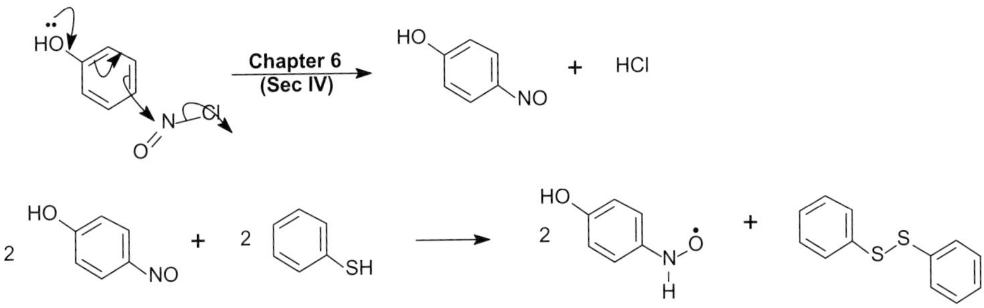

Scheme 10.11. The nitrosation of phenol (C_6H_5OH) with nitrosyl chloride, NOCl (see Chapter 8, Section IV (b) D) to produce *p*-nitrosophenol (*p*-HOC_6H_4NO) and its subsequent reduction by thiophenol to the corresponding *p*-hydroxyphenylnitrosyl radical (*p*-$HOC_6H_4NHO\cdot$). The oxidation product is the corresponding diphenyldithio ether ($C_6H_5SSC_6H_5$).

Interestingly, nitroxyl radicals can also be produced by reduction of nitroso compounds. So, for example, nitrosation of hydroxybenzene (phenol)* (Chapter 8, Section IV, b) with, for example, nitrosyl chloride (NOCl) produces the corresponding *p*-hydroxynitrosobenzene and reduction of the latter with, for example, a thiol such as thiophenol, yields the corresponding nitroxyl radical (and dithiophenol) (Scheme 10.11).

Numerous such nitroxyl radicals have been bound to biological macromolecules where they serve as probes for structural parameters. They provide information

*Nitrosation is generally much more difficult than nitration (as the nitrosonium cation [NO^+] is apparently much less reactive than the nitronium cation [NO_2^+]), and so activated rings undergo nitrosation much more easily than inactivated rings. The hydroxyl group is, however, not only necessary for a facile nitrosation, but it can also be used to form derivatives desirable for incorporation into biological systems.

about the orientation of the radical (called the "spin label") by examining the electron spin resonance (ESR) spectrum (Chapter 2, VI, b) of the labeled material as a function of its position in the magnetic field. For example, if the radical is fixed in a membrane or crystalline lattice, the ESR signal observed will be, in part, a function of how the labeled material sits in the magnetic field. Then, when the lattice melts or is otherwise disrupted, the signal will change since the nitroxyl radical or its bound counterpart is now more free to move about.

The products obtained on oxidation of aminobenzene (aniline) and other aminoaryl compounds with oxygen and/or peroxide are somewhat dependent on the specific conditions used. For example, freshly distilled aminobenzene (aniline) is colorless, but in the presence of air, it readily darkens and small quantities of crude "anilinium dyes" result. The same materials, which have been used since the middle of the nineteenth century to color fabric, also result when other oxidizing agents are used, and indeed, the dye industry of that time was built on the oxidation of aniline and related compounds (this chapter). With both air and peroxides, aminobenzene (aniline) has been reported to undergo oxidation to yield, among others, phenylhydroxylamine, nitrosobenzene, p-hydroxynitrosobenzene, benzophenone, and nitrobenzene (Scheme 10.12).

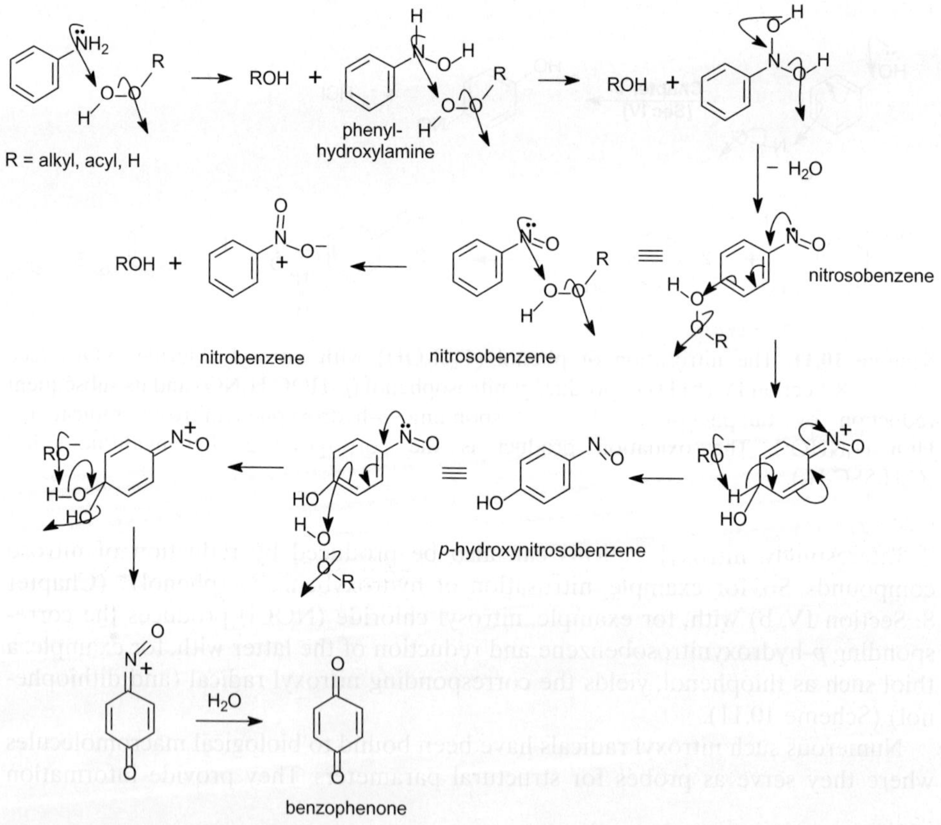

Scheme 10.12. A representation of a path to some products (phenylhydroxylamine, nitrosobenzene, p-hydroxynitrosobenzene, benzophenone, and nitrobenzene) obtained on peroxide oxidation of aminobenzene (aniline).

II. Other Oxidizing Agents

a. General. It is likely that alkylamines bearing a primary amino group ($-NH_2$) on a primary or secondary carbon react with almost any oxidizing agent to initially produce an imine. It is generally presumed that these imines form in much the same way an aldehyde or ketone forms on reaction of a primary or secondary alcohol, respectively, with an oxidizing agent. When the primary amino group is on a primary carbon, imines are seldom isolated since they may oxidize further to nitriles, dimerize (and sequester the metal of the oxidizing agent), or form other intractable materials (Scheme 10.13). Except for the oxidation to a nitrile (which clearly cannot happen without carbon–carbon bond cleavage), similar chemistry, along with azine formation (Equation 10.35), is obtained when the primary amino group is on a secondary carbon and is appropriately substituted.

$$(C_6H_5)_2CH-NH_2 \xrightarrow{\textbf{Scheme 10.13}} (C_6H_5)_2C=NH \longrightarrow \underset{N=C(C_6H_5)_2}{\overset{(C_6H_2)_2C=N}{\diagdown}} \quad (10.35)$$

However, when the primary amino group is on a suitably substituted tertiary carbon, nitro group formation (Equation 10.36) and degradation by carbon–carbon bond cleavage occur.

$$(CH_3)_3C-NH_2 + KMnO_4/H_2O \longrightarrow (CH_3)_3C-NO_2 + MnO_2 \quad (10.36)$$

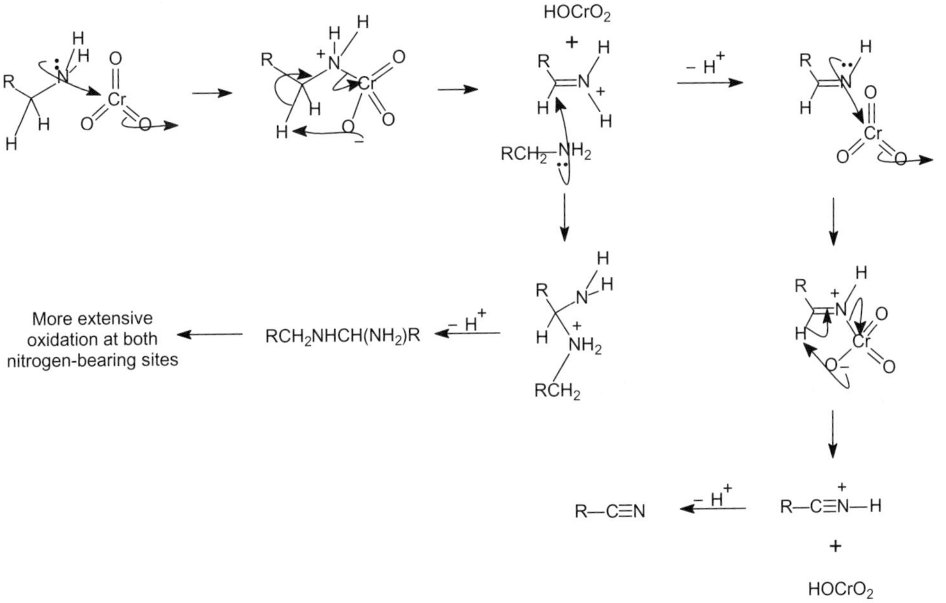

Scheme 10.13. An imaginary process showing a possible route from a primary amine attached to a primary carbon undergoing oxidation with chromic anhydride to an imine, which (a) reacts further with the original amine substrate yielding an adduct containing two nitrogens and (b) with additional oxidant yielding nitrile.

The oxidation of the primary aromatic amine, aminobenzene (aniline), has been the subject of countless lectures and at least one book.* It has been claimed on a number of occasions that this oxidation reaction, which served (in theory if not in practice) as the basis for the establishment of the chemical dye industry (beginning in about 1856), can therefore be thought of as serving as the parent reaction for the entire chemical industry.

The story is told that a bright young man, William Henry Perkin (1838–1907), entering the new Royal (now Imperial) College of Chemistry in London, over family opposition at age 15, began to study under its young director August Wilhelm von Hofmann ([Chapter 7, Scheme 7.24] discoverer of the **Hofmann reaction**, *vide infra*). Having established a small chemistry laboratory in his own home and being pressed for time in Hofmann's laboratory, it is claimed that during the Easter recess of 1856, Perkin undertook his own investigation[†] on the oxidation of an aminobenzene (aniline) derivative, *N*-allyl-*p*-methylaniline (allyltoluidine) (Figure 10.1). Perkin (apparently following a vague suggestion of Hofmann) considered that the addition of oxygen (O_2 or two "O") to two equivalents of "allyltoluidine" ($C_{10}H_{12}N$) would produce quinine ($C_{20}H_{24}N_2O_2$), an alkaloid isolated from the bark of various *Cinchona* species and appreciated as a medicinally valuable palliative in the battle against malaria, a major problem in maintaining the empire.

It is interesting to note that the information provided in Figure 10.1, which represents, in outline, the creation of an industry, is a clear affirmation of previous comments (see, e.g., Chapter 6, Scheme 6.93). As shown in Scheme 10.14, nitration of impure (contaminated by toluene) benzene, followed by iron and acetic acid nitro group reduction, produces both aminobenzene (aniline) and 2- and 4-methylaminobenzene (*o*- and *p*-toluidine, respectively).

b. Oxidation by Halogen. Treatment of amines with a source of positive halogen, for example, sodium hypochlorite (HOCl), bromine (Br_2), or iodine (I_2), in the presence of base results in the formation of the corresponding *N*-halogen and *N,N*-dihalogen derivatives. Interestingly, *N*-fluorocompounds can be formed if fluorine (F_2) diluted in an inert medium (e.g., nitrogen, N_2) is added to an amine. It will be recognized that it is "positive" halogen that is adding and the *N*-halo compounds formed are thus oxidizing agents. That this is so is readily seen in Scheme 10.15 where halogenation of cyclohexylamine produces the corresponding *N*-chloro

---

*Garfield, S. *Mauve*, Norton, New York, **2000**.

[†]Perkin, it is said, left school, patented "mauvine," and set up a dyeworks in West London. He was highly successful in this and other dye agents, including alizarin and (it is believed) malachite green, which were attributed to his efforts. Later, losing out to the dynamic German dye industry, he experimented again on his own and found the reaction bearing his name (Chapter 9, Scheme 9.83).

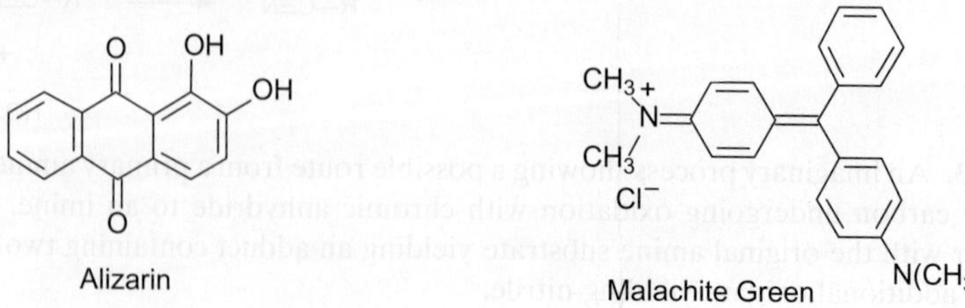

Alizarin

Malachite Green

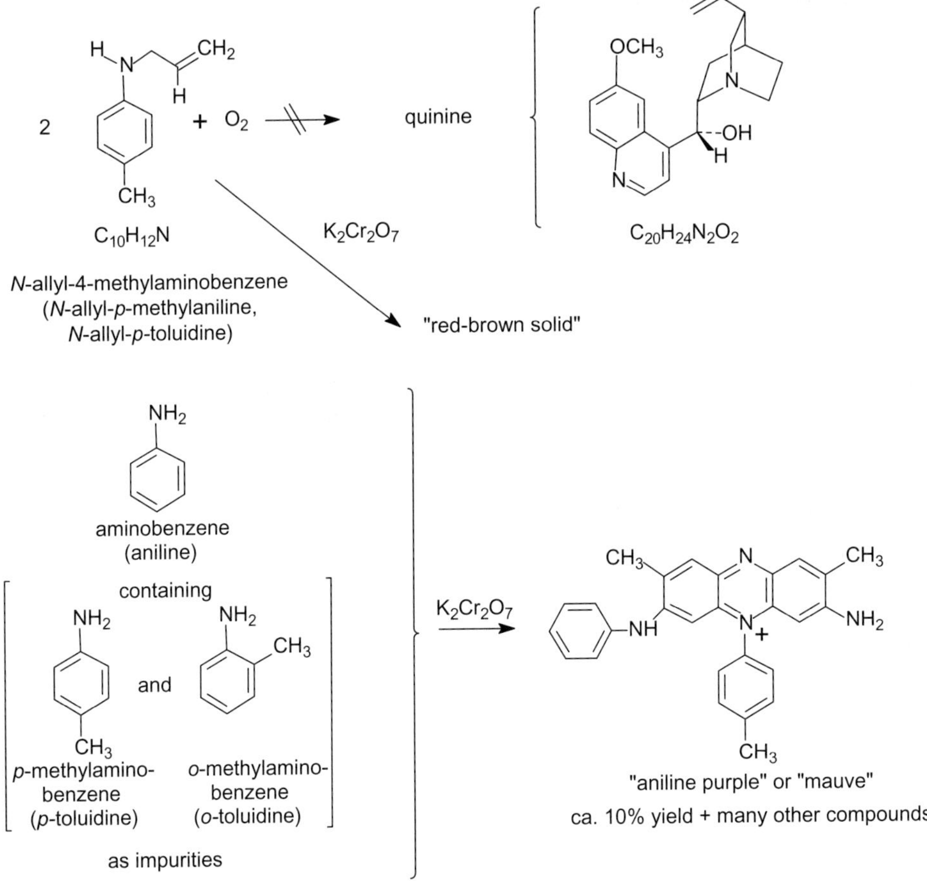

Figure 10.1. The mistaken assumption that quinine could be produced on oxidation of allyltoluidine. The subsequent oxidation of aniline(s) to "mauve."

Scheme 10.14. The formation of aminobenzene (aniline) and *o*- and *p*-methylaminobenzenes (*o*- and *p*-toluidines) from the nitration of a mixture of benzene and methylbenzene (toluene) and subsequent reduction of the mixture with iron (Fe) in acetic acid (CH₃CO₂H).

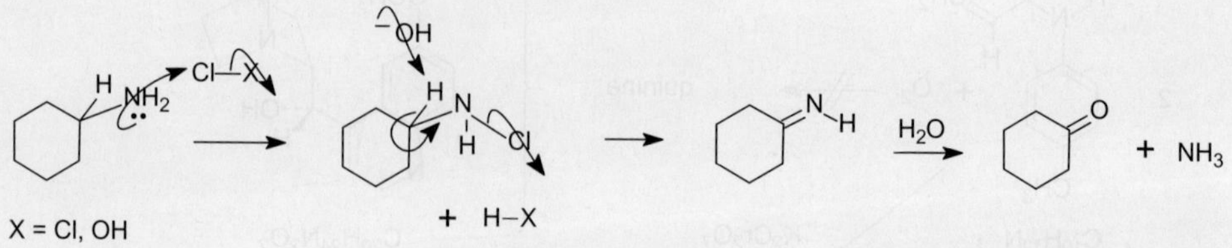

X = Cl, OH

+ H–X

Scheme 10.15. A depiction of the N-chlorination of cyclohexylamine and the subsequent loss of hydrogen chloride (HCl) with imine formation followed by hydrolysis to cyclohexanone and ammonia.

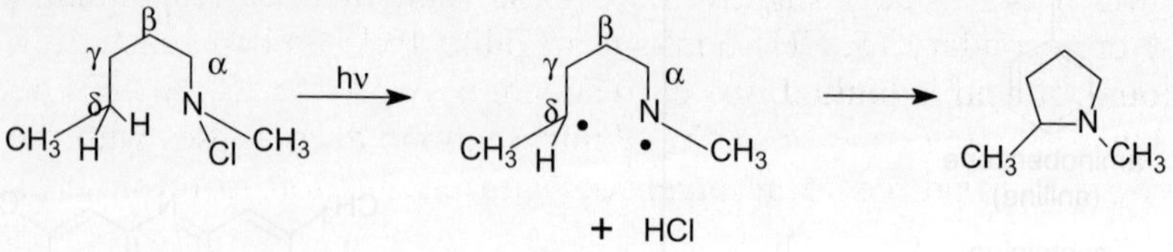

+ HCl

Scheme 10.16. The formation of 1,2-dimethylazacyclopentane (1,2-dimethylpyrrolidine) by radical cleavage and loss of HCl from the corresponding N-chloroamine.

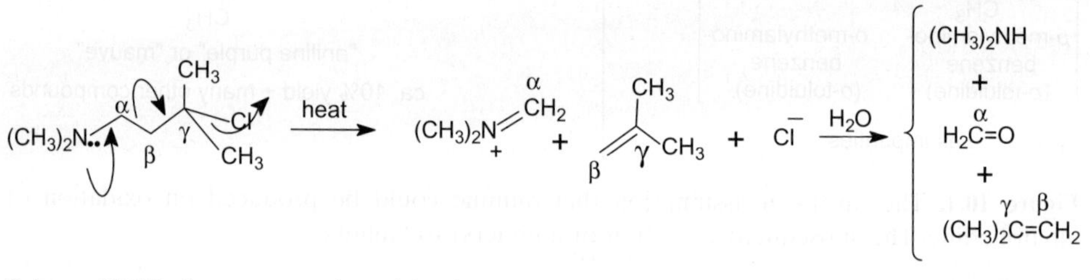

Scheme 10.17. A representation of the thermal decomposition of α-chloro-N,N-dimethylamine with bond fragmentation to produce N,N-dimethylamine [$(CH_3)_2NH$], methanal (formaldehyde, $H_2C=O$), and an alkene (2-methylpropene). The process is called the **Hofmann–Löffler–Freytag** reaction (see Hofman, A. W. *Chem. Ber.*, 1883, *16*, 558 and Löffler, K.; Freytag, C. *Chem. Ber.*, 1909, *42*, 3427).

derivative which, on treatment with aqueous base, loses hydrogen chloride (HCl) and produces the corresponding imine, the latter hydrolyzing to the ketone in the aqueous environment.

Interestingly, some alkyl N-chloroamines possessing one or more hydrogens on carbon δ to the nitrogen and capable of undergoing a cyclization reaction to form a five-membered ring (azacyclopentane or pyrrolidine) do so on irradiation. It is held that this is a free radical reaction, and since a radical initiator might be used in place of UV light, the suggestion is reasonable (Scheme 10.16).

On the other hand, if the halogen is γ to a fully alkylated nitrogen and on a fully alkylated carbon, thermal decomposition (Scheme 10.17) can intrude (the **Hofmann–Löffler–Freytag** reaction) with (after hydrolysis) formation of a ketone and an alkene.

c. Oxidation with Nitrous Acid. Reaction of alkyl amines with nitrous acid (HONO or HNO_2) or its equivalent,* might, in principle, be expected to produce the corresponding *N*-nitroso (R–NH–N=O) derivative, and indeed, under carefully controlled conditions such compounds can be prepared.

While nitrosation can often be effected through the use of nitrosyl chloride (Cl–N=O), nitrogen sesquioxide (N_2O_3, *vide supra*), or dinitrogen tetroxide (N_2O_4), it is much more common to use an aqueous acidic solution of a nitrite salt, for example, sodium nitrite ($NaNO_2$). As might be expected (*vide infra*), the results of the reactions utilizing different nitrosating agents (and usually in different solvents) are frequently slightly different.

Using acidic solutions of nitrite salt, it is generally found that tertiary amines (R_3N), which are usually slightly more basic than their corresponding primary (RNH_2) or secondary (R_2NH) analogues (Table 10.1), fail to react. However, if the amount of acid is limited, the corresponding *N*-nitroso (R–N=NO) derivative forms. Under forcing conditions, the *N*-nitrosoamine decomposes with the loss of an alkyl group (oxidized to an aldehyde, with elimination to form an alkene or simple hydrolysis to the corresponding alcohol) and formation of nitrosated secondary amine. A cartoon representation of these processes is shown in Scheme 10.18.

Secondary *N*-nitrosamines are frequently isolable and are formed reversibly from secondary amines and nitrous acid (HONO) and other nitrosating agents.

Primary amines on reaction with nitrous acid undergo deamination to generate nitrogen (N_2), alcohols, and alkenes.

With methylamine (methanamine, CH_3NH_2) oxidation by nitrous acid (HONO) or its anhydride (dinitrogentrioxide, N_2O_3) apparently produces diazomethane (CH_2N_2) (Scheme 10.19), which decomposes as it is formed. Under other conditions, diazomethane (CH_2N_2) can be formed and used as a reagent for organic synthesis.

Diazomethane (CH_2N_2), which is toxic and known to explode in the presence of sharp objects or rough surfaces (**Caution!** Do not work with ground glass flasks or boiling stones, etc.), is generally prepared and used as diethyl ether (ethyl ether,

---

*Nitrous acid, HONO (or HNO_2) is known only in the form of its salts, for example, sodium nitrite ($NaNO_2$), and in solution where it is in equilibrium with its corresponding anhydride, N_2O_3. The latter is also in equilibrium with nitric oxide (NO) and nitrogen dioxide (NO_2).

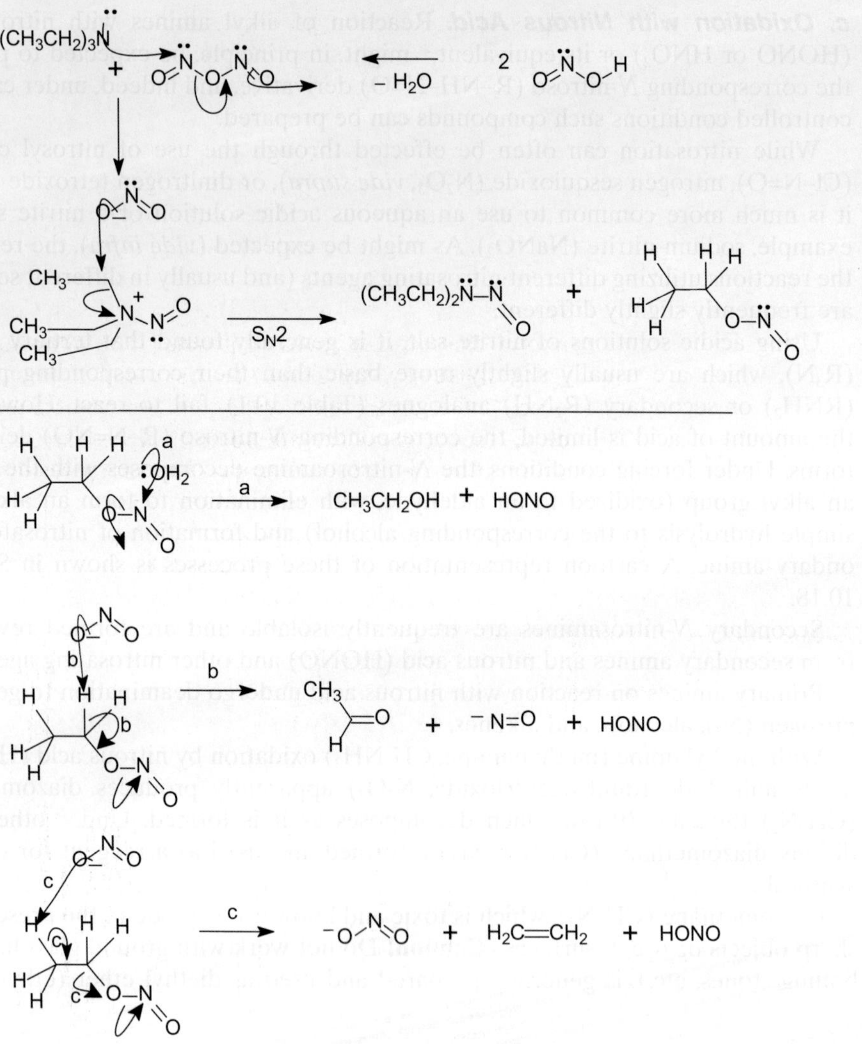

Scheme 10.18. A cartoon representation of the decomposition of *N*-nitroso-*N*,*N*,*N*-triethylamine by various pathways to *N*-nitroso-*N*,*N*-diethylamine [(CH$_3$CH$_2$)$_2$N–NO] and then to (a) ethanol (CH$_3$CH$_2$OH), (b) ethanal (acetaldehyde, CH$_3$CHO), and (c) ethene (ethylene, H$_2$C=CH$_2$) along with a variety of oxides of nitrogen.

CH$_3$CH$_2$OCH$_2$CH$_3$) solutions. The solutions are used directly after separating the aqueous phase and drying the ether (ethyl ether, CH$_3$CH$_2$OCH$_2$CH$_3$) phase. The preparation uses base-induced decomposition of *N*-nitrosomethylurea (Scheme 10.20), which, as shown in the same scheme, is produced by the reaction of aqueous solutions of methylamine hydrochloride (CH$_3$NH$_3^+$ Cl$^-$) with urea (H$_2$NCONH$_2$) (Scheme 10.20).

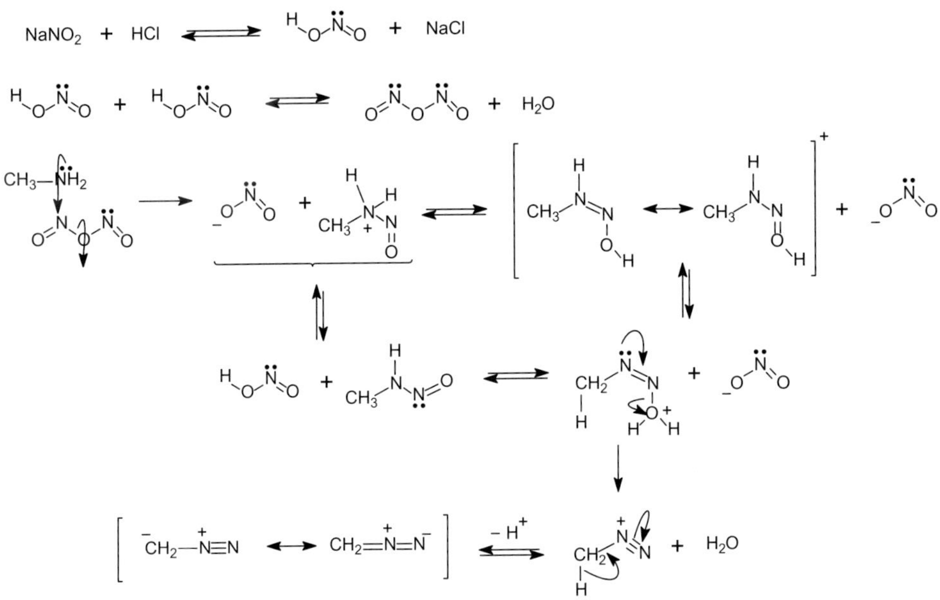

Scheme 10.19. A representation of a pathway to diazomethane (CH_2N_2) by nitrosation of methylamine (methanamine, CH_3NH_2) with nitrous acid (HONO). Generally, any diazomethane (CH_2N_2) formed under these conditions decomposes before it can be used in subsequent reactions.

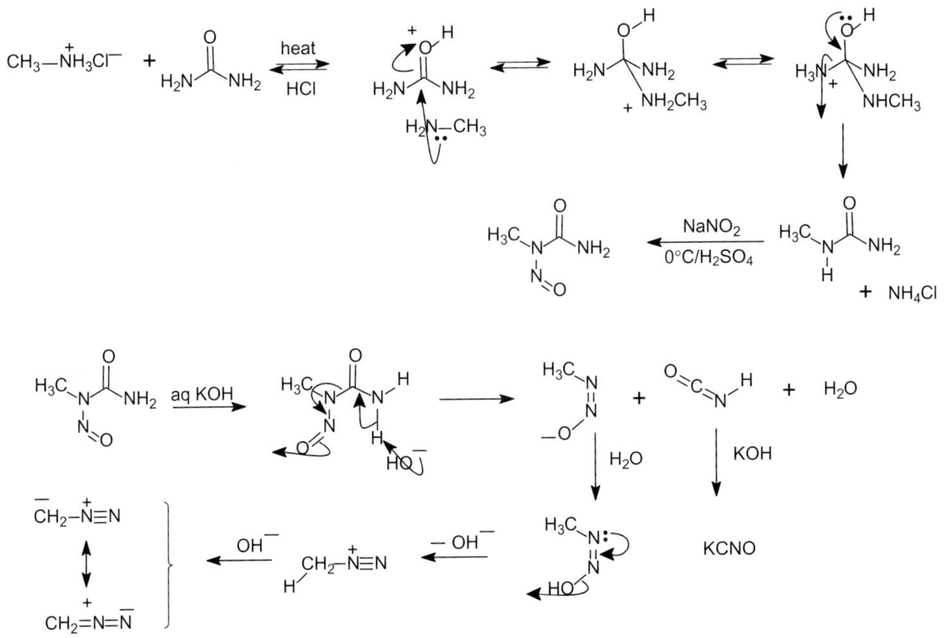

Scheme 10.20. A representation of the formation and base-induced decomposition of N-nitrosomethylurea to produce diazomethane (CH_2N_2) and potassium isocyanate (KCNO).

D. REDUCTION OF AMINES

Reduction of groups attached to amines (aromatic rings and carbon–carbon double and triple bonds) can be effected by methods previously discussed. These include (a) catalytic hydrogenation with hydrogen gas (H_2) over a metal catalyst such as platinum (Pt), nickel (Ni), or palladium (Pd) (Equation 10.37) and (b) dissolving metals such as sodium (Na) in an alcohol solvent/reactant (Equation 10.38). Some nitrogenous materials have been known to poison some catalysts and thus experimentation is often necessary.

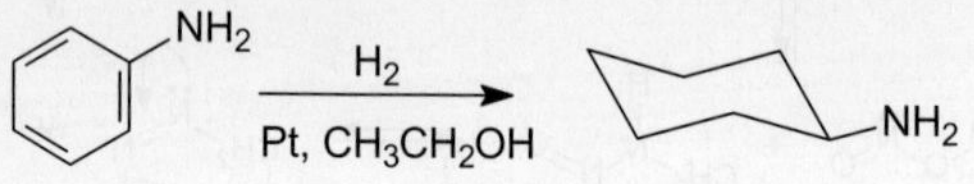

$$(10.37)$$

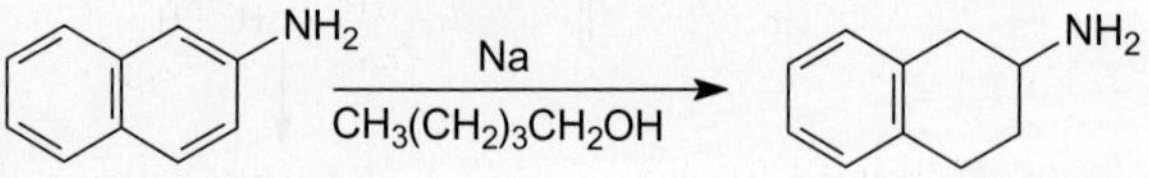

$$(10.38)$$

Hydrogenolysis (reduction with bond breaking) of benzylamines resulting in the formation of toluene and debenzylated amine is formally a reduction of the amine. Thus, as shown in Equation 10.39, treatment of N-benzyl-1,2,3,4-tetrahydroisoquinoline dissolved in ethanol with hydrogen (H_2) gas at about 1 atm pressure in the presence of aqueous hydrochloric acid ($HCl_{(aq)}$) and a platinum (Pt) catalyst, results in formation of the tetrahydroisoquinoline base (present as is hydrochloric acid salt) and toluene.

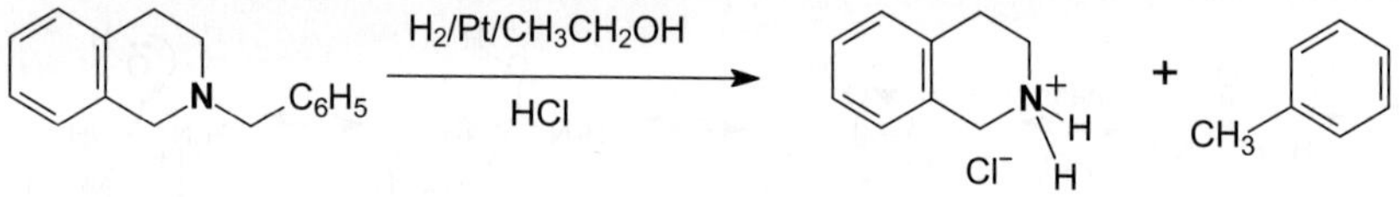

$$(10.39)$$

Similarly, it has been reported that hydrocarbons can be obtained from primary amines by treatment of the amine with aqueous sodium hydroxide and hydroxylamine-O-sulfonic acid (Scheme 10.21) and enamines, obtained from ketones and secondary amines, can be converted to alkenes by reduction with alane (AlH_3) (see Coulter, J. M.; Lewis, J. W.; Lynch, P. P. *Tetrahedron*, **1968**, *24*, 4489) (Equation 10.40).

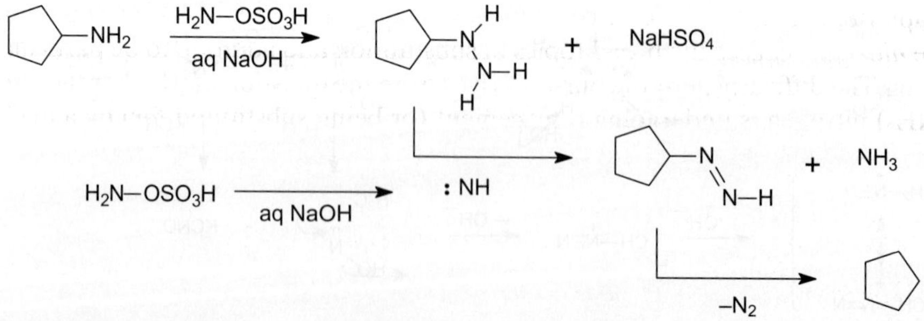

Scheme 10.21. A representation of a possible series of compounds lying between aminocyclopentane and cyclopentane. The reaction is shown to involve hydroxylamine O-sulfonic acid in the presence of base (see Nickon, A.; Hill, A. *J. Am. Chem. Soc.*, **1964**, *86*, 1152).

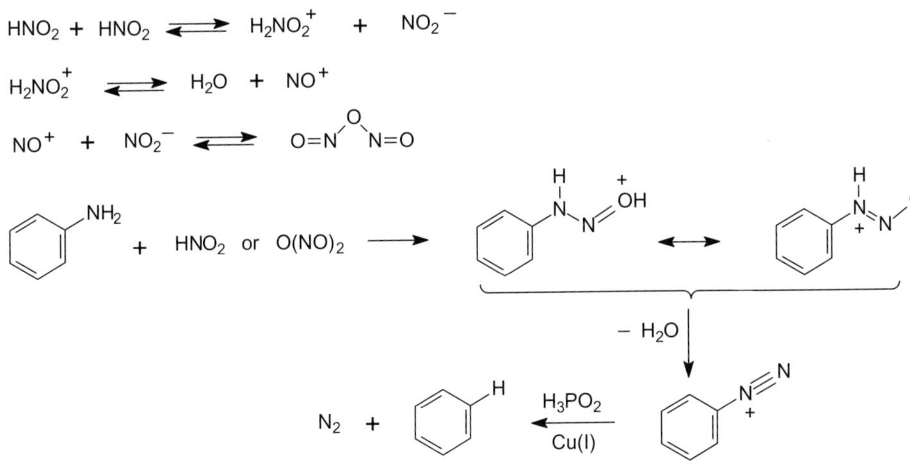

$$HNO_2 + HNO_2 \rightleftharpoons H_2NO_2^+ + NO_2^-$$

$$H_2NO_2^+ \rightleftharpoons H_2O + NO^+$$

$$NO^+ + NO_2^- \rightleftharpoons O=N{-}O{-}N=O$$

Scheme 10.22. A representation of the conversion of aminobenzene ($C_6H_5NH_2$, aniline) to benzene (C_6H_6) through diazotization of the former with nitrous acid ($HONO_2$) followed by loss of nitrogen (as N_2) and hydrogen transfer from hypophosphorus acid (H_3PO_2) in the presence of cuprous copper [Cu(I)] to the aromatic ring.

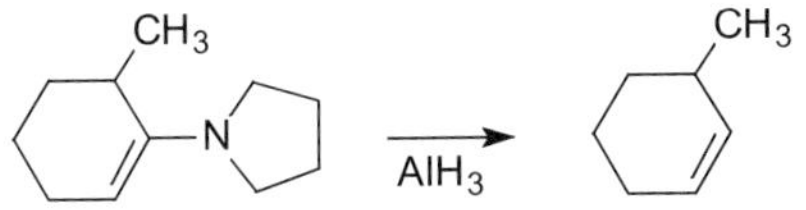

$$(10.40)$$

Finally, in this vein, amino groups attached to aromatic rings can be replaced by hydrogen via diazotization* and reduction of the diazotized species with hypophosphorus (H_3PO_2) acid and copper(I) salts (Scheme 10.22).

E. ADDITION AND SUBSTITUTION REACTIONS OF AMINES

I. General Introduction

As pointed out earlier when considering alcohols (Chapter 5, Section C, III and Chapter 8, Part I, E), the electron-rich heteroatom (in that case *oxygen* and in this case *nitrogen*) causes the broad topics of substitution and addition to be particularly vexing. The difficulty here, of course, derives from the question of whether the amino ($-NH_2$) nitrogen is undergoing replacement (or being substituted *for*) by a nucleophile or whether it is serving *as* a nucleophile and substituting itself for some leaving

---

*The process of diazotization of aromatic amines as noted in Scheme 10.22 (and as will be seen subsequently in this chapter) by treatment of the amine with nitrous acid (or its equivalent) has traditionally been used to effect *substitution* of the amino group by other functional groups. In that vein, replacement of the diazo group by hydrogen, while a reduction, may also be thought of as a "substitution" in which the hydrogen is substituting for the amino group. As will also be seen (*vide infra*), the diazotization process when carried out on alkylamines results in carbocation formation and subsequent elimination or skeletal rearrangement.

group. For example, substitution of oxygen for nitrogen resulting from the hydrolysis of an amide (Equation 10.41 and Chapter 9, Table 9.10, item 1) and nucleophilic substitution of sulfur for nitrogen (Equation 10.42 and Chapter 8, Part III) by displacement are examples of the first kind, whereas formation of the amide from, for example, an acid chloride and an alkylamine (Equation 10.43 and Chapter 9, Table 9.7, item 4) or nucleophilic substitution on nitrogen (Equation 10.44 and Chapter 7, Table 7.7, item h) are examples of the latter. It should now be clear that many of the reactions to be discussed here and which might be called either substitution or addition have been seen before. Some examples are provided in Table 10.2.

$$(10.41)$$

$$(10.42)$$

$$(10.43)$$

$$(10.44)$$

Just as alcohols (or phenols, enols, etc.) can be considered as derived from water (H_2O) by replacement of one proton by an alkyl (or aryl, or alkenyl, etc.) group and ethers can be considered as derived from alcohol (or phenols, enols, etc.) by replacement of both protons by identical (or different) alkyl (or aryl, or alkenyl, etc.) groups, so amines, as noted earlier, can be considered as derived from ammonia (NH_3) by replacement of one or more hydrogens by alkyl or aryl groups.

The substitution of an amino group for a hydroxyl has been of interest for many years. Early workers found that treatment of alcohols (primary alcohols worked best) with ammonia gas (NH_3) in the presence of a thorium oxide (ThO_2) catalyst at high temperature ($>300°C$) and pressure (about 100 atm) for about 2 h resulted in the formation of amines (Equation 10.45).

$$CH_3CH_2CH_2OH + NH_3 \xrightarrow[\text{300°C/100 atm}]{ThO_2} CH_3CH_2CH_2NH_2 + OH_2 \qquad (10.45)$$

TABLE 10.2. Reactions of Amines Encountered in Earlier Chapters

Chapter 5

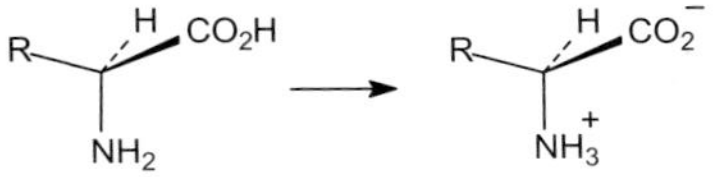

Equation 5.16. The addition of the proton to nitrogen and its loss from the carboxylic acid to form the zwitterion represents the neutral species.

Chapter 6

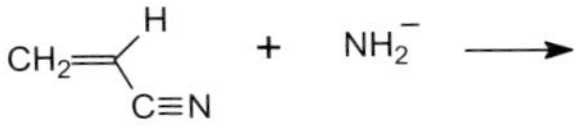

Anionic polymerization; presumably via addition to the alkene, i.e.,

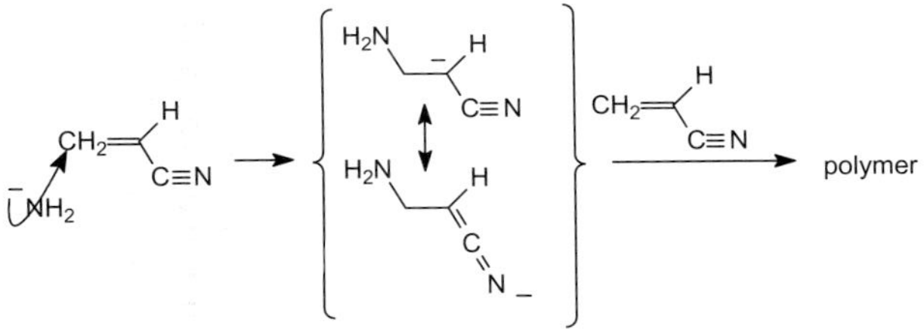

polymer

Chapter 7

CH_3CH_2—Br + :$N(CH_3)_3$ $\longrightarrow$ $[CH_3CH_2N(CH_3)_3]^+$ Br^- S_N2-type displacement reaction Table 7.7, item h

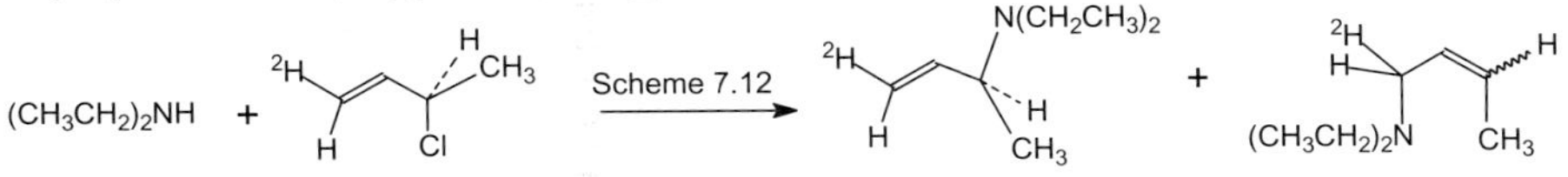

$(CH_3CH_2)_2NH$ +   Scheme 7.12   + S_N2-type displacement reaction accompanied by an apparent S_N2'-type process

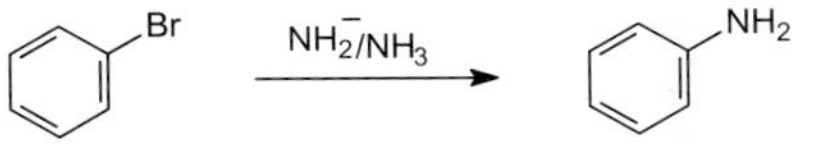

(*E*-) and (*Z*-)

Equation 7.64. The aromatic "substitution" process is shown to occur via a benzyne intermediate and is thus "elimination" and subsequent "addition."

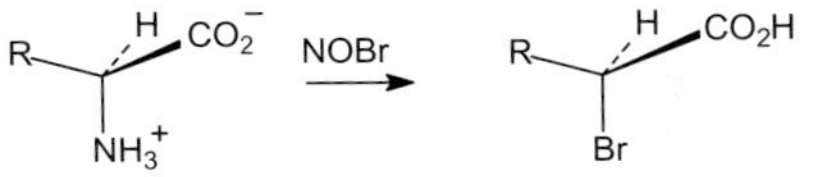

See the footnote on the Walden inverstion, Chapter 7. Presumably, a lactone intrudes to maintain stereochemistry.

Chapter 8

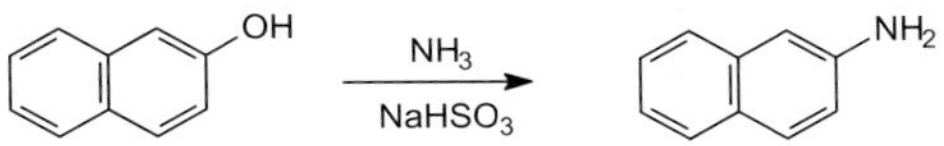

Scheme 8.35

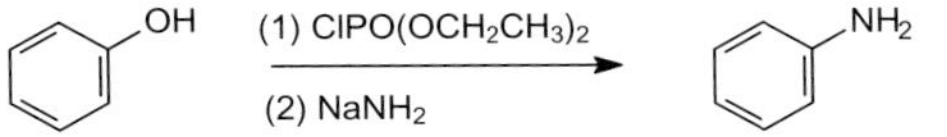

Scheme 8.36

(*Continued*)

TABLE 10.2. *Continued*

Chapter 9

$H_2C=O$ + NH_3 $\longrightarrow$ (structure) + $4H_2O$ — Table 9.4

(acetaldehyde) + H_2N-CH_3 $\longrightarrow$ (structure) an imine + H_2O — Table 9.4

(cyclohexanone) + $H-N$(pyrrolidine) $\longrightarrow$ (structure) an enamine + H_2O — Table 9.4

(alanine) + (phenylpyruvate structure) $\underset{\text{transaminase enzyme}}{\overset{\text{transamination}}{\rightleftharpoons}}$ (pyruvate) + (phenylalanine) —

(benzophenone) + NH_3 $\longrightarrow$ (imine) + H_2O — Equation 9.51

(aniline) + (cyclohexanone) $\longrightarrow$ (imine) + (aminal structure) —

(structure) $\longrightarrow$ (2-methylquinoline) + H_2O — Equation 9.53

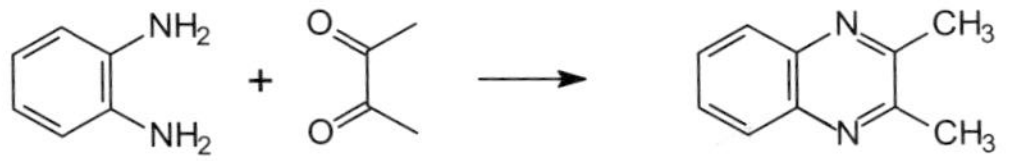

Equation 9.54

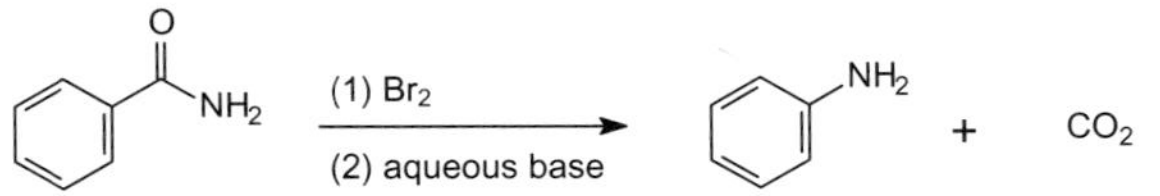

Scheme 9.103 (the Hofmann rearrangement)

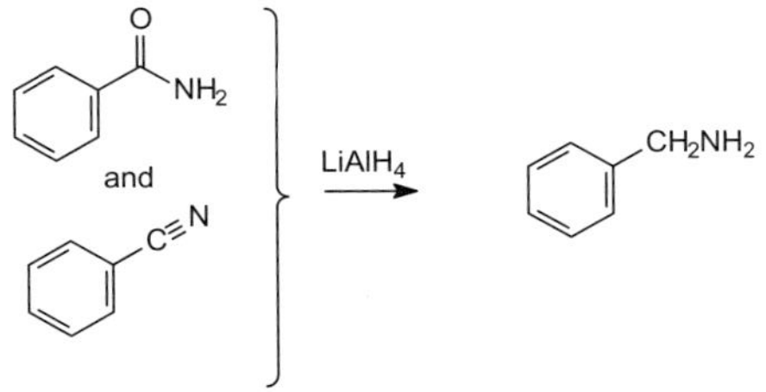

Scheme 9.110

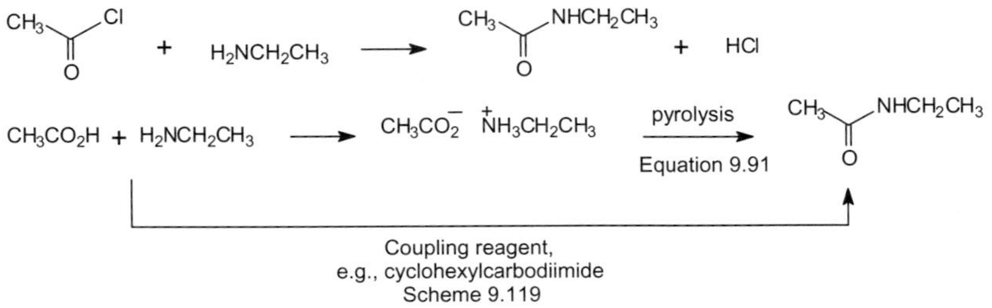

Table 9.7

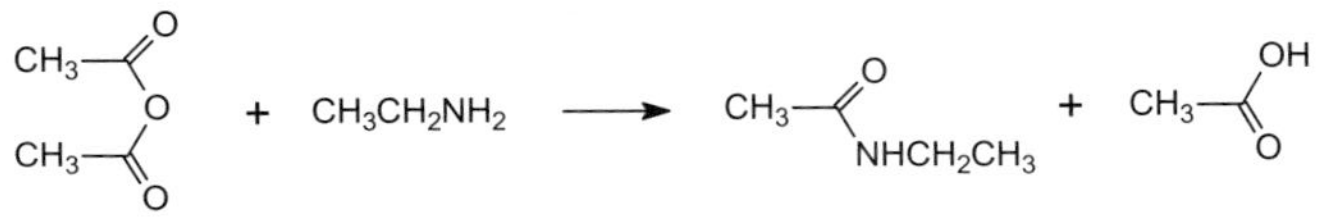

Table 9.8

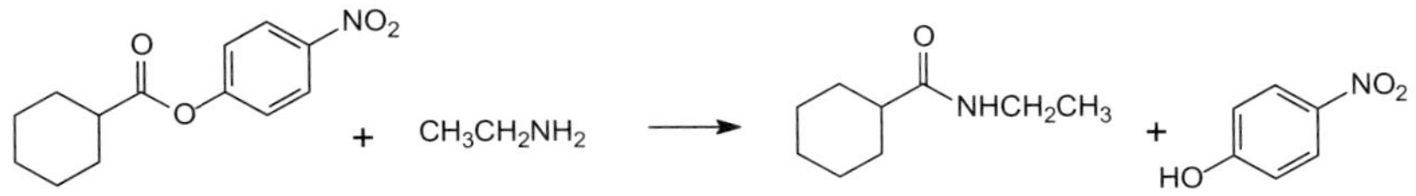

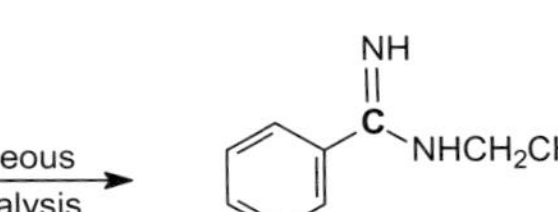

Table 9.9

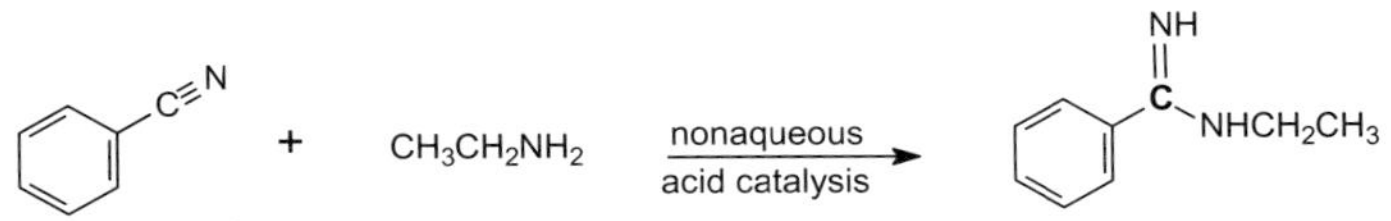

An amidine, by analogy with material in Chapter 9

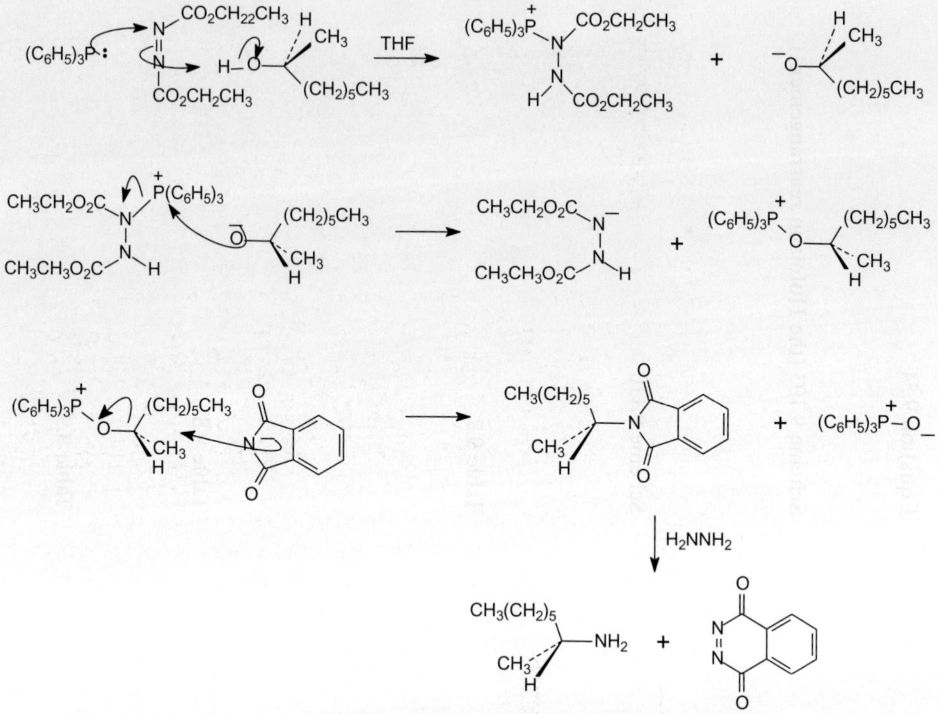

Scheme 10.23. Use of the **Mitsunobu protocol** (alcohol, triphenylphosphine, diethyl azodicarboxylate, and phthalimide in THF) in reaction with (S)-$(+)$-2-octanol to produce (R)-$(-)$-2-octylamine. See Corelli, F.; Summa, V.; Brogi, A.; Monteagudo, E.; Botta, M. *J. Org. Chem.*, **1995**, *60*, 2008.

More recently, the **Mitsunobu protocol** (Chapter 8, Scheme 8.32) has been widely used to effect the substitution as, under the proper conditions, diethyl azodicarboxylate can be shown to react with, for example, secondary alcohols, to form products with inverted absolute stereochemistry. Thus, as shown in Scheme 10.23 (and the details of Scheme 8.32), treatment of (S)-$(+)$-2-octanol, $[\alpha]_D = +14.3°$ (heptane) with diethyl azodicarboxylate, triphenylphosphine, and phthalimide in THF at room temperature for several hours produced the corresponding N-2-octylphthalimide in high yield. Then, treatment of the latter with hydrazine hydrate yielded the desired, optically pure, (R)-$(-)$-2-octylamine in about 50% overall yield (and with complete inversion of configuration).

Addition of ammonia and amines across the carbon–carbon double bond of alkenes was considered restricted to activated unsaturated compounds. However, some cases of addition to inactivated compounds in the presence of alkali metals at high temperatures and pressures had been found. More recently, using selective catalysts and in the presence of carbon monoxide (CO), hydroaminomethylation of terminal olefins has proven itself to be attractive for the synthesis of specific secondary and tertiary aliphatic amines (Equation 10.46, after Ahmed et al., 2003).

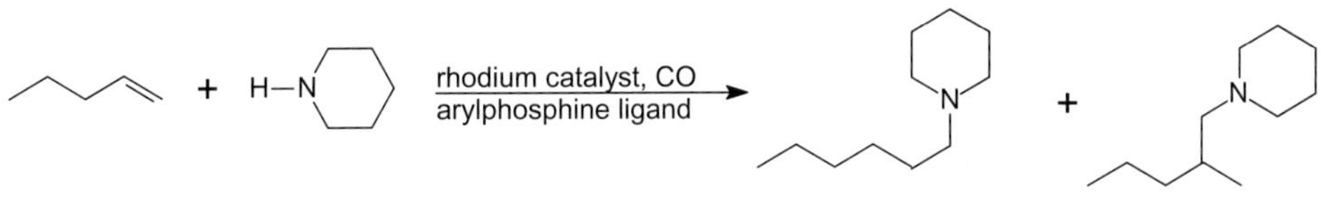

$$(10.46)$$

In an interesting but somewhat more specialized environment, which was initially without precedent but which has, nonetheless, been used in a variety of similar situations subsequently, it proved possible to produce an anion (or radical) by reductive detosylation of an N-tosyl amide and then to allow that reactive species to add to a carbon–carbon double bond. Subsequent quenching of the resulting radical (or anion) with an alcohol or other species then produces an adduct (Scheme 10.24).

As expected, the addition of amines (pK_a of NH_3 [Table 5.16] ≈ 36) to alkynes (pK_a of $HC{\equiv}CH$ [Table 5.16] ≈ 23) can also be effected (Equation 10.47). However, as shown in the equation, the addition does not stop at the vinylamine but rather an acetylenic amine forms due to a second addition of alkyne.

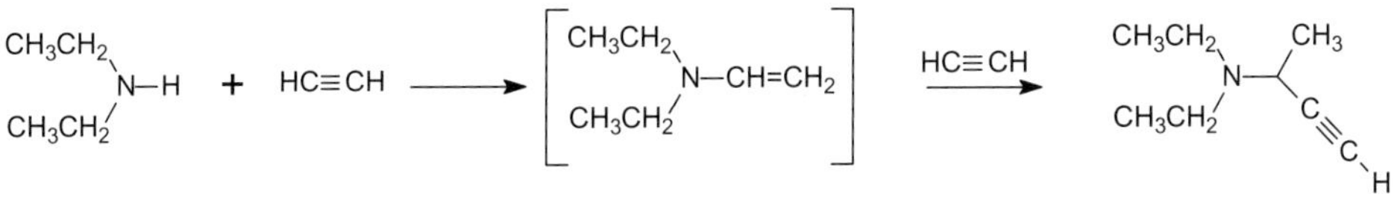

$$(10.47)$$

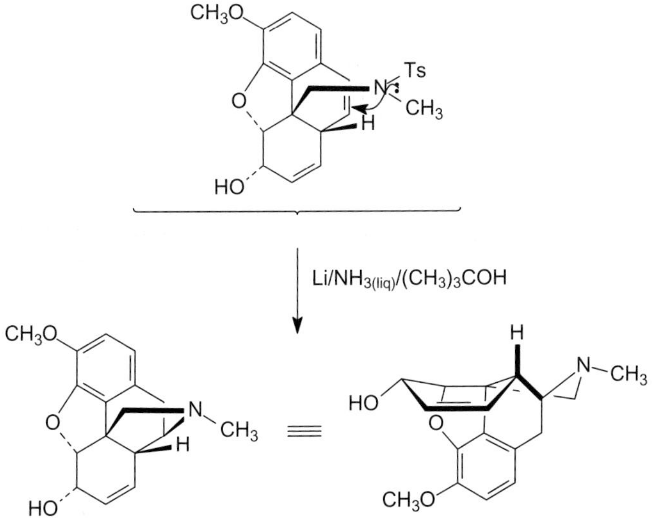

Scheme 10.24. Reductive addition of an amine to a double bond, after Parker, K. A.; Foka, D. *J. Am. Chem. Soc.*, **1992**, *114*, 9688.

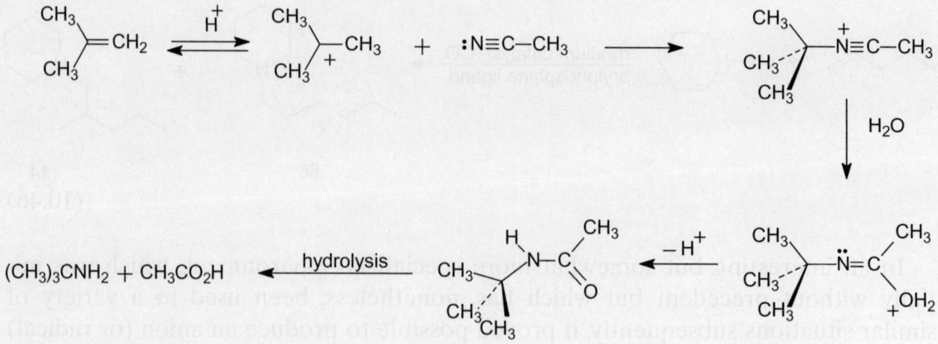

Scheme 10.25. A representation of the reaction between 2-methylpropene, strong acid (e.g., H_2SO_4), and acetonitrile to produce the corresponding acetamide which, on hydrolysis, yields 2-methyl-2-propanamine (*t*-butylamine) and ethanoic acid.

Returning to alkenes and as an aside, it is interesting to note that in specialized circumstances, the nitrogen of hydrogen cyanide (HCN) ($H{\equiv}CN$) or an alkyl nitrile ($R{\equiv}CN$) is capable of undergoing addition to a protonated carbon–carbon double bond. This process, followed by hydrolysis generates, first, an amide and then the corresponding amine (Scheme 10.25).

In a more general sense and despite the pioneering work of Parker (*vide supra*) on the amination of a carbon–carbon double bond in the morphine synthesis, a continuing effort has been under way to find high-yield hydroamination reactions. Various approaches including the use of alkali metals and early- and late-transition metals as well as actinide and lanthanide complexes have been tried and, depending on the specific circumstances, addition to either the more- or less-substituted carbon of an unsymmetrically substituted alkene can be obtained. The generality of the process, while established in principle, continues to require extensive investigation for specific cases.

As noted earlier, the replacement of halogen by an amino group (a substitution reaction at carbon and at nitrogen [!]) is generally not useful for the preparation of primary (RNH_2) or secondary (R_2NH) amines since the alkylamines are stronger bases than ammonia. However, the substitution reaction is useful for tertiary (R_3N) amines, for example, the *N*-methylation of pyridine with methyl iodide (Equation 10.48) and the formation of other quaternary ammonium salts (cf. the **Menschutkin** reaction, Table 7.7, item h).

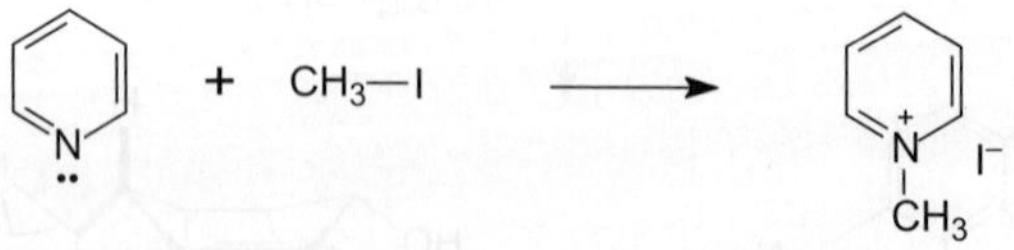

$$\text{pyridine} \; + \; CH_3{-}I \; \longrightarrow \; \text{N-methylpyridinium } I^- \tag{10.48}$$

As noted in Table 10.2 and in the footnote following Figure 7.3, a portion of the expanded **Walden inversion** involved treatment of (S)-(+)-2-aminosuccinic acid (aspartic acid) with NOBr (nitrosyl bromide) to produce (S)-(−)-2-bromosuccinic acid (Equation 10.49). We now know that NOBr acts as an oxidizing agent (see Section I [c], Oxidation with Nitrous Acid and Schemes 10.18–10.20), which causes

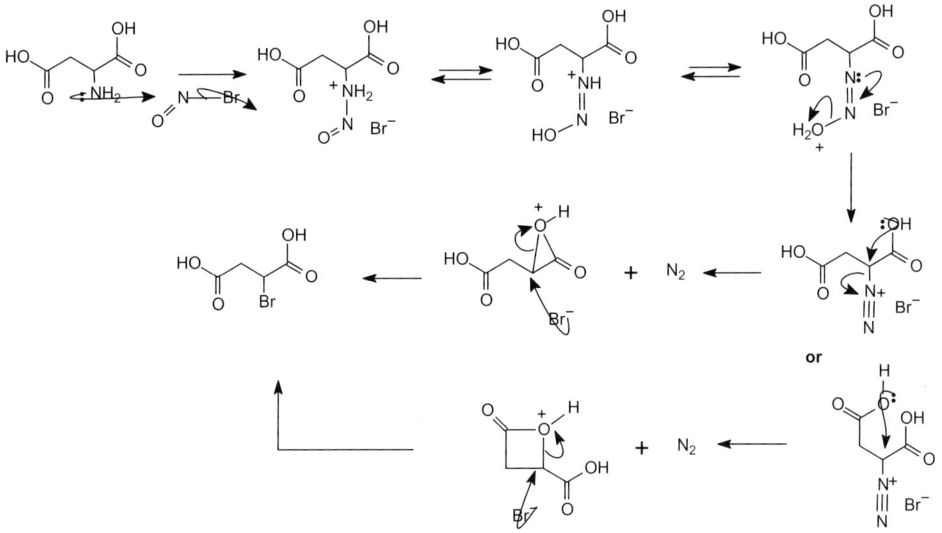

Scheme 10.26. A portion of the scheme (1895) of the original **Walden inversion** sequence of reactions. This part involves treatment of (S)-(+)-2-aminosuccinic acid (aspartic acid) with NOBr (nitrosyl bromide) to produce (S)-(–)-2-bromosuccinic acid and presumably involves formation of a diazonium ion, an α- or β-lactone, and opening of the latter (with overall retention) by bromide anion. Given the bond angles involved, the β-lactone is considered more likely. See Clough, G. W. *J. Chem. Soc. Trans.*, **1918**, *113*, 526.

deamination of the amino acid and results in the formation of a carbocation—which might be expected to undergo attack by bromide anion on *either* face. Nonetheless, it appears that this part of the Walden inversion process proceeds with retention of configuration at carbon (presumably with intrusion of a neighboring carboxylate to form either an α- or a β-lactone) as shown in Scheme 10.26.

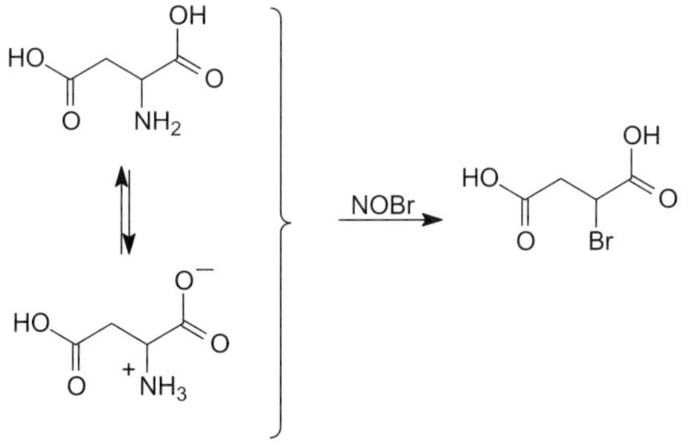

$$(10.49)$$

In the material from Chapter 8 provided in Table 10.2, it should be clear that there are two fundamentally different processes occuring. Thus, in the case of 2-aminonaphthalene (Equation 10.50 and Scheme 8.35), the initial process is one in which the position α to the phenolic –OH undergoes substitution and the resulting adduct subsequently is converted to a carbinolamine, which undergoes dehydration.

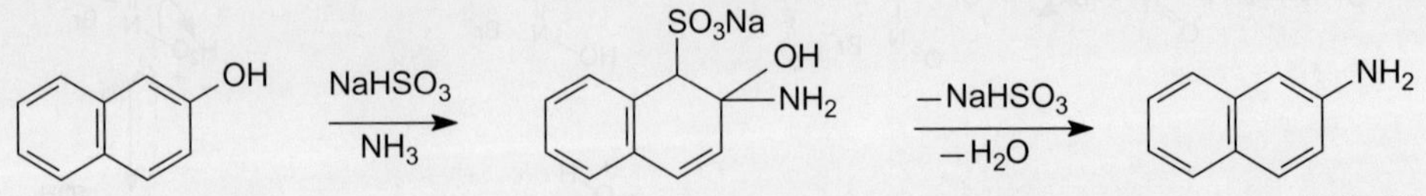

$$(10.50)$$

In contrast, the replacement of the –OH in the second example from Chapter 8 shown in Table 10.2, the conversion of the poor leaving group on the aromatic ring into a good leaving group and treatment with strong base (the amide anion, NH_2^-) in ammonia presumably involves the intermediacy of benzyne (Equation 10.51 and Scheme 8.36).

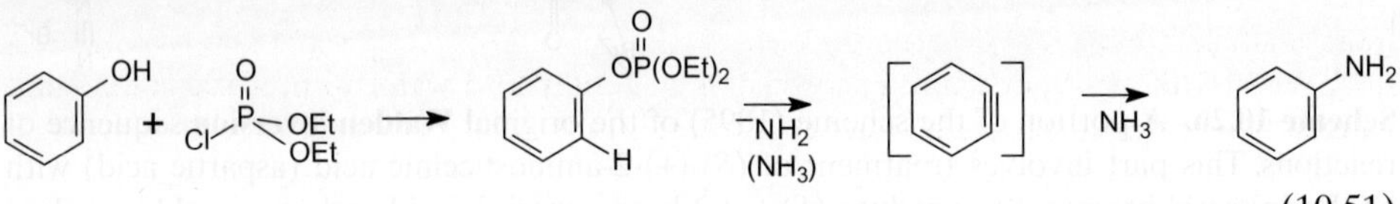

$$(10.51)$$

Interestingly, other benzyne sources accomplish the same end. Thus, as shown in Scheme 10.27, when the *ortho*-trifluoromethanesulfonate (triflate, –OTf) derivative of phenyltrimethylsilane [i.e., 2-(trimethylsilyl)-phenyl triflate] is treated with cesium fluoride in acetonitrile (cyanomethane) at room temperature in the presence of *N*-phenyltrifluoroacetamide, 2,2,2-trifluoro-1-[2-(phenylamino)phenyl]ethanone results. Presumably, elimination occurs to produce benzyne, and this is followed by addition and rearrangement.

Almost without exception, the material from Chapter 9 and Table 9.4, shown again in Table 10.2, hinges on the addition of the electron-rich amino function to the carbon of the carbonyl group. It is not uncommon to find that this process is accelerated (as noted before, Chapter 9, Figure 9.10) by slightly acidic media (although oxime formation is an exception) and that the angle of attack is given by the Bürgi–Dunitz angle (Scheme 10.28). Subsequently, elimination occurs (Scheme 10.29).

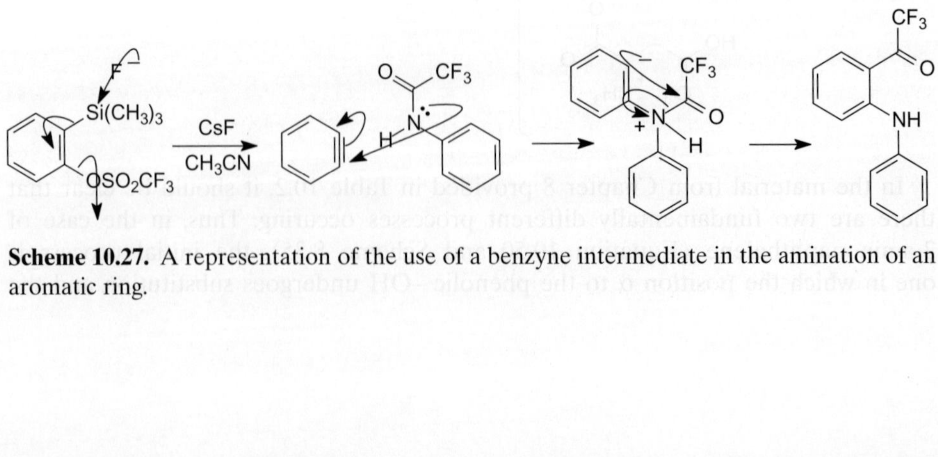

Scheme 10.27. A representation of the use of a benzyne intermediate in the amination of an aromatic ring.

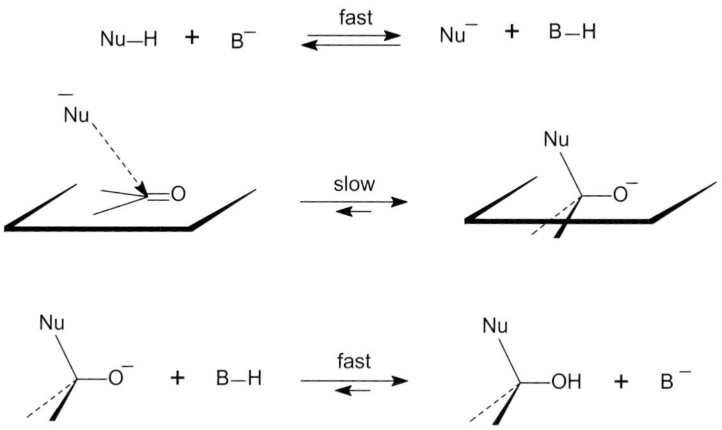

Scheme 10.28. A repeat of Scheme 9.35. Here, Nu represents an amino function and the attack of the incoming group is drawn to be greater than 90° as suggested by Bürgi et al. (Bürgi, H. B.; Dunitz, J. D.; Shefter, E. *J. Am. Chem. Soc.*, **1973**, *95*, 5065). Affirming the data in Table 10.4, the carbonylamines shown here generally lose water to produce compounds with carbon–nitrogen double bonds. See Scheme 10.29.

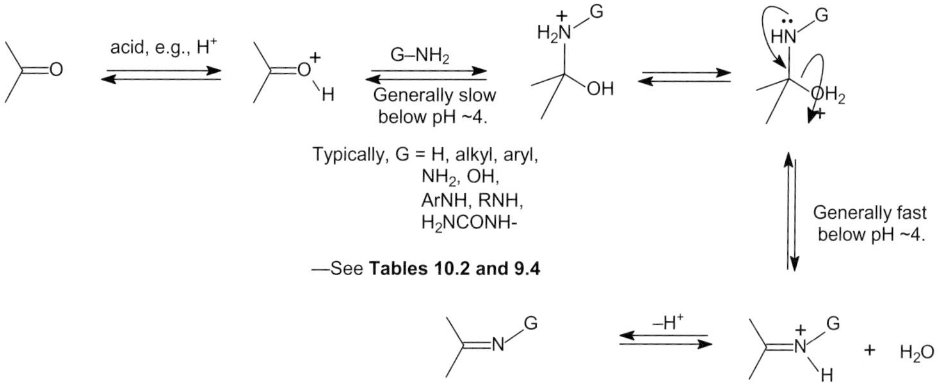

Typically, G = H, alkyl, aryl,
NH$_2$, OH,
ArNH, RNH,
H$_2$NCONH-

—See **Tables 10.2 and 9.4**

Scheme 10.29. A cartoon representation of the overall pathway for the conversion of compounds containing a carbon–oxygen double bond (C=O) to the corresponding compound containing a carbon–nitrogen double bond (C=N).

F. ADDITION AND REARRANGEMENT REACTIONS OF AMINES

As already seen, substitution reactions at nitrogen can be difficult because reaction can occur more than once (*vide supra*). Nonetheless, it is possible to create aziridines (azacyclopropanes), azetidines (azacyclobutanes), and pyrrolidines (azacyclopentanes) this way. Thus, 4-chloro-1-aminobutane can undergo base-catalyzed cyclization to produce pyrrolidine (azacyclopentane) (Equation 10.52) and 2-chloro-2-aminoethane yields aziridine (azacyclopropane) the same way (Equation 10.53). Although azetidine (azacyclobutane) cannot be prepared the same way, treatment of 1,3-dibromopropane with *p*-toluenesulfonamide in the presence of base (*vide supra*, Scheme 10.3 and the discussion accompanying the **Hinsberg reaction**) does produce the azetidine-*N*-*p*-toluenesulfonamide and the *N*-substituent can be removed by sodium in alcohol reduction (Equation 10.54).

$$(10.52)$$

$$(10.53)$$

$$(10.54)$$

Interestingly, simply heating the hydrochloride salts of 1,4-diaminobutane (putrescine) and 1,5-diaminobutane (cadaverine) also provides the corresponding cyclic derivatives. Thus, the former leads to pyrrolidine (azacyclopentane) (Equation 10.55) while the latter produces piperidine (azacyclohexane) (Equation 10.56). In the same vein, the unsaturated heterocyclic compound pyrrole (aza-2,4-cyclopentadiene) can also be made by a substitution reaction. Thus, if furan (oxa-2,4-cyclopentadiene, Chapter 8, Scheme 8.102) and ammonia (NH_3) or furan and a primary amine (RNH_2) are passed in the vapor phase over an alumina catalyst, pyrrole (aza-2,4-cyclopentadiene) (or an *N*-substituted pyrrole) forms (Equation 10.57).

putrescine
hydrochloride

$$(10.55)$$

cadaverine
hydrochloride

$$(10.56)$$

R = H or alkyl

$$(10.57)$$

The corresponding six-membered heterocycle, pyridine (azabenzene) was initially obtained by the dry distillation of bones, and much of it is obtained from pyrolysis of coal or oil shale. There is a commercial synthesis that utilizes the vapor-phase reaction between formaldehyde (methanal, $H_2C{=}O$), acetaldehyde (ethanal,

H_3CCHO), and ammonia over a proprietary catalyst at high temperature. The details of this process are not well worked out and some 4-methylpyridine (4-methylazabenzene) accompanies the parent unsubstituted product and must be separated by distillation (Equation 10.58).

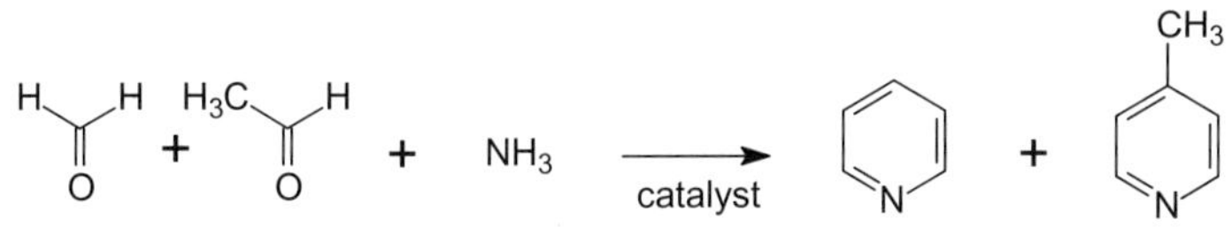

$$(10.58)$$

Additional members of the vast general class of heterocyclic compounds (containing nitrogen as well as other "hetero" atoms) will be discussed in Part III.

However, it is worth returning to the chemistry of hexamethylenetetramine (1,3,5,7-tetraazaadamantane or 1,3,5,7-tetraazatricyclo[3.3.1.1^{3,7}]decane, $C_6H_{12}N_4$, Chapter 9 and Table 10.2) that is formed by the condensation of ammonia (NH_3) with methanal (formaldehyde, $H_2C=O$). As was noted as early as 1895 by M. Delépine, in a reaction now bearing his name, alkyl halides react with hexamethylenetetramine to produce N-substituted derivatives of the hexamethylenetetramine, which can then be hydrolyzed to aldehyde, ammonia, and substituted amine. Not unexpectedly, it has become clear that benzylic and allylic halides react best (Scheme 10.30).

Problem 10.2. Write a pathway for the hydrolysis of the substituted N-benzylhexamethylene-tetramine to methanal (formaldehyde, H_2CO), the ammonium ion, and the benzylammonium salt shown in Scheme 10.30.

As already noted, the amino group can affect displacement reactions on halogen (*vide supra*) and, in the presence of activating reagents (see, e.g., the **Mitsunobu protocol**, Chapter 8, Scheme 8.32 and this chapter, Scheme 10.23), can also displace oxygen. In that vein, it is worthwhile noting that β-aminoalcohols can be prepared by attack of amines on relatively unhindered epoxides (oxiranes) (Equation 10.59).

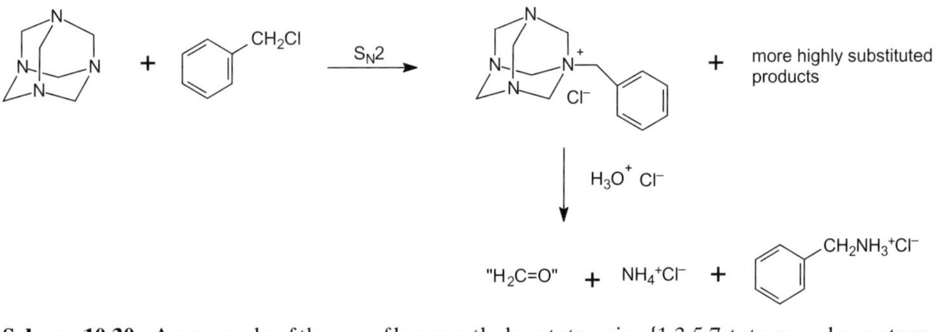

Scheme 10.30. An example of the use of hexamethylenetetramine {1,3,5,7-tetraazaadamantane or 1,3,5,7-tetraazatricyclo[3.3.1.1^{3,7}]decane}, followed by hydrolysis to effect replacement of a halogen by nitrogen (the **Delépine reaction**, see Delépine, M. *C. R. Séances Acad. Sci.*, **1895**, *120*, 501).

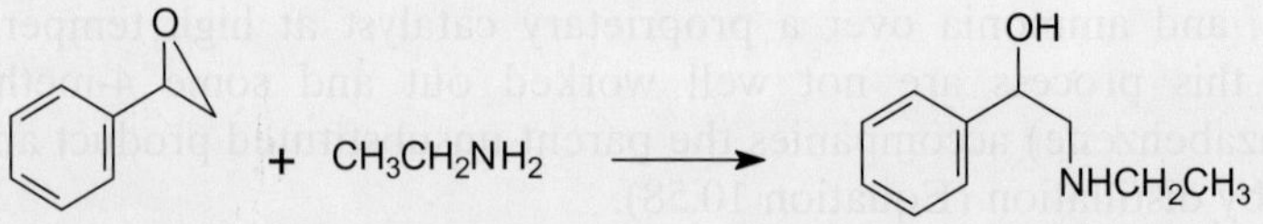

$$(10.59)$$

This type of displacement can, of course also be effected by other nitrogenous reactants. Indeed, there is a family of such species, which, although not amines themselves can, be converted to amines by reductive or hydrolytic methods. Thus, amides, can be alkylated either upon nitrogen or oxygen—the former yielding a new amine on hydrolysis or reduction (Scheme 10.31), and the latter yielding an imidate (frequently isolated as a salt), which, on aminolysis, can produce an amidine (Scheme 10.32). The alkylation (on nitrogen or oxygen and sometimes on both) can be accomplished by adjusting the conditions. For alkylation on nitrogen (Scheme 10.31), the amide anion is generated with, for example, potassium hydride, KH, in benzene solvent, while for O-alkylation (Scheme 10.32), a potent alkylating agent [e.g., triethyloxonium tetrafluoroborate, $(CH_3CH_2)_3O^+BF_4^-$] is employed in dichloromethane (CH_2Cl_2) or chloroform ($CHCl_3$) solution.

In addition to the "usual" substitution reactions directly on the nitrogen atom of the amino or amido group, there are (as noted earlier, e.g., Table 7.7) substitution reactions with other nucleophiles that can be *converted* to the amino (or substituted amino) group. These include incorporation of the azido function (produced by, e.g., a nucleophilic substitution reaction of azide anion [N_3^-] on an alkyl halide) and its subsequent reduction (with lithium aluminum hydride, $LiAlH_4$) (Equation 10.60) or triphenylphosphine [$(C_6H_5)_3P$] (Equation 10.61) to the corresponding amine, as well as a similar displacement reaction with isocyanate ($O=C=N^-$) (Equation

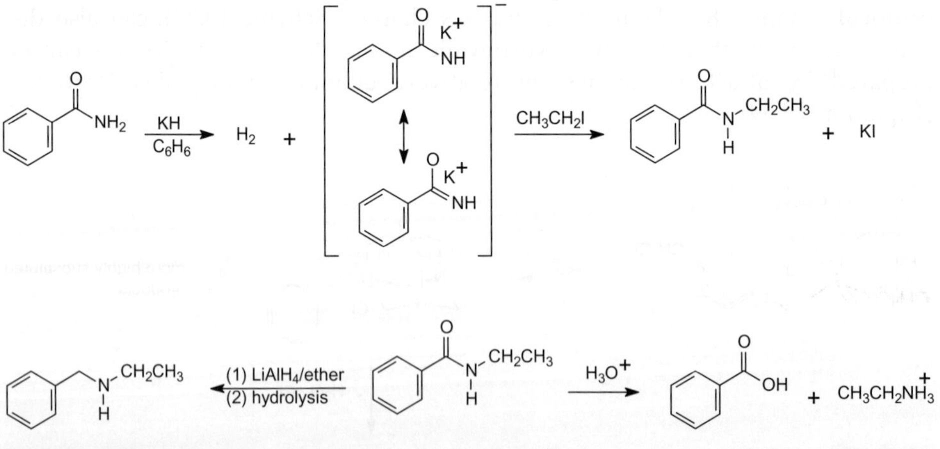

Scheme 10.31. Alkylation on nitrogen. Benzamide (benzenecarboxamide) is alkylated by iodoethane (ethyl iodide, CH_3CH_2I) on nitrogen after generation of the amide anion with potassium hydride (KH). Subsequently, the *N*-ethylbenzamide (*N*-ethylbenzenecarboxamide) can be hydrolyzed by aqueous acid to produce benzoic acid (benzenecarboxylic acid) and the corresponding ethylammonium salt. Alternatively, the *N*-ethylbenzamide (*N*-ethylbenzene-carboxamide) can be reduced with, for example, lithium aluminum hydride ($LiAlH_4$) in ether [$(CH_3CH_2)_2O$], and, after hydrolysis, *N*-benzylethanamine is produced.

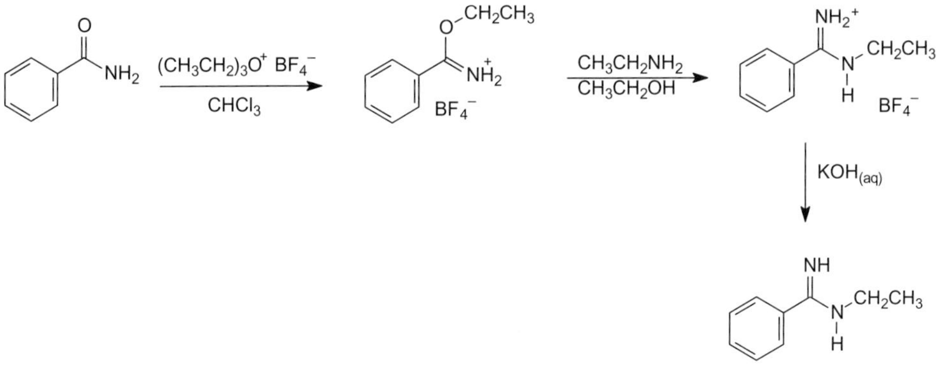

Scheme 10.32. Alkylation on oxygen. Benzamide (benzenecarboxamide) is alkylated by triethyloxonium tetrafluoroborate $[(CH_3CH_2)_3O^+ \ BF_4^-]$ on oxygen in chloroform $(CHCl_3)$ solution. Subsequently, the benzimidate tetrafluoroborate salt can be aminated by treatment with ethanolic ethanamine (ethylamine, $CH_3CH_2NH_2$) to produce the corresponding amidine.

10.62) or isothiocyanate $(S=C=N^-)$ (Equations 10.64 and 10.65) anions. As shown in Equation 10.62, reaction with the isocyanate anion produces an alkyl isocyanate,* which can subsequently undergo alcoholysis to yield a urethane (a carboxyamic acid ester, i.e., a carbamate, $RNHCO_2R'$) or aminolysis (Equation 10.63) to yield an *N*-alkylurea, $RNHCONH_2$. As shown in Equation 10.64, both isothiocyanates (R–N=C=S) and thiocyanates (R–S–C≡N) (Equation 10.65) can be isolated from the reaction between the alkyl halide and the ambident nucleophilic isothiocyanate (thiocyanate) anion. Reactions of these nitrogen- and sulfur-containing species parallel those of the oxygen analogues.

*Interestingly, isocyanic acid, H–N=C=O, is written as if it were in equilibrium with its proton tautomer cyanic acid, N≡C–OH, although there is no spectroscopic evidence for the presence of the latter. Isocyanic acid, H–N=C=O, which might be expected to arise by the acidification of a salt such as potassium isocyanate, $K^{+ \, -}N=C=O$, is unstable and treatment of the salt with aqueous acid results in the formation of ammonia (NH_3) and carbon dioxide (CO_2). In the same vein, when urea (H_2NCONH_2) is heated above its melting point, it decomposes into ammonia (NH_3) and isocyanic acid (HNCO) and the latter, in the absence of a reacting partner reacts with itself to produce cyanuric acid (below) and the linear polymer "cyamelide." Heating cyanuric acid regenerates isocyanic acid.

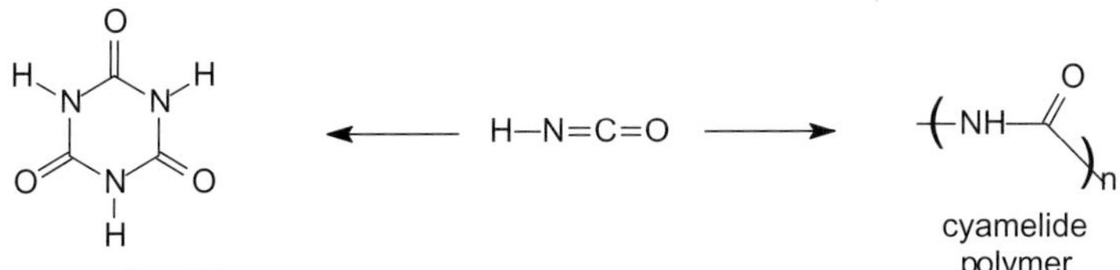

cyanuric acid

H–N=C=O

cyamelide
polymer

The other potential isomer of HOCN (in addition to isocyanic acid [H–N=C=O] and cyanic acid [H–O–C≡N]) is fulminic acid (H–O–N$^+$≡C$^-$), which is also incapable of isolation as the free acid (since attempts to do so result in polymerization). However, salts of fulminic acid are known and mercury fulminate, prepared by treatment of a mixture of mercury and ethanol with nitric acid, is used commercially as a detonator (blasting cap) for explosives.

$$\text{PhCH}_2\text{Br} \xrightarrow[\text{acetone}]{\text{NaN}_3} \text{PhCH}_2\text{-N}=\overset{+}{\text{N}}=\text{N}^- \xrightarrow[\text{(2) aqueous work-up}]{\text{(1) LiAlH}_4 \text{ in ether}} \text{PhCH}_2\text{NH}_2 + \text{H}_2\text{NNH}_2$$

$$(10.60)$$

$$\text{PhCH}_2\text{-N}=\overset{+}{\text{N}}=\bar{\text{N}} \xrightarrow[\text{toluene}]{(C_6H_5)_3P} N_2 + \text{PhCH}_2\text{-N}=\text{P}(C_6H_5)_3 \xrightarrow{H_3O^+} \text{PhCH}_2\text{-}\overset{+}{\text{NH}}_3 + (C_6H_5)_3\text{PO}$$

$$(10.61)$$

$$\text{PhCH}_2\text{Br} \xrightarrow[\text{acetone}]{\text{KOCN}} \text{PhCH}_2\text{-N}=\text{C}=\text{O} \xrightarrow{CH_3OH} \text{PhCH}_2\text{-N(H)-C(=O)-OCH}_3 \quad (10.62)$$

$$\text{PhCH}_2\text{Br} \xrightarrow[\text{acetone}]{\text{KOCN}} \text{PhCH}_2\text{-N}=\text{C}=\text{O} \xrightarrow{NH_3} \text{PhCH}_2\text{-N(H)-C(=O)-NH}_2 \quad (10.63)$$

$$\text{PhCH}_2\text{Br} \xrightarrow[\text{acetone}]{\text{KSCN}} \text{PhCH}_2\text{-N}=\text{C}=\text{S} \xrightarrow{NH_3} \text{PhCH}_2\text{-N(H)-C(=S)-NH}_2 \quad (10.64)$$

$$\text{PhCH}_2\text{Br} \xrightarrow[\text{acetone}]{\text{KSCN}} \text{PhCH}_2\text{-S-C}\equiv\text{N} + \text{PhCH}_2\text{-N}=\text{C}=\text{S} \quad (10.65)$$

It is worthwhile reconsidering (as a reprise) the **Beckmann, Curtius, Hofmann, Lossen,** and **Schmidt rearrangements**, which, although the first is distinctly different from the last four, are often lumped together in amine-type rearrangements. The similarity derives from, in all cases, the migration of a group from carbon to nitrogen **with** retention of stereochemistry at the migrating carbon. Furthermore, as noted below, there is a variant of the **Schmidt rearrangement** that bears a striking resemblance to the **Beckmann** process.

In the **Beckmann rearrangement**, it will be recalled, the hydroxyl of an oxime (derived from an aldehyde or ketone) is converted into a leaving group and, as it leaves, the alkyl group attached to the carbon bearing the doubly bound nitrogen migrates. As shown in Scheme 10.33, if the (*E*)-oxime derived from optically active 3-methyl-2-pentanone undergoes the rearrangement, the amide that results retains the sense of handedness at the carbon that has moved.

The synthesis of chiral amines is important because many pharmaceuticals and other materials contain the amino (–NH$_2$) functional group and current good practice requires that a single enantiomer be tested and sold. If synthesis of an amine where the nitrogen is attached to an asymmetrically substituted carbon is wanted, special techniques can sometimes be employed.

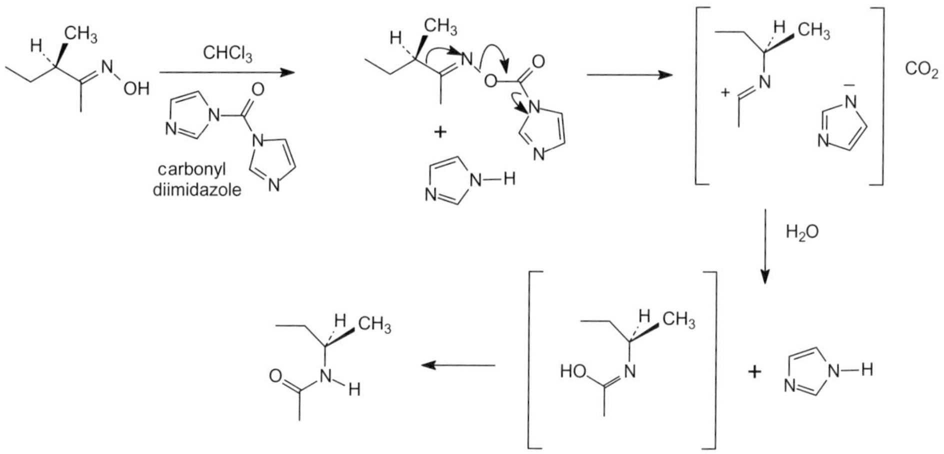

Scheme 10.33. An example of the **Beckmann rearrangement** to produce an amide (that can be hydrolyzed to an amine) showing migration of the alkyl group with retention of configuration.

Among the more successful techniques has been to make use of addition reactions to the carbon of the carbon–nitrogen double bond found in imines that bear suitable substituents.* For example, *tert*-butyl disulfide can be oxidized by hydrogen peroxide in the presence of **both** a vanadium [a vanadium(IV) complex with acetylacetone (pentane-2,4-dione), i.e, VO(acac)$_2$, was used] catalyst in chloroform (CHCl$_3$) solution **and** a chiral iminoalcohol [viz. (*S*)-2-(*N*-3,5-di-*t*-butylsalicylidene)amino-3,3-dimethyl-1-butanol]. The latter is capable of exchanging with the acetylacetonate ("acac," pentane-2,4-dione) to produce the chiral sulfinate, (*S*)-*t*-butyl-*t*-butanethiosulfinate (Scheme 10.34). Then, as also shown in the scheme, addition of the sulfinate in THF (oxocyclopentane, THF) to a suspension of lithium amide (LiNH$_2$) in liquid ammonia (NH$_{3(l)}$) produced (*R*)-*t*-butanesulfinamide. When the optically active sulfinamide was allowed to react with an aldehyde (e.g., propionaldehyde), the corresponding sulfinimine resulted. Reaction of the latter with a Grignard reagent (e.g., phenylmagnesium bromide, C$_6$H$_5$MgBr) in ether to yield *N*-(1-phenylpropyl)-*t*-butanesulfinamide, followed by hydrolysis of the product in methanolic HCl generated the corresponding (1*R*)-1-phenylpropylamine hydrochloride in high yield and high diastereomeric excess. A cyclic transition state was proposed to account for the stereochemical outcome (Scheme 10.34).

It should be clear that this method generates the chirality rather than maintains it. In the **Curtius rearrangement**[†] (Scheme 10.35), an acid chloride, such as that from **chiral** 2-methylbutanoic acid [CH$_3$CH$_2$CH(CH$_3$)CO$_2$H] is first converted to the corresponding acid chloride and the latter treated with sodium azide. Decomposition of the acyl azide so formed generates the corresponding **chiral** 2-butylisocyanate, which, on hydrolysis, produces carbon dioxide (CO$_2$) and the **chiral** 2-aminobutane (Scheme 10.35).

---

*A review of this process can be found in Ellman, J. A.; Owens, T. D.; Tang, T. P. *Acc. Chem. Res.*, **2002**, *35*, 984.

[†]Curtius, T. *Chem. Ber.*, **1890**, *23*, 3023.

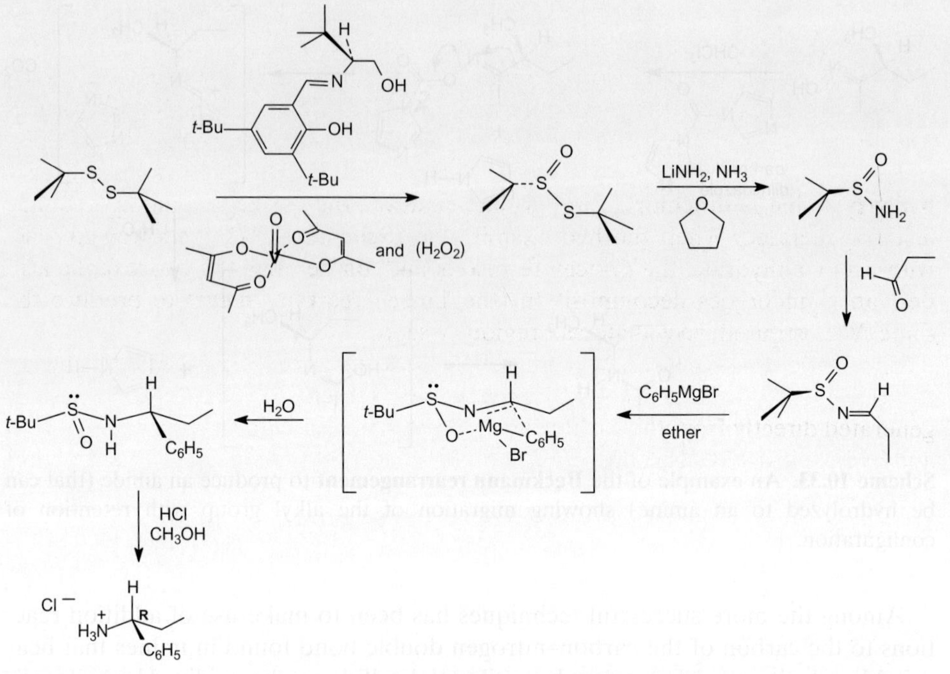

Scheme 10.34. The synthesis of (1*R*)-1-phenylpropylamine hydrochloride. Beginning with *tert*-butyl disulfide, oxidation with hydrogen peroxide in the presence of VO(acac)$_2$ and (*S*)-2-(*N*-3,5-di-*t*-butylsalicylidene)amino-3,3-dimethyl-1-butanol produces the chiral sulfinate (*S*)-*t*-butyl-*t*-butanethiosulfinate. Then, addition of the sulfinate in THF (oxocyclopentane, THF) to a suspension of lithium amide (LiNH$_2$) in liquid ammonia (NH$_{3(l)}$) generates (*R*)-*t*-butanesulfinamide. The optically active sulfinamide reacts with propanal to form the corresponding sulfinimine. Reaction of the latter with phenylmagnesium bromide (C$_6$H$_5$MgBr) in ether yields *N*-(1-phenylpropyl)-*t*-butanesulfinamide and hydrolysis in methanolic HCl generates the corresponding (1*R*)-1-phenylpropylamine hydrochloride. The cartoon drawing of the presumed cyclic transition state is used to account for the stereochemical outcome. But see Hose, D. R. J.; Mahon, M. F.; Molloy, K. C.; Raynham, T.; Wills, M. *J. Chem. Soc. Perkin Trans. I*, **1996**, *7*, 691.

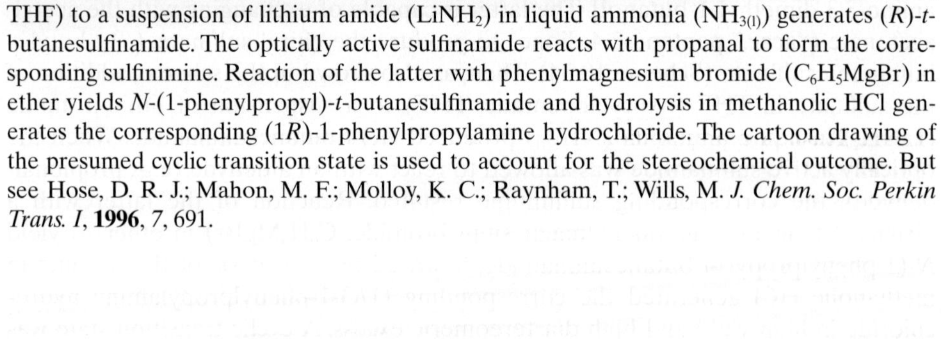

Scheme 10.35. A representation of the **Curtius rearrangement** showing that the amine prepared from the corresponding carboxylic acid retains the absolute configuration.

The same isocyanate derivative is also seen in the **Hofmann rearrangement** (Scheme 10.36). That is, if the carboxylic acid chloride from the 2-methylbutanoic acid is first converted to the corresponding amide by reaction with ammonia (NH_3) and the latter treated with bromine (Br_2) and sodium hydroxide (i.e., NaOBr formed from reaction of the two reagents), the corresponding *N*-bromo compound forms.

If the acid chloride prepared as in Schemes 10.35 and 10.36 is treated with hydroxylamine hydrochloride in pyridine solution, the corresponding hydroxamic acid is generated. When the hydroxamic acid (Scheme 10.37) is allowed to react with acetic anhydride, the O-acetate results and, on heating, the hydroxamic acid derivative undergoes decomposition (the **Lossen rearrangement**) to produce the same *N*-substituted isocyanate as previously seen.

The **Schmidt rearrangement** is essentially the same as the **Curtius rearrangement** (as indeed are the **Hofmann** and **Lossen rearrangements**) in that the acid azide is generated directly from the carboxylic acid on treatment with sulfuric acid (H_2SO_4) and hydrazoic acid (HN_3). So, formation of the acid chloride is not necessary, but it

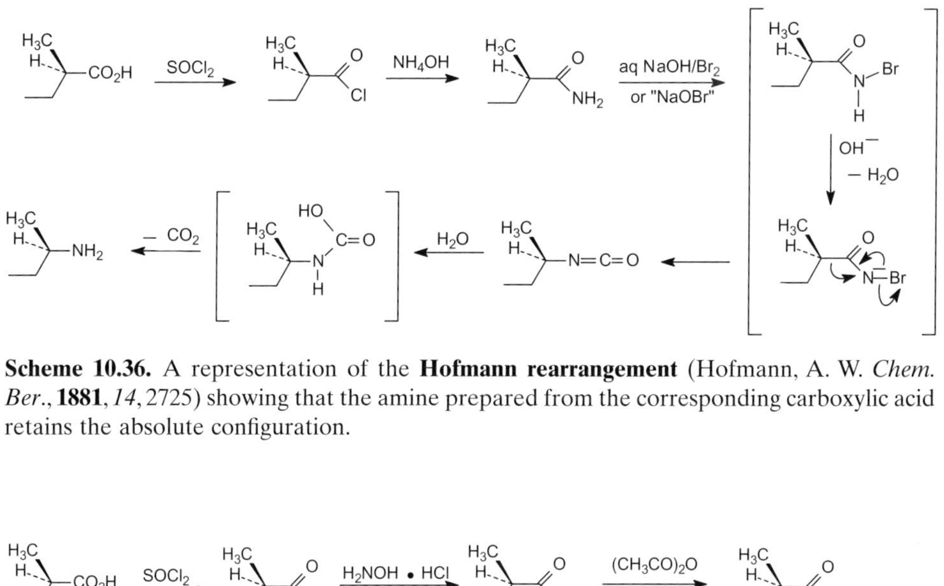

Scheme 10.36. A representation of the **Hofmann rearrangement** (Hofmann, A. W. *Chem. Ber.*, **1881**, *14*, 2725) showing that the amine prepared from the corresponding carboxylic acid retains the absolute configuration.

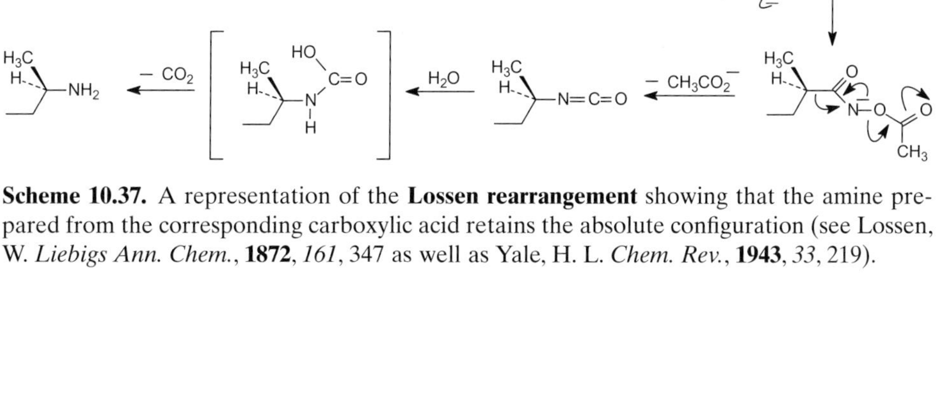

Scheme 10.37. A representation of the **Lossen rearrangement** showing that the amine prepared from the corresponding carboxylic acid retains the absolute configuration (see Lossen, W. *Liebigs Ann. Chem.*, **1872**, *161*, 347 as well as Yale, H. L. *Chem. Rev.*, **1943**, *33*, 219).

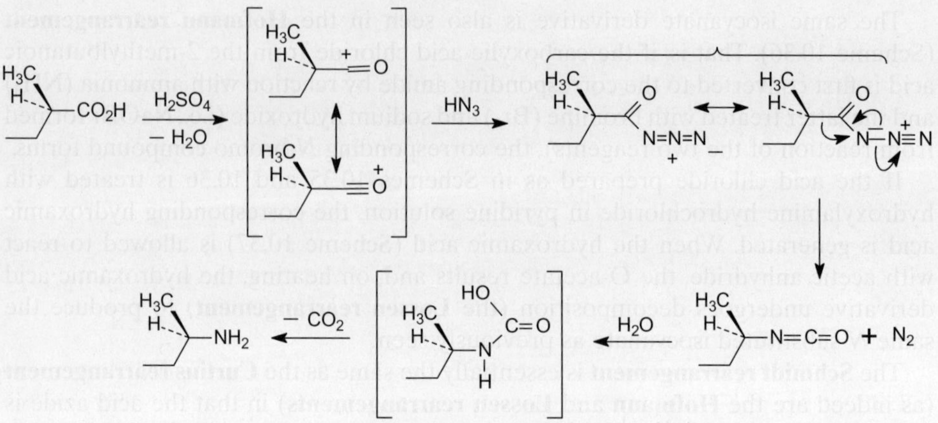

Scheme 10.38. A representation of the **Schmidt rearrangement** showing that the amine prepared from the corresponding carboxylic acid retains the absolute configuration of the acid (see Schmidt, R. F. *Angew. Chem.*, **1923**, *36*, 511; Schemit, R. F. *Chem. Ber.*, **1924**, *57*, 704; Benson, R. F. *Chem. Rev.*, **1947**, *41*, 1).

is necessary that the alkyl (or aryl) group that is to migrate be stable to the rather vigorous conditions (Scheme 10.38).

Finally, in this regard, it is worthwhile to recall that the substituted isocyanate need not be converted to the amine. That is, reaction of the isocyanate with an alcohol yields the corresponding carboxyamic acid ester (the urethane) and reaction with an amine generates an *N*-substituted urea. Because water is ubiquitous, the urea is often isolated as a "side product" (Equation 10.66).

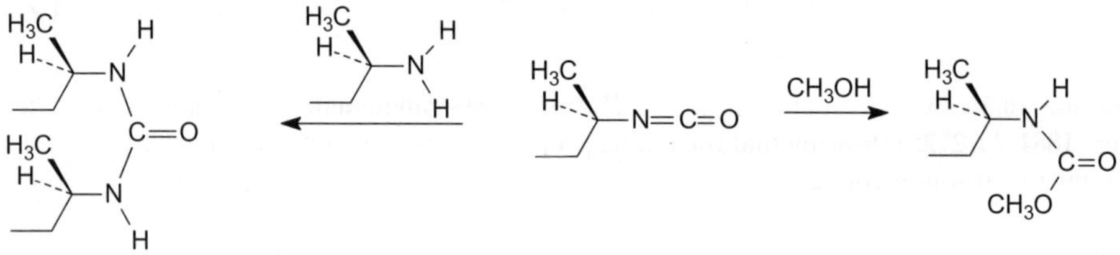

$$(10.66)$$

Interestingly, the **Schmidt rearrangement** can also be carried out on ketones (Scheme 10.39). As shown in the scheme, cyclohexanone, on treatment with hydrazoic acid (generated from sodium azide [NaN_3] and sulfuric acid [H_2SO_4]), undergoes rearrangement with ring expansion that leads directly to the corresponding amide, hexahydro-2H-azepin-2-one (2-oxohexamethyleneimine, ε-caprolactam).[*]

Addition of **Grignard** reagents to the carbon of the carbonyl group of dialkyl formamides (the **Bouveault reaction**)[†] can follow either of two possible paths. Depending on the Grignard reagent and the formamide, either (a) deformylation of the amide with the generation of the dialkyl amine and the corresponding aldehyde occurs or (b) simple addition occurs. The latter is followed by loss of water to

---

[*]See, additionally, Hutchison, G. I.; Prager, R.; Ward, A. D. *Aust. J. Chem.*, **1980**, *33*, 2477.
[†]Bouveault, L.; Blanc, G. *Compt. Rend. Séances Acad. Sci.*, **1903**, *136*, 1676.

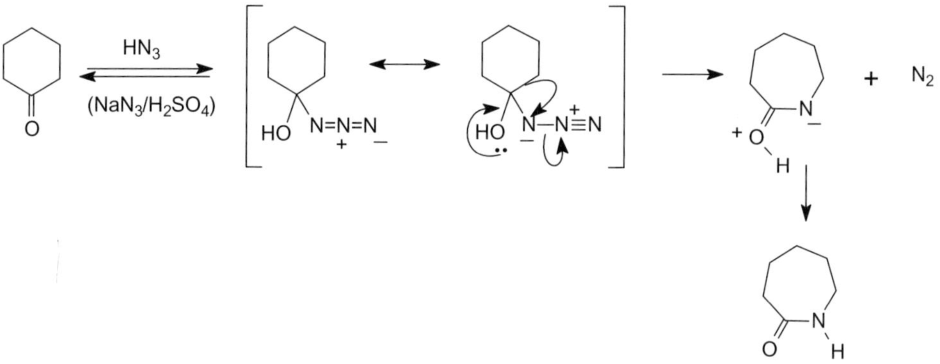

Scheme 10.39. The **Schmidt rearrangement** as carried out on cyclohexanone to yield the corresponding amide, hexahydro-2H-azepin-2-one (2-oxohexamethyleneimine, ε-caprolactam).

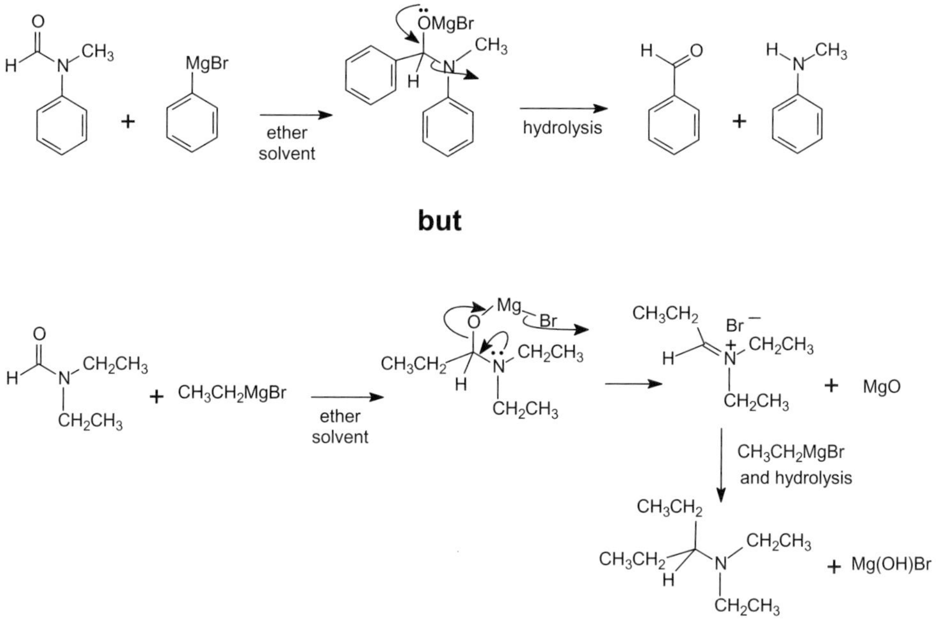

but

Scheme 10.40. The **Bouveault reaction**. Addition of phenylmagnesium bromide in ether to N-methylformanilide followed by hydrolysis yields, as the major products, benzaldehyde (benzenecarboxyaldehyde) and N-methylaniline (N-methylbenzenamine). However, addition of ethylmagnesium bromide in ether to N,N-diethylformamide produces (N,N-diethyl)-3-amino-pentane (after Smith, L. I.; Bayless, M. *J. Org. Chem.*, **1941**, *6*, 437).

an imine and then a second addition of Grignard reagent to yield, ultimately, a tertiary amine (Scheme 10.40).

Interestingly, in the same vein, other aryl amines (often simply referred to as anilides as they are clearly related to aminobenzene [aniline]) are frequently found to undergo specific rearrangements as a consequence of the presence of the

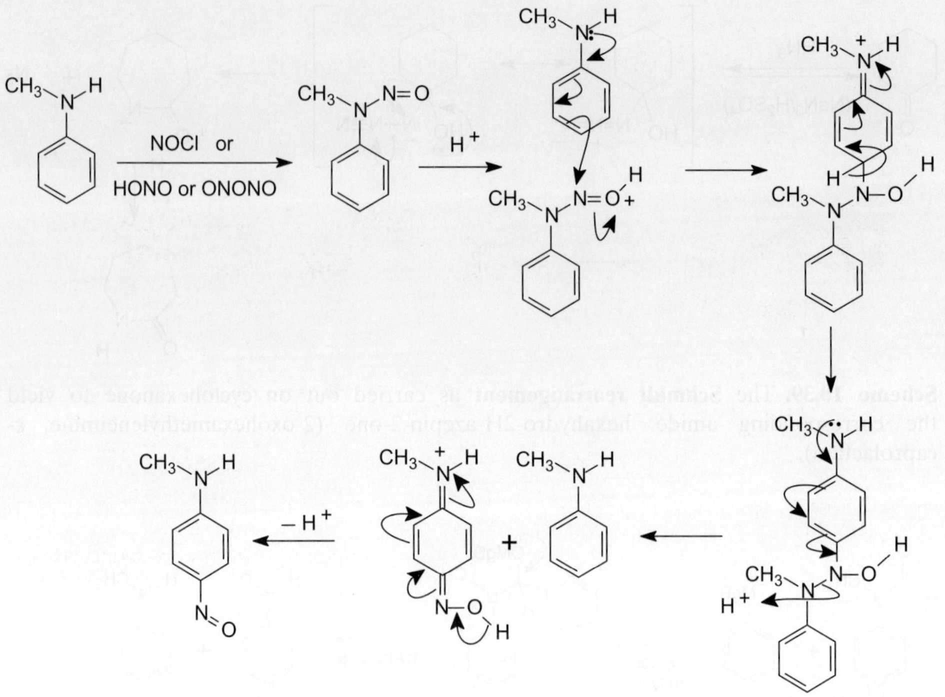

Scheme 10.41. The **Fischer–Hepp** rearrangement. A representation of nitrosation (with either nitrous acid [HONO] or nitrosyl chloride [NOCl]) (*vide supra*, this chapter) of *N*-methylaminobenzene (*N*-methylaniline) produces the corresponding *N*-nitroso derivative and the latter, on treatment with hydrogen chloride, undergoes intramolecular rearrangement through attack by *N*-methylaminobenzene (*N*-methylaniline). 4-Nitroso-*N*-methylaminobenzene (4-nitroso-*N*-methylaniline) results. For the Fischer–Hepp process here and in the next scheme see Fischer, O.; Hepp, E. *Chem. Ber.*, **1886**, *19*, 2991 and Johal, S. S.; Williams, D. L. H.; Buncel, E. *J. Chem. Soc. Perkin Trans. II*, **1980**, 165.

aromatic ring. Thus, nitrosation (with either nitrous acid [HONO] or nitrosyl chloride [NOCl]) of *N*-methylaminobenzene (*N*-methylaniline) produces the corresponding *N*-nitroso derivative and the latter, on treatment with hydrogen chloride, undergoes intramolecular rearrangement to 4-nitroso-*N*-methylaminobenzene (4-nitroso-*N*-methylaniline) (Scheme 10.41). This process is called the **Fischer–Hepp** rearrangement. The corresponding aryldiazo derivative undergoes the same reaction (Scheme 10.42).

As indicated in Scheme 10.42, treatment of aminobenzene with nitrous acid produces the corresponding diazonium species. Variously substituted aryl amines can also be diazotized and some diazonium salts can be isolated (often as their tetrafluoroborates). Either directly as formed or after isolation and reintroduction, these salts undergo reaction with a variety of aryl compounds to produce what are often brightly colored materials. Many of these diazo derivatives have found use as dyes (Figure 10.2 and Table 10.3) and some have been found to undergo loss of nitrogen in the presence of metals (e.g., copper) to produce either aryl radicals or aryl carbocations that are capable of reacting to produce other aryl derivatives (Figure 10.3).

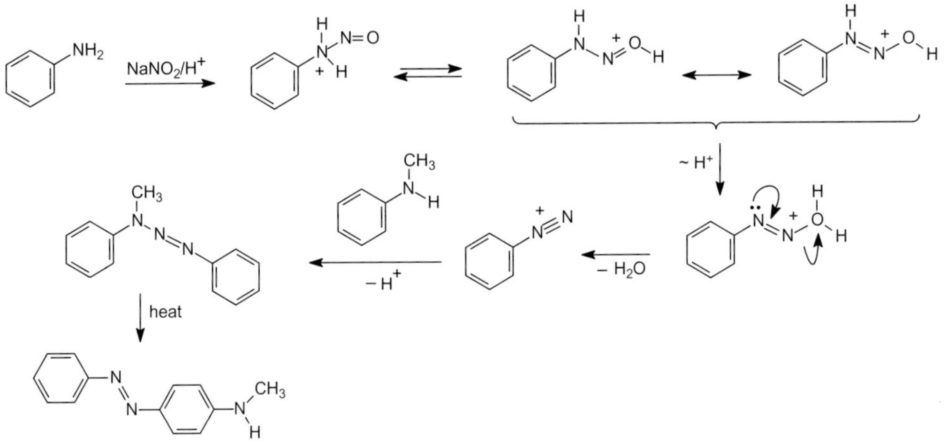

Scheme 10.42. The **Fischer–Hepp** rearrangement. A representation of diazotization of aminobenzene (aniline) with nitrous acid and subsequent reaction of the diazonium ion with N-methylaminobenzene (N-methylaniline) to yield a nitrogen-substituted derivative, which then undergoes intramolecular rearrangement to the corresponding 4-substituted azo derivative.

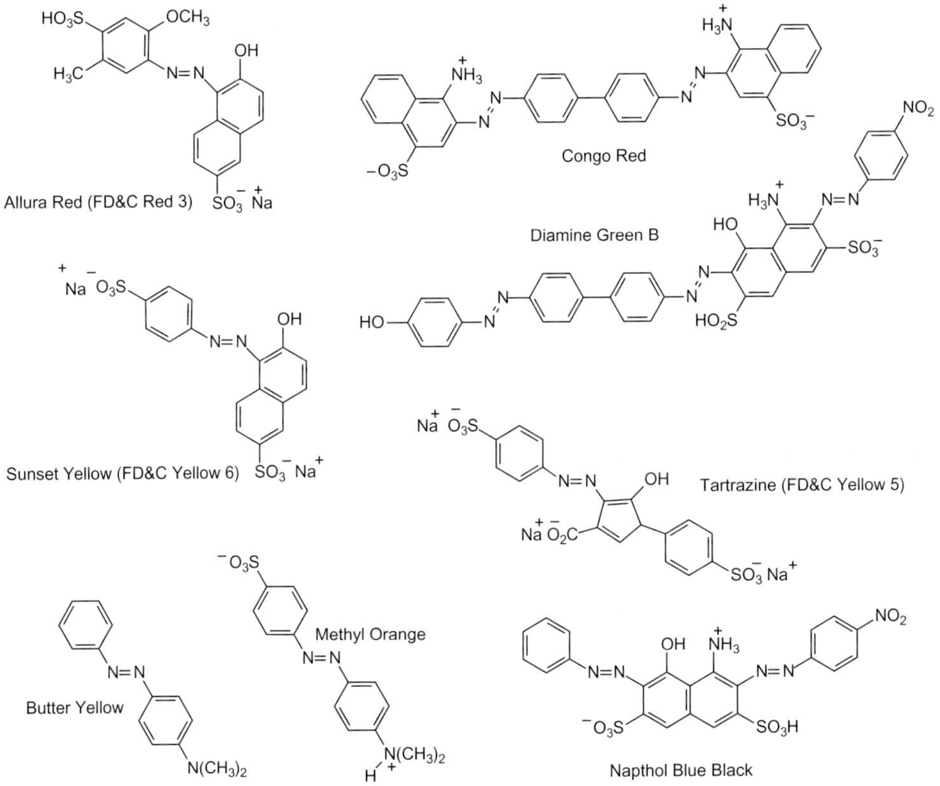

Figure 10.2. Commercially available dyestuffs containing the diazo group. The letters FD&C refer, respectively, to Food, Drug and Cosmetic and mean that these materials were or are used in media serving in any of those capacities as regulated in the United States by the Food and Drug Administration (FDA) and specified in the Code of Federal Regulations (CFR).

TABLE 10.3. Currently (2005) Approved Food and Drug Administration (FDA) Food Coloring Materials and Their Colors and Uses

FD&C blue no. 1
Brilliant blue FCF (bright blue): used in beverages, dairy products powders, jellies, confections, condiments, icings, syrups, and extracts
FD&C blue no. 2
Indigotine (royal blue): used in baked goods, cereals, snack foods, ice cream, confections, and cherries
FD&C green no. 3
Fast green FCF (sea green): used in beverages, puddings, ice cream, sherbet, cherries, confections, baked goods, and dairy products
FD&C red no. 40
Allura red AC (orange-red): used in gelatins, puddings, dairy products, confections, beverages, and condiments
FD&C red no. 3
Erythrosine (cherry red): used in coloring cherries in fruit cocktail and in canned fruits for salads, confections, baked goods, dairy products, and snack foods
FD&C yellow no. 5
Tartrazine (lemon yellow): used in custards, beverages, ice cream, confections, preserves, and cereals
FD&C yellow no. 6
Sunset yellow (orange): used in cereals, baked goods, snack foods, ice cream, beverages, dessert powders, and confections

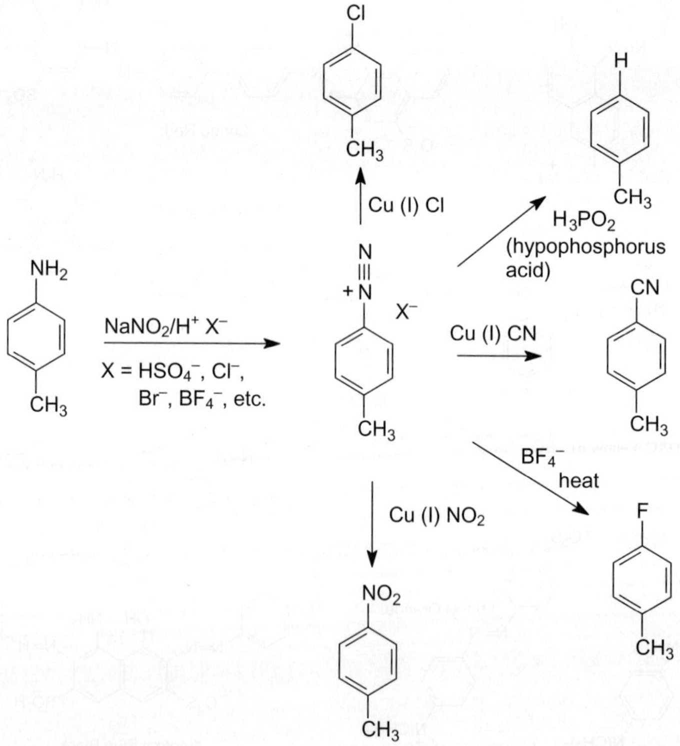

Figure 10.3. Aryl derivatives (clockwise from the top): 4-chloro-1-methylbenzene (4-chlorotoluene), methylbenzene (toluene), 4-cyano-1-methylbenzene (4-cyanotoluene), 4-fluoro-1-methylbenzene (4-fluorotoluene), 4-nitro-1-methylbenzene (4-nitrotoluene) prepared by copper(I) salt-catalyzed decomposition of the diazonium salt generated on treatment of 4-amino-1-methylbenzene, (4-aminotoluene) with sodium nitrite ($NaNO_2$), and protic acid.

Problem 10.3. The structures for FD&C yellow numbers 5 and 6 (tartrazine and sunset yellow, respectively) and FD&C red 3 are provided. Utilizing search engines on the world wide web and any other source, find the structures for the other allowed coloring materials. **Note: not all of these materials are azo dyes.**

Aliphatic amines can also be diazotized by treatment with sodium nitrite in the presence of acid. However, these diazo derivatives are not stable and immediately undergo loss of nitrogen to form the corresponding carbocations. Even primary carbocations can be generated in this way because nitrogen is such a good leaving group. Thus, as shown in Scheme 10.43, optically active 1-[^{2}H]-1-aminobutane on treatment with sodium nitrite in ethanoic acid (acetic acid, CH_3CO_2H) yields the corresponding 1-acetate (87%), which is $69 \pm 7\%$ inverted. The remainder of the hydrocarbon was found as 1-[^{2}H]-2-acetoxybutane (presumably arising via hydride migration followed by capture of the secondary carbocation by solvent) and a small amount of 1-[^{2}H]-1-nitrobutane. Because the reaction did not proceed with complete inversion of configuration, it is argued that a primary carbocation formed and that this was followed by collapse of the solvent shell. Interestingly, other potential

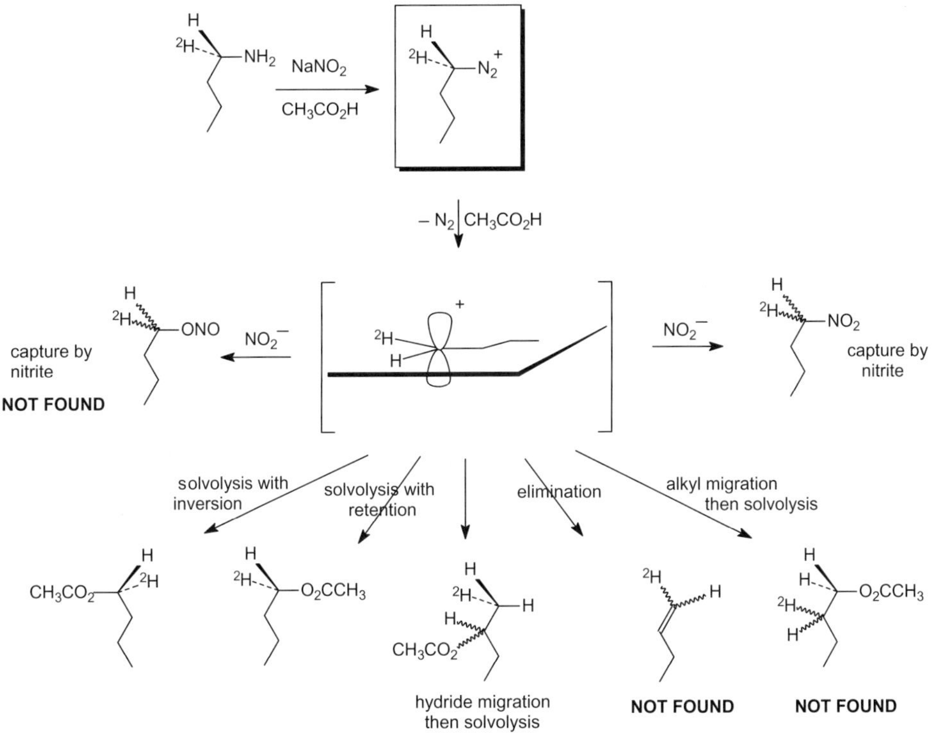

Scheme 10.43. A representation of the formation and decomposition of the diazonium ion produced from optically active (but of unknown absolute stereochemistry!) 1-^{2}H-1-aminobutane in acetic acid. It has been reported that 87% of the product is the primary acetate and, of that, $69 \pm 7\%$ had inverted configuration. Some hydride migration occurred and the secondary carbocation underwent acetolysis. A small amount of 1-[^{2}H]-1-nitrobutane was also detected (see Streitwieser, A. Jr.; Schaeffer, W. D. *J. Am. Chem. Soc.*, **1957**, *79*, 2888).

products that might have formed in the solvent shell, for example, arising from proton loss or even alkyl migration (Scheme 10.43), were not observed.

Problem 10.4. Using curved arrows as appropriate, account for all of the products (potential and actually isolated) shown in Scheme 10.43.

Rearrangements such as that seen in the **Fischer–Hepp** variant seen in Scheme 10.42 are also common for N-halo derivatives. Thus, chlorination of N-acetylaminobenzene (N-acetylaniline) produces the corresponding N-chloro derivative (Scheme 10.44). Then, when the N-acetyl-N-chloroaminobenzene (N-acetyl-N-chloroaniline) is treated with HCl, the **Orton rearrangement** occurs and N-acetyl-4-chloroaminobenzene (N-acetyl-4-chloroaniline) results. It has been suggested, as shown in the scheme, that "free" chlorine is the chlorinating agent.

The **Passerini** and **Ugi reactions** are closely related. As shown in Scheme 10.45 (for the **Passerini reaction**), when an isonitrile (isocyanide) is allowed to react with

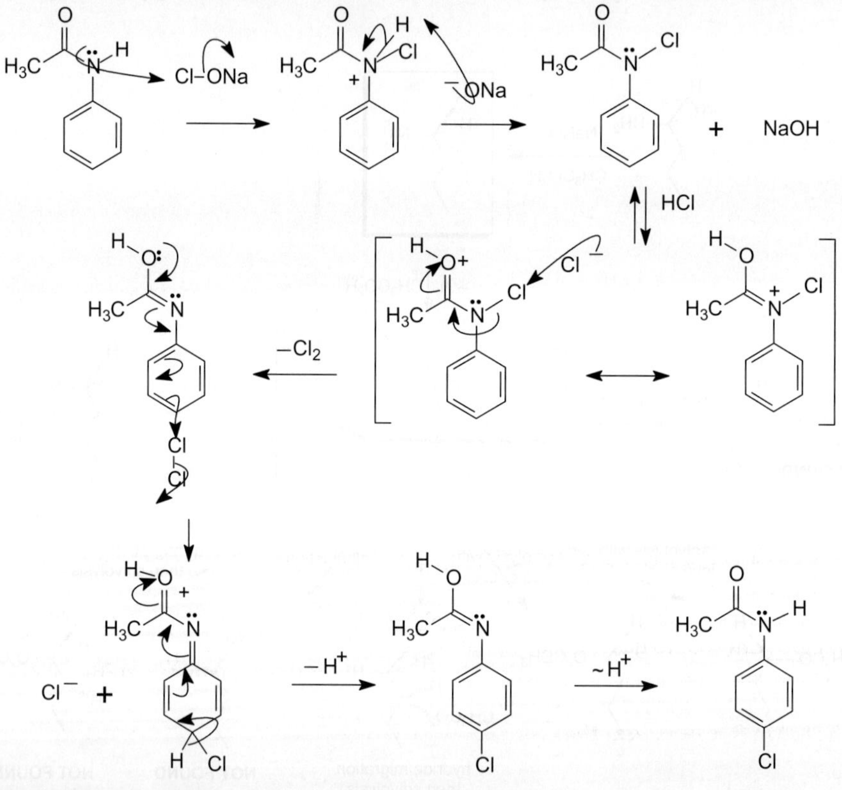

Scheme 10.44. The **Orton rearrangement**. Chlorination of N-acetylaminobenzene (N-acetylaniline) with sodium hypochlorite (NaOCl) produces the corresponding N-chloro derivative. Then, when the N-acetyl-N-chloroaminobenzene (N-acetyl-N-chloroaniline) is treated with HCl N-acetyl-4-chloroaminobenzene (N-acetyl-4-chloroaniline) results. It has been suggested that "free" chlorine is the chlorinating agent. See Bender, G. *Chem. Ber.*, **1886**, *19*, 2272; Chattaway, F. D.; Orton, K. J. P. *J. Chem. Soc.*, **1899**, *75*, 2046; Onoda, R.; Kawaguchi, H. *Kagaku Kyoiku*, **1981**, *29*, 220 (CA 95:203032).

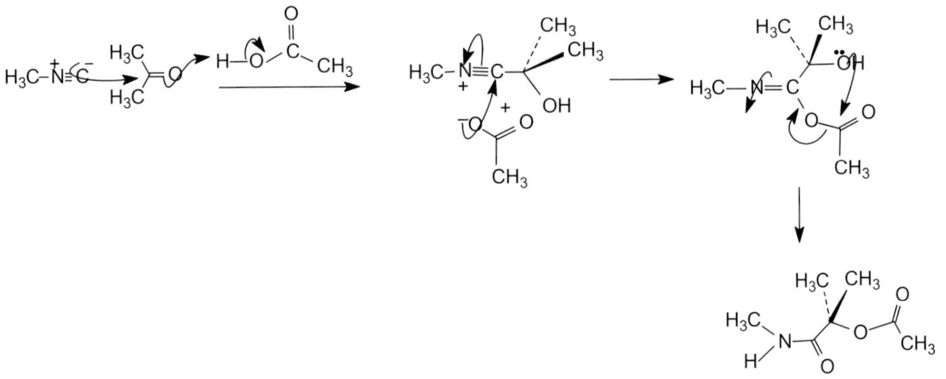

Scheme 10.45. The **Passerini reaction**. Methyl isocyanide (CH₃NC) is shown undergoing reaction with propanone (acetone, CH₃COCH₃) in ethanoic acid (acetic acid, CH₃CO₂H). The product, after rearrangement is the acetate ester of the *N*-methylamide of 2-hydroxy-2-methylpropanoic acid (see Passerini, M. *Gazz. Chim. Ital.*, **1921**, *51*, 126; Ugi, I. *Angew. Chem. Int. Ed. Engl.*, **1962**, *1*, 8).

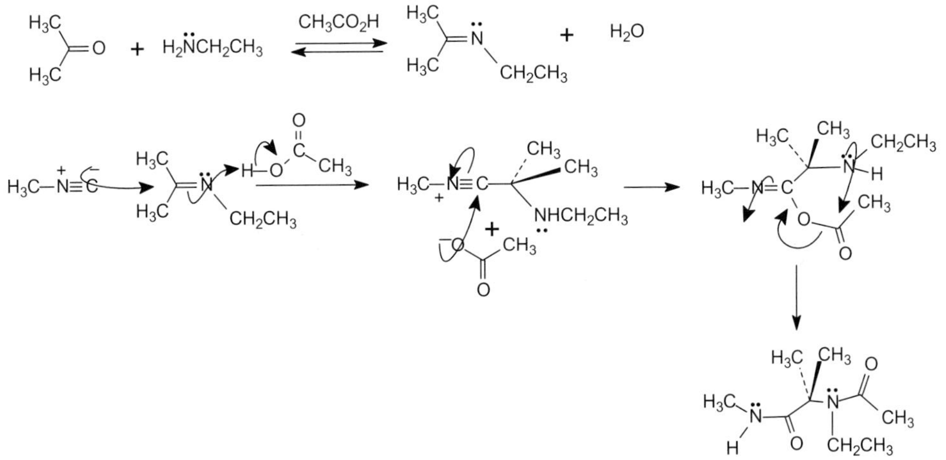

Scheme 10.46. Ugi reaction. A four-component variant of the **Passerini** process, where, in this example, ethanamine (ethyl amine, CH₃CH₂NH₂) first undergoes reaction with propanone (acetone, CH₃COCH₃) to yield the corresponding imine. The imine then reacts with methyl isocyanide (CH₃NC) to generate an amino-imine, and it is this that reacts with the carboxylic acid. The product, after rearrangement is the acetamide of the *N*-methylamide of 2-(*N*-ethyl)-amino-2-methylpropanoic acid (see Baker, R. H.; Stanonis, D. *J. Am. Chem. Soc.*, **1951**, *73*, 699; Ugi, I.; Meyr, R. *Chem. Ber.*, **1961**, *94*, 2229; Ugi, I. *Angew. Chem. Int. Ed. Engl.*, **1962**, *1*, 8).

an aldehyde or ketone in the presence of a carboxylic acid, the initial adduct between the isocyanide and ketone (aldehyde) is attacked by the acid and an amidoacetate results.

Interestingly, as seen in Scheme 10.46, the **Ugi reaction** variant of this process adds one more component to make it a four-component process. In the **Ugi** process, an amine is initially added to the reaction mixture so that the ketone or aldehyde

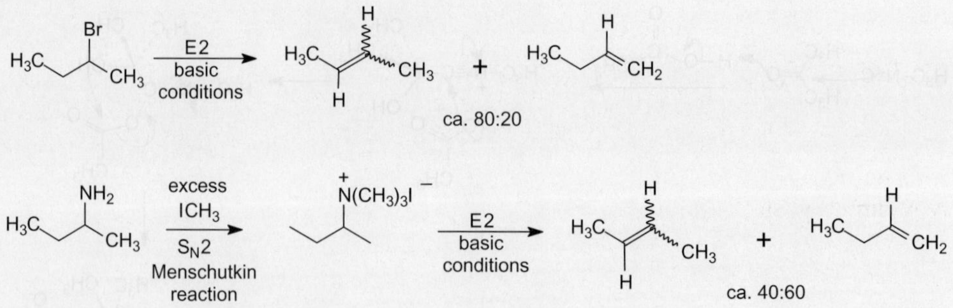

ca. 80:20

ca. 40:60

Scheme 10.47. Elimination to form 1-butene and (*E*)- and (*Z*)-2-butenes. E2 elimination from 2-bromobutane yields mainly 2-butenes while elimination from the 2-aminobutane *N,N,N*-trimethylamonium iodide yields mainly 1-butene (see Hofmann, A. W. *Liebigs Ann. Chem.*, **1851**, *78*, 253; Hofmann, A. W. *Chem. Ber.*, **1881**, *14*, 494).

will first react with the amine to form an imine. It is the imine that then reacts with the isocyanide to generate an amino-imine, and it is this that reacts with the carboxylic acid. The process is shown in Scheme 10.46.

Finally, as foreshadowed in Chapter 7 (Scheme 7.24 ff., the **Hofmann elimination)** and Chapter 8, Part I, Section F, A. W. Hofmann (1818–1895) pointed out that elimination from ammonium ions yielded the least highly substituted alkene rather than the most highly substituted alkenes (the **Saytzeff elimination**). Thus, as shown in Scheme 10.47, when 2-aminobutane is exhaustively methylated with iodomethane (ICH$_3$) and treated with base, there is more elimination to yield 1-butene (about 60%) than when, for example, the same elimination reaction is effected on 2-bromobutane (about 20%). It has been argued that this result comes about because the base is closely associated with the positively charged ion and the hindered environment favors loss of the proton from the primary carbon.

Interestingly, this reaction has been closely investigated by J. Sicher (1972), and, when carrying out the elimination on suitably deuterated 1,1,4,4-tetramethyl-7-cyclodecyltrimethylammonium chloride, the striking observation made that the elimination occurs in such a way that *anti*-elimination produces (*Z*)-alkene while *syn*-elimination produces (*E*)-alkene (Scheme 10.48a)! While it is clear that base should associate with the ammonium ion and that this would produce significant *syn*-elimination, the reason for the dichotomy and why the preference is so pronounced is not yet clear.

The reaction of such ammonium ions can also be affected even if there is no β-hydrogen. Under such circumstances, a facile rearrangement occurs if there is an acidic proton on the carbon bearing the quaternary nitrogen. Thus, as shown in Scheme 10.48b, a rearrangement seen in the analogous sulfur compounds (the **Stevens rearrangement**, Scheme 8.108) occurs when the phenacylbenzyldimethylammonium bromide is treated with base in aqueous solution. While a carbanion (shown) might be presumed to be involved, it is argued there is some evidence that, with the appropriate substituents present, a diradical is formed and recombines in the rearranged configuration.

Finally, in circumstances where there are neither acidic α-protons nor acidic β-protons so that neither the Stevens rearrangement nor the Hofmann elimination can occur, heating the ammonium ion nonetheless results in rearrangement if *an aryl group is present*. Thus, for the dibenzyldimethylammonium bromide shown in Scheme 10.48c, treatment with strong base {e.g., lithium diisopropylamide [$(CH_3)_2CH]_2N^-$ Li^+], LDA} results in carbanion formation. The anion is capable, as shown for this **Sommelet–Hauser** reaction, of attacking the neighboring ring to form the corresponding diphenylmethane in which one of the rings still bears the *N,N*-dimethyl substitution.

PART II. SOME ORGANOPHOSPHORUS CHEMISTRY

Phosphorus compounds were introduced in Chapter 5 (Section VII). That material could be profitably reviewed at this time.

Table 10.4 contains some of the compounds already seen in various portions of this work. Some organophosphorus compounds not discussed previously will also be found in Table 10.4. The phosphate backbones of deoxyribonucleic acid and ribonucleic acid (**DNA** and **RNA**, respectively) will be considered in Chapter 14.

The discovery of phosphorus (white phosphorus, P_4) is reported to have occurred as a consequence of the reductive pyrolysis of phosphate (PO_4^{-3}) found in human (and other animal) urine; the carbon-containing components being oxidized in the process to CO_2 and CO. Phosphate-containing minerals are currently used with charcoal to produce the same material.

Reduction of phosphorus (P_4) in the presence of hydrogen (H_2) produces phosphine (PH_3) (Equation 10.67), while treatment with halogens, for example, chlorine, generates phosphorus trihalides (phosphorus trichloride) (Equation 10.68).

$$P_4 + 6H_2 \longrightarrow 4PH_3 \tag{10.67}$$

$$P_4 + 6Cl_2 \longrightarrow 4PCl_3 \tag{10.68}$$

Interestingly, white phosphorus (P_4) will also react with alkyl and aryl lithium compounds in oxacyclopentane (THF) to give dark red solutions of organophosphides, which, on hydrolysis, produce modest yields of primary phosphines (Equation 10.69).

$$\text{(C}_6\text{H}_5)\text{-Li} + P_4 \longrightarrow \text{"red solution"} \xrightarrow{H_2O} \text{(C}_6\text{H}_5)\text{-PH}_2 \tag{10.69}$$

(ca. 35%)

At high temperatures (>ca. 250°C), phosphorus reacts with organic halides to produce aryl (alkyl) dihalophosphines and diaryl (dialkyl) halophosphines (Equation 10.70).

$$\text{(C}_6\text{H}_5)\text{-Br} + P_4 \xrightarrow{heat} \text{(C}_6\text{H}_5)\text{-PBr}_2 + [\text{(C}_6\text{H}_5)]_2\text{PBr} \tag{10.70}$$

44% 23%

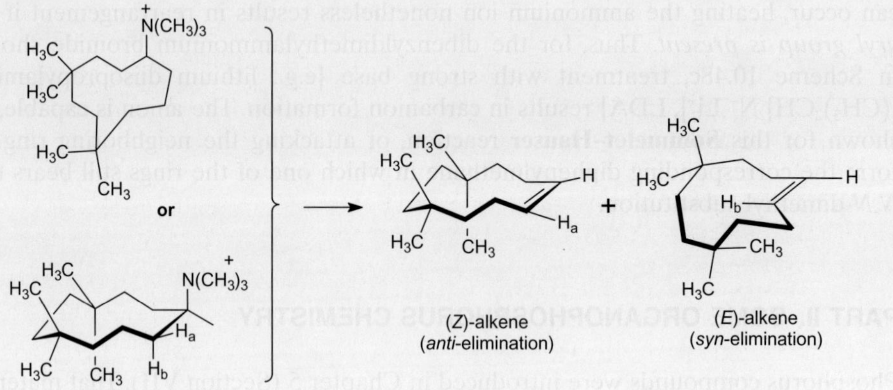

(Z)-alkene
(anti-elimination)

(E)-alkene
(syn-elimination)

(a)

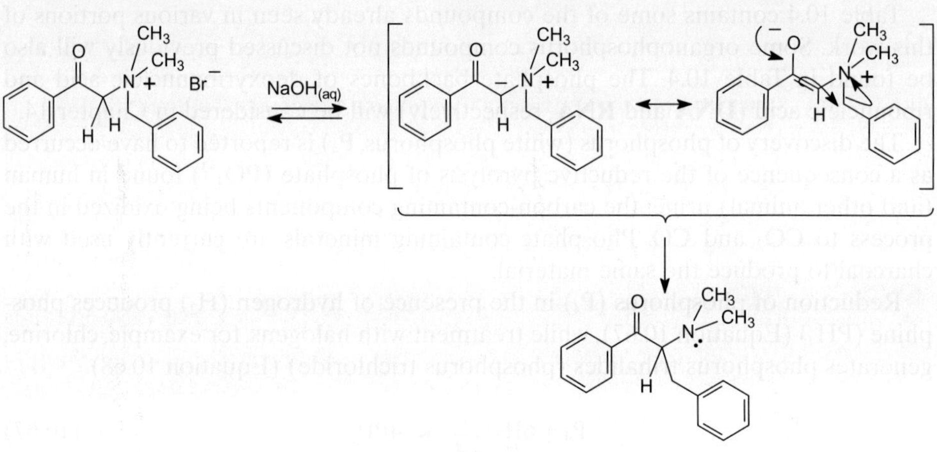

NaOH$_{(aq)}$

(b)

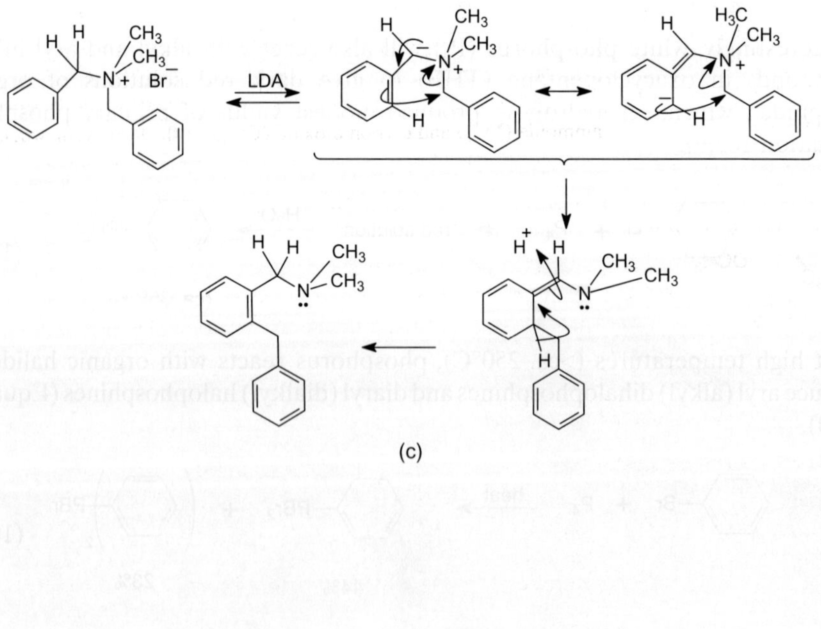

LDA

(c)

Scheme 10.48. (a) After Sicher, J. *Angew. Chem. Int. Ed. Engl.*, **1972**, *11*, 200. The Hofmann elimination from the 1,1,4,4-tetramethyl-7-cyclodecyltrimethylammonium ion, showing that *anti*-elimination produces (*Z*)-alkene and *syn*-elimination produces (*E*)-alkene. (b) The **Stevens rearrangement** of ammonium salts. See Scheme 8.108 for an example of the sulfur analogue. Phenacylbenzyldimethylammonium bromide is treated with base in aqueous solution, a carbanion (shown) is generated α to the carbonyl, and migration of the benzyl group occurs. It has also been argued that, with the appropriate substituents present, a pair of radicals is formed and remains solvent caged for recombination (Stevens, T. S.; Creighton, E. M.; Gordon, A. B.; MacNicol, M. *J. Chem. Soc.*, **1928**, 3193; Wittig, G. *Angew. Chem.*, **1954**, *66*, 14). (c) A representation of the **Sommelet–Hauser** reaction showing the attack of an aromatic ring on a neighboring benzylic carbon. The Hofmann elimination cannot occur (there are no β protons for elimination) and the Stevens rearrangement cannot occur as there is no carbonyl to act as a stabilizing influence and "electron sink." (Sommelet, M. *C. R. Séances Acad. Sci.*, **1937**, *205*, 56; Kantor, S. W.; Hauser, C. R. *J. Am. Chem. Soc.*, **1951**, *73*, 4122; Jones, G. C.; Hauser, C. R. *J. Org. Chem.*, **1962**, *27*, 3572).

TABLE 10.4. Some Phosphorus-Containing Compounds Previously Encountered

Chapter 5	
PH_3	Phosphine
$(CH_3)PH_2$	Methylphosphine
$(CH_3)_2PH$	Dimethylphosphine
$(CH_3)_3P$	Trimethylphosphine
$(CH_3)_3P{=}O$	Trimethylphosphine oxide
$(CH_3O)_3P$	Trimethylphosphite
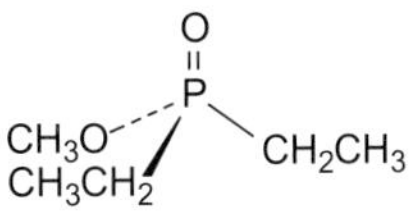	Methyl diethylphosphinate
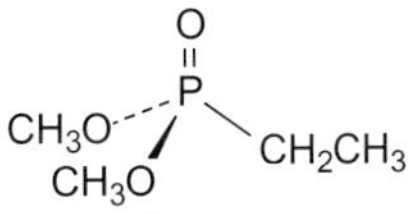	Dimethyl ethylphosphonate
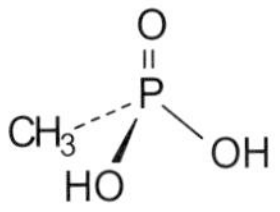	Methylphosphonic acid
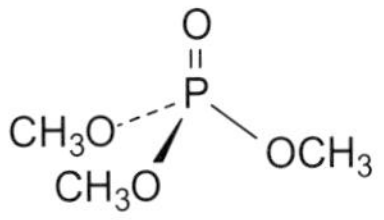	Trimethyl phosphate
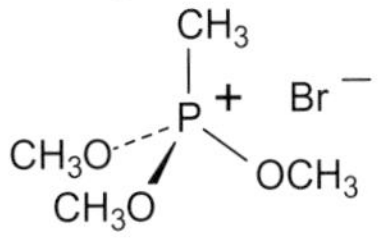	Methyl trimethoxyphosphonium bromide
Chapter 7	
PCl_5	Phosphorus pentachloride
$(C_6H_5)_3P$	Triphenylphosphine
$[(C_6H_5)_3PCH_2CH_3]^+ Br^-$	Ethyl triphenylphosphonium bromide
Chapter 8	
$(C_6H_5)_3PO$	Triphemylphosphine oxide

TABLE 10.4. *Continued*

$(C_6H_5)_3P$	Triphenylphosphine in the Mitsunobu reaction
	Geranyl diphosphate
	Phosphomevalonate [(R)-3-methyl,3,5-dihydroxypentanoic acid-5-phosphate]
	Isopentenyl diphosphate (4-hydroxy-2-methyl-1-pentene diphosphate)
	Diethylchlorophosphate
$POCl_3$	Phosphorus oxychloride
PBr_3	Phosphorus tribromide
85% H_3PO_4	Phosphoric acid
$(CH_3CH_2CH_2CH_2)_3P$	tri-n-Butylphosphine
$[(CH_3)_2N]_3P{=}O$	Hexamethylphosphoramidate [tris(dimethylamino)phosphine oxide]
Chapter 9	
$[(C_6H_5)_3PCH_3]^+\,Br^-$	Triphenylmethylphosphonium bromide
$(C_6H_5)_3P{=}CH_2$	Triphenylphosphinemethylene (methylenetriphenylphosphorane) (a **Wittig** reagent)
$(C_6H_5)_3P{=}CHCO_2CH_3$	Methyl (triphenylphosphoranylidene) acetate
	Methyl (diphenylphosphine oxide) acetate
	Methyl (diethylphosphono) acetate

If ethanol is allowed to react with phosphorus in the presence of oxygen, the corresponding diethyl phosphite forms in about 50% yield (Equation 10.71).

$$CH_3CH_2OH \;+\; P_4 \;+\; O_2 \;\longrightarrow\; \underset{CH_3CH_2O}{\overset{CH_3CH_2O}{{>}}}\!\!P\!\!\underset{}{\overset{O}{\|}}\!H \qquad (10.71)$$

Phosphorus halides generally react, in oxacyclopentane (THF) with **Grignard** reagents to produce triaryl or alkyl phosphines. Thus, as shown in Equation 10.72, when four equivalents of phenylmagnesium bromide are allowed to react with phosphorus trichloride, triphenylphosphine results. Trivinylphosphine can be prepared in a similar fashion (Equation 10.73) by simply changing the nucleophile.

$$4C_6H_5MgBr + PCl_3 \longrightarrow (C_6H_5)_3P + MgBrCl \qquad (10.72)$$

$$4CH_2=CHMgCl + PCl_3 \longrightarrow (CH_2=CH)_3P + MgBrCl \qquad (10.73)$$

As might therefore be imagined (correctly), substituted phosphonous dihalides (e.g., dibromophenylphosphine, the product of Equation 10.70) can also react with **Grignard** reagents to produce unsymmetrical alkaryl-substituted phosphines (Equation 10.74). This reaction, like those of **Grignard** reactions in general, is usually exothermic and is preformed at low temperature and under nitrogen (to avoid the production of phosphine oxides).

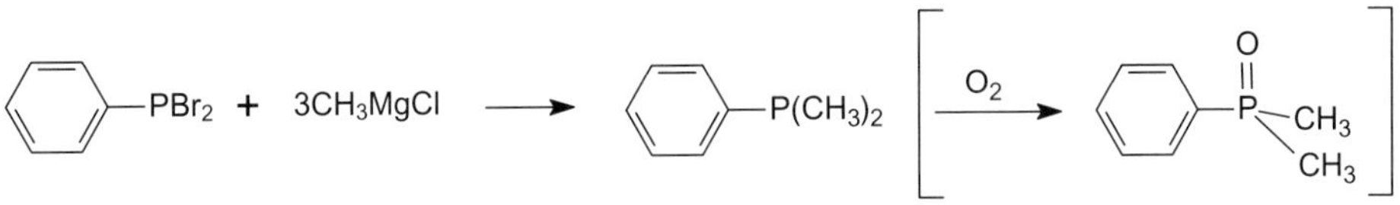

$$(10.74)$$

In the same vein, cyclic phosphines such as 1-phenylphosphocyclohexane can be prepared by the reaction of the corresponding bifunctional **Grignard** reagent, generated from, for example, 1,5-dibromopentane, with the corresponding dichlorophenylphosphine (Scheme 10.49).

Similar nucleophilic substitution at phosphorus with the loss of alkoxide instead of halide (*vide supra*) also occurs readily. Thus, under a nitrogen atmosphere (to avoid oxidation at phosphorus) when four or more equivalents of phenylmagnesium bromide are treated with trimethyl phosphite (Equation 10.75), triphenylphosphine is cleanly obtained. However, if three or fewer equivalents of the same Grignard reagent are used, then dimethyl phenylphosphonite, methyl diphenylphosphinite, and triphenylphosphine are all found (Equation 10.76) with the exact ratio being,

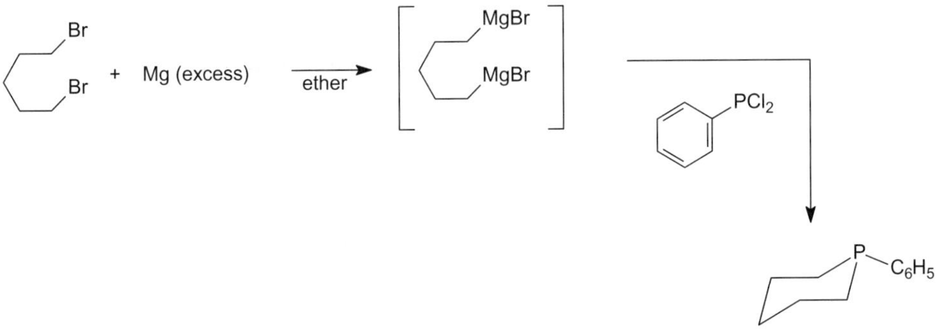

Scheme 10.49. A bifunctional **Grignard** reagent undergoing reaction with a dihalophosphine to produce a heterocycle phosphine (1-phenylphosphocyclohexane).

in part, determined by the rate at which the addition of trimethyl phosphite is effected.

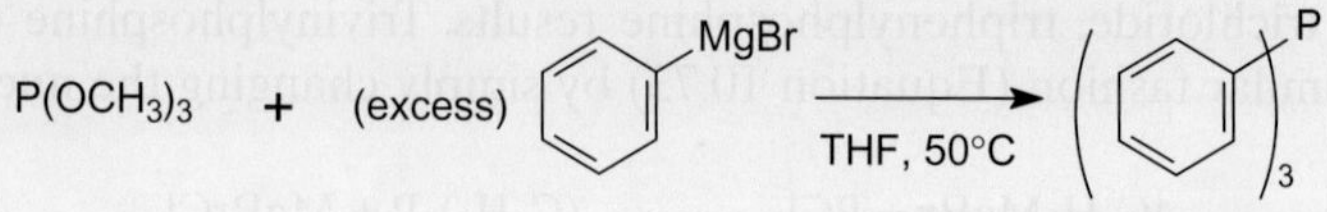

$$P(OCH_3)_3 \quad + \quad (excess) \quad C_6H_5MgBr \xrightarrow[\text{THF, 50°C}]{} (C_6H_5)_3P \qquad (10.75)$$

$$C_6H_5MgBr \quad + \quad (excess)\ P(OCH_3)_3 \xrightarrow[\text{THF, 50°C}]{} C_6H_5P(OCH_3)_2 \quad + \quad (C_6H_5)_2POCH_3 \quad + \quad (C_6H_5)_3P \qquad (10.76)$$

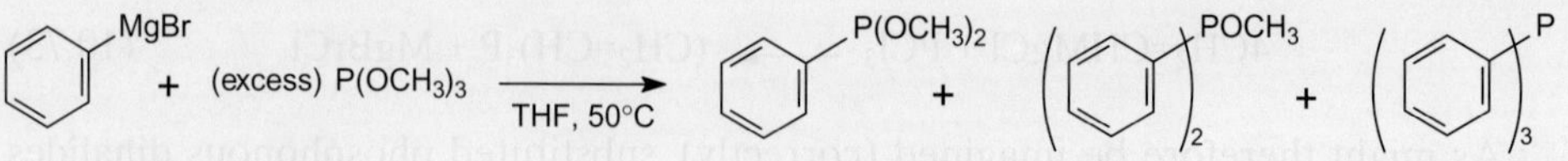

Interestingly, the reaction of trimethylphosphate $[(CH_3O)_3PO]$ under the same conditions fails. Even at higher temperatures, an intractable mixture of products is obtained. In part, the oversimplification of the concept of the Grignard reagent as a negatively charged carbon species attached to a positively charged magnesium is to blame (please see the discussions in Chapters 5, C, VIII and 7, D, I) since it is clear that the simple orbital picture for first row elements does not obtain here. Nonetheless, nucleophilic substitutions reactions readily occur on the carbon atoms attached to oxygen.

For example, in a typical process, methyl iodide reacts with ethyl dimethyl phosphite to produce iodomethane and ethyl methyl methylphosphonate. Presumably, the alkylation occurs on phosphorus and the liberated iodide anion then subsequently attacks the methyl group and displace the phosphonate (Scheme 10.50) in a **Michaelis–Arbuzov** reaction.

Although the Michaelis–Arbuzov* reaction has been widely used for the preparation of phosphonates and the path shown in Scheme 10.50 has been generally accepted, it appears that if the alkyl group attached to oxygen is capable of leaving without the help of the nucleophile, some portion of the reaction may proceed by an S_N1-type process.

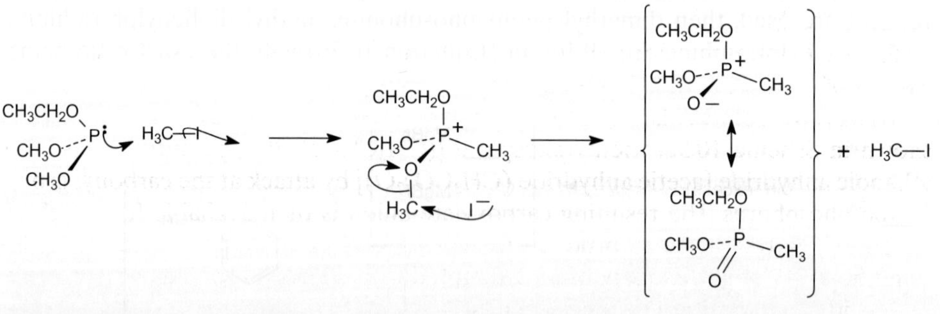

Scheme 10.50. A representation of the formation of a phosphonate (ethyl methyl methylphosphonate) by an S_N2 reaction at one of the methyl groups of ethyl dimethyl phosphite (a **Michaelis–Arbuzov** process).

*Arbuzov, A. E. (1877–1968), student (1896–1900) and then faculty member (1911–1930), University of Kazan, Russia. Michaelis, A. (1847–1916), Chemical Institute, University of Rostock, Germany. See Michaelis, A.; Kaaehne, R. *Chem. Ber.*, **1898**, *31*, 1048; Arbuzov, A. E. *J. Russ. Phys. Chem.*, **1906**, *38*, 687.

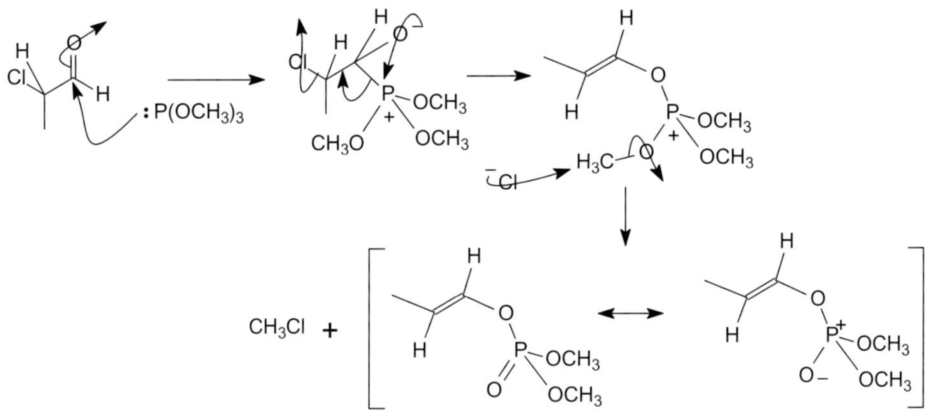

Scheme 10.51. A representation of the **Perkow** reaction where 2-chloropropanal undergoes reaction with trimethylphosphite to produce dimethylvinyl phosphate and chloromethane.

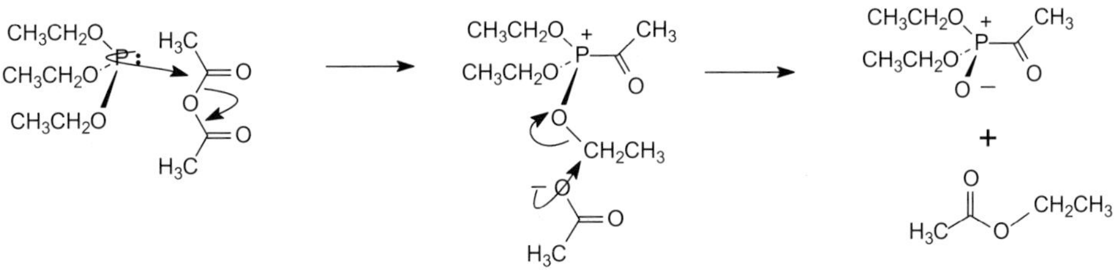

Scheme 10.52. A representation of the reaction of triethylphosphite $[(CH_3CH_2O)_3P]$ with ethanoic anhydride [acetic anhydride $(CH_3CO)_2O)$] to produce the corresponding acyl diethylphosphonate and ethyl ethanoate (ethyl acetate $[CH_3CO_2CH_2CH_3]$).

Nearly 50 years after the initial reports of Arbuzov and Michaelis, and despite early reports affirming the contrary, it was found that α-haloaldehydes (and, later, some α-haloketones and α-haloesters) underwent reaction with phosphites to produce vinylphosphates rather than the expected phosphonates. The path (Scheme 10.51, the **Perkow** reaction)* is presumed to involve attack of phosphorus on the carbon of the carbonyl and subsequent rearrangement.

Trialkyl phosphites also react with anhydrides in an analogous reaction. Thus, as shown in Scheme 10.52, triethylphosphite $[(CH_3CH_2O)_3P]$ apparently reacts with ethanoic anhydride [acetic anhydride $(CH_3CO)_2O)$] by attack at the carbonyl group by the phosphorus. The resulting carboxylate anion is then available for attack at an ethoxy carbon atom to produce the corresponding acyl phosphonate. On the other hand, a similar reaction, when attempted with the corresponding acyl chloride results in decomposition of the reactants.

When triphenylphosphine $[(C_6H_5)_3P]$ (see, e.g., Equation 10.72) is allowed to react with carbon tetrachloride (see Scheme 8.28), the corresponding chlorotriphenylphosphonium trichloromethide salt $[(C_6H_5)_3PCl^+\ CCl_3^-]$ forms, as seen again here

*Perkow, W.; Ullerich, K.; Meyer, F. *Naturwissenschaften*, **1952**, *39*, 353; Perkow, W. *Chem. Ber.*, **1952**, *87*, 755; Lichtenthaler, F. W. *Chem. Rev.*, **1961**, *61*, 607.

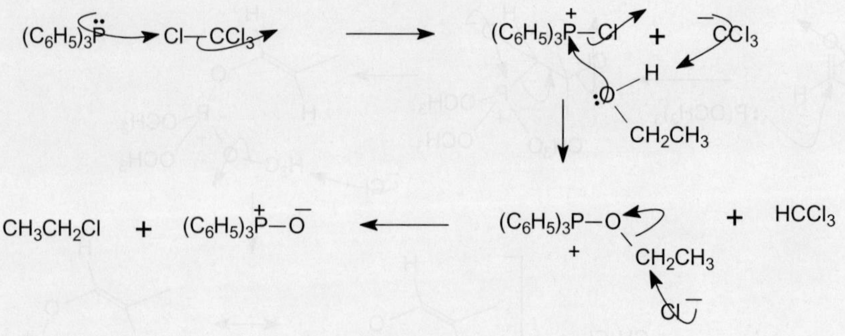

Scheme 10.53. The reaction of triphenylphosphine, $(C_6H_5)_3P$, with tetrachloromethane (carbon tetrachloride, CCl_4) to produce the chlorotriphenylphosphonium trichloromethide salt $[(C_6H_5)_3PCl^+ CCl_3^-]$, which, as seen before (Chapter 8), is capable of reaction with alcohols (shown here as ethanol, CH_3CH_2OH) to generate the corresponding chloride (via an apparent S_N2-type process) (see Appel, R. *Angew. Chem. Int. Ed. Engl.*, **1975**, *14*, 801).

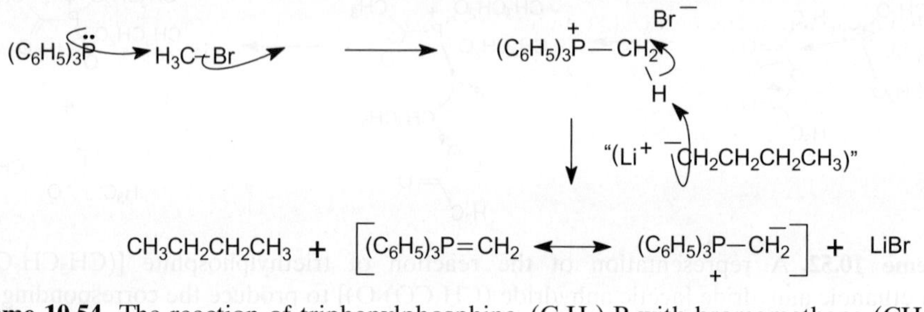

Scheme 10.54. The reaction of triphenylphosphine, $(C_6H_5)_3P$, with bromomethane (CH_3Br) to form the phosphonium bromide salt and its subsequent treatment with *n*-butyllithium (written here in quotation marks since it should be realized now that *n*-butyllithium is used in solution and it may actually not be the monomeric species shown) to produce butane, lithium bromide, and the phosphorane.

as Scheme 10.53. And in that vein, as pointed out earlier (Table 10.4 and Chapter 9), triphenylphosphine undergoes reaction with a variety of alkyl halides to produce the corresponding triphenylphosphonium halide, which on treatment with strong base produces the corresponding triphenylphosphinemethylene (methylenetriphenylphosphorane) ylid (Wittig reagent) (Scheme 10.54).

When phosphorus is substituted with four different substituents, the resulting tetrahedral phosphonium salt is, of course, chiral. When (*S*)-benzylethylmethylphenylphosphonium iodide is treated with phenyllithium (Scheme 10.55), the corresponding benzylidenium ion results. Because this ion, on reaction with benzaldehyde, generates (*R*)-ethylmethylphenylphosphine oxide and (*E*)-stilbene, it is argued that the reaction proceeds with retention of configuration.

Interestingly, trigonal bipyramidal structures shown as chiral intermediates in Scheme 10.55 and common in pentacoordinate phosphorus compounds enjoy a small energy difference between the trigonal bipyramid and tetragonal pyramid structures. The consequence is that axial and equatorial groups (unless linked as in

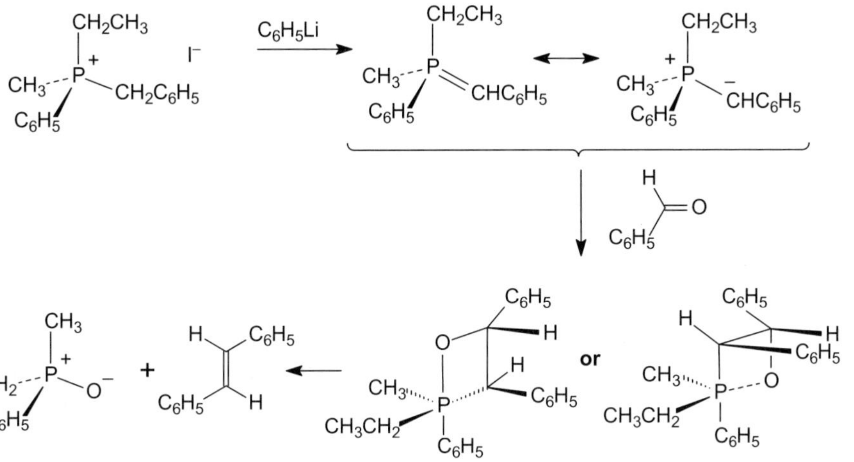

Scheme 10.55. A representation of the Wittig reaction utilizing chiral benzylethylmethyl-phenylphosphonium iodide and phenyllithium to generate the ylid and reaction of that intermediate with benzenecarboxyaldehyde (benzaldehyde) to produce chiral ethylmethyl-phenylphosphine oxide and (*E*)-stilbene. As shown, the number of possible intermediates is limited by the requirements of retention of chirality and production of a single stilbene (Wittig, G.; Rieber, M. *Liebigs Ann. Chem.*, **1949**, *562*, 187).

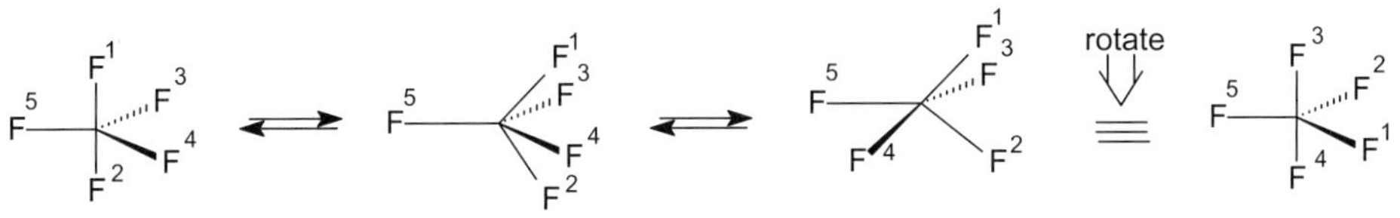

Figure 10.4. PF_5 (phosphorus pentafluoride): a symmetrical tetragonal pyramid. A representation (using phosphorus pentafluoride) of the process by which axial and equatorial substituents are interchanged by intramolecular tunneling (called a **Berry pseudorotation**).

Scheme 10.55) can interconvert. The pathway is called a **Berry pseudorotation*** and is illustrated in Figure 10.4.

Phosphorus nuclear magnetic resonance (NMR) (Tables 5.7 and 10.5) rose to importance as the presence of phosphates in biological systems became amenable to study.

The only naturally occurring isotope of phosphorus is ^{31}P, and it has nonzero spin ($I = 1/2$). In a 300 MHz (for ^1H) or 7.05 T NMR system, ^{31}P comes into resonance at about 121.48 MHz (^{13}C would be at about 75.47 MHz in the same system). Some NMR chemical shifts (relative to 85% phosphoric acid) are provided in Table 10.5. Efforts to correlate chemical shift values of ^{31}P phosphorus chemical shifts over its nearly 500 ppm range (as it has been for carbon over its 100 ppm range) have been made. Results are shown in Table 10.6.

---

*Berry, R. S., a professor at the University of Michigan, proposed the intramolecular tunneling process that bears his name in 1960. Subsequently, in 1981, while at the University of Chicago, he noted how widely quoted the work had become. (See Berry, R. S. *J. Chem. Phys.*, **1960**, *32*, 933.)

TABLE 10.5. Representative Examples of Organophosphorus Compounds

Compound	Name	mp (°C)	bp (°C)	^{31}P–^{13}C coupling (J_{PC})	^1H NMR data [δ, ppm TMS = 0.00]	^{13}C and ^{31}PNMR data [δ, ppm TMS = 0.00 (C): 85% H_3PO_4 = 0.00 (P)]
CH_3PH_2	Methylphosphine			9.3 Hz		(-4.6 $^{13}CH_3$); [(+163 ^{31}P)]
$(CH_3)_2PH$	Dimethylphosphine			11.6 Hz		(6.9 $^{13}CH_3$); [(+98 ^{31}P)]
$(CH_3)_3P$	Trimethylphosphine	40		13.5 Hz	0.90	(17.1 $^{13}CH_3$); [+62 (^{31}P)]
$(CH_3CH_2CH_2CH_2O)_3PO$	tri-*n*-Butylphosphate			-5.9 (C_1), 6.5 (C_2), 0.0 (C_3,C_4)		76.5 ($^{13}CH_2$) C_1; 38.2 ($^{13}CH_2$) C_2; 19.1 ($^{13}CH_2$) C_3; 13.7 ($^{13}CH_2$) C_4
$(C_6H_5)_3P$	Triphenylphosphine					(^{13}C); [+6.0 (^{31}P)]
$(C_2H_5O)_3P$	Triethylphosphite	158			3.85 (2H), 1.19 (3H)	(^{13}C); [-139.0 (^{31}P)]
$(CH_3O)_3P$	Trimethylphosphite	111			3.43	(^{13}C); [-141 (^{31}P)]
$(C_6H_5)_3PO$	Triphenylphosphine oxide					(^{13}C); [-28 (^{31}P)]
$(C_6H_5)_2PCH_3$	Methyldiphenylphosphine					(^{13}C); [+28 (^{31}P)]
P–H (phosphacyclopentane ring)	Phosphacyclopentane			5.4 Hz (C_2); 9.95 Hz (C_3)		31.6 (CH_2); 21.2 (CH_2)
$(C_6H_5)_2P(O)CH_3$	Methyldiphenylphosphine oxide					(^{13}C); [-27 (^{31}P)]
$(CH_3)_3PO$	Trimethylphosphine oxide					(^{13}C); [-36 (^{31}P)]
$(C_6H_5O)_3P$	Triphenylphosphite					(^{13}C); [-128 (^{31}P)]
PCl_3	Phosphorus trichloride					-220 (^{31}P)
$(C_6H_5)PCl_2$	Dichlorophenylphosphine					(^{13}C); [-166 (^{31}P)]
$(C_6H_5O)_3PO$	Triphenylphosphate					(^{13}C); [+18 (^{31}P)]
$(CH_3O)_2HPO$	Dimethyl phosphite					(^{13}C); [+18 (^{31}P)]
$(C_6H_5)PO(OH)_2$	Phenylphosphonic acid					(^{13}C); [-18 (^{31}P)]
PCl_5	Phosphorus pentachloride					+80 (^{31}P)

The information for this table is derived from experiment. See Denney, D. B.; Denney, D. Z.; Chang, B. C.; Marsi, K. L. *J. Am. Chem. Soc.*, **1969**, *91*, 5243; Weigert, F. J.; Roberts, J. D. *Inorg. Chem.*, **1973**, *12*, 313; Jones. C. E.; Coskran, K. J. *Inorg. Chem.*, **1971**, *10*, 1536; and Van Wazer, J. R.; Callis, C. F.; Shoolery, J. N.; Jones, R. C. *J. Am. Chem. Soc.*, **1956**, *78*, 5715.

TABLE 10.6. Increments in the Position of the Chemical Shift of ^{13}C (Tetracoordinate Carbon) and ^{31}P (Tricoordinate Phosphorus) as a Function of Substitution

	X = H	X = R′	X = NR$_2$	X = OR′	X = F	X = Cl	
^{13}C	0	9	28	58	70	31	
^{31}P	−69	−20	47	55	25	74	

From Dransfeld, A.; Schleyer, P. V. R. *Mag. Res. Chem.*, **1998**, *36*, S29.

PART III. SOME ORGANOSILICON CHEMISTRY

Silicon lies directly beneath carbon in the periodic table.

Although all the allotropes of carbon (graphite, diamond, buckeyball) demonstrate that carbon enjoys the ability to self-link into chains and rings, that property fails for the single allotrope of silicon, which appears to be octahedral. Although more than one reason might be put forth for the inability of silicon to concatenate, it is clear that bond strength plays an important role.

As pointed out earlier (Chapter 1, Table 1.1), a typical carbon–carbon single bond (e.g., in ethane, H_3C–CH_3, length 154 pm, rotational barrier ca. 12 kJ/mol [2.7 kcal/mol]) has a bond dissociation energy (BDE) of about 380 kJ/mol (90 kcal/mol) (Chapter 1, Table 1.1). A silicon–silicon bond (e.g., in disilane, H_3Si–SiH_3, length 234 pm, rotational barrier ca. 4 kJ/mol [0.95 kcal/mol]) has a reported BDE of about 230 kJ/mol (55 kcal/mol); that is, the bond between the two silicons in H_3Si–SiH_3 is just over half as strong as the bond between the two carbons in H_3C–CH_3. A typical C–Si bond, e.g., in CH_3SiH_3, has a length of 186 pm and a BDE of 318 kJ/mol [79.5 kcal/mol] and a rotational barrier of 7.1 kJ/mol [1.7 kcal/mol]. However, bonds between silicon and elements more electronegative than carbon tend to be stronger than the same bonds between carbon and those same elements that are more electronegative than carbon. For example, the Si–F bond at about 580 kJ/mol (140 kcal/mol) is about 130 kJ/mol (30 kcal/mol) stronger than the comparable C–F bond (450 kJ/mol [110 kcal/mol]). The most obvious consequence of the strong Si–F bond is that fluoride ion can be used to effect removal of a silicon substituent on carbon or oxygen when such a substituent is used, for example, as a **protecting group**.

Interestingly, the silicon–hydrogen bond (Si–H) is relatively weak (323 kJ/mol [77 kcal/mol]) compared with the carbon–hydrogen (C–H) bond (418 kJ/mol [100 kcal/mol]) so that compounds containing the Si–H bond can serve as reducing agents.

Elemental silicon is obtained by reduction of silicon dioxide with carbon in an electric arc (Equation 10.77). Subsequent reaction of silicon with chlorine produces the corresponding silicon tetrachloride (Equation 10.78). Silicon tetrafluoride, on the other hand, is prepared by treating silica (SiO_2) and a source of fluoride ion (e.g., sodium hexafluorosilane, Na_2SiF_6) with sulfuric acid (H_2SO_4) (Equation 10.79) or simply treating phosphate rock (i.e., the minerals apatite [$Ca_5(PO_4)_3OH$] and the isomorphous fluorapatite [$Ca_5(PO_4)_3F$] among others) with silica (SiO_2) and sulfuric acid (H_2SO_4) (Equation 10.80).

$$2C + SiO_2 \rightarrow 2CO + Si \tag{10.77}$$

$$2Cl_2 + Si \rightarrow SiCl_4 \tag{10.78}$$

$$Na_2SiF_6 + H_2SO_4 \rightarrow Na_2SO_4 + 2HF + SiF_6 \tag{10.79}$$

$$SiO_2 + 4Ca_5(PO_4)_3F + 4H_2SO_4 \rightarrow SiF_4 + 4Ca_5(PO_4)_3(HSO_4) + 2H_2O \tag{10.80}$$

Classically, silanes (SiH_4, Si_2H_6, etc.) have been separated by fractional distillation from the reaction mixture generated when magnesium silicide (Mg_2Si), formed by heating magnesium and silicon in the absence of air, is treated with hydrochloric acid in the presence of hydrogen gas (Equation 10.81).*

$$Mg_2Si + 2HCl \xrightarrow{\;H_2\;} MgCl_2 + SiH_4 + Si_2H_6 + \ldots \tag{10.81}$$

Now, it had been clear for some time that reaction of organometallic reagents, with silicon tetrachloride ($SiCl_4$) would produce the corresponding alkylated species.† Thus, as shown in Equation 10.82, treatment of silicon tetrachloride ($SiCl_4$) with methylmagnesium bromide (CH_3MgBr), prepared in the usual way from methyl bromide (CH_3Br) and magnesium metal (Mg) in diethyl ether [$(CH_3CH_2)_2O$] produced the corresponding trichloromethylsilane (CH_3SiCl_3) and, with excess Grignard reagent (CH_3MgBr), dichlorodimethylsilane [$(CH_3)_2SiCl_2$].

$$SiCl_4 + CH_3MgBr \rightarrow MgBrCl + CH_3SiCl_3 \xrightarrow{\;CH_3MgBr\;} MgBrCl + (CH_3)_2SiCl_2 \tag{10.82}$$

Treatment of dichlorodimethylsilane [$(CH_3)_2SiCl_2$] with yet a third equivalent of organometallic reagent such as methylmagnesium bromide (CH_3Br) or *tert*-butylmagnesium bromide in the presence of copper reagents (e.g., CuCl/KCN) produces the corresponding trialkylchlorosilane [Equation 10.83, R = CH_3 or R = $(CH_3)_3C$].

$$R-MgBr + (CH_3)_2SiCl_2 \xrightarrow[\substack{hydrocarbon \\ solvent}]{CuCl/KCN} R-Si(CH_3)_2Cl + MgBrCl \tag{10.83}$$
$$R = CH_3, (CH_3)_3C \qquad\qquad\qquad R = CH_3, (CH_3)_3C$$

Furthermore, it has been reported that any and all of these silyl halide derivatives can be converted to the corresponding silanes (i.e., compounds bearing at least one Si–H) by treatment (often at high temperature in hydrocarbon solvents) with lithium aluminum hydride ($LiAlH_4$) and that such silanes can be used to generate the chlorides in the presence of a suitable Lewis acid (e.g., $AlCl_3$) on treatment with a second (usually lower boiling) chlorosilane (Scheme 10.56).

Trialkylchlorosilanes (R_3SiCl), as already seen in a few instances in previous chapters and shown again in Table 10.7, are quite valuable synthetic tools as they have been made to serve to preserve enols from tautomerization to their corresponding aldehydes or ketones as well as to serve in double-bond formation

---

*Mulla, I. S.; Choube, A. C.; Dongare, M. K.; Sinha, A. P. B. *Indian J. Chem.*, **1988**, *27A*, 756.
†Tao, T.; Maciel, G. E. *J. Am. Chem. Soc.*, **2000**, *122*, 3118.

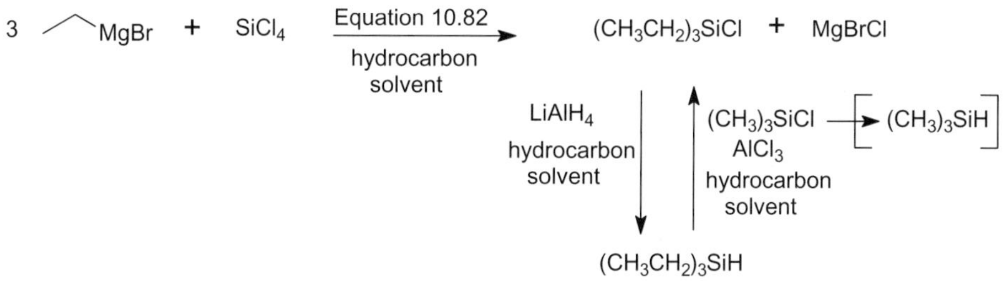

$$3 \ \text{\textbackslash}MgBr \ + \ SiCl_4 \xrightarrow[\substack{\text{hydrocarbon} \\ \text{solvent}}]{\text{Equation 10.82}} (CH_3CH_2)_3SiCl \ + \ MgBrCl$$

$$\downarrow \substack{LiAlH_4 \\ \text{hydrocarbon} \\ \text{solvent}} \qquad \uparrow \substack{(CH_3)_3SiCl \longrightarrow (CH_3)_3SiH \\ AlCl_3 \\ \text{hydrocarbon} \\ \text{solvent}}$$

$$(CH_3CH_2)_3SiH$$

Scheme 10.56. A representation of the formation of chlorotriethylsilane [(CH₃CH₂)₃SiCl] (via a Grignard reaction) between tetrachlorosilane (SiCl₄) and ethylmagnesium bromide (CH₃CH₂MgBr) and the conversion of that trialkylchlorosilane [(CH₃CH₂)₃SiCl] to the corresponding triethylsilane [(CH₃CH₂)₃SiH]. The reaction of the triethylsilane [(CH₃CH₂)₃SiH] with aluminum trichloride (AlCl₃) and chlorotrimethylsilane [(CH₃)₃SiCl] to convert the dehalogenated material back to the chlorosilane from which it came is also shown. In the latter, the chlorotrimethylsilane [(CH₃)₃SiCl] reactant is, in turn, converted to trimethylsilane [(CH₃)₃SiH] (Fry, J. L.; Rahaim, R. J. Jr.; Maleczka, R. E. Jr. Triethylsilane, in Crich, D. (ed.), e-*EROS Encyclopedia of Reagents for Organic Synthesis*, **2007**, John Wiley & Sons, Ltd, DOI 10.1002/047084289X.rt226.pub2).

(**Peterson olefination**, item 7, Table 10.7)* in a process analogous to the Wittig reaction.

And, with regard to the Peterson olefination reaction, the Grignard reagent used (derived from chloromethyltrimethylsilane [ClCH₂Si(CH₃)₃]) could only be prepared after it was found (in the early 1950s) that direct chlorination of tetramethylsilane [(CH₃)₄Si] could be effected (Scheme 10.57). Interestingly, the same material had been prepared earlier by chlorination of chlorotrimethylsilane [(CH₃)₃SiCl] to give chloromethyldimethylchlorosilane [ClCH₂Si(CH₃)₂Cl] and then reaction of the latter with methylmagnesium bromide (CH₃MgBr) (Scheme 10.57). But the direct chlorination was important since it could also be used (under free radical conditions) to chlorinate the related benzyltrimethylsilane [C₆H₅CH₂Si(CH₃)₃ to the α-chlorobenzyl derivative [C₆H₅CH(Cl)Si(CH₃)₃] (Equation 10.84) with sulfuryl chloride (SO₂Cl₂) in the presence of benzoyl peroxide [(C₆H₅CO)₂O₂].†

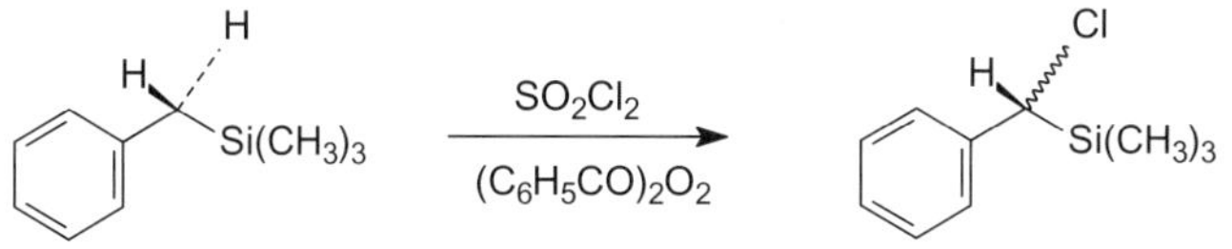

$$\xrightarrow[\text{(C}_6\text{H}_5\text{CO)}_2\text{O}_2]{\text{SO}_2\text{Cl}_2} \tag{10.84}$$

Similar free radical **substitution reactions** can be effected with bromine (using, e.g., *N*-bromosuccinimide [NBS]) and, once the halogen is in place, **nucleophilic substitution** at carbon α to silicon (or indeed, elsewhere) occurs with relative ease.‡

---

*Peterson, D. J. *J. Org. Chem.*, **1968**, *33*, 780 and Corey, E. J.; Enders, D.; Bock, M. G. *Tetrahedron Lett.*, **1976**, 7. Petersen, R. C.; Ross, S. D. *J. Am. Chem. Soc.*, **1963**, *85*, 3164.

†Baney, R. H.; Krager, R. J. *Inorg. Chem.*, **1964**, *3*, 1657.

‡Kulicke, K. J.; Chatgilialoglu, C.; Kopping, B.; Giese, B. *Helv. Chim. Acta*, **1992**, *75*, 935 and references therein and Holmes, R. R. *Chem. Rev.*, **1990**, *90*, 17.

TABLE 10.7. Silanes and Derivatives as used in Earlier Chapters

Substrate	Reagent Added to Substrate	Result of Addition of Reagent to Substrate Alcohol	Reference
1. CH_3CH_2OH	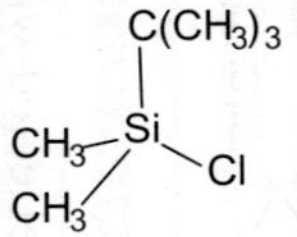 $C(CH_3)_3$ / $CH_3-Si-Cl$ / CH_3	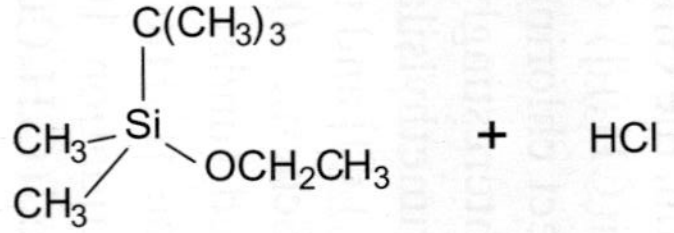 $C(CH_3)_3$ / $CH_3-Si-OCH_2CH_3$ / CH_3 $+$ HCl	Chapter 8: Table 8.6
2. 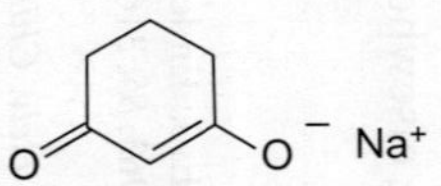	$(CH_3)_3SiCl$	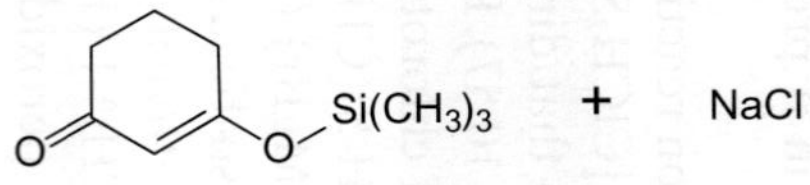 $Si(CH_3)_3$ $+$ NaCl	Chapter 8: Table 8.6
3. 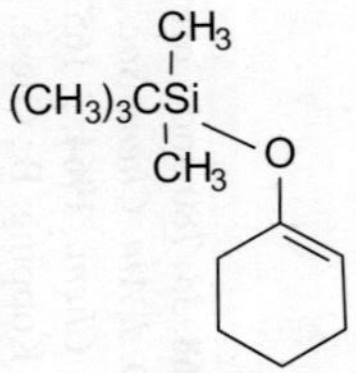	$[:CH_2]$	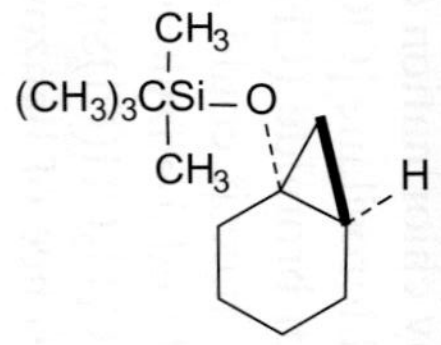	Chapter 8: Scheme 8.74
4. 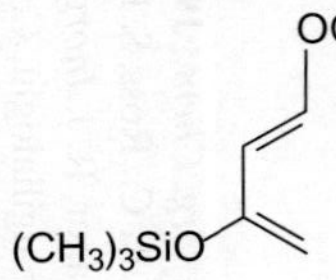	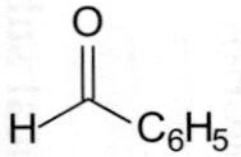	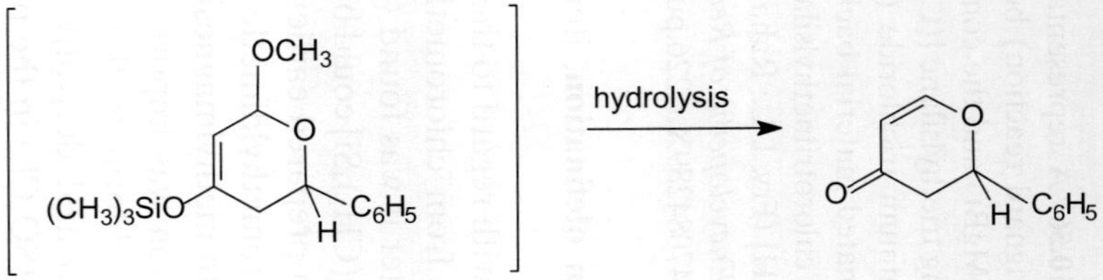 hydrolysis	Chapter 9: Equation 9.31

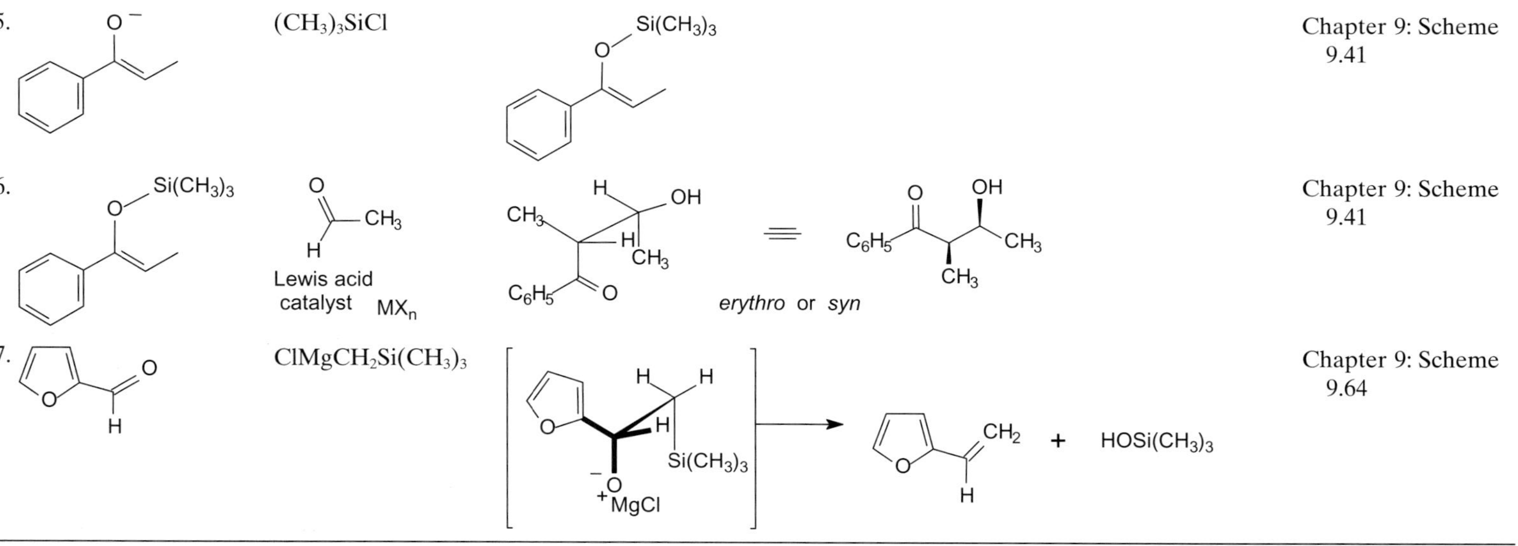

5.
(CH3)3SiCl
Chapter 9: Scheme
9.41
6.
Lewis acid catalyst MXn
erythro or syn
Chapter 9: Scheme
9.41
7.
ClMgCH2Si(CH3)3
CH2 + HOSi(CH3)3
Chapter 9: Scheme
9.64

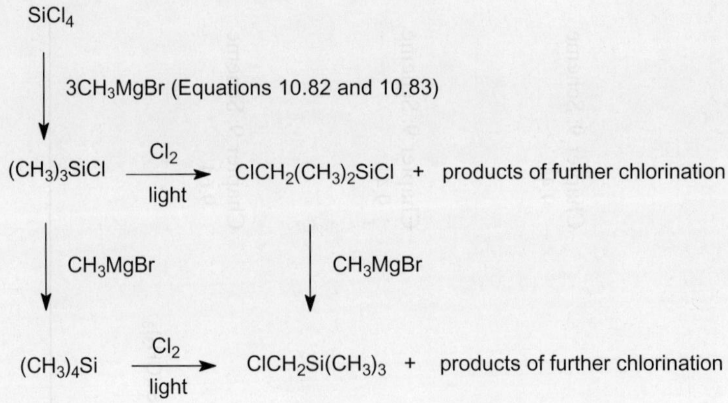

SiCl$_4$

3CH$_3$MgBr (Equations 10.82 and 10.83)

(CH$_3$)$_3$SiCl $\xrightarrow[\text{light}]{\text{Cl}_2}$ ClCH$_2$(CH$_3$)$_2$SiCl   +   products of further chlorination

CH$_3$MgBr   CH$_3$MgBr

(CH$_3$)$_4$Si $\xrightarrow[\text{light}]{\text{Cl}_2}$ ClCH$_2$Si(CH$_3$)$_3$ +   products of further chlorination

Scheme 10.57. A representation of two paths to chloromethyltrimethylsilane [ClCH$_2$Si(CH$_3$)$_3$]. In both, free radical chlorination in solution (by passage of chlorine gas [Cl$_2$] into pure alkylsilane while irradiating with a UV lamp) is used successfully.

As already indicated in Table 10.7 and earlier, a major use for trialkylchlorosilanes (R$_3$SiCl) (which include, but of course are not limited to, chlorotrimethylsilane [(CH$_3$)$_3$SiCl] and t-butyldimethylchlorosilane [(CH$_3$)$_3$CSi(CH$_3$)$_2$Cl]) is to block a specific site (temporarily) from reaction and then free it at will. Given (*vide supra*) the strength of the Si–F bond, the strategy that has developed is to (a) generate an anion, for example, at oxygen or carbon, at the site to be blocked, (b) allow the anion to react with the silylating reagent, (c) isolate and purify the silylated intermediate, (d) continue with the reaction(s) for which the blocked site was created, and (e) remove the silylated blocking group with fluoride (F$^-$) anion.*

Consider, for example, **item 3**, Table 10.7, originally described in Chapter 8, Scheme 8.80 and shown again as Scheme 10.58. Here, the t-butyldimethylsilyl (t-BDMS) derivative of the enol of cyclohexanone (prepared by, e.g., treatment of cyclohexanone with butyllithium at low temperature followed by reaction with t-butyldimethylchlorosilane) is allowed to react with methylene (or carbene [:CH$_2$]) generated from the corresponding diiodide (CH$_2$I$_2$) in the **Simmons–Smith** reaction. The corresponding cyclopropane [1-(t-butyldimethylsilyloxy)bicyclo[4.1.0]heptane] results. Subsequent treatment with base results in ring opening and formation of 2-methylcyclohexanone and t-butyldimethylsilanol [(CH$_3$)$_3$C(CH$_3$)$_2$SiOH] in one step or, by using, for example, tetrabutylammonium fluoride (TBAF) [(CH$_3$CH$_2$CH$_2$CH$_2$)$_4$N$^+$F$^-$], into the corresponding alcohol, bicyclo[4.1.0]heptan-1-ol, ready for further reaction.†

In Scheme 10.58, it is important to notice that the silicon intermediate is, as a consequence of d-orbital participation, shown with five bonds to silicon. Thus, the reaction is **not** S$_N$2 type but rather an addition–elimination. The subsequent reaction of the enol-t-BDMS derivative to produce the bicyclo[4.1.0]heptane system is then

---

*Petersen, R. C.; Ross, S. D. *J. Am. Chem. Soc.*, **1963**, *85*, 3164.

†Simmons, H. E.; Smith, R. D. *J. Am. Chem. Soc.*, **1959**, *81*, 4256; Boyer, J.; Breliere, C.; Carre, F.; Corriu, R. J. P.; Kpoton, A.; Poirier, M.; Royo, G.; Young, J. C. *J. Chem. Soc. Dalton Trans.*, **1989**, *1*, 43; Holmes, R. R. *Chem. Rev.*, **1990**, *90*, 17.

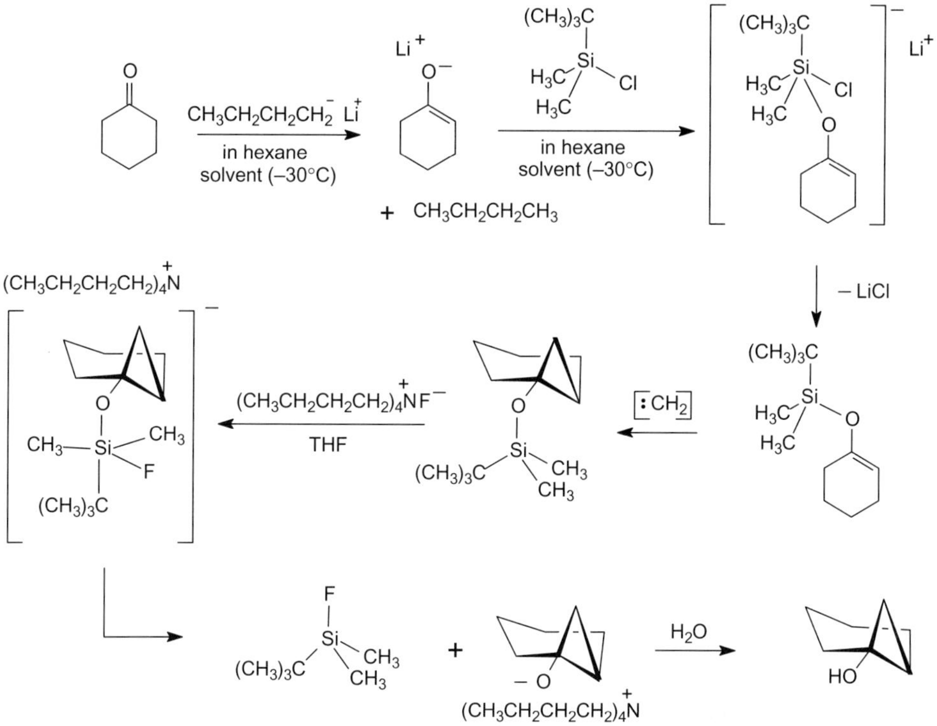

Scheme 10.58. An elaboration of Scheme 8.80. The formation of the anion of cyclohexanone and its reaction with *t*-butyldimethylchlorosilane [$(CH_3)_3CSi(CH_3)_2Cl$] (TBDMS).

followed by deprotection of the oxygen with TBAF. Again, pentacovalent silicon is invoked.

Silylation in a "protecting" or stabilizing role is also effective at carbon. For example (Scheme 10.59), treatment of 2-propyne-1-ol (propargyl alcohol) with dihydropyran (oxacyclohex-2-ene) in the presence of an acid catalyst (Scheme 8.60) produces the corresponding masked alcohol. When the protected alcohol is treated with *n*-butyllithium in oxacyclopentane (THF) at $-30°C$, the proton on the triple bond is lost and the resulting carbanion is capable of reaction with *t*-butyldimethylchlorosilane [$(CH_3)_3CSi(CH_3)_2Cl$] after the temperature is lowered to $-78°C$ to produce the corresponding silylated product. Then, treatment with methanolic acid generates the corresponding free alcohol. Now, that primary hydroxyl is capable of the plethora of reactions to be found in Chapter 8, and ultimately, the alkyne is capable of liberation on treatment with a source of fluoride anion.*

It has been argued that the Nazarov reaction, following the rules laid down in Chapter 4, Section G on electrocyclic pathways, has a counterpart found in reactions of α, β-unsaturated silanes (Scheme 10.60). Thus, as shown in the scheme, when cyclopentenoyl chloride was treated with silver tetrafluoroborate (other Lewis acids

---

*Rajagopalan, S.; Zweifel, G. *Synthesis*, **1984**, *2*, 111; Dunogues, J.; Bourgeois, P.; Pillot, J. P.; Merault, G.; Calas, R.; Lapouyade, P. *J. Organomet. Chem.*, **1975**, *87*, 169.

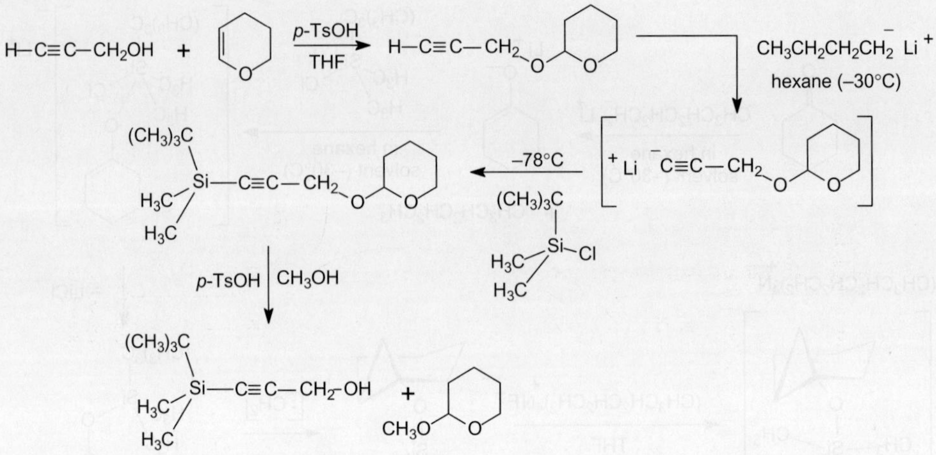

Scheme 10.59. A scheme showing a representation of what occurs on treatment of 2-propyne-1-ol (propargyl alcohol) with dihydropyran (oxacyclohex-2-ene) in the presence of an acid catalyst followed by *n*-butyllithium in hexane at –30°C and then with *t*-butyldimethylchlorosilane [(CH$_3$)$_3$CSi(CH$_3$)$_2$Cl] after the temperature is lowered to –78°C. The silylated product on treatment with methanolic acid generates the corresponding free alcohol with the trimethylsilyl group in place.

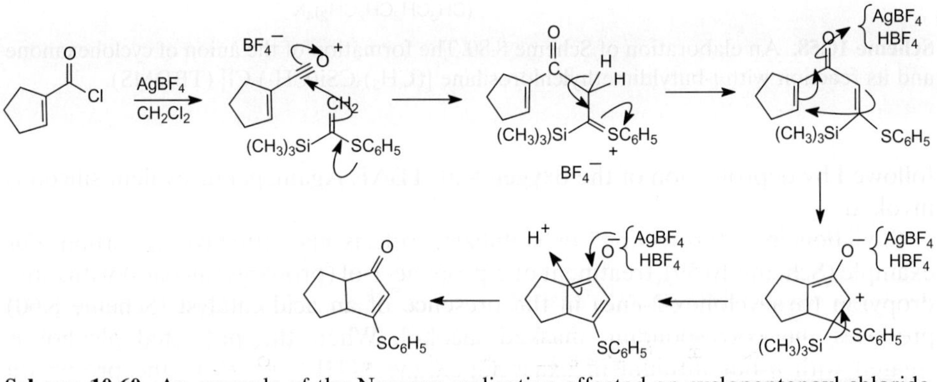

Scheme 10.60. An example of the Nazarov cyclization effected on cyclopentenoyl chloride. The example shows a potential rationalization for the observation that treatment of the acid chloride with silver tetrafluoroborate in dichloromethane at –50°C followed by addition of 1-(trimethylsilyl)-1-(phenylthio)ethene results in the formation of 4-(thiophenyl)bicyclo[3.3.0] oct-3-ene-2-one. It is argued that the cyclization step is a concerted process (see Habermas, K. L.; Denmark, S. E.; Jones, T. K. The Nazarov cyclization, in Paquette, L. A. (ed.), *Organic Reactions*, Vol. 45, Wiley, Hoboken, NJ, **1994**; Magnus, P.; Quagliato, D. *J. Org. Chem.*, **1985**, *50*, 1621).

fail) in dichloromethane at –50°C and 1-(trimethylsilyl)-1-(phenylthio)ethene added, 4-(thiophenyl)bicyclo[3.3.0]oct-3-ene-2-one resulted (in about 30% yield).

Interestingly, however, if the thiophenyl substituent is absent, a different course must be taken as a different product results.

It is argued that silicon substituents stabilize positive charges on the carbon β to the one bearing the silicon. The stabilization is said to result from electron donation

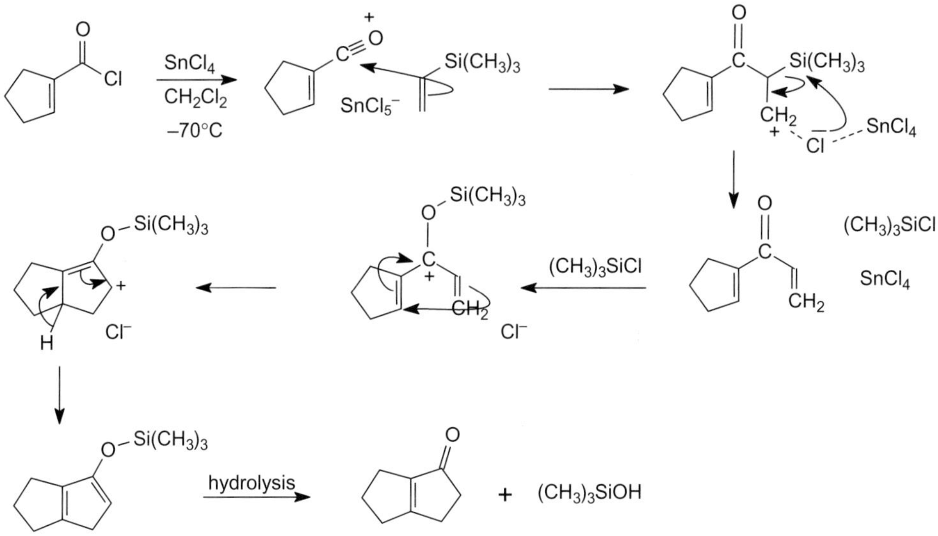

Scheme 10.61. A representation of a Nazarov-type cyclization occurring as a result of the addition of trimethylsilylethene (vinyltrimethylsilane [CH_2=$CHSi(CH_3)_3$]) to cyclopentenoyl chloride in the presence of tin(IV) chloride. It is important to note that the putative primary carbocation is presumed to be stabilized by overlap of the empty p- type orbital on carbon with the bond between the silicon and the carbon to which it is directly attached, that is, "the β-effect" as well as partial bonding to $SnCl_5^-$. (see Habermas, K. L.; Denmark, S. E.; Jones, T. K. The Nazarov cyclization, in Paquette, L. A. (ed.), *Organic Reactions*, Vol. 45, Wiley, Hoboken, NJ, **1994**; Magnus, P.; Quagliato, D. *J. Org. Chem.*, **1985**, *50*, 1621).

from the filled orbital between the silicon and the carbon to which it is attached and the neighboring empty orbital. Thus, as shown in Scheme 10.61, which should be compared with Scheme 10.60, when cyclopentenoyl chloride is treated with tin(IV) chloride (stannic chloride, $SnCl_4$) in dichloromethane at −70°C and trimethylsilylethene (vinyltrimethylsilane [CH_2=$CHSi(CH_3)_3$]) is added, bicyclo[3.3.0]oct-7-ene-1-one results (in about 50% yield). Again, a Nazarov-type cyclization is suggested.

When chlorotrimethylsilane [$(CH_3)_3SiCl$] is treated with potassium cyanide (KCN) in *N*-methylpyrrolidone solvent, trimethylsilylcyanide [$(CH_3)_3SiCN$)] is obtained. As shown in Scheme 10.62, this cyano compound has found significant use as a reagent for the introduction of the nitrile function at the carbonyl (compare with Scheme 9.37). The reversible nature of the addition reaction typical of such carbonyl additions with, for example, HCN (Scheme 9.37), is dramatically diminished when using trimethylsilylcyanide [$(CH_3)_3SiCN$] since the otherwise free hydroxyl is now derivatized.

Also, as shown in Scheme 10.62, trimethylsilylcyanide [$(CH_3)_3SiCN$] readily converts acid chlorides to the corresponding ketonitriles.

As pointed out earlier (Chapter 9, Part II), ethers (generically R–O–R, R–O–Ar, and Ar–O–Ar, where R = alkyl and Ar = aryl) can be induced to undergo cleavage reactions with Lewis acids. In particular, hydrogen iodide (HI) found utility as a reagent suited (albeit in low yield and with many side reactions) for

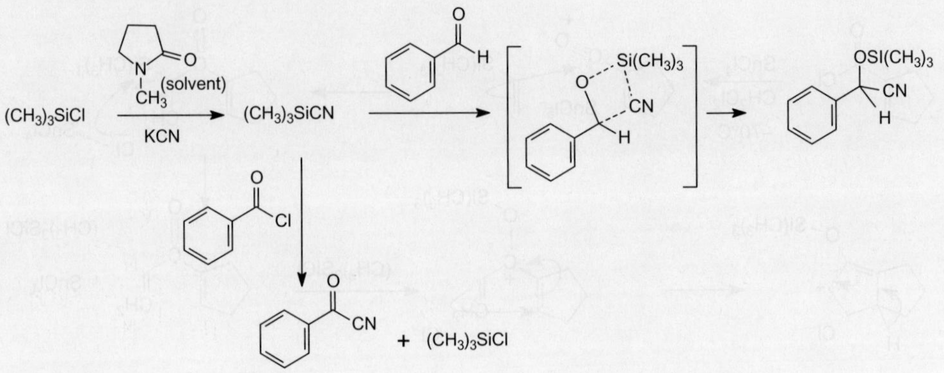

Scheme 10.62. A representation of the formation of carbonyl adducts through the use of trimethylsilylcyanide [(CH$_3$)$_3$SiCN] (see Sundermeier, M.; Mutyala, S.; Zapf, A.; Spannenberg, A.; Beller, M. *J. Organomet. Chem.*, **2003**, *684*, 50).

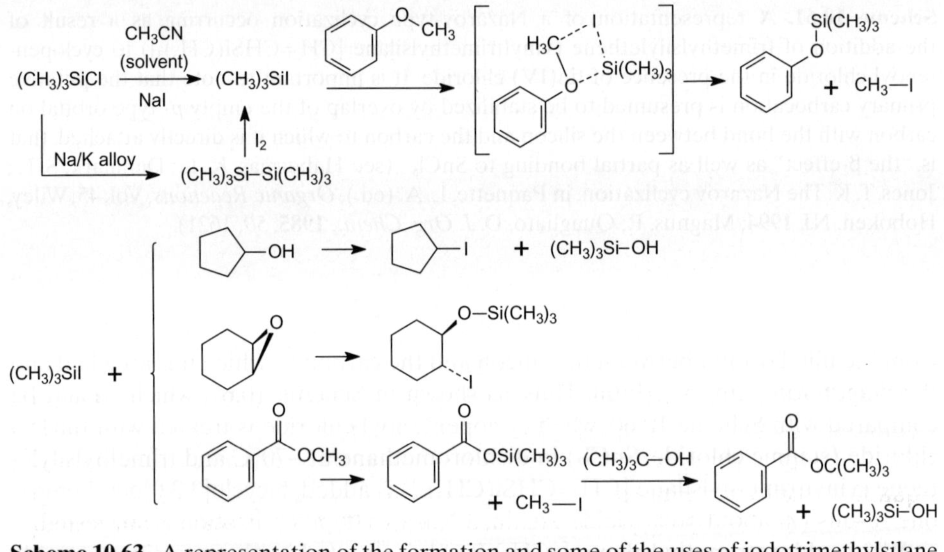

Scheme 10.63. A representation of the formation and some of the uses of iodotrimethylsilane [(CH$_3$)$_3$SiI]. See Pray, B. O.; Sommer, L. H.; Goldberg, G. M.; Kerr, G. T.; Di Giorgio, P. A.; Whitmore, F. C. *J. Am. Chem. Soc.*, **1948**, *70*, 433; Fleming, I.; Dunogues, J.; Smithers, R. *Organic Reactions*, Vol. 37, Kende, A. S., ed. Wiley, Hoboken, NJ, **1989**.

the cleavage of, in particular, methyl ethers. The reaction using iodotrimethylsilane [(CH$_3$)$_3$SiI)] is much more facile, and given the ready formation of the silicon–oxygen bond (*vide supra*), this reagent has found favor when replacement of oxygen by iodine is sought (Scheme 10.63). Thus, as shown in Scheme 10.63, when chlorotrimethylsilane [(CH$_3$)$_3$SiCl] is treated with sodium iodide in acetonitrile, iodotrimethylsilane [(CH$_3$)$_3$SiI] is produced.

Interestingly, and as an aside, the initial preparation of this iodide was, as shown, accomplished by treatment of chlorotrimethylsilane [(CH$_3$)$_3$SiCl] with

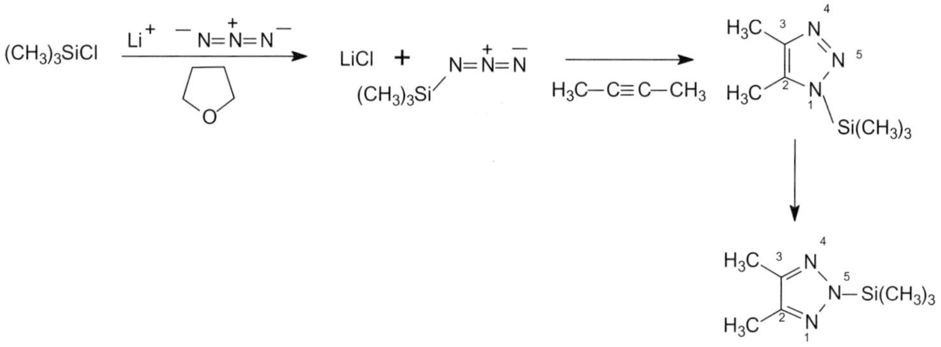

Scheme 10.64. A cycloaddition (and rearrangement) reaction of trimethylsilylazide with 2-butyne (dimethylacetylene, $CH_3C{\equiv}CCH_3$). See Birkofer, L.; Wegner, P. *Chem. Ber.*, **1966**, *99*, 2512; Hoffmann, S.; Hartung, K. J.; Nguyen, T. H.; Mewes, R.; Baluzow, W. *Zeit Chemie*, **1986**, *26*, 105; Rickborn, B. *Organic Reactions*, Vol. 52, Paquett, L. A., ed. Wiley, Hoboken, NJ, **1998**.

sodium–potassium alloy to produce hexamethyldisilane, which, in turn, underwent reaction with iodine (I_2) to yield the iodotrimethylsilane $[(CH_3)_3SiI]$.

A variety of useful reactions, avoiding strong acid usually needed (Chapter 8) to convert an oxygen-bearing substituent (–OH, –OR, etc.) from a poor leaving group into a good leaving group, is now be undertaken with iodotrimethylsilane $[(CH_3)_3SiI]$ as seen in Scheme 10.63.

When chlorotrimethylsilane $[(CH_3)_3SiCl]$ is treated with lithium azide (LiN_3) in oxacyclopentane (THF), trimethylsilylazide $[(CH_3)_3SiN_3]$ results. Treatment of the latter with 2-butyne (dimethylacetylene, $CH_3C{\equiv}CCH_3$) results in a cycloaddition reaction (Scheme 10.64), and, what is of particular interest, the 1-trimethylsilyl-1,2,3-triazole that results undergoes a 1,5-sigmatropic rearrangement to generate the more stable 2-trimethylsilyl-1,2,5-triazole.

Vinylsilanes have proved to be of some small interest because, as noted earlier, it appears that the silyl group stabilizes carbocations β to the position bearing the silicon. Trimethylvinylsilane $[(CH_3)_3CH{=}CH_2]$ is generally prepared, as shown in Equation 10.85 by the reaction between trimethylsilane $[(CH_3)_3SiH]$ and ethene (ethylene, $H_2C{=}CH_2$) over an iron(V) pentacarbonyl $[Fe(CO)_5]$ catalyst. Interestingly, the corresponding trichlorovinylsilane ($Cl_3SiCH{=}CH_2$) can also be made in a condensation reaction. In this case (Equation 10.86), the reaction is between ethyne (acetylene, $HC{\equiv}CH$) and trichlorosilane in the presence of a ruthenium dichloride ($RuCl_2$) catalyst.

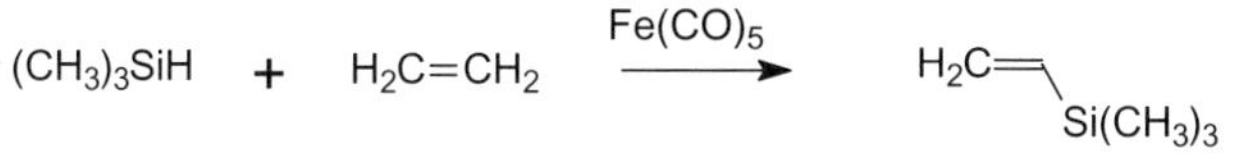

$$(CH_3)_3SiH \;+\; H_2C{=}CH_2 \;\xrightarrow{Fe(CO)_5}\; H_2C{=}\!\!\diagdown_{Si(CH_3)_3} \qquad (10.85)$$

$$Cl_3SiH \;+\; H{-}C{\equiv}C{-}H \;\xrightarrow{RuCl_2}\; H_2C{=}\!\!\diagdown_{SiCl_3} \qquad (10.86)$$

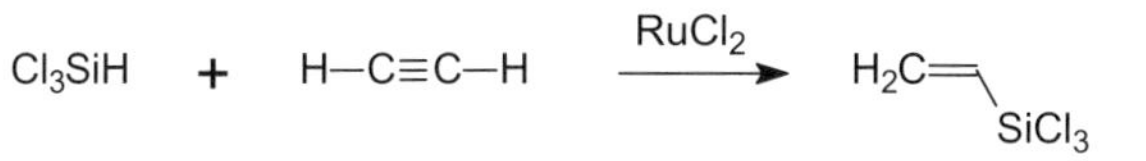

That the trimethylsilyl group actively participates at the β position was demonstrated as shown in Scheme 10.65. So, when the 1,1-dideutero-2-trimethylsilyl ethanol $[(CH_3)_3SiCH_2C^2H_2OH]$ shown in the scheme was treated with phosphorus

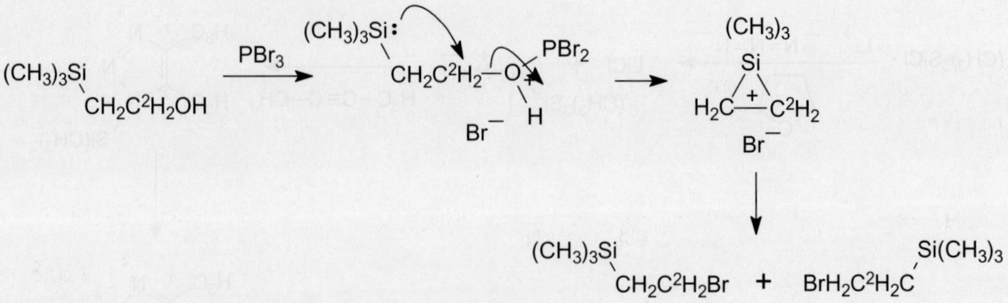

Scheme 10.65. The scrambling of the trimethylsilyl group across the two carbon atoms via neighboring group participation as the –OH group is lost from ethanol. A demonstration of the β-effect. Schmidbaur, H.; Tronich, W. *Angew. Chem. Int. Ed. Engl.*, **1968**, *7*, 220.

tribromide in dichloromethane (CH_2Cl_2), the corresponding dideutero-2-trimethyl-silylbromoethanes resulted. As expected for a "bridging" trimethylsilyl group, substitution occurred equally well at each end.

Interestingly, silyl-substituted aryl compounds frequently undergo *ipso* substitution. It is argued that this may be the result of *d*-orbital participation in the substitution process. As shown in Equation 10.87, treatment of (4-methoxyphenyl) trimethylsilane with aqueous acid results in the slow addition of a proton to the carbon bearing the trimethylsilyl group and eventual loss of hydroxytrimethylsilane [$(CH_3)_3SiOH$] (see Bott et al., 1967).

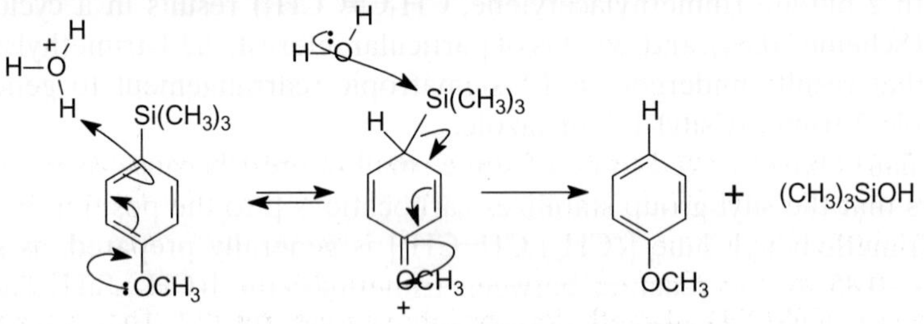

$$(10.87)$$

The aryltrimethylsilane itself can be prepared by a simple reaction between the corresponding Grignard reagent and chlorotrimethylsilane in oxacyclopentane (THF) as shown in Equation 10.88 (see Effenberger and Haebich, 1979).

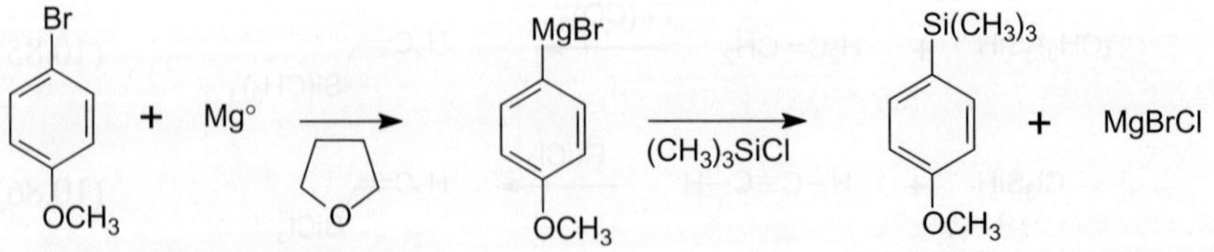

$$(10.88)$$

Allylic silanes can be prepared by the reaction between 3-chloro-1-propene (allyl chloride, $CH_2=CHCH_2Cl$) and trichlorosilane ($HSiCl_3$) in the presence of triethyl-amine [$(CH_3CH_2)_3N$] and a copper(I) chloride catalyst (Equation 10.89) to yield allyltrichlorosilane ($CH_2=CHCH_2SiCl_3$). Then, reaction of the latter with a suitable Grignard reagent (e.g., methylmagnesium bromide [CH_3MgBr]) or with zinc (Zn) and chloromethane (H_3CCl) yields the corresponding allyltrimethylsilane [$CH_2=CHCH_2Si(CH_3)_3$] (Equation 10.90) (see Chelucci et al., 2008).

$$CH_2=\overset{\displaystyle\diagup\hspace{-0.3em}\diagdown Cl}{} \; + \; HSiCl_3 \quad \xrightarrow[\text{(CH}_3\text{CH}_2)_3\text{N}]{\text{CuCl}} \quad CH_2=\overset{\displaystyle\diagup\hspace{-0.3em}\diagdown SiCl_3}{} \; + \; (CH_3CH_2)_3\overset{+}{N}HCl^-$$

$$(10.89)$$

$$CH_2=\overset{\displaystyle\diagup\hspace{-0.3em}\diagdown SiCl_3}{} \; + \; 3CH_3MgBr \quad \xrightarrow[\text{(CH}_3\text{CH}_2)_2\text{O}]{} \quad CH_2=\overset{\displaystyle\diagup\hspace{-0.3em}\diagdown Si(CH_3)_3}{} \; + \; 3MgBrCl$$

$$(10.90)$$

With hydrogen bromide (HBr), allyltrimethylsilane [$CH_2=CHCH_2Si(CH_3)_3$] yields 2-bromo-1-(trimethylsilyl)propane (Equation 10.91); however, with hydrogen chloride (HCl), the same substrate yields chlorotrimethylsilane and propene (Equation 10.92). This has led some observers to conclude that the reaction with hydrogen bromide (HBr) is not an ionic process beginning, for example, with proton transfer from HBr to the substrate, but rather that the reaction is a radical reaction (of hydrogen and bromine atoms), which simply adds in what appears to be a "Markownikoff fashion" as a result of stabilization by silicon (see Russell and Nagpal, 1961).

$$HBr_{(aq)} \; + \; CH_2=\overset{\displaystyle\diagup\hspace{-0.3em}\diagdown Si(CH_3)_3}{} \quad \longrightarrow \quad H_3C\overset{\displaystyle\diagup\hspace{-0.3em}\diagdown SiCl_3}{\underset{Br}{\diagdown}} \qquad (10.91)$$

$$HCl_{(aq)} \; + \; CH_2=\overset{\displaystyle\diagup\hspace{-0.3em}\diagdown Si(CH_3)_3}{} \quad \longrightarrow \quad H_2C=\overset{\displaystyle\diagup}{} \; + \; ClSi(CH_3)_3 \qquad (10.92)$$

Thus, it comes as no surprise (Scheme 10.66) that when the silane resulting from the reaction between chlorotrimethylsilane and the Grignard reagent prepared from 1-methoxy-4-chloromethylbenzene and magnesium is reduced under Birch conditions ($Li/NH_{3\,(l)}$, CH_3CH_2OH) and worked up with hydrogen chloride (HCl) in methanol (CH_3OH), 4-methylenecyclohexanone results. That is, apparently, protodesilylation and enol hydrolysis co-occur.

Allyltrimethylsilane [$CH_2=CHCH_2Si(CH_3)_3$] also undergoes reaction with ketones in the presence of aluminum trichloride ($AlCl_3$). Thus, as shown in Scheme 10.67, when propanone (acetone) is treated with aluminum trichloride ($AlCl_3$), the polarized carbonyl group is, arguably, capable of undergoing attack by the alkene. The resulting carbocation then undergoes fragmentation to yield a tertiary alcohol, which is isolated as the trimethysilyl derivative. Aldehydes behave similarly.

The allyltrimethylsilyl group [$CH_2=CHCH_2Si-$] is, of course, capable of participating in Claisen-type processes. Thus, as shown in Scheme 10.68, enol exchange with 1-ethoxycyclohexene and 3-(trimethylsilyl)-1-propanol in the presence of *p*-toluenesulfonic acid catalyst produces a trimethylsilylallyl vinyl ether, which undergoes rearrangement to the corresponding α-(1-trimethylsilylallyl) ketone.

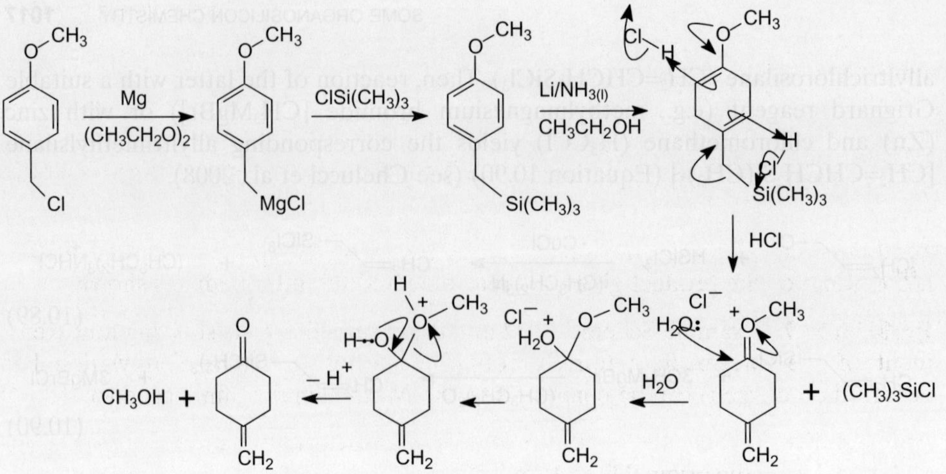

Scheme 10.66. A potential pathway for protodesilylation and enol hydrolysis of the Birch reduction product of 1-methoxy-4-(trimethylsilyl)methylbenzene to generate 4-methylene-cyclohexanone (see Rabideau, P. W.; Marcinow, Z. *Organic Reactions*, Paquette, L. A., ed. Vol. 42, Wiley, Hoboken, NJ, 1992).

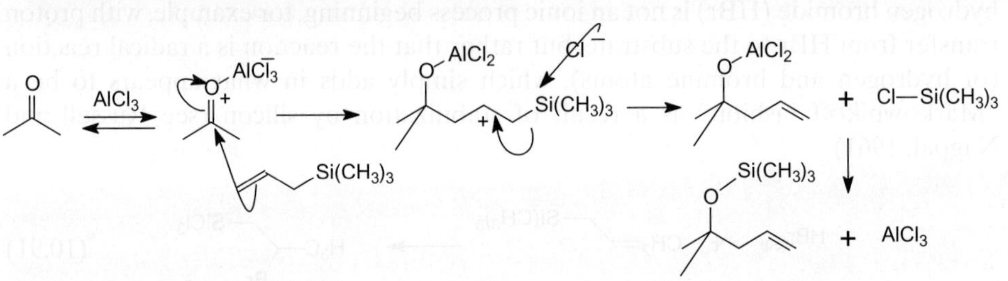

Scheme 10.67. A representation of the reaction of propanone [acetone (CH_3COCH_3)] with vinyltrimethylsilane [CH_2=$CHSi(CH_3)_3$] in the presence of aluminum trichloride ($AlCl_3$). The fragmentation of the cation to yield a tertiary alcohol, which is isolated as the trimethysilyl derivative, is worthy of note.

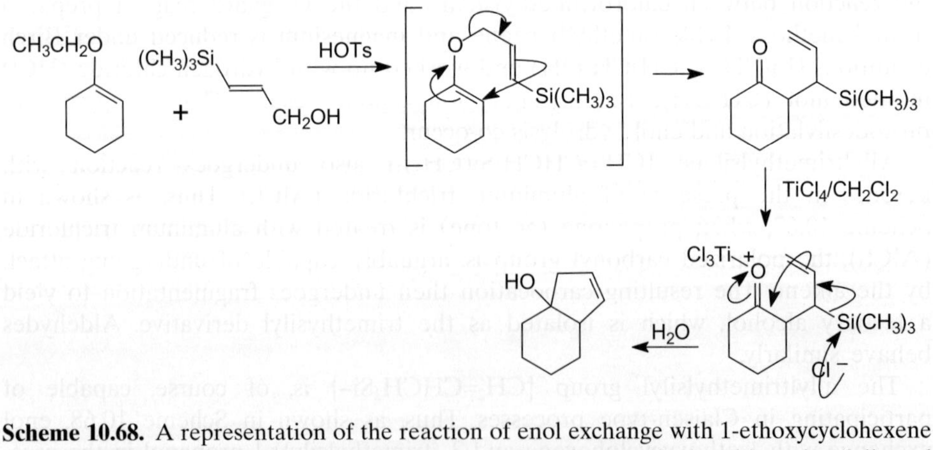

Scheme 10.68. A representation of the reaction of enol exchange with 1-ethoxycyclohexene and 3-(trimethylsilyl)-1-propanol in the presence of *p*-toluenesulfonic acid catalyst followed by a Claisen-type rearrangement. Subsequent ring closure under the influence of titanium tetrachloride is also represented (see, e.g., Trost, B. M.; Dong, G.; Vance, J. A. *J. Am. Chem. Soc.*, **2007**, *129*, 4540). HOTs = toluenesulfonic acid.

1018

ADDITIONAL PROBLEMS

Problem 10.5. Consider the processes of Equation 10.50. Write curved arrows with the shown substrate using the shown reactants that account for both steps of the reaction.

Problem 10.6. Consider the processes of Equation 10.41. Write curved arrows accounting for the product given the reactants and the intermediates shown.

Problem 10.7. Examine Scheme 10.32 and draw curved arrows showing how treatment of the intermediate tetrafluoroborate species (i.e., where alkylation has occurred on oxygen) can be converted to the *N*-ethylamidine final product.

Problem 10.8. In Scheme 10.40, the **Bouveault reaction**, two different results are obtained. Using the original literature (reference given) or any other source, attempt to justify the outcome.

Problem 10.9. How might you decide (what experiment might be done) to determine if the version of the Fischer–Hepp process shown in Scheme 10.42 produced the diaryl diazo product via an intra- or intermolecular pathway?

REFERENCES

Ahmed, M.; Seaya, A. M.; Jackstell, R.; Beller, M. *J. Am. Chem. Soc.*, **2003**, *125*, 10311.

Bott, R. W.; Eaborn, C.; Jackson, P. M. *J. Organomet. Chem.*, **1967**, *7*, 79.

Chelucci, G.; Baldino, S.; Pinna, G. A.; Benaglia, M.; Buffa, L.; Guizzetti, S. *Tetrahedron*, **2008**, *64*, 7574.

Effenberger, F.; Haebich, D. *Liebigs Ann. Chem.*, **1979**, *6*, 842.

Russell, G. A.; Nagpal, K. L. *Tetrahedron Lett.*, **1961**, 421.

Sicher, J. *Angew. Chem. Int. Ed. Engl.*, **1972**, *11*, 200.

FOREGROUND

The future is a world limited by ourselves …
—M. Maeterlinck
Jyzelle, Act I

INTRODUCTION

The parts of this work that have preceded this section dealt with some of the compounds constituting the body of organic chemistry, some methods of preparation of some of those compounds and some transformations within and between them. The same theme continues here but now with somewhat more complicated systems, somewhat more detail, and with particular attention to many of the molecules that are found in nature. The material in this part will build on what you have already learned.

For example, as discussed in Chapter 7, Walden carried out studies involving the chiral dicarboxylic acid, 2-hydroxybutane-1,4-dioic acid (2-hydroxysuccinic acid; malic acid), which, as noted there, he was able to obtain from the chiral dicarboxylic acid, 2-aminobutane-1,4-dioic acid (2-aminosuccinic acid; aspartic acid). These dioic acids were readily available to Walden because they are naturally occurring. The first occurs as a species found in the citric acid cycle (also called "the tricarboxylic acid cycle" or the "Krebs cycle") and will be discussed in more detail in Chapter 11. The second, an amino acid, will be discussed with other members of the same family in Chapter 12.

2-hydroxybutanedioic acid
(malic acid)

2-aminobutanedioic acid
(adipic acid)

It will not have escaped notice that these compounds, and indeed, most of those to be studied here are chiral, and some of the reactions they undergo are found to be enzyme (Chapters 8, 12, and 14) catalyzed. But it is also frequently possible to carry out similar reactions in the laboratory and occasionally to be able to suggest that some of the details of the laboratory processes mimic what is found in nature. And so it is that the particular transformations of *oxidation, reduction, substitution, addition, elimination,* and *rearrangement* are, once again, found to be important.

Furthermore, in formulating the transformations of *starting materials to products,* it is often the case that evidence has accumulated that the pathways by which the former are converted to the latter involves the intermediacy of *carbocations, carbanions,* and *free radicals* and kinetic and thermodynamic measurements have been made.

So, an understanding of the general principles can be applied to what appear to be, on first examination, unfamiliar cases.

Chapter 11, with which this part begins, is an introduction to carbohydrates and other carbonaceous materials (e.g., fatty acids, terpenes, steroids Figure III-A) derived, initially, from the absorption of carbon dioxide (CO_2) by plants and passing through acetate ($CH_3CO_2^-$) or its equivalent. Where appropriate, some references to enzymes involved in these conversions will be indicated.

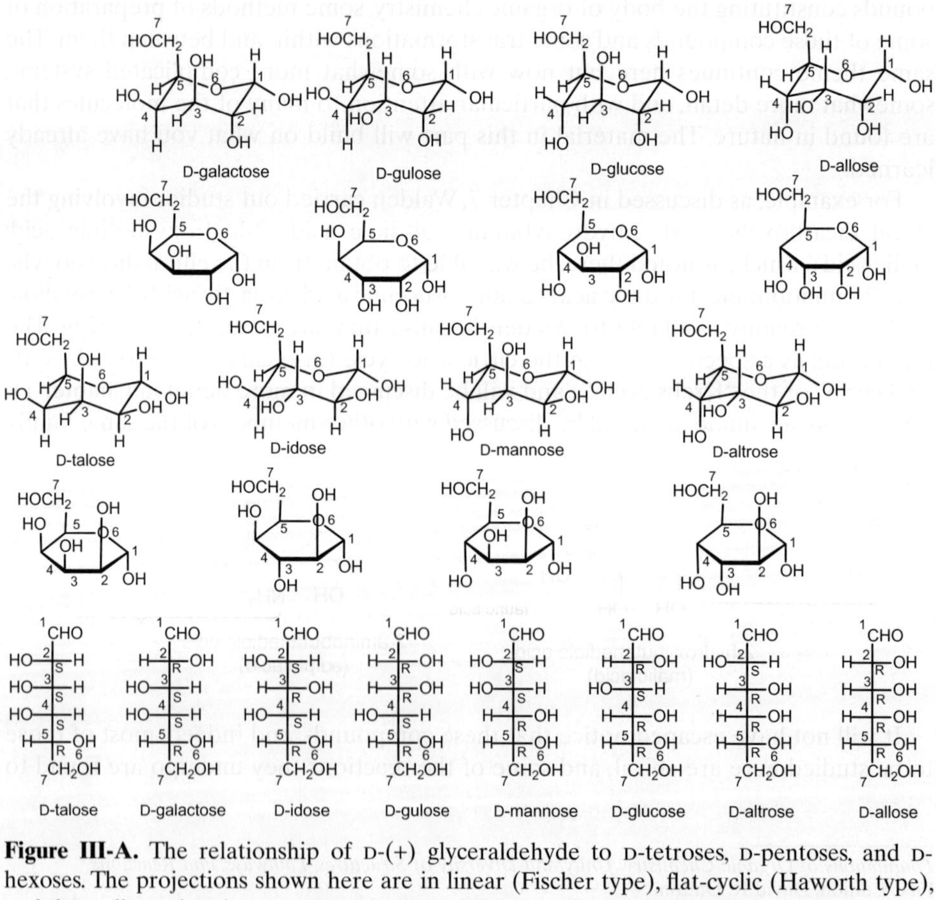

D-galactose   D-gulose   D-glucose   D-allose

D-talose   D-idose   D-mannose   D-altrose

D-talose   D-galactose   D-idose   D-gulose   D-mannose   D-glucose   D-altrose   D-allose

Figure III-A. The relationship of D-(+) glyceraldehyde to D-tetroses, D-pentoses, and D-hexoses. The projections shown here are in linear (Fischer type), flat-cyclic (Haworth type), and three-dimensional type representations. All are used interchangeably as needed.

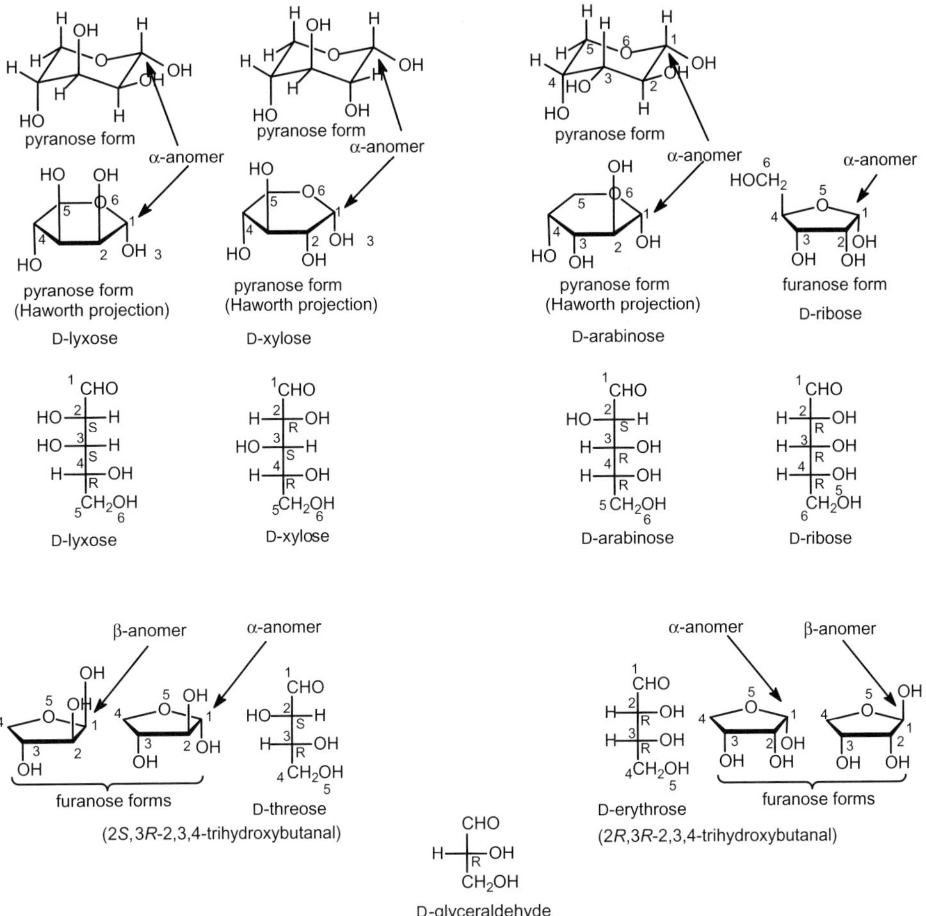

Figure III-A. *Continued.*

Chapter 12 will utilize some of the compounds developed in Chapter 11 to produce amino acids and the peptides and related materials that can be, at least in principle, synthesized by reactions between suitably substituted amino acids. This will reasonably lead to a discussion of enzymes, coenzymes, and metabolic processes, which, of course, can only be covered briefly (Figure III-B).

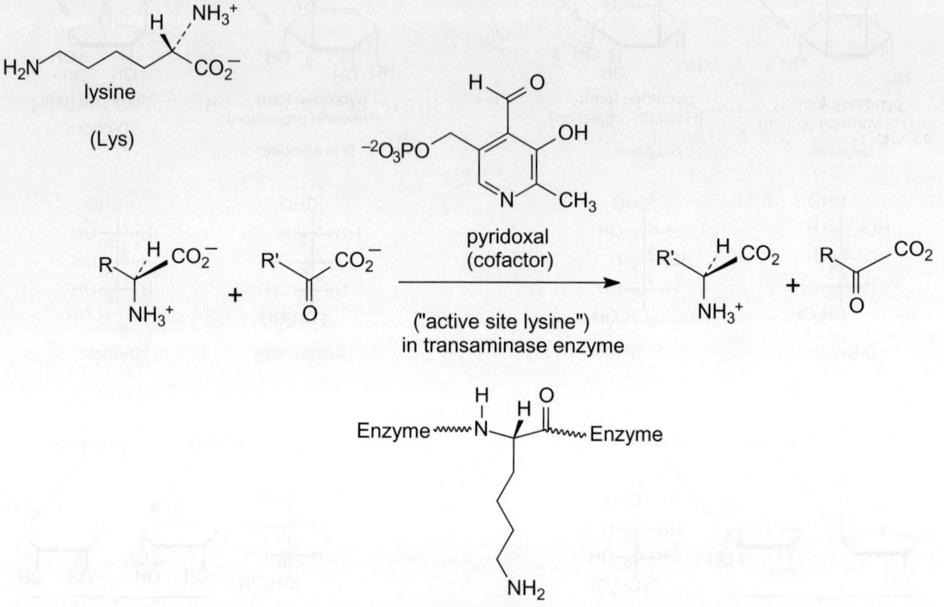

Figure III-B

Plants and animals also use amino acids to synthesize nitrogen-containing materials (most of which contain the nitrogen in a ring and are thus "heterocycles") of what has been referred to as secondary metabolism because obvious use of these materials in or to the plant/animal is not apparent to us. Some of these compounds are discussed in Chapter 13 under the general heading "Alkaloids."

Finally, in what follows, the Enzyme Commission (EC) numbers lead to the most current references for the information cited and are used in place of lists of references. Details can be found at http://www.chem.qmul.ac.uk/iubmb/enzyme/search.html.

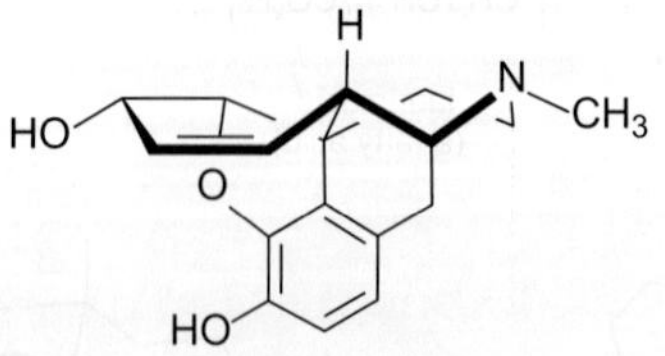

morphine
(an alkaloid)

In the first part of the last chapter (Chapter 14), heterocyclic materials, called nucleic acids as well as some of their derivatives that encode (in DNA where A, G, C, and T are found) and transmit (in RNA where A, G, C, and U are found) the stuff of life, will be briefly covered (Figure III-C).

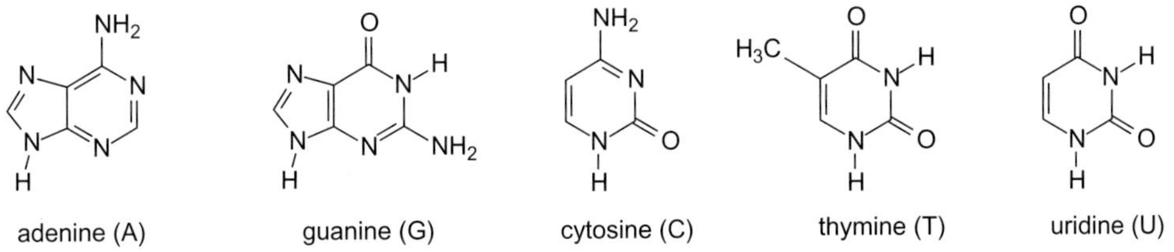

adenine (A)   guanine (G)   cytosine (C)   thymine (T)   uridine (U)

Figure III-C

In the second part of Chapter 14, a brief introduction to the uroporphyrinogens as derived from δ-aminolevulinic acid is provided.

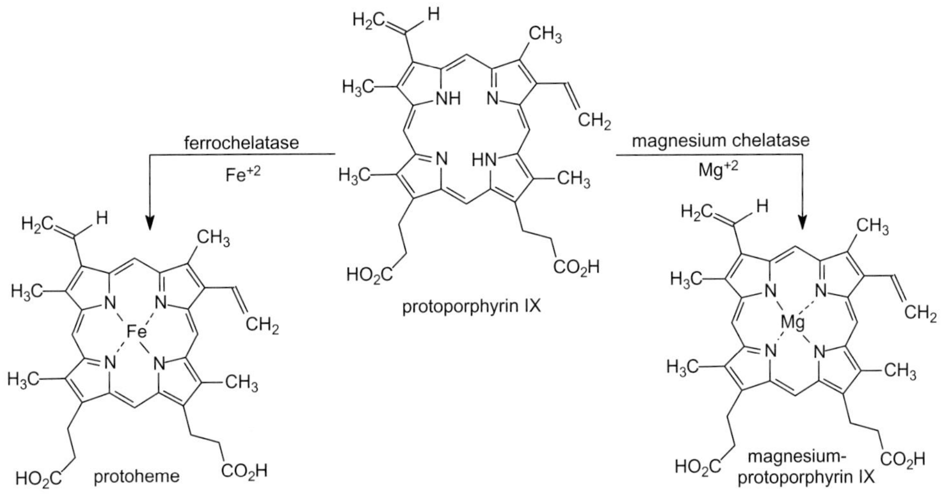

protoporphyrin IX

ferrochelatase
Fe^{+2}

magnesium chelatase
Mg^{+2}

protoheme

magnesium-
protoporphyrin IX

CHAPTER 11

An Introduction to Carbohydrates, Acetogenins, and Steroids

And the earth brought forth grass, herb yielding seed after its kind and tree bearing fruit, wherein is the seed thereof, after its kind; and God saw that it was good. And there was evening and there was morning, a third day.

—Genesis

ותצא הארץ דשא עשב מזריע זרע למינהו ועץ עשה־פרי אשר
זרעו־בו למינהו וירא אלהים כי־טוב: ויהי־ערב ויהי־בקר יום שלישי:
בראשית

A. INTRODUCTION

Although there are some forms of life on Earth that do not, directly, need sunlight to prosper,* plants are not among them, as solar energy appears to be required to convert carbon dioxide $(CO_2)^\dagger$ into **carbohydrates**[‡] $[C(H_2O)_n$; as phosphate derivatives] in the process called **photosynthesis**. The details of photosynthetic processes are beyond the scope of this work.

---

*These so-called extremophiles are usually unicellular species found growing under conditions that were, for many years, considered inimical to life.

[†]Carbon dioxide (CO_2) is processed through the Calvin cycle ([or the Calvin–Benson–Bassham cycle] which is also referred to as the RPP cycle) *vide infra*.

Although some plants, such as those called C_4 and those known as using Crassulacean acid metabolism (CAM) photosynthesis (i.e., xerophytes) initially capture the carbon dioxide (CO_2) differently than the C_3 plants do, they too subsequently release it to the Calvin cycle (M. Calvin [1911–1997], professor, University of California, Berkeley, since 1937, Nobel Prize in Chemistry 1961; Bassham, J.; Benson, A.; Calvin, M. [1950]. *J. Biol. Chem.*, **1950**, *185*, 781).

[‡]**Carbohydrates** (i.e., hydrates of carbon) made from the fragments (carbon dioxide [CO_2] and water [H_2O]) are put together (**synthesis**) with light (**photo**) ultimately providing the energy.

Foundations of Organic Chemistry: Unity and Diversity of Structures, Pathways, and Reactions, First Edition. David R. Dalton.
© 2011 John Wiley & Sons, Inc. Published 2011 by John Wiley & Sons, Inc.

However, it is possible to begin with the Calvin or reductive pentose phosphate (RPP) cycle to discuss some of the organic chemistry involved in the enzyme-catalyzed processes of the cycle.*

B. THE CALVIN CYCLE (Bassham et al., 1950)

The overall picture of the Calvin cycle is shown in Figure 11.1.

The grand picture, requiring three turns of the cycle, is that 3 equivalents of carbon dioxide (CO_2) (i.e., 3 C_1) react with 3 equivalents of the five-carbon carbohydrate derivative, ribulose-1,5-bisphosphate[†] (i.e., 3 C_5) to yield three (hypothetical) six-carbon adducts, 2-carboxy-3-ketoribitol-1,5-bisphosphate (3 C_1 + 3 C_5 = 3 C_6). The three C_6 species then undergo what can be written as a retroaldol Chapter 9, Part 1, IV, C, 2 and Part 2, E to yield 6 equivalents of the monophosphate ester of the three-carbon dihydroxy monocarboxylate, 3-phosphoglycerate (PGA) (i.e., 3 C_6 = 6 C_3). Of the 6 equivalents of PGA produced, five are used to regenerate 3 equivalents of ribulose-1,5-bisphosphate (i.e., 5 C_3 = 3 C_5), while the sixth is now available to combine with another to make a six-carbon sugar (2 C_3 = 1 C_6) such as glucose or for other ends, such as decarboxylation to acetate.

So, ribulose-1,5-bisphosphate (abbreviated as RuBP) using the enzyme ribulose-1,5-bisphosphate carboxylase (EC 4.1.1.39, carboxydismutase, Rubisco) catalyzes the Mg^{+2}-dependent conversion of the 1,5-bisphosphate ester of the carbohydrate ribulose with carbon dioxide (CO_2) to produce 2 equivalents of PGA (Equation 11.1). As shown in Scheme 11.1, the six-carbon intermediate, 2-carboxy-3-ketoribitol-1,5-bisphosphate, is often written since it conforms to what little data are available— but it has not been isolated.

$$\text{RuBP} + CO_2 \xrightarrow[Mg^{+2}]{\text{Rubisco}} 2\,\text{PGA} \qquad (11.1)$$

It is important to keep in mind that we want the PGA for other purposes, although, as noted above, three turns of the cycle are necessary to produce six PGAs and five of them are reused in making the three ribulose-1,5-bisphosphates necessary to turn the cycle three times.

So, as shown in Figure 11.1, after the absorption of CO_2 and the formation of 2 equivalents of glycerate 3-phosphate (3-phospho-D-glycerate), a further phosphorylation occurs. The 3-phospho-D-glycerate is now phosphorylated on the carboxyl oxygen to make a mixed carboxylate-phosphate anhydride, 1,3-bisphospho-D-glycerate (3-phospho-D-glyceroyl phosphate). The phosphate exchange involves the removal of a phosphate group from adenosine triphosphate (ATP), to form the

---

*A cycle cannot have a beginning or an end. If it did, it would not be a cycle. So, for the cycles to be discussed, chemical species will be added and removed, tangentially as the cycle turns, with energy for the turning being provided (ultimately) by the sunlight. It is found that catalysis can be used to lower the barriers for reactions to occur. These catalysts are frequently proteins (Chapter 12) called **enzymes** that help organize the reactants (called **substrates**) for processing.

[†]A note about the use of "bis" rather than "di" is needed. Generally, the former is utilized when there are two independent phosphate esters (here, one at C1 and one at C5) where the latter is used when the two phosphates are cojoined (as $H_2O_3P\text{-}O\text{-}PO_3H_2$ or adenine diphosphate [ADP]).

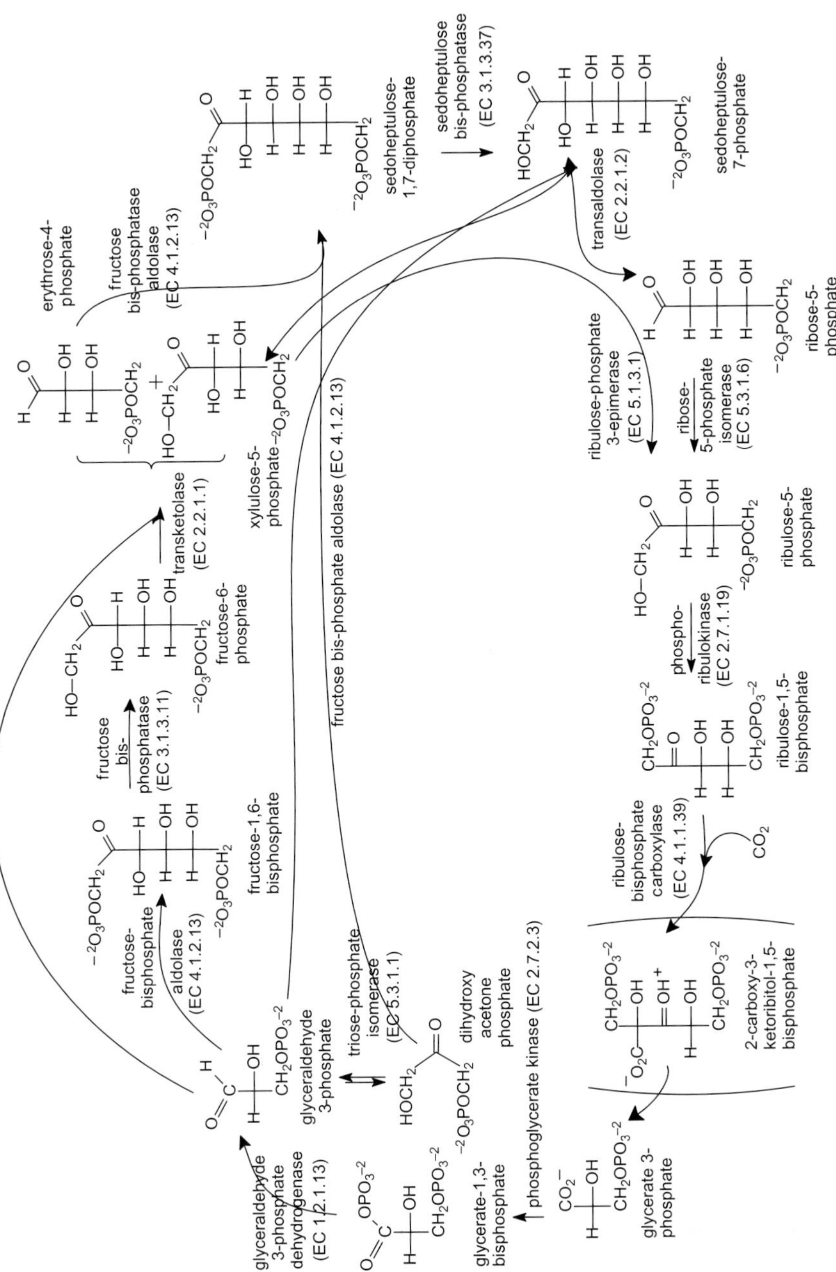

Figure 11.1. A representation of the Calvin cycle.

1029

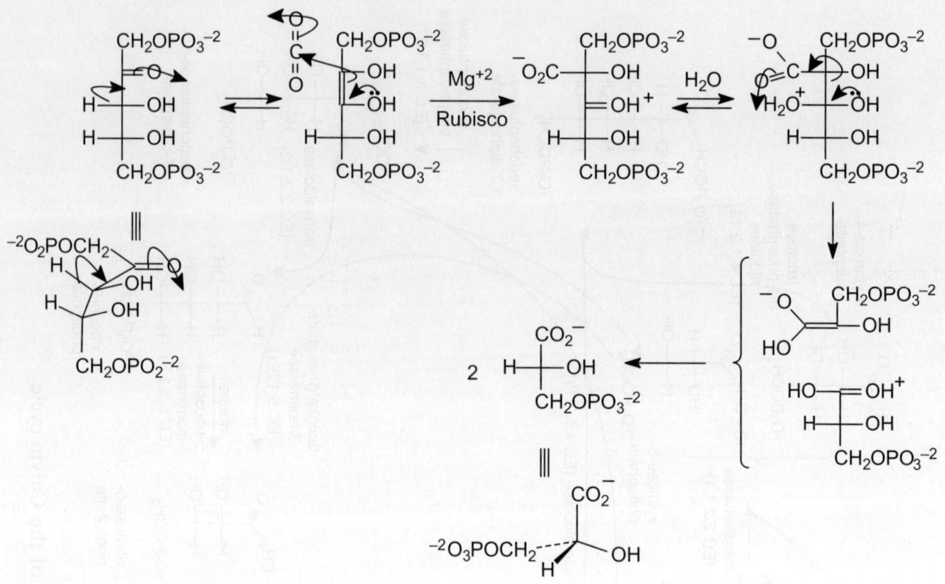

Scheme 11.1. A cartoon representation of the addition of carbon dioxide (CO_2) to the five-carbon ketosugar diphosphate, ribulose-1,5-bisphosphate, in an imagined aldol-type process in the presence of Mg^{+2} (which might further be imagined to coordinate with the enediol as in pinacol formation, Chapter 8) and the enzyme ribulose-1,5-bisphosphate carboxylase (EC 4.1.1.39). Since 2-carboxy-3-ketoribitol-1,5-bisphosphate is not isolable, the sense of addition (*re* or *si*) cannot be established. Retroaldol-type fragmentation is presumed to occur to yield, after suitable proton tautomeric adjustments, 2 equivalents of 3-phosphoglycerate ($C_1 + C_5 = 2\ C_3$).

bisphosphoglycerate and adenosine diphosphate (ADP) (Scheme 11.2). The process is catalyzed by the enzyme phosphoglycerate kinase (EC 2.7.2.3).*

Now, with a good leaving group attached to a carbonyl oxygen, reduction can be affected to produce 3-phosphoglyceraldehyde (glyceraldehyde 3-phosphate). The reduction is accomplished enzymatically with glyceraldehyde 3-phosphate dehydrogenase (EC 1.2.1.13) by using the reduced form of the cofactor nicotinamide adenine

---

*A kinase (or kinases in general) belongs to a subclass of enzymes known as transferases. The function of a kinase is to participate as a catalyst in the transfer of a phosphate group. The phosphate anhydride (or phosphoanhdyride) linkages (i.e., O–P–O–P–O, as in two or more equivalents of phosphoric acid, H_3PO_4, from which water has been removed) are considered to be "high-energy bonds" in that their hydrolysis (not $D°$) is accompanied by a relatively large $-\Delta G$ (ca. $-30\,kJ\,mol^{-1}$). Nonetheless, as they are generally not solvated well, hydrolysis is not easy, and so the phosphoglycerate kinase lowers the barrier.

The Enzyme Commission (EC) numbers are part of a well-defined (although occasionally modified) system worked out (and on) by the *Nomenclature Committee of the International Union of Biochemistry and Molecular Biology (NC-IUBMB)* in consultation with the *International Union of Pure and Applied Chemistry (IUPAC)*. Details can be found through those bodies, but, in general, all enzymes beginning EC 1 refer to **Oxidoreductases**, EC 2 to **Transferases**, EC 3 to **Hydrolayses**, EC 4 to **Lyases**, EC 5 **Isomerases**, and EC 6 to **Ligases**. So, all enzyme-catalyzed reactions involve (1) oxidation–reduction, (2) transfer of an atom or functional group from one species to another, (3) hydrolysis, (4) breaking bonds in elimination processes with (commonly) double-bond formation, (5) isomerization of one species to another, or (6) making bonds (often with ATP hydrolysis). Some overlap occurs.

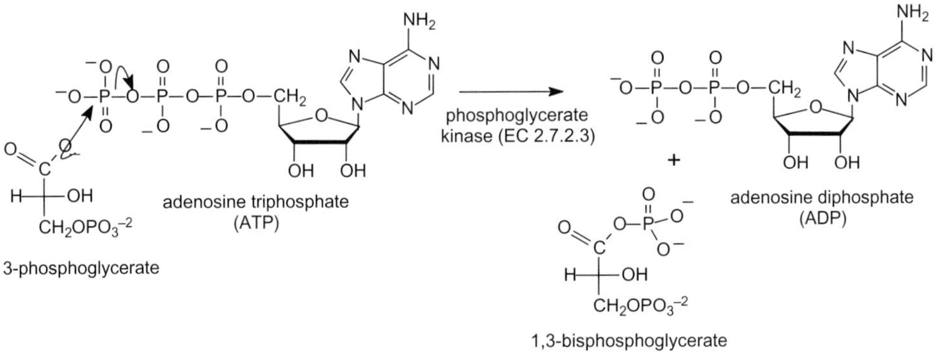

Scheme 11.2. The phosphorylation of the carboxylate group of 3-phosphoglycerate to produce a mixed carboxylate-phosphate anhydride, 1,3-bisphosphoglycerate. The phosphoglycerate kinase-catalyzed (EC 2.7.2.3) formation of the anhydride involves the removal of a phosphate group from adenosine triphosphate (ATP), yielding adenosine diphosphate (ADP) and the bisphosphoglycerate. The stereochemistry of the attack at phosphorus is backside to the leaving group.

dinucleotide phosphate (NADPH), which is oxidized in the process to the oxidized form nicotinamide adenine dinucleotide phosphate ($NADP^+$) (Scheme 11.3).[*][†]

[*]As seen in the Scheme 11.3, NADPH and $NADP^+$ both bear a phosphate group linked to the C2 of that ribosyl group attached to the adenine portion of the dinucleotide rather than the ribosyl group attached to the pyridine ring of the nicotinamide portion. While to the chemist the change from these species to the corresponding NADH and NAD^+, which lack the phosphate substitution, is, overall, slight and, further, the change clearly does not involve the site where the reduction occurs, the enzymes dedicated to the cause find that $NADP^+$ and NAD^+ are not interchangeable! The couple, $NADPH/NADP^+$, is generally found in low concentration and is used in reductive biosynthetic processes, while the much more common $NADH/NAD^+$ is used in energy metabolism.

[†]As seen in Scheme 11.3, hydride is transferred from NADPH to the 1,3-bisphosphoglycerate resulting in the formation of glyceraldehyde 3-phosphate, phosphate, and $NADP^+$. To the organic chemist, this would appear as an addition–elimination process with the hydride adding to the carbon of the carbonyl, the formation of a tetrahedral intermediate, and the subsequent elimination of the phosphate leaving group. The presence of the enzyme complicates the issue since its role is that of a catalyst and forces might be brought into play, which would change the course of the "normal" addition–elimination. Second, as shown below, the two faces of the reduced pyridine ring (NADPH) are not identical. Thus, hydrogen transfer will occur from either the *re*- or *si*-face (depending on which the enzyme steers to the substrate). While it will be specific, there are about as many enzymes that use the *re* hydrogen as those that use *si*.

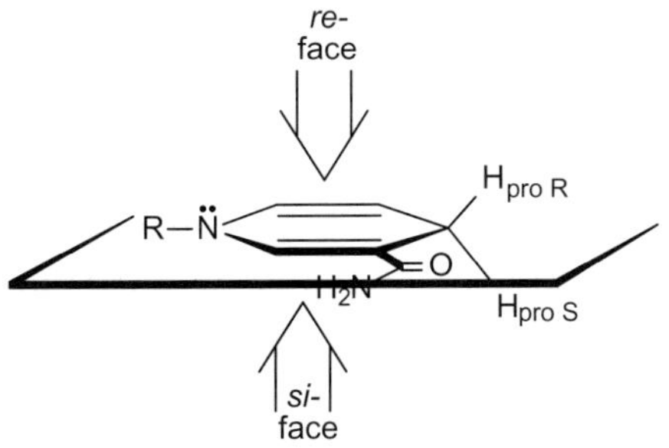

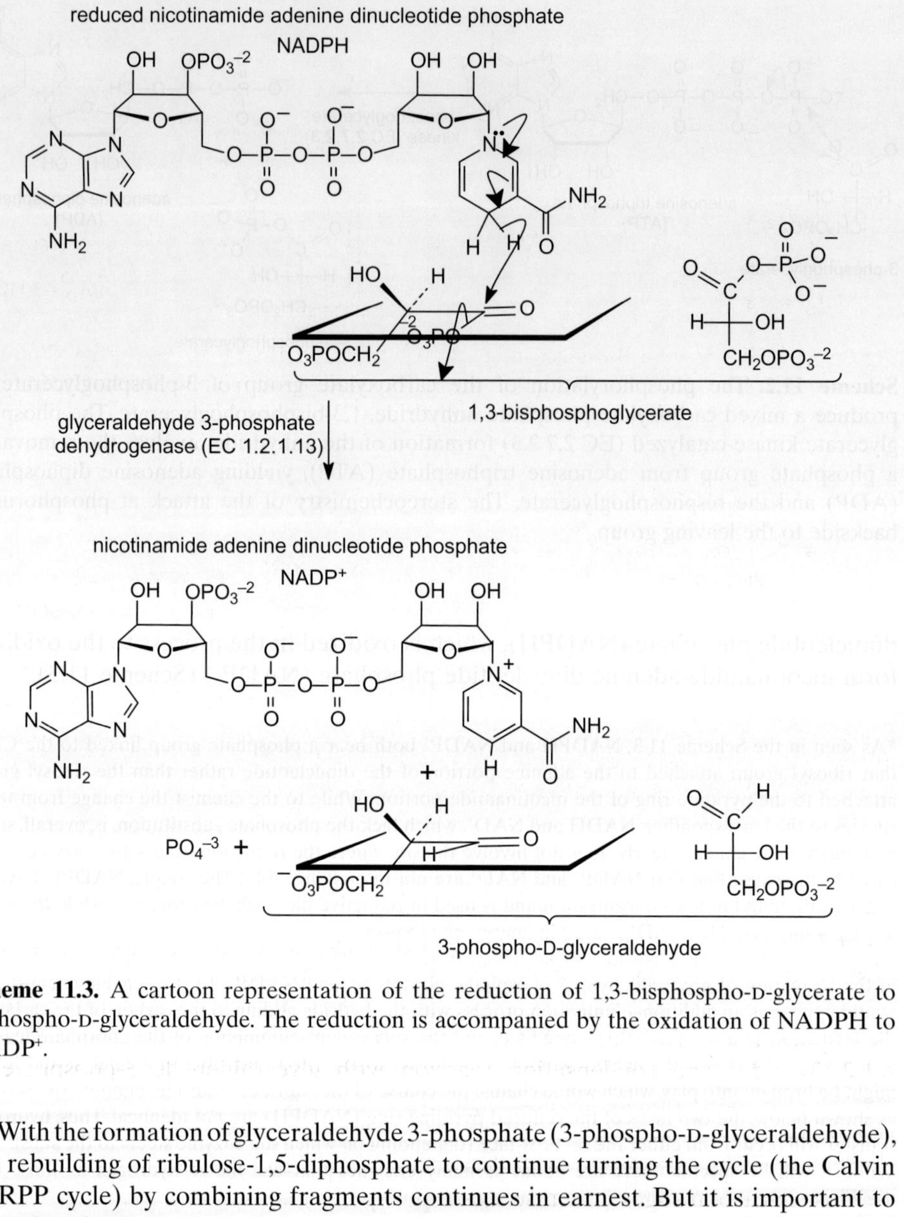

reduced nicotinamide adenine dinucleotide phosphate
NADPH

1,3-bisphosphoglycerate

glyceraldehyde 3-phosphate
dehydrogenase (EC 1.2.1.13)

nicotinamide adenine dinucleotide phosphate
NADP⁺

3-phospho-D-glyceraldehyde

Scheme 11.3. A cartoon representation of the reduction of 1,3-bisphospho-D-glycerate to 3-phospho-D-glyceraldehyde. The reduction is accompanied by the oxidation of NADPH to NADP⁺.

With the formation of glyceraldehyde 3-phosphate (3-phospho-D-glyceraldehyde), the rebuilding of ribulose-1,5-diphosphate to continue turning the cycle (the Calvin or RPP cycle) by combining fragments continues in earnest. But it is important to note that the formation of glyceraldehyde 3-phosphate cannot be underestimated.

Thus, **glyceraldehyde 3-phosphate will be used to produce hexoses**, materials used for the "storage of energy" as well as for structural purposes.

Furthermore, **phospho-3-glycerate**, precursor to glyceraldehyde 3-phosphate, will be seen again for it is from here that complex molecule formation arises. An isomerase converts it to **phospho-2-glycerate** and thence by dehydration to **phosphoenolpyruvate** and on to **pyruvate**, acetate (**acetyl coenzyme [CoA]**), and products built therefrom.

Now, continuing with Figure 11.2, glyceraldehyde 3-phosphate isomerizes to dihydroxyacetone phosphate (Scheme 11.4). While this is simply the isomerization of an

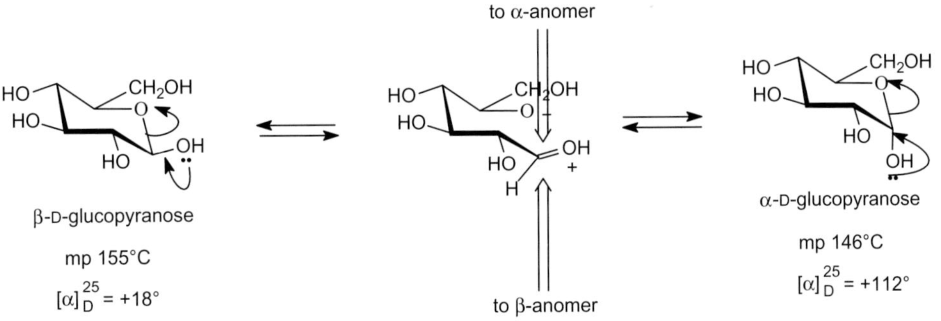

Figure 11.2. The α- and β-D-glucose (α- and β-D-glucopyranose to emphasize their six-membered rings) forms and their equilibration. The equilibrium mixture at about 36:64 (α:β) has an optical rotation of +52.5°.

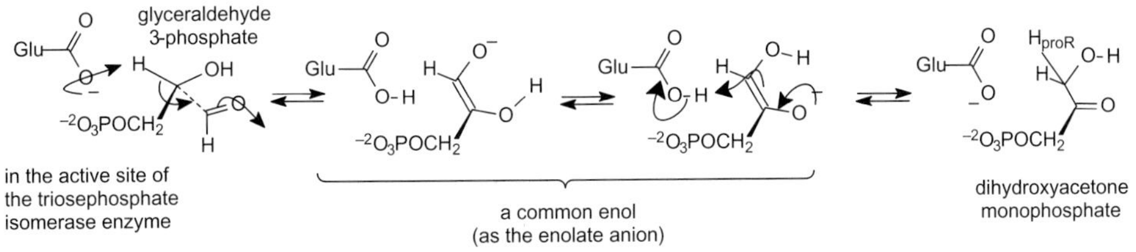

Scheme 11.4. The isomerization of glyceraldehyde 3-phosphate to dihydroxyacetone phosphate catalyzed by triosephosphate isomerase (EC 5.3.1.1) and through a "common" enol. It is presumed that proton abstraction (and readdition as the (*pro-R*) hydrogen) is affected by an active site glutamate anion.

α-hydroxyaldehyde to the corresponding α-hydroxyketone via the common enol, it is catalyzed by a triosephosphate isomerase (EC 5.3.1.1), where, presumably, proton abstraction (and readdition as the (*pro-R*) hydrogen) is affected by an active site glutamate anion [$^-O_2CCH_2CH_2(NH_3^+)CO_2^-$]. And then, the dihydroxyacetone monophosphate formed undergoes a catalyzed (fructose-bisphosphate aldolase, EC 4.1.2.13) aldol-type condensation reaction with glyceraldehyde 3-phosphate to produce the six-carbon ketosugar fructose-1,6-bisphosphate. It is held that an active site lysine Lys-NH₂ [$^+H_3NCH_2CH_2CH_2CH_2CH(NH_3^+)CO_2^-$] serves as the catalyst through addition at the carbonyl followed by proton tautomerization (Scheme 11.5).

This six-carbon sugar (fructose 1,6-bisphosphate) is used as a building block to return to ribulose 6-phosphate as well as to produce other fragments **that will be used subsequently to build amino acids, terpenes, steroids, acetogenens, and peptides and proteins** as well as glucose for storage as starch or cellulose. The next step in that process is the hydrolysis of the phosphate at C1 to generate fructose 6-phosphate. Once again, an enzyme (fructose-bisphosphatase, EC 3.1.3.11) assures that hydrolysis occurs at only one of the two phosphate linkages. Clearly, the two phosphates are different and the difference is easily seen from the vantage point of the cyclic form of the fructose 1,6-bisphosphate in Schemes 11.5 and 11.6.

Now, with the intervention of the enzyme transketolase (EC 2.2.1.1), an enzyme that utilizes thiamine diphosphate (vitamin B₁, written as the ylid, $pK_a \approx 18$) cofac-

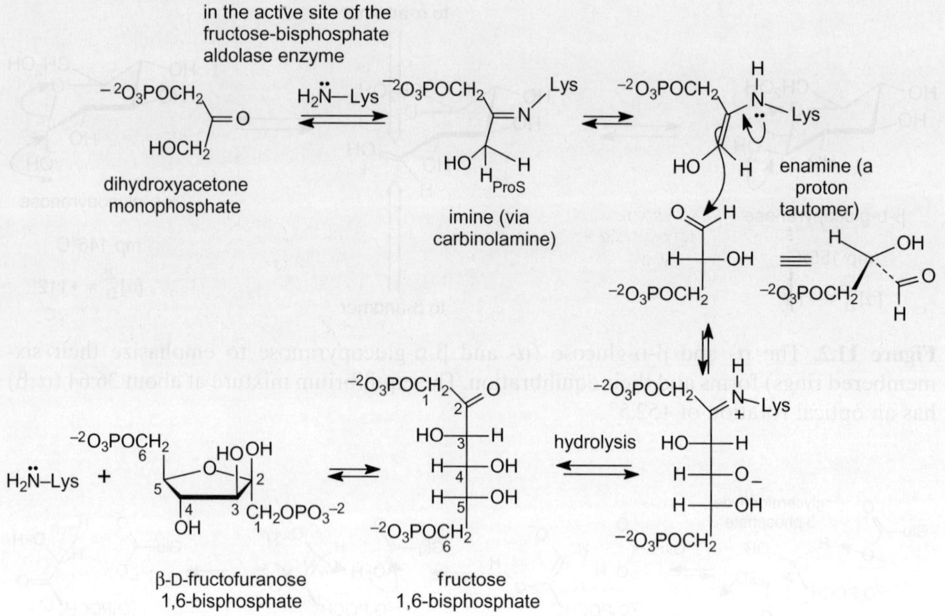

Scheme 11.5. A cartoon representation of the catalyzed (fructose-bisphosphate aldolase, EC 4.1.2.13) aldol-type condensation between glyceraldehyde 3-phosphate and dihydroxyacetone monophosphate to produce the six-carbon ketosugar fructose-1,6-bisphosphate. An active site lysine Lys-NH$_2$ [$^+$H$_3$NCH$_2$CH$_2$CH$_2$CH$_2$CH(NH$_3$$^+$)CO$_2$$^-$] apparently serves as the catalyst through addition at the carbonyl followed by proton tautomerization.

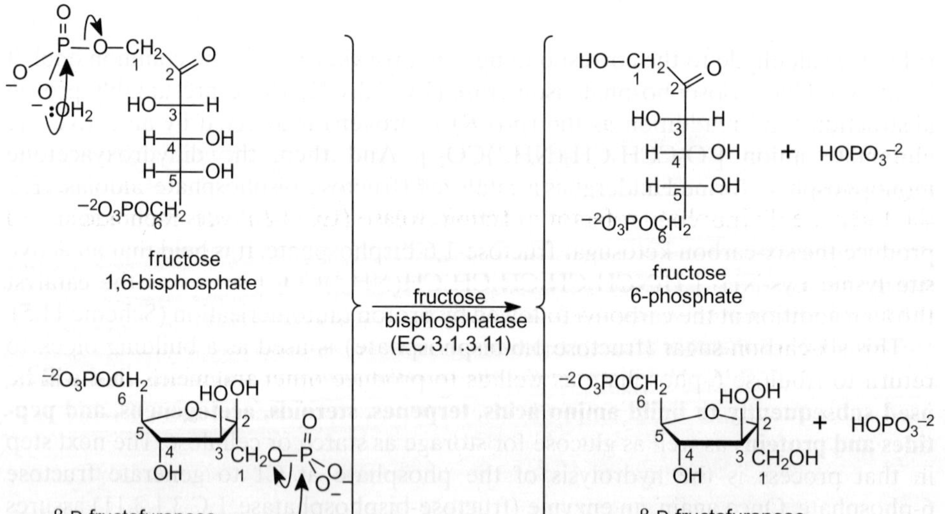

Scheme 11.6. A representation of the hydrolysis of the phosphate at C1 of fructose 1,6-bis-phosphate to generate fructose 6-phosphate. The enzyme (fructose-bisphosphatase, EC 3.1.3.11) assures the hydrolysis occurs only at C1.

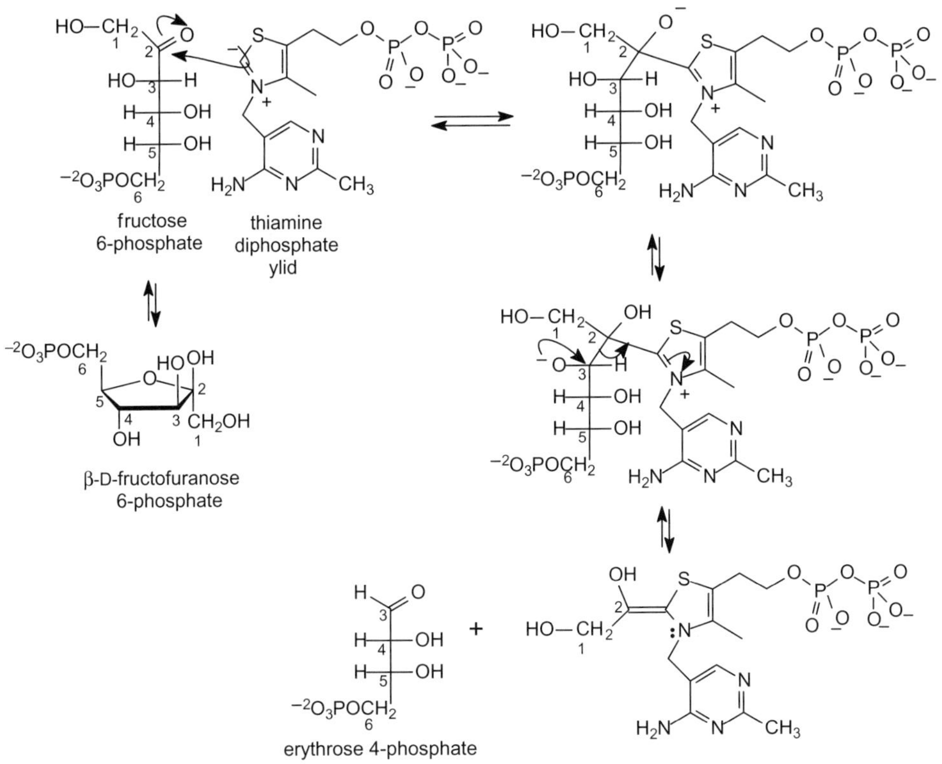

fructose
6-phosphate

thiamine
diphosphate
ylid

β-D-fructofuranose
6-phosphate

erythrose 4-phosphate

Scheme 11.7. A representation of fructose 6-phosphate undergoing a retroaldol-type reaction with the ketolase enzyme cofactor thiamine diphosphate to yield erythrose 4-phosphate and a two-carbon fragment that has remained attached to the thiamine cofactor of transketolase (EC 2.2.1.1).

tor, *two carbons are removed from fructose 6-phosphate* (C1 and C2) by what we write as a retroaldol-like process ($C_6 - C_2 = C_4$), thus leaving behind the four-carbon sugar erythrose 4-phosphate (Scheme 11.7). That two-carbon unit, still attached to the thiamine diphosphate cofactor of transketolase (EC 2.2.1.1), is then added in a supposed aldol-like process, to one of the PGAs (glyceraldehyde 3-phosphate) made available initially (*vide supra*, Scheme 11.3) from reaction of carbon dioxide (CO_2) with ribulose 1,6-bisphosphate ($C_2 + C_3 = C_5$) (Scheme 11.8). The product, xylulose 5-phosphate, is a diastereomer of ribulose 5-phosphate as a consequence of the opposite stereochemistry *at the site α to the carbonyl* and which, in the presence of the enzyme ribulose-phosphate 3-epimerase (EC 5.1.3.1), can therefore be isomerized to the latter (Scheme 11.9) through (presumably) a common enol.

The erythrose 4-phosphate generated in the same step (Scheme 11.7) as the derivative of thiamine diphosphate is then available for enzyme-catalyzed aldol-type condensation with dihydroxyacetone monophosphate just as shown in Scheme 11.5 for the analogous reaction with glyceraldehyde 3-phosphate and using the same fructose-bisphosphate aldolase (EC 4.1.2.13).

In the event, as there is now one more carbon in erythrose 4-phosphate than in glyceraldehyde 3-phosphate, a seven-carbon sugar, sedoheptulose 1,7-bisphosphate,

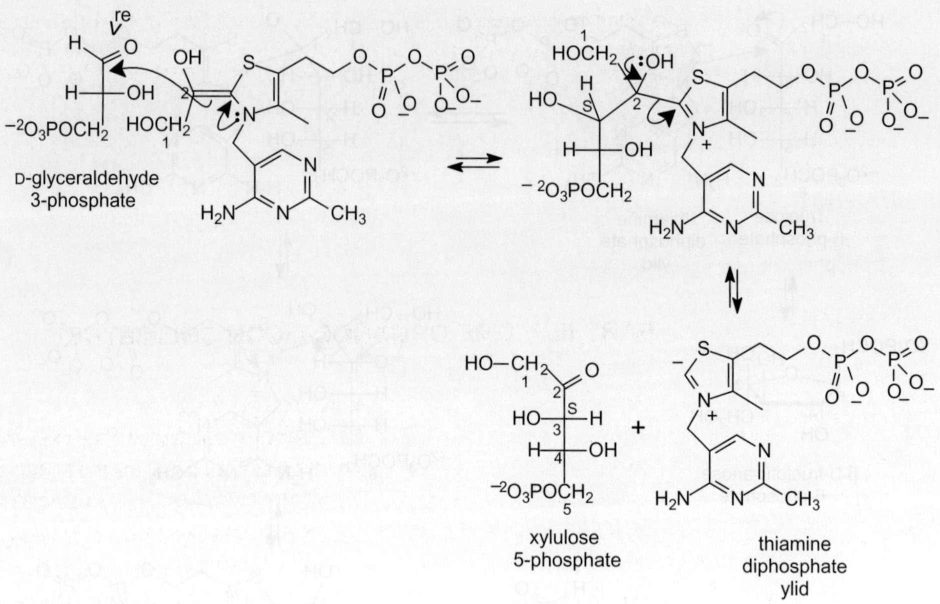

Scheme 11.8. The addition of the two-carbon fragment attached to the thiamine diphosphate cofactor of transaldolase (EC 2.2.1.2) to 3-phosphoglycerates (glyceraldehyde 3-phosphate) forming xylulose 5-phosphate and the regeneration of thiamine diphosphate ylid.

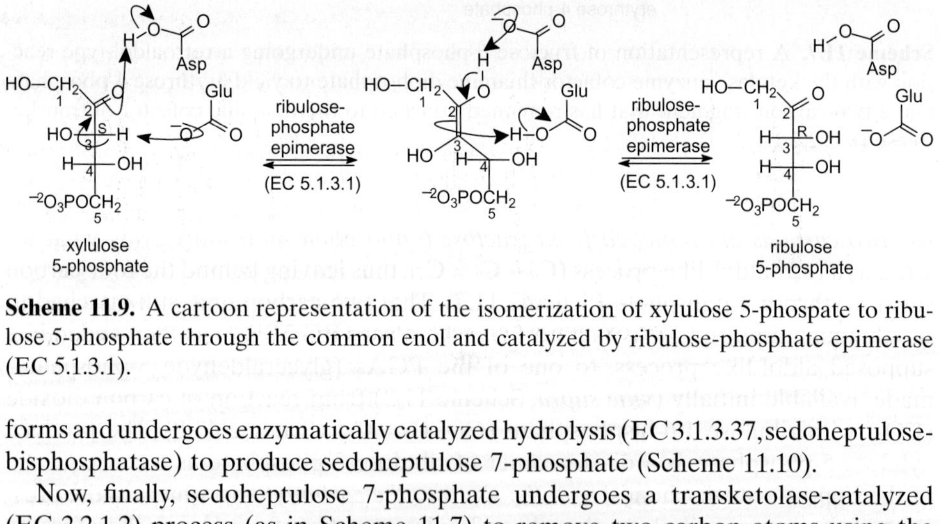

Scheme 11.9. A cartoon representation of the isomerization of xylulose 5-phospate to ribulose 5-phosphate through the common enol and catalyzed by ribulose-phosphate epimerase (EC 5.1.3.1).

forms and undergoes enzymatically catalyzed hydrolysis (EC 3.1.3.37, sedoheptulose-bisphosphatase) to produce sedoheptulose 7-phosphate (Scheme 11.10).

Now, finally, sedoheptulose 7-phosphate undergoes a transketolase-catalyzed (EC 2.2.1.2) process (as in Scheme 11.7) to remove two carbon atoms using the enzyme cofactor thiamine diphosphate to yield ribose 5-phosphate and a two-carbon fragment that has remained attached to the thiamine cofactor of transketolase (EC 2.2.1.1, sedoheptulose 7-phosphate) (Scheme 11.11). When the two-carbon fragment is added to glyceraldehyde 3-phosphate, the material of Scheme 11.8 again is applied and xylulose 5-phosphate results. The xylulose 5-phosphate isomerizes to ribulose 5-phosphate as in Scheme 11.9 (with intervention of ribulose phosphate 3-epimerase (EC 5.1.3.1). And, the ribose 5-phosphate, an aldose, isomerizes (an aldose–ketose isomerase, EC 5.3.1.6, ribose 5-phosphate isomerase) to ribulose 5-phosphate.

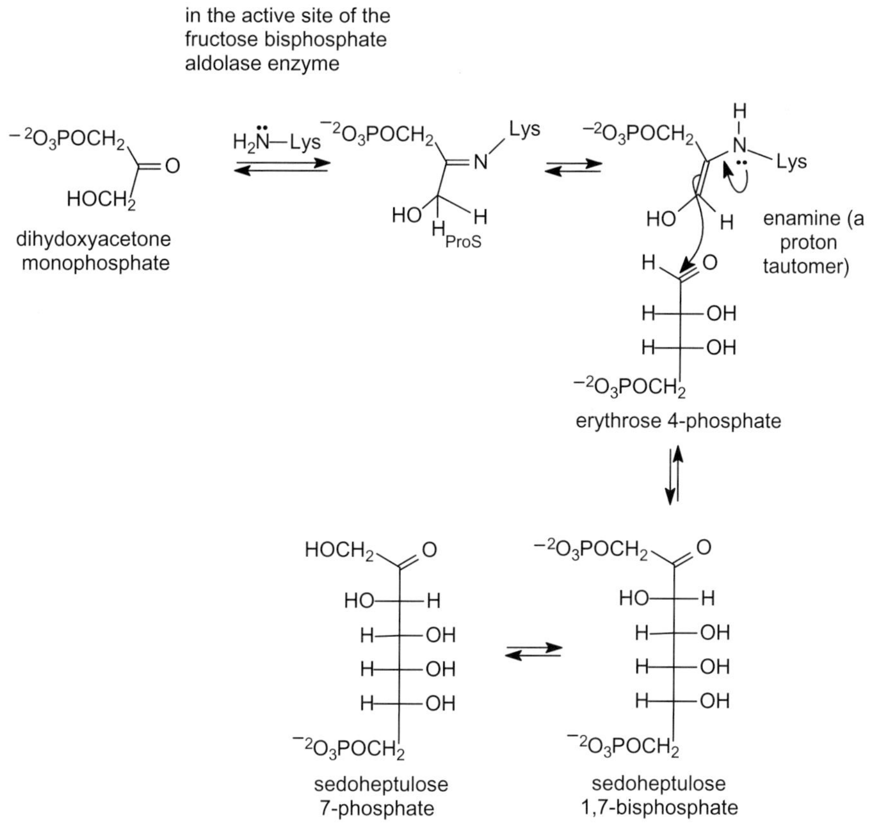

in the active site of the
fructose bisphosphate
aldolase enzyme

$-^2O_3POCH_2$ =O
$HOCH_2$
dihydoxyacetone
monophosphate

$H_2\ddot{N}-Lys$ $-^2O_3POCH_2$ =N—Lys
HO H
$_{ProS}$

$-^2O_3POCH_2$ N—H—Lys
HO H enamine (a
H O   proton
tautomer)

H——OH
H——OH
$-^2O_3POCH_2$
erythrose 4-phosphate

$HOCH_2$ O
HO——H
H——OH
H——OH
H——OH
$^-2O_3POCH_2$
sedoheptulose
7-phosphate

$-^2O_3POCH_2$ O
HO——H
H——OH
H——OH
H——OH
$^-2O_3POCH_2$
sedoheptulose
1,7-bisphosphate

Scheme 11.10. A representation of the aldol-type condensation of erythrose 4-phosphate with dihydroxyacetone monophosphate using fructose bisphosphate aldolase (EC 4.1.2.13) to produce a seven-carbon sugar, sedoheptulose 1,7-bisphosphate. The enzyme-catalyzed hydrolysis (EC 3.1.3.37) yielding sedoheptulose 7-phosphate is also shown.

So, as the circle turned, two ribulose 5-phosphates have been regenerated from two xylulose 5-phosphates and one ribulose 5-phosphate has come from ribose 5-phosphate, and in the overall turning (as shown in Figure 11.1), some molecular species were created that will prove useful in further discussions.

C. CARBOHYDRATES

I. Biosynthesis

D-Glyceraldehyde,* the "simplest sugar" (as the 3-phosphate), has been produced in the turning of the Calvin cycle (Scheme 11.3). Hydrolysis of the phosphate gener-

---

*D, because when the Fischer projection structure is drawn vertically, with the most highly oxidized carbon at the "top," the –OH on the penultimate carbon is on the right. It will be recalled that the Fischer projection is a representation in which the groups that lie on the horizontal lines are to be thought of as lying above the plane of the paper and the groups on the vertical lines as those lying below the plane of the paper, thus defining a tetrahedron view "edge on."

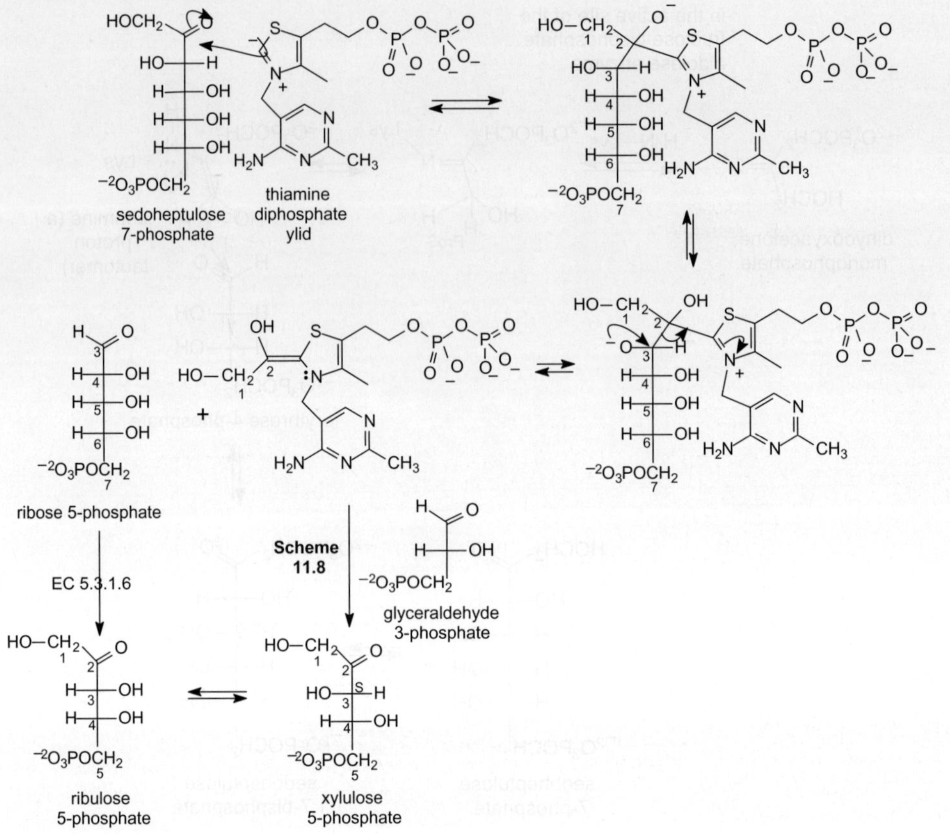

Scheme 11.11. A representation of the conversion of sedoheptulose 7-phosphate to ribose 5-phosphate and xylulose 5-phosphate. The initial loss of the two-carbon fragment by an enzyme using transketolase (EC 2.2.1.2) with the enzyme cofactor thiamine diphosphate yields ribose 5-phosphate and the two-carbon fragment that has remained attached to the thiamine cofactor (Scheme 11.11). When the two-carbon fragment is added to glyceraldehyde 3-phosphate, the material of Scheme 11.8 is seen once again and xylulose 5-phosphate results. The xylulose 5-phosphate isomerizes to ribulose 5-phosphate as in Scheme 11.9 (with intervention of ribulose phosphate 3-epimerase [EC 5.1.3.1]). And, the ribose 5-phosphate, an aldose, isomerizes (an aldose-ketose isomerase, EC 5.3.1.6, ribose 5-phosphate isomerase) to ribulose 5-phosphate.

ates the free alcohol. Similarly, D-erythrose 4-phosphate (Scheme 11.7) was also formed in the Calvin cycle, and hydrolysis leads to D-erythrose. As shown in Equation 11.2, epimerization of the hydroxyl at C-2 via the common enol, produces D-threose.

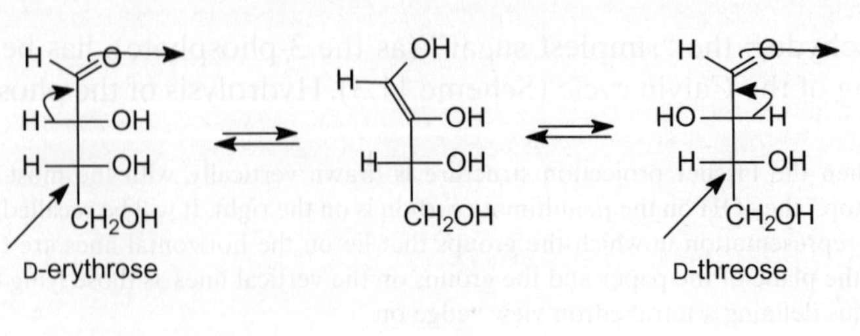

$$\text{(11.2)}$$

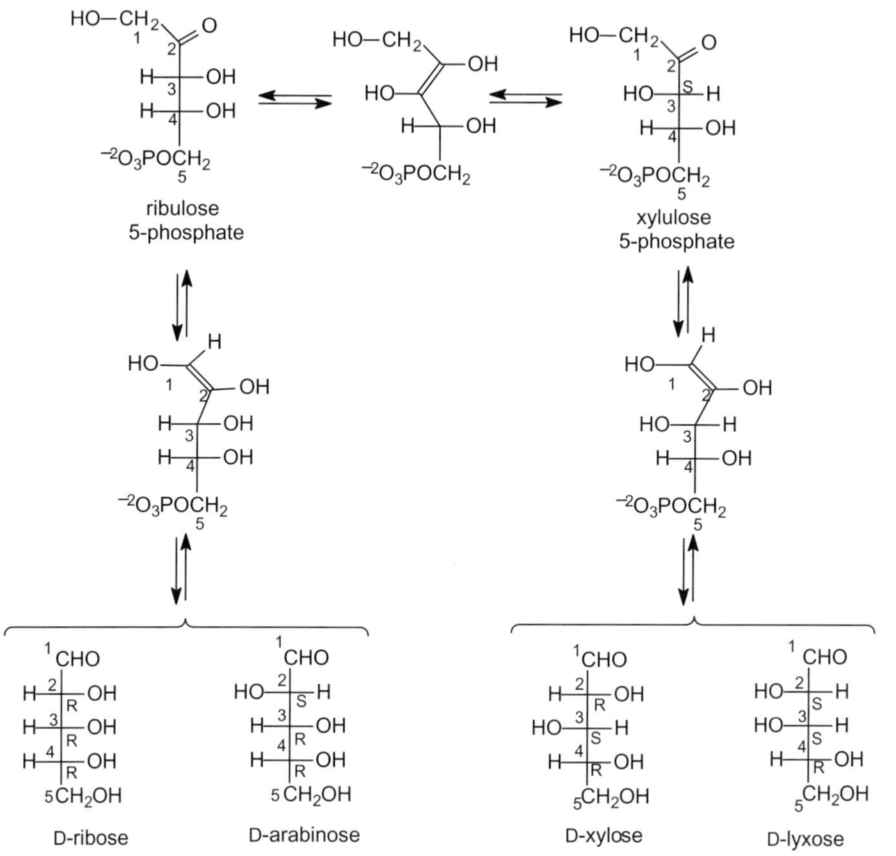

ribulose
5-phosphate

xylulose
5-phosphate

D-ribose   D-arabinose   D-xylose   D-lyxose

Scheme 11.12. A representation of the interconversion of ribulose 5-phosphate and xylulose 5-phosphate through the common enol and their subsequent hydrolysis and epimerization (or vice versa) to ribose and arabinose on the one hand and xylose and lyxose on the other.

Not surprisingly, xylulose 5-phosphate on hydrolysis and epimerization (or vice versa) of the common enol can be seen as equivalent to xylose on the one hand and to lyxose (Scheme 11.12) on the other. Furthermore, as already seen, xylulose 5-phosphate can be epimerized through the common enol to ribulose 5-phosphate (Scheme 11.11) and the latter, on hydrolysis and epimerization (or vice versa), to ribose and arabinose (Scheme 11.12).

The synthesis of glucose (gluconeogenesis), which is then stored as starch (a mixture of α-glucose polymers, *vide infra*) and/or cellulose (a mixture of β-glucose polymers, *vide infra*), occurs by condensation of 3-phosphoglyceraldehyde (glyceraldehyde 3-phosphate) with dihydroxyacetone monophosphate (Scheme 11.13).

II. Chemistry

Early laboratory work (before 1900) showed that D-arabinose could be converted to a mixture of D-glucose and D-mannose (Scheme 11.14, the Kiliani–Fischer

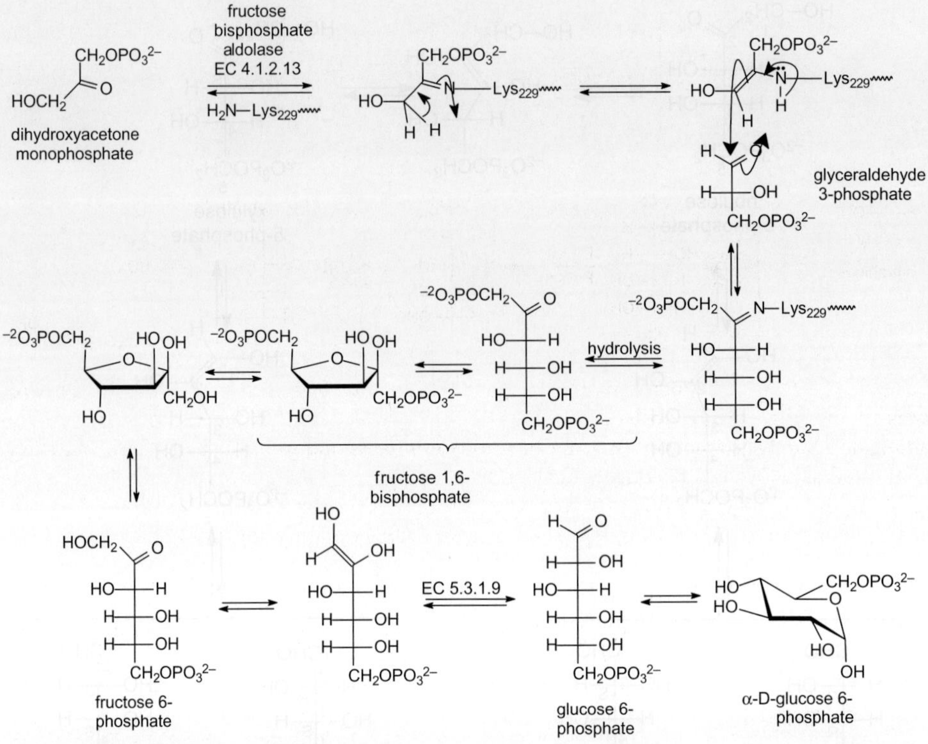

dihydroxyacetone monophosphate

fructose bisphosphate aldolase EC 4.1.2.13

glyceraldehyde 3-phosphate

hydrolysis

fructose 1,6-bisphosphate

fructose 6-phosphate

EC 5.3.1.9

glucose 6-phosphate

α-D-glucose 6-phosphate

Scheme 11.13. A representation of a portion of the pathway of gluconeogenesis. The 6-phosphate of glucose is considered capable of hydrolysis to glucose with the appropriate hydrolyase (EC 3.1.3.9) catalyst. The anomer shown is α (because the hydroxyl on the anomeric carbon is *cis* or *syn* to the hydroxyl at C5, Scheme 11.14). A caution concerning nomenclature is needed. In the acyclic form, it is clear which carbon is number six. Cyclization to the six-membered ring pyranose (oxane) (or furanose [oxolane] when the five-membered ring forms) leaves us with a nomenclature problem since the carbon that was C-1 (the anomeric carbon) is often referred to as C-1 in both structures that involve formation of bonds to that carbon as well as in reactions that may occur at that carbon. However, it will be recalled (Chapter 5) that the nomenclature of cyclic ethers (and other heterocycles) generally requires that the heteroatom (or a heteroatom if more than one is present) receive the lowest number. So, the IUPAC and the IUBMB numbering system is used for archiving purposes, while the trivial nomenclature is commonly used in the literature. Both systems will be found here as appropriate.

synthesis)* and both of the latter reconverted to the former (Scheme 11.15, the Wohl degradation).[†]

   As pointed out early in Chapter 9, aldehydes are easily oxidized by aqueous ammoniacal silver solutions [i.e., $Ag(NH_3)_2^+$, Tollen's reagent] as well as Fehling's reagent (Cu^{+2} in aqueous sodium tartrate) and Benedict's reagent (Cu^{+2} in aqueous

---

*Heinrich Kiliani (1855–1945), professor, Freiberg, Germany. Emil Fischer (1852–1919), professor, Berlin, Germany, Nobel Prize 1902. See Killiani, H. *Chem. Ber.*, **1885**, *18*, 3066 and Fischer, E. *Chem. Ber.*, **1889**, *22*, 2204.

[†]Alfred Wohl (1863–1933), professor, University of Danzig. See Wohl, A. *Chem. Ber.*, **1893**, *26*, 730.

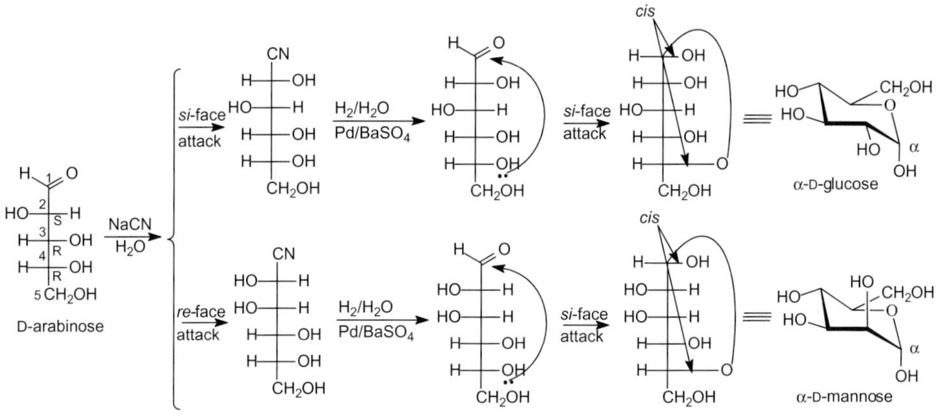

Scheme 11.14. A modified Kiliani–Fischer synthesis. The original synthesis involved hydrolysis of the nitrile function to the corresponding carboxylic acid and subsequent sodium-amalgam (Na/Hg) reduction. This modification eliminates one step. Since the "new" hydroxyl generated as a result of hemiacetal formation is *cis* or *syn* to the hydroxyl at C5, the configuration at the anomeric carbon is α. Were the hemiacetal hydroxyl *trans* or *anti* to the hydroxyl at C5, the anomeric –OH would be equatorial and the configuration would be β.

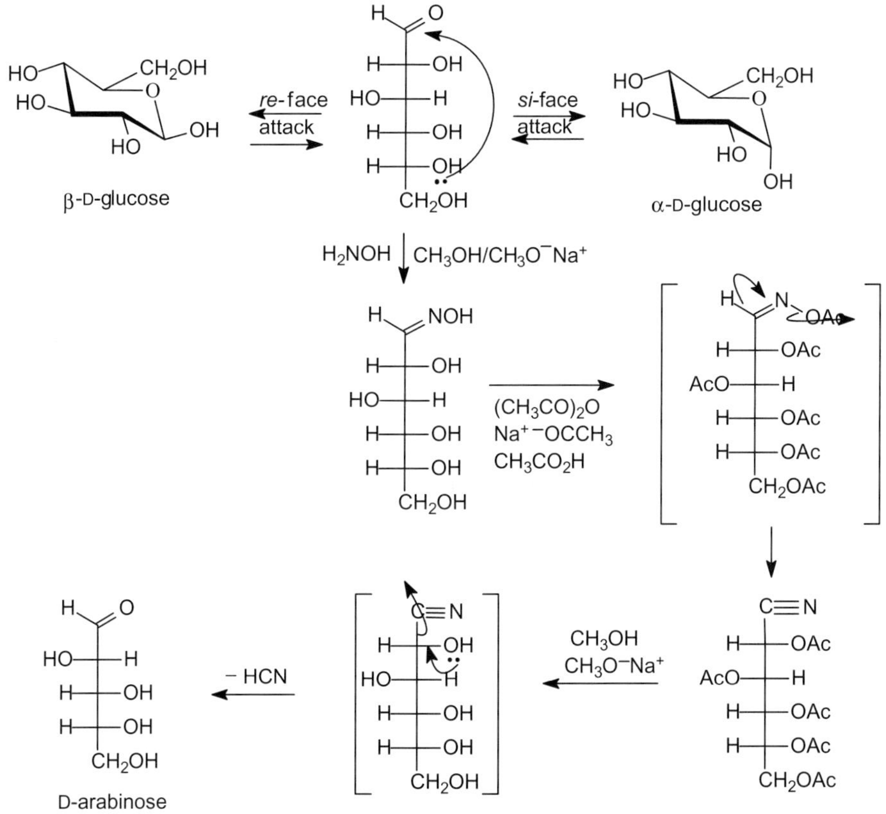

Scheme 11.15. An example of the Wohl degradation. The anomers of D-glucose (i.e., the isomers that are epimeric at the anomeric carbon) can be considered as the aldehyde rather than the hemiacetals. Formation of the oximes (presumably both (*E*)- and (*Z*)-oxime isomers form) followed by acetylation results in elimination to the nitrile. Hydrolysis of the remaining acetates is accompanied by elimination of hydrogen cyanide and formation of an aldehyde with *one carbon less* than the aldehyde with which the sequence began. Thus, both mannose and glucose yield arabinose.

sodium citrate) and other oxidizing agents that are less mild. With these gentle oxidizing agents, however, the carbohydrates mentioned so far give the corresponding "aldonic" acids. Thus, glucose yields gluconic acid (Equation 11.3).

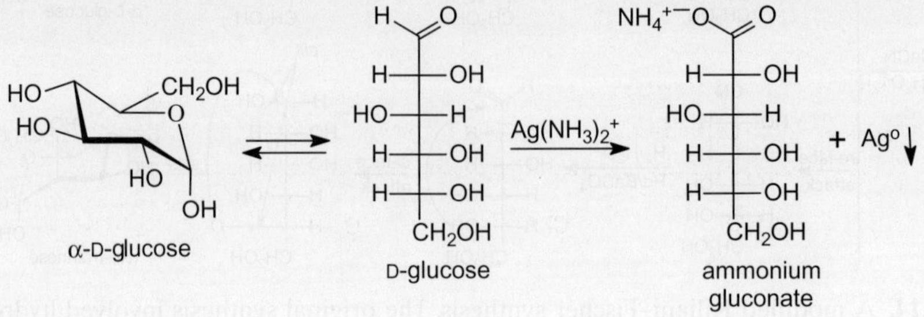

$$(11.3)$$

But in a reaction that has not hitherto been encountered, that is, with nitric acid, the carbohydrates yield dicarboxylic acids called aldaric acids, compounds in which both termini have been oxidized! So, D-glucose yields glucaric acid (Equation 11.4).

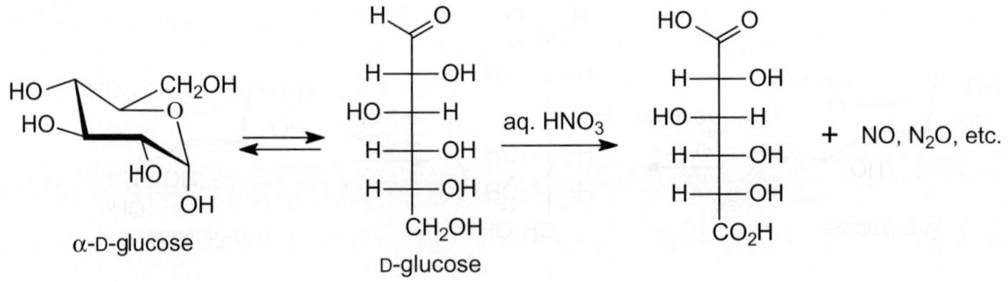

$$(11.4)$$

As noted in Scheme 11.15 and earlier schemes, closure of the pyranose ring by attack of the hydroxyl at C5 on the carbon of the carbonyl can occur on either face (*re* or *si*) of the carbonyl. As a consequence, the two anomers, when in solution, are in equilibrium. As already noted, the α-anomer has the –OH of the hemiacetal *cis* or *syn* to the hydroxyl at C5 (*trans* or *anti* to the –CH$_2$OH group and "down" in the Haworth projection *vide infra*), while same hydroxyl has the opposite configuration in the β-anomer. The crystalline isomers, α- and β-D-glucopyranose are stable entities (Figure 11.2), but when dissolved, they apparently enter into equilibrium via the ring-opened species. The phenomenon which brings the isomers (at the anomeric carbon) into an equilibrium α:β of about 36:64 in aqueous solution is called "mutarotation (Figure 11.3)."

The structure determination of glucose was undertaken by Fischer (*vide supra*). It was known that on treatment with nitric acid the aldopentose called arabinose yielded an aldaric acid that was "optically active." As shown in Figure 11.4, two of the four aldopentoses could therefore be eliminated!

Therefore, it was clear (Figure 11.5) that one of the two aldopentoses that gave rise to the dicarboxylic acids that retained "optical activity" had to be arabinose. Furthermore, when the Kiliani–Fischer process (Scheme 11.14, *vide supra*) was undertaken on both of those aldopentoses, only one of them gave hexoses that produced *two six-membered aldaric acids*, **both of which were chiral**. It was known

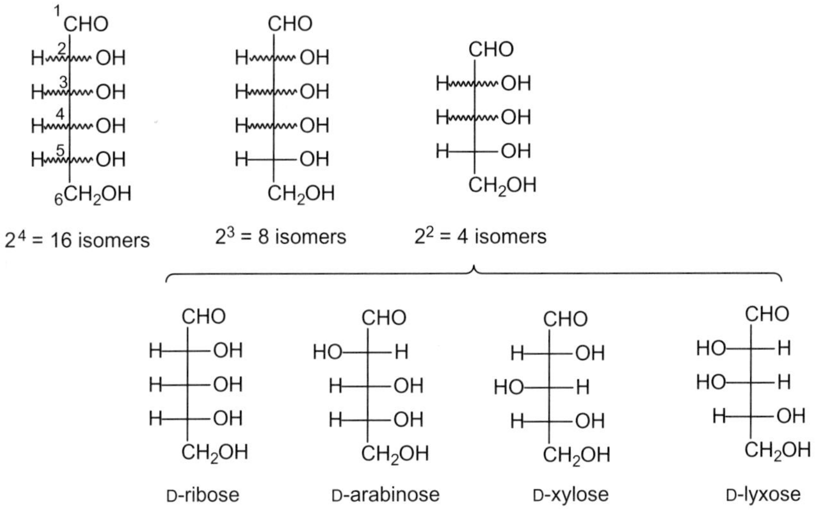

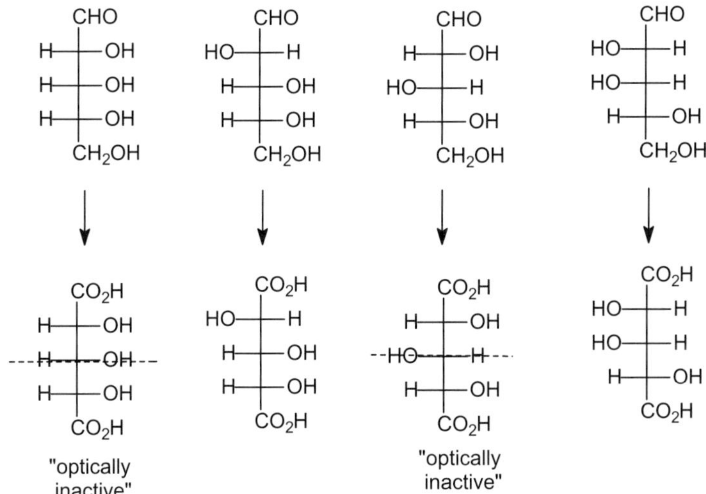

Figure 11.3. Part I of the Fischer determination of the structure of D-glucose. While in principle it might be any one of 16 possible isomers, once the penultimate carbon is written "to the right," only eight possible isomers remain. And, since it was known that the Kiliani synthetic method could be used on pentoses of which there can be only four (once the penultimate carbon is written "to the right"), the chemistry of the pentoses could be used to help define the structure of glucose.

Figure 11.4. Formation of aldaric (dicarboxylic) acids that are "optically active" on oxidation of pentoses.

that both glucose and mannose gave aldaric acids that were chiral! Thus, as shown in Figure 11.6, the structures of mannose and glucose are known but it is not known which is which.

The final piece of the puzzle fell into place and thus the structure of glucose was determined when it was found that one of the other six remaining hexoses yielded

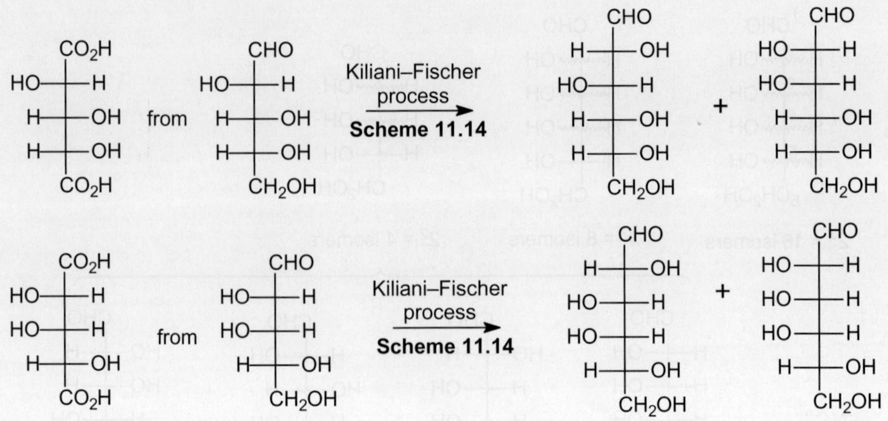

Figure 11.5. Of the two pentoses shown, only the first yields two hexoses that, on oxidation, produce chiral aldaric acids.

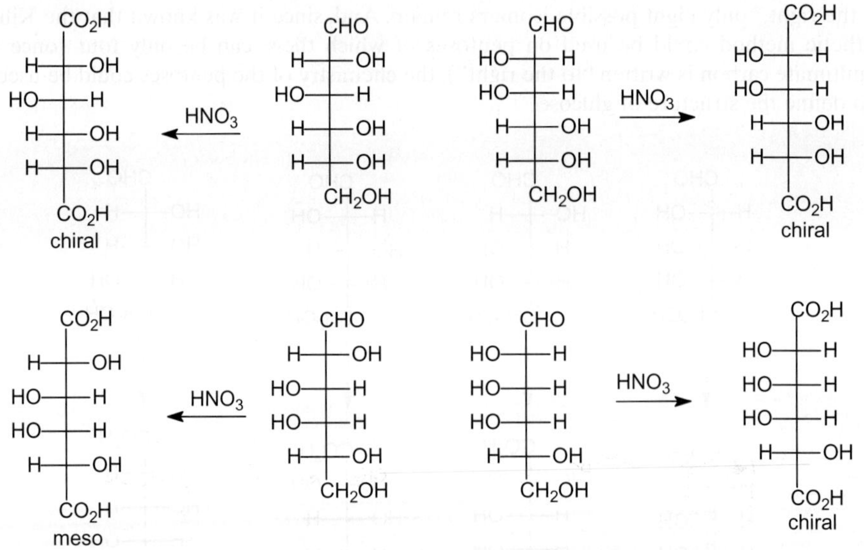

Figure 11.6. Oxidation of the hexoses of Figure 11.5. Only the two hexoses at the top yield aldaric acids that are chiral. On the bottom, one yields a meso aldaric acid.

the same aldaric acid as that one derived from glucose. The eight aldoses are provided in Figure 11.7.

Problem 11.1. Examine the structures of the aldohexoses in Figure 11.7. Two are crossed off and one of those must be glucose and one must be mannose (as shown on the left). Identify the "other" aldohexose among the remaining six that, on oxidation, gives the **same** aldaric acid as that derived from glucose.

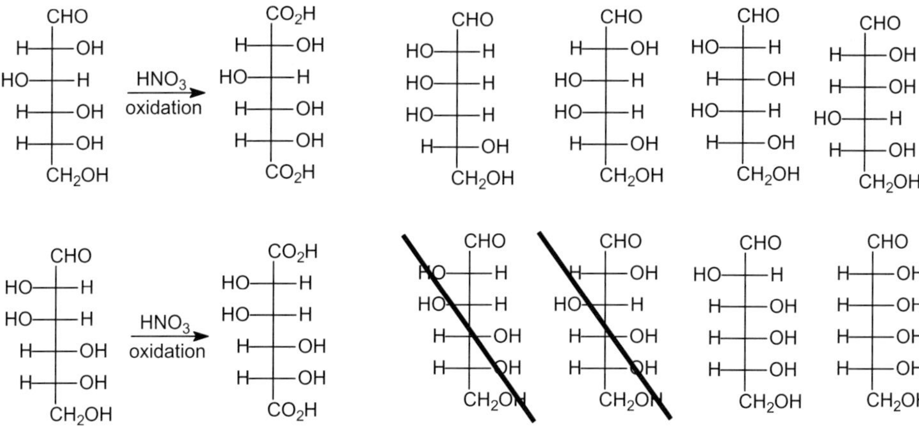

Figure 11.7. The aldohexoses (required for the solution to Problem 11.1). The two on the right of the figure that are crossed out, one of which is mannose and the other glucose, are shown, along with their respective glucaric acids on the left of the figure.

In the early part of the twentieth century, extensive efforts were undertaken by, among others, Sir Walter Norman Haworth* to affect the synthesis of the carbohydrate vitamin C, which had been long recognized as necessary to avoid scurvy[†] by British seamen. The Haworth synthesis (that barely preceded that of Reichstein, *vide infra*) is provided as Scheme 11.16.

Interestingly, in his Nobel lecture (Haworth, 1966, pp. 415–432), Haworth pointed out that "the synthesis of natural L-ascorbic acid (vitamin C) proceeds … from L-xylosone, which is obtainable from D-galactose as the outcome of the following series of transformations: D-galactose → D-galactose-1:2,3:4-diacetonide → D-galacturonic acid-1:2,3:4-diacetonide → D-galacturonic acid → L-galactonic acid → L-galactonamide → L-lyxose → L-xylosazone → L-xylosone." These processes are presented in Scheme 11.17.

Another synthesis of vitamin C (that of Reichstein[‡] and coworkers) followed that of Haworth et al. by **1 year**.

---

*Sir Walter Norman Haworth (1883–1950), professor and department head, University of Birmingham, Nobel Prize 1937, Chemistry, shared with Professor Paul Karrer (about whom more later). The Nobel Prize in Chemistry 1937 was divided equally between Walter Norman Haworth *for his investigations on carbohydrates and vitamin C* and Paul Karrer *for his investigations on carotenoids, flavins, and vitamins A and B₂*.

[†]A mammalian vitamin deficiency disease (resulting from lack of vitamin C) that results in the failure to synthesize the protein collagen. Collagen is the main protein of connective tissue and its loss is severely debilitating. In the middle of the eighteenth century (about 1753), a Scottish surgeon, James Lind, sailing with the British Navy, demonstrated that ingestion of citrus fruit could prevent scurvy.

[‡]Tadeus Reichstein (1897–1996), professor of Chemistry and director of the Pharmaceutical Institute, University of Basel (Switzerland), Nobel Prize in Physiology or Medicine 1950. As a major figure in the elucidation of the chemistry of steroids, his work will be encountered again (*vide infra*). The Haworth work was published in 1934 (Ault, R. G.; Baird, D. K.; Carrington, H. C.; Haworth, W. N.; Herbert, R.; Hirst, E. L.; Percival, E. G. V.; Smith, F.; Stacey, M. *J. Chem. Soc.*, **1933**, 1419 while the full synthesis of Reichstein, T.; Grussner, A. *Helv. Chim. Acta*, **1934**, *17*, 311 and then Reichstein, T.; Grüssner, A.; Oppenauer, R. appeared in *Helv. Chim. Acta*, **1934**, *17*, 510).

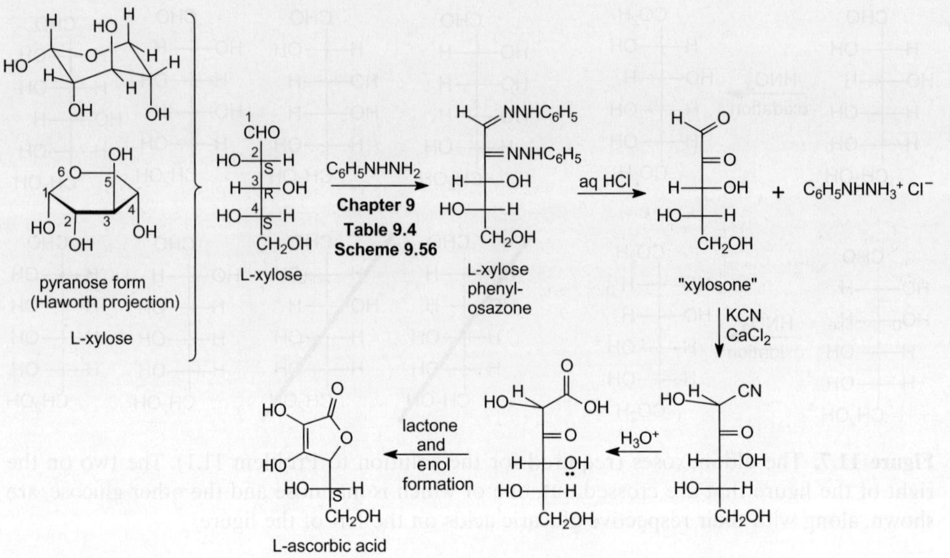

Scheme 11.16. The Haworth synthesis of ascorbic acid. It was argued that use of calcium chloride dramatically improved the yield (Haworth, W. N.; Hirst, E. L.; Jones, J. K. N.; Smith, F. *J. Chem. Soc.*, **1934**, 1192 and work cited therein).

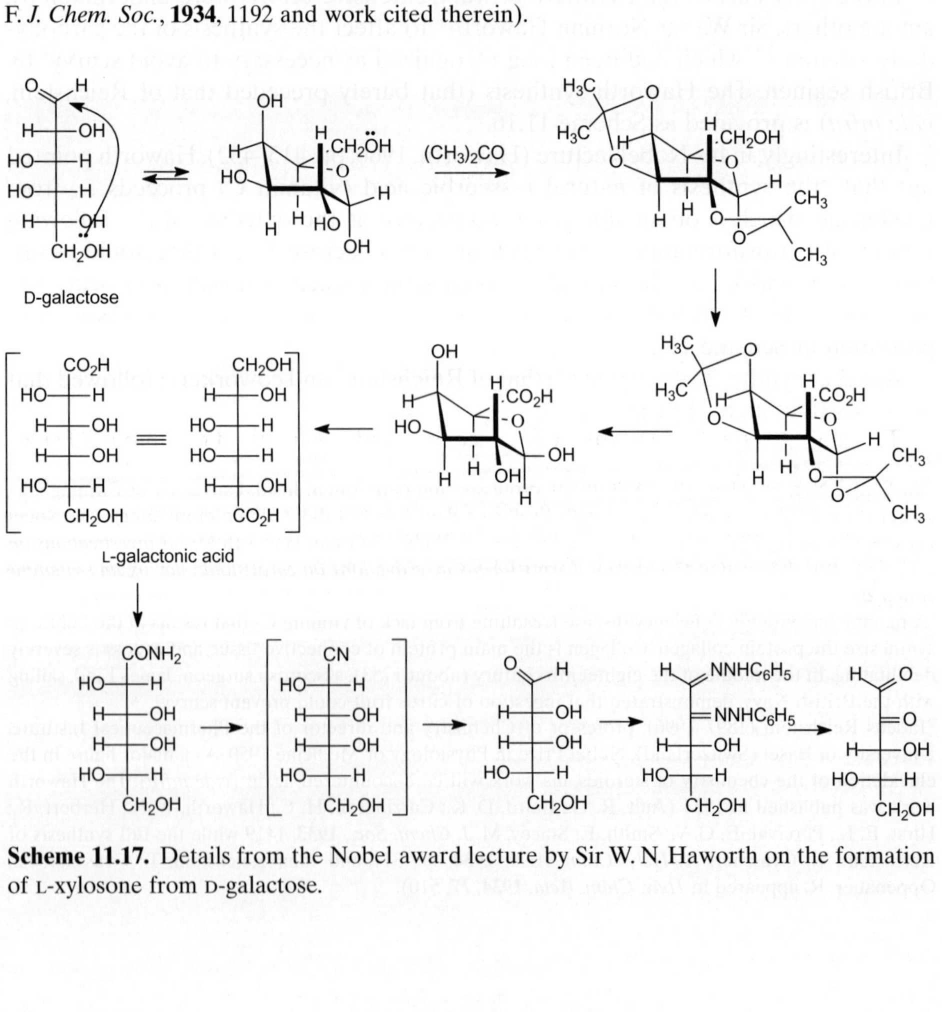

Scheme 11.17. Details from the Nobel award lecture by Sir W. N. Haworth on the formation of L-xylosone from D-galactose.

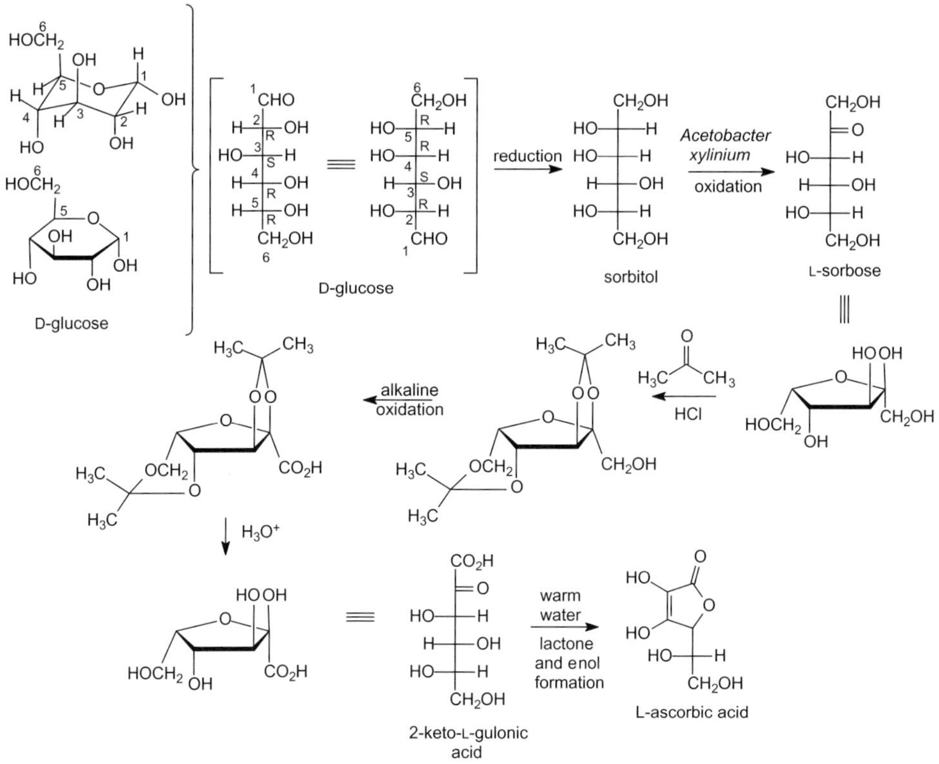

Scheme 11.18. A representation of the Reichstein synthesis of L-ascorbic acid (Reichstein, T.; Grussner, A. *Helv. Chim. Acta*, **1934**, *17*, 311). *Note with regard to the structure of* D-*glucose as shown in the brackets*: when the structure is "flipped" 180°, the Fischer representation (where the groups on the horizontal are "above the plane of the paper") becomes reversed and would be "below the plane." Thus, to preserve the absolute stereochemistry, the groups appear reversed *in the drawing*.

The main biosynthetic pathway for ascorbic acid was worked out about 20 years later (also in the United Kingdom) by Mapson, Isherwood, and Chen at the University of Cambridge, but many others have contributed subsequently) and, just as Reichstein synthesis (Scheme 11.18) began from the readily available D-glucose, so it is that the biosynthesis is found to proceed from D-glucose (as well as from D-galactose) as shown in Scheme 11.19.

Problem 11.2. With regard to Scheme 11.18, (a) suggest how the reduction of D-glucose to sorbitol might have been carried out. (b) Although it is difficult to compete with *Acetobacter xylinium* (who "work for food"), suggest an alternative oxidation to sorbose from sorbitol. (c) Why is alkaline oxidation used to convert the bisacetonide to the corresponding carboxylic acid. Suggest a suitable reagent. (d) Both here and in Scheme 11.16, the 2-keto-L-gulonic acid intermediate are shown to convert to L-ascorbic acid by "lactone and enol formation." Write curved arrows to account for those transformations.

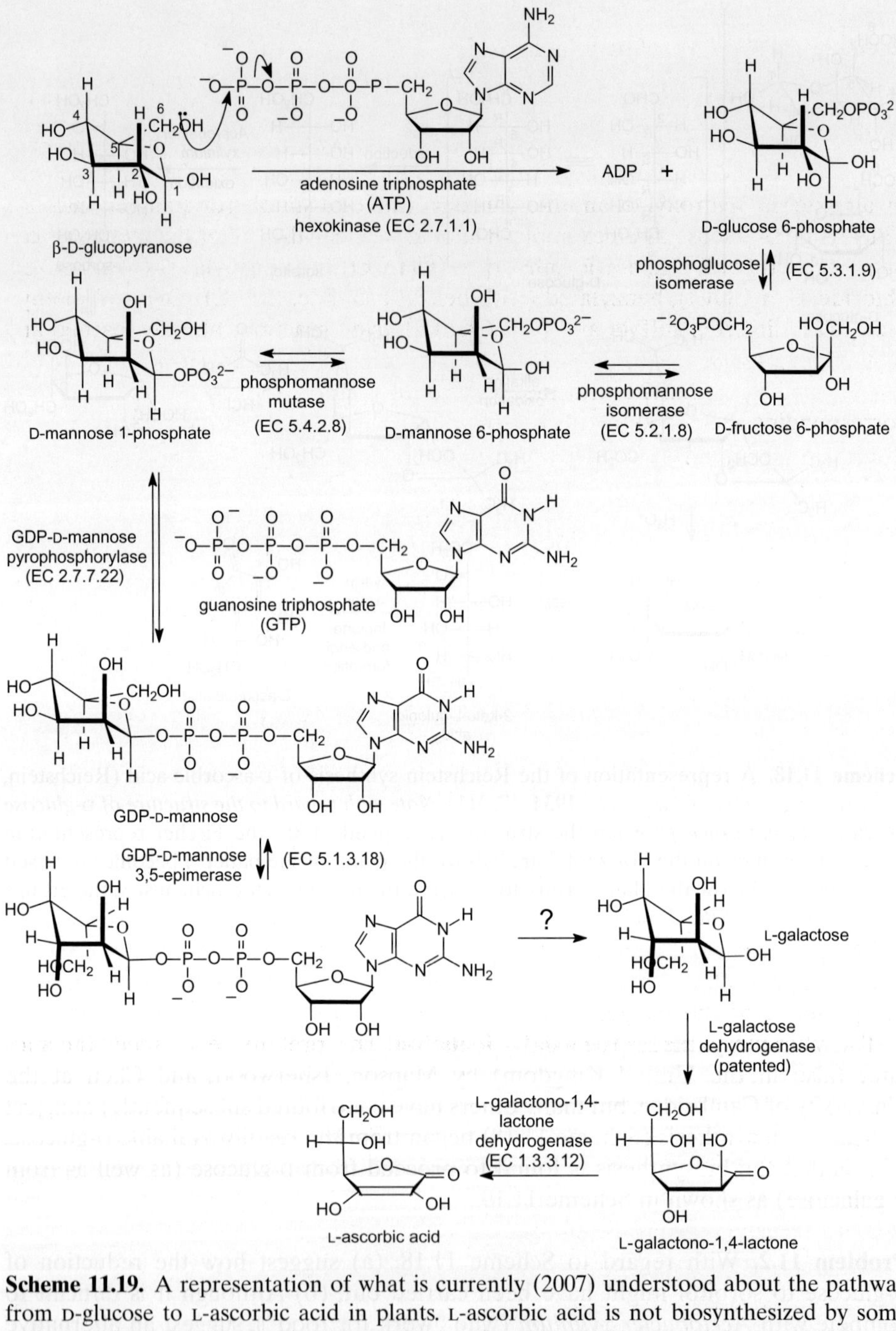

Scheme 11.19. A representation of what is currently (2007) understood about the pathway from D-glucose to L-ascorbic acid in plants. L-ascorbic acid is not biosynthesized by some mammals (including humans). EC numbers and some graphic materials provided in this scheme have been taken with permission from appropriate links in a URL starting with http://www.chem.qmul.ac.uk/iubmb/enzyme/.

Based on Figure 11.2 and the chemistry discussed so far, it should be clear that the hydroxyl (–OH) groups, primary and secondary, present in the cyclic and acyclic forms of carbohydrates, and the one hydroxyl (–OH) on the anomeric carbon are different from each other.

Indeed, the hydroxyl on the anomeric carbon, arising as it does as a consequence of cyclization, to a hemiacetal (or hemiketal in, e.g., fructose) can easily be replaced by an external hydroxyl of another carbohydrate or a simple (or complex) alcohol. Thus, using D-glucose as an example, it should be clear that all of the hydroxyls can be acetylated [with e.g., acetic anhydride, $(CH_3CO)_2O$], benzoylated (with benzoyl chloride, C_6H_5COCl), benzylated (with benzyl bromide, $C_6H_5CH_2Br$), and methylated [with dimethyl sulfate and $(CH_3O)_2SO_2$] using reactions already learned and that involve *replacing the proton* on oxygen (Equation 11.5). However, with the same example, treatment with methyl alcohol under acidic conditions yields the corresponding acetal where only the anomeric carbon has been replaced (Equation 11.6).

β-D-glucopyranose $\xrightarrow{(CH_3CO)_2O}$ β-D-glucopyranose pentaacetate

$$(11.5)$$

α-D-glucopyranose $\xrightarrow[HCl]{CH_3OH}$ 1-(O-methyl)-α-D-glucopyranose + 1-(O-methyl)-β-D-glucopyranose

$$(11.6)$$

The path by which hemiacetals are converted to acetals has already been discussed, in general, in Chapter 9. However, it is of some small interest to note that the presence of the oxygen *in the six-membered ring* induces the hydroxyl (or methoxy) group to **prefer the axial orientation**! This preference is called "the anomeric effect." As shown in Scheme 11.20, it has been argued that there is significant overlap between the nonbonded electron pair on the heterocyclic oxygen and the C-1 carbon, and thus, the direction of attack on this species should be from the axial direction. An alternative explanation arguing that the axial and equatorial species are in equilibrium and that dipole–dipole repulsion destabilizes the equatorial isomer has also gained some currency.

Species, such as those of Scheme 11.20, in which a sugar group (the D-glucopyranose in the scheme) is attached through the anomeric carbon via a heteroatom bond (the $O–CH_3$) are called **glycosides**. So the 1-(O-methyl)-α-D-glucopyranose in the scheme is commonly referred to as "the α-anomer of glucose O-methylglycoside."

If the α-anomer of glucose O-methylglycoside (or the β-anomer for that matter) is treated with an acetylating agent such as acetic anhydride, the corresponding tetraacetyl ester results. Treatment of the 1-(O-methyl)tetraacetate with dilute aqueous acid then results in demethylation, that is, the conversion of an acetal to a

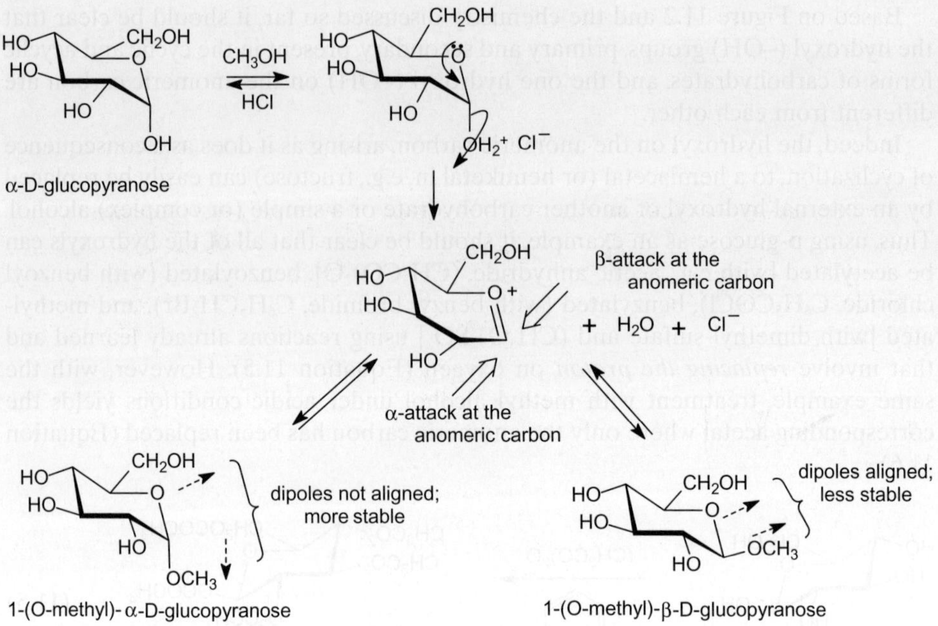

α-D-glucopyranose

β-attack at the anomeric carbon

+ H₂O + Cl⁻

α-attack at the anomeric carbon

dipoles not aligned; more stable

dipoles aligned; less stable

1-(O-methyl)-α-D-glucopyranose

1-(O-methyl)-β-D-glucopyranose

Scheme 11.20. A representation of the anomeric effect. One argument for the preference of axial attack on the anomeric carbon of the oxonium ion invokes the orbitals aligned in that ion for attack from beneath the ring (as drawn). A second argues that as the anomers are in equilibrium, it is the alignment of dipoles that disfavors the β-anomer. See Lemieux, R. U.; Hayami, J. *Can. J. Chem.*, **1965**, *43*, 2162 and Wolfe, S.; Whangbo, M.-H.; Mitchell, D. J. *Carbohydr. Res.*, **1979**, *69*, 1.

hemiacetal. Now, the hemiacetal undergoes reaction with a variety of acids (both Brønsted and Lewis) to produce the corresponding C-1 halides (Scheme 11.21).

By allowing the C-1-substituted halides to undergo substitution reactions with the anions of appropriate nucleophiles, families of glycosides are generated. Indeed, it is very common to find that biologically active substances isolated from plants and animals are composed of two parts, one of which is a carbohydrate. Figure 11.8 presents only three of a legion of such compounds and more will be encountered later (e.g., as cardiac glycosides) after polysaccharides have been introduced.

In the laboratory, the classical process of combining such fragments is called the Koenigs–Knorr reaction and the product one obtains depends, in part, on the conditions of the reaction. However, since the carbon α to the seat of substitution is generally acetylated, neighboring group participation insures the incoming group has the β-position when equilibration cannot be effected, that is, as shown in Scheme 11.22. The Koenigs–Knorr reaction incorporates the use of a silver salt (usually silver carbonate, Ag₂CO₃) and the leaving group (bromide anion [Br⁻]) cannot return.

In living systems where acetylation (presumably) does not occur, the sense of addition to the anomeric carbon clearly involves enzymatic control, and, of course,

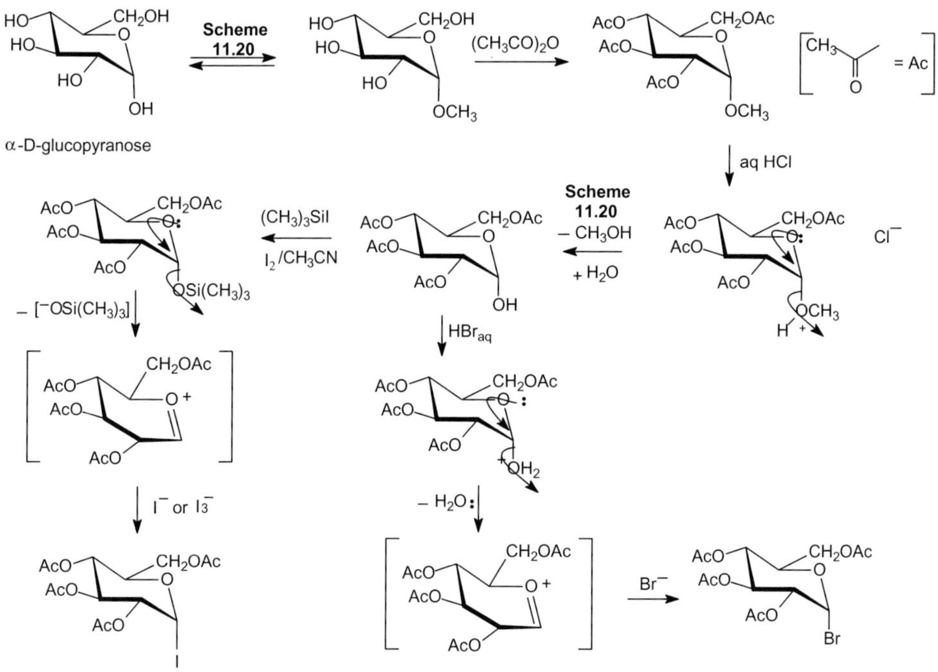

α-D-glucopyranose

Scheme 11.21. The conversion of the tetraacetate of the O-methylglycoside of glucose into the corresponding α-C-1 iodide (with trimethylsilyl iodide [(CH$_3$)$_3$SiI]) and bromide (with aqueous hydrogen bromide, HBr).

the "preformed" hexose (as pyranose [oxane] or furanose [oxolane]) need not actually be involved.

So, the process shown in Figure 11.9 that leads to sucrose need not actually occur by a nucleophilic substitution reaction at the anomeric carbon. Indeed, the derivative of galactose (shown as guanosine diphosphate [GDP]) may be the uridine diphosphate (UDP) or ADP instead, and, of course, the "displacement" is enzyme mediated.

Nonetheless, it should be clear that such processes that yield dimers could easily be extended to include more monomeric species. There are several other considerations to be introduced:

(a) First, although the bonding shown in Figure 11.9 is β1 → 4, there is no need that it always be so, and both the geometry at the anomeric carbon (α or β) and the site of bonding to the second monomer might be at any of the other hydroxyl groups (including the anomeric carbon itself).

(b) Second, if the bonding between the two (or more) monomeric carbohydrates does **not** involve the hydroxyl on the anomeric carbon on **both** (or more) carbohydrates (i.e., there is at least one hemiacetal), the resulting dimer will be capable of reducing Tollens' (or Benedict's) reagent. These sugars are traditionally called "reducing sugars." If both anomeric carbons are tied up as acetals, the sugar is a "nonreducing sugar."

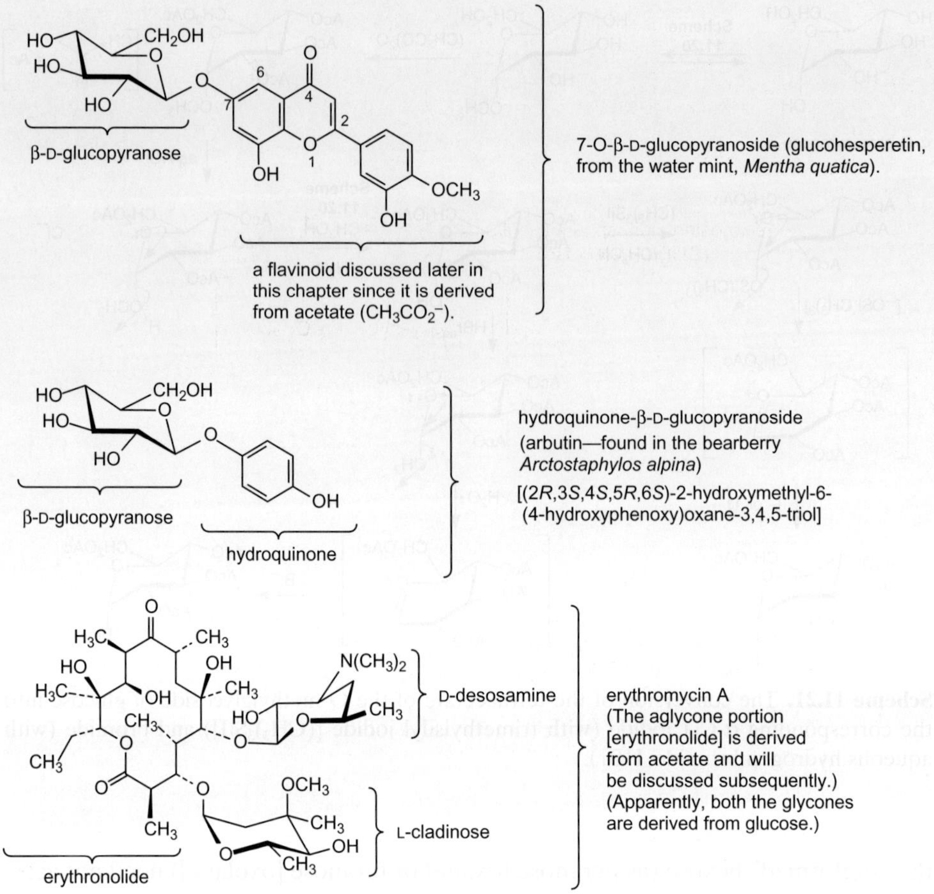

Figure 11.8. Some examples of glycosides.

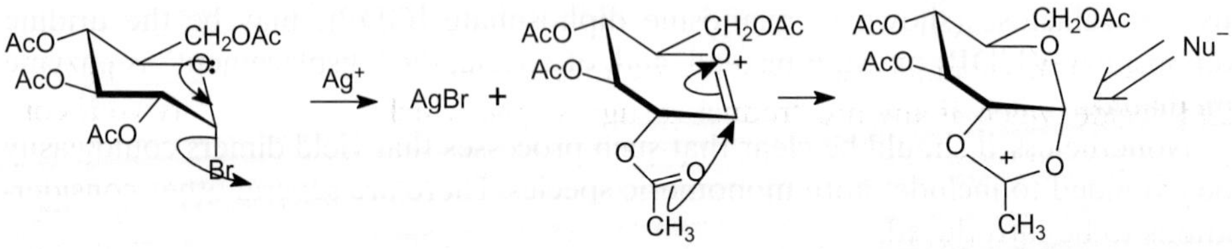

Scheme 11.22. A representation of a Koenigs–Knorr process. The incoming nucleophile will orient itself β on the glucose tetraacetate because of neighboring group participation by the acetate at C-2. See Koenigs, W.; Knorr, E. *Chem. Ber.*, **1901**, *34*, 957 and Stazi, F.; Palmisano, G.; Turconi, M.; Clini, S.; Santagostino, M. *J. Org. Chem.*, **1994**, *69*, 1097.

III. Oligosaccharides

Oligosaccharides are considered to consist of that group of polysaccharides that contain two to five or so monomeric units. As described above, they can be reducing or nonreducing sugars. Figures 11.9 and 11.10 present some examples of such sugars.

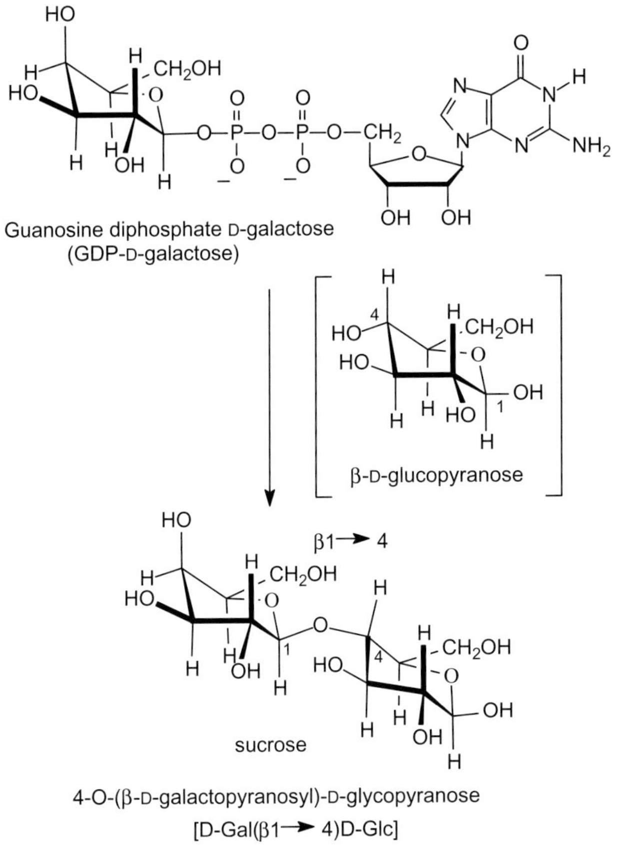

Figure 11.9. A cartoon depiction of the formation of sucrose from GDP-D-galactose and β-D-glucopyranose. The linkage between the carbohydrates is from C4 of the latter to the anomeric carbon of the former and is β, that is, a β1 → 4 linkage.

Problem 11.3. Examine the structures of the disaccharides in Figures 11.9 and 11.10 and decide which, if any, are "reducing sugars" and which are not. Justify your conclusion with a drawing using curved arrows.

It has become clear that cell recognition of other cells is frequently mediated by the carbohydrates on the surface. In part, it should already be clear that these highly functionalized molecules can be connected to one another in many different ways. So, with enough carbohydrates, the variation that can be created is very large, and thus, cell recognition can be finely tuned to allow, for example, "self" to be distinguished from "nonself." Thus, as shown in Figure 11.11, blood types are defined by using three or four surface carbohydrates that are attached to cell wall protein, and it is not uncommon to find 10–12 carbohydrates woven in specific patterns on cell surfaces to promote recognition to other ends. The study of glycobology is an exciting new field.

As shown in Figure 11.11, there are carbohydrates with nitrogen substituents. *N*-acetyl-D-glucosamine is a component of cell wall envelope in bacteria and, *vide*

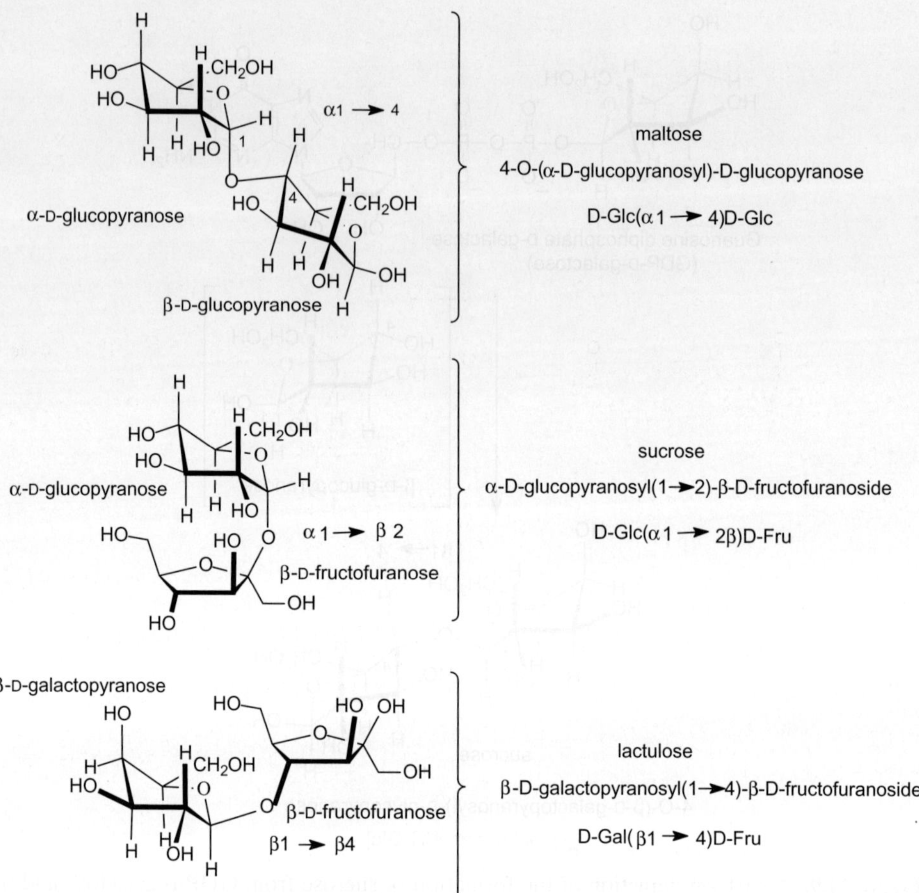

Figure 11.10. Three common disaccharides: lactose is a sugar found in mammalian milk; sucrose is the standard natural sweetening agent in many foods and obtained from sugar cane; lactulose is a disaccharide prepared from lactose and which is not absorbed by the gastrointestinal tract. It is used as a laxative.

infra, polymerized, is the principal component of the exoskeletons of invertebrates (insects, crustaceans, etc.). It appears that a biosynthesis of both *N*-acetyl-D-glucosamine and *N*-acetylgalactosamine starts from D-fructose 6-phosphate (Scheme 11.13) and, although not of plant origin, is produced here for completeness as Scheme 11.23. There are several interesting steps on which more discussion will follow in later chapters.

First, there is the problem of transamination, shown in the first step, and for which a transaminase enzyme (EC 2.6.1.16) is needed. It appears that the process, in this case, is not reversible, and, in a backward reaction (catalyzed by a deaminase, EC 3.5.99.6), an equivalent of ammonia (but almost certainly not as "NH₃") is lost and the nitrogen is replaced by an –OH group. The transaminase obtains the nitrogen in the (effective) hydrolysis of an amide (glutamine). Furthermore, there is probably

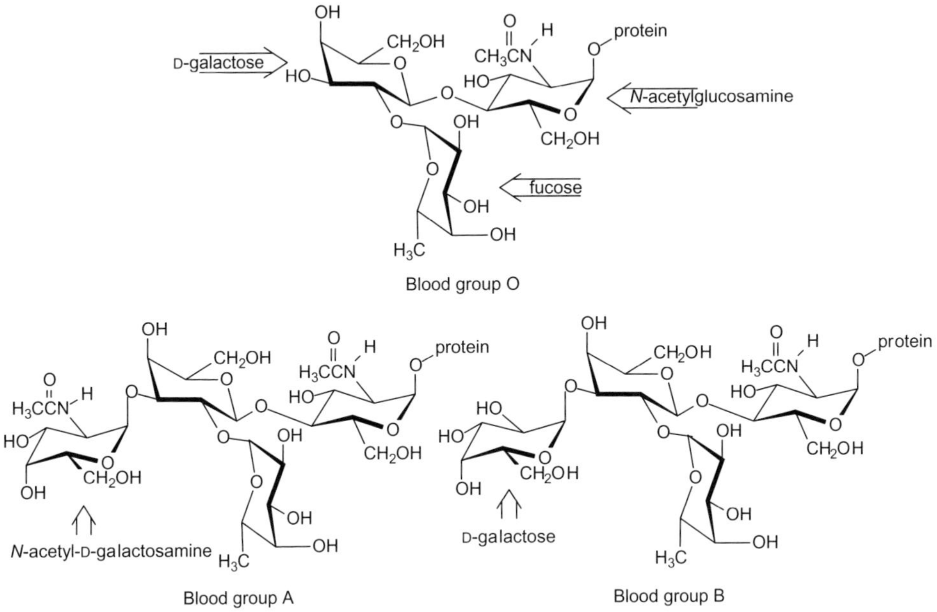

Figure 11.11. The cell surface carbohydrates required by blood cells to insure a specific "blood type."

an imine involved, the reduction of which dictates the absolute chemistry shown. The subject of transamination will be discussed from several different points of view in this and the remaining chapters but, for the present case, it is presumed that a carbonyl group is involved at C_2 to accept the equivalent of "NH_3" from the enzyme.

Second, there is the problem of acetylation by acetyl-CoA ($CH_3COSCoA$). The formation of acetyl-CoA follows on the heels of that of acetate and the acetogenins in the next portion of this chapter.

Third, there is the necessity to utilize the UDP ester to consummate the isomerization (epimerization) of a single hydroxyl group in the conversion of a derivative of D-glucose into one of D-galactose. It can be argued that this is because the epimerase (EC 5.1.3.7) needed just that ester.

Lastly, there is the presumed hydrolysis step to produce N-acetyl-D-galactosamine, which is drawn only for the sake of convenience. There is no evidence for this "free" species, and it may be used directly as the UDP ester or otherwise modified before its ultimate cellular fate.

Finally, among the oligosaccharides to be discussed here and also important in cell–cell recognition is the blood group antigen sialyl Lewis[X] (SLeX) (Figure 11.12). This carbohydrate tetramer is displayed at the termini of glycolipids (*vide infra*) present on the surface of white blood cells. The complex between SLeX and E-selectin, a glycoprotein (*vide infra*) that belongs to the family of cell adhesion molecules that mediate binding processes appears important in a variety of pathological states.

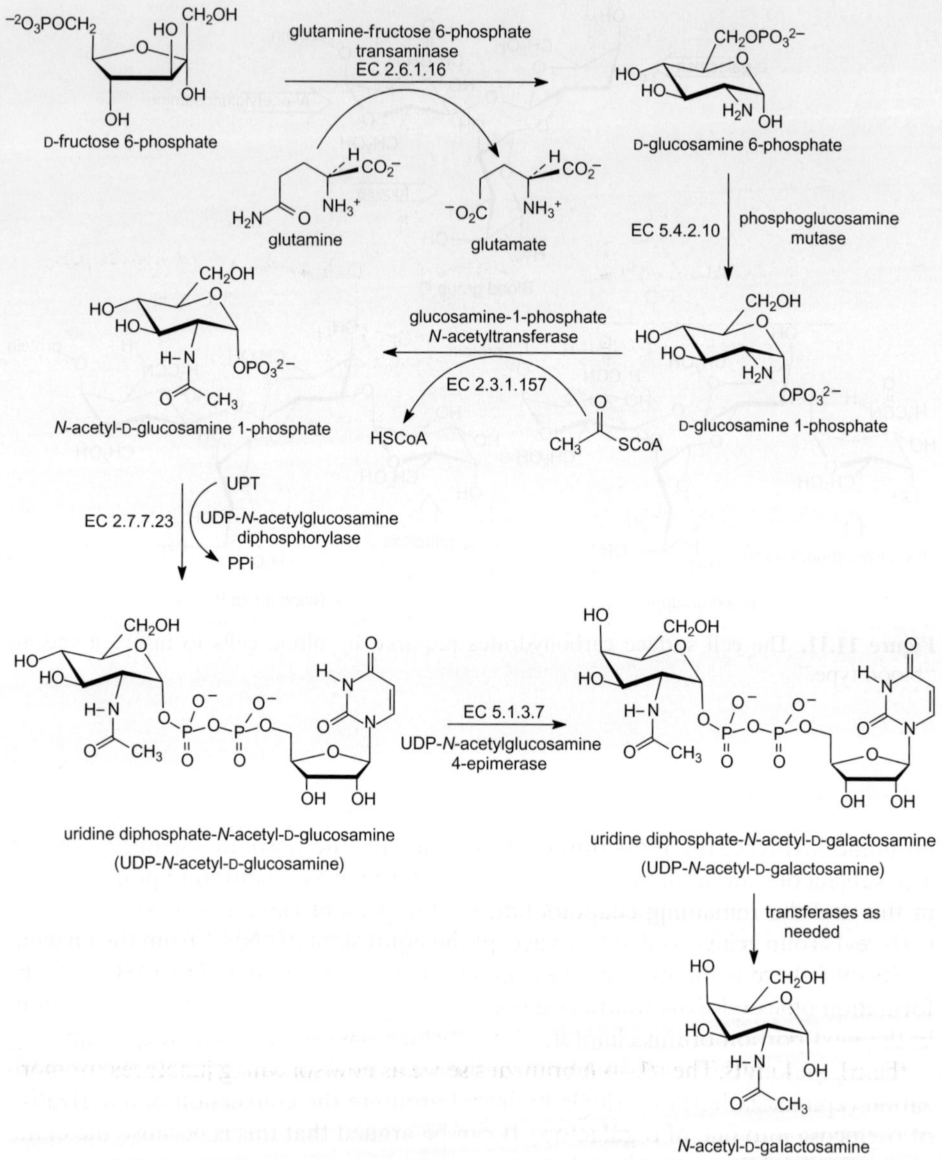

Scheme 11.23. Pathways to *N*-acetylglucosamine and *N*-acetylgalactosamine. The transamination process shown in the first step is presumed to involve an oxidation to a ketone, transamination, enamine formation, isomerization, and ring closure. Transamination, use of acetyl-CoA, and uridine will be discussed subsequently (*vide infra*). EC numbers and some graphic materials provided in this scheme have been taken with permission from appropriate links in a URL starting with http://www.chem.qmul.ac.uk/iubmb/enzyme/.

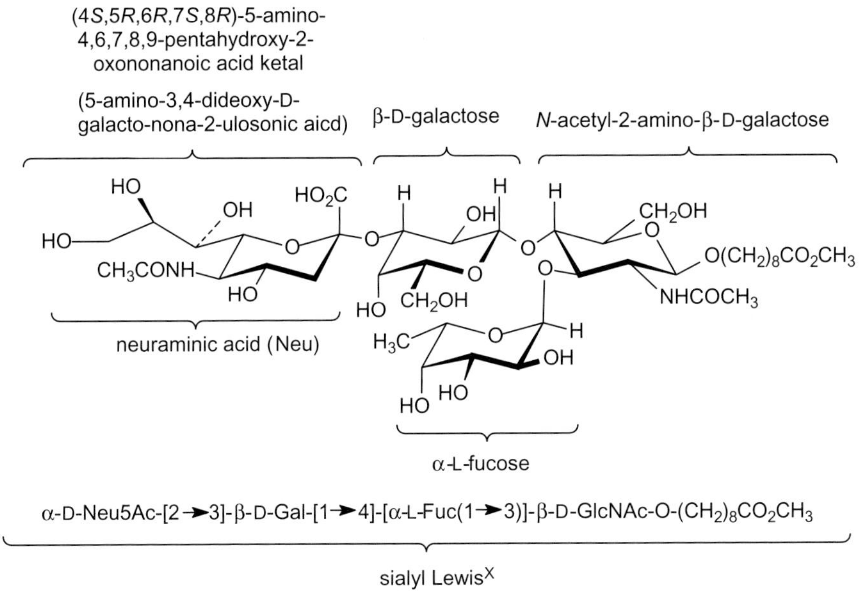

α-D-Neu5Ac-[2→3]-β-D-Gal-[1→4]-[α-L-Fuc(1→3)]-β-D-GlcNAc-O-(CH₂)₈CO₂CH₃

sialyl Lewis^X

Figure 11.12. The tetrasaccharide sialyl Lewis$^\text{X}$. This bioactive oligosaccharide is found on the termini of glycolipids (one shown here as the methyl ester of 9-hydroxyoctanoic acid) and is believed to play an important role in cell–cell recognition.

IV. Polysaccharides

It is widely held that polysaccharides (frequently thousands of saccharide mono-mers all linked together) serve two functions. Not only do polysaccharides serve plants as structural components (stems, trunks, etc.), but they also serve as a food reserve (e.g., starch).

Starch is composed of two types of polymer, viz. amylose, a linear polymer of 1000–2000 (ca. 10^3) glucopyranose units linked $\alpha 1 \rightarrow 4$ and amylopectin, with about a thousand times more (ca. 10^6) units, and which has $\alpha 1 \rightarrow 6$ branches in addition to the $\alpha 1 \rightarrow 4$ units. The $\alpha 1 \rightarrow 6$ branches serve as new spawning junctures for more $\alpha 1 \rightarrow 4$ units to join. Since the $\alpha 1 \rightarrow 6$ branches occur about once every 20 or so glucopyranose units, a thick network of insoluble polymer quickly builds up. Interestingly, mammals store the same carbohydrate (glucopyranose) as glycogen, a somewhat more branched version of amylopectin (branching about every 10 or so units). Representations for amylose and amylopectin–glycogen are shown in Figure 11.13.

Cellulose is probably the most abundant organic material on the planet as it is the main constituent of plant cell walls. In cellulose, the glucopyranose units are linked $\beta(1 \rightarrow 4)$ in their linear chain rather than $\alpha(1 \rightarrow 4)$ as in amylose and amy-lopectin. Interestingly, the structure of chitin, noted (*vide supra*) for its important role in shells and carapace of crustaceans and insect exoskeletons, is essentially the same as cellulose save that $\beta(1 \rightarrow 4)$ linked N-acetylglucosamine residues are involved. Representations of cellulose and chitin are shown in Figure 11.14.

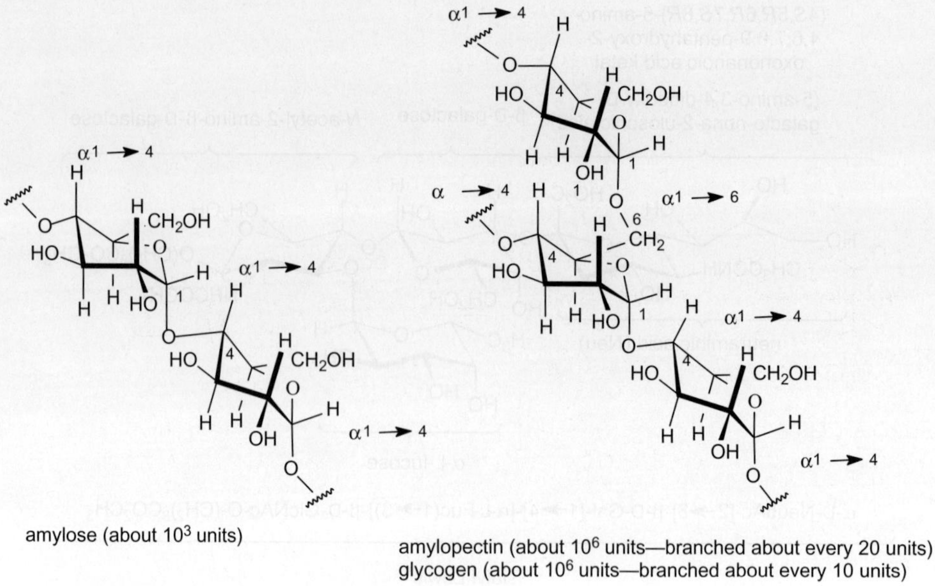

amylose (about 10^3 units)

amylopectin (about 10^6 units—branched about every 20 units)
glycogen (about 10^6 units—branched about every 10 units)

Figure 11.13. Representations of fragments of amylose, amylopectin–glycogen.

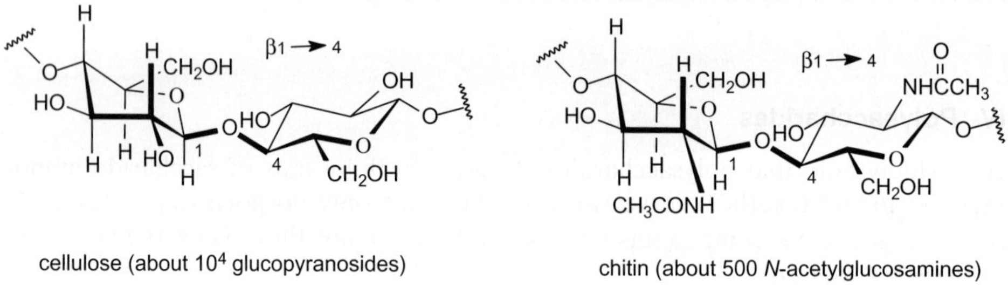

cellulose (about 10^4 glucopyranosides)

chitin (about 500 *N*-acetylglucosamines)

Figure 11.14. Representations of two units (each) of cellulose and chitin polymers.

D. ACETOGENINS

I. Acetyl-CoA (CH₃COSCoA)

Those compounds derived by condensation of acetyl units (effectively the two-carbon "CH₃C=O" unit) in a linear or branched fashion are often referred to as acetogenins. Given the wealth of materials so derived and the numbers of new compounds reported annually, it will seem that almost endless combinations are possible. The beginning of the discussion of the "two-carbon unit" requires a return to the Calvin cycle to pick up one of the PGA units produced as the cycle turned (Figure 11.1).

As shown in Scheme 11.24, action of phsophoglycerate mutase (EC 5.4.2.1) on PGA produces the corresponding isomer, 2-phosphoglycerate. In plants, it appears that the phosphate group is actually transferred to an active site histidine from the oxygen at C3 and then transferred back again to the oxygen at C2.

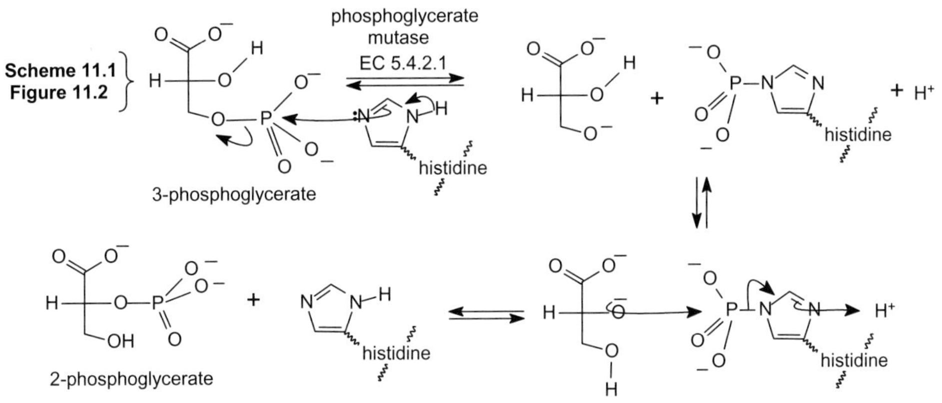

Scheme 11.24. A representation of the conversion of 3-phosphoglycerate into 2-phosphoglycerate via participation of the enzyme phosphoglycerate mutase (EC 5.4.2.1).

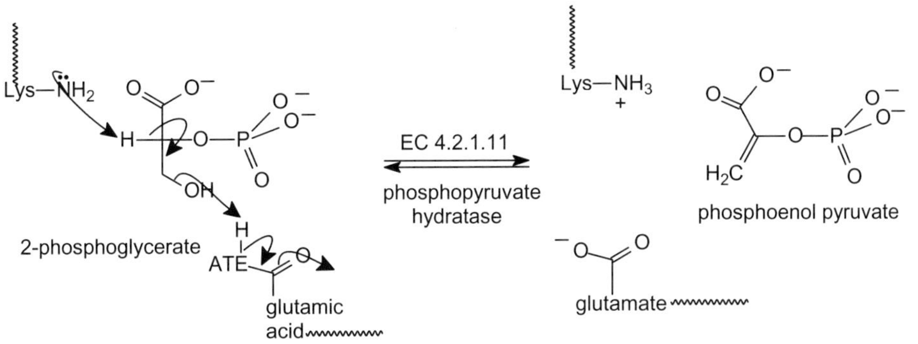

Scheme 11.25. A representation of the conversion of 2-phosphoglycerate to phosphoenolpyruvate with catalysis by phosphopyruvate hydratase (EC 4.2.1.11).

Then, in the presence of the enzyme enolase (phosphopyruvate hydratase, EC 4.2.1.11) and as shown in Scheme 11.25, dehydration of the 2-phospho-D-glycerate (2-phosphoglycerate) occurs to produce the phosphate enol of pyruvic acid, phosphoenolpyruvate (PEP).

Pyruvate, as its enol, is generated from phosphoenolpyruvate (Scheme 11.26) when the phosphoryl group is transferred to ADP with catalysis by the enzyme pyruvate kinase (EC 2.7.1.40). The enol then tautomerizes to the ketone.

Pyruvate can be formed, as indicated (Schemes 11.24 and 11.25), from glyceraldehyde 3-phosphate produced in the Calvin cycle as well as by catabolism of glucose, which, since it serves as a food storage material (*vide supra*, starch), is probably its main source.* However, it is the further uses of pyruvate to which attention must be turned.

As shown, pyruvate can be reduced to lactic acid, decarboxylated and reduced to ethanol, and (effectively) decarbonylated to acetate and, while all three are

---

*The catabolism of glucose has been well studied, and many of the reactions are similar to those shown as reversible and portrayed in Figure 11.2.

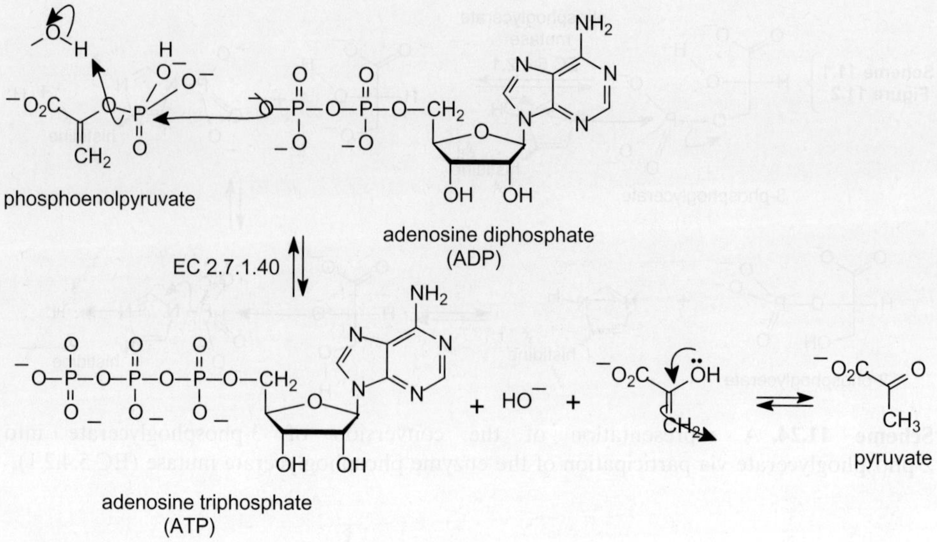

phosphoenolpyruvate

adenosine diphosphate
(ADP)

EC 2.7.1.40

adenosine triphosphate
(ATP)

pyruvate

Scheme 11.26. A representation of the enzyme-catalyzed (EC 2.7.1.40) reversible dephosphorylation of phosphoenolpyruvate to produce pyruvate and the ADP–ATP conversion accompanying the process.

elaborated on (Schemes 11.27–11.29), it is the latter that allows the formation of fatty acids; polyketide antibiotics and related compounds; and finally, mevalonic acid and then terpenes, sesquiterpenes, triterpenes, and steroids.

The conversion of pyruvate ($CH_3COCO_2^-$) to ethanol (CH_3CH_2OH) requires two steps. First, the pyruvate undergoes decarboxylation with participation of the enzyme pyruvate decarboxylase (EC 4.1.1.1). The enzyme uses thiamine diphosphate as a cofactor. As was the case earlier (Scheme 11.7), the nucleophilic ylid adds to the carbon of the carbonyl. In this case, the addition product corresponds to a β,γ-unsaturated carboxylic acid, ready for decarboxylation to produce an enol. Protonation followed by elimination of the thiamine diphosphate ylid then yields ethanal (acetaldehyde, CH_3CHO). This process is shown in Scheme 11.28.

Then, in the second step, a process essentially identical to that of Scheme 11.27, the ethanal (acetaldehyde, CH_3CHO) is reduced to ethanol (CH_3CH_2OH) by hydride transfer from the reduced form of nicotinamide adenine dinucleotide (NADH). The latter being oxidized to NAD^+ (Scheme 11.29).

Finally (for the moment), the third product derived from phosphoenolpyruvate is acetyl-CoA, written as **$CH_3COSCoA$** to emphasize the carbon–sulfur bond. That is, the (terminal) thiol of CoA forms the bond to the carbonyl group. The process forming $CH_3COSCoA$ from pyruvate is catalyzed by what appears to be a three-enzyme complex. The enzymes of the complex are pyruvate dehydrogenase complex (containing an oxoreductase for which the EC number has not yet been assigned), the enzyme N(6)-(S-acetyldehydrolipoyl)lysine (EC 2.3.1.12), and dihydrolipoamide dehydrogenase (EC 1.8.1.1.4). It is important to note (and it should be clear) that it is necessary to effect an oxidation subsequent to the decarboxylation of pyruvate since the thioester is in a higher oxidation state at the carbonyl carbon

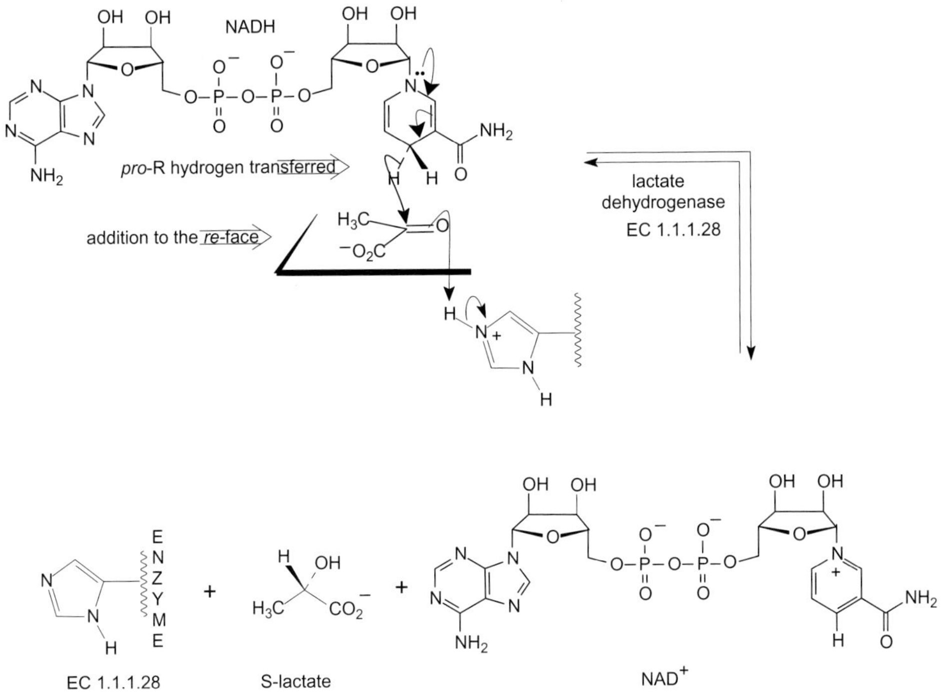

Scheme 11.27. A representation of the reduction of pyruvate to S-lactate (L-lactate) by action of the enzyme lactate dehydrogenase and the cofactor NADH (reduced nicotinamide adenine dinucleotide, which is oxidized to NAD$^+$ [nicotinamide adenine dinucleotide]) in the process.

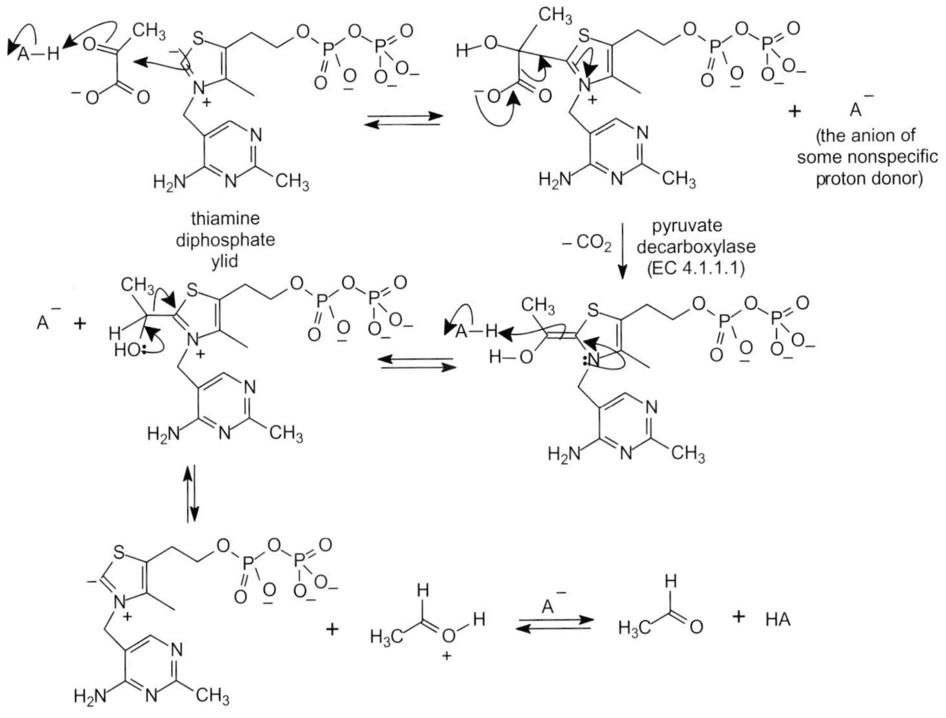

Scheme 11.28. The decarboxylation of pyruvate catalyzed by pyruvate decarboxylase (EC 4.1.1.1) whose cofactor is thiamine diphsophate. The product is ethanal (acetaldehyde, CH$_3$CHO) and the thiamine diphosphphate cofactor is regenerated. The acid AH is an unidentified proton donor. EC numbers and some graphic materials provided in this scheme have been taken with permission from appropriate links in a URL starting with http://www.chem.qmul.ac.uk/iubmb/enzyme/.

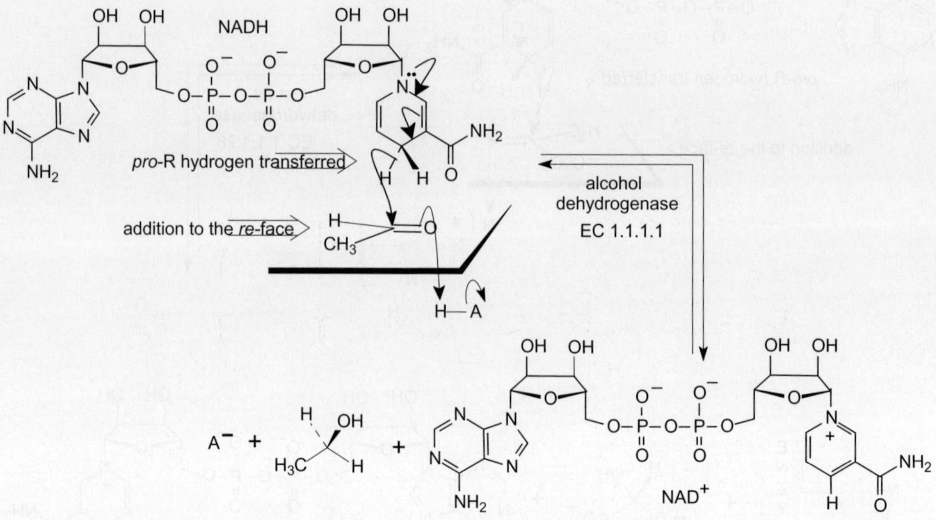

Scheme 11.29. A representation of the reduction of ethanal (acetaldehyde, CH_3CHO) by the enzyme alcohol dehydrogenase (EC 1.1.1.1) using the reduced form of nicotinamide adenine dinucleotide (NADH) as cofactor to ethanol (CH_3CH_2OH). The cofactor is oxidized to NAD^+. EC numbers and some graphic materials provided in this scheme have been taken with permission from appropriate links in a URL starting with http://www.chem.qmul.ac.uk/iubmb/enzyme/.

than was pyruvic acid itself. The oxidation occurs through the use of lipoic acid [linked in this case through an amide bond to a lysine in the enzyme $N(6)$-(S-acetyldehydrolipoyl)lysine (EC 2.3.1.12)] (Scheme 11.30) after which CoA replaces the lipoyl substituent via an addition–elimination on the carbonyl of the acetate unit (Scheme 11.31). Then, the enzyme lipoamide dehydrogenase (EC 1.8.1.4) is reduced by the dihydrolipoamide so that lipoamide can be regenerated for further use (Scheme 11.32). The lipoamide dehydrogenase (EC 1.8.1.4) is itself reoxidized with the reduction being effected by flavin adenine dinucleotide (FAD) (Scheme 11.33). $FADH_2$ formed in that process is itself reoxidized to FAD by the $NADH/NAD^+$ couple.

II. Acetyl-CoA ($CH_3COSCoA$) to Fatty Acids and Related Compounds

As foreshadowed in Chapter 8, Scheme 8.70, it is now clear that in many species, fatty acids (and as will become apparent later, polyketide antibiotics and related materials) are derived from acetyl-CoA and, occasionally, other starter CoA (propanoyl, butryl, etc.) units. Overall, and in a most general sense, the assembly of fatty acids and related compounds on a polyketide synthase (PKS) enzyme begins with an acetyl (or related) starter unit (as abbreviated in Scheme 11.34) transferred to the synthase enzyme from an acetyl-CoA (or related) species (Scheme 11.35). Once attached to the enzyme synthase complex, the addition of a two-carbon unit via (what appears to be) a Claisen-type condensation takes place with a malonyl

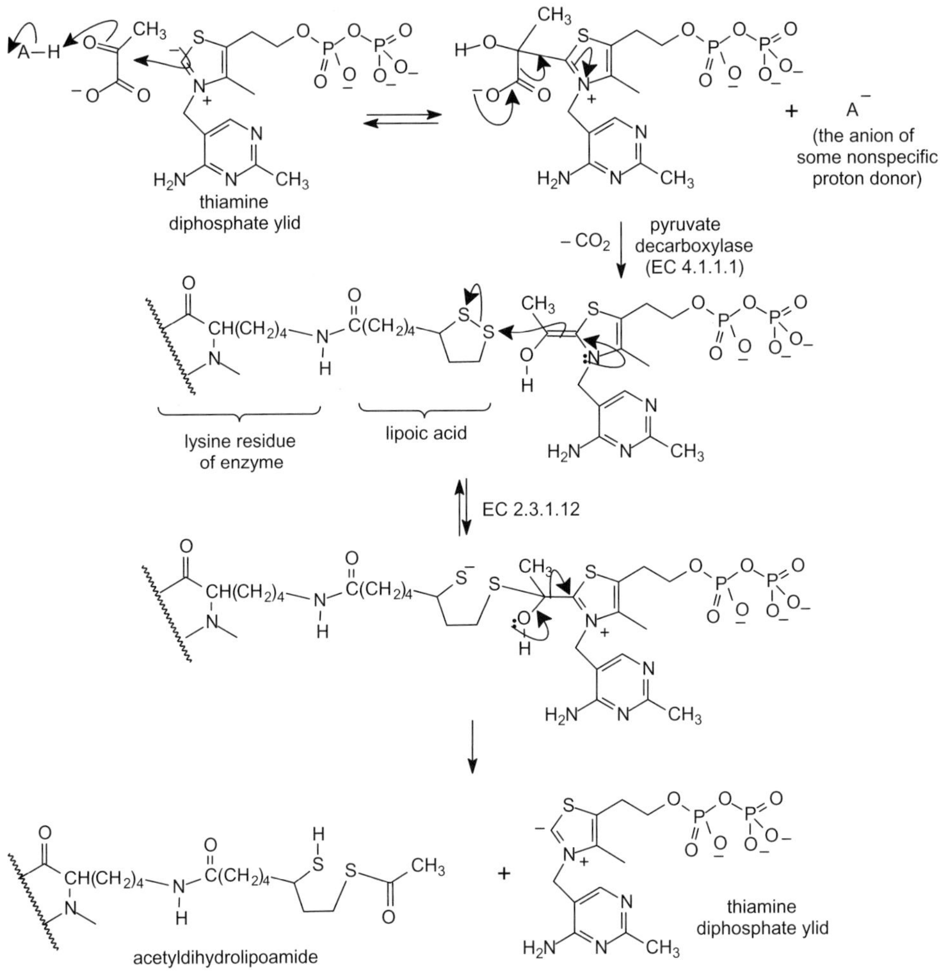

thiamine
diphosphate ylid

+   A$^-$
(the anion of
some nonspecific
proton donor)

$-CO_2$ | pyruvate
decarboxylase
(EC 4.1.1.1)

lysine residue
of enzyme

lipoic acid

EC 2.3.1.12

acetyldihydrolipoamide

thiamine
diphosphate ylid

Scheme 11.30. The formation of acetyldihydrolipoamide from pyruvate. Decarboxylation is mediated by the cofactor thiamine diphosphate. EC numbers and some graphic materials provided in this scheme have been taken with permission from appropriate links in a URL starting with http://www.chem.qmul.ac.uk/iubmb/enzyme/.

unit (three carbons followed by decarboxylation) also attached to the acyl carrier protein synthase. The malonyl unit is apparently synthesized on or near the same complex by an aldol-type reaction of carbon dioxide with the carbanion generated from the methyl group of acetyl-CoA! The enzyme cofactor that helps this carboxylation is biotin carboxylase (EC 6.4.1.2) and a representation of the pathway to accomplish the formation of malonate end is shown in Scheme 11.36. A representation of a pathway showing the condensation of synthase attached malonyl CoA with the acetyl (or similar) unit attached to the PKS is presented in Scheme 11.37. The addition may occur prior to, subsequent to, or concomitant with decarboxylation.

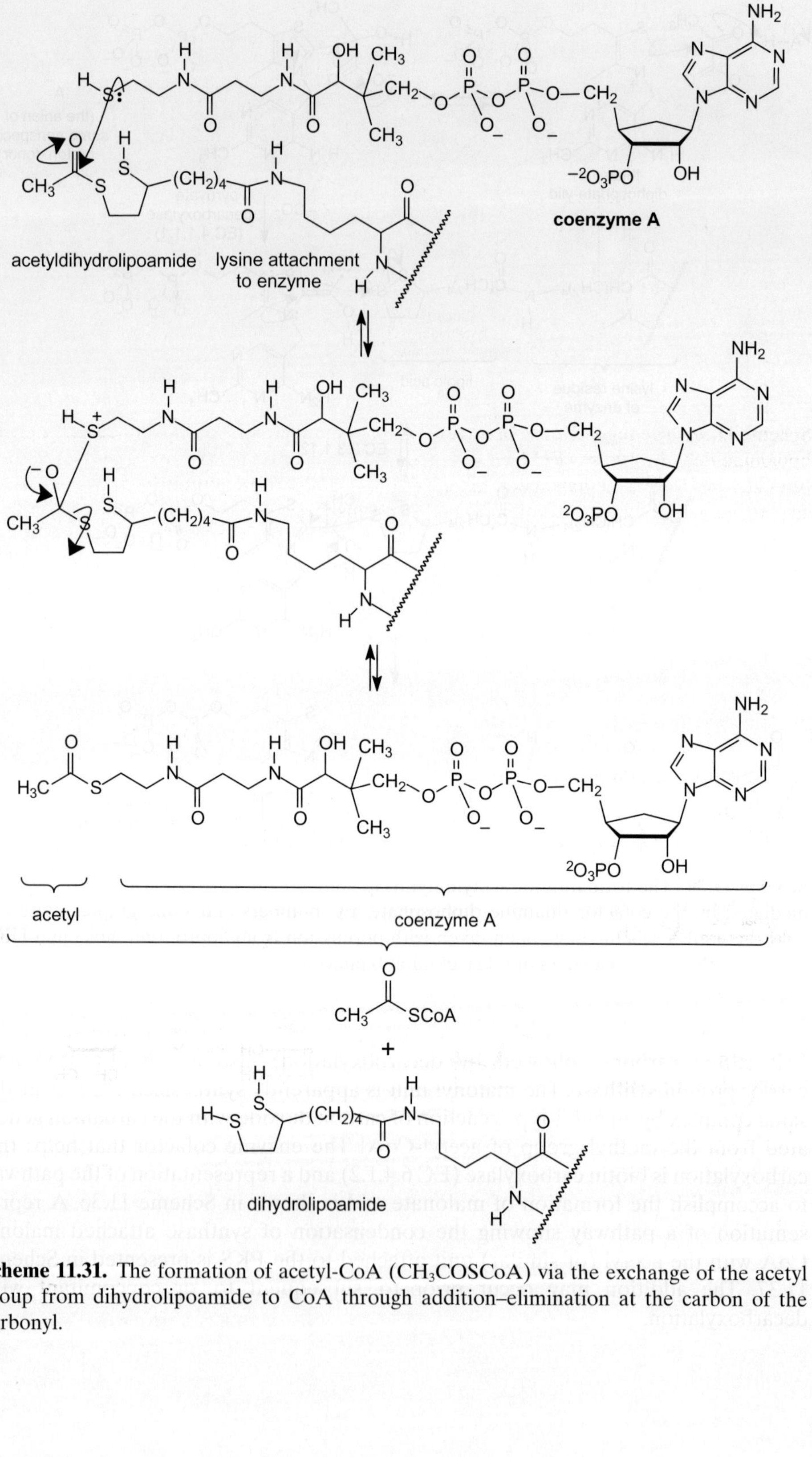

coenzyme A

acetyldihydrolipoamide   lysine attachment
to enzyme

acetyl   coenzyme A

$$CH_3 \overset{O}{\underset{}{\mathrm{C}}} SCoA$$

+

dihydrolipoamide

Scheme 11.31. The formation of acetyl-CoA ($CH_3COSCoA$) via the exchange of the acetyl group from dihydrolipoamide to CoA through addition–elimination at the carbon of the carbonyl.

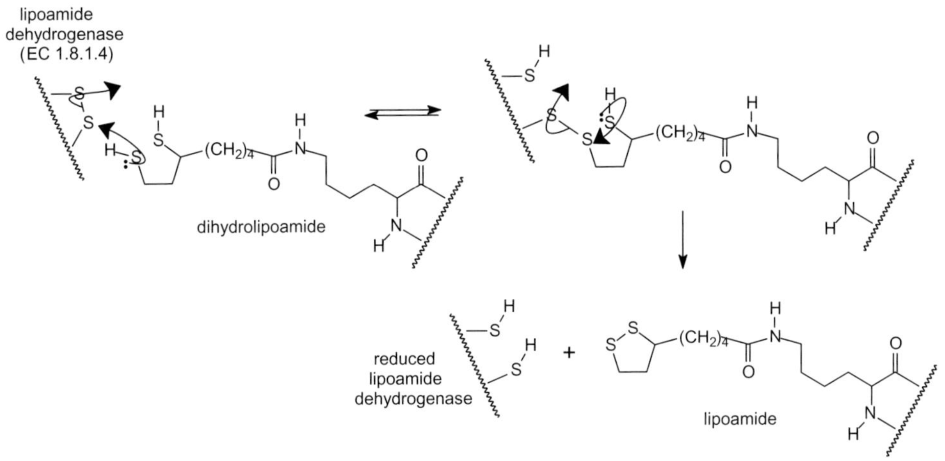

Scheme 11.32. The regeneration of lipoamide by oxidation of dihydrolipoamide. The enzyme lipoamide dehydrogenase (EC 1.8.1.4) is reduced. EC numbers and some graphic materials provided in this scheme have been taken with permission from appropriate links in a URL starting with http://www.chem.qmul.ac.uk/iubmb/enzyme/.

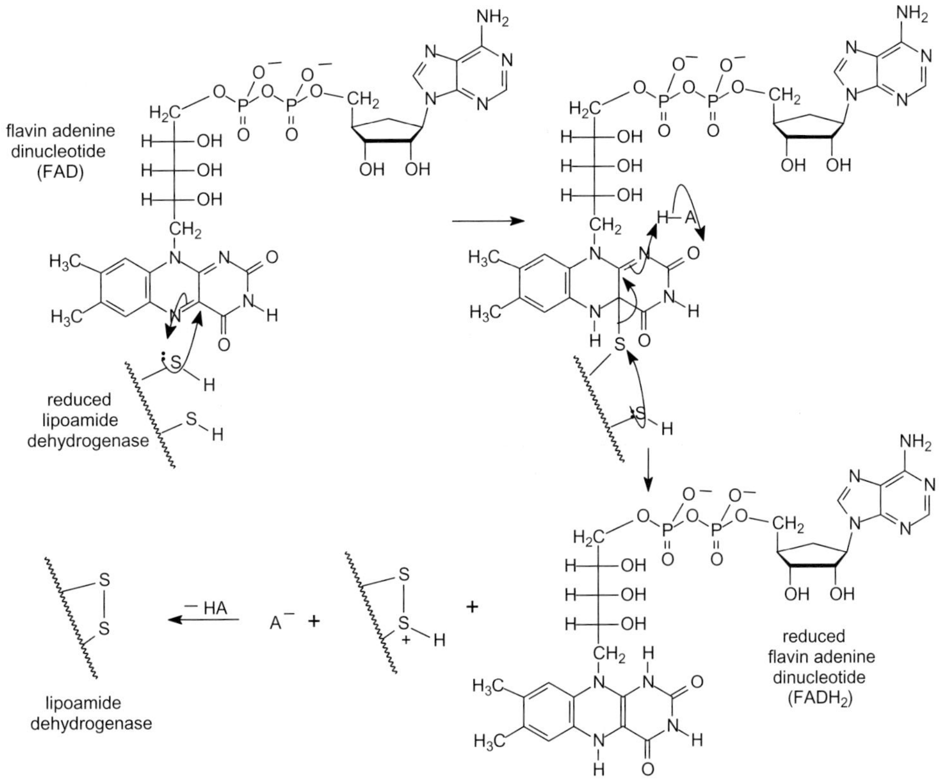

Scheme 11.33. The oxidative regeneration of the enzyme lipoamide dehydrogenase (EC 1.8.1.4). Flavin adenine dinucleotide (FAD) is reduced in the process to $FADH_2$. EC numbers and some graphic materials provided in this scheme have been taken with permission from appropriate links in a URL starting with http://www.chem.qmul.ac.uk/iubmb/enzyme/.

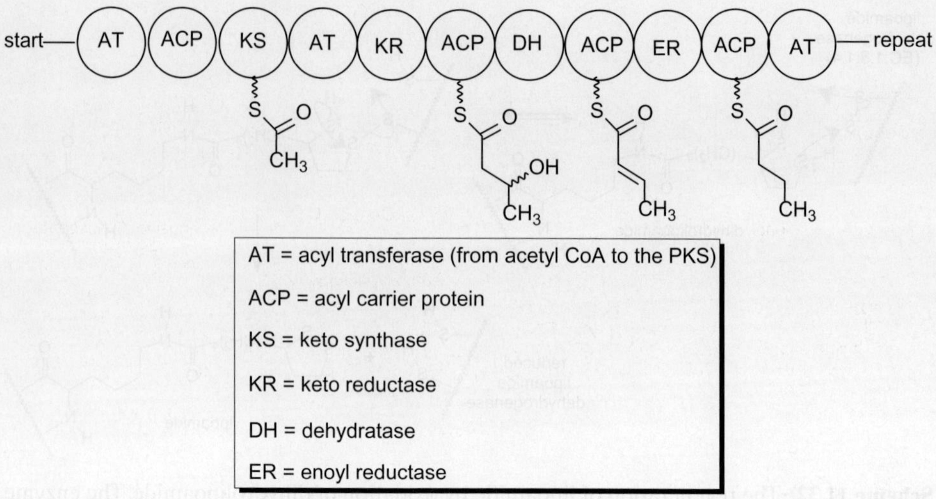

AT = acyl transferase (from acetyl CoA to the PKS)

ACP = acyl carrier protein

KS = keto synthase

KR = keto reductase

DH = dehydratase

ER = enoyl reductase

Scheme 11.34. A cartoon representation of a polyketide synthase that might be used to produce fatty acids or polyketides.

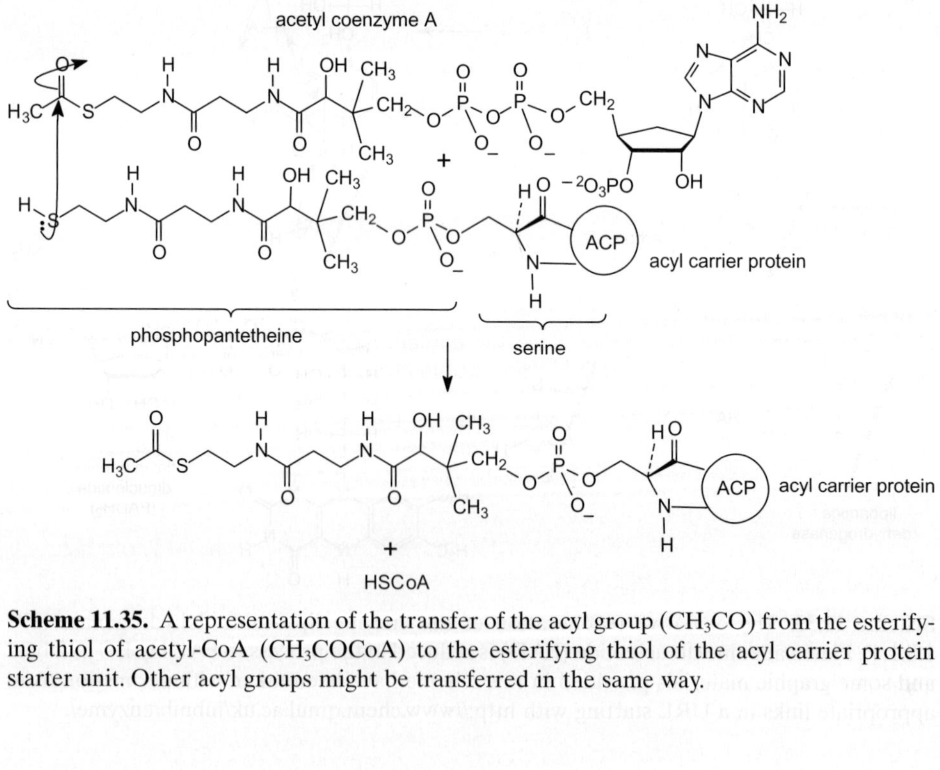

Scheme 11.35. A representation of the transfer of the acyl group (CH_3CO) from the esterifying thiol of acetyl-CoA (CH_3COCoA) to the esterifying thiol of the acyl carrier protein starter unit. Other acyl groups might be transferred in the same way.

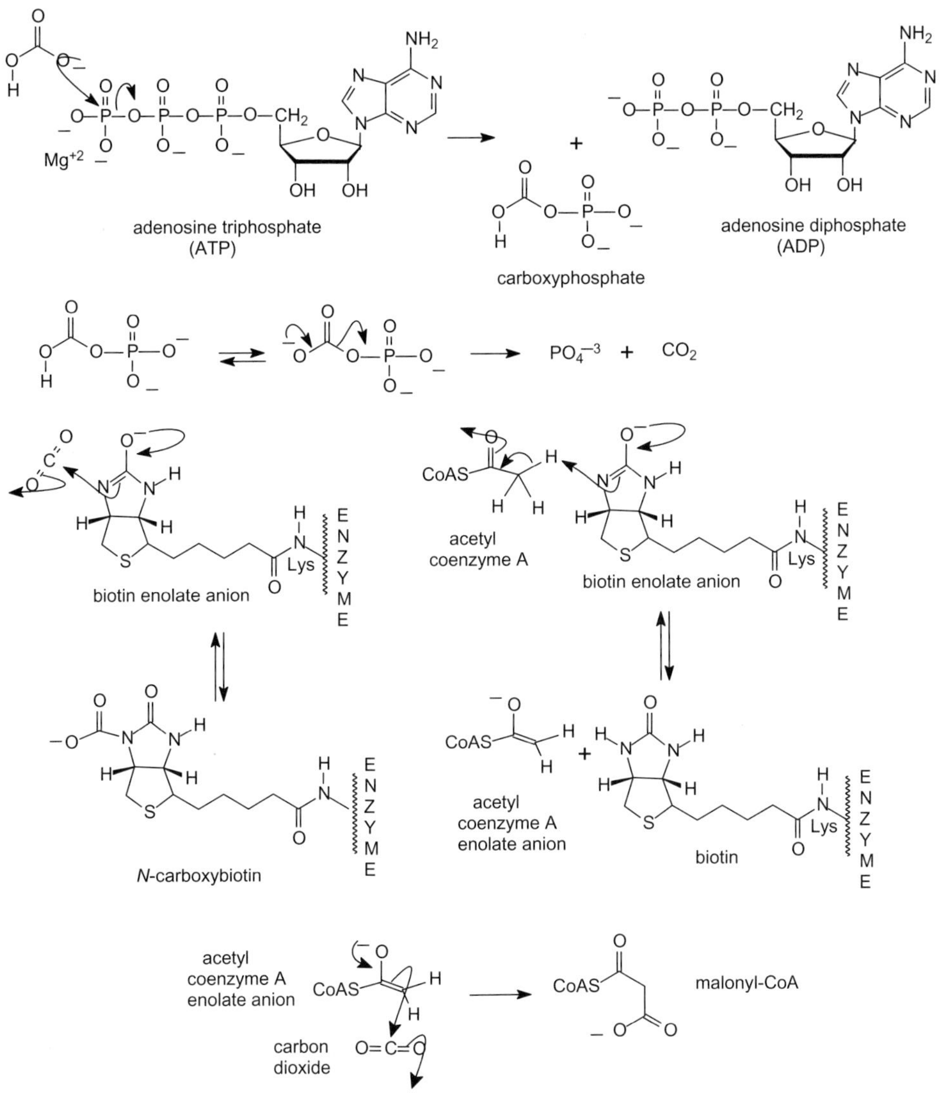

Scheme 11.36 diagram with labels:

adenosine triphosphate
(ATP)

carboxyphosphate

adenosine diphosphate
(ADP)

PO_4^{-3} + CO_2

biotin enolate anion

acetyl
coenzyme A

biotin enolate anion

N-carboxybiotin

acetyl
coenzyme A
enolate anion

biotin

acetyl
coenzyme A
enolate anion

carbon
dioxide

malonyl-CoA

Scheme 11.36. Potential pathways in concert with evidence for the formation of malonyl-CoA. The cofactor biotin helps produce the enolate anion of acetyl-CoA and to deliver carbon dioxide to that enolate anion to produce malonyl-CoA.

After the addition and decarboxylation, the chain can continue to grow through the addition of another malonyl fragment (and decarboxylation) forming keto groups interrupted by methylenes until told to "stop" as dictated by the particular synthase enzyme. Alternatively, reduction at each stage of carbonyl to hydroxyl at each stage of malonyl addition, elimination of water to an (*E*)- or (*Z*)-alkene, and

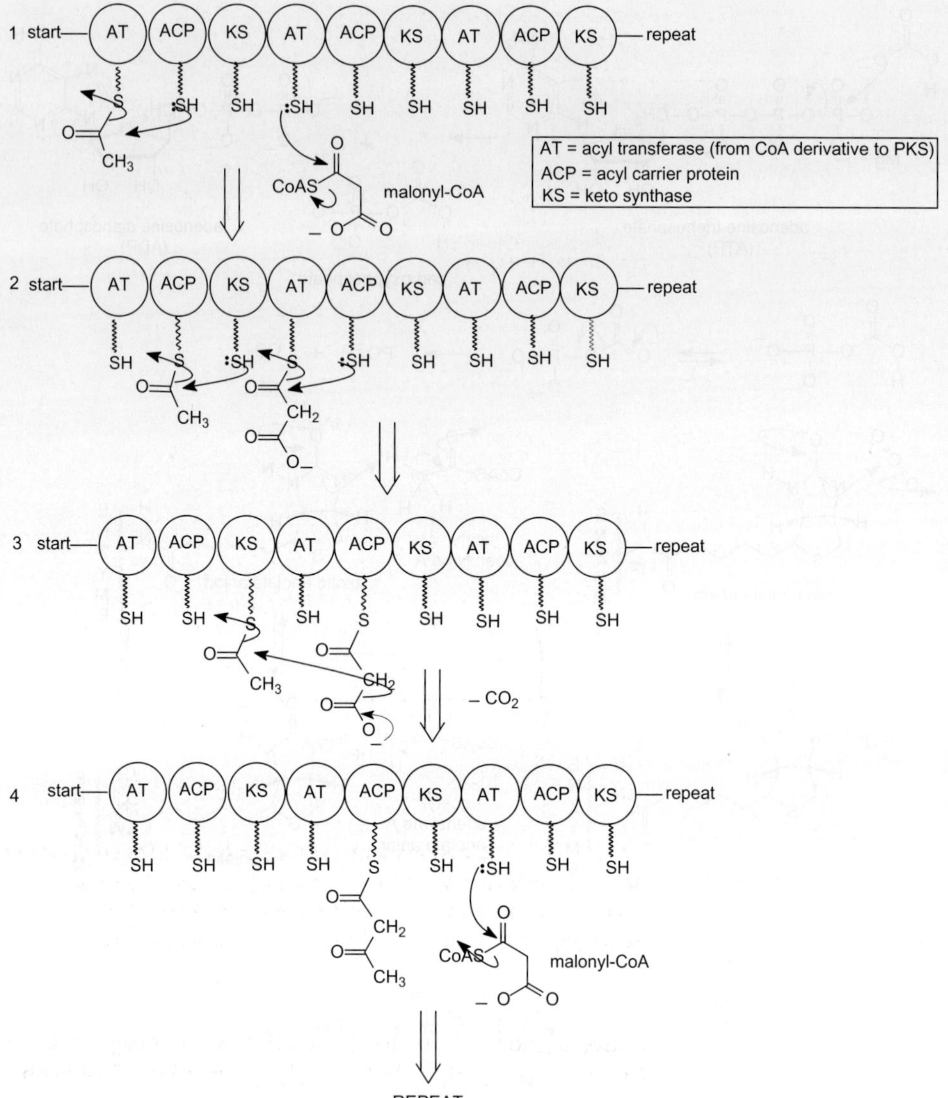

Scheme 11.37. A cartoon representation of the steps in a single coupling reaction on the way to formation of a polyketide. As the process moves "down the chain," two carbon units are added.

then, perhaps, further reduction to an alkyl group could follow. The former (Figure 11.15) produces more than 10,000 known polyketides **and derivatives** while the latter (Figure 11.18) produces thousands of saturated and unsaturated fatty acids **and derivatives**.

Not only can different starter units, for example, propionyl CoA, and butryl CoA, be used in polyketide formation but different extenders (e.g., methylmalonyl CoA) are also found. A now classical example is the bacterial system *Saccharopolyspora*

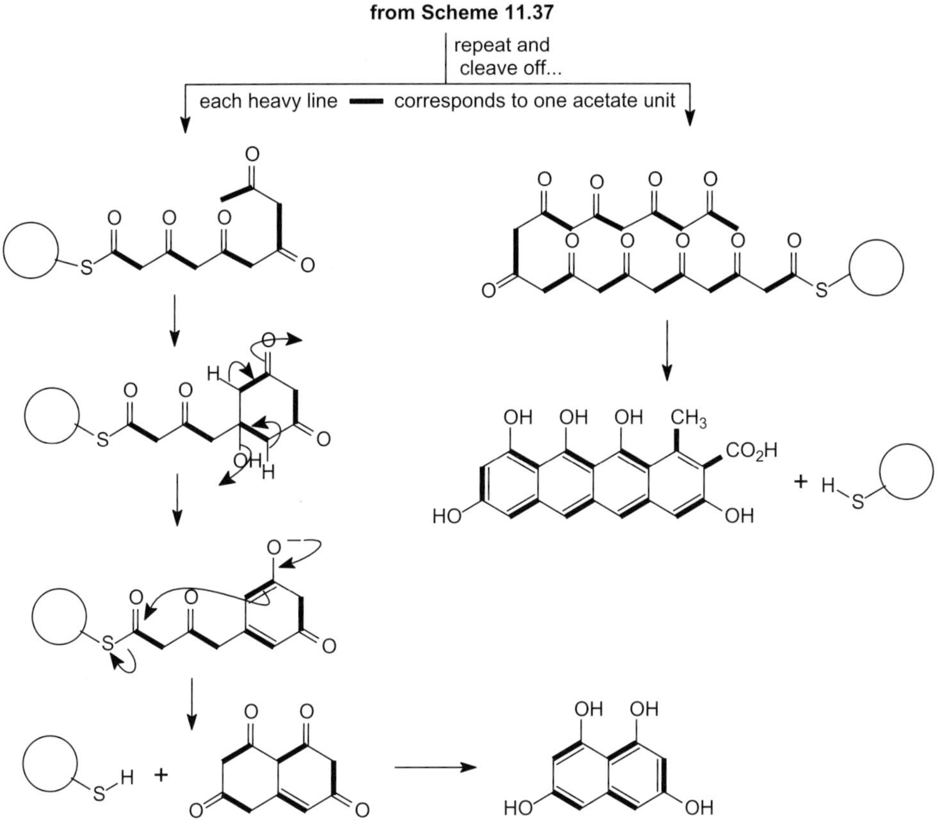

Figure 11.15. Polyketide formation from malonyl addition (and decarboxylation) to a starter acetyl-CoA on a polyketide synthase. Two examples of thousands possible are shown. The heavy lines in each represent two carbons of an "acetate" unit, and in many cases, the patterns are confirmed by isotopic (^{14}C) labeling experiments.

erythraea that synthesizes erythromycin and utilizes a PKS that has been extensively investigated.*

As shown in Scheme 11.38, the biosynthesis of erythromycin uses PKS consisting of a complex of three units that are named after the 6-deoxyerythronolide B product eventually elaborated into the antibiotic erythromycin. The three units are deoxyerythronolide B synthase DEBS1, DEBS2, and DEBS3 and, as noted in Scheme 11.38, involve beginning with propanoyl CoA[†] and adding methylmalonyl

---

*See Pfeifer, B. A.; Admiraal, S. J.; Gramago, H.; Cane, D.; Khosla, C. *Science*, **2001**, *291*, 1790 for a leading reference to this elegant work.

[†]One of the pathways to propanoyl-CoA is from catabolism of the amino acid threonine (Chapter 12). Thus, threonine (threonine dehydratase, EC 4.3.1.19, cofactor pyridoxal phosphate) undergoes deamination to give 2-oxobutanoate (α-ketobutyrate) as shown below. Then, 2-oxobutanoate (α-ketobutyrate) undergoes decarboxylation (perhaps as shown in Scheme 11.30) with formation of propanoyl dihydrolipoamide in a (cofactor) thiamine diphosphate mediated step. Finally, as in Scheme 11.31, propanoyl-CoA is formed. An alternative pathway uses a ferrodoxin to effect the decarboxylation of 2-oxobutanoate (α-ketobutyrate) Ferredoxins are small proteins containing iron and sulfur atoms in iron–sulfur clusters.

CoA* units. In Scheme 11.38, the abbreviations are KS, ketosynthase; AT, acyl transferase; ACP, acyl carrier protein; KR, ketoreductase; ER, enoyl reductase; DH, dehydratase; and TE, thioesterase (to remove the completed chain prior to or concomitant with) cyclization. Finally, as also seen in Scheme 11.38, the formation of erythromycin A requires some small oxidative modifications (and attachment of two sugars at the anomeric carbon) for completion.

In addition to the saturated fatty acids derived from "acetate" such as dodecanoic acid (lauric acid, C_{12}, Figure 11.16, $n = 10$), tetradecanoic acid (myristic acid, C_{14}, Figure 11.16, $n = 12$), hexadecanoic acid (palmitic acid, C_{16}, Figure 11.16, $n = 14$), and octadecanoic acid (stearic acid, C_{18}, Figure 11.16, $n = 16$) by a pathway such as that in Figure 11.16 (an abbreviated version of Schemes 11.37 and 11.38), many unsaturated fatty acids also occur naturally.

As shown in Figure 11.17, these include such monoalkenes as (Z)-9-octadecanoic acid (oleic acid) and (Z,Z)-9,12-octadecanoic acid (linoleic acid) and 20 carbon 5,8,11,14-eicosatetraenoic acid (arachidonic acid), the tetraene from which the prostaglandins are derived.

In yet another variation on the same theme, PKSs have been linked to the family of polyether ionophore (Chapter 8) antibiotics that are widely found in bacteria and, interestingly, in a number of toxic marine species. These polyketide-based species are apparently capable of encircling cations and moving them across cellular membranes or inhibiting that motion. A typical member, monensin, is provided in Figure 11.18, and it is assumed that along with the usual array of modules, there are oxidases (epoxidases) that perform so as to allow oxygen (O_2) to participate in construction of the acid (as cleaved from the synthase).

*(S)-methylmalonyl CoA is formed from propanoyl-CoA with a biotin cofactor-mediated carboxylation as in Scheme 11.36 (EC 6.4.1.3).

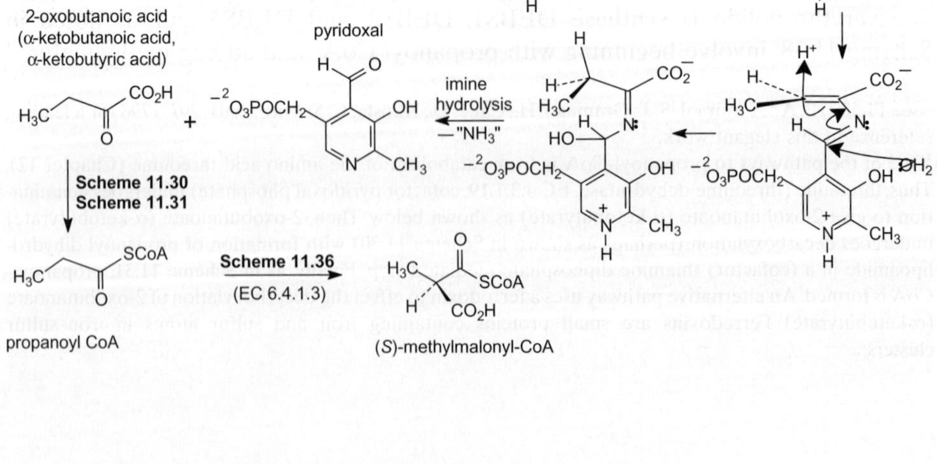

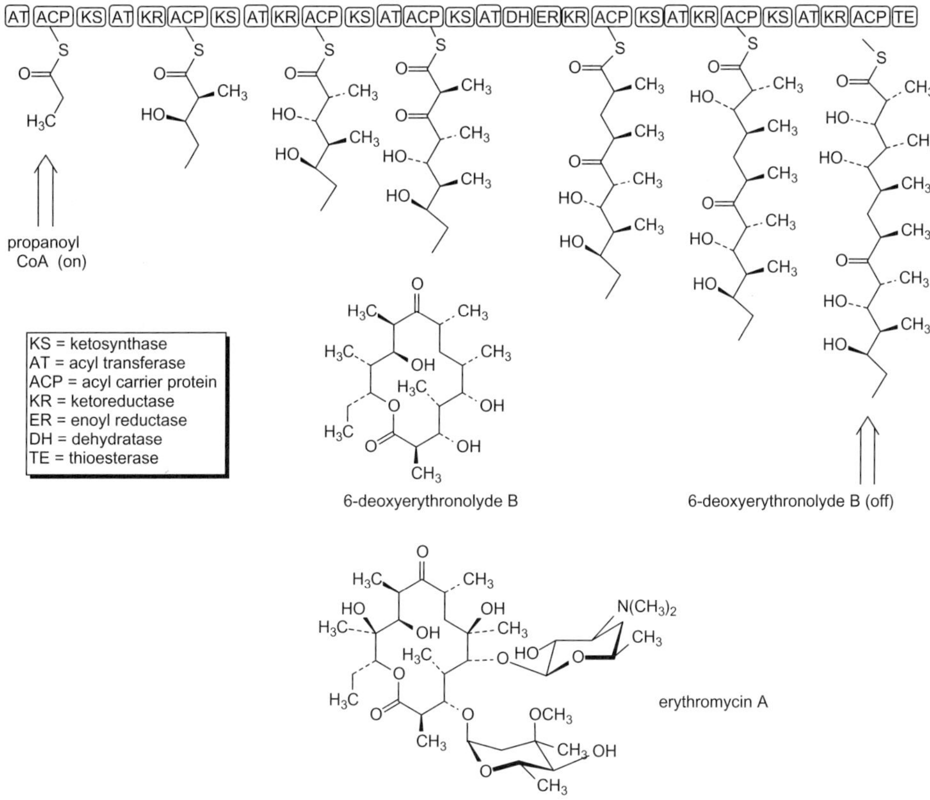

Scheme 11.38. A representation of the polyketide synthase that is found to produce 3-deoxyerythronolyde B on the way to erythromycin A. See Pfeifer, B. A.; Admiraal, S. J.; Gramago, H.; Cane, D.; Khosla, C. *Science*, **2001**, *291*, 1790 for a leading reference to this elegant work and Chen, A. Y.; Schnarr, N. A.; Kim, C.-Y.; Cane, D. E.; Khosla, C. *J. Am. Chem. Soc.*, **2006**, *128*, 3067.

III. Isoprenoides: To Dimethylallyl Diphosphate and Beyond

a. Dimethylallyl Diphosphate from Acetyl-CoA via Mevalonate. Mevalonate (Scheme 11.39) can be biosynthesized in more than one way. The more recently discovered path from 1-deoxy-D-xylulose 5-phosphate will be considered second.

As shown in Scheme 11.39, acetyl-CoA is transferred to the enzyme acetyl-CoA C-acetyltransferase (EC 2.3.1.9) where it is attacked (Claisen condensation) by the anion of acetyl-CoA to produce acetoacetyl-CoA. Then, the aldol-like condensation between acetoacetyl-CoA and acetyl-CoA is catalyzed by hydroxymethylglutaryl-CoA synthase (EC 2.3.3.10). It is held that the acetoacetyl-CoA binds to the enzyme (with the CoAS being replaced by a sulfhydryl bond from the enzyme) and that the addition to the carbonyl occurs on the *re*-face to yield (3*S*)-3-hydroxy-3-methylglutaryl CoA.

The next two steps (Scheme 11.40) occur on the enzyme hydroxymethylglutaryl-CoA reductase (EC 1.1.1.34), require 2 equivalents of NADPH, and involve, first, the reduction of the thioester to aldehyde and then, second, the reduction of

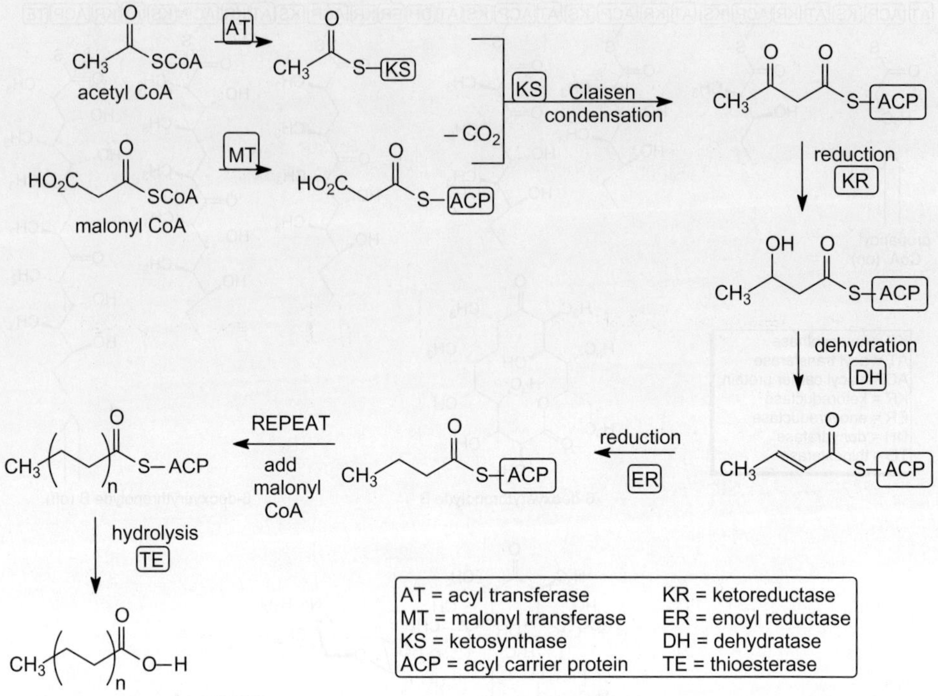

AT
acetyl CoA
MT
malonyl CoA
KS
Claisen condensation
– CO2
reduction
KR
dehydration
DH
reduction
ER
REPEAT
add malonyl CoA
hydrolysis
TE
AT = acyl transferase
MT = malonyl transferase
KS = ketosynthase
ACP = acyl carrier protein
KR = ketoreductase
ER = enoyl reductase
DH = dehydratase
TE = thioesterase

Figure 11.16. Saturated fatty acid production using a modular ketosynthase to build the fatty acids two carbons at a time. For $n = 10$, dodecanoic acid (lauric acid); for $n = 12$, tetradecanoic acid (myristic acid); for $n = 14$, hexadecanoic acid (palmitic acid); and for $n = 16$, octadecanoic acid (stearic acid).

the aldehyde to the corresponding alcohol, (*R*)-mevalonate [(*R*)-3,5-dihydroxy-3-methylpentanoic acid].

Then (Scheme 11.41) mevalonate [(*R*)-3,5-dihydroxy-3-methylpentanoic acid] is phosphorylated by the enzyme mevalonate kinase (EC 2.7.1.36) at the primary hydroxyl. The phosphate that is added is obtained (an addition–elimination reaction or a displacement reaction) from the terminal phosphate group of ATP forming ADP in turn. Another phosphorylation, again using ATP yields mevalonate diphosphate, this time by the enzyme phosphomevalonate kinase (EC 2.7.4.2) and then, after one more phosphorylation, this time on the tertiary hydroxyl, decarboxylation and dehydration co-occur, catalyzed by the enzyme mevalonate diphosphate decarboxylase (EC 4.1.1.43). The products are inorganic phosphate, carbon dioxide, and isopentenyl diphosphate (diphosphoric acid mono[3-methylbut-3-enyl] ester). Isopentenyl diphosphate isomerase (EC 5.3.3.2) catalyzes the isomerization, via loss of the C_2 *pro-R* hydrogen, between isopentenyl diphosphate and dimethylallyl diphosphate.

b. Dimethylallyl Diphosphate from Pyruvate and Glyceraldehyde. Within the last several decades, it has been found that a major pathway in plants for the formation of isoprenoids relies on the reaction between pyruvate and D-glyceraldehyde 3-phosphate (D-phosphoglyceraldehyde).

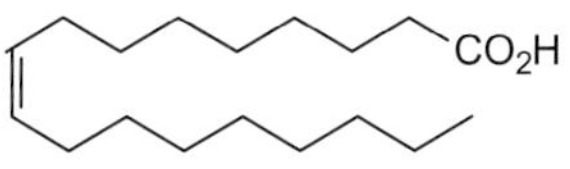

oleic acid (C$_{18}$)
(9Z)-octadecanoic acid

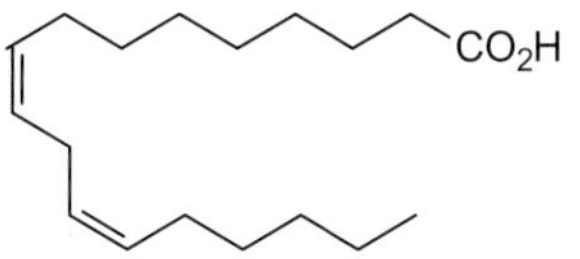

linoleic acid (C$_{18}$)
(9Z,12Z)-octadecadienoic acid

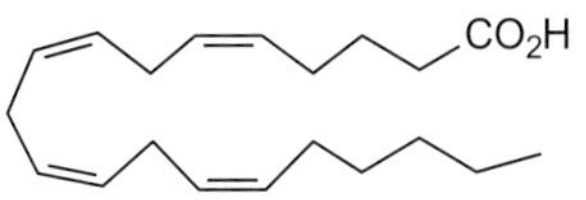

arachidonic acid (C$_{20}$)
(5Z,8Z,11Z,14Z)-eicosatatraenoic acid

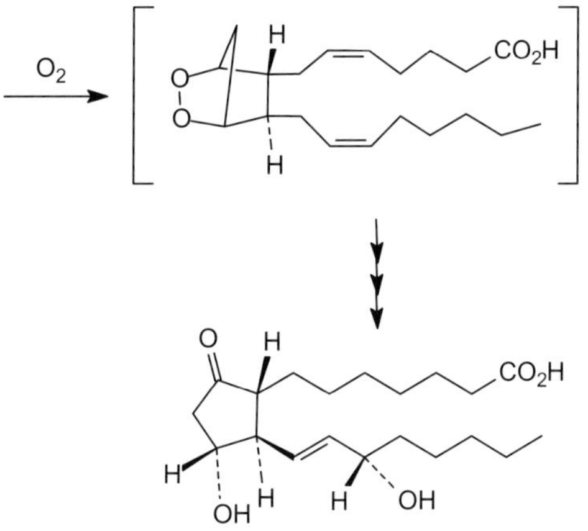

O$_2$

prostaglandin E$_1$ (PGE$_1$)

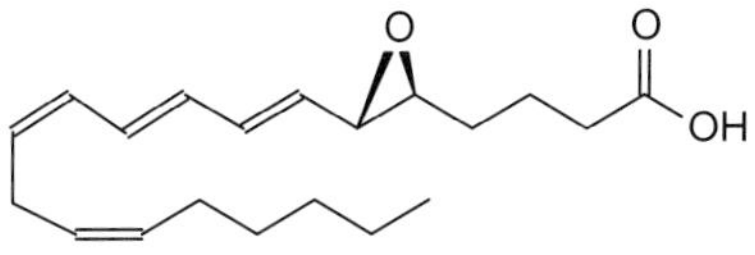

leukotriene A4

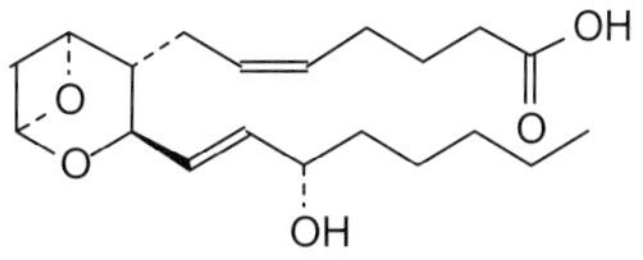

thromboxane A2

Figure 11.17. Examples of unsaturated fatty acids. Large numbers of such alkenes both as (*E*)- and (*Z*)-isomers exist. The intermediate in the oxidation of arachidonic acid to (only one shown member of the) prostaglandins is part of a cascade of reactions to an entire family of compounds that vary in oxidation state. Leukotrienes (three conjugated double bonds and no five-membered ring) and thromboxanes with six-membered oxygen-containing rings (oxacyclohexane, pyran, oxane) are also derived from C$_{20}$ polyalkenes.

As already seen in Schemes 11.28 (pyruvate to ethanol) and 11.30 (pyruvate to acetyl-CoA), the three-carbon unit of pyruvate is a good substrate for reaction with the thiamine diphosphate cofactor. Indeed, as shown in Scheme 11.42, the same two-carbon unit found in Scheme 11.30 is generated, but now, in place of a lipoic acid fragment, attack occurs on the carbonyl of glyceraldehyde 3-phosphate under the influence of 1-deoxy-D-xylulose synthase (EC 2.2.1.7).

Then, 1-deoxy-D-xylulose undergoes an alkyl migration and that isomerization is accompanied by reduction (NADPH $\rightarrow$ NADP$^+$) under the influence of 1-

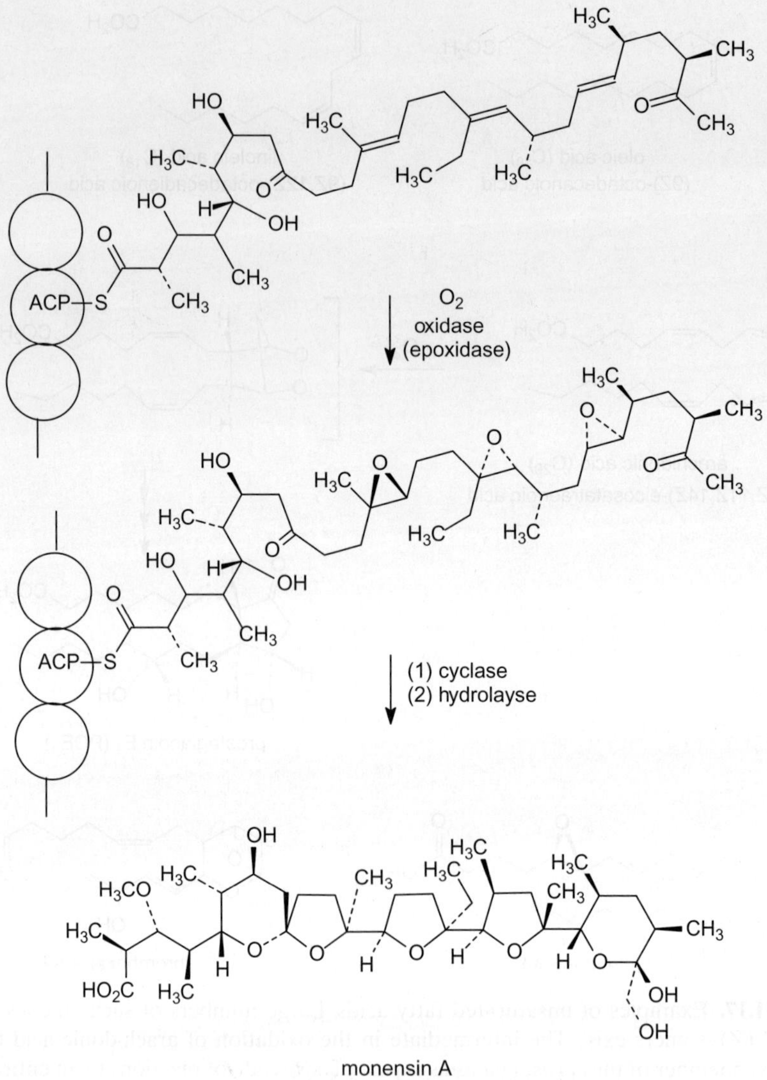

Figure 11.18. A potential pathway to the polyether ionophore, monensin A as built on a modular polyketide synthase that includes an oxidase that utilizes oxygen (O_2).

deoxy-D-xylulose 5-phosphate reductoisomerase (EC 1.1.1.267) to generate 2-C-methyl-D-erythritol 4-phosphate (Scheme 11.43). The *pro-S* hydrogen on NADPH adds to the *re*-face of the aldehyde. After additional phosphorylation, it appears that 2C-methyl-D-erythritol-2,4-cyclodiphosphate forms before isopentenyl diphosphate is generated. But details are lacking.

Nonetheless, it is interesting to note* that some of the dimethylallyl diphosphate formed in this way is converted to 2-methyl-1,3-butadiene ("isoprene") and is directly released by plants into the atmosphere. It appears that "those plants that

*Lerdau, M. *Science*, **2007**, *316*, 212.

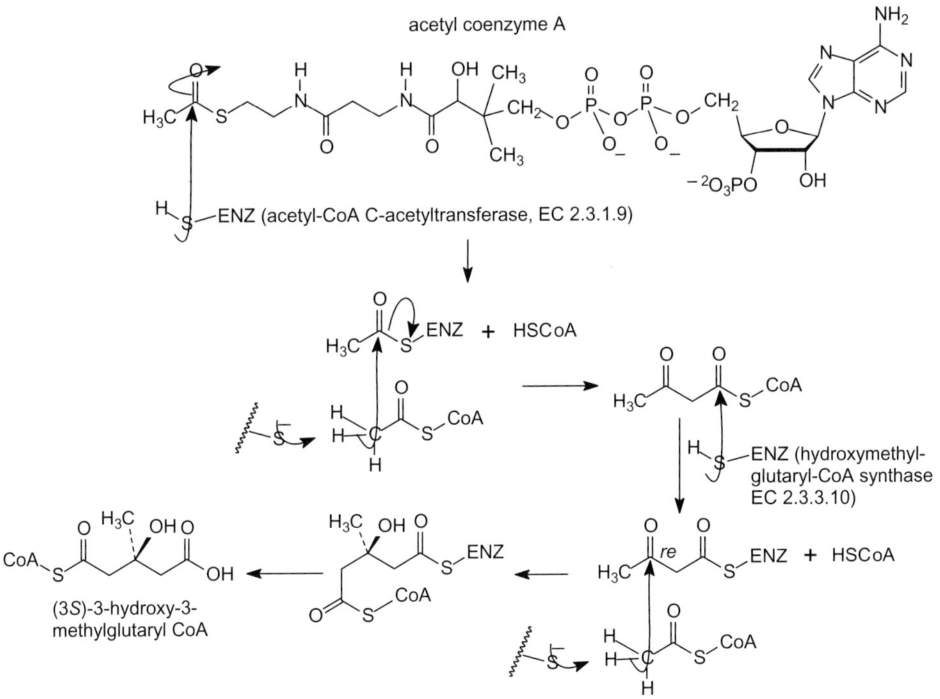

Scheme 11.39. A representation of a pathway for the conversion of acetyl-CoA to (3*S*)-3-hydroxy-3-methylgluraryl CoA on the way to mevalonate. EC numbers and some graphic materials provided in this scheme have been taken with permission from appropriate links in a URL starting with http://www.chem.qmul.ac.uk/iubmb/enzyme/.

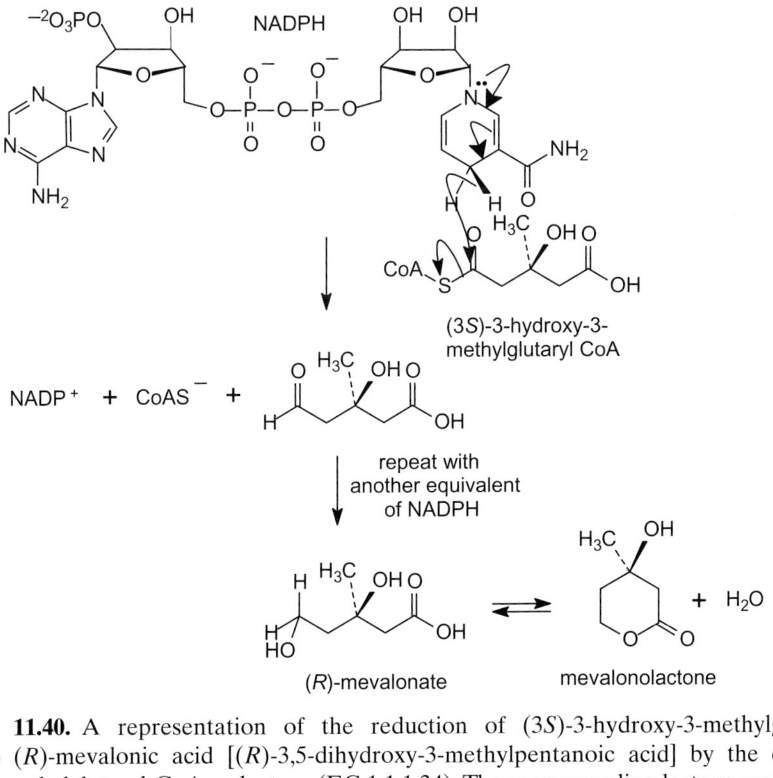

Scheme 11.40. A representation of the reduction of (3*S*)-3-hydroxy-3-methylglutaryl CoA to (*R*)-mevalonic acid [(*R*)-3,5-dihydroxy-3-methylpentanoic acid] by the enzyme hydroxymethylglutaryl-CoA reductase (EC 1.1.1.34). The corresponding lactone mevalonolactone is also shown.

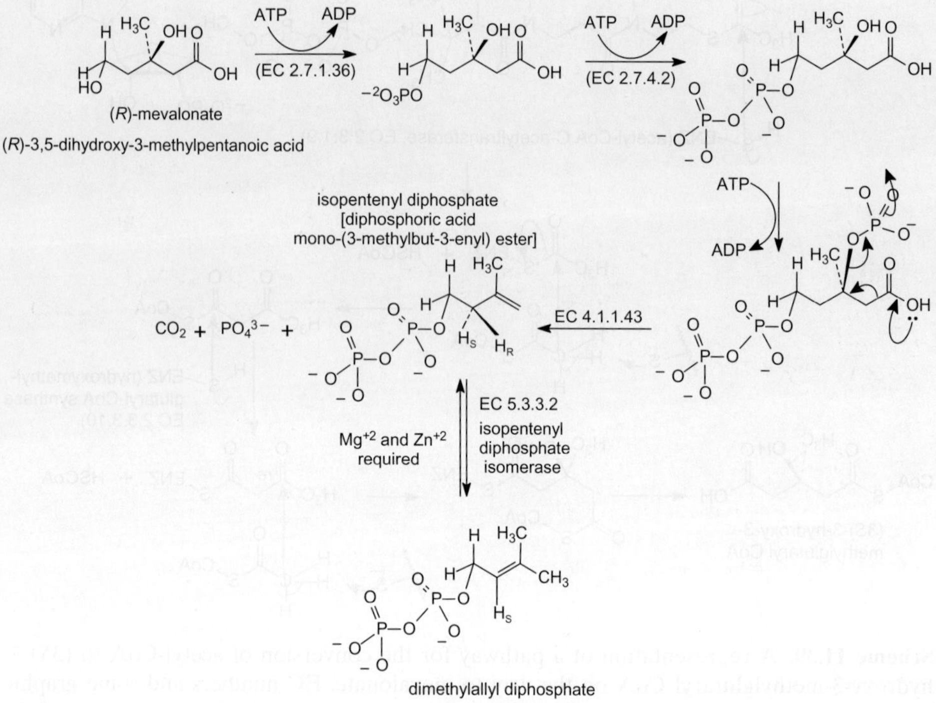

(R)-mevalonate

(R)-3,5-dihydroxy-3-methylpentanoic acid

isopentenyl diphosphate
[diphosphoric acid
mono-(3-methylbut-3-enyl) ester]

CO_2 + PO_4^{3-} +

EC 4.1.1.43

EC 5.3.3.2

Mg^{+2} and Zn^{+2}
required

isopentenyl
diphosphate
isomerase

dimethylallyl diphosphate

Scheme 11.41. The enzyme-catalyzed phosphorylation (ATP → ADP) and decarboxylation (with elimination of inorganic phosphate) of (R)-mevalonate [(R)-3,5-dihydroxy-3-methylpentanoic acid] to isopentenyl diphosphate (diphosphoric acid mono-[3-methylbut-3-enyl] ester) and the isomerization of the latter to dimethylallyl diphosphate (EC 5.3.3.2). It is important to note that the isomerization involves the specific removal of the *pro-R* hydrogen of C_2 of the former. EC numbers and some graphic materials provided in this scheme have been taken with permission from appropriate links in a URL starting with http://www.chem.qmul.ac.uk/iubmb/enzyme/.

emit isoprene … are better protected against oxidative stress" and that this process can alter forest ecosystems.

Beyond this direct effect played by isoprene, the five carbons of dimethylallyl diphosphate and isopentenyl diphosphate serve as realized precursors to the idealized concept of thousands of naturally occurring compounds whose structures are multiples of C_5 and with the particular branching seen there. Although not originally recognized as such, they may be considered as the parent of 2-methyl-1,3-butadiene or "isoprene," and it is from here that the concept of an "isoprene rule" begins.*

c. Terpenes. With the serious study of odoriferous oils derived from plants, for example, camphor (known since biblical times), pine oil, and citrus oil ("essential oils") and their chemical analysis by combustion (Chapter 1) in the nineteenth century, came the clear evidence that many of these otherwise unrelated materials were made up of a number of carbons that occurred in multiples of five. In 1887, O.

*Ruzicka, L. *Experientia*, **1953**, *9*, 357.

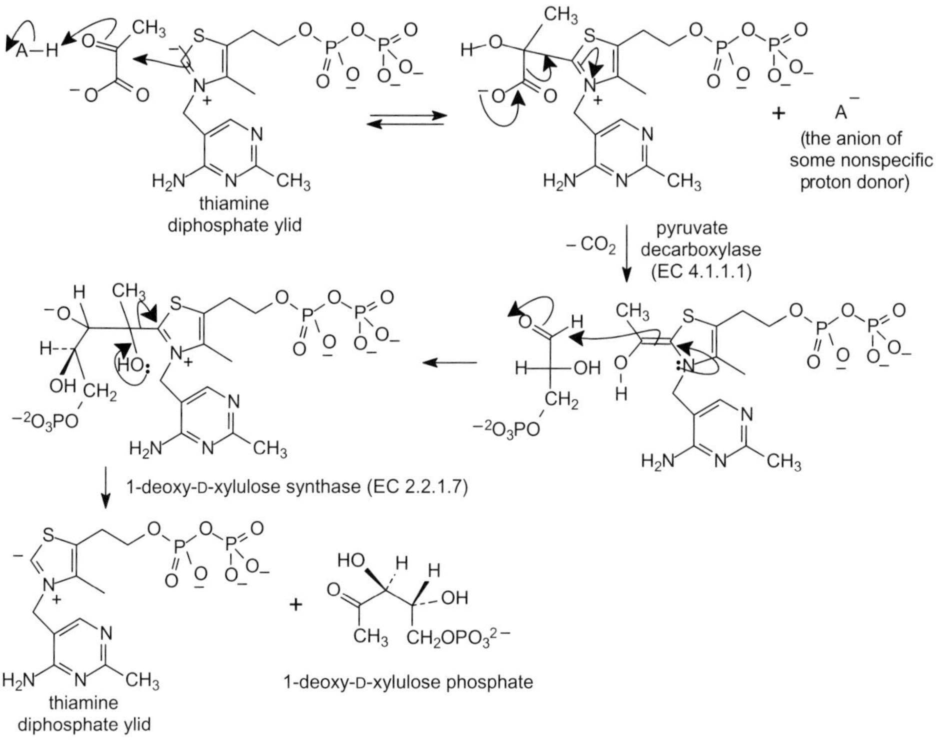

thiamine
diphosphate ylid

pyruvate
decarboxylase
(EC 4.1.1.1)

$-CO_2$

(the anion of
some nonspecific
proton donor)

1-deoxy-D-xylulose synthase (EC 2.2.1.7)

thiamine
diphosphate ylid

+

1-deoxy-D-xylulose phosphate

Scheme 11.42. A representation of the formation of 1-deoxy-D-xylulose phosphate from the reaction of thiamine diphosphate ylid with pyruvic acid and D-glyceraldehyde 3-phosphate in the presence of the synthase EC 2.2.1.7.

Wallach[*] proposed that one "isoprene" (2-methyl-1,3-butadiene) unit of five carbon atoms could serve as this function. This idea was eventually elaborated on by Ruzicka[†] and Bloch[‡] into the "isoprene rule." Eventually, as we will see, the family of terpenes (C_{10}), sesquiterpenes (C_{15}), diterpenes (C_{20}), sesterpenes (C_{25}), triterpenes (C_{30}), and truly large compounds (polymers) found in latex and gutta-percha comprises tens of thousands of individual, unique compounds all of which may be considered as derived from "isoprene" (2-methyl-1,3-butadiene) and that the pair of C5 isomers, isopentenyl diphosphate and dimethylallyl diphosphate, compounds linked by interconversion (Scheme 11.41) under the influence of the isopentenyl isomerase (EC 5.3.3.1) serve as the biochemical parents to the family.[§]

As shown in Scheme 11.44, the attack of isopentenyl diphosphate, with the loss of the C_2 *pro-R* hydrogen on to the primary carbon of dimethylallyl phosphate in

[*]Wallach, O. *Liebigs Ann. Chem.*, **1887**, *238*, 78.
[†]Ruzicka, L. *Experientia*, **1953**, *9*, 357 where the "isoprene rule" is defined.
[‡]Bloch, K. *J. Biol. Chem.*, **1959**, *234*, 2595. "With the characterization of phosphomevalonic acid and the two pyrophosphate esters described in this paper ... the active 'isoprenoid' capable of condensing ... (is described). The transformation of mevalonic acid to the isoprenoid-condensing units involves, apart from the decarboxylation, the elimination of two hydroxyl groups. It is a reasonable assumption that esterification with ... phosphate serves to facilitate eliminations..."
[§]See end of chapter, Problem 11.17.

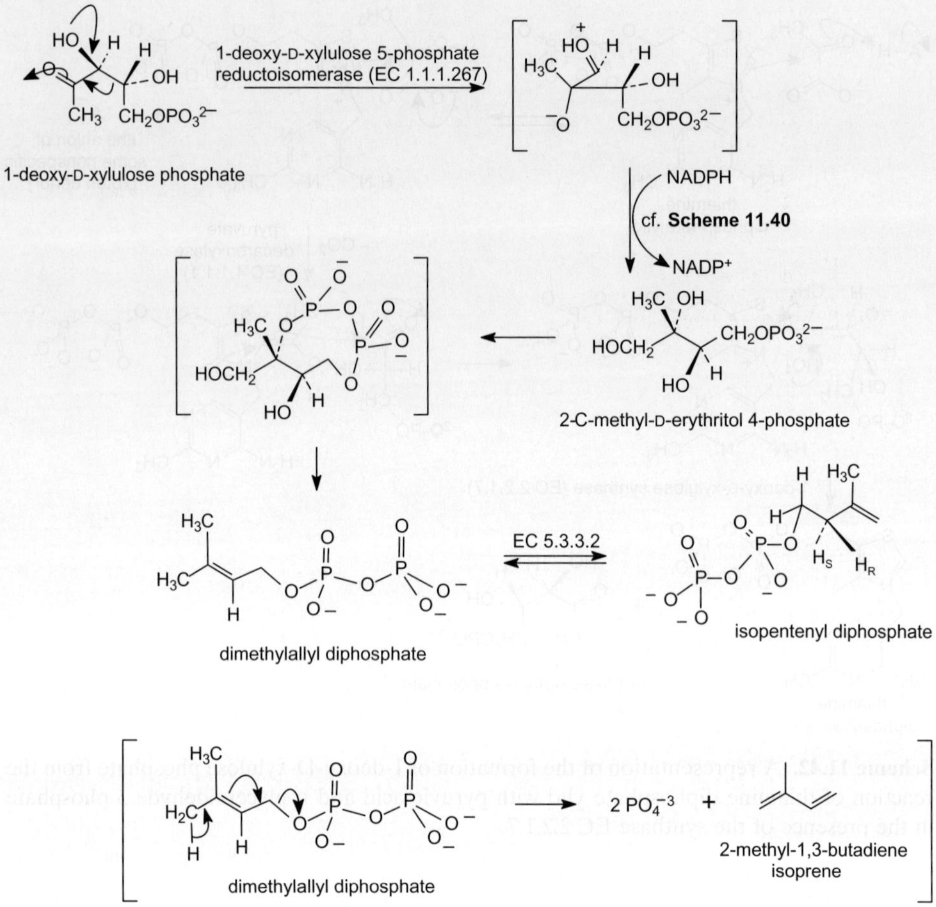

1-deoxy-D-xylulose phosphate

2-C-methyl-D-erythritol 4-phosphate

dimethylallyl diphosphate

isopentenyl diphosphate

dimethylallyl diphosphate

2-methyl-1,3-butadiene
isoprene

Scheme 11.43. A depiction of the conversion of 1-deoxy-D-xylulose phosphate to dimethylallyl diphosphate and thence to isopentenyl diphosphate and to "isoprene," which is in concert with what is known. Details for the steps after 2-C-methyl-D-erythritol-4-phosphate are less clear than those preceding it. EC numbers and some graphic materials provided in this scheme have been taken with permission from appropriate links in a URL starting with http:// www.chem.qmul.ac.uk/iubmb/enzyme/.

an S_N1- or S_N2-like process produces (with dimethylallyl*trans*transferase, EC 2.5.1.1) geranyl diphosphate as well as (with dimethylallyl*cis*transferase, EC 2.5.1.28) neryl diphosphate. It has been argued that both geranyl diphosphate and neryl diphosphate are capable of isomerization to linalyl diphosphate and that both the $(3R)$- and $(3S)$-isomers form as a function of the progenitor.

With the recent ability to isolate and clone various monoterpene synthases (e.g., the family of carbon–oxygen lyases acting on phosphates (EC 4.2.3.1 through EC 4.2.3.27) has come the realization that it may be possible to actually link the structures identified through a small family of highly reactive presumed carbocation intermediates (Figure 11.19). Thus, if linalyl diphosphate undergoes cyclization, as shown in Equation 11.7, the α-terpinyl carbocation generated can be (easily) seen to give rise to the structures of Figure 11.19.

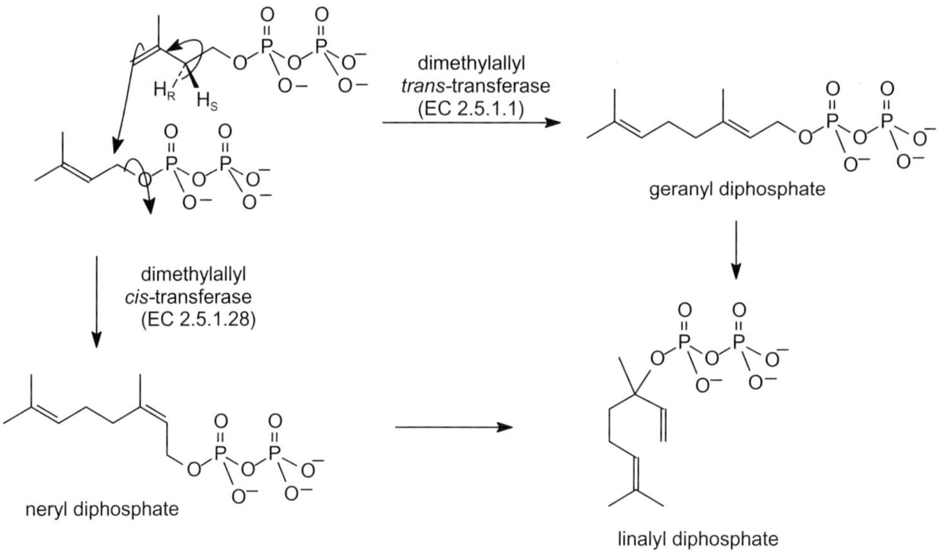

Scheme 11.44. The formation of linalyl diphosphate from neryl diphosphate and geranyl-diphosphate. Both of the latter arise from coupling between dimethylallyl diphosphate and isopentenyl diphosphate. Both (3*R*)- and (3*S*)-isomers form.

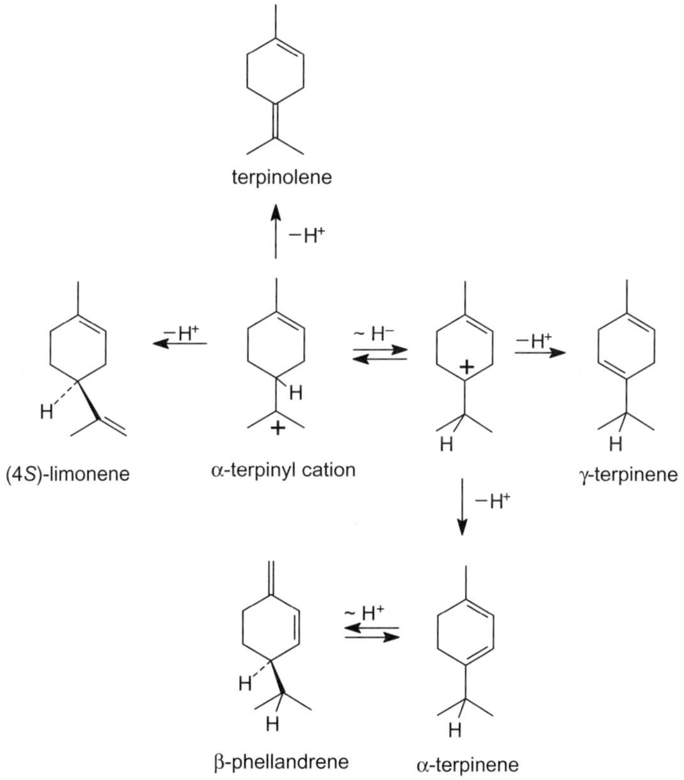

Figure 11.19. Some of the alkenes derived by simple proton loss that are suggested to arise from the common, enzyme-associated α-terpinyl carbocation.

$$\text{(11.7)}$$

linalyl diphosphate   α-terpinyl cation

Furthermore, in addition to those structures resulting from proton loss (and double-bond formation) in Figure 11.18 and those structures that could be argued as resulting from carbocation trapping by water (H_2O) followed by proton loss (Equation 11.8), there is that subset that clearly result from Wagner–Meerwein-type rearrangements (Scheme 11.45) and then proton loss or trapping as well as those that result from further oxidative processes and for which the enzymes (noted in the figure) have been isolated (Scheme 11.46).

$$\text{(11.8)}$$

α-terpinyl cation   α-terpineol

Finally, along the lines of the simple monoterpenes, once the allylic oxidations have occurred (or in the case of the carbocations, water has added to produce alcohols), additional oxidation is possible and the products are commonly found in nature. Some of these compounds are presented in Figure 11.20.

As shown in Scheme 11.47, isopentenyl diphosphate can (again in principle) attack geranyl diphosphate (or, indeed, in principle neryl diphosphate or linalyl diphosphate) and, in this way, account for the formation of farnesyl diphosphate. Indeed, under the influence of geranyl*trans*transferase (EC 2.5.1.10), such a process occurs!

Farnesyl diphosphate serves as the precursor of those thousands of C_{15} derivatives known as sesquiterpenes (L. sesqui = one and one-half), and they are often considered to be categorized under the general rubrics of (1) acyclic, (2) monocyclic, (3) bicyclic, and (4) tricyclic as a function of the number of rings (not including those containing only three members) present. While many works have been written on the general subject of sesquiterpenes detailing their isolation, properties, and syntheses, most of that will not be considered here.

As shown in Figure 11.21, the acyclic sesquiterpenes include the parent alcohol, farnesol, the isomer nerolidol, and many oxidation products such as juvenile hormone III.*

*The family of hormones known as "juvenile hormones" play a critical role in regulating the development of a variety of insects in and beyond the larval stage. As long as the juvenile hormones are present, development cannot proceed properly, and under normal circumstances, their gradual diminution allows molting. Synthetic hormones, for example, methoprene, can be used to prevent maturation in insects.

Methoprene

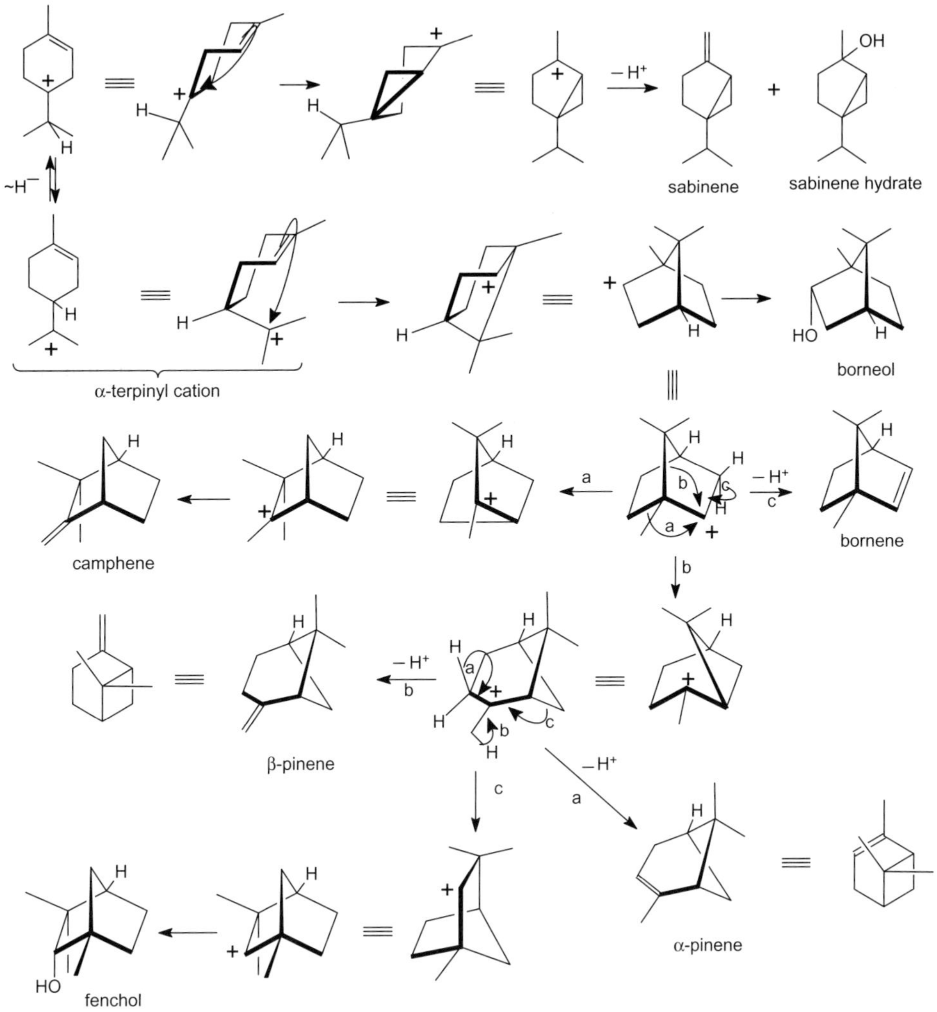

Scheme 11.45. Some of the monoterpene rearrangement products for which some specific synthases have been found [e.g., (–)-*endo*-fenchol synthase (EC 4.2.3.10), sabinene hydrate synthase (EC 4.2.3.11), pinene synthase (EC 4.2.3.14), and (*R*)-limonene synthase (EC 4.2.3.20), Figure 11.19]. The carbocation rearrangements shown here are largely speculative (some evidence does exist for these rearrangements outside of living systems) with regard to the enzymes themselves.

Problem 11.4. Examine the structure of (*S*)-(+)-solanone in Figure 11.21. It is clear that it cannot follow the pattern dictated by the "isoprene rule," which has sets of five carbons added in a linear fashion from isopentenyl diphosphate to geranyl diphosphate to farnesyl diphosphate as shown earlier. Of course, (*S*)-(+)-solanone also has "only" 13 carbon atoms. How might it have, nevertheless, been formed?

There are thousands of monocyclic sesquiterpenes. Figure 11.22 contains a small sample.The particular compounds chosen in Figure 11.22 have been chosen because they are typical of the many variations on the theme of cyclization, rearrangement, and oxidation. These compounds are very widespread in the environment.

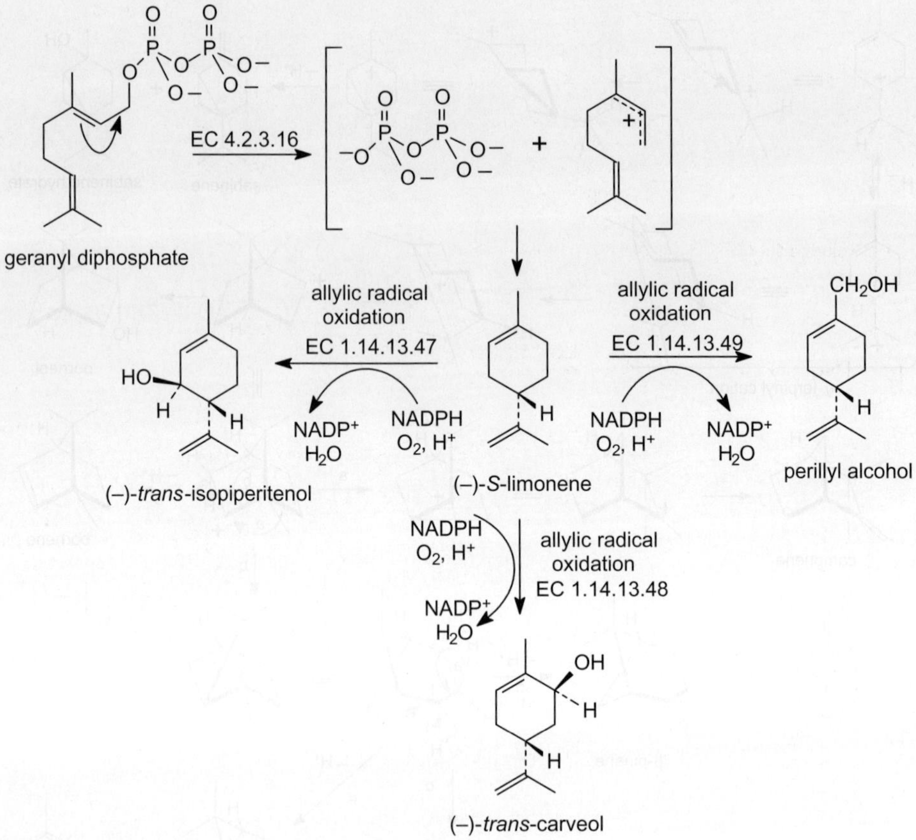

Scheme 11.46. The results of allylic oxidation around the endocyclic double bond in (−)-*S*-limonene by the family of oxidases, (*S*)-limonene 3-monooxygenase (EC 1.14.13.47), (*S*)-limonene 6-monooxygenase (EC 1.14.13.48), and (*S*)-limonene 7-monooxygenase (EC 1.14.13.49). EC numbers and some graphic materials provided in this scheme have been taken with permission from appropriate links in a URL starting with http://www.chem.qmul.ac.uk/iubmb/enzyme/.

Nuciferol (Sakai et al., 1965) is a constituent of *Torreya nucifera*, the Japanese nutmeg-yew, and bisabolangelone is a constituent of the small shrub *Angelica silvestris* and a few other members of the **Umbelliferae**. Bisabolangelone has been reported to be an antifeedant, that is, a material that discourages foraging of leaves on (or in) which it is found by herbivores.

α-Elemene (Paknikar and Bhattacharyya, 1962) is a novel anticancer drug isolated from the Chinese medicinal herb *Rhizoma zedoariae* (the dried rhizome of the flowering plant, *Curcuma phaeocaulis*, a member of the **Zingiberaceae**, from which the spice "turmeric" is also found).

The leaves of the Amazonian tree *Siparuma guianensis* contain cruzerene, which (in an example of an "ethnoveterinary remedy") is reported to be used by certain hunters indigenous to its environs to treat snakebites and scorpion stings received by their hunting dogs.

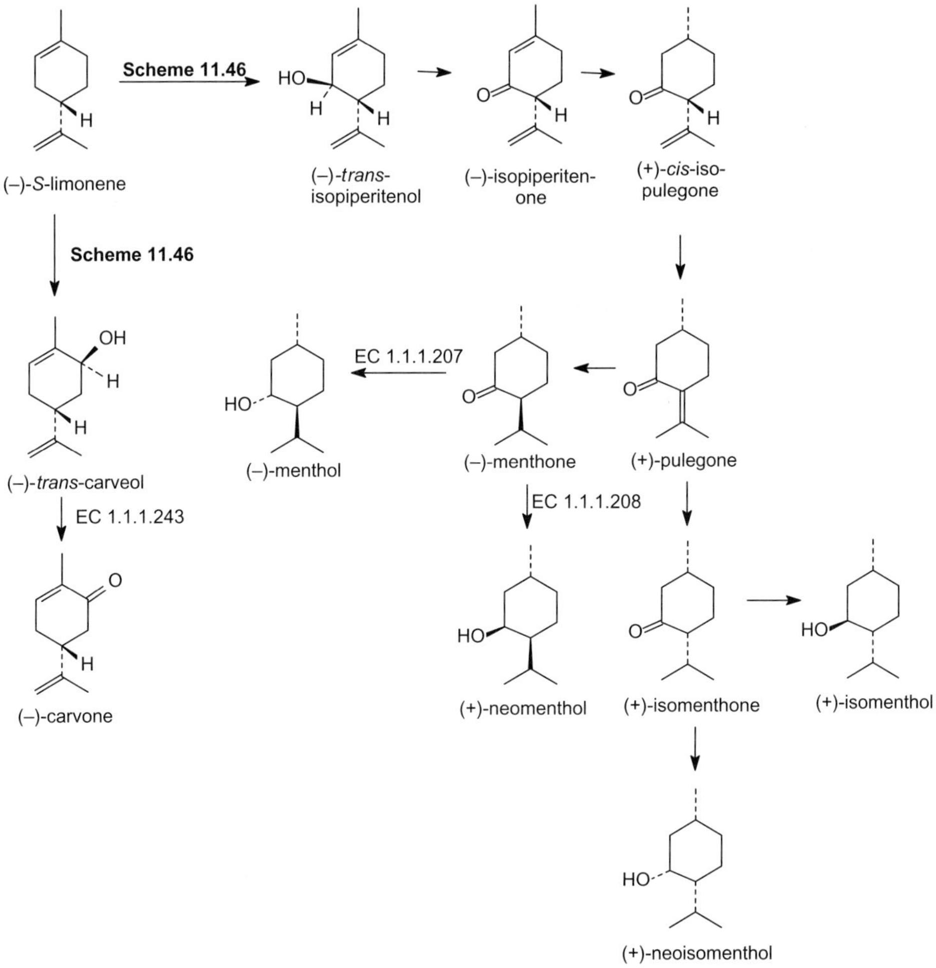

Figure 11.20. Further oxidation products derived from simple terpenes following cyclization and hydration (of carbocations and/or double bonds). In a few cases, the enzymes have been identified (as noted), and NAD$^+$ or NADP$^+$ are generally involved as the oxidizing agents.

Artemisinin,* isolated from the shrub *Artemisia annua*, has long been used for a variety of purposes in traditional Chinese medicine. Now, it is considered as an archetypical member of a class of dioxygen-containing compounds that are actively being studied as drugs to treat multidrug-resistant strains of *Falciparum* malaria.

Aristolactone (Martin-Smith et al., 1964) is an "essential oil" obtained along with limonene (*vide supra*) from the aerial parts of the woody vine *Aristolochia gibertii* (Aristolochiaceae), a plant with large calabash-shaped spotted flowers with an unpleasant odor.

Deoxytrisporone is a metabolite of the plant fungus *Choanephora trispora*.

*See end of chapter, Problem 11.17.

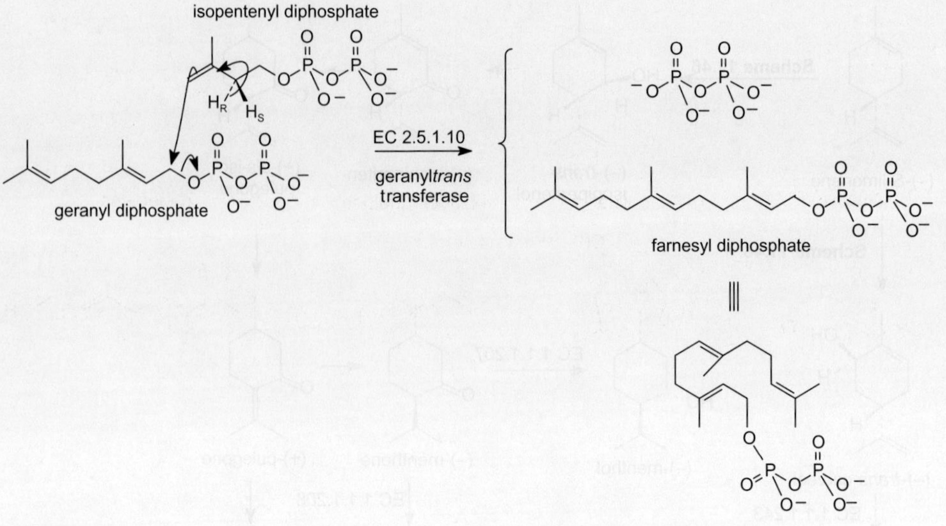

Scheme 11.47. A representation of the formation of farnesyl diphosphate from geranyl diphosphate and isopentenyl diphosphate as catalyzed by geranyl*trans*transferase (EC 2.5.1.10).

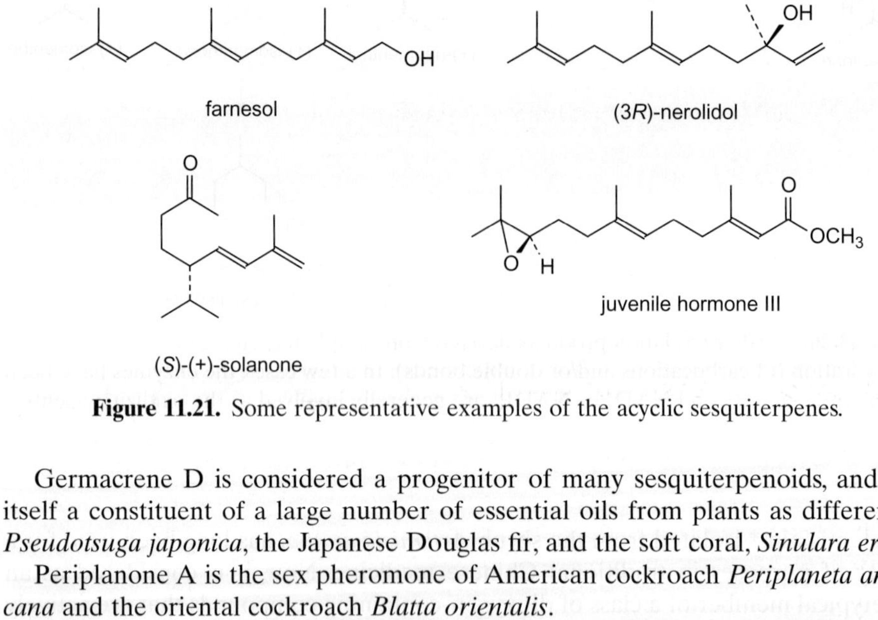

farnesol

(3*R*)-nerolidol

(*S*)-(+)-solanone

juvenile hormone III

Figure 11.21. Some representative examples of the acyclic sesquiterpenes.

Germacrene D is considered a progenitor of many sesquiterpenoids, and it is itself a constituent of a large number of essential oils from plants as different as *Pseudotsuga japonica*, the Japanese Douglas fir, and the soft coral, *Sinulara erecta*.

Periplanone A is the sex pheromone of American cockroach *Periplaneta americana* and the oriental cockroach *Blatta orientalis*.

Humulene (better referred to as α-humulene) is also a constitutent of many essential oils and is found in hops (*Humulus lupulus*) and cloves (*Syzygium aromaticum*).

Zerumbone (Dev, 1960) is reported to be an antifungal, anti-inflammatory, anti-tumor member of the sesquiterpene family found in the rhizomes of wild ginger (*Zingiber zerumbet*) and the lilac, *Syringa pinnatifolia*.

It is interesting to examine (Scheme 11.48) some of the thinking that relates these compounds to the parent farnesyl diphosphate and to understand that many years

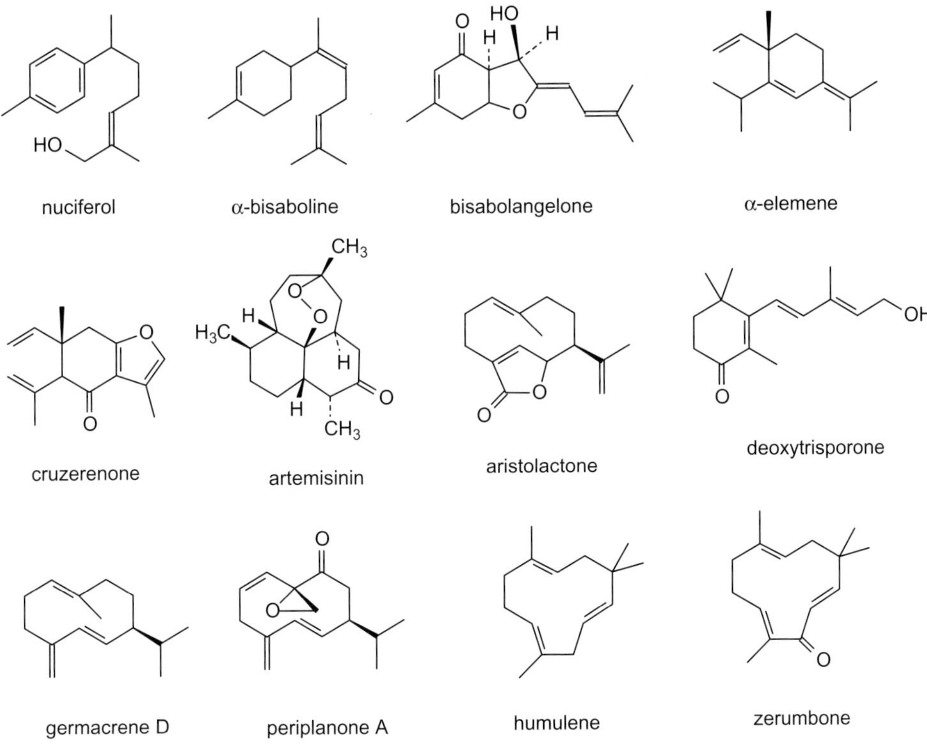

Figure 11.22. A sample of naturally occurring monocyclic sesquiterpenes.

Scheme 11.48. A representation of potential pathways to monocyclic sesquiterpenes α-elemene and α-bisaboline as examples for other sesquiterpenes in Figure 11.22. Cyclizations and proton migrations are readily accommodated to produce specific products.

of experimental "feeding" of precursors (e.g., by adding suitable material to the water given to the plant) labeled with ^{14}C and ^{3}H followed by extensive degradative procedures to show the location of label in the product were necessary to convince workers of the reasonable nature of their ideas.

Problem 11.5. Given the patterns encountered in Schemes 11.45 and 11.48, suggest curved arrow representations of reasonable pathways from farnesyl diphosphate to (a) germacrene D and (b) humulene.

Some of the chemistry surrounding the synthesis of these compounds is very interesting. For example, as shown in Scheme 11.49, one of the syntheses* of periplanone A begins with dihydrolimonene. Analysis of the synthesis, using reactions studied previously, is found in Problem 11.6.

Problem 11.6. Examine each step in Scheme 11.49 and decide exactly what processes are occurring. Use curved arrows to approximate movement of electrons to account for the way changes are occurring.

It should be clear that bicyclic sesquiterpenes can also be formed. Figure 11.23 provides a few examples of this large number of compounds.

β-Eudesmol, isolated from "So-jutsu" (*Atractylodis lanceae* rhizomas), a perennial growing to the height of about 1 m and widely used in traditional Chinese medicine, is known to have various unique effects on the nervous system.

α-Santonin occurs widely in plants, especially *Artemisia* spp. (from which the liquor absinthe is prepared) and was widely used in the past as an anthelmintic. But it has fallen out of use with the development of safer drugs. Interestingly, some deep-seated rearrangements occur in this system. Thus, as shown in Equation 11.9 (and set as Problem 11.7), treatment of santonin with alkali yields santonic acid and santonic acid is converted into santonide by heating in acetic acid (followed by pyrolysis).

$$(11.9)$$

*Both because of its interest as a synthetic target and because there is a belief that humans and cockroaches cannot coexist peacefully and because sex is considered to be such a fundamental driving force, sex lures have been sought to control cockroaches. The synthesis described here is from Hauptmann, H.; Mühlbauer, G.; Sass, H. *Tetrahedron Lett.*, **1986**, 6189.

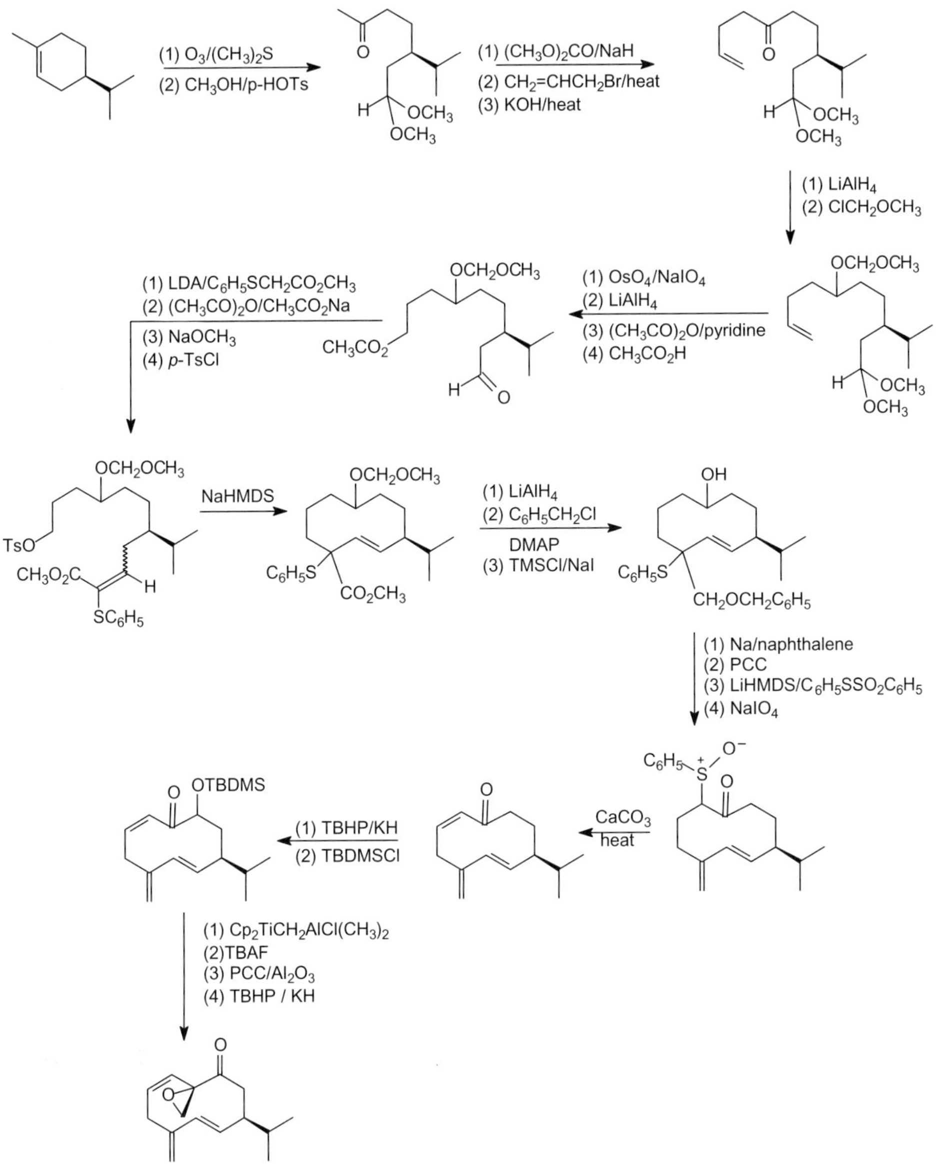

Scheme 11.49. A synthesis of periplanone A (after Hauptmann, H.; Mühlbauer, G.; Sass, H. *Tetrahedron Lett.*, **1986**, 6189). LDA = lithium diisopropyl amide; *p*-TsCl = *para*-toluenesulfonyl chloride; NaHMDS = sodium hexamethyldisilazide; DMAP = 4-dimethyl-aminopyridine; TMSCl = trimethylchlorosilane; PCC = pyridinium chlorochromate; TBHP = *t*-butyl hydrogen peroxide; TBDMSCl = *t*-butyldimethylchlorosilane; TBAF = tetrabutylammonium fluoride; Cp = cyclopentenyl.

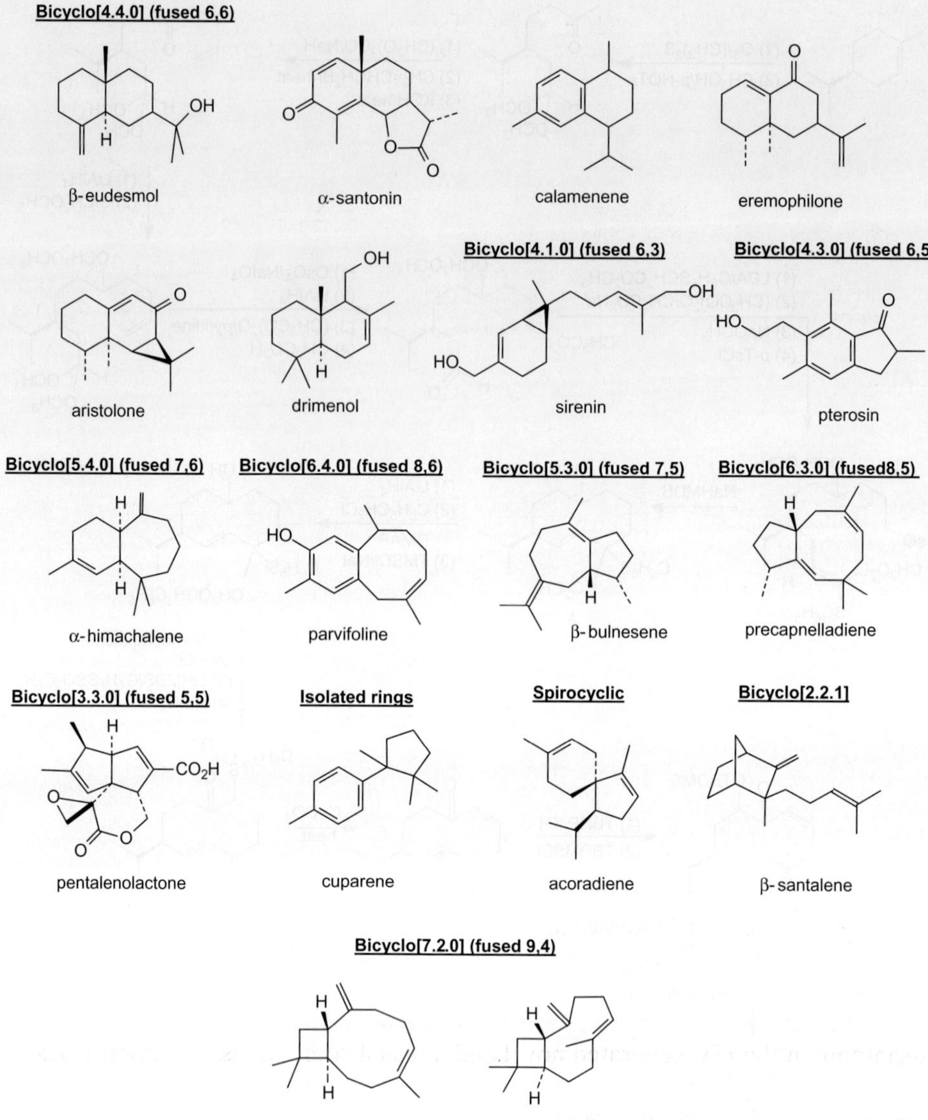

Figure 11.23. A few typical examples of the thousands of bicyclic sesquiterpenes found in nature.

Problem 11.7. Using the "curved arrow" formalism, write a suitable pathway from α-santonin to santonic acid and then to santonide (Woodward, 1950). **Answered.** (Note: the details of the mechanism are suggestions!)

The pattern of substitution of both β-eudesmol and α-santonin can be accounted for by assuming that the cyclization of farnesyl diphosphate, followed by (or preceded by) suitable oxidation changes at allylic positions, is as shown in Scheme 11.50.

The results of leaving α-santonin in sunlight and eventually, actually irradiating α-santonin in the UV, generated novel and unusual compounds. Structural elucidation of these compounds, understanding their interrelationships, and determining their mode of formation consumed the interests and efforts of some of the leading chemists of the twentieth century.

Some of that material is shown in Scheme 11.51.

Calamenene (Simpson and McQuilkin, 1976) and related oxygen-containing sesquiterpenes are widely found in nature. It is included here as a variant on the cyclization pathway that gave rise to α-santonin and β-eudesmol.

Problem 11.8. Draw farnesyl diphosphate in a manner that suggests how calamenene might form. Hint: examine the pathways in Scheme 11.50.

Although eremophilone was isolated from an Australian shrub (*Eremophila mitchelli*) in the 1930s, it took nearly a decade before its structure was elucidated. The problem, in part, was that it does not obey the isoprene rule (a methyl group

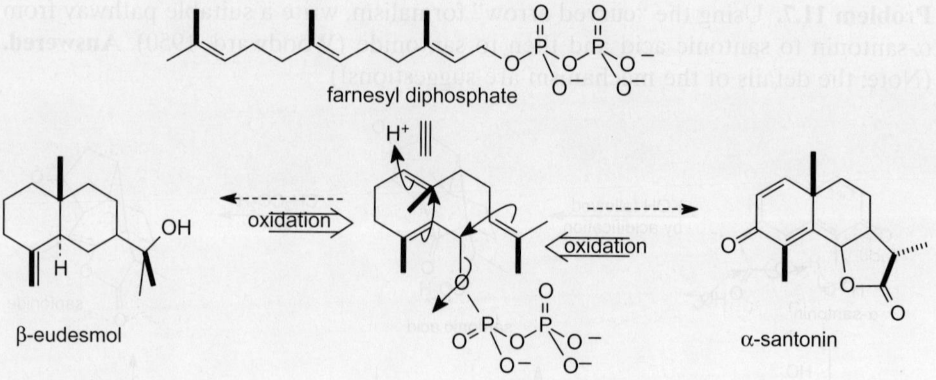

farnesyl diphosphate

β-eudesmol

oxidation

H⁺

oxidation

α-santonin

Scheme 11.50. A cartoon representation of the organization of farnesyl diphosphate that, on oxidation and cyclization (or cyclization and oxidation), produces both β-eudesmol and α-santonin.

is in the wrong place!), and it was the first sesquiterpene found that violated that rule. As shown in Scheme 11.52, one rationalization for the outcome is a methide shift, which, as will be seen throughout the sesquiterpenes, is a fairly common occurrence. Interestingly, eremophilone oil (from the shrub) is approved as an active ingredient in a commercial product designed for termite control. It is claimed to be nontoxic to humans.

As shown in Scheme 11.53, a minor change in the direction of oxidation can account for the formation of aristolone, isolated from both the root of the Chinese fairy vine and the rhizome of Canadian ginger and widely used in traditional medicines. It is of some interest that a few sesquiterpene (and diterpene, *vide infra*) alcohols are found in these vines along with one of the few nitro group containing natural products, aristolochic acid.

Drimenol (Appel et al., 1959) is another member of a large group of sesquiterpenes (the drimanes) and has been isolated from, among other sources, the bark of the Chilean canelo tree. It has been suggested that its formation can be accounted for by a path such as that shown in Scheme 11.54.

Sirenin (Machlis et al., 1966) is the chemotactic (i.e., causing movement along a chemical concentration gradient) sexual hormone isolated from a water mold (*Allomyces*) and which is produced by female gametes (ovum) and that serves to attract male gametes (sperm cells). A representation for the possible formation of sirenin is provided in Scheme 11.55.

A brief synthesis of sirenin (Scheme 11.56) has been produced and is of some interest because of the stereochemical outcome (not discussed here but found in Gant et al., 1995) and because of the path itself, which closely resembles what might be the biosynthetic path as shown in Scheme 11.55.

The substituted bicyclo[4.3.0] system found in pterosin is of particular interest since it is clear that there are only 14 carbons (!) in the economically important class of sesquiterpenes of which this is a member. There are a number of pterosins isolated from the brachen fern *Pteridum aquilinium* and similar plants that result in poisoning when ingested by browsing cattle.

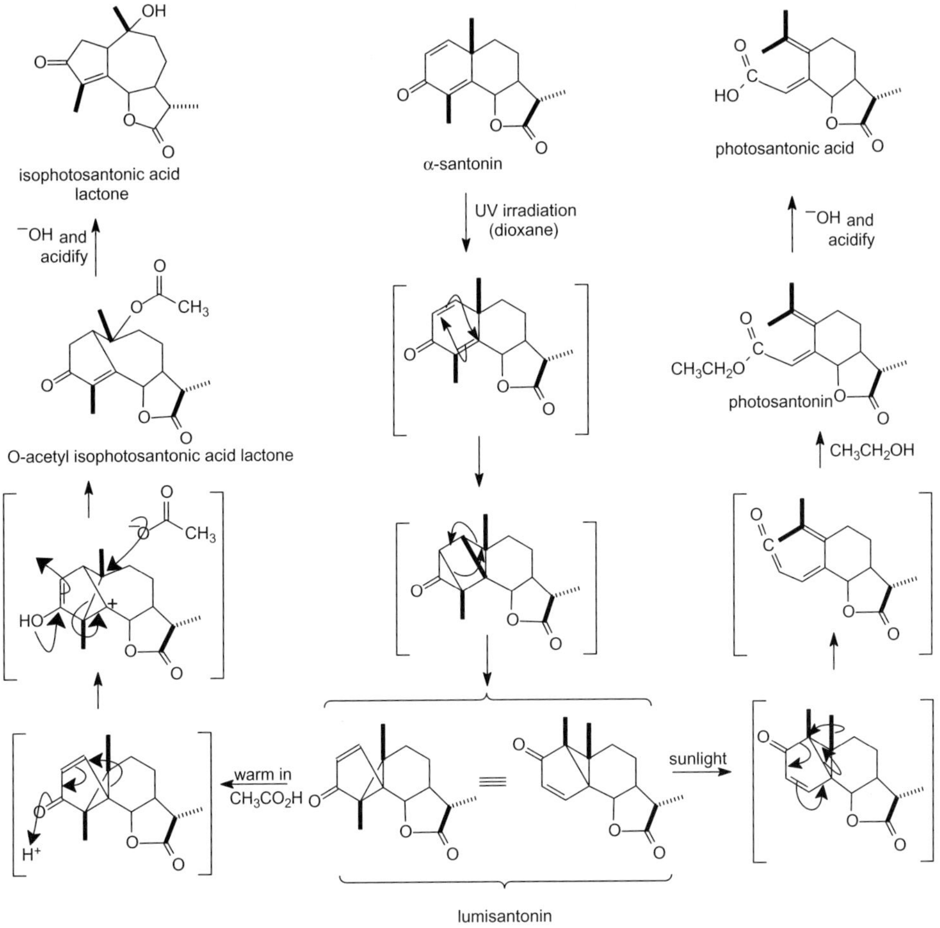

Scheme 11.51. Some of the interrelationships between santonin and its photochemically derived products. For the pieces of the puzzle of the various compounds formed and how they were assembled as structural work was performed, see, for example, (a) Woodward, R. B.; Kovach, E. G. *J. Am. Chem. Soc.*, **1950**, *72*, 1009; (b) Arigoni, D.; Bosshard, H.; Bruderer, H.; Buchi, G.; Jeger, O.; Krebaum, L. J. *Helv. Chim. Acta*, **1957**, *40*, 1732; and (c) Barton, D. H. R. *Helv. Chim. Acta*, **1959**, *4*, 2604 and the references in these works. Interestingly, additional products continue to be found; see Natarajan, A.; Tsai, C. K.; Khan, S. I.; McCarren, P.; Houk, K. N.; Garcia-Garibay, M. A. *J. Am. Chem. Soc.*, **2007**, *129*, 9846. The latter have reported mazdasantonin and dimers resulting from a lattice-controlled furan reaction.

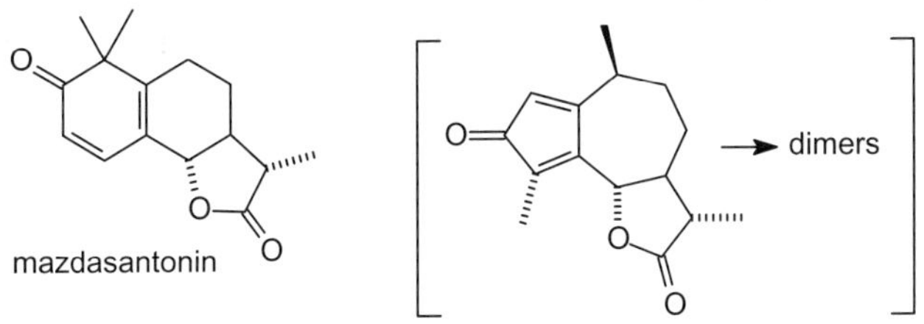

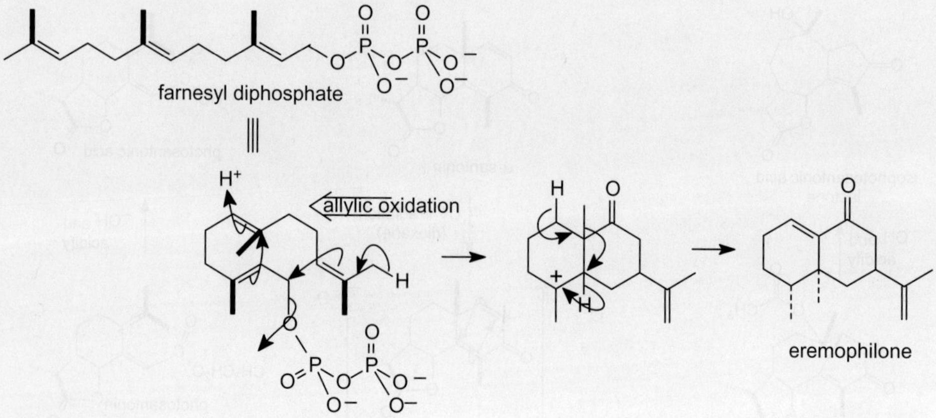

Scheme 11.52. A cartoon representation of a potential pathway to eremophilone from farnesyl diphosphate. The pattern of methylation anticipated has changed due to (presumably) a series of hydride and methide shifts. See Zalkow, L. H.; Markley, F. X. M.; Djerasi, C. *J. Am. Chem. Soc.*, **1980**, *82*, 6354.

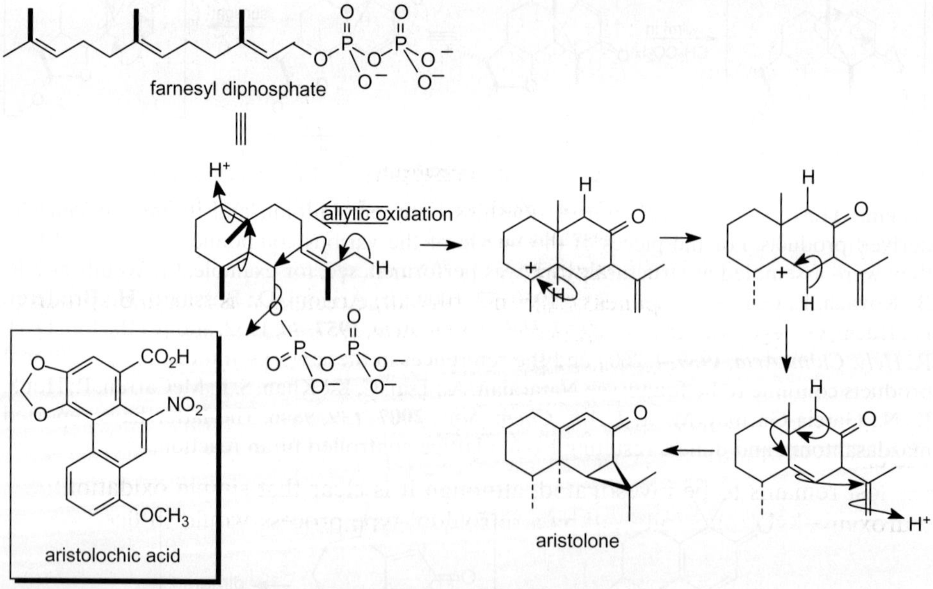

Scheme 11.53. A cartoon representation of a potential pathway to aristolone from farnesyl diphosphate. The pattern of methylation anticipated has changed due to (presumably) a series of hydride and methide shifts. The structure of aristolochic acid, one of the few nitro-containing natural products known is also shown since some sesquiterpene alcohols are found as esters.

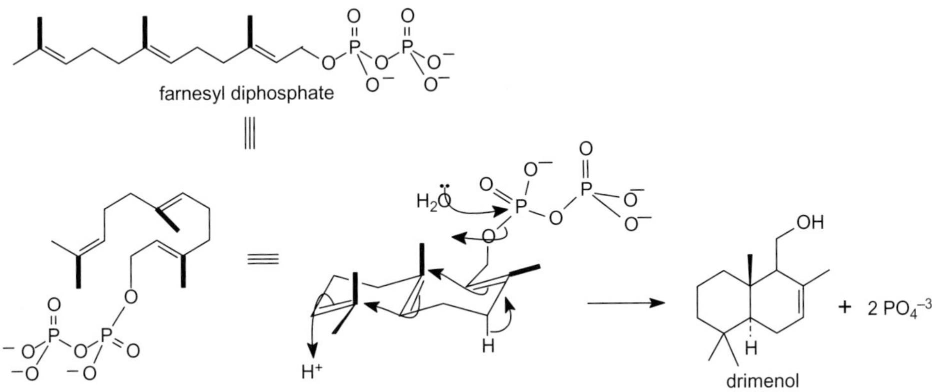

Scheme 11.54. A representation of a pathway to the sesquiterpene drimenol involving cyclization and hydrolysis of farnesyl diphosphate.

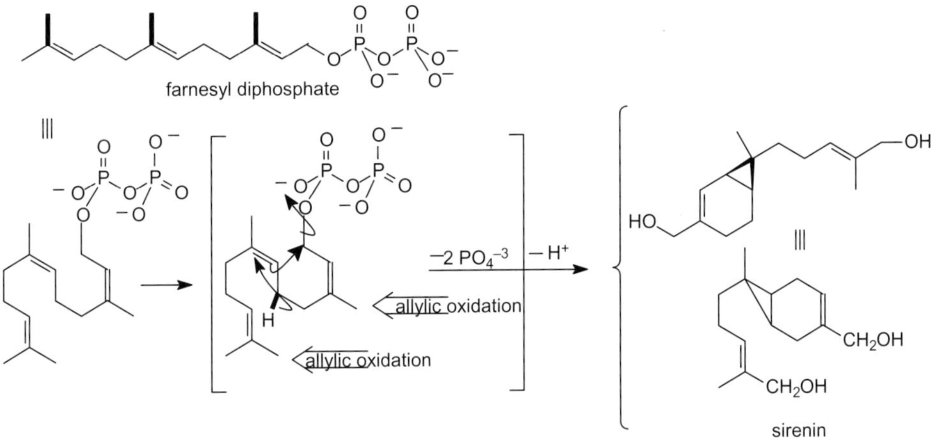

Scheme 11.55. A representation of a pathway accounting for the formation of serenin from farnesyl diphosphate.

It can be argued that the pathway (Scheme 11.57) to pterosin involves initial formation of a four-membered ring, which is then opened to form the hydroxyethyl side chain. The timing of the loss of the methyl group and the exact means by which it is lost remains to be investigated, although it is clear that simple oxidation to a hydroxymethyl group, followed by a retroaldol-type process, would suffice.

The sesquiterpene α-himachalene* has been isolated from the New World sandfly *Lutzomyia longipalpis*, a carrier of a protozoan that is transmitted in the bite of the fly. The protozoan infection results in the disease leishmaniasis, one form of which can cause enlargement of the spleen and, if untreated, death. Scheme 11.58 presents a cartoon of a biosynthetic rationalization of its formation.

*See end of chapter, Problem 11.17.

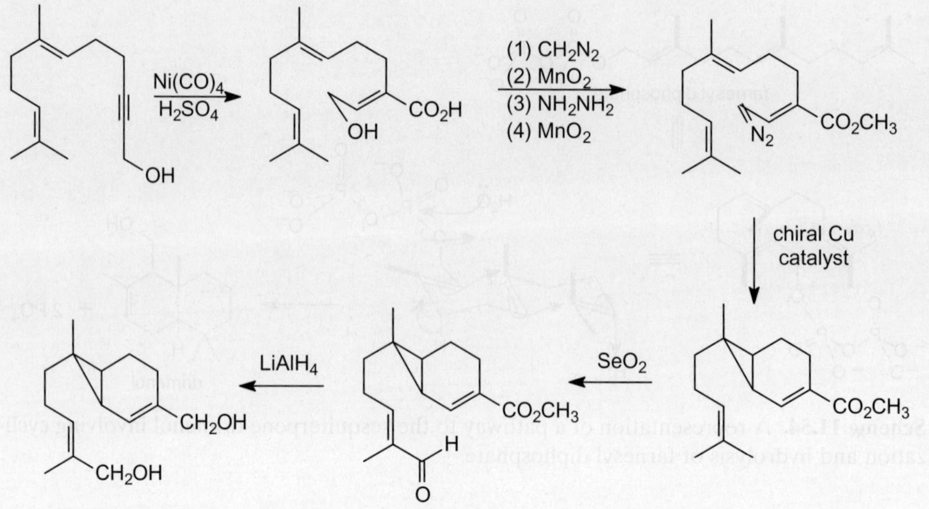

Scheme 11.56. A synthesis of sirenin as set forth in Gant, T. G.; Noe, M. C.; Corey, E. J. *Tetrahedron Lett.*, **1995**, 8745. The synthesis, using a chiral catalyst, produces chiral material.

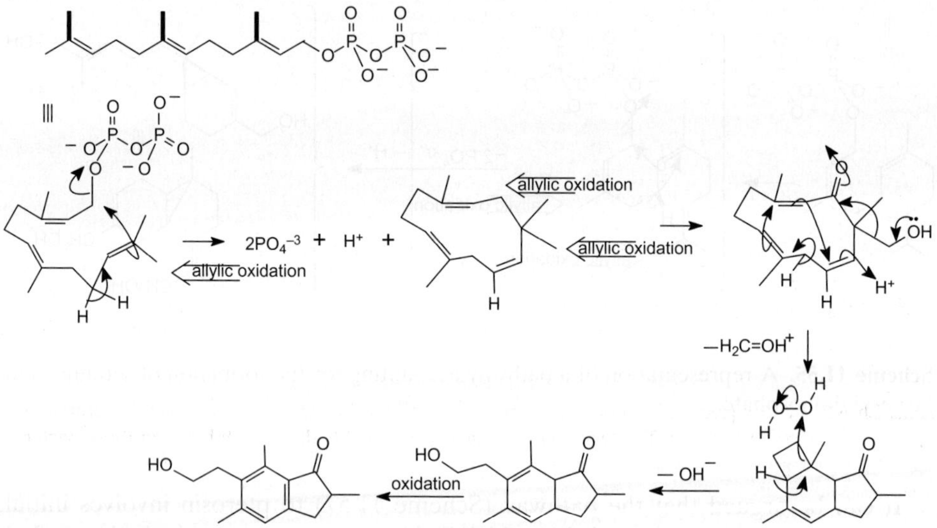

pterosin

Scheme 11.57. A representation of the formation of pterosin that suggests that a four-membered ring forms, and it is subsequently opened to form the hydroxyethyl side chain. The timing of the loss of the methyl group is uncertain, and the retroaldol is without proof. The nature of the oxidizing agents is not known, and the use of hydrogen peroxide as an oxygen donor is unlikely.

Problem 11.9. In Figure 11.23, structures are given for the following: (a) parvifoline, reported to be cytotoxic and used in Chinese folk medicine for reducing swelling, is isolated from the plant *isodon parvifolius*; (b) β-bulnesene, reported to inhibit platelet aggregation and serve as an anti-inflammation agent, is obtained from *pog-*

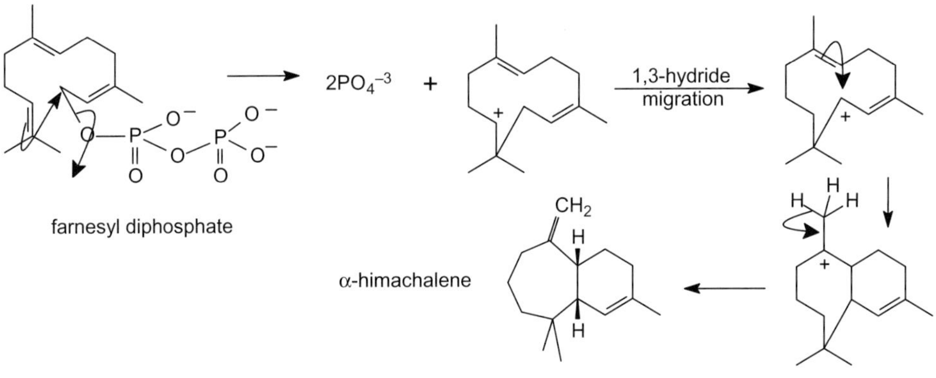

farnesyl diphosphate

$2PO_4^{-3}$ +

1,3-hydride migration

CH_2

α-himachalene

Scheme 11.58. A pathway to account for the formation of α-himachalene from farnesyl diphosphate.

ostemom cablin (patchouli), a flowering aromatic herb; and (c) precapnelladiene, from the coral *Capnella imbricate*.

(i) Suggest a suitable biosynthetic path for the production of these sesquiterpenes beginning with farnesyl diphosphate.

(ii) Using the open literature, find a synthesis of each of these and show how the sesquiterpene was prepared.

More than 30 species of the genus *Streptomyces* produce the sesquiterpenoid antibiotic pentalenolactone. This antibiotic is active against both gram-positive* and gram-negative bacteria, as well as other biota and apparently functions by inactivating glyceraldehyde 3-phosphate dehydrogenase. A variety of experiments indicate that pentalenolactone is generated from the sesquiterpene pentalenene, and Scheme 11.59 reflects this observation.[†]

*Gram-positive and gram-negative refer to the empirical method (the work of the Danish bacteriologist, Hans Christian Gram [1853–1938]) of staining bacteria with the dye crystal violet to help determine the pathogenic nature of bacteria. The technique is designed to quickly distinguish between those bacteria that have outer polysaccharide membranes and those that have peptidoglycan membranes.
[†]An excellent discussion of the early and current work on pentalenolactone can be found in You, Z.; Omura, S.; Ikeda, H.; Cane, D. E. *Arch. Biochem. Biophys.*, **2007**, *459*, 233.

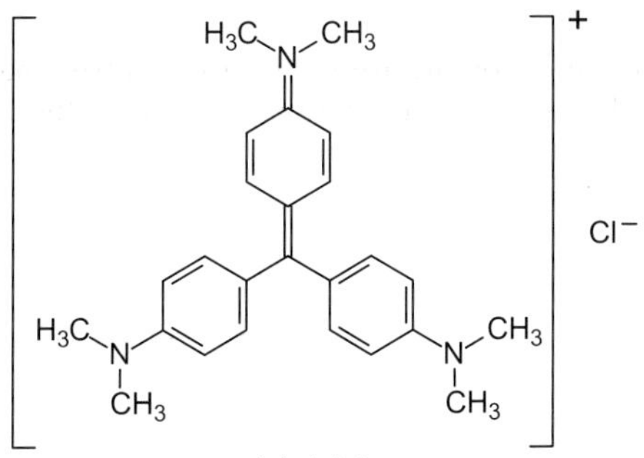

crystal violet

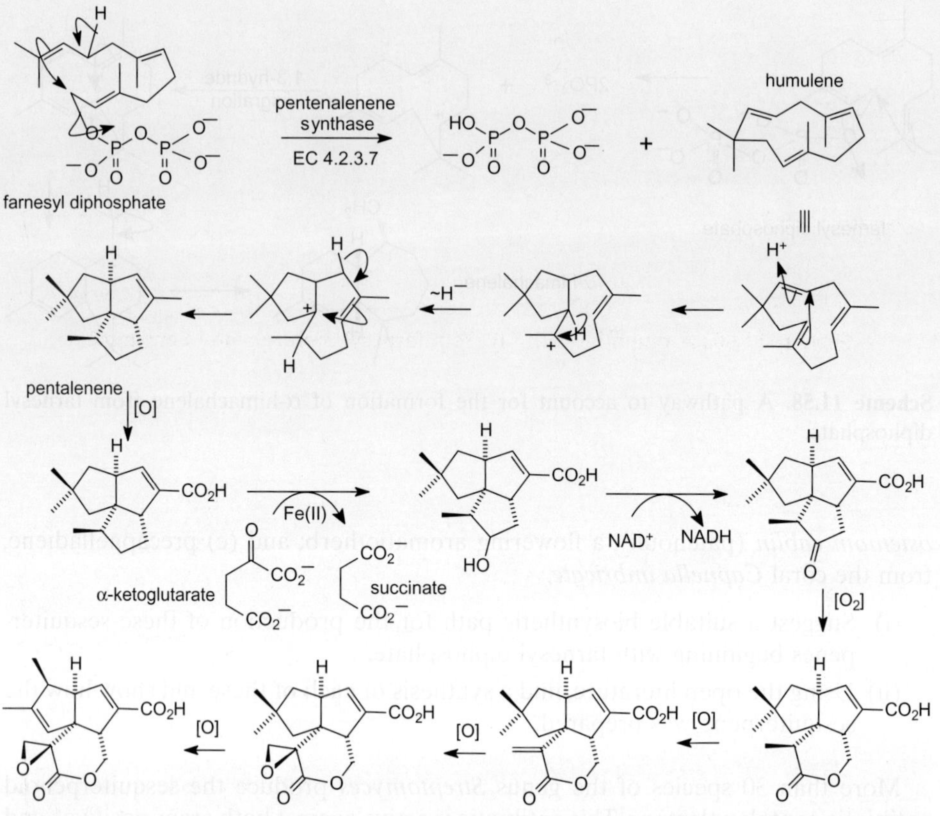

Scheme 11.59. An abbreviated biosynthetic pathway to pentalenene and pentalenolactone as suggested by Lesburg, C. A.; Zhai, G.; Cane, D. E.; Christianson, D. W. *Science*, **1997**, *277*, 1820 and You, Z.; Omura, S.; Ikeda, H.; Cane, D. E. *Arch. Biochem. Biophys.*, **2007**, *459*, 233.

Cuparene (Enzell and Erdtman, 1958) has been found in fossil resins and today is among the hydrocarbons (terpene, sesquiterpene, diterpenes, etc.) present in cedarwood oil (from, e.g., *Juniperes virginiana* L.). Interestingly, halogenated relatives are also found in red algae (*Laurencia* spp.). A few biosynthetic studies have been undertaken, and it is clear that a pattern of cyclization processes, such as that shown in Scheme 11.60, obtains.

Acoradiene and its isomers are widespread, the isomer shown is isolated from vetiver oil but one of its double-bond isomers is reported to be the aggregation pheromone of the Broad-horned flour beetle (*Gnatocerus cornutus*). A pathway to acoradiene is shown in Scheme 11.61.

Yellow wormwood (*Artemisia xantaphora*) is reported to contain, among other sesqui- and diterpenes, β-santalene (Corey et al., 1962). This bicyclic compound is presumably derived from farnesyl diphosphate by a somewhat more extensive series of rearrangements including those typical of other norbornane-type skeletons. Presumably, a process such as that shown in Scheme 11.62 is involved.

The sesquiterpenes caryophyllene (*E*) and isocaryophyllene (*Z*) (Corey et al., 1964) frequently occur together in essential oils derived from cloves and black

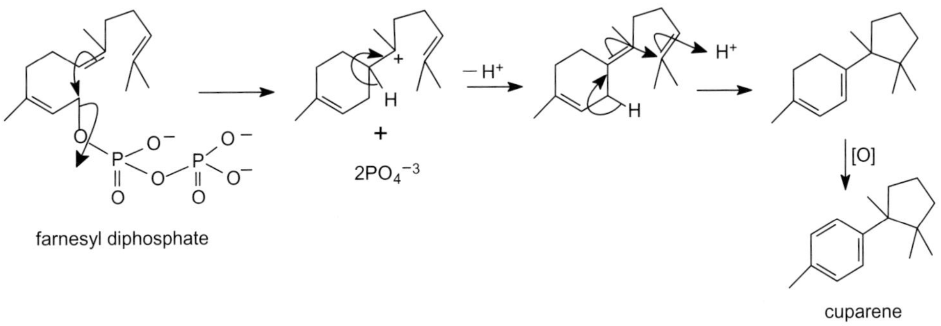

farnesyl diphosphate

cuparene

Scheme 11.60. A potential pathway from farnesyl diphosphate to cuparene.

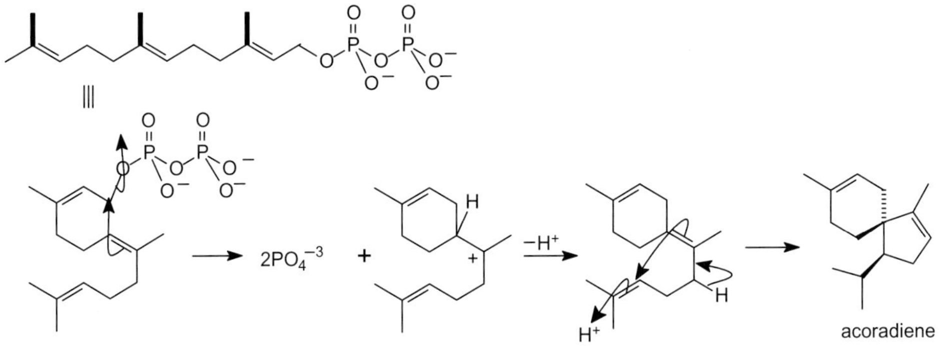

acoradiene

Scheme 11.61. A potential pathway to acoradiene from farnesyl diphosphate (see end of chapter Problem 11.17).

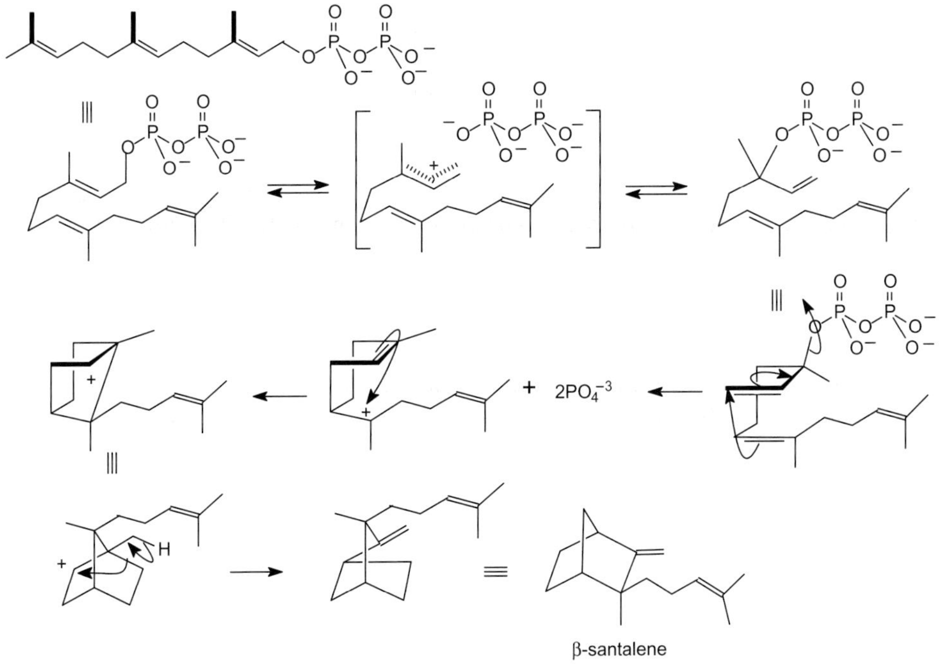

β-santalene

Scheme 11.62. A representation of a possible route to β-santalene from farnesyl diphosphate.

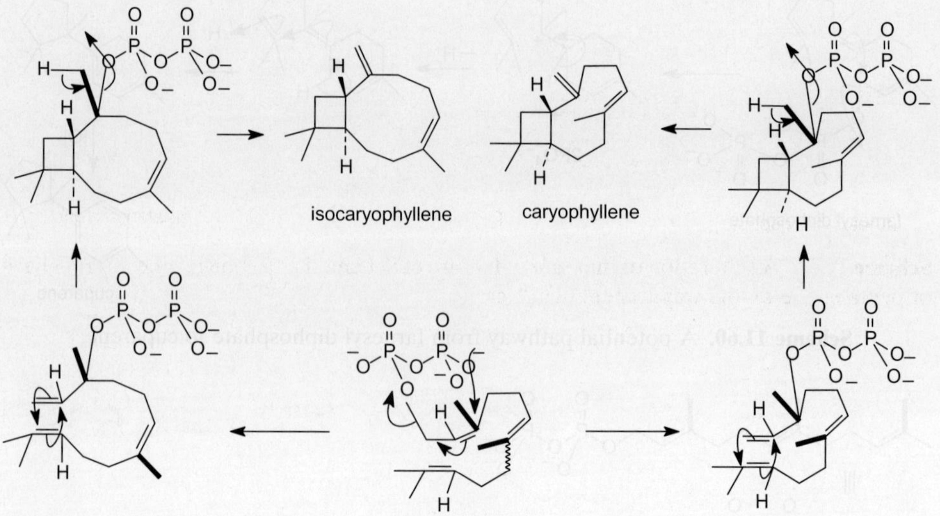

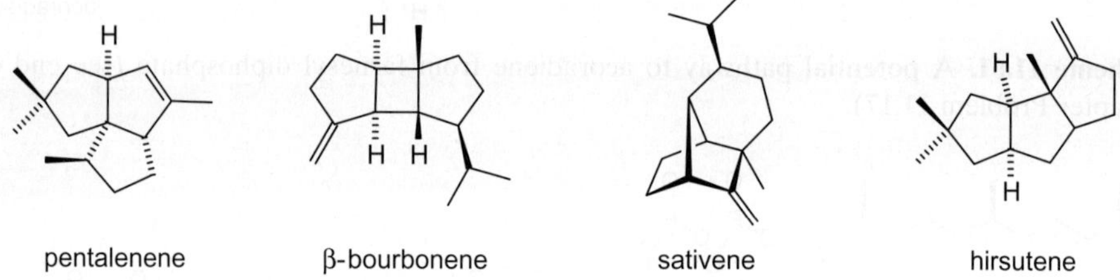

Scheme 11.63. Potential pathways to isocaryophyllene [(*Z*)-double bond in the eight-membered ring] and caryophyllene [(*E*)-double bond in the eight-membered ring] from farnesyl diphosphate. The double-bond isomerization is suggested to arise before formation of the four-membered ring.

Figure 11.24. Examples of three-ring sesquiterpene carbocycles.

pepper. Presumably, the cavity of the enzyme in which the cyclization of farnesyl diphosphate takes place so constrains that precursor as to introduce the (rare) four-membered ring. A potential cyclization process is provided as Scheme 11.63.

Sesquiterpenes with three rings are the final group to be considered. Indeed, one has already been seen in Scheme 11.59. Thus, it will be recalled that, on the way to pentalenolactone (which itself contains three rings albeit one is a lactone and not carbocyclic), the three-ring carbocycle pentalenene was found to be a biosynthetic intermediate. Figure 11.24 presents pentalenene again along with a few other examples of three-ring carbocycles. So, Scheme 11.64 reiterates the intermediacy from Scheme 11.59, of pentalenene on the way from farnesyl diphosphate to pentalenolactone (Quaderer et al., 2006).

Several syntheses of pentalenene have been reported. One, by T. Hudlicky* and coworkers is outlined in Scheme 11.65. In this scheme, the dimethylcyclopentene

*(a) Hudlicky, T.; Sinai-Zingde, G.; Natchus, M. G.; Ranu, B. C.; Papadopolous, P. *Tetrahedron*, **1987**, *43*, 5685. (b) Hudlicky, T.; Natchus, M. G.; Sinai-Zingde, G. *J. Org. Chem.*, **1987**, *52*, 4641.

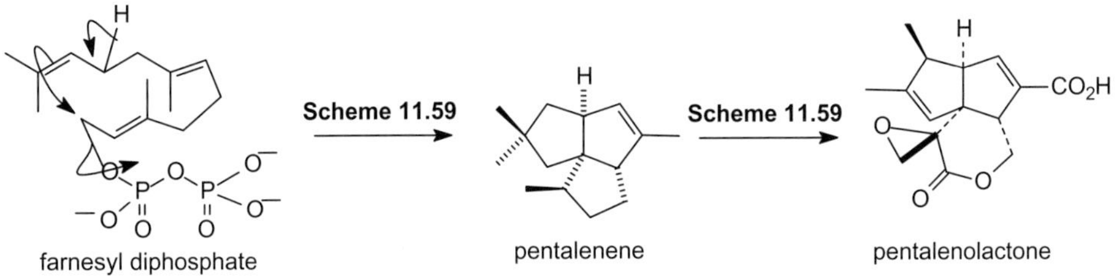

Scheme 11.64. A reiteration of that salient feature of Scheme 11.59 showing the intermediacy of pentalenene on the way to pentalenolactone.

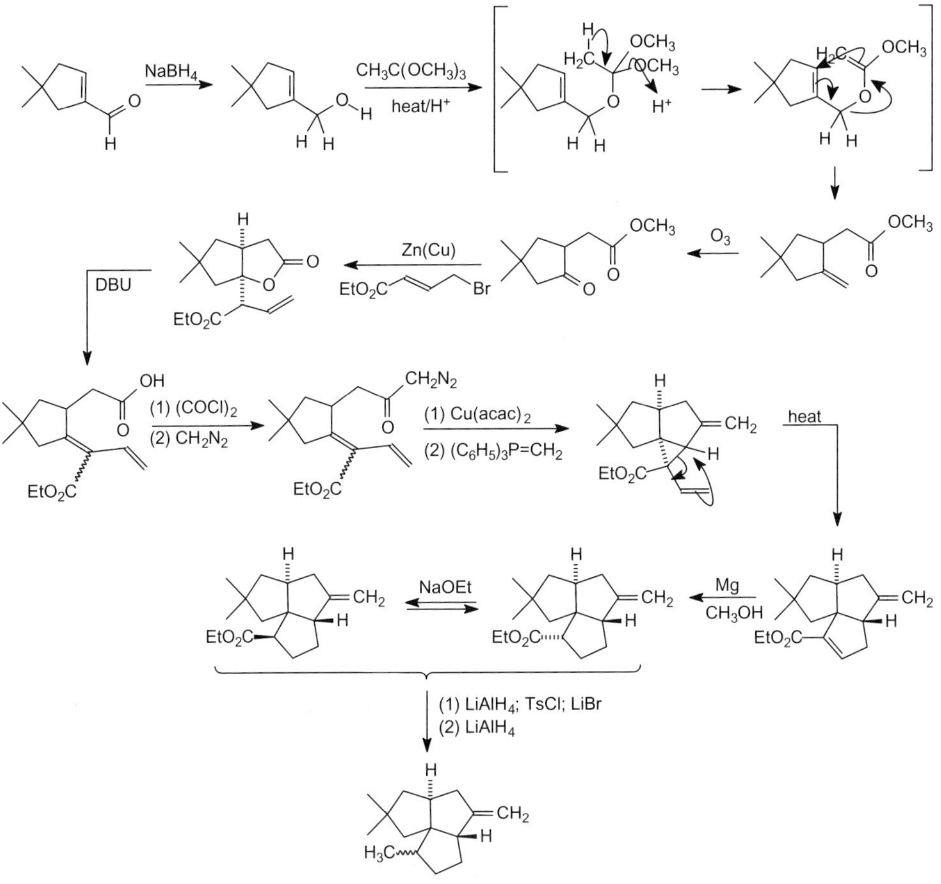

Scheme 11.65. A synthesis of pentalenene. See the text for comments. The work has been published. See (a) Hudlicky, T.; Sinai-Zingde, G.; Natchus, M. G.; Ranu, B. C.; Papadopolous, P. *Tetrahedron*, **1987**, *43*, 5685; (b) Hudlicky, T.; Natchus, M. G.; Sinai-Zingde, G. *J. Org. Chem.*, **1987**, *52*, 4641. DBU = 1,8-Diazabicyclo[5.4.0]undec-7-ene.

carboxyaldehyde shown was reduced with sodium borohydride, the alcohol treated with triethyl orthoacetate and the resulting ester rearranged. Ozonolyis regenerated the carbonyl group. Then, condensation in a Reformatsky-type process (Chapter 9) was followed by lactonization and treatment with base resulted in elimination.

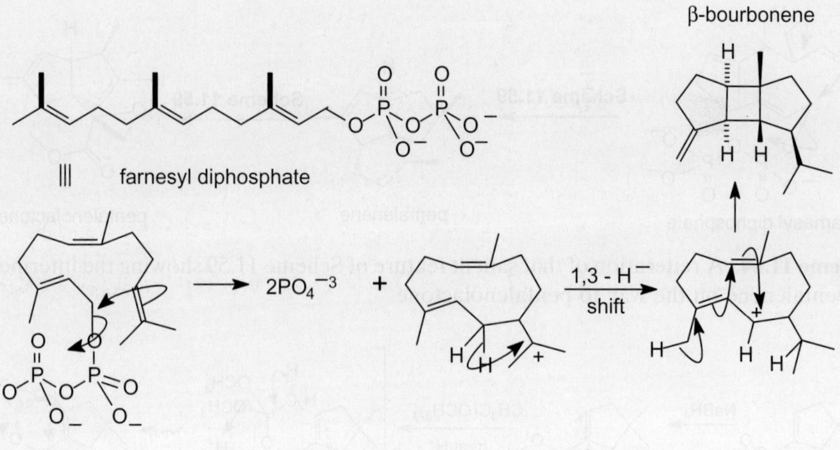

β-bourbonene

||| farnesyl diphosphate

2PO$_4$$^{-3}$ +

1,3 - H$^-$
shift

Scheme 11.66. A representation of a potential pathway to β-bourbonene from farnesyl diphosphate.

Then the carboxylic acid was converted (via the acid chloride) to the diazoketone (Chapter 9), and decomposition of the latter via the carbene and insertion produced the substituted cyclopropane. Next, the keto group was converted to a methylene unit via the Wittig, and thermolysis generated the third cyclopentyl ring. Dissolving metal reduction and epimerization in sodium ethoxide set the stage for conversion of the ester to an alkyl group.

So, first the reduction of the ester to the corresponding primary alcohol with lithium aluminum hydride was effected, and the alcohol was converted to the toluenesulfonate (tosylate). Displacement by bromide anion (LiBr) to produce the primary bromide was followed by another lithium aluminum hydride reduction to the final hydrocarbon (Scheme 11.65).

Many synthetic endeavors have been undertaken to produce some of the tricyclic sesquiterpenes, and some biosynthetic efforts, such as that shown in Scheme 11.59, for pentalenene are under way, to attempt to find the path from farnesyl diphosphate to them.

However, most of the pathways are speculative. Such speculative schemes are shown for β-bourbonene* (Scheme 11.66), sativene (Scheme 11.67), and hirsutene (Scheme 11.68).

It will not come as a surprise that yet another isopentenyl diphosphate can attack farnesyl diphosphate and, in this way, account for the formation of the diterpene precursor, geranylgeranyl diphosphate. Indeed, under the influence of farnesyltransferase (EC 2.5.1.29), such a process occurs (Scheme 11.69)!

Phytol is an alcohol with which chlorophyll (Chapter 12) is esterified. After the phytol is isolated by hydrolysis, it is subsequently used in the manufacture of vitamin E (of which a variety of similar compounds are categorized). Interestingly, the particular vitamin E derivative commonly prepared (α-tocopherol) is obtained by heating phytol and 2,3,5-trimethylhydroquinone (a compound derived in the plant from shikimic acid, *vide infra*) in the presence of an acid catalyst (Scheme 11.70).

*See end of chapter, Problem 11.17.

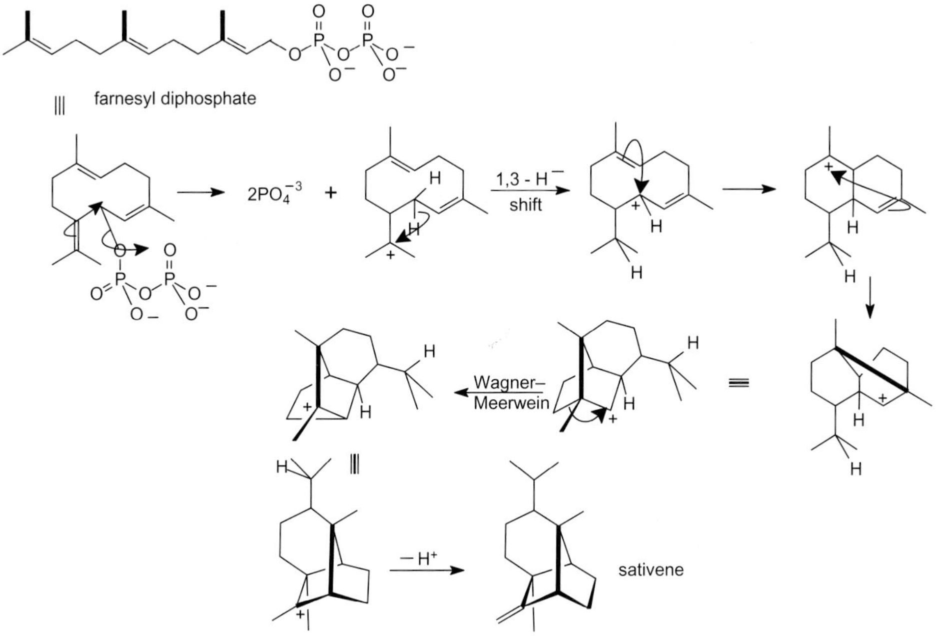

Scheme 11.67. A representation of a potential pathway to sativene from farnesyl diphosphate.

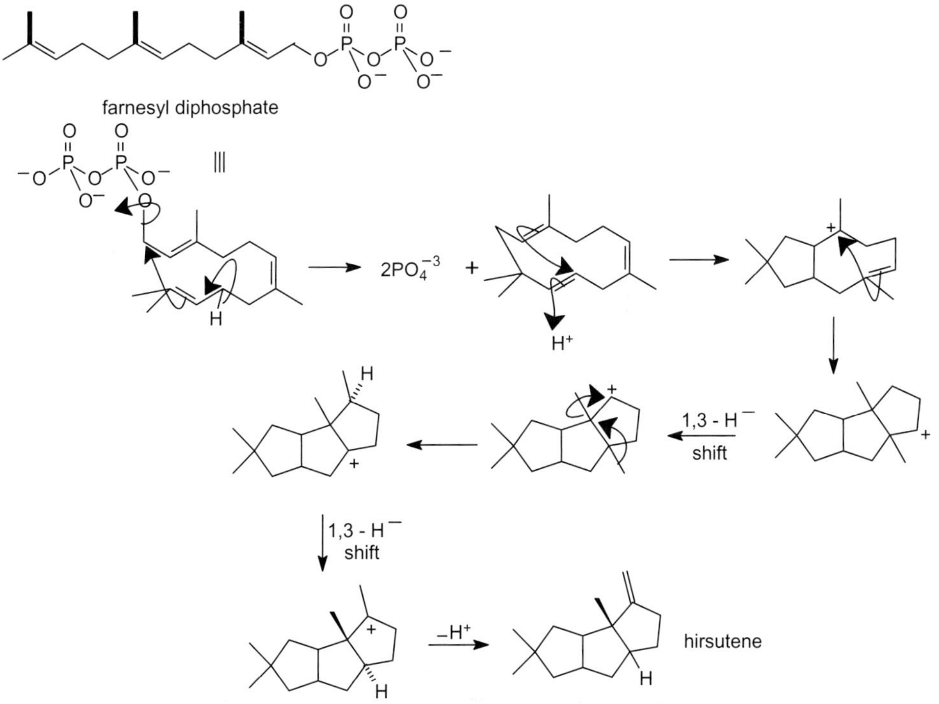

Scheme 11.68. A representation of a potential pathway to hirsutene from farnesyl diphosphate.

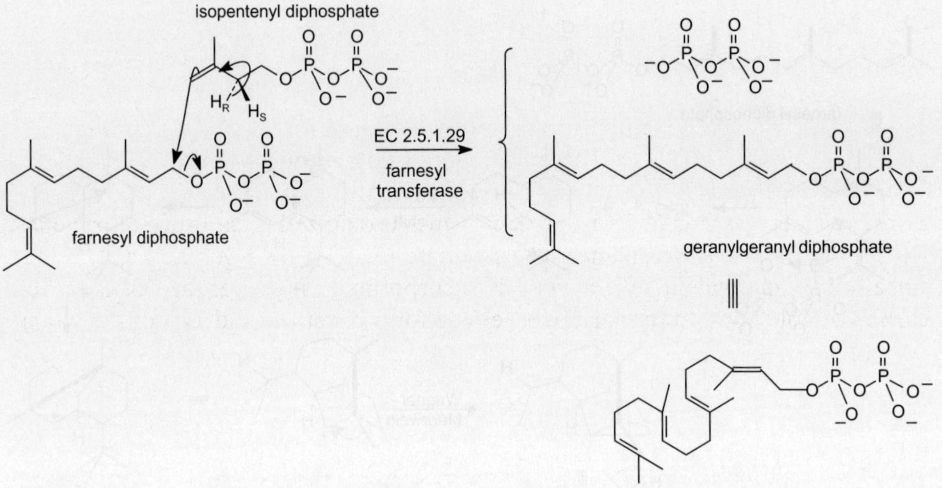

Scheme 11.69. A representation of the formation of geranylgeranyl diphosphate from farnesyl diphosphate and isopentenyl diphosphate as catalyzed by farnesyltransferase (EC 2.5.1.58 and 2.5.2.59). EC numbers and some graphic materials provided in this scheme have been taken with permission from appropriate links in a URL starting with http://www.chem.qmul.ac.uk/iubmb/enzyme/.

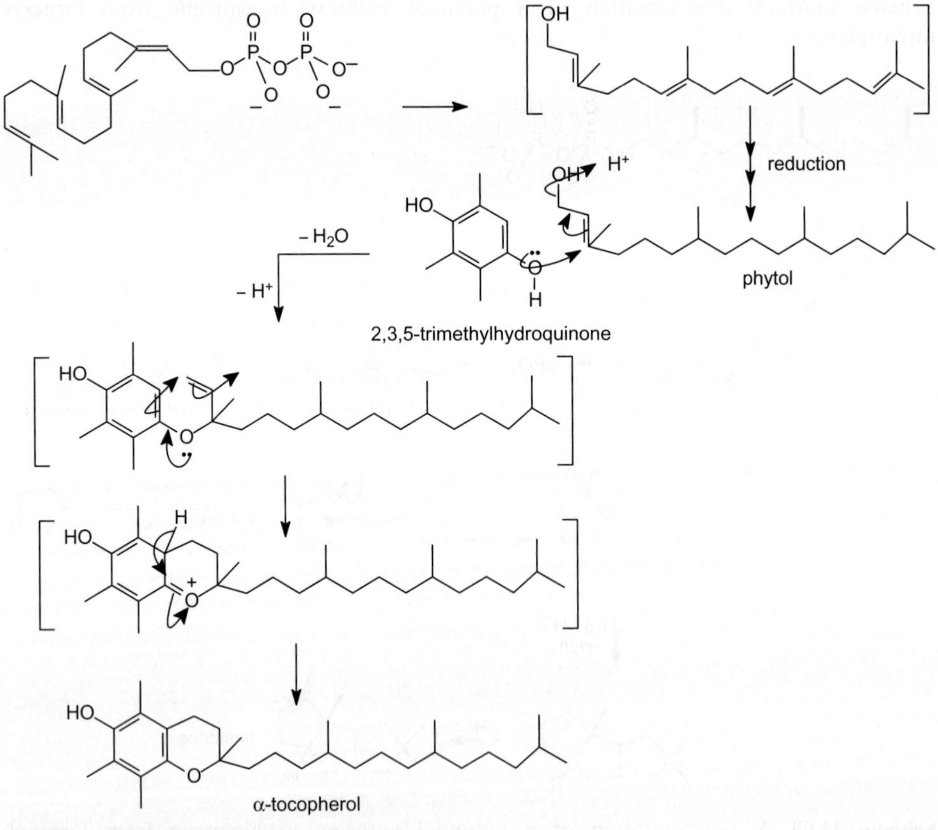

Scheme 11.70. A common synthesis of α-tocopherol. The phytol and 2,3,5-trimethylhydroquinone are brought together in the presence of an acid catalyst to produce vitamin E.

Given the structure of geranylgeranyl diphosphate, it should be clear that reduction of the double bonds and hydrolysis (to remove the diphosphate group) produces the primary alcohol.

Similarly, vitamin A, important in bone growth and vision, can be accounted for by a simple cyclization of the parent geranylgeranyl diphosphate and oxidation to introduce additional double bonds. However, it appears that the equivalent of eight isoprene units combine to form carotene, which is oxidized (β-carotene dioxygenase, EC 1.13.11.21, has been renamed as β-carotene 15,15′-monoxygenase, EC 1.14.99.36 since only 1 equivalent of dioxygen is incorporated) with cleavage of a carbon–carbon double bond to retinal, the aldehyde form of retinol and part of the vitamin A group.

In that vein, it has been proposed that geranylgeranyl diphosphate and its isomer, geranyllinalyl diphosphate (Scheme 11.71) combine to generate the corresponding tetraterpene (C_{40}) derivative (phytoene), and the latter is isomerized to a carotene and then cleaved (presumably) after cyclization (as noted above) to 2 equivalents of either the aldehyde or the alcohol.

Cembrene, a termite trail-marking pheromone* is thought of as a simple diterpene that serves as a "backbone" representative for many other diterpenes. Scheme 11.72 presents a potential pathway to cembrene from geranylgeranyl diphosphate. The same scheme (Scheme 11.72) also provides a path from the same starting material to taxadiene. Taxadiene is the hydrocarbon core of the antitumor drug taxol (paclitaxol), which has proved of some use in treatment of breast, ovarian, and lung cancer. Since taxol (paclitaxol) is obtained in only small quantities from the bark of the Pacific yew (*Taxus brevifolia*) and removal of the bark to obtain the drug kills the tree, a synthetic pathway from more readily available materials was sought. Many years of work were necessary before a pathway was found.[†]

Problem 11.10. Consider the two remaining diterpenes of Figure 11.25 viz. labdadienyl diphosphate and aphidicolin. Based on what has been shown regarding biosynthetic hypotheses for the other four diterpenes in the same figure, provide reasonable possible pathways for their formation from geranylgeranyl diphosphate.

Although names and properties of thousands of hydrocarbons and their variously oxidized and rearranged derivatives have been compiled and catalogued (variously "complete" catalogues that are incomplete when published), the exact number of terpenes (C_{10}), sesquiterpenes (C_{15}), diterpenes (C_{20}), sesterpenes (C_{25}), triterpenes (C_{30}), carotenoids (C_{40}), and varieties of rubber (e.g., laytex and gutta-percha) is unknown. Similarly, tomes have been written on the subject, and the discussion of these compounds here will end with an introduction to the triterpenes because these serve as sources of steroids (animal) and sterols (plants).

As shown in Scheme 11.73, two molecules of (E,E)-farnesyl diphosphate combine to form a head-to-head dimer containing a cyclopropane ring, viz. presqualene

*A pheromone is any of a group of naturally occurring chemical species produced by one member of the species to alert other members of the same species about specific circumstances. Some pheromones alert conspecifics to danger, to sexual receptivity, to nearby food, and so on.
[†]Holton, R. A.; Somoza, C.; Kim, H. B.; Liang, F.; Biediger, R. J.; Boatman, P.D.; Shindo, M.; Smith, C. C.; Kim, S. *J. Am. Chem. Soc.*, **1994**, *116*, 1597 and following papers.

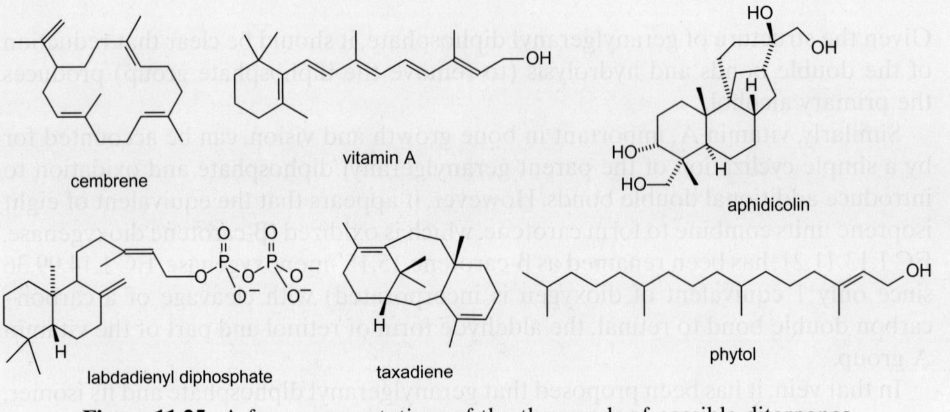

Figure 11.25. A few representatives of the thousands of possible diterpenes.

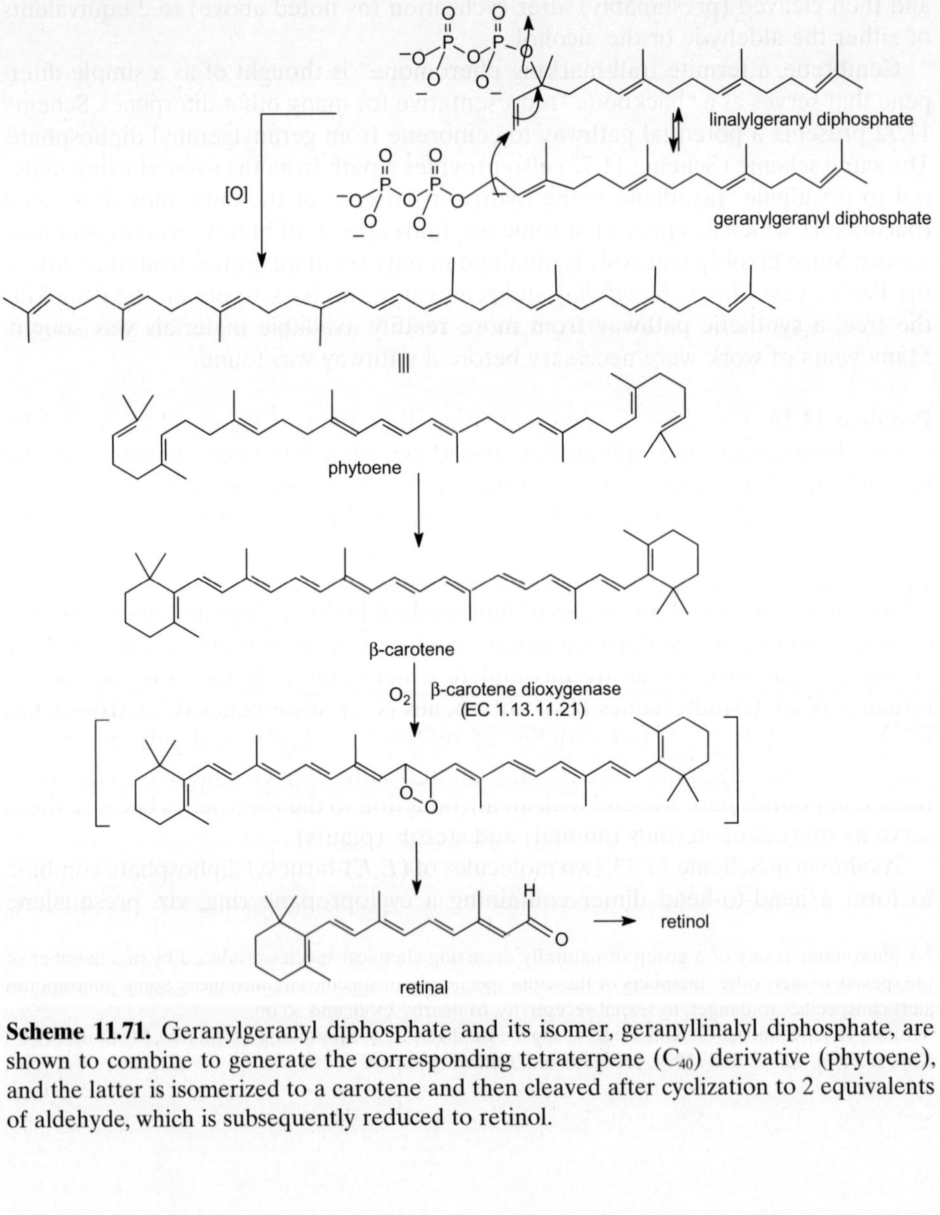

Scheme 11.71. Geranylgeranyl diphosphate and its isomer, geranyllinalyl diphosphate, are shown to combine to generate the corresponding tetraterpene (C_{40}) derivative (phytoene), and the latter is isomerized to a carotene and then cleaved after cyclization to 2 equivalents of aldehyde, which is subsequently reduced to retinol.

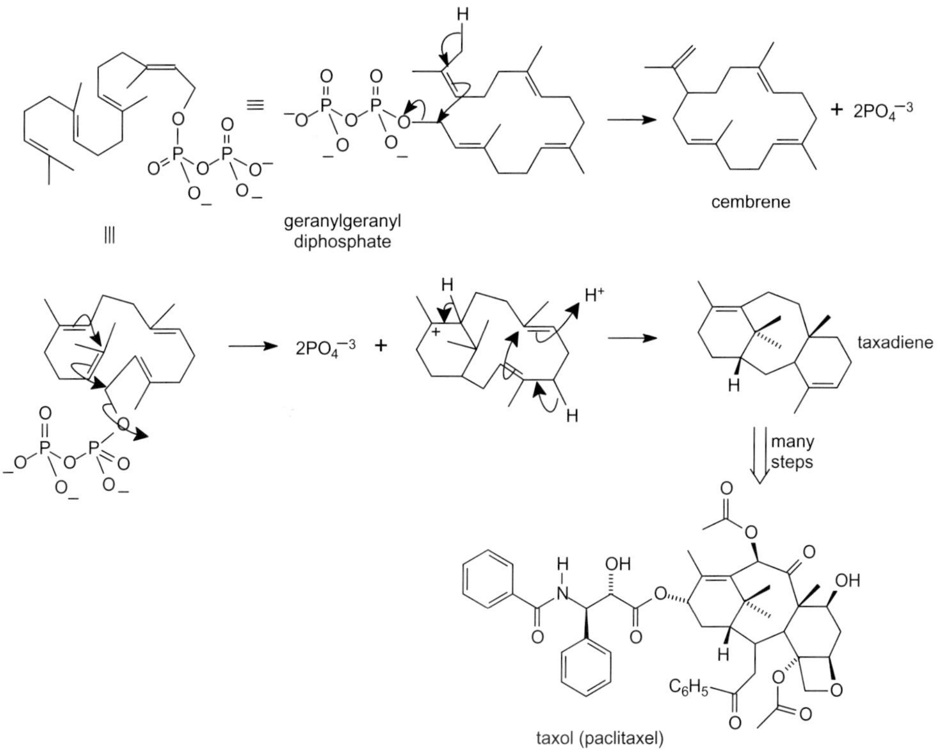

Scheme 11.72. Potential pathways from geranylgeranyl diphosphate to cembrene and to taxadiene. Taxadiene is presumably a biochemical precursor to taxol (paclitaxel).

diphosphate, in a process catalyzed by squalene synthase (EC 2.5.1.2). The cyclopropane ring opens to squalene.

The road from squalene to lanosterol, the fully cyclized precursor to cholesterol,* and other steroids passes through (3S)-2,3-oxidosqualene (squalene oxide). The

---

*At this writing, there is a widely available group of medicines utilized to affect the concentration of cholesterol in the bloodstream. Cholesterol is one of the steroids implicated in heart disease in humans and other primates. The drugs affecting the concentration of bloodstream cholesterol are called "statins," and it has been found that these compounds appear to function by inhibiting the enzyme hydroxymethylglutaryl-CoA (HMG-CoA) reductase (see Scheme 11.40). It will not have escaped the astute reader that inhibition at that level will have other consequences on all of the compounds "downstream" and so careful dosing is required.

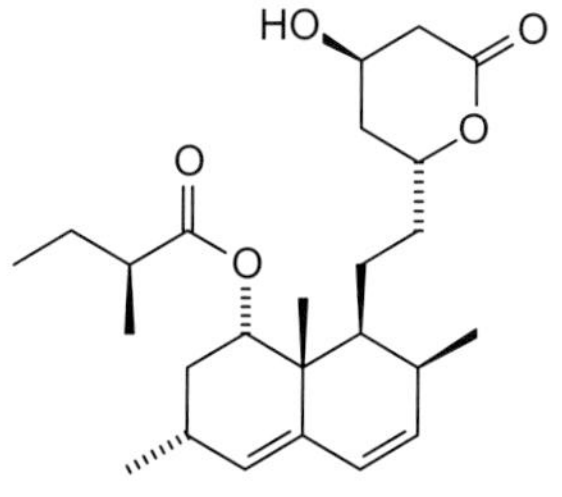

Lovastatin, Mevacor ™ (isolated from a strain of *Aspergillus terreus*).

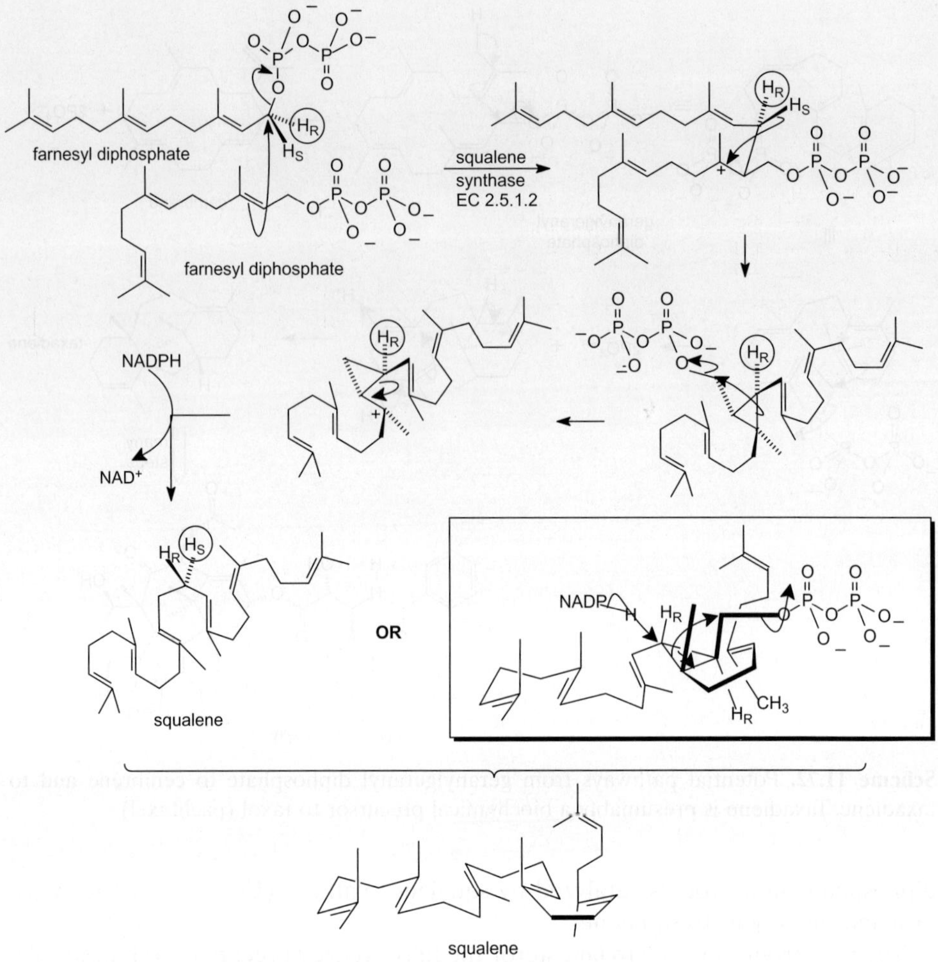

Scheme 11.73. The squalene synthase-catalyzed (EC 2.5.1.21) conversion of farnesyl diphosphate to presqualene diphosphate (containing a cyclopropane ring) and thence to squalene. The details confirming much of this work was initially set forth and then summarized by Popjak, G. J.; Cornforth, S. J. *Biochem. J.*, **1966**, *101*, 553 (Nobel Prize, 1975). EC numbers and some graphic materials provided in this scheme have been taken with permission from appropriate links in a URL starting with http://www.chem.qmul.ac.uk/iubmb/enzyme/.

oxidation of squalene to the epoxide requires oxygen (O_2) and a flavin (i.e., $FADH_2$) along with the enzyme squalene monoxygenase (EC 1.14.99.7). Squalene picks up one oxygen atom in forming the epoxide and the $FADH_2$ on the other. Then NADH reduces the oxidized flavin and is converted to NAD^+ (Scheme 11.74). The cyclization to lanosterol goes forth in the presence of the enzyme oxidosqualene cyclase (EC 5.4.99.7) from the epoxide (Scheme 11.75).

Thorough investigation has delineated 19 steps in the conversion of lanosterol to cholesterol. An abbreviated version of the path is provided in Scheme 11.76 where it is intended to show that the conversion requires a series of methyl group oxidations in the sequence expected (viz. methyl [$-CH_3$] → hydroxymethyl

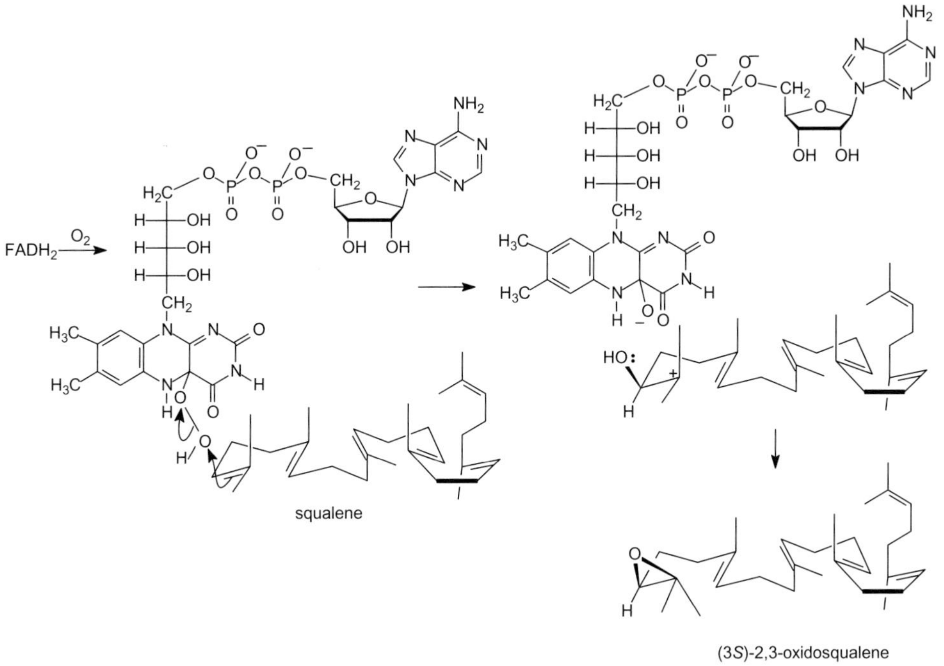

Scheme 11.74. The formation of (*S*)-2,3-oxidosqualene from squalene in the presence of FADH$_2$ and squalene monoxygenase (EC 1.14.99.7). After McMurry, J.; Begley, T. *The Organic Chemistry of Biological Pathways*, Roberts & Co., Englewood, CO, **2005**, 148 ff.

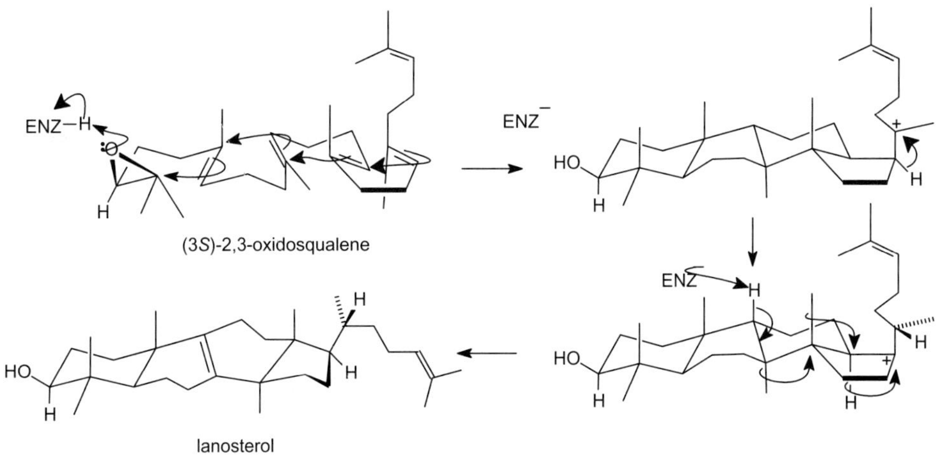

Scheme 11.75. The cyclization to lanosterol from (3*S*)-2,3-oxidosqualene in the presence of the enzyme oxidosqualene cyclase (EC 5.4.99.7). The details (demonstrated with labeled materials and specific chemical degradation!) of the movement (migration) of the hydride and methide units were carefully documented. See Popjak, G. J.; Cornforth, S. J. *Biochem. J.*, **1966**, *101*, 553 (Nobel Prize, 1975). ENZ = oxidosqualene cyclase.

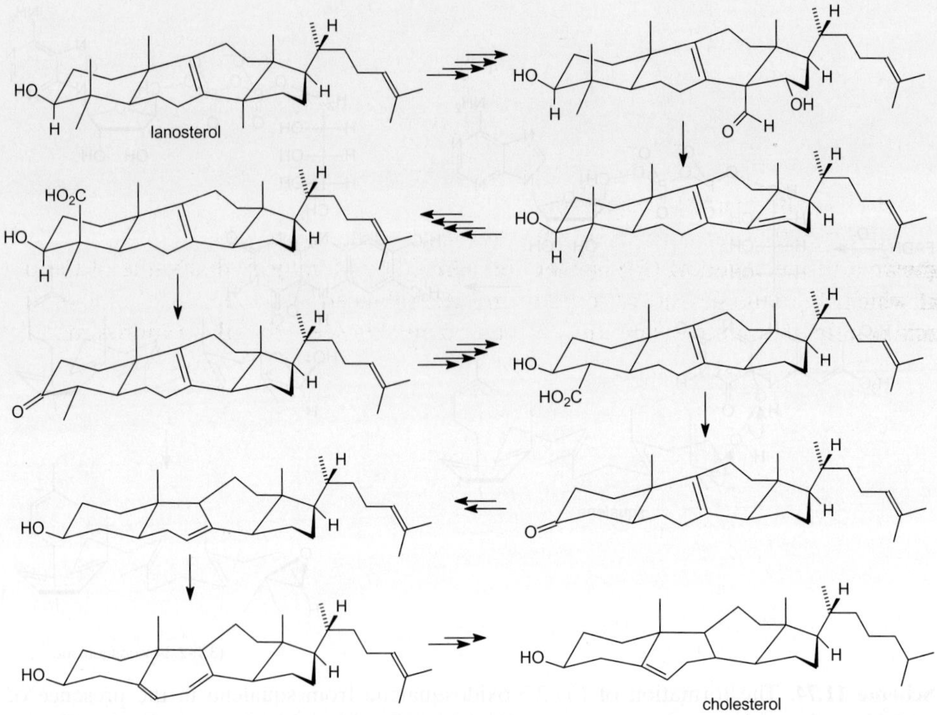

Scheme 11.76. The path from lanosterol to cholesterol. After Rilling, H. C.; Chayet, L. T., in Danielsson, H; Sjövall, J. (eds.), *Sterols and Bile Acids*, Elsevier, New York, **1985**, 33.

$[CH_2OH] \rightarrow$ aldehyde $[O=CH] \rightarrow$ carboxylic acid $[CO_2H] \rightarrow$ carbon dioxide $[CO_2]$ and replacement of the lost carbon by a proton) and a few other cosmetic changes.

Cholesterol is the parent of both estrogenic (female) and androgenic (male) steroids. In Scheme 11.77, a biogenetic path from cholesterol to testosterone, typical of the latter (via pregnenolone and progesterone, both typical of the former), is shown. Furthermore, testosterone itself (Scheme 11.77) undergoes aromatization (with the "aromatase" enzyme, EC 1.14.14.1) to estradiol, an estrogenic steroid.

Although steroid syntheses were reported as early as 1951,[*] more recent work[†] follows a biogenetic pathway and is shown in Scheme 11.78.

Problem 11.11. Examine the chemistry of Scheme 11.78. Describe (with suitable curved arrow formalism) what is happening in each step.

d. Loose Ends

1. Iridoids (Loganin and Secologanin). The relatively small group of richly oxidized terpenes resulting from a cyclization giving rise to a fused pyran-cyclopentane ring system is widely known as iridoids. All of those isolated so far have a *cis*-fused

[*]Woodward, R. B.; Sondheimer, F.; Taub, D.; Heusler, K.; McLamore, W. M. *J. Am. Chem. Soc.*, **1951**, *73*, 2403.
[†]Gravestock, M. B.; Johnson, W. S.; McCarry, B. E.; Parry, R. J.; Ratcliffe, B. E. *J. Am. Chem. Soc.*, **1978**, *100*, 4274.

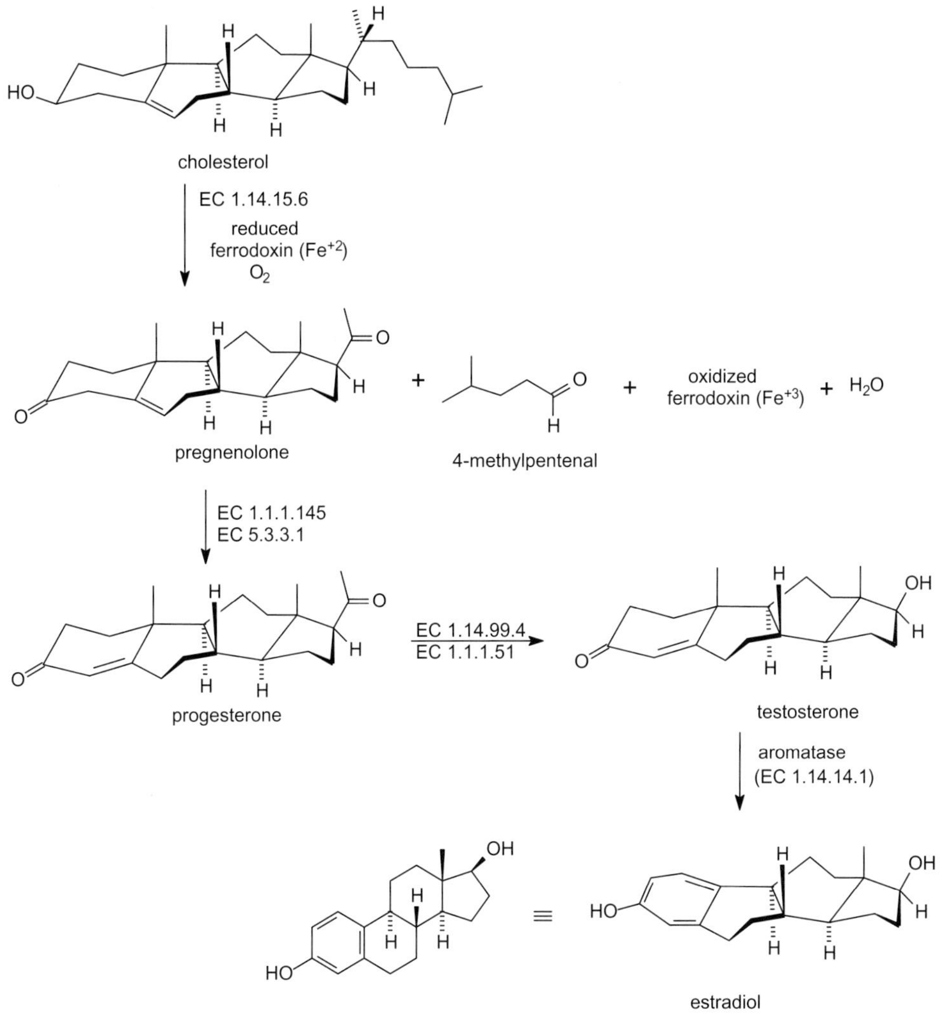

cholesterol

EC 1.14.15.6
reduced
ferrodoxin (Fe^{+2})
O$_2$

pregnenolone

4-methylpentenal

+ oxidized ferrodoxin (Fe^{+3}) + H$_2$O

EC 1.1.1.145
EC 5.3.3.1

progesterone

EC 1.14.99.4
EC 1.1.1.51

testosterone

aromatase
(EC 1.14.14.1)

≡

estradiol

Scheme 11.77. A representation of a biogenetic path from cholesterol to testosterone (via pregnenolone and progesterone) and then to estradiol.

ring system. Although not widely distributed, one iridoid, secologanin, and its progenitor, loganin, have sparked the interest of several generations of chemists since they were implicated in the biosynthesis of alkaloids, valued for antitumor drugs, derived from the Madagascar periwinkle, *Catharanthus roseus** and some other indole alkaloids as well as some ipacec alkaloids.

Scheme 11.79 presents a potential pathway to loganin and secologanin from geranyl diphosphate. Only a few of the enzymes necessarily involved are known with any certainty, and while descriptive names have been given to them, details of

---

*While this particular member of the Apocynaceae has proved most valuable for the study, other plants will be considered too (Chapter 12). See (a) Oudin, A.; Courtois, M.; Rideau, M.; Clastre, M. *Phytochem. Rev.*, **2007**, *6*, 259; (b) Sampaio-Santos, M. I.; Kaplan, M. A. C. *J. Braz. Chem. Soc.*, **2001**, *12*, 144.

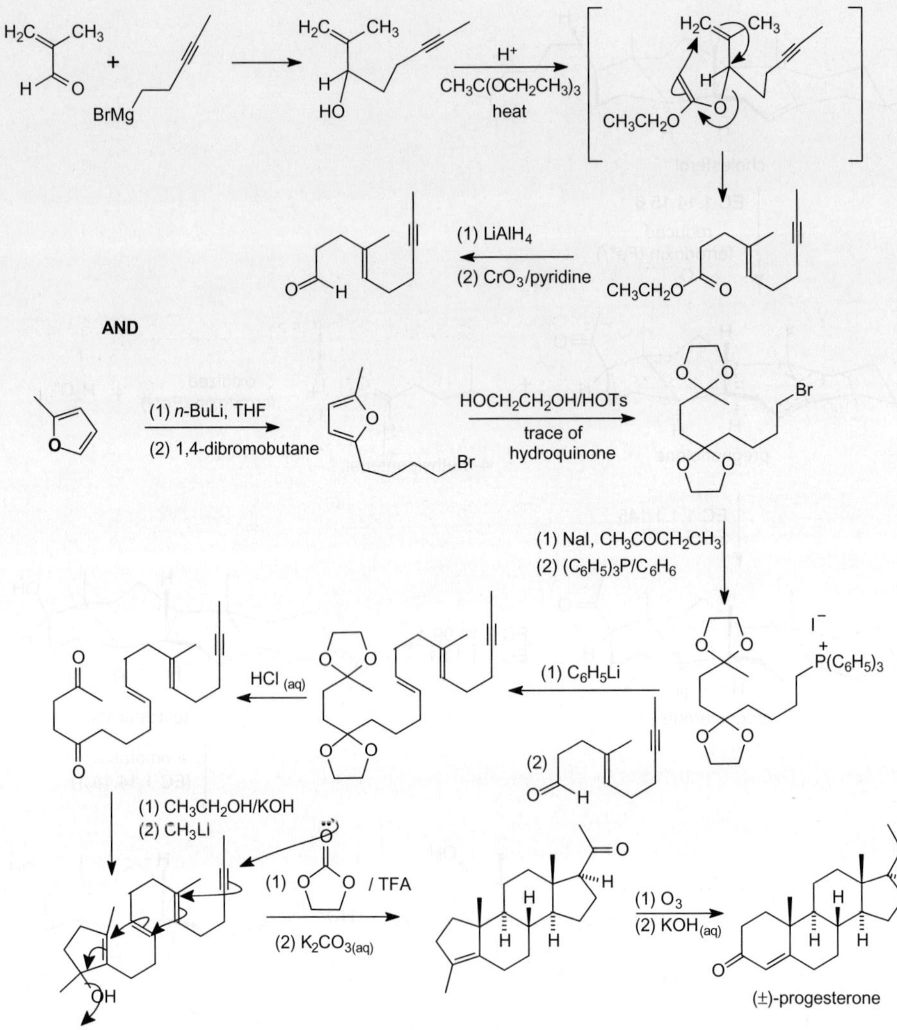

Scheme 11.78. A synthesis of racemic progesterone. See Gravestock, M. B.; Johnson, W. S.; McCarry, B. E.; Parry, R. J.; Ratcliffe, B. E. *J. Am. Chem. Soc.*, **1978**, *100*, 4274.

their structures are wanting. Thus, it is clear that geranyl diphosphate must almost certainly be converted to the 10-hydroxygeraniol and thence to a dialdehyde. Cyclization followed by methyl group oxidation, esterification, and hemiacetal addition of a glucose unit would then lead to a deoxyloganin, which needs only a subsequent oxidation to loganin and an oxidative ring cleavage to yield secologanin.

2. Shikimic Acid, Isoshikimic Acid, and Prephenic Acid. As will be seen in the next chapter (Chapter 12), the amino acids phenylalanine, tyrosine, and tryptophan all contain aryl rings. The biosynthesis of the aromatic rings of these amino acids passes through the same seven-carbon carboxylic acid, shikimic acid, which itself is derived from erythrose and phosphoenolpyruvate. As shown in Scheme 11.10, as part of the

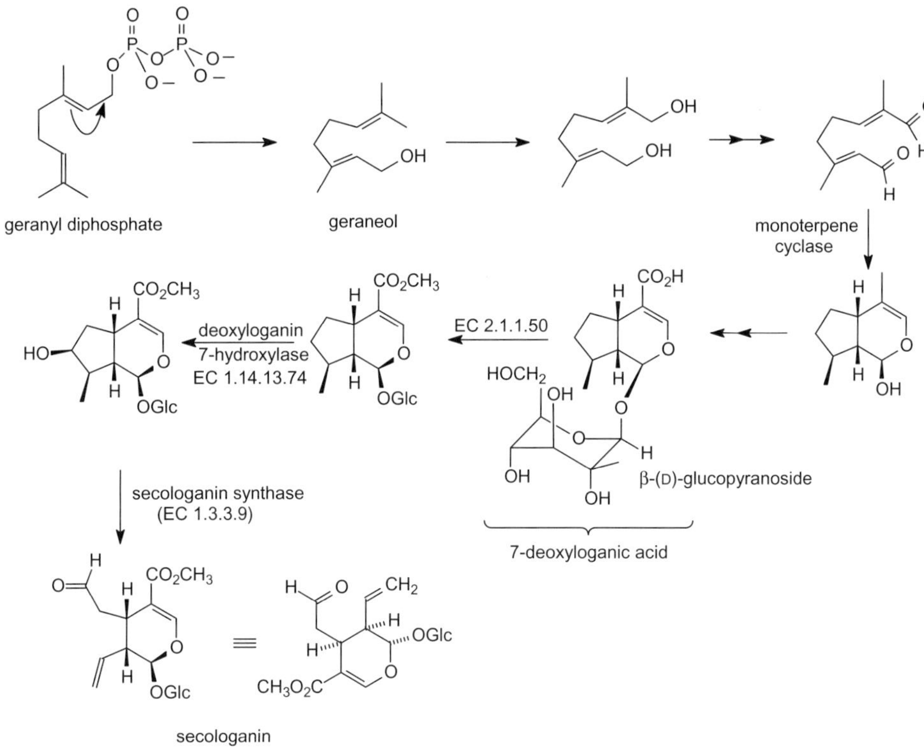

geranyl diphosphate

geraneol

monoterpene cyclase

deoxyloganin 7-hydroxylase EC 1.14.13.74

EC 2.1.1.50

β-(D)-glucopyranoside

7-deoxyloganic acid

secologanin synthase (EC 1.3.3.9)

secologanin

Scheme 11.79. A proposed pathway to secologanin from geranyl diphosphate.

Calvin–Benson–Bassham cycle, a seven-carbon sugar arises from the reaction of dihydroxyacetone monophosphate with erythrose 4-phosphate.

However, if the erythrose 4-phosphate reacts with phosphoenoyl pyruvate instead of dihydroxyacetone monophosphate, a seven-carbon sugar lacking a hydroxyl at one carbon, a dehydroheptanoate (2-dehydro-3-deoxy-7-phospho-D-*arabino* heptanoate) results. Cyclization of this heptanoate and rearrangement under the influence of 3-dehydroquinate synthase (EC 4.2.3.4) as shown in Scheme 11.80 results in the formation of 3-dehydroquinate. It is particularly important to recognize that this rationalization (1) both uses and forms $NADH/H^+$ in the process, (2) apparently requires divalent cations in the synthase enzyme (not shown in the path of Scheme 11.80), and (3) requires that the six-membered oxygen-containing pyran ring open and then reclose as a cyclohexane!

Labeling studies are in accord with the process shown in Scheme 11.80.

Next, the 3-dehydroquinate needs to undergo dehydration to produce 3-dehydroshikimic acid. While, in principle, it appears that one might simply expect the protons α to the carbonyl to be acidic enough to be lost to an active site amine and dehydration to occur (perhaps after involving phosphorylation of the tertiary hydroxyl) via an enolate anion, it has been suggested that the enzyme 3-dehydroquinate dehydratase (EC 4.2.1.10) appears to utilize an active site amine to form an imine, then an enamine, and then a second imine before hydrolyzing to the product. The matter remains to be resolved. The process utilizing an enamine, with loss of the pro-R hydrogen, is shown in Scheme 11.81.

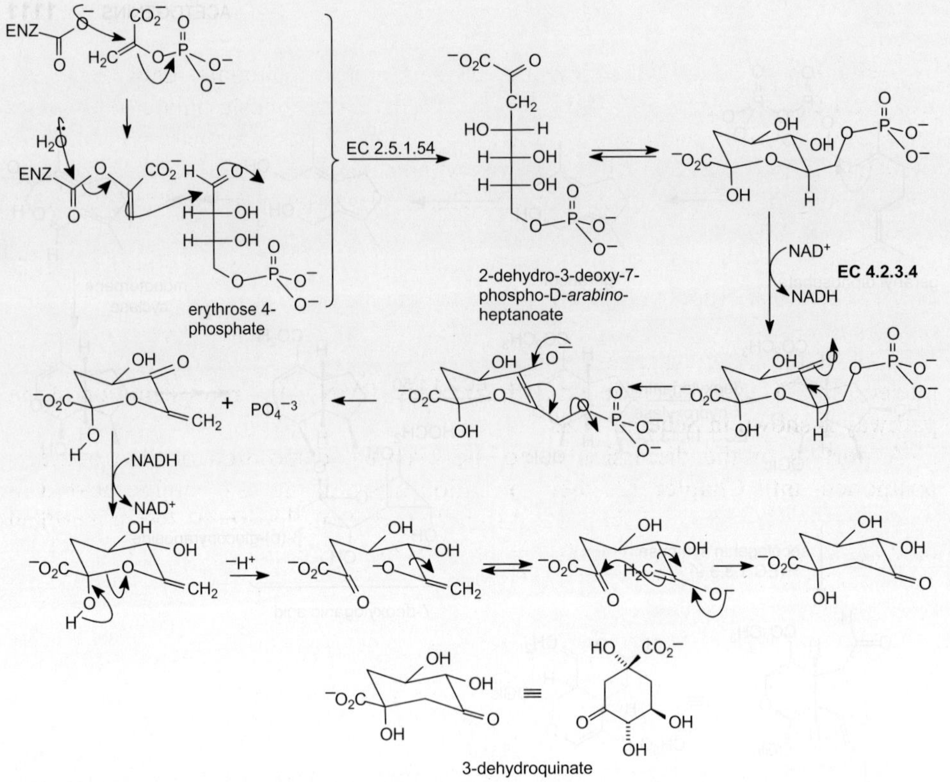

erythrose 4-phosphate

2-dehydro-3-deoxy-7-phospho-D-*arabino*-heptanoate

EC 2.5.1.54

EC 4.2.3.4

3-dehydroquinate

Scheme 11.80. A rationalization for the formation of 3-dehydroquinate from 2-dehydro-3-deoxy-7-phospho-D-*arabino*-heptanoate. After Bender, S. L.; Mehdi, S; Knowles, J. R. *Biochemistry*, **1989**, *28*, 7555.

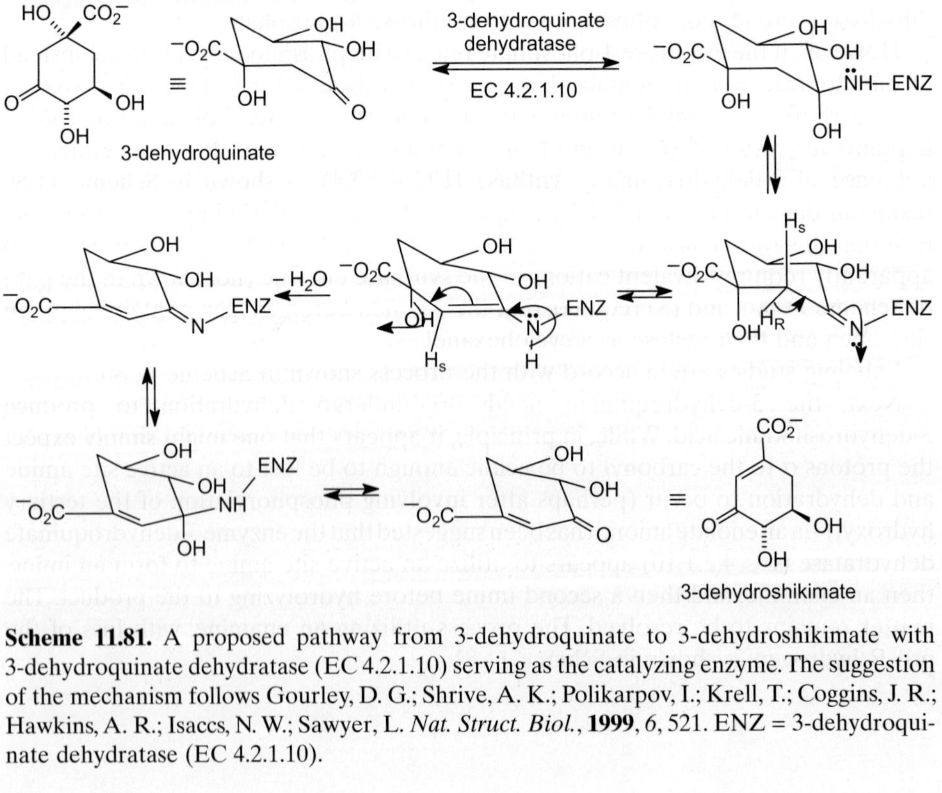

3-dehydroquinate

3-dehydroquinate dehydratase

EC 4.2.1.10

3-dehydroshikimate

Scheme 11.81. A proposed pathway from 3-dehydroquinate to 3-dehydroshikimate with 3-dehydroquinate dehydratase (EC 4.2.1.10) serving as the catalyzing enzyme. The suggestion of the mechanism follows Gourley, D. G.; Shrive, A. K.; Polikarpov, I.; Krell, T.; Coggins, J. R.; Hawkins, A. R.; Isaccs, N. W.; Sawyer, L. *Nat. Struct. Biol.*, **1999**, *6*, 521. ENZ = 3-dehydroquinate dehydratase (EC 4.2.1.10).

Reduction of 3-dehydroshikimate to shikimate (shikimate dehydrogenase, EC 1.1.1.25) followed by phosphorylation to 3-phosphoshikimate (shikimate kinase, EC 2.7.1.71) sets the stage for the reaction of the latter with phosphoenol pyruvate (3-phosphoshikimate 1-carboxyvinyltransferase, EC 2.5.1.19) to yield 5-(1-carboxyvinyl)-3-phosphoshikimate. These processes are shown in Scheme 11.82.

Now, in the presence of chorismate synthase (EC 4.2.3.5), 5-(1-carboxyvinyl)-3-phosphoshikimate loses phosphate and the H_R proton at C_6 to form the conjugated diene, chorismate.

It has been argued that this is a free radical-type process using a flavin mononucleotide (FMN) that is reduced in the process (FMNH$_2$). A representation of the pathway is shown in Scheme 11.83.

A portion of the discussion as to the eventual disposition of chorismate is postponed until Chapter 12 when formation and utilization of amino acids will be discussed. However, it is important to note here that chorismate lies at a critical biochemical junction. Chorismate is linked to the formation of the aromatic

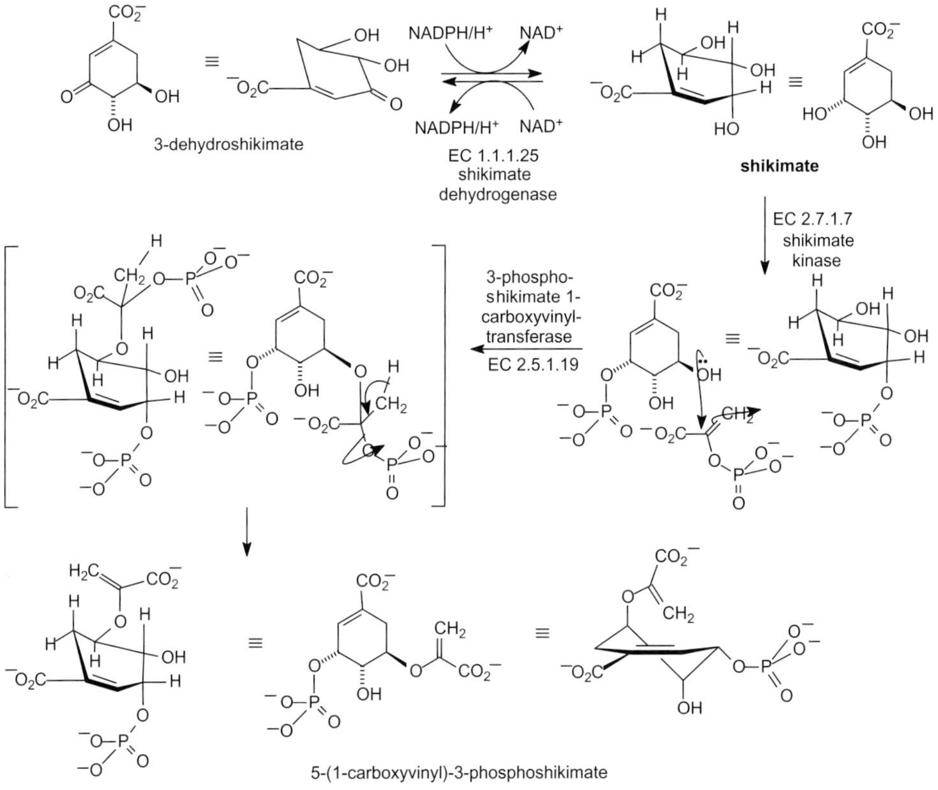

Scheme 11.82. A representation of a pathway to 5-(1-carboxyvinyl)-3-phosphoshikimate from 3-dehydroshikimate and via shikimate. After Lewis, J.; Johnson, K. A.; Anderson, K. S. *Biochemistry*, **1999**, *38*, 7372. EC numbers and some graphic materials provided in this scheme have been taken with permission from appropriate links in a URL starting with http://www.chem.qmul.ac.uk/iubmb/enzyme/.

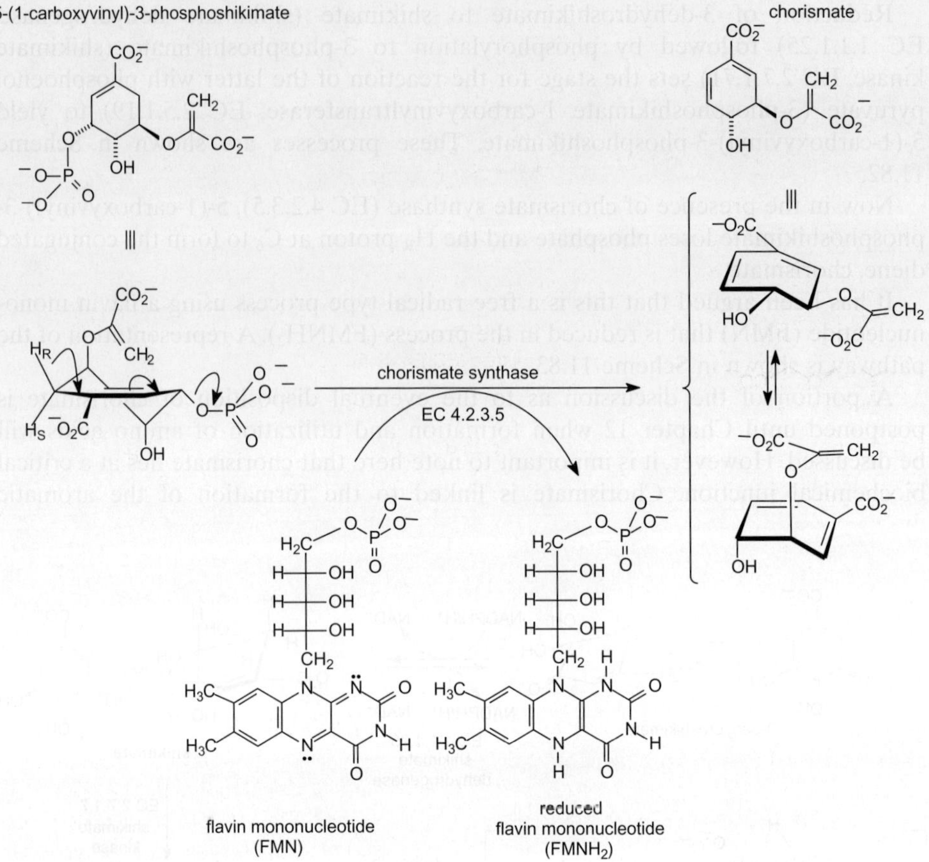

Scheme 11.83. The conversion of 5-(1-carboxyvinyl)-3-phosphoshikimate to the conjugated diene, chorismate, in the presence of chorismate synthase (EC 4.2.3.5). It has been argued that this is a free radical-type process using a flavin mononucleotide (FMN) that is reduced in the process (FMNH₂). After Osborne, A.; Thorneley, R. N.; Abell, C.; Bornemann, S. *J. Biol. Chem.*, **2000**, *275*, 35825.

rings of the amino acids phenylalanine and tyrosine (Chapter 12) via prephenic acid (Scheme 11.84), and tryptophan (Chapter 12) via anthranilic acid and folate (Chapter 13) via *p*-aminobenzoic acid (Scheme 11.85) as well as ubiquinone (Scheme 11.86).

Isochorismate is linked to the formation of *o*-hydroxybenzoic acid (salicylate), as well as 2,3-dihydroxybenzoic acid required for enterobactin biosynthesis (Scheme 11.87). Isochorismate is also a precursor to the vitamins K phylloquinone and menaquinone (Scheme 11.88).

Problem 11.12. Using any suitable source, look up the structure of enterobactin and write a suitable pathway for its formation from 2,3-dihydroxybenzoic acid.

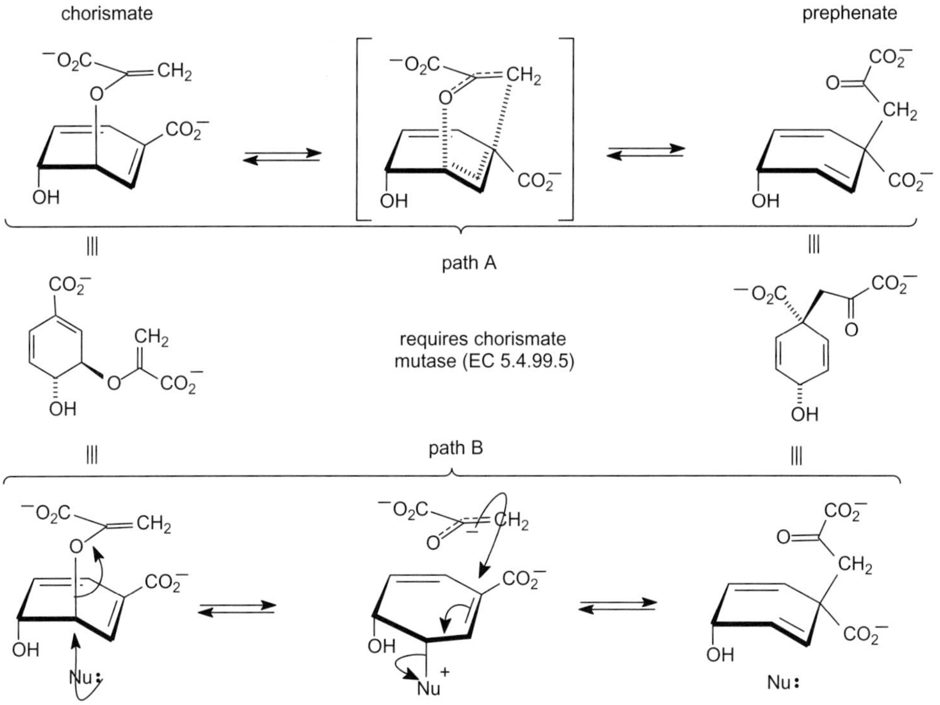

chorismate

prephenate

path A

requires chorismate
mutase (EC 5.4.99.5)

path B

Scheme 11.84. The conversion of chorismate to prephenate as catalyzed by chorismate mutase (EC 5.4.99.5). Two paths are shown. In path A, the proximity of the groups and the well-known Claisen rearrangement suggests a pathway that is synchronic or one that is nearly so with bond breaking and making occurring simultaneously (or nearly so). The role of the enzyme is then largely organizational. In path B, some nucleophilic group on the chorismate mutase enzyme leads the process and causes fragmentation of the carbon–oxygen bond. The group is subsequently expelled. For more detail, see Walsh, C. T.; Liu, J.; Fusnak, F.; Sakaitani, M. *Chem. Rev.*, **1990**, *90*, 1105.

3. The Citric Acid Cycle (or the Tricarboxylic Acid [TCA] Cycle or the Krebs Cycle).* It will be recalled (Schemes 11.24–11.26) that the enzyme phosphoglycerate mutase (EC 5.4.2.1) acts on PGA (obtained from, e.g., fructose-1,6-bisphosphate) to produce the isomeric, 2-phosphoglycerate and that phosphopyruvate hydratase (EC 4.2.1.11) then converts the 2-phosphoglycerate to phosphoenolpyruvate.

After the conversion of phosphoenolpyruvate to pyruvate (Scheme 11.26) with the aid of the enzyme pyruvate kinase (EC 2.7.1.40), pyruvate is decarboxylated (thiamine diphosphate) and oxidized (lipoic acid), and the thioester (acetyl-CoA) is produced (EC 1.2.4.1) (Schemes 11.30 and 11.31). The fate of some of the acetyl-CoA ($CH_3COSCoA$) produced this way has been discussed above.

After Sir Hans Adolf Krebs (Nobel Prize, 1953, "for his discovery of the citric acid cycle") who identified and studied the cycle. See Krebs, H. A. *Perspect. Biol. Med.*, **1970, *14*, 154.*

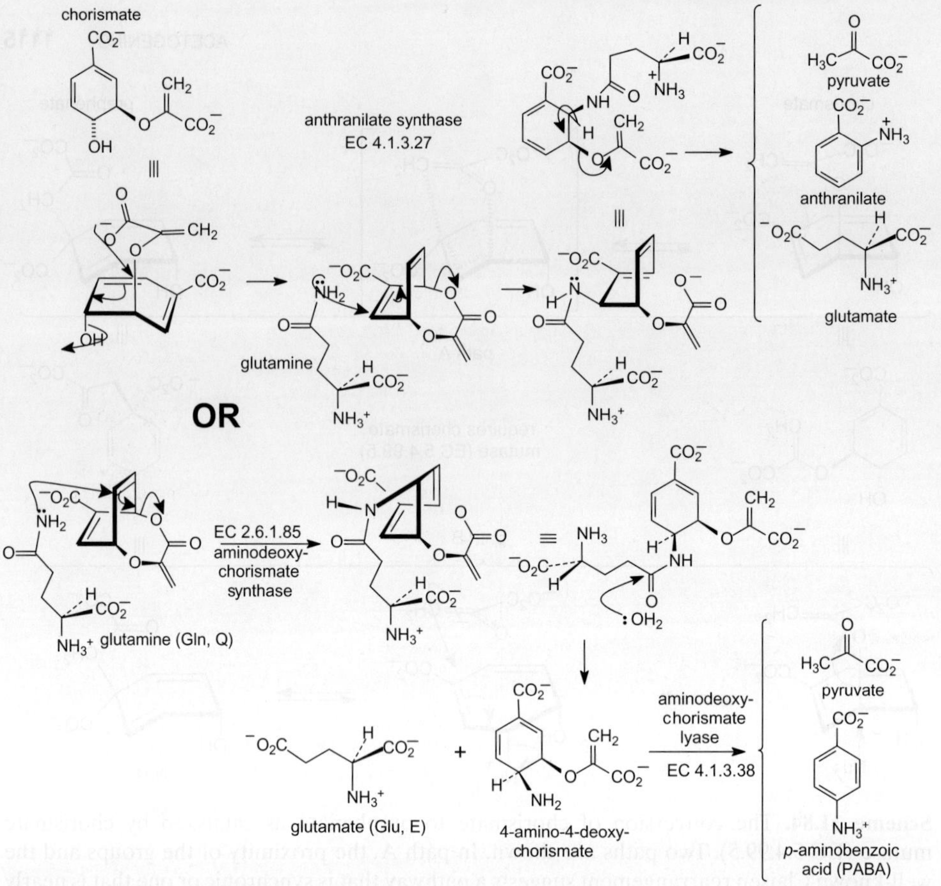

Scheme 11.85. A representation of the conversion of chorismate to anthranilate (under the control of anthranilate synthase, EC 4.1.3.27) and the conversion of chorismate to *p*-aminobenzoic acid (PABA), under the control of aminodeoxychorismate synthase (EC 2.6.1.85), to 4-amino4-deoxychorismate and thence to 4-aminobenzoic acid and pyruvate. The attractive common bicyclic lactone intermediate shown, although not required, is based on a suggestion of Walsh, C. T.; Liu, J.; Rusnak, F.; Sakaitani, M. *Chem. Rev.*, **1990**, *90*, 1105. EC numbers and some graphic materials provided in this scheme have been taken from appropriate links in a URL starting with http://www.chem.qmul.ac.uk/iubmb/enzyme/.

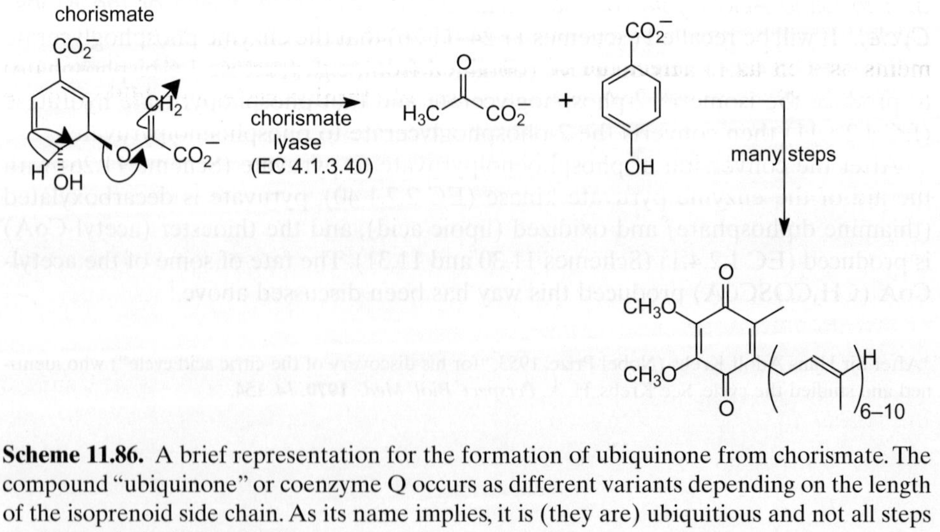

Scheme 11.86. A brief representation for the formation of ubiquinone from chorismate. The compound "ubiquinone" or coenzyme Q occurs as different variants depending on the length of the isoprenoid side chain. As its name implies, it is (they are) ubiquitious and not all steps are known in its biosynthesis.

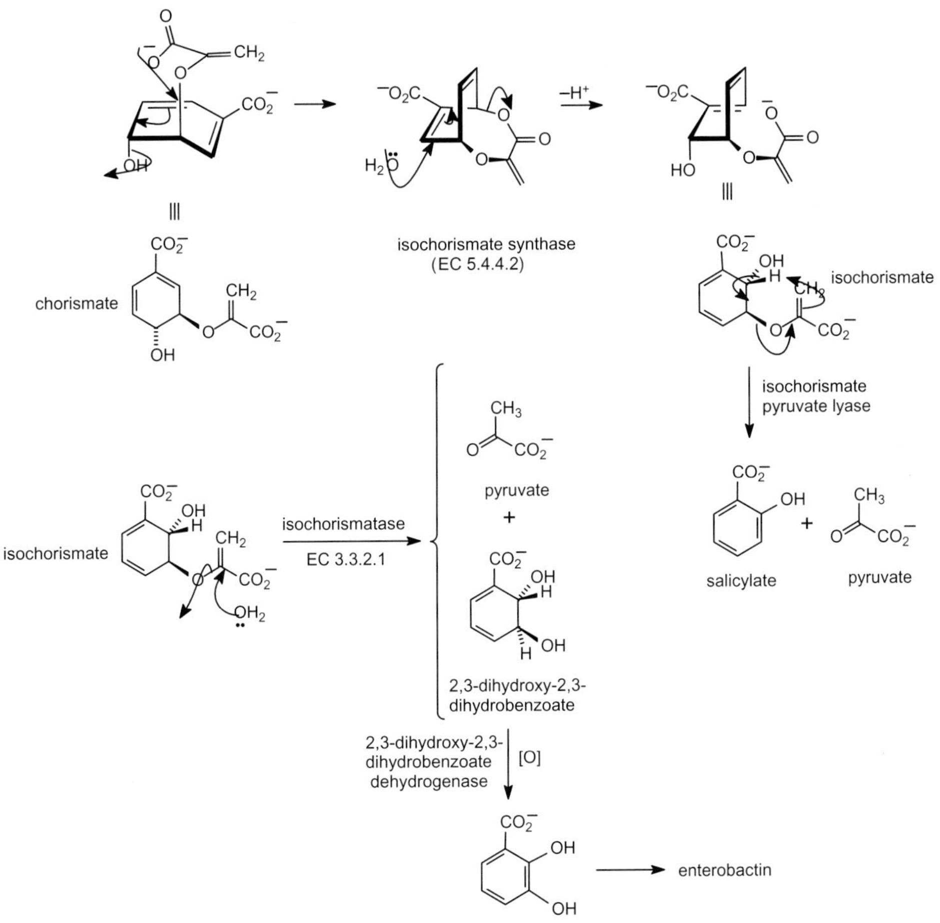

Scheme 11.87 reaction diagram with labeled structures: chorismate, isochorismate synthase (EC 5.4.4.2), isochorismate, isochorismate pyruvate lyase, isochorismatase EC 3.3.2.1, pyruvate, 2,3-dihydroxy-2,3-dihydrobenzoate, salicylate, pyruvate, 2,3-dihydroxy-2,3-dihydrobenzoate dehydrogenase, [O], enterobactin.

Scheme 11.87. A representation of the conversion of isochorismate to *o*-hydroxybenzoic acid (salicylate) and to 2,3-dihydroxybenzoic acid (required for enterobactin biosynthesis).

In addition to polyketides and mevalonic acid-derived species, acetyl-CoA moves through the TCA (or the citric acid cycle, or the Krebs cycle), a cycle (which by definition cannot have a beginning or end) that serves to produce many useful fragments as well as to effect the complete oxidation of pyruvic acid. As shown in Scheme 11.89, as acetyl-CoA enters the already turning cycle, an addition to the carbonyl group of oxaloacetate occurs (citrate synthase, EC 2.3.3.1) to produce citrate. In the addition process, the methyl group of the acetyl-CoA becomes the *pro-S* carbon in citrate.

Problem 11.13. As shown in Scheme 11.89, if the methyl group of acetyl-CoA becomes the *pro-S* carbon in citrate, has the addition occurred to the *re-* or *si*-face of the carbonyl?

Then, continuing in Scheme 11.89, isomerization of citrate (aconitase, aconitate hydratase, EC 4.2.1.3) to isocitrate occurs. This isomerization passes through

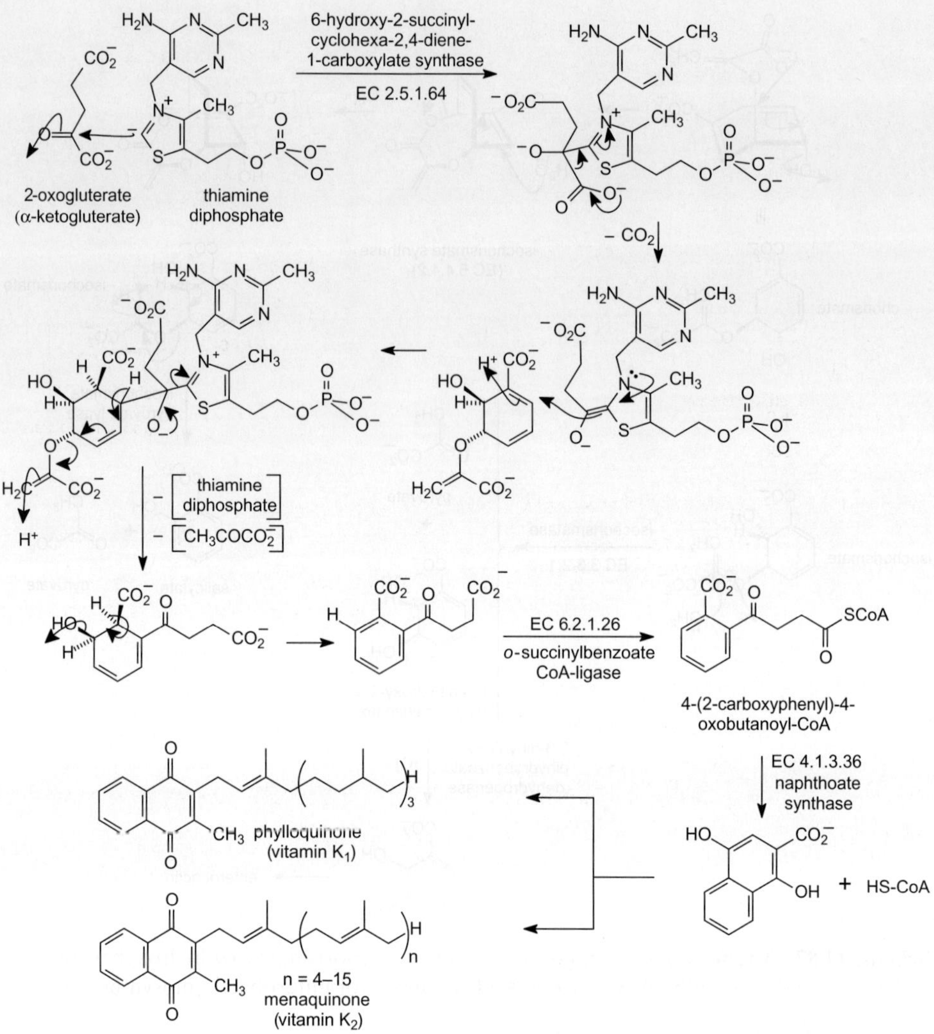

Scheme 11.88. Although the enzymes are (apparently) not yet known for the addition of the terpene fragments onto the napthoate (that is subsequently oxidized to the quinone), some of the enzymes for the earlier steps have been identified. The early steps in the biosyntheses of the vitamins K are from a suggestion of Emmons, G. T.; Campbell, I. M.; Bentley, R. *Biochem. Biophys. Res. Commun.*, **1958**, *131*, 956.

cis-aconitate. In the dehydration, water is eliminated from the subunit that was originally the oxalacetate moiety in a *trans*-manner (or antarafacial) from the conformer in which the carboxyl on the oxalacetate-derived end is *anti* to the carboxymethyl group derived from the introduced acetate. In the hydration (aconitate hydratase, EC 4.2.1.3) process that converts *cis*-aconitate to *erythro*-(2*R*,3*S*)-isocitrate, the hydroxyl adds to the position from which the hydrogen was lost in the preceding step and vice versa. This appears to be a *trans* (or antarafacial) addition.

Isocitrate (*erythro*-(2*R*,3*S*)-isocitrate) is oxidized (isocitrate dehydrogenase, EC 1.1.1.41) to oxalosuccinate and the latter undergoes decarboxylation to α-

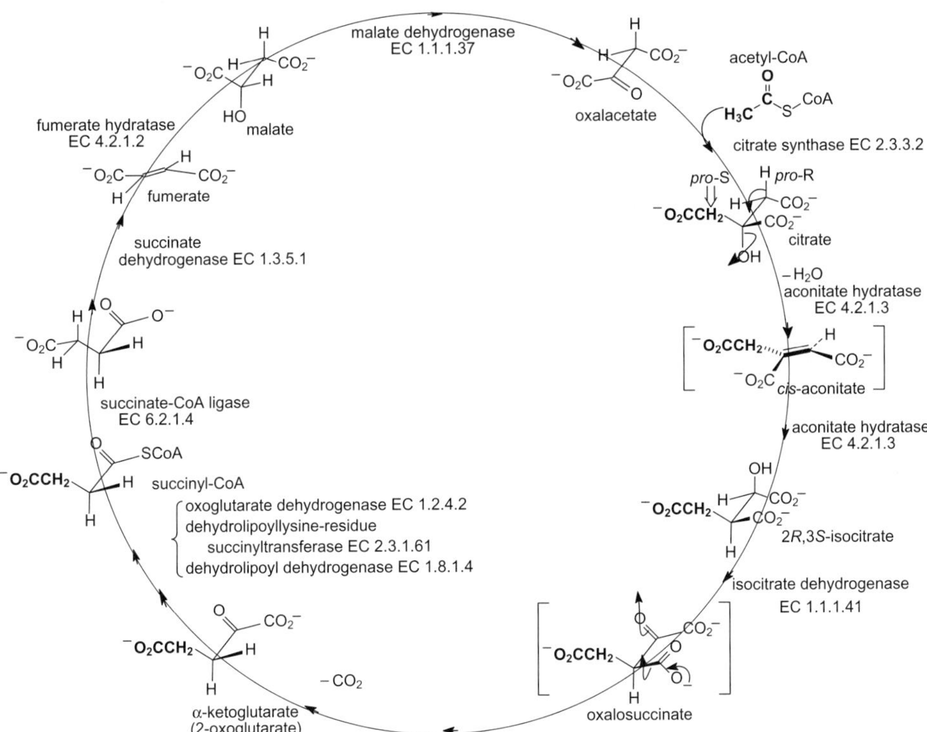

Scheme 11.89. A representation of the tricarboxylic acid (TCA) (or citric acid or Krebs cycle, after Sir Hans Adolf Krebs, Nobel Prize, 1953, who identified and studied the cycle) for the complete oxidation of pyruvate.

ketoglutarate (2-oxoglutarate) by loss of the carboxyl group (as CO_2). This is the decarboxylation of a β-ketoacid, and it may occur concomitant with the oxidation in which it forms. The carboxyl is replaced with retention by a hydrogen from exogenous water and which is now the *pro-S* hydrogen and, after conversion of glutarate to succinate—see below—will be the *pro-S* hydrogen in succinate.

So, α-ketoglutarate (2-oxoglutarate) undergoes a decarboxylation by a path that is similar to the decarboxylation of pyruvate to acetyl-CoA (CH_3CO-CoA, acetyl-CoA) as seen in Schemes 11.30 and 11.31 (*vide supra*). That is, both thiamine diphosphate and a lipoic acid residue are needed as well as CoA itself to effect the conversion to the thioester of CoA. The conversion of α-ketoglutarate (2-oxoglutarate) to succinyl-CoA requires a multienzyme complex that consists of (1) oxoglutarate dehydrogenase, EC 1.2.4.2; (2) dihydrolipoyllysine-residue succinyltransferase, EC 2.3.1.61; (3) dihydrolipoyl dehydrogenase, EC 1.8.1.4.

Succinyl-CoA is hydrolyzed (succinate-CoA ligase, EC 6.2.1.4) to succinate, which then undergoes dehydrogenation (involving the conversion of ubiquinone to ubiquinol, succinate dehydrogenase, EC 1.3.5.1) to fumarate in a process that involves the loss of one *pro-R* and one *pro-S* hydrogen. Fumarate is then hydrated (fumarate hydratase, EC 4.2.1.2) to malate. The addition of water occurs stereospecifically to yield only S-malate. On the adjacent carbon, the *pro-R* hydrogen is derived from exogenous sources and the *pro-S* hydrogen is from the fumarate.

S-malate is oxidized (malate dehydrogenase, EC 1.1.1.37) to oxaloacetate. And the cycle continues.

ADDITIONAL PROBLEMS

Problem 11.14. Compare and contrast the reducing agents NADH (or NADPH) and $FADH_2$. Describe how they differ and how they are similar.

Problem 11.15. Show how the Kiliani–Fischer synthesis can be used to prepare all four D-pentoses from the corresponding D-erythrose and D-threose.

Problem 11.16. Show how the Wohl degradation can be used to convert each of the four D-pentoses to the corresponding tetrose (either erythrose or threose).

Problem 11.17. Use the information provided in Appendix II to find information for compounds about which you want to learn more.

REFERENCES

Appel, H. H.; Brooks, C. J. W.; Overton, K. H. *J. Chem. Soc.*, **1959**, 3322.

Bassham, J.; Benson, A.; Calvin, M. *J. Biol. Chem.*, **1950**, *185*, 781.

Corey, E. J.; Hartmann, R.; Vatakencherry, P. A. *J. Am. Chem. Soc.*, **1962**, *84*, 2611.

Corey, E. J.; Mitra, R. B.; Uda, H. *J. Am. Chem. Soc.*, **1964**, *86*, 485,

Dev, S. *Tetrahedron*, **1960**, *8*, 171.

Enzell, C.; Erdtman, H. *Tetrahedron*, **1958**, *4*, 361.

Gant, T. G.; Noe, M. C.; Corey, E. J. *Tetrahedron Lett.*, **1995**, 8745.

Haworth, W. N. *Nobel Lectures, Chemistry 1922–1941*, Elsevier, Amsterdam, **1966**, pp. 415–432.

Machlis, L.; Nutting, W. H.; Williams, M. W.; Rapoport, H. *Biochemistry*, **1966**, *5*, 2147.

Martin-Smith, M.; de Mayo, P.; Smith, S. J.; Stenlake, J. B.; Williams W. D. *Tetrahedron Lett.*, **1964**, 2391.

Paknikar, S. K.; Bhattacharyya, S. C. *Tetrahedron*, **1962**, *18*, 1509.

Quaderer, R.; Omura, S.; Ikeda, H.; Cane, D. E. *J. Am. Chem. Soc.*, **2006**, *128*, 13036.

Sakai, T.; Nishimura, K.; Hirose, Y. *Bull. Chem. Soc. Jpn.*, **1965**, *38*, 381.

Simpson, R. F.; McQuilkin, R. M. *Phytochemistry*, **1976**, *15*, 328.

An Introduction to Amino Acids, Peptides and Proteins, Enzymes, Coenzymes, and Metabolic Processes

Let philosophers dabble in science and babble
Bout Oxy-gin, Hydro-gin, Nitro-gin's fame
For their gin, to my thinking, is not worth the drinking
Their labour's all lost and their learning a drame [dream]
—"The Humors of Whisky" (Traditional)

A. INTRODUCTION

All of the amino acids, peptides and proteins, enzymes, and coenzymes, materials with which this chapter deals, contain carbon, hydrogen, oxygen, and nitrogen. Some contain sulfur, phosphorus, selenium, and other metals (e.g., iron, magnesium, molybdenum, and copper), but these metals, as well as the other elements, are always found with C–, H–, N–, and O-containing species.

In the previous chapter (Chapter 11), cycles in which carbon-, hydrogen-, and oxygen-containing fragments underwent oxidation, reduction, and rearrangement were considered. Further, Chapter 9 provided some information about carboxylic acids and Chapter 10 about amines. The main concern of this chapter is to consider the role of those compounds where the acidic carboxylic acid and the basic amino functionalities coexist in the same molecules.

There are 20 "common" α-amino acids that are found in all living systems (Table 12.1), and each enjoys a more or less unique biosynthesis from fragments found in Chapter 11 and a more or less unique degradation back to those same fragments. Generally, as shown in Table 12.1, the amino acids have the "L" (in the Fischer projection nomenclature, Chapter 7) or "S" configuration, although the "R" enantiomers are also found—but rarely.

Since the amino acids have both amino (basic) and carboxylic acid (acidic) groups, they are commonly written as zwitterions with both charges shown. However, it will be appreciated that in basic solution, the proton on the amino group (and the carboxylic acid too) is lost and the species that results is an anion. Contrarily, in

Foundations of Organic Chemistry: Unity and Diversity of Structures, Pathways, and Reactions,
First Edition. David R. Dalton.
© 2011 John Wiley & Sons, Inc. Published 2011 by John Wiley & Sons, Inc.

TABLE 12.1. The 20 Common L-Amino Acids (Names, Abbreviations, Structures, pK$_a$'s)*

Name (Abbreviations)	Structure	pK$_a$ (–COOH)	pK$_a$ (NH$_3$)	pK$_a$ (Side Chain)	Pl	NMR (D$_2$O)
Glycine (Gly, G)		2.35	9.58		5.97	^{1}H 3.542 (D$_2$O) ^{13}C 44.11 (CH$_2$)
Alanine (Ala, A)		2.35	9.67		6.01	^{1}H 3.77 (CH), 1.47 (CH$_3$) ^{13}C 18.85 (CH$_3$), 53.2 (CH), 178.58 (CO$_2$)
Valine (Val, V)		2.30	9.62		5.96	^{1}H 3.59 (α-CH), 2.26 (β-CH), 1.03, 0.98 (CH$_3$) ^{13}C 19.33, 20.66 (CH$_3$), 31.8 (β-CH), 63.1 (α-CH), 177.03 (CO$_2$)
Leucine (Leu, L)		2.33	9.64		5.98	^{1}H 3.72 (α-CH), 1.71 (β-CH$_2$), 1.71 (CH), 0.95 (CH$_3$) ^{13}C 23.58, 24.72 (CH$_3$), 26.86 (γ-CH), 42.52 (β-CH$_2$), 56.11 (α-CH), 178.33 (CO$_2$)
Isoleucine (Ile, I)		2.36	9.60		6.02	^{1}H 3.66 (α-CH), 1.46 (γ-CH$_2$), 1.96 (β-CH), 0.99, 0.93 (CH$_3$) ^{13}C 13.8, 17.4 (CH$_3$), 27.2 (γ-CH2), 38.6 (β-CH), 62.25 (α-CH), 176.9 (CO$_2$)
Methionine (Met, M)		2.28	9.21		5.74	^{1}H 2.12 (CH$_3$), 2.63 (S-CH$_2$), 2.15 (β-CH$_2$), 3.85 (α-CH) ^{13}C 16.63 (β-CH$_2$), 31.51 (CH$_3$), 32.40 (S-CH$_2$), 56.60 (α-CH), 177.0 (CO$_2$)
Proline (Pro, P)		1.95	1C.64		6.30	^{1}H 2 (γ-CH$_2$), 2.33 (β-CH$_2$), 3.33, 3.39 (N–CH$_2$), 4.12 (α-CH) ^{13}C 31.69 (β-CH$_2$), 26.45 (γ-CH$_2$), 48.76 (N–CH$_2$), 63.92 (α-CH), 177.4 (CO$_2$)

Amino acid	Structure	pK_1	pK_2	pK_3	pI	NMR data
Phenylalanine (Phe, F)	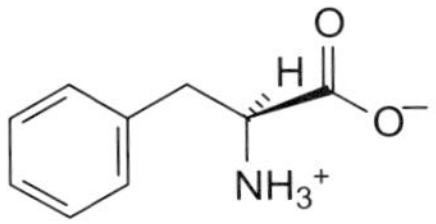	1.80	9.16		5.48	^{1}H 7.41, 7.38, 7.32 (Ar–H), 3.97 (CH), 3.20 (CH_2) ^{13}C 39.10 (CH_2), 58.75 (CH), 130.40, 131.83, 132.08, 137.82 (Ar–C), 176.76 (CO_2)
Tryptophan (Trp, W)	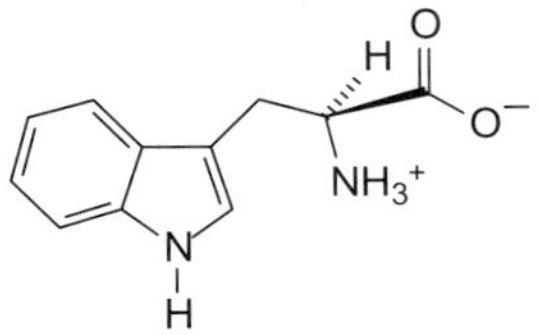	2.86	9.41		5.89	^{1}H 7.72, 7.57, 7.30, 7.27, 7.19 (Ar–H), 4.04 (CH), (CH_2), 3.20 (CH_2) ^{13}C 29.10 (CH_2), 57.76 (CH), 114.66, 121.16, 122.16, 124.83, 127.74, 129.36, 139.03 (Ar–C), 177.26 (CO_2)
Serine (Ser, S)	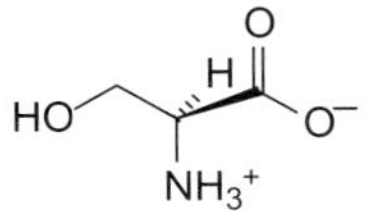	2.19	9.21		5.68	^{1}H 3.96 (CH_2), 3.83 (CH) ^{13}C 59.08 (CH_2), 62.89 (CH), 175.17 (CO_2)
Threonine (Thr, T)	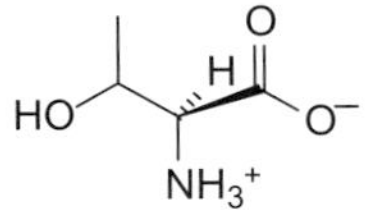	2.09	9.10		5.60	^{1}H 1.32 (CH_3), 3.57 (α-CH), 4.24 (β-CH) ^{13}C 22.14 (CH_3), 63.14 (α-CH), 69.64 (β-CH), 175.63 (CO_2)
Asparagine (Asn, N)	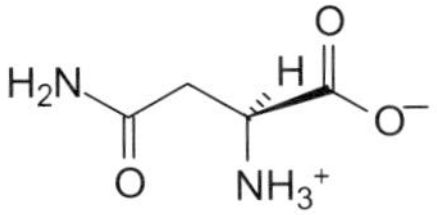	2.02	8.82		5.41	^{1}H 3.99 (CH), 2.94, 2.84 (CH_2) ^{13}C 37.3 (CH_2), 53.99 (CH), 176.21 (C=O), 177.17 (CO_2)
Glutamine (Gln, Q)	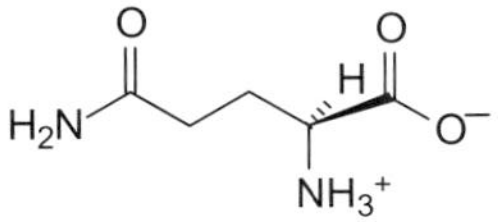	2.17	9.13		5.65	^{1}H 2.14 (β-CH_2), 2.46 (γ-CH_2), 3.76 (CH) ^{13}C 28.95 (β-CH_2), 33.53 (γ-CH_2), 56.83 (CH), 176.83 (C=O), 180.37 (CO_2)
Tyrosine (Tyr, Y)	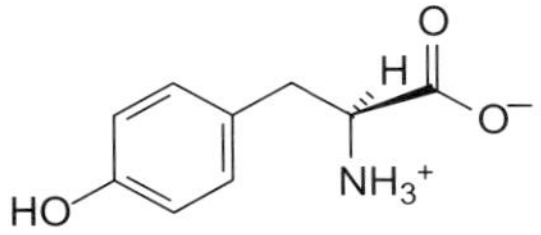	2.20	9.21	10.06	5.66	^{1}H 7.04, 6.67 (Ar–H), 3.49 (CH), 2.82 (CH_2) ^{13}C 41.83 (CH_2), 60.04 (CH), 120.30, 128.05, 133.40, 163.79 (Ar–C), 184.22 (CO_2)

(Continued)

TABLE 12.1. *Continued*

Name (Abbreviations)	Structure	pK_a (–COOH)	pK_a (NH_3)	pK_a (Side Chain)	PI	NMR (D_2O)
Cysteine (Cys, C)		1.92	10.30	8.17	5.07	^{1}H 3.94 (CH), 3.04 (CH_2) ^{13}C 27.69 (CH_2), 58.73 (CH), 175.56 (CO_2)
Lysine (Lys, K)		2.16	9.06	10.54	9.74	^{1}H 1.44–1.90 (CH_2's), 2.99–3.0 (N–CH_2), 3.75 (CH) ^{13}C 24.13 (CH_2), 29.11 (CH_2), 32.61 (CH_2), 41.76 (N–CH_2), 57.20 (CH), 177.41 (CO_2)
Arginine (Arg, R)		2.17	8.99	12.48	10.76	^{1}H 1.66 (CH_2), 1.90 (CH_2), 3.27 (N–CH_2), 3.76 (CH) ^{13}C 26.57 (CH_2), 30.29 (CH_2), 43.2 (CH_2), 57.03 (N–CH), 159.5 (CN_3), 177.21 (CO_2)
Histidine (His, H)		1.80	9.23	6.04	7.59	^{1}H 3.23, 3.26 (CH_2), 3.98 (CH), 7.05, 7.79 (Het–CH) ^{13}C 30.78 (CH_2), 57.47 (CH), 119.55 (HCN_2), 134.67 (NCH), 139.97 (NCC), 176.65 (CO_2)
Aspartic acid (Asp, D)		1.89	9.70	3.70	2.77	^{1}H 2.64, 2.80 (CH_2), 3.89 (CH) ^{13}C 39.26 (CH_2), 54.92 (CH), 177 (CO_2), 180.3 (CO_2)
Glutamic acid (Glu, E)		2.20	9.57	4.27	3.22	^{1}H 2.07, 2.38 (CH_2), 3.75 (CH) ^{13}C 29.69,36.17 (CH_2), 57.34 (CH), 177.29 (CO_2), 184.04 (CO_2)

*The data in this table are taken from Dawson, R. M. C.; Elliott, D. C.; Elliott, W. H.; Jones, K. M. *Data for Biochemical Research*, 3rd edition, Oxford Science Publication, Oxford, **1986**, pp.1–31. Nuclear magnetic resonance (NMR) data are from the Biological NMR Data Bank (http://www.bmrb.wisc.edu).
^{15}N values, although widely available, are not reported because they are pH dependent.

acidic solution, both the carboxylic acid and the amino group are protonated and the cation results. Of course, for those five common amino acids (aspartic acid, glutamic acid, arginine, histidine, and lysine) that have additional carboxylic acid (aspartic acid and glutamic acid) or additional amino (arginine, histidine, and lysine) functional groups, the situation is slightly different in that dianions and/or dications result, but, in general, the principle applies that both cations and anions result as a function of the pH of the medium.

Interestingly, therefore, in addition to the pK$_a$'s shown in Table 12.1, there is also shown a value called **pI** (for **isoelectric point**). The pI value is that value of the pH at which the anionic and cationic forms are balanced and the zwitterionic amino acid can be thought of as neutral. As can be quickly seen, the pI is (approximately) the average of the two acid dissociation constants (pK$_a$'s) all determined under the same conditions for those amino acids with neutral side chains. Those with basic side chains (lysine, histidine, and arginine) have pIs that are approximated by the average of the two highest pK$_a$'s, and those with acidic side chains have pI's approximated by the average of the two lowest pK$_a$'s.

Now, if an aqueous solution of a mixture of amino acids at some fixed pH is placed on absorbent paper and the paper is placed between a pair of electrodes, when current is applied, cationic species will migrate toward the cathode and anionic species toward the anode. The distance through which the amino acid migrates is a function of the electrical current and the pH. At its pI, an amino acid remains stationary! This powerful tool for separating amino acids is called **electrophoresis**.

The individual α-amino acids joined together make peptides (two yielding a dipeptide, three a tripeptide, and many, a polypeptide). In peptides, the α-amino group of one is joined to the carboxylate of the next, thus (with the loss of water) forming an amide bond.

The specific sequence of amino acids of which the polypeptide is made is called the **primary structure**, and its formation is controlled by that piece of deoxyribonucleic acid (**DNA**) (Chapter 14) that dictates, through the genetic "code" (*vide infra*), which amino acid the machinery (e.g., the **ribosome** [about which more will be said later]) that joins them is to use and in what order the amino acids are to be combined. In principle, the ribosome could build a linear chain from "right to left" or "left to right," so, by convention, peptides are written with the N-terminal amino acid to the left and the carboxylate terminus on the right.

Additionally, it is now clear that the primary structure is derivatized by other processes such as phosphorylation, alkylation, and glycosylation as it leaves the ribosome and that these processes help govern the folding of the growing peptide as the backbone of amino acids increases. In this process, a secondary structure is generated.

The **secondary structure** of the peptide refers to the three-dimensional aspect of various domains of the protein. These domains may be helical or sheetlike and may also include the polypeptide threadlike connectors of the helices and sheets (Figure 12.1). Interestingly, given the many possibilities for folding patterns that can be imagined for any peptide of any reasonable length (e.g., 100–1000 amino acids long), only one appears to form spontaneously as the chain leaves the ribosome, and that one is the "correct" one for the protein's function.

When the end sequence is read and fully grown protein has finally separated from the ribosome, the separated, domain-organized, partially derivatized giant molecule

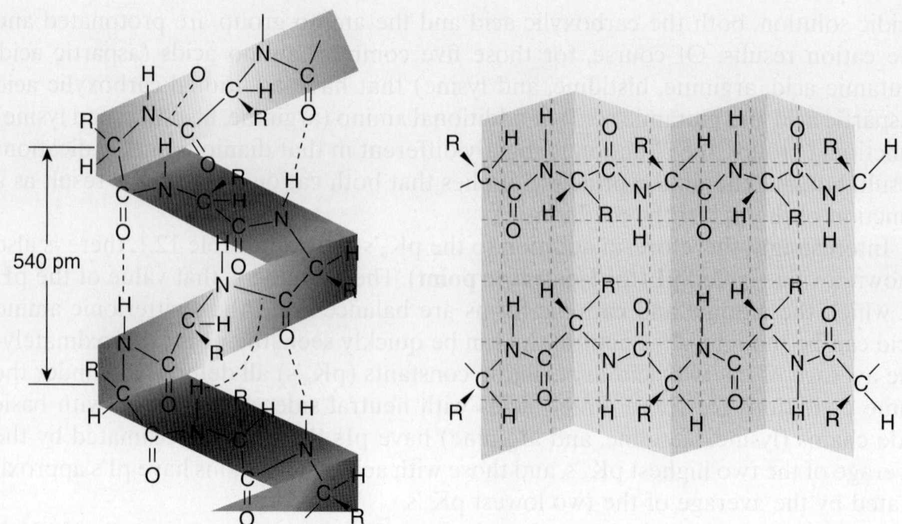

Figure 12.1. Cartoon representations of α-helix and β-sheet structures of polypeptides (proteins).

assumes a unique shape as a result, in part, of overall intramolecular interactions. This "final" shape, remaining subject to substrate-induced changes, is called the **tertiary structure**.

An aggregation of these proteins after assembly is the **quaternary structure**, and it is not unknown for the aggregate to be needed for a specific function.

Static versions of these common forms, having been crystallized, are readily seen in the Protein Data Bank (PDB) (http://www.wwpdb.org) for a variety of proteins.

The large proteins (peptides) that catalyze intra- and intercellular reactions are called **enzymes**. The enzymes (as pointed out in Chapter 11) serve to catalyze the reactions between simple substrates by providing an environment in which the reactants can be brought into the appropriate positions for reaction. Generally, these catalysts are very specific and, as will be recalled (Chapter 11), they are organized by the type of reaction that they affect (e.g., see the information to be found through the International Union of Biochemistry and Molecular Biology at http://www.iubmb.org).

The small molecules that assist in the catalytic processes enzymes promote and that are not themselves proteins are called **coenzymes**. Again, it will be recalled that these molecules are used to help consummate the reactions between substrates. So, for example, the coenzyme pyridoxal (vitamin B$_6$, Table 12.2) can (among other functions already discussed in Chapter 11) be utilized by a transaminase enzyme to help move a nitrogen from an amino acid (via initial reaction at an active-site lysine *in the active site of the enzyme*) to the α-carbonyl of an α-ketocarboxylic acid and vice versa (the process is called **transamination**). The pyridoxal (Table 12.2) has served as a catalytic "nitrogen carrier." In some cases where pyridoxal (vitamin B$_6$) is used as a coenzyme, it is altered in the process. Thus, for example, in amino acid decarboxylation, pyridoxamine (HC=O replaced by H$_2$CNH$_2$) can result and an additional step is required to reconvert it to pyridoxal. Indeed, such changes are common among coenzymes.

Name	Structure	Function
Pyridoxal phosphate	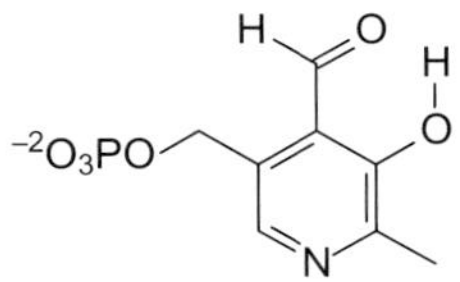	Amino acid metabolism, aldo-type reactions
Lipoic acid	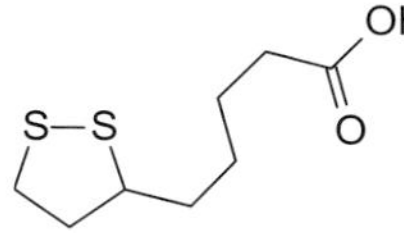	Acyl transfer reactions
Thiamine diphosphate	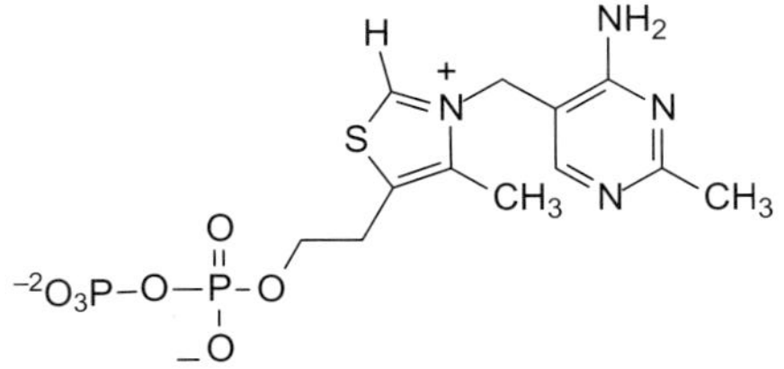	Decarboxylation
Biotin	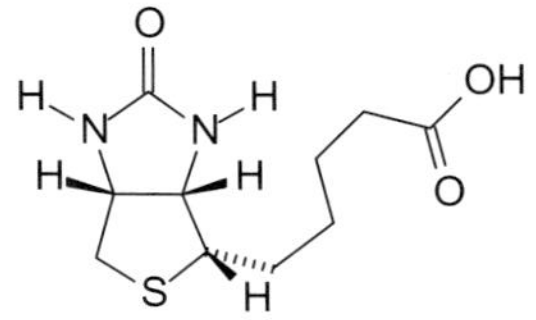	Carboxylation
Adenosine triphosphate (ATP)	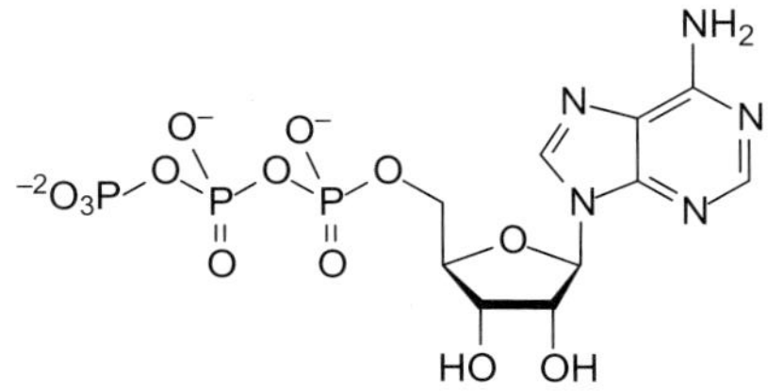	Phosphorylation
Nicotinamide adenine dinucleotide (NAD$^+$) and (NADP$^+$)	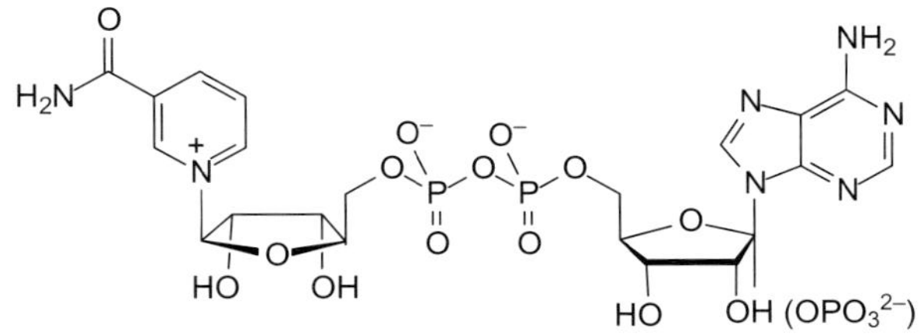	Oxidation and reduction

(Continued)

TABLE 12.2. *Continued*

Name	Structure	Function
Coenzyme A (CoA-SH)	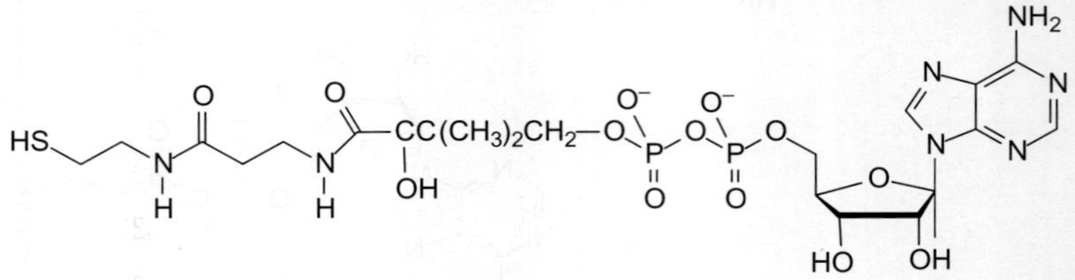	Acyl transfer
Flavin adenine dinucleotide (FAD)	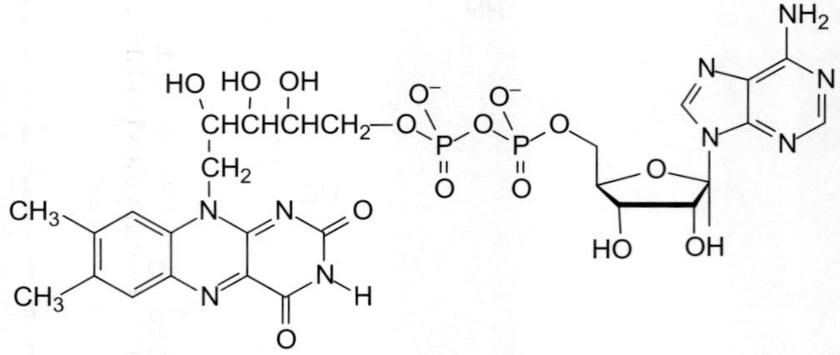	Oxidation and reduction
S-Adenosylmethionine	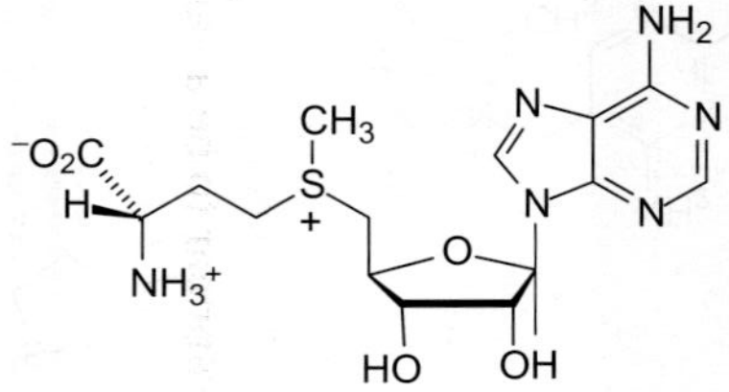	Methyl transfer
Tetrahydrofolate	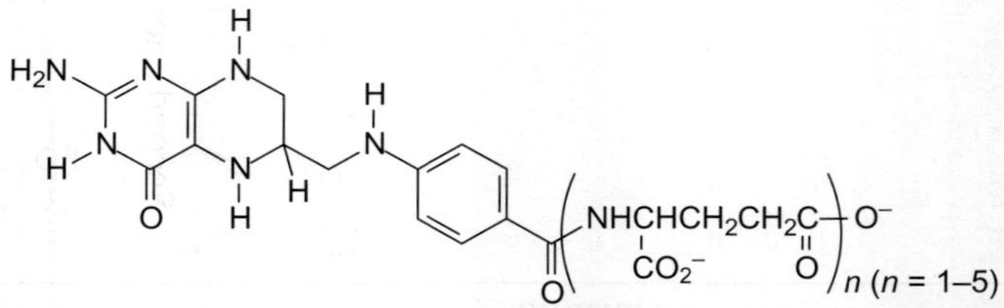	One–carbon (C_1) transfer agent, Often as an equivalent of "$H_2C{=}O$"

Table 12.2 provides a list of some of the common coenzymes, many of which have already been encountered in Chapter 11.

Finally, for this introduction, **metabolic processes** are, in short, all of the reactions that must occur for life to be maintained. The processes themselves are generally divided into **catabolic** (or **exergonic** [but not necessarily in the form of work] or **energy releasing**) paths as in the breakdown of food in cellular respiration and **anabolic** (or **endergonic** [but not necessarily in the form of work] or **energy utilizing**) paths that are used to construct proteins and nucleic acids, cell walls, and so on.

The metabolic processes utilize the building blocks discussed in Chapter 11 (carbohydrates, acetogenins, and steroids) and amino acids (discussed here) along paths mediated by enzymes and coenzymes so that thermodynamically unfavorable reactions can be coupled to favorable ones. Finally, although a discussion of more than only a small fraction of the total number of reactions involved in metabolic processes (some of which have already been met in Chapter 11) is beyond the scope of this work, it is hoped that an appreciation of the details of the common elements of metabolism across animal and plant kingdoms and of the building blocks used throughout will serve to affirm the unity of all life.

B. AMINO ACIDS

I. Biosynthesis

Because all of the amino acids contain nitrogen and because carbon metabolism was treated (at least in part) in Chapter 11, the initial question in the biosynthesis of amino acids deals with nitrogen metabolism. The knotty question of the genesis of some form of nitrogenous compound(s) so as to allow interaction with the fragments met in Chapter 11 must be faced.

First, ferredoxin-nitrate reductase (EC 1.7.7.2) is a molybdenum–iron–sulfur protein that converts nitrate to nitrite while undergoing oxidation. Its structure is not yet well-defined. Then, ferridoxin-nitrite reductase (EC 1.7.7.1), an iron (as heme and at least one iron sulfur complex, *vide infra*) can continue the reduction to ammonia (NH_3).

Second, there appear to be a variety of enzymes (in the general, EC 1.7 reductase rubric) with more or less well-defined structures that reduce a variety of other oxides of nitrogen. However, the real challenge, given the composition of the earth's atmosphere, is the reduction of nitrogen (N_2) to ammonia (NH_3), and, to that end, there appear to be a number of different enzymes in the generalized rubric "nitrogenase" as found in the PDB.

The bacteria (*diazotrophs*) from which the enzymes are isolated are found living in symbiosis with root nodule cells of legumes, and all appear to promote the same reaction (Equation 12.1), which, boldly put, is the conversion of atmospheric nitrogen to ammonia.

$$N_2 + 8H^+ + 8e^- + 16ATP + 16H_2O \rightarrow 2NH_3 + H_2 + 16ADP + 16PO_4^{-3} \quad (12.1)^*$$

*Although the charges in Equation 12.1 *appear* to be unbalanced, the full number of charges on the adenosine triphosphate (ATP), which is converted to adenosine diphosphate (ADP), is not noted. As a consequence, PO_4^{-3} is often written as P_i to avoid the problem of "missing charges."

In the nitrogenase enzyme isolated from *Azotobacter vinelandii* (e.g., see PDB 1MIN, 2MIN, and 3MIN entries), it appears that there are three sites required for the process given by Equation 12.1 to be consummated. One site is called the "4Fe-4S cluster," a second site is referred to as the "P-cluster," and the third site is the "MoFe cofactor" (sometimes called the "M-center") or "MoFe protein" where, some have argued, the actual reduction occurs.

It has been appreciated for some time that protein-bound iron–sulfur clusters are important in biological electron transfer reactions and, reasonably, they can be found in a variety of proteins associated with redox systems. Although data are incomplete, the extent to which they are effective in any particular system (i.e., their measured reduction potential) appears to be a function of their position (e.g., buried vs. exposed) in the protein. However, in general, as shown in Figure 12.2, the 4Fe-4S cluster is surrounded by protein-bound cysteine residues, forming sulfur–iron bonds external to the cluster, holding it in place.

Now, it appears that the actual reduction occurs in the so-called MoFe protein. This site has the other two metal groups, the P-cluster and the MoFe cofactor. It is often argued that the P-cluster serves the role as electron transfer agent from the 4Fe-4S site to the MoFe cofactor, lying as it does at the interface between subunits (nitrogenase appears to be a tetramer that has two different pairs of subunits) (PDB 1MIN, 2MIN, and 3MIN). A representation of the P-cluster is provided in Figure 12.3 and one of the MoFe cofactor in Figure 12.4. Details of the reduction and the actual use of these three interrelated iron–sulfur domains remain unknown at this time.

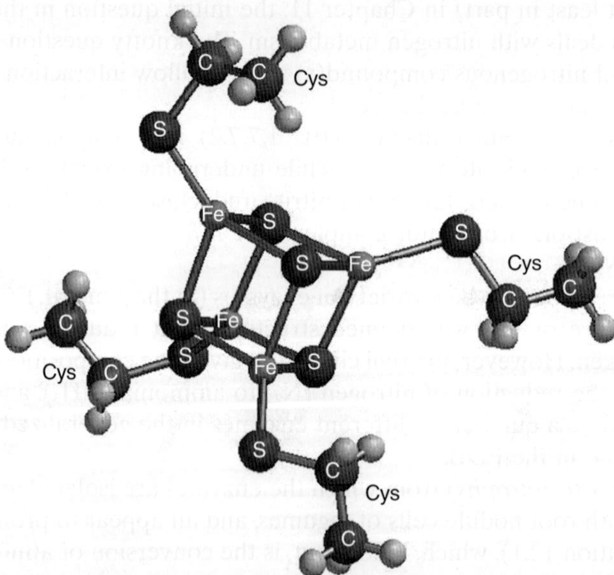

Figure 12.2. A representation of the 4Fe-4S cluster of the nitrogenase enzyme isolated from *A. vinelandii*. It is argued that the cysteine residues of the protein (shown) that form sulfur–iron bonds hold it in place in the enzyme. (The figure is repeated here so that it is clear that a source of sulfur is available. It can be argued that radicals developed as a result of the abstraction of hydrogen by the 5′-deoxyadenosin-5′-yl radical encourage sulfur capture from the 4Fe-4S complex.)

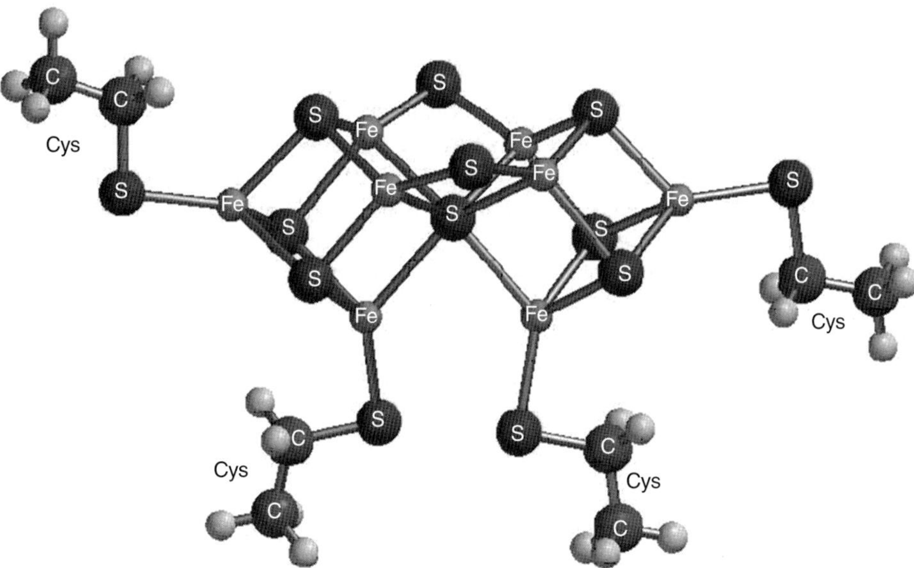

Figure 12.3. A representation of the P-cluster of the nitrogenase enzyme isolated from *A. vinelandii*. It is argued that the cysteine residues of the protein (shown) that form sulfur–iron bonds hold it in place in the enzyme.

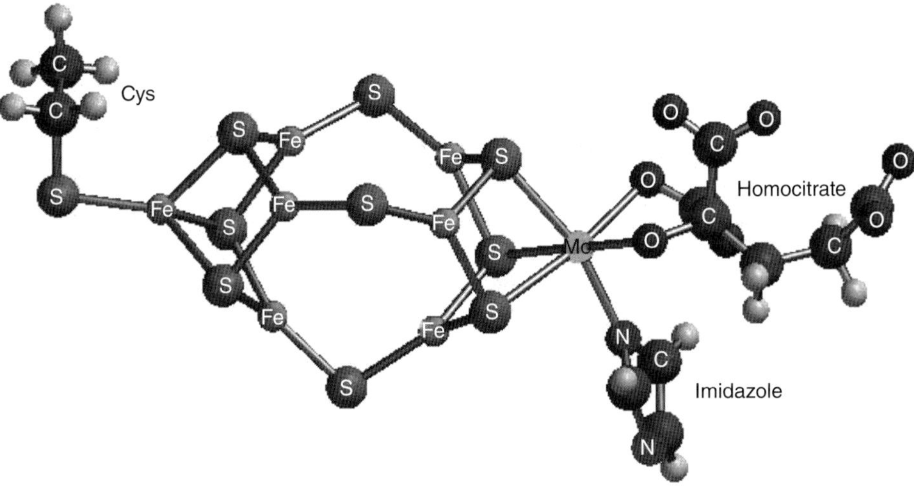

Figure 12.4. A representation of the MoFe cofactor of the nitrogenase enzyme isolated from *A. vinelandii*. It is argued that the cysteine residue on the one end and the imidazole on the other (which also bears a homocitrate bound to molybdenum) immobilize this in the active site where the reduction of nitrogen actually occurs.

Then, after nitrogen is reduced to ammonia, the latter must be absorbed, transported, and converted into nitrogenous materials. Here, the latter are the amino acids.

The transport of ammonia appears to occur through cell wall "ammonia membrane channels" called ammonia transporters (**Amt proteins**). Ammonia, having been transported, is, at least in principle, capable of being utilized by, for example, the enzyme aspartase (aspartate ammonia-lyase, EC 4.3.1.1) that catalyzes the reversible deamination of **aspartate** (**Asp, D**) to fumarate (Equation 12.2).

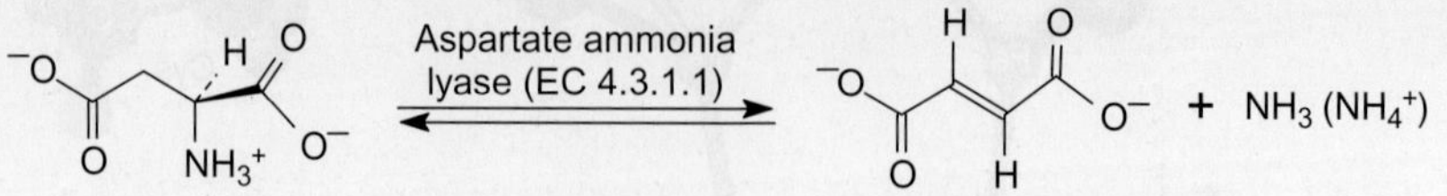

$$\text{(aspartate)} \quad \xrightarrow[\text{lyase (EC 4.3.1.1)}]{\text{Aspartate ammonia}} \quad \text{(fumarate)} \; + \; NH_3 \, (NH_4^+) \qquad (12.2)$$

Interestingly, although a number of such reversible ammonia lyases are known, only a few seem to need pyridoxal as a cofactor, for example, threonine ammonia lyase (EC 4.3.1.19) (Scheme 12.1), where the ammonia is presumably lost via a hydrolytic mechanism to generate the corresponding carbonyl compound or (as is not shown in the Scheme 12.1) the corresponding pyridoxamine but **not** the alkene.

So, the reversible formation of alkenes currently appears limited to a handful of ammonia lyases that include the already noted aspartate ammonia-lyase that produces fumarate (Equation 12.2), methylaspartate ammonia-lyase that yields mesaconate (EC 4.3.1.2) (Equation 12.3), the urocanate-producing histidine

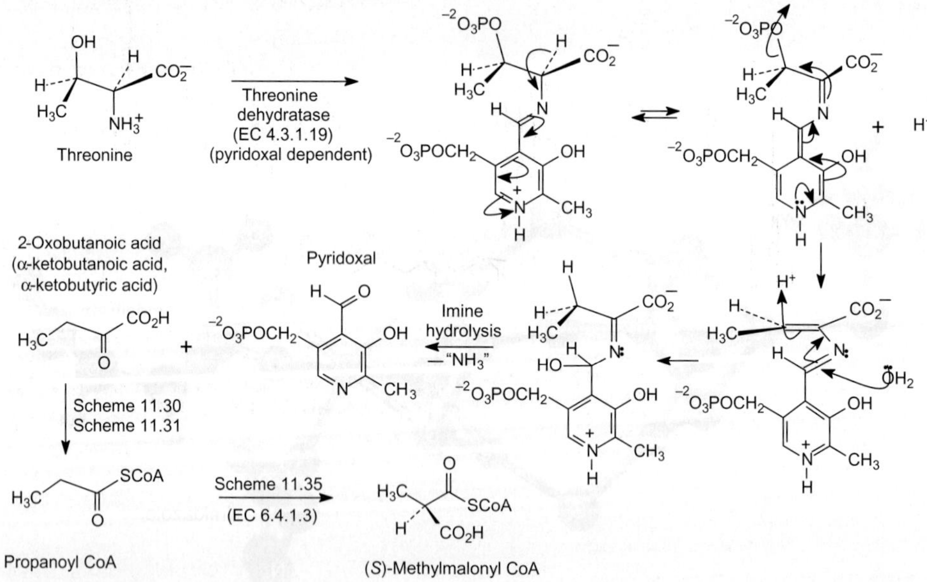

Scheme 12.1. A cartoon representation of a pathway for the deamination of L-**threonine** (**Thr, T**) using pyridoxal as a cofactor. Note that 2-oxobutanoic acid and pyridoxal form (along with, in principle, ammonia, NH₃). The ammonia may actually be retained by the pyridoxal fragment to produce pyridoxamine (after reduction), which can then be used for further transaminations.

ammonia-lyase (EC 4.3.1.3) (Equation 12.4), and phenylalanine ammonia-lyase (EC 4.3.1.5) that generates (E)-cinnamate (Equation 12.5).*

$$\text{Methylaspartate ammonia lyase (EC 4.3.1.2)} \qquad + \ NH_3 \ (NH_4^+) \qquad (12.3)$$

$$\text{Histidine ammonia lyase (EC 4.3.1.3)} \qquad + \ NH_3 \ (NH_4^+) \qquad (12.4)$$

$$\text{Phenylalanine ammonia lyase (EC 4.3.1.5)} \qquad + \ NH_3 \ (NH_4^+) \qquad (12.5)$$

Now, with aspartate (Asp, D) in hand, for example, it should be clear that transamination will allow production of both essential and nonessential amino acids.[†]

Alanine (Ala, A), as shown in Scheme 12.2, can be formed when pyruvate (derived, e.g., from phosphoenolpyruvate via 2-phosphoglycerate, as shown *inter alia* Scheme 11.25) undergoes transamination from aspartate (Asp, D) (produced as in Equation 12.3 from fumarate). Fumarate, it will be remembered (Scheme 11.89), is produced in the tricarboxylic acid cycle. So, as shown in Scheme 12.2, the aminotransferase cofactor, pyridoxal, serves to deaminate aspartate (Asp, D) to produce oxaloacetate. The amino group is then transferred to pyruvate to yield alanine (Ala, A) and the pyridoxal cofactor is regenerated.[‡]

In a similar way, transamination from aspartate (Asp, D) to α-ketoglutarate, which, like fumarate, is produced in the tricarboxylic acid cycle (Scheme 11.89), then yields **glutamate (Glu, E)** and, exactly as in Scheme 12.2, oxalacetate (Equation 12.6).

---

*As shown in Scheme 11.84, chorismate yields prephenate. As will be seen here, *vide infra*, prephenate can yield both phenylpyruvate and p-hydroxyphenylpyruvate. Although these phenylpropanoids can simply cyclize to the corresponding lactones, reductions to the respective alcohols and subsequent dehydrations lead to the respective cinnamates. Amino acids (**phenylalanine [Phe, F]** and **tyrosine [Tyr, Y]**) result from the transamination (respectively) of these ketoacids.

[†]Humans and other animals cannot, it seems, biosynthesize all of the amino acids they need. So, in an anthropomorphic sense, some amino acids have been defined as "essential"; that is, those amino acids that we cannot biosynthesize on our own (or not enough of on our own) and must be obtained from our diet are called essential. Those we can biosynthesize on our own (or with the aid of intestinal biota) are "nonessential." Of the 20 most common amino acids, **11** (viz., alanine [Ala, A], arginine [Arg, R], aspartate [Asp, D], asparagine [Asn, N], cysteine [Cys, C], glutamate [Glu, E] glutamine [Gln, Q], glycine [Gly, G], proline [Pro, P], serine [Ser, S], and tyrosine [Tyr, Y]) are considered **nonessential**, while **9** (viz., histidine [His, H], isoleucine [Ile, I], leucine [Leu, L], lysine [Lys, K], methionine [Met, M], phenylalanine [Phe, F], threonine [Thr, T], tryptophan [Trp, W], and valine [Val, V]) are **essential** and must be obtained from our diet. These amino acids are all **S** (or **L**) and stereoisomers, and other modifications of them, as well as 300 related compounds, that is, also called amino acids (containing both carboxylate and amino functions), are found naturally.

[‡]A pathway to pyridoxal itself from 1-deoxy-D-xylulose-5-phosphate and 4-hydroxy-L-threonine 4-phosphate will be elaborated upon in the discussion of cofactors.

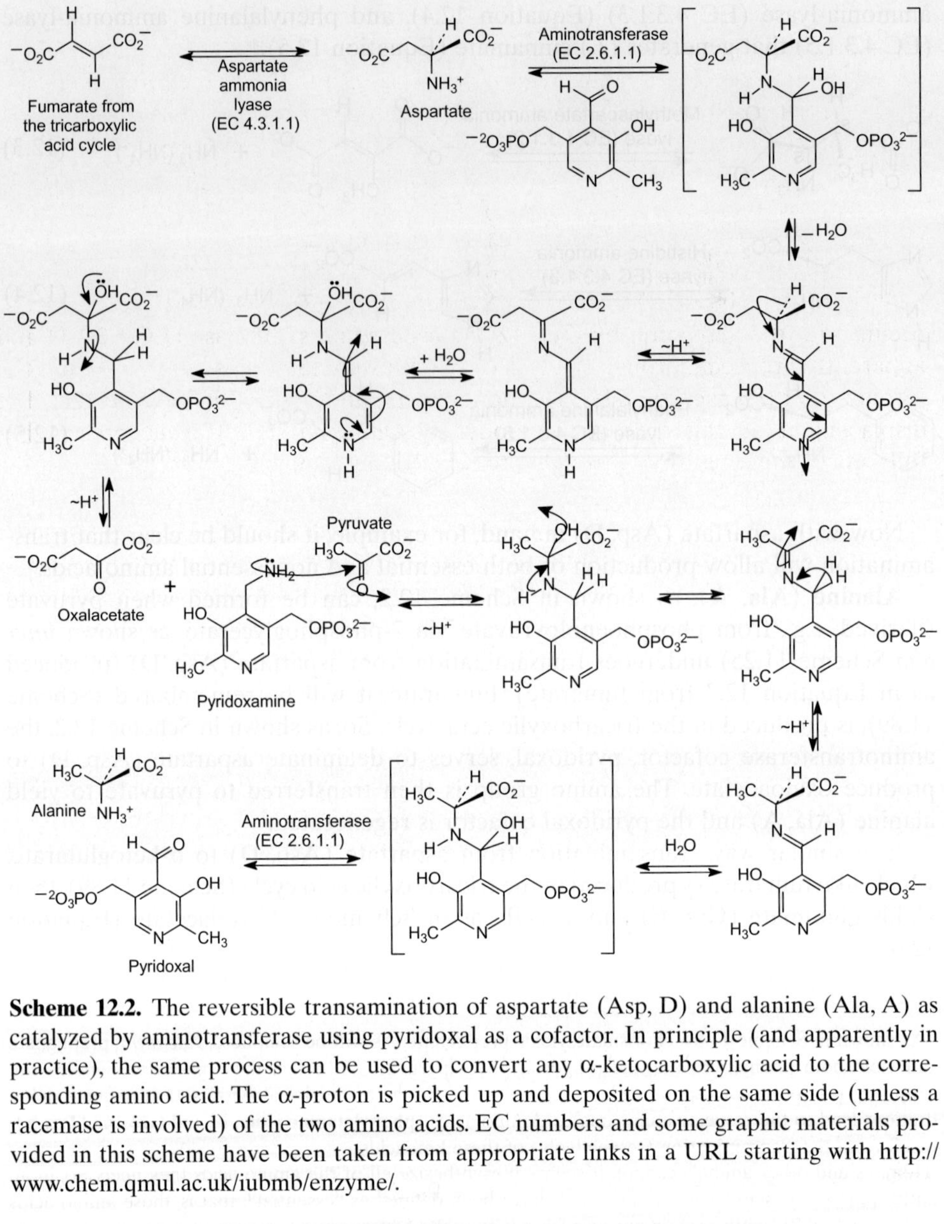

Fumarate from
the tricarboxylic
acid cycle

Aspartate
ammonia
lyase
(EC 4.3.1.1)

Aspartate

Aminotransferase
(EC 2.6.1.1)

$-H_2O$

Oxalacetate

Pyruvate

Pyridoxamine

Alanine

Aminotransferase
(EC 2.6.1.1)

Pyridoxal

Scheme 12.2. The reversible transamination of aspartate (Asp, D) and alanine (Ala, A) as catalyzed by aminotransferase using pyridoxal as a cofactor. In principle (and apparently in practice), the same process can be used to convert any α-ketocarboxylic acid to the corresponding amino acid. The α-proton is picked up and deposited on the same side (unless a racemase is involved) of the two amino acids. EC numbers and some graphic materials provided in this scheme have been taken from appropriate links in a URL starting with http://www.chem.qmul.ac.uk/iubmb/enzyme/.

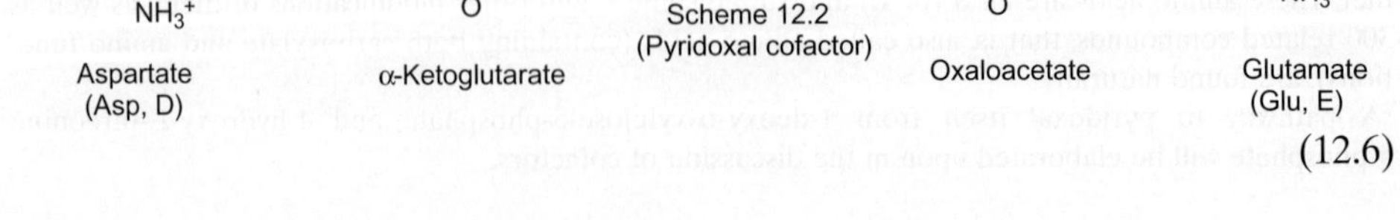

Aspartate
(Asp, D)

α-Ketoglutarate

Aminotransferase
(EC 2.6.1.1)

Scheme 12.2
(Pyridoxal cofactor)

Oxaloacetate

Glutamate
(Glu, E)

$$(12.6)$$

Interestingly, however, the formation of **glutamine (Gln, Q)** from glutamate (Glu, E) requires another enzyme (glutamine synthetase [glutamate-ammonia ligase], EC 6.3.1.2) controlled step to convert the carboxylate to the corresponding amide. As shown in Scheme 12.3, glutamate (Glu, E) undergoes phosphorylation on the carboxylate (C5) to glutamate 5-phosphate by ATP (which is converted to ADP). Ammonia (NH_3) (presumably produced and transmitted through an appropriate channel) displaces the inorganic phosphate. Of course, this is simply the activation of a carboxylic acid for amide formation as seen in Chapter 9.

The conversion of aspartate (Asp, D) to **asparagine (Asn, N)** follows a similar pathway for activation. However, here, the attack of aspartate (Asp, D) onto ATP occurs with the activation by the enzyme asparagine synthetase (EC 6.3.1.1) and apparently proceeds through the loss of two phosphate groups as ($P_2O_7^{-4}$) and the formation of a β-aspartyl adenosyl monophosphate as shown in Scheme 12.4. The displacement on this activated carbonyl by ammonia (NH_3) again, presumably, follows the same pathway through a tetrahedral intermediate.

Glutamate (Glu, E) also gives rise to both **arginine (Arg, R)** and **proline (Pro, P)**.

The direct reduction of glutamate 5-phosphate (Scheme 12.3) with the nicotinamide adenine diphosphate (NADPH)/NADP$^+$-dependent glutamate-5-semialdehyde dehydrogenase (EC 1.2.1.41) as shown in Scheme 12.5 produces glutamate 5-semialdehyde. The aldehyde spontaneously undergoes cyclization to (S)-3,4-dihydro-2H-pyrrole-2-carboxylate, which is then reduced to L-proline (Pro, P). The reduction is accomplished again with the phosphorylated nicotinamide adenine dinucleotide being oxidized while in the presence of the enzyme pyrroline-5-carboxylate reductase (EC 1.5.1.2).

The formation of arginine (Arg, R) from glutamate (Glu, E) takes a different pathway. As shown in Scheme 12.5, an acetyl group is transferred to the α-nitrogen of glutamate (Glu, E) from acetyl coenzyme A (acetyl-CoA) (e.g., see Scheme 11.39)

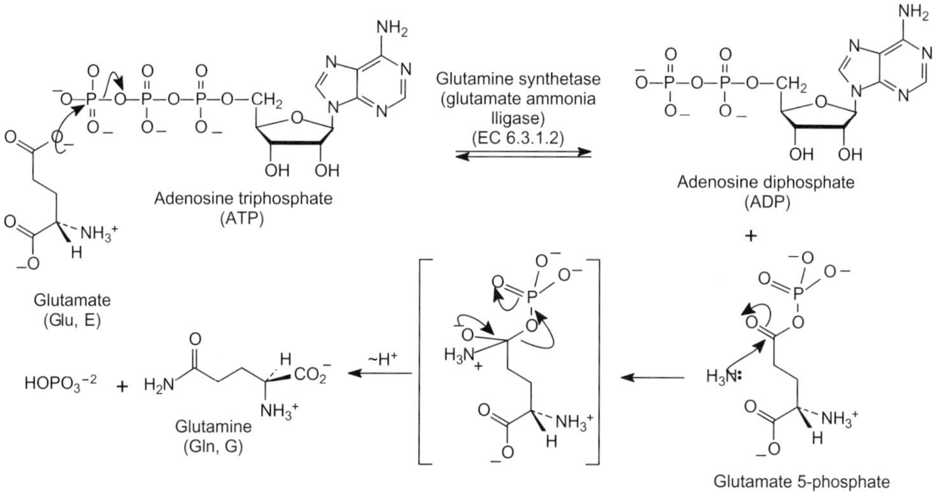

Scheme 12.3. A representation of a pathway for the phosphorylation of glutamate (Glu, E) for the activation of the carboxylate toward substitution and conversion to the corresponding amide, glutamine (Gln, Q). Although "free" ammonia (NH_3) is shown as attacking the carbonyl, it is unlikely that this is the actual process.

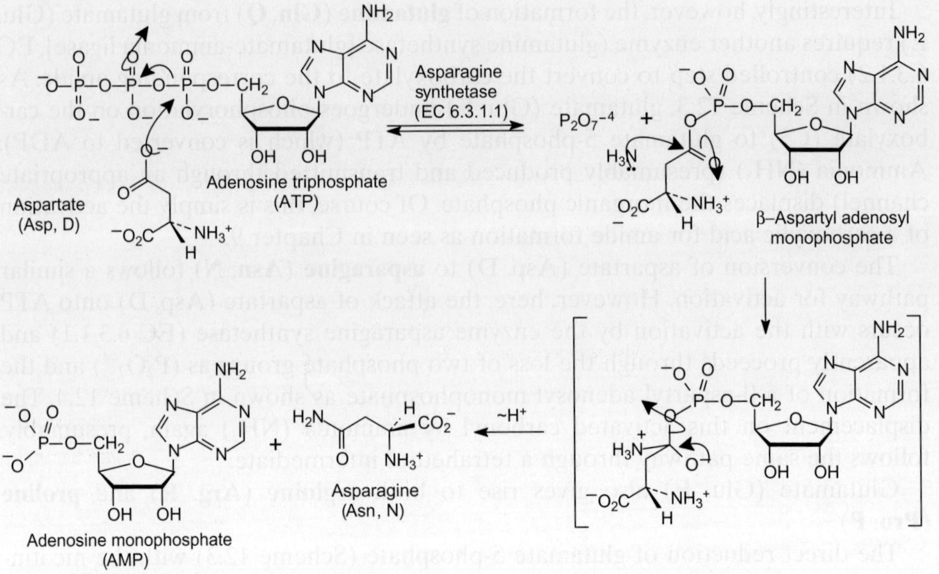

Scheme 12.4. A representation of a pathway for the conversion of aspartate (Asp, D) to the corresponding amide, asparagine (Asn, N). Although "free" ammonia (NH_3) is shown as attacking the carbonyl, it is unlikely that this is the actual process.

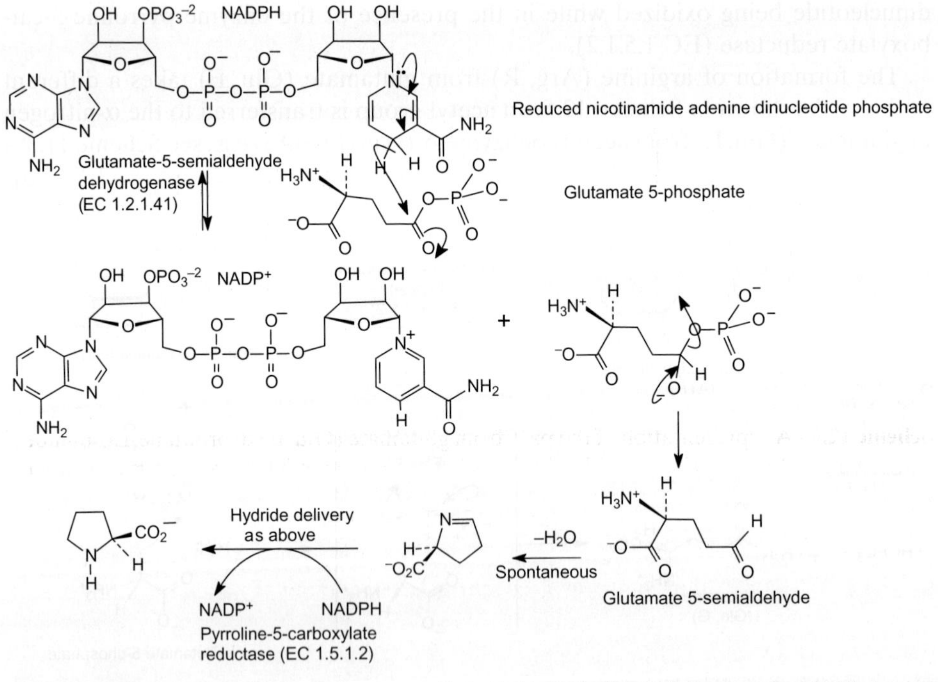

Scheme 12.5. The formation of glutamate 5-semialdehyde from glutamate 5-phosphate and its conversion to L-proline (Pro, P). EC numbers and some graphic materials provided in this scheme have been taken from appropriate links in a URL starting with http://www.chem.qmul.ac.uk/iubmb/enzyme/.

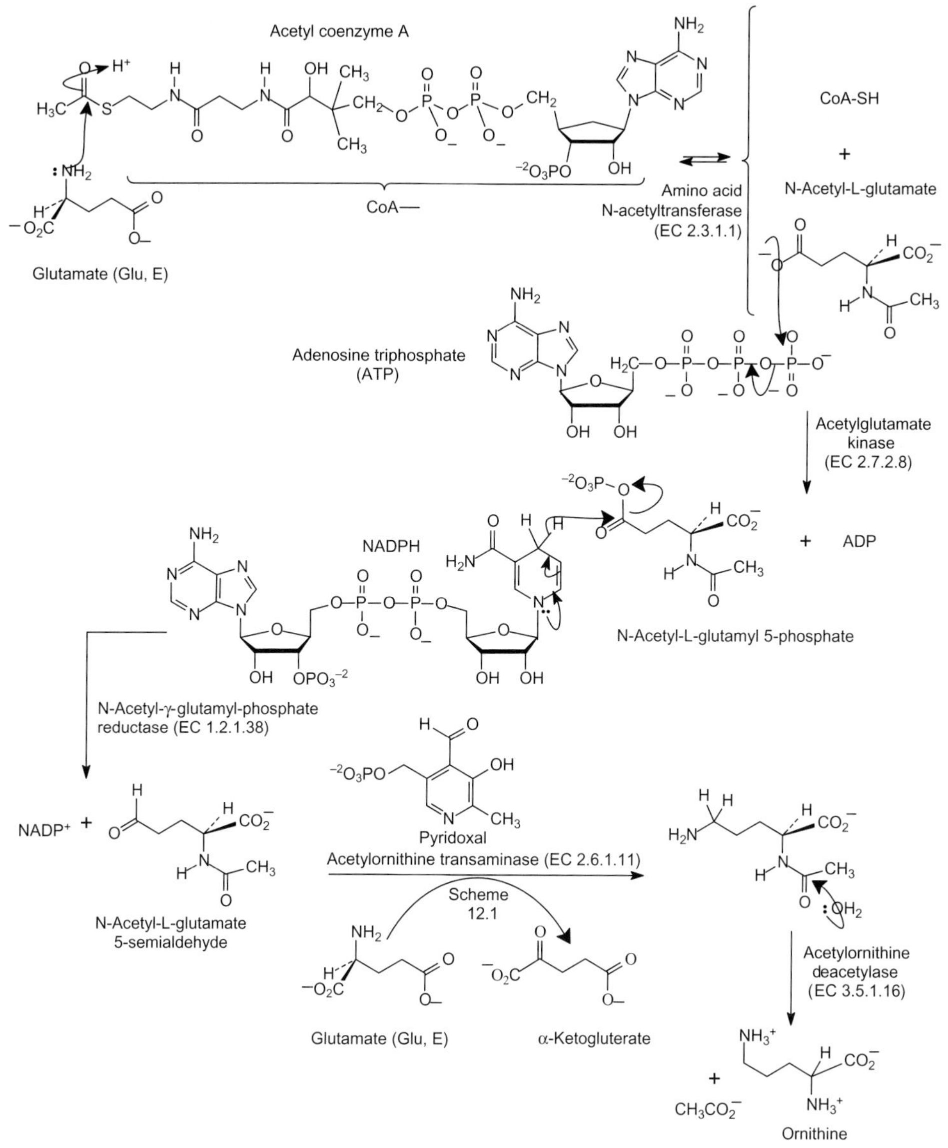

Scheme 12.6. A representation of the path from glutamate (Glu, E) to ornithine. EC numbers and some graphic materials provided in this scheme have been taken from appropriate links in a URL starting with http://www.chem.qmul.ac.uk/iubmb/enzyme/.

to yield N-acetyl-L-glutamate in a reaction catalyzed by the enzyme amino acid N-acetyltransferase (EC 2.3.1.1). Then, phosphorylation on oxygen by acetylglutamate kinase (EC 2.7.2.8) in a process similar to that seen in Scheme 12.3, with ATP being converted to ADP, produces N-acetyl-L-glutamyl 5-phosphate. On reduction with NADPH, which is converted in the process to $NADP^+$ (viz., Scheme 12.6), and in the presence of N-acetyl-γ-glutamyl-phosphate reductase (EC 1.2.1.38), the phosphate at C-5 is replaced by hydrogen and N-acetyl-L-glutamate 5-semialdehyde is

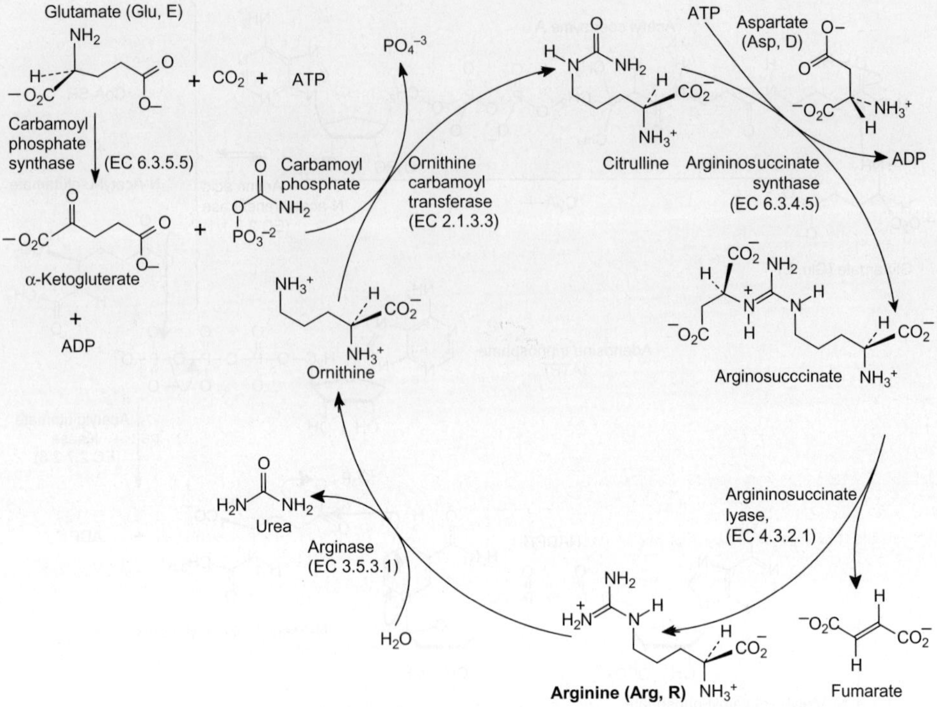

Scheme 12.7. The urea cycle producing both fumarate (cf. the tricarboxylic acid cycle; Chapter 11, Scheme 11.89) as well citrulline and the nonessential amino acid arginine (Arg, R). EC numbers and some graphic materials provided in this scheme have been taken from appropriate links in a URL starting with http://www.chem.qmul.ac.uk/iubmb/enzyme/.

produced. Transamination (the amino group coming from glutamate, which goes to 2-oxoglutarate, as in Scheme 12.1) catalyzed by acetylornithine transaminase (EC 2.6.1.11) then produces N^2-acetylornithine, and deacetylation, a hydrolysis by water, is catalyzed by acetylornithine deacetylase (EC 3.5.1.16). The product is ornithine.

Ornithine is a member of the urea cycle and, as shown in Scheme 12.7, turning the cycle results in the formation of the nonessential (as we synthesize it) amino acid arginine (Arg, R).

As seen in Scheme 12.7, carbamoyl phosphate, generated from glutamine (Glu, Q) (converted to α-ketoglutarate), ATP, and carbon dioxide with the enzyme carbamoyl phosphate synthase* (EC 6.3.5.5) carbamylates the terminal amino group of ornithine using ornithine carbamoyl transferase (EC 2.1.3.3) to generate citrulline. Then, aspartate **adds** to the carbonyl of citrulline to generate argininosuccinate, a process catalyzed by the enzyme argininosuccinate synthase (EC 6.3.4.5) while ATP goes to ADP and inorganic phosphate (PO_4^{-3}) to account for the water (H_2O)

---

*Alternatively, ammonia (NH_3) and carbon dioxide (CO_2), along with two adenosine triphosphate (2ATP) and water (H_2O), can combine in the presence of carbamoyl phosphate synthase (ammonia) (EC 6.3.4.16) to produce carbamoyl phosphate, two adenosine diphosphates (2ADP) and inorganic phosphate (P_i).

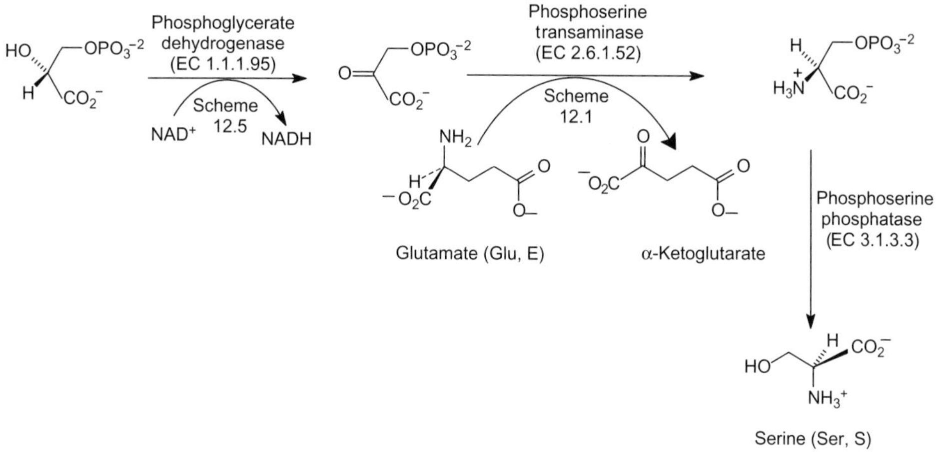

Scheme 12.8. A path to the formation of serine (Ser, S) from 3-phosphoglycerate. EC numbers and some graphic materials provided in this scheme have been taken from appropriate links in a URL starting with http://www.chem.qmul.ac.uk/iubmb/enzyme/.

generated as the imine forms. Elimination of fumarate (argininosuccinate lyase, EC 4.3.2.1) produces arginine (Arg, R) ready for hydrolysis to urea and ornithine. The hydrolysis is catalyzed by arginase (EC 3.5.3.1).

As discussed in Chapter 11, the ubiquitous 3-phosphoglycerate is formed in the Calvin cycle, and this three-carbon fragment is then used throughout nature in the synthesis of larger fragments. The formation of **serine (Ser, S)** is also dependent upon 3-phosphoglycerate. As shown in Scheme 12.8, 3-phosphoglycerate is oxidized to 3-phosphohydroxypyruvate while nicotinamide adenine dinucleotide (NAD^+) is reduced to NADH (as in Scheme 12.6) as catalyzed by the enzyme phosphoglycerate dehydrogenase (EC 1.1.1.95). Then, phosphoserine transaminase (EC 2.6.1.52), a pyridoxal phosphate protein, catalyzes the transfer of the amino group from glutamate (Glu, E) as seen in Schemes 12.1 and 12.5 to produce 3-phosphoserine, and hydrolysis of the latter with phosphoserine phosphatase (EC 3.1.3.3) generates serine (Ser, S).

Glycine (Gly, G) is generated from serine (Ser, S) by the loss of the "elements of formaldehyde (CH_2O)"; that is, while it is possible that formaldehyde is actually formed *in situ*, it is considered unlikely as a free species and so the enzyme catalyzing this process, glycine hydroxymethyltransferase (EC 2.1.2.1), has both pyridoxal and tetrahydrofolate in close proximity in the active site. As shown in Scheme 12.9, pyridoxal-coordinated serine (Ser, S) loses the equivalent of formaldehyde ("CH_2O") to the cofactor tetrahydrofolate to produce 5,10-methylenetetrahydrofolate, a species utilized as a source of methylene units for methylation of other intermediates.

Interestingly, there is an alternate pathway (Scheme 12.10) for the formation of glycine (Gly, G) that involves the transamination (using pyridoxamine, Scheme 12.2) of glyoxylate ($OCHCO_2^-$). Glyoxylate ($OCHCO_2^-$) is formed as part of the "glyoxylate cycle" (Scheme 12.11), a pathway that is found in germinating plants and is generally absent in animals.

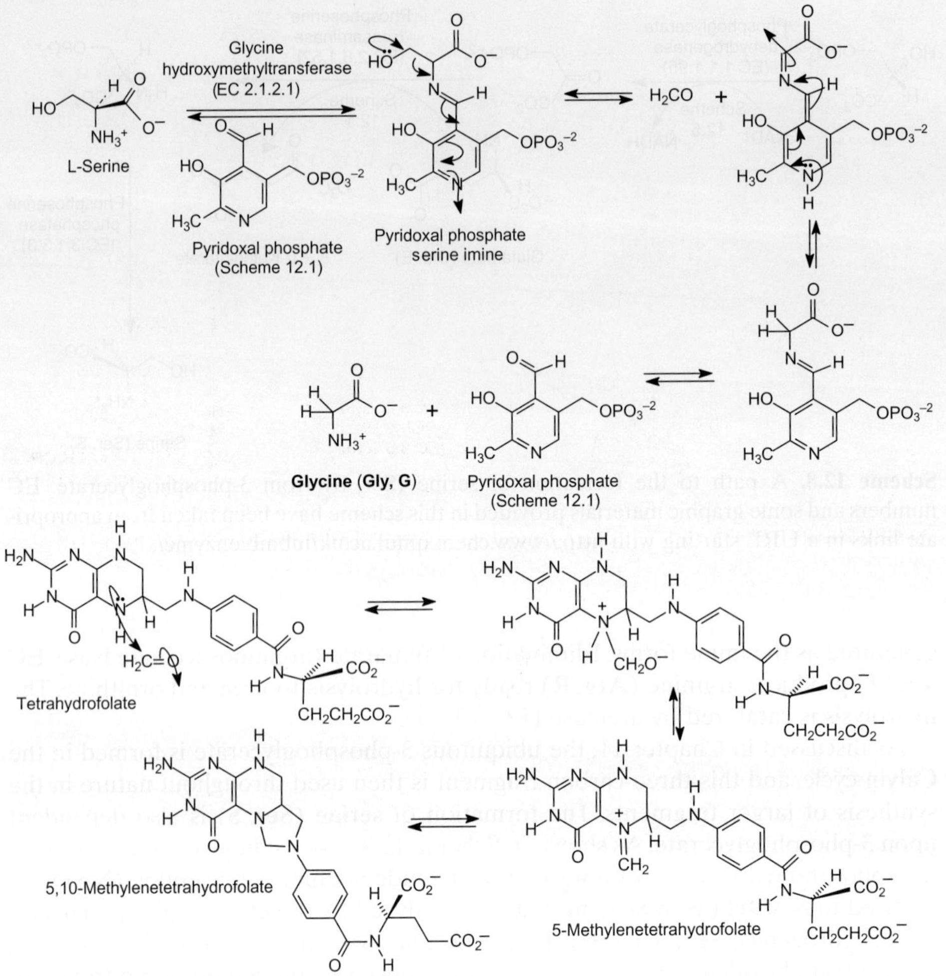

Scheme 12.9. A retroaldol-like cleavage for the removal of a one-carbon unit as an equivalent of formaldehyde CH_2O from serine (Ser, S) to generate the amino acid glycine (Gly, G). The enzyme cavity contains the cofactor pyridoxal as well as the tetrahydrofolate. The latter is converted to 5,10-methylenetetrahydrofolate in the process.

While the initial stages of the citric acid cycle and the glyoxylate ($OCHCO_2^-$) cycle are the same, they differ after the hydration of *cis*-aconitate to isocitrate. In the citric acid cycle (Chapter 11, Scheme 11.89), isocitrate undergoes oxidation (isocitrate dehdrogenase, EC 1.1.1.41) to oxalosuccinate, which then decarboxylates to α-ketoglutarate. In the glyoxylate cycle, isocitrate undergoes cleavage into glyoxylate and succinate (isocitrate lyase, EC 4.1.3.1). Succinate can be moved directly into the citric acid cycle on the way to fumarate (succinate dehydrogenase, EC 1.3.5.1) and eventually to form oxaloacetate. Glyoxylate, in addition to undergoing transamination to produce glycine (Gly, G) (Scheme 12.10), is available to condense with acetyl-CoA (Scheme 12.11) to yield malate (malate synthase, EC 2.3.3.9), which, again, is part of the citric acid cycle (Scheme 11.89).

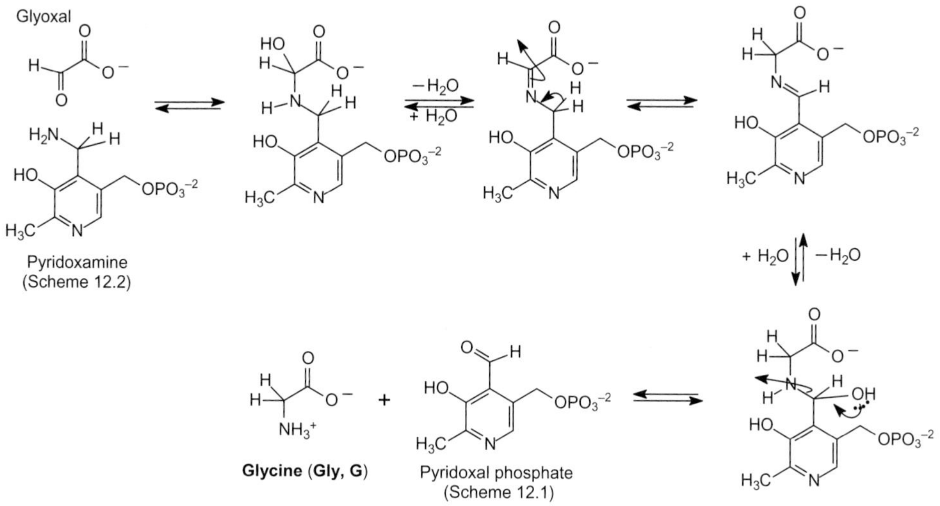

Scheme 12.10. The formation of glycine (Gly, G) by transamination from pyridoxamine to glyoxal.

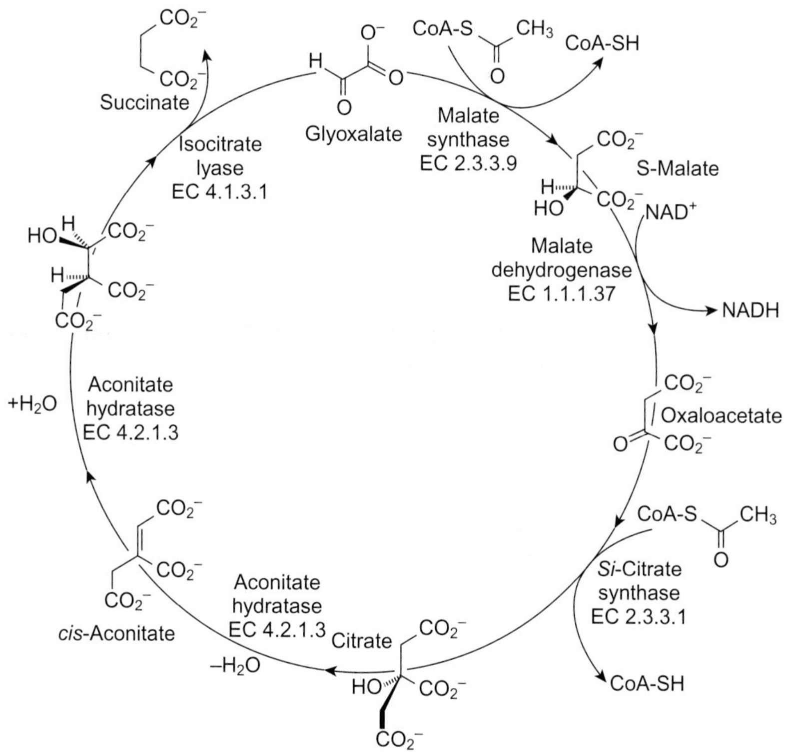

Scheme 12.11. The glyoxylate cycle. In addition to glyoxylate, fragments common to the tricarboxylic acid cycle are also present and can be utilized in that cycle as appropriate.

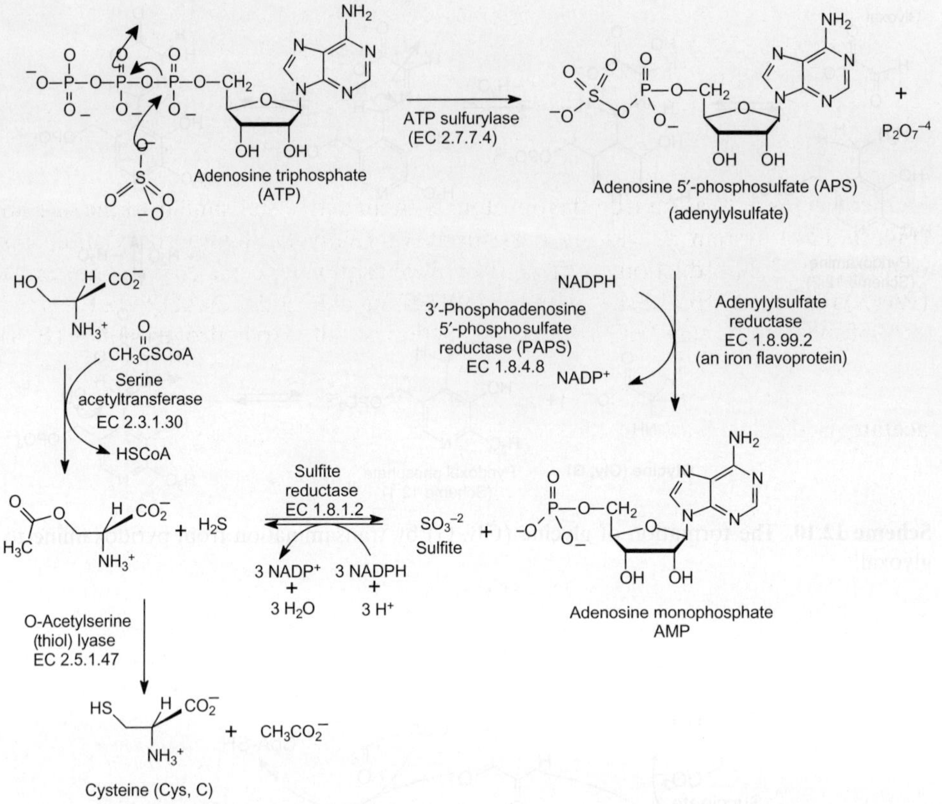

Scheme 12.12. The formation of cysteine (Cys, C) from sulfate (SO_4^{-2}).* EC numbers and some graphic materials provided in this scheme have been taken from appropriate links in a URL starting with http://www.chem.qmul.ac.uk/iubmb/enzyme/.

The last amino acid in the group of "nonessential" amino acids to be considered here (**tyrosine [Tyr, Y]** will be considered along with **phenylalanine [Phe, F]**) is **cysteine (Cys, C)**. The discussion of this member of the class has been postponed until the end because of the necessity of digression into the fate of sulfur in the environment and its subsequent necessary incorporation into the fabric of life. While sulfur (as S_8) is found naturally occurring and while hydrogen sulfide (H_2S) also abounds, it is the highest oxidation state of sulfur (as sulfate, SO_4^{-2}) that is finally found in soils and that is ready for incorporation into cysteine (Cys, C). Thus, as shown in Scheme 12.12, ATP is converted into adenosine 5′-phosphosulfate

*Some reports utilize glutathione and a 2Fe-2S electron transport system for the reduction of sulfate to sulfite.

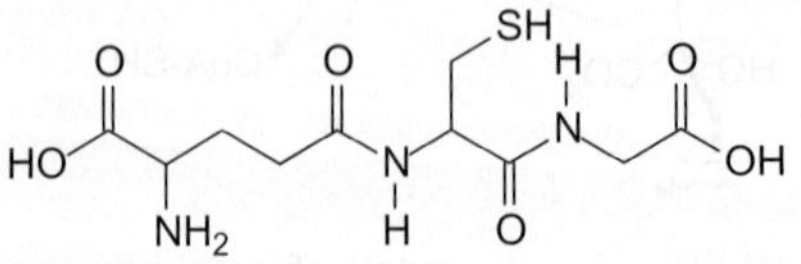

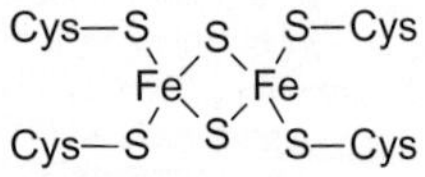

(adenylylsulfate, APS) by ATP sulfurylase (EC 2.7.7.4). Then (perhaps after additional phosphorylation at 3′), reduction of sulfate (SO_4^{-2}) to sulfite (SO_3^{-2}) occurs with NADPH being oxidized to $NADP^+$. Several enzymes have been found capable of carrying out this important reaction. If phosphorylation at C 3′ has *not* occurred, the reduction is effected by adenylylsulfate reductase (EC 1.8.99.2) and $FADH_2$ conversion to flavin adenine dinucleotide (FAD) (cf. Chapter 11, Scheme 11.33) may be involved as well as an iron flavoprotein with an active site similar to the 4Fe-4S (Figure 12.2) system previously described. Alternatively, if phosphorylation has occurred at C 3′, reduction is effected by 3′-phosphoadenosine 5′-phosphosulfate (PAPS) reductase (EC 1.8.4.8) with NADPH being oxidized to $NADP^+$. In the presence of sulfite reductase (EC 1.8.1.2), a reduction of sulfite to hydrogen sulfide (H_2S) with concomitant oxidation of NADPH to $NADP^+$ occurs. Then, in what appears to be an enzyme (O-acetylserine [thiol] lyase, EC 2.5.1.47)-mediated S_N2 reaction, acetate is displaced by the hydrogen sulfide anion (HS^-) to generate cysteine (Cys, C).

Although cysteine (Cys, C) is considered "nonessential" in animals, it is biosynthesized from the "essential" amino acid **methionine** (**Met, M**), which, as it is essential, must be ingested.

So, the biosynthesis of methionine (Met, M), the first of the essential amino acids to be considered (Scheme 12.13), begins by the conversion of aspartate (Asp, D) to aspartate semialdehyde in the same way glutamate (Glu, E) was converted to glutamate semialdehyde (*vide supra*, Scheme 12.6). Phosphorylation on the terminal carboxylate of aspartate (Asp, D) by ATP in the presence of aspartate kinase (EC 2.7.2.4) and subsequent reduction of the aspart-4-yl phosphate by NADPH in the presence of aspartate semialdehyde dehydrogenase (EC 1.2.1.11) yields the aspartate semialdehyde. The aspartate semialdehyde is further reduced to homoserine (homoserine oxoreductase, EC 1.1.1.3) and the latter is succinylated by succinyl-CoA with the liberation of coenzyme A (CoA-SH) in the presence of homoserine O-succinyl-transferase (EC 2.3.1.46). Then, reaction with cysteine (Cys, C) in the presence of cystathionine γ-synthase (EC 2.5.1.48) produces cystathionine and succinate. In the presence of the pyridoxal phosphate protein cystathionine β-lyase (EC 4.4.1.8), *both* ammonia and pyruvate are lost from cystathionine and homocysteine is produced. Finally, methylation on sulfur to generate methionine (Met, M) occurs by the donation of the methyl from 5-methyltetrahydrofolate in the presence of methonine synthase (EC 2.1.1.13).

The biosynthesis of threonine (Thr, T) as shown in Scheme 12.14 also begins with aspartate semialdehyde, which is reduced, as before (homoserine dehydrogenase, EC 1.1.1.3), to homoserine (Scheme 12.13). Now, however, for threonine (Thr, T), the phosphorylating enzyme homoserine kinase (EC 2.7.1.39) effects the transfer of phosphate to the primary hydroxyl of homoserine from ATP, forming ADP and 4-phosphohomoserine. The latter, with threonine synthase (EC 4.2.3.1) and using pyridoxal as cofactor, generates threonine (Thr, T).

With threonine (Thr, T) in hand, the path to the biosynthesis of **isoleucine** (**Ile, I**) is clear. The first part of the biosynthesis is shown in Scheme 12.15 where, in the presence of the pyridoxal (cofactor) protein (enzyme) threonine ammonia lyase (EC 4.3.1.19), threonine (Thr, T) is converted to 2-oxobutanoate. Then, with thiamine diphosphate (cf. Chapter 11, Scheme 11.7) as a cofactor for the enzyme acetolactate synthase (EC 2.2.1.6), an acetyl group is added at C2 of the 2-oxobutanoate

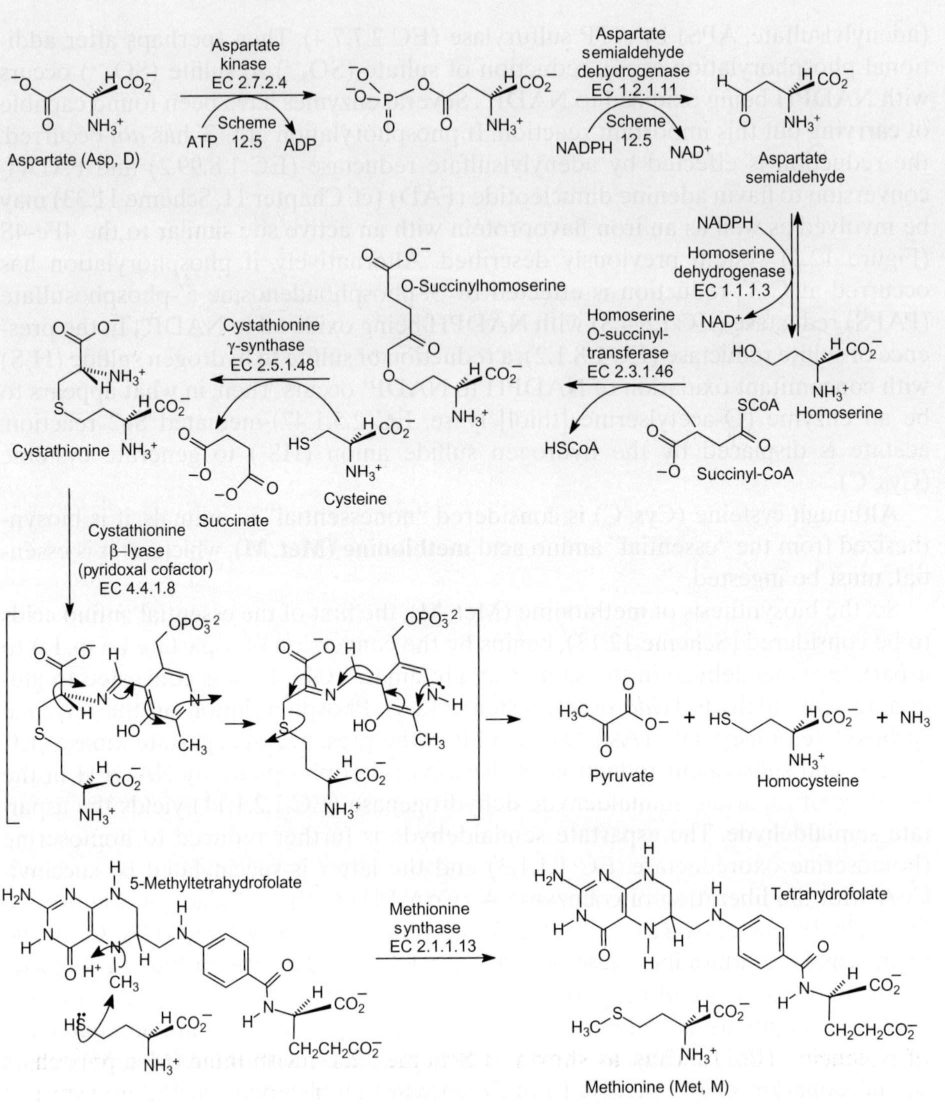

Aspartate kinase EC 2.7.2.4 (Scheme 12.5) ATP → ADP

Aspartate (Asp, D)

Aspartate semialdehyde dehydrogenase EC 1.2.1.11 (Scheme 12.5) NADPH → NAD$^+$

Aspartate semialdehyde

NADPH

Homoserine dehydrogenase EC 1.1.1.3 NAD$^+$

Homoserine

Homoserine O-succinyl-transferase EC 2.3.1.46 SCoA HSCoA Succinyl-CoA

O-Succinylhomoserine

Cystathionine γ-synthase EC 2.5.1.48

Cysteine

Cystathionine

Cystathionine β–lyase (pyridoxal cofactor) EC 4.4.1.8

Succinate

OPO$_3^{2-}$ CH$_3$ HO N

Pyruvate + Homocysteine + NH$_3$

5-Methyltetrahydrofolate

Methionine synthase EC 2.1.1.13

Tetrahydrofolate

Methionine (Met, M)

Scheme 12.13. A pathway for the biosyntheis of the essential amino acid methionine (Met, M). It is important to note that cysteine (Cys, C) is required to introduce the sulfur. EC numbers and some graphic materials provided in this scheme have been taken from appropriate links in a URL starting with http://www.chem.qmul.ac.uk/iubmb/enzyme/.

to produce (S)-2-hydroxy-2-ethyl-3-oxobutanoate. In the presence of reduced NADPH and the enzyme ketol acid reductoisomerase (EC 1.1.1.86) (Scheme 12.16), the ethyl group migrates from C2 to C3 and the carbonyl generated in that process is reduced to produce (2R,3R)-2,3-dihydroxy-3-methyl-pentanoate. With the enzyme dihydroxy-acid dehydratase (EC 4.2.1.9), the hydroxyl at C3 is lost and the hydroxyl at C2 is converted to the corresponding ketone. Presumably, this dehydration occurs through the enol, which then tautomerizes.

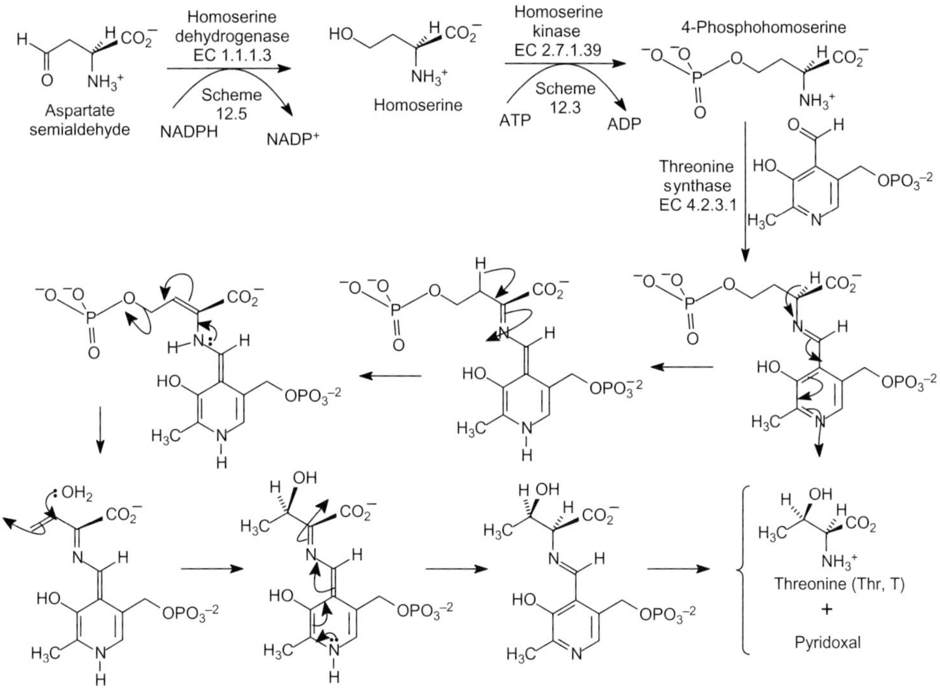

Scheme 12.14. A representation of the biosynthesis of threonine (Thr, T). EC numbers and some graphic materials provided in this scheme have been taken from appropriate links in a URL starting with http://www.chem.qmul.ac.uk/iubmb/enzyme/.

Finally, transamination with the branched-chain amino acid transaminase (EC 2.6.1.42), the coenzyme for which is pyridoxal and the source of the amino group from which pyridoxal transfers the nitrogen is glutamate (Glu, E), occurs. 2-Oxoglutarate forms in the process.

Interestingly, the biosynthesis of **valine** (**Val, V**) is almost identical with that of isoleucine (Ile, I). Thus, as shown in Scheme 12.17, with thiamine diphosphate (cf. Chapter 11, Scheme 11.7) as a cofactor for the enzyme acetolactate synthase (EC 2.2.1.6), an acetyl group derived from pyruvate is added to the carbonyl of a second pyruvate to produce (*S*)-2-hydroxy-2-*methyl*-3-oxobutanoate (rather than the (*S*)-2-hydroxy-2-*ethyl*-3-oxobutanoate produced in the biosynthesis of isoleucine (Ile, I). As before, in the presence of reduced NADPH and the enzyme ketol acid reductoisomerase (EC 1.1.1.86) (Scheme 12.16), there is a group migration from C2 to C3 (a *methyl* rather than an *ethyl*), and the carbonyl group generated in that process is reduced to produce (2*R*, 3*R*)-2,3-dihydroxy-3-methyl-butanoate. With the enzyme dihydroxy acid dehydratase (EC 4.2.1.9), the hydroxyl at C3 is lost and the hydroxyl at C2 is converted to the corresponding ketone. As before, it is presumed that this dehydration occurs through the enol, which then tautomerizes to 3-methyl-2-oxobutanoate.

Finally, the major difference is that transamination with the valine pyruvate transaminase (EC 2.6.1.66) occurs. Pyridoxal is the coenzyme again, but now the source of the nitrogen is alanine (Ala, A) (thus reforming pyruvate in the process).

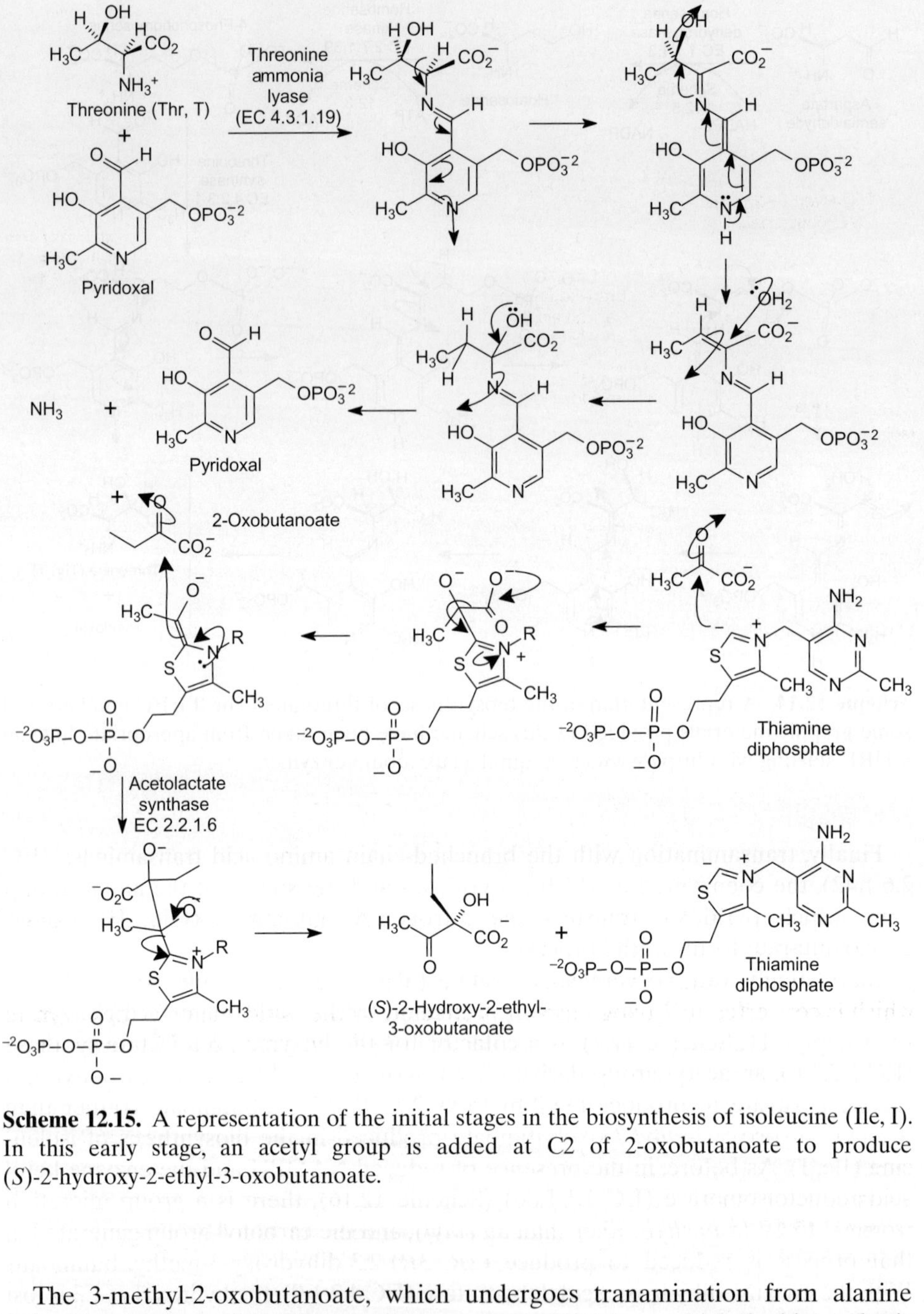

Threonine (Thr, T)

Threonine ammonia lyase (EC 4.3.1.19)

Pyridoxal

Pyridoxal

NH_3 +

2-Oxobutanoate

Thiamine diphosphate

Acetolactate synthase EC 2.2.1.6

(*S*)-2-Hydroxy-2-ethyl-3-oxobutanoate

Thiamine diphosphate

Scheme 12.15. A representation of the initial stages in the biosynthesis of isoleucine (Ile, I). In this early stage, an acetyl group is added at C2 of 2-oxobutanoate to produce (*S*)-2-hydroxy-2-ethyl-3-oxobutanoate.

The 3-methyl-2-oxobutanoate, which undergoes tranamination from alanine (Ala, A) to lead to valine (Val, V), also serves as a point from which the biosynthesis of **leucine** (**Leu**, **L**) can begin to be discussed. Thus, as shown in Scheme 12.18, 3-methyl-2-oxobutanoate reacts with acetyl-CoA ($CH_3COSCoA$) in the presence of 2-isopropylmalate synthase (EC 2.3.3.13) to generate (2*S*)-2-isopropylmalate, which then loses water (isopropylmalate dehydratase, EC 4.2.1.33) to generate 2-isopropylmaleate. Now, water readds under the influence of the same enzyme to

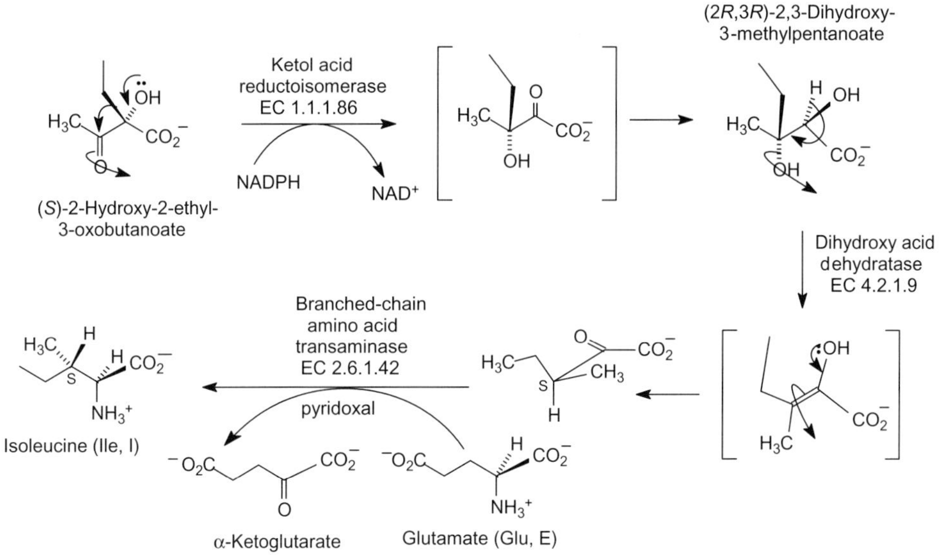

Scheme 12.16. A representation of the final stages in the biosynthesis of isoleucine (Ile, I). In the late stages shown here, the migration of the ethyl group is accompanied by the reduction of the carbonyl generated in that migration (the enzyme is a "reductoisomerase" (EC 1.1.1.86). Dehydration to an enol (a "dihydroxy acid dehydratase," EC 4.2.1.9), tautomerization to the corresponding ketone, and a final transamination ("branched-chain amino acid transaminase" with pyridoxal as a cofactor) from glutamate (Glu, E) produces isoleucine (Ile, I). EC numbers and some graphic materials provided in this scheme have been taken from appropriate links in a URL starting with http://www.chem.qmul.ac.uk/iubmb/enzyme/.

produce the isomeric $(2R,3S)$-3-isopropylmalate. Oxidation at C2 (3-isopropylmalate dehydrogenase, EC 1.1.1.85, NAD^+ reduced to NADH) provides $(2S)$-2-isopropyl-3-oxosuccinate, a β-ketoacid that spontaneously decarboxylates to 4-methyl-2-oxopentanoate. Then, transamination (the nitrogen coming from glutamate [Glu, E] which is converted to 2-oxoglutarate) is effected by the pyridoxal cofactor enzymes leucine transaminase (EC 2.6.1.6) and/or branched-chain amino acid transaminase (EC 2.6.1.42) to produce leucine (Leu, L).

Aspartate 4-semialdehyde, seen, for example, in Scheme 12.13, which provided a pathway for the biosynthesis of the essential amino acid methionine (Met, M) and in Scheme 12.14, which holds a representation of the biosynthesis of threonine (Thr, T), is also a place to begin to describe a pathway to **lysine (Lys, K)**. As shown in Scheme 12.19, aspartate 4-semialdehyde undergoes an aldol-type reaction with pyruvate $(CH_3COCO_2^-)$ in the presence of dihydropicolinate synthase (EC 4.2.1.52) to produce a series of intermediates that, it is presumed, lead to (S)-2,3-dihydropyridine-2,6-dicarboxylate. Then, dihydrodipicolinate reductase (EC 1.3.1.26) working with NADPH produces the tetrahydropyridine, (S)-2,3,4,5-tetrahydropyridine-2,6-dicarboxylate. This heterocycle, in the presence of glutamate (Glu, E) and water, is capable of transamination directly to 2-oxoglutarate and $(2S, 6S)$-2,3-diaminopimelate in the presence of LL-diaminopimelate aminotransferase (EC 2.6.1.83), while the latter, in the presence of the pyridoxal dependent racemase

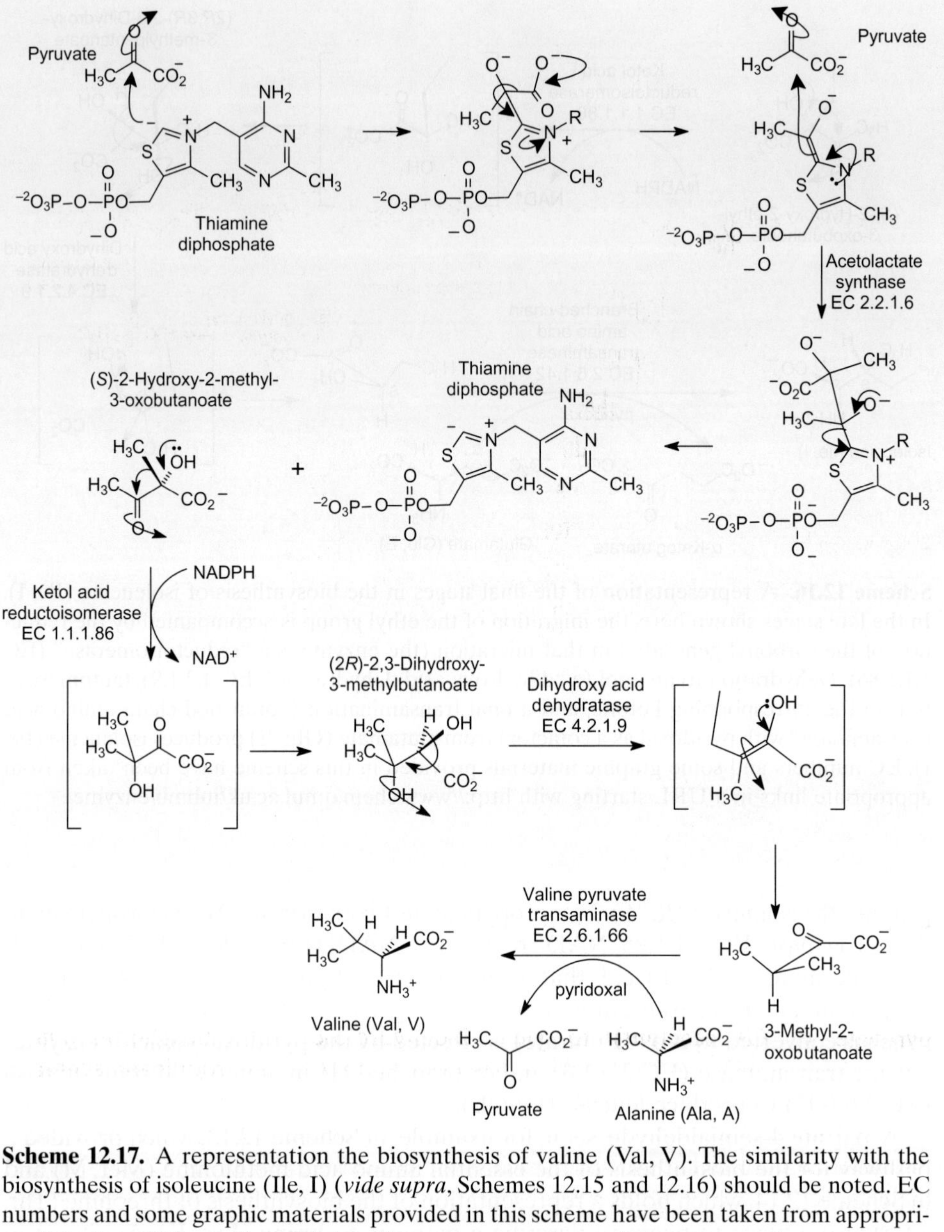

Scheme 12.17. A representation the biosynthesis of valine (Val, V). The similarity with the biosynthesis of isoleucine (Ile, I) (*vide supra*, Schemes 12.15 and 12.16) should be noted. EC numbers and some graphic materials provided in this scheme have been taken from appropriate links in a URL starting with http://www.chem.qmul.ac.uk/iubmb/enzyme/.

pimelate epimerase (EC 5.1.1.7), generates the corresponding *meso*-2,6-diaminopimelate. Then, finally, in the presence of the pyridoxal-phosphate protein diaminopimelate decarboxylase (EC 4.1.1.20), decarboxylation to lysine (Lys, K) occurs.

In Chapter 11, Part IV, D, d (ii), shikimic acid (Scheme 12.20) was shown to arise from the reaction between erythrose-4-phosphate (a C4 reactant) with phosphoenol

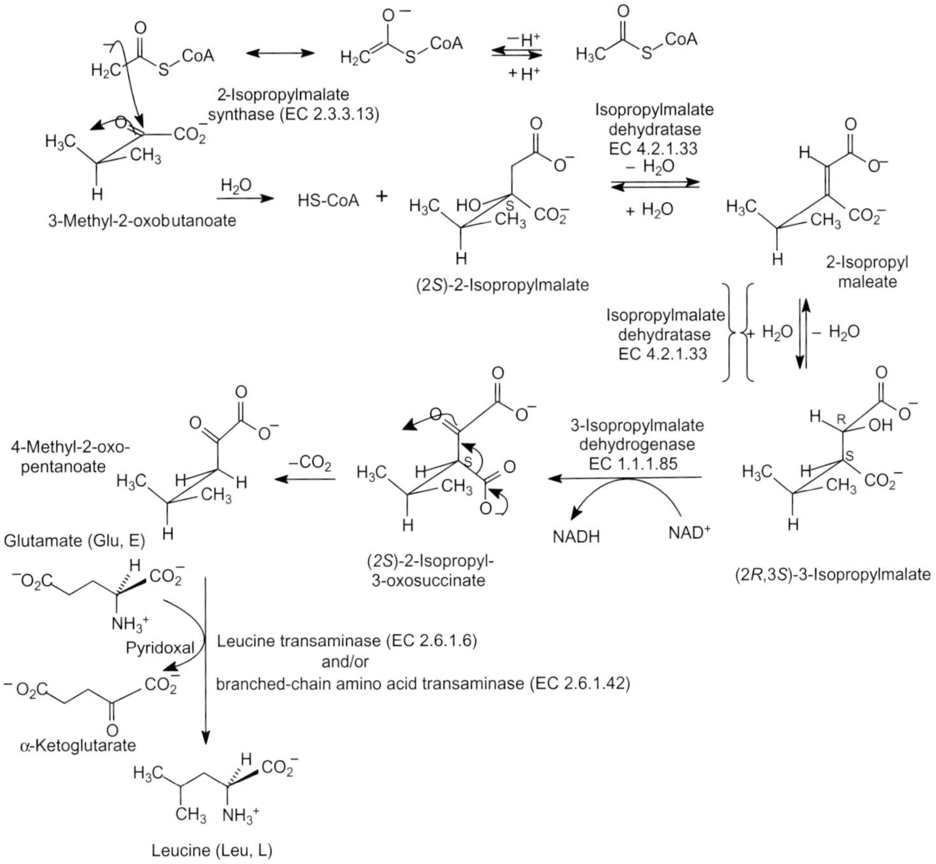

2-Isopropylmalate
synthase (EC 2.3.3.13)

$-H^+$ / $+H^+$

H₂C ... S–CoA   H₂C ... S–CoA   H₃C ... S–CoA

Isopropylmalate
dehydratase
EC 4.2.1.33
$-H_2O$ / $+H_2O$

3-Methyl-2-oxobutanoate

H_2O → HS-CoA + (2S)-2-Isopropylmalate

2-Isopropyl
maleate

Isopropylmalate
dehydratase
EC 4.2.1.33
$+H_2O$ / $-H_2O$

4-Methyl-2-oxo-
pentanoate

$-CO_2$

3-Isopropylmalate
dehydrogenase
EC 1.1.1.85

NADH   NAD⁺

(2S)-2-Isopropyl-
3-oxosuccinate

(2R,3S)-3-Isopropylmalate

Glutamate (Glu, E)

Pyridoxal

Leucine transaminase (EC 2.6.1.6)
and/or
branched-chain amino acid transaminase (EC 2.6.1.42)

α-Ketoglutarate

Leucine (Leu, L)

Scheme 12.18. A representation of the biosynthesis of leucine (Leu, L). EC numbers and some graphic materials provided in this scheme have been taken from appropriate links in a URL starting with http://www.chem.qmul.ac.uk/iubmb/enzyme/.

pyruvate (a C3 reactant) to yield 2-dehydro-3-deoxy-7-phospho-D-*arabino*-heptanoate (a C7 product), which, on cyclization (3-dehydroquinate synthase, EC 4.2.3.4) resulted in the formation of 3-dehydroquinate (cf. Scheme 11.80).

Next (cf. Scheme 11.81), the 3-dehydroquinate was shown to undergo dehydration to produce 3-dehydroshikimic acid and reduction (cf. Scheme 11.82) to shikimate (shikimate dehydrogenase, EC 1.1.1.25). In the same scheme, a depiction of phosphorylation to 3-phosphoshikimate (shikimate kinase, EC 2.7.1.71) followed and was shown to set the stage for the reaction of the latter with phosphoenol pyruvate under the influence of the enzyme (3-phosphoshikimate 1-carboxyvinyl transferase, EC 2.5.1.19) to yield 5-(1-carboxyvinyl)-3-phosphoshikimate. Finally, with chorismate synthase (EC 4.2.3.5) (Scheme 11.83), the penultimate progenitor of phenylalanine (Phe, F), tyrosine (Tyr, Y), and **tryptophan (Trp, W)**, viz., chorismate, is produced. These processes are shown in detail in Schemes 11.80–11.83 and, in less detail, in Scheme 12.20.

Now, as shown in Scheme 12.21, chorismate rearranges (chorismate mutase, EC 5.4.99.5) in what might be an internal Cope-type process (see Scheme 11.84 and the

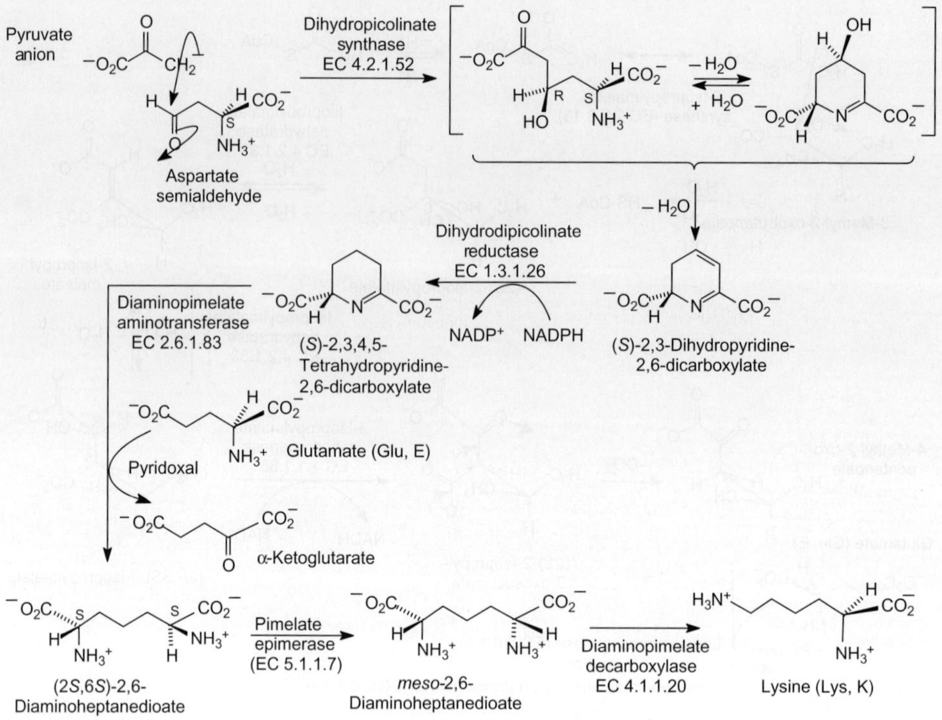

Pyruvate anion

Aspartate semialdehyde

Dihydropicolinate synthase
EC 4.2.1.52

$-H_2O$
$+ H_2O$

$-H_2O$

(S)-2,3-Dihydropyridine-2,6-dicarboxylate

Dihydrodipicolinate reductase
EC 1.3.1.26

$NADP^+$ NADPH

(S)-2,3,4,5-Tetrahydropyridine-2,6-dicarboxylate

Diaminopimelate aminotransferase
EC 2.6.1.83

Pyridoxal

Glutamate (Glu, E)

α-Ketoglutarate

(2S,6S)-2,6-Diaminoheptanedioate

Pimelate epimerase
(EC 5.1.1.7)

meso-2,6-Diaminoheptanedioate

Diaminopimelate decarboxylase
EC 4.1.1.20

Lysine (Lys, K)

Scheme 12.19. A representation of a pathway from aspartate semialdehyde to lysine (Lys, K). EC numbers and some graphic materials provided in this scheme have been taken from appropriate links in a URL starting with http://www.chem.qmul.ac.uk/iubmb/enzyme/.

discussion accompanying it in Chapter 11) to produce prephenate, the actual precursor to phenylalanine (Phe, F) and tyrosine (Tyr, Y).

Thus, for phenylalanine (Phe, F), decarboxylation and dehydration (prephenate dehydratase, EC 4.2.1.51) to phenylpyruvate is followed by transamination with either of the enzymes tyrosine transaminase (EC 2.6.1.5) or aromatic amino acid transaminase (EC 2.6.1.57). Both of these use pyridoxal as cofactor and derive the nitrogen for the amino function from glutamate (Glu, E).

Similarly, for tyrosine (Tyr, Y), the hydroxyl group in prephenate is oxidized using the enzyme prephenate dehydrogenase (NAD^+ to NADH [EC 1.3.1.12] or $NADP^+$ to NADPH [EC 1.3.1.13]) to the corresponding carbonyl (a β, γ unsaturated δ-ketocarboxylic) acid that then loses CO_2 and generates 4-hydroxyphenylpyruvate. Transamination with the aromatic amino acid transaminase (EC 2.6.1.57) (using pyridoxal as a cofactor) where the nitrogen for the amino function comes from glutamate (Glu, E) completes the biosynthesis of tyrosine (Tyr, Y).

It is important at this point to digress because the amino acid phenylalanine (Phe, F) is the progenitor of a vast class of *nonnitrogenous* plant products: the **phenylpropanoids**.

First, it is necessary to note that, as pointed out above (Equation 12.5), phenylalanine ammonia lyase (EC 4.3.1.5) serves both to generate phenylalanine (Phe, F) by the direct aminolysis of *trans-* or (*E*)-cinnamate and the reverse. Thus,

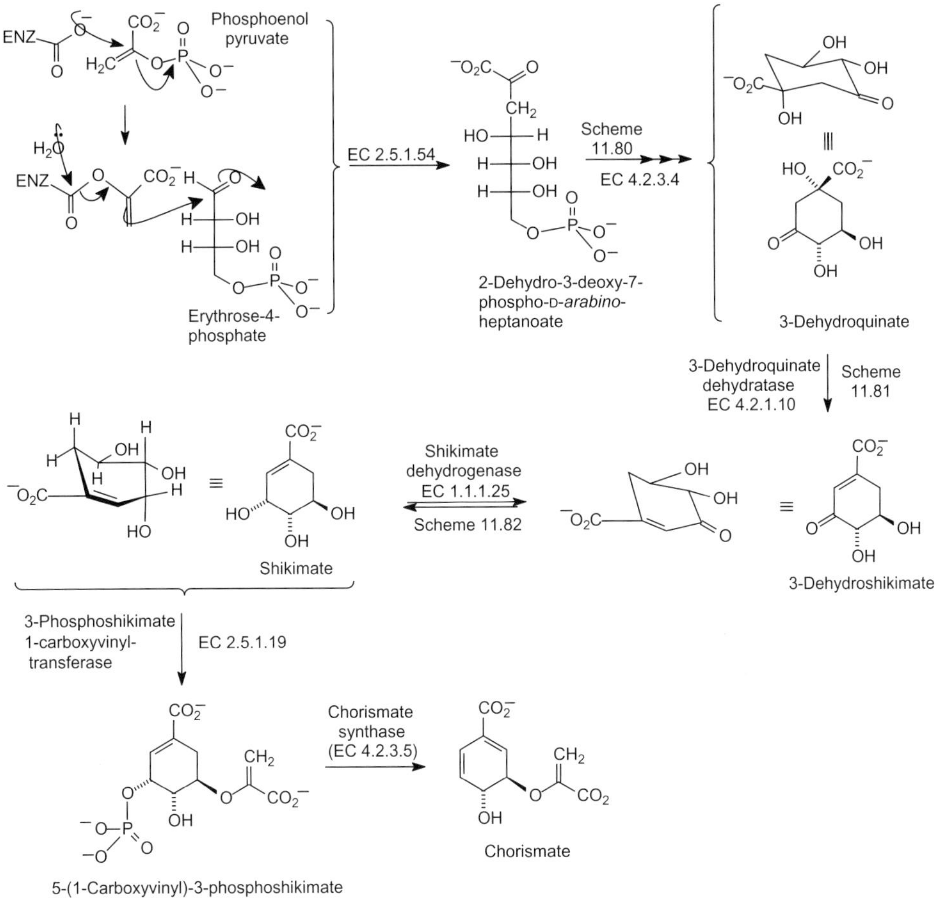

Scheme 12.20. An abbreviation of Schemes 11.80–11.83 presenting the overall transformation of erythrose-4-phosphate and phosphoenol pyruvate to chorismate.

(E)-cinnamic acid formed from phenylalanine (Phe, F) serves as the parent of the large family of C6-C3 compounds known as phenylpropanoids. As shown in Scheme 12.22, (*trans*) or (E)-cinnamic acid, with oxygen and NADPH is converted, in the presence of *trans*-cinnamate 4-monooxygenase (EC 1.14.13.11), to 4-hydroxycinnamate (*p*-coumaric acid) and then, with other common (and not necessarily specific) oxygenases such as one of the group of copper proteins known as monophenol monooxygenase (EC 1.14.18.1), to caffeic acid (3,4-dihydroxycinnamic acid). Now, monomethylation, with the methyl group being transferred from S-adenosylmethionine (SAM) using an enzyme such as catechol *O*-methyltransferase (EC 2.1.1.6), produces ferulic acid (4-hydroxy-3-methoxy)cinnamic acid. Further oxidation to introduce additional hydroxyl groups and methylation of some or all of them in different patterns are common. These species, either by themselves or as esters of carbohydrates (Chapter 11), occur widely in nature.

Continuing, briefly, with the digression, cinnamic acid itself and the various oxidized derivatives of cinnamic acid are capable of reduction to their respective

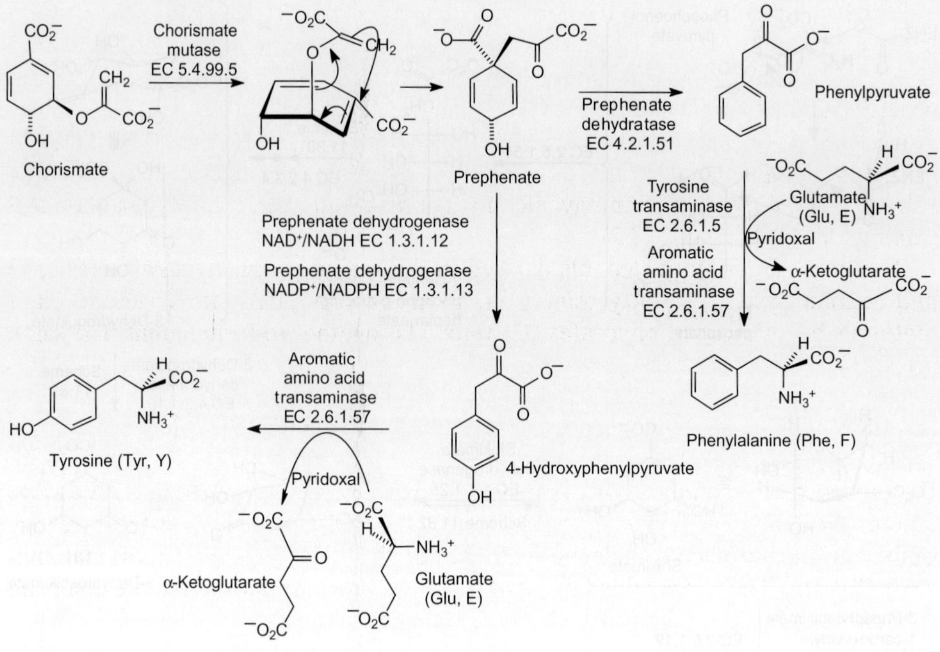

Scheme 12.21. An abbreviated version (Chapter 11, Scheme 11.84) showing the formation of prephenate as precursor to phenylalanine (Phe, F) and tyrosine (Tyr, Y). For the former, decarboxylation and dehydration to phenylpyruvate is followed by transamination to the amino acid. For the latter, the hydroxyl group in prephenate is oxidized to the corresponding carbonyl to avoid its expulsion, and this is followed by decarboxylation and transamination. EC numbers and some graphic materials provided in this scheme have been taken from appropriate links in a URL starting with http://www.chem.qmul.ac.uk/iubmb/enzyme/.

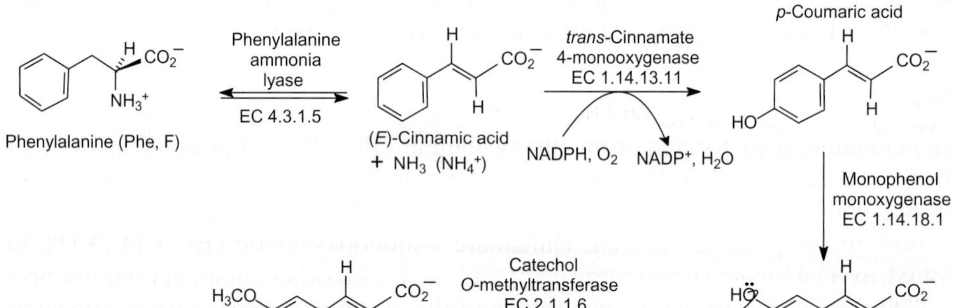

Scheme 12.22. A representation of the deamination of phenylalanine (Phe, F) to (*E*)-cinnamic acid and the conversion of the latter into more highly oxidized derivatives. The caffeic acid and ferulic acid serve as introductory compounds to phenylpropanoids. EC numbers and some graphic materials provided in this scheme have been taken from appropriate links in a URL starting with http://www.chem.qmul.ac.uk/iubmb/enzyme/.

aldehydes and from there to the corresponding alcohols. It is likely that this process involves initial phosphorylation of the carboxylate (as already shown in Schemes 12.6 and 12.13) by the appropriate (as yet unidentified) kinase (a member of the class of EC 2.7.2.X). Further reduction (again the enzyme is not known but, presumably, with an alcohol dehydrogenase of the EC 1.1.1.1 class) with NADH (or NADPH) to NAD^+ (or $NADP^+$) as likely cofactor yields the family of corresponding primary alcohols such as cinnamyl alcohol, coumaryl alcohol, coniferyl alcohol, and sinaphyl alcohol (Figure 12.5)

The phenolic primary alcohols derived in this way from phenylalanine (Phe, F) and in some systems from tyrosine (Tyr, Y) are incorporated into structural plant materials by linking carbohydrates (Chapter 11) together to form lignins. The same alcohols give rise to phenylpropenes, such as eugenol and safrole, which are commonly referred to as "essential oils," while further oxidation and double-bond isomerization of *p*-coumaric acid leads to umbelliferone (Figure 12.6).

Finally, in this vein, as shown in Scheme 12.23, *p*-coumaric acid is converted (in the presence of 4-coumarate-CoA ligase, EC 6.2.1.12, coenzyme A, and ATP) into its corresponding coenzyme A thioester, 4-coumaroyl-CoA. Then, in the presence of the enzyme naringenin chalcone synthase (EC 2.3.1.74), 3 equivalents of malonyl-CoA (Scheme 11.36) are added sequentially, a cyclization follows, and the chalcone (naringenin chalcone) is produced. This is the beginning of the pathway for flavonoid (as derived from this progenitor chalcone) biosynthesis.

We return now to the biosynthesis of tryptophan (Trp, W).

As seen in Scheme 12.20, chorismate is generated from erythrose-4-phosphate and phosphoenol pyruvate. Further, as seen before in Scheme 11.85 and again here in Scheme 12.24, chorismate, obtaining an amino function from glutamine (Gln, Q), is converted to anthranilate (and glutamate [Glu, E] and pyruvate) using anthranilate synthase (EC 4.1.3.27). Then, the reaction of anthranilate with

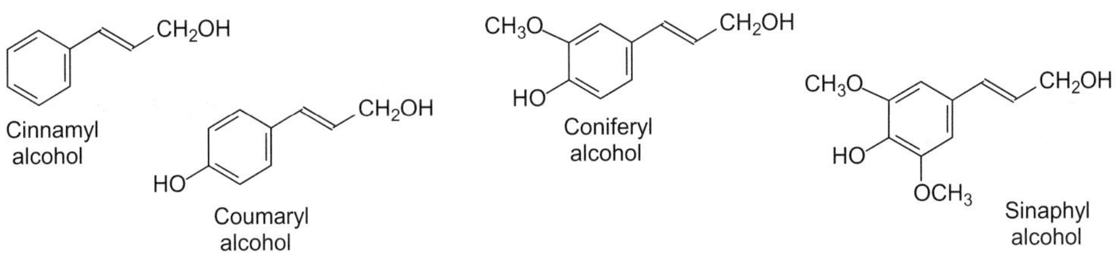

Figure 12.5. Some of the alcohols derived from substituted cinnamic acids. The oxidation pattern is common to phenylalanine- (Phe, F) and tyrosine- (Tyr, Y) derived phenols.

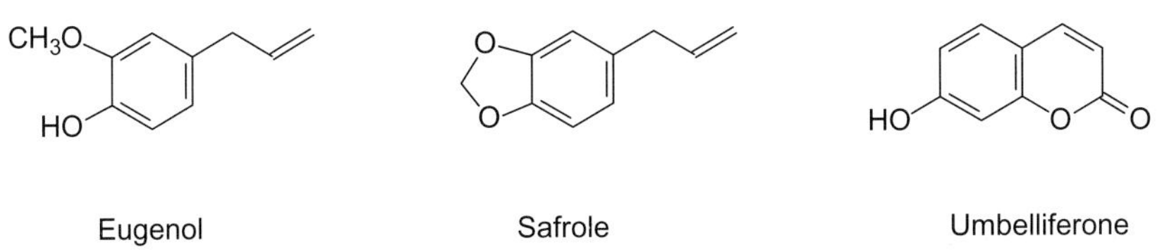

Figure 12.6. The phenylpropenes, eugenol and safrole ("essential oils"), and the lactone umbelliferone, generated on oxidation (and double-bond isomerization) of *p*-coumaric acid.

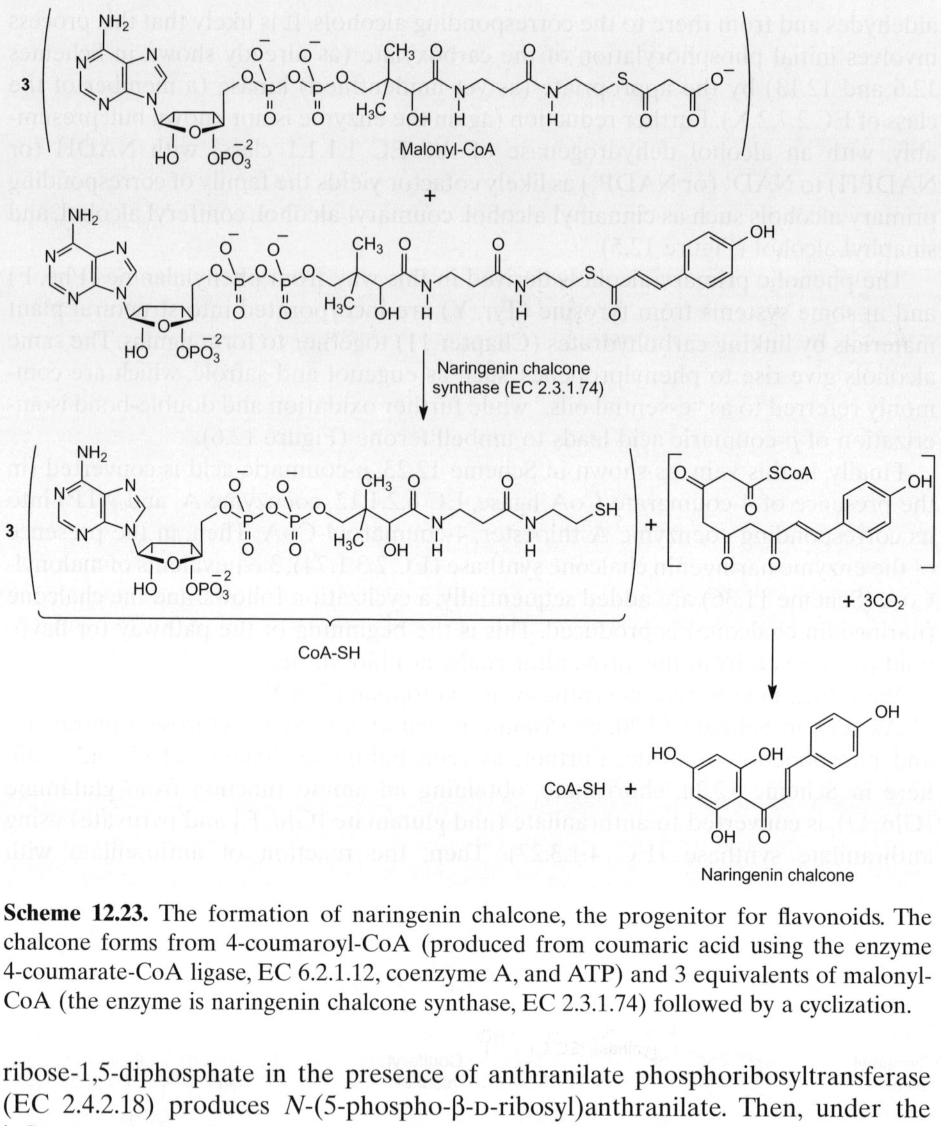

Malonyl-CoA

+

Naringenin chalcone
synthase (EC 2.3.1.74)

CoA-SH

+ $3CO_2$

CoA-SH +

Naringenin chalcone

Scheme 12.23. The formation of naringenin chalcone, the progenitor for flavonoids. The chalcone forms from 4-coumaroyl-CoA (produced from coumaric acid using the enzyme 4-coumarate-CoA ligase, EC 6.2.1.12, coenzyme A, and ATP) and 3 equivalents of malonyl-CoA (the enzyme is naringenin chalcone synthase, EC 2.3.1.74) followed by a cyclization.

ribose-1,5-diphosphate in the presence of anthranilate phosphoribosyltransferase (EC 2.4.2.18) produces *N*-(5-phospho-β-D-ribosyl)anthranilate. Then, under the influence of phosphoribosylanthranilate isomerase (EC 5.3.1.24), the ribosyl ring opens to the corresponding enol and tautomerization to the keto alcohol 1-(2-carboxyphenylamino)-1-deoxy-D-ribulose-5-phosphate takes place. Then, through the action of indole-3-glycerolphosphate synthase (EC 4.1.1.48), cyclization occurs with the loss of carbon dioxide (CO_2) and the formation of indole-3-glycerolphosphate.

Finally, the three-carbon side chain of the indole-3-glycerolphosphate is removed to produce indole itself and glyceraldehyde-3-phosphate (Scheme 12.25). The indole moves from one subunit of the enzyme, tryptophan synthase (EC 4.2.1.20), to another where it encounters the result of dehydration of a pyridoxal-bound serine (Ser, S) to which it adds. Then, hydrolytic cleavage of the resulting pyridoxal derivative of tryptophan (Trp, W) yields pyridoxal, ready for another reaction with, for example, serine (Ser, S) and tryptophan (Trp, W).

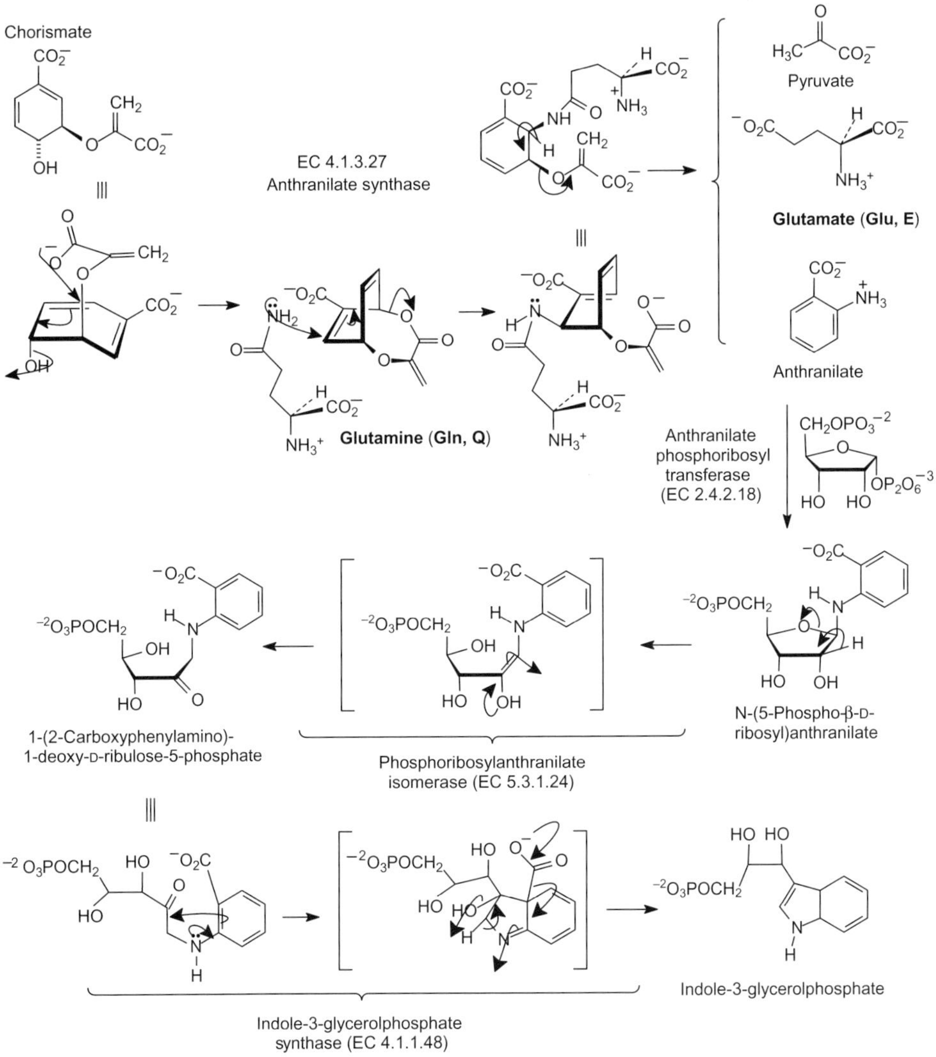

Chorismate

EC 4.1.3.27
Anthranilate synthase

Pyruvate

Glutamate (Glu, E)

Anthranilate

Glutamine (Gln, Q)

Anthranilate
phosphoribosyl
transferase
(EC 2.4.2.18)

N-(5-Phospho-β-D-
ribosyl)anthranilate

1-(2-Carboxyphenylamino)-
1-deoxy-D-ribulose-5-phosphate

Phosphoribosylanthranilate
isomerase (EC 5.3.1.24)

Indole-3-glycerolphosphate

Indole-3-glycerolphosphate
synthase (EC 4.1.1.48)

Scheme 12.24. Chorismate is converted to anthranilate (and glutamate [Glu, E] and pyruvate) using anthranilate synthase (EC 4.1.3.27). Anthranilate and ribose-1,5-diphosphate, in the presence of anthranilate phosphoribosyltransferase (EC 2.4.2.18), produces N-(5-phospho-β-D-ribosyl) anthranilate. Finally, indole-3-glycerolphosphate is formed via 1-(2-carboxyphenyl-amino)-1-deoxy-D-ribulose-5-phosphate with loss of carbon dioxide (CO_2).

The final amino acid to be considered here is **histidine (His, H)**. As shown in Scheme 12.26, the initial steps begin with the reaction of ribose-1,5-diphosphate with ATP in the presence of ATP-phosphoribosyl transferase (EC 2.4.2.17); then, after phosphate is lost (phosphoribosyl-ATP diphosphatase [EC 3.6.1.31]), the pyrimidine ring is hydrolytically cleaved (phosphoribosyl-adenosine 5′-phosphate [AMP] cyclohydrolase [EC 3.5.4.19]), a process that is followed by the opening of

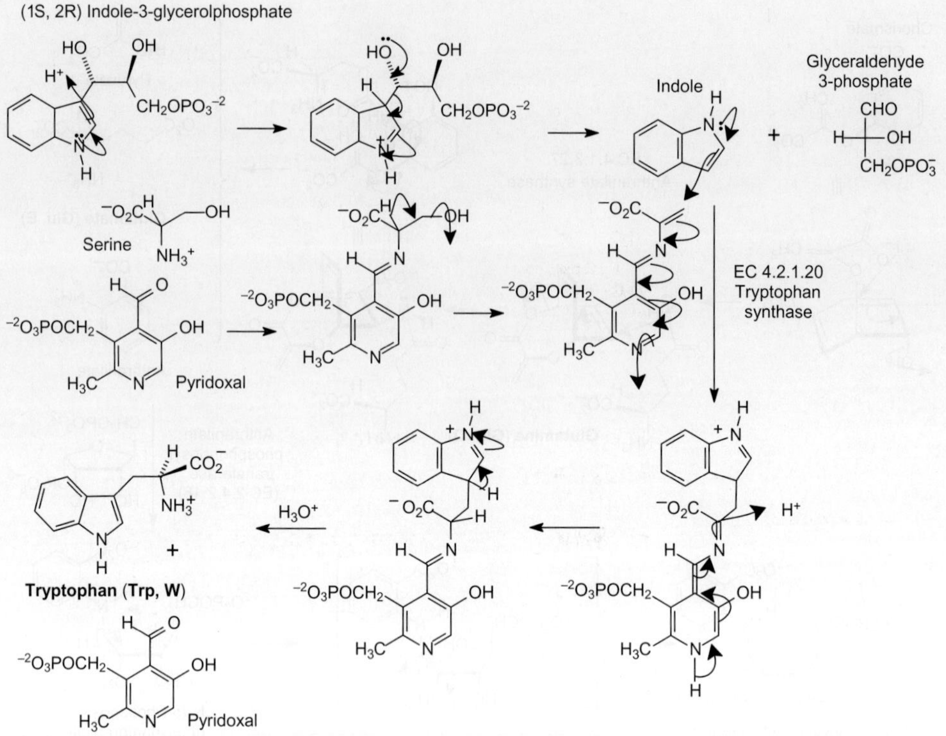

Scheme 12.25. A pathway for the formation of tryptophan (Trp, W). The side chain of the indole is derived from serine (Ser, S).

the furan ring to which the heterocycle was attached (1-(5-phosphoribosyl)-5-[(5-phosphoribosylamino)methylideneamino]imidazole-4-carboxamide isomerase [EC 5.3.1.16]).

Continuing in Scheme 12.27, cleavage into fragments occurs by hydrolysis of the imine, producing a formamide to be used for the synthesis of the histidine (His, H) and a second imidazole-containing fragment to be subsequently used in purine biosynthesis. Then, transfer of the amino function from glutamine (Gln, Q) presumably generating glutamate (Glu, E) (the transferase does not seem to have specifically been identified) and cyclization of the amide to the newly introduced amine generates the required imidazole functionality. Dehydration (imidazoleglycerol phosphate dehydratase [EC 4.2.1.19]) followed by rearrangement produces the carbonyl functionality required for the introduction of the histidine (His, H) side-chain amino group. The amino function is introduced, using glutamine (Gln, Q) to glutamate (Glu E) with histidinol-phosphate transaminase (EC 2.6.1.9). Then, finally, dephosphorylation (histidinolphosphatase [EC 3.1.3.15]) and oxidation by NAD^+, which is reduced in the process (histidinol dehydrogenase [EC 1.1.1.23]), converts the alcohol to the corresponding carboxylic acid found in histidine (His, H).

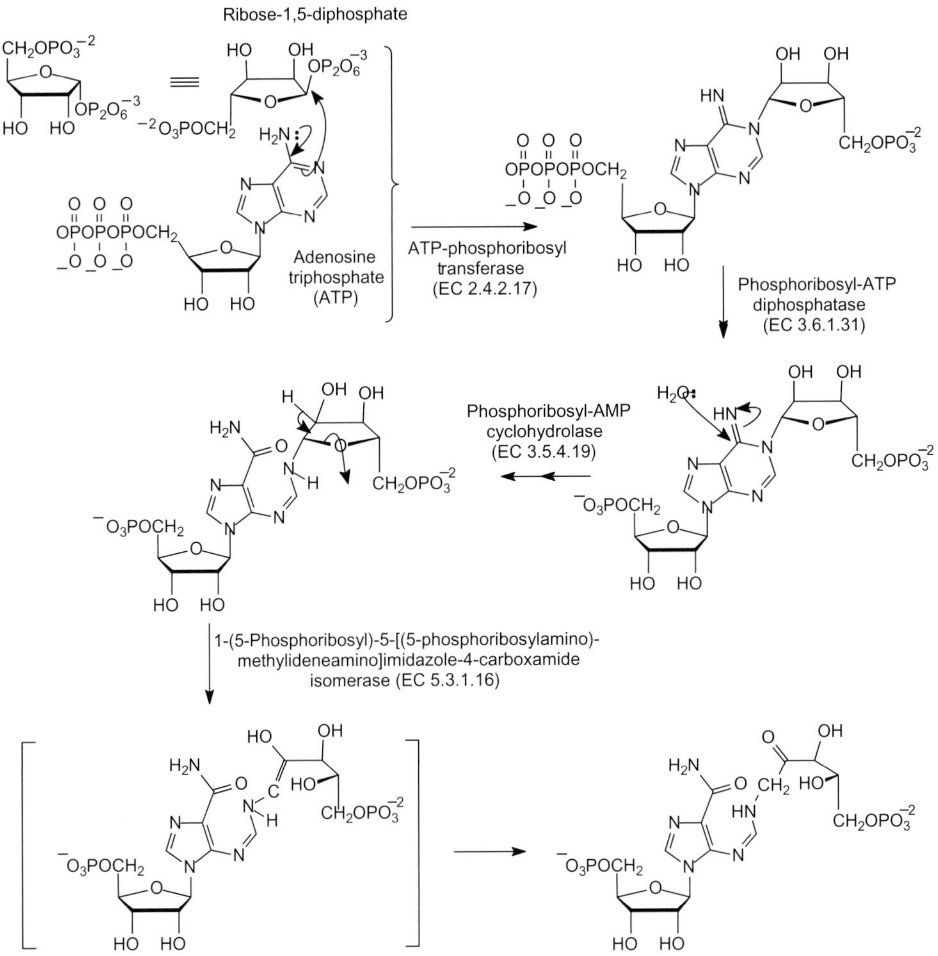

Scheme 12.26. The initial steps in the biosynthesis of histidine (His, H) correspond to a building up of the appropriate network and results in the degradation of an equivalent of ribose (as the 1,5-diphosphate) and adenosine triphosphate (ATP). EC numbers and some graphic materials provided in this scheme have been taken from appropriate links in a URL starting with http://www.chem.qmul.ac.uk/iubmb/enzyme/.

II. Synthesis

The variety of syntheses available for the preparation of amino acids is large because of, at least in part, the desire to produce peptides by linking the amino acids from one to another to make proteins (*vide infra*). Therefore, while it is possible to utilize a different method for each amino acid discussed in Part I, the variety is, at least in principle, applicable to each of them. The particular method used in the synthesis of any amino acid is often a function of other substituents that might be present and would require masking or derivatization.

As noted earlier (see Chapter 8, Scheme 8.49, and Chapter 9, Scheme 9.104), chlorine and bromine are introduced with relative ease into carboxylic acids at the

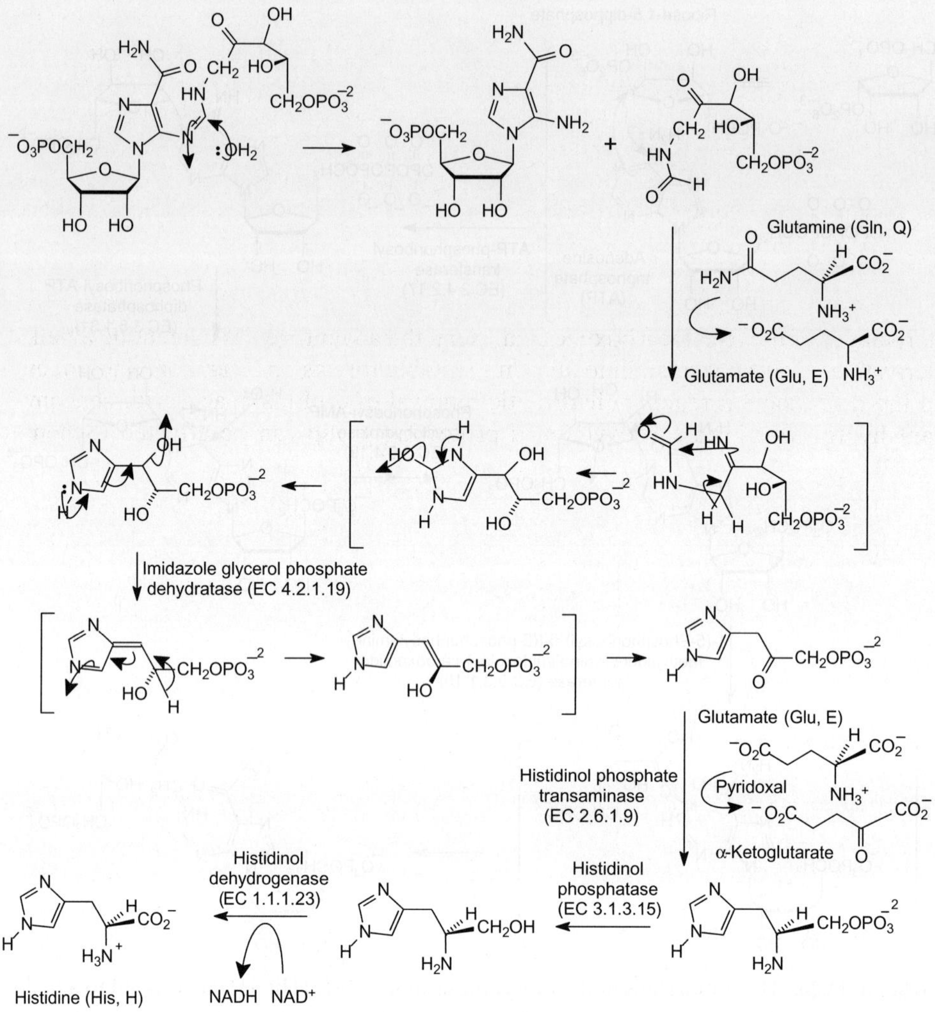

Scheme 12.27. The final stages in the biosynthesis of histidine (His, H) correspond to the introduction of the appropriate nitrogens and cyclization and oxidation to the final amino acid. EC numbers and some graphic materials provided in this scheme have been taken from appropriate links in a URL starting with http://www.chem.qmul.ac.uk/iubmb/enzyme/.

position alpha to the carbonyl. Then, on treatment with 3 equivalents of ammonia in a sealed tube at elevated temperature the ammonium salt of the corresponding amino acid and ammonium chloride (NH_4Cl) result. Equation 12.7 shows the conversion of chloroacetic acid to the corresponding glycine (Gly, G) ammonium salt.

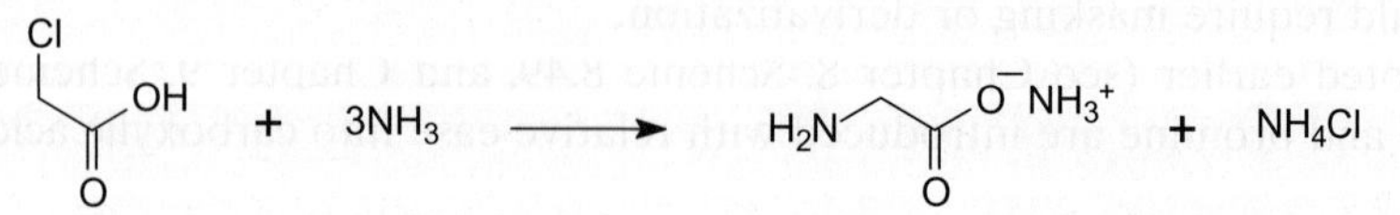

$$\text{(12.7)}$$

An interesting variant to produce glycine (Gly, G) or, indeed, any other α-amino acid from the corresponding halide requires the use of sodium azide. Thus, reaction of chloroacetic acid with sodium azide in acetone solvent produces the corresponding α-azido acid, which, in a subsequent step, can be reduced with hydrogen gas using a palladium on carbon (Pd/C) catalyst (Scheme 12.28). The advantage of the sodium azide method lies in the ability to use α-halo nitriles and amides as well as the carboxylic acid and to avoid the necessity of blocking the amino function (while carrying out other reactions); since until the azido group is reduced, there is no amino function present.

It is not uncommon to utilize the anion of phthalimide to the same purpose on derivatives of the α-halocarboxylic acid. Here, the advantage lies in creating a bulky derivative, invariably crystalline and thus suitable for easy purification. Subsequent to the displacement reaction, hydrolytic removal of *o*-phthalic acid with concomitant hydrolysis of the carboxylic acid protecting group can be effected (Scheme 12.29).

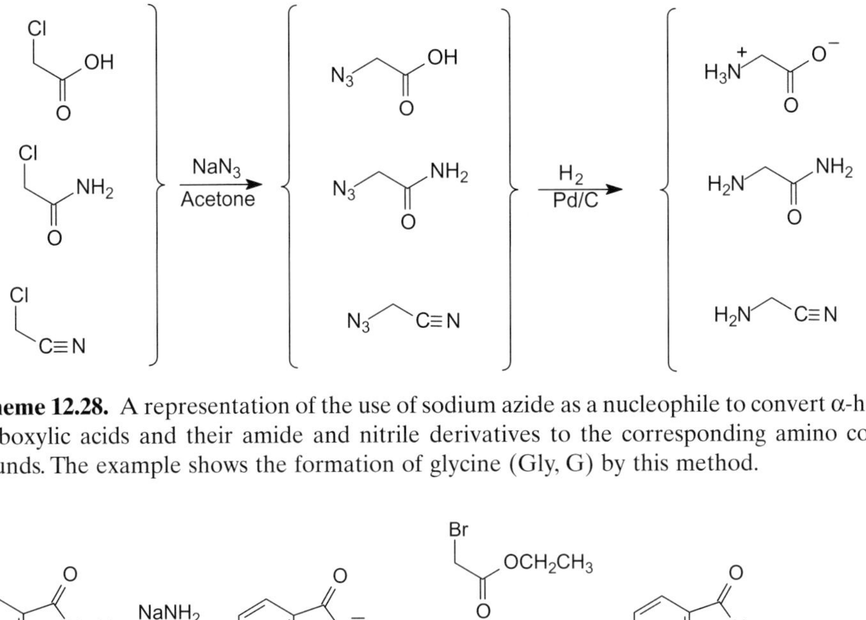

Scheme 12.28. A representation of the use of sodium azide as a nucleophile to convert α-halo carboxylic acids and their amide and nitrile derivatives to the corresponding amino compounds. The example shows the formation of glycine (Gly, G) by this method.

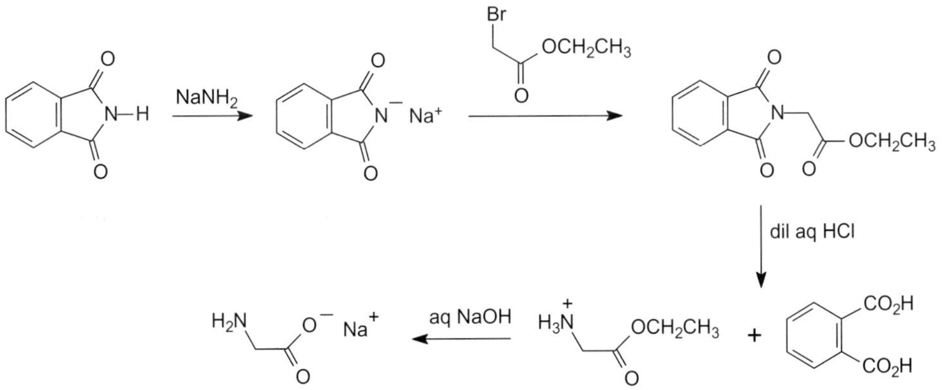

Scheme 12.29. A representation of the use of phthalimide to produce protected amino acids. The corresponding amides and nitriles could also be used in this process. The formation of the sodium salt of glycine (Gly, G) is shown.

Displacement reactions can also be useful in the synthesis of amino acids when more than one halogen is present. Thus, for example, proline (Pro, P) is readily prepared from 1,3-dibromopropane. On treatment of the dibromide with sodium diethylmalonate, a single displacement occurs to generate a monobromodiester. Bromination of that diester produces the corresponding 2,5-dibromodicarboxylate ester. The latter, on treatment, in a sealed tube with methanolic ammonia at 140°C undergoes cyclization and conversion to the bis-amide. Acidic (HCl) hydrolysis generates proline (Scheme 12.30).

The synthesis of ^{15}N-labeled glutamic acid (glutamate [Glu, E]) has been reported to be effectively accomplished by mixing 2-ketoglutaric acid (α-ketoglutaric acid) in aqueous solution with ^{15}N-enriched ammonium chloride and dilute sodium hydroxide. Then, as shown in Scheme 12.31, when that mixture is added dropwise over a period of about 5 h to a suspension of reduced platinum oxide in water,

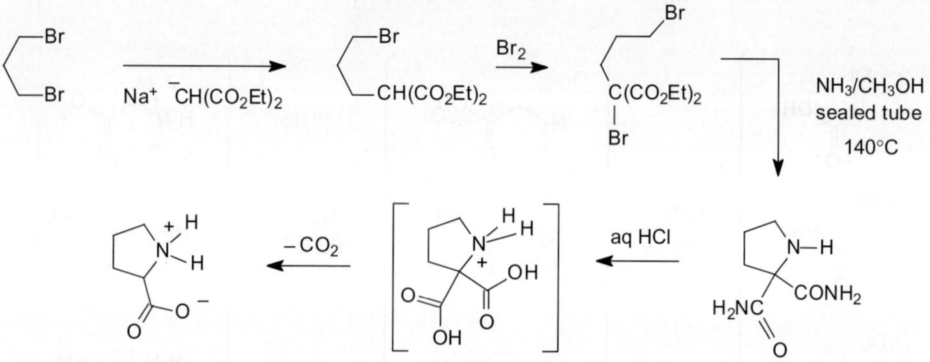

Scheme 12.30. A representation of the use of a dihalide, the anion of diethyl malonate and two substitution reactions, followed by an elimination of carbon dioxide (CO_2) to produce the amino acid proline (Pro, P).

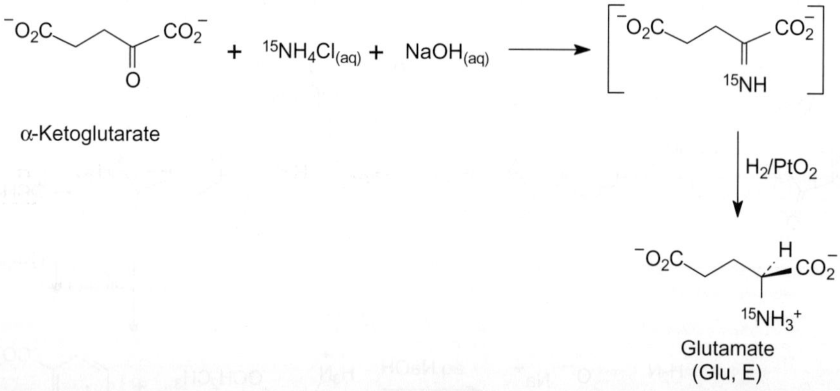

Scheme 12.31. A scheme for the synthesis of ^{15}N-glutamate (Glu, E) by the formation of an imine from α-ketoglutarate and ^{15}N-labeled ammonium chloride ($^{15}NH_4Cl$) followed by reduction with hydrogen (H_2) in the presence of a platinum oxide (PtO_2) catalyst. See, for example, Kragl, U.; Gödde, A.; Wandry, C.; Kinzy, W.; Cappon, J. J.; Lugtenburg, J. *Tetrahedron Assymetry*, **1993**, *6*, 1193.

additional hydrogen is taken up and the presumed imine intermediate is reduced to the corresponding amine. Removal of the catalyst by filtration and acidification with dilute HCl produces the corresponding ^{15}N-enriched glutamate (Glu, E).

Interestingly, given the pK$_a$ differences in the two carboxylic acid units, it is possible (Equation 12.8) to simply dissolve the glutamate (Glu, E) in absolute ethanol and warm the solution in the presence of concentrated hydrochloric acid (HCl$_{(aq)}$) and to obtain a good yield of the corresponding glutamic acid γ-monoethylester hydrochloride. This ester can be converted to the corresponding *N*-carbobenzoxy monoester with benzylchloroformate in pyridine and the latter (Equation 12.9) on aminolysis with liquid ammonia and catalytic reduction to effect debenzylation produces glutamine (Gln, Q).

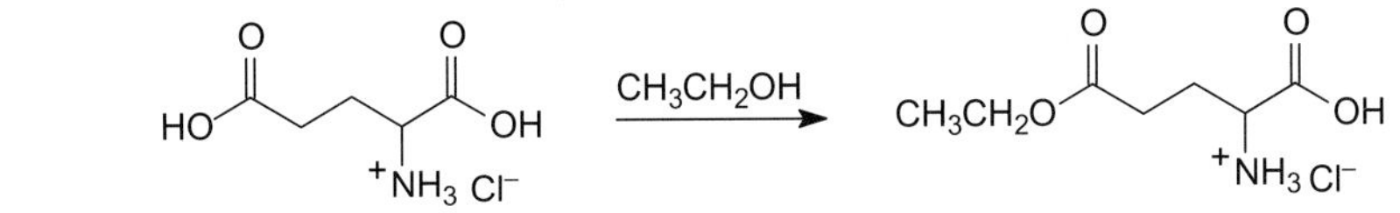

$$(12.8)$$

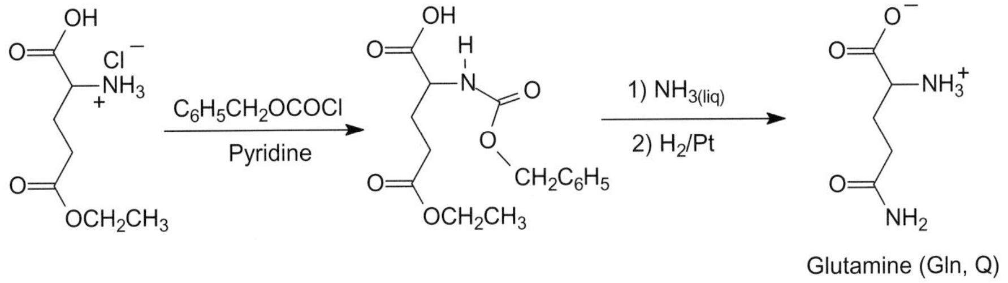

$$(12.9)$$

Beginning in about 1850, Adolph Strecker (Chapter 9, Scheme 9.68) undertook experiments on the reaction between aldehydes, cyanide anion [CN$^-$] (or hydrogen cyanide [HCN] itself) and ammonia [NH$_3$] (commonly in the form of ammonium salts such as ammonium chloride [NH$_4$Cl]). The reaction that bears his name, the Strecker reaction, has been used for the preparation of a wide variety of racemic amino acids, and Scheme 12.32 shows how it can be used in the synthesis of alanine

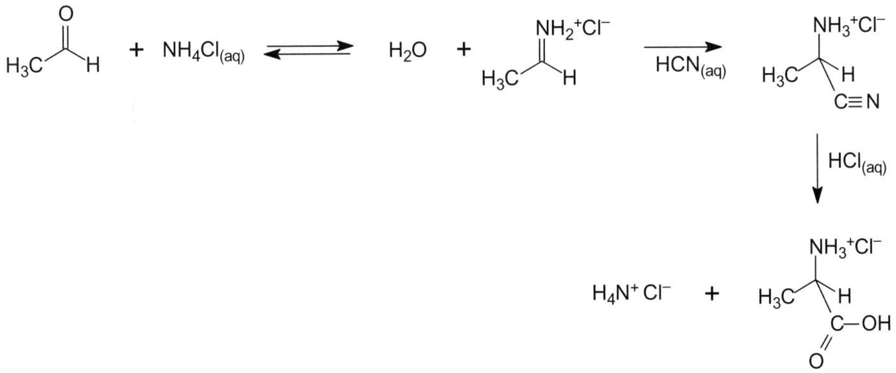

Scheme 12.32. A representation of the Strecker amino acid synthesis utilizing the synthesis of racemic alanine (Ala, A) as a typical example.

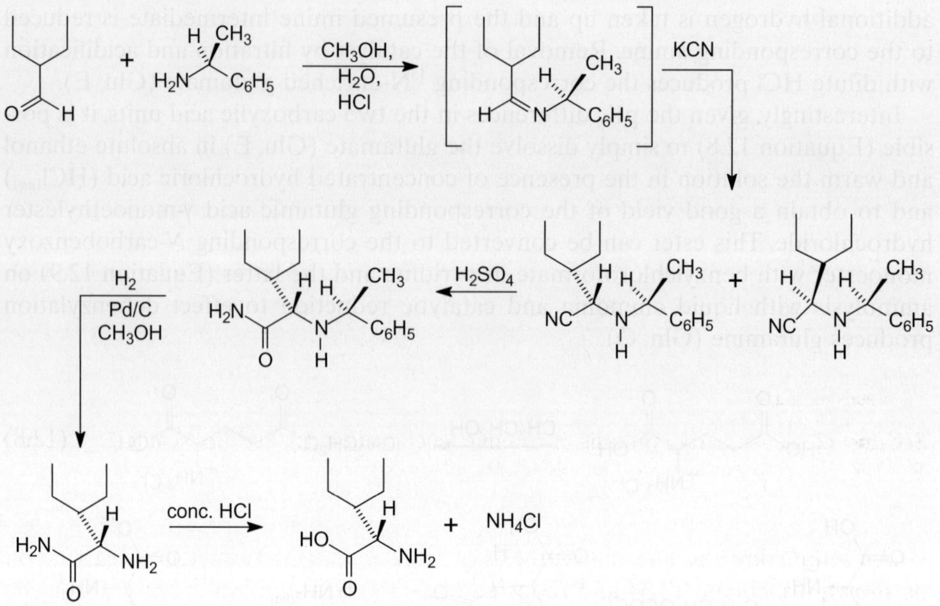

Scheme 12.33. A synthesis of the specific enantiomer, 3-ethyl-L-norvaline, as described in Resnick, L.; Galante, R. J. *Tetrahedon Asymmetry*, **2006**, *17*, 846.

(Ala, A). Thus, reaction of ethanal (acetaldehyde, CH_3CHO) with ammonium chloride ($NH_4Cl_{(aq)}$) and HCN leads to the production of racemic 2-aminopropionitrile, which, on hydrolysis with aqueous HCl, produces racemic alanine (Ala, A).

Interestingly, while the Strecker synthesis has proved to be a very powerful tool for the general synthesis of many racemic amino acids, a number of modifications have been developed to either (a) provide higher yields or, more recently, (b) produce a chiral product.

In the first case, the typical problem of aldehyde oxidation and general instability has been ameliorated by the formation of the aldehyde bisulfite derivative (Chapter 9, IVb). Cleaner products result when the bisulfite derivative is formed first.

In the second case, a nonnatural amino acid of commercial value was synthesized in a Strecker reaction using a chiral amine to help induce chirality in the product. In this case,* the authors were interested in the synthesis of 3-ethyl-L-norvaline. So, as shown in Scheme 12.33, a mixture of (S)-(-)-α-methylbenzylamine, potassium cyanide (KCN), and 2-ethylbutyraldehyde in dilute aqueous methanolic HCl was allowed to stir for several days at room temperature. During this time, the hydrochloride salt of the Strecker adduct precipitated and, under these conditions, the precipitate was found to be enriched in the levorotatory isomer (>30:1 by chromatographic separation of the diastereomers). Hydrolysis of the appropriate amino nitrile with cold concentrated sulfuric acid produced 3-ethyl-N^2-[(1S)-phenylethyl]-L-norvalinamide and debenzylation was accomplished by hydrogenation at 100 psi of hydrogen (H_2) over a 5% Pd/C catalyst. Finally, the amide was hydrolyzed in concentrated HCl to produce 3-ethyl-L-norvaline hydrochloride.

*Taken from Resnick, L.; Galante, R. J. *Tetrahedon Asymmetry*, **2006**, *17*, 846.

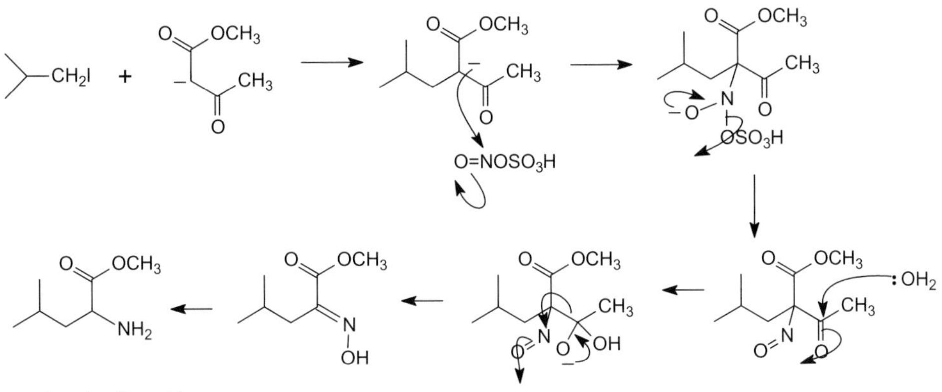

Leucine (**Leu, L**)

Scheme 12.34. An early synthesis of racemic leucine (Leu, L). See Bouveault, L.; Locquin, R. *C. R. Hebd. Seances Acad. Sci.*, **1905**, *141*, 115.

An interesting and unusual synthesis of racemic leucine (Leu, L) was carried out as shown in Scheme 12.34 and reported early in the twentieth century. As shown, 2-methyl-1-propanol was converted to the corresponding iodide with hydrogen iodide (HI) and the primary alkyl iodide reacted with sodium salt derived from methyl acetoacetate to yield 5-methyl-3-carboxymethyl-2-hexanone. This, treated with nitrosyl sulfate (O=N-OSO$_3$H), yields the decarbonylated methyl 5-methyl-pentanoate-2-oxime. Reduction of the latter produced racemic leucine (Leu, L).

Although some very elegant schemes for the synthesis of chiral materials have been developed, early work is also quite instructive. Consider the case of arginine (Arg, R), whose total synthesis was first reported by Sörensen in 1910. At that time, it was already known that bromination of diethyl malonate with bromine (Br$_2$) in carbon tetrachloride (CCl$_4$) solvent produced the corresponding monobromo diester, which could be treated with the potassium salt of phthalimide to generate phthalimidomalonic ester. The latter, when treated in turn with sodium metal in toluene, yielded the sodium salt, which was easily alkylated to any number of monoalkyl or monoarylphthalimido malonic esters (Scheme 12.35). In particular, for racemic arginine (Arg, R) and as shown in Scheme 12.35, the sodium salt of the phthalimidomalonic ester was allowed to react with γ-bromopropylphthalimide to produce the corresponding γ-phthalimidopropylphthalimidomalonic ester. Then, treatment of the latter with aqueous sodium hydroxide (NaOH) produced a hydrolysis mixture containing tetrasodium γ-phthaloylaminopropylphthaloylaminomalonic acid; the hydrolysis was completed with dilute aqueous hydrogen chloride (HCl). The product is racemic ornithine dihydrochloride.

When the racemic ornithine (Scheme 12.35) was treated with benzoyl chloride in the presence of pyridine, the bisbenzoate was formed, and after purification, it was subjected to hydrolysis using aqueous barium hydroxide [Ba(OH)$_2$]. The benzoyl group attached to the primary amine underwent hydrolysis to produce α-benzoylamino-δ-aminopentanoate, which, after acidification, was treated with cyanamide (N≡C–NH$_2$) to form α-benzoylamino-δ-guanidopentanoic acid. A final hydrolysis of the benzoyl group with dilute aqueous hydrochloric acid (HCl) generated arginine (Arg, R) (Scheme 12.36).

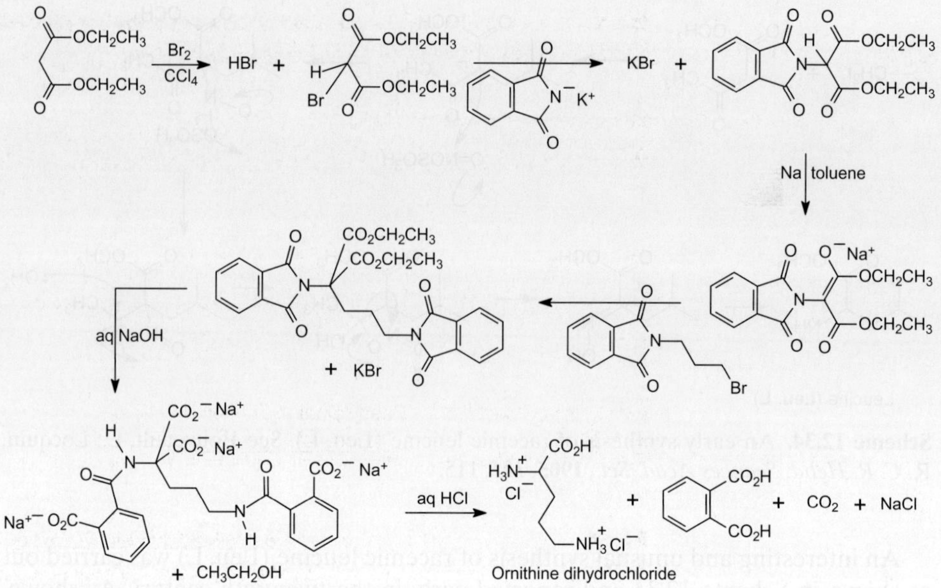

Scheme 12.35. An early synthesis of ornithine (after Sörensen, S. P. L. *Chem. Ber.*, **1910**, *43*, 643) making use of phthalimide. Ornithine was prepared on the way to arginine (Arg, R) as shown in Scheme 12.36.

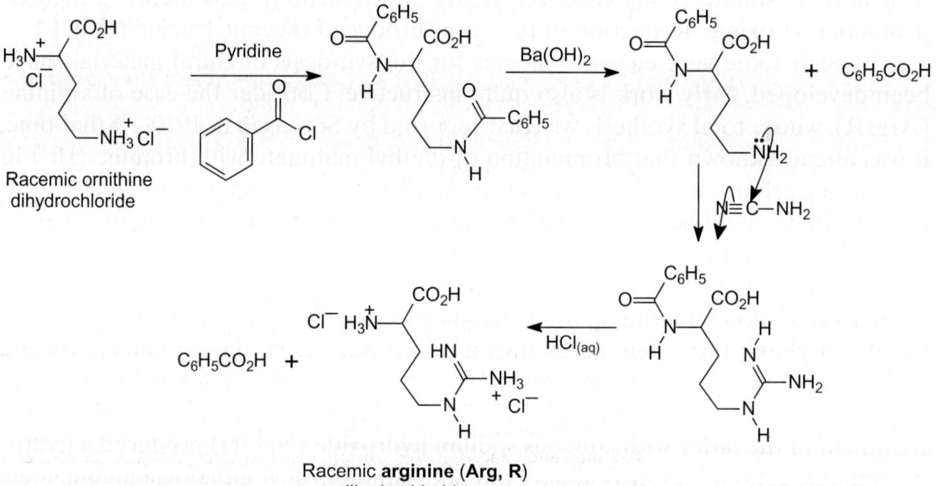

Scheme 12.36. The completion of the synthesis of arginine (Arg, R) from ornithine (after Sörensen, S. P. L. *Chem. Ber.*, **1910**, *43*, 643).

One of the earliest and most interesting syntheses of racemic aspartate (Asp, D) was reported in 1933 and is shown in Scheme 12.37.*

When diethyl fumarate is treated with 4 equivalents of ammonia (NH_3) in ethanol (CH_3CH_2OH), the diazoheterocyclic compound 3,6-diacetamino-2,5-diketopiperazine results.

*Dunn, M. S.; Fox, S. W. *J. Biol. Chem.*, **1933**, *101*, 493.

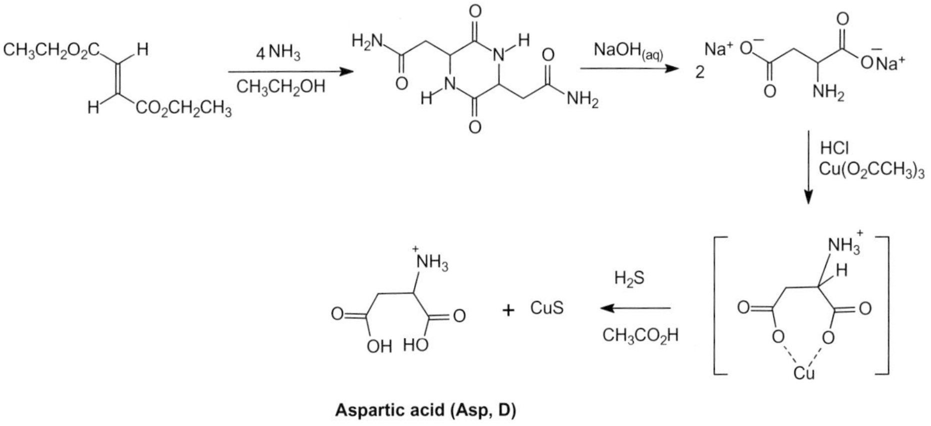

Aspartic acid (Asp, D)

Scheme 12.37. A synthesis of aspartate (Asp, D) (after Dunn, M. S.; Fox, S. W. *J. Biol. Chem.*, **1933**, *101*, 493) via a diketopiperzaine.

Saponification of the latter with aqueous sodium hydroxide (NaOH) resulted in cleavage to 2 equivalents of disodium aspartate (Asp, D). The product was purified by treatment with copper acetate [$Cu(O_2CCH_3)_2$] (which produced a crystalline trihydrate) and the latter, on careful treatment with hydrogen sulfide (H_2S) in acetic acid (CH_3CO_2H), yielded (Scheme 12.37) racemic aspartate (Asp, D).

With aspartate (Asp, D) in hand, the conversion to asparagine (Asn, N) has been shown to be straightforward. As expressed in U.S. Patent 5326908,* when a slurry of **aspartic acid (Asp, D)** in methanol (CH_3OH) and sulfuric acid (H_2SO_4) is prepared, the 4-carboxylate (that furthest from the amino group) is preferentially methylated and, without isolation, treatment of the reaction mixture with at least a fivefold excess of ammonia (NH_3) produces asparagine (Asn, N). After the removal of excess ammonia, adjustment of the pH to the isoelectric point of the amino acid (pI ≈ 5.4) yielded a crystalline material (Scheme 12.38).

Before the dawn of the twentieth century, the reaction between hypochlorous acid (HOCl) and propenoic acid was under investigation. It soon became clear that two different chlorohydrins, separable on the basis of solubility, were present. One of the isomers could be reduced with loss of chlorine to lactic acid and the other to an isomer of lactic acid. Both of the isomers could be converted to the same epoxide. Interestingly, as a consequence of the latter observation, treatment of *either isomer* with ammonia produces β-amino-α-hydroxypropanoic acid, the structural isomer of serine (Ser, S) (Scheme 12.39).

Years later, it was found that the reaction of ethyl formate with the ethyl ester of hippuric acid (ethyl *N*-benzoylglycine) in the presence of sodium ethoxide would lead to an aldol adduct, the reduction and hydrolysis of which generates racemic serine (Ser, S) (Scheme 12.40).

Problem 12.1. Serine (Ser, S) can also be prepared using the technique shown in Scheme 12.35 except that condensation with formaldehyde or an equivalent that can be reduced to the primary hydroxyl of serine (Ser, S). Sketch such a synthesis paying attention to reagents and conditions.

*Erickson, R.; Bray, R.; Johnson, M.; Klein, L.; Seagle, D. A. U. S. Patent 5326908, 1994.

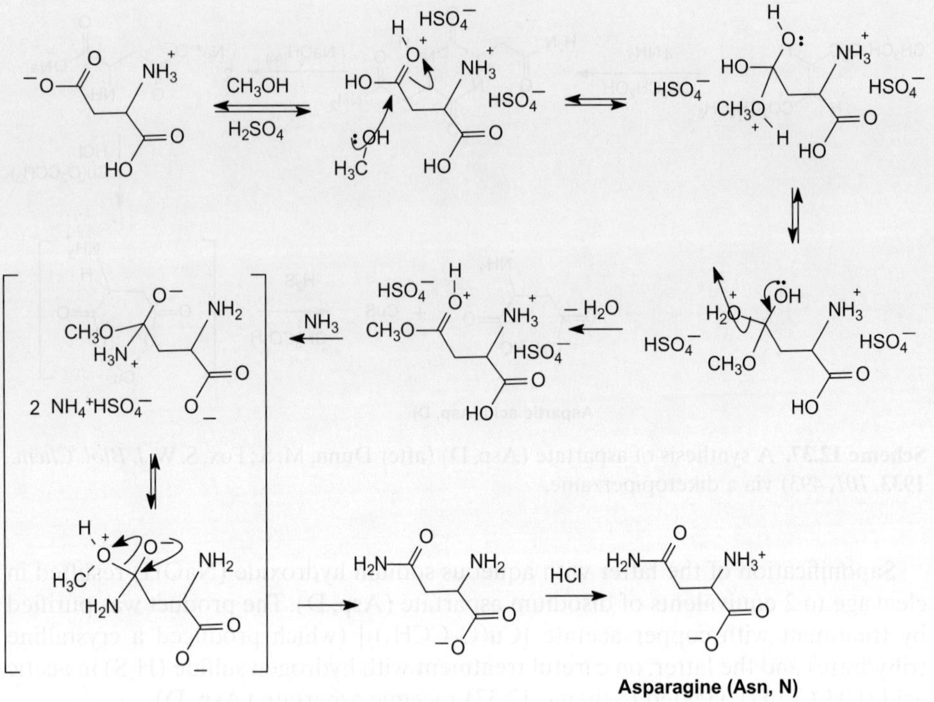

Asparagine (Asn, N)

Scheme 12.38. A synthesis of asparagine (Asn, N) (after Erickson, R.; Bray, R.; Johnson, M.; Klein, L.; Seagle, D. A. U.S. Patent 5326908, 1994).

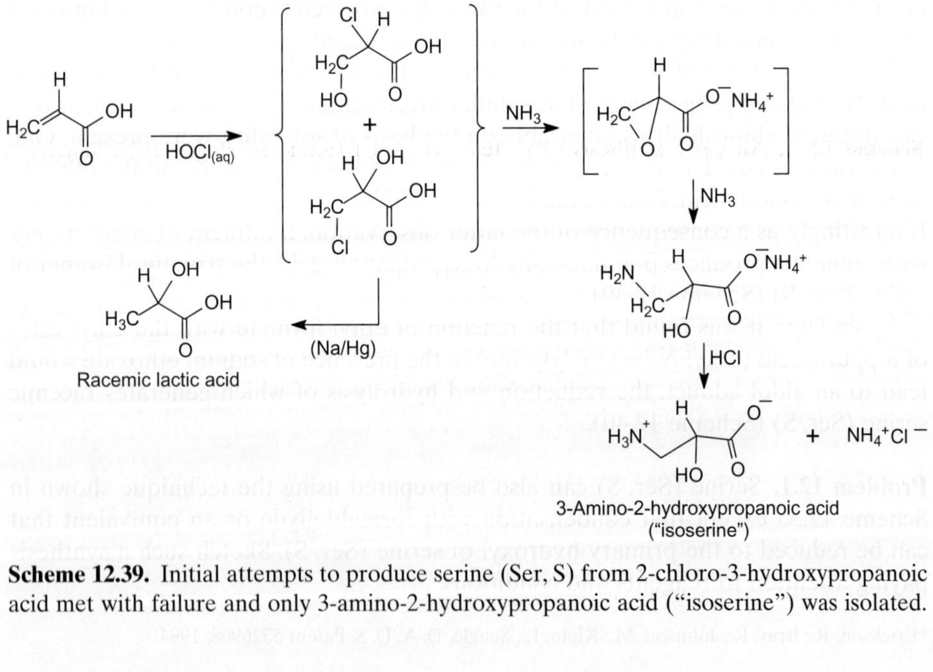

Racemic lactic acid

3-Amino-2-hydroxypropanoic acid
("isoserine")

Scheme 12.39. Initial attempts to produce serine (Ser, S) from 2-chloro-3-hydroxypropanoic acid met with failure and only 3-amino-2-hydroxypropanoic acid ("isoserine") was isolated.

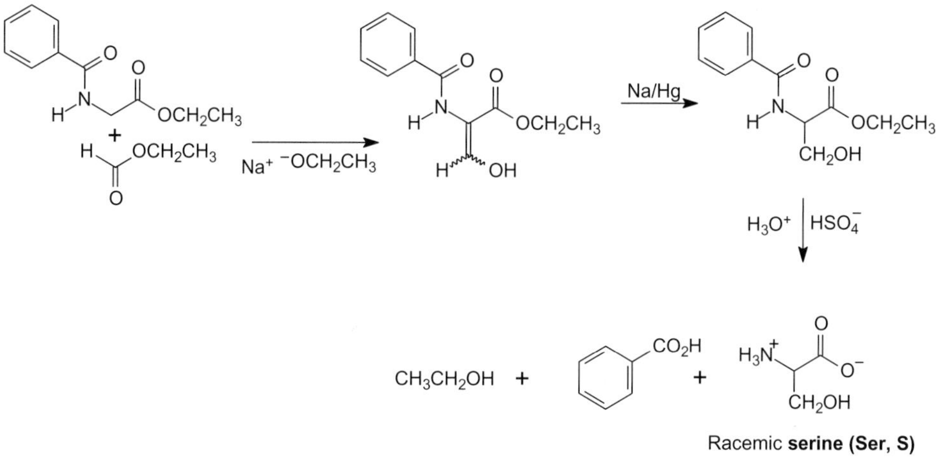

Scheme 12.40. An early successful synthesis of racemic serine (Ser, S) (after Erlenmeyer, E.; Stoop, F. *Liebigs Ann. Chem.*, **1904**, *337*, 236).

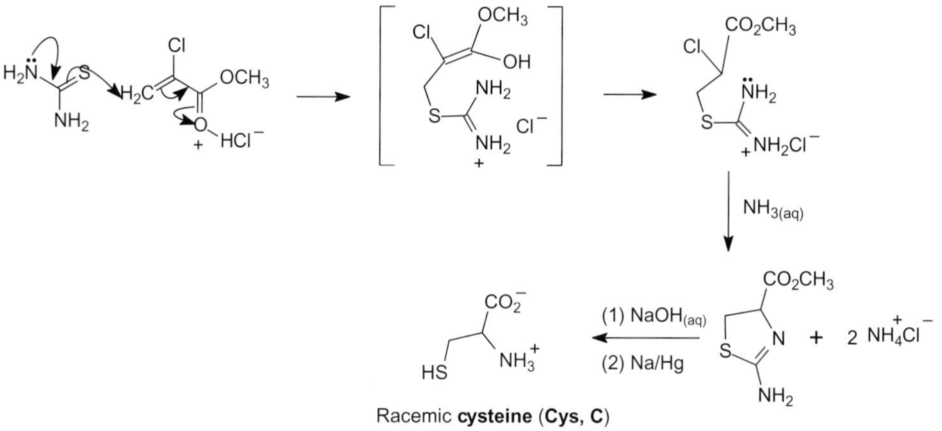

Racemic **serine (Ser, S)**

Racemic **cysteine (Cys, C)**

Scheme 12.41. An early synthesis of cysteine (Cys, C) (after Behringer, H.; Zillikens, P. *Liebigs Ann. Chem.*, **1951**, *574*, 140).

Although early cysteine (Cys, C) preparations were fraught with difficulty due to oxidation at sulfur and the absence of a good preparation of serine (Ser, S), the starting material from which all of the early workers began, some clever paths were found. Among them* was the acid-catalyzed (HCl) conjugate addition of thiourea to the methyl ester of 2-chloropropenoic acid (α-chloroacrylic acid ethyl ester) to yield the corresponding isothiuronium salt. Then, as shown in Scheme 12.41, cyclization to the 2-aminothiazoline carboxylic acid ester was easily effected by treatment with aqueous ammonia and, after base hydrolysis to the carboxylic acid, sodium amalgam (Na/Hg) reduction produced cysteine (Cys, C).

*The work is that of Behringer, H.; Zillikens, P. *Liebigs Ann. Chem.*, **1951**, *574*, 140.

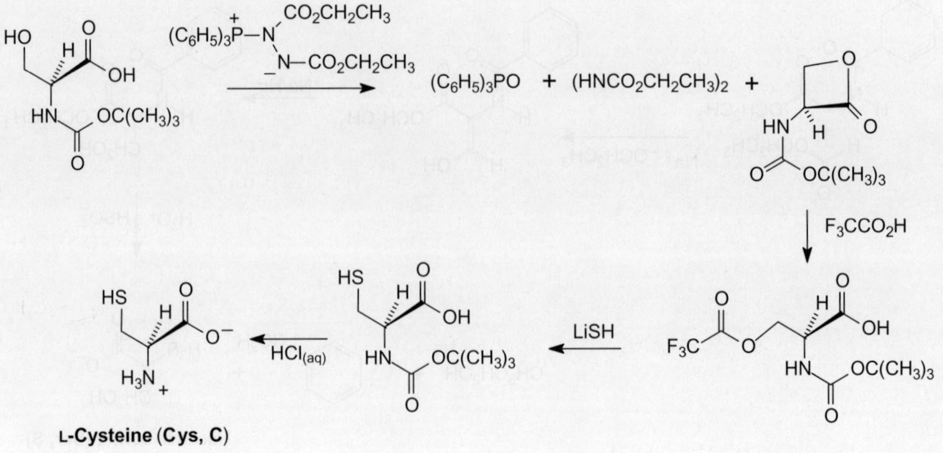

Scheme 12.42. A synthesis of L-cysteine (Cys, C) (after Arnold, L. D.; May, R. G.; Vederas, J. C. *J. Am. Chem. Soc.*, **1988**, *110*, 2237).

A more recent method,* also available for the synthesis of other similar compounds, makes use of the substituted four-membered ring heterocycle α-amino-β-propiolactone.

As shown in Scheme 12.42, when a *t*-butoxycarbonyl (*t*-BOC) or other N-protected L-serine (Ser, S) derivative was treated with preformed Mitsunobu (triphenylphosphine and dimethyl azodicarboxylate) reagent, the corresponding N-protected α-amino-β-lactone was formed in good yield. Then, when the *N-t*-butoxycarbonyl α-amino-β-lactone was treated with anhydrous trifluoroacetic acid (CF$_3$CO$_2$H), the corresponding trifluoroacetate ester resulted and, after pumping off the excess trifluoroacetic acid, the residue in acetonitrile/tetrahydrofuran (CH$_3$CN/THF) solvent produced protected L-cysteine (Cys, C) on reaction with lithium hydrogen sulfide (LiSH). Hydrolysis with dilute aqueous HCl generated L-cysteine (Cys, C) itself.

The last of the nonessential amino acids to be considered here is tyrosine (Tyr, Y). Although a number of different methods have been used in the synthesis of this amino acid, the one shown here, which presumably requires the intermediacy of an azlactone, has a generality (viz., the condensation of an aldehyde with hippuric acid) that allows similar syntheses of other amino acids too.

So, when *p*-methoxybenzaldehyde is allowed to react with hippuric acid (*N*-benzoylglycine), not only does aldol condensation occur but the product of that condensation is also subject to cyclization to an azlactone. Reduction and hydrolysis of the azlactone (oxazolone) produces an amino acid and benzoic acid, and, since a methoxy group is present in this particular synthesis, its cleavage is required and thus reduction here was effected with red phosphorus and HI. These reactions are shown in Scheme 12.43.

Beginning in the early part of the twentieth century, F. L. Pyman reported two different syntheses of histidine (His, H). Interestingly, he was also able to affect a resolution into the enantiomers. The initial synthesis was reported in 1911. As shown

*Arnold, L. D.; May, R. G.; Vederas, J. C. *J. Am. Chem. Soc.*, **1988**, *110*, 2237.

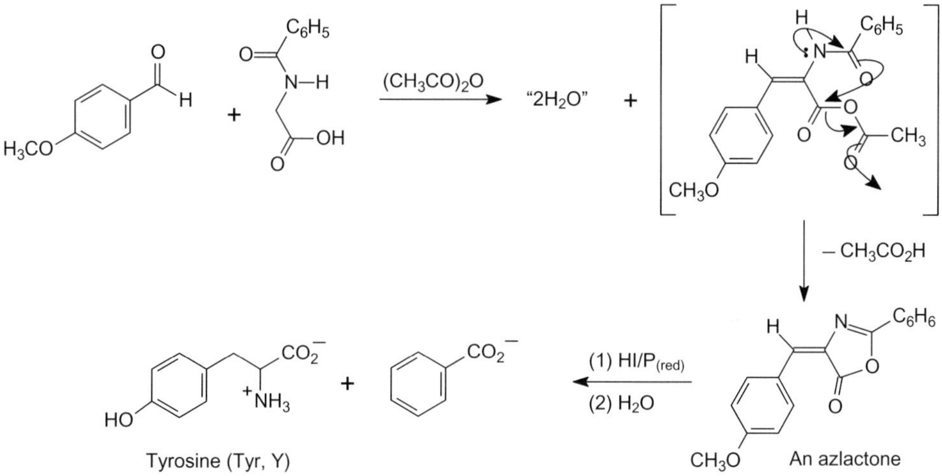

Tyrosine (Tyr, Y)   An azlactone

Scheme 12.43. A synthesis of tyrosine (Tyr, Y) via the condensation of *p*-methoxybenzaldehyde with hippuric acid (*N*-benzoylglycine) in the presence of acetic anhydride. The reaction is presumed to involve the intermediacy of an azlactone (oxazolone), and the reduction (and ether cleavage) is accomplished with hydrogen iodide (HI) and red phosphorus (see Badshah, A.; Khan, N. H.; Kidwai, A. R. *J. Org. Chem.*, **1972**, *37*, 2916 and references therein.)

in Scheme 12.44, Pyman,* following the older German literature, knew that treatment of citric acid with sulfuric acid (H_2SO_4) resulted in oxidative decarboxylation to yield acetone-1,3-dicarboxylic acid, and, further, that treatment of the latter with nitrous acid (HONO) yielded carbon dioxide (CO_2) and the corresponding bis-oxime. Reduction of the bis-oxime with zinc and hydrochloric acid (Zn/HCl) generated the corresponding dihydrochloride of 1,3-diaminoacetone (propanone).

Then, in the new work, Pyman found that the bis-amine hydrochloride, when heated with potassium thiocyanate (KSCN), yielded 2-thiol-4-aminomethylgloxaline as the major product. Further, when oxidized with nitric acid ($HONO_2$), the thiol sulfur was removed and the corresponding glyoxaline (imidazole) resulted, in which the primary amino function had also been replaced by a hydroxyl (presumably via nitrosation with nitrous acid [HONO] generated by a reduction of nitrate [NO_3^-] in the oxidation of the sulfur of the thiol). Then, the immediately useful precursor to histidine (His, H) was generated by treatment of the hydrochloride salt of the primary alcohol, 4-hydroxylmethylglyoxaline (i.e., 4-hydroxymethylimidazole) with phosphorus pentachloride (PCl_5) in chloroform ($CHCl_3$). To complete the synthesis of histidine (His, H), the 4-chloromethylimidazole (Scheme 12.45) was brought into reaction with the sodium salt of diethylchloromalonate in ethanol, and the resulting diethyl 4-methylchloromalonate imidazole derivative was isolated as its hydrogen oxalate salt (through the addition of oxalic acid ($HO_2C)_2$ during crystallization). The oxalic acid was replaced with hydrogen chloride in aqueous HCl and, on heating a 20% HCl aqueous solution of the hydrochloride, the corresponding α-chloro-β-4-glyoxaline-propanoic acid resulted and a single enantiomer spontaneously crystallized. Treatment with ammonia yielded histidine (His, H).

*Pyman, F. L. *J. Chem. Soc.*, **1911**, *99*, 1389.

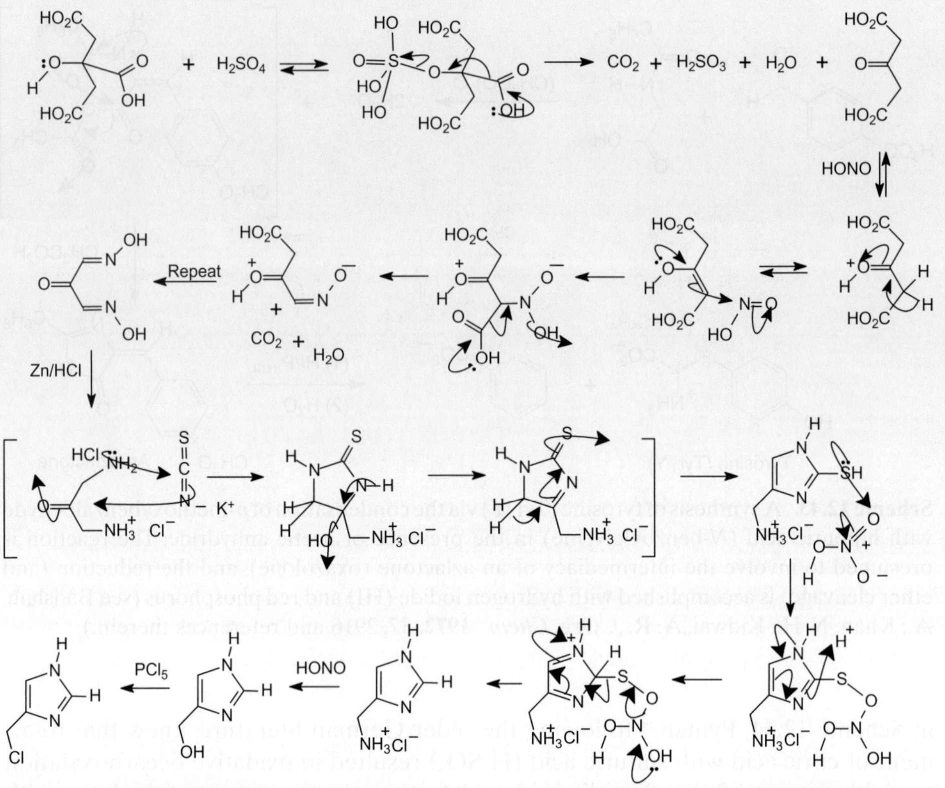

Scheme 12.44. The synthesis of 4-chloromethylglyoxaline (i.e., 4-chloromethylimidazole) from citric acid, as carried out by Pyman (Pyman, F. L. *J. Chem. Soc.*, **1911**, *99*, 1389), and on the way to histidine (His, H).

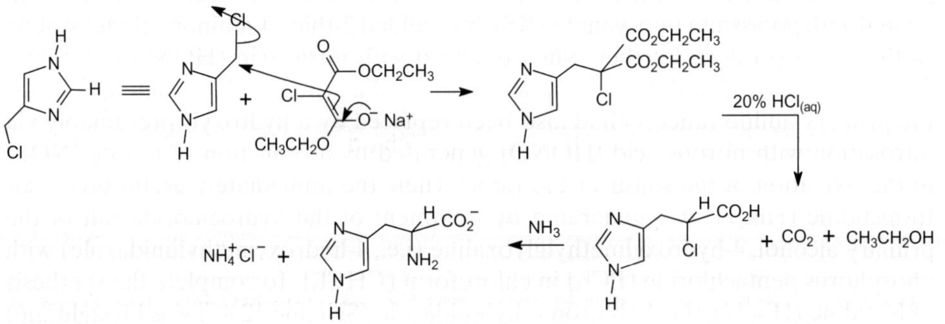

Scheme 12.45. The conversion of 4-chloromethylglyoxaline from citric acid (Scheme 12.44) to histidine (His, H) as carried out by Pyman (Pyman, F. L. *J. Chem. Soc.*, **1911**, *99*, 1389).

In the second synthesis (also undertaken by Pyman and published some 5 years later,* the same 4-hydroxylmethylimidazole utilized in the first synthesis and shown in Scheme 12.44 was used again, but this time, it was oxidized with nitric acid to the corresponding aldehyde, the so-called glyoxaline formaldehyde (Scheme 12.46). The

*Pyman, F. L. *J. Chem. Soc.*, **1916**, *109*, 186.

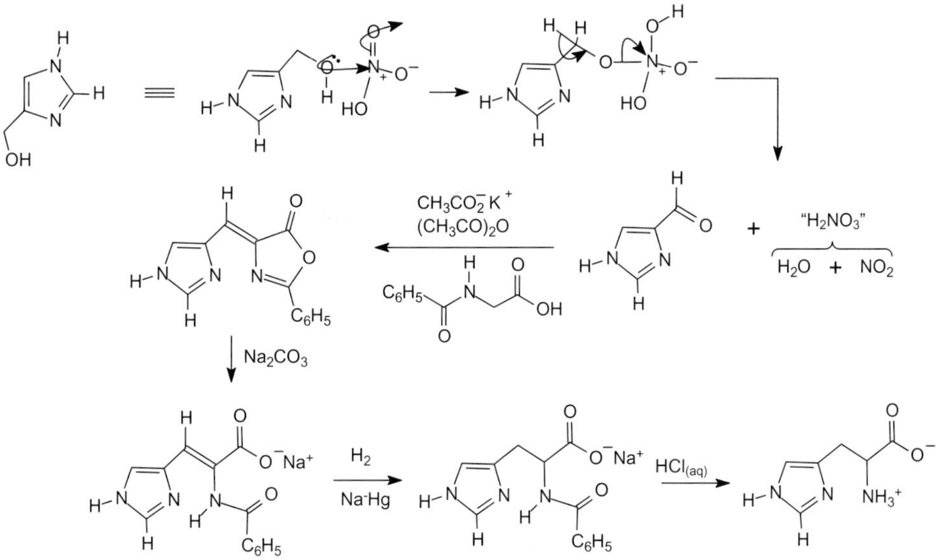

Scheme 12.46. A synthesis of histidine (His, H) after Pyman (Pyman, F. L. *J. Chem. Soc.*, **1916**, *109*, 186) utilizing the 4-hydroxylmethylimidazole shown in Scheme 12.44 and effecting its oxidation and an azlactone synthesis with hippuric acid and acetic anhydride–sodium acetate.

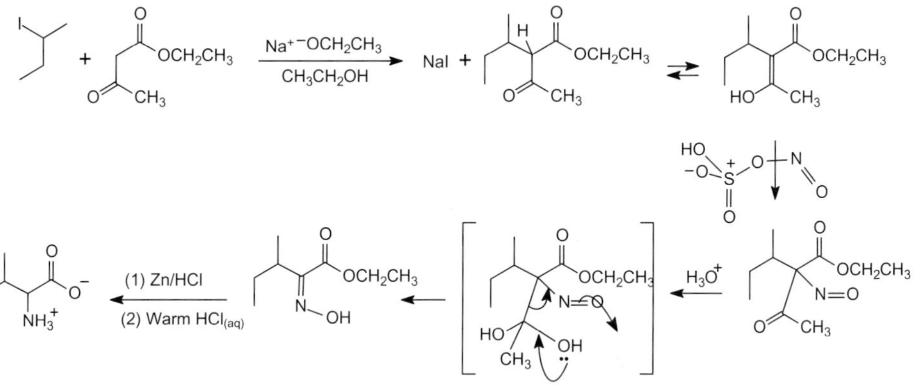

Scheme 12.47. An early synthesis of isoleucine (Ile, I) from 2-iodobutane and acetoacetic ester.

reaction of that aldehyde with hippuric acid (*N*-benzoylglycine) in the presence of acetic anhydride and sodium acetate generated the corresponding azlactone, α-benzoylamino-β-4-(1-acetylimidazole) acrylic acid lactimide. When the azlactone was treated with sodium carbonate, the sodium salt of α-benzoylamino-β-4-imidazole acrylic acid resulted and reduction (hydrogen [H_2] and sodium amalgam [Na/Hg]) produced the corresponding α-*N*-benzoylhistidine, the hydrolysis of which (aqueous HCl) led to histidine (His, H) itself.

Isoleucine (Ile, I) was first synthesized about the same time by allowing 2-iodobutane to react with ethyl acetoacetate in ethanol in the presence of sodium ethoxide (Scheme 12.47) and then by treating the resulting 2-butyl derivative of the

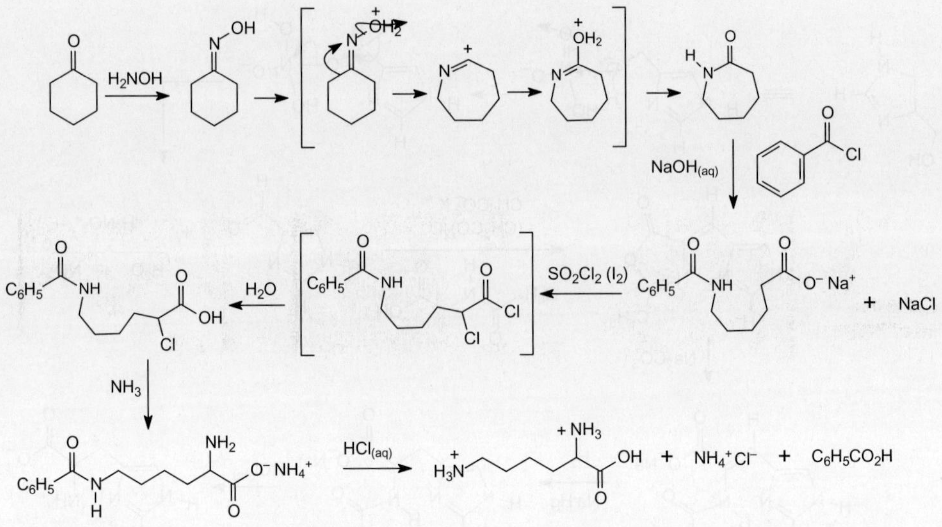

Scheme 12.48. A representation of a synthesis of lysine (Lys, K) based upon the ready availability of ε-caprolactam (after Galat, A. *J. Am. Chem. Soc.*, **1947**, *69*, 86).

acetoacetic ester with nitrosyl sulfate ($ONSO_4H$), which resulted in the formation of the corresponding oxime derivative. Reduction of the latter with zinc (Zn) and HCl, the ethyl ester of isoleucine, resulted and, finally, the amino acid itself was readily obtained by a basic hydrolysis of the ester followed by reacidification.

A number of syntheses of racemic lysine (Lys, K) have been reported. However, as pointed out by Galat,* with the advent of the ready availability of ε-caprolactam,† which can serve as a starting material, a simple method (Scheme 12.48) is derived from that lactam. Thus, as shown in Scheme 12.48, cyclohexanone was converted to the corresponding oxime and the Beckmann rearrangement was effected with sul-

---

*Galat, A. *J. Am. Chem. Soc.*, **1947**, *69*, 86.

†The polyamide polymer nylon-6 became available for nonmilitary use at the end of World War II. It is produced by the catalyzed ring opening and the subsequent polymerization of ε-caprolactam. The number "6" refers to the number of carbon atoms lying between nitrogens in the polymer, that is,

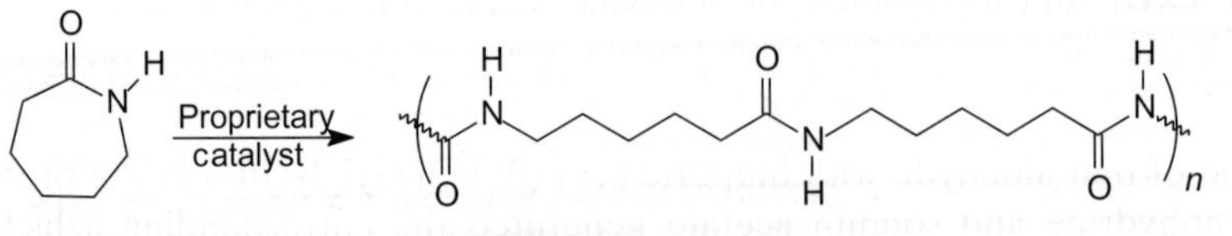

The related polyamide polymer nylon-6,6 is formed by the copolymerization of hexamethylenediamine (1,6-diaminohexane) and adipic acid (hexanedioic acid), that is,

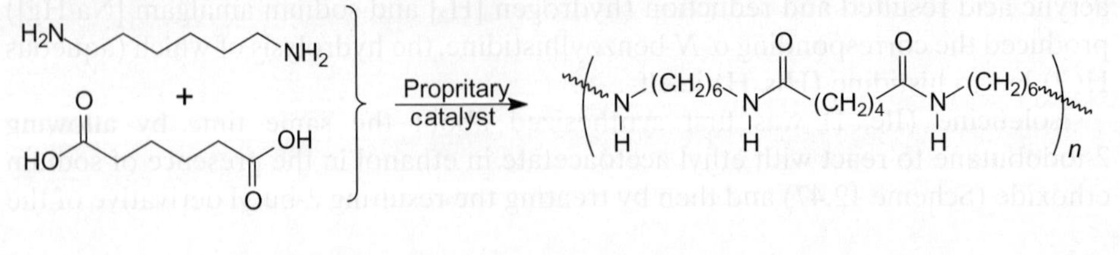

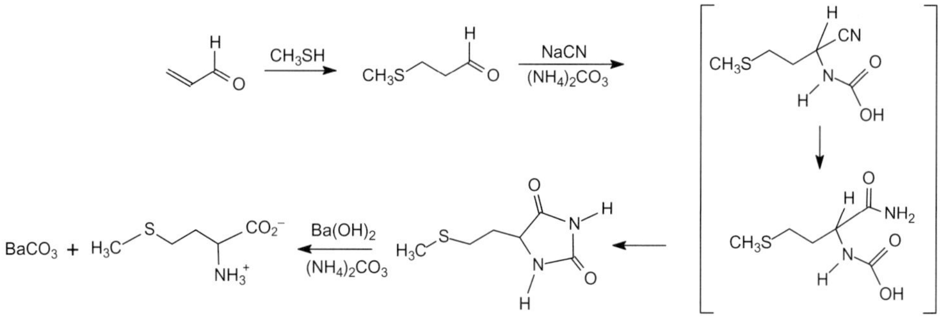

Scheme 12.49. A synthesis of methionine (Met, M) via the corresponding hydantoin (after Bucherer, H. T.; Steiner, W. *J. Prakt. Chem.*, **1934**, *140*, 291).

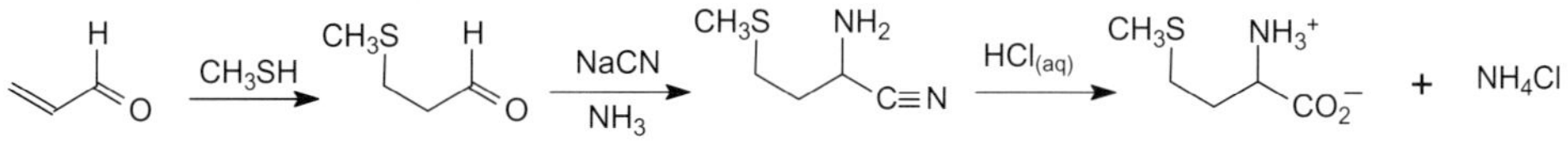

Scheme 12.50. A synthesis of methionine (Met, M) via the Strecker process (after Catch, J. R.; Cook, A. H.; Graham, A. R.; Heilbron, I. M. *J. Chem. Soc.*, **1947**, 1609).

furic acid (H_2SO_4). The resulting ε-caprolactam was then treated with sodium hydroxide (NaOH) and benzoyl chloride (C_6H_5COCl) in a Schotten–Bauman reaction to produce the *N*-benzoyl derivative (i.e., ε-benzoylaminocaproic acid [6-benzoylaminohexanoic acid]). Chlorination of the latter with sulfuryl chloride (SO_2Cl_2) resulted in the formation of the corresponding α-chlorocarboxylic acid (a Hell–Volhard–Zelinsky reaction), and aminolysis followed by acid-catalyzed debenzoylation produced racemic lysine (Lys, K).

In the early part of the twentieth century, a good synthesis of racemic methionine (Met, M) was published by Bucherer and Steiner.* It was found that acrolein (2-propenal) underwent Michael addition with methane thiol (CH_3SH) (Scheme 12.49) to yield the corresponding 3-thio-methylpropanal. When this aldehyde was added to an aqueous solution of sodium cyanide (NaCN), ammonium carbonate [$(NH_4)_2CO_3$], ethanol (CH_3CH_2OH), and triethylamine [$(CH_3CH_2)_3N$] and was allowed to stand, heated briefly, evaporated to dryness, and treated with sulfuric acid (H_2SO_4), methionine hydantoin resulted in greater than 90% yield. The amino acid itself was then obtained by heating the hydantoin in aqueous barium hydroxide followed by treatment of the residue with a solution of ammonium carbonate [$(NH_4)_2CO_3$] (Scheme 12.49).

Interestingly, only a few years later, Heilbron and coworkers[†] found that equally good results could be obtained by a simple Strecker-type synthesis. Thus, when the same thioaldehyde was treated with potassium cyanide (KCN) and ammonia, the corresponding α-aminonitrile resulted and hydrolysis of the cyano group with aqueous hydrochloric acid (HCl) produced methionine (Met, M) (Scheme 12.50).

Valine (Val, V) has also been synthesized by the Strecker process beginning with isobutyraldehyde and by adding cyanide in the presence of ammonia (analogously

*Bucherer, H. T.; Steiner, W. *J. Prakt. Chem.*, **1934**, *140*, 291.
[†]Catch, J. R.; Cook, A. H.; Graham, A. R.; Heilbron, I. M. *J. Chem. Soc.*, **1947**, 1609.

to the process shown in Scheme 12.50). However, it is reported that better yields of the racemic amino acid are obtained on the aminolysis of α-bromoisovaleric acid (Scheme 12.51), the latter generated via the Hell–Volhard–Zelenski reaction (bromine and red phosphorus) on isovaleric acid itself.

Phenylalanine (Phe, F) is most conveniently synthesized by the azlactone (oxazolone) method. Thus, when hippuric acid (*N*-benzoylglycine) is heated with benzaldehyde in the presence of anhydous sodium acetate (NaO_2CCH_3) in glacial acetic acid (CH_3CO_2H), the corresponding azlactone (oxazolone) is obtained. Treatment of the latter with cold HI and red phosphorus followed by hot water produces the amino acid phenylalanine (Phe, F), which, after the removal of the benzoic acid by extraction, is isolated by adjusting the pH to between 5 and 6 with concentrated ammonia (Scheme 12.52).

As shown in Scheme 12.53, threonine (Thr, T) can be synthesized readily by the addition of mercury(II) acetate (mercuric acetate, $Hg(O_2CCH_3)_2$) in methanol

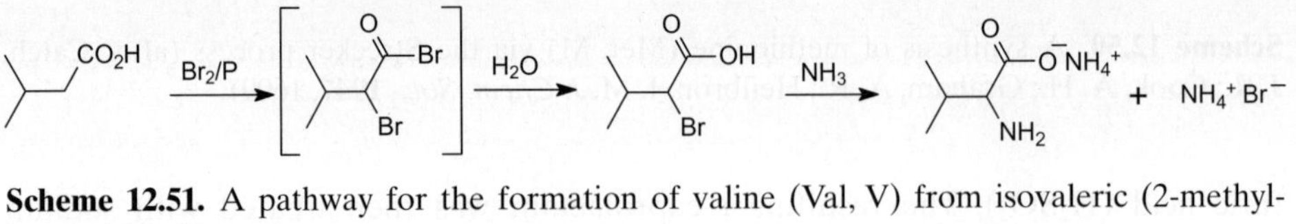

Scheme 12.51. A pathway for the formation of valine (Val, V) from isovaleric (2-methylbutanoic) acid.

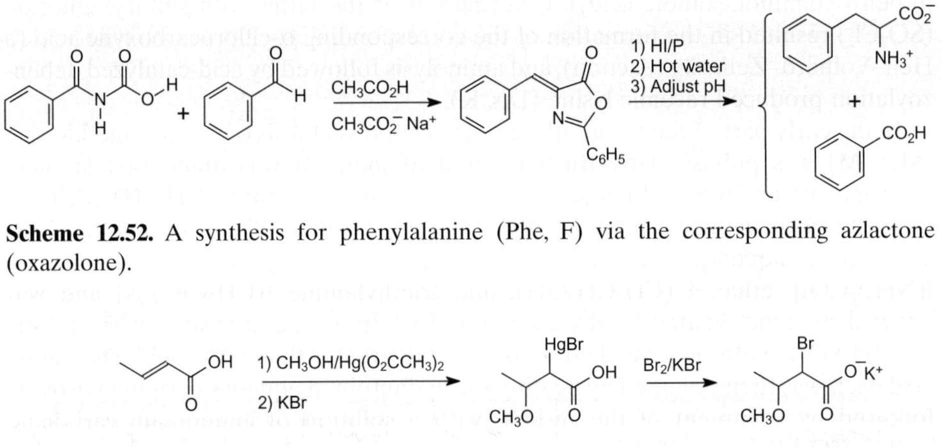

Scheme 12.52. A synthesis for phenylalanine (Phe, F) via the corresponding azlactone (oxazolone).

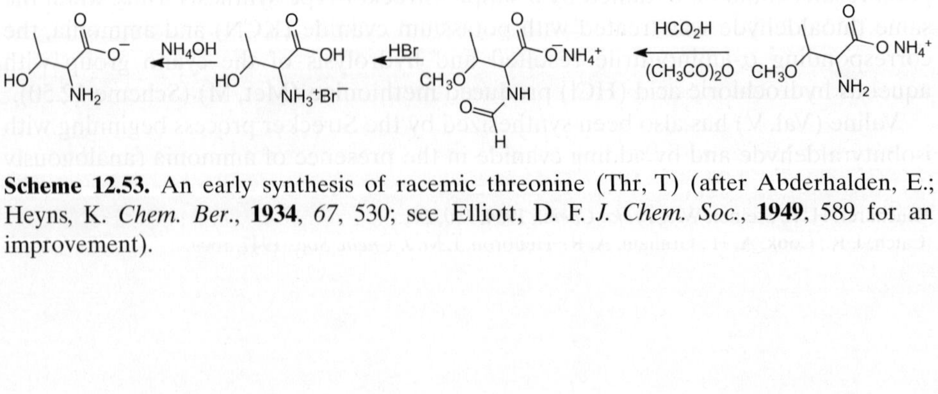

Scheme 12.53. An early synthesis of racemic threonine (Thr, T) (after Abderhalden, E.; Heyns, K. *Chem. Ber.*, **1934**, *67*, 530; see Elliott, D. F. *J. Chem. Soc.*, **1949**, 589 for an improvement).

(CH$_3$OH) across the carbon–carbon double bond of crotonic acid [(E)-2-butenoic acid] followed by potassium bromide treatment of the initial adduct to yield the corresponding mercuric bromide and then the formation of 2-bromo-3-methoxybutanoic acid by the addition of bromine (Br$_2$) to the latter. Acidification to the free acid, followed by amination (NH$_3$), formylation with the equivalent of "acetic-formic anhydride," hydrogen bromide (HBr) ether cleavage, and a final pH adjustment with ammonium hydroxide produced the desired amino acid.

A synthesis of tryptophan (Trp, W) traditionally begins with the synthesis of the indole nucleus. Although a variety of pathways to variously substituted indole nuclei have been developed over the years (using the "Fischer indole synthesis," Chapter 13) the simple unsubstituted system has generally proved more difficult than more highly substituted derivatives. Nonetheless, a decent synthesis of indole itself was worked out by A. Reissert and was reported as early as 1897.

As shown in Scheme 12.54, when *ortho*-nitrotoluene is allowed to react with diethyl oxylate in the presence of sodium (or potassium) ethoxide, a condensation occurs and ethoxide is eliminated. Then, with zinc (Zn) in acetic acid, reduction of the nitro group and cyclization to the keto carbonyl occurs. Hydrolysis results in the formation of the indole-2-carboxylic acid and pyrolysis leads to decarboxylation with the formation of the free base and carbon dioxide (CO$_2$).

Nearly a century later, in 1976, Hoffmann–La Roche Ltd. was awarded patents (U.S. Patents 3,732,245 and 3,976,639) largely through the work of Willy Leimbgruber on a simpler high-yield method that, again, begins (Scheme 12.55) with *ortho*-nitrotoluene. In this version, *N,N*-dimethylaminoacetaldehyde dimethylacetal is allowed to react in the presence of pyrrolidine with the nitro substrate. Exchange of the pyrrolidine for the dimethylamino group is presumed to occur, and then condensation at the methyl group follows to produce a styrene derivative, which, on reduction, undergoes cyclization and elimination to indole itself.

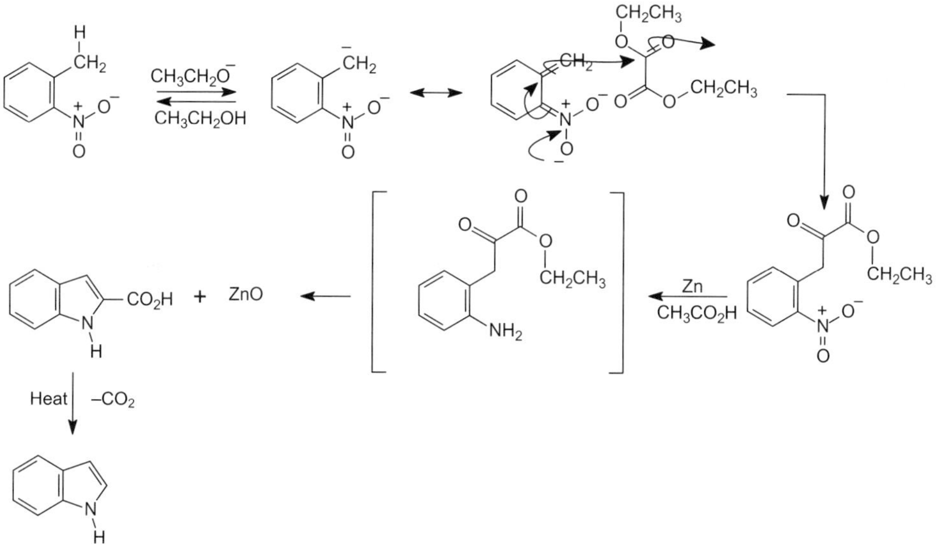

Scheme 12.54. An early synthesis of indole from *o*-nitrotoluene (after Reissert, A. *Chem. Ber.*, **1897**, *30*, 1030).

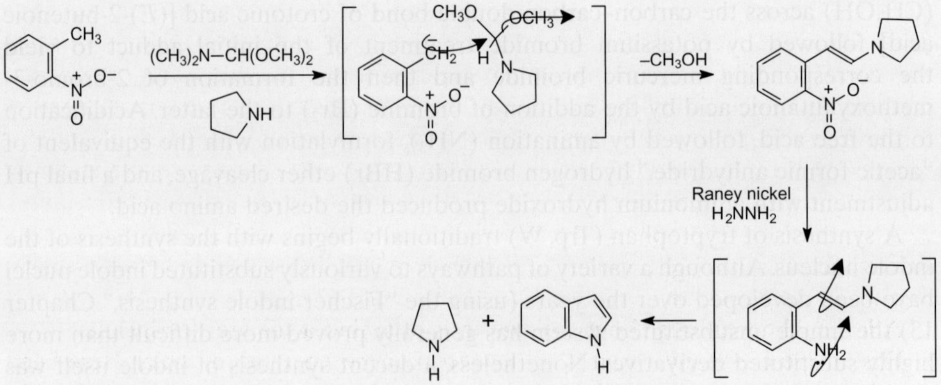

Scheme 12.55. An early synthesis of indole from *o*-nitrotoluene (after U.S. Patents 3,732,245 and 3,976,639 to Willy Leimbgruber and Hoffmann–La Roche Ltd.).

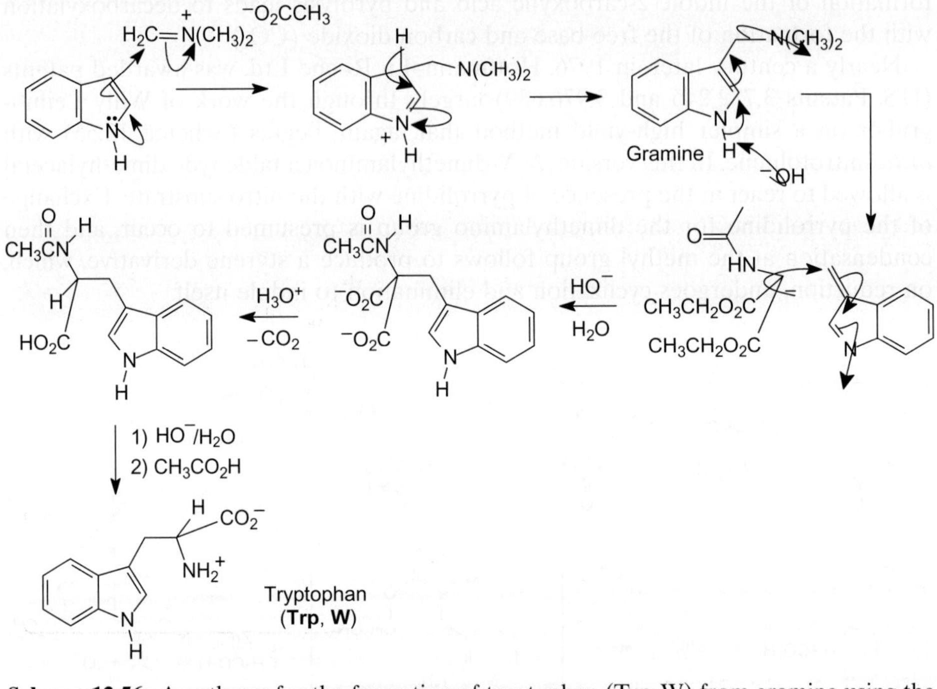

Scheme 12.56. A pathway for the formation of tryptophan (Typ, W) from gramine using the diethyl ester of *N*-acetylmalonate.

Now, for the synthesis of tryptophan (Trp, W), the intermediate gramine was first generated. As shown in Scheme 12.56, when a cold solution of dimethylamine, acetic acid, and aqueous formaldehyde (40%) was added to indole, rapid warming occurred and after standing, addition of an aqueous base resulted in the crystallization of gramine. The latter was dried over potassium hydroxide (KOH) and, under nitrogen and in refluxing toluene containing powdered sodium hydroxide (NaOH), was treated with the diethyl ester of *N*-acetylmalonic acid. After several hours, dimethylamine [(CH$_3$)$_2$NH] was evolved and, after filtering the hot solution, cooling resulted in the crystallization of the ethyl ester of α-acetylamino-α-carboethoxy-β-3-indolpropanoic acid. Hydrolysis of the esters was accomplished with aqueous sodium hydroxide (NaOH) and, on acidification and heating, decarboxylation occurred to yield *N*-acetyltryptophan. With the completion of the decarboxylation, deacetylation was accomplished by heating the *N*-acetylcarboxylic acid in aqueous sodium hydroxide (NaOH) for 20 h. Tryptophan (Trp, W) was isolated on acidification with acetic acid. The synthesis is outlined in Scheme 12.56.

Additional comments about the synthesis of amino acids are worthwhile.

First, there are many variants of the Strecker method, two of which have been noted above, viz., in the syntheses of alanine (Ala, A) and methionine (Met, M). However, the Strecker variants all involve the addition of cyanide (CN⁻) to an imine (or equivalent); the latter formed *in situ* from a carbonyl compound and a suitable equivalent of ammonia (e.g., ammonia [NH$_3$], ammonium chloride [NH$_4$Cl], and ammonium acetate [NH$_4$O$_2$CCH$_3$]). The generalized pathway is shown in Scheme 12.57.

A second tried-and-true method, already seen above in the synthesis of both tyrosine (Tyr, Y) and phenylalanine (Phe, F), involves the intermediacy of an azlactone (oxazolone). There are two choices here: either the azlactone (oxazolone) can be formed (as was the case for both phenylalanine [Phe, F] and tyrosine [Tyr, Y]) or a "generalized" azlactone (oxazolone) can be formed from glycine (Gly, G) and any appropriate anhydride as shown in Scheme 12.58. Then, the azlactone (oxazolone) having formed, reaction with an aldehyde of choice (that becomes the amino acid side chain) followed by reduction (e.g., with hydrogen [H$_2$] in the presence of an appropriate catalyst) and finally, hydrolysis produces the corresponding α-amino acid (Scheme 12.58).

Third, there is the hydantoin (glycolylurea, imidazolidine-2,4-dione) method. Here, the reactive intermediate is again formed from glycine (Gly, G), which, on reaction with potassium cyanate (KOCN), generates the hydantoin. The resulting

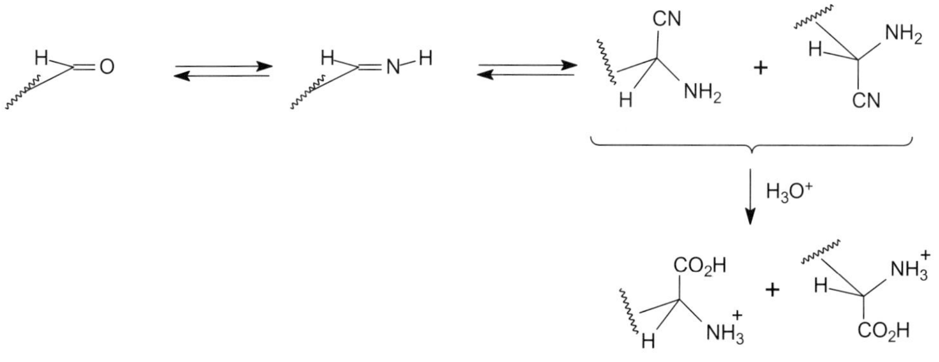

Scheme 12.57. A generalized representation of the Strecker process. The source of the nitrogen required to form the imine is not specified as many different sources are available (see text). A significant effort has been (and continues, e.g., see Scheme 12.33) made to find paths to single enantiomers.

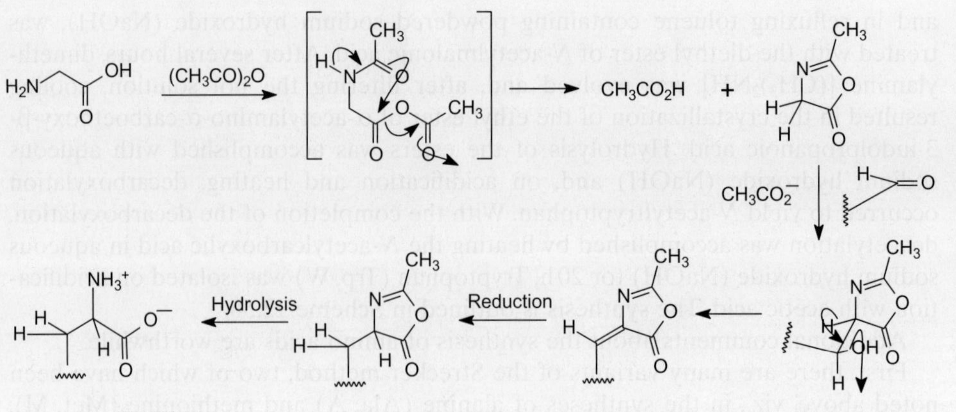

Scheme 12.58. A representation of a pathway to an amino acid using the azlactone (oxazolone) synthetic route. Glycine (Gly, G) is allowed to react with acetic anhydride so that it undergoes both N-acetylation and conversion to the mixed anhydride. On loss of acetic acid, the mixed anhydride produces the azlactone (oxazolone), which can then react with a choice of aldehydes (in the presence of acetate anion—to serve as a proton sink). Dehydration, reduction, and hydrolysis then generates the amino acid.

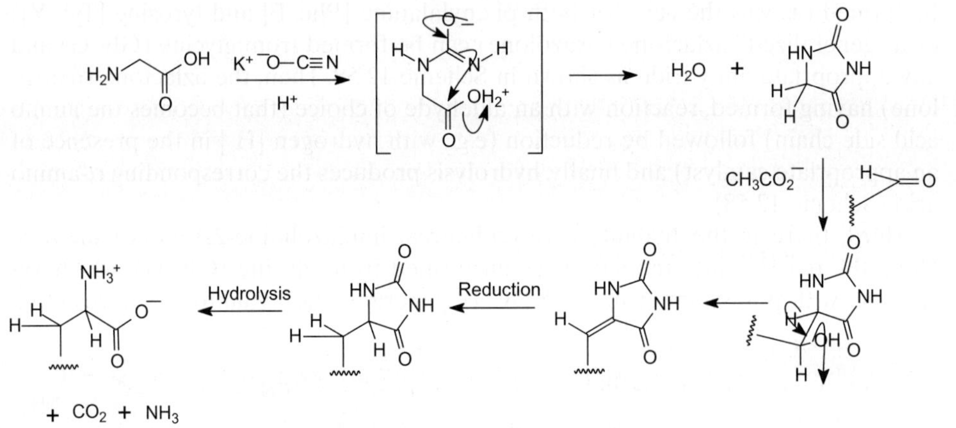

Scheme 12.59. A representation of a pathway to an amino acid using the hydantoin (glycolyl urea, imidazolidine-2,4-diene) synthetic route. Glycine (Gly, G) is allowed to react with potassium cyanate in the presence of an acid catalyst. Addtion occurs at nitrogen and cyclization with loss of water produces the hydantoin (glycolyl urea, imidazolidine-2,4-diene). The hydantoin is then available to react with a choice of aldehydes (in the presence of acetate anion—to serve as a proton sink). Dehydration, reduction, and hydrolysis then generate the amino acid.

heterocyclic system can, as was the case with the azlactone (oxazolone), react with an aldehyde that is destined to become the amino acid side chain after reduction (Scheme 12.59).

In addition to the classical general methods outlined above, there are a number of newer methods that deserve mention.

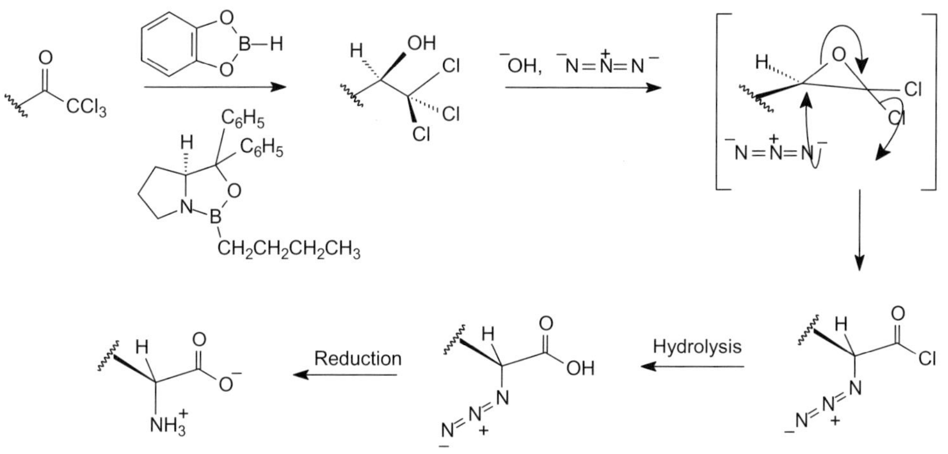

Scheme 12.60. The trichloromethyl ketone is reduced with catecholborane in the presence of the (S)-oxazaborolidine catalyst. The resulting (R)-secondary alcohol (produced in high enantioselectivity) is treated with sodium hydroxide and sodium azide to yield the corresponding (S)-α-azido carboxylic acid, and reduction then produces the amino acid (after Corey, E. J.; Link, J. O. *J. Am. Chem. Soc.*, **1992**, *114*, 1906).

It has been pointed* out that trichloromethyl ketones are generally available by the reaction of aldehydes with nucleophilic trichloromethide reagents followed by oxidation as well as in other ways. So, as shown in Scheme 12.60, when the trichloromethyl ketone is reduced with catecholborane in the presence of the (S)-oxazaborolidine catalyst, the corresponding (R)-secondary alcohol is produced in high enantioselectivity. Then, when the alcohol is treated with sodium hydroxide and sodium azide, the corresponding (S)-α-azidocarboxylic acid results and reduction (hydrogen [H_2] in the presence of a palladium [Pd] catalyst) produces the amino acid.

An exciting method for the synthesis of either (or both) enantiomers of amino acids is based upon the use of pseudoephedrine, a generally available commodity chemical.† However, an excellent synthesis of pseudoephedrine is also available.‡ For the latter (Scheme 12.61), the synthesis began with the N-carbobenzoxy derivative (N-Cbz) of alanine (Ala, A) (shown here as L-alanine, Ala, A), which was then treated with paraformaldehyde in the presence of a trace of p-toluenesulfonic acid (PTSA) catalyst to form the corresponding N-carbobenzoxyoxazolidinone. Treatment of the latter with sodium borohydride in methanol produced the corresponding lactol, which was allowed to react with phenylmagnesium bromide to yield an inseparable mixture of diastereomeric aminoalcohols in a 95:5 ratio. When the mixture was converted to the corresponding mixture of oxazolines on treatment with paraformaldehyde, separation could be effected by chromatography. Treatment of the major isomer with sodium cyanoborohydride/trimethylchlorosilane (NaCNBH₃/TMSCl) in acetonitrile then reductively opened the heterocycle with

*Corey, E. J.; Link, J. O. *J. Am. Chem. Soc.*, **1992**, *114*, 1906.
†Reddy, G. B.; Rao, G. V.; Sreevani, B.; Iyengar, D. S. *Tetrahedron Lett.*, **2000**, *41*, 953.
‡Myers, A. G.; Gleason, J. L.; Yoon, T. *J. Am. Chem. Soc.*, **1995**, *117*, 8488.

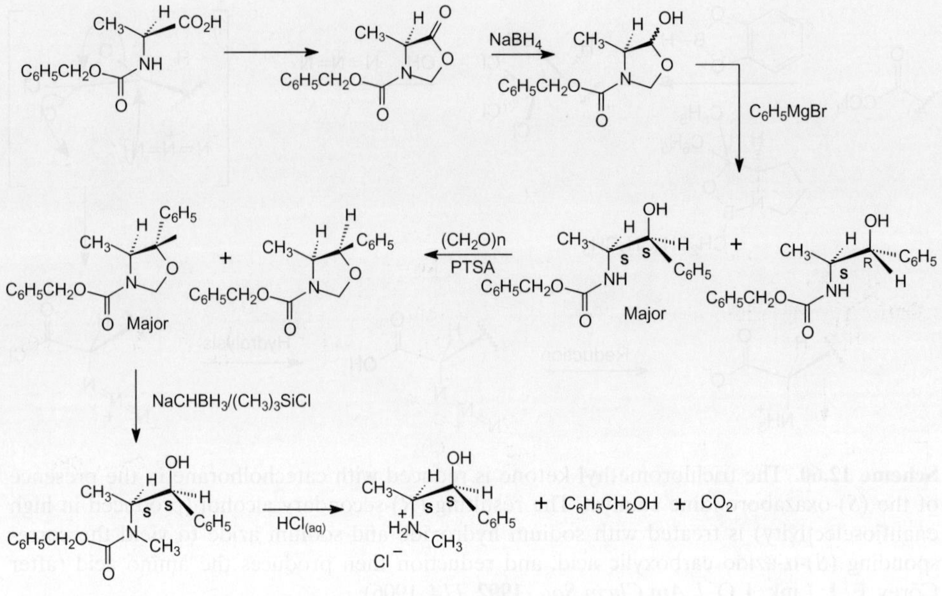

Scheme 12.61. A synthesis of (1S, 2S)-pseudoephedrine (after Reddy, G. B.; Rao, G. V.; Sreevani, B.; Iyengar, D. S. *Tetrahedron Lett.*, **2000**, *41*, 953).

the formation of the carbobenzoxy derivative of pseudoephedrine, and the hydrolysis of the latter by heating with concentrated hydrochloric acid (HCl) generated pseudoephedrine hydrochloride itself.

Now, when *N-tert*-butoxycarbonyl (N-BOC) glycine (Gly, G) is treated with 2,2-dimethylpropanoic acid chloride (pivaloyl chloride)—to form the mixed anhydride—and the mixture is used to acetylate pseudoephedrine, the resulting bis-amide can be hydrolyzed in acid to generate a chiral pseudoephedrine glycinamide (Scheme 12.62). The absolute stereochemistry of the pseudoephedrine remains unaffected. Then, when the pseudoephedrine glycinamide is treated with lithium chloride (LiCl) and just less than 2 equivalents of lithium diisopropylamide (LDA) in THF at −78°C, both acidic protons (on O and on N) are lost, but, on warming that dianion to 0°C followed by the addition of an electrophile (such as an alkyl halide), alkylation *occurs on carbon*, not on nitrogen or oxygen. It is argued that the (Z)-enolate must be an intermediate because of the diasteromeric excess of one isomer in the resulting product mixture and that the C-methyl of the pseudoephedrine influences the sense of the attack (Figure 12.7). Finally, aqueous acid hydrolysis removes the pseudoephedrine for reuse and produces the appropriate amino acid (Scheme 12.62).

In yet another interesting new method, which cannot only give rise to either enantiomer of any desired amino acid but can also produce α,α-disubstituted amino acids, it has been found* (Scheme 12.63) that S-glycidol (2-oxiranemethanol, 2,3-epoxy-1-propanol) will undergo a Mitsunobu reaction (Scheme 8.33) with phthalimide in the presence of triphenylphosphine and di-*tert*-butyl

*Harding, C. I.; Dixon, D. J.; Ley, S. V. *Tetrahedron*, **2004**, *60*, 7679.

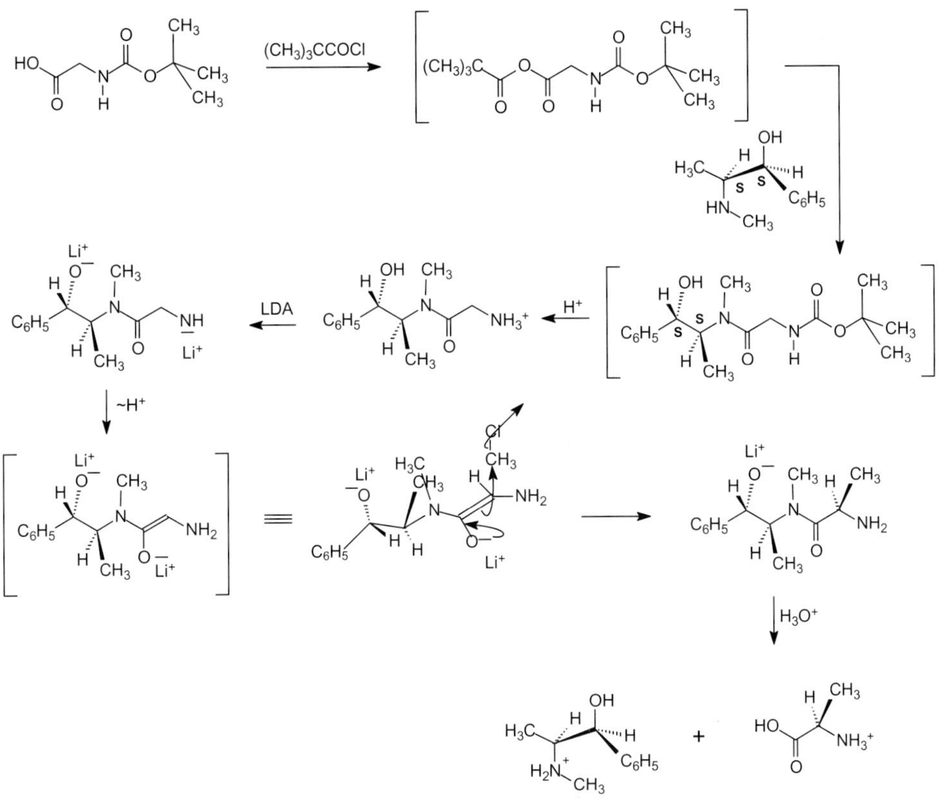

Scheme 12.62. A generalized scheme for the preparation of any amino acid whose alkyl side chain can be introduced by nucleophilic substitution. The synthesis, after Myers, A. G.; Gleason, J. L.; Yoon, T. *J. Am. Chem. Soc.*, **1995**, *117*, 8488, makes use of pseudoephedrine as a chiral auxiliary group to guide in the placement of the incoming group.

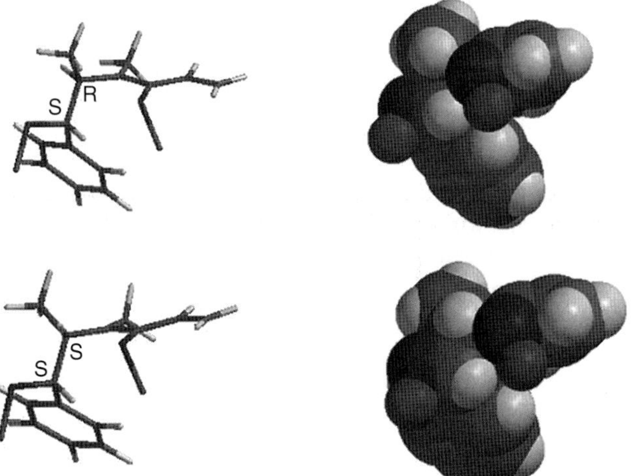

Figure 12.7. The (S, R)- and (S, S)-(Z)-enolates of Scheme 12.62. The C-methyl group helps determine the position of the aryl ring and the sense of the direction of alkylation.

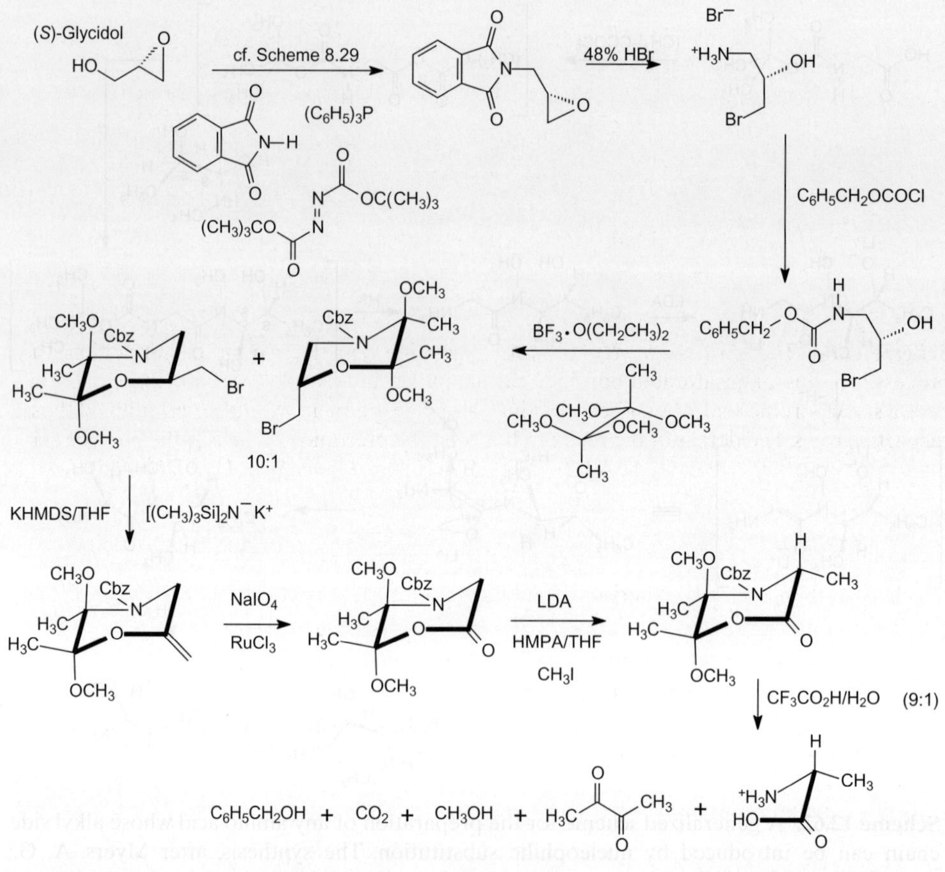

Scheme 12.63. A synthesis of (R)-alanine (Ala, A) based upon a path utilizing the formation of morpholine derivatives and after Harding, C. I.; Dixon, D. J.; Ley, S. V. *Tetrahedron*, **2004**, *60*, 7679. Additional alkyl groups can be added to form α,α-disubstituted derivatives and, beginning with the opposite glycidol derivative, the opposite enantiomer can be obtained. HMPA = hexamethylphosphoramide.

azidodicarboxylate to produce the 2,3-epoxy-1-(N-phthalimido)propane, which, with hydrogen bromide (HBr), underwent both ring opening and hydrolytic cleavage of the phthalimido group to produce 3-bromo-2-hydroxy-1-aminopropane hydrobromide. The absolute stereochemistry of the hydroxyl at C-2 was preserved. Then, the amino group was protected as the N-carbobenzoxy derivative, by the reaction with benzylchloroformate and the reaction with the diketal, 2,2,3,3-tetramethoxybutane, effected in the presence of borontrifluoride etherate to produce the corresponding morpholino derivatives in a 10:1 mixture of diastereomers.

Treatment of the major isomer with potassium hexamethyldisilizane (KHMDS, $[(CH_3)_3Si]_2N^-\,K^+$) caused the elimination of hydrogen bromide (HBr), and the oxidation of the resulting alkene with sodium periodate–ruthenium trichloride couple ($NaIO_4/RuCl_3$) yielded the desired lactone, where the carbonyl group was appropriately positioned to activate the methylene α to the nitrogen for subsequent

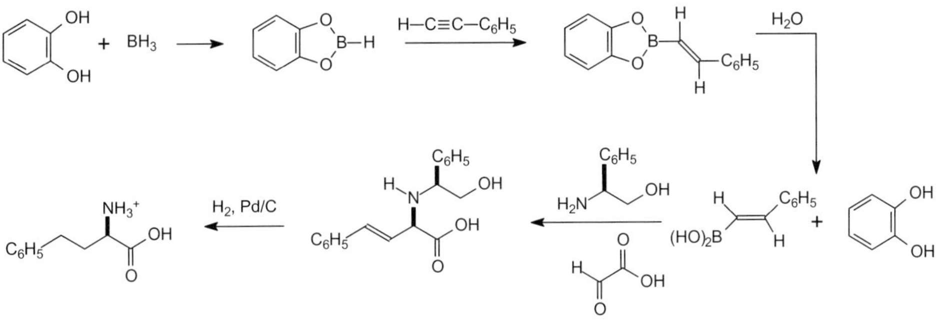

Scheme 12.64. The synthesis of (R)-homophenylalanine via a three-component reaction. The process involves an unsaturated boronic acid, an amine, and an α-ketocarboxylic acid. In this case, a styrylboronic acid, (S)-2-phenylglycinol, and glyoxylic acid ($CHOCO_2H$) fulfilled those respective roles. The details of the reaction between the presumed imine and the boronic acid remain obscure (after Petasis, N. A.; Zavialov, I. A. *J. Am. Chem. Soc.*, **1997**, *119*, 445).

alkylation. With the introduction of an alkyl group destined to become the defining feature of the amino acid and hydrolytic cleavage of the morpholine ring, the *N*-carbobenzoxy protected amino acid was liberated (Scheme 12.63).

Although new and clever methods of synthesis of variously substituted α-amino acids continue to be found and development of various older techniques continues to evolve, and thus review articles and modestly large volumes describing these advances are issued frequently, this section on synthesis will close with the recently found practical synthesis of β,γ-unsaturated α-amino acids by Petasis et al.* and a reminder by Schölkopf[†] on the value of diketopiperazines.

The Petasis method involves the use of organoboron compounds to produce a side chain on a Mannich product derived from the imine condensation product between an amine and an α-ketocarboxylic acid. Thus, as shown in Scheme 12.64, as an example of this three-component reaction, when the styrylboronic acid derived from the reaction of phenylacetylene with catecholborane was allowed to react with glyoxylic acid ($CHOCO_2H$) and (S)-2-phenylglycinol, a single diastereomer was isolated, which gave an enantiomerically pure (R)-homophenylalanine hydrochloride on reduction. This procedure is limited only by the availability of the components.

Finally, as shown in Scheme 12.65, it is valuable to remember that simple amino acids can often easily be converted (by simply heating them) into the corresponding diketopiperazines, which can be further manipulated. Thus, L-alanine dimerizes on heating and when the dimer is treated with trimethyloxyonium tetrafluoroborate, methylation on oxygen ensues. Treatment of the dimethoxy bis-enol ether generates a monoanion that can be benzylated on carbon. Alkylation with benzyl chloride followed by hydrolysis regenerates L-alanine and the methyl substituted phenylalanine. Such a pattern of reactivity can be widely employed.

*Petasis, N. A.; Zavialov, I. A. *J. Am. Chem. Soc.*, **1997**, *119*, 445.
†Schölkopf, U. *Tetrahedron*, **1983**, *39*, 1983.

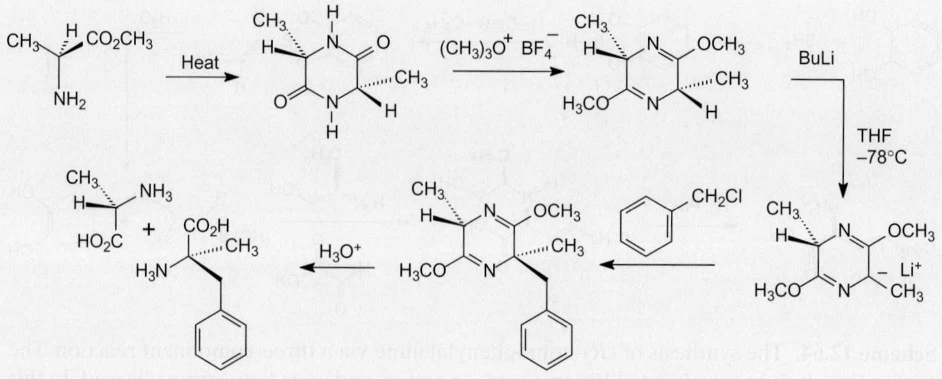

Scheme 12.65. A demonstration of the value of diketopiperazines in the synthesis of α-amino acids (after Schölkopf, U. *Tetrahedron*, **1983**, *39*, 1983).

C. PEPTIDES AND PROTEINS—INTRODUCTION

Amino acids, by synthesis, as described in Section B of this chapter, and by degradation of proteins (i.e., polypeptides, *vide infra*), are available for the construction of new proteins.*

When two amino acids ("A" and "B") are covalently joined (forming an amide bond or, in a biochemical sense, a "peptide bond") by using the α-amino group of one and the α-carboxylate of the other, a dipeptide unit is created. Since each amino acid can, in principle, utilize either its α-amino group or its α-carboxylate group, respectively, there are two possibilities for the dipeptide (either A–B or B–A, unless the two amino acids are identical). Equation 12.10 provides a generalized view of this procedure, without specifying how the condensation (with loss of water) or the hydrolysis (with gain of water) occurs.

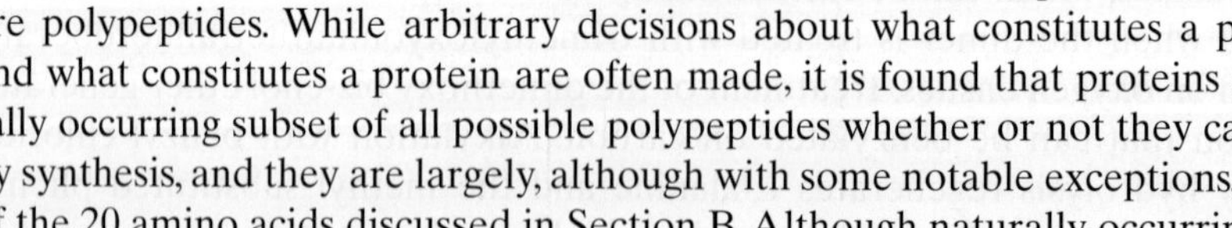

$$(12.10)$$

When three amino acids are so cojoined, a tripeptide is formed. As the number of amino acids increases, small and then large oligopeptides and then polypeptides result. Proteins, some of which are enzymes that act on or help build other proteins, are polypeptides. While arbitrary decisions about what constitutes a polypeptide and what constitutes a protein are often made, it is found that proteins are a naturally occurring subset of all possible polypeptides whether or not they can be made by synthesis, and they are largely, although with some notable exceptions, composed of the 20 amino acids discussed in Section B. Although naturally occurring proteins, when isolated, will have the amino acid "backbone" as dictated by their polypeptide

---

*The word "protein" is derived from the Greek work *prota*, meaning "of first [primary] form or quality," whereas the word "peptide," from the same source, appears to mean "easily digested or dissolved."

nature, they may have been further modified by reactions such as acetylation or methylation. In the naturally occurring proteins, it is generally held that these modifications occur after the protein is synthesized.

Both biochemical (i.e., *in vivo*) and chemical (i.e., *in vitro*) pathways for the decomposition of proteins (polypeptides) to amino acids and for the formation of peptides (proteins) from amino acids are known. Having shown some biosynthetic and synthetic pathways to some amino acids in Section B, this part begins with obtaining the same amino acids by the decomposition of the polypeptides (proteins) by biochemical and chemical pathways and concludes with using those amino acids to produce the peptides and polypeptides.

I. Amino Acids from Peptides

Acid hydrolysis of peptides yields the amino acids from which the peptide was assembled and a few other fragments arising from postassembly modifications. This is simply the acid-catalyzed hydrolysis of the peptide, that is, amide (Chapter 9) bonds. Hundreds, and in some cases thousands, of copies of amino acids are obtained on complete hydrolysis (sometimes called "digestion") of many peptides (proteins), and although any given one might be focused upon as exemplary, only one, α-chymotrypsin, because it has been studied so extensively, is used here.

Chymotrypsin (a protein that is an enzyme that acts on other proteins and cleaves them in specific places)* is found in the pancreas of mammals and other vertebrates. It is classified as an endopeptidase because it catalyzes the cleavage of the peptide on which it acts on internal—in contrast to terminal—amide bonds. The PDB contains a number of examples of this peptide as well as a significant amount of detailed information. The quaternary[†] structure of a sample of α-chymotrypsin can be seen in **PDB 5cha**[‡] (EC 3.4.21.1), reproduced here as Figure 12.8. The figure shows the result of an X-ray crystal structure determination (at a resolution of 167 pm) of the dimeric peptide, at ph 3.5, from domestic cow (*Bos taurus*). The primary structure of each half of the 245 amino acid dimer (molecular mass = 25,666 Da) begins with **Cys-Gly-Val-Pro-Ala-Ile-Gln-Pro-Val** or (**CGVPAIQPV**) as seen in Figure 12.9 and which, as can be seen from the complete primary sequence in Table 12.3, describes the beginning of the peptide chain.

Although it is not clear from the drawing, disulfide bridges (Chapter 8, Scheme 8.105) exist between various cysteines (Cys, C). A tube representation of the top (from the perspective of Figure 12.8) of the dimeric enzyme is shown in Figure 12.9

---

*Chymotrypsin serves as a catalyst that promotes the proteolytic (it is therefore a "protease") cleavage on the carbonyl (rather than the amino) side of peptide bonds to aromatic ring-containing amino acids (viz., phenylalanine [Phe, F], tyrosine [Tyr, Y], and tryptophan [Trp, W]). Of course, other enzymes promote other cleavages.

[†]An ordered list of the amino acids from the N-terminus (where an amino group remains) to the carboxylic acid terminus (where a carboxylate remains) that also includes bridges across the structure created by sulfur–sulfur bonds from cysteine to cysteine is called the primary structure. The secondary structure of a peptide usually refers to repeating units that give rise to coils or ribbons or other repeating structural variants, while the tertiary structure consists of all of the contributions to the three-dimensional representation. When two or more polypeptide subunits are present in the protein, their relationship is called a quaternary structure.

[‡]Blevins, R. A.; Tulinsky, A. *J. Biol. Chem.*, **1985**, *260*, 4264.

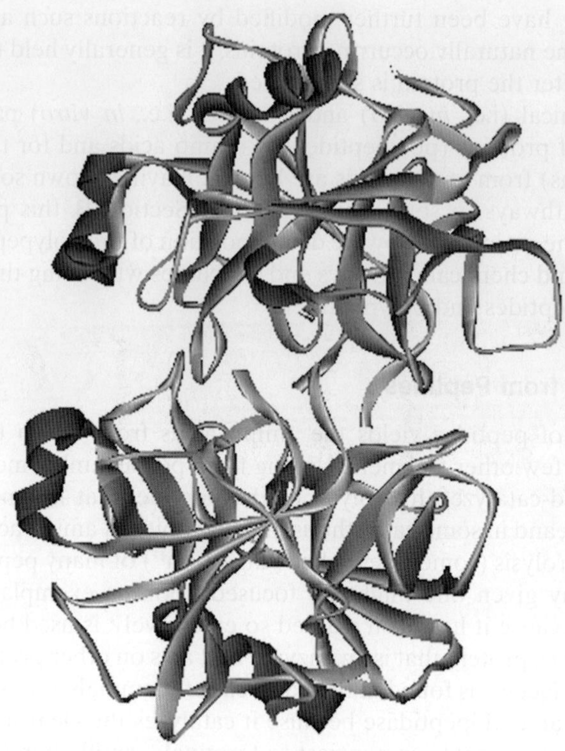

Figure 12.8. A ribbon representation of the result of an X-ray crystal structure determination (at a resolution of 167 pm) of the dimeric chymotrypsin peptide, at pH 3.5, from domestic cow (*B. taurus*). The primary structure of each half of the 245 amino acid dimer (molecular mass = 25,666 Da) begins with **Cys-Gly-Val-Pro-Ala-Ile-Gln-Pro-Val** or (**CGVPAIQPV**) (after Blevins, R. A.; Tulinsky, A. *J. Biol Chem..* **1985**, *260*, 4264).

with the amino acids labeled with their single-letter labels and all five of the sulfur bridges (viz., **Cys** 1–**Cys** 122, **Cys** 42–**Cys** 58, **Cys** 136–**Cys** 201, **Cys** 168–**Cys** 182, and **Cys** 192–**Cys** 220) also shown.

While it is not critical to the discussion of how this enzyme works on proteins, it will not have escaped the astute reader that while chymotrypsin cleaves peptide bonds on the carbonyl side of the bonds to phenylalanine (Phe, F), tyrosine (Tyr, Y), and tryptophan (Trp, W) amino acids in the chain, *those same amino acids are found in chymotrypsin itself*. Is it reasonable to inquire why chymotrypsin does not cleave itself?*

The general answer appears to be that chymotrypsin and, indeed, many other peptidases, are synthesized in the pancreas as inactive "proenzymes" (or zymogens), which are incapable of carrying out their catalytic function, until they are secreted into the intestine. Once in the intestine, the blocking (usually terminal) groups are hydrolyzed off and the active enzyme is produced.

*__From the nursery rhyme__, "There once were two cats of Kilkenny; Each thought there was one cat too many; So they fought and they fit; And they scratched and they bit; Till excepting their nails; And the tips of their tails; Instead of two cats; There weren't any."

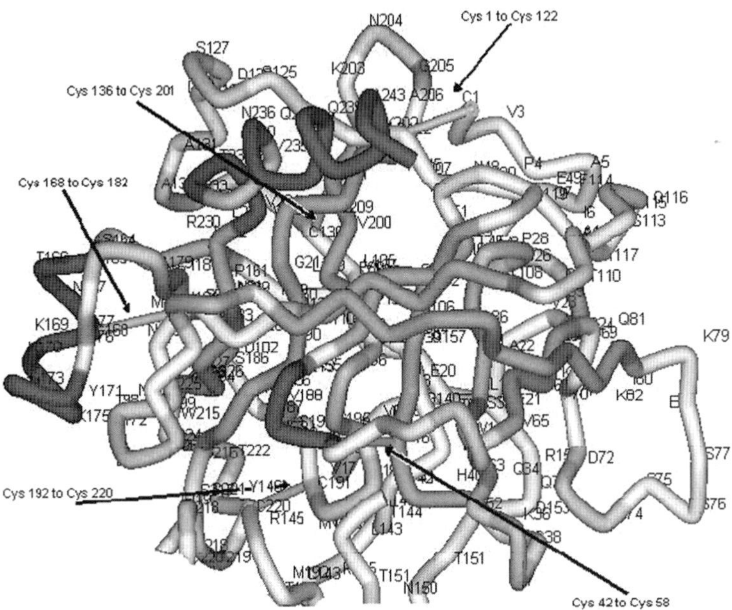

Figure 12.9. A tube representation of the top (from the perspective of Figure 12.8) of the dimeric chymotrypsin with the amino acids labeled with their single-letter labels and all five of the sulfur bridges (viz., **Cys** 1–**Cys** 122, **Cys** 42–**Cys** 58, **Cys** 136–**Cys** 201, **Cys** 168–**Cys** 182, and **Cys** 192–**Cys** 220) also shown (after Blevins, R. A.; Tulinsky, A. *J. Biol. Chem.*, **1985**, *260*, 4264).

In its active form, the mechanism of action has been studied extensively. Many potential pathways have been excluded, and it appears (the consensus mechanism) that the enzyme utilizes a catalytic triad of **Asp (D)**-102, **His (H)**-57, and **Ser (S)**-195 (Figure 12.10a,b).

The use of the catalytic triad is shown in Scheme 12.66.

Chymotrypsin is only one member of the family of serine proteases (enzymes that utilize an active-site serine (Ser, S) to cleave the peptide chain). Trypsin, also an endopeptidase obtained from bovine pancreas, is another member of the same family of serine (Ser, S) proteases. Interestingly, however, trypsin cleaves peptides on the carbon side of lysine (Lys, K) and arginine (Arg, R) groups. The catalytic triad discussed above for chymotrypsin also appears to apply here, while still other peptidases have other conserved units that allow them to catalyze the cleavage of peptides at specific sites. While it is not appropriate to provide an exhaustive list of peptidases that have been found to effect protein cleavage, it is important to be aware that in addition to serine (Ser, S) proteases, there are cysteine (Cys, C) proteases and aspartate (Asp, D) proteases and, in addition to endopeptidases, there are aminopeptidases that act on N-termini (N-terminal exopeptidases) and carboxypeptidases (carboxyl-terminus exopeptidases) that act at the corresponding carboxylic acid termini.

Utilizing protein cleaving enzymes, peptides can be broken into smaller pieces and, as noted above, the termini determined (or the termini determined chemically,

TABLE 12.3. The Amino Acid Sequence Found in PDB 5cha (EC 3.4.21.1), Chymotrypsin from (*Bos taurus*)

10	20	30	40	50	60
CGVPAIQPVL	SGLSRIVNGE	EAVPGSWPWQ	VSLQDKTGFH	FCGGSLINEN	WVVTAAHCGV
70	80	90	100	110	120
TTSDVVVAGE	FDQGSSSEKI	QKLKIAKVFK	NSKYNSLTIN	NDITLLKLST	AASFSQTVSA
130	140	150	160	170	180
VCLPSASDDF	AAGTTCVTTG	WGLTRYTNAN	TPDRLQQASL	PLLSNTNCKK	YWGTKIKDAM
190	200	210	220	230	240
ICAGASGVSS	CMGDSGGPLV	CKKNGAWTLV	GIVSWGSSTC	STSTPGVYAR	VTALVNWVQQ
TLAA					

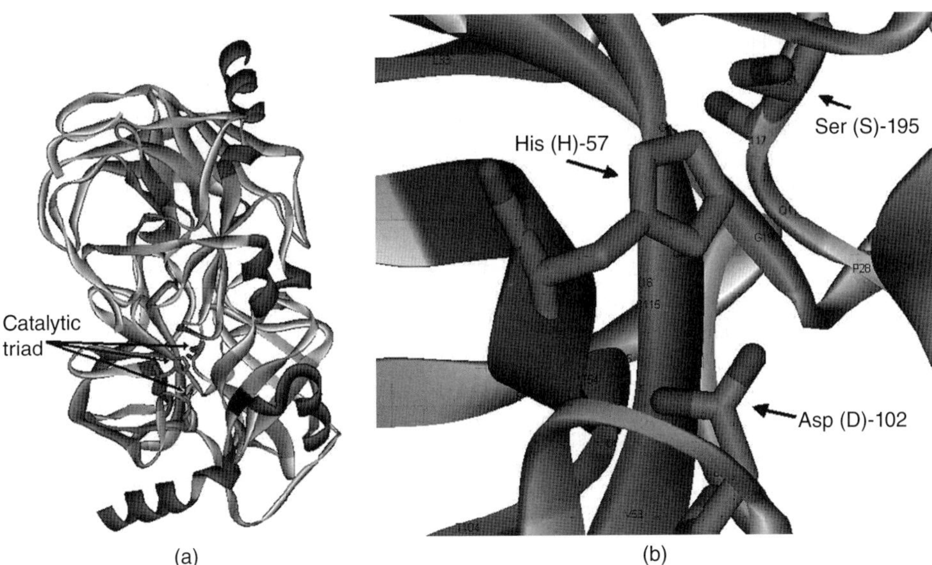

Figure 12.10. (a) On the left, a view of chymotrypsin showing the general placement of the catalytic triad of **Asp (D)**-102, **His (H)**-57, and **Ser (S)**-195; (b) an expanded view of the marked area of (a). The graphic is from **PDB** file **5cha**.

vide infra) and then the smaller pieces isolated and broken further. Thus, for example, as already discussed, chymotrypsin cleaves the peptide on the carbon side of amino acids with aryl side chains (phenylalanine [Phe, F], tyrosine [Tyr, Y], and tryptophan [Trp, W]), while trypsin cleaves peptides on the carbon side of lysine (Lys, K) and arginine (Arg, R) groups. Pepsin (from any of the species of the porcine genus *Sus* [pig]) also cleaves peptides with aryl side chains (phenylalanine [Phe, F], tyrosine [Tyr, Y], and tryptophan [Trp, W]) but *on the nitrogen side*. Further hydrolysis and separation of the amino acids by thin layer or column chromatography allows all of the amino acids in the peptide to be isolated and characterized.

Additionally, information can be obtained by end group analysis of the peptide as well as by different mass spectrometric techniques.

With regard to the former, there appears to be only one old and not too reliable method for determining which amino acid lies at the carboxyl terminus (barring the use of a carboxyl-terminus exopeptidase specific for the amino acid or acids it is to excise). The difficulty is that methods that might be used are frequently too vigorous to identify; only the carboxyl end and exopeptidases and mass spectrometric techniques (*vide infra*) have been developed. The (now old) Bergmann method required converting the carboxylic acid at the carboxyl terminus into the corresponding acid azide (Scheme 12.67). As shown in Scheme 12.67, where there is a carboxyl terminal alanine (Ala, A), treatment of the carboxylic acid with diazomethane (CH_2N_2) (a useful process for small molecules but difficult with large peptides) produced the corresponding methyl ester, which, on reaction with hydrazoic acid (HN_3) (a material with which it is difficult to work), lost methanol and allowed isolation of the corresponding acid azide. Then, heating the acid azide in benzyl alcohol resulted in

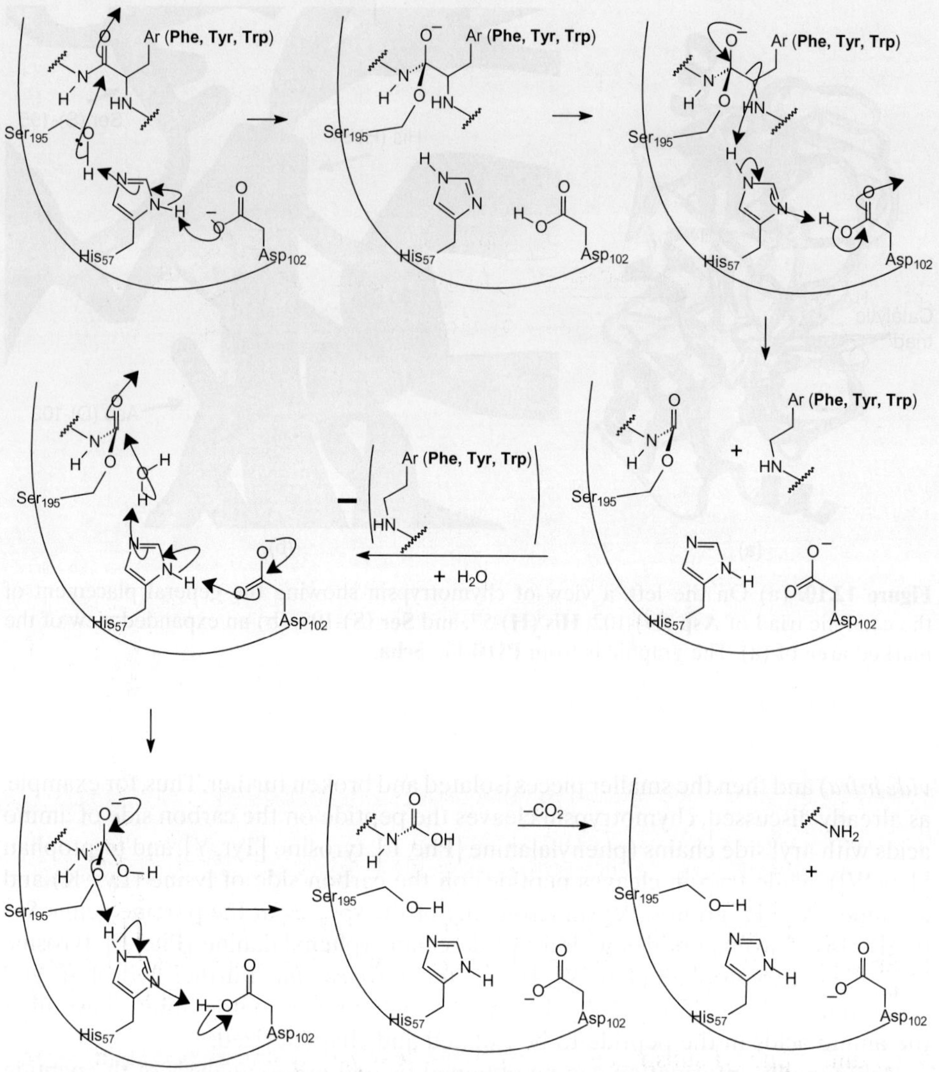

Scheme 12.66. A cartoon representation of the chymotrypsin catalytic triad in action. Cleavage of a peptide on the carbonyl side of the bond (to phenylalanine [Phe, F], tyrosine [Tyr, Y], or tryptophan) by the triad of Asp (D)-102, His (H)-57, and Ser (S)-195.

the intrusion of a Curtius rearrangement to yield an isocyanate, which, without isolation, produced the corresponding benzyl alcohol ester of the carbamic acid. Purification of the latter followed by aqueous acid treatment, resulted in hydrolysis of the ester, decarboxylation, and decomposition of the resulting bisamine to the shortened peptide chain and an aldehyde (acetaldehyde in this case). The latter derived from the terminal carboxylic acid. In contrast, there are several methods for determining which amino acid lies at the N-terminus of the peptide.

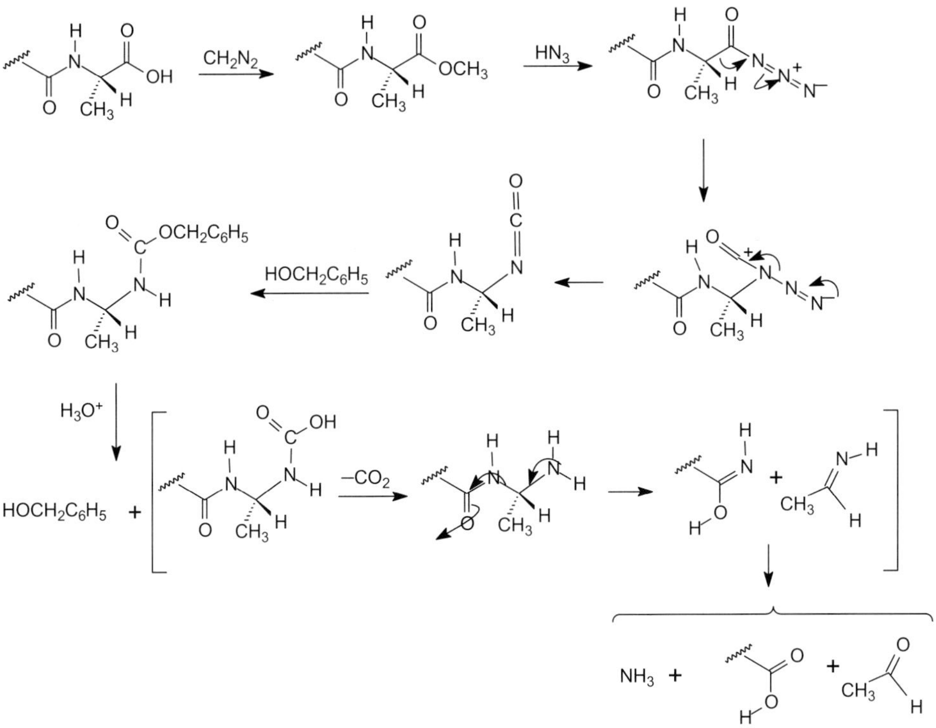

Scheme 12.67. An example of the Bergmann method (Bergmann, M. *Science*, **1934**, *79*, 439) for the determination of the amino acid at the carboxylate end of a peptide chain. The details are described in the text.

Among the earliest successful methods for deciding which amino acids lay at the N-terminus is that of Sanger.* When a peptide is treated with 2,4-dinitrofluorobenzene (Scheme 12.68), attack at the ipso carbon occurs and the nucleophilic aromatic substitution followed by hydrolysis of the peptide results in the formation of the *N*-2,4-dinitrophenyl substituted amino acid. Clearly, only amino acids lying at the N-terminus can be so labeled.

In addition to the Sanger method, it has become common to use the Edman[†] degradation.

The Edman degradation is often chosen in place of the Sanger method because of the stability and ease of isolation of the hydantoin derivative of the amino acid—the final product of the degradation method. Using the same model as in Scheme 12.68, a representation of the Edman degradation is presented in Scheme 12.69.

When the peptide is treated with phenylisothiocyanate, reaction at the amino terminus occurs to produce the corresponding thiourea derivative of the peptide.

*Frederick Sanger, who invented the method, is among the few individuals to have won two Nobel Prizes. He received the Nobel Prize in Chemistry in 1958 for determination of the complete structure of the protein "insulin" and, in 1980, he received the Chemistry prize again for the discovery of the chain termination method of DNA sequencing (Chapter 14).

[†]Pehr Edman, Karolinska Institutet, Stockholm, Sweden, first published on the technique in 1950.

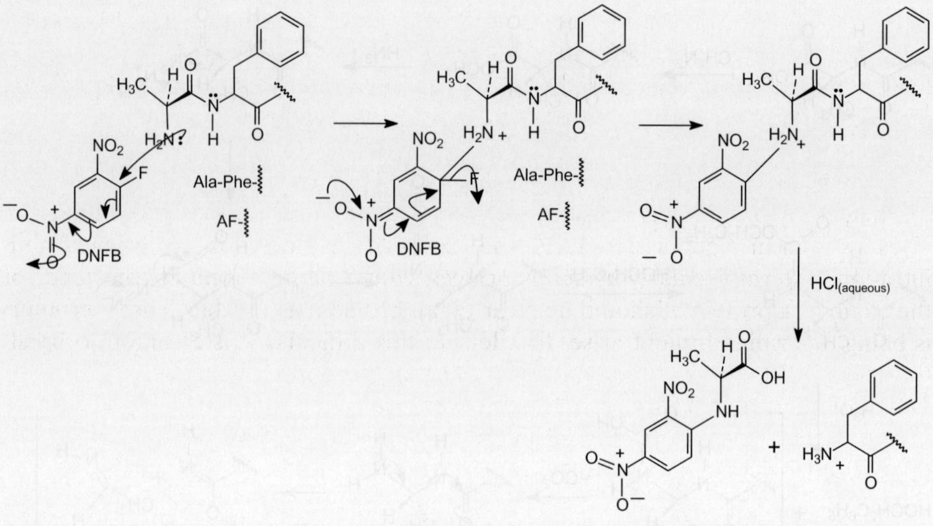

Scheme 12.68. A representation of the Sanger method of degradation of a peptide at the N-terminus. The example chosen shows the reaction between the N-terminus amino acid alanine (Ala, A) and 2,4-dinitrofluorobenzene (DNFB). A phenylalanine (Phe, F) is shown adjacent to the terminal alanine (Ala, A) as the second from the N-terminus. The remainder of the peptide is not shown. The products of the reaction are shown as the *N*-(2,4-dinitrophenyl) alanine and the peptide chain where the phenylalanine (Phe, F) is now the new terminal amino acid.

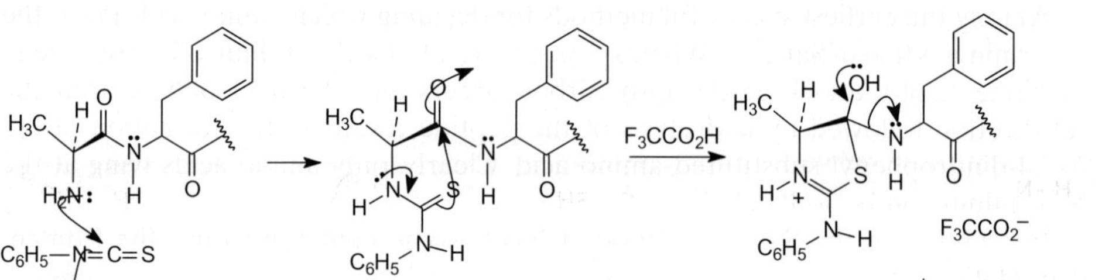

Scheme 12.69. A cartoon representation of the reaction between the Ala-Phe terminus amino acid from Scheme 12.68 with phenylisothiocyanate. After the initial thioamide forms, heat and acid convert the adduct into an anilinothiazolinone. The latter rearranges to the methyl *N*-phenylthiohydantoin.

Then, on treatment of that derivative with trifluoroacetic acid (F_3CCO_2H), protonation followed by cyclization takes place and an anilinothiazolinone is generated. On warming, the thiazolinone rearranges to the corresponding phenylthiohydantoin of alanine (Ala, A) and the peptide chain is left with a phenylalanine (Phe, F) at the terminus.

All three of the degradation methods suffer two flaws.

First, many proteins, as isolated, are composed of more than one peptide chain. Thus, as seen in Figures 12.8–12.10, the chains of chymotrypsin are bridged both intra- and intermolecularly by sulfur bridges. Thus, when determining the "end" of the chain of a protein, it should be clear that a problem as to which chain terminus is being determined might arise. To alleviate this difficulty, it is common to break the sulfur–sulfur linkages *before* undertaking amino acid sequencing.

The sulfur–sulfur bridges can be destroyed oxidatively or reductively. Treatment of the protein containing such disulfide bridges with performic acid $[(HCO)_2O]$ (Equation 12.11) converts both sulfur atoms of the disulfide to sulfonic acids.

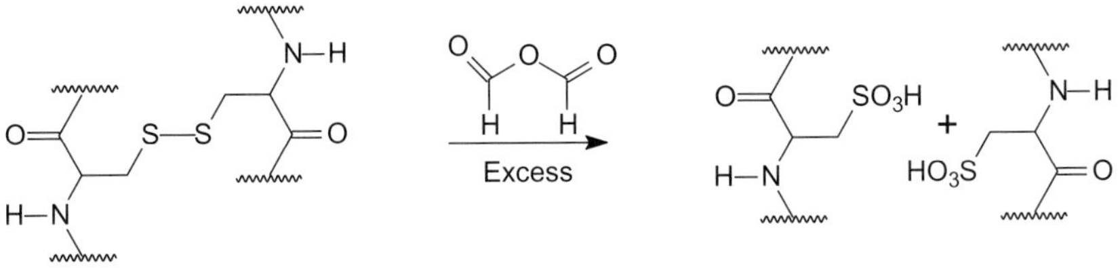

$$(12.11)$$

Alternatively, treatment with either dithiothritol or dithioerythritol (Equation 12.12) results in reductive cleavage to the corresponding thiols, which then require derivitization so that reformation does not occur, as well as to insure that the free thiols do not intrude in further reactions. The latter is accomplished by treating the thiols (without isolation) with iodoacetic acid (Equation 12.13).

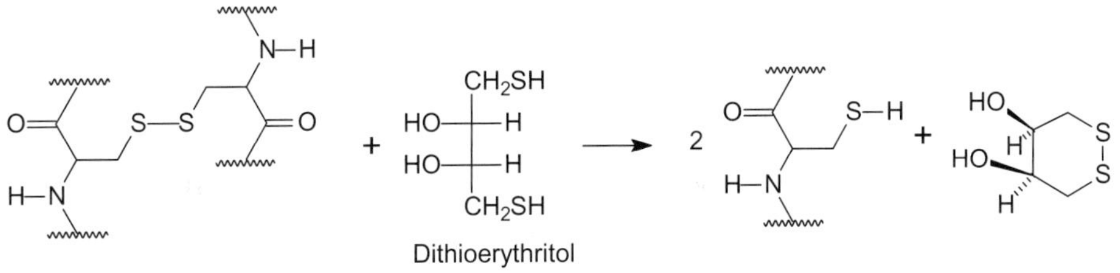

$$(12.12)$$

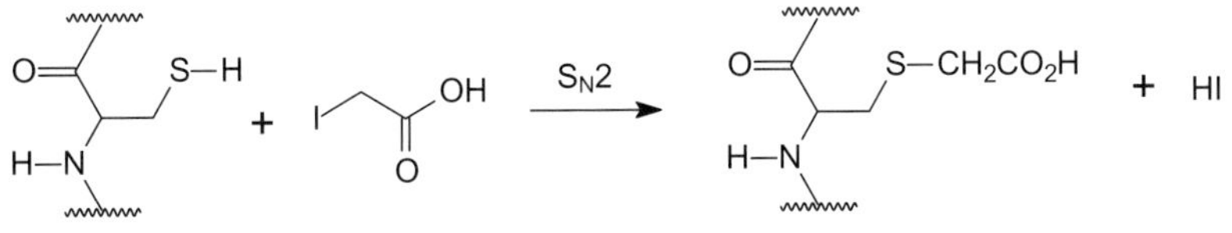

$$(12.13)$$

The second drawback is that, at least in the early work, it was not uncommon to have only trace quantities of protein for structural analysis. Thus, isolation of the product derived from N-terminus excision was not always trivial and chromatographic separations with careful comparison of R_f (Rate of flow) values were needed. With similar alkyl side chains, the analysis was not always straightforward. So, fluorescent indicators were sought and, in the mid-1950s, "dansyl chloride"

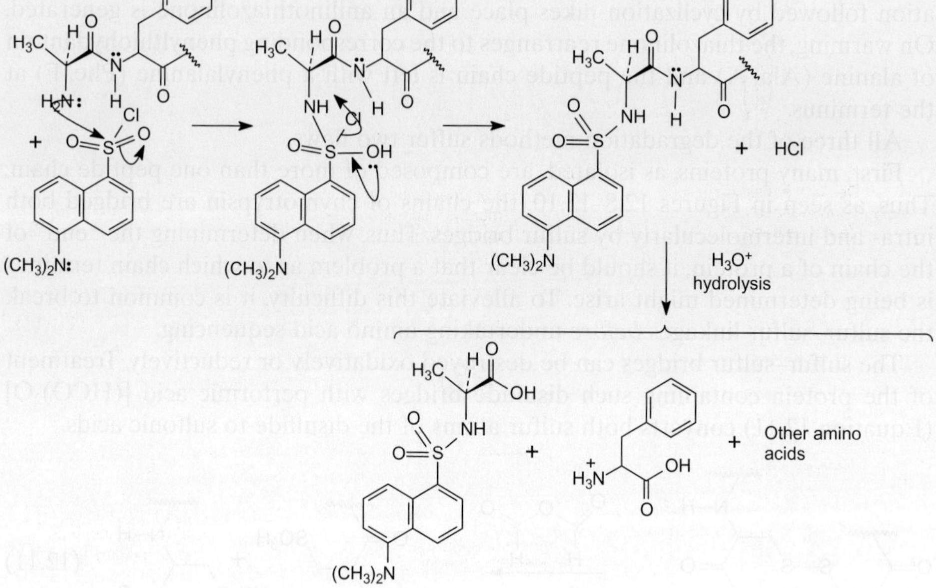

Scheme 12.70. A cartoon representation of the reaction between the Ala-Phe terminus amino acid from Schemes 12.68 and 12.69 with 1-dimethylaminonaphthalene-5-sulfonyl chloride (dansyl chloride). After the initial sulfonamide forms, acid hydrolysis is used to destroy the entire peptide. The easily detected (fluorescent) dimethylsulfonamide can be readily identified.

(1-dimethylaminonaphthalene-5-sulfonyl chloride) was developed by the California Foundation for Biochemical Research. The advantage to making this sulfonic acid derivative of the N-terminus amino acid lies in the ability to detect about 10^{-10} mol of the brightly blue-to-blue green fluorescent product (Scheme 12.70), well beyond the ability to detect other derivatives.

Finally, from the point of view of the organic chemist, obtaining amino acids from proteins, which is done by either acid or base hydrolysis (after cleaving the sulfur–sulfur bonds) rather than by structural analysis, is of concern. Indeed, at this writing, a significant amount of structural analysis is accomplished by specialized computer-coupled mass spectrometric (Chapter 2) techniques. The special nature of these techniques derives from the need to put the very large molecules in the gas phase and so desorption from various materials—without decomposition—becomes important and exotic technology, such as matrix-assisted laser desorption ionization (**MALDI**) MS becomes necessary.

II. Peptides from Amino Acids—*In Vivo*

The millennia needed for the establishment of pathways in living systems for the synthesis of peptides (proteins) have resulted in dynamic processes that are wondrous indeed, and all of the necessary events have not yet been exposed! Much work remains!

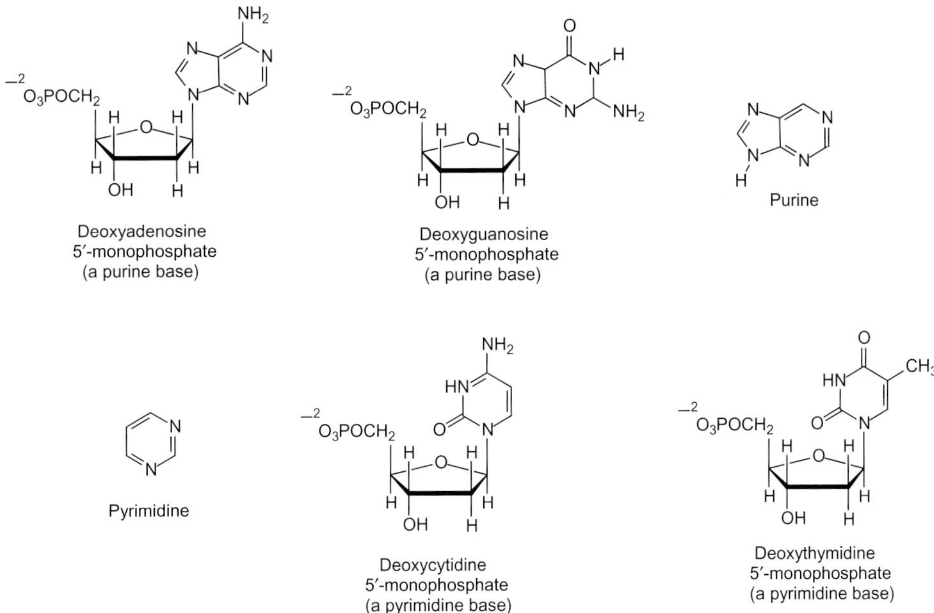

Figure 12.11. The four bases, adenine (A), guanine (G), thymine (T), and cytosine (C), attached to the anomeric carbon of 2-deoxyribose and esterified with phosphoric acid at C5 of the deoxyribose unit. The purine and pyrimidine ring systems are also shown.

In bacterial cells and in yeast, which are, for obvious reasons, much more commonly studied than other animals and plants, amino acids, synthesized de novo or by degradation of proteins of various kinds and sizes, are found in the cytosol. The amino acids from these proteins are used for the synthesis of peptides. Much of the synthesis occurs either on/in the ribosomes of the endoplasmic reticulum (ER) or through nonribosomal pathways using acyl carrier protein-like condensation reactions.

For the former, **the ribosomal pathway**, the place to start, is with the genetic code and the carrier of the code, DNA. For the most general case, and as will be seen subsequently (Chapter 14), DNA is composed of four deoxyribonucleotides (i.e., monophosphate ester derivatives of deoxynucleosides) (Figure 12.11) held in chains (Figure 12.12) by the formation of phosphate diesters, linearly from one deoxyribose (5′) to the next (3′). The chains are paired, in the classical helical structure, by hydrogen bonding between the purine and pyrimidine bases (A–T and C–G) on opposite chains (Figure 12.12).*

The pattern of the base pairs over thousands of bases constitutes our respective genomes, and it is in the genome that the instructions (genes) for peptide synthesis

---

*There are three levels of complexity. At the lowest level, there are four bases: two purines (adenine [A] and guanine [G]) and two pyrimidines (thymine [T] and cytosine [C]). At the next level, each of these is attached to the anomeric carbon of 2-deoxyribose, and they are now called deoxyribonucleosides and their names are deoxyadenosine, deoxyguanosine, deoxythymidine, and deoxycytidine. The addition of phosphate esters at C5 of the deoxyribose units changes the name to deoxyribonucleotide, for example, deoxyadenosine 5′-monophosphate.

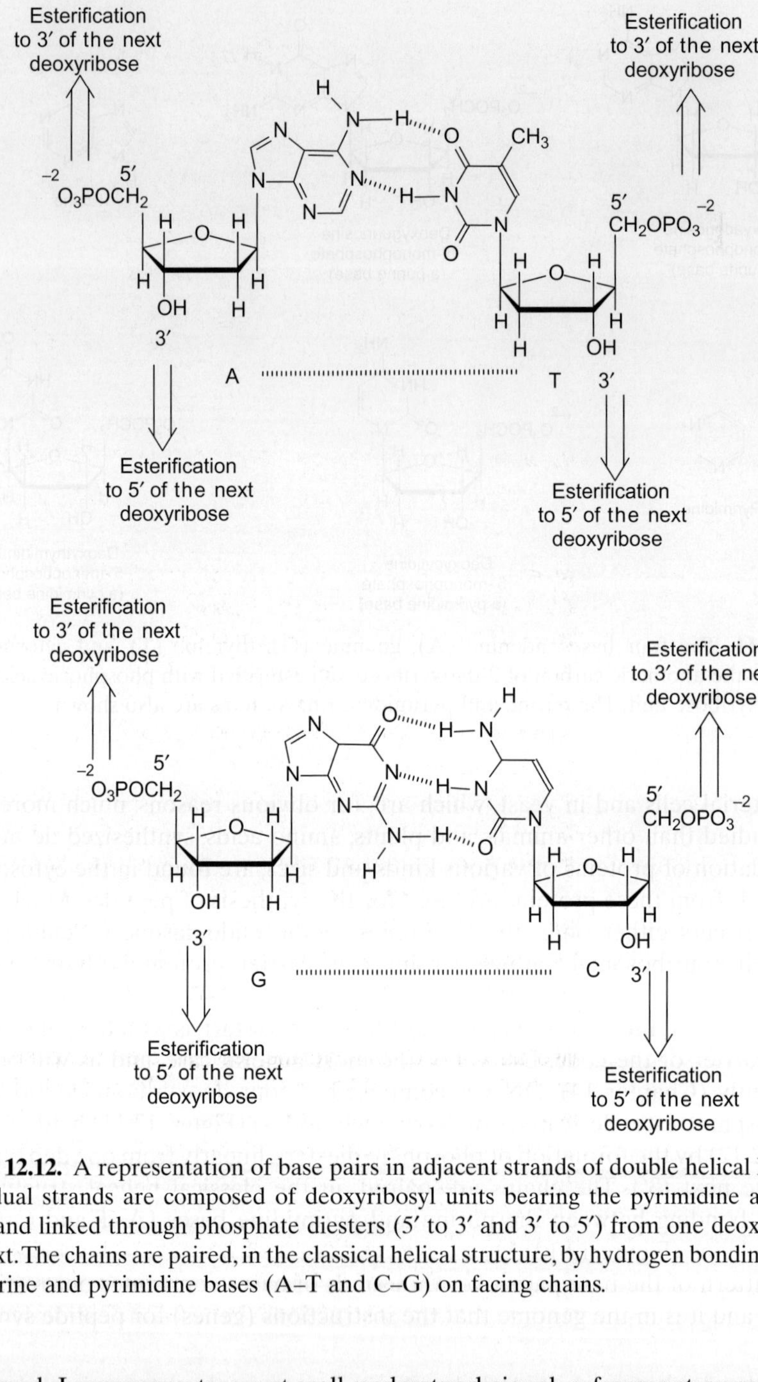

Esterification to 3′ of the next deoxyribose

Esterification to 3′ of the next deoxyribose

$^{-2}O_3POCH_2$ 5′

$CH_2OPO_3^{-2}$ 5′

CH₃

A ⋯⋯⋯⋯⋯⋯⋯⋯ T   3′

OH   3′

Esterification to 5′ of the next deoxyribose

Esterification to 5′ of the next deoxyribose

Esterification to 3′ of the next deoxyribose

Esterification to 3′ of the next deoxyribose

$^{-2}O_3POCH_2$ 5′

$CH_2OPO_3^{-2}$ 5′

G ⋯⋯⋯⋯⋯⋯⋯⋯ C   3′

Esterification to 5′ of the next deoxyribose

Esterification to 5′ of the next deoxyribose

Figure 12.12. A representation of base pairs in adjacent strands of double helical DNA. The individual strands are composed of deoxyribosyl units bearing the pyrimidine and purine bases and linked through phosphate diesters (5′ to 3′ and 3′ to 5′) from one deoxyribose to the next. The chains are paired, in the classical helical structure, by hydrogen bonding between the purine and pyrimidine bases (A–T and C–G) on facing chains.

are found. In response to a not-well-understood signal, a fragment or segment of the double strand unravels (over a short distance) with the aid of a member of the family of helicase enzymes to create an open reading frame that allows the encoded information to be transcribed. The process, transcription, takes the information encoded in the DNA and converts it into transportable form as ribonucleic acids

(RNAs) to allow the information to be put to use. The information can be used to make peptides (proteins) as well as for other purposes, not all of which are well understood.

As the name implies, RNAs have a ribose rather than a deoxyribose as the carbohydrate portion of the chain. Additionally, many RNAs have only a single strand rather than a double helical strand. Finally, although base pairing again obtains when there are two strands, and adenine (A), guanine (G), and cytosine (C) are present in the RNAs, the base thymine (T) is replaced by uracil (U), which, as seen in Figure 12.13, is the same as thymine, T, except for the methyl group, which is missing in uracil, U.

Now, as the DNA unravels over its short genome distance (catalyzed by helicase), RNA polymerase, using the DNA template strand, transcribes the single DNA strand in the $3' \rightarrow 5'$ direction with the RNA elongation taking place in the $5' \rightarrow 3'$ direction with base pairing guiding the sequence. Thus, every G in the DNA is paired to a C in the RNA being formed, and every C in the DNA is paired to a G in the RNA being formed. Every T in the DNA is paired to an A in the RNA being formed, and every A in the DNA is paired to a U in the RNA being formed since the base thymine, T, is replaced by uracil, U.

Interestingly, although a wide variety of RNAs are created in this way and are used to an equally wide variety of ends, it is held that to make peptides (proteins), three different RNAs must be generated. These are ribosomal RNA (rRNA),

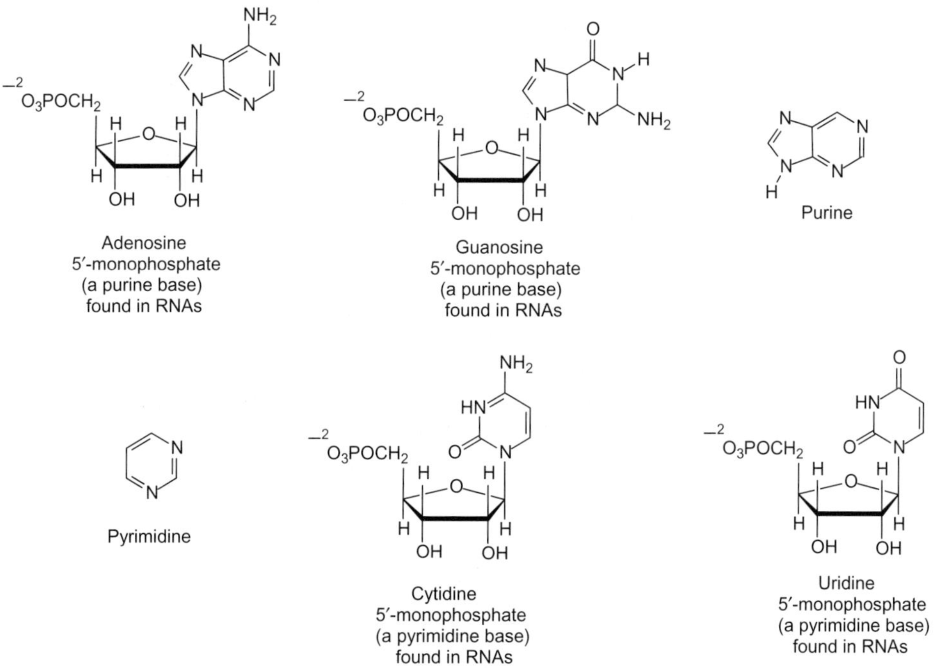

Figure 12.13. The four bases, adenine (A), guanine (G), cytosine (C), and uracil (U) attached to the anomeric carbon of ribose and esterified with phosphoric acid at C5 of the ribose unit. The purine and pyrimidine ring systems are also shown.

messenger RNA (mRNA), and transfer RNA (tRNA), and each is produced separately.

rRNA is incorporated into the machinery of the ribosome that is used to produce proteins, and the information helps define the specific protein to be synthesized.

mRNA carries the three-letter genetic code (the codon) (Figure 12.14) that determines the start of the protein synthesis, the specific amino acids to be added, the sequence of the addition, and it carries the codon required for the termination of the peptide chain. This mRNA binds to the ribosome with the codon exposed to allow the tRNA to bring the amino acids in the appropriate sequence into the growing peptide (protein) chain.

The amino acids to be brought into the peptide (protein) are attached to the tRNA on what is called the "amino acid arm" of the "crosslike" structure of the

The RNA codons

	Second nucleotide				
	U	**C**	**A**	**G**	
U	UUU **Phenylalanine** (Phe)	UCU **Serine** (Ser)	UAU **Tyrosine** (Tyr)	UGU **Cysteine** (Cys)	U
	UUC Phe	UCC Ser	UAC Tyr	UGC Cys	C
	UUA **Leucine** (Leu)	UCA Ser	UAA **STOP**	UGA **STOP**	A
	UUG Leu	UCG Ser	UAG **STOP**	UGG **Tryptophan** (Trp)	G
C	CUU **Leucine** (Leu)	CCU **Proline** (Pro)	CAU **Histidine** (His)	CGU **Arginine** (Arg)	U
	CUC Leu	CCC Pro	CAC His	CGC Arg	C
	CUA Leu	CCA Pro	CAA **Glutamine** (Gln)	CGA Arg	A
	CUG Leu	CCG Pro	CAG Gln	CGG Arg	G
A	AUU **Isoleucine** (Ile)	ACU **Threonine** (Thr)	AAU **Asparagine** (Asn)	AGU **Serine** (Ser)	U
	AUC Ile	ACC Thr	AAC Asn	AGC Ser	C
	AUA Ile	ACA Thr	AAA **Lysine** (Lys)	AGA **Arginine** (Arg)	A
	AUG **Methionine** (Met) or **START**	ACG Thr	AAG Lys	AGG Arg	G
G	GUU **Valine** Val	GCU **Alanine** (Ala)	GAU **Aspartic acid** (Asp)	GGU **Glycine** (Gly)	U
	GUC (Val)	GCC Ala	GAC Asp	GGC Gly	C
	GUA Val	GCA Ala	GAA **Glutamic acid** (Glu)	GGA Gly	A
	GUG Val	GCG Ala	GAG Glu	GGG Gly	G

Figure 12.14. The RNA **codons** for mRNA read from DNA and are installed on the ribosome to be read by tRNA bringing the amino acids to the ribosome for peptide (protein) synthesis. Note that AUG is both the start codon (and thus requires the tRNA **anticodon** of UAC for the first amino acid, methionine [Met, M]) as well as the codon for methionine (Met, M) that might be found elsewhere, and that UAA, UGA, and UAG are stop codons. Note, too, that this 4 × 16 array, with 64 combinations, is redundant for the 20 amino acids, and thus some amino acids are specified in more than one way, while the start codon is unique. Information in this table was obtained from Nirenberg, M; Leder, P.; Bernfield, M.; Brimacombe, R.; Trupin, J.; Rottman, F.; O'Neal, C. *Nat. Acad. Sci. U.S.*, **1965**, *53*, 1161 and subsequent work.

tRNA and, as traditionally drawn, across from the anticodon arm, the arm read by the codon of the mRNA attached to the ribosome (Figure 12.15).

As shown in Scheme 12.71, and as affirmed by labeling experiments, an amino acid (shown using phenylalanine [Phe, F] as an example in the scheme) to be added to the growing peptide (protein) chain is activated for reaction by attachment at the carboxylate to adenosine triphosphate (ATP) to make an anhydride of the amino acid with adenosine monophosphate (and the loss of inorganic phosphate). Then, in the next step, the activated amino acid is esterified by the C-2 or the C-3 hydroxyl of a ribosyl unit of AMP, which is attached via phosphate at the 5′ carbon to the aminoacyl transfer end of tRNA. If attachment is to C-2, rearrangement to C-3 follows and the aminoacyl-tRNA is activated and ready to be added to the growing peptide chain at the synthesis site on the ribosome.

Although there may be more than one way to grow the peptide chain, a common way (bacteria) requires three sites. The first site is called "P" for the "peptidyl" site. Here, the growing peptide end has the carbonyl to which the addition will occur, that is, the **receiving** site where the incoming amino group of the amino acid binds. The second site is the "A" site of the process where the recognition and binding of

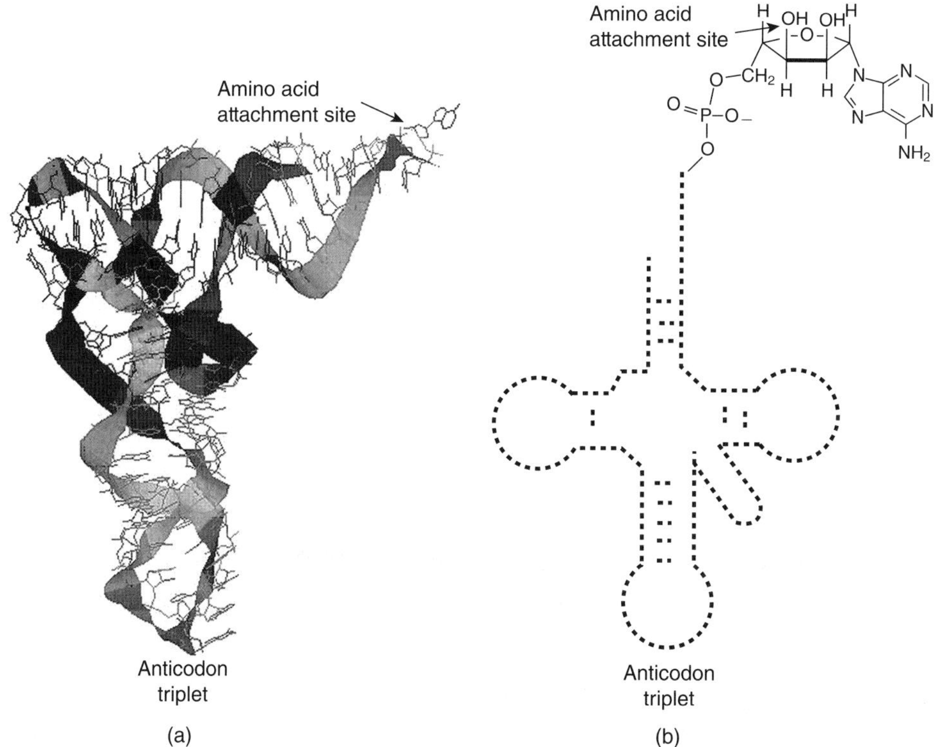

Figure 12.15. Cartoon representations of tRNA. The "amino acid arm" (shown here bearing an adenosine monophosphate to which the amino acid will be added) as traditionally drawn is across from the anticodon arm, the arm read by the codon of the mRNA attached to the ribosome. The drawing on the left (a) is from PDB 1TRA (after Shi, H.; Moore, P. B. *RNA* **2000**, *6*, 1091). The drawing on the right (b) is a simplified version of that on the left.

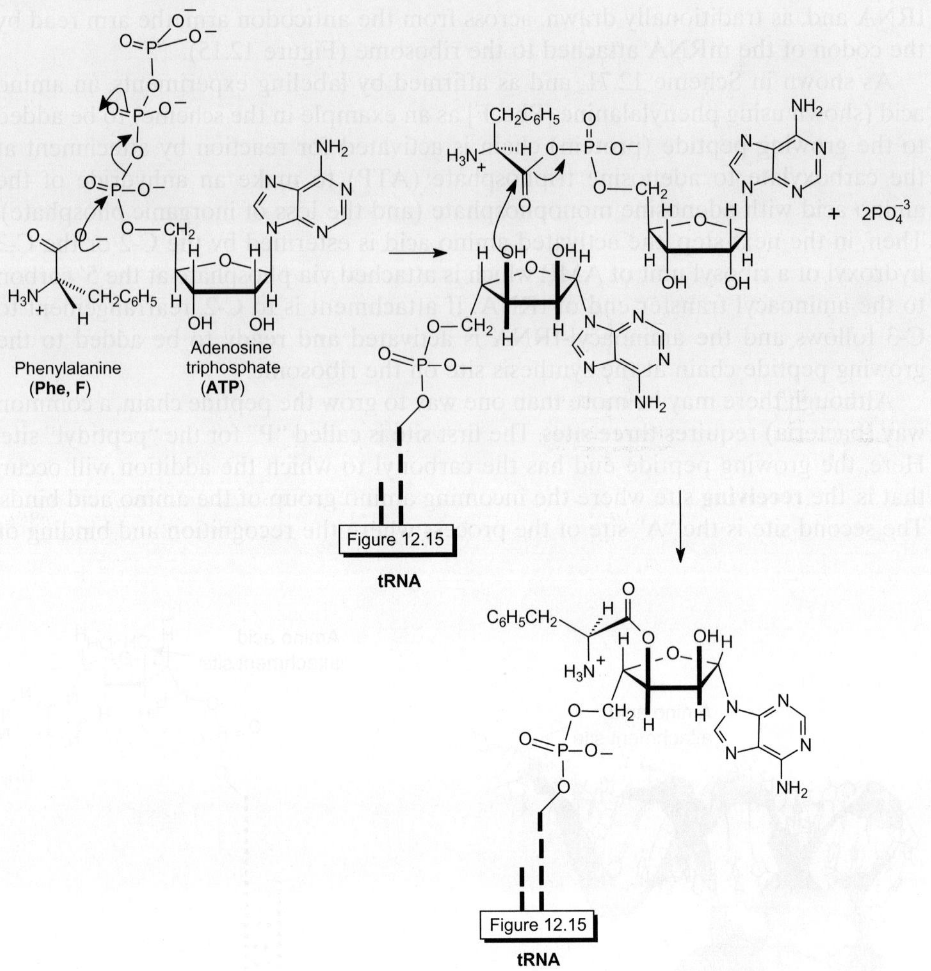

Phenylalanine
(Phe, F)

Adenosine
triphosphate
(ATP)

$+ \ 2PO_4^{-3}$

Figure 12.15

tRNA

Figure 12.15

tRNA

Scheme 12.71. A cartoon representation of phenylalanine (Phe, F) as an example added to ATP for activation followed by esterification at the C-3 hydroxyl of a ribosyl unit of AMP, which is attached via phosphate at the 5′ carbon to the aminoacyl transfer end of tRNA.

the **incoming** amino acid "aminoacyl" group is attached for reaction with the growing peptide. The final site is called "E" for the "exit" site; it is the empty tRNA fragment attached to the site on the mRNA, which was "P" until the coupling of the incoming amino acid occurred. In this process, the previously empty A site has become the new P site. The space—in a now new, empty A site on the mRNA—is for the next amino acid in the growing peptide chain (Scheme 12.72), which remains attached to the P site. Note that in Scheme 12.72, the AUG codon that begins the translocation motion along the ribosome uses *N*-formylmethionine (fMet). The same codon recognizes methionine (Met, M) subsequently. The one-carbon unit of the formyl group is derived from folate (*vide infra* and Chapter 11). Posttranscriptional modification can account for either hydrolytic deformylation if the peptide is to begin with methionine, reduction if it is to begin with *N*-methylmethionine, or cleavage of

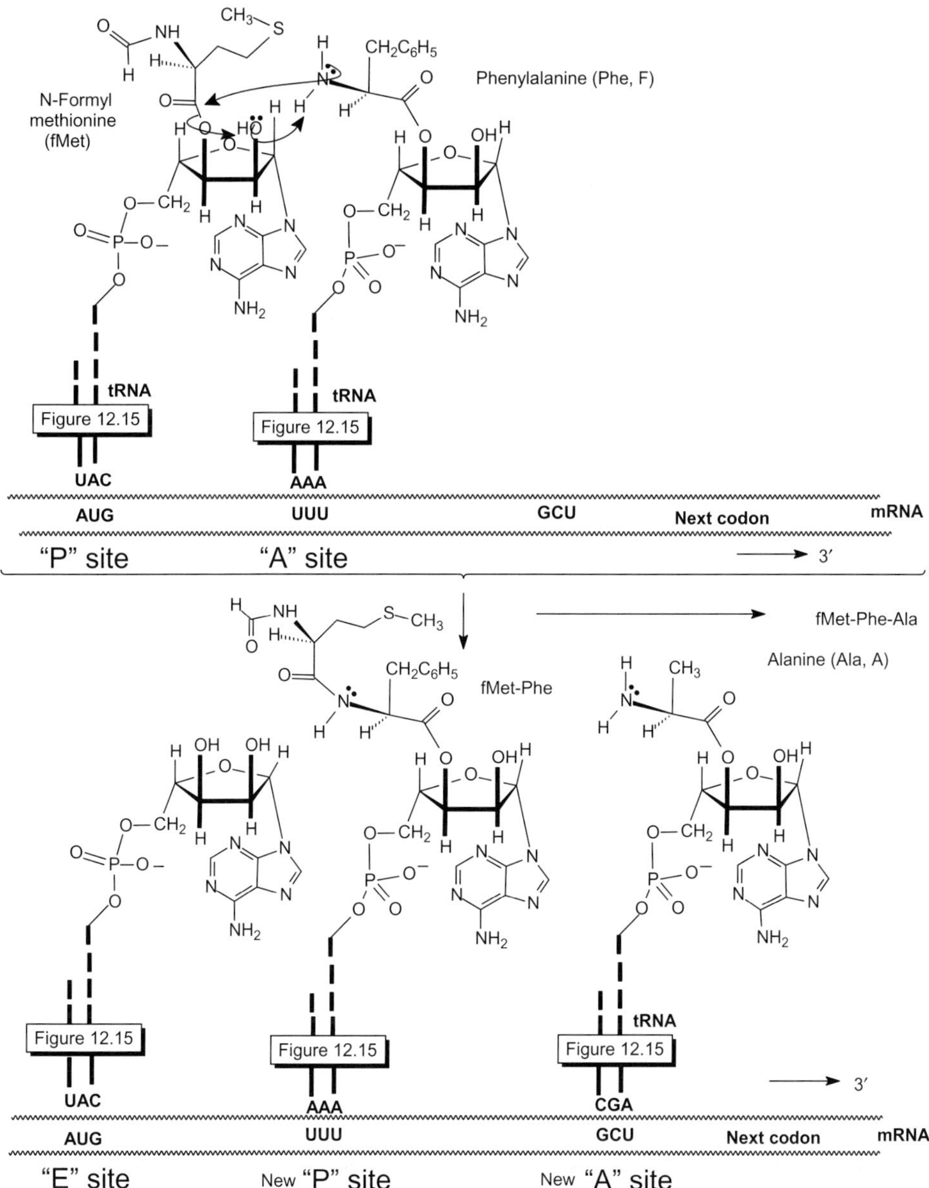

Scheme 12.72. Cartoon of a method of peptide bond formation as catalyzed by RNA that is consonant with experimental data (after Strobel, S. A.; Cochrane, J. C. *Curr. Opin. Chem. Biol.*, **2007**, *11*, 636).

the initiating *N*-formylmethionine (or a longer sequence including the initiating *N*-formylmethionine) to produce the active peptide (*vide infra*).

The **nonribosomal pathway** may be operational alongside the ribosomal pathway outlined above, but it is **independent of mRNA** and, rather than producing proteins (peptides) obviously required for life, a different set of peptides (called secondary metabolites) results. Thus, it is frequently found that if these secondary metabolites are not produced or if they are modified by changing growth medium or other factors, life continues regardless. Additionally, it is proving possible to modify some of these nonribosomal peptides because they appear to be produced and used in much the same way as the polyketides discussed in Chapters 8 (Scheme 8.70), 9 (Table 9.1), and 11 (Schemes 11.34, 11.37, and 11.38, Figure 11.16). The difference is that in place of polyketide synthases (PKSs), there are nonribosomal peptide synthases (NRPSs) that play the critical role. The nonribosomal peptides, usually "small," cyclic peptides such as cyclosporine A (Figure 12.16), are produced by participation of several nonribosomal peptide synthetase enzymes that are organized in modules (as with PKSs) for the insertion of the small units into the growing peptide as well as enzymes that start the small peptide, enzymes that induce cyclization, enzymes that promote addition of methyl groups, carbohydrates, and so on, and, of course, enzymes that terminate the synthesis process.

Again, in a vein similar to that of the PKSs (Chapters 8 [Scheme 8.70], 9 [Table 9.1], and 11 [Schemes 11.34, 11.37, and 11.38, Figure 11.16]), there appears to be a relatively small set of modules that operate. A major difference between PKSs and the NRPSs is that the modules (enzymes) in the latter that elongate (extend) the peptide chain are different for each amino acid (unlike the repetitive nature of the PKSs for loading, e.g., propionate, to make 6-deoxyerythronolide B, Scheme 11.38), but the similarity allows some crossover so that propionate, for example, is occasionally incorporated into NRPSs.

In abbreviated form, a synthetic sequence is shown in Scheme 12.73.

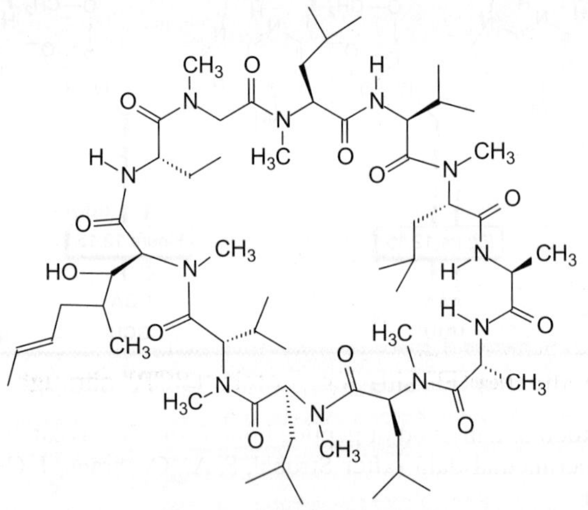

Cyclosporin A

Figure 12.16. Cyclosporin A is a naturally occurring cyclic nonribosomal polypeptide useful in organ transplants (since it suppresses the immune system, a system that would normally reject the organ transplanted). It is reported that it was originally isolated from fungus (*Tolypocladium inflatum* Gams) found in a sample of Norwegian soil. See Ruegger, A.; Kuhn, M.; Lichti, H.; Loosli, H. R.; Huguenin, R.; Quiquerez, C.; Von Wartburg, A. *Helv. Chim. Acta,* **1976,** *59,* 1075.

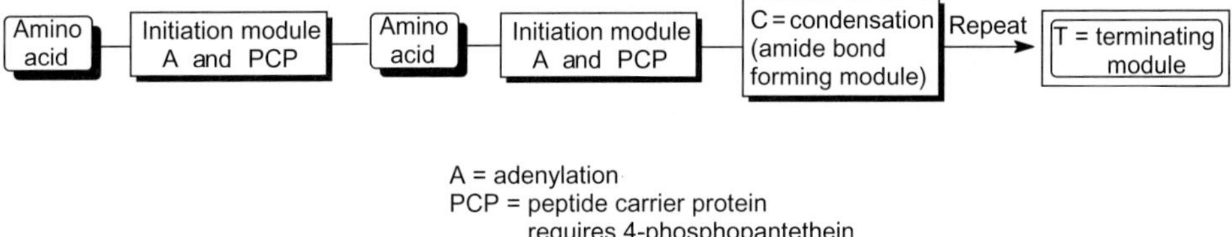

Scheme 12.73. A flow diagram-type scheme showing the overall use of the modules present and required to effect peptide synthesis utilizing a nonribosomal pathway.

Scheme 12.74. The adenylation of isoleucine (Ile, I) used only as an example followed by its conversion to the corresponding 4′-phosphopantetheine* derivative for condensation.

Some details of the pieces in Scheme 12.73 are considered in Schemes 12.74 and 12.75.

An amino acid (Scheme 12.74) such as isoleucine (Ile, I) is adenylated and is then attached to the peptide carrier protein (PCP) through a sulfur linkage (4-phosphopantetheine) as shown in the block diagram of Scheme 12.73. A second

---

*The biosynthesis of 4′-phosphopantetheine is quite interesting. This transfer agent (also, of course, part of coenzyme A) is derived from the pathway leading to and from valine (Val, V). As shown in the next figure and in Scheme 12.17, 3-methyl-2-oxobutanoate (2-ketoisovalerate) is the immediate precursor/product (transamination, EC 2.6.1.42) to/from valine (Val, V). When 2-ketoisovalerate (3-methyl-2-oxobutanoate) abstracts the one-carbon unit from 5,10-methylenetetrahydrofolate (below and Scheme 12.9) in an aldol-type condensation, 2-dehydropantoate results (EC 2.1.2.11). Reduction of the latter (NADPH/H⁺, EC 1.1.1.169) generates (*R*)-pantoate. After activation of the carboxylate by adenylation, reaction with β-alanine produces pantothenate (EC 6.3.2.1), which is phosphorylated at the 4-position

(Continued on next page)

(EC 2.7.1.33) to N-(R)-4′-phosphopantothenate. Activation of the carboxylate results from the addition of β-alanine (β-alanine is formed on decarboxylation of aspartate [Asp, D], EC 4.1.1.11), by cytidylate formation (cytosine triphosphate [CTP] is followed by coupling to L-cysteine (Cys, C) (EC 6.3.2.5) to produce N-[(R)-4′-phosphopantothienoyl]-L-cysteine. Oxidation to the thioaldehyde with flavin mononucleotide (FMN → FMNH$_2$) allows decarboxylation of the latter with the formation of the corresponding enol. Then, reduction with nicotinamide adenine dinucleotide phosphate (NADPH/H$^+$ → NADP$^+$) (EC 4.1.1.36) leads to 4-phosphopantetheine (pantetheine 4′-phosphate).

(Continued on next page)

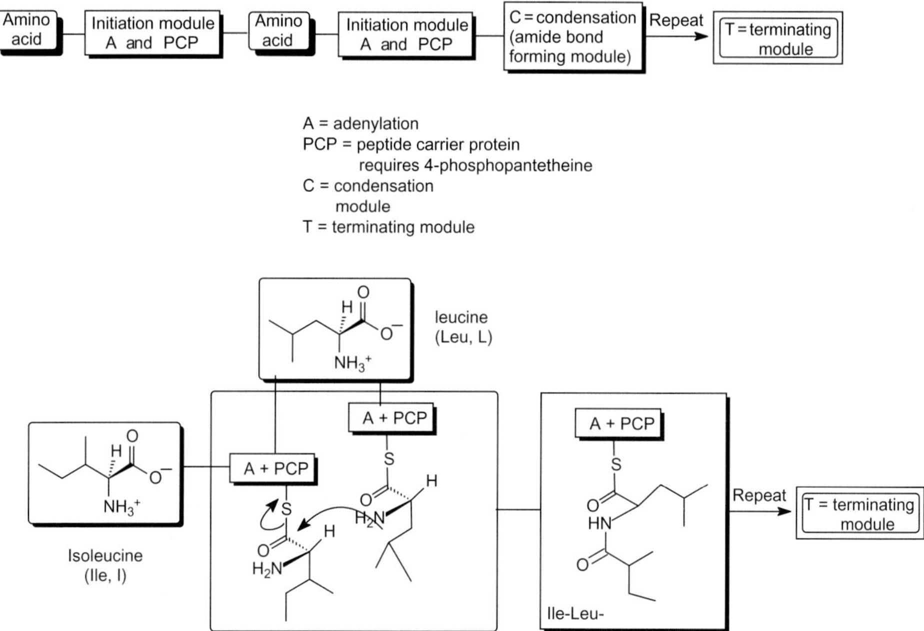

Scheme 12.75. A repetition of Scheme 12.73 with the amino acids isoleucine (Ile, I) and leucine (Leu, L) shown as examples of the process of the pathway to nonribosomal peptides.

amino acid, for example, leucine (Leu, L), is similarly activated. Then, they are allowed to condense and the product is ready for the next (A, PCP) module.

III. Peptides from Amino Acids—*In Vitro*

At this writing, the synthesis of peptides (proteins) falls into two very broad classes, which, by their breadth, leaves room for overlap and combination.

The first method, used for simple and unusual (because of the amino acid makeup) peptides, consists of combining the appropriately protected amino acids in the desired sequence in the laboratory, literally adding one to the next, until the sequence is prepared. It should be clear that small sequences might be prepared by different individuals and then the short pieces combined in the right order to make the final product. The drawback is that this is a time-consuming process as each step must be optimized in order to produce enough material to pay the salaries of the men and women engaged in the synthesis.

The second method is to build a machine that accomplishes what the individuals themselves did for each of the steps in a synthesis. Such machines exist, they are very valuable for peptide synthesis, and, while the capital outlay may be large, other concerns about the treatment of workers do not apply. Further, the machines can be programmed to operate "around the clock." However, the drawbacks are not easily overcome. Thus, small errors multiply and purity decreases, unusual amino acids that might need special treatment do not get it, and yields at each step may not be (and usually are not) optimal. However, the merits of this system, utilizing a solid phase on which syntheses could be repeated with new amino acid units in a "mechanical" fashion, resulted in the award of the Nobel Prize in Chemistry to R. Bruce Merrifield in 1984.

In brief, the Merrifield method (Scheme 12.76) involves an insoluble polymeric bead bearing a benzyl halide (chlorides work well) "linker." Of course, there is more than one per bead, but the principle is the same. On reaction with an N-protected amino acid, the benzyl halide is displaced by the carboxylate anion and an ester linkage is formed. Then, the amino protecting group is removed and a carboxylate-activated N-protected amino acid is added. The amide link from the activated carboxylic acid to the unprotected but resin linked amino acid is then forged, the unused reagents and side products washed away from the insoluble bead now bearing two amino acids with the last added retaining an N-protection. The entire sequence is then repeated, treating the last amino acid added as if it were that one first added to the bead.

At the end, the benzylic ester is hydrolzyed and the last N protection is removed to provide the desired product. An example of the process is provided in Scheme 12.76.

In the absence of peptide synthesizing equipment and for special cases where it is either inappropriate (for reasons of economy) or where the mechanical method will not work because the methods that are used for carboxylic acid activation, protection, and or coupling fail with a particular compound or set of compounds, the fallback position is to carry out the coupling reaction at the bench. To this end, protecting groups for amines and carboxylic acids are put in place so that, as with the Merrifield method, the free amino group of one amino acid or peptide fragment, protected at the carboxylate, is available to react with the activated carboxylate of

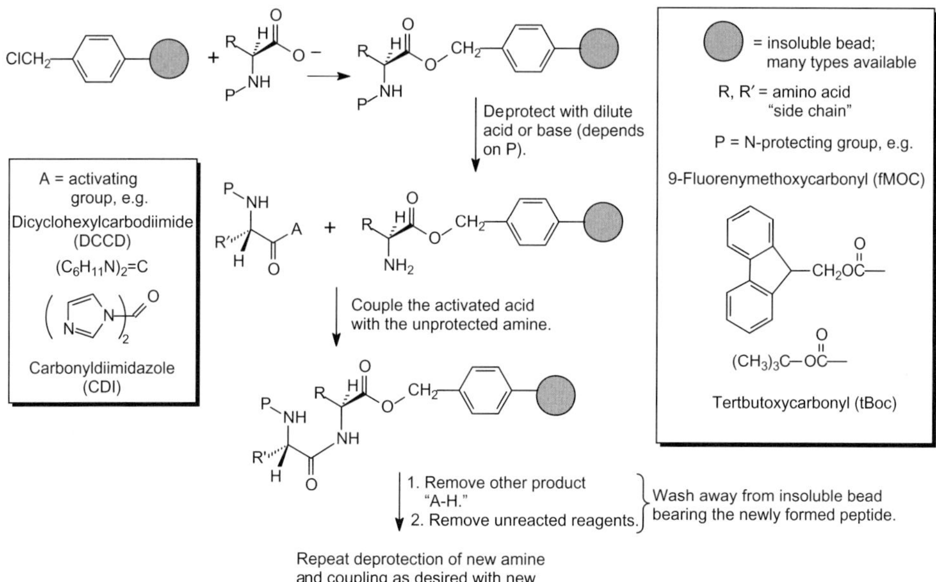

Scheme 12.76. An abbreviated outline of the Merrifield method for the synthesis of peptides. The details of the processes in each step are the same as those in a sequence of individual steps to be considered subsequently. The method can be mechanized and automatic peptide synthesizers are commercially available.

another, N-protected amino acid or peptide fragment to forge the new amide (peptide) bond. The major novelty is the coupling reagent.

Consider the Merrifield method described above and the use of an N-protected serine (Ser, S) reactant that is to be reacted by coupling the carboxylate of this N-protected serine (Ser, S) at the amine of an amino acid with a protected ester group. It should be clear, based on the chemistry discussed in Chapter 8 (Alcohols) that both of the coupling agents discussed in conjunction with the Merrifield method above, viz., dicyclohexylcarbodiimide (DCCD) and carbonyldiimidazole (CDI) (Scheme 12.76) cannot be used because they both are known to react with alcohols. Of course, in principle, the alcohol can be protected too, but a separate step is then required and the standard Merrifield system as outlined above will need modification.

Examples of the coupling reagents are shown in Figure 12.17. Examples of the alcohols and phenols that have been found particularly appropriate for carboxylate protection are provided in Figure 12.18. Interestingly, only a few protecting groups are routinely used on the amino functionality, and they are often esters of carbamic acids, compounds easily hydrolyzed to alcohol, carbon dioxide, and the corresponding "free" amino function. These are shown in Figure 12.19. The special case of hexafluoroacetone will be dealt with separately.

Examples of the use of the coupling reagents will be provided utilizing various amine (nitrogen) and carboxylic acid (carboxylate) protecting groups.

As shown in Scheme 12.77, when serine (Ser, S) is allowed to react with 2-(*tert*-butoxycarbonyloxyimino)-2-phenylacetonitrile (BOC-ON) in methylene chloride,

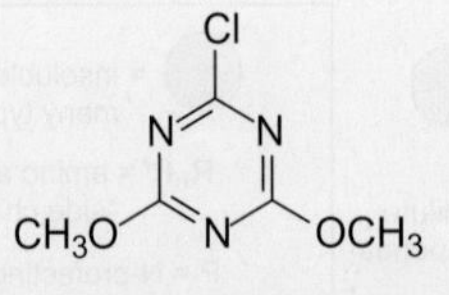

6-Chloro-2,4-dimethoxy-s-triazine
(CDMT)

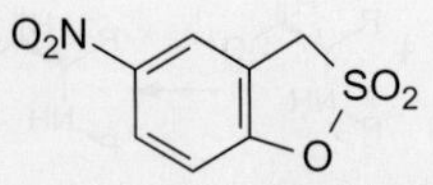

5-Nitro-3H-1,2-benzoxathiole
S,S-dioxide

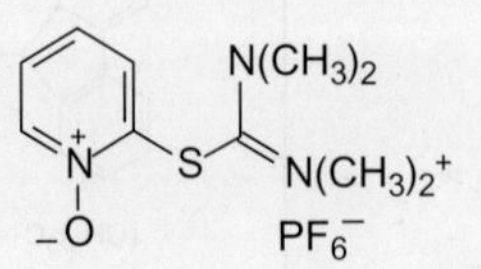

S-(1-Oxido-2-pyridinyl)-1,1,3,3-tetramethyl-
thiouronium hexafluorophosphate (HOTT)

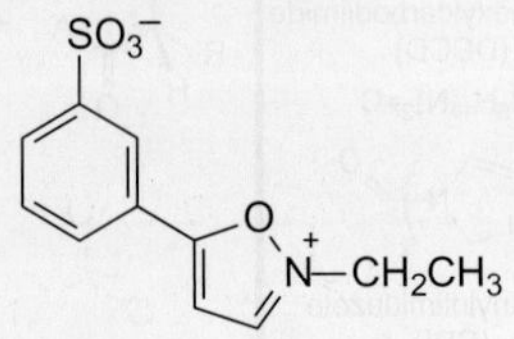

N-Ethyl-5-phenylisoxazolium-3′-sulfonate
(Woodward's reagent K)

Figure 12.17. Examples of peptide coupling reagents chosen from several hundred similar compounds. Carbonyldiimidazole (CDI) and dicyclohexylcarbodiimide (DCCD) are not shown here since they are provided as part of the Merrifield process described in Scheme 12.76 and have been discussed earlier (Chapter 9, Scheme 9.102).

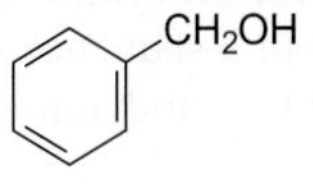

Benzyl alcohol

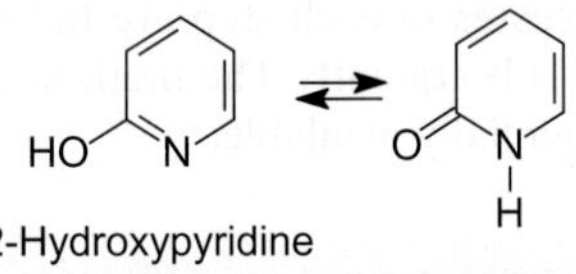

2-Hydroxypyridine

HOCH$_2$CCl$_3$

2,2,2-Trichloroethanol

Pentafluorophenol

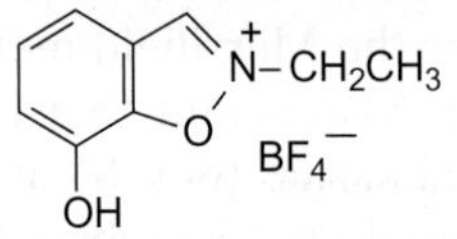

2-Ethyl-7-hydroxy-1,2-benzisoxazolium
tetrafluoroborate

Figure 12.18. Examples of alcohols and phenols that are used to block the carboxylic acid portion of amino acids so that amide-forming reactions can occur at the nitrogen. Simple alcohols that include methanol (CH$_3$OH), ethanol (CH$_3$CH$_2$OH), and *tert*-butyl alcohol (*t*-butanol, (CH$_3$)$_3$COH) are omitted intentionally. Methyl esters or many amino acids, in particular, are often commercially available and are frequently used as received.

the major product after washing with dilute aqueous acid and base is the corresponding *t*-BOC N-protected amino acid. Reaction at nitrogen occurs in preference to oxygen, and what little oxygen substitution does occur can generally be hydrolyzed with dilute acid in preference to the hydrolysis reaction at nitrogen. Then reaction of this N-BOC-protected amino acid with benzyl ester of glycine (Gly, G) in the presence of the coupling agent 6-chloro-2,4-dimethoxy-s-triazine (CDMT), a reagent with which the hydroxyl group of serine (Ser, S) does not react, produces the corresponding N- and carboxy-protected dipeptide. The amino and carboxylate protecting groups can be removed selectively.

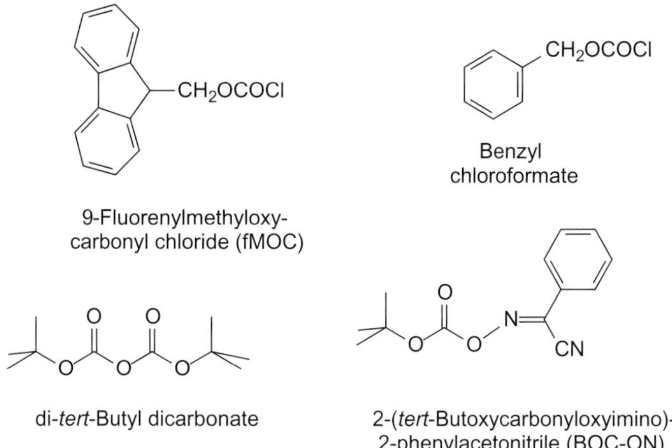

9-Fluorenylmethyloxy-
carbonyl chloride (fMOC)

Benzyl
chloroformate

di-*tert*-Butyl dicarbonate

2-(*tert*-Butoxycarbonyloxyimino)-
2-phenylacetonitrile (BOC-ON)

Figure 12.19. Examples of chloroformates and carbonates that are used to block the amine portion of amino acids so that amide-forming reactions can occur at the carboxylic acid. Phthalic anhydride (Scheme 10.2) is omitted intentionally since an amino acid prepared by the Gabriel synthesis using phthalimide would retain the phthalimide protecting group subsequently. Phthalic anhydride is not generally used as a protecting group with preformed amino acids.

Alternatively, as shown in Scheme 12.78, when N-BOC phenylalanine (Phe, F), prepared by treatment of the amino acid with BOC-ON as with serine (Ser, S) in Scheme 12.77, is treated with DCCD and 2,2,2-trichloroethanol, the corresponding trichloroethanol ester is formed. Removal of the butoxycarbonyl (BOC) group with dilute acid is easily effected and the ester remains intact. When this ester is allowed to react with the same N-BOC-protected serine (Ser, S) of Scheme 12.77, but now in the presence of *N*-ethyl-5-phenylisoxazolium-3′-sulfonate (Woodward's reagent K), coupling occurs. Again, this coupling reagent, unlike DCCD or CDI, can be used in the presence of serine (Ser, S). Then, the *t*-BOC protecting group can be removed by acid hydrolysis, while the trichloroethyl ester only comes apart in the presence of zinc dust and acetic acid. These reactions can, once again, be completed separately so that either or both of the protecting groups can be removed. The sulfonate coupling reagent as well as the amidosulfonate product are water soluble and are easily removed from the reaction by extraction.

In yet another example (Scheme 12.79), if the 9-fluorenylmethyloxycarbonyl (fMOC) derivative of glycine (Gly, G) were allowed to react with the ethyl ester of valine (Val, V) in the presence of 5-nitro-3H-2,3-benzoxathiole S,S-dioxide, the peptide would again form and the nitrophenol generated from the coupling reagent is easily removed as a consequence of its acidic properties.

At this point, the pattern of these synthetic efforts should be clear. All of the methods utilize an amino acid protected from reaction at one or the other functional group and, with an appropriate coupling agent, produce a reaction with another amino acid oppositely protected to form an amide. The choice in each case, in addition to the amino acids, simply involves protecting groups and coupling agent.

Finally, for this section, the last case to be considered involves the use of hexafluoroacetone (CF_3COCF_3) as a protecting group. As shown in Scheme 12.80, when

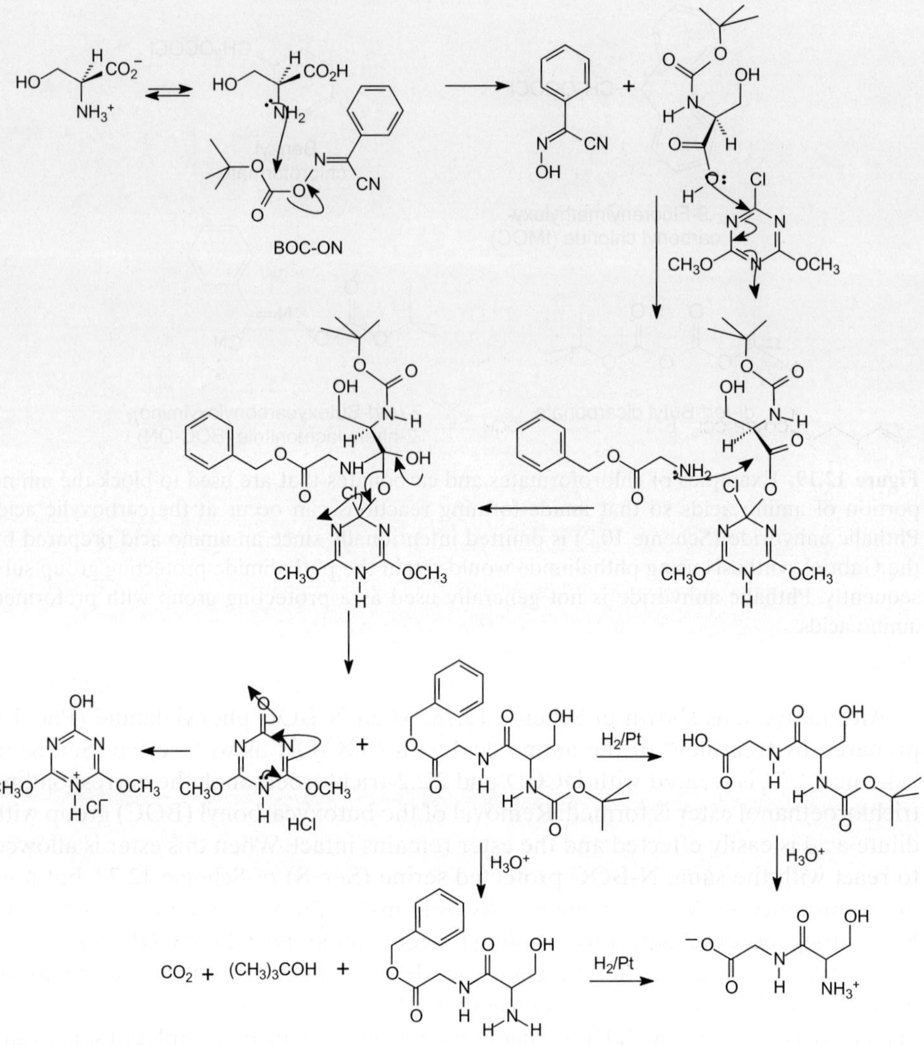

Scheme 12.77. Peptide coupling using 6-chloro-1,4-dimethoxy-2-triazine as the coupling reagent between the benzyl ester of glycine and N-BOC-protected serine.

aspartic acid (Asp, D) is allowed to react with hexafluoroacetone (CH_3COCF_3), the corresponding bistrifluoromethyloxazolone forms in over 85% yield. Then, simple treatment of this derivative with the methyl ester of phenylalanine (methyl phenyl-alaninate) at room temperature results in a 72% conversion to the amide dimer, Asp-Phe-OCH_3, sold under the commercial name aspartame.

D. THE COENZYMES

As noted in the introduction to this chapter, there are 10 coenzymes that are "common." The coenzymes in question have been utilized throughout the portion

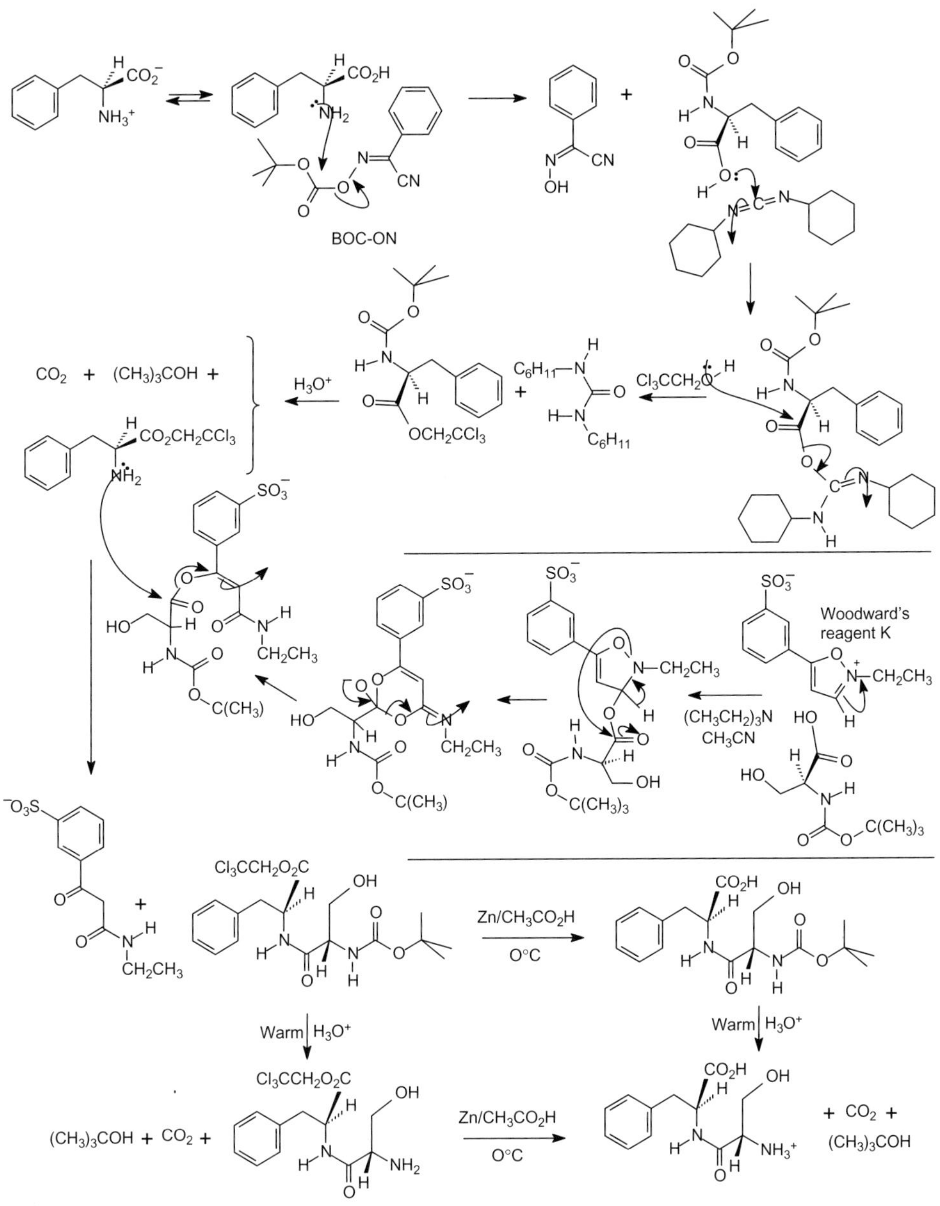

Scheme 12.78. Peptide formation using Woodward's reagent K and a trichloroethanol ester.

of the work called "Foreground" prior to this and will be utilized again subsequently as needed in the reactions. Here, each of the "common coenzymes" is examined on its own; that is, (a) the current thinking on the biosynthesis of the coenzyme itself and the enzymes that are used in its biosynthesis will be elaborated upon, and (b) at least one chemical synthesis of the coenzyme or its immediate biosynthetic precursor, where stability is a question, will be provided.

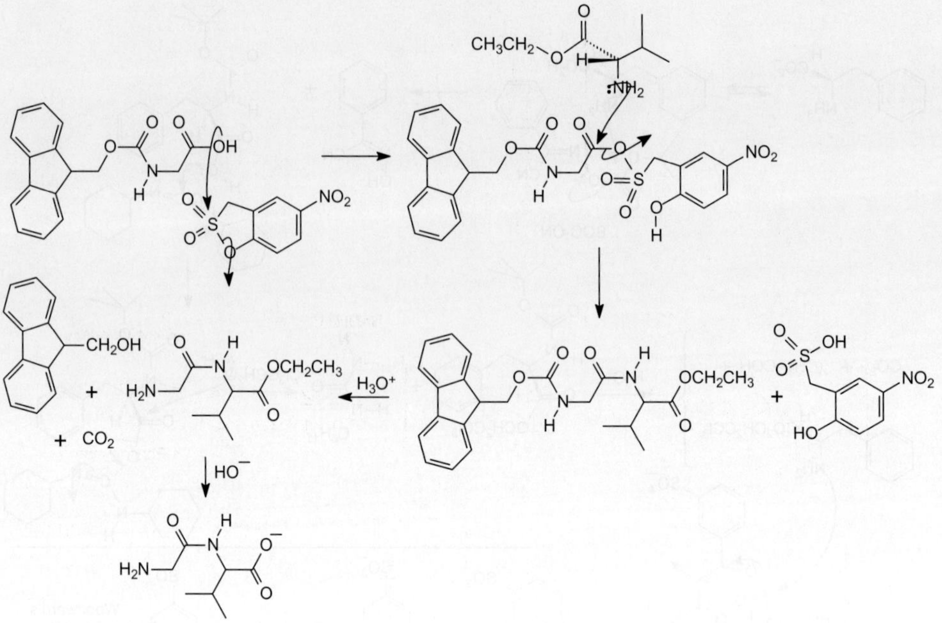

Scheme 12.79. Coupling between fMOC-glycine (Gly, G) and the ethyl ester of valine (Val, V) utilizing 5-nitro-3H-2,3-benzoxathiole S,S-dioxide as the coupling reagent.

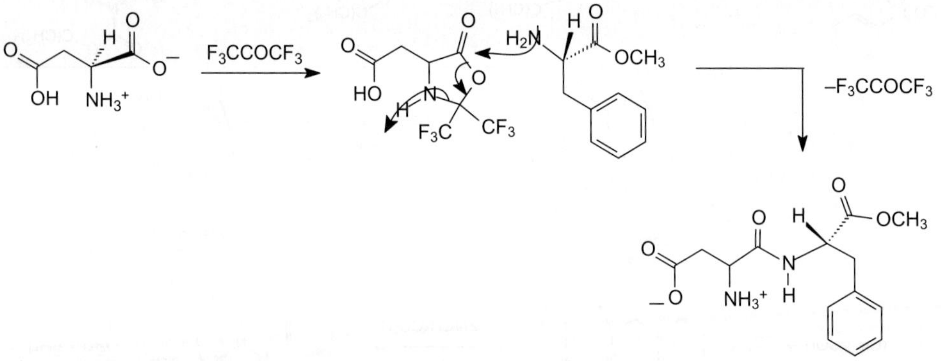

Scheme 12.80. A representation of the synthesis of aspartame utilizing hexafluoroacetone to block both the amino and carboxylate functionalities in aspartic acid (Asp, D) and thus to allow attack by methyl phenylalaninate.

I. Pyridoxal Phosphate

As shown in Scheme 12.81, the biosynthesis of pyridoxal begins with erythrose-4-phosphate. Oxidation of the aldehyde (EC 1.2.1.72, NAD^+ to $NADH/H^+$) produces, first, 4-phosphoerythronate, and a second oxidation with the same cofactor (EC 1.1.1.290) yields the corresponding α-ketocarboxylic acid $(3R)$-3-hydroxy-2-oxo-4-phosphonooxybutanoate. Transamination (glutamate [Glu, E $\rightarrow$ α-ketoglutarate],

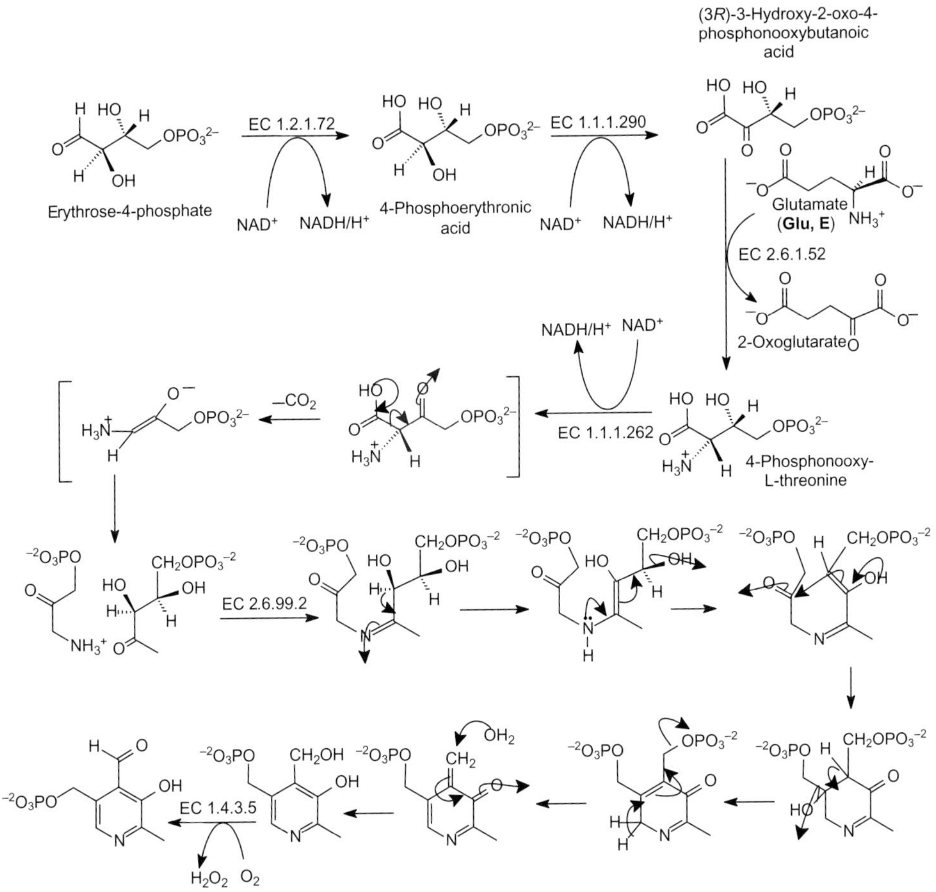

Scheme 12.81. The biosynthesis of pyridoxal. The latter parts shown, including the incorporation of exogenous water, are taken from Cane, D. E.; Du, S.; Spenser, I.D. *J. Am. Chem. Soc.*, **2000**, *122*, 4213. EC numbers and some graphic materials provided in this scheme have been taken from appropriate links in a URL starting with http://www.chem.qmul.ac.uk/iubmb/enzyme/.

EC 2.6.1.52) generates 4-phosphonooxy-L-threonine, and another oxidation with NAD^+ going to $NADH/H^+$ (EC 1.1.1.262) converts the hydroxyl to a keto group and yields (2*S*)-2-amino-3-oxo-4-phosphonooxybutanoate, which spontaneously decarboxylates to the unstable 3-amino-2-oxopropylphosphate. This amine then condenses with 1-deoxy-D-xylulose-5-phosphate (Chapter 11, Scheme 11.42, generated from pyruvic acid and D-glyceraldehyde-3-phosphate, EC 2.2.1.7) to the corresponding imine, which undergoes cyclization to pyridoxine 5′-phosphate (EC 2.6.99.2). Oxidation (EC 1.4.3.5, $O_2 \rightarrow H_2O$) yields pyridoxal 5′-phosphate.

With regard to the synthesis of pyridoxal, the situation is a bit more complex. The oxidation in the last step is particularly facile. As a consequence, syntheses are directed toward pyridoxine, the corresponding benzylic alcohol present in the penultimate step in the above biosynthesis.

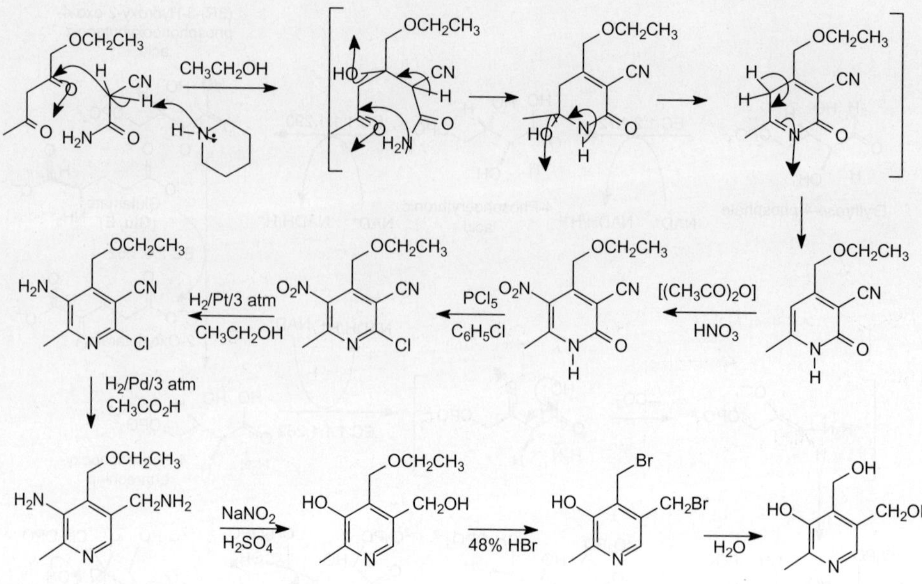

Scheme 12.82. The early synthesis of pyridoxine (vitamin B_6), a reduced form of pyridoxal. The synthesis shown here and described in the text is after Harris, S. A., Folkers, K. *J. Am. Chem. Soc.*, **1939**, *61*, 1245.

The initial classic synthesis, completed in 1939, is shown in Scheme 12.82. The condensation reaction between 1-ethoxypentane-2,4-dione and cyanoacetamide in ethanol (CH_3CH_2OH) using piperidine as a catalyst generated 3-cyano-4-ethoxymethyl-6-methyl-2-pyridone, which, in acetic anhydride [$(CH_3CO)_2O$] with fuming nitric acid (HNO_3), produced 3-cyano-4-ethoxymethyl-5-nitro-6-methyl-2-pyridone. Then, with phosphorus pentachloride (PCl_5) in chlorobenzene, aromaticity was introduced and the oxygen at C2 was replaced by chlorine. After reduction of the nitro group to the corresponding aryl amine with hydrogen and platinum (Pt) in aqueous ethanol (CH_3CH_2OH) at 3 atm of hydrogen (H_2), a second reduction in acetic acid with palladium (Pd) and hydrogen (H_2), also at high pressure, was utilized to reduce the nitrile and to dehalogenate the aromatic ring. The dihydrochloride of the bis-amino-pyridine derivative (2-methyl-3-amino-4-ethoxymethyl-5-aminomethylpyridine), on treatment with aqueous nitrous acid ($NaNO_2 + H_2SO_4$), underwent two diazotization reactions so that basification yielded 2-methyl-3-hydroxy-4-ethoxymethyl-5-hydroxymethylpyridine. Treatment of the hydrochloride of the dihydroxyethoxymethypyridine with 48% aqueous hydrogen bromide (HBr) resulted in the replacement of hydroxyl of the primary alcohol by bromine as well as cleavage of the O-ethyl ether to yield 2-methyl-3-hydroxy-4,5-di(bromomethyl) pyridine as the hydrobromide; finally, boiling this dibromide in water replaced the benzylic-type bromines with hydroxyl groups and the bromide anions were removed from the solution by treatment with silver chloride (AgCl). The residue, after evaporation, was the hydrochloride of 2-methyl-3-hydroxy-4,5-di(hydroxymethyl) pyridine (pyridoxine, vitamin B_6).

Interestingly, somewhat more than two decades later, an almost trivial synthesis (by comparison) was reported (Scheme 12.83). In this newer method, the ethyl ester

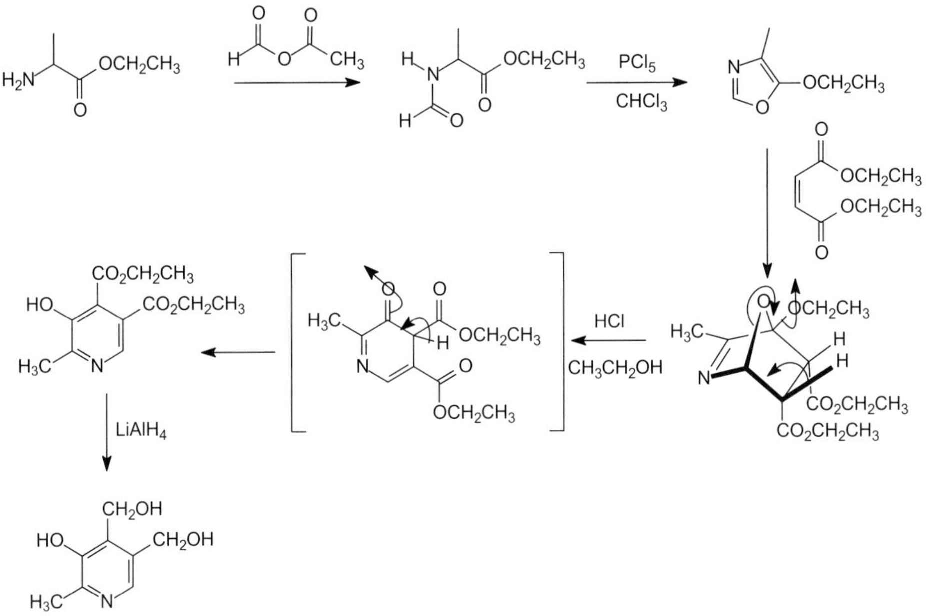

Scheme 12.83. An improved synthesis of pyridoxine (vitamin B_6) (after Harris, E. E.; Firestone, R. A.; Pfister, K.; Boettcher, R. R.; Cross, F. J.; Currie, R. B.; Monaco, M.; Peterson, E. R.; Reuter, W. *J. Org. Chem.*, **1962**, *27*, 2605).

of racemic alanine (Ala, A) was N-formylated by the use of formic-acetic anhydride and the amide was treated with phosphorous pentachloride (PCl_5) in chloroform ($CHCl_3$) to produce 3-methyl-4-ethoxyoxazole. Treatment of the oxazole with diethyl maleate at 100°C for several hours yielded a Diels–Alder adduct, which was readily cleaved with ethanolic hydrogen chloride (HCl) to give the diethyl ester of 2-methyl-3-hydroxypyridine-4,5-dicarboxylic acid hydrochloride. When the diester was reduced with lithium aluminum hydride ($LiAlH_4$), the diol pyridoxine resulted.

II. Lipoic Acid

The details of the biosynthesis of lipoic acid as brought about by the lipoyl synthase polypeptide enzyme (EC 2.8.1.8) remain somewhat obscure. It is argued that the enzyme belongs to the class of the "radical SAM" family, all members of which catalyze by using an electron from the reduced form of the 4Fe-4S (this chapter, Section B, I) cluster in the enzyme to split the SAM into the 5′-deoxyadenosyl radical and methionine (Scheme 12.84). It is argued further that two 5′-deoxyadenosyl radicals formed in this way go on to abstract the hydrogens from the eight-carbon saturated acyl chain* (which is bound to the enzyme via an auspiciously placed lysine [Lys, K]) at carbons 6 and 8, and then the two sulfur atoms are introduced.

*The saturated acyl chain, while lysine (Lys, K) bound, has presumably been biosynthesized in the usual way (see Chapter 11, Scheme 11.34, et seq.) and then transferred (lipoyl[octanoyl] transferase, EC 2.3.1.181) to the active-site lysine (Lys, K) of the lipoyl synthase enzyme (EC 2.8.1.8).

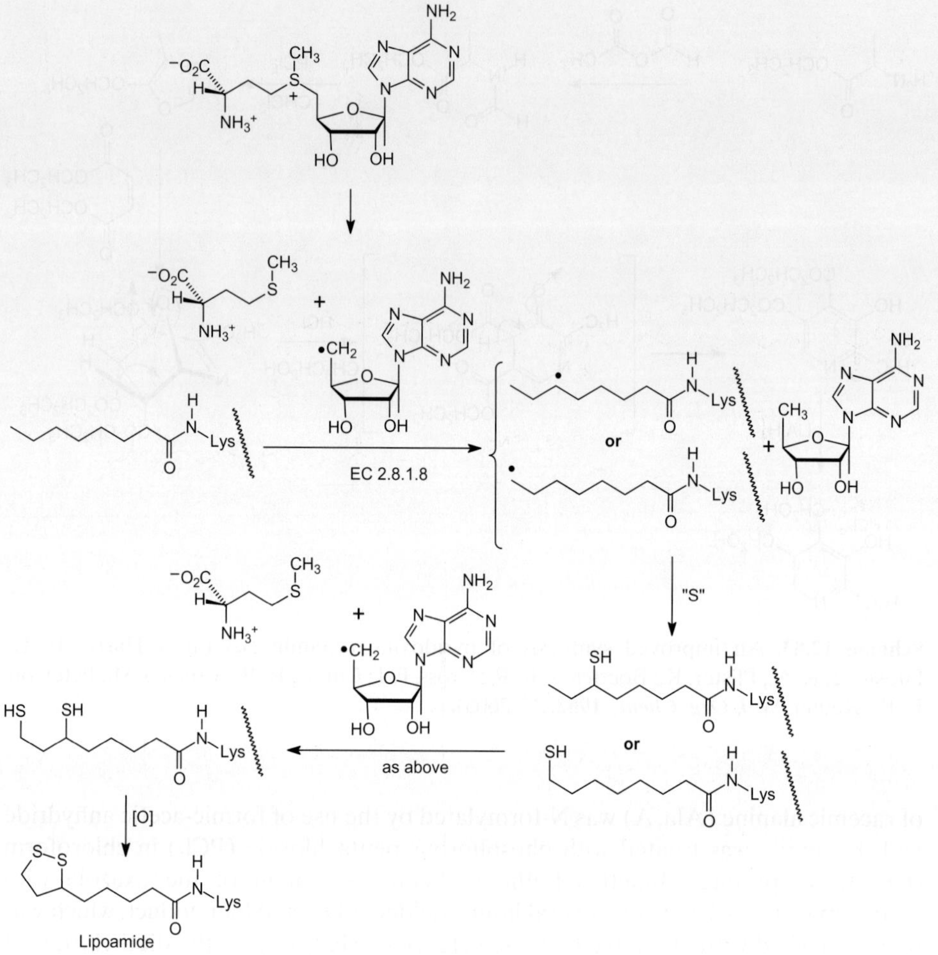

Scheme 12.84. Some information in accord with what is known about the biosynthesis of lipoic acid (after Cicchillo, R. B.; Booker, S. J. *J. Am. Chem. Soc.*, **2005**, *127*, 2860).

It is unlikely that both abstractions and sulfur insertions occur simultaneously. However, it has been shown in *Escherichia coli* (Cicchillo and Booker, 2005) using sulfur labeling, that both sulfur atoms come from the same enzyme and that, therefore, one is not added, the substrate released, and then the second is added. It is presumed that the sulfur may come from a 2Fe-2S source on the enzyme, although this may be fiction.

One of the syntheses (Scheme 12.85) of lipoic acid began with the half-acid chloride of the ethyl ester of adipic acid [1,6-hexanedioic acid, $HO_2C(CH_2)_4CO_2H$], which, in nitrobenzene solvent, at 45°C, in the presence of a large excess of aluminum trichloride ($AlCl_3$) was added to ethene (ethylene, $H_2C=CH_2$) to yield ethyl 6-keto-Δ^7-octenoate in about 50% yield (after distillation). Thioacetic acid underwent conjugate addition to the double bond of the ketoalkene to generate ethyl 8-acetylthio-6-ketooctanoate, which was readily reduced with sodium borohydride

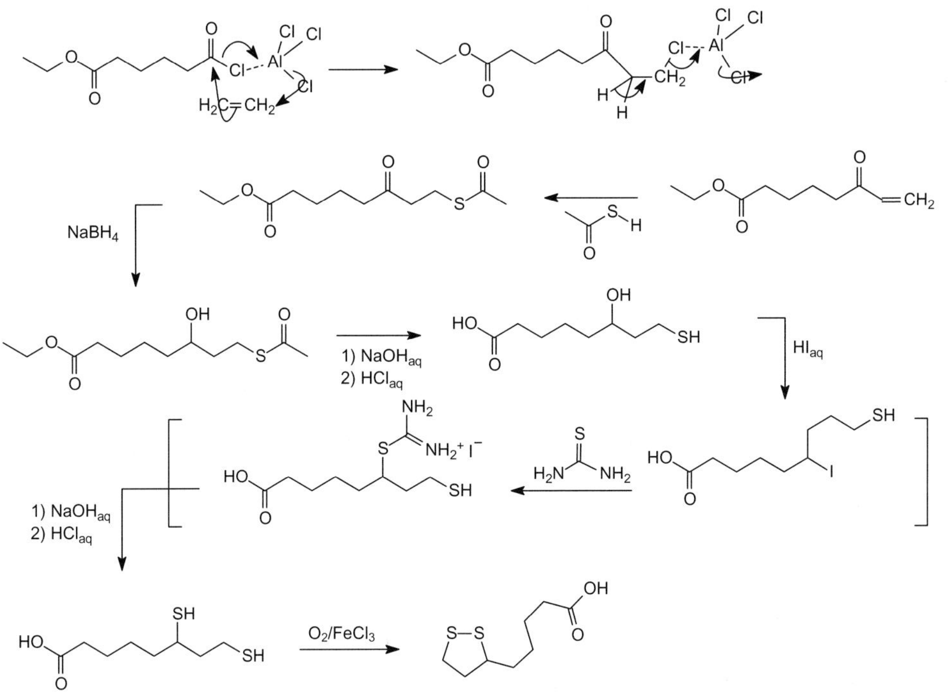

Scheme 12.85. A synthesis of lipoic acid (after Bullock, M. W.; Brockman, J. A.; Patterson, E. L.; Pierce, J. V.; von Saltza, M. H.; Sanders, F.; Stokstad, L. R. *J. Am. Chem. Soc.*, **1954**, *76*, 1828).

in methanol to the corresponding alcohol, 8-acetylthio-6-hydroxyoctanoate. Hydrolysis with aqueous sodium hydroxide generated the corresponding 8-thiol-6-hydroxyoctanoic acid. Then, heating the hydroxy acid with 50% aqueous hydrogen iodide and thiourea for 10 h at reflux, followed by basification with aqueous sodium hydroxide (NaOH) and reacidification with hydrochloric acid, resulted in the introduction of the second sulfur to yield 6,8-dithiooctanoic acid. Distillation of the crude acid under vacuum and oxidation (O_2) in the presence of ferric chloride ($FeCl_3$) produced (in 30% yield from 8-thio-6-hydroxyoctanoic acid) lipoic acid.

III. Thiamine Diphosphate

The story of the biosynthesis of the cofactor thiamine diphosphate (vitamin B_1), the anion of which is necessary for its condensation reactions (Scheme 12.86), is still being written. Not only is there some evidence that this cofactor may have different origins in prokaryotes and eukaryotes but also, not all of the steps in plants or animals have been made clear. However, while it is an active area of research, some pieces are fairly well established, and it is generally agreed that a convergent pathway, bringing together two phosphorylated and independently biosynthesized fragments, obtains. The exposition of those pieces is carried out below in three parts. The first part deals with the biosynthesis of the pyrimidine ring (i.e., the presumed genesis of 4-amino-5-hydroxymethyl-2-methylpyrimidine), which,

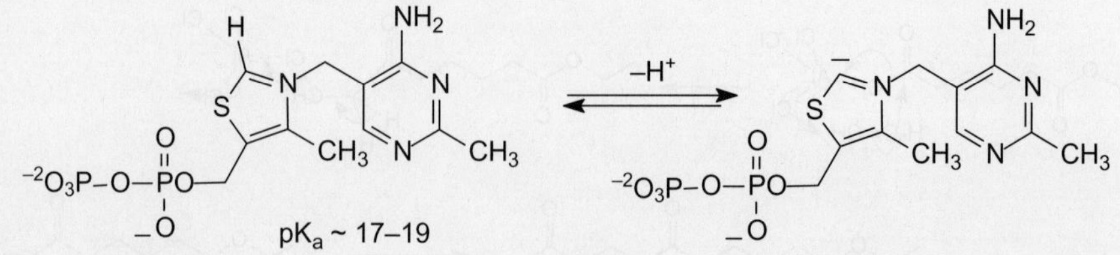

Scheme 12.86. Formation of the anion of thiamine (Washabaugh, M. W.; Jenks, W. P. *Biochemistry*, **1988**, *27*, 5044).

after phosphorylation, is added to the thiamine thiazole [4-methyl-5-(2-phosphonooxyethyl) thiazole]. Work on the biosynthesis of the thiazole ring (i.e., 5-(2-hydroxy-ethyl)-4-methylthiazole) is the second part. The combination of the two pieces into thiamine diphosphate is the third part.

a. 4-Amino-5-hydroxymethyl-2-methylpyrimidine. At the beginning of Chapter 11 (Figure 11.2), there is a description of the Calvin cycle. It will be recalled that this cycle performs the overall function of showing how 3 equivalents of carbon dioxide (CO_2) are converted to one useful (for other purposes) equivalent of 3-phosphoglycerate. Along the way, as the cycle turns, many exciting compounds are generated, including D-ribulose-5-phosphate (Scheme 12.87)* and glyceraldehyde 3-phosphate.

Glyceraldehyde 3-phosphate, it will be recalled (Scheme 11.13), is a progenitor of glucose-6-phosphate, which, via the pentose phosphate pathway, is also (i.e., in addition to the Calvin cycle) on the path to ribulose-5-phosphate. Now, via the common enol, ribulose-5-phosphate is readily converted to ribose-5-phosphate (EC 5.3.1.6), which is then bisphosphorylated (EC 2.7.6.1) to the α-diphosphate so that a leaving group is in place that will allow replacement by the terminal amido function of glutamine (Gln, Q) on its way to glutamate (Glu, E) with the formation of 2-aminoribose-5-phosphate.

Following conversion of the carboxylate of glycine (Gly, G) to a phosphate anhydride (ATP→ADP + P$_i$) displacement of the phosphate by the amino group of the 2-aminoribose-5-phosphate (EC 6.3.4.13) generates the corresponding amido product, which is formylated on the primary amine (EC 2.1.2.2) by CHO–C$_{10}$ substituted tetrahydrofolate.

Now, aminolysis of the carbonyl function of the original amide generated by the addition of glycine (Gly, G) to 2-aminoribose-5-phosphate using, once again, the terminal amido function of glutamine (Gln, Q) (which is again converted to glutamate [Glu, E]) results in imine formation at that carbonyl. Finally, a cyclization with dehydration (ATP → ADP + P$_i$, EC 6.3.3.1) produces the desired (5-aminoimidazol) ribose-5-phosphate [5-amino-1-(5-phospho-D-ribosyl)imidazole, 5-aminoimidazole ribotide] (Scheme 12.87).

While a significant amount of solid labeling studies have confirmed the chemistry outlined in Scheme 12.87, the same kinds of studies have shed less firm information

---

*As shown in Scheme 12.87, D-ribulose-5-phosphate can also be derived from D-glucose-6-phosphate, which, as shown in Scheme 11.13, is generated from dihydroxyacetone monophosphate (i.e., an isomer of 3-phosphoglycerate—from the Calvin cycle).

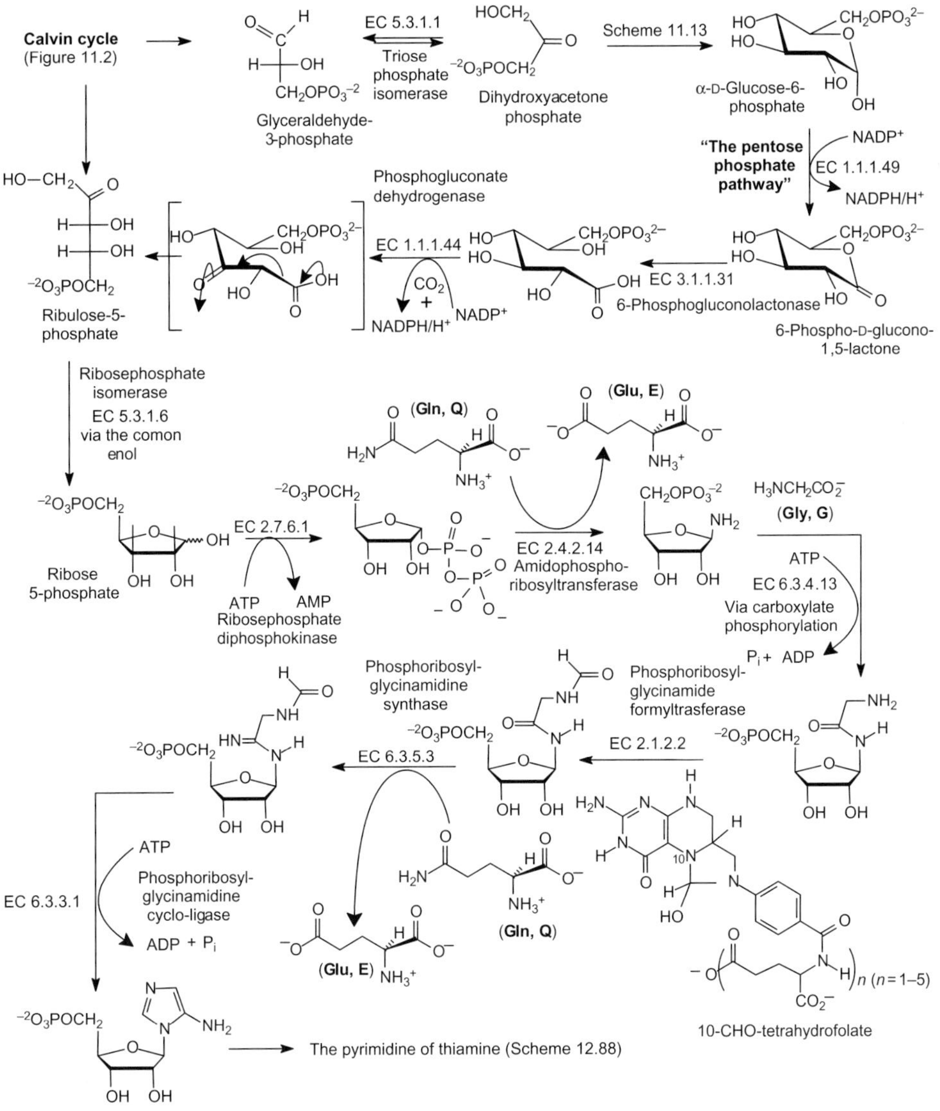

(5-Aminoimidazol)ribose [5-amino-1-(5-phospho-D-ribosyl)imidazole]

Scheme 12.87. Paths to 5-aminoimidazolribotide, the precursor to the 4-amino-5-hydroxymethyl-2-methylpyrimidine portion of thiamine. EC numbers and some graphic materials provided in this scheme have been taken from appropriate links in a URL starting with http://www.chem.qmul.ac.uk/iubmb/enzyme/.

on the conversion of the aminoimidazolribotide 5-phosphate to 4-amino-5-hydroxymethylpyrimidine phosphate. Indeed, the studies are less than clear in defining a pathway. Nonetheless, an excellent start has been made as shown in Scheme 12.88.

b. 4-Methyl-5-(2-phosphonooxyethyl)thiazole. While the biosynthesis of the thiamine thiazole is actively under investigation, some of the enzymes involved

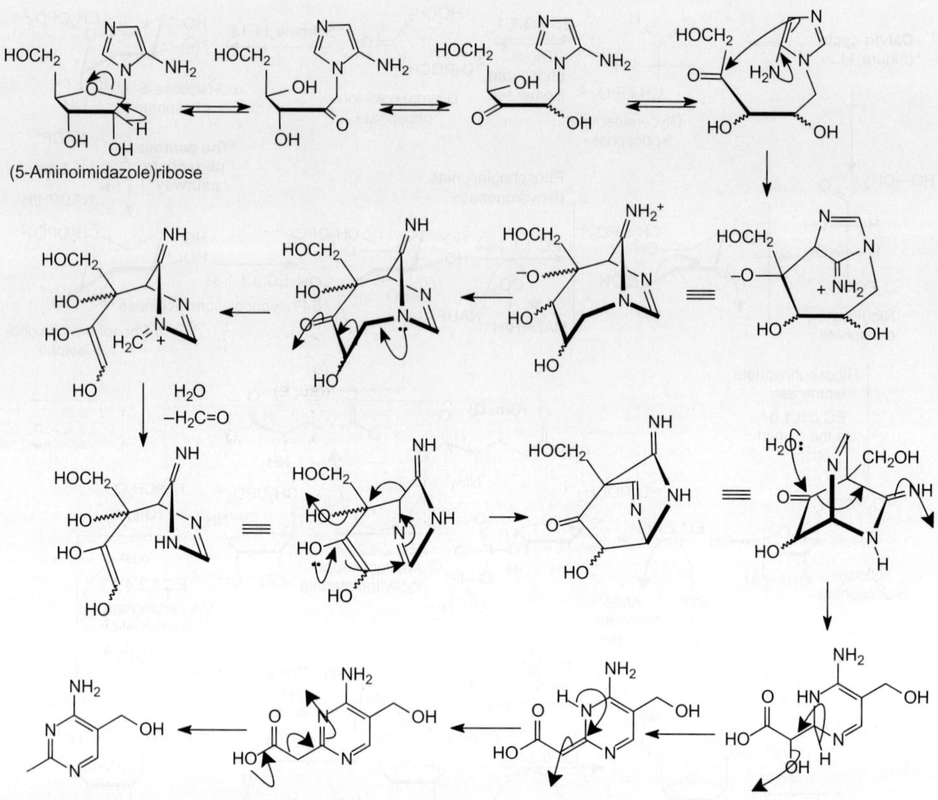

Scheme 12.88. A proposed pathway from 5-aminoimidazoleribotide to 4-amino-5-hydroxy-methylpyrimidine (Spenser, I. D.; White, R. L. *Angew. Chem. Int. Ed. Engl.*, 1997, *36*, 1032).

(largely in *Bacillus subtilis*) have been identified.* As shown in Scheme 12.89, 1-deoxy-D-xylulose-5-phosphate (Chapter 11, Scheme 11.42, and this chapter, Scheme 12.81) forms an imine with Lys96 of a protein in the thiamine thiazole biosynthetic complex. This imine tautomerizes to the corresponding aminoketone, which reacts with a protein thiocarboxylate carrier at the carbonyl, which then undergoes an S to O acyl shift and elimination of water to the corresponding thiocarbonyl derivative. Enolization of the thioketone followed by elimination of the neighboring carboxylate generates a thioenol, which undergoes addition to the imine of glycine (formed in a separate oxidative step from glycine [Gly, G]). The resulting aminoimine undergoes cyclic transimination and decarboxylation to the thiazole product, 4-methyl-5-(2-phosphonooxyethyl)thiazole.

c. Thiamine Diphosphate. Finally, as shown in Equation 12.14, 4-amino-5-hydroxymethyl-2-methylpyrimidine undergoes phosphorylation to the corresponding diphosphate, and displacement on that benzylic-type carbon of the diphosphate (EC 2.5.1.3) results in the formation of thiamine phosphate and diphosphate.

---

*This description is distilled from a more thorough précis. Chatterjee, A.; Han, X.; McLafferty, F. W.; Begley, T. P. *Angew. Chem.*, **2006**, *118*, 3587.

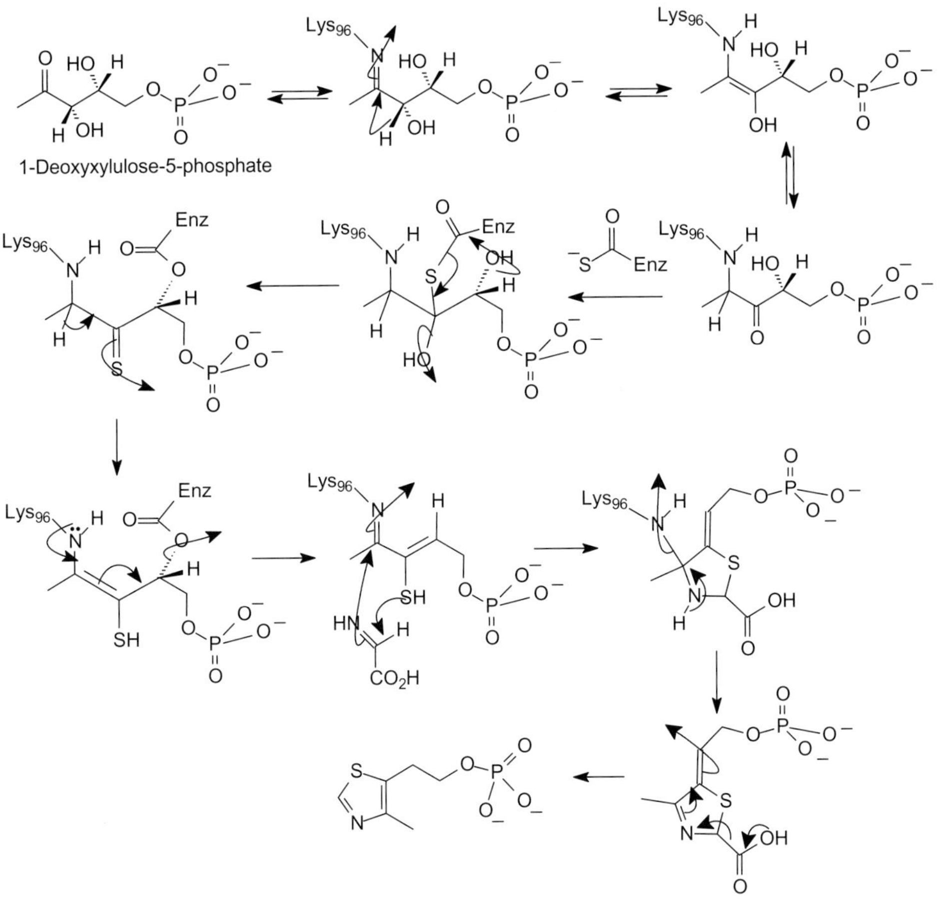

Scheme 12.89. A possible path for the formation of 4-methyl-5-(2-phosphonooxyethyl)thiazole from 1-deoxyxylulose-5-phosphate (after Chatterjee, A.; Han, X.; McLafferty, F. W.; Begley, T. P. *Angew. Chem.*, **2006**, *118*, 3587).

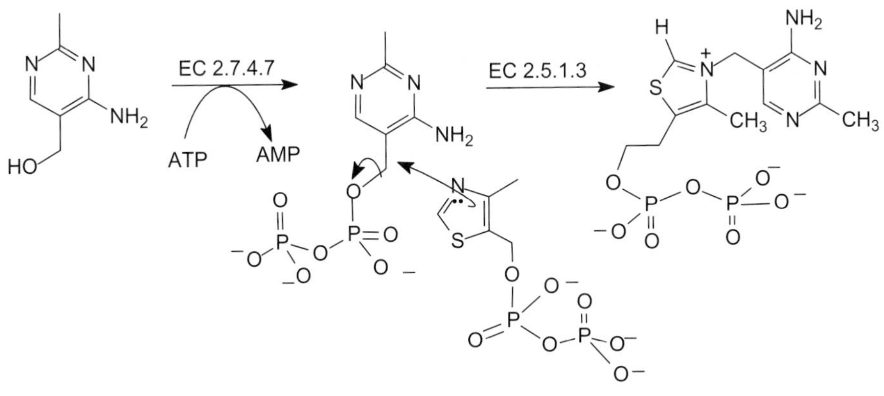

$$(12.14)$$

The synthesis of thiamine itself was accomplished relatively early by a group at the laboratories of Merck, Inc. who utilized a brief (then unpublished!) synthesis of 4-methyl-5-(β-hydroxyethyl)thiazole to combine with 4-amino-5-bromomethyl-2-methylpyrimidine in an S_N2-type process. The synthesis of the 4-methyl-5-(β-hydroxyethyl)thiazole was published a month after it was used. The synthesis of the 4-methyl-5-(β-hydroxyethyl)thiazole is shown in Scheme 12.90 and that using it in combination with 4-amino-5-bromomethyl-2-methylpyrimidine to produce the vitamin is shown in Scheme 12.91.

For the thiazole (Scheme 12.90), ethyl acetoacetate was treated with ethylene oxide (oxirane) in the presence of sodium ethoxide to produce the corresponding furanone, which yielded a chloride on treatment with thionyl chloride. The chlorine served to remove the only active proton remaining and introduced the future leaving group. Thus, hydrolysis of the lactone yielded 3-chloro-5-hydroxy-2-pentanone, which, on treatment with thioformamide, provided a modest yield of the 4-methyl-5-(β-hydroxyethyl)thiazole.

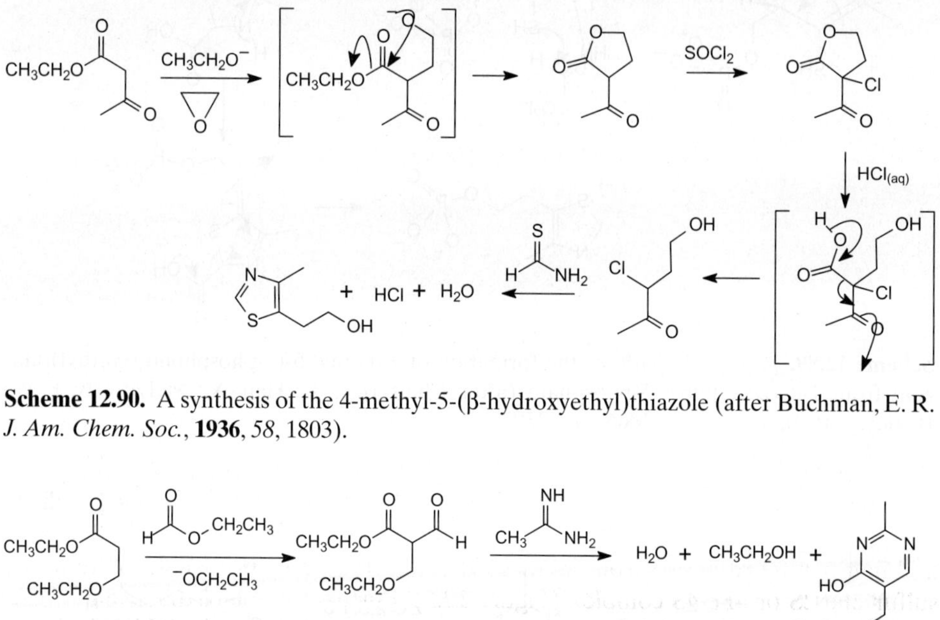

Scheme 12.90. A synthesis of the 4-methyl-5-(β-hydroxyethyl)thiazole (after Buchman, E. R. *J. Am. Chem. Soc.*, **1936**, *58*, 1803).

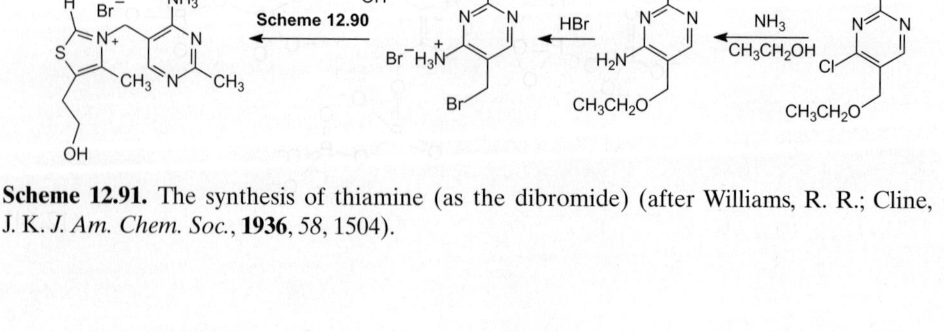

Scheme 12.91. The synthesis of thiamine (as the dibromide) (after Williams, R. R.; Cline, J. K. *J. Am. Chem. Soc.*, **1936**, *58*, 1504).

The group at Merck, Inc. formylated ethyl 3-ethoxypropanoate at the α-carbon with ethyl formate in the presence of ethoxide and then treated the resulting aldehydic ester with the amidine from acetamide. The resulting 4-hydroxy-5-(β-ethoxyethyl)-2-methylpyrimidine yielded the corresponding 4-chloroderivative with phosphorus oxychloride, which, with ethanolic ammonia produced the 4-amino-5-(β-ethoxyethyl)-2-methylpyrimidine. Ether cleavage to the β-bromohydrobromide produced the immediate precursor to the thiamine hydrobromide (Scheme 12.91).

IV. Biotin

At this writing, there appears to be a general consensus that oxidation of a long-chain fatty acid (or long-chain fatty acids) by some as yet unspecified cytochrome P450 enzyme or enzymes* (using oxygen, O_2) serves to generate 1,7-heptanedioic acid ([$HO_2C(CH_2)_5CO_2H$], pimelic acid). The acid, derivatized as pimeloyl–SCoA, is activated for reaction with alanine (Ala, A) to produce 7-keto-8-aminonanoate (7-keto-8-aminopellargonic acid) by a pyridoxal phosphate-catalyzed process (EC 2.3.1.47) (Scheme 12.92). The aminoketone then undergoes transamination with SAM as an imine derived from pyridoxal, serving as the amino transfer agent, producing S-adenosyl-4-methylthio-2-oxobutanoate and 7,8-diaminononanoate (7,8-diaminoplargonic acid, EC 2.6.1.62). This appears to be one of the few cases that the amino group of methionine (Met, M) is used as a source of nitrogen.

Then the 7,8-diaminononanoate (7,8-diaminopelargonic acid) (Scheme 12.93) is converted to the bisamide dethiobiotin (an imidazolidone) by carboxylation with carbon dioxide (CO_2) in the presence of ATP in a reaction requiring magnesium (Mg^{+2}) and catalyzed by dethiobiotin synthase (EC 6.3.3.3). The well-known reversible reaction of amines with carbon dioxide (CO_2) and an X-ray crystal structure showing the 8-amino group as a mixed carbonic phosphate anhydride has lent support to this picture (**PDB 1a82**). Finally, in the presence of biotin synthase (EC 2.8.1.6), an enzyme that appears to be used only once in each synthesis, reaction occurs with 2 equivalents (but almost certainly one at a time) of SAM and sulfur to produce biotin, 2 equivalents of methionine (Met, M), and 2 equivalents of 5'-deoxyadenosine.

It is argued that biotin synthase (EC 2.8.1.6) is a member of the "radical SAM" family. The members of this family produce the 5'-deoxyadenosin-5'-yl radical and methionine (Met, M) from SAM by the addition of an electron from an iron–sulfur 2Fe-2S or 4Fe-4S complex (Figure 12.2). Further, although the origin of the sulfur in biotin is uncertain, it is known that its insertion into dethiobiotin occurs with the retention of configuration at carbon and that it is not directly derived from SAM.

---

*The enzymes in the family called cytochrome P450 are among the most versatile and ubiquitous catalysts for biological oxidation. It is common that many different (sometimes hundreds of) genes for different cytochrome P450s occur in every plant and animal. The cytochrome P450 proteins generally have a high-valence iron in a heme or hemelike structure that is held in place by a cysteine or cysteines and utilize oxygen (as O_2) in a series of one-electron processes. Overall, one of the two oxygens in O_2 is found in water and the other is attached to carbon; that is,

$$R–H + NAD(P)H + H^+ + O_2 = R–OH + NAD(P)^+ + H_2O.$$

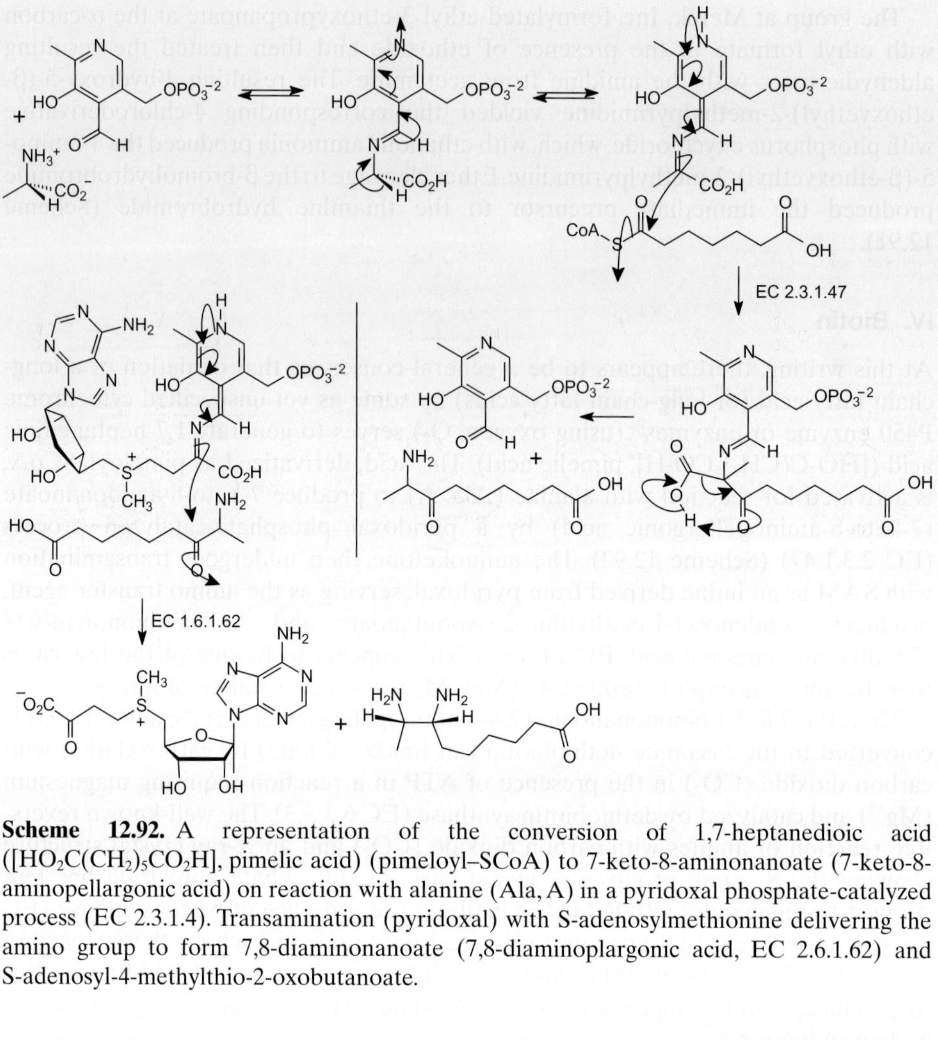

Scheme 12.92. A representation of the conversion of 1,7-heptanedioic acid ([$HO_2C(CH_2)_5CO_2H$], pimelic acid) (pimeloyl–SCoA) to 7-keto-8-aminonanoate (7-keto-8-aminopellargonic acid) on reaction with alanine (Ala, A) in a pyridoxal phosphate-catalyzed process (EC 2.3.1.4). Transamination (pyridoxal) with S-adenosylmethionine delivering the amino group to form 7,8-diaminonanoate (7,8-diaminoplargonic acid, EC 2.6.1.62) and S-adenosyl-4-methylthio-2-oxobutanoate.

An early synthesis of biotin (1975)* is particularly interesting since by beginning with readily available L-(+)-cysteine (Cys, C) and thus by incorporating the appropriate chiralty at the beginning, a late-stage resolution was not necessary.

So, as shown in Scheme 12.94, on reaction with benzaldehyde, L-(+)-cysteine (Cys, C) was converted to (4R)-carboxy-(2S)-phenylthiazolidone. Further protection of the nitrogen as the N-carbomethoxy derivative was effected with methyl chloroformate and reduction of the latter with diborane yielded the corresponding alcohol. The latter was oxidized to the aldehyde with chromium trioxide-pyridine. The use of vinylmagnesium chloride at low temperature succeeded in adding a vinyl group, which, on treatment with trimethyl orthoformate, in the presence of an acid catalyst, underwent ester exchange and then a Claisen rearrangement to produce the corresponding δ,γ-unsaturated carboxylic acid ester as the *trans*- or (*E*)-isomer exclusively.

*Confalone, P. J.; Pizzolato, G.; Baggiolini, E. G.; Lollar, D.; Uskoković. M. *J. Am. Chem. Soc.*, **1975**, *97*, 5936.

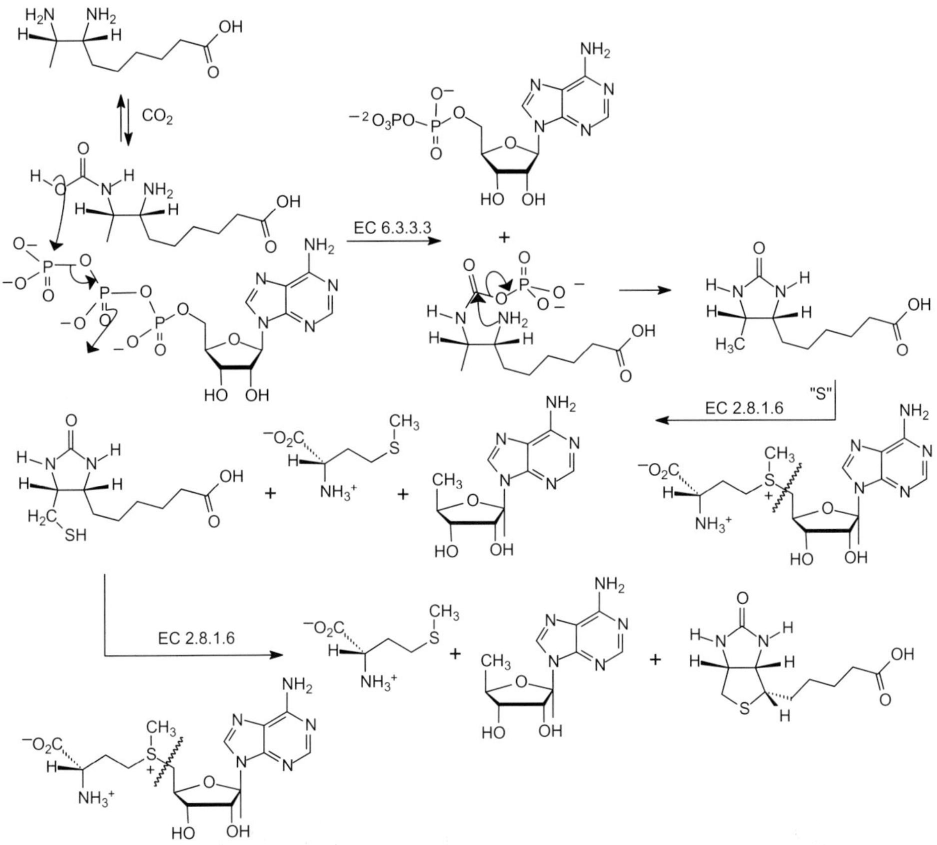

Scheme 12.93. A representation of the conversion of 7,8-diaminononanoate (7,8-diaminopelargonic acid) to the bisamide dethiobiotin (an imidazolidone) by carboxylation with carbon dioxide (CO_2) in the presence of ATP magnesium (Mg^{+2}) and dethiobiotin synthase (EC 6.3.3.3). X-ray of a mixed carbonic phosphate anhydride has lent support to this picture (PDB 1a82) (after Käck, H; Gibson K. J.; Lindqvist, Y; Schneider, G. *Proc. Natl. Acad. Sci. U.S.A.*, **1998**, *95*, 5495). With biotin synthase (EC 2.8.1.6), biotin results. EC numbers and some graphic materials provided in this scheme have been taken from appropriate links in a URL starting with http://www.chem.qmul.ac.uk/iubmb/enzyme/.

Then, bromination of the double bond resulted in a rearrangement to a trisubstituted tetrahydrothiophene with the expulsion of benzaldehyde. Sulfur participation in bromonium ion opening is suspected.

As shown in Scheme 12.95, the methylcarbamate was removed with hydrogen bromide (HBr) in acetic acid to form hydrobromide salt. When the amino hydrobromide salt was heated in acetic acid, a bromolactam was generated from the presumed aziridine intermediate and then treatment of the bromolactam with lithium azide in dimethylformamide produced a *cis* azidolactam, which then yielded the corresponding amine on reduction. Hydrolysis of the lactam with aqueous barium hydroxide ([Ba(OH)$_2$] then produced the diamino acid, and addition of phosgene (COCl$_2$) in aqueous sodium bicarbonate (NaHCO$_3$) produced bisnorbiotin, which was isolated as the corresponding methyl ester.

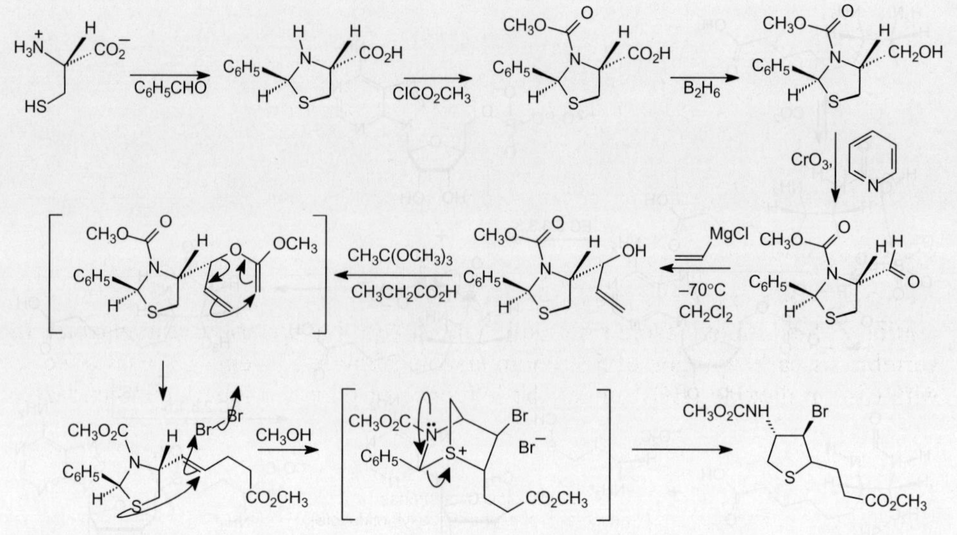

Scheme 12.94. The beginning of a synthesis of biotin from L-(+)-cysteine (Cys, C) (after Confalone, P. J.; Pizzolato, G.; Baggiolini, E. G.; Lollar, D.; Uskoković, M. *J. Am. Chem. Soc.*, **1975**, *97*, 5936).

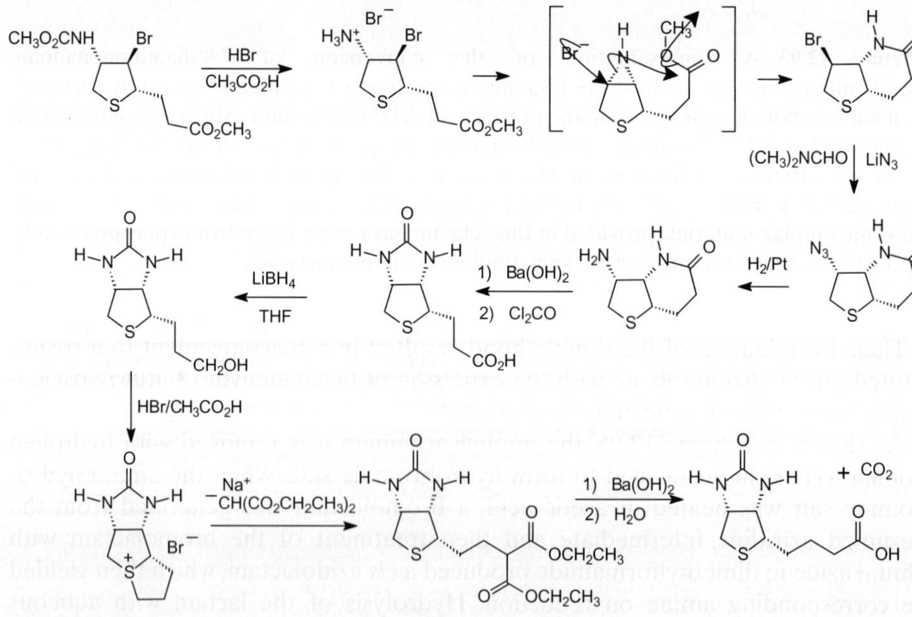

Scheme 12.95. The completion of a synthesis of biotin from L-(+)-cysteine (Cys, C) (after Confalone, P. J.; Pizzolato, G.; Baggiolini, E. G.; Lollar, D.; Uskoković, M. *J. Am. Chem. Soc.*, **1975**, *97*, 5936).

The final push to biotin itself required a reduction of the methyl ester with lithium borohydride ($LiBH_4$), treatment with hydrogen bromide (HBr) to generate a thiophanium bromide salt, which underwent reaction with diethyl malonate to a diester. The latter was not isolated but was hydrolyzed directly to the corresponding dicarboxylic acid (barium hydroxide), which, on acidification, underwent decarboxylation to yield biotin.

V. Adenosine

As noted in Scheme 12.87, the early stages of purine biosynthesis involve the generation of 5-aminoimidazole ribotide [5-amino-2-(5′-phosphoribosyl)imidazole]. In vertebrates, carboxylation of 5-aminoimidazole ribotide (Scheme 12.96) is effected with carbon dioxide (CO_2) under the influence of phosphoribosylaminoimidazole

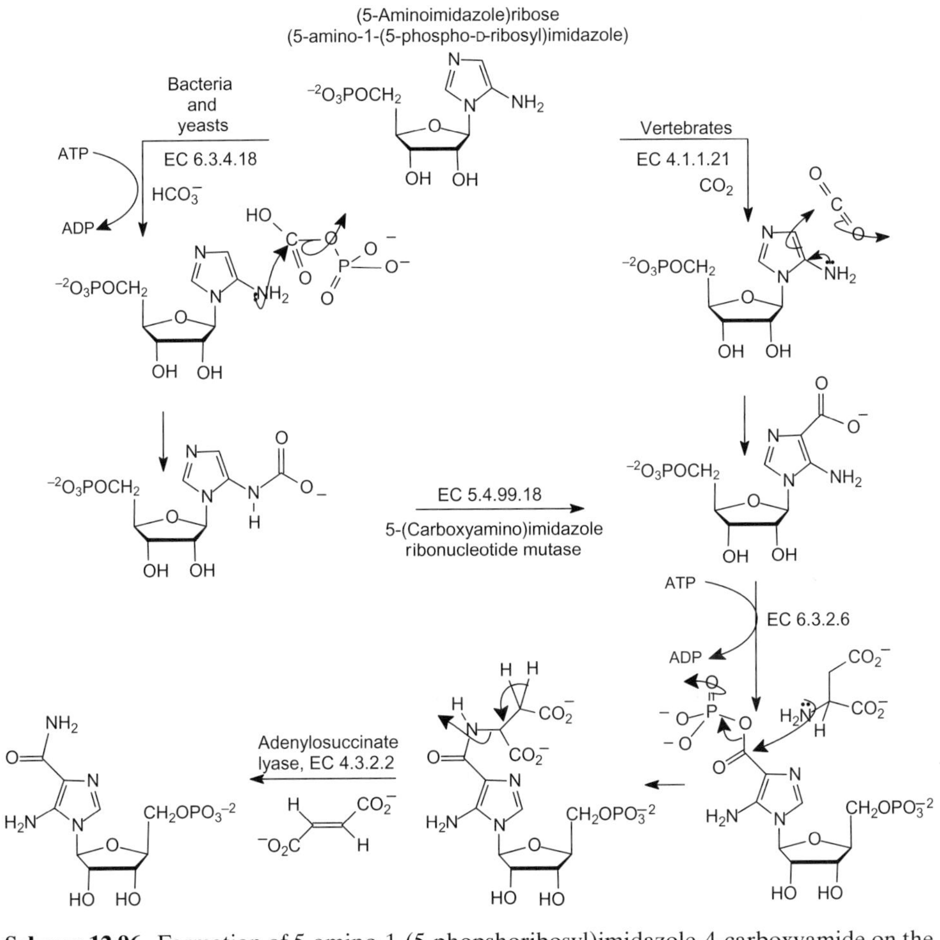

Scheme 12.96. Formation of 5-amino-1-(5-phopshoribosyl)imidazole-4-carboxyamide on the way to inosine and thence to adenosine. EC numbers and some graphic materials provided in this scheme have been taken from appropriate links in a URL starting with http://www.chem.qmul.ac.uk/iubmb/enzyme/.

carboxylase (EC 4.1.1.21) to produce the corresponding 5-aminoimidazole ribotide-4-carboxylate [5-amino-1-(5″-phosphoribosyl)imidazole-4-carboxylate]. In contrast, in bacteria and yeast, it appears that N-carboxylation using bicarbonate occurs first (5-(carboxyamino) imidazole ribonucleotide synthase, EC 6.3.4.18) to produce 5-carboxyamino-1-(5′-phosphoriboxyl) imidazole and that this N-carboxy-derivative rearranges [5-(carboxyamino)imidazole ribonucleotide mutase, EC 5.2.99.18] *to the same 5-aminoimidazole ribotide-4-carboxylate* found in vertebrates.

Phosphorylation at the carboxylate (ATP → ADP) followed by reaction with aspartate (Asp, D) in the presence of phosphoribosylaminoimidazolesuccinocarbox-amide synthase (EC 6.3.2.6) produces the corresponding (S)-succinate derivative. This is followed by loss of fumarate (adenylosuccinate lyase, EC 4.3.2.2) to the corresponding amide (last seen in the biosynthesis of histidine (His, H) (Scheme 12.26). Then, as shown in Scheme 12.97, N-formylation (as in Scheme 12.87) on the C-5 amino group is effected with 10-CHO-H_4 folate (EC 2.1.2.3) to yield the N-formyl derivative (5-formamido-1-(5′-phosphoribosyl) imidazole-4-carboxamide). Loss of water accompanies cyclization (EC 3.5.2.10) to yield inosine 5′-phosphate (IMP).

The conversion of IMP to AMP requires amination at C-6 of the purine system, and the nitrogen for this process is derived from aspartate (Asp, D) (adenylosuccinate synthase, EC 6.3.4.4). It appears that the driving force for the loss of water in this process that yields N^6-1,2-dicarboxyl adenosine monophosphate (adenosine 5′-phosphate, AMP) is the conversion of guanosine triphosphate (GTP) to guanosine diphosphate (GDP). Adenosine monophosphate (adenosine 5′-phosphate, AMP) is formed from the N^6-derivative by loss of fumarate (catalyzed by adenylosuccinate lyase, EC 4.3.2.2).

The synthesis of adenosine can be conveniently divided into two parts. The first part is the synthesis of the purine system, while the second part is the attachment of the ribose to the appropriate position of the former. An interesting description of the issue of the phosphorylation (needed for the synthesis of adenosine mono-, di-, and triphosphates) is provided by Sir Alexander Todd in his lecture (December 11, 1957) on the occasion of his acceptance of the Nobel Prize and, in principle, might be considered a third part of the overall synthesis problem.

It is widely held that the early work on purine synthesis was unified and codified by W. Traube in a reaction that bears his name. As outlined in Scheme 12.98, a typical Traube synthesis* begins with either urea or thiourea and a condensation reaction with a malonic acid derivative. For adenine itself (Scheme 12.98), the work began with thiourea. Thus, heating an ethanol solution of thiourea with sodium metal and malononitrile produced 4,6-diamino-2-thiopyrimidine. Nitrosation with nitrous acid followed by reduction of the nitroso group with ammonium sulfide yielded 4,5,6-triamino-2-thiopyrimidine. When the triamine was warmed with formic acid, cyclization with loss of water occurred and 6-amino-2-thiopurine was isolated as its potassium salt. Then, warming the salt with a solution of 20% sulfuric acid containing 3% hydrogen peroxide produced adenine itself. Interestingly, when the same potassium salt was warmed with nitric acid, hypoxanthine (6-hydroxypurine) was generated.

There are several other syntheses that are noteworthy. Todd[†] and coworkers found that formamidine (formed by treating HCN in ethanol with HCl to produce

---

*See (a) Traub, W. *Chem. Ber.*, **1900**, *33*, 1371, 3035; (b) Traub, W. *Liebigs Ann. Chem.*, **1904**, *331*, 64.
[†]Baddiley, J.; Lythgoe, B.; Todd, A. R. *J. Chem. Soc.*, **1943**, 386.

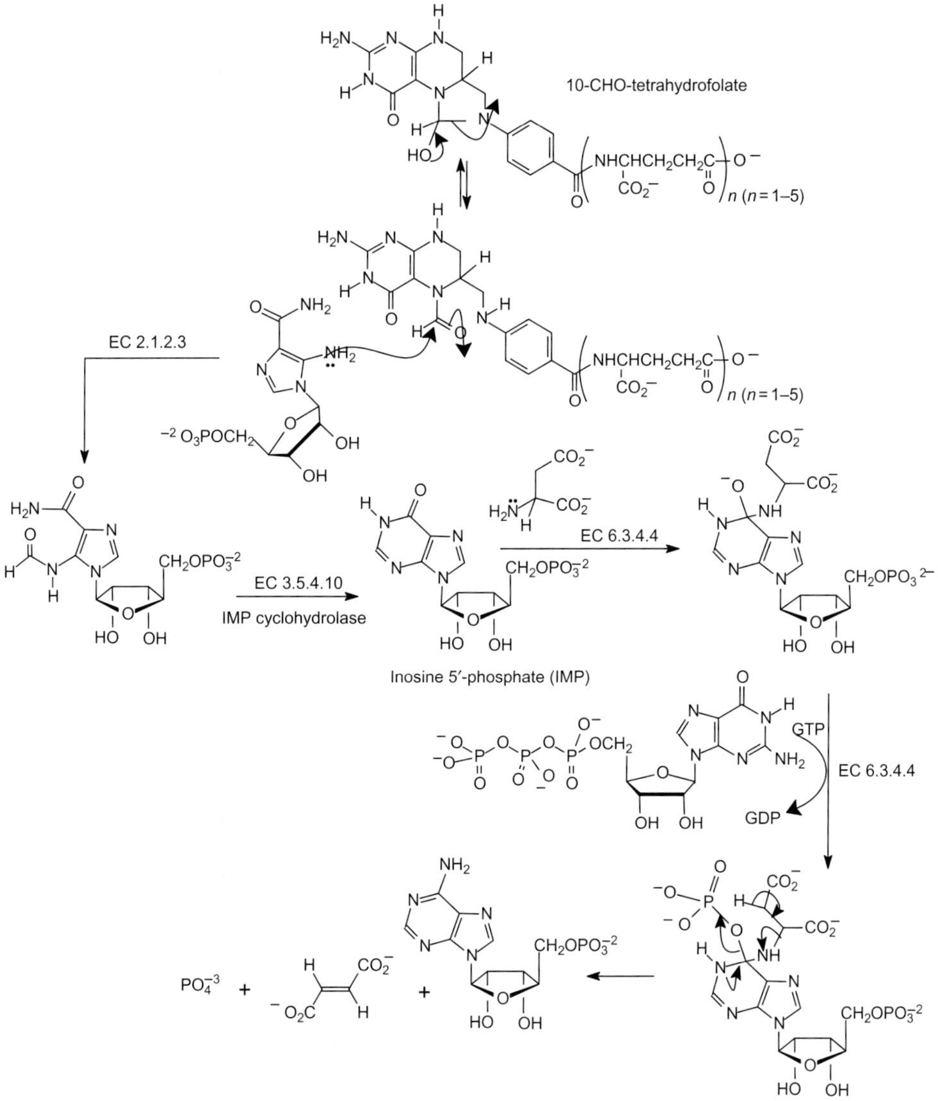

Scheme 12.97. Formation of adenosine phosphate from inosine. EC numbers and some graphic materials provided in this scheme have been taken from appropriate links in a URL starting with http://www.chem.qmul.ac.uk/iubmb/enzyme/.

the ethoxyimine followed by aminolysis) reacted with benzoazomalononitrile (formed by treating malonic dinitrile with diazotized aniline) to produce 4,6-diamino-5-benzeneazopyrimidine, which could be reduced with hydrogen (100°C, 6 atm) in the presence of Raney nickel to 4,5,6-triaminopyrimidine (Scheme 12.99). When the pyrimidine was treated with the sodium salt of dithioformic acid, 4,6-diamino-5-thioforamidopyrimidine formed and simply heating the latter resulted in cyclization, with loss of hydrogen sulfide (H_2S), to adenine.

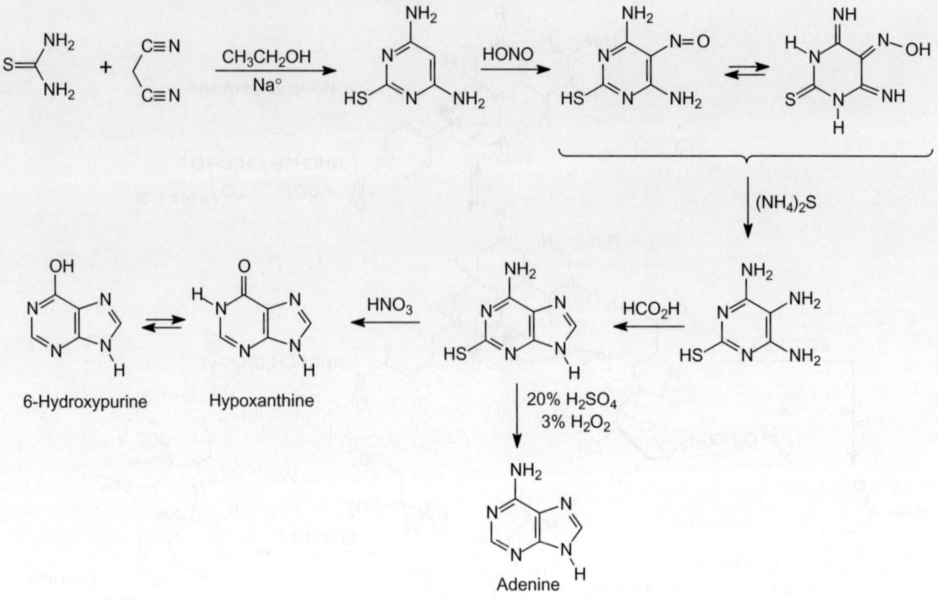

6-Hydroxypurine   Hypoxanthine

Adenine

Scheme 12.98. A representation of an early synthesis of adenine (and hydroxanthine) (after Traub, W. *Liebigs Ann. Chem.*, **1904**, *331*, 64).

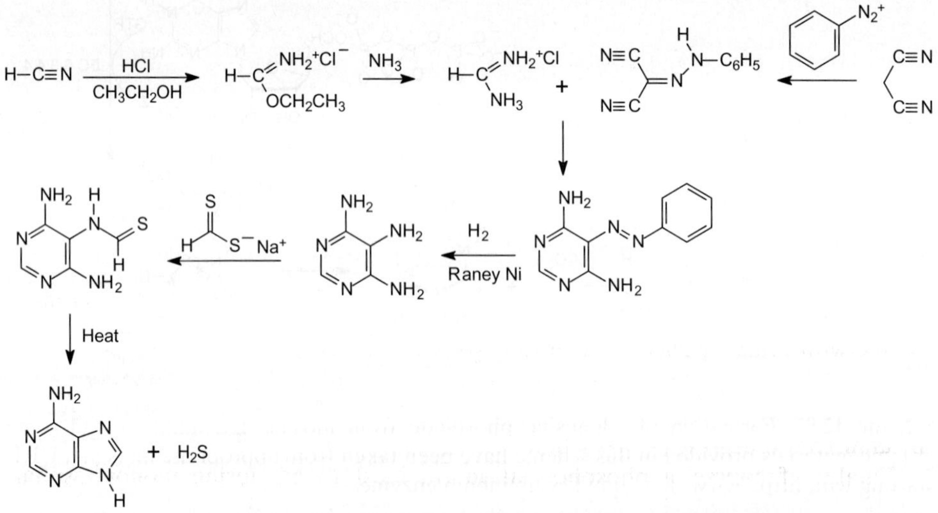

Scheme 12.99. A representation of an efficient synthesis of adenine after Baddiley, J.; Lythgoe, B.; Todd, A.R. *J. Chem. Soc.*, **1943**, 386.

As impressive as this relatively simple synthesis was, it was subsequently found* that, as shown in Equation 12.15, simply heating formamide ($HCONH_2$) with phosphorous oxychloride ($POCl_3$) in a sealed tube yielded adenine (43.5%, based on formamide).

*Ochiai, M.; Marumoto, R.; Kobayashi, S.; Shimazu, H.; Morita, K. *Tetrahedron*, **1968**, *24*, 5731.

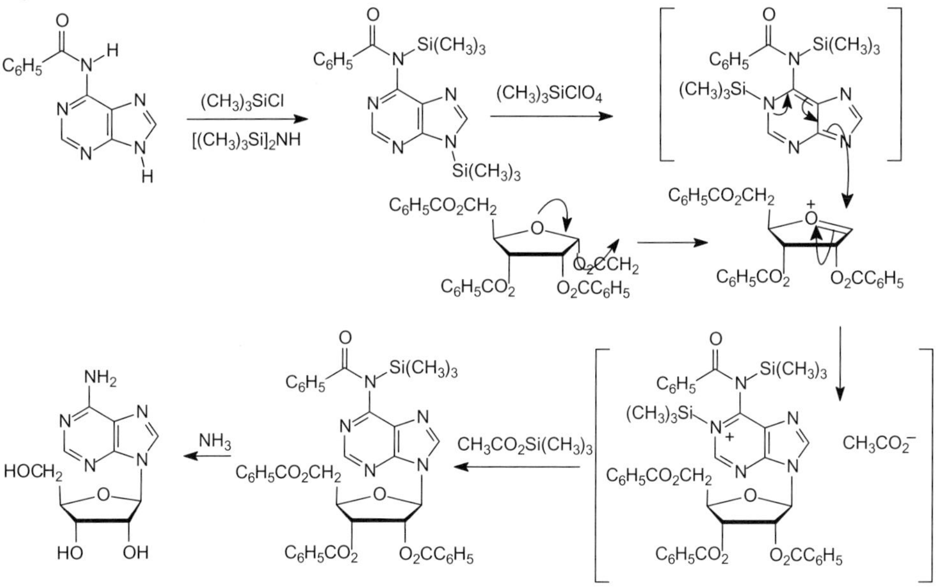

Scheme 12.100. A synthesis of adenosine from adenine (after Vorbrüggen, H.; Kroliklewicz, K.; Bennua, B. *Chem. Ber.*, **1981**, *114*, 1234). A version of the Hilbert–Johnson reaction.

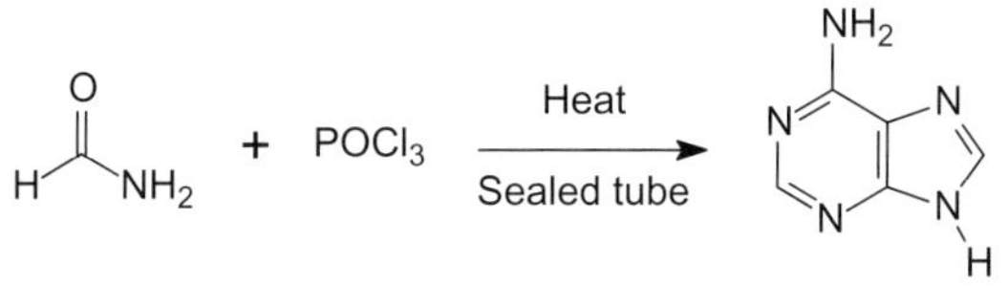

$$\text{HCONH}_2 + POCl_3 \xrightarrow[\text{Sealed tube}]{\text{Heat}} \text{adenine} \qquad (12.15)$$

One of the better processes for the conversion of adenine to adenosine involves the initial conversion of adenine to the corresponding N^6-benzoyl derivative with benzoyl chloride in pyridine.[*] Then, as shown in Scheme 12.100, silylation with trimethylchlorosilane in hexamethyldisilazane $\{[(CH_3)_3Si]_2NH\}$ solvent at reflux yielded a silyl derivative that was heated in 1,2-dichloroethane $(ClCH_2CH_2Cl)$ at reflux with trimethylsilyl perchlorate $[(CH_3)_3SiClO_4]$ and 1-O-acetyl-2,3,5-tri-O-benzoyl-β-D-ribofuranose for 12 h. The resulting crude, protected adenosine was converted to unprotected material by standing at room temperature with methanolic ammonia followed by recrystallization from aqueous methanol.

Finally, of course, a phosphorylation (at least to adenosine monophosphate, although the di- and triphosphates have also been prepared) is required. As noted earlier, the issue of phosphorylation was discussed by A. R. Todd on his receipt of the Nobel Prize in Chemistry in 1957, but the initial work on the preparation of adenosine mono- and diphosphates was reported in 1947[†] and the triphosphate in 1949.[‡]

In the event, the ribosyl unit of adenosine was protected as its acetone ketal by the treatment of adenosine with acetone and zinc chloride. The resulting

[*]Vorbrüggen, H.; Kroliklewicz, K.; Bennua, B. *Chem. Ber.*, **1981**, *114*, 1234.
[†]Baddiley, J.; Todd, A. R. *J. Chem. Soc.*, **1947**, 648.
[‡]Baddiley, J.; Michelson, A. M.; Todd, A. R. *J. Chem. Soc.*, **1949**, 582.

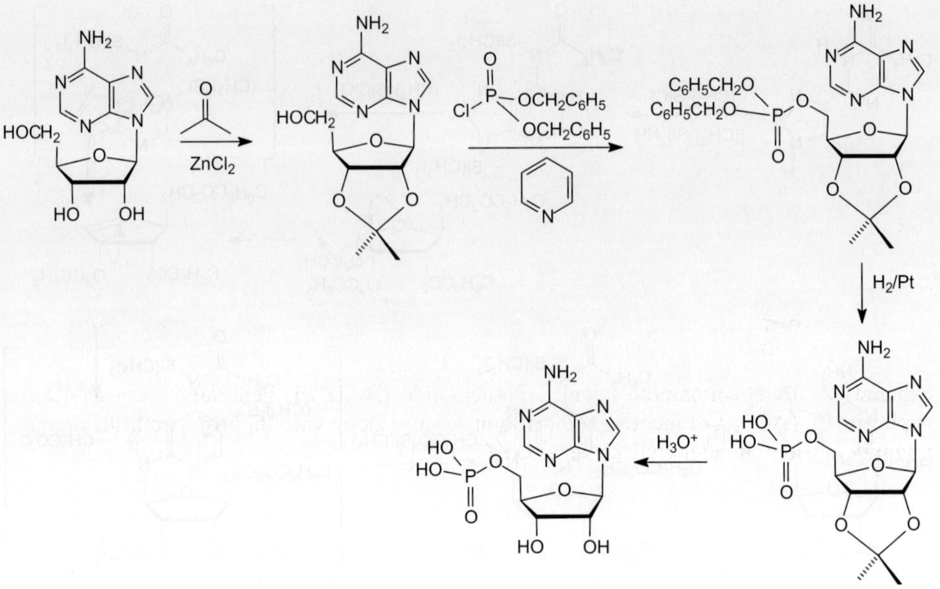

Scheme 12.101. A representation of the synthesis of adenosine monophosphate (after Baddiley, J.; Todd, A. R. *J. Chem. Soc.*, **1947**, 648 and Baddiley, J.; Michelson, A. M.; Todd, A. R. *J. Chem. Soc.*, **1949**, 582).

2′,3′-isopropylidene adenosine reacted with dibenzyl chlorophosphonate in the presence of pyridine to yield the corresponding 2′,3′-isopropylidene adenosine-5′-dibenzylphosphate. The benzyl groups were removed by hydrogenolysis and "mild" acid treatment opened the ketal to yield adenosine-5′-phosphate (Scheme 12.101). Then, in a modification of this procedure, the 2′,3′-isopropylidene adenosine-5′-dibenzylphosphate was carefully hydrolyzed to yield a monobenzylphosphate derivative, adenosine 5′-benzylphosphate, which could be phosphorylated again to produce what was presumed to be adenosine 5′-tribenzyldiphosphate (since on hydrogenation in alcohol, adenosine 5′-diphosphate resulted). The ATP was synthesized by partial hydrolysis of the tribenzyl diphosphate and subsequent phosphorylation and hydrogenolysis.

VI. NAD⁺

The dinucleotide NAD⁺ (and NADH, NAD(P)⁺, and NAD(P)H) is composed of ADP and a second ribosyl unit to which a nicotinamide is appended at the anomeric carbon. The biosynthesis of adenosine monophosphate (adenosine 5′-phosphate, AMP) was outlined above, and AMP can be converted to ADP by the action of adenylate kinase (EC 2.7.4.3). The enzymes called ATP synthases (EC 3.6.3.14 and 3.6.3.15) are complicated membrane- and nonmembrane-associated species that convert the energy associated with proton electrochemical gradients into ATP from ADP and inorganic phosphate (P_i) (Chapter 14).

NAD⁺ is generated from ATP and nicotinamide ribonucleotide (Scheme 12.102) along with diphosphate by utilizing nicotinamide nucleotide adenylyltransferase

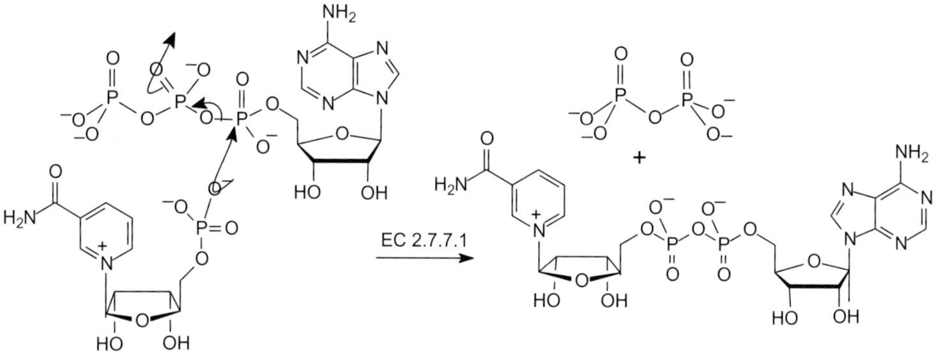

Scheme 12.102. Nicotinamide adenine dinucleotide (NAD$^+$) is generated from adenosine triphosphate (ATP) and nicotinamide ribonucleotide along with diphosphate utilizing nicotinamide nucleotide adenylyltransferase (EC 2.7.7.1).

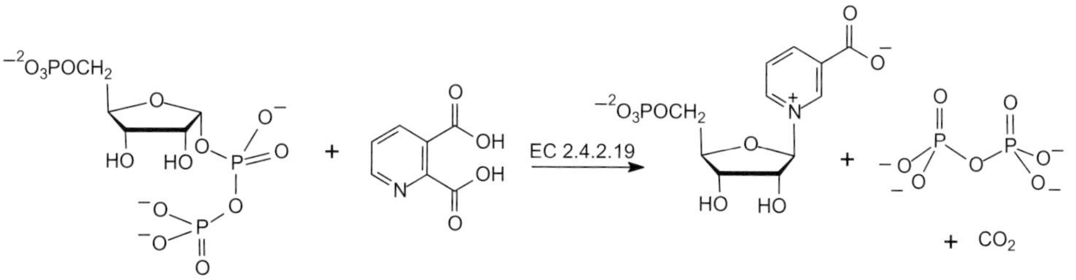

5-Phospho-α-D-ribose-
1-diphosphate

Quinolinic acid
(pyridine 1,3-dicarboxylic acid)

Nicotinate
ribonucleotide

Scheme 12.103. Nicotinate nucleotide formed by the phosphoribosyltransferase quinolinate phosphoribosyltransferase (EC 2.4.2.19).

(EC 2.7.7.1). The nicotinamide nucleotide itself can be produced with a phosphoribosyltransferase such as the quinolinate phosphoribosyltransferase (EC 2.4.2.19) (Scheme 12.103) to nicotinate ribonucleoide, which is subsequently converted to the amide. Other transferases, for example, EC 2.4.2.1, purinenucleoside phosphorylase, have also been implicated in the same process. It is generally assumed that phosphorylation of the carboxylate precedes amination and that a suitable source of nitrogen is found. Details remain unavailable at this writing.

So, only the biosynthesis of quinolinate itself is required to complete the biosynthesis of NAD$^+$.

Quinolinic acid is generated by one of two different pathways. In one pathway, by utilizing the enzyme quinolinate synthase (EC 1.4.3.16), whose crystal structure has recently been obtained,* only the broadest outlines of a possible pathway arising from a reaction between dihydroxyacetone monophosphate and iminoaspartate have been adumbrated (Scheme 12.104),

---

*Sakuraba, H.; Tsuge, H.; Yoneda, K.; Katumuma, N.; Ohshima, T. *J. Biol. Chem.*, **2005**, *280*, 26645.

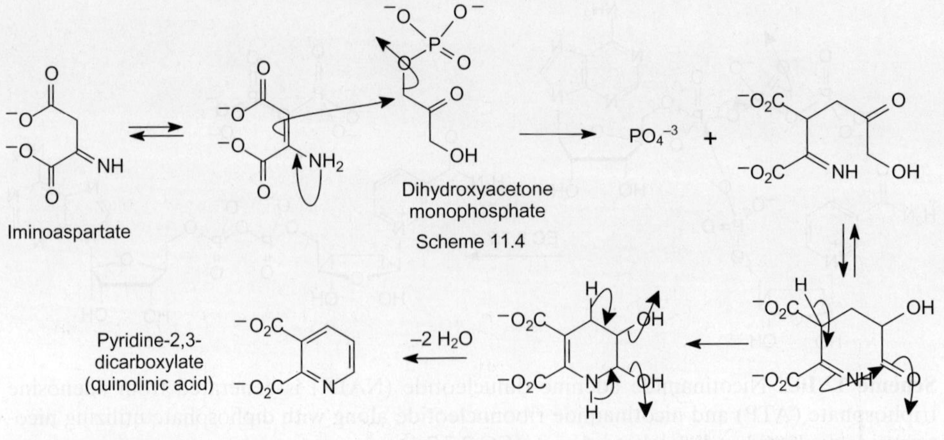

Scheme 12.104. A proposed pathway from aspartate (Asp, D) and dihydroxyaetone monophosphate to pyridine-2,3-dicarboxylate (quinolinic acid) (after Sakuraba, H.; Tsuge, H.; Yoneda, K.; Katumuma, N.; Ohshima, T. *J. Biol. Chem.*, **2005**, *280*, 26645).

In the other pathway* (the "eukaryotic pathway"), tryptophan (Trp, W) is catabolized to quinolinic acid (Scheme 12.105). Here, tryptophan (Trp, W) undergoes oxidation (O_2) in the presence of tryptophan-2,3-dioxygenase (EC 1.13.11.11) to produce N-formyl-L-kynurenine, which, in turn, is hydrolyzed with the loss of formate to yield L-kynurenine (arylformamidase, EC 3.5.1.9). A second oxidation (kynurenine 3-monooxygenase, EC 1.14.13.9) produces 3-hydroxykynurenine, which, with kynureninase (EC 3.7.1.3), undergoes cleavage to alanine and 3-hydroxyanthranilate. Then, 3-hydroxyanthranilate, with 3-hydroxyanthranilate-3,4-dioxygenase (EC 1.13.11.6) and oxygen (O_2), undergoes ring cleavage to 2-amino-3-carboxymuconate semialdehyde, which spontaneously cyclizes to quinolinic acid (Scheme 12.105).

At this writing, both quinoline (obtained from coal tar) and quinolininc acid are materials of commerce. Nonetheless, both have been synthesized and production of the latter from the former is common.

As shown in Equation 12.16, the classical Skraup[†] reaction that entails the treatment of aryl amines, such as aniline, with glycerol (1,2,3-propanetriol) and sulfuric acid (H_2SO_4) in nitrobenzene solvent readily produces quinoline (or substituted quinolines). Details of the pathway remain unclear. Then, oxidation of the aromatic ring can be effected to produce quinolinic acid. The oxidation has been reported to be successful utilizing permanganate (MnO_4^-) or nitric acid ($HONO_2$) as oxidizing agents, but, for some time, the method of Stix and Bulgatsh,[‡] which makes use of hydrogen peroxide (H_2O_2) and metal salts, for example, Cu(II), has been preferred (Equation 12.17).

*(a) Begley, T. P.; Kinsland, C.; Mehl, R. A.; Osterman, A.; Dorrestein, P. *Vitam. Horm.*, **2001**, *61*, 103; (b) Kurnasov, O.; Goral, V.; Colabroy, K.; Gerdes, S.; Anantha, S.; Osterman, A.; Begley, T. P. *Chem. Biol.*, **2003**, *10*, 1195; (c) Begley, T. P. *Nat. Prod. Rep.*, **2006**, *23*, 15; (d) Webb, M. E.; Marquet, A.; Mendel, R. R.; Rébeillé, F.; Smith, A. G. *Nat. Prod. Rep.*, **2007**, *24*, 988.
†(a) Skraup, Z. H. *Chem. Ber.*, **1880**, *13*, 2086; (b) Clarke, H. T.; Davis, A. W., *Organic Synthesis*, Coll. Vol. 1, Wiley, New York, 1941, p. 478.
‡Stix, W.; Bulgatsh, S. A. *Chem. Ber.*, **1832**, *65B*, 11.

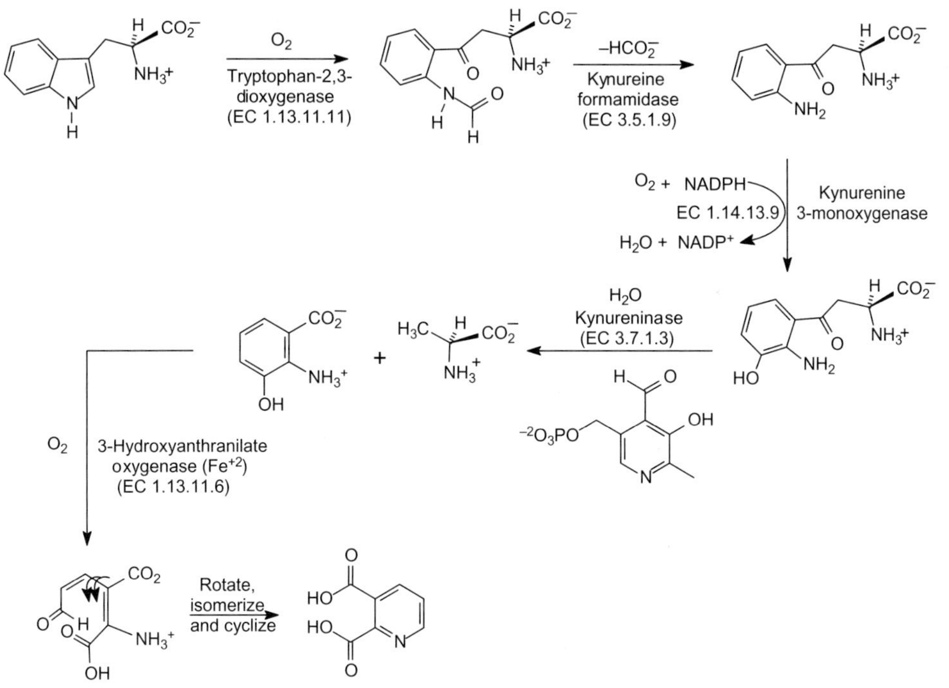

Scheme 12.105. A path from tryptophan (Trp, W) to 2-amino-3-carboxymuconate semialdehyde, which spontaneously cyclizes to quinolinic acid (Begley, T. P. *Nat. Prod. Rep.*, **2006**, *23*, 15).

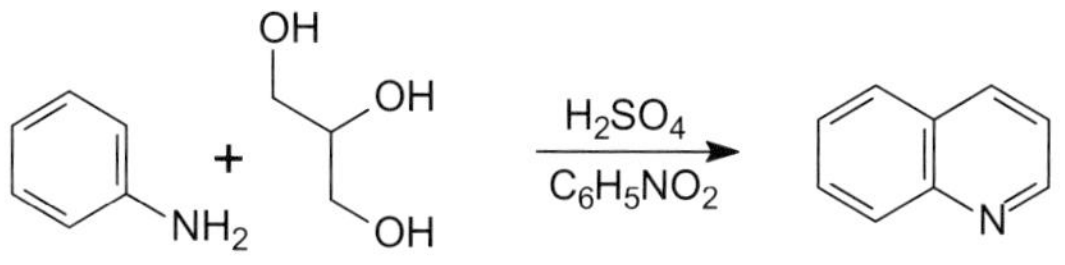

$$(12.16)$$

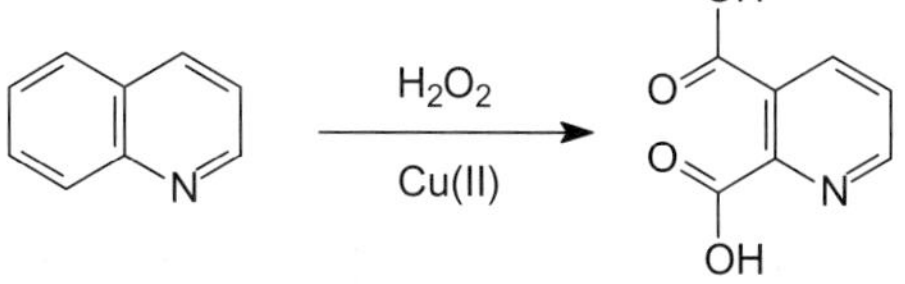

$$(12.17)$$

Finally, as will be seen later (Chapter 13), nicotinic acid is readily obtained as described in an early study* by the oxidation of nicotine (Equation 12.18) with nitric acid.

$$(12.18)$$

*Weidel, H. *Liebigs Ann. Chem.*, **1873**, *165*, 349.

VII. Coenzyme A (CoA-SH)

Coenzyme A (CoA-SH) is visualized here as an adenosine unit attached to an elaborated (i.e., an additional amino acid, cysteine [Cys, C], participates) pantothenic acid side chain.

As shown in Scheme 12.106, deamination (by transamination using 2-oxoglutarate aminotransferase, EC 2.6.1.42, an enzyme requiring pyridoxal) of valine (Val, V) leads to 3-methyl-2-oxobutanoate (e.g., see Scheme 12.17). Introduction of a hydroxymethyl group from 5,10-methylenetretahydrofolate (3-methyl-2-oxobutanoate hydroxymethyltransferase, EC 2.1.2.11) in an aldol-like process produces 2-dehydropantoate (4-hydroxy-3,3,-dimethyl-2-oxobutanoate). Stereospecific

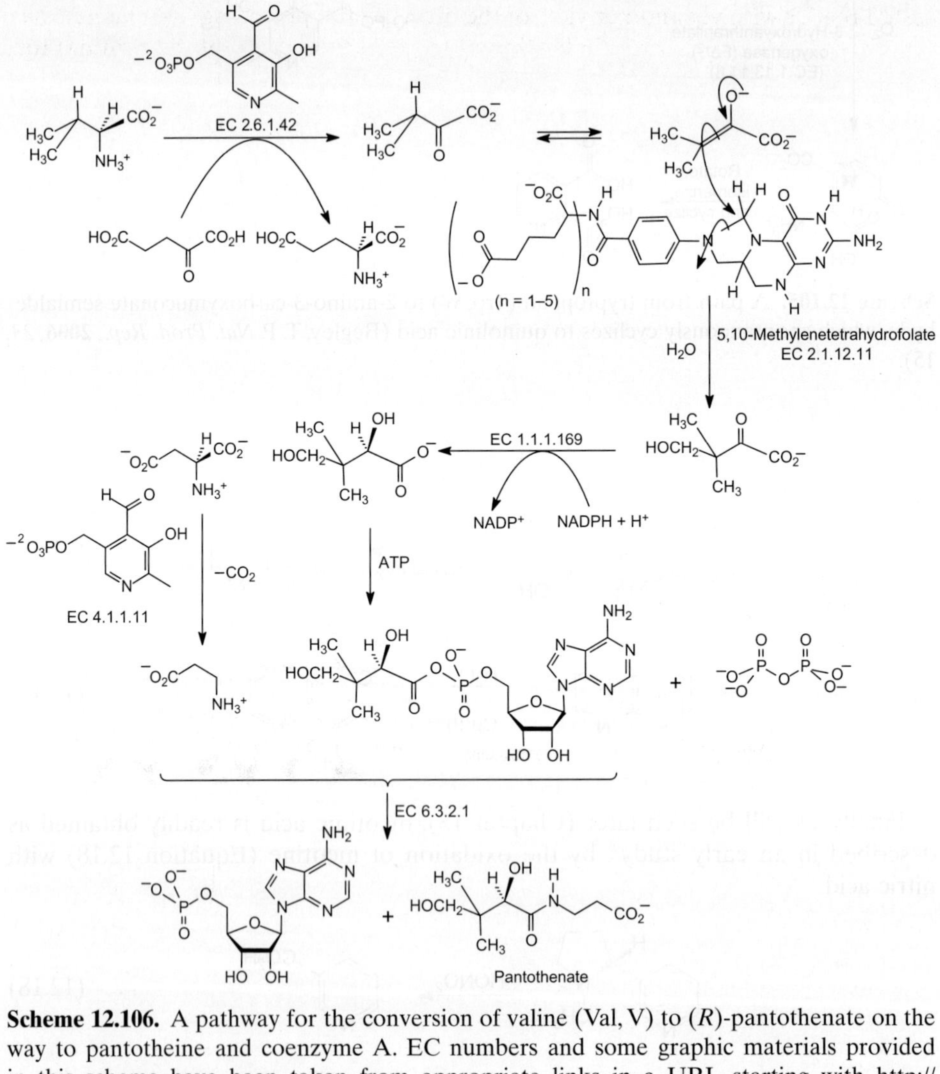

Scheme 12.106. A pathway for the conversion of valine (Val, V) to (*R*)-pantothenate on the way to pantotheine and coenzyme A. EC numbers and some graphic materials provided in this scheme have been taken from appropriate links in a URL starting with http://www.chem.qmul.ac.uk/iubmb/enzyme/.

reduction of the α-keto group (2-dehydropantoate 2-reductase, EC 1.1.1.169) yields (*R*)-pantoate. Activation of the carboxylate by esterification with ATP (with the formation of inorganic phosphate) is then followed by amide formation using the amino group of β-alanine (from the pyridoxal-catalyzed decarboxylation of aspartate [Asp, D] with aspartate 1-decarboxylase, EC 4.1.1.11) and the enzyme pantoate-β-alanine ligase (EC 6.3.2.1) with the formation of pantothenate. Then, as shown in Scheme 12.107, the path to coenzyme A is clear.

The primary hydroxyl group of pantothenate is phosphorylated (ATP → ADP, EC 2.7.1.33) and then the carboxylate of pantothenate is activated for displacement by conversion to the monophosphate ester of cytosine (from CTP with the

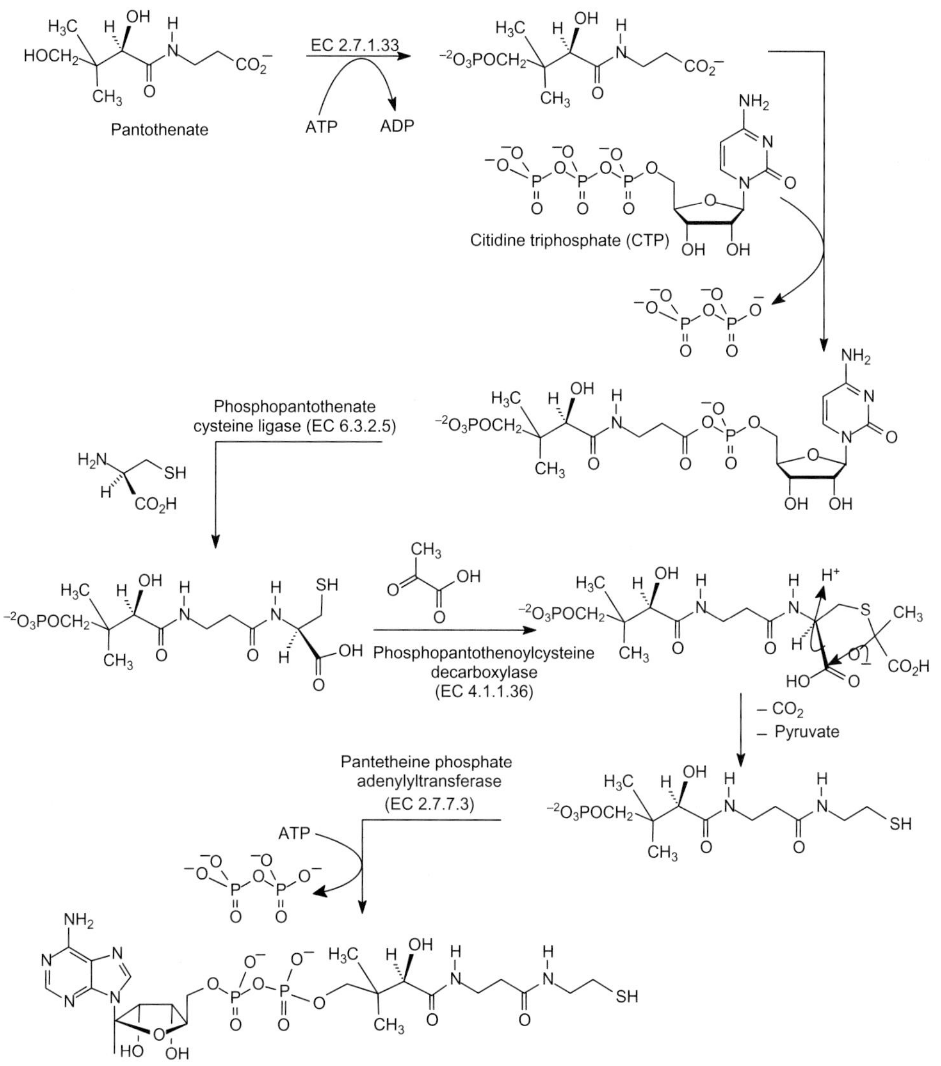

Scheme 12.107. A pathway for the conversion of (*R*)-pantothenate on the way to coenzyme A. EC numbers and some graphic materials provided in this scheme have been taken from appropriate links in a URL starting with http://www.chem.qmul.ac.uk/iubmb/enzyme/.

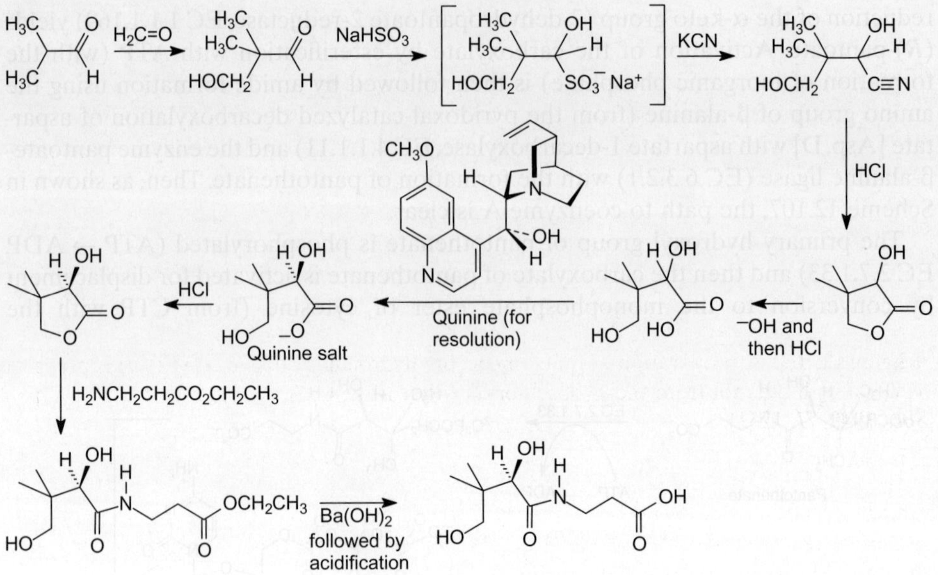

Scheme 12.108. A synthesis of pantothenic acid (after Stiller, E. T.; Stanton, A. H.; Finkelsetin, J.; Keresztesy, J. C.; Folkers, K. *J. Am. Chem. Soc.*, **1940**, *62*, 1785).

simultaneous production of inorganic phosphate). Displacement of cytosine monophosphate by L-cysteine (Cys, C) produces N-[(R)-4-phosphopantothenoyl]–L-cysteine. It appears that both the carboxylate activation and the subsequent displacement are catalyzed by the same enzyme, phosphopantothenate cysteine ligase (EC 6.3.2.5).

Decarboxylation of the newly added cysteine (Cys, C) residue utilizing pyruvate (CH_3COCO_2H), a cofactor with phosphopantothenoylcysteine decarboxylase (EC 4.1.1.36), generates carbon dioxide (CO_2) and pantotheine phosphate. Finally, coenzyme A is formed by the reaction of the pantotheine phosphate with ATP in the presence of pantotheine phosphate adenylyltransferase (EC 2.7.7.3). Inorganic diphosphate is formed at the same time (Scheme 12.107).

The synthesis of coenzyme A can be divided into the synthesis of adenosine (adumbrated above) and the synthesis of pantotheine phosphate. The latter is then conveniently divided into the preparation of pantothenic acid and its subsequent conversion to the desired (pantetheine) thiol.

Although there are a number of syntheses of pantothenic acid, that of Folkers et al.* is among the earliest. As shown in Scheme 12.108, a clean synthesis of α-hydroxy-β,β-dimethyl-γ-butyrolactone was needed. First, 2-methylpropanal (isobutyraldehyde) underwent an aldol condensation with formaldehyde (formalin, $H_2C=O$) to produce α,α-dimethyl-β-hydroxypropanal. Then, the bisulfite adduct of this aldehyde underwent reaction with potassium cyanide (KCN) and the separated cyanohydrin was hydrolyzed with concentrated hydrochloric acid (HCl), and the product was carefully made basic. The racemic α-hydroxy-β,β-dimethyl-γ-butyrolactone was separated, while the lactone underwent ring opening in the

---

*Stiller, E. T.; Stanton, A. H.; Finkelsetin, J.; Keresztesy, J. C.; Folkers, K. *J. Am. Chem. Soc.*, **1940**, *62*, 1785.

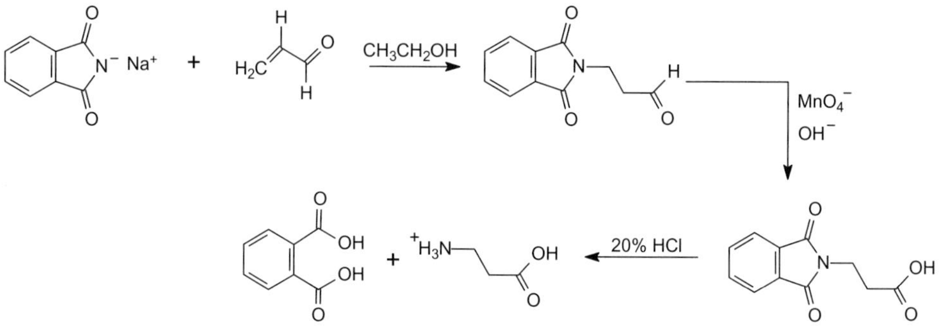

Scheme 12.109. A preparation of β-alanine as the hydrochloride. The ethyl ester is prepared as usual with ethanol from the hydrochloride (after Moe, O. A.; Warner, D. T. *J. Am. Chem. Soc.*, **1949**, *71*, 1251).

presence of excess sodium hydroxide. Then, neutralization of the excess base with dilute acid and resolution with quinine was effected.

Finally, reaction of the butyrolactone with the ethyl ester of β-alanine* $H_2NCH_2CH_2CO_2CH_2CH_3$), a preparation of which is shown in Scheme 12.109, led to the preparation of the ethyl ester of pantothenic acid. The acid itself formed on the basic hydrolysis of the ester.

Now, pantetheine itself (Scheme 12.110) was synthesized from the lactone; that is, as shown in Scheme 12.110, when the easily oxidized thiol *N*-β-alanyl-2-mercaptoethylamine reacted with the chiral pantolactone in the absence of solvent at 100°C, pantetheine formed. The formation of the thiol *N*-β-alanyl-2-mercaptoethylamine and the reaction to form pantetheine[†] began with the reaction of 2-bromoethylamine hydrobromide with benzylthiol.

The resulting 2-benzylthioethylamine could be debenzylated by treatment with sodium in liquid ammonia. However, when 2-benzylthioethylamine was treated with carbobenzyloxy-β-alanine azide (prepared from carbobenzyloxy-β-alanylhydrazide by nitrosation), 2-benzylthio-*N*-(carbobenzyloxy-β-alanyl)ethylamine formed. Reduction with sodium in liquid ammonia was sufficient to remove both the benzyl and carbobenzyloxy protecting groups and, as noted above, reaction with pantolactone yielded pantetheine. Phosphorylation to the mono- and diphosphates of pantetheine has been effected with the corresponding dibenzylphosphonates (*vide supra*, ATP).

VIII. FAD

For the FAD cofactor composed of two nucleotides, only one (adenine) has already been described. The flavin portion is one of current interest and of some excitement since "the formation of the isoalloxazine ring ... is (considered to be) one of the most remarkable reactions in cofactor biosynthesis."[‡]

*Moe, O. A.; Warner, D. T. *J. Am. Chem. Soc.*, **1949**, *71*, 1251.
[†]Baddiley, J.; Thain, E. M. *J. Chem. Soc.*, **1952**, 800.
[‡]Webb, M. E.; Marquet, A.; Mendel, R. R.; Rébeillé, F.; Smith, A. G. *Nat. Prod. Rep.*, **2007**, *24*, 988.

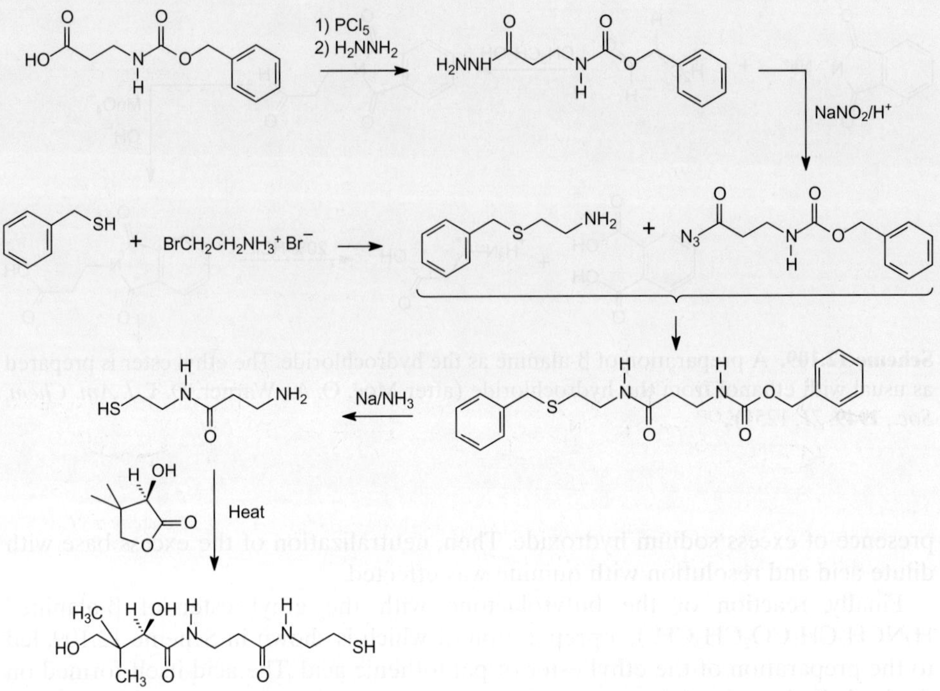

Scheme 12.110. A representation of the synthesis of pantetheine from pantothenic acid lactone (after Baddiley, J.; Thain, E. M. *J. Chem. Soc.*, **1952**, 800).

As shown in Scheme 12.111, the biosynthesis of flavin can be picked up with Schemes 12.87 and 12.97 (*vide supra*) in which the formation of adenosine monophosphate (adenosine 5′-monophosphate, AMP) from 5-amino-1-(5-phosphoribosyl)imidazole-4-carboxyamide via inosine (IMP) was outlined. Inosine monophosphate (inosine 5′-phosphate, IMP) is hydrated (H_2O) and then oxidized ($NAD^+ \rightarrow NADH$, IMP dehydrogenase, EC 1.1.1.205) forming xanthosine 5′-phosphate. Then, xanthosine 5′-phosphate is converted to guanosine 5′-phosphate. The oxygen is replaced via initial derivatization with ATP to produce the AMP derivative and inorganic phosphate. Attachment of ammonia (EC 6.3.4.1, xanthosine-5′-phosphate–ammonia ligase) or glutamine (Gln, Q) (EC 6.3.5.2, xanthosine 5′-phosphate amidotransferase) generates the guanosine 5′-phosphate.*

GTP, in the presence of GTP cyclohydrolase II (EC 3.5.4.25) and water (Equation 12.19), loses formate and produces the corresponding substituted tri-amine, 2,5-diamino-6-keto-4-(5-phosphoribosylamino) pyrimidine and diphosphate. Presumably, this is a "simple" hydrolysis.

---

*In the latter, hydrolysis of the amide to generate glutamic acid (Glu, E) from the glutamine (Gln, Q) also occurs.

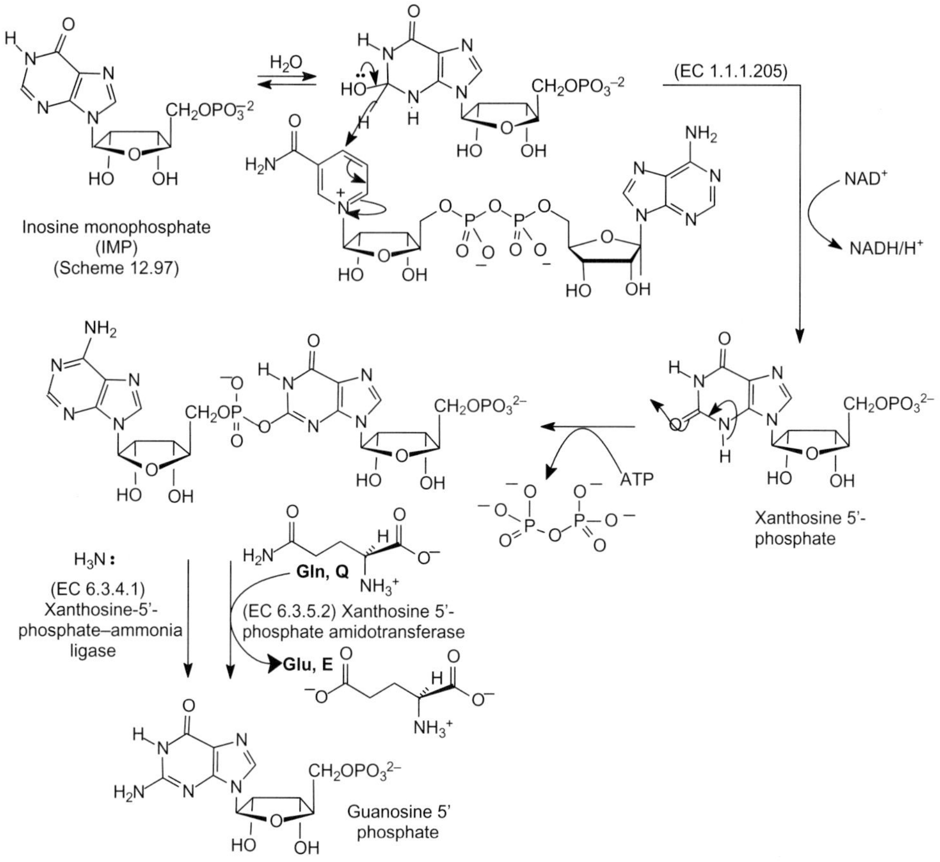

Scheme 12.111. A representation of the biosynthesis of guanosine 5′-phosphate from xanthine 5′-phosphate. EC numbers and some graphic materials provided in this scheme have been taken from appropriate links in a URL starting with http://www.chem.qmul.ac.uk/iubmb/enzyme/.

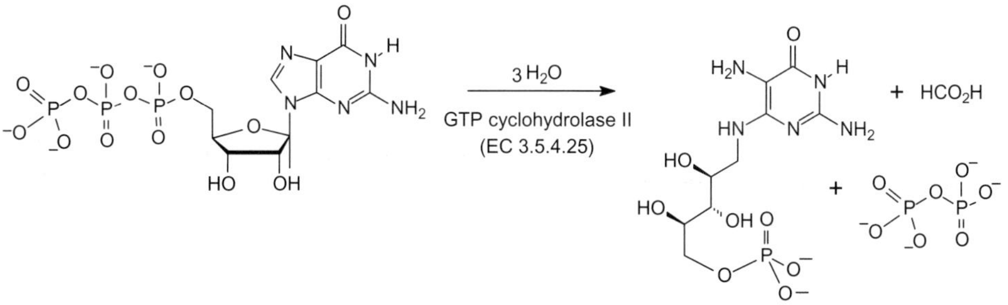

$$(12.19)$$

A second catalyzed hydrolysis then converts the 2-amino group to the corresponding keto function with the generation of ammonia. This is effected by the diaminoketophosphoribosylaminopyrimidine deaminase (EC 3.5.4.26) as shown in Equation 12.20.

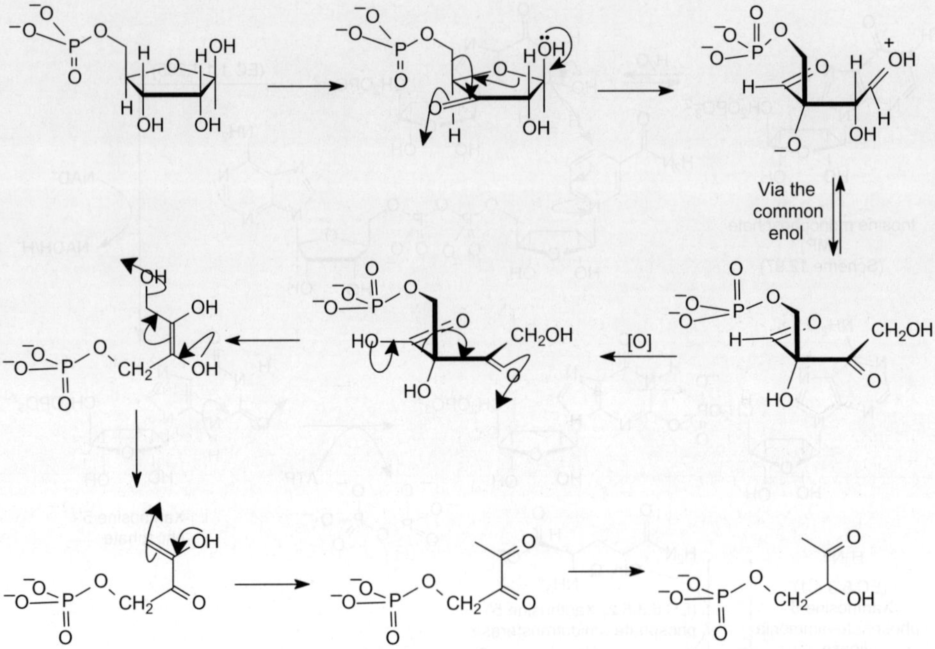

Scheme 12.112. A pathway to a four-carbon fragment derived from ribose-5-phosphate with the loss of that one carbon (C4) as suggested by labeling studies (after Nielsen, P.; Neuberger, G.; Fujii, I.; Brown, D. H.; Keller, P. J.; Floss, H. G.; Bacher, A. *J. Biol. Chem.*, **1986**, *261*, 3661).

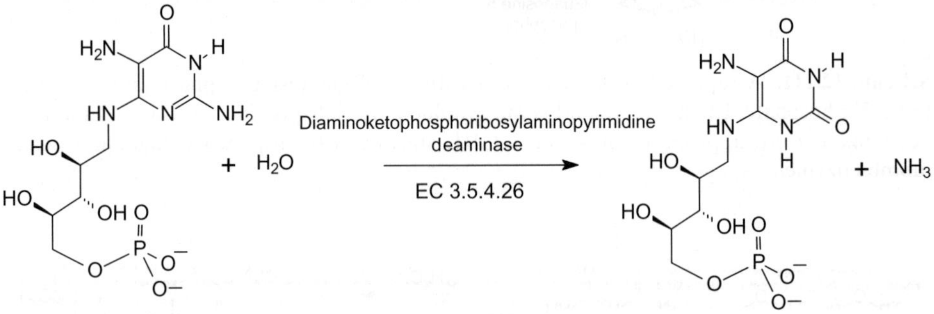

$$(12.20)$$

Now, it has been affirmed* that the 5-amino-6-ribitylamino-2,4(1H, 3H)-pyrimidinedione 5'-phosphate makes use of a four-carbon ribose-5-phosphate derived unit to produce 6,7-dimethyl-8-ribityllumazine (6,7-dimethyl-8-ribityllumazine synthase). Scheme 12.112 provides a pathway to a four-carbon fragment derived from ribose-5-phosphate and Scheme 12.113 provides one from the four-carbon fragment through a reaction with 5-amino-6-ribitylamino-2,4(1H,3H)-pyrimidinedione 5'-phosphate to produce 6,7- dimethyl-8-ribityllumazine (6,7-dimethyl-8-ribityllumazine synthase, EC 2.5.1.9).

*Nielsen, P.; Neuberger, G.; Fujii, I.; Brown, D. H.; Keller, P. J.; Floss, H. G.; Bacher, A. *J. Biol. Chem.*, **1986**, *261*, 3661.

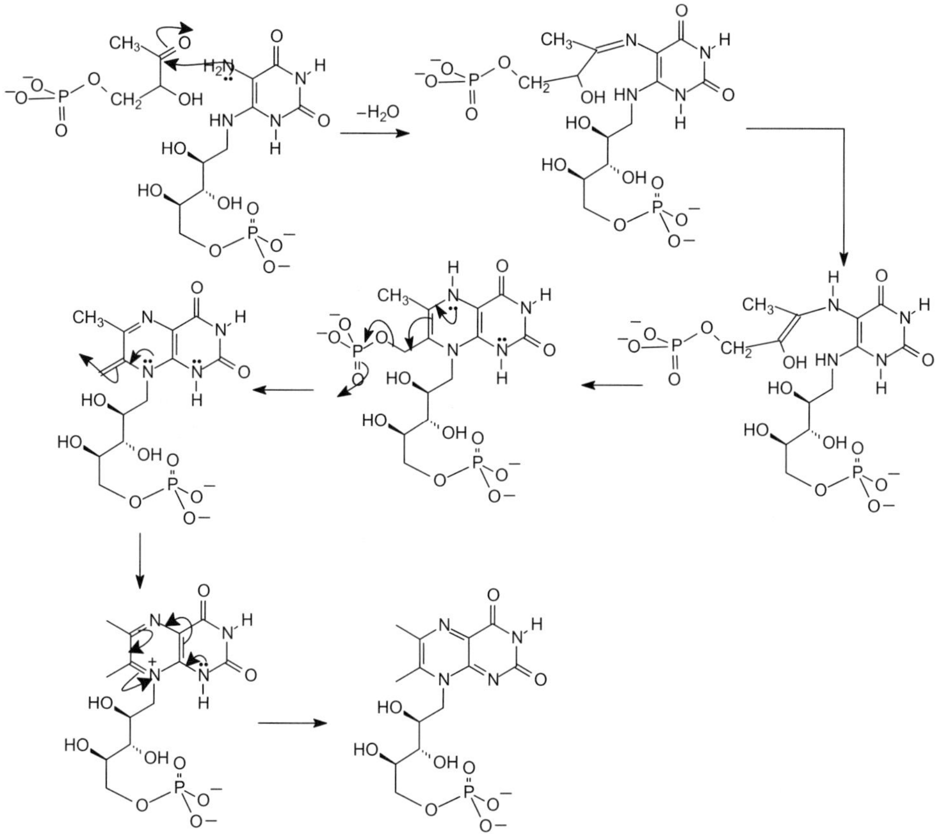

Scheme 12.113. A representation of a potential pathway to 6,7- dimethyl-8-ribityllumazine from a four-carbon ribose derivative and 5-amino-6-ribitylamino-2,4(1H,3H)-pyrimidinedione 5′-phosphate utilizing 6,7-dimethyl-8-ribityllumazine synthase (EC 2.5.1.9) (after Zylberman, V.; Klinke, S; Haase, I.; Bacher, A.; Fischer, M.; Goldbaum F. A. *J. Bacteriol.*, **2006**, *188*, 6135).

Then, in what has been called "one of the most remarkable reactions in cofactor biosynthesis," the isoalloxazine ring of riboflavin is produced by a "dismutation" of two 6,7-dimethyl-8-ribityllumazines. So, as shown in Scheme 12.114, the process removes four carbons of one of the 6,7-dimethyl-8-ribityllumazines and transfers them to the other, leaving a 5-amino-6-ribitylamino-2,4(1H,3H)-pyrimidinedione 5′-phosphate behind*—again ready for conversion to a new 6,7-dimethyl-8-ribityllumazine.

The synthesis of riboflavin begins (Scheme 12.115) with the formation of barbituric acid by the condensation of diethyl malonate [$(CH_3CH_2O_2C)_2CH_2$] with urea [$(NH_2)_2C=O$] in ethanolic sodium ethoxide.[†] Then, the barbituric acid can either (a) be oxidized directly to the hydrate of alloxan or (b) first be converted to the ben-zylidine adduct and then oxidized to the same compound. Both oxidations are

*Begley, T. P. *Nat. Prod. Rep.*, **2006**, *23*, 15.
[†]Dickey, J. B.; Gray, A. R. *Organic Synthesis*, Coll. Vol. 2, Wiley, New York, 1943, p. 60 ff.

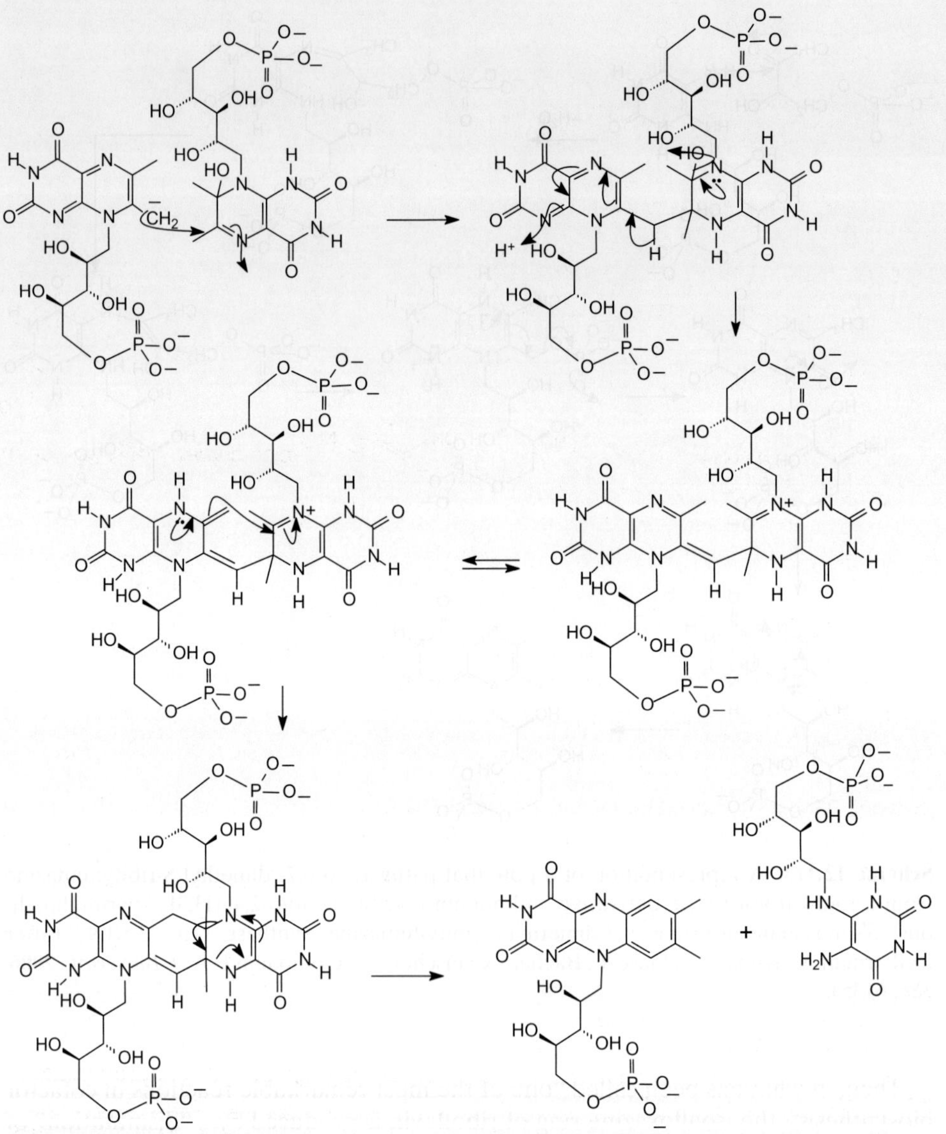

Scheme 12.114. A representation of a potential pathway to riboflavin and 5-amino-6-ribitylamino-2,4(1H,3H)-pyrimidinedione 5′-phosphate utilizing two 6,7-dimethyl-8-ribityllumazines and riboflavin synthase (EC 2.5.1.9) (after Begley, T. P. *Nat. Prod. Rep.*, **2006**, *23*, 15).

reported to be best carried out with chromic acid (CrO_3).* Interestingly, oxidation of uric acid with nitric acid is also reported to produce alloxan.† Then, nitration of *o*-xylene produces the corresponding mono- and then dinitroxylenes. Reduction of

***Method A**: Holmgren, A. V.; Wenner, W. *Organic Synthesis*, Coll. Vol. 4, Wiley, New York, 1952, p. 32 ff.;
Method B: Speer, J. H.; Dabovich, T. C. *Organic Synthesis*, Coll. Vol. 4, Wiley, New York, 1952, p. 37 ff.
†Santos, C. X. C.; Anjos, E. I.; Augosto, O. *Arch. Biochem. Biophys.*, **1999**, *372*, 285.

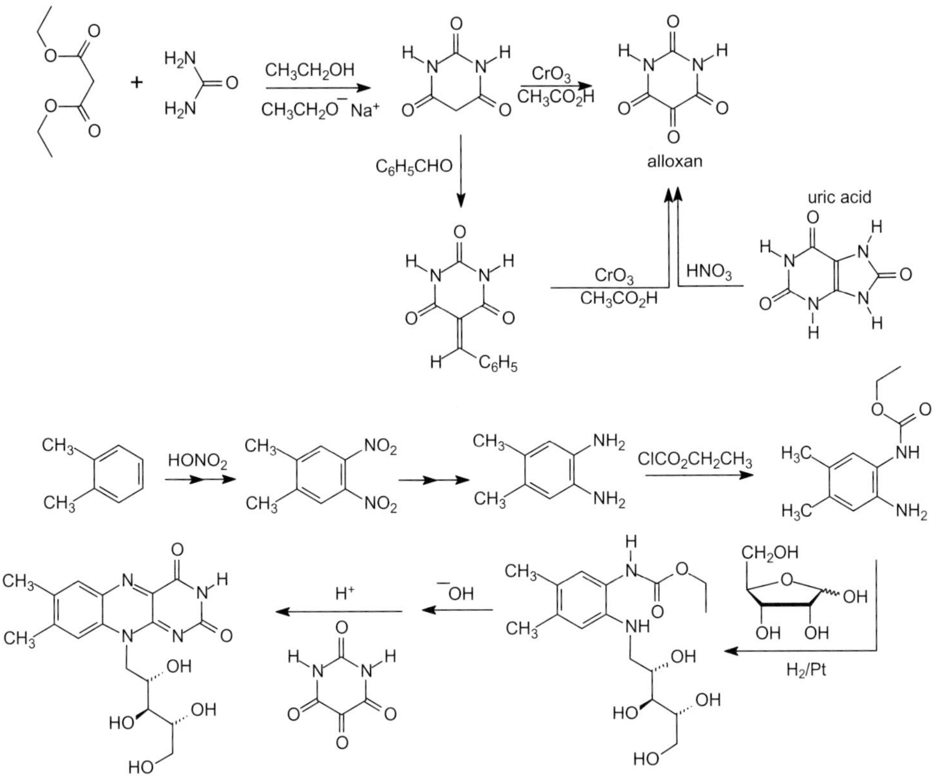

Scheme 12.115. A representation of the synthesis of riboflavin (after Dickey, J. B.; Gray, A. R. *Organic Synthesis*, Coll. Vol. 2, Wiley, New York, 1943, p. 60 ff.; Method A: Holmgren, A. V.; Wenner, W. *Organic Synthesis*, Coll. Vol. 4, Wiley, New York, 1952, p. 32 ff.; Method B. Speer, J. H.; Dabovich, T.C. *Organic Synthesis*, Coll. Vol. 4, Wiley, New York, 1952, p. 37 ff.; Santos, C. X. C.; Anjos, E. I.; Augosto, O. *Arch. Biochem. Biophys.*, **1999**, *372*, 285; Karrer, P. *Organic Chemistry*, Elsevier, New York, 1950, p. 730).

the latter produces the corresponding diamine and treatment with ethyl chloroformate produces the corresponding 1-amino-2-carbethoxyamino-4,5-dimethylbenzene. Treatment of the latter with an equivalent of ribose and hydrogenation of the resulting imine yields 2-carboethoxyamino-4,5-dimethylphenylribamine. Basic hydrolysis of that product followed by acid-catalyzed condensation with alloxan generates riboflavin.*

Interestingly, the synthesis of riboflavin *phosphate* has taken a slightly different course. As reported in 1950[†] (Scheme 12.116), when the reduced *N*-ribosyl derivative of 2,3-dimethylaniline is phosphorylated with phosphorus oxychloride and then carefully hydrolyzed, the corresponding monophosphate is capable of isolation. Then, treatment of that ester with diazotized aniline yielded 1-D-1'-ribitylamino-6-phenylazo-3,4-dimethylbenzene. Reduction to the corresponding amine followed by condensation with alloxan then yielded riboflavin monophosphate.

*Karrer, P. *Organic Chemistry*, Elsevier, New York, 1950, p. 730.

[†]Flexser, L. A. U. S. Patent 2,610,176, 1950.

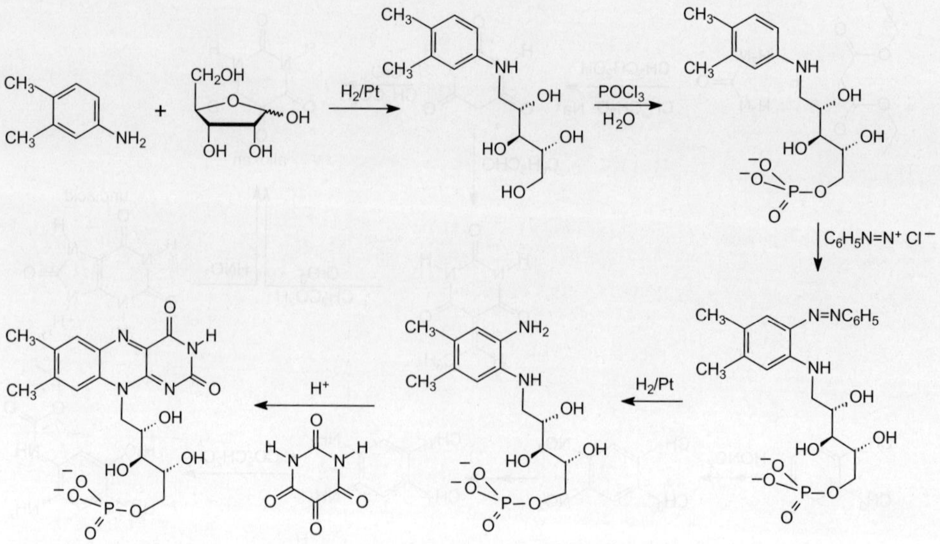

Scheme 12.116. A representation of the synthesis of riboflavin phosphate (after Flexser, L. A. U.S. Patent 2,610,176, 1950).

IX. SAM

The biosynthesis of SAM, as shown in Equation 12.21 (similar to material in a URL starting with http://www.chem.qmul.ac.uk/iubmb/enzyme/), relies on methionine adenosyltransferase (EC 2.5.1.6) to effect the condensation between L-methionine (Met, M) and ATP, a process in which all three phosphate residues initially bound to ATP are lost (as inorganic phosphate). The biosynthesis of methionine (Met, M) and ATP have been previously considered (*vide supra*).

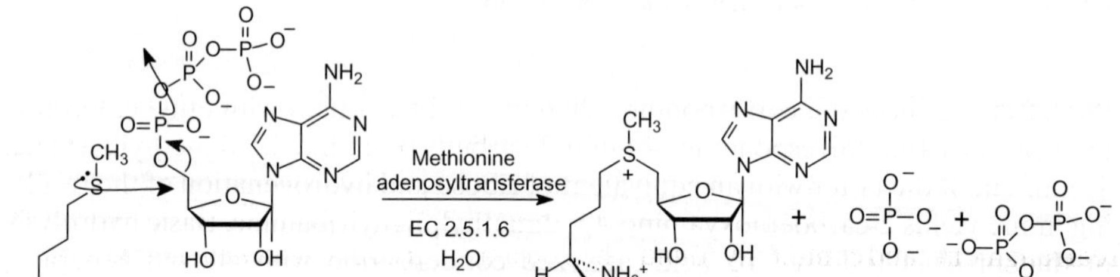

$$(12.21)$$

One synthesis of SAM* involves the initial conversion of adenosine (the synthesis of which has been described earlier, *vide supra*) to the corresponding primary chloride hydrochloride (Scheme 12.117) with thionyl chloride ($SOCl_2$) in pyridine, demethylation of methionine (Met, M) with sodium metal ($Na°$) in liquid ammonia (NH_3) to produce homocysteine (the synthesis of which has been described earlier,

*Deshpande, P. B.; Senthilkumar, U. P.; Padmanabhan, R. U. S. Patent 6,881,837, 2005.

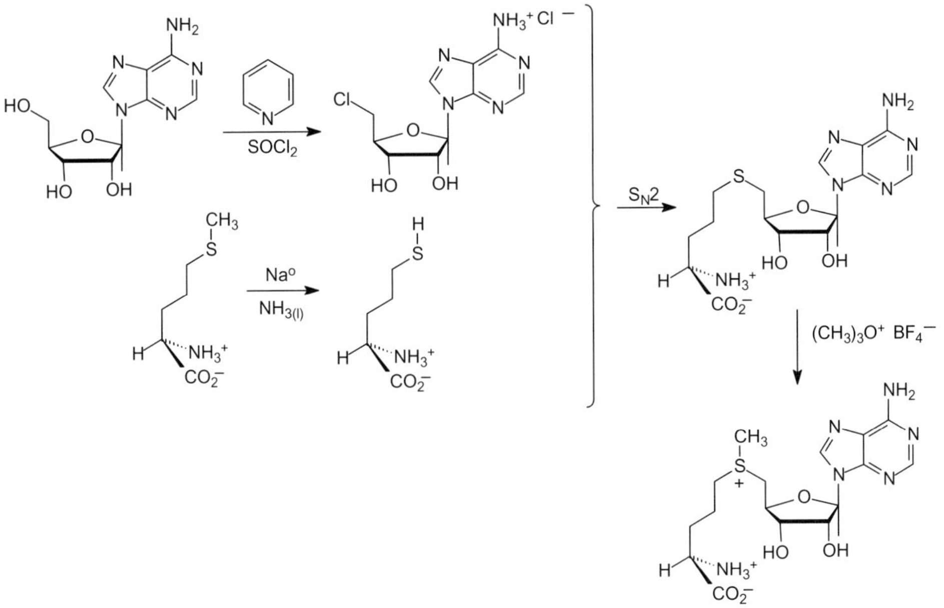

Scheme 12.117. A practical synthesis of S-adenosylmethionine (after Deshpande, P. B.; Senthilkumar, U. P.; Padmanabhan, R. U.S. Patent 6,881,837, 2005).

vide supra), condensing the homocysteine with the 5′-chloromethyladenosine hydrochloride in a typical S_N2 process, and finally, methylating the resulting thioether with trimethyloxonium tetrafluoroborate to yield SAM.

X. Tetrahydrofolate

The early stages of tetrahydrofolate (and folate) biosynthesis begin with GTP undergoing loss of formic acid and thus ring opening to an N-substituted triaminopyrimidine (GTP cyclohydrolase I, EC 3.5.4.16) much as seen earlier (Scheme 12.111) when ring opening was also accompanied by phosphate hydrolysis (GTP cyclohydrolase II). As shown in Scheme 12.118, loss of formate is followed by ring opening, imine-enol to amine-ketone rearrangement (a process called the "Amadori rearrangment" and commonly found when an amino group replaces the hemiacetal— or acetal—hydroxyl of cyclic carbohydrates), and then **recyclization** to form a six-membered ring. Phosphate hydrolysis to 7,8-dihydroneopterin follows.

As shown in Scheme 12.119, the 7,8-dihydroneopterin undergoes a retroaldol with the liberation of α-hydroxyacetaldehyde to 6-hydroxymethyl-7,8-dihydropterin (dihydroneopterin aldolase, EC 4.1.2.25), which is, in turn, phosphorylated, with ATP being converted to AMP (2-amino-4-hydroxy-6-hydroxymethyldihydropteridine diphosphokinase, EC 2.7.6.3) to yield 6-hydroxymethyl-7,8-dihydropterin diphosphate.

Meanwhile, as shown in Scheme 12.120, chorismate is converted to 4-amino-4-deoxychorismate (aminodeoxychorismate synthase, EC 2.6.1.85) and thence to 4-aminobenzoate and pyruvate (aminodeoxychorismate lyase, EC 4.1.3.38).

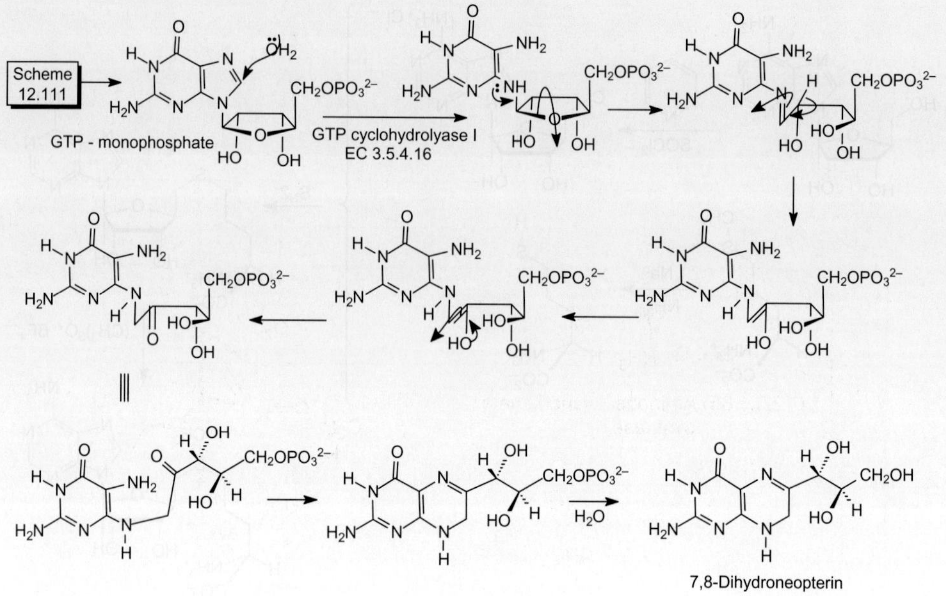

Scheme 12.118. The conversion of guanosine triphosphate (GTP) to 7,8-dihydroneopterin under the influence of GTP cyclohydrolase I (EC 3.5.4.16). EC numbers and some graphic materials provided in this scheme have been taken from appropriate links in a URL starting with http://www.chem.qmul.ac.uk/iubmb/enzyme/.

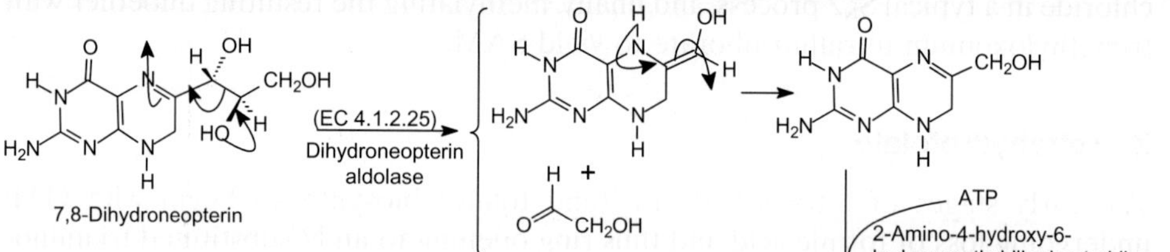

Scheme 12.119. A representation of the formation of 6-hydroxymethyl-7,8-dihydropterin diphosphate from 7,8-dihydroneopterin. EC numbers and some graphic materials provided in this scheme have been taken from appropriate links in a URL starting with http://www.chem.qmul.ac.uk/iubmb/enzyme/.

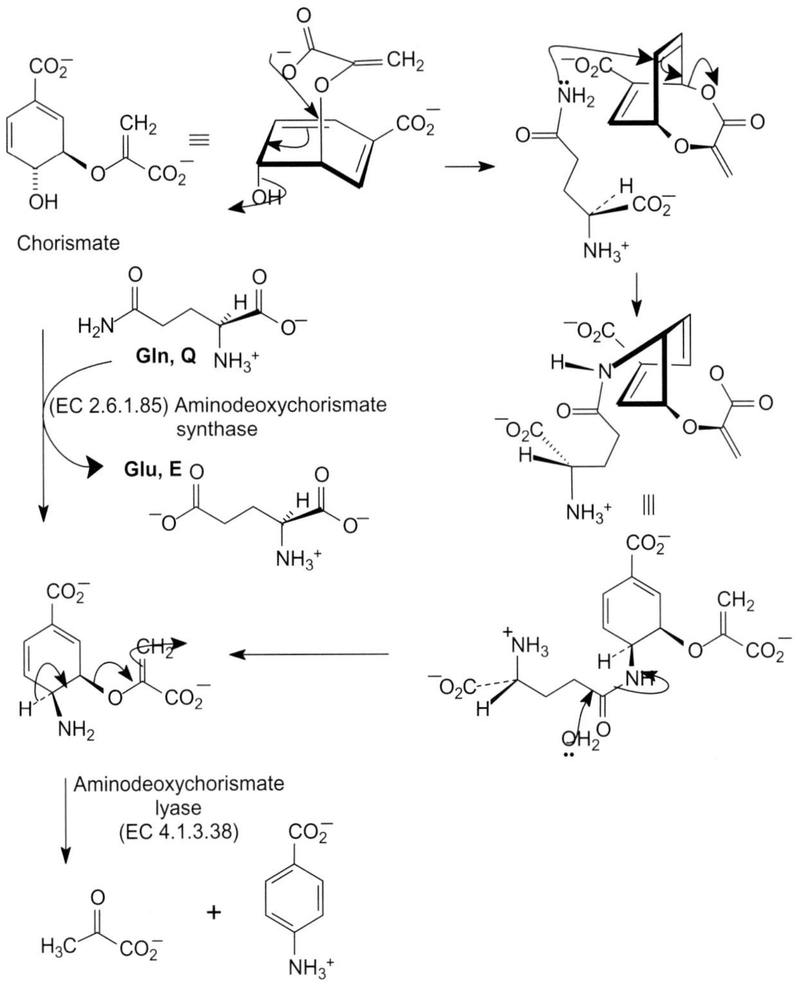

Scheme 12.120. A representation of the conversion of chorismate to 4-amino-4-deoxychorismate (aminodeoxychorismate synthase, EC 2.6.1.85) and thence to 4-aminobenzoate and pyruvate (aminodeoxychorismate lyase, EC 4.1.3.38). The attractive common bicyclic lactone intermediate shown, although not required, is based on a suggestion of Walsh, C. T.; Liu, J.; Rusnak, F.; Sakaitani, M. *Chem. Rev.*, **1990**, *90*, 1105. EC numbers and some graphic materials provided in this scheme have been taken from appropriate links in a URL starting with http://www.chem.qmul.ac.uk/iubmb/enzyme/.

The combination of 6-hydroxymethyl-7,8-dihydropterin diphosphate with 4-aminobenzoate in a simple displacement reaction (dihydropteroate synthase, EC 2.5.1.15) produces 7,8-dihydropteroate (Equation 12.22).

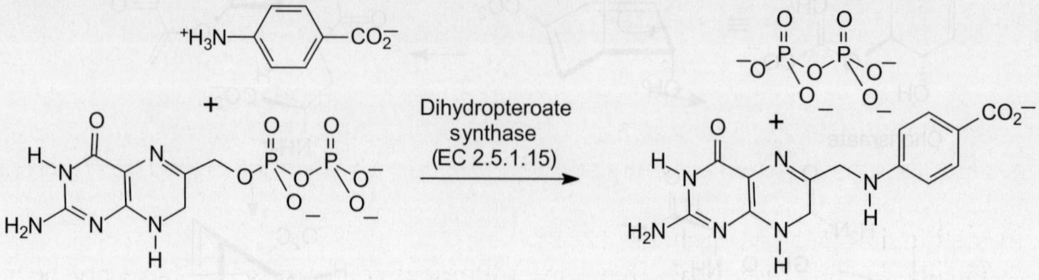

$$(12.22)$$

Finally, as shown in Scheme 12.121, the carboxylate is phosphorylated (while ATP is converted to AMP) and this is followed by amide formation with glutamate (Glu,

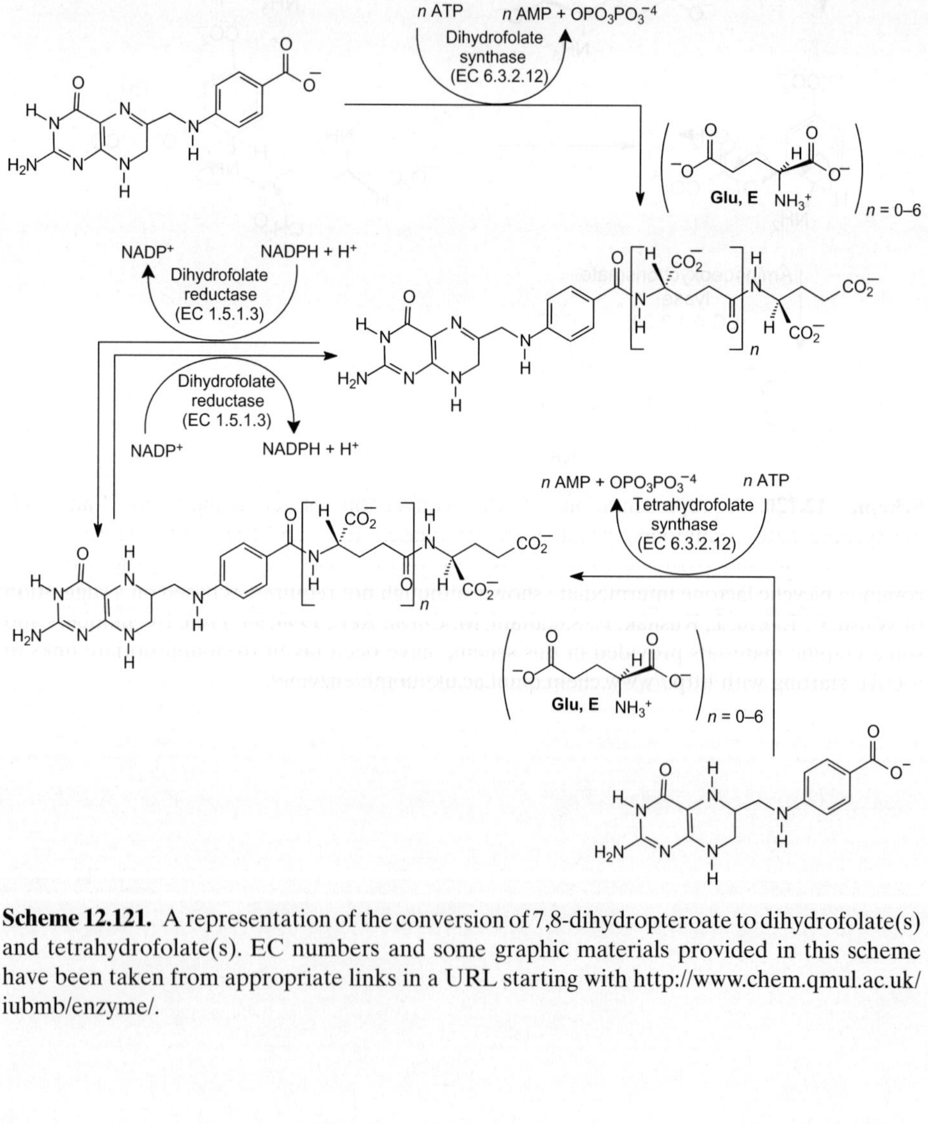

Scheme 12.121. A representation of the conversion of 7,8-dihydropteroate to dihydrofolate(s) and tetrahydrofolate(s). EC numbers and some graphic materials provided in this scheme have been taken from appropriate links in a URL starting with http://www.chem.qmul.ac.uk/iubmb/enzyme/.

E) under the influence of dihydrofolate synthase (EC 6.3.2.12). Repetition of the same sequence of phosphorylation and amide formation serves to extend the amide side chain. Reduction (NADPH, dihydrofolate reductase, EC 1.5.1.3) leads to tetrahydrofolate, which, itself, can undergo changes in the number of glutamate (Glu, E) side chains (tetrahydrofolate synthase, EC 6.3.2.17).

Although not shown, there is a further group of folate coenzymes that, in particular, allow for methylation, hydroxymethylation, and formylation. Some of these processes have been shown as part of other biosynthetic pathways earlier and in general account for the introduction and "movement" of single-carbon atoms (viz., the conversion of serine [Ser, S] to glycine [Gly, G], Scheme 12.9, and homosysteine to methionine [Met, M], Scheme 12.13).

Finally, for this discussion, there are additional coenzymes that have not been discussed. They have been omitted only because of limited space. The interested reader is directed to the biochemical literature. A good place to begin is with *Natural Product Reports*, published by the Royal Society of Chemistry, since reviews of biosynthetic pathways regularly appear in that journal.

REFERENCE

Cicchillo, R. B.; Booker, S. J. *J. Am. Chem. Soc.*, **2005**, *127*, 2860.

An Introduction to Alkaloids and Some Other Heterocyclic Compounds

> Our present and ... humbler ... task is to study the
> molecular architecture of some plant products and ...
> to note just a few of the mechanistic findings of
> classical biochemistry and of modern research ...
>
> —Robert Robinson
> *The Structural Relations of Natural Products*
> Clarendon Press, Oxford, 1955, p. 2

A. INTRODUCTION

Heterocyclic compounds have one or more noncarbon atoms in one or more rings (cycles), and many such compounds with boron, oxygen, sulfur, phosphorus and nitrogen have already been encountered. One such very large group, singled out here where most members contain nitrogen in a ring or rings, are called "alkaloids". On the one hand, alkaloids might be defined as "a highly diverse group of compounds that are related only by the occurrence of a nitrogen atom in a heterocyclic ring."* However, there are some compounds that contain nitrogen and are generally regarded as alkaloids but do not contain nitrogen in a heterocyclic ring.

Sometimes, alkaloids are considered simply as nitrogen-containing basic compounds (i.e., "alkali bodies" or amines) of secondary metabolism," a group of compounds not clearly recognized as being required as part of "primary metabolism." However, it is clear that some are chemically neutral (not basic) and, for most, their place in metabolism is not known.

Exceptions aside, the general definitions given above do appear to apply since most alkaloids are basic and most do have nitrogen in a ring.

Furthermore, although alkaloids have traditionally been considered products of plant secondary metabolism, we now know that compounds that are strikingly similar to those found in plants and which may have arisen by similar pathways from the same small family of fundamental building blocks are also found in animals. In some cases, it is held that the alkaloids or alkaloid precursors are ingested by the

*Ziegler, J.; Facchini, P. J. *Ann. Rev. Plant Biol.*, **2008**, *59*, 735.

animals from the plants on which the animals feed (or even further down the food chain, animals the animals eat) and thus are not actually biosynthesized by the animal in which they are found. In other cases, the source is clear.

There are now over 20,000 alkaloids of known structures, and it should be obvious that only a few can be considered here. Indeed, it could be asked why any of them are considered here.

The rationale for their discussion derives, first, from the observation that plant bases were among the earliest compounds isolated in pure crystalline form by chemists (and their forbearers) who were seeking information about the extracts of plants that were used as medications. Second, alkaloids are, today, among those compounds whose biosynthesis in well-known plants is being most closely examined gene by gene. Third, and finally, many of these compounds serve as templates for exploration of new and clever synthetic methods. Thus, their inclusion here provides the interested reader with yet another view of the diversity the preceding two chapters have emphasized as well as some indication of some of the newer (at this writing) tools of those who practice synthetic organic chemistry.

So, from the hundreds of families of alkaloids and the thousands of alkaloids in those families, only a few of the families and, within them, only some (Figure 13.1) of the typical bases are considered in this chapter. The choice of what to consider is based on two criteria.

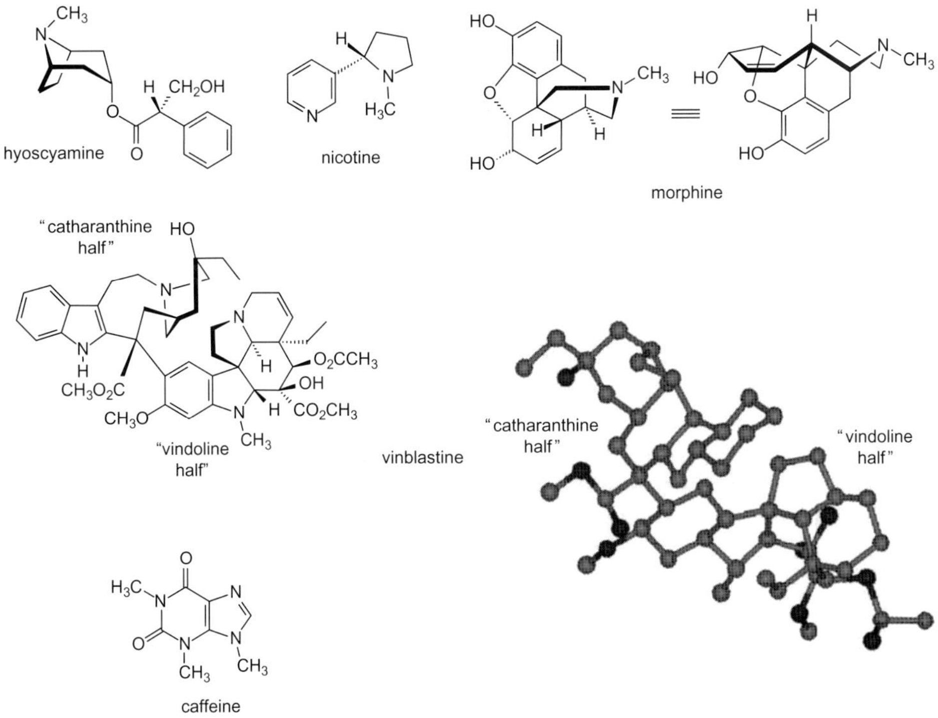

Figure 13.1. Representatives of a few families of alkaloids to be discussed in this chapter. The ball-and-stick representation is from reported X-ray crystallographic information. See Bau, R.; Jin, K. K. *J. Chem. Soc. Perkin Trans. I*, **2000**, 2079.

First, several groups of compounds have enjoyed lengthy exploration along degradative, synthetic, and biosynthetic paths. This combination resulted from their early isolation (and they are plentiful) and their structures (they were intriguingly elaborate but, save for the most recent, not wildly so) lent themselves to the chemistry that was available for degradation studies, and as the tools became available, adventurous investigators applied (largely) radiolabeling to growing plants to demonstrate the viability of what became known as "biogenetic theory." Biogenetic theory is an intellectual undertaking that attempted (with some major successes) to map the simpler compounds of primary metabolism already discussed to these more elaborate formations.

Second, despite the elaborate structures of these thousands of compounds, it is generally true that they are derived from an amino acid or amino acids. As a consequence, the relatively small number of amino acids, products of primary metabolism, could reasonably be used as biosynthetic precursors once prepared in labeled (radioactive or other isotopic) form and incorporation into a plant demonstrated.

Third, in addition to amino acids, it has become clear that other simple compounds, shown to be closely related to already elaborated carbohydrate, polyketide, and terpenoid fragments, are incorporated into alkaloids.

Finally, therefore, only a few examples need to be provided to demonstrate the power of the application of the laboratory tools and intellectual vigor that has led to our current understanding of these naturally occurring heterocycles. The interested reader is directed to the current literature and the various compendia of these compounds.

Thus, from the Solanaceae, hyoscyamine, a tropane alkaloid, which we now know to be derived from ornithine and/or **arginine (Arg, R**, Scheme 12.7), and nicotine, derived from a combination of nicotinic acid (see Scheme 12.103, et seq.) and ornithine, serve as examples of alkaloids based on nonaromatic amino acids. In the benzylisoquinoline alkaloids, only morphine, arguably the first alkaloid isolated in purified form and known to be derived from **tyrosine (Tyr, T**), itself a prephenic acid derivative, is discussed. For the indole alkaloids, the single example of this rich family to be illustrated is vinblastine, a (relatively) recently isolated base of some medicinal value and which is shown to be composed of two different expressions of **tryptophan (Trp, W**) and mevalonate (Scheme 11.40). Finally, caffeine is chosen as the member of the purine alkaloids, a relatively small family of compounds but one of major commercial import and one whose biosynthesis is closely tied to the library of life.

Finally, in this vein, plants in these families appear to have trained us to cultivate them and to find ways to insure their health. Thus, it can be argued that alkaloid production has survival value to the plant.

B. TROPANE ALKALOIDS

I. Chemistry of Hyoscyamine

In 1879, W. Ladenburg, furthering the "beautiful investigations" (*schönen Untersuchungen*) of Kraut and Lossen who had shown that hyoscyamine ($C_{17}H_{23}NO_3$) could be decomposed into tropine ($C_8H_{15}NO$) and tropic acid ($C_9H_{10}O_3$), succeeded

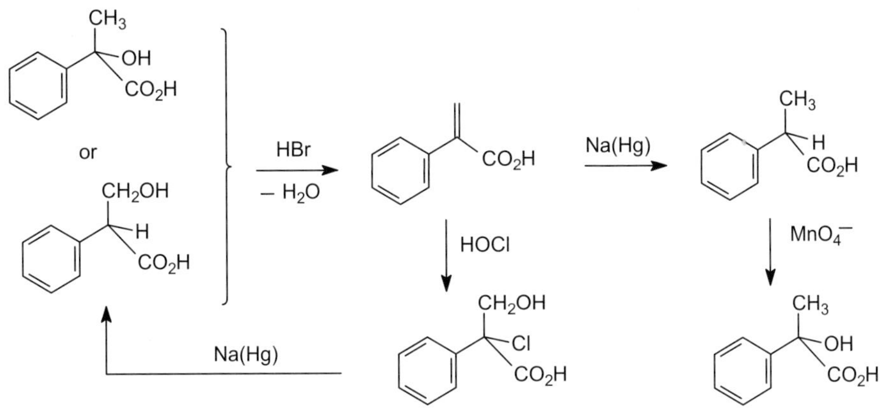

Scheme 13.1. Degradation and partial synthesis determining the structure of tropic acid. After Ladenburg, A.; Rügheimer, L. *Chem. Ber.*, **1880**, *13*, 373.

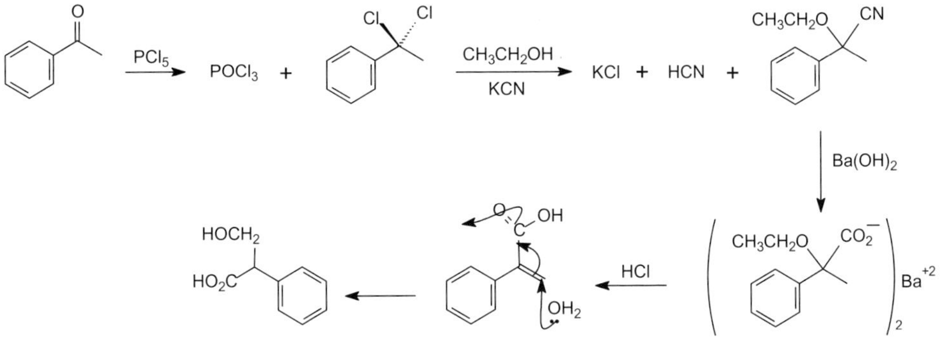

Scheme 13.2. A synthesis of racemic tropic acid. After Ladenburg, A.; Rügheimer, L. *Chem. Ber.*, **1880**, *13*, 2041.

in reconstituting the racemic hyoscyamine by treating tropine and tropic acid with an excess of dilute (überschüssiger, verdünnter) hydrochloric acid.[*]

The structure of tropic acid was worked out as shown in Scheme 13.1,[†] and the synthesis (Scheme 13.2) was accomplished by Ladenburg and Rügheimer later the same year.[‡]

With regard to the structure, it was known that phenyllactic acid [$C_6H_5CH(OH)CO_2H$] was different from tropic acid, but it was not clear if tropic acid was a tertiary or a primary alcohol. But Ladenburg and Rügheimer knew that a carbon directly attached to the aromatic ring could be (gently) oxidized by basic permanganate to the corresponding alcohol. The carboxy-alcohol they obtained by this method was different from tropic acid. Furthermore, the dehydration product of tropic acid on treatment with hypochlorous acid yielded a chlorohydrin, which, on reduction, afforded tropic acid. They concluded that the 3-hydroxy-2-phenylpropanoic acid represented tropic acid.

[*]Ladenburg, A. *Chem. Ber.*, **1879**, *12*, 941.
[†]Ladenburg, A.; Rügheimer, L. *Chem. Ber.*, **1880**, *13*, 373.
[‡]Ladenburg, A.; Rügheimer, L. *Chem. Ber.*, **1880**, *13*, 2041.

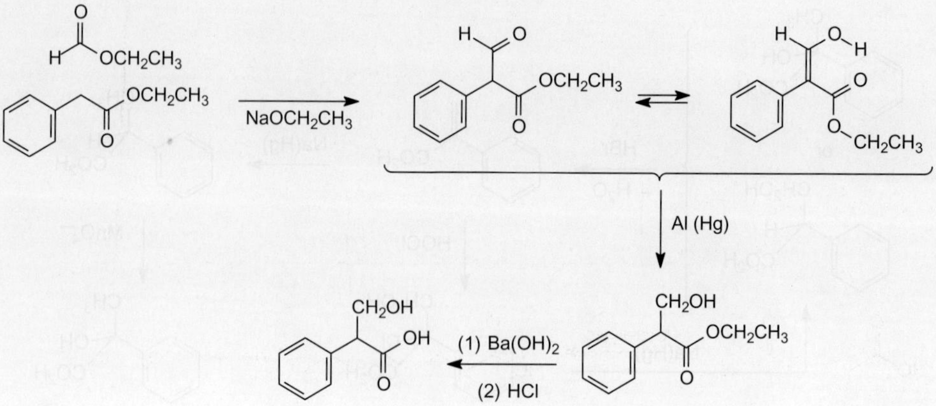

Scheme 13.3. A later synthesis of racemic tropic acid. After Müller, E. *Chem. Ber.*, **1918**, *51*, 252.

For the synthesis, as shown in Scheme 13.2, when acetophenone was treated with phosphorus pentachloride, first with cooling and then warming to 40°C to complete the reaction, the corresponding benzylic dichloride (1,1-dichloro-1-phenylethane) was obtained in good yield. Dissolution of the dichloride in ethanol followed by addition of ethanolic potassium cyanide (KCN) and standing for 48 h at room temperature resulted in a solution from which potassium chloride (KCl) precipitated and which had a strong odor of hydrogen cyanide (HCN). The ethanol was removed by distillation, and the residue was treated at room temperature for 8 h with barium hydroxide [Ba(OH)$_2$]. Acidification resulted in the isolation of tropic acid in low yield.

Nearly 40 years later, when additional amounts of tropic acid were sought, an alternative synthesis was developed by Müller. As shown in Scheme 13.3, when the ethyl ester of phenylacetic acid was allowed to condense with ethyl formate in ether and in the presence of sodium metal and the resulting condensation products was reduced with aluminum amalgam, the corresponding ethyl ester of tropic acid was isolated. Hydrolysis of the ester was accomplished (again) with barium hydroxide (Scheme 13.3). The question of stereochemistry was not addressed.

With regard to tropine, the nitrogenous fragment of hyoscyamine, a series of papers, by Richard Willstätter, perhaps best described as a *tour de force*, began with the structure determination of tropine (in 1898) and concluded with its synthesis (in 1903). There resulted a masterful body of work that stands, even today, as truly remarkable.* First, it had been known for a few years that chromic acid oxidation of tropine produced a dicarboxylic acid (tropinic acid, C$_8$H$_{13}$NO$_4$).[†] Willstätter, carrying out two Hofmann-type elimination reactions on this acid showed that the product was piperylenedicarboxylic acid (which yielded pimelic acid [heptanedioic acid] on reduction). Willstätter concluded that the nitrogen must be present in the form of a bridge. He further concluded that this must be a bridge across a

---

*Willstätter, R. *Chem. Ber.*, **1898**, *31*, 1534; Willstätter, R. *Liebigs Ann. Chem.*, **1901**, *317*, 204, 267, 307; Willstätter, R. *Liebigs Ann. Chem.*, **1903**, *326*, 1, 23.
[†]Merling, G. *Liebigs Ann. Chem.*, **1888**, *216*, 329.

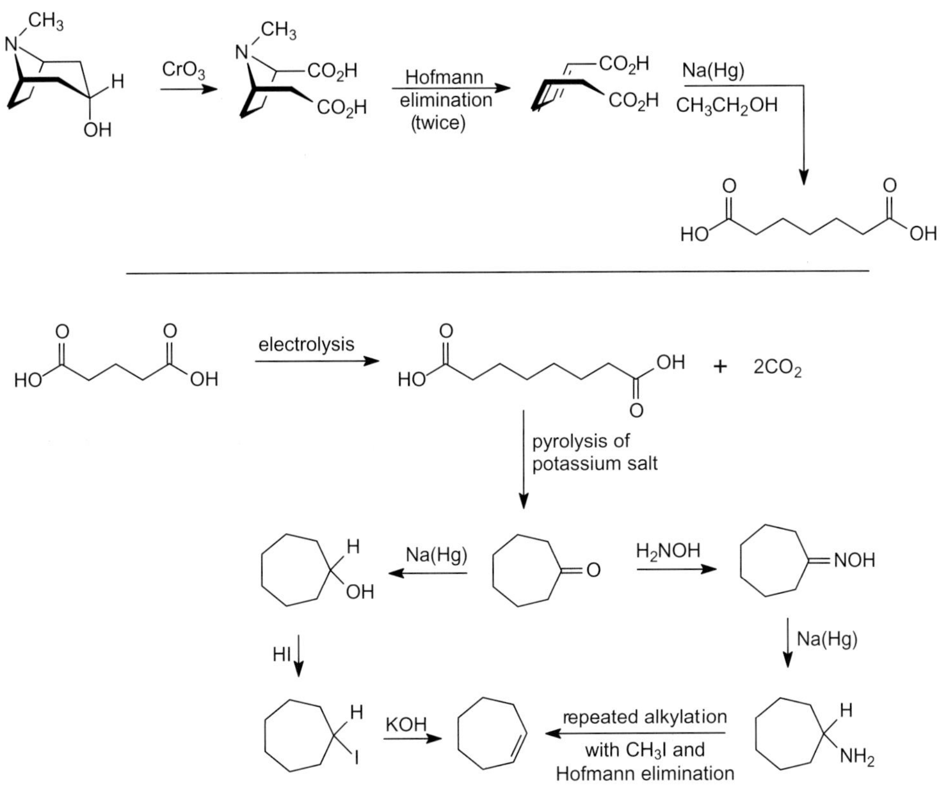

Scheme 13.4. In the upper part, a representation of the oxidation of tropine to an *N*-methyl-pyrrolidinedioic acid (tropinic acid) and the subsequent production of piperylenedicarboxylic acid and pimelic acid (heptanedioic acid). In the lower part, a representation of the electrolytic conversion of glutaric acid (pentanedioic acid) to suberic acid (octanedioic acid) and carbon dioxide (CO_2) and thence to cycloheptanone and, in two different ways, to cycloheptene.

seven-membered ring, which he set out to synthesize. Cycloheptanone was the goal, and this was built by taking the newly found electrolytic procedure of Crum Brown[*] to convert 2 equivalents of glutaric acid (pentanedioic acid) to suberic acid (octane-dioic acid) and carbon dioxide (CO_2) and then transforming that by pyrolysis of the potassium salt to cycloheptanone. Two different paths were then used to transform the ketone into cycloheptene. The chemistry to accomplish these ends is outlined in Scheme 13.4.

Continuing (Scheme 13.5), bromination of cycloheptene to the corresponding dibromide and dehydrohydrohalogenation yielded cycloheptadiene. A second bromination to the corresponding 1,4-dibromide (not the only product) and another dehydrohalogenation produced cycloheptatriene. Now, addition of bromine to cycloheptadiene in the presence of dimethylamine yielded the corresponding *N,N*-dimethylaminocycloheptadiene (presumably via an addition and elimination) and then reduction (sodium in alcohol) provided 5-(*N,N*-dimethyl)aminocycloheptene.

[*]Brown, A. C.; Walker, J. *Liebigs Ann. Chem.*, **1891**, *261*, 107.

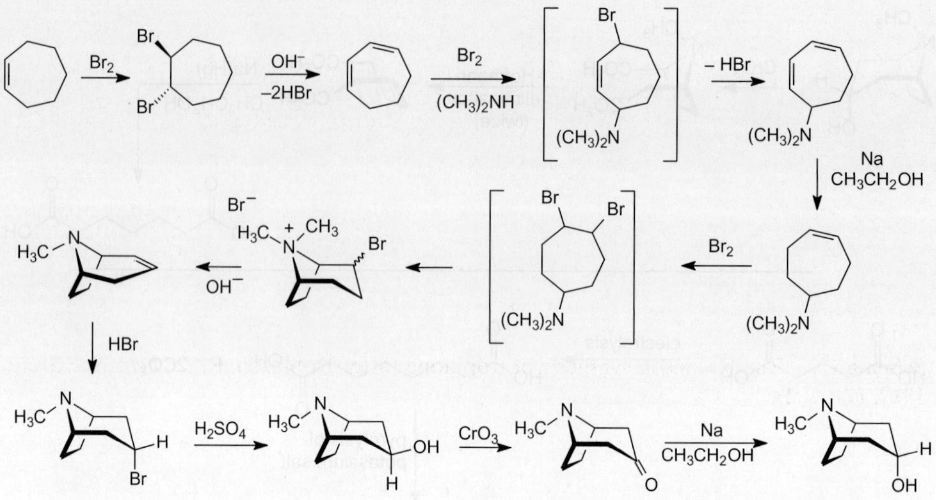

Scheme 13.5. The final stages in the Willstätter tropine synthesis (*vide supra*).

Bromination of the latter produced a transient dibromide that underwent cyclization to the desired 8-(*N*,*N*-dimethyl)aza-2-bromobicyclo[3.2.1]octane hydrobromide (i.e., tropane nucleus). Base treatment effected dehydrohalogenation and then
N-demethylation and subsequent addition of hydrogen bromide (and neutralization) afforded the corresponding (presumed) *endo*-3-bromide. Strikingly, sulfuric
acid treatment of the bromide was apparently required to replace the bromine with
(after basification) a hydroxyl group, and it is reported that the wrong isomer
(hydroxyl *exo*) resulted!

So, oxidation to tropinone and another reduction (sodium in ethanol) was
required (Scheme 13.5).

As pointed out by Sir Robert Robinson with regard to tropanone, almost 15 years
after the work outlined above was published,* "… it appears to be of too complicated a character to admit of development into an economical process." The
critique was accompanied by a new synthesis, which required only that succinic
dialdehyde react with acetonedicarboxylate and methylamine in ethanol solution
(Scheme 13.6). As simple as this reaction appears, however, it clearly serves more
as a proof of principle than a good preparative method as side reactions frequently
intrude. Significantly more work has been carried out on tropinone since these early
days, and a more recent synthesis (also containing references to other methods) is
shown in Scheme 13.7.

A recent synthesis of tropine and the isomeric pseudotropine (ψ-tropine) by
Noyori et al.[†] exemplifies the use of a generalized "3 + 4" addition pathway to
produce bicyclic compounds. Indeed, this synthesis allows for the production of a
large family of tropine-related bases since a double bond, ready for further elaboration, is retained in the bicyclic system.

*Robinson, R. *J. Chem. Soc. Trans.*, **1917**, *111*, 762.
[†]Hayakawa, Y.; Baba, Y.; Makino, S.; Noyori, R. *J. Am. Chem. Soc.*, **1978**, *100*, 1786.

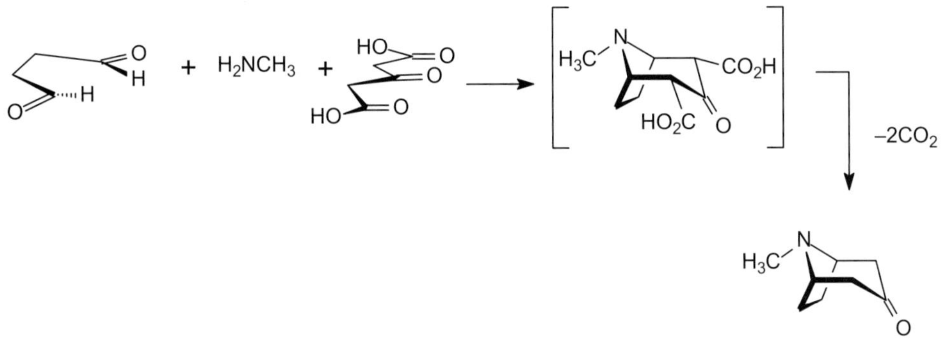

Scheme 13.6. The Robinson synthesis of tropinone (after Robinson, R. *J. Chem. Soc. Trans.*, **1917**, *111*, 762).

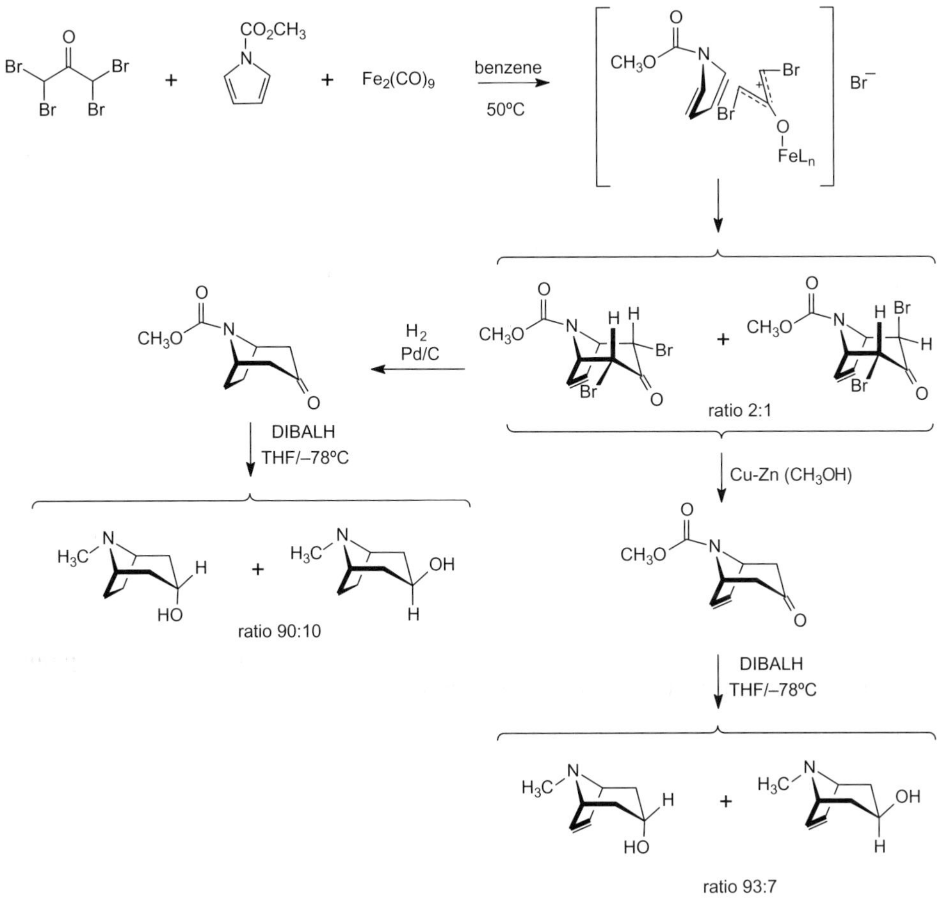

Scheme 13.7. Preparation of tropine and ψ-tropine as well as the corresponding dehydro derivatives from *N*-carbomethoxypyrrole, diiron enneacarbonyl [Fe$_2$(CO)$_9$] and α,α,α′,α′-tetrabromoacetone. After Hayakawa, Y.; Baba, Y.; Makino, S.; Noyori, R. *J. Am. Chem. Soc.*, **1978**, *100*, 1786.

In this synthesis, $\alpha,\alpha,\alpha',\alpha'$-tetrabromoacetone is allowed to react with N-carbomethoxy-pyrrole in the presence of diiron enneacarbonyl [$Fe_2(CO)_9$] (3:1:3 ratio) to aid the process. As shown in Scheme 13.7, this addition produces a mixture of cis- and $trans$-dibromocycloadducts, which can, with or without separation, be converted to their dehalogenated counterparts on reduction with hydrogen (H_2) in the presence of palladium (Pd). A further reduction with diisobutylaluminum hydride (DIBAL-H) in tetrahydrofuran (THF) at $-78°C$ then generates a 90:10 mixture of tropine and pseudotropine (ψ-tropine). In an interesting aside, the authors point out that α,α'-dibromoacetone could not be used as the presumed intermediate cation is too reactive. It is argued that the additional two bromine substituents in the tetrabromide shown in the scheme ($\alpha,\alpha,\alpha',\alpha'$-tetrabromoacetone) stabilize the required addend.

Aside from the brief introduction to synthetic efforts, outlined above, and before going on with an introduction to the biosynthetic studies on hyoscyamine, there are two questions the interested reader might, at this point, be asking. First, how was it known which of the two tropanone reduction products was tropine and which was pseudotropine (ψ-tropine), and second, what was the absolute stereochemistry of tropic acid and how was that determined?

With regard to the first, as already mentioned, in 1888, G. Merling (*vide supra*) had examined the oxidation of tropine with chromic acid and had isolated a dicarboxylic acid. Additionally, he also examined the oxidation of tropine with basic permanganate solution and found that under very carefully controlled conditions (dilute aqueous solution and low temperature), a compound was isolated that had lost a methyl group. Methylation to produce the original (with methyl iodide) showed that demethylation had been the only change, and it was presumed that the methyl group that had been lost was on nitrogen. The work was repeated by others (notably by Willstätter),[*] and finally, in 1953, Fodor and Nádor[†] made use of the N-des-methyl isomers and showed (Scheme 13.8) that, whereas N-benzoylnor-ψ-tropine on treatment with hydrogen chloride in dioxane solution underwent rearrangement to the corresponding O-benzoyl derivative, N-benzoylnortropine did not. Thus, the hydroxyl must be on the same side of the ring as the nitrogen bridge in the former and on the opposite side in the latter.

The second question, that of the absolute stereochemistry [(R) or (S)], of the naturally occurring (–)-tropic acid was answered by Fodor and Csepreghy.[‡] Thus, as shown in Scheme 13.9, it was already known[§] that the absolute stereochemistry of (+)-alanine was (S). It had also been determined[¶] that (S)-(+)-α-phenylpropanoic acid underwent Curtius reaction to yield (S)-(–)-1-phenylethylamine. Since the Curtius reaction occurs with retention of stereochemistry within the migrating group, the relative geometry of the (S)-(–)-1-phenethylamine was known to be the

[*]Willstätter, R. *Chem. Ber.*, **1896**, *29*, 1636.
[†]Fodor, G.; Nádor, K. *J. Chem. Soc.*, **1953**, 721.
[‡]Fodor, G.; Csepreghy, G. *J. Chem. Soc.*, **1962**, 3222.
[§]Fischer, E.; Raske, K. *Chem. Ber.*, **1907**, *40*, 3717 demonstrated the interrelationship between (–)-serine, (+) alanine, and (+) glyceric acid. Among other ways, the **absolute stereochemistry** of (–)-serine has been demonstrated by the X-ray (with anomalous dispersion) of ferrichrome A by Zalkin, A.; Forrester, J. D.; Templeton, D. H. *J. Am. Chem. Soc.*, **1966**, *88*, 1810.
[¶]Bernstein, H. I.; Whitmore, F. C. *J. Am. Chem. Soc.*, **1939**, *61*, 1324.

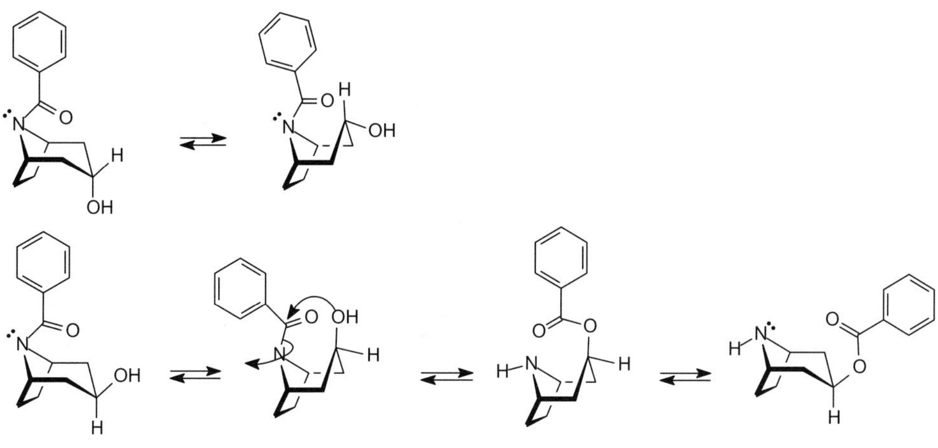

Scheme 13.8. *N*-benzoylnortropine (top) cannot undergo N to O (N → O) benzoyl migration as the –OH does not approach closely enough to the nitrogen. However, for *N*-benzoylnor-ψ-tropine (bottom), the rearrangement is quite facile. Thus, the hydroxyl must be on the same side of the ring as the nitrogen bridge in the latter and on the opposite side in the former. After Fodor, G.; Nádor, K. *J. Chem. Soc.*, **1953**, 721.

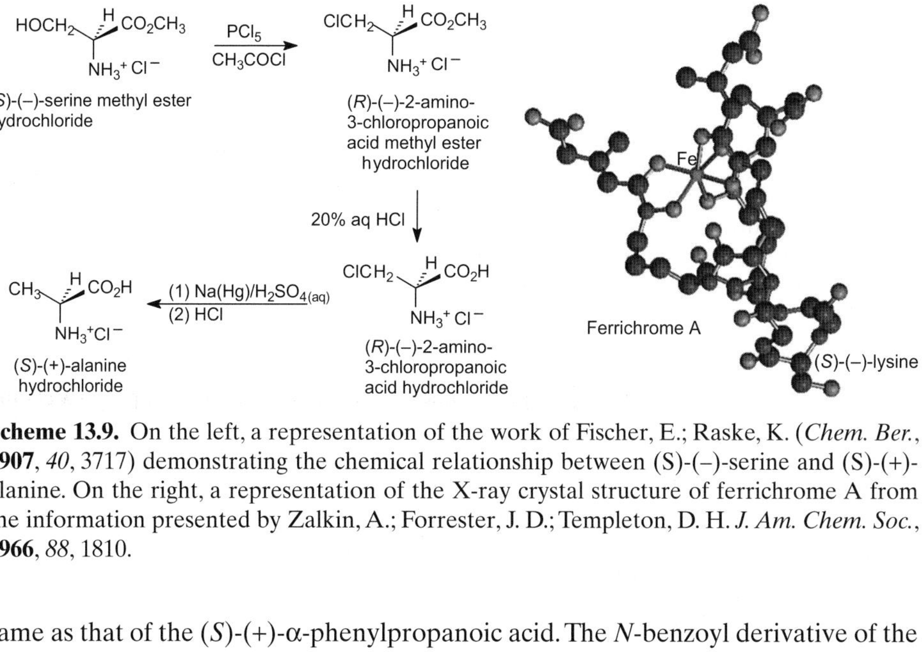

Scheme 13.9. On the left, a representation of the work of Fischer, E.; Raske, K. (*Chem. Ber.*, **1907**, *40*, 3717) demonstrating the chemical relationship between (S)-(–)-serine and (S)-(+)-alanine. On the right, a representation of the X-ray crystal structure of ferrichrome A from the information presented by Zalkin, A.; Forrester, J. D.; Templeton, D. H. *J. Am. Chem. Soc.*, **1966**, *88*, 1810.

same as that of the (*S*)-(+)-α-phenylpropanoic acid. The *N*-benzoyl derivative of the (*S*)-(–)-1-phenylethylamine had been correlated with (*S*)-(+)-alanine (Scheme 13.10).*

*This correlation (Scheme 13.10) had originally been performed by Leithe, W. *Chem. Ber.*, **1931**, *64*, 2827 by benzoylation of the (–)-1-phenylethylamine, nitration of the aromatic ring of the resulting *N*-benzoyl derivative, reduction of the nitro group to the amine, diazotization of the amine and conversion of the diazo compound to the phenol. Oxidation of the phenol with chromic anhydride in acetic acid yielded the *N*-benzoyl derivative of (+)-alanine—by direct comparison to that compound.

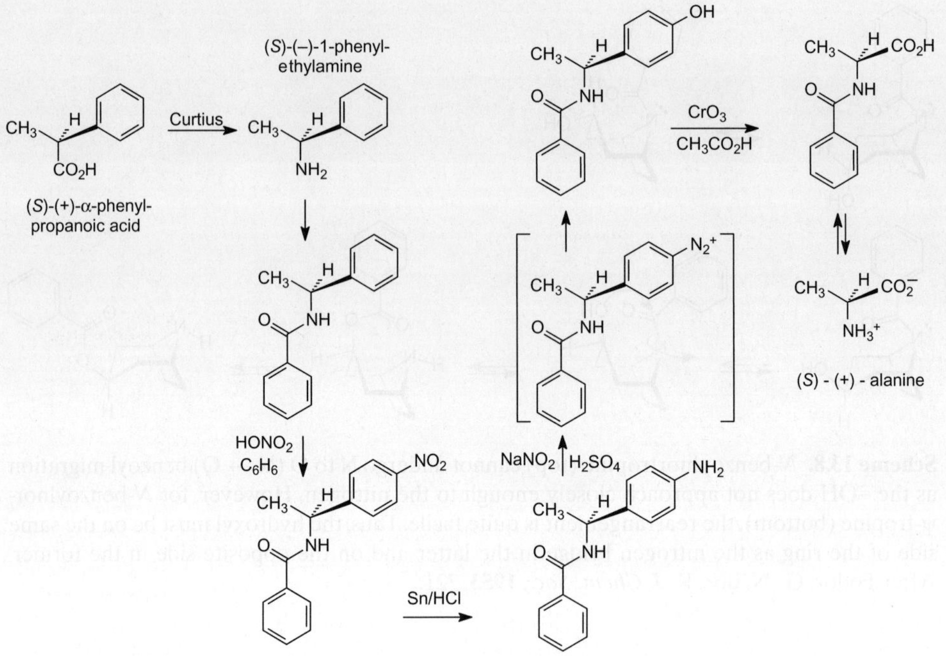

Scheme 13.10. An outline of the chemistry demonstrating the relationship between (S)-(+)-α-phenylpropanoic and (S)-(+)-alanine. After Leithe, W. *Chem. Ber.*, **1931**, *64*, 2827 and Bernstein, H. I.; Whitmore, F. C. *J. Am. Chem. Soc.*, **1939**, *61*, 1324.

Even earlier,* the enantiomers of tropic acid had been resolved by treating a synthetic mixture of the enantiomers with quinine (Scheme 13.11). The synthetic mixture had been prepared by simply forming the cyanohydrin of acetophenone, dehydrating the cyanohydrin by simple distillation and adding dry hydrogen chloride (HCl) gas into a solution of the alkene in ether. Hydrolysis of the resulting primary (!) chloride [(±)-3-chloro-2-phenylpropanoic acid] yielded (±)-tropic acid. The quinine resolution was affected in ethanol.

Now, Fodor and Csepreghy resolved (±)-3-chloro-2-phenylpropanoic acid with codeine and dehalogenated the resolved, separated enantiomers by hydrogenolysis (H_2) over palladium on charcoal. They found that (S)-(−)-3-chloro-2-phenylpropanoic acid yielded (R)-(−)-α-phenyl-propanoic acid and concluded, therefore, that the naturally occurring (−)-tropic acid was (S) (Scheme 13.12).

II. Chemistry of Nicotine

In an exciting report in 1843[†] (stated[‡] to be nearly 300 years after Jean Nicot sent tobacco seed from Portugal to Paris), Melsens pointed out that the product of condensation of fumes of tobacco acted on insects ("… sont rapidement asphyxiés dans

---

*McKenzie, A.; Wood, J. K. *J. Chem. Soc.*, **1919**, *115*, 828.
[†]Melsens, L. H. F. *Ann. Chim.*, **1843**, *9*, 465.
[‡]Jackson, K. E. *Chem. Rev.*, **1941**, *29*, 123.

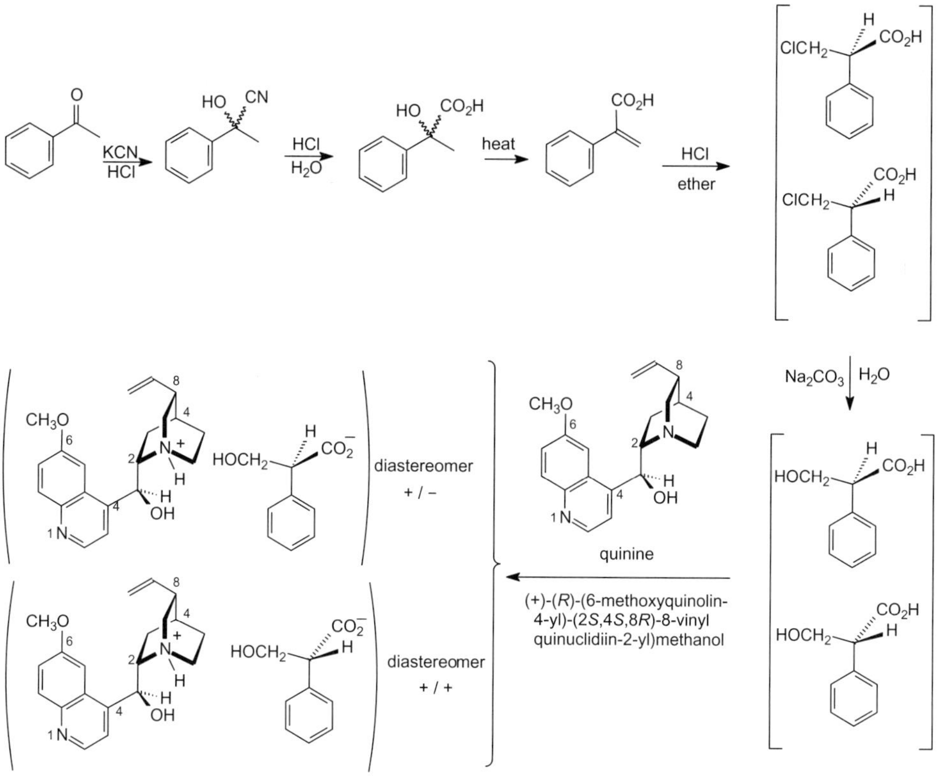

Scheme 13.11. The synthesis and resolution of the enantiomers of tropic acid. After McKenzie, A.; Wood, J. K. *J. Chem. Soc.*, **1919**, *115*, 828.

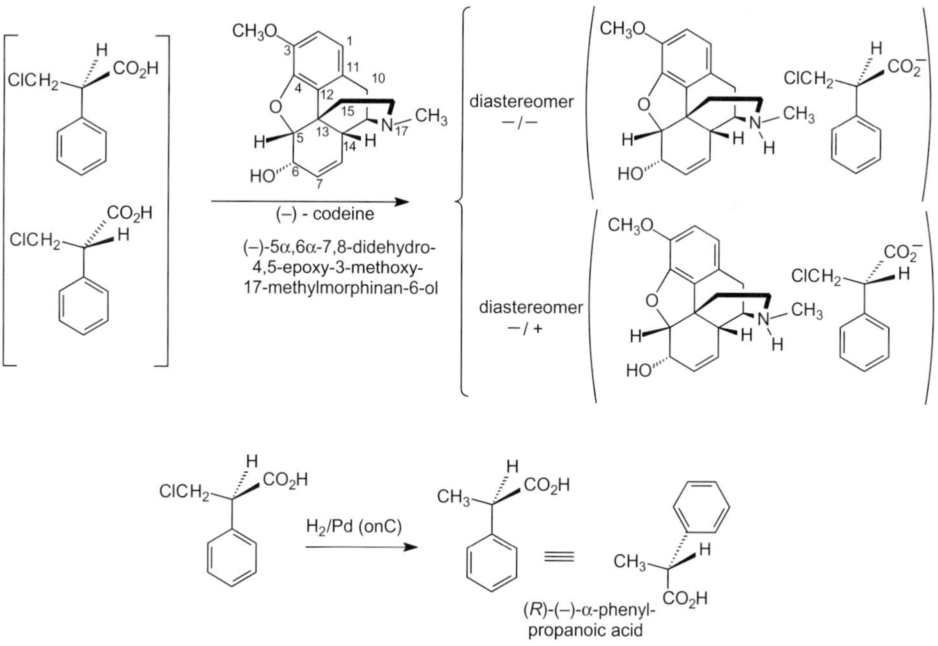

Scheme 13.12. Resolution of (±)-3-chloro-2-phenylpropanoic acid with codeine and hydrogenolytic dehalogenation of (*S*)-(−)-3-chloro-2-phenylpropanoic acid to (*R*)-(−)-α-phenylpropanoic acid. After Fodor, G.; Csepreghy, G. *J. Chem. Soc.*, **1962**, 3222.

une atmosphère de fume …") and on how he subsequently followed the work of Barral* to isolate nicotine from tobacco. Melsens reported that the empirical formula of nicotine corresponded to $C_{10}H_{14}N_2$.

It was clear that some chemists had begun to examine the chemical properties of this base, and in 1870, Huber[†] reported in a brief communication ("Vorläufige Mittheilung") that the acid he isolated ($C_6H_5NO_2$) some 3 years earlier through the action of potassium dichromate ($K_2Cr_2O_7$) and sulfuric acid (H_2SO_4) on nicotine was not, as reported in the earlier work, an amino acid but rather a pyridine carboxylic acid. He stated a more complete work would follow; however, it appears that he was overtaken by Weidel,[‡] among others. Indeed, a much more complete description of nicotine, its properties, a large number of derivatives, and of nicotinic acid, which was identified as pyridine 3-carboxylic acid by studies on naphthoquinolines, appeared shortly thereafter (Scheme 13.13).[§]

The excitement derived from the identification of 6 of the 10 carbons, 4 of the 5 hydrogens, and 1 of the 2 nitrogen atoms in nicotine did not wane, but finding additional information about the remainder appears to have been fraught with difficulty. In part, this arose from intractable mixtures and dearth of separation techniques. Indeed, it was not until nearly a decade later that Pinner,[¶] after correcting

*Barral, M. *Comptes Rendus*, **1835**, *1*, 224.
[†]Huber, C. *Chem. Ber.*, **1870**, *3*, 849.
[‡]Weidel, H. *Liebigs Ann. Chem.*, **1873**, *165*, 328.
[§]Skraup, Z. H.; Cobenzl, A. *Monatsh. Chem.*, **1883**, *1*, 436. As shown in Scheme 13.13, the logic is as follows: there are two aminonaphthalines (prepared by reduction of the corresponding nitronaphthalines, which, in turn, came about on nitration of naphthalene). One aminonaphthalene on reaction with sulfuric acid and glycerol in, for example, nitrobenzene solvent (the **Skraup** reaction) yields a single naphthoquinoline. The other aminonaphthalene yields two different naphthoquinolines. Now, if the amino group is on position 1 (or α), then only one naphthoquinoline is possible, and if it is on position 2 (or β), then there are two such isomers. Oxidation (with permanganate in base and then chromic acid) of *both* of the β-naphthoquinolines can yield only quinolinic acid, the decarboxylation of which can yield only nicotinic acid—with which it was compared.

The **Skraup** reaction itself involves the reaction between an aryl amine (such as aniline [aminobenzene]) and glycerol (1,2,3-propanetriol) or some other similar three-carbon fragment in the presence of sulfuric acid (H_2SO_4) to effect a condensation and cyclization. The sulfuric acid (H_2SO_4) can also act as an oxidizing agent to consummate the reaction, although other oxidizing agents are used from time to time.

[¶]Pinner, A. *Chem. Ber.*, **1893**, *26*, 292.

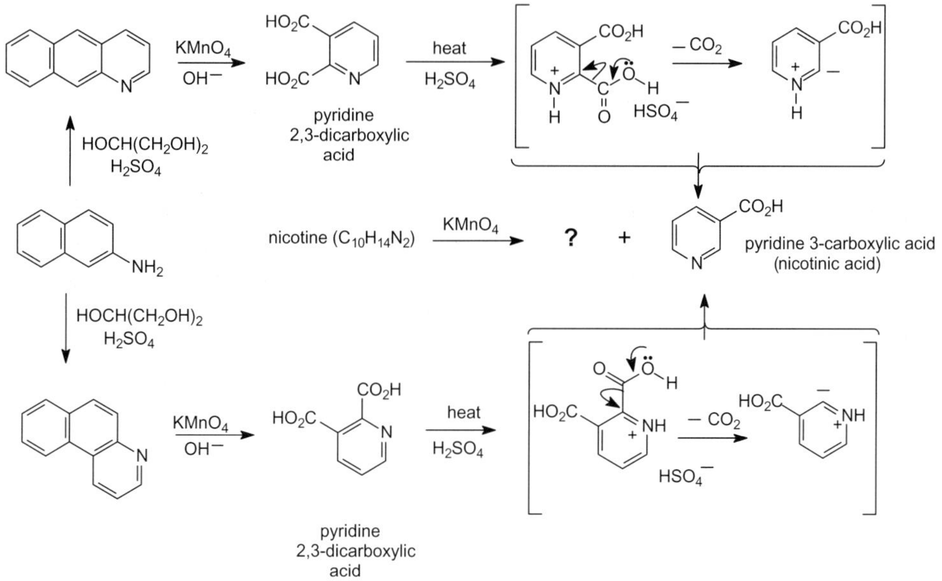

Scheme 13.13. A set of oxidative reactions on quinoline and isoquinoline from which dicarboxylic acids are initially obtained. Both pyridine-2,3-dicarboxylic acid and pyridine-3,4-dicarboxylic acid undergo decarboxylation to pyridine-3-carboxlyic acid. The latter is identical to the acid obtained on oxidation of nicotine. After Skraup, Z. H.; Cobenzl, A. *Monatsh. Chem.*, **1883**, *1*, 436.

misinterpretations (including some of his own) and describing careful work on bromination of nicotine in acetic acid (CH_3CO_2H) and aqueous hydrogen bromide ($HBr_{(aq)}$) solutions, was able to make some progress.

In brief, Pinner found (Scheme 13.14) that two oxygen (and bromine) compounds could be isolated. He called them dibromcotinine ($C_{10}H_{10}N_2OBr_2$) and dibromticonine ($C_{10}H_8N_2O_2Br_2$), and he studied their decomposition in basic (OH^-) solution. As shown in the scheme, he found that the former, $C_{10}H_{10}N_2OBr_2$, underwent decomposition to yield methanamine (CH_3NH_2), methyl β-pyridyl ketone and, presumably, glyoxylic acid, which was further oxidized to the isolated oxalic acid [$(HO_2C)_2$]. The latter, $C_{10}H_8N_2O_2Br_2$, under the same conditions, produced methanamine (CH_3NH_2), nicotinic acid, and malonic acid [$CH_2(CO_2H)_2$]. And so, Pinner concluded that the structure of nicotine consisted of the five-membered *N*-methylpyrrolidine ring fused at its α-position to the C-3 position of pyridine ring. Of course, the question of stereochemistry did not yet arise.

The issue of the stereochemistry of (–)-nicotine was initially addressed by Karrer and Widmer* who found that the hydrogen iodide (HI) salt of nicotine could be

*The naturally occurring alkaloid "hygrine" is easily racemized and, as normally isolated, is only slightly levorotatory. The acids (+)-hygrinic acid and (–)-hygrinic acid obtained from hygrine are not readily racemized.

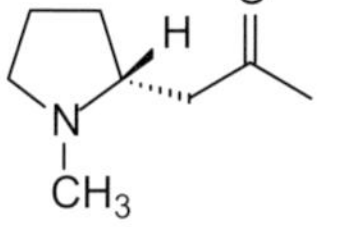

(R)-(+)-hygrine

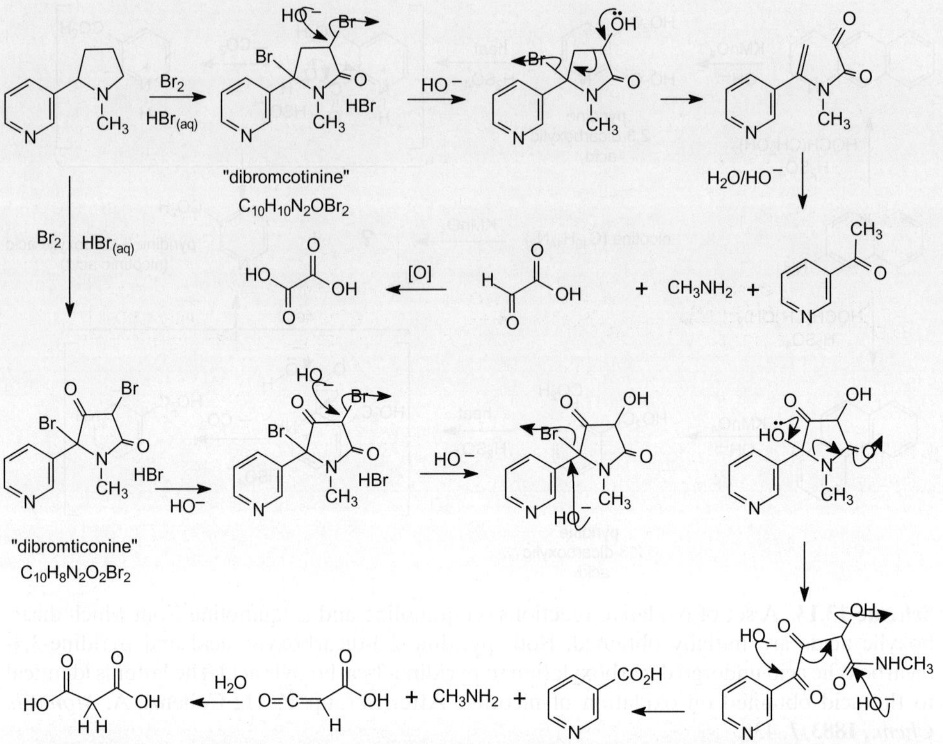

Scheme 13.14. The results of the investigations of Pinner on the bromination of nicotine and the subsequent hydrolysis of the two isolable products. After Pinner, A. *Chem. Ber.*, **1893**, 26, 292.

methylated (Scheme 13.15) and then oxidized, first with ferricyanide [$K_3Fe(CN)_6$] and then with chromic anhydride (CrO_3) in sulfuric acid (H_2SO_4) to (–)-hygrinic acid [a derivative of the naturally occurring alkaloid (–)-hygrine*]. They further found that treatment of (–)-hygrinic acid with methyl iodide (CH_3I) and silver oxide (Ag_2O) produced (–)-stachydrine, the zwitterion produced on methylation of (–)-**proline** (**Pro, P**). This clearly demonstrated that the stereochemistry at the stereogenic carbon was the same for (–)-**proline** (**Pro, P**), (–)-stachydrine, (–)-hygrinic acid, and (–)-nicotine.

In 1969, Buckingham and coworkers,[†] examined the X-ray crystal structures of two diastereoisomeric (–)-proline (**Pro, P**) cobalt complexes and affirmed that (–)-proline (**Pro, P**) is (*S*)-(–)-proline (**Pro, P**). Interestingly the complexes (Figure 13.2) are best described as "catoptric," that is, they are diastereoisomeric but with the same configuration of the (*S*)-(–)-proline (**Pro, P**) in each.

*Karrer, P.; Widmer, R. *Helv. Chim. Acta*, **1925**, 8, 364.
[†]Buckingham, D. A.; Marzilli, L. G.; Maxwell, I. E.; Sargeson, A. M. *Chem. Commun.*, **1969**, 583.

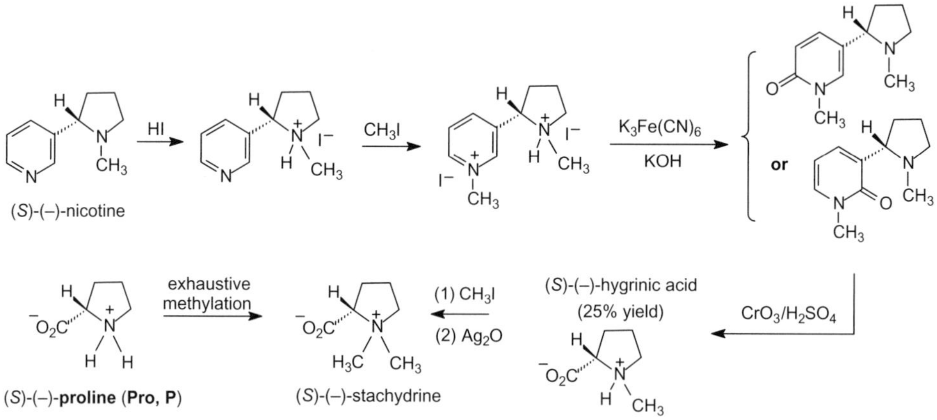

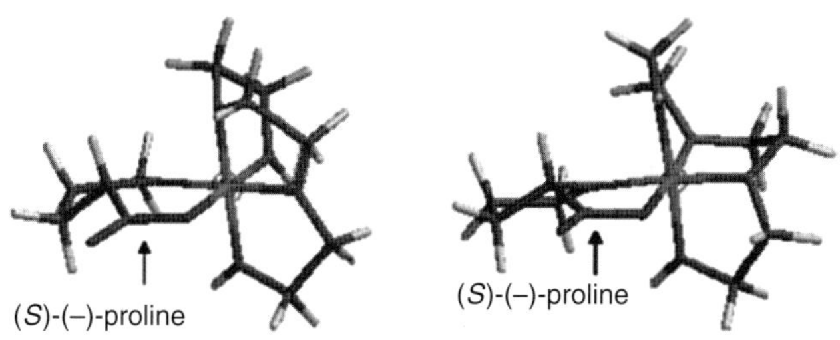

Scheme 13.15. The chemical correlation of the absolute stereochemistry of (S)-$(-)$-nicotine with (S)-$(-)$-proline (Pro, P). After Karrer, P.; Widmer, R. *Helv. Chim. Acta*, **1925**, *8*, 364.

Figure 13.2. On the left, a representation of $L(-)_{589}$-β_2-RRS-[Co trien(S-Pro)]$^{2+}$ ion and on the right, a representation of $D(+)_{589}$-β_2-SSS-[Co trien(S-Pro)]$^{2+}$ ion as determined by X-ray crystallography. After Buckingham, D. A.; Marzilli, L. G.; Maxwell, I. E.; Sargeson, A. M. *Chem. Commun.*, **1969**, 583.

III. Biosynthesis of Hyoscyamine and Nicotine

a. The Common Feature. It has become clear after many years of work* that the intermediacy of an N-methyl-Δ^1-pyrrolium cation lies on the path both to hyoscyamine and to nicotine. The intermediate can be generated from both ornithine and arginine (**Arg, R**) via butane-1,4-diamine (putrescine). The paths from the amino acids to N-methylputrescine are shown in Scheme 13.16.

Pyridoxal phosphate-catalyzed decarboxylation (ornithine decarboxylase, EC 4.1.1.17) yields butane-1,4-diamine (putrescine) directly. **Arginine (Arg, R)** first undergoes pyridoxal phosphate-catalyzed decarboxylation (**arginine [Arg, R]** decarboxylase, EC 4.1.1.19) to agmatine and then hydrolytic loss of urea (agmatinase, EC 3.5.3.11) to produce the same diamine. Ornithine and **arginine (Arg, R)** are members

*Ziegler, J.; Facchini, P. J. *Ann. Rev. Plant Biol.*, **2008**, *59*, 735.

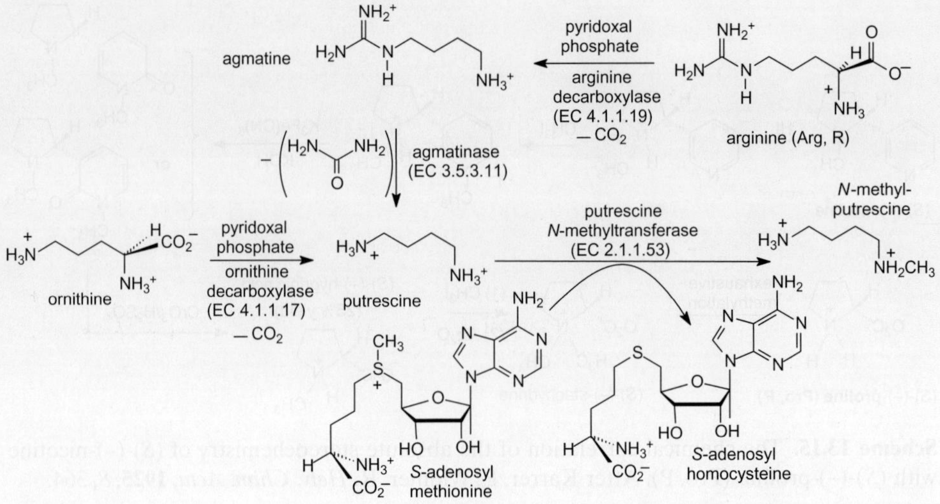

Scheme 13.16. The initial steps in the formation of an N-methyl-Δ^1-pyrrolium cation. Pyridoxal phosphate-catalyzed decarboxylation of ornithine (ornithine decarboxylase, EC 4.1.1.17) yields butane-1,4-diamine (putrescine). **Arginine (Arg, R)** also undergoes pyridoxal phosphate-catalyzed decarboxylation (**arginine [Arg, R]** decarboxylase, EC 4.1.1.19) to agmatine and then hydrolytic loss of urea (agmatinase, EC 3.5.3.11) to produce the same diamine. Methylation on nitrogen by S-adenosylmethionine is catalyzed by putrescine N-methyltransferase (EC 2.1.1.53). EC numbers and some graphic materials provided in this scheme have been taken from appropriate links in a URL starting with http://www.chem.qmul.ac.uk/iubmb/enzyme/.

of the urea cycle (Scheme 12.7). Finally, in Scheme 13.16, methylation on nitrogen is accomplished by transfer of the methyl from S-adenosylmethionine (the latter forming S-adenosylhomocysteine in the process). This process, generating N-methylputrescine, is catalyzed by putrescine N-methyltransferase (EC 2.1.1.53).

In the next step toward the formation of the N-methyl-Δ^1-pyrrolium cation, the N-methylputrescine undergoes oxidation to 4-(N-methyl)aminobutanal. This process is catalyzed by a primary amine oxidase (EC 1.4.3.21). The oxidase is a member of a group of copper-containing enzymes that apparently utilize a dihydroxytyrosine-type (or trihydroxyphenylalanine) quinone to effect the reaction. As shown in Scheme 13.17, a tyrosine residue within the oxidase undergoes posttranslational modified copper-catalyzed oxidation to the corresponding topaquinone (TPQ). It is argued that only oxygen (O_2), copper, and the protein are required. Thus, the oxidation of the amine to aldehyde involves the initial conversion of oxygen (O_2), a proton source, and tyrosine to TPQ and hydrogen peroxide, while it is in the second step that the reduction of TPQ back to a trisphenol brings about the oxidation of the amine to the imine. Hydrolysis of the imine provides the aldehyde 4-(N-methyl) aminobutanal, the spontaneous cyclization of which produces water and the N-methyl-Δ^1-pyrrolium cation. Scheme 13.17 shows the formation of TPQ as a series of two-electron processes. It is far more likely that a series of one-electron processes are involved in its formation.

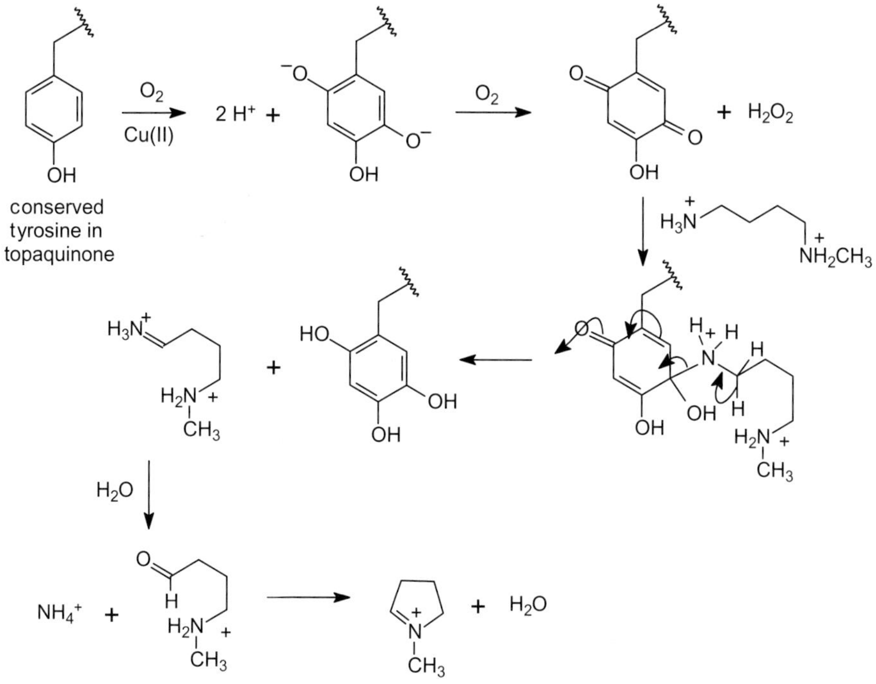

Scheme 13.17. A cartoon showing the possible formation of the N-methyl-Δ^1-pyrrolium cation from a topaquinone-bound oxidant and the N-methylputrescine cation. This process is catalyzed by a primary amine oxidase (EC 1.4.3.21). Additional information can be found at Dooley, D. M. *J. Biol. Inorg. Chem.*, **2008**, *4*, 1 and Heim, W. G.; Sykes, K. A.; Hildreth, S. B.; Sun, J.; Lu, R.-H.; Jelsko, J. G. *Phytochemistry*, **2007**, *68*, 454.

b. The Biosynthesis of Nicotine. It will be recalled that nicotinic acid is produced by decarboxylation of quinolinic acid as shown in Scheme 12.103. Scheme 13.18 reproduces part of Scheme 12.103 and adds to that earlier material a hydrolysis process involving removal of the carbohydrate from the nicotinic acid to free the latter.

The reaction of nicotinic acid with the N-methyl-Δ^1-pyrrolium cation has been shown to involve exchange at C-6 as shown in Scheme 13.19.

As recently pointed out, the joining of the two pieces (i.e., the N-methyl-Δ^1-pyrrolium cation to the pyridine ring) to produce nicotine occurs with the participation of enzymes whose nature remains "poorly characterized." Indeed, the condensation process itself with labeled materials is not, even at this writing, as clean an experiment as is desirable.* Nonetheless, it appears that it is widely accepted that a hydride source donates a hydrogen, with its pair, to C-6 of the nicotinic acid, and a ^{3}H (tritium) label originally placed there is lost as the condensation reaction is consummated.

*Ziegler, J.; Facchini, P. J. *Ann. Rev. Plant Biol.*, **2008**, *59*, 735 and Leete, E.; Liu, Y.-Y. *Phytochemistry*, **1973**, *12*, 593.

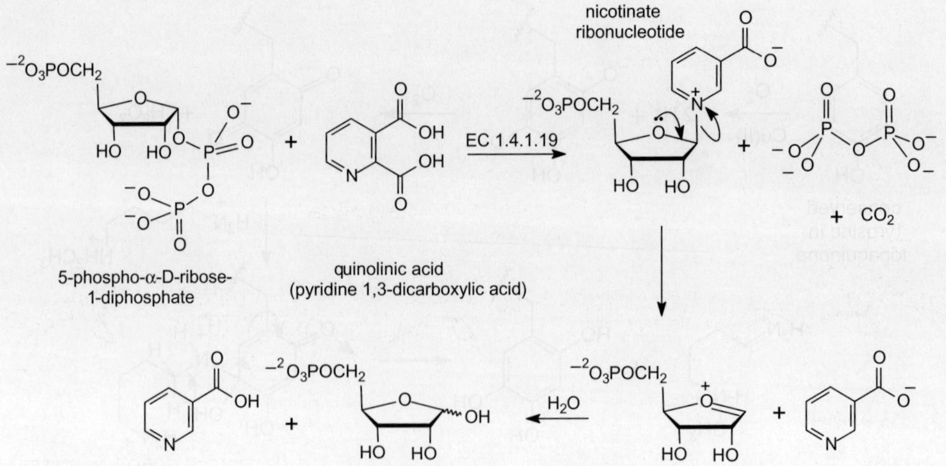

Scheme 13.18. An expansion of Scheme 12.103 allowing the formation of "free" nicotinic acid and ribose-5-phosphate from a nicotinate ribonucleotide.

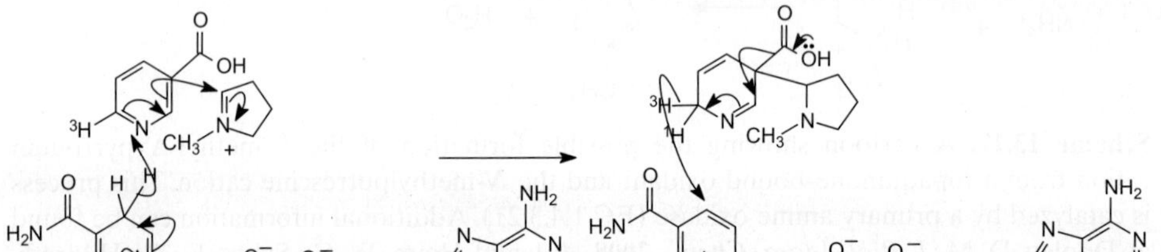

Scheme 13.19. A cartoon representation of a possible path to (–)-nicotine from a combination of nicotinic acid with an *N*-methyl-Δ^1-pyrrolium cation. There is no evidence that the source of hydride is NADH or NAD(P)H, and it is shown to be so here only as a matter of convenience. As noted earlier, the enzymes involved are currently "poorly characterized."

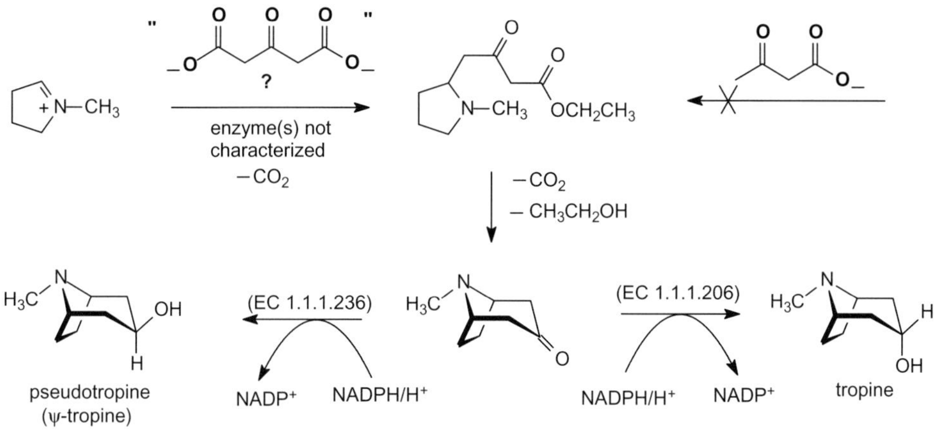

Scheme 13.20. Some indications of the pathways available for the conversion of N-methyl-Δ^1-pyrrolium cation to tropine and pseudotropine (ψ-tropine) by reaction with an acetate-derived representative fragment.

c. The Biosynthesis of Hyoscyamine.

Some current ideas on the biosynthesis of hyoscyamine are affirmed and some are confused by numerous labeling experiments with both ^{14}C and ^3H isotopes. Some enzymatic information is available but even that is not in complete accord with any simple pathway. Thus, it is quite clear that the N-methyl-Δ^1-pyrrolium cation is acted upon by an acetate-derived fragment to generate the remainder of the bicyclic system. But it is not clear exactly what that acetate-derived fragment must be! Thus, as shown in Scheme 13.20, it has been reported* that the ethyl ester of (R,S)-[2,3-$^{13}C_2$, 3-^{14}C]-4-(1-methyl-2-pyrrolidinyl)-3-oxobutanoate is well incorporated into hyoscyamine. This hyoscyamine precursor is clearly derived from the N-methyl-Δ^1-pyrrolium cation on the one hand and what appears to be an equivalent of the half-acid ester of acetone dicarboxylic acid on the other (interestingly, however, acetoacetate [$CH_3COCH_2CO_2H$] does not appear to be utilized intact). Second, as just noted, *both* the R- and S-enantiomers seem to be accepted with about equal facility, implying some sort of subsequent epimerization. Third, it appears that there is scrambling of deuterium label from [2-^2H]-N-methylpyrrolidine into C-1 *and* C-5 (i.e., the bridgehead carbons) of the tropane ring.

These details aside however, the pieces needed on the path to the tropone ring are now approximately understood, while the details concerning the process (or processes) and the enzyme (or enzymes) responsible remain to be characterized.

Scheme 13.20 also shows that the action of tropone reductase I (EC 1.1.1.206) (requiring nicotinamide adenine dinucleotide [NADPH]) on tropone produces tropine, while tropone reductase II (EC 1.1.1.236) (also requiring NADPH) generates pseudotropine (ψ-tropine).

*Robins, R. J.; Abraham, T. W.; Parr, A. J.; Eagles, J.; Walton, N. J. *J. Am. Chem. Soc.*, **1997**, *119*, 10929.

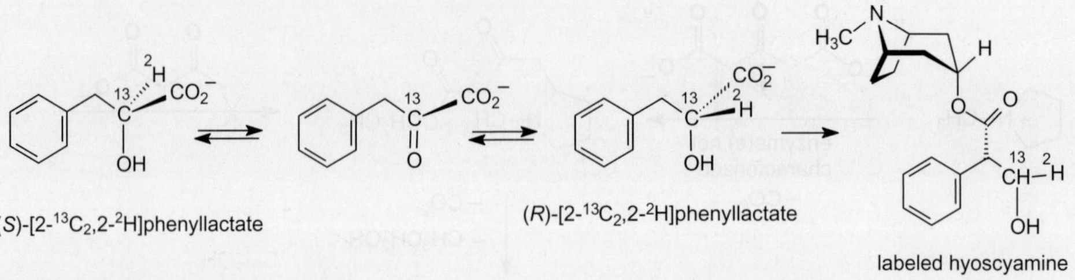

(S)-[2-$^{13}C_2$,2-^{2}H]phenyllactate

(R)-[2-$^{13}C_2$,2-^{2}H]phenyllactate

labeled hyoscyamine

Scheme 13.21. Incorporation of labeled (R)-[2-^{13}C,2-^{2}H]phenyllactate but not (S)-[2-^{13}C, 2-^{2}H]phenyllactate into hyoscyamine.

d. The Biosynthesis of Tropic Acid. The reasonable suggestion that (–)-tropic acid was likely related to a phenyl propanoid or phenylalanine was made very early, but it was not until recently, using ^{13}C nuclear magnetic resonance (NMR) spectroscopy, that it was found that phenylalanine* and the presumed products of transamination of phenylalanine, viz. phenylpyruvic acid and phenyllactic acid, would also serve.[†] Indeed, in the event, it appears clear that (R)-phenyllactate is esterified with tropine to produce the alkaloid (R)-littorine and it is the latter, in the presence of a cytochrome P_{450} oxidase, that undergoes rearrangement to (S)-hyoscyamine. Thus, combined with the isolation of a gene to silence the cytochrome P_{450} (CYP80F1),[‡] the presence of which then resulted in inhibiting the formation of hyoscyamine and the accumulation of littorine, and the lack of any clear indications that the rearrangement of phenyllactic acid to tropic acid could proceed without the initial formation of littorine, it appears that efforts that might have shown if that rearrangement could occur were abandoned.

The details of the process, shown in Scheme 13.21, involved a series of experiments[§] that showed that carbon from both (S)-[2-^{13}C,2-^{2}H]phenyllactate and $(R)(-)$-[2-^{13}C,2-^{2}H]phenyllactate is incorporated into (–)-tropic acid but that *only* the deuterium from $(R)(-)$-[2-^{13}C,2-^{2}H]phenyllactate made it into the alkaloid. The deuterium was found on the carbinol carbon (C-3′). These data were interpreted as indicating that (S)-[2-^{13}C,2-^{2}H]phenyllactate must have undergone oxidation to phenylpyruvate and then reduction to the (R)-isomer of phenyllactate acid in order to be incorporated.

As noted above, the ability to use labeled synthetic intermediates has led to further understanding of the process by which the (–)-tropic acid was produced from littorine. First, it was shown that multiply labeled littorine was converted to hysocyamine.[¶] Thus, [^{2}H$_3$-N-methyl, 1′-3′-^{13}C$_2$]littorine was converted into [^{2}H$_3$-N-methyl, 1′-2′-^{13}C$_2$]hyoscyamine *in vivo*, showing that the rearrangement occurred from one intact alkaloid into the other (Equation 13.1).

*Leete, E.; Kowanko, N.; Newmark, R. A. *J. Am. Chem. Soc.*, **1975**, *97*, 6825.

[†]Chesters, N. C. J. E.; O'Hagan, D.; Robins, R. J. *J. Chem. Soc. Chem. Commun.*, **1995**, 127.

[‡]Li, R.; Reed, D. W.; Lie, E.; Nowak, J.; Pelcher, L. E.; Page, J. E.; Covello, P. S. *Chem. Biol.*, **2006**, *13*, 513.

[§]Robins, R. J.; Chesters, N. C. J. E.; O'Hagan, D.; Paar, A. J.; Walton, N. J.; Woolley, J. G. *J. Chem. Soc. Perkin Trans. I*, **1995**, 481.

[¶]Robins, R. J.; Bachmann, P.; Woolley, J. G. *J. Chem. Soc. Perkin Trans. I*, **1994**, 615.

$[^2H_3\text{-}N\text{-methyl, }1'\text{-}3'\text{-}^{13}C_2]$littorine $[^2H_3\text{-}N\text{-methyl, }1'\text{-}2'\text{-}^{13}C_2]$hyoscyamine

(13.1)

Now, as shown in Scheme 13.22, both $(2R,3S)[3\text{-}^2H_1]$phenyllactate and $(2R,3R)$ $[3\text{-}^2H_1]$phenyllactate were fed to *Datura stramonium*. On harvest, it was found that the (–)-hyoscyamine formed from the $(2R,3R)[[3\text{-}^2H_1]$phenyllactate had lost deuterium while that from $(2R,3S)[3\text{-}^2H_1]$phenyllactate had retained the deuterium. This result requires that the migration of the carboxylate group, and its appended tropinol (in littorine, the obligatory intermediate) occurs with inversion of configuration at the benzylic carbon.*

Furthermore (also shown in Scheme 13.22), when a mixture of $(\pm)\text{-}[2\text{-}^3H]\text{-}$phenyllactate (only the $2R$-isomer is shown in the scheme) is fed (as above), it was found that the tritium appeared on the hydroxymethyl group in (–)-hyoscyamine. The migration was stereospecific and occurred with inversion of configuration at the seat of migration. This was demonstrated (shown in Scheme 13.22) by the degradation of the (–)-hyoscyamine and isolation of the chiral acetic acid.[†‡]

---

*Chesters, N. C. J. E.; Walker, K.; O'Hagen, D.; Floss, H. G. *J. Am. Chem. Soc.*, **1996**, *118*, 925.

[†]Chesters, N. C. J. E.; O'Hagen, D.; Robins, R. J.; Kastelle, A.; Floss, H. G. *J. Chem. Soc. Chem. Commun.*, **1995**, 129.

[‡]Although the details differ in minor ways, two works, published in the same journal in 1969 laid out the details of the analysis of chiral methyl groups. The work was described by Cornforth, J. W.; Redmond, J. W.; Eggerer, H.; Buckel, W.; Gutschow, C. *Nature*, **1969**, *221*, 1212 and Lüthy, J.; Rétey, J.; Arigoni, D. *Nature*, **1969**, *221*, 1213. In the first case, Cornforth et al. converted phenylacetylene to the corresponding 1-deuterio compound and reduced the mono-deuterio alkyne with diimide to the alkene. Epoxidation yielded two epoxides, which on reduction with lithium borotritide gave a mixture of $(1R,2R)[^2H_1,^3H_1]\text{-}1\text{-}$phenylethanol and $(1S,2S)\text{-}[^2H_1,^3H_1]\text{-}1\text{-}$phenylethanol. The alcohols were converted to their corresponding acid phthalates, and the acid phthalates resolved as their brucine salts.

(Continued on next page)

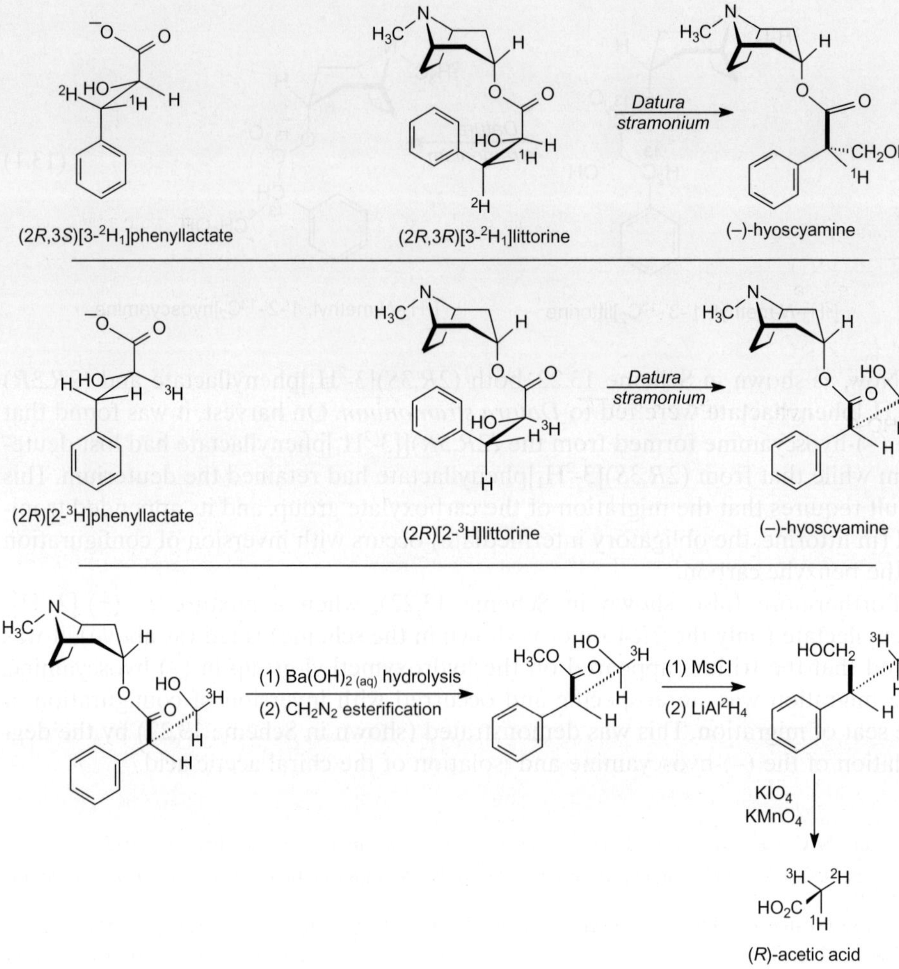

(2R,3S)[3-^{2}H$_1$]phenyllactate

(2R,3R)[3-^{2}H$_1$]littorine

(−)-hyoscyamine

(2R)[2-^{3}H]phenyllactate

(2R)[2-^{3}H]littorine

(−)-hyoscyamine

(1) Ba(OH)$_2$ (aq) hydrolysis

(2) CH$_2$N$_2$ esterification

(1) MsCl

(2) LiAl^{2}H$_4$

KIO$_4$
KMnO$_4$

(R)-acetic acid

Scheme 13.22. A representation of experiments showing the results of deuterium- and tritium-labeled phenyllactate feedings to *Datura stramonium*. The results shown must be accounted for in any subsequently formulated mechanism (yet forthcoming). After Chesters, N. C. J. E.; Walker, K.; O'Hagen, D.; Floss, H. G. *J. Am. Chem. Soc.*, **1996**, *118*, 925 and Chesters, N. C. J. E.; O'Hagen, D.; Robins, R. J.; Kastelle, A.; Floss, H. G. *J. Chem. Soc. Chem. Commun.*, **1995**, 129.

The observation that acetate shown in Scheme 13.22 was actually $(R)[^2H_1,{}^3H_1]$ acetate was demonstrated by the method of Cornforth and Arigoni to which reference was made above. That is, the acetate was allowed to react with glyoxylate in the presence of malate synthase (EC 2.3.3.9), adenosine triphosphate (ATP), acetate

Oxidation with chromic acid destroyed the asymmetry at the benzylic carbon and yielded the corresponding acetophenones, which were induced to undergo the Baeyer–Villiger oxidation to generate the corresponding phenylacetates. Hydrolysis of the phenylacetates yielded the corresponding $(R)[^2H_1,{}^3H_1]$ acetate and (S)-$[^2H_1,{}^3H_1]$acetate.

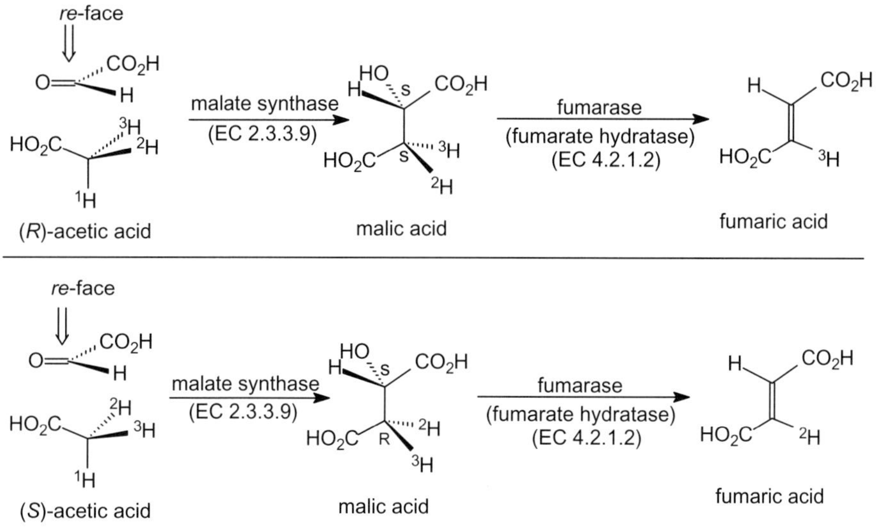

Scheme 13.23. Conversion of acetate to fumarate via malate. After Cornforth, J. W.; Redmond, J. W.; Eggerer, H.; Buckel, W.; Gutschow, C. *Nature*, **1969**, *221*, 1212 and Lüthy, J.; Rétey, J.; Arigoni, D. *Nature*, **1969**, *221*, 1213.

kinase (EC 2.7.2.1), phosphate acetyltransferase (EC 2.3.1.8), and coenzyme A (CoA) (glyoxylate cycle). Each provided a sample of malate and a sufficient isotope effect was shown so that ^{1}H loss would be preferential.

In the presence of fumarase (fumarate hydratase, EC 4.2.12) malate undergoes reversible dehydration to generate fumarate (citric acid cycle). In this process (a *trans*-elimination), the hydroxyl at C-2 and the *pro-R* hydrogen at C-3 are lost (Scheme 13.23).

It was in this way that the acetic acid depicted as (*R*)-acetate in Scheme 13.22 was shown to be (*R*) and, since reduction of the mesylate with lithium borotritide [LiB(^{3}H)$_4$] occurred with inversion of configuration, the original rearrangement must itself have gone with inversion.

C. MORPHINE (AND CODEINE AND THEBAINE)

I. Chemistry of Morphine (and Codeine and Thebaine)

Figure 13.3 provides a representation of the close relationship between morphine, its O-methyl ether, codeine, and the enol ether, thebaine. These three are shown despite the thrust of this portion toward a discussion of morphine since they are frequently isolated together and their chemistry is intertwined.

The viscous crude latex-like exudate of the unripe seed pods of the opium poppy, *Papaver somniferum*, is "opium." This material has a long history of human use, and a wide variety of literature sources can be consulted for more interesting aspects of what is known.

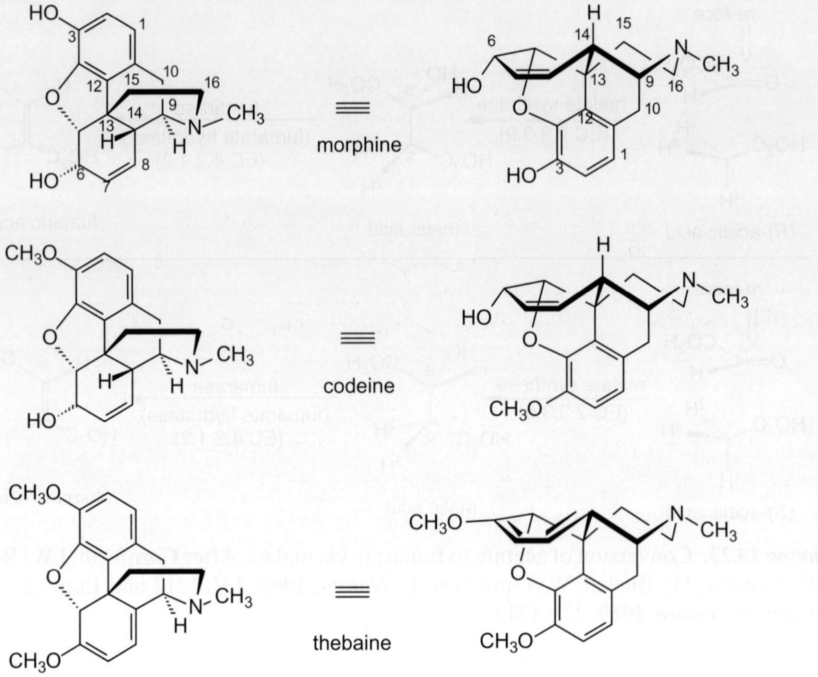

Figure 13.3. Representations of the structures of morphine, codeine, and thebaine.

While there is a significant amount of interesting chemistry that was developed as a consequence of efforts to prove the structure of these (and related) alkaloids,* it can be argued that highlights in the story of morphine are (1) the isolation of the pure alkaloid morphine by Sertürner in 1805, (2) the deduction of what is now recognized as the correct structure by Gulland and Robinson in 1925, (3) the determination of the absolute configuration by Mackay and Hodgkin in 1955, and (4) the first total synthesis by Tschudi and Gates in 1952.[†]

It had been found quite early that acid hydrolysis of thebaine yielded codeinone, which had, of course, been obtained by the oxidation of codeine with chromic acid in acetic acid. Methylation of morphine with, for example, trimethylphenylammonium hydroxide, yielded codeine.

The basic ring structure of these systems was determined by the observation (Scheme 13.24) that the methiodide of thebaine when refluxed in acetic anhydride [(CH₃CO)₂O] underwent decomposition to the ester of a new phenol (called "thebaol"), which yielded phenanthrene on zinc dust distillation. The ester of

*Much of the early work concerning the chemistry and biosynthetic hypotheses on morphine and related alkaloids has been summarized in (a) Holmes, H. L. *The Alkaloids*, Vol. 2, Academic Press, New York, **1952**, pp. 1–159; (b) Holmes, H. L.; Stork, G. *The Alkaloids*, Vol. 2, Academic Press, New York, **1952**, pp. 161–217; (c) Dalton, D. R. *The Alkaloids*, Dekker, New York, **1979**, pp. 234–235 and 347–365; (d) Hesse, M. *Alkaloids*, Wiley-VCH, New York, **2002**, pp. 260–279.

[†](1) Sertürner, F. W. *J. Pharm.*, **1805**, *13*, 234. (2) Gulland, J. M.; Robinson, R. *Mem. Proc. Manchester Lit. Phil. Soc.*, **1925**, *69*, 79. (3) Mackay, M.; Hodgkin, D. C. *J. Chem. Soc.*, **1955**, 3261. (4) Gates, M.; Tschudi, G. *J. Am. Chem. Soc.*, **1952**, *74*, 1109.

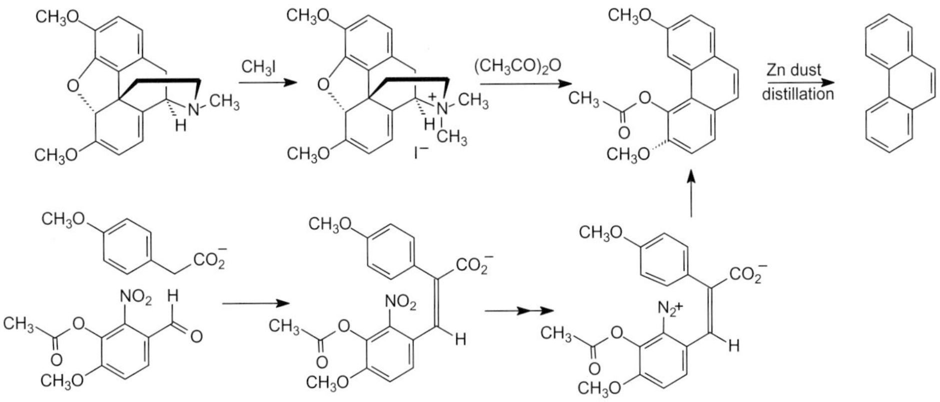

Scheme 13.24. The demonstration of the location of oxygen substituents in thebaine (and hence morphine and codeine) and the presence of a phenanthrene (barring rearrangement) system. The degradative work is from Vongerichten, E.; Dittmer, O. *Chem. Ber.*, **1906**, *39*, 1718. The synthesis, using the sodium salt of the 4-methoxyphenylacetic acid reacting with the nitrobenzaldehyde derivative (in acetic acid for 3 days at 100°C) followed by reduction [iron(II) sulfate], diazotization (sodium nitrite in methanolic sulfuric acid), cyclization, and pyrolysis (loss of carbon dioxide) was effected by Pschorr, H. *Liebigs Ann. Chem.*, **1912**, *391*, 40.

thebaol was synthesized (as shown) by the condensation between the sodium salt of 4-methoxyphenylacetic acid and 3-acetoxy-4-methoxy-2-nitrobenzaldehyde followed by reduction, diazotization, cyclization, and pyrolysis (loss of carbon dioxide).

The correct formulation of the structure of these bases by J. M. Gulland and R. Robinson as noted above and the determination of the structure by X-ray crystallography by M. Mackay and D. C. Hodgkin, as well as the absolute configuration by chemical means (once the structure was on firm ground), also led to understanding of a variety of puzzles.

The absolute stereochemistry at C-13 and C-14 was determined as shown in Schemes 13.25 and 13.26.

Thus, as shown in Scheme 13.25, when dihydrocodeinone (produced by catalytic hydrogenation of codeine followed by chromic acid oxidation) is subjected to two successive Hofmann degradations (with reduction after the first), a single alkene is obtained. Conversion of the carbonyl group in this alkene to the corresponding ketal then permits osmium tetroxide–lead acetate oxidation of the alkene to the corresponding aldehyde, which, under Wolff–Kishner (Huang–Minlon modification) reduction conditions, is converted to a methyl group. Then, removal of the ketal protecting group, hydrogenolysis of the ether bridge, and methylation of the phenol generated in that process produced a ketomethyl ether. Finally, formation of the thioketal of the ketone and its reduction followed by benzylic oxidation with chromic anhydride (CrO_3) in sulfuric acid (H_2SO_4) produced a new ketone, which underwent oxidative ozonolysis to (1S,2S)-(–)-2-carboxy-2-methylcyclohexane acetic acid.

The absolute configuration of the dicarboxylic acid, (1S,2S)-(–)-2-carboxy-2-methylcyclohexane acetic acid had been laboriously established over a period of

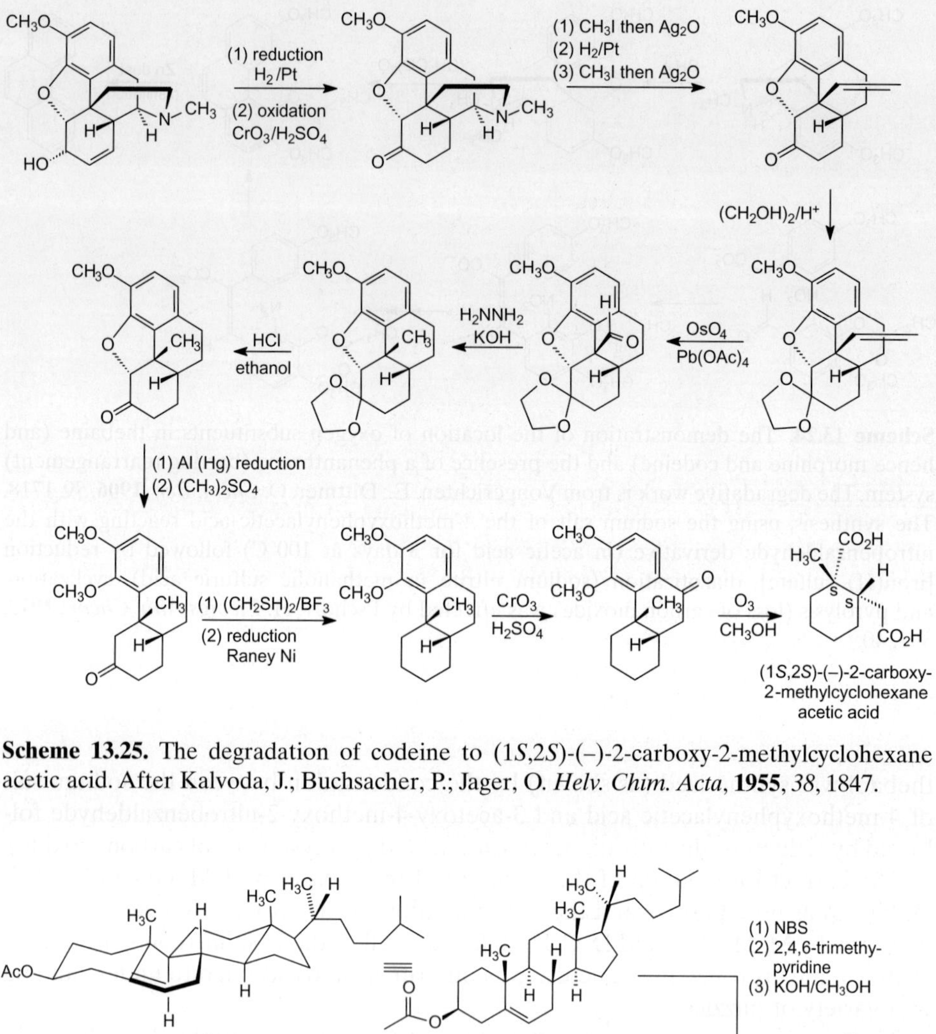

Scheme 13.25. The degradation of codeine to (1S,2S)-(−)-2-carboxy-2-methylcyclohexane acetic acid. After Kalvoda, J.; Buchsacher, P.; Jager, O. *Helv. Chim. Acta*, **1955**, *38*, 1847.

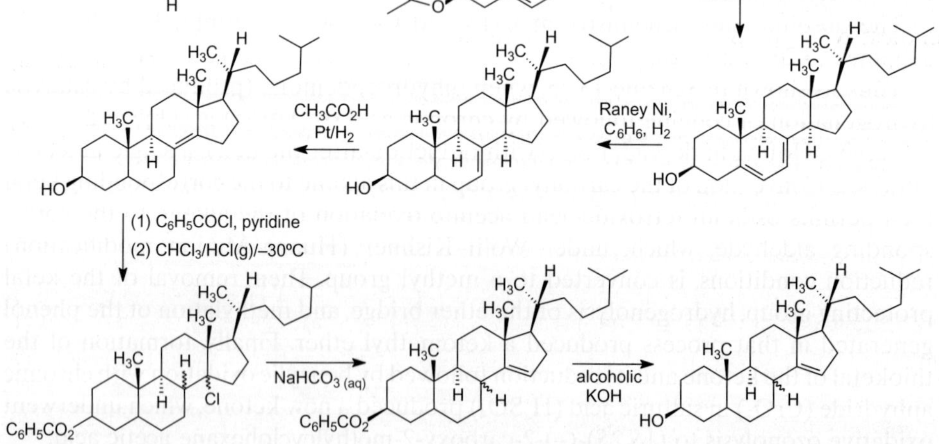

Scheme 13.26. A representation of the formation of cholest-$\Delta^{14(15)}$en-3β-ol from cholesterol acetate. After Cornforth, J. W.; Gore, I. Y.; Popjak, G. *Biochem. J.*, **1957**, *65*, 94. NBS = N-bromosuccinimide.

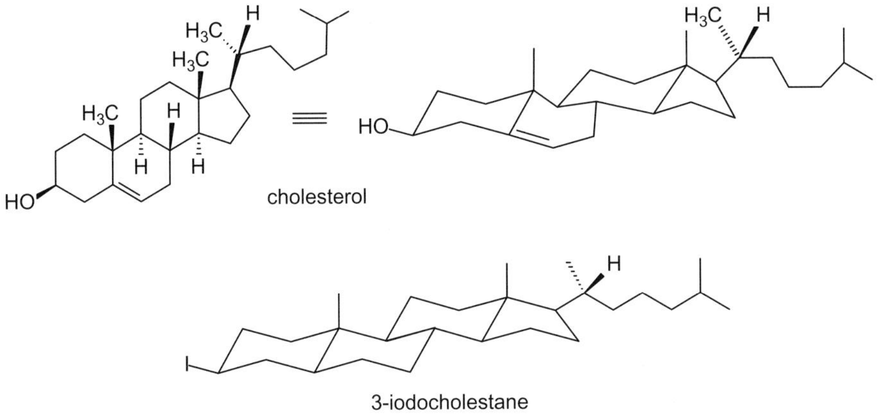

Figure 13.4. Representations of the structures of cholesterol and 3-iodocholestane. See Bernal, J. D.; Crowfoot, D.; Fankuchen, I. *Proc. R. Soc.*, **1940**, *239*, 17; Carlisle, C. H.; Crowfoot, D. *Proc. R. Soc. A*, **1945**, *184*, 64; and Fieser, L. F.; Fieser, M. *Steroids*, Reinhold, New York, 1959.

years. It is a diastereomer of (1*R*,2*S*)-(–)-2-carboxy-2-methylcyclohexane acetic acid obtained from steroid degradation.

First, it was necessary to establish the absolute stereochemistry of the steroid nucleus.

As shown in Figure 13.4, and as reported in 1940, an extensive survey of the structures of steroids as determined by X-ray crystallography was undertaken[*] and shortly thereafter, the X-ray crystal structure of cholesteryl iodide (3-iodocholestane) was reported.[†] The structure was largely that anticipated on the basis of a vast amount of chemical evidence (that was subsequently summarized).[‡] This established the relative stereochemistry of the side chain and rings.

Almost a decade after the X-ray crystal structure was determined, it was reported that, as part of "more extensive work on the degradation of cholesterol," cholest-$\Delta^{14(15)}$en-3β-ol underwent ozonolysis to cleave the double bond[§] and that pyrolysis of the ozonolysis product gave an unsaturated aldehyde. The work on cholesterol and the ozonolysis experiment are shown in Schemes 13.26 and 13.27. Scheme 13.27 also shows the relationship between the **chiral** aldehyde isolated on pyrolysis of the ozonolysis product from cholest-$\Delta^{14(15)}$en-3β-ol and optically active natural (*R*)-(+)-citronellal.

The absolute stereochemistry of (+)-citronellal had, in turn, been determined by its degradation to (+)-methyl succininic acid (Scheme 13.28). The rather circuitous route begins with the treatment of (+)-citronellal with acetic anhydride [$(CH_3CO)_2O$]. The treatment results in the formation of the acetate ester of an isopulegol. Hydrolysis of the ester followed by chromic acid (CrO_3) oxidation produces a

[*]Bernal, J. D.; Crowfoot, D.; Fankuchen, I. *Proc. R. Soc.*, **1940**, *239*, 17.

[†]Carlisle, C. H.; Crowfoot, D. *Proc R. Soc. A*, **1945**, *184*, 64.

[‡]Fieser, L. F.; Fieser, M. *Steroids*, Reinhold, New York, 1959.

[§](1) Cornforth, J. W.; Youhotsky, I.; Popják, G. *Nature*, **1954**, *173*, 536. (2) Cornforth, J. W.; Gore, I. Y.; Popjak, G. *Biochem. J.*, **1957**, *65*, 94.

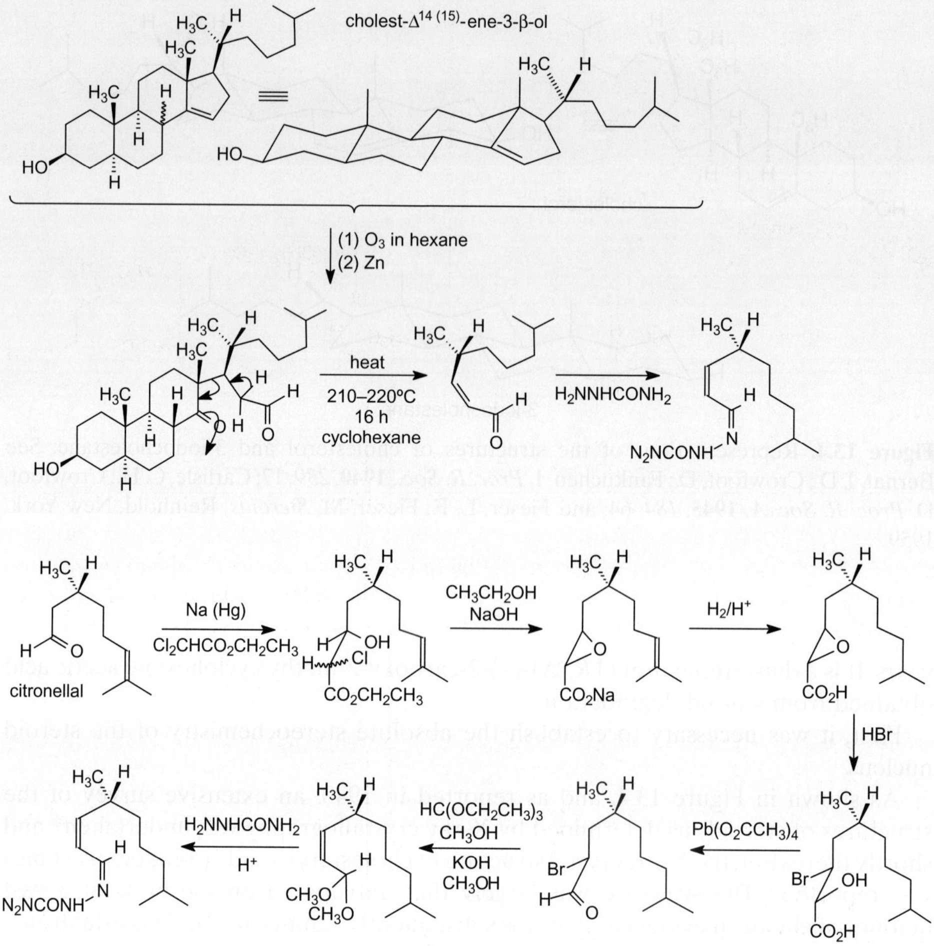

Scheme 13.27. The relationship between the **chiral** aldehyde isolated on pyrolysis of the cholest-$\Delta^{14(15)}$en-3β-ol ozonolysis product and (R)-(+)-citronellal as demonstrated by the identity of their semicarbazone derivatives. After Cornforth, J. W.; Youhotsky, I.; Popják, G. *Nature*, **1954**, *173*, 536 and Cornforth, J. W.; Gore, I. Y.; Popjak, G. *Biochem. J.*, **1957**, *65*, 94.

ketone, isopulegone. Then, treatment of the isopulegone with barium hydroxide [Ba(OH)₂] induces rearrangement to (+)-pulegone [and passing (+)-pulegone over glass wool at 600°C regenerates citronellal!].* Oxidation of (+)-pulegone with permanganate (KMnO₄) produces both (+)-β-methyladipic acid and (+)-methylsuccinic acid.†

Although it is accepted that the method of quasi-racemates (as defined by Fredga‡ and which correlates the absolute stereochemistry of amino acids to the

---

*Tiemann, F.; Schmidt, R. *Chem. Ber.*, **1895**, *28*, 2137 and Karrer, P. *Organic Chemistry*, 1950, Elsevier, New York, p. 684.
†von Braun, J.; Jostes, F. *Chem. Ber.*, **1926**, *59*, 1091, 1444.
‡Fredga, A. *Tetrahedron*, **1960**, *8*, 126.

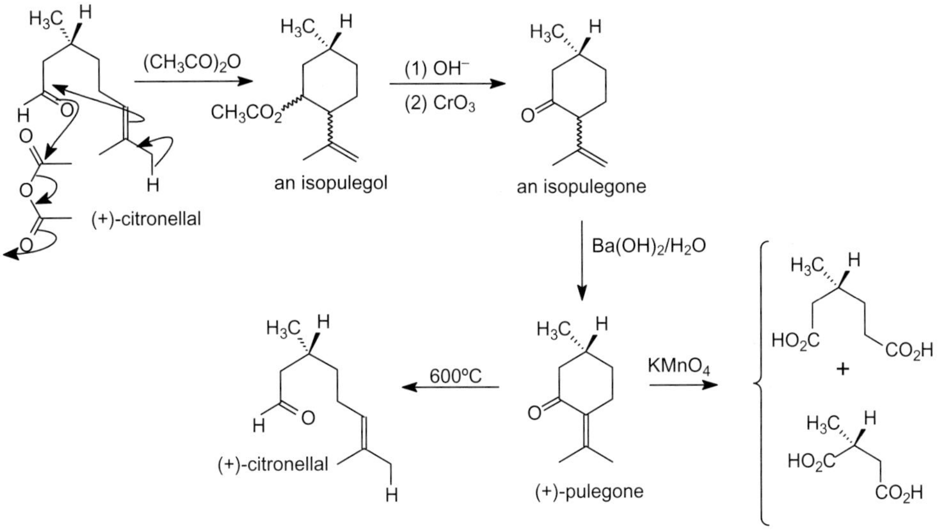

Scheme 13.28. The conversion of (+)-citronellal to (+)-β-methyl adipic acid and (+)-methyl-succinic acids. After Tiemann, F.; Schmidt, R. *Chem. Ber.*, **1895**, *28*, 2137 and von Braun, J.; Jostes, F. *Chem. Ber.*, **1926**, *59*, 1091, 1444.

terpenes and glyceraldehyde) is satisfactory to define the absolute stereochemistry of (+)-methylsuccinic acid,* it appears that Fischer,[†] using the work of Walden,[‡] was able to relate (−)-methylsuccinic acid directly to (+)-alanine (Scheme 13.29). As indicated in the scheme, (+)-alanine (which we know as *S*) can be converted to (*S*)-(−)-2-bromopropanoic acid with nitrosyl bromide (NOBr) (see Chapter 7), and the result is the retention of configuration (presumably, the α-lactone is involved). Then, treatment of the latter with the anion of ethyl ester of malononitrile results in the displacement of the bromide with inversion of configuration. Hydrolysis of the nitrile and decarboxylation leads to (−)-methylsuccinic acid, the optical antipode of that material obtained from (+)-citronellal. The method is not considered particularly valuable since some racemization occurs at the tricarboxylic acid stage.

The establishment of the absolute stereochemistry of the side chain of the steroid nucleus, coupled with the knowledge of the relative stereochemistry of the remainder of the steroid system and the wealth of information derived from decades of the study of the chemistry of steroids was then used to define the absolute stereochemistry of (1*S*,2*S*)-(−)-2-carboxy-2-methylcyclohexane acetic acid. Although it is

*Kaneko, T.; Wakabayashi, K.; Katsura, H. *Bull Chem. Soc. Jpn.*, **1962**, *35*, 1149. Additionally, Cornforth (Cornforth, J. W.; Youhotsky, I.; Popják, G. *Nature*, **1954**, *173*, 536) suggests that the method of Fredga is suitable.
[†]Fischer, E.; Flatau, E. *Liebigs Ann. Chem.*, **1909**, *365*, 13. It is noteworthy that Fischer comments that the method used for the demonstration that (−)-bromopropanoic acid has the same configuration as (−)-methylsuccinic acid is not a good method to consider for the synthesis of the latter (i.e., "Aus der Verwandlung der 1-Brompropiosäure in 1-Methylbernsteinsäure könnte man den Schluss ziehen das beide Verbindungen die gleiche Configuration besitzen … Für die praktische Darstellung der 1-Methyl-bernsteinsäure kommt die Synthese auch nicht in Betracht…").
[‡]Walden, P. *Chem. Ber.*, **1895**, *28*, 1287.

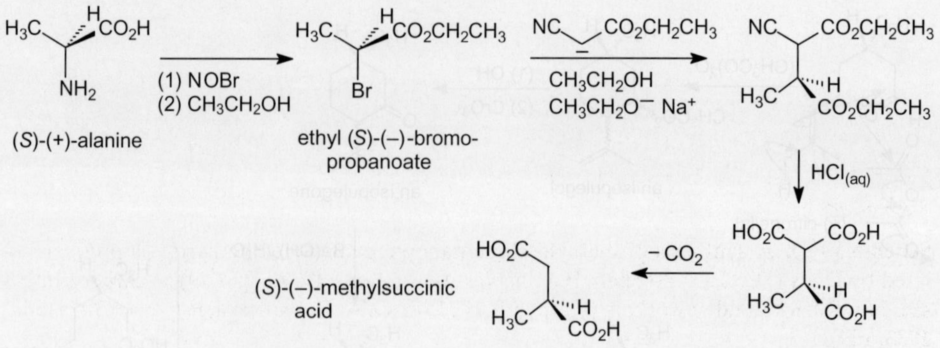

(S)-(+)-alanine

ethyl (S)-(–)-bromo-propanoate

(S)-(–)-methylsuccinic acid

Scheme 13.29. A relationship established between (S)-(+)-alanine and (S)-(–)-methylsuccinic acid, the antipode of the methylsuccinic acid obtained from citronellal, as demonstrated by Fischer, E.; Flatau, E. *Liebigs Ann. Chem.*, **1909**, *365*, 13. It is argued that some racemization occurs at the tricarboxylic acid stage.

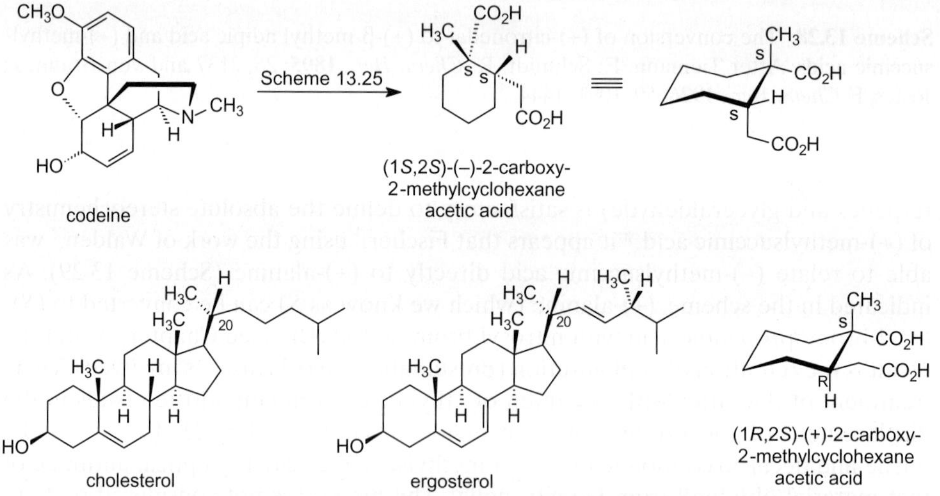

codeine

(1S,2S)-(–)-2-carboxy-2-methylcyclohexane acetic acid

cholesterol

ergosterol

(1R,2S)-(+)-2-carboxy-2-methylcyclohexane acetic acid

Figure 13.5. A representation of the degradation of codeine to (1S,2S)-(–)-2-carboxy-2-methyl-cyclohexane acetic acid (see Scheme 13.25) and representations of cholesterol (the absolute configuration of the side chain at C20 is known [Schemes 13.26 through 13.29]) and its relationship to ergosterol, which has the same absolute configuration at C20 and which has been degraded to (1R,2S)-(+)-2-carboxy-2-methylcyclohexane acetic acid.

somewhat moot given the number of structures whose absolute stereochemistry has subsequently been elucidated by the Bijvoet* method, the degradative chemistry linking cholesterol with ergosterol and the latter with (1R,2S)-(+)-2-carboxy-2-methylcyclohexane acetic acid (Figure 13.5) remains exciting!

Otto Diels[†] had shown quite early that pyrolysis of both cholesterol and ergosterol in the presence of selenium (Se) produced the same methylcyclopentanophen-

*Bijvoet, J. M.; Peerdeman, A. F.; van Bommel, A. J. *Nature*, **1951**, *168*, 271.
[†]Diels, O.; Stephan, H. J. *Liebigs Ann. Chem.*, **1937**, *527*, 279.

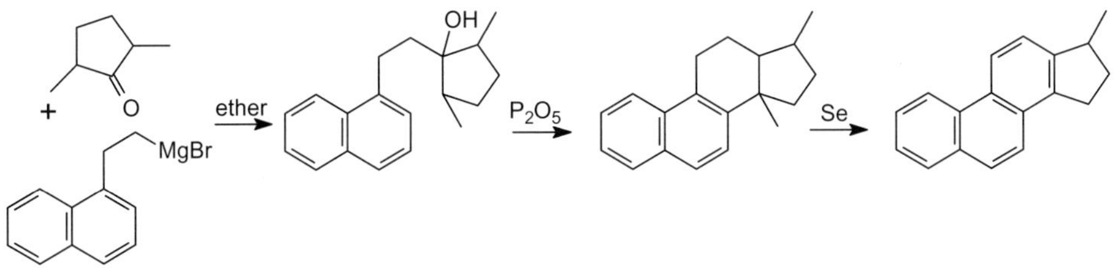

Scheme 13.30. A synthesis of methylcyclopentanophenanthrene, the pyrolysis product isolated by Diels (Diels, O.; Stephan, H.J. *Liebigs Ann. Chem.*, **1937**, *527*, 279) as earlier synthesized by Ruzicka and coworkers (Harper, S. H.; Kon, G. A. R.; Ruzicka, F. C. J.*J. Chem. Soc.*, **1933**, 124).

anthrene as synthesized several years earlier as shown in Scheme 13.30 by Ruzicka[*] and coworkers (Equation 13.2).

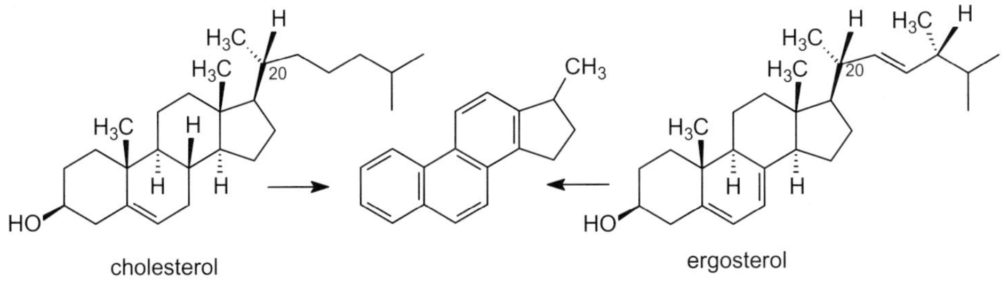

$$(13.2)$$

In 1931, it was shown by Reindel and Kipphan[†] that ozonolysis (O_3) of ergosterol acetate in petroleum ether–ethyl acetate produced a mixture containing 2,3-dimethylbutanal, thus identifying the side chain and the position of the double bond therein (Equation 13.3).

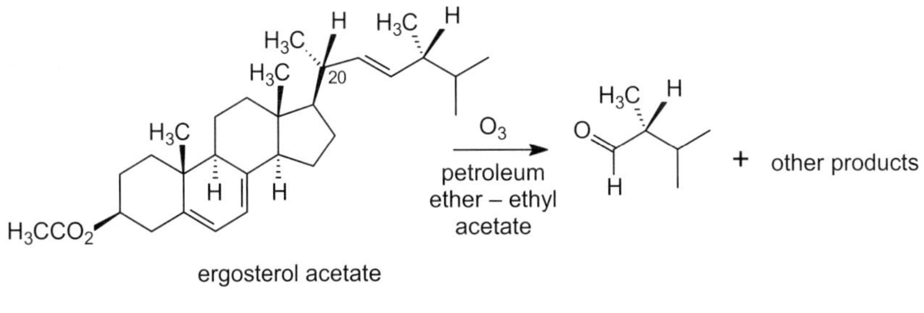

$$(13.3)$$

A year later, Inhoffen[‡] and coworkers made an extensive study of the oxidation products derived from ergosterol, and in 1933, Chuang[§] using the information of both earlier groups noted that after reduction of ergosterol with hydrogen (H_2) over

[*]Harper, S. H.; Kon, G. A. R.; Ruzicka, F. C. J.*J. Chem. Soc.*, **1933**, 124.
[†]Reindel, F.; Kipphan, H. *Liebigs Ann. Chem.*, **1932**, *493*, 181.
[‡]Guiteras, A.; Nakamiya, Z.; Inhoffen, H. H. *Liebigs Ann. Chem.*, **1932**, *494*, 117.
[§]Chuang, C. K. *Liebigs Ann. Chem.*, **1933**, *500*, 270.

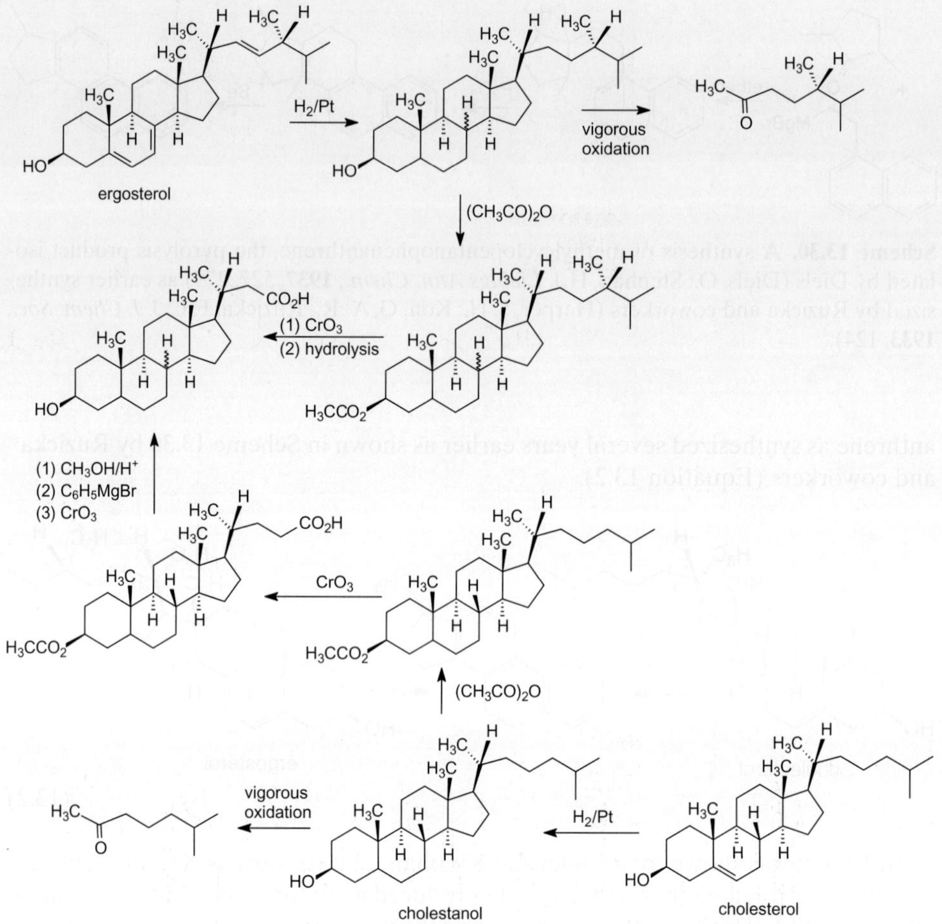

Scheme 13.31. A summary of some of the exciting work taking place during the 1932–1934 period on the relationship between plant (ergosterol) and animal (cholesterol) steroids. See (a) Reindel, F.; Kipphan, H. *Liebigs Ann. Chem.*, **1932**, *493*, 181; (b) Guiteras, A.; Nakamiya, Z.; Inhoffen, H. H. *Liebigs Ann. Chem.*, **1932**, *494*, 117; (c) Chuang, C. K. *Liebigs Ann. Chem.*, **1933**, *500*, 270; (d) Fernholz, E.; Chakravorty, P. N. *Chem. Ber.*, **1934**, *67*, 2021; (e) Windaus, A.; Inhoffen, H. H.; von Reichel, S. *Liebigs Ann. Chem.*, **1934**, *510*, 248.

a platinum (Pt) catalyst, the resulting ergostanol, on "energischen oxidation" with chromic anhydride (CrO_3), gave an aliphatic ketone, which proved to be 5,6-dimethylheptan-2-one, one carbon more than the 6-methylheptan-2-one found on identical treatment of cholestanol. Chuang also showed that among the chromic anhydride oxidation products of the acetate ester of ergostanol, there was a carboxylic acid best fit by keeping the four rings intact and diminishing the side chain so that only four carbons remained (Scheme 13.31).

In the following year, Fernholz and Chakravorty* succeeded in oxidizing the acetate of cholestanol (prepared by acetylation of cholesterol in which the carbon–

*Fernholz, E.; Chakravorty, P. N. *Chem Ber.*, **1934**, *67*, 2021.

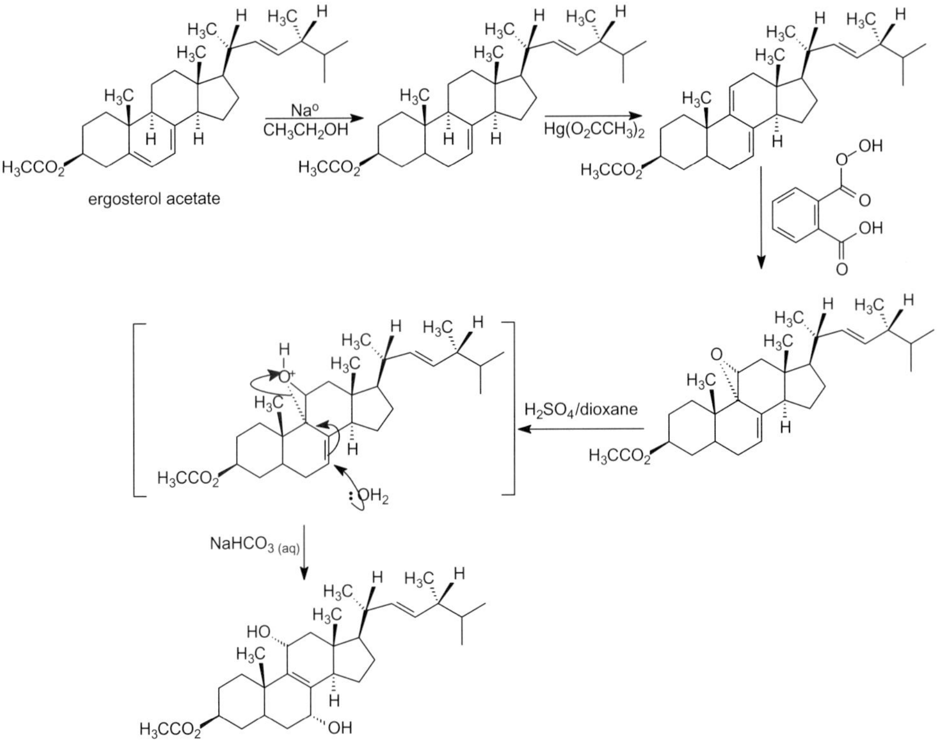

Scheme 13.32. A representation of the path to the formation of $\Delta^{8(9);22}$-3β-acetoxy-7α,11α-dioxyergostadiene from the acetate of ergosterol after Heusser, H.; Eichenberger, K.; Kurath, P.; Dällenbach, H. R.; Jeger, O. *Helv. Chim. Acta*, **1951**, *34*, 2106.

carbon double bond had been reduced* with platinum [Pt] black and hydrogen [H$_2$] in ether) with chromic acid (CrO$_3$) to a monocarboxylic acid (isolated in less than 5% yield). As noted above, ergostanol acetate (the endocyclic double-bond system having been deductively reasoned through in detail in the same year)[†] had given a low yield of carboxylic acid whose side chain was one methylene unit shorter than that found from cholestanol. So Fernholz and Chakravorty prepared the methyl ester of cholestanol and treated it with the Grignard reagent (the method of Wieland[‡]) from bromobenzene. The resulting diphenylcarbinol was then reoxidized with chromic anhydride (CrO$_3$) to yield a carboxylic acid from that was the same as that derived from ergostanol, thus establishing the identity of the two catalytically reduced and chemically oxidized systems (Scheme 13.31).

In 1951, Jeger and coworkers[§] reported (as shown in Scheme 13.32) that they had converted ergosterol acetate (by now clearly written as the $\Delta^{5;7;22}$-triene acetate) into $\Delta^{7;22}$-3β-acetoxyergostadiene (a so-called α-dihydroergosterol acetate) by reduction of the acetate of ergosterol with sodium in ethanol. When the diene was oxidized

*Willstatter, R; Mayer, E. W. *Chem. Ber.*, **1908**, *41*, 2199.

[†]Windaus, A.; Inhoffen, H. H.; von Reichel, S. *Liebigs Ann. Chem.*, **1934**, *510*, 248.

[‡]Wieland, H.; Schlichting, O.; Jacobi, R. *Z. Physiol. Chem.*, **1926**, *161*, 80.

[§]Heusser, H.; Eichenberger, K.; Kurath, P.; Dällenbach, H. R.; Jeger, O. *Helv. Chim. Acta*, **1951**, *34*, 2106.

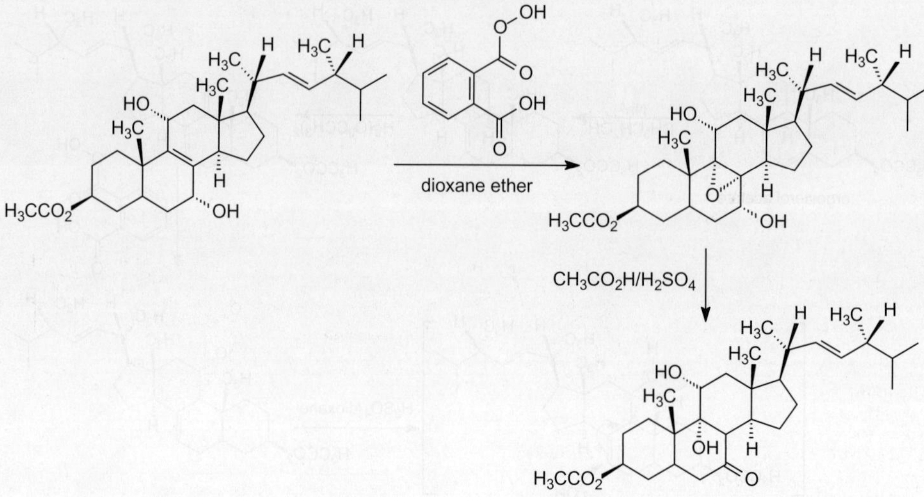

Scheme 13.33. A representation of the conversion of the dienediol, $\Delta^{8(9);22}$-3β-acetoxy-7α,11α-dioxyergostadiene to the corresponding ketone Δ^{22}-3β-acetoxy-9α,11α-dioxy-7-ketoergostene. After Heuser, H.; Anliker, R.; Eichenberger, K.; Jeger, O. *Helv. Chim. Acta*, **1952**, *35*, 936.

with mercury(II) acetate, a new triene, $\Delta^{7;9;22}$-3β-acetoxyergostatriene, was formed. Infrared and ultraviolet spectra were in accord with the conjugated diene formulation. Further oxidation of the $\Delta^{7;9;22}$-3β-acetoxy-ergostatriene with 1 equivalent of monoperoxyphthalic acid in ether generated an epoxide, formulated as $\Delta^{7;22}$-3β-acetoxy-9,11-α-dioxyergostadiene. Amazingly, the epoxide, on brief treatment with sulfuric acid in dioxane, followed by aqueous bicarbonate workup, yielded $\Delta^{8(9);22}$-3β-acetoxy-7α,11α-dioxyergostadiene!

As reported a year later* (and shown in Scheme 13.33), the acetoxydienol was then taken up in dioxane and oxidized again with monoperoxyphthalic acid in ether, and after standing for 3 days at room temperature, the reaction mixture was diluted with ether to generate a new epoxide, Δ^{22}-3β-acetoxy-7α,11α-dioxy-8α,9α-oxidoergostene. This epoxide, on standing in for a day in acetic acid with a trace of sulfuric acid, underwent rearrangement to the ketone Δ^{22}-3β-acetoxy-9α,11α-dioxy-7-ketoergostene.

Finally, as shown in Scheme 13.34 and reported by the same group,† the mono-acetate, Δ^{22}-3β-acetoxy-9α,11α-dioxy-7-ketoergostene, was acetylated with acetic anhydride to yield the corresponding diacetate Δ^{22}-3β,11α-diacetoxy-9α-hydroxy-7-ketoergostene, and treatment of the latter, in dioxane, with aqueous hydrochloric acid (HCl) for an hour resulted in partial hydrolysis and formation of a different monoacetate, that is, Δ^{22}-11α-acetoxy-3β, 9α-dihydroxy-7-ketoergostene. Oxidation of the two bare hydroxyl groups with chromic anhydride (CrO_3) in this diol resulted in the formation of the corresponding dione, Δ^{22}-11α-acetoxy-9α-hydroxyergostene-3,7-dione. After hydrolysis of the remaining acetate ester group, the resulting alcohol, in benzene solution, was oxidized with lead(IV) acetate, and the ether

*Heuser, H.; Anliker, R.; Eichenberger, K.; Jeger, O. *Helv. Chim. Acta*, **1952**, *35*, 936.
†Heusser, H.; Beriger, E; Anliker, R.; Jeger, O.; Ruzicka, L. *Helv. Chim. Acta*, **1953**, *36*, 1918.

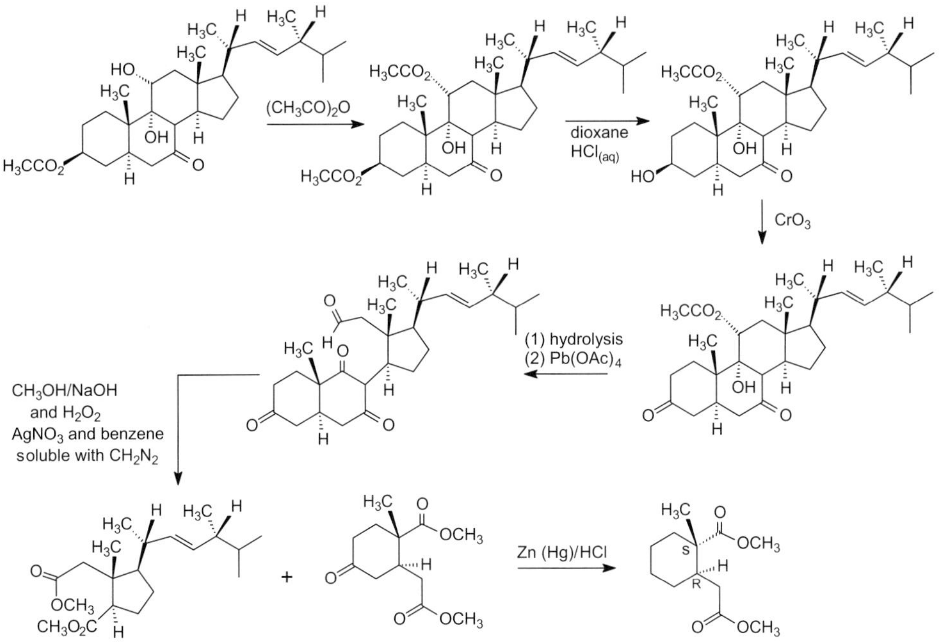

Scheme 13.34. A representation of the final steps in the degradation of ergosterol via Δ^{22}-3β-acetoxy-9α,11α-dioxy-7-ketoergostene to (1R,2S)-(+)-2-carboxy-2-methylcyclohexane acetic acid, confirming the absolute stereochemistry of codeine (and thus morphine). After Heusser, H.; Beriger, E.; Anliker, R.; Jeger, O.; Ruzicka, L. *Helv. Chim. Acta*, **1953**, *36*, 1918.

soluble residue from that oxidation was treated with methanolic sodium hydroxide, hydrogen peroxide, and then silver nitrate. After basification and removal of the silver oxide, the benzene soluble residue was treated with diazomethane and the mixture of esters was chromatographed on activity grade III alumina. The desired ketoester was reduced under Clemmensen conditions to (1R,2S)-(+)-2-carboxy-2-methylcyclohexane acetic acid, which was, as expected from the stereochemistry of the ring junction in the steroids, diastereomeric with the material isolated from morphine.

Since that time, a vast amount of NMR[*] and configurational correlations[†] by X-ray analysis has become available to confirm the conclusions drawn.

Before leaving the discussion of ergosterol, it is worthwhile examining its place in the family of vitamins called the "D" vitamins (apparently because the first members were isolated after A, B, and C!).

Ergosterol was first isolated from the ergot fungus but has since been found in other fungi, including yeast, as well as in vegetable oils. It is directly related to vitamin D_2 (as well as to vitamin D_3 which is missing the side chain double bond, and other, otherwise modified, D vitamins) by an electrocyclic ring opening and a bond rotation. A representation currently accepted for this process is provided in Scheme 13.35.

[*]Reich, H. J.; Jautelat, M.; Messe, M. T.; Weigert, F. J.; Roberts, J. D. *J. Am. Chem. Soc.*, **1969**, *91*, 7445.
[†]Klyne, W.; Buckingham, J. *Atlas of Stereochemistry*, Oxford University Press, New York, 1974.

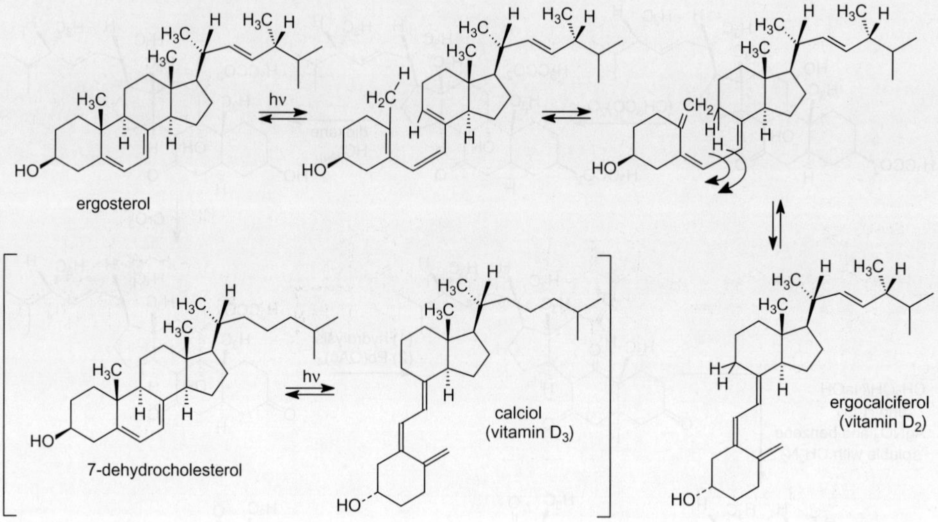

Scheme 13.35. A representation of the photochemical electrocyclic transformation between ergosterol (a "plant sterol") and "ergocalciferol" or vitamin D_2 as well as the relationship between $\Delta^{5,7}$-cholesteroldiene (an "animal sterol") and "calciol" or vitamin D_3.

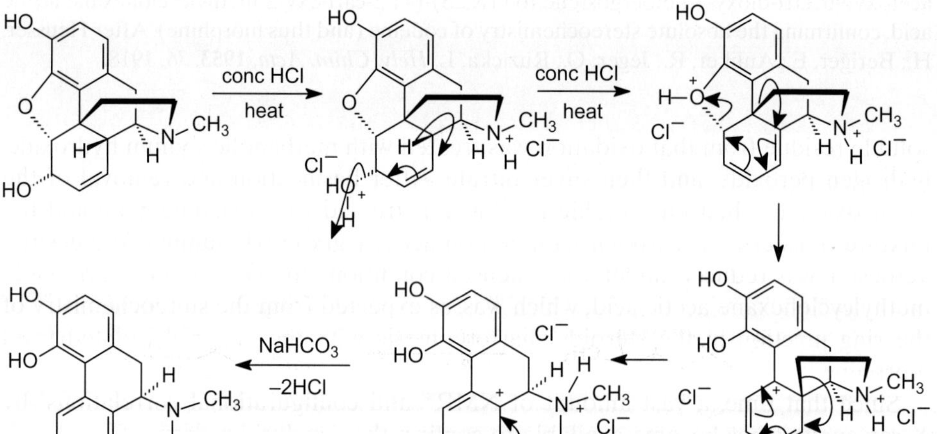

Scheme 13.36. A representation of a pathway to apomorphine from morphine. After Holmes, H. L.; Stork, G. *The Alkaloids*, Vol. 2, Academic, New York, 1952, p. 195.

Among the more interesting reactions that have been found while investigating the chemistry of morphine is the transformation that occurs when morphine is heated with concentrated hydrochloric acid at 130–140°C in a sealed tube for 3 h and the resulting reaction mixture is neutralized with sodium bicarbonate. As shown in Scheme 13.36, there has been a deep-seated transformation (for which a proposed pathway is provided) to apomorphine.

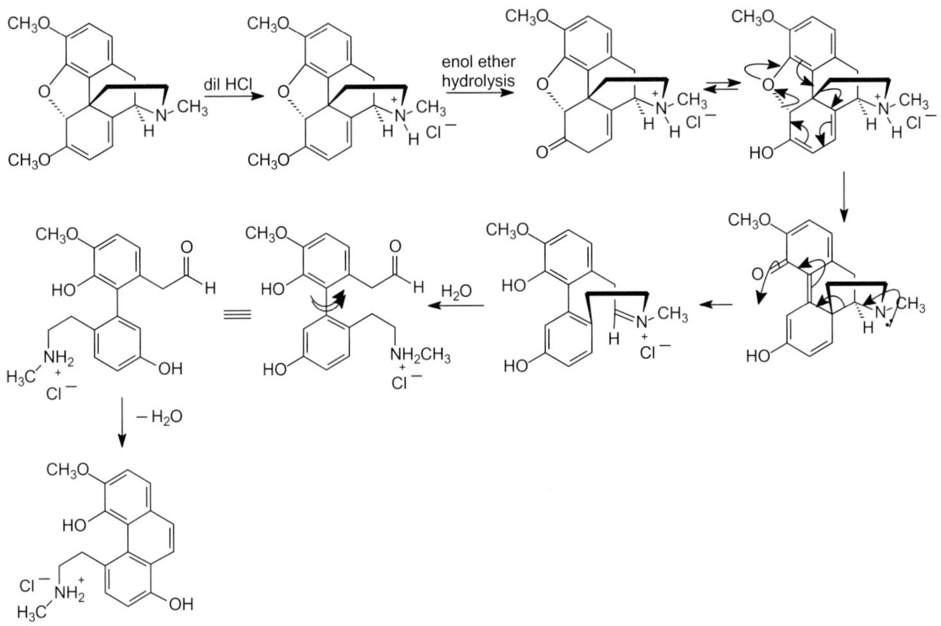

Scheme 13.37. A representation of a pathway from thebaine to thebenine. It is suggested that the dilute acid permits utilization of the electron pair on nitrogen. After Holmes, H. L.; Stork, G. *The Alkaloids*, Vol. 2, Academic, New York, 1952, p. 197.

A similar result, noted quite early, was obtained when thebaine is treated with 40% HCl for about 3 h on the steam bath. Morphothebaine is produced (Equation 13.4).*

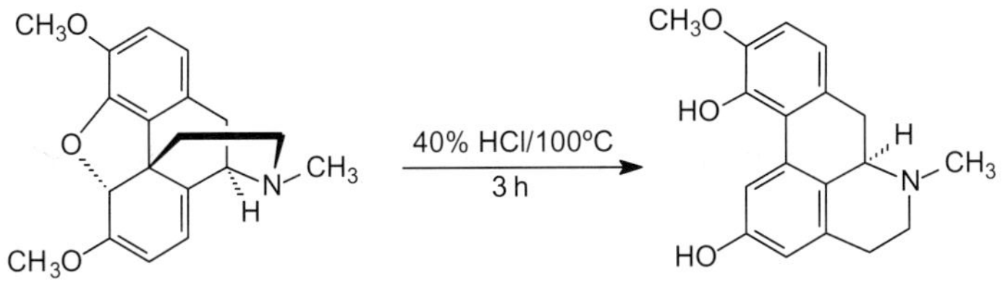

$$\text{(13.4)}$$

Problem 13.1. Utilizing Scheme 13.36 (or something similar as a model), provide a potential pathway for the conversion of thebaine to morphothebaine.

Interestingly, however, when thebaine is warmed with dilute hydrochloric acid (HCl) for only 2 min, the hydrochloride of an amorphous base, thebenine results. A potential pathway for the formation of this base is shown in Scheme 13.37.

Yet another interesting rearrangement of thebaine is encountered when it is treated with phenylmagnesium bromide in ether. As shown in Scheme 13.38, the treatment results in the formation of phenyldihydrothebaine as the major product.

*Freund, M.; Holthof, C. *Chem. Ber.*, **1899**, *32*, 168.

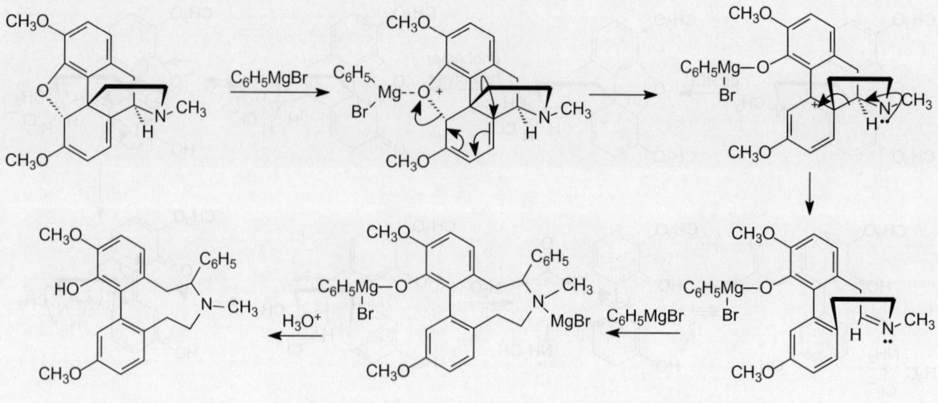

Scheme 13.38. A representation of a pathway from thebaine to phenyldihydrothebaine, the product resulting from the addition of phenylmagnesium bromide to thebaine. After Holmes, H. L.; Stork, G. *The Alkaloids*, Vol. 2, Academic, New York, 1952, p. 198.

The scheme shows a potential path to that product, but it is clear that other paths might also be drawn.

II. The Biosynthesis of Morphine (and Codeine and Thebaine)

It has been argued* that the "first major contribution to biogenetic hypothesis" was the visual dissection of the benzylisoquinoline skeleton found in laudanosine and paperverine (Figure 13.6) by Winterstein and Trier.[†] The argument continues that the subsequent insight of Sir Robert Robinson,[‡] utilizing the idea of more-or-less intact amino acid (or logically obtained derivatives of amino acids, e.g., amines, acids, and aldehydes, that retain the basic skeleton of the amino acid) implicit in biogenetic hypothesis, reasonably led to the correct structure of morphine. The basic idea has, of course, enjoyed considerable expansion and has been affirmed by many subsequent experiments. Indeed, as will be seen here, current work strives not only to identify the enzymes and enzyme cofactors required to effect the reactions adumbrated by early workers but also to identify the specific genes involved.

The early work, then, was in accord with two "units" corresponding to C_6-C_2 each being incorporated into these bases (and many other related bases not considered here). Thus, the suggestion that they were derived from a phenethylamine, such as tyrosine (having lost CO_2), or from a hydroxylated phenylalanine (having lost CO_2), was readily accepted.[§] Feeding experiments with radioactively labeled substrates were in accord with this hypothesis.

*Spenser, I. D. in Florkin, M.; Stotz, E. H. (eds.), *Comprehensive Biochemistry*, Vol. 20, Elsevier, New York, **1968**, p. 231 ff.

[†]Winterstein, E.; Trier, G. *Die Alkaloide*, Bornträger, Berlin, 1910.

[‡]Robinson, R. *J. Chem. Soc.*, **1917**, *111*, 876.

[§]Among the more exciting and valuable expositions on the early adventures in speculation that have been, more-or-less, realized, the interested reader is directed to (a) the Fifth Pedler Lecture as provided by Robinson, R. *J. Chem. Soc.*, **1936**, 1079; (b) Robinson, R. *The Structural Relations of Natural Products*, Oxford University Press, London, 1955; (c) Barton, D. H. R.; Cohen, T. *Festschrift Arthur Stoll*, Birkhäuser Verlag, Basel, **1957**, p. 117 ff.

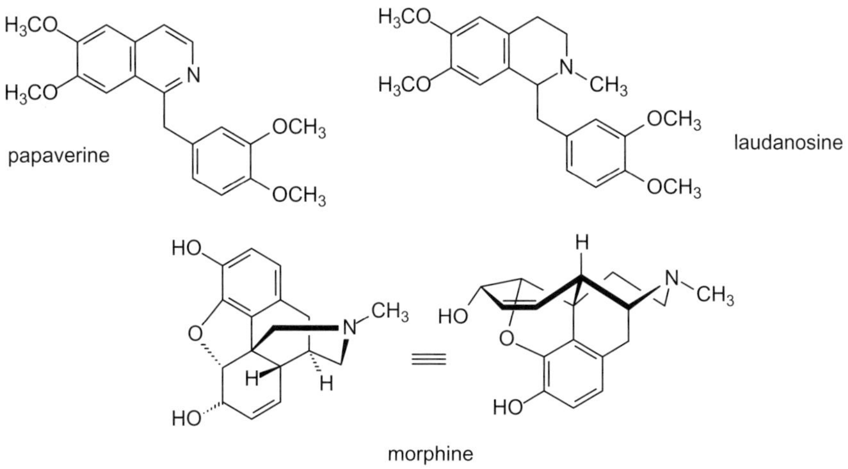

papaverine

laudanosine

morphine

Figure 13.6. Structures of papaverine and laudanosine (after Winterstein, E.; Trier, G. *Die Alkaloide*, Bornträger, Berlin, 1910) and morphine (as suggested—but without the stereochemistry shown—by Robinson, R. *J. Chem. Soc.*, **1917**, *111*, 876).

However, it was clearly recognized that, barring a symmetrical intermediate or incorporation of two identical phenethylamine portions, identical labeling of both C_6-C_2 units could not be obtained. And indeed, subsequent studies showed that two tyrosine-derived C_6-C_2 units differed.[*]

At this writing, a significant amount of work has been undertaken and more is under way utilizing genomic approaches. Thus, for example, the oxidation of **tyrosine (Tyr, Y)**, to dihydroxyphenylalanine is well established across a variety of plant bases and the subsequent decarboxylation of **tyrosine (Tyr, Y)** to tyramine as well as the oxidation of **phenylalanine (Phe, F)** to 3,4-dihydroxyphenylalanine (dopa) and its subsequent decarboxylation to 3,4-dihydroxyphenethylamine (dopamine), while catalyzed by different enzymes, involves processes that occur in members of a large gene family all of which are in the opium poppy.[†]

These oxidation reactions require oxygen (O_2) and tetrahydrobiopterin as a cofactor. Thus, as shown in Scheme 13.39, 7,8-dihydroneopterin 3′-triphosphate (generated from guanosine triphosphate [GTP] as seen in Scheme 12.118) is converted to 6-pyruvoyl-5,6,7,8-tetrahydropterin by an elimination reaction and two keto-enol isomerizations. The process is catalyzed by the enzyme 6-pyruvoyltetrahydropterin synthase (EC 4.2.3.12). Then, via an intermediate, written as an equilibrium between α-hydroxyketones (named dihydrosepiapterin) linked by a common enol, reduction to tetrahydrobiopterin is effected (in two separate steps) by 2 equivalents of NADPH used by the enzyme sepiapterin reductase (EC 1.1.1.153). Tetrahydrobiopterin is the cofactor involved in the "National Institutes of Health (NIH) shift" (cf. Chapter 6) pathway used by the iron-containing enzyme phenylalanine 4-monooxygenase (EC 1.14.16.1) to convert **phenylalanine (Phe, F)** to **tyrosine (Tyr, Y)** and is converted to (6*R*)-6-(L-erythro-1,2-dihydroxypropyl)-5,6,7,8-tetrahydro-4a-hydroxypterin in the process.

[*]Battersby, A. R.; Francis, R. J. *J. Chem. Soc.*, **1963**, 4071.
[†]Facchini, P. J.; De Luca, V. *J. Biol. Chem.*, **1994**, *269*, 26684.

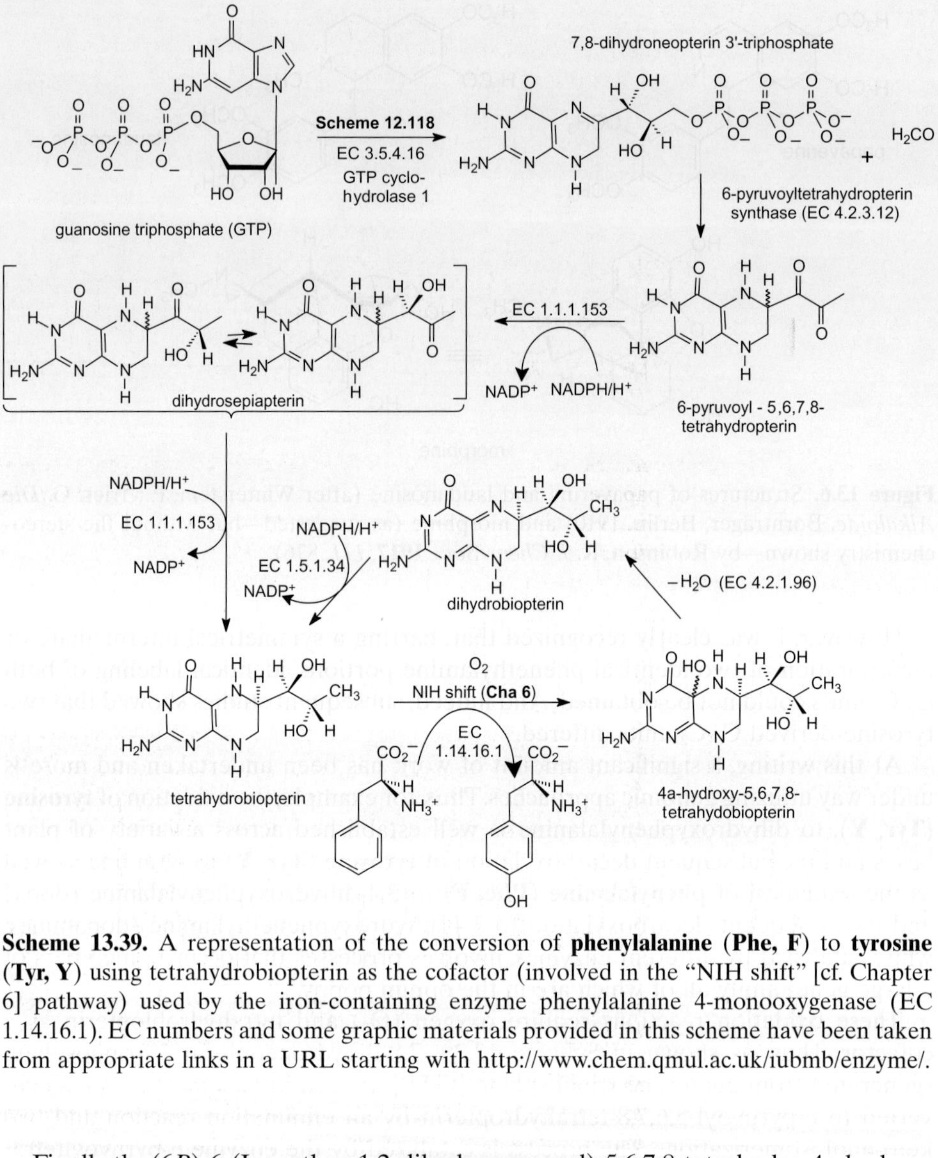

Scheme 13.39. A representation of the conversion of **phenylalanine (Phe, F)** to **tyrosine (Tyr, Y)** using tetrahydrobiopterin as the cofactor (involved in the "NIH shift" [cf. Chapter 6] pathway) used by the iron-containing enzyme phenylalanine 4-monooxygenase (EC 1.14.16.1). EC numbers and some graphic materials provided in this scheme have been taken from appropriate links in a URL starting with http://www.chem.qmul.ac.uk/iubmb/enzyme/.

Finally, the (6R)-6-(L-erythro-1,2-dihydroxypropyl)-5,6,7,8-tetrahydro-4a-hydroxypterin undergoes dehydration either spontaneously or with the aid of the 4a-hydroxytetrahydrobiopterin dehydratase enzyme (EC 4.2.1.96) to yield dihydrobiopterin, which can be recycled by reduction back to tetrahydrobiopterin with NADPH acting as a cofactor for 6,7-dihydropteridine reductase (EC 1.5.1.34).

Interestingly, this cofactor is a member of a family of enzymes that include tyrosine 3-monooxygenase (EC 1.14.16.2) that participates in the conversion of **tyrosine (Tyr, Y)** to dihydroxyphenylanaline and tryptophan 5-monooxygenase (EC 1.14.16.4) that participates in the conversion of **tryptophan (Trp, W)** to 5-hydroxytryptophan.

As shown in Schemes 13.40 and 13.41, the biosynthesis of morphine (via thebaine and codeine) has been shown to involve the conversion of **tyrosine (Tyr, Y)** into

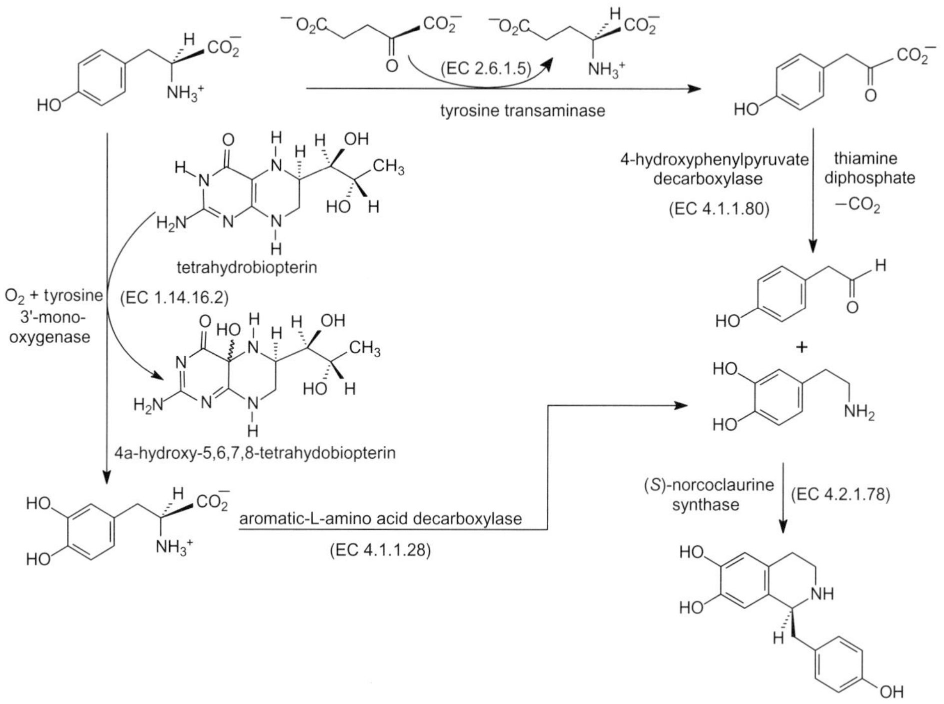

Scheme 13.40. A representation of the formation of (S)-norcoclaurine by a combination of dopamine [3,4-dihydroxyphenethylamine, 2-(3,4-dihydroxyphenyl)ethanamine] with 4-hydroxyphenylacetaldehyde [2-(4-hydroxyphenyl)acetaldehyde] in the presence of (S)-norcoclaurine synthase (EC 4.2.1.78). EC numbers and some graphic materials provided in this scheme have been taken from appropriate links in a URL starting with http://www.chem.qmul.ac.uk/iubmb/enzyme/.

two different C_6-C_2 fragments (viz. 4-hydroxyphenylacetaldehyde and 3,4-dihydroxyphenethylamine [dopamine]) that subsequently recombine to form an isolable intermediate, (S)-norcoclaurine.

In Scheme 13.40 and as noted above, the action of the iron-containing enzyme tyrosine 3-monooxygenase (EC 1.14.16.2) is shown to effect the conversion of **tyrosine (Tyr, Y)** and oxygen (O_2) to 3,4-dihydroxyphenylalanine (L-dopa), while the cofactor tetrahydrobiopterin undergoes oxidation to 4a-hydroxytetrahydrobiopterin. Then, the general aromatic-L-amino acid decarboxylase (EC 4.1.1.28), an enzyme that uses pyridoxal as a cofactor, effects the decarboxylation of the bisphenolic acid to the corresponding amine, dopamine [3,4-dihydroxyphenethylamine, 2-(3,4-dihydroxyphenyl)ethanamine].

In the same scheme, a representation of the action of tyrosine transaminase (EC 2.6.1.5) acting on **tyrosine (Tyr, Y)** is shown. This enzyme utilizes pyridoxal phosphate to remove the amino group from **tyrosine (Tyr, Y)** and transfer it to α-ketoglutarate (2-oxoglutarate) with formation of **glutamate (Glu, E)** from the latter and 4-hydroxyphenylpyruvate from the former. Then, 4-hydroxyphenylpyruvate decarboxylase (EC 4.1.1.80), which appears to require thiamine diphosphate and

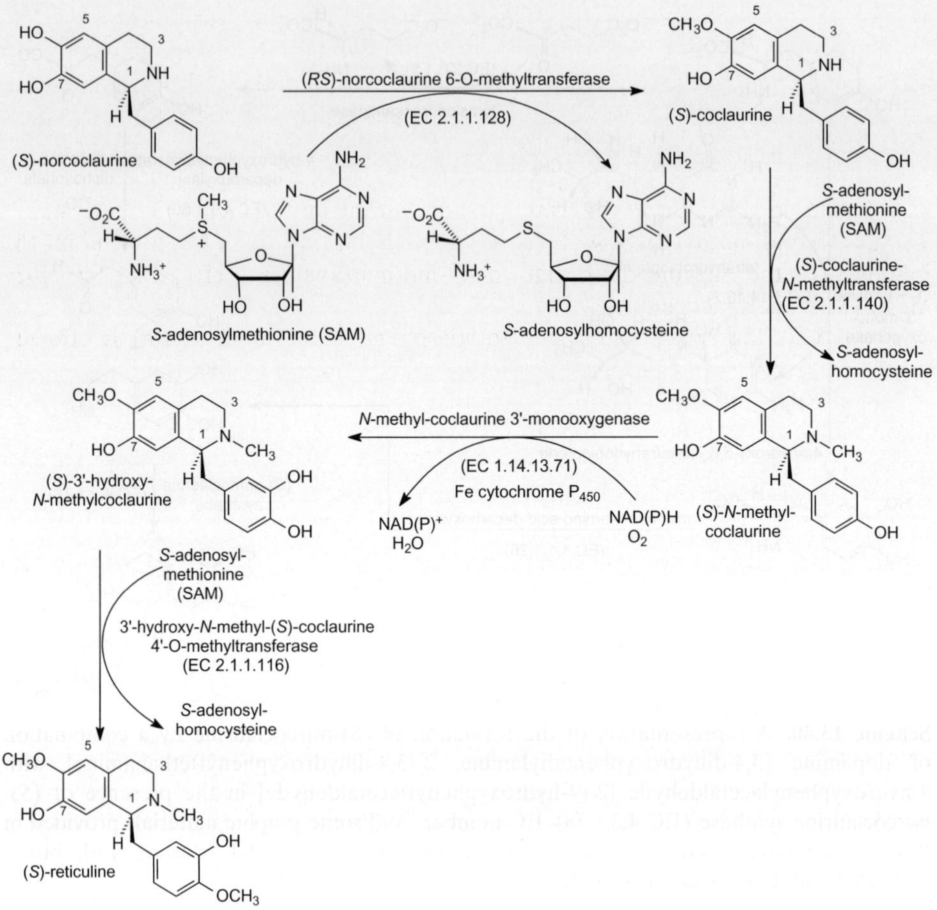

(S)-norcoclaurine

(RS)-norcoclaurine 6-O-methyltransferase

(EC 2.1.1.128)

(S)-coclaurine

S-adenosylmethionine (SAM)

S-adenosylhomocysteine

S-adenosyl-methionine (SAM)

(S)-coclaurine-N-methyltransferase (EC 2.1.1.140)

S-adenosyl-homocysteine

(S)-3'-hydroxy-N-methylcoclaurine

N-methyl-coclaurine 3'-monooxygenase

(EC 1.14.13.71) Fe cytochrome P_{450}

$NAD(P)^+$ H_2O

$NAD(P)H$ O_2

(S)-N-methyl-coclaurine

S-adenosyl-methionine (SAM)

3'-hydroxy-N-methyl-(S)-coclaurine 4'-O-methyltransferase (EC 2.1.1.116)

S-adenosyl-homocysteine

(S)-reticuline

Scheme 13.41. A representation of the conversion of (S)-coclaurine into (S)-reticuline. EC numbers and some graphic materials provided in this scheme have been taken from appropriate links in a URL starting with http://www.chem.qmul.ac.uk/iubmb/enzyme/.

the magnesium (Mg^{+2}) cation, effects the conversion of 4-hydroxyphenylpyruvate to 4-hydroxyphenylacetaldehyde [2-(4-hydroxyphenyl)acetaldehyde].

The combination of the dopamine [3,4-dihydroxyphenethylamine, 2-(3,4-dihydroxyphenyl)ethanamine] with 4-hydroxyphenylacetaldehyde [2-(4-hydroxyphenyl)acetaldehyde] in the presence of (S)-norcoclaurine synthase (EC 4.2.1.78), a modified form of which has recently been crystallized (PDB 2vq5), results in the production of (S)-norcoclaurine.

It is interesting to note that, using complementary DNA (cDNA) (Chapter 14), that is, DNA obtained by synthesis from a messenger RNA (mRNA) template in a reaction catalyzed by reverse transcriptase enzyme, the norcoclaurine synthase enzyme has been characterized, and the pathway from the precursors has been shown to most likely involve imine formation and then cyclization in a process resembling the Pictet–Spengler reaction.*

*This material and much of what follows has recently been thoroughly reviewed in Ziegler, J.; Facchini, P. J. *Ann. Rev. Plant Biol.*, **2008**, *59*, 735.

The conversion of (*S*)-norcoclaurine to (*R*)-reticuline, the immediate precursor to oxidative cyclization, clearly requires three methylations (!), an oxidation, and an isomerization (the latter effected by an oxidation and then a reduction). Thus, as shown in Scheme 13.41, it has been found that (*RS*)-norcoclaurine 6-O-methyltransferase (EC 2.1.1.128) effects the methylation of (*S*)-norcoclaurine through transfer of a methyl group from *S*-adenosylmethionine to phenolic hydroxyl at C-6 of (*S*)-norcoclaurine generating (*S*)-coclaurine and *S*-adenosylhomocysteine. Then, a second methylation, with the same methylation source but now under the influence of the enzyme (*S*)-coclaurine-*N*-methyltransferase (EC 2.1.1.140) *produces* (*S*)-*N*-methylcoclaurine.

Oxidation (the introduction of another phenolic hydroxyl group) is effected by *N*-methylcoclaurine 3′-monooxygenase (EC 1.14.13.71), the gene for which (CYP80B1) produces a heme-thiolate cytochrome P_{450}-dependent monooxygenase protein that uses reduced NADPH as a cofactor and oxygen (!) to produce water, oxidized cofactor (nicotinamide adenine dinucleotide [$NADP^+$]), and (*S*)-3′-hydroxy-*N*-methylcoclaurine.

The final methylation, generating (*S*)-reticuline, utilizes the enzyme 3′-hydroxy-*N*-methyl-(*S*)-coclaurine 4′-O-methyltransferase (EC 2.1.1.116) to convert 3′-hydroxy-*N*-methyl-(*S*)-coclaurine to (*S*)-reticuline, while *S*-adenosylmethionine goes to *S*-adenosylhomocysteine.

Now, it is known (Scheme 13.42) that 1,2-dehydroreticuline is reduced by 1,2-dehydroreticulinium reductase (EC 1.5.1.27) with the reduced form of NADPH being oxidized to $NADP^+$. However, the enzyme required for the oxidation of (*S*)-reticuline to 1,2-dehydroreticuline has not yet been characterized.

It is then argued (Scheme 13.42) that (*R*)-reticuline undergoes oxidation to a diradical, which couples intramolecularly to yield salutaridine. The coupling process is catalyzed by salutaridine synthase (EC 1.14.21.4) and, in the process, nicotinamide adenine dinucleotide [NAD(P)H] is oxidized to $NADP^+$ and oxygen is converted to water.* Then, use of another equivalent of nicotinamide adenine dinucleotide [NAD(P)H → $NAD(P)^+$] as a cofactor for salutaridine reductase (EC 1.1.1.248) is used to effect reduction of salutaridine to salutaridinol.

In the next step, acetylation of the hydroxyl at C7 utilizing acetyl-CoA (acetyl-CoA is the cofactor that is used by the enzyme salutaridinol 7-O-acetyltransferase [EC 2.3.1.150]) *and* CoA-SH and 7-O-acetylsalutaridinol results. Spontaneous cyclization accompanies loss of acetic acid, and thebaine results.

As shown in Scheme 13.43, the conversion of thebaine to codeine and thence to morphine is effected by enzymes that have not yet been completely characterized. It is clear that methyl groups must be removed and a reduction [codeinone reductase, EC 1.1.1.247, NAD(P)H → NAD(P)] must be effected. The sequence is not clear, although there does seem to be a general consensus that the scheme shown probably applies.

---

*The suggestion that this process obtains here is based on the early suggestions of phenolic coupling already outlined above and, as noted there, in Robinson, R. *The Structural Relations of Natural Products*, **1955**, Oxford University Press, London, and Barton, D. H. R.; Cohen, T. *Festschrift Arthur Stoll*, **1957**, Birkhäuser Verlag, Basel, p. 117 ff.

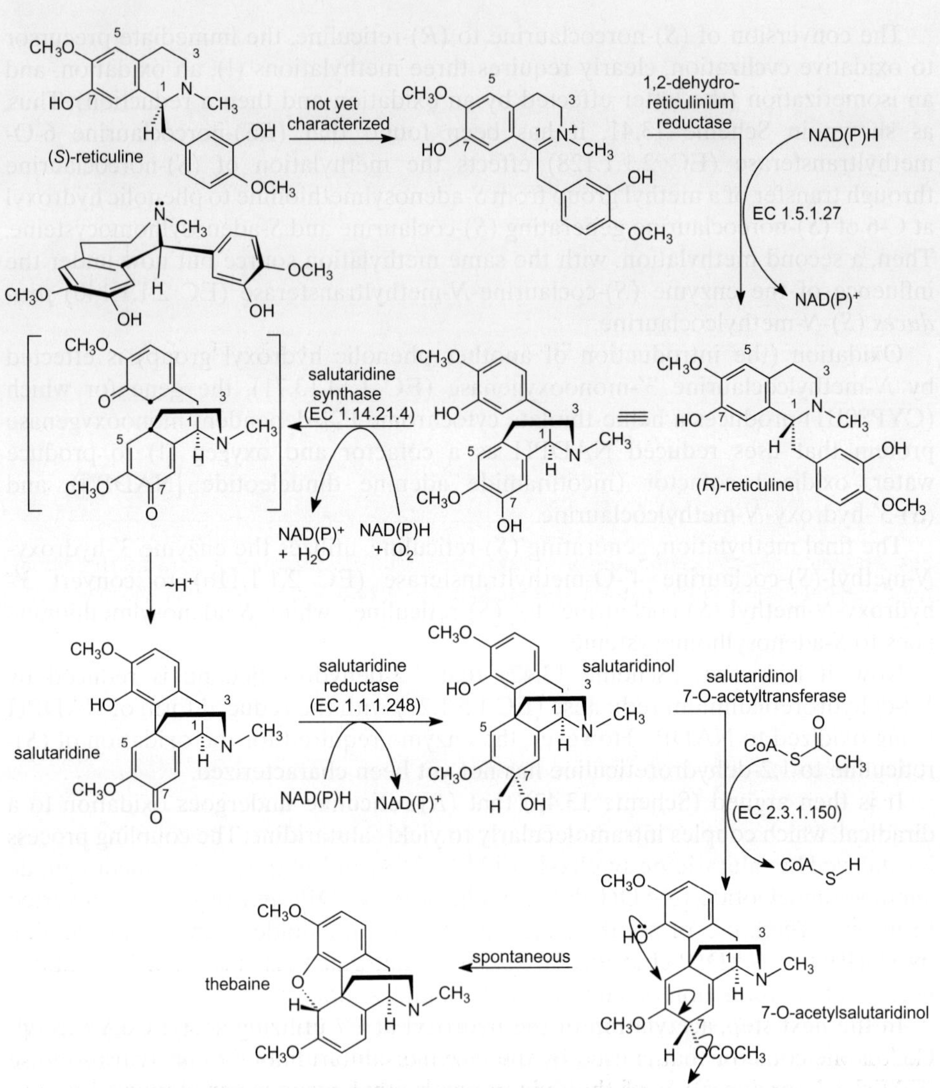

Scheme 13.42. A representation of the formation of thebaine from (*S*)-reticuline via an initial isomerization to (*R*)-reticuline, cyclization to salutaridine, reduction to salutaridinol, acetylation, and a final (spontaneous) elimination. EC numbers and some graphic materials provided in this scheme have been taken from appropriate links in a URL starting with http://www.chem.qmul.ac.uk/iubmb/enzyme/.

III. The Synthesis of Morphine

The literature of organic chemistry is replete with attempts to synthesize morphine and codeine and a number of successes have been achieved. The first total synthesis of codeine and morphine, completed in 1952 (that of Gates and Tschudi*), is shown in Scheme 13.44.

*Gates, M.; Tschudi. G. *J. Am. Chem. Soc.*, **1950**, *72*, 228; Gates, M.; Tschudi. G. *J. Am. Chem. Soc.*, **1952**, *74*, 1109.

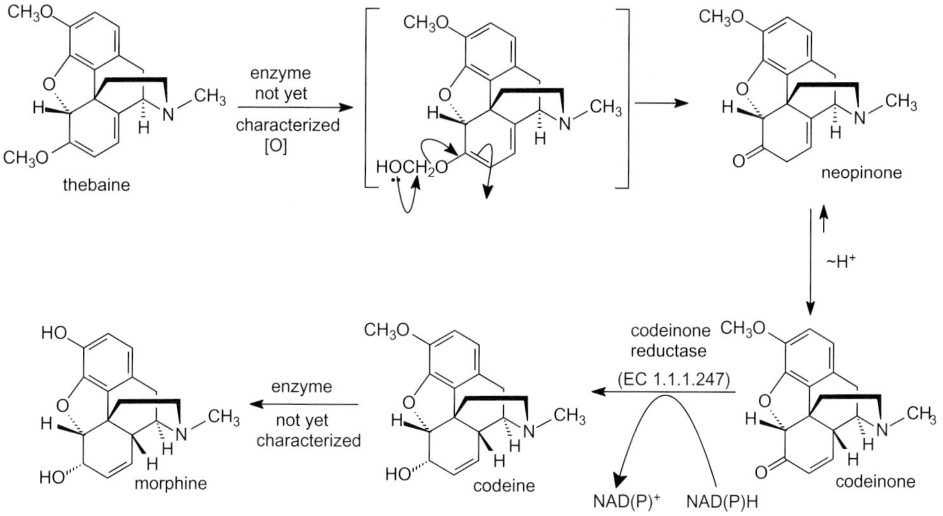

Scheme 13.43. A representation of a path from thebaine to codeine and morphine. EC numbers and some graphic materials provided in this scheme have been taken from appropriate links in a URL starting with http://www.chem.qmul.ac.uk/iubmb/enzyme/.

Although more syntheses have followed that early work* (and it remains likely that yet even more will be developed because these alkaloids serve as templates for testing a variety of experimental tools, ideas and techniques), that of Kenner Rice[†] (Scheme 13.45) continues to stand as one of the cleanest and highest yielding.

The synthesis of Gates and Tschudi (Scheme 13.44) began with 2,6-dihydroxynaphthalene, which was converted to its monobenzoate and nitrosated. Reduction of the nitroso (H_2, Pd/C) yielded the corresponding α-aminophenol, the oxidation of which, with iron(III) chloride, produced a quinone, and reduction of the quinone with sodium hydrosulfite followed by methylation (dimethyl sulfate [$(CH_3)_2SO_4$]) yielded a dimethoxybenzoate. Removal of the benzoate protecting group and repetition of the entire sequence outlined above produced a dimethoxyquinone.

Michael condensation of the quinone with ethyl cyanoacetate followed by ferricyanide oxidation (the latter serving to regenerate the quinone) yielded an ester, which, on hydrolysis and acidification, underwent decarboxylation to the corresponding nitrile. Serving as the dienophile, this "ene" portion underwent the Diels–Alder reaction with butadiene to produce a hydrophenanthrene. The reduction of the phenanthrene, in ethanol, with hydrogen over a copper chromite catalyst led to, among others, a lactam, the keto group of which could be removed by the Wolff–Kishner procedure, and which could be methylated with methyl iodide and reduced with lithium hydride to a tetracyclic amine. Resolution of the amine with dibenzoyltartaric acid yielded an enantiomer (dextrorotatory) that was identical to a degradation product obtained from thebaine! Thebaine was then used as a relay material

---

*All of the syntheses reported in the 10-year span 1992–2002 have been reviewed. See Mascavage, L. M.; Wilson, M. L.; Dalton, D. R. *Curr. Org. Synth.*, **2006**, *3*, 175.
[†]Rice, K. J. *Org. Chem.*, **1980**, *45*, 3135.

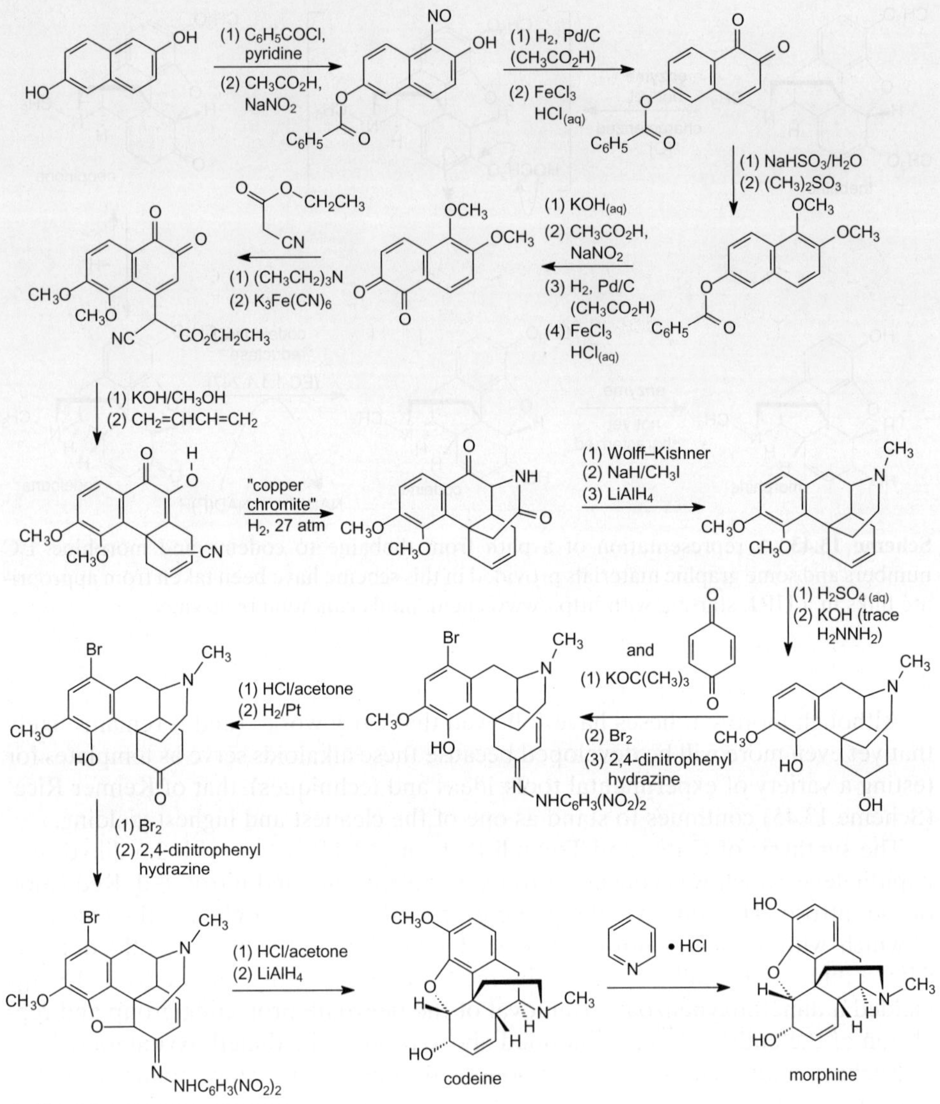

Scheme 13.44. A representation of the first total synthesis of codeine and morphine (after Gates, M.; Tschudi, G. *J. Am. Chem. Soc.*, **1950**, *72*, 228; Gates, M.; Tschudi. G. *J. Am. Chem. Soc.*, **1952**, *74*, 1109).

since it was easier to obtain that alkaloid and convert it to the degradation product than it was to synthesize the intermediate!

Hydration of the double bond led to a mixture of alcohols, which were separated and heated with potassium hydroxide in diethylene glycol at 225°C in the presence of a trace of hydrazine (H_2NNH_2) to cause partial demethylation. That was followed by a modified Oppenauer (potassium *t*-butoxide-benzophenone) oxidation. Then, bromination of the arylketone produced a dibromide that formed a 2,4-dinitrophenylhydrazone (DNPH) derivative identical to one obtained from 1-bromothe-

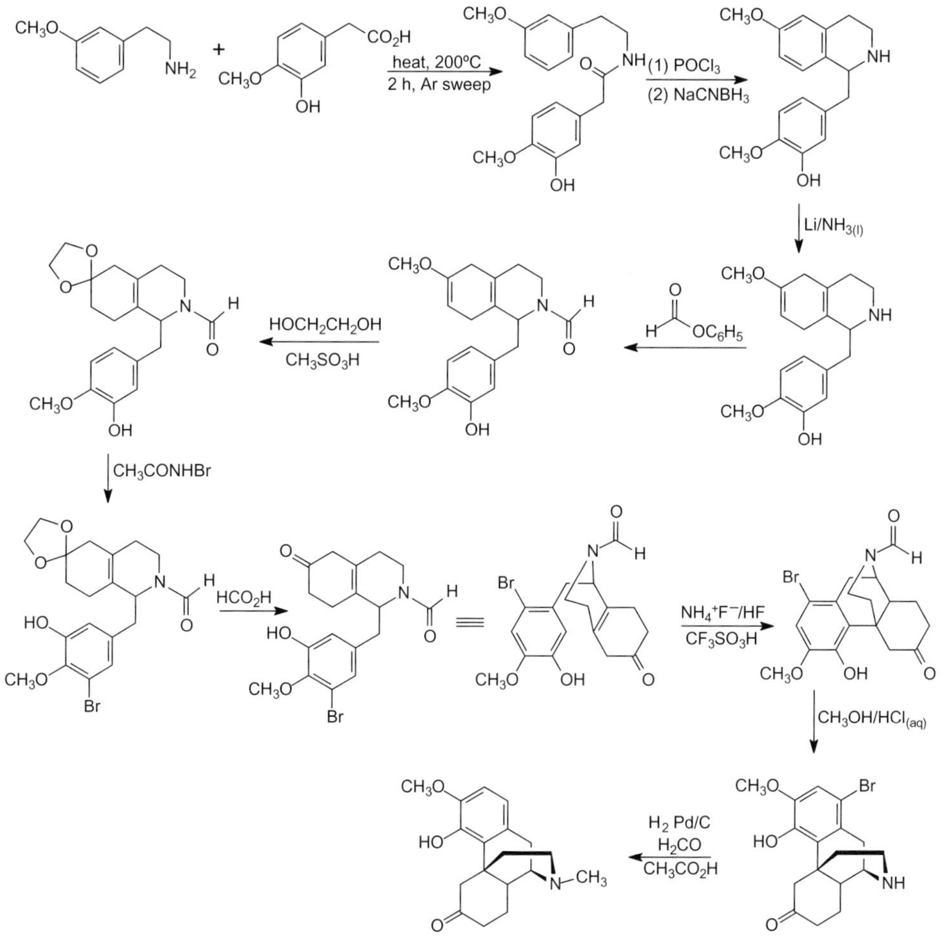

Scheme 13.45. Representation of a high-yield synthesis of racemic dihydrothebainone. The product was subsequently converted to thebaine, codeine, and morphine (see Scheme 13.46). After Rice, K. *J. Org. Chem.*, **1980**, *45*, 3135.

baine, indicating that epimerization at C-14 had occurred. At this point, exchange of DNPH with acetone (HCl catalysis) produced a new ketone, reduction of which yielded 1-bromodihydrothebainone.

Now, dibromination of the 1-bromodihydrothebainone followed by treatment with DNPH produced the DNP derivative of 1-bromocodeinone, which, again, was converted to the corresponding ketone by treatment with acetone and aqueous hydrochloric acid.

Finally, "energetic treatment" of the bromoketone product with lithium aluminum hydride ($LiAlH_4$) led to codeine. Codeine was cleaved to morphine with pyridinium hydrochloride.

The synthesis devised by Kenner Rice (*vide supra*) (Scheme 13.45) was actually directed toward dihydrothebainone since the conversion of that intermediate to thebaine and codeine had been detailed several years earlier.* The Rice protocol began with heating 2-(3-methoxyphenyl)ethylamine with (3-hydroxy-4-methyoxy-phenyl)acetic acid. While it is generally the case that an acid–base reaction might be expected and little else would occur, at a temperature of 200°C (for 2 h under an argon [Ar] atmosphere while water that formed was swept away), the corresponding amide formed. Treatment of the amide with phosphorus oxychloride (POCl$_3$) in acetonitrile (CH$_3$CN) resulted in cyclization and subsequent reduction of the imine formed in that reaction with sodium cyanoborohydride (NaCNBH$_3$) in methanol (CH$_3$OH) provided the corresponding tetrahydroisoquinoline.

Then, Birch reduction with lithium (Li) in liquid ammonia (NH$_3$) produced the corresponding dihydroamine where, as anticipated, the double bonds were not in conjugation. The amine was next protected as the corresponding formamide by refluxing the "unpurified" Birch reduction product with phenyl formate (C$_6$H$_5$O$_2$CH) in ethyl acetate (CH$_3$CO$_2$CH$_2$CH$_3$) to produce the expected pair of rotational isomers. When the N-formyl enol ether was stirred in THF-containing methanesul-fonic acid (CH$_3$SO$_3$H) and ethylene glycol [(HOCH$_2$)$_2$], the enol ether underwent hydrolysis to the corresponding ketone, which, in turn, was converted to the ethyl-ene ketal. Bromination of the aromatic ring *para-* to the free hydroxyl was accom-plished with N-bromoacetamide (CH$_3$CONHBr), which was effectively deketalized with aqueous formic acid (HCO$_2$H).

The resulting bromoketone underwent Grewe-type cyclization to racemic 1-bromo-N-formylnordihydrothebainone when it was treated with ammonium flu-oride–hydrogen fluoride (NH$_4$-HF) in trifluoromethylsulfonic acid (CF$_3$SO$_3$H). Then, racemic dihydrothebainone was obtained when the formyl group was removed by hydrolysis in aqueous methanolic hydrogen chloride (CH$_3$OH/HCl) and the residue from that hydrolysis was hydrogenated (H$_2$) in acetic acid (CH$_3$CO$_2$H) containing formaldehyde (H$_2$CO) and sodium acetate (Na$^+$ $^-$O$_2$CCH$_3$).

As seen in Scheme 13.46, dihydrothebainone could be brominated (2 equivalents of bromine, Br$_2$, were used) in acetic acid (CH$_3$CO$_2$H) and basification yielded 1-bromodihydrocodeinone. The bromine (Br) from the latter was removed by hydrogenolysis (H$_2$) over palladium on carbon (Pd/C). This resulted in the formation of dihydrocodeinone. The dimethylketal of dihydrocodeinone was produced by treating the latter with methyl orthoformate [HC(OCH$_3$)$_3$] and sulfuric acid (H$_2$SO$_4$), and then, on heating in chloroform with toluenesulfonic acid (p-CH$_3$C$_6$H$_5$SO$_3$H), the ketal lost methanol (CH$_3$OH) to produce the enol ether, Δ^6-dihydrothebaine. Treatment of this enol ether with aqueous hydrogen bromide (HBr) followed by methanolic N-bromoacetamide (CH$_3$CONHBr) regenerated the ketal but now with a bromine on the adjacent carbon (7-bromodihydrocodeinone)dimethyl ketal. Then, treatment of this bromo-ketal with potassium *tert*-butoxide in dimethyl sulfoxide, while heating for 1 h at 120°C induced elimination of both hydrogen bromide (HBr) and methanol (CH$_3$OH) and generated thebaine. However, if the same bromoketal was heated at only 60°C for longer (7 h), the elimination did not involve both the hydrogen bromide (HBr) and methanol (CH$_3$OH) but rather just hydrogen bromide (HBr) so that the dimethylketal of codeinone (hydrolyzed to codeinone with acetic acid) was produced. Codeinone was converted to codeine by reduction with sodium borohydride (NaBH$_4$) in methanol (CH$_3$OH) and the latter converted to morphine

---

*Weller, D. D.; Rapaport, H. *J. Med. Chem.*, **1976**, *19*, 1171.

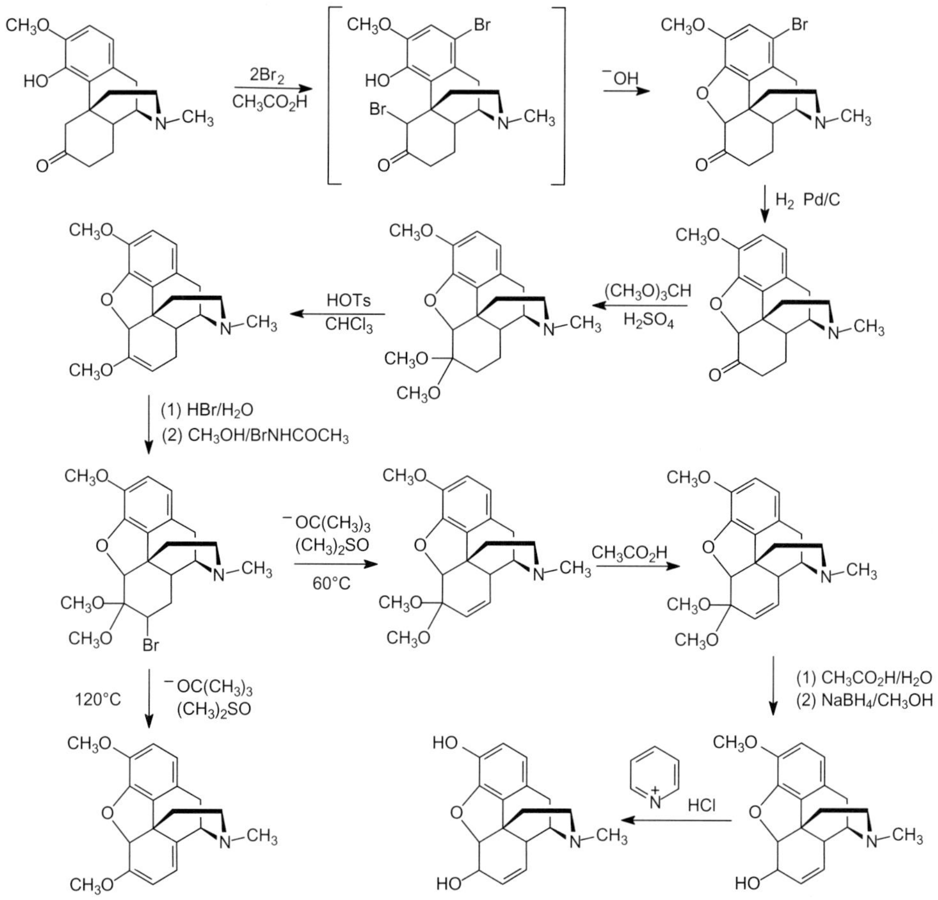

Scheme 13.46. A representation of the conversion of dihydrothebainone to thebaine, codeine, and morphine. After Weller, D. D.; Rapaport, H. *J. Med. Chem.*, **1976**, *19*, 1171. HOTs = toluenesulfonic acid.

as worked out by Gates and Tschudi and shown above (i.e., with pyridinium hydrochloride).

D. VINBLASTINE

I. Chemistry of Vinblastine

In 1958, a most curious paper appeared in the *Annals of the New York Academy of Science* in which even the authors suggested that its appearance was "unorthodox."* The work was entitled "Role of Chance Observations in Chemotherapy: *Vinca rosea*" and needs to be read to be enjoyed. In this work, the authors describe following activity in extracts from *Vinca rosea* Linn. (now called *Catharanthus roseus*) through a series of purifications cycles until crystalline material of plant alkaloid is obtained.

*Noble, R. L.; Beer, C. T.; Cutts, J. H. *Ann. N. Y. Acad. Sci.*, **1958**, *76*, 882.

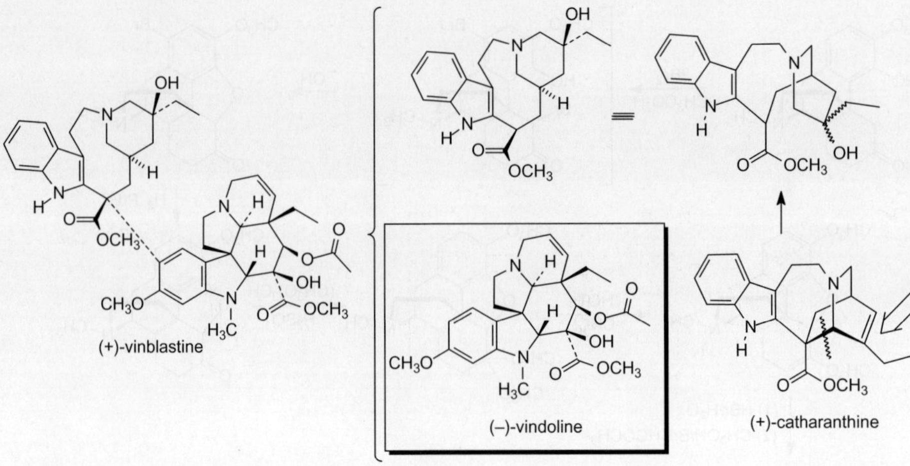

Scheme 13.47. A cartoon representation of the potential coupling of fragments resembling (–)-vindoline and (modified) (+)-catharanthine to produce vinblastine.

Because the crystalline material (called, at that time, vincaleucoblastine) had what were reported to be "interesting" properties, further examination was undertaken, physical data were collected, and suggestions for the structure based on spectroscopic analysis were set forth.[*] Finally, the X-ray crystal structure, using the Bijvoet method of anomalous dispersion (*vide supra*), succeeded in establishing the structure and stereochemistry beyond reasonable doubt.[†‡]

The first synthesis of vinblastine was reported by Potier and coworkers more than a decade later[§] and three more, again, after about another decade had passed.[¶] The syntheses all used a naturally occurring fragment (viz. vindoline) to complete the work. These works are worthy of study. However, a recent synthesis by Fukuyama and coworkers[#] has captured the imagination and is reported below.

As shown in Scheme 13.47, and in concert with the syntheses that had preceded this one, it was recognized that vinblastine could be looked at as derived from a coupling between species related to the "simpler" fragments vindoline and (a modified) catharanthine. For the latter, it is clear that the "modification" (were the alka-

[*]Neuss, N.; Gorman, M.; Svoboda, G. H.; Maciak, G.; Beer, C. T. *J. Am. Chem. Soc.*, **1959**, *81*, 4754; Neuss, N.; Gorman, M.; Boaz, H. D.; Cone, N. J. *J. Am. Chem. Soc.*, **1962**, *84*, 1509; Neuss, N.; Gorman, M.; Hargrove, W.; Cone, J. H.; Beimann, K.; Buchi, G.; Manning, R. E. *J. Am. Chem. Soc.*, **1964**, *86*, 1440.

[†]Moncrief, J. W.; Libscomb, W. N. *Acta Crystallogr.*, **1966**, *21*, 322.

[‡]A significant amount of degradative and synthetic work on analogous compounds was known. Much of this has been summarized in Kutney, J. P. in Hey, D. H.; Wiesner, K. F. (eds.), *MTP International Review of Science*, Vol. 9, Butterworths, London, 1973, p. 27 ff.

[§]Mangeney, P.; Andriamialisoa, R. Z.; Langlois, N.; Langlois, Y.; Potier, P. *J. Am. Chem. Soc.*, **1979**, *101*, 2243.

[¶](a) Kutney, J. P.; Choi, L. S. S.; Nakano, J.; Tsukamoto, H.; McHugh, M.; Boulet, C. A. *Heterocycles*, **1988**, *27*, 1845. (b) Kuehne, M. E.; Matson, P. A.; Bornmann, W. G. *J. Org. Chem.*, **1991**, *56*, 513. (c) Magnus, P.; Mendoza, J. S.; Stamford, A.; Ladlos, M.; Willis, P. *J. Am. Chem. Soc.*, **1992**, *114*, 10232.

[#]Yokoshima, S.; Ueda, T.; Kobayashi, S.; Sato, A.; Kuboyama, T.; Tokuyama, H.; Fukuyama, T. *Pure Appl. Chem.*, **2003**, *75*, 29.

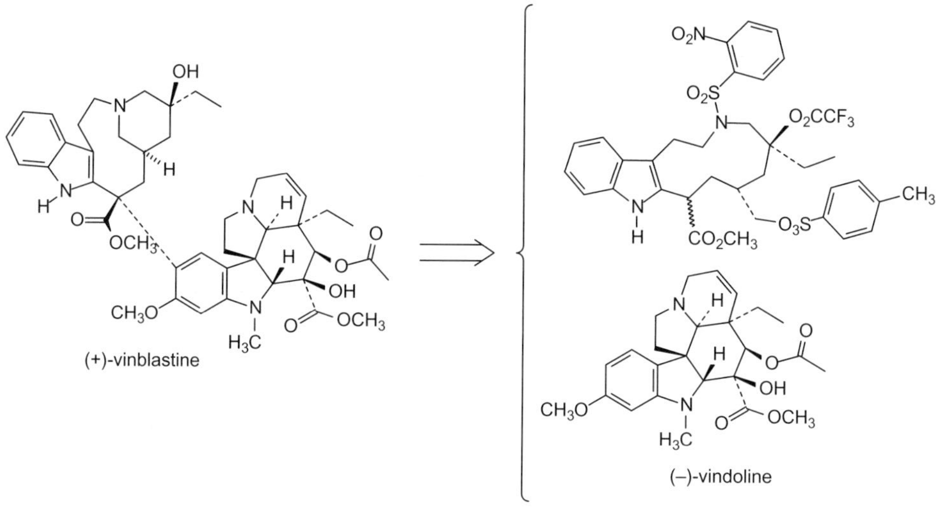

Scheme 13.48. The fragments (vindoline) and catharanthine-like needed to produce (+)-vinblastine. See Yokoshima, S.; Ueda, T.; Kobayashi, S.; Sato, A.; Kuboyama, T.; Tokuyama, H.; Fukuyama, T. *Pure Appl. Chem.*, 2003, *75*, 29.

loid itself be formed rather than a precursor that was suitably directed) that the azabicyclo[2.2.2]octene ring would require oxidation and cleavage either before or, for cleavage, concomitant with the two systems becoming cojoined. As will be seen, at this writing, details of the *in vivo* process are lacking.

Indeed, as shown in Scheme 13.48, the plan of the synthesis to be described, that is, that of Fukuyana et al. (*vide supra*) uses a similar understanding.

The synthesis of (–)-vindoline as effected by Fukuyama et al. began with the preparation of a substituted indole and is shown in Schemes 13.49–13.53.

When 3-aminophenol is treated with toluenesulfonyl chloride at low temperature in pyridine, selective tosylation at nitrogen occurs. Then, reaction with acrolein (propenal) in the presence of ethanamine (ethylamine) produces the corresponding β-aminoaldehyde, which undergoes cyclization to an *N*-tosylated dihydoquinoline when treated with hydrogen chloride (HCl) in warm THF. Elimination of toluenesulfinic acid by treatment with potassium hydroxide (KOH) yields the corresponding quinoline, which undergoes reaction with methanesulfonyl chloride (CH_3SO_2Cl) to yield the mesylate of the phenol. Treatment of the mesylate with thiophosgene ($CSCl_2$) in the presence of sodium carbonate in wet THF solvent resulted in ring opening to the aldehydic thiocyanate, which was reduced with sodium borohydride ($NaBH_4$) to the corresponding primary alcohol and the latter then protected as the tetrahydropyranyl acetal (THP) by reaction with dihydropyran.

Then, as shown in Scheme 13.50, when the protected thiocyanate was treated with the anion of benzyl methyl malonate (generated from the diester with sodium hydride) in THF, the corresponding thioanilide was formed. Then, using the Fukuyama reaction, treatment of the thioamide with tri-*n*-butyltin hydride [(*n*-$CH_3CH_2CH_2CH_2)_3SnH$] in the presence of azobisisobutyronitrile (AIBN) resulted in cyclization to a 2,3-disubstituted indole nucleus, which was protected at nitrogen by formation of the *t*-butylcarboxyamide (Boc) through reaction with the corre-

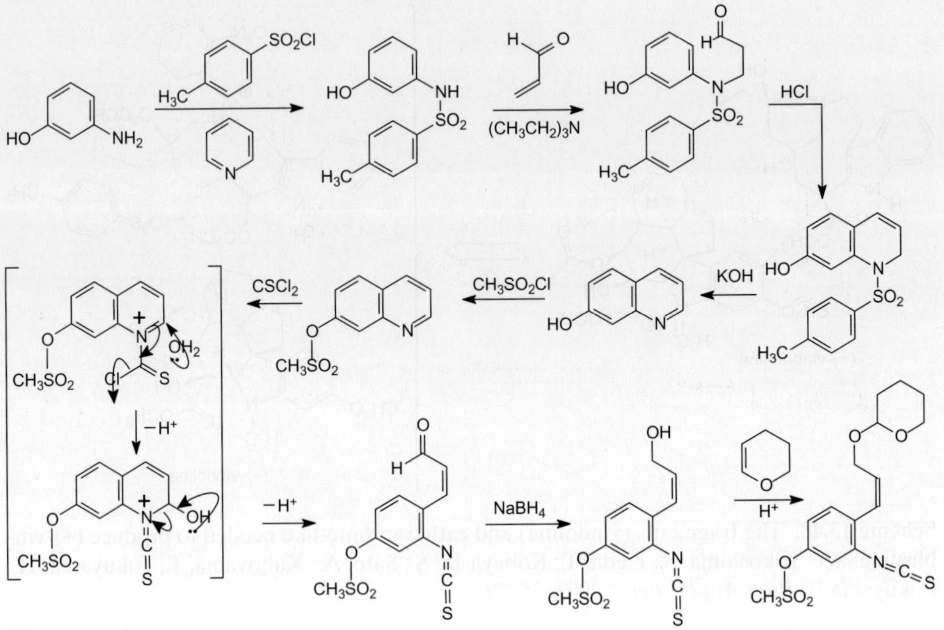

Scheme 13.49. The beginning of the path to (–)-vindoline. After Yokoshima, S.; Ueda, T.; Kobayashi, S.; Sato, A.; Kuboyama, T.; Tokuyama, H.; Fukuyama, T. *Pure Appl. Chem.*, **2003**, *75*, 29.

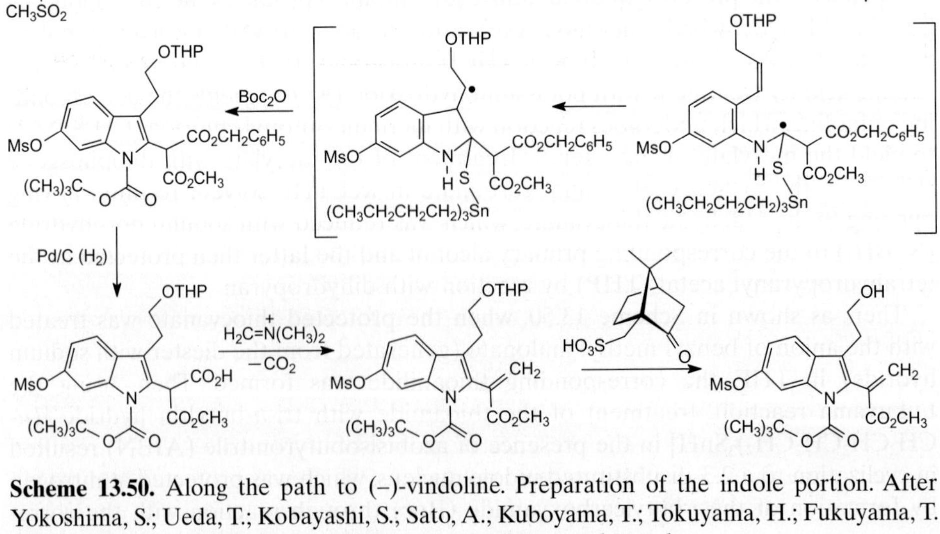

Scheme 13.50. Along the path to (–)-vindoline. Preparation of the indole portion. After Yokoshima, S.; Ueda, T.; Kobayashi, S.; Sato, A.; Kuboyama, T.; Tokuyama, H.; Fukuyama, T. *Pure Appl. Chem.*, **2003**, *75*, 29. OTHP = tetrahydropyranyl acetal.

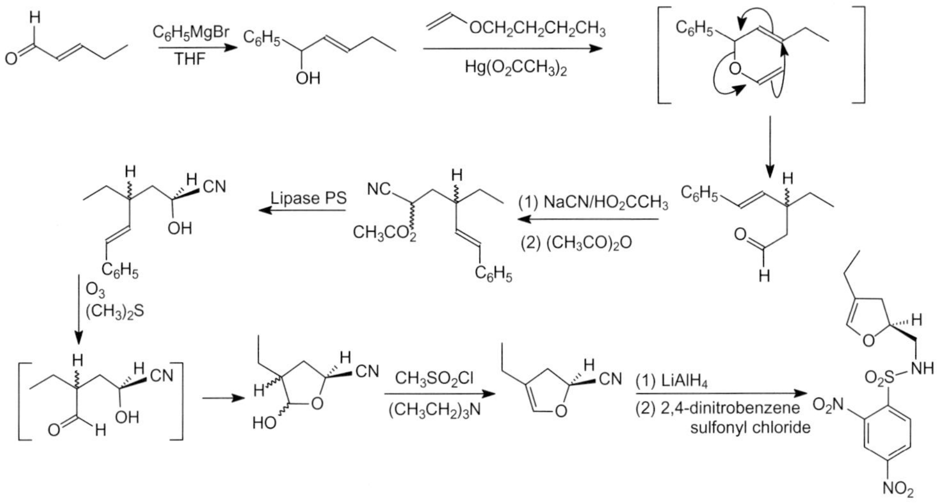

Scheme 13.51. Along the path to (–)-vindoline. Preparation of the basic nitrogen portion destined to become the dienophile. After Yokoshima, S.; Ueda, T.; Kobayashi, S.; Sato, A.; Kuboyama, T.; Tokuyama, H.; Fukuyama, T. *Pure Appl. Chem.*, **2003**, *75*, 29.

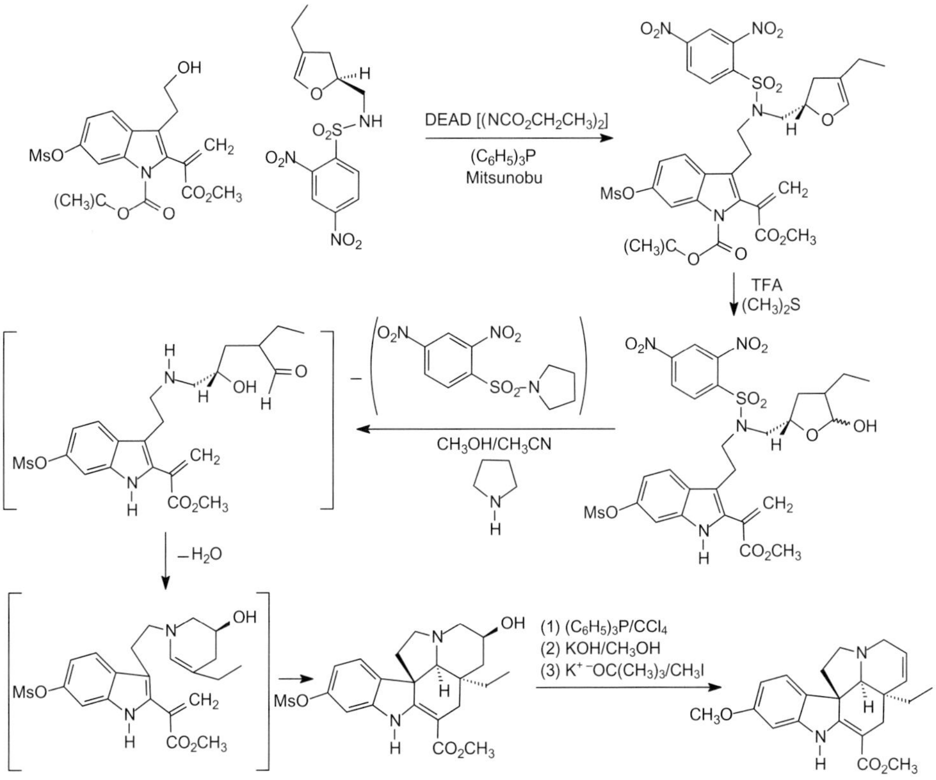

Scheme 13.52. Along the path to (–)-vindoline. Preparation of (–)-11-methoxytabersonine. After Yokoshima, S.; Ueda, T.; Kobayashi, S.; Sato, A.; Kuboyama, T.; Tokuyama, H.; Fukuyama, T. *Pure Appl. Chem.*, **2003**, *75*, 29.

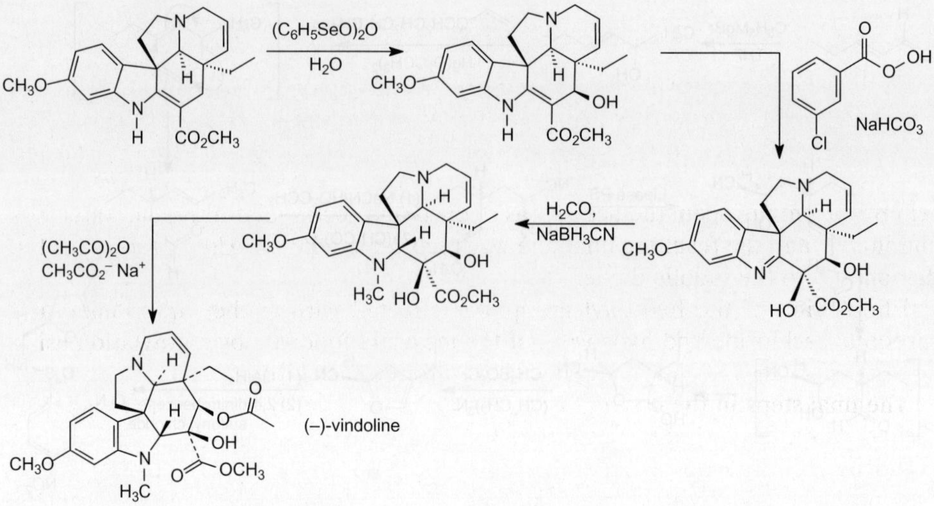

Scheme 13.53. The final steps in the conversion of (–)-11-methoxytabersonine to (–)-vindoline. After Yokoshima, S.; Ueda, T.; Kobayashi, S.; Sato, A.; Kuboyama, T.; Tokuyama, H.; Fukuyama, T. *Pure Appl. Chem.*, **2003**, *75*, 29.

sponding *t*-butylcarboxy anhydride (Boc$_2$O). Hydrogenolysis (H$_2$ with Pd/C) permitted the removal of the benzyl protecting group, and the half-acid ester underwent decarboxylation. A Mannich reaction with formaldehyde in the presence of *N,N*-dimethylamine followed by deprotection (removal of the THP protecting group) resulted in hydroxydiene formation.

With the diene portion in hand, it was now necessary to produce a dienophile that would allow construction of the complete system. The dienophile would contain the basic nitrogen in vindoline.

As shown in Scheme 13.51, 2-pentenal underwent Grignard reaction with phenylmagnesium bromide and the alcohol that resulted was treated with vinyl butyl ether in the presence of mercury(II) acetate [Hg(O$_2$CCH$_3$)$_2$] to produce a new ether that underwent Claisen rearrangement to the corresponding aldehyde. A Strecker reaction (sodium cyanide, NaCN, in acetic acid [CH$_3$CO$_2$H]) successfully converted the aldehyde to the corresponding mixture of isomeric cyanohydrins, which was acetylated with acetic anhydride [(CH$_3$CO)$_2$O]. Then the mixture was treated with Amano lipase polystyrene (lipase PS [*Pseudomonas cepacia*]) in aqueous THF to effect hydrolysis of one diastereomer preferentially and to generate mixture of (*S*)-cyanohydrins in high enantiomeric purity (97% ee).* Ozonolysis of the double bond yielded the corresponding aldehyde, which cyclized to the hemiacetal, and could be converted to the corresponding mesylate. Elimination to the dihydrofuran and reduction of the nitrile to the corresponding amine was followed by conversion to the 2,4-dinitrosulfonamide by treatment with dinitrobenzene-sulfonyl chloride.

When the indolic diene was allowed to react with the dinitrosulfonamide derivative of the aminodihydrofuran (Scheme 13.52) under Mitsunobu conditions [diethyl

*Faber, K. *Biotransformation in Organic Chemistry: A Textbook*, Springer-Verlag, New York, 2004, pp. 29–176.

azodicarboxylate, DEAD, in the presence of triphenylphosphine, $(C_6H_5)_3P$], the fragments coupled. Then, treatment of the coupled product with trifluoroacetic acid (TFA) resulted in removal of the Boc protecting group on the indolic nitrogen and, on workup with water, (re)hydration of the dihydrofuran to the acetal. Now, when the acetal was (briefly, 5 min!) treated with pyrrolidine in acetonitrile-methanol and the resulting mixture was warmed for 2 h at 60°C, the dinitrosulfonamide was removed from nitrogen, the acetal was opened and reclosed at nitrogen, water was eliminated, and the resulting enamine was cyclized in a Diels–Alder process (as the dienophile) to the indolic diene.

Elimination of the hydroxyl group was effected with triphenylphosphine and carbon tetrachloride and hydrolysis of the mesylate followed by methylation using potassium *t*-butoxide and methyl iodide to produce (–)-11-methoxytabersonine.

The final steps in the conversion of (–)-11-methoxytabersonine to (–)-vindoline (Scheme 13.53) involved benzeneseleninic anhydride $[(C_6H_5SeO)_2O]$ oxidation followed by a second oxidation with *meta*-chloroperbenzoic acid (MCPBA) in the presence of sodium bicarbonate $(NaHCO_3)$. Without isolation, but with adjustment of the pH, reductive *N*-methylation of the indoline was effected with formaldehyde (H_2CO) addition to the imine, followed by sodium cyanoborohydride $(NaBH_3CN)$ reduction. Finally, selective acetylation of the secondary hydroxyl necessary to produce (–)-vindoline (in the presence of the tertiary alcohol) was accomplished with acetic anhydride-containing sodium acetate.

As shown in Schemes 13.54–13.56, the "catharanthine-like" piece was next prepared.

When the mixed anhydride between 4-ethyl-4-pentenoic acid and pivalic acid (prepared using pivaloyl chloride on the pentenoic acid) was treated with the lithium salt of (*R*)-4-benzyl-2-oxazolidinone an imide was formed where one (the "bottom" as drawn) face is encumbered by the bulky benzyl group. Thus, cyanoethylation occurred with diastereoselectivity, and a single isomer was isolated. Reduction of the imide with sodium borohydride $(NaBH_4)$ then resulted in the formation of the primary alcohol, and the latter was protected as the *t*-butyldiphenylsilyl (TBDPS)

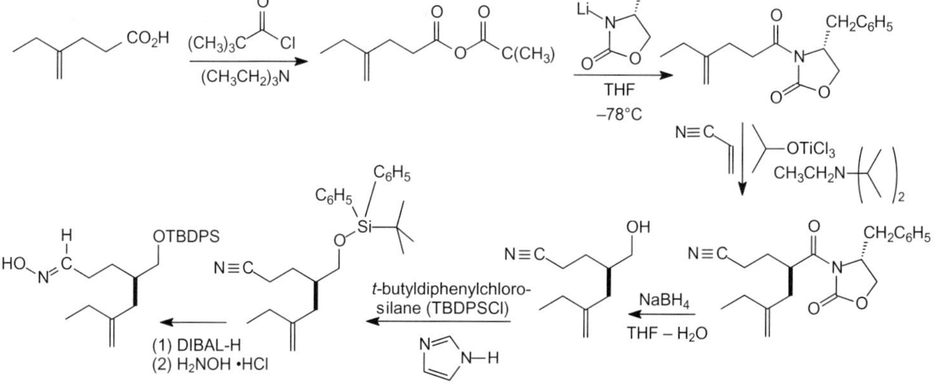

Scheme 13.54. The initial processes in building the "catharanthine-like" piece of (+)-vinblastine. After Yokoshima, S.; Ueda, T.; Kobayashi, S.; Sato, A.; Kuboyama, T.; Tokuyama, H.; Fukuyama, T. *Pure Appl. Chem.*, **2003**, *75*, 29. OTBDPS = *t*-butyldiphenylsilyl ether.

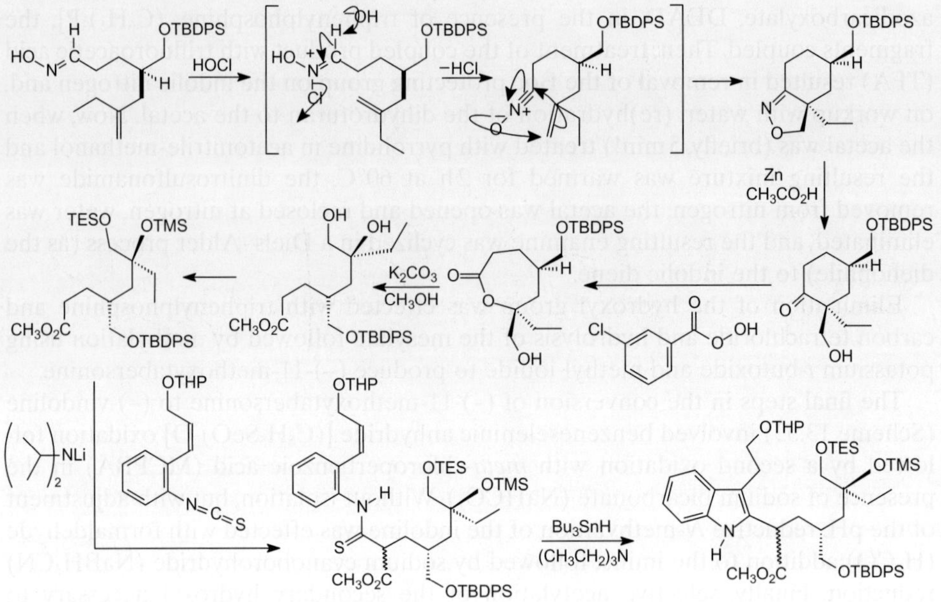

Scheme 13.55. Additional work along to path to building the "catharanthine-like" piece of (+)-vinblastine. After Yokoshima, S.; Ueda, T.; Kobayashi, S.; Sato, A.; Kuboyama, T.; Tokuyama, H.; Fukuyama, T. *Pure Appl. Chem.*, **2003**, *75*, 29. OTBDPS = *t*-butyldiphenylsilyl ether; TESO = triethylsilyl ether; OTMS = trimethylsilyl ether; OTHP = tetrahydropyranyl acetal; OTES = triethylsilyl ether.

derivative by treatment with *t*-butyldiphenylchlorosilane (TBDPSCl) in the presence of imidazole. Reduction of the nitrile to aldehyde with diisobutylaluminum hydride (DIBAL-H) was followed by treatment with hydroxylamine hydrochloride to produce the oxime.

The oxime (Scheme 13.54), written here as the (*E*)-isomer (but there is not enough data to be specific), was oxidized (Scheme 13.55) with sodium hypochlorite to produce the corresponding nitrile oxide, which spontaneously underwent intramolecular 1,3-dipolar cycloaddition to the corresponding isoxazoline. On treatment with zinc (Zn) in acetic acid (CH_3CO_2H), the N–O bond of the isoxazoline was cleaved and the resulting iminoaclohol was hydrolyzed to the corresponding ketoalcohol. The Baeyer–Villiger oxidation of the latter with *meta*-chloroperbenzoic acid in acetic acid (CH_3CO_2H) produced a lactone. Methanolysis of the lactone generated the corresponding methyl ester of a diol, the hydroxyl groups of which could be differentially protected first, at the primary hydroxyl, with triethylchlorosilane (TES) and imidazole and then, at the secondary hydroxyl, with trimethylchlorosilane.

Treatment of the enolate anion, generated with lithium diisopropylamide (LDA) α to the carbonyl of the ester of the derivatized triolester, with a THP-protected thiocyanate similar to that of Scheme 13.50 (*vide supra*) then led to a new indole (the Fukuyama indole synthesis).

As shown in Scheme 13.56, protection of the newly formed indole was accomplished by treatment with Boc anhydride in the presence of dimethylaminopyridine (DMAP) and the deprotection of the primary alcohol that previously bore the triethylsilyl group, the secondary alcohol that bore the trimethylsilyl group and the

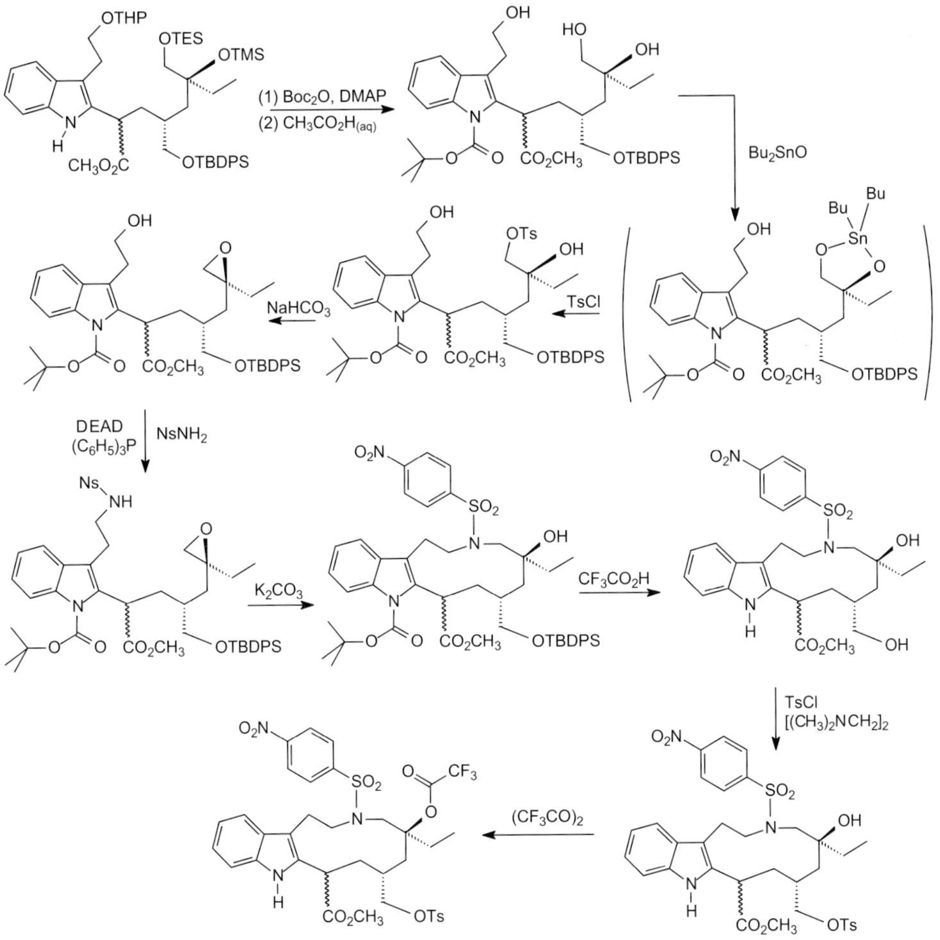

Scheme 13.56. A path showing the formation of the piece of (+)-vinblastine representing the "catharanthine"-like fragment needed for the total synthesis. After Yokoshima, S.; Ueda, T.; Kobayashi, S.; Sato, A.; Kuboyama, T.; Tokuyama, H.; Fukuyama, T. *Pure Appl. Chem.*, **2003**, *75*, 29. OTBDPS = *t*-butyldiphenylsilyl ether; OTMS = trimethylsilyl ether; OTHP = tetrahydropyranyl acetal; OTES = triethylsilyl ether; OTs = toluenesulfonate ester.

primary alcohol that was protected by the THP was effected with aqueous acetic acid (CH_3CO_2H). Then, treatment of this triol with di-*n*-butyltin oxide (Bu_2SnO), a reagent that complexes with the vicinal diol, activating that one which is primary,* followed by addition of toluenesulfonyl chloride, resulted in the formation of a tosylate ester of the primary alcohol. Bicarbonate treatment of this tosylate resulted in epoxide formation and when, under Mitsunobu conditions, the epoxide was treated with diethyl azodicarboxylate and *para*-nitrobenzenesulfonamide ($NsNH_2$), the only remaining hydroxyl function was converted to the corresponding *para*-nitrobenzenesulfonyl derivative of a primary amine.

*Martinelli, M. J.; Nayyar, N. K.; Moher, E. D.; Dhokte, U. P.; Pawlak, J. M.; Vaidyanathan, R. *Org. Lett.*, **1999**, *1*, 447.

With bicarbonate, the acidic sulfonamide proton was removed and the primary carbon end of the epoxide was attacked, thus completing the cyclization of the required 11-membered ring.

The final steps in the preparation of the "catharanthine" fragment required that the last remaining protecting group (the *t*-butyldiphenylsilyl) be removed. This was accomplished with trifluoroacetic acid (TFA). The newly liberated primary alcohol was tosylated (toluensulfonyl chloride in the presence of tetramethylpropylenediamine), and the tertiary alcohol was protected as the trifluoroacetate derivative by treatment with trifluoroacetic anhydride in the presence of pyridine.

As shown in Scheme 13.57, consummation of the synthesis of (+)-vinblastine required activation of the "catharanthine" fragment (from Scheme 13.56) and reaction of that activated piece with the (−)-vindoline fragment from Scheme 13.53. Then, once the two fragments had been cojoined, the final protecting groups were removed, and the last ring was put in place.

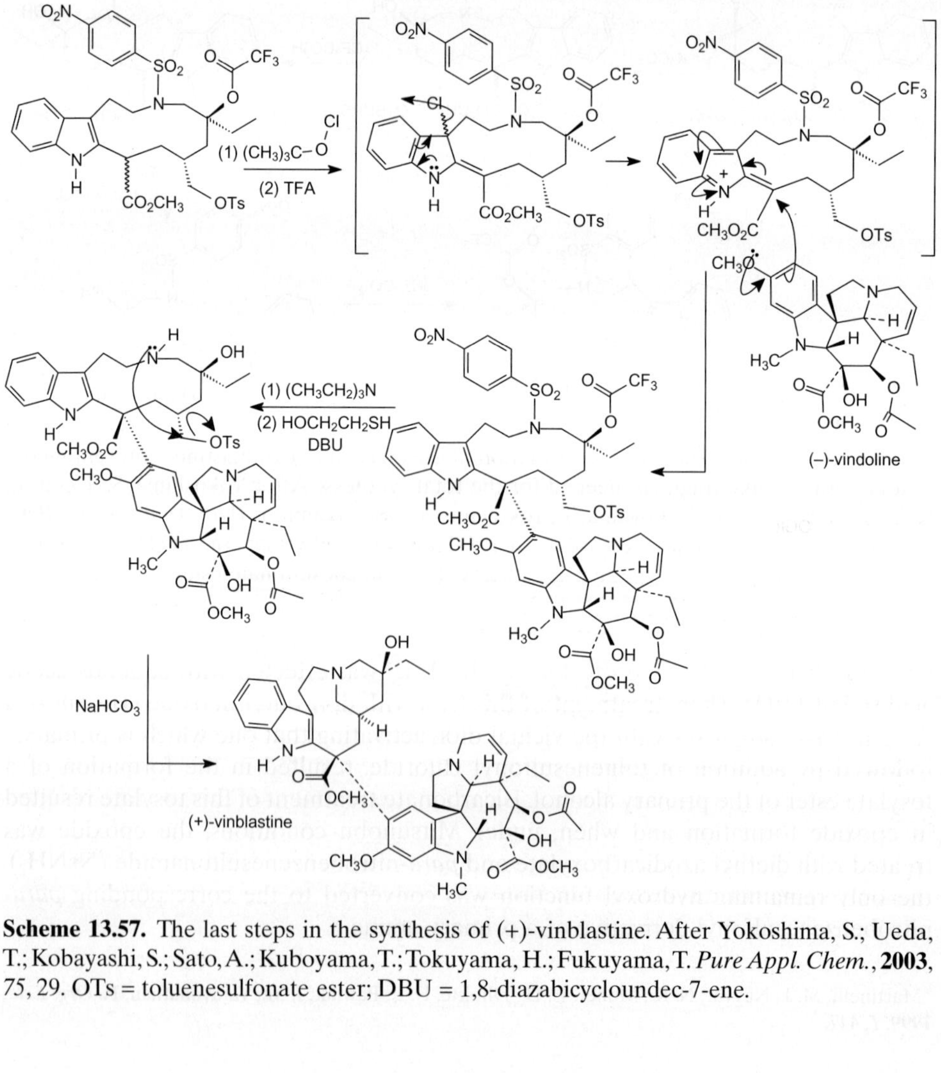

Scheme 13.57. The last steps in the synthesis of (+)-vinblastine. After Yokoshima, S.; Ueda, T.; Kobayashi, S.; Sato, A.; Kuboyama, T.; Tokuyama, H.; Fukuyama, T. *Pure Appl. Chem.*, **2003**, 75, 29. OTs = toluenesulfonate ester; DBU = 1,8-diazabicycloundec-7-ene.

II. Biosynthesis of Vinblastine

In earlier chapters (Chapters 11 and 12), some of the materials needed to bring important intermediates forward in the biosynthesis of vinblastine have been discussed. Thus, the Schemes, beginning with Scheme 11.79 that provides a path to secologanin (repeated here as Scheme 13.58) and continuing through Schemes 11.80–11.83 where a pathway to chorismate (produced from shikimic acid and pyruvate) was shown, are now needed. Schemes 12.24 and 12.25 (repeated here as Schemes 13.59 and 13.60) presenting a pathway (first) from chorismate to anthranilate and (second) to **tryptophan (Trp, W)** and tryptamine are also required, as it is the combination of the pyridoxal-catalyzed decarboxylation product (EC 4.1.1.28, the L-amino acid decarboxylase) of **tryptophan (Trp, W)**, that is, tryptamine, with secologanin that gives rise to the vast array of products collectively known as the "monoterpene indole alkaloids."

As shown in Scheme 13.61, a Pictet–Spengler-type condensation and cyclization between tryptamine and secologanin forms the first committed step with the generation of strictosidine. The process is effected by the enzyme strictosidine synthase (EC 4.3.3.2), and, interestingly, strictosidine synthase (the complementary DNA) is

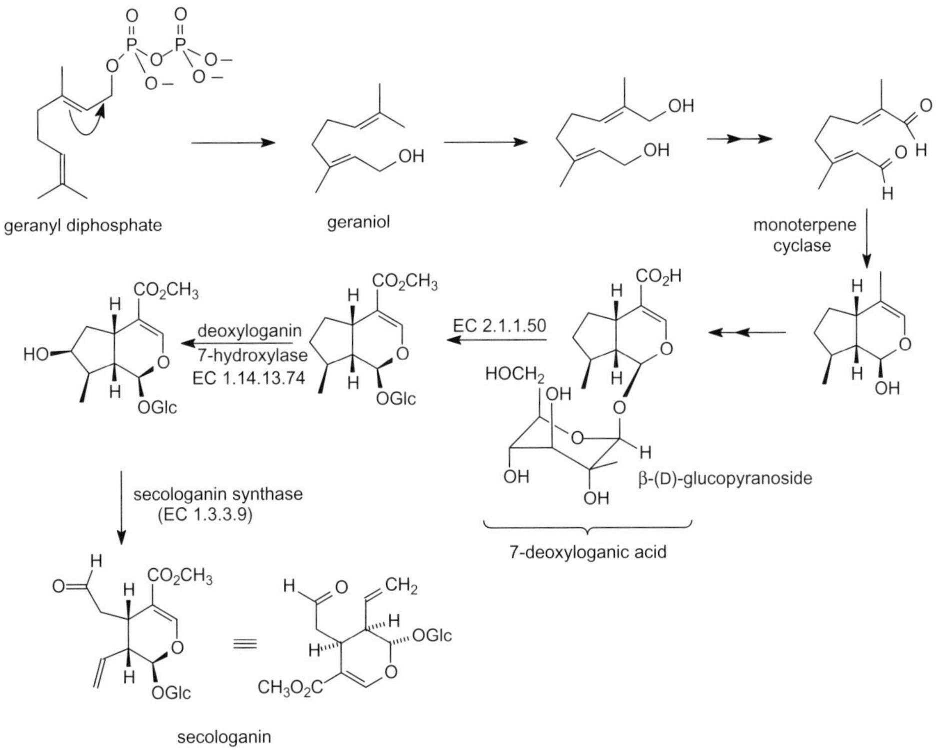

Scheme 13.58. A repetition of Scheme 11.79, providing a pathway for the formation of secologanin from geraniol. EC numbers and some graphic materials provided in this scheme have been taken from appropriate links in a URL starting with http://www.chem.qmul.ac.uk/iubmb/enzyme/.

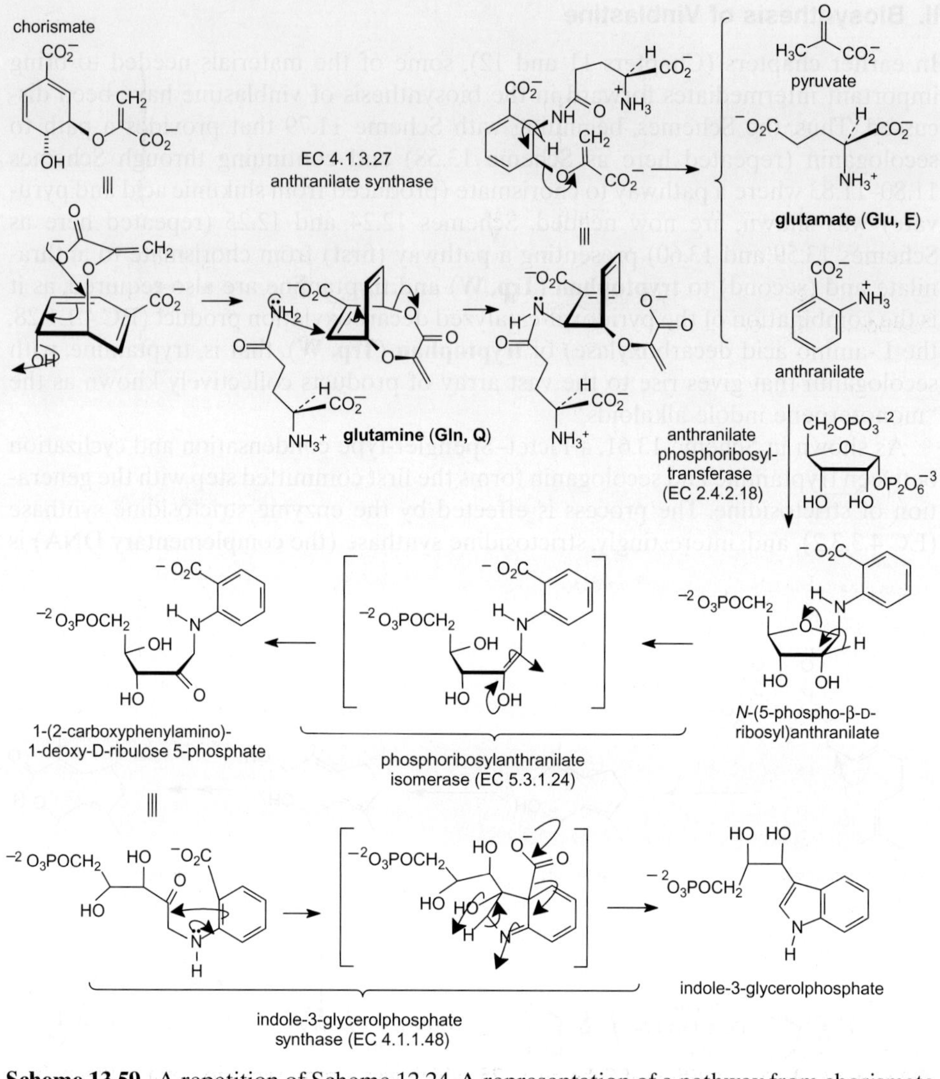

chorismate

EC 4.1.3.27
anthranilate synthase

pyruvate

glutamate (Glu, E)

glutamine (Gln, Q)

anthranilate

anthranilate
phosphoribosyl-
transferase
(EC 2.4.2.18)

N-(5-phospho-β-D-
ribosyl)anthranilate

1-(2-carboxyphenylamino)-
1-deoxy-D-ribulose 5-phosphate

phosphoribosylanthranilate
isomerase (EC 5.3.1.24)

indole-3-glycerolphosphate

indole-3-glycerolphosphate
synthase (EC 4.1.1.48)

Scheme 13.59. A repetition of Scheme 12.24. A representation of a pathway from chorismate to indole-3-glycerolphosphate (on the way to **tryptophan [Trp, W]**).

reported to have been "the first cDNA involved in alkaloid biosynthesis to be cloned and functionally expressed in microorganisms."*

After the condensation has been consummated, the glucose (Glc) side chain is hydrolytically removed with the aid of strictosidine β-D-glucohydrolase (EC 3.2.1.105), and it appears that a cDNA has also been characterized for this step.

The resulting aglycone apparently undergoes spontaneous dehydration to 4,21-dehydrogeissoschizine, which is then reduced (EC 1.3.1.36, geissoschizine dehydrogenase) to geissoschizine. At this juncture (Scheme 13.62), imagination runs wild

*Ziegler, J.; Facchini, P. J. *Ann. Rev. Plant Biol.*, **2008**, *59*, 735. See Chapter 14 regarding complementary DNA (cDNA).

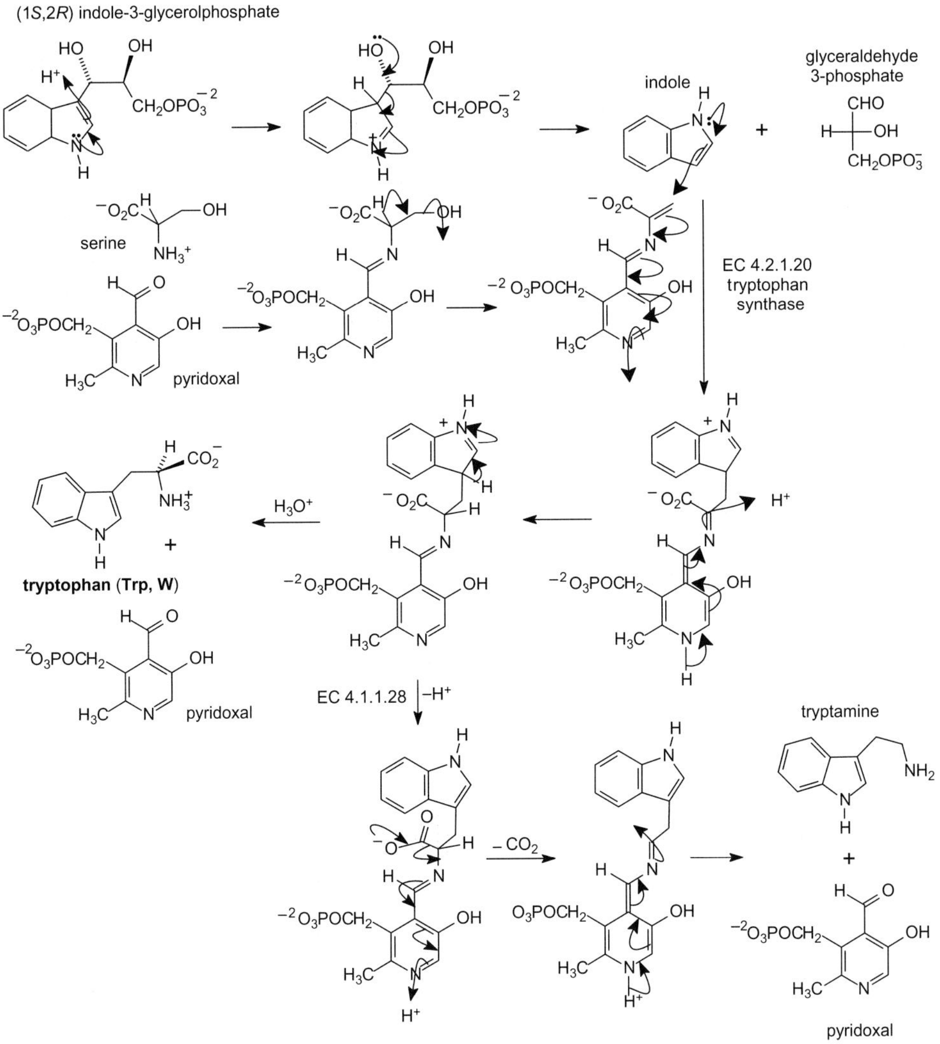

Scheme 13.60. A repetition of Scheme 12.25, which was a representation of a pathway from indole-3-glycerolphosphate to **tryptophan (Trp, W)**. The path has been modified with addition of steps (using the amino acid deaminase EC 4.1.1.28) to carry **tryptophan (Trp, W)** to tryptamine.

since there is little evidence about what occurs in going to catharanthine (or a "catharanthine"-like fragment) and the base tabersonine from geissoschizine. However, tabersonine is known (Scheme 13.63) to lead to vindoline.

As shown in Scheme 13.63, tabersonine, in the presence of tabersonine 16-hydroxylase (EC 1.14.13.73), is converted to 16-hydroxytabersonine. The reaction requires both NADPH and oxygen (O_2), and $NADP^+$ and water (H_2O) are produced. Tabersonine 16-hydroxylase is a heme-thiolate protein (a P_{450}). The phenolic hydroxyl group on the aromatic ring undergoes methylation (tabersonine 16-O-methyltransferase, EC 2.1.1.94), while S-adenosylmethionine is converted to

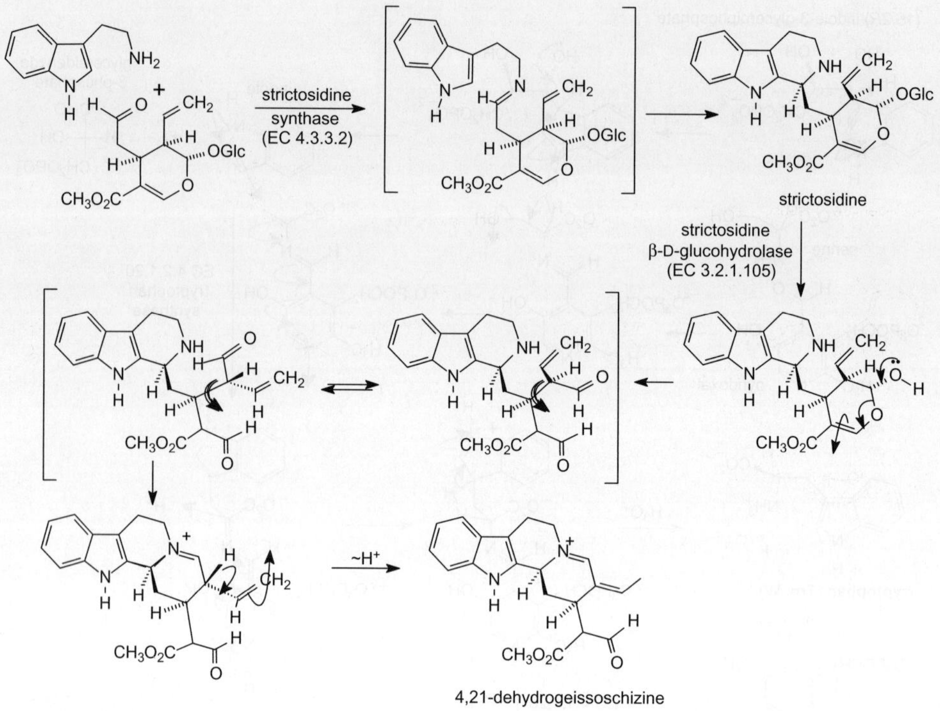

strictosidine synthase
(EC 4.3.3.2)

strictosidine

strictosidine β-D-glucohydrolase
(EC 3.2.1.105)

~H⁺

4,21-dehydrogeissoschizine

Scheme 13.61. A path, involving a Pictet–Spengler ring closure of the imine formed from the reaction of tyramine with secologanin to strictosidine and thence to dehydrogeissoschizine. The process is supported by a large number of feeding experiments with labelled compounds. (Much of this chemistry has been reviewed and, at any rate, is too extensive to repeat here. See, for example, Scott, A. I. in Hey, D. H.; Wiesner, K. F. (eds.), *MTP International Review of Science*, Vol. 9, Butterworths, London, 1973, pp. 105 ff.) The final cyclization and ring closure processes appear to be spontaneous.

S-adensoylhomocysteine. Then, the 16-methoxytabersonine is converted either by reduction of the double bond and subsequent oxidation or in a single hydroxylation step to 3-hydroxy-16-methoxy-2,3-dihydrotabersonine and another methylation, this time at the dihydroindolic nitrogen (again using *S*-adenosylmethionine but now with the enzyme 3-hydroxy-16-methoxy-2,3-dihydrotabersonine *N*-methyltransferase, EC 2.1.1.99). The resulting desacetoxyvindoline is oxidized to deacetylvindoline in the presence of deacetoxyvindoline 4-hydroxylase (EC 1.14.11.20), an enzyme that requires 2-oxoglutarate (α-ketoglutarate). The α-ketoacid is decarboxylated (to succinate and carbon dioxide [CO_2]) and ferrous iron (Fe^{2+}) and ascorbic acid also appear to be necessary. Finally, deacetylvindoline is acetylated by acetyl-CoA in the presence of deacetylvindoline O-acetyltransferase (EC 2.3.1.107), and vindoline is generated.

Of course, the final step in the discussion of the biosynthesis and in the production of vinblastine in *C. roseus* requires an oxidative coupling of vindoline and catharanthine (or their respective equivalents) as shown earlier in Scheme 13.47. At this writing, it appears that the coupling of the fragments is accomplished with the help of a nonspecific peroxidase.

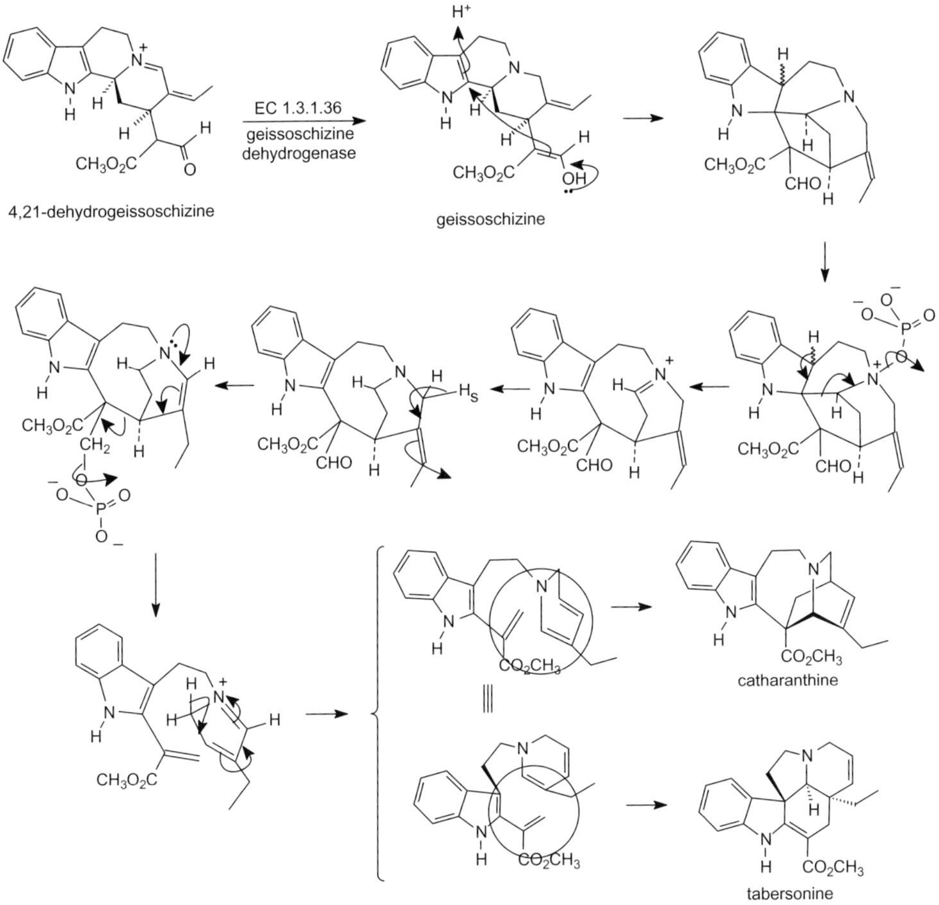

Scheme 13.62. Possible pathways (involving different Diels–Alder reactions) to catharanthine and to tabersonine (the latter known to be converted to vindoline) from geissoschizine. At this writing, the pathway shown is largely substantiated by labeling experiments. See, for example, Scott, A. I. in Hey, D. H.; Wiesner, K. F. (eds.), *MTP International Review of Science*, Vol. 9, Butterworths, London, **1973**, pp. 105 ff.

E. CAFFEINE

I. Some History and the Synthesis of Caffeine

In their seminal work dealing with the history of caffeine-containing beverages (mainly coffee, tea, and cocoa), Weinberg and Bealer* argued that the original isolation of caffeine from coffee was by Runge as a consequence of, and subsequent to, a meeting with Goethe in 1819. Indeed, Kränzlein, in his celebratory article on the centenary of the coal–tar–dye industry in Germany,[†] the founding of which is attrib-

---

*Weinberg, B. A.; Bealer, B. K. *The World of Caffeine*, Routledge, New York, **2001**.
[†]Kränzlein, G. *Angew. Chem.*, **1935**, *48*, 1.

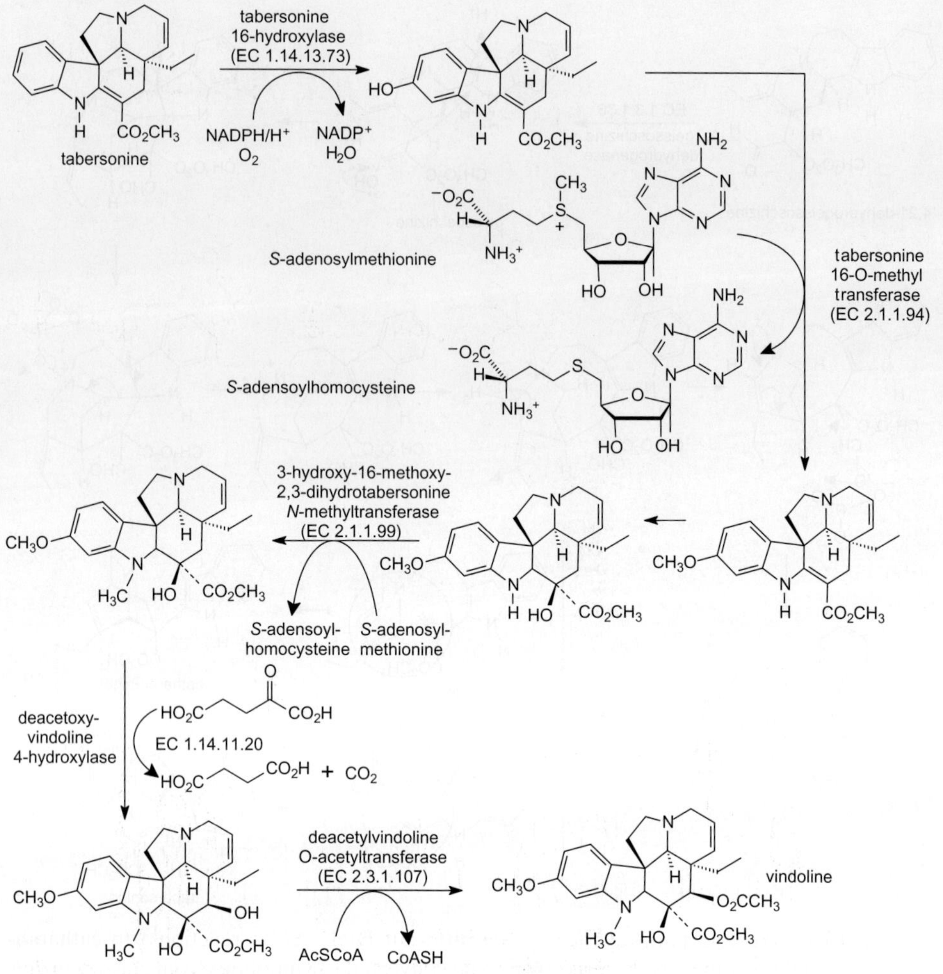

Scheme 13.63. A pathway for the formation of vindoline from tabersonine. EC numbers and some graphic materials provided in this scheme have been taken from appropriate links in a URL starting with http://www.chem.qmul.ac.uk/iubmb/enzyme/.

uted to Runge, made a similar claim. Atta-Ur-Rahman and Choudhary, in their full report on "Purine Alkaloids"[*] noted that Runge did make the claim in 1820,[†] and they argued that this report is that of the "first purine base isolated from any plant source." The claim was ignored by Pelletier and Caventou in their larger work in 1822.[‡]

Much of the early chemistry highlighted the relationships among urea, allantoic acid, allentoin, and the fully elaborated purines, uric acid, caffeine, theobromine, theophylline, and xanthine (Figure 13.7).

[*]Atta-Ur-Rahman; Choudhary, M. I. in Brossi, A. (ed.), *The Alkaloids*, Vol. 38, Academic Press, New York, **1990**, pp. 226–313.

[†]Runge, F. F. *Neuste Phytochemische Entdeckungen*, **1820**, *1*, 144.

[‡]Pelletier, J.; Caventou, J. B. *Dictionnaire de medicine*, Vol. IV, Paris, France: Béchet Jeune, **1822**, p. 35.

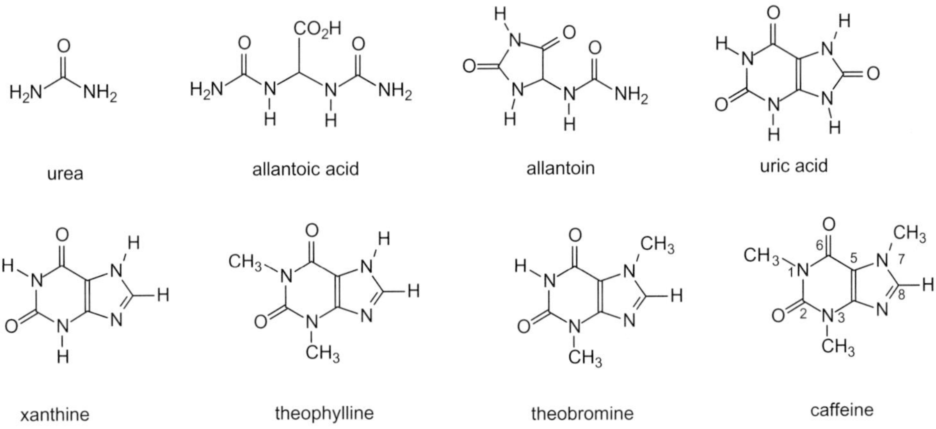

Figure 13.7. Structures of some of the purines and their degradation products.

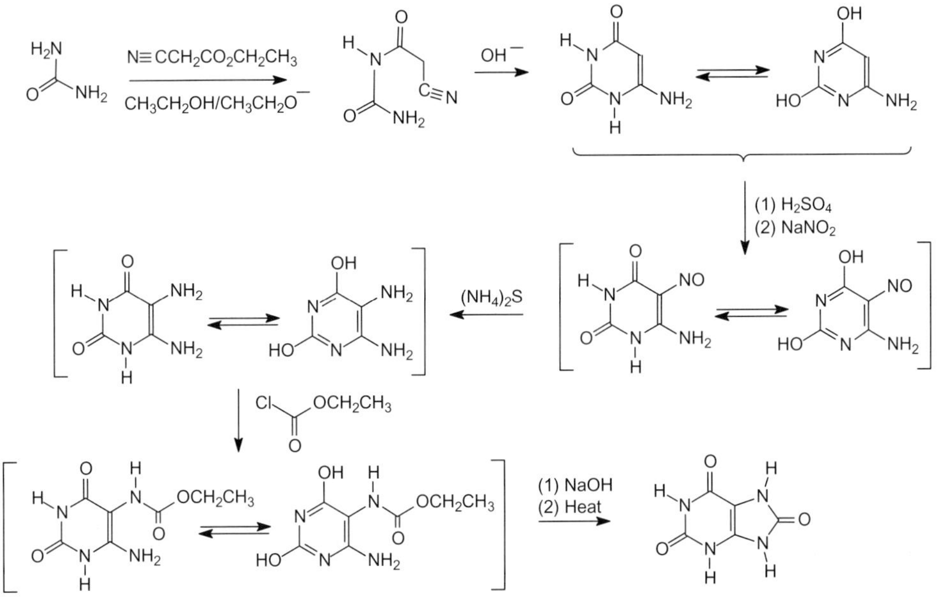

Scheme 13.64. A synthesis of uric acid. After Traube, W. *Chem. Ber.*, **1900**, *33*, 3035.

That early work was summarized in a series of three papers in 1881–1882 entitled *Uber das Caffein* (i.e., *About Caffeine*) by Emil Fischer* and a fourth paper, also in 1882, detailing the conversion of xanthine into theobromine and, separately, 8-hydroxycaffeine ("coffein") and its keto-tautomer.

This work ultimately resulted in a synthesis of caffeine by Traube.† As shown in Scheme 13.64, Traube began his work with the synthesis of uric acid. He then went on to the methylated derivatives.

*Fischer, E. *Chem. Ber.*, **1881**, *14*, 637, 1905; Fischer, E. *Chem. Ber.*, **1882**, *15*, 29, 453.
†Traube, W. *Chem. Ber.*, **1900**, *33*, 3035.

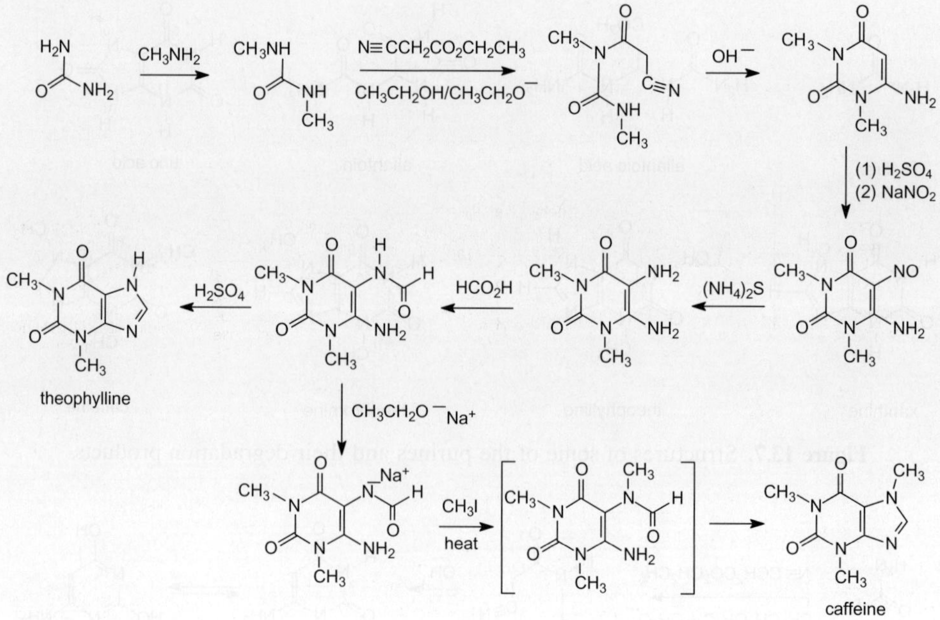

Scheme 13.65. Syntheses of theophylline and caffeine. After Traube, W. *Chem. Ber.*, **1900**, *33*, 3035.

The work began with the condensation of urea with ethyl cyanoacetate. The product, cyanoacetylurea on treatment with dilute, warm sodium hydroxide (NaOH) generated 4-aminouracil (4-amino-2,6-dioxypyrimidine). Nitrosation of the sulfate salt with sodium nitrite produced the corresponding isonitroso derivative, which underwent easy reduction (with ammonium sulfide) to the corresponding 5,6-diaminouracil (4,5-diamino-2,6-dioxypyrimidine).

When the bisamine was treated with a basic aqueous solution of ethylchloroformate, the corresponding urethane formed and a sodium salt of the urethane on heating to 180–190°C lost ethanol with cyclization to uric acid.

Then, as shown in Scheme 13.65, repeating the reaction sequence with *N,N*-dimethylurea (prepared by simply heating urea with *N*-methylamine) yielded the corresponding *N,N*-dimethyl-4,5-diamino-2,6-dioxypyrimidine and heating that bisamine, in 90% formic acid yielded the corresponding *N*-formyl derivative, which was cyclized to theophylline with sulfuric acid.

The sodium salt of the *N*-formyl derivative (formed with sodium ethoxide) was treated with methyl iodide and the resulting mixture of methylated *N*-formylpyrimidine and sodium iodide heated to induce cyclization and produce caffeine.

II. Biosynthesis of Caffeine

Much of the biosynthetic pathway for the formation of caffeine has already been discussed in the context of purine biosynthesis (Chapter 12). Thus, as shown in Scheme 12.87 (reproduced here as Scheme 13.66), following a more comprehensive

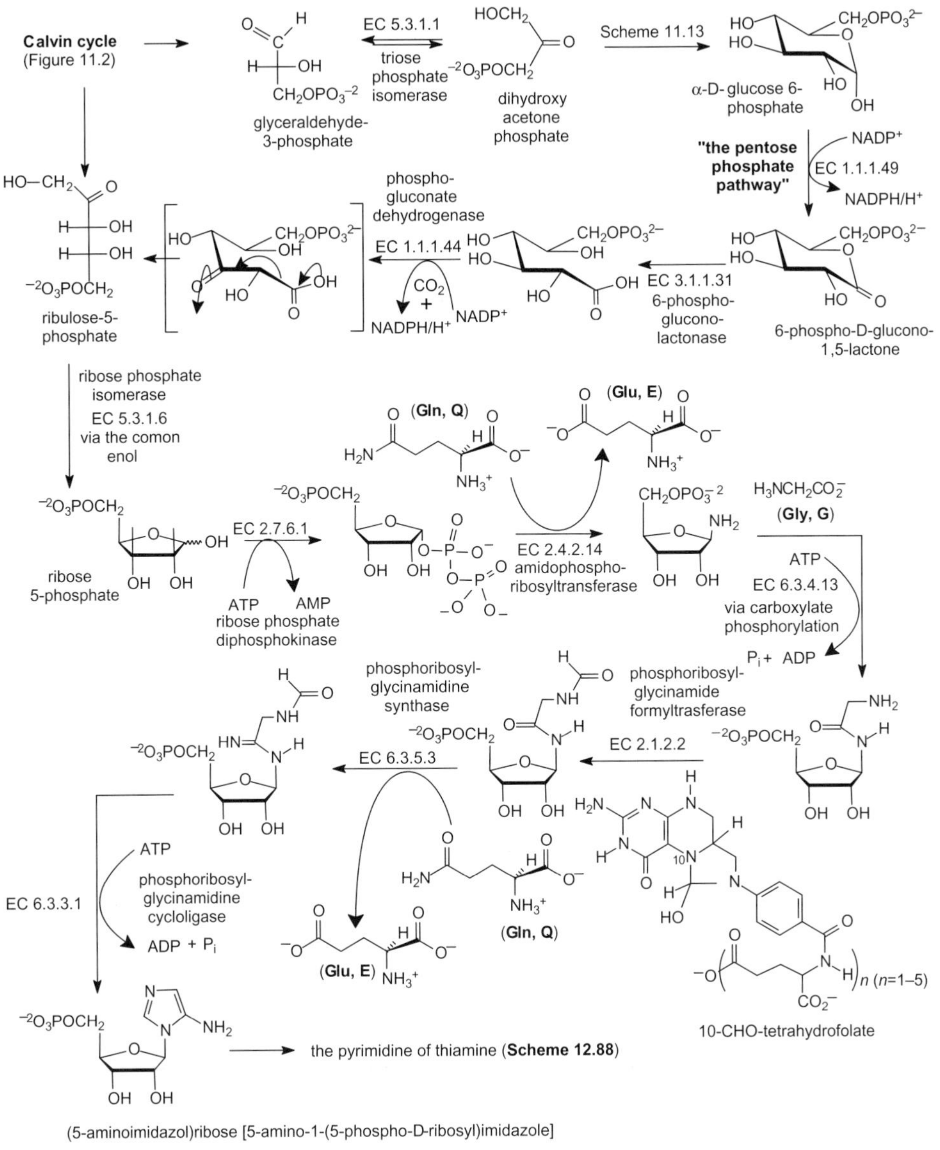

Scheme 13.66. Paths to 5-aminoimidazol ribotide, the precursor to the 4-amino-5-hydroxy-methyl-2-methylpyrimidine portion of thiamine. EC numbers and some graphic materials provided in this scheme have been taken from appropriate links in a URL starting with http://www.chem.qmul.ac.uk/iubmb/enzyme/.

discussion than that presented here, ribulose 5-phosphate is converted to 2-aminoribose 5-phosphate where the 2-amino group is derived from **glutamine (Gln, Q)** on its way to **glutamate (Glu, E)**. Then, amide formation using the 2-aminoribose 5-phosphate amine and the carbonyl of O-phosphorylated **glycine (Gly, G)** leads to an amide with another free amine, and the latter is converted to the corresponding terminal *N*-formyl amide with tetrahydrofolate donating the one carbon. Finally, another amino group derived from **glutamine (Gln, Q)** is utilized to produce the required imine from the carbonyl group of the original glycine adduct, and this is followed by cyclization to yield 5-aminoimidazol ribose 5-phosphate [5-amino-1-(5-phospho-D-ribosyl)-imidazole, 5-aminoimidazole ribotide].

Then as shown in Scheme 13.67 (a reproduction of Scheme 12.96), 5-aminoimidazole ribose 5-phosphate is carboxylated (either with CO_2 or with bicarbonate) to 5-carboxyamino-1-(5′-phosphoribosyl)imidazole, which rearranges to 5-aminoimidazole ribotide-4-carboxylate, and the latter undergoes amination from **aspartate (Asp, D)** to yield 5-amino-1-(5-phosphoribosyl)imidazole-4-carboxamide after the loss of fumarate. Finally (Scheme 13.8), N-formylation is effected with 10-CHO-H_4 folate to yield the N-formyl derivative (5-formamido-1-(5′-phosphoribosyl)imidazole-4-carboxyamide) and cyclization yields inosine 5′-phosphate (IMP).

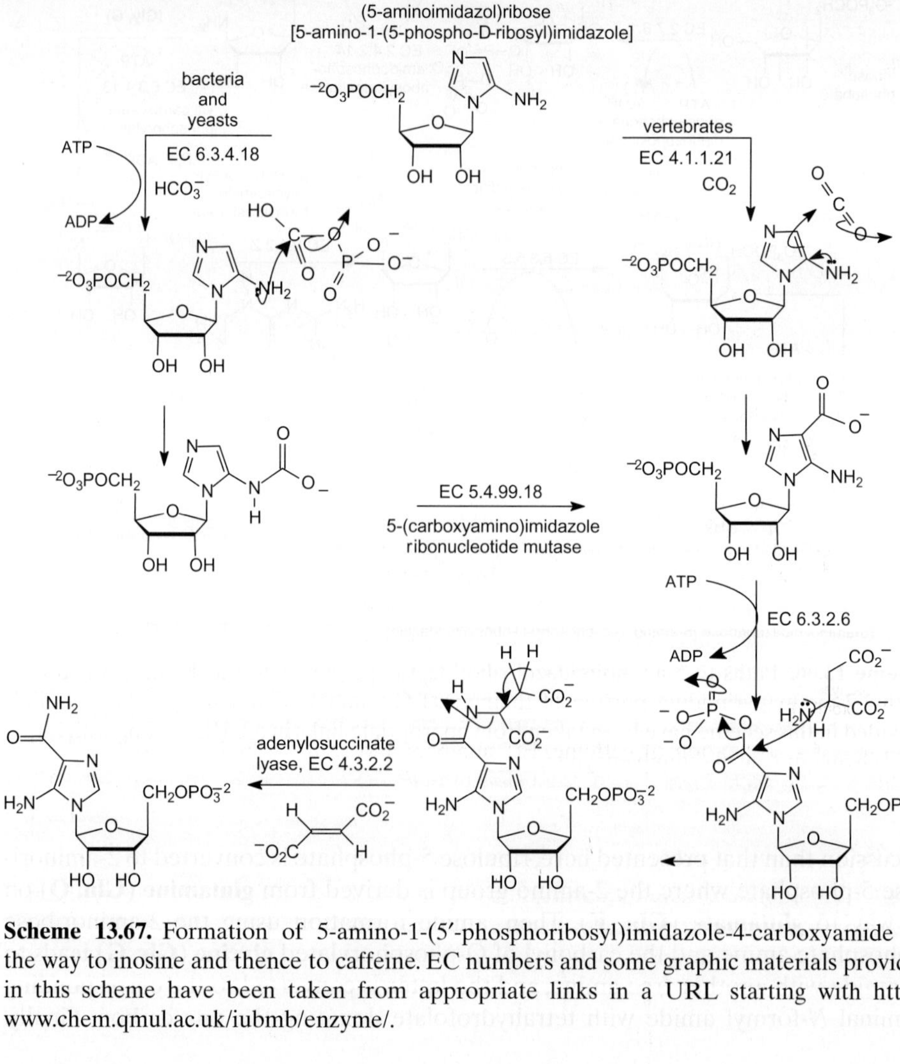

Scheme 13.67. Formation of 5-amino-1-(5′-phosphoribosyl)imidazole-4-carboxyamide on the way to inosine and thence to caffeine. EC numbers and some graphic materials provided in this scheme have been taken from appropriate links in a URL starting with http://www.chem.qmul.ac.uk/iubmb/enzyme/.

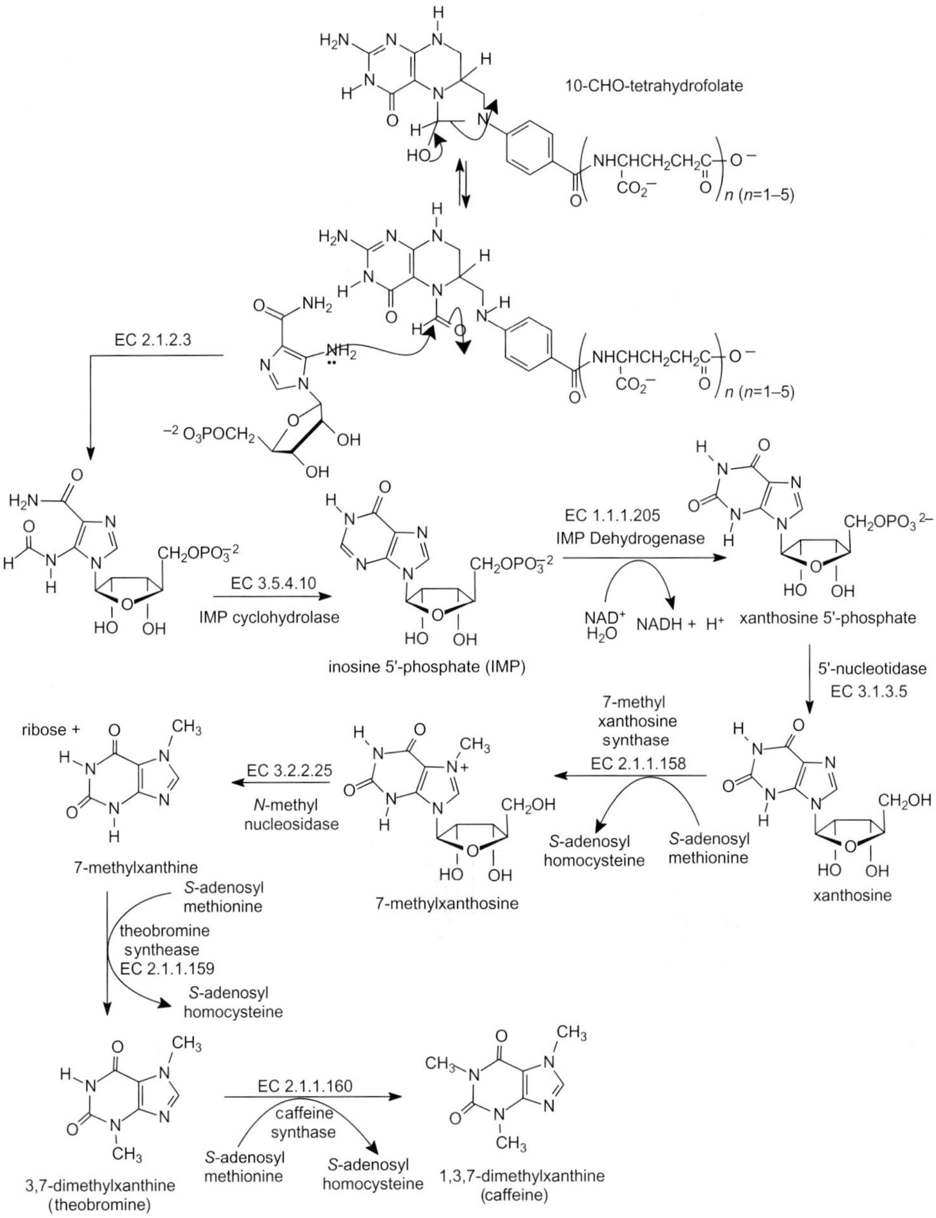

Scheme 13.68. Formation of caffeine. EC numbers and some graphic materials provided in this scheme have been taken from appropriate links in a URL starting with http://www.chem.qmul.ac.uk/iubmb/enzyme/.

IMP (Scheme 13.68) undergoes hydration and oxidation (EC 1.1.1.205) to xanthosine 5′-phosphate. Then, dephosphorylation (EC 3.1.3.5), methylation at N_7 (EC 2.1.1.58), and hydrolytic loss of ribose to 7-methylxanthine (EC 3.2.2.25) follow. Finally, a second methylation (EC 2.1.1.159 and/or EC 2.1.1.160) yields 3,7-dimethylxanthine (theobromine) and the third methylation (EC 2.1.1.160) yields 1,3,7-trimethylxanthine (caffeine).

CHAPTER 14

Part I. On the Genetic Code: Unity and Diversity
Part II. The Tetrapyrrolic Cofactors: Unity and Diversity

PART I. ON THE GENETIC CODE: UNITY AND DIVERSITY

> ... the idea that evolution has a main track or
> privileged axis is unsupported by scientific evidence.
>
> —Sir Peter Medawar, *Mind*, **1961**, *70*, 99

A. INTRODUCTION (THE GENETIC CODE)*

It should be clear to the interested reader that at the time of this writing, vast swaths of genomic information are rapidly becoming available and correlations of genomic, proteomic, and even metabolomic information, aided by high-speed computational techniques, are in the process of clarifying our understanding of living systems.

It should also be clear that this volume is not a biochemistry text.

Given the vast changes in the landscape of biology at the molecular level that are occurring at ever-increasing speeds, it seems clear that the general subject of what is now called "biochemistry" will fragment even further from what is now found.

Nonetheless, organic chemistry has lent its fabric to the background of the molecular interactions occurring among biological systems. These systems are extended portions of the weave since the reactions occurring are, in detail, the familiar bond making and breaking processes of organic chemistry, albeit within larger bodies. So, following this introduction, this chapter is divided into two broad parts. The first

---

*"It is one of the more striking generalizations of biochemistry—which surprisingly is hardly ever mentioned in the biochemical textbooks—that twenty amino acids and the four bases, are, with minor reservations, the same throughout Nature ..." (Sir Francis Crick, Nobel lecture, December 11, 1962).

---

part finishes the discussion, broached earlier, of the nitrogenous bases (adenine [A], guanine [G], thymine [T], cytosine [C], and uracil [U]) making up part of the information code (nucleotides) of the polymers deoxyribonucleic acid (DNA) and ribonucleic acid (RNA). Here, a bit of history, synthetic chemistry, and biosynthesis are brought forward.

In the same part, it is made clear again that the nitrogenous bases are attached to a ribose or deoxyribose unit at the anomeric carbon of the carbohydrate unit. Phosphate fragments are also affixed to the carbohydrate portion of the nitrogenous base–carbohydrate couple. Interestingly, as already seen in the discussion of the biosynthesis of both adenine and guanine (Chapter 12), the attachment of *at least one* phosphate unit to the terminal methylene of the ribosyl (deoxyribosyl) (the 5′-end) unit appears to be necessary during these biosynthetic processes. The conversion of adenosine diphosphate (ADP) to adenosine triphosphate (ATP), the reverse of which (ATP to ADP and inorganic phosphate, (PO_4^{-3})) has been used as the major currency to drive reactions is briefly discussed. It is worthwhile to consider how to get the ATP back again for reuse. This part also includes an examination of some of the chemistry of DNA and RNA as chemical, rather than biochemical, entities.

In the second part of this chapter, a very brief introduction to the ongoing studies of tetrapyrrolic derivatives is provided. These compounds are common to living systems, for example, cobalt-containing cobalamin (vitamin B_{12} derivatives), heme, which holds an iron in place of cobalt, and finally, chlorophyll a, where there is a magnesium centrally held in a more or less common core of four cojoined pyrrole (or pyrrole derived) rings. It will be abundantly clear that much remains to be done.

B. PART A

I. The Bases of DNA and RNA

As shown in earlier chapters, there are five bases that will be considered. Their structures are shown in Figure 14.1. All of these, when found in DNA (A, G, C, and T) and RNA (A, G, C, and U), occur attached to their respective (ribose [RNA]or deoxyribose [DNA]) carbohydrate esters of phosphoric acid. These derivatives have also been introduced in Chapter 12 and are discussed further below.

Finally, as selected works in history are prowled through for material suitable to an organic chemistry text, it is worthwhile to keep in mind, as pointed out by Arthur

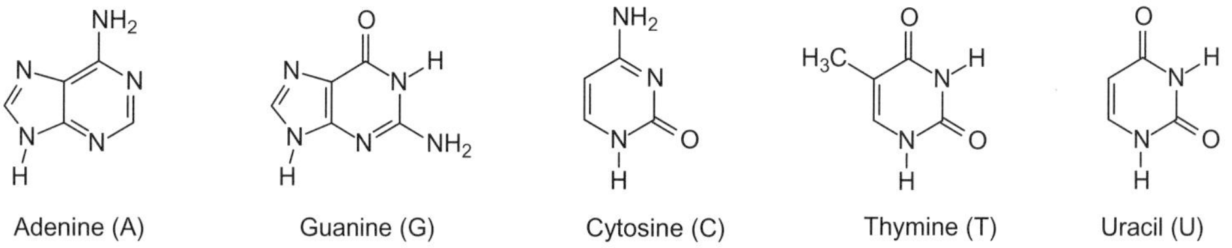

Figure 14.1. Representations of the bases (A, G, C, and T) of deoxyribonucleic acid (DNA) and (A, G, C, and U) ribonucleic acid (RNA).

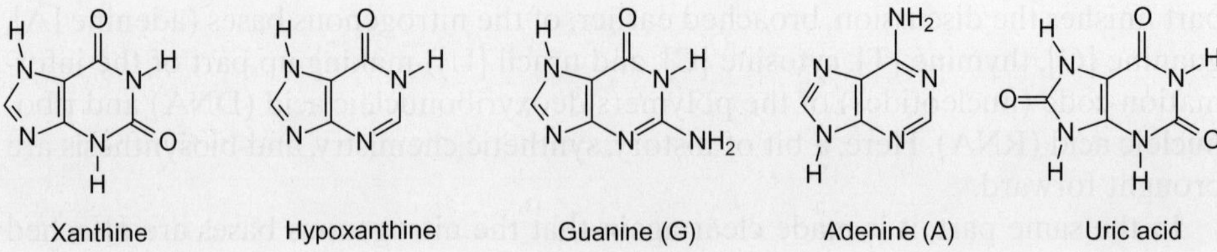

Figure 14.2. Representations of bases isolated by Kossel (Kossel, A. *Chem. Ber.*, 1885, *18*, 79) and a representation of uric acid as determined by Wöhler and Liebig (Wöhler. F.; Liebig, J. *Ann. Pharm.*, 1838, *26*, 256) and worked on by Fischer (Fischer, E. *Liebigs Ann.*, 1882, *215*, 253, et seq.).

Kornberg,* "… in the 1930's DNA was still called thymus nucleic acid and widely believed to occur only in animal cells; RNA had been isolated only from plant cells and was called yeast nucleic acid. In fact, plant and animal cells were sometimes distinguished on the basis of this chemical feature."

a. Adenine (A). **Adenine (A)** was discussed in Chapter 12, Section E, where a biosynthetic pathway by which it is formed, as well as several syntheses, was provided. As already noted, it is present in cofactors including adenosine mono-, di-, and triphosphates (adenosine 5′-phosphate [AMP], ADP and ATP, respectively), the oxidized and reduced forms of nicotinamide adenine dinucleotide (NAD$^+$ and NADH), acetyl coenzyme A (CoASCOCH$_3$), the oxidized and reduced forms of flavin adenine dinucleotide (FAD and FADH$_2$), and S-adenosylmethonine (SAM).

In 1885, A. Kossel[†] reported the isolation of xanthine, guanine, and hypoxanthine from albumin as well as a new base, which he named "adenine" (Figure 14.2). It was thought that these bases were part of the material derived from the cellular nuclei since Kossel isolated them from material similar to that used by (and following the procedures of) Friedrich Miescher,[‡] who had reported on the phosphorus-rich ("phosphorreiche") nuclear material he called "nucleïn" (and which we now call DNA), which could be broken down into a carbohydrate, phosphate, and a base on acid treatment. Although it is clear that elemental analysis (Chapter 1) played a key role in structure assignment, it is only in the context of the vast amount of work generated in the same time period by E. Fischer and coworkers (as outlined in Chapter 13) that the structure was solved.

b. Guanine (G). Guanine (originally mistaken for xanthine) was initially isolated by Unger from Peruvian bird guano (manure) deposits.[§] The structure was uncertain until the rich series of experiments were carried out by Fischer[¶] (already referred

---

*Kornberg, A. *DNA Synthesis*, W. H. Freeman & Co, San Francisco, 1974, p. 2.

[†]Kossel, A. *Chem. Ber.*, **1885**, 18, 79.

[‡]Miescher, F. *Chem. Ber.*, **1874**, 7, 376. It is interesting to speculate that if Berzelius had not worked as assiduously on perfecting the analytical balance of Lavoisier, this work could not have been done (cf. Chapter 1).

[§]Unger, B. *Ann. Chem. Phys. (Poggendorff)*, **1844**, 62, 158; Unger, B. *Ann. Chem. Phys. (Poggendorff)*, **1845**, 65, 222; Unger, B. *Ann. Chem. Pharm.*, **1846**, 58, 18; Unger, B. *Ann. Chem. Pharm.*, **1846**, 59, 58.

[¶]Fischer, E. *Chem. Ber.*, **1881**, 14, 637; Fischer, E. *Chem. Ber.*, **1881**, 14, 1905; Fischer, E. *Chem. Ber.*, **1882**, 15, 29; Fischer, E. *Chem. Ber.*, **1882**, 15, 453.

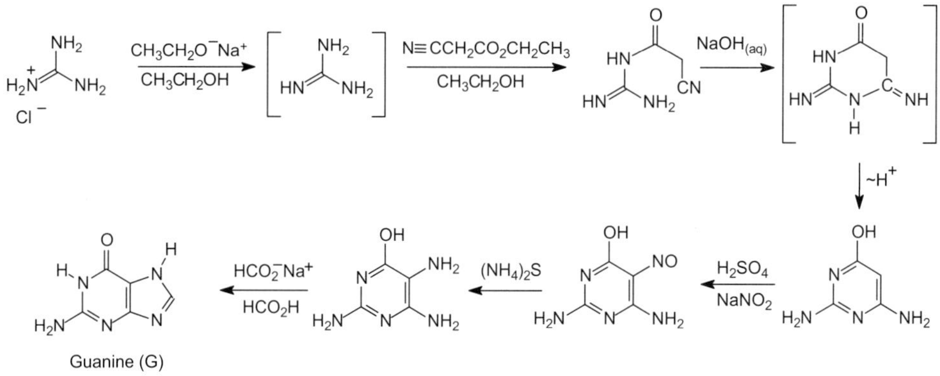

Scheme 14.1. A representation of the synthesis of guanine (G) (after Traub, W. *Chem. Ber.*, **1900**, *33*, 1371).

to in Chapter 13), which culminated in a synthesis, by Fischer, in 1897.* The Fischer synthesis was subsequently improved upon (only 3 years later) by Traub.† The Traub synthesis is shown in Scheme 14.1.

As shown in Scheme 14.1, guanidine hydrochloride [$H_2NC(NH_2)NH_2^+$ Cl$^-$] was converted to the free base with sodium ethoxide in ethanol. After the precipitated sodium chloride was removed, the solution was treated with ethyl cyanoacetate. The product of the condensation reaction, on heating (and also on treatment with aqueous base), underwent cyclization to produce 2,4-diamino-6-oxypyrimidine. On nitrosation, followed by reduction (with ammonium sulfide, $(NH_4)_2S$), the corresponding 2,4,5-triamino-8-oxopyrimidine resulted and the reaction of the pyrimidine with sodium formate and formic acid yielded guanine (G), indistinguishable from the authentic material.

It was shown in Scheme 12.97 (which is partially repeated here as Scheme 14.2) that inosine 5′-phosphate (IMP) underwent amination by the addition of **aspartate** (**Asp, D**) to the carbon of the carbonyl to produce N^6-(*S*)-1,2-dicarboxyethyl adenosine monophosphate. The reaction was aided by adenylosuccinate synthase (EC 6.3.4.4). The phosphate was derived from guanosine triphosphate (GTP). Then, after the elimination of fumarate (adenylosuccinate lyase, EC 4.3.2.2), AMP resulted.

However, as shown in Scheme 14.3, some of which has been seen earlier as Scheme 12.111 in Section H of Chapter 12) inosine phosphate, also can undergo hydration at C2. Oxidation of the resulting carbinolamine in the presence of inosine monophosphate dehydrogenase (EC 1.1.1.205) produces xanthosine 5′-phosphate. The xanthosine 5′-phosphate is then converted to guanosine 5′-monophosphate (GMP).

Because GMP is so important, there are two different pathways for its production. In one, amination is effected using ammonia (EC 6.3.4.1, GMP synthase) as the source of nitrogen, and loss of water is aided, preceding its elimination, by phosphorylation by ATP. The alternative path to GMP is to utilize the nitrogen in **glutamine (Gln, Q)** (which then goes to **glutamic acid [Glu, E]**) (EC 6.3.5.2, GMP

---

*Fischer, E. *Chem. Ber.*, **1897**, *30*, 2226.
†Traub, W. *Chem. Ber.*, **1900**, *33*, 1371.

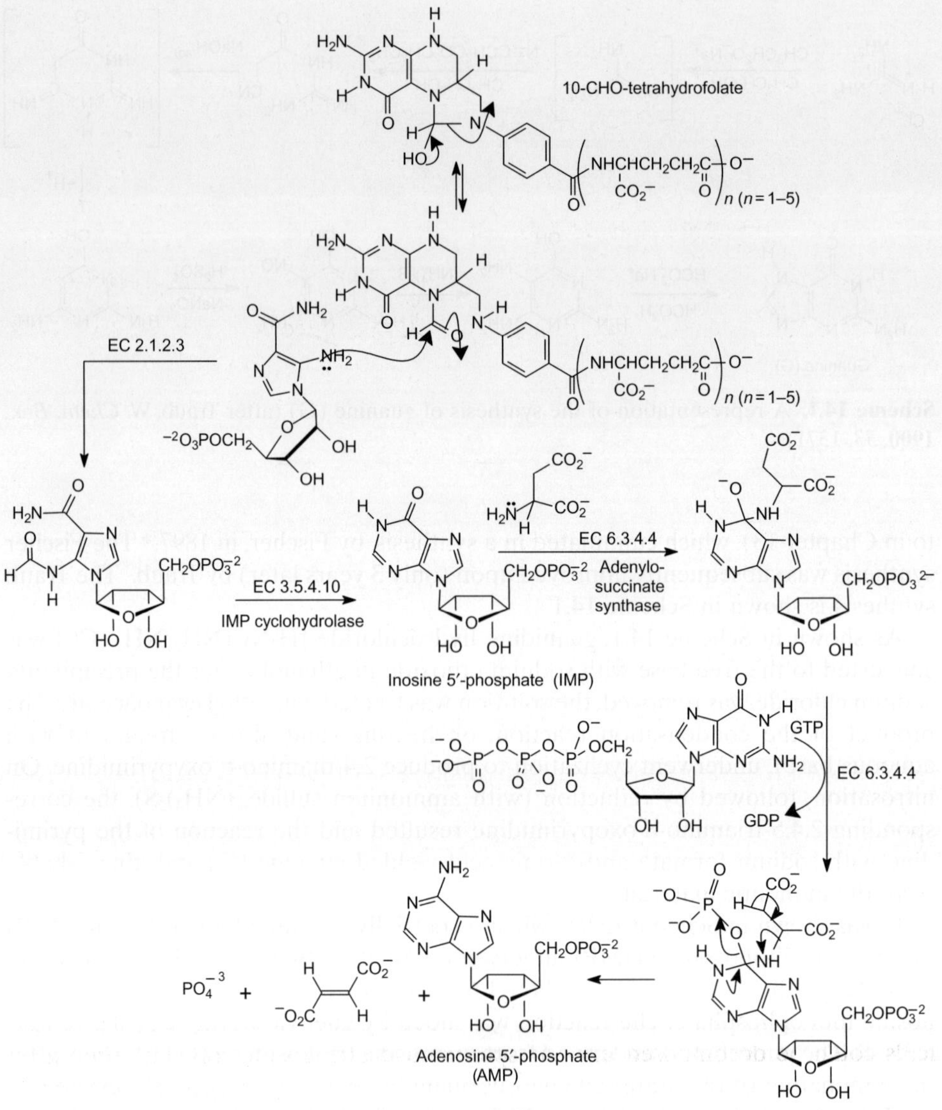

Scheme 14.2. A repetition of Scheme 12.97, serving to generate adenosine 5′-phosphate (AMP) via inosine 5′-phosphate (IMP). EC numbers and some graphic materials in this scheme have been taken from appropriate links in a URL starting with http://www.chem.qmul.ac.uk/iubmb/enzyme.

synthase) and with, again, loss of water being aided by phosphorylation of the oxygen by ATP before it is eliminated.

c. Uracil (U) and Thymine (T). In 1893, Kossel and Neumann reported* the remarkably prescient observation that nucleic acids were widespread in both plant

*Kossel, A.; Neumann, A. *Chem. Ber.*, **1893**, *26*, 2753.

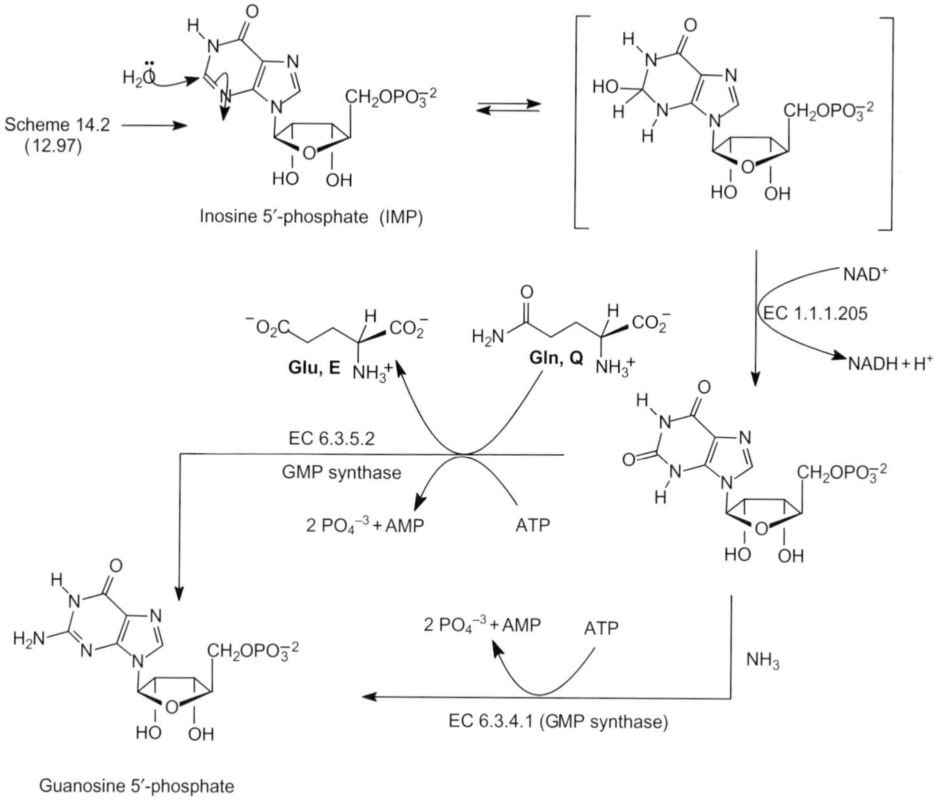

Scheme 14.2 ⟶
(12.97)

Inosine 5′-phosphate (IMP)

NAD⁺

EC 1.1.1.205

NADH + H⁺

Glu, E

Gln, Q

EC 6.3.5.2

GMP synthase

2 PO₄⁻³ + AMP   ATP

2 PO₄⁻³ + AMP   ATP

NH₃

EC 6.3.4.1 (GMP synthase)

Guanosine 5′-phosphate

Scheme 14.3. A representation of a pathway for the conversion of inosine 5′-monophosphate (IMP) to guanosine 5′-monophosphate. Some of this material has been presented previously in Scheme 12.111. EC numbers and some graphic materials in this scheme have been taken from appropriate links in a URL starting with http://www.chem.qmul.ac.uk/iubmb/enzyme.

and animal kingdoms and that lengthy investigation had demonstrated that these acids could be decomposed into phosphoric acid and a series of "Nucleïnbasen," that is, the bases of nucleic acids. The bases mentioned are adenine, hypoxanthine, guanine, and xanthine (Figure 14.2, *vide supra*). The authors then went on to describe careful fractionation of the adenine portion of the isolate from calf thymus and defined what, to them, appeared to be a dimeric material from which a new base, thymine, was isolated. It was not, however, clear that the same series of bases of nucleic acids existed in both animal and plant kingdoms, so, for example, the bases of calf thymus and yeast might be different. For example, Ascoli* reported, in 1900, that he was able to isolate a simple pyrimidine derivative from yeast, which he called uracil (following along the lines set out by Kossel as noted above).

A year later (1901), Fischer[†] described a continuation of the research he was then favoring. He and Roeder had undertaken the examination of the reactions between derivatives of acrylic acid (propenoic acid) and urea. As shown in Scheme 14.4, they

*Ascoli, A. *Z. Physiol. Chem.*, **1900**, *31*, 162.
[†]Fischer, E.; Roeder, G. *Chem. Ber.*, **1901**, *34*, 3751.

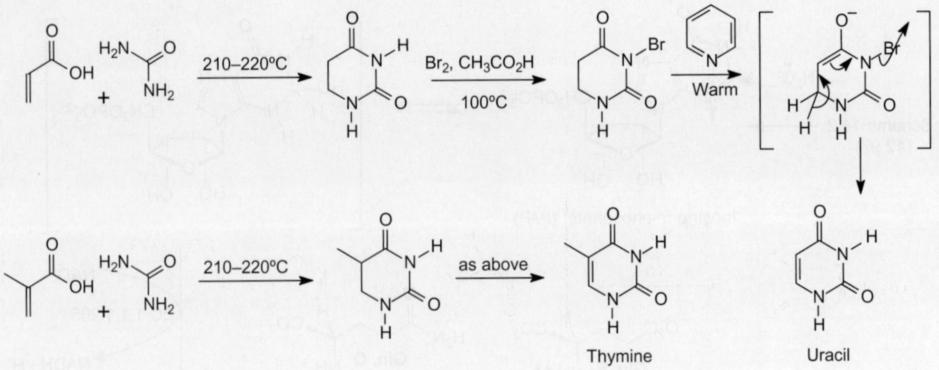

Thymine

Uracil

Scheme 14.4. A representation of an original synthesis of uracil and thymine developed shortly after the isolation of the former (after Fischer, E.; Roeder, G. *Chem. Ber.*, **1901**, *34*, 3751).

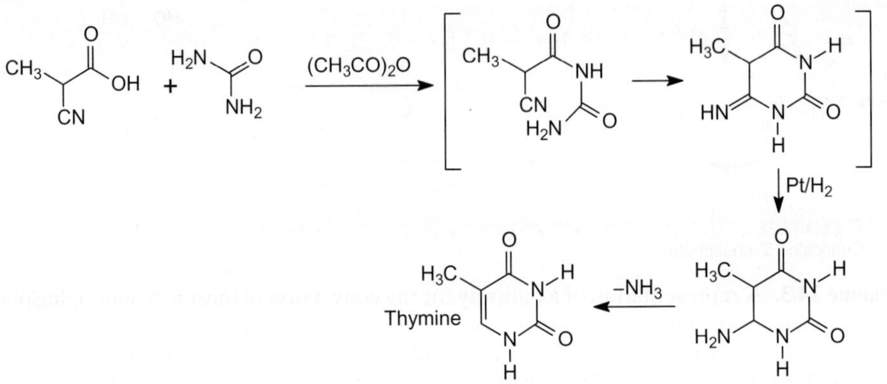

Thymine

Scheme 14.5. A representation of the formation of thymine by the reaction between 2-cyanoprop-anoic acid and urea followed by partial hydrogenation (after Bergmann, W.; Johnson, T. B. *J. Am. Chem. Soc.*, **1933**, *55*, 1733).

had found that bromination of the initially formed product produced a monobromide that underwent elimination to yield uracil. The use of methylacrylic acid (2-methylpropenoic acid), which was treated similarly, yielded thymine.

A variety of additional synthetic studies have been undertaken to produce these simple pyrimidines. The work of Bergmann and Johnson,* some 30 years later, to produce thymine takes advantage of the cyanoacetic acid idea originally outlined by Traube (*vide supra*). Thus, as shown in Scheme 14.5, the reaction of "methylcyanacetic acid" (2-cyanopropanoic acid) with urea and acetic anhydride generated the corresponding ureide, which, when dissolved in water and stirred in an atmosphere of hydrogen at 70°C in the presence of platinum black (Pt) until half the theoretical quantity of hydrogen (H_2) was absorbed, lost ammonia and yielded thymine directly.

*Bergmann, W.; Johnson, T. B. *J. Am. Chem. Soc.*, **1933**, *55*, 1733.

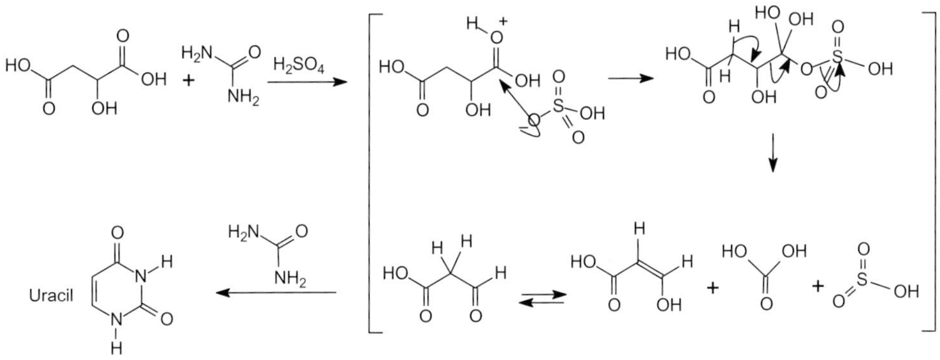

Scheme 14.6. A representation of the formation of uracil bythe reaction between malic acid and urea in the presence of sulfuric acid (after Davidson, D.; Baudisch, O. *J. Am. Chem. Soc.*, **1926**, *48*, 2379).

Baudisch and Davidson* found an even simpler way to prepare uracil (Scheme 14.6).

They found that uracil formed from the reaction of urea with malonic semialdehyde. Thus, when urea was dissolved in cold sulfuric acid, malic acid added and the reaction mixture heated on the steam bath for an hour, carbon monoxide (CO), carbon dioxide (CO_2), and sulfur dioxide (SO_2) were evolved. On cooling and quenching in water, uracil separated and, routinely, 50–55% yields were obtained.

d. Cytosine (C). In 1903, Wheeler and Johnson[†] wrote the following:

> … a new discovery of Kossel and Stendel. Working with a large quantity of sturgeons' testicles they obtained, in the histidine fraction, a basic substance agreeing in composition with the formula $C_4H_5ON_3$. Kossel and Stendel (*Zeit. Physiol Chem.*, **1902**, *37*, 179) state their belief that this compound is an aminooxypyrimidine, since hitherto all of the nitrogen compounds obtained from nucleic acid have been found to contain the pyrimidine ring. This new substance was found to closely resemble **cytosine**, which Kossel and Neumann (*Chem. Ber.*, **1894**, *27*, 1894) obtained in the year 1894 as a cleavage-product of thymus nucleic acid.

Since Kossel and Stendel had also shown that cytosine (isolated with sulfuric acid), on treatment with nitrous acid, gave a substance "having the properties of uracil," Wheeler and Johnson (also) concluded that cytosine was most likely an aminooxypyrimidine and considered the possibilities. They decided that there were seven possibilities (ignoring tautomers) for cytosine (Figure 14.3) based on the formation of uracil (Equation 14.1).

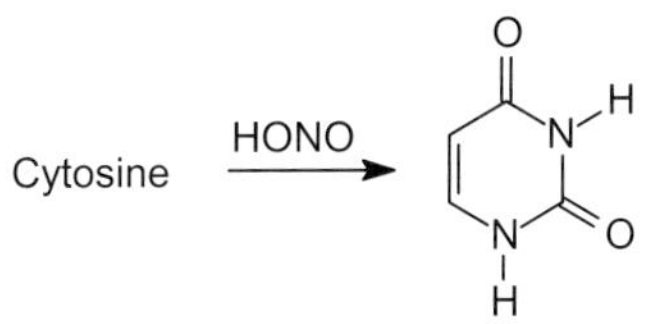

$$\text{Cytosine} \xrightarrow{\text{HONO}} \text{uracil} \tag{14.1}$$

*Davidson, D.; Baudisch, O. *J. Am. Chem. Soc.*, **1926**, *48*, 2379.
[†]Wheeler, H. L.; Johnson, T. B. *Am. Chem. J.*, **1903**, *29*, 492.

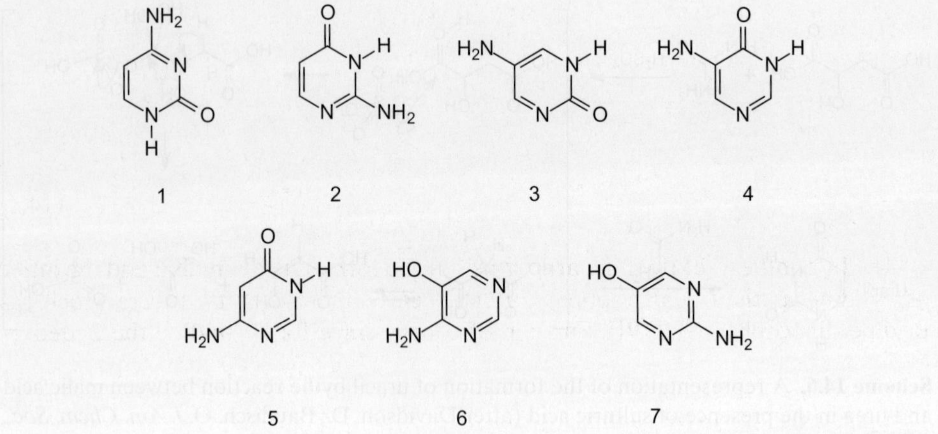

Figure 14.3. A representation of the seven possibilities considered for the structure of cytosine by Wheeler and Johnson (after Wheeler, H. L.; Johnson, T. B. *Am. Chem. J.* **1903**, *29*, 492).

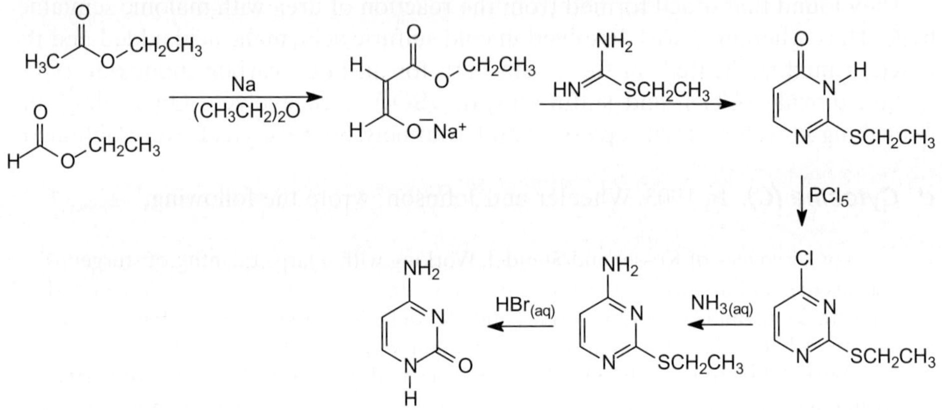

Scheme 14.7. A representation of the synthesis of cytosine (after Wheeler, H. L.; Johnson, T. B. *Am. Chem. J.*, **1903**, *29*, 492).

Wheeler and Johnson concluded that only structures representing compounds 1 and 6 in Figure 14.3 were possible and set about synthesizing them.

Thus, as shown in Scheme 14.7, the reaction of ethyl formate with ethyl acetate in ether in the presence of sodium metal yielded ethyl sodium formyl acetate. Then, the addition product of ethyl bromide with thiourea was treated with aqueous base and the ethyl sodium formyl acetate was added so that, after standing for "a number" of hours, acidification with acetic acid yielded 2-ethylmercapto-6-oxypyrimidine. Treatment of the latter with phosphorus pentachloride yielded the corresponding 2-ethylmercapto-6-chloropyrimidine; subsequent alcoholic ammonolysis generated 2-ethylmercapto-6-aminopyrimidine; and boiling aqueous hydrobromic acid resulted in the production of 6-amino-2-oxypyrimidine (cytosine).

The isomeric 2-amino-6-oxypyrimidine was prepared (Equation 14.2) by the reaction of ethyl sodium formyl acetate with guanidine.

$$(14.2)$$

The biosynthesis of thymine and cytosine (the former as thymidine and the latter as cytidine as they bear 5′-phosphorylated carbohydrate units) utilize uracil (as uridine diphosphate [UDP]). Further, of course, since for thymidine the 2′-deoxy-sugar must form first, the carbohydrate reduction must also be considered.

The formation of uridine 5′-monophosphate (UMP) (already shown, in part, in Schemes 9.98 and 12.7) is shown in greater detail in Scheme 14.8.

First, shown as part of the biosynthesis of **arginine** (**Arg**, **R**) in Chapter 12 (Scheme 12.7), carbamoylphosphate can be produced either from ammonia (NH_3) and carbon dioxide (CO_2) in the presence of carbamoylphosphate synthase (EC 6.3.4.16) where two equivalents of ATP and water (H_2O) are also required or it can be generated from glutamine (Gln, Q) and carbon dioxide (CO_2), where, again, two equivalents of ATP and water (H_2O) are required, **glutamate** (**Glu**, **E**) is formed, and a different carbamoylphosphate synthase enzyme (EC 6.3.3.5) is employed. Aspartate (Asp, D) is added with the aid of aspartate carbamoyltransferase (EC 2.1.3.2) to generate *N*-carbamoylaspartate, which cyclizes to dihydroorotate (dihydroorotase, EC 3.5.2.3). Oxidation of dihydroorotate (dihydroorotate oxidase, EC 1.3.3.1 utilizing the cofactors flavin mononucleotide $FMNH_2$ and flavin dinucle-otide [$FADH_2$]) with oxygen (O_2) while also producing hydrogen peroxide (H_2O_2) or its equivalent, then follows (cf. Scheme 14.9). Then, on reaction with 5-phospho-α-D-ribose-1-diphosphate in the presence of orotate phosphoribosyl-transferase (EC 2.4.2.10), the bond between carbon and nitrogen at the anomeric carbon of the ribosyl unit is made and orotidine 5′-monophosphate results. Decarboxylation of the latter in the presence of orotidine-5′-phosphate decarboxylase (EC 4.1.1.23) results in the formation of UMP.

Uridine triphosphate (UTP) is generated by the phosphorylation of UMP in separate steps through the transfer of phosphate from ATP. The first step is the transfer of phosphate *using UMP* kinase (EC 2.7.4.22) to yield ADP and UDP. Then, with the aid of nucleoside kinases such as nucleoside diphosphate kinase (EC 2.7.4.6), UDP is converted to UTP, while another ADP is produced from an ATP.

Cytidine triphosphate (CTP) is formed from UTP using either glutamine (Gln, Q), which is converted to glutamate (Glu, E), or ammonia (NH_3) as the source of the amino group and CTP synthase (EC 6.3.4.2) (Scheme 14.10).

II. Deoxynucleotides

The deoxynucleotides, that is, the nucleotides where the hydroxyl at C2 of the ribose unit is missing (including thymidine phosphate), are generated from the nucleotides where the hydroxyl group is present using a thioredoxin disulfide on the diphos-phates (ribonucleoside diphosphate reductase, EC 1.17.4.1). In this way, ribonucleo-

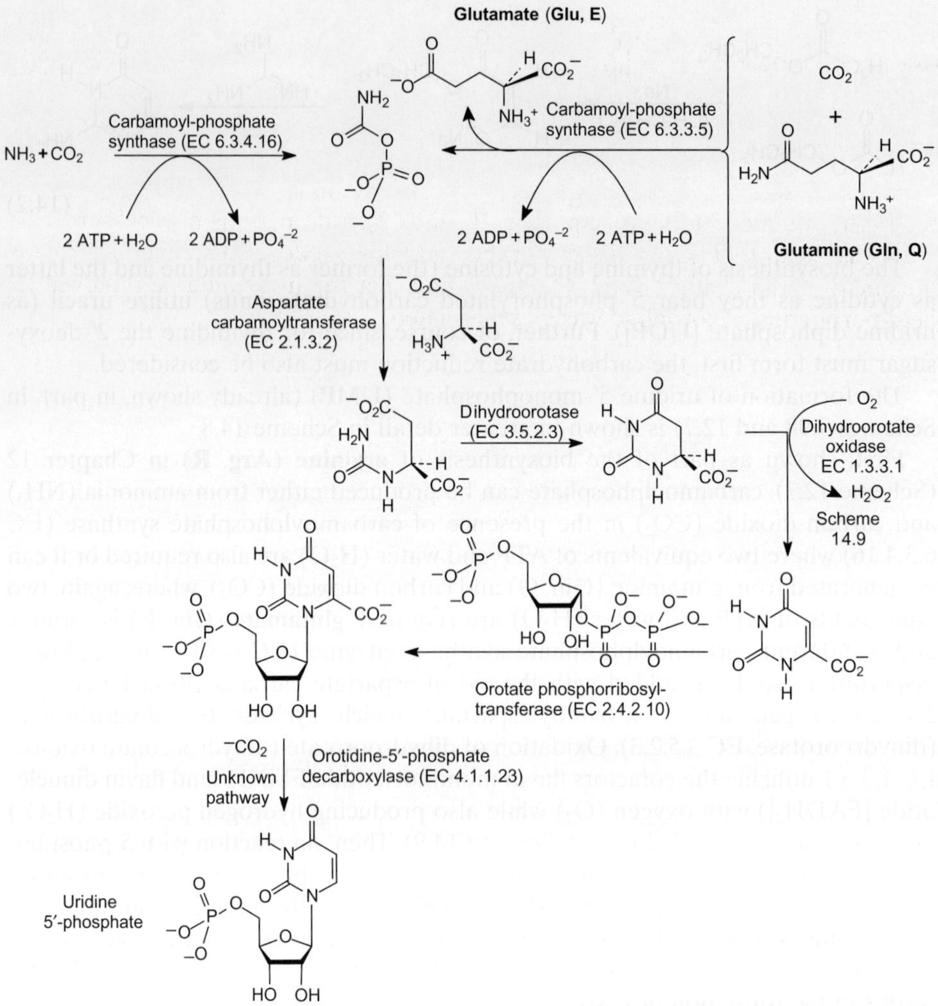

Scheme 14.8. A representation of a pathway to uridine 5′-monophosphate (UMP). EC numbers and some graphic materials in this scheme have been taken from appropriate links in a URL starting with http://www.chem.qmul.ac.uk/iubmb/enzyme.

side diphosphate reductases are also known as cytidine diphosphate (CDP) reductase, UDP reductase, ADP reductase, and GDP reductase, and the processes that occur are to convert CDP to deoxycytidine diphosphate (dCDP), UDP to deoxyuridine diphosphate (dUDP), ADP to deoxyadenosine diphosphate (dADP), and GDP to deoxyguanosine 5′-diphosphate (dGDP). A pathway has been proposed* (Scheme 14.11), and the structures of a number of such ribonucleoside reductases (ribonucleotide reductases) are available in the Protein Data Bank (PDB), access to which can be found at http://www.pdb.org/pdb/ and http://www.ebi.ac.uk/thornton-srv/databases/pdbsum/.

*Stubbe, J. *J. Biol. Chem.*, **1990**, *265*, 5330.

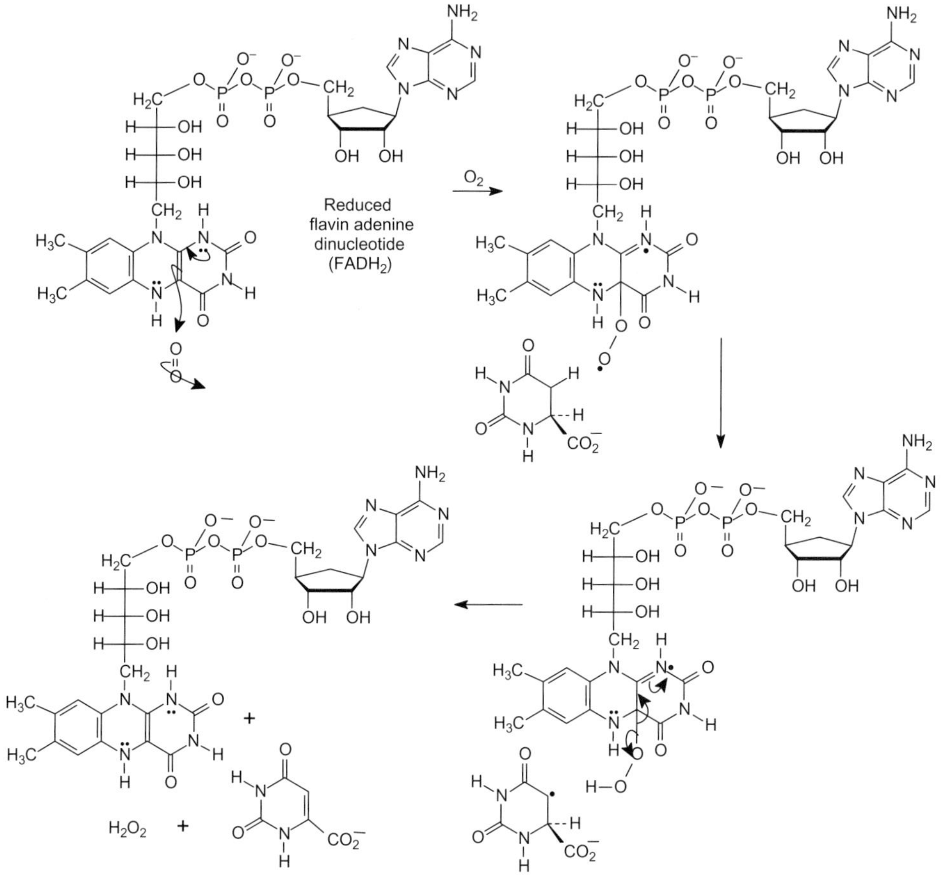

Scheme 14.9. A cartoon representation using a "one-electron pathway" (although a similar pathway using two electrons might be written) to convert dihydroorotidine to orotidine with reduced flavin adenine dinucleotide and oxygen. The other product is hydrogen peroxide (or its equivalent).

As can be seen in Scheme 14.11, it is argued that a radical species initiates the process by the removal of a hydrogen atom from C 3′ of the ribosyl unit. The loss of water after protonation from the thiol (or any other adventitious acid) is presumably facilitated by the adjacent "free" electron as a radical cation* is formed, and, at least in principle, the positive charge that must develop can be stabilized by the adjacent electron and its distribution onto the remaining hydroxyl oxygen. The ribonucleoside reductases (ribonucleotide reductases) possess thiol pairs as shown generically in Scheme 14.11. Generally, these are present (as shown) as cysteine (**Cys**, **C**) portions of one of the appropriately located peptide chains involved in the reductase. Since the amount and kind of deoxyribonucleotide (dADP, dCDP, dUDP, and dGDP) (deoxythymidine diphosphate [dTDP] is generated from deoxyuridine

---

*Radical cations have been observed in the gas phase. Steill, J.; Zhao, J.; Siu, C-K.; Ke, Y.; Verkerk, U. H.; Oomens, J.; Dunar, R. C.; Hopkinson, A. C.; Siu, K. W. M. *Angew. Chem. Int. Ed.*, **2008**, *50*, 9666.

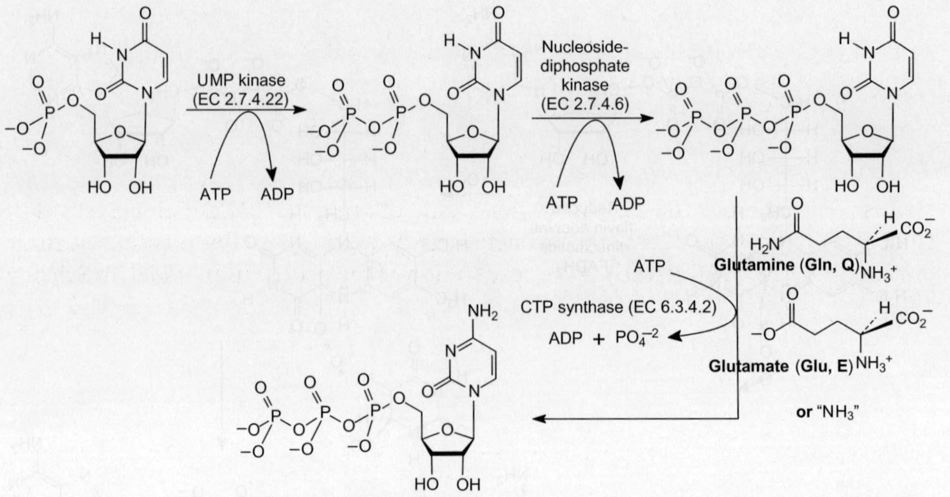

Scheme 14.10. A representation of the pathway from uridine monophosphate (UMP) to uridine triphosphate (UTP) and to cytidine triphosphate (CTP). EC numbers and some graphic materials in this scheme have been taken from appropriate links in a URL starting with http://www.chem.qmul.ac.uk/iubmb/enzyme.

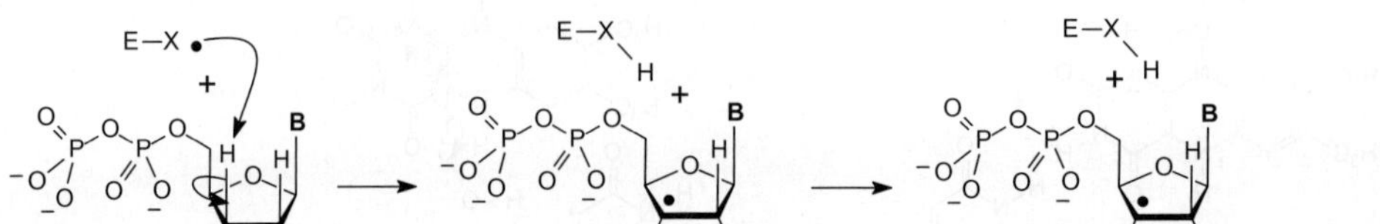

Scheme 14.11. A proposal for the use of ribonucleoside diphosphate reductase (ribonucleotide diphosphate reductase, EC 1.17.4.1) to convert adenosine diphosphate (ADP), uridine diphosphate (UDP), and cytidine diphosphate (CDP) into their respective 2′-deoxy analogues deoxyadenosine diphosphate (dADP), deoxyuridine diphosphate (dUDP), and deoxycytidine diphosphate (dCDP), or, in general, ribonucleic acids (RNAs) into the corresponding deoxyribonucleic acids (DNAs) (after Stubbe, J. *J. Biol. Chem.*, **1990**, *265*, 5330).

phosphate, *vide infra*, Scheme 14.12) as well as that of the respective nucleotides themselves are absolutely essential for normal growth and reproduction, it is held that the quantities needed are controlled by elaborate feedback systems. Both inhibition and activation need continuous monitoring.

Thymidine monophosphate (deoxythymidine monophosphate [dTMP]) is generated from deoxyuridine monophosphate (dUMP). The methyl group added to the ring is derived from the C10 methylene unit of tetrahydrofolate as described in Chapter 12 (Scheme 12.9), and reduction is brought about by nicotine adenine dinucleotide (NADH, NAD$^+$). A representation of the process is provided in Scheme 14.12.

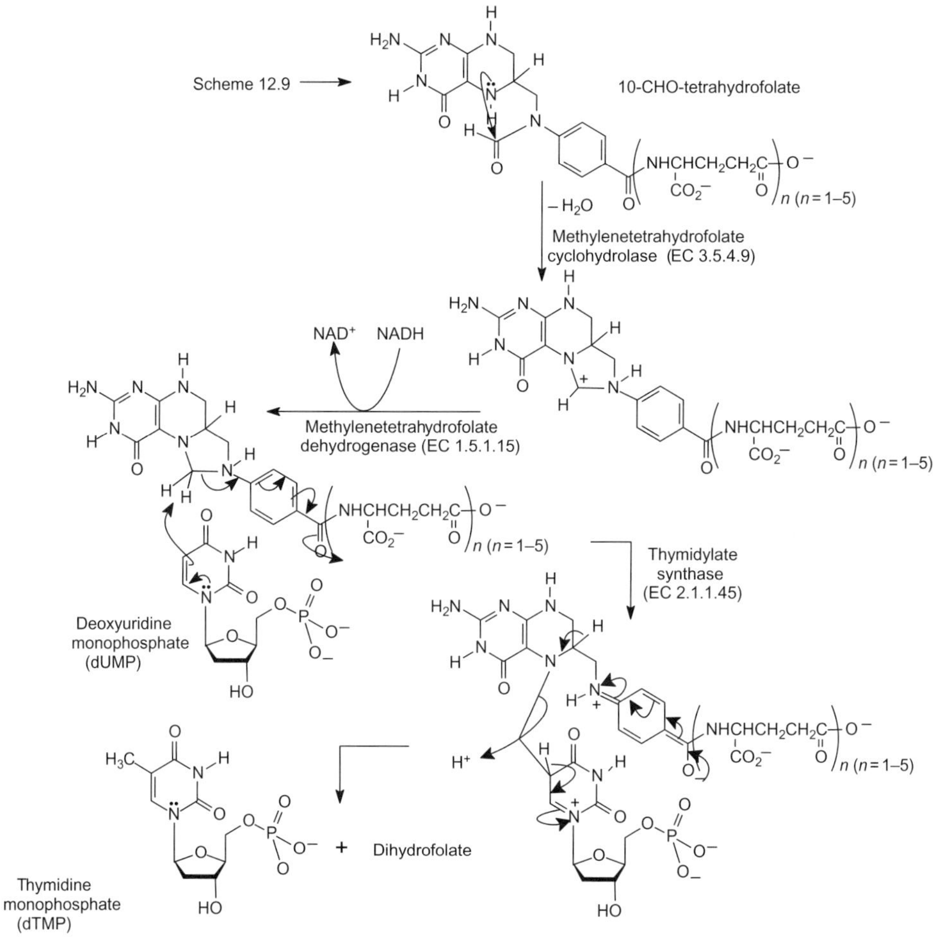

Scheme 14.12. A representation of a potential pathway for the conversion of deoxyuridine monophosphate (dUMP) to deoxythymidine monophosphate (dTMP) while tetrahydrofolate is oxidized to dihydrofolate using nicotinamide adenine dinucleotide (NAD$^+$/NADH). EC numbers and some graphic materials in this scheme have been taken from appropriate links in a URL starting with http://www.chem.qmul.ac.uk/iubmb/enzyme.

III. The Role of Phosphate

Although it is clear that the energy driving the systems with which we have dealt has its origin in our sun, and although it is clear that the laws of thermodynamics remain inviolate,[*] there is a special role played by phosphate both as a structural contributor and as a portion of the currency that is used to pay the price required by the various transformations in the biological systems functioning throughout life.[†]

Westheimer's thought-provoking report (i.e., "Why Nature Chose Phosphates"), despite is potential as an interrogative, is an exposition rather than a question. It is clear that were it an argument it must be *ex post facto*; that is, phosphate (PO_4^{-3}) "works," and our justification of its apt use as a linker between nucleotides that resists hydrolysis under physiological conditions and as a storehouse for energy in the form of its anhydrides (considering that diphosphates and triphosphates are anhydrides of the corresponding acids) derives from our ability to make those judgments. Enzymes and phosphates (as in DNA and RNA) have coevolved.

Coevolution suggests that the simplest pathway might not always be favored and elegant appearing solutions are a consequence of observer bias. Nonetheless, the "rotary engine" of ATP synthase, used to regenerate ATP, the currency of energy expenditure in all plant and animal cells, appears to be an "elegant" solution to recovering the price paid for carrying out the chemistry.

In the regeneration of ATP from ADP and inorganic phosphate, water must be lost as an anhydride is forming. The details of that process on the enzyme (ATP synthase or ATPase, e.g., EC 3.6.3.14) surface remain unclear at this writing.

Although there are minor differences, ATPase is highly conserved[‡] across living systems. It is found in the plasma membranes of bacteria as well as in the thylakoid membranes of chloroplasts and mitochondria of animals (e.g., PDB code **1bmf**).

In 1960, Racker and coworkers[§] reported that the oxidative phosphorylation system found in beef heart mitochondria could be separated into two fragments or factors (F_0 and F_1). One fragment, membrane associated and insoluble (F_0), catalyzed electron transport without phosphorylation. The other fraction (F_1, the "soluble component"), when added to the particulate fraction, "restored oxidative phosphorylation."

A year later, Mitchell[¶] proposed (in the Nobel prize-winning work) that phosphorylation of ADP to ATP (and the reverse) was tightly coupled to proton transfer across a membrane.

The Mitchell hypothesis was affirmed in 1971 by Kagawa and Racker[#] by reconstituting the ATPase, and then, in 1977, Kagawa and coworkers[||] showed that phosphorylation occurred when a proton gradient was established.

ATP synthases are highly conserved very large proteins (molecular masses ~500,000 Da) (e.g., PDB code **1bmf**). The F_1 fraction (above the membrane) section is linked to the F_0 section (Figure 14.4).

[*]Atkins. P. W. *Four Laws That Drive the Universe*, Oxford University Press, New York, 2007.
[†]Westheimer, F. H. Why nature chose phosphates, *Science*, **1987**, *235*, 1173.
[‡]Noji, H.; Yoshida, M. *J. Biol. Chem.*, **2001**, *276*, 1665.
[§]Pullman, M. E.; Penefsky, H. S.; Datta, A.; Racker, E. *J. Biol. Chem.*, **1960**, *235*, 3322.
[¶]Mitchell, P. *Nature*, **1961**, *191*, 144; Nobel Prize in Chemistry, 1978.
[#]Kagawa, Y.; Racker, E. *J. Biol. Chem.*, **1971**, *246*, 5477.
[||]Sone, N.; Yoshida, M.; Hirata, H.; Kagawa, Y. *J. Biol. Chem.*, **1977**, *252*, 2956.

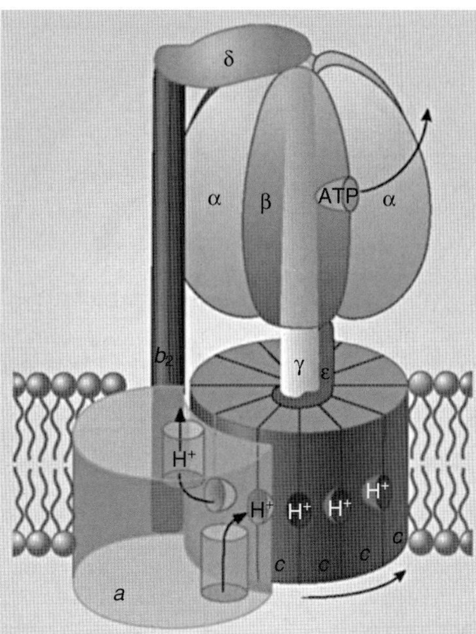

Figure 14.4. A representation of an F_0F_1 ATPase viewed from the side. It is currently held that the a-subunit in F_0 contains two partial channels. For a proton to traverse the membrane, it moves from one channel to the center where it binds to one c-subunit of the c-ring. Then, the c-ring subunit rotates and the proton moves out. The c-subunits are anchored to the γ-subunit (part of a rotor), whereas the a-subunit is anchored through the $b_2\delta$ to the $\alpha_3\beta_3$ hexamer in F_1. So, rotation of the c-ring relative to the a-subunit in F_0 will drive the rotation of the γ-subunit relative to the $\alpha_3\beta_3$ hexamer. This rotation induces conformational changes in the proteins making up the α- and β-subunits where ADP, ATP, and phosphate are bound. Adapted with permission from Macmillan Publishers Ltd: Cross, R. L. *Nature*, **2004**, *427*, 407. This has since been elaborated upon in greater detail (see Junge, W.; Sielaff, H.; Engelbrecht, S. *Nature*, **2009**, *459*, 354.

The operation of the synthase is suggested to be as follows: a proton enters a channel in the a-subunit F_0 and it moves from that channel on the outside to the center where it binds to one c-subunit of the membrane unit c-ring. Then, the c-ring subunit rotates and the proton moves out. The c-subunits are anchored to the γ-subunit (part of a rotor), whereas the a-subunit is anchored through the $b_2\delta$-fragment to the $\alpha_3\beta_3$ hexamer in F_1. So, rotation of the c-ring relative to the a-subunit in F_0 will drive the rotation of the γ-subunit relative to the $\alpha_3\beta_3$ hexamer. This rotation induces conformational changes in the proteins making up the α- and β-subunits where ADP, ATP, and phosphate are bound.

The data are consistent with three states of the three $\alpha\beta$ pairs making up the $\alpha_3\beta_3$ hexamer where, largely in the β strand, the adenine and ribosyl units are held. As shown in Figure 14.5, these three states are referred to as "open" (O) or "loose" (L) and "tight" (T). As the γ-subunit turns, the three states in the three $\alpha\beta$ pairs alternate and ATP is formed from ADP + PO_4^{-3} as it turns in one direction or ADP + PO_4^{-3} is formed from ATP as it turns in the opposite direction. In forming ATP then,

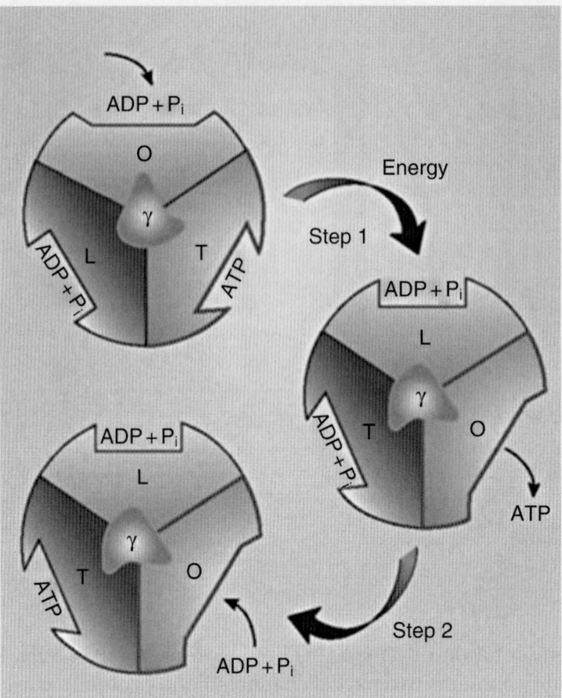

Figure 14.5. A representation of the formation of ATP from ADP in the F$_1$ portion of ATPase. As the γ-subunit turns, the three states (O, T, and L) in the three αβ pairs alternate and ATP is formed from ADP + PO$_4^{-3}$. Adapted with permission from Macmillan Publishers Ltd: Cross, R. L. *Nature*, **2004**, *427*, 407.

ADP + PO$_4^{-3}$ enter into the open state, are brought together to react in the loose state, and the reaction is consummated in the tight state (Equation 14.3). The next rotation into the open state allows the ATP to leave and be recharged with additional ADP + PO$_4^{-3}$ for the next round. In the hydrolysis of ATP to ADP + PO$_4^{-3}$, the process is reversed (Figure 14.5).

$$(14.3)$$

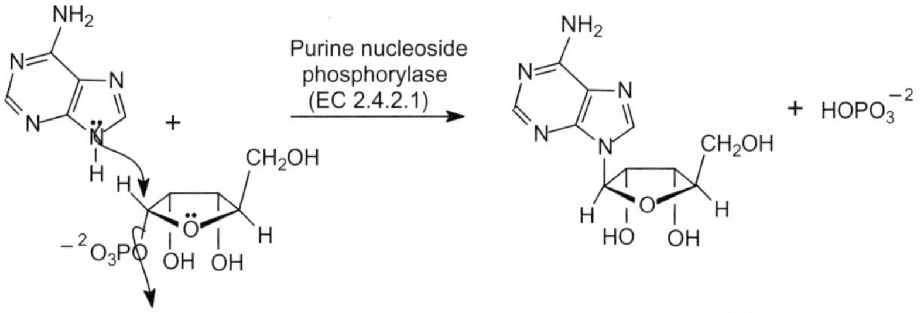

Scheme 14.13. A representation of a "salvage" reaction in which a purine (adenine, A) undergoes reaction with α-D-ribose-1-phosphate in what is represented as an S_N2-type process but which almost certainly involves participation by the ring oxygen as well as purine nucleoside phosphorylase (EC 2.4.2.1). NB: The name of the enzyme as a "phosphorylase" refers to its potential to catalyze the reverse of the reaction shown.

Although ATP (and a few other nucleotide triphosphates) has been used widely in this chapter and in previous chapters as an energy source to facilitate reactions, that use has been largely limited to cases where a reasonable pathway can be written. But it is important to note for the future that our current understanding of more elaborate biochemical processes that use ATP might best be described as vague. For example, the use of ATP by helicases remains poorly understood.

Helicases are proteins, such as ribonuclease E (EC 3.1.26.12), which is a bacterial ribonuclease used in the processing of ribosomal RNA, the chemical degradation of bulk cellular RNA, the decay of specific regulatory, messenger, and structural RNAs, and the control of plasmid DNA replication.[*] Although distinctive DNA sequence dependence is known to change the rate (and presumably the use of ATP) of unwinding of DNA, the details are not known,[†] nor is it clear how ATP plays the role it must in the cleavage of tightly wound DNA by using protein-bound tyrosine to attack the scissile phosphate bond.[‡]

The exciting work that resulted in a second Nobel Prize for Sanger[§] relied heavily on the formation of DNA in the presence of labelled (generally ^{31}P phosphate) nucleotides with suitable primers, on such restriction enzymes to cleave them, and on the intellectual skill required to put the pieces back together in the right order.

On the other hand, as seen in this chapter and in earlier chapters, the formation of phosphates of adenine (e.g., AMP, ADP, and ATP), guanidine (e.g., GTP), cytosine (e.g., cytidine monophosphate [CMP]), uracil (e.g., uridine monophosphate [UMP]), and dTMP have all involved the carbohydrate scaffold as a building block for the formation of the finished heterocyclic base (purine or pyrimidine). It is also important to realize that, as part of nucleotide salvage pathways, it has been found that a family of enzymes collectively known as phosphorylases serves to catalyze reactions between "free" bases and phosphate esters of carbohydrates (and related compounds). For example, as shown in Scheme 14.13, the generalized enzyme, purine nucleoside phosphorylase (EC 2.4.2.1), catalyzes the conversion of a purine with

[*]Feng, Y.; Vickers, T. A.; Cohen, S. N. *Proc. Natl. Acad. Sci. U.S.A.*, **2002**, *99*, 14, 746.
[†]Johnson, D. S.; Bai, L.; Smith, B. Y.; Patel, S. S.; Wang, M. D. *Cell*, **2007**, *129*, 1299.
[‡]Krogh, B. O.; Schuman, S. *J. Biol. Chem.*, **2002**, *277*, 5711.
[§]Sanger, F. Determination of nucleotide sequences in DNA, Nobel lecture, 1980.

α-D-ribose-1-phosphate to produce the purine nucleoside and phosphate. In principle, the reaction might be thought of as a simple substitution, but, of course, occurring at the anomeric carbon, participation by the oxygen of the ring is likely.

Although it is outside the scope of this chapter to even begin to expose the wealth of information concerning the formation of polymeric DNA and RNA beyond that which has already been mentioned, it is, nonetheless, important to recognize that the overall reaction catalyzed by DNA polymerase enzymes is to take a growing deoxyribonucleotide monophosphate polymer and to add to it a deoxyribonucleotide triphosphate and to eliminate two linked phosphate units, that is, $P_2O_7^{-4}$ or its equivalent (commonly written as PP_i). Further, it is important to note that in principle and in practice, the polymerization process is reversible. And, both in principle and in practice, it is common to find the phosphate "units" scrambled. Indeed, it is presumed that this is one of the ways of introducing radioactive phosphate into deoxyribonucleotide triphosphates.*

C. PART B

I. The Sequencing of DNA

As outlined in the Nobel lecture, Sanger[†] and his coworkers developed a technique that allowed the determination of the sequence of nucleic acids in a fragment of DNA. It is critical to acknowledge that separation techniques, including gel electrophoresis (i.e., the movement of charged fragments within a gel such as polyacrylamide as a function of their mass and the applied electrical field), were important to the analysis. Simply, it is assumed that one has a DNA polymerase suitable to the task and a primer that will allow the particular DNA fragment to begin growing if appropriate deoxynucleotides and suitable growth and temperature requirements are met. Given those common and commonly understood conditions, the principle is straightforward. The nucleoside triphosphates, deoxycytidine triphosphate (dCTP), deoxyguanosine triphosphate (dGTP), deoxyadenosine triphosphate (dATP), and deoxythymidine triphosphate (dTTP), are provided along with an identical set of dideoxynucleoside triphosphates (i.e., those missing the 2′- and 3′-ribosyl hydroxy groups), dideoxycytidine triphosphate (ddCTP), dideoxyguanosine triphosphate (ddGTP), dideoxyadenosine triphosphate (ddATP), and ddeoxythymidine triphosphate (ddTTP). Since the incorporation of the dideoxynucleoside triphosphate will cause the polymer to terminate, the examination of the fragments at the end of the process by electrophoresis and comparison to fragments whose structure is known will define the structure of the piece of DNA produced. Of course, fragments of known structure are necessary, and the synthesis of such fragments has been undertaken and, much as with the Merrifield (Chapter 12) synthesis of peptides, has also been automated.

II. Chemical Synthesis of DNA

Again, beginning with the work of Khorana (*vide supra*) and continuing, largely with Letsinger and Caruthers,[‡] the synthetic details were fined-tuned. In principle,

---

*Kornberg, A. *Science,* **1969,** *163,* 1410.
[†]Sanger, F. Determination of nucleotide sequences in DNA, Nobel lecture, 1980.
[‡]Caruthers, M. H. *Acc. Chem. Res.,* **1991,** *24,* 278.

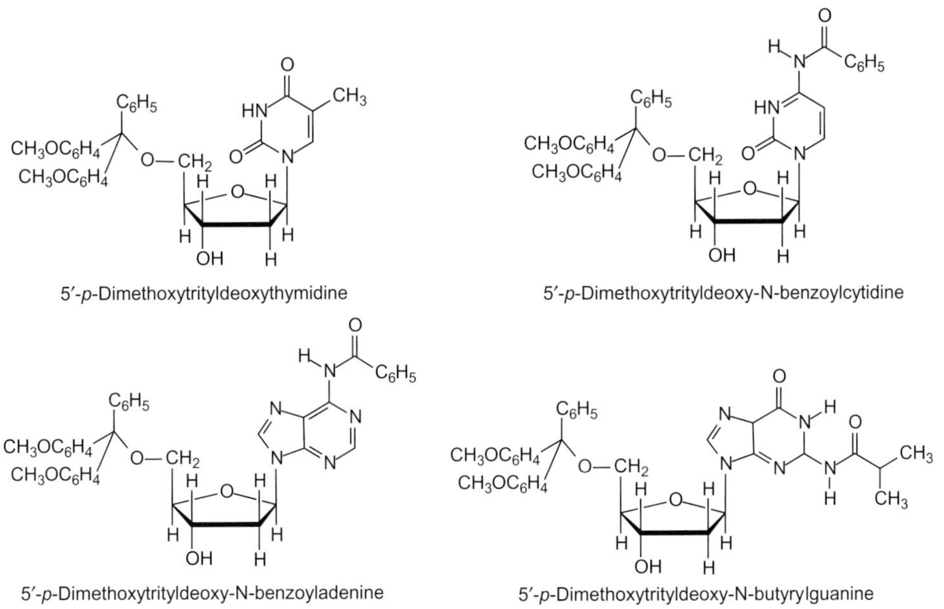

5′-*p*-Dimethoxytrityldeoxythymidine

5′-*p*-Dimethoxytrityldeoxy-N-benzoylcytidine

5′-*p*-Dimethoxytrityldeoxy-N-benzoyladenine

5′-*p*-Dimethoxytrityldeoxy-N-butyrylguanine

Figure 14.6. A set of functional group protected DNA nucleosides ready for synthesis.

of course, it is clear what must be done: (a) each base must be (reversibly) protected so that it does not interfere with the addition reactions occurring at the 3′ and 5′ hydroxyl groups; (b) one hydroxyl must be blocked while reaction occurs at the other; and then, (c) the blocked hydroxyl in the species just created must be unblocked so that it is ready for reaction again. Over the years, beginning in earnest in the last several decades of the twenty-first century, tuning of the various blocking groups and the process best suited for the condensation were laboriously studied and worked out.

A set of 5′-*p*-dimethoxytrityldeoxynucleosides was prepared* (Figure 14.6), and then those from cytidine, adenosine, and guanosine were protected at their respective reactive primary nitrogens. This yielded 5′-*p*-dimethoxytrityldeoxy-thymidine, for which no blocking group was necessary, 5′-*p*-dimethoxytrityl-*N*-benzoyldeoxycytidine, 5′-*p*-dimethoxytrityl-*N*-benzoyldeoxyadenosine, and 5′-*p*-dimethoxytrityl-*N*-isobutyryldeoxyguanosine.

With the protected nucleosides in hand, the synthesis is begun at what will become the 3′-end of the fragment. For example, if the DNA fragment is to have a deoxythymidine at the 3′-end, then the 5′-*p*-dimethoxytrityldeoxythymidine is first added to a polymeric or, following the Caruthers (*vide supra*) protocol, silica support. If the latter is to be followed, as shown in Scheme 14.14, then (3-aminopropyl) triethoxysilane is heated in refluxing toluene with the suitable silica matrix. As anticipated, at least one of the ethoxy substituents is replaced by an active oxygen (presumably an –OH initially) on the silica surface. Then, the protected deoxynu-cleoside, having been reacted with succinic anhydride followed by *p*-nitrophenol in the presence of dicyclohexylcarbodiimide (DCCD), was added and the nitrophenol,

*Schaller, H.; Weimann, G.; Lerch, B.; Khorana, H. G. *J. Am. Chem. Soc.*, **1963**, *85*, 3821.

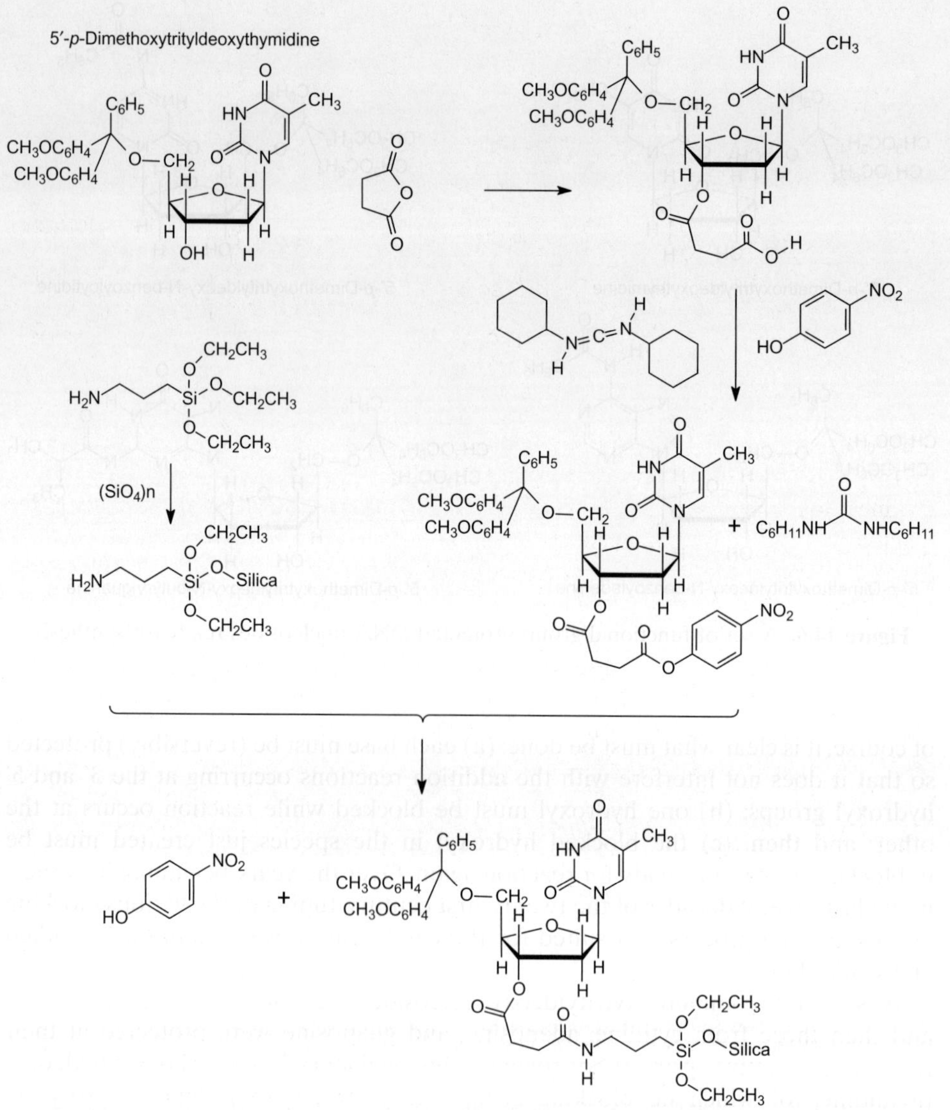

Scheme 14.14. Attachment of a 5′-protected deoxynucleoside (deoxythymidine) to a silica surface as a starter for the synthesis of a DNA fragment (after Caruthers, M. H. *Acc. Chem. Res.*, **1991**, *24*, 278).

the better leaving group, was displaced by the amino function to yield the corresponding amide, which was, as expected, attached to the silica matrix. Surface-attached unreacted primary amines were passivated by treatment with acetic anhydride.

The dimethoxytrityl group was next removed, in an S_N1-type process, from the 5′-hydroxyl with a trace of dichloroacetic acid in dichloromethane and the next deoxynucleoside was readied for addition (Equation 14.4).

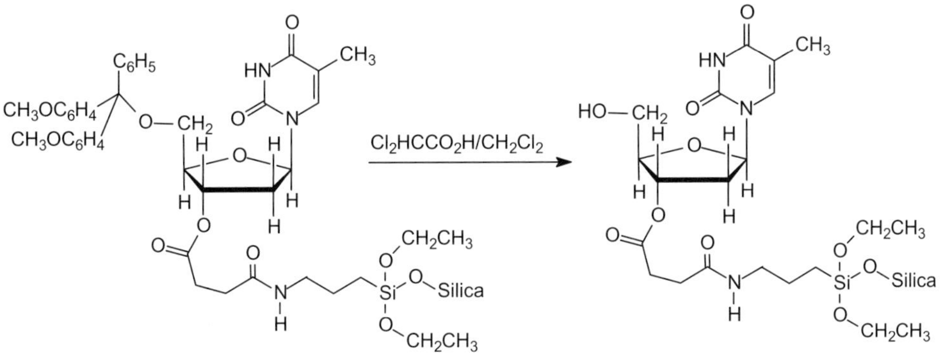

$$(14.4)$$

The second (and subsequent nucleosides) was activated for the addition as follows (Scheme 14.15): phosphorus trichloride was treated with β-cyanoethanol to produce the corresponding β-cyanoethoxyphosphorus dichloride, and the latter reacted with *N,N*-diisopropylamine in the presence of pyridine to produce the corresponding β-cyanoethoxy-*N,N*-diisopropylamidophosphorus chloride, and the last was treated with 3′-unprotected nucleoside in the presence of 2,4,6-trimethylpyridine (collidine).

In the next (simple) step, the 5′-unprotected (but silica-attached) deoxynucleoside was treated with the 3′-phosphoramidite substituted deoxynucleoside in the presence of tetrazole in acetonitrile to affect the coupling. Then, with acetic anhydride in the presence of *N,N*-dimethyl-4-aminopyridine, any unreacted hydroxyl groups that remain are capped as acetates for subsequent disposal, and the remaining attached and protected dimer is oxidized with iodine (in oxacyclopentane, tetrahydrofuran, THF/2,6-dimethylamino-pyridine [lutidine]) to convert it to the corresponding (protected) phosphate (Scheme 14.16).

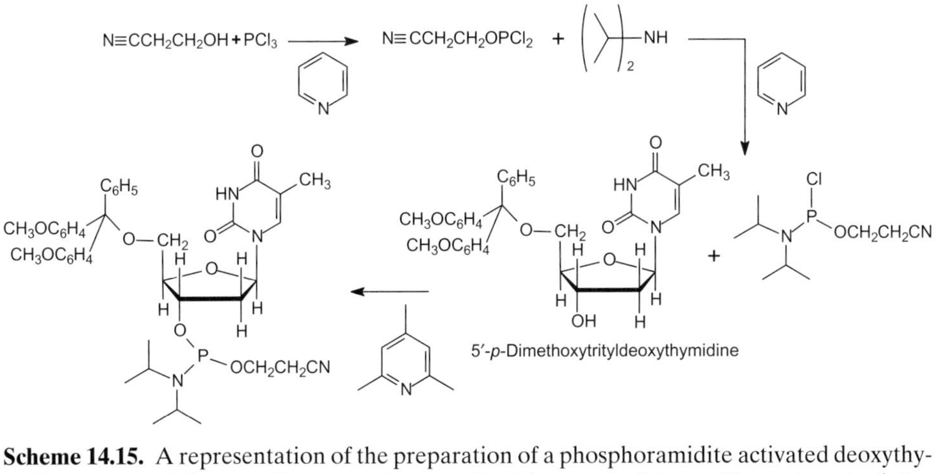

Scheme 14.15. A representation of the preparation of a phosphoramidite activated deoxythymidine prepared for coupling to a deoxynucleoside attached to a silica substrate (after Caruthers, M. H. *Acc. Chem. Res.*, **1991**, *24*, 278).

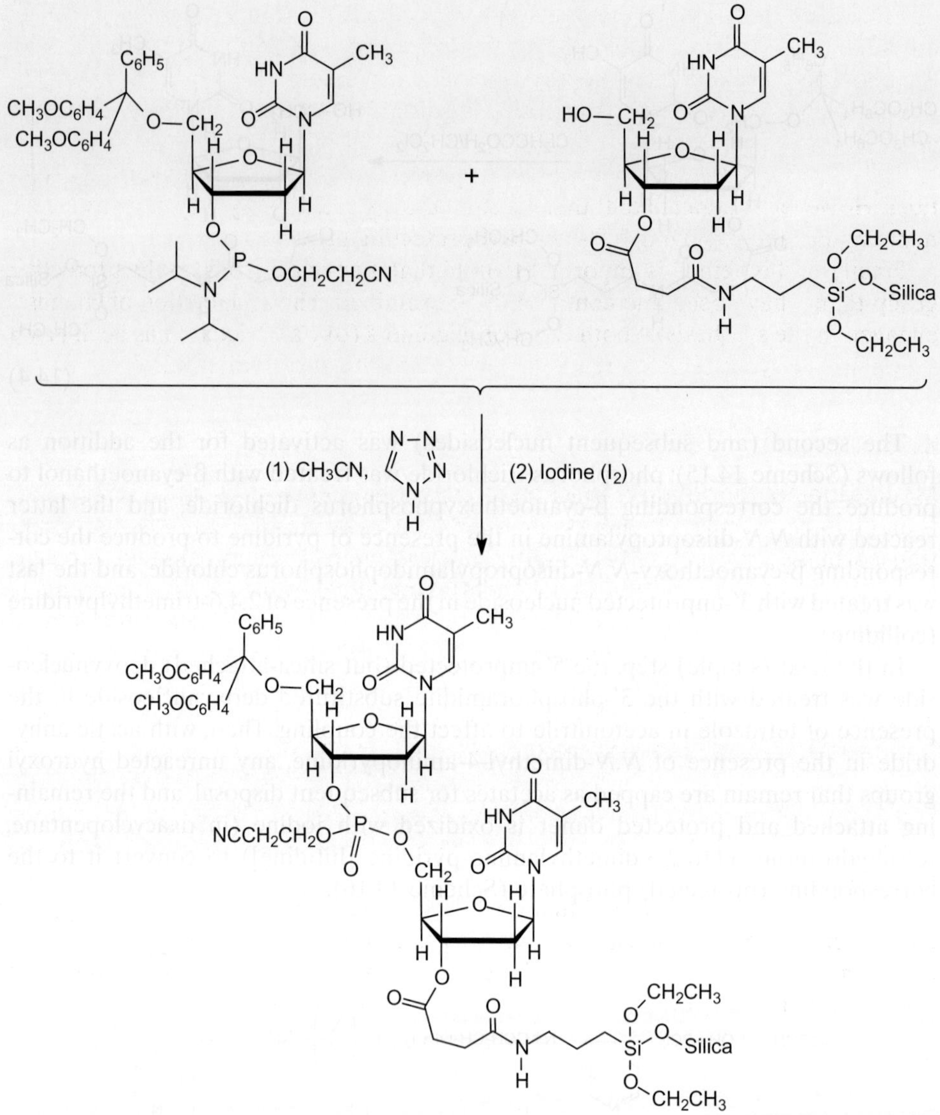

Scheme 14.16. A representation of the coupling of a protected, silica-attached (at the 3′-hydroxyl) deoxythymidine nucleoside to a second 5′-dimethoxytritylthymidine protected nucleoside. Oxidation of the trivalent phosphorus is effected by iodine (I_2) (after Caruthers, M. H. *Acc. Chem. Res.*, **1991**, *24*, 278).

It should be clear that the 5′-dimethoxytrityl of the dimer created by the coupling needs to be removed for the next (suitably protected) deoxynucleoside addition. Tens of deoxynucleosides have been coupled in this way. However, a simple calculation, assuming at least 90% yield of the desired product is obtained at each step, shows that only very small amounts of (protected) DNA fragments of any reasonable size can be produced in this fashion.

Protecting groups that were mounted on the deoxynucleosides, it will be recalled, are all amides (*N*-benzoyl and *N*-butryl), and, along with the one that remains on the phosphate backbone (cyanoethyl), are all easily removed when the synthesis of the completed oligodeoxyribonucleotide is finished, by treatment with concentrated ammonium hydroxide.

The same fundamental approach is also used in the synthesis of oligoribonucleotides, However, it appears that there is still no truly satisfactory way of introducing, maintaining, and removing a 2′-hydroxyl protecting group.

Finally, in this vein, it is important to note that many variations on the protecting group theme have been and continue to be examined. The imagination of chemists engaged in the synthesis of both types of oligomers (oxy and deoxy) has been given free play, and the interested reader is directed to the current literature.

III. Modification to DNA

Modification of the "library" function of DNA (and hence its eventual replication) is an ongoing process in living systems. First, it should be clear that the process of transporting deoxyribonucleotide phosphates across cell membranes and into cells for whole cell studies is not a trivial undertaking. If cell walls are made permeable to such charged species (consider the phosphate backbone), it is not uncommon that the cells do not survive. Further, disrupting a cell wall requires (if the cell is to survive) repair processes to begin. The obvious question as to whether the cellular activity observed is in response to the need to repair the cell or whether the normal cell processes are occurring is often raised. Clear answers are not always forthcoming.

The modifications themselves can, of course, be naturally occurring (mutations) or they can be induced by chemical synthesis of the modified species. Both kinds are known.

Thus, within cells, it is found that deamination of cytosine to uracil occurs. As a consequence, the cytosine cannot hydrogen bond appropriately to guanine in the double helix (Figure 14.7). The flaw is recognized by the repair system and the uracil is excised and replaced.

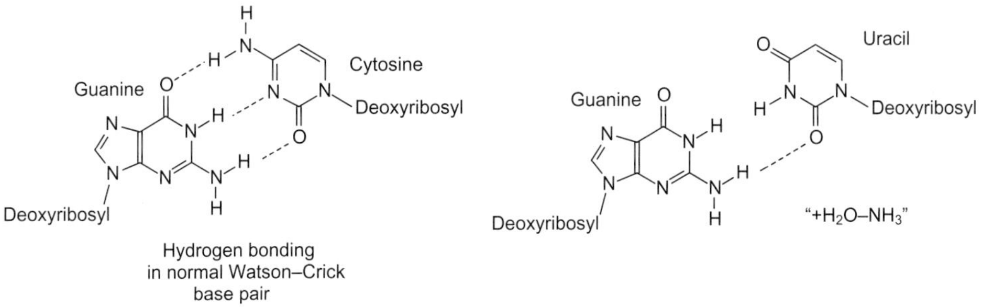

Guanine — Cytosine — Deoxyribosyl — Deoxyribosyl

Hydrogen bonding in normal Watson–Crick base pair

Uracil — Guanine — Deoxyribosyl — Deoxyribosyl — "+H₂O–NH₃"

Figure 14.7. A representation of the formal loss of ammonia (gain of water) in the deamination of a normal hydrogen-bonded G–C DNA base pair to a G, U set which cannot appropriately hydrogen bond.

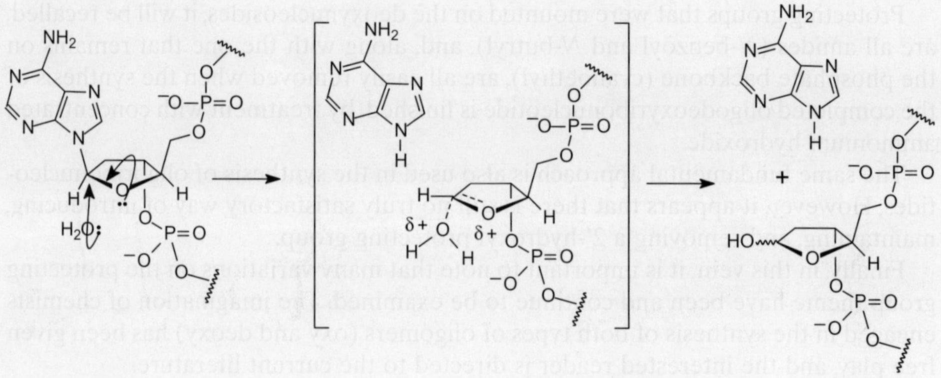

Scheme 14.17. A cartoon representation of the "depurination" of a fragment of DNA by hydrolysis.

As an interesting aside, the reader may recall that the conversion of cytosine to uracil (Equation 14.1) is accomplished by deamination through nitrosation and that the deamination shown in Figure 14.7 might raise questions about the role of nitrites in food preservation. It is difficult to establish whether the deamination process occurred as commonly prior to the widespread use of nitrites.

It is interesting to speculate that the presence of the methyl group in thymine, a clear and obvious characteristic distinguishing it from uracil, is critical for just that reason. Thus, to the extent that cytosine is deaminated (a rate estimated as frequent as 1 in 10^7 per day or about 10^2),* the product cannot "fit" in place of thymine in its pairing to adenine.

On the other hand, it is argued that as many as 10^5 purines are lost every day* as a consequence of hydrolysis of the N-β-glycosyl bond at the anomeric carbon (i.e., similar to the reverse of what is seen in Scheme 14.13) where water rather than phosphate acts as the nucleophile on DNA and with the participation of the ring oxygen (Scheme 14.17).

As already indicated, nitrites, used as preservatives, largely in proteinacous products such as preserved meats, sausages, and fishes, promote deamination via nitrosation, but more stable N-nitrosamines (such as N-nitrosodimethylamine) and N-nitrosourea also participate. Methylating agents such as SAM and dimethylsulfate (both endogenous and from the environment) induce DNA damage by methylation on nitrogen, which will inhibit hydrogen bonding, and by methylating phosphate linkers. Bisulfite,[†] added as a preservative to food and beverages, is considered an antioxidant (it is oxidized to sulfate and it inhibits oxidation of aldehydes by forming bisulfite adducts [Chapter 9]), and of course it works to that end. However, it does not discriminate between oxidations that are necessary to function and those that are not.

*Nelson, D. L.; Cox, M. M. *Principles of Biochemistry*, 5th edition, W. H. Freeman & Co., New York, **2008**, p. 289 ff.
†Frommer, M.; McDonald, L. D.; Millar, D. S.; Collis, C. M.; Watt, F.; Grigg, G. W.; Molloy, P. L.; Paul, C. L. *Proc. Natl. Acad. Sci. U.S.A.*, **1992**, *89*, 1827.

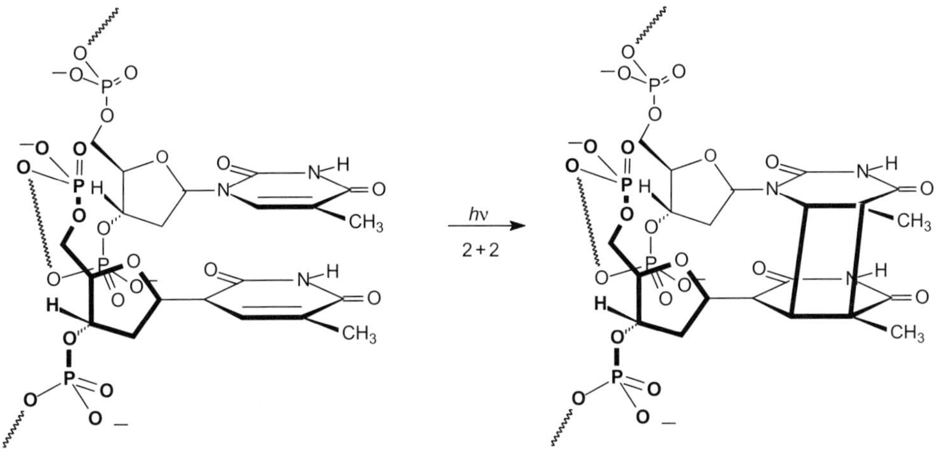

Scheme 14.18. A representation of the photochemical dimerization of thymine to yield a thymine dimer. The dimer is repaired by excision using a repair enzyme and replacement with an undimerized nucleoside.

Also, there is a variety of other agents that can affect DNA and that are not naturally occurring, so called sulfur and nitrogen mustards [$S(CH_2CH_2Cl)_2$ and $CH_3N(CH_2CH_2Cl)_2$], which serve to cross-link DNA strands. The cross-links cannot be removed.

Finally, there are the photochemical dimers resulting from absorption of UV radiation and bond formation between nearby thymine bases (Scheme 14.18).

PART II. THE TETRAPYRROLIC COFACTORS: UNITY AND DIVERSITY

... A Survey of the Last Four Billion Years.
—A. I. Scott, *Angew. Chem. Int. Ed.*, **1993**, *32*, 1223

A. THE TETRAPYRROLIC COFACTORS

I. Introduction

Adenosylcobalamin, cyanocobalamin (both active forms of vitamin B_{12}), heme, and chlorophyll a all enjoy the structural similarity of possessing a metal (cobalt in B_{12}, iron in heme, and magnesium in chlorophyll a) centrally held in a core of four cojoined pyrrole (or pyrrole derived) rings. That these tetrapyrrolic systems are individually adorned with other substitution elements allows them and a large number of related materials, which are also composed of the same four pyrrole (or pyrrole-derived) rings, their respective functions.

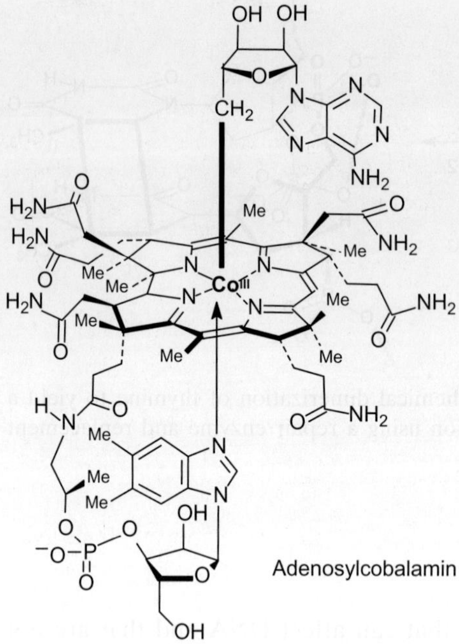

Cyanocobalamin

Adenosylcobalamin

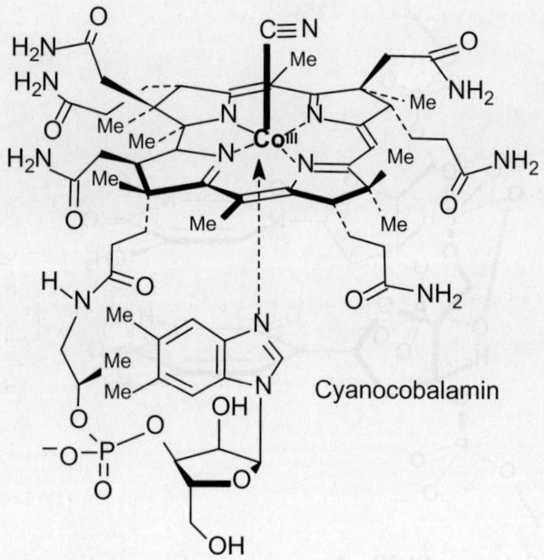

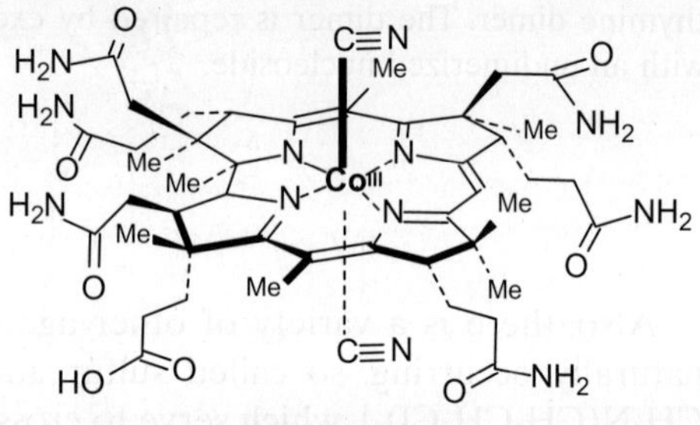

Cobyric acid

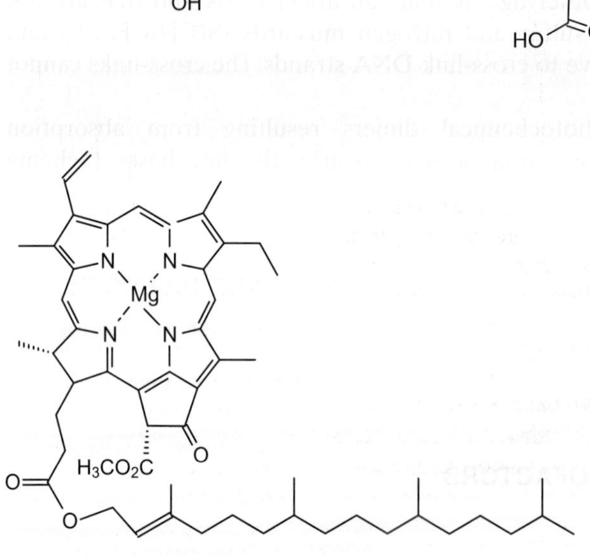

Chlorophyll a

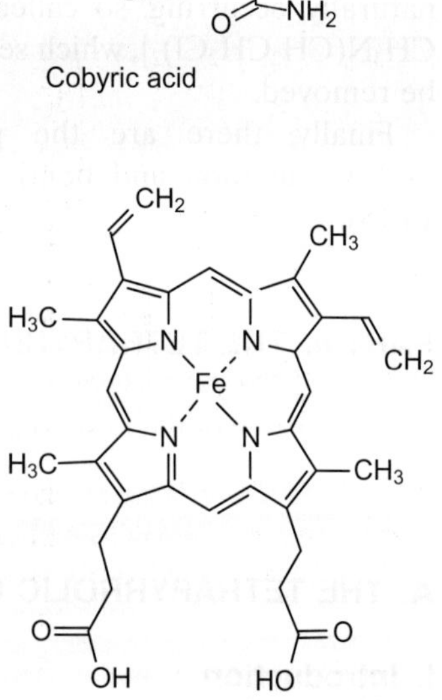

Heme

Although it has been argued that the pyrrole ring system itself can be produced in a "prebiotic" soup from simpler compounds and, indeed, there is some evidence that even assembled tetrapyrrolic systems might have arisen quite early, there is also a very large body of work showing that, today, other paths are followed in the biosynthesis of these compounds.

II. Some Early Chemistry

The year 1930 was a watershed year in the study of the macrocyclic tetrapyrrolic natural products found in plants and animals for it was then that the Nobel Prize in Chemistry was awarded to Hans Fischer for the work leading to the structures of heme and chlorophyll.

Although some discussion of pyrrole (and other heterocyclic compounds) has been touched upon briefly (Chapters 8, 10, and 13), a bit more is needed here.

It was clear to Fischer, his coworkers, and his predecessors that the isolate "hemin" had the formula $C_{34}H_{32}O_4N_4FeCl$ and that the iron could be removed to yield porphyrins. Chlorophyll was closely related.

The early degradative studies relied on oxidation and reduction reactions of the porphyrins and some basic chemistry of pyrroles, including the observations that

(a) the distillation of succinimide with zinc dust yielded pyrrole (Equation 14.5);

$$(14.5)$$

(b) many 1,4-dicarbonyl compounds on treatment with amines and their derivatives yielded pyrroles (e.g., oximes after reduction, Equation 14.6) and pyrrole (having been converted to its anion with potassium metal) could be alkylated and acetylated on nitrogen (Equation 14.7);

$$(14.6)$$

$$(14.7)$$

(c) hematinic acid [3-(4-methyl-2,5-dioxo-2,5-dihydropyrrol-3-yl)propanoic acid, Equation 14.8] was among the oxidation products obtained from a hematin chloride (and related materials) with chromic acid;

3-(4-Methyl-2,5-dioxo-2,5-dihydropyrrol-3-yl)propanoic acid

Hematinic acid

$$(14.8)$$

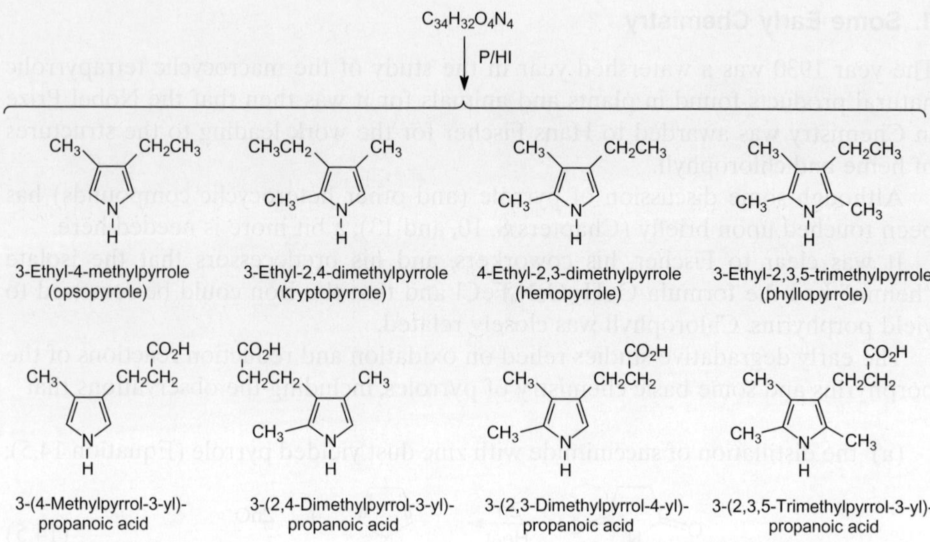

$C_{34}H_{32}O_4N_4$

P/HI

3-Ethyl-4-methylpyrrole
(opsopyrrole)

3-Ethyl-2,4-dimethylpyrrole
(kryptopyrrole)

4-Ethyl-2,3-dimethylpyrrole
(hemopyrrole)

3-Ethyl-2,3,5-trimethylpyrrole
(phyllopyrrole)

3-(4-Methylpyrrol-3-yl)-
propanoic acid

3-(2,4-Dimethylpyrrol-3-yl)-
propanoic acid

3-(2,3-Dimethylpyrrol-4-yl)-
propanoic acid

3-(2,3,5-Trimethylpyrrol-3-yl)-
propanoic acid

Figure 14.8. Some of the products obtained on the degradation of heme and, where the propanoic acid side chain has been replaced by an allyl group, from chlorophyll.

(d) and, as shown in Figure 14.8, when heme and chlorophyll (from which the metals had been removed) were cleaved under reducing conditions (phosphorus/hydrogen iodide milieu), a wide variety of pyrroles resulted.

As reported in his Nobel lecture* and in a variety of subsequent publications by others,[†] the large number of reactions that had been undertaken to synthesize various pyrroles and the techniques of linking them to form porphyrins led to a clear understanding of the structures of heme and chlorophyll and their various degradation products.

Typical synthetic processes are provided in Schemes 14.19–14.23 for the synthesis of pyrroles and their assembly into a porphyrin. Many variations are possible.

As shown in Scheme 14.19, nitrosation of ethyl acetoacetate (ethyl 3-oxobutanoate) with sodium nitrite and hydrogen chloride generates the corresponding 2-nitroso derivative, which is in equilibrium with the oxime. Reduction of the nitroso (or oximino) derivative with zinc and hydrogen chloride yields the corresponding **labile** amine (ethyl 2-amino-3-oxobutanoate) derivative. The amine cannot, generally, be stored, as it undergoes self-condensation; however, it can be used at once to react with appropriate carbonyl compounds to form pyrroles.

As shown in Scheme 14.20, treatment of ethyl acetoacetate (ethyl 3-oxobutanoate) with methyl iodide in the presence of sodium hydride produces the corresponding

*Fischer, H. On haemin and the relationships between haemin and chlorophyll, http://nobelprize.org/nobel_prizes/chemistry/laureates/1930/fischer-lecture.pdf.
[†]There is a truly large volume of work with various synthetic steps explained in greater or lesser detail scattered through the literature and summarized in the Nobel lecture. Additional interesting material can be found in Jackson, A. H.; Kenner, G. W. *Nature*, **1967**, *215*, 1126; Frydman, B.; Reil, S.; Despuy, M. E.; Rapoport, H. *J. Am. Chem. Soc.*, **1969**, *91*, 2338; and de Almeida, J. A. P. B.; Kenner, G. W.; Rimmer, J.; Smith, K. M. *Tetrahedron*, **1976**, *32*, 1793, as well as references found in these publications.

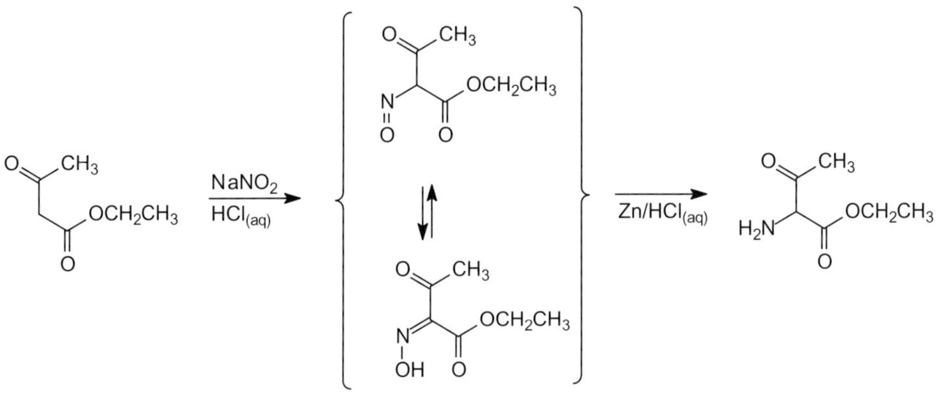

Scheme 14.19. A pathway to the labile ethyl 2-amino-3-oxobutanoate.

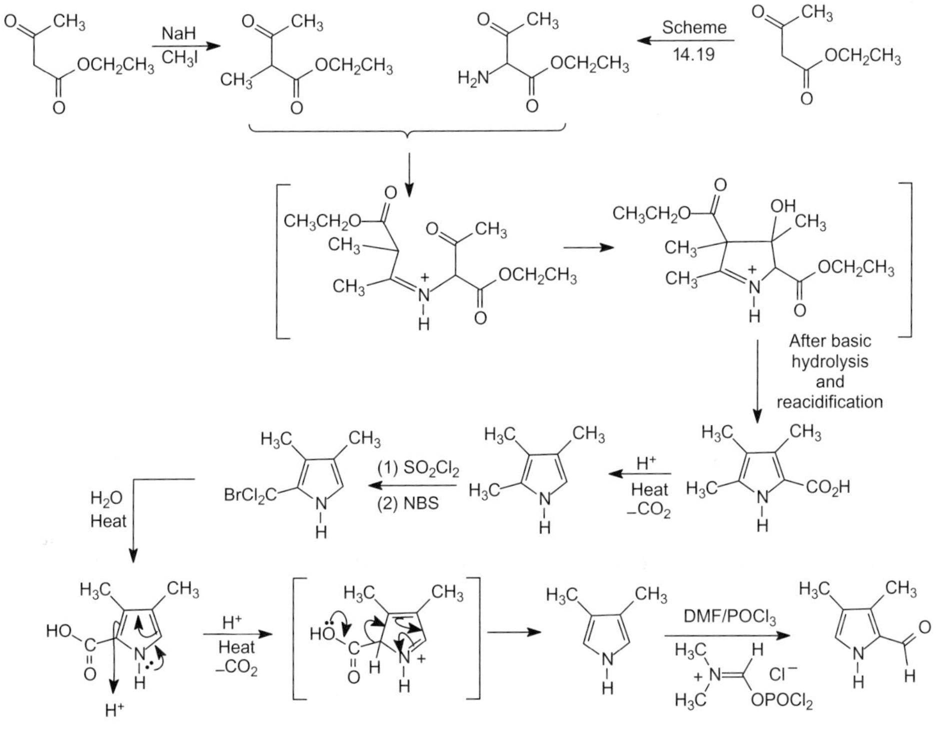

Scheme 14.20. A pathway for the formation of 2,3-dimethylpyrrole-2-carboxyaldehyde from ethyl acetoacetate (ethyl 3-oxobutanoate). DMF = *N,N*-dimethylformamide.

ethyl 2-methyl-3-oxobutanoate, which, on condensation with ethyl 2-amino-3-oxobutanoate followed by hydrolysis, produces a dicarcarboxylic acid that undergoes cyclization and hydrolysis (with base) and results in the formation of 3,4,5-trimethylpyrrole-2-carboxylic acid when acidified. Further heating of the latter under acidic conditions yields 2,3,4-trimethylpyrrole and CO_2.

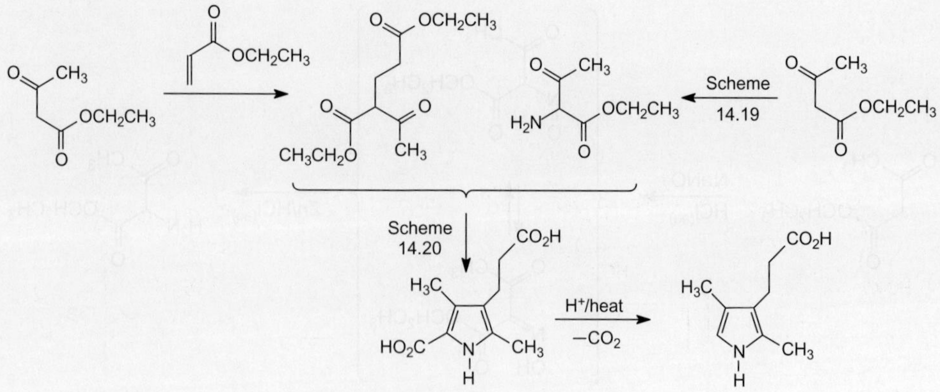

Scheme 14.21. Utilization of the condensation chemistry of ethyl acetoacetate (ethyl 3-oxopropanoate) to produce yet another pyrrole derivative; 3-(2,4-dimethyl-3-yl) propanoic acid.

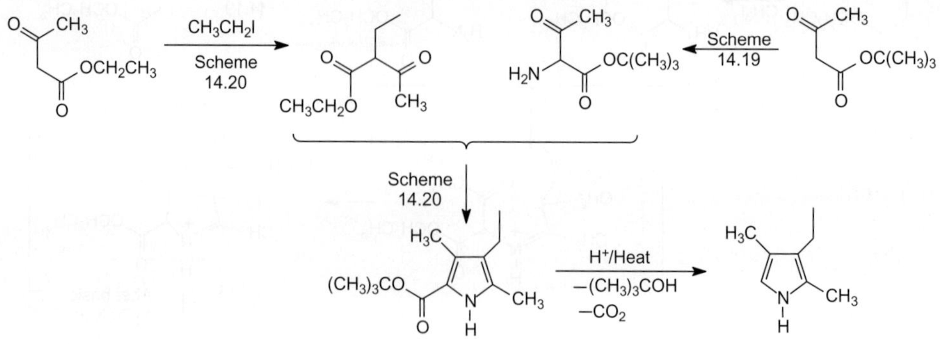

Scheme 14.22. Utilization of condensation chemistry of ethyl acetoacetate (ethyl 3-oxopropanoate) with ethyl iodide and the t-butyl ester of acetoacetic acid to produce yet another pyrrole derivative; 3-ethyl-2,4-dimethylpyrrole.

Chlorination of 2,3,4-trimethylpyrrole with sulfuryl chloride (SO_2Cl_2) yields 2-dichloromethyl-3,4-dimethylpyrrole, which, on bromination with *N*-bromosuccinimide, generates 2-bromodichloromethyl-3,4-dimethylpyrrole. Hydrolysis of the trihalide produces an intermediate carboxylic acid that undergoes decarboxylation to yield 2,3-dimethylpyrrole. Vilsmeier–Haack formulation generates the corresponding 2,3-dimethylpyrrole-2-carboxyaldehyde.

In an analogous way (Scheme 14.21), condensation of ethyl acetoacetate (ethyl 3-oxobutanoate) with the ethyl ester of acrylic acid (ethyl acrylate, ethyl propenoate) followed by a second condensation with ethyl 2-amino-3-oxobutanoate (as above) produces the dicarboxylic acid 3-(3,5-dimethyl-2-carboxy-4-yl)propanoic acid. On decarboxylation, the monoacid 3-(2,4-dimethyl-3-yl)propanoic acid results.

As shown in Scheme 14.22, other variations can also be effected.

Scheme 14.22 utilizes ethyl iodide (iodoethane) and the t-butyl ester of acetoacetic acid (t-butyl 3-oxobutanoate) in a similar set of condensation reactions. Here, however, the t-butyl ester is not hydrolyzed under basic conditions, and

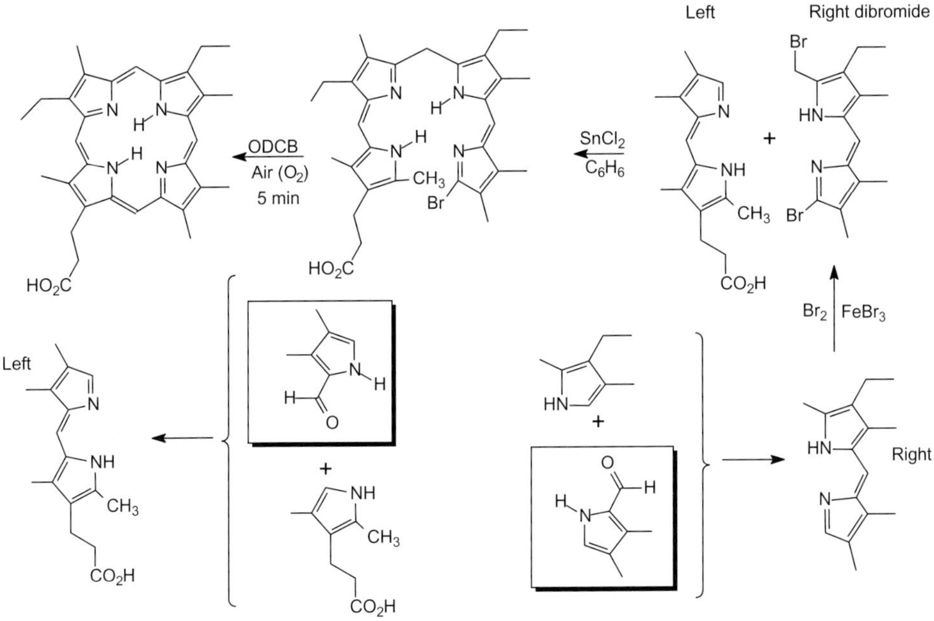

Scheme 14.23. Formation of a porphyrin from the fragments of Schemes 14.20–14.22. ODCB = *ortho*-dichlorobenzene.

the final acid treatment effects both hydrolysis and decarboxylation to produce 3-ethyl-2,4-dimethylpyrrole.

Finally, as shown in Scheme 14.23, all of the fragments described in Scheme 14.20–Scheme 14.22 can be brought together to form a porphyrin nucleus.

Thus, condensation between pyrrole-3,4-dimethyl-2-carboxaldehyde and the 3-(2,4-dimethyl-3-yl)propanoic acid yields a dipyrrole fragment, and condensation between the same aldehyde and the 3-ethyl-2,4-dimethylpyrrole forms a second such dipyrrole. Bromination of the latter in the presence of iron(III) tribromide introduces a bromine into the *only* vacant 2-position as well as onto the only α-methyl available. This derivative can then be coupled to the only available α-position on the first fragment utilizing tin(II) chloride. Finally, heating in *ortho*-dichlorobenzene (ODCB) at reflux (180°C) in the presence of air for 5 min completes the cyclization to the porphyrin.

It is interesting to note that the same kinds of reactions were utilized in the total synthesis of chlorophyll a as reported in communication form in 1960 and expanded upon subsequently.*

The synthesis is undertaken to produce chlorophyll a in the usual pattern; that is, the early work on pyrrole chemistry† is utilized extensively, and the four required

---

*Woodward, R. B.; Ayer, W. A.; Beaton, J. M.; Bickelhaupt, F.; Bonnet, R.; Buchschacher, P.; Closs, G. L.; Dutler, H.; Hannah, J.; Hauck, F. P.; Itô, S.; Langemann, A.; Le Goff, E.; Leimgruber, W.; Lwowski, W.; Saure, J.; Valenta, Z.; Volz, H. *J. Am. Chem. Soc.*, **1960**, *82*, 3800; *Tetrahedron*, **1990**, *22*, 7599; Woodward, R. B.; *Pure Appl. Chem.*, **1961**, *2*, 383.

†The synthesis begins with the preparation of Knorr's pyrrole (diethyl 3,5-dimethylpyrrole-2,4-dicarboxylate), which is made by the reaction of ethyl acetoacetate (ethyl 3-oxobutanoate) with ethyl 2-amino-3-oxobutanoate. The latter is made by nitrosation and subsequent reduction of ethyl acetoacetate.

rings are synthesized independently and linked, pairwise, at the end. Further, it will have been noted that chlorophyll a is chiral, and so it follows that chirality is introduced during the synthesis through the use of chiral reagents or a resolution is required. In this case, a resolution was required.

In contrast, the total synthesis of vitamin B_{12} (cyanocobalamin) was undertaken in a different way and that work is summarized in a handsome series of three publications[*] authored by Robert B. Woodward on behalf of the Eschenmoser-Woodward cooperative effort and spanning 5 of the 12 years that ended with the synthesis of cobyric acid. The latter had earlier been converted to vitamin B_{12}.[†]

III. Current Biosynthetic Understanding

The creation of all of the macrocyclic ring systems comes about starting with the same substituted pyrrole, viz., porphobilinogen (PBG) [5-(aminomethyl)-4-(carboxymethyl)-1H-pyrrole-3-propanoic acid]. However, there is more than one pathway to get to this starting place.

In plants, as shown in Scheme 14.24, the chloroplast tRNA activates glutamate by ligation (EC 6.1.1.17, glutamate-tRNA ligase, using ATP and Mg^{+2}), and then reduction to glutamate-1-semialdehyde occurs (EC 1.2.1.70, glutamyl-tRNA reductase, using NADPH).[‡]

Transposition of the amino and carbonyl groups follows (Scheme 14.24).

A different pathway is obtained in mammals where, in the mitochondrial matrix, glycine and succinyl CoA, in the presence of 5-aminolevulinate synthase, EC 2.3.1.37, and using pyridoxal as a cofactor, direct formation of 5-aminolevulinate occurs (Scheme 14.25).

PBG synthase (EC 4.2.1.24) acts on two equivalents of 5-aminolevulinate to produce the condensation product, prophobilinogen (PBG) [5-(aminomethyl)-4-(carboxy-methyl)-1H-pyrrole-3-propanoic acid] and two equivalents of water (Scheme 14.26). The pathway has been examined in some detail and it appears that imines are formed with two separate lysine residues. The imines undergo condensation.[§]

Now, although details are lacking,[¶] it appears that PBG molecules condense under the influence of PBG deaminase (hydroxymethylbilane synthase [HMBS], EC 2.5.1.61) and a growing chain of porphyrinogens evolves. Interestingly, it has been reported that acute intermittent porphyria (AIP) is an autosomal dominant inherited disease caused by a decreased activity of HMBS.[#] The disease, porphyria,

[*]Woodward, R. B. *Pure Appl. Chem.*, **1968**, *17*, 519; Woodward, R. B. *Pure Appl. Chem.*, **1971**, *25*, 283; Woodward, R. B. *Pure Appl. Chem.*, **1973**, *33*, 145. See also Eschenmoser, A. *Pure Appl. Chem.*, **1963**, *7*, 279 and Eschenmoser, A. *Angew. Chem. Int. Ed.*, **1998**, *27*, 5; Eschenmoser, A.; Wintner, C. E. *Science*, **1977**, *196*, 1410.

[†]Friedrich, W.; Gross, G.; Bernhauer, K.; Zeller, P. *Helv. Chim. Acta*, **1960**, *43*, 704.

[‡]Kannangara, C. G.; Gough, S. P.; Bruyant, P.; Hoober, J. K.; Kahn, A.; von Wettstein, D. *Trends Biochem. Sci. (TIBS)*, **1988**, *13*, 139.

[§]Goodwin, C. E.; Leeper, F. J. *Org. Biomol. Chem.*, **2003**, *1*, 1443.

[¶]Battersby, A. R. *Nat. Prod. Rep.*, **2000**, *17*, 507.

[#]Maeda, N.; Horie, Y.; Adachi, K.; Nanba, E.; Kawasaki H.; Daimon, M.; Kudo, Y.; Kondo, M. *J. Hum. Genet.*, **2000**, *45*, 263.

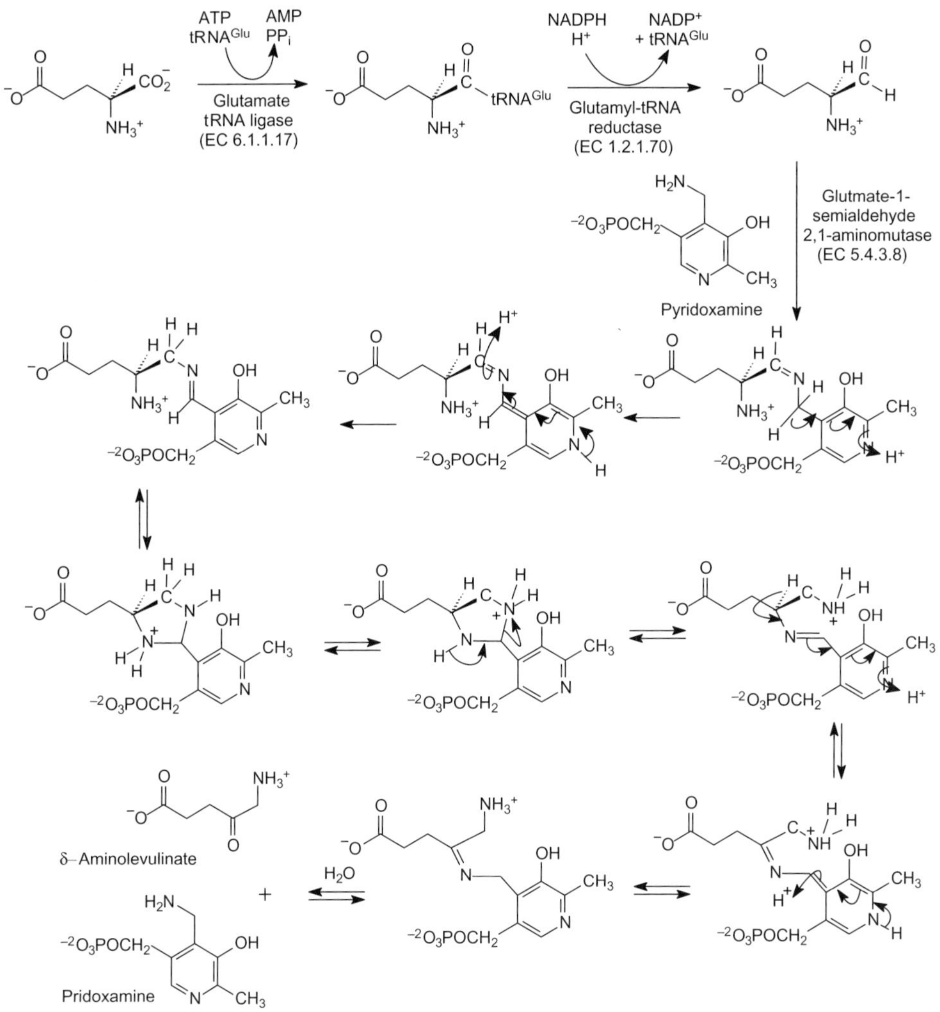

Scheme 14.24. A representation of a pathway for the formation of 5-aminolevulinic acid (δ-aminolevulinic acid) from glutamate in plants using a tRNA ligase and, subsequently, amino to carbonyl transposition aided by pyridoxamine. EC numbers and some graphic materials in this scheme have been taken from links in http://www.chem.qmul.ac.uk/iubmb/enzyme.

is also called "the royal madness" because it is held that it was transmitted, by marriage, through some of the royal houses of Europe.

As shown in Scheme 14.27, it appears that there is a growing chain of pyrrole units attached to a starter unit of two dipyrromethanes (an "enz-dipyrromethane") on hydroxymethylbilane synthase (EC 2.5.1.61). When the chain reaches a total of six units, a cleavage of the four latest added occurs and species such as hydroxymethylbilane or its equivalent, which has evolved, are released. Whatever the process occurring, it is convenient to think of the growing chain as occurring by a series of electrophilic aromatic substitution reactions.

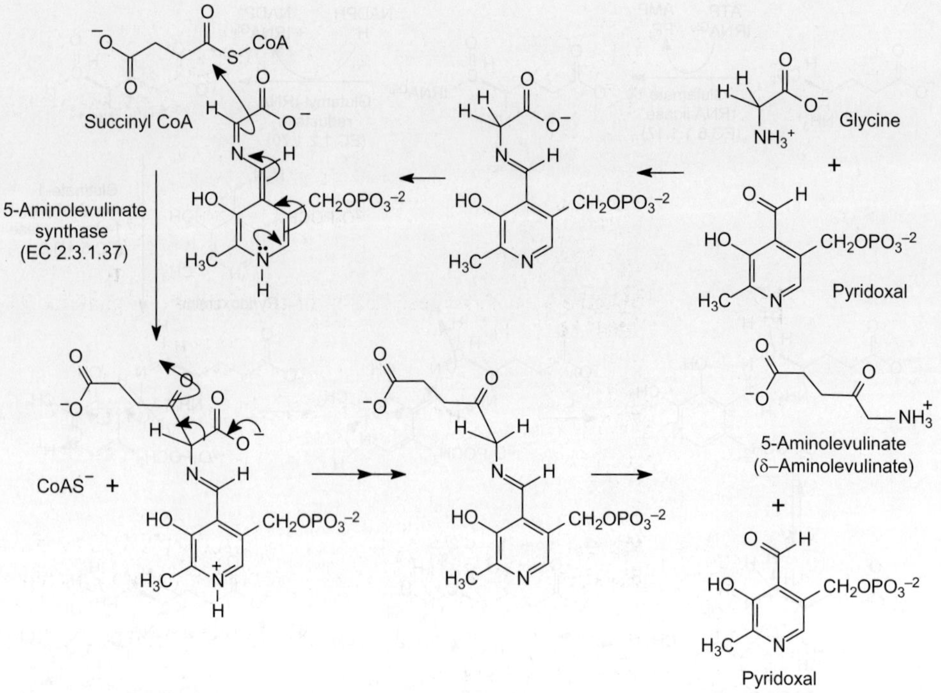

Scheme 14.25. A representation of a pathway for the formation of 5-aminolevulinate (δ-aminolevulinate) from succinyl CoA and glycine (Gly, G) in the presence of 5-aminolevulinate synthase (EC 2.3.1.37).

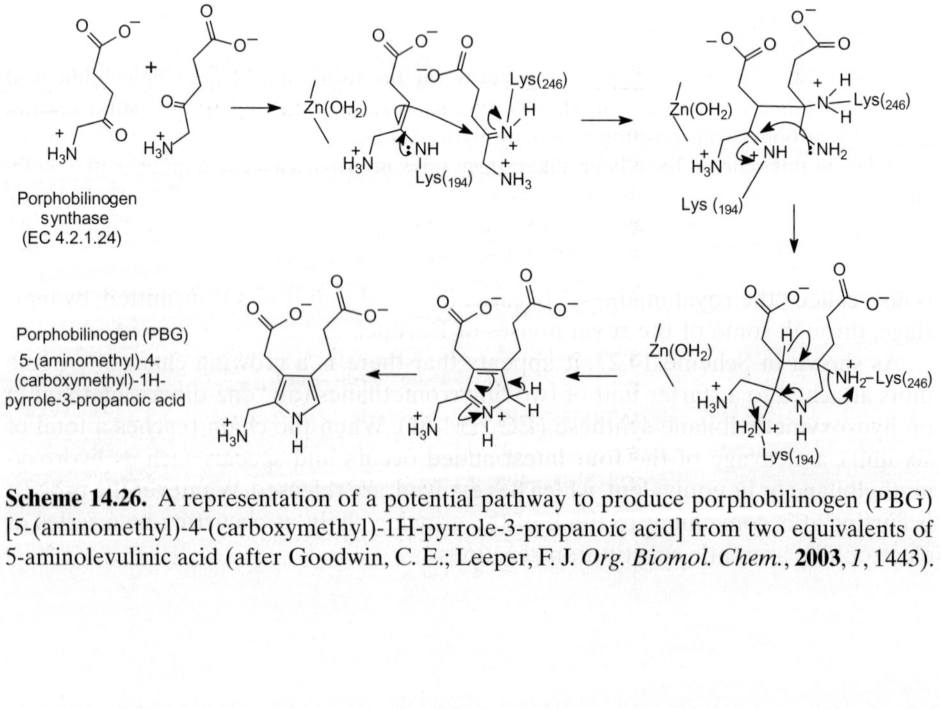

Scheme 14.26. A representation of a potential pathway to produce porphobilinogen (PBG) [5-(aminomethyl)-4-(carboxymethyl)-1H-pyrrole-3-propanoic acid] from two equivalents of 5-aminolevulinic acid (after Goodwin, C. E.; Leeper, F. J. *Org. Biomol. Chem.*, **2003**, *1*, 1443).

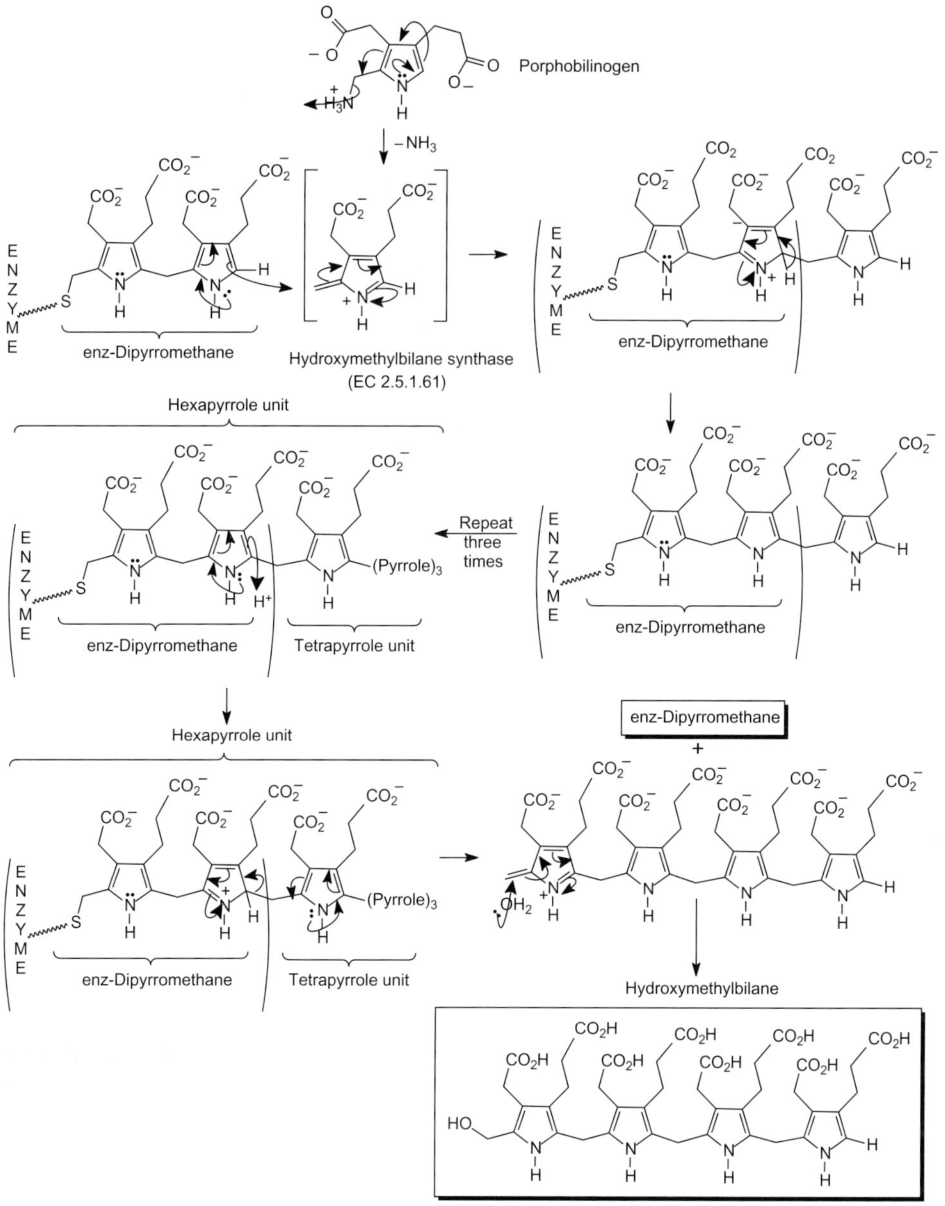

Scheme 14.27. A representation of a growing chain of pyrrole units attached to a "starter unit" of two dipyrromethanes. At a total of six units, cleavage of the four last added occurs.

Then, catalyzed by uroporphyrogen III cyclase (EC 4.2.1.75), the 20 carbon ring forms.

That this is the case was studied in some detail by Frydman and coworkers.*

*Sburlati, A.; Frydman, R. B.; Valasinas, A.; Rosé, S.; Priesta, H. A.; Frydman, B. *Biochemistry*, **1983**, *22*, 4006.

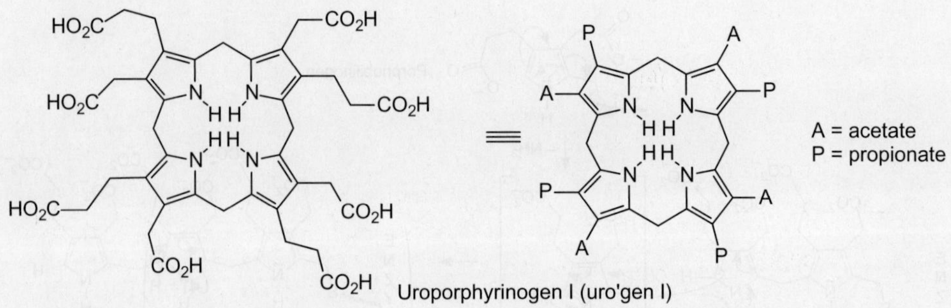

Uroporphyrinogen I (uro'gen I)

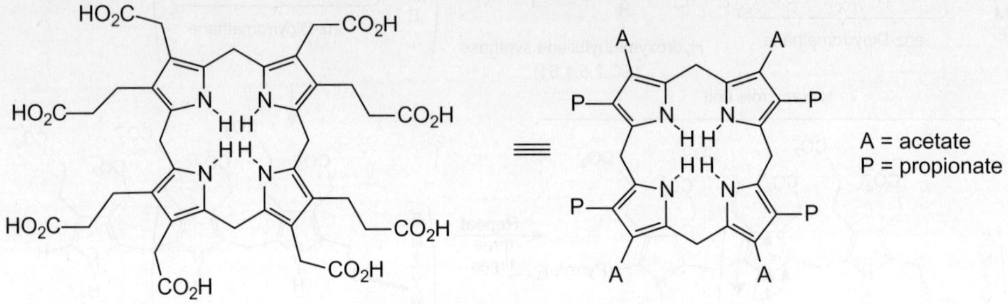

A = acetate
P = propionate

Uroporphyrinogen II (uro'gen II)

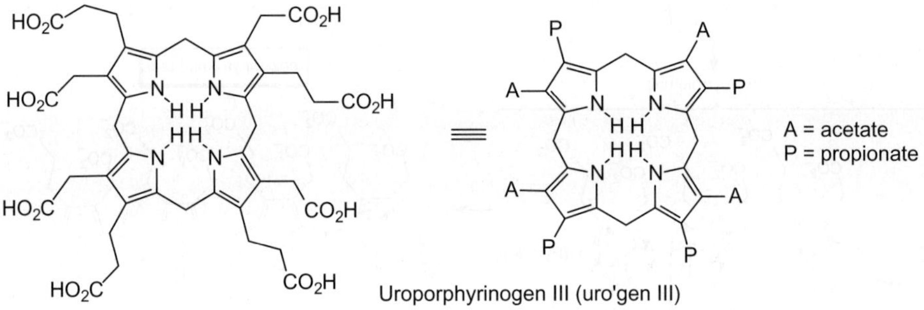

A = acetate
P = propionate

Uroporphyrinogen III (uro'gen III)

A = acetate
P = propionate

Figure 14.9. Three uroporphyrinogens to be considered. Uroporphyrinogen I forms in the presence of porphobilinogen deaminase (EC 2.5.1.61). The reaction does not go on. When both porphobilinogen deaminase (EC 2.5.1.61) and uroporphyrogen III cyclase (EC 4.2.1.75) are present, uroporphyrinogen III forms. Uroporphyrinogen II is not formed.

As shown in Figure 14.9 (where $A = -CH_2CO_2H$ and $P = -CH_2CH_2CO_2H$), a family of tetrameric pyrroles, the "bilanes" was created. These 2-(aminomethyl) bilanes were then studied in some detail, and it was found that the 2-(aminomethyl) bilane formally derived from the head-to-tail condensation of four units of PBG yielded a cyclic product (a uroporphyrinogen [or uro'gen] called uroporphyrinogen I [or uro'gen I]). Uro'gen I could be formed either by chemical means or by taking the same 2-(aminomethyl) bilane and effecting the cyclization in the presence of PBG deaminase (EC 2.5.1.61) *and in the absence of uroporphyrinogen III synthase* (EC 4.2.1.75). In the presence of both enzymes, uroporphyrinogen III (uro'gen III) resulted.*

*Sburlati, A.; Frydman, R. B.; Valasinas, A.; Rosé, S.; Priesta, H. A.; Frydman, B. *Biochemistry*, **1983**, *22*, 4006.

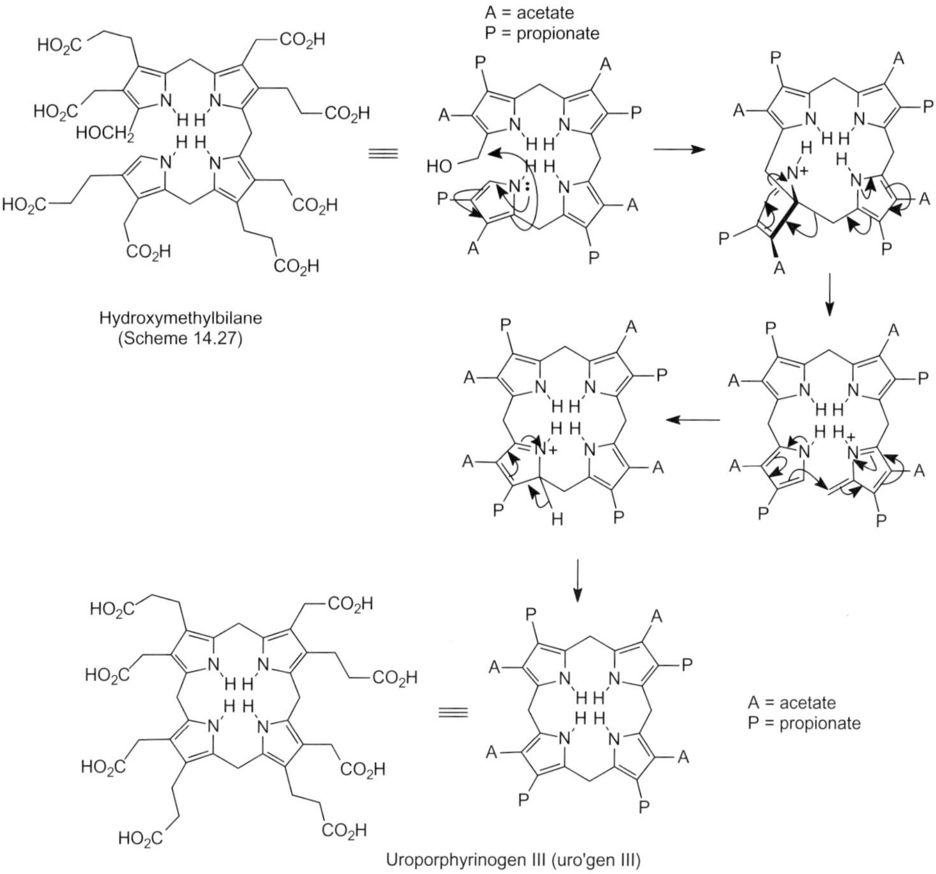

Scheme 14.28. A representation of a pathway that accounts for the formation of uroporphyrinogen III from hydroxymethylbilane. Presumably, the hydroxyl group is activated for displacement, although the details of that activation are not yet known.

Uro'gen II was produced by the chemical cyclization of the head-to-head polymer. The three uroporphyrinogens (uro'gens) are shown in Figure 14.9.

The question as to how uro'gen III formed was subsequently elegantly elucidated by A. I. Scott and A. R. Battersby* and their coworkers, who demonstrated that the first step in the process was for the PBG deaminase (EC 2.5.1.61) to convert 2-(aminomethyl) bilane to hydroxymethylbilane. As indicated above, the hydroxymethylbilane (the product of the action of HMBS [EC 2.5.1.61]) was the substrate for the cosynthetase.

Scheme 14.28 provides a representation of the pathway that is accounted for by labeling studies and provides uroporphyrinogen III from the starting hydroxymethylbilane.

Uroporphyrinogen III (uro'gen III) lies at the branching point of porphyrin and corrin biosynthesis. Down one path, following initial decarboxylation, both chloro-

*Battersby, A. R. *Nat. Prod. Rep.*, **2000**, *17*, 207. See also Battersby, A. R.; Fookes, C. J. R.; Gustafson-Potter, K. E.; McDonald, E.; Matcham, G. W. J. *J. Chem. Soc., Perkin Trans.*, **1982**, *1*, 2427.

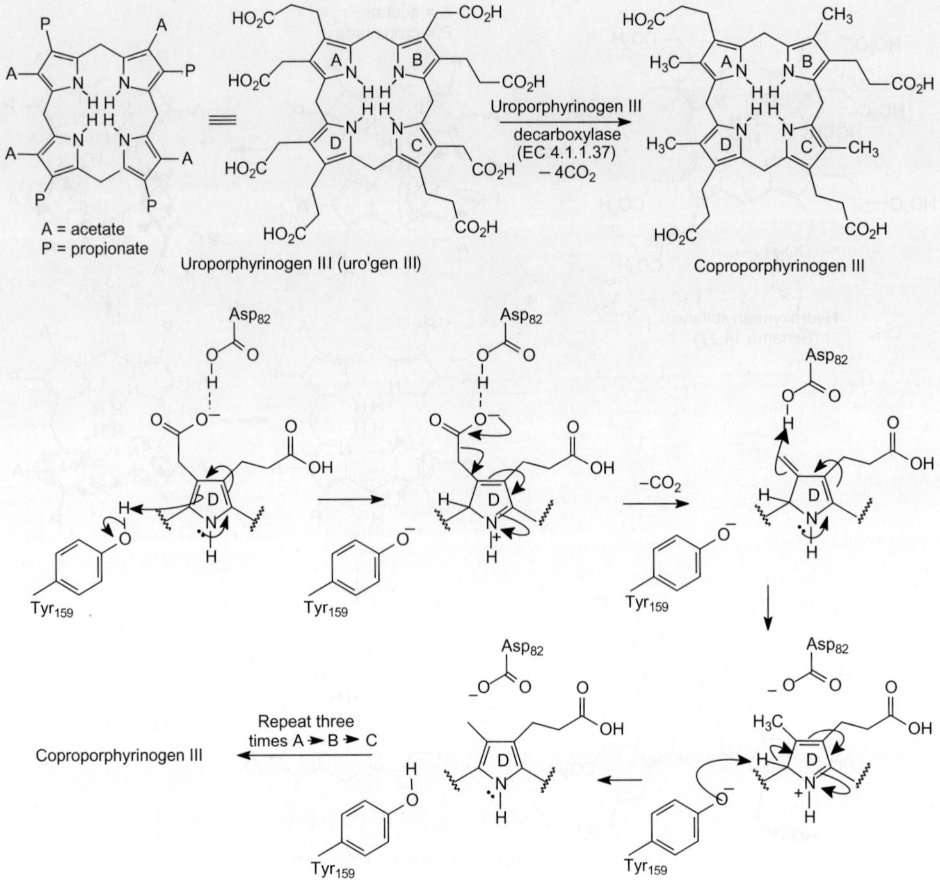

Scheme 14.29. A representation of a potential pathway for uroporphyrinogen III decarboxylase catalysis to form coproporphyrinogen III (after Martins, B. M.; Grimm, B.; Mock, H.-P.; Huber, R.; Messerschmidt, A. *J. Biol Chem.*, **2001**, *276*, 44108).

phyll and heme can be found. Down the other path, following initial methylation on carbon, cobalamin (vitamin B_{12}) and siroheme result.

For the former, and based on crystallographic studies with uroporphyrinogen III decarboxylase (EC 4.1.1.37) from *Nicotiana tabacum* (Scheme 14.29), the proposal has been made* that, beginning with pyrrole ring D and proceeding clockwise, loss of carbon dioxide (CO_2) occurs, sequentially, to generate the coproporphyrinogen III. The involvement of aspartate (Asp, D) 82 and tyrosinate (Tyr, Y) 159 groups has been suggested as shown.

To convert coproporphyrinogen III to protoporphyrin IX, the ultimate precursor to heme and chlorophyll, additional decarboxylations and an oxidation are necessary (Scheme 14.30). The former results in the replacement of the propionate side chains of rings "A" and "B" with vinyl substituents, and the latter fully aromatizes the macrocyle.

*Martins, B. M.; Grimm, B.; Mock, H.-P.; Huber, R.; Messerschmidt, A. *J. Biol. Chem.*, **2001**, *276*, 44108.

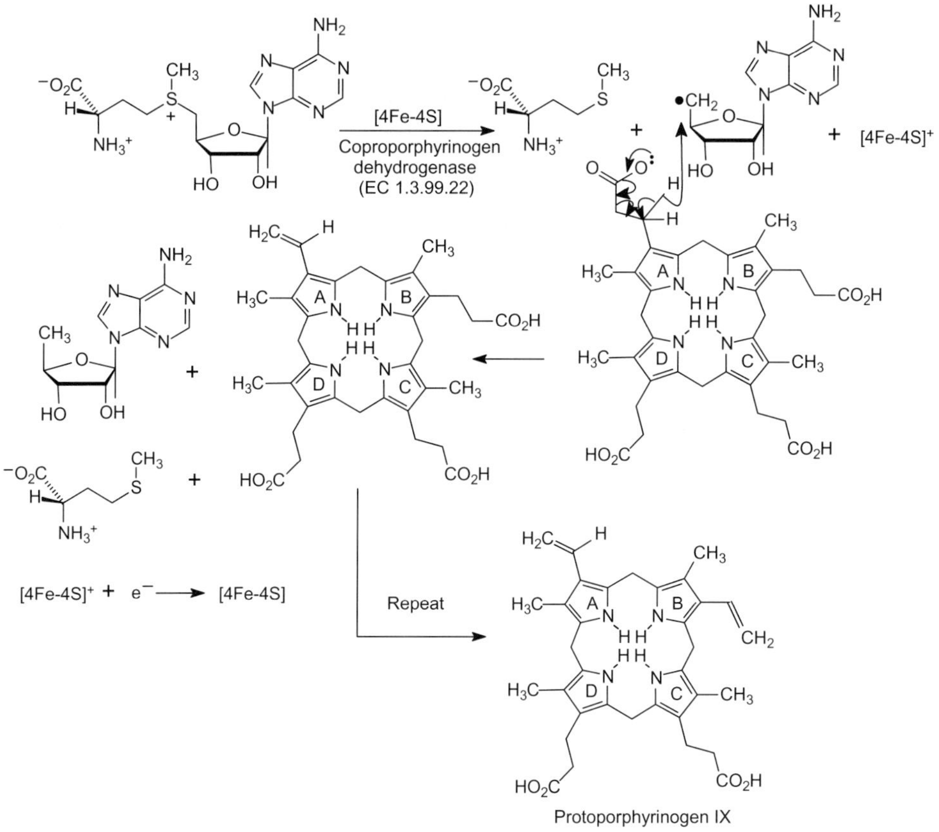

Scheme 14.30. A representation of a pathway in bacteria using coproporphyrinogen dehydrogenase (EC 1.3.99.22) to produce protoporphyrinogen IX from coproporphyrinogen III.

To introduce the vinyl substituent in bacteria, SAM is cleaved with the aid of the [4Fe-4S] cluster of the coproporphyrinogen dehydrogenase (EC 1.3.99.22) to yield methionine and a 5-deoxyadenosine-5′-yl radical. The radical, on reaction with coproporphyrinogen III, abstracts an electron from a carboxylate anion and the product undergoes loss of CO_2. This occurs twice, producing the two vinyl side chains in protoporphyrinogen IX. The electron generated at each returns to the cluster.[*]

For the production of the vinyl side chains in eukaryotes, coproporphyrinogen oxidase (EC 1.3.3.3) catalyzes the conversion of coproporphyrinogen III to protoporphyrinogen IX. Except for the observation that the enzyme is not a metalloprotein,[†] its structure remains unknown.

Regardless of the source of protoporphyrinogen IX, that is, whether by oxidation with oxygen and coproporphyrinogen oxidase (EC 1.3.3.3) in eurkaryotes or with SAM via a radical pathway and using coproporphyrinogen dehydrogenase (EC 1.3.99.22), subsequent oxidation using protoporphyrinogen oxidase (EC 1.3.3.4)

[*]Layer, G.; Moser, J.; Heinz, D. W.; Jahn, D.; Schubert, W. D. *EMBO J.*, **2003**, *22*, 6214.
[†]Medlock, A. E.; Dailey, H. A. *J. Biol. Chem.*, **1996**, *271*, 32507.

ensues. It appears that several different isoenzymes (enzymes differing in amino acid sequence but performing the same function—albeit with different kinetics) of protoporphyrinogen oxidase (EC 1.3.3.4) occur. The cofactor FAD appears to be utilized throughout for the oxidation and $FADH_2$ results.

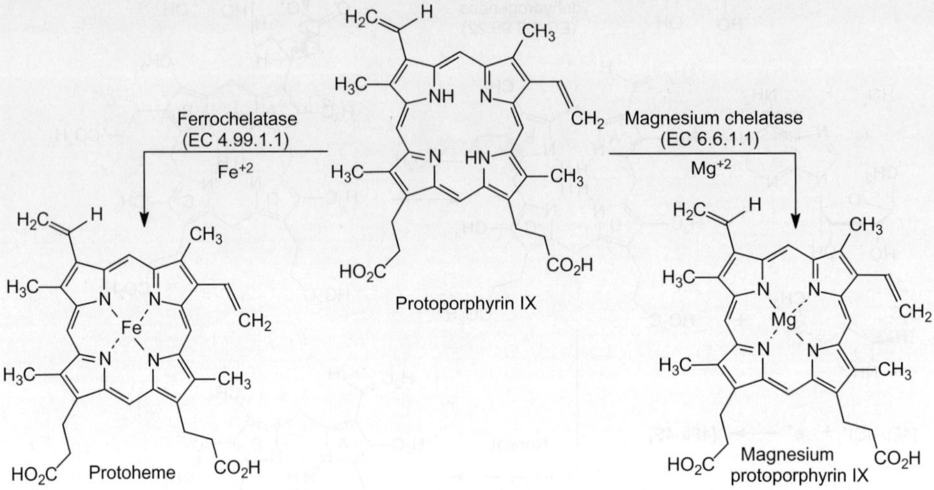

Ferrochelatase (EC 4.99.1.1) Fe^{+2}

Magnesium chelatase (EC 6.6.1.1) Mg^{+2}

Protoporphyrin IX

Protoheme

Magnesium protoporphyrin IX

Scheme 14.31. A representation of the incorporations of metals (Fe^{+2} and Mg^{+2}) into protoporphyrin IX to produced heme (protoheme) and magnesium protoporphyrin IX, respectively. The latter is the precursor for chlorophyll. EC numbers and some graphic materials in this scheme have been taken from links in http://www.chem.qmul.ac.uk/iubmb/enzyme.

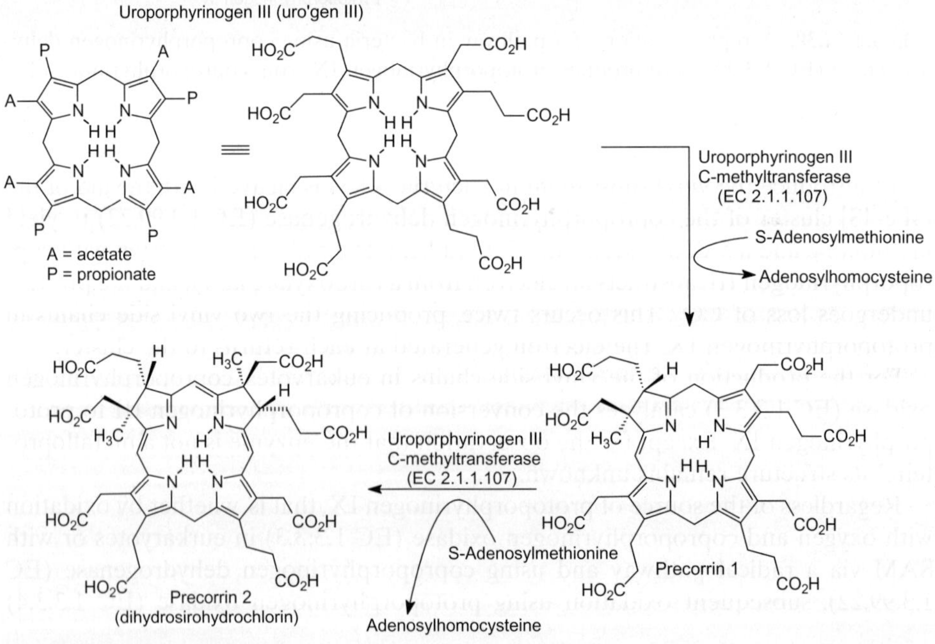

Uroporphyrinogen III (uro'gen III)

A = acetate
P = propionate

Uroporphyrinogen III C-methyltransferase (EC 2.1.1.107)

S-Adenosylmethionine

Adenosylhomocysteine

Precorrin 1

Uroporphyrinogen III C-methyltransferase (EC 2.1.1.107)

S-Adenosylmethionine

Adenosylhomocysteine

Precorrin 2 (dihydrosirohydrochlorin)

Scheme 14.32. The initial steps on the pathway from uroporphyrinogen III to cobalamin (vitamin B_{12}) and siroheme.

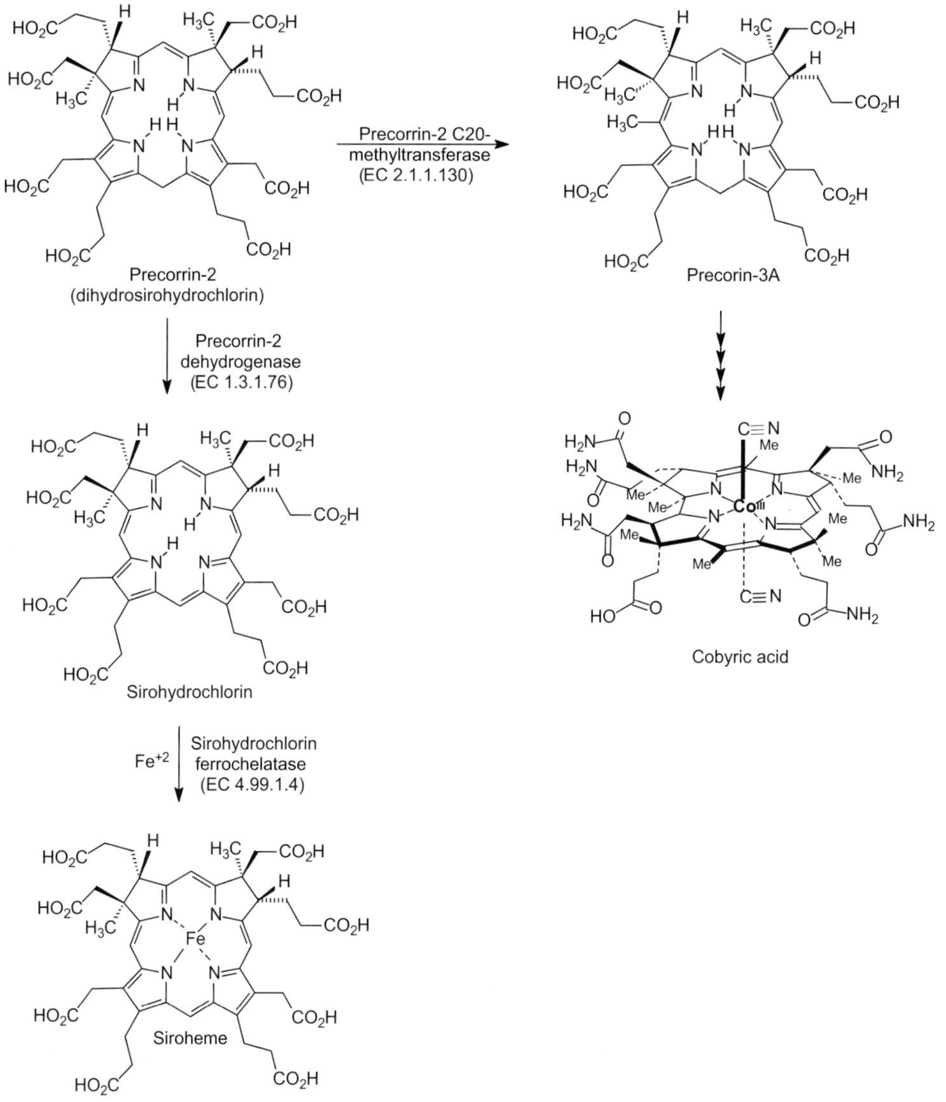

Scheme 14.33. A representation of pathways to siroheme and to precorrin-3A cobalamin. EC numbers and some graphic materials in this scheme have been taken from links in http://www.chem.qmul.ac.uk/iubmb/enzyme.

The product of the oxidation, protoporphyrin IX (Equation 14.9), is the jumping-off point for the biosyntheses of heme and chlorophyll (Scheme 14.31).

Protoporphyrinogen oxidase (EC 1.3.3.4) FAD/FADH$_2$

O_2 H_2O_2

Protoporphyrinogen IX

Protoporphyrin IX

$$(14.9)$$

As shown in Scheme 14.31, incorporation of Fe(II) (ferrous) iron into protoporphyrin IX with the aid of the enzyme ferrochelatase (EC 4.99.1.1) produces protoheme or heme (the particular variant labeled protoheme here is called heme b in some classifications; substituents derived from those shown in the protoheme give rise to other variants).

Also, as shown in Scheme 14.31, incorporation of magnesium with the aid of magnesium chelatase (EC 6.6.1.1.) leads to magnesium protoporphyrin IX, now committed to the biosynthesis of chlorophyll.

The second branch from uroporphyrinogen III (uro'gen III) referred to above leads to corrin rather than heme and requires initial methylation on carbon (Scheme 14.32). This pathway, which eventually leads to cobalamin (vitamin B$_{12}$) and siroheme, requires the use of SAM (which is converted to S-adenosylhomocysteine) and the enzyme uroporphyrinogen III C-methyltransferase (EC 2.1.1.107). The first methylation occurs on the A ring, forming precorrin-1. Repeating the process with the same enzyme and the same methylating agent yields precorrin-2 (dihydrosirohydrochlorin).

Finally, in Scheme 14.33, oxidation of precorrin-2 (NAD$^+$/NADH is the cofactor for the use of the enzyme precorrin-2 dehydrogenase, EC 1.3.1.76) yields sirohydrochlorin, which, with ferrous (Fe^{+2}) iron and the enzyme sirohydrochlorin ferrochelatase (EC 4.99.1.4), produces siroheme. If, in place of oxidation, yet another methylation were to occur, precorrin-2 would be converted to precorrin-3A (precorrin-2 C20-methyltransferase, EC 2.1.1.130) and then, after addition methylation, and modification, to cobalamin.

Epilogue

It is difficult to pen this epilogue.

Your author regrets that so little of the rich fabric of organic chemistry was provided and that you might be disappointed that so little current material, replacing the classical studies that were given, is present.

With regard to the former, the tapestry is so large that only a few threads, designed (hopefully) to provide the overall picture (but with loss of some detail), could be given in a single volume.

With regard to the latter, it appears that modern search engines do not easily allow for substantive inquiry, and so your author decided that some flavor of the early work of those whose shoulders we stand upon was needed. But hints of how to use the solid foundation to build new structures abound, and it is now your duty to use that hard-won material to increase the substance of humanity.

Good luck.

The Schrödinger Equation

Consider a simple harmonic sine function such as that shown in Figure AI.1, which might, for example, represent a plucked string.

The frequency (ν) is the **the number of vibrations per unit of time** and is the reciprocal of the **period (τ) of the motion** (i.e., $\tau = 1/\nu$), the **time** required for a single vibration. Just as τ is a measure of the periodicity of the wave in time, the wavelength (λ) is the periodicity of the wave in **space**. The numerical value, n, of the time (t) or the distance (x) is such that the amplitude (A) of the wave is zero (the wave is said to cross the x-axis) at any integral multiple of n. The amplitude (A) of the wave at $n/2$, where n is an odd integer, is the maximum displacement of the curve in the y-direction, and at these values, the sense of motion of the wave is reversed. At $n/2$ (n is odd) then, the kinetic energy (E_k) is zero, while the potential energy (E_p) is at a maximum. Since the total energy (E_T) is simply the sum of the potential (E_p) and kinetic energies (E_k), when y is at a maximum, the total energy (E_T) equals the potential energy ($E_T = E_p$).

Now, consider that the entire wave depicted in Figure AI.1 has moved. Everything written about it above remains the same. However, if the wave in Figure AI.1 is moving with some *velocity*, c, in the x-direction (i.e., it is said to be propagating), then at some later time, $t + \Delta t$ (where Δt is some small increment in time, t), the new picture of the wave appears as the dotted line in Figure AI.2, while the solid line is that originally seen in Figure AI.1.

As pointed out by de Broglie (L. De Broglie, 1892–1987, Nobel Prize in Physics, 1929), if an electron is moving as a wave, around a nucleus, a more general description of a wave is needed than that given by a sine function.

In that vein, regardless of the exact form of the wave, if there is a wave moving in the xy plane (as in Figure AI.2), then a plot of y as a function of x corresponds to the wave. The **slope of any curve over any given small incremental value of y and x is $\Delta y/\Delta x$**. If the increments Δy and Δx are made small enough (to get an instantaneous picture, i.e., the slope of the curve at some point x, y), the slope is better represented as the first derivative (i.e., dy/dx). However, to describe a **wave**, it is the **curvature** at any given point that is necessary rather than the slope. The curvature at the point is given by the *second derivative, d^2y/dx^2*.

If, however, the wave is propagating as in Figure AI.2, it is necessary to consider both the direction (x) and the time (t), and thus *partial derivatives* $(\partial^2y/\partial x^2)_t$ and $(\partial^2y/\partial t^2)_x$ are required.

Foundations of Organic Chemistry: Unity and Diversity of Structures, Pathways, and Reactions,
First Edition. David R. Dalton.
© 2011 John Wiley & Sons, Inc. Published 2011 by John Wiley & Sons, Inc.

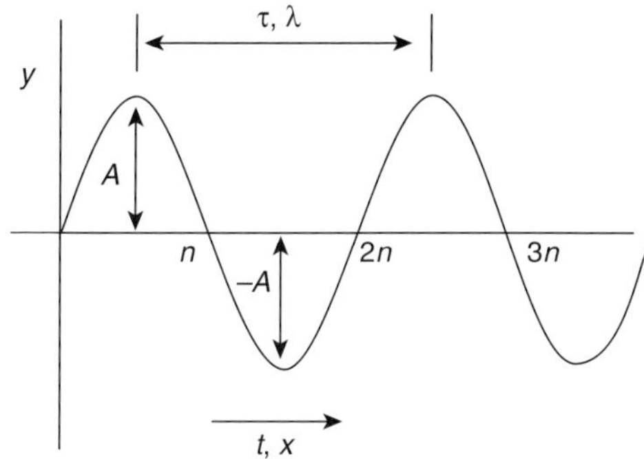

Figure AI.1. Representation of a simple harmonic sine function.

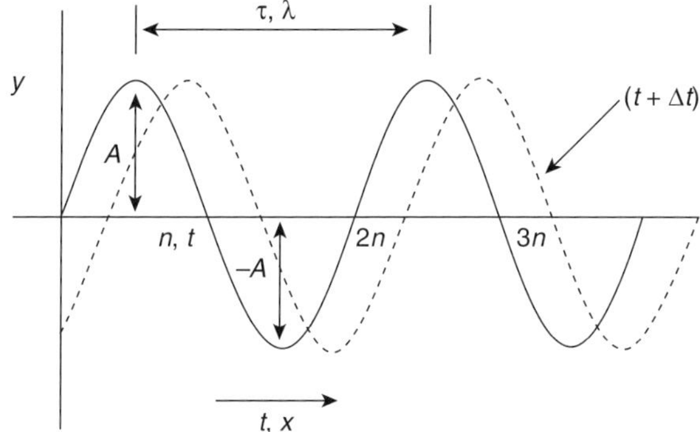

Figure AI.2. Representation of a simple harmonic sine function as in Figure AI.1 before and after its motion along the horizontal time (t) or distance (x) axis.

Overall then, for a moving (with velocity c) wave form, as shown in Figure AI.2, Equation AI.1 describes what is seen:

$$\partial^2 y/\partial x^2 = (1/c^2)\cdot(\partial^2 y/\partial t^2). \tag{AI.1}$$

Now, although it cannot be derived (it is the leap that most of us who plod can only envy), Schrödinger (E. Schrödinger, 1887–1961, University of Vienna, Nobel Prize in Physics, 1933), in 1926, following de Broglie's suggestion that electrons around the nucleus moved as waves, replaced y in Equation AI.1 with a function for the wave (in which a W became a Ψ), such that $\Psi(x, t)$ would describe the electron's behavior and the velocity of the wave (c) with the velocity of light (c) to give

$$\partial^2 \Psi/\partial x^2 = (1/c^2)\cdot(\partial^2 \Psi/\partial t^2). \tag{AI.2}$$

One solution to the expression of Equation AI.2 is

$$\Psi(x,\ t) = \text{constant}\ (e^{i\alpha}), \tag{AI.3}$$

where $i = \sqrt{-1}$, and

$$\alpha = 2\pi[(x/\lambda) - vt]. \tag{AI.4}$$

The general expression of Equation AI.4 correlates with the particular case of the sine wave shown in Figures AI.1 and AI.2 since the wave in Figure AI.1 can be described as

$$\alpha = A\sin(2\pi/\lambda)x \tag{AI.5}$$

if it is stationary, while if it is moving, as in Figure AI.2,

$$\alpha = A\sin[(2\pi/\lambda)x - 2\pi vt], \tag{AI.6}$$

where v is the **frequency** (in cycles per seconds) with which it is moving.

Combining constants and substituting Equation AI.6 into the more general expression of Equation AI.4 (which will apply not only to sine waves),

$$\Psi(x, y) = \text{constant}\ (e^{i[(2\pi/\lambda)x - 2\pi vt)]}) \tag{AI.7}$$

or

$$\Psi(x, y) = \text{constant}\ (e^{2\pi ix/\lambda})(e^{-2\pi ivt}). \tag{AI.8}$$

Then, following Planck (M. Planck, 1879–1947, Nobel Prize in Physics, 1918), who pointed out that $E = hv$ (where h is Planck's constant) and Einstein (A. Einstein, 1879–1955, Nobel Prize in Physics, 1921), who related energy to mass and the speed of light ($E = mc^2$), Schrödinger noted that

$$mc^2 = hv = hc/\lambda, \tag{AI.9}$$

or

$$\lambda = h/mc. \tag{AI.10}$$

If a particle, instead of a photon, is considered, following de Broglie, so that velocity is v instead of c, then

$$\lambda = h/mv = h/p_z, \tag{AI.11}$$

where p_x is the momentum of the particle in the x-direction, and thus

$$\Psi(x, y) = \text{constant}\ (e^{2\pi ixp/h})(e^{-2\pi iEt/h}). \tag{AI.12}$$

Then, if the *constant* in the above expressions is abbreviated as C,

$$\partial\Psi/\partial t = -(2\pi iEC/h)(e^{2\pi ixp/h})(e^{-2\pi iEt/h}) \tag{AI.13}$$

or

$$\partial\Psi/\partial t = -(2\pi i E/h)\Psi \tag{AI.14}$$

or

$$(-h/2\pi i)\cdot\partial\Psi/\partial t = E\Psi. \tag{AI.15}$$

Similarly

$$\partial\Psi/\partial x = -(2\pi i p_x C/h)\Psi \tag{AI.16}$$

so that, by analogy,

$$(-h/2\pi i)\cdot\partial\Psi/\partial t = p_x\Psi \tag{AI.17}$$

or

$$p_x = (-h/2\pi i)\cdot\partial/\partial x. \tag{AI.18}$$

Schrödinger's next argument was to introduce the total energy of the system into the wave equation. Thus, as noted at the beginning of this discussion, the total energy of the system must be the sum of the kinetic and potential energies (i.e., $E_T = E_k + E_p$). Thus, since $E_k = \frac{1}{2}mv^2$,

$$E_T = (p_x^2/2m) + E_p \tag{AI.19}$$

or

$$E_T = [(h^2/8\pi^2 mv)\cdot\partial^2/\partial x^2 + E_p] \tag{AI.20}$$

and, therefore, by Equation AI.15,

$$(-h/2\pi i)\cdot\partial\Psi/\partial t = [(-h^2/8\pi^2 m)\cdot\partial^2/\partial x^2 + E_p]\Psi. \tag{AI.21}$$

If this is for a **stationary wave** (and thus *time* independent) and, since for such a case, $E = (h/2\pi i)$, we may then replace Ψ by a *time-independent* function, ψ, rearrange, and write

$$(\partial^2\psi/\partial x^2) + (8\pi^2 m/h^2)(E + E_p)\psi = 0. \tag{AI.22}$$

If x is the distance r of the electron from the nucleus, Ze is the charge on the nucleus, and e is the charge on the electron, then E_p is (Ze^2/r) and, for one dimension,

$$(\partial^2\psi/\partial x^2) + (8\pi^2 m/h^2)[E + (Ze^2/r)]\psi = 0, \tag{AI.23}$$

and for three dimensions (x, y, and z),

$$(\partial^2\psi/\partial x^2 + \partial^2\psi/\partial y^2 + \partial^2\psi/\partial z^2) + (8\pi^2 m/h^2)[E + (Ze^2/r)]\psi = 0. \qquad \text{(AI.24)}$$

This is the "famous" Schrödinger equation.

The energies allowed by solution of the equation are called **eigenvalues**, and the corresponding wave functions (ψ) are called **eigenfunctions**. With a few exceptions, the Schrödinger equation can be solved only approximately.

As pointed out by Born (M. Born, 1882–1970, Nobel Prize in Physics in 1954), if the eigenfuction is *real* (i.e., does not involve i), then its square (ψ^2) multiplied by a volume element (dx, dy, dz) is proportional *to the probability that the electron is present in the volume element.*

The Literature

The current (2010) state of the literature in chemistry and her sister sciences is in a state of flux.

Books, such as this one, as well as specialist treatises on almost every topic discussed in this volume, continue to be produced and housed in libraries (although distribution on the World Wide Web [WWW]), with or without hard copy production, is becoming common). These materials are part of the "secondary" literature.

The "primary" literature consists of journals, to which many references in this and other secondary materials can be found and which contain the fruits of original thought and research as well as references to prior art. Journals are largely produced by scientific societies and related publishers. For the most part, the work found in journals has been examined by "referees" for its likely veracity before publication. Journals, which continue to proliferate as more and more men and women become involved in scientific endeavors, are, in many cases, no longer produced in hard copy or, if so produced, are no longer distributed. The distribution of the information is now, largely, through the WWW.

The volume of material is no longer easily mastered by any individual.

To attempt to provide an indication of the substance of the research itself, societies such as the International Union of Pure and Applied Chemistry (IUPAC) in Switzerland, the American Chemical Society (ACS) in the United States, and the Royal Society of Chemistry (RCS) in England set standards for the style and content of what they publish. It is unusual to find these standards violated.

In addition, most publishing organizations require authors to begin their work with an "abstract," which is subsequently entered into a database. Databases, using these abstracts and key words from the work, are then sold (although some are publicly available, e.g., *Public Library of Science [PLoS]*) to commercial organizations (pharmaceutical companies, chemical producers, and so on) and universities.

The production of databases has become very expensive and many earlier compendia, such as *Beilsteins Handbuch der Organischen Chemie*, have been essentially driven out of competition as hardbound volumes (although at this writing much of it remains available through a commercial online data base, **Reaxys** (www.reaxys. com) owned and operated by Elsevier B.V. in Amsterdam, The Netherlands). Similarly, other specialist databases such as *Methoden der Organischen Chemie (Houben–Weyl)* have been transformed into Science of Synthesis: Houben–Weyl

Foundations of Organic Chemistry: Unity and Diversity of Structures, Pathways, and Reactions,
First Edition. David R. Dalton.
© 2011 John Wiley & Sons, Inc. Published 2011 by John Wiley & Sons, Inc.

Methods of Molecular Transformations owned and operated by Thieme Publishing Group, Stuttgart, Germany (www. thieme.com) as online sources of information. The current major compendium through which most searches for chemical information are directed is that of the Chemical Abstracts (CA) (http://cas.org) Division of the American Chemical Society (ACS) (www.acs.org) and its SciFinder online service.

Librarians and other information technology-sensitive authorities on searching the literature can often help identify suitable online sources of information.

In addition to the above, three major sources of material used in Part III of this book need to be recognized. First, there is the Protein Data Bank (PDB) (whose current address on the WWW is www.pdb.org). Second, there is the International Union of Biochemistry and Molecular Biology (IUBMB) (whose current address on the WWW is www.iubmb.org) and the access to the enzyme nomenclature graciously hosted by Queen Mary University of London and searchable through www.chem.qmul.ac.uk/iubmb/enzyme. Finally, there is the extensive set of pathways that interlink life to be found through the Kyoto Encyclopedia of Genes and Genomes (KEGG), where the Pathway Database can be found at www.genome.jp/kegg. These entry points provided a wealth of reliable data and appropriate literature references as well as links to other worthwhile sources.

Foundations of Organic Chemistry: Unity and Diversity of Structures, Pathways, and Reactions,
First Edition. David R. Dalton.
© 2011 John Wiley & Sons, Inc. Published 2011 by John Wiley & Sons, Inc.